Teacher's Edition

Mathematical Connections

A Bridge to Algebra and Geometry

Francis J. Gardella

Patricia R. Fraze

Joanne E. Meldon

Marvin S. Weingarden

Cleo Campbell, CONTRIBUTING AUTHOR

TEACHER CONSULTANTS

Alma Cantu Aguirre

Cheryl Arevalo

Douglas Denson

Jackie G. Piver

Jocelyn Coleman Walton

Houghton Mifflin Company • BOSTON

Atlanta • Dallas • Geneva, Ill. • Palo Alto • Princeton • Toronto

AUTHORS

Francis J. Gardella, Supervisor of Mathematics and Computer Studies, East Brunswick Public Schools, East Brunswick, New Jersey

Patricia R. Fraze, former Mathematics Department Chairperson, Huron High School, Ann Arbor, Michigan

Joanne E. Meldon, Mathematics Teacher, Taylor Allderdice High School, Pittsburgh, Pennsylvania

Marvin S. Weingarden, Supervisor of Secondary Mathematics, Detroit Public Schools, Detroit, Michigan

Cleo Campbell, Coordinator of Mathematics, Anne Arundel County Public Schools, Annapolis, Maryland

TEACHER CONSULTANTS

Alma Cantu Aguirre, Mathematics Department Chairperson, Thomas Jefferson High School, San Antonio, Texas

Cheryl Arevalo, Supervisor of Mathematics, Des Moines Public Schools, Des Moines, Iowa

Douglas Denson, Principal, Twin Bridges Elementary School, Twin Bridges, Montana

Jackie G. Piver, Mathematics Department Chairperson, Enloe High School, Raleigh, North Carolina

Jocelyn Coleman Walton, Supervisor of Mathematics, Plainfield High School, Plainfield, New Jersey

The authors wish to thank Linda Dritsas, Instructional Coordinator of Secondary Mathematics for the Fresno, California, Unified School District, for her contribution to this Teacher's Edition.

ISBN: 0-395-47020-X   BCDEFGHIJ-VH-9987654321

Contents of the Teacher's Edition

Program Highlights

Problem Solving/Decision Making

- *Problem Solving Lessons* throughout help students develop a variety of problem solving strategies. (See page 119.)
- *Decision Making Lessons* help students select appropriate materials and methods for solving problems. (See page 28.)
- *Problem Solving/Application* exercise sets ask students to choose appropriate strategies to solve problems related to real-world situations. (See pages 104 and 426.)

Technology

- *Calculator Usage* is integrated into lesson exposition and exercises wherever it is appropriate. (See pages 108 and 114.)
- *Computer Applications* are included in every chapter. (See pages 174 and 596).
- *Focus on Computers* features introduce students to the use of various kinds of software. (See pages 70 and 220.)

Applications

- *Real-world Applications* lead into the concept being presented in many lessons. (See pages 4 and 151.)
- *Focus on Applications* features introduce such topics as map reading, business uses of graphs, and packaging of consumer products. (See pages 132, 224, and 658.)
- *Chapter Openers* suggest many ways in which mathematics is used in the world today. (See pages 92 and 186.)

Connections

- *Connections* between the various branches of mathematics are emphasized throughout, in lesson introductions and exercises. (See pages 33 and 142.)
- *Data Analysis* lesson introductions and exercises contribute to students' understanding of the importance of data analysis in the real world. (See pages 170 and 172.)
- *Connection/Mixed Review* features connect mathematics with other subject areas—history, science, music, and language arts—and with the contributions of various cultures and ethnic groups to our society. (See pages 160 and 360.)
- *Historical Notes and Career/Application* features emphasize the role of mathematics in society, past and present. (See pages 175 and 178.)

Reasoning

- *Focus on Explorations* lessons use concrete models and manipulatives to help students discover new concepts and develop their reasoning power. (See pages 51 and 300.)
- *Exploration* lesson openers encourage discovery-based learning. (See pages 148 and 340.)
- *Thinking Skills* and *Logical Reasoning* exercises enhance students' ability to think critically and to reason both inductively and deductively. (See page 144.)

Communication

- *Check Your Understanding* and *Guided Practice* questions and the *Exercises* promote communication and understanding of math concepts. (See pages 162–164.)
- *Communication: Reading, Writing, and Discussion* exercises help students to read with understanding and to communicate mathematical ideas orally and in writing, whether working independently or in cooperative groups. (See pages 31 and 120.)
- *Writing Word Problems* and *Writing About Mathematics* exercises help students acquire competence in writing mathematics. (See pages 121 and 159.)
- *Group Activities* help students to work and communicate effectively to solve problems. (See pages 191 and 377.)

Review/Testing/Reteaching

- *Spiral Review* exercises in every lesson. (See page 367.)
- *Self-Tests* at convenient checkpoints in each chapter give students an opportunity to assess their own progress. (See page 203.)
- *Chapter Reviews and Chapter Tests* cover all the essential concepts in that chapter. (See pages 40 and 42.)
- *Cumulative Reviews* in a standardized test format include topics previously presented. (See page 336.)
- *Mixed Reviews* relate concepts previously taught to other subject areas. (See pages 122 and 123.)
- *Extra Practice* and *Toolbox Skills Practice* provide exercises for reteaching and exercises for covering skills presented in earlier courses. (See pages 730 and 753.)

CONTENTS

photo from page 85

v

photo from page 355

photo from page 331

photo from page 278

photo from page 523

ix

photo from page 698

photo from page 466

photo from page 679

xi

Support for all Teaching and Learning Needs

Teacher's Edition

- Professional essays that discuss teaching and learning strategies
- Special pages preceding each chapter include:
  Chapter Planning Material
  Alternate Teaching Approaches
  Support Material Facsimiles
- Student text pages with answers annotated in red
- Side columns with a complete teaching plan for every lesson

Overhead Visuals

Color transparencies, with suggestions and scripts for enhanced presentations:
- Visual presentations of abstract concepts
- Graphing aids
- Geometric models
- Probability and statistics help

Resource Book

- Lesson Starter Masters
- Practice Masters
- Nonroutine Practice
- Enrichment Masters
- Chapter Projects
- Teacher Aids

T12

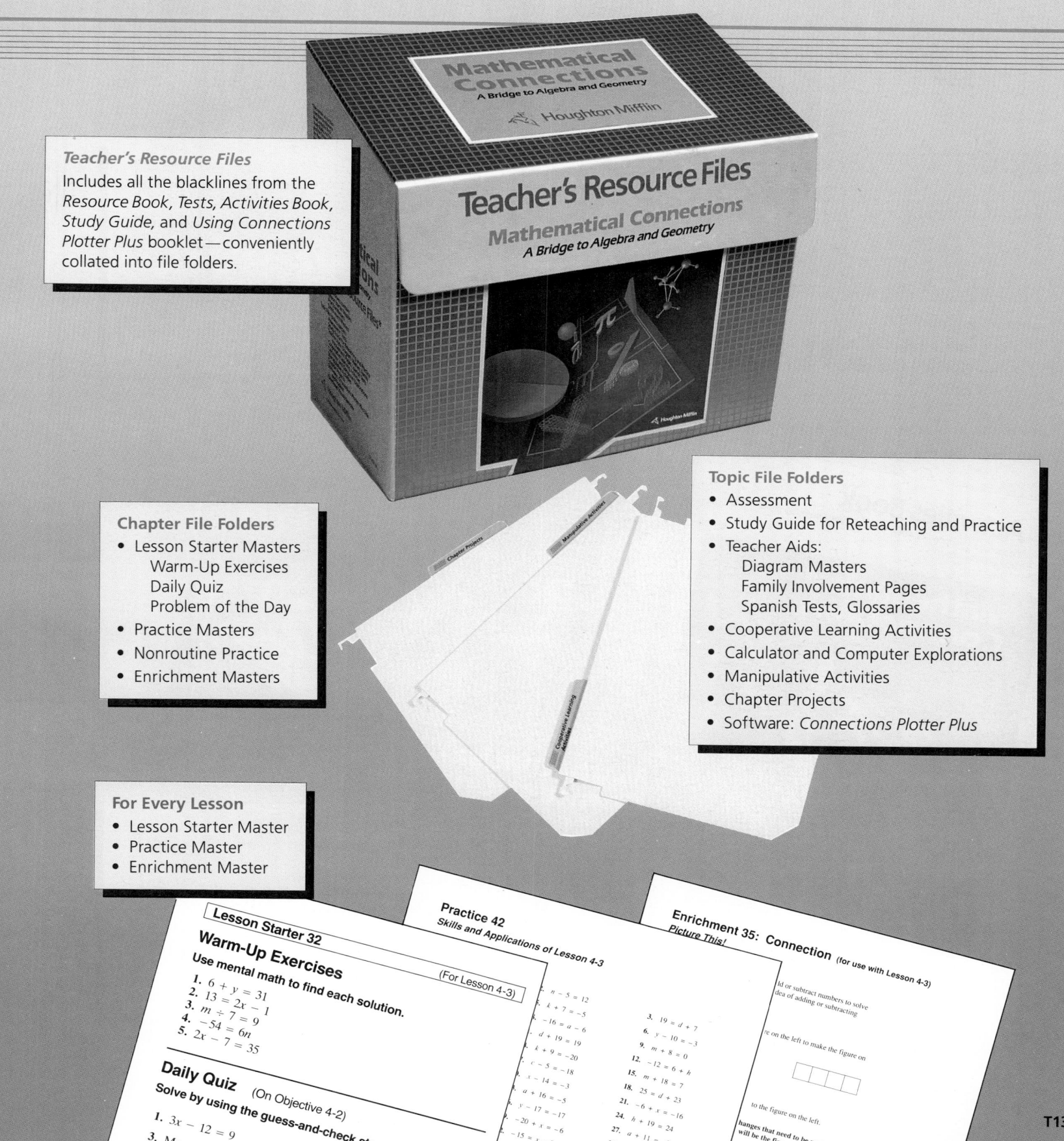

Teacher's Resource Files
Includes all the blacklines from the *Resource Book, Tests, Activities Book, Study Guide,* and *Using Connections Plotter Plus* booklet—conveniently collated into file folders.

Chapter File Folders
- Lesson Starter Masters
  - Warm-Up Exercises
  - Daily Quiz
  - Problem of the Day
- Practice Masters
- Nonroutine Practice
- Enrichment Masters

Topic File Folders
- Assessment
- Study Guide for Reteaching and Practice
- Teacher Aids:
  - Diagram Masters
  - Family Involvement Pages
  - Spanish Tests, Glossaries
- Cooperative Learning Activities
- Calculator and Computer Explorations
- Manipulative Activities
- Chapter Projects
- Software: *Connections Plotter Plus*

For Every Lesson
- Lesson Starter Master
- Practice Master
- Enrichment Master

Lesson Starter 32
Warm-Up Exercises
Use mental math to find each solution.
(For Lesson 4-3)
1. $6 + y = 31$
2. $13 = 2x - 5$
3. $m \div 7 = 9$
4. $-54 = 6n$
5. $2x - 7 = 35$

Daily Quiz   (On Objective 4-2)
Solve by using the guess-and-check strategy.
1. $3x - 12 = 9$
2. $19 = x$
3. Mary Kwon spent $12...
   at $1.50 per...

Practice 42
Skills and Applications of Lesson 4-3

. $n - 5 = 12$
. $k + 7 = -5$
. $-16 = a - 6$
. $d + 19 = 19$
. $k + 9 = -20$
. $c - 5 = -18$
. $x - 14 = -3$
$a + 16 = -5$
. $y - 17 = -17$
. $-20 + x = -6$
. $-15 = x - 8$
...mber x, the result is...

3. $19 = d + 7$
6. $y - 10 = -3$
9. $m + 8 = 0$
12. $-12 = 6 + h$
15. $m + 18 = 7$
18. $25 = d + 23$
21. $-6 + x = -16$
24. $h + 19 = 24$
27. $a + 11 = -7$
30. $m + 4 = 14$
33. $a - 14$

Enrichment 35: Connection (for use with Lesson 4-3)
Picture This!

...ld or subtract numbers to solve
...dea of adding or subtracting

...e on the left to make the figure on

...to the figure on the left.

...hanges that need to be made in
...will be the figure on the right.

Support for Teaching, Testing, Reteaching, and Technology

Activities Book
- Cooperative Learning Activities
- Calculator and Computer Explorations
- Manipulative Activities

Available in blackline master format.

Tests
- Quizzes at each Self-Test checkpoint
- Chapter Tests (in two forms, A and B)
- Cumulative Tests

Available in blackline master format.

A separate Answer Key is provided.

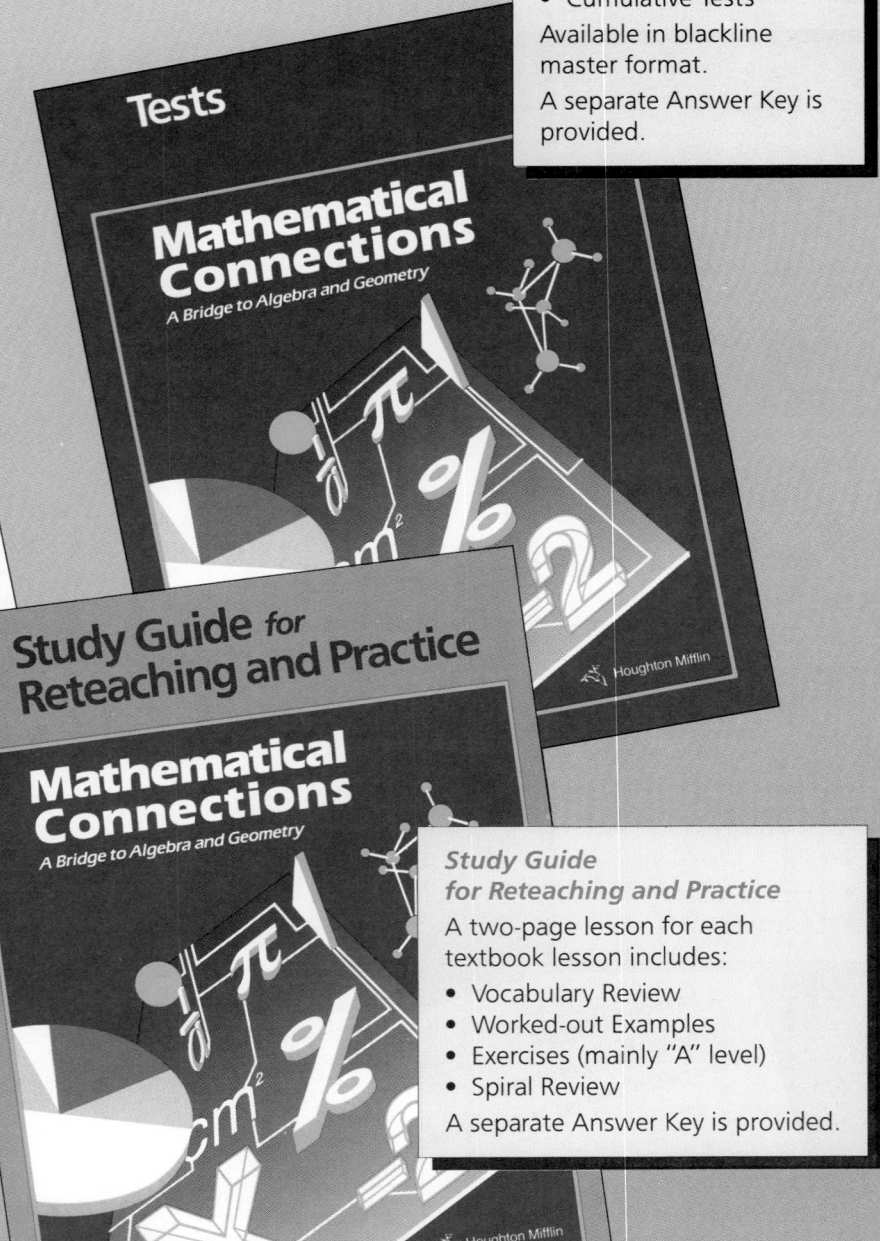

Study Guide for Reteaching and Practice

A two-page lesson for each textbook lesson includes:
- Vocabulary Review
- Worked-out Examples
- Exercises (mainly "A" level)
- Spiral Review

A separate Answer Key is provided.

A *Solution Key* with step-by-step solutions, including diagrams, for all written exercises in the student book is also available.

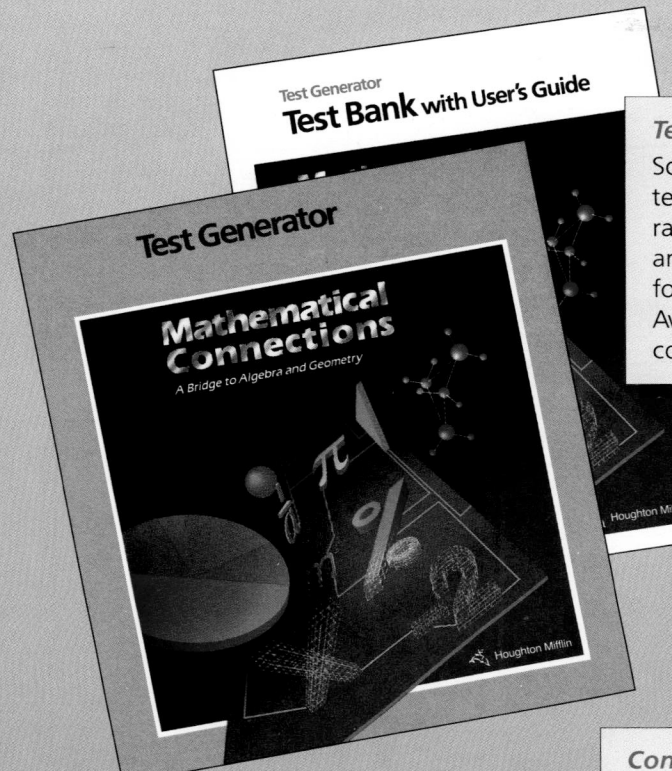

Test Generator
Test Bank with User's Guide

Mathematical Connections
A Bridge to Algebra and Geometry

Houghton Mifflin

Test Generator

Software for generating your own tests, either manually or by random selection, with diagrams and options for customizing the format and level of difficulty. Available for Macintosh and IBM computers.

Using Connections Plotter Plus

Mathematical Connections
A Bridge to Algebra and Geometry

Houghton Mifflin

Connections Plotter Plus

Software for:
- Drawing and measuring geometric figures
- Graphing equations
- Displaying data

Available for Apple II, Macintosh, and IBM computers.

Texas Instruments Calculators

All the support you need for teaching problem solving in *Mathematical Connections* with these TI calculators available through Houghton Mifflin:

- TI-Math Explorer, with fraction capability
- TI-Challenger, a basic scientific calculator
- TI-34 Calculator, a full-range scientific calculator with fraction capability
- TI-81 Graphics Calculator, for graphing equations

Special Planning Pages for Every Chapter

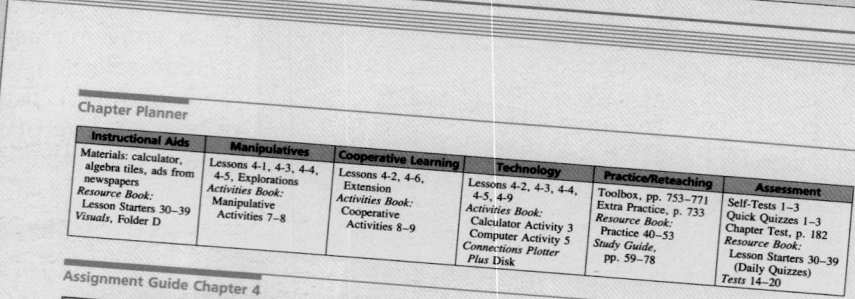

Chapter Planner

Instructional Aids	Manipulatives	Cooperative Learning	Technology	Practice/Reteaching	Assessment
Materials: calculator, algebra tiles, ads from newspapers *Resource Book:* Lesson Starters 30–39 *Visuals,* Folder D	Lessons 4-1, 4-3, 4-4, 4-5, Explorations *Activities Book:* Manipulative Activities 7–8	Lessons 4-2, 4-6, Extension *Activities Book:* Cooperative Activities 8–9	Lessons 4-2, 4-3, 4-4, 4-5, 4-9 *Activities Book:* Calculator Activity 3 Computer Activity 5 *Connections Plotter Plus* Disk	Toolbox, pp. 753–771 Extra Practice, p. 733 *Resource Book:* Practice 40–53 *Study Guide,* pp. 59–78	Self-Tests 1–3 Quick Quizzes 1–3 Chapter Test, p. 182 *Resource Book:* Lesson Starters 30–39 (Daily Quizzes) *Tests* 14–20

Assignment Guide Chapter 4

Day	Skills Course	Average Course	Advanced Course
1	**4-1:** 1–20, 21–37 odd, 43–46	**4-1:** 1–20, 21–41 odd, 43–46	**4-1:** 1–19 odd, 21–46
2	**4-2:** 1–15 odd, 16–19	**4-2:** 1–19	**4-2:** 1–19
3	**4-3:** 1–21, 23, 24, 32–35	**4-3:** 1–26, 32–35	**4-3:** 1–13 odd, 15–35
4	**4-4:** 1–31 odd, 33–37 **Exploration:** Activities I–III	**4-4:** 1–31 odd, 33–37 **Exploration:** Activities I–III	**4-4:** 1–31 odd, 33–37 **Exploration:** Activities I–III
5	**4-5:** 1–17, 19–23 odd, 30–33	**4-5:** 1–17, 19–27 odd, 30–33	**4-5:** 1–17 odd, 18–33
	Mixed Review **4-6:** 1–18, 21–24	**Mixed Review** **4-6:** 1–13 odd, 15–24	
... odd, 21–25	**4-7:** 1–10, 11–19 odd, 21–25	**Mixed Review** **4-6:** 1–13 odd, 15–24	
		4-7: 1–7 odd, 9–25, Challenge	
	4-8: 1–14, 16–21	**4-8:** 1–21	
...d, 27–32	**4-9:** 1–16, 17–21 odd, 23–25, 27–32	**4-9:** 1–15 odd, 17–32, Historical Note	
...3–30	**4-10:** 1–15, 17, 18, 23–30	**4-10:** 1–7 odd, 8–30	
...st:	*Prepare For Chapter Test:* Chapter Review	*Prepare for Chapter Test:* Chapter Review	
...t	*Administer Chapter 4 Test;* Cumulative Review	*Administer Chapter 4 Test; Chapter Extension;* Cumulative Review	

140B

Teacher's Resources

Resource Book
Chapter 4 Project
Lesson Starters 30–39
Practice 40–53
Enrichment 33–43
Diagram Masters
Chapter 4 Objectives
Family Involvement 4
Spanish Test 4
Spanish Glossary

Activities Book
Cooperative Activities 8–9
Manipulative Activities 7–8
Calculator Activity 3
Computer Activity 5

Study Guide, pp. 59–78

Tests
Tests 14–20

Visuals
Folder D

Connections Plotter Plus Disk

Planning Chapter 4

Chapter Overview

Chapter 4 connects and extends the skills learned in the first three chapters by having students learn to solve one- and two-step equations. The chapter gives students an opportunity to sharpen their skills in working with integers, to apply the distributive property, and to simplify expressions. It also serves as an introduction to solving equations, a skill used extensively in algebra and in more advanced mathematics courses.

Chapter 4 begins by introducing students to equations that can be solved by substitution and mental math. *Lesson 2* continues the development of problem-solving skills by introducing the guess-and-check method. *Lessons 3 and 4* focus on solving one-step equations using inverse operations. A manipulative approach to modeling and solving equations by using algebra tiles is presented in the *Focus on Explorations.* This approach is used to demonstrate how to solve the two-step equations in *Lesson 5.* Writing variable expressions and equations from word phrases and sentences is the topic of *Lessons 6 and 7.* This helps ease the way into *Lesson 8* on using equations to solve word problems. *Lesson 9* extends the concepts from previous lessons by presenting equations that must be solved by combining like terms or by using the distributive property. Formulas, a logical extension of solving equations and a connecting link to many areas such as medicine, science, and business, are introduced in *Lesson 10.* The *Chapter Extension* presents equations with variables on both sides of the equation.

Background

Chapter 4 presents students with many opportunities to use previously learned skills. These include substituting a number for a variable, working with integers, combining like terms, using the distributive property, and using estimation and mental math skills. Writing has an important role in the chapter as students are asked to write variable expressions and equations for word phrases and sentences.

The use of calculators in doing mathematics is further developed in this chapter. Students are shown how to use a calculator to check solutions to equations and also how to use a calculator to help students solve an equation. Spreadsheets are used to help students solve certain equations that they cannot yet solve by using paper and pencil alone.

Manipulatives are used to aid students' understanding of the processes involved in solving equations. The first few lessons use the balance-scale model, an effective classroom demonstration device. The *Focus on Explorations* shows students how to model and solve equations by using algebra tiles, and this technique is then used to demonstrate solving two-step equations in the lesson that follows.

Problem-solving concepts and skills are developed in two lessons in Chapter 4. *Lesson 2* presents the strategy of guess-and-check. This strategy is very useful in the real world because many problems can be solved by using it. *Lesson 8* presents the strategy of using an equation, a strategy that students will need to employ throughout their study of mathematics. Since reading a word problem and writing an equation for it tends to be difficult for many students, the two lessons preceding *Lesson 8* provide instruction on how to write variable expressions and equations for word phrases and sentences.

Objectives

4-1	To find solutions of equations by substitution and mental math.
4-2	To solve problems by guessing and checking.
4-3	To solve equations using addition or subtraction.
4-4	To solve equations using multiplication or division.
4-4	FOCUS ON EXPLORATIONS To use algebra tiles to model equations.
4-5	To solve equations using two steps.
4-6	To write variable expressions for word phrases.
4-7	To write equations for sentences.
4-8	To solve problems by using equations.
4-9	To solve equations by simplifying expressions involving combining like terms and the distributive property.
4-10	To work with formulas.
	CHAPTER EXTENSION To solve equations having variables on both sides.

140A

Planning Information

- Chapter Overview
- Chapter Background
- Objectives
- Chapter Planner Chart
- Assignment Guide
- Teacher's Resources Chart

Using Technology

CALCULATORS

This chapter is an introduction to linear equations. Calculators can be used by students to check guesses and solutions found algebraically. After students are adept at solving linear equations with positive integral solutions, it is useful to give them some problems that do not work out so nicely. Students will appreciate fully the power of the algebraic methods if they see that these methods work regardless of the complexity of the numbers involved.

COMPUTERS

Computers can be used in the same way as calculators to help students solve equations containing "ugly" numbers. A spreadsheet program can be used to provide insight that is quite different from the balance model of equation-solving presented in the text. Adventurous students can learn to solve linear equations using function-graphing software, although they will not yet appreciate fully the connections between linear graphs and linear equations.

Lesson 4-2

CALCULATORS OR COMPUTERS

The guess-and-check strategy can be made easier and can be seen in a different light if a calculator or computer is used to perform the computations involved. Punching keys on a calculator, or typing formulas into a spreadsheet or a BASIC program, will reinforce understanding of the operations involved in a solution process. The most helpful way to use a calculator or a computer for the guess-and-check method is to produce a table of calculations (checks) for a range of guesses. To solve Guided Practice Exercises 1 through 4, for example, you (or a student) might write the following BASIC program:

```
10 FOR T = 1 TO 10
20 FOR C = 1 TO 20
30 PRINT T, "TRUCKS: ";C, "CARS: "
40 PRINT "AMOUNT EARNED: ";15 * T + 7 * C
50 NEXT C
60 NEXT T
```

The same kind of table can be produced, with a bit more work, by using a spreadsheet program or a calculator.

140C

Lessons 4-4, 4-5

COMPUTERS

The simplest use of computers in this and the following lessons is to have students write a program in BASIC or a formula for a spreadsheet that solves equations of the form $ax + b = c$, perhaps with special cases at first, such as $a = 1$, or $b = 0$. For example, the line 100 LET X = (C − B)/A might appear in a BASIC program of this type. This process is empowering because the student is in effect teaching the computer how to solve the equation. Writing a program or formula requires understanding at a deeper level than simply carrying out the solution algorithm itself.

A spreadsheet program can be used to give students an alternative way of looking at the algebraic process of solving linear equations. Essentially, the spreadsheet can perform each operation, one step at a time. Using the spreadsheet, students can see quickly that the two-step algorithm on page 157 involves inverting the operations contained in the equation. The order in which the two steps are carried out, which might otherwise seem arbitrary, will be clear immediately if students try the wrong order.

A graphing calculator or a computer program that graphs functions can be used to solve linear equations. For example, to solve $5x + 3 = 2x + 7$, simply graph the two functions $y = 5x + 3$ and $y = 2x + 7$. The x-value of the intersection of the two graphs provides the solution. Although students will be better equipped to understand graphing later in the course, more capable students will find this intriguing.

Using Manipulatives

Lesson 4-1

Algebra tiles can be used to demonstrate how to solve an equation. Example 1 on page 142 could be modeled as follows:

$$x \quad +1 \quad = \quad 6$$

Use inverse operations.

$$x \quad +1-1 \quad = \quad 6-1$$

Notice that 1 was subtracted from both sides of the equation.

Alternate Teaching Approaches

- Using Technology
- Using Manipulatives
- Reteaching/Alternate Approach
- Cooperative Learning

...e pictured here. See the Teacher's Resources chart
...isting of all materials available for this chapter.

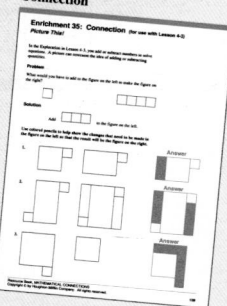

4-1 Enrichment
Data Analysis

4-2 Enrichment
Problem Solving

4-3 Enrichment
Connection

4-4 Enrichment
Communication

4-5 Enrichment
Application

4-6 Enrichment
Exploration

140E

Teacher's Resources

Facsimiles of the Enrichment Masters available for use with each lesson in the chapter are pictured.

Facsimiles of the *Practice Masters* and *Tests* appear in the side columns next to the lesson or test that they accompany.

A Teaching Plan for Every Lesson

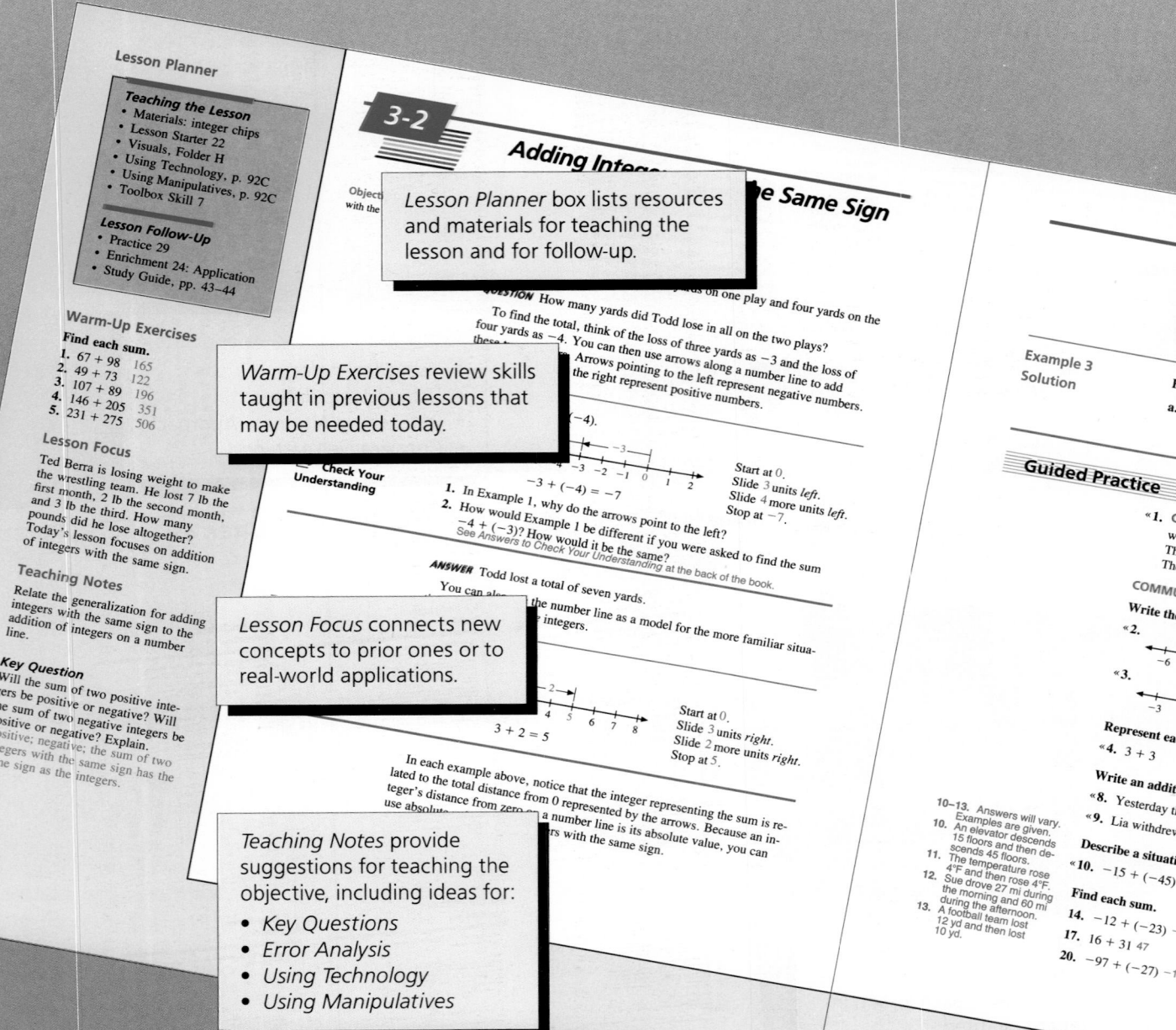

Lesson Planner

Teaching the Lesson
- Materials: integer chips
- Lesson Starter 22
- Visuals, Folder H
- Using Technology, p. 92C
- Using Manipulatives, p. 92C
- Toolbox Skill 7

Lesson Follow-Up
- Practice 29
- Enrichment 24: Application
- Study Guide, pp. 43–44

Warm-Up Exercises
Find each sum.
1. 67 + 98 165
2. 49 + 73 122
3. 107 + 89 196
4. 146 + 205 351
5. 231 + 275 506

Lesson Focus
Ted Berra is losing weight to make the wrestling team. He lost 7 lb the first month, 2 lb the second month, and 3 lb the third. How many pounds did he lose altogether? Today's lesson focuses on addition of integers with the same sign.

Teaching Notes
Relate the generalization for adding integers with the same sign to the addition of integers on a number line.

Key Question
Will the sum of two positive integers be positive or negative? Will the sum of two negative integers be positive or negative? Explain. Positive; negative; the sum of two integers with the same sign has the same sign as the integers.

Lesson Planner box lists resources and materials for teaching the lesson and for follow-up.

Warm-Up Exercises review skills taught in previous lessons that may be needed today.

Lesson Focus connects new concepts to prior ones or to real-world applications.

Teaching Notes provide suggestions for teaching the objective, including ideas for:
- *Key Questions*
- *Error Analysis*
- *Using Technology*
- *Using Manipulatives*

98

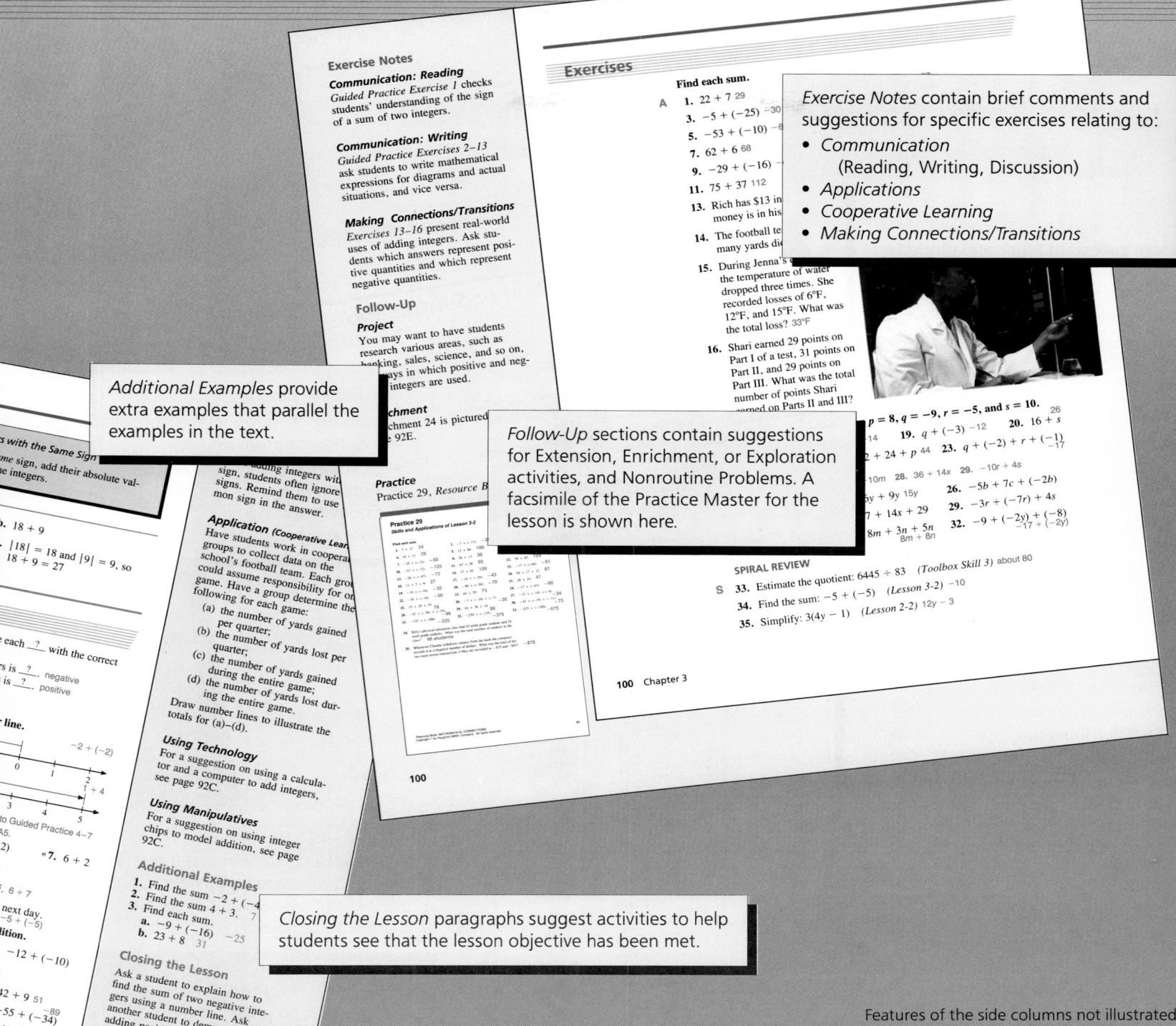

Exercise Notes

Communication: Reading
Guided Practice Exercise 1 checks students' understanding of the sign of a sum of two integers.

Communication: Writing
Guided Practice Exercises 2–13 ask students to write mathematical expressions for diagrams and actual situations, and vice versa.

Making Connections/Transitions
Exercises 13–16 present real-world uses of adding integers. Ask students which answers represent positive quantities and which represent negative quantities.

Follow-Up

Project
You may want to have students research various areas, such as banking, sales, science, and so on, ____ays in which positive and neg- ____ integers are used.

___chment
___chment 24 is pictured ___ 92E.

Practice
Practice 29, *Resource B*___

Practice 29
Skills and Applications of Lesson 3-2

____

___adding integers wit____ sign, students often ignore ____ signs. Remind them to use ___ mon sign in the answer.

Application (Cooperative Lear___
Have students work in coopera___ groups to collect data on the school's football team. Each gro___ could assume responsibility for on___ game. Have a group determine the___ following for each game:
(a) the number of yards gained per quarter;
(b) the number of yards lost per quarter;
(c) the number of yards gained during the entire game;
(d) the number of yards lost during the entire game.
Draw number lines to illustrate the totals for (a)–(d).

Using Technology
For a suggestion on using a calcula- tor and a computer to add integers, see page 92C.

Using Manipulatives
For a suggestion on using integer chips to model addition, see page 92C.

Additional Examples
1. Find the sum −2 + (−4___
2. Find the sum 4 + 3. 7
3. Find each sum.
   a. −9 + (−16)   −25
   b. 23 + 8   31

Closing the Lesson
Ask a student to explain how to find the sum of two negative inte- gers using a number line. Ask another student to demonstrate adding positive integers using a number line.

Suggested Assignments
Skills: 1–16, 17–31 odd, 33–35
Average: 1–16, 17–31 odd, 33–35
Advanced: 1–15 odd, 17–35

(Left-edge partial page)

___rs with the Same Sign___
___ame sign, add their absolute val-___
___ the integers.

   b. 18 + 9
___b. |18| = 18 and |9| = 9, so
      18 + 9 = 27

___ce each __?__ with the correct

___rs is __?__. negative
___rs is __?__. positive

___r line.

                    −2 + (−2)

   0   1   2
            2
            1 + 4
   3   4   5
___ to Guided Practice 4–7
   A5.
   −2)   «7. 6 + 2

___F. 6 + 7
___e next day.
      −5 + (−5)
___dition.
___. −12 + (−10)

   42 + 9 51
   −55 + (−34) ⁻⁸⁹
   −44 + (−19)

___ntegers 99

Exercises

Find each sum.

A 1. 22 + 7 29
   3. −5 + (−25) −30
   5. −53 + (−10) −6___
   7. 62 + 6 68
   9. −29 + (−16) ___
   11. 75 + 37 112
   13. Rich has $13 in ___
        money is in his ___
   14. The football te___
        many yards di___
   15. During Jenna's ___
        the temperature of water dropped three times. She recorded losses of 6°F, 12°F, and 15°F. What was the total loss? 33°F
   16. Shari earned 29 points on Part I of a test, 31 points on Part II, and 29 points on Part III. What was the total number of points Shari ___ed on Parts II and III?

$p = 8, q = −9, r = −5,$ and $s = 10.$
   ___14   19. $q + (−3)$ −12   20. $16 + s$ 26
   ___2 + 24 + p 44   23. $q + (−2) + r + (−1)$ ⁻¹⁷
   ___−10m   28. 36 + 14x   29. −10r + 4s
   ___5y + 9y 15y   26. −5b + 7c + (−2b)
   ___7 + 14x + 29   29. −3r + (−7r) + 4s
   ___8m + 3n + 5n   32. −9 + (−2y) + (−8)
        8m + 8n   ⁻¹⁷ + (−2y)

SPIRAL REVIEW

S 33. Estimate the quotient: $6445 \div 83$ *(Toolbox Skill 3)* about 80
   34. Find the sum: $−5 + (−5)$ *(Lesson 3-2)* −10
   35. Simplify: $3(4y − 1)$ *(Lesson 2-2)* 12y − 3

100   Chapter 3

Exercise Notes contain brief comments and suggestions for specific exercises relating to:
- *Communication*
  (Reading, Writing, Discussion)
- *Applications*
- *Cooperative Learning*
- *Making Connections/Transitions*

Additional Examples provide extra examples that parallel the examples in the text.

Follow-Up sections contain suggestions for Extension, Enrichment, or Exploration activities, and Nonroutine Problems. A facsimile of the Practice Master for the lesson is shown here.

Closing the Lesson paragraphs suggest activities to help students see that the lesson objective has been met.

Suggested Assignments are provided for a skills course, an average course, and an advanced course.

Features of the side columns not illustrated here include: *Calculator* and *Computer* commentary, *Reteaching/Alternate Approach* ideas, *Quick Quizzes*, and *Alternative Assessment* exercises. *Critical Thinking, Life Skills, Estimation* and *Number Sense* ideas are also identified.

A Lesson Format that Gets Students Involved

Objectives and Vocabulary

Objectives are stated at the start of the lesson. Important terms and properties are listed in a *Terms to Know* box and are printed **boldface** in the text for easy reference.

4-3

Using Addition or Subtraction

Objective: To solve equations using addition or subtraction.

 EXPLORATION

Terms to Know
- *solve*
- *inverse operations*

1. What operation could you use to undo the addition on the left pan of this balance scale?

2. What happens to the balance scale if you subtract 17 from the expression on the left pan?

3. What happens to the balance scale if you subtract 17 from both the left and the right pans at the same time?

4. What is in each pan after you simplify the expressions in Step 3?

To **solve** an equation, you find all values of the variable that make the equation true. Recall that each of these values is called a *solution*. When you solve an equation, you get the variable alone on one side of the equals sign. To do this, you need to undo any operations on the variable.

Motivating Lesson Introduction

Various lesson introductions encourage active student involvement:
- *Explorations*
- *Applications*
- *Connections*
- *Data Analysis*

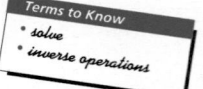

Example 1
Solution

Solve $6 + k = 31$. Check the solution.

$6 + k$ | 31

$6 + k - 6$ | $31 - 6$ ← To undo the addition of 6, subtract 6 from both sides.

k | 25

$6 + k = 31$

$6 + k - 6 = 31 - 6$

Examples and Models

Worked-out *Examples* and concrete models illustrate and reinforce concepts and skills presented in the text.

 **Check Your Understanding**

$6 + 25$ | 31 ← To check the solution, substitute 25 for k in the original equation.

1. In Example 1, what is the purpose of subtracting 6 from the equation?

Check for Understanding

Check Your Understanding exercises help students solidify their understanding of processes and concepts developed in worked-out Examples.

148 Chapter

Helpful Student Annotations

Brief annotations provide explanations for the more difficult steps in a solution.

Just as you used subtraction to undo addition, you can use addition to undo subtraction.

Example 2
Solution

Solve $-29 = s - 15$. Check the solution.

$-29 = s - 15$
$-29 + 15 = s - 15 + 15$
$-14 = s$

The solution is -14.

 **Check Your Understanding**

2. How would Example 2 b
   $s + 15$?

Generalization Boxes
Important properties, concepts, and procedures are highlighted in tinted boxes for easy reference.

Because addition and subtraction undo each other, they are called **inverse operations.** Recognizing inverse operations can help you to solve equations.

Generalization: *Solving Equations Using Addition or Subtraction*

If a number has been *added* to the variable, subtract that number from both sides of the equation.
If a number has been *subtracted* from the variable, add that number to both sides of the equation.

Technology
Calculator usage is integrated into lessons where appropriate.

345

on   off

You can use a calculator to check the solution of an equation. For instance, to check if -508 is the solution of $n + 163 = -345$, substitute -508 for n and use this key sequence.

508. +/- + 163. =

Guided Practice

« **1. COMMUNICATION** « *Reading*   Replace each __?__ with the correct word.

To solve an equation, you undo any operation on the variable by using a(n) __?__ operation. To undo the addition of 8, for example, you would __?__ 8 from both sides of the equation.

Explain how you would solve each equation.

2. $x - 16 = 18$ 3. $y + 6 = 27$
4. $-41 = 14 + w$ 5. $-62 = r$

Guided Practice
Guided Practice Exercises for students to do under the teacher's guidance.

Solve. Check each solution.

6. $w + 8 = 6$ 7. $16 = a -$
9. $12 + t = -6$ 10. $-11 = z$

Exercises

Solve. Check each solution.

1. $a + 2 = 11$ 2. $5 +$
4. $45 = d + 8$ 5. $1 =$
7. $h - 10 = -21$ 8. $-$
10. $-9 = a - 9$ 11. a

13. The sum of 16 and a number

14. When nine is subtracted fro
    the value of c?

CALCULATOR

Solve. Then use a calculator to check the s

15. $c + 463 = 859$ 16.
17. $-1026 = 554 + d$ 18.
19. $562 = 999 + t$ 20.

THINKING SKILLS

Complete.

21. If $|x| = 5$, then $x = $ __?__ or $x = $ __?__
22. If $|x + 2| = 7$, then $x + 2 = $ __?__ or

Analyze your answers to Exercises 21
of each of the following equations. If t

solution.

23. $|x| = 3$ 24. $|m| =$
26. $|c + 4| = 6$ 27. $|h| =$
29. $|p| + 5 = 25$ 30. $|n| -$

Relevant Exercises
- *Opening exercises* relate directly to the Examples in the lesson.
- *Labeled exercises* encourage critical thinking and reasoning, provide for the use of technology, and connect the mathematics to related topics, to other subject areas, and to the real world through applications.
- *Spiral Review* exercises provide a mixed review of previously taught skills and concepts.

A, B, and *C* labels indicating the difficulty level of the *Exercises* and an *S* label denoting *Spiral Review* are included in the Teacher's Edition.

Not shown are these special end-of-lesson features: *Historical Notes* and *Challenge* exercises

SPIRAL REVIEW

32. Find the next three numbers: 4, 8, 12, __?__ , __?__ , __?__   *(Lesson 2*
33. Solve: $m - 19 = 24$ *(Lesson 4-3)*
    (3y)(2x)   *(Lesson 2-1)*

Strategies for Solving Problems and Making Decisions

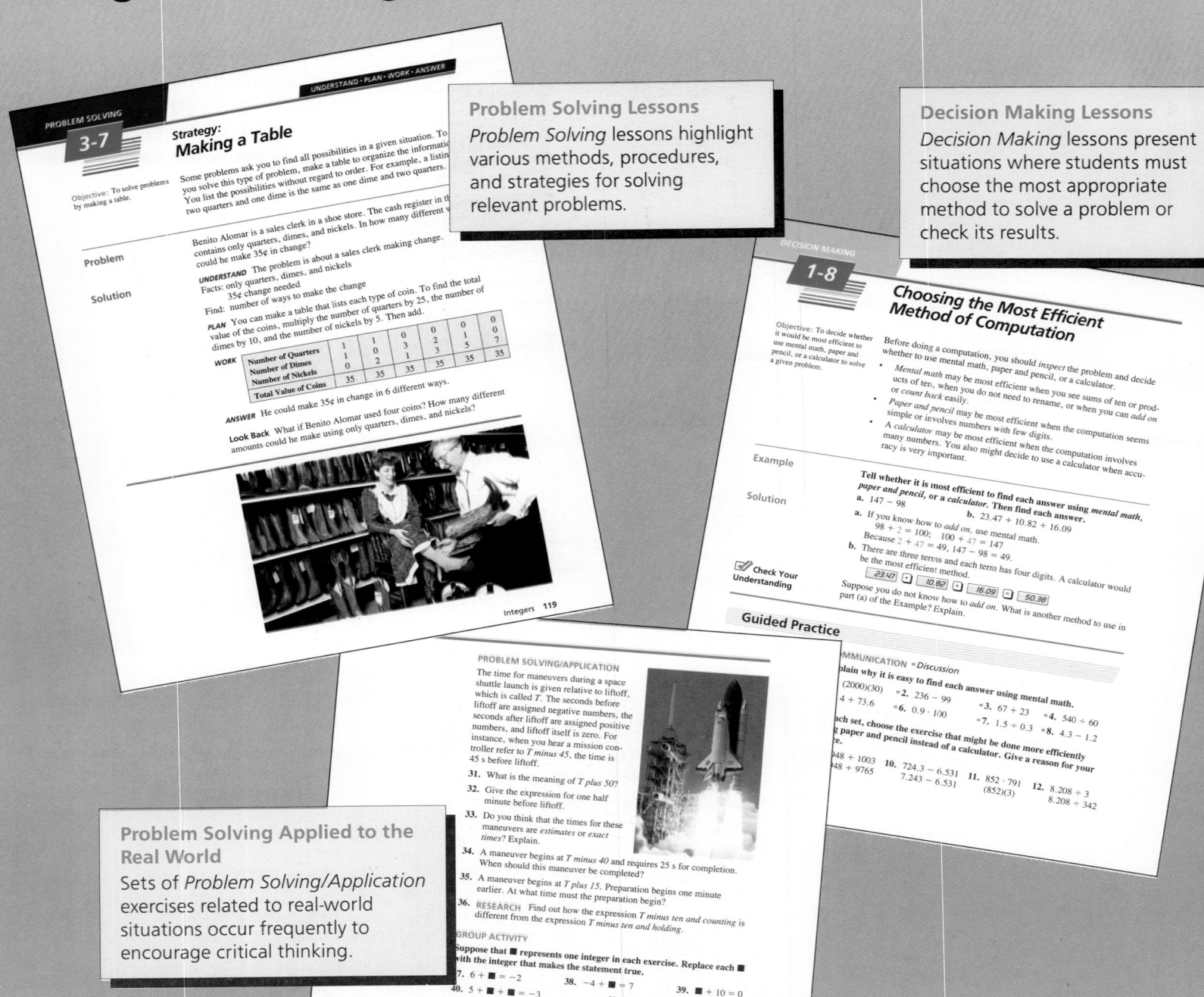

Problem Solving Lessons
Problem Solving lessons highlight various methods, procedures, and strategies for solving relevant problems.

Decision Making Lessons
Decision Making lessons present situations where students must choose the most appropriate method to solve a problem or check its results.

PROBLEM SOLVING
3-7

UNDERSTAND · PLAN · WORK · ANSWER

Strategy:
Making a Table

Objective: To solve problems by making a table.

Some problems ask you to find all possibilities in a given situation. To help you solve this type of problem, make a table to organize the information. You list the possibilities without regard to order. For example, a listing of two quarters and one dime is the same as one dime and two quarters.

Problem Benito Alomar is a sales clerk in a shoe store. The cash register in the store contains only quarters, dimes, and nickels. In how many different ways could he make 35¢ in change?

Solution **UNDERSTAND** The problem is about a sales clerk making change.
Facts: only quarters, dimes, and nickels
35¢ change needed
Find: number of ways to make the change

PLAN You can make a table that lists each type of coin. To find the total value of the coins, multiply the number of quarters by 25, the number of dimes by 10, and the number of nickels by 5. Then add.

WORK

Number of Quarters	1	1	0	0	0	0
Number of Dimes	1	0	3	2	1	0
Number of Nickels	0	2	1	3	5	7
Total Value of Coins	35	35	35	35	35	35

ANSWER He could make 35¢ in change in 6 different ways.

Look Back What if Benito Alomar used four coins? How many different amounts could he make using only quarters, dimes, and nickels?

Integers **119**

DECISION MAKING
1-8

Choosing the Most Efficient Method of Computation

Objective: To decide whether it would be most efficient to use mental math, paper and pencil, or a calculator to solve a given problem.

Before doing a computation, you should *inspect* the problem and decide whether to use mental math, paper and pencil, or a calculator.

- *Mental math* may be most efficient when you see sums of ten or products of ten, when you do not need to rename, or when you can *add on* or *count back* easily.
- *Paper and pencil* may be most efficient when the computation seems simple or involves numbers with few digits.
- A *calculator* may be most efficient when the computation involves many numbers. You also might decide to use a calculator when accuracy is very important.

Example Tell whether it is most efficient to find each answer using *mental math, paper and pencil,* or a *calculator.* Then find each answer.
a. 147 − 98   b. 23.47 + 10.82 + 16.09

Solution a. If you know how to *add on,* use mental math.
98 + 2 = 100; 100 + 47 = 147
Because 2 + 47 = 49, 147 − 98 = 49.
b. There are three terms and each term has four digits. A calculator would be the most efficient method.

[23.47] [+] [10.82] [+] [16.09] [=] [50.38]

☑ **Check Your Understanding**
Suppose you do not know how to *add on.* What is another method to use in part (a) of the Example? Explain.

Guided Practice

COMMUNICATION «*Discussion*
Explain why it is easy to find each answer using mental math.
1. (2000)(30)   «2. 236 − 99   «3. 67 + 23   «4. 540 ÷ 60
5. 4 + 73.6   «6. 0.9 · 100   «7. 1.5 ÷ 0.3   «8. 4.3 − 1.2

each set, choose the exercise that might be done more efficiently paper and pencil instead of a calculator. Give a reason for your
948 + 1003   10. 724.3 − 6.531   11. 852 · 791   12. 8.208 ÷ 3
948 + 9765   7.243 − 6.531   (852)(3)   8.208 ÷ 342

Problem Solving Applied to the Real World
Sets of *Problem Solving/Application* exercises related to real-world situations occur frequently to encourage critical thinking.

PROBLEM SOLVING/APPLICATION
The time for maneuvers during a space shuttle launch is given relative to liftoff, which is called *T.* The seconds before liftoff are assigned negative numbers, the seconds after liftoff are assigned positive numbers, and liftoff itself is zero. For instance, when you hear a mission controller refer to *T minus 45,* the time is 45 s before liftoff.

31. What is the meaning of *T plus 50?*
32. Give the expression for one half minute before liftoff.
33. Do you think that the times for these maneuvers are *estimates* or *exact times*? Explain.
34. A maneuver begins at *T minus 40* and requires 25 s for completion. When should this maneuver be completed?
35. A maneuver begins at *T plus 15.* Preparation begins one minute earlier. At what time must the preparation begin?
36. RESEARCH Find out how the expression *T minus ten and counting* is different from the expression *T minus ten and holding.*

GROUP ACTIVITY
Suppose that ■ represents one integer in each exercise. Replace each ■ with the integer that makes the statement true.
37. 6 + ■ = −2   38. −4 + ■ = 7   39. ■ + 10 = 0
40. 5 + ■ + ■ = −3   41. −8 + ■ + ■ = −12
42. ■ + (−1,000,000) = 1   43. ■ + 1,000,000,000 = −1

Suppose that ■ and ▲ represent different integers.
44. List four pairs of integers for which ■ + ▲ = −1 is true. Compare your list with other lists in the group. Are all lists the same?
45. For how many different pairs of integers is ■ + ▲ = −1 a true statement? Give a convincing argument to support your answer.

SPIRAL REVIEW
46. Find the sum mentally: 27 + 21 + 43 (Lesson 1-6)
47. Continue the pattern: 1, 5, 9, 13, _?_, _?_, _?_ (Lesson 2-5)
48. Write 2,700,000 ... (Lesson 2-5?)

T22

Tools for Discovery and Problem Solving

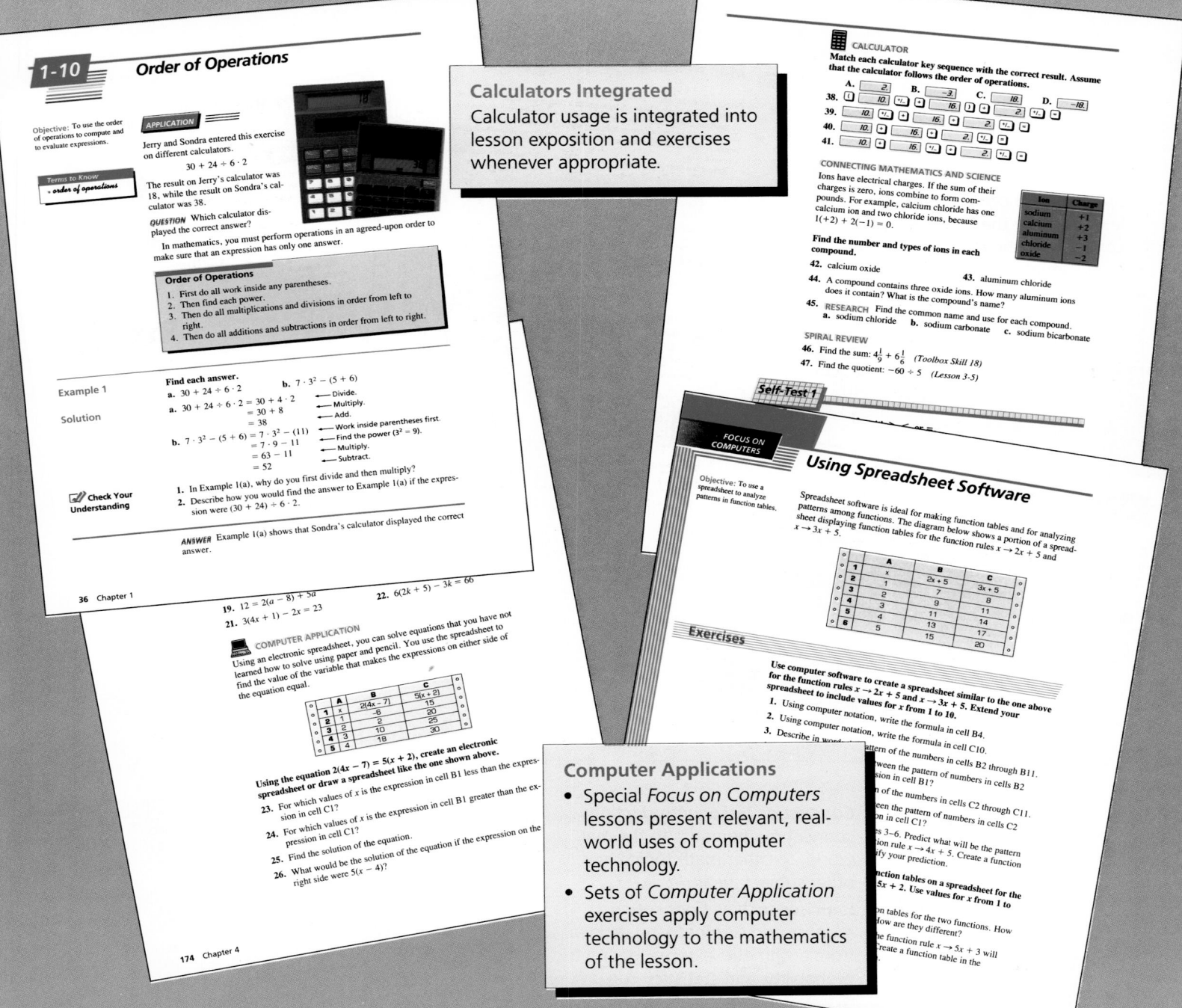

Calculators Integrated
Calculator usage is integrated into lesson exposition and exercises whenever appropriate.

Computer Applications
- Special *Focus on Computers* lessons present relevant, real-world uses of computer technology.
- Sets of *Computer Application* exercises apply computer technology to the mathematics of the lesson.

Real-World Applications and Data Analysis

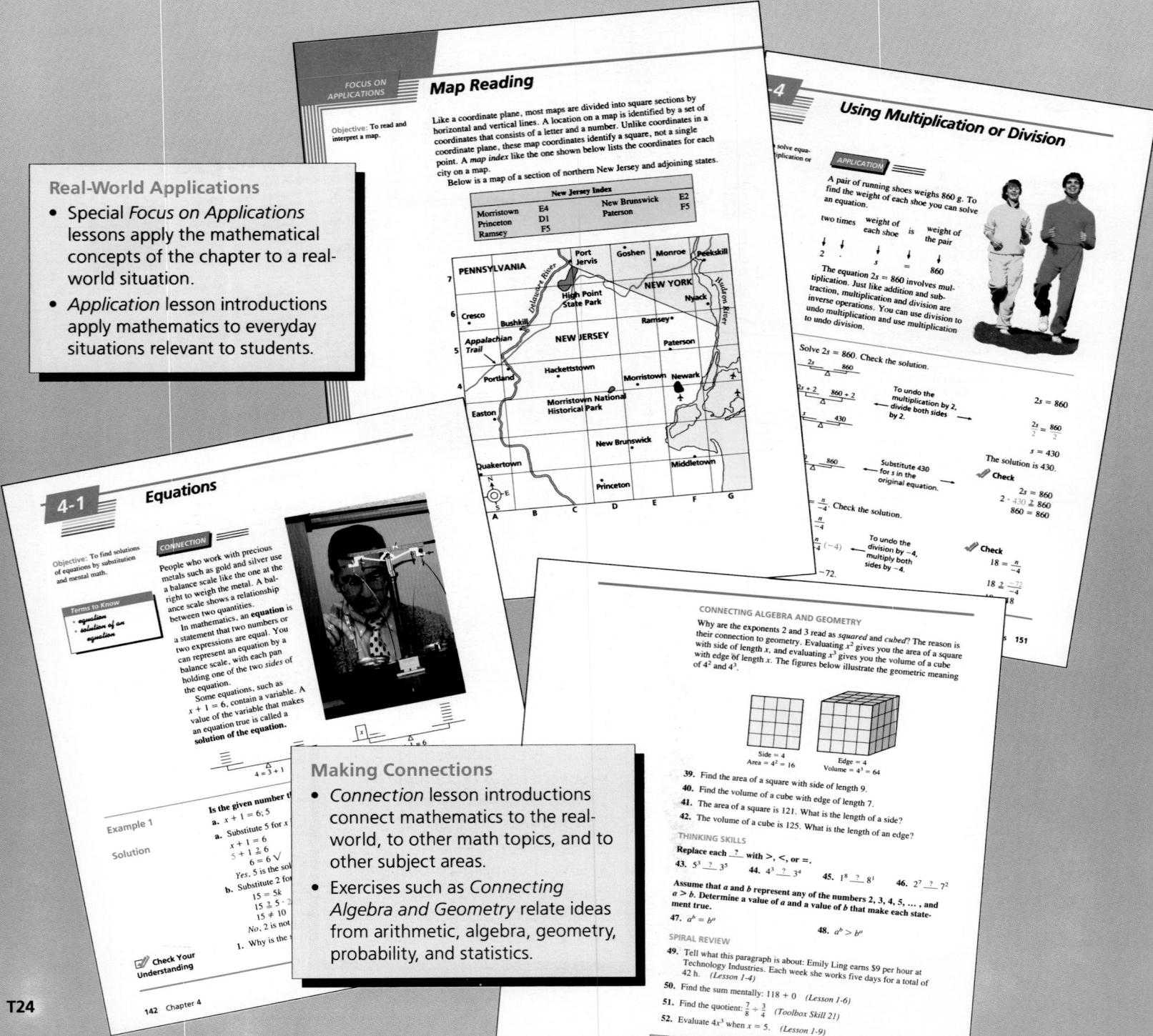

Real-World Applications
- Special *Focus on Applications* lessons apply the mathematical concepts of the chapter to a real-world situation.
- *Application* lesson introductions apply mathematics to everyday situations relevant to students.

Making Connections
- *Connection* lesson introductions connect mathematics to the real-world, to other math topics, and to other subject areas.
- Exercises such as *Connecting Algebra and Geometry* relate ideas from arithmetic, algebra, geometry, probability, and statistics.

Map Reading

Like a coordinate plane, most maps are divided into square sections by horizontal and vertical lines. A location on a map is identified by a set of coordinates that consists of a letter and a number. Unlike coordinates in a coordinate plane, these map coordinates identify a square, not a single point. A *map index* like the one shown below lists the coordinates for each city on a map.

Below is a map of a section of northern New Jersey and adjoining states.

New Jersey Index

Morristown	E4	New Brunswick	E2
Princeton	D1	Paterson	F5
Ramsey	F5		

Using Multiplication or Division

APPLICATION

A pair of running shoes weighs 860 g. To find the weight of each shoe you can solve an equation.

two times weight of is weight of
each shoe the pair
2 · s = 860

The equation $2s = 860$ involves multiplication. Just like addition and subtraction, multiplication and division are inverse operations. You can use division to undo multiplication and use multiplication to undo division.

Solve $2s = 860$. Check the solution.

$\frac{2s}{2} = \frac{860}{2}$ To undo the multiplication by 2, divide both sides by 2. $2s = 860$

$\frac{2s}{2} = \frac{860}{2}$

$s = 430$

The solution is 430.

Check

$2s = 860$
$2 \cdot 430 \stackrel{?}{=} 860$
$860 = 860$ ✓

$\frac{n}{-4}$. Check the solution.

To undo the division by -4, multiply both sides by -4.

Check

$18 = \frac{n}{-4}$

$18 \stackrel{?}{=} \frac{-72}{-4}$

$18 = 18$ ✓

151

4-1 Equations

CONNECTION

Objective: To find solutions of equations by substitution and mental math.

Terms to Know
- equation
- solution of an equation

People who work with precious metals such as gold and silver use a balance scale like the one at the right to weigh the metal. A balance scale shows a relationship between two quantities.

In mathematics, an **equation** is a statement that two numbers or two expressions are equal. You can represent an equation by a balance scale, with each pan holding one of the two *sides* of the equation.

Some equations, such as $x + 1 = 6$, contain a variable. A value of the variable that makes an equation true is called a **solution of the equation.**

$4 = 3 + 1$

Example 1 Is the given number t...
  a. $x + 1 = 6$; 5
  a. Substitute 5 for x
  $x + 1 = 6$
  $5 + 1 \stackrel{?}{=} 6$
  $6 = 6$ ✓
  Yes, 5 is the sol...

Solution
  b. Substitute 2 for...
  $15 = 5k$
  $15 \stackrel{?}{=} 5 \cdot 2$
  $15 \neq 10$
  No, 2 is not...

  1. Why is the s...

☑ **Check Your Understanding**

CONNECTING ALGEBRA AND GEOMETRY

Why are the exponents 2 and 3 read as *squared* and *cubed*? The reason is their connection to geometry. Evaluating x^2 gives you the area of a square with side of length x, and evaluating x^3 gives you the volume of a cube with edge of length x. The figures below illustrate the geometric meaning of 4^2 and 4^3.

Side = 4
Area = $4^2 = 16$

Edge = 4
Volume = $4^3 = 64$

39. Find the area of a square with side of length 9.
40. Find the volume of a cube with edge of length 7.
41. The area of a square is 121. What is the length of a side?
42. The volume of a cube is 125. What is the length of an edge?

THINKING SKILLS

Replace each ? with $>$, $<$, or $=$.
43. 5^3 _?_ 3^5 **44.** 4^3 _?_ 3^4 **45.** 1^8 _?_ 8^1 **46.** 2^7 _?_ 7^2

Assume that a and b represent any of the numbers 2, 3, 4, 5, … , and $a > b$. Determine a value of a and a value of b that make each statement true.
47. $a^b = b^a$
48. $a^b > b^a$

SPIRAL REVIEW

49. Tell what this paragraph is about: Emily Ling earns $9 per hour at Technology Industries. Each week she works five days for a total of 42 h. *(Lesson 1-4)*
50. Find the sum mentally: $118 + 0$ *(Lesson 1-6)*
51. Find the quotient: $\frac{7}{8} \div \frac{3}{4}$ *(Toolbox Skill 21)*
52. Evaluate $4x^3$ when $x = 5$. *(Lesson 1-9)*

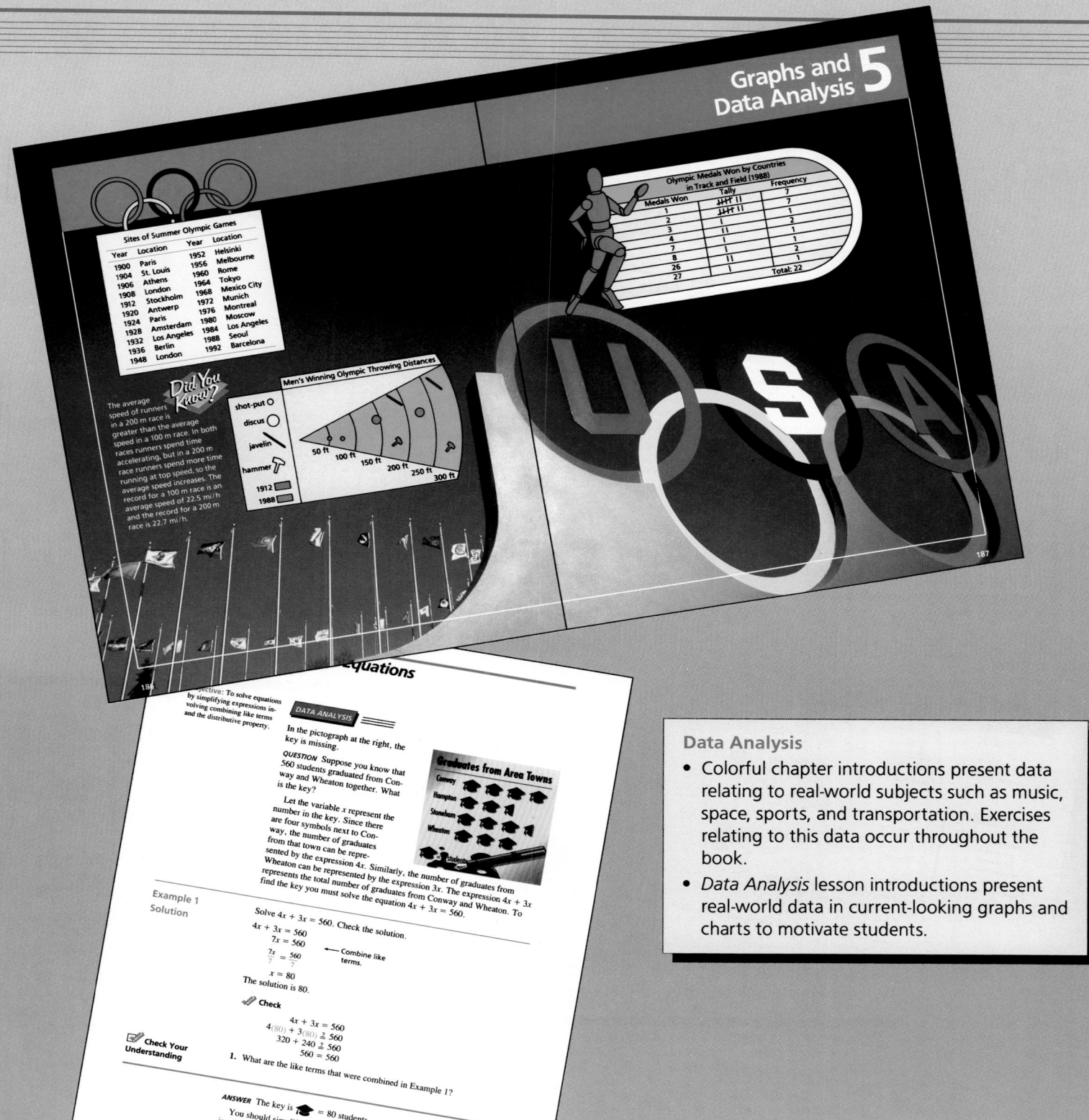

Graphs and Data Analysis 5

Sites of Summer Olympic Games

Year	Location	Year	Location
1900	Paris	1952	Helsinki
1904	St. Louis	1956	Melbourne
1906	Athens	1960	Rome
1908	London	1964	Tokyo
1912	Stockholm	1968	Mexico City
1920	Antwerp	1972	Munich
1924	Paris	1976	Montreal
1928	Amsterdam	1980	Moscow
1932	Los Angeles	1984	Los Angeles
1936	Berlin	1988	Seoul
1948	London	1992	Barcelona

Did You Know?

The average speed of runners in a 200 m race is greater than the average speed in a 100 m race. In both races runners spend time accelerating, but in a 200 m race runners spend more time running at top speed, so the average speed increases. The record for a 100 m race is an average speed of 22.5 mi/h and the record for a 200 m race is 22.7 mi/h.

Men's Winning Olympic Throwing Distances

shot-put
discus
javelin
hammer

50 ft 100 ft 150 ft 200 ft 250 ft 300 ft

1912
1988

Olympic Medals Won by Countries in Track and Field (1988)

Medals Won	Tally	Frequency
1	̄HT II	7
2	̄HT II	7
3	II	2
4	I	1
7	I	1
8	II	2
26	I	1
27		1
	Total: 22	

186 / 187

Equations

Objective: To solve equations by simplifying expressions involving combining like terms and the distributive property.

DATA ANALYSIS

In the pictograph at the right, the key is missing.

QUESTION Suppose you know that 560 students graduated from Conway and Wheaton together. What is the key?

Let the variable x represent the number in the key. Since there are four symbols next to Conway, the number of graduates from that town can be represented by the expression $4x$. Similarly, the number of graduates from Wheaton can be represented by the expression $3x$. The expression $4x + 3x$ represents the total number of graduates from Conway and Wheaton. To find the key you must solve the equation $4x + 3x = 560$.

Graduates from Area Towns

Conway
Hampton
Stoneham
Wheaton

Students

Example 1
Solution

Solve $4x + 3x = 560$. Check the solution.

$$4x + 3x = 560$$
$$7x = 560 \quad \leftarrow \text{Combine like terms.}$$
$$\frac{7x}{7} = \frac{560}{7}$$
$$x = 80$$

The solution is 80.

Check

$$4x + 3x = 560$$
$$4(80) + 3(80) \stackrel{?}{=} 560$$
$$320 + 240 \stackrel{?}{=} 560$$
$$560 = 560$$

Check Your Understanding

1. What are the like terms that were combined in Example 1?

ANSWER The key is 🎓 = 80 students.

You should simplify the sides of an equation before solving. This may involve combining like terms, as in Example 1. It may also involve using the distributive property.

172 Chapter 4

Data Analysis

- Colorful chapter introductions present data relating to real-world subjects such as music, space, sports, and transportation. Exercises relating to this data occur throughout the book.
- *Data Analysis* lesson introductions present real-world data in current-looking graphs and charts to motivate students.

Activities to Promote Critical Thinking

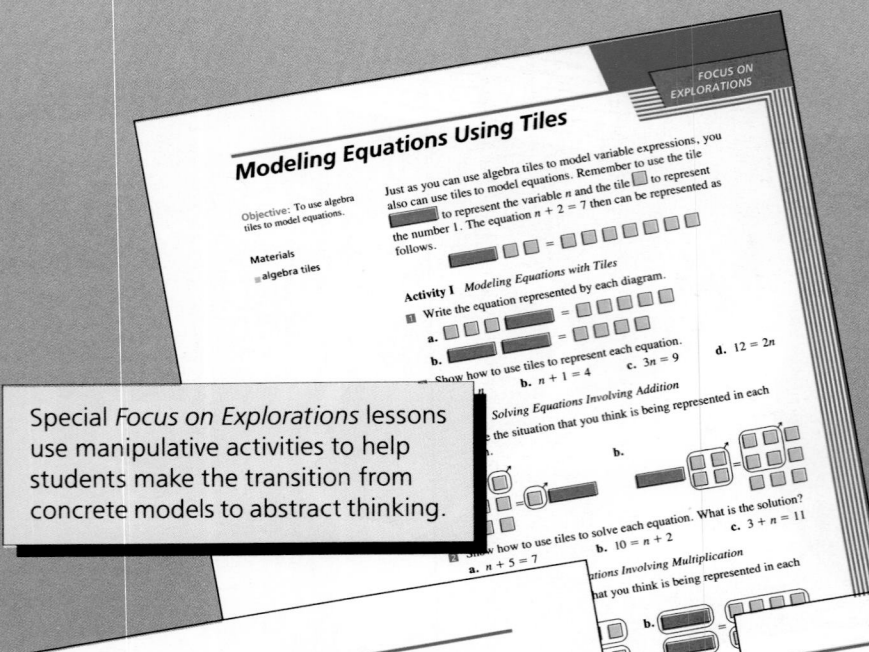

Special *Focus on Explorations* lessons use manipulative activities to help students make the transition from concrete models to abstract thinking.

Logical Reasoning exercises encourage inductive and deductive thinking to solve problems.

Thinking Skills exercises help develop students' higher order thinking processes.

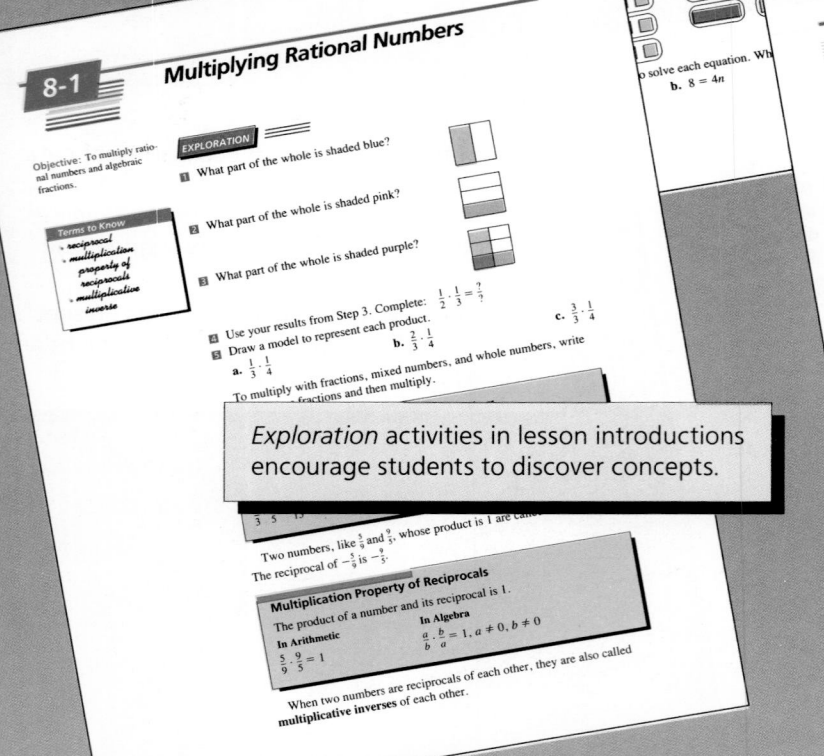

Exploration activities in lesson introductions encourage students to discover concepts.

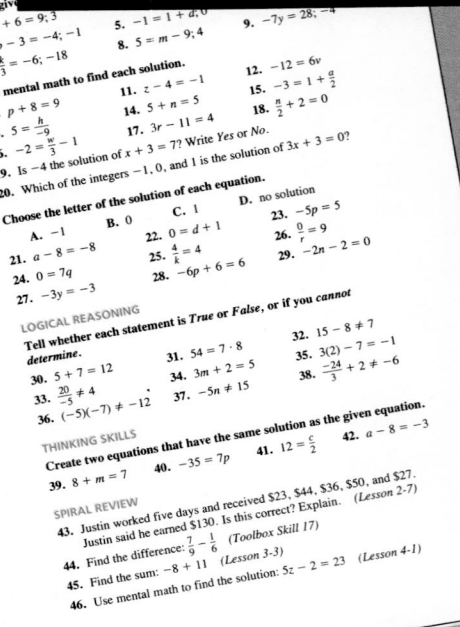

Opportunities to Read, Write, and Discuss Math

Communication: Reading, Writing, and *Discussion* exercises help students read and comprehend lesson concepts and encourage the verbalization of mathematical ideas.

... expressions that contain exponents.

... 4.   b. $(3n)^2$

Solution   a. $3n^2 = 3$ ...
b. $(3n)^2 = (3n)(3n) = (3 \cdot 4)(\ldots) = (12)(12) = 144$

Some calculators have a $\boxed{y^x}$ key. If you have this key on your calculator, you may wish to use it when you work with exponents. For example, you can use the following key sequence to find 8^6.
$\boxed{8}$ $\boxed{y^x}$ $\boxed{6}$ $\boxed{=}$

Guided Practice
COMMUNICATION « *Reading*

Each word is followed by four correct meanings. Choose the meaning that is used in this lesson.

«1. base
   a. the lowest point
   b. a supporting layer
   c. a number that is raised to a power
   d. a corner of the infield in softball

«2. power
   a. strength
   b. authority
   c. electricity
   d. exponent

COMMUNICATION « *Writing*

Write an expression for each phrase.

«3. three to the fifth power
«4. a number x to the sixth power
«5. a number x cubed multiplied by six
«6. six times a number x, cubed

Write a phrase for each expression.
«7. 8^3 «8. a^4 «9. $3x^5$ «10. $(3x)^5$

Give the exponential form of each expression.
11. $(7)(7)(7)(7)(7)$
12. $x \cdot x \cdot x \cdot x$
13. $5 \cdot d \cdot d \cdot d$
14. $4y \cdot 4y \cdot 4y$

Give the multiplication that each expression represents.
15. 6^7 16. c^5 17. $5x^4$ 18. $(2c\ldots$

Connecting Arithmetic and Alg...

Group Activities help students to work and communicate effectively in cooperative groups to better understand concepts and solve problems.

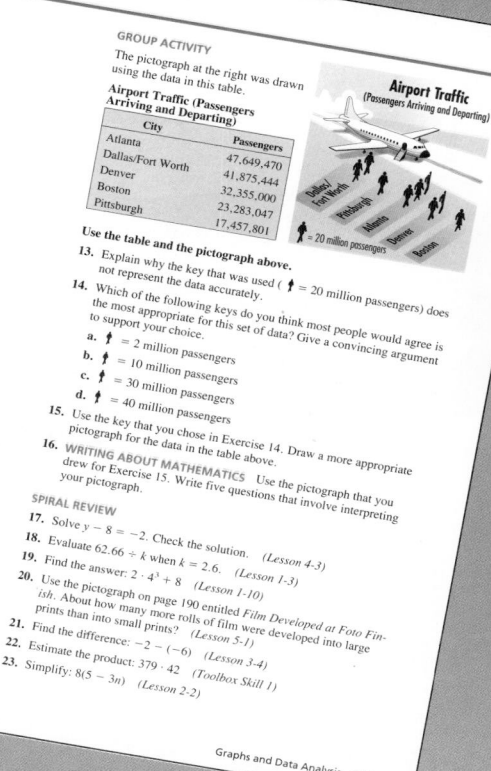

GROUP ACTIVITY

The pictograph at the right was drawn using the data in this table.

Airport Traffic (Passengers Arriving and Departing)

City	Passengers
Atlanta	47,649,470
Dallas/Fort Worth	41,875,444
Denver	32,355,000
Boston	23,283,047
Pittsburgh	17,457,801

Use the table and the pictograph above.
13. Explain why the key that was used (✈ = 20 million passengers) does not represent the data accurately.
14. Which of the following keys do you think most people would agree is the most appropriate for this set of data? Give a convincing argument to support your choice.
   a. ✈ = 2 million passengers
   b. ✈ = 10 million passengers
   c. ✈ = 30 million passengers
   d. ✈ = 40 million passengers
15. Use the key that you chose in Exercise 14. Draw a more appropriate pictograph for the data in the table above.
16. WRITING ABOUT MATHEMATICS   Use the pictograph that you drew for Exercise 15. Write five questions that involve interpreting your pictograph.

SPIRAL REVIEW
17. Solve $y - 8 = -2$. Check the solution.   *(Lesson 4-3)*
18. Evaluate $62.66 \div k$ when $k = 2.6$.   *(Lesson 1-3)*
19. Find the answer: $2 \cdot 4^3 + 8$ *(Lesson 1-10)*
20. Use the pictograph on page 190 entitled *Film Developed at Foto Finish.* About how many more rolls of film were developed into large prints than into small prints?   *(Lesson 5-1)*
21. Find the difference: $-2 - (-6)$ *(Lesson 3-4)*
22. Estimate the product: $379 \cdot 42$ *(Lesson 3-4)*
23. Simplify: $8(5 - 3n)$ *(Lesson 2-2)*

Writing About Mathematics exercises help students improve their problem solving abilities by translating symbolic representations into real-world situations.

Graphs and Data Analysis   **191**

Extensive Support to Assess Student Performance

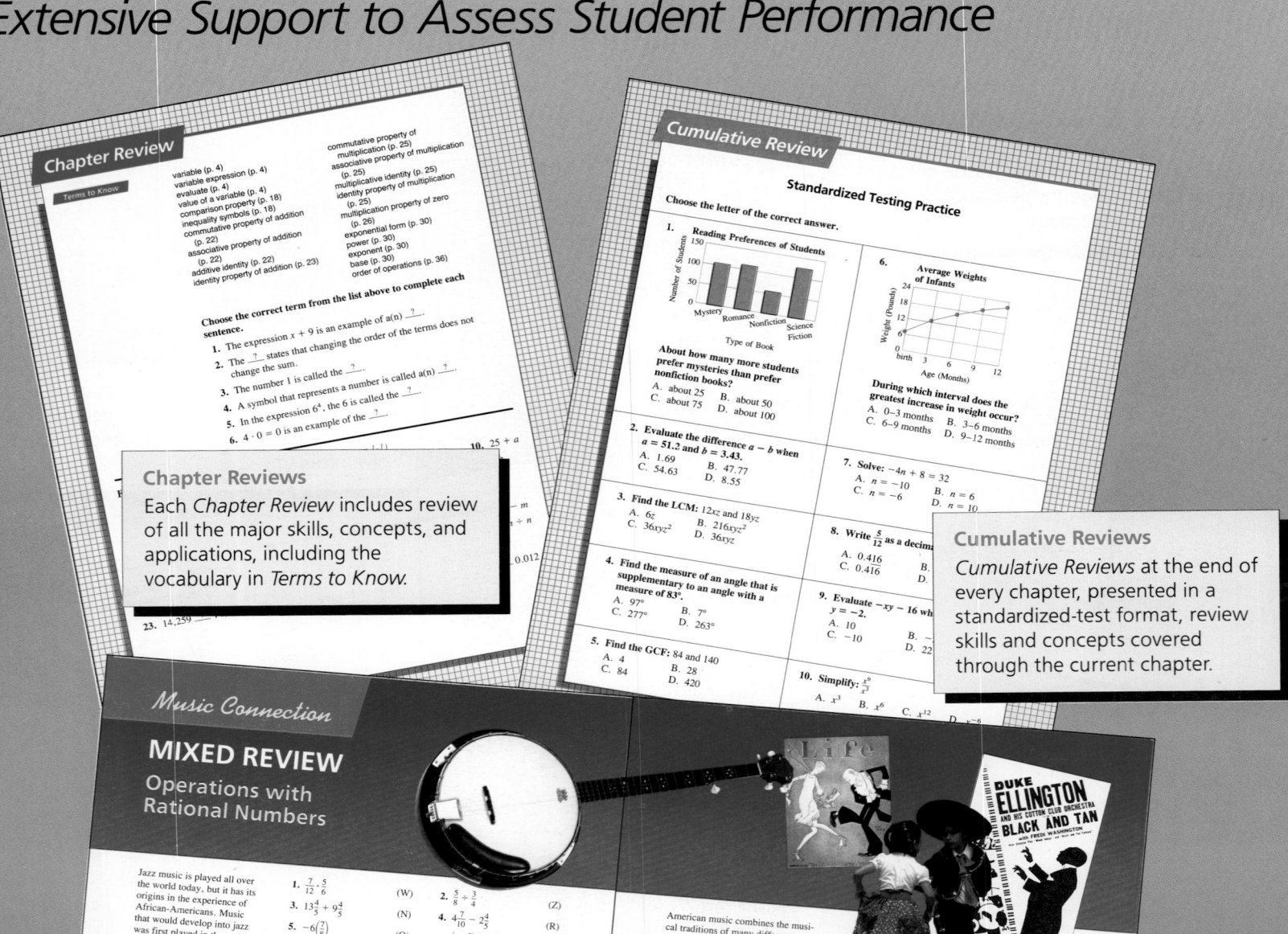

Chapter Reviews

Each *Chapter Review* includes review of all the major skills, concepts, and applications, including the vocabulary in *Terms to Know*.

Cumulative Reviews

Cumulative Reviews at the end of every chapter, presented in a standardized-test format, review skills and concepts covered through the current chapter.

Mixed Reviews

Special *Connection/Mixed Reviews* present skill practice in an innovative format that illustrates connections to other subject areas and the contributions of various cultures to our society.

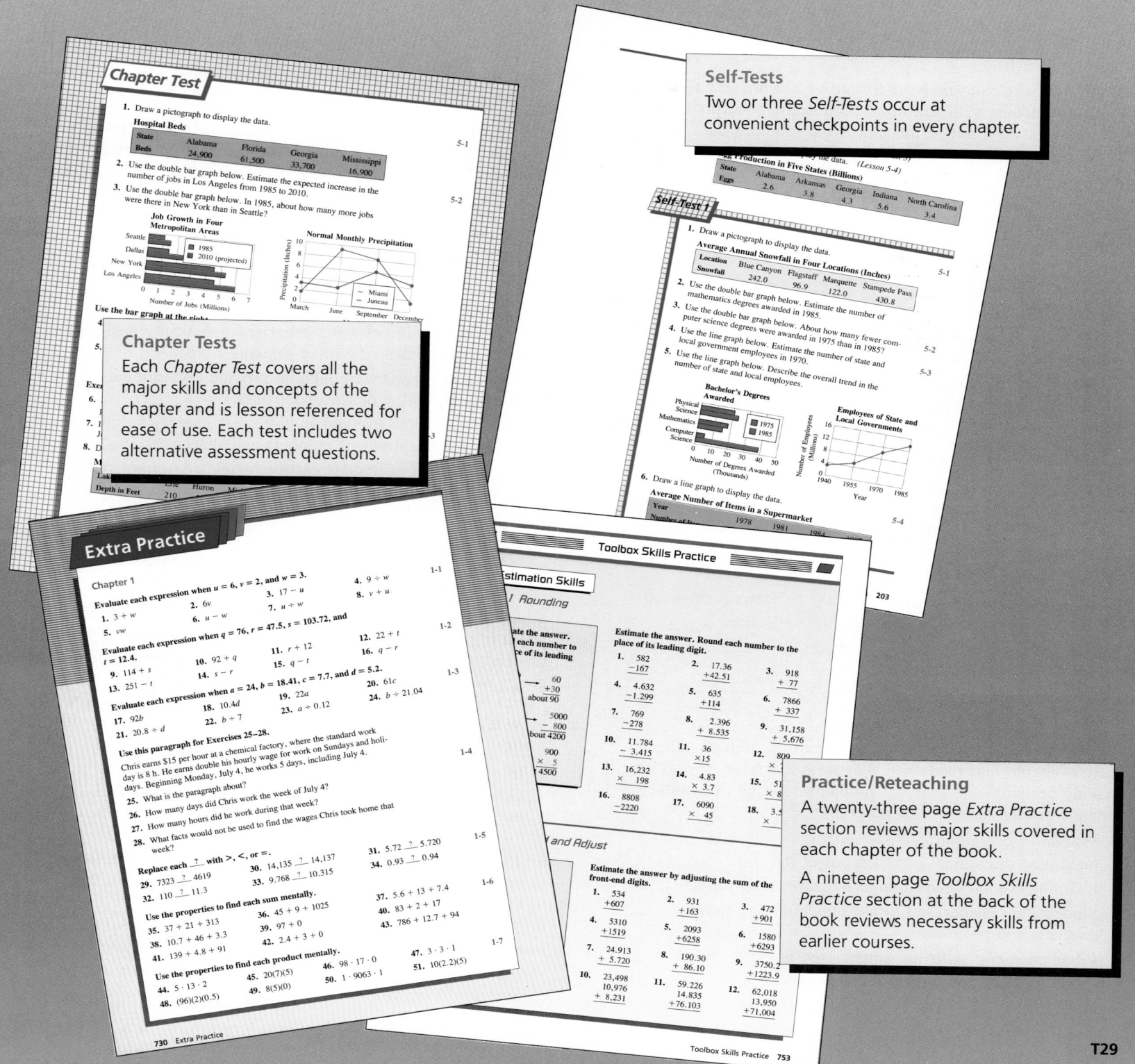

Chapter Test

1. Draw a pictograph to display the data.

Hospital Beds

State	Alabama	Florida	Georgia	Mississippi
Beds	24,900	61,500	33,700	16,900

5-1

2. Use the double bar graph below. Estimate the expected increase in the number of jobs in Los Angeles from 1985 to 2010.

3. Use the double bar graph below. In 1985, about how many more jobs were there in New York than in Seattle?

5-2

Job Growth in Four Metropolitan Areas

□ 1985
■ 2010 (projected)

Seattle
Dallas
New York
Los Angeles

0 1 2 3 4 5 6 7
Number of Jobs (Millions)

Normal Monthly Precipitation

— Miami
— Juneau

March June September December

Use the bar graph at the right.

4.

5.

Exer
6.

7.

8. D

M

Lak...	Erie	Huron	Mi...
Depth in Feet	210		

Chapter Tests

Each *Chapter Test* covers all the major skills and concepts of the chapter and is lesson referenced for ease of use. Each test includes two alternative assessment questions.

Self-Tests

Two or three *Self-Tests* occur at convenient checkpoints in every chapter.

...play the data. *(Lesson 5-4)*

Egg Production in Five States (Billions)

State	Alabama	Arkansas	Georgia	Indiana	North Carolina
Eggs	2.6	3.8	4.3	5.6	3.4

Self-Test 1

1. Draw a pictograph to display the data.

Average Annual Snowfall in Four Locations (Inches)

Location	Blue Canyon	Flagstaff	Marquette	Stampede Pass
Snowfall	242.0	96.9	122.0	430.8

5-1

2. Use the double bar graph below. Estimate the number of mathematics degrees awarded in 1985.

3. Use the double bar graph below. About how many fewer computer science degrees were awarded in 1975 than in 1985?

5-2

4. Use the line graph below. Estimate the number of state and local government employees in 1970.

5. Use the line graph below. Describe the overall trend in the number of state and local employees.

5-3

Bachelor's Degrees Awarded

Physical Science
Mathematics
Computer Science

□ 1975
■ 1985

0 10 20 30 40 50
Number of Degrees Awarded (Thousands)

Employees of State and Local Governments

16
12
8
4
0
1940 1955 1970 1985
Year

6. Draw a line graph to display the data.

Average Number of Items in a Supermarket

Year	1978	1981	1984	
Number of It...				

5-4

Extra Practice

Chapter 1

Evaluate each expression when $u = 6$, $v = 2$, and $w = 3$.

1-1

1. $3 + w$
2. $6v$
3. $17 - u$
4. $9 \div w$
5. vw
6. $u - w$
7. $u \div w$
8. $v + u$

Evaluate each expression when $q = 76$, $r = 47.5$, $s = 103.72$, and $t = 12.4$.

1-2

9. $114 + s$
10. $92 + q$
11. $r + 12$
12. $22 + t$
13. $251 - t$
14. $s - r$
15. $q - t$
16. $q - r$

Evaluate each expression when $a = 24$, $b = 18.41$, $c = 7.7$, and $d = 5.2$.

1-3

17. $92b$
18. $10.4d$
19. $22a$
20. $61c$
21. $20.8 \div d$
22. $b \div 7$
23. $a \div 0.12$
24. $b \div 21.04$

Use this paragraph for Exercises 25–28.

Chris earns $15 per hour at a chemical factory, where the standard work day is 8 h. He earns double his hourly wage for work on Sundays and holidays. Beginning Monday, July 4, he works 5 days, including July 4.

1-4

25. What is the paragraph about?
26. How many days did Chris work the week of July 4?
27. How many hours did he work during that week?
28. What facts would not be used to find the wages Chris took home that week?

Replace each __?__ **with** >, <, **or** =.

1-5

29. 7323 __?__ 4619
30. $14,135$ __?__ $14,137$
31. 5.72 __?__ 5.720
32. 110 __?__ 11.3
33. 9.768 __?__ 10.315
34. 0.93 __?__ 0.94

Use the properties to find each sum mentally.

1-6

35. $37 + 21 + 313$
36. $45 + 9 + 1025$
37. $5.6 + 13 + 7.4$
38. $10.7 + 46 + 3.3$
39. $97 + 0$
40. $83 + 2 + 17$
41. $139 + 4.8 + 91$
42. $2.4 + 3 + 0$
43. $786 + 12.7 + 94$

Use the properties to find each product mentally.

1-7

44. $5 \cdot 13 \cdot 2$
45. $20(7)(5)$
46. $98 \cdot 17 \cdot 0$
47. $3 \cdot 3 \cdot 1$
48. $(96)(2)(0.5)$
49. $8(5)(0)$
50. $1 \cdot 9063 \cdot 1$
51. $10(2.2)(5)$

730 Extra Practice

Toolbox Skills Practice

203

Estimation Skills

/ Rounding

...ate the answer.
...each number to
...ce of its leading

60
+30
about 90

5000
− 800
bout 4200

900
× 5
4500

Estimate the answer. Round each number to the place of its leading digit.

1. $\begin{array}{r} 582 \\ -167 \end{array}$
2. $\begin{array}{r} 17.36 \\ +42.51 \end{array}$
3. $\begin{array}{r} 918 \\ + 77 \end{array}$
4. $\begin{array}{r} 4.632 \\ -1.299 \end{array}$
5. $\begin{array}{r} 635 \\ +114 \end{array}$
6. $\begin{array}{r} 7866 \\ + 337 \end{array}$
7. $\begin{array}{r} 769 \\ -278 \end{array}$
8. $\begin{array}{r} 2.396 \\ + 8.535 \end{array}$
9. $\begin{array}{r} 31,158 \\ + 5,676 \end{array}$
10. $\begin{array}{r} 11.784 \\ - 3.415 \end{array}$
11. $\begin{array}{r} 36 \\ \times 15 \end{array}$
12. $\begin{array}{r} 809 \\ \times 1 \end{array}$
13. $\begin{array}{r} 16,232 \\ \times 198 \end{array}$
14. $\begin{array}{r} 4.83 \\ \times 3.7 \end{array}$
15. $\begin{array}{r} 51 \\ \times 8 \end{array}$
16. $\begin{array}{r} 8808 \\ -2220 \end{array}$
17. $\begin{array}{r} 6090 \\ \times 45 \end{array}$
18. $\begin{array}{r} 3.5 \\ \times \end{array}$

/ and Adjust

Estimate the answer by adjusting the sum of the front-end digits.

1. $\begin{array}{r} 534 \\ +607 \end{array}$
2. $\begin{array}{r} 931 \\ +163 \end{array}$
3. $\begin{array}{r} 472 \\ +901 \end{array}$
4. $\begin{array}{r} 5310 \\ +1519 \end{array}$
5. $\begin{array}{r} 2093 \\ +6258 \end{array}$
6. $\begin{array}{r} 1580 \\ +6293 \end{array}$
7. $\begin{array}{r} 24.913 \\ + 5.720 \end{array}$
8. $\begin{array}{r} 190.30 \\ + 86.10 \end{array}$
9. $\begin{array}{r} 3750.2 \\ +1223.9 \end{array}$
10. $\begin{array}{r} 23,498 \\ 10,976 \\ + 8,231 \end{array}$
11. $\begin{array}{r} 59.226 \\ 14.835 \\ +76.103 \end{array}$
12. $\begin{array}{r} 62,018 \\ 13,950 \\ +71,004 \end{array}$

Practice/Reteaching

A twenty-three page *Extra Practice* section reviews major skills covered in each chapter of the book.

A nineteen page *Toolbox Skills Practice* section at the back of the book reviews necessary skills from earlier courses.

Toolbox Skills Practice 753

Teaching Strategies

Philosophy of the Program

Mathematical Connections, A Bridge to Algebra and Geometry provides students with an introduction to the concepts of algebra and geometry, and at the same time solidifies their grasp of arithmetic concepts and procedures. It also provides exploratory experiences in data analysis and probability. A student who is successful with this program should be ready to move on into algebra and geometry or into an integrated mathematics curriculum.

Challenge of the Future

As we stand on the threshold of a new century, we are approaching an exciting new era in mathematics education. Advances in technology not only have reaffirmed mathematics as an alive and dynamic science in itself, but also have made it an integral part of our daily lives. Now, more than at any time in the past, educating students for life requires that we provide them with a solid background in mathematics.

The greatest challenge of this new era is clear: mathematics can, and must, be made accessible to all students. *Mathematical Connections, A Bridge to Algebra and Geometry* was written to help you meet this challenge. Its philosophy is rooted in these three basic principles.

- To learn mathematics, students must be actively involved.
- The learning of mathematics requires reflection.
- The mathematics that students learn must be meaningful.

Active Involvement

Simply stated, knowing mathematics requires *doing* mathematics—exploring, modeling, making conjectures, researching, making decisions, and solving problems. Activities of this nature form the heart of this program.

- *Problem solving* and *decision making* experiences are integrated throughout most lessons of the text.
- Learning is motivated through frequent *explorations*.
- Students are encouraged to use *manipulatives* to model concepts and procedures.
- *Communication skills* are emphasized in every lesson.
- *Group activities* are suggested for every chapter.
- Each chapter contains suggestions for independent *research*.

Reflection

The answer to a textbook exercise cannot be considered an end in itself. Rather, students must learn to reflect on a result—to develop a sense of its reasonableness, to think critically about its implications, to consider alternative ways to approach it, and to generalize from it. This program encourages the process of reflection in several ways.

- In every lesson, *Check Your Understanding* questions provide students with a means of assessing their comprehension of the mathematics being presented.
- Every chapter incorporates exercises specifically designed to develop and enhance *critical thinking skills*.
- *Estimation* exercises throughout provide opportunities to explore the notion of reasonableness.
- Sets of *number sense* and *spatial sense* exercises sharpen students' perceptions of numbers and shapes.
- Special sets of exercises practice *logical reasoning* skills.
- A *challenge* in every chapter requires students to create their own approaches to nonroutine problems.

Meaningful Mathematics

Students truly appreciate mathematics when they see its utility, and it is this appreciation that can lead to understanding. To this end, students must observe how the need for mathematics grows out of real-life situations, and they must see how mathematics is used in resolving real-world problems. This program provides a variety of contexts in which students can see mathematics applied to the real world.

- In every chapter, sets of *Problem Solving/Application*, *Career/Application*, and *Connecting Mathematics and . . .* exercises show students how mathematics relates to their everyday lives and to their future plans.
- Throughout the text, students learn how current *technology*, both *calculator* and *computer*, relates to the mathematics they are learning.
- Each chapter is introduced through a display of relevant data, and practice in *data analysis* is integrated throughout.
- *Historical Notes* and *Connection/Mixed Review* features explore the human side in the development of mathematics.

Making Connections

Many people truly enjoy mathematics. Some find an inherent beauty in the structure of the subject, particularly during those ''Aha!'' moments when a new idea suddenly, almost miraculously, slips into its place within that structure. Others are drawn to the seemingly boundless utility of mathematics, whether it is being used to unlock the mysteries of the universe or simply to speed up the line at the supermarket checkout. In either case, a powerful appeal of mathematics lies in its connections, both within mathematics itself and to the real world.

Connections Within Mathematics

Within the structure of mathematics itself, there are two types of connections that students must make.

- *Connections among different representations of a single concept or procedure* In general, there are five ways by which mathematical ideas can be represented: concrete objects, pictures, spoken language, written symbols, and real-world situations. It is important for students to understand how the various representations are connected if they are to master a given concept or procedure.

- *Connections among different topics in mathematics* Students must learn to view mathematics as an integrated whole rather than as a series of isolated, unrelated topics. To prepare for high school mathematics, it is particularly important for students to form links between arithmetic and algebra and between algebra and geometry. It is also important for students to understand how topics in arithmetic, algebra, and geometry are related to data analysis and probability.

Connections Outside the Mathematics Class

For students to value mathematics, they need to see its relevance in settings other than the mathematics classroom. This type of connection can be found in and out of school.

- *Connections to other disciplines* Students who have been turned off to mathematics often show renewed interest when they see how mathematics is related to other subjects. Students in a pre-algebra class may be motivated especially by connections to art, music, science, and social studies.

- *Connections to daily life* Students should understand that mathematics pervades our entire society and, consequently, is connected intrinsically to our daily lives. In some ways, real-world connections are the most critical to nurture, because they will have a profound effect on a student's ability to function as a productive member of society.

Maintaining a Multicultural Perspective

All the connections that have been discussed here have little meaning if students do not appreciate mathematics in its historical context. Students should have some awareness of the importance of mathematics throughout history, and they should have a sense of the evolution of mathematical thought over time. Most importantly, they should come to understand that the body of mathematical knowledge we have today is not the work of a select few, but rather the result of a vast and culturally diverse group of men and women.

In the Teacher's Edition, the side columns accompanying each chapter-opening photograph contain multicultural notes with suggestions for research activities and resource material.

As its title implies, this program is committed to helping your students make connections in mathematics.

- Many lessons are introduced by way of a *Connection* or a real-world *Application*.
- Whenever appropriate, connections between representations of a concept are highlighted. These appear under the labels *In Arithmetic/In Algebra* and *In Words/In Symbols*.
- In every chapter, a set of exercises focuses on *Connecting Algebra and Geometry, Connecting Algebra and Data Analysis,* and so on.
- In every chapter, a set of exercises labeled *Connecting Mathematics and . . .* relates mathematics to another subject.
- Special *Connection/Mixed Review* features connect mathematics with other subject areas—history, science, music, and language arts—and with the contributions of various cultures and ethnic groups.

> *See student pages 4, 52, 127, 159, 208, and 360–361 and side-column commentary on pages 92, 237, and 354.*

Teaching Strategies

Problem Solving

Problem solving is the heart of mathematics. Without problems to solve, there is little need for mathematics. Without problem-solving methods, there is no way for mathematics to grow and flourish. To be a teacher of mathematics, then, is to be a teacher of problem solving.

Problem Solving Strategies

A problem-solving strategy is a plan of action. When you choose a strategy for solving a problem, you are identifying the method of solution that you intend to use.

Here is a list of strategies that good problem solvers tend to use time and time again. These strategies are given special emphasis in this textbook.

- making a table
- guess and check
- using equations
- identifying a pattern
- drawing a diagram
- supplying missing facts
- using logical reasoning
- using a Venn diagram
- using a simpler problem
- making a model
- working backward
- using proportions

Problem Solving Applications

Not all problems are applications. For instance, determining how many prime numbers there are between 100 and 200 is a problem that students might solve—and many students will enjoy—but the result has no meaning within the realm of their experience. For students to comprehend the importance of problem-solving skills, they need to apply these skills to problems that arise from real-world situations. On page T33, you will find some suggestions for teaching applications of mathematics and a description of the way applications are incorporated into this program.

Some Tips on Teaching Problem Solving

Just as there is no one correct strategy for solving a problem, there is no one correct way to teach problem solving. However, you may find the following suggestions helpful in assuring students a broad range of problem-solving experiences.

- *Teach students to approach problem solving in a systematic, orderly manner*   In this program, students are encouraged to organize their efforts around a four-step plan: understand, plan, work, answer.
- *Encourage students to look back*   At the most basic level, looking back involves checking the reasonableness of an answer, and students should do this for every problem. However, there are several other important aspects of looking back: considering an alternative strategy, changing the conditions of the problem, extending the problem, and generalizing the results of the problem.
- *Provide experiences in solving nonroutine problems*   When problems become routine, or predictable, problem-solving skills can no longer develop. Try to challenge students with a nonroutine "brainteaser" problem at least once a week.
- *Give students practice in problem formulation*   A good indicator of how well your students understand the problem-solving process is whether they can write meaningful problems. Writing problems—and solving each others' problems—can be an excellent group activity.
- *Emphasize decision making*   Students often are unaware of the decisions that are theirs to make—what method of computation to use, whether an estimate rather than an exact answer is sufficient, and so on. Class discussions periodically should focus on this facet of problem solving.

Throughout this program, problem solving is considered to be an integral part of the mathematics being taught.

- *Problems* appear in the exercise sets of all lessons.
- In each chapter, one or more *Problem Solving* lessons focus on a problem-solving strategy.
- In each chapter, sets of *Problem Solving/Application* and *Career/Application* exercises give students a chance to apply a range of problem-solving skills to real-world situations.
- *Decision Making* lessons give students guidance in making intelligent choices when solving problems.
- In each chapter, the *Challenge* provides an opportunity for nonroutine problem solving.

See student pages 20, 28–29, 121–122, 164, and 216 and side-column commentary on pages 50 and 268.

Applications and Life Skills

Few if any students would question the importance of learning to read and write. True, they may object to a given assignment, but their resistance often is a simple matter of likes and dislikes. Most students see the written word everywhere in the world around them, and they perceive the ability to read and write as a basic survival skill. Society has coined the word *illiterate* to describe a person who cannot read or write, and most students recognize that being illiterate is a stigma.

When you change the subject to mathematics, however, the situation becomes very different. Quite simply, many students view mathematics as an alien topic; they suppose it's useful to someone, somewhere, but they feel it has no relevance in their world. It probably would be difficult to find a mathematics teacher who has not come face-to-face with one of these students and the question: *When am I ever going to use this?*

The answer is that, in our rapidly changing society, students probably will have to use mathematics every day of their lives. As consumers, they will enter a world where they are confronted with a vast array of facts and figures—weights and measures, discounts, markups, taxes, loans, investments, newspaper graphs, charts, and statistics, just to name a few. As workers, they will enter a changed job market where applicants are required more and more to demonstrate skill in mathematics and problem solving. The fact is that the ability to do mathematics is becoming a basic survival skill, so much so that a new word, *innumeracy,* has been coined to describe the mathematical equivalent of illiteracy.

Mathematics in Everyday Life

How can you teach students to appreciate mathematics as a basic life skill? At the pre-algebra level, your students probably will be motivated by exploring applications of mathematics to their immediate environment. Real-world applications are incorporated into this program in several ways.

- Many lessons are introduced with a brief discussion of an *Application* of the mathematics of the lesson.
- Whenever appropriate, the problems that are integrated into each exercise set discuss *applications.*
- In each chapter, a brief set of *Problem Solving/Application* exercises explores how the mathematics of the chapter applies to a real-world situation.

- Periodic *Focus on Applications* lessons give students the chance to investigate an application in greater depth.
- In this Teacher's Edition, further support is to be found in the side-column commentary that accompanies each lesson. The *Lesson Focus* often furnishes a real-world application that can be used to motivate the lesson, and a section called *Life Skills* describes how a particular facet of the lesson cultivates your students' practical life skills.

Looking Ahead: Mathematics on the Job

At the pre-algebra level, few students realize that turning off mathematics in high school automatically turns off a number of career options in the future. Many students believe that the only occupations requiring expertise in mathematics are oriented towards engineering and the sciences. To help your students understand how mathematics might impact on their career choices, each chapter of the student text includes a set of *Career/Application* exercises.

Textbook exercises, of course, can provide only a brief glimpse of any one occupation. If your students' curiosity is piqued by a particular occupation, you may wish to have them do further research about the uses of mathematics in that field. As a long-range project, you might have your students work collaboratively to present the results of their research in a mathematics career day.

A final word on life skills: The skills you teach in your mathematics class are not necessarily restricted to mathematical content. The very way that you structure your class—the sort of activities you choose, the type of interaction you encourage, and so on—can help your students develop other, equally important life skills. For instance, a cooperative learning activity in your class not only will strengthen your students' mathematical skills, but also will develop their skills in working collaboratively and will sharpen their social interaction skills. In other words, the life skills that you are teaching your students extend outside of mathematics as well and have a profound impact on many other aspects of their lives.

See student pages 4, 54, 104, and 132 and side-column commentary on pages 189 and 346.

Teaching Strategies

Critical Thinking Skills

Problem solving, decision making, logical reasoning—these aspects of mathematics all require a student to handle complex situations, to exercise judgment, to cope with uncertainty, and to exercise a degree of independence. In other words, a student of mathematics must learn to be a critical thinker. It follows that, as a teacher of mathematics, an important objective should be to teach students to think for themselves.

The Levels of Critical Thinking

Most educators agree that *critical thinking* occurs on several levels. These levels range from the very simple, such as direct recall of facts, to the more complex *higher order* thinking skills. However, there is no universal agreement among educators about how these levels of thinking should be classified. In this program, you will find the levels identified according to the following hierarchy, first outlined by Benjamin Bloom in the *Taxonomy of Educational Objectives*.

- *Knowledge* is defined as the simple recall of information in the same form, or nearly the same form, in which it was presented. Reciting a definition or a multiplication fact from memory is an example of this level of thinking.
- *Comprehension* is the simplest level of understanding. At this level students are able to translate information into a different form and make use of it, although they still might not see how pieces of information relate to each other. For instance, students at this level are able to read data from a table and to use the data as instructed.
- *Application* is the level of thinking at which a student begins to see relationships among things. A student functioning at this level is able to use an appropriate skill or concept in a new situation without being specifically instructed to do so. For example, students at this level are able to choose the arithmetic operation needed to solve a word problem.
- *Analysis* is the level of thinking at which students are able to break down ideas into smaller parts and to identify the relationships between parts. Comparing the subsets of the real number system is an example of this type of thinking.
- *Synthesis* is the ability to combine skills and concepts already mastered to discover new ideas. At this level, students are able to observe patterns, make connections, and generalize.

- *Evaluation* is the highest level of thinking. At this level, students make judgments about the usefulness of an idea or a method in a particular situation. For instance, a student at this level is able to explain whether a problem is better solved using mental math, paper and pencil, or a calculator.

The levels of critical thinking are considered to be cumulative. That is, each successive level incorporates all the levels beneath it in the hierarchy.

Evaluating Critical Thinking

By its very nature mathematics involves thinking, and so your students practice thinking skills at some level every time they enter your classroom. However, it is a good idea periodically to focus on problems that are specifically designed to develop the higher order thinking skills. In this textbook, you will find such problems under the label *Thinking Skills*. Each chapter contains one or more sets of thinking-skills problems, and an annotation next to the problems in this Teacher's Edition alerts you to the levels of thinking that are required. In addition, the side-column commentary in this Teacher's Edition frequently highlights the ways in which the basic objective of a lesson relates to the development of critical thinking skills.

How can you evaluate your students' progress in critical thinking? A person's approach to critical thinking is highly individual, so there is no "best" method of evaluation. However, you may want to watch for the following behaviors that usually indicate that a student is making progress in developing critical thinking skills.

- improved performance on quizzes and tests, particularly on those items that require more than simple arithmetic calculation or algebraic manipulation
- increased participation in classroom discussions
- improved quality of oral and written responses
- decreased tendency to respond to questions impulsively, without thinking
- increased tendency to ask reflective, thought-provoking questions

See student pages 27 and 252 and side-column commentary on pages 82 and 146.

Communication Skills

Today's students must assimilate a vast amount of information from a wide variety of sources. These sources include teachers, textbooks, reference books, and videos, to name just a few. To master a subject, though, a student must also be able to *convey* information: to both answer and formulate questions, to clarify a point of view, to express opinions, and to develop hypotheses. It is this entire process—learning facts, asking questions, stating convictions, making conjectures, expressing feelings, and so on—that constitutes the communication of knowledge. Without it, learning cannot take place.

Each lesson in this text includes a set of exercises under the label *Communication*. These exercises, usually incorporated into the Guided Practice section, are designed to provide you with lesson-specific ideas for developing communication skills. These exercises focus on one of the three basic forms of communication: reading, writing, and discussion.

Reading Mathematics

Perhaps the most fundamental way to assimilate knowledge is through reading. Students must learn to appreciate mathematics as a language to be read, with its own special vocabulary, symbols, and syntax. The following are some suggestions that might help you develop your students' reading skills.

- *Vocabulary-building* Whenever possible, encourage students to connect mathematical terms to everyday language. Train them to examine new words routinely for familiar prefixes, suffixes, and root words.
- *Reading for comprehension* For each textbook lesson, ask students to identify the main idea and the major supporting details. Periodically check your students' understanding of a direction line such as "Simplify" by asking them to describe what they are to do in their own words.
- *Using other resources* Show your students how to use a mathematics dictionary or glossary.

Writing Mathematics

Composing an essay, report, or even a simple paragraph requires the writer to organize and analyze thoughts. For this reason, there are few vehicles more powerful than the written word to help you assess students' grasp of basic concepts.

Here are a few suggestions for ways that you might incorporate writing into your pre-algebra class.

- *Keeping a journal* Have students keep a daily mathematics journal that is more than a simple notebook. Encourage them to record questions, observations, and conjectures.
- *Letter-writing* Have students write responses to questions about mathematics from classmates or younger students.
- *Interviews* Have students write a script of an imaginary interview with a famous mathematician, or report on a real interview with a person who uses mathematics on the job.

Discussing Mathematics

In some ways, the most important of the communication skills is discussion. In a discussion, information flows in two directions as students talk with you and with each other about their perceptions of mathematics. In the process, students learn to view mathematics less as a lifeless body of facts and exercises and more as a complex, dynamic, and very human endeavor.

The following are some ways that you can help your students develop effective verbal communication skills.

- In whole-class discussions, encourage students not only to answer your questions, but also to raise their own questions.
- Plan some activities that occur in small groups, where many students feel freer to express opinions and uncertainties.
- Have a student or group of students prepare a class presentation on an enrichment topic.

Multicultural Considerations

Learning about the contributions of various cultures to the development of mathematics can lead students to a better understanding of mathematical concepts, as well as an appreciation for the cultures involved. Topics such as different number systems (for example, Mayan, Babylonian, African) or the contributions of mathematicians from different countries can serve as the basis for presentations, discussions, and written reports.

See student pages 31, 96, and 152 and side-column commentary on pages 73, 122, and 304.

Teaching Strategies

Using Technology

The most effective way to introduce students to computers and calculators is to have them use these tools in meaningful ways in their school work. Throughout this textbook, numerous opportunities are provided for students to read about and use computers and calculators. A special *Using Technology* feature, which students should read at the beginning of the course, is located on pages xiv–xviii. Pages entitled *Focus on Computers* are devoted to this topic and can be found in many chapters of the book. Also, in every chapter, there are *Computer Application* exercises and *Calculator* exercises. When appropriate, calculators also are used to introduce and develop new ideas and approaches to the content of some lessons.

In the Teacher's Edition, the pages preceding each chapter contain a *Using Technology* section that outlines suggestions for integrating technology into the chapter. In the Activities Book, there are two technology activities for each chapter in a blackline master format.

What's Available?

The calculator and computer technology available today is very powerful and developing rapidly. Scientific calculators, many now capable of dealing with fractions exactly, are readily available at low cost. Recently, calculators have been developed that are capable of graphing virtually any function, of solving equations approximately at the touch of a button, and of evaluating expressions that are typed in normal mathematical notation. A great deal of useful mathematical software is available for the older computers already installed in many schools. Prices of increasingly powerful, fast, and easy-to-use computers are steadily decreasing. Mathematics students can make use of Houghton Mifflin's *Connections Plotter Plus* or commercial software programs for spreadsheet, statistics, function graphing, and general drawing programs. As computer software becomes better known, easier to use, and less expensive, it has a good chance of revolutionizing the teaching of mathematics. Just as calculators are making tedious pencil-and-paper calculations and trigonometry tables obsolete, these software programs will help shift the emphasis of mathematics education from learning to do accurate algebraic manipulations to exploring patterns and solving problems.

Technology in the Classroom

The introduction of calculators and computers into the classroom presents a decision-making situation for students. Some problems can best be solved by simply using paper and a pencil, while others can be solved expeditiously using a calculator or computer. Students need to learn how and when to apply technology to solve problems and when to use other means.

Calculators and computers have several important uses in the pre-algebra classroom. At the most fundamental level, they facilitate numerical calculations. The ease and accuracy of these tools allow teachers to pose more interesting problems and allow students to concentrate on the noncomputational aspects of the material. In addition, students using these tools can quickly test many guesses for the solution of a problem, which can lead them to a correct solution or to an insight that will allow an algebraic solution.

At a higher level, these technologies provide students with an opportunity to discover mathematics for themselves. By examining many cases, students can observe patterns that lead to important generalizations. For example, a carefully designed set of exercises can lead them to graph many straight lines and thereby discover the role of m and b in equations of the form $y = mx + b$. This kind of activity is valuable not only because it varies the usual classroom routine, but also because of the excitement of discovery that students can experience. Even if students use spreadsheets, graphing calculators, or software simply to check their answers to routine problems, using the technology can reinforce their understanding and intuition by providing another pathway to learning.

Technology and Communication

One of the goals of this book is to develop students' communication skills in mathematical language. In addition to the normal notation printed in books, mathematical ideas can be expressed by calculator keypresses, by spreadsheets, or by computer programs. All three modes of expression are included in appropriate places in the text and supplementary material.

> See student pages xiv–xviii, 174, 270, 341, and 393 and Teacher's Edition pages 234C and 288C.

Using Manipulatives

What is a manipulative? For many people, the very word conjures up the image of catalogs describing expensive sets of balance scales, fraction bars, geoboards, and algebra tiles. These indeed are manipulatives, but the fact is that *any* object a student handles physically to investigate a concept is a manipulative. Using this definition, a manipulative can be something as simple as a piece of paper or a cardboard box.

In the past, manipulatives often were equated with toys and, as such, were relegated to the primary grades. If students were ready for pre-algebra, the reasoning went, then surely it was time to dispense with the toys and get on with the serious business of learning how to manipulate symbols. So why do we now see manipulatives appearing in so many secondary, and even post-secondary, classrooms?

The answer is that, in recent years, we have come to understand a great deal more about how people learn mathematics. Research has shown that for students to truly comprehend mathematics, they must actively construct their own knowledge of mathematical concepts. At all levels, one of the most powerful ways of doing this is to model concepts with manipulatives.

This program is designed to help you incorporate the use of manipulatives into your pre-algebra class at virtually any point you choose. A *Focus on Explorations* lesson in every chapter of the student book presents a set of activities that model the content of the chapter with manipulatives. In the Teacher's Edition, the pages preceding each chapter contain a *Using Manipulatives* section that outlines several additional suggestions for integrating manipulatives into the chapter. In the Activities Book, there are two manipulative activities for each chapter in a blackline master format.

When to Use Manipulatives

Many teachers are concerned about the amount of classroom time consumed when students use manipulatives. Manipulatives, however, are not really additional content. Rather, they are just one of a number of optional methods you might consider when planning how to teach your curriculum. When you regard manipulatives in this way, then it is important to know when they are the best method of choice, that is, the method that will best serve you and your students. The following are two situations in which you may find manipulatives to be particularly effective.

- *Introducing new concepts* When students use manipulatives to construct their own knowledge, the knowledge truly becomes their own. You probably will discover that less time is needed for review.
- *Remediation* Students who have difficulty with a concept seldom are helped by additional worksheets of written exercises—they usually make the same mistakes over and over again. These students need to spend additional time manipulating a physical model before they can work with the symbols on paper.

Some Tips on Working with Manipulatives

- Buying sets of manipulatives from catalogs can become expensive. If your budget is limited, you may want to buy only those manipulatives that you will use frequently throughout the year, such as algebra tiles. For manipulatives that you will use less often, consider paper-and-pencil alternatives—for instance, use dot paper or graph paper to do geoboard activities.

- Be resourceful—many types of materials can be purchased in bulk from business suppliers. For instance, inexpensive squares of waxed paper for paper folding might be available from a nearby restaurant supplier.

- Package materials as needed for a given class period. Considerable class time may be lost if students have to first gather or assemble the materials needed for an activity.

- If possible, allow students some "warm-up" time to explore a new manipulative. A brief period of exploration will satisfy their natural curiosity, and your students may even provide a key question or observation to motivate the main objective of the lesson.

- In general, manipulative activities are best suited to a cooperative learning environment. If you can, have students work in pairs or in small groups of three or four.

- Be sure to plan sufficient time at the end of class to collect and account for any materials that must be returned to you.

See student pages 51 and 450–451 and Teacher's Edition pages 140C and 140D.

Teaching Strategies

Cooperative Learning

It is widely recognized today that students need to have many different kinds of experiences to learn mathematics effectively. The traditional classroom structure of teaching a large group of 20 or more students is effective for some topics or lessons but not for all instruction. Having students participate in small cooperative learning groups of two to six students to learn mathematics has many desirable results.

What is the Teacher's Role?

At first, the teacher needs to discuss with students how groups will be formed and the purposes of group work. You may wish to assign students to groups or have students form their own groups. Over time, however, try to make sure that the members of the groups change so that all students have an opportunity to work with one another in developing academic and social skills. When students work in groups, you can move from group to group, providing help as needed, answering or asking questions, making informal assessments of how well students are doing, and, in general, guiding students toward accomplishing the objectives of the lesson.

What Purposes Do Groups Serve?

The main purpose of having students work in groups is to increase their understanding of mathematics. Students working in small groups are less inhibited to discuss their ideas with classmates, ask questions, or present solutions to problems. Group work, then, can be used to foster an attitude of ''let's try it,'' which is essentially an experimental and problem-solving attitude that is fundamental to achieving success in mathematics.

In a pre-algebra classroom, there may be some students who have not been particularly successful in studying mathematics. Group work provides opportunities for the teacher to understand those difficulties students may be having and then to provide individual assistance tailored to specific problem areas.

Less capable students need both individual help with their problems and emotional support and encouragement to try doing mathematics. Their progress at times is slow and fragile, but through small group work, you have a good chance of providing the assistance and support necessary for success.

Team Work

Cooperative learning groups can function as teams that work together to improve all team members' skills and attitudes. The more capable students can help the less capable ones, so the whole team's performance is improved. Thus, the stronger students sharpen their own understanding of the material being studied, while the weaker students achieve success and feel good about contributing to their team's efforts.

Team work stimulates a positive attitude toward mathematics and motivates students to take responsibility for their own learning and that of other team members. Students gain in their understanding of mathematical skills and concepts, and in their ability to work together and to communicate with one another.

Through teams, students can compete with other teams instead of with one another. This approach adds an element of fun and excitement to the class environment and is a positive experience for students.

What Can I Tell Parents?

In talking with parents about cooperative learning groups, you can cite the following advantages of group activities.

- Students take responsibility for their own learning.
- Students actively do mathematics with other students.
- Students improve their skills in using mathematical language as they work with others in the group.
- Students share their ideas and improve their social skills.
- Students become less fearful of making errors.
- Students become more confident in their own abilities by successful group participation.
- Students learn that a group can often solve problems that an individual cannot solve.
- Students learn how to work in a group environment.
- Students learn more mathematics, enjoy learning it more, and develop a positive attitude toward mathematics and school.

See student pages 164 and 659, Teacher's Edition pages 186D and 338D, and side column commentary on pages 61 and 291.

Questioning Techniques

Suppose that a lesson about types of triangles begins this way.

A triangle with one right angle is called a right triangle. Which of these three triangles is a right triangle?

I

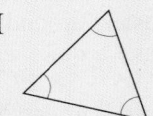

Now consider the following very different approach to the same topic.

II

One of these triangles is called a right triangle. Which one do you think it is?

Why do you think it is triangle II?

III

Why do you think that triangles I and III are not right triangles?

What do you think would be a good definition of a right triangle?

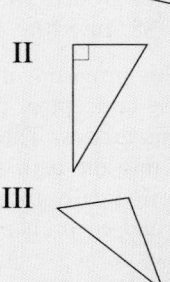

In the first approach, the teacher is acting as a *dispenser* of knowledge. A definition of right triangle is given and students are expected merely to apply it. In contrast, the second approach has the teacher acting as a *facilitator*, guiding students to construct their own knowledge of right triangles. Although both approaches elicit the same fact—namely, that triangle II is the right triangle—the second will be more productive. After completing this lesson, students are more likely to retain their knowledge of right triangles and to develop a solid understanding of other types of triangles. The strength of the second approach lies in asking good questions, not in the use of any elaborate manipulative or audio-visual aid.

What Constitutes a Good Question?

In order to describe a good question, it may be easier to first consider what makes a question *not good*. In general, a question is not good for class discussion if students can answer it by rote. Therefore, a good question can be defined as one that requires students to think.

Good questions may arise spontaneously in class discussion, and often they come from the students themselves. However, the teacher sets the tone for the types of questions that are asked, and so the roots of good questioning actually lie in your lesson plan. To help you incorporate good questions into your plan, you may find the following guidelines especially helpful.

- *Plan a series of questions*   Good questions elicit not just the correct solution to a problem, but also some information as to why that solution is correct. Often this is best done through a carefully planned *sequence* of questions.
- *Plan open-ended questions*   By their very nature, open-ended questions are more insightful than single-response questions. For instance, asking "What are ten numbers between 0 and 0.1?" will stimulate a more productive discussion than asking "Is 0.03 between 0 and 0.1?"

How to Ask Questions

As important as knowing what questions to ask is knowing how to ask them. Class discussions can be unpredictable, of course, so there are no fixed rules as to how you approach a question on a given day. Nonetheless, here are a few general suggestions that you may find useful.

- When you initially pose a question, address it to the entire group rather than to an individual. If one student is singled out from the start, other students may assume that they have no need to think about the question at all.
- After a question has been posed, give students some time to think about it. Three to five seconds usually is sufficient for this "wait time," although you may need to vary it according to the difficulty level of the question and the ability level of the students in your class.
- If students are unable to answer a question, try to avoid answering it yourself. Instead, find a different way to phrase the question, or perhaps go back and ask a few questions about intermediate steps.

Suggestions for questions that you can use with your pre-algebra class appear throughout this program. In each lesson of the student text, a *Check Your Understanding* or *Look Back* section poses questions that encourage students to reflect on or extend the worked examples. In the Teacher's Edition, the side columns adjacent to each lesson contain one or more *Key Questions* that are designed to help you guide your students through some critical aspect of the lesson.

See student pages 36 and 121 and side-column commentary on pages 246 and 376.

Teaching Strategies

Estimation Skills

When do you estimate? Research has shown that, in everyday life, people estimate far more often than they find exact answers. In fact, an estimate frequently *is* the answer, as when you determine about how much paint you need for a room. Even when you perform an exact calculation—as in balancing a checkbook—you estimate to check for reasonableness. Clearly, by teaching your students to become good estimators, you not only pave the way for greater success in mathematics, you also provide them with a valuable life skill.

Estimation Strategies

All too often, estimation is erroneously equated with rounding. However, rounding is only one of a number of techniques used by good estimators. In fact, there are times that rounding is actually inefficient or unproductive as an estimation strategy. Therefore, one of your goals should be to expose your students to a variety of other ways to estimate. In particular, pre-algebra students should have some experience with these methods.

- *Front-end-and-adjust* This type of estimation may be more efficient than rounding when working with sums and differences of whole numbers and decimals. Interestingly, it is the estimation technique that seems to come most naturally to young children. Here is an example.

$$\begin{aligned}&\$8.56\\&\ \ 1.92\\&+\ 3.61\end{aligned} \longrightarrow$$

First, add the *front-end* digits:
$$\$8 + \$1 + \$3 = \$12$$
Now *adjust* this sum:
$.92 is about $1
$.56 + $.61 is about $1
A good estimate is
$12 + $1 + $1, or $14.

- *Compatible numbers* Using compatible numbers—sometimes called "nice" numbers—may be more productive than rounding if you need to estimate a quotient of whole numbers or decimals. (Dividing the rounded numbers may be nearly as difficult as the original division.) You also use compatible numbers to estimate products and quotients involving fractions and percents. Here are two examples.

$$8\overline{)5719} \longrightarrow \overset{\text{about}}{\underset{}{\ \ 700}}\ 8\overline{)5600}$$

$$24\% \text{ of } 371$$
$$\downarrow \qquad \downarrow$$
$$\tfrac{1}{4} \times \ 360, \text{ or about } 90$$

- *Clustering* Numbers that are all near a particular value are said to *cluster* around that value. When a group of addends clusters around one number, you can use that number to make a quick estimate. For instance, in the addition $7.95 + $8.27 + $8.23 + $7.89 + $7.79, the addends cluster around $8, so a reasonable estimate of the sum is $5 \times \$8$, or $40.

In this program, it is assumed that students are comfortable with the concept of estimation as a result of their work in previous grades. Therefore, the discussion of estimation is integrated into the text whenever it is appropriate. Worked examples periodically focus on estimation in order to further develop skills, and each chapter contains at least one set of estimation exercises to help maintain skills. For those students who may be unfamiliar with some techniques, or who may need additional practice, four brief sections on basic estimation skills are provided in the *Toolbox Skills Practice* at the back of the book.

The Language of Estimation

When asked to find an estimate, many students mistakenly believe that there is only one correct answer. In an interesting paradox, these students persist in their attempts to obtain the "exact estimate" for a given computation. All students need to be aware that there is generally a *range* of acceptable estimates for any given problem. One way to build this awareness is to encourage your students to phrase their estimates in the "language of estimation." For instance, here are some ways that you might prompt students to respond when they estimate.

The sum $193 + 414$ is *about* 600.

The product 7×95 is *between* 630 and 700.

The quotient $294 \div 5$ is *a little less* than 60.

The fraction $\frac{19}{36}$ is *close to* one half.

Thirty-five percent is *roughly* equal to one third.

Once you have taught your students the language of estimation, encourage them to use it not only in classroom discussions, but also in their written responses.

> *See student pages 74, 154, 341, 753, and 754 and side-column commentary on pages 341 and 515.*

Alternative Assessment

Mathematics educators are aware of the need to make the mathematics classroom an environment that fosters exploration, investigation, discussion, writing, computational proficiency, and an emphasis on higher order thinking. Students should be encouraged to reason, make and test conjectures, show connections, justify conclusions, and so on. The reality is that our current assessment practices quite often do not support this approach to learning mathematics.

Many efforts at the national, state, and district level are already reflecting major changes in test design and format, with the major goal being able to determine students' ability to reason in mathematics. These tests include assessment approaches that should be modeled in the classroom.

Performance Tasks

One means for assessment is that of performance tasks, which is a logical approach for mathematics instruction. With performance tasks, a student is presented a problem or situation that requires several steps before arriving at a conclusion. Throughout the problem, the student is required to explain, make conjectures, write descriptions, justify decisions, and so on; the major emphasis is on higher order thinking skills with a demonstration of computational proficiency being a suboutcome. The mastery of content is a means to the end, not an end in itself. Performance task assessment is an effective means for determining students' ability to exhibit mathematical knowledge and apply it in a meaningful context. This approach to assessment also helps to promote connections to topics within mathematics, other disciplines, and to the real-world.

More Alternatives for Assessment

Other alternative forms of assessment include: Open-ended questions, observations, interviews, oral questions and explanations, demonstrations, take-home tests, journal writing, and portfolios. It is impossible to discuss all of these forms fully here; the important point to mention is that these alternatives are exciting, authentic, interesting ways to make assessment an integral part of the learning process. Students will be provided with more opportunities requiring application, analysis, and synthesis instead of just simple recall and computation. A sample alternative test item for a traditional question is given above.

Traditional item:
Find the perimeter of quadrilateral ABCD.

Alternative item:
Draw an irregular quadrilateral that has a perimeter of 17 cm.

Which is the more authentic assessment?

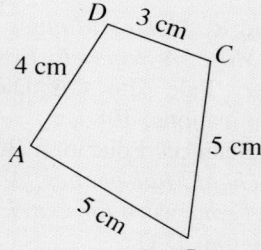

Teaching to the Test

Some teachers worry about "teaching to the test." Simply put, student learning objectives should be made very clear and the instruction should be designed to help students achieve those delineated objectives. Students' assessment, both in content and procedure, should model the instructional practices. To give insufficient attention to the alignment of content and assessment usually results in poor test scores.

What About Multiple-Choice and Short-Answer Tests?

"Alternative assessment," as the term implies, does not suggest that traditional paper-and-pencil tests be eliminated. Obviously these types of tests are, and will remain, an integral part of the assessment framework. However, students learn in many different ways and should have varied modes of assessment opportunities. Effective assessment should allow and enable students to show what they do know and what they can do, with the emphasis on these positive accomplishments rather than on what is not known.

For your convenience, this program has included alternative assessment questions in each Chapter Test in the student book and in the Tests booklet, and after each Quick Quiz in the side columns of the Teacher's Edition.

See student pages 136 and 334 and side-column commentary on pages 433 and 704.

Bibliography

General References

Johnson, David R. *Making Minutes Count Even More: A Sequel to Every Minute Counts*. Palo Alto, CA: Dale Seymour Publications, 1986.
Mathematical Sciences Education Board. *Reshaping School Mathematics: A Philosophy and Framework for Curriculum*. Washington, D.C.: National Academy Press, 1990.
National Council of Teachers of Mathematics. *Curriculum and Evaluation Standards for School Mathematics*. Reston, VA: NCTM, 1989.
———. *Professional Standards for Teaching Mathematics*. Reston, VA: NCTM, 1991.
National Research Council. *Everybody Counts: A Report to the Nation on the Future of Mathematics Education*. Washington, D.C.: National Academy Press, 1989.

Applications and Life Skills

American Association for the Advancement of Science. *Project 2061: Science for All Americans–Summary*. Washington, D.C.: AAAS, 1989.
Mathematical Sciences Education Board. *Mathematics Education: Wellspring of U.S. Industrial Strength*. Washington, D.C.: MSEB, 1989.

Communication Skills

Mumme, Judith, and Nancy Shepherd. "Communication in Mathematics." *Arithmetic Teacher* 38 (Sept., 1990): 18–22.
Texas Education Agency. *Learning Partners: Reading and Mathematics*. Austin, TX: TEA, 1979.

Cooperative Learning

Burns, Marilyn, and Cathy McLaughlin. *A Collection of Math Lessons from Grades 6–8*. New Rochelle, NY: Cuisenaire Company of America, Inc., 1990.
Davidson, Neil, ed. *Cooperative Learning in Mathematics: A Handbook for Teachers*. Menlo Park, CA: Addison-Wesley, 1990.

Critical Thinking Skills

Bloom, Benjamin, ed. *Taxonomy of Educational Objectives*. New York: David McKay Company, Inc., 1956.
Resnick, Lauren. *Education and Learning to Think*. Washington, D.C.: National Academy Press, 1987.
Seymour, Dale, and Ed Beardslee. *Critical Thinking Skills in Patterns, Imagery, Logic: Grades 7–12*. Palo Alto, CA: Dale Seymour, 1990.

Estimation Skills

Reys, Barbara J., and Robert E. Reys. *Guide to Using Estimation Skills and Strategies (GUESS)*. Palo Alto, CA: Dale Seymour, 1983.

Making Connections

Baumgart, John K. *Historical Topics for the Mathematics Classroom*. Reston, VA: The National Council of Teachers of Mathematics, 1969, 1989.
Glatzer, David J., and Joyce Glatzer. *Math Connections: Middle-School Activities*. Palo Alto, CA: Dale Seymour, 1988.

Krause, Marina C. *Multicultural Mathematics Materials*. Reston, VA: The National Council of Teachers of Mathematics, Inc., 1983.

Questioning Techniques

Johnson, David R. *Every Minute Counts: Making Your Math Class Work*. Palo Alto, CA: Dale Seymour, 1982.
McCullough, Dorothy, and Edye Findley. "How to Ask Effective Questions." *Arithmetic Teacher* 30 (March, 1983): 8–9.

Using Manipulatives

Howden, Hilde. *Algebra Tiles for the Overhead Projector*. New Rochelle, NY: Cuisenaire Company of America, 1985.
Joyner, Jeane M. "Using Manipulatives Successfully." *Arithmetic Teacher* 38 (Oct., 1990): 6–7.
National Council of Teachers of Mathematics. "Arithmetic Teacher: Focus Issue on Manipulatives." *Arithmetic Teacher* 33 (Feb., 1986).

Using Technology

Bloom, Marjorie W., and Grace K. Galton. *Estimate! Calculate! Evaluate!: Calculator Activities for the Middle Grades*. New Rochelle, NY: Cuisenaire Company of America, 1990.
Coburn, Terrence G., Shirley Hoogeboom, and Judy Goodnow. *The Problem Solver with Calculators*. Sunnyvale, CA: Creative Publications, Inc., 1989.
Papert, Seymour. *Mindstorms: Children, Computers and Powerful Ideas*. New York: Basic Books, Inc., 1980.

Pacing Chart/Assignment Guide

Pacing Chart

A yearly *Pacing Chart* and daily assignments are provided for three levels of courses—a skills course, an average course, and an advanced course. All three levels provide for 160 days, including review and testing days. The *Pacing Chart* at the right shows the number of days allotted for each chapter of the three courses. Semester and trimester divisions are indicated by a red rule and blue rule, respectively.

Chapter	1	2	3	4	5	6	7	8	9	10	11	12	13	14	15
Skills Course	12	12	11	12	12	11	13	12	12	13	11	10	10	9	0
Average Course	12	12	11	12	12	11	13	12	10	11	10	10	10	9	5
Advanced Course	12	12	11	12	12	10	13	11	8	11	9	10	11	9	9

trimester   semester   trimester

Skills Course

The *skills course* is intended for students who need practice and review of arithmetic skills in order to study algebra successfully. The course covers many of the lessons in the first 14 chapters, usually allowing one day for each lesson. The suggested daily assignments include the A-level skill exercises, some B-level exercises, and all of the Spiral Review exercises.

Average Course

The *average course* is intended for students who, while they may need some practice and review of arithmetic skills, can apply these skills to problem-solving situations. The course covers, in a more in-depth manner than the skills course, many of the lessons in Chapters 1–15. The suggested daily assignments include many of the A-level and B-level exercises, plus some C-level exercises, and all of the Spiral Review exercises.

Advanced Course

The *advanced course* is an accelerated course intended for students who need little practice or review of arithmetic skills. These students master and apply concepts quickly. The course covers nearly all of the lessons in Chapters 1–15. The suggested daily assignments include some A-level exercises, most of the B-level and C-level exercises, and all of the Spiral Review exercises. If time permits, the enrichment lessons in Looking Ahead may be covered as well.

Assignment Guide

The *Assignment Guide* for each chapter is located on the interleaved pages preceding the chapter. A part of the Assignment Guide for Chapter 11 is shown here. A key describing the numbering system for the assignments is shown below.

key

Exercises

Lesson Number
1–7: 1–29 odd, 31–40
Exploration: Activity I

Focus on Explorations
Focus on Applications
Focus on Computers
Chapter Extension

Day	Skills Course	Average Course	Advanced Course
1	**11-1:** 1–11	**11-1:** 1–7 odd, 8–11 **11-2:** 1–20	**11-1:** 1–7 odd, 8–11, Challenge **11-2:** 1–20, Historical Note
2	**11-2:** 1–20	**11-3:** 1–19, 22–26	**11-3:** 1–26
3	**11-3:** 1–17, 22–26	**11-4:** 1–11, 13–16	**11-4:** 1–16
4	**11-4:** 1–8, 13–16	**11-5:** 1–16, 21–28	**11-5:** 1–28
5	**11-5:** 1–16, 23–28	**11-6:** 1–26	**11-6:** 1–17 odd, 19–26 **11-7:** 1–7 odd, 9–18
6	**11-6:** 1–18, 23–26	**11-7:** 1–12, 16–18	**Focus on Computers** **11-8:** 1–16

Diagnostic Test on Toolbox Skills

Test on Estimation Skills

Skill 1 — *Rounding*

Estimate the answer. Round each number to the place of its leading digit.

1. $\begin{array}{r} 212 \\ +186 \\ \hline \end{array}$

2. $\begin{array}{r} 372.9 \\ -168.2 \\ \hline \end{array}$

3. $\begin{array}{r} 85.6 \\ \times 21.2 \\ \hline \end{array}$

4. $\begin{array}{r} 75.659 \\ +13.346 \\ \hline \end{array}$

Skill 2 — *Front-end and Adjust*

Estimate the answer by adjusting the sum of the front-end digits.

5. $\begin{array}{r} 476 \\ +521 \\ \hline \end{array}$

6. $\begin{array}{r} 157.13 \\ +\ 38.24 \\ \hline \end{array}$

7. $\begin{array}{r} 456.218 \\ 21.004 \\ +\ \ 18.219 \\ \hline \end{array}$

8. $\begin{array}{r} 16,340 \\ 73,281 \\ +50,962 \\ \hline \end{array}$

Skill 3 — *Compatible Numbers*

Use compatible numbers to estimate the quotient.

9. $38\overline{)791}$

10. $9.1\overline{)281.6}$

11. $21.64\overline{)5986.95}$

12. $593\overline{)25,284}$

Skill 4 — *Clustering*

Use clustering to estimate the sum.

13. $41.3 + 40.1 + 39.9$

14. $63 + 55 + 61 + 59$

15. $127 + 129 + 133 + 131 + 135 + 132$

16. $10.65 + 9.59 + 10.03 + 10.34$

Test on Numerical Skills

Skill 5 — *Place Value; Writing Numbers*

Give the place and value of the under-lined digit.

17. $19,7\underline{2}8$

18. $6.05\underline{1}$

Write the word form of each number.

19. 537

20. 209.03

Skill 6 — *Rounding Whole Numbers and Decimals*

Round to the nearest hundred.

21. 757

22. 9633

Round to the nearest tenth.

23. 15.92

24. 543.063

Adding Whole Numbers

Add.

25. 72
  +85

26. 983
  +714

27. 517
  + 96

28. 4736
  208
  + 592

Skill 8   *Subtracting Whole Numbers*

Subtract.

29. 579
  −461

30. 86,329
  −35,104

31. 193
  − 57

32. 43,716
  −25,419

Skill 9   *Multiplying Whole Numbers*

Multiply.

33. 903
  × 31

34. 53,104
  ×   22

35. 78
  ×35

36. 609
  ×147

Skill 10   *Dividing Whole Numbers*

Divide. If there is a remainder, show it in fraction form.

37. $47\overline{)106}$

38. $16\overline{)5396}$

Divide. Round the quotient to the nearest tenth.

39. $28\overline{)514}$

40. $79\overline{)9234}$

Skill 11   *Adding Decimals*

Add.

41. 832.27
  +366.59

42. 428.461
  19.035
  +53.197

43. 36.9
  +18.23

44. 4.008
  59.26
  + 0.9

Skill 12   *Subtracting Decimals*

Subtract.

45. 800.11
  −673.19

46. 753.895
  −203.776

47. 27.4
  − 9.13

48. 0.1597
  −0.06

Skill 13   *Multiplying Decimals*

Multiply.

49.   8.7
      ×4.1

50.   15.06
      ×   9.3

51.   0.38
      × 0.2

52.   0.059
      ×   0.8

Skill 14   *Dividing Decimals*

Divide. If necessary, round the quotient to the nearest hundredth.

53. $7.6\overline{)34.2}$

54. $4.1\overline{)33.333}$

55. $16.9\overline{)58.61}$

56. $0.13\overline{)3.6847}$

Skill 15   *Multiplying and Dividing by 10, 100, and 1000*

Multiply by 1000.

Divide by 100.

57. 0.3

58. 172.5124

59. 120.34

60. 0.00483

Skill 16   *Divisibility Tests*

Test the number for divisibility. Write *Yes* or *No*.

61. Is 736 divisible by 2?

62. Is 609 divisible by 9?

63. Is 1934 divisible by 4?

64. Is 3360 divisible by 8?

Skill 17   *Adding and Subtracting Fractions*

Add or subtract. Write the answer in lowest terms.

65. $\dfrac{5}{14}$
   $+\dfrac{3}{14}$

66. $\dfrac{5}{12}$
   $+\dfrac{1}{3}$

67. $\dfrac{12}{25}$
   $-\dfrac{7}{25}$

68. $\dfrac{9}{10}$
   $-\dfrac{1}{15}$

Skill 18   *Adding Mixed Numbers*

Add. Write each sum in lowest terms.

69. $8\dfrac{1}{9}$
   $+4\dfrac{5}{9}$

70. $11\dfrac{5}{7}$
   $+23\dfrac{3}{7}$

71. $2\dfrac{4}{7}$
   $+9\dfrac{11}{14}$

72. $2\dfrac{11}{16}$
   $+16\dfrac{1}{5}$

Subtracting Mixed Numbers

Subtract. Write each difference in lowest terms.

73. $26\frac{25}{27}$
    $-21\frac{16}{27}$

74. 15
    $-\ 8\frac{5}{9}$

75. $27\frac{6}{7}$
    $-12\frac{3}{14}$

76. $19\frac{1}{2}$
    $-\ 8\frac{7}{8}$

Skill 20 *Multiplying Fractions and Mixed Numbers*

Multiply. Write each product in lowest terms.

77. $\frac{2}{15} \times \frac{3}{10}$

78. $12 \times \frac{5}{24}$

79. $18\frac{3}{4} \times \frac{1}{15}$

80. $3\frac{3}{4} \times 9\frac{3}{5}$

Skill 21 *Dividing Fractions and Mixed Numbers*

Divide. Write each quotient in lowest terms.

81. $\frac{17}{38} \div \frac{2}{19}$

82. $17\frac{1}{3} \div \frac{13}{15}$

83. $33 \div \frac{3}{11}$

84. $2\frac{6}{7} \div 1\frac{1}{4}$

Skill 22 *U.S. Customary System of Measurement*

Complete.

85. $9\frac{1}{4}$ yd = _?_ ft = _?_ in.

86. 11 gal = _?_ qt = _?_ pt

87. 160,000 oz = _?_ lb = _?_ t

88. 29 ft = _?_ yd _?_ ft

Test on Graphing Skills

Skill 23 *Reading Pictographs*

Use the pictograph at the right.

89. In which month were the fewest team banners sold?

90. Was the number of team banners sold greater in July or in August?

91. About how many team banners were sold in June?

92. About how many team banners were sold from May through September?

Sales of Team Banners

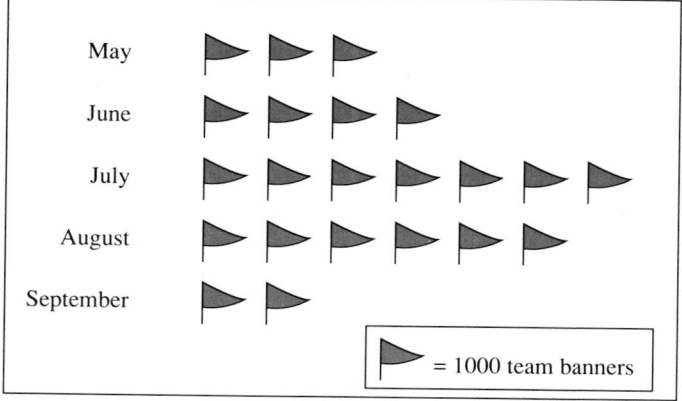

Skill 24   *Reading Bar Graphs*

Use the bar graph at the right.

93. Which city is the farthest from Denver?

94. Which city is farther from Denver, New Orleans or Seattle?

95. Estimate the distance from Dallas to Denver.

96. About how much farther is it from Miami to Denver than from Seattle to Denver?

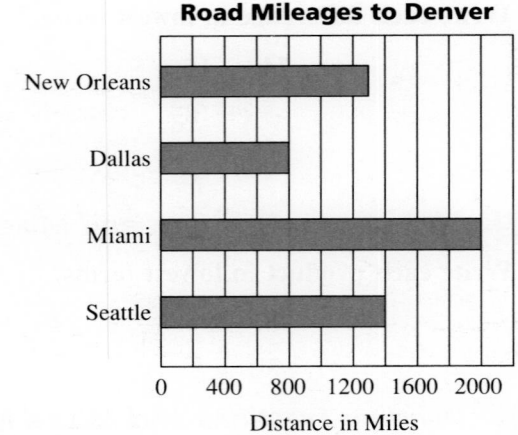

Road Mileages to Denver

Distance in Miles

Skill 25   *Reading Line Graphs*

Use the line graph at the right.

97. About how much money did the Hsu family spend on home heating oil in October?

98. During which month did the Hsu family pay the most money for home heating oil?

99. During which month did the Hsu family pay the least money for home heating oil?

100. About how much more money did the Hsu family pay for home heating oil in January than in December?

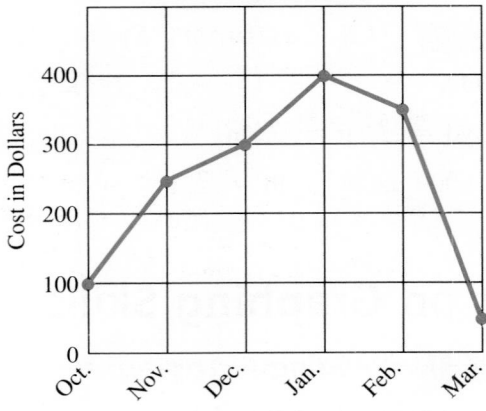

Home Heating Oil Costs for the Hsu Family

Cost in Dollars

Answers to Diagnostic Test on Toolbox Skills are on p. A46.

Mathematical Connections

A Bridge to Algebra and Geometry

Francis J. Gardella

Patricia R. Fraze

Joanne E. Meldon

Marvin S. Weingarden

Cleo Campbell, CONTRIBUTING AUTHOR

TEACHER CONSULTANTS

Alma Cantu Aguirre

Cheryl Arevalo

Douglas Denson

Jackie G. Piver

Jocelyn Coleman Walton

Houghton Mifflin Company • BOSTON

Atlanta • Dallas • Geneva, Ill. • Palo Alto • Princeton • Toronto

AUTHORS

Francis J. Gardella, Supervisor of Mathematics and Computer Studies, East Brunswick Public Schools, East Brunswick, New Jersey

Patricia R. Fraze, former Mathematics Department Chairperson, Huron High School, Ann Arbor, Michigan

Joanne E. Meldon, Mathematics Teacher, Taylor Allderdice High School, Pittsburgh, Pennsylvania

Marvin S. Weingarden, Supervisor of Secondary Mathematics, Detroit Public Schools, Detroit, Michigan

Cleo Campbell, Coordinator of Mathematics, Anne Arundel County Public Schools, Annapolis, Maryland

TEACHER CONSULTANTS

Alma Cantu Aguirre, Mathematics Department Chairperson, Thomas Jefferson High School, San Antonio, Texas

Cheryl Arevalo, Supervisor of Mathematics, Des Moines Public Schools, Des Moines, Iowa

Douglas Denson, Principal, Twin Bridges Elementary School, Twin Bridges, Montana

Jackie G. Piver, Mathematics Department Chairperson, Enloe High School, Raleigh, North Carolina

Jocelyn Coleman Walton, Supervisor of Mathematics, Plainfield High School, Plainfield, New Jersey

The authors wish to thank Robert H. Cornell, Mathematics Teacher, Milton Academy, Milton, Massachusetts, and James M. Sconyers, Mathematics Teacher, East Preston High School, Terra Alta, West Virginia, for their contributions to this textbook.

ISBN: 0-395-46150-2

LETTER TO STUDENTS

Dear student,

You're invited . . .

Mathematical Connections is a bridge that will take you from where you are in your study of mathematics to algebra and geometry. Since topics in mathematics are connected, this course will also lead you to data analysis and probability. We have written this textbook because we want you to enjoy your continuing journey into mathematics.

to explore . . .

In this course you're expected to get actively involved in learning: explore, ask questions, discuss alternatives, and make connections between what's new and what's known. Exploration can be done by using mathematical models, by using a calculator or a computer, and by using your active, open mind.

together . . .

If you have difficulty, and you may at times, don't get discouraged. Your teacher wants you to succeed and will be working hard to help you. Your classmates are there with you, and you may be able to solve a problem more easily by working cooperatively in a group. We've also built help for you into the textbook. We encourage you to read the explanations in your book, study the *Examples*, answer the *Check Your Understanding* questions, and compare your answers with those at the back of your book.

for your future . . .

Don't close off any roads to the future at this stage of your journey. In tomorrow's world, the new situations you'll face as a citizen may call for decision-making skills you're building now, and your job may be one that you can't even imagine today. Through the study of mathematics, prepare yourself to cross any bridge along the way.

We wish you a successful, enjoyable journey.

Francis J. Gardella

Cleo Campbell

Patricia R. Fraze

Joanne E. Meldon

Marvin Weingarden

CONTENTS

photo from page 85

CHAPTER 1

CHAPTER 2

v

photo from page 355

photo from page 331

photo from page 278

photo from page 523

photo from page 698

photo from page 466

photo from page 679

TABLE OF SYMBOLS

		Page			Page
$\cdot$	$\times$ (times)	12	$\parallel$	is parallel to	249
$\dfrac{x}{y}$	$\div$ (division)	12	$\leq$	is less than or equal to	325
$>$	is greater than	18	$\geq$	is greater than or equal to	325
$<$	is less than	18	$\dfrac{b}{a}$	reciprocal of $\dfrac{a}{b}$	340
$=$	equals, is equal to	18	$a:b$	ratio of a to b	388
( )	parentheses—a grouping symbol	22	%	percent	404
$-$	negative	94	π	pi, a number approximately equal to 3.14 and $\dfrac{22}{7}$.	441
$+$	positive	94			
$-n$	opposite of n	94	$\cong$	is congruent to	462
$\lvert n \rvert$	absolute value of n	95	$\triangle ABC$	triangle ABC	463
(x, y)	ordered pair of numbers	124	$\sim$	is similar to	466
$\neq$	is not equal to	142	$\sqrt{\ }$	positive square root	475
$\approx$	is approximately equal to	213	$P(E)$	probability of event E	541
$\overleftrightarrow{AB}$	line AB	236	$n!$	n factorial	559
$\overline{AB}$	line segment AB	237	$_nP_r$	number of permutations of n items taken r at a time	560
$\overrightarrow{AB}$	ray AB	237	$_nC_r$	number of combinations of n items taken r at a time	560
$\angle A$	angle A	237	$f(x)$	f of x, the value of f at x	613
$^\circ$	degree(s)	240	$\sin A$	sine of angle A	716
$m\angle A$	measure of angle A	240	$\cos A$	cosine of angle A	716
$\perp$	is perpendicular to	249	$\tan A$	tangent of angle A	716

TABLE OF MEASURES

Time

60 seconds (s) = 1 minute (min)	365 days
60 minutes = 1 hour (h)	52 weeks (approx.) } = 1 year
24 hours = 1 day	12 months
7 days = 1 week	10 years = 1 decade
4 weeks (approx.) = 1 month	100 years = 1 century

Metric	United States Customary

Length

10 millimeters (mm) = 1 centimeter (cm)

$\left.\begin{array}{l}100 \text{ cm} \\ 1000 \text{ mm}\end{array}\right\}$ = 1 meter (m)

1000 m = 1 kilometer (km)

Length

12 inches (in.) = 1 foot (ft)

$\left.\begin{array}{l}36 \text{ in.} \\ 3 \text{ ft}\end{array}\right\}$ = 1 yard (yd)

$\left.\begin{array}{l}5280 \text{ ft} \\ 1760 \text{ yd}\end{array}\right\}$ = 1 mile (mi)

Area

100 square millimeters = 1 square centimeter (mm^2)   (cm^2)

10,000 cm^2 = 1 square meter (m^2)

10,000 m^2 = 1 hectare (ha)

Area

144 square inches (in.^{2}) = 1 square foot (ft^2)

9 ft^2 = 1 square yard (yd^2)

$\left.\begin{array}{l}43,560 \text{ ft}^2 \\ 4840 \text{ yd}^2\end{array}\right\}$ = 1 acre (A)

Volume

1000 cubic millimeters = 1 cubic centimeter (mm^3)   (cm^3)

1,000,000 cm^3 = 1 cubic meter (m^3)

Volume

1728 cubic inches (in.^{3}) = 1 cubic foot (ft^3)

27 ft^3 = 1 cubic yard (yd^3)

Liquid Capacity

1000 milliliters (mL) = 1 liter (L)

1000 L = 1 kiloliter (kL)

Liquid Capacity

8 fluid ounces (fl oz) = 1 cup (c)

2 c = 1 pint (pt)

2 pt = 1 quart (qt)

4 qt = 1 gallon (gal)

Mass

1000 milligrams (mg) = 1 gram (g)

1000 g = 1 kilogram (kg)

1000 kg = 1 metric ton (t)

Weight

16 ounces (oz) = 1 pound (lb)

2000 lb = 1 ton (t)

Temperature–Degrees Celsius (°C)

0°C = freezing point of water

37°C = normal body temperature

100°C = boiling point of water

Temperature–Degrees Fahrenheit (°F)

32°F = freezing point of water

98.6°F = normal body temperature

212°F = boiling point of water

USING TECHNOLOGY

TECHNOLOGY IN OUR DAILY LIVES

Today's World. Over the past 20 years, technology has dramatically changed the way that we live and work. From hand-held calculators to microwave ovens and VCRs to large-scale computers that process financial data or help plan space explorations, calculators and computers have become an important part of our lives.

Computers are well known for their ability to perform numerical calculations quickly. An even more important contribution of computers may be their ability to transmit information (words, pictures, sounds) electronically. By means of computers, information can be shared almost instantly with people throughout the world.

USING CALCULATORS

Calculators for Everyday Use. Since calculators are useful in a wide variety of different situations, specialized calculators have been designed to meet particular home, school, or job needs. For example, a calculator that does arithmetic operations is useful for everyday needs such as checking an itemized bill, balancing a checkbook, or preparing a tax form.

Tomorrow's World. In the 21st century, even more jobs than today will involve the use of calculators or computers. Becoming familiar with these technological tools will help you prepare for your future work. It will also help you become a well-informed citizen. In order to make effective decisions about the public and private uses of technology, every citizen needs a general understanding of what calculators and computers can—and cannot—do.

Scientific Calculators.
A calculator that can perform more advanced mathematical operations, such as those studied in advanced high school or college courses, is particularly useful to scientists, engineers, and others who use mathematics in their daily work.

A calculator that can perform financial and statistical operations is especially useful to bankers, financial analysts, and statisticians.

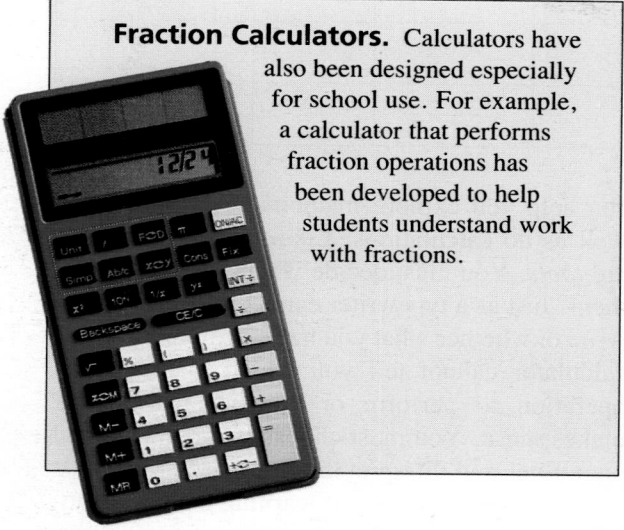

Fraction Calculators. Calculators have also been designed especially for school use. For example, a calculator that performs fraction operations has been developed to help students understand work with fractions.

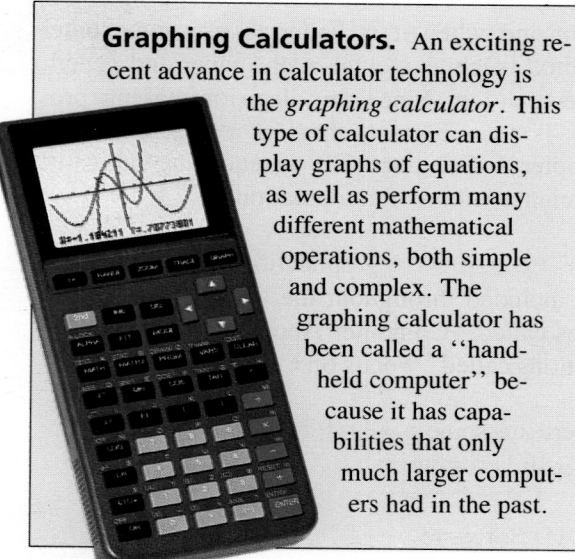

Graphing Calculators. An exciting recent advance in calculator technology is the *graphing calculator*. This type of calculator can display graphs of equations, as well as perform many different mathematical operations, both simple and complex. The graphing calculator has been called a "hand-held computer" because it has capabilities that only much larger computers had in the past.

USING COMPUTERS

Computer Software. Programmers can write special programs to solve particular problems. However, many programs used in schools and businesses are written in a general form, with instructions for inserting formulas or values. Such programs may be supplied on disks to be transferred to the computer's memory. Here are some general categories:

Function Graphing and Automatic Drawing. Function graphing software allows the user to quickly plot graphs of functions by entering an equation or a formula.

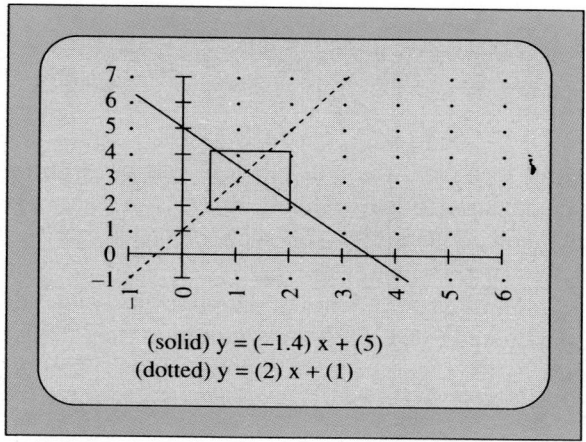

(solid) $y = (-1.4) x + (5)$
(dotted) $y = (2) x + (1)$

Most function graphing programs contain options for changing the scale, or "zooming," to better investigate small portions of the graph, such as the region where the graph crosses an axis. Automatic drawing software can be used to draw and measure geometric figures.

Word Processing. A word processing program accepts typed text and makes corrections as requested, adjusting the lines to fit. The text may be divided into paragraphs and pages, and words may be made italic or boldface. The methods of page make-up can produce pages ready to be printed directly. This process is called "desktop publishing."

Manipulating databases. With database software, collections of data, such as membership lists with addresses, telephone numbers, and other information, can be sorted and analyzed in various ways. For instance, mailings can be sorted by zip codes.

Spreadsheets. With an electronic spreadsheet like the one pictured on the next page, a user can enter a number, a formula, or a label (word) into each compartment, or *cell*. Such a program can be used to produce financial statements, solve equations, or determine function values. An especially useful feature is that by

Spreadsheet program provides a network of cells.

	A	B	C	D	E	F	G
1							
2							
3							

using formulas, the entire layout can be recomputed automatically whenever a given value is changed. The effects of such changes can be determined immediately.

Simulations. Simulation programs can be constructed to take certain assumptions and compute what the results would be. These results

can then be used to predict what will happen in the real-world situation. Simulation techniques are currently used in a wide range of fields, including financial forecasting, astronaut training, consumer-buying patterns, and building and city planning. Automobile manufacturers often use wind-tunnel simulations to investigate the effects of different design modifications. Simulations are particularly useful when the situation under consideration is difficult, impossible, costly, or very dangerous to observe directly.

USING TECHNOLOGY IN THIS COURSE

Tools for Learning. Calculators and computers can help you learn mathematics. They can help you explore mathematical concepts as well as do calculations. Like a typewriter, they are *tools*. You must decide when and how to use them. Just as a typewriter cannot tell you what to write or whether what you have written is good, a calculator cannot tell you which mathematical operation to perform or whether the answer makes sense. You must choose the mathematical operations and interpret the result.

An important part of learning to use technological tools is deciding when it is efficient to use them and when it is better to use some other method, such as mental math, paper and pencil, or estimation. Since this decision-making process is so important, there is a lesson right in Chapter 1 (see page 28) on choosing the most efficient method of computation.

Computers. Applications of computers are included throughout the book—in the exercises called "Computer Application" and in the sections called "Focus on Computers."

Here are some page references:

BASIC Programs	6, 298, 485, 567, 688
CAD Software	631
Database Software	220, 349
Geometric Drawing Software	270, 470
Graphing Software	
Function Graphing	596, 598
Geometric Graphing	131
Statistical Graphing	212, 512, 521
Logo Programs	265
Spreadsheets	55, 70, 174, 411

Calculators. Calculators are featured in many places in this course—in the lessons, in the exercises, and in special explorations. The following chart lists some topics and the associated calculator keys that you will learn about in this course.

Topics	Keys	Explained, Pages
Using Memory	[M+] [M−] [MR] [MC]	xviii
Basic Operations	[+] [−] [×] [÷] [=]	7–8, 11–12
Exponents	[y^x] [x^2]	31, 453
Order of Operations	[(] [)]	37
Repeated Operations	[K]	63
Scientific Notation	[EE] [EXP]	81
Operations with Integers	[+/−]	108
Finding Statistics	[Σx] [n] [$\bar{x}$]	231
Fraction Operations	[$a^b/_c$]	341
Finding Reciprocals	[1/X]	345
Finding Percents	[%]	409
Geometric Measurement	[π]	441
Finding Square Roots	[$\sqrt{x}$]	476
Finding Permutations	[x!]	561
Graphing Functions	[Y=] [X|T] [GRAPH] [(−)] [∧]	586, 617
Trigonometric Functions	[SIN] [COS] [TAN] [INV]	719, 723

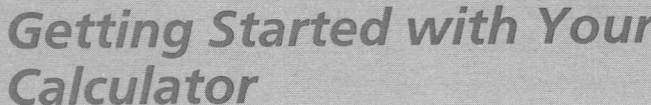

Getting Started with Your Calculator

Calculators can help you to do complicated computations easily. In order to use a calculator effectively, you should become familiar with its basic keys. Some particularly useful keys are those that control the memory.

[M+] adds the number displayed to the number in the memory.

[M−] subtracts the number displayed from the number in the memory.

[MR] or [MRC] recalls the number stored in the memory.

[MC] clears the memory by setting it to zero.

5. The 6-digit product will be the 3-digit number written twice.

Activity I   *Using the Memory Keys*

1 Turn on your calculator and be sure that the display and memory are clear.

2 Store the number 101 in the memory using the following key sequence:

[1] [0] [1] [M+]

3 Clear the display, then multiply the number in the memory by 53 using the following key sequence:

[5] [3] [×] [MR] [=]

Record the result on a piece of paper. 5353

4 Repeat Step 3 using other two-digit numbers. What pattern do you notice in the results? You can form the 4-digit product of a 2-digit number and 101 by writing the 2-digit number twice.

5 Predict what might happen when you multiply a three-digit number by the number 1001. Test your prediction using your calculator.

6 What number should you store in memory to obtain a similar pattern when you multiply a four-digit number? Use a calculator to test your prediction. 10001

Activity II   *Looking for Patterns*

1. Turn on your calculator and be sure that the display and memory are clear.

2. Store the number 11 in the memory.

3. Clear the display. Then divide 1 by the number in the memory using the following key sequence.

   Record the result on a piece of paper. 0.090909

4. Repeat Step 3 using the digits from 2 through 9. What pattern do you notice in the results? The quotient is a repeating decimal formed by repeating the 2-digit product of the dividend and 9.

1–3. Answers will vary. Examples are given.
   1. 89 + 6 − 75
   2. 5 + 6 + 9 − 7 − 8
   3. 5 + 8 + 9 − 6 − 7

Activity III   *Problem Solving*

Using the key sequence shown below, you will obtain the displayed result.

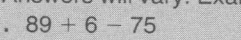

1. Copy the diagram shown below on your paper. Using each of the digits from 5 through 9, the plus key, and the minus key, fill in the boxes to obtain the displayed result. Keep a record of your attempts.

   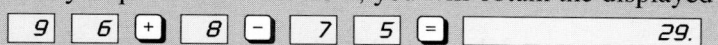

2. Repeat Step 1 using the diagram shown below.

   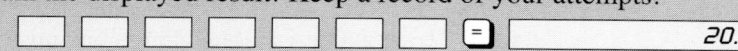

3. Repeat Step 2, but this time obtain the number 9.

4. Using each of the digits from 5 through 9, the plus key, and minus key, what is the greatest possible result you can obtain using the diagram shown below? 988

   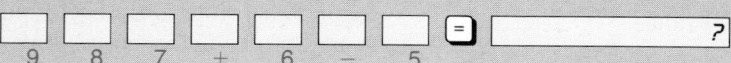
   9   8   7   +   6   −   5

Activity IV   *Problem Solving*

1. Copy the diagram at the right on your paper.

2. Using each of the digits from 5 through 9, fill in the boxes to obtain the greatest possible product. Do the computations on your calculator. Make a table to record your attempts. 96 × 875 = 84,000

3. Repeat Step 2, but this time obtain the least possible product. 68 × 579 = 39,372

Planning Chapter 1

Chapter Overview

Chapter 1 connects the arithmetic students should know with the algebra they will be expected to learn. In a sense, this chapter begins the journey from the concrete to the abstract. However, to begin this journey, students should have certain skills. To check these skills, we suggest you use the Diagnostic Test on page T44. If students are weak in any area, you can assign some exercises from the appropriate Toolbox Skills Practice that begins on page 753. Point out to students that the skills on the Diagnostic Test are important for success in Chapter 1 and subsequent chapters in the book.

Chapter 1 begins by introducing students to variables and variable expressions by using whole numbers as replacements for the variables. *Lessons 2 and 3* take a more in-depth look at variable expressions, using both whole numbers and decimals. Problem solving begins in *Lesson 4*, as students are asked to read a paragraph and then analyze what they have read. *Lessons 5–7* present properties of real numbers. *Lesson 8* begins the decision-making strand by requiring students to decide the most efficient method for doing a computation. *Lessons 9 and 10* present positive exponents and the order of operations.

Background

Chapter 1 introduces algebra early. This gives students a sense that they are learning new mathematics and not just reviewing previously learned skills. Basic arithmetic skills are reviewed, however, as many exercises in the chapter have students evaluate variable expressions for given values of the variables.

Problem solving is introduced early in Chapter 1. Students are asked to analyze what they have read and to identify information that is given or not given. The important first step in problem solving is to think about what has been read.

Chapter 1 also contains the first decision-making lesson. Students often do not take the time to decide the most efficient method to do a problem; instead, they reach for a calculator. *Lesson 8* shows that in many cases using either paper and pencil or mental math can be more efficient than a calculator.

Calculator usage is encouraged in Chapter 1. A number of lessons show how to use a calculator. Sample key sequences are provided, and students should be encouraged to try them. Calculator usage is reinforced by the inclusion of calculator exercises. These exercises are not merely "push the button" exercises. They are designed to make students think about whether a calculator answer is correct or about the appropriateness of a particular key sequence. Estimation is a key skill when students use calculators. Checking for the reasonableness of an answer is important because it is easy to make a mistake when entering numbers on a calculator.

The *Focus on Explorations* uses paper folding to explore powers of two. This manipulative activity demonstrates to students how quickly powers of numbers increase. Activity III encourages the use of logical reasoning, as students have to unfold sheets of paper mentally to create designs and then physically create designs by folding and cutting sheets of paper.

Objectives

1-1	To evaluate variable expressions involving whole numbers.
1-2	To evaluate variable expressions involving addition and subtraction of whole numbers and decimals.
1-3	To evaluate variable expressions involving multiplication and division of whole numbers and decimals.
1-4	To read problems for understanding.
1-5	To use the comparison property to compare and to order numbers.
1-6	To recognize the properties of addition and to use them to find sums mentally.
1-7	To recognize the properties of multiplication and to use them to find products mentally.
1-8	To decide whether it would be most efficient to use mental math, paper and pencil, or a calculator to solve a given problem.
1-9	To compute with exponents and to evaluate expressions involving exponents.
	FOCUS ON EXPLORATIONS To use paper folding to model powers of two.
1-10	To use the order of operations to compute and to evaluate expressions.
	CHAPTER EXTENSION To create and to use new operations.

Chapter Planner

Instructional Aids	Manipulatives	Cooperative Learning	Technology	Practice/Reteaching	Assessment
Materials: paper clips, cups, calculators, scissors *Resource Book:* Lesson Starters 1–10	Lessons 1-1, 1-2, 1-3, Focus on Explorations *Activities Book:* Manipulative Activities 1–2	Lessons 1-3, 1-4, Chapter Extension *Activities Book:* Cooperative Activities 1–4	Lessons 1-1, 1-2, 1-3, 1-5, 1-8, 1-9, 1-10 *Activities Book:* Calculator Activity 1 Computer Activity 1 *Connections Plotter Plus* Disk	Toolbox, pp. 753–771 Extra Practice, p. 730 *Resource Book:* Practice 1–13 *Study Guide,* pp. 1–20	Self-Tests 1–2 Quick Quizzes 1–2 Chapter Test, p. 42 *Resource Book:* Lesson Starters 1–10 (Daily Quizzes) *Tests 1–4*

Assignment Guide Chapter 1

Day	Skills Course	Average Course	Advanced Course
1	**1-1:** 1–18, 21–37 odd, 39–44	**1-1:** 1–20, 21–37 odd, 39–44	**1-1:** 1–17 odd, 19–44
2	**1-2:** 1–12, 13–17 odd, 21–27 odd, 30–36	**1-2:** 1–27 odd, 28–36	**1-2:** 1–11 odd, 13–36, Historical Note
3	**1-3:** 1–12, 13–39 odd, 42–46	**1-3:** 1–39 odd, 40–46	**1-3:** 1–11 odd, 13, 14, 15–39 odd, 40–46
4	**1-4:** 1–16	**1-4:** 1–16	**1-4:** 1–16
5	**1-5:** 1–12, 13–35 odd, 45–52	**1-5:** 1–14, 15–43 odd, 45–52	**1-5:** 1–11 odd, 13, 14, 15–31 odd, 33–52
6	**1-6:** 1–16, 17–25 odd, 29, 30	**1-6:** 1–25 odd, 26–30	**1-6:** 1–25 odd, 26–30
7	**1-7:** 1–18, 19–29 odd, 35–40	**1-7:** 1–29 odd, 31–40	**1-7:** 1–29 odd, 31–40
8	**1-8:** 1–12, 13, 15, 17–20	**1-8:** 1–20	**1-8:** 1–20
9	**1-9:** 1–20, 21–37 odd, 39, 40, 43, 45, 49–52 **Exploration:** Activity I and Activity II	**1-9:** 1–20, 21–37 odd, 39–46, 49–52 **Exploration:** Activities I–III	**1-9:** 1–17 odd, 19, 20, 21–37 odd, 39–52, Challenge **Exploration:** Activities I–III
10	**1-10:** 1–23, 25–35 odd, 38–41	**1-10:** 1–21 odd, 22, 23–37 odd, 38–41	**1-10:** 1–21 odd, 22–41
11	*Prepare for Chapter Test:* Chapter Review	*Prepare for Chapter Test:* Chapter Review	*Prepare for Chapter Test:* Chapter Review
12	*Administer Chapter 1 Test*	*Administer Chapter 1 Test;* Cumulative Review	*Administer Chapter 1 Test;* Chapter Extension; Cumulative Review

Teacher's Resources

Resource Book
Chapter 1 Project
Lesson Starters 1–10
Practice 1–13
Enrichment 1–11
Diagram Masters
Chapter 1 Objectives
Family Involvement 1
Spanish Test 1
Spanish Glossary

Activities Book
Cooperative Activities 1–4
Manipulative Activities 1–2
Calculator Activity 1
Computer Activity 1

Study Guide, pp. 1–20

Tests
Tests 1–4

Connections Plotter Plus **Disk**

Using Technology

CALCULATORS

This chapter introduces algebraic notation and variables. One of the principal goals of the chapter is to give students practice reading and evaluating algebraic and arithmetic expressions. Students will benefit from these activities most fully if they try them in several different ways. Calculators not only provide an alternative to paper and pencil and mental calculations, but also expand the types of problems that are feasible for students to tackle. In this chapter and throughout the book, students are asked to experiment with arithmetic expressions in order to find patterns that they can generalize by using algebra. Calculators allow for a greater range and depth of experimentation than paper and pencil or mental calculations.

COMPUTERS

Computers can provide yet another means to understand algebraic expressions. Students can gain insight by writing expressions in BASIC (or another computer language) and having the computer evaluate them. Spreadsheet programs, which use quite a different and helpful model of variables, can be used for the same purpose.

Lesson 1-2

COMPUTERS

Have students evaluate some algebraic expressions by using BASIC in immediate mode. For instance, they can type:

X = 4.96

Y = 3.77

PRINT X + 21.08 + Y

Students can verify the computer's results by using paper and pencil or a calculator.

Lesson 1-3

COMPUTERS

A spreadsheet program uses locations on the screen to store numerical values and variable expressions. For instance, you might store the number 16 in location A3, the number 27 in location A4, and the formula A3 × 19 × A4 in location A5 (the notation used depends on the program). When the program performs its calculations, the value of the expression will appear in A5. This use of a location as a *place* where numbers may be stored is a useful model for students to know.

Lesson 1-5

COMPUTERS

A spreadsheet program could be used to enter the data in Exercises 33–36. Help students write formulas for obtaining the total population of the country, for the change in population in each region, and for the country as a whole. You might also explore in a class discussion what other formulas, or what additional data, would give further insight into how the population of the United States has grown and shifted.

Lesson 1-9

CALCULATORS

Have students use a calculator to discover some of the laws of exponents. Then have them write the laws both in words and in algebraic notation. For example, if students calculate 3^4, 3^2, and 3^6, they should notice that the product of the first two numbers equals the third. After several examples, they can express their observations in words and then in an algebraic form such as

$$b^m b^n = b^{m+n}.$$

To solidify students' understanding, have them predict the result of $(7.1)^5(7.1)^3$ by calculating the single power of $(7.1)^8$.

Lesson 1-10

CALCULATORS AND COMPUTERS

It would be useful to have students explore the order of operations used on their calculators, in BASIC, and in spreadsheet programs. Any discrepancies they find will provide a useful lesson; that is, algebraic order of operations is an arbitrary convention, but it is an important one to learn.

Using Manipulatives

Paper cups and paper clips are very useful in demonstrating the idea of a variable expression.

Lesson 1-1

Give each student three paper cups and tell them that they are going to evaluate the expression $3n$ for various values of n. First, suppose that $n = 4$. You can demonstrate by placing four paper clips in each of the cups. The number of clips in the three cups is 3×4, or 12. Tell students that the variable, n, was replaced each time with a 4. The value of the expression is 12.

Repeat the above activity, but let $n = 5$. Fill each cup with five clips. There are three cups each containing five clips. The value of the expression is 15. It is easy to see that the value of the expression depends on the value of the variable.

The cup and paper clip idea can be used for addition and subtraction expressions by letting the cup represent the variable. For example, to evaluate $x + 3$, use a cup and three clips. Tell students to let $x = 2$. Fill the cup with two clips. Ask students: How many clips are there in all? 5 Thus, the expression $x + 3$ equals 5 when $x = 2$. Choose different values for x and repeat the process.

To evaluate subtraction expressions such as $12 - x$, first count out 12 clips. Now, suppose $x = 4$. Remove four clips and put them in the cup. Remove the cup. How many clips are left? 8 Thus, the value of $12 - x$ is 8 when $x = 4$.

Lessons 1-2, 1-3

If students are having difficulty with either lesson, you may wish to modify and repeat the manipulative activities given for Lesson 1-1.

Reteaching/Alternate Approach

Here are some alternate approach ideas for teaching or reteaching concepts in Lessons 1-4 through 1-8. The approach suggested for Lesson 1-4 works particularly well as a cooperative activity.

Lesson 1-4

COOPERATIVE LEARNING

Divide students into small groups of three or four. Ask each group to make up two paragraphs similar to those in this lesson. For each paragraph, the group should develop a list of four to six questions. Have the groups exchange paragraphs and answer the questions.

Lesson 1-5

To help students remember the meaning of the $<$ and $>$ symbols, there are several possible interpretations that you might use.

1. Tell students to think of each inequality symbol as an alligator's mouth and that the alligator eats the larger fish.
2. Tell students that each inequality symbol is a hand reaching out to grab the larger object.
3. Tell students that the large side of the symbol is by the larger number and that the small side of the symbol is by the smaller number.

Lessons 1-6, 1-7

To help students learn the properties of addition and multiplication, you may wish to make a poster. The properties can be grouped as in the text, or you may wish to group the commutative properties together, the associative properties together, and the identity properties together. Students can then give In Arithmetic–In Algebra examples as in the lessons. When other properties, such as the distributive property, are introduced later in the course, the poster can be expanded.

Lesson 1-8

Many students think that using a calculator is always the quickest way to do a computation. To demonstrate that this is not true, put three exercises on the chalkboard that can be done mentally. Tell students that you will race them to the answers, but that they must use their calculators. While students are still pushing buttons, give the answers to the three exercises.

Enrichment masters from the Resource Book are pictured here. See the Teacher's Resources chart on page 2B for a complete listing of all materials available for this chapter.

1-1 Enrichment
Connection

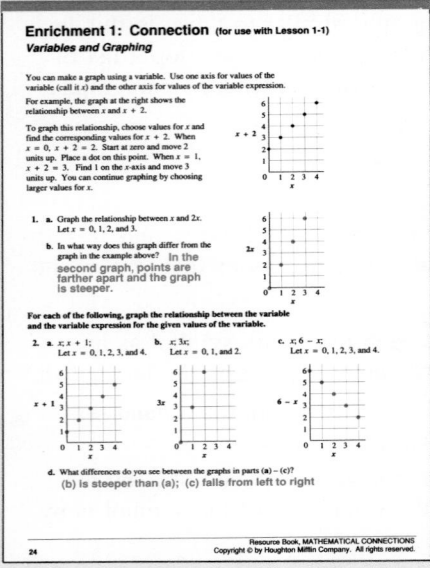

Enrichment 1: Connection (for use with Lesson 1-1)
Variables and Graphing

You can make a graph using a variable. Use one axis for values of the variable (call it x) and the other axis for values of the variable expression.

For example, the graph at the right shows the relationship between x and $x + 2$.

To graph this relationship, choose values for x and find the corresponding values for $x + 2$. When $x = 0$, $x + 2 = 2$. Start at zero and move 2 units up. Place a dot on this point. When $x = 1$, $x + 2 = 3$. Find 1 on the x-axis and move 3 units up. You can continue graphing by choosing larger values for x.

1. a. Graph the relationship between x and $2x$. Let $x = 0, 1, 2,$ and 3.

   b. In what way does this graph differ from the graph in the example above? In the second graph, points are farther apart and the graph is steeper.

For each of the following, graph the relationship between the variable and the variable expression for the given values of the variable.

2. a. x; $x + 1$; Let $x = 0, 1, 2, 3,$ and 4.
   b. x; $3x$; Let $x = 0, 1,$ and 2.
   c. x; $6 - x$; Let $x = 0, 1, 2, 3,$ and 4.

   d. What differences do you see between the graphs in parts (a) – (c)? (b) is steeper than (a); (c) falls from left to right

1-2 Enrichment
Application

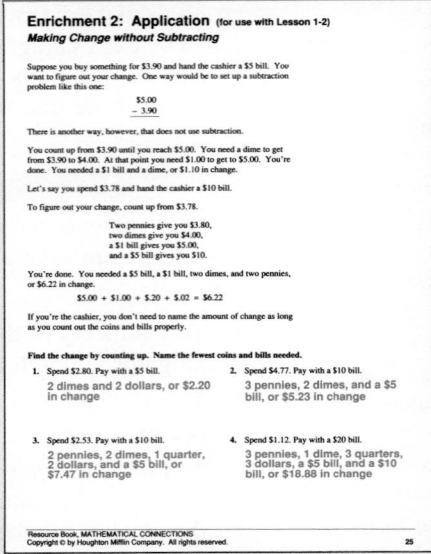

Enrichment 2: Application (for use with Lesson 1-2)
Making Change without Subtracting

Suppose you buy something for $3.90 and hand the cashier a $5 bill. You want to figure out your change. One way would be to set up a subtraction problem like this one:

$$\begin{array}{r} \$5.00 \\ -\ 3.90 \end{array}$$

There is another way, however, that does not use subtraction.

You count up from $3.90 until you reach $5.00. You need a dime to get from $3.90 to $4.00. At that point you need $1.00 to get to $5.00. You're done. You needed a $1 bill and a dime, or $1.10 in change.

Let's say you spend $3.78 and hand the cashier a $10 bill.

To figure out your change, count up from $3.78.

Two pennies give you $3.80,
two dimes give you $4.00,
a $1 bill gives you $5.00,
and a $5 bill gives you $10.

You're done. You needed a $5 bill, a $1 bill, two dimes, and two pennies, or $6.22 in change.

$$\$5.00 + \$1.00 + \$.20 + \$.02 = \$6.22$$

If you're the cashier, you don't need to name the amount of change as long as you count out the coins and bills properly.

Find the change by counting up. Name the fewest coins and bills needed.

1. Spend $2.80. Pay with a $5 bill.
   2 dimes and 2 dollars, or $2.20 in change

2. Spend $4.77. Pay with a $10 bill.
   3 pennies, 2 dimes, and a $5 bill, or $5.23 in change

3. Spend $2.53. Pay with a $10 bill.
   2 pennies, 2 dimes, 1 quarter, 2 dollars, and a $5 bill, or $7.47 in change

4. Spend $1.12. Pay with a $20 bill.
   3 pennies, 1 dime, 3 quarters, 3 dollars, a $5 bill, and a $10 bill, or $18.88 in change

1-3 Enrichment
Thinking Skills

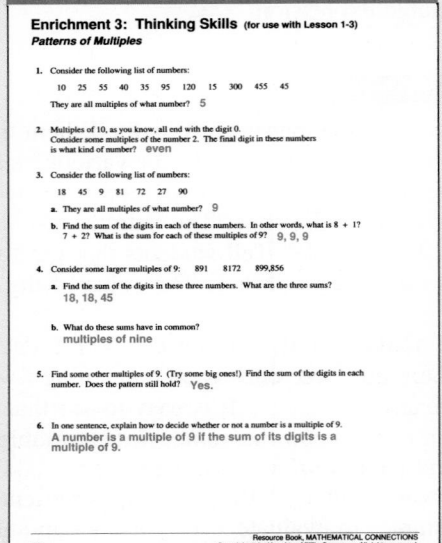

Enrichment 3: Thinking Skills (for use with Lesson 1-3)
Patterns of Multiples

1. Consider the following list of numbers:
   10   25   55   40   35   95   120   15   300   455   45
   They are all multiples of what number? 5

2. Multiples of 10, as you know, all end with the digit 0. Consider some multiples of the number 2. The final digit in these numbers is what kind of number? even

3. Consider the following list of numbers:
   18   45   9   81   72   27   90
   a. They are all multiples of what number? 9
   b. Find the sum of the digits in each of these numbers. In other words, what is 8 + 1? 7 + 2? What is the sum for each of these multiples of 9? 9, 9, 9

4. Consider some larger multiples of 9:   891   8172   899,856
   a. Find the sum of the digits in these three numbers. What are the three sums? 18, 18, 45
   b. What do these sums have in common? multiples of nine

5. Find some other multiples of 9. (Try some big ones!) Find the sum of the digits in each number. Does the pattern still hold? Yes.

6. In one sentence, explain how to decide whether or not a number is a multiple of 9.
   A number is a multiple of 9 if the sum of its digits is a multiple of 9.

1-4 Enrichment
Problem Solving

Enrichment 4: Problem Solving (for use with Lesson 1-4)
What's the Problem?

Sometimes one situation can create several different problems. A clear understanding of the situation will allow you to solve many of these problems. Studying the problems — by reading, looking, and questioning — is always the best way to start. There are times, however, when some problems just cannot be solved.

Consider this situation:

Sarah invited five friends, Jorge, Celeste, Jody, Jason, and Sonya, to go to the movies at 7:15. Before the movie they are all going to meet for dinner. The restaurant is a 10-minute walk from the movie theater, and dinner will take about 20 minutes. Dinner will cost about $6 for each person, and the movie will cost exactly $6.50. Sarah told everybody to bring $20. Sarah lives 5 minutes from the restaurant.

Problems

1. What is the latest time Sarah can leave her home to go to the restaurant? 6:40

2. Explain why you *cannot* tell how long it will take Sarah to walk home after the movie. Her home could be between the restaurant and the theater, or not.

3. Is there anything you can tell about the time it will take Sarah to walk home? The longest it could take her would be 15 min.

4. After reading the paragraph carefully, write two *different* questions that can be answered with the given information. Identify what information is used and explain how to solve each problem. Answers will vary. About how much change will be left from $20 after dinner and a movie?

5. Write two questions that *cannot* be answered with the given information. Explain what other information would be needed to solve the problem. Answers will vary. When should Jorge leave his home to go to the restaurant? You need the time it takes him to get to the restaurant from his home.

6. Write a description of a *problem situation* different from the one above. Identify some problems that can be answered and some that cannot. Answers will vary. Students should write a story similar to the one above.

1-5 Enrichment
Communication

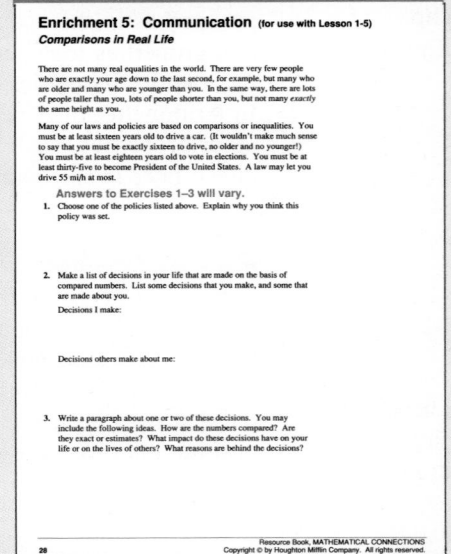

Enrichment 5: Communication (for use with Lesson 1-5)
Comparisons in Real Life

There are not many real equalities in the world. There are very few people who are exactly your age down to the last second, for example, but many who are older and many who are younger than you. In the same way, there are lots of people taller than you, lots of people shorter than you, but not many *exactly* the same height as you.

Many of our laws and policies are based on comparisons or inequalities. You must be at least sixteen years old to drive a car. (It wouldn't make much sense to say that you must be exactly sixteen to drive, no older and no younger!) You must be at least thirty-five to become President of the United States. A law may let you drive 55 mi/h at most.

Answers to Exercises 1–3 will vary.

1. Choose one of the policies listed above. Explain why you think this policy was set.

2. Make a list of decisions in your life that are made on the basis of compared numbers. List some decisions that you make, and some that are made about you.
   Decisions I make:

   Decisions others make about me:

3. Write a paragraph about one or two of these decisions. You may include the following ideas. How are the numbers compared? Are they exact or estimates? What impact do these decisions have on your life or on the lives of others? What reasons are behind the decisions?

1-6 Enrichment
Exploration

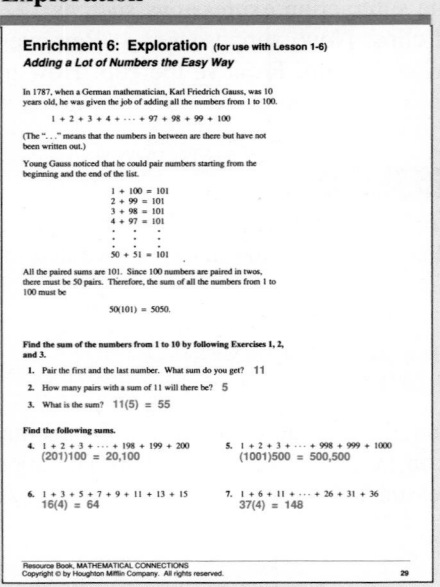

Enrichment 6: Exploration (for use with Lesson 1-6)
Adding a Lot of Numbers the Easy Way

In 1787, when a German mathematician, Karl Friedrich Gauss, was 10 years old, he was given the job of adding all the numbers from 1 to 100.

$$1 + 2 + 3 + 4 + \cdots + 97 + 98 + 99 + 100$$

(The "..." means that the numbers in between are there but have not been written out.)

Young Gauss noticed that he could pair numbers starting from the beginning and the end of the list.

$$\begin{array}{l} 1 + 100 = 101 \\ 2 + 99 = 101 \\ 3 + 98 = 101 \\ 4 + 97 = 101 \\ \qquad \cdot \\ \qquad \cdot \\ \qquad \cdot \\ 50 + 51 = 101 \end{array}$$

All the paired sums are 101. Since 100 numbers are paired in twos, there must be 50 pairs. Therefore, the sum of all the numbers from 1 to 100 must be

$$50(101) = 5050.$$

Find the sum of the numbers from 1 to 10 by following Exercises 1, 2, and 3.

1. Pair the first and the last number. What sum do you get? 11

2. How many pairs with a sum of 11 will there be? 5

3. What is the sum? 11(5) = 55

Find the following sums.

4. $1 + 2 + 3 + \cdots + 198 + 199 + 200$
   (201)100 = 20,100

5. $1 + 2 + 3 + \cdots + 998 + 999 + 1000$
   (1001)500 = 500,500

6. $1 + 3 + 5 + 7 + 9 + 11 + 13 + 15$
   16(4) = 64

7. $1 + 6 + 11 + \cdots + 26 + 31 + 36$
   37(4) = 148

1-7 Enrichment
Connection

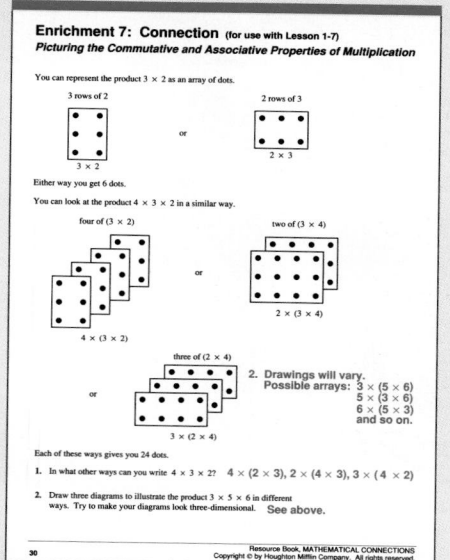

Enrichment 7: Connection (for use with Lesson 1-7)
Picturing the Commutative and Associative Properties of Multiplication

You can represent the product 3 × 2 as an array of dots.

3 rows of 2   2 rows of 3

or

3 × 2   2 × 3

Either way you get 6 dots.

You can look at the product 4 × 3 × 2 in a similar way.

four of (3 × 2)   two of (3 × 4)

or

4 × (3 × 2)   2 × (3 × 4)

three of (2 × 4)

or

3 × (2 × 4)

2. Drawings will vary.
Possible arrays: 3 × (5 × 6)
5 × (3 × 6)
6 × (5 × 3)
and so on.

Each of these ways gives you 24 dots.

1. In what other ways can you write 4 × 3 × 2?   4 × (2 × 3), 2 × (4 × 3), 3 × (4 × 2)

2. Draw three diagrams to illustrate the product 3 × 5 × 6 in different ways. Try to make your diagrams look three-dimensional.   See above.

1-8 Enrichment
Data Analysis

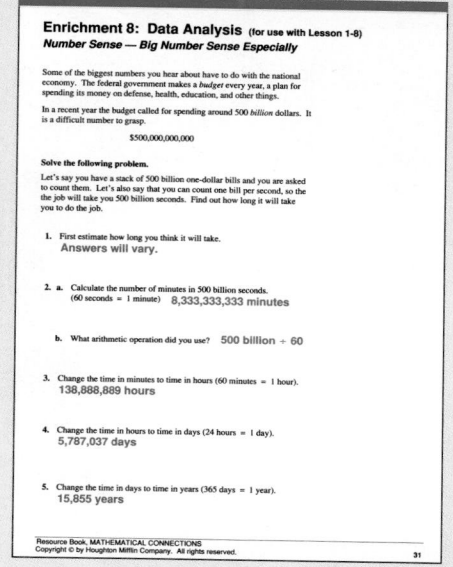

Enrichment 8: Data Analysis (for use with Lesson 1-8)
Number Sense — Big Number Sense Especially

Some of the biggest numbers you hear about have to do with the national economy. The federal government makes a *budget* every year, a plan for spending its money on defense, health, education, and other things.

In a recent year the budget called for spending around 500 *billion* dollars. It is a difficult number to grasp.

$500,000,000,000

Solve the following problem.

Let's say you have a stack of 500 billion one-dollar bills and you are asked to count them. Let's also say that you can count one bill per second, so the job will take you 500 billion seconds. Find out how long it will take you to do the job.

1. First estimate how long you think it will take.
   Answers will vary.

2. a. Calculate the number of minutes in 500 billion seconds.
      (60 seconds = 1 minute)   8,333,333,333 minutes

   b. What arithmetic operation did you use?   500 billion ÷ 60

3. Change the time in minutes to time in hours (60 minutes = 1 hour).
   138,888,889 hours

4. Change the time in hours to time in days (24 hours = 1 day).
   5,787,037 days

5. Change the time in days to time in years (365 days = 1 year).
   15,855 years

1-9 Enrichment
Exploration

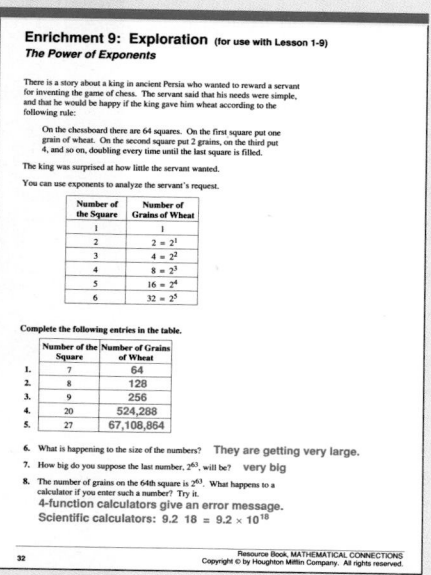

Enrichment 9: Exploration (for use with Lesson 1-9)
The Power of Exponents

There is a story about a king in ancient Persia who wanted to reward a servant for inventing the game of chess. The servant said that his needs were simple, and that he would be happy if the king gave him wheat according to the following rule:

On the chessboard there are 64 squares. On the first square put one grain of wheat. On the second square put 2 grains, on the third put 4, and so on, doubling every time until the last square is filled.

The king was surprised at how little the servant wanted.

You can use exponents to analyze the servant's request.

Number of the Square	Number of Grains of Wheat
1	1
2	2 = 2^1
3	4 = 2^2
4	8 = 2^3
5	16 = 2^4
6	32 = 2^5

Complete the following entries in the table.

	Number of the Square	Number of Grains of Wheat
1.	7	64
2.	8	128
3.	9	256
4.	20	524,288
5.	27	67,108,864

6. What is happening to the size of the numbers?   They are getting very large.

7. How big do you suppose the last number, 2^{63}, will be?   very big

8. The number of grains on the 64th square is 2^{63}. What happens to a calculator if you enter such a number? Try it.
   4-function calculators give an error message.
   Scientific calculators: 9.2 18 = 9.2×10^{18}

1-10 Enrichment
Communication

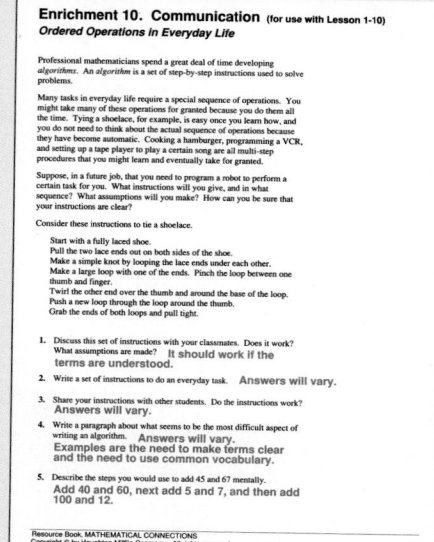

Enrichment 10. Communication (for use with Lesson 1-10)
Ordered Operations in Everyday Life

Professional mathematicians spend a great deal of time developing *algorithms*. An *algorithm* is a set of step-by-step instructions used to solve problems.

Many tasks in everyday life require a special sequence of operations. You might take many of these operations for granted because you do them all the time. Tying a shoelace, for example, is easy once you learn how, and you do not need to think about the actual sequence of operations because they have become automatic. Cooking a hamburger, programming a VCR, and setting up a tape player to play a certain song are all multi-step procedures that you might learn and eventually take for granted.

Suppose, in a future job, that you need to program a robot to perform a certain task for you. What instructions will you give, and in what sequence? What assumptions will you make? How can you be sure that your instructions are clear?

Consider these instructions to tie a shoelace.

Start with a fully laced shoe.
Pull the two lace ends out on both sides of the shoe.
Make a simple knot by looping the lace ends under each other.
Make a large loop with one of the ends. Pinch the loop between one thumb and finger.
Twirl the other end over the thumb and around the base of the loop.
Push a new loop through the loop around the thumb.
Grab the ends of both loops and pull tight.

1. Discuss this set of instructions with your classmates. Does it work? What assumptions are made?   It should work if the terms are understood.

2. Write a set of instructions to do an everyday task.   Answers will vary.

3. Share your instructions with other students. Do the instructions work?
   Answers will vary.

4. Write a paragraph about what seems to be the most difficult aspect of writing an algorithm.   Answers will vary.
   Examples are the need to make terms clear and the need to use common vocabulary.

5. Describe the steps you would use to add 45 and 67 mentally.
   Add 40 and 60, next add 5 and 7, and then add 100 and 12.

End of Chapter Enrichment
Extension

Enrichment 11: Extension (for use after Chapter 1)
Further Explorations of Multiplication and Exponents

When you multiply numbers, the products tend to be larger than the factors. For example, 3 × 6 = 18. In the same way, raising numbers to powers usually gives even larger numbers. For example $4^2 = 16$, $10^3 = 1000$.

Does this pattern always hold true in multiplication? Complete the following tables and look for patterns. Use your calculator, recalling that

$$21^3 = 21 \times 21 \times 21.$$

1. Find each answer.

a.
$(0.1)^2$	0.01
$(0.1)^3$	0.001
$(0.1)^4$	0.0001

b.
$(0.9)^2$	0.81
$(0.9)^3$	0.729
$(0.9)^4$	0.6561

c.
$(1.1)^2$	1.21
$(1.1)^3$	1.331
$(1.1)^4$	1.4641

d. Describe the patterns that you see in the three columns.
   column (a) is getting close to 0 (or getting smaller), column (b) is getting smaller, column (c) is getting larger

e. Try the same operations with 0.2, 0.8, and 1.2. What patterns appear?
   The same patterns hold.

2. Is $(0.8)^2 < (0.9)^2$ or is $(0.8)^2 > (0.9)^2$? $(0.8)^2 < (0.9)^2$

3. What happens when you raise the number 1 to any power?   You get 1.

4. If x can stand for any number, which one is larger, x or x^2? Explain.
   $x = x^2$ if $x = 0$ or 1; $x > x^2$ if $0 < x < 1$; $x^2 > x$ if $x > 1$

5. If x and y are any numbers, when will xy (the product of x and y) be larger than both x and y? When will xy be smaller than both x and y? To answer these questions, try some numbers. Experiment!
   xy will be larger than both x and y if $x > 1$ and $y > 1$;
   xy will be smaller if $0 < x < 1$ and $0 < y < 1$.

About the Chapter Openers

The data items that appear in the tables and graphs with each of the chapter-opening photographs are used throughout the book in exercises labeled **DATA**.

Multicultural Notes

The multicultural notes in the side columns accompanying each of the chapter-opening photographs provide additional information about the theme of each chapter opener and about other topics covered in the lessons. The information focuses on contributions of various cultural and ethnic groups to our society. Research activities and other resources are also suggested.

For Chapter 1 Opener

Students may be interested to learn that the first gold record was awarded in 1942 to Glenn Miller. The big band leader received a gold-plated album as part of a promotional campaign when his recording of "Chattanooga Choo Choo" brought in more than one million dollars in sales.

In 1958, the Recording Industry Association of America began to certify gold record status for records with million dollar sales. But the association changed this practice in 1975, mainly because higher-priced, double-record, long-playing albums so easily reached sales figures of one million dollars. Gold and platinum records are now awarded based on the number of copies sold, not on sales figures.

Even before the first gold record was awarded, many musicians had hit songs that either sold one million copies or made over one million dollars. For example, in 1918, the opera singer Alma Gluck became the first woman to make a record that sold one million copies, with her rendition of a song by African-American songwriter James Bland.

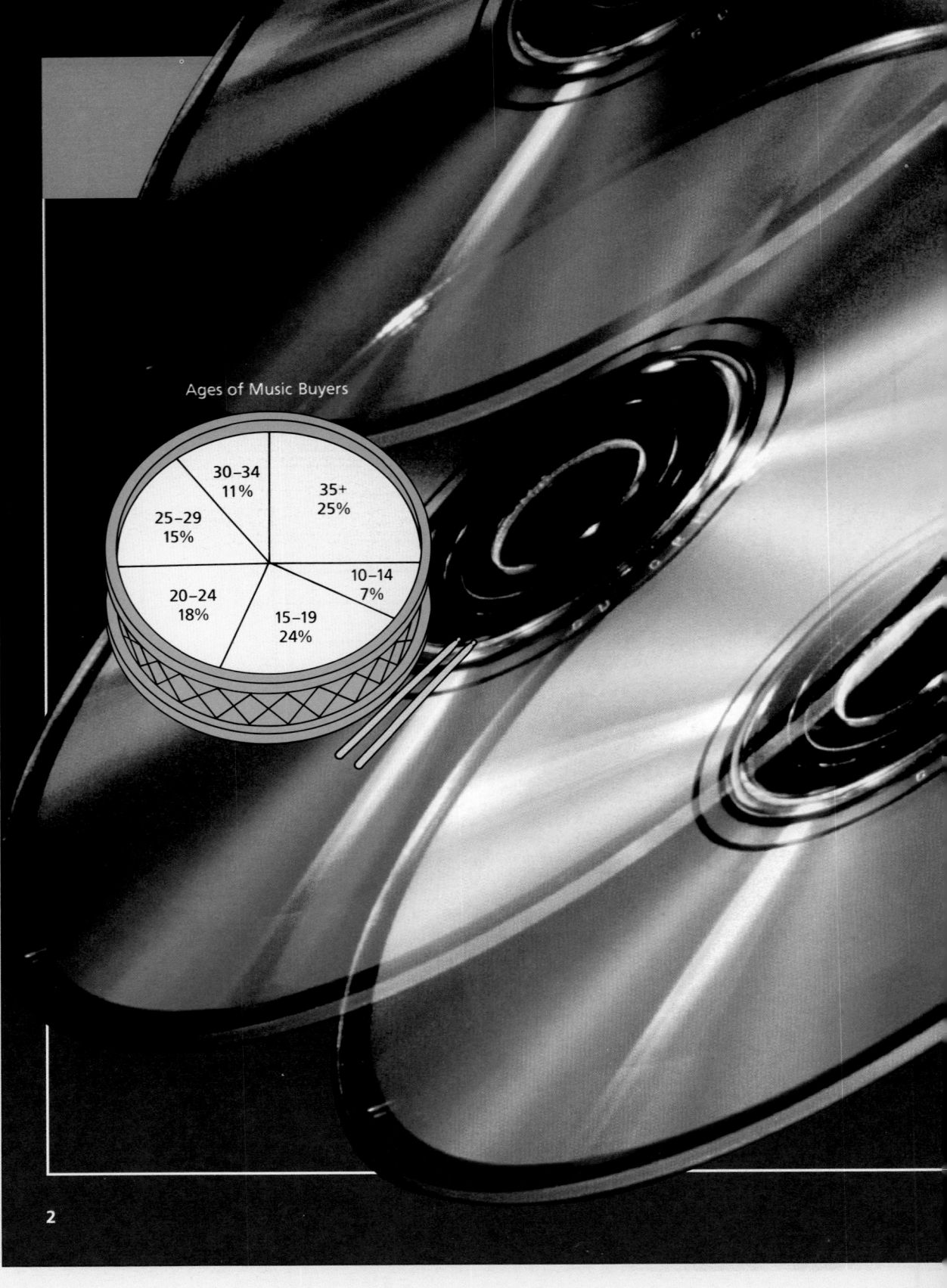

Ages of Music Buyers

30–34 11%

35+ 25%

25–29 15%

10–14 7%

20–24 18%

15–19 24%

Connecting Arithmetic and Algebra 1

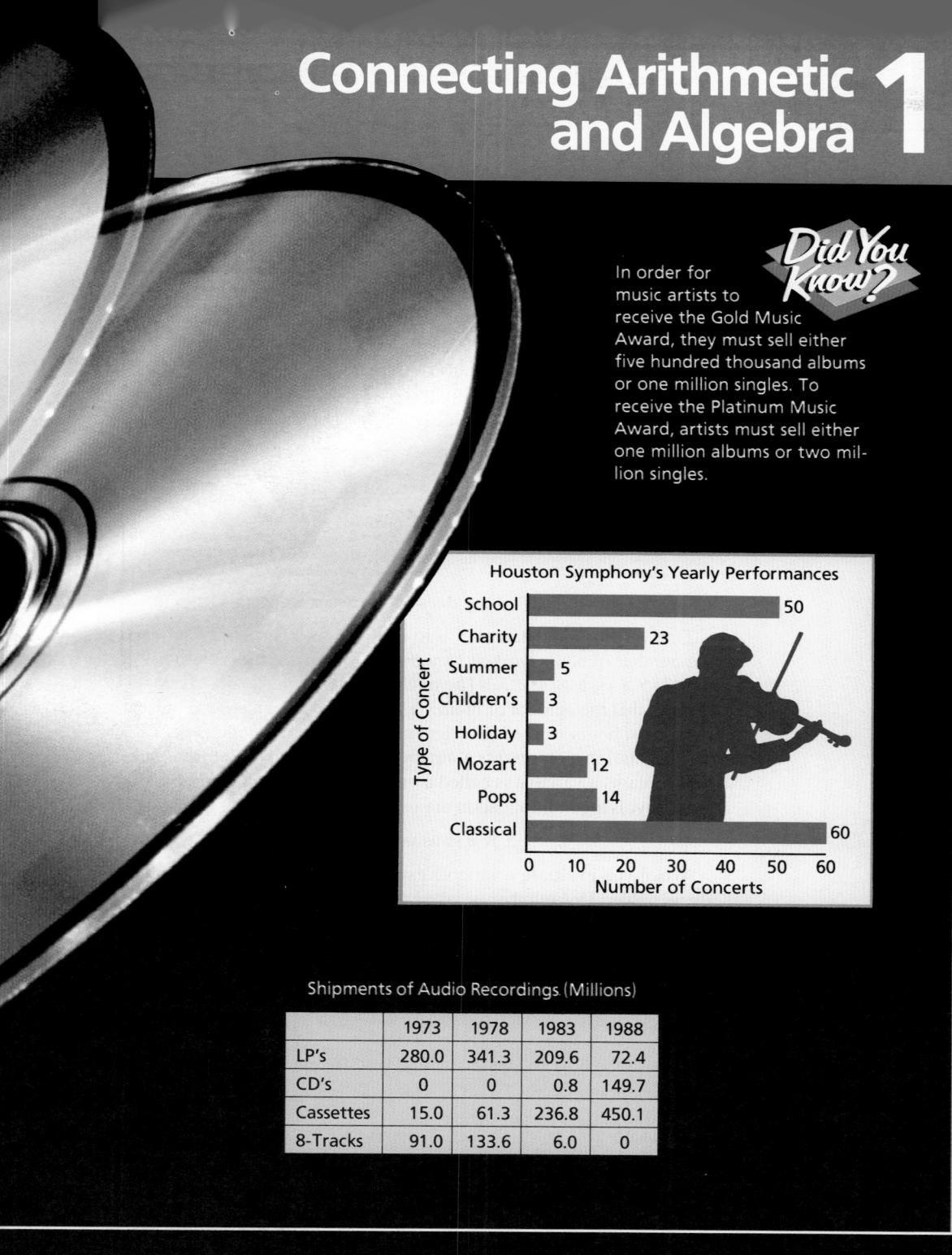

In order for music artists to receive the Gold Music Award, they must sell either five hundred thousand albums or one million singles. To receive the Platinum Music Award, artists must sell either one million albums or two million singles.

Houston Symphony's Yearly Performances

Type of Concert	Number of Concerts
School	50
Charity	23
Summer	5
Children's	3
Holiday	3
Mozart	12
Pops	14
Classical	60

Shipments of Audio Recordings (Millions)

	1973	1978	1983	1988
LP's	280.0	341.3	209.6	72.4
CD's	0	0	0.8	149.7
Cassettes	15.0	61.3	236.8	450.1
8-Tracks	91.0	133.6	6.0	0

Multicultural Notes
(continued)

In 1930, the Mills Brothers issued a record that made them the first vocal group to earn over one million dollars for one recording.

Research Activity

Students may enjoy researching, either individually or in cooperative groups, other musicians or groups who have made top-selling records. By calculating how much money is made from sales of a gold or platinum album at current record prices, students can get a feel for how much a million really is. Students can also calculate how many copies of a record must be sold in order to reach sales of one million dollars.

Suggested Resources

1991 Guinness Book of World Records. New York: Sterling Publishing Company, 1990.

Sanders, Dennis. *The First of Everything*. New York: Delacorte Press, 1981.

3

Teaching the Lesson
- Materials: paper clips, cups
- Lesson Starter 1
- Using Manipulatives, p. 2D
- Toolbox Skills 7–10

Lesson Follow-Up
- Practice 1
- Enrichment 1: Connection
- Study Guide, pp. 1–2
- Cooperative Activity 1

Warm-Up Exercises

Find each answer.
1. 124 + 17 + 706   847
2. 300 − 174   126
3. 25 × 17 × 4   1700
4. 308 ÷ 7   44
5. 288 ÷ 9   32

Lesson Focus

Ask students if they have a part-time job and if so, how do they figure out how much they will be paid each week. Ask if the amount they earn *varies* based on the number of hours worked. Today's lesson focuses on mathematical expressions that vary.

Teaching Notes

Work through the examples with the students stressing the importance of substituting first, and then performing the indicated operations.

Key Question

Why is an expression containing a variable called a *variable expression*?

Because the value of the expression varies as the value of the variable changes.

Critical Thinking

You may wish to have students *analyze* the relationship given by the chart, the pattern, and the variable expression in the lesson opener.

Variables and Variable Expressions

Objective: To evaluate variable expressions involving whole numbers.

APPLICATION

Jan earns $5 per hour working at The Music Shop. The amount of money she earns changes, or *varies,* with the amount of time she works. The table below shows how much money Jan earns for working 1, 2, 3, 4, and 5 hours.

Terms to Know
- *variable*
- *variable expression*
- *evaluate*
- *value of a variable*

Hours Worked	Money Earned ($)
1	5 × 1 = 5
2	5 × 2 = 10
3	5 × 3 = 15
4	5 × 4 = 20
5	5 × 5 = 25

QUESTION How much money does Jan earn for working 11 hours?

One way to find the answer is to continue the chart until you reach 11 for the number of hours worked. Another way is to use the pattern 5×1, 5×2, 5×3, 5×4, ... (The three dots mean *and so on.*) If you do, you can say that the amount of money earned is $5 \times n$, where n represents the number of hours worked. The letter n is called a *variable*.

A symbol that represents a number is called a **variable.** An expression that contains a variable is called a **variable expression.** Variable expressions involving multiplication are usually written without the $\times$ sign.

$$5 \times n \text{ is usually written as } 5n.$$

When you **evaluate** a variable expression, you substitute a number for the variable. This number is called the **value** of the variable.

Example 1

Solution

Evaluate the expression $5n$ when $n = 11$.

$5n = 5 \times 11$ ⟵ Substitute 11 for n.
$\quad = 55$

 **Check Your Understanding**

1. Describe how Example 1 would be different if Jan works 20 hours.
2. Describe how the variable expression in Example 1 would be different if Jan earned $6 per hour.
   See *Answers to Check Your Understanding* at the back of the book.

ANSWER Jan earns $55 for working 11 hours.

Some variable expressions contain more than one variable. To evaluate these expressions you substitute the value given for each variable.

Example 2

Solution

Evaluate the expression $y + 9 + z$ when $y = 2$ and $z = 8$.

$$\begin{aligned} y + 9 + z &= 2 + 9 + 8 \qquad \longleftarrow \text{Substitute 2 for } y \\ &= 11 + 8 \qquad\qquad\;\; \text{and 8 for } z. \\ &= 19 \end{aligned}$$

Guided Practice

COMMUNICATION « *Reading*

« **1.** Replace each __?__ with the correct word.

variable   A __?__ is a symbol that represents a number. An expression that con-
variable   tains a __?__ is called a variable expression. Any number that you substitute for a variable is called a __?__ of the variable. value

« **2.** Explain the meaning of the word *variable* in the following sentence.

The weather report today calls for variable cloudiness.
The weather is likely to vary between cloudy and sunny.

Assume that $a = 3$, $b = 9$, and $c = 5$. Replace each __?__ with a number that makes the statement true.

3. $7a = 7 \times$ __?__ $=$ __?__   3; 21

4. $c + 13 =$ __?__ $+ 13 =$ __?__   5; 18

5. $c - a =$ __?__ $-$ __?__ $=$ __?__   5; 3; 2

6. $b \div a =$ __?__ $\div$ __?__ $=$ __?__   9; 3; 3

Evaluate each expression when $x = 6$, $y = 8$, and $z = 4$.

7. $8x$ 48

8. $11 + z$ 15

9. $16 \div z$ 4

10. $y - 7$ 1

11. $y + 2 + z$ 14

12. $y - x$ 2

13. $2xz$ 48

14. $y \div z$ 2

15. $x + 4 + x$ 16

16. $5zx$ 120

17. $z + z + 1$ 9

18. $x + z + y$ 18

Exercises

Evaluate each expression when $a = 3$, $b = 12$, and $c = 4$.   12. 9

A   **1.** $10a$ 30

2. $12c$ 48

3. $b - 2$ 10

4. $27 - c$ 23

5. $8 + c$ 12

6. $a + 13$ 16

7. $c \div 2$ 2

8. $b \div 6$ 2

9. $5ba$ 180

10. $3ac$ 36

11. $a + 5 + b$ 20

12. $c + 2 + a$

13. $b \div a$ 4

14. $b \div c$ 3

15. $b - c$ 8

16. $c - a$ 1

17. $b + c + a$ 19

18. acb 144

Connecting Arithmetic and Algebra   **5**

Error Analysis
When replacing a variable in an expression that does not show a multiplication sign, students may forget to multiply. If $n = 32$, then $5n$ may be erroneously written as 532. Remind students that when a number and a variable are written side by side, it means that they should be multiplied.

Using Manipulatives
For a suggestion on using manipulatives to model variable expressions, see page 2D.

Additional Examples
1. Evaluate the expression $12n$ when $n = 3$.   36
2. Evaluate the expression $a + 17 + b$ when $a = 12$ and $b = 14$.   43

Closing the Lesson
Check students' understanding of the concepts in the lesson by having them work the following application.

Application
Michelle bought some compact disks (CDs) at $12 each.
1. How much did she spend if she bought 4 CDs? $48
2. If she bought n CDs, what variable expression can be used to represent the cost? $12n$
3. Evaluate $12n$ when $n = 5$.   60

Suggested Assignments
Skills: 1–18, 21–37 odd, 39–44
Average: 1–20, 21–37 odd, 39–44
Advanced: 1–17 odd, 19–44

Exercise Notes
Communication: Reading
Guided Practice Exercises 1 and 2 check students' understanding of the vocabulary presented in the lesson.

Making Connections/Transitions

The connection between a numerical pattern and a mathematical interpretation of the pattern can be developed by having students work *Exercises 21–26.*

Reasoning

Exercises 35–38 require students to make a conjecture about the relative size of two variables, to test their conjectures, and to make additional conjectures based upon the results.

Follow-Up

Extension

In Exercises 21–24, ask students to determine the 9th term in each pattern and to find its value.

Exploration

Have students use a calculator to explore what happens to the value of the variable expression $1 \div n$ as n becomes greater and greater. The value approaches 0.

Enrichment

Enrichment 1 is pictured on page 2E.

Practice

Practice 1, *Resource Book*

Practice 1
Skills and Applications of Lesson 1-1

Evaluate each expression when $x = 2, y = 8,$ and $z = 10.$

1. $10z$ 100
2. $x + y$ 10
3. $25 - y$ 17
4. $7y$ 56
5. $x + y + z$ 20
6. $x + y - z$ 0
7. $33 + y$ 41
8. $y - 7 + x$ 3
9. $z - 8$ 2
10. $y + 3 - 2$ 9
11. $y + 4$ 2
12. $2xy$ 32
13. $z + 15 - y$ 17
14. $2yz$ 160
15. $18 - y + 10$ 20
16. $20 - z + 5$ 15
17. $z + 5$ 2
18. $z + x$ 5

Evaluate each expression when $a = 5, b = 7, c = 12,$ and $d = 4.$

19. $a + c + d$ 21
20. $3d$ 12
21. $8 + d$ 12
22. $a + b - 8$ 4
23. $a + 10$ 15
24. $c + 3$ 4
25. $b + c + d$ 23
26. $c + d$ 3
27. $9a$ 45
28. ab 35
29. bd 28
30. $14 + b$ 2

31. In Central High School there are six ninth grade homerooms. Each homeroom has n students. Write a variable expression that represents the total number of students in the ninth grade. $6n$
32. Todd counted t items in his mother's shopping cart. She then bought seven more items. Write a variable expression to show the total number of items. $t + 7$
33. In major league baseball there are twenty-six teams and each team has p players. Write a variable expression to show the total number of players. $26p$
34. The latest count of eighth grade students shows a total of b male students and g female students. Write a variable expression showing the total number of eighth grade students. $b + g$

---

19. Rosita was born on her brother's sixth birthday. Write a variable expression that represents how many years old Rosita is when her brother is b years old. $b - 6$

20. A ream of paper contains five hundred sheets. Write a variable expression that represents the number of reams in s sheets of paper. $s \div 500$

PATTERNS

Write a variable expression for each pattern. Use n as the variable.

B 21. $7 \times 1, 7 \times 2, 7 \times 3, 7 \times 4, \ldots$ $7n$

22. $4 + 1, 4 + 2, 4 + 3, 4 + 4, \ldots$ $4 + n$

23. $1 - 1, 2 - 1, 3 - 1, 4 - 1, \ldots$ $n - 1$

24. $9 - 1, 9 - 2, 9 - 3, 9 - 4, \ldots$ $9 - n$

25. $60 \div 1, 60 \div 2, 60 \div 3, 60 \div 4, \ldots$ $60 \div n$

26. $1 \div 5, 2 \div 5, 3 \div 5, 4 \div 5, \ldots$ $n \div 5$

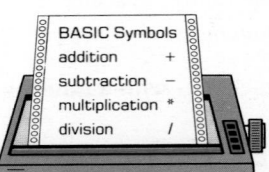

COMPUTER APPLICATION

BASIC Symbols
addition   +
subtraction   −
multiplication *
division   /

In a computer, the term *variable* indicates a location in the memory. When you write a BASIC computer program, you use capital letters for variables. To perform operations on variables, you create BASIC expressions using the operation symbols shown at the left.

Evaluate each BASIC expression when $A = 2$, $B = 6$, and $C = 12$.

27. $A + 5$ 7
28. $3 * C$ 36
29. $A + B + C$ 20
30. C/A 6

Write a BASIC expression for each variable expression.

31. $6m$
$6 * M$
32. $p \div q$
P/Q
33. $j + 12 + k$
$J + 12 + K$
34. rst
$R * S * T$

LOGICAL REASONING

Find a value of r and a value of s that satisfy the given conditions.

C 35. $rs = 96$ and $r - 4 = s$ $r = 12; s = 8$

36. $rs = 64$ and $r \div s = 16$ $r = 32; s = 2$

37. $r \div 2 = s$ and $r + s = 45$ $r = 30; s = 15$

38. $r + s = 17$ and $rs = 60$ $r = 12; s = 5$ or $r = 5; s = 12$

SPIRAL REVIEW

39. nine and two ten-thousandths
42. Yes; No; No; Yes; Yes

S 39. Write the word form of 9.0002. *(Toolbox Skill 5)*

40. Estimate the product: 3185×79 *(Toolbox Skill 1)* about 240,000

41. Evaluate $4cd$ when $c = 5$ and $d = 9$. *(Lesson 1-1)* 180

42. Is 430 divisible by 2? by 3? by 4? by 5? by 10? *(Toolbox Skill 16)*

43. Find the product: $\frac{1}{2} \times \frac{4}{5}$ *(Toolbox Skill 20)* $\frac{2}{5}$

44. Round 69.553 to the nearest tenth. *(Toolbox Skill 6)* 69.6

6   Chapter 1

1-2 Addition and Subtraction Expressions

Objective: To evaluate variable expressions involving addition and subtraction of whole numbers and decimals.

APPLICATION

John Carter works in a shoe store. He earns $287 per week in base pay plus a bonus, or commission, on his total sales.

QUESTION Last week John's commission was $116.49. What was his total pay for the week?

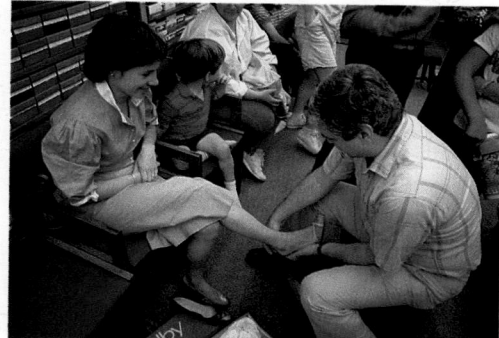

You can find John's total pay for any week by evaluating the addition expression $287 + n$, where n represents his commission for the week.

Example 1
Solution

Evaluate $287 + n$ when $n = 116.49$.

First substitute 116.49 for n. $287 + n = 287 + 116.49$

Next estimate the sum by adjusting the sum of the front-end digits.

$$287.00$$
$$+116.49$$ about 100
$$300 \quad + 100 \longrightarrow \text{about } 400$$

Then add. Remember to line up the decimal points.

$$\begin{array}{r} \overset{1\ 1}{287.00} \\ +116.49 \\ \hline 403.49 \end{array}$$

The answer 403.49 is close to the estimate of 400, so 403.49 is reasonable.

Check Your Understanding

1. In Example 1, explain why you estimate before adding.
2. In Example 1, explain how you get the estimate *about 100*.
3. Why is 287 written as 287.00 in Example 1?
   See *Answers to Check Your Understanding* at the back of the book.

You might decide to use a calculator to find the sum in Example 1. This is the key sequence you would use.

 287. [+] 116.49 [=]

ANSWER John's total pay last week was $403.49.

Connecting Arithmetic and Algebra   **7**

Lesson Planner

> **Teaching the Lesson**
> - Materials: paper clips, cups, calculator
> - Lesson Starter 2
> - Using Technology, p. 2C
> - Using Manipulatives, p. 2D
> - Toolbox Skills 7, 8, 11, 12
>
> **Lesson Follow-Up**
> - Practice 2
> - Enrichment 2: Application
> - Study Guide, pp. 3–4

Warm-Up Exercises

Find each answer.

1. $23.95 + 42.83$ 66.78
2. $104.80 + 58.75$ 163.55
3. $82.63 - 47.87$ 34.76
4. $106.28 - 77.47$ 28.81
5. $321.45 - 300.45$ 21.00

Lesson Focus

Ask students to imagine that they are shopping. Suppose they bought a cassette for $11.95. How much would the students have spent if they also had bought something else? The amount would vary depending upon what else was bought. In today's lesson, the focus is on variable expressions involving addition or subtraction.

Teaching Notes

Work through the examples with the students. Stress the three key steps of *substitute, estimate,* and *add* (or *subtract*).

Key Question

Why is it important to include an estimation step when evaluating an expression? An estimation step provides a check on the reasonableness of the actual answer.

Critical Thinking

Checking the reasonableness of an answer, as shown in the Examples, requires a student to *compare* the actual and estimated answers and then to *draw a conclusion* about whether his or her actual answer is reasonable.

Error Analysis

When adding or subtracting decimals with paper and pencil, students may forget to line up the decimal points and thus get an incorrect answer. Remind students to check the alignment of all decimal points before they add or subtract.

Using Technology

For information on deciding when to use a calculator, paper and pencil, or mental math, refer to Lesson 1-8 on decision making.

For a suggestion on how to use a simple BASIC program with this lesson, see page 2C.

Using Manipulatives

For a suggestion on using manipulatives to model variable expressions, see page 2D.

Additional Examples

1. Evaluate $x + 89$ when $x = 207.63$. 296.63
2. Evaluate $x - y$ when $x = 96.72$ and $y = 64.35$. 32.37

Closing the Lesson

Have students write an addition or subtraction expression and pass it to another student. Then ask various students to describe the three-step process they would follow to evaluate the other students' expressions.

Suggested Assignments

Skills: 1–12, 13–17 odd, 21–27 odd, 30–36
Average: 1–27 odd, 28–36
Advanced: 1–11 odd, 13–36, Historical Note

To evaluate subtraction expressions, you use a similar procedure.

Example 2
Solution

Evaluate $x - y$ when $x = 67.8$ and $y = 48.63$.

First substitute 67.8 for x and 48.63 for y.

$$x - y = 67.8 - 48.63$$

Next estimate the difference by rounding each number to the place of the leading digit.

$$\begin{array}{r} 67.8 \longrightarrow 70 \\ -48.63 \longrightarrow -50 \\ \hline \text{about } 20 \end{array}$$

Then subtract. Remember to line up the decimal points.

$$\begin{array}{r} \overset{5\ 17\ 7\ 10}{\cancel{6}\cancel{7}.\cancel{8}\cancel{0}} \\ -\ 4\ 8.6\ 3 \\ \hline 1\ 9.1\ 7 \end{array}$$

The answer 19.17 is close to the estimate of 20, so 19.17 is reasonable.

If you decide to use a calculator to find the difference in Example 2, you would use this key sequence.

 $\boxed{67.8}$ $\boxed{-}$ $\boxed{48.63}$ $\boxed{=}$

Guided Practice

COMMUNICATION « *Reading*

« **1.** Replace each __?__ with the correct word.
When you add, the result is called the __?__. sum
When you subtract, the result is called the __?__. difference

« **2.** What is the objective of this lesson? Name two ways that it is different from the objective of Lesson 1-1.

Answers to Guided Practice 2-6 are on p. A1.

COMMUNICATION « *Writing*

Describe how to obtain a reasonable estimate of each answer.

« **3.** $42.1 + 37.7$ « **4.** $714.6 + 81$

« **5.** $5.14 - 2.81$ « **6.** $6.9037 - 0.751$

Evaluate each expression when $x = 26.4$, $y = 163.5$, and $z = 39$.

7. $z + 88$ 127 **8.** $103 - z$ 64 **9.** $y + 92$ 255.5

10. $83 - x$ 56.6 **11.** $x - 7.6$ 18.8 **12.** $x + y$ 189.9

Evaluate each expression when $a = 94$, $b = 21.4$, $c = 12.86$, **and** $d = 106.4$.

6. 79.2   9. 200.4

A **1.** $17.7 + b$ 39.1

2. $c + 9.37$ 22.23

3. $d +$ 204.21
97.81

4. $c + 43.9$ 56.76

5. $102.5 - b$ 81.1

6. $a - 14.8$

7. $114.5 - c$ 101.64

8. $d - 44.6$ 61.8

9. $a + d$

10. $c + b$ 34.26

11. $d - b$ 85.0

12. $a - c$ 81.14

13. Bill Mihalko spent $149.95 for a CD player, $227 for an amplifier, and $199.49 for a pair of speakers. How much did he spend for this audio equipment? $576.44

14. DATA, *pages 2–3* How many more classical concerts than school concerts does the Houston Symphony perform yearly? 10

PROBLEM SOLVING/APPLICATION

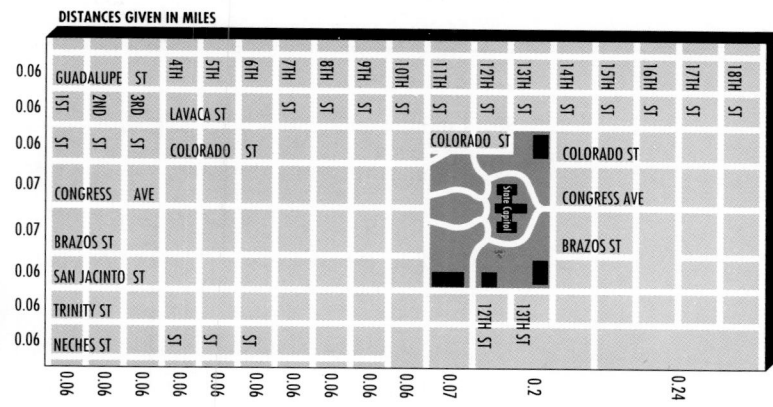

Use the map of a part of Austin, Texas, above.

B **15.** What is the length of the part of Trinity Street shown on the map?
1.11 mi

16. How much longer is the part of San Jacinto Street shown on the map than the part of Neches Street? 0.57 mi

17. You go from the corner of 2nd Street and Brazos Street to the corner of 2nd Street and Guadalupe Street. How far have you gone? 0.26 mi

18. You go from the corner of Trinity Street and 4th Street to the corner of Trinity Street and 11th Street to the corner of 11th Street and Lavaca Street. How far have you gone? 0.74 mi

19. ESTIMATION Estimate the length of 12th Street from Guadalupe Street to Colorado Street. about 0.09 mi

Write a variable expression for each phrase.

«**20.** the sum of 7.9 and a number n $7.9 + n$

«**21.** 43.61 added to a number p $p + 43.61$

«**22.** a number x minus 270.5 $x - 270.5$

«**23.** a number z subtracted from 8.7 $8.7 - z$

Write a phrase for each variable expression. Answers may vary.

«**24.** $x + 6.2$ a number x plus 6.2   «**25.** $15.8 + z$ the sum of 15.8 and a number z

«**26.** $31.7 - a$ 31.7 minus a number a «**27.** $b - 16.4$ a number b minus 16.4

THINKING SKILLS **28.** Evaluation **29.** Analysis; synthesis

C **28.** Determine a value of p and a value of q for which $p + q = p - q$ is a true statement. Answers may vary; p can be any number. q must be zero.

29. Compare your answer to Exercise 28 to your classmates' answers. Make a generalization about the statement $p + q = p - q$.
The statement $p + q = p - q$ is true only when $q = 0$.

SPIRAL REVIEW

S **30.** Find the quotient: $5 \div \frac{1}{5}$ *(Toolbox Skill 21)* 25

31. Estimate the sum: $6.9 + 7.2 + 6.7 + 6.9 + 7.4$ *(Toolbox Skill 4)* about 35

32. Evaluate $a + b$ when $a = 6.3$ and $b = 8.9$. *(Lesson 1-2)* 15.2

33. Find the difference: $\frac{7}{8} - \frac{1}{2}$ *(Toolbox Skill 17)* $\frac{3}{8}$

34. Evaluate $x \div y$ when $x = 48$ and $y = 4$. *(Lesson 1-1)* 12

35. Find the sum: $69,483 + 35,670$ *(Toolbox Skill 7)* 105,153

36. Find the product: 4.93×1000 *(Toolbox Skill 15)* 4930

Historical Note

I n the past, mathematicians have used many different forms of variables. Records show that before 1600 B.C., Egyptians used a word meaning *heap* in much the same way that a variable is used today. The French mathematician François Viète, who is pictured at the left, used letters as variables. Viète lived from 1540 to 1603.

See answer on p. A1.

Research

Both the ancient Greeks and Hindus used symbols for unknown quantities. Find out what these symbols were and how they originated.

1-3

Multiplication and Division Expressions

Objective: To evaluate variable expressions involving multiplication and division of whole numbers and decimals.

DATA ANALYSIS

The pictograph at the right shows the value of home sales in Leeville.

QUESTION What was the value of home sales in 1990?

You can find the value of home sales in a given year by evaluating the multiplication expression $2.5n$, where n represents the number of symbols on the graph for that year.

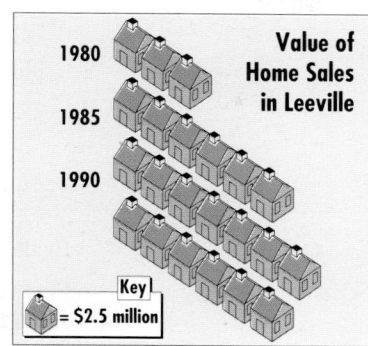

Value of
Home Sales
in Leeville

1980

1985

1990

Key
= $2.5 million

Example 1

Solution

Evaluate $2.5n$ when $n = 13$.

First substitute 13 for n. $\qquad 2.5n = 2.5 \times 13$

Next estimate the product by rounding each factor to the place of the leading digit.

$$2.5 \longrightarrow 3$$
$$\times 13 \longrightarrow \times 10$$
$$\text{about } 30$$

Then multiply. Remember to include the correct number of decimal places in the product.

$$\begin{array}{r} 2.5 \\ \times 13 \\ \hline 75 \\ 250 \\ \hline 32.5 \end{array}$$

The answer 32.5 is close to the estimate of 30, so 32.5 is reasonable.

 **Check Your Understanding**

1. In Example 1, how do you know the correct number of decimal places to show in the product?
   See Answers to Check Your Understanding at the back of the book.

You might decide to use a calculator to find the product in Example 1. This is the key sequence you would use.

ANSWER The value of home sales in 1990 was $32.5 million.

Connecting Arithmetic and Algebra **11**

The division expression 32,500,000 ÷ *n*, where *n* is the number of homes sold in Leeville in 1990, can be used to find the average cost of a home. Ask students to find the average cost of a home in Leeville if the number of homes sold were (a) 200, (b) 250, (c) 260. $162,500; $130,000; $125,000

Key Question

Why is it important to estimate an answer when using a calculator? By pushing an incorrect key, you can get a wrong answer. Estimation helps determine whether or not an answer is reasonable.

Reasoning

Ask students to evaluate xy, yx, $x \div y$, and $y \div x$ when $x = 2.4$ and $y = 0.6$. Does the order of the variables matter? Yes, it matters when dividing. Without giving examples, ask students whether order matters in addition and subtraction.

Error Analysis

Students often misplace the decimal point in a quotient or product. Remind students of the procedures they should follow for the correct placement of the decimal point.

Using Technology

For a suggestion on how to use a spreadsheet program with this lesson, see page 2C.

Using Manipulatives

For a suggestion on using manipulatives to model variable expressions, see page 2D.

Additional Examples

1. Evaluate $2.5x$ when $x = 3.3$.
   8.25
2. Evaluate $a \div b$ when $a = 16.8$ and $b = 7$. 2.4

In algebra, it is important to realize that you can show both multiplication and division in a number of different ways. For example, each of these symbols represents multiplication.

times sign	*raised dot*	*parentheses*
7×5	$7 \cdot 5$	(7)5 or 7(5) or (7)(5)

Similarly, these expressions all represent division.

division sign	*division house*	*fraction bar*
$42 \div 7$	$7\overline{)42}$	$\frac{42}{7}$

Example 2 Evaluate $\frac{y}{x}$ when $y = 17.4$ and $x = 6$.

Solution First substitute 17.4 for y and 6 for x. $\qquad \frac{y}{x} = \frac{17.4}{6}$

Next estimate the quotient using compatible numbers. $\qquad 6\overline{)17.4} \longrightarrow \begin{array}{r} \text{about } 3 \\ 6\overline{)18} \end{array}$

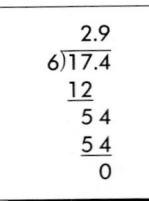

Then use the division process: divide, multiply, subtract, bring down the next digit

$$\begin{array}{r} 2.9 \\ 6\overline{)17.4} \\ \underline{12} \\ 5\,4 \\ \underline{5\,4} \\ 0 \end{array}$$

The answer 2.9 is close to the estimate of 3, so 2.9 is reasonable.

 **Check Your Understanding**

2. In Example 2, how do you know where to place the decimal point in the quotient?
See *Answers to Check Your Understanding* at the back of the book.

If you decide to use a calculator to find the quotient in Example 2, you would use this key sequence.

Guided Practice

COMMUNICATION «*Reading*

«**1.** Choose all the words that are associated with *multiplication*.
   (**a.**) factor **b.** sum **c.** quotient (**d.**) product

«**2.** Choose all the words that are associated with *division*.
   a. addend (**b.**) divisor (**c.**) quotient **d.** difference

COMMUNICATION « *Writing* Answers to Guided Practice 3-10 are on p. A1.

Write each expression in two other ways.

«**3.** $27 \cdot 15$ 　　«**4.** $884 \div 22$ 　　«**5.** $6.5\overline{)124.2}$ 　　«**6.** $(12.4)(26.9)$

Describe how to obtain a reasonable estimate of each answer.

«**7.** $(3.7)81$ 　　«**8.** $4.3(5.1)$ 　　«**9.** $2.2\overline{)12.32}$ 　　«**10.** $14.73 \div 2.8$

Evaluate each expression when $a = 18$, $b = 3.2$, and $c = 0.8$.

11. $17a$ 306 　　**12.** $22b$ 70.4 　　**13.** $385.2 \div a$ 21.4 　　**14.** $27.2 \div c$ 34

15. ab 57.6 　　**16.** bc 2.56 　　**17.** $\dfrac{b}{c}$ 4 　　**18.** $\dfrac{a}{c}$ 22.5

Exercises

Evaluate each expression when $w = 63$, $x = 1.6$, $y = 62.72$, and $z = 18.27$.

A

1. $87x$ 139.2 　　**2.** $12.4w$ 781.2 　　**3.** $27y$ 1693.44 　　**4.** $3.4z$ 62.118

5. $30.87 \div w$ 0.49 　　**6.** $z \div 30$ 0.609 　　**7.** $\dfrac{y}{32}$ 1.96 　　**8.** $\dfrac{35.28}{x}$ 22.05

9. wx 100.8 　　**10.** xy 100.352 　　**11.** $z \div w$ 0.29 　　**12.** $y \div x$ 39.2

13. Sandy Haig drives 37.4 mi round trip to work and back home. Sandy works five days each week. How many miles does she drive to work and back home each week? 187 mi

14. Martin Chun purchased nine tickets to a dance festival. He paid a total of $139.50 for the tickets. How much did each ticket cost? $15.50

 CALCULATOR

Estimate to place the decimal point in each calculator answer.

B **15.** $(5.92)(7.15)$ 42.328 | 42328. | 　　**16.** $92 - 79.65$ 12.35 | 1235. |

17. $48.678 \div 915$ 0.0532 | 000532. | 　　**18.** $(32.5)(8.64)$ 280.8 | 2808. |

19. $862.37 + 1158.93$ 2021.3 | 20213. | 　　**20.** $21.973 \div 0.73$ 30.10 | 3010. |

Estimate to tell whether each calculator answer is *reasonable* or *unreasonable*. If an answer is unreasonable, find the correct answer.

21. $(3.32)(21.6)$ | 71712. | 　　**22.** $17.4 - 9.38$ | 8.02 |

23. $59.78 \div 98$ | 6.1 | 　　**24.** $1121 \div 19$ | 590. |

25. $0.35 + 1.9$ | 2.25 | 　　**26.** $(52.4)(2.3)$ | 120.52 |

21. unreasonable; 71.712 　　**22.** reasonable 　　**23.** unreasonable; 0.61
24. unreasonable; 59 　　**25.** reasonable 　　**26.** reasonable

Connecting Arithmetic and Algebra 　**13**

Closing the Lesson

Ask students to explain how estimation can be used to provide a quick check of solutions when using decimals.

Suggested Assignments

Skills: 1–12, 13–39 odd, 42–46
Average: 1–39 odd, 40–46
Advanced: 1–11 odd, 13, 14, 15–39 odd, 40–46

Exercise Notes

Communication: Reading

As a follow-up to *Guided Practice Exercises 1 and 2*, ask students to list all the words that they associate with each operation.
Addition: add, plus, sum, addends
Subtraction: difference, minus, minuend, subtrahend
Multiplication: factor, product
Division: divisor, dividend, quotient, remainder

Estimation

In *Guided Practice Exercises 7–10*, discuss the methods students used in their descriptions. Different methods may lead to different estimates. The purpose of the discussion is to encourage students to verify their thinking and to discover that there may be more than one "correct" answer to an estimation problem.

Calculator

Exercises 15–26 offer reinforcement of the Key Question presented in the Teaching Notes.

Problem Solving

Checking for reasonableness of answers is a key step in problem solving. *Exercises 21–26* reinforce this concept.

14

COMMUNICATION « *Writing*

Write a variable expression for each phrase.

« **27.** 15.4 times a number z 15.4z

« **28.** the product of a number x and 1.01   1.01x

« **29.** 986.4 divided by a number n 986.4 ÷ n

« **30.** a number y divided by 2.4 y ÷ 2.4

Write a phrase for each variable expression. Answers may vary.

« **31.** 12.3a « **32.** 9.7b « **33.** x ÷ 5.4   « **34.** 26.8 ÷ z
12.3 times a number a 9.7 times a number b a number x divided by 5.4   26.8 divided by a number z

GROUP ACTIVITY

Suppose that numbers are assigned to the letters of the alphabet as follows:

A = 1   B = 2   C = 3   D = 4   E = 5

and so on to Z = 26.

Using this code, you find the value of a name by multiplying the values of its digits. For example:

ED = 5 · 4 = 20

Find the numerical value of each name.

35. BOB   60

36. AMY   325

37. SUE   1995

38. JUAN   2940

39. Find the value of your name. Compare it with the values of the names of others in your group. Are there two names with the same value?
Answers will vary.

C   **40.** Find each of the following. Answers will vary.
a. a man's name that has three letters and a value of 140   BEN
b. a woman's name that has three letters and a value of 140   JAN
c. a person's name that has four letters and a value of 700   JEAN
d. a person's name that has four letters and a value of 1400   ANDY

41. Find a value other than 140, 700, and 1400 that can represent two different names. Give two names that have this value.
Answers will vary. Example: 840; LEN and DON

SPIRAL REVIEW

S   **42.** Find the product: $35 \times \frac{3}{5}$ *(Toolbox Skill 20)*   21

43. Estimate the quotient: $2674 \div 9$ *(Toolbox Skill 3)*   about 300

44. Evaluate $16n$ when $n = 5.4$.   *(Lesson 1-3)*   86.4

45. Give the place of the underlined digit: 13,207.9586   *(Toolbox Skill 5)*
hundredths

46. Find the difference: $6\frac{3}{4} - 4\frac{1}{2}$ *(Toolbox Skill 19)* $2\frac{1}{4}$

1-4

Reading for Understanding

Objective: To read problems for understanding.

To solve a problem, you must first understand it. This means that you may need to read the problem several times to determine what information is given, what you must find, and whether any facts are not needed.

Problem

Use this paragraph for parts (a)–(d).

Aaron had $234 in his savings account on December 31. During the next two months, he made deposits of $53, $65, and $40. He also made withdrawals of $25 and $37. The account earned $1.95 in interest during the same period.

a. What is the paragraph about?

b. How much interest did the account earn during the two months?

c. Does the paragraph tell how much money Aaron spent during the two months?

d. Identify any facts that are not needed to find the total amount of money added to Aaron's account during the two months.

Solution

a. The paragraph is about the money in Aaron's savings account.

b. The account earned $1.95 interest during the two months.

c. No. Aaron withdrew $25 and $37, but the paragraph does not tell how much he spent.

d. The following facts are not needed.
   the amount of money in his account [$234]
   the withdrawals [$25 and $37]

Look Back What if Aaron had deposited $17 instead of withdrawing $25? How would the amount in his savings account be changed?
There would be $42 more in his account.

Guided Practice

Use this paragraph for Exercises 1–3. Choose the letter of the correct answer.

When Keisha filled the gas tank of her car on June 5, the odometer showed 7251.3 mi. She bought 11.7 gal of gasoline. On June 18, Keisha filled the gas tank with 14.2 gal of gasoline and the odometer showed 7588.7 mi.

1. How many gallons of gasoline did Keisha buy on June 5?
   a. 11.7 gal   b. 14.2 gal   c. $(14.2 - 11.7)$ gal

2. How many miles did the car's odometer show on June 18?
   a. 7251.3 mi   b. 7588.7 mi   c. $(7251.3 + 7588.7)$ mi

Connecting Arithmetic and Algebra   **15**

Teaching Notes *(continued)*

Making Connections/Transitions

Students need to recognize that many different problems may arise from a single situation. They must be able to identify a problem within a given situation and then make the transition from the verbal or written situation to the mathematical representation.

Reteaching/Alternate Approach

For a cooperative learning activity to use with this lesson, see page 2D.

Additional Example

Amar and Sara went bowling. Shoes cost $1.25 to rent. Bowling costs 85¢ per game. Amar bowled 7 games. Sara bowled three games less than Amar. Amar did not rent shoes.
1. What is the paragraph about? the costs and number of games when Amar and Sara went bowling
2. How many games did Sara bowl? 4
3. Who spent more on bowling? Amar
4. Identify any facts that are not needed to find the amount of money Sara spent. Amar did not rent shoes.

Closing the Lesson

Check students' understanding of the concepts of the lesson by having them work the following application.

Application

Manny, Moe, and Jack are going to lunch. Manny spends $3.78. Moe spends $2 more than Manny. Jack spends half the amount that Moe spent. Ask a student to create a problem from this information. Continue this process until at least four different problems have been created.

Suggested Assignments

Skills: 1–16
Average: 1–16
Advanced: 1–16

16

3. Identify any facts that are not needed to find the number of miles traveled from June 5 to June 18.
   a. the amounts of gasoline bought on June 5 and on June 18
   b. the number of miles shown on the odometer on June 5
   c. the number of miles shown on the odometer on June 18

Use this paragraph for Exercises 4–7. Answers to Guided Practice 4–7 are on p. A1.

On August 23, Brett bought two pairs of jeans for $24.99 each, a shirt for $15, and a sweater for $34.49. Brett paid with five $20 bills.

4. What is the paragraph about?
5. How many items of clothing did Brett buy?
6. Identify any facts not needed to find the total cost of the clothes.
7. Describe how you would find the amount of change Brett received.

Problem Solving Situations

Use this paragraph for Exercises 1–4. Answers to Problems 1–12 are on p. A1.

Cathy earns $6.50 per hour working part-time as a cashier. She works one hour more on Tuesday than on Thursday. On Wednesday she works 4 h. On Thursday she works half as long as on Wednesday.

1. What is the paragraph about?
2. How many hours does Cathy work on Wednesday?
3. Identify any facts that are not needed to find the number of hours Cathy works during the week.
4. Describe how you would find the number of hours Cathy works on Tuesday.

Use this paragraph for Exercises 5–8.

Paul is saving money to buy a video game system. The game system costs $105 and comes with one free game cartridge. Game cartridges cost $35.49 each. Paul saved $34 in May, $45 in June, $61 in July, and $33 in August.

5. What is the paragraph about?
6. How much did Paul save in July?
7. Does the paragraph tell how many game cartridges Paul plans to buy?
8. Identify any facts that are not needed to find the month in which Paul will be able to buy the game system.

16 Chapter 1

Use this paragraph for Exercises 9–12.

Carol Jigargian bought a stereo system for $600. The tax on the stereo system was $30. She made a down payment of $100 and agreed to pay the remainder in 10 equal payments.

9. What is the paragraph about?

10. How much tax did Carol Jigargian pay?

11. Identify any facts that are not needed to find the total cost of the stereo system.

12. Describe how you would find the amount of each payment.

SPIRAL REVIEW

S 13. Find the difference: $16.53 - 0.5319$ *(Toolbox Skill 12)* 15.9981

14. Evaluate $a + b$ when $a = 7.65$ and $b = 12.4$. *(Lesson 1-2)* 20.05

15. Find the product: 13.87×1000 *(Toolbox Skill 15)* 13,870

16. Identify any facts that are not needed to solve this problem: Last baseball season Todd's batting average was 0.350 with 21 hits. This season his batting average was 0.331 with 19 hits. How many hits did Todd have during the two baseball seasons? *(Lesson 1-4)* batting averages of 0.350 and 0.331

Self-Test 1

Evaluate each expression when $a = 2$, $b = 5$, and $c = 8$.

1. $16a$ 32	2. $65 \div b$ 13	3. $a + 6 + a$ 10	4. $4ac$ 64	**1-1**

Evaluate each expression when $w = 13.67$, $x = 42$, $y = 75.4$, and $z = 9.04$.

5. $97 + y$ 172.4   6. $w + z$ 22.71   7. $w - 8.9$ 4.77   8. $y - z$ 66.36   **1-2**

9. $52w$ 710.84   10. xz 379.68   11. $x \div 0.6$ 70   12. $268.8 \div x$ 6.4   **1-3**

Use this paragraph for Exercises 13–16. Answers to Self-Test 13–16 are on p. A1.

Jill rented 48 movies and 15 video games last year. Each movie rental cost $2.99 and each video game rental cost $1.50.

13. What is the paragraph about?   **1-4**

14. How many video games did Jill rent last year?

15. Identify any facts that are not needed to find the amount of money Jill spent on movie rentals last year.

16. Describe how you would find the total amount Jill spent on video game rentals last year.

Connecting Arithmetic and Algebra   **17**

Exercise Note

Life Skills
All the problems involve situations that students might encounter in the real world: determining the amount of gasoline purchased, buying clothes, computing wages, saving money, and buying on an installment plan.

Follow-Up

Extension
Have students bring to class some paragraphs from newspapers or magazines that contain numerical data. Make up four to six questions that relate to each paragraph.

Enrichment
Enrichment 4 is pictured on page 2E.

Practice
Practice 4 is shown below.

Quick Quiz 1
See page 44.

Alternative Assessment
See page 44.

Practice 4, *Resource Book*

Practice 4
Skills and Applications of Lesson 1-4

Use this paragraph for Exercises 1-4.

Karen, Cosmo, Kevin, and Melba were on the committee for the ninth grade Halloween dance. Expenses for the dance totaled $187.60. Eighty-six tickets were sold at $2.50 per ticket.

1. What is the paragraph about?   the Halloween dance

2. What were the total expenses? $187.60

3. Identify any facts that are not needed to find the profit.   The members of the dance committee.

4. Describe how you would find the amount of money collected from the sale of the tickets.   Multiply the number of tickets sold times the price per ticket.

Use this paragraph for Exercises 5-9.

Daphne attended a college that was 240 miles from home. Her expenses at the college bookstore were $240.50 for three textbooks, $4.95 for three notebooks, $6.50 for a college banner, and $3.95 for five pens.

5. What is the paragraph about?   Daphne's expenses at the bookstore.

6. How much did Daphne spend on notebooks? $4.95

7. How much did each pen cost? $.79

8. Identify any facts that are not needed to find Daphne's total expenses at the bookstore.   Daphne attended a college 240 miles from home.

9. Describe how you would find Daphne's total expenses.   Total the expenses.

Use this paragraph for Exercises 10-13.

Tom's mother bought a new car. The car was white with blue interior. She paid $4000 as a down payment and $325.20 per month for three years.

10. What is the paragraph about?   The cost of a new car.

11. How much is the monthly payment for the car? $325.20

12. Identify the facts that are not needed to find the total cost of the car.   The car was new, white with blue interior.

13. Describe how you would find the total cost of the car.   Multiply the cost per month by 36 months (3 years) and add the down payment.

Teaching the Lesson
- Lesson Starter 5
- Using Technology, p. 2C
- Reteaching/Alternate
  Approach, p. 2D
- Toolbox Skill 5

Lesson Follow-Up
- Practice 6
- Enrichment 5:
  Communication
- Study Guide, pp. 9–10
- Calculator Activity 1

Warm-Up Exercises

Tell which number is greater.

1. 30; 50   50
2. 90; 80   90
3. 509; 59   509
4. 301; 103   301
5. 890; 980   980

Lesson Focus

Tell students to imagine that the school photographer is trying to take a picture of the entire class. To get everyone into the picture, the photographer has to arrange people by height—shorter people in front and taller people in back. Today's lesson focuses on comparing and ordering numbers by size.

Teaching Notes

Point out that the boxes in the Examples were put around numbers with like place values to make the comparison easier.

1-5 The Comparison Property

Objective: To use the comparison property to compare and to order numbers.

Terms to Know
- *comparison property*
- *inequality symbols*

EXPLORATION

1 Al's height is 5 ft 2 in. and Bob's height is 4 ft 11 in. What is the relationship between their heights?
- **a.** Al is taller than Bob.
- **b.** Al is shorter than Bob.
- **c.** Al is the same height as Bob.
- **3.** Al weighs more than Bob; Al weighs less than Bob; Al weighs the same as Bob.

2 Is there a different way to express the relationship between their heights? Explain. Yes; Bob is shorter than Al.

3 List all the possible relationships between their *weights*.

4 How many possible relationships are there between their *ages*? 3

When you compare any two measurements, such as heights, weights, or ages, there are only three possible relationships between them. The **comparison property** of numbers summarizes these relationships.

Comparison Property

For any two numbers a and b, exactly one of the following is true.

In Words	In Symbols
a is greater than b	$a > b$
a is less than b	$a < b$
a is equal to b	$a = b$

The symbols $>$ and $<$ are called **inequality symbols**.

Example 1

Replace each __?__ with >, <, or =.

a. $15{,}824$ __?__ $15{,}794$ **b.** 3.059 __?__ 3.51

Solution

Compare the numbers place-by-place from left to right. Find the first place in which the digits are different.

a. 1 5 , [8] 2 4
   1 5 , [7] 9 4
   $8 > 7$
   So $15{,}824 > 15{,}794$.

b. 3 . [0] 5 9
   3 . [5] 1
   $0 < 5$
   So $3.059 < 3.51$.

Check Your Understanding

1. In Example 1(a), why is there a box around the digits 8 and 7?
2. Describe two ways that Example 1(b) is different from Example 1(a).
   See *Answers to Check Your Understanding* at the back of the book.

18   Chapter 1

Usually there are two ways to express an order relationship.

$a > b$ has the same meaning as $b < a$.

You can use what you know about comparing two numbers when you need to put three or more numbers in numerical order.

Example 2
Solution

Write 0.47, 0.4, 0.247 in order from least to greatest.

$0 . \boxed{4} 7$ $2 < 4$, so $0 . 4 \boxed{7}$ $7 > 0$, so
$0 . \boxed{4}$ 0.247 is the $0 . 4 \boxed{0}$ 0.47 is the
$0 . \boxed{2} 4 7$ *least* number.   *greatest* number.

From the least to greatest, the numbers are: 0.247; 0.4; 0.47
You write this with two inequality symbols as $0.247 < 0.4 < 0.47$.

Check Your Understanding

3. In the Solution of Example 2, why is 0.4 written as 0.40?
4. Write the numbers in Example 2 in order from *greatest* to *least*. Use two inequality symbols.
   See *Answers to Check Your Understanding* at the back of the book.

There are two ways to read a statement that contains two inequality symbols. For example, you can read $0.247 < 0.4 < 0.47$ as follows.

0.247 is less than 0.4, and 0.4 is less than 0.47,

or

0.4 is between 0.247 and 0.47.

Guided Practice

COMMUNICATION « *Writing*

Write each sentence in symbols.

« **1.** Ninety-five is greater than seventeen. 95 > 17

« **2.** Six and fifty-nine hundredths is less than eight and one tenth.   6.59 < 8.1

« **3.** Twelve and forty hundredths is equal to twelve and four tenths.   12.40 = 12.4

« **4.** Five is greater than four and nine thousandths, and four and nine thousandths is greater than four.   5 > 4.009 > 4

Write each statement in words. Answers to Guided Practice 5–8 are on p. A1.

« **5.** 5001 < 5100   « **6.** 0.6 = 0.600

« **7.** 9.03 > 9.007   « **8.** 550 < 581 < 600

Express each relationship in another way.

Tom is younger than Sue.
9. Sue is older than Tom.   **10.** 8.9 < 10.5   10.5 > 8.9

11. 14.0 = 14   14 = 14.0   **12.** 0 < 0.861 < 1   1 > 0.861 > 0

Connecting Arithmetic and Algebra   **19**

Exercise Notes

Communication: Discussion
In their answers to *Guided Practice Exercise 19*, students may have a number of ideas on how to remember the symbols. For more ideas, see page 2D.

Number Sense
Exercises 27–32 require students to make conjectures about the size of products when multiplying by numbers less than, greater than, or equal to 1 without actually multiplying. Students can test their conjectures by using a calculator.

Making Connections/Transitions
Exercises 33–36 may be related to the census process that occurs every ten years. Ask students how a change in population affects the building industry, transportation industry, marketing of products, and representation in Congress.

Reasoning
Exercises 37–44 require students to examine mathematical statements and draw conclusions based on logical reasoning.

Nonroutine Problem
The comparison property is often used in simple logic problems. You may wish to have students solve the following problem.
1. Alvin, Bob, Carol, and Diane are friends.
2. Carol is the taller of the girls.
3. Alvin gets better grades than Diane, but she is taller.
4. Bob is not the shortest, but Diane is taller than he is.

Arrange the students in order from shortest to tallest.   Alvin, Bob, Diane, Carol

Replace each __?__ with >, <, or =.

13. 80,492 __?__ 80,942 <   **14.** 14.69 __?__ 14.690 =   **15.** 16.40 __?__ 16.04 >

Write in order from least to greatest. Use two inequality symbols.

16. 3674; 372; 3476
372 < 3476 < 3674

17. 5.01; 0.51; 0.015
0.015 < 0.51 < 5.01

18. 2.09; 0.209; 2
0.209 < 2 < 2.09

« **19.** COMMUNICATION «*Discussion*   Describe a way to remember which inequality symbol represents *is less than* and which represents *is greater than*. See answer on p. A1.

Exercises

Replace each __?__ with >, <, or =.

A   **1.** 11,388 __?__ 11,614 <   **2.** 27,459 __?__ 26,721 >   **3.** 6.40 __?__ 6.4 =

4. 5 __?__ 0.54 >   **5.** 18.073 __?__ 13.562 >   **6.** 45.09 __?__ 45.090 =

Write in order from least to greatest. Use two inequality symbols.

7. 0.26; 0.2; 0.238
0.2 < 0.238 < 0.26

8. 0.57; 0.5; 0.519
0.5 < 0.519 < 0.57

9. 14.36; 12.03; 14
12.03 < 14 < 14.36

10. 36; 31.86; 36.93
31.86 < 36 < 36.93

11. 25.60; 25.08; 25.04
25.04 < 25.08 < 25.60

12. 69.43; 69.48; 69.45
69.43 < 69.45 < 69.48

13. In a recent year, the population of Canada was 26.1 million and the population of Mexico was 84 million. Which country had the greater population? Mexico

14. The best three times in a 100-m dash were as follows:
Julie → 13.24 s; Anne → 12.92 s; Sonya → 13.41 s.
Which woman finished first? second? third? Anne; Julie; Sonya

Let $w = 6$, $x = 12$, $y = 6.3$, and $z = 0.85$. Replace each __?__ with >, <, or =.

B   **15.** w __?__ 21 <   **16.** 16 __?__ x >   **17.** 6.21 __?__ y <

18. z __?__ 0.805 >   **19.** z __?__ 8.5 <   **20.** 6.30 __?__ y =

21. w __?__ y <   **22.** z __?__ y <   **23.** $x - 6$ __?__ w =

24. $z + 5$ __?__ w <   **25.** $2y$ __?__ x >   **26.** $\frac{w}{6}$ __?__ z >

NUMBER SENSE

***Without multiplying*, replace each __?__ with >, <, or =.**

27. 57 __?__ 57(0.3) >   **28.** 4.8 __?__ (4.8)(0.91) >

29. 5.42 __?__ (5.42)1 =   **30.** 33 __?__ (33)(1.08) <

31. 2.05 __?__ (2.05)(1.899) <   **32.** 84 __?__ 84(13.004) <

CAREER/APPLICATION

A *statistician* collects, organizes, and interprets numerical facts, called *data*. Often the statistician must compare data in order to identify trends and analyze the characteristics of a population.

Answers to Exercises 35 and 36 are on p. A2.

Use the data on the clipboard at the right.

Population of the United States
(Millions)

	1960	1980
Northeast	44.7	49.1
Midwest	51.6	58.9
South	55.0	75.4
West	28.1	43.2

33. Which region had the least population in 1960? in 1980? West; West

34. Which region had the least change in population from 1960 to 1980? Northeast

35. Find the total population of the United States in 1960 and in 1980. About how many more people were there in 1980 than in 1960?

36. **RESEARCH** Find the region with the greatest change in population from 1960 to 1980. Find two probable reasons for this change.

LOGICAL REASONING

Assume that x represents any of the numbers 1, 2, 3, 4, 5,... . Tell whether each statement is *always*, *sometimes*, or *never* true.

C 37. $x + 1 > x$ always

38. $x + 3 > x + 1$ always

39. $x + 1 = 2x$ sometimes

40. $3x = 6$ sometimes

41. $x < 2x$ always

42. $x \cdot x > 2x$ sometimes

43. $1 > x \div x$ never

44. $x \div 1 < x \cdot 1$ never

SPIRAL REVIEW

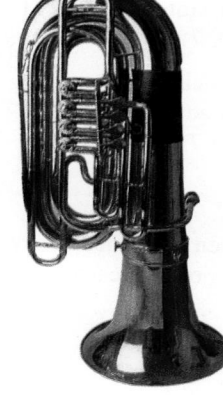

S 45. Evaluate the expression $2yz$ when $y = 3$ and $z = 1.2$. *(Lesson 1-3)* 7.2

46. Find the quotient: $389,760 \div 96$ *(Toolbox Skill 10)* 4060

47. Estimate the sum: $657.2 + 194 + 34.91$ *(Toolbox Skill 2)* about 900

48. Round 9.975 to the nearest hundredth. *(Toolbox Skill 6)* 9.98

49. Find the sum: $\frac{6}{7} + \frac{4}{7}$ *(Toolbox Skill 17)* $1\frac{3}{7}$

50. Replace the __?__ with >, <, or =: 1.72 __?__ 1.072 *(Lesson 1-5)* >

51. Find the product: 462×709 *(Toolbox Skill 9)* 327,558

52. **DATA**, *pages 2–3* How many pops concerts does the Houston Symphony perform each year? *(Toolbox Skill 24)* 14

Connecting Arithmetic and Algebra **21**

Extension

Ask students what process they would use to compare fractions with like denominators. For instance, ask how they would order $\frac{3}{7}$, $\frac{1}{7}$, and $\frac{2}{7}$ from least to greatest.

Enrichment

Enrichment 5 is pictured on page 2E.

Practice

Practice 6 is shown below.

Practice 6, *Resource Book*

Practice 6
Skills and Applications of Lesson 1-5

Replace each __?__ with >, <, or =.

1. 378 __?__ 873 <
2. 14,853 __?__ 14,358 >
3. 3.84 __?__ 38.4 <
4. 0.54 __?__ 154 <
5. 18.537 __?__ 12.989 >
6. 0.654 __?__ 65.4 <
7. 48.7 __?__ 5.37 >
8. 0.732 __?__ 0.276 >
9. 53.09 __?__ 53.090 =
10. 0.00639 __?__ 63900 <
11. 15.81 __?__ 1.581 >
12. 27.007 __?__ 35.007 <
13. 45.08 __?__ 45.8 <
14. 0.009 __?__ 0.001 >
15. 36.07 __?__ 36.071 <

Write in order from least to greatest. Use two inequality symbols. 16–27. See below.

16. 48, 35.9, 39.87
17. 3.5, 0.35, 35.0
18. 0.0017, 0.17, 0.017
19. 23.780, 23.870, 23.087
20. 22.4, 0.224, 224
21. 0.8, 0.008, 0.08
22. 93, 93.1, 93.01
23. 1.14, 11.4, 0.114
24. 432, 324, 243
25. 32.5, 11.89, 21.365
26. 0.01, 0.011, 0.0101
27. 78.30, 78.03, 783

28. The distance from Earth to Mars is approximately 48.5 million miles. The distance from Earth to Mercury is approximately 56.9 million miles. Which of the two planets is farther from Earth? **Mercury**

29. In his major league baseball career, Hank Aaron hit 755 home runs. Babe Ruth hit 714 home runs in his career. Who hit more home runs? **Hank Aaron**

30. The Sears Tower in Chicago is 1454 feet high, the Empire State Building is 1350 feet high, and the First Bank Tower in Toronto is 935 feet high. Which of the three is the tallest? List their heights in order from shortest to tallest. Use two inequality symbols. **Sears Tower;** 935 < 1250 < 1454

16. 35.9 < 39.87 < 48
17. 0.35 < 3.5 < 35.0
18. 0.0017 < 0.017 < 0.17
19. 23.087 < 23.780 < 23.870
20. 0.224 < 22.4 < 224
21. 0.008 < 0.08 < 0.8
22. 93 < 93.01 < 93.1
23. 0.114 < 1.14 < 11.4
24. 243 < 324 < 432
25. 11.89 < 21.365 < 32.5
26. 0.01 < 0.0101 < 0.011
27. 78.03 < 78.30 < 783

Lesson Planner

Teaching the Lesson
- Lesson Starter 6
- Reteaching/Alternate Approach, p. 2D
- Toolbox Skills 7, 11

Lesson Follow-Up
- Practice 7
- Enrichment 6: Exploration
- Study Guide, pp. 11–12

Warm-Up Exercises

Find each sum.

1. $235 + 43 + 65$ 343
2. $2.7 + 6.6 + 3.4$ 12.7
3. $13.5 + 34 + 23.5$ 71
4. $0.5 + 1.5 + 3$ 5
5. $3.6 + 4.4 + 3$ 11

Lesson Focus

Ask students if they ever use short-cut methods for adding. For example, have they ever first added numbers whose sum is 10 or 100? These short-cut methods are based on the properties of addition that are the focus of today's lesson.

Teaching Notes

Work through the Examples with the students. Point out how the properties are used, especially in Example 2 where they make addition easier.

Key Question

Why should you use the properties when solving an addition exercise mentally?   The properties allow you to rearrange the numbers so as to make addition easier.

Reteaching/Alternate Approach

For a suggestion on making and using a poster, see page 2D.

Properties of Addition

Objective: To recognize the properties of addition and to use them to find sums mentally.

CONNECTION

Diego Sanchez travels 16 mi from his home to his office. After work he travels the same 16 mi from his office to his home. Reversing the order in which he *commutes* does not change the distance that he commutes. This idea is similar to the **commutative property** of addition.

Terms to Know
- *commutative property*
- *associative property*
- *additive identity*
- *identity property of addition*

Commutative Property of Addition

Changing the order of the terms does not change the sum.

In Arithmetic	In Algebra
$36 + 10 = 10 + 36$	$a + b = b + a$

You can sit between two friends and say the same thing first to one and then to the other. The result is the same no matter which friend you speak to first. This idea of *associating* first with one friend and then with the other is similar to the **associative property** of addition.

Associative Property of Addition

Changing the grouping of the terms does not change the sum.

In Arithmetic	In Algebra
$(36 + 10) + 5 = 36 + (10 + 5)$	$(a + b) + c = a + (b + c)$

Parentheses show you how to group the numbers in an expression. Do the work within the parentheses first.

Example 1

Replace each _?_ with the number that makes the statement true.

a. $29 + 46 = 46 + \underline{?}$

b. $20 + (18 + 7) = (\underline{?} + 18) + 7$

Solution

a. Use the commutative property of addition.
$29 + 46 = 46 + 29$

b. Use the associative property of addition.
$20 + (18 + 7) = (20 + 18) + 7$

The number 0 has a special addition property. When 0 is added to any number, the sum is *identical* to the original number. For this reason, the number 0 is called the **additive identity.**

22   Chapter 1

Identity Property of Addition

The sum of any number and zero is the original number.

In Arithmetic
$14 + 0 = 14$

In Algebra
$a + 0 = a$

You can use the properties of addition to help you find sums mentally.

Example 2

Use the properties to find each sum mentally.

a. $24 + 0$

b. $76 + 29 + 14$

Solution

a. $24 + 0 = 24$

b. $76 + 29 + 14 = (76 + 14) + 29$ ⟵ Group numbers whose sum is 10, 20, 30, and so on.
$= 90 + 29$
$= 119$

 **Check Your Understanding**

1. Which property was used to find the sum in Example 2(a)?
2. In Example 2(b), explain why it is helpful to group 76 and 14.
   See *Answers to Check Your Understanding* at the back of the book.

Guided Practice

COMMUNICATION « *Reading*

Refer to the text on pages 22–23.

«**1.** State the identity property of addition in words. The sum of any number and zero is the original number.

«**2.** Give an everyday word that is related to *commutative*. commute

Name the property shown by each statement.

3. $17 + 22 = 22 + 17$

4. $89 + 0 = 89$

5. $(8 + 4) + 5 = 8 + (4 + 5)$

6. $n + 0 = n$

7. $b + c = c + b$

8. $(x + y) + z = x + (y + z)$

Answers to Guided Practice 3–8 are on p. A2.

Replace each ? with the number that makes the statement true.

9. $17 + 56 = \underline{?} + 17$ 56

10. $\underline{?} + 43 = 43 + 29$ 29

11. $\underline{?} + (2.3 + 9) = (31 + 2.3) + 9$ 31

12. $(42 + 15) + \underline{?} = 42 + (15 + 0.29)$ 0.29

Use the properties to find each sum mentally.

13. $36 + 48 + 14$ 98

14. $0 + 147$ 147

15. $5.9 + 3.7 + 3.1$ 12.7

16. $104 + 47 + 53$ 204

17. $15 + 67 + 55$ 137

18. $0 + 2.8 + 11.2$ 14

Exercises

Replace each __?__ with the number that makes the statement true.

A **1.** $37 + 15 = 15 + $ __?__ 37   **2.** $62 + 47 = $ __?__ $ + 62$ 47

3. $(7 + 5) + 2 = 7 + ($ __?__ $ + 2)$ 5   **4.** $0 + 3.2 = 3.2 + $ __?__ 0

5. __?__ $ + 8.3 = 8.3 + 29$ 29   **6.** $(6 + $ __?__ $) + 7 = 6 + (8 + 7)$ 8

Use the properties to find each sum mentally.

7. $29 + 14 + 71$ 114   **8.** $25 + 31 + 35$ 91   **9.** $4.4 + 1.9 + 3.6$ 9.9

10. $4.8 + 2.7 + 4.2$ 11.7   **11.** $0 + 155$ 155   **12.** $23.4 + 0$ 23.4

13. $41 + 15 + 9 + 50$ 115   **14.** $23 + 12 + 7 + 68$ 110

15. $5.5 + 2.4 + 3.6 + 7.5$ 19   **16.** $9.1 + 7.4 + 2.9 + 4.6$ 24

Use the properties to evaluate each expression when $a = 1.3$, $b = 4.7$, and $c = 26$.

B **17.** $c + 49 + 64$ 139   **18.** $1.7 + 2.2 + a$ 5.2   **19.** $0 + c + 54$ 80

20. $b + 0 + 3.3$ 8   **21.** $a + 6.2 + b$ 12.2   **22.** $c + a + b$ 32

23. $11.2 + c + 8.8 + 14$ 60   **24.** $c + 0 + 24 + 12$ 62   **25.** $a + 3.1 + 6.9 + b$ 16

C **26.** **WRITING ABOUT MATHEMATICS** Suppose that your friend is having difficulty understanding the commutative and associative properties of addition. Write a paragraph that describes for your friend at least two ways in which the properties are different. See answer on p. A2.

CONNECTING MATHEMATICS AND LANGUAGE ARTS

Once you understand the meaning of a mathematical term like *commutative* or *associative*, you may be able to figure out the meaning of other, unfamiliar words that share the same root word.

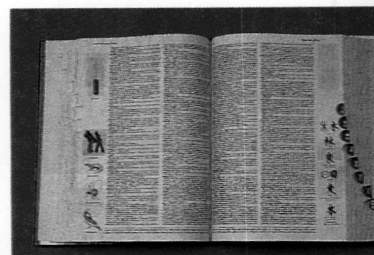

Choose the correct word for each definition. Then use a dictionary to check your answer.

27. to reverse the direction of an electric current
   a. comminute   **(b.)** commutate   **c.** concatenate   **d.** communicate

28. capable of being joined in a relationship
   a. assessable   **b.** accessible   **(c.)** associable   **d.** asocial

SPIRAL REVIEW

S **29.** Find the quotient: $66.69 \div 5.4$ *(Toolbox Skill 14)* 12.35

30. Find the sum mentally: $54 + 32 + 46$ *(Lesson 1-6)* 132

24   Chapter 1

1-7

Properties of Multiplication

Objective: To recognize the properties of multiplication and to use them to find products mentally.

EXPLORATION

Replace each __?__ with the number or word that makes the statement true.

1 The products $15 \times 30 = $ __?__ and $30 \times 15 = $ __?__ suggest that multiplication is __?__. 450; 450; commutative

2 The products $(15 \times 30) \times 2 = $ __?__ and $15 \times (30 \times 2) = $ __?__ suggest that multiplication is __?__. 900; 900; associative

Multiplication, like addition, has commutative and associative properties.

Commutative Property of Multiplication

Changing the order of the factors does not change the product.

In Arithmetic	In Algebra
$15 \times 30 = 30 \times 15$	$ab = ba$

Associative Property of Multiplication

Changing the grouping of the factors does not change the product.

In Arithmetic	In Algebra
$(15 \times 30) \times 2 = 15 \times (30 \times 2)$	$(ab)c = a(bc)$

Example 1

Replace each __?__ with the number that makes the statement true.

a. $18 \cdot 31 = 31 \cdot $ __?__

b. $10 \cdot (28 \cdot 5) = ($ __?__ $\cdot 28) \cdot 5$

Solution

a. Use the commutative property of multiplication.
$18 \cdot 31 = 31 \cdot 18$

b. Use the associative property of multiplication.
$10 \cdot (28 \cdot 5) = (10 \cdot 28) \cdot 5$

The number 1 has a special multiplication property. When any number is multiplied by 1, the product is *identical* to the original number. For this reason, the number 1 is called the **multiplicative identity.**

Identity Property of Multiplication

The product of any number and 1 is the original number.

In Arithmetic	In Algebra
$14 \times 1 = 14$	$a \cdot 1 = a$

Connecting Arithmetic and Algebra **25**

Reteaching/Alternate Approach
For a suggestion on making and
using a poster, see page 2D.

Additional Examples

1. Replace each $\underline{?}$ with the number that makes the statement
true.
   a. $23 \cdot 45 = 45 \cdot \underline{?}$ 23
   b. $10 \cdot (14 \cdot 3) =$
      $(10 \cdot \underline{?}) \cdot 3$ 14
2. Use the properties to find the
products mentally.
   a. $(7)(9)(0)(6)$ 0
   b. $8 \cdot 1 \cdot 30$ 240
   c. $20 \cdot 34 \cdot 5$ 3400

Closing the Lesson

Use the cooperative learning activity
below to check students' understanding of the properties of multiplication.

Cooperative Learning

In cooperative groups, students
should make up three exercises
whose solutions are facilitated by
using the (a) associative, (b) commutative, and (c) identity properties
of multiplication. Each group
should present the exercises orally
to the class, including an explanation of why the stated property simplifies the solution.

Suggested Assignments

Skills: 1–18, 19–29 odd, 35–40
Average: 1–29 odd, 31–40
Advanced: 1–29 odd, 31–40

Exercise Notes

Communication: Reading

Reading for the main idea is critical
in many areas such as social studies, language arts, mathematics, or
even in reading a news article.
Guided Practice Exercises 1 and 2
check students' ability to elicit key
points from information given.

The number 0 also has a special multiplication property.

> ### Multiplication Property of Zero
>
> The product of any number and zero is zero.
>
> | **In Arithmetic** | **In Algebra** |
> |---|---|
> | $14 \times 0 = 0$ | $a \cdot 0 = 0$ |

You can use the properties of multiplication to find products mentally.

MENTAL MATH

Example 2

Use the properties to find each product mentally.
   a. $3(0)(8)(4)$ b. $15 \cdot 1 \cdot 3$ c. $25 \cdot 13 \cdot 4$

Solution
   a. $3(0)(8)(4) = 0$
   b. $15 \cdot 1 \cdot 3 = 15 \cdot 3 = 45$
   c. $25 \cdot 13 \cdot 4 = (25 \cdot 4) \cdot 13$ ← Group numbers whose
      $ = 100 \cdot 13$ product is easy to find.
      $ = 1300$

✔ **Check Your Understanding**

1. Which property was used to find the product in Example 2(a)?
   Example 2(b)?
2. In Example 2(c), explain why you find the product of 25 and 4 first.
   See *Answers to Check Your Understanding* at the back of the book.

Can you divide a number, such as 8, by zero? If so, you could find a
number n such that $8 \div 0 = n$. But this means that $n \cdot 0 = 8$. However, the
multiplication property of zero states that $n \cdot 0 = 0$. So you cannot divide
by zero, and you say that the quotient $8 \div 0$ is *undefined*.

Guided Practice

COMMUNICATION «*Reading*   Answers to Guided Practice 1–6
                              are on p. A2.

Refer to the text on pages 25–26.

«**1.** What is the main idea of the lesson?

«**2.** List four major points that support the main idea of the lesson.

Name the property shown by each statement.

   3. $12 \cdot 46 = 46 \cdot 12$ **4.** $(7.3)(1) = 7.3$

   5. $29(0) = 0$ **6.** $(6 \cdot 9) \cdot 3 = 6 \cdot (9 \cdot 3)$

Replace each $\underline{?}$ with the number that makes the statement true.

   7. $54 \cdot 63 = \underline{?} \cdot 54$ 63   **8.** $(\underline{?})(8.2) = (8.2)(2.6)$ 2.6

   9. $\underline{?} \cdot (18 \cdot 7) = (71 \cdot 18) \cdot 7$ 71   **10.** $(23 \cdot 5) \cdot 32 = 23 \cdot (\underline{?} \cdot 32)$ 5

Use the properties to find each product mentally.

11. $(2)(4.2)(1)$ **12.** $2 \cdot 78 \cdot 5$ **13.** $5(1.2)(20)$ **14.** $25(11)(4)(0)$
8.4   780   120   0

Exercises

Replace each __?__ with the number that makes the statement true.

A   **1.** $26 \cdot 38 = 38 \cdot \underline{\ ?\ }$ 26   **2.** $(27)(1.2) = (\underline{\ ?\ })(27)$ 1.2

3. $(9 \cdot 7) \cdot 3 = 9 \cdot (\underline{\ ?\ } \cdot 3)$ 7   **4.** $6 \cdot (8 \cdot 9) = (6 \cdot 8) \cdot \underline{\ ?\ }$ 9

5. $(\underline{\ ?\ })(52) = (52)(2.1)$ 2.1   **6.** $(5 \cdot \underline{\ ?\ }) \cdot 4 = 5 \cdot (6 \cdot 4)$ 6

Use the properties to find each product mentally.

7. $2 \cdot 32 \cdot 5$ 320   **8.** $2(8)(50)$ 800   **9.** $1 \cdot 89$ 89   **10.** $56(1)$ 56

11. $24(0)(5)$ 0   **12.** $0 \cdot 22 \cdot 5$ 0   **13.** $(24)(1)(5)$ **14.** $33 \cdot 2 \cdot 1$ 66
                                                 120

15. $(0.6)(1.1)(5)$ **16.** $(0.5)(3.7)(2)$ **17.** $(4.4)(0)(5)$ 0 **18.** $(1)(0.5)(8)$ 4
3.3   3.7

Use the properties to evaluate each expression when $x = 5$, $y = 12$, and $z = 3$.

B   **19.** $(20)(43)(x)$ **20.** $(25)(z)(4)$ 300   **21.** $x \cdot 7 \cdot y$ 420 **22.** $x \cdot z \cdot 1$ 15
4300

23. $y \cdot 1 \cdot z$ 36   **24.** $z \cdot 0 \cdot y$ 0   **25.** $x \cdot y \cdot 0$ 0   **26.** xzy 180

27. $(20)(y)(0.2)$ 48 **28.** $(0.5)(x)(14)$ 35 **29.** $0.5zy$ 18   **30.** $0.2zx$ 3

THINKING SKILLS   **31.** Knowledge   **32.** Analysis; synthesis
                          33. Knowledge   **34.** Analysis; synthesis

In Exercises 31 and 33, do the work in parentheses first.

C   **31.** Find $49 - (27 - 14)$. Then find $(49 - 27) - 14$. 36; 8

32. Examine the results of Exercise 31. Is subtraction associative? No Create another example to support your answer. See answer below.

33. Find $24 \div (6 \div 2)$. Then find $(24 \div 6) \div 2$. 8; 2

34. Examine the results of Exercise 33. Is division associative? No Create another example to support your answer. Answers will vary.

                32. Answers will vary. Example: Example:
SPIRAL REVIEW $(12 - 8) - 2 = 2, 12 - (8 - 2) = 6$ $(48 \div 12) \div 2 = 2,$
                                                          $48 \div (12 \div 2) = 8$

S   **35.** Evaluate $a \div b$ when $a = 10$ and $b = 2$. *(Lesson 1-1)* 5

36. Complete: 6 ft 4 in. $= \underline{\ ?\ }$ in. *(Toolbox Skill 22)* 76

37. Find the product mentally: $12(5)(0)(8)$ *(Lesson 1-7)* 0

38. Find the sum: $3\frac{2}{3} + 4\frac{5}{6}$ *(Toolbox Skill 18)* $8\frac{1}{2}$

39. Evaluate $b - a$ when $a = 4.6$ and $b = 11.2$. *(Lesson 1-2)* 6.6

40. Is 6240 divisible by 2? by 3? by 4? by 5? by 8? by 9? by 10? *(Toolbox Skill 16)* Yes; Yes; Yes; Yes; Yes; No; Yes

Connecting Arithmetic and Algebra   **27**

Critical Thinking

Exercises 31–34 are designed to help students discover that the associative and commutative properties are not true for all operations. To achieve this, students are asked to *analyze* given examples, *draw conclusions* from their analyses, and then support their conclusions by *creating* new examples.

Follow-Up

Extension

You may want to have students use the addition and multiplication properties with fractions. For addition, you may want to use fractions with common denominators to simplify computation.

Enrichment

Enrichment 7 is pictured on page 2F.

Practice

Practice 8 is shown below.

Practice 8, *Resource Book*

Practice 8
Skills and Applications of Lesson 1-7

Replace each __?__ with the number that makes the statement true.

1. $14 \cdot 19 = 19 \cdot \underline{\ ?\ }$ 14   **2.** $\underline{\ ?\ } \cdot 33 = 33 \cdot 29$ 29

3. $(7.8)(\underline{\ ?\ }) = (9.3)(7.8)$ 9.3   **4.** $7 \cdot (9 \cdot 5) = (\underline{\ ?\ } \cdot 9) \cdot 5$ 7

5. $(7.3)(8.4) = (\underline{\ ?\ })(7.3)$ 8.4   **6.** $(19 \cdot 4) \cdot 7 = 19 \cdot (4 \cdot \underline{\ ?\ })$ 7

7. $(\underline{\ ?\ } \cdot 12) \cdot 5 = 20 \cdot (12 \cdot 5)$ 20   **8.** $(16.3)(14.8) = (\underline{\ ?\ })(16.3)$ 14.8

9. $(7 \cdot \underline{\ ?\ }) \cdot 10 = 7 \cdot (14 \cdot 10)$ 14   **10.** $(0.07)(0.13) = (\underline{\ ?\ })(0.07)$ 0.13

11. $(11 \cdot 13) \cdot 20 = 11 \cdot (\underline{\ ?\ } \cdot 20)$ 13   **12.** $(\underline{\ ?\ })(32) = (32)(38)$ 38

13. $\underline{\ ?\ } \cdot (13 \cdot 17) = (10 \cdot 13) \cdot 17$ 10   **14.** $(16.8)(\underline{\ ?\ }) = (11.3)(16.8)$ 11.3

15. $(73)(39) = (39)(\underline{\ ?\ })$ **16.** $(25 \cdot 0.32) \cdot \underline{\ ?\ } = 25 \cdot (0.32 \cdot 100)$
             73   100

Use the properties to find each sum mentally.

17. $(32.5) \cdot 0$ 0   **18.** $5 \cdot 14 \cdot 2$ 140   **19.** $33.5(1)$ 33.5

20. $(5)(14.8)(2)$ 148   **21.** $(16.5)(0)(42.8)$ 0   **22.** $(2)(1)(50)$ 100

23. $(20)(38)(5)$ 3800   **24.** $(4)(27.5)(25)$ 2750   **25.** $(50)(2)(0)$ 0

26. $(0.037)(1)$ 0.037   **27.** $(0.004)(0.013)(0)$ 0   **28.** $(0.5)(4.8)(2)$ 4.8

29. $(2.5)(13.7)(4)$ 137   **30.** $(2.5)(44.6)(0.4)$ 44.6   **31.** $(16.3)(0)(0.07)$ 0

32. $(16.5)(46.5)(0)(13.8)$ 0   **33.** $(25)(50)(4)(2)$ 10,000   **34.** $(0)(4.38)(1)$ 0

Use the properties to evaluate each expression when $x = 2$, $y = 14$, and $z = 5$.

35. $x \cdot y \cdot 88 \cdot 0$ 0   **36.** $(0.6)(3.3)(z)$ 9.9   **37.** $16.82xz$ 168.2

38. $(50)(92.89)(x)$ 9289   **39.** $(0.20)(y)(5)$ 14   **40.** $(y)(0)(16.9)$ 0

Teaching the Lesson
- Materials: calculator
- Lesson Starter 8
- Reteaching/Alternate Approach, p. 2D
- Toolbox Skill 15

Lesson Follow-Up
- Practice 9
- Enrichment 8: Data Analysis
- Study Guide, pp. 15–16

Warm-Up Exercises

Find each answer.

1. $228 + 56 + 172$ 456
2. $36.6 + 3.56 + 44.4$ 84.56
3. $34 \cdot 5 \cdot 2$ 340
4. $56(34.5)(0)(2)$ 0
5. $(3.25)(4.14)$ 13.455

Lesson Focus

Ask students if they have ever raced another person to see who could compute faster. Using a calculator to compute is sometimes slower than computing mentally. Today's lesson focuses on choosing the most efficient method of computation.

Teaching Notes

Key Question
How do you decide on a method of computation?   Answers will vary.

Reteaching/Alternate Approach
For a suggestion on a hands-on activity using calculators with this lesson, see page 2D.

Additional Examples

Tell whether it is most efficient to find each answer using *mental math, paper and pencil*, or a *calculator*. Then find each answer using the method that you chose.

1. $57 + 49$ pp; 106
2. $1.9 + 7.6 + 6.1$ mm; 15.6
3. $(3.42)(0.567)$ c; 1.93914

DECISION MAKING

1-8 Choosing the Most Efficient Method of Computation

Objective: To decide whether it would be most efficient to use mental math, paper and pencil, or a calculator to solve a given problem.

Before doing a computation, you should *inspect* the problem and decide whether to use mental math, paper and pencil, or a calculator.

- *Mental math* may be most efficient when you see sums of ten or products of ten, when you do not need to rename, or when you can *add on* or *count back* easily.

- *Paper and pencil* may be most efficient when the computation seems simple or involves numbers with few digits.

- A *calculator* may be most efficient when the computation involves many numbers. You also might decide to use a calculator when accuracy is very important.

Example

Tell whether it is most efficient to find each answer using *mental math*, *paper and pencil*, or a *calculator*. Then find each answer.

a. $147 - 98$ b. $23.47 + 10.82 + 16.09$

Solution

a. If you know how to *add on*, use mental math.
$98 + 2 = 100; \quad 100 + 47 = 147$
Because $2 + 47 = 49$, $147 - 98 = 49$.

b. There are three terms and each term has four digits. A calculator would be the most efficient method.

| 23.47 | 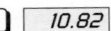 | 10.82 | | 16.09 | | 50.38 |

✓ Check Your Understanding

Suppose you do not know how to *add on*. What is another method to use in part (a) of the Example? Explain.
See *Answers to Check Your Understanding* at the back of the book.

Guided Practice

COMMUNICATION «*Discussion* Answers to Guided Practice 1–12 are on p. A2.

Explain why it is easy to find each answer using mental math.

«1. $(2000)(30)$ «2. $236 - 99$ «3. $67 + 23$ «4. $540 \div 60$

«5. $4 + 73.6$ «6. $0.9 \cdot 100$ «7. $1.5 \div 0.3$ «8. $4.3 - 1.2$

In each set, choose the exercise that might be done more efficiently using paper and pencil instead of a calculator. Give a reason for your choice.

9. $948 + 1003$ 10. $724.3 - 6.531$ 11. $852 \cdot 791$ 12. $8.208 \div 3$
$948 + 9765$ $7.243 - 6.531$ $(852)(3)$ $8.208 \div 342$

Tell whether it is most efficient to find each answer using *mental math,* *paper and pencil,* **or a** *calculator.* **Then find each answer using the method that you chose.** Answers may vary. Likely answers are given.

13. $49.5 \div 3$
pp; 16.5

14. $100 \cdot 2 \cdot 3$
mm; 600

15. $5414 - 678$
c; 4736

16. $2.5 + 7$
mm; 9.5

Exercises

1–16. Answers may vary. Likely answers are given.
Tell whether it is most efficient to find each answer using *mental math,* *paper and pencil,* **or a** *calculator.* **Then find each answer using the method that you chose.**

A
1. $13 \cdot 6$
pp; 78

2. $165 - 45$
mm; 120

3. $380 + 135$
pp; 515

4. $84 \div 6$ pp; 14

5. $783 \cdot 35$
c; 27,405

6. $1.2 \div 0.4$
mm; 3

7. $521(0.9)$
pp; 468.9

8. $10.93 - 2.981$
c; 7.949

9. $6.6 \div 2.75$
c; 2.4

10. $143.9 + 2.3$
pp; 146.2

11. $(0.4)(0.7)$
mm; 0.28

12. $45.7 \div 100$
mm; 0.457

Tell whether it is most efficient to solve each problem using *mental math, paper and pencil,* **or a** *calculator.* **Then solve each problem using the method that you chose.**

B
13. Greg spent $2.70 on 3 packages of seeds. How much did each package of seeds cost? mm; $.90

14. Sharla has $12.50. She wants to buy as many subway tokens as possible. Tokens cost 75¢ each. How many can she buy? c; 16

15. Alan bought items priced at $.89, $2.49, $5.45, and $1.49 at the grocery store. Estimate the total cost of the items. mm; about $10.50

16. Mai purchased seven flower arrangements for a party. Each flower arrangement cost $19.85. About how much was the total cost?
mm; about $140

SPIRAL REVIEW

S
17. Find the product: 0.04×0.05
(Toolbox Skill 13) 0.002

18. Is it most efficient to find the product $(4.82)(5.64)$ using *mental math, paper and pencil,* or a *calculator*? *(Lesson 1-8)*
calculator

19. Replace the __?__ with $>$, $<$, or $=$: 2.04 __?__ 2.040
(Lesson 1-5) $=$

20. Find the quotient: $8.48 \div 100$
(Toolbox Skill 15) 0.0848

Connecting Arithmetic and Algebra   **29**

Teaching the Lesson
- Materials: calculator
- Lesson Starter 9
- Using Technology, p. 2C
- Toolbox Skill 9

Lesson Follow-Up
- Practice 10
- Enrichment 9: Exploration
- Study Guide, pp. 17–18
- Computer Activity 1

Warm-Up Exercises

Evaluate each expression when $a = 3$ and $b = 5$.

1. $3 \cdot a \cdot a$ 27
2. $a \cdot a \cdot a \cdot a \cdot a$ 243
3. $5 \cdot b \cdot b$ 125
4. $b \cdot b \cdot b \cdot b \cdot b$ 3125

Lesson Focus

Tell students to imagine that they are climbing the stairs in a very tall building. They would bring one foot up a step and repeat the process with the other foot and so on, over and over. In mathematics, we often deal with repetitive patterns. Today's lesson focuses on multiplication factors that repeat and the concept of exponents.

Teaching Notes

Stress that an exponent represents a shortened form for expressing a number of like factors. If students do not have a calculator with a $\boxed{y^x}$ key, they may find 4^5 by pressing $\boxed{4}$ $\boxed{\times}$ $\boxed{=}$ $\boxed{=}$ $\boxed{=}$ $\boxed{=}$. The display should show 1024.

Key Question

What is the difference between $5x^2$ and $(5x)^2$? In $5x^2$, only the x is squared. In $(5x)^2$, both the 5 and the x are squared.

1-9 Exponents

Objective: To compute with exponents and to evaluate expressions involving exponents.

Terms to Know
- *exponential form*
- *power*
- *exponent*
- *base*

EXPLORATION

1 Enter each key sequence on a calculator. Record the results.

a. $\boxed{3.}$ $\boxed{=}$ 3

b. $\boxed{3.}$ $\boxed{\times}$ $\boxed{3.}$ $\boxed{=}$ 9

c. $\boxed{3.}$ $\boxed{\times}$ $\boxed{3.}$ $\boxed{\times}$ $\boxed{3.}$ $\boxed{=}$ 27

d. $\boxed{3.}$ $\boxed{\times}$ $\boxed{3.}$ $\boxed{\times}$ $\boxed{3.}$ $\boxed{\times}$ $\boxed{3.}$ $\boxed{=}$ 81

2 Suppose that n represents the number of times you entered 3 as a factor. Find the value of n that makes each statement true.

a. $3^n = 3$ 1   b. $3^n = 9$ 2   c. $3^n = 27$ 3   d. $3^n = 81$ 4

3 What number is represented by 3^{10}? 59,049

4 Find the value of x that makes this statement true: $3^x = 1,594,323$ 13

You can write a multiplication expression in which all the factors are the same in a shortened form called **exponential form.**

$$\underbrace{3 \cdot 3 \cdot 3 \cdot 3 \cdot 3 \cdot 3}_{6 \text{ factors}} = 3^{\overset{\longleftarrow \text{ exponent}}{6}}$$

base

The number 3^6, or 729, is called a **power** of 3. The **exponent** 6 shows that the **base** 3 is used as a factor six times.

You read 3^6 as *three to the sixth power* or *the sixth power of three.*
You read 3^2 as *three to the second power* or *three squared.*
You read 3^3 as *three to the third power* or *three cubed.*

Any number to the first power is equal to that number, as in $3^1 = 3$. The number 1 to any power equals 1, as in $1^7 = 1$.

Example 1

Solution

☑ **Check Your Understanding**

Find each answer: a. 6^3 b. 1^4 c. 10^1

a. $6^3 = 6 \cdot 6 \cdot 6 = 216$ b. $1^4 = 1$ c. $10^1 = 10$

1. Describe how 6^3 is different from $6 \cdot 3$.
2. Describe how 6^3 is different from 3^6.
3. In Example 1(b), explain why $1^4 = 1$.
   See *Answers to Check Your Understanding* at the back of the book.

You can use exponents with variables as well as with numbers. For instance, you can write $x \cdot x \cdot x \cdot x \cdot x$ as x^5. You read x^5 as *a number x to the fifth power* or *the fifth power of a number x.*

30   Chapter 1

You may need to evaluate variable expressions that contain exponents.

Example 2

Evaluate each expression when $n = 4$.

a. $3n^2$ **b.** $(3n)^2$

Solution

a. $3n^2 = 3 \cdot n \cdot n = 3 \cdot 4 \cdot 4 = 48$

b. $(3n)^2 = (3n)(3n) = (3 \cdot 4)(3 \cdot 4) = (12)(12) = 144$

 Some calculators have a $\boxed{y^x}$ key. If you have this key on your calculator, you may wish to use it when you work with exponents. For example, you can use the following key sequence to find 8^6.

Guided Practice

COMMUNICATION «*Reading*

Each word is followed by four correct meanings. Choose the meaning that is used in this lesson.

«**1.** base
    a. the lowest point
    b. a supporting layer
    (c.) a number that is raised to a power
    d. a corner of the infield in softball

«**2.** power
    a. strength
    b. authority
    c. electricity
    (d.) exponent

COMMUNICATION «*Writing*

Write an expression for each phrase.

«**3.** three to the fifth power 3^5

«**4.** a number x to the sixth power x^6

«**5.** a number x cubed multiplied by six $6x^3$

«**6.** six times a number x, cubed $(6x)^3$

7. eight to the third power or eight cubed

8. a number a to the fourth power

9. three times the fifth power of a number x

10. the fifth power of the product of three and a number x

Write a phrase for each expression.

«**7.** 8^3 «**8.** a^4 «**9.** $3x^5$ «**10.** $(3x)^5$

Give the exponential form of each expression.

11. $(7)(7)(7)(7)(7)$ 7^5 **12.** $x \cdot x \cdot x \cdot x$ x^4

13. $5 \cdot d \cdot d \cdot d$ $5d^3$ **14.** $4y \cdot 4y \cdot 4y$ $(4y)^3$

Give the multiplication that each expression represents.

15. 6^7 **16.** c^5 **17.** $5x^4$ **18.** $(2c)^3$
$6 \cdot 6 \cdot 6 \cdot 6 \cdot 6 \cdot 6 \cdot 6$ $c \cdot c \cdot c \cdot c \cdot c$ $5 \cdot x \cdot x \cdot x \cdot x$ $(2c)(2c)(2c)$

Connecting Arithmetic and Algebra   **31**

31

Reasoning

To help students learn how to reason, you may wish to ask the following questions.

1. Can you determine an exact value for 1^x? Explain.
    Yes; $1^x = 1$, since 1 to any power equals 1.

2. Can you determine an exact value for x^1? Explain. No; $x^1 = x$, so the value depends upon the value of x.

Error Analysis

A common error students make is to multiply the base by the exponent. For example, students may find 3^5 as follows: $3^5 = 3 \times 5$, or 15. Point out that the exponent tells how many times the base is used as a factor. Thus, $3^5 = 3 \times 3 \times 3 \times 3 \times 3$, or 243.

Using Technology

For a suggestion on how to use a calculator to discover some laws of exponents, see page 2C.

Additional Examples

1. Find each answer.
    a. 3^4 81
    b. 10^3 1000
    c. 1^5 1

2. Evaluate each expression when $n = 3$.
    a. $2n^4$ 162
    b. $(2n)^4$ 1296

Closing the Lesson

Ask students to explain in their own words why $2^3 \neq 6$. Also, ask them to explain how to evaluate variable expressions involving exponents.

Suggested Assignments

Skills: 1–20, 21–37 odd, 39, 40, 43, 45, 49–52
Average: 1–20, 21–37 odd, 39–46, 49–52
Advanced: 1–17 odd, 19, 20, 21–37 odd, 39–52, Challenge

Exercise Notes

Communication: Reading
Guided Practice Exercises 1 and 2 reinforce the fact that many mathematical terms relate to commonly used words in English.

Calculator
Exercises 35–38 help students recognize the correct key sequence needed to raise a given number to a given power.

Making Connections/Transitions
Exercises 39–42, and the paragraph before them, show a connection between algebra and geometry. This connection can be made even more clear by using tiles and blocks to illustrate the concept. *Exercises 41 and 42* can be used to show students how to form square and cube roots.

Application
In the paragraph preceding *Exercises 39–42*, the diagrams can be related to real-world applications. Area, for example, is used to determine the amount of carpet needed in a house. What other items in a house are purchased based on their area? tile for bathrooms, kitchens or floors, wallpaper, paint, roofing materials
The capacity of heating and air-conditioning units is based on the number of cubic feet of space. What other common item(s) identify cubic feet or inches? refrigerators, freezers, car trunks

Follow-Up

Extension
Is a number raised to a whole number power always larger than the original number? No; the value of a positive decimal number less than 1 raised to a whole number power becomes smaller as the power increases. For example, $(0.5)^1 = 0.5$, $(0.5)^2 = 0.25$, $(0.5)^3 = 0.125$, $(0.5)^4 = 0.0625$, and so on.

Find each answer.

19. 4^3 64 **20.** 1^5 1 **21.** 2^5 32 **22.** 10^4 10,000 **23.** 11^2 121 **24.** 2^8 256

Evaluate each expression when $n = 3$.

25. n^5 243 **26.** $5n^2$ 45 **27.** $3n^3$ 81 **28.** $(8n)^1$ 24 **29.** $(4n)^2$ 144 **30.** $(3n)^4$ 6561

Exercises

Find each answer.

A **1.** 7^3 343 **2.** 1^{14} 1 **3.** 9^1 9 **4.** 12^2 144 **5.** 10^3 1000 **6.** 6^4 1296

Evaluate each expression when $n = 2$.

7. n^4 16   **8.** n^6 64   **9.** n^7 128   **10.** n^9 512

11. $6n^3$ 48   **12.** $10n^5$ 320   **13.** $4n^4$ 64   **14.** $7n^2$ 28

15. $(2n)^2$ 16   **16.** $(9n)^2$ 324   **17.** $(4n)^3$ 512   **18.** $(5n)^3$ 1000

19. The memory of a computer is measured in a unit called a byte. The letter K represents 2^{10} bytes. Write this number without an exponent. Then write the number of bytes represented by 40K of memory. 1024; 40,960

20. DATA, *pages 666–667* Calculate 11^1, 11^2, 11^3, and 11^4. Where can these powers of 11 be found in Pascal's triangle? 11; 121; 1331; 14,641; they are formed by the numbers in the second through fifth rows of the triangle.

Evaluate each expression for the given value of the variable.

B **21.** 5^x, when $x = 2$ 25   **22.** 3^v, when $v = 5$ 243

23. 4^m, when $m = 4$ 256   **24.** 9^w, when $w = 3$ 729

25. 12^x, when $x = 1$ 12   **26.** 1^c, when $c = 12$ 1

Find each answer.

27. $(0.9)^2$ 0.81   **28.** $(0.7)^2$ 0.49   **29.** $(0.5)^3$ 0.125   **30.** $(0.6)^3$ 0.216

31. $(0.2)^3$ 0.008   **32.** $(0.3)^3$ 0.027   **33.** $(0.1)^3$ 0.001   **34.** $(0.1)^4$ 0.0001

CALCULATOR

Match each expression with the correct calculator key sequence.

35. $(1.2)^5$ D   A. [0.5] [y^x] [12.] [=]

36. 12^5 C   B. [5.] [y^x] [12.] [=]

37. 5^{12} B   C. [12.] [y^x] [5.] [=]

38. $(0.5)^{12}$ A   D. [1.2] [y^x] [5.] [=]

CONNECTING ALGEBRA AND GEOMETRY

Why are the exponents 2 and 3 read as *squared* and *cubed*? The reason is their connection to geometry. Evaluating x^2 gives you the area of a square with side of length x, and evaluating x^3 gives you the volume of a cube with edge of length x. The figures below illustrate the geometric meaning of 4^2 and 4^3.

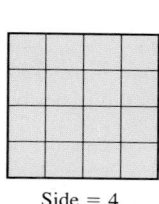
Side = 4
Area = 4^2 = 16

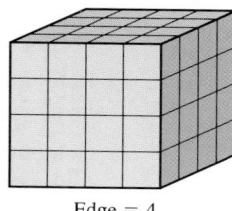
Edge = 4
Volume = 4^3 = 64

39. Find the area of a square with side of length 9. 81

40. Find the volume of a cube with edge of length 7. 343

41. The area of a square is 121. What is the length of a side? 11

42. The volume of a cube is 125. What is the length of an edge? 5

THINKING SKILLS 43–46. Knowledge   **47–48.** Evaluation

Replace each __?__ with >, <, or =.

C **43.** 5^3 __?__ 3^5 < 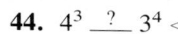   **44.** 4^3 __?__ 3^4 <   **45.** 1^8 __?__ 8^1 <   **46.** 2^7 __?__ 7^2 >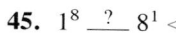

Assume that a and b represent any of the numbers 2, 3, 4, 5, ... , and $a > b$. Determine a value of a and a value of b that make each statement true.

47. $a^b = b^a$ $a = 4; b = 2$

48. $a^b > b^a$ $a = 3; b = 2$

SPIRAL REVIEW

The paragraph is about the amount of money Emily Ling earns and the number of hours that she works.

S **49.** Tell what this paragraph is about: Emily Ling earns \$9 per hour at Technology Industries. Each week she works five days for a total of 42 h. *(Lesson 1-4)*

50. Find the sum mentally: $118 + 0$ *(Lesson 1-6)* 118

51. Find the quotient: $\frac{7}{8} \div \frac{3}{4}$ *(Toolbox Skill 21)* $1\frac{1}{6}$

52. Evaluate $4x^3$ when $x = 5$.   *(Lesson 1-9)* 500

Challenge

Describe how to find this sum using only mental math. See answer on p. A2.

$$20 + 21 + 22 + 23 + 24 + 25 + 26 + 27 + 28 + 29 + 30$$

Connecting Arithmetic and Algebra   **33**

33

Activities I and II are designed to give students an idea of how quickly powers of a number increase. Activity III helps students to think spatially, a useful skill for studying the geometry in Chapters 6, 10, and 14.

Activity Notes

Activity I

Before beginning *Activity I,* ask students to estimate the number of times that they can fold a sheet of letter-size paper in half. After the activity, ask them to compare their results with their original estimates.

Could additional folds be made if the paper were larger or thinner? Ask students to make a conjecture, repeat the activity, and compare the results.

Activity II

Problem-solving strategies that might be used in *Activity II* are make a table, organize the data, and look for a pattern.

Activity III

The software program *Factory* by Sunburst supports the concepts in this activity.

FOCUS ON EXPLORATIONS

Objective: To use paper folding to model powers of two.

Materials
- letter-size paper
- scissors

Modeling Powers of Two

You can use a sheet of paper to show how quickly powers of numbers increase. To do this, you simply keep folding the paper again and again.

Activity I *Collecting Data*

1 Fold a sheet of paper in half. Unfold the paper. Into how many parts has the paper been separated? 2

2 Refold the paper. Now fold it in half again. Unfold the paper. Into how many parts has the paper been separated this time? 4

3 Continue the process of folding and unfolding the paper. After each new fold, open the paper and count the number of parts into which it has been separated. To record your results, copy and complete the table below.

Number of Folds	1	2	3	4	5
Number of Parts	2	?	?	?	?

                                               4   8   16   32

Activity II *Problem Solving*

1 Look at the data in the table from Activity I. Describe the numbers that appear in the row labeled "Number of Parts." The numbers are powers of 2.

2 At this point, you may not be able to fold the paper anymore. However, you can use the data in the table from Activity I to predict the number of parts that the paper would have if you could fold it six times. How many parts would the paper have? 64

3 How many parts would the paper have if you could fold it seven times? eight times? nine times? ten times? 128; 256; 512; 1024

4 What do the numbers you found in the previous step represent mathematically? 2^7; 2^8; 2^9; 2^{10}

Origami, the art or process of folding paper into shapes, originated in Japan.

5 How many folds would you have to make for the paper to have 4096 parts? 12

6 After how many folds would the number of parts be more than one million? 20

Activity III *Visual Thinking* Answers to Activity III are on p. A2.

1 Each piece of paper has been folded twice. The dark lines indicate cuts that have been made through the folded paper. Sketch how each piece of paper would look when unfolded.

a.

first fold

b.
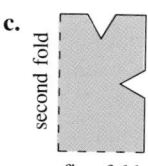
first fold

c.
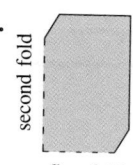
first fold

d.
first fold

e.
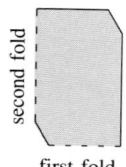
first fold

f.
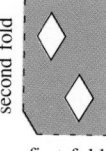
first fold

g.

first fold

h.
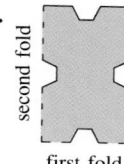
first fold

2 Fold a letter-size piece of paper twice. Then cut the folded paper so that you can create each design when it is unfolded.

a.
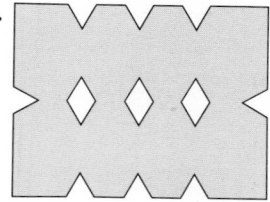

b.
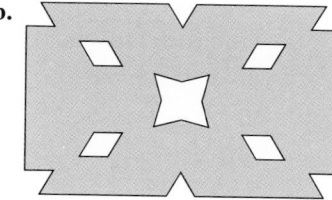

c.
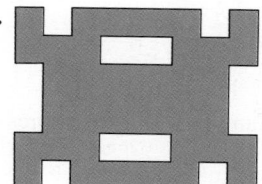

d.
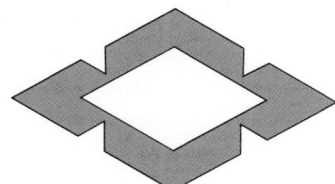

Connecting Arithmetic and Algebra **35**

Making Connections/Transitions
Students can compare *Activity III* to the activity of making snowflakes from folded paper. Such designs could be incorporated into tile or wallpaper designs. Students may wish to create their own design which could be used for wallpaper.

Follow-Up

Cooperative Learning
Divide students into cooperative groups. Ask students to tear a sheet of paper in half and pile the pieces on top of each other. Then ask them to tear this stack in half and again pile the pieces on top of each other. Ask students to do this twice more. Now ask the students what they are modeling. powers of 2 Tell students to continue this process until they get a stack tall enough to measure. Five tears, or 32 pieces, will be about one-eighth inch. Ask students to predict the height of the stack after 10 tears, 4 in. after 15 tears, 128 in. and after 20 tears. 4096 in. Ask students to explain how they arrived at their answers.

Teaching the Lesson
- Lesson Starter 10
- Using Technology, p. 2C
- Toolbox Skills 7–14

Lesson Follow-Up
- Practice 11
- Practice 12: Nonroutine
- Enrichment 10: Communication
- Study Guide, pp. 19–20
- Manipulative Activity 2
- Cooperative Activity 3
- Test 2

Warm-Up Exercises

Find each answer. Perform the operations in order from left to right.

1. $19 + 27 + 56$ 102
2. $83 - 14 - 26$ 43
3. $5 \times 9 + 11$ 56
4. $64 \div 4 - 9$ 7

Lesson Focus

Tell students that when they prepare to go to school in the morning, they do things in a certain order. For instance, they first get up before they get dressed, and they get dressed before they leave their homes. In mathematics, operations must also be performed in a specified order. Today's lesson focuses on order of operations.

Teaching Notes

A mnemonic device that may help students remember the order of operations is "Please Please My Dear Aunt Sally." The underlined letters of the phrase represent Parentheses, Power, Multiplication, Division, Addition, and Subtraction.

Order of Operations

Objective: To use the order of operations to compute and to evaluate expressions.

APPLICATION

Terms to Know
- *order of operations*

Jerry and Sondra entered this exercise on different calculators.

$$30 + 24 \div 6 \cdot 2$$

The result on Jerry's calculator was 18, while the result on Sondra's calculator was 38.

QUESTION Which calculator displayed the correct answer?

In mathematics, you must perform operations in an agreed-upon order to make sure that an expression has only one answer.

Order of Operations

1. First do all work inside any parentheses.
2. Then find each power.
3. Then do all multiplications and divisions in order from left to right.
4. Then do all additions and subtractions in order from left to right.

Example 1

Find each answer.

a. $30 + 24 \div 6 \cdot 2$ b. $7 \cdot 3^2 - (5 + 6)$

Solution

a. $30 + 24 \div 6 \cdot 2 = 30 + 4 \cdot 2$ ◄— Divide.
$= 30 + 8$ ◄— Multiply.
$= 38$ ◄— Add.

b. $7 \cdot 3^2 - (5 + 6) = 7 \cdot 3^2 - (11)$ ◄— Work inside parentheses first.
$= 7 \cdot 9 - 11$ ◄— Find the power ($3^2 = 9$).
$= 63 - 11$ ◄— Multiply.
$= 52$ ◄— Subtract.

Check Your Understanding

1. In Example 1(a), why do you first divide and then multiply?
2. Describe how you would find the answer to Example 1(a) if the expression were $(30 + 24) \div 6 \cdot 2$.
See *Answers to Check Your Understanding* at the back of the book.

ANSWER Example 1(a) shows that Sondra's calculator displayed the correct answer.

The fraction bar acts like parentheses in the order of operations.

Example 2

Find the answer: $\dfrac{24 + 12}{13 - 4}$

Solution

$\dfrac{24 + 12}{13 - 4} = \dfrac{36}{9}$ ← Work above the fraction bar, then below the fraction bar.

        $= 4$ ← Divide.

Check Your Understanding

3. Write the expression in Example 2 with parentheses instead of with a fraction bar.
   See *Answers to Check Your Understanding* at the back of the book.

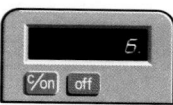

You can use a calculator to find an answer when order of operations is involved, but the key sequence that you use depends on the type of calculator you have. For instance, if your calculator has parenthesis keys, you can find $18 \div (7 - 4)$ using this key sequence.

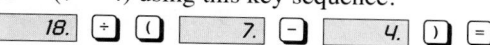

If your calculator does not have parenthesis keys, you will need to use the memory keys.

You must also use the order of operations when you evaluate variable expressions.

Example 3

Evaluate each expression when $a = 5$ and $b = 4$.

a. $(8 + b^2) \div 3 + a$ b. $\dfrac{43 + a}{10 - b}$

Solution

a. $(8 + b^2) \div 3 + a = (8 + 4^2) \div 3 + 5$
                 $= (8 + 16) \div 3 + 5$
                 $= 24 \div 3 + 5$
                 $= 8 + 5$
                 $= 13$

b. $\dfrac{43 + a}{10 - b} = \dfrac{43 + 5}{10 - 4}$

          $= \dfrac{48}{6}$

          $= 8$

Check Your Understanding

4. Describe how you would evaluate the expression in Example 3(a) if it were $(8 + b)^2 \div 3 + a$.
5. Describe how you would evaluate the expression in Example 3(a) if it were $(8 + b)^2 \div (3 + a)$.
   See *Answers to Check Your Understanding* at the back of the book.

Suggested Assignments

Skills: 1–23, 25–35 odd, 38–41
Average: 1–21 odd, 22, 23–37 odd, 38–41
Advanced: 1–21 odd, 22–41

Exercise Notes

Communication: Reading and Writing

In *Guided Practice Exercise 1*, students are asked to clarify their understanding of the order of operations by restating the rules in their own words. In *Guided Practice Exercise 2*, students are asked to analyze the rules and to create their own exercises to demonstrate their understanding.

Critical Thinking

Exercises 30–37 require students to *evaluate* whether a given statement is true or false under the order of operations. For any false statements, students are required to insert parentheses to *create* a true statement.

Guided Practice

«**1.** COMMUNICATION «*Reading* Without referring to the book, write the correct order of operations on a sheet of paper. Then check your answer against the text on pages 36–37. See answer on p. A2.

«**2.** COMMUNICATION «*Writing* Write an expression in which you can work from left to right to find the answer. Write an expression in which you *cannot* work from left to right to find the answer.
Answers will vary. Examples are: $12 \cdot 5 - 2$; $12 - 5 \cdot 2$.

List the operations in the order you would perform them.

3. $34 - 10 + 4 \cdot 2$

4. $(53 - 8) \div 5$

5. $4 \cdot 5^2 + 7$
Answers to Guided Practice 3–6 are on p. A2.

6. $\dfrac{16 + 14}{10 - 5}$

Find each answer.

7. $7 + 45 \div 9$ 12

8. $4^3 + (15 - 7)$ 72

9. $8 + 32 \div 4 \cdot 5$ 48

10. $\dfrac{10 \cdot 4}{14 - 9}$ 8

Evaluate each expression when $a = 2$, $b = 6$, and $c = 12$.

11. $3b + 7$ 25

12. $(2a + c) \div 2$ 8

13. $3(b - a^2) + 16$ 22

14. $\dfrac{36 + c}{b - a}$ 12

Exercises

Find each answer.

A **1.** $9^2 + 3 \cdot 5$ 96

2. $4^3 - 30 \div 2$ 49

3. $9 + 45 \div 9 \cdot 8$ 49

4. $7^2 - 14 + 5 \cdot 2$ 45

5. $8^2 \div (8 - 4)$ 16

6. $11 + 3(19 + 2)$ 74

7. $2^2 + 6(3 + 10)$ 82

8. $14 + (3^3 - 7)$ 34

9. $\dfrac{18 + 10}{7 - 3}$ 7

10. $\dfrac{51 - 9}{11 - 4}$ 6

11. $\dfrac{9 \cdot 5}{3 + 6}$ 5

12. $\dfrac{54 + 18}{2 \cdot 4}$ 9

Evaluate each expression when $a = 2$, $b = 5$, and $c = 4$.

13. $9 + 7a$ 23

14. $8c - 13$ 19

15. $(14 - a) \cdot c$ 48

16. $b + c^2 \div a + 4$ 17

17. $81 - (b + 9) + a^2$ 71

18. $65 \div b - a + 6$ 17

19. $\dfrac{b + 19}{a + 4}$ 4

20. $\dfrac{15c}{1 + b}$ 10

21. $b^2 - a^3$ 17

22. June runs 37 mi each week, twelve miles of which she runs on the weekend. How many miles does June run each weekday? 5 mi

23. DATA, *pages 2–3* The Rolling Pebbles had five gold albums and two platinum albums last year. What is the least number of albums they could have sold last year? 4,500,000 albums

Find each answer. Work inside the parentheses first. Then work inside the square brackets.

B **24.** $[(12 - 4) \cdot 2 + 11] \div 3$ 9

25. $[(6 + 8) \cdot 3 - 12] + 6$ 36

26. $110 - [20 + (4^2 - 9)]$ 83

27. $45 + [6^2 - (12 + 6)]$ 63

28. $34 + [(16 + 12) \div 4] - 16$ 25

29. $48 - [36 \div (4 + 5)] + 11$ 55

THINKING SKILLS 30–37. Evaluation; synthesis
Answers to Exercises 30–37 are on p. A2.
Classify each statement as *True* or *False*. Then make each false statement true by inserting parentheses where necessary.

C **30.** $4 \cdot 5 + 6 = 44$

31. $24 - 4 \cdot 2 = 40$

32. $24 \div 3 + 5 \cdot 2 = 6$

33. $6 \cdot 12 - 5 \cdot 8 = 32$

34. $4 \cdot 6 + 3 \cdot 3 = 33$

35. $3 \cdot 4 - 2 \cdot 3 = 18$

36. $4 + 4^2 \div 2 = 32$

37. $12 - 2^2 \div 4 = 2$

SPIRAL REVIEW

S **38.** Find the answer: 5^5 *(Lesson 1-9)* 3125

39. Find the difference: $40{,}007 - 6218$ *(Toolbox Skill 8)* 33,789

40. Find the sum: $8.635 + 27.9$ *(Toolbox Skill 11)* 36.535

41. Evaluate $5a - 18 \div b$ when $a = 9$ and $b = 6$.   *(Lesson 1-10)* 42

Self-Test 2

Replace each __?__ with >, <, or =.

1. 2874 __?__ 2957
<

2. 7.02 __?__ 7.020
=

3. 4.19 __?__ 4.0187
>

1-5

Use the properties of addition and multiplication to find each answer mentally.

4. $81 + 42 + 19$ 142

5. $2.3 + 3.9 + 4.7$ 10.9

6. $106 + 0$ 106

1-6

7. $(0.5)(1.7)(20)$ 17

8. $15 \cdot 1 \cdot 10$ 150

9. $7(3)(0)(4)$ 0

1-7

Tell whether it is most efficient to find each answer using *mental math*, *paper and pencil*, or a *calculator*. Then find each answer using the method that you chose.
Answers may vary. Likely answers are given.

10. $3.487 + 69.88$
c; 73.367

11. $84 - 18$
pp; 66

12. $(59.1)(10)$
mm; 591

1-8

Evaluate each expression when $a = 5$, $b = 2$, and $c = 7$.

13. a^4 625

14. $3b^6$ 192

15. $(2c)^2$ 196

1-9

16. $17 - 8 + b^2$ 13

17. $48 \div (16 - 4) \cdot b$ 8

18. $\dfrac{28 - c}{a + b}$ 3

1-10

Connecting Arithmetic and Algebra   **39**

Follow-Up

Extension
Insert grouping symbols in the expression $6 \cdot 5^2 - 7 + 9$ so that the value is:
1. greater than 800. $(6 \cdot 5)^2 - 7 + 9$
   or $(6 \cdot 5)^2 - (7 + 9)$
2. between 125 and 140.
   $6 \cdot 5^2 - (7 + 9)$
3. less than 125. $6 \cdot (5^2 - 7) + 9$
4. between 150 and 175.
   $6 \cdot (5^2 - 7 + 9)$

Enrichment
Enrichment 10 is pictured on page 2F.

Practice
Practice 11 is shown below.

Quick Quiz 2
See page 45.

Alternative Assessment
See page 45.

Practice 11, *Resource Book*

Practice 11
Skills and Applications of Lesson 1-10

Find each answer.

1. $3^2 + 5 - 10$ 4	2. $(6 + 9) + 5 - 2$ 1	3. $4 + 7^2 - 9$ 44
4. $10 + 2 \cdot 6 + 3$ 10	5. $2(7 - 9) + 3 \cdot 4$ 16	6. $5 \cdot 7 - 4 + 2$ 33
7. $17 + 12 - 10 + 4$ 23	8. $4 - 3^2 + 9$ 3	9. $5(3 + 2) + 5$ 5
10. $23 - 3(11 - 7)$ 11	11. $(4^2 - 6) \cdot 8 - 5$ 75	12. $5^2 - 3^2 + 4^2$ 32
13. $8(9 - 6) + 4$ 6	14. $(4^3 - 7^2) + 15$ 1	15. $18(9 - 3^2) + 11$ 11
16. $(9 - 4) + (6 - 5)$ 5	17. $5(19 - 11) - 4(7 - 2)$ 20	18. $12 + 3(7 - 5)$ 2
19. $\frac{18 + 10}{9 - 5}$ 7	20. $\frac{49 - 9}{7 - 3}$ 10	21. $\frac{37 + 8}{5 + 4}$ 5
22. $\frac{8 + 4}{9 + 7}$ 2	23. $\frac{5(4 + 3)}{12 - 7}$ 7	24. $\frac{64 + 8}{7 + 1}$ 1

Evaluate each expression when $a = 3$, $b = 7$, and $c = 5$.

25. $8a - b$ 17	26. $12a + 3b$ 57	27. $(a \cdot b + 4) + 5$ 5
28. $b^2 - c + 5$ 48	29. $4(7 + ac) + 4$ 22	30. $a^2 - b + c$ 7
31. $(c^2 + a^2) + 2$ 26	32. $105 + 3c - b$ 0	33. $c^2 + b - b(2)$ 18
34. $\frac{7 + 3a}{c - 1}$ 4	35. $\frac{c^2 - 5}{3a - 5}$ 5	36. $\frac{a + b + c}{ac}$ 1
37. $10 + 3c - c^2$ 0	38. $c^2 - 3b + 4a$ 16	39. $b^2 - a^3 - 4c$ 2

40. Seven of the algebra classes at Eliot School have 24 students. The eighth class has 26 students. What is the total number of students? 194

41. During a paper drive, four teams collected 2 tons of paper the first week and 3 tons each week for the next six weeks. How many tons of paper did each team collect if they each collected equal amounts? 5

Chapter Review

Terms to Know

variable (p. 4)
variable expression (p. 4)
evaluate (p. 4)
value of a variable (p. 4)
comparison property (p. 18)
inequality symbols (p. 18)
commutative property of addition (p. 22)
associative property of addition (p. 22)
additive identity (p. 22)
identity property of addition (p. 23)

commutative property of multiplication (p. 25)
associative property of multiplication (p. 25)
multiplicative identity (p. 25)
identity property of multiplication (p. 25)
multiplication property of zero (p. 26)
exponential form (p. 30)
power (p. 30)
exponent (p. 30)
base (p. 30)
order of operations (p. 36)

Choose the correct term from the list above to complete each sentence.

1. The expression $x + 9$ is an example of a(n) __?__. variable expression

2. The __?__ states that changing the order of the terms does not change the sum. commutative property of addition

3. The number 1 is called the __?__. multiplicative identity

4. A symbol that represents a number is called a(n) __?__. variable

5. In the expression 6^4, the 6 is called the __?__. base

6. $4 \cdot 0 = 0$ is an example of the __?__. multiplication property of zero

---

Evaluate each expression when $a = 3$ and $b = 8$. *(Lesson 1-1)*

7. $7a$ 21	**8.** $4b$ 32	**9.** $39 - b$ 31	**10.** $25 + a$ 28
11. $27 \div a$ 9	**12.** $b \div 2$ 4	**13.** $b + b$ 16	**14.** ab 24

Evaluate each expression when $x = 5.6$, $y = 85$, $m = 1.32$, and $n = 24$.
(Lessons 1-2, 1-3)

15. $31.9 + x$ 37.5	**16.** xm 7.392	**17.** $m \div 0.3$ 4.4	**18.** $x - m$ 4.28
19. $2.3n$ 55.2	**20.** $y - 4.7$ 80.3	**21.** $x + y$ 90.6	**22.** $m \div n$ 0.055

Replace each __?__ with >, <, or =. *(Lesson 1-5)*

23. $14{,}259$ __?__ $14{,}312$ <	**24.** 4.86 __?__ 4.860 =	**25.** 0.12 __?__ 0.012 >

Use this paragraph for Exercises 26–29. *(Lesson 1-4)* Answers to Chapter Review 26–29 are on p. A2.

Yvonne earns $7 per hour typing term papers. Last week, she typed 6 h on Monday, 5 h on Tuesday, and 7 h on Wednesday. On Thursday, Yvonne typed the same number of hours as on Monday.

26. What is the paragraph about?

27. How many hours did Yvonne type on Thursday?

28. Identify any facts that are not needed to find the total number of hours Yvonne typed.

29. Describe how you would find the total number of hours that Yvonne typed last week.

Write in order from least to greatest. Use two inequality symbols.
(Lesson 1-5)

30. 12,456; 5642; 12,375
$5642 < 12{,}375 < 12{,}456$

31. 0.62; 0.078; 0.102
$0.078 < 0.102 < 0.62$

32. 6.10; 6; 0.611
$0.611 < 6 < 6.10$

Replace each __?__ with the number that makes the statement true.
(Lessons 1-6, 1-7)

33. $25 + 0.72 = $ __?__ $+ 25$
0.72

34. __?__ $\cdot 31 = 31 \cdot 44$
44

35. __?__ $\cdot (5 \cdot 14) = (12 \cdot 5) \cdot 14$
12

Use the properties of addition and multiplication to find each answer mentally. *(Lessons 1-6, 1-7)*

36. $37 + 49 + 63$ 149

37. $6(3)(0)(8)$ 0

38. $(4)(0.7)(2.5)$ 7

39. $2.9 + 0 + 4.1$ 7

Tell whether it is most efficient to find each answer using *mental math*, *paper and pencil*, or a *calculator*. Then find each answer using the method that you chose. *(Lesson 1-8)* Answers may vary. Likely answers are given.

40. $56 + 87$ pp; 143

41. $98.25 - 3.667$ c; 94.583

42. $(79)(6.32)$ c; 499.28

43. $1200 \div 6$ mm; 200

Find each answer. *(Lessons 1-9, 1-10)*

44. 3^5 243

45. 2^7 128

46. 1^8 1

47. 4^5 1024

48. 10^4 10,000

49. 8^3 512

50. $7^2 - 8 \cdot 3 + 6$ 31

51. $\dfrac{3 \cdot 14}{10 - 4}$ 7

52. $3^2 + 5(2 + 9)$ 64

Evaluate each expression when $a = 3$, $b = 5$, and $c = 6$.
(Lessons 1-9, 1-10)

53. b^2 25

54. a^5 243

55. $4c^3$ 864

56. $2a^5$ 486

57. $(4b)^2$ 400

58. $(2a)^3$ 216

59. $25 - 3c$ 7

60. $64 - (c + 9) + a^2$ 58

61. $\dfrac{b + a}{c - 2}$ 2

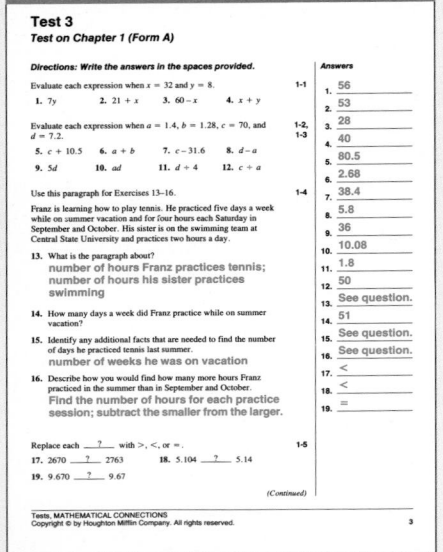

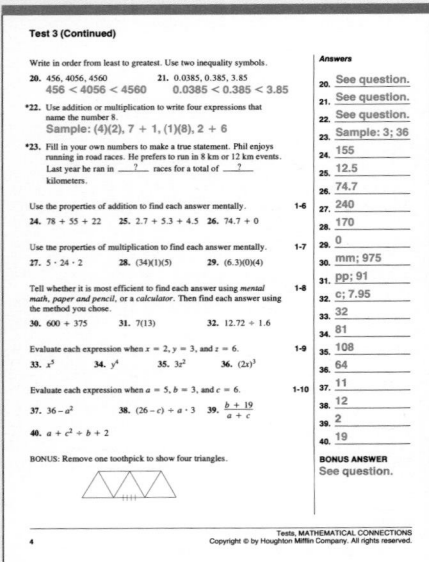

Chapter Test

Evaluate each expression when $x = 45$ and $y = 9$.

1. $8y$ 72   **2.** $23 + x$ 68   **3.** $57 - x$ 12   **4.** $x \div y$ 5   1-1

Evaluate each expression when $a = 1.6$, $b = 1.44$, $c = 48$, and $d = 8.1$.

5. $c + 12.7$ 60.7   **6.** $a + b$ 3.04   **7.** $c - 26.4$ 21.6   **8.** $d - a$ 6.5   1-2

9. $7b$ 10.08   **10.** ad 12.96   **11.** $d \div 3$ 2.7   **12.** $b \div a$ 0.9   1-3

Use this paragraph for Exercises 13–16. Answers to Chapter Test 13–16 are on p. A3.

Dale bought two cassette tapes for $9.99 each and a compact disc for $13.95. An album costs $8.95. Dale gave the clerk two $20 bills.

13. What is the paragraph about?   1-4

14. How much did Dale pay for the compact disc?

15. Identify any facts that are not needed to find the total amount of Dale's purchase.

16. Describe how you would find the amount of change Dale received.

Replace each $\underline{\ ?\ }$ with >, <, or =.

17. $2324 \underline{\ ?\ } 2243$ >   **18.** $3.16 \underline{\ ?\ } 3.106$ >   **19.** $2.50 \underline{\ ?\ } 2.5$ =   1-5

Write in order from least to greatest. Use two inequality symbols.

20. 7634; 779; 7073 $779 < 7073 < 7634$ **21.** 8.65; 0.0865; 0.865 $0.0865 < 0.865 < 8.65$

★**22.** Name four numbers that are between 3.1 and 3.15. Answers will vary. Examples are: 3.11; 3.12; 3.13; 3.14.

★**23.** How many numbers in all are there between 3.1 and 3.15? Answers will vary. Accept any answer that expresses the idea that there are infinitely many.

Use the properties of addition and multiplication to find each answer mentally.

24. $76 + 38 + 24$ 138   **25.** $3.5 + 5.2 + 4.5$ 13.2   **26.** $0 + 12.8$ 12.8   1-6

27. $8(12)(1)$ 96   **28.** $(50)(9)(0.2)$ 90   **29.** $15 \cdot 8 \cdot 0$ 0   1-7

Tell whether it is most efficient to find each answer using *mental math*, *paper and pencil*, or a *calculator*. Then find each answer using the method that you chose. Answers may vary. Likely answers are given.

30. $800 + 755$ mm; 1555   **31.** $8(17)$ pp; 136   **32.** $13.58 \div 1.4$ c; 9.7   1-8

Evaluate each expression when $x = 2$, $y = 5$, and $z = 9$.

33. y^3 125   **34.** x^5 32   **35.** $2z^2$ 162   **36.** $5x^4$ 80   **37.** $(3y)^2$ 225   1-9

38. $45 - y^2$ 20   **39.** $(33 - z) \div x \cdot 6$ 72   **40.** $\dfrac{y + 39}{x + z}$ 4   1-10

Other Operations

Objective: To create and to use new operations.

You have used the operations of addition, subtraction, multiplication, and division. By combining these four basic operations, it is possible to create other operations. For instance, the example below shows an operation that might be called "diamond."

Example

For all numbers a and b, $a \blacklozenge b = ab - b$. Find each of the following.
a. $7 \blacklozenge 3$ 　　　　　　　　　　b. $3 \blacklozenge 7$

Solution

a. $7 \blacklozenge 3 = (7)(3) - 3$ 　　　　b. $3 \blacklozenge 7 = (3)(7) - 7$
$ = 21 - 3$ 　　　　　　　　$ = 21 - 7$
$ = 18$ 　　　　　　　　　$ = 14$

Exercises

For all numbers a and b, $a \heartsuit b = a^2 + b^2$. Find each of the following.

1. $2 \heartsuit 3$ 13 　　　**2.** $3 \heartsuit 2$ 13 　　　**3.** $1 \heartsuit 6$ 37 　　　**4.** $6 \heartsuit 1$ 37

5. Does $a \heartsuit b = b \heartsuit a$? Explain. Yes. Addition is commutative.

6. Does $a \heartsuit (b \heartsuit c) = (a \heartsuit b) \heartsuit c$? Give an example.
No. Answers will vary. An example is $2 \heartsuit (3 \heartsuit 4) = 629$, $(2 \heartsuit 3) \heartsuit 4 = 185$.

For all numbers x and y, $x \blacksquare y = x + y^2$. Find each of the following.

7. $5 \blacksquare 4$ 21 　　　**8.** $4 \blacksquare 5$ 29 　　　**9.** $6 \blacksquare 8$ 70 　　　**10.** $8 \blacksquare 6$ 44

11. Is $\blacksquare$ commutative? Give an example to support your answer.
No. Answers will vary. An example is $5 \blacksquare 4 = 21$, $4 \blacksquare 5 = 29$.

12. Is $\blacksquare$ associative? Give an example to support your answer.
No. Answers will vary. An example is $2 \blacksquare (3 \blacksquare 4) = 363$, $(2 \blacksquare 3) \blacksquare 4 = 27$.

13. Write a variable expression for the operation $\star$. Use a and b as the variables.

$$1 \star 2 = 5$$
$$2 \star 3 = 8$$
$$7.1 \star 5 = 17.1$$
$a \star b = a + 2b$ 　　$12 \star 10 = 32$

14. **GROUP ACTIVITY** Make up an operation, $\triangle$. List the results when you perform your operation on several pairs of numbers. Trade results with a partner and determine the operation your partner created. Write your partner's operation in the form "For all numbers a and b, $a \triangle b = \ldots$." Answers will vary.

Additional Example

For all numbers x and y, $x \odot y = xy + y^2$. Find the following.
1. $5 \odot 4$ 　$5 \cdot 4 + 4^2 = 36$
2. $4 \odot 5$ 　$4 \cdot 5 + 5^2 = 45$

Exercise Note

Cooperative Learning

You may want to extend *Exercise 14* and turn it into a game. Each group should create an operation and then make a chart showing a, b, and $\triangle$. Groups should exchange charts. The winner is the first group to identify correctly the operation $\triangle$ and to tell whether it is associative, commutative, or both.

Evaluate each expression when $x = 3$, $y = 4$, and $z = 7$.

1. $14x$ 42
2. $76 \div y$ 19
3. $z + 9 + z$ 23
4. $4xz$ 84

Evaluate each expression when $a = 12.56$, $b = 32$, $c = 83.1$, and $d = 8.06$.

5. $87 + c$ 170.1
6. $a + d$ 20.62
7. $a - 7.6$ 4.96
8. $c - d$ 75.04
9. $63a$ 791.28
10. bd 257.92
11. $b \div 0.4$ 80
12. $172.8 \div b$ 5.4

Use this paragraph for Exercises 13–16.

Inez bought 17 compact disks and 10 cassette tapes last year. Each compact disk cost \$12.98 and each cassette tape cost \$8.95.

13. What is the paragraph about?
    the number and cost of compact disks and cassette tapes that Inez bought last year
14. How many compact disks did Inez buy last year?   17
15. Identify any facts that are not needed to find the amount of money Inez spent on cassette tapes last year.   17 compact disks at \$12.98 each
16. Describe how you would find the total amount Inez spent on compact disks last year.
    multiply 17 by 12.98

Alternative Assessment

Have students bring some real-world information to class, write a paragraph containing at least four pieces of numerical information, and develop four related problems. At least two of the related problems should include writing and/or evaluating a variable expression.

Cumulative Review

Standardized Testing Practice

Choose the letter of the correct answer.

1. **What information is not needed to solve this problem?**
   Joe earns \$9 per hour and works 8 h per day. He works 40 h per week. How much does Joe earn per week?
   A. earns \$9 per hour
   B. works 8 h per day
   C. works 40 h per week
   D. all the information is needed

2. **Evaluate $45.97 + x$ when $x = 32.5$.**
   A. 13.47   B. 49.22
   C. 4922   D. 78.47

3. **Evaluate $54.4 - a$ when $a = 17.9$.**
   A. 72.3
   B. 46.5
   C. 36.5
   D. 37.5

4. **Tran, Kay, Jarreau, and Lea live 0.61 mi, 0.061 mi, 0.601 mi, and 0.16 mi from school, respectively. Who lives closest to school?**
   A. Tran   B. Kay
   C. Jarreau   D. Lea

5. **Evaluate $247.04 \div a$ when $a = 6.4$.**
   A. 38.6   B. 253.44
   C. 240.64   D. 386

6. **Which of the following do you know is true for all numbers x, y, and z?**
   I. $(xy)z = x(yz)$
   II. $(xyz)^2 = xyz^2$
   III. $(x \div y) \cdot z = x \div y \cdot z$
   A. I only
   B. II only
   C. I and II
   D. I and III

7. **Evaluate $16m$ when $m = 4.3$.**
   A. 688   B. 6.88
   C. 0.688   D. 68.8

8. **List 0.847, 0.0847, 8.47, 0.1847 in order from least to greatest.**
   A. 8.47, 0.847, 0.1847, 0.0847
   B. 0.1847, 0.0847, 0.847, 8.47
   C. 0.0847, 0.1847, 0.847, 8.47
   D. 8.47, 0.1847, 0.0847, 0.847

9. **$1 + 0 = 1$ is an example of which property?**
   A. commutative property of addition
   B. associative property of addition
   C. identity property of addition
   D. identity property of multiplication

10. **Which is a statement of the commutative property of addition?**
    A. $ab = ba$ B. $a + b = b + a$
    C. $a = b$ D. $a(bc) = (ab)c$

11. Use the properties of addition to find $68 + 0 + 42$ mentally.
- A. 100
- B. 120
- C. 0
- (D.) 110

12. $7(0) = 0$ is an example of which property?
- A. comparison property
- B. identity property of addition
- (C.) multiplication property of zero
- D. identity property of multiplication

13. Which is a statement of the associative property of multiplication?
- A. $ab = ba$ B. $a + b = b + a$
- C. $a > b$ (D.) $a(bc) = (ab)c$

14. Use the properties of multiplication to find $4(1)(6)(25)$ mentally.
- A. 1600
- B. 0
- (C.) 600
- D. 2400

15. Write in exponential form.
$4 \cdot 4 \cdot 4 \cdot 4 \cdot 4$
- A. 5^4 (B.) 4^5
- C. 20   D. 1024

16. The volume of a cube is 8^3 cubic feet. How many cubic feet is that?
- A. 24
- B. 83
- C. 6561
- (D.) 512

17. Evaluate a^4 when $a = 6$.
- A. 24
- B. 4096
- (C.) 1296
- D. 10

18. Choose the best estimate.
$716.54 - 388.16$
- (A.) about 300
- B. about 400
- C. about 500
- D. about 1100

19. Find the answer.
$24 \div 8 + 4 \times 3^2$
- A. 63
- B. 147
- (C.) 39
- D. 441

20. Evaluate $a^4 - (33 + a)$ when $a = 3$.
- (A.) 45
- B. 51
- C. 118
- D. 70

Quick Quiz 2 (1-5 through 1-10)
Replace each ? with >, <, or =.
1. $13{,}784$? $13{,}847$ <
2. 8.06 ? 8.060 =
3. 314.28 ? 314.0276 >

Use the properties of addition and multiplication to find each answer mentally.
4. $73 + 36 + 17$ 126
5. $3.4 + 2.9 + 4.6$ 10.9
6. $709 + 0$ 709
7. $(0.4)(2.8)(25)$ 28
8. $16 \cdot 1 \cdot 10$ 160
9. $8(4)(0)(5)$ 0

Tell whether it is most efficient to find each answer using *mental math, paper and pencil*, or a *calculator*. Then find each answer using the method that you chose.
10. $2.786 + 79.88$ c; 82.666
11. $76 - 18$ pp; 58
12. $(67.2)(10)$ mm; 672

Evaluate each expression when $x = 4$, $y = 3$, and $z = 7$.
13. x^4 256
14. $4y^6$ 2916
15. $(2z)^2$ 196
16. $16 - 7 + y^2$ 18
17. $56 \div (15 - 7) \cdot y$ 21
18. $\dfrac{20 + x}{z - y}$ 6

Alternative Assessment
Have students make up three problems that require using the order of operations. They must use the digits 1, 2, 4, 6, and 9 in each problem. Students must also use either three or more operations, or two or more operations and parentheses in each of their problems. Each solution must have a different value.   Answers may vary. Possible solutions are:
$1 + 2 \cdot 4 - 6 + 9 = 12$
$(9 \cdot 6 + 2) \div 4 + 1 = 15$
$6^2 \div 9 - 4 + 1 = 1$

Planning Chapter 2

Chapter Overview

Chapter 2 begins the study of algebra in earnest. A number of concepts important in first-year algebra are presented in a way that students should find relatively simple to follow: numerical examples first, followed by examples involving variables. To further aid students' understanding, some examples are illustrated by using algebra tiles, which provide a pictorial representation of the algebraic concept being presented. Although the tile illustration can be considered optional, we do suggest its use as it will help many students. As a preparation for working with tiles, use the *Focus on Explorations* on page 51. As with Chapter 1, if you find any students still having difficulty with basic skills, we suggest assigning some exercises from the appropriate Toolbox Skills at the back of the text.

Chapter 2 begins by introducing the concept of simplifying expressions. In *Lessons 1–3*, students learn to simplify multiplication expressions, use the distributive property, and combine like terms. *Lesson 4* presents a four-step plan to solve problems. Patterns and functions are discussed in *Lessons 5 and 6*, while *Lesson 7* presents some problem-solving methods students can use to check their answers. The decision-making strand, which began in Chapter 1, is continued in *Lesson 8*, where students are asked to determine whether an exact or an estimated answer is needed to solve a problem. The chapter concludes on a scientific note with discussions of the metric system and scientific notation in *Lessons 9 and 10*.

Background

Chapter 2 continues the introduction to algebra by building on the concepts presented in Chapter 1. Students once again work with exponents, this time to simplify multiplication expressions by using the product of powers rule. Parentheses occur again in the lesson on the distributive property, where they are used as a multiplication symbol and as a grouping symbol. Students continue their work with variables as they learn how to combine like terms. The concept of function is introduced through the lesson on patterns by having students work with function tables, either completing them or using them to find the function rule.

The problem-solving strand that began in Chapter 1 asked students to read a problem for understanding. The next logical step is to have students read a problem and solve it. This is done in the first of two problem-solving lessons in this chapter where students solve problems using a four-step plan.

The second problem-solving lesson focuses on the last step of the plan by presenting a number of methods students can use to check their answers.

Chapter 2 connects mathematics to the real world. The decision-making lesson asks students to decide whether an exact answer or an estimate is needed to solve a problem, an important skill that is used frequently in the real world. Lessons on the metric system and on scientific notation continue this connection, as does the *Focus on Applications* on page 84. The *Focus on Computers* on page 70 has students use spreadsheets, a skill that is becoming more important. Students also are helped to become better calculator users by learning about the constant key and the scientific notation key.

Objectives

2-1 To simplify multiplication expressions using the product of powers rule and the properties of multiplication.

FOCUS ON EXPLORATIONS To use algebra tiles to model expressions.

2-2 To use the distributive property to compute mentally and to simplify expressions.

2-3 To simplify expressions by combining like terms.

2-4 To use a four-step plan to solve problems.

2-5 To recognize and to extend patterns.

2-6 To complete function tables and to find function rules using tables.

FOCUS ON COMPUTERS To use a spreadsheet to analyze patterns in function tables.

2-7 To check the answer to a problem.

2-8 To decide whether an estimate or an exact answer is needed to solve a problem.

2-9 To recognize and to use metric units of measure.

2-10 To write numbers in scientific notation.

FOCUS ON APPLICATIONS To convert customary units to metric units.

CHAPTER EXTENSION To use the laws of exponents.

Instructional Aids	Manipulatives	Cooperative Learning	Technology	Practice/Reteaching	Assessment

Materials: algebra tiles, colored beads, metric measuring devices, calculators
Resource Book:
  Lesson Starters 11–20
Visuals, Folder C
	Lessons 2-2, 2-3,
	  Focus on Explorations
	Activities Book:
	  Manipulative
	    Activities 3–4
		Lessons 2-2, 2-5, 2-6
		Activities Book:
		  Cooperative
		    Activity 5
			Lessons 2-2, 2-4, 2-6,
			  2-8, 2-10, Focus on
			  Computers
			Activities Book:
			  Calculator Activity 2
			  Computer Activity 2
			Connections Plotter Plus Disk
				Toolbox, pp. 753–771
				Extra Practice, p. 731
				Resource Book:
				  Practice 14–26
				Study Guide, pp. 21–40
					Self-Tests 1–3
					Quick Quizzes 1–3
					Chapter Test, p. 88
					Resource Book:
					  Lesson Starters 11–20
					    (Daily Quizzes)
					Tests 5–9

Assignment Guide Chapter 2

Day	Skills Course	Average Course	Advanced Course

1	**2-1:** 1–15, 17–31 odd,
	  37–40
		2-1: 1–17, 19–31 odd, 33,
		  34, 37–40
			2-1: 1–15 odd, 16–40

2	**Exploration:** Activities I, II
	2-2: 1–18, 19–23 odd,
	  27–32
		Exploration: Activities I, II
		2-2: 1–20, 21–25 odd,
		  27–32
			Exploration: Activities I, II
			2-2: 1–17 odd, 19–32,
			  Historical Note

3	**2-3:** 1–18, 19–29 odd,
	  36–40
		2-3: 1–18, 19–35 odd,
		  36–40
			2-3: 1–17 odd, 19–40

4	**2-4:** 1–6, 7–11 odd, 13–16	**2-4:** 1–16	**2-4:** 1–16

5	**2-5:** 1–15, 17, 19–22,
	  28–31
		2-5: 1–16, 17–25 odd,
		  28–31
			2-5: 1–13 odd, 15–31,
			  Challenge

6	**2-6:** 1–7, 9, 10, 18–21	**2-6:** 1–8, 9–17 odd, 18–21	**2-6:** 1–21, **Computers:** 1–9
7	**2-7:** 1, 2, 3–9 odd, 11–14	**2-7:** 1–14	**2-7:** 1–14
8	**2-8:** 1–12	**2-8:** 1–12	**2-8:** 1–12

9	**2-9:** 1–20, 21–41 odd, 42,
	  43, 52–56
		2-9: 1–20, 21–45 odd,
		  52–56
			2-9: 1–19 odd, 20, 21–39
			  odd, 40–56

10	**2-10:** 1–18, 19–27 odd,
	  39–44
	Application: 1–12
		2-10: 1–18, 19–33 odd,
		  35–37, 39–44
		Application: 1–17 odd
			2-10: 1–21 odd, 23–44
			Application: 9–19

11	*Prepare for Chapter Test:* Chapter Review	*Prepare for Chapter Test:* Chapter Review	*Prepare for Chapter Test:* Chapter Review
12	*Administer Chapter 2 Test*	*Administer Chapter 2 Test;* Cumulative Review	*Administer Chapter 2 Test;* Chapter Extension; Cumulative Review

Teacher's Resources

Resource Book
  Chapter 2 Project
  Lesson Starters 11–20
  Practice 14–26
  Enrichment 12–22
  Diagram Masters
  Chapter 2 Objectives
  Family Involvement 2
  Spanish Test 2
  Spanish Glossary

Activities Book
  Cooperative Activity 5
  Manipulative Activities 3–4
  Calculator Activity 2
  Computer Activity 2

Study Guide, pp. 21–40

Tests
  Tests 5–8

Visuals
  Folder C

Connections Plotter Plus **Disk**

Alternate Approaches Chapter 2

Using Technology

CALCULATORS
Calculators provide a means of exploring the algebra properties introduced in this chapter and also allow students to attempt problems having realistic numbers.

COMPUTERS
BASIC programs, as well as spreadsheet programs, also give students practice using realistic numbers, but they have another important role. By using variables that look similar or identical to those used in algebra, these programs provide a bridge from purely arithmetic problem solving to the use of variables in algebra. On computers, the connection between numbers and variables is immediate.

Lesson 2-2

COMPUTERS
A spreadsheet program can be used to explore the distributive property, as suggested in Exercises 25 and 26. A somewhat similar setup would involve three or four cells, along with an appropriate label for each of them. For example, cell B2 might contain a value labeled X, cell C2 might contain the formula $5 * (B2-2)$, and cell D2 might contain the formula $5 * B2-10$. (Be aware that some complex expressions such as $5 * B3-5 * A3$ may not be evaluated correctly by some spreadsheets; the order of operations may be strictly from left to right.)

Lesson 2-4

COMPUTERS
A spreadsheet program can be used to ease the transition from the arithmetic word problems presented in this lesson to the algebraic word problems that students will encounter later. Once students have solved a problem using pencil and paper or a calculator, they can translate the solution into a spreadsheet, which will allow automatic solution of any problem of the same form. For example, in the problem on page 59, you could put 35 in cell C2, 31 in cell D2, and the formula $7.15 * (C2 + D2)$ in cell E2, with appropriate labels, such as "HRS, WK 1," "HRS, WK 2," "EARNINGS," in cells C1, D1, and E1, respectively. This kind of activity will accustom students to using algebra-like notation to represent quantities in word problems.

Lesson 2-6

COMPUTERS
BASIC programs or spreadsheets can be used to give students practice in finding the function rule for a given set of number pairs. For example, you could have students RUN the following simple BASIC program:

```
10 INPUT "X = "; X
20 PRINT "RESULT = "; 2 * X + 1
30 GOTO 10
```

Students could be asked to come up with a rule that generates the numbers the program gives them. They can test their solution by stopping the program and LISTing it. A spreadsheet program can be used in the same way; this will be most effective if the program allows you to hide formulas, so that the answer is not immediately available.

Lesson 2-8

CALCULATORS
Use calculators to explore the size of errors introduced by rounding. Individually or in groups, students can add a set of four decimals, round the decimals to the nearest integer, add the results, and then compare their answers. By varying the decimals, students can try to find the largest possible series obtainable. They can then repeat the process using multiplication. Do the errors tend to be larger or smaller for multiplication than for addition? Is there a maximum possible error? (It might be possible to guide a discussion into relative, or percent error, for the two operations.)

Using Manipulatives

Algebra tiles may be used to demonstrate a variety of algebraic concepts.

Focus on Explorations
Algebra tiles are used to model expressions with like terms, such as $n + 3n$, and expressions with unlike terms, such as $2n + 3$. As a further check on whether students understand how to use the tiles, you may wish to have them model some more expressions, such as $2x + 3x$, $3a + 4$, and so on.

Lesson 2-2

Algebra tiles can be used to demonstrate the distributive property. If students are having difficulty understanding this property, you may wish to have them build or draw additional models. For example, you could present the following:

Build a model to simplify $2(3x + 2)$.

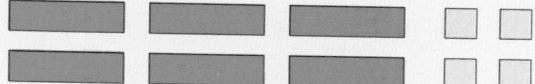

a. Do you have two groups of $3x + 2$? yes

b. Regroup like terms to find the solution by drawing circles on the diagram.

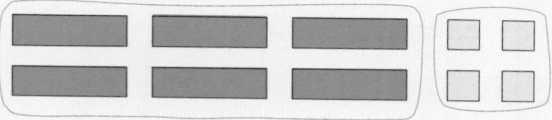

c. How many x tiles do you have? 6

d. How many 1 tiles do you have? 4

e. What is the product of $2(3x + 2)$? $6x + 4$

Lesson 2-3

Algebra tiles can be used to facilitate the understanding of combining like terms. On page 56, Example 1(b) uses an arrow to indicate the term being subtracted or "taken away." If you do not have actual tiles, an alternative approach is to cross out the pictorial representation of the terms being subtracted. For example:

Simplify $5n - 2n$.
  a. Draw $5n$.

  b. Cross out $2n$.

  c. Read the answer. Each box represents n. Three boxes are not crossed out. Thus, $5n - 2n = 3n$.

Reteaching/Alternate Approach

Lesson 2-2
COOPERATIVE LEARNING

In cooperative groups, students should design a spreadsheet to keep a record of their grades, including quizzes and homework. The spreadsheet should make use of the distributive property in a manner similar to its use on page 55.

Lesson 2-4

There are a variety of strategies that can be used to solve problems. To see how many of these strategies your students are familiar with, ask them to write down the ones they know. You may wish to discuss the following problem that can be solved in more than one way.

Six friends ordered a pizza. Each person made one straight cut across the pizza. What is the greatest number of pieces into which they could have cut the pizza? Explain how you determined your answer.

Lesson 2-5
COOPERATIVE LEARNING

Divide the class into cooperative groups of three or four. Give to each group a sheet of visual patterns, such as the one below.

Ask the groups to determine the next three figures in each pattern you provided. After the groups have worked with the visual patterns, have them work some numerical patterns and variable patterns similar to those presented in the text.

Lesson 2-6
COOPERATIVE LEARNING

Divide the class into cooperative groups and have each group develop a function, listing at least five of its terms. Have the groups display the terms pictorially as a sequence of numbers from left to right and in a function table. Each group should explain its functions to the class, describe how it could be extended, and discuss which method of representation is easiest to understand.

46D

Teacher's Resources Chapter 2

Enrichment masters from the Resource Book are pictured here. See the Teacher's Resources chart on page 46B for a complete listing of all materials available for this chapter.

2-1 Enrichment
Connection

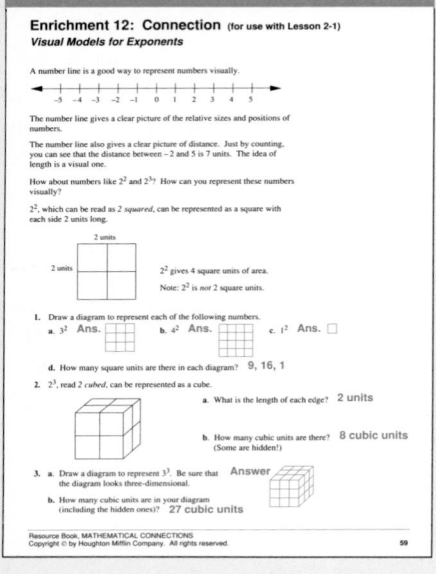

2-2 Enrichment
Connection

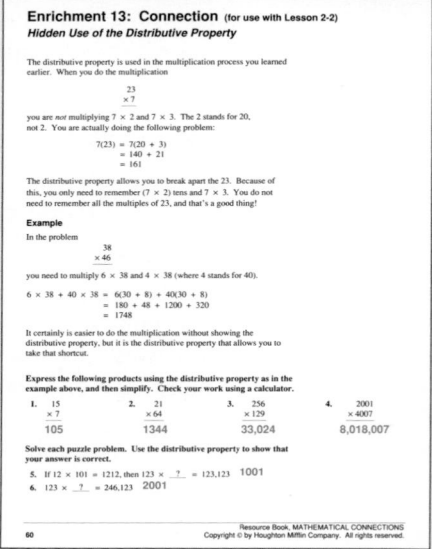

2-3 Enrichment
Data Analysis

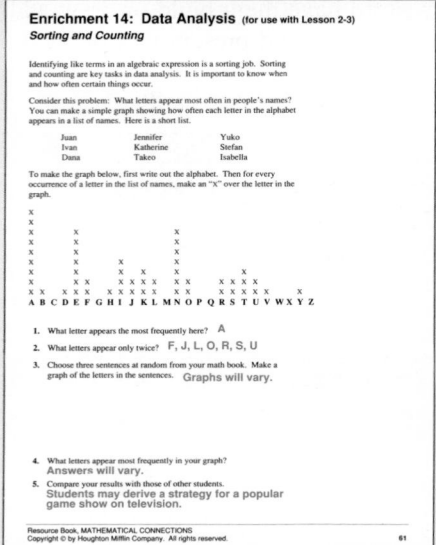

2-4 Enrichment
Problem Solving

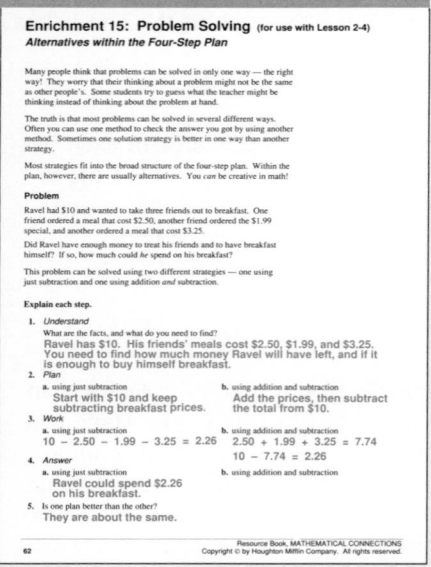

2-5 Enrichment
Thinking Skills

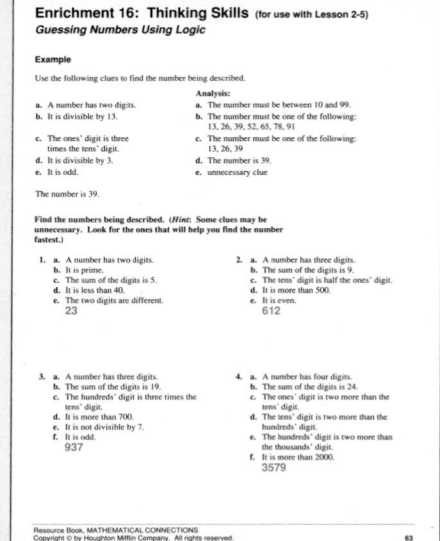

2-6 Enrichment
Communication

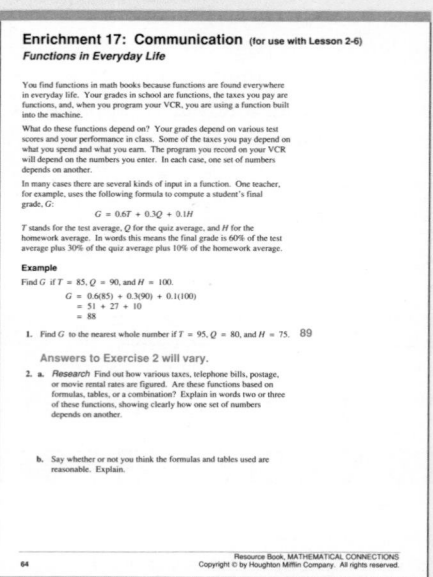

2-7 Enrichment
Problem Solving

Checking the Check-Out

Sooner or later you will have to go shopping at a supermarket. If you buy a shopping cart full of small items, you will probably spend between $50 and $100. You will get a receipt when you go through the check-out line. Many receipts show both the name and the price of the item. Consider the following items from a register tape.

apples	3.25	gal milk	2.31	p butter	2.89
cereal	2.59	juice	2.21	frz broccoli	1.89
paper towel	0.89	5 lb flour	3.78	1 lb mayo	2.67
chicken	4.78	light bread	1.49	am cheese	3.21
pickles	2.13	celery	1.89	tomatoes	2.79
pencils	2.11	lined paper	3.05	compass	1.99

The receipt lists the total at $45.92. With so many small items on the list, it can be a bit difficult to tell whether the sum is reasonable or not. A good way to check for reasonableness is to round each of the prices to the nearest dollar, and add them together. The next time you go shopping, try to keep a running sum by rounding all the prices to the nearest dollar. You should be able to do that mentally.

1. Run your finger down the columns, rounding off each price to the nearest dollar and adding it to a running sum. (A running sum gets bigger and bigger.) Try to do this without writing anything down. What is your estimate?   **$46**

2. Does the total $45.92 seem reasonable?   **Yes.**

3. The cash register arrived at $45.92 by addition. Check its work by subtracting all of the prices on the list from $45.92. You can do so by entering 45.92 on your calculator, hitting the (–) key, entering 3.25 (for apples), hitting (–), entering the next price, and so on until you finish the list. What is your conclusion?   **The total is correct.**

4. Make up a shopping list (along with prices), read off the names and prices, and see if your class can make a good estimate. Repeating this exercise will help you become better at estimating and better at checking the check-out.   **Answers will vary.**

2-8 Enrichment
Application

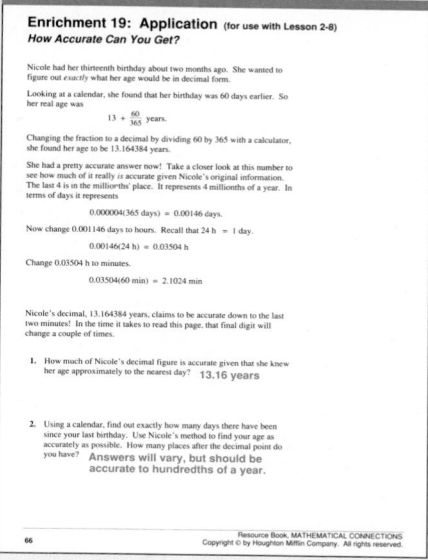

How Accurate Can You Get?

Nicole had her thirteenth birthday about two months ago. She wanted to figure out *exactly* what her age would be in decimal form.

Looking at a calendar, she found that her birthday was 60 days earlier. So her real age was

$$13 + \frac{60}{365} \text{ years.}$$

Changing the fraction to a decimal by dividing 60 by 365 with a calculator, she found her age to be 13.164384 years.

She had a pretty accurate answer now! Take a closer look at this number to see how much of it really *is* accurate given Nicole's original information. The last 4 is in the millionths' place. It represents 4 millionths of a year. In terms of days it represents

$$0.000004(365 \text{ days}) = 0.00146 \text{ days.}$$

Now change 0.00146 days to hours. Recall that 24 h = 1 day.

$$0.00146(24 \text{ h}) = 0.03504 \text{ h}$$

Change 0.03504 h to minutes.

$$0.03504(60 \text{ min}) = 2.1024 \text{ min}$$

Nicole's decimal, 13.164384 years, claims to be accurate down to the last two minutes! In the time it takes to read this page, that final digit will change a couple of times.

1. How much of Nicole's decimal figure is accurate given that she knew her age approximately to the nearest day?   **13.16 years**

2. Using a calendar, find out exactly how many days there have been since your last birthday. Use Nicole's method to find your age as accurately as possible. How many places after the decimal point do you have?   **Answers will vary, but should be accurate to hundredths of a year.**

2-9 Enrichment
Exploration

A Measurement Activity

You should try to develop a good *feel* for the metric system because it is being used more and more. The following activity will help you visualize distances in meters.

Materials

You will need a metric measuring tape — a 10 m tape would be excellent, although a meter stick can be used if you do not have a tape.

What to do

1. Find a room or a hallway where you can take at least ten steps, or paces, in a straight line.

2. Mark a starting point where the back of your heel will rest when you begin taking steps.

3. Using your usual walking step, take ten paces in a straight line from the starting point. Mark where the toe of your leading shoe is when you have finished.

4. Measure the distance from the starting point to the endpoint to the nearest centimeter. You might have a measure like 6.12 m or 7.92 m. (Recall that 92 cm = 0.92 m.)

5. Compute the length of your average pace in meters:

   $$\frac{\text{actual distance in meters}}{10 \text{ paces}} = \underline{\hspace{2cm}} \text{ meters/pace}$$

   Answers should be in the 0.6 – 0.8 m range.

6. "Pace off" a building whose length you can look up. (Perhaps your teacher can identify one for you.) Record the number of paces you made.   **Answers will vary.**

7. Multiply the number of paces you made by the number of meters in your average pace (recorded in Exercise 5 above). This product is your measurement of the length of the building in meters.   **Answers will vary.**

8. Compare your results to those of other students and to the actual length of the building. How accurate were you?   **Students should be able to measure the lengths to the nearest meter or so.**

2-10 Enrichment
Data Analysis

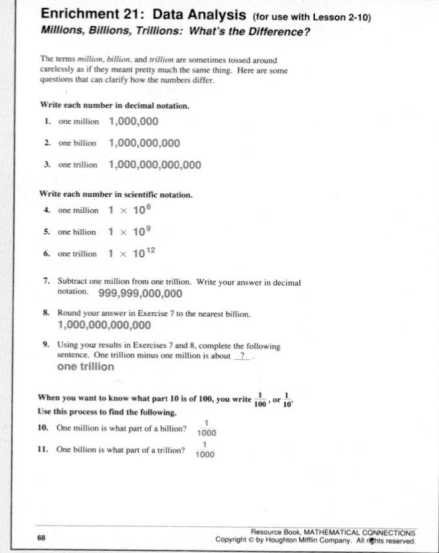

Millions, Billions, Trillions: What's the Difference?

The terms *million*, *billion*, and *trillion* are sometimes tossed around carelessly as if they meant pretty much the same thing. Here are some questions that can clarify how the numbers differ.

Write each number in decimal notation.

1. one million   **1,000,000**

2. one billion   **1,000,000,000**

3. one trillion   **1,000,000,000,000**

Write each number in scientific notation.

4. one million 1×10^6

5. one billion 1×10^9

6. one trillion 1×10^{12}

7. Subtract one million from one trillion. Write your answer in decimal notation.   **999,999,000,000**

8. Round your answer in Exercise 7 to the nearest billion.
   1,000,000,000,000

9. Using your results in Exercises 7 and 8, complete the following sentence. One trillion minus one million is about   _?_ .
   one trillion

When you want to know what part 10 is of 100, you write $\frac{1}{100}$, or $\frac{1}{10}$.
Use this process to find the following.

10. One million is what part of a billion? $\frac{1}{1000}$

11. One billion is what part of a trillion? $\frac{1}{1000}$

End of Chapter Enrichment
Extension

World and National Numbers

Materials
You will need an almanac.

Activity

1. What is the approximate world population?
   about 5 billion

2. a. What is the approximate population of the United States?
   about 250 million

   b. In a recent year the government spent approximately $149 billion on public education. According to this figure, how much money was spent per person in the United States?   **$600**

3. a. Recently, the federal government made a *budget*, a plan for spending its money. In that year the federal budget was $500 billion. According to this budget, how much money would be spent per person in the United States?   **$2000**

   b. Suppose that the budget was revised to cut out $2 million worth of spending. Subtract $2 million from $500 billion. Rounding to the nearest billion, what would the budget be now?   **$500 billion**

4. *Research* Use an almanac to write a problem like the ones above. Then let a classmate try to solve it.   **Answers will vary.**

Multicultural Notes

For Chapter 2 Opener

Women have made many important contributions to the exploration of space. Two years after the first manned space flight took place, Soviet cosmonaut Valentina Tereshkova became the first woman to fly in space. She orbited the earth 48 times. The next space flight by a woman, Svetlana Savitskaya, took place in 1982. On another mission in 1984, Savitskaya became the first woman to walk in space. The first American woman to fly in space was Sally Ride, who took part in a space shuttle mission in 1983. She flew on a second shuttle mission in 1984.

For Page 55

Students following the direction for research under the Historical Note may want to learn about other ancient number systems as well. Both the Mayan and Babylonian systems, for example, used place value and had symbols for zero.

For Page 62

Leonardo Fibonacci was born in the city of Pisa, in what is now Italy, and grew up on the north coast of Africa. He traveled to Greece, Syria, Sicily, and Egypt, where he became familiar with Hindu-Arabic mathematics. The books on mathematics that he published helped spread the use of the Hindu-Arabic number system into Europe.

For Page 76

Some students may not be aware that all scientists and many countries have adopted the metric system. You may want to find out if any students are familiar with the metric system or have observed it being used in other countries.

Notable Firsts in Space

1959	Space vehicle lands on the moon.
1961	Manned flight orbits Earth.
1965	Closeup pictures of Mars.
1965	Humans "walk in space."
1968	Space vehicle orbits the sun.
1968	Live TV transmissions from orbit.
1969	Space vehicle lands on Venus.
1969	Manned flight lands on the moon.
1973	Closeup pictures of Jupiter.
1974	Closeup pictures of Mercury.
1979	Closeup pictures of Saturn.
1986	Closeup pictures of Uranus.
1989	Closeup pictures of Neptune.

46

Introduction to Algebra 2

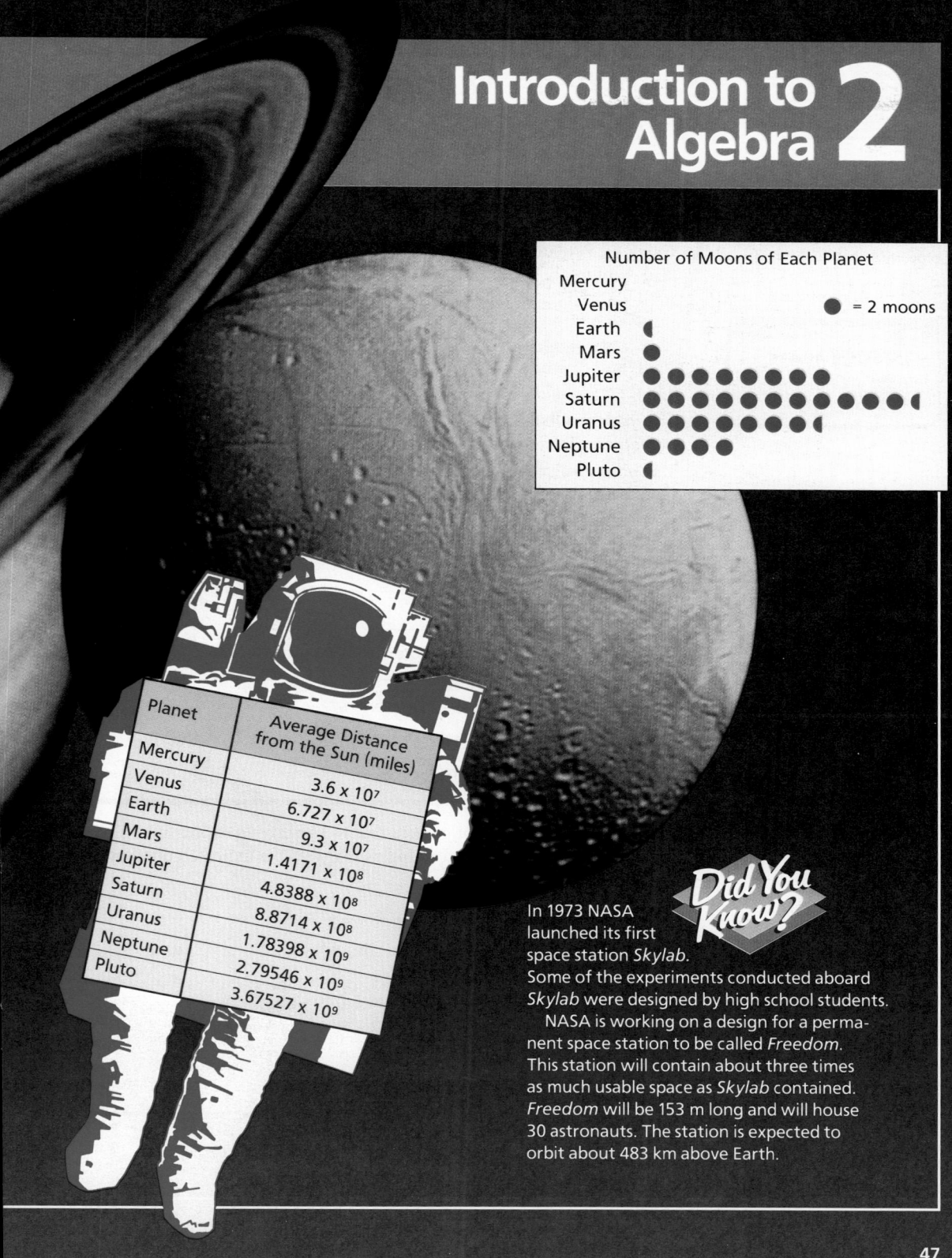

Number of Moons of Each Planet

● = 2 moons

Planet	Number of Moons
Mercury	
Venus	
Earth	◖
Mars	●
Jupiter	●●●●●●●
Saturn	●●●●●●●●●●●◖
Uranus	●●●●●●◖
Neptune	●●●●
Pluto	◖

Planet	Average Distance from the Sun (miles)
Mercury	3.6×10^7
Venus	6.727×10^7
Earth	9.3×10^7
Mars	1.4171×10^8
Jupiter	4.8388×10^8
Saturn	8.8714×10^8
Uranus	1.78398×10^9
Neptune	2.79546×10^9
Pluto	3.67527×10^9

Did You Know?

In 1973 NASA launched its first space station *Skylab*. Some of the experiments conducted aboard *Skylab* were designed by high school students.

NASA is working on a design for a permanent space station to be called *Freedom*. This station will contain about three times as much usable space as *Skylab* contained. *Freedom* will be 153 m long and will house 30 astronauts. The station is expected to orbit about 483 km above Earth.

47

Research Activities

For Chapter 2 Opener

The space shuttle and space stations have made it easier for scientists to conduct experiments in space, and many of the men and women now involved in space exploration are scientists as well as astronauts. You may want to have students research the different experiments—biological, medical, physical—that have been conducted in space.

For Page 76

Students can research the history of the development and acceptance of the metric system and compare it with the history of the customary, or English, system.

Suggested Resource

Summerlin, Lee B., ed. *Skylab, Classroom in Space*. Washington, D.C.: National Aeronautics and Space Administration, 1977.

Teaching the Lesson
• Lesson Starter 11
• Toolbox Skills 7, 9

Lesson Follow-Up
• Practice 14
• Enrichment 12: Connection
• Study Guide, pp. 21–22

Warm-Up Exercises

Find each answer.

1. 7^3 343
2. 4^4 256
3. $2(3^5)$ 486
4. $2(5^3)$ 250
5. $(2 \cdot 5)^3$ 1000

Lesson Focus

Ask students if they ever had to multiply the same factor by itself a number of times. Ask if they actually counted the factors. Multiplication expressions that have the same factors can be simplified because the number of factors is known. This is the focus of today's lesson.

Teaching Notes

Use the Exploration at the beginning of the lesson to help students discover the product of powers rule. Work through the Examples, pointing out how the product of powers rule and the properties of multiplication are used to simplify each expression. Remind students that variables are usually written in alphabetical order.

Key Questions

1. How do you simplify the expression $(2x)(3x^5)$? What property or rule is used to perform each step?
$(2x)(3x^5) = 2 \cdot 3 \cdot x \cdot x^5 =$
$(2 \cdot 3)(x \cdot x^5) = 6x^6$; commutative property, associative property, product of powers rule

---

2-1 Simplifying Multiplication Expressions

Objective: To simplify multiplication expressions using the product of powers rule and the properties of multiplication.

Terms to Know
• *product of powers rule*
• *simplify*

EXPLORATION

1 Using a calculator, replace each __?__ with the number that makes the statement true.

a.
$4^3 = $ __?__ 64
$4^2 = $ __?__ 16
$4^3 \cdot 4^2 = $ __?__ 1024
$4^5 = $ __?__ 1024
$4^3 \cdot 4^2 = 4^?$ 5

b.
$5^2 = $ __?__ 25
$5^4 = $ __?__ 625
$5^2 \cdot 5^4 = $ __?__ 15,625
$5^6 = $ __?__ 15,625
$5^2 \cdot 5^4 = 5^?$ 6

c.
$2^3 = $ __?__ 8
$2^6 = $ __?__ 64
$2^3 \cdot 2^6 = $ __?__ 512
$2^9 = $ __?__ 512
$2^3 \cdot 2^6 = 2^?$ 9

2 Use your results from Step 1. Complete this statement: To multiply powers that have the same base, you __?__ the exponents. add

3 Use your result from Step 2. Find the value of n that makes each statement true. Use a calculator to check your answers.
a. $6^3 \cdot 6^7 = 6^n$ 10
b. $3^5 \cdot 3^9 = 3^n$ 14
c. $7^5 \cdot 7^4 = 7^n$ 9

You can use exponents to count factors when finding a product of powers.

$$7^5 \cdot 7^4 = \underbrace{(7 \cdot 7 \cdot 7 \cdot 7 \cdot 7)}_{\text{5 factors}} \cdot \underbrace{(7 \cdot 7 \cdot 7 \cdot 7)}_{\text{4 factors}} = \overset{\uparrow}{7^9}_{\text{9 factors}}$$

Generalization: *Product of powers rule*

To multiply powers having the same base, add the exponents.

$$a^m \cdot a^n = a^{m+n}$$

You can use the product of powers rule to *simplify* an expression. You **simplify** an expression by performing as many of the indicated operations as possible.

Example 1
Solution

Simplify: a. $x^3 \cdot x^5$ b. $w^6 \cdot w$ c. $c^5 \cdot c^2 \cdot c^4$

a. $x^3 \cdot x^5 = x^{3+5}$ b. $w^6 \cdot w = w^6 \cdot w^1$ c. $c^5 \cdot c^2 \cdot c^4 = c^{5+2+4}$
       $= x^8$ $= w^{6+1}$ $= c^{11}$
                              $= w^7$

✏️ **Check Your Understanding**

In Example 1(b), why is w rewritten as w^1?
See *Answers to Check Your Understanding* at the back of the book.

To simplify some expressions, you might need to use the product of powers rule together with the commutative and associative properties.

Example 2
Solution

Simplify: **a.** $6a^2 \cdot 4a^3$ **b.** $(5n)(7n^2)$

a. $6a^2 \cdot 4a^3 = (6 \cdot 4)(a^2 \cdot a^3)$
$= (24)(a^{2+3})$
$= 24a^5$

b. $(5n)(7n^2) = (5 \cdot 7)(n \cdot n^2)$
$= (5 \cdot 7)(n^1 \cdot n^2)$
$= (35)(n^{1+2})$
$= 35n^3$

You can also simplify multiplication expressions that involve more than one variable.

Example 3
Solution

Simplify: **a.** $(7x)(2y)$ **b.** $5y \cdot 3x \cdot 2y$

a. $(7x)(2y) = (7 \cdot 2)(x \cdot y)$
$= 14xy$

b. $5y \cdot 3x \cdot 2y = (5 \cdot 3 \cdot 2)(y \cdot x \cdot y)$
$= (5 \cdot 3 \cdot 2)(x \cdot y \cdot y)$
$= 30xy^2$

Guided Practice

COMMUNICATION «*Reading*

Refer to the text on pages 48–49.

1. To multiply powers having the same base, add the exponents.

«**1.** Describe the product of powers rule in words.
«**2.** What does it mean to *simplify* an expression?

2. You simplify an expression by performing as many of the indicated operations as possible.

Choose the letter of the correct answer.

3. $q^6 \cdot q^2$ **a.** q^{12} **b.** q^8 **c.** q^4 **d.** q^3

4. $t \cdot t^4$ **a.** $2t^4$ **b.** $2t^5$ **c.** t^4 **d.** t^5

5. $(5c^4)(3c^6)$ **a.** $8c^{10}$ **b.** $15c^{10}$ **c.** $8c^{24}$ **d.** $15c^{24}$

6. $9w \cdot 4z$ **a.** $36wz$ **b.** $13wz$ **c.** $94wz$ **d.** $36w^2z^2$

Simplify.

7. $y^2 \cdot y^3$ y^5 **8.** $x^5 \cdot x$ x^6 **9.** $w^4 \cdot w^5 \cdot w^4$ w^{13}

10. $d \cdot d^3 \cdot d^6$ d^{10} **11.** $3k^4 \cdot 4k^2$ $12k^6$ **12.** $(2c^3)(10c^4)$ $20c^7$

13. $(6z^3)(6z)$ $36z^4$ **14.** $5p \cdot 9p^7$ $45p^8$ **15.** $(5p)(4q)$ $20pq$

16. $3a \cdot 9b$ $27ab$ **17.** $4w \cdot 2z \cdot 6w$ $48w^2z$ **18.** $(7a)(5b)(4b)$
                                                     $140ab^2$

Introduction to Algebra   **49**

---

2. When multiplying powers of the same number, why are the exponents added rather than multiplied?   The exponents represent the number of times the number is used as a factor. To get the total, you add the exponents.

Error Analysis
When simplifying multiplication expressions using the product of powers rule, students may multiply the exponents and get an incorrect answer. Remind students to *add* exponents when finding the product of powers of the same number. Also, remind students that x means x^1 in order to avoid errors such as $x^3 \cdot x = x^3$.

Additional Examples
Simplify.

1. a. $n^2 \cdot n^7$ n^9
    b. $x^4 \cdot x$ x^5
    c. $y^6 \cdot y^3 \cdot y^5$ y^{14}

2. a. $5m^3 \cdot 2m^4$ $10m^7$
    b. $(6x)(9x^4)$ $54x^5$

3. a. $(3a)(4b)$ $12ab$
    b. $(7s)(8r)(s)$ $56rs^2$

Closing the Lesson
Use the cooperative learning activity below to check students' understanding of simplifying multiplication expressions.

Cooperative Learning
Have students work in cooperative groups to solve each exercise. Ask one member from each of three groups to explain the group's solution to the class.

1. $2x^4 \cdot 3x^3$ $6x^7$
2. $x^6 \cdot \underline{\ ?\ } = x^{11}$ x^5
3. $y \cdot y \underline{\ ?\ } = y^7$ 6

Suggested Assignments
Skills: 1–15, 17–31 odd, 37–40
Average: 1–17, 19–31 odd, 33, 34, 37–40
Advanced: 1–15 odd, 16–40

49

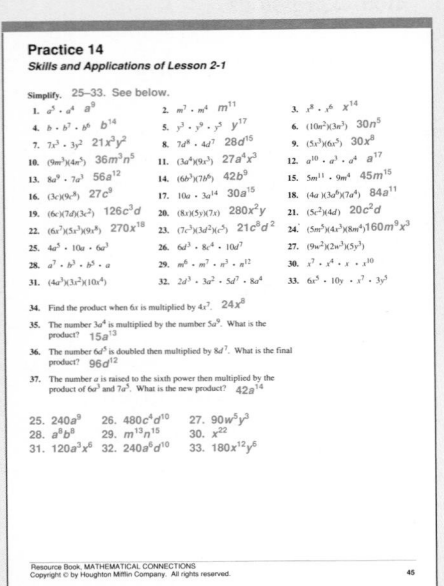

Exercises

Simplify.

A 1. $c^6 \cdot c^4$ c^{10} 2. $b^8 \cdot b^3$ b^{11} 3. $n^2 \cdot n$ n^3

4. $y \cdot y^8$ y^9 5. $x^3 \cdot x^7 \cdot x^3$ x^{13} 6. $a^5 \cdot a^8 \cdot a^6$ a^{19}

7. $4d^2 \cdot 3d^3$ $12d^5$ 8. $5k^6 \cdot 3k^2$ $15k^8$ 9. $(6b^2)(5b)$ $30b^3$

10. $(8c)(4c^3)$ $32c^4$ 11. $(2c)(14d)$ $28cd$ 12. $(12q)(4r)$ $48qr$

13. $(7w)(4w)(2y)$ $56w^2y$ 14. $5b \cdot 3a \cdot 2b$ $30ab^2$ 15. $(5w)(6x)(8x)$ $240wx^2$

16. Find the product when $2a$ is multiplied by $3a^5$. $6a^6$

17. Find the product of a number z squared and the same number z cubed. z^5

Replace each ___?___ with the expression that makes the statement true.

B 18. $x^5 \cdot \underline{\ ?\ } = x^{13}$ x^8 19. $c^7 \cdot \underline{\ ?\ } = c^{11}$ c^4 20. $\underline{\ ?\ } \cdot n^{12} = n^{19}$ n^7

21. $(5c)(\underline{\ ?\ }) = 15c^2$ $3c$ 22. $8p \cdot \underline{\ ?\ } = 48p^2$ $6p$ 23. $(\underline{\ ?\ })(4a) = 32ab^2$ $8b^2$

Find the value of n that makes each statement true.

24. $z^3 \cdot z^n = z^{12}$ 9   25. $x^n \cdot x^2 = x^{10}$ 8   26. $d^2 \cdot d^4 \cdot d^n = d^{18}$ 12

27. $y^5 \cdot y^n \cdot y^3 = y^{10}$ 2   28. $q^n \cdot q = q^4$ 3   29. $t \cdot t^n = t^7$ 6

30. $a^n \cdot a^n = a^{10}$ 5   31. $k^n \cdot k^n \cdot k^n = k^6$ 2   32. $y^n \cdot y \cdot y^3 \cdot y^n = y^{18}$ 7

LOGICAL REASONING Answers to Exercises 35–37 are on p. A3.

Exercises 33–36 refer to these statements:
    I. $2^m \cdot 2^0 = 2^{m+0} = 2^m$ **II.** $2^m \cdot 1 = 2^m$

C 33. Which property or rule is illustrated by Statement I? product of powers rule

34. Which property or rule is illustrated by Statement II? identity property of multiplication

35. Explain why this statement must be true: $2^m \cdot 2^0 = 2^m \cdot 1$

36. What number do you think is represented by 2^0? Give a convincing argument to support your answer.

SPIRAL REVIEW

37. Tell what this paragraph is about: Amy Gold bought six tickets to a basketball game and five tickets to a hockey game. The basketball tickets cost $67.50 and the hockey tickets cost $53.75. *(Lesson 1-4)*

38. Find the answer: $4^3 - (12 + 9) \div 7$ *(Lesson 1-10)* 61

39. Simplify: $(5c^2)(6c)$ *(Lesson 2-1)* $30c^3$

40. **DATA,** *pages 46–47* How many moons does Jupiter have? *(Toolbox Skill 23)* 16

Modeling Expressions Using Tiles

Objective: To use algebra tiles to model expressions.

Materials

■ algebra tiles

You can use algebra tiles to visualize expressions. For example, let the tile [ ] represent the variable n and the tile [ ] represent the number 1. You can represent the expression $3n + 2$ as shown in the diagram below. Group the tiles by circling them.

Activity I *Expressions Involving Similar Tiles* Answers to Activity I, Steps 2-3 are on p. A3.

1 Write the expression represented by each diagram.

a. (diagram) $2 + 6$

b. (diagram) $3n + 2n$

2 Show how to use tiles to represent each expression.

a. $4 + 2$ b. $5 + 3$ c. $1 + 6$

d. $3n + 3n$ e. $4n + n$ f. $2n + 7n$

3 Show two different ways to separate each set of tiles into two groups. Write the expression represented by each grouping.

a. (diagram)

b. (diagram)

Activity II *Expressions Involving Unlike Tiles* Answers to Activity II, Steps 2-3 are on p. A3.

1 Write the expression represented by each diagram.

a. (diagram) $3n + 4$

b. (diagram) $2n + 5$

2 Show how to use tiles to represent each expression.

a. $n + 4$ b. $6 + n$ c. $5n + 2$

d. $4n + 1$ e. $3 + 2n$ f. $5 + 6n$

3 Show two different ways to separate this set of tiles into two groups. Write the expression represented by each grouping.

a. (diagram)

b. (diagram)

Teaching Notes

Algebra tiles can be used to model a number of algebraic concepts. Students who have difficulty learning abstractly may be helped by the use of concrete models. Before students attempt the activities, go over what each tile represents and why $3n + 2$ is pictured the way it is.

Using Manipulatives

For a suggestion on how to extend these activities, see page 46C.

Activity Notes

The activities are designed to introduce students to algebra tiles. *Activity I* has students work with similar tiles only, while *Activity II* uses unlike tiles.

Error Analysis

Some students may think that the expressions in Activity II, Exercise 1 represent $7n$ or 7. Have students model $7n$ and 7 to see that $3n + 4$, $2n + 5$, $7n$, and 7 are all different.

Follow-Up

Extension

To extend the activities, and to give students a preview of combining like terms, you may wish to have students group similar tiles and then write the expression that the grouping represents.

Introduction to Algebra **51**

51

Warm-Up Exercises

Find each answer.
1. $42 + 26 \div 2 - 7$ 48
2. $6 \cdot 7 + 6 \cdot 4$ 66
3. $6 \cdot (7 + 4)$ 66
4. $8 \cdot 11 - 8 \cdot 4$ 56
5. $8 \cdot (11 - 4)$ 56

Lesson Focus

To motivate today's lesson on the distributive property, you may want to use the application below.

Application

A salesperson sold 24 pairs of basketball shoes and 8 pairs of jogging shoes to each of 10 customers today. How many pairs of shoes were sold in all?   The solution can be found by using the distributive property. For example, $10(24 + 8) = 10 \cdot 24 + 10 \cdot 8 = 320$.

Teaching Notes

Tell students that a good way to remember the distributive property is to think about distributing newspapers. The newscarrier distributes a paper to each customer just as the factor outside the parentheses distributes itself to each term inside the parentheses.

The Distributive Property

Objective: To use the distributive property to compute mentally and to simplify expressions.

APPLICATION

A parking garage has three levels. On each level, there are 10 parking spaces for the disabled and 80 other parking spaces. To determine the total number of parking spaces in the garage, you can perform the calculations in two different ways.

$$3(80 + 10) = 3(90) = 270$$
$$3(80) + 3(10) = 240 + 30 = 270$$

The fact that you get the same answer either way is an application of the *distributive property*.

The **distributive property** allows you to multiply each term inside a set of parentheses by a factor outside the parentheses. You say that multiplication is *distributive* over addition and over subtraction.

The Distributive Property

In Arithmetic	In Algebra
$3(80 + 10) = 3(80) + 3(10)$	$a(b + c) = ab + ac$
$3(80 - 10) = 3(80) - 3(10)$	$a(b - c) = ab - ac$

The distributive property can help you to perform some calculations mentally.

Example 1   Use the distributive property to find each answer mentally.

a. $8 \cdot 36 - 8 \cdot 16$ b. $7(108)$

Solution

a. $8 \cdot 36 - 8 \cdot 16 = 8(36 - 16)$
$\qquad\qquad\qquad = 8(20)$
$\qquad\qquad\qquad = 160$

b. $7(108) = 7(100 + 8)$
$\qquad\qquad = 7(100) + 7(8)$
$\qquad\qquad = 700 + 56$
$\qquad\qquad = 756$

✓ **Check Your Understanding**

1. In Example 1(b), why is 108 written as $100 + 8$ and not as $110 - 2$?
2. Suppose Example 1(b) had been $7(98)$. Describe a method to find this product mentally.
   See *Answers to Check Your Understanding* at the back of the book.

You can use the distributive property to simplify variable expressions.

Example 2

Solution

Simplify $3(n + 2)$.

Let  represent n and □ represent 1.

← 3 groups of $n + 2$ → $3(n + 2)$

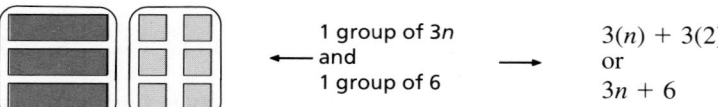

1 group of $3n$
and $3(n) + 3(2)$
1 group of 6   or
                       $3n + 6$

So $3(n + 2) = 3n + 6$.

Example 3

Simplify.

a. $5(6 - m)$ **b.** $4(2x - 7)$ **c.** $2(9 + 3y)$

Solution

a. $5(6 - m) = 5(6) - 5(m)$
$\qquad\qquad = 30 - 5m$

b. $4(2x - 7) = 4(2x) - 4(7)$
$\qquad\qquad = 8x - 28$

c. $2(9 + 3y) = 2(9) + 2(3y)$
$\qquad\qquad = 18 + 6y$

 **Check Your Understanding**

3. In Example 3, what is the meaning of the curved arrows?
See *Answers to Check Your Understanding* at the back of the book.

Guided Practice

COMMUNICATION «*Reading*

«**1.** Replace each __?__ with the correct word or phrase.

The __?__ allows you to multiply each term inside a set of parentheses by a __?__ outside the parentheses. The distributive property involves two operations. The operations are multiplication and either __?__ or subtraction. distributive property; factor; addition

«**2.** Explain the meaning of the word *distributed* in the following sentence.

The teacher distributed the tests to the students.
The teacher gave one test to each student.

Introduction to Algebra   **53**

Key Questions

1. How does the distributive property make the solutions to Example 1 easier? The numbers can be rewritten to simplify the multiplication.

2. Why cannot exercises such as $9 \cdot 54 - 9 \cdot 26$ and $12(237)$ be done mentally using the distributive property? The numbers cannot be easily rewritten in a way that simplifies the exercise.

Reasoning
The Check Your Understanding asks students to examine different ways to rewrite problems and then to draw conclusions as to which way is the best.

Error Analysis
When using the distributive property, students may apply the property to only one of the terms inside the parentheses, or they may multiply all three terms. Remind students that the number outside the parentheses multiplies each term inside.

Using Technology
For a suggestion on how to use a spreadsheet program to demonstrate the distributive property, see page 46C.

Using Manipulatives
For a suggestion on using algebra tiles to demonstrate the distributive property, see page 46D.

Reteaching/Alternate Approach
For a suggestion on using a cooperative learning activity with this lesson, see page 46D.

Calculator
The distributive property can be verified experimentally by using a calculator.

Additional Examples

1. Use the distributive property to find each answer mentally.
   a. $7 \cdot 74 - 7 \cdot 24$ 350
   b. $8(106)$ 848
2. Simplify: $2(n + 1)$ $2n + 2$
3. Simplify.
   a. $3(2x - 5)$ $6x - 15$
   b. $2(8 + 4y)$ $16 + 8y$

Closing the Lesson

Have students write an exercise that can be solved using the distributive property and then pass it to another student. Then ask some students to describe how to solve the exercise they have been given.

Suggested Assignments

Skills: 1–18, 19–23 odd, 27–32
Average: 1–20, 21–25 odd, 27–32
Advanced: 1–17 odd, 19–32,
        Historical Note

Exercise Notes

Communication: Reading
Guided Practice Exercises 1 and 2 ask students to interpret the text material and to explain the common usage of the word distribute.

Communication: Writing
Guided Practice Exercises 3 and 4 ask students to represent a picture by mathematical symbols and vice versa.

Life Skills
Exercises 25 and 26 require students to work with spreadsheets, an important skill in business and industry.

Follow-Up

Extension
Have students write a word problem that uses the distributive property in its solution. Have students explain how the distributive property is used in the solution.

COMMUNICATION « *Writing*

«**3.** Let 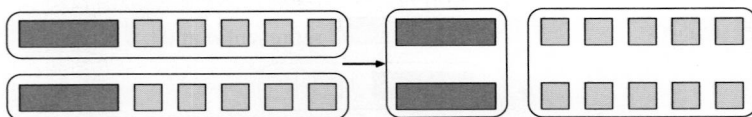 represent n and ▢ represent 1. Write the statement that is represented by this diagram. $2(n + 5) = 2n + 10$

«**4.** Draw a diagram similar to the one in Exercise 3 to represent this statement: $3(2n + 1) = 6n + 3$ See answer on p. A4.

Replace each __?__ with the number that makes the statement true.

5. $8(43) + 8(7) = 8(\underline{?})$ 50

6. $7(199) = 7(\underline{?}) - 7(1)$ 200

7. $4(n + 6) = (\underline{?})n + 4(6)$ 4

8. $2(3c - 12) = 2(3c) - \underline{?}(12)$ 2

Use the distributive property to find each answer mentally.

9. $6(62) - 6(12)$ 300

10. $9 \cdot 13 + 9 \cdot 7$ 180

11. $4(307)$ 1228

Simplify.

12. $6(x + 4)$ $6x + 24$

13. $7(3 - n)$ $21 - 7n$

14. $2(5 - 12a)$ $10 - 24a$

Exercises

Use the distributive property to find each answer mentally.

A **1.** $7(68) + 7(12)$ 560

2. $8(31) + 8(19)$ 400

3. $15(8) - 15(6)$ 30

4. $6(13) - 6(7)$ 36

5. $4(109)$ 436

6. $7(206)$ 1442

7. $9(197)$ 1773

8. $5(296)$ 1480

9. $8(398)$ 3184

Simplify.

10. $5(n + 12)$ $5n + 60$

11. $9(10 + a)$ $90 + 9a$

12. $6(8 - y)$ $48 - 6y$

13. $5(x - 9)$ $5x - 45$

14. $7(7 + 2c)$ $49 + 14c$

15. $8(6m + 9)$ $48m + 72$

16. $4(4w - 6)$ $16w - 24$

17. $3(9 - 4a)$ $27 - 12a$

18. $5(3c + 7)$ $15c + 35$

19. Peter bought two pairs of pants for $29.99 each and two shirts for $19.99 each. How much more did he spend on pants than on shirts? $20

20. Last week Lauren ran 5 mi three times. This week she ran 5 mi four times. Did she run more than 40 mi during the two weeks? Explain. No; in all, she ran $5(3 + 4) = 5(7) = 35$ mi, which is less than 40 mi.

Replace each __?__ with the number that makes the statement true.

B **21.** $4(2n - \underline{?}) = 8n - 16$ 4

22. $2(5t + \underline{?}) = 10t + 2$ 1

23. $\underline{?}(2n + 1) = 6n + 3$ 3

24. $\underline{?}(6z - 5) = 12z - 10$ 2

COMPUTER APPLICATION

Like the distributive property, the spreadsheet below finds an amount in two ways. It calculates weekly earnings both by finding the product of the total hours and the hourly rate and by finding the sum of the daily earnings.

	A	B Hours	C Earnings	
1		Hours	Earnings	
2	Monday	6	$54	
3	Tuesday	5	$45	
4	Wednesday	8	$72	
5	Thursday	7	$63	
6	Friday	5	$45	
7				
8	Total Hours	31		
9	Hourly Rate	$9		
10				
11	Weekly Earnings	$279	$279	

Use an electronic spreadsheet or draw a spreadsheet similar to the one shown to find each person's weekly earnings in two ways.

25. Leotie earns $8/h. She worked 7 h on Monday, 4 h on Tuesday, 8 h on Wednesday, 5 h on Thursday, and 6 h on Friday. $240

26. Ellen earns $7/h. She worked 6 h on Monday, 3 h on Tuesday, 6 h on Wednesday, 8 h on Thursday, and 7 h on Friday. $210

SPIRAL REVIEW 30. No; Yes; No; No; No

S **27.** Evaluate ab when $a = 7$ and $b = 9$. *(Lesson 1-1)* 63

28. Find the sum: $1.63 + 0.96$ *(Toolbox Skill 11)* 2.59

29. Simplify: $5(3x + 5)$ *(Lesson 2-2)* $15x + 25$

30. Is 177 divisible by 2? by 3? by 4? by 5? by 9? *(Toolbox Skill 16)*

31. Evaluate $4n$ when $n = 2.9$. *(Lesson 1-3)* 11.6

32. Estimate the quotient: $46,287 \div 59$ *(Toolbox Skill 3)* about 800

The word *algebra* was not used by Europeans until the twelfth century. The word first appears in the Latin translation of the title of a ninth-century book by the Arabian mathematician al-Khowarizmi. This work by al-Khowarizmi popularized the use of the Hindu-Arabic number system.

Research — The features are place value and the use of zero.

Find out two features of the Hindu-Arabic number system that make it different from the Roman numeral system.

Introduction to Algebra **55**

55

Teaching the Lesson
- Materials: algebra tiles
- Lesson Starter 13
- Using Manipulatives, p. 46D
- Toolbox Skills 7, 8

Lesson Follow-Up
- Practice 16
- Practice 17: Nonroutine
- Enrichment 14: Data Analysis
- Study Guide, pp. 25–26
- Manipulative Activity 4
- Cooperative Activity 5
- Test 5

Warm-Up Exercises

Evaluate each expression when $x = 4$.

1. $2x + x$ 12
2. $3x$ 12
3. $11x - 6x$ 20
4. $5x$ 20
5. $x + 4x + 3x$ 32
6. $8x$ 32

Lesson Focus

Tell students to imagine the kitchens in their houses if plates, forks, knives, spoons, and glasses were put away randomly—for example, a glass in one cupboard and the other glasses on different shelves. In addition to not looking very neat, it would be difficult to locate objects. Today's lesson focuses on a way to arrange mathematical terms. We are going to combine like terms, just as at home, you combine like objects in the kitchen.

Teaching Notes

Be certain students understand that like terms must have *identical variable* parts. Examples 1(c) and 2(b) can be used to stress the fact that terms such as ac and ab or n and n^2 are not like terms.

Combining Like Terms

Objective: To simplify expressions by combining like terms.

CONNECTION

Barry simplified the process of locating items in his music collection by combining like items. He put CDs in one box, cassettes in another box, and record albums in a third box.

In mathematics, sometimes you can simplify an expression by *combining like terms*. The expression $2n + 5m + 4n$ contains three *terms*: $2n$, $5m$, and $4n$. The terms $2n$ and $4n$ have identical variable parts, so they are called **like terms.** The terms $2n$ and $5m$ have different variable parts, so they are called **unlike terms.**

The process of adding or subtracting like terms is often called **combining like terms.** Unlike terms cannot be combined.

> **Terms to Know**
> - *like terms*
> - *unlike terms*
> - *combining like terms*

Example 1

Simplify: **a.** $2n + 4n$ **b.** $4n - n$ **c.** $3n + 4 + 2n$

Solution

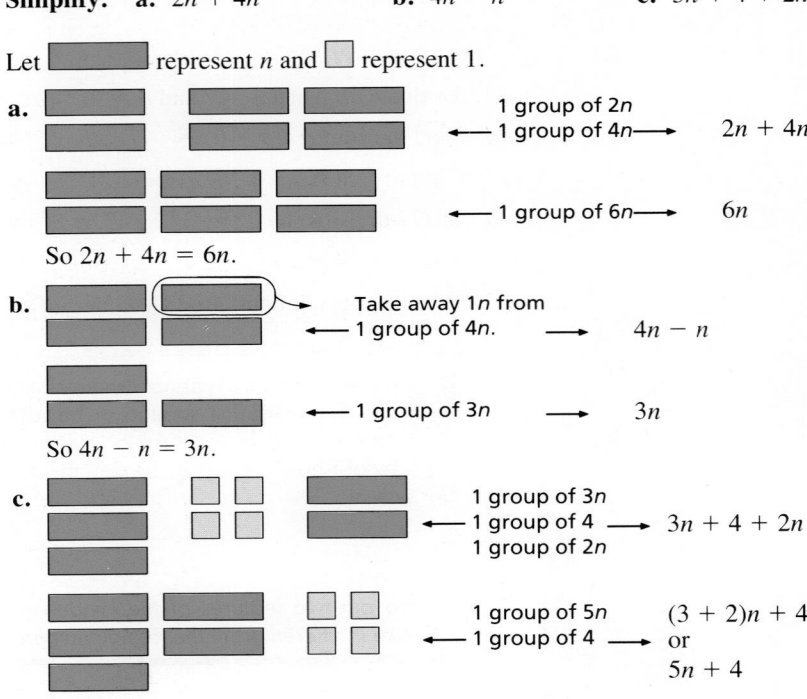

Let ▭ represent n and ▫ represent 1.

a. 1 group of $2n$ / 1 group of $4n$ → $2n + 4n$

1 group of $6n$ → $6n$

So $2n + 4n = 6n$.

b. Take away $1n$ from 1 group of $4n$. → $4n - n$

1 group of $3n$ → $3n$

So $4n - n = 3n$.

c. 1 group of $3n$ / 1 group of 4 / 1 group of $2n$ → $3n + 4 + 2n$

1 group of $5n$ / 1 group of 4 → $(3 + 2)n + 4$ or $5n + 4$

So $3n + 4 + 2n = 5n + 4$.

When you combine like terms, you are applying the distributive property.

Example 2

Solution

Simplify: **a.** $9x + 7x$ **b.** $11c + c - 8b$

a. $9x + 7x = (9 + 7)x$
$= 16x$

b. $11c + c - 8b = 11c + 1c - 8b$ ◄—— Recall that $c = 1 \cdot c$.
$= (11 + 1)c - 8b$
$= 12c - 8b$

 **Check Your Understanding**

1. In Example 2(b), why do you combine only $11c$ and c?
2. What property is used in Example 2(b) to write c as $1c$?
   See *Answers to Check Your Understanding* at the back of the book.

Guided Practice

« **1.** COMMUNICATION « *Reading*   Replace each __?__ with the correct word.

In an expression, terms that can be combined are called __?__ terms. like
Terms that cannot be combined are called __?__ terms. unlike

Tell whether the following are *like terms* or *unlike terms*.

2. $4a$; $6a$ **3.** $5d$; $8b$ **4.** $6xy$; $6x$ **5.** $2w$; 2 **6.** $54r$, r
like terms   unlike terms   unlike terms   unlike terms   like terms

Simplify.

7. $7a + 4a$ 11a   **8.** $c + 3c$ 4c   **9.** $12x - 4x$ 8x

10. $10y - y$ 9y   **11.** $3a + 9 + 11a$ 14a + 9   **12.** $5b + 8b - 7$
                                                                                              13b − 7
13. $c + 4n + 6n$ c + 10n   **14.** $8d - d - 5m$ 7d − 5m   **15.** $6z + z + 9z$
                                                                                              16z

Exercises

Simplify.

12. $7x - 5$

A **1.** $6x + 8x$ 14x   **2.** $7a + 10a$ 17a   **3.** $5m + m$ 6m

4. $c + 21c$ 22c   **5.** $12w - w$ 11w   **6.** $14p - 5p$ 9p

7. $8x - 3x$ 5x   **8.** $9y - y$ 8y   **9.** $2z + 7 + 6z$
                                                                                     8z + 7
10. $6n + 9n + 4$ 15n + 4   **11.** $6k + 3k - 6$ 9k − 6   **12.** $5x + 2x - 5$

13. $4w + 4r - 3r$ 4w + r   **14.** $8y - 7y + 4$ y + 4   **15.** $2a + 4b + 5b$
                                                                                              2a + 9b
16. $3x + 7x + 9y$ 10x + 9y   **17.** $5n + 12n + n$ 18n   **18.** $m + 4m + 6m$
                                                                                              11m

Introduction to Algebra   **57**

Key Question

Why does $5n + n = 6n$? Since $n = 1 \cdot n$, the coefficient of n is 1. So, $5n + n = 5n + 1n = (5 + 1)n = 6n$.

Critical Thinking

When combining like terms, students must *examine* the terms for similarities. They must also *evaluate* any similarities in the terms and then *make a judgment* as to which terms are like and which are unlike.

Error Analysis

Students may mistakenly combine unlike terms by multiplying the variables. This is especially likely to happen in situations such as $5x + 5y$, for which students may write $5xy$. Remind students that terms must have identical variable parts in order to be combined.

Using Manipulatives

For a suggestion on using algebra tiles to illustrate combining like terms, see page 46D.

Closing the Lesson

Have students make up expressions that can be simplified by combining like terms. Ask some students to put their expressions on the chalkboard and explain how they would simplify them.

Suggested Assignments

Skills: 1–18, 19–29 odd, 36–40
Average: 1–18, 19–35 odd, 36–40
Advanced: 1–17 odd, 19–40

Exercise Notes

Communication: Writing

Exercise 31 asks students to compare four previously learned methods, discuss these methods in their own words, and create new examples that reflect the points in their summary.

Problem Solving

Exercise 36 asks students to identify extraneous information. To do this, they must identify (1) the problem to be solved, and (2) the information needed to solve the problem.

Follow-Up

Nonroutine Problem

Using a phone book, randomly choose 5 to 10 telephone numbers. Let the hyphen in the number represent an equals sign. Students are to insert mathematical symbols to make an equation. For example, 225-6780 could represent $2 - 2 + 5 = 6 + 7 - 8 + 0$ or $2 + 2^5 = 6 \cdot 7 - 8 - 0$.

Enrichment

Enrichment 14 is pictured on page 46E.

Practice

Practice 16 is shown below.

Quick Quiz 1

See page 90.

Alternative Assessment

See page 90.

Practice 16, *Resource Book*

Practice 16
Skills and Applications of Lesson 2-3

Simplify.
1. $5x + 7x$ 12x
2. $3a + 12a$ 15a
3. $9d - 6d$ 3d
4. $18m - 4m$ 14m
5. $16w + 5w$ 21w
6. $8k - 8k$ 0
7. $t + 15t$ 16t
8. $5b + 3b + 4b$ 12b
9. $10x - 6x$ 4x
10. $13x - x$ 12x
11. $14c + 7c - c$ 20c
12. $7w - 5w + 2w$ 4w
13. $6z + 10 + 3z$ 9z + 10
14. $k + 23k$ 24k
15. $7 + 14n + 3n$ 7 + 17n
16. $8x - 5 + 3x$ 11x − 5
17. $2a + 7a + 3a$ 12a
18. $5x + 10w - 2x$ 3x + 10w
19. $6y + 9 - y$ 5y + 9
20. $7m - 7 + 2m$ 9m − 7
21. $8t - 8s + t$ 9t − 8s
22. $14x - 3 - 3x$ 11x − 3
23. $6k + 9 - 6x$ 9
24. $13a + 9a + 7a$ 29a
25. $5m + 8m - 3t$ 13m − 3t
26. $11a + 7b + 3b$ 11a + 10b
27. $22x + 17x - 2x$ 37x
28. $9u - 5 - 5u$ 4a − 5
29. $15k - k + 8$ 14k + 8
30. $98m - 25m + 2m$ 75m
31. $25 + 6w + 8w$ 25 + 14w
32. $17x + 28x - 7x$ 38x
33. $28a + 37a + 2a$ 67a
34. $6x + 12 + 3a$ 9a + 12
36. $6(a + 2) + 3a$ 9a + 12
35. $3(2x - 5) - 2x$ 4x − 15

37. On the first day of the month, a car dealership had an inventory (number of new cars and number of used cars) of eighteen new cars and fourteen used cars. Twelve new cars were delivered to the dealer and nine used cars were added. How would you report the dealer's new inventory? **30 new cars and 23 used cars**

38. Mario's music collection included seventeen cassettes, fourteen tapes, and nine compact discs. Luis's music collection included twenty-two cassettes, eleven tapes, and five compact discs. How would you report their combined collection in a simplified way? **39 cassettes, 25 tapes, and 14 compact discs**

Simplify, if possible. If not possible, write *cannot be simplified*.

19. $10a + 13b$
20. cannot be simplified
21. $11c + 4w$
22. $7y + 6z$
23. cannot be simplified
24. $8n + 4$
25. cannot be simplified
26. $12h + 17k$
27. $5a + 13b + 3x$
28. cannot be simplified
29. $14c + 8$
30. $2x + 2w - 6z$

B 19. $4a + 5b + 6a + 8b$
20. $b + 8c + 2x + 5z$
21. $7c + 6w - 2w + 4c$
22. $4y + 3y + 12z - 6z$
23. $a + 9m - n + 4x$
24. $7n - 2n + 4 + 3n$
25. $d - 4n + 3w - 7x$
26. $2h + 8k + 10h + 9k$
27. $5a + 6b + 3x + 7b$
28. $4a + 8b + 9c + z$
29. $12c - 3c + 8 + 5c$
30. $2x + 7w - 5w - 6z$

31. **WRITING ABOUT MATHEMATICS** In Lessons 2-1, 2-2, and 2-3 you learned four methods for simplifying expressions. Write a brief report that summarizes these methods. Be sure to include an example for each method. (*Hint:* Refer to the objective of each lesson.)
See answer on p. A4.

Simplify.

32. $2(x + 7) + 8x$ 10x + 14
33. $7y + 9(4 + 9y)$ 88y + 36
34. $6(2a + 1) + 3(2 + 3a)$ 21a + 12
35. $3(6 + b) + 5(b + 1)$ 8b + 23

SPIRAL REVIEW

36. Amy drove 255 mi in four days and used 10 gal of gasoline. Identify any facts that are not needed to find miles per gallon. (*Lesson 1-4*) the time (four days)
37. Evaluate x^4 when $x = 5$. (*Lesson 1-9*) 625
38. Simplify: $4c + 7c - 5d$ (*Lesson 2-3*) 11c − 5d
39. Find the sum: $\frac{3}{4} + \frac{7}{8}$ (*Toolbox Skill 17*) $1\frac{5}{8}$
40. Find the product: $(0.37)(1000)$ (*Toolbox Skill 15*) 370

Self-Test 1

Simplify.

1. $x^4 \cdot x^3$ x^7
2. $c \cdot c^7$ c^8
3. $a^3 \cdot a^8 \cdot a^5$ a^{16} 2-1
4. $3w^2 \cdot 9w^5$ $27w^7$
5. $(6x)(15z)$ 90xz
6. $(4a)(7b)(3a)$ $84a^2b$

Use the distributive property to find each answer mentally.

7. $6 \cdot 12 + 6 \cdot 8$ 120
8. $5(107)$ 535
9. $8(196)$ 1568   2-2

Simplify.

10. $2(x + 9)$ 2x + 18
11. $8(7 - 5x)$ 56 − 40x
12. $2(11 + 7c)$ 22 + 14c
13. $3n + 4n$ 7n
14. $14x - x$ 13x
15. $7x - 4x - 5y$ 3x − 5y
16. $4a + 2b + 9b$ 4a + 11b
17. $6z + 1 + 2z$ 8z + 1
18. $w + 3w + 8w$ 12w   2-3

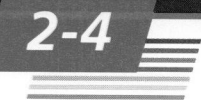

Using a Four-Step Plan

Objective: To use a four-step plan to solve problems.

In Lesson 1-4 you learned how to read a problem for understanding. Understanding a problem is the first step to solving the problem. Below are all four steps you should carry out when solving a problem.

Four-Step Plan

UNDERSTAND	**Read and understand the problem.** Know what is given and what you have to find.
PLAN	**Make a plan.** Choose a problem solving strategy.
WORK	**Carry out the plan.** Use the strategy and do any necessary calculations.
ANSWER	**Check any calculations and answer the problem.** Interpret the answer, if necessary.

One strategy you might decide to use in the second step—making a plan—is *choosing the correct operation*. Sometimes it takes more than one operation to solve a problem. In that case, you must decide not only which operations are needed, but also in which order to use them.

Problem

Two weeks ago Michael worked 35 h. Last week he worked 31 h. Michael earns an hourly wage of $7.15. How much did Michael earn during these two weeks?

Solution

UNDERSTAND The problem is about Michael's earnings.
Facts: worked 35 h
       worked 31 h
       earns $7.15 per hour
Find: the total amount Michael earned during two weeks

PLAN First add to determine the number of hours Michael worked. Then multiply the total number of hours by the hourly wage.

WORK $35 + 31 = 66$
$7.15(66) = 471.9$

ANSWER Check the calculations: $7.15(35 + 31) = 7.15(66) = 471.9$
Michael earned $471.90 during the two weeks.

Look Back An alternative method for solving the problem is to first find the amount earned each week and then add. Solve the problem again using this alternative method. $471.90

Introduction to Algebra   **59**

Lesson Planner

> ### Teaching the Lesson
> - Lesson Starter 14
> - Using Technology, p. 46C
> - Reteaching/Alternate Approach, p. 46D
> - Toolbox Skills 7–14
>
> ### Lesson Follow-Up
> - Practice 18
> - Enrichment 15: Problem Solving
> - Study Guide, pp. 27–28

Warm-Up Exercises

Use this paragraph for Exercises 1–3.

Jan bought three loaves of bread that cost $1.09 each, two pounds of peaches that cost $1.29 per pound, and a bottle of water that cost $.87.

1. What is the paragraph about? what Jan bought and how much it cost

2. What information is not needed to find how much Jan spent on bread? the cost of the peaches and the water

3. How much did Jan spend on bread? $3.27

Lesson Focus

In the real world, problems are not well organized. Situations must be examined, information sorted, and problems formulated. Today's lesson focuses on a four-step plan to solve problems.

Teaching Notes

Work through the Example with students. Stress each step of the problem solving plan, making certain students realize that all four steps must be followed to solve a given problem.

Key Question

Why is it important to include a step to check the calculations and interpret the answer?

Checking calculations helps avoid errors, and interpreting the answers helps avoid impossible answers, such as 4.5 radios.

Critical Thinking

Formulating a plan requires students to understand the problem and to *develop* a method to use to find the solution. Students may use the success of their plan to validate their thinking process.

Using Technology

For a suggestion on how to use a spreadsheet program with this lesson, see page 46C.

Error Analysis

Upon arriving at an answer, students may forget to check it against the original problem to see that it is reasonable. This leads to answers such as 6.5 buses. Encourage students to check their answers.

Reteaching/Alternate Approach

For a suggestion on a class discussion about problem solving, see page 46D.

Additional Example

Margo works as a purchasing agent. Last week she purchased 23 boxes of computer disks for her company. This week she purchased 21 more boxes. The price per box is $8.45. How much did Margo spend on the computer disks?

8.45(23 + 21) = 371.80
She spent $371.80 on computer disks.

Closing the Lesson

Give students a problem that can be solved in more than one way. Work through the four-step plan with students listing the various methods of solution. Have students use each method to solve the problem.

Guided Practice

COMMUNICATION «*Reading*

Use this paragraph for Exercises 1–3.

In ice hockey a team earns two points for a win, one point for a tie, and no points for a loss. The Hornets won 36, tied 12, and lost 16 games.

«**1.** How many points does a team earn for a win? 2 points

«**2.** Identify any facts that are not needed to find the number of games the Hornets played. one point for a tie game; two points for a win; no points for a loss

«**3.** Describe how you would find the number of points the Hornets earned. multiply 36 by 2; add 12

Choose the letter of the key sequence needed to solve each problem.

4. Keisha bought a dress for $36.98 and two blouses for $20 each. What was the total cost of her purchases? B

   A. ` 2. ` × ` 20. ` − ` 36.98 ` =
   B. ` 2. ` × ` 20. ` + ` 36.98 ` =
   C. ` 2. ` × ` 36.98 ` + ` 20. ` =

5. Kay bought a sweater for $36.98. She gave the clerk two $20 bills. How much change did she receive? A

Solve.

6. Larry bought six pairs of socks for $3.99 per pair and three shirts for $19.95 each. What was the total cost of the socks? $23.94

7. ABC Company bought 145 new office chairs. The office manager put 26 new chairs in each of four small conference rooms. How many new chairs remained for the manager to put in the large conference room? 41 chairs

Problem Solving Situations

Solve.

1. Ron has a 30-ft sailboat. He sailed 7 mi from Kingsport to Elk Island. After sailing back to Kingsport, Ron then sailed 12 mi to Cape West. How many miles did Ron sail in all? 26 mi

---

PROBLEM SOLVING

CHECKLIST

Keep this in mind:
Using a Four-Step Plan

Consider this strategy:
Choosing the Correct Operation

---

2. Gail's bus trip from home to work is 3 mi and usually takes 20 min. Bus tokens cost $.75. Gail bought twelve tokens. How much did she spend? $9

3. Jamie bought two loaves of bread for $1.29 each and three heads of lettuce for $.95 each. What was the total cost? $5.43

4. King High School has budgeted $4350 for new desks. Each desk costs $115 including tax. How many desks can the school buy? 37 desks

5. The student council bought 400 sweatshirts, 650 T-shirts, and 1100 notebooks to sell during the school year. At the end of the year the council had 96 sweatshirts, 139 T-shirts, and 227 notebooks left. How many items did the student council sell during the school year?
1688 items

6. Jill Panov bought a television that cost $299 and a VCR that cost $349.98. The sales tax was $38.94. She made a down payment of $140 and agreed to pay the rest in eight equal payments. What was the amount of each payment? $68.49

DATA ANALYSIS

Use the diagram at the right.

7. How many more one-room schools are there in Pennsylvania than in South Dakota? 163

8. About how many times the number of one-room schools in Montana are there in Nebraska? about 3

9. The total number of one-room schools in the United States is 1279. How many one-room schools are not in one of the four states shown? 495

One-Room Schools in Four States

SOUTH DAKOTA	94
MONTANA	111
PENNSYLVANIA	257
NEBRASKA	322

10. There are 84 one-room schools in Ohio. Is the number of one-room schools in Nebraska *greater than* or *less than* the total number in Ohio, South Dakota, and Montana? by how many? greater than; 33

WRITING WORD PROBLEMS

Using the information given, write a word problem for each exercise. Then solve the problem. Answers to Problems 11–12 are on p. A4.

11. $600 - 180 = 420; 420 \div 12 = 35$

12. $(10.75)42 = 451.50; (451.50)52 = 23,478$

SPIRAL REVIEW

S 13. Simplify: $4(3a - 8)$ *(Lesson 2-2)* 12a − 32

14. Carlos bought six greeting cards and six stamps. Each card cost $1.25 and each stamp cost $.25. How much did he spend?   *(Lesson 2-4)*
$9

15. Find the sum mentally: $73 + 0 + 27 + 19$ *(Lesson 1-6)* 119

16. **DATA,** *pages 2–3* How many summer concerts does the Houston symphony perform each year?   *(Toolbox Skill 24)* 5 concerts

Introduction to Algebra   **61**

Suggested Assignments
Skills: 1–6, 7–11 odd, 13–16
Average: 1–16
Advanced: 1–16

Exercise Note

Data Analysis
Problem Solving Situations 7–10 check students' ability to read and use information from a bar graph.

Follow-Up

Extension (Cooperative Learning)
Have students work in cooperative groups in order to develop a problem solving situation. From each situation, ask the groups to make up as many problems as they can. This might be done as a competition with points awarded for:
(a) the group that makes up the most problems;
(b) the group with the most unusual problems;
(c) the shortest situation with at least five related problems.

Enrichment
Enrichment 15 is pictured on page 46E.

Practice
Practice 18, *Resource Book*

2-5 Patterns

Objective: To recognize and to extend patterns.

APPLICATION

Many patterns occur naturally in the real world. In 1202, Leonardo Fibonacci wrote about a pattern from nature that is called the *Fibonacci sequence*. These are the first ten numbers of that pattern.

$$1, 1, 2, 3, 5, 8, 13, 21, 34, 55$$

Beginning with the third number, 2, each number in the pattern is the sum of the two numbers immediately preceding it. You can find this pattern in the spirals of the seeds on most sunflowers and in the spirals of the scales on many pineapples.

Recognizing a pattern is a useful way to solve some problems in mathematics. Many mathematical patterns involve addition, subtraction, multiplication, and division.

Example 1

Solution

Find the next three numbers in each pattern.

a. $3, 6, 12, 24, \underline{\ ?\ }, \underline{\ ?\ }, \underline{\ ?\ }$

b. $5, 7, 10, 14, \underline{\ ?\ }, \underline{\ ?\ }, \underline{\ ?\ }$

a. Each number is 2 times the preceding number. The pattern is *multiply by 2*.

$$3 \quad 6 \quad 12 \quad 24 \quad 48 \quad 96 \quad 192$$
$${\times 2}\quad{\times 2}\quad{\times 2}\quad{\times 2}\quad{\times 2}\quad{\times 2}$$

The next three numbers are 48, 96, and 192.

b. The second number is 2 more than the first number. The third number is 3 more than the second number. The fourth number is 4 more than the third number. The pattern is *add 2, add 3, add 4, and so on*.

$$5 \quad 7 \quad 10 \quad 14 \quad 19 \quad 25 \quad 32$$
$${+2}\quad{+3}\quad{+4}\quad{+5}\quad{+6}\quad{+7}$$

The next three numbers are 19, 25, and 32.

 **Check Your Understanding**

1. Suppose you extend the pattern in Example 1(a). What would be the next two numbers after 192?

2. Suppose you extend the pattern in Example 1(b). What would you add to 32 to get the next number? What would be the next number?
   See Answers to Check Your Understanding at the back of the book.

Example 2

Find the next three expressions in each pattern.

a. $x, x + 3, x + 6, \underline{\ ?\ }, \underline{\ ?\ }, \underline{\ ?\ }$

b. $12a + 5, 10a + 5, 8a + 5, \underline{\ ?\ }, \underline{\ ?\ }, \underline{\ ?\ }$

Solution

a. Each expression is 3 more than the preceding expression. The pattern is *add 3*.

$$x \quad x + 3 \quad x + 6 \quad x + 9 \quad x + 12 \quad x + 15$$
$$\underset{+3}{\nearrow}\ \underset{+3}{\nearrow}\ \underset{+3}{\nearrow}\ \underset{+3}{\nearrow}\ \underset{+3}{\nearrow}$$

The next three expressions are $x + 9$, $x + 12$, and $x + 15$.

b. Each expression is $2a$ less than the preceding expression. The pattern is *subtract 2a*.

$$12a + 5 \quad 10a + 5 \quad 8a + 5 \quad 6a + 5 \quad 4a + 5 \quad 2a + 5$$
$$\underset{-2a}{\nearrow}\ \underset{-2a}{\nearrow}\ \underset{-2a}{\nearrow}\ \underset{-2a}{\nearrow}\ \underset{-2a}{\nearrow}$$

The next three expressions are $6a + 5$, $4a + 5$, and $2a + 5$.

☑ **Check Your Understanding**

3. Suppose the pattern in Example 2(b) continued. What would be the next expression after $2a + 5$?

See *Answers to Check Your Understanding* at the back of the book.

You can use a calculator to work with patterns that involve a repeated operation. For instance, you can use a calculator to find the pattern 6, 9, 12, 15, 18. If your calculator has a *constant key*, [K], you can use this key sequence.

$$\boxed{3.}\ \boxed{+}\ \boxed{K}\ \boxed{\ }\ \boxed{6.}\ \boxed{=}\ \boxed{=}\ \boxed{=}\ \boxed{=}$$

If your calculator does not have a constant key, you might be able to use the equals key.

$$\boxed{6.}\ \boxed{+}\ \boxed{\ }\ \boxed{3.}\ \boxed{=}\ \boxed{=}\ \boxed{=}\ \boxed{=}$$

To find out how the calculator you are using handles constants, consult the user's manual.

Guided Practice

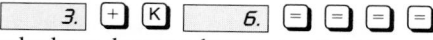

COMMUNICATION «*Writing* **3.** 128, 113, 98, 83, 68
6. $x - 2, 5x - 2, 9x - 2, 13x - 2, 17x - 2$

Write the first five numbers or expressions in each pattern.

«**1.** 8; *add 6*
8, 14, 20, 26, 32

«**2.** 2; *multiply by 2*
2, 4, 8, 16, 32

«**3.** 128; *subtract 15*

«**4.** 400; *divide by 2*
400, 200, 100, 50, 25

«**5.** $8x$; *multiply by 5*
8x, 40x, 200x, 1000x, 5000x

«**6.** $x - 2$; *add 4x*

Describe each pattern in words. **7.** divide by 2
8. add 3, add 4, add 5, and so on

«**7.** 240, 120, 60, 30, 15

«**8.** 5, 8, 12, 17, 23

«**9.** $c, 3c, 9c, 27c$
multiply by 3

«**10.** $a + 22, a + 18, a + 14, a + 10$
subtract 4

Introduction to Algebra   **63**

Key Question
How does working with patterns that contain variables differ from working with number patterns?
Patterns with variables may involve the variable, the constant, or both.

Critical Thinking
Analyzing a pattern requires students to look for similarities and differences among the terms, to *make a conjecture* as to the relationship that is present, and to *check* the conjecture by testing it for accuracy.

Reteaching/Alternate Approach
For a suggestion on using a cooperative activity, see page 46D.

Additional Examples

1. Find the next three numbers in each pattern.
a. 2, 6, 18, $\underline{\ ?\ }, \underline{\ ?\ }, \underline{\ ?\ }$
54, 162, 486
b. 4, 7, 11, 16, $\underline{\ ?\ }, \underline{\ ?\ }, \underline{\ ?\ }$
22, 29, 37

2. Find the next three expressions in each pattern.
a. $n, n + 2, n + 4, n + 6,$
$\underline{\ ?\ }, \underline{\ ?\ }, \underline{\ ?\ }$
$n + 8, n + 10, n + 12$
b. $24x + 7, 21x + 7, 18x + 7,$
$\underline{\ ?\ }, \underline{\ ?\ }, \underline{\ ?\ }$
$15x + 7, 12x + 7, 9x + 7$

Closing the Lesson
Have a few students explain what they look for when working with patterns. Then ask a student to name the next three numbers of the pattern 4, 5, 7, 10, 14, 19. 25, 32, 40 Ask another student to identify the pattern. Add 1, add 2, add 3, and so on.

Find the next three numbers or expressions in each pattern.

11. 23, 27, 31

12. 8, 4, 2

13. 48, 240, 1440

14. 12x, 15x, 18x

11. 7, 11, 15, 19, _?_, _?_, _?_   **12.** 128, 64, 32, 16, _?_, _?_, _?_

13. 2, 2, 4, 12, _?_, _?_, _?_   **14.** 3x, 6x, 9x, _?_, _?_, _?_

15. $2x + 11, 2x + 9, 2x + 7,$ _?_, _?_, _?_ $2x + 5, 2x + 3, 2x + 1$

16. $7c + 1, 6c + 1, 5c + 1,$ _?_, _?_, _?_ $4c + 1, 3c + 1, 2c + 1$

Exercises

1. 52, 62, 72
2. 76, 73, 70
3. 625, 3125, 15,625
4. 9, 3, 1
5. 20, 26, 33
6. 44, 53, 63
7. 13, 8, 2
8. 120, 720, 5040
9. 56m, 112m, 224m
10. 128y, 512y, 2048y
15. January
16. 2187 people
17. 89, 144
18. $\frac{1}{64}, \frac{1}{128}$

Find the next three numbers or expressions in each pattern.

A

1. 12, 22, 32, 42, _?_, _?_, _?_   **2.** 88, 85, 82, 79, _?_, _?_, _?_

3. 1, 5, 25, 125, _?_, _?_, _?_   **4.** 729, 243, 81, 27, _?_, _?_, _?_

5. 6, 8, 11, 15, _?_, _?_, _?_   **6.** 18, 23, 29, 36, _?_, _?_, _?_

7. 23, 22, 20, 17, _?_, _?_, _?_   **8.** 1, 2, 6, 24, _?_, _?_, _?_

9. $7m, 14m, 28m,$ _?_, _?_, _?_   **10.** $2y, 8y, 32y,$ _?_, _?_, _?_

11. $a, a + 5, a + 10,$ _?_, _?_, _?_ $a + 15, a + 20, a + 25$

12. $n + 12, n + 15, n + 18,$ _?_, _?_, _?_ $n + 21, n + 24, n + 27$

13. $n + 3, 2n + 3, 3n + 3,$ _?_, _?_, _?_ $4n + 3, 5n + 3, 6n + 3$

14. $13x - 2, 11x - 2, 9x - 2,$ _?_, _?_, _?_ $7x - 2, 5x - 2, 3x - 2$

15. John needs $550 to buy a television. He had $150 in May, $200 in
June, $250 in July, and $300 in August. If this pattern continues, in
what month will he be able to buy the television?

16. Suppose that you told a secret to three friends. On Monday, each
friend told the secret to three other people. On Tuesday, each of these
people told the secret to three other people, and so on. How many
people are told the secret on Saturday?

B   **17.** Find the next two numbers in the Fibonacci sequence on page 62.

18. Find the next two numbers in the pattern $\frac{1}{2}, \frac{1}{4}, \frac{1}{8}, \frac{1}{16}, \frac{1}{32}$.

CONNECTING ARITHMETIC AND GEOMETRY

Some number patterns are associated with geometric figures. For instance, the figure below represents the first four *triangular numbers*. This number pattern has been used since ancient times.

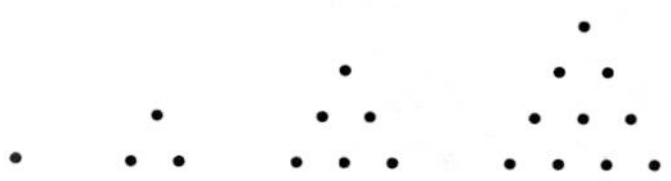

19. Draw the figures that represent the fifth and sixth triangular numbers.
See answer on p. A4.

20. Without drawing the figures, find the seventh and eighth triangular numbers.
28, 36

21. Describe the pattern of the triangular numbers in words.
Start with 1, add 2, add 3, add 4, and so on.

22. The figure at the right represents the first four *square numbers*. Use the pattern to make a list of the first ten square numbers.
1, 4, 9, 16, 25, 36, 49, 64, 81, 100

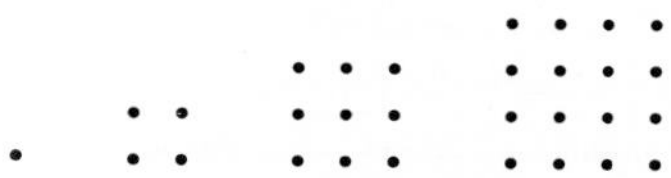

GROUP ACTIVITY

24. multiply by 2; two to the first power, two to the second power, two to the third power, and so on

Give two different ways to describe each pattern.

C **23.** 1, 4, 9, 16, 25, 36, 49, . . .
add 3, add 5, add 7, and so on; square 1, square 2, square 3, and so on

24. 2, 4, 8, 16, 32, 64, 128, . . .

Find two different ways to continue each pattern.

25. 2, 4, 8, 10, ___?___, ___?___, ___?___
14, 16, 20; 20, 22, 44

26. 1, 3, 9, 11, ___?___, ___?___, ___?___
17, 19, 25; 33, 35, 105

27. Make up a number pattern, then trade patterns with a partner.
- **a.** Describe your partner's pattern in words. Answers will vary.
- **b.** Find the next three numbers in the pattern.
- **c.** Determine if there is more than one way to continue the pattern.

SPIRAL REVIEW

S **28.** Evaluate $2a + 3b$ when $a = 6$ and $b = 4$. *(Lesson 1-10)* 24

29. Find the next three numbers in the pattern:
1, 6, 36, 216, ___?___, ___?___, ___?___ *(Lesson 2-5)* 1296, 7776, 46,656

30. Find the product: (6.7)(3.2) *(Toolbox Skill 13)* 21.44

31. Find the quotient: $4\frac{3}{8} \div 1\frac{2}{5}$ *(Toolbox Skill 21)* $3\frac{1}{8}$

Challenge

Find the next three items in each pattern.

a. J, F, M, A, M, J, J, ___?___, ___?___, ___?___ A, S, O

b. O, T, T, F, F, S, S, ___?___, ___?___, ___?___ E, N, T

c.  ___?___, ___?___, ___?___

d.  ___?___, ___?___, ___?___

Answers to Challenge (c) and (d) are on p. A4.

Follow-Up

Project

Puzzles and brain teasers are often based on patterns.

1. Have students collect some puzzles to share with the class or to use as a bulletin board display.
2. Have students develop some of their own brain teasers based on well-known information.

Enrichment

Pascal's triangle has many different patterns. You might supply students with a portion of Pascal's triangle or show them how to build it. Ask them to find as many different number patterns as they can.

Enrichment

Enrichment 16 is pictured on page 46E.

Practice

Practice 19 is shown below.

Practice 19, *Resource Book*

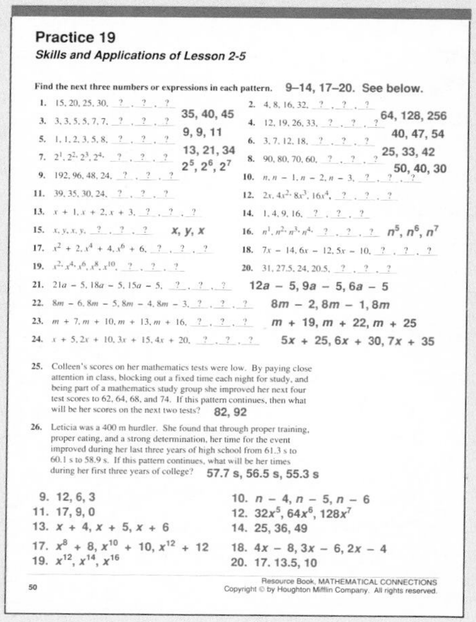

Practice 19
Skills and Applications of Lesson 2-5

Find the next three numbers or expressions in each pattern. 9–14, 17–20. See below.

1. 15, 20, 25, 30, ?, ?, ? 35, 40, 45
2. 4, 8, 16, 32, ?, ?, ? 64, 128, 256
3. 3, 3, 5, 5, 7, 7, ?, ?, ? 9, 9, 11
4. 12, 19, 26, 33, ?, ?, ? 40, 47, 54
5. 1, 1, 2, 3, 5, 8, ?, ?, ? 13, 21, 34
6. 3, 7, 12, 18, ?, ?, ? 25, 33, 42
7. $2^1, 2^2, 2^3, 2^4$, ?, ?, ? $2^5, 2^6, 2^7$
8. 90, 80, 70, 60, ?, ?, ? 50, 40, 30
9. 192, 96, 48, 24, ?, ?, ?
10. $n, n - 1, n - 2, n - 3$, ?, ?, ?
11. 39, 35, 30, 24, ?, ?, ?
12. $2x, 4x^2, 8x^3, 16x^4$, ?, ?, ?
13. $x + 1, x + 2, x + 3$, ?, ?, ?
14. 1, 4, 9, 16, ?, ?, ?
15. x, y, x, y, ?, ?, ? x, y, x
16. n^1, n^2, n^3, n^4, ?, ?, ? n^5, n^6, n^7
17. $x^2 + 2, x^4 + 4, x^6 + 6$, ?, ?, ?
18. $7x - 14, 6x - 12, 5x - 10$, ?, ?, ?
19. $x^2, x^4, x^6, x^8, x^{10}$, ?, ?, ?
20. 31, 27.5, 24, 20.5, ?, ?, ?
21. $21a - 5, 18a - 5, 15a - 5$, ?, ?, ? $12a - 5, 9a - 5, 6a - 5$
22. $8m - 6, 8m - 5, 8m - 4, 8m - 3$, ?, ?, ? $8m - 2, 8m - 1, 8m$
23. $m + 7, m + 10, m + 13, m + 16$, ?, ?, ? $m + 19, m + 22, m + 25$
24. $x + 5, 2x + 10, 3x + 15, 4x + 20$, ?, ?, ? $5x + 25, 6x + 30, 7x + 35$

25. Colleen's scores on her mathematics tests were low. By paying close attention in class, blocking out a fixed time each night for study, and being part of a mathematics study group she improved her next four test scores to 62, 64, 68, and 74. If this pattern continues, then what will be her scores on the next two tests? 82, 92

26. Leticia was a 400 m hurdler. She found that through proper training, proper eating, and a strong determination, her time for the event improved during her last three years of high school from 61.3 s to 60.1 s to 58.9 s. If this pattern continues, what will be her times during her first three years of college? 57.7 s, 56.5 s, 55.3 s

9. 12, 6, 3
10. $n - 4, n - 5, n - 6$
11. 17, 9, 0
12. $32x^5, 64x^6, 128x^7$
13. $x + 4, x + 5, x + 6$
14. 25, 36, 49
17. $x^8 + 8, x^{10} + 10, x^{12} + 12$
18. $4x - 8, 3x - 6, 2x - 4$
19. x^{12}, x^{14}, x^{16}
20. 17. 13.5, 10

50

Resource Book, MATHEMATICAL CONNECTIONS

Teaching the Lesson
- Lesson Starter 16
- Using Technology, p. 46C
- Reteaching/Alternate Approach, p. 46D
- Cooperative Learning, p. 46D
- Toolbox Skills 7–10

Lesson Follow-Up
- Practice 20
- Enrichment 17: Communication
- Study Guide, pp. 31–32
- Computer Activity 2

Warm-Up Exercises

Evaluate each expression when $x = 2$, $y = 6$, and $z = 8$.

1. $3x - 4$ 2
2. $xz + y$ 22
3. $x^2 + z^2$ 68
4. $(x + y)^2$ 64
5. $3xy + 2yz$ 132

Lesson Focus

Today's lesson focuses on relationships in which one quantity depends on another. To motivate this lesson on functions, you may want to use the application below.

Application

The Chun family is on vacation and they wish to rent a car. The sign at the rental agency states that the rental fees are $25 per day plus 13¢ per mile traveled. How much will it cost to rent the car? *The answer depends on how far the car travels.*

Teaching Notes

Functions may be a new concept for many students. Stress that each number in the first column is paired with *exactly* one number in the second column. Also, emphasize the terminology: *function*, *function rule*, *arrow notation*, and *function table*.

2-6   *Functions*

Objective: To complete function tables and to find function rules using tables.

Terms to Know
- *function*
- *function rule*
- *arrow notation*
- *function table*

EXPLORATION

1 Emily Scott bought a bouquet of flowers for $4. Can you tell how much change she received from the florist? Why or why not? *No; you do not know how much money she gave the florist.*

2 How much change would Emily receive if she gave the florist a $5 bill? a $10 bill? a $20 bill? *$1; $6; $16*

3 Complete this statement: The amount of change that Emily receives depends on ___?___. *the amount of money that she gives the florist*

There are many times when one quantity depends on another. For instance, the amount of change that you receive when you make a purchase depends on the amount of money that you give the salesperson. In mathematics, it is said that the amount of change is a *function* of the amount of money you give the salesperson. A **function** is a relationship that pairs each number in a given set of numbers with *exactly one* number in a second set of numbers.

Often you can describe a function by a **function rule.** For instance, suppose your purchase costs $4. If you give the salesperson x dollars, the amount of change you receive is $(x - 4)$ dollars. You can use the variable expression $x - 4$ to create a function rule.

You say: *x is paired with x − 4*
You use **arrow notation** to write: $x \rightarrow x - 4$

You can also use the variable expression to make a **function table,** like the one at the right.

x	$x - 4$
5	1
10	6
20	?
50	?
100	?

Example 1

Complete the function table shown above.

Solution

The function rule is $x \rightarrow x - 4$.
Substitute 20, 50, and 100 for x in the expression $x - 4$.

$x - 4 = 20 - 4 = 16$
$x - 4 = 50 - 4 = 46$
$x - 4 = 100 - 4 = 96$

x	$x - 4$
5	1
10	6
20	16
50	46
100	96

✓ **Check Your Understanding**

1. In Example 1, what number does the function pair with 50?
2. What number would be paired with 500 using the function rule in Example 1?

See *Answers to Check Your Understanding* at the back of the book.

You can sometimes use a function table to find a pattern that can help you write the function rule.

Example 2 Write a function rule for the function table shown at the right.

x	?
1	8
2	16
5	40
6	48
10	80

Solution Notice the pattern: Each number in the second column is eight times the number in the first column.

$8(1) = 8$ $\quad 8(2) = 16$ $\quad 8(5) = 40$ $\quad 8(6) = 48$ $\quad 8(10) = 80$

The function rule is $x \rightarrow 8x$.

Guided Practice

Is the first quantity a function of the second quantity? Write *Yes* or *No*.

1. the cost of a long-distance phone call; the time of day you place the call Yes.
2. the cost of a long-distance phone call; the color of your phone No.
3. the cost of mailing a package; the temperature No.
4. the cost of mailing a package; the weight of the package Yes.

COMMUNICATION « *Writing*

Write each function rule using arrow notation.

«5. x is paired with $2x + 15$ $\quad x \rightarrow 2x + 15$

«6. x is paired with $3x - 1$ $\quad x \rightarrow 3x - 1$

Write each function rule in words.

«7. $x \rightarrow x + 7$ $\quad x$ is paired with $x + 7$

«8. $x \rightarrow 7x$ $\quad x$ is paired with $7x$

«9. $x \rightarrow \frac{x}{3}$ $\quad x$ is paired with $\frac{x}{3}$

«10. $x \rightarrow 6x - 9$ $\quad x$ is paired with $6x - 9$

Choose the letter of the correct function rule for each function table.

A. $x \rightarrow x + 2$ $\quad$ **B.** $x \rightarrow 4x - 4$ $\quad$ **C.** $x \rightarrow 4x$

D. $x \rightarrow x + 4$ $\quad$ **E.** $x \rightarrow x^2$ $\quad$ **F.** $x \rightarrow 2x$

11. F

x	?
2	4
3	6
4	8
5	10
6	12

12. A

x	?
2	4
4	6
6	8
8	10
10	12

13. B

x	?
2	4
3	8
4	12
5	16
6	20

Key Question

How do the numbers in the second column of a function table depend on the numbers in the first column? The numbers in the second column are found by substituting each number in the first column into the function rule.

Critical Thinking

Ask students to *compare* the process of completing the right-hand side of a function table with the process of evaluating a variable expression.

Application

Functions are used often in areas other than mathematics. For example, library or rental agency fees can be represented by functions.

Using Technology

For a suggestion on how to use a BASIC program with this lesson, see page 46C.

Reteaching/Alternate Approach

For a suggestion on using a cooperative learning activity with this lesson, see page 46D.

Additional Examples

1. Complete the function table shown.

n	$n + 7$	
5	12	
10	17	
25	?	32
50	?	57
100	?	107

2. Write a function rule for the function table shown. $x \rightarrow 2x$

x	?
1	2
2	4
5	10
6	12
10	20

Closing the Lesson

Have students make up exercises similar to those in Examples 1 and 2 and then give them to their classmates. Ask a few students to describe how they would find the missing numbers and the rules.

Suggested Assignments

Skills: 1–7, 9, 10, 18–21
Average: 1–8, 9–17 odd, 18–21
Advanced: 1–21

Exercise Notes

Reasoning

In *Guided Practice Exercises 1–4*, students examine two quantities in order to determine whether or not one is a function of the other.

Communication: Writing

Guided Practice Exercises 5–10 require students to translate written expressions into mathematical symbols and vice versa.

Making Connections/Transitions

Exercises 9–12 illustrate the usefulness of the function concept in understanding a particular situation from science.

Problem Solving

Exercises 13–17 require students to write function rules to solve problems. Using a function rule to solve a problem is a problem-solving strategy that is used often.

14. Complete the function table.

x	x − 6	
10	4	
12	6	
14	?	8
16	?	10
18	?	12

15. Use the table to find the function rule. $x \rightarrow \frac{x}{7}$

x	?
7	1
14	2
21	3
28	4
35	5

Exercises

Complete each function table.

A **1.**

x	10x	
1	10	
2	20	
4	?	40
6	?	60
8	?	80

2.

x	$\frac{x}{6}$	
6	1	
12	2	
18	?	3
24	?	4
30	?	5

3.

x	2x + 9
2	13
4	17
21 6	?
25 8	?
29 10	?

Use each function table to find the function rule.

4. $x \rightarrow x + 3$

x	?
5	8
6	9
7	10
8	11
9	12

5. $x \rightarrow 5x$

x	?
1	5
2	10
3	15
4	20
5	25

6. $x \rightarrow x - 7$

x	?
8	1
10	3
12	5
14	7
16	9

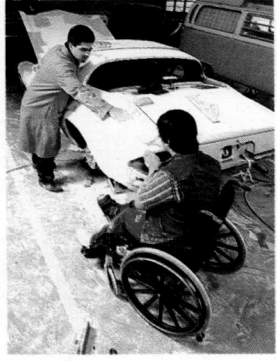

7. A person's weekly pay often depends on the number of hours the person works that week. Daniel earns $8.25 per hour. Use the function rule $t \rightarrow 8.25t$ to complete the function table at the right. Find Daniel's weekly pay for weeks that he works 7, 9, 12, 13, and 16 hours. $57.75, $74.25, $99, $107.25, $132

	t	8.25t
57.75	7	?
74.25	9	?
99	12	?
107.25	13	?
132	16	?

8. Suppose Daniel has received a raise and now earns $9.75 per hour. Make a function table to find his weekly pay for weeks that he works 5, 8, 12, 15, and 20 hours. $48.75, $78, $117, $146.25, $195

68   Chapter 2

CONNECTING MATHEMATICS AND SCIENCE

The number of times a cricket chirps per minute is thought to be a function of the air temperature. If c is the air temperature in degrees Celsius (°C), then the function rule for the number of chirps per minute is $c \rightarrow 7c - 30$.

B

9. The air temperature is 20°C. How many times per minute does a cricket chirp? 110 times

10. The air temperature is 5°C. How many times does a cricket chirp in 1 h? 300 times

11. How many more times per minute does a cricket chirp at 28°C than at 25°C? 21 times

12.

c	$7c - 30$
8	26
10	40
12	54
14	68
16	82
18	96

12. Make a function table for a cricket's chirps per minute when the air temperature is 8°C, 10°C, 12°C, 14°C, 16°C, and 18°C.

PROBLEM SOLVING/APPLICATION

The Center Town Library charges $.15 for each day a book is overdue.

C

13. What does it mean for a library book to be *overdue*?

13. The book has been kept out of the library past the day that its return was requested.

14. Write the function rule that describes the relationship between the number of days a book is overdue and the amount charged by the library. Represent the number of days by d and write the function rule using arrow notation. $d \rightarrow 0.15d$

15. Use the function rule from Exercise 14 to find the charge for a book that is three days overdue. $.45

16. A book that was due on Monday is returned on Friday. Use the function rule from Exercise 14 to find the overdue charge. $.60

17. RESEARCH Find out the amount that your school library or public library charges for overdue books. If possible, write a function rule to represent the overdue charge. Answers will vary.

SPIRAL REVIEW

S

18. Complete the function table at the right. *(Lesson 2-6)*
25; 37; 49

x	$6x - 5$
1	1
3	13
5	?
7	?
9	?

19. Find the quotient: $146.94 \div 6.2$ *(Toolbox Skill 14)* 23.7

20. Find the answer: 13^2 *(Lesson 1-9)* 169

21. Simplify: $8x - 3x + 5y$ *(Lesson 2-3)* $5x + 5y$

Introduction to Algebra **69**

Extension *(Cooperative Learning)*
Have students work in cooperative groups. Ask each group to develop a problem involving functions in real life. Problems could include such things as car rentals, taxi rides, car repairs, service repair calls, and so on. Each group can present its problem to the entire class, including the method used to research the information used in the problem.

Enrichment
Enrichment 17 is pictured on page 46E.

Practice
Practice 20 is shown below.

Practice 20, *Resource Book*

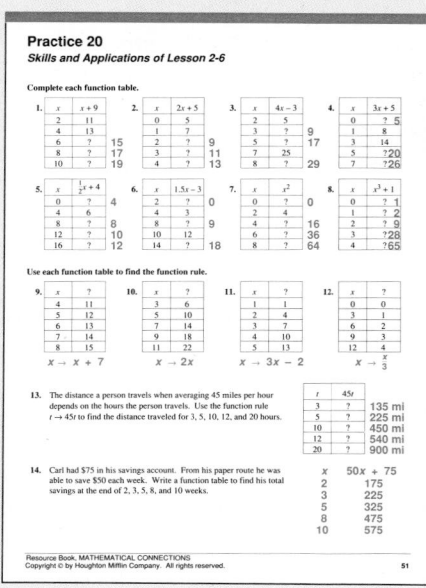

Spreadsheet software is becoming more and more prevalent in today's world, especially in business and industry. This Focus on Computers will help to familiarize students with the use of spreadsheet software.

Exercise Note

Critical Thinking

By using spreadsheets, students are asked to *analyze* a pattern of numbers and then to *formulate* a relationship between that pattern and an expression in another cell. Using their results, students are asked to *predict* the pattern for a related function and then to check their predictions by *creating* new function tables.

Suggested Assignment

Advanced: 1–9

Follow-Up

Enrichment

Have students create a table using the function rule $x \rightarrow ax + b$ for $x = 0, 1, 2, 3, 4,$ and 5. The entries in the second column will be $b, a + b, 2a + b, 3a + b, 4a + b,$ and $5a + b$. Ask students to find the differences between the entries in the second column. a

Have students repeat the activity for functions where a and b are known. Then reverse the process and have students find the function rule given the entries in the two columns.

Using Spreadsheet Software

Objective: To use a spreadsheet to analyze patterns in function tables.

Spreadsheet software is ideal for making function tables and for analyzing patterns among functions. The diagram below shows a portion of a spreadsheet displaying function tables for the function rules $x \rightarrow 2x + 5$ and $x \rightarrow 3x + 5$.

	A	B	C
1	x	2x + 5	3x + 5
2	1	7	8
3	2	9	11
4	3	11	14
5	4	13	17
6	5	15	20

Exercises

x	2x + 5	3x + 5
6	17	23
7	19	26
8	21	29
9	23	32
10	25	35

Answers to Exercises 4, 6–9 are on p. A4.

Use computer software to create a spreadsheet similar to the one above for the function rules $x \rightarrow 2x + 5$ and $x \rightarrow 3x + 5$. Extend your spreadsheet to include values for x from 1 to 10.

1. Using computer notation, write the formula in cell B4. 2 * A4 + 5

2. Using computer notation, write the formula in cell C10. 3 * A10 + 5

3. Describe in words the pattern of the numbers in cells B2 through B11.
add 2

4. What is the relationship between the pattern of numbers in cells B2 through B11 and the expression in cell B1?

5. Describe in words the pattern of the numbers in cells C2 through C11.
add 3

6. What is the relationship between the pattern of numbers in cells C2 through C11 and the expression in cell C1?

7. Use your results from Exercises 3–6. Predict what will be the pattern in a function table for the function rule $x \rightarrow 4x + 5$. Create a function table in your spreadsheet to verify your prediction.

Use computer software to create function tables on a spreadsheet for the function rules $x \rightarrow 5x + 1$ and $x \rightarrow 5x + 2$. Use values for x from 1 to 10.

x	5x + 1	5x + 2
1	6	7
2	11	12
3	16	17
4	21	22
5	26	27
6	31	32
7	36	37
8	41	42
9	46	47
10	51	52

8. Describe the patterns in the function tables for the two functions. How are the two function tables alike? How are they different?

9. Predict how the function table for the function rule $x \rightarrow 5x + 3$ will compare with the given functions. Create a function table in the spreadsheet to verify your prediction.

70 Chapter 2

2-7

Checking the Answer

Objective: To check the answer to a problem.

When you find an answer to a problem, it is important to check that your answer is correct, or at least reasonable. There are several ways to do this.

- Check your calculations. Compare your work to the wording of the problem to make sure you performed the operations correctly.
- Solve the problem again using an alternative method. Compare the new answer to your original answer to see if they match.
- Solve the problem again using estimation. Compare the estimate to your answer to see if your answer is reasonable.

Problem

Kelly earns $9.75 per hour. Last week he worked 8 h on Monday, 7 h on Tuesday, 5 h on Friday, and 9 h each on Wednesday and Thursday. Kelly said he earned $282.75 for the week. Is this correct? Explain.

Solution

You can use estimation to check Kelly's answer. Round the hourly wage to the leading digit. Then multiply by the total number of hours.

Hourly wage: $9.75 → $10

Number of hours: $8 + 7 + 5 + 9 + 9 = 38$

Multiply: $\begin{array}{r} 38 \\ \times 10 \\ \hline 380 \end{array}$ → about $380

Because $282.75 is not close to $380, the answer is not correct.

The hours worked on Thursday were not included in the total number of hours for the week.

Look Back Check the answer by using an alternate method to solve the problem. First multiply the number of hours for each day by $9.75. Then add these amounts to find the total pay for the week.
$370.50; so Kelly's answer of $282.75 is not correct.

Guided Practice

COMMUNICATION «*Discussion* Answers to Guided Practice 1–4 are on p. A4.

«**1.** Why is it important to check your answer?

«**2.** List two ways of checking an answer.

«**3.** What types of errors might someone make when solving a problem?

«**4.** Why should you answer a problem with a complete sentence?

Introduction to Algebra **71**

Lesson Planner

Teaching the Lesson
- Lesson Starter 17
- Toolbox Skills 7–14

Lesson Follow-Up
- Practice 21
- Practice 22: Nonroutine
- Enrichment 18: Problem Solving
- Study Guide, pp. 33–34
- Test 6

Warm-Up Exercises

Round to the leading digit.
1. $28.36 $30
2. $11.13 $10
3. $9.17 $9
4. $40.49 $40
5. $125.55 $100

Lesson Focus

Today's lesson uses estimation and other methods to check if an answer to a problem is reasonable. To motivate this lesson, you may want to use the application below.

Application

Erica spent $13.78 on Monday, $3.86 on Tuesday, and $14.19 on Wednesday. She said she had spent more than $30. Does the total amount seem reasonable?
Yes; the question can be answered by making an estimate.

Teaching Notes

After working through the Example with students, have them check the answer by using the method that is described in Look Back.

5. Bob bought two pairs of jeans for $24.90 each and one sweatshirt for
$32.95. Bob said that he spent $115.70. This is not correct. What error did Bob make?
   a. He multiplied $32.95 by 2.
   b. He forgot to multiply $24.90 by 2.
   c. He added before multiplying.

6. Lisa had $243.50 in her checking account on June 4. During June she
deposited $80 and wrote checks for $42.10 and $27.80. Lisa said her
new balance at the end of June was $93.60. Is this correct? Explain.
No; Lisa subtracted rather than added when she deposited $80.

Problem Solving Situations

1. Yes. Estimating gives
   20 + 7 + 3(3) = 36,
   which is close to $37.42.
2. No; Pearl did not include
   the one-point baskets.
3. $20.30
4. 13 students
5. $45.10
6. $469.50

1. Daniel spent $23.40 at the su-
permarket, $6.52 at the drug
store, and rented three video
tapes for $2.50 each. Daniel
said he spent $37.42. Is this
correct? Explain.

2. Pearl scored seven two-point
baskets in a basketball game.
In the next game, she scored
five two-point baskets and two
one-point baskets. Pearl said
she scored 24 points. Is this
correct? Explain.

Solve. Check your answer.

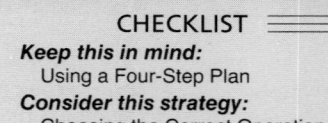

PROBLEM SOLVING

CHECKLIST

Keep this in mind:
Using a Four-Step Plan
Consider this strategy:
Choosing the Correct Operation

3. Chris exchanged two pairs of
shoes that cost $36.85 each for
one pair of boots that cost $65
and one pair of sneakers that
cost $29. How much more
money did Chris need to pay?

4. Ninety-one students volunteered to clean parks and visit nursing
homes. The students were evenly assigned to five parks and two
nursing homes. How many students were assigned to each place?

5. On March 2 Sue spent $11.60 for gasoline. She paid $13.25 for gasoline on March 10 and $87.43 for a tune-up on March 13. On March 21
Sue spent $9.75 for gasoline. She paid $10.50 for gasoline on March
29. How much did Sue spend for gasoline during March?

6. Paul earns $12.25 for each new cable TV customer he recruits, plus a
bonus of $8.50 for each subscriber to the arts channel. Last week Paul
recruited 30 new customers, 12 of whom subscribed to the arts channel. How much did Paul earn last week?

7. A math test contained two parts and was worth 100 points. Each item on Part A was worth eight points and each item on Part B was worth six points. Rosa got four items correct on Part A and eight items correct on Part B. What was Rosa's score on the test? 80 points

8. Trang wants to budget money for telephone service. His telephone bills for the last six months were $21.50, $18.78, $17.52, $22.15, $23.60, and $19.20. He assumes these amounts are typical. About how much should Trang expect to pay for telephone service for a year?
about $240

WRITING WORD PROBLEMS Answers to Problems 9–10 are on p. A4.

Using the information given, write a word problem for each exercise. Then solve the problem and check your answer.

9. 250 mi; 5 days **10.** a cassette costs $9.95; a compact disc costs $14.50

SPIRAL REVIEW

S 11. Replace the __?__ with >, <, or =: 0.309 __?__ 0.39   *(Lesson 1-5)* <

12. Estimate the difference: $76.81 - 39.5$ *(Toolbox Skill 1)* about 40

13. Jamal bought nine packs of baseball cards. Each pack cost $.79. How much did Jamal spend? Check your answer.   *(Lesson 2-7)* $7.11

14. Simplify: $(8a)(3a)$ *(Lesson 2-1)* $24a^2$

Self-Test 2

1. The price of a sofa is $389.99. It can be bought on an install-ment plan for a $50 down payment and 18 payments of $25. How much more does the sofa cost on the installment plan?   2-4
$110.01

Find the next three numbers in each pattern.

2. 5, 8, 11, 14, __?__, __?__, __?__   3. 1, 4, 16, 64, __?__, __?__, __?__   2-5
   17, 20, 23   256, 1024, 4096

4. Complete the function table.   5. Find the function rule. $x \rightarrow 7x$ 2-6

x	2x + 6
1	8
2	10
5	?	16
9	?	24
11	?	28

x	?
3	21
5	35
7	49
9	63
10	70

No; Marvin multiplied his 5 extra hours by $7.50 instead of $10.75.

6. Marvin earns $7.50 per hour. If he works more than 40 h per week, he earns $10.75 per hour for each extra hour. Last week Marvin worked 45 h. He says he earned $337.50. Is this cor-rect? Explain.   2-7

Introduction to Algebra   **73**

Suggested Assignments
Skills: 1, 2, 3–9 odd, 11–14
Average: 1–14
Advanced: 1–14

Exercise Notes

Communication: Discussion
Guided Practice Exercises 1–4 can be used as the basis for a class dis-cussion on answers to problems.

Communication: Writing
Problem Solving Situations 9 and 10 require students to use their own experiences and the given formula-tions to create a problem.

Follow-Up

Enrichment
Enrichment 18 is pictured on page 46F.

Practice
Practice 21 is shown below.

Quick Quiz 2
See page 91.

Alternative Assessment
See page 91.

Practice 21, *Resource Book*

Lesson Planner

Warm-Up Exercises

Estimate each answer.

1. $23.78 + 39.98$ about 60
2. $45.67 - 16.06$ about 30
3. 3×20.22 about 60
4. 5×18.32 about 100
5. 53.42×4 about 200

Lesson Focus

Today's lesson focuses on deciding when an exact answer is needed and when an estimate is appropriate.

Teaching Note

Key Question

Why is an estimate sufficient to determine how much money Ellen should take to the store in the Example?   Ellen needs to bring more money than the total amount of her purchase. This can be estimated.

Additional Example

Decide whether an *estimate* or an *exact answer* is needed. Then solve the problem.

Pacific West is having a sale on cassette tapes. The sale price is $6.98 each. Grace finds two tapes she wants to buy and has $17 in her pocket.

1. Does Grace have enough money with her to buy both tapes?
Estimate; $6.98 → about $7, $7 \cdot 2 = $14. Grace has enough money.

2-8 Finding an Estimate or an Exact Answer

Objective: To decide whether an estimate or an exact answer is needed to solve a problem.

Some real-life problems require only an *estimate* for a solution. Others require an *exact answer*. Before attempting to solve a problem, you should decide whether an estimate or an exact answer is needed.

Example

Decide whether an *estimate* or an *exact answer* is needed for parts (a) and (b). Then solve.

Best Clothes is having a sale on sweaters. The sale price is $17.95 each. Ellen wants to buy three sweaters.

a. How much money should Ellen take to the store to make sure that she has enough to pay for the sweaters?

b. What will be the total cost of the three sweaters excluding tax?

Solution

a. Ellen needs to *estimate* the amount of money to bring. To be sure that she has enough money, Ellen should *round up*.
cost of one sweater = $17.95 → about $20
$20 \cdot 3 = $60
Ellen should bring about $60.

b. Finding the total cost of the sweaters requires an *exact answer*.
$17.95 \cdot 3 = $53.85
The total cost of the sweaters excluding tax is $53.85.

 **Check Your Understanding**

In part(a) of the Example, why did Ellen round $17.95 up to $20 instead of up to $18?
See *Answers to Check Your Understanding* at the back of the book.

Guided Practice

1. estimate   2. exact answer   3. exact answer   4. estimate

Decide whether an *estimate* or an *exact answer* is needed in each situation.

1. the number of hours a trip will take

2. the amount an employee is paid

3. the width of a new window shade

4. the number of books in a library

Decide whether an *estimate* or an *exact answer* is needed. Then solve.

5. Ana Rivera earns $14.25 per hour as a researcher. She works 40 h per week. How much does she earn each week? exact answer; $570

« 6. COMMUNICATION « *Discussion*   Describe a recent situation in which you needed only an estimate to make a decision. Explain why you did not need an exact answer. Answers will vary.

Exercises

Decide whether an *estimate* or an *exact answer* is needed. Then solve.

A **1.** Eli wants to plant lettuce seeds in a garden. Packets of seeds cost $.85 each. How much money should Eli take to the store to buy six packets?
estimate; $6

2. Helen pumped 11.5 gal of gasoline into her car's gas tank. Gasoline costs $1.10 per gallon. How much did Helen pay for the gasoline?
exact answer; $12.65

3. On Monday, Sky Airlines flew 120 flights. Of these, 111 arrived on time. How many flights did not arrive on time?
exact answer; 9 flights

4. It is recommended that restaurant customers leave a tip of $.15 for each dollar spent. Cy's bill is $19.97. How much tip should he leave?
estimate; $3

5. The coach of a baseball team has $400 with which to buy equipment. Including tax, a shirt costs $8.95, a cap costs $4.95, and a bat costs $12.75. Can the coach buy a shirt and a cap for each of the 15 players?
estimate; Yes, the coach will spend about $210.

6. Use the information in Exercise 5. Can the coach buy a shirt, a cap, and a bat for each of the 15 players? exact answer; Yes, the coach will spend $399.75.

GROUP ACTIVITY 7–10. Answers will vary.

Your class decides to remodel your classroom by painting the walls and retiling the floor. Paint costs $12.50 per gallon, and each gallon covers 400 ft². Floor tile costs $1.27 per square foot.

B **7.** Find the length, width, and height of the classroom in feet. Use these values to find the areas of the floor and walls. Remember to subtract the areas of the doors and windows.

8. How many square feet of tile will be needed? What will this cost?

9. How many gallons of paint will be needed? What will this cost?

10. Give examples of when your group used estimates and when your group used exact answers.

SPIRAL REVIEW

S **11.** Find the product mentally: (25)(1)(6)(4) *(Lesson 1-7)* 600

12. Find the sum: $5\frac{2}{3} + 4\frac{3}{4}$ *(Toolbox Skill 18)* $10\frac{5}{12}$

Introduction to Algebra **75**

2. What will be the cost of the two tapes? Exact answer;
$6.98 · 2 = $13.96. The total cost, excluding tax, is $13.96.

Closing the Lesson

List the prices of a number of items students might buy on the chalkboard. Ask students to list what they think they can buy with $20. Then have the students find the actual costs to see if they spent more than $20.

Suggested Assignments

Skills: 1–12
Average: 1–12
Advanced: 1–12

Exercise Note

Critical Thinking

In *Guided Practice Exercises 1–5*, students need to read carefully and understand a problem before *deciding* whether an exact or estimated answer is appropriate.

Follow-Up

Enrichment

Enrichment 19 is pictured on page 46F.

Practice

Practice 23, *Resource Book*

Practice 23
Skills and Applications of Lesson 2-8

Decide whether an *estimate* or an *exact answer* is needed. Then solve.

1. At the Royal Pizza, Chen ordered a large pizza and had three different toppings added. The plain pizza cost $6.75 and each topping cost $.50. The tax was $.58. What did Chen pay? **exact; $8.83**

2. One of the tall buildings in the center of the city has 22 stories. The height of each story is 11 ft. How tall is the building? **estimate; about 220 ft**

3. Sheila and Todd emptied five baskets of golf balls during their practice swings at the golf range. Todd recalled that the first basket had sixty-four golf balls. How many golf balls did they hit during practice? **estimate; about 300**

4. Cosmo's bedroom has dimensions 12 ft by 15 ft. From floor to ceiling the measure is 8 ft. If one gallon of paint covers 325 ft², then how much should be budgeted to paint the walls of the room if a gallon of paint costs $18.95? **estimate; about $40.00**

5. Yoko's glasses cost $84. The second pair was at half price. What did Yoko pay for the two pairs of glasses? **exact; $126**

6. A new mathematics book costs $26.50. The school ordered sixty-five books. How much will these new books cost? **exact; $1722.50**

7. If ground beef sells for $1.89 per pound, can John buy $5\frac{1}{2}$ pounds of ground beef for $10.00? **exact; No.**

8. The distance from Brian's house to school is 1.4 mi. If Brian walks at a rate of 3.5 mi/h, can he walk to school in 30 min? **estimate; Yes.**

Teaching the Lesson
- Materials: metric measuring devices
- Lesson Starter 19
- Visuals, Folder C
- Toolbox Skill 15

Lesson Follow-Up
- Practice 24
- Enrichment 20: Exploration
- Study Guide, pp. 37–38

Warm-Up Exercises

Multiply.

1. 23×100 2300
2. 4.5×1000 4500
3. 3.6×0.01 0.036
4. 37×0.1 3.7
5. 4.8×0.001 0.0048

Lesson Focus

Ask students if they have ever used a 35 mm camera or bought a one liter bottle of juice. These are metric measurements. Today's lesson focuses on recognizing and using metric units of measure.

Teaching Notes

Point out the most frequently used metric units. Stress that the relationships among the prefixes are the same whether working with length, liquid capacity, or mass.

Key Question

How can you use the metric system chart to determine how many places to move the decimal when changing from one measure to another? The number of places you move the decimal point is the same as the number of rows in the chart from the given measure to the new measure.

2-9 The Metric System of Measurement

Objective: To recognize and to use metric units of measure.

Terms to Know
- *meter*
- *liter*
- *gram*

DATA ANALYSIS

Sam Foster has a recipe that calls for 0.25 liter of milk. However, Sam's measuring cup is marked in milliliters.

QUESTION How many milliliters is 0.25 liter?

To find the answer, Sam used the table below, which he found in a reference book.

Metric System of Measurement

Prefix and Meaning		Length	Liquid Capacity	Mass
kilo-	1000	kilometer (km)	kiloliter (kL)	kilogram (kg)
hecto-	100	hectometer (hm)	hectoliter (hL)	hectogram (hg)
deka-	10	dekameter (dam)	dekaliter (daL)	dekagram (dag)
	1	meter (m)	liter (L)	gram (g)
deci-	0.1	decimeter (dm)	deciliter (dL)	decigram (dg)
centi-	0.01	centimeter (cm)	centiliter (cL)	centigram (cg)
milli-	0.001	millimeter (mm)	milliliter (mL)	milligram (mg)

Note: The most frequently used metric units are in red.

The table shows that the three basic units of measure in the metric system are the **meter** (m) for length, the **liter** (L) for liquid capacity, and the **gram** (g) for mass. Units beginning with *kilo-* are the largest units in the table and units beginning with *milli-* are the smallest.

The table lists enough data so that you can change one unit of measure to another. Each unit in the table is 10 times as large as the unit immediately below it. For example, 1 cm is equal to 10 mm. Therefore, to change from a *larger* metric unit to a *smaller* metric unit, you multiply by 10, 100, 1000, and so on.

Example 1

Solution

Write 0.25 L in mL.

Liters are larger than milliliters. Multiply by 1000.
$0.25(1000) = 250$ ← You can use mental math.
$0.25 \text{ L} = 250 \text{ mL}$

 **Check Your Understanding**

1. Why do you multiply by 1000 in Example 1?
   See *Answers to Check Your Understanding* at the back of the book.

ANSWER 0.25 liter is equal to 250 milliliters.

In order to change from a *smaller* metric unit to a *larger* metric unit, you multiply by 0.1, 0.01, 0.001, and so on.

Example 2
Solution

a. Write 48 mm in cm.

a. Multiply by 0.1.
48(0.1) = 4.8
48 mm = 4.8 cm

b. Write 37.5 g in kg.

b. Multiply by 0.001.
37.5(0.001) = 0.0375
37.5 g = 0.0375 kg

 **Check Your Understanding**

2. In Example 2(b), how many places do you move the decimal point?
See *Answers to Check Your Understanding* at the back of the book.

Error Analysis
When converting units of measure, students may move the decimal in the wrong direction. Remind students that they need many small units to make one large unit.

Additional Examples
1. Write 0.43 m in cm.   43 cm
2. Write 34 g in kg.   0.034 kg
3. Write 47.6 mL in L.   0.0476 L

Closing the Lesson
Have a student give a measurement. Ask another student to describe how to convert the measurement to a different given unit. Then ask a third student to do the conversion.

Suggested Assignments
Skills: 1–20, 21–41 odd, 42, 43, 52–56
Average: 1–20, 21–45 odd, 52–56
Advanced: 1–19 odd, 20, 21–39 odd, 40–56

Exercise Notes

Communication: Reading
In *Guided Practice Exercises 1 and 2*, students are asked to connect words and their definitions.

Guided Practice

COMMUNICATION « *Reading*

Choose the correct phrase to complete each definition.

«1. A milligram is:
   a. one thousand grams
   b. one million grams
   c. one thousandth of a gram
   d. one millionth of a gram

«2. A kilometer is:
   a. one hundred meters
   b. one thousand meters
   c. one hundredth of a meter
   d. one thousandth of a meter

Multiply.

3. (2.43)(10) 24.3
4. (31.5)(100) 3150
5. (6.83)(1000) 6830

6. (4.4)(1000) 4400
7. (13.7)(0.1) 1.37
8. (11.8)(0.01) 0.118

9. (12.5)(0.01) 0.125
10. (6.19)(0.001) 0.00619
11. (58.6)(0.001) 0.0586

Write each measure in the unit indicated.

12. 350 cm; m 3.5 m
13. 6.2 L; mL 6200 mL
14. 0.75 kg; g 750 g

15. 12.5 mm; cm 1.25 cm
16. 735 mL; L 0.735 L
17. 65 m; km 0.065 km

Exercises

Write each measure in the unit indicated.

A
1. 30 mm; cm 3 cm
2. 4150 mL; L 4.15 L
3. 450 m; km 0.45 km

4. 2000 mg; g 2 g
5. 25 g; mg 25,000 mg
6. 400 cm; m 4 m

7. 615 mm; m 0.615 m
8. 1.2 m; mm 1200 mm
9. 2345 mL; L 2.345 L

10. 25 m; cm 2500 cm
11. 3.4 L; mL 3400 mL
12. 45.2 g; kg 0.0452 kg

13. 0.74 m; cm 74 cm
14. 0.07 L; mL 70 mL
15. 0.098 km; m 98 m

16. 21.6 cm; mm 216 mm
17. 0.88 km; m 880 m
18. 5.6 g; mg 5600 mg

Introduction to Algebra   77

Data Analysis
Exercise 20 asks students to find
information from a table and write
it using a designated unit of measure.

Mental Math
Exercises 27–35 can be done easily
using mental math even though they
involve multiplying decimals. When
multiplying by 0.1, 0.01, 0.001,
and so on, mental math is an effective technique.

Number Sense
Exercises 36–41 are designed to
help students develop their number
sense in relation to metric measures.

Application
Exercises 42–45 illustrate a realistic
use of the metric system. You may
wish to have students list some
other careers in which the metric
system is used.

Research
Exercises 46–51 present some very
large and very small metric units.
Further research may be done by
students to find the fields in which
these terms are used.

19. The mass of a hummingbird is about 0.002 kg. About how many grams is the mass of a hummingbird? 2 g

B **20.** **DATA** pages 46–47 The proposed space station *Freedom* will orbit Earth at a height of how many meters? 483,000 m

Replace each __?__ with >, <, or =.

21. 540 mL __?__ 0.54 L =   **22.** 0.87 km __?__ 87 m >

23. 0.25 g __?__ 2500 mg <   **24.** 34 cm __?__ 0.034 m >

25. 907 g __?__ 0.0907 kg >   **26.** 125 mm __?__ 12.5 cm =

MENTAL MATH

You can use mental math to multiply numbers by 0.1, 0.01, 0.001, and so on. You simply move the decimal point of the number being multiplied to the left the same number of decimal places as there are in the number by which you are multiplying. For instance, 0.01 has two decimal places, so $(231.4)(0.01) = 2.314$.

Find each product mentally.

27. $(43.6)(0.1)$ 4.36   **28.** $(764.4)(0.01)$ 7.644   **29.** $(57.8)(0.01)$ 0.578

30. $(891.3)(0.001)$ 0.8913 **31.** $(8.09)(0.1)$ 0.809   **32.** $(0.56)(0.1)$ 0.056

33. $(1.36)(0.01)$ 0.0136   **34.** $(3.08)(0.001)$ 0.00308 **35.** $(24.5)(0.001)$ 0.0245

NUMBER SENSE

It is sometimes necessary to estimate measures. You can make more reasonable estimates using familiar measures as references. For instance, a
dime is about one millimeter thick, a fingernail is about one centimeter
wide, and the length of twelve city blocks is about one kilometer.

Select the most reasonable measure for each item.

36. length of a soccer field
   a. 100 cm   **b.** 100 m   **c.** 100 km

37. height of a person
   a. 175 mm   **b.** 175 cm   **c.** 175 m

38. width of a computer screen
   a. 23 cm   **b.** 23 m   **c.** 23 km

39. distance from New York to London
   a. 5567 mm   **b.** 5567 cm   **c.** 5567 km

40. The normal mass of an adult male is about 70 kg. Use this fact to estimate the normal mass of a five-year-old child. A reasonable answer is 17.5 kg.

41. The capacity of a bucket is about 8 L. Use this fact to estimate the capacity of a drinking glass. A reasonable answer is 0.25 L.

CAREER/APPLICATION

Automotive technicians maintain and repair cars. They must be familiar with metric units because many cars are manufactured using parts with metric measures.

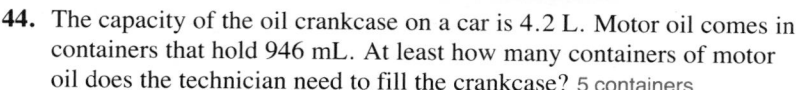

42. The measure of the head of the oil drain plug on a car is 2.2 cm. Write this measure in millimeters. 22 mm

43. A radiator holds 17.1 L of coolant. Write this measure in milliliters. 17,100 mL

44. The capacity of the oil crankcase on a car is 4.2 L. Motor oil comes in containers that hold 946 mL. At least how many containers of motor oil does the technician need to fill the crankcase? 5 containers

45. The tread on an old tire is 185 mm wide. The tread on a new tire is 20.5 cm wide. How many centimeters wider is the tread on the new tire than on the old tire? 2 cm

RESEARCH **46.** one billion meters **47.** one-millionth of a meter
49. one-billionth of a meter
Find the meaning of the metric unit used in each exercise.

C **46.** In June 1989, an asteroid passed within 16 gigameters of Earth.

47. Scientists can make a scratch in aluminum 1 micrometer wide.

48. A transistor can open and close a pathway of electrons in 0.01 nanosecond. one-billionth of a second

49. Ultraviolet light has a wavelength of 300 nanometers.

50. Laser pulses last about 1 picosecond. one-trillionth of a second

51. Soil surrounding the roots of one giant redwood tree can hold 0.5 megaliter of water. one million liters

SPIRAL REVIEW

S **52.** Complete the function table at the right. *(Lesson 2-6)*

53. Write 15 lb in oz. *(Toolbox Skill 22)* 240 oz

54. Round to the nearest hundredth: 12.8728 *(Toolbox Skill 6)* 12.87

55. Simplify: $5(2x - 4)$ *(Lesson 2-2)* $10x - 20$

56. Write 187.4 mm in cm. *(Lesson 2-9)* 18.74 cm

x	$x + 5$
2	7
4	9
6	?
8	?
10	?

11
13
15

Introduction to Algebra **79**

Extension

A metric scavenger hunt may be used to further solidify students' understanding of the metric system. Give students different metric measures, such as 1 m, 25 cm, 47 mm, 3 g, 6 kg, and so on. Ask them to name an object in the room that they think fits the measure. After completing all estimates, students are to measure each of the objects that they named and find the difference between the estimated measure and the actual measure. This activity can be reversed. Objects can be named first. Students are then asked to estimate different measures and then follow the process above.

Enrichment

Enrichment 20 is pictured on page 46F.

Practice

Practice 24 is shown below.

Practice 24, *Resource Book*

Practice 24
Skills and Applications of Lesson 2-9

Write each measure in the unit indicated.

1. 100 cm; m 1 m
2. 1 kL; L 1000 L
3. 0.50 L; mL 500 mL
4. 100 mg; g 0.1 g
5. 5000 L; kL 5 kL
6. 525 mm; m 0.525 m
7. 75 g; kg 0.075 kg
8. 12.5 cm; mm 125 mm
9. 5 L; mL 5000 mL
10. 0.9 m; cm 90 cm
11. 400 cm; mm 4000 mm
12. 8.5 g; mg 8500 mg
13. 10 km; m 10,000 m
14. 0.03 kg; g 30 g
15. 458 mm; cm 45.8 cm
16. 6000 mL; L 6 L
17. 3 kg; g 3000 g
18. 15 m; cm 1500 cm
19. 1 L; kL 0.001 kL
20. 35 cm; m 0.35 m
21. 4.1 kg; g 4100 g
22. 24,580 mm; m 24.58 m
23. 5000 mL; L 5 L
24. $\frac{1}{2}$ g; mg 500 mg
25. 6587 mg; g 6.587 g
26. 0.001 kL; L 1 L
27. 4 m; mm 4000 mm
28. 15 m; km 15,000 m
29. 860,000 mm; m 860 m
30. 5876 mg; g 5.876 g
31. 4 L; mL 4000 mL
32. 585 cm; m 5.85 m
33. 432 m; km 0.432 km
34. 0.05 g; mg 50 mg
35. 9435 mm; m 9.435 m
36. 795 L; kL 0.795 kL

37. The width of a paper clip is about one centimeter. How many millimeters is the width of a paper clip? 10 mm

38. If four encyclopedias placed end to end measure one meter, then how many centimeters does each encyclopedia measure? 25 cm

39. Your mathematics textbook has a mass of approximately 0.75 kg. About how many grams is the mass of the book? 750 g

Teaching the Lesson
- Materials: calculator
- Lesson Starter 20
- Toolbox Skill 15

Lesson Follow-Up
- Practice 25
- Practice 26: Nonroutine
- Enrichment 21: Data Analysis
- Study Guide, pp. 39–40
- Test 7

Warm-Up Exercises

Find each answer.
1. 10^3 1000
2. 10^5 100,000
3. 4^5 1024
4. 7^3 343
5. 10^7 10,000,000

Lesson Focus

Ask students how they would find the product (450,000)(12,000,000), or (12,500,000)(356,000,000). The answer might not fit on a standard calculator. When scientists and engineers calculate with greater numbers, they use scientific notation, the focus of today's lesson.

Teaching Notes

Have students work through the Exploration to discover the relationship between the exponent and the number of places the decimal point moves.

Key Question

How is a number written in scientific notation?   The number is written as a number that is at least 1 but less than 10 multiplied by a power of ten.

2-10

Scientific Notation

Objective: To write numbers in scientific notation.

Terms to Know
- *scientific notation*
- *decimal notation*

EXPLORATION

Look at the following pattern.

$$9.3 \times 10^1 = 9.3 \times 10 = 93$$
$$9.3 \times 10^2 = 9.3 \times 100 = 930$$
$$9.3 \times 10^3 = 9.3 \times 1000 = 9300$$
$$9.3 \times 10^4 = 9.3 \times 10,000 = 93,000$$

1 Replace each __?__ with the number that makes the statement true.
   a. $9.3 \times 10^5 = 9.3 \times 100,000 = $ __?__ 930,000
   b. $9.3 \times 10^6 = 9.3 \times$ __?__ $= 9,300,000$ 1,000,000
   c. $9.3 \times 10^? = 9.3 \times 10,000,000 = 93,000,000$ 7

2 Write the number represented by 9.3×10^8. What is the relationship between the exponent of 10 and the number of places the decimal point in 9.3 is moved to the right? 930,000,000; They are both 8.

Greater numbers can be difficult to read and to write. Scientists and other people who use these numbers often write them in *scientific notation*. A number is written in **scientific notation** when it is written as a number that is at least one but less than ten multiplied by a power of ten.

$$6 \times 10^5$$
at least 1, but less than 10 ⟷ 2.3×10^8 ⟷ a power of 10
$$1.52 \times 10^9$$

Example 1

Write each number in scientific notation.
   a. 13,000   b. 24,500,000

Solution

Move the decimal point to get a number that is at least 1, but less than 10.
   a. $13,000 = 1.3 \times 10^4$

   4 places

   b. $24,500,000 = 2.45 \times 10^7$

   7 places

☑ **Check Your Understanding**

1. In Example 1(a), why do you multiply by 10^4?
   See *Answers to Check Your Understanding* at the back of the book.

A number such as 24,500,000 is said to be in **decimal notation.** Numbers written in scientific notation can be rewritten in decimal notation.

Example 2

Solution

Write each number in decimal notation.

a. 7.16×10^5 **b.** 5.2×10^6

Move the decimal point to the right.

a. $7.16 \times 10^5 = 716{,}000$ **b.** $5.2 \times 10^6 = 5{,}200{,}000$

          5 places   6 places

☑ **Check Your Understanding**

2. In Example 2(b), why do you move the decimal point 6 places?

3. Describe how Example 2(a) would be different if the number were 7.16×10^2.

See *Answers to Check Your Understanding* at the back of the book.

 Some calculators display greater numbers in scientific notation. For instance, a calculator will display 480,000,000,000 as $\boxed{4.8 \ 11}$.

Many calculators have an $\boxed{EE}$ or $\boxed{EXP}$ key. You use this key to enter numbers in scientific notation. For instance, to enter 4.8×10^{11} you use this key sequence.

$\boxed{4.8}$ 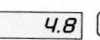 $\boxed{11}$

Guided Practice

COMMUNICATION «*Reading*

1. A number that is written as a number that is at least one but less than ten multiplied by a power of ten.

Refer to the text on pages 80–81.

«**1.** Describe the form of a number written in scientific notation.

«**2.** List at least two reasons for writing numbers in scientific notation.
Greater numbers can be difficult to read and to write.

Is each number written in scientific notation? Write *Yes* or *No*.

3. 5.8×10^9 **4.** 1.07×5^{10} **5.** 0.64×10^7 No.   **6.** 12.7×10^6
   Yes.   No.   No.

Replace each __?__ with the number that makes the statement true.

7. $560 = 5.6 \times 10^?$ 2   **8.** $7000 = 7 \times 10^?$ 3

9. $67{,}500 = 6.75 \times 10^?$ 4   **10.** $\underline{\ ?\ } \times 10^4 = 24{,}000$ 2.4

11. $\underline{\ ?\ } \times 10^3 = 2758$ 2.758   **12.** $\underline{\ ?\ } \times 10^5 = 900{,}000$ 9

Write each number in scientific notation.

13. 34,000   **14.** 150,000   **15.** 600,000 6×10^5 **16.** 1,420,000
   3.4×10^4 1.5×10^5 1.42×10^6

Write each number in decimal notation.

17. 4.2×10^3 **18.** 1.65×10^5 **19.** 2.173×10^8 **20.** 8×10^9
   4200   165,000   217,300,000   8,000,000,000

Exercise Notes *(continued)*

Number Sense
Exercises 29–34 help students to develop their number sense when using scientific notation. These exercises require students to examine two numbers in scientific notation and then draw a conclusion as to the relative size of the numbers.

Critical Thinking
Exercises 35–38 require students to use previously acquired *knowledge* to complete and *compare* the results of two computations. Students must then *formulate* a rule for multiplying numbers in scientific notation.

Follow-Up

Extension
You may want to have students research subjects that use greater numbers. Have students write a report on their findings, identifying the source.

Enrichment
Enrichment 21 is pictured on page 46F.

Practice
Practice 25, *Resource Book*

Practice 25
Skills and Applications of Lesson 2-10

Write each number in scientific notation. 1–16. See below.
1. 8000   2. 5600   3. 10,000   4. 8300
5. 3,000,000   6. 900,000   7. 760,000   8. 46,000,000
9. 63,500   10. 480,000   11. 3,780,000   12. 637,000
13. 47,000,000   14. 6,350,000   15. 695,000   16. 14,580,000

Write each number in decimal notation. 17–36. See below.
17. 7×10^3 18. 6.5×10^4 19. 5.8×10^5 20. 8.35×10^6
21. 1.15×10^6 22. 9.2×10^7 23. 6.25×10^6 24. 7.35×10^4
25. 3.5×10^9 26. 4.853×10^8 27. 1.05×10^5 28. 9.345×10^6
29. 2.85×10^7 30. 5.65×10^6 31. 2.985×10^7 32. 1.095×10^9
33. 1.11×10^4 34. 5.55×10^6 35. 9.05×10^5 36. 8.235×10^7

37. A Wall Street report listed the selling price of a "hot stock." The total shares of the stock sold were 2,295,000. Write this number in scientific notation. 2.295×10^6

38. One of the leading money winners in stock car racing had earned $783,000 with the season only at the halfway point. Write his winnings in scientific notation. 7.83×10^5

39. The diameter of Earth's orbit is about 186,000,000 mi. Write this figure in scientific notation. 1.86×10^8

1. 8×10^3 2. 5.6×10^3 3. 1×10^4 4. 8.3×10^3
5. 3×10^6 6. 9×10^5 7. 7.6×10^5 8. 4.6×10^7
9. 6.35×10^4 10. 4.8×10^5 11. 3.78×10^6 12. 6.37×10^5
13. 4.7×10^7 14. 6.35×10^6 15. 6.95×10^5 16. 1.458×10^7

17. 7000   18. 65,000   19. 580,000   20. 8,350,000
21. 1,150,000   22. 92,000,000   23. 6,250,000   24. 73,500
25. 3,500,000,000   26. 485,300,000   27. 105,000   28. 9,345,000
29. 28,500,000   30. 5,650,000   31. 29,850,000   32. 1,095,000,000
33. 11,100   34. 5,550,000   35. 905,000   36. 82,350,000

Exercises

1. 1.2×10^6

Write each number in scientific notation.

A 1. 1,200,000   2. 6000 6×10^3 3. 5700 5.7×10^3 4. 254,000 2.54×10^5

5. 45,200 4.52×10^4 6. 95,000,000 9.5×10^7 7. 851,400,000 8.514×10^8 8. 3,680,000 3.68×10^6

Write each number in decimal notation.

9. 3.8×10^3 3800   10. 7×10^5 700,000   11. 1.52×10^6 1,520,000   12. 7.5×10^4 75,000

13. 5×10^8 500,000,000   14. 6.15×10^9 6,150,000,000   15. 3.425×10^5 342,500   16. 1.06×10^7 10,600,000

17. Scientists use solar telescopes to study the sun. The diameter of the sun is approximately 1,392,000 km. Write this number in scientific notation. 1.392×10^6 km

18. The distance that light travels in one year is approximately 9.46×10^{15} m. Write this number in decimal notation. 9,460,000,000,000,000 m

DATA, *pages 46–47*

Write in decimal notation the average distance from each planet to the sun.

19. Mercury 36,000,000 mi   20. Venus 67,270,000 mi   21. Saturn 887,140,000 mi   22. Pluto 3,675,270,000 mi

CALCULATOR

For each calculator display, write the number first in scientific notation and then in decimal notation. 7.9×10^9; 7,900,000,000

23. | 3.5 11 | 3.5×10^{11}; 350,000,000,000   24. | 7.9 09 |

25. | 4.853 07 | 4.853×10^7; 48,530,000   26. | 1.963 10 | 1.963×10^{10}; 19,630,000,000

Choose the key sequence you would use to enter each number on a calculator.

27. 1.2×10^{12} D   A. | 1.4 | [EE] | 14 |

28. 1.4×10^{14} A   B. | 12 | [EE] | 1.2 |

C. | 14 | [EE] | 1.4 |

D. | 1.2 | [EE] | 12 |

NUMBER SENSE

Without writing in decimal notation, **replace each __?__ with >, <, or =.**

29. 4×10^5 __?__ 4×10^2 >   30. 6×10^6 __?__ 6×10^8 <

31. 1.7×10^5 __?__ 3.2×10^5 <   32. 8×10^3 __?__ 7×10^3 >

33. 0×10^7 __?__ 2×10^4 <   34. 0×10^4 __?__ 0×10^6 =

B

Kitt Peak National Observatory, Arizona

82 Chapter 2

THINKING SKILLS 35–36. Knowledge   **37.** Analysis   **38.** Synthesis

Replace each ? with the number that makes the statement true.

C **35.** $(2 \times 10^2)(4 \times 10^3) = (2 \times \underline{\ ?\ })(10^2 \times 10^3) = 8 \times 10^?$ 4; 5

    36. $(3 \times 10^2)(5 \times 10^4) = (3 \times 5)(10^? \times 10^4)$
                          $= 15 \times 10^6 = 1.5 \times 10^7$ 2

    37. Compare Exercises 35 and 36. How are they alike? How are they different? See answer on p. A4.

    38. Develop a method for multiplying numbers that are written in scientific notation. Use your method to find each product.
        a. $(3 \times 10^4)(2 \times 10^7)$ 6×10^{11}
        b. $(6 \times 10^2)(8 \times 10^5)$ 4.8×10^8

SPIRAL REVIEW

S **39.** Find the product: $2\frac{1}{3} \times 3\frac{1}{4}$ *(Toolbox Skill 20)* $7\frac{7}{12}$

    40. Evaluate $3.65 + m$ when $m = 4.8$.   *(Lesson 1-2)* 8.45

    41. Write 42,540,000 in scientific notation.   *(Lesson 2-10)* 4.254×10^7

    42. Is it most efficient to find the quotient $1400 \div 20$ by using *paper and pencil, mental math*, or a *calculator*?   *(Lesson 1-8)* mental math

    43. Simplify: $5a - 2a + 7b$ *(Lesson 2-3)* $3a + 7b$

    44. Lou works five days per week, eight hours per day. She earns $11.50 per hour. How much does Lou earn per week?   *(Lesson 2-4)* $460

Self-Test 3

Decide whether an *estimate* or an *exact answer* is needed. Then solve.

  1. Grapes cost $1.49 per pound. Sergei needs to buy 12 lb of grapes for a fruit platter. He has $20. Does he have enough money to buy 12 lb of grapes? estimate; Yes he needs about $18.   2-8

Write each measure in the unit indicated.

  2. 480 mm; cm 48 cm   **3.** 3.2 L; mL 3200 mL   **4.** 0.5 kg; g   500 g   2-9

Write each number in scientific notation.

  5. 44,000 4.4×10^4 **6.** 9,870,000   **7.** 21,300,000   2-10
                               9.87×10^6 2.13×10^7

Write each number in decimal notation.

  8. 5.6×10^3 5600   **9.** 3.78×10^5 378,000 **10.** 6.43×10^7 64,300,000

Introduction to Algebra **83**

Converting Customary Units to Metric Units

Objective: To convert customary units to metric units.

Because most other countries use the metric system, many United States companies convert the customary measurements of their products into metric units before exporting them. You can do this type of conversion using a chart like the one below.

When You Know	Multiply By	To Find
inches	2.54	centimeters
feet	0.31	meters
yards	0.91	meters
miles	1.61	kilometers
ounces	28.35	grams
pounds	0.454	kilograms
fluid ounces	29.573	milliliters
pints	0.473	liters
quarts	0.946	liters
gallons	3.785	liters

Example

Anderson Architects designs buildings worldwide. They recently designed a house that might be built in both the United States and Canada. The United States plans show that the length of the house is 57 ft and the width of the house is 33 ft. What are the length and width in meters that should be shown on the Canadian plans?

Solution

To convert from feet to meters, multiply by 0.31.

$$57(0.31) = 17.67$$
$$33(0.31) = 10.23$$

Estimate to check the reasonableness of these answers.

$$57(0.31) \rightarrow 60(0.3) = 18$$
$$33(0.31) \rightarrow 30(0.3) = 9$$

The answers 17.67 and 10.23 are close to the estimates of 18 and 9, so 17.67 and 10.23 are reasonable.

On the Canadian plans, the length should be 17.67 m and the width should be 10.23 m.

Exercises

Write each measure in the unit indicated.

1. 6 pt; L 2.838 L
2. 27 mi; km 43.47 km
3. 62.5 in.; cm 158.75 cm
4. 9.4 qt; L 8.8924 L

ESTIMATION 5. about 210 mL

Estimate each measure in the unit indicated.

5. 7 fl oz; mL   6. 5 ft; m   7. 98 lb; kg   8. 21 oz; g

about 50 kg   about 630 g

about 1.5 m

9. Which metric measure is approximately equal to one yard? meter

10. Which metric measure is approximately equal to one quart? liter

11. Which customary measure is approximately equal to four liters? gallon

12. Which customary measure is approximately equal to two and one half centimeters? inch

PROBLEM SOLVING/APPLICATION   Answers will vary, depending on method of solution and conversion factor used.

13. According to the owner's manual, a car has a capacity for 4.3 qt of oil and 6.9 pt of transmission fluid. The car is to be exported to Mexico. What capacities in liters should be listed in the owner's manual?
4.0678 L; 3.2637 L

14. A set of plans for a house indicates that its width is 28 ft and its length is 42 ft. The length of the driveway is 19 yd. What are these dimensions in meters? 8.68 m; 13.02 m; 17.29 m

15. The plans for a new medical building show that the dimensions of the lobby will be 16 ft 9 in. by 14 ft 3 in. What dimensions should the plans show in meters? 5.1925 m; 4.4175 m

16. The label on a jar of fruit juice indicates that the capacity is 2 qt 10 fl oz. What is the capacity of the jar in milliliters? 2188.402 mL

17. An adjustable wrench is advertised for use with bolts that measure from 0.25 in. through 1.75 in. The wrench is being exported to South America. How should the sizes be advertised in millimeters?
6.35 mm; 44.45 mm

18. The label on a package of modeling clay indicates that the weight is 12 lb 6 oz. What is the mass of the package in kilograms? 5.61825 kg

19. RESEARCH   When was the metric system created? Name two advantages of using the metric system. See answer on p. A4.

Introduction to Algebra   85

Closing the Lesson
Have students measure objects using customary units. Then have them convert these measures to metric units. Have students check their conversions by measuring the objects in metric units.

Suggested Assignments
Skills: 1–12
Average: 1–17 odd
Advanced: 9–19

Exercise Notes

Estimation
Exercises 5–12 give students a better understanding of the size of some metric measures by having them make estimates. You may wish to mention that 7 fl oz is about a glass of water, that 5 ft is about some students' height, and that 98 lb is about some students' weight.

Problem Solving
Exercises 13–18 require students to solve problems by using information from the table on page 84. Using a table or chart is a common problem-solving strategy.

Research
Exercise 19 requires students to research the history of the metric system and to explore the advantages of its use.

Follow-Up

Extension
Have students create a table to convert metric units to customary units. The students should use their tables to convert some metric units to customary units.

Chapter Review

Terms to Know

product of powers rule (p. 48)
simplify (p. 48)
distributive property (p. 52)
like terms (p. 56)
unlike terms (p. 56)
combining like terms (p. 56)
function (p. 66)
function rule (p. 66)

arrow notation (p. 66)
function table (p. 66)
meter (p. 76)
gram (p. 76)
liter (p. 76)
scientific notation (p. 80)
decimal notation (p. 80)

Choose the correct term from the list above to complete each sentence.

1. A relationship that pairs each number in a given set of numbers with exactly one number in a second set of numbers is called a(n) __?__. function

2. The __?__ allows you to multiply each term inside parentheses by a factor outside the parentheses. distributive property

3. The expressions $7x$ and $4z$ are called __?__. unlike terms

4. The basic unit of mass in the metric system is the __?__. gram

5. The number 3.6×10^5 is written in __?__. scientific notation

6. $x \rightarrow 2x + 1$ is written in __?__. arrow notation

---

Simplify. *(Lessons 2-1, 2-2, 2-3)*   12. $17a$ 15. b^{12} 18. $24c - 72$ 21. $48wx^2$

7. $a^5 \cdot a$ a^6

8. $c^4 \cdot c^2 \cdot c^7$ c^{13}

9. $6(4x + 5)$ $24x + 30$

10. $2(12 + x)$ $24 + 2x$

11. $8n + 7m + 16n$ $24n + 7m$

12. $11a + 6a$

13. $9x + 12 + x$ $10x + 12$

14. $(5a)(11b)$ $55ab$

15. $b^5 \cdot b \cdot b^6$

16. $x^7 \cdot x^9$ x^{16}

17. $9(6 - z)$ $54 - 9z$

18. $8(3c - 9)$

19. $3x^2 \cdot 5x^6$ $15x^8$

20. $17c - 8c$ $9c$

21. $(8w)(3x)(2x)$

22. $15 + 13z - z$ $15 + 12z$

23. $18d - 11d - 5c$ $7d - 5c$

24. $6c \cdot 7c^3$ $42c^4$

Use the distributive property to find each answer mentally. *(Lesson 2-2)*

25. $6(23) + 6(7)$ 180

26. $4 \cdot 48 - 4 \cdot 8$ 160

27. $8(104)$ 832

28. $3(94)$ 282

Solve. *(Lesson 2-4)*

29. Laurie bought two sets of screwdrivers at $19.49 each, three boxes of screws at $4.98 each, and a drill for $39.95. What was the total cost of Laurie's purchase? $93.87

Find the next three numbers or expressions in each pattern.
(Lesson 2-5)

30. 3, 10, 17, 24, __?__, __?__, __?__ 31, 38, 45

31. 2, 6, 18, 54, __?__, __?__, __?__ 162, 486, 1458

32. $a + 6, a + 7, a + 8,$ __?__, __?__, __?__ $a + 9, a + 10, a + 11$

33. $38m, 35m, 32m,$ __?__, __?__, __?__ $29m, 26m, 23m$

Complete each function table. *(Lesson 2-6)*

34.

x	x − 8
12	4
16	8
20	? 12
24	? 16
28	? 20

35.

x	9x
2	18
4	36
6	? 54
8	? 72
10	? 90

Find each function rule. *(Lesson 2-6)*

36.

x	?
3	1
6	2
9	3
12	4
15	5

$x \to \frac{x}{3}$

37.

x	?
5	10
10	15
15	20
20	25
25	30

$x \to x + 5$

Tell whether the final statement is correct. Explain. *(Lesson 2-7)*

38. The mortgage loan, property tax, and maintenance fee on Myeesha's condominium total $1135 per month. The maintenance fee is $110 per month and the property tax is $75 per month. Myeesha said that her mortgage loan is $950 per month. Yes.

Decide whether an *estimate* or an *exact answer* is needed. Then solve.
(Lesson 2-8)

39. Tim charges $1.50 per page to type term papers. How much will he charge to type a 33-page paper? exact answer; $49.50

Write each measure in the unit indicated. *(Lesson 2-9)*

40. 1700 g; kg
1.7 kg

41. 750 mL; L
0.75 L

42. 0.76 km; m
760 m

43. 1.9 m; cm
190 cm

Write each number in scientific notation. *(Lesson 2-10)*

44. 350,000
3.5×10^5

45. 12,700
1.27×10^4

46. 6,550,000
6.55×10^6

47. 48,000,000
4.8×10^7

Write each number in decimal notation. *(Lesson 2-10)*

48. 1.3×10^5
130,000

49. 9.7×10^7
97,000,000

50. 2.64×10^4
26,400

51. 5.88×10^8
588,000,000

Exercises 13 and 14 are marked with stars to indicate that they represent alternative forms of assessment. Both exercises test students' understanding of unlike terms and thus, indirectly, of like terms.

Chapter Test (Form A)

Test 8, *Tests*

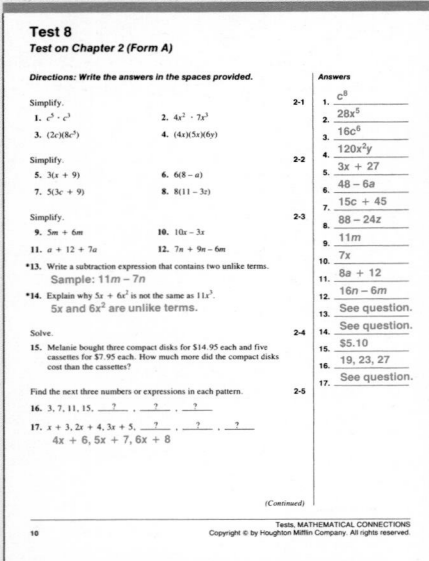

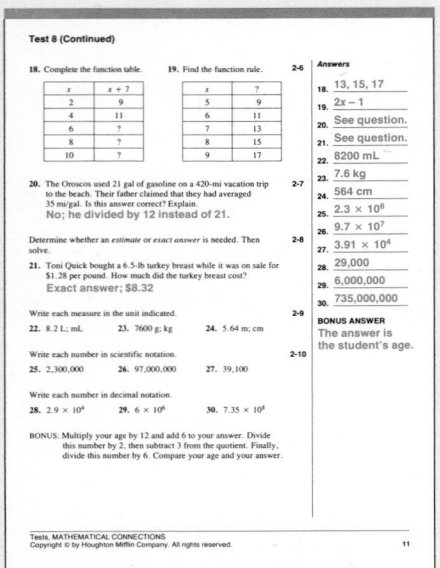

Chapter Test

Simplify. 8. $45z - 72$ 11. $9c + 14$ 12. $2n + 3m$

1. $a^7 \cdot a^6$ a^{13} 2. $6x \cdot 7x^3$ $42x^4$ 3. $(5m)(8n)$ $40mn$ 4. $(4b)(7c)(3b)$ $84b^2c$ **2-1**

5. $5(x + 11)$ $5x + 55$ 6. $3(9 - a)$ $27 - 3a$ 7. $6(7 + 4c)$ $42 + 24c$ 8. $9(5z - 8)$ **2-2**

9. $6m + 8m$ $14m$ 10. $12x - 5x$ $7x$ 11. $c + 14 + 8c$ 12. $6n - 4n + 3m$ **2-3**

★13. Write an addition expression that contains two unlike terms.
Answers will vary. Example: $5x + 2y$

★14. Explain why $5x$ and $5x^2$ are unlike terms.
They do not have identical variable parts.

15. Michi bought two hardcover books for $17.98 each and four paperback books for $3.95 each. How much more did the hardcover books cost than the paperback books? $20.16   **2-4**

Find the next three numbers or expressions in each pattern.

16. $1, 3, 9, 27, \underline{\ ?\ }, \underline{\ ?\ }, \underline{\ ?\ }$ 81, 243, 729   17. $x - 1, 3x - 1, 5x - 1, \underline{\ ?\ }, \underline{\ ?\ },$ $\underline{\ ?\ }$ $7x - 1, 9x - 1, 11x - 1$ **2-5**

18. Complete the function table.   19. Find the function rule.   **2-6**

x	$x - 3$
3	0
6	3
9	? 6
12	? 9
15	? 12

x	?	$x \rightarrow x + 7$
5	12	
6	13	
7	14	
8	15	
9	16	

20. Levi lives 12 mi from work. He works five days each week. When asked how many miles he commutes each week, Levi said he commutes 60 mi. Is this answer correct? Explain.   **2-7**
No; Levi did not include the miles both to and from work.

Decide whether an *estimate* or *exact answer* is needed. Then solve.

21. Ground beef sells for $2.09 per pound. Sue needs to buy 36 lb for a picnic. How much money should she take to the store? estimate; $80   **2-8**

Write each measure in the unit indicated.

22. 655 mL; L 0.655 L   23. 8.3 kg; g 8300 g   24. 56.4 cm; mm 564 mm   **2-9**

25. Write 2,400,000 in scientific notation. 2.4×10^6 26. Write 453,000 in scientific notation. 4.53×10^5 **2-10**

27. Write 4.7×10^5 in decimal notation. 470,000   28. Write 1.62×10^7 in decimal notation. 16,200,000

Other Rules for Exponents

Warm-Up Exercises

Simplify.

1. $a^2 \cdot a^4$ a^6
2. $c^5 \cdot c^7$ c^{12}
3. $x \cdot x^3$ x^4
4. $z^3 \cdot z^7 \cdot z^6$ z^{16}
5. $y^2 \cdot y^2 \cdot y^2$ y^6

Objective: To use the rules of exponents.

EXPLORATION

1. Replace each __?__ with the number that makes the statement true.
   a. $(x^5)^3 = x^5 \cdot x^5 \cdot x^5 = x^{5+5+5} = x^?$ 15
   b. $(a^3)^4 = a^3 \cdot a^3 \cdot a^3 \cdot a^3 = a^{3+3+3+3} = a^?$ 12

2. Use your results from Step 1. Complete this statement:
   To find a power of a power, you __?__ the exponents. multiply

3. Use your result from Step 2. Find the value of n that makes each statement true. Use a calculator to check your answer.
   a. $(5^2)^3 = 5^n$ 6
   b. $(3^2)^7 = 3^n$ 14
   c. $(2^4)^3 = 2^n$ 12

When you simplify expressions involving exponents, there are a number of rules that may make your work easier. For instance, in Lesson 2-1 you learned the *product of powers rule*.

$$a^m \cdot a^n = a^{m+n}$$

Another helpful rule for exponents is the *power of a power rule*.

> **Power of a Power Rule**
>
> To find the power of a power, multiply the exponents.
> $$(a^m)^n = a^{mn}$$

Teaching Note

Students may confuse the product of powers rule and the power of a power rule. Stress the difference between these two rules of exponents. After students have created the power of a product rule in Exercise 9, you may wish to put all three rules on the chalkboard and have students discuss ways to remember them.

Exercise Note

Critical Thinking
Exercises 7–10 require students to use their previous knowledge of exponents to create a new rule for simplifying an expression of the form $(ab)^m$.

Example

$$(c^2)^7 = c^{2 \cdot 7} = c^{14}$$

Exercises

Simplify.

1. $(b^2)^5$ b^{10}
2. $(x^5)^3$ x^{15}
3. $(c^7)^4$ c^{28}
4. $(m^3)^9$ m^{27}
5. $(x^6)^6$ x^{36}
6. $(z^4)^8$ z^{32}

THINKING SKILLS 7–8. Knowledge **9–10.** Synthesis

Find each answer.

7. a. $(3 \cdot 4)^2$ 144 b. $3^2 \cdot 4^2$ 144
8. a. $(2 \cdot 3)^4$ 1296 b. $2^4 \cdot 3^4$ 1296

9. Power of a product rule: To find the power of a product, you find the power of each factor and then multiply. $(ab)^m = a^m b^m$

9. Examine the results of Exercises 7 and 8. Create a rule for simplifying $(ab)^m$. Call it the *power of a product rule*.

10. Simplify each expression by combining the rules for exponents.
    a. $(xy)^5$ $x^5 y^5$ b. $(2k)^3$ $8k^3$ c. $(a^2 b)^3$ $a^6 b^3$ d. $(m^3 n^2)^4$ $m^{12} n^8$ e. $(3c^3 d^5)^2$ $9c^6 d^{10}$

Introduction to Algebra **89**

Quick Quiz 1 (2-1 through 2-3)

Quick Quiz 1 (2-1 through 2-3)

Simplify.

1. $b^5 \cdot b^3$ b^8
2. $n \cdot n^8$ n^9
3. $x^4 \cdot x^7 \cdot x^3$ x^{14}
4. $y^3 \cdot y^6 \cdot y^4$ y^{13}
5. $2m^3 \cdot 8m^7$ $16m^{10}$
6. $(3c)(7c^2)$ $21c^3$
7. $(7a)(12b)$ $84ab$
8. $(5x)(7y)(4z)$ $140xyz$

Use the distributive property to find each answer mentally.

9. $8(45) - 8(35)$ 80
10. $7 \cdot 13 + 7 \cdot 7$ 140
11. $5(409)$ 2045
12. $7(194)$ 1358

Simplify.

13. $3(n + 7)$ $3n + 21$
14. $7(x - 5)$ $7x - 35$
15. $8(9 - 4y)$ $72 - 32y$
16. $3(9 + 5k)$ $27 + 15k$
17. $4m + 6m$ $10m$
18. $24x - x$ $23x$
19. $8x + 3y + 5x$ $13x + 3y$
20. $5x + 3y + 8y$ $5x + 11y$

Alternative Assessment

Have students make up 4 exercises that meet the following criteria:
1. one that requires multiplication of variable expressions with exponents;
2. one that can be solved mentally using the distributive property;
3. one with variables that uses the distributive property;
4. one that requires combining like terms.

Use the exercises to develop a test for the class.

Cumulative Review

Standardized Testing Practice

Choose the letter of the correct answer.

1. **How can you use the distributive property to find 8(105) mentally?**
   A. $8(10 + 5) = 8 \times 15 = 120$
   B. $8(100) + 8(5) = 800 + 40 = 840$
   C. $5(105) + 3(105) = 525 + 315 = 840$
   D. $8(100) - 8(5) = 800 - 40 = 760$

2. **Find the next number in the pattern:**
   3, 6, 10, 15, __?__
   A. 30   B. 18
   C. 21   D. 20

3. **Evaluate $412.5 + n$ when $n = 86$.**
   A. 498.5
   B. 326.5
   C. 421.1
   D. 403.9

4. **Write 76,500 in scientific notation.**
   A. 7.65×10^4
   B. 76.5×10^3
   C. 7.65×10^3
   D. 765×10^2

5. **Which expression has a value of 216 when $n = 3$?**
   A. $2n^3$ B. $(2n)^3$
   C. $3n^2$ D. $(3n)^2$

6. **Complete:** The exact weight of a package rather than the estimated weight is needed to __?__.
   A. store the package on a shelf
   B. carry the package on a bike rack
   C. mail the package
   D. all of the above

7. **Evaluate $a + b + 2$ when $a = 4$ and $b = 8$.**
   A. 64   B. 12
   C. 34   D. 14

8. **Simplify:** x^2y^3
   A. $(xy)^5$
   B. xy^5
   C. $(xy)^6$
   D. already simplified

9. **Which property enables you to find $(4 + 1)(0)$ mentally?**
   A. commutative property of addition
   B. multiplication property of zero
   C. identity property of addition
   D. identity property of multiplication

10. **Choose the correct relationship:**
    $x = 6(2b + 3); y = 2(9 + 6b)$
    A. $x < y$ B. $x = y$
    C. $x > y$ D. cannot determine

11. During the last 3 days, Ruth drove 120 mi, 380 mi, and 250 mi. Gas costs $1.10 per gallon. Her car used 30 gal of gas. Which of the following cannot be determined?

A. number of mi/gal car averages
B. number of miles driven
C. capacity of gas tank
D. total cost of gas used

(C is circled)

12. Which number is greatest?

A. 0.2346   B. 0.3246
C. 0.3264   D. 0.3624

(D is circled)

13. Jorge bought 3 lb of apples at $.89/lb and 2 lb of grapes at $2.49/lb. Find the total cost.

A. $7.65   B. $9.25
C. $3.38   D. $8.45

(A is circled)

14. Evaluate the expression
$5 + 3(x - y^2)$ when $x = 10$ and $y = 2$.

A. 197
B. 48
C. 512
D. 23

(D is circled)

15. Write 6.45 kg in g.

A. 645 g
B. 64.5 g
C. 6450 g
D. 64,500 g

(C is circled)

16. Find the function rule.

A. $x \to x + 1$
B. $x \to 3x - 3$
C. $x \to 2x - 1$
D. $x \to 4x + 5$

(C is circled)

x	?
2	3
3	5
4	7
5	9
6	11

17. Simplify: $3x + 5y + 4x$

A. $12xy$ B. $7x + 5y$
C. $12(x + y)$ D. $5x + 7y$

(B is circled)

18. Evaluate the quotient $7.5 \div b$ when $b = 1.5$.

A. 6   B. 5
C. 9   D. 0.2

(B is circled)

19. A bill for two $38-sweaters and one $24-shirt came to $62. Find the error.

A. $38 was not multiplied by 2.
B. $24 was multiplied by $38.
C. $24 was multiplied by 2.
D. No error was made.

(A is circled)

20. Write 43.5 mm in cm.

A. 435 cm
B. 0.435 cm
C. 4.35 cm
D. 4350 cm

(C is circled)

Quick Quiz 2 *(2-4 through 2-7)*

1. The price of a stereo is $685. The sales tax is $41.10. The manufacturer is offering a rebate of $75. What is the final cost of the stereo? $651.10

Find the next three numbers in each pattern.

2. 3, 9, 15, 21, _?_, _?_, _?_
27, 33, 39

3. 1, 3, 9, 27, _?_, _?_, _?_
81, 243, 729

4. Complete the function table.

x	4x − 1	
2	7	
4	15	
6	?	23
8	?	31
10	?	39

5. Find the function rule.

x	?	$x \to 8x$
2	16	
3	24	
5	40	
7	56	
11	88	

6. Maria earns $6.50 per hour plus tips. Yesterday, she worked 7 hours and earned $18.50 in tips. Maria said she earned $45.50 yesterday. Is this correct? Explain.   No; Maria forgot to add in the $18.50 in tips.

Alternative Assessment

Sam bought a stereo for $75 down and monthly payments of $35 each. The price of the stereo is $550.

1. Find the total amount Sam has paid by the end of each of the first three months. $180

2. Use the given information to complete the function table.

n	?	$35n + 75$
1	?	110
2	?	145
5	?	250
7	?	320
11	?	460

3. How many payments will be required to pay off the stereo? Explain.   14; after 13 months, Sam has paid $530, so he must make one more payment.

Planning Chapter 3

Chapter Overview

Chapter 3 introduces students to the integers and thus extends the number system they have been using. Absolute value is introduced in the first lesson and then used to explain addition of integers. The other basic operations with integers are addressed, sometimes with the use of integer chips to provide a concrete model for the operation. Expressions are evaluated again, this time with integers. Problem solving is continued with the presentation of a new strategy, making a table. The coordinate plane is introduced and then used to graph functions.

The chapter begins by introducing positive numbers, negative numbers, zero, opposites, and absolute value by means of the number line. *Lessons 2 and 3* cover addition of integers: *Lesson 2* concentrates on integers with the same sign and *Lesson 3* concentrates on integers with different signs. The *Focus on Explorations* shows students how to work with integer chips, which are used in *Lessons 4 and 5* to provide concrete models for subtraction, multiplication, and division of integers. *Lesson 6* returns to the important concept of evaluating variable expressions, this time with integers. *Lesson 7* continues the problem-solving strand by introducing a new strategy, that of making a table. *Lessons 8 and 9* introduce the coordinate plane and show students how to graph points and functions on it. The *Focus on Applications,* which immediately follows these lessons, ties the lessons to real life by having students work with the grid system found on many maps.

Background

Chapter 3 extends the algebraic concepts presented in the first two chapters to a new set of numbers, the integers. Negative numbers are introduced, and the first half of the chapter teaches students how to perform the operations of addition, subtraction, multiplication, and division with them. Absolute value is presented by using a visual, geometric approach; it is defined as the distance from the number to zero on a number line. The remainder of the chapter extends the concept of evaluating a variable expression and the function concept to integers, introducing and using the coordinate plane in the last two lessons.

The problem-solving strand is expanded once again by the inclusion of a new strategy, that of making a table. Many students try to solve problems by using an operation only or, if they know how, by using an equation. However, there are a whole range of problems that can be solved more easily by using a table. *Lesson 7* points this out explicitly to students.

Technology is featured in Chapter 3, as students are shown how the change-sign key can be used to add, subtract, multiply, and divide integers. The purpose of this instruction is to make students better calculator users; users less likely to make errors and more likely to see them if they do. The power of computers to graph figures in the coordinate plane is shown in the Computer Application on page 131.

Students sometimes learn operations on integers more easily if they can see them being performed concretely. The *Focus on Explorations* on pages 105 and 106 introduces students to integer chips. The manipulatives are then used to illustrate some examples in the lessons following this section. The chips, of course, can also be used to illustrate addition, which is presented before the Focus section.

Objectives

3-1	To recognize and compare integers and to find opposites and absolute values.
3-2	To add integers with the same sign.
3-3	To add integers with different signs.
☰	**FOCUS ON EXPLORATIONS** To use integer chips to model integers.
3-4	To find the difference of two integers.
3-5	To find products and quotients of integers.
3-6	To evaluate expressions involving integers.
3-7	To solve problems by making a table.
3-8	To find coordinates and graph points on a coordinate plane.
3-9	To graph functions on a coordinate plane.
☰	**FOCUS ON APPLICATIONS** To read and interpret a map.
☰	**CHAPTER EXTENSION** To determine whether a set of numbers is closed under a given operation.

Chapter Planner

Instructional Aids	Manipulatives	Cooperative Learning	Technology	Practice/Reteaching	Assessment
Materials: integer chips, decks of cards, calculator *Resource Book:* Lesson Starters 21–29 *Visuals,* Folders H, I	Lessons 3-1, 3-2, 3-3, 3-4, 3-5, Focus on Explorations *Activities Book:* Manipulative Activities 5–6	Lessons 3-3, 3-5, Focus on Applications *Activities Book:* Cooperative Activities 6–7	Lessons 3-2, 3-3, 3-4, 3-5, 3-7, 3-9 *Activities Book:* Computer Activities 3–4 *Connections Plotter Plus* Disk	Toolbox, pp. 753–771 Extra Practice, p. 732 *Resource Book:* Practice 28–39 *Study Guide,* pp. 41–58	Self-Tests 1–2 Quick Quizzes 1–2 Chapter Test, p. 136 *Resource Book:* Lesson Starters 21–29 (Daily Quizzes) *Tests* 10–13

Assignment Guide Chapter 3

Day	Skills Course	Average Course	Advanced Course
1	**3-1:** 1–15, 18–26 even, 35–38	**3-1:** 1–16, 17–33 odd, 35–38	**3-1:** 1–21 odd, 22–38
2	**3-2:** 1–16, 17–31 odd, 33–35	**3-2:** 1–16, 17–31 odd, 33–35	**3-2:** 1–15 odd, 17–35
3	**3-3:** 1–20, 21–35 odd, 46–49 **Exploration:** Activities I–IV	**3-3:** 1–20, 21–43 odd, 46–49 **Exploration:** Activities I–IV	**3-3:** 1–19 odd, 21–49 **Exploration:** Activities I–IV
4	**3-4:** 1–20, 21–37 odd, 39–44	**3-4:** 1–20, 21–37 odd, 39–44	**3-4:** 1–17 odd, 19–44
5	**3-5:** 1–20, 21–41 odd, 46, 47	**3-5:** 1–17 odd, 19–41, 46, 47	**3-5:** 1–19 odd, 21–47
6	**3-6:** 1–24, 26, 28, 33, 35, 51–58	**3-6:** 1–28, 29–43 odd, 51–58	**3-6:** 1–27 odd, 29–58, Challenge
7	**3-7:** 1–16 **Mixed Review**	**3-7:** 1–16 **Mixed Review**	**3-7:** 1–16 **Mixed Review**
8	**3-8:** 1–28, 37, 38, 41–44	**3-8:** 1–20, 21–39 odd, 41–44	**3-8:** 1–19 odd, 21–44, Historical Note
9	**3-9:** 1–10, 17–20 **Application:** 1–14	**3–9:** 1–10, 11–15 odd, 17–20 **Application:** 1–14	**3–9:** 1–20 **Application:** 1–16
10	*Prepare for Chapter Test:* Chapter Review	*Prepare for Chapter Test:* Chapter Review	*Prepare for Chapter Test:* Chapter Review
11	*Administer Chapter 3 Test*	*Administer Chapter 3 Test;* Cumulative Review	*Administer Chapter 3 Test;* Chapter Extension; Cumulative Review

Teacher's Resources

Resource Book
- Chapter 3 Project
- Lesson Starters 21–29
- Practice 28–39
- Enrichment 23–32
- Diagram Masters
- Chapter 3 Objectives
- Family Involvement 3
- Spanish Test 3
- Spanish Glossary

Activities Book
- Cooperative Activities 6–7
- Manipulative Activities 5–6
- Computer Activities 3–4

Study Guide, pp. 41–58

Tests
- Tests 10–13

Visuals
- Folders H, I

Connections Plotter Plus **Disk**

Alternate Approaches Chapter 3

Using Technology

Lessons 3-2, 3-3, 3-4

CALCULATORS

Students can explore how algebraic computations work on their calculators, using the $\boxed{+/-}$ key to find the negative of an entered number. They also can be asked to extend the rules they have learned to decimals by doing problems such as $2.5 - 3.8$.

COMPUTERS

Students can investigate how BASIC and spreadsheet formulas deal with positive and negative numbers. In addition, a simple spreadsheet can provide a model of addition similar to that of integer chips. In this model, positive chips are represented by entries of 1 and negative chips are represented by -1. If the positive entries and the negative entries are lined up next to each other in columns, and the spreadsheet program is directed to add them cumulatively, students will be able to see how $+/-$ pairs cancel each other and the excess in either the positive or negative column gives the final answer. A similar model may be used for subtraction by adding the opposite.

Lesson 3-7

COMPUTERS

The tables used in this lesson could be constructed using a spreadsheet program. This approach would provide a bridge to algebraic solutions of problems like the money problems found in the Example and exercises. In solving the sample problem, which involves enumeration of all the possible ways to make 35¢ in change, a formula giving the total value of each column will require multiplying the number of quarters by 25, the number of dimes by 10, and the number of nickels by 5. The resulting formula, which follows from a few specific examples, is a step toward constructing equations of this type in algebra.

Lesson 3-9

COMPUTERS

As mentioned in the text, you can use geometric graphing software to do Exercises 15 and 16. A graphing calculator or a computer running function-graphing software also can be used to allow students to explore graphs. Such software almost always graphs function values for every x-value on the screen, not just for integer values. This important difference needs to be explained to students.

Using Manipulatives

Lesson 3-1

Use Activity I on page 105 to model integers. As further examples, use the following:

Draw models to represent each integer.

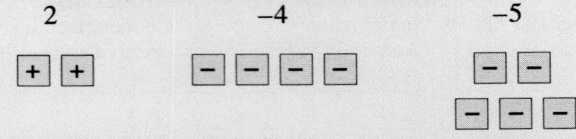

What integer is represented by each of the following?

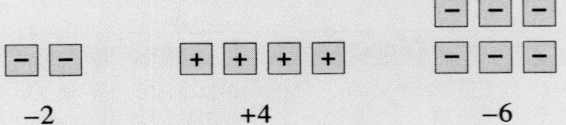

Lesson 3-2

Adding integers with the same sign can be illustrated by combining groups. You can compare addition of integers to combining the contents of two containers.

Use chips to find the following sums:

$3 + 2 = \underline{?}$

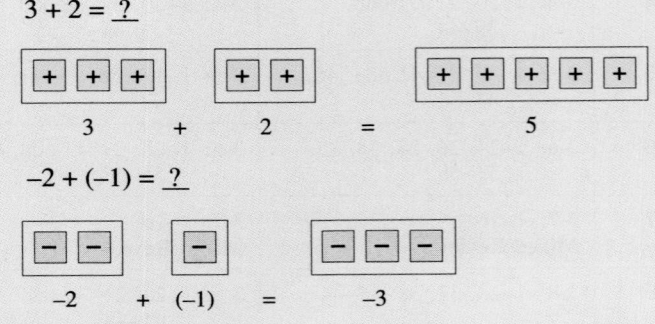

$$3 \quad + \quad 2 \quad = \quad 5$$

$-2 + (-1) = \underline{?}$

$$-2 \quad + \quad (-1) \quad = \quad -3$$

Lesson 3-3

The addition property of opposites may be demonstrated by pairing a positive chip with a negative chip.

Find the sum of $3 + (-3)$ using chips.

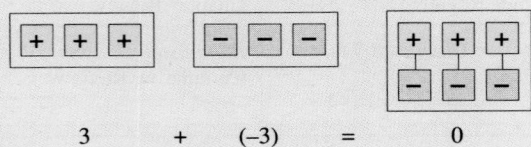

$$3 \quad + \quad (-3) \quad = \quad 0$$

Lesson 3-4

Subtraction can be modeled as the addition of the opposite by using integer chips. For instance, the solution to Example 2 can be shown as follows:

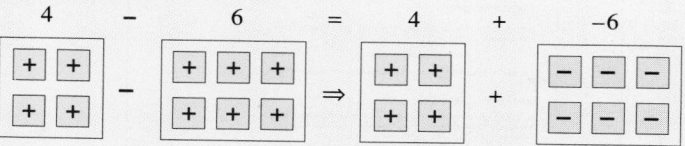

The last model may be combined as follows to show -2.

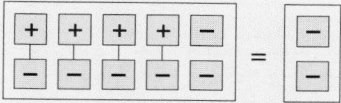

Lesson 3-5

Integer chips may be used instead of a pattern to model the product of two negative integers. Begin by reminding students that the symbol "$-$" means "the opposite of." Then proceed as follows:

Since $-(2)(-3)=(-2)(-3)$, then $(-2)(-3) = 6$.

Reteaching/Alternate Approach

Lesson 3-5

COOPERATIVE LEARNING
Divide students into groups of three or four. Give each group a deck of cards and have them remove all the kings, queens, and jacks. The remaining cards represent their face values, (ace = 1), with red cards being negative and black cards being positive. Have each group shuffle its cards and turn the top card face up. This card is the goal number. Now have each group draw four more cards and try to use them to create an expression equal to the goal number. For example:

Goal number: 2 of hearts (-2)

Cards drawn: ace of diamonds, 3 of hearts, 8 of diamonds, 10 of hearts. (All red cards were drawn. Each number is negative.)

Possible solution: $-2 = -1 + (-3) + (-8) - (-10)$

Should you wish, this activity can be performed as a competition. Impose a time limit for each draw and award 3 points to the first group to get a correct expression, 2 points to the second group, and 1 point to the third group. The competition ends when one group gets a certain number of points, say 11 or 15.

Lesson 3-7

Students sometimes make up tables with little regard to order. Tell students that it is helpful to look at the greatest quantity first. For instance, in the Example on page 119, the greatest coin value is 25¢, or a quarter. Students should ask themselves the following questions: What is the greatest number of quarters I can use? If I use one quarter, how many dimes do I need? If I have one quarter and one dime, do I need any nickels? If I have one quarter but do not use any dimes, how many nickels do I need? Students should continue this reasoning until all possibilities have been exhausted.

Lesson 3-8

Put a coordinate grid on the chalkboard and label the vertical lines as 0 St., 1st St., 2nd St., and so on. Label the horizontal lines as 0 Ave., 1st Ave., 2nd Ave., and so on. Ask students a series of questions that will move them around the grid, always starting at (0 St., 0 Ave.) and always ending on a "corner." Point out that the order in which the street corners are named makes a difference. This activity will help students to understand the concept of an ordered pair.

Teacher's Resources Chapter 3

Enrichment masters from the Resource Book are pictured here. See the Teacher's Resources chart on page 92B for a complete listing of all materials available for this chapter.

3-1 Enrichment
Communication

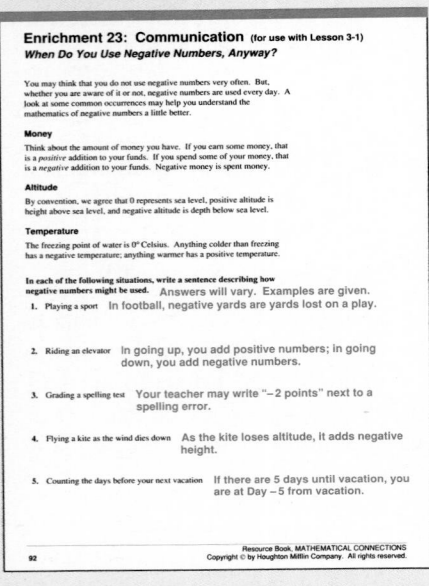

Enrichment 23: Communication (for use with Lesson 3-1)
When Do You Use Negative Numbers, Anyway?

You may think that you do not use negative numbers very often. But, whether you are aware of it or not, negative numbers are used every day. A look at some common occurrences may help you understand the mathematics of negative numbers a little better.

Money
Think about the amount of money you have. If you earn some money, that is a *positive* addition to your funds. If you spend some of your money, that is a *negative* addition to your funds. Negative money is spent money.

Altitude
By convention, we agree that 0 represents sea level, positive altitude is height above sea level, and negative altitude is depth below sea level.

Temperature
The freezing point of water is 0° Celsius. Anything colder than freezing has a negative temperature; anything warmer has a positive temperature.

In each of the following situations, write a sentence describing how negative numbers might be used. Answers will vary. Examples are given.
1. Playing a sport   In football, negative yards are yards lost on a play.

2. Riding an elevator   In going up, you add positive numbers; in going down, you add negative numbers.

3. Grading a spelling test   Your teacher may write "−2 points" next to a spelling error.

4. Flying a kite as the wind dies down   As the kite loses altitude, it adds negative height.

5. Counting the days before your next vacation   If there are 5 days until vacation, you are at Day −5 from vacation.

3-2 Enrichment
Application

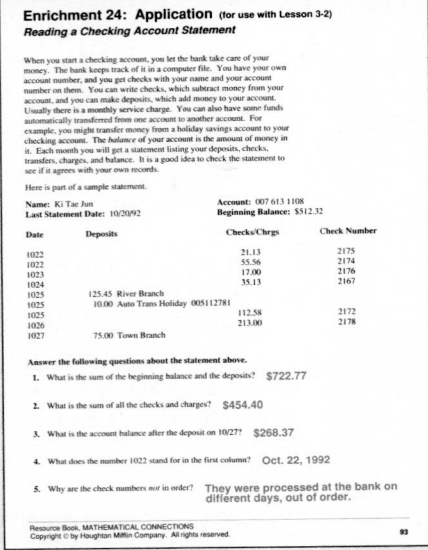

Enrichment 24: Application (for use with Lesson 3-2)
Reading a Checking Account Statement

When you start a checking account, you let the bank take care of your money. The bank keeps track of it in a computer file. You have your own account number, and you get checks with your name and your account number on them. You can write checks, which subtract money from your account, and you can make deposits, which add money to your account. Usually there is a monthly service charge. You can also have some funds automatically transferred from one account to another account. For example, you might transfer money from a holiday savings account to your checking account. The *balance* of your account is the amount of money in it. Each month you will get a statement listing your deposits, checks, transfers, charges, and balance. It is a good idea to check the statement to see if it agrees with your own records.

Here is part of a sample statement.

Name: Ki Tae Jun   **Account:** 007 613 1108
Last Statement Date: 10/20/92   **Beginning Balance:** $512.32

Date	Deposits	Checks/Chrgs	Check Number
1022		21.13	2175
1022		55.56	2174
1023		17.00	2176
1024		35.13	2167
1025	125.45 River Branch		
1025	10.00 Auto Trans Holiday 005112781		
1025		112.58	2172
1026		213.00	2178
1027	75.00 Town Branch		

Answer the following questions about the statement above.
1. What is the sum of the beginning balance and the deposits? $722.77

2. What is the sum of all the checks and charges? $454.40

3. What is the account balance after the deposit on 10/27? $268.37

4. What does the number 1022 stand for in the first column?   Oct. 22, 1992

5. Why are the check numbers *not* in order?   They were processed at the bank on different days, out of order.

3-3 Enrichment
Exploration

Enrichment 25: Exploration (for use with Lesson 3-3)
Quick Averaging

Let's say that you have the following test scores in math and want to know your average.

93, 86, 94, 82, 96, 97, 89, 85

You could certainly find the average by adding the numbers and dividing by 8.

Let's explore a quicker way.
By looking at the numbers you can see that the average will be about 90. You can call 90 the *reference number*. Now, 93 is 3 more than 90, so you record that number as +3. 86 is 4 less than 90; you record it as −4. So, instead of writing down the original numbers, write down the following sum.

+3 + (−4) + (+4) + (−8) + (+6) + (+7) + (−1) + (−5)

Answer the following questions about the sum above.
1. Explain where the (+7) comes from.   97 is 7 more than 90, so you record it as (+7).

2. Explain where the (−5) comes from.   85 is 5 less than 90, so you record it as (−5).

3. What is the sum of this new list of numbers?   2

4. What is the average of this list?   0.25

5. Add this average to 90 to get the average of the original list.   90.25

6. Why does it make sense to add the new average to 90?   Because all the numbers in the new list refer to 90.

7. Find the average by adding the original numbers and dividing by 8.   90.25; Yes.
Do you get the same average?

8. Could we use 80 as a reference number? What would be different about the second list of numbers?   Yes; All the numbers would be positive.

Use the method described above to find the averages of the following sets of numbers. Try to do some of these mentally.

9. 78, 82, 83, 85   82
10. 81, 85, 86, 83, 85   84
11. 71, 75, 81, 92, 88   81.4
12. 90, 94, 87, 93, 89, 99   92
13. 67, 72, 68, 82, 78, 91   76.3
14. 130, 150, 141, 143, 145   141.8
15. 67, 70, 71, 82, 89, 99, 100   82.6
16. 90, 90, 90, 68, 94, 94   87.7

3-4 Enrichment
Application

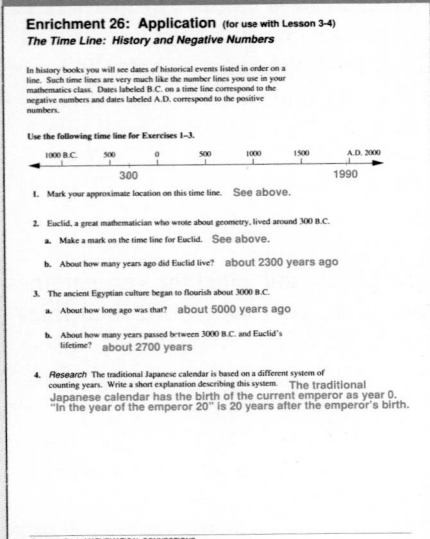

Enrichment 26: Application (for use with Lesson 3-4)
The Time Line: History and Negative Numbers

In history books you will see dates of historical events listed in order on a line. Such time lines are very much like the number lines you use in your mathematics class. Dates labeled B.C. on a time line correspond to the negative numbers and dates labeled A.D. correspond to the positive numbers.

Use the following time line for Exercises 1–3.

1000 B.C.   500   0   500   1000   1500   A.D. 2000
                    300   1990

1. Mark your approximate location on this time line.   See above.

2. Euclid, a great mathematician who wrote about geometry, lived around 300 B.C.
   a. Make a mark on the time line for Euclid.   See above.
   b. About how many years ago did Euclid live?   about 2300 years ago

3. The ancient Egyptian culture began to flourish about 3000 B.C.
   a. About how long ago was that?   about 5000 years ago
   b. About how many years passed between 3000 B.C. and Euclid's lifetime?   about 2700 years

4. **Research** The traditional Japanese calendar is based on a different system of counting years. Write a short explanation describing this system.   The traditional Japanese calendar has the birth of the current emperor as year 0. "In the year of the emperor 20" is 20 years after the emperor's birth.

3-5 Enrichment
Data Analysis

Enrichment 27: Data Analysis (for use with Lesson 3-5)
What Do You Expect?

The following activity will show you some things about positive and negative numbers. It will also show you something about *expected value*, an important idea in data analysis. If you toss a coin a number of times, you expect heads about half the time. If you make T-shirts, you expect most of your customers to be certain sizes.

Materials
You will need paper, pencil, and a cube whose sides are numbered 1 through 6.

What to do
Keep a running sum according to the following rule: Start with a sum of 0. Roll the cube. If an odd number (1, 3, 5) comes up, write that number down and add it to your sum. If an even number comes up, add the opposite of the number to your sum. For example, if you roll a four, add a −4 to your sum. Keep a record like this one.

Roll	Add	Sum
		0 (Start)
3	3	3
2	−2	1 (3 + (−2) = 1)
4	−4	−3
6	−6	−9
5	5	−4
3	3	−1

Answers to Exercises 1–4 will vary.
1. Before you roll your cube, decide what you would *expect* the sum to be after six rolls. Explain your thinking.   Since all numbers are equally likely to come up, you might expect the sum 1 + (−2) + 3 + (−4) + 5 + (−6) = −3 after a large number of trials.

2. What would you expect the sum to be after 30 rolls? Why?   −15; Since 30 = 5 × 6, you would expect 5 × (−3) = −15.

3. Try 30 rolls and see what your sum is. Add your sum to those of your classmates. What is the new sum?

4. Devise rules for a game where the expected sum will be 0.   An example of a game could be tossing a single coin. If a head comes up, you get 1, and if a tail comes up, you get −1

3-6 Enrichment
Thinking Skills

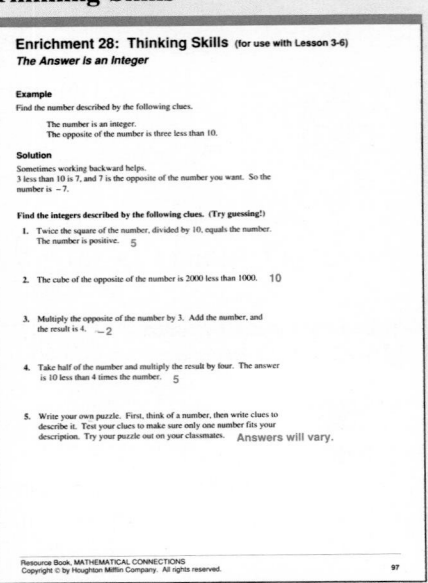

Enrichment 28: Thinking Skills (for use with Lesson 3-6)
The Answer is an Integer

Example
Find the number described by the following clues.
   The number is an integer.
   The opposite of the number is three less than 10.

Solution
Sometimes working backward helps.
3 less than 10 is 7, and 7 is the opposite of the number you want. So the number is −7.

Find the integers described by the following clues. (Try guessing!)
1. Twice the square of the number, divided by 10, equals the number. The number is positive.   5

2. The cube of the opposite of the number is 2000 less than 1000.   10

3. Multiply the opposite of the number by 3. Add the number, and the result is 4.   −2

4. Take half of the number and multiply the result by four. The answer is 10 less than 4 times the number.   5

5. Write your own puzzle. First, think of a number, then write clues to describe it. Test your clues to make sure only one number fits your description. Try your puzzle out on your classmates.   Answers will vary.

3-7 Enrichment
Problem Solving

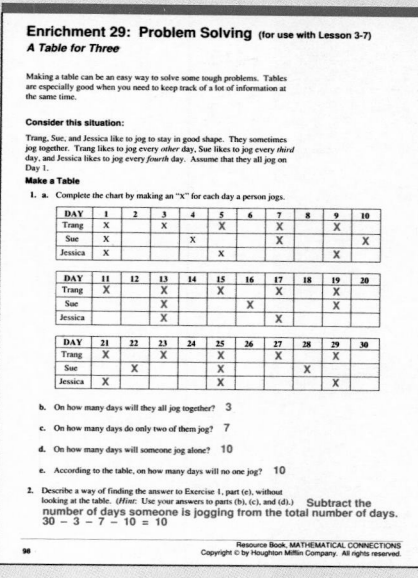

Enrichment 29: Problem Solving (for use with Lesson 3-7)
A Table for Three

Making a table can be an easy way to solve some tough problems. Tables are especially good when you need to keep track of a lot of information at the same time.

Consider this situation:

Trang, Sue, and Jessica like to jog to stay in good shape. They sometimes jog together. Trang likes to jog every *other* day, Sue likes to jog every *third* day, and Jessica likes to jog every *fourth* day. Assume that they all jog on Day 1.

Make a Table

1. a. Complete the chart by making an "X" for each day a person jogs.

DAY	1	2	3	4	5	6	7	8	9	10
Trang	X		X		X		X		X	
Sue	X			X			X			X
Jessica	X				X				X	

DAY	11	12	13	14	15	16	17	18	19	20
Trang	X		X		X		X		X	
Sue			X			X			X	
Jessica			X				X			

DAY	21	22	23	24	25	26	27	28	29	30
Trang	X		X		X		X		X	
Sue		X			X			X		
Jessica	X				X				X	

b. On how many days will they all jog together? 3

c. On how many days do only two of them jog? 7

d. On how many days will someone jog alone? 10

e. According to the table, on how many days will no one jog? 10

2. Describe a way of finding the answer to Exercise 1, part (e), without looking at the table. (*Hint:* Use your answers to parts (b), (c), and (d).) Subtract the number of days someone is jogging from the total number of days.
30 − 3 − 7 − 10 = 10

3-8 Enrichment
Connection

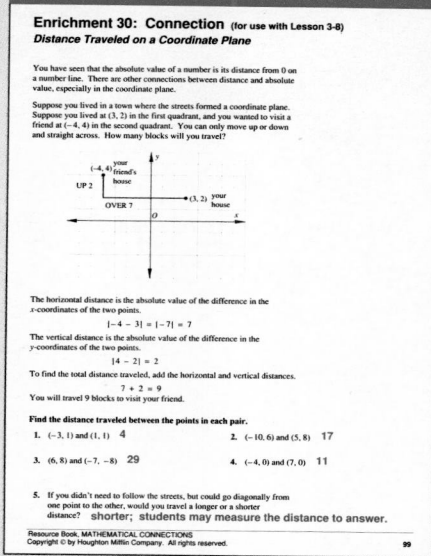

Enrichment 30: Connection (for use with Lesson 3-8)
Distance Traveled on a Coordinate Plane

You have seen that the absolute value of a number is its distance from 0 on a number line. There are other connections between distance and absolute value, especially in the coordinate plane.

Suppose you lived in a town where the streets formed a coordinate plane. Suppose you lived at (3, 2) in the first quadrant, and you wanted to visit a friend at (−4, 4) in the second quadrant. You can only move up or down and straight across. How many blocks will you travel?

The horizontal distance is the absolute value of the difference in the x-coordinates of the two points.

$$|-4 - 3| = |-7| = 7$$

The vertical distance is the absolute value of the difference in the y-coordinates of the two points.

$$|4 - 2| = 2$$

To find the total distance traveled, add the horizontal and vertical distances.

$$7 + 2 = 9$$

You will travel 9 blocks to visit your friend.

Find the distance traveled between the points in each pair.

1. (−3, 1) and (1, 1) 4

2. (−10, 6) and (5, 8) 17

3. (6, 8) and (−7, −8) 29

4. (−4, 0) and (7, 0) 11

5. If you didn't need to follow the streets, but could go diagonally from one point to the other, would you travel a longer or a shorter distance? shorter; students may measure the distance to answer.

3-9 Enrichment
Communication

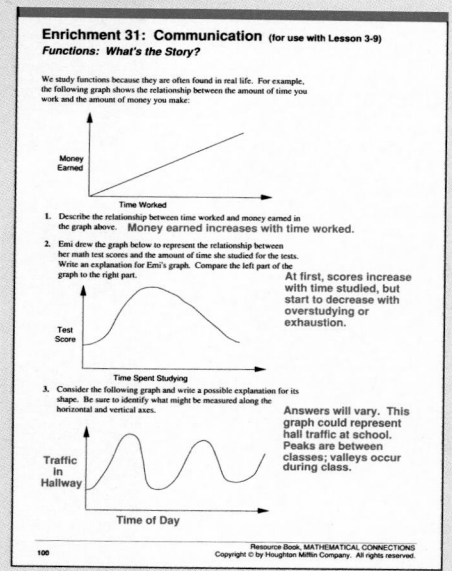

Enrichment 31: Communication (for use with Lesson 3-9)
Functions: What's the Story?

We study functions because they are often found in real life. For example, the following graph shows the relationship between the amount of time you work and the amount of money you make:

1. Describe the relationship between time worked and money earned in the graph above. Money earned increases with time worked.

2. Emi drew the graph below to represent the relationship between her math test scores and the amount of time she studied for the tests. Write an explanation for Emi's graph. Compare the left part of the graph to the right part. At first, scores increase with time studied, but start to decrease with overstudying or exhaustion.

3. Consider the following graph and write a possible explanation for its shape. Be sure to identify what might be measured along the horizontal and vertical axes. Answers will vary. This graph could represent hall traffic at school. Peaks are between classes; valleys occur during class.

End of Chapter Enrichment
Extension

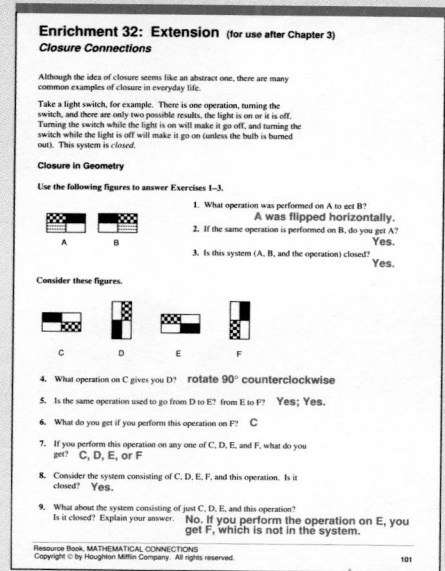

Enrichment 32: Extension (for use after Chapter 3)
Closure Connections

Although the idea of closure seems like an abstract one, there are many common examples of closure in everyday life.

Take a light switch, for example. There is one operation, turning the switch, and there are only two possible results, the light is on or it is off. Turning the switch while the light is on will make it go off, and turning the switch while the light is off will make it go on (unless the bulb is burned out). This system is *closed*.

Closure in Geometry

Use the following figures to answer Exercises 1–3.

1. What operation was performed on A to get B? A was flipped horizontally.

2. If the same operation is performed on B, do you get A? Yes.

3. Is this system (A, B, and the operation) closed? Yes.

Consider these figures.

4. What operation on C gives you D? rotate 90° counterclockwise

5. Is the same operation used to go from D to E? from E to F? Yes; Yes.

6. What do you get if you perform this operation on F? C

7. If you perform this operation on any one of C, D, E, and F, what do you get? C, D, E, or F

8. Consider the system consisting of C, D, E, F, and this operation. Is it closed? Yes.

9. What about the system consisting of just C, D, E, and this operation? Is it closed? Explain your answer. No. If you perform the operation on E, you get F, which is not in the system.

For Chapter 3 Opener

The words tornado and hurricane have both come to English via Spanish. Tornado comes from the Spanish word *tronada*, meaning thunderstorm. Tornado has also been influenced by the Spanish verb *tornar*, meaning to turn or return. Spanish explorers adopted the word *huracán* from the Carib language. Typhoon is derived both from an Arabic word for hurricane and two Cantonese words that mean "big wind." A British ship captain coined the word cyclone in 1846. The word is similar to a Greek word that means circle and implies the coil of a snake.

For Page 94

Encourage students to locate Mount Everest and the Marianas Trench on a world map; one that shows elevation would be particularly useful for this lesson. Students may be interested to learn that the mountains in many countries are not as high as the bottoms of certain valleys in Nepal and Tibet.

For Page 104

In recent years, it has become more common for countries to cooperate with each other in the exploration of space. The first joint U.S.-Soviet space mission took place in 1975. In 1983, the space shuttle carried the European-built Spacelab into orbit.

Research Activities

For Chapter 3 Opener

Many newspapers give daily high and low temperature readings for cities in different parts of the world. Many almanacs also contain temperature data. You may want to have students compare and contrast temperature readings for cities below, near, and above the equator. Discuss with students how temperature influences people's lives in the various cities.

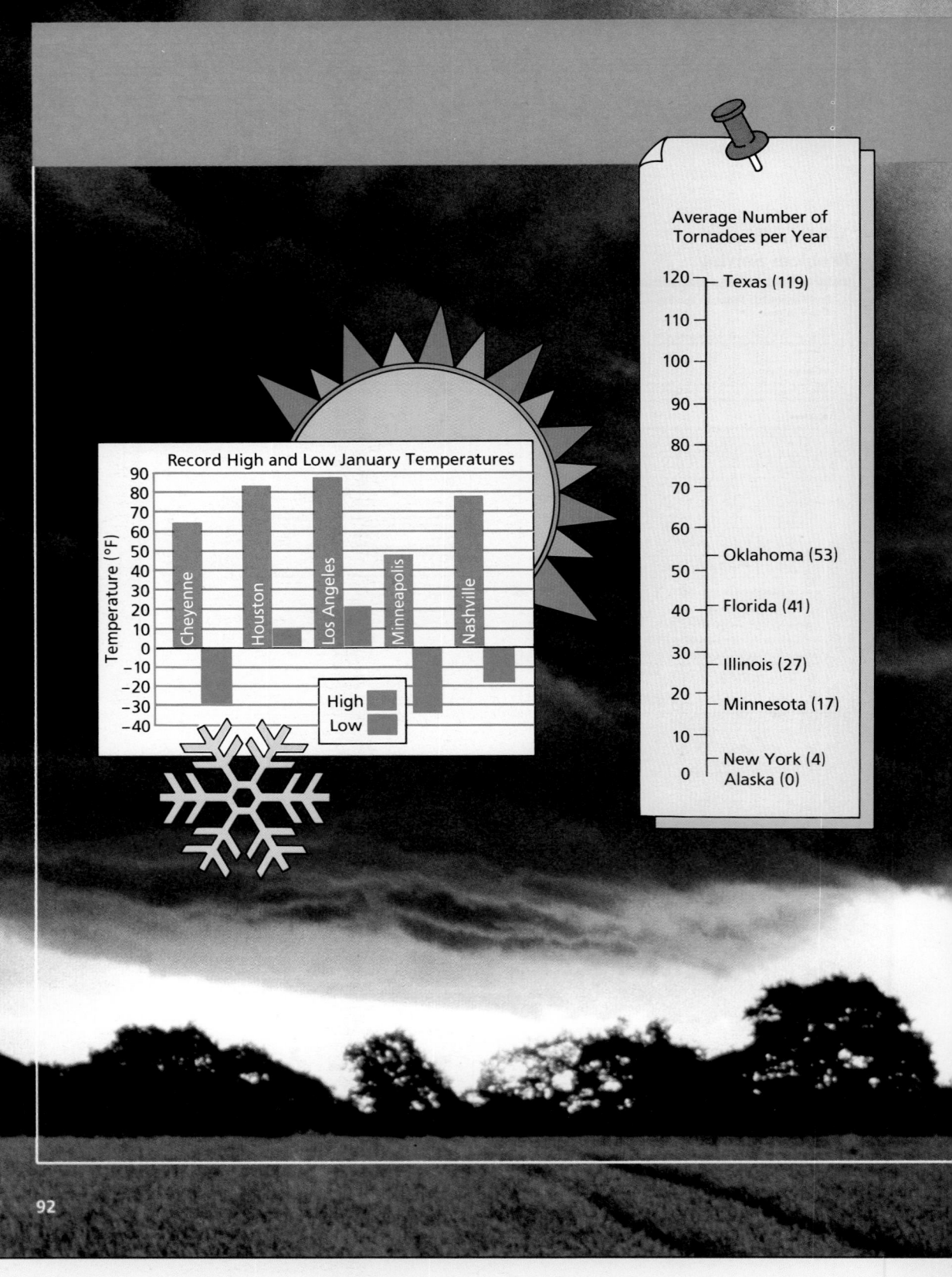

Integers 3

Normal Low January Temperatures (°F)

50 40   30 20   10   0 −5   −10   −5   0   10

60

70
   60
      50   40
           30

                                          20

                                    40
                                    50
                                    60
                          40 50   70
                             60
                              70

Did You Know?

A tropical storm has a wind speed that is greater than 39 mi/h. Storms in the Atlantic and eastern Pacific oceans are identified by a person's first name. These names have an international flavor because storms are tracked by many nations.

When the wind speed of a tropical storm reaches 74 mi/h, the storm is upgraded. An upgraded storm is called a *hurricane* in the Atlantic and eastern Pacific oceans, a *typhoon* in the western Pacific Ocean, and a *cyclone* in the Indian Ocean.

93

Research Activities
(continued)

For Page 104
You may want to have students work in cooperative groups to create timelines showing the history of international cooperation in space.

Suggested Resources

Cassutt, Michael. *Who's Who in Space: The First 25 Years*. Boston: G. K. Hall and Company, 1987.

NASA, The First 25 Years, 1958–1983: A Resource for Teachers. Washington, D.C.: National Aeronautics and Space Administration, 1983.

The 1991 Information Please Almanac. Boston: Houghton Mifflin Company, 1990.

Teaching the Lesson
- Materials: integer chips
- Lesson Starter 21
- Visuals, Folder I
- Using Manipulatives, p. 92C
- Toolbox Skill 5

Lesson Follow-Up
- Practice 28
- Enrichment 23: Communication
- Study Guide, pp. 41–42

Warm-Up Exercises

Replace each ? with >, <, or =.

1. 10,974 ? 10,794   >
2. 6935 ? 6635   >
3. 1.04 ? 1.40   <
4. 2.071 ? 2.71   <
5. 0.50 ? 0.5   =

Lesson Focus

Ask students if they have ever heard a weather forecaster say that the temperature for the day would go from a high of 27°F above zero to a low of 6°F below zero. Today's lesson focuses on numbers that can be used to represent positive and negative quantities.

Teaching Notes

Stress the terminology *integers, positive, negative, opposites,* and *absolute value*.

Key Questions

1. Can the absolute value of a number ever be negative? Explain.   No; absolute value is defined as the distance a number is from zero on a number line and distance is never negative.

Integers on a Number Line

Objective: To recognize and compare integers and to find opposites and absolute values.

Terms to Know
- *integers*
- *positive integers*
- *negative integers*
- *opposites*
- *absolute value*

DATA ANALYSIS

The diagram at the right shows information about the highest and lowest points on Earth. The positive sign on $^+29{,}028$ tells you that the top of Mount Everest is 29,028 ft *above* sea level. The negative sign on $^-38{,}635$ tells you that the bottom of the Marianas Trench is 38,635 ft *below* sea level. Sea level is represented by 0. The numbers $^+29{,}028$, $^-38{,}635$, and 0 are examples of *integers*.

An **integer** is any number in the following set.

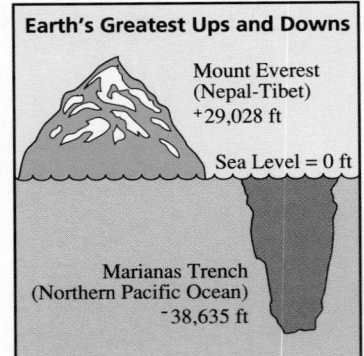

Earth's Greatest Ups and Downs

Mount Everest
(Nepal-Tibet)
+29,028 ft

Sea Level = 0 ft

Marianas Trench
(Northern Pacific Ocean)
$^-38{,}635$ ft

$$\{\ldots, {}^-4, {}^-3, {}^-2, {}^-1, 0, {}^+1, {}^+2, {}^+3, {}^+4, \ldots\}$$ ⟵ The braces { } mean the set that contains.

Integers greater than zero are called **positive integers.** Integers less than zero are called **negative integers.** Zero is neither positive nor negative. To make notation simpler, you generally write positive integers without the positive sign.

Another way to show the integers is to locate them as points on a number line. On a horizontal number line, *positive integers* are to the right of zero and *negative integers* are to the left.

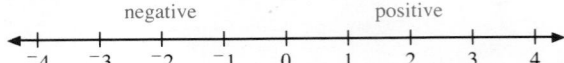

Numbers that are the same distance from zero, but on opposite sides of zero, are called **opposites.** To indicate the opposite of a number n, you write $-n$. You read $-n$ as "the opposite of n."

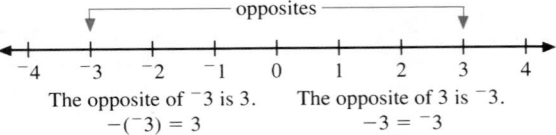

The opposite of $^-3$ is 3.   The opposite of 3 is $^-3$.
$-(^-3) = 3$ $-3 = {}^-3$

On the number line above, you see that the symbols -3 and $^-3$ represent the same number, negative three. To make notation simpler, this textbook will use the *lowered* sign to indicate a negative number. From this point on, you will see negative three written as -3.

The distance that a number is from zero on a number line is the **absolute value** of the number. You use the symbol | | to indicate absolute value. You read $|n|$ as "the absolute value of n."

Example 1

Find each absolute value.

a. $|3|$ **b.** $|-4|$

Solution

a.

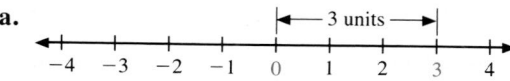

3 is 3 units from 0, so $|3| = 3$.

b.

4 is 4 units from 0, so $|-4| = 4$.

 **Check Your Understanding**

1. Name another integer that has an absolute value of 3.
2. What is the absolute value of 0?
   See *Answers to Check Your Understanding* at the back of the book.

When you compare numbers, you may want to picture them on a number line. On a horizontal number line, numbers increase in order from left to right.

Example 2

Replace each __?__ with >, <, or =.

a. $1 \underline{\ ?\ } -3$ **b.** $-4 \underline{\ ?\ } -2$

Solution

a.

1 is *to the right of* -3, so $1 > -3$.

b.

-4 is *to the left of* -2, so $-4 < -2$.

Guided Practice

COMMUNICATION «*Reading*

Replace each __?__ with the correct phrase.

«**1.** The __?__ of a number is its distance from zero on a number line.
   absolute value
«**2.** The expression $-b$ is read as __?__. the opposite of b

Integers   **95**

2. If n represents an integer, when does $-n$ represent a positive number? The number $-n$ represents a positive integer when n represents a negative integer. For example, if $n = -6$, then $-n = -(-6) = 6$.

Error Analysis
When ordering negative integers, students may associate the size of the numbers without considering the sign and thus get an incorrect answer. Remind students to imagine the numbers on a number line.

Application
People routinely use absolute value to refer to quantities even when the quantity is negative. For instance, a temperature of $-12°$ is referred to as "12° below zero." Ask students for other examples.

Using Manipulatives
For a suggestion on using integer chips to model integers, see page 92C.

Additional Examples

1. Find each absolute value.
   a. $|5|$ 5
   b. $|-2|$ 2
2. Replace each __?__ with >, <, or =.
   a. $-3 \underline{\ ?\ } -5$ >
   b. $-2 \underline{\ ?\ } 3$ <

Closing the Lesson
Have students order a set of integers from least to greatest by placing them on a number line. Ask some students to describe what absolute value means and to give an example of its use.

Suggested Assignments
Skills: 1–15, 18–26 even, 35–38
Average: 1–16, 17–33 odd, 35–38
Advanced: 1–21 odd, 22–38

Communication: Writing

Guided Practice Exercises 3–10 have students make connections between real-world information and positive and negative integers. *Exercise 21* presents the algebraic definition of absolute value and asks students to compare it to the geometric definition given in the text.

Data Analysis

Exercises 16 and 22–26 show integers being used in diagrams and graphs, a valuable real-world connection. As a further connection to geography, have students find the names of the 18 states referred to in *Exercise 16.*

Reasoning

Exercises 27–34 require students to examine abstract statements and to draw conclusions about relationships that exist. Students should then use numerical examples to support their reasoning.

COMMUNICATION «*Writing*

Write an integer that represents each situation.

«**3.** At night, temperatures on Mars can reach 130°F below zero. −130

«**4.** Jacksonville, Florida, is located at sea level. 0

«**5.** Linda deposited $25 in her savings account. 25

«**6.** The Bay City Bengals football team lost a total of 80 yd in one game. −80

Describe a situation that can be represented by each integer.

«**7.** 44   «**8.** −6   «**9.** −15   «**10.** 0

7–10. Answers will vary. An example is John deposited $44 in his savings account.

Write the opposite of each integer.

11. 5 −5   **12.** −9 9   **13.** −11 11   **14.** 0 0

For each pair of integers, tell which integer is farther to the right on a horizontal number line.

15. −6, 2 2   **16.** 0, −1 0   **17.** 10, −10 10   **18.** −11, −8 −8

Find each absolute value.

19. $|-8|$ 8   **20.** $|33|$ 33   **21.** $|0|$ 0   **22.** $|-23|$ 23

Replace each ? with >, <, or =.

23. 0 ? 6 <   **24.** 5 ? −2 >   **25.** −10 ? −12 >   **26.** −13 ? −9 <

Exercises

Find each absolute value.

A   **1.** $|-5|$ 5   **2.** $|7|$ 7   **3.** $|13|$ 13

      4. $|-6|$ 6   **5.** $|-1|$ 1   **6.** $|10|$ 10

Replace each ? with >, <, or =.

7. 8 ? −9 >   **8.** −4 ? 4 <

9. ⁻18 ? −18 =   **10.** 12 ? 0 >

11. −11 ? −7 <   **12.** −1 ? −8 >

13. 0 ? −6 >   **14.** −5 ? 6 <

15. Two different depths in the Carlsbad Caverns are 754 ft below sea level and 900 ft below sea level. Which depth is closer to sea level? −754 ft

16. DATA, *pages 92–93* How many states have regions with normal low January temperatures less than or equal to 0°F? 17

Write in order from least to greatest. Use two inequality symbols.

B **17.** 4, −3, 9 $-3 < 4 < 9$ **18.** 1, 0, −1 $-1 < 0 < 1$ **19.** −10, −8, −6 $-10 < -8 < -6$ **20.** 2, −2, 0 $-2 < 0 < 2$

21. WRITING ABOUT MATHEMATICS In some textbooks, *absolute value* is defined as follows.

If a is positive, $|a| = a$.
If a is negative, $|a| = -a$.
If a is zero, $|a| = 0$.

Write a paragraph that compares this definition to the definition given in this lesson.

21. The definition given in the lesson refers to distance from zero. This definition uses a variable to represent any integer. This definition specifically includes zero.

DATA ANALYSIS

Use the graph at the right.

22. How much profit does the Value Company show? $20,000

23. How much loss does the Do Right Company show? $10,000

24. How much loss does the Hearty Company show? $30,000

25. Which company shows a greater loss? Hearty Company

26. List the companies in order from the company with the greatest profit to the company with the greatest loss.
Buyers, Value, Essential, Do Right, Hearty

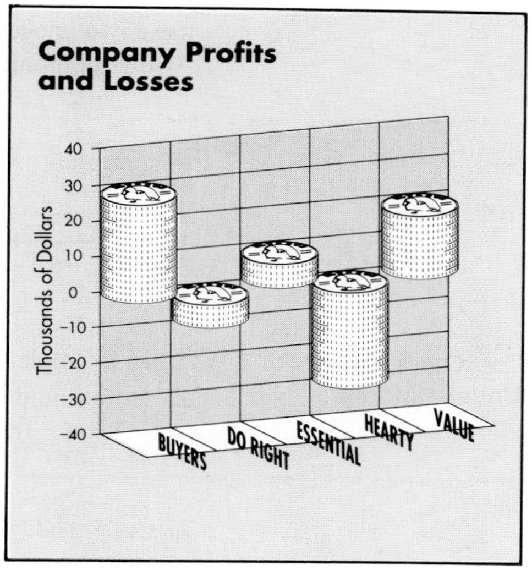

LOGICAL REASONING

Assume that *n* represents a negative integer and *p* represents a positive integer. Tell whether each statement is *always, sometimes,* or *never* true.

C **27.** $n > p$ never **28.** $p > n$ always

29. $0 > p$ never **30.** $-n < 0$ never

31. $|p| = p$ always **32.** $|n| < 0$ never

33. $|n| = |p|$ sometimes **34.** $|n| < |p|$ sometimes

SPIRAL REVIEW

S **35.** Find the sum: 1529 + 674 *(Toolbox Skill 7)* 2203

36. Evaluate $8qr$ when $q = 5$ and $r = 4$. *(Lesson 1-1)* 160

37. Find the absolute value: $|-2|$ *(Lesson 3-1)* 2

38. Evaluate $51.3 \div p$ when $p = 3$. *(Lesson 1-2)* 17.1

Follow-Up

Extension

Have students research the yearly average high and low temperatures in degrees Celsius for six major cities. Then have them find the difference between the high and low temperatures for each city.

Enrichment

Another way of writing the definition of absolute value is as follows:

$$|a| = \begin{cases} a; & a \geq 0 \\ -a; & a < 0 \end{cases}$$

Have students relate this definition to the one given in Exercise 21 on page 97.

Enrichment

Enrichment 23 is pictured on page 92E.

Practice

Practice 28 is shown below.

Practice 28, *Resource Book*

Practice 28
Skills and Applications of Lesson 3-1

Find each absolute value.

1. |10| 10	2. |−7| 7	3. |−14| 14
4. |0| 0	5. |−25| 25	6. |15| 15
7. |22| 22	8. |−9| 9	9. |16| 16
10. |33| 3	11. |−19| 19	12. |−45| 45
13. |27| 27	14. |−17| 17	15. |−34| 34
16. |37| 37	17. |−89| 89	18. |125| 125
19. |−350| 350	20. |−75| 75	21. |6| 6

Replace each ? with >, <, or =.

22. −7 ? 7 <	23. −1 ? 3 <	24. −11 ? −10 <
25. −6 ? 6 <	26. 11 ? 5 >	27. −5 ? −3 <
28. 0 ? −6 >	29. −13 ? 8 <	30. −11 ? −11 =
31. 0 ? 5 <	32. −25 ? −36 >	33. 28 ? −17 >
34. 14 ? −14 >	35. −8 ? −19 >	36. 11 ? −2 >
37. −9 ? 9 <	38. −23 ? −29 >	39. −33 ? −53 >
40. −35 ? −21 <	41. 12 ? −32 >	42. 17 ? −17 >
43. −45 ? −40 <	44. |5| ? |5| =	45. |−8| ? |8| =

46. Katie read the outdoor temperature on the thermometer on Wednesday and recorded −6° C. The next day James read the temperature at −9° C. Which of the two is the colder temperature? −9° C

47. The sports announcer gave one golfer's score as −7 (7 below par) and the second golfer's score as −12 (12 below par). The winner is the golfer with a score farthest below par (zero would be a score of par). Which golfer has the winning score? the second golfer

80 Resource Book, MATHEMATICAL CONNECTIONS
Copyright © by Houghton Mifflin Company. All rights reserved.

Teaching the Lesson
- Materials: integer chips
- Lesson Starter 22
- Visuals, Folder H
- Using Technology, p. 92C
- Using Manipulatives, p. 92C
- Toolbox Skill 7

Lesson Follow-Up
- Practice 29
- Enrichment 24: Application
- Study Guide, pp. 43–44

Warm-Up Exercises

Find each sum.

1. $67 + 98$ 165
2. $49 + 73$ 122
3. $107 + 89$ 196
4. $146 + 205$ 351
5. $231 + 275$ 506

Lesson Focus

Ted Berra is losing weight to make the wrestling team. He lost 7 lb the first month, 2 lb the second month, and 3 lb the third. How many pounds did he lose altogether? Today's lesson focuses on addition of integers with the same sign.

Teaching Notes

Relate the generalization for adding integers with the same sign to the addition of integers on a number line.

Key Question

Will the sum of two positive integers be positive or negative? Will the sum of two negative integers be positive or negative? Explain.
Positive; negative; the sum of two integers with the same sign has the same sign as the integers.

Adding Integers with the Same Sign

Objective: To add integers with the same sign.

In a football game, Todd lost three yards on one play and four yards on the next play.

QUESTION How many yards did Todd lose in all on the two plays?

To find the total, think of the loss of three yards as -3 and the loss of four yards as -4. You can then use arrows along a number line to add these two integers. Arrows pointing to the left represent negative numbers. Arrows pointing to the right represent positive numbers.

Example 1

Solution

Find the sum $-3 + (-4)$.

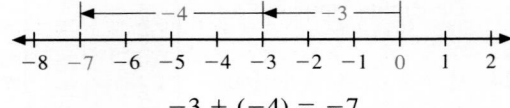

$$-3 + (-4) = -7$$

Start at 0.
Slide 3 units *left*.
Slide 4 more units *left*.
Stop at -7.

☑ **Check Your Understanding**

1. In Example 1, why do the arrows point to the left?
2. How would Example 1 be different if you were asked to find the sum $-4 + (-3)$? How would it be the same?
See *Answers to Check Your Understanding* at the back of the book.

ANSWER Todd lost a total of seven yards.

You can also use the number line as a model for the more familiar situation of adding *positive* integers.

Example 2

Solution

Find the sum $3 + 2$.

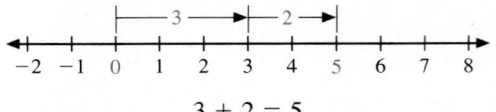

$$3 + 2 = 5$$

Start at 0.
Slide 3 units *right*.
Slide 2 more units *right*.
Stop at 5.

In each example above, notice that the integer representing the sum is related to the total distance from 0 represented by the arrows. Because an integer's distance from zero on a number line is its absolute value, you can use absolute value to add integers with the same sign.

98 Chapter 3

Error Analysis

When adding integers with the same sign, students often ignore the signs. Remind them to use the common sign in the answer.

> **Generalization:** *Adding Integers with the Same Sign*
>
> To add two integers that have the *same* sign, add their absolute values. Then give the sum the sign of the integers.

Example 3
Solution

Find each sum: **a.** $-10 + (-14)$ **b.** $18 + 9$

a. $|-10| = 10$ and $|-14| = 14$
$10 + 14 = 24$, so
$-10 + (-14) = -24$

b. $|18| = 18$ and $|9| = 9$, so
$18 + 9 = 27$

Guided Practice

«**1.** **COMMUNICATION** «*Reading*   Replace each ___?___ with the correct word.
The sign of the sum of two negative integers is ___?___. negative
The sign of the sum of two positive integers is ___?___. positive

COMMUNICATION «*Writing*

Write the addition represented on each number line.

«**2.**

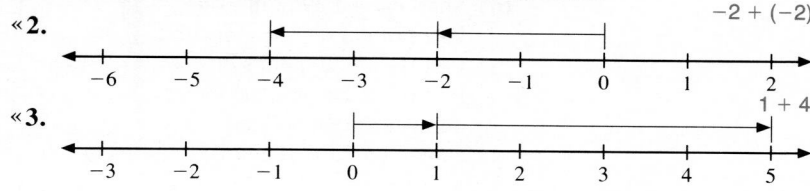

$-2 + (-2)$

«**3.**

$1 + 4$

Represent each addition on a number line.

Answers to Guided Practice 4–7 are on p. A5.

«**4.** $3 + 3$ «**5.** $-4 + (-4)$ «**6.** $-1 + (-2)$ «**7.** $6 + 2$

Write an addition that represents each situation.

«**8.** Yesterday the temperature rose 6°F and then rose 7°F. $6 + 7$

«**9.** Lia withdrew $5 from her account one day and $5 the next day.
$-5 + (-5)$

Describe a situation that can be represented by each addition.

«**10.** $-15 + (-45)$ «**11.** $4 + 4$ «**12.** $27 + 60$ «**13.** $-12 + (-10)$

Find each sum.

14. $-12 + (-23)$ −35   **15.** $-7 + (-15)$ −22   **16.** $42 + 9$ 51

17. $16 + 31$ 47   **18.** $39 + 86$ 125   **19.** $-55 + (-34)$ −89

20. $-97 + (-27)$ −124   **21.** $48 + 53$ 101   **22.** $-44 + (-19)$ −63

10–13. Answers will vary. Examples are given.
10. An elevator descends 15 floors and then descends 45 floors.
11. The temperature rose 4°F and then rose 4°F.
12. Sue drove 27 mi during the morning and 60 mi during the afternoon.
13. A football team lost 12 yd and then lost 10 yd.

Integers   **99**

Application (*Cooperative Learning*)

Have students work in cooperative groups to collect data on the school's football team. Each group could assume responsibility for one game. Have a group determine the following for each game:
  (a) the number of yards gained per quarter;
  (b) the number of yards lost per quarter;
  (c) the number of yards gained during the entire game;
  (d) the number of yards lost during the entire game.
Draw number lines to illustrate the totals for (a)–(d).

Using Technology

For a suggestion on using a calculator and a computer to add integers, see page 92C.

Using Manipulatives

For a suggestion on using integer chips to model addition, see page 92C.

Additional Examples

1. Find the sum $-2 + (-4)$. -6
2. Find the sum $4 + 3$. 7
3. Find each sum.
  a. $-9 + (-16)$ -25
  b. $23 + 8$ 31

Closing the Lesson

Ask a student to explain how to find the sum of two negative integers using a number line. Ask another student to demonstrate adding positive integers using a number line.

Suggested Assignments

Skills: 1–16, 17–31 odd, 33–35
Average: 1–16, 17–31 odd, 33–35
Advanced: 1–15 odd, 17–35

Exercise Notes

Communication: Reading
Guided Practice Exercise 1 checks students' understanding of the sign of a sum of two integers.

Communication: Writing
Guided Practice Exercises 2–13 ask students to write mathematical expressions for diagrams and actual situations, and vice versa.

Making Connections/Transitions
Exercises 13–16 present real-world uses of adding integers. Ask students which answers represent positive quantities and which represent negative quantities.

Follow-Up

Project
You may want to have students research various areas, such as banking, sales, science, and so on, for ways in which positive and negative integers are used.

Enrichment
Enrichment 24 is pictured on page 92E.

Practice
Practice 29, *Resource Book*

Exercises

Find each sum.

A
1. 22 + 7   29
2. −14 + (−13)   −27
3. −5 + (−25)   −30
4. 8 + 34   42
5. −53 + (−10)   −63
6. −72 + (−8)   −80
7. 62 + 6   68
8. 42 + 52   94
9. −29 + (−16)   −45
10. 68 + 49   117
11. 75 + 37   112
12. −32 + (−18)   −50

13. Rich has $13 in his savings account. He deposits $25. How much money is in his savings account after the deposit? $38

14. The football team lost 11 yd, 11 yd, and 12 yd on three plays. How many yards did the team lose in all on the plays?   34 yd

15. During Jenna's experiment, the temperature of water dropped three times. She recorded losses of 6°F, 12°F, and 15°F. What was the total loss?   33°F

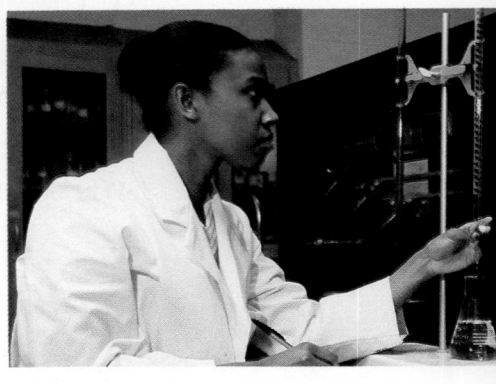

16. Shari earned 29 points on Part I of a test, 31 points on Part II, and 29 points on Part III. What was the total number of points Shari earned on Parts II and III?   60 points

Evaluate each expression when $p = 8$, $q = -9$, $r = -5$, and $s = 10$.

B
17. $p + s$ 18
18. $r + q$ −14
19. $q + (-3)$ −12
20. $16 + s$ 26
21. $-2 + r + (-15)$ −22
22. $12 + 24 + p$ 44
23. $q + (-2) + r + (-1)$ −17

Simplify. 26. $-7b + 7c$ 27. $-10m$ 28. $36 + 14x$ 29. $-10r + 4s$

24. $-3z + (-5z)$ −8z
25. $6y + 9y$ 15y
26. $-5b + 7c + (-2b)$
27. $-8m + (-2m)$
28. $7 + 14x + 29$
29. $-3r + (-7r) + 4s$
30. $4d + 2c + 11c$ 13c + 4d
31. $8m + 3n + 5n$ 8m + 8n
32. $-9 + (-2y) + (-8)$ −17 + (−2y)

SPIRAL REVIEW

S
33. Estimate the quotient: $6445 \div 83$ *(Toolbox Skill 3)*   about 80
34. Find the sum: $-5 + (-5)$ *(Lesson 3-2)*   −10
35. Simplify: $3(4y - 1)$ *(Lesson 2-2)*   12y − 3

100   Chapter 3

3-3 Adding Integers with Different Signs

Objective: To add integers with different signs.

Terms to Know
- *addition property of opposites*
- *additive inverse*

APPLICATION

Carl enters an elevator at a garage level that is two floors below the lobby level. He leaves the elevator after going up five floors.

QUESTION At which floor does Carl leave the elevator?

To find the answer, you can represent the garage level as −2 and the number of floors Carl goes up as +5. Then add the integers. Sometimes it is helpful to use a number line to add integers that have different signs.

Example 1 Find each sum.

a. −2 + 5   **b.** 1 + (−3)

Solution

a.

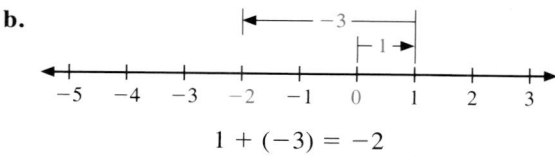

Start at 0.
Slide 2 units *left*.
Slide 5 units *right*.
Stop at 3.

−2 + 5 = 3

b.

Start at 0.
Slide 1 unit *right*.
Slide 3 units *left*.
Stop at −2.

1 + (−3) = −2

ANSWER Part (a) shows that Carl leaves the elevator at the third floor.

In Example 1, notice that the *difference* between the distances the arrows represent is the absolute value of the sum of the two integers. You can use the difference of absolute values to add integers with different signs.

> **Generalization: Adding Integers with Different Signs**
>
> To add two integers with different signs, first find their absolute values. Then subtract the lesser absolute value from the greater absolute value. Give the result the sign of the integer with the greater absolute value.

Integers   **101**

Lesson Planner

Teaching the Lesson
- Materials: integer chips
- Lesson Starter 23
- Visuals, Folder H
- Using Technology, p. 92C
- Using Manipulatives, p. 92C
- Toolbox Skill 7

Lesson Follow-Up
- Practice 30
- Enrichment 25: Exploration
- Study Guide, pp. 45–46
- Cooperative Activity 6

Warm-Up Exercises

Find each sum.

1. −4 + (−7)   −11
2. 5 + 17   22
3. −13 + (−8)   −21
4. |−6| + |7| 13
5. |−6| + |−8| 14

Lesson Focus

To motivate today's lesson, you may want to use the application problem below. To solve the problem, students must add integers with different signs, the focus of today's lesson.

Application

At midnight on a winter night, the temperature was −8°F. By noon, the temperature had risen 37°. What was the temperature at noon? 29°F

Teaching Notes

Students may have difficulty adding integers with different signs. Use number lines as models and discuss thoroughly the generalization for adding integers given at the bottom of this page.

Key Question

How do you determine whether to add or subtract absolute values when adding integers? same signs: add; different signs: subtract

Error Analysis

Students often do not take the absolute values when adding integers. Stress that when adding integers, you always take the absolute value first and then add or subtract according to whether the integers have the same or different signs.

Using Technology

For a suggestion on using a calculator and a computer to add integers, see page 92C.

Using Manipulatives

For a suggestion on using integer chips to model addition, see page 92C.

Application

Have students make up problems with checking accounts. Writing a check is the same as using a negative number; money is being withdrawn from an account. Making a deposit represents a positive number. Balancing a checkbook requires using the concepts presented in this lesson.

Additional Examples

Find each sum.

1. **a.** $-3 + 4$ 1
   b. $1 + (-4)$ -3
2. **a.** $8 + (-15)$ -7
   b. $-8 + 13$ 5

Closing the Lesson

Ask a student to explain how to add integers with different signs. Ask another student to demonstrate $-7 + 9$ by using a number line. Ask another student to explain how to solve the same problem by using absolute values.

Example 2
Solution

Find each sum: **a.** $10 + (-16)$ **b.** $-7 + 12$

a. $|10| = 10$ and $|-16| = 16$
Subtract: $16 - 10 = 6$
The negative integer has the greater absolute value, so the sum is negative.
$10 + (-16) = -6$

b. $|-7| = 7$ and $|12| = 12$
Subtract: $12 - 7 = 5$
The positive integer has the greater absolute value, so the sum is positive.
$-7 + 12 = 5$

☑ Check Your Understanding

1. Describe how Example 1(a) would be different if you found the sum $-10 + 16$.

2. Describe how to find the sum $3 + (-3)$.
   See *Answers to Check Your Understanding* at the back of the book.

In the case of adding opposites, the sum will always be zero. This fact is so useful in algebra that it is identified as a *property* of opposites.

> **Addition Property of Opposites**
>
> The sum of a number and its opposite is zero.
> $$a + (-a) = 0 \text{ and } -a + a = 0$$

Because the sum of a number and its opposite is zero, the opposite of a number is sometimes called the **additive inverse** of the number.

Guided Practice

COMMUNICATION «*Reading*

Refer to the text on pages 98–99 and 101–102.

1. The generalizations are alike in that they both use absolute value. They are different in that when the integers have the same sign, you add the absolute values and when the integers have different signs, you subtract the lesser absolute value from the greater.

«**1.** Compare the generalizations for adding integers with the same sign and adding integers with different signs. How are the generalizations alike? How are they different?

«**2.** Describe the addition property of opposites in words.
When you add a number and its opposite, the sum is zero.

COMMUNICATION «*Writing*

Write the addition represented on each number line.

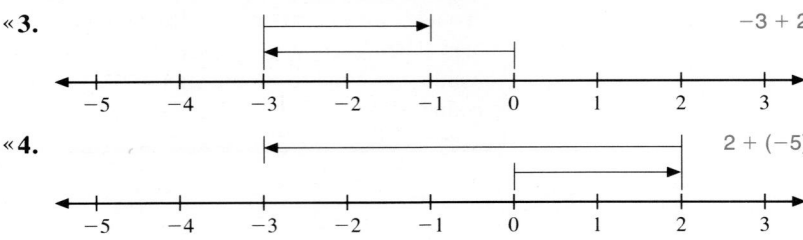

«**3.** $-3 + 2$

«**4.** $2 + (-5)$

Represent each addition on a number line.

« **5.** $-1 + 4$ « **6.** $-4 + 6$ « **7.** $5 + (-1)$ « **8.** $4 + (-7)$

Answers to Guided Practice 5–8 are on p. A5.

Tell whether each sum is *positive*, *negative*, or *zero*.

9. $88 + (-67)$ positive   **10.** $-22 + 19$ negative   **11.** $25 + (-26)$ negative   **12.** $14 + (-14)$ zero

Find each sum.

13. $-13 + 2$ −11   **14.** $4 + (-9)$ −5   **15.** $8 + (-8)$ 0

16. $-4 + 4$ 0   **17.** $3 + (-7) + 9$ 5   **18.** $-2 + 6 + (-10)$ −6

Exercises

Find each sum.

A
1. $-12 + 12$ 0   **2.** $19 + (-19)$ 0   **3.** $5 + (-26)$ −21

4. $-1 + 27$ 26   **5.** $9 + (-9)$ 0   **6.** $-6 + 6$ 0

7. $-2 + 7$ 5   **8.** $-25 + 18$ −7   **9.** $-8 + 18$ 10

10. $26 + (-15)$ 11   **11.** $-13 + 6$ −7   **12.** $-28 + 24$ −4

13. $14 + (-40)$ −26   **14.** $41 + (-45)$ −4   **15.** $20 + (-6) + (-7)$ 7

16. $-8 + 23 + (-14)$ 1   **17.** $-8 + (-13) + 7$ −14   **18.** $-25 + 11 + 5$ −9

19. Sue deposited $45 in her savings account and then withdrew $53. After the withdrawal, did she have more or less money than when she started? How much more or less? less money; $8 less

20. Pete lost three yards on one play and then gained eight yards on the next play. Find the total number of yards gained or lost. 5 yd gained

Evaluate each expression when $a = 4$, $b = -6$, and $c = -3$.

B
21. $-1 + a + c$ 0   **22.** $a + c + b$ −5   **23.** $b + a + b$ −8

24. $-7 + a + c$ −6   **25.** $b + 10 + a$ 8   **26.** $c + b + 9$ 0

MENTAL MATH

To add integers mentally, it is helpful to look for opposites. You can also group positive and negative integers.

Find each sum mentally.

27. $-3 + 5 + (-8) + (-6) + 9 + 3$ 0

28. $-10 + (-7) + 12 + (-12) + 8 + 8$ −1

29. $-2 + (-11) + 5 + 11 + (-7) + 16$ 12

30. $6 + 10 + (-4) + (-10) + (-2)$ 0

Integers   **103**

Suggested Assignments

Skills: 1–20, 21–35 odd, 46–49
Average: 1–20, 21–43 odd, 46–49
Advanced: 1–19 odd, 21–49

Exercise Notes

Critical Thinking
Determining how generalizations are alike and how they are different, as in *Guided Practice Exercise 1*, requires students to *compare* the generalizations and *draw conclusions* about their use.

Communication: Writing
Guided Practice Exercises 3–8 ask students to translate pictorial representations as mathematical expressions, and vice versa.

Mental Math
Exercises 27–30 show how mental math can be used to add integers. These exercises also require the use of the associative and commutative properties.

Problem Solving

In *Exercises 31–36*, negative integers are used in the launch of a space shuttle.

Reasoning

Exercises 44 and 45 lead students to the discovery that an equation may have many different solutions that are correct. The argument developed in *Exercise 45* should indicate that there are an infinite number of possible solutions to ■ + ▲ = −1.

Follow-Up

Extension *(Cooperative Learning)*

A sample bank statement could be given to students with a list of checks that have been written and deposits that have been made but not yet credited to the account. Working in cooperative groups, students should reconcile the balance and explain how they arrived at their answers.

Enrichment

Enrichment 25 is pictured on page 92E.

Practice

Practice 30, *Resource Book*

PROBLEM SOLVING/APPLICATION

The time for maneuvers during a space shuttle launch is given relative to liftoff, which is called *T*. The seconds before liftoff are assigned negative numbers, the seconds after liftoff are assigned positive numbers, and liftoff itself is zero. For instance, when you hear a mission controller refer to *T minus 45*, the time is 45 s before liftoff.

31. 50 s after liftoff
32. T minus 30
33. exact times; each maneuver begins at an exact time before liftoff.
34. T minus 15
35. T minus 45
36. *T minus ten and counting* means that the countdown is continuing; *T minus ten and holding* means that the countdown has stopped 10 s before liftoff.

31. What is the meaning of *T plus 50*?

32. Give the expression for one half minute before liftoff.

33. Do you think that the times for these maneuvers are *estimates* or *exact times*? Explain.

34. A maneuver begins at *T minus 40* and requires 25 s for completion. When should this maneuver be completed?

35. A maneuver begins at *T plus 15*. Preparation begins one minute earlier. At what time must the preparation begin?

36. **RESEARCH** Find out how the expression *T minus ten and counting* is different from the expression *T minus ten and holding*.

GROUP ACTIVITY

Suppose that ■ represents one integer in each exercise. Replace each ■ with the integer that makes the statement true.

37. 6 + ■ = −2 −8

38. −4 + ■ = 7 11

39. ■ + 10 = 0 −10

40. 5 + ■ + ■ = −3 −4

41. −8 + ■ + ■ = −12 −2

42. ■ + (−1,000,000) = 1 1,000,001

43. ■ + 1,000,000,000 = −1 −1,000,000,001

Suppose that ■ and ▲ represent different integers.

44. Lists will vary. **C**
Example: 5, −6; 4, −5; 3, −4; −3, 2.

45. infinitely many pairs; for every integer, you can add the opposite of one greater than the integer.

44. List four pairs of integers for which ■ + ▲ = −1 is true. Compare your list with other lists in the group. Are all lists the same?

45. For how many different pairs of integers is ■ + ▲ = −1 a true statement? Give a convincing argument to support your answer.

SPIRAL REVIEW

S

46. Find the sum mentally: 27 + 21 + 43 *(Lesson 1-6)* 91

47. Continue the pattern: 1, 5, 9, 13, __?__, __?__, __?__ *(Lesson 2-5)* 17, 21, 25

48. Write 2,700,000 in scientific notation. *(Lesson 2-10)* 2.7×10^6

49. Find the sum: −44 + 16 *(Lesson 3-3)* −28

Modeling Integers Using Integer Chips

Objective: To use integer chips to model integers.

Materials

■ integer chips

A $\boxed{+}$ integer chip represents positive 1. A $\boxed{-}$ integer chip represents negative 1. Because $1 + (-1) = 0$, the pair of integer chips $\boxed{+}\boxed{-}$ represents 0.

Activity I *Representing Integers*

1 Name the integer represented in each diagram.

a. $\boxed{+}\ \boxed{+}\ \boxed{+}$
    $\boxed{+}\ \boxed{+}$ 5

b. $\boxed{-}\ \boxed{-}\ \boxed{-}$ −3

c. $\boxed{-}$ −1

d. $\boxed{+}\ \boxed{+}$ 2

2 Show how to use integer chips to represent each integer. See answer on p. A5.
   a. 4   b. −2   c. −5   d. 3

3 Show how to represent the number 0 using the number of chips indicated. See answer on p. A5.
   a. 2   b. 4   c. 6   d. 8

Activity II *Combining Chips*

The combination of $\boxed{+}$ and $\boxed{-}$ chips in the diagram below represents −2.

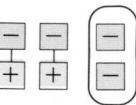

1 Name the integer represented by each diagram.

a. 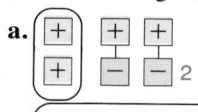 2

b. 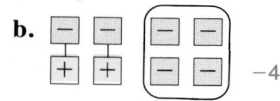 −4

c. 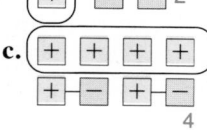 4

d. 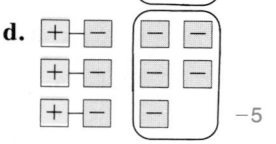 −5

2 Show how to combine as many integer chips as possible in each diagram. What integer does the diagram represent?

a. 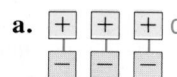 0

b. 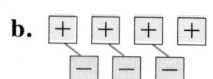 1

c. 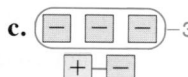 −3

d. 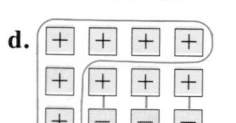 6

Integers **105**

Teaching Note

Making a model and drawing a diagram are common problem-solving strategies. The use of integer chips may help some students to solidify their understanding of the topics covered in the first three lessons of this chapter.

Activity Notes

Activity I

Activity I relates to Lesson 3-1. The meaning of positive integers, negative integers, and zero is reinforced through the use of a model.

Activities II and III

Activities II and III allow students to practice combining chips and using zero pairs, thus providing a concrete representation of the addition property of opposites.

Activity IV

In *Activity IV* chips are used to represent addition of integers with both the same and with different signs. The addition property of opposites is used again to add integers with different signs.

Activity III *Using Pairs of Zero*

1 Use the diagram below. Find the minimum number of ⊞ or ⊟ chips you would have to add to the diagram to obtain each integer.

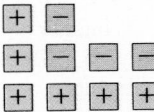

a. 3 one ⊞   **b.** 1 one ⊟   **c.** −3 five ⊟   **d.** 5 three ⊞

2 Use the diagram from Step 1. Explain how you could add two integer chips without changing the integer that is represented.

2. Add one ⊞ chip and one ⊟ chip.

3 Show how to represent the integer −3 using the number of chips indicated. See answer on p. A5.

a. 3   **b.** 5   **c.** 7   **d.** 9

4 Is it possible to represent the integer −3 using an even number of chips? Explain. No; you always need 3 ⊟ chips along with pairs of ⊞⊟ chips.

Activity IV *Adding Integers*

To add integers using integer chips, bring all the chips together. Combine any pairs of ⊞ and ⊟ chips. The remaining integer chips represent the sum.

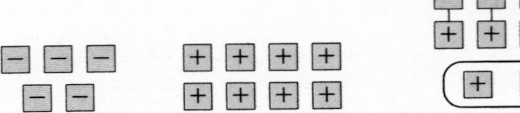

5 negative chips   8 positive chips   After combining 5 pairs of 0, you have 3 positive chips.

$$-5 \quad + \quad 8 \quad = \quad 3$$

1 Write the addition that is represented by each diagram. Combine the chips to find the sum. −1 + (−8) = −9

a.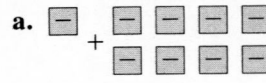

b. 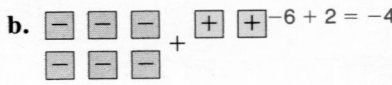 −6 + 2 = −4

c. 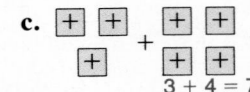 3 + 4 = 7

d. 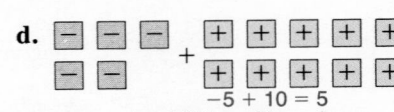 −5 + 10 = 5

2 Show how to use integer chips to represent each addition. Show how you would combine the chips to find the sum. What is the sum?

a. 4 + 6   **b.** −2 + (−7)   **c.** −8 + 4   **d.** 6 + (−5)

See answer on p. A5.

3 Suppose you are trying to find a sum using integer chips. Explain how you can tell what the sign of the sum will be. The sign of the sum will be the sign of the greater number of chips.

3-4

Subtracting Integers

Objective: To find the difference of two integers.

APPLICATION

An atom is made up of tiny particles: electrons, protons, and neutrons. An electron has a charge of -1, a proton has a charge of $+1$, and a neutron has no charge. During chemical reactions, only electrons move to other atoms. If an atom loses or gains electrons, its charge changes.

QUESTION During a chemical reaction, an atom with a charge of -3 loses one electron. What is the charge of the atom after the reaction?

To find the charge, represent the electron by -1. Then subtract -1 from the atom's charge of -3.

Example 1
Solution

Find the difference $-3 - (-1)$.

One way to subtract integers is to use integer chips.

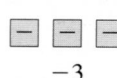
Start with 3
negative chips.
-3

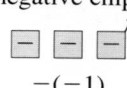
Take away 1
negative chip.
$-(-1)$

$=$

2 negative
chips remain.
-2

ANSWER The atom has a charge of -2 after the reaction.

Example 2
Solution

Find the difference $4 - 6$.

Start with 4
positive chips.

4

In order to take away 6 positive
chips, add 2 sets of 0.

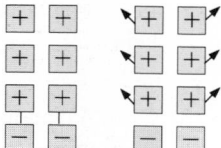

-6

$=$

2 negative
chips remain.

-2

✓ **Check Your**
Understanding

1. In Example 2, why is the answer negative?
See *Answers to Check Your Understanding* at the back of the book.

Integers   **107**

Teaching the Lesson
- Materials: integer chips, calculator
- Lesson Starter 24
- Visuals, Folder H
- Using Technology, p. 92C
- Using Manipulatives, p. 92D
- Toolbox Skill 8

Lesson Follow-Up
- Practice 31
- Enrichment 26: Application
- Study Guide, pp. 47–48
- Computer Activity 3
- Manipulative Activity 5

Warm-Up Exercises

Find each answer.
1. $-3 + (-7)$ -10
2. $3 + (-7)$ -4
3. $7 - 3$ 4
4. $-3 + 7$ 4
5. $-7 + (-3)$ -10

Lesson Focus

Ask students if they have ever borrowed money from a friend and then repaid a portion of it. If so, while still in debt, they did owe less money. This situation involves subtracting with negative numbers, which is the focus of today's lesson.

Teaching Notes

Reinforce the generalization that subtracting an integer is the same as adding its opposite by using many examples. Integer chips can be used to model the subtraction of integers.

Teaching Notes *(continued)*

Key Questions

1. Why is it important to rewrite a subtraction problem as an addition problem before working it? You can use the addition rules presented earlier.

2. Why does subtracting a negative integer result in a greater answer? Subtracting a negative integer is the same as adding a positive integer, so the answer becomes greater.

Error Analysis

Students often become confused when subtracting integers. Remind students to rewrite subtraction problems as addition problems and then apply the generalizations for finding the sum of two integers.

Using Technology

For a suggestion on using calculators and computers to subtract integers, see page 92C.

Using Manipulatives

For a suggestion on using integer chips to model subtraction, see page 92D.

Additional Examples

Find the difference.

1. $-5 - (-2)$ -3
2. $3 - 5$ -2
3. $-4 - 6$ -10
4. $-6 - (-7)$ 1

Closing the Lesson

Ask students to do the following.

1. Draw a diagram to illustrate subtracting a positive integer from a negative integer.
2. Draw a diagram to illustrate subtracting a negative integer from a negative integer.
3. Rewrite a subtraction problem as an addition problem and solve it.
4. Explain the generalization that states how to subtract integers.

When you find the sum $-3 + 1$, the answer is -2. In Example 1 you found that the difference $-3 - (-1)$ also is -2. You can see that subtracting -1 is the same as adding 1.

> **Generalization:** *Subtracting Integers*
>
> Subtracting an integer is the same as adding its opposite.
> $$a - b = a + (-b)$$

Example 3

Find each difference: **a.** $-3 - 4$ **b.** $-7 - (-8)$

Solution

a. $-3 - 4 = -3 + (-4)$
       $= -7$

b. $-7 - (-8) = -7 + 8$
           $= 1$

✏️ **Check Your Understanding**

2. In Example 3(b), why is $-7 - (-8)$ written as $-7 + 8$?
See *Answers to Check Your Understanding* at the back of the book.

Some calculators have a change-sign key, [+/-] . Pressing this key changes the sign of the number that is displayed. You use this key to enter negative numbers. For example, you can use this key sequence to add -49 and -78.

Guided Practice

«**1. COMMUNICATION** «*Reading*   Replace the __?__ with the correct word.

To subtract an integer, add its __?__. opposite

COMMUNICATION «*Writing*

$-5 - (-3) = -2$

«**2.** Let ⊟ represent -1. Write the subtraction represented by the diagram below.

«**3.** Draw a diagram similar to the one in Exercise 2 to illustrate this subtraction: $-3 - (-2) = -1$ ⊟⊟⊟ → ⊟

Replace each __?__ with the integer that makes the statement true.

4. $6 - (-8) = 6 + $ __?__ 8   **5.** $-12 - 14 = -12 + $ __?__ -14

Find each difference.

6. $12 - (-3)$ 15   **7.** $-4 - 14$ -18 **8.** $32 - 40$ -8

9. $-11 - (-15)$ 4   **10.** $-24 - (-16)$ -8 **11.** $19 - 7$ 12

108 Chapter 3

Exercises

Suggested Assignments
Skills: 1–20, 21–37 odd, 39–44
Average: 1–20, 21–37 odd, 39–44
Advanced: 1–17 odd, 19–44

Find each difference. 12. -20

A
1. $-5 - 4$ -9
2. $-21 - 3$ -24
3. $-12 - (-18)$ 6
4. $-3 - (-3)$ 0
5. $10 - (-15)$ 25
6. $7 - (-14)$ 21
7. $9 - 9$ 0
8. $16 - 5$ 11
9. $8 - 11$ -3
10. $6 - 35$ -29
11. $-13 - (-9)$ -4
12. $-28 - (-8)$
13. $-18 - (-18)$ 0
14. $-24 - (-32)$ 8
15. $-15 - 37$ -52
16. $38 - 47$ -9
17. $11 - 7$ 4
18. $23 - 23$ 0

19. The record low temperature in Denver for March is $-11°F$. The normal low temperature for March is $25°F$. How much greater is the normal than the record temperature? $36°F$

20. The peak of Mount Everest is 8848 m above sea level. The shore of the Dead Sea is 400 m below sea level. How much higher is the peak of Mount Everest than the shore of the Dead Sea? 9248 m

Evaluate each expression when $c = -3$ and $d = 9$.

B
21. $c - (-6)$ 3
22. $-4 - d$ -13
23. $7 - d$ -16 -14
24. $5 - 19 - c$ -11
25. $c - d$ -12
26. $d - c$ 12

 CALCULATOR

Match each subtraction with the correct calculator key sequence.

27. $87 - (-469)$ B
28. $87 - 469$ D
29. $-87 - (-469)$ C
30. $-87 - 469$ A

A. $\boxed{87.}$ $\boxed{+/-}$ $\boxed{-}$ $\boxed{469.}$ $\boxed{=}$

B. $\boxed{87.}$ $\boxed{-}$ $\boxed{469.}$ $\boxed{+/-}$ $\boxed{=}$

C. $\boxed{87.}$ $\boxed{+/-}$ $\boxed{-}$ $\boxed{469.}$ $\boxed{+/-}$ $\boxed{=}$

D. $\boxed{87.}$ $\boxed{-}$ $\boxed{469.}$ $\boxed{=}$

PATTERNS

Find the next three integers in each pattern.

31. $16, 11, 6, 1, \underline{?}, \underline{?}, \underline{?}$ $-4, -9, -14$
32. $2, 0, -3, -7, \underline{?}, \underline{?}, \underline{?}$ $-12, -18, -25$
33. $-23, -19, -15, -11, \underline{?}, \underline{?}, \underline{?}$ $-7, -3, 1$
34. $-1, -3, -5, -7, \underline{?}, \underline{?}, \underline{?}$ $-9, -11, -13$

Exercise Notes

Communication: Writing
Guided Practice Exercises 2 and 3 require students to interpret a model and to make a pictorial representation of a mathematical statement.

Calculator
Exercises 27–30 check students' understanding of the use of the $\boxed{+/-}$ key.

Number Sense
Exercises 31–34 extend the study of patterns, begun in Lesson 2-5, to the integers.

Integers 109

Application

Exercises 35–38 relate mathematics to the field of meteorology. Determining the wind-chill factor is a skill used by meteorologists.

Estimation

Exercise 38 requires students to estimate a wind-chill factor by using two values from the chart.

Follow-Up

Enrichment

Enrichment 26 is pictured on page 92E.

Practice

Practice 31 is shown below.

Practice 31, *Resource Book*

CAREER/APPLICATION

Meteorologists analyze weather data and forecast the weather. They report the *wind-chill factor* on windy days. The wind-chill factor is the temperature that results from the air temperature combined with the wind speed. Meteorologists use a chart like the one below to find the wind-chill factor.

Wind Speed in Miles per Hour	Air Temperature (°F)			
	20	10	0	−10
10	3	−9	−22	−34
20	−10	−24	−39	−53
30	−18	−33	−49	−64

35. The air temperature is 10°F and the wind speed is 30 mi/h. What is the wind-chill factor? −33°F

36. The air temperature is −10°F. How much colder does it feel when the wind speed is 10 mi/h than when there is no wind? 24°F

37. The air temperature is 0°F. The wind speed increases from 10 mi/h to 20 mi/h. By how much does the temperature seem to drop? 17°F

38. **ESTIMATION** The wind speed is 10 mi/h and the air temperature is 15°F. Estimate the wind-chill factor. about −3°F

SPIRAL REVIEW

S 39. **DATA,** *pages 92–93* What is the record low January temperature for Houston? *(Toolbox Skill 24)* about 10°F

40. Find the difference: $5 - 9$ *(Lesson 3-4)* −4

41. Write 107 cm in mm.   *(Lesson 2-9)* 1070 mm

42. Find the quotient: $\frac{9}{10} \div \frac{2}{5}$ *(Toolbox Skill 21)* $2\frac{1}{4}$

43. Estimate the sum: $7.9 + 8.4 + 8.1 + 7.8$ *(Toolbox Skill 4)* about 32

44. Evaluate $a + 10.9$ when $a = 17.5$.   *(Lesson 1-2)* 28.4

110   Chapter 3

Multiplying and Dividing Integers

Lesson Planner

Teaching the Lesson
- Materials: integer chips, calculator, decks of cards
- Lesson Starter 25
- Using Manipulatives, p. 92D
- Reteaching/Alternate Approach, p. 92D
- Cooperative Learning, p. 92D
- Toolbox Skills 9, 10

Lesson Follow-Up
- Practice 32
- Practice 33: Nonroutine
- Enrichment 27: Data Analysis
- Study Guide, pp. 49–50
- Test 10

Objective: To find products and quotients of integers.

DATA ANALYSIS

Companies that make a profit are said to be operating "in the black." A company that loses money is operating "in the red." The pictograph at the right uses black symbols for profits and red symbols for losses. For 1988, you can calculate the loss in millions of dollars by finding $2(-3)$.

Markets, Inc.			
1988	$	$	
1989	$		
1990		$	$
1991		$	$ $

$ = -\$3 million $ = \$3 million

Example 1

Solution

Find the product $2(-3)$.

You can use integer chips to multiply integers.

2 groups of -3 1 group of -6

$2(-3)$ = -6

Example 1 shows that, when you multiply a negative integer by a positive integer, the product is *negative*. What is the result when you multiply a positive integer by a negative integer? From Example 1 and the commutative property of multiplication, you know these two facts.

$$2(-3) = -6$$
$$2(-3) = (-3)2$$

You can conclude that $(-3)2$ must also equal -6. So, when you multiply a positive integer by a negative integer, the product is *negative*.

What is the result when you multiply two negative integers? Using Example 1 as a starting point, you can generate the pattern at the right. Notice that, when -3 is multiplied by a negative integer, the product is *positive*.

$$2(-3) = -6$$
$$1(-3) = -3$$
$$0(-3) = 0$$
$$(-1)(-3) = 3$$
$$(-2)(-3) = 6$$

Generalization: *Multiplying Integers*

The product of two integers with the same sign is positive.
The product of two integers with different signs is negative.

Warm-Up Exercises

Find each answer.
1. (26)4   104
2. 7(33)   231
3. $256 \div 4$ 64
4. $108 \div 6$ 18

Lesson Focus

Tell students that a high-speed elevator descends 20 ft every second. This can be represented as -20. After 8 seconds, the elevator will have moved $8(-20)$, or -160 ft. That is, it will have descended 160 ft. Today's lesson focuses on how to multiply and divide integers.

Teaching Notes

Stress that the product or quotient of two numbers with like signs is positive, and that the product or quotient of two numbers with unlike signs is negative.

Integers   111

Teaching Notes *(continued)*

Key Questions

1. How are the generalizations for multiplication and division of integers alike? Both generalizations state that if the signs of the integers being multiplied or divided are the same, the answer is positive; if the signs are different, then the answer is negative.

2. What property of numbers have you already learned that involves the order of the factors in a multiplication exercise? How can you use that property to explain why $3(-2)$ is the same as $(-2)(3)$? Commutative property of multiplication; the property applies to all real numbers, so $3(-2) = (-2)(3) = 6$.

Using Manipulatives

For a suggestion on using integer chips to model multiplication, see page 92D.

Reteaching/Alternate Approach

For a suggestion on using cooperative learning, see page 92D.

Additional Examples

1. Find the product of $3(-4)$. -12
2. Find each product.
   a. $(-12)(5)$ -60
   b. $(-8)(-11)$ 88
   c. $(-5)(6)(-4)$ 120
3. Find each quotient.
   a. $-35 \div 7$ -5
   b. $\frac{-56}{-7}$ 8

Closing the Lesson

Have students write a multiplication or division problem using integers and pass it to another student. Ask some students to describe the process they would use to find the answer.

Suggested Assignments

Skills: 1–20, 21–41 odd, 46, 47
Average: 1–17 odd, 19–41, 46, 47
Advanced: 1–19 odd, 21–47

Example 2

Find each product: a. $(-11)(4)$ b. $(-7)(-10)$ c. $(-5)(4)(-2)$

Solution

a. $(-11)(4)$
   $= -44$

b. $(-7)(-10)$
   $= 70$

c. $(-5)(4)(-2)$
   $= (-20)(-2)$
   $= 40$

 **Check Your Understanding**

1. In Example 2(c), why is the product positive?
   See *Answers to Check Your Understanding* at the back of the book.

You can use the relationship between multiplication and division to divide integers.

$$3(-7) = -21 \quad \rightarrow \quad -21 \div (-7) = 3$$
$$(-3)7 = -21 \quad \rightarrow \quad -21 \div 7 = -3$$
$$(-3)(-7) = 21 \quad \rightarrow \quad 21 \div (-7) = -3$$

> **Generalization:** *Dividing Integers*
>
> The quotient of two integers with the same sign is positive.
> The quotient of two integers with different signs is negative.

Keep in mind that division by zero is *undefined*.

Example 3

Find each quotient: a. $-15 \div 5$ b. $\frac{-48}{-12}$

Solution

a. $-15 \div 5 = -3$

b. $\frac{-48}{-12} = 4$

 **Check Your Understanding**

2. In Example 3(a), how do you know the quotient is negative?
   See *Answers to Check Your Understanding* at the back of the book.

Guided Practice

«1. **COMMUNICATION** «*Reading*   Replace each ___?___ with the correct word.

The product of two integers is positive if both factors are ___?___ or both factors are ___?___. The quotient of two integers is ___?___ if the two integers have different signs. positive; negative; negative

Tell whether each answer is *positive*, *negative*, *zero*, or *undefined*.

2. $(-7)14$ negative

3. $(-19)(0)(-3)$ zero

4. $8 \div (-2)$ negative

5. $\frac{0}{-5}$ zero

6. $-6 \div 0$ undefined

7. $\frac{-24}{-6}$ positive

Find each answer.

8. $2(-5)$ -10

9. $(-8)(9)$ -72

10. $4(-32)(-3)$ $\overset{384}{}$

11. $0(-2)(-5)$ 0

12. $-63 \div (-9)$ 7

13. $-36 \div 6$ -6

14. $\dfrac{39}{-13}$ -3

15. $\dfrac{-84}{-4}$ 21

16. $(-7)(-11)(-2)$ -154

Exercises

Find each answer.

A

1. $(-4)(-1)$ 4

2. $2(-30)$ -60

3. $-77 \div 7$ -11

4. $-42 \div (-6)$ 7

5. $(6)(0)(-15)$ 0

6. $-70 \div (-5)$ 14

7. $32 \div (-8)$ -4

8. $-10(12)$ -120

9. $(-7)(-22)$ 154

10. $144 \div (-3)$ -48

11. $-250 \div 25$ -10

12. $3(3)(-7)$ -63

13. $(-4)(5)(-5)$ 100

14. $(-6)(3)(0)$ 0

15. $(-7)(-6)(-2)$ -84

16. $\dfrac{75}{-5}$ -15

17. $\dfrac{-48}{3}$ -16

18. $\dfrac{-169}{-13}$ 13

19. Tina Landon noticed an unusual temperature change of $-20°F$ in just 4 h. Assume that the temperature changed at a steady rate. What was the change in temperature each hour? $-5°F$

20. Maria Savino's watch lost 2 min every day. How many minutes did the watch lose over a period of three days? 6 min

Use the order of operations to find each answer.

B

21. $-18 \div 3 + 5(-6)$ -36

22. $16 \div 4 + 2(-8)$ -12

23. $-3(1-8) + 2^3$ 29

24. $\dfrac{2}{8-10}$ -1

25. $\dfrac{-39+3}{-4}$ 9

Evaluate each expression when $x = -12$.

26. $2x + 1$ -23

27. $3x - 5$ -41

28. $8 - 5x$ 68

29. $\dfrac{x}{-2} + 5$ 11

30. $\dfrac{x}{-3} - 6$ -2

31. $9 - \dfrac{x}{4}$ 12

Simplify.

32. $2(5d - 2)$ $10d - 4$

33. $4(-3c + 6)$ $-12c + 24$

34. $-2(-3a + 5)$ $6a - 10$

35. $-7(-2y - 8)$ $14y + 56$

36. $6(-7 + x)$ $-42 + 6x$

37. $-5(4 - 4b)$ $-20 + 20b$

Integers **113**

Exercise Notes

Mental Math
Many of *Exercises 1–18* can be solved using mental math.

Calculator
Exercises 38–41 test students' ability to correctly perform complicated key sequences involving multiplication and division.

Making Connections/Transitions
Exercises 42–45 have students apply mathematical thinking to solve problems that arise in chemistry.

Follow-Up

Enrichment
You may wish to ask students to examine several manuals for administering standardized tests. On a test where there is a penalty for an incorrect answer, how is it determined what to deduct? Are the same number of points deducted for each item missed? Is the deduction based on the number of items?

Enrichment
Enrichment 27 is pictured on page 92E.

Practice
Practice 32, *Resource Book*

Quick Quiz 1 (3-1 through 3-5)

Replace each __?__ with >, <, or =.

1. $4 \underline{\;?\;} -5$ >
2. $-7 \underline{\;?\;} 4$ <
3. $-8 \underline{\;?\;} -5$ <
4. $-6 \underline{\;?\;} -6$ =

Find each answer.

5. $-3 + (-3)$ -6
6. $-4 + (-19)$ -23
7. $-5 + (-13)$ -18
8. $-6 + 9$ 3
9. $8 + (-14)$ -6
10. $-23 + 23$ 0
11. $-28 - 4$ -32
12. $16 - 29$ -13
13. $-2 - (-9)$ 7
14. $9(-5)$ -45
15. $5(-8)(-2)$ 80
16. $45 \div (-9)$ -5

Alternative Assessment

Use the cooperative activity on page 92D as an alternative method to assess whether students understand operations with integers completely.

 CALCULATOR

Match each calculator key sequence with the correct result. Assume that the calculator follows the order of operations.

A. | 2. | B. | -3. | C. | 18. | D. | -18. |

38. (| 10. | +/- | + | 16. |) | ÷ | 2. | +/- | =   B

39. | 10. | +/- | + | 16. | + | 2. | +/- | =   D

40. | 10. | + | 16. | ÷ | 2. | +/- | =   A

41. | 10. | + | 16. | +/- | ÷ | 2. | +/- | =   C

CONNECTING MATHEMATICS AND SCIENCE

Ions have electrical charges. If the sum of their charges is zero, ions combine to form compounds. For example, calcium chloride has one calcium ion and two chloride ions, because $1(+2) + 2(-1) = 0$.

Ion	Charge
sodium	+1
calcium	+2
aluminum	+3
chloride	-1
oxide	-2

Find the number and types of ions in each compound.

C 42. calcium oxide
1 calcium, 1 oxide

43. aluminum chloride
1 aluminum, 3 chloride

44. A compound contains three oxide ions. How many aluminum ions does it contain? What is the compound's name? 2; aluminum oxide

45. **RESEARCH** Find the common name and use for each compound.
a. sodium chloride   b. sodium carbonate   c. sodium bicarbonate
table salt; food seasoning   washing soda; detergents   bicarbonate of soda; stomach antacids or baking soda; baking

SPIRAL REVIEW

S 46. Find the sum: $4\frac{1}{9} + 6\frac{1}{6}$ (Toolbox Skill 18) $10\frac{5}{18}$

47. Find the quotient: $-60 \div 5$ (Lesson 3-5) -12

Self-Test 1

Replace each __?__ with >, <, or =.

1. $5 \underline{\;?\;} -6$ 2. $-9 \underline{\;?\;} -4$ 3. $-3 \underline{\;?\;} -3$ 3-1
   >   <   =

Find each answer.

4. $-2 + (-2)$ -4 5. $-5 + (-17)$ -22 6. $-4 + (-14)$ -18 3-2

7. $-4 + 9$ 5 8. $7 + (-13)$ -6 9. $-16 + 16$ 0 3-3

10. $-26 - 5$ -31 11. $14 - 28$ -14 12. $-1 - (-7)$ 6 3-4

13. $8(-4)$ -32 14. $(-7)(5)(-2)$ 70 15. $36 \div (-3)$ -12 3-5

3-6 Evaluating Expressions Involving Integers

Objective: To evaluate expressions involving integers.

CONNECTION

In previous chapters you learned how to evaluate expressions. In this chapter you learned how to perform the four basic operations with integers. You can combine these skills to evaluate expressions in which the values of the variables are integers. These expressions might also involve exponents, absolute value, and opposites.

Example 1

Solution

Evaluate each expression when $y = -4$:   **a.** y^2 **b.** $2y^3 + 18$

a. $y^2 = (-4)^2$
$= 16$ ⟵ $(-4)^2 = (-4)(-4) = 16$

b. $2y^3 + 18 = 2(-4)^3 + 18$
$= 2(-64) + 18$
$= -128 + 18$
$= -110$

 **Check Your Understanding**

1. In Example 1(b), why does $(-4)^3$ equal -64?
See *Answers to Check Your Understanding* at the back of the book.

Absolute value signs have the same priority as parentheses in the order of operations. When evaluating expressions involving absolute value, you evaluate any expression within absolute value signs first.

Example 2

Solution

Evaluate each expression when $c = -9$ **and** $d = 4$.
a. $|c + d|$ **b.** $|c| + |d|$

a. $|c + d| = |-9 + 4|$
$= |-5|$
$= 5$

b. $|c| + |d| = |-9| + |4|$
$= 9 + 4$
$= 13$

 **Check Your Understanding**

2. In Example 2(b), why do you add 9 and 4 instead of -9 and 4?
See *Answers to Check Your Understanding* at the back of the book.

You will sometimes find it necessary to use the *multiplication property of −1* when evaluating expressions involving integers.

Integers **115**

Warm-Up Exercises

Evaluate each expression when
$a = 5$, $b = 2$, and $c = 6$.
1. $7a$ 35
2. $14 + c$ 20
3. $(a + 19) \div c$ 4
4. b^4 16
5. $c^2 \div b$ 18

Lesson Focus

Mary Azia has $22 per week deducted from her pay for medical insurance. This deduction can be represented as -22. To find how much is deducted for any number of weeks, students can use the expression $-22n$, where n is the number of weeks. Today's lesson focuses on evaluating expressions involving integers.

Teaching Notes

Point out to students that evaluating expressions involving integers is no more difficult than evaluating expressions involving only positive numbers. The same rules apply.

Key Question

What does the multiplication property of −1 state? A number multiplied by −1 is the opposite of the number.

Error Analysis

Many students think that $-n$ is always negative. Stress that $-n$ should be read as the *opposite of n*. If n is negative, $-n$ is positive, since the opposite of a negative number is positive.

Additional Examples

Evaluate each expression when $n = -3$.

1. n^2 9
2. $2n^3 + 26$ -28

Evaluate each expression when $a = -8$ and $b = 6$.

3. $|a + b|$ 2
4. $|a| + |b|$ 14

Evaluate each expression when $m = -4$ and $n = 5$.

5. $-mn$ 20
6. $-m + 12$ 16

Closing the Lesson

Ask students to list the similarities and the differences between evaluating expressions that involve integers and those that involve positive numbers only.

Suggested Assignments

Skills: 1–24, 26, 28, 33, 35, 51–58
Average: 1–28, 29–43 odd, 51–58
Advanced: 1–27 odd, 29–58, Challenge

Exercise Notes

Making Connections/Transitions

These exercises combine skills learned in several previous lessons. This helps students to see mathematics as an integrated whole.

Communication: Reading

Guided Practice Exercises 1 and 2 give students the chance to clarify their understanding of the multiplication property of -1 and of how absolute value fits in the order of operations.

> **Multiplication Property of -1**
>
> The product of any number and -1 is the opposite of the number.
>
> $$-1n = -n \quad \text{and} \quad -n = -1n$$

Example 3

Evaluate each expression when $p = -3$ and $q = 5$.

a. $-q + 9$ **b.** $-pq$

Solution

a. $\begin{aligned} -q + 9 &= -1q + 9 \\ &= (-1)(5) + 9 \\ &= -5 + 9 \\ &= 4 \end{aligned}$

b. $\begin{aligned} -pq &= -1pq \\ &= (-1)(-3)(5) \\ &= 3(5) \\ &= 15 \end{aligned}$

 **Check Your Understanding**

3. In Example 3(b), why can $-pq$ be written as $-1pq$?
See *Answers to Check Your Understanding* at the back of the book.

Guided Practice

COMMUNICATION « *Reading*

Refer to the text on pages 115–116.

« **1.** State the multiplication property of -1 in words. The product of any number and -1 is the opposite of the number.

« **2.** Where does absolute value fit in the order of operations? Absolute value has the same priority as parentheses.

Find each answer.

3. $(-7)^2$ 49   **4.** $(-1)^5$ -1 **5.** $|-16 + 23|$ 7   **6.** $|-5| - |-4|$ 1

Evaluate each expression when $a = -5$, $b = -10$, and $c = 7$.

7. b^3 -1000 **8.** $4a^2$ 100   **9.** $b^2 + (-6)$ 94

10. $|a - b|$ 5   **11.** $|a| - |b|$ -5 **12.** $-10 - |c|$ -17

13. $-ac$ 35   **14.** $-b - 18$ -8 **15.** $-ab + 12$ -38

Exercises

Evaluate each expression when $m = -3$, $n = 8$, and $s = -6$.

A   **1.** s^4 1296   **2.** m^5 -243 **3.** $5m^3$ -135

     4. $-10s^2$ -360 **5.** $s^3 + 50$ -166 **6.** $m^4 - 2$ 79

     7. $-9 + 4m^2$ 27   **8.** $30 - 4s^2$ -114 **9.** $m^2 - n$ 1

     10. $2s^2 + n$ 80   **11.** $|m - n|$ 11   **12.** $|m + s|$ 9

Evaluate each expression when $m = -3$, $n = 8$, and $s = -6$.

13. $|n| + |s|$ 14

14. $|m| + |n|$ 11

15. $|m + 3| - 6$ $^{-6}$

16. $7 - |n - 8|$ 7

17. $|m| - |n|$ -5

18. $|s| - |n|$ $^{-2}$

19. $-mn$ 24

20. $-ns$ 48

21. $-s + (-7)$ $^{-1}$

22. $-n - 11$ -19

23. $-ns - 6$ 42

24. $-mn + 14$ ^{38}

25. Find the sum of 10 squared and -4 cubed.   36

26. Find the difference when -11 is subtracted from -3 squared.   20

27. Find the difference when the absolute value of 24 is subtracted from the opposite of -9. -15

28. Find the sum when the absolute value of -8 is added to the opposite of 5.   3

LOGICAL REASONING

Choose the correct relationship between the expressions in columns I and II. Assume that each variable can represent any integer except zero.

	I	II	
B 29.	pq	$-pq$ D	**A.** I > II
30.	$(3m)^2$	$(-3m)^2$ C	**B.** I < II
31.	$|c|$	$-|c|$ A	**C.** I = II
32.	$-5a^2$	$6a^2$ B	**D.** cannot determine

FUNCTIONS

Complete each function table.

33.

x	x^2
-6	36
-3	9
-2	? 4
-1	? 1
4	? 16

34.

x	$-x + 2$
-5	7
-4	6
-1	? 3
0	? 2
2	? 0

Use each table to find the function rule.

35.

x	?
-6	24
-5	20
-4	16
1	-4
3	-12

$x \rightarrow -4x$

36.

x	?
-4	-10
-1	-7
0	-6
3	-3
8	2

$x \rightarrow x - 6$

Integers   **117**

Reasoning
Exercises 29–32 require students to analyze mathematical situations and then make conjectures based on their analyses.

Making Connections/Transitions
Exercises 33–36 extend the concept of function, first introduced in Lesson 2-6, to the integers.

Critical Thinking
Exercises 37–44 require students to *evaluate* six expressions, *analyze* their relationship to one another, and *make generalizations* based on their analyses.

Nonroutine Problem
The first eight powers of 3 are:
$3^1 = 3$; $3^2 = 9$; $3^3 = 27$; $3^4 = 81$;
$3^5 = 243$; $3^6 = 729$; $3^7 = 2187$;
$3^8 = 6561$. Without multiplying, find the units digit in each of the following.
1. 3^{16} 1
2. 3^{23} 7
3. 3^{46} 9
4. 3^{100} 1
The units digits are found by using the product of powers rule. For example, since $3^{16} = 3^8 \cdot 3^8$ and $3^8 = 6561$, then the product 6561×6561 must have 1 as its units digit.

Enrichment

Since absolute value has the same priority as parentheses in the order of operations, students may wonder if the distributive property holds for absolute value. Have students explore this idea by creating and analyzing examples such as $-5|3 + 4|$ and $|(-5)3 + (-5)4|$.

Enrichment

Enrichment 28 is pictured on page 92E.

Practice

Practice 34 is shown below.

Practice 34, *Resource Book*

Practice 34
Skills and Applications of Lesson 3-6

NUMBER SENSE

Evaluate each expression when $x = -1$.

37. x^2 1   **38.** x^3 −1   **39.** x^{10} 1   **40.** x^{15} −1   **41.** x^{47} −1   **42.** x^{100} 1

43. Assume that n is a positive integer. Explain how you know whether the value of $(-1)^n$ is 1 or -1. If n is even, $(-1)^n$ is 1. If n is odd, $(-1)^n$ is -1.

C **44.** Assume that n is a positive integer and that a is any negative integer. Explain how you know whether the value of a^n is positive or negative. If n is even, a^n is positive. If n is odd, a^n is negative.

Without computing, tell whether each answer is *positive* or *negative*.

45. $(-2)^{14}$ positive   **46.** $(-3)^9$ negative   **47.** $(-68)^3$ negative

48. $(-122)^2$ positive   **49.** $(-7)^{24}$ positive   **50.** $(-8)^{15}$ negative

SPIRAL REVIEW

51. 150,000,000
52. about 1200
53. 16
54. −48
55. No; Julia did not include the amount that either her brother or her sister paid.

S **51.** Use the line graph at the right. What was the population of the United States in 1950? *(Toolbox Skill 25)*

52. Estimate the sum: $547 + 298 + 369$ *(Toolbox Skill 2)*

53. Evaluate $-2x + 6$ when $x = -5$.   *(Lesson 3-6)*

54. Find the difference: $-34 - 14$ *(Lesson 3-4)*

55. Julia rode her bicycle in a charity bike-a-thon. She collected $.10/mile, $.75/mile, and $1/mile from each of three friends. Two others paid $.25/mile each. Her brother and sister paid $.50/mile each. Julia rode 25 mi. She said that she raised $71.25 for charity. Is this correct? Explain.   *(Lesson 2-7)*

56. Replace the ___?___ with >, <, or =: -9 ___?___ 4   *(Lesson 3-1)* <

57. Multiply mentally: $10(6)(0)(5)$ *(Lesson 1-7)* 0

58. Find the difference: $4\frac{2}{5} - 2\frac{3}{10}$ *(Toolbox Skill 19)* $2\frac{1}{10}$

United States Population

Challenge

There are many ways to write an expression equal to -2. For example, $-6 \div 2 + 5 \cdot 1 - 4 = -2$. Find two other expressions that equal -2. Use only the integers from -9 to 9, do not repeat any integer, and use all four operations exactly once.
Answers will vary. Examples: $-5 + 7 - 8 \div 4 \cdot 2$ and $-8 \div 2 + 5 - 1 \cdot 3$

118   Chapter 3

3-7

Strategy:
Making a Table

Objective: To solve problems by making a table.

Some problems ask you to find all possibilities in a given situation. To help you solve this type of problem, make a table to organize the information. You list the possibilities without regard to order. For example, a listing of two quarters and one dime is the same as one dime and two quarters.

Problem

Benito Alomar is a sales clerk in a shoe store. The cash register in the store contains only quarters, dimes, and nickels. In how many different ways could he make 35¢ in change?

Solution

UNDERSTAND The problem is about a sales clerk making change.
Facts: only quarters, dimes, and nickels
        35¢ change needed
Find: number of ways to make the change

PLAN You can make a table that lists each type of coin. To find the total value of the coins, multiply the number of quarters by 25, the number of dimes by 10, and the number of nickels by 5. Then add.

WORK

Number of Quarters	1	1	0	0	0	0
Number of Dimes	1	0	3	2	1	0
Number of Nickels	0	2	1	3	5	7
Total Value of Coins	35	35	35	35	35	35

ANSWER He could make 35¢ in change in 6 different ways.

Look Back What if Benito Alomar used four coins? How many different amounts could he make using only quarters, dimes, and nickels?
14 amounts

Integers **119**

Lesson Planner

Teaching the Lesson
• Lesson Starter 27
• Using Technology, p. 92C
• Reteaching/Alternate Approach, p. 92D

Lesson Follow-Up
• Practice 35
• Enrichment 29: Problem Solving
• Study Guide, pp. 53–54

Warm-Up Exercises

The Booters won 17 soccer matches, lost 6, and tied 2. A win is worth 2 points, a tie is worth 1, and a loss is worth 0.
1. What is the paragraph above about? wins, losses, and ties of the Booters and how many points each is worth
2. How many points is a tie worth? 1
3. How many games did the Booters lose? 6

Lesson Focus

Tell students that a hostess in a restaurant has a group of 24 people to be seated. The restaurant contains tables that will seat 3, 4, or 6 people each. In how many different ways can the group be seated so that there are no empty seats at a table? One way to solve the problem is to write a table showing all the possible arrangements. Today's lesson focuses on the problem-solving strategy of making a table.

Teaching Notes

Work through the sample problem with students, stressing the importance of organizing information in a table in a systematic way.

Teaching Notes *(continued)*

Key Question

What system was used to arrange the information in the table on page 119? *The coins were listed in order from greatest value to least. Then all possibilities for each value were considered.*

Reasoning

The Look Back section on page 119 asks students to consider different questions involving four coins. If the four coins were not restricted to quarters, nickels and dimes, how would the response change? *You would have to include all possibilities involving pennies and half-dollars.*

Using Technology

For a suggestion on using a spreadsheet program to construct tables, see page 92C.

Reteaching/Alternate Approach

For a suggestion on how to format tables, see page 92D.

Additional Example

Bill wants to borrow 30¢ from Sally. Sally has only quarters, dimes, and pennies. In how many different ways can Sally give Bill 30¢?

Number of Quarters	1	0	0	0	0
Number of Dimes	0	3	2	1	0
Number of Pennies	5	0	10	20	30
Total Value of Coins	30	30	30	30	30

There are 5 different ways Sally can give Bill 30¢.

Closing the Lesson

Use the cooperative activity below to check students' understanding of the strategy of making a table.

Cooperative Learning

Have students work in groups to create a problem that can be solved by making a table. Each group should give its problem to another group. Each group should solve the problem and explain its solution.

Guided Practice

COMMUNICATION «*Reading*

Use the problem below for Exercises 1–4.

Last weekend Pets, Inc. cared for 12 animals, all of which were either dogs or birds. The workers at Pets, Inc. counted the number of animal legs. How many different totals are possible?

«**1.** How many animals were at Pets, Inc. last weekend? 12 animals

«**2.** What was the greatest number of dogs possible at Pets, Inc. last weekend? 12 dogs

«**3.** What was the least number of birds possible at Pets, Inc. last weekend? 0 birds

4. Solve the problem by making a table. 13 totals

5. Solve by making a table: Basketball shots are worth one, two, or three points. In a basketball game Mei Lin scored 7 points. In how many different ways could she have scored the 7 points? 8 ways

«**6.** COMMUNICATION «*Discussion* There are several ways to organize information in a table. Discuss these ways. Which seems most efficient to you? Explain. See answer on p. A5.

Problem Solving Situations

Solve by making a table.

1. Lewanda Warner has to pay $27 for food at the grocery store. In how many different ways can she pay the $27 with bills of $10, $5, and $1? 12 ways

2. Three darts are thrown at the target at the right. All three darts hit the target. How many different point totals are possible? 9 totals

3. Carpenter Joe Gibbs makes three-legged tables and four-legged tables. Both kinds of tables have the same type of leg. Last month Joe made 14 tables and then totaled the number of legs he made. How many different totals are possible? 15 totals

4. Stanley Weldman scored 11 points in a basketball game. Basketball shots are worth one, two, or three points. In how many different ways could Stanley have scored the 11 points? 16 ways

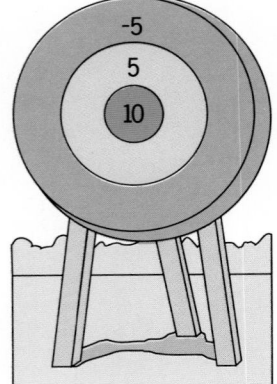

Solve using any problem solving strategy.

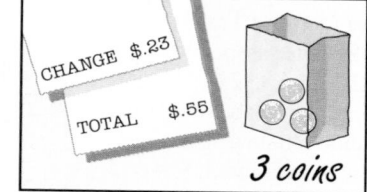

PROBLEM SOLVING
CHECKLIST
Keep this in mind:
Using a Four-Step Plan
Consider these strategies:
Choosing the Correct Operation
Making a Table

5. Four darts are thrown at the target shown on page 120. All four darts hit the target. How many different point totals are possible?

6. Daniel Kuo bought two gallons of milk for $1.99 each, a dozen eggs for $1.19, and a loaf of bread for $1.09. He gave the cashier a $10 bill. How much change did Daniel receive?

7. Angela Marini wants to buy 8 tickets to a concert. Tickets cost $20 each in advance and $24 each on the day of the concert. There is a $2 discount for sales of 10 tickets or more. How much will Angela save if she buys the tickets in advance?

8. Juniors at Bentley High School held a car wash and earned $100. They charged $6 per truck and $4 per car. In how many different ways could the juniors have earned the $100? making a table; 9 ways

9. In ice hockey, a team earns 2 points for a win, 1 point for a tie game, and no points for a loss. The Ice Pirates earned 12 points and lost 3 games in January. In how many different ways could the team have earned the 12 points? making a table; 7 ways

10. A restaurant contains two-person tables and four-person booths. There are 15 tables and 28 booths in the restaurant. How many people can dine at the restaurant at one time? choosing the correct operation; 142 people

WRITING WORD PROBLEMS

For each exercise, use one piece of information at the right. Write a word problem that you would solve by making a table. Then solve the problem.

CHANGE $.23

TOTAL $.55

3 coins

11. Use quarters, dimes, and nickels in the problem.

12. Use dimes, nickels, and pennies in the problem.
Answers to Problems 11–12 are on p. A5.

SPIRAL REVIEW

S **13.** Find the sum: $-8 + (-4)$ *(Lesson 3-2)* −12

14. Bill Blanchette has to pay $43. In how many different ways could he pay the $43 with bills worth $20, $10, and $1? *(Lesson 3-8)* 9 ways

15. Evaluate $x - |y|$ when $x = -3$ and $y = -8$. *(Lesson 3-6)* −11

16. Find the product: $(11.5)(3.2)$ *(Toolbox Skill 13)* 36.8

Integers **121**

Exercise Notes
Communication: Reading
Guided Practice Exercises 1–3 ask students to read and analyze the given problem. The idea of limits is addressed in the second and third exercises.

Communication: Discussion
Guided Practice Exercise 6 promotes group discussion and clarification of the ideas of the lesson. Logical reasoning is needed to determine the most efficient method and to support why it was chosen.

Follow-Up
Enrichment
Enrichment 29 is pictured on page 92F.

Practice
Practice 35 is shown below.

Practice 35, *Resource Book*

Practice 35
Skills and Applications of Lesson 3-7

Solve by making a table.

1. Yoko flipped a penny, then recorded whether it fell with a head showing or a tail showing. How many different outcomes were there if she flipped the coin twice? 4

2. Each face of a cube is assigned a different number from one to six. The cube is tossed on a table and the number on the face which is facing upward is recorded. How many different ways can you get a sum of nine if the cube is tossed twice? 4

3. Andres had a three-question quiz. Each question had choice A and choice B from which to choose the answer. In how many different ways can the choices be made for the three-question quiz? 8

4. Assume a football team can only score points in the following way: Six points for a touchdown, one point for an extra point (only following a touchdown), and three points for a field goal. In how many different ways could a football team score 19 points? 3

5. To pay for an item costing $9.45, Ken paid with a $10 bill. How many different ways could he receive change if he received no pennies and at most one nickel? 3

6. The sum of the digits of a three-digit number is four. How many different three-digit numbers are possible? 10

7. Alicia selected a chip from a container that had one red, one blue, and one green chip. She recorded the color and returned the chip to the container. After shaking the container, she selected a second chip and recorded the color. In how many different ways could she end up with two different colors? 6

MIXED REVIEW
Operations with Integers

Background

By the mid 1500s, the land that Spain claimed in North and South America was divided into two viceroyalties, one of which was called New Spain. New Spain included Venezuela and all Spanish-claimed lands north of the Isthmus of Panama. California and New Mexico were both part of New Spain.

Spanish settlement of New Spain was aided by the king's granting of large estates (*haciendas*) to nobles. The establishment of Spanish missions also encouraged settlers to move into new areas. Los Angeles was one such mission town. In 1769, a Spanish expedition found a pleasant area on the coast of California. They called the place *Nuestra Señora la Reina de Los Angeles* (Our Lady, Queen of the Angels) because their landing took place on the feast day of Our Lady of the Angels of Porciuncúla. Two years later, Franciscan monks built a mission in the area, San Gabriel Arcángel. El Pueblo de Los Angeles—the city of Los Angeles—was founded near the mission ten years later.

On September 4, 1781, forty-four pioneers ended a seven-month journey from Sonora, Mexico, to the site of a new town in Alta California. The group—made up of Africans, Native Americans, Spaniards, and people of mixed ancestry—had come to California to found what would become a famous American city.

To find out the name of the new settlement, complete the code boxes below. Begin by solving Exercise 1. The answer to Exercise 1 is −9, and the letter associated with Exercise 1 is *U*. Write *U* in the box above −9 as shown below. Continue in this way with Exercises 2–24.

1. $-4 + (-5)$ −9 (U)
2. $12 + 7$ 19 (A)
3. $35 + (-97)$ −62 (L)
4. $-28 - (-5)$ −23 (D)
5. $8(-4)$ −32 (E)
6. $(-7)(-9)$ 63 (G)
7. $\frac{55}{5}$ 11 (E)
8. $\frac{-96}{12}$ −8 (E)
9. $728 \div (-14)$ −52 (O)
10. $-69 \div (-3)$ 23 (B)
11. $5(-7)$ −35 (L)
12. $(-8)(10)$ −80 (P)
13. $-17 - 17$ −34 (O)
14. $-31 - (-31)$ 0 (A)
15. $(-15)(6)$ −90 (T)
16. $(-5)(-7)$ 35 (S)
17. $-11 + 19$ 8 (A)
18. $14 + (-27)$ −13 (L)
19. $\frac{-64}{16}$ −4 (O)
20. $\frac{-48}{-8}$ 6 (A)
21. $46 - 83$ −37 (D)
22. $33 - (-21)$ 54 (U)
23. $57 + (-5)$ 52 (A)
24. $(-9)(3)$ −27 (L)

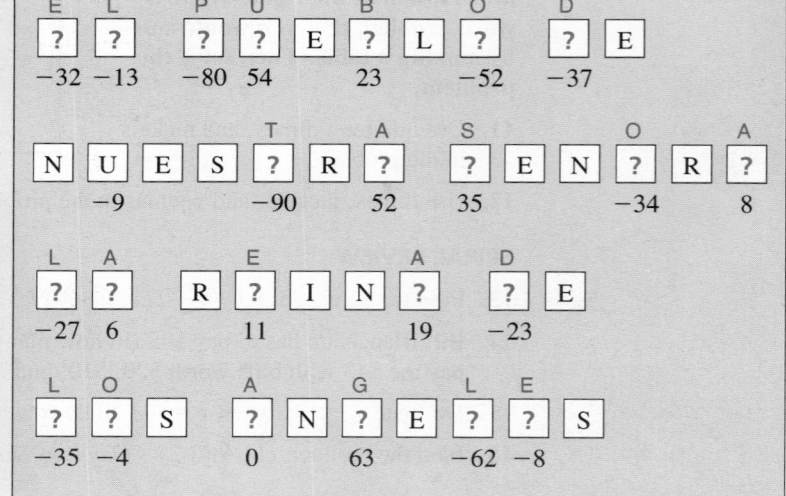

E	L		P	U		B		O		D	
?	?		?	?	E	?	L	?		?	E
−32	−13		−80	54		23		−52		−37	

				T		A		S			O		A
N	U	E	S	?	R	?		?	E	N	?	R	?
	−9			−90		52		35			−34		8

L	A			E			A		D	
?	?		R	?	I	N	?		?	E
−27	6			11			19		−23	

L	O			A		G		L	E	
?	?	S		?	N	?	E	?	?	S
−35	−4			0		63		−62	−8	

4-7 Enrichment
Thinking Skills

Enrichment 39: Thinking Skills (for use with Lesson 4-7)
An Equation Chain

The following is an equation chain. The solution to each puzzle will help you solve the next puzzle in the sequence.

Write each puzzle as an equation. Solve for the variable and record the value on the right. Then use that value to solve the next puzzle. All the values are integers.

1. 3 more than twice a equals 15. $2a + 3 = 15$; 6

2. 5 more than 3 times a equals b. $3a + 5 = b$; 23

3. 4 times c plus 7 equals b. $4c + 7 = b$; 4

4. d less than 6 times c equals 17. $6c - d = 17$; 7

5. Nine less than d times d equals twice e. $(d)(d) - 9 = 2e$; 20

6. e plus f equals 5 times 5. $e + f = (5)(5)$; 5

7. f times g equals 30. $(f)(g) = 30$; 6

8. g divided by 2 equals h. $\frac{g}{2} = h$; 3

9. h less than c equals 1. Check: $c - h \overset{?}{=} 1$
   $c - h = 1$; $4 - 3 = 1$ ✓

If you did not get 1 in Exercise 9, then go back and check your equation chain!

4-8 Enrichment
Problem Solving

Enrichment 40: Problem Solving (for use with Lesson 4-8)
Use an Equation on That Telephone Bill!

Miguel wanted to have his own telephone. He had enough money in his savings account to cover the installation fee of $95.35. The monthly rate for the phone service would be $19.55, and each local call would be charged at 2 cents ($.02) per minute. Miguel planned to make only local calls to his friends.

After the first month he got a bill for $123.58 that included charges for installation, monthly service, and the calls he had made. He was surprised by the total bill, and wanted to know how much time his calls had taken that month. This is what he did.

Understand
Miguel knew that his phone bill should be equal to the sum of the installation fee, the monthly rate, and the cost of the local calls. He wanted to find the time taken by his local calls.

Plan
Miguel wanted to write an equation about his phone bill. On one side he would add the installation charge, the monthly charge, and the local call cost. The cost of the local calls depended on the number of minutes he was on the phone. Each minute cost $.02. He let x be the number of minutes.

1. Write an expression with x for the cost of his local calls. $0.02x$

2. Add the local call cost, the installation charge, and the monthly charge. (There should be three terms in the expression.) $0.02x + 95.35 + 19.55$

3. Set this sum equal to the amount of the telephone bill. (Write an equation.) $0.02x + 95.35 + 19.55 = 123.58$

Work

4. To solve the equation in Exercise 3, what do you subtract from both sides? $(95.35 + 19.55)$, or 114.90

5. Your new equation should have $0.02x$ equal to a number. What does that number represent? Write your new equation.
   the total cost of local calls; $0.02x = 8.68$

Answer

6. Solve for x by dividing both sides by 0.02. What was the total length of Miguel's local calls in minutes? 434 min

---

4-9 Enrichment
Thinking Skills

Enrichment 41: Thinking Skills (for use with Lesson 4-9)
Find the Mystery Message!

To find what the mystery message is, solve Exercises 1–17. On the line above each answer, write the letter that goes with that exercise as many times as the answer appears.

1. $2x - 1 = 15$ (E)   8
2. $2x = 12$ (D)   6
3. $5x + 1 = 36$ (Y)   7
4. $40x - 1 = 39$ (T)   1
5. $5x - 8 = 37$ (S)   9
6. $100x - 900 = 300$ (G)   12
7. $\frac{x}{2} = 30$ (M)   60
8. $6x - 3x = 6$ (I)   2
9. $7x + 5 = 82$ (N)   11
10. $2(x - 1) = 30$ (R)   16
11. $45x + 23 = 23$ (A)   0
12. $96x = 384$ (P)   4
13. $3(x + 1) = 48$ (W)   15
14. $6x + 2x - 7 = 33$ (F)   5
15. $4x - 3 = 49$ (C)   13
16. $3(x + 5) = 66$ (O)   17
17. $2x - 1 = 59$ (H)   30

```
 H    Y    P    A    T    I    A         I    S
30    7    4    0    1    2    0         2    9

 R    E    G    A    R    D    E    D         A    S
16    8   12    0   16    6    8    6         0    9

 T    H    E         F    I    R    S    T
 1   30    8         5    2   16    9    1

 W    O    M    A    N
15   17   60    0   11

 M    A    T    H    E    M    A    T    I    C    I    A    N
60    0    1   30    8   60    0    1    2   13    2    0   11
```

4-10 Enrichment
Application

Enrichment 42: Application (for use with Lesson 4-10)
Rules of Thumb, Formulas, and Cricket Chirps

Many people use *rules of thumb*, practical formulas useful in everyday life. If you are cooking a turkey, for example, a rule of thumb might be to cook it 20 minutes per pound. You could translate that rule into the formula
$$t = 20w,$$
where w is the weight of the turkey and t is the cooking time in minutes. A 12-pound turkey should be cooked $(12)(20)$, or 240, minutes, which is 4 hours.

Another interesting rule of thumb has to do with crickets and how fast they chirp. As the weather gets warmer in the spring and summer, the crickets chirp faster. As the weather gets cooler, the chirping slows down. The following formula can approximate the number of chirps per minute depending on the temperature:
$$c = 4t - 150,$$
where c is the number of chirps per minute and t is the temperature in degrees Fahrenheit. For example, at 50°F, there will be $(4)(50) - 150 = 50$ chirps per minute. At 80°F, there will be $(4)(80) - 150 = 320 - 150 = 170$ chirps per minute.

1. How many chirps are there per minute at 60°F? 90

2. How many chirps are there per minute at 70°F? 130

3. How many chirps are there per minute at 100°F? 250

4. a. How many chirps are there at 30°F, according to the formula? −30

   b. Does this number make sense? No.

   c. What does your answer to part (b) imply about this rule of thumb?
   It makes sense only for temperatures above 37.5°F.

End of Chapter Enrichment
Extension

Enrichment 43: Extension (for use after Chapter 4)
Exploring Alternatives

There may be more than one way to solve an equation. When you are about to solve an equation, consider the alternative methods of solution that are available. Choose the method that will be easiest for you.

Example

Solve the equation $56(x + 42) = 5600$.

Solution 1

Use the distributive property.
$$56(x + 42) = 5600$$
$$56x + 2352 = 5600$$
$$56x + 2352 - 2352 = 5600 - 2352$$
$$56x = 3248$$
$$\frac{56x}{56} = \frac{3248}{56}$$
$$x = 58$$

Solution 2

Notice that both sides of the equation can be divided evenly by 56. Let's solve the equation by dividing each side by 56 first.
$$56(x + 42) = 5600$$
$$\frac{56(x + 42)}{56} = \frac{5600}{56}$$
$$x + 42 = 100$$
$$x + 42 - 42 = 100 - 42$$
$$x = 58$$

Use the example above to answer the following.

1. How many steps were necessary to solve the equation in Solution 1? in Solution 2? 4; 3

2. Which method of solving the equation involved the least amount of work? Explain. Dividing each side by 56 first; the calculations were much easier to do.

Solve. Check each solution.

3. $2(y + 195) = 400$ 5

4. $25(x + 4) = 125$ 1

5. $314(a - 15) = 628$ 17

6. $24(5 + 2z) = 264$ 3

Multicultural Notes

For Chapter 4 Opener

Many countries are far more dependent on railroads for public transportation than the United States, which relies heavily on the automobile. Japan, for example, has been a leader in developing fast, quiet, and safe trains. Hundreds of thousands of passengers a day ride high speed trains called *shinkansen* (new trunk line). These trains have cut travel time drastically over the last thirty years, making it possible for people to commute from far suburbs to jobs in the city.

The countries of Western Europe have also developed extremely fast trains. Such trains have already substantially increased travel and commerce between European countries. Future plans to upgrade old tracks and build new ones will make possible an international network of fast speed railways.

For Page 179

In Exercises 19–22, you may want to have students learn more about Georg Simon Ohm and Albert Einstein. Ohm (1787–1854) was a German physicist who discovered the law relating electromotive force to current and resistance in 1826. His work was neglected for a number of years, until he became a professor at the University of Nuremberg. Einstein (1879–1955) was born in Germany and became a Swiss citizen in 1905. He first framed his famous equation relating matter and energy, $E = mc^2$, in 1905.

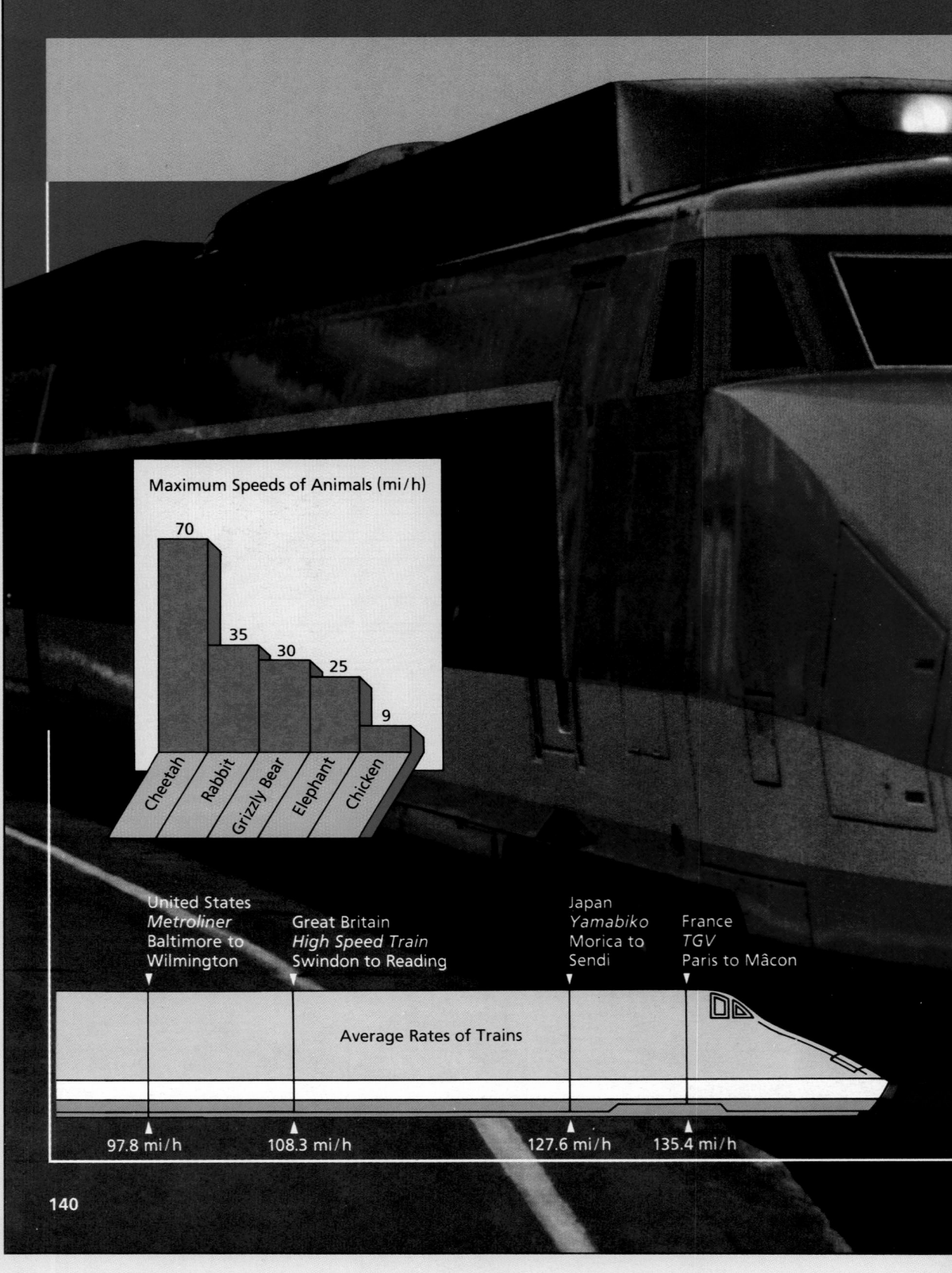

Maximum Speeds of Animals (mi/h)

Cheetah	Rabbit	Grizzly Bear	Elephant	Chicken
70	35	30	25	9

United States
Metroliner
Baltimore to
Wilmington
97.8 mi/h

Great Britain
High Speed Train
Swindon to Reading
108.3 mi/h

Japan
Yamabiko
Morica to
Sendi
127.6 mi/h

France
TGV
Paris to Mâcon
135.4 mi/h

Average Rates of Trains

140

Equations 4

Frequently Used Rates		
		Formula
	Distance	Distance = rate x time
	Price	Price = rate x weight
	Earnings	Earnings = rate x time

Did You Know?
Light travels at a rate of 186,282 mi/s. Electric impulses are slower and travel at a rate of 126,263 mi/s. Sound is much slower. At 32°F and sea level, sound travels at a rate of only 1086 ft/s.

141

Research Activity

For Chapter 4 Opener

Students may find it interesting to compare average rates of trains in the United States to average rates of trains in other countries. For example, they can figure out how much travel time could be saved if a train similar to France's TGV, an acronym for *train à grande vitesse* (high speed train), were to carry passengers from Chicago to New York. Students can also research the arguments for and against developing a fast speed railway system in the United States.

Suggested Resource

Turnbull, Lucia S. "Lightning on the Rails," *1985 Yearbook of Science and the Future*, pp. 10–27. Chicago: Encyclopaedia Britannica, Inc., 1984.

Teaching the Lesson
- Materials: algebra tiles, balance scales
- Lesson Starter 30
- Visuals, Folder D
- Using Manipulatives, p. 140C
- Toolbox Skills 7–10

Lesson Follow-Up
- Practice 40
- Enrichment 33: Data Analysis
- Study Guide, pp. 59–60

Warm-Up Exercises

Find each answer.

1. $4 \cdot 6 - 7 + 2 \cdot (-3)$ 11
2. $-7 - (-3)(-3)$ -16
3. $32 \div 4 + 7(2 - 9)$ -41

Evaluate.

4. $n \div 4 + 3n$ if $n = -8$. -26
5. $3(y + 7) - 2y$ if $y = 4$.   25

Lesson Focus

Ask students how they evaluated expressions when given a value for the variable. The same technique will be used in today's lesson to determine if a given number is a solution of an equation.

Teaching Notes

Key Question
What is the procedure used to check if a number is a solution of an equation or not?   Substitute the number for the variable in the equation. Perform the calculations. If the two sides of the equation are equal, the number is a solution.

Application
Pharmacists may "count" pills by using a finely calibrated balance scale. The variable would represent the mass of one pill.

Equations

Objective: To find solutions of equations by substitution and mental math.

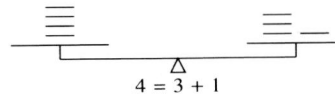   CONNECTION

Terms to Know
- equation
- solution of an equation

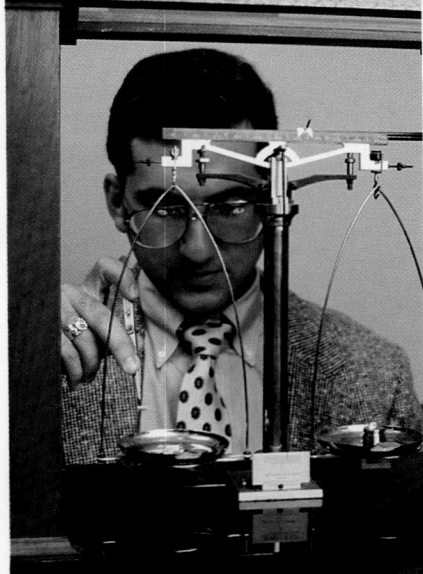

People who work with precious metals such as gold and silver use a balance scale like the one at the right to weigh the metal. A balance scale shows a relationship between two quantities.

In mathematics, an **equation** is a statement that two numbers or two expressions are equal. You can represent an equation by a balance scale, with each pan holding one of the two *sides* of the equation.

Some equations, such as $x + 1 = 6$, contain a variable. A value of the variable that makes an equation true is called a **solution of the equation.**

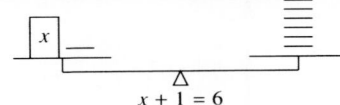

$4 = 3 + 1$ $x + 1 = 6$

Example 1

Is the given number the solution of the equation? Write *Yes* or *No*.

a. $x + 1 = 6; 5$ b. $15 = 5k; 2$

Solution

a. Substitute 5 for x in the equation.

$x + 1 = 6$
$5 + 1 \overset{?}{=} 6$
$6 = 6 \checkmark$

Yes, 5 is the solution of $x + 1 = 6$.

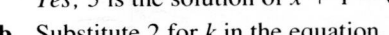
$5 + 1$ 6

b. Substitute 2 for k in the equation.

$15 = 5k$
$15 \overset{?}{=} 5 \cdot 2$ The symbol $\neq$ means
$15 \neq 10$ ⟵ *is not equal to.*

No, 2 is not the solution of $15 = 5k$.

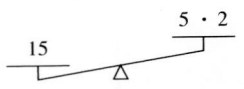
$5 \cdot 2$
15

 **Check Your Understanding**

1. Why is the scale in Example 1(b) not in balance?
See *Answers to Check Your Understanding* at the back of the book.

---

142 Chapter 4

Sometimes you can use mental math to find a solution. State the equation as a question and see if you can find the answer easily.

Example 2

Use mental math to find the solution of $2n - 6 = 14$.

Solution

First, think about the value of $2n$:
What number minus six equals fourteen?
$20 - 6 = 14$, so $2n = 20$.

Then think about the value of n:
Two times *what number* equals 20?
$2 \cdot 10 = 20$, so $n = 10$.

The solution of $2n - 6 = 14$ is 10.

See *Answers to Check Your Understanding* at the back of the book.

☑ Check Your Understanding

2. What questions would you ask if the equation in Example 2 were $4n - 6 = 14$?

3. Describe how Example 2 would be different if the equation were $2n + 6 = 14$.

Guided Practice

COMMUNICATION « *Reading*

Does each of the following fit the definition of an equation? Write *Yes* or *No*. If *No*, explain.

« 1. $3 + 4 = 7$ Yes.

« 2. $a + 12$ No; no = sign.

« 3. $x + 19 - 17$ No; no = sign.

« 4. $18 = 3b - 6$ Yes.

COMMUNICATION « *Writing*

Write the equation represented by each balance scale.

« 5. $6 = 4 + 2$

« 6. $5 = 2n + 3$

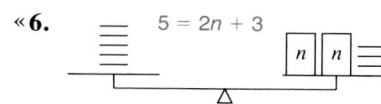

Draw a balance scale similar to those in Exercises 5 and 6 to represent each equation. Answers to Exercises 7–9 are on p. A9.

« 7. $5 = 4 + 1$

« 8. $x + 6 = 9$

« 9. $7 = 3n + 1$

Is the given number a solution of the equation? Write *Yes* or *No*.

10. $n - 8 = 16$; 8 No.

11. $-2 = x + 5$; -7 Yes.

12. $6y = 20$; -2 No.

Use mental math to find each solution.

13. $5 + r = 37$ 32

14. $-18 = z - 5$ -13

15. $\frac{c}{5} = 10$ 50

16. $4m + 10 = -30$ -10

17. $11 = 2t - 1$ 6

18. $-1 = \frac{w}{3} - 1$ 0

Equations **143**

Exercise Notes (continued)

Reasoning
Exercises 30–38 use numerical and algebraic expressions and connect with skills learned in previous lessons, as well as in this one.

Critical Thinking
Exercises 39–42 require students to *create* equations that have the same solution as a given equation. This leads to the concept of equivalent equations.

Cooperative Learning
Exercises 39–42 can be solved by dividing students into cooperative groups and asking each group to follow the directions. Have the groups share their results with the entire class. List the different equations that yield the same results on the chalkboard.

Follow-Up

Enrichment
Enrichment 33 is pictured on page 140E.

Practice
Practice 40, *Resource Book*

Exercises

Is the given number a solution of the equation? Write *Yes* or *No.*

A 1. $x + 6 = 9$; 3 Yes.   2. $18 = 3n$; 15 No.   3. $-20 = \frac{c}{-4}$; -5 No.

4. $b - 3 = -4$; -1 Yes.   5. $-1 = 1 + d$; 0 No.   6. $12w = 12$; 1 Yes.

7. $\frac{k}{3} = -6$; -18 Yes.   8. $5 = m - 9$; 4 No.   9. $-7y = 28$; -4 Yes.

Use mental math to find each solution.

10. $p + 8 = 9$ 1   11. $z - 4 = -1$ 3   12. $-12 = 6v$ -2

13. $5 = \frac{h}{-9}$ -45   14. $5 + n = 5$ 0   15. $-3 = 1 + \frac{a}{2}$ -8

16. $-2 = \frac{w}{3} - 1$ -3   17. $3r - 11 = 4$ 5   18. $\frac{n}{2} + 2 = 0$ -4

19. Is -4 the solution of $x + 3 = 7$? Write *Yes* or *No.* No.

20. Which of the integers -1, 0, and 1 is the solution of $3x + 3 = 0$? -1

Choose the letter of the solution of each equation.

A. -1   B. 0   C. 1   D. no solution

B 21. $a - 8 = -8$ B   22. $0 = d + 1$ A   23. $-5p = 5$ A

24. $0 = 7q$ B   25. $\frac{4}{k} = 4$ C   26. $\frac{0}{r} = 9$ D

27. $-3y = -3$ C   28. $-6p + 6 = 6$ B   29. $-2n - 2 = 0$ A

LOGICAL REASONING

Tell whether each statement is *True* or *False,* or if you *cannot determine*.

30. $5 + 7 = 12$ True   31. $54 = 7 \cdot 8$ False   32. $15 - 8 \neq 7$ False

33. $\frac{20}{-5} \neq 4$ True   34. $3m + 2 = 5$ cannot determine   35. $3(2) - 7 = -1$ True

36. $(-5)(-7) \neq -12$ True   37. $-5n \neq 15$ cannot determine   38. $\frac{-24}{3} + 2 \neq -6$ False

THINKING SKILLS   39–42. Synthesis

Answers to Exercises 39–42 are on p. A9.

Create two equations that have the same solution as the given equation.

C 39. $8 + m = 7$ 40. $-35 = 7p$ 41. $12 = \frac{c}{2}$ 42. $a - 8 = -3$

SPIRAL REVIEW

S 43. Justin worked five days and received $23, $44, $36, $50, and $27. Justin said he earned $130. Is this correct? Explain. *(Lesson 2-7)* No; the amount $50 was not included in the total.

44. Find the difference: $\frac{7}{9} - \frac{1}{6}$ *(Toolbox Skill 17)* $\frac{11}{18}$

45. Find the sum: $-8 + 11$ *(Lesson 3-3)* 3

46. Use mental math to find the solution: $5z - 2 = 23$ *(Lesson 4-1)* 5

Strategy:
Guess and Check

Objective: To solve problems by guessing and checking.

One method for solving a problem is to use the guess-and-check strategy. Guess an answer and check it against the information in the problem to see if it is correct. If your first guess is not correct, use the results to make a second guess. Keep guessing and checking until you find the correct answer.

Problem

Denzel spent $17.53 on pet food for his cats and dogs. The cat food cost $.89 per can and the dog food cost $1.19 per can. How many cans of each type of pet food did he buy?

Solution

UNDERSTAND The problem is about buying cans of pet food.
Facts: cat food cans at $.89 each
        dog food cans at $1.19 each
        total value of $17.53
Find: the number of cans of each type of pet food bought

PLAN First guess the number of cans of pet food, say 8 cat food cans and 8 dog food cans. Check the total value. Continue guessing and checking until you find the correct value of $17.53.

WORK Make a table to organize the work.

	Cat	Dog	Total Value Spent	
First Guess	8	8	$.89(8) + $1.19(8) = $16.64	too low
Second Guess	9	9	$.89(9) + $1.19(9) = $18.72	too high
Third Guess	9	8	$.89(9) + $1.19(8) = $17.53	correct

ANSWER Denzel bought 9 cans of cat food and 8 cans of dog food.

Look Back What if Denzel spent $20.80 on pet food? How many cans of each type of pet food did he buy? 10 cans of cat food, 10 cans of dog food

---

You can also use the guess-and-check strategy to find the solution of an equation. Check your guesses by substitution.

Suppose you
guess that 12 is ——➤
the solution.

$3x - 14 = 13$
$3(12) - 14 \stackrel{?}{=} 13$
$36 - 14 \stackrel{?}{=} 13$
$22 \neq 13$, so 12 is incorrect.

Because 22 is greater than 13, the guess of 12 is too high. The next guess should be less than 12. Continue guessing and checking until you find the solution, 9.

Equations **145**

Lesson Planner

Teaching the Lesson
- Materials: calculator, newspaper advertisements
- Lesson Starter 31
- Using Technology, p. 140C
- Reteaching/Alternate Approach, p. 140D
- Cooperative Learning, p. 140D
- Toolbox Skills 7–10

Lesson Follow-Up
- Practice 41
- Enrichment 34: Problem Solving
- Study Guide, pp. 61–62
- Calculator Activity 3

Warm-Up Exercises

Find each answer.
1. 4($.79 + $1.29) $8.32
2. 5($.86) + 6($2.34) $18.34
3. $6 \cdot 78 - 5 \cdot 97$ -17

Use mental math to find each solution.
4. $3m - 8 = 19$ 9
5. $11 + 4y = 15$ 1

Lesson Focus

Ask students if they have ever guessed at an answer. Ask them if they tested their guess to see whether it was correct or not. Today's lesson encourages guessing *and* checking.

Teaching Notes

Key Question
Why is organizing information in a table useful when using the guess-and-check strategy? The strategy allows for quick comparisons between results, prevents using the same guess twice, and keeps track of whether each guess is too high or too low.

Reasoning

Students need to use logical reasoning to make a reasonable first guess. For instance, the first guess in the Problem was 8 cat food cans and 8 dog food cans since each can costs about $1, and the total amount spent was about $17.

Critical Thinking

The guess-and-check strategy requires students to *analyze* their guesses, *interpret* the results, and to *create* a new, more accurate guess based on their previous one.

Application

Travel agencies work with guess-and-check type situations when setting up travel arrangements. Clients indicate where they would like to travel, about how much they are willing to spend, and how long they want to vacation. Agents then look at a variety of variables, such as hotel and transportation costs, to find an itinerary that fits the client's needs.

Using Technology

For a suggestion on using calculators and a BASIC program to use the guess-and-check method, see page 140C.

Reteaching/Alternate Approach

For a suggestion on using cooperative learning groups to solve real-world problems by the guess-and-check method, see page 140D.

Additional Example

Denise spent $13.27 for batteries and tape. The tape cost $.79 per roll and the batteries cost $1.29 each. How many of each item did she buy?
$7(\$.79) + 6(\$1.29) = \$13.27$;
7 rolls of tape and 6 batteries.

Guided Practice

COMMUNICATION «*Reading*

Use the problem below for Exercises 1–4.

Sarah charges $7 per truck and $4 per car to wash her neighbors' vehicles. She earned $41 last weekend. How many vehicles of each type did she wash?

1. The paragraph is about Sarah earning money.

« **1.** What is the paragraph about?

« **2.** How much did Sarah earn last weekend? $41

3. Suppose you guess that Sarah washed 4 cars and 4 trucks. Did she earn *less than* or *more than* $41? more than; $44

4. Solve the problem using the guess-and-check strategy.
5 cars, 3 trucks

Solve using the guess-and-check strategy.

5. $14v - 8 = 90$ 7
6. $13 + 2m = 57$ 22

7. Rene Valmont collected fees from entrants in a race. He collected $5.50 each from runners who registered before the day of the race and $8 each from those who registered on the day of the race. Rene collected $177. How many runners registered before the day of the race and how many on the day of the race? (There is more than one answer.) before: 22, on: 7; or before: 6, on: 18

Problem Solving Situations

Solve using the guess-and-check strategy.

1. $5n - 18 = 102$ 24
2. $23 + \frac{x}{8} = 48$ 200
3. $94 = \frac{b}{12} - 16$ 1320
4. $87 = 15 + 6y$ 12

5. 5 sweaters, 7 shirts

5. Pedro Sonez bought sweaters at $26 each and shirts at $16 each. He paid a total of $242. How many of each did Pedro buy?

6. In a two-day bike race, Rachel Davis traveled 30 mi. She biked 6 mi farther on the first day than she did on the second day. How many miles did Rachel travel each day? first: 18 mi, second: 12 mi

8. 3 bunches of tulips, 14 bunches of daffodils; or 8 bunches of tulips, 6 bunches of daffodils

7. Linda Chang has fifteen quarters and dimes in all. Their total worth is $2.55. How many of each type of coin does she have? 7 quarters, 8 dimes

8. A bunch of tulips costs $4 and a bunch of daffodils costs $2.50. Paula Wilcox bought some tulips and daffodils for prom decorations. She spent $47. How many bunches of each type of flower did Paula buy? (There is more than one answer.)

9–13. Strategies may vary. Likely strategies are given.

Solve using any problem solving strategy.

9. Rose Perrini has only quarters and dimes. In how many different ways can she pay a $1.35 tip? making a table; 3 ways

PROBLEM SOLVING

CHECKLIST

Keep this in mind:
  Using a Four-Step Plan
Consider these strategies:
  Choosing the Correct Operation
  Making a Table
  Guess and Check

10. A child's bank contains 39 nickels and pennies in all. Their total worth is $.87. How many of each type of coin are in the bank? guess and check; 12 nickels, 27 pennies

11. Michael Olin bought several baseball cards for $10 each. When the price increased to $17 per card, he sold all of the cards except one. His profit was $46. How many baseball cards did Michael buy? guess and check; 9

12. Bradley Ellis went on a trip to visit some friends in another state. She traveled 350 mi on each of the first two days and 400 mi on the third day. How far did she travel? choosing the correct operation; 1100 mi

13. The top three students in a creative writing contest shared a $10,000 college scholarship award. The second-place student received $4000 and the third-place student received $1500. How much did the first-place student receive? choosing the correct operation; $4500

WRITING WORD PROBLEMS Answers to Problems 14 and 15 are on p. A9.

Using the information given, write a word problem that you could solve by using the guess-and-check strategy. Then solve the problem.

14. apples at $.50 each and bananas at $.75 each

15. sweaters at $17 each and shirts at $11 each

SPIRAL REVIEW

S **16.** Find the difference: $11.06 - 7.98$ *(Toolbox Skill 12)* 3.08

17. DATA, *pages 2–3* How many charity concerts does the Houston Symphony perform each year? *(Toolbox Skill 24)* 23

18. Find the answer: $-7 + 5 \times 6 \div (-2)$ *(Lesson 3-6)* −22

19. Sanjee bought compact discs at $14 each and cassette tapes at $9 each. He paid a total of $83. How many of each did Sanjee buy? *(Lesson 4-2)* 4 compact discs, 3 cassette tapes

Equations **147**

Closing the Lesson

Ask a student to explain how the guess-and-check method may be used to solve a problem. Ask another student to explain how the four-step problem-solving plan relates to the guess-and-check method.

Suggested Assignments

Skills: 1–15 odd, 16–19
Average: 1–19
Advanced: 1–19

Exercise Note

Cooperative Learning

Exercises 14 and 15 may be done in groups. Each group should create problems as instructed in the text. One problem could be written to identify only the total amount spent, and the other one could identify the number of items purchased and the total amount spent. Groups can exchange problems and solve them.

Follow-Up

Enrichment

Enrichment 34 is pictured on page 140E.

Practice

Practice 41, *Resource Book*

Practice 41
Skills and Applications of Lesson 4-2

Solve using the guess-and-check strategy.

1. The combined ninth grade and tenth grade classes total 605 students. The tenth grade class has 45 more students than the ninth grade class. How many students are in each class? 9th grade 280, 10th grade 325

2. Jonah's school bought two types of calculators. The regular calculators cost $8 each and the scientific calculators cost $20 each. The bill to the school was $500. How many of each were purchased if a total of 40 calculators were delivered? 25 regular calculators, 15 scientific calculators

3. Sonia scored 40 points in a basketball game. She made twenty-six baskets. Some were two-point baskets and some were one-point baskets (foul shots). How many baskets of each type did she make? 12 foul shots, 14 baskets

4. Tickets for the ninth grade show were $2.50 each for students and $5.00 each for adults. The 225 tickets sold resulted in a gross amount of $875. How many of each type of ticket were sold? 100 student tickets, 125 adult tickets

5. Virgil's junior varsity football team scored 33 points in one of their games. They scored both seven-point touchdowns (all extra points were made) and three-point field goals. How many touchdowns and how many field goals did they score to total 33 points? 3 touchdowns, 4 field goals

6. Eva counted her quarters and dimes. She had a total of 107 coins with a value of $17.45. How many of each type of coin did she have? 62 dimes, 45 quarters

7. Clara's father is the manager of the local shopping market. He told Clara that a truckload of peaches and plums was delivered. The cost to the store was $.50/lb for peaches and $.60/lb for plums. The total bill was $760. How many pounds of each were delivered if the total weight was 1400 lb? 800 lb of peaches, 600 lb of plums

Teaching the Lesson
• Materials: algebra tiles,
  balance scale, calculator
• Lesson Starter 32
• Using Manipulatives,
  p. 140D
• Toolbox Skills 7, 8

Lesson Follow-Up
• Practice 42
• Enrichment 35: Connection
• Study Guide, pp. 63–64

Warm-Up Exercises

Use mental math to find each solution.

1. $6 + y = 31$ 25
2. $13 = 2x - 1$ 7
3. $m \div 7 = 9$ 63
4. $-54 = 6n$ -9
5. $2x - 7 = 35$ 21

Lesson Focus

Ask students if they have ever used a balance scale. What happens when weights are added to only one pan of the scale? Does the scale remain balanced? Today's lesson focuses on solving equations, a mathematical model of balance scales.

Teaching Notes

Emphasize the importance of keeping an equation balanced. Whatever is done to one side of an equation must be done to the other side.

Key Question

Why is it important to check a solution? If a mistake has been made in solving the equation, checking the proposed solution will reveal the error.

Using Addition or Subtraction

Objective: To solve equations using addition or subtraction.

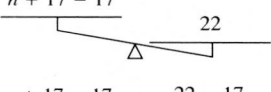

EXPLORATION

1 What operation could you use to undo the addition on the left pan of this balance scale? subtraction

$$n + 17 \qquad 22$$

2 What happens to the balance scale if you subtract 17 from the expression on the left pan? The scale is not in balance.

$$n + 17 - 17 \qquad 22$$

3 What happens to the balance scale if you subtract 17 from both the left and the right pans at the same time? The scale stays in balance.

$$n + 17 - 17 \qquad 22 - 17$$

4 What is in each pan after you simplify the expressions in Step 3? n; 5

$$n \qquad 5$$

Terms to Know
• *solve*
• *inverse operations*

To **solve** an equation, you find all values of the variable that make the equation true. Recall that each of these values is called a *solution*. When you solve an equation, you get the variable alone on one side of the equals sign. To do this, you need to undo any operations on the variable.

Example 1

Solution

Solve $6 + k = 31$. Check the solution.

$$6 + k \qquad 31$$

$$6 + k - 6 \qquad 31 - 6$$

$$k \qquad 25$$

To undo the addition of 6, subtract 6 from both sides.

$$6 + k = 31$$

$$6 + k - 6 = 31 - 6$$

$$k = 25$$

The solution is 25.

$$6 + 25 \qquad 31$$

To *check* the solution, substitute 25 for k in the original equation.

✓ Check

$$6 + k = 31$$
$$6 + 25 \stackrel{?}{=} 31$$
$$31 = 31$$

 **Check Your Understanding**

1. In Example 1, what is the purpose of subtracting 6 from *both* sides of the equation?
See *Answers to Check Your Understanding* at the back of the book.

Just as you used subtraction to undo addition, you can use addition to undo subtraction.

Example 2
Solution

Solve $-29 = s - 15$. Check the solution.

$$-29 = s - 15$$
$$-29 + 15 = s - 15 + 15$$
$$-14 = s$$

To undo the subtraction of 15, add 15 to both sides.

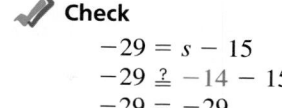 **Check**
$$-29 = s - 15$$
$$-29 \stackrel{?}{=} -14 - 15$$
$$-29 = -29$$

The solution is -14.

 **Check Your Understanding**

2. How would Example 2 be different if the right side of the equation were $s + 15$?
See Answers to Check Your Understanding at the back of the book.

Because addition and subtraction undo each other, they are called **inverse operations.** Recognizing inverse operations can help you to solve equations.

> **Generalization:** *Solving Equations Using Addition or Subtraction*
>
> If a number has been *added* to the variable, subtract that number from both sides of the equation.
>
> If a number has been *subtracted* from the variable, add that number to both sides of the equation.

 You can use a calculator to check the solution of an equation. For instance, to check if -508 is the solution of $n + 163 = -345$, substitute -508 for n and use this key sequence.

Guided Practice

« **1. COMMUNICATION** «*Reading* Replace each __?__ with the correct word.

To solve an equation, you undo any operation on the variable by using a(n) __?__ operation. To undo the addition of 8, for example, you would __?__ 8 from both sides of the equation. inverse; subtract

Explain how you would solve each equation.

2. $x - 16 = 18$ add 16

3. $y + 6 = 27$ subtract 6

4. $-41 = 14 + w$ subtract 14

5. $-62 = r - 37$ add 37

Equations **149**

Using Manipulatives
The Exploration at the beginning of the lesson could be illustrated visually by using a balance scale. Adding or removing an amount from only one pan shows that the equation becomes unbalanced.

For a suggestion on using algebra tiles to model equation solving, see page 140D.

Calculator
Calculators may be used to check the results of a problem. They also may be used to perform computations when using inverse operations. Exercises 15–20 encourage the use of a calculator in checking solutions.

Error Analysis
When solving an equation, students may use the operation indicated rather than its inverse to solve for the variable. Remind students that the goal is to get the variable alone on one side of the equation. This is done by using inverse operations to "undo" any operations on the variable.

Additional Examples
1. Solve $7 + n = 12$. Check the solution. 5
2. Solve $-34 = v - 16$. Check the solution. -18

Closing the Lesson
Give students a list of ten equations. Ask them to identify the operation used in each equation and to state its inverse. Then have them solve the equations.

Suggested Assignments
Skills: 1–21, 23, 24, 32–35
Average: 1–26, 32–35
Advanced: 1–13 odd, 15–35

Exercise Notes
Communication: Reading
Guided Practice Exercise 1 requires students to interpret information they have read in the lesson.

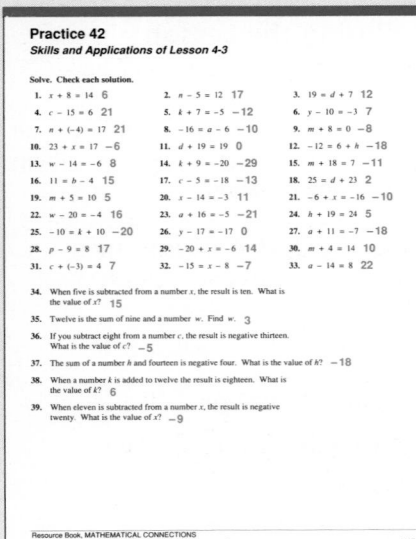

Solve. Check each solution.

6. $w + 8 = 6$ −2   7. $16 = a - 17$ 33   8. $4 + n = 9$ 5

9. $12 + t = -6$ −18   10. $-11 = z - 5$ −6   11. $m - 8 = -8$ 0

Exercises

Solve. Check each solution.

A   1. $a + 2 = 11$ 9   2. $5 + b = -19$ −24   3. $-2 = 7 + c$ −9

4. $45 = d + 8$ 37   5. $1 = r - 3$ 4   6. $w - 15 = -40$ −25

7. $h - 10 = -21$ −11   8. $-10 = c - 20$ 10   9. $a - 9 = 9$ 18

10. $-9 = a - 9$ 0   11. $a + 9 = 9$ 0   12. $-9 = a + 9$ −18

13. The sum of 16 and a number *k* is 5. Find *k*. −11

14. When nine is subtracted from a number *c*, the result is twelve. What is the value of *c*? 21

 CALCULATOR

Solve. Then use a calculator to check the solution.

B   15. $c + 463 = 859$ 396   16. $n - 862 = 98$ 960

17. $-1026 = 554 + d$ −1580   18. $-523 = 146 + w$ −669

19. $562 = 999 + t$ −437   20. $h - 772 = 3480$ 4252

THINKING SKILLS   21, 22. Comprehension   23–31. Analysis

Complete.

21. If $|x| = 5$, then $x = $ __?__ or $x = $ __?__. 5; −5

22. If $|x + 2| = 7$, then $x + 2 = $ __?__ or $x + 2 = $ __?__. 7; −7

Analyze your answers to Exercises 21 and 22. Then find the solution(s) of each of the following equations. If there is no solution, write *no solution*.

C   23. $|x| = 3$ 3; −3   24. $|m| = 0$ 0   25. $|d - 2| = 0$ 2

26. $|c + 4| = 6$ 2; −10   27. $|h| = -5$ no solution   28. $|y - 4| = -14$ no solution

29. $|p| + 5 = 25$ 20; −20 30. $|n| - 8 = 0$ 8; −8   31. $|z| + 7 = 0$ no solution

SPIRAL REVIEW

S   32. Find the next three numbers: 4, 8, 12, __?__, __?__, __?__   *(Lesson 2-5)* 16, 20, 24

33. Solve: $m - 19 = 24$ *(Lesson 4-3)* 43

34. Simplify: $(4x)(3y)(2x)$ *(Lesson 2-1)* $24x^2y$

35. Find the product: $(-9)(13)$ *(Lesson 3-5)* −117

4-4 Using Multiplication or Division

Lesson Planner

Teaching the Lesson
• Materials: algebra tiles, calculator
• Lesson Starter 33
• Using Technology, p. 140C
• Using Manipulatives, p. 140D
• Toolbox Skills 9, 10

Lesson Follow-Up
• Practice 43
• Practice 44: Nonroutine
• Enrichment 36: Communication
• Study Guide, pp. 65–66
• Cooperative Activity 8
• Test 14

Objective: To solve equations using multiplication or division.

APPLICATION

A pair of running shoes weighs 860 g. To find the weight of each shoe you can solve an equation.

two times	weight of each shoe	is	weight of the pair
↓	↓	↓	↓
2 ·	s	=	860

The equation $2s = 860$ involves multiplication. Just like addition and subtraction, multiplication and division are inverse operations. You can use division to undo multiplication and use multiplication to undo division.

Example 1

Solution

Solve $2s = 860$. Check the solution.

$$2s = 860$$

To undo the multiplication by 2, divide both sides by 2.

$$\frac{2s}{2} = \frac{860}{2}$$

$$s = 430$$

The solution is 430.

✓ **Check**

Substitute 430 for s in the original equation.

$$2s = 860$$
$$2 \cdot 430 \stackrel{?}{=} 860$$
$$860 = 860$$

Example 2

Solution

Solve $18 = \frac{n}{-4}$. Check the solution.

$$18 = \frac{n}{-4}$$

To undo the division by −4, multiply both sides by −4.

$$18(-4) = \frac{n}{-4}(-4)$$

$$-72 = n$$

The solution is −72.

✓ **Check**

$$18 = \frac{n}{-4}$$
$$18 \stackrel{?}{=} \frac{-72}{-4}$$
$$18 = 18$$

Warm-Up Exercises

Solve each equation.
1. $m + 12 = 5$ −7
2. $n - 4 = 7$ 11
3. $p \div (-6) = 8$ −48
4. $-7x = 91$ −13
5. $2m - 11 = 19$ 15

Lesson Focus

Ask students if they have ever eaten at a restaurant with several friends and then split the bill equally. Today's lesson focuses on equations that arise from this type of problem, namely equations that can be solved using multiplication or division.

Teaching Notes

Stress the inverse relationship of multiplication and division. Remind students that the "answer" to a division problem may be checked by using multiplication.

Key Question

What is the procedure used to solve equations of the form $-x = k$, where k is a number? Divide both sides by −1.

Equations   **151**

Error Analysis

As with addition and subtraction equations, students may use the operation indicated rather than the inverse, and thus get an incorrect answer. Remind students to use inverse operations when solving equations.

Using Technology

For a suggestion on using a BASIC program or a spreadsheet to solve equations, see page 140C.

Using Manipulatives

For a suggestion on using algebra tiles to model equation solving, see page 140D.

Additional Examples

1. Solve $3x = 807$. Check the solution. 269
2. Solve $-28 = v \div 3$. Check the solution. -84
3. Solve $-x = 11$. Check the solution. -11

Closing the Lesson

Have students make a poster listing the rules for solving all four types of equations. The first rule could be: To solve an addition equation, subtract the number being added to the variable from both sides of the equation. The poster will be a useful reminder to students on how to solve equations. Also, as students learn to solve more types of equations, the poster can be expanded.

Suggested Assignments

Skills: 1–31 odd, 33–37
Average: 1–31 odd, 33–37
Advanced: 1–31 odd, 33–37

Exercise Notes

Making Connections/Transitions

Guided Practice Exercise 1 helps students connect the concept of inverse operation with everyday tasks.

As you have seen in Examples 1 and 2, you can use inverse operations to help you decide whether to multiply or divide to solve an equation.

> **Generalization:** *Solving Equations Using Multiplication or Division*
>
> If a variable has been *multiplied* by a nonzero number, divide both sides by that number.
>
> If a variable has been *divided* by a number, multiply both sides by that number.

Before multiplying or dividing to solve an equation, you sometimes might need to use the multiplication property of -1.

Example 3

Solution

Solve $-n = 9$. Check the solution.

$$-n = 9$$
$$-1n = 9 \quad \longleftarrow \text{Use the multiplication property of } -1.$$
$$\frac{-1n}{-1} = \frac{9}{-1}$$
$$n = -9$$

 **Check**

$$-n = 9$$
$$-1n = 9$$
$$(-1)(-9) \overset{?}{=} 9$$
$$9 = 9$$

The solution is -9.

 **Check Your Understanding**

1. Why is $-n$ rewritten as $-1n$ in Example 3?
2. Describe how Example 3 would be different if the equation were $-3n = -9$.

   See *Answers to Check Your Understanding* at the back of the book.

Guided Practice

«**1.** COMMUNICATION «*Discussion* Many everyday actions, like tying and untying a shoelace, can be thought of as inverse operations. List other everyday actions that can be thought of as inverse operations.
Answers will vary.

Tell whether you would *multiply* or *divide* to solve each equation.

2. $-8b = -64$ **3.** $\frac{y}{7} = 14$ **4.** $19 = \frac{t}{-5}$ **5.** $-81 = 9d$
   divide   multiply   multiply   divide

Replace each __?__ with the number that makes the statement true.

6. $3h = 45$

$$\frac{3h}{3} = \frac{45}{?} \quad 3$$
$$h = \underline{\ ?\ } \quad 15$$

7. $-54 = 2v$

$$\frac{-54}{?} = \frac{2v}{?} \quad 2; 2$$
$$\underline{\ ?\ } = v \quad -27$$

8. $\frac{z}{-3} = 32$

$$\frac{z}{-3}(\underline{\ ?\ }) = 32(\underline{\ ?\ })$$
$$\quad\quad\quad\quad -3; -3$$
$$z = \underline{\ ?\ } \quad -96$$

COMMUNICATION «*Reading*

Use this paragraph for Exercises 9 and 10.

The term *nonzero* is very important in mathematics. All numbers except zero are nonzero numbers. When you use division in a computation or when you use division to solve an equation, you can divide only by non-zero numbers. Division by zero is said to be *undefined*.

«**9.** What is a nonzero number? any number except zero

«**10.** Name two instances in mathematics in which you may use only non-zero numbers. using division in a computation and using division to solve an equation

Solve. Check each solution.

11. $12n = 24$ 2 　　**12.** $-15 = -5c$ 3 　　**13.** $11 = -y$ −11 　　**14.** $\dfrac{-15}{-a} = 15$

15. $\dfrac{m}{-7} = -4$ 28 　　**16.** $-100 = \dfrac{d}{10}$ −1000 　　**17.** $12 = \dfrac{r}{-8}$ −96 　　**18.** $\dfrac{n}{2} = 25$ 50

Exercises

Solve. Check each solution.

A

1. $2b = 30$ 15 　　　　**2.** $-6c = 108$ −18 　　　　**3.** $-64 = 8d$ −8

4. $-55 = -5v$ 11 　　　**5.** $-q = 0$ 0 　　　　　**6.** $-6 = -k$ 6

7. $-14 = \dfrac{x}{4}$ −56 　　　**8.** $\dfrac{y}{5} = -20$ −100 　　　**9.** $\dfrac{a}{-3} = 63$ −189

10. $\dfrac{b}{-7} = 7$ −49 　　　**11.** $-31 = \dfrac{r}{-2}$ 62 　　　**12.** $3 = \dfrac{n}{12}$ 36

13. $-7t = -105$ 15 　　　**14.** $-10u = 120$ −12 　　　**15.** $17 = -w$ −17

16. $-19 = -h$ 19 　　　**17.** $-9y = 9$ −1 　　　　**18.** $1 = \dfrac{x}{-5}$ −5

19. If $2r = 0$, what is the value of r? 0

20. The quotient $-63 \div m$ is equal to -9. Find the value of m. 7

 CALCULATOR

The solution of each equation is -12. Match each equation with the calculator key sequence needed to check the solution.

B

21. $\dfrac{-72}{n} = 6$ B

22. $-72 = \dfrac{864}{n}$ D

23. $864 = -72n$ A

24. $864n = -10{,}368$ C

A. 72. [+/−] [×] 12. [+/−] [=]

B. 72. [+/−] [÷] 12. [+/−] [=]

C.  864. [×] 12. [+/−] [=]

D. 864. [÷] 12. [+/−] [=]

Equations **153**

Communication: Reading
Guided Practice Exercises 9 and 10 check students' ability to read a short paragraph and respond correctly to some related questions.

Calculator
Exercises 21–24 require students to identify the correct key sequence for checking a solution involving negative numbers. Calculators can be very useful in checking solutions.

Number Sense
Some of the estimates given in *Exercises 25–28* differ by multiples of ten. Determining the size of an answer promotes the development of number sense in students.

Estimation
Estimation skills are used in *Exercises 25–32*. *Exercises 29–32* require students to estimate the solution to an equation rather than solving it.

Follow-Up

Enrichment
Enrichment 36 is pictured on page 140E.

Practice
Practice 43, *Resource Book*

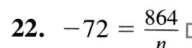

Practice 43
Skills and Applications of Lesson 4-4

Solve. Check each solution.

1. $3c = 24$ 8 　　2. $5y = -15$ −3 　　3. $\frac{x}{4} = -7$ 28

4. $36 = -9a$ −4 　　5. $\frac{x}{-7} = 3$ −21 　　6. $\frac{d}{6} = -6$ −36

7. $-12h = 60$ −5 　　8. $-8w = 72$ −9 　　9. $-25 = -x$ 25

10. $-5d = -85$ 17 　　11. $-6y = 84$ −14 　　12. $\frac{n}{11} = -4$ −44

13. $-64 = -8n$ 8 　　14. $\frac{b}{-18} = 1$ −18 　　15. $-56 = 7x$ −8

16. $-9a = 108$ −12 　　17. $\frac{a}{8} = -4$ −32 　　18. $16 = \frac{c}{-4}$ −64

19. $-11m = -99$ 9 　　20. $\frac{w}{-7} = 0$ 0 　　21. $\frac{k}{-6} = 7$ −42

22. $-45 = -5a$ 9 　　23. $-h = 0$ 0 　　24. $8b = -96$ −12

25. $56 = -m$ −56 　　26. $-108 = -420$ 42 　　27. $-17x = 0$ 0

28. $18 = -18n$ −1 　　29. $-144 = -12a$ 12 　　30. $\frac{w}{4} = -14$ −56

31. $7k = -49$ −7 　　32. $9m = 81$ 9 　　33. $\frac{x}{8} = -13$ −104

34. The product of 9 and a number w is zero. What is the value of w? 0

35. When a number k is divided by 7 the quotient is negative five. What is the value of k? −35

36. The quotient $-72 \div c$ is equal to 8. What is the value of c? −9

37. The product of six and a number d is -36. What is the value of d? −6

38. Negative twenty is equal to the product of five and a number m. What is the value of m? −4

Use mental math to find each solution.

1. $m + 5 = -3$ -8
2. $12 = r - 7$ 19
3. $g \div 4 = -4$ -16

Solve by using the guess-and-check strategy.

4. Sandy bought a CD and a clock radio for her desk. The total cost was $50. The clock radio cost $14 more than three times the CD. How much did each item cost?
   CD, $9; clock radio, $41

Solve. Check each solution.

5. $15 = c + 8$ 7
6. $n - 7 = -11$ -4
7. $17 = 14 + w$ 3
8. $-4x = 76$ -19
9. $-5 = b \div 7$ -35
10. $-c = 9$ -9

Alternative Assessment

Ask each student to make up four equations, one for each of the four basic operations, and a word problem that can be solved by using the guess-and-check strategy. Have students exchange papers and solve the equations and word problem. Check to see how students are doing by observing their work and asking questions informally.

ESTIMATION

To estimate the solution of an equation, first decide which inverse operation to use. Then use an appropriate estimation method for that operation.

Choose the letter of the best estimate for the solution of each equation.

25. $346 = -49m$ a. -5 (b.) -7 c. -70 d. -50

26. $\frac{n}{67} = 205$ a. 3   b. 30   c. 1400   (d.) 14,000

27. $x - 402 = 234$ a. 200   b. 80,000   c. -2 (d.) 600

28. $2070 = 486 + d$ (a.) 1500   b. 2500   c. -1500 d. 4

Estimate the solution of each equation.

29. $\frac{b}{88} = 17$ about 1800

30. $314 = 21p$ about 15

31. $1597 = z + 206$ about 1400

32. $n - 482 = -695$ about -200

SPIRAL REVIEW

33. DATA, *pages 46–47*   How many moons does Saturn have? *(Toolbox Skill 23)* 23

34. Solve: $-8x = 72$ *(Lesson 4-4)* -9

35. Graph points $A(2, 4)$, $B(-3, 2)$, and $C(-1, -3)$ on a coordinate plane.   *(Lesson 3-8)* See answer on p. A9.

36. Estimate the difference: $903 - 719$ *(Toolbox Skill 1)* about 200

37. Estimate the sum: $3.8 + 4.1 + 4.4 + 3.7$ *(Toolbox Skill 4)* about 16

Self-Test 1

Use mental math to find each solution.

1. $r + 4 = -3$ -7 2. $\frac{h}{-3} = 3$ -9 3. $13 = 3w - 5$ 4-1
   6

Solve by using the guess-and-check strategy.

4. Lucia bought birthday cards at $1.50 each and thank-you notes at $1.10 each. She paid a total of $10.80. How many of each did she buy? 5 birthday cards, 3 thank you notes   4-2

Solve. Check each solution.

5. $c + 12 = 9$ -3 6. $m - 8 = -10$ -2 7. $15 = 13 + w$ 2   4-3

8. $-3z = 90$ -30 9. $-4 = \frac{a}{6}$ -24 10. $-n = 7$ -7 4-4

Modeling Equations Using Tiles

Objective: To use algebra tiles to model equations.

Materials

■ algebra tiles

Just as you can use algebra tiles to model variable expressions, you also can use tiles to model equations. Remember to use the tile  to represent the variable n and the tile □ to represent the number 1. The equation $n + 2 = 7$ then can be represented as follows.

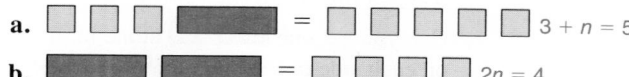

Activity I *Modeling Equations with Tiles*

1 Write the equation represented by each diagram.

a. 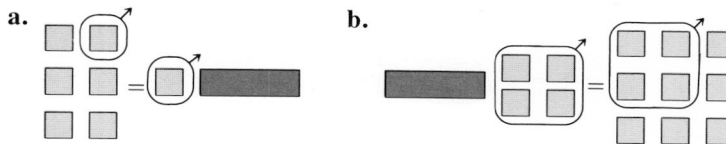 $3 + n = 5$

b. $2n = 4$

2 Show how to use tiles to represent each equation.

   a. $6 = n$ **b.** $n + 1 = 4$ **c.** $3n = 9$ **d.** $12 = 2n$

Answers to Activity I, Step 2, and Activities II and III are on p. A9.

Activity II *Solving Equations Involving Addition*

1 Describe the situation that you think is being represented in each diagram.

a.   **b.**

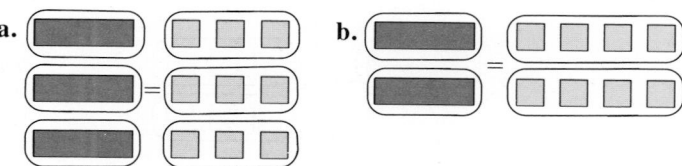

2 Show how to use tiles to solve each equation. What is the solution?

   a. $n + 5 = 7$ **b.** $10 = n + 2$ **c.** $3 + n = 11$

Activity III *Solving Equations Involving Multiplication*

1 Describe the situation that you think is being represented in each diagram.

a.   **b.**

2 Show how to use tiles to solve each equation. What is the solution?

   a. $2n = 10$ **b.** $8 = 4n$ **c.** $6n = 6$

Teaching Note

Using Manipulatives

This entire exploration focuses on using algebra tiles. Tiles offer a tactile way for students to solve equations.

Activity Notes

Activity I

This activity allows students to visualize equations prior to solving them. A mat with an equal sign in the center, such as the one shown below, might be used to represent the two sides of an equation.

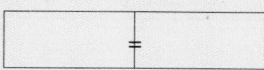

It is important for students to experience and understand Activity I before moving to Activity II.

Activity II

This activity allows students to manipulate tiles to show that the same number is being subtracted from each side of the equation.

Activity III

This activity allows students to visualize division by 2. The matching of sets reinforces the idea that whatever is done to one side of the equation must be done to the other side.

Equations **155**

Teaching the Lesson
• Materials: algebra tiles, calculator
• Lesson Starter 34
• Using Technology, p. 140C
• Toolbox Skills 7–10

Lesson Follow-Up
• Practice 45
• Enrichment 37: Application
• Study Guide, pp. 67–68
• Computer Activity 5
• Manipulative Activity 7

Warm-Up Exercises

Solve each equation.
1. $m + 16 = 85$ 69
2. $x - 29 = 61$ 90
3. $y \div (-7) = 18$ -126
4. $-7n = 91$ -13
5. $6x = 66$ 11

Lesson Focus

Ask students to think about shopping for groceries. If a person buys several of one item and only one of a different item, how can that person find the total cost of all the items? Is it necessary to use more than one step to solve the problem? Today's lesson focuses on solving two-step equations.

Teaching Notes

Review the order of operation rules with students. The steps for solving equations are the reverse of those rules. Work through the examples and stress the generalization on page 157.

Key Question

What steps would you follow to solve an equation involving both addition and multiplication? First undo the addition by using subtraction. Then undo the multiplication by using division.

Two-Step Equations

Objective: To solve equations using two steps.

EXPLORATION   See answers on p. A10.

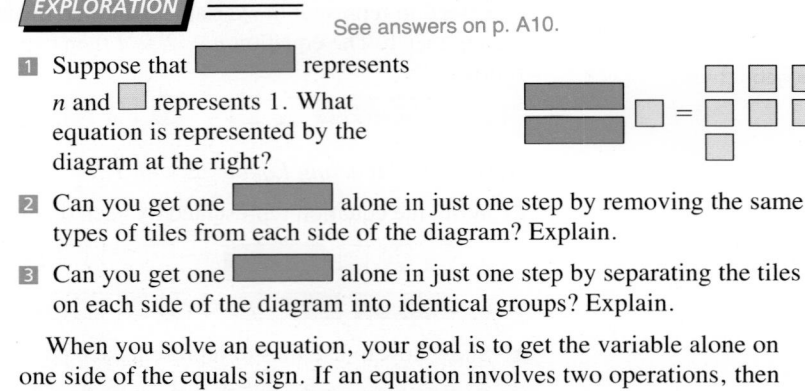

1 Suppose that ▭ represents n and ▢ represents 1. What equation is represented by the diagram at the right?

2 Can you get one ▭ alone in just one step by removing the same types of tiles from each side of the diagram? Explain.

3 Can you get one ▭ alone in just one step by separating the tiles on each side of the diagram into identical groups? Explain.

When you solve an equation, your goal is to get the variable alone on one side of the equals sign. If an equation involves two operations, then you need to use two steps to achieve that goal.

Example 1   Solve $2n + 1 = 7$.

Solution   Let ▭ represent n and ▢ represent 1.

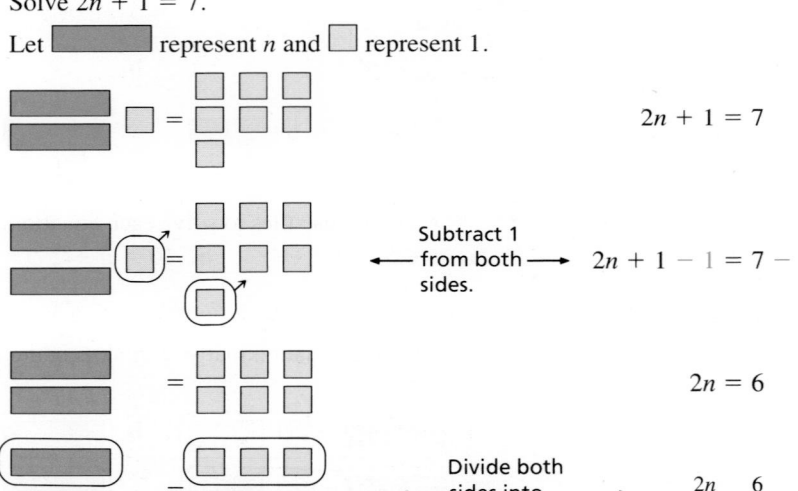

$2n + 1 = 7$

Subtract 1 from both sides. $2n + 1 - 1 = 7 - 1$

$2n = 6$

Divide both sides into two identical groups. $\dfrac{2n}{2} = \dfrac{6}{2}$

$n = 3$

The solution is 3.

✓ **Check Your Understanding**

1. How would you check that 3 is the solution of the equation in Example 1?
See *Answers to Check Your Understanding* at the back of the book.

156   Chapter 4

156

Example 2   Solve $-20 = \frac{t}{3} - 4$. Check the solution.

Solution

$$-20 = \frac{t}{3} - 4$$

$$-20 + 4 = \frac{t}{3} - 4 + 4 \longleftarrow \text{Add 4 to both sides.}$$

$$-16 = \frac{t}{3}$$

$$-16 \cdot 3 = \frac{t}{3} \cdot 3 \longleftarrow \text{Multiply both sides by 3.}$$

$$-48 = t$$

The solution is -48.

✔ **Check**

$$-20 = \frac{t}{3} - 4$$

$$-20 \overset{?}{=} \frac{-48}{3} - 4$$

$$-20 \overset{?}{=} -16 - 4$$

$$-20 = -20$$

 **Check Your Understanding**

2. What two operations are involved in the equation in Example 2?

3. In Example 2, what two operations were used to solve the equation?
See *Answers to Check Your Understanding* at the back of the book.

To solve equations involving two steps, such as the equations in Examples 1 and 2, you need to use *two* inverse operations.

> **Generalization:** *Solving Two-Step Equations*
>
> First undo the addition or subtraction, using the inverse operation.
> Then undo the multiplication or division, using the inverse operation.

 You can use a calculator to solve an equation. Use inverse operations as you normally would. To solve $11m - 44 = -121$, use this key sequence.

Guided Practice

COMMUNICATION «*Writing*

«**1.** Let ▭ represent n and ▯ represent 1. Write the equation that is represented by the diagram below. $2n + 7 = 9$

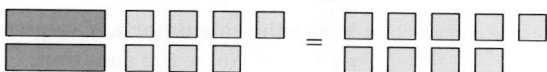

«**2.** Draw a diagram similar to the diagram in Exercise 1 to represent this equation: $8 = 3n + 5$ See answer on p. A10.

Choose the letter of the equation that has the same solution as the given equation.

3. $3x - 12 = 18$ **a.** $x - 12 = 6$ **b.** $3x = 6$ **c.** $3x = 30$

Equations   **157**

157

Sidebar:

Using Manipulatives
Algebra tiles demonstrate the process used to solve two-step equations. You may want to give students a few equations similar to Example 1 and have them solve the equations by using tiles.

Using Technology
For a suggestion on using a BASIC program or a spreadsheet to solve equations, see page 140C.

Application
Chemists, physicists, and other scientists often use mathematical equations in their work.

Error Analysis
Students may use inverse operations in the wrong order. If multiplication is involved, they may divide both sides, excluding any term that is added or subtracted. Remind students to first undo any addition or subtraction, then undo the multiplication or division.

Additional Examples
1. Solve $2x + 3 = 9$.   3
2. Solve $-30 = \frac{d}{4} - 6$. -96

Closing the Lesson
Give students a list of ten equations, some one-step and some two-step equations. Have students identify the type of equation and then describe how to solve it.

Suggested Assignments
Skills: 1–17, 19–23 odd, 30–33
Average: 1–17, 19–27 odd, 30–33
Advanced: 1–17 odd, 18–33

Exercise Notes

Communication: Writing
Guided Practice Exercises 1 and 2 have students write an equation represented by a diagram and vice versa.

Calculator
Calculators can be useful in solving
equations. *Exercises 18–23 focus*
on the key sequences needed to
solve various types of equations.

Critical Thinking
Exercise 25 requires students to
compare two equations and to iden-
tify any differences in the procedures used to solve them.

Making Connections/Transitions
Exercises 26–29 connect mathemat-
ics and physical science. Chemical
reactions depend upon the valences
of the elements.

Choose the letter of the equation that has the same solution as the given equation.

4. $\frac{w}{5} + 10 = 20$ (**a.**) $\frac{w}{5} = 10$ **b.** $w + 10 = 100$ **c.** $\frac{w}{5} = 2$

Replace each __?__ with the number that makes the statement true.

5.
$$6n + 3 = -39$$
$$6n + 3 - \underline{\ ?\ } = -39 - \underline{\ ?\ }$$
$$6n = -42 \qquad 3; 3$$
$$\frac{6n}{?} = \frac{-42}{?} \qquad 6; 6$$
$$n = \underline{\ ?\ } \ -7$$

6.
$$25 = \frac{h}{2} - 9$$
$$25 + \underline{\ ?\ } = \frac{h}{2} - 9 + \underline{\ ?\ } \qquad 9; 9$$
$$34 = \frac{h}{2}$$
$$34 \cdot \underline{\ ?\ } = \frac{h}{2} \cdot \underline{\ ?\ } \ 2; 2$$
$$\underline{\ ?\ } = h \ 68$$

Solve. Check each solution.

7. $4c - 2 = 10 \ 3$

8. $-28 = 12 + 2z \ -20$

9. $15 = \frac{t}{4} - 1 \ 64$

10. $4 + \frac{x}{-2} = 16$

Exercises

Solve. Check each solution.

A   **1.** $6n + 4 = 28 \ 4$ **2.** $3t + 1 = -8 \ -3$ **3.** $7w - 5 = -19 \ ^{-2}$

4. $8b - 5 = 35 \ 5$ **5.** $37 = -5c + 2 \ -7$ **6.** $-21 = -9b + 6 \ 3$

7. $28 = -3x - 2 \ -10$ **8.** $35 = 9m - 10 \ 5$ **9.** $\frac{r}{5} - 1 = 7 \ 40$

10. $\frac{n}{-6} - 4 = 6 \ -60$ **11.** $7 = \frac{a}{3} - 6 \ 39$ **12.** $\frac{w}{-3} + 8 = 14 \ -18$

13. $\frac{t}{6} - 6 = -32 \ -156$ **14.** $-53 = \frac{k}{4} - 53 \ 0$ **15.** $86 = \frac{u}{-2} + 86 \ 0$

16. The sum $6 + 5n$ is equal to 16. Find the value of n. 2

17. If $3x - 11 = 4$, what is the value of x? 5

 CALCULATOR Answers to Exercises 18–23 are on p. A10.

Write the calculator key sequence you would use to solve each equation. Solve the equation.

B   **18.** $16c - 71 = 153$ **19.** $187 = \frac{k}{8} + 135$

20. $\frac{x}{13} - 32 = 58$ **21.** $-42z + 161 = 1505$

22. $1480 + 7w = 2040$ **23.** $\frac{b}{-15} + 112 = -88$

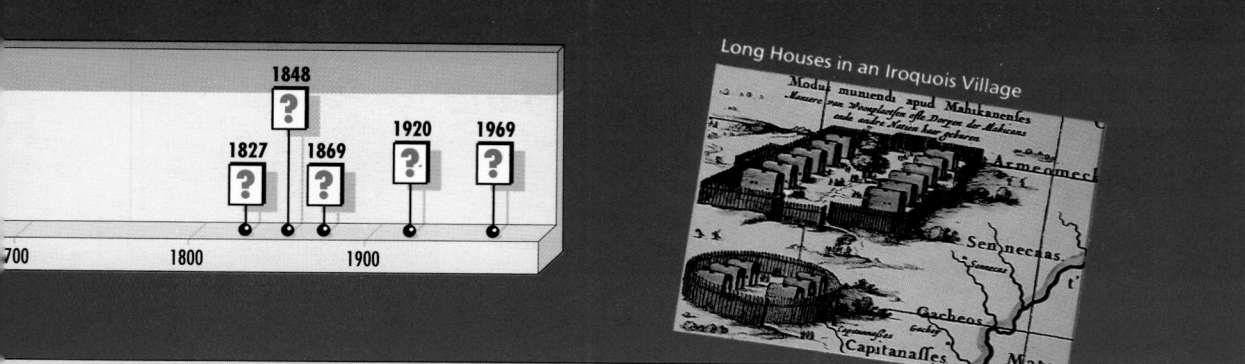

1848 **?**

1827 **?** 1869 **?** 1920 **?** 1969 **?**

700   1800   1900

Long Houses in an Iroquois Village

The timeline on this page gives seven important dates in American history. Each date is associated with one of the events listed next to Exercises 25–31. To match each date with the correct event, first solve all seven exercises. Then put your answers in order from least to greatest. The order of your answers will correspond to the order of the dates on the timeline, from 1500 to 1969. For example, the event associated with the smallest answer took place first.

25. −20 + (−4) + 38   14 (1920)
Women win voting rights with the passage of the 19th Amendment to the Constitution.

27. 5(−3)(4)   −60 (about 1500)
The Cayuga, Mohawk, Oneida, Onondaga, and Seneca nations form the Iroquois Confederacy, whose system of representative government later inspires the drafters of the United States Constitution.

29. $\frac{-36}{-9}$ 4 (1848)
Men and women meet at Seneca Falls, New York, to draw up a declaration of rights for women.

26. −5 + 17 + (−3)   9 (1869)
Workers, most of them Chinese and European immigrants, complete the first transcontinental railroad.

28. (−2)(0)(−5)   0 (1827)
The United States' first newspaper run by African-Americans, *Freedom's Journal*, is founded in New York.

30. 19 + (−28) + 37   28 (1969)
American pioneers in space first walk on the moon.

31. −8 + (−15) + (−3)   −26 (1609)
Hispanic settlers from Mexico establish Santa Fe as the capital of the colony of New Mexico.

Suggested Activities

Though timelines are useful for telling when events took place, they cannot reveal why or how they took place, or what effect they had on people's lives. You may want to ask students to come up with questions about events on the timeline on page 123. Such questions might include: Why did the nations of the Iroquois Confederacy band together? What kind of articles appeared in *Freedom's Journal*? How long did it take to cross the country once the transcontinental railroad was completed? Students can work in cooperative groups to generate questions for particular events. They can conduct research to find answers to their questions and share their findings with the class.

Suggested Resources

Meltzer, Milton, ed. *The Black Americans: A History in Their Own Words, 1619-1983*. New York: Thomas Y. Crowell, 1984.

_____. *The Chinese Americans*. New York: Thomas Y. Crowell, 1980.

Waldman, Carl. *Encyclopedia of Native American Tribes*. New York: Facts on File Publications, 1988.

Supporters of Women's Voting Rights

Oldest House, Santa Fe, New Mexico

First Transcontinental Railroad

Integers   **123**

Teaching the Lesson
- Lesson Starter 28
- Visuals, Folder I
- Reteaching/Alternate Approach, p. 92D

Lesson Follow-Up
- Practice 36
- Enrichment 30: Connection
- Study Guide, pp. 55–56
- Cooperative Activity 7

Warm-Up Exercises

Use a number line to find each sum.
1. $-2 + 6$ 4
2. $-3 + (-4)$ -7
3. $7 + (-5)$ 2
4. $3 + (-5)$ -2
5. $-2 + (-6)$ -8

Lesson Focus

Tell students that cities on a world map can be located by using longitude and latitude. The longitude and latitude provide coordinates for a city. Today's lesson focuses on using coordinates to find points in a plane.

Teaching Notes

Much of the content in this lesson may be new to students. Emphasize the terms *coordinate plane, axes, origin, quadrant, ordered pair,* and *coordinates.*

Key Question

How can you graph a point on a coordinate plane?   Start at the origin. Look at the x-coordinate and move that many units left (negative) or right (positive). Then look at the y-coordinate and move that many units up (positive) or down (negative). Graph a point at this location.

3-8

The Coordinate Plane

Objective: To find coordinates and graph points on a coordinate plane.

The grid at the right is called a **coordinate plane.** A coordinate plane is formed by two number lines called **axes.** The horizontal number line is the **x-axis,** and the vertical number line is the **y-axis.** The point where the axes meet is called the **origin.** The axes separate the coordinate plane into four sections called **quadrants.**

You can assign an **ordered pair** of numbers to any point on the plane. The first number in an ordered pair is the **x-coordinate.** The second number is the **y-coordinate.** The origin has coordinates $(0, 0)$.

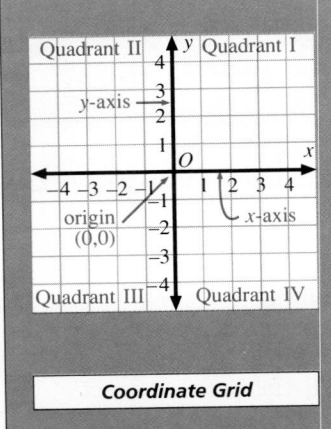

Coordinate Grid

Example 1

Use the coordinate plane at the right. Write the coordinates of each point.

a. *E*   b. *F*

Solution

a. Start at the origin. Point *E* is 2 units left (negative) and 4 units up (positive). The coordinates are $(-2, 4)$.

b. Start at the origin. Point *F* is 3 units right (positive) and 0 units up or down. The coordinates are $(3, 0)$.

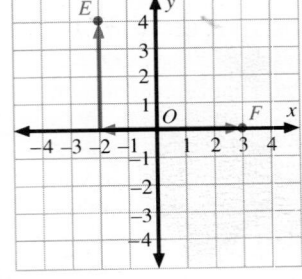

 **Check Your Understanding**

1. In Example 1(a), why is the x-coordinate negative?
2. In Example 1(b), why is the y-coordinate zero?
   See *Answers to Check Your Understanding* at the back of the book.

When you **graph a point** $A(x, y)$ on a coordinate plane, you show the point that is assigned to the ordered pair (x, y).

Generalization: *Graphing a point A(x, y)*

1. Start at the origin.
2. Move x units horizontally along the x-axis.
3. Then move y units vertically.
4. Draw the point and label it *A.*

124   Chapter 3

Example 2

Graph each point on a coordinate plane.

a. $A(4, 1)$ **b.** $B(1, 4)$

c. $C(-3, 0)$ **d.** $D(0, -4)$

Solution

In each case, start at the origin.

a. Move 4 units to the right and then 1 unit up.

b. Move 1 unit to the right and then 4 units up.

c. Move 3 units to the left and then 0 units up or down.

d. Move 0 units to the left or right and then 4 units down.

☑️ **Check Your Understanding**

3. In Example 2(d), how do you know point D is on the y-axis?

4. Describe how the graph of point A would be different if the coordinates were $(4, -1)$.

See *Answers to Check Your Understanding* at the back of the book.

Guided Practice

COMMUNICATION « *Reading* **3(b).** Any of the four areas into which the axes separate the coordinate plane.

Explain the meaning of the underlined word in each sentence.

Answers to Guided Practice 1–2 are on p. A6.

« **1.** The members of the team agreed to <u>coordinate</u> their efforts.

« **2.** The <u>origin</u> of the Mississippi River is Lake Itasca, Minnesota.

« **3.** The prefix *quadr-*, as in *quadrant*, means "four."

     a. What is the meaning of the familiar term *quadruplet*? group of four
     b. What is the meaning of the mathematical term *quadrant*?
     c. How can knowing the meaning of *quadruplet* help you remember the meaning of *quadrant*? A group of four will help you remember that the coordinate plane is separated into four quadrants.

To graph each point, tell how many units and in what direction from the origin you would move for the underlined coordinate.

4. $Q(\underline{2}, 3)$ 2 units to the right   **5.** $R(-8, \underline{8})$ 8 units up

6. $S(\underline{-5}, -9)$ 5 units to the left   **7.** $T(4, \underline{-2})$ 2 units down

8. $U(\underline{0}, 6)$ 0 units to the left or right   **9.** $V(-3, \underline{0})$ 0 units up or down

Answers to Guided Practice 10–15 are on p. A6.

Describe how you would move from the origin to graph each point.

10. $A(-7, 2)$ **11.** $B(5, -3)$

12. $C(0, 2)$ **13.** $D(-4, 0)$

14. $E(-2, -4)$ **15.** $F(3, 5)$

Integers   **125**

Error Analysis

In graphing an ordered pair (x, y), students may reverse the coordinates. Remind them that the first number, or x-coordinate, of an ordered pair indicates the number of units to move left or right; the second number, or y-coordinate, indicates the number of units to move up or down.

Reteaching/Alternate Approach

For a suggestion on relating a street map and a coordinate plane, see page 92D.

Application

An application of a coordinate grid to maps is given on student pages 132 and 133.

Additional Examples

1. Use the coordinate plane below. Write the coordinates of each point.
     a. M $(-3, 4)$
     b. P $(2, 0)$

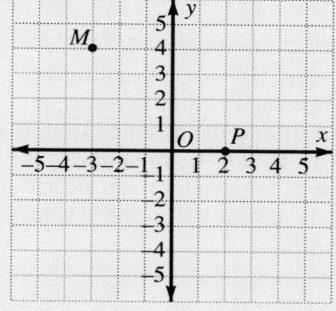

2. Graph each point on a coordinate plane.
     a. Q $(3, 2)$ **b.** R $(2, 3)$
     c. S $(-4, 0)$ **d.** T $(0, -2)$

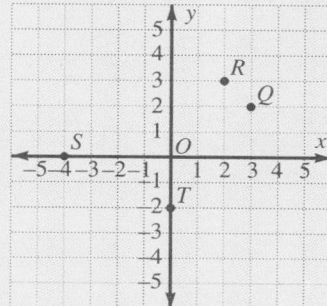

Use the coordinate plane at the right. Write the coordinates of each point.

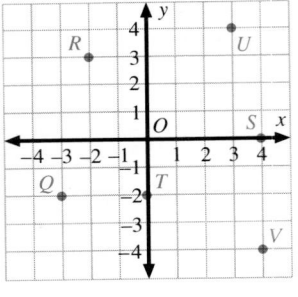

16. Q (−3, −2)   17. R (−2, 3)

18. S (4, 0)   19. T (0, −2)

20. U (3, 4)   21. V (4, −4)

Graph each point on a coordinate plane.

22. $J(3, 5)$ 23. $K(3, −2)$

24. $L(0, −1)$ 25. $M(2, 0)$

Answers to Guided Practice 22–25 are on p. A6.

Exercises

Use the coordinate plane at the right. Write the coordinates of each point.

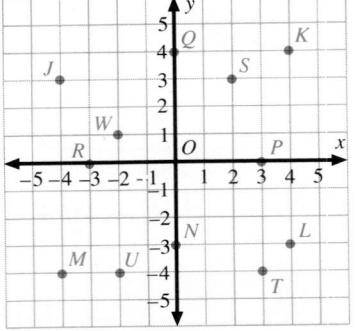

A   1. J (−4, 3)   2. K (4, 4)

     3. L (4, −3)   4. M (−4, −4)

     5. N (0, −3)   6. P (3, 0)

     7. R (−3, 0)   8. Q (0, 4)

     9. S (2, 3)   10. T (3, −4)

     11. U (−2, −4)   12. W (−2, 1)

Graph each point on a coordinate plane.

13. $A(5, 1)$ 14. $B(3, 3)$ 15. $C(6, −2)$ 16. $O(0, 0)$

17. $E(−3, −3)$ 18. $F(−2, 4)$ 19. $G(−6, 1)$ 20. $H(−4, −1)$

Answers to Exercises 13–20 are on p. A6.

Tell in which quadrant or on which axis each point lies.

B   21. $P(1, 9)$ I   22. $Q(5, −14)$ IV   23. $R(−12, 6)$ II   24. $S(−8, −3)$ III

     25. $A(0, −3)$ 26. $B(−12, 0)$ 27. $C(1, 0)$ 28. $D(0, 9)$
        y-axis   x-axis   x-axis   y-axis

THINKING SKILLS 29–36. Synthesis

Develop a rule that describes what is true of all points in the given part of the coordinate plane.

29. Quadrant I   30. Quadrant II

31. Quadrant III   32. Quadrant IV

33. Quadrants II and III   34. Quadrants III and IV

35. the *x*-axis   36. the *y*-axis

CONNECTING ALGEBRA AND GEOMETRY

In *coordinate geometry*, you represent geometric figures on a coordinate plane. For example, at the right you see a square on a coordinate plane. The points $A(-2, -2)$, $B(-2, 2)$, $C(2, 2)$, and $D(2, -2)$ are called the *vertices* of the square.

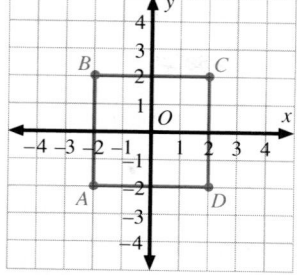

37. What is the shape of the figure that has vertices $Q(-3, 5)$, $R(-3, 1)$, $S(4, 1)$, and $T(4, 4)$? quadrilateral

38. Figure *RSTV* is a square. Its vertices are $R(-1, 4)$, $S(-1, 1)$, $T(2, 1)$ and a point V. What are the coordinates of V? (2, 4)

C **39.** Triangle *ABC* has vertices $A(4, -3)$, $B(2, 1)$, and $C(0, -1)$. Add 2 to the *x*-coordinate and 3 to the *y*-coordinate of all the vertices. Call the new points *D*, *E*, and *F*. Compare the shape and location of triangle *DEF* to the shape and location of triangle *ABC*.

40. Rectangle *JKLM* has vertices $J(2, 2)$, $K(2, 5)$, $L(8, 5)$, and $M(8, 2)$. Multiply the *x*-coordinate and *y*-coordinate of all the vertices by -1. Call the new points *N*, *P*, *Q*, and *R*. Compare the shape and location of rectangle *NPQR* to the shape and location of rectangle *JKLM*.
Answers to Exercises 39–40 are on p. A6.

SPIRAL REVIEW

S **41.** Simplify: $4a - 7b + 3a$ *(Lesson 2-3)* 7a – 7b

42. Find the product: $(-16)(7)$ *(Lesson 3-6)* –112

43. Graph the point $A(-5, 4)$ on a coordinate plane.   *(Lesson 3-8)*
See answer on p. A6.

44. Do you need an *estimate* or an *exact answer* to find the amount of money you should bring to school to buy lunch?   *(Lesson 2-8)*
estimate

Historical Note
The Nile River flooded every year, so it was necessary to resurvey the land each year.

Historical Note

I n the ancient world, the Egyptians and the Romans used coordinates in surveying. The idea of using a coordinate plane to graph points assigned to ordered pairs came much later. René Descartes, a French mathematician, developed it in a book published in 1637. This idea revolutionized mathematics because it tied together geometry and algebra.

Research

Find out why it was important to the Egyptians to learn how to use coordinates to survey the land.

Integers   **127**

Follow-Up

Extension
Ask students to draw a triangle in any quadrant of the coordinate plane. Label the coordinates of each vertex. Find three new vertices by writing the opposite of each number in the ordered pairs. Graph the new vertices. What relationship does the new graph have to the original graph? same shape but turned one half-turn   Try drawing a triangle in a different quadrant. Repeat the same process as above. Does the relationship you observed hold true? Yes

Enrichment
Enrichment 30 is pictured on page 92F.

Practice
Practice 36 is shown below.

Practice 36, *Resource Book*

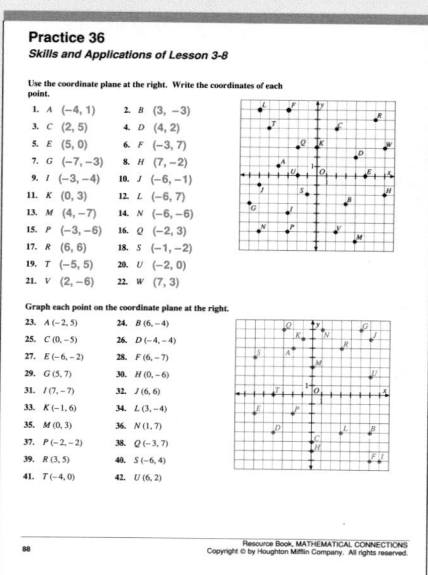

Teaching the Lesson
- Lesson Starter 29
- Using Technology, p. 92C

Lesson Follow-Up
- Practice 37
- Practice 38: Nonroutine
- Enrichment 31: Communication
- Study Guide, pp. 57–58
- Computer Activity 4
- Test 11

Warm-Up Exercises

Complete the function table.

x	$3x - 4$	
2	2	
4	8	
6	?	14
8	?	20
10	?	26

Lesson Focus

Ask students if they have ever seen line graphs in magazines or newspapers. Line graphs are examples of graphs of functions. Today's lesson begins the study of graphing functions by graphing ordered pairs of the functions.

Teaching Notes

Show students how to create ordered pairs from a function table. Remind students that the x-coordinate is given first.

Key Question

How can the coordinate plane be used to represent the information given in a function table? The information in the table can be used to create ordered pairs. These can be graphed on a coordinate plane.

Graphing Functions

Objective: To graph FUNC-TIONS on a coordinate plane.

Terms to Know
- graph of a function

1. $(-4, -3)$, $(-3, -2)$, $(-1, 0)$, $(0, 1)$, $(3, 4)$

3.

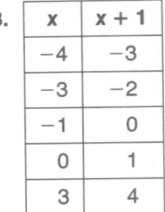

x	$x + 1$
-4	-3
-3	-2
-1	0
0	1
3	4

EXPLORATION

1️⃣ Make a list of the ordered pairs that are graphed at the right.

2️⃣ Complete this statement: In this list, each x-coordinate is paired with exactly one y-coordinate, so the relationship is a __?__. function

3️⃣ Use the ordered pairs in your list to make a function table.

4️⃣ Write the function rule. $x \rightarrow x + 1$

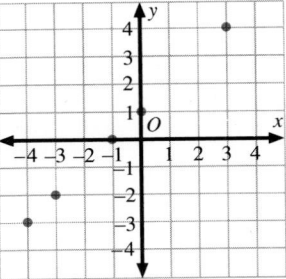

In Lesson 2-6, a function was defined as a relationship that pairs each number in a given set of numbers with exactly one number in a second set of numbers. You learned to represent this pairing of numbers using either a *function table* or a *function rule*. For instance, the function shown in the table at the right can be represented by the rule $x \rightarrow x - 2$.

x	$x - 2$
-3	-5
-2	-4
0	-2
1	-1
5	3

Because it is a pairing of numbers, a function can also be represented by a set of ordered pairs. Therefore, you can draw the **graph of the function** by graphing the points that correspond to all the ordered pairs.

Example 1

Solution

Graph the function shown in the table above.

- Use the table to get a set of ordered pairs.
- Graph the ordered pairs on a coordinate plane.

$(-3, -5)$
$(-2, -4)$
$(0, -2)$ ⟶
$(1, -1)$
$(5, 3)$

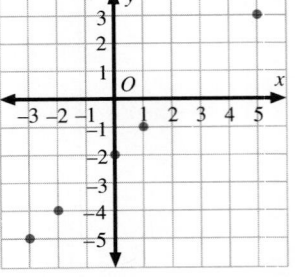

✓ **Check Your Understanding**

1. In Example 1, how do you determine the x-coordinates of the ordered pairs? the y-coordinates?
See *Answers to Check Your Understanding* at the back of the book.

Example 2

Solution

Graph the function $x \rightarrow x^2 - 1$ when $x = -2, -1, 0, 2,$ and 3.

- Make a function table.
- Use the table to get a set of ordered pairs.
- Graph the ordered pairs on a coordinate plane.

x	$x^2 - 1$
-2	$(-2)^2 - 1 = 3$
-1	$(-1)^2 - 1 = 0$
0	$0^2 - 1 = -1$
2	$2^2 - 1 = 3$
3	$3^2 - 1 = 8$

$(-2, 3)$
$(-1, 0)$
$(0, -1)$
$(2, 3)$
$(3, 8)$

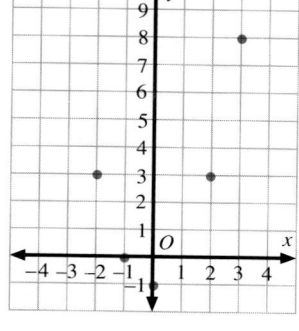

$\longrightarrow$

✎ Check Your Understanding

2. In Example 2, how do you determine which numbers to enter in the first column of the table?

See *Answers to Check Your Understanding* at the back of the book.

Guided Practice

COMMUNICATION «*Reading*

Refer to the text on pages 128–129.

«**1.** What is the main idea of the lesson?

«**2.** What is the main difference between Example 1 and Example 2?

«**3.** Name three different ways to represent a function. function table, function rule, set of ordered pairs

1. The main idea of the lesson is that functions can be represented by sets of ordered pairs that can be graphed on a coordinate plane.

2. The main difference is that a function table is given in Example 1, and a function rule is given in Example 2. You must make a function table in Example 2.

Match each graph with the correct function rule.

4. B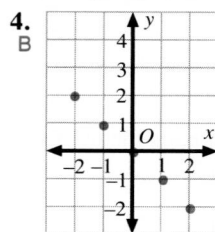

5. D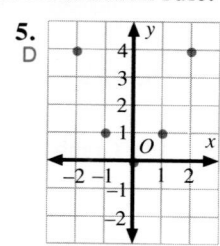

6. C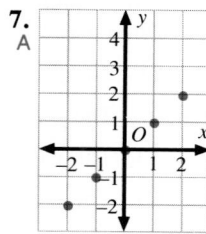

7. A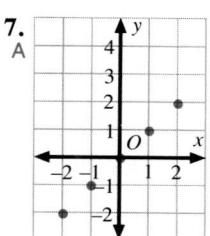

A. $x \rightarrow x$

B. $x \rightarrow -x$

C. $x \rightarrow |x|$

D. $x \rightarrow x^2$

Error Analysis
When graphing functions from tables, students may reverse the order of the coordinates. Remind students that the first column contains the x-coordinate and the second column contains the y-coordinate.

Using Technology
For a suggestion on using a graphing calculator or graphing software to graph functions, see page 92C.

Additional Examples

1. Graph the function given in the table below.

x	$x - 3$
-3	-6
-2	-5
0	-3
1	-2
3	0

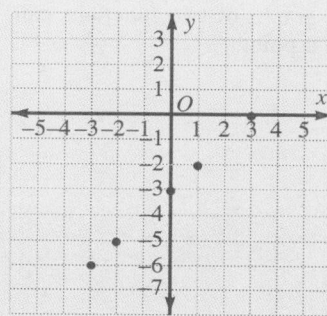

2. Graph the function $x \rightarrow x^2 + 1$ when $x = -2, -1, 0, 1, 2.$

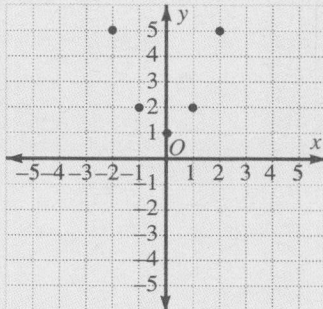

Closing the Lesson

Reverse the procedure given in the lesson. Give students a graph of a function and ask them to make a function table and to give the function rule.

Suggested Assignments

Skills: 1–10, 17–20
Average: 1–10, 11–15 odd, 17–20
Advanced: 1–20

Exercise Notes

Communication: Reading

Guided Practice Exercises 1–3 ask students to think about functions and how they are represented.

Computer

Exercises 15 and 16 make use of a computer to graph functions. Exercise 15 shows that the graph of a function such as $x \rightarrow x + 3$ is a line. Exercise 16 presents the idea that the graphs of two functions can meet in a point.

Graph each function. Answers to Guided Practice 8–11 are on p. A6.

8.

x	$-2x$
-2	4
-1	2
0	0
3	-6
4	-8

9.

x	$\lvert x \rvert + 1$
-5	6
-2	3
0	1
1	2
3	4

10. $x \rightarrow 5 - x$, when $x = -4, -2, 0, 3,$ and 5

11. $x \rightarrow -x - 3$, when $x = -2, -1, 0, 3,$ and 6

Exercises

Graph each function. Answers to Exercises 1–14 are on pp. A6–7.

A **1.**

x	$x + 3$
-4	-1
-3	0
0	3
1	4
2	5

2.

x	$x \div (-2)$
-6	3
-4	2
0	0
2	-1
4	-2

3. $x \rightarrow \dfrac{6}{x}$, when $x = -3, -2, 1, 2,$ and 6

4. $x \rightarrow -5 + x$, when $x = -3, -1, 0, 2,$ and 5

5.

x	$2x - 3$
-2	-7
-1	-5
0	-3
2	1
3	3

6.

x	$x^2 + 2$
-2	6
-1	3
0	2
2	6
3	11

7. $x \rightarrow -x + 2$, when $x = -5, -3, 0, 2,$ and 4

8. $x \rightarrow \lvert x - 2 \rvert$, when $x = -3, -1, 0, 4,$ and 5

9. Graph the function that satisfies these conditions: The values of x are $-4, -2, -1, 2,$ and 5, and the function rule is *x is paired with* $\lvert x \rvert$.

10. Graph the function that pairs the number x with $x - 4$. The values of x are $-5, -1, 0, 2,$ and 4.

Using $-2, -1, 0, 1,$ and 2 as values for x, graph five ordered pairs that satisfy each condition. Use the ordered pairs to make a function table. Then give the rule for the function.

B **11.** The y-coordinate is twice the x-coordinate.

12. The y-coordinate is six more than the x-coordinate.

13. The y-coordinate is the opposite of the x-coordinate.

14. The y-coordinate is the third power of the x-coordinate.

COMPUTER APPLICATION Answers to Exercises 15–17 are on p. A7.

A type of computer software called *geometric graphing software* allows you to use a computer to graph points on a coordinate plane. For Exercises 15 and 16, use this type of software if it is available, or use graph paper.

15. Graph the function $x \rightarrow x + 5$ when $x = -4, 0,$ and 2. Draw a line through the points that you graphed.
  a. Name three other points that appear to lie on this line.
  b. Does the same function rule represent the relationship between the coordinates of these three points?

16. Graph the function $x \rightarrow x - 3$ when $x = -1, 0,$ and 6. Draw a line through the points that you graphed. On the same coordinate plane, graph the function $x \rightarrow -x + 1$ when $x = -4, 0,$ and 3. Draw a line through these three points.
  a. Name the point where the two lines meet.
  b. Which function rule represents the relationship between the coordinates of this point?

SPIRAL REVIEW

S **17.** Graph the function shown in the table at the right.   *(Lesson 3-9)*

x	$3x - 1$
-2	-7
-1	-4
0	-1
1	2
2	5

18. Find the sum: $13.6 + 9.4 + 7.13$
(Toolbox Skill 11) 30.13

19. Ed bought three shirts for $13.50 each and a pair of pants for $27.50. What was the total cost?   *(Lesson 2-4)* $68

20. Find the difference: $-9 - (-12)$
(Lesson 3-4) 3

Self-Test 2

Evaluate each expression when $a = -5$ and $b = 2$.

1. $a^2 + 3b$ 31   **2.** $|a| - b$ 3   **3.** $-ab$ $\overset{10}{}$ 3-6

4. Solve by making a table: Pilar has quarters and dimes. She wants to buy a sandwich that costs $2.95. In how many different ways could Pilar pay for the sandwich? 6 ways   3-7

Answers to Self-Test 5–7 are on p. A8.
Graph each point on a coordinate plane.

5. $D(-2, 3)$ **6.** $E(0, -4)$ 3-8

7. Graph the function $x \rightarrow |x + 2|$ when $x = -3, -2, 0, 1,$ and 4.   3-9

Integers   **131**

Follow-Up

Enrichment
Have students graph pairs of functions such as those given in Exercises 15 and 16. Ask them to try to find functions that: (a) intersect; (b) are parallel; and (c) are the same line.

Enrichment
Enrichment 31 is pictured on page 92F.

Practice
Practice 37 is shown below.

Quick Quiz 2
See pages 138 and 139.

Alternative Assessment
See page 139.

Practice 37, *Resource Book*

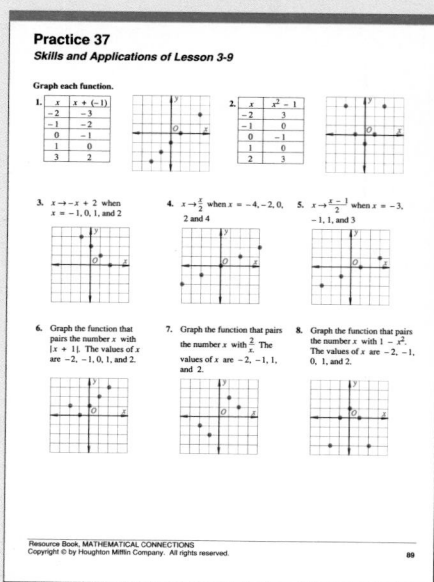

Map Reading

Objective: To read and interpret a map.

Like a coordinate plane, most maps are divided into square sections by horizontal and vertical lines. A location on a map is identified by a set of coordinates that consists of a letter and a number. Unlike coordinates in a coordinate plane, these map coordinates identify a square, not a single point. A *map index* like the one shown below lists the coordinates for each city on a map.

Below is a map of a section of northern New Jersey and adjoining states.

New Jersey Index			
Morristown	E4	New Brunswick	E2
Princeton	D1	Paterson	F5
Ramsey	F5		

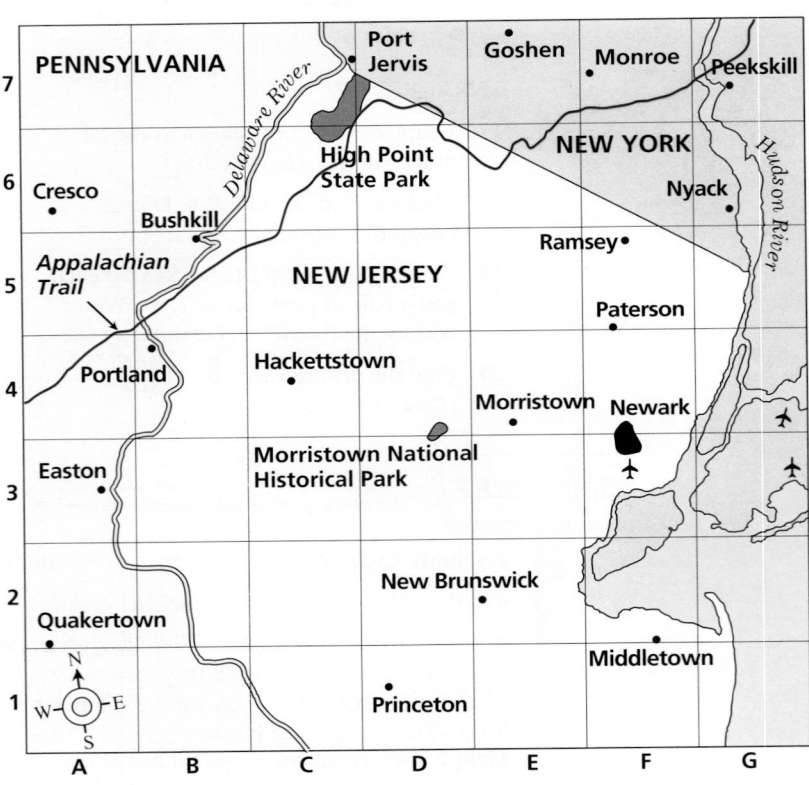

Example

Solution

Write the coordinates of Middletown, New Jersey.

Find the square that contains Middletown.
Move *down* the column to find the letter: F.
Move *left* on the row to find the number: 2.
Combine the letter and number to give the coordinates: F2

Exercises

Use the map shown on page 132.

1. Write the coordinates of Bushkill, Pennsylvania. B5

2. Write the coordinates of Hackettstown, New Jersey. C4

3. Name two New Jersey cities in Column F.

4. Name two New York cities in Row 7.

5. In which squares is Newark, New Jersey located?

6. In which squares do Pennsylvania and New Jersey share a border?

7. In which squares is Morristown National Historical Park located?

8. Name the squares in Pennsylvania shown on the map that include the Appalachian Trail. A4, A5, B5

9. Which rivers are shown on the map? Delaware and Hudson rivers

10. How many airports are indicated on the map? In which squares are they located? three; F3, G3, G4

11. Which town is farther west, Morristown or Newark? Morristown

12. Which town is farther north, Paterson or Ramsey? Ramsey

13. Make a map index for the five Pennsylvania towns shown.

14. Make a map index for the five New York towns shown.

15. GROUP ACTIVITY   Make a map of your school and the area around your school. Label points of interest. Include a map index of the points of interest. Answers will vary.

16. WRITING ABOUT MATHEMATICS   Write a paragraph that describes three similarities and three differences between a map and a coordinate plane. See answer on p. A8.

3. Answers will vary. Example: Newark and Ramsey
4. Answers will vary. Example: Goshen and Monroe
5. F3 and F4
6. C7, C6, B6, B5, B4, A4, A3, A2, B2, B1, C1
7. D3 and D4

13.
Pennsylvania Index	
Bushkill	B5
Cresco	A6
Easton	A3
Quakertown	A2
Portland	B4

14.
New York Index	
Goshen	E7
Monroe	F7
Nyack	G6
Peekskill	G7
Port Jervis	C7

Integers   **133**

Additional Example

Using the map on page 132, write the coordinates of Cresco, Pennsylvania.   A6

Exercise Notes

Life Skills

Exercises 1–14 involve finding locations on a map and interpreting the relative locations, a practical life skill.

Cooperative Learning

Exercise 15 gives students an opportunity to create their own maps. Groups could compare maps and note any similarities or differences.

Communication: Writing

Exercise 16 encourages students to look for similarities and differences between a map and a coordinate plane and to formalize their thinking by writing a paragraph.

Suggested Assignments

Skills: 1–14
Average: 1–14
Advanced: 1–16

Chapter Review

Terms to Know

integers (p. 94)
positive integers (p. 94)
negative integers (p. 94)
opposites (p. 94)
absolute value (p. 95)
addition property of opposites
   (p. 102)
additive inverse (p. 102)
multiplication property of −1
   (p. 116)
coordinate plane (p. 124)

axes (p. 124)
origin (p. 124)
x-axis (p. 124)
y-axis (p. 124)
quadrant (p. 124)
ordered pair (p. 124)
x-coordinate (p. 124)
y-coordinate (p. 124)
graph a point (p. 124)
graph of a function
   (p. 128)

Choose the correct term from the list above to complete each sentence.

1. On a number line, numbers that are the same distance from zero but on opposite sides of zero are called __?__. opposites

2. On a number line, the distance of a number from zero is the __?__ of the number. absolute value

3. The __?__ are the numbers in the set $\{\ldots, -3, -2, -1, 0, 1, 2, 3, \ldots\}$. integers

4. The __?__ states that the sum of a number and its opposite is zero.

5. The __?__ is the point in a coordinate plane where the axes intersect. origin

6. In an ordered pair, the first number is called the __?__. x-coordinate

   4. addition property of opposites

Find each absolute value. *(Lesson 3-1)*

7. $|-11|$ 11   8. $|5|$ 5   9. $|0|$ 0   10. $|-2|$ 2   11. $|4|$ 4   12. $|-3|$ 3

Replace each __?__ with >, <, or =. *(Lesson 3-1)*

13. -12 __?__ 3 <   14. -8 __?__ -8 =   15. -2 __?__ 20 <   16. -5 __?__ 0 <

Solve by making a table. *(Lesson 3-7)*

17. Corey Watson, a florist, sells daisies for 50¢ each, carnations for 75¢ each, and roses for $1 each.

   a. In how many different ways can he make a bouquet that sells for $3? 7 ways

   b. A customer wants to buy four flowers. How many different price totals are possible? 9 totals

Find each answer.   *(Lessons 3-2, 3-3, 3-4, 3-5)* **29.** −16

18. $-2 + (-3)$ −5

19. $14 + 8$ 22

20. $48 - (-28)$ 76

21. $-16 - (-4)$ −12

22. $4(-5)$ −20

23. $(-6)(-8)$ 48

24. $\frac{88}{8}$ 11

25. $\frac{-84}{12}$ −7

26. $1001 \div (-13)$ −77

27. $-72 \div (-6)$ 12

28. $3(-6)(-6)$ 108

29. $(-4)(-1)(-4)$

30. $-23 - 23$ −46

31. $-24 - (-24)$ 0

32. $(-9)(5)$ −45

33. $(-2)(7)(0)$ 0

34. $-5 + 16$ 11

35. $12 + (-32)$ −20

36. $\frac{-32}{4}$ −8

37. $\frac{-63}{-9}$ 7

38. $26 - 49$ −23

39. $38 - 74$ −36

40. $26 + (-26)$ 0

41. $-45 + 3$ −42

Evaluate each expression when $a = 6$, $b = -2$, and $c = -7$.
(Lesson 3-6)

42. b^2 4

43. a^3 216

44. $3b^3$ −24

45. $-5a^2$ −180

46. $c^2 - 9$ 40

47. $b^4 + 12$ 28

48. $|a| + |c|$ 13

49. $|b| - |c|$ −5

50. $|c| - b$ 9

51. $a + |b|$ 8

52. $|a + b|$ 4

53. $|c - 2|$ 9

Use the coordinate plane at the right.
Write the coordinates of
each point.   *(Lesson 3-8)*

54. Q (−2, 4)

55. R (1, 0)

56. S (3, 3)

57. T (−1, −2)

58. U (4, −3)

59. V (0, −4)

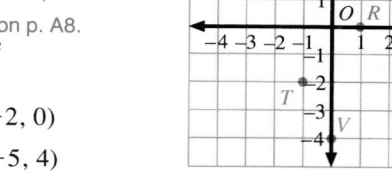

Answers to Chapter Review 60–67 are on p. A8.
Graph each point on a coordinate
plane.   *(Lesson 3-8)*

60. $D(5, -3)$

61. $E(-2, 0)$

62. $F(1, 3)$

63. $G(-5, 4)$

Graph each function.   *(Lesson 3-9)*

64.

x	$x - 3$
−3	−6
−2	−5
0	−3
3	0
5	2

65.

x	$3x + 1$
−3	−8
−1	−2
0	1
2	7
3	10

66. $x \to -2 + x$,
when $x = -3, -1, 0, 3$, and 5

67. $x \to x^2 + 1$,
when $x = -2, -1, 0, 1$, and 2

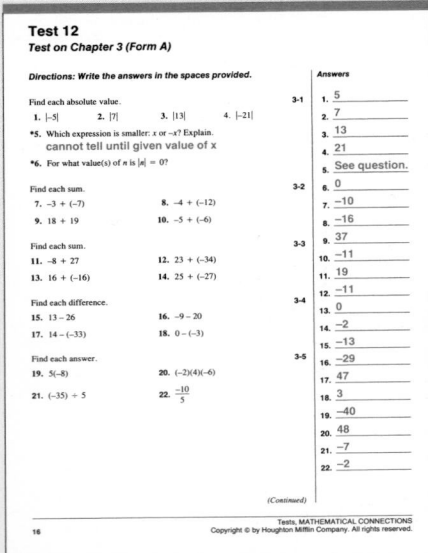
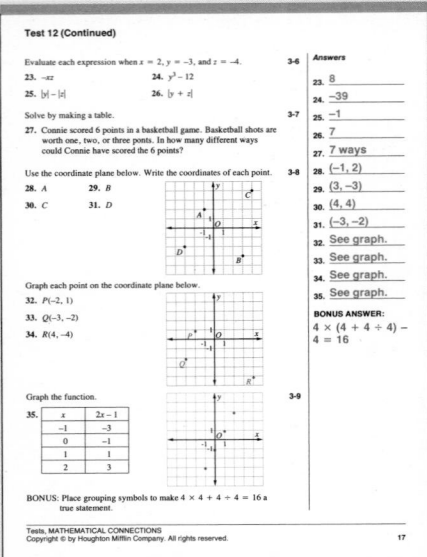
Chapter Test

Find each absolute value.

1. $|-2|$ 2   **2.** $|9|$ 9   **3.** $|21|$ 21   **4.** $|-14|$ 14   3-1

★ **5.** Which expression is greater: x or $-x$? Explain. See answer on p. A8.

★ **6.** For what values of n is $|n| = -n$? 0 and all negative numbers

Find each answer. 10. −43 12. 0 13. −18 16. 67 18. −6

7. $-3 + (-5)$ −8	**8.** $4 + 26$ 30	**9.** $18 + 32$ 50	**10.** $-7 + (-36)$	3-2
11. $-18 + 72$ 54	**12.** $49 + (-49)$	**13.** $12 + (-30)$	**14.** $-7 + 35$ 28	3-3
15. $-6 - 25$ −31	**16.** $14 - (-53)$	**17.** $29 - 30$ −1	**18.** $-12 - (-6)$	3-4
19. $(-7)(-11)$ 77	**20.** $5(-2)(9)$ −90	**21.** $-64 \div 8$ −8	**22.** $\dfrac{-6}{-2}$ 3	3-5

Evaluate each expression when $v = 3$, $w = -2$, and $y = -4$.

23. $-vy$ 12   **24.** $y^3 - 10$ −74   **25.** $|w| - |v|$ −1   **26.** $|w + y|$ 6   3-6

Solve by making a table.

27. Dwayne has only nickels, dimes, and quarters. The *Weekly Times* costs 50¢. In how many different ways could he pay the 50¢ to buy a copy of the *Weekly Times*? 10 ways   3-7

Use the coordinate plane at the right. Write the coordinates of each point.

28. A **29.** B **30.** C **31.** D 3-8
$(-3, -2)$ $(-2, 3)$ $(1, 0)$ $(0, -3)$

Graph each point on a coordinate plane.

32. $P(1, -1)$ **33.** $Q(0, 3)$ **34.** $R(-3, 5)$
Answers to Chapter Test 32–38 are on p. A8.

Graph each function.

35.

x	$x + 4$
−4	0
−2	2
0	4
1	5
3	7

36.

x	$3x - 1$
−2	−7
−1	−4
0	−1
1	2
2	5

37. $x \rightarrow -x - 1$ 3-9
when $x = -4, -2, -1, 0,$ and 1

38. $x \rightarrow |x + 3|$
when $x = -6, -4, 0, 2,$ and 3

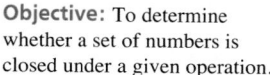
Closure

Objective: To determine whether a set of numbers is closed under a given operation.

Terms to Know
- *whole numbers*
- *closure*

The numbers in the set {0, 1, 2, 3, . . .} are called **whole numbers.** You already know that, when you add any two whole numbers, the sum is also a whole number.

$$1 + 2 = 3$$
$$35 + 35 = 70$$
$$146 + 0 = 146$$

For this reason, the set of whole numbers is said to have *closure* under addition.

A set of numbers has **closure** under a given operation when performing the operation on any numbers in the set results in a number that is also in the set. You describe a set of numbers as being either *closed* or *not closed* under a given operation.

Example

Tell whether the set of whole numbers is *closed* or *not closed* under the operation of subtraction.

Solution

Look for a whole-number difference that is *not* a whole number.
$$8 - 5 = 3 \qquad 25 - 25 = 0 \qquad 7 - 12 = -5$$
Because $7 - 12 = -5$, and -5 is not a whole number, the whole numbers are *not closed* under subtraction.

There are, of course, many other examples of whole-number differences that are negative integers. However, to show that the set of whole numbers is not closed under subtraction, you need to find only one of these differences.

Exercises

Tell whether the given set of numbers is *closed* or *not closed* under each operation. Answers to Exercises 1–8 are on pp. A8–9.
a. addition
b. subtraction
c. multiplication
d. division

1. the set of whole numbers
2. the set of integers
3. the set of positive integers
4. the set of negative integers
5. {0, 1}
6. {1, 3, 5, . . .}
7. {2, 4, 6, . . .}
8. {. . ., −5, −3, −1}

Integers   **137**

Teaching Note

Closure may be a new concept to many students. Often students think that if you perform an operation on two numbers, you always get a number of the same type. Use the Example to show that this is not so.

Additional Example

Tell whether the set of whole numbers is closed or not closed under multiplication.   You test to see if the whole numbers are closed under multiplication by looking for a product of whole numbers that is not a whole number. The rule for multiplying integers with the same signs tells you that the product is positive, and a positive integer is a whole number. The multiplication property of zero tells you that zero times a number is zero, and zero is a whole number. Since the product of any two whole numbers is always a whole number, the whole numbers are closed under multiplication.

Exercise Note

Critical Thinking

Exercises 1–8 require students to analyze the given sets of numbers to see if they are closed or not. In some cases, students may have to create an example to show that a set is not closed under a given operation. In other cases, students have to apply properties and rules to determine that a set is closed.

Using the coordinate plane below, write the coordinates of each point.

1. *A* (−3, 4)
2. *B* (2, −4)
3. *C* (3, 0)

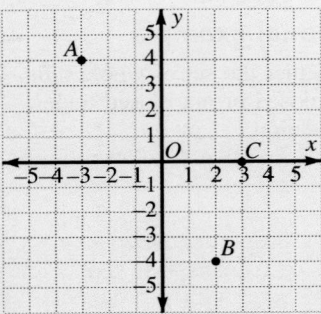

Graph each point on a coordinate plane.

4. *D* (−1, 4)
5. *E* (0, −3)

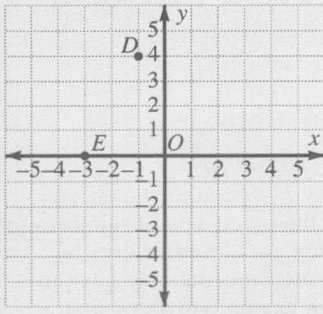

6. Graph the function *x* → 2*x* when *x* = −2, −1, 0, 1, 2.

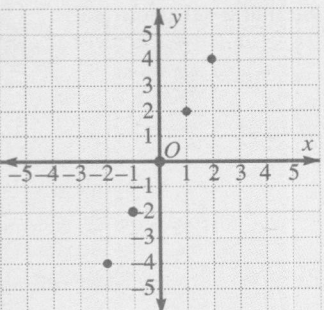

Cumulative Review

Standardized Testing Practice

Choose the letter of the correct answer.

1. Write 53 m in km.
   A. 5300 km   B. 53,000 km
   C. 0.53 km   D. 0.053 km ✓

2. Find the coordinates of point *D*.
   A. (−2, 1)
   B. (2, −1) ✓
   C. (−1, 2)
   D. (1, −2)

3. In 24 h the temperature went from −12°C to 10°C. Find the change in temperature.
   A. 22°C ✓
   B. −22°C
   C. 2°C
   D. −2°C

4. Tyson bought two ties at $12.99 each. The total tax was $1.30. Tyson got a discount of $2.60. Find the total amount he paid.
   A. $16.89   B. $24.68 ✓
   C. $29.88   D. $27.28

5. Find the answer mentally.
   7(28) + 7(52)
   A. 1480   B. 560 ✓
   C. 700   D. 490

6. Simplify: 3(*x* + 5*y*) + 2*x* + *y*
   A. 5*x* + 16*y* ✓   B. 5*x* + 6*y*
   C. 9*x* + 18*y*   D. 21*xy*

7. Express the relationship 6 < 8 < 10 in another way.
   A. 6 > 8 > 10
   B. 6 < 10 < 8
   C. 10 > 8 > 6 ✓
   D. 10 > 6 > 8

8. Amy has $2.05 in dimes and quarters. It is *not* possible for Amy to have which grouping of coins?
   A. 7 quarters, 3 dimes
   B. 5 quarters, 8 dimes
   C. 3 quarters, 13 dimes
   D. 6 quarters, 5 dimes ✓

9. Find the function rule.
   A. *x* → *x* − 5
   B. *x* → 3*x* − 13
   C. *x* → 3*x* − 1 ✓
   D. *x* → −*x* + 5

x	?
−2	−7
−1	−4
0	−1
3	8
5	14

10. Evaluate *b* + *a* when *a* = 43.87 and *b* = 116.5.
    A. 55.52   B. 555.2
    C. 160.37 ✓   D. 72.63

11. Find the sum: $-42 + 18$

A. -60
B. -24
C. 60
D. 24

12. Find the quotient: $\frac{-32 + 8}{-4}$

A. 10
B. -10
C. 6
D. -6

13. Evaluate $3 - 2c^2$ **when** $c = -5$.

A. -47
B. -97
C. 53
D. -22

14. Find the answer: $\frac{64 - 48}{8 - 6}$

A. 0
B. -2
C. 24
D. 8

15. Write in order from least to greatest:
$|12|, |0|, |-15|, |-1|$

A. $|0|, |-1|, |12|, |-15|$
B. $|-15|, |-1|, |0|, |12|$
C. $|-15|, |12|, |-1|, |0|$
D. $|12|, |0|, |-15|, |-1|$

16. Simplify: $-4(-2t + 3)$

A. $-8t + 12$
B. $8t - 12$
C. $8t + 12$
D. $-8t - 12$

17. Evaluate $xy + z$ **when** $x = 13$, $y = 4$, **and** $z = 2$.

A. 19
B. 78
C. 104
D. 54

18. Which word or phrase, used twice, correctly completes "The product of any number and __?__ is __?__"?

A. one
B. zero
C. itself
D. its inverse

19. Find the sum: $-32 + (-18)$

A. -14
B. 14
C. -50
D. 50

20. Complete: The cost of three tickets at $3 each and two tickets at $9.50 each is __?__.

A. $28
B. $12.50
C. $38
D. $34.50

7. Graph the function $x \rightarrow |x - 2|$ **when** $x = -3, -2, 0, 1, 3$.

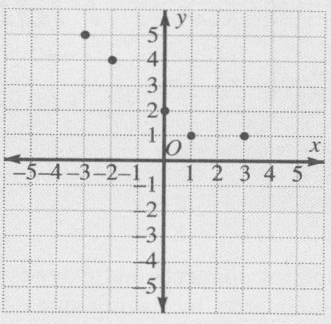

Alternative Assessment
Have students graph the following points: $(0, 2)$, $(6, 0)$, $(3, 10)$, $(18, 8)$, $(15, 14)$, $(18, 3)$, $(0, 7)$, $(6, 5)$. Tell students to connect the points in the given order and to describe the figure they see.   house

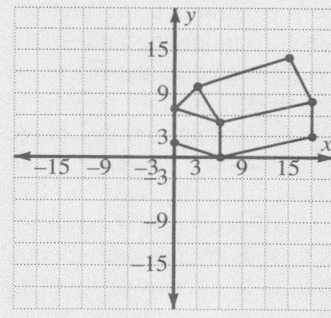

Integers   139

139

Planning Chapter 4

Chapter Overview

Chapter 4 connects and extends the skills learned in the first three chapters by having students learn to solve one- and two-step equations. The chapter gives students an opportunity to sharpen their skills in working with integers, to apply the distributive property, and to simplify expressions. It also serves as an introduction to solving equations, a skill used extensively in algebra and in more advanced mathematics courses.

Chapter 4 begins by introducing students to equations that can be solved by substitution and mental math. *Lesson 2* continues the development of problem-solving skills by introducing the guess-and-check method. *Lessons 3 and 4* focus on solving one-step equations using inverse operations. A manipulative approach to modeling and solving equations by using algebra tiles is presented in the *Focus on Explorations*. This approach is used to demonstrate how to solve the two-step equations in *Lesson 5*. Writing variable expressions and equations from word phrases and sentences is the topic of *Lessons 6 and 7*. This helps ease the way into *Lesson 8* on using equations to solve word problems. *Lesson 9* extends the concepts from previous lessons by presenting equations that must be solved by combining like terms or by using the distributive property. Formulas, a logical extension of solving equations and a connecting link to many areas such as medicine, science, and business, are introduced in *Lesson 10*. The *Chapter Extension* presents equations with variables on both sides of the equation.

Background

Chapter 4 presents students with many opportunities to use previously learned skills. These include substituting a number for a variable, working with integers, combining like terms, using the distributive property, and using estimation and mental math skills. Writing has an important role in the chapter as students are asked to write variable expressions and equations for word phrases and sentences.

The use of calculators in doing mathematics is further developed in this chapter. Students are shown how to use a calculator to check solutions to equations and also how to use a calculator to solve an equation. Spreadsheets are used to help students solve certain equations that they cannot yet solve by using paper and pencil alone.

Manipulatives are used to aid students' understanding of the processes involved in solving equations. The first few lessons use the balance-scale model, an effective classroom demonstration device. The *Focus on Explorations* shows students how to model and solve equations by using algebra tiles, and this technique is then used to demonstrate solving two-step equations in the lesson that follows.

Problem-solving concepts and skills are developed in two lessons in Chapter 4. *Lesson 2* presents the strategy of guess-and-check. This strategy is very useful in the real world because many problems can be solved by using it. *Lesson 8* presents the strategy of using an equation, a strategy that students will need to employ throughout their study of mathematics. Since reading a word problem and writing an equation for it tends to be difficult for many students, the two lessons preceding *Lesson 8* provide instruction on how to write variable expressions and equations for word phrases and sentences.

Objectives

4-1 To find solutions of equations by substitution and mental math.

4-2 To solve problems by guessing and checking.

4-3 To solve equations using addition or subtraction.

4-4 To solve equations using multiplication or division.

FOCUS ON EXPLORATIONS To use algebra tiles to model equations.

4-5 To solve equations using two steps.

4-6 To write variable expressions for word phrases.

4-7 To write equations for sentences.

4-8 To solve problems by using equations.

4-9 To solve equations by simplifying expressions involving combining like terms and the distributive property.

4-10 To work with formulas.

CHAPTER EXTENSION To solve equations having variables on both sides.

Chapter Planner

Instructional Aids	Manipulatives	Cooperative Learning	Technology	Practice/Reteaching	Assessment
Materials: calculator, algebra tiles, ads from newspapers *Resource Book:* Lesson Starters 30–39 *Visuals,* Folder D	Lessons 4-1, 4-3, 4-4, 4-5, Explorations *Activities Book:* Manipulative Activities 7–8	Lessons 4-2, 4-6, Extension *Activities Book:* Cooperative Activities 8–9	Lessons 4-2, 4-3, 4-4, 4-5, 4-9 *Activities Book:* Calculator Activity 3 Computer Activity 5 *Connections Plotter Plus* Disk	Toolbox, pp. 753–771 Extra Practice, p. 733 *Resource Book:* Practice 40–53 *Study Guide,* pp. 59–78	Self-Tests 1–3 Quick Quizzes 1–3 Chapter Test, p. 182 *Resource Book:* Lesson Starters 30–39 (Daily Quizzes) *Tests* 14–20

Assignment Guide Chapter 4

Day	Skills Course	Average Course	Advanced Course
1	**4-1:** 1–20, 21–37 odd, 43–46	**4-1:** 1–20, 21–41 odd, 43–46	**4-1:** 1–19 odd, 21–46
2	**4-2:** 1–15 odd, 16–19	**4-2:** 1–19	**4-2:** 1–19
3	**4-3:** 1–21, 23, 24, 32–35	**4-3:** 1–26, 32–35	**4-3:** 1–13 odd, 15–35
4	**4-4:** 1–31 odd, 33–37 **Exploration:** Activities I–III	**4-4:** 1–31 odd, 33–37 **Exploration:** Activities I–III	**4-4:** 1–31 odd, 33–37 **Exploration:** Activities I–III
5	**4-5:** 1–17, 19–23 odd, 30–33	**4-5:** 1–17, 19–27 odd, 30–33	**4-5:** 1–17 odd, 18–33
6	**Mixed Review** **4-6:** 1–15, 21–24	**Mixed Review** **4-6:** 1–18, 21–24	**Mixed Review** **4-6:** 1–13 odd, 15–24
7	**4-7:** 1–10, 11–19 odd, 21–25	**4-7:** 1–10, 11–19 odd, 21–25	**4-7:** 1–7 odd, 9–25, Challenge
8	**4-8:** 1–13, 16–21	**4-8:** 1–14, 16–21	**4-8:** 1–21
9	**4-9:** 1–16, 17–21 odd, 27–32	**4-9:** 1–16, 17–21 odd, 23–25, 27–32	**4-9:** 1–15 odd, 17–32, Historical Note
10	**4-10:** 1–12, 17, 18, 23–30	**4-10:** 1–15, 17, 18, 23–30	**4-10:** 1–7 odd, 8–30
11	*Prepare for Chapter Test:* Chapter Review	*Prepare For Chapter Test:* Chapter Review	*Prepare for Chapter Test:* Chapter Review
12	*Administer Chapter 4 Test*	*Administer Chapter 4 Test;* Cumulative Review	*Administer Chapter 4 Test;* Chapter Extension; Cumulative Review

Teacher's Resources

Resource Book
Chapter 4 Project
Lesson Starters 30–39
Practice 40–53
Enrichment 33–43
Diagram Masters
Chapter 4 Objectives
Family Involvement 4
Spanish Test 4
Spanish Glossary

Activities Book
Cooperative Activities 8–9
Manipulative Activities 7–8
Calculator Activity 3
Computer Activity 5

Study Guide, pp. 59–78

Tests
Tests 14–20

Visuals
Folder D

Connections Plotter Plus Disk

Alternate Approaches Chapter 4

Using Technology

CALCULATORS

This chapter is an introduction to linear equations. Calculators can be used by students to check guesses and solutions found algebraically. After students are adept at solving linear equations with positive integral solutions, it is useful to give them some problems that do not work out so nicely. Students will appreciate fully the power of the algebraic methods if they see that these methods work regardless of the complexity of the numbers involved.

COMPUTERS

Computers can be used in the same way as calculators to help students solve equations containing "ugly" numbers. A spreadsheet program can be used to provide insight that is quite different from the balance model of equation-solving presented in the text. Adventurous students can learn to solve linear equations using function-graphing software, although they will not yet appreciate fully the connections between linear graphs and linear equations.

Lesson 4-2

CALCULATORS OR COMPUTERS

The guess-and-check strategy can be made easier and can be seen in a different light if a calculator or computer is used to perform the computations involved. Punching keys on a calculator, or typing formulas into a spreadsheet or a BASIC program, will reinforce understanding of the operations involved in a solution process. The most helpful way to use a calculator or a computer for the guess-and-check method is to produce a table of calculations (checks) for a range of guesses. To solve Guided Practice Exercises 1 through 4, for example, you (or a student) might write the following BASIC program:

```
10 FOR T = 1 TO 10
20 FOR C = 1 TO 20
30 PRINT T, "TRUCKS: ";C, "CARS: "
40 PRINT "AMOUNT EARNED: ";15 * T + 7 * C
50 NEXT C
60 NEXT T
```

The same kind of table can be produced, with a bit more work, by using a spreadsheet program or a calculator.

Lessons 4-4, 4-5

COMPUTERS

The simplest use of computers in this and the following lessons is to have students write a program in BASIC or a formula for a spreadsheet that solves equations of the form $ax + b = c$, perhaps with special cases at first, such as $a = 1$, or $b = 0$. For example, the line 100 LET X = (C − B)/A might appear in a BASIC program of this type. This process is empowering because the student is in effect teaching the computer how to solve the equation. Writing a program or formula requires understanding at a deeper level than simply carrying out the solution algorithm itself.

A spreadsheet program can be used to give students an alternative way of looking at the algebraic process of solving linear equations. Essentially, the spreadsheet can perform each operation, one step at a time. Using the spreadsheet, students can see quickly that the two-step algorithm on page 157 involves inverting the operations contained in the equation. The order in which the two steps are carried out, which might otherwise seem arbitrary, will be clear immediately if students try the wrong order.

A graphing calculator or a computer program that graphs functions can be used to solve linear equations. For example, to solve $5x + 3 = 2x + 7$, simply graph the two functions $y = 5x + 3$ and $y = 2x + 7$. The x-value of the intersection of the two graphs provides the solution. Although students will be better equipped to understand graphing later in the course, more capable students will find this intriguing.

Using Manipulatives

Lesson 4-1

Algebra tiles can be used to demonstrate how to solve an equation. Example 1 on page 142 could be modeled as follows:

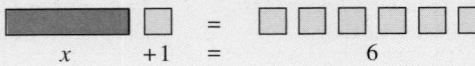

Use inverse operations.

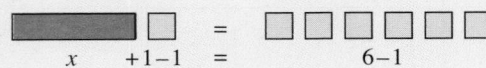

Notice that 1 was subtracted from both sides of the equation.

Read the solution.

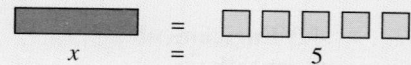

$$x \quad = \quad 5$$

Lesson 4-3

Addition and subtraction equations can be solved using algebra tiles. Addition may be shown as in the example for Lesson 4-1 above. Shaded tiles can be used to indicate subtraction. Remind students that subtraction is the same as adding the opposite.

Build a model.

$$k \quad -4 \quad = \quad 5$$

Use inverse operations.

$$k \quad -4+4 \quad = \quad 5+4$$

Simplify by removing zero pairs.

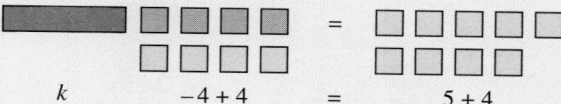

$$k \quad -4+4 \quad = \quad 5+4$$

Read the solution.

$$k \quad = \quad 9$$

Lesson 4-4

Multiplication equations can be solved using algebra tiles. For example, to solve $2n = 8$, you would do the following:

Build a model.

$$2n \quad = \quad 8$$

Use inverse operations.

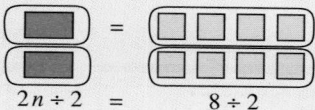

$$2n \div 2 \quad = \quad 8 \div 2$$

Read the solution.

$$n \quad = \quad 4$$

Reteaching/Alternate Approach

Lesson 4-2

COOPERATIVE LEARNING

Divide students into groups. Using an advertisement from a catalog or newspaper, circle two items having different prices for each group. Give each group three different amounts of money they have to spend on the circled items. Ask the groups to use the guess-and-check method to see how many items of each they can purchase. You may wish to repeat the process using three different items and different amounts of money.

Lesson 4-6

COOPERATIVE LEARNING

Students usually enjoy number puzzles, such as the following:

- Start with the number of the month in which you were born.
- Triple it.
- Subtract the number of days in a week.
- Add the number of quarters in $3.
- Multiply the result by 3.
- Add 5 less than the original number.
- The result should equal one more than the original number followed by a zero.

Do the puzzle with students, correctly guessing their numbers. Then divide students into groups. Have the groups translate each step of the puzzle into a variable expression. Discuss the meaning of each variable expression and replace the variables by numbers to see why the puzzle works.

Teacher's Resources Chapter 4

Enrichment masters from the Resource Book are pictured here. See the Teacher's Resources chart on page 140B for a complete listing of all materials available for this chapter.

4-1 Enrichment
Data Analysis

Enrichment 33: Data Analysis (for use with Lesson 4-1)
How Can We Make These Sides Equal?

The following activity is built around equations with one variable such as:

$$4 + x = 6$$

In order for this equation to be true, x has to be 2.

Materials
You will need pencil, paper, and two cubes whose sides are numbered 1 through 6.

What to do
Roll the cubes one at a time. Call the number that comes up on the first cube A, and the number on the second cube B. What number will you have to add to A to get B? In other words, solve the equation $A + x = B$.

Example
Let's solve the equation $A + x = B$ for two rolls of the cubes.

Roll	A	B	$A + x = B$	x
1	4	6	$4 + x = 6$	2
2	5	3	$5 + x = 3$	-2

1. Complete this table for five rolls of the cubes. **Answers to Exercises 1–4 will vary. Examples are given.**

Roll	A	B	$A + x = B$	x
1	4	2	$4 + x = 2$	-2
2	6	5	$6 + x = 5$	-1
3	1	2	$1 + x = 2$	1
4	3	3	$3 + x = 3$	0
5	4	1	$4 + x = 1$	-3

2. What is the average x-value for your five rolls? **-1**

3. What would you expect the average x-value to be if you repeated this experiment many times? **0**

4. Roll the cubes ten times to make ten more equations and find your average x-value. Compare your results with those of other students.

4-2 Enrichment
Problem Solving

Enrichment 34: Problem Solving (for use with Lesson 4-2)
Guess and Check, Then Guess Again

Many people feel that guessing isn't using mathematics to solve a problem. This is simply not true. Guessing and checking is a good strategy for solving certain problems. Feel free to guess on problems and to develop your guessing skills.

Problem
Estelle's mother, who ran the family farm, needed to know how many horses and how many chickens were out in the field behind the barn. She sent Estelle out to count the animals. Estelle went out to the field, returned, and told her mother, "There are 37 heads and 118 feet in the field, Mom." "Thanks a lot," said her mother.

Solution
1. *Understand* Explain what the problem is about. **Estelle's mother needs to translate the given information into the number of horses and the number of chickens.**

2. *Plan* Her mother could guess the number of horses and check her answer. She could continue guessing and checking until she found the correct number of horses and chickens. What is the maximum number of horses that there could be? **37**

Answers to Exercises 3–4 will vary.

3. *Work* a. Guess the number of horses. **Try 20.**
   b. How many legs would that many horses have? **80**
   c. With that many horses, how many chickens would there have to be? **17**
   d. How many legs would that many chickens have? **34**

4. *Answer* Is your guess about the number of horses too high or too low? How can you tell? **too low; 80 + 34 < 118**

5. Repeat Exercises 1 through 4 until you have a solution.
   number of horses = ___?___ **22**   number of chickens = ___?___ **15**

6. What might be an alternative to the guess-and-check strategy for this problem? **Make a table.**

4-3 Enrichment
Connection

Enrichment 35: Connection (for use with Lesson 4-3)
Picture This!

In the Exploration in Lesson 4-3, you add or subtract numbers to solve equations. A picture can represent the idea of adding or subtracting quantities.

Problem
What would you have to add to the figure on the left to make the figure on the right?

Solution
Add [ ] to the figure on the left.

Use colored pencils to help show the changes that need to be made in the figure on the left so that the result will be the figure on the right.

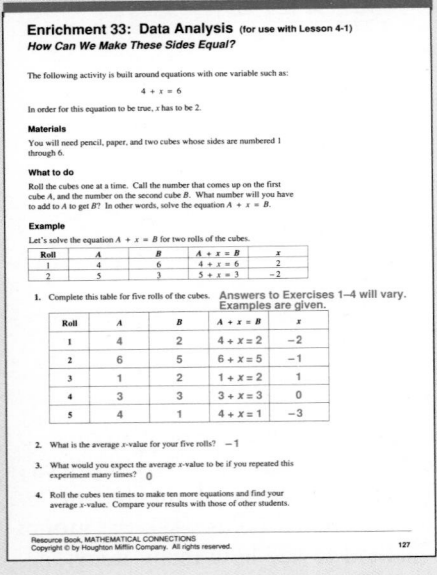

1. **Answer**
2. **Answer**
3. **Answers may vary.**   **Answer**

4-4 Enrichment
Communication

Enrichment 36: Communication (for use with Lesson 4-4)
What Do Equations Mean, Anyway?

Many people wonder, "When am I ever going to use the equation $3x = 9$?" It's a good question and it deserves a good answer. An equation is a sentence stating that one thing is the same as another. "$3x = 9$" states that "three x's are the same as 9."

Not All Equations Are True
It is easy to write a sentence that is not true. "I am standing on the moon," is a perfectly good sentence grammatically, but it is clearly false. You could change it to "Neil Armstrong was standing on the moon," and it would be true! He was the first person to step on the moon, back in 1969. If x happens to be 7, then "$3x = 9$" is false, too. (This is because $(3)(7) = 21$, not 9.)

Making an Equation True
When you solve an equation like "$3x = 9$," you find the values of x that will make it true. For this equation, there is only one such value. For most values of x this equation is false!

1. Look at the "standing on the moon" sentences above. Then write five sentences of your own that are false. **Answers will vary.**

2. Rewrite your five sentences so that they are true. **Answers will vary.**

Translate the following equations into sentences. For example, "$2x = 14$" means "Two x's are the same as 14."

3. $4x = 13$ **Four x's are the same as 13.**

4. $15,000x = 17$ **Fifteen thousand x's are the same as 17.**

5. $\frac{x}{4} = 24$ **One fourth of x is the same as 24.**

6. $Ax = B$ **A x's are the same as B.**

4-5 Enrichment
Application

Enrichment 37: Application (for use with Lesson 4-5)
An Average Equation

Problem 1
Jeff bought three apples at a fruit market and paid $1.19 for them. They were priced at $1.09 per pound. He wanted to figure out how much a single apple would cost, so he wrote down the equation $3x = 1.19$. He reasoned that x would represent the cost of each apple. He used his calculator to divide 1.19 by 3, and saw the result, 0.3966. . . . "That's how much one apple would cost," he thought.

1. What is wrong with Jeff's reasoning? **Answers may vary. For example: $.3966 . . .$ is not the cost of each apple because they aren't necessarily the same weight.**

2. What does the value of x, 0.39666. . . , really represent? **the average cost of an apple**

3. In what situation would Jeff happen to be correct? **if all the apples were the same weight**

Problem 2
The telephone bills for Ramon's family totaled $429.79 for the year.

4. Write an equation that will help you find their average monthly telephone bill. **$12x = 429.79$**

5. What is their average monthly cost? **$35.82**

6. Use the same method to find the average daily cost of their telephone. **$1.18**

Problem 3
Juanita has a total of 365 points on her mathematics tests this term. She has had 4 tests.

7. Write and solve an equation to find her average test score. **$4x = 365$ $x = 91.25$**

8. Does Juanita have to have a test score equal to her average score? Explain why or why not. **No; the average of a set of numbers does not need to be one of the numbers.**

4-6 Enrichment
Exploration

Enrichment 38: Exploration (for use with Lesson 4-6)
Patterns in Equations

1. Solve each equation for x.

   a. $3x - 7 = 8$ **5**   b. $5x + 3 = 8$ **1**
   $3x - 7 = 11$ **6** $5x + 3 = 13$ **2**
   $3x - 7 = 14$ **7** $5x + 3 = 18$ **3**
   $3x - 7 = 17$ **8** $5x + 3 = 23$ **4**

   c. In part (a), as the value of x is increased by 1, what happens to the value of $3x - 7$? **It goes up by 3.**

   d. In part (b), as the value of x is increased by 1, what happens to the value of $5x + 3$? **It goes up by 5.**

2. a. Complete the following table, evaluating the expression $8x - 7$ for each value of x.

x	$8x - 7$
1	1
2	9
3	17
4	25
400	3193
401	3201

   b. As the value of x is increased by 1, what happens to $8x - 7$? **It goes up by 8.**

   c. Suppose you use the expression $ax - 7$ instead of $8x - 7$. What happens to the value of $ax - 7$ when the value of x is increased by 1? **It goes up by a.**

WRITING ABOUT MATHEMATICS Answers to Exercises 24–27 are on p. A10.

24. Describe in your own words the process of solving an equation.

25. Tell how solving an equation like $x + 6 = 10$ is different from solving an equation like $2x + 3 = 15$.

CONNECTING MATHEMATICS AND PHYSICAL SCIENCE

A chemical reaction has a natural balance: there are the same number of atoms of each element after the reaction as there were before. For this reason, *chemical equations* are used to represent reactions. For example, the diagram below shows how hydrogen and oxygen react to form water.

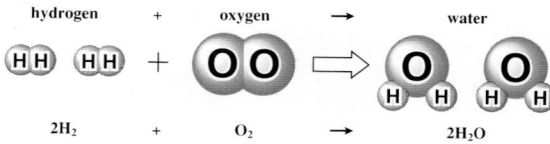

26. What does it mean to say that a chemical reaction *balances*?

27. If there are two oxygen atoms before a reaction, how many must there be after the reaction? Explain.

C   28. Use the diagram below to complete the chemical equation.

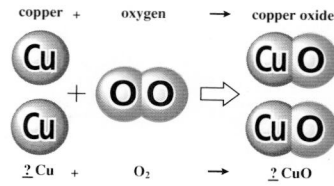

2; 2

29. The elements sodium and oxygen combine to form sodium oxide. Complete the chemical equation for this reaction.

$\underline{\ ?\ }$ Na + O_2 = $\underline{\ ?\ }$ Na_2O 4; 2

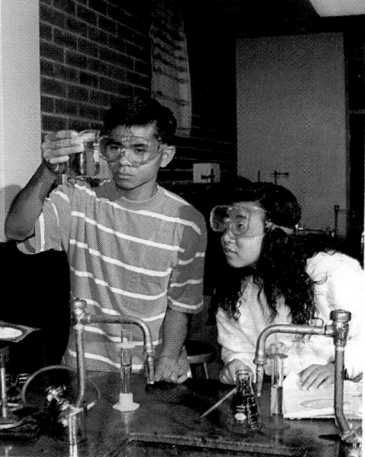

SPIRAL REVIEW

S   30. Find the sum: $6\frac{5}{8} + 4\frac{2}{3}$ *(Toolbox Skill 18)* $11\frac{7}{24}$

31. Write 8.32×10^9 in decimal notation.   *(Lesson 2-10)*   8,320,000,000

32. Solve: $\frac{x}{4} - 15 = -12$ *(Lesson 4-5)*   12

33. Find the product mentally: $(9)(6)(0)(5)(2)$ *(Lesson 1-7)*   0

Equations   **159**

Project

You may wish to ask students to make a list of occupations that use mathematics routinely. They may also find it interesting to research the math requirements for working in various occupations.

Enrichment

Enrichment 37 is pictured on page 140E.

Practice

Practice 45 is shown below.

Practice 45, *Resource Book*

Practice 45
Skills and Applications of Lesson 4-5

Solve. Check each solution.

1. $4x + 3 = 7$ 1
2. $3n - 8 = 10$ 6
3. $2n - 15 = 5$ 10
4. $\frac{a}{-4} + 14 = 20$ −24
5. $0 = 7n - 21$ 3
6. $-8 = \frac{c}{5} - 15$ 35
7. $42 = -6t - 12$ −9
8. $56 = -9x + 56$ 0
9. $8b - 15 = -31$ −2
10. $15w + 15 = 0$ −1
11. $-12x - 21 = 15$ −3
12. $8 + \frac{u}{3} = 28$ 60
13. $4k + 32 = 4$ −7
14. $\frac{c}{7} + 20 = 7$ −91
15. $44 = 3x - 31$ 25
16. $-7w - 19 = 9$ −4
17. $\frac{k}{8} - 21 = -13$ 64
18. $16 = \frac{a}{-4} + 37$ 84
19. $28 = -5s + 8$ −4
20. $112 = -4c + 84$ −7
21. $-6n + 15 = 75$ −10
22. $24 + 7a = -32$ −8
23. $40 = \frac{k}{11} + 32$ 88
24. $9x - 16 = 65$ 9
25. $47 = 10m + 87$ −4
26. $\frac{c}{7} + 48 = 48$ 0
27. $-4x + 27 = -1$ 7
28. $24 = -3x - 9$ −11
29. $\frac{x}{-8} - 9 = 3$ −96
30. $42 + \frac{n}{6} = 41$ −6
31. $9m - 13 = 68$ 9
32. $47 = 15 + 4c$ 8
33. $-8x + 63 = 63$ 0
34. When $5n$ is added to 34 the result is 24. What is the value of n? −2
35. If $10x - 23 = 17$, what is the value of x? 4
36. The quotient $\frac{n}{7}$ is added to 19 and the result is 11. What is the value of n? −56
37. Seventeen is added to the product of a number w and 6. The result is 23. What is the value of w? 1
38. The sum of $9x$ and 50 is −13. What is the value of x? −7

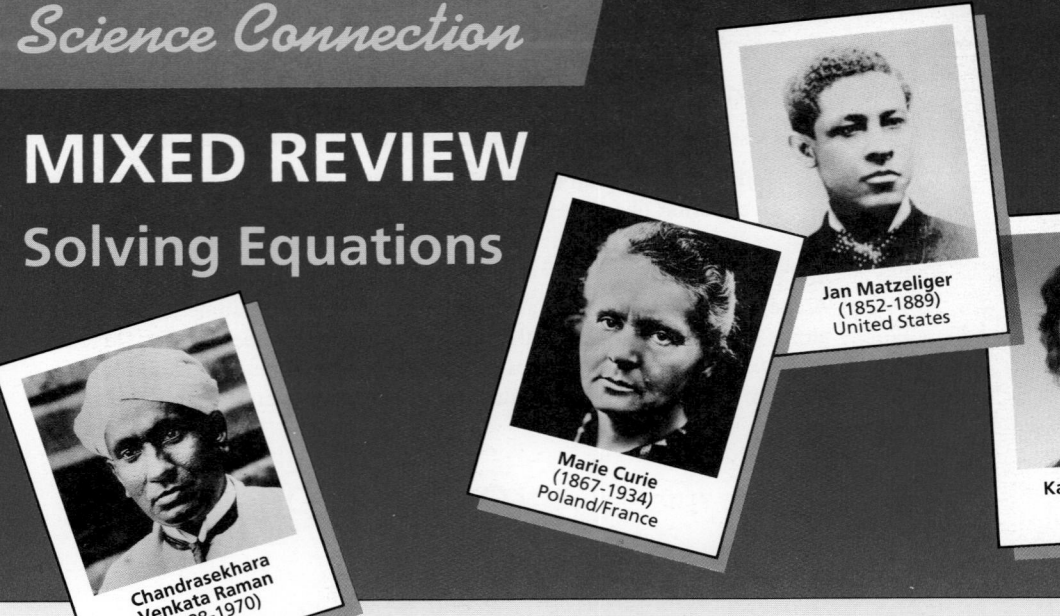

MIXED REVIEW
Solving Equations

Chandrasekhara Venkata Raman (1888-1970) India

Marie Curie (1867-1934) Poland/France

Jan Matzeliger (1852-1889) United States

Katherine Johnson (b. 1918) United States

Men and women from around the world have made important scientific and technical contributions. To find out what Chandrasekhara Venkata Raman and Marie Curie have in common, complete the code boxes below. Begin by solving Exercise 1. The answer to Exercise 1 is −3, and the letter associated with Exercise 1 is N. Write N in the box above −3 as shown below. Continue in this way with Exercises 2–24.

1. $6n = -18$ −3 (N)

2. $-a = 0$ 0 (E)

3. $14 = -3x + 2$ −4 (E)

4. $\frac{b}{4} = 7$ 28 (N)

5. $5 = t + 7$ −2 (Y)

6. $13 + x = 16$ 3 (E)

7. $-15 + 9d = 21$ 4 (R)

8. $8 = \frac{m}{6} + 2$ 36 (P)

9. $-y - 18 = -13$ −5 (I)

10. $12t = 108$ 9 (W)

11. $11 = \frac{a}{-4}$ −44 (T)

12. $-21 = \frac{-n}{2}$ 42 (O)

13. $10c - 18 = 42$ 6 (H)

14. $54 = -6g$ −9 (N)

15. $\frac{d}{7} = -9$ −63 (B)

16. $-8 = 16 - 3w$ 8 (Z)

17. $x + 12 = 38$ 26 (A)

18. $\frac{n}{5} - 2 = 0$ 10 (R)

19. $32 = b - 18$ 50 (I)

20. $-a - 19 = 27$ −46 (E)

21. $3w - 8 = 13$ 7 (S)

22. $\frac{m}{-8} - 2 = 0$ −16 (L)

23. $-7c = -84$ 12 (E)

24. $4 + g = 56$ 52 (R)

T	H	E	Y		A	R	E		N	O	B	E	L
?	?	?	?		?	?	?		N	?	?	?	?
−44	6	12	−2		26	4	−46		−3	42	−63	0	−16

P	R	I	Z	E		W	I	N	N	E	R	S
?	?	?	?	?		?	?	?	?	?	?	?
36	52	−5	8	−4		9	50	28	−9	3	10	7

Juan de la Cierva
(1895-1936)
Spain

Chien-Shiung Wu
(b. 1912)
United States

To find out more about the contributions of the scientists and inventors on these pages, solve the equations and match your solutions to the answers at the bottom of the page.

25. $\frac{x}{3} + 1 = 5$

Marie Curie f

27. $-6 = a + 4$

Chandrasekhara Venkata Raman c

29. $55 = -11g$

Juan de la Cierva d

26. $2n - 6 = 8$

Jan Matzeliger a

28. $7t = -21$

Chien-Shiung Wu e

30. $7 + 6n = 61$

Katherine Johnson b

---

a. 7
Inventor who patented a machine that made possible the mass production of shoes.

d. -5
Inventor who designed the autogiro, a cross between a helicopter and an airplane.

b. 9
Aerospace engineer who helped design systems used in tracking space missions.

e. -3
Physicist whose innovative experiment disproved the law that like nuclear particles always act alike.

c. -10
Physicist who made new observations about the properties of light frequencies.

f. 12
Chemist who discovered the elements polonium and radium.

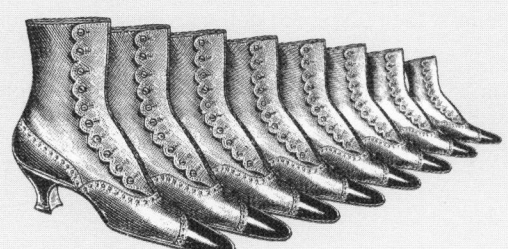

Equations **161**

Teaching the Lesson
- Lesson Starter 35
- Reteaching/Alternate Approach, p. 140D
- Cooperative Learning, p. 140D

Lesson Follow-Up
- Practice 46
- Enrichment 38: Exploration
- Study Guide, pp. 69–70

Warm-Up Exercises

Write a variable expression for each phrase.

1. the sum of a number x and 12
$x + 12$

2. 18 minus a number z $18 - z$

3. 24 times a number c $24c$

4. a number b divided by 14 $\dfrac{b}{14}$

5. a number n squared n^2

Lesson Focus

Ask students if they have ever heard the French expression *C'est la vie!* It means *That's life!* in English. Today's lesson focuses on translating word phrases into symbols, the language of mathematics.

Teaching Notes

Ask students to name some terms for mathematical operations. As you teach the lesson, emphasize how key words can be used to unite variable expressions.

Key Question

What is the first step in writing a variable expression?
Choose a variable and decide what it represents.

Writing Variable Expressions

Objective: To write variable expressions for word phrases.

CONNECTION

When you visit another country, you may need to ask directions in a language other than your own. You can translate what you want to say by using words and phrases from the other language that have similar meanings.

Mathematics has a language of its own, the language of symbols. In many problem solving situations you need to translate from a word phrase to a variable expression.

Example 1

Write a variable expression that represents the phrase *eight increased by five times a number n.*

Solution

Increased by suggests addition. *Times* suggests multiplication.

$$8 + 5n$$

Check Your Understanding

1. Does the expression $5n + 8$ also represent the phrase in Example 1? Explain.
See *Answers to Check Your Understanding* at the back of the book.

When writing a variable expression that represents a word phrase, you first choose a variable to represent the unknown number.

Example 2

Write a variable expression that represents the phrase *$35 less than twice Mary's salary.*

Solution

Choose a variable to represent the unknown number.

Let s = Mary's salary.
Then $2s$ = twice Mary's salary.
So $2s - 35$ = $35 less than twice Mary's salary.

If Mary's salary is s dollars, then $(2s - 35)$ dollars represents $35 less than twice that salary.

Check Your Understanding

2. Does the expression $35 - 2s$ also represent the phrase in Example 2? Explain.
See *Answers to Check Your Understanding* at the back of the book.

Guided Practice

COMMUNICATION *«Reading*

Choose all the terms that are associated with each operation.

*«***1.** addition C, F, G

*«***2.** subtraction D, E, J

*«***3.** multiplication A, I

*«***4.** division B, H

A. times

B. quotient

C. combined

D. fewer than

E. difference

F. increased by

G. sum

H. shared equally

I. product

J. decreased by

Replace each __?__ with the correct number or variable expression.

5. $r - \underline{\ ?\ }$ represents *one less than a number r.* 1

6. $\underline{\ ?\ }n$ represents *twice a number n.* 2

7. $\underline{\ ?\ } + \underline{\ ?\ }$ represents *six more than three times a number m.* 3m; 6

8. $\underline{\ ?\ } + \underline{\ ?\ }$ represents *the sum of seven and a number w divided by two.* 7; $\frac{w}{2}$

Write a variable expression that represents each phrase. If necessary, choose a variable to represent the unknown number.

9. a number *v* divided by 30 $\frac{v}{30}$

10. $4 less than last paycheck $p - 4$

11. six more than twice as many hits 2h + 6

12. the sum of four times a number *r* and two 4r + 2

Exercises

Write a variable expression that represents each phrase. If necessary, choose a variable to represent the unknown number.

Choices of variables may vary.

A

1. five more than a number *x* $x + 5$

2. seven times a number *y* 7y

3. twice as old as Walter 2w

4. six divided by a number *n* $\frac{6}{n}$

5. the total of three times a number *c* and eight 3c + 8

6. a number *x* decreased by three $x - 3$

7. four less than six times a number *d* 6d − 4

8. seven more sunny days than in May m + 7

9. twelve fewer apples on the tree than yesterday a − 12

10. three inches shorter than Francesca f − 3

11. students separated into six equal teams $\frac{s}{6}$

12. double the number of points Derek scored 2d

13. seventeen points more than Tina scored t + 17

14. grapes shared equally by four people $\frac{g}{4}$

Equations   **163**

Critical Thinking

Translating word phrases into variable expressions requires students to *identify* the variables and operation(s) involved before writing the variable expression.

Error Analysis

In writing variable expressions, students often confuse the meaning of "less than" and "is less than." Eight *less than* a number *x* means 8 is subtracted from *x*, that is, $x - 8$. Eight *is less than x* means that 8 is a smaller number than *x*, which is written $8 < x$.

Reteaching/Alternate Approach

For a suggestion on using a cooperative learning activity to develop number puzzles, see page 140D.

Additional Examples

1. Write a variable expression that represents the phrase *twelve less than three times a number k.* $3k - 12$

2. Write a variable expression that represents the phrase *$65 more than three times the cost of a book.* Let *b* = cost of a book. $3b + 65$

Closing the Lesson

Ask a student to give a word or phrase used to designate a mathematical operation. Ask another student to give the symbol for the operation.

Suggested Assignments

Skills: 1–15, 21–24
Average: 1–18, 21–24
Advanced: 1–13 odd, 15–24

Cooperative Learning

Exercises 15 and 16 present an opportunity for students to discuss mathematics. Giving a convincing argument requires that students clarify their thinking.

Problem Solving

Exercises 17–19 begin the process of writing equations to solve problems. *Exercise 20* carries the process one step further by asking for a function rule of the form $x \rightarrow ax + b$.

Application

Functions occur in a variety of real-world situations. In *Exercises 17–20*, a telephone call is used as the basis for a series of problems. Rates vary with the time of day, the duration of the call, the day of the week, the distance involved, and whether or not the call is operator-assisted.

Follow-Up

Enrichment

Enrichment 38 is pictured on page 140E.

Practice

Practice 46, *Resource Book*

Practice 46
Skills and Applications of Lesson 4-6

Write a variable expression that represents each phrase. If necessary, choose a variable to represent the unknown number.   Choices of variables may vary.

1. seven more than a number *n* $n + 7$
2. eight less than a number *h* $h - 8$
3. four times a number *y* $4y$
4. a number *x* squared x^2
5. three more than a number *n* $n + 3$
6. a number *a* divided into three equal parts $\frac{a}{3}$
7. seven crayons more than Jan has $n + 7$
8. a class equally divided into four reading groups $\frac{c}{4}$
9. a pizza shared equally by six people $\frac{z}{6}$
10. triple the number of points Sue scored $3n$
11. four times the age of Meg $4a$
12. a collection of books shared equally by 36 students. $\frac{n}{36}$
13. double the number of points Ray scored $2x$
14. two seconds less than Yoshi's time to run the 440 hurdles $y - 2$
15. six years older than Tina $t + 6$
16. three more apples than Rob has $x + 3$
17. a ruler divided into twelve equal parts $\frac{x}{12}$
18. a task completed in half the time $\frac{t}{2}$
19. five more than twice a number *x* $2x + 5$
20. seven less than a number *c* squared $c^2 - 7$
21. Mike's salary if he works *h* hours at $4.50 per hour $4.50h$
22. five years older than twice Kim's age $2k + 5$
23. the square of the sum of a number *s* and five $(x + 5)^2$
24. triple the result of adding seven to a number *x* $3(x + 7)$

B **15.** Make a chart with four columns. Write the name of one of the four arithmetic operations at the top of each column. Under each operation, list all the words and phrases you associate with the operation.
See answer on p. A10.

16. Is the expression *four less than a number x* different from *four is less than a number x*? Give a convincing argument to support your answer.
Yes; the second phrase includes the verb *is*. Expressions: $x - 4$; $4 < x$

PROBLEM SOLVING/APPLICATION

Use this information for Exercises 17–20.

At the weekday rate, the cost of a long-distance telephone call from Centerville to Newtown is 26¢ for the first minute and 12¢ for each additional minute.

17. Let *x* represent the number of additional minutes. Write a variable expression that represents the cost in cents of a telephone call made at the weekday rate. $26 + 12x$

18. How would the variable expression that you wrote for Exercise 17 be different if it represented the cost in dollars rather than the cost in cents? $0.26 + 0.12x$

C **19.** Suppose you make a call that lasts *x* additional minutes at the weekday rate, then extends *y* additional minutes into the evening period. At the evening rate, the cost of each additional minute is 7¢. Write a variable expression that represents the cost in cents of your call. $26 + 12x + 7y$

20. FUNCTIONS   Make a function table that shows the cost in cents of a long-distance telephone call at the weekday rate for each whole number of minutes from 1 min through 10 min. What is the function rule? See answer on p. A10.

SPIRAL REVIEW

S **21.** Complete the function table at the right.
(Lesson 2-6)

22. Estimate the quotient: $38\overline{)2749}$
(Toolbox Skill 3) about 70

23. Write a variable expression that represents this phrase: the sum of seven and three times a number *x*   *(Lesson 4-6)* $7 + 3x$

24. Evaluate $x \div 12$ when $x = 1.44$.   *(Lesson 1-3)*
0.12

x	$4x - 1$	
2	7	
4	15	
6	?	23
8	?	31
10	?	39

4-7 Writing Equations

Objective: To write equations for sentences.

CONNECTION

In the English language, you can form simple sentences by using a verb to join phrases. For instance, you join the phrases "My art class" and "in Room 112" by a linking verb such as "is" to form a complete thought: "My art class is in Room 112." Similarly, in mathematics you form an equation by using an equals sign to join mathematical expressions.

Example 1

Write an equation that represents each sentence.

a. Nine more than a number x is 12.

b. Twenty-four is a number t divided by 3.

Solution

a. $x + 9 = 12$ b. $24 = \frac{t}{3}$

Check Your Understanding

1. In Example 1, what verb is translated into an equals sign?
See *Answers to Check Your Understanding* at the back of the book.

In problem solving situations, you may need to translate a sentence into an equation. To do this, first find the verb that you can represent by an equals sign. Then translate the remaining phrases into two mathematical expressions and join them by the equals sign.

Example 2

Write an equation that represents the relationship in the following sentence.

A financial software package costs $115, which is $25 more than the cost of a game software package.

Solution

Choose a variable to represent the unknown number.

The cost of the financial package is $115.
Let g = cost of the game package.
Then $g + 25$ = cost of the financial package.
So $g + 25 = 115$.

The equation $g + 25 = 115$ represents the relationship between the costs of the financial package and the game package.

Check Your Understanding

2. In Example 2, why would you decide to let the variable represent the cost of the game package rather than the financial package?
See *Answers to Check Your Understanding* at the back of the book.

Equations   **165**

Lesson Planner

Teaching the Lesson
• Lesson Starter 36

Lesson Follow-Up
• Practice 47
• Enrichment 39: Thinking Skills
• Study Guide, pp. 71–72

Warm-Up Exercises

Write a mathematical expression for each of the following.

1. four more than a number y
   $y + 4$
2. eight less than two times a number c $2c - 8$
3. a number x decreased by 5
   $x - 5$
4. 17 more than four times a number t $4t + 17$
5. five more than g divided by 4
   $\frac{g}{4} + 5$

Lesson Focus

Ask students if they have ever thought of mathematics as a language. There are many terms and symbols that are unique to mathematics. Today's lesson focuses on the connection between English sentences and mathematical equations.

Teaching Notes

Point out to students that just as word sentences always contain a verb, mathematical sentences always contain a relational symbol. This may be an equals sign or another symbol, such as one meaning "is less than" or "is greater than."

Key Question

Why is an equation easier to solve than a problem that uses only words?   The relationships have been identified.

Critical Thinking

Translating word sentences into equations requires students to *interpret* words as mathematical symbols.

Reteaching/Alternate Approach

For a suggestion on using problem-solving diagrams to model equations, see page 140D.

Error Analysis

Students often make errors in writing an equation for a sentence. Remind them to examine a sentence phrase by phrase, and then to connect the phrases with an equals sign.

Additional Examples

1. Write an equation that represents each sentence.
   a. Seven more than a number y is 15. $y + 7 = 15$
   b. Thirty-six is equal to a number n divided by 4.
      $36 = \frac{n}{4}$
2. Write an equation that represents the relationship in the following sentence. A combination telephone/answering machine costs $105, which is $56 more than the cost of a telephone.
   Let t = cost of the telephone.
   $105 = t + 56$

Closing the Lesson

Ask a student what verb is often used to indicate an equals sign. Ask other students to explain how they choose and identify a variable. Ask them to explain how to solve an equation.

Suggested Assignments

Skills: 1–10, 11–19 odd, 21–25
Average: 1–10, 11–19 odd, 21–25
Advanced: 1–7 odd, 9–25,
   Challenge

Guided Practice

COMMUNICATION « *Reading* Answers to Guided Practice 1–2 are on p. A10.

« **1.** What is the objective of this lesson? How is it different from the objective of Lesson 4-6?

« **2.** In this lesson, how is Example 1 different from Example 2?

Use this information for Exercises 3–5.

Karen has $46, which is twice as much money as Sal has.

3. Let s = number of dollars Sal has. Then Karen has __?__ · s dollars. 2

4. The actual amount of money Karen has is __?__. $46

5. An equation that represents this situation is __?__ = __?__. 2s; 46

Write an equation that represents each sentence.

6. Three times a number x is 18.
   $3x = 18$

7. A number t more than 9 is 17.
   $9 + t = 17$

8. Sixteen is a number m divided by 3. $16 = \frac{m}{3}$

9. A number z decreased by 3 is 39. $z - 3 = 39$

Write an equation that represents the relationship in each sentence.

10. Eve sold 45 tickets, which is three times as many tickets as Ben sold.
   $45 = 3b$

11. At the elementary school there are 22 teachers, which is nine fewer teachers than at the high school. $h - 9 = 22$

Exercises

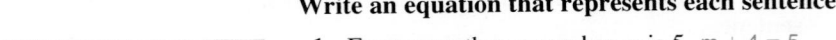

A

Write an equation that represents each sentence.

1. Four more than a number m is 5. $m + 4 = 5$

2. Sixty-three is 7 times a number n. $63 = 7n$

3. A number a divided by 6 is 12. $\frac{a}{6} = 12$

4. Thirty-four is a number t increased by 7. $34 = t + 7$

5. The product of 15 and a number k is 105. $15k = 105$

6. Two subtracted from a number b is 9. $b - 2 = 9$

Write an equation that represents the relationship in each sentence.

7. In his locker Alexander has seven books, which is the number of books Joanne has in her locker divided by two. $7 = \frac{j}{2}$

8. At South High School there are 534 girls, which is 89 more than the number of boys. $534 = b + 89$

9. The low temperature on Monday was 10°F, which is 15°F less than the low temperature on Sunday. $10 = s - 15$

10. Last week Manuel earned $297, which is twice the amount that Tony earned. $297 = 2t$

Each of the situations in Exercises 9–14 can be represented by one of the equations below. Choose the letter of the correct equation.

  A. $x + 8 = 24$ **B.** $x - 8 = 24$ **C.** $8x = 24$ **D.** $\frac{x}{8} = 24$

B **11.** Rich scored 8 more points than Teresa. Rich scored 24 points. A

12. Mario is 8 times as old as his sister. Mario is 24 years old. C

13. Paula has 24 albums, which is the number of cassettes Richard has divided by 8. D

14. Christine worked a total of 24 h during the last two weeks. The first week she worked 8 h. A

15. The total rainfall in Warwick last year was 8 times the amount of rain that fell during March. Warwick had 24 cm of rainfall last year. C

16. Raoul's score of 24 on today's quiz is 8 points lower than his score on last week's quiz. B

COMMUNICATION « *Writing*

Write a sentence that could be represented by each equation.
Answers to Exercises 17–20 are on p. A10.

C «**17.** $m - 8 = 16$ «**18.** $y + 9 = 32$

  «**19.** $4w = 48$ «**20.** $\frac{u}{9} = 14$

SPIRAL REVIEW

S **21.** Is it most efficient to find the quotient $1800 \div 90$ using *mental math, paper and pencil,* or a *calculator?* *(Lesson 1-8)* mental math

22. Write an equation to represent this sentence: The product of 5 and a number x is 42. *(Lesson 4-7)* $5x = 42$

23. Graph the function $x \rightarrow 2x$ when $x = -3, -1, 0, 1,$ and 3. *(Lesson 3-9)* See answer on p. A10.

24. Find the product: $2\frac{6}{7} \times 3\frac{3}{5}$ *(Toolbox Skill 20)* $10\frac{2}{7}$

25. Is 4 a solution of $6x = -24$? Write *Yes* or *No.* *(Lesson 4-1)* No.

Challenge

Use four of these weights to balance the scale.

    11   19   47   56   92

$11 + 92$ $47 + 56$

Can you balance the scale using all five weights? No

Equations   **167**

Reasoning
Exercises 11–16 require students to read each problem and interpret its meaning. Since all of the problems use the same numbers, careful distinctions must be made.

Nonroutine Problem
In the Challenge problem, five weights cannot be used to balance the scale. If all five weights were used, the sum on one side would be an even number; on the other side, the sum would be odd.

Follow-Up

Exploration
Ask students to look for puzzles or brain teasers that can be solved by using mathematics. Place them on the bulletin board in the classroom.

Enrichment
Enrichment 39 is pictured on page 140F.

Practice
Practice 47 is shown below.

Practice 47, *Resource Book*

Practice 47
Skills and Applications of Lesson 4-7

Write an equation that represents each sentence.
1. Six less than a number n is 8. $n - 6 = 8$
2. Twice a number m is 28. $2m = 28$
3. The product of a number a and 5 is 35. $5a = 35$
4. The sum of 12 and a number w is 32. $w + 12 = 32$
5. The quotient of a number x and 6 is -4. $\frac{x}{6} = -4$
6. The difference when eight is subtracted from a number x is 18. $x - 8 = 18$
7. The product of -9 and a number d is 36. $-9d = 36$
8. A number n added to 14 is 29. $14 + n = 29$
9. The product of 12 and a number t is 48. $12t = 48$
10. The sum of 24 and a number n is 52. $24 + n = 52$
11. The square of a number x is 36. $x^2 = 36$
12. The quotient of a number k and -9 is -6. $\frac{k}{-9} = -6$
13. The sum of a number n and 14 is 4. $n + 14 = 4$
14. The product of 8 and a number c is -56. $8c = -56$
15. The difference of a number m and 14 is 11. $m - 14 = 11$
16. Four added to three times a number x is 28. $3x + 4 = 28$
17. Twelve subtracted from the product of a number x and 2 results in a difference of 8. $2x - 12 = 8$

Write an equation that represents the relationship in each sentence.
18. In Maria's mathematics class there are 14 boys. The number of boys is three more than the number of girls in the class. $g + 3 = 14$
19. Clyde jogged for 16 miles last week. His distance was 4 miles less than the miles jogged by Jefferson. $J - 4 = 16$
20. Jana scored nine goals during last year's soccer season. Jana's total was four more than Rhonda's total. $R + 4 = 9$
21. If all of the library books were divided equally among 6 classrooms, then each room would have 45 books. $\frac{x}{6} = 45$

4-8

Strategy:
Using Equations

Objective: To solve problems by using equations.

Some problems describe a relationship between two or more numbers. To solve this type of problem, choose a variable to represent one of the unknown numbers in the problem. Use that variable to write expressions for the other unknown numbers. Then use the facts of the problem to write an equation. You solve the problem by solving this equation and finding the unknown numbers.

Problem

A parking garage charges $3 for the first hour and $2 for each additional hour. On a recent day, a motorist paid $17 to park a car in the garage. How many hours was the car parked in the garage?

Solution

UNDERSTAND The problem is about the cost of parking a car in a parking garage.

Facts: $3 for the first hour
$2 for each additional hour
total of $17 paid

Find: the number of hours the car was parked

PLAN Choose a variable and decide what the variable will represent. Use the variable to write expressions and then an equation for the problem. Solve the equation to answer the question.

WORK Let h = the number of additional hours the car was parked. Then $2h$ = the cost of parking for the additional hours.

The cost for the first hour plus the cost for the additional hours is 17.

$$\underbrace{\quad\quad\quad}\;\;\downarrow\;\underbrace{\quad\quad\quad\quad}\;\;\downarrow\downarrow$$
$$3 \qquad\qquad + \qquad\qquad 2h \qquad\qquad = 17$$

$$3 + 2h = 17$$
$$3 + 2h - 3 = 17 - 3$$
$$2h = 14$$
$$\frac{2h}{2} = \frac{14}{2}$$
$$h = 7$$

✓ **Check**
$$3 + 2h = 17$$
$$3 + 2(7) \stackrel{?}{=} 17$$
$$3 + 14 \stackrel{?}{=} 17$$
$$17 = 17$$

ANSWER Because $h = 7$, the car was parked for 7 additional hours. So, including the first hour, the car was parked for a total of 8 h.

Look Back An alternative method for solving the problem is to use the guess and check strategy. Solve the problem again using this alternative method. 8 h

Warm-Up Exercises

Write an equation that represents each sentence.

1. The product of 7 and a number k is 18. $7k = 18$
2. Four more than a number m is 21. $m + 4 = 21$
3. Seventy-five is 3 more than 4 times a number j. $75 = 4j + 3$
4. Mitch has twelve CD's, which is 3 times the number of tapes Joan has. $12 = 3t$
5. Write a sentence that the equation $n - 2 = 22$ might represent. Answers will vary. Sally had $22 left after spending $2 for stamps.

Lesson Focus

Ask students if they have ever tried to solve number puzzles. Books, magazines, and newspapers often have puzzles that use simple equations. Today's lesson focuses on the problem-solving strategy of using equations.

Teaching Notes

Review briefly the generalizations about solving equations and the four-step problem-solving plan. Work through the lesson with students, stressing the importance of *looking back* at a solution once one is found.

168 Chapter 4

Guided Practice

COMMUNICATION «*Reading*

Use the problem below for Exercises 1–7.

A construction site has carpenters and electricians. The number of carpenters is eight fewer than twice the number of electricians. If there are twelve carpenters on the job site, how many electricians are there?

«**1.** What do you need to find out? the number of electricians

«**2.** Use the variable n. What will you let n represent?
the number of electricians

«**3.** Write a variable expression that represents the number of carpenters.
$2n - 8$

«**4.** How many carpenters are there? 12

«**5.** Use your answers in Exercises 3 and 4 to write an equation.
$2n - 8 = 12$

6. Solve the equation. Check the solution. $n = 10$

7. Use the solution of the equation to write an answer to the problem.
There are 10 electricians.

Solve using an equation.

8. The number of drummers in a band is two more than three times the number of tuba players. If there are eight drummers in the band, how many tuba players are there? 2

9. A number is divided by two. Then five is added to the result. If the answer is -2, what was the original number? -14

10. Hector bought a computer system for $989. He made a $125 down payment and paid the remaining amount in twelve equal payments. What was the amount of each payment? $72

Problem Solving Situations

Solve using an equation.

1. At her new job Mary earns $100 more than twice the salary she earned as a college intern. If she now earns $2100, how much did she earn as an intern? $1000

2. The greater of two numbers is nine less than four times the other number. If the greater number is 71, find the lesser number. 20

3. Fran's second bowling score was 72 points less than twice her first score. If her second score was 156, what was her first score? 114

4. On his tenth birthday, Tom's monthly allowance was doubled. When he turned twelve, his allowance increased by $10 to $34 per month. What was Tom's allowance before his tenth birthday? $12

Equations **169**

Critical Thinking
Checking the reasonableness of an answer requires students to *compare* the answer with the original problem and to *draw a conclusion* as to whether it is correct or not.

Life Skills
The problems in this lesson relate to real-world skills. Students are allowed to use a variety of strategies to solve them.

Error Analysis
Errors in problem solving with equations usually result when the wrong equation is used. Remind students to identify what the variable represents and what the relationships within the problem are.

Additional Example
Airport parking is $1 for the first hour and 50¢ for each additional hour. Paul paid $3.50 to park his car. How long was Paul's car parked at the airport?
Let h = number of additional hours parked.
Let $0.5h$ = the cost of parking for additional hours.
$1 + 0.5h = 3.50; h = 5$
Paul's car was parked for 6 hours.

Closing the Lesson
Ask a student to summarize the four-step problem-solving plan. Ask another student to explain how the plan is used when working with equations.

Suggested Assignments

Skills: 1–13, 16–21
Average: 1–14, 16–21
Advanced: 1–21

Exercise Notes

Communication: Reading
Guided Practice Exercises 1–7 guide students step-by-step through the process necessary to write an equation to solve a problem.

Problem Solving
Problem Solving Situations 1–4 should be solved by using the four-step plan and an equation. Problem Solving Situations 5–11 may be solved by using any strategy introduced thus far in the textbook.

Making Connections/Transitions
Problem Solving Situations 12–15 connect mathematics with the type of data analysis used by dieticians or by others interested in nutrition.

Cooperative Learning
Problem Solving Situation 15 could be explored as a group activity. What are the minimum daily Recommended Dietary Allowances for your age group? What foods would meet those requirements? How does a fast food meal compare?

Communication: Writing
Problem Solving Situations 16 and 17 require students to create a word problem for a given mathematical equation.

Solve using any problem solving strategy. 5–11. Strategies may vary. Likely strategies are given.

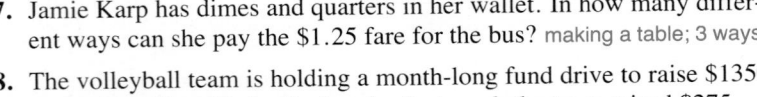

PROBLEM SOLVING
CHECKLIST
Keep this in mind:
Using a Four-Step Plan
Consider these strategies:
Choosing the Correct Operation
Making a Table
Guess and Check
Using Equations

5. Scott Bradley bought five shirts at $14 each and six pairs of socks at $3.50 each. What was the total cost of his purchase? choosing the correct operation; $91

6. Nineteen more than twice a number is 17. What is the number? using equations; −1

7. Jamie Karp has dimes and quarters in her wallet. In how many different ways can she pay the $1.25 fare for the bus? making a table; 3 ways

8. The volleyball team is holding a month-long fund drive to raise $1350 for a trip to New York. During the first week the team raised $275. The second week the team raised $490. How much does the team still need to raise? choosing the correct operation; $585

9. Niabi Thomas bought a used car for $4800. The tax on the car was $216. Niabi made a down payment of $1200 and agreed to pay the remainder in 24 equal payments. What was the amount of each payment? using equations; $159

10. Alfonso Gomez went on a cross-country bus trip. On the first day he traveled 250 mi. He traveled 320 mi on each of the next two days. During the fourth day he traveled 280 mi. Alfonso says that by the end of the fourth day he had traveled a total of 850 mi. Is this correct? Explain. No; he forgot to include one of the 320-mi days.

11. Samantha Marshall bought some paperback books for $5 each and some magazines for $3 each. She paid a total of $32. How many of each did she buy? guess and check; 4 books, 4 magazines

DATA ANALYSIS

Use the table at the right. Solve using any problem solving strategy.

Food	Calcium (mg)
1 egg	27
1 orange	54
1 tomato	24
1 glass (240 mL) whole milk	288

12. How many more milligrams of calcium are in a 240-mL glass of milk than in one egg? 261 mg

13. How many milligrams of calcium are in three tomatoes? 72 mg

14. The Recommended Daily Dietary Allowance (RDA) of calcium for Melania's age group is 800 mg. Melania has already had two oranges and one 240-mL glass of whole milk today. How many more glasses of milk must she have to meet her RDA for calcium? 2

15. **RESEARCH** What is the RDA of calcium for your age group? Answers will vary.

WRITING WORD PROBLEMS

Write a word problem that you could solve by using each equation. Then solve the problem. Answers to Problems 16 and 17 are on p. A10.

16. $16 = x + 5$

17. $3x - 1 = 26$

SPIRAL REVIEW

S **18.** Complete: 64 oz = __?__ lb   *(Toolbox Skill 22)* 4

19. Is 426 divisible by 2? by 3? by 4? by 5? by 8? by 9? by 10? *(Toolbox Skill 16)* Yes; Yes; No; No; No; No; No.

20. Find the difference: $4\frac{7}{12} - 2\frac{5}{6}$ *(Toolbox Skill 19)* $1\frac{3}{4}$

21. The sum of three times a number and seven is 55. Find the number. *(Lesson 4-8)* 16

Solve. Check each solution.

1. $6m + 3 = -15$ -3   **2.** $\frac{n}{-4} + 8 = 2$ 24   **3.** $89 = 10q - 11$ 10   **4-5**

Write a variable expression that represents each phrase. If necessary, choose a variable to represent the unknown number.

4. a number z divided by fourteen $\frac{z}{14}$ **4-6**

5. three times as many ripe tomatoes as yesterday 3t

Write an equation that represents the relationship in each sentence.

6. Seven more than a number n is 35. $n + 7 = 35$ **4-7**

7. Sandra has twenty-two video cassettes, which is nine fewer than Thomas has. $t - 9 = 22$

Solve using an equation.

8. Lynn's second test score was 60 points less than twice her first score. If her second score was 86, what was Lynn's first test score? 73   **4-8**

9. Pat Burns is planning to compete in a 300-km bicycle race. He plans to travel 90 km on the first day and then to travel the same distance on each of the next three days. How far does he plan to travel on the fourth day? 70 km

Equations **171**

Teaching the Lesson
- Lesson Starter 38
- Toolbox Skills 7–10

Lesson Follow-Up
- Practice 50
- Enrichment 41: Thinking Skills
- Study Guide, pp. 75–76
- Manipulative Activity 8
- Cooperative Activity 9

Warm-Up Exercises

Solve.

1. $x + 12 = 8$ -4
2. $z - 7 = -15$ -8
3. $2n + 17 = 37$ 10
4. $3x - 12 = 42$ 18
5. $5c + 17 = -13$ -6

Lesson Focus

Today's lesson focuses on equations that involve combining like terms or the distributive property. To motivate the lesson, you may wish to use the application below.

Application

Sam bought two short-sleeve shirts and four long-sleeve shirts on sale. All the shirts were the same price. Sam spent a total of $84. Use the equation $2x + 4x = 84$ to find the cost of each shirt. Each shirt cost $14.

Teaching Notes

Point out that this lesson extends the previously learned skills of solving an equation, combining like terms, and using the distributive property.

Key Question

Why were the terms inside the parentheses not combined in Example 2? $4x$ and 3 are not like terms.

More Equations

Objective: To solve equations by simplifying expressions involving combining like terms and the distributive property.

DATA ANALYSIS

In the pictograph at the right, the key is missing.

QUESTION Suppose you know that 560 students graduated from Conway and Wheaton together. What is the key?

Graduates from Area Towns

Let the variable x represent the number in the key. Since there are four symbols next to Conway, the number of graduates from that town can be represented by the expression $4x$. Similarly, the number of graduates from Wheaton can be represented by the expression $3x$. The expression $4x + 3x$ represents the total number of graduates from Conway and Wheaton. To find the key you must solve the equation $4x + 3x = 560$.

Example 1

Solve $4x + 3x = 560$. Check the solution.

Solution

$$4x + 3x = 560$$
$$7x = 560 \quad \longleftarrow \text{ Combine like terms.}$$
$$\frac{7x}{7} = \frac{560}{7}$$
$$x = 80$$

The solution is 80.

✓ **Check**

$$4x + 3x = 560$$
$$4(80) + 3(80) \overset{?}{=} 560$$
$$320 + 240 \overset{?}{=} 560$$
$$560 = 560$$

Check Your Understanding

1. What are the like terms that were combined in Example 1?
   See *Answers to Check Your Understanding* at the back of the book.

ANSWER The key is 🎓 = 80 students.

You should simplify the sides of an equation before solving. This may involve combining like terms, as in Example 1. It may also involve using the distributive property.

Example 2

Solution

Solve $2(4x + 3) = 46$. Check the solution.

$$2(4x + 3) = 46$$
$$8x + 6 = 46 \quad \longleftarrow \text{Use the distributive property.}$$
$$8x + 6 - 6 = 46 - 6$$
$$8x = 40$$
$$\frac{8x}{8} = \frac{40}{8}$$
$$x = 5$$

The solution is 5.

✔ **Check**
$$2(4x + 3) = 46$$
$$2(4[5] + 3) \stackrel{?}{=} 46$$
$$2(20 + 3) \stackrel{?}{=} 46$$
$$2(23) \stackrel{?}{=} 46$$
$$46 = 46$$

 **Check Your Understanding**

2. Why was the distributive property used in Example 2?
See *Answers to Check Your Understanding* at the back of the book.

Guided Practice

COMMUNICATION « *Reading*

Refer to the text on pages 172–173. Answers to Guided Practice 1–2 are on p. A11.

« **1.** What is the main idea of the lesson?

« **2.** List two major points that support the main idea of the lesson.

Match each equation with the equation that has the same solution.

3. $3(t + 2) = 30$ B

4. $2(t + 3) = 30$ D

5. $3t + 2t = 30$ C

6. $3t - 2t = 30$ A

A. $t = 30$

B. $3t + 6 = 30$

C. $5t = 30$

D. $2t + 6 = 30$

Solve. Check each solution.

7. $8n + 5n = 39$ 3

8. $2p - 6p = 40$ −10

9. $5(3x + 4) = -10$ −2

10. $6 = 2(7t - 4)$ 1

11. $3a - 6 + 4a = 22$ 4

12. $9b + 7 - b = 71$ 8

Exercises

Solve. Check each solution.

A **1.** $7n + 4n = 132$ 12

2. $12a + 3a = 60$ 4

3. $2(3v + 4) = -40$ −8

4. $5(2x + 7) = 45$ 1

5. $-5c + 9c = -20$ −5

6. $14g - 17g = -21$ 7

Equations **173**

Problem Solving
To find the missing key to the pictograph shown on page 172, students might use the strategy of *working backwards*.

Making Connections/Transitions
Data analysis frequently requires that calculations are made based on information presented in a graph or chart. The introductory section of the lesson connects the concepts of graphing, solving equations, and analyzing data.

Additional Examples

1. Solve $4x + 2x = 900$. Check the solution. 150

2. Solve $3(2x + 5) = 135$. Check the solution. 20

Closing the Lesson
Ask a student to explain the steps needed to solve an equation that involves combining like terms. Ask another student to explain the steps needed to solve an equation that involves the distributive property.

Suggested Assignments
Skills: 1–16, 17–21 odd, 27–32
Average: 1–16, 17–21 odd, 23–25, 27–32
Advanced: 1–15 odd, 17–32, Historical Note

Exercise Notes

Communication: Reading
Guided Practice Exercises 1 and 2 use the language arts approach of looking for the main idea. Students have to identify the main idea of the lesson and provide evidence to support this idea.

Exercise Notes *(continued)*

Computer
*Exercises 23–26 direct students to
use a spreadsheet to solve an equation that they have not yet learned
to solve algebraically. Spreadsheets
are used in the accounting departments of many businesses.*

Solve. Check each solution.

7. $3(2q - 7) = 3$ 4

8. $6(7k - 10) = 24$ 2

9. $-12 = 3(2x - 10)$ 3

10. $36 = 4(z + 11)$ −2

11. $4y + 7 + 8y = 43$ 3

12. $36 = 6b - 6 + b$ 6

13. The sum of twice a number and four times the number is 54. Find the number. 9

14. A number subtracted from five times the number is −28. Find the number. −7

15. Twice the difference of 7 subtracted from a number is −70. Find the number. −28

16. Three times the sum of a number and 4 is 108. Find the number. 32

Solve. Check each solution.

B 17. $5(v + 1) + 5v = 35$ 3

18. $4(g + 2) + 8g = 56$ 4

19. $12 = 2(a - 8) + 5a$ 4

20. $24 = 7(n - 3) + 8n$ 3

21. $3(4x + 1) - 2x = 23$ 2

22. $6(2k + 5) - 3k = 66$ 4

 COMPUTER APPLICATION

Using an electronic spreadsheet, you can solve equations that you have not
learned how to solve using paper and pencil. You use the spreadsheet to
find the value of the variable that makes the expressions on either side of
the equation equal.

	A	B	C
1	x	2(4x − 7)	5(x + 2)
2	1	−6	15
3	2	2	20
4	3	10	25
5	4	18	30

**Using the equation $2(4x - 7) = 5(x + 2)$, create an electronic
spreadsheet or draw a spreadsheet like the one shown above.**

C 23. For which values of x is the expression in cell B1 less than the expression in cell C1? all values of x less than 8

24. For which values of x is the expression in cell B1 greater than the expression in cell C1? all values of x greater than 8

25. Find the solution of the equation. 8

26. What would be the solution of the equation if the expression on the
right side were $5(x - 4)$? −2

SPIRAL REVIEW

S **27.** On his first day of a new job, Steve Rosadini worked for 7 h. On the next three days he worked for 8.5 h, 7 h, and 2 h. His hourly wage is $9.25. *(Lesson 1-4)*

    a. What is the paragraph about? The paragraph is about Steve working at a job.

    b. How long did Steve work on the first day? 7 h

    c. Identify any facts that are not needed to find the total number of hours that Steve worked for the four days. the hourly wage

    d. Describe how you would find the total amount that Steve earned for the four days. Add 7, 8.5, 7, and 2. Multiply the sum by 9.25.

28. Use the bar graph below. Find the height of the waterfall named Tugela. *(Toolbox Skill 24)* 3000 ft

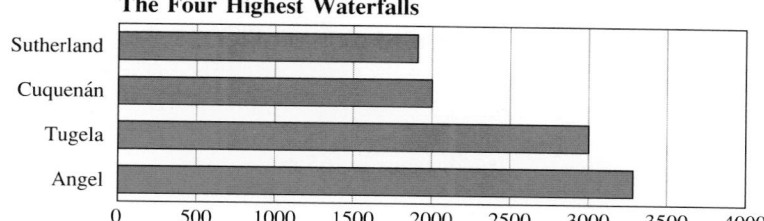

The Four Highest Waterfalls

Height in Feet

29. Estimate the sum: $156 + 203 + 242$ *(Toolbox Skill 2)* about 600

30. Solve: $2(4x + 5) = 66$ *(Lesson 4-9)* 7

31. PATTERNS Find the next three numbers in the pattern: 18, 15, 12, 9, __?__, __?__, __?__   *(Lesson 2-5)* 6, 3, 0

32. Ed has only $10, $5, and $1 bills. In how many different ways could he pay $26?   *(Lesson 3-7)* 12 ways

Historical Note

S tars that change in brightness are called *variable stars*. Cecilia Payne-Gaposchkin (1903–1979), an astronomer from the United States, perfected the technique of studying variable stars photographically. Her findings provided important clues to the structure of our galaxy, the Milky Way.

See answer on p. A11.

RESEARCH

Recent astronomical studies indicate that there may be a black hole at the center of the Milky Way. Find out what a black hole is.

Equations **175**

Follow-Up

Extension

Ask students to develop word problems that could be solved by using the equations in Guided Practice Exercises 3–12.

Enrichment

Enrichment 41 is pictured on page 140F.

Practice

Practice 50 is shown below.

Practice 50, *Resource Book*

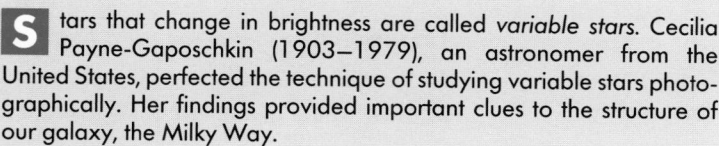

Practice 50
Skills and Applications of Lesson 4-9

Solve. Check each solution.

1. $5x + 4x = 27$ 3   2. $7w - 2w = 25$ 5   3. $11a + 4a = 60$ 4
4. $4(3s + 5) = 44$ 2   5. $15y - 3y = 48$ 4   6. $7(6x - 8) = 70$ 3
7. $4a + 7 + 3a = -14$ -3 8. $72 = -8(4x - 5)$ -1 9. $18x - 11x = 49$ 7
10. $5(3w - 11) = 65$ 8   11. $6d + 11 + 3d = 47$ 4   12. $6c - 9c + 14 = 20$ -2
13. $7(4b - 4) = 112$ 5   14. $-15t + 9t = 0$ 0   15. $176 = 8(5x - 3)$ 5
16. $12(a + 7) = 96$ 1   17. $18y - 15y + 19 = 25$ 2   18. $100 = 48 + 3w + 10w$ 4
19. $8d - 3d - 12 = 53$ 13   20. $4h - 27 + 3h = 43$ 10   21. $209 = 11(5a - 6)$ 5
22. $9(x - 5) = 54$ 11   23. $3(6x + 10) = 120$ 5   24. $19x - 4x + 15 = -15$ -2
25. $2a + 88 + 6a = 88$ 0   26. $5(-6c + 7) = 35$ 0   27. $9(3d - 8) = 36$ 4
28. $71 = 16w - 8w - 9$ 10   29. $-11(6y - 12) = 0$ 2   30. $4(3t + 30) = -24$ -12
31. $17(4 - 11) = 0$ 11   32. $13m - 120 + 5m = 60$ 10   33. $-15n + 8n - 48 = -7$ -7
34. $4(a - 2) = 36$ 11   35. $9(6d - 42) = 0$ 7   36. $70 = -10x + 8x + 54$ -8
37. $12 = 12c - 10c + 10$ 1   38. $5(-5n + 5) = 75$ -2 39. $22 + 5x - 5x + 4x = 34$ 3

40. The sum of four times a number and three times the number is 35. Find the number. 5
41. The sum of six times a number and five is tripled. The result is -21. Find the number. -2
42. Fourteen is subtracted from six times a number and then twice the number is added. The result is 10. Find the number. 3
43. The difference when twenty is subtracted from three times a number is multiplied by five. The result is 35. Find the number. 9

Resource Book, MATHEMATICAL CONNECTIONS
Copyright © by Houghton Mifflin Company. All rights reserved.   123

175

Teaching the Lesson
- Lesson Starter 39
- Toolbox Skills 7–10

Lesson Follow-Up
- Practice 51
- Practice 52: Nonroutine
- Enrichment 42: Application
- Study Guide, pp. 77–78
- Test 16

Warm-Up Exercises

Solve. Then check each equation.

1. $9x + 3x = 108$ 9
2. $-54 = 3(7n - 4)$ -2
3. $4a - 9 - 5a = 11$ -20
4. The sum of twice a number n and five equals 49. Find the number. $2n + 5 = 49; 22$
5. Four less than the number c divided by five is equal to 9. Find the number. $\frac{c}{5} - 4 = 9; 65$

Lesson Focus

Ask students if they have heard the pilot of an airplane announce the flying time to the destination. Today's lesson works with formulas, including the distance formula which can be used to determine flying time by substituting values for the plane's speed and distance from the destination.

Teaching Notes

Remind students of formulas they might have used in previous courses, such as those for the perimeter or area of a rectangle. Stress that substituting values for a variable in a formula is similar to evaluating expressions. A specific value replaces each variable.

Formulas

Objective: To work with formulas.

Terms to Know
• *formula*

A **formula** is an equation that states a relationship between two or more quantities. The quantities are usually represented by variables. The variable used is often the first letter of the word it represents.

For instance, the *distance formula* represents a relationship between distance (D), rate (r), and time (t).

$$\text{distance} = \text{rate} \times \text{time}$$
$$D = rt$$

Example 1

Use the formula $D = rt$. Find the distance when $r = 55$ mi/h and $t = 3$ h.

Solution

Substitute 55 for r and 3 for t. Solve.
$$D = rt$$
$$D = 55(3)$$
$$D = 165$$

The distance is 165 mi.

Sometimes you might need to use inverse operations to find an unknown value in a formula.

Example 2

Use the formula $D = rt$. Find the time when $D = 240$ mi and $r = 40$ mi/h.

Solution

Substitute 240 for D and 40 for r. Solve.
$$D = rt$$
$$240 = 40t$$
$$\frac{240}{40} = \frac{40t}{40}$$
$$6 = t$$

The time is 6 h.

 **Check Your Understanding**

1. In Example 2, how do you get the equation $240 = 40t$?
2. How would Example 2 be different if a value for t, not r, were given?
   See *Answers to Check Your Understanding* at the back of the book.

Guided Practice

COMMUNICATION «*Reading*

Explain the meaning of the word *formula* in each sentence.

«**1.** Melissa was fed baby formula every four hours. Melissa was fed a specially mixed liquid food for babies.
«**2.** Andrew's formula for success on tests included reviewing his notes. Andrew had a systematic plan for success on tests.

Choose the letter of the equation that reflects the given information.

3. $D = rt$; $r = 20$ mi/h and $D = 600$ mi

    a. $20 = 600t$ b. $600 = \frac{t}{20}$

    (c.) $600 = 20t$ d. $600 = 20 + t$

4. $P = 2l + 2w$; $w = 3$ m and $P = 24$ m

    a. $3 = 2l + 48$ (b.) $24 = 2l + 6$

    c. $24 = 6 + 2w$ d. $3 = 48 + 2w$

Use the formula $C = p - d$, where C represents cost, p represents price, and d represents discount. Replace each ? with the correct value.

5. $p = \$50$, $d = \$5$, $C = \$$? 45   6. $p = \$240$, $d = \$60$, $C = \$$?^{180}

7. $C = \$604$, $d = \$151$, $p = \$$? 755   8. $C = \$76$, $d = \$4$, $p = \$$? 80

Exercises

The amount of force (F) applied to move an object a certain distance (d) is defined as work (W). When a force is applied in the direction that the object moves, the relationship between these three variables is $W = Fd$.

Use the formula $W = Fd$. Replace each ? with the correct value.

	Work (W)	Force (F)	Distance (d)
A 1.	? ft-lb 2200	40 lb	55 ft
2.	520 ft-lb	13 lb	? ft 40
3.	2400 ft-lb	? lb 40	60 ft

4. An elevator does 62,400 ft-lb of work lifting a load 120 ft. How heavy is the load? 520 lb

The speed that an airplane travels in the air depends upon the speed of the wind. The relationship between ground speed (g), air speed (a), and head wind speed (h) is given by the formula $g = a - h$.

Use the formula $g = a - h$. Replace each ? with the correct value.

	Ground Speed (g)	Air Speed (a)	Head Wind Speed (h)
5.	? km/h 610	620 km/h	10 km/h
6.	575 mi/h	? mi/h 600	25 mi/h

7. An airplane travels into a 15 mi/h head wind at a ground speed of 460 mi/h. What is the airplane's air speed? 475 mi/h

8. An airplane traveling into a 35 mi/h head wind has an air speed of 325 mi/h. What is the airplane's ground speed? 290 mi/h

Equations   **177**

Making Connections/Transitions

Exercises 9–12 connect two branches of mathematics, algebra and geometry. Many geometric relationships can be expressed by using formulas.

Application

Exercises 13–16 show an application of formulas to the medical field. Formulas are used in a variety of ways; for example, finding the appropriate dosages for medicine.

Follow-Up

Extension

Give students a short list of formulas from various fields. Ask them to try to determine what the variables in each formula could represent.

Enrichment

Enrichment 42 is pictured on page 140F.

Practice

Practice 51 is shown below.

Practice 51, *Resource Book*

Practice 51
Skills and Applications of Lesson 4-10

Use the formula $D = rt$. Replace each _?_ with the correct value.

	Distance (D)	Rate (r)	Time (t)
1.	_?_ mi 150	30 mi/h	5 h
2.	44 ft	_?_ ft/s 22	2 s
3.	21 mi	7 mi/h	_?_ h 3
4.	_?_ mi 165	55 mi/h	3 h

5. Kim and her parents drove 114 mi in 3 h. What was the average rate they traveled? **38 mi/h**

6. The high school band director estimated that the distance to be traveled by bus to a band competition was 540 mi. Assuming the bus would travel at an average rate of 30 mi/h, how long will the band be on the bus? **18 h**

The amount of current, I, passing through a wire with resistance R produces a voltage, E. The relationship between these three variables can be represented by $E = IR$ (Ohm's Law).

Use the formula $E = IR$. Replace each _?_ with the correct value.

	Volts (E)	Current (I)	Resistance (R)
7.	_?_ v 126	6 amp	21 ohms
8.	12 v	_?_ amp 3	4 ohms
9.	30 v	5 amp	_?_ ohms 6
10.	110 v	_?_ amp 22	5 ohms

11. If the heating element in a water heater draws 15 amp of current from a 210 v line, then what is the resistance of the system? **14 ohms**

12. What voltage is required to operate a 15 amp electric dryer if the resistance is 2 ohms? **30 volts**

CONNECTING ALGEBRA AND GEOMETRY

Geometry is a branch of mathematics that involves many formulas. These include the formulas for perimeter (P) and area (A). In order to do problems in geometry, you will often have to work with formulas.

Use the given formula. Replace each _?_ with the correct value.

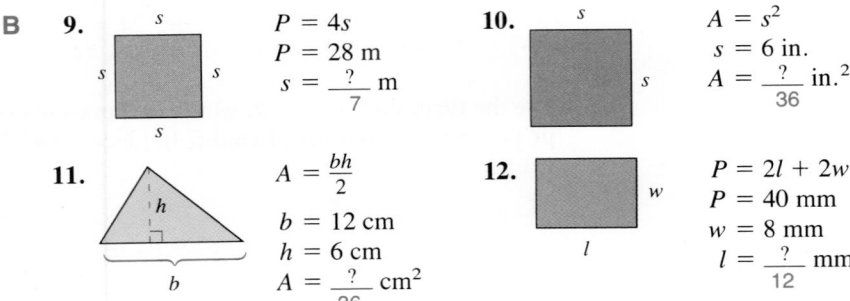

B   9. $P = 4s$
     $P = 28$ m
     $s = \dfrac{?}{7}$ m

10. $A = s^2$
     $s = 6$ in.
     $A = \dfrac{?}{36}$ in.^{2}

11. $A = \dfrac{bh}{2}$
     $b = 12$ cm
     $h = 6$ cm
     $A = \dfrac{?}{36}$ cm^2

12. $P = 2l + 2w$
     $P = 40$ mm
     $w = 8$ mm
     $l = \dfrac{?}{12}$ mm

CAREER/APPLICATION

A *nurse* or *health practitioner* might determine blood pressure. This is a measure of the pressure of the blood within the arteries. *Systolic* pressure is the highest level in the pressure cycle; *diastolic* pressure is the lowest level.

13. The normal systolic blood pressure (P) for a person of a given age (a) is 110 more than the quotient when that person's age is divided by two. Write a formula for systolic blood pressure. $P = \dfrac{a}{2} + 110$

14. Use the formula in Exercise 13. What would you expect the normal systolic blood pressure to be for an 18-year-old person? 119

15. Use the formula in Exercise 13. If a person has a normal systolic blood pressure of 150, how old might you expect that person to be? 80 years old

16. **RESEARCH** Blood pressure is often reported as a relationship between the systolic and diastolic pressures. Find out how this relationship is expressed. See answer on p. A11.

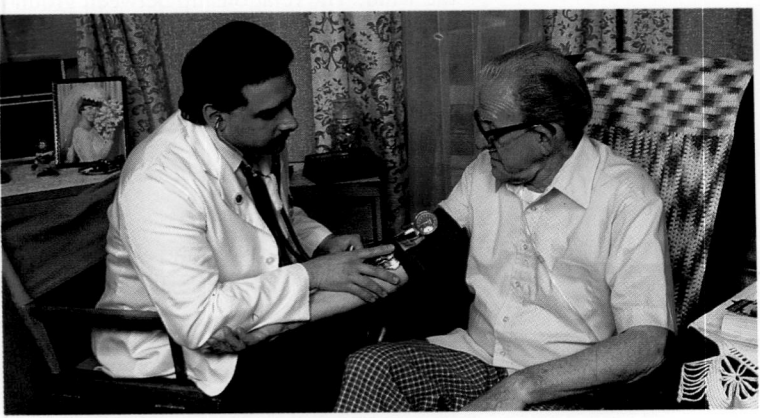

DATA, pages 140–141

Use the formula $D = rt$.

17. How far will the *Metroliner* normally travel in 3 h? 293.4 mi

18. How long will it take an impulse of electricity to travel 631,315 mi?
5 s

RESEARCH 21. E = force (in volts), I = current (in amperes),
R = resistance (in ohms).

C **19.** Find out the formula representing Ohm's law. $E = IR$

20. Find out the Einstein equation. $E = mc^2$

21. Find out what the variables represent in the formula for Ohm's law.

22. Find out what the variables represent in the Einstein equation.
E = energy, m = mass, c = speed of light

SPIRAL REVIEW

S **23.** Find the quotient: $31.08 \div 3.7$ *(Toolbox Skill 14)* 8.4

24. Use the formula $W = Fd$. Let $W = 70$ ft-lb and $F = 14$ lb. Find d.
(Lesson 4-10) 5 ft

25. Evaluate $|x + y|$ when $x = 7$ and $y = -9$. *(Lesson 3-6)* 2

26. Complete: 1.23 km = $\underline{\ ?\ }$ m *(Lesson 2-9)* 1230

27. Simplify: $9a + 5b - b$ *(Lesson 2-3)* $9a + 4b$

28. Solve: $\frac{x}{9} = 14$ *(Lesson 4-4)* 126

29. Evaluate n^5 when $n = 4$. *(Lesson 1-9)* 1024

30. Solve: $x + 9 = -6$ *(Lesson 4-3)* -15

Self-Test 3

Solve. Check each solution.

1. $3a + 6a = 108$ 12   **2.** $8y - 15y = 42$ -6 4-9

3. $4(2n - 7) = 68$ 12   **4.** $76 = 4(h + 9)$ 10

5. $-19 = 6x - 5 + x$ -2 **6.** $-3c - 4 + 7c = 8$ 3

Use the formula $D = rt$.   4-10

7. Let $r = 45$ mi/h and $t = 4$ h. Find D. 180 mi

8. Let $r = 50$ mi/h and $t = 7$ h. Find D. 350 mi

9. Let $D = 180$ mi and $r = 15$ mi/h. Find t. 12 h

10. Let $D = 2106$ mi and $t = 9$ h. Find r. 234 mi/h

Equations   **179**

Chapter Review

Terms to Know

equation (p. 142)
solution of an equation (p. 142)
solve (p. 148)

inverse operations (p. 149)
formula (p. 176)

Choose the correct term from the list above to complete each sentence.

1. Multiplication and division are __?__ because each can be used to undo the other in solving an equation. inverse operations

2. A(n) __?__ is a statement that two numbers or two expressions are equal. equation

3. To __?__ an equation, you find all values of the variable that make the equation true. solve

4. A(n) __?__ is an equation that states a relationship between two or more quantities that are usually represented by variables. formula

5. A value of the variable that makes an equation true is a(n) __?__.
   solution of an equation

Is the given number a solution of the equation? Write *Yes* or *No*.
(Lesson 4-1)

6. $x - 3 = 7$; 10   Yes.

7. $\frac{n}{8} = -2$; 6 No.

8. $9 = t + 7$; 16 No.

9. $36 = -4a$; -9 Yes.

Use mental math to find each solution.   *(Lesson 4-1)*

10. $11 = r + 6$ 5

11. $8m - 4 = -60$ -7

12. $13 = \frac{b}{5} + 3$ 50

13. $x - 2 = -2$ 0

Solve using the guess-and-check strategy.   *(Lesson 4-2)*

14. The prices at Salons Now are \$11 for a haircut and \$34 for a perm. On Tuesday Salons Now collected \$279 from customers who were given haircuts or perms. How many of each were given on Tuesday?
    13 haircuts, 4 perms

15. Malik bought some pens for \$1.25 each. He also bought some notebooks for \$2.75 each. He spent \$14.75. How many of each item did he buy? 3 pens, 4 notebooks

Solve. Check each solution.   *(Lessons 4-3, 4-4, and 4-5)*

16. $r + 12 = 15$ 3

17. $b - 4 = -6$ -2

18. $8 = 13 + y$ -5

19. $14 = a - 9$ 23

20. $-q = 16$ -16

21. $\frac{v}{7} = -2$ -14

22. $15 = \frac{m}{3}$ 45

23. $11x = 132$ 12

24. $1 + 4n = 9$ 2

25. $5 = \frac{w}{3} + 2$ 9

26. $-13 = 5 + 2t$ -9

27. $\frac{a}{-8} - 5 = 6$ -88

Write a variable expression that represents each phrase. If necessary, choose a variable to represent the unknown number. *(Lesson 4-6)*

28. 6 less than 8 times a number q $8q - 6$

29. a number h divided by 19 $\frac{h}{19}$

30. 7 times as many books as Ann read last year $7b$

31. 4 more than twice the number of tickets sold yesterday $2t + 4$

Write an equation that represents the relationship in each sentence. *(Lesson 4-7)*

32. A number z decreased by 15 is 33. $z - 15 = 33$

33. The school debating team has 8 members, which is 5 fewer than the number of students on the school mathematics team. $m - 5 = 8$

34. Jack has 48 trophies, which is three times as many as Mark has. $3x = 48$

Solve using an equation. *(Lesson 4-8)*

35. Beth MacCurdy paid a total of $355 for two identical wool coats, including tax of $17. What was the price of each coat? $169

36. Andrew Morrison takes a group of students on a fishing trip each year. This year they caught a record 27 fish, which is one less than four times as many fish as the group caught last year. How many fish were caught last year? 7 fish

Solve. Check each solution. *(Lesson 4-9)*

37. $-2x - 7x = 18$ -2

38. $6m + 4 + 5m = -29$ -3

39. $4(v - 10) = 24$ 16

40. $110 = 5(1 + 3k)$ 7

Use the formula $C = np$, where C is the total cost, n is the number of items purchased, and p is the price per item. *(Lesson 4-10)*

41. Let $C = \$182$ and $p = \$13$. Find n. 14

42. Let $n = 24$ and $p = \$1.99$. Find C. $47.76

43. Lucas bought 12 posters at $10.50 per poster. How much did he spend? $126

44. Robin spent $60 for 8 tickets. What was the price of each ticket? $7.50

Chapter Test (Form A)
Test 17, *Tests*

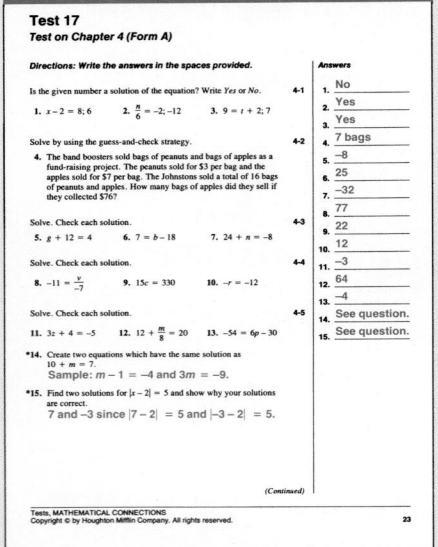

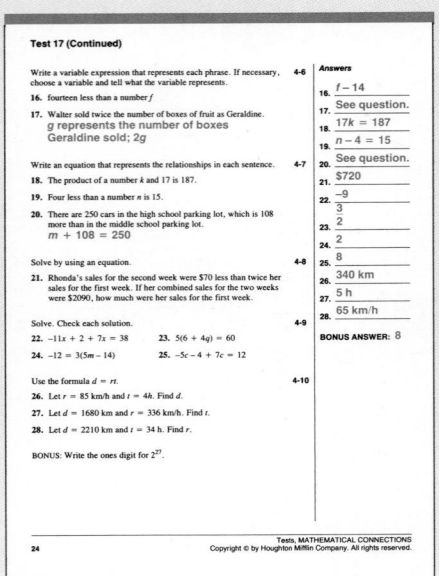

Chapter Test

Is the given number a solution of the equation? Write *Yes* or *No*.

1. $-6x = 18; -3$ Yes. 2. $9 = u - 4; 5$ No. 3. $\frac{a}{7} = -7; 49$ No. **4-1**

Solve using the guess-and-check strategy.

4. Florina has eighteen dimes and nickels in all. Their total worth is $1.40. How many of each type of coin does she have? 10 dimes, 8 nickels **4-2**

Solve. Check each solution.

5. $g + 15 = 7$ -8 6. $9 = b - 17$ 26 7. $28 + n = -10$ -38 **4-3**

8. $-13 = \frac{v}{-5}$ 65 9. $18c = 180$ 10 10. $-r = 20$ -20 **4-4**

11. $3z + 5 = -4$ -3 12. $15 + \frac{m}{8} = 20$ 40 13. $-72 = 6p - 30$ -7 **4-5**

★14. Explain the meaning of *inverse operations*. Give two examples.
    See answer on p. A11.

Write a variable expression that represents each phrase. If necessary, choose a variable to represent the unknown number.

15. eighteen more than four times a number j $4j + 18$ **4-6**

16. ten times the number of boxes Jo sold $10b$

Write an equation that represents the relationship in each sentence.

17. A number n divided by 8 is 90. $\frac{n}{8} = 90$ **4-7**

18. Salim has 15 model cars in his collection, which is 8 fewer model cars than Franco has. $f - 8 = 15$

★19. Explain the difference between an equation and an expression.
    See answer on p. A11.

Solve using an equation.

20. Ali's mathematics textbook contains 690 pages. This is 248 pages less than twice the number of pages in Ali's science textbook. How many pages are in the science textbook? 469 **4-8**

Solve. Check each solution.

21. $-12x + 3 + 5x = 38$ -5 22. $6(8 + 3q) = 66$ 1 23. $-10 = 5(7m - 16)$ 2 **4-9**

Use the formula $g = a - h$.

24. Let $a = 720$ km/h and $h = 15$ km/h. Find g. 705 km/h **4-10**

25. Let $g = 650$ mi/h and $h = 25$ mi/h. Find a. 675 mi/h

Equations Having Variables on Both Sides

Objective: To solve equations having variables on both sides.

Some equations have the variable on both sides. To solve, you need to get the variable alone on one side.

Example

Solution

Solve $5a - 9 = 2a - 3$. Check the solution.

Subtract a variable term from both sides.

$$5a - 9 = 2a - 3$$
$$5a - 9 - 2a = 2a - 3 - 2a$$
$$3a - 9 = -3$$
$$3a - 9 + 9 = -3 + 9$$
$$3a = 6$$
$$\frac{3a}{3} = \frac{6}{3}$$
$$a = 2$$

The solution is 2.

 **Check**

$$5a - 9 = 2a - 3$$
$$5(2) - 9 \stackrel{?}{=} 2(2) - 3$$
$$10 - 9 \stackrel{?}{=} 4 - 3$$
$$1 = 1$$

Exercises

Solve. Check the solution.

1. $7n + 10 = 3n + 2$ -2
2. $5h - 7 = 2h + 2$ 3
3. $4v - 9 = 2v + 1$ 5
4. $9x - 4 = 6x + 23$ 9
5. $3t + 12 = 2t - 14$ -26
6. $2y + 2 = y - 8$ -10
7. $5a = 2a + 18$ 6
8. $8u = 6u - 20$ -10
9. $-5 + 12v = 11v - 7$ -2
10. $-9 + 16g = 5g + 13$ 2
11. $1 + 9h = 4h + 11$ 2
12. $8 + 32t = 31t + 17$ 9
13. $-5p = 2p + 14$ -2
14. $-7g = 2g + 36$ -4
15. $5a - 7 = -3a + 17$ 3
16. $9b + 7 = -6b + 52$ 3
17. $2v + 7 = 4v - 19$ 13
18. $8x + 17 = 9x - 8$ 25

GROUP ACTIVITY Answers to Exercises 19–20 are on p. A11.

19. The first step in solving the Example above involved subtracting a variable term from both sides of the equation. What other first step could have been taken to solve the equation?

20. Describe four different ways to begin solving the equation $3x + 5 = 8x - 10$.

Equations **183**

Quick Quiz 2 (4-5 through 4-8)

Solve. Check each solution.

1. $4 + 5m = -21$ $m = -5$
2. $x \div 24 = 3$ $x = 72$
3. $56 = 10b - 44$ $b = 10$

Write a variable expression that represents each phrase.

4. a number c divided by seventeen
$\frac{c}{17}$

5. six times as high as the department store $6d$

Write an equation that represents the relationships in each sentence.

6. The sum of a number d and six is 47. $d + 6 = 47$

7. Eduardo has eighty-eight photos, which is twenty-four fewer than Sylvia took on the trip.
$88 = s - 24$

Solve by using an equation.

8. Suzanne is 13 cm shorter than Tony. Suzanne is 167 cm tall. How tall is Tony?
$167 = t - 13; t = 180;$
Tony is 180 cm tall.

9. Tom is planning to visit his grandmother. She lives 700 mi away. The family plans to drive 300 mi the first day. They plan to travel an equal distance on four other days so that they can sightsee. How many miles does the family plan to travel on the third day? $700 = 300 + 4d;$
$d = 100;$ the family plans to travel 100 mi on the third day.

Alternative Assessment

Have each student make up two word problems that can be solved by using equations. Have students exchange their problems. Ask a few students to explain how to solve each problem using the four-step plan. Then have all students solve the problems.

Standardized Testing Practice

Choose the letter of the correct answer.

1. **Which equation could you use to solve the problem?**
   The Tigers' score was 35 points less than twice that of the Colonials. The Tigers scored 74 points. How many points did the Colonials score?
   A. $2c - 35 = 74$
   B. $2c + 35 = 74$
   C. $2c - 74 = 35$
   D. $74 - 2c = 35$

2. **Solve:** $16 = \frac{t}{4} + 4$
   A. $t = 80$ B. $t = 60$
   C. $t = 192$ D. $t = 48$

3. **Which number is to the left of -6 on a number line?**
   A. -8 B. 0
   C. $|-6|$ D. 7

4. **Write the phrase *7 less than 4 times a number* as a variable expression.**
   A. $7 - 4n$
   B. $7 < 4n$
   C. $4n - 7$
   D. $4n < 7$

5. **Evaluate $24 - a$ when $a = -6$.**
   A. 18   B. 30
   C. -18 D. -4

6. **Eldon bought 3 apples at 45¢ each, 2 pears at 38¢ each, and a juice drink for 65¢. He gave the cashier $5. Which question can be answered?**
   A. How much did he spend?
   B. How much change did he get?
   C. How many items did he buy?
   D. all of the above

7. **Solve:** $-13n = 52$
   A. $n = -4$ B. $n = 4$
   C. $n = -676$ D. $n = 676$

8. **Evaluate the quotient $41.6 \div z$ when $z = 8$.**
   A. 52   B. 5.2
   C. 332.8   D. 33.28

9. **How would you move the decimal point to change 47.5 mm to m?**
   A. 2 places to the right
   B. 2 places to the left
   C. 3 places to the right
   D. 3 places to the left

10. **Evaluate $(2b)^3$ when $b = 4$.**
    A. 512   B. 128
    C. 32   D. 216

11. In four weeks, Emma worked 35 h, 42 h, 22 h, and 18 h. Choose the best estimate of her total hours.

   A. about 80 h
   B. about 100 h
   C. about 120 h
   D. about 140 h

16. Complete: If $x + a = b$, then $x =$ ___?___.

   A. $b - a$
   B. $a - b$
   C. ab
   D. $\frac{b}{a}$

12. Solve: $2(y + 6) = -24$

   A. $y = -15$ B. $y = -18$
   C. $y = 18$ D. $y = -6$

17. Solve: $\frac{x}{-16} = 8$

   A. $x = 128$ B. $x = -128$
   C. $x = 2$ D. $x = -2$

13. Find the function rule.

   A. $x \rightarrow 5x - 4$
   B. $x \rightarrow 3x + 6$
   C. $x \rightarrow 5x + 1$
   D. $x \rightarrow 4x + 2$

x	?
1	6
2	10
3	14
5	22
6	26

18. A rectangle has perimeter 28 cm and width 3 cm. Use the formula $P = 2l + 2w$ to find the length of the rectangle.

   A. 25 cm   B. 22 cm
   C. 17 cm   D. 11 cm

14. Find the sum: $-12 + (-18)$

   A. -6 B. 6
   C. -30 D. 30

19. Solve: $15 - 4y = 3$

   A. $y = 3$ B. $y = -3$
   C. $y = 4.5$ D. $y = -4.5$

15. Of which equation(s) is -5 a solution?

   I. $x - 1 = 4$ II. $12 - x = 17$
          III. $3x = -15$

   A. I and II only
   B. I and III only
   C. II and III only
   D. I, II, and III

20. Which statement shows the commutative property of addition?

   A. $a + (b + c) = (a + b) + c$
   B. $a(b + c) = ab + ac$
   C. $a + (b + c) = a + (c + b)$
   D. $a(b + c) = (b + c)a$

Planning Chapter 5

Chapter Overview

Chapter 5 serves as an introduction to the study of probability and statistics, topics that will be examined in greater depth later in the book. The content of the chapter provides an interesting connection between mathematics and the real world; graphs are used in many places, such as magazines, newspapers, research journals, and daily news reports. Bulletin board displays or graphical information that is readily available can motivate students to learn the content of the chapter.

Lesson 1 begins by introducing students to pictographs. *Lessons 2 and 3* focus on the interpretation of bar graphs and line graphs. *Lesson 4* extends those skills by having students draw their own graphs. In *Lessons 5 and 6*, students must rely more strongly on their reasoning ability and problem-solving skills: in *Lesson 5*, they choose the most appropriate type of graph to represent data, and in *Lesson 6*, they analyze situations for too much or not enough information. In *Lesson 7*, students study misleading graphs. *Lesson 8* presents the ideas of mean, median, mode, and range. The decision-making strand continues in *Lesson 9*, as students choose the most appropriate statistic to represent a set of data. In *Lesson 10*, students construct frequency tables. Technology and applications in the business world are addressed in the *Focus on Computers* and *Focus on Applications*.

Background

Chapter 5 introduces graphs immediately. Students often feel a sense of security when working with graphical information. Basic arithmetic skills are included in the context of the lesson, as students determine intervals and find the measures of central tendency. Logical reasoning is developed as students analyze data and draw conclusions.

One problem-solving strategy is to create a table or graph. The content of this chapter promotes the development of this strategy. The decision-making strand is further developed by requiring students to analyze data, determine whether the data items are independent, or whether they represent information that is changing over time. This analysis is used to help choose the most appropriate type of graph as a representation of the data. Students also examine data for extremes that distort the measures of central tendency, which leads to a decision as to the measure that is most representative of the data.

Technology is included through the use of database software on the computer and the use of calculators when finding measures of central tendency. If computers and appropriate software are available, you may want to demonstrate how spreadsheets can be used to generate graphs of varying types and how changing the parameters impacts the presentation of data.

The *Focus on Explorations* uses a paper ruler and paper clips to explore the definition of the mean, median, and mode. This manipulative activity demonstrates how changing a single number affects the results.

Objectives

5-1 To interpret and draw pictographs.

5-2 To interpret single and double bar graphs.

5-3 To interpret single and double line graphs.

5-4 To draw a bar or line graph for a given set of data.

5-5 To decide whether it would be more appropriate to draw a bar graph or a line graph to display a set of data.

5-6 To solve problems involving too much or not enough information.

5-7 To recognize how bar graphs and line graphs can be misleading.

5-8 To find the mean, median, mode, and range of a set of data.

FOCUS ON EXPLORATIONS To build a concrete model to represent the mean of a set of numbers.

5-9 To decide whether a given measure of central tendency is appropriate for a set of data.

FOCUS ON COMPUTERS To use database software to analyze a set of data.

5-10 To calculate statistical measures for data in a frequency table.

FOCUS ON APPLICATIONS To analyze profits and losses using a double line graph.

CHAPTER EXTENSION To recognize and use statistical symbols associated with the mean.

Chapter Planner

Instructional Aids	Manipulatives	Cooperative Learning	Technology	Practice/Reteaching	Assessment
Materials: cardboard, ruler, hole punch, string, tape, paper clips, colored candy or cereal, calculator *Resource Book:* Lesson Starters 40–49	Lessons 5-2, 5-8, Focus on Explorations *Activities Book:* Manipulative Activities 9–10	Lessons 5-1, 5-4, 5-6, 5-7, 5-9, Focus on Applications *Activities Book:* Cooperative Activities 10–11	Lessons 5-4, 5-5, 5-7, 5-8, Focus on Computers *Activities Book:* Calculator Activity 4 Computer Activity 6 *Connections Plotter Plus* Disk	Toolbox, pp. 753–771 Extra Practice, p. 735 *Resource Book:* Practice 54–67 *Study Guide,* pp. 79–98	Self-Tests 1–3 Quick Quizzes 1–3 Chapter Test, p. 229 *Resource Book:* Lesson Starters 40–49 (Daily Quizzes) *Tests 21–25*

Assignment Guide Chapter 5

Day	Skills Course	Average Course	Advanced Course
1	**5-1:** 1–8, 10, 12, 17–23	**5-1:** 1–12, 17–23	**5-1:** 1–7 odd, 9–23
2	**5-2:** 1–12, 13–21 odd, 25, 26	**5-2:** 1–22, 25, 26	**5-2:** 1–11 odd, 13–26
3	**5-3:** 1–24, 33–36	**5-3:** 1–20, 21–29 odd, 33–36	**5-3:** 1–19 odd, 21–36
4	**5-4:** 1–5, 11–14	**5-4:** 1–8, 11–14	**5-4:** 1–14
5	**5-5:** 1–8	**5-5:** 1–8	**5-5:** 1–8
6	**5-6:** 1–14	**5-6:** 1–14	**5-6:** 1–14, Historical Note
7	**5-7:** 1–9, 14–17	**5-7:** 1–9, 14–17	**5-7:** 1–17
8	**5-8:** 1–6, 7–19 odd, 23–25 **Exploration:** Activities I–II	**5-8:** 1–6, 7–19 odd, 23–25 **Exploration:** Activities I–II	**5-8:** 1–11 odd, 13–25, Challenge **Exploration:** Activities I–II
9	**5-9:** 1–11 **Computers:** 1–7	**5-9:** 1–11 **Computers:** 1–9	**5-9:** 1–11 **Computers:** 1–9
10	**5-10:** 1–4, 9–12 **Application:** 1–10	**5-10:** 1–6, 9–12 **Application:** 1–14	**5-10:** 1–12 **Application:** 1–14
11	*Prepare for Chapter Test:* Chapter Review	*Prepare for Chapter Test:* Chapter Review	*Prepare for Chapter Test:* Chapter Review
12	*Administer Chapter 5 Test*	*Administer Chapter 5 Test; Cumulative Review*	*Administer Chapter 5 Test; Chapter Extension; Cumulative Review*

Teacher's Resources

Resource Book
Chapter 5 Project
Lesson Starters 40–49
Practice 54–67
Enrichment 44–54
Diagram Masters
Chapter 5 Objectives
Family Involvement 5
Spanish Test 5
Spanish Glossary

Activities Book
Cooperative Activities 10–11
Manipulative Activities 9–10
Calculator Activity 4
Computer Activity 6

Study Guide, pp. 79–98

Tests
Tests 21–25

Connections Plotter
Plus Disk

Alternate Approaches Chapter 5

Using Technology

CALCULATORS

Calculators can be used most profitably in this chapter to explore the elementary statistics introduced in Lesson 8.

COMPUTERS

Spreadsheet and statistics programs, as well as some database programs, can be used to explore bar and line graphs. The difference between the graphs produced by powerful spreadsheet programs and those that statistics programs usually generate can lead to fruitful discussions with students.

Lessons 5-4, 5-5

COMPUTERS

There are many computer programs that can produce bar and line graphs. In particular, many of the more powerful spreadsheet programs include options for drawing graphs of the data in a user-selected range of cells. Even if the spreadsheet programs available to you do not have this feature, there are many business-oriented graphics programs that will draw graphs based on data imported from a spreadsheet program. The ease with which graphs are produced by these programs provides students with an opportunity to explore the advantages and disadvantages of each type of graph.

Lesson 5-8

COMPUTERS

Statistics programs, including such general mathematical exploration software as Houghton Mifflin's *Connections Plotter Plus*, allow students to discover some of the properties of the mean, median, mode, and range. These programs allow the user to input a sequence of numbers (for example, the heights of the members of a class) and then calculate elementary statistics for the data. They usually include the standard deviation, which students could explore, although details might be difficult for them at this time. These programs often will draw graphs based on the data and in a form different from that used by spreadsheet programs. Rather than plotting each data point as the height of a bar or of a point on a line graph, they can draw frequency diagrams, which show the number of data points that fall into each range of values. This is an important difference, and appreciating it can help students understand more about statistics.

CALCULATORS

Calculators can provide good practice with simple statistics and discovery of simple properties. Students can explore what happens to the mean, median, mode, and range, for example, when each piece of data is added to, or multiplied by, the same constant.

Using Manipulatives

Lesson 5-2

Bar graphs can be made by counting cereal or candies which come in a variety of colors. Have students work in cooperative groups. Give each group one package of candy or a specified amount, such as one-half cup, of cereal. Each group should develop a bar graph so that the bars compare like colors.

After making the bar graph, each group should develop three questions to share with the class. At least one of the questions should apply to the results of the entire class, such as "Which color appears most (or least) often?"

Lesson 5-8

Grid paper and cereal may be used to provide a representation of the mean. Make a graph using grid paper, placing one piece of cereal in each square of the graph to make bars of varying lengths. The mean, or average length, can be found by removing cereal from the longer bars and adding them to the shorter bars. This process is repeated until the bars are as close to the same length as possible. The mean can be read directly from the graph. The median can be found by arranging the bars in order from least to greatest and selecting the "middle" length. The mode can be identified by selecting the length(s) that appear(s) most often.

Reteaching/Alternate Approach

Lesson 5-1

Students enjoy creating pictographs based on their own data. For example, they could create "birthday" pictographs based on the birthdays of the class. Other topics might include the colors of family cars, number of brothers and sisters, and so on.

Lesson 5-4

COOPERATIVE LEARNING

Divide students into cooperative groups. Ask each group to examine the data presented in Guided Practice Exercises 2 or 5, or Exercises 1–4. Each group should develop four questions that can be answered by using the graph and two problems that cannot be answered from the graph. For example, in a bar graph relating the number of adults participating in leisure-time activities (Exercise 1), the question "Which sport has fewer adult participants than in prior years?" cannot be answered.

Lesson 5-5

Show the class a variety of bar and line graphs you have collected from newspapers and magazines. Discuss the similarities and differences among the data presented. Ask questions such as the following: What common characteristics do you see on the bar graphs? When comparing different values on line graphs, does the intermediate value have any meaning? What differences can you find between the information presented on a bar graph and that presented on a line graph? After the class discussion, you may wish to stress the similarities and differences noted in the text.

Lesson 5-6

COOPERATIVE LEARNING

Have students bring in articles from magazines or newspapers that contain a variety of facts and figures. Divide the students into cooperative learning groups and ask each group to develop a series of problems from the information in their articles. The development of a graph should facilitate the solution to the problem(s) posed. Students should present a table of facts, a graph, problem(s), and identification of extraneous information or missing information.

Lesson 5-7

COOPERATIVE LEARNING

Have students bring in graphs from newspapers and magazines. In cooperative groups, students can analyze the graphs as to whether they are misleading or not. Groups should explain why they think a graph is misleading and then redraw any misleading graphs so as to present the data correctly.

Lesson 5-8

Arrange students in order of height from shortest to tallest. You can then visually pick out the student who represents the median height of the class. The range is determined by the difference in heights between the tallest and the shortest students. If several people are the same height, that most frequent measure would represent the mode. Students should find the mean of the heights of the class and then check to see if anyone in the class is that height.

Lesson 5-9

COOPERATIVE LEARNING

Working in cooperative groups, students should collect data about the members of the group. This could include heights, eye colors, favorite foods, number of brothers and sisters, and so on. Students should then decide which measure of central tendency best describes each set of data.

Lesson 5-10

Students can use the data collected from their cooperative groups in Lesson 5-9 to create frequency tables for the class as a whole.

Teacher's Resources Chapter 5

Enrichment masters from the Resource Book are pictured here. See the Teacher's Resources chart on page 186B for a complete listing of all materials available for this chapter.

5-1 Enrichment
Communication

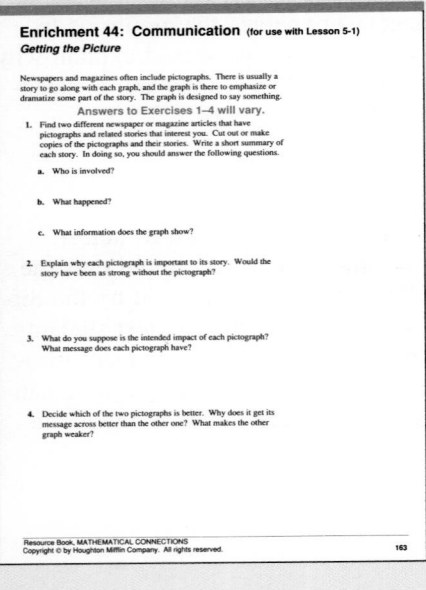

Enrichment 44: Communication (for use with Lesson 5-1)
Getting the Picture

Newspapers and magazines often include pictographs. There is usually a story to go along with each graph, and the graph is there to emphasize or dramatize some part of the story. The graph is designed to say something.

Answers to Exercises 1–4 will vary.

1. Find two different newspaper or magazine articles that have pictographs and related stories that interest you. Cut out or make copies of the pictographs and their stories. Write a short summary of each story. In doing so, you should answer the following questions.
   a. Who is involved?
   b. What happened?
   c. What information does the graph show?

2. Explain why each pictograph is important to its story. Would the story have been as strong without the pictograph?

3. What do you suppose is the intended impact of each pictograph? What message does each pictograph have?

4. Decide which of the two pictographs is better. Why does it get its message across better than the other one? What makes the other graph weaker?

Resource Book, MATHEMATICAL CONNECTIONS
Copyright © by Houghton Mifflin Company. All rights reserved.
163

5-2 Enrichment
Data Analysis

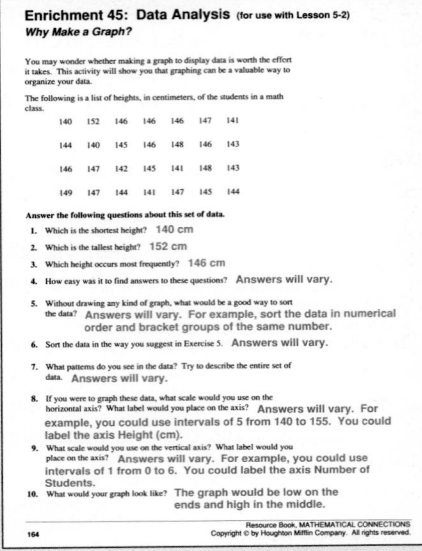

Enrichment 45: Data Analysis (for use with Lesson 5-2)
Why Make a Graph?

You may wonder whether making a graph to display data is worth the effort it takes. This activity will show you that graphing can be a valuable way to organize your data.

The following is a list of heights, in centimeters, of the students in a math class.

140	152	146	146	146	147	141
144	140	145	146	148	146	143
146	147	142	145	141	148	143
149	147	144	141	147	145	144

Answer the following questions about this set of data.

1. Which is the shortest height? **140 cm**
2. Which is the tallest height? **152 cm**
3. Which height occurs most frequently? **146 cm**
4. How easy was it to find answers to these questions? **Answers will vary.**
5. Without drawing any kind of graph, what would be a good way to sort the data? **Answers will vary. For example, sort the data in numerical order and bracket groups of the same number.**
6. Sort the data in the way you suggest in Exercise 5. **Answers will vary.**
7. What patterns do you see in the data? Try to describe the entire set of data. **Answers will vary.**
8. If you were to graph these data, what scale would you use on the horizontal axis? What label would you place on the axis? **Answers will vary. For example, you could use intervals of 5 from 140 to 155. You could label the axis Height (cm).**
9. What scale would you use on the vertical axis? What label would you place on the axis? **Answers will vary. For example, you could use intervals of 1 from 0 to 6. You could label the axis Number of Students.**
10. What would your graph look like? **The graph would be low on the ends and high in the middle.**

164
Resource Book, MATHEMATICAL CONNECTIONS
Copyright © by Houghton Mifflin Company. All rights reserved.

5-3 Enrichment
Exploration

Enrichment 46: Exploration (for use with Lesson 5-3)
How Long Does a Penny Last?

Many people collect pennies. You get a lot of pennies in change when you buy things, but there is very little you can buy with a penny. When you empty your pockets at night the pennies might go into one pile and the more valuable coins into another. In the morning you might take the quarters, the dimes, and the nickels with you, and leave the pennies at home. The pile of pennies would get bigger every day. **Answers to Exercises 1–10 will vary.**

Project

1. Gather all the pennies you can find. (You might borrow some from your parents or combine your pennies with a classmate's.) The more you have, the better.

2. Sort the pennies according to the year they were produced, or *minted*.

3. Make a pile for each year, and put the piles in order, from the earliest on the left to the latest on the right.

4. Make a sketch of your row of stacks.

5. Make a list of all the years represented in your collection, from the earliest to the latest.

6. For each year in your list, write down the number of pennies from that year.

7. What year has the most pennies? (There may be a tie!) The number of pennies in that year is your *maximum*.

8. In what year (or years) do the fewest pennies appear? The number in that year is your *minimum*.

9. In what year (or years) is the number of pennies about one half of the *maximum*?

10. Compare your results with those of your classmates. Are your sketches the same? In what years do most students find their maximum, minimum, and half-maximum values? On a separate piece of paper, write a paragraph about your results.

Resource Book, MATHEMATICAL CONNECTIONS
Copyright © by Houghton Mifflin Company. All rights reserved.
165

5-4 Enrichment
Application

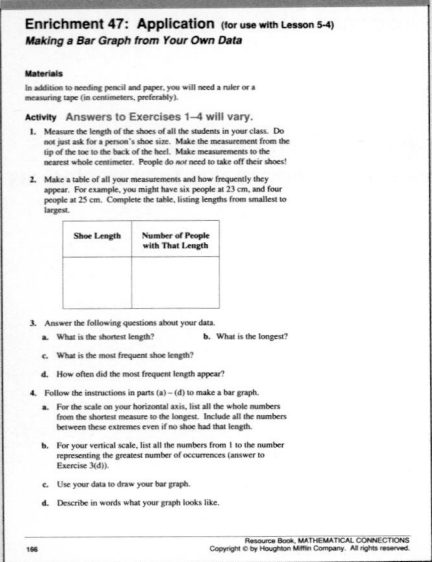

Enrichment 47: Application (for use with Lesson 5-4)
Making a Bar Graph from Your Own Data

Materials

In addition to needing pencil and paper, you will need a ruler or a measuring tape (in centimeters, preferably).

Activity Answers to Exercises 1–4 will vary.

1. Measure the length of the shoes of all the students in your class. Do not just ask for a person's shoe size. Make the measurement from the tip of the toe to the back of the heel. Make measurements to the nearest whole centimeter. People do *not* need to take off their shoes!

2. Make a table of all your measurements and how frequently they appear. For example, you may have six people at 23 cm, and four people at 25 cm. Complete the table, listing lengths from smallest to largest.

Shoe Length	Number of People with That Length

3. Answer the following questions about your data.
   a. What is the shortest length? b. What is the longest?
   c. What is the most frequent shoe length?
   d. How often did the most frequent length appear?

4. Follow the instructions in parts (a) – (d) to make a bar graph.
   a. For the scale on your horizontal axis, list all the whole numbers from the shortest measure to the longest. Include all the numbers between these extremes even if no shoe had that length.
   b. For your vertical scale, list all the numbers from 1 to the number representing the greatest number of occurrences (answer to Exercise 3d).
   c. Use your data to draw your bar graph.
   d. Describe in words what your bar graph looks like.

166
Resource Book, MATHEMATICAL CONNECTIONS
Copyright © by Houghton Mifflin Company. All rights reserved.

5-5 Enrichment
Communication

Enrichment 48: Communication (for use with Lesson 5-5)
When There Isn't Just One Correct Answer

There are times in mathematics when there is not just one answer to a problem. What is the best price to charge for a school dance, for example? This question could be hotly debated, with many possible answers each justifiable for different reasons.

The best way to display data can also be debatable. Your textbook suggests, correctly, that bar graphs are best when data "fall into distinct categories and you want to compare the totals." A line graph is best when "you want to emphasize trends in data that change continuously over time." These are good general guidelines.

There may be times, however, when it is best to disregard these guidelines for a particular reason. There may be an aspect of the data that you want to emphasize to make your point. There may be a special audience that has an interest in one way of looking at the data. One group of people may be interested in general trends, while another group may want to compare categories. If you graphed enrollment in your school by class, for example, students would probably want to know exactly how many students were in each class. Parents and school board members, who have to plan for changes in the school, might be more interested in general trends.

In any case, when you want to go against generally accepted guidelines, you need to justify your case. You need to explain yourself fully.

Activity

1. Choose a graph in Chapter 5 of your textbook. Cut out a graph that interests you in a newspaper or magazine. For both of these graphs answer the following questions.
   a. What audience might be most interested in the graph as it is?
   b. Think of a different way of displaying the data. Is this a reasonable presentation of the data? Justify your answer.
   c. Draw your new graph. **Answers will vary. Students should clearly justify their ideas by identifying target audiences and their interests.**

Resource Book, MATHEMATICAL CONNECTIONS
Copyright © by Houghton Mifflin Company. All rights reserved.
167

5-6 Enrichment
Problem Solving

Enrichment 49: Problem Solving (for use with Lesson 5-6)
Too Much or Too Little Information in a Graph

Graphs in newspapers and magazines can have more information than you need to answer your questions. Graphs can also raise questions that they do not answer. It is a good idea to be aware of what graphs do and do not tell you. **Answers to Exercises 1–8 will vary.**

What To Do

Look in a newspaper or magazine to find a graph that interests you. Cut it out or make a copy of it.

1. Explain in a general way what the graph is about.

2. Write a question that you *can* answer with the information in the graph.

3. Identify the facts in the graph that you will need to answer your question.

4. What is the answer to the question?

5. Identify any information in the graph that you did *not* use.

6. Write a question that is related to the graph but would require more information.

7. What are some possible answers to your question?

8. What other information would you need?

168
Resource Book, MATHEMATICAL CONNECTIONS
Copyright © by Houghton Mifflin Company. All rights reserved.

5-7 Enrichment
Application

Enrichment 50: Application (for use with Lesson 5-7)
When Numbers Can Be Misleading

Graphs can sometimes be misleading because pictures can be distorted and give the wrong impression about the data they represent.

Numbers can be misleading, too. That may be hard to believe, but it can easily happen.

Many packaged foods are labeled with nutritional information about the vitamins, calories, fat, cholesterol, and other aspects of the food in the package. People use this information to compare the nutritional values of various brands of the same food products. It is sometimes hard to understand what the numbers really mean. Here are partial listings for two breakfast foods.

Food A (per serving)	
calories	240
protein	24 grams
carbohydrates	3 grams
fat	15 grams
cholesterol	45 milligrams
serving size	3 oz (84 grams)

Food B (per serving)	
calories	210
protein	9 grams
carbohydrates	6 grams
fat	16 grams
cholesterol	0 milligrams
serving size	2 tablespoons

1. Which product, Food A or Food B, appears to be the one with more fat and calories? **The products appear to contain about the same amount of fat and calories.**

2. Which one has more cholesterol? **Food A**

3. Why is it difficult to compare these two products fairly? **The serving sizes differ.**

4. Given that 1 tablespoon of Food B weighs 16 grams, how many calories are there per gram of Food A? Food B? **6.6 cal/g; 2.9 cal/g**

5. Do your answers to Exercises 1 and 4 seem to agree or disagree? Why do you think this is? **disagree; the serving size for Food A is three times as much as for Food B.**

6. Find tables similar to the ones above on the labels of two food products you use. What do you notice about the serving sizes of the products? Are they reasonable? Explain your answer. **Answers will vary. Serving sizes are probably small.**

169

5-8 Enrichment
Connection

Enrichment 51: Connection (for use with Lesson 5-8)
A Visual Look at Mean, Median, Mode, and Range

This activity will give you a physical model of the ideas of mean, median, mode, and range.

Materials
You will need scissors and strips of paper. (The edges pulled off computer paper are excellent for this.)

What To Do
Cut nine strips of paper into varying lengths. Try to make some of them the same length. Some of your strips might look like this:

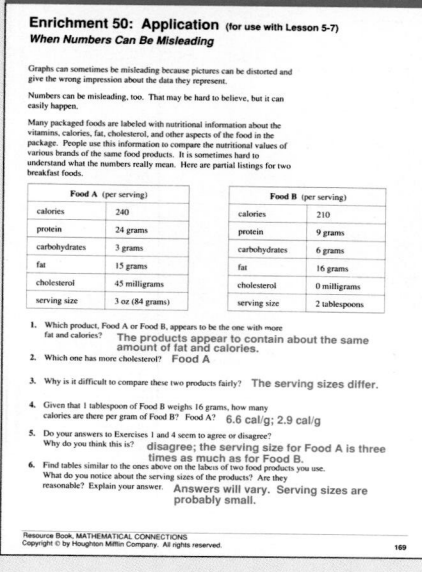

Now arrange the strips in order from smallest to largest.

1. How can you tell what the range is? **The range is the difference in length between the longest piece and the shortest.**

2. How can you tell the mode? Is there more than one mode? **The mode is the most frequent length; Answers will vary.**

3. How can you tell the median? **It is the length of the strip in the middle.**

4. Tape all the pieces together end to end. What does the length of your new long strip represent? **the sum of the lengths**

5. Fold the long strip so there are nine equal lengths. What does each length represent? **the mean**

6. If you had started with an even number of pieces, how would you have found the median? **by taping the two middle pieces together and folding the resulting strip in half**

5-9 Enrichment
Thinking Skills

Enrichment 52: Thinking Skills (for use with Lesson 5-9)
When the Mean Doesn't Mean Much

As you saw in Lesson 5-9, the mean can often give a distorted picture of the data it is supposed to represent. For example, consider the following numbers:

30	4	21	56	1,800,000

1. What is the mean? **360,022.2**

2. What number in the list is *closest* to the mean? **56**

3. What is the range? **1,799,996**

4. How much information does the mean give you? **very little information**

An Example from Business
Here is the salary scale for Sisitzky Sales Candy Company.

Owner	$120,000	Sales Clerk	$19,000	Secretary	$16,000
Sales Manager	$32,000	Sales Clerk	$19,000	Part-time	$9000
Office Manager	$32,000	Sales Clerk	$19,000	Part-time	$9000
Sales Clerk	$19,000	Secretary	$16,000	Part-time	$9000
Sales Clerk	$19,000	Secretary	$16,000	Part-time	$9000

5. Find the mean salary in this company. **$24,200**

6. What is the median salary? **$19,000**

7. What is the mode? **$19,000**

8. Tell which measure of central tendency (mean, mode, or median) describes these data best. **the median and the mode**

171

5-10 Enrichment
Data Analysis

Enrichment 53: Data Analysis (for use with Lesson 5-10)
This Week's Top Five

You may have wondered how the "TOP 40" songs for a particular week are selected. The process involves looking at record sales and asking radio stations about requests they receive. It is a fairly complicated process.

You can do your own poll of your classmates. Ask your classmates to choose five of their favorite songs. You will get some interesting and reliable results.

Ask your teacher to make enough copies of this form for your class.

The Top Five Songs of the Week

Song Title	Singer or Group Name
1.	
2.	
3.	
4.	
5.	

1. Have everyone in your class fill out a form with his or her favorite songs in order of preference. Encourage people not to discuss their choices with one another. You will get more reliable data that way. **Answers to Exercises 1–3 will vary.**

2. Make a list of all the songs that are named on the forms. Give a song 5 points for each time it gets first place, 4 points for each second place, and so forth. Each fifth place will be worth 1 point.

3. Add the points for each song and put the songs in numerical order. The songs with the highest five scores are your TOP 5. You can repeat this in a few weeks if there is interest in doing so.

End of Chapter Enrichment
Extension

Enrichment 54: Extension (for use after Chapter 5)
Give Me Five!

Find five numbers for each of the following puzzles. All of the numbers are integers. (Remember that integers can be negative!)

1. The range is 40.
   There is no mode.
   The median is also the mean.
   When put in order, the numbers increase by 10 every time.
   The smallest number is one fifth of the largest number.

   ☐ ☐ ☐ ☐ ☐ **10; 20; 30; 40; 50**

2. The median is the mode.
   The sum of the numbers is 25.
   The range is 0.

   ☐ ☐ ☐ ☐ ☐ **5; 5; 5; 5; 5**

3. The median is twice the mode.
   The mean is one more than the median.
   The sum of all the numbers is 15.
   The mean is one of the five numbers.

   ☐ ☐ ☐ ☐ ☐ **1; 1; 2; 3; 8**

4. The median and the mean are both 0.
   The mode is greater than the mean.
   The sum of the absolute values of all the numbers is 20.
   The smallest number is 4 less than the next smallest.

   ☐ ☐ ☐ ☐ ☐ **−7; −3; 0; 5; 5**

173

For Chapter 5 Opener

Students may be aware that our modern Olympic games are named after similar contests held at Olympia in ancient Greece. Games were held in other places as well—Delphi, Corinth, and Argolis—but the most important were those held every four years at Olympia. The Olympic games were so important that, in about the year 300 B.C., the Greeks began to base their calendar on their occurrence. In this calendar system, the four-year period between Olympic games was called an Olympiad. The first Olympiad was dated from the first recorded Olympic games, which were held in 776 B.C. People marked time by Olympiads even after the Olympic games were discontinued in 394 A.D. After the 304th Olympiad (about 440 A.D.), the Greek calendar system fell out of use.

For Page 193

In Exercises 2–11, students can point out on a world map each of the countries shown in the graph. Students can also identify the most and least populous areas in each country and conjecture as to why certain areas have either large or small populations.

For Page 208

For the Historical Note, students can consider other women who have made contributions to the fields of health and medicine. For example, Alice Hamilton (1869–1970) was a physician who led early efforts to protect workers from hazardous materials in the workplace. The physician Helen Taussig (1898–1986) discovered that a condition affecting many babies (an oxygen deficiency in the blood that gave the skin a bluish cast) was due to a congenital heart defect. She suggested an operation that has saved the lives of many infants.

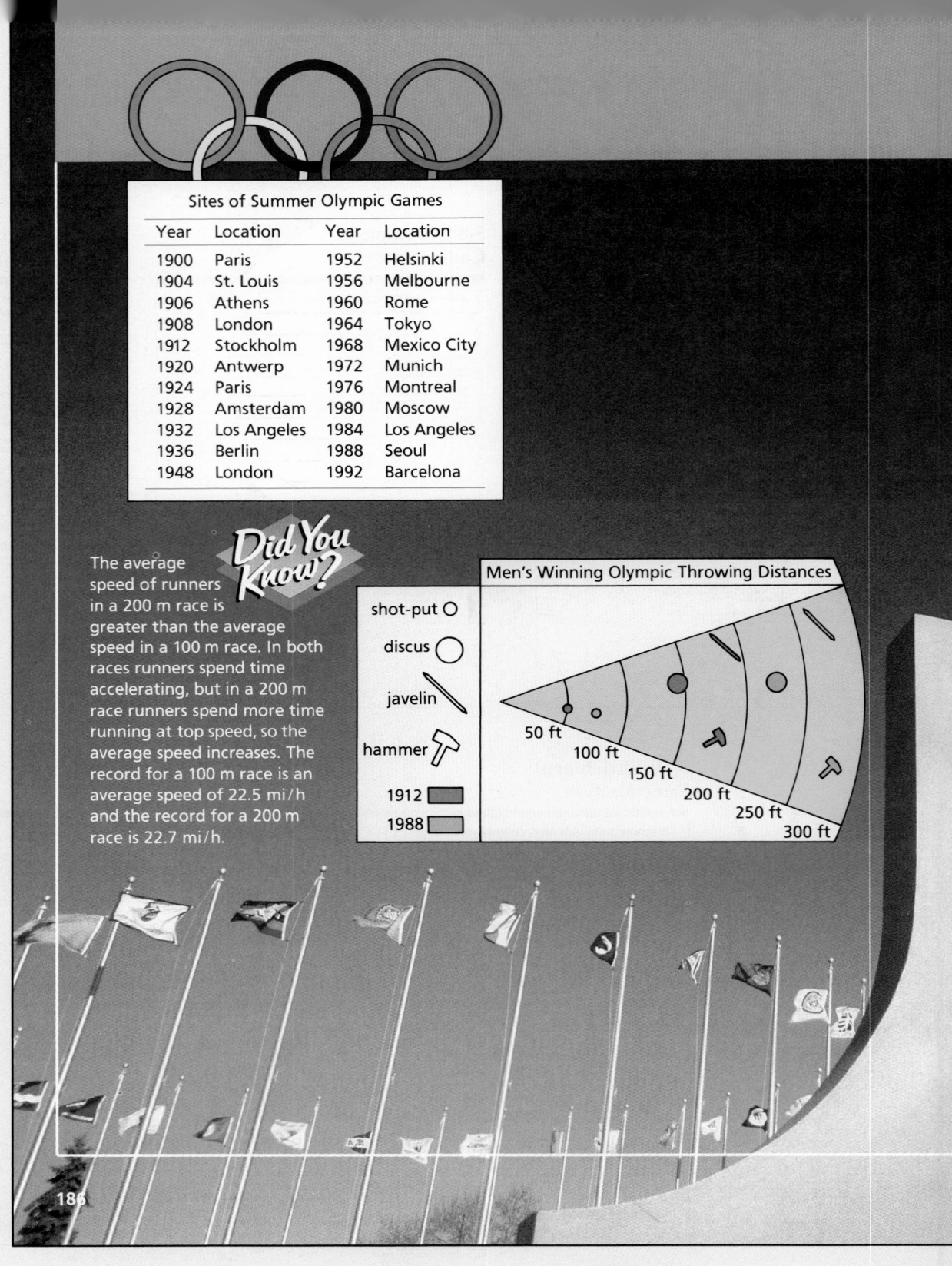

Sites of Summer Olympic Games

Year	Location	Year	Location
1900	Paris	1952	Helsinki
1904	St. Louis	1956	Melbourne
1906	Athens	1960	Rome
1908	London	1964	Tokyo
1912	Stockholm	1968	Mexico City
1920	Antwerp	1972	Munich
1924	Paris	1976	Montreal
1928	Amsterdam	1980	Moscow
1932	Los Angeles	1984	Los Angeles
1936	Berlin	1988	Seoul
1948	London	1992	Barcelona

Did You Know?

The average speed of runners in a 200 m race is greater than the average speed in a 100 m race. In both races runners spend time accelerating, but in a 200 m race runners spend more time running at top speed, so the average speed increases. The record for a 100 m race is an average speed of 22.5 mi/h and the record for a 200 m race is 22.7 mi/h.

Men's Winning Olympic Throwing Distances

shot-put
discus
javelin
hammer
1912
1988

50 ft
100 ft
150 ft
200 ft
250 ft
300 ft

Graphs and Data Analysis 5

Olympic Medals Won by Countries in Track and Field (1988)		
Medals Won	Tally	Frequency
1	ЖЖ II	7
2	ЖЖ II	7
3	I	1
4	II	2
7	I	1
8	I	1
26	II	2
27	I	1
	Total: 22	

187

Teaching the Lesson
- Lesson Starter 40
- Reteaching/Alternate Approach, p. 186C
- Toolbox Skill 23

Lesson Follow-Up
- Practice 54
- Enrichment 44: Communication
- Study Guide, pp. 79–80
- Manipulative Activity 9

Warm-Up Exercises

Solve.

1. (3.5)225   787.5
2. (7.3)350   2555
3. (2.9)475   1377.5

Round to the nearest thousand.

4. 4760   5000
5. 26,089   26,000

Lesson Focus

Today's lesson focuses on pictographs. Ask students if they have ever seen pictures used to display numerical facts in magazines. Pictures of shoes, for example, could be used to illustrate the number of baseball cleats, basketball shoes, or jogging shoes sold.

Teaching Notes

Ask students to make up some questions that could be asked about the pictograph on page 189. Make sure they understand the approximate nature of the numbers represented in the pictograph and how to read the key.

5-1 Pictographs

Objective: To interpret and draw pictographs.

A **graph** is a picture that displays numerical facts, called **data.** One type of graph that you often see in newspapers and magazines is a **pictograph.** In this type of graph, a symbol is used to represent a given number of items. A **key** on the graph tells you how many items the symbol represents. It is important to note that pictographs often show only *approximations* of the data.

Example 1

Use the pictograph above.

a. About how many lime cones were sold?

b. About how many more chocolate cones were sold than orange cones?

Solution

a. Each symbol represents 150 cones.
There are $2\frac{1}{2}$, or 2.5, symbols on the lime cone.
Multiply.
2.5(150) = 375
About 375 lime cones were sold.

b. Compare the numbers of symbols.
The chocolate cone has 4 more symbols than the orange cone.
Multiply.
4(150) = 600
About 600 more chocolate cones were sold than orange cones.

 **Check Your Understanding**

1. Describe how Example 1(a) would be different if each symbol represented 500 cones.

2. Describe a different way to find the answer to Example 1(b).
See *Answers to Check Your Understanding* at the back of the book.

In order to draw a pictograph from a set of data like the one below, you must first choose a symbol and find an appropriate amount for that symbol to represent.

Farms in the United States (Millions)

Year	1900	1920	1940	1960	1980
Farms	5.7	6.5	6.1	4.0	2.4

188   Chapter 5

Example 2

Solution

Draw a pictograph to display the data in the table at the bottom of page 188.

- Choose a symbol to represent 1 million farms. Half a symbol represents 0.5 million farms.

- Round each number to the nearest half-million to find the number of symbols for each year.

  5.7 ⟶ 5.5 symbols
  6.5 ⟶ 6.5 symbols
  6.1 ⟶ 6 symbols
  4.0 ⟶ 4 symbols
  2.4 ⟶ 2.5 symbols

Farms in the United States

1900
1920
1940
1960
1980

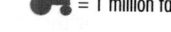 = 1 million farms

- Draw the graph. Include a title and the key.

 **Check Your Understanding**

3. In Example 2, explain why 5.7 is rounded to 5.5 instead of 6.
See *Answers to Check Your Understanding* at the back of the book.

Guided Practice

« **1.** COMMUNICATION « *Reading* Replace each ? with the correct word.   A graph in which a symbol is used to represent a certain number of items is called a ? . In this type of graph, the ? tells you how many items each symbol represents.

pictograph; key

Use the table below. Assume that an hourglass is the symbol used in a pictograph displaying the data.

2. How many years might an hourglass represent?

3. How many years would half of the hourglass represent?

4. To what number would you round the life spans?

5. Draw a pictograph to display the data.
Answers to Guided Practice 2–5 are on p. A11.

Use the pictograph on page 188.

6. About how many banana cones were sold? about 450

7. About how many more strawberry cones were sold than lime cones?
about 225

Maximum Life Spans of Certain Animals

Animal	horse	sea lion	goat	fox	chipmunk
Years	46	28	18	14	8

Key Question
How do you choose a symbol and an appropriate amount for a pictograph symbol? The symbol should be related to the subject being graphed. The amount each symbol represents is determined by finding a common unit for all the data.

Critical Thinking
To *determine* the amount that each symbol represents in a pictograph, students must *compare* and *analyze* the numerical data.

Making Connections/Transitions
Pictographs offer a visual representation of data and allow for quick comparisons and interpretations.

Error Analysis
Students may incorrectly interpret the key for a pictograph, thus getting an incorrect answer. Stress that reading the key accurately is an important part of using a pictograph.

Life Skills
Pictographs are often used in magazines and newspapers to present data. Information presented in such a way can be interpreted quickly and easily.

Reteaching/Alternate Approach
For a suggestion on topics for students to draw pictographs, see page 186C.

Additional Examples
1. Use the pictograph on page 188.
  a. About how many banana cones were sold? 450
  b. About how many more strawberry cones were sold than lime cones? 225

190

Additional Examples
(continued)

2. Draw a pictograph to display the data in the table below.

Median Family Income

Year	Dollars
1970	$9,867
1975	$13,719
1980	$21,023
1985	$27,144

Median Family Income

1970	$	$			
1975	$	$	$		
1980	$	$	$	$	
1985	$	$	$	$	$

$ = $5000

Closing the Lesson

Provide students with a pictograph from a magazine or newspaper. Have them make a table of data using the graph and then analyze the data to see if a more accurate pictograph can be made.

Suggested Assignments

Skills: 1–8, 10, 12, 17–23
Average: 1–12, 17–23
Advanced: 1–7 odd, 9–23

Exercise Notes

Mental Math
Exercises 9–12 have students use mental math to answer questions about data from a pictograph. This can be done with most pictographs whose keys are powers of 10.

Cooperative Learning
Exercises 13–15 allow students to discuss the accuracy of a pictograph and its key while working in cooperative groups.

Communication: Writing
Exercise 16 gives students an opportunity to analyze their pictographs as they develop appropriate questions.

Exercises

1. about 3,500,000 **2.** about 1,750,000 **3.** about 1,000,000

Use the pictograph at the right. About how many passengers were there?

A **1.** in 1992   **2.** in 1984

3. About how many more passengers were there in 1988 than in 1980?

4. In which years were there fewer than 2,000,000 passengers?
1980, 1984

5. Name the year that there were about twice as many passengers as in 1984. 1992

6. How many symbols represent 5,000,000 passengers?
10 symbols

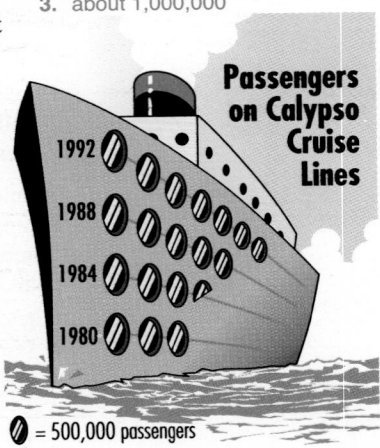

Passengers on Calypso Cruise Lines

⬭ = 500,000 passengers

Draw a pictograph to display the data. Answers to Exercises 7 and 8 are on p. A11.

7. Airport Limousine Company Earnings per Quarter

Quarter	first	second	third	fourth
Earnings	$80,477	$99,812	$86,040	$51,252

8. Number of Women in State Legislatures

Year	1978	1980	1982	1984	1986	1988
Women	767	871	911	996	1101	1175

MENTAL MATH 11. about $3600   **12.** about $4000

Use the pictograph below. Find each answer mentally.

B **9.** About how many rolls of film were developed altogether? about 1550

10. About how many more rolls were developed into textured prints than into glossy prints? about 450

Foto Finish received a set amount of money for each roll developed into prints. About how much was received altogether?

11. small glossy prints: $12 per roll   **12.** large glossy prints: $16 per roll

Film Developed at Foto Finish   📷 = 100 rolls

Small Textured Prints   Small Glossy Prints   Large Textured Prints   Large Glossy Prints

GROUP ACTIVITY

The pictograph at the right was drawn using the data in this table.

Airport Traffic (Passengers Arriving and Departing)

City	Passengers
Atlanta	47,649,470
Dallas/Fort Worth	41,875,444
Denver	32,355,000
Boston	23,283,047
Pittsburgh	17,457,801

Airport Traffic
(Passengers Arriving and Departing)

= 20 million passengers

Use the table and the pictograph above.
Answers to Exercises 13–15 are on p. A11.

C **13.** Explain why the key that was used (= 20 million passengers) does not represent the data accurately.

14. Which of the following keys do you think most people would agree is the most appropriate for this set of data? Give a convincing argument to support your choice.

   a. = 2 million passengers

   b. = 10 million passengers

   c. = 30 million passengers

   d. = 40 million passengers

15. Use the key that you chose in Exercise 14. Draw a more appropriate pictograph for the data in the table above.

16. WRITING ABOUT MATHEMATICS Use the pictograph that you drew for Exercise 15. Write five questions that involve interpreting your pictograph. Answers will vary.

SPIRAL REVIEW

S **17.** Solve $y - 8 = -2$. Check the solution. *(Lesson 4-3)* 6

18. Evaluate $62.66 \div k$ when $k = 2.6$. *(Lesson 1-3)* 24.1

19. Find the answer: $2 \cdot 4^3 + 8$ *(Lesson 1-10)* 136

20. Use the pictograph on page 190 entitled *Film Developed at Foto Finish*. About how many more rolls of film were developed into large prints than into small prints? *(Lesson 5-1)* about 150

21. Find the difference: $-2 - (-6)$ *(Lesson 3-4)* 4

22. Estimate the product: $379 \cdot 42$ *(Toolbox Skill 1)* about 16,000

23. Simplify: $8(5 - 3n)$ *(Lesson 2-2)* $40 - 24n$

Graphs and Data Analysis **191**

191

Teaching the Lesson
* Materials: multicolored candy or cereal
* Lesson Starter 41
* Using Manipulatives, p. 186C
* Toolbox Skill 24

Lesson Follow-Up
* Practice 55
* Enrichment 45: Data Analysis
* Study Guide, pp. 81–82

Warm-Up Exercises

Use the bar graphs on the pages indicated to answer each question.

1. How many Pops concerts does the Houston Symphony perform each year? (pp. 2–3)   14

2. What is the record low January temperature in Cheyenne? (pp. 92–93)   −30° F

3. What is the maximum speed a rabbit can run? (pp. 140–141) 35 mi/h

Lesson Focus

Ask students if they have ever participated in a survey. Results of a political survey may compare the President's popularity each month of this year with the corresponding month of last year. Dentists may be interested in a survey on the type of toothpaste preferred. Today's lesson focuses on bar graphs, which can be used to show comparisons.

Teaching Notes

While students may be familiar with single bar graphs, the double bar graph may be new to them. Point out the parts of the graph at the bottom of this page, especially the legend and its importance in understanding the graph. Stress that a numerical scale is on one of the axes and the categories are on the other.

5-2   Interpreting Bar Graphs

Objective: To interpret single and double bar graphs.

APPLICATION

Terms to Know
* *bar graph*
* *scale*
* *legend*

The results of a survey about music preferences were published in the Alltown High School newspaper in the form of a *bar graph*. This made it easier for students to compare preferences.

A **bar graph** has two axes. One axis is labeled with a numerical **scale**. The other is labeled with the categories. When reading a bar graph, you might find it is often necessary to estimate where the bars end.

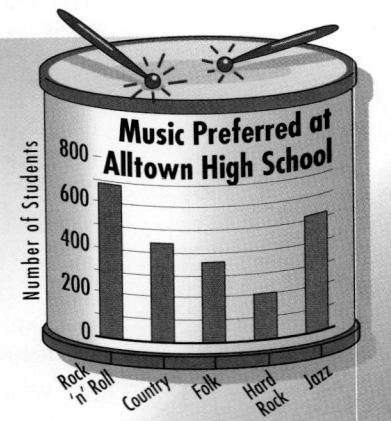

Example 1

Use the bar graph above.

a. Estimate the number of students who prefer folk music.

b. About how many more students prefer jazz than prefer hard rock?

Solution

a. Locate the bar for *folk* music.
It ends about halfway between 300 and 400, near 350.
About 350 students prefer folk music.

b. Compare the heights of the bars for *jazz* and *hard rock*.
The jazz bar is about 3 intervals higher than the hard rock bar.
Multiply: 3(100) = 300
About 300 more students prefer jazz than prefer hard rock.

 **Check Your Understanding**

Describe a different way to find the answer to Example 1(b).
See *Answers to Check Your Understanding* at the back of the book.

There are many kinds of bar graphs. The graph at the right is a *double bar graph*. In this type of bar graph, two bars appear in each category. A **legend** is included on the graph to identify the two types of bars.

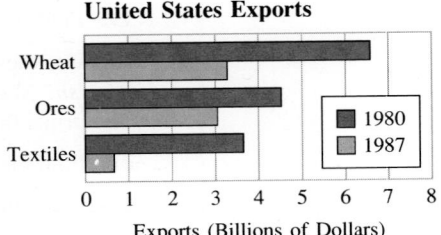

192   Chapter 5

Example 2

Use the double bar graph on page 192. About how much greater was the value of ores exported in 1980 than in 1987?

Solution

Locate the bars for ores. The 1980 bar is *red*. It ends at about 4.5. The 1987 bar is *blue*. It ends at about 3.
Subtract: 4.5 billion − 3 billion = 1.5 billion
The value of ores exported was about $1.5 billion greater in 1980 than in 1987.

Guided Practice

« 1. **COMMUNICATION** « *Reading* Choose the definition of the word *scale* that is used in this lesson.

scale (skāl) *noun* **1.** A dry, thin flake or crust. **2.** A series of marks placed at fixed distances, used for measuring. **3.** The relationship between the actual size of something and the size of a model or drawing that represents it. **4.** A series of musical tones that goes up or down in pitch. **5.** An instrument for weighing. definition 2

Use the double bar graph below. 8, 10, 11. Estimates may vary.

2. What does each interval on the scale represent? 0.1 billion people

3. What do the red bars represent? 1988 population

4. Which country is expected to have the greatest population in 2100?
United States
5. Which country is expected to have the least increase in population?
Japan
6. Tell whether the bar for Brazil in 1988 ends closer to 0.1, 0.15, or 0.2.
0.15
7. Tell whether the bar for Mexico in 2100 ends closer to 0.1, 0.15, or 0.2. 0.2

8. Estimate the population of the United States in 1988.
about 0.25 billion
9. Which country is expected to have a population of about 0.2 billion in 2100? Mexico

10. Estimate the projected increase in population for Brazil.
about 0.15 billion
11. For the year 2100, about how much greater is the projected population for the United States than for Mexico?
about 0.1 billion

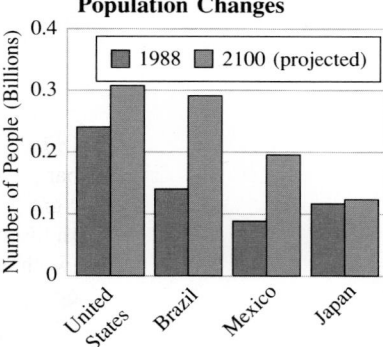

Population Changes

■ 1988 ■ 2100 (projected)

Number of People (Billions): 0.4, 0.3, 0.2, 0.1, 0

United States, Brazil, Mexico, Japan

Graphs and Data Analysis **193**

Key Question
Are the data in the bar graphs on page 192 exact or approximate?
approximate

Life Skills
Being able to read and interpret a bar graph is an important life skill. These graphs occur frequently in newspapers and magazines.

Critical Thinking
To use a bar graph, students must *analyze* and *interpret* information presented visually and *translate* it into numerical data.

Using Manipulatives
For a suggestion on using cereal or candy to determine a bar graph, see page 186C.

Additional Examples

1. Cinema 5 took a survey of the number of people who bought tickets for each type of movie currently being shown.

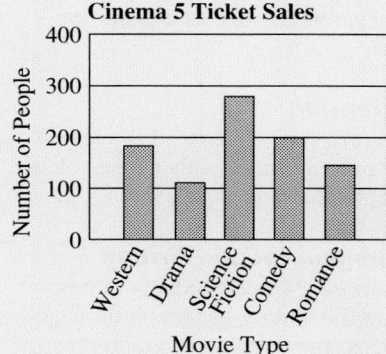

Cinema 5 Ticket Sales

Number of People: 400, 300, 200, 100, 0

Western, Drama, Science Fiction, Comedy, Romance

Movie Type

Use the bar graph above to answer the following.
a. Estimate the number of people who prefer romance movies.   about 150
b. About how many more people prefer science fiction to westerns?   about 100
2. Use the double bar graph on page 192. About how much greater was the value of wheat exported in 1980 than in 1987?   about $3.5 billion

194

Closing the Lesson

Ask students when they would use a single bar graph and when they would use a double bar graph.

Suggested Assignments

Skills: 1–12, 13–21 odd, 25, 26
Average: 1–22, 25, 26
Advanced: 1–11 odd, 13–26

Exercise Notes

Communication: Reading

Guided Practice Exercise 1 presents definitions of the word "scale," thus showing how a mathematical term can be used in everyday life.

Estimation

Estimation is used to interpret data from bar graphs. Many of the exercises reinforce this notion, as students are asked "to estimate" or find "about how many."

Data Analysis

Exercises 13–16 require students to interpret a double bar graph that has a negative as well as a positive scale.

Extension

Exercises 17–22 extend the lesson by presenting two other types of bar graphs that are often used.

Communication: Writing

Exercise 24 asks students to compare the sliding bar graph on page 195 to the double bar graph they created in *Exercise 23*.

Exercises

1. about 2 million 2. about 2.5 million 3. about 3.5 million 4. about 1 million

Use the bar graph at the right. Estimate the number of balls sold for each sport.

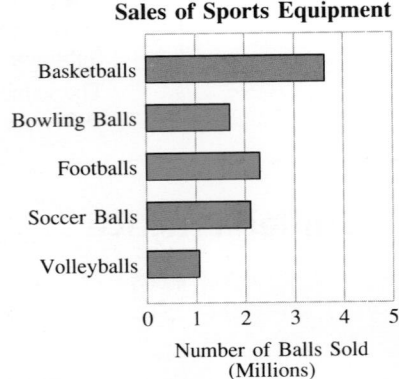
Sales of Sports Equipment

A
1. soccer   2. football

3. basketball   4. volleyball

5. Name all the sports for which fewer than 2 million balls were sold.

6. About how many times as many soccer balls were sold as volleyballs?

7. About how many more basketballs were sold than bowling balls?

8. About how many more footballs were sold than volleyballs?

5. bowling, volleyball 6. about 2 times 7. about 2 million 8. about 1.5 million

Use the double bar graph at the right.

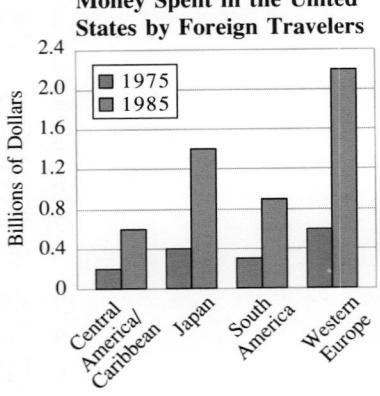
Money Spent in the United States by Foreign Travelers

9. In 1975, travelers from which region spent about $0.6 billion in the United States?
Western Europe

10. In 1985, travelers from which region spent about $0.6 billion in the United States?
Central America/Caribbean

11. About how much more did travelers from Japan spend in the United States in 1985 than in 1975? Estimate may vary; about $1 billion

12. Travelers from which region increased their spending in the United States the most between 1975 and 1985? Western Europe

DATA, *pages 92–93* **13–16.** Estimates may vary.

B 13. Estimate the record low temperature in Minneapolis. about −35°F

14. About how much warmer is the record high temperature in Cheyenne than the record low temperature? about 95°F

15. About how much warmer is the record high temperature in Los Angeles than the record high temperature in Minneapolis? about 40°F

16. About how much colder is the record low temperature in Nashville than the record low temperature in Houston? about 25°F

Newspapers and magazines often put the numerical fact at the end of each bar on a bar graph and omit the scale. This type of graph is called an *annotated* bar graph.

Use the annotated bar graph at the right.

17. How many more calculators were shipped in 1987 than in 1975? 15.4 million

18. In which four-year interval did the number of calculators shipped increase by less than 3 million? 1983–1987

19. Estimate the income from the sales of all the calculators shipped in 1975.

20. Estimate the income from the sales of all the calculators shipped in 1987.

   19. about $1159.2 million   **20.** about $845 million

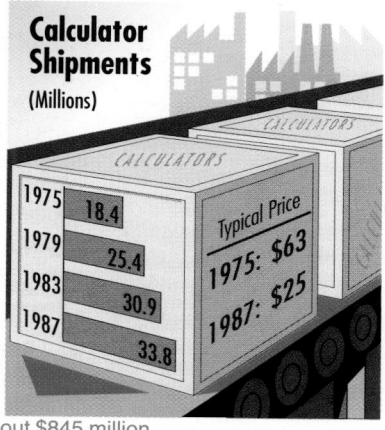

Calculator Shipments (Millions)

1975	18.4
1979	25.4
1983	30.9
1987	33.8

Typical Price
1975: $63
1987: $25

Data given in pairs, such as wins and losses, are sometimes graphed on a *sliding* bar graph. On this type of graph there are two scales. One scale extends to the right of zero and the other extends to the left.

Use the sliding bar graph at the right.

21. About how many games did the Cincinnati team lose? Estimate may vary; about 90.

22. About how many more games did the San Francisco team win than the Atlanta team? Estimate may vary; about 30.

C 23. Make a double bar graph that displays the same set of data as is shown in this sliding bar graph. See answer on p. A11.

24. **WRITING ABOUT MATHEMATICS** Write a paragraph that compares the graph you drew for Exercise 23 to the graph shown at the right. Answers will vary.

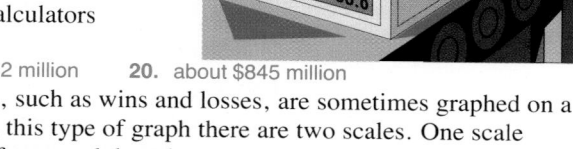

National League Standings

Western Division

Games Lost   Games Won

Team   100 80 60 40 20 0 20 40 60 80 100

San Francisco
San Diego
Houston
Los Angeles
Cincinnati
Atlanta

SPIRAL REVIEW

S 25. Simplify: $6u + 5v + 11u$ (Lesson 2-3) $17u + 5v$

26. Use the bar graph at the top of page 194. About how many more basketballs were sold than soccer balls?   (Lesson 5-2)
Estimate may vary; about 1.5 million.

Graphs and Data Analysis   **195**

Follow-Up

Extension (*Cooperative Learning*)
Ask students to bring in a bar graph from a newspaper or magazine. Working in groups, students should create five questions that can be answered from the graph. At least one of the questions should require drawing a conclusion. A written explanation of the group's conclusion should be part of the work.

Enrichment
Enrichment 45 is pictured on page 186E.

Practice
Practice 55 is shown below.

Practice 55, *Resource Book*

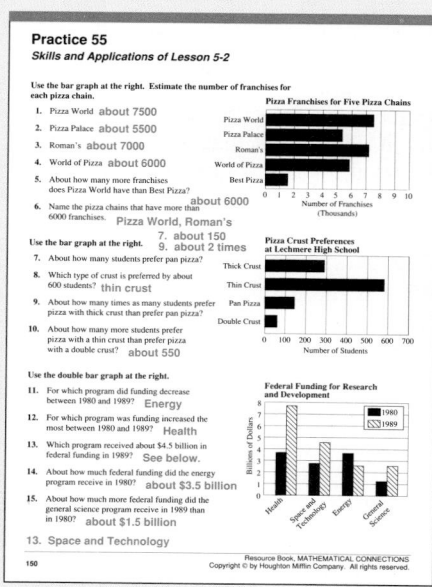

Teaching the Lesson
• Lesson Starter 42
• Toolbox Skill 25

Lesson Follow-Up
• Practice 56
• Enrichment 46: Exploration
• Study Guide, pp. 83–84

Warm-Up Exercises

Write as a number.

1. 0.3 million   300,000
2. 500 thousand   500,000
3. 1.1 million   1,100,000
4. 2.25 million   2,250,000
5. 0.26 million   260,000

Lesson Focus

Ask students to imagine that they are sales representatives in a meeting with their manager. At the front of the room is a chart with a line graph showing sales in recent months. They see that the line on the graph moves down from left to right, showing that sales are decreasing. Today's lesson focuses on line graphs and how to interpret them.

Teaching Notes

Students have probably seen line graphs, but may not have thought much about the purposes of such graphs. Point out that line graphs are used to indicate trends or changes in data.

Key Question

What is the purpose of a line graph?   A line graph shows an amount and a direction of a change in data over a period of time.

5-3   *Interpreting Line Graphs*

Objective: To interpret single and double line graphs.

A **line graph** shows an *amount* and a *direction* of change in data over a period of time. In a line graph the data are represented by points. These points are connected by line segments.

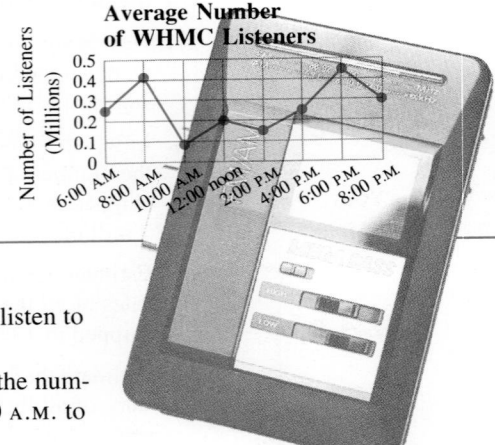

Average Number of WHMC Listeners

Example 1

Use the line graph above.

a. About how many people listen to WHMC at 5:00 P.M.?

b. Estimate the decrease in the number of listeners from 8:00 A.M. to 10:00 A.M.

Solution

a. Locate 5:00 P.M. halfway between 4:00 P.M. and 6:00 P.M. Move up to the red line, then left to the vertical axis.
   The number for 5:00 P.M. is between 0.3 million and 0.4 million, at about 0.35 million.
   About 0.35 million people listen at 5:00 P.M.

b. Estimate the number of listeners at each of the two hours.

   8:00 A.M. ⟶ about 0.4 million
   10:00 A.M. ⟶ about 0.1 million

   Subtract: $0.4 - 0.1 = 0.3$
   The number of listeners decreases by about 0.3 million from 8:00 A.M. to 10:00 A.M.

Terms to Know
• *line graph*
• *increasing trend*
• *decreasing trend*

If a series of segments on a line graph slopes upward over a given interval, there is an **increasing trend** in the data over that interval. If a series of segments slopes downward, there is a **decreasing trend** over that interval.

Double line graphs such as the one shown at the right are useful for comparing trends in two sets of data. This particular double line graph has a gap in the scale, marked with a jagged line (⚡).

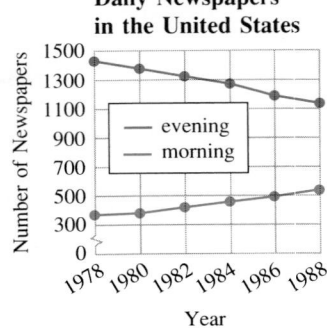

Daily Newspapers in the United States

— evening
— morning

196   Chapter 5

Example 2

Use the double line graph at the bottom of page 196.

a. Estimate the number of morning newspapers in 1982.

b. Describe the overall trend in the number of evening newspapers.

Solution

a. Locate the point for 1982 on the *blue* line.
The number for 1982 is between 300 and 500, at about 400.
There were about 400 morning newspapers in 1982.

b. Each *red* segment slopes downward. Overall, there was a decreasing trend in the number of evening newspapers.

 **Check Your Understanding**

Describe how Example 2(a) would be different if you were asked to find the number of evening newspapers in 1982.
See *Answers to Check Your Understanding* at the back of the book.

Guided Practice

1. Reading popular magazines can help a person keep up with the latest fads and fashions.

«**1.** COMMUNICATION «*Reading* Explain the meaning of the word *trends* in the following sentence.

Reading popular magazines can help a person keep up with the latest trends.

Use the line graph below. Tell whether there was an *increase* or a *decrease* in the data over each interval.

2. 1960–1965 decrease   **3.** 1980–1985 increase

4. What does each interval on the horizontal axis represent? 5 years

5. What does each interval on the vertical axis represent? 20 million acres

6. Which interval along the vertical axis contains the point for 1965?
760–780

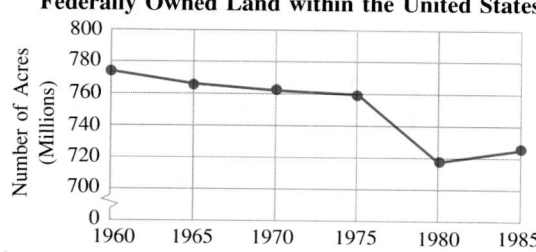

Federally Owned Land within the United States

Use the double line graph at the bottom of page 196.

7. In which given year were there about 1100 evening newspapers? 1988

8. About how many more evening newspapers than morning newspapers were there in 1986? Estimate may vary; about 700.

9. Describe the overall trend in the number of morning newspapers.
increasing

Graphs and Data Analysis **197**

Exercise Notes

Communication: Reading
Guided Practice Exercise 1 connects the concept of trend presented in this lesson to the trends that students find interesting, such as clothing fashions or movies.

Making Connections/Transitions
Exercises 25–32 connect line graphs to previously learned algebraic skills and functions.

1–4. Estimates may vary. **1.** about 8 million **2.** about 7.5 million **3.** about 9.5 million **4.** about 9 million

Exercises

Use the line graph at the right. Estimate the number of members in each year.

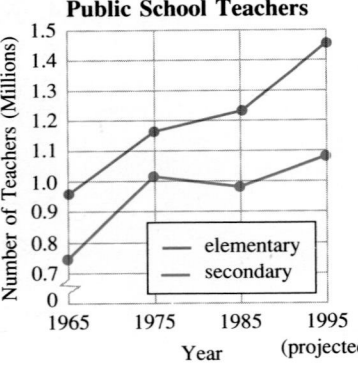

American Bowling Congress Membership

A **1.** 1985 **2.** 1969

3. 1979 **4.** 1975

5. In which given years were there about 9.5 million members in the American Bowling Congress?
1977, 1981

6. In which given year were there about 7 million members in the American Bowling Congress?
1989

7. Estimate the decrease in membership from 1981 to 1989.
Estimate may vary; about 2.5 million

8. During which four-year interval did membership decrease by about one million people? 1985–1989

9. During which four-year interval did the greatest increase in membership occur? 1973–1977

10. Describe the trend in membership from 1969 to 1981. increasing
Answers to Exercises 11–14 are on p. A12.

Use the double line graph at the right. Estimate the number of public school teachers in each year at each level.

11. 1975 **12.** 1995

13. 1970 **14.** 1990

15–19. Estimates may vary.
15. About how many more elementary than secondary public school teachers were there in 1965? about 0.2 million

16. About how many more elementary than secondary public school teachers were there in 1985? about 0.3 million

Public School Teachers

17. Estimate the total number of public school teachers in 1980. about 2.2 million

18. Estimate the total number of public school teachers in 1970.
about 2 million
19. Estimate the increase in the number of secondary public school teachers from 1965 to 1995. about 0.35 million

20. Describe the overall trend in the number of elementary public school teachers. increasing

198 Chapter 5

Line graphs are often shown without labels on the axes. To interpret these graphs, you must find the values of the intervals.

Use the graph at the right.

B 21. What does each interval on the horizontal axis represent?
5 years

22. What does each interval on the vertical axis represent?
2 million trucks

23. In which year were there about 4 million fewer trucks from 3 to 5 years old than there were trucks 12 years and over? 1985

24. Estimate the year in which there were as many trucks from 3 to 5 years old as there were trucks 12 years and over.
Estimate may vary; about 1982.

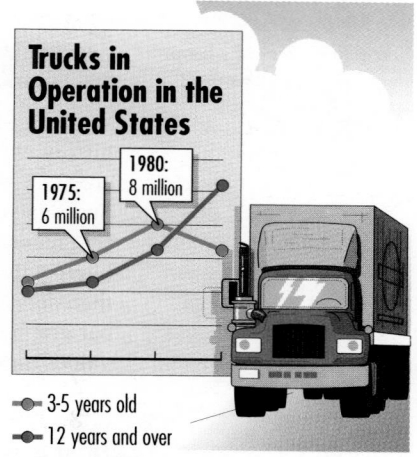

Trucks in Operation in the United States

1975: 6 million

1980: 8 million

● 3-5 years old
● 12 years and over

CONNECTING DATA ANALYSIS AND ALGEBRA

Each table and graph in this chapter represents a relationship between two sets of data. Each item in the first set of data is paired with exactly one item in the second set, so each relationship is a function.

25. 50 million   **27.** 30 million   **29.** 50 million

Use the line graph below to complete each ordered pair.

25. (1910, ___?___)   **26.** (___?___, 10 million)   1985   **27.** (1955, ___?___)

28. (1925, ___?___)   **29.** (1940, ___?___)   1970 **30.** (___?___, 20 million)
40 million

C 31. Use the ordered pairs from Exercises 25–30 to make a function table for the data.
See answer on p. A12.

32. Use the function table that you made for Exercise 31 to complete this statement: The number of sheep on farms in the United States is a function of ___?___. the year

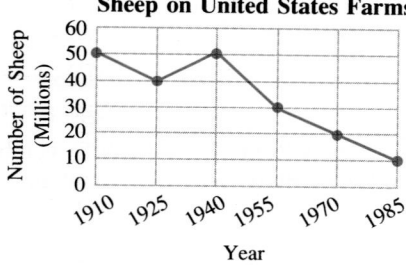

Sheep on United States Farms

Number of Sheep (Millions)

Year

SPIRAL REVIEW

S 33. Simplify: $(4x)(60x)$ *(Lesson 2-1)* $240x^2$

34. Find the product: $8(-6)(-3)$ *(Lesson 3-5)* 144

35. Estimate the sum: $0.59 + 0.61 + 0.627 + 0.584$ *(Toolbox Skill 4)*
2.4

36. Use the line graph on page 197. In which five-year interval did the greatest decrease occur?   *(Lesson 5-3)* 1975–1980

Graphs and Data Analysis   **199**

Follow-Up

Project
Have students research a topic relating to health-care statistics, such as diseases, hospital costs, number of physicians by area, and so on. After researching the topic, students should organize the data, prepare and label a graph, write a report on their sources, and explain why they organized the data as they did. Students should report conclusions and predictions that they can make from their graphs.

Enrichment
Enrichment 46 is pictured on page 186E.

Practice
Practice 56 is shown below.

Practice 56, *Resource Book*

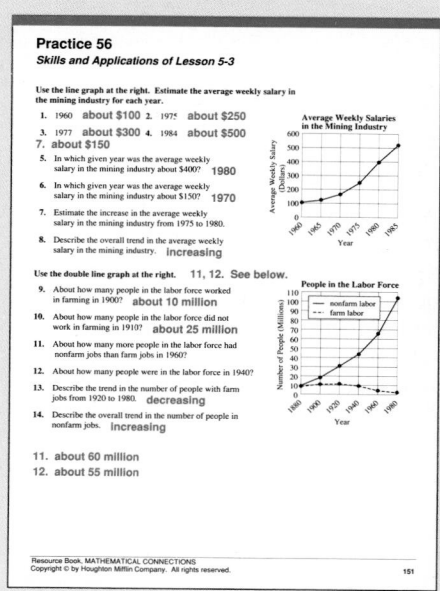

199

Teaching the Lesson
- Lesson Starter 43
- Using Technology, p. 186C
- Reteaching/Alternate Approach, p. 186D
- Cooperative Learning, p. 186D
- Toolbox Skills 24 and 25

Lesson Follow-Up
- Practice 57
- Practice 58: Nonroutine
- Enrichment 47: Application
- Study Guide, pp. 85–86
- Test 21

Warm-Up Exercises

Draw a pictograph to display the data in the table.

Music Sales

Compact Disks	500
Cassettes	625
Cassette Singles	370
45's	127
Albums	315

Music Sales

Compact Disks	⬭⬭⬭⬭
Cassettes	⬭⬭⬭⬭⬭
Cassette Singles	⬭⬭⬭
45's	⬭
Albums	⬭⬭⊂
	Key ⬭ = 125 Records

Lesson Focus

Ask students how often they have read an article filled with facts and figures. Did they find the information difficult to remember? Today's lesson focuses on displaying data in the form of a bar or line graph. Information may be compared quickly and easily by using a visual display.

5-4 Drawing Bar and Line Graphs

Objective: To draw a bar or line graph for a given set of data.

APPLICATION

While doing a research project for her history class, Lynne Greyson discovered that some counties in the United States have a president's name. She first recorded the information in a table like the one at the right. Lynne then decided to display these data in a bar graph so that she could compare values more easily.

County Name	Number of Counties
Washington	32
Adams	12
Jefferson	25
Madison	20
Monroe	17

Example 1

Solution

Draw a bar graph to display the data in the table above.

- Draw the axes. Position the names of the presidents on the vertical axis. This will make the names easier to read.

- Choose a scale. The greatest number is 32, so draw and label a scale from 0 to 35, using intervals of five.

- Draw a bar for each entry in the table. Use the scale to decide where to end the bars.

- Label the horizontal axis and title the graph.

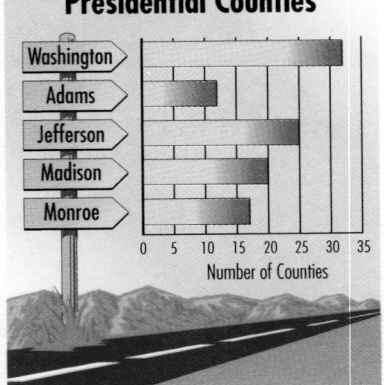

Check Your Understanding

Why is it better to use intervals of five rather than intervals of ten on the graph in Example 1?
See *Answers to Check Your Understanding* at the back of the book.

As part of his science project, John Nodlem wanted to compare his growth to that of the average male. To make it easier to compare the data and analyze trends, he drew a line graph.

Growth Record for John Nodlem

Age (years)	birth	4	8	12	16
Recorded Height (cm)	55	85	108	147	182
Height of Average Male (cm)	51	99	126	147	169

Example 2
Solution

Draw a line graph to display the data at the bottom of page 200.

- Draw the axes, then write the ages at evenly spaced intervals on the horizontal axis.

- Choose a scale. Notice that the heights range from 51 cm to 182 cm. Start with a gap in the scale. Then, using intervals of twenty-five, number the scale from 50 to 200.

- Place a point on the graph for John's height at each age. Connect the points from left to right with solid line segments. Repeat this process for the average heights, using dashes or a different color.

- Include a legend on the graph. Label both axes and title the graph.

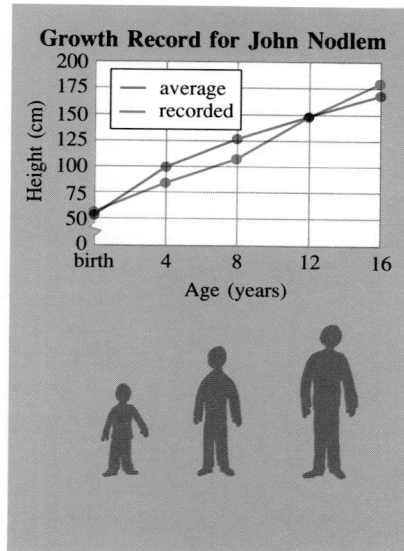

Growth Record for John Nodlem

Guided Practice

Answers to Guided Practice 1, 4, and 5 are on p. A12.

«**1. COMMUNICATION** «*Reading*   Refer to the text on pages 200–201. Using your own words, develop two outlines for your notebook: *How to Draw a Bar Graph* and *How to Draw a Line Graph*.

Consider drawing a line graph to display the data below.

In-State Tuition and Fees at Public Two-Year Colleges

Year	1979	1982	1985	1988
Cost	$327	$434	$584	$690

Answers may vary. Example: gap from 0 to 300 followed by intervals of 50 up to 700.
2. Describe the scale you would use.

3. What labels would you place on the axes? *Year* on horizontal axis; *Cost in Dollars* on vertical axis

4. Draw a line graph to display the data.

5. Draw a bar graph to display the data below.

Appliance Shipments (Millions)

Year	1975	1979	1983	1987
Electric Ranges	2.1	3.0	2.8	3.6
Microwave Ovens	0.8	2.8	6.0	12.7

Graphs and Data Analysis   **201**

Teaching Notes

While students may have some idea on how to draw bar and line graphs, they should follow the four steps presented in the Examples.

Key Question

What are the four steps in drawing a graph?   Draw the axes, choose a scale, plot the data, and provide all necessary labels.

Critical Thinking

When drawing a graph, students must *assess* the data before *creating* a scale for the graph.

Error Analysis

Students may plot data incorrectly when creating a graph. Remind them to check that they have drawn the correct size bar or plotted the points correctly.

Using Technology

For a suggestion on using computers to generate bar and line graphs, see page 186C.

Reteaching/Alternate Approach

For a suggestion on using cooperative learning to examine bar and line graphs, see page 186D.

Additional Examples

1. Draw a bar graph to display the data in the table below.

Mobile Homes Shipped

Year	Number (Thousands)
1970	401
1975	213
1980	222
1985	284

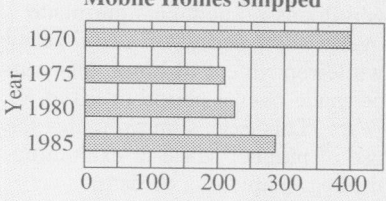

Mobile Homes Shipped

2. Draw a line graph to display the data in the table below.

Farm Income

Year	Dollars (Millions)
1970	$54
1975	$90
1980	$141
1985	$152

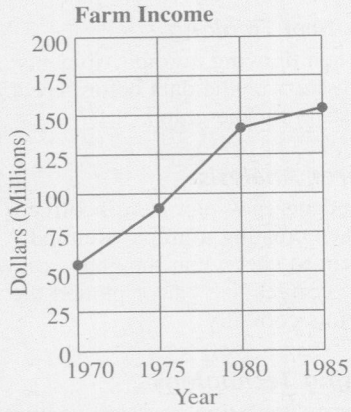

Farm Income

Closing the Lesson

Ask a student to explain how to label the axes on a bar graph that shows the ages of the Presidents when they took office. Ask another student how to label the scale. Ask a third student to explain when bar graphs are used.

Suggested Assignments

Skills: 1–5, 11–14
Average: 1–8, 11–14
Advanced: 1–14

Exercise Notes

Number Sense

To extend *Exercise 3*, ask students why loans on new cars are made over a short period of 4 to 5 years, while repayment of home loans may be made over a period of 20 to 30 years. This extension promotes the use of number sense in examining the reasonableness of data.

Exercises

Answers to Exercises 1–10 are on pp. A12–A13.

Draw a bar graph to display the data.

A   1. **Adults Participating in Leisure-Time Activities (Millions)**

Activity	bicycling	softball	swimming	volleyball
Adults	60	39	75	34

2. **Tourism in the United States and Overseas (Millions)**

Year	1981	1983	1985	1987
Tourists in the U.S.	9.1	7.9	7.5	10.4
U.S. Tourists Overseas	8.0	9.6	12.3	13.7

Draw a line graph to display the data.

3. **Average Payment Period, Finance Company Loans on New Cars**

Year	1984	1985	1986	1987	1988	1989
Number of Months	48.3	53.5	50.0	53.5	56.2	54.2

4. **Recreational Boats in the United States (Millions)**

Year	1975	1979	1983	1987
Outboard Motor Boats	5.7	6.7	7.2	7.8
Canoes and Rowboats	2.4	2.7	3.4	4.1

B   5. DATA, *pages 92–93* Draw a bar graph to display the data in the diagram entitled *Average Number of Tornadoes per Year*.

6. DATA, *pages 2–3* Draw a bar graph to display the data in the table entitled *Shipments of Audio Recordings*.

CONNECTING MATHEMATICS AND SOCIAL STUDIES

7. Draw a line graph to display the data in the table at the right.

Use your graph from Exercise 7.

C   8. Do the retail and farm prices of oranges always change by the same amount? Explain.

9. Describe the overall relationship between the retail and farm prices of oranges.

10. THINKING SKILLS Compare the graph to the table. Explain why an economist might find the graph more useful than the table.

California Orange Prices
Cents per Pound

Year	Retail Price	Farm Price
1979	44	13
1981	39	8
1983	39	5
1985	53	12
1987	55	10
1989	56	11

202   Chapter 5

SPIRAL REVIEW

S **11.** Solve $15 = \frac{c}{11} + 17$. Check the solution. *(Lesson 4-5)* −22

12. Write 4,900,000 in scientific notation. *(Lesson 2-10)* 4.9×10^6

13. Estimate the quotient: $31.69 \div 4.2$ *(Toolbox Skill 3)* about 8

14. Draw a bar graph to display the data. *(Lesson 5-4)* See answer on p. A13.

Egg Production in Five States (Billions)

State	Alabama	Arkansas	Georgia	Indiana	North Carolina
Eggs	2.6	3.8	4.3	5.6	3.4

Self-Test 1

1. Draw a pictograph to display the data. See answer on p. A13. **5-1**

Average Annual Snowfall in Four Locations (Inches)

Location	Blue Canyon	Flagstaff	Marquette	Stampede Pass
Snowfall	242.0	96.9	122.0	430.8

2-4. Estimates may vary.

2. Use the double bar graph below. Estimate the number of mathematics degrees awarded in 1985. about 15 thousand

3. Use the double bar graph below. About how many fewer computer science degrees were awarded in 1975 than in 1985? **5-2**
about 35 thousand

4. Use the line graph below. Estimate the number of state and local government employees in 1970. about 10 million **5-3**

5. Use the line graph below. Describe the overall trend in the number of state and local employees. increasing

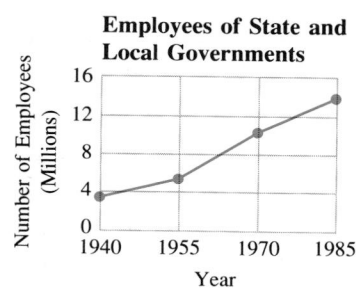

Bachelor's Degrees Awarded

Employees of State and Local Governments

6. Draw a line graph to display the data. **5-4**
See answer on p. A13.

Average Number of Items in a Supermarket

Year	1978	1981	1984	1987
Number of Items	11,767	12,877	15,379	24,531

Graphs and Data Analysis **203**

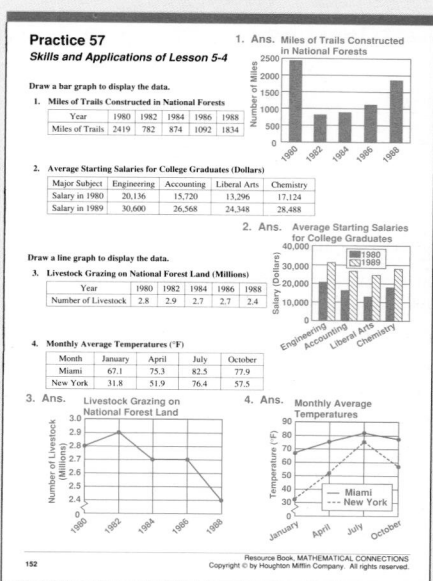

Teaching the Lesson
- Lesson Starter 44
- Using Technology, p. 186C
- Reteaching/Alternate Approach, p. 186D
- Toolbox Skills 23–25

Lesson Follow-Up
- Practice 59
- Enrichment 48: Communication
- Study Guide, pp. 87–88

Warm-Up Exercise

Describe the scale you would use to draw a bar graph of the data below. Answers will vary. For example: intervals of 1000 miles

Lengths of U.S. Coastlines

Atlantic	2069 mi
Pacific	7623 mi
Gulf	1631 mi
Arctic	1060 mi

Lesson Focus

Today's lesson focuses on choosing the appropriate type of graph to display data.

Teaching Notes

Key Question

What data would be appropriate to display on a line graph? *changes in temperature, population, price, and growth, for example*

Additional Example

Decide whether it would be more appropriate to draw a bar graph or line graph to display the data. Then draw the graph. *line graph*

Daily Temperature (°F)

4:00 A.M.	52
8:00 A.M.	64
4:00 P.M.	71
8:00 P.M.	62

DECISION MAKING

5-5

Choosing the Appropriate Type of Graph

Objective: To decide whether it would be more appropriate to draw a bar graph or a line graph to display a set of data.

Before displaying a set of data, you should first inspect it to decide whether to draw a bar graph or a line graph.

- A *bar graph* may be most appropriate when the data fall into distinct categories and you want to compare the totals.
- A *line graph* may be most appropriate when you want to emphasize trends in data that change continuously over time.

Example

Decide whether it would be more appropriate to draw a *bar graph* or a *line graph* to display the data. Then draw the graph.

National League Home Runs

Year	1985	1986	1987	1988	1989
Home Runs	1424	1523	1824	1279	1365

Solution

Bar graph. Yearly totals are distinct amounts. They are not carried over from one year to the next. In other words, yearly totals do not change continuously over time. For this reason, a line graph is not appropriate.

National League Home Runs

☑ **Check Your Understanding**

Suppose the data in the Example represented the average attendance per game in a given stadium. Would it be appropriate to draw a line graph to display the data? Explain.

See *Answers to Check Your Understanding* at the back of the book.

Guided Practice

COMMUNICATION «*Discussion*

Decide whether each type of data changes continuously. Explain.

« **1.** the number of graduates from a certain high school
No; this is a yearly total, which is a distinct amount.

« **2.** the enrollment at a certain high school
Yes; throughout the year, students move and change schools.

« **3.** the amount of money in a savings account
Yes; over a period of time, the amount increases and decreases.

« **4.** the amount of money deposited to a savings account
No; this is a distinct amount that is added at a particular time.

bar graph; see graph on p. A13.

5. Decide whether it would be more appropriate to draw a *bar graph* or a *line graph* to display the data. Then draw the graph.

Sales in Foreign Countries of Goods from the United States (Billions of Dollars)

Product	aircraft	auto parts	computers	autos	gold
Billions	$20.3	$13.2	$11.6	$9.1	$5.2

Exercises

Graphs for Exercises 1–3 are on pp. A13–A14.
Decide whether it would be more appropriate to draw a *bar graph* or a *line graph* to display the data. Then draw the graph.

A **1. Money Households Spend Annually on Reading Materials** bar graph

Metropolitan Area	Houston	Boston	Chicago	Anchorage
Dollars	164	202	178	246

2. Women and the Labor Force (Millions) line graph

Year	1965	1970	1975	1980	1985
Women in the Labor Force	26	32	37	45	51
Women out of the Labor Force	41	41	43	43	43

3. Theater Box Office Receipts (Millions) bar graph

Year	1980	1982	1984	1986
Broadway Receipts	$143	$221	$227	$191
Road Show Receipts	$181	$250	$206	$236

B **4. THINKING SKILLS** Explain why someone might decide to draw a pictograph rather than a bar graph to display the data.
See answer on p. A14.
Pieces of Mail Handled Annually in Major Cities (Millions)

City	Atlanta	Denver	Louisville	Tampa
Pieces of Mail	2340	1660	948	1470

SPIRAL REVIEW 7. Line graph; the population changes continuously.

S **5.** Simplify: $5(3n + 4)$ *(Lesson 2-2)* 15n + 20

6. Write as a variable expression: 5 less than twice the number *a*
(Lesson 4-6) 2a − 5

7. Decide whether it would be more appropriate to draw a *bar graph* or a *line graph* to display data about the population of the United States over a period of years. Explain.   *(Lesson 5-5)*

8. Estimate the sum: $213 + 548 + 194 + 362$ *(Toolbox Skill 2)*about 1300

Graphs and Data Analysis   **205**

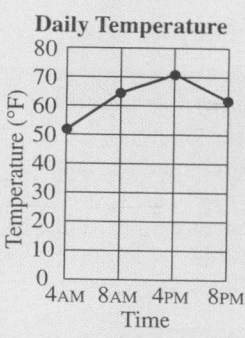

Daily Temperature

Closing the Lesson

Present different types of data to students. Ask what type of graph should be used to represent the data.

Suggested Assignments

Skills: 1–8
Average: 1–8
Advanced: 1–8

Follow-Up

Enrichment

Enrichment 48 is pictured on page 186E.

Practice

Practice 59, *Resource Book*

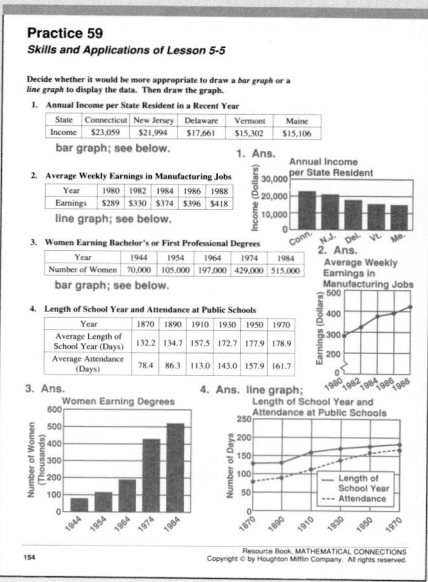

Lesson Planner

Teaching the Lesson
- Lesson Starter 45
- Reteaching/Alternate Approach, p. 186D
- Cooperative Learning, p. 186D
- Toolbox Skills 23–25

Lesson Follow-Up
- Practice 60
- Enrichment 49: Problem Solving
- Study Guide, pp. 89–90

Warm-Up Exercises

Solve.

1. Luis earns $10.50 per hour. Last week he worked 35 h. This week he worked 32 h. How much did Luis earn last week? $367.50

2. Anna delivers 42 newspapers each weekday and also on Saturday. On Sunday, Anna delivers 57 newspapers. How many newspapers does Anna deliver on weekends? 99

Lesson Focus

Ask students if they have ever solved a problem in real life for which all the facts were available. In real life, the needed facts often are supplied by the problem solver. Today's lesson focuses on solving problems involving too much or not enough information.

Teaching Notes

Ask students to look at the Problem on this page and interpret the information.

Key Question

Where can you find the answer to the question in the Problem? You can find the information in an almanac or other reference book.

5-6

Too Much or Not Enough Information

Objective: To solve problems involving too much or not enough information.

Sometimes you need to use data from a graph to solve a problem. Graphs might contain more information than you need. They also might not contain enough information to solve a problem. When trying to solve a problem, you should always read graphs carefully.

Problem

Use the diagram below. Which state in the United States has the greatest number of daily newspapers?

Number of Daily Newspapers and Their Circulation

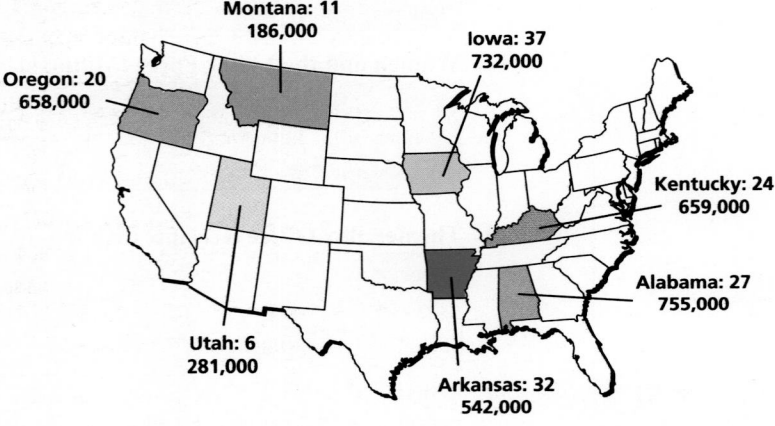

Montana: 11
186,000

Iowa: 37
732,000

Oregon: 20
658,000

Kentucky: 24
659,000

Alabama: 27
755,000

Utah: 6
281,000

Arkansas: 32
542,000

Solution

UNDERSTAND The problem is about daily newspapers.
Facts: the number of daily newspapers in seven states
the circulation of daily newspapers in seven states
Find: the state in the United States with the greatest number of daily newspapers

PLAN Compare the number of daily newspapers in every state in the United States.

WORK Not enough information. There is information for only seven states.

Look Back What if you are asked to find how much greater daily newspaper circulation is in Iowa than in Montana? What information about those states should you ignore? the number of daily newspapers

206 Chapter 5

Guided Practice

COMMUNICATION « *Reading*

Tell whether each problem contains *too much* or *not enough* information. Identify any extra or missing information.

not enough; cost of each notebook
« **1.** Carla wants to buy five notebooks. How much will she spend?

« **2.** Michael mows lawns for six clients. He earns $72 each week. In how many weeks will Michael earn $360? too much; six clients

4. the expected flights over a jet's lifetime and the number of jets in the world fleet

Use the table at the right. Solve, if possible. If there is not enough information, tell what facts are needed.

3. How many types of jets are listed? 5

4. What facts are given about the jets?

5. Does the diagram show the actual number of flights made by each type of jet? No

6. How many DC-10's are there?

7. How many more 727's are there than 737's? 90

6. not enough information; the number of DC-10's in the world fleet

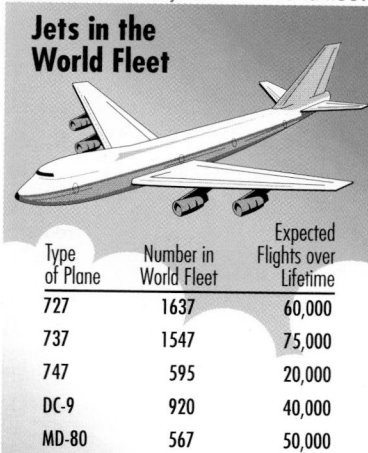

Jets in the World Fleet

Type of Plane	Number in World Fleet	Expected Flights over Lifetime
727	1637	60,000
737	1547	75,000
747	595	20,000
DC-9	920	40,000
MD-80	567	50,000

Problem Solving Situations

Use the double bar graph below. Solve, if possible. If there is not enough information, tell what facts are needed.

1. How many more grocery stores in Midtown had a bakery in 1990 than in 1980? 7

2. not enough information; the number of grocery stores in Midtown with video departments in 1985

2. How many more grocery stores in Midtown had a video department in 1990 than in 1985?

3. not enough information; the number of books sold in Midtown in 1980 and 1990

3. How many more books were sold in Midtown grocery stores in 1990 than in 1980?

4. By how much more did the number of deli departments increase than the number of catering departments in Midtown grocery stores from 1980 to 1990? 3

Specialty Departments at Grocery Stores in Midtown

Number of Stores (y-axis: 0, 2, 4, 6, 8, 10, 12, 14, 16, 18)

Legend: ■ 1980 ■ 1990

Categories: Deli, Bakery, Video, Catering, Book

Graphs and Data Analysis **207**

Exercise Notes

Communication: Reading
Reading for understanding is very important in this lesson. Students must analyze the problems for extraneous or missing information. *Guided Practice Exercises 1 and 2* have students identify extra or missing information before being asked to solve problems.

Life Skills
Being able to read a table or graph is an important life skill. *Guided Practice Exercises 3–7* and *Problem Solving Situations 1–4* help students to practice reading tables and graphs.

Making Connections/Transitions
The previously learned problem-solving strategies and mathematical skills of making a table, guess and check, and using equations are integrated in *Problem Solving Situations 5–8*.

Follow-Up

Enrichment
Enrichment 49 is pictured on page 186E.

Practice
Practice 60, *Resource Book*

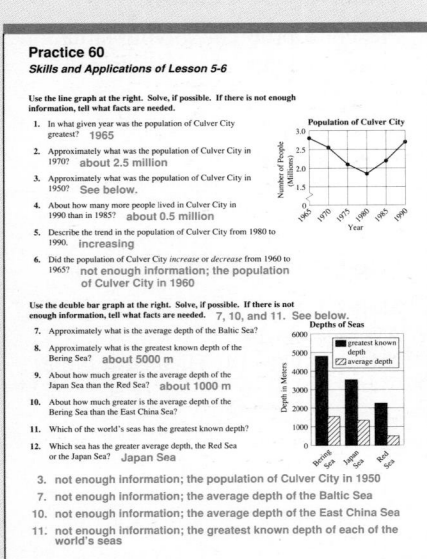

Solve using any problem solving strategy.

5. Sarah has only dimes and quarters. In how many different ways could she make $1.45?
making a table; 3

6. Jeremy paid $10.80 for a phone call. The rates were $2.40 for the first minute, and $.60 for each additional minute. For how many additional minutes was Jeremy charged?
guess and check; 14 min

7. A soccer team sold 496 raffle tickets and collected $372.50. Expenses totaled $75.98. How much money did the team make after expenses?
choosing the correct operation; $296.52

8. A clothing manufacturer distributed 300 suits to several department stores. The manufacturer sent each store the same number of suits. How many suits did each store receive?
not enough information; the number of stores is needed.

WRITING WORD PROBLEMS Answers to Problems 9 and 10 are on p. A14.

For each exercise, write two word problems. One problem should contain not enough information. Then solve the problems, if possible.

9. Use the bar graph at the top of page 192.

10. Use the line graph at the top of page 196.

SPIRAL REVIEW

S 11. Find the difference: $5\frac{5}{6} - 3\frac{5}{12}$ *(Toolbox Skill 19)* $2\frac{5}{12}$

12. Replace the __?__ with >, <, or =: -3 __?__ -7 *(Lesson 3-1)* >

13. Use the diagram at the top of page 207. How many more DC-9's are there than MD-80's in the world fleet? *(Lesson 5-6)* 353

14. Is -3 a solution of the equation $x - 5 = -8$? *(Lesson 4-1)* Yes.

W orking as a nurse in British hospitals, Florence Nightingale (1820–1910) was shocked by the crude treatment patients received. To convince people of the need for reform, she developed new types of statistical graphs that highlighted the poor conditions. Her efforts led to substantial improvements in hospital health care.

Answers will vary.

Research

In a recent health or biology textbook, find two graphs that illustrate data related to health or fitness.

Historical Note

5-7 *Misleading Graphs*

Objective: To recognize how bar graphs and line graphs can be misleading.

1, 2. Estimates may vary.
1. about 55,000; about 60,000; about 50,000
2. about 55,000; about 60,000; about 50,000
3. *Graph B* is visually misleading because its scale has a gap.

1 Use *Graph A*. Estimate the attendance at the rock concert in Franklin, in Midway, and in Sunville.

2 Repeat Step 1, but use *Graph B* instead.

3 Describe how *Graph A* and *Graph B* are different.

Bar graphs illustrate data in a way that makes it easy to compare values. This is one of the advantages of displaying data in bar graphs rather than in tables. A bar graph is misleading, however, when it creates a false visual impression.

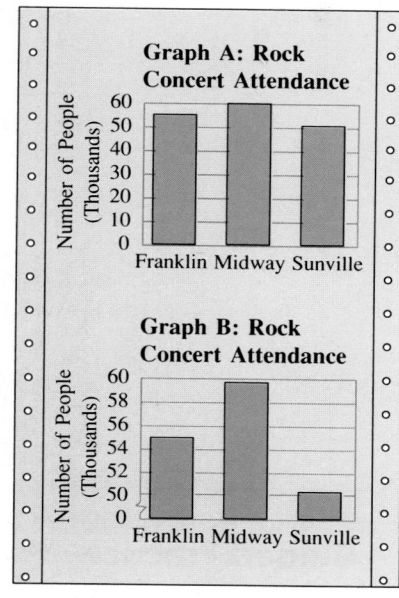

Example 1

Use *Graph B* above.

a. Explain why the graph is visually misleading.

b. Explain why someone who wants to attract performers to Midway might use this graph.

Solution

a. Because there is a gap in the scale, the graph shows only a part of each bar. This creates the false impression that there were great differences in attendance in the three cities.

b. The graph gives a visual impression that attendance in Midway was not only much greater than attendance in either Franklin or Sunville, but was also greater than the combined attendance in the two cities.

✓ Check Your Understanding

In *Graph B*, describe the visual impression of the relationship between the attendance in Midway and the attendance in Sunville.
See *Answers to Check Your Understanding* at the back of the book.

Line graphs are useful for analyzing changes in data. However, a line graph is misleading when it makes these changes appear more dramatic than they really are.

Graphs and Data Analysis **209**

Lesson Planner

Teaching the Lesson
• Lesson Starter 46
• Reteaching/Alternate Approach, p. 186D
• Cooperative Learning, p. 186D
• Toolbox Skills 23–25

Lesson Follow-Up
• Practice 61
• Practice 62: Nonroutine
• Enrichment 50: Application
• Study Guide, pp. 91–92
• Test 22

Warm-Up Exercises

Use the graph on page 207 on Specialty Departments at Grocery Stores in Midtown to answer the following questions.

1. How many specialty departments were represented altogether for 1990? 60

2. Which type of specialty department had the greatest increase from 1980 to 1990? Video

3. How many grocery stores are there in Midtown? not enough information

4. If the bar on the graph falls between two of the labeled lines on the graph, can you determine the exact or an approximate number of specialty departments represented? Explain. Exact; the labeled lines are in increments of two; a bar extending between two labeled lines would represent one additional store more than the previously labeled line.

Lesson Focus

Graphs may present data factually, but they also can be misleading. Today's lesson focuses on recognizing misleading graphs.

209

Key Question

How can you recognize a misleading graph? *gaps in the scale; distorted scales; unnumbered scales*

Reteaching/Alternate Approach

For a suggestion on using cooperative learning to determine whether graphs are misleading, see page 186D.

Additional Examples

1. Use the graphs below.

Graph A: Average Visitors per Day

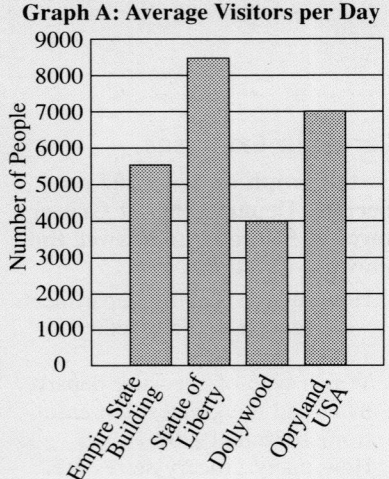

Graph B: Average Visitors per Day

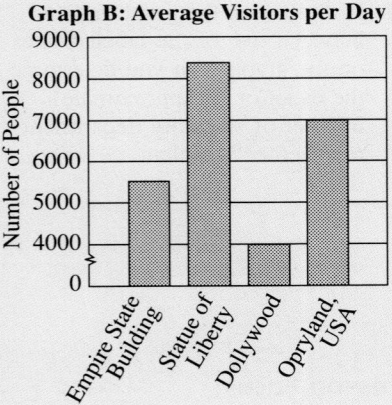

a. Which graph is visually misleading? Why? *Graph B; the gap in the scale gives the impression that the Statue of Liberty has many more visitors than the other places.*

Example 2

Graph C: Employment at Carter's Department Store

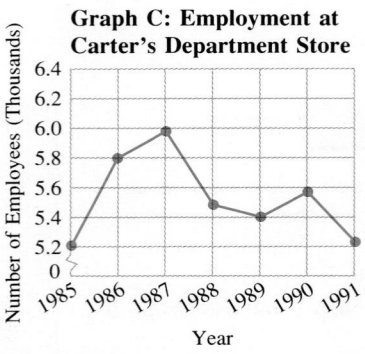

Graph D: Employment at Carter's Department Store

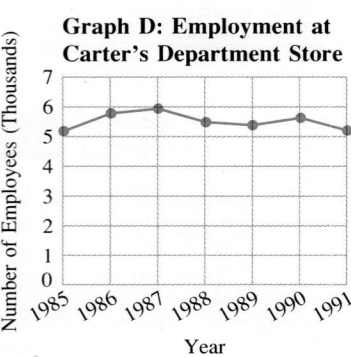

a. Contrast the visual impressions given by the graphs above.

b. Which graph might the department store present to new employees to show them that the store provides job security? Explain.

Solution

a. *Graph C* makes it appear that the employment level changed dramatically between 1985 and 1991. *Graph D* gives the opposite impression.

b. *Graph D*. It makes the employment level appear fairly stable.

Guided Practice

COMMUNICATION « *Reading*

Refer to the text on pages 209–210.

« **1.** State the main idea of the lesson.

« **2.** List two major points that support the main idea.
Answers to Guided Practice 1 and 2 are on p. A14.

Use the bar graph at the right. **3, 4.** Estimates may vary.

3. Do not use the scale. About how many times as high as bar C does bar F appear to be? *about 3 times*

4. Use the scale. About how much greater is the value represented by bar F than by bar C? *about 0.4*

5. *Graph A gives the impression that concert attendance in the three towns was about the same.*

5. Use the graphs on page 209. Explain why someone who wants to attract performers to Franklin would use *Graph A* rather than *Graph B*.

6. Use the graphs in Example 2 above. Which graph might employees use to show that they lack job security? Explain. *Graph C; it makes it appear that the employment level varies dramatically.*

Exercises

Answers to Exercises 2, 4, and 5 are on p. A14.

Use the bar graph at the right.

A

1. Describe the visual impression of the relationship between production levels at Plants C and D.
   about the same

2. Explain why Plant B might use this graph.

3. Explain how the graph might be redrawn to give a better impression of Plant A's performance.
   The gap could be eliminated and intervals of 1 from 0 to 4 could be used.

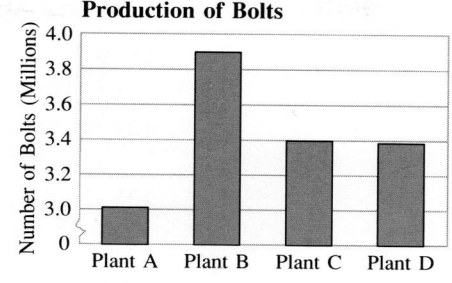

Production of Bolts

Use the line graphs below.

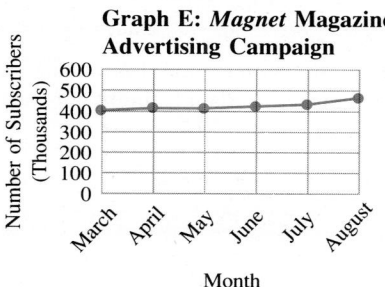

Graph E: *Magnet* Magazine Advertising Campaign

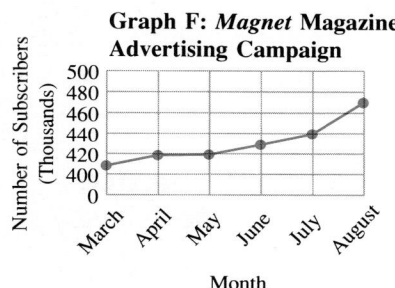

Graph F: *Magnet* Magazine Advertising Campaign

4. Contrast the visual impressions given by the two graphs.

5. Which graph might the magazine owner use to show displeasure at the results of the advertising campaign? Explain.

6. Which graph might the advertising agency use to attract new clients? Explain. Graph F; it may give the client the impression that the agency's advertising campaigns are very successful.

THINKING SKILLS 7–9. Evaluation

Use the bar graph at the right. Classify each statement as *accurate* or *misleading*.

B

7. VCRs cost four times as much at Video Center as they do at Movie House.

8. Prices at the stores vary widely.

9. Video Center charges about $30 more than Movie House charges for a VCR.

   7. misleading   8. misleading   9. accurate

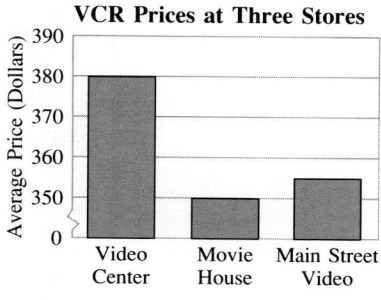

VCR Prices at Three Stores

Graphs and Data Analysis **211**

---

b. Which graph would someone promoting Dollywood want to use? Graph A; it gives a better impression of daily attendance.

2. Use the graphs below.

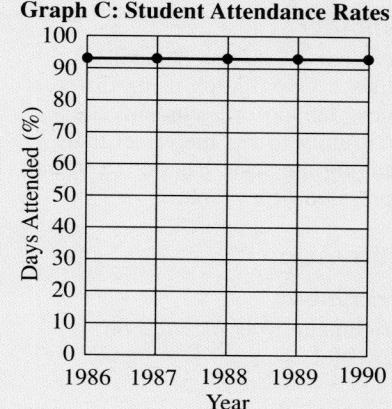

Graph C: Student Attendance Rates

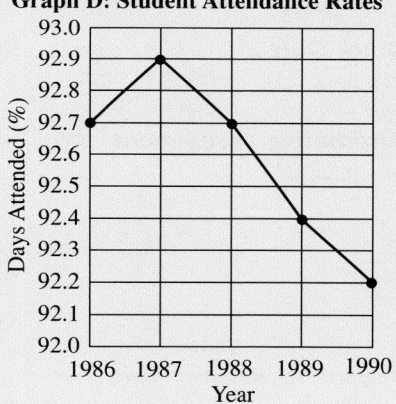

Graph D: Student Attendance Rates

a. Contrast the visual impressions given by graphs C and D. Graph C gives the appearance that attendance is fairly constant at a high rate. Graph D gives the appearance that the attendance rate has declined sharply since 1987.

b. Which graph might the school use if it wanted to improve student attendance? Graph D

Closing the Lesson

Discuss with students why certain graphs may be misleading or not.

211

Suggested Assignments

Skills: 1–9, 14–17
Average: 1–9, 14–17
Advanced: 1–17

Exercise Note

Computer

Exercises 10–13 ask students to redraw a given graph using different scales. This affords students the opportunity to see the effect that changing the scale has on the visual impression of a graph.

Follow-Up

Enrichment

Enrichment 50 is pictured on page 186F.

Practice

Practice 61 is shown below.

Quick Quiz 2

See page 232.

Alternative Assessment

See page 233.

Practice 61, *Resource Book*

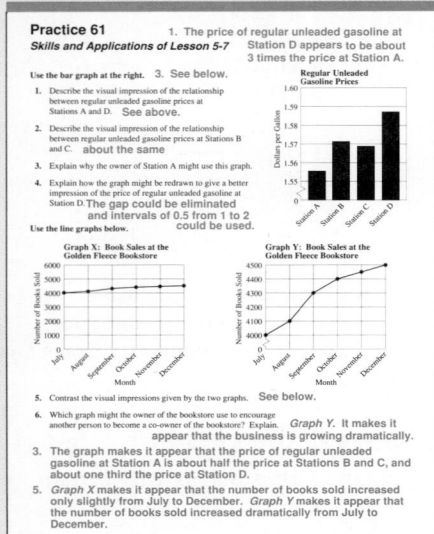

 COMPUTER APPLICATION Answers to Exercises 10–13 are on p. A14.

Use graphing software or graph paper. Refer to *Graph F* on page 211.

C   10. Copy the graph as it is shown. (If you are using a computer, you may not be able to show the gap in the scale.)

11. Redraw the graph, using a scale from 350 to 550 with intervals of 50. Keep the dimensions of the graph the same as in Exercise 10.

12. Repeat Exercise 11, this time using a scale from 200 to 600 with intervals of 100.

13. Use the graphs you drew for Exercises 10–12 to explain how changing the scale on a line graph affects its visual impact.

17. It gives the false visual impression that the level of
SPIRAL REVIEW   production at Plant C was about half that at Plant B.

S   14. Write 36 mL in L.   *(Lesson 2-9)* 0.036 L

15. Find the sum: $2\frac{3}{10} + 7\frac{3}{5}$ *(Toolbox Skill 18)* $9\frac{9}{10}$

16. Solve $-20 = 8q + 6q + 8$. Check the solution.   *(Lesson 4-9)* -2

17. Use the bar graph on page 211 entitled *Production of Bolts*. Explain why the manager of Plant C might not want to use this graph. *(Lesson 5-7)*

Self-Test 2

line graph; see graph on p. A15.
1. Decide whether it would be more appropriate to draw a *bar graph* or a *line graph* to display the data. Then draw the graph.   5-5

Median Prices of New One-Family Homes in the Northeast

Year	1976	1978	1980	1982	1984	1986
Price	$47,300	$58,100	$69,500	$78,200	$88,600	$125,000

3. not enough; the number of miles showing on the odometer before the trip
Solve, if possible. If there is not enough information, tell what facts are needed.

2. Sabrina had $987.63 in her checking account. She made two deposits of $472.38 each. How much did Sabrina deposit altogether? $944.76   5-6

3. Miguel drove 379 mi to visit a friend in Chicago. How many miles did the odometer show when he finished his trip?

4. Use the bar graph on page 211 entitled *Production of Bolts*. Describe the visual impression of the relationship between the levels of production at Plants A and B.   5-7

4. The level of production at Plant B appears to be five times that at Plant A.

Mean, Median, Mode, and Range

Objective: To find the mean, median, mode, and range of a set of data.

Terms to Know
- *statistics*
- *mean*
- *median*
- *mode*
- *range*

The branch of mathematics that deals with collecting, organizing, and analyzing data is called **statistics.** Statisticians use graphs and a variety of *statistical measures* to describe a set of data.

The **mean,** or *average*, of a set of data is the sum of the data items divided by the number of items.

The **median** of a set of data is the middle item when the data are listed in numerical order. If there is an even number of items, the median is the average of the two middle items.

The **mode** of a set of data is the item that appears most often. There can be more than one mode. There can also be no mode, if each item appears only once.

The **range** of a set of data is the difference between the greatest and least values of the data.

Example

Find the mean, median, mode(s), and range of the data.

Marie's Bowling Scores: 149, 183, 149, 193, 147, 193

Solution

- Find the sum of Marie's scores.
  149 + 183 + 149 + 193 + 147 + 193 = 1014

  $mean = \frac{1014}{6} = 169$ ◄—— Divide by the number of scores.

- List the scores in numerical order.
  147, 149, 149, 183, 193, 193

  $median = \frac{149 + 183}{2} = 166$ ◄—— Find the average of the two middle scores.

- modes = 149 and 193   ◄—— Both 149 and 193 appear twice.

- range = 193 − 147 = 46   ◄—— Subtract the least score from the greatest score.

 **Check Your Understanding**

1. In the Example, what does the 6 appearing in the first denominator represent?
2. In the Example, why is it necessary to average 149 and 183 to find the median?
   See *Answers to Check Your Understanding* at the back of the book.

When you divide to find the mean, you may get an answer with many decimal places. When this happens, you should round the answer to the nearest tenth. Indicate that your answer is rounded by using the approximation symbol ≈.

Graphs and Data Analysis **213**

Warm-Up Exercises

1. Use the information in Exercise 6 on page 215 to make a graph of typical temperatures in the Mountain states in February.

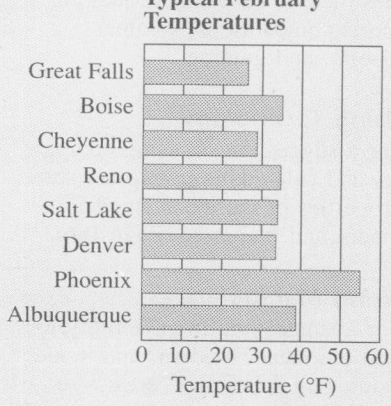

Typical February Temperatures

Great Falls, Boise, Cheyenne, Reno, Salt Lake, Denver, Phoenix, Albuquerque

0 10 20 30 40 50 60
Temperature (°F)

2. Which city had the highest typical temperature? Phoenix
3. What was the difference between the highest and lowest typical temperature of the cities shown? about 28°
4. Which two cities had the most similar typical temperature? Salt Lake City and Denver

Lesson Focus

Ask students if they have ever watched a gymnastics event or a diving competition. Gymnastics and diving are scored by a number of judges. The score received by a gymnast or diver is an average. Today's lesson focuses on statistical measures used to describe a set of data.

Teaching Notes

Students tend to confuse the meanings of statistical terms. Stress the meanings of *mean, median, mode, range,* and *statistics*.

Key Question

How can you find the mean and the median of a set of data? Mean: Find the sum of the data items and divide by the number of items. Median: Arrange the data in numerical order and find the middle term. If there is an even number of terms, find the average of the two middle terms.

Life Skills

Statistical measures are used in many different areas of life; examples include advertising, judging sports competitions, weather reports, and grading.

Using Technology

For a suggestion on using computers and calculators to explore some properties of the mean, median, mode, and range, see page 186C.

Using Manipulatives

For a suggestion on an activity to find the mean, median, and mode using grid paper and cereal, see page 186C.

Reteaching/Alternate Approach

For a suggestion on using student data to find statistical measures, see page 186D.

A calculator can be used to find the mean of a set of data. You would use this key sequence to find the mean of 234, 175, and 488.

234 [+]   175 [+]   488 [=] [÷]   3 [=]

Guided Practice

COMMUNICATION « *Reading*

Identify the statistical measure most closely related to each term.

« **1.** average — mean   « **2.** difference — range   « **3.** middle — median   « **4.** most often — mode

Find the average of the two numbers.

  5. 8 and 14 11   **6.** −15 and 21 3   **7.** 85 and 88 86.5   **8.** 0.9 and 1.7 1.3

List the data in order, then find the middle number(s).

  9. 3.2, 2.7, 2.3, 3.8, 0.3, 3.0 0.3, 2.3, 2.7, 3.0, 3.2, 3.8; 2.7, 3.0

  10. 304, 403, −3040, 3004, 4003, −340, 3400
     −3040, −340, 304, 403, 3004, 3400, 4003; 403

Find the mean, median, mode(s), and range of the data. Answers are given in the order mean, median, mode, and range.

  11. **Typical Number of Prime Time TV Viewers Each Day (Millions):**
     101.1, 94.5, 93.9, 97.0, 89.4, 87.1, 105.8
     ≈95.5; 94.5; none; 18.7

  12. **Points Scored by Steve during a Week of Basketball Practice:**
     2, 7, 7, 10, 12 7.6; 7; 7; 10

  13. **Deposits to and Withdrawals from a Savings Account:**
     $35, $80, −$25, $95, −$50, $60, −$105, −$50, $80, $20
     $14; $27.50; −$50 and $80; $200

Exercises

Find the mean, median, mode(s), and range of the data.

A **1.** **Luggage Space in Large Cars (Cubic Feet):**
     20, 16, 22, 16, 15, 20, 21, 16 18.25, 18; 16; 7

  2. **Jay's Charitable Contributions in a Recent Year:**
     $35, $50, $25, $55, $30, $35, $25, $70, $50, $30, $25, $50
     $40; $35; $25 and $50; $45

  3. **Profits and Losses from School Plays:**
     −$79, $24, −$118, $37, $349, $48 $43.50; $30.50; none; $467

  4. **Base Prices for Two-Door Hatchbacks:**
     $5500, $8300, $6300, $6400, $9900, $7000, $5850, $4350, $6700
     $6700; $6400; none; $5550

5. ≈ −4.2°C; −4°C; none; 7°C   6.   ≈36.6°F; 35.1°F; none; 29.4°F

Find the mean, median, mode(s), and range of the entire set of data in each diagram.

5. **Typical February Temperatures in New England**

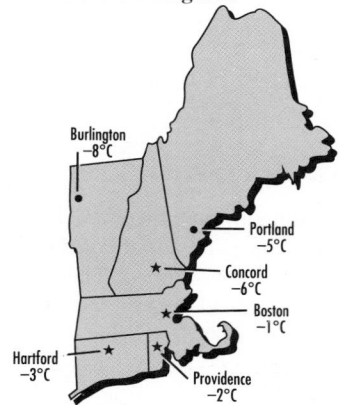

6. **Typical February Temperatures in the Mountain States**

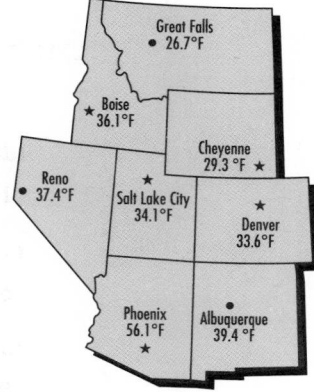

B 7.

75; 60; 60; 135

8.

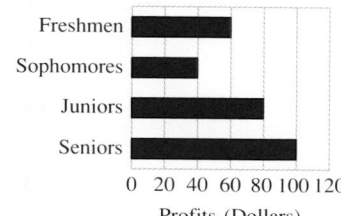

70; 70; none; 60

9. Jack's scores on the first four days of a golf tournament were 72, 76, 71, and 74. What score must he receive on the last day of the tournament in order to have a mean score of 73? 72

10. Palladin spent a total of $100 for five shirts. Later he bought another shirt. He spent an average of $18.78 per shirt for the six shirts. What did Palladin pay for the sixth shirt? $12.68

11. The youngest person in an audience of 400 people is 26 years old. The range of ages in the audience is 38 years. Find the age of the oldest person in the audience. 64 years

12. In May the average of the agents' commissions at Country Realty was $2680.35. Find the agents' total commissions for May. not enough information; the number of agents is needed.

216

Exercise Notes (continued)

Nonroutine Problem

Divide students into cooperative groups. Give each group four test scores and a grading scale. Assuming there are two tests left in this grading period, have each group determine what must be scored on the two tests to (a) maintain the current grade, (b) raise the current grade to the next higher letter grade, and (c) drop the current grade to the next lower letter grade.

Follow-Up

Exploration

Ask students to research some decisions that were made based on statistical data. These could be decisions to market certain products in an area, to terminate employees because sales were not as high as expected, and so on. Students should write a brief report that summarizes the data, the decision that was reached, and an analysis of their own that supports or refutes the decision.

Enrichment

Enrichment 51 is pictured on page 186F.

Practice

Practice 63, Resource Book

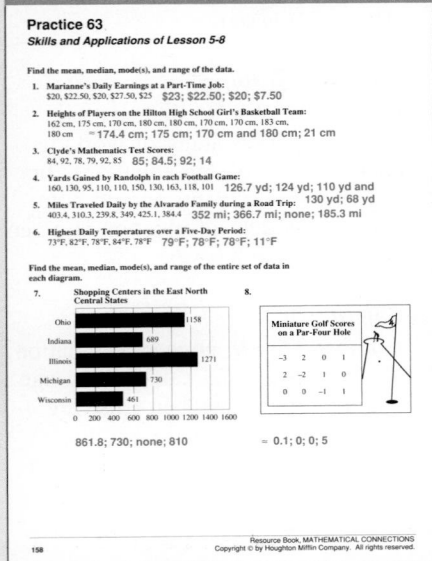

 CALCULATOR

Tell if each calculator mean is *reasonable* or *unreasonable*. If a mean is unreasonable, use your calculator to find the correct mean.

13. 5, 7, 9, 8, 5, 6, 5 ⬛ 4. unreasonable; ≈ 6.4

14. 116, 84, 123, 97, 65 ⬛ 150. unreasonable; 97

15. 12.4, 15.7, 11.9, 14.6 ⬛ 13.65 reasonable

16. 1.25, 3.6, 8.12, 6.9 ⬛ 0.78 unreasonable; ≈ 5.0

PROBLEM SOLVING/APPLICATION

In gymnastics competitions, there are usually four judges and a head judge. They rate performances on a scale of 0 to 10. The highest and lowest scores from the four judges are dropped. The final score is the mean of the two remaining scores. The head judge's score is used only if there is some question about the fairness of the scoring.

	Head Judge	Judge 1	Judge 2	Judge 3	Judge 4
Sachi	8.8	8.4	8.9	8.5	9.0
Janice	9.1	9.3	7.5	7.4	9.2

Use the table above.

17. What score did the head judge give to Sachi? 8.8

18. Which two of Sachi's scores will be dropped? 8.4 and 9.0

19. Using the method described above, what will be Sachi's final score? 8.7

20. Using the method described above, what will be Janice's final score? 8.35

C 21. Explain how the method of scoring explained above compares with finding the median of the four judges' scores. It is the same as finding the median of the four judges' scores.

22. **THINKING SKILLS** If you were the head judge, how would you rule on the scores that the other four judges gave to Janice? Explain.
Evaluation; Answers may vary. The scores given by Judge 2 and Judge 3 are much lower than the others. The head judge's score could be used in this case to obtain a more reasonable score of 8.6.

SPIRAL REVIEW

S 23. Is 8460 divisible by 2? by 3? by 4? by 5? by 8? by 9? by 10? (*Toolbox Skill 16*) yes; yes; yes; yes; no; yes; yes.

24. Find the mean, median, mode(s), and range: 4, −4, −18, 26 (*Lesson 5-8*) 2; 0; none; 44

25. Evaluate $2n^3$ when $n = 5$. (*Lesson 1-9*) 250

Challenge

Find a set of five numbers with a mean of 24, a median of 15, a mode of 10, and a range of 41. 10, 10, 15, 34, and 51

Modeling the Mean

Objective: To build a concrete model to represent the mean of a set of numbers.

Materials

- strip of cardboard 1" × 12"
- ruler
- hole punch
- string
- tape
- 7 paper clips

Mark off a strip of cardboard as shown in the diagram at the left below. Punch holes one inch apart and one-eighth inch from the edge. Using string and tape, suspend the strip of cardboard over the edge of a desk.

1b, 1d, 1e. Answers will vary. Examples: **1. b.** 77, 77   **d.** 77, 78
     e. 77, 77, 77

Activity I *Exploring the Mean*

1   Hang paper clips in the holes for the given numbers. To reach a balance, hang in the appropriate holes as many additional paper clips as there are blanks. Record the results. You may not use the hole in the center.

     a. 71, 74, 81, __?__ 78
     b. 71, 74, 81, __?__, __?__
     c. 71, 74, 80, __?__ 79
     d. 71, 74, 80, __?__, __?__
     e. 71, 74, 80, __?__, __?__, __?__

2   Find the mean of the data in each part of Step 1. How do the means compare with the number at the center of the strip of cardboard?
Each mean is 76 and equals the number at the center.

3   Would any of the means in Step 2 change if you hung a paper clip in the hole in the center of the strip of cardboard? Explain.
No; it would not change the balance.

Activity II *Exploring the Median and the Mode*

1   Find the median and the mode of the data:
71, 73, 74, 76, 78, 78, 79   76; 78

2   Hang a paper clip in each hole corresponding to a number in Step 1. If a number appears more than once, use a paper clip for each time it appears. Do the data balance around the median? No.

3   Do you think the data balance around the mode found in Step 1?
No.

4   Under what condition(s) do you think a set of data would balance around the mean, the median, and the mode? Use the strip of cardboard to produce such a set of data.
See answer on p. A15.

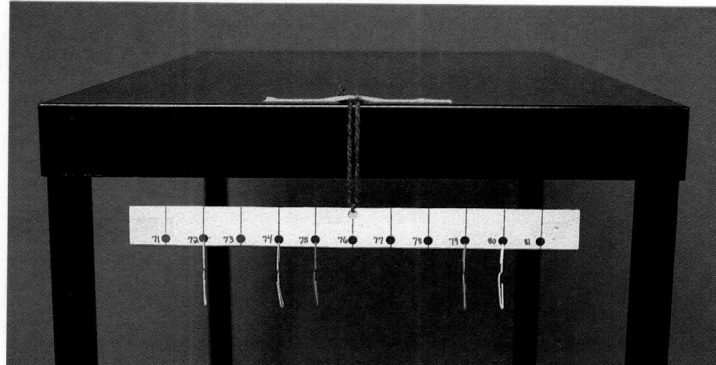

Teaching Notes

Using Manipulatives

This lesson uses a manipulative technique to reinforce the concepts of mean, median, and mode through visual representations.

Number Sense

The moving of the paper clips helps students visualize the relationship of the data to the statistical measurement.

Extension

To extend *Activities I and II*, have students show that the sum of the distances to data items less than the mean equals the sum of the distances to data items greater than the mean. Have them show that this is not necessarily true for the median.

Teaching the Lesson
- Lesson Starter 48
- Reteaching/Alternate Approach, p. 186D
- Cooperative Learning, p. 186D

Lesson Follow-Up
- Practice 64
- Enrichment 52: Thinking Skills
- Study Guide, pp. 95–96

Warm-Up Exercises

Use the data below.

Michael had test scores of 78, 72, 85, 91, 85, and 87.

1. What is the mean? 83

2. What is the mode? 85

3. What is the median? 85

4. What is the range? 19

Lesson Focus

To motivate today's lesson on choosing the appropriate statistic, use the application below.

Application

At Burger Barn, eight employees each earn $15,000 per year. The assistant manager earns $40,000, and the manager earns $69,000. The manager tells a new employee that the average salary is $23,000. Is the manager using an appropriate statistic to describe the data? No; the $69,000 distorts the average.

Teaching Notes

Key Question

Which statistical measure best describes data under each condition?

a. extreme value present in data median

b. data of yogurt flavor preferences mode

c. data grouped close together mean

DECISION MAKING

5-9

Choosing the Appropriate Statistic

Objective: To decide whether a given measure of central tendency is appropriate for a set of data.

Terms to Know

- *measure of central tendency*

The mean, median, and mode are sometimes called **measures of central tendency** because they describe how data are *centered*. Each of these measures describes a set of data in a slightly different way.

- The *mean* is the most familiar statistical measure. It is an appropriate measure of central tendency when the data are reasonably centered around it. However, the mean can be distorted by an *extreme* value, one much greater or less than the other values.

- When there is an extreme value that distorts the mean, the *median* is an appropriate measure of central tendency.

- When data cannot be averaged or listed in numerical order, the *mode* is an appropriate measure of central tendency.

Example

Ace Mechanics has seven employees. Their salaries are $18,000, $19,000, $19,000, $22,000, $24,000, $30,000, and $120,000. Decide whether the mean describes these data well. Explain.

Solution

No. The mean is $36,000 and is greater than six of the seven salaries. It is distorted by the extreme value of $120,000.

✒ **Check Your Understanding**

In the Example, how do you find that the mean is $36,000?
See *Answers to Check Your Understanding* at the back of the book.

Guided Practice

COMMUNICATION «*Discussion*

Which measure of central tendency do you think is generally used in each situation? Explain.
Answers to Guided Practice 1–3 are on p. A15.

«**1.** reporting yearly incomes for company employees

«**2.** summarizing scores on a math test

«**3.** determining the size of shoe in greatest demand

4. Manuel reports that people prefer Brand X more often than any other brand. Which measure of central tendency is he using? mode

5. Sharon says that half her employees earn more than $30,000 per year and half earn less. Which measure of central tendency is she using? median

6. The breeds of the six dogs in a certain neighborhood were retriever, poodle, collie, retriever, bulldog, and beagle. Decide which measure of central tendency is appropriate for these data. Explain.
mode; the data cannot be averaged or listed in numerical order.

218 Chapter 5

Answers to Exercises 1–7, 9, and 10 are on p. A15.

Exercises

A **1.** The heights of the five starting players on a basketball team are 75 in., 74 in., 73 in., 70 in., and 68 in. Decide whether the mean describes these data well. Explain.

2. The monthly car payments for Ted Nelson's last eight customers were $266, $285, $285, $285, $285, $315, $325, and $344. Decide whether the mode describes these data well. Explain.

3. At the end of the winter, the seven sweaters left in stock at Angie's Outlet Store were brown, orange, orange, green, brown, orange, and orange. Decide which measure of central tendency is appropriate for these data. Explain.

4. The profits at five schools that sold Gems & Jewels products were $318.22, $440.79, $607.16, $1090.38, and $4790.15. Decide whether the mean or the median is a better measure of central tendency for these data. Explain.

5. The sizes of the thirteen families living on Wood Road are 2, 3, 3, 3, 4, 4, 5, 5, 6, 6, 7, 8, and 9. Decide whether the median or the mode is a better measure of central tendency for these data. Explain.

6. The favorite vegetables of the six members of the Walker family are peas, carrots, beans, peas, carrots, and carrots. Decide which measure of central tendency is appropriate for these data. Explain.

B **7.** The numbers of hours worked by the seven nurses at Dale General Hospital in one week were 20, 20, 20, 29, 48, 52, and 56. Decide which measure of central tendency might encourage a nurse who wants to work at least 35 h per week to apply for a job at Dale General. Explain.

SPIRAL REVIEW

S **8.** Solve $13 = \dfrac{b}{12}$. Check the answer. *(Lesson 4-4)* 156

9. Graph the points on a coordinate plane: $A(3, 0)$, $B(2, -4)$, $C(-3, -2)$ *(Lesson 3-8)*

10. A student's scores on five tests were 84, 78, 89, 94, and 74. Decide whether the mean describes these data well. Explain. *(Lesson 5-9)*

11. PATTERNS Find the next three numbers in the pattern: 5, 8, 12, 17, __?__, __?__, __?__ *(Lesson 2-5)* 23, 30, 38

Graphs and Data Analysis **219**

Reteaching/Alternate Approach
For a suggestion on using cooperative learning to study statistical measures, see page 186D.

Additional Example
The ages of the people in a room are 92, 23, 1, 2, 1, and 1. Decide whether the mean best describes the data. Explain. No; the mean is 20. Only one of the six people is near 20. The data are distorted by the extreme value.

Closing the Lesson
Ask students to explain how an extreme value can distort data, and when the use of the mode and median is most appropriate.

Suggested Assignments
Skills: 1–11
Average: 1–11
Advanced: 1–11

Follow-Up

Enrichment
Enrichment 52 is pictured on page 186F.

Practice
Practice 64, *Resource Book*

Practice 64
Skills and Applications of Lesson 5-9

Ex. 1–6. See below.
1. The ages of the children in the Mendoza family are 17, 15, 14, and 2. Decide whether the mean describes these data well. Explain.

2. Paula took the SAT twice in her junior year and twice in her senior year in high school. Her verbal scores were 480, 520, 540, and 590. Decide whether the mean describes these data well. Explain.

3. The students in the ninth grade at Winston High School were asked to select their favorite lunch. Of the 100 students in the class, 32 prefer pizza, 29 prefer hamburgers, 28 prefer spaghetti, and 11 prefer chicken. Decide which measure of central tendency is appropriate for these data. Explain.

4. The Clinton School District hired five new teachers. Their salaries are $22,500, $23,500, $25,500, $26,000, and $48,500. Decide whether the mean or the median is a better measure of central tendency for these data. Explain.

5. Paloma's mathematics test scores were 70, 88, 70, 78, 94, 70, and 98. Decide whether the median or the mode is a better measure of central tendency for these data. Explain.

6. A group of business travelers were asked which type of transportation they usually use. Their answers were automobile, automobile, airplane, train, airplane, airplane, automobile, airplane, airplane, and train. Decide which measure of central tendency is appropriate for these data. Explain.

1. No; the mean is 12. It is distorted by the extreme value 2.

2. Yes; the mean is 532.5. It is not distorted by an extreme value.

3. mode; the data cannot be averaged or listed in numerical order.

4. median; the median is $25,500 and the mean is $29,200. The mean is distorted by the extreme value $48,500.

5. median; the median is 78 and the mode is 70. The mode is less than all the other test scores.

6. mode; the data cannot be averaged or listed in numerical order.

159

220

Teaching Notes

Database management systems simplify working with large amounts of data. Point out, for example, how much simpler it is for a computer to list all metropolitan areas with transportation costs of less than $4500 than it is to find them by hand.

Key Question

How can you use the information in the given database to compare expenditures by region? The data can be grouped by region and then measures of central tendency can be found and compared.

Life Skills

Database software is used regularly in many businesses, including those dealing with credit, accounting, stocks, and financial planning.

Suggested Assignments

Skills: 1–9
Average: 1–9
Advanced: 1–9

FOCUS ON
COMPUTERS

Using Database Software

Objective: To use database software to analyze a set of data.

People who have to analyze great amounts of data will often use a **database management system.** They can then more easily sort through tables of data like the one below. In some cases, database software can also be used to calculate statistical measures.

Average Annual Household Expenditures (Dollars)

Metropolitan Area	Region	Food	Housing	Transportation
Atlanta	South	3472	9462	5578
Buffalo	Northeast	3956	6087	4048
Cleveland	Midwest	4012	7462	6084
Dallas/Fort Worth	South	4006	8953	6384
Honolulu	West	5126	9386	6023
Milwaukee	Midwest	3982	8488	4343
New York City	Northeast	4484	10191	4677
Philadelphia	Northeast	4262	8942	4700
Portland	West	3939	7961	4231
San Francisco	West	4422	11174	5600
St. Louis	Midwest	3757	7584	4590
Washington, D.C.	South	4388	12069	6751

Exercises

Answers to Exercises 8 and 9 are on p. A15.

Use database software to create a database containing the information in the table above.

1. What names did you give the fields in the database?
   Metropolitan Area, Region, Food, Housing, Transportation

2. How many records did you enter in the database? 12

3. Find all the records whose region is *South*. Which metropolitan areas do they represent? Atlanta, Dallas/Fort Worth, Washington, D.C.

4. Find all the records with food expenditures less than $4000. Which metropolitan areas do they represent?
   Atlanta, Buffalo, Milwaukee, Portland, St. Louis

5. Sort the records in decreasing order of transportation expenditures. Which record appears first? Washington, D.C.

6. Sort the records alphabetically both by region and by metropolitan area. Which record appears first? Cleveland

7. Find the mean of the food expenditures for all the metropolitan areas.
   $4150.50

8. Describe how the database software can help you find the median of the housing expenditures for all the metropolitan areas.

9. Which metropolitan area do you think has the greatest total costs for food, housing, and transportation? Explain how you made your decision.

Data Collection and Frequency Tables

Objective: To calculate statistical measures for data in a frequency table.

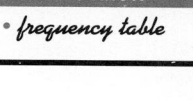 **APPLICATION**

Mrs. Fischer's first period gym class was doing a unit on health. The students were asked to measure their pulse rates. They organized the data in a **frequency table** like the one shown below.

| Pulse Rates (beats/min) | | |
Rate	Tally	Frequency
69	卌 I	
70	卌 I	6
71	I	6
72	卌 卌	1
73	卌 I	10
		6
Total:		29

Example

Find the mean, median, mode(s), and range of the data in the frequency table above.

Solution

• Multiply each *rate* by its *frequency*:

$69 \cdot 6 = 414$ ← In finding the mean, a calculator may be helpful.
$70 \cdot 6 = 420$
$71 \cdot 1 = 71$
$72 \cdot 10 = 720$
$73 \cdot 6 = \underline{438}$
Add: 2063

Divide by the total of the frequencies:

$2063 \div 29 \approx 71.1$

The mean is about 71.1 beats/min.

• There are twenty-nine items. The median is the middle item, so look for the fifteenth item.

The median is 72 beats/min.

• Look for the rate with the most tally marks.

The mode is 72 beats/min.

• Subtract: $73 - 69 = 4$

The range is 4 beats/min.

 **Check Your Understanding**

1. In the Example, why was the number 29 used to find the mean?
2. In the Example, how do you determine that 72 is the fifteenth item?
See *Answers to Check Your Understanding* at the back of the book.

Warm-Up Exercises

Use the information on page 220.

1. Create a double bar graph to compare the annual household expenditures in Atlanta and Portland.

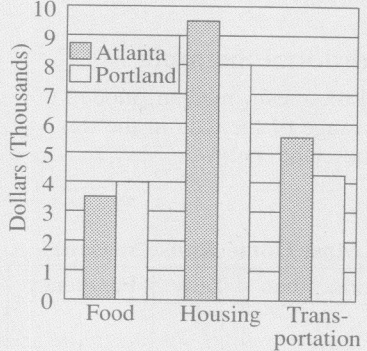

Average Annual Household Expenditures

2. Which of the two cities is less expensive? Explain. Portland; food costs more in Portland, but the total costs are less.
3. Find the average cost for food in the Western region. $4495.67
4. Find the median cost of transportation in the South. $6384
5. What is the range of the average housing costs in the Northeast? $10,191 − $6,087 = $4,104

Guided Practice

«**1.** COMMUNICATION «*Writing* Make a frequency table for the data below.
Hours Worked: 5, 6, 6, 5, 7, 8, 6, 5, 7, 5, 5, 6, 6, 8, 5, 6, 8, 6
See answer on p. A15.

Use the frequency table at the right.

Soup Prices	Tally	Frequency
$.69	IIII	4
$.79	I	1
$.89	II	2
$.99	III	3
	Total:	10

2. What is the total of the frequencies? 10

3. What was the total price for all of the soup? $8.30

4. Find the mean, median, mode(s), and range of the data.
$.83; $.84; $.69; $.30

Exercises

Find the mean, median, mode(s), and range of the data.

A **1.**

Miles per Gallon	Tally	Frequency
30	HHI	5
31	III	3
32	III	3
33	I	1
	Total:	12

31; 31; 30; 3

2.

Bat Lengths (Inches)	Tally	Frequency
30	III	3
31	III	3
32	HHI II	7
34	I	1
	Total:	14

31.5; 32; 32; 4

CAREER/APPLICATION

A *market researcher* collects data from consumers. These data are used to help people who provide goods and services decide how best to satisfy the needs of consumers.

Number of Movies Seen by Students in a Month

4	2	4	4	3	5
5	3	4	12	4	5
4	5	3	4	2	4
3	4	4	3	3	15

Use the data at the right.

B **3.** Make a frequency table for the data. See answer on p. A15.

4. Use the frequency table that you made for Exercise 3. Find the mean, median, mode(s), and range of the data. ≈ 4.5; 4; 4; 13

5. Decide whether the mean describes the data well. Explain.
No; the mean, 4.5, is distorted by the extreme values 12 and 15.

C **6.** WRITING ABOUT MATHEMATICS Ask ten of your classmates how many movies they have seen in the past month. Write a paragraph comparing your data to the data given in the table. Answers will vary.

7–8. Answers may vary due to
RESEARCH possible changes in state laws.

7. Use an almanac to find the minimum age requirement for obtaining a moped license in each of the fifty states. Make a frequency table for the data.

8. Use the frequency table that you made for Exercise 7. Find the mean, median, mode(s), and range of the data.

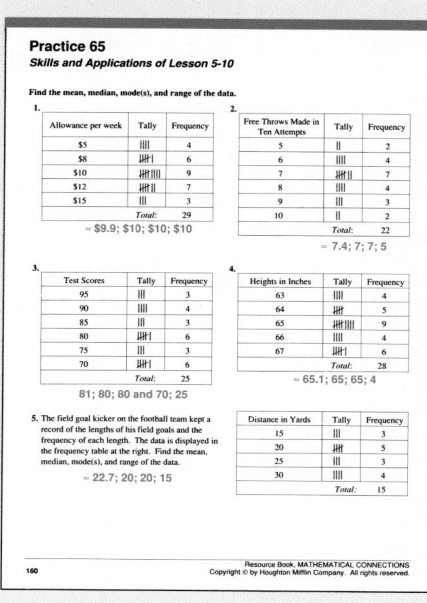

SPIRAL REVIEW

S 9. Use the formula $g = a - h$. Let $g = 756$ km/h and $h = 14$ km/h. Find a.   *(Lesson 4-10)* 770 km/h

10. Find the quotient: $4\frac{1}{2} \div \frac{3}{4}$ *(Toolbox Skill 21)* 6

11. The number of laps that Olga swam on Friday was four more than twice the number of laps she swam on Monday. If Olga swam 76 laps on Friday, how many laps did she swim on Monday?   *(Lesson 4-8)*
36 laps

12. **DATA,** *pages 186–187* Find the mean, median, mode(s), and range of the data in the frequency table about Olympic medals.   *(Lesson 5-10)*
≈ 5.7; 2; 1 and 2; 26

Self-Test 3

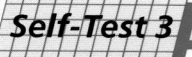

$43.75; $40; $50 and $20; $90

1. The amounts withdrawn in one hour from an automatic teller machine were $20, $10, $50, $70, $30, $50, $100, and $20. Find the mean, median, mode(s), and range of these data.   5-8

2. The ages of the twelve students in a dance class were 14, 13, 15, 14, 15, 13, 16, 15, 16, 15, 15, and 13. Decide whether the mean describes these data well. Explain. Answers to Self-Test 2 and 3 are on p. A15.   5-9

3. The five residents in a dormitory suite were from New York, Maine, New York, Iowa, and New York. Decide which measure of central tendency is appropriate for these data. Explain.

4. Find the mean, median, mode(s), and range of the data.   5-10

7; 7; 8; 3

Hours of Sleep	Tally	Frequency
5					3
6	ЖН		6
7	ЖН ||||	9
8	ЖН ЖН ||	12
	Total:	30

Graphs and Data Analysis   **223**

Exercise Note

Life Skills

Exercises 7 and 8 require students to research actual data and then to make and use a frequency table containing the data.

Follow-Up

Enrichment

Enrichment 53 is pictured on page 186F.

Practice

Practice 65 is shown below.

Quick Quiz 3

See page 230.

Alternative Assessment

See page 230.

Practice 65, *Resource Book*

Using Graphs in Business

Objective: To analyze profits and losses using a double line graph.

Veronica studied cosmetology at South Vocational High School and graduated as a licensed beautician. She opened her own manicure salon at a nearby mall. She kept careful records of the amount of money she collected, the amount of money she spent, and the number of manicures. After two months she drew the double line graph shown at the right to display her progress.

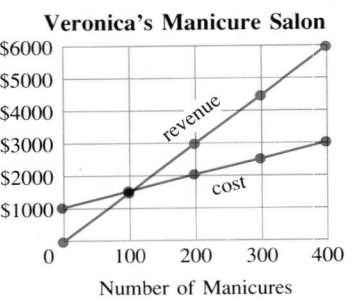

Veronica's Manicure Salon

The amount of money that a company collects is called **revenue**. The amount of money that the company spends is called **cost.** When revenue is greater than cost, the company makes a *profit*.

$$\text{profit} = \text{revenue} - \text{cost}$$

When cost is greater than revenue, the company experiences a *loss*.

$$\text{loss} = \text{cost} - \text{revenue}$$

When the company's cost and revenue are equal, the company is said to *break even*.

Example

Use the double line graph above.

a. What was the revenue from 400 manicures?

b. During which interval did Veronica lose money?

c. How many manicures had Veronica done when she broke even?

d. Estimate the profit from 300 manicures.

Solution

a. Locate the point for 400 on the revenue line, which is *blue*. The revenue from 400 manicures was $6000.

b. Veronica lost money during the interval where the line representing cost is *above* the line representing revenue. She lost money during the interval from 0 up to 100.

c. Veronica broke even at the point where the two lines meet. She had done 100 manicures when she broke even.

d. The revenue from 300 manicures was about $4500. The cost for 300 manicures was about $2500.
Subtract: 4500 − 2500 = 2000
The profit from 300 manicures was about $2000.

Exercises

Jack opened a photo shop specializing in passport photos. He monitored his business for three months. Then he displayed his progress by drawing the double line graph at the right. **1, 2, 5, 6, 8–10.** Estimates may vary.

Jack's Passport Photo Shop

Use the double line graph at the right.

1. Estimate the cost for 2000 customers. about $14,000

2. Estimate the revenue from 1800 customers. about $18,000

3. During which interval did Jack lose money? 0 up to 800

4. During which interval did Jack make a profit? 801 to 2000

5. Estimate Jack's loss from photographing 400 people. about $2000

6. Estimate Jack's profit from photographing 1600 people. about $4000

7. How many people had Jack photographed when he broke even? 800

8. Estimate Jack's revenue at the point when he broke even. about $8000

9. Estimate Jack's profit per customer for 1200 customers. about $1.67

10. Estimate Jack's profit per customer for 2000 customers. about $3

11. Jack's cost for 2800 customers was $18,000. His revenue was $28,000. What was his profit? $10,000

12. Jack's cost for 2400 customers was $16,000. His profit was $8,000. What was his revenue? $24,000

GROUP ACTIVITY **13.** Answers will vary. Examples: camera equipment, film, rent for shop, office furniture

13. Make a list of four things that might be included in the initial cost of opening a passport photo shop.

14. Make a list of four things that might be included in the initial cost of starting any business. Compare this list to the list you made for Exercise 13.

    See answer on p. A15.

Exercise Notes

Estimation

Exercises 1, 2, 5, 6, and 8–10 require students to estimate answers from the data presented in the graph.

Cooperative Learning

When comparing their lists from *Exercises 13 and 14,* groups should become aware that there are common items on both.

Follow-Up

Extension

Companies often make projections based on costs and revenues. If profits are not possible, a decision may be made to close an operation. For example, suppose each manicure requires one hour of time and Veronica is working alone. What is the maximum amount of profit she can earn in four 40-hour weeks? about $600 Is it reasonable to expect that there will be a customer every hour and that there will be no cancellations? No What are some options that Veronica might explore to increase her profit? hire an assistant, raise prices, work more hours

Quick Quiz 1 *(5-1 through 5-4)*

1. Draw a bar graph to display the data in the table below.

Average Price Paid to U.S. Farmers

Type of Meat	Price per 100 lb
Hogs	$42.30
Beef	$66.60
Veal	$89.20
Sheep	$25.60
Lambs	$69.10

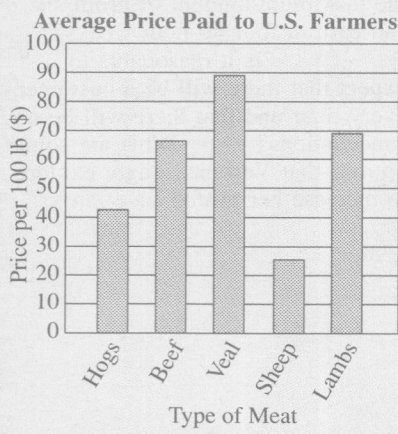

Average Price Paid to U.S. Farmers

2. Use the double bar graph below. Estimate the increase in direct aid.   about $160,000,000

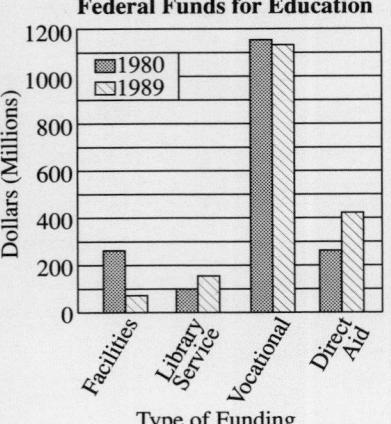

Federal Funds for Education

Chapter Review

Terms to Know

graph (p. 188)
data (p. 188)
pictograph (p. 188)
key (p. 188)
bar graph (p. 192)
scale (p. 192)
legend (p. 192)
line graph (p. 196)
increasing trend (p. 196)

decreasing trend (p. 196)
statistics (p. 213)
mean (p. 213)
median (p. 213)
mode (p. 213)
range (p. 213)
measure of central tendency (p. 218)
frequency table (p. 221)

Choose the correct term from the list above to complete each sentence.

1. The __?__ of a set of data is the item that appears most often. mode

2. A(n) __?__ is used to show data that change continuously over time. line graph

3. A branch of mathematics that deals with collecting, organizing, and analyzing data is called __?__. statistics

4. A group of numerical facts is often referred to as __?__. data

5. In a double bar graph, the __?__ is included to help you distinguish between the two types of bars. legend

6. The __?__ on a line graph allows you to estimate the values represented by the points on the line. scale

---

9. not enough information; the amount of corn produced in 1980

Use the pictograph at the right. Solve, if possible. If there is not enough information, tell what facts are needed. *(Lessons 5-1, 5-6)*

7. About how many bushels of soybeans were produced in 1980?
   about 1800 million

8. About how many more bushels of soybeans were produced in 1985 than in 1970? about 800 million

9. In 1980, about how much more corn was produced than soybeans?

10. Estimate the income from the sale of all the soybeans produced in 1975.
    about $7872 million

Soybean Production on United States Farms

	Average Price per Bushel
1970	$2.85
1975	$4.92
1980	$7.57
1985	$5.05

= 400 million bushels

11. Draw a pictograph to display the data. *(Lesson 5-1)* See answer on p. A16.

Parks in Four Major Cities

City	Fort Worth	Minneapolis	New Orleans	Omaha
Parks	136	153	250	99

12–13. Estimates may vary. Accept reasonable estimates.

Use the bar graph at the right. *(Lesson 5-2)*

12. About how much money was spent for advertising on television?
about $25 billion

13. About how much more money was spent for advertising in newspapers than for advertising in magazines?
about $25 billion

14. For which method of advertising was the amount of money spent about $20 billion? direct mail

Money Spent for Advertising

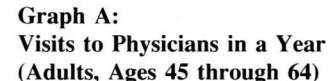

Use Graph A below. *(Lesson 5-3)*

15. During which five-year interval did the greatest increase in the average number of visits occur? 1980–1985

16. Estimate the average number of visits in 1975. Estimate may vary; about 5.5 visits.

17. In which given year was the average number of visits about 6? 1985

Use Graph A and Graph B below. *(Lesson 5-7)*

18. Contrast the visual impressions given by the two graphs. See answer on p. A16.

19. Which graph might someone use to show that the average number of visits to physicians did not change very much between 1970 and 1985? Graph A

Graph A:
Visits to Physicians in a Year
(Adults, Ages 45 through 64)

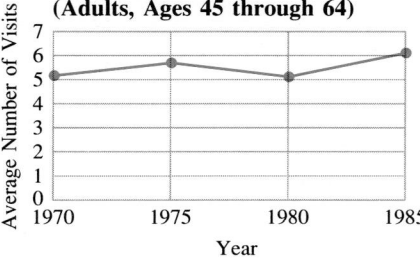

Graph B:
Visits to Physicians in a Year
(Adults, Ages 45 through 64)

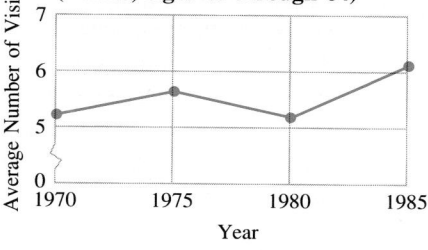

Use the line graph below.

3. Estimate the number of recreational vehicles in 1980. about 180,000

4. Describe the overall trend in the number of recreational vehicles. increasing

Number of Recreational Vehicles

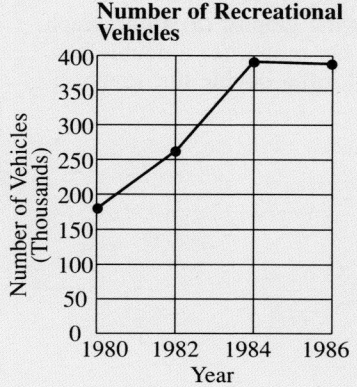

5. Draw a line graph to display the data.

Average Number of Truck Campers

Year	Truck Campers (Thousands)
1970	95.9
1975	44.3
1980	5.0
1985	6.9

Number of Truck Campers

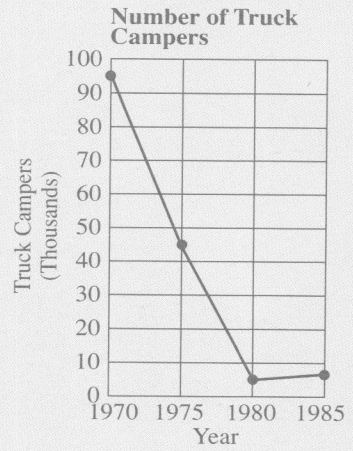

Alternative Assessment
See next page.

228

Chapter 5

Decide whether it would be more appropriate to draw a *bar graph* or a *line graph* to display the data. Then draw the graph.
(Lessons 5-4, 5-5)

20. **Attendance at Professional Basketball Playoffs (Millions)** bar graph

Year	1975	1979	1983	1987
Attendance	0.7	0.9	0.6	1.1

21. **Class I Locomotives in Service (Thousands)** line graph

Year	1975	1979	1983	1987
Locomotives	28	28	26	20

Solve, if possible. If there is not enough information, tell what facts are needed. *(Lesson 5-6)*

22. The price per pound for apples is $.99. How much will a bag of 12 apples cost? not enough information; the weight of the bag of apples

23. The 16 members of a swim team swam a total of 1000 laps to raise money for charity. They raised $250. How much money did they raise per lap? $.25

Find the mean, median, mode(s), and range of the data.
(Lessons 5-8, 5-10)

24.

Sweater Prices

$25.49 $28.99 $34.60 $47.99 $79.48

$43.31; $34.60; none; $53.99

25.

Number of Siblings	Tally	Frequency
0	卌 I	6
1	卌	5
2	卌 III	8
3	I	1
	Total:	20

1.2; 1; 2; 3

26. Typical February temperatures in the West North-Central States are −10°C, −7°C, −8°C, −2°C, −4°C, 1°C, and 2°C. Find the mean, median, mode(s), and range of these data. *(Lesson 5-8)* −4°C; −4°C; none; 12°C

27. The seven faculty members in an art department specialized in photography, painting, sculpting, printmaking, painting, painting, and graphic design. Decide which measure of central tendency is appropriate for these data. Explain. *(Lesson 5-9)*
mode; the data cannot be averaged or listed in numerical order.

Chapter Test

2, 3. Estimates may vary.

1. Draw a pictograph to display the data. See answer on p. A16.   5-1

Hospital Beds

State	Alabama	Florida	Georgia	Mississippi
Beds	24,900	61,500	33,700	16,900

2. Use the double bar graph below. Estimate the expected increase in the number of jobs in Los Angeles from 1985 to 2010. about 1.5 million   5-2

3. Use the double bar graph below. In 1985, about how many more jobs were there in New York than in Seattle? about 3.5 million

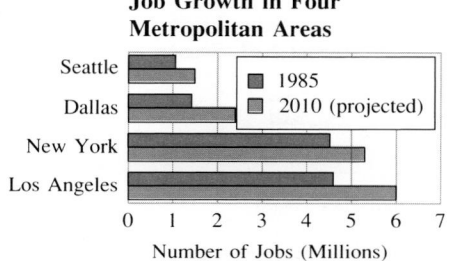

Job Growth in Four Metropolitan Areas

Seattle, Dallas, New York, Los Angeles
■ 1985 ■ 2010 (projected)
Number of Jobs (Millions) 0 1 2 3 4 5 6 7

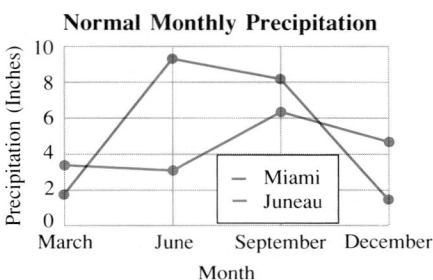

Normal Monthly Precipitation

Precipitation (Inches) — Miami — Juneau
March June September December
Month

Use the bar graph at the right.
Answers to Chapter Test 4 and 5 are on p. A16.

★ **4.** What do you think the scale could represent?

★ **5.** Use your answer to Exercise 4. What do you think is an appropriate title for the graph?

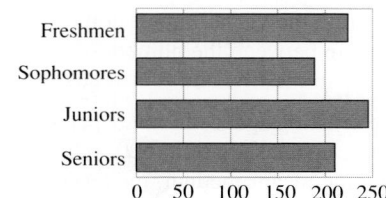

Freshmen
Sophomores
Juniors
Seniors
0   50   100   150   200   250

Exercises 6 and 7 refer to the double line graph above.

6. In which given month does Juneau normally get about six inches of precipitation? September   5-3

7. In June, about how much greater is precipitation in Miami than in Juneau? Estimates may vary; about 6 in.

8. Draw a bar graph to display the data. See answer on p. A16.   5-4

Maximum Depths of the Great Lakes

Lake	Erie	Huron	Michigan	Ontario	Superior
Depth in Feet	210	750	923	802	1333

Alternative Assessment

Exercises 4 and 5 are marked with stars to indicate that they represent alternative forms of assessment. Both exercises test students' abilities to finish a partially completed graph.

Chapter Test (Form A)

Test 24, *Tests*

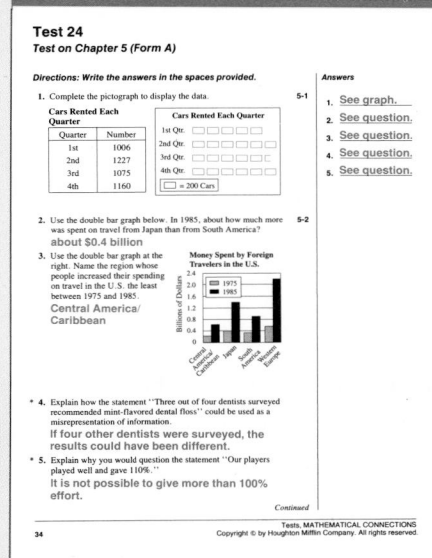

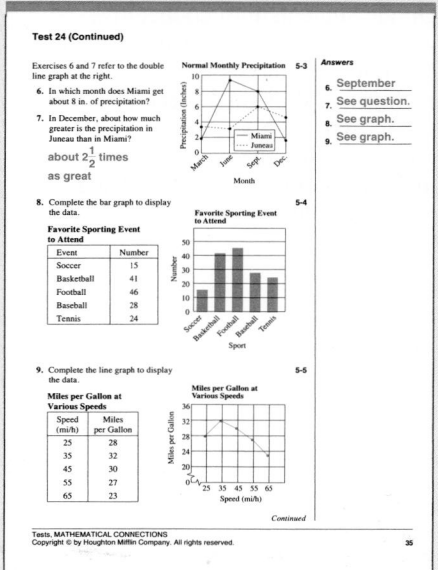

230

1. The amounts deposited in a savings account during one year were $50, $30, $60, $90, $50, $60, $100, and $40. Find the mean, median, mode(s), and range of the data. $60; $55; $50 and $60; $70

2. The bowling scores during the first twelve games of a league were 124, 123, 125, 124, 125, 123, 126, 125, 126, 125, 125, and 123. Decide whether the mean best describes the data. Explain. Yes; the scores are not distorted by an extreme value.

3. The people ordering the newest environmental book were from California, Nevada, California, New York, and California. Decide which measure of central tendency is appropriate for the data. Explain. Mode; the data are not numerical.

4. Find the mean, median, mode(s), and range of the data. 5; 5; 4; 3

Hours of TV	Tally	Frequency
4	⊥⊥⊥ ⊥⊥⊥ ||	12
5	⊥⊥⊥ ||||	9
6	⊥⊥⊥		6
7					3

Alternative Assessment

Divide students into cooperative groups. Write mean, median, mode, and range on pieces of paper. Have each group draw one piece of paper. Ask each group to make up a problem using the stated statistical measure. Groups should exchange problems and solve them.

Chapter 5

Graphs for Chapter Test 9 and 10 are on p. A16.

Decide whether it would be more appropriate to draw a *bar graph* or a *line graph* to display the data. Then draw the graph.

9. **Median Age of United States Citizens** line graph — 5-5

Year	1920	1940	1960	1980
Median Age	25.3	29.0	29.5	30.0

10. **Money Spent for Radio and Television Repair (Billions)** bar graph

Year	1970	1975	1980	1985
Dollars	1.4	2.2	2.6	3.1

Use the bar graph at the right. Solve, if possible. If there is not enough information, tell what facts are needed.

Points Scored in a Recent NFL Season

11. About how many more points were scored in the second quarter than in the first quarter?
Estimate may vary; about 1000 points. — 5-6

12. About how many points were scored in overtime? not enough information; the number of points scored in overtime

13. Describe the visual impression of the relationship between the number of points scored in the first and fourth quarters. — 5-7
It appears that there were about twice as many points scored in the fourth quarter as in the first quarter.

14. The cruising speeds in miles per hour of three-engine jets used by domestic airlines are 622, 600, 622, 620, 615, 580, 608, 615, and 615. Find the mean, median, mode(s), and range of these data. — 5-8
≈ 610.8 mi/h; 615 mi/h; 615 mi/h; 42 mi/h

15. The hourly wages for Mrs. Doyle's six employees are $7.50, $10.50, $11.00, $7.50, $12.50, and $29.00. Decide whether the mean describes these data well. Explain. No; the mean is $13 and is distorted by the extreme value of $29. — 5-9

16. The favorite colors of the seven people who worked in an office were red, blue, purple, blue, red, blue, and blue. Decide which measure of central tendency is appropriate for these data. Explain. See answer on p. A16. — 5-10

17. Find the mean, median, mode(s), and range of the data in the frequency table at the right.
24; 24; 25; 5

Class Sizes	Tally	Frequency
22	⊥⊥⊥ ||	7
24	⊥⊥⊥		6
25	⊥⊥⊥ |||	8
27				2
	Total:	23

Using Variables in Statistics

Objective: To recognize and use statistical symbols associated with the mean.

Statistical measures are used so frequently that mathematicians have developed special symbols for them. Certain scientific calculators even have a *statistical mode* and show statistical symbols directly on the key pad. Three of these symbols are associated with the mean.

n represents the *number* of items in a set of data.

Σx read "the sum of all x," represents the *sum* of the data.

$\bar{x}$ read "x bar," represents the *mean* of the data.

Example

Solution

Find n, Σx, and $\bar{x}$ for the data: 30, 28, 36, 45, 48, 38

- n = the number of items = 6

- $\Sigma x = 30 + 28 + 36 + 45 + 48 + 38 = 225$

- $\bar{x} = \dfrac{\Sigma x}{n} = \dfrac{225}{6} = 37.5$

Exercises

Find n, Σx, and $\bar{x}$ for each set of data. Use a calculator with a statistical mode if you have one. (Refer to the manual that accompanies the calculator to determine the correct key sequence.)
Answers are given in the order n, Σx, and $\bar{x}$.

1. **New England's Electoral College Votes:**

   8, 4, 13, 4, 4, 3 6; 36; 6

2. **Ages of Presidents from Ohio at Inauguration:**

   46, 54, 49, 55, 54, 51, 55 7; 364; 52

3. **Areas of the Great Lakes (Thousands of Square Miles):**

   31.8, 23.0, 22.4, 9.9, 7.5 5; 94.6; 18.92

4. **Park and Recreation Area Visitors in Pacific States (Millions):**

   46.7, 37.2, 72.9, 5.3, 20.2 5; 182.3; 36.46

Graphs and Data Analysis   **231**

Quick Quiz 2 (5-5 through 5-7)

1. Decide whether it would be more appropriate to draw a bar graph or a line graph to display the data. Then draw the graph. line graph

Tuition and Fees at 4-Year Public Universities

Year	Cost
1970	$427
1975	$599
1980	$840
1985	$1386

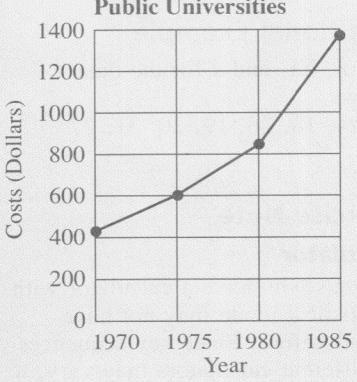

Tuition and Fees at 4-Year Public Universities

Solve if possible. If there is not enough information, tell what facts are needed.

2. Michelle had $426.87 in her savings account. She made two deposits of $326.17 each. How much does she now have in her savings account? $1079.21

3. Jeff spent $43.41 for groceries this week. How much did he spend this month for groceries? not enough information; the amount he spent on groceries for the other three weeks

4. Use the bar graph on page 211 entitled ''VCR Prices at Three Stores.'' Describe the visual impression of the relationship between the prices at the three stores. It appears that the prices at Video Center are four times those at Movie House and almost three times those at Main Street Video.

Cumulative Review

Standardized Testing Practice

Choose the letter of the correct answer.

1. $a + (b + c) = (a + b) + c$ is an example of which property?
   A. commutative property of addition
   B. associative property of addition
   C. distributive property
   D. identity property of addition

2. Write 3.45×10^4 in decimal notation.
   A. 138 B. 3,450,000
   C. 34,500 D. 0.000345

3. Chernak's Chicken Pies sold 5500 pies in March, 5000 in April, and 3000 in May. If the symbol on a pictograph of these data represents 1000 pies, how many symbols will there be for March?
   A. 5 B. 3
   C. 11 D. 5.5

4. Write the coordinates of the point 3 units to the left of the y-axis and 4 units up from the x-axis.
   A. $(-3, -4)$ B. $(-3, 4)$
   C. $(3, -4)$ D. $(3, 4)$

5. Find the answer.
   $4^2 \cdot 3 - (5 - 2)$
   A. 0 B. 45
   C. 21 D. 41

6. Choose the most appropriate graph to display a patient's temperature over a period of twelve hours.
   A. bar graph
   B. pictograph
   C. line graph
   D. double bar graph

7. Continue the pattern.
   2, 9, 16, 23, __?__, __?__, __?__
   A. 24, 26, 29 B. 28, 33, 38
   C. 30, 37, 44 D. 32, 42, 53

8. An auto magazine listed the wheelbases of 4 cars. They were the Archer (261.5 cm), the Bella (248.72 cm), the Cara (238 cm), and the Dante (230.8 cm). Which car has the longest wheelbase?
   A. Archer B. Bella
   C. Cara D. Dante

9. Choose the correct relationship.
   $x = (-12)(3)(-5)(0)(2)$
   $y = (-7)(20)(-3)(2)$
   A. $x > y$ B. $x < y$
   C. $x = y$ D. can't determine

10. Write the phrase as a variable expression: 7 less than $5n$
    A. $7 - 5n$ B. $5n - 7$
    C. $7 < 5n$ D. $5n < 7$

11.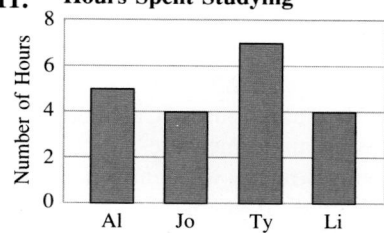

Hours Spent Studying

Who studied for about 5 h?

A. Al   B. Jo
C. Ty   D. Li

16. Cherise bought a tennis racket and some tennis balls. What information is needed to find the total amount she spent?

A. the price of the tennis racket
B. the price of each tennis ball
C. the number of tennis balls
D. all of the above

12. Simplify: $x^4 \cdot x^3$

A. x^{12} B. x^{24}
C. x^7 D. $2x^7$

17. Solve: $24 = -3t$

A. $t = -8$ B. $t = 8$
C. $t = -72$ D. $t = 72$

13. Find the mean of the data.
10, 27, 10, 15

A. 17   B. 12.5
C. 10   D. 15.5

18. Decide which statistical measure best describes the data.
62, 86, 88, 94, 94

A. the mean   B. the median
C. the range   D. the mode

14. Solve: $y - 9 = -12$

A. $y = 3$ B. $y = -3$
C. $y = -21$ D. $y = 21$

19. Solve: $21 = 3m + 6$

A. $m = 9$ B. $m = 1$
C. $m = 45$ D. $m = 5$

15. The temperature increased from −5°F to 12°F in 4 h. Find the change in temperature.

A. 7°F   B. −7°F
C. −17°F   D. 17°F

20. Write an equation for the situation.
Ann has 32 tapes. She has 6 fewer tapes than Ted. How many tapes does Ted have?

A. $t + 6 = 32$
B. $6t = 32$
C. $t - 6 = 32$
D. $t + 32 = 6$

Planning Chapter 6

Chapter Overview

Chapter 6 begins the geometry strand of the course. Students are introduced to basic geometric figures and their properties. *Lesson 1* presents the building blocks of geometry: points, lines, planes, segments, rays, and angles. *Lesson 2* instructs students in the use of the protractor, both to measure and to draw angles. The two *Focus on Explorations* in this chapter, on pages 243 and 244 and on pages 254 and 255, show students how to use a compass and a straightedge to do various constructions. *Lesson 3* presents the many types of angles that one encounters in geometry, and *Lesson 4* discusses parallel and perpendicular lines. *Lesson 5* begins the presentation of plane figures with a general discussion of polygons. *Lessons 6 and 7* continue the presentation by discussing the two simplest polygons, namely, triangles and quadrilaterals. The *Focus on Computers* shows how geometric drawing software can be used to make and test conjectures. *Lesson 8* covers the problem-solving strategy of identifying a pattern, and *Lesson 9* closes the chapter with a discussion of the ideas of symmetry.

Background

Some of the concepts in Chapter 6 may be familiar to students from previous courses; however, the material in this chapter is presented in a more formal and in-depth fashion. Throughout the chapter, connections are made between geometry and algebra, as students use equations and formulas to find missing angle measures. Additional connections are made to the application of geometry in the real world and to other subject areas such as language arts. Logical reasoning is encouraged to determine whether statements are true or false and to make and test conjectures. Communication of mathematical ideas is stressed once again through numerous reading, writing, and discussion exercises.

Since geometry is both visual and physical, manipulatives can be used effectively in this chapter. Geoboards, an overhead projector, and such everyday objects as straws and string can be used to demonstrate and clarify concepts, as shown in the Using Manipulatives section on page 234C. Furthermore, both *Focus on Explorations* in this chapter demonstrate geometric constructions, an obvious topic for manipulative activities.

Computers and calculators can play an important role in the study of geometry. Geometric drawing software can be used to draw and measure figures and to make and test conjectures, as shown in the *Focus on Computers* on page 270. The computer language Logo can be used to draw plane figures. Calculators can be used to help compute missing angle measures.

In Chapter 6, the problem-solving strategy of identifying a pattern is used frequently. This strategy offers another connection between geometry and algebra, as is demonstrated in the Look Back question at the bottom of page 271, where students use an algebraic expression to find the number of diagonals in a geometric figure.

Objectives

6-1 To identify the basic geometric figures.

6-2 To measure and draw angles using a protractor.

FOCUS ON EXPLORATIONS To use a straightedge and a compass to construct and bisect line segments and angles.

6-3 To find measures of acute, right, obtuse, straight, complementary, supplementary, vertical, and adjacent angles.

6-4 To identify perpendicular and parallel lines and calculate the measures of the angles formed when a transversal intersects parallel lines.

FOCUS ON EXPLORATIONS To construct a line perpendicular to a given line and to construct a line parallel to a given line.

6-5 To identify types of polygons and find the measures of their angles.

6-6 To classify triangles by their sides and angles and to use the triangle inequality.

6-7 To identify special quadrilaterals and apply their properties.

FOCUS ON COMPUTERS To use geometric drawing software to test conjectures.

6-8 To solve problems by identifying a pattern.

6-9 To recognize and find lines of symmetry in plane figures.

CHAPTER EXTENSION To recognize and name regular and semiregular tessellations.

Chapter Planner

Instructional Aids	Manipulatives	Cooperative Learning	Technology	Practice/Reteaching	Assessment
Materials: overhead projector, craft sticks, board compass, geoboards, string, geo-dot paper, wax paper, calculator, protractor, straightedge *Resource Book:* Lesson Starters 50–58 *Visuals,* Folders A, B	Lessons 6-2, 6-3, 6-4, 6-5, 6-6, 6-7, 6-8, Focus on Explorations *Activities Book:* Manipulative Activities 11–12	Lessons 6-3, 6-4, 6-7, Chapter Extension *Activities Book:* Cooperative Activity 12	Lessons 6-1, 6-2, 6-4, 6-5, 6-6, 6-9 *Activities Book:* Calculator Activity 5 Computer Activity 7 *Connections Plotter Plus* Disk	Toolbox, pp. 753–771 Extra Practice, p. 737 *Resource Book:* Practice 68–79 *Study Guide,* pp. 99–116	Self-Tests 1–2 Quick Quizzes 1–2 Chapter Test, pp. 283–284 *Resource Book:* Lesson Starters 50–58 (Daily Quizzes) *Tests 26–30*

Assignment Guide Chapter 6

Day	Skills Course	Average Course	Advanced Course
1	**6-1:** 1–22, 25–28	**6-1:** 1–22, 25–28	**6-1:** 1–28, Historical Note
2	**6-2:** 1–16 **Exploration:** Activities I–IV	**6-2:** 1–16 **Exploration:** Activities I–IV	**6-2:** 1–16 **Exploration:** Activities I–IV **6-3:** 1–7 odd, 9–27
3	**6-3:** 1–18, 25–27	**6-3:** 1–22, 25–27	**6-4:** 1–28 **Exploration:** Activities I–IV
4	**6-4:** 1–16, 23–28 **Exploration:** Activities I–IV	**6-4:** 1–18, 23–28 **Exploration:** Activities I–IV	**6-5:** 1–18
5	**6-5:** 1–12, 15–18	**6-5:** 1–12, 15–18	**6-6:** 1–19 odd, 20–57, Challenge
6	**6-6:** 1–20, 21–39 odd, 54–57	**6-6:** 1–47 odd, 48–57	**6-7:** 1–25 **Focus on Computers**
7	**6-7:** 1–12, 13–19 odd, 22–25 **Focus on Computers**	**6-7:** 1–20, 22–25 **Focus on Computers**	**6-8:** 1–16
8	**6-8:** 1–6, 11–16	**6-8:** 1–16	**6-9:** 1–24
9	**6-9:** 1–11, 13, 22–24	**6-9:** 1–15, 22–24	*Prepare for Chapter Test:* Chapter Review
10	*Prepare for Chapter Test:* Chapter Review	*Prepare for Chapter Test:* Chapter Review	*Administer Chapter 6 Test; Chapter Extension; Cumulative Review*
11	*Administer Chapter 6 Test*	*Administer Chapter 6 Test; Cumulative Review*	

Teacher's Resources

Resource Book
  Chapter 6 Project
  Lesson Starters 50–58
  Practice 68–79
  Enrichment 55–64
  Diagram Masters
  Chapter 6 Objectives
  Family Involvement 6
  Spanish Test 6
  Spanish Glossary

Activities Book
  Cooperative Activity 12
  Manipulative Activities 11–12
  Calculator Activity 5
  Computer Activity 7

Study Guide, pp. 99–116

Tests
  Tests 26–30

Visuals
  Folders A, B

Connections Plotter Plus **Disk**

Using Technology

Lessons 6-1, 6-2

COMPUTERS

Have students use a geometric drawing program such as *Connections Plotter Plus* to draw segments and angles. Then have them measure these segments and angles. This will give students practice naming these objects. Students can learn easily what a bisector of a segment or of an angle does, as well as the fact that two intersecting lines form two pairs of congruent angles and four pairs of supplementary angles.

Lesson 6-4

COMPUTERS

A geometric drawing program can be used to discover the fact that a transversal and two parallel lines form congruent pairs of corresponding and alternate interior angles. Students can discover important theorems such as: A line perpendicular to one of two coplanar parallel lines is also perpendicular to the other one; or, the shortest distance from a point to a line lies along the line through the point, perpendicular to the given line.

Lesson 6-5

CALCULATORS OR COMPUTERS

A spreadsheet program or a calculator can be used to explore the formula giving the sum of the angles of a polygon. For example, students can be given a sum of angles and then be asked to find the number of sides that a polygon with that sum has. Students can discover easily by example that the sum of the exterior angles of a polygon is 360°. (This result also can be discovered by exploring shapes in Logo.)

Lesson 6-6

COMPUTERS

Geometric drawing software can be used to discover the meanings and properties of altitudes, angle bisectors, and medians of triangles. Students can discover that these three objects can be the same in the case of equilateral triangles, but are distinct from each other in general. A sequence of exercises in this lesson refers to Logo, which allows a very active exploration of geometric shapes. For example, a simple drawing loop such as REPEAT 20 [FD 100 RT 150] can be used to explore star shapes, in particular such questions as: Given a certain turning angle, how can you predict the number of points on the resulting star?

Lesson 6-9

COMPUTERS

Software exists that makes it easy to apply transformations, including reflections, on a drawn geometric shape. It is natural to use such a program to explore *symmetry*. (A figure is symmetric about a line if its reflection in the line looks the same as the original figure.)

Using Manipulatives

Lesson 6-2

To introduce the lesson, give groups of students a large paper circle and have them fold the circle into four equal angles, each 90°. Then ask the groups to fold the circle again to make eight equal angles, each 45°. Students should notice that if the folding continues, the angles become smaller and smaller. Lead the class to discover that the standard unit of measure, called a degree, is $\frac{1}{360}$th of the circle.

An overhead projector can be used effectively with manipulatives. A plastic transparent protractor can be used on the overhead projector to demonstrate how to draw and measure angles. Also, you can project an image of an acute angle on the screen and ask students which angle is larger, the one on the overhead or the one on the screen. Have a student measure both angles with a protractor to show that they are equal. Stress that the opening between the sides of an angle determines the measure of an angle.

Lesson 6-3

A chalkboard compass can be used to demonstrate different types of angles. Students can create their own angles by using wooden craft sticks.

Lesson 6-4

Make a model to show two lines cut by a transversal. Use paper fasteners or magnetic tape so that the lines and angles are movable. When the lines are parallel, the angles can be made to "fit" in other locations. This shows that when the lines are parallel there are only two different sizes for angles. Make the lines "not parallel" and show, for instance, that corresponding angles are no longer equal in measure.

Lesson 6-5

Students can use geoboards to form various polygons.

To demonstrate how to find the sum of the angles of a polygon, use a model of a pentagon that is divided into three triangles by diagonals from a given vertex. Separate the triangles and multiply $3 \times 180°$ to obtain the sum of the angles of the three triangles. Reassemble the triangles to form the pentagon. The sum of the angles remains the same, 540°. Repeat this process for other polygons. Ask students to form conjectures about finding the sum of the angles of any polygon:

$$\text{sum} = (n - 2)180°$$

Lesson 6-6

Have students use straws to construct quadrilaterals and triangles. Students can then contrast the stability (rigidity) of these two types of figures. Use this in the Application of building a gate.

Lesson 6-7

Give each student a string 12 to 14 inches long with the ends tied together. Using the loop (and holding it up so all can see), ask them to form different types of quadrilaterals.

Lesson 6-9

Transparent mirrors are very useful for helping students understand symmetry. Each student can draw a figure, place the mirror along an edge, and draw its reflection. Students will soon realize that the mirror acts as a line of symmetry.

As a related activity, give students a piece of wax paper and have them fold it in half and draw a figure on one side. When they unfold it, the reflection of the figure will be on the other side.

Reteaching/Alternate Approach

Lesson 6-3

COOPERATIVE LEARNING

Have students work in groups to develop a drawing that will incorporate all of the types of angles covered in this lesson. Each group is to share its drawing for discussion with the entire class. The drawings should be on large sheets of paper.

Lesson 6-4

COOPERATIVE LEARNING

Have students work in groups to construct parallel lines with transversals, perpendicular lines, and so on, on a geoboard. Each group should explore the properties of each figure created.

Lesson 6-5

COOPERATIVE LEARNING

Students may enjoy discovering the formula for finding the sum of the angles of a polygon. Have students work in groups to draw polygons having four to ten sides. Each group should find the sum of the angles of each polygon by cutting it into as many nonoverlapping triangles as possible. Instruct the groups to use their results to make a generalization. Each group should share its generalization with the class.

Lesson 6-6

COOPERATIVE LEARNING

The Exploration at the beginning of the lesson can be done with students working in cooperative groups. Each group can extend the Exploration by testing various lengths of its own choosing. The groups can present their conclusions to the class.

Lesson 6-7

You can use geoboards to do preliminary identification and comparisons of various quadrilaterals. Geo-dot paper can be used to examine properties of specific figures. For example, students can draw a parallelogram and then use it to list three facts about parallelograms.

Teacher's Resources Chapter 6

Enrichment masters from the Resource Book are pictured here. See the Teacher's Resources chart on page 234B for a complete listing of all materials available for this chapter.

6-1 Enrichment
Connection

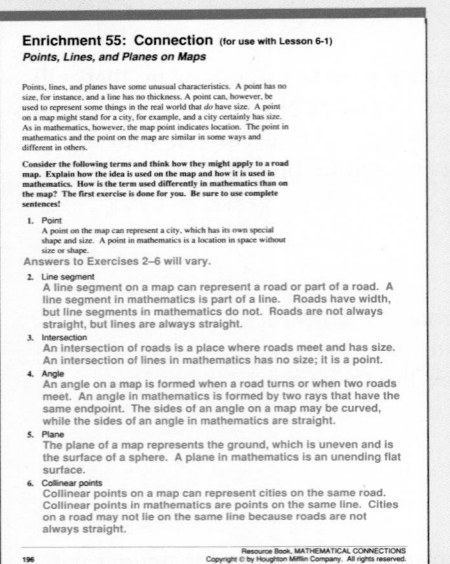

Enrichment 55: Connection (for use with Lesson 6-1)
Points, Lines, and Planes on Maps

Points, lines, and planes have some unusual characteristics. A point has no size, for instance, and a line has no thickness. A point can, however, be used to represent some things in the real world that *do* have size. A point on a map might stand for a city, for example, and a city certainly has size. As in mathematics, however, the map point indicates location. The point in mathematics and the point on the map are similar in some ways and different in others.

Consider the following terms and think how they might apply to a road map. Explain how the idea is used on the map and how it is used in mathematics. How is the term used differently in mathematics than on the map? The first exercise is done for you. Be sure to use complete sentences.

1. **Point**
A point on the map can represent a city, which has its own special shape and size. A point in mathematics is a location in space without size or shape.
Answers to Exercises 2–6 will vary.

2. **Line segment**
A line segment on a map can represent a road or part of a road. A line segment in mathematics is part of a line. Roads have width, but line segments in mathematics do not. Roads are not always straight, but lines are always straight.

3. **Intersection**
An intersection of roads is a place where roads meet and has size. An intersection of lines in mathematics has no size; it is a point.

4. **Angle**
An angle on a map is formed when a road turns or when two roads meet. An angle in mathematics is formed by two rays that have the same endpoint. The sides of an angle on a map may be curved, while the sides of an angle in mathematics are straight.

5. **Plane**
The plane of a map represents the ground, which is uneven and is the surface of a sphere. A plane in mathematics is an unending flat surface.

6. **Collinear points**
Collinear points on a map can represent cities on the same road. Collinear points in mathematics are points on the same line. Cities on a road may not lie on the same line because roads are not always straight.

6-2 Enrichment
Exploration

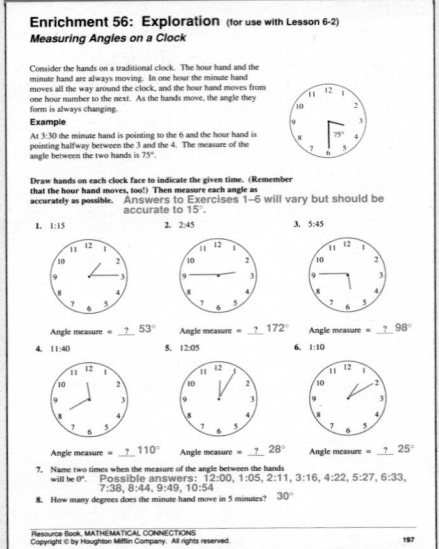

Enrichment 56: Exploration (for use with Lesson 6-2)
Measuring Angles on a Clock

Consider the hands on a traditional clock. The hour hand and the minute hand are always moving. In one hour the minute hand moves all the way around the clock, and the hour hand moves from one hour number to the next. As the hands move, the angle they form is always changing.

Example
At 3:30 the minute hand is pointing to the 6 and the hour hand is pointing halfway between the 3 and the 4. The measure of the angle between the two hands is 75°.

Draw hands on each clock face to indicate the given time. (Remember that the hour hand moves, too!) Then measure each angle as accurately as possible. *Answers to Exercises 1–6 will vary but should be accurate to 15°.*

1. 1:15   Angle measure = __?__ 53°
2. 2:45   Angle measure = __?__ 172°
3. 5:45   Angle measure = __?__ 98°
4. 11:40 Angle measure = __?__ 110°
5. 12:05 Angle measure = __?__ 28°
6. 1:10   Angle measure = __?__ 25°

7. Name two times when the measure of the angle between the hands will be 0°. *Possible answers:* 12:00, 1:05, 2:11, 3:16, 4:22, 5:27, 6:33, 7:38, 8:44, 9:49, 10:54

8. How many degrees does the minute hand move in 5 minutes? 30°

6-3 Enrichment
Application

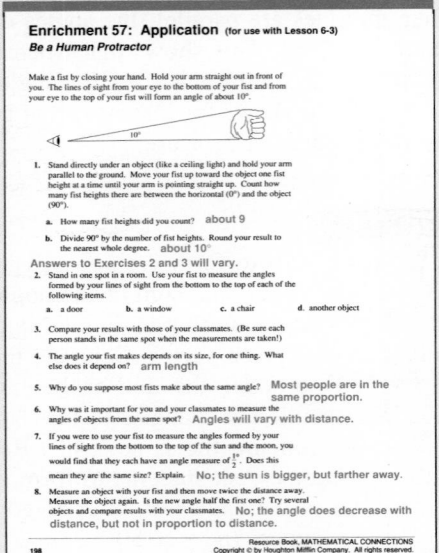

Enrichment 57: Application (for use with Lesson 6-3)
Be a Human Protractor

Make a fist by closing your hand. Hold your arm straight out in front of you. The lines of sight from your eye to the bottom of your fist and from your eye to the top of your fist will form an angle of about 10°.

1. Stand directly under an object (like a ceiling light) and hold your arm parallel to the ground. Move your fist up toward the object one fist height at a time until your arm is pointing straight up. Count how many fist heights there are between the horizontal (0°) and the object (90°).
   a. How many fist heights did you count? about 9
   b. Divide 90° by the number of fist heights. Round your result to the nearest whole degree. about 10°
   Answers to Exercises 2 and 3 will vary.

2. Stand in one spot in a room. Use your fist to measure the angles formed by your lines of sight from the bottom to the top of each of the following items.
   a. a door   b. a window   c. a chair   d. another object

3. Compare your results with those of your classmates. (Be sure each person stands in the same spot when the measurements are taken!)

4. The angle your fist makes depends on its size, for one thing. What else does it depend on? arm length

5. Why do you suppose most fists make about the same angle? Most people are in the same proportion.

6. Why was it important for you and your classmates to measure the angles of objects from the same spot? Angles will vary with distance.

7. If you were to use your fist to measure the angles formed by your lines of sight from the bottom to the top of the sun and the moon, you would find that they each have an angle of $\frac{1}{2}$°. Does this mean they are the same size? Explain. No; the sun is bigger, but farther away.

8. Measure an object with your fist and then move twice the distance away. Measure the object again. Is the new angle half the first one? Try several objects and compare results with your classmates. No; the angle does decrease with distance, but not in proportion to distance.

6-4 Enrichment
Communication

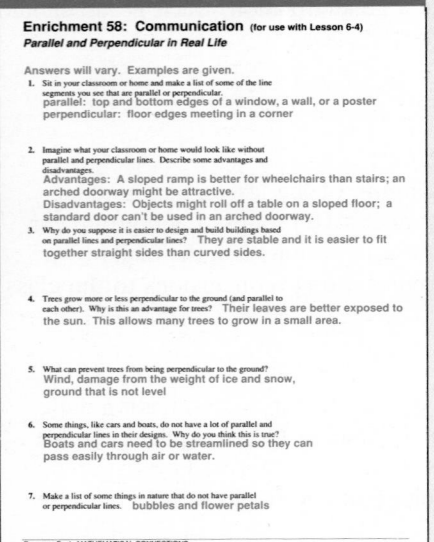

Enrichment 58: Communication (for use with Lesson 6-4)
Parallel and Perpendicular in Real Life

Answers will vary. Examples are given.

1. Sit in your classroom or home and make a list of some of the line segments you see that are parallel or perpendicular.
   parallel: top and bottom edges of a window, a wall, or a poster
   perpendicular: floor edges meeting in a corner

2. Imagine what your classroom or home would look like without parallel and perpendicular lines. Describe some advantages and disadvantages.
   Advantages: A sloped ramp is better for wheelchairs than stairs; an arched doorway might be attractive.
   Disadvantages: Objects might roll off a table on a sloped floor; a standard door can't be used in an arched doorway.

3. Why do you suppose it is easier to design and build buildings based on parallel lines and perpendicular lines? They are stable and it is easier to fit together straight sides than curved sides.

4. Trees grow more or less perpendicular to the ground (and parallel to each other). Why is this an advantage for trees? Their leaves are better exposed to the sun. This allows many trees to grow in a small area.

5. What can prevent trees from being perpendicular to the ground? Wind, damage from the weight of ice and snow, ground that is not level

6. Some things, like cars and boats, do not have a lot of parallel and perpendicular lines in their designs. Why do you think this is true? Boats and cars need to be streamlined so they can pass easily through air or water.

7. Make a list of some things in nature that do not have parallel or perpendicular lines. bubbles and flower petals

6-5 Enrichment
Connection

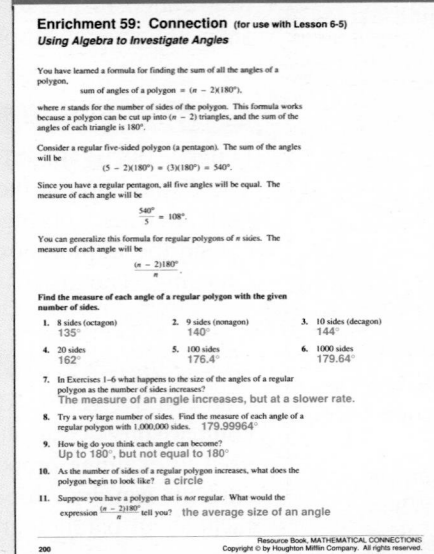

Enrichment 59: Connection (for use with Lesson 6-5)
Using Algebra to Investigate Angles

You have learned a formula for finding the sum of all the angles of a polygon,

sum of angles of a polygon = $(n - 2)(180°)$,

where n stands for the number of sides of the polygon. This formula works because a polygon can be cut up into $(n - 2)$ triangles, and the sum of the angles of each triangle is 180°.

Consider a regular five-sided polygon (a pentagon). The sum of the angles will be

$$(5 - 2)(180°) = (3)(180°) = 540°.$$

Since you have a regular pentagon, all five angles will be equal. The measure of each angle will be

$$\frac{540°}{5} = 108°.$$

You can generalize this formula for regular polygons of n sides. The measure of each angle will be

$$\frac{(n - 2)180°}{n}.$$

Find the measure of each angle of a regular polygon with the given number of sides.

1. 8 sides (octagon) 135°
2. 9 sides (nonagon) 140°
3. 10 sides (decagon) 144°
4. 20 sides 162°
5. 100 sides 176.4°
6. 1000 sides 179.64°

7. In Exercises 1–6 what happens to the size of the angles of a regular polygon as the number of sides increases? The measure of an angle increases, but at a slower rate.

8. Try a very large number of sides. Find the measure of each angle of a regular polygon with 1,000,000 sides. 179.99964°

9. How big do you think each angle can become? Up to 180°, but not equal to 180°

10. As the number of sides of a regular polygon increases, what does the polygon begin to look like? a circle

11. Suppose you have a polygon that is *not* regular. What would the expression $\frac{(n - 2)180°}{n}$ tell you? the average size of an angle

6-6 Enrichment
Data Analysis

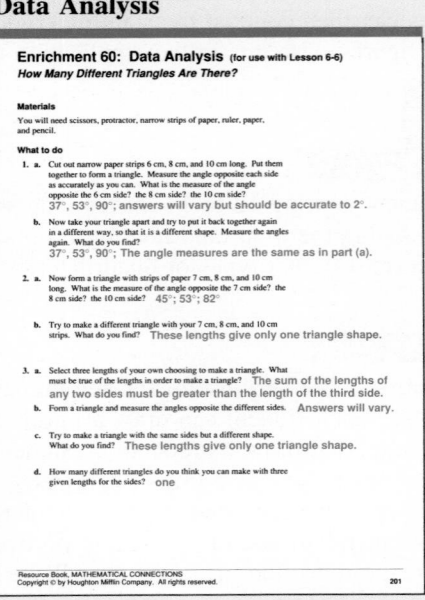

Enrichment 60: Data Analysis (for use with Lesson 6-6)
How Many Different Triangles Are There?

Materials
You will need scissors, protractor, narrow strips of paper, ruler, paper, and pencil.

What to do
1. a. Cut out narrow paper strips 6 cm, 8 cm, and 10 cm long. Put them together to form a triangle. Measure the angle opposite each side as accurately as you can. What is the measure of the angle opposite the 6 cm side? the 8 cm side? the 10 cm side? 37°, 53°, 90°; answers will vary but should be accurate to 2°.
   b. Now take your triangle apart and try to put it back together again in a different way, so that it is a different shape. What do you find? 37°, 53°, 90°; The angle measures are the same as in part (a).

2. a. Now form a triangle with strips of paper 7 cm, 8 cm, and 10 cm long. What is the measure of the angle opposite the 7 cm side? the 8 cm side? the 10 cm side? 45°; 53°; 82°
   b. Try to make a different triangle with your 7 cm, 8 cm, and 10 cm strips. What do you find? These lengths give only one triangle shape.

3. a. Select three lengths of your own choosing to make a triangle. What must be true of the lengths in order to make a triangle? The sum of the lengths of any two sides must be greater than the length of the third side.
   b. Form a triangle and measure the angles opposite the different sides. Answers will vary.
   c. Try to make a triangle with the same sides but a different shape. What do you find? These lengths give only one triangle shape.
   d. How many different triangles do you think you can make with three given lengths for the sides? one

234E

6-7 Enrichment
Thinking Skills

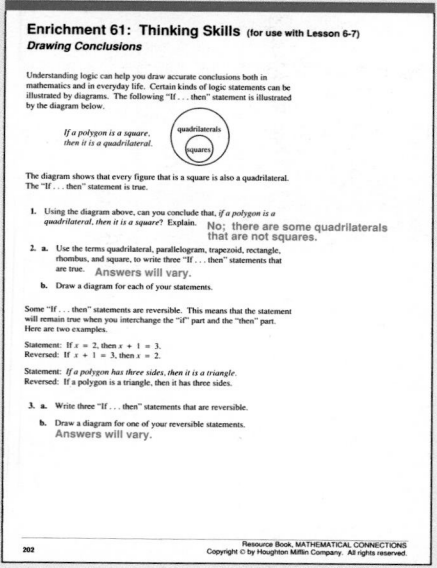

Enrichment 61: Thinking Skills (for use with Lesson 6-7)
Drawing Conclusions

Understanding logic can help you draw accurate conclusions both in mathematics and in everyday life. Certain kinds of logic statements can be illustrated by diagrams. The following "If . . . then" statement is illustrated by the diagram below.

If a polygon is a square, then it is a quadrilateral.

quadrilaterals
squares

The diagram shows that every figure that is a square is also a quadrilateral. The "If . . . then" statement is true.

1. Using the diagram above, can you conclude that, *if a polygon is a quadrilateral, then it is a square?* Explain.   No; there are some quadrilaterals that are not squares.

2. a. Use the terms quadrilateral, parallelogram, trapezoid, rectangle, rhombus, and square, to write three "If . . . then" statements that are true.   Answers will vary.

   b. Draw a diagram for each of your statements.

Some "If . . . then" statements are reversible. This means that the statement will remain true when you interchange the "if" part and the "then" part. Here are two examples.

Statement: If $x = 2$, then $x + 1 = 3$.
Reversed: If $x + 1 = 3$, then $x = 2$.

Statement: *If a polygon has three sides, then it is a triangle.*
Reversed: If a polygon is a triangle, then it has three sides.

3. a. Write three "If . . . then" statements that are reversible.

   b. Draw a diagram for one of your reversible statements.
   Answers will vary.

6-8 Enrichment
Problem Solving

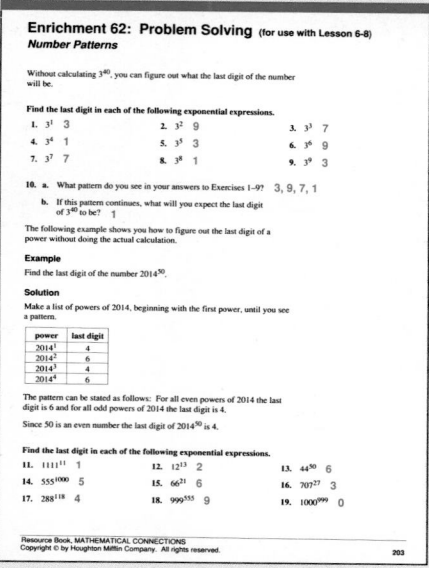

Enrichment 62: Problem Solving (for use with Lesson 6-8)
Number Patterns

Without calculating 3^{40}, you can figure out what the last digit of the number will be.

Find the last digit in each of the following exponential expressions.

1. 3^1 3
2. 3^2 9
3. 3^3 7
4. 3^4 1
5. 3^5 3
6. 3^6 9
7. 3^7 7
8. 3^8 1
9. 3^9 3

10. a. What pattern do you see in your answers to Exercises 1–9?   3, 9, 7, 1

   b. If this pattern continues, what will you expect the last digit of 3^{40} to be?   1

The following example shows you how to figure out the last digit of a power without doing the actual calculation.

Example
Find the last digit of the number 2014^{50}.

Solution
Make a list of powers of 2014, beginning with the first power, until you see a pattern.

power	last digit
2014^1	4
2014^2	6
2014^3	4
2014^4	6

The pattern can be stated as follows: For all even powers of 2014 the last digit is 6 and for all odd powers of 2014 the last digit is 4.

Since 50 is an even number the last digit of 2014^{50} is 4.

Find the last digit in each of the following exponential expressions.

11. 1111^{11} 1
12. 12^{13} 2
13. 44^{50} 6
14. 555^{1000} 5
15. 66^{21} 6
16. 707^{27} 3
17. 288^{118} 4
18. 999^{555} 9
19. 1000^{999} 0

6-9 Enrichment
Connection

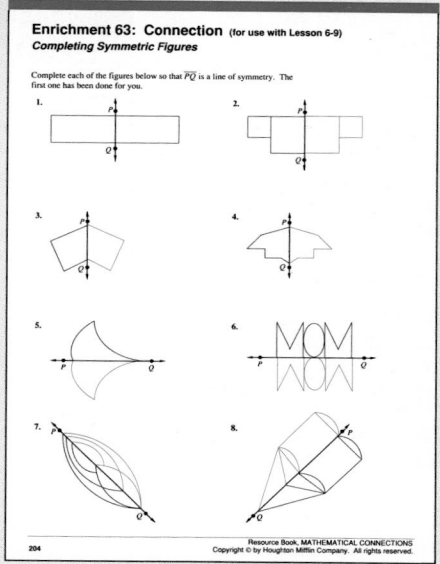

Enrichment 63: Connection (for use with Lesson 6-9)
Completing Symmetric Figures

Complete each of the figures below so that $\overleftrightarrow{PQ}$ is a line of symmetry. The first one has been done for you.

1.   2.   3.   4.   5.   6.   7.   8.

End of Chapter Enrichment
Extension

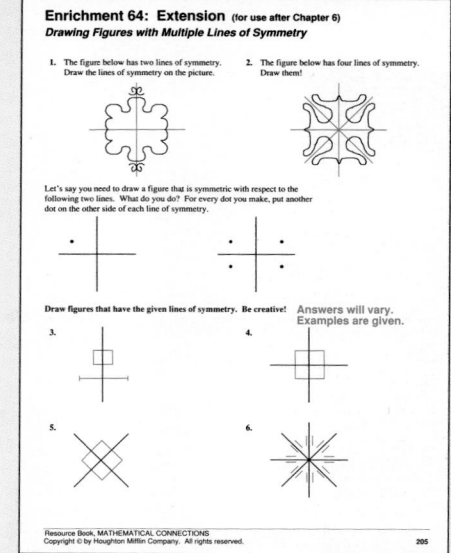

Enrichment 64: Extension (for use after Chapter 6)
Drawing Figures with Multiple Lines of Symmetry

1. The figure below has two lines of symmetry. Draw the lines of symmetry on the picture.

2. The figure below has four lines of symmetry. Draw them!

Let's say you need to draw a figure that is symmetric with respect to the following two lines. What do you do? For every dot you make, put another dot on the other side of each line of symmetry.

Draw figures that have the given lines of symmetry. Be creative!   Answers will vary. Examples are given.

3.   4.   5.   6.

Multicultural Notes

For Chapter 6 Opener

The first woman to obtain an airplane pilot's license was the Baroness Raymonde de Laroche of France, in 1910. She was followed by Harriet Quimby, an American, in 1911. By 1917, there were eleven licensed women pilots in the United States, and many more besides who flew without licenses. (In the early days of aviation, one did not necessarily need a license to fly a plane.)

Many of these early women aviators lobbied to be given the opportunity to fly planes for the Air Service in World War I. Though their lobbying efforts were unsuccessful, these aviators played an important role in the Air Service's recruiting efforts. Many of them were also active in the Red Cross.

Throughout the 1920s and 1930s, women continued to prove themselves to be capable, professional pilots. In World War II, women flew transport planes in Europe.

For Page 236

The ancient Greeks developed the study of geometry to a high level of sophistication. Students should be aware that many principles of geometry and other branches of mathematics were previously known to the Egyptians, the Babylonians, and other groups as well.

For Page 278

The woman pictured on this page is weaving a traditional Zapotec design. From about 1500 B.C. to 750 A.D. the Zapotec empire flourished in what is now the state of Oaxaca in southern Mexico. Zapotecs still live in this part of Mexico.

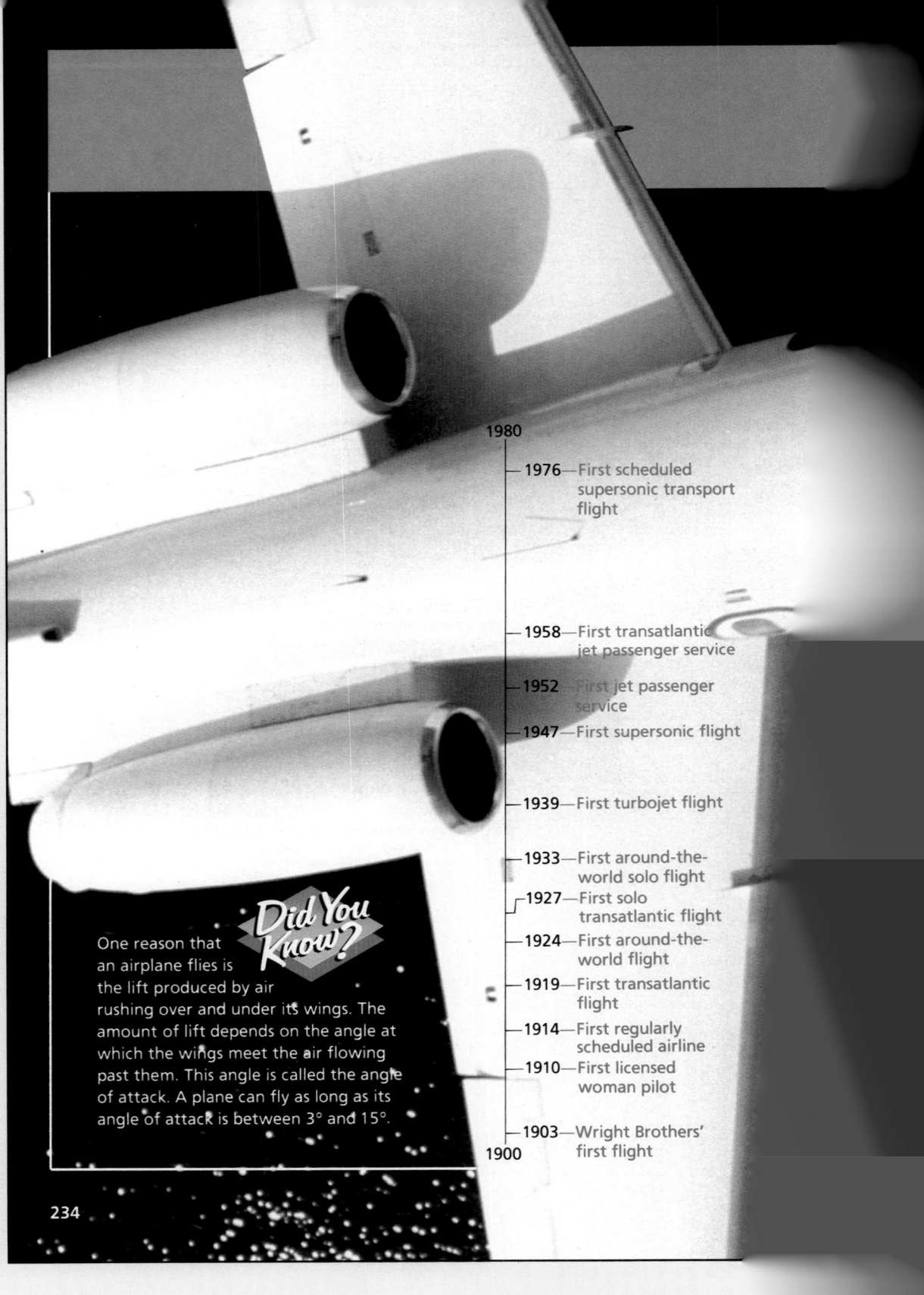

One reason that an airplane flies is the lift produced by air rushing over and under its wings. The amount of lift depends on the angle at which the wings meet the air flowing past them. This angle is called the angle of attack. A plane can fly as long as its angle of attack is between 3° and 15°.

Did You Know?

234

1980

1976—First scheduled supersonic transport flight

1958—First transatlantic jet passenger service

1952—First jet passenger service

1947—First supersonic flight

1939—First turbojet flight

1933—First around-the-world solo flight

1927—First solo transatlantic flight

1924—First around-the-world flight

1919—First transatlantic flight

1914—First regularly scheduled airline

1910—First licensed woman pilot

1903—Wright Brothers' first flight

1900

Introduction to Geometry 6

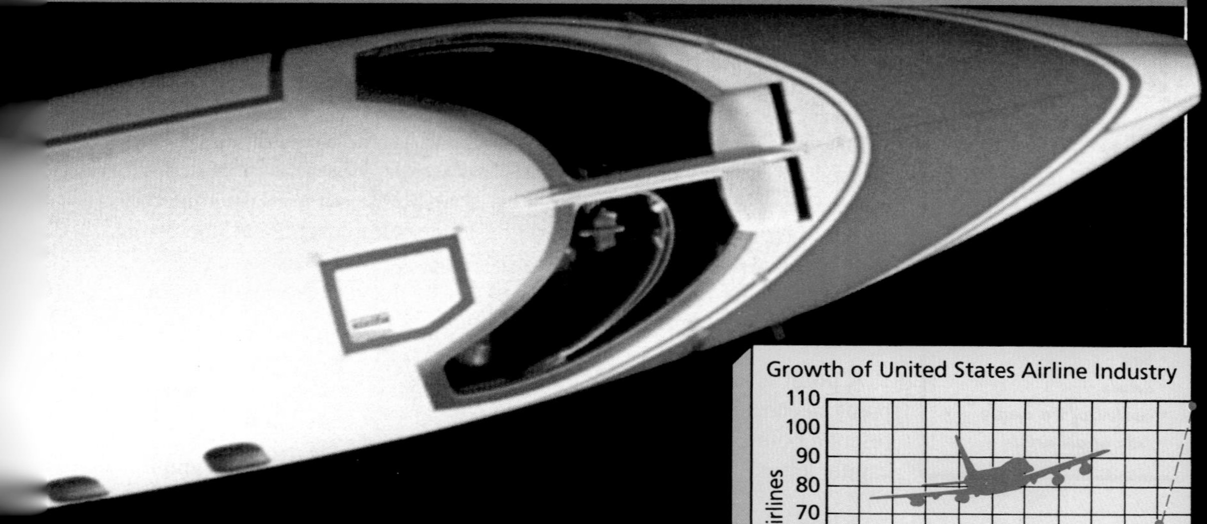

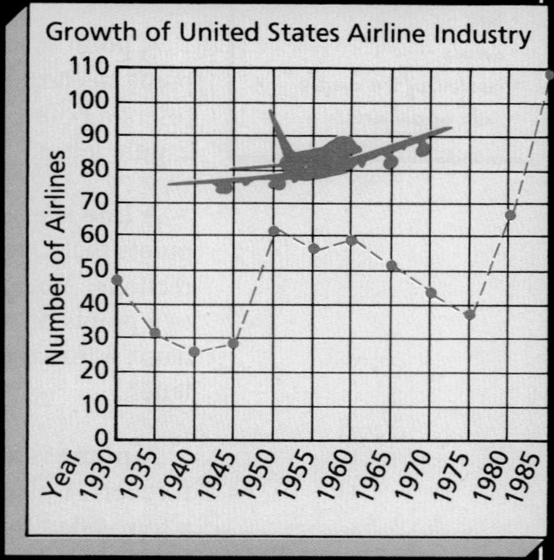

Research Activities

For Chapter 6 Opener
Studying about the history of women in aviation can be both lively and inspiring for students. You may want to have students prepare oral or written biographies of women aviators to share with the class.

For Page 278
For the Native American designs researched in Exercise 16, students can also report on which Native American groups or nations developed the designs.

Suggested Resources

Brooks-Pazmany, Kathleen. *United States Women in Aviation, 1919–1929*. Washington, D.C.: Smithsonian Institution Press, 1985.

Naylor, Maria, ed. *Authentic Indian Designs*. New York: Dover Publications, Inc., 1975.

Twelve Busiest United States Airports

Airport	Location	Passengers (Millions)
O'Hare	Chicago	56.3
Hartsfield	Atlanta	47.6
Los Angeles	Los Angeles	44.9
Dallas/Fort Worth	Dallas/Fort Worth	41.9
Stapleton	Denver	32.4
Kennedy	New York	30.2
San Francisco	San Francisco	29.8
LaGuardia	New York	24.2
Miami	Miami	24.0
Newark	Newark	23.5
Logan	Boston	23.3
Honolulu	Honolulu	20.4

Growth of United States Airline Industry

235

Teaching the Lesson
- Lesson Starter 50
- Using Technology, p. 234C

Lesson Follow-Up
- Practice 68
- Enrichment 55: Connection
- Study Guide, pp. 99–100

Warm-Up Exercises

What geometric terms are represented by each of the following descriptions?

1. the tip of a pencil   point
2. one edge of the front cover of a book   line segment
3. the top of a desk   plane
4. where one wall of the classroom meets the floor   line segment
5. a corner of a book cover   angle

Lesson Focus

Calling a person by his or her right name is important. Also, we recognize one another by our features. The importance of recognizing geometric figures and naming them correctly is the focus of today's lesson.

Teaching Notes

Approach the meanings of the various geometric terms by using many physical examples, as in the Warm-Up Exercises. Students should be able to identify physical examples of each term and provide a sketch for each with labels.

6-1 Points, Lines, and Planes

Objective: To identify the basic geometric figures.

CONNECTION

When you write a report, an essay, or even a letter to a friend, your work is based on ideas that you developed over many years. For instance, the words you write are created using the alphabet that you learned as a child, and sentences and paragraphs flow from the rules of grammar and composition that you learned in school. You will find that geometry is similar because it involves organizing several basic ideas into useful and interesting mathematical structures.

The geometry that you will study in this course has its foundation in three undefined terms: *point, line,* and *plane.*

Terms to Know
- *point*
- *line*
- *plane*
- *collinear*
- *intersect*
- *line segment*
- *endpoint*
- *ray*
- *angle*
- *vertex of an angle*
- *side of an angle*

A **point** is an exact location in space. A point has no size, but you use a dot to represent a point. You name a point by a capital letter.

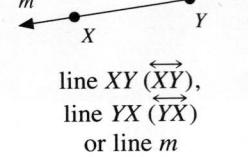

points A and B

A **line** is a straight arrangement of points that extends forever in opposite directions. You can name a line using any two points on the line. Another way to name a line is to use a single lowercase letter.

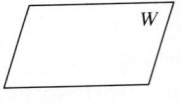

line XY ($\overleftrightarrow{XY}$), line YX ($\overleftrightarrow{YX}$) or line m

A **plane** is a flat surface that extends forever. A plane has no edges, but you use a four-sided figure to represent a plane. You can name a plane using a capital letter.

plane W

Points that lie on the same line are called **collinear** points. In the figure below, points $P, Q, R, S,$ and T are collinear.

Two lines that meet at one point are said to **intersect** in that point.

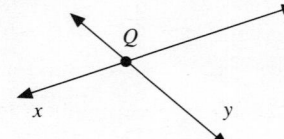

When two distinct planes intersect, their intersection is a line.

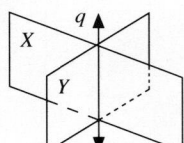

A **line segment** is a part of a line that consists of two **endpoints** and all the points between. You name a line segment using its endpoints.

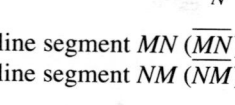

line segment MN ($\overline{MN}$)
line segment NM ($\overline{NM}$)

A **ray** is a part of a line that has one endpoint and extends forever in one direction. You name a ray by writing the endpoint first, then writing one other point on the ray.

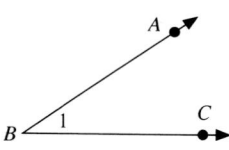

ray YZ ($\overrightarrow{YZ}$)

When two rays share a common endpoint, the figure that is formed is an **angle**. The endpoint is called the **vertex** of the angle, and the rays are called the **sides**. In the angle at the right, point B is the vertex. The sides are $\overrightarrow{BA}$ and $\overrightarrow{BC}$.

The symbol for an angle is $\angle$. You name an angle using letters or numbers, usually depending on what is most appropriate in the given situation. For the angle above, any of these is an appropriate name.

$$\angle ABC \qquad \angle CBA \qquad \angle B \qquad \angle 1$$

To name an angle using three letters, the vertex letter must be in the center. It would *not* be correct to refer to the angle above as $\angle BAC$.

Example

Write the name of each figure.

a.  b. 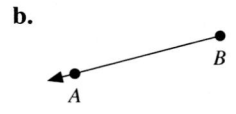 c.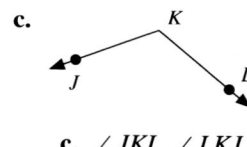

Solution

a. $\overline{DC}$ or $\overline{CD}$

b. $\overrightarrow{BA}$

c. $\angle JKL$, $\angle LKJ$, or $\angle K$

Check Your Understanding

1. In part (b) of the Example, why isn't $\overrightarrow{AB}$ a correct answer?
2. In part (c) of the Example, would $\angle KJL$ be an appropriate answer?
   See *Answers to Check Your Understanding* at the back of the book.

Guided Practice

COMMUNICATION «*Reading*

Refer to the text on pages 236–237.

«**1.** Which three undefined terms form the foundation of geometry?
point, line, plane

«**2.** How are a ray and a line segment alike? How are they different?
Each is a part of a line. A ray has one endpoint; a line segment has two endpoints.

Introduction to Geometry **237**

Key Questions
1. What are the differences among a line, a line segment, and a ray? A line is of infinite length; a line segment has a definite length with two endpoints; a ray has one endpoint and extends infinitely in one direction.
2. How is an angle formed and named? An angle is formed by two rays with a common endpoint and is named by three letters, one letter, or a number.

Making Connections/Transitions
Use a model of a rectangular prism (shoe box, cereal box, commercial model) to discuss points, line segments, and planes. Have students sketch a box, label it, and identify points, line segments, and planes from their sketches. Ask students why a ray or line is not represented.

Reasoning
Students should observe the progression from zero dimension to three dimensions as points, lines, planes, and solids are considered. This should help them to understand certain units of measurement (linear, square, cubic).

Using Technology
For a suggestion on using geometric drawing software, see page 234C.

Additional Examples
Write the name of each figure.

1.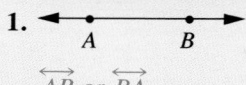

$\overleftrightarrow{AB}$ or $\overleftrightarrow{BA}$

2.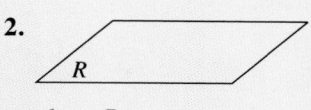

plane R

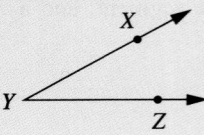
Choose the letters of all the correct names for each figure.

3.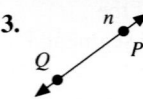

a. line n
b. $\overleftrightarrow{PQ}$
c. $\overleftrightarrow{QP}$
d. line Q

4.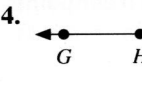

a. $\overleftrightarrow{GH}$
b. $\overrightarrow{GH}$
c. $\overrightarrow{HG}$
d. $\overleftrightarrow{HG}$

5.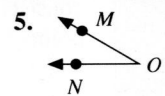

a. $\angle ONM$
b. $\angle NOM$
c. $\angle O$
d. $\angle NMO$

6.

a. point T
b. plane T
c. angle T
d. line T

Write the name of each figure.

7. $\angle RAS$, $\angle SAR$, or $\angle A$

7.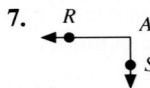

8.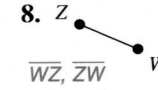

$\overline{WZ}$, $\overline{ZW}$

9.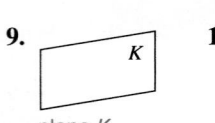

plane K

10.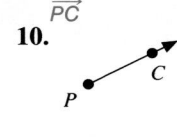

$\overrightarrow{PC}$

COMMUNICATION «*Discussion*

«**11.** Make a list of all the line segments that you can name in the figure below.

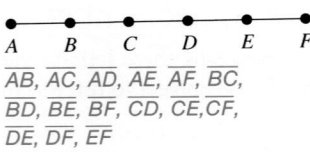

$\overline{AB}$, $\overline{AC}$, $\overline{AD}$, $\overline{AE}$, $\overline{AF}$, $\overline{BC}$, $\overline{BD}$, $\overline{BE}$, $\overline{BF}$, $\overline{CD}$, $\overline{CE}$, $\overline{CF}$, $\overline{DE}$, $\overline{DF}$, $\overline{EF}$

«**12.** Explain why $\angle M$ is not an appropriate name for the figure below. More than one angle has M as a vertex.

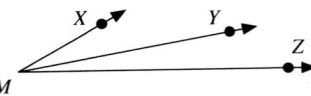

Exercises

Write the name of each figure. $\angle TRV$, $\angle VRT$, or $\angle R$ $\overleftrightarrow{CB}$, $\overleftrightarrow{BC}$

A

1.

R

plane R

2.

Z   N

$\overrightarrow{ZN}$

3. T

V

R

4. C

B

5. Explain why the symbols $\overrightarrow{AG}$ and $\overrightarrow{GA}$ do not represent the same geometric figure. They have different endpoints.

6. Give four different names for the angle shown at the right.
$\angle CTM$, $\angle MTC$, $\angle T$, $\angle 3$

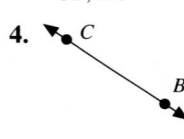

Make a sketch of each figure. Answers to Exercises 7–12 are on p. A16.

B

7. $\angle RXM$ **8.** $\overline{RS}$ **9.** $\overrightarrow{ZN}$ **10.** plane A

11. lines j and k, which intersect at point D

12. $\angle CBA$ and $\angle ABD$, which share $\overrightarrow{BA}$ as a common side

SPATIAL SENSE

Use the figure at the right. Name another
point that lies in the same plane as the
given points.

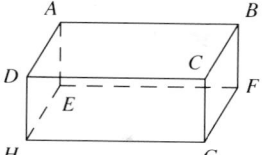

13. points *A, C, D* B   14. points *D, C, H* G

15. points *A, B, F* E   16. points *D, H, E* A

Use the figure at the right above. Name three
line segments that intersect at the given point.

17. point *C*
$\overline{BC}, \overline{DC}, \overline{GC}$

18. point *A*
$\overline{DA}, \overline{BA}, \overline{EA}$

19. point *F*
$\overline{BF}, \overline{GF}, \overline{EF}$

20. point *E*
$\overline{HE}, \overline{FE}, \overline{AE}$

MENTAL MATH

First answer each question by picturing the geometric figure in your
mind. Then check if your mental picture was accurate by making a
sketch of the figure on paper.

21. In angle *AMT*, which point is the vertex? M

22. What are the sides of angle *SYQ*? $\overrightarrow{YS}, \overrightarrow{YQ}$

C 23. Points *A, B,* and *C* are noncollinear. What three different line
segments have these points as their endpoints? $\overline{AB}, \overline{BC}, \overline{AC}$

24. What figure is formed by the intersection of ray *GH* and ray *HG*? $\overline{HG}$

SPIRAL REVIEW

S 25. Find the sum: $\frac{7}{8} + \frac{9}{16}$ *(Toolbox Skill 17)* $1\frac{7}{16}$

26. Write the name of the figure shown at the
right. *(Lesson 6-1)* $\overrightarrow{AZ}$

27. Sue sold tickets for a choral performance. Adult tickets cost $5.75
each and student tickets cost $2 each. Sue collected $72. How many
of each type of ticket did she sell? *(Lesson 4-2)*
8 adult tickets, 13 student tickets

28. Find the answer: $-2 + 8 \div (-2) - 10$ *(Lesson 3-6)* −16

Historical Note

Archie Alexander, one of the first African-American technicians to
be internationally famous, was an outstanding structural engineer. After graduating from the University of Iowa in 1912, he studied
bridge design at the University of London. He then became a specialist
in the design and construction of concrete and steel bridge spans.
Geometry played a very important role in his work.

Research

Find out two ways that geometry is used in the design and
construction of bridges. See answer on p. A17.

Introduction to Geometry **239**

Practice

Practice 68, *Resource Book*

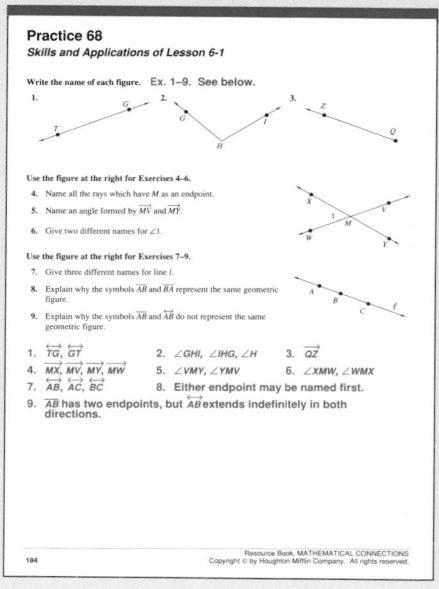

Follow-Up

Extension *(Cooperative Learning)*
Have students work in groups to
draw four rays, each starting at
point *D* and having one other point,
E, F, G, or *H,* on them. Ask the
groups to identify all the angles in
the sketches.

Enrichment
Enrichment 55 is pictured on
page 234E.

239

Teaching the Lesson
• Materials: protractor, paper circles, overhead projector
• Lesson Starter 51
• Visuals, Folder A
• Using Technology, p. 234C
• Using Manipulatives, p. 234C

Lesson Follow-Up
• Practice 69
• Enrichment 56: Exploration
• Study Guide, pp. 101–102

Warm-Up Exercises

Draw each of the following.

1. ray *AB*

2. line segment *CD*

3. line *XY*

4. ∠*ABC*

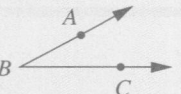

5. plane *R*

Lesson Focus

Is it important that airplanes follow specific flight paths? How do ships travel in the right direction during periods of heavy fog or darkness? This lesson should help answer questions such as these as it focuses on angle measurements.

6-2

Measuring and Drawing Angles

Objective: To measure and draw angles using a protractor.

APPLICATION

Terms to Know
• *degree*
• *protractor*

The architect who plans a building usually presents the plan in a blueprint. A blueprint shows not only the sizes of pieces such as walls and built-in cabinets, but also their positions in relation to each other. To show positions accurately, the architect indicates the size of the angle formed where these pieces meet.

The unit that is commonly used to measure the size of an angle is the **degree.** The number of degrees in an angle's measure indicates the amount of openness between the sides of the angle. To measure an angle, you use the geometric tool called a **protractor.**

Example 1

Use a protractor to measure ∠*ABC*.

Solution

Put the center mark of the protractor on the vertex of the angle, which is *B*. Place the 0° mark on one side of the angle, $\overrightarrow{BC}$. Then read the number where the other side, $\overrightarrow{BA}$, crosses the scale. The measure of ∠*ABC* is 75°.

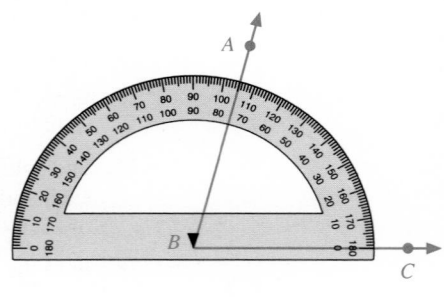

☑ **Check Your Understanding**

1. In Example 1, why do you read the measure of the angle from the bottom scale of numbers rather than the top scale?
See *Answers to Check Your Understanding* at the back of the book.

To indicate the degree measure of an angle, you use a small letter *m*.

In Words	In Symbols
The measure of angle *ABC* is seventy-five degrees.	$m\angle ABC = 75°$

240 Chapter 6

Example 2

Solution

Use a protractor to draw ∠RST with measure 160°.

Draw a ray, $\overrightarrow{ST}$, to represent one side of the angle. Put the center of the protractor on the endpoint, point S, and line up the 0° mark along the ray. Make a mark at 160° and remove the protractor. Draw a ray, $\overrightarrow{SR}$, through the mark.

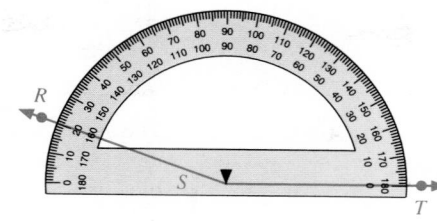

 **Check Your Understanding**

2. How would Example 2 be different if $\overrightarrow{ST}$ were drawn in the opposite direction?

See *Answers to Check Your Understanding* at the back of the book.

Guided Practice

1. a measure of accuracy
2. a measure of temperature
3. a measure of an angle

COMMUNICATION «*Reading*

Explain the meaning of the word *degree* in each sentence.

«**1.** Jerrold can compute mentally with a high degree of accuracy.

«**2.** The temperature increased ten degrees in just one hour.

«**3.** The pilot made a two-degree correction in the airplane's flight path.

«**4.** Lawanda earned a master's degree in English literature.

a measure of education

Refer to the sentences in Exercises 1–4.

Degree is a measurement.

«**5.** What do all these meanings of the word *degree* have in common?

«**6.** In which sentence is the meaning of *degree* closest to the meaning used in this lesson? 3

Use a protractor to measure each angle.

7. 126°
L
M   N

8. 45°
R
P   Q

Answers to Guided Practice 9–12 are on p. A17.

Use a protractor to draw an angle of the given measure.

9. 35°   **10.** 155°   **11.** 90°   **12.** 67°

Introduction to Geometry   **241**

Teaching Notes

Students may have difficulty in using a protractor. As you work the Examples with students, be certain they understand how to place the protractor correctly.

Key Question

What determines the number of degrees in an angle? the amount of opening between the sides of the angle

Using Technology

For a suggestion on using geometric drawing software, see page 234C.

Using Manipulatives

For suggestions on using a paper-folding activity and an activity with a protractor on the overhead projector, see page 234C.

Error Analysis

Students may use the wrong scale on a protractor for measuring or drawing angles. Comparing angles to 90° (a corner) can help students avoid this problem.

Additional Examples

1. Use a protractor to measure ∠DEF.   115°

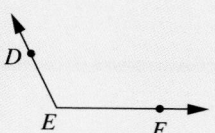

2. Use a protractor to draw ∠PQR with measure 40°.

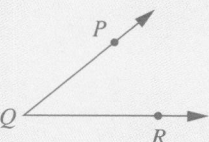

Closing the Lesson

Ask a student to explain what is meant by "the measure of an angle."

Skills: 1–16
Average: 1–16
Advanced: 1–16

Suggested Assignments

Skills: 1–16
Average: 1–16
Advanced: 1–16

Exercise Note

Estimation

Exercises 9–12 help students develop an intuitive sense for the size of an angle.

Follow-Up

Extension *(Cooperative Learning)*
Have students work in groups and use a protractor to draw the face of a clock. Each group should create three questions such as: How many degrees did the hands of the clock move while you were in school today? Groups should exchange questions and discuss answers.

Enrichment

Enrichment 56 is pictured on page 234E.

Practice

Practice 69, *Resource Book*

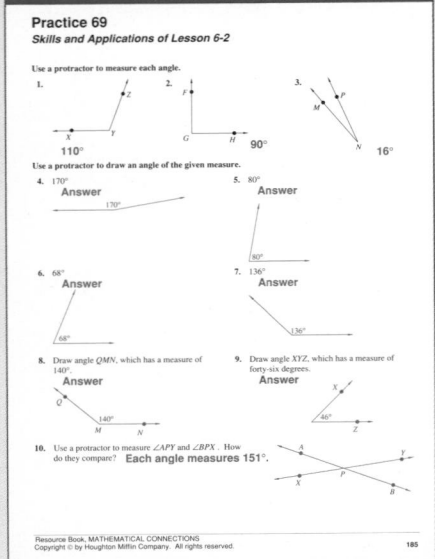

Exercises

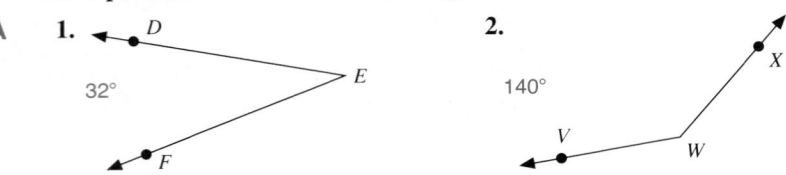

Use a protractor to measure each angle.

A   **1.**   32°   **2.**   140°

Use a protractor to draw an angle of the given measure. Answers to Exercises 3–8 are on p. A17.

3. 115°   **4.** 20°   **5.** 88°   **6.** 164°

7. Draw angle *JKL*, which has a measure of sixty degrees.

8. The measure of the angle formed by joining rays *XY* and *XZ* is 172°. Draw the angle.

ESTIMATION

Match each angle with the best estimate of its measure. Then use a protractor to check your answer.

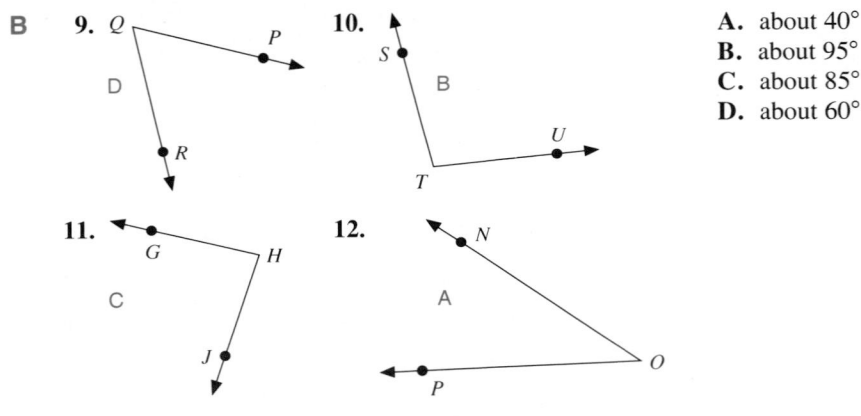

B   **9.**   **10.**   **A.** about 40°

                                              B. about 95°

                                              C. about 85°

                                              D. about 60°

11.   **12.**

SPIRAL REVIEW

S   **13.** Solve: $x - 12 = -3$ *(Lesson 4-3)* 9

14. Estimate the sum: $47 + 51 + 44 + 58$ *(Toolbox Skill 4)* about 200

15. Draw an angle that measures 104°.   *(Lesson 6-2)* See answer on p. A17.

16. The weekly salaries at Office Rite Company are $315, $340, $295, $305, $850, $800, and $345. Decide whether the mean describes the data well. Explain.   *(Lesson 5-9)* No; extreme values of $850 and $800 distort the mean.

242   Chapter 6

Constructing and Bisecting Line Segments and Angles

Objective: To use a straightedge and a compass to construct and bisect line segments and angles.

Materials
- straightedge
- compass

In this exploration, you will *construct* geometric figures using only two tools, a *straightedge* and a *compass*. You may use a ruler as a straightedge, but you may not use the marks on the ruler for measuring. You will use a compass to draw a circle, or part of a circle called an *arc*.

Activity I *Constructing a Line Segment the Same Length as a Given Line Segment*

Check students' drawings.

1. Use a straightedge to draw any segment $\overline{AB}$.
   a. Use a straightedge to draw a line. Call it ℓ.
   b. Choose any point on ℓ and label it C.
   c. Open your compass so that the sharp tip is on A and the writing tip is on B. Keeping the same compass opening, place the sharp tip on C and draw an arc crossing line ℓ. Label the point of intersection D. $\overline{CD}$ is the same length as $\overline{AB}$.

2. a. Draw any line segment $\overline{EF}$. Then use a straightedge and a compass to construct $\overline{MN}$ so that the length of $\overline{MN}$ is twice the length of $\overline{EF}$. See answer on p. A17.
   b. Describe how you would construct $\overline{MN}$ so that the length of $\overline{MN}$ is three times the length of $\overline{EF}$. See answer on p. A17.

Activity II *Constructing an Angle with the Same Measure as a Given Angle*

1. Draw any $\angle B$. Check students' drawings.
   a. To copy $\angle B$, begin by using a straightedge to construct any ray with endpoint E.
   b. Using any compass opening, place the sharp tip of your compass on point B and draw an arc intersecting the sides of $\angle B$. Label the points of intersection A and C.
   c. Using the same compass opening, place the sharp tip of your compass on E and draw an arc. Label the point of intersection F.
   d. Open your compass so that the sharp tip is on C and the writing tip is on A. Using the same compass opening, place the point of your compass on F and draw an arc intersecting the first arc. Label the point of intersection D. Draw $\overrightarrow{ED}$. $m\angle E$ equals $m\angle B$.

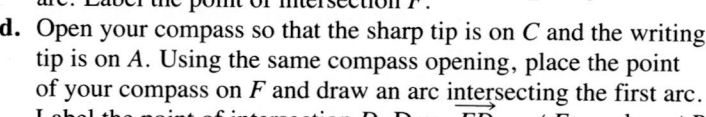

2. Draw any obtuse angle. Use a straightedge and a compass to construct an angle with the same measure. See answer on p. A17.

Introduction to Geometry **243**

Teaching Notes

Constructions are an important manipulative activity in geometry. The activities in this Focus on Explorations introduce students to some of the less complicated types of constructions.

Activity Notes

Activity I
Activity I has students construct a line segment with the same length as a given line segment. Students are then asked to apply this construction to create line segments that are two and three times as long as a given segment.

Activity II
Activity II has students construct an angle with the same measure as a given acute angle. Students are then asked to apply this construction to create an angle equal in measure to a given obtuse angle.

Activity III
Activity III has students construct the bisector of a given line segment. Students are then asked to apply this construction to divide a given line segment into four segments of equal length.

Activity IV
Activity IV has students construct the bisector of a given angle. Students are then asked to consider how the measures of the parts of the bisected angle would compare if the measure of the original angle increased or decreased. Finally, students are asked to apply this construction to bisect each of two given angles.

You can also use a straightedge and a compass to *bisect* a geometric figure, that is, to divide it into two parts of equal measure.

Activity III   *Bisecting a Line Segment*

1 Draw any line segment $\overline{AB}$. Check students' drawings.

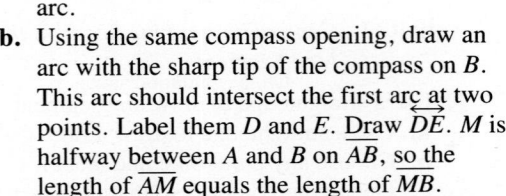

   a. Choose a point C on $\overline{AB}$, so that the length of $\overline{AC}$ clearly is more than half the length of $\overline{AB}$. Set the compass so that the sharp tip is on A and the writing tip is on C. Draw an arc.

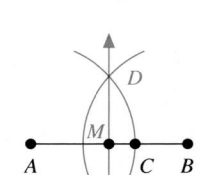

   b. Using the same compass opening, draw an arc with the sharp tip of the compass on B. This arc should intersect the first arc at two points. Label them D and E. Draw $\overleftrightarrow{DE}$. M is halfway between A and B on $\overline{AB}$, so the length of $\overline{AM}$ equals the length of $\overline{MB}$.

   c. Which line segment bisects $\overline{AB}$: $\overline{AC}$, $\overline{CB}$, or $\overline{DE}$? $\overline{DE}$

2 Trace the segment $\overline{FG}$ below onto your paper. Then use a straightedge and a compass to divide the segment into four segments of equal length.
See answer on p. A17.

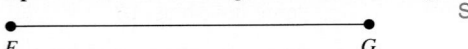

F   G

Answers to Activity IV, Steps 2 and 3, are on p. A17.

Activity IV   *Bisecting an Angle*

1 Draw any $\angle B$. Check students' drawings.

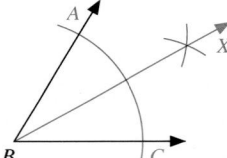

   a. Using any compass opening, place the sharp tip of your compass on point B and draw an arc intersecting the sides of $\angle B$. Label the points of intersection A and C.

   b. Then place the sharp tip of the compass on A and draw an arc. Using the same compass opening, place the sharp tip of your compass on C and draw an arc that intersects the arc you just drew. Label the point of intersection X. Draw $\overrightarrow{BX}$. $m\angle ABX$ equals $m\angle XBC$.

2 How would the measures of $\angle ABX$ and $\angle XBC$ compare if $m\angle ABC$ increased? if $m\angle ABC$ decreased?

3 Trace each angle onto your paper. Bisect each angle.

   a.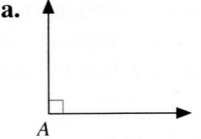

   A

   b.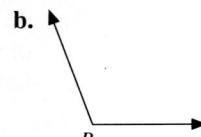

   B

6-3

Types of Angles

Objective: To find measures of acute, right, obtuse, straight, complementary, supplementary, vertical, and adjacent angles.

Terms to Know
- *acute angle*
- *right angle*
- *obtuse angle*
- *straight angle*
- *complementary*
- *supplementary*
- *adjacent*
- *vertical*

EXPLORATION

1 In medicine, a pain that is very sharp is called *acute* pain. What do you think an *acute angle* looks like? a sharp angle

2 In sailing, you *right* a capsized boat when you set it upright again. What do you think a *right angle* looks like? one side is upright

3 In botany, a leaf with a blunt or rounded tip is called an *obtuse* leaf. What do you think an *obtuse angle* looks like? a blunt angle

4 In art, a line that is not curved or bent is called a *straight* line. What do you think a *straight angle* looks like? a straight line

Angles are often classified by their measures. The table below gives a summary to help you identify and compare the basic types.

Angle	Name	Measure
	acute angle	greater than 0°, less than 90°
	right angle	equal to 90° (*Note:* A small square indicates a right angle.)
	obtuse angle	greater than 90°, less than 180°
	straight angle	equal to 180°

Names are also given to special pairs of angles. For instance, some pairs are classified by the sum of their measures.

Two angles are **complementary** when the sum of their measures is 90°.

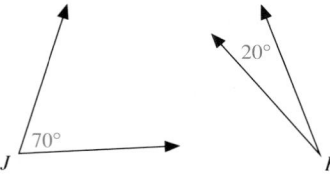

$m\angle J + m\angle K = 70° + 20°$
$= 90°$
$\angle J$ and $\angle K$ are complementary.

Two angles are **supplementary** when the sum of their measures is 180°.

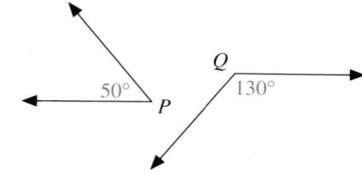

$m\angle P + m\angle Q = 50° + 130°$
$= 180°$
$\angle P$ and $\angle Q$ are supplementary.

Introduction to Geometry **245**

Lesson Planner

Teaching the Lesson
- Materials: board compass, wooden craft sticks
- Lesson Starter 52
- Visuals, Folder A
- Using Manipulatives, p. 243C
- Reteaching/Alternate Approach, p. 234D
- Cooperative Learning, p. 234D

Lesson Follow-Up
- Practice 70
- Enrichment 57: Application
- Study Guide, pp. 103–104

Warm-Up Exercises

1. Draw two lines intersecting at point *A*.

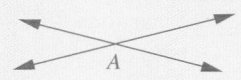

2. Name two rays and two line segments in this figure.

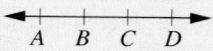

Answers may vary. Examples:
$\overrightarrow{AD}, \overrightarrow{BD}; \overline{AB}, \overline{BC}$

3. Sketch an angle that has a measure of approximately 30°.

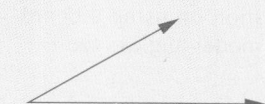

245

Lesson Focus

Choose individual students to follow instructions such as these:

1. Take 2 steps and turn *right*.
2. Start at your desk and continue *straight* ahead.
3. Place another book *adjacent* to your math book.

Such activities demonstrate the need to understand each italicized term. Today's lesson focuses on these and other terms related to angles.

Teaching Notes

Spend adequate time discussing the terms in the Exploration section at the beginning of the lesson. Have students point out several examples of each type of angle within the classroom.

Key Question

Can an obtuse angle have a complement? Explain.   No; the sum of two complementary angles must be 90°, and the measure of an obtuse angle is already greater than 90°.

Reasoning

The Check Your Understanding questions for Example 1 require students to explain the changes in the problems if the terms supplement and complement are interchanged.

You may also want to ask students if two complementary angles or two supplementary angles must be adjacent.   No; they may be, but it is not a necessary condition.

Using Manipulatives

For a suggestion on using a board compass to model angles, see page 234C.

Reteaching/Alternate Approach

For a suggestion on using a cooperative learning activity to explore angles, see page 234D.

246

Example 1

a. Find the measure of an angle complementary to a 23° angle.

b. Find the measure of an angle supplementary to a 170° angle.

Solution

a. Subtract the given measure from 90°.
$90° - 23° = 67°$

b. Subtract the given measure from 180°.
$180° - 170° = 10°$

✎ Check Your Understanding

1. How would part (a) of Example 1 be different if you were asked to find the measure of a *supplement* of an angle that measures 23°?

2. In part (b) of Example 1, would it be possible to find a *complement* of an angle that measures 170°? Explain.
   See Answers to Check Your Understanding at the back of the book.

Two angles that share a common side, but do not overlap each other, are called **adjacent** angles. In the figure below at the left, ∠*ABX* and ∠*XBC* are adjacent, but ∠*ABC* and ∠*XBC* are not adjacent. Two special situations occur when complementary and supplementary angles are adjacent.

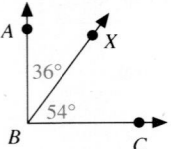   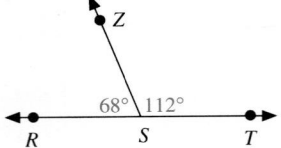

$m\angle ABX + m\angle CBX = 90°$,
so ∠*ABC* is a right angle.

$m\angle RSZ + m\angle TSZ = 180°$,
so ∠*RST* is a straight angle.

When two lines intersect, the angles that are *not* adjacent to each other are called **vertical** angles. Vertical angles are always equal in measure. In the figure at the right, ∠1 and ∠2 are vertical angles.

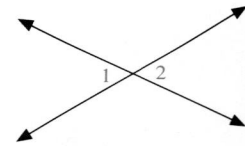

Example 2

Use the figure at the right.
Find the measure of each angle.

a. ∠2   b. ∠3

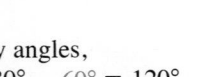

Solution

a. ∠2 and ∠4 are vertical angles, so $m\angle 2 = m\angle 4 = 60°$.

b. ∠3 and ∠4 are supplementary angles, so $m\angle 3 = 180° - m\angle 4 = 180° - 60° = 120°$.

✎ Check Your Understanding

3. Name the two pairs of vertical angles in the figure for Example 2.

4. In the figure for Example 2, name three pairs of supplementary angles other than ∠3 and ∠4.
   See Answers to Check Your Understanding at the back of the book.

Guided Practice

COMMUNICATION « *Reading*

« **1.** Explain the meaning of *supplement* in the following sentence.

Elena works part time to earn money to supplement her college scholarship. Elena earns money to add to her scholarship.

2. In mathematics, *complementary* means that the sum of the measures of two angles is 90°. *Complimentary* means expressing a compliment or praise.

« **2.** How is the meaning of the word *complementary* different from the meaning of *complimentary*?

Find each difference.

3. 90 − 74 16

4. 90 − 15 75

5. 90 − 53 37

6. 180 − 166 14

7. 180 − 82 98

8. 180 − 59 121

Tell whether two angles with the given measures are *complementary*, *supplementary*, or *neither*.

9. 40°, 50° complementary

10. 125°, 55° supplementary

11. 12°, 88° neither

12. 90°, 90° supplementary

Find the measure of a complement of an angle of the given measure.

13. 15° 75°

14. 79° 11°

15. 60° 30°

16. 33° 57°

Find the measure of a supplement of an angle of the given measure.

17. 41° 139°

18. 145° 35°

19. 88° 92°

20. 27° 153°

Use the figure at the right. Give the measure of each angle.

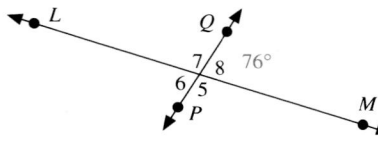

21. ∠5 104°

22. ∠6 76°

23. ∠7 104°

24. ∠8 76°

Exercises

Find the measure of a complement of an angle of the given measure.

A **1.** 30° 60°

2. 5° 85°

3. 29° 61°

4. 87° 3°

Find the measure of a supplement of an angle of the given measure.

5. 10° 170°

6. 65° 115°

7. 102° 78°

8. 90° 90°

Use the figure at the right. Give the measure of each angle.

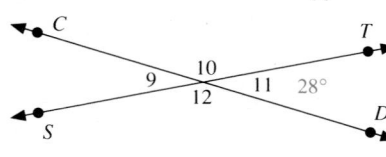

9. ∠9 28°

10. ∠10 152°

11. ∠11 28°

12. ∠12 152°

Additional Examples

1. Find the measure of an angle complementary to a 35° angle. 55°
2. Find the measure of an angle supplementary to a 135° angle. 45°
3. Find the measure of each angle in the figure if *m*∠2 is 50°.

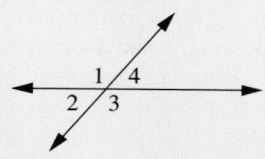

$m\angle 2 = m\angle 4 = 50°$
$m\angle 1 = m\angle 3 = 130°$

Closing the Lesson

Choose five students to go to the chalkboard. Have five other students ask them to do tasks related to the lesson. For example:

1. Sketch vertical angles.
2. Sketch complementary angles.
3. Sketch supplementary angles.
4. Find the complement of 40°.
5. Find the supplement of 110°.

Suggested Assignments

Skills: 1–18, 25–27
Average: 1–22, 25–27
Advanced: 1–7 odd, 9–27

Exercise Notes

Communication: Reading

Guided Practice Exercises 1 and 2 require students to relate two mathematical terms introduced in this lesson to the usage of words in everyday life.

Exercise Notes (continued)

Reasoning
Exercises 15–18 require students to determine whether certain geometric statements are always true, never true, or sometimes true.

Making Connections/Transitions
In order to solve *Exercises 19–24*, students are required to apply an algebraic process to the geometric figures.

Follow-Up

Enrichment
Have students develop a "proof" that vertical angles are equal. They can use either paragraph form or a sketch and statements of their own choosing.

Enrichment
Enrichment 57 is pictured on page 234E.

Practice
Practice 70, *Resource Book*

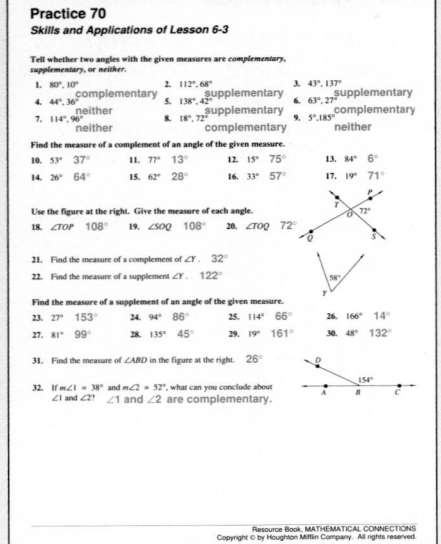

---

13. Angle *XYZ* has the same measure as its complement. Find $m\angle XYZ$. 45°

14. **DATA,** *pages 234–235* Determine whether an airplane's angle of attack is an *acute*, *right*, or *obtuse* angle. acute

LOGICAL REASONING

Replace each __?__ with *always*, *sometimes*, or *never* to make a true statement.

B 15. A supplement of an obtuse angle is __?__ an acute angle. always

16. A complement of an acute angle is __?__ an obtuse angle. never

17. The measure of an angle is __?__ equal to the measure of its supplement. sometimes

18. Two vertical angles that are supplementary are __?__ right angles. always

CONNECTING GEOMETRY AND ALGEBRA

Sometimes the measures of angles in a figure are given only as variable expressions. If you are given enough information, though, you can determine what the exact measures must be by using an equation.

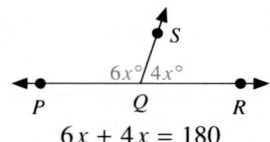

$$6x + 4x = 180$$

Exercises 19–22 refer to the figure at the right. Answers to Exercises 19 and 20 are on p. A17.

19. Why was the number 180 used as the right side of the equation?

20. Why are 6x and 4x added on the left side of the equation?

21. What value of *x* is the solution of the equation? 18

22. What are the measures of $\angle PQS$ and $\angle RQS$? 108°, 72°

In each figure, find $m\angle ABC$.

C 23. 60°

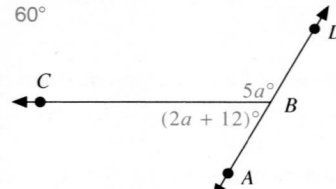

24. 72°

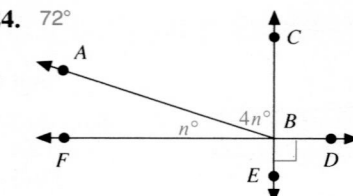

SPIRAL REVIEW

$$2n + 11 = -31$$

S 25. Write an equation that represents the sentence: The sum of twice a number *n* and eleven is negative thirty-one. *(Lesson 4-7)*

26. Find the measure of an angle that is supplementary to a 56° angle. *(Lesson 6-3)* 124°

27. Find the sum: $-17 + (-14)$ *(Lesson 3-2)* −31

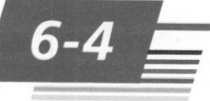

6-4 Perpendicular and Parallel Lines

Objective: To identify perpendicular and parallel lines and calculate the measures of the angles formed when a transversal intersects parallel lines.

APPLICATION

Many cities were planned so that the streets lie in an orderly grid pattern. At the point where streets intersect, they form right angles and are said to be *perpendicular*. Streets that do not intersect are *parallel*.

New Orleans, circa 1765

Mississippi River

Two lines that intersect to form right angles are **perpendicular** lines. In the figure at the right, you see that $\angle AMY$ is a right angle, and so you know that $\overleftrightarrow{AB}$ and $\overleftrightarrow{XY}$ are perpendicular. The symbol for *is perpendicular to* is $\perp$.

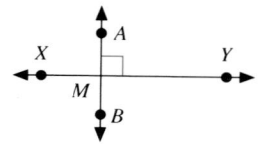

In Words	In Symbols
Line AB is perpendicular to line XY.	$\overleftrightarrow{AB} \perp \overleftrightarrow{XY}$

Two lines in the same plane that do not intersect are **parallel** lines. In the figure at the right, $\overleftrightarrow{PQ}$ and $\overleftrightarrow{RS}$ will always remain the same distance apart, and so they are parallel. The symbol for *is parallel to* is $\parallel$.

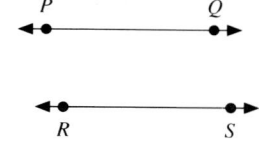

In Words	In Symbols
Line PQ is parallel to line RS.	$\overleftrightarrow{PQ} \parallel \overleftrightarrow{RS}$

The words *parallel* and *perpendicular* also can be used to refer to line segments and rays.

Introduction to Geometry **249**

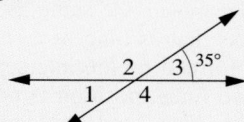

Lesson Focus

Ask students if they have ever watched a carpenter build a house. Is it important that the walls be "straight up and down," or *vertical*, to the foundation? What is the relationship of the *opposite* walls? These concepts are the focus of today's lesson.

Teaching Notes

Use the Application at the beginning of the lesson to further students' understanding of the meaning of the terms *perpendicular* and *parallel*.

Key Questions

1. Why, in a plane, must two lines be either parallel or intersecting? Any two lines that are not always the same distance apart will eventually intersect.
2. Are corresponding angles always equal in measure? No; only if the lines cut by the transversal are parallel. The same is true for alternate interior angles.

Application

Some students may have had experience with blueprints in industrial arts or vocational education classes. Relate the concepts of parallel and perpendicular to their work.

Error Analysis

Students sometimes think that perpendicular means the same as vertical. Stress that as long as a right angle is formed by the intersection of the two lines, position is not important. Also, students tend to think that pairs of alternate interior and corresponding angles are always equal, even if the lines cut by the transversal are not parallel. Stress that the lines must be parallel for the angles to be equal.

Example 1

Use the figure at the right. Tell whether each statement is *True* or *False*.

a. $\overleftrightarrow{AB} \parallel \overleftrightarrow{CD}$ **b.** $\overleftrightarrow{MN} \perp \overleftrightarrow{CD}$

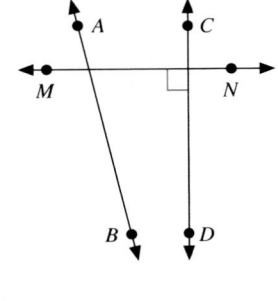

Solution

a. False. If $\overleftrightarrow{AB}$ and $\overleftrightarrow{CD}$ were extended downward, they would intersect.

b. True. A right angle is formed at the point where $\overleftrightarrow{MN}$ intersects $\overleftrightarrow{CD}$.

✏️ **Check Your Understanding**

1. In Example 1, how many right angles are formed at the point in the figure where $\overleftrightarrow{MN}$ intersects $\overleftrightarrow{CD}$? Explain.

See *Answers to Check Your Understanding* at the back of the book.

A **transversal** is a line that intersects two or more lines in the same plane at different points. In the figure at the right, line ℓ is a transversal that intersects $\overleftrightarrow{JK}$ and $\overleftrightarrow{MN}$.

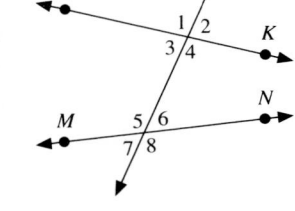

Alternate interior angles are *interior* to the two lines, but on *alternate* sides of the transversal. In the figure, these are the pairs of alternate interior angles.

$\angle 3$ and $\angle 6$ $\angle 4$ and $\angle 5$

Corresponding angles are in the same position with respect to the two lines and the transversal. In the figure above, these are the pairs of corresponding angles.

$\angle 1$ and $\angle 5$ $\angle 2$ and $\angle 6$ $\angle 3$ and $\angle 7$ $\angle 4$ and $\angle 8$

There are special properties when the lines intersected by a transversal are parallel.

The angles in a pair of alternate interior angles are equal in measure.
The angles in a pair of corresponding angles are equal in measure.

Example 2

In the figure at the right, $\overleftrightarrow{EF} \parallel \overleftrightarrow{GH}$. Find the measure of each angle.

a. $\angle 6$ **b.** $\angle 8$

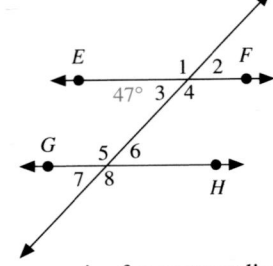

Solution

a. $\angle 3$ and $\angle 6$ are alternate interior angles, so $m\angle 6 = m\angle 3 = 47°$.

b. $\angle 8$ and $\angle 6$ are supplementary:
$$m\angle 8 = 180° - m\angle 6$$
$$= 180° - 47° = 133°$$

✏️ **Check Your Understanding**

2. In Example 2, which angle in the figure forms a pair of corresponding angles with $\angle 3$? What is the measure of this angle?

See *Answers to Check Your Understanding* at the back of the book.

---

250 Chapter 6

Guided Practice

COMMUNICATION « *Writing*

Write each sentence in symbols.

« **1.** Line *ST* is perpendicular to line *YZ*. $\overleftrightarrow{ST} \perp \overleftrightarrow{YZ}$

« **2.** Line segment *PQ* is parallel to line segment *MN*. $\overline{PQ} \parallel \overline{MN}$

Write each statement in words. Answers to Guided Practice 3–6 are on p. A17.

« **3.** $\overleftrightarrow{AB} \parallel \overleftrightarrow{ST}$ « **4.** $\overrightarrow{FG} \perp \overleftrightarrow{UV}$ « **5.** $\overline{XY} \perp \overline{ZW}$

« **6.** Draw a diagram that represents the statement $\overrightarrow{MN} \perp \overline{GH}$.

Use the figure at the right. Tell whether the given angles are a pair of corresponding angles or alternate interior angles.

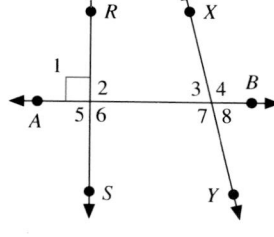

7. ∠1 and ∠3
corr.

8. ∠2 and ∠7
alt. int.

9. ∠6 and ∠3
alt. int.

10. ∠4 and ∠2
corr.

Use the figure above at the right. Tell whether each statement is *True* or *False*.

11. $\overleftrightarrow{RS} \parallel \overleftrightarrow{XY}$ False   **12.** $\overleftrightarrow{AB} \perp \overleftrightarrow{RS}$ True

In the figure at the right, $\overleftrightarrow{EF} \parallel \overleftrightarrow{GH}$ and $m\angle 9 = 75°$. Find the measure of each angle.

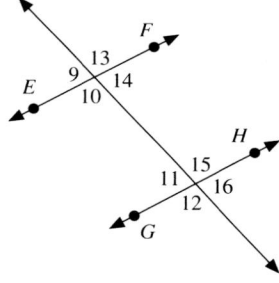

13. ∠10 105° **14.** ∠11 75° **15.** ∠12 105°

16. ∠13 105° **17.** ∠14 75° **18.** ∠15 105°

« **19.** **COMMUNICATION** « *Discussion*
Why do you think so many cities are laid out as a grid of parallel and perpendicular streets? See answer on p. A17.

Exercises

Use the figure at the right. Tell whether each statement is *True* or *False*.

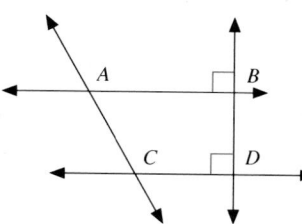

A **1.** $\overleftrightarrow{AB} \perp \overleftrightarrow{CD}$ False   **2.** $\overleftrightarrow{AB} \perp \overleftrightarrow{DB}$ True

3. $\overleftrightarrow{CD} \parallel \overleftrightarrow{AB}$ True   **4.** $\overleftrightarrow{AC} \parallel \overleftrightarrow{BD}$ False

5. $\overleftrightarrow{AC} \perp \overleftrightarrow{CD}$ False   **6.** $\overleftrightarrow{AC} \parallel \overleftrightarrow{CD}$ False

7. $\overleftrightarrow{AB} \parallel \overleftrightarrow{AC}$ False   **8.** $\overleftrightarrow{CD} \perp \overleftrightarrow{BD}$ True

Introduction to Geometry **251**

Using Technology
For a suggestion on using geometric drawing software, see page 234C.

Using Manipulatives
For a suggestion on using a model to study lines cut by a transversal, see page 234D.

Reteaching/Alternate Approach
For a suggestion on using a cooperative learning activity to study types of intersecting lines, see page 234D.

Additional Examples

Use the figure to tell whether each statement is *true* or *false*.

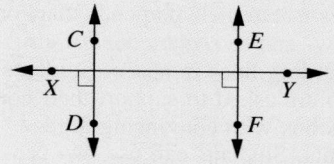

1. $\overleftrightarrow{EF} \perp \overleftrightarrow{XY}$ True
2. $\overleftrightarrow{EF} \parallel \overleftrightarrow{CD}$ True

Use the figure to find the measure of each angle. $\overleftrightarrow{AB} \parallel \overleftrightarrow{CD}$

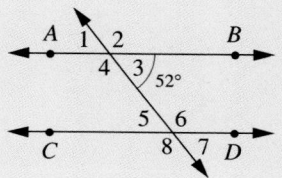

3. ∠5   52°
4. ∠8   128°

Closing the Lesson

Draw two parallel lines cut by two nonparallel transversals on the chalkboard. Give students categories of angles, such as supplementary, alternate interior, equal, and so on, and ask them to identify all the angles or pairs of angles that fit into the category.

Suggested Assignments

Skills: 1–16, 23–28
Average: 1–18, 23–28
Advanced: 1–28

Exercise Notes

Communication: Writing

Guided Practice Exercises 1–5 ask students to translate sentences into geometric symbols, and vice versa. *Guided Practice Exercise 6* asks students to draw a diagram that represents a symbolic statement.

Critical Thinking

In *Exercises 19–22*, students are asked to *compare* the definitions of skew and parallel lines, to *classify* lines as parallel, perpendicular, or skew, and to *reach conclusions* regarding lines in space. Students also are asked to support their conclusions with convincing arguments. Students can use two pencils, for example, to show examples of parallel, perpendicular, and skew lines.

17. $m\angle 1 = 96°$
    $m\angle 2 = 84°$
    $m\angle 3 = 96°$
    $m\angle 4 = 90°$
    $m\angle 5 = 90°$
    $m\angle 6 = 90°$
    intersecting
18. $m\angle 7 = 78°$
    $m\angle 8 = 102°$
    $m\angle 9 = 78°$
    $m\angle 10 = 102°$
    $m\angle 11 = 78°$
    $m\angle 12 = 102°$
    intersecting
19. See answer on p. A17.

In the figure at the right,
$\overleftrightarrow{JK} \parallel \overleftrightarrow{QR}$ **and** $m\angle 7 = 143°$.
Find the measure of each angle.

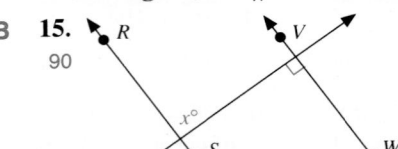
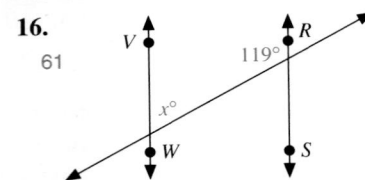

9. $\angle 1$ 37°　　　　10. $\angle 2$ 143°

11. $\angle 3$ 143°　　　12. $\angle 4$ 37°

13. $\angle 5$ 37°　　　　14. $\angle 6$ 143°

In each figure, $\overrightarrow{RS} \parallel \overrightarrow{VW}$. **Find the value of** *x.*

B 15.
    90

16.
    61

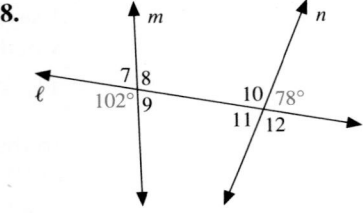

In each figure, first find the measures of all the numbered angles. Then tell whether lines *m* **and** *n* **are parallel or intersecting lines.**

17.　　　　　　　　　　18.

THINKING SKILLS 19. Analysis　　**20–22.** Evaluation

Lines that do not lie in the same plane are called **skew** lines.

19. Compare the definition of skew lines, given above, to the definition of parallel lines given on page 249. How are skew lines and parallel lines different? How do you think they are alike?

20. The figure at the right represents a box, similar to an ordinary packing carton. Classify each pair of line segments as *parallel, perpendicular,* or *skew.*

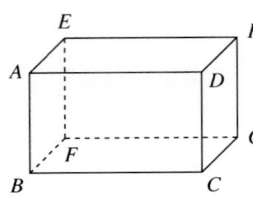

  parallel   a. $\overline{AB}$ and $\overline{DC}$　　　b. $\overline{EF}$ and $\overline{DH}$ skew
perpendicular c. $\overline{CG}$ and $\overline{FG}$　　　d. $\overline{AB}$ and $\overline{GH}$
                                      parallel
21. Is it possible for two skew lines to be perpendicular? Give a convincing argument to support your answer. See answer on p. A17.

C 22. Determine whether the following statement is true or false. True
Any two lines in space are related in exactly one of these ways: the lines either intersect, they are parallel, or they are skew.

S **23.** Estimate the quotient: 3495 ÷ 38
(*Toolbox Skill 3*) about 90

24. In the figure at the right,
$\overleftrightarrow{AB} \parallel \overleftrightarrow{CD}$. Find the measure of
∠8. (*Lesson 6-4*) 140°

25. Evaluate $-4a^2b$ when $a = 5$ and
$b = -2$. (*Lesson 3-6*) 200

26. Complete: 148 mm = <u>?</u> m
(*Lesson 2-9*) 0.148

27. Draw a bar graph to display the
data in the table at the right.
(*Lesson 5-4*) See answer on p. A17.

28. This week the Panthers scored
37 points. This is 5 points less
than twice the number of
points they scored last week.
How many points did the
Panthers score last week?
(*Lesson 4-8*) 21

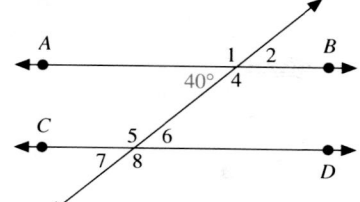

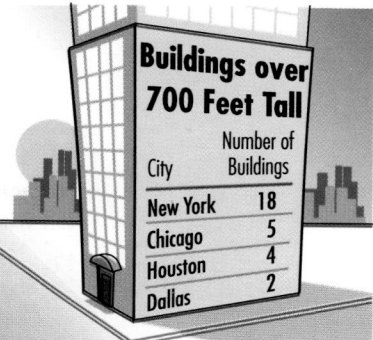

Buildings over
700 Feet Tall

City	Number of Buildings
New York	18
Chicago	5
Houston	4
Dallas	2

Self-Test 1

Write the name of each figure.

1.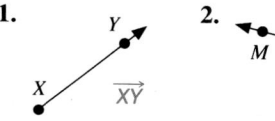
$\overrightarrow{XY}$

2. $\overleftrightarrow{MN}, \overleftrightarrow{NM}$

3. ∠TSR, ∠RST, or ∠S

6-1

Answers to Self-Test 4–6 are on p. A18.
Use a protractor to draw an angle of the given measure.

4. 55° **5.** 100° **6.** 82° 6-2

7. Find the measure of an angle complementary to a 49° angle. 6-3
41°

8. Find the measure of an angle supplementary to a 77° angle.
103°

In the figure at the right,
$\overleftrightarrow{ST} \parallel \overleftrightarrow{UV}$ and $m\angle 6 = 65°$.
Find the measure of each
angle.

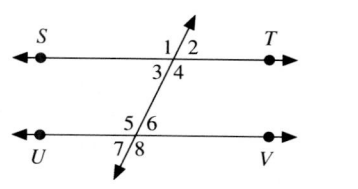

9. ∠7 65° **10.** ∠5 115°

11. ∠2 65° **12.** ∠4 115°

6-4

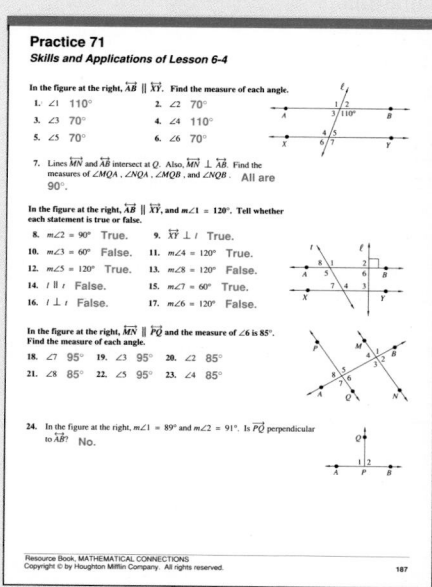

254

Teaching Notes

This Focus on Explorations continues the introduction to constructions begun on pages 243 and 244. Students are shown how to construct perpendicular lines and parallel lines and then are asked to use these constructions to reach conclusions about figures that contain both types of lines in them.

Activity Notes

Activity I

Activity I has students construct a line perpendicular to a given line through a point on the line. You may wish to mention how this activity is similar to Activity III on page 244.

Activity II

Activity II has students construct a line perpendicular to a given line through a point not on the line. You may wish to compare this activity with Activity I, pointing out both the similarities and differences.

Activity III

Activity III has students construct a line parallel to a given line through a given point. Students are then asked to explain why the construction works and to reach a conclusion about two lines parallel to the same line.

Activity IV

Activity IV combines what students have learned in Activities I through III. Students are asked to apply the constructions from these activities to create a number of figures containing both perpendicular and parallel lines. Using these figures, students are asked to reach some conclusions regarding perpendicularity, parallelism, and the types of angles and figures formed by perpendicular and parallel lines.

Objective: To construct a line perpendicular to a given line and to construct a line parallel to a given line.

Materials

■ straightedge
■ compass

Constructing Perpendicular and Parallel Lines

Graphic designers, architects, and engineers often draw perpendicular and parallel lines in their work. The construction of these lines can be done by computer-aided design programs. You can also construct perpendicular and parallel lines with a straightedge and a compass.

Activity I   *Constructing a Line Perpendicular to a Given Line through a Point on the Line*

■ Draw a line ℓ. Check students' drawings.
  a. Choose any point P on the line. Put the sharp tip of your compass on P and draw arcs that intersect the line in two points. Label the points of intersection A and B.
  b. Does $AP = BP$? Yes
  c. Widen the opening of your compass. With the sharp tip on A, draw an arc.
  d. Without changing the compass setting, place the sharp tip on B and draw an arc that intersects the arc you just drew. Label this point of intersection Q. Draw $\overleftrightarrow{PQ}$. $\overleftrightarrow{PQ} \perp$ line ℓ.

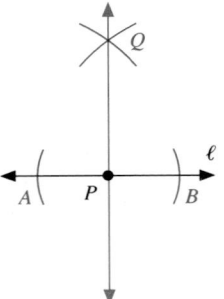

Activity II   *Constructing a Line Perpendicular to a Given Line through a Point not on the Line*

■ Draw a line ℓ. Check students' drawings.
  a. Draw a point N, not on ℓ. With the sharp tip of your compass on N, draw arcs that intersect line ℓ in two points. Label the points of intersection S and T.
  b. Widen the opening of your compass. With the sharp tip on S, draw an arc not on the line.
  c. Without changing the compass setting, put the sharp tip of your compass on T and draw an arc that intersects the arc you just drew. Label the point of intersection R. Draw $\overleftrightarrow{NR}$. $\overleftrightarrow{NR} \perp$ line ℓ.

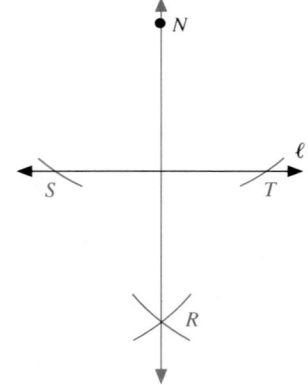

Answers to Activity III, Steps 2–3, and Activity IV are on p. A18.

Activity III *Constructing a Line through a Given Point and Parallel to a Given Line*

1. Draw a line *c*. Check students' drawings.

   a. Through any point *P* not on line *c*, draw any line *ℓ* that intersects line *c*. Label the angle formed ∠1 as shown.

   b. Construct angle ∠2, corresponding to ∠1, at *P*. Extend the side of ∠2 through *P* and label that line *d*.

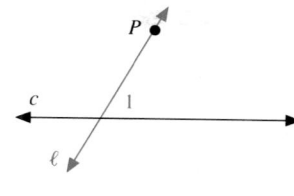

2. Explain why line *d* is parallel to line *c*.

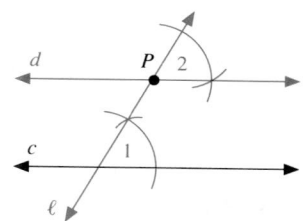

3. Draw a point *Q* on line *ℓ*, far below line *c*. Construct line *e* that is parallel to line *c*. Is line *e* parallel to line *d*? Explain.

Activity IV *More Constructions*

1. Draw a line *ℓ*. Construct a line *m* that is perpendicular to line *ℓ* through a point *A* that is not on *ℓ*.

2. Construct a line *q* that is perpendicular to line *ℓ* through a point *B* that is not on *ℓ* or *m*. Is line *m* parallel to line *q*? Explain.

3. Make a general statement about two lines that are perpendicular to the same line.

4. Construct a line *r* parallel to line *l* through point *B*. Is line *r* perpendicular to line *q*? line *m*?

5. If a line is perpendicular to one of two parallel lines, is it also perpendicular to the other parallel line? Explain.

6. Make a generalization about the kind of angles formed by the intersection of parallel lines with lines perpendicular to them.

7. Trace triangle *ABC*. Construct a line parallel to $\overline{AB}$ at *C*. Construct a line parallel to $\overline{BC}$ at *A*. Construct a line parallel to $\overline{AC}$ at *B*.

   a. What geometric shape do these new lines form?

   b. What do you notice about the relationship between the sizes of the new shape and triangle *ABC*?

8. Trace triangle *DEF*. Construct lines parallel to $\overline{DE}$ at *F*, parallel to $\overline{EF}$ at *D*, and parallel to $\overline{DF}$ at *E*. Make two observations about the largest new shape you have constructed.

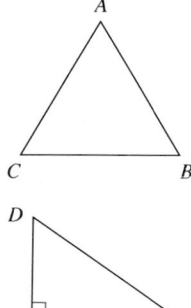

Introduction to Geometry **255**

Teaching the Lesson
- Materials: geoboards
- Lesson Starter 54
- Visuals, Folder B
- Using Technology, p. 234C
- Using Manipulatives, p. 234D
- Reteaching/Alternate Approach, p. 234D
- Cooperative Learning, p. 234D

Lesson Follow-Up
- Practice 73
- Enrichment 59: Connection
- Study Guide, pp. 107–108
- Calculator Activity 5

Warm-Up Exercises

Use the figure below.

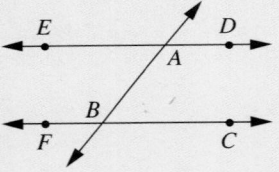

Answers will vary. Examples are given.

1. Name two rays. $\overrightarrow{AD}$, $\overrightarrow{AE}$
2. Name two line segments. $\overline{ED}$, $\overline{FC}$
3. Name two lines. $\overleftrightarrow{ED}$, $\overleftrightarrow{FC}$
4. Name an obtuse angle. $\angle ABF$
5. Name a pair of alternate interior angles. $\angle ABC$ and $\angle EAB$

Lesson Focus

Show students a picture of the Pentagon. Ask them to observe the shape and relate it to its name. The identification of polygons and their angles is the focus of today's lesson.

Polygons

Objective: To identify types of polygons and find the measures of their angles.

Terms to Know
- *polygon*
- *side of a polygon*
- *vertex of a polygon*
- *diagonal*
- *regular polygon*

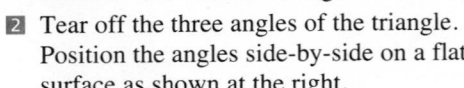

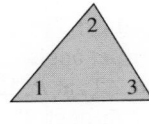

1. Using a pencil and a straightedge, draw a triangle on a piece of paper. Then use scissors to cut out the triangle.
2. Tear off the three angles of the triangle. Position the angles side-by-side on a flat surface as shown at the right.
3. Compare your results from Step 2 with those of other students in your class. What do all the triangles seem to have in common? They form a straight angle.
4. Make a guess about triangles by completing this statement: The sum of the measures of the angles of a triangle is __?__. Your guess is called a *conjecture*. 180°
5. Now test your conjecture: Draw a different triangle on a piece of paper and measure each of the three angles with a protractor. What is the sum of the measures? Was your conjecture correct? 180°; yes

A **polygon** is a closed figure formed by joining three or more line segments in a plane at their endpoints, with each line segment joining exactly two others. Each line segment is called a **side of the polygon.** Each point where two sides meet is a **vertex of the polygon.** (The plural of *vertex* is *vertices.*) A **diagonal** is a line segment that joins two nonconsecutive vertices.

These figures are all examples of polygons.

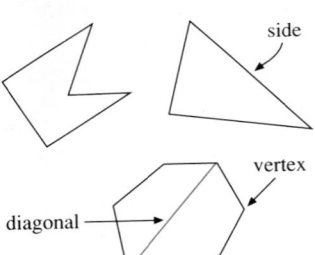

These figures are *not* polygons.

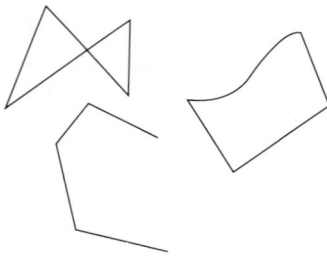

Polygons are identified by the numbers of their sides, using names given to them by the early Greek mathematicians. The table on the next page lists the names for some polygons. In each name, the prefix is underlined.

Name of Polygon	Meaning of Prefix	Number of Sides
triangle	three	3
quadrilateral	four	4
pentagon	five	5
hexagon	six	6
heptagon	seven	7
octagon	eight	8
nonagon	nine	9
decagon	ten	10

A polygon in which all sides have the same length and all angles have the same measure is called a **regular polygon.** In the figure at the right, the red marks show that all sides are equal in length and all angles are equal in measure, so the figure is a regular pentagon.

To name a polygon, you list its vertices in order. Two of the many names for the figure at the right are pentagon *PQRST* and pentagon *RSTPQ*. It should *not* be called pentagon *PRQST*.

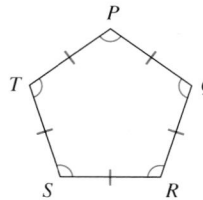

Example 1

Identify each polygon. Then list all its diagonals.

a.

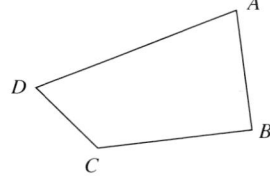

b.

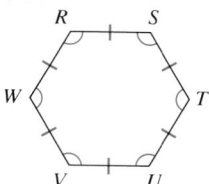

Solution

a.

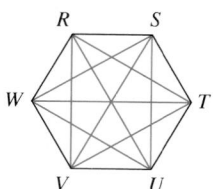

Polygon *ABCD* is a quadrilateral.
Its diagonals are
$\overline{AC}$ and $\overline{BD}$.

b.

Polygon *RSTUVW* is a regular hexagon.
Its diagonals are
$\overline{RT}$, $\overline{RU}$, $\overline{RV}$, $\overline{SU}$, $\overline{SV}$, $\overline{SW}$, $\overline{TV}$, $\overline{TW}$, and $\overline{UW}$.

Introduction to Geometry   **257**

Teaching Notes

Allow ample time for the Exploration at the beginning of the lesson. By doing this activity, students will discover the fact that the sum of the measures of the angles of a triangle is 180°.

Key Questions

1. Why are the figures shown at the bottom right of page 256 not polygons?   One figure has a side that is not a line segment; another is not closed; another has sides that intersect at a point that is not an endpoint.
2. Why is the sum of the angles of a polygon equal to $(n - 2)180°$?   The diagonals from any vertex divide the polygon into $(n - 2)$ triangles.

Error Analysis

Students sometimes count the same diagonal twice when determining the total number of diagonals for a given polygon. Placing a small tick mark on each diagonal will help to avoid this error.

Using Technology

For a suggestion on using a spreadsheet program or a calculator with this lesson, see page 234C.

Using Manipulatives

For a suggestion on using models to show that the sum of the angles of a polygon is given by $(n - 2)180°$, see page 234D.

Reteaching/Alternate Approach

For a suggestion on using a cooperative learning activity to discover the formula for the sum of the angles of a polygon, see page 234D.

Reasoning

The development of the formula for the sum of angles of a polygon, $(n - 2)180°$, requires students to generalize the relationship after observing a pattern.

Additional Examples

1. Identify each polygon. Then list all its diagonals.

   a.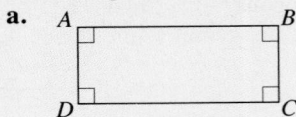

   ABCD is a rectangle; its diagonals are $\overline{AC}$ and $\overline{BD}$.

   b.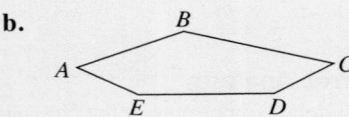

   ABCDE is a pentagon; its diagonals are $\overline{AD}$, $\overline{AC}$, $\overline{EB}$, $\overline{EC}$, and $\overline{BD}$.

2. Find the unknown angle measure in the pentagon below.
   140°

   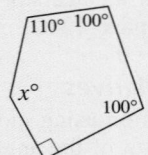

Closing the Lesson

Review with students the prefixes and names for polygons and the procedure for finding the measures of the angles of any polygon.

The simplest of all polygons is the triangle. The word *triangle* means "three angles." The Exploration on page 256 leads to the following conclusion about the three angles in a triangle.

The sum of the measures of the angles of a triangle is 180°.

You can use this fact to find the sum of the measures of the angles of any polygon. Simply use the diagonals from one vertex to separate the polygon into triangles.

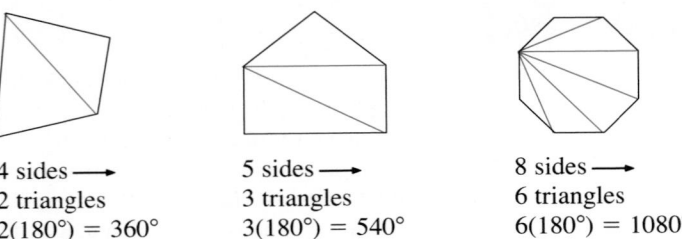

4 sides →	5 sides →	8 sides →
2 triangles	3 triangles	6 triangles
$2(180°) = 360°$	$3(180°) = 540°$	$6(180°) = 1080°$

In each case, the number of triangles is two fewer than the number of sides of the polygon. This leads to the following conclusion.

> **Generalization:** *Finding the Sum of the Angles of a Polygon*
>
> To find the sum of the measures of the angles of a polygon with n sides, subtract two from n and multiply the result by 180°.
> $$\text{sum of angles of a polygon} = (n - 2)(180°)$$

Example 2

Find the unknown angle measure in the polygon at the right.

Solution

First find the sum of the measures of all the angles. The polygon has 6 sides.
$$(n - 2)(180°) = (6 - 2)(180°)$$
$$= 4(180°) = 720°$$

Add the known measures.
$$110° + 107° + 140° + 120° + 132° = 609°$$

Subtract this sum from 720°.
$$720° - 609° = 111°$$

The unknown angle measure is 111°.

 **Check Your Understanding**

1. In Example 2, what number was substituted for n in the expression $(n - 2)(180°)$? Why?

2. What is represented by 609° in Example 2?
   See *Answers to Check Your Understanding* at the back of the book.

Guided Practice

COMMUNICATION «*Reading*

Match the name of each polygon with the everyday word that has the same prefix. Then give the meaning of the everyday word.

«**1.** triangle D; three-wheeled cycle   **A.** quadruple

«**2.** quadrilateral A; four times   **B.** octopus

«**3.** octagon B; animal with eight tentacles   **C.** decade

«**4.** decagon C; ten years   **D.** tricycle

Evaluate the expression $(n - 2)(180)$ for each value of the variable.

5. when $n = 7$ 900   **6.** when $n = 9$ 1260   **7.** when $n = 12$ 1800

Find the sum of the measures of the angles of each type of polygon.

8. pentagon 540°   **9.** decagon 1440°   **10.** octagon 1080°

Identify each polygon. Then list all its diagonals.

11. pentagon; $\overline{JL}$, $\overline{JM}$, $\overline{KM}$, $\overline{KN}$, $\overline{LN}$
12. triangle; no diagonals
13. regular pentagon; $\overline{YV}$, $\overline{YW}$, $\overline{ZX}$, $\overline{VX}$, $\overline{ZW}$

11. 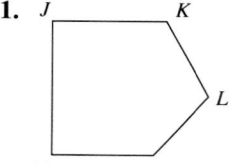   **12.** 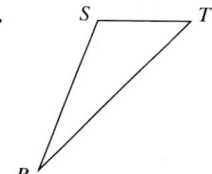   **13.**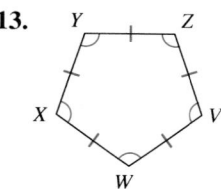

Find each unknown angle measure.

14. 30° 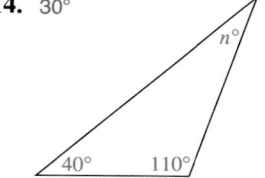   **15.** 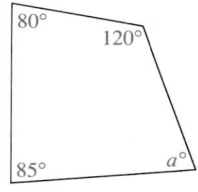   **16.**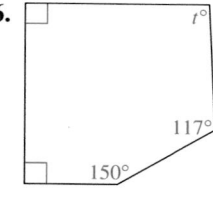

75°   93°

Exercises

Identify each polygon. Then list all its diagonals.

1. regular triangle; no diagonals
2. hexagon; $\overline{AC}$, $\overline{AD}$, $\overline{AE}$, $\overline{BD}$, $\overline{BE}$, $\overline{BF}$, $\overline{CE}$, $\overline{CF}$, $\overline{DF}$
3. quadrilateral; $\overline{LN}$, $\overline{OM}$

A 1. 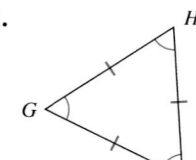   **2.** 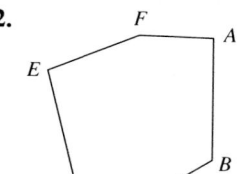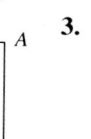   **3.**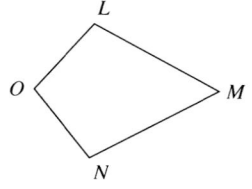

Introduction to Geometry   **259**

Suggested Assignments

Skills: 1–12, 15–18
Average: 1–12, 15–18
Advanced: 1–18

Exercise Notes

Communication: Reading
Guided Practice Exercises 1–4
relate geometric terms to the
students' vocabularies.

Making Connections/Transitions
Exercises 9–12 demonstrate the
strong vocabulary linkage of mathematics and language arts. This is
also evident in the chart that relates
the name of the polygon to the
number of sides.

Critical Thinking
In *Exercises 13 and 14*, students
must *generalize* the results of an
exercise to *develop* a formula.

Extension

Have students fill in the chart below.

Regular polygon	No. of sides	Sum of the interior angles	Measure of one interior angle
	3		
	4		
	5		
	• • • • • • • n		

Enrichment

Enrichment 59 is pictured on page 234E.

Practice

Practice 73, *Resource Book*

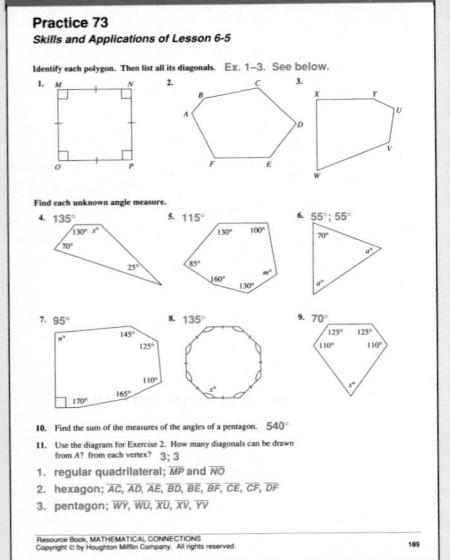

Practice 73
Skills and Applications of Lesson 6-5

Identify each polygon. Then list all its diagonals. Ex. 1–3. See below.

Find each unknown angle measure.

10. Find the sum of the measures of the angles of a pentagon. 540°
11. Use the diagram for Exercise 2. How many diagonals can be drawn from A? from each vertex? 3; 3

1. regular quadrilateral; $\overline{MP}$ and $\overline{NO}$
2. hexagon; $\overline{AC}$, $\overline{AD}$, $\overline{AE}$, $\overline{BD}$, $\overline{BE}$, $\overline{BF}$, $\overline{CE}$, $\overline{CF}$, $\overline{DF}$
3. pentagon; $\overline{WY}$, $\overline{WU}$, $\overline{XU}$, $\overline{XV}$, $\overline{YV}$

Find each unknown angle measure.

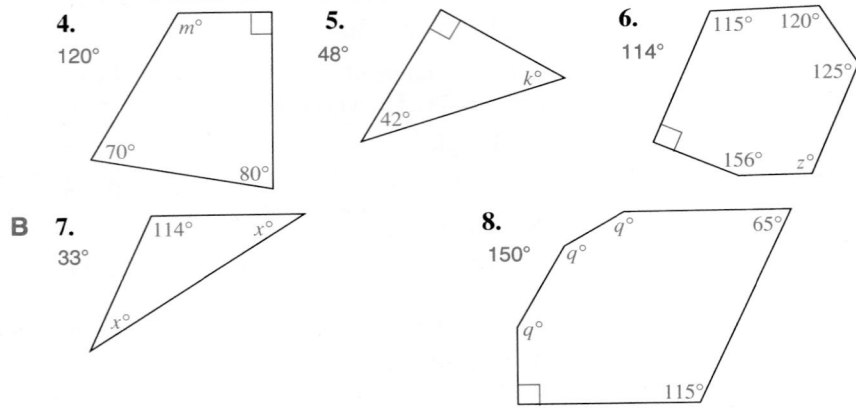

4. 120° 70° 80° $m°$

5. 48° 42° $k°$

6. 114° 115° 120° 125° 156° $z°$

B 7. 33° 114° $x°$ $x°$

8. 150° $q°$ $q°$ $q°$ 65° 115°

CONNECTING MATHEMATICS AND LANGUAGE ARTS

The prefixes *quadr-* and *quadri-*, as in quadrilateral, mean "four." Remembering this fact might help you figure out the meanings of other words with the same prefix.

Match each word with its correct meaning.

9. quadraphonic B
10. quadrennial D
11. quadrant A
12. quadruped C

A. one of four major areas of the coordinate plane
B. having four channels that reproduce sound
C. a four-footed animal
D. occurring every four years

THINKING SKILLS 13. Comprehension 14. Synthesis

13. Replace each __?__ with the number that makes the sentence true.
   10; A regular decagon has __?__ angles. The sum of the measures of the an-
   1440; gles is __?__ degrees. To find the measure of just one angle, find the
   1440; 10; 144 quotient __?__ ÷ __?__. The measure of each angle is __?__ degrees.

C 14. Generalize the results of Exercise 13. Develop a formula to find the measure of one angle of a regular polygon that has n sides. $\frac{(n-2)(180°)}{n}$

SPIRAL REVIEW

S 15. Solve: $3x - 8 = 25$ *(Lesson 4-5)* 11

16. Find the unknown angle measure in the polygon at the right.   *(Lesson 6-5)* 114°

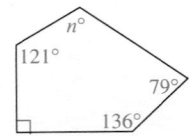
121° $n°$ 79° 136°

17. Find the difference: $7\frac{3}{8} - 5\frac{3}{4}$ *(Toolbox Skill 19)* $1\frac{5}{8}$

18. Find the mean, median, mode(s), and range for the data:
   27, 31, 26, 31, 24, 27, 32, 27, 31, 32, 26   *(Lesson 5-8)* 28.5, 27; 27 and 31; 8

6-6 Triangles

Objective: To classify triangles by their sides and angles and to use the triangle inequality.

Terms to Know
- *scalene triangle*
- *isosceles triangle*
- *equilateral triangle*
- *Triangle Inequality*
- *acute triangle*
- *right triangle*
- *obtuse triangle*

EXPLORATION

1. Cut three narrow strips of paper so that their lengths are 4 in., 5 in., and 8 in. Then lay the papers on a flat surface and try to form a triangle. Is it possible? Yes

2. Now repeat Step 1, this time using papers of lengths 3 in., 5 in., and 8 in. Is it possible to make a triangle? No

3. Without actually cutting them out, predict whether strips of paper of the given measures would form a triangle.
   - **a.** 2 in., 3 in., 5 in. No
   - **b.** 2 in., 4 in., 5 in. Yes
   - **c.** 3 in., 4 in., 8 in. No
   - **d.** 4 in., 5 in., 7 in. Yes

4. Compare your answers to Step 3 with those of other students in your class. Are your answers the same? If they are not the same, can you give a convincing argument to support your answers? See answer on p. A18.

5. Complete this statement: In a triangle, the sum of the lengths of any two sides must be __?__ the length of the third side. greater than

Triangles are often classified by the measures of their sides.

scalene triangle	isosceles triangle	equilateral triangle
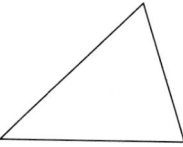		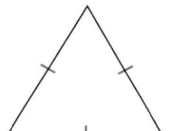
No sides have the same length.	At least two sides have the same length.	All three sides have the same length.

An equilateral triangle is also *equiangular*. This means that all the angles have the same measure. Because all sides have the same length and all angles have the same measure, the term *equilateral triangle* is really a special name for a *regular* triangle.

For three line segments to be the sides of a triangle, there must be a specific relationship among their lengths. This relationship among the sides of a triangle is important in so many applications of geometry that it is given the special name of *The Triangle Inequality*.

The Triangle Inequality

In any triangle, the sum of the lengths of any two sides is greater than the length of the third side.

Introduction to Geometry **261**

Lesson Planner

Teaching the Lesson
- Materials: straws, calculator
- Lesson Starter 55
- Using Technology, p. 234C
- Using Manipulatives, p. 234D
- Reteaching/Alternate Approach, p. 234D
- Cooperative Learning, p. 234D

Lesson Follow-Up
- Practice 74
- Enrichment 60: Data Analysis
- Study Guide, pp. 109–110
- Computer Activity 7

Warm-Up Exercises

Sketch an irregular pentagon *ABCDE*. Draw the diagonals from vertex *A*.

1. How many triangles are formed? 3
2. What is the sum of the measures of the angles in the pentagon? 540°
3. If the pentagon were regular, what would be the measure of each angle? 108°
4. Sketch in the remaining diagonals. What is the total number of diagonals? 5

Lesson Focus

To motivate today's lesson, the focus of which is the study of triangles, you may wish to use the following application.

Application

If a carpenter builds a gate, it is usually a rectangle. What else is usually a part of the construction? cross-strip or diagonal Why is this diagonal piece there? to strengthen the construction

Example 1

Tell whether line segments of the given lengths *can* or *cannot* be the sides of a triangle. If they can, tell whether the triangle would be *scalene*, *isosceles*, or *equilateral*.

a. 8 ft, 7 ft, 9 ft   **b.** 9 m, 3 m, 4 m

Solution

Compare each sum of two lengths to the third length.

a. $8 + 7 \underline{\ ?\ } 9$ $8 + 9 \underline{\ ?\ } 7$ $7 + 9 \underline{\ ?\ } 8$
      $15 > 9$ $17 > 7$ $16 > 8$

Each sum of two lengths is greater than the third, so the line segments *can* be the sides of a triangle. No lengths are the same, so the triangle is *scalene*.

b. $9 + 3 \underline{\ ?\ } 4$ $9 + 4 \underline{\ ?\ } 3$ $3 + 4 \underline{\ ?\ } 9$
      $12 > 4$ $13 > 3$ $7 < 9$

The sum $3 + 4$ is less than the third length, 9, so the line segments *cannot* be the sides of a triangle.

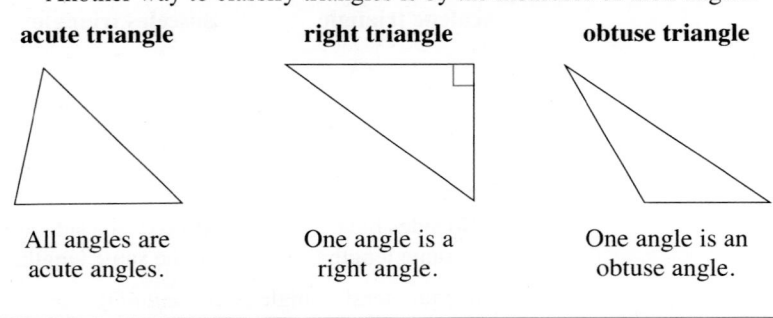

✏️ **Check Your Understanding**

1. In part (b) of Example 1, describe a way to form a triangle by changing just one of the given lengths.
   See *Answers to Check Your Understanding* at the back of the book.

Another way to classify triangles is by the measures of their angles.

acute triangle	right triangle	obtuse triangle
All angles are acute angles.	One angle is a right angle.	One angle is an obtuse angle.

Example 2

The measures of two angles of a triangle are 28° and 40°. Tell whether the triangle is *acute*, *right*, or *obtuse*.

Solution

Find the measure of the third angle.
   Add the known measures: $28° + 40° = 68°$
   Subtract this sum from 180°: $180° - 68° = 112°$
The third angle measure is 112°, so the triangle is *obtuse*.

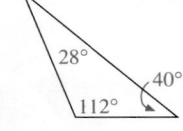

✏️ **Check Your Understanding**

2. What does 180° represent in the solution of Example 2?
3. In Example 2, how do you know that the triangle is obtuse?
   See *Answers to Check Your Understanding* at the back of the book.

262   Chapter 6

 You can find the measure of the third angle of a triangle using the memory keys on a calculator. Here is a key sequence that you might use for Example 2.

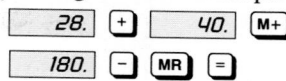

Guided Practice

Answers to Guided Practice 4–6 are on p. A18.

COMMUNICATION « *Writing*

Classify each triangle first by its sides, then by its angles.

1. scalene; obtuse
2. isosceles; right
3. equilateral; acute

« **1.** 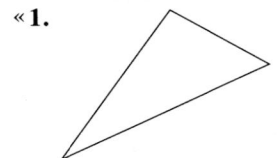 « **2.** 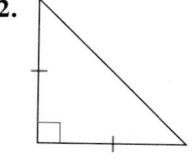 « **3.**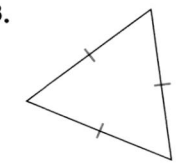

Use a pencil and straightedge to sketch an example of a triangle of the given classifications. Answers to Guided Practice 4–6 are on p. A18.

« **4.** scalene, right « **5.** isosceles, acute « **6.** equilateral, acute

Tell whether line segments of the given lengths *can* or *cannot* be the sides of a triangle. If they can, tell whether the triangle would be *scalene*, *isosceles*, or *equilateral*.

7. 12 ft, 3 ft, 11 ft can; scalene

8. 4 m, 1.3 m, 2.7 m cannot

The measures of two angles of a triangle are given. Tell whether the triangle is *acute*, *right*, or *obtuse*.

9. 48°, 16° obtuse

10. 61°, 29° right

Exercises

Tell whether line segments of the given lengths *can* or *cannot* be the sides of a triangle. If they can, tell whether the triangle would be *scalene*, *isosceles*, or *equilateral*.

A

1. 3 km, 1 km, 4 km cannot

2. 1 ft, 1 ft, 1 ft can; equilateral

3. 5 in., 10 in., 13 in.
  can; scalene

4. 24 yd, 13 yd, 5 yd cannot

5. 8 m, 3.1 m, 3.1 m cannot

6. 6 cm, 4.5 cm, 4.5 cm
  can; isosceles

7. 7.1 cm, 6.4 cm, 7.1 cm
  can; isosceles

8. 9.81 m, 16 m, 6.19 m cannot

9. 1.5 mi, 1.5 mi, 1.5 mi
  can; equilateral

10. 6 mm, 9 mm, 4 mm
  can; scalene

Introduction to Geometry **263**

Closing the Lesson

Have one student ask another student a question such as "What is the difference between an isosceles triangle and a scalene triangle?" Have students continue with similar questioning until the main points of the lesson have been reviewed.

Suggested Assignments

Skills: 1–20, 21–39 odd, 54–57
Average: 1–47 odd, 48–57
Advanced: 1–19 odd, 20–57, Challenge

Exercise Notes

Communication: Writing
Guided Practice Exercises 1–6 ask students to classify triangles from sketches, and then to make their own sketches given the classification.

Reasoning
Exercises 40–47 help students learn how to reason by understanding that justifications must be given for some conclusions.

Computer
Exercises 48–53 introduce students to Logo and show how the language can be used to draw triangles.

The measures of two angles of a triangle are given. Tell whether the triangle is *acute*, *right*, or *obtuse*.

11. 27°, 141° obtuse   **12.** 90°, 52° right   **13.** 69°, 53° acute

14. 24°, 26° obtuse   **15.** 67°, 23° right   **16.** 39°, 66° acute

17. 50°, 50° acute   **18.** 34°, 56° right   **19.** 7°, 39° obtuse

20. No; the sum of the angles would be greater than 180°.

20. Is it possible for a triangle to have two right angles? Explain.

21. Can an isosceles triangle have sides that measure 8 cm, 4 cm, and 6 cm? Explain. No; two sides must be equal in length.

Use the figure at the right. Name all the triangles of the given classification.

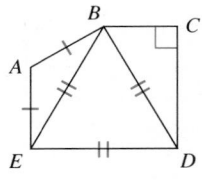

B **22.** acute *EDB*   **23.** obtuse *BAE*

24. scalene *BCD*   **25.** isosceles *BAE, BED*

26. right *BCD*   **27.** equilateral *BED*

In the figure at the right, $\overleftrightarrow{AB} \parallel \overleftrightarrow{XY}$. Find the measure of each angle.

28. ∠1 38°   **29.** ∠2 38°

30. ∠3 52°   **31.** ∠4 90°

32. ∠5 38°   **33.** ∠6 52°

34. ∠7 128°   **35.** ∠8 52°

36. ∠9 128°   **37.** ∠10 142°

38. ∠11 38°   **39.** ∠12 142°

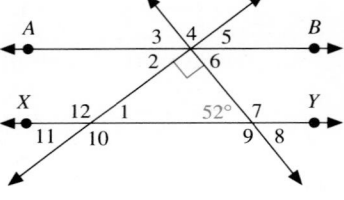

LOGICAL REASONING

Tell whether each statement is *True* or *False*. If the statement is false, give a reason that supports your answer.

40. An obtuse triangle can have a right angle. False; the sum of the angles would be greater than 180°.

41. An equilateral triangle is also an isosceles triangle. True

42. A right triangle can be a scalene triangle. True

43. An obtuse triangle can have more than one obtuse angle. False; the sum of the angles would be greater than 180°.

44. An equilateral triangle can be a scalene triangle. False; an equilateral triangle has three equal sides but a scalene triangle has no equal sides.

45. An isosceles triangle can be a right triangle. True

46. A scalene triangle can never be an isosceles triangle. True

47. An acute triangle can never be an equilateral triangle. False; an acute triangle can have three equal sides.

COMPUTER APPLICATION

The program at the right is written in the Logo computer language. It will instruct a computer to draw an equilateral triangle like the one shown.

FD 40 RT 120
FD 40 RT 120
FD 40 RT 120

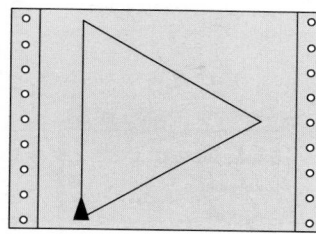

In this program, the command FD instructs the Logo "turtle" to go forward a specified distance. The command RT turns the turtle to the right a specified number of degrees. Notice that the angles used in this program must be the *supplements* of the angles of the triangle.

48. Write the Logo command that instructs the turtle to move forward a distance of 65. FD 65

49. Write the Logo command that creates a 45° angle for the triangle in the program above. FD 40 RT 135

Use what you know about the properties of triangles. Explain why each of these Logo programs will *not* output a triangle.

51. The three angles are each 65°. Their sum is greater than 180°.

50.
FD 20 RT 100   Because
FD 50 RT 100   20 + 20 < 50
FD 20 RT 160

51.
FD 62 RT 115
FD 62 RT 115
FD 62 RT 115

Sketch the triangle that you think will be output when each program is run. Then, if you have access to a computer and the Logo language, use them to run the program and check your answer.

Answers to Exercises 52 and 53 are on p. A18.

C **52.**
FD 40 RT 90
FD 30 RT 127
FD 50 RT 143

53.
FD 40 RT 103
FD 90 RT 154
FD 90 RT 103

SPIRAL REVIEW

S **54.** Use the formula $P = 2l + 2w$. Let $l = 11$ cm and $P = 34$ cm. Find w. *(Lesson 4-10)* 6 cm

55. The measures of two angles of a triangle are 73° and 17°. Tell whether the triangle is *acute, right,* or *obtuse.* *(Lesson 6-6)* right

56. Find the quotient: $\frac{6}{7} \div \frac{2}{3}$ *(Toolbox Skill 21)* $1\frac{2}{7}$

57. Find the difference: $-13 - (-24)$ *(Lesson 3-4)* 11

Challenge

How many times each day are the hands of a clock perpendicular? 44

Follow-Up

Extension (Cooperative Learning)
Students working in groups can discuss the design of easels or "tripods." Each group should write a paragraph explaining why three legs are used instead of four. The groups should share their ideas with the class. This activity should lead to the conclusion that three points always determine a plane.

Enrichment
Enrichment 60 is pictured on page 234F.

Practice
Practice 74, *Resource Book*

Practice 74
Skills and Applications of Lesson 6-6

Tell whether line segments of the given lengths *can* or *cannot* be the sides of a triangle. If they can, tell whether the triangle would be *scalene, isosceles,* or *equilateral*.
6. can; scalene
9. can; equilateral
12. can; scalene

1. 10 in., 14 in., 19 in. can; scalene
2. 7 ft, 3 ft, 10 ft cannot
3. 8.5 cm, 12.5 cm, 8.5 cm can; isosceles
4. 6.8 m, 6.8 m, 6.8 m can; equilateral
5. 6 ft, 10 ft, 8 ft can; scalene
6. 11.85 in., 20.36 in., 27.04 in. See above.
7. 10.5 cm, 10.5 cm, 25.5 cm cannot
8. 18.3 ft, 20.1 ft, 20.1 ft can; isosceles
9. 16.5 in., 16.5 in., 16.5 in. See above.
10. 15.3 cm, 25 cm, 7.8 cm cannot
11. 17.03 ft, 14.16 ft, 5 ft can; scalene
12. 19.03 in., 19.003 in., 19.3 in. See above.

The measures of two angles of a triangle are given. Tell whether the triangle is *acute, right,* or *obtuse*.

13. 40°, 50° right
14. 63°, 109° obtuse
15. 63°, 54° acute
16. 74°, 49° acute
17. 60°, 60° acute
18. 150°, 20° obtuse
19. 58°, 84° acute
20. 63°, 27° right
21. 35°, 43° obtuse
22. 37°, 90° right
23. 52°, 28° obtuse
24. 74°, 15° obtuse
25. 18°, 64° obtuse
26. 74°, 68° acute
27. 11°, 85° acute

28. What is true about the two acute angles of a right triangle?
29. Give the lengths of three line segments that cannot be sides of a triangle.
30. If the lengths of two sides of a triangle are 10 in. and 14 in., can the third side be 26 in. long? Why or why not?

28. The sum of their measures is 90°.
29. Answers will vary. Example: 3 cm, 5 cm, 9 cm
30. No; the length of the third side must be less than 24 in.

Teaching the Lesson
- Materials: string, geoboards
- Lesson Starter 56
- Using Manipulatives, p. 234D
- Reteaching/Alternate Approach, p. 234D

Lesson Follow-Up
- Practice 75
- Enrichment 61: Thinking Skills
- Study Guide, pp. 111–112

Warm-Up Exercises

Match each set of measurements with the type of triangle it describes: (a) obtuse, (b) scalene, (c) isosceles, (d) right, (e) not a triangle

1. 3 ft, 4 ft, 6 ft (b)
2. 44°, 46°, 90° (d)
3. 30°, 45°, 105° (a)
4. 5 m, 9 m, 15 m (e)
5. 8 in., 6 in., 8 in. (c)

Lesson Focus

Quadrilaterals are used frequently in building, making furniture, and so on. Have students give examples of other real-world uses of quadrilaterals. Special quadrilaterals and their properties are the focus of today's lesson.

Teaching Notes

Discuss the organizational chart to help students see the relationships of the various quadrilaterals and the progression from general to specific.

Quadrilaterals

Objective: To identify special quadrilaterals and apply their properties.

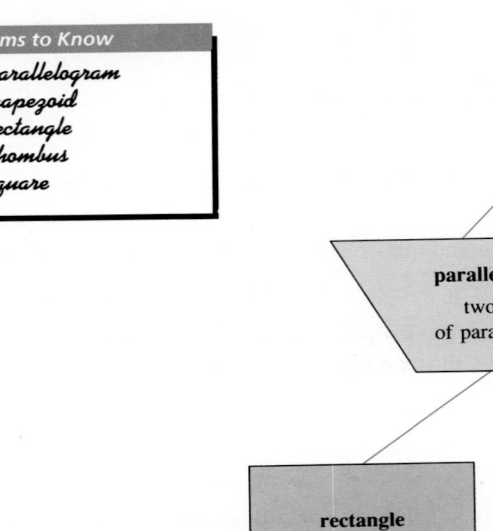

DATA ANALYSIS

Many mathematics texts contain a reference chart similar to this one.

Terms to Know
- *parallelogram*
- *trapezoid*
- *rectangle*
- *rhombus*
- *square*

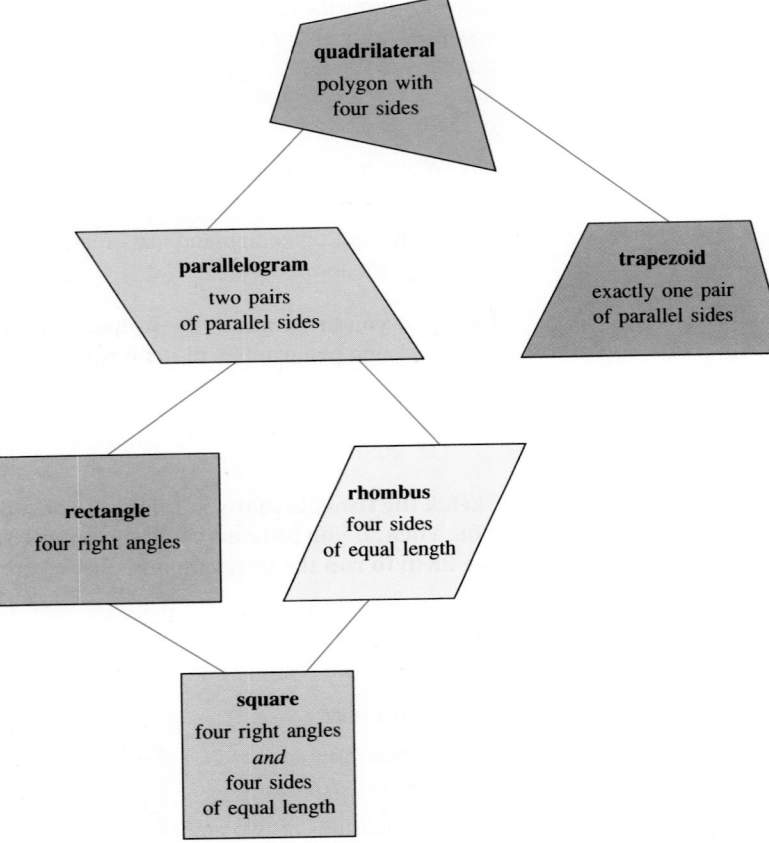

The chart contains data about quadrilaterals. Here is how you read the definition of a *rectangle* from the chart.

A rectangle is a quadrilateral with four right angles.

You can also use the chart to determine relationships among the special types of quadrilaterals. Here is an example.

A rectangle is a type of parallelogram.

266   Chapter 6

A chart like the one on the previous page can contain only a limited amount of data. Many quadrilaterals have other properties that are not listed. Here are two important properties of the parallelogram.

The opposite sides of a parallelogram have the same length.
The opposite angles of a parallelogram have the same measure.

Example

In the figure at the right, *ABCD* is a parallelogram.
a. Find the length of $\overline{AB}$.
b. Find $m\angle 1$.

Solution

a. Opposite sides of a parallelogram have the same length, so $\overline{AB}$ is equal in length to $\overline{CD}$. The length of $\overline{AB}$ is 7 in.
b. Opposite angles of a parallelogram have the same measure, so
$m\angle B = m\angle D = 115°$.
In triangle *ABC*, the sum of the measures of the angles must be 180°. The measures of $\angle B$ and $\angle ACB$ are known, so subtract the sum of their measures from 180°.
$$m\angle B + m\angle ACB = 115° + 43° = 158°$$
$$m\angle 1 = 180° - 158° = 22°$$

✓ Check Your Understanding

1. In part (a) of the Example, is it possible to find $m\angle DAC$? Explain.
2. Explain why 180° was used in part (b) of the Example.
See *Answers to Check Your Understanding* at the back of the book.

Guided Practice

COMMUNICATION «*Reading*

Refer to the chart on page 266.

«**1.** What is the definition of a *square*? a quadrilateral with four right angles and four sides of equal lengths
«**2.** What is the difference between a parallelogram and a trapezoid?
«**3.** Is a rectangle a type of rhombus? No.

Refer to the terms defined in the chart on page 266. List *all* the names that apply to each figure.

4.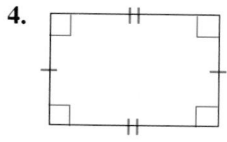

quadrilateral, parallelogram, rectangle

5.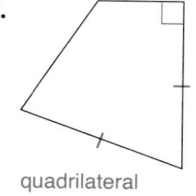

quadrilateral

6.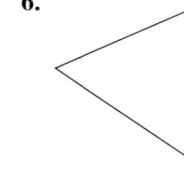

quadrilateral, trapezoid

Introduction to Geometry   **267**

2. A parallelogram has two pairs of parallel sides; a trapezoid has exactly one pair of parallel sides.

Key Questions
1. What are the special types of quadrilaterals? parallelogram, trapezoid, rectangle, rhombus, square
2. What is the difference between (a) a parallelogram and a rectangle, and (b) a rhombus and a square? (a) A parallelogram does not necessarily have right angles; a rectangle is a parallelogram that does have right angles. (b) A square is a special type of rhombus that must have right angles; a rhombus does not necessarily have right angles.

Error Analysis
Students may have difficulty identifying the properties of quadrilaterals. To overcome these difficulties, direct students to use the chart on page 266 and to use geoboards and geo-dot paper.

Using Manipulatives
For a suggestion on using string to model quadrilaterals, see page 234D.

Reteaching/Alternate Approach
For a suggestion on using geoboards and geo-dot paper, see page 234D.

Additional Example
In the figure below, *DEFG* is a rhombus.

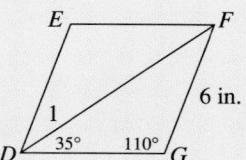

a. Find the sum of the lengths of the four sides. 24 in.
b. Find $m\angle 1$. 35°

Closing the Lesson

Ask a student to select a quadrilateral and list one of its characteristics. Ask another student to list a second characteristic. Continue the process using different students and quadrilaterals.

Suggested Assignments

Skills: 1–12, 13–19 odd, 22–25
Average: 1–20, 22–25
Advanced: 1–25

Exercise Notes

Communication: Reading

Guided Practice Exercises 1–3 help students to understand the definitions of the quadrilaterals and their distinguishing properties.

Problem Solving

Exercises 17 and 18 require that students analyze the drawings and the given information and then select the appropriate strategies for solving the problems. Several deductions are inherent in the solution process.

Cooperative Learning

Exercises 19–21 ask students working in cooperative groups to sketch, analyze, and categorize the various quadrilaterals according to specific characteristics.

In the figure at the right, *RSTU* is a trapezoid.

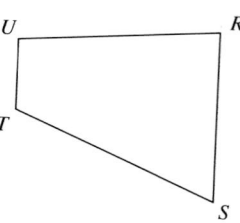

7. Name all pairs of parallel sides. $\overline{UT}$ and $\overline{RS}$

8. Name the side that is opposite $\overline{RU}$. $\overline{TS}$

9. Name all pairs of opposite angles. $\angle U$ and $\angle S$; $\angle R$ and $\angle T$

10. Are opposite angles equal in measure? No

In the figure at the right, *JKLM* is a rhombus. Find each measure.

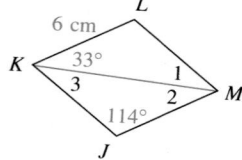

11. the length of $\overline{JM}$ 6 cm
12. $m\angle L$ 114°

13. the length of $\overline{LM}$ 6 cm
14. $m\angle 1$ 33°

15. $m\angle 2$ 33°
16. $m\angle 3$ 33°

Exercises

In the figure at the right, *XYZW* is a rectangle. Find each measure.

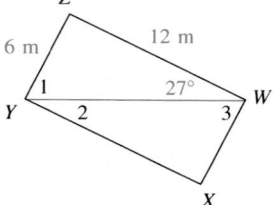

A   1. the length of $\overline{WX}$ 6 m
2. $m\angle Z$ 90°

3. the length of $\overline{XY}$ 12 m
4. $m\angle 1$ 63°

5. $m\angle 2$ 27°
6. $m\angle 3$ 63°

In the figure at the right, *ACDF* is a parallelogram and *ABDE* is a square. Find each measure.

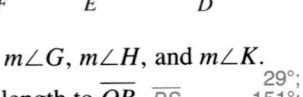

7. the length of $\overline{BD}$ 4 in.
8. $m\angle BAE$ 90°

9. the length of $\overline{DE}$ 4 in.
10. $m\angle F$ 53°

11. the length of $\overline{CD}$ 5 in.
12. $m\angle 1$ 37°

B   13. the length of $\overline{DF}$ 7 in.
14. $m\angle FDC$ 127°

15. In parallelogram *GHJK*, $m\angle J = 29°$. Find $m\angle G$, $m\angle H$, and $m\angle K$. 29°; 151°; 151°

16. In rectangle *PQRS*, name the side equal in length to $\overline{QR}$. $\overline{PS}$

17. $m\angle 1 = 102°$
    $m\angle 2 = 33°$
    $m\angle 3 = 57°$
18. $m\angle 1 = 20°$
    $m\angle 2 = 48°$
    $m\angle 3 = 20°$
    $m\angle 4 = 112°$
    $m\angle 5 = 68°$
    $m\angle 6 = 112°$

17. *MNOP* is a trapezoid. Find the measures of all the numbered angles.

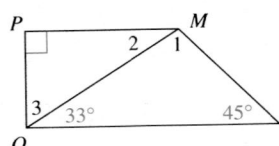

18. *QRST* is a parallelogram. Find the measures of all the numbered angles.

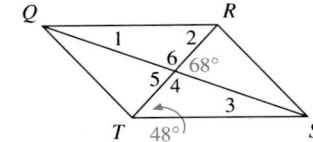

DATA ANALYSIS/GROUP ACTIVITY

Rectangle			
Definition: a quadrilateral with four right angles			
Parallel sides:	none	one pair	(two pairs)
Opposite sides equal:		(yes)	no
Opposite angles equal:		(yes)	no
Four right angles:		(yes)	no
Four equal sides:		yes	(no)

Answers to Exercises 19–21 are on p. A18.

The index card shown above catalogs the major characteristics of a rectangle. For Exercises 19–21, you can use index cards similar to the one shown, or you might have access to computer database software.

19. Begin a quadrilateral database by copying the card above. Then create a card similar to it for each of the following figures: quadrilateral, parallelogram, trapezoid, rhombus, square.

20. Sort through the database that you created in Exercise 19. List all the figures that have the given characteristic.
    a. two pairs of parallel sides   **b.** four right angles
    c. opposite sides equal   **d.** opposite angles equal

C   **21.** In a quadrilateral, two angles that are not opposite are *consecutive angles*. Add the following entry to each card in your file.
    Consecutive angles supplementary: none   two pairs   four pairs
    Working with others in your group, determine how to complete this entry for each figure in your database.

SPIRAL REVIEW

S   **22.** Estimate the difference: $887 - 519$
    (Toolbox Skill 1) about 400

23. Solve: $\frac{x}{3} = -17$ *(Lesson 4-4)* −51

24. In the figure at the right, *CDEF* is a rhombus. Find the length of $\overline{EF}$ and $m\angle 1$.
    (Lesson 6-7) 9.8 cm; 36°

25. *True or False: $\overline{XY}$ is the symbol for line XY.*   *(Lesson 6-1)*
    False

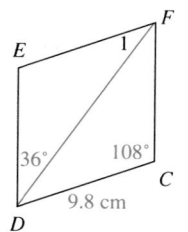

Extension (*Cooperative Learning*)
Have students work in groups to develop "Description Response Cards" for quadrilaterals. One set of cards should contain descriptions of quadrilaterals and a second set should contain the names of quadrilaterals. These cards can then be used for class drills or reviews. For example, the cards could be dealt so that each group receives an equal number of each type of card. One student begins by reading a description of a quadrilateral. The person who answers is the one whose "Description" card is the response. This process continues until the last card describes the first one, thus completing the cycle.

Project
Have students assist in the development of a bulletin board display on quadrilaterals. Assign different responsibilities based on interest and ability.

Enrichment
Enrichment 61 is pictured on page 234F.

Practice
Practice 75, *Resource Book*

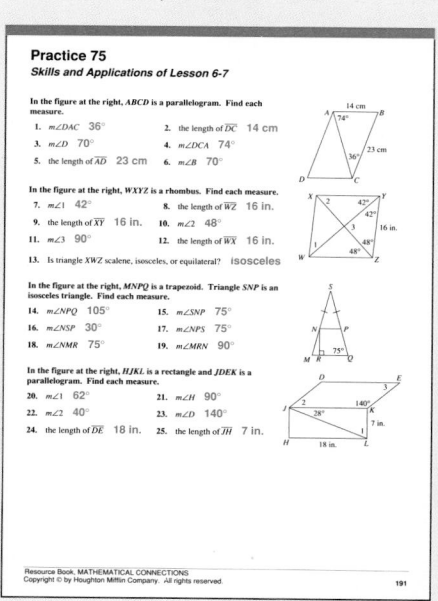

Using Geometric Drawing Software

With **geometric drawing software,** you can use a computer to make accurate drawings of plane geometric figures. You also can instruct the computer to measure line segments and angles in the figures. Because the computer performs these tasks so quickly, you can use this software to test conjectures through experimentation.

Exercises

In Exercises 1–4, use geometric drawing software to help you test this conjecture: *The diagonals of a rectangle are equal in measure.*

Experiment Number	$\overline{AC}$	$\overline{BD}$
1	?	?
2	?	?
3	?	?
4	?	?

1. In your notebook, make a table like the one shown at the right. Check students' papers.

2. Draw a rectangle *ABCD* on the computer screen. Find the measure of diagonals $\overline{AC}$ and $\overline{BD}$. Record the results in your table as Experiment 1. Answers will vary.

3. Repeat Exercise 2 for three different rectangles. Record these results in your table as Experiments 2, 3, and 4. Answers will vary.

4. Arrive at a conclusion: Is the conjecture *true* or *false*? True

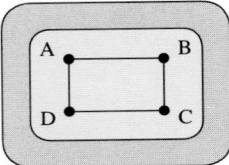

In Exercises 5 and 6, use geometric drawing software to test the conjecture. Then tell whether the conjecture is *true* or *false*.

5. The diagonals of a parallelogram are equal in measure. False

6. The diagonals of a rhombus are perpendicular. True

Answers to Exercises 7 and 8 are on p. A19.

7. To **bisect** a line segment means to separate it into two line segments that are equal in measure. Using the following statement, make a conjecture by replacing the _?_ with the name of one of the special quadrilaterals.

   The diagonals of a _?_ bisect each other.

   Now test your conjecture. Was it true or false?

8. In an **isosceles trapezoid,** the nonparallel sides are equal in measure. Make a conjecture about the diagonals of an isosceles trapezoid. Now test your conjecture. Was it true or false?

6-8

Strategy:
Identifying a Pattern

Objective: To solve problems by identifying a pattern.

A useful strategy for solving many problems in mathematics is to look for a pattern in the given information. This strategy may be especially helpful when the problem is one that involves geometric figures.

Problem

The figures below show all the diagonals of a triangle, quadrilateral, pentagon, and hexagon. How many diagonals does a decagon have?

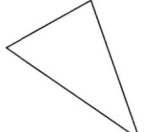 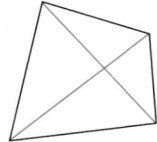 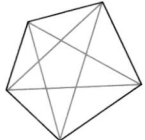 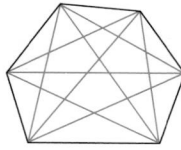

Solution

UNDERSTAND The problem is about the diagonals of polygons.
Facts: all diagonals are shown for 3, 4, 5, and 6 sides
Find: the number of diagonals when there are 10 sides

PLAN Count the diagonals in each of the given figures. Make a table that pairs the number of diagonals with the number of sides for each figure. Then look for a pattern among the numbers in the table.

WORK

Number of Sides	3	4	5	6
Number of Diagonals	0	2	5	9

+2   +3   +4

The pattern of the number of diagonals is *add 2, add 3, add 4,* and so on. Using this pattern, extend the table until the number of sides is 10.

Number of Sides	3	4	5	6	7	8	9	10
Number of Diagonals	0	2	5	9	14	20	27	35

+2   +3   +4   +5   +6   +7   +8

ANSWER A decagon has 35 diagonals.

Look Back Often you can *generalize* a pattern by using a variable expression. For instance, to find the number of diagonals of a polygon with n sides, you can use the expression $\frac{n(n-3)}{2}$. Show how to use this expression to find how many diagonals a decagon has. $\frac{10(7)}{2} = 35$

Lesson Planner

Teaching the Lesson
• Lesson Starter 57

Lesson Follow-Up
• Practice 76
• Enrichment 62: Problem Solving
• Study Guide, pp. 113–114

Warm-Up Exercises

1. Explain the difference between a regular and an irregular polygon. A regular polygon has equal angles and equal sides; an irregular polygon does not.
2. What is the sum of the measures of the angles of a quadrilateral? 360°

Give the next two terms in each pattern.

3. 1, 1, 2, 3, 5, _?_, _?_ 8, 13
4. 2, 4, 8, 16, _?_, _?_ 32, 64

Lesson Focus

Remind students of their work in Chapter 2 on identifying patterns. Today's lesson focuses on how to use this strategy to solve problems involving geometric patterns.

Teaching Notes

Stress the importance of analyzing the information in a problem and trying to identify a pattern. Also, students need to realize that more than one strategy is often used to solve a problem. Making a table or an organized list frequently is helpful in identifying a pattern.

Key Questions

1. How are number patterns related to geometric figures?
Various characteristics of geometric figures form number patterns.

2. Is there always only one correct strategy for solving a problem?
Not always; many problems can be solved by using more than one strategy.

Critical Thinking

Students have to *analyze* information to identify patterns in drawings or other situations. Each conjecture or hypothesis has to be *tested*. Is the conjecture or hypothesis true for all situations? Can a generalization be stated?

Problem Solving

Point out to students that mathematics is "full of patterns," and therefore identifying patterns is one of the most useful strategies to learn.

Additional Example

Find the next figure in the pattern below.

Closing the Lesson

Have students summarize what is meant by "identifying a pattern," and how patterns can be used to solve problems.

Suggested Assignments

Skills: 1–6, 11–16
Average: 1–16
Advanced: 1–16

Guided Practice

Use this problem for Exercises 1–6.

The figures below show a pattern of acute angles formed by 2, 3, 4, and 5 rays with a common endpoint. In this pattern, how many acute angles are formed by 8 rays with a common endpoint?

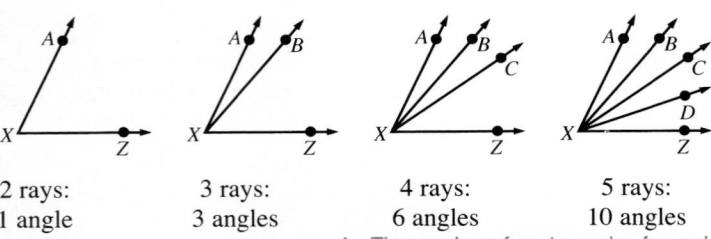

2 rays:	3 rays:	4 rays:	5 rays:
1 angle	3 angles	6 angles	10 angles

COMMUNICATION «*Reading*

1. The number of acute angles formed by rays with a common endpoint.
2. How many acute angles are formed by 8 rays with a common endpoint?

«**1.** What is the problem about?

«**2.** What is the question that has to be answered?

3. Complete the following statement: In the figure that shows 3 rays with a common endpoint, the 3 acute angles are ∠*AXB*, ∠*BXZ*, and ∠__?__.
    AXZ

4. In the figure that shows 4 rays with a common endpoint, name the 6 acute angles. ∠*AXB*, ∠*AXC*, ∠*AXZ*, ∠*BXC*, ∠*BXZ*, ∠*CXZ*

5. Copy and complete this table.

Number of Rays	_?_	2	3	4	5	_?_	6	_?_	7	8
Number of Acute Angles		1	3	_?_	6	10	_?_	15	_?_	21	_?_ 28

6. What is the solution of the problem? 28

The figures below represent the first four *oblong numbers*.

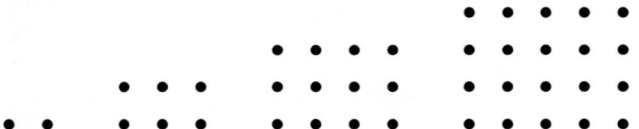

7. What is the eighth oblong number? 72

«**8.** **COMMUNICATION** «*Discussion*   What variable expression can be used to represent the *n*th oblong number? $n(n + 1)$

Problem Solving Situations

Solve by identifying a pattern.

1. The figures below show the first four figures in a pattern of squares. How many squares are in the tenth figure in this pattern? 21

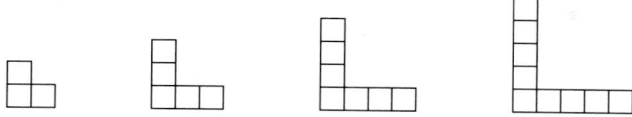

2. The figures below show the number of line segments that can be drawn between 2 points and between 3, 4, and 5 noncollinear points. How many line segments can be drawn between 9 noncollinear points? 36

Solve using any problem solving strategy.   3–6. Strategies may vary; likely strategies are given.

3. When a number is decreased by six, the result is −27. What is the number? using an equation; −21

4. Jennifer Ng bought a bicycle helmet that cost $31.29. She gave the clerk two $20 bills. How much change did Jennifer receive? choosing the correct operation; $8.71

5. Darnel Jefferson has three $1 bills, two $5 bills, and one $20 bill. How many different amounts of money can Darnel make using these bills? making a table; 23

6. The figures below show all the diagonals that can be drawn from one vertex of a quadrilateral, pentagon, hexagon, and heptagon. How many diagonals can be drawn from one vertex of a decagon? identifying a pattern; 7

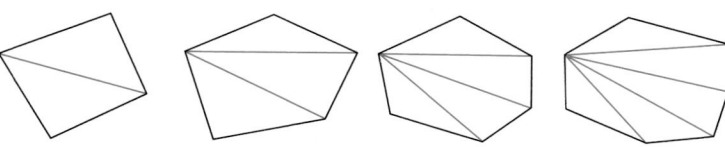

> **PROBLEM SOLVING**
>
> **CHECKLIST**
>
> **Keep these in mind:**
> Using a Four-Step Plan
> Too Much or Not Enough Information
>
> **Consider these strategies:**
> Choosing the Correct Operation
> Making a Table
> Guess and Check
> Using Equations
> Identifying a Pattern

Project

Have students research the patterns found in the Fibonacci sequence and give some examples of where they occur in nature.

Enrichment

Enrichment 62 is pictured on page 234F.

Practice

Practice 76, *Resource Book*

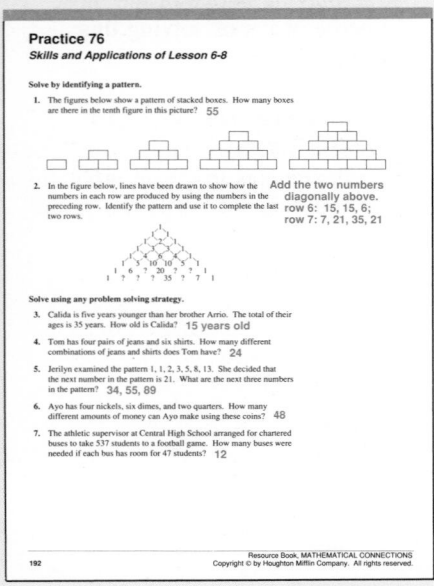

PROBLEM SOLVING/APPLICATION Answers to Problems 7–12 are on p. A19.

Often you can use a geometric problem to model a real-life situation. For example, Exercises 7–10 refer to the following problem.

> Eight people meet at a party. Each person shakes hands with each of the others exactly once. How many handshakes are exchanged?

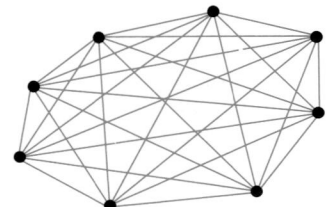

7. A geometric model for this problem is shown at the right. What do the black dots and the red line segments represent?

8. Which geometry problem in this lesson seems most closely related to the handshake problem?

9. Describe how to solve the handshake problem by identifying a pattern. What is the solution?

10. Describe how to solve the handshake problem using a strategy other than identifying a pattern.

WRITING WORD PROBLEMS

Write a problem that you could solve by identifying each pattern. Then solve the problem.

11.
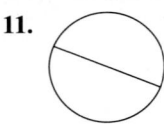 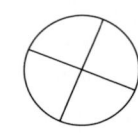 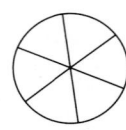 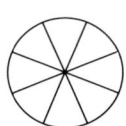

12.
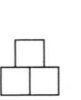 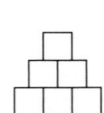 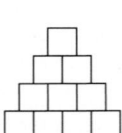 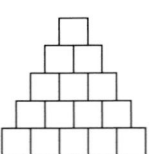

SPIRAL REVIEW

S 13. Estimate the sum: $238 + 177 + 315 + 150$ *(Toolbox Skill 2)*
about 890

14. Solve: $3(2x - 5) = 15$ *(Lesson 4-9)* 5

15. Draw an angle that measures 82°.   *(Lesson 6-2)* See answer on p. A19.

16. How many small triangles (△) are in the ninth figure in this pattern of triangles?   *(Lesson 6-8)* 81

274   Chapter 6

6-9 Symmetry

Objective: To recognize and find lines of symmetry in plane figures.

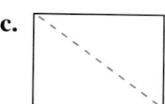

1 Fold a rectangular sheet of paper that is not a square three different times. Each time, make the fold as indicated by the dashed lines in the figures below. In each case, does one half of the folded paper fit exactly over the other half? **a.** Yes **b.** Yes **c.** No

a. [rectangle with horizontal dashed line]   b. [rectangle with vertical dashed line]   c. [rectangle with diagonal dashed line]

2 Suppose that your sheet of paper were shaped like a square. How would the results of Step 1 be different? The answer to part (c) would be Yes.

3 Suppose you had a sheet of paper shaped like the parallelogram in the figure at the right. Do you think you could fold this paper so that one half fits exactly over the other? No

4 Suppose your paper were shaped like a circle. How many ways could you fold it so that one half fits exactly over the other? infinitely many ways

The figure at the right shows a butterfly with its wings in two different positions. When the butterfly flaps its wings, it is as if the butterfly "folds" itself along $\overleftrightarrow{AB}$. One half of the butterfly seems to fit exactly over the other half. Mathematically, the butterfly's shape has **line symmetry**, and $\overleftrightarrow{AB}$ is called a **line of symmetry.**

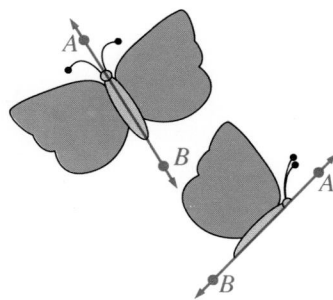

Example 1

Is $\overleftrightarrow{AB}$ a line of symmetry? Write *Yes* or *No*.

a. [triangle]   b. 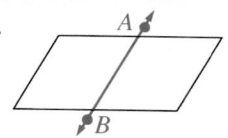 [parallelogram]   c. 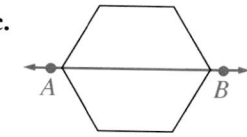 [hexagon]

Solution

a. Yes.   b. No.   c. Yes.

Check Your Understanding

1. In part (b) of Example 1, why isn't $\overleftrightarrow{AB}$ a line of symmetry?
2. In part (c) of Example 1, is $\overleftrightarrow{AB}$ the only line of symmetry?
   See *Answers to Check Your Understanding* at the back of the book.

Introduction to Geometry **275**

Warm-Up Exercises

Draw a line that divides these figures in half.

1.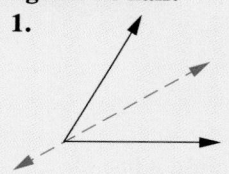

2.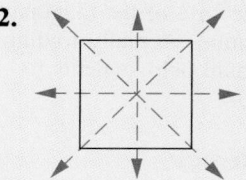

3.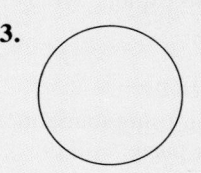

infinitely many (through the center)

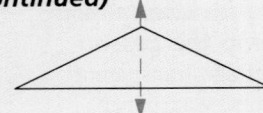
Lesson Focus

When you look in a mirror, you see a reflection of yourself. A "mirror image" or reflection is used as the basis for many designs. The mathematical idea involved is called *symmetry*, the focus of today's lesson.

Teaching Notes

Use the Exploration at the beginning of the lesson to review the concept of fitting one half of a figure over the other half.

Key Question

How can you determine if a figure has line symmetry? Check to see if there is any way the figure can be folded so that the "fold" is a line of symmetry.

Error Analysis

Students may confuse a line of symmetry with a line that simply divides the figure into two equal areas, as in the case of the diagonal of a parallelogram. Actually folding the figures should help students to avoid this error.

Using Technology

For a suggestion on using geometric drawing software to explore symmetry, see page 234C.

Using Manipulatives

For a suggestion on using transparent mirrors or wax paper with this lesson, see page 234D.

Some geometric figures have many lines of symmetry, while others have no lines of symmetry.

Example 2

Draw all the lines of symmetry in each figure.

a. 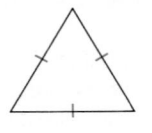   b.   c.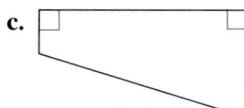

Solution

a.   b.   c. There are no lines of symmetry in the figure.

Guided Practice

COMMUNICATION « *Reading*

« **1.** What is the main idea of this lesson? To recognize and find lines of symmetry.

« **2.** What is the main difference between Example 1 and Example 2? (*Hint:* It may help to refer to the objective of the lesson.) In Ex. 1, lines of symmetry must be recognized; in Ex. 2, lines of symmetry must be drawn.

Does each pictured item have line symmetry? Explain.

3.
Yes

4.
Yes

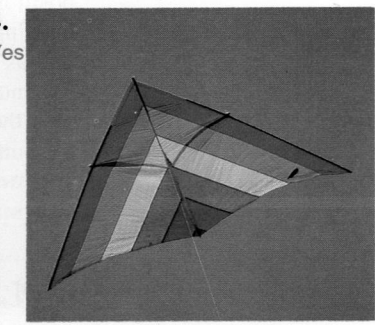

Is $\overleftrightarrow{XY}$ a line of symmetry? Write *Yes* or *No*.

5.
Yes

6.
No

7. Yes

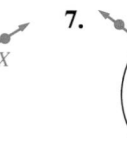

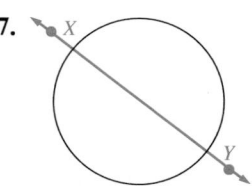

Answers to Guided Practice 8 and 10 are on p. A19.

Trace each figure onto a piece of paper. Then draw all the lines of symmetry. If there are no lines of symmetry, write *None*.

8.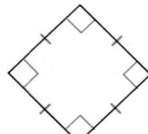

9. 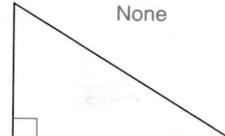 None

10.

Exercises

Is $\overleftrightarrow{MN}$ a line of symmetry? Write *Yes* or *No*.

A

1. No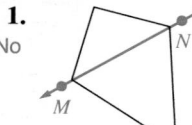

2. Yes

3. No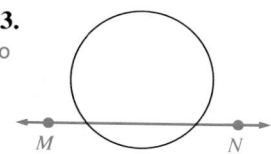

Trace each figure onto a piece of paper. Then draw all the lines of symmetry. If there are no lines of symmetry, write *None*.

4. None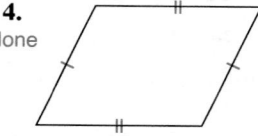

5.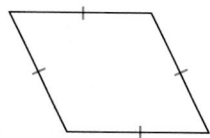

6.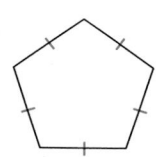

Answers to Exercises 5 and 6 are on p. A19.

7. Identify the type of quadrilateral that has four lines of symmetry.
square

8. What type of triangle has exactly one line of symmetry? an isosceles triangle that is not equilateral

Trace each figure onto a piece of paper. Then complete the figure so that $\overleftrightarrow{PQ}$ is a line of symmetry. Answers to Exercises 9 and 10 are on p. A19.

B

9.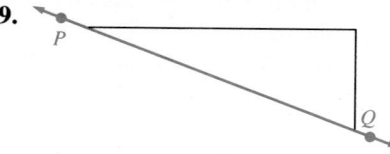

10.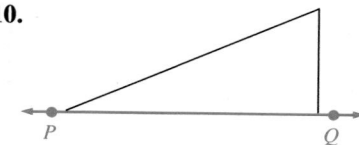

11. PATTERNS Complete this table for regular polygons.

Number of Sides	3	4	5	6	7	8
Number of Lines of Symmetry	?	?	?	?	?	?
	3	4	5	6	7	8

12. FUNCTIONS Write a function rule to represent the relationship between the number of sides of a regular polygon and the number of lines of symmetry. Use *n* to represent the number of sides. (*Hint:* You might want to use the results from Exercise 11.) $n \rightarrow n$

Introduction to Geometry **277**

 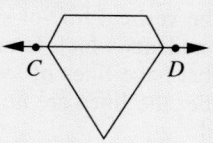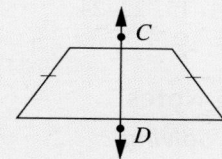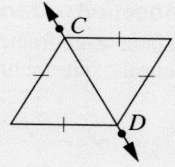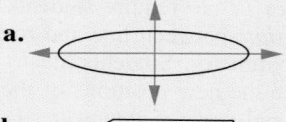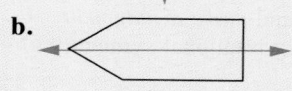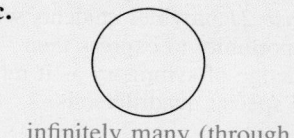

CAREER/APPLICATION

A *craftsperson* is skilled in work that is done by hand rather than by machine. Practicing a craft often requires a knowledge of the line symmetry of traditional designs. Pottery, woodworking, sewing, and quilting are examples of the many crafts that maintain cultural heritages.

Each of the following is a sketch of the basic *template* for a quilt design. How many lines of symmetry does each template have?

13.
4
14.
8
15.
14

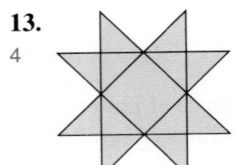

16. RESEARCH Find a Native American design that has line symmetry. Trace the design and draw in all the lines of symmetry.

Check students' drawings.

THINKING SKILLS 17–20. Evaluation

A figure has **rotational symmetry** if it fits exactly over its original position after being rotated less than a complete turn. A figure does not have to have line symmetry in order to have rotational symmetry. For instance, the figure at the right illustrates the rotational symmetry of a parallelogram. Given a half-turn, with point *A* as the center of the turn, the entire parallelogram *WXYZ* fits exactly over its original position.

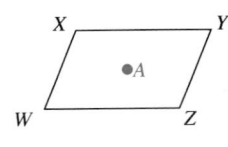

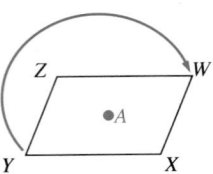

Determine whether each figure has *line symmetry*, *rotational symmetry*, or *both*.

C **17.** **18.** **19.** **20.**

rotational   line   both   both

21. WRITING ABOUT MATHEMATICS Suppose that you are a reporter for a high school mathematics newsletter. Write a feature article entitled *Symmetry and the Special Quadrilaterals*. See answer on p. A20.

S **22.** Find the product: $(-21)(8)$ *(Lesson 3-5)* −168

23. In the figure at the right, is $\overleftrightarrow{RS}$ a line of symmetry?   *(Lesson 6-9)* No

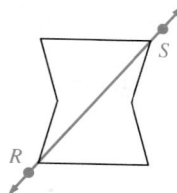

24. Decide whether it would be more appropriate to draw a *bar graph* or a *line graph* to display the data. Then draw the graph.   *(Lesson 5-5)*

Evening TV Viewers   line graph; see answer on p. A20.

Time	6:00 P.M.	8:00 P.M.	10:00 P.M.	12:00 A.M.
Viewers	45,000	65,000	50,000	10,000

Self-Test 2

Identify each polygon. Then list all its diagonals.

regular pentagon; $\overline{VS}, \overline{VT}, \overline{RU}, \overline{RT}, \overline{SU}$ 6-5

1.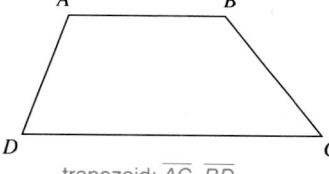

trapezoid; $\overline{AC}, \overline{BD}$

2.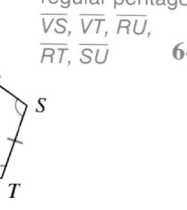

3. Tell whether line segments of lengths 9 ft, 7 ft, and 2 ft *can* or *cannot* be the sides of a triangle. If they can, tell whether the triangle would be *scalene*, *isosceles*, or *equilateral*. cannot   6-6

4. The measures of two angles of a triangle are 57° and 44°. Tell whether the triangle is *acute*, *right*, or *obtuse*. acute

In the figure at the right, *ABCD* is a rectangle.

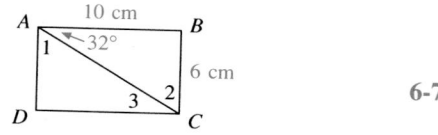

5. Find the length of $\overline{CD}$. 10 cm   6-7

6. Find $m\angle 2$. 58°

7. How many line segments are there if there are 8 points? 28   6-8

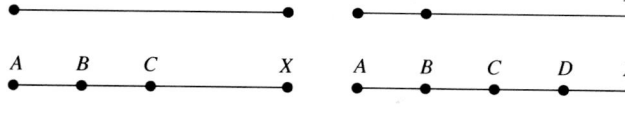

8. Draw all the lines of symmetry in the figure at the right. See answer on p. A20.   6-9

Introduction to Geometry   **279**

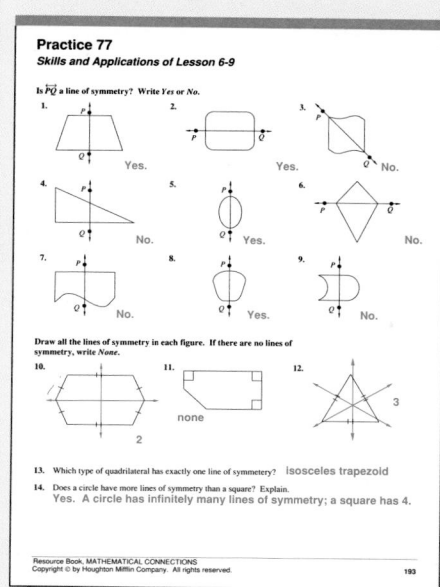

Chapter Review

Terms to Know

point (p. 236)
line (p. 236)
plane (p. 236)
collinear (p. 236)
intersect (p. 236)
line segment (p. 237)
endpoint (p. 237)
ray (p. 237)
angle (p. 237)
vertex of an angle (p. 237)
side of an angle (p. 237)
degree (p. 240)
protractor (p. 240)
acute angle (p. 245)
right angle (p. 245)
obtuse angle (p. 245)
straight angle (p. 245)
complementary (p. 245)
supplementary (p. 245)
adjacent (p. 246)
vertical (p. 246)
perpendicular (p. 249)

parallel (p. 249)
transversal (p. 250)
alternate interior angles (p. 250)
corresponding angles (p. 250)
polygon (p. 256)
side of a polygon (p. 256)
vertex of a polygon (p. 256)
diagonal (p. 256)
regular polygon (p. 257)
scalene triangle (p. 261)
isosceles triangle (p. 261)
equilateral triangle (p. 261)
The Triangle Inequality (p. 261)
acute triangle (p. 262)
right triangle (p. 262)
obtuse triangle (p. 262)
parallelogram (p. 266)
trapezoid (p. 266)
rectangle (p. 266)
rhombus (p. 266)
square (p. 266)
line symmetry (p. 275)
line of symmetry (p. 275)

Choose the correct term from the list above to complete each sentence.

1. A part of a line with two endpoints is called a(n) __?__. line segment

2. Two lines in the same plane that do not intersect are __?__. parallel

3. A(n) __?__ has a measure of 180°. straight angle

4. A rectangle with four sides that are the same length is a(n) __?__. square

5. A quadrilateral with exactly one set of parallel sides is a(n) __?__. trapezoid

6. A(n) __?__ has three sides of the same length. equilateral triangle

7. Two angles whose measures have a sum of 90° are called __?__. complementary angles

8. A(n) __?__ is an instrument used to measure angles. protractor

9. A(n) __?__ has a measure greater than 90°, but less than 180°. obtuse angle

10. A triangle all of whose angles are less than 90° is a(n) __?__. acute triangle

Write the name of each figure. *(Lesson 6-1)*

11.
$\overline{GH}, \overline{HG}$

12.
$\angle PQR, \angle RQP,$ or $\angle Q$

13.
Plane J

Use a protractor to measure each angle. *(Lesson 6-2)*

14.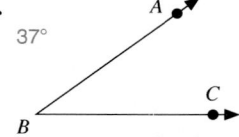
37°

15.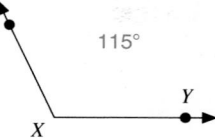
115°

Use a protractor to draw an angle of the given measure. *(Lesson 6-2)*
Answers to Chapter Review 16–19 are on p. A20.

16. 25°   **17.** 95°   **18.** 165°   **19.** 62°

Find the measure of a complement of an angle of the given measure.
(Lesson 6-3)

20. 24° 66°   **21.** 45° 45°   **22.** 73° 17°   **23.** 88° 2°

Find the measure of a supplement of an angle of the given measure.
(Lesson 6-3)

24. 30° 150°   **25.** 75° 105°   **26.** 85° 95°   **27.** 122° 58°

In the figure at the right, $\overleftrightarrow{AB} \parallel \overleftrightarrow{CD}$. Find the measure of each angle. *(Lessons 6-3 and 6-4)*

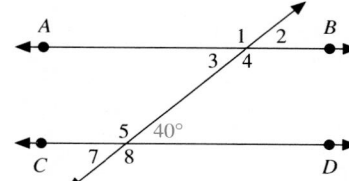

28. ∠2 40°   **29.** ∠5 140°

30. ∠4 140°   **31.** ∠1 140°

32. ∠7 40°   **33.** ∠8 140°

Use the figure at the right. Tell whether each statement is *True* or *False*. *(Lesson 6-4)*

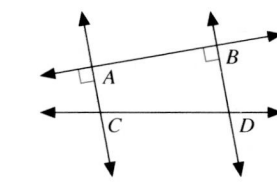

34. $\overleftrightarrow{AB} \parallel \overleftrightarrow{CD}$ False   **35.** $\overleftrightarrow{AC} \parallel \overleftrightarrow{BD}$ True

36. $\overleftrightarrow{BD} \perp \overleftrightarrow{AB}$ True   **37.** $\overleftrightarrow{CD} \perp \overleftrightarrow{AB}$ False

38. $\overleftrightarrow{AC} \perp \overleftrightarrow{AB}$ True   **39.** $\overleftrightarrow{AC} \parallel \overleftrightarrow{AB}$ False

Identify each polygon. Then list all its diagonals. *(Lesson 6-5)*

40.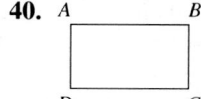
quadrilateral; $\overline{AC}, \overline{BD}$

41.
triangle; no diagonals

42.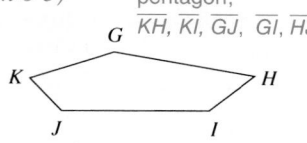
pentagon; $\overline{KH}, \overline{KI}, \overline{GJ}, \overline{GI}, \overline{HJ}$

Find each unknown angle measure. *(Lesson 6-5)*

43.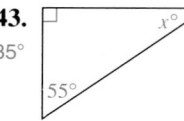
35°
55°

44.
129°
130°
101°
n°

45.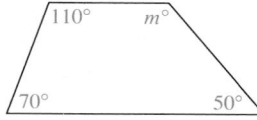
110° m°
130°
70° 50°

281

Quick Quiz 1 *(6-1 through 6-4)*

Write the name of each figure.

1.

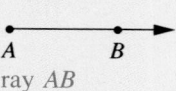

ray *AB*

2.

line segment *CD*

3.

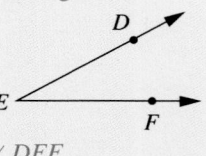

∠*DEF*

Use a protractor to draw an angle of the given measure.

4. 65°   5. 110°

6. 72°

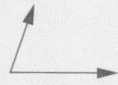

7. Find the measure of an angle complementary to a 63° angle.
   27°

8. Find the measure of an angle supplementary to a 115° angle.
   65°

In the figure, $\overleftrightarrow{AB} \parallel \overleftrightarrow{CD}$ and $m\angle 4 = 52°$. Find the measure of each angle.

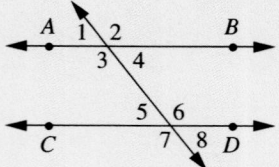

9. ∠1   52°   10. ∠3   128°
11. ∠5   52°   12. ∠7   128°

Chapter 6

Tell whether line segments of the given lengths *can* or *cannot* be the sides of a triangle. If they can, tell whether the triangle would be *scalene*, *isosceles*, or *equilateral*.   *(Lesson 6-6)*

46. 8 cm, 8 cm, 8 cm   47. 7 in., 6 in., 11 in.   48. 7 mm, 24 mm, 25 mm
    can; equilateral   can; scalene   can; scalene

The measures of two angles of a triangle are given. Tell whether the triangle is *acute*, *right*, or *obtuse*.   *(Lesson 6-6)*

49. 15°, 52° obtuse   50. 45°, 45° right   51. 38°, 67° acute   52. 75°, 29° acute

In the figure at the right, *ABCD* is a rhombus. Find each measure.   *(Lesson 6-7)*

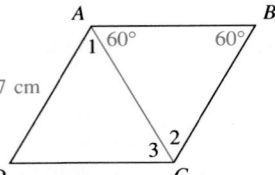

53. $m\angle 2$ 60°   54. $m\angle 3$ 60°

55. $m\angle 1$ 60°   56. the length of $\overline{BC}$ 7 cm

57. $m\angle D$ 60°   58. the length of $\overline{CD}$ 7 cm

Solve by identifying a pattern.   *(Lesson 6-8)*

59. The figures below show the first four figures in a pattern of squares. How many small squares are in the ninth figure of this pattern? 18

   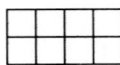

Is $\overleftrightarrow{XY}$ a line of symmetry? Write *Yes* or *No*.   *(Lesson 6-9)*

60.   61.   62.

No   Yes   Yes

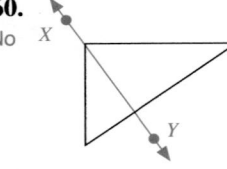 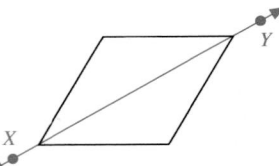 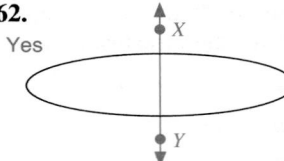

Trace each figure onto a piece of paper. Then draw all the lines of symmetry in each figure. If there are no lines of symmetry, write *none*.
(Lesson 6-9)   Answers to Chapter Review 63 and 65 are on p. A20.

63.   64.   65.

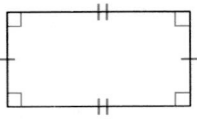 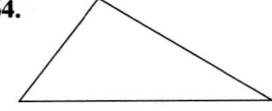 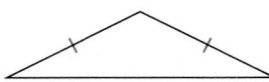

None

Write the name of each figure.

1. $\overleftrightarrow{CD}$, $\overrightarrow{DC}$

2. 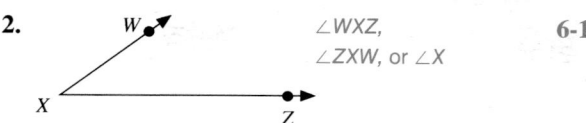 ∠WXZ, ∠ZXW, or ∠X   6-1

Use a protractor to measure each angle.

3.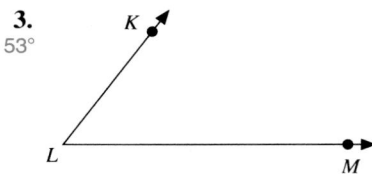
53°

4.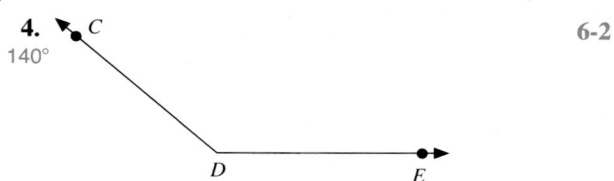
140°

6-2

Use a protractor to draw an angle of the given measure. Answers to Chapter Test 5–8 are on p. A20.

5. 50°   6. 85°   7. 125°   8. 67°

Find the measure of a complement of an angle of the given measure.

9. 25° 65°   10. 55° 35°   11. 38° 52°   12. 79° 11°   6-3

Find the measure of a supplement of an angle of the given measure.

13. 45° 135°   14. 70° 110°   15. 94° 86°   16. 112° 68°

★ 17. What is the difference in the measure of a complement and a supplement of any given acute angle? 90°

In the figure at the right, $\overleftrightarrow{ST} \parallel \overleftrightarrow{UV}$. Find the measure of each angle.

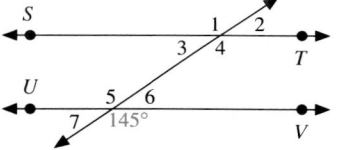

18. ∠1 145°   19. ∠4 145°   6-4

20. ∠6 35°   21. ∠7 35°

Identify each polygon. Then list all its diagonals.

22.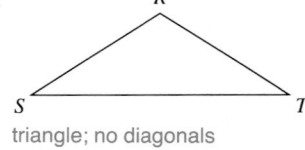
triangle; no diagonals

23.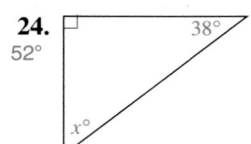
quadrilateral; $\overline{MO}$, $\overline{NP}$

6-5

Find each unknown angle measure.

24.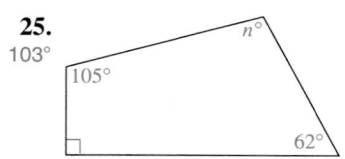
52°   38°
x°

25.
103°
105°   n°
62°

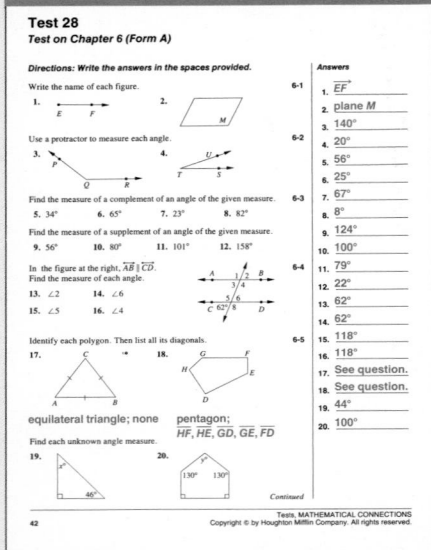

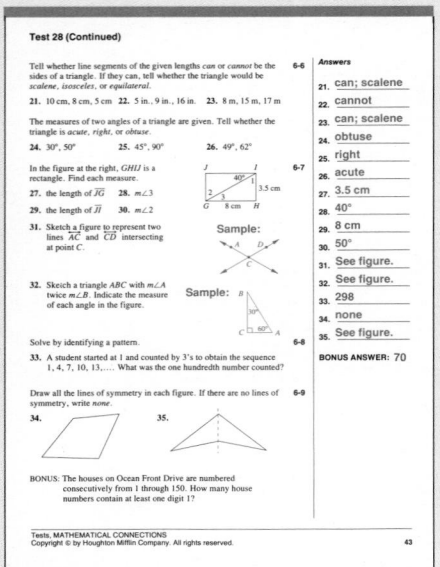

Chapter 6

Tell whether line segments of the given lengths *can* or *cannot* be the sides of a triangle. If they can, tell whether the triangle would be *scalene, isosceles,* or *equilateral.*

26. 3 ft, 5 ft, 9 ft
cannot

27. 6 m, 8 m, 10 m
can; scalene

28. 8 cm, 13 cm, 8 cm
can; isosceles   6-6

The measures of two angles of a triangle are given. Tell whether the triangle is *acute, right,* or *obtuse.*

29. 26°, 75° acute

30. 31°, 44° obtuse

31. 62°, 28° right

★**32.** How many acute angles can there be in a triangle? how many obtuse angles? Explain. See answer on p. A20.

In the figure at the right, *WXYZ* is a square. Find each measure.

33. $m\angle 1$ 45°

34. the length of $\overline{XY}$ 5 in.   6-7

35. $m\angle 2$ 45°

36. the length of $\overline{ZY}$ 5 in.

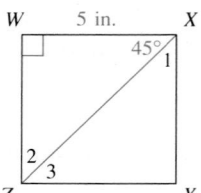

Solve by identifying a pattern.

37. The figures below show the first four figures in a pattern of circles. How many parts is the eighth circle divided into? 9   6-8

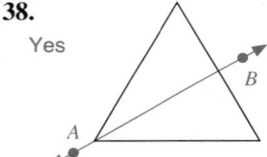

Is $\overleftrightarrow{AB}$ a line of symmetry? Write *Yes* or *No.*

38.

Yes

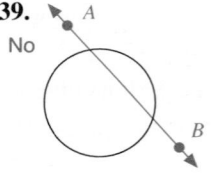

39.

No   6-9

Trace each figure onto a piece of paper. Then draw all the lines of symmetry in each figure. If there are no lines of symmetry, write *none.*

40.

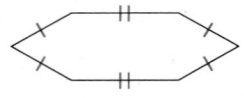

See answer on p. A20.

41.

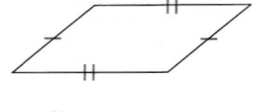

None

Tessellations

Teaching Note

Initially, students may confuse designs with tessellations. The precise definition of a tessellation should be stressed and illustrated. Have some examples available to show the class, such as Escher prints, tile patterns, and so on.

Exercise Note

Critical Thinking

Students are required to *create* tessellations and *determine* the codes. In order to form combinations that will tessellate, students need to experiment with different shapes.

Objective: To recognize and name regular and semiregular tessellations.

Terms to Know
* *tessellation*

A **tessellation** is a pattern in which identical copies of a figure cover a plane without gaps or overlaps. When the figure is a regular polygon, the pattern is a *regular* tessellation. A pattern formed by two or more types of regular polygons is a *semiregular* tessellation. For a pattern to be a regular or semiregular tessellation, the arrangement of regular polygons at every vertex must be identical.

Example

Solution

Write the code for the tessellation in the honeycomb above.

The arrangement at every vertex is hexagon-hexagon-hexagon. Because a hexagon has 6 sides, the code is 6-6-6.

Exercises

Write the code for each tessellation.

1.

4-4-4-4

2.

6-3-6-3

Use the regular polygons at the right. Make a *template* for each figure by first tracing it onto heavy paper or cardboard, then cutting it out.

3. Draw a tessellation of equilateral triangles. Write the code for your tessellation.

4. Make a sketch that demonstrates why a tessellation of regular pentagons is impossible.
Answers to Exercises 3 and 4 are on p. A20.

GROUP ACTIVITY

5. The codes for the combinations:
6-3-6-3, 3-4-6-4,
4-4-3-3-3,
3-3-4-3-4,
6-3-3-3-3
6. Code: 4-8-8

5. List the five combinations of regular polygons shown that will form a semiregular tessellation.

6. When you add a regular octagon to the set of regular polygons above, one other semiregular tessellation is possible. Create a template for a regular octagon. Then find the other tessellation.

Identify each polygon. Then list all its diagonals.

1.

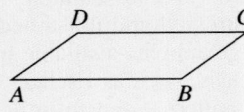

parallelogram *ABCD*; diagonals $\overline{AC}$, $\overline{BD}$

2.

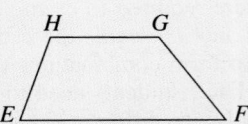

trapezoid *EFGH*; diagonals $\overline{EG}$, $\overline{FH}$

3. Tell whether line segments of 6 ft, 7 ft, and 8 ft *can* or *cannot* be the sides of a triangle. If they can, tell whether the triangle would be *scalene, isosceles,* or *equilateral*. can; scalene

4. The measures of two angles of a triangle are 42° and 33°. Tell whether the triangle is *acute, right,* or *obtuse*. obtuse

In the figure below, *ABCD* is a rhombus.

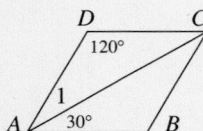

5. Find $m\angle B$. 120°

6. Find $m\angle 1$. 30°

7. How many line segments can connect five noncollinear points? Study the pattern in the figures below. 10

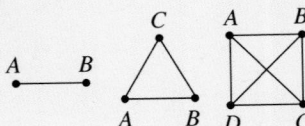

Cumulative Review

Standardized Testing Practice

Choose the letter of the correct answer.

1. Average Monthly Temperatures

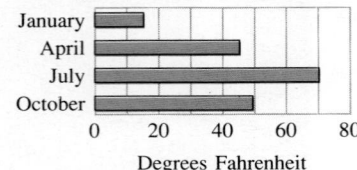

What information can be read from the graph?

A. the high temperature in May
B. the rainfall in July
C. the average temperature in July
D. the annual temperature range

2. The commutative property of multiplication states that changing the ___?___ of the factors does not change the product.

A. grouping   B. order
C. identities   D. size

3. Simplify: $4x + 3y + 7x$

A. $14x$ B. $7y + 7x$
C. $11x + 3y$ D. $14xy$

4. How many lines of symmetry does an equilateral triangle have?

A. 1   B. 3
C. 6   D. 0

5. Evaluate $q^2 - r$ when $q = 4$ and $r = -7$.

A. 9   B. 23
C. 11   D. 45

6.

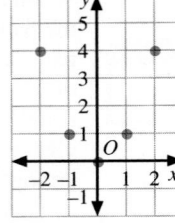

Which function rule does this graph represent?

A. $x \rightarrow 2x$ B. $x \rightarrow x$
C. $x \rightarrow x^2$ D. $x \rightarrow |x|$

7. Write the phrase *11 more than twice a number* as a variable expression.

A. $11 > 2x$ B. $11x + 2$
C. $2x > 11$ D. $2x + 11$

8. Evaluate: 5^3

A. 125   B. 15
C. 25   D. 5

9. Solve: $-44 = 4(2x - 7)$

A. $x = -2$ B. $x = -9$
C. $x = 9$ D. $x = -4\frac{5}{8}$

10. Complete: $z^5 \cdot \underline{} = z^{15}$

A. 3   B. z^3
C. z^{10} D. 10

11. What does $\overleftrightarrow{WX} \parallel \overleftrightarrow{YZ}$ mean?
I. The lines are parallel.
II. The lines are perpendicular.
III. The lines do not intersect.
A. I only B. I and II
C. I and III D. II only

12. Evan started practicing piano at 3 P.M. and played for 2 h. How many songs did he practice?
A. 5 B. 6
C. $1\frac{1}{2}$ D. not enough information

13. At 6 P.M. the temperature is −11°F. By 8 P.M. it has fallen 7°F. What is the temperature at 8 P.M.?
A. −4°F B. 4°F
C. 18°F D. −18°F

14. Choose the data most appropriate for a line graph.
A. voters in areas of a city
B. school enrollment over 10 years
C. total sales of three magazines
D. price per pound of apples in six grocery stores

15. The length of a book is about 230 mm. About how many centimeters long is the book?
A. 0.23 B. 2.3
C. 2300 D. 23

16. Which figure has no lines of symmetry?
A. regular octagon
b. square
C. scalene triangle
D. rhombus

17.

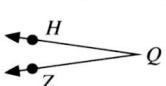

What is a correct name for this figure?
A. $\angle HZQ$ B. $\angle QHZ$
C. $\angle 4$ D. $\angle ZQH$

18. The area of a parallelogram is 42 cm² and the height is 6 cm. Use the formula $A = bh$ to find the base.
A. 7 cm B. 9 cm
C. 252 cm D. 3 cm

19. A pictograph shows that 900 people bought tapes and 600 people bought CDs. If 6 symbols represent the people who bought tapes, how many people does one symbol represent?
A. 250 B. 150
C. 100 D. 300

20. Two parallel lines are intersected by a transversal. What is the relationship between the alternate interior angles?
A. they are equal in measure
B. they are complementary
C. they are adjacent
D. they are vertical

---

8. Draw all the lines of symmetry in the figure below.

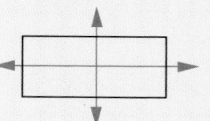

Alternative Assessment

1. What is the relationship of complements of angles with equal measures? Their measures are equal.

2. In the figure, *ABCD* is a rectangle. Explain why $\overleftrightarrow{EF}$ is a line of symmetry and why $\overleftrightarrow{AC}$ is not.

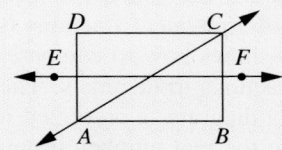

If the figure were folded along $\overleftrightarrow{EF}$, the parts would coincide. If $\overleftrightarrow{AC}$ were used as a fold, this would not occur.

3. The sum of the measures of the angles of a quadrilateral is equal to the number of degrees in _?_ straight angles. two

Introduction to Geometry **287**

Planning Chapter 7

Chapter Overview

Chapter 7 presents some elementary topics from number theory as well as fraction concepts that are essential for success in algebra. *Lesson 1* introduces prime and composite numbers and discusses ways to factor the latter. Factoring is used in *Lessons 2 and 3* to find the GCF and LCM of both numbers and algebraic expressions. The *Focus on Explorations* models fractions using fraction bars, which are then used in *Lesson 4* to demonstrate visually the concept of equivalent fractions. In *Lesson 5*, the idea of equivalent fractions is applied to algebraic fractions. *Lesson 6* shows how to compare fractions, while *Lesson 7* relates fractions to decimals. The problem solving strategy of drawing a diagram is presented in *Lesson 8*. In *Lesson 9*, the subsets of the real number system are introduced. *Lesson 10* shows students how to graph the solutions to equations and inequalities on a number line. The chapter concludes with *Lesson 11* extending the concept of exponents to both zero and the negative numbers.

Background

Lesson 1 begins by using a geometric model to explain factors and primes. This approach relates factoring to the number of different arrangements that can be made with rectangular tiles. Exponents, which were introduced in Chapter 1, are used again in *Lessons 1 and 2*. *Lesson 3* uses the greatest common power of each prime to determine the greatest common factor.

When students respond to the Thinking Skills and Logical Reasoning Exercises found in *Lessons 1–5*, they are developing their sense of numbers.

Connections made throughout the chapter link various areas of mathematics and relate concepts to real-world applications. In *Lesson 4*, arithmetic and geometry are joined through the use of symmetry and fractions. In *Lesson 6*, the determination of shutter speed and data analysis are used to relate fractions to the real world.

The problem solving in *Lesson 7* focuses on applications. *Lesson 8* introduces the problem solving strategy of drawing a diagram, reinforces the four-step process, and uses strategies that have been introduced previously in the problem set.

Lesson 10 expands the concept of inequality, introduced in Chapter 1, by graphing inequalities. *Lesson 11* builds on the concept of scientific notation introduced in Lesson 2-10. In that lesson, positive powers of ten were used to express large measures. Lesson 7-11 uses negative exponents to express very small numbers in scientific notation. Science applications frequently use scientific notation.

Manipulatives are very useful in teaching and learning fraction concepts. The *Focus on Explorations* shows how fractions can be modeled by using fraction bars. Fraction bars are then used to illustrate equivalent fractions and the comparison of fractions, thus providing students with a clear visual representation of these two concepts.

Technology is addressed by a computer application and the appropriate use of calculators throughout the chapter.

Objectives

7-1	To find the prime factorization of a number.
7-2	To find greatest common factors.
7-3	To find least common multiples.
≡	**FOCUS ON EXPLORATIONS** To use fraction bars to model fractions.
7-4	To find equivalent fractions and to write fractions in lowest terms.
7-5	To simplify algebraic fractions.
7-6	To compare fractions.
7-7	To write fractions as decimals and decimals as fractions.
7-8	To solve problems by drawing a diagram.
7-9	To recognize rational numbers.
7-10	To graph solutions of equations and inequalities on a number line.
7-11	To simplify expressions involving negative and zero exponents, and to use negative exponents to write numbers in scientific notation.
≡	**CHAPTER EXTENSION** To write repeating decimals as fractions.

Chapter Planner

Instructional Aids	Manipulatives	Cooperative Learning	Technology	Practice/Reteaching	Assessment
Materials: fraction bars, clothesline, clothespins, index cards, calculator *Resource Book:* Lesson Starters 59–69 *Visuals,* Folder C	Lessons 7-1, 7-4, 7-6, 7-9, Focus on Explorations *Activities Book:* Manipulative Activities 13–14	Lessons 7-1, 7-2, 7-3, 7-10, Focus on Explorations *Activities Book:* Cooperative Activities 13–15	Lessons 7-1, 7-2, 7-3, 7-6, 7-7, 7-10, 7-11 *Activities Book:* Calculator Activities 6–7 *Connections Plotter Plus* Disk	Toolbox, pp. 753–771 Extra Practice, p. 739 *Resource Book:* Practice 80–94 *Study Guide,* pp. 117–138	Self-Tests 1–3 Quick Quizzes 1–3 Chapter Test, p. 334 *Resource Book:* Lesson Starters 59–69 (Daily Quizzes) *Tests* 31–35

Assignment Guide Chapter 7

Day	Skills Course	Average Course	Advanced Course
1	**7-1:** 1–33, 36, 37–42	**7-1:** 1–33, 36, 37–42	**7-1:** 1–25 odd, 27–42
2	**7-2:** 1–24, 31–34	**7-2:** 1–26, 31–34	**7-2:** 1–19 odd, 20–34
3	**7-3:** 1–20, 27, 28, 35–38 **Exploration:** Activities I–IV	**7-3:** 1–20, 27–32, 35–38 **Exploration:** Activities I–IV	**7-3:** 1–17 odd, 19–38 **Exploration:** Activities I–IV
4	**7-4:** 1–22, 31–37, 39–42	**7-4:** 1–42	**7-4:** 1–19 odd, 21–42, Challenge
5	**7-5:** 1–17, 19–25 odd, 32–35	**7-5:** 1–29, 32–35	**7-5:** 1–15 odd, 16–35, Historical Note
6	**7-6:** 1–20, 29–34	**7-6:** 1–34	**7-6:** 1–34
7	**7-7:** 1–30, 49–52	**7-7:** 1–30, 31–45 odd, 49–52	**7-7:** 1–19 odd, 21–52
8	**7-8** 1–19	**7-8:** 1–19	**7-8:** 1–19
9	**7-9:** 1–28, 33–36	**7-9:** 1–36	**7-9:** 1–13 odd, 15–36
10	**7-10:** 1–19, 33–39	**7-10:** 1–28, 33–39	**7-10:** 1–13 odd, 15–39
11	**7-11:** 1–24, 31–34	**7-11:** 1–27, 31–34	**7-11:** 1–19 odd, 21–34
12	*Prepare for Chapter Test:* Chapter Review	*Prepare for Chapter Test:* Chapter Review	*Prepare for Chapter Test:* Chapter Review
13	*Administer Chapter 7 Test*	*Administer Chapter 7 Test; Cumulative Review*	*Administer Chapter 7 Test; Chapter Extension; Cumulative Review*

Teacher's Resources

Resource Book
Chapter 7 Project
Lesson Starters 59–69
Practice 80–94
Enrichment 65–76
Diagram Masters
Chapter 7 Objectives
Family Involvement 7
Spanish Test 7
Spanish Glossary

Activities Book
Cooperative Activities 13–15
Manipulative Activities 13–14
Calculator Activities 6–7

Study Guide, pp. 117–138

Tests
Tests 31–35

Visuals, Folder C

Connections Plotter Plus **Disk**

Alternate Approaches Chapter 7

Using Technology

Lesson 7-1

CALCULATORS

Students can use calculators to determine whether one integer is divisible by another, if the numbers involved are not too large. They simply need to divide and check whether the quotient is an integer.

COMPUTERS

A simple BASIC program can test easily whether a relatively small INPUT integer is prime or not. Here is an example:

```
10 INPUT "INTEGER: ";N
20 FOR I = 2 TO SQR(N)
30 IF INT(N/I) = (N/I) THEN 100 : REM FOUND DIVISOR
40 NEXT I
50 PRINT "A PRIME"
60 GO TO 110
100 PRINT "NOT A PRIME"
110 END
```

Lesson 7-2

CALCULATORS

The Euclidean Algorithm for finding greatest common factors is efficient and easy to carry out, especially using a calculator. Not only does this method produce a result quickly, but it also provides students with an opportunity to use calculators in a slightly different way (finding integer quotients and remainders).

COMPUTERS

The Euclidean Algorithm can be programmed on a computer. Students using such a program can explore divisibility rules and properties of greatest common factors and least common multiples.

Lesson 7-3

CALCULATORS

Since the least common multiple of a pair of numbers equals their product divided by the greatest common divisor, any work done previously on greatest common divisors is applicable to the least common multiple.

Lesson 7-6

CALCULATORS

Calculators can be used to compare fractions rather simply by dividing and comparing decimal approximations. Using this method, students can explore what happens when the numerator and denominator of a fraction are multiplied by the same integer, or when an integer is added to both the numerator and denominator.

Lesson 7-7

CALCULATORS AND COMPUTERS

Students can use a calculator or a BASIC program to determine repeating decimal equivalents for rational numbers by dividing the numerator by the denominator. If the number of digits in the repeating pattern is large, however, the entire pattern may not be displayed, and this method will fail. Students also should be aware that calculators may round or truncate repeating decimals.

Lesson 7-10

COMPUTERS

A simple BASIC program or spreadsheet program can be used to find points in the graph of an open sentence. For instance, students can run the following program, typing in different values for X (be sure they do not restrict the input to integers). On a number line, students can make a dot at each X for which the program returns "TRUE." As enrichment, complex inequalities such as $|x - 3| < 4$ also can be solved in this way.

```
10 INPUT "X = ";X
20 IF (X > 3) THEN 50
30 PRINT "FALSE"
40 GOTO 10
50 PRINT "TRUE"
60 GOTO 10
```

Using Manipulatives

Lesson 7-1

The Exploration at the beginning of the lesson can be done as a manipulative activity with students working in cooperative groups. Give each group tiles or cubes with which to build rectangular arrays. Ask each group to build as many rectangles as possible using 24 tiles. Each group should sketch its results on a sheet of graph paper and discuss how the results relate to multiplication. To further solidify students' understanding of prime and composite numbers, you may wish to have the groups repeat this activity with numbers other than 24.

Lesson 7-6

Fraction bars can be used to help students order three or more fractions, rather than just two as is done in the Examples. Have students randomly select four or five bars and then arrange them in order. (Bars that are equal in value should be placed on top of each other.) Then have students write the relationships that they observe.

Another highly effective manipulative is a "fraction line." Place a clothesline across the front of the room and then clip an index card with 0 on it at the left end of the line and an index card with 1 on it at the right end. These cards can be held with clothespins. Put clothespins and a pile of index cards with fractions written on them on your desk. Have each student randomly select a fraction card and place it on the line in its correct position. The position of incorrectly placed fractions can be adjusted. This procedure gives students a visual representation of the relative size of fractions.

Lesson 7-9

The "fraction line" discussed above can be used with rational numbers. Place the 0 card at the center of the clothesline and then place cards with positive and negative integers on them in their correct positions to create a number line. Students then select cards with positive and negative fractions and decimal numbers written on them and place the cards in their relative positions on the line.

Reteaching/Alternate Approach

Lesson 7-6

Cross multiplication can be used as a shortcut to compare fractions. For example, to compare $\frac{1}{5}$ and $\frac{2}{7}$, set up the following.

$$\frac{7}{\Box} = \frac{1}{5} \qquad \frac{2}{7} = \frac{10}{\Box}$$

Notice that the numerator of each of the new equivalent "fractions" is equal to the numerator of one of the original fractions multiplied by the denominator of the other original fraction; in this case, $1 \times 7 = 7$ and $2 \times 5 = 10$. The numerators of the new "fractions" can then be compared ($7 < 10$) to conclude that $\frac{1}{5} < \frac{2}{7}$.

Lesson 7-10

COOPERATIVE LEARNING

Place students in cooperative groups. Have each group create a set of 20 to 25 numbers consisting of positive and negative integers, fractions, and decimals. Each number should be written on a 3" x 5" index card. Additional cards should be labeled with a variable, such as x, and the symbols $<$, $>$, $\leq$, and $\geq$.

Place the numbered cards face down in one pile and the symbol cards face down in another pile so that they may be selected at random. One member of a group selects two cards, one from each pile, and then forms an inequality by using the variable, the number selected, and the symbol selected. The group works together to draw a graph of the inequality. Then each member of the group selects one number card at random and determines whether the number selected is a solution of the inequality. Each person must also show where on the number line the point would be graphed.

Teacher's Resources Chapter 7

Enrichment masters from the Resource Book are pictured here. See the Teacher's Resources chart on page 288B for a complete listing of all materials available for this chapter.

7-1 Enrichment
Connection

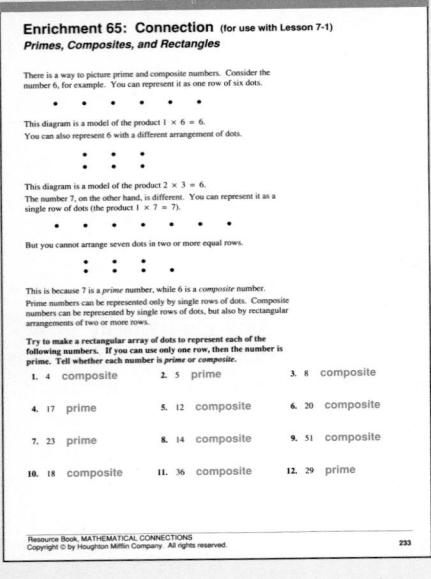

Enrichment 65: Connection (for use with Lesson 7-1)
Primes, Composites, and Rectangles

7-2 Enrichment
Thinking Skills

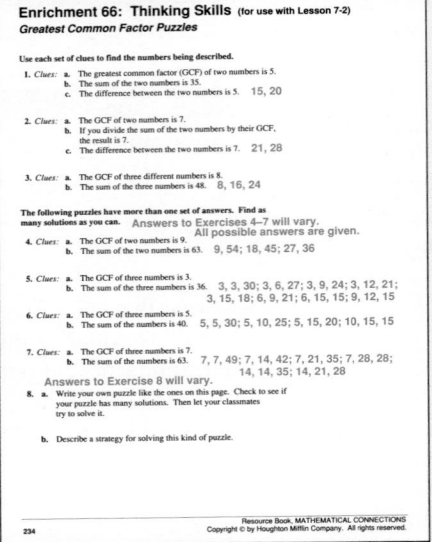

Enrichment 66: Thinking Skills (for use with Lesson 7-2)
Greatest Common Factor Puzzles

7-3 Enrichment
Application

Enrichment 67: Application (for use with Lesson 7-3)
Common Multiples

7-4 Enrichment
Connection

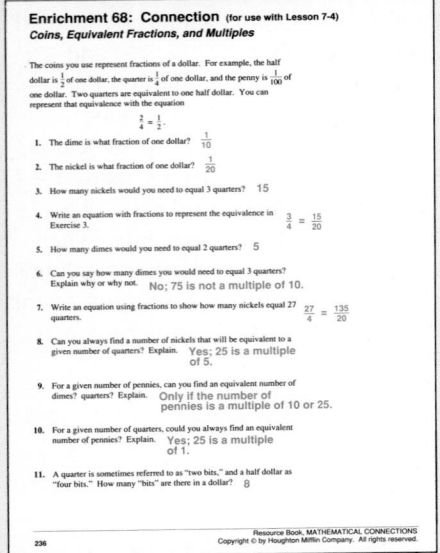

Enrichment 68: Connection (for use with Lesson 7-4)
Coins, Equivalent Fractions, and Multiples

7-5 Enrichment
Communication

Enrichment 69: Communication (for use with Lesson 7-5)
A Common Fraction Mistake

7-6 Enrichment
Exploration

Enrichment 70: Exploration (for use with Lesson 7-6)
Why You Can't Divide by Zero

7-7 Enrichment
Data Analysis

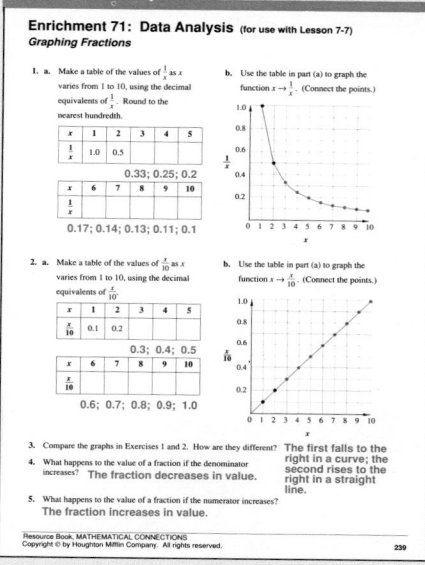

Enrichment 71: Data Analysis (for use with Lesson 7-7)
Graphing Fractions

1. **a.** Make a table of the values of $\frac{1}{x}$ as x varies from 1 to 10, using the decimal equivalents of $\frac{1}{x}$. Round to the nearest hundredth.

   b. Use the table in part (a) to graph the function $x \rightarrow \frac{1}{x}$. (Connect the points.)

x	1	2	3	4	5
$\frac{1}{x}$	1.0	0.5			

   0.33; 0.25; 0.2

x	6	7	8	9	10
$\frac{1}{x}$					

   0.17; 0.14; 0.13; 0.11; 0.1

2. **a.** Make a table of the values of $\frac{x}{10}$ as x varies from 1 to 10, using the decimal equivalents of $\frac{x}{10}$.

   b. Use the table in part (a) to graph the function $x \rightarrow \frac{x}{10}$. (Connect the points.)

x	1	2	3	4	5
$\frac{x}{10}$	0.1	0.2			

   0.3; 0.4; 0.5

x	6	7	8	9	10
$\frac{x}{10}$					

   0.6; 0.7; 0.8; 0.9; 1.0

3. Compare the graphs in Exercises 1 and 2. How are they different? **The first falls to the right in a curve; the second rises to the right in a straight line.**

4. What happens to the value of a fraction if the denominator increases? **The fraction decreases in value.**

5. What happens to the value of a fraction if the numerator increases? **The fraction increases in value.**

7-8 Enrichment
Problem Solving

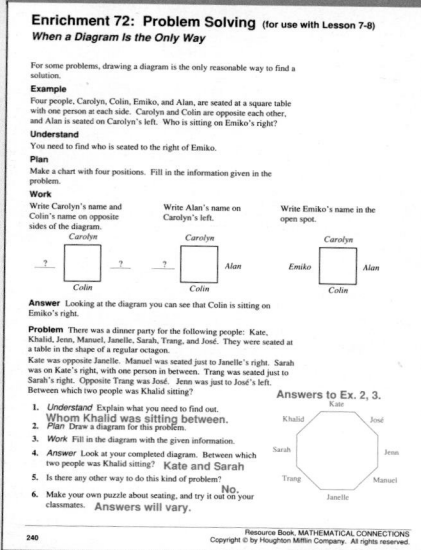

Enrichment 72: Problem Solving (for use with Lesson 7-8)
When a Diagram Is the Only Way

For some problems, drawing a diagram is the only reasonable way to find a solution.

Example
Four people, Carolyn, Colin, Emiko, and Alan, are seated at a square table with one person at each side. Carolyn and Colin are opposite each other, and Alan is seated on Carolyn's left. Who is sitting on Emiko's right?

Understand
You need to find who is seated to the right of Emiko.

Plan
Make a chart with four positions. Fill in the information given in the problem.

Work
Write Carolyn's name and Colin's name on opposite sides of the diagram.

Write Alan's name on Carolyn's left.

Write Emiko's name in the open spot.

Answer Looking at the diagram you can see that Colin is sitting on Emiko's right.

Problem There was a dinner party for the following people: Kate, Khalid, Jenn, Manuel, Janelle, Sarah, Trang, and José. They were seated at a table in the shape of a regular octagon.
Kate was opposite Janelle. Manuel was seated just to Janelle's right. Sarah was on Kate's right, with one person in between. Trang was seated just to Sarah's right. Opposite Trang was José. Jenn was just to José's left. Between which two people was Khalid sitting?

1. *Understand* Explain what you need to find out. **Whom Khalid was sitting between.**
2. *Plan* Draw a diagram for this problem.
3. *Work* Fill in the diagram with the given information.
4. *Answer* Look at your completed diagram. Between which two people was Khalid sitting? **Kate and Sarah**
5. Is there any other way to do this kind of problem? **No.**
6. Make your own puzzle about seating, and try it out on your classmates. **Answers will vary.**

Answers to Ex. 2, 3.

7-9 Enrichment
Thinking Skills

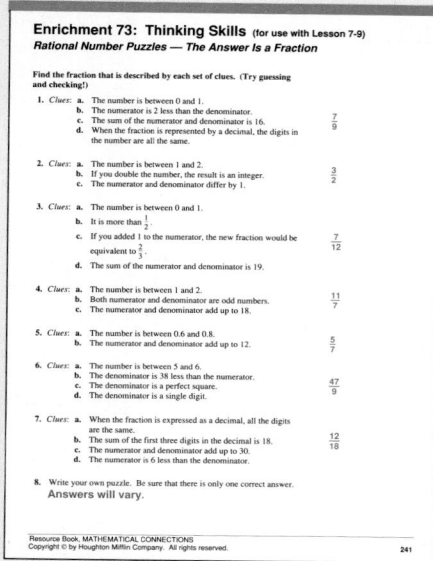

Enrichment 73: Thinking Skills (for use with Lesson 7-9)
Rational Number Puzzles — The Answer Is a Fraction

Find the fraction that is described by each set of clues. (Try guessing and checking!)

1. *Clues:* **a.** The number is between 0 and 1.
   b. The numerator is 2 less than the denominator.
   c. The sum of the numerator and denominator is 16.
   d. When the fraction is represented by a decimal, the digits in the number are all the same. $\frac{7}{9}$

2. *Clues:* **a.** The number is between 1 and 2.
   b. If you double the number, the result is an integer.
   c. The numerator and denominator differ by 1. $\frac{3}{2}$

3. *Clues:* **a.** The number is between 0 and 1.
   b. It is more than $\frac{1}{2}$.
   c. If you added 1 to the numerator, the new fraction would be equivalent to $\frac{2}{3}$. $\frac{7}{12}$
   d. The sum of the numerator and denominator is 19.

4. *Clues:* **a.** The number is between 1 and 2.
   b. Both numerator and denominator are odd numbers.
   c. The numerator and denominator add up to 18. $\frac{11}{7}$

5. *Clues:* **a.** The number is between 0.6 and 0.8.
   b. The numerator and denominator add up to 12. $\frac{5}{7}$

6. *Clues:* **a.** The number is between 5 and 6.
   b. The denominator is 38 less than the numerator.
   c. The denominator is a perfect square. $\frac{47}{9}$
   d. The denominator is a single digit.

7. *Clues:* **a.** When the fraction is expressed as a decimal, all the digits are the same.
   b. The sum of the first three digits in the decimal is 18. $\frac{12}{18}$
   c. The numerator and denominator add up to 30.
   d. The numerator is 6 less than the denominator.

8. Write your own puzzle. Be sure that there is only one correct answer. **Answers will vary.**

7-10 Enrichment
Exploration

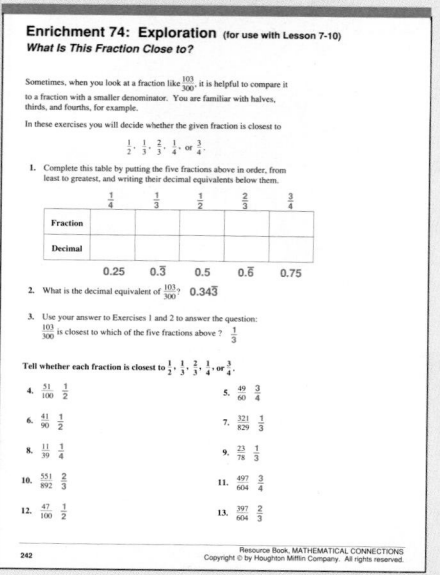

Enrichment 74: Exploration (for use with Lesson 7-10)
What Is This Fraction Close to?

Sometimes, when you look at a fraction like $\frac{103}{300}$, it is helpful to compare it to a fraction with a smaller denominator. You are familiar with halves, thirds, and fourths, for example.

In these exercises you will decide whether the given fraction is closest to

$$\frac{1}{4}, \quad \frac{1}{3}, \quad \frac{1}{2}, \quad \frac{2}{3}, \quad \text{or} \quad \frac{3}{4}.$$

1. Complete this table by putting the five fractions above in order, from least to greatest, and writing their decimal equivalents below them.

Fraction	$\frac{1}{4}$	$\frac{1}{3}$	$\frac{1}{2}$	$\frac{2}{3}$	$\frac{3}{4}$
Decimal					

   0.25 $0.\overline{3}$ 0.5 $0.\overline{6}$ 0.75

2. What is the decimal equivalent of $\frac{103}{300}$? **$0.34\overline{3}$**

3. Use your answer to Exercises 1 and 2 to answer the question: $\frac{103}{300}$ is closest to which of the five fractions above? **$\frac{1}{3}$**

Tell whether each fraction is closest to $\frac{1}{2}, \frac{1}{3}, \frac{2}{3}, \frac{1}{4}$, or $\frac{3}{4}$.

4. $\frac{51}{100}$ $\frac{1}{2}$ 5. $\frac{49}{60}$ $\frac{3}{4}$

6. $\frac{41}{90}$ $\frac{1}{2}$ 7. $\frac{321}{829}$ $\frac{1}{3}$

8. $\frac{11}{39}$ $\frac{1}{4}$ 9. $\frac{23}{78}$ $\frac{1}{3}$

10. $\frac{551}{892}$ $\frac{2}{3}$ 11. $\frac{497}{604}$ $\frac{3}{4}$

12. $\frac{47}{100}$ $\frac{1}{2}$ 13. $\frac{397}{604}$ $\frac{2}{3}$

7-11 Enrichment
Exploration

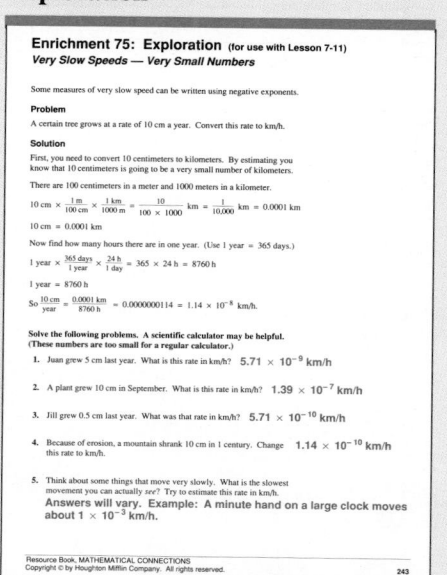

Enrichment 75: Exploration (for use with Lesson 7-11)
Very Slow Speeds — Very Small Numbers

Some measures of very slow speed can be written using negative exponents.

Problem
A certain tree grows at a rate of 10 cm a year. Convert this rate to km/h.

Solution
First, you need to convert 10 centimeters to kilometers. By estimating you know that 10 centimeters is going to be a very small number of kilometers.
There are 100 centimeters in a meter and 1000 meters in a kilometer.

$$10 \text{ cm} \times \frac{1 \text{ m}}{100 \text{ cm}} \times \frac{1 \text{ km}}{1000 \text{ m}} = \frac{10}{100 \times 1000} \text{ km} = \frac{10}{1,000,000} \text{ km} = 0.0001 \text{ km}$$

10 cm = 0.0001 km

Now find how many hours there are in one year. (Use 1 year = 365 days.)

$$1 \text{ year} \times \frac{365 \text{ days}}{1 \text{ year}} \times \frac{24 \text{ h}}{1 \text{ day}} = 365 \times 24 \text{ h} = 8760 \text{ h}$$

1 year = 8760 h

So $\frac{10 \text{ cm}}{\text{year}} = \frac{0.0001 \text{ km}}{8760 \text{ h}} = 0.0000000114 = 1.14 \times 10^{-8}$ km/h.

Solve the following problems. A scientific calculator may be helpful. (These numbers are too small for a regular calculator.)

1. Juan grew 5 cm last year. What is this rate in km/h?   **5.71×10^{-9} km/h**

2. A plant grew 10 cm in September. What is this rate in km/h?   **1.39×10^{-7} km/h**

3. Jill grew 0.5 cm last year. What was that rate in km/h?   **5.71×10^{-10} km/h**

4. Because of erosion, a mountain shrank 10 cm in 1 century. Change this rate to km/h.   **1.14×10^{-10} km/h**

5. Think about some things that move very slowly. What is the slowest movement you can actually *see*? Try to estimate this rate in km/h. **Answers will vary. Example: A minute hand on a large clock moves about 1×10^{-3} km/h.**

End of Chapter Enrichment
Extension

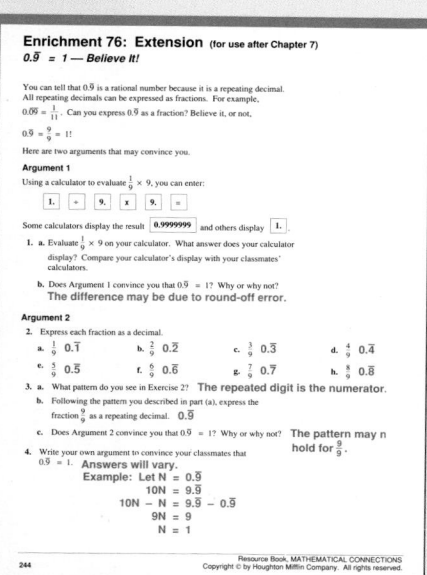

Enrichment 76: Extension (for use after Chapter 7)
$0.\overline{9} = 1$ — Believe It!

You can tell that $0.\overline{9}$ is a rational number because it is a repeating decimal.
All repeating decimals can be expressed as fractions. For example,
$0.\overline{06} = \frac{1}{11}$. Can you express $0.\overline{9}$ as a fraction? Believe it, or not,

$$0.\overline{9} = \frac{9}{9} = 1!$$

Here are two arguments that may convince you.

Argument 1
Using a calculator to evaluate $\frac{1}{9} \times 9$, you can enter:

1.	÷	9.	×	9.	=

Some calculators display the result $\boxed{0.9999999}$ and others display $\boxed{1.}$

1. **a.** Evaluate $\frac{1}{9} \times 9$ on your calculator. What answer does your calculator display? Compare your calculator's display with your classmates' calculators.

   b. Does Argument 1 convince you that $0.\overline{9} = 1$? Why or why not? **The difference may be due to round-off error.**

Argument 2

2. Express each fraction as a decimal.
   a. $\frac{1}{9}$ **$0.\overline{1}$**   **b.** $\frac{2}{9}$ **$0.\overline{2}$**   **c.** $\frac{3}{9}$ **$0.\overline{3}$**   **d.** $\frac{4}{9}$ **$0.\overline{4}$**
   e. $\frac{5}{9}$ **$0.\overline{5}$**   **f.** $\frac{6}{9}$ **$0.\overline{6}$**   **g.** $\frac{7}{9}$ **$0.\overline{7}$**   **h.** $\frac{8}{9}$ **$0.\overline{8}$**

3. **a.** What pattern do you see in Exercise 2? **The repeated digit is the numerator.**
   b. Following the pattern you described in part (a), express the fraction $\frac{9}{9}$ as a repeating decimal. **$0.\overline{9}$**
   c. Does Argument 2 convince you that $0.\overline{9} = 1$? Why or why not? **The pattern may not hold for $\frac{9}{9}$.**

4. Write your own argument to convince your classmates that $0.\overline{9} = 1$. **Answers will vary.**
   Example: Let $N = 0.\overline{9}$
   $$10N = 9.\overline{9}$$
   $$10N - N = 9.\overline{9} - 0.\overline{9}$$
   $$9N = 9$$
   $$N = 1$$

Multicultural Notes

For Chapter 7 Opener

In today's world, it is difficult to think of an occupation that does not in some way depend upon computers. For example, librarians use computers to store and retrieve information about library holdings and to gain access to information from other organizations' databases. Meteorologists use computers to describe and predict changing atmospheric conditions around the world. Supermarket cashiers use computers to scan packages for codes that translate into prices. Medical technicians use a computerized tomography scan (CT scan, for short) to create computer images of body organs from x-ray data. Doctors can then use these images to diagnose disease.

For Page 303

You may want to discuss with students working on Guided Practice Exercises 1 and 2 the different etymologies of the words *equal* and *equivalent*. *Equal* comes from the Latin *aequalis*, meaning even, level, or equal. *Equivalent* comes from the Latin verb *aequivalere*, meaning to have equal force of value. *Aequivalere* is a compound of *aequus* (equal) and *valere* (to be well or strong).

For Page 320

Exercise 17 offers a good opportunity to build students' knowledge of world geography and climate patterns. Discuss with students what other major cities in the world are at Houston's latitude (30° N) and what their climates are like, and, if their climates are different from Houston's, how those differences might be explained. Among the cities that can be discussed are Jacksonville, Florida; Chongqing, China; Hangzhou, China; Cairo, Egypt; New Delhi, India; and Abadan, Iran.

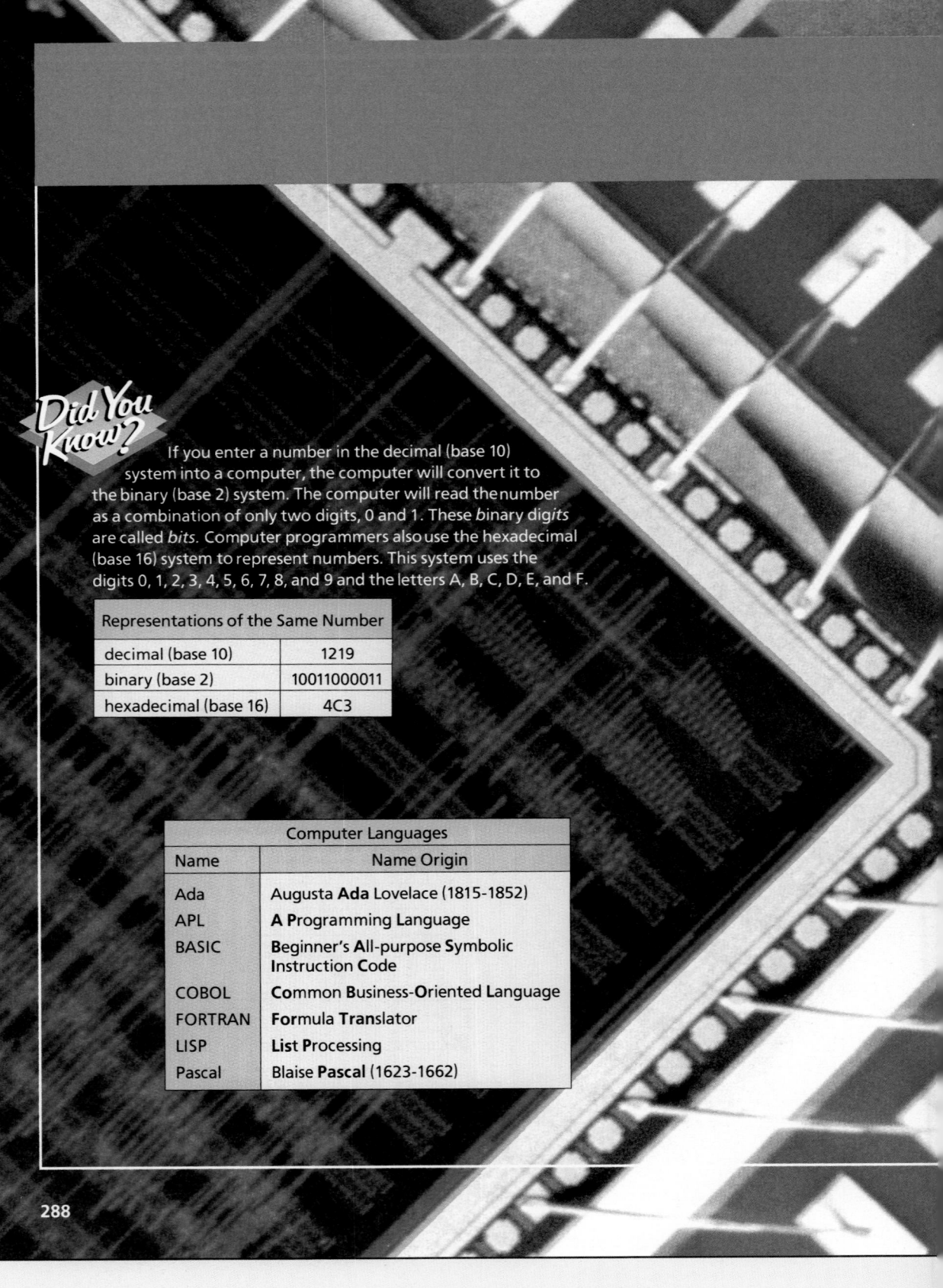

Did You Know?

If you enter a number in the decimal (base 10) system into a computer, the computer will convert it to the binary (base 2) system. The computer will read the number as a combination of only two digits, 0 and 1. These *binary digits* are called *bits*. Computer programmers also use the hexadecimal (base 16) system to represent numbers. This system uses the digits 0, 1, 2, 3, 4, 5, 6, 7, 8, and 9 and the letters A, B, C, D, E, and F.

Representations of the Same Number	
decimal (base 10)	1219
binary (base 2)	10011000011
hexadecimal (base 16)	4C3

Computer Languages	
Name	Name Origin
Ada	Augusta **Ada** Lovelace (1815-1852)
APL	**A** **P**rogramming **L**anguage
BASIC	**B**eginner's **A**ll-purpose **S**ymbolic **I**nstruction **C**ode
COBOL	**Co**mmon **B**usiness-**O**riented **L**anguage
FORTRAN	**For**mula **Tran**slator
LISP	**Lis**t **P**rocessing
Pascal	Blaise **Pascal** (1623-1662)

Number Theory and Fraction Concepts 7

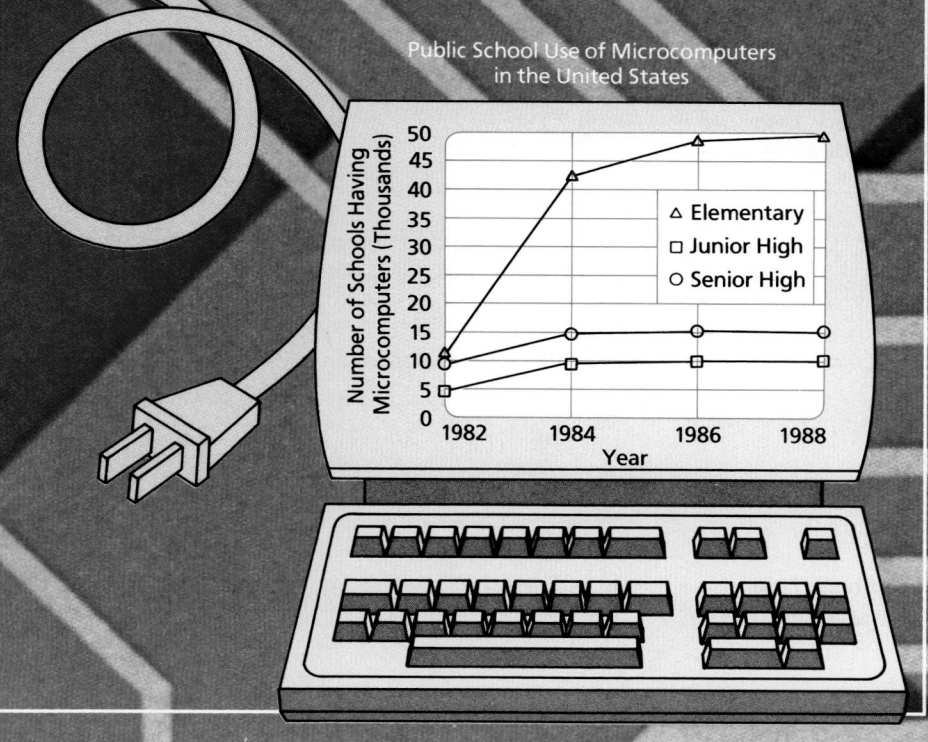

1946   The first electronic digital computer, called ENIAC (Electronic Numerical Integrator and Calculator), uses vacuum tubes for power. It weighs about 60,000 lbs and can do 5000 additions per second.

1958   Transistors replace vacuum tubes in computers. Transistors make computers smaller, less expensive, and able to do 200,000 additions per second.

1964   Integrated circuits replace transistors, making computers even smaller and able to do 1,250,000 additions per second.

1978   The computer chip reduces thousands of circuits to an area about the size of a pencil eraser, making computers able to do 10,000,000 additions per second.

1989   Supercomputers are capable of performing billions of operations per second with 20-digit accuracy.

Public School Use of Microcomputers in the United States

Number of Schools Having Microcomputers (Thousands)

△ Elementary
□ Junior High
○ Senior High

Year: 1982, 1984, 1986, 1988

289

Research Activity

For Chapter 7 Opener

For students to gain an appreciation of the many ways that computers are used today, ask them to interview people in the workplace about how they use computers. Students can brainstorm in cooperative groups to come up with a variety of occupations to explore. Cooperative groups can help students develop good strategies for interviewing, organizing information, and reporting to the class. If students have access to database software, you may want to have them create a database of the different uses of computers in various occupations.

Suggested Resource

Hunter, Beverly. ''Problem Solving With Databases,'' in Terence R. Cannings and Stephen W. Brown, eds., *The Information Age Classroom: Using the Computer As A Tool* (Irvine, CA: Franklin, Beedle and Associates, 1986), pp. 157–164.

Teaching the Lesson
- Materials: tiles or cubes, calculator
- Lesson Starter 59
- Using Technology, p. 288C
- Using Manipulatives, p. 288D
- Toolbox Skill 16

Lesson Follow-Up
- Practice 80
- Enrichment 65: Connection
- Study Guide, pp. 117–118
- Cooperative Activity 13

Warm-Up Exercises

Write each number as a product of two whole numbers in as many different ways as possible.

1. 12 $1 \cdot 12, 2 \cdot 6, 3 \cdot 4$
2. 15 $1 \cdot 15, 3 \cdot 5$
3. 13 $1 \cdot 13$
4. 27 $1 \cdot 27, 3 \cdot 9$
5. 40 $1 \cdot 40, 2 \cdot 20, 4 \cdot 10, 5 \cdot 8$

Lesson Focus

Ask students to imagine that they are arranging eight small rectangular tables into one large rectangular table for a party. The eight tables could be arranged to form two different rectangles; a two table by four table rectangle or a one table by eight table rectangle. However, if there were only seven tables, they could only be arranged in a single rectangle of one table by seven tables. Today's lesson focuses on prime and composite numbers, which can be used to explain the solution to the table problem.

Teaching Notes

Emphasize the relationship of the rectangular model to prime and composite numbers. Prime numbers allow only one rectangular model; composite numbers may be arranged in two or more rectangular models.

Factors and Prime Numbers

Objective: To find the prime factorization of a number.

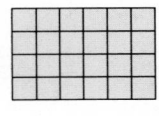

EXPLORATION

Terms to Know
- *factor*
- *prime number*
- *composite number*
- *prime factorization*

1 At the right you see a 4 by 6 rectangular arrangement of 24 tiles. Describe all the other rectangular arrangements that can be made using exactly 24 tiles. 1 by 24; 2 by 12; 3 by 8

2 Describe all the rectangular arrangements that can be made if you have only 23 tiles. 1 by 23

A given number of tiles can be arranged in more than one way only when the given number has more than two *factors*. Recall that when one whole number is divisible by a second whole number, the second number is a **factor** of the first. A whole number greater than 1 with exactly two factors, 1 and the number itself, is called a **prime number**. A **composite number** has more than two factors.

Example 1

Tell whether each number is *prime* or *composite*.

a. 11   b. 21   c. 31

Solution

a. Prime. The only factors of 11 are 1 and 11.

b. Composite. The factors of 21 are 1, 3, 7, and 21.

c. Prime. The only factors of 31 are 1 and 31.

✎ Check Your Understanding

1. In Example 1, how many factors does each of the prime numbers have?
See *Answers to Check Your Understanding* at the back of the book.

When you write a number as a product of prime numbers, you are writing the **prime factorization** of the number.

Example 2

Solution

Write the prime factorization of 140.

Use divisibility rules to help you make a *factor tree*.

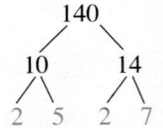

$$140 = 2 \cdot 5 \cdot 2 \cdot 7$$
$$= 2 \cdot 2 \cdot 5 \cdot 7$$
$$= 2^2 \cdot 5 \cdot 7$$

✎ Check Your Understanding

2. In Example 2, how do you know that 140 is divisible by 10?
See *Answers to Check Your Understanding* at the back of the book.

Guided Practice

COMMUNICATION « *Reading*

Replace each __?__ with the correct phrase.

« **1.** A whole number whose only factors are 1 and itself is a(n) __?__.
prime number

« **2.** A number with more than two factors is a(n) __?__.
composite number

Rewrite each statement using exponents.

3. $450 = 2 \cdot 3 \cdot 3 \cdot 5 \cdot 5$
$450 = 2 \cdot 3^2 \cdot 5^2$

4. $2000 = 2 \cdot 2 \cdot 5 \cdot 2 \cdot 5 \cdot 2 \cdot 5$
$2000 = 2^4 \cdot 5^3$

Find all the factors of each number.

5. 48 1, 2, 3, 4, 6, 8, 12, 16, 24, 48 **6.** 37 1, 37 **7.** 64 1, 2, 4, 8, 16, 32, 64 **8.** 100 1, 2, 4, 5, 10, 20, 25, 50, 100

Tell whether each number is *prime* or *composite*.

9. 2 prime **10.** 10 composite **11.** 28 composite **12.** 17 prime **13.** 51 composite **14.** 63 composite

Write the prime factorization of each number.

15. 49 7^2 **16.** 15 $3 \cdot 5$ **17.** 18 $2 \cdot 3^2$ **18.** 44 $2^2 \cdot 11$ **19.** 90 $2 \cdot 3^2 \cdot 5$ **20.** 144 $2^4 \cdot 3^2$

COMMUNICATION « *Discussion*

« **21.** List all the divisibility rules that you remember. See list in *Toolbox Skills Practice*, Skill 16.

« **22.** How do the divisibility rules help you find the prime factorization of 5448? The rules help to find the factors. For instance, 3 is a factor of 5448 because the digits add to 21 and 8 is a factor of 5448 because 448 is a multiple of 8.

Exercises

Tell whether each number is *prime* or *composite*.

A **1.** 27 composite **2.** 32 composite **3.** 19 prime **4.** 41 prime
5. 100 composite **6.** 61 prime **7.** 47 prime **8.** 96 composite

Write the prime factorization of each number. 20. $2^2 \cdot 3^2 \cdot 5^2$

9. 27 3^3 **10.** 32 2^5 **11.** 22 $2 \cdot 11$ **12.** 69 $3 \cdot 23$
13. 20 $2^2 \cdot 5$ **14.** 36 $2^2 \cdot 3^2$ **15.** 28 $2^2 \cdot 7$ **16.** 52 $2^2 \cdot 13$
17. 108 $2^2 \cdot 3^3$ **18.** 132 $2^2 \cdot 3 \cdot 11$ **19.** 500 $2^2 \cdot 5^3$ **20.** 900
21. 1008 $2^4 \cdot 3^2 \cdot 7$ **22.** 624 $2^4 \cdot 3 \cdot 13$ **23.** 2625 $3 \cdot 5^3 \cdot 7$ **24.** 2808 $2^3 \cdot 3^3 \cdot 13$

25. List all the different ways to write 60 as a product of two whole-number factors. (*Hint:* There are six ways.)
$1 \cdot 60, 2 \cdot 30, 3 \cdot 20, 4 \cdot 15, 5 \cdot 12, 6 \cdot 10$

26. What is the prime factorization of 1764? $2^2 \cdot 3^2 \cdot 7^2$

Number Theory and Fraction Concepts **291**

Exercise Notes

Communication: Discussion
Guided Practice Exercises 21 and 22 have students recall the divisibility rules and then apply them to finding the prime factorization of a number. Should students not remember any of the rules, refer them to Toolbox Skill 16 at the back of the book.

Critical Thinking
In *Exercises 27–32*, students have to *apply* their knowledge of factoring whole numbers to variable expressions and then *formulate* a rule for factoring such expressions.

Nonroutine Problem
Find all the numbers between 1 and 100 inclusive that have an odd number of factors. What do these numbers have in common?
1, 4, 9, 16, 25, 36, 49, 64, 81, 100. The numbers are all squares of other numbers.

Follow-Up

Enrichment
Enrichment 65 is pictured on page 288E.

Practice
Practice 80, *Resource Book*

THINKING SKILLS 27–29. Knowledge 30. Application
31. Synthesis 32. Application

B **Replace each __?__ with the expression that makes the statement true.**

27. $y^5 = 1 \cdot \underline{\ ?\ } \ y^5$

28. $y^5 = y \cdot \underline{\ ?\ } \ y^4$

29. $y^5 = y^2 \cdot \underline{\ ?\ } \ y^3$

30. Use your answers to Exercises 27–29 to list all the factors of y^5.
   $1, y, y^2, y^3, y^4, y^5$

31. Make a rule for finding all the factors of any given power a^n.
   The factors are $a^n, a^{n-1}, a^{n-2}, \ldots, a^1, 1$

32. Use the rule you made in Exercise 31 to find all the factors of x^8.
   $1, x, x^2, x^3, x^4, x^5, x^6, x^7, x^8$

GROUP ACTIVITY

C 33. Find all the prime numbers between 0 and 100. (*Hint:* There are 25 of these prime numbers.) See answer on p. A20.

34. **a.** Find all the prime numbers between 100 and 200. See answer on p. A20.
   b. How many prime numbers are there in part (a)? 21
   c. Are there *more than* or *fewer than* 25 prime numbers in part (a)?
   fewer

35. **a.** How many prime numbers do you think there are between 200 and 300? Explain. fewer than 21; the number of primes decreases.
   b. Find all the prime numbers between 200 and 300. See answer on p. A20.
   c. How many prime numbers are there in part (b)? 16
   d. How does your answer to part (c) compare with your answer to part (a)? It supports the answer to (a).

36. **RESEARCH** The *sieve of Eratosthenes*, a method for finding prime numbers, has been used for more than 2000 years. Find how the sieve is used and report your findings to your group. See answer on p. A20.

SPIRAL REVIEW

S 37. Find the mean, median, mode(s), and range: 84, 96, 72, 77, 91
   (*Lesson 5-8*) 84; 84; none; 24

38. Find the sum: $\frac{9}{16} + \frac{3}{4}$ (*Toolbox Skill 17*) $1\frac{5}{16}$

39. Write the prime factorization of 96. (*Lesson 7-1*) $2^5 \cdot 3$

40. Draw a bar graph for the data below. (*Lesson 5-4*)

   Cars Registered (Thousands) See answer on p. A20.

State	Maine	Utah	Alaska	Idaho	Nevada
Number of Cars	698	760	225	625	624

41. Find the answer: $-12 - (-12)$ (*Lesson 3-4*) 0

42. Find the next three expressions in the pattern: $20n + 5$, $18n + 5$, $16n + 5$, __?__, __?__, __?__ (*Lesson 2-5*)
   $14n + 5, 12n + 5, 10n + 5$

7-2

Greatest Common Factor

Objective: To find greatest common factors.

Terms to Know
- *common factor*
- *greatest common factor (GCF)*

John has 28 seedlings to plant and Henrietta has 40. They want to plant two rectangular gardens side by side using the same number of rows in each garden.

QUESTION What is the greatest number of rows they can have in their gardens?

This question can be answered using *common factors*. A number that is a factor of two numbers is called a **common factor** of those two numbers. The greatest number in a list of common factors is called the **greatest common factor (GCF)**. The greatest number of rows that John and Henrietta can have in their gardens will equal the GCF of 28 and 40.

Example 1

Find the GCF of 28 and 40.

Solution 1

The factors of 28 are 1, 2, 4, 7, 14, and 28.
The factors of 40 are 1, 2, 4, 5, 8, 10, 20, and 40.
The common factors of 28 and 40 are 1, 2, and 4.
The GCF of 28 and 40 is 4.

Solution 2

Write the prime factorization of each number.

$28 = 2^2 \cdot 7$
$40 = 2^3 \cdot 5$ The common prime factor is 2,
$GCF = 2^2 = 4$ ⟵ and the lesser power of 2 is 2^2.

 **Check Your Understanding**

1. In Solution 1 of Example 1, why was 2 not chosen as the GCF?
2. In Solution 2 of Example 1, why were 5 and 7 not used as factors in the GCF?

See *Answers to Check Your Understanding* at the back of the book.

ANSWER The greatest number of rows that John and Henrietta can have in their gardens is 4.

As shown in Example 1, you can use prime factorizations to find a GCF. To do this, find all the common prime factors. Then form a product using the least power that appears for each factor.

Warm-Up Exercises

Write the prime factorization of each number.

1. 30 $2 \cdot 3 \cdot 5$
2. 60 $2^2 \cdot 3 \cdot 5$
3. 100 $2^2 \cdot 5^2$
4. 54 $2 \cdot 3^3$
5. 132 $2^2 \cdot 3 \cdot 11$

Lesson Focus

Ask students if they remember how to write a fraction in lowest terms. They may have asked themselves a question such as ''What is the largest number that divides both the numerator and the denominator of the fraction?'' This question uses the concept of the greatest common factor. Although today's lesson does not deal with fractions, it does teach students how to find the greatest common factor for a set of numbers.

Teaching Notes

Use examples to show that factors of a number are always less than or equal to half the number. Emphasize the meaning of the two terms *common factor* and *greatest common factor*.

294

Teaching Notes (continued)

Key Questions

1. Is it possible for an odd number and an even number to have common factors? Yes; many even numbers have odd numbers as factors. For example, $10 = 2 \cdot 5$ and $15 = 3 \cdot 5$.

2. Can the GCF of two numbers ever be as large as the greater of the two numbers? No; the GCF is a factor of both numbers; thus, it cannot be greater than the smaller of the two numbers.

Error Analysis

When selecting common factors to find the GCF, students may erroneously choose the greatest power, rather than the least power. Remind students to choose the least power of the common factors.

Using Technology

For a suggestion on using a calculator and a computer to find the GCF, see page 288C.

Additional Examples

Find the GCF.

1. 27 and 45   9
2. $16x^5$ and $10x^7$ $2x^5$
3. $32bc$ and $80c$ $16c$

Closing the Lesson

You may wish to use the following cooperative learning activity to close the lesson.

Cooperative Learning

Use the cards that you made for Lesson 7-1. Show two cards to students working in groups. Each group should create a list of factors of the numbers, name the common factors, and find the GCF. If there are no common factors other than 1, the groups should state this.

Suggested Assignments

Skills: 1–24, 31–34
Average: 1–26, 31–34
Advanced: 1–19 odd, 20–34

To find the GCF of variable expressions, include in it the least power that appears for each common variable factor.

Example 2

Find the GCF.

a. $12a^4$ and $27a^6$

b. $24xy$ and $84x$

Solution

a. $12a^4 = 2^2 \cdot 3 \cdot a^4$
$27a^6 = 3^3 \cdot a^6$
$GCF = 3 \cdot a^4 = 3a^4$

b. $24xy = 2^3 \cdot 3 \cdot x \cdot y$
$84x = 2^2 \cdot 3 \cdot 7 \cdot x$
$GCF = 2^2 \cdot 3 \cdot x = 12x$

☑ **Check Your Understanding**

3. In Example 2(a), why was the power a^4 used instead of a^6?
4. In Example 2(b), why was y not used as a factor in the GCF?
See *Answers to Check Your Understanding* at the back of the book.

Guided Practice

«1. COMMUNICATION «*Reading*   Replace each __?__ with the correct phrase.

A __?__ of two numbers is a number that is a factor of the two numbers. The __?__ is the greatest number in a list of common factors.
common factor; GCF

Identify the common prime factor(s).

2. $2^3 \cdot 5$ 5
$3 \cdot 5^4$

3. $2 \cdot 3^6$ 3
$3 \cdot 7^2$

4. $3^2 \cdot 5 \cdot 7$
$3 \cdot 7^3$ 3, 7

5. $5 \cdot 7^2 \cdot 11$
$5^3 \cdot 11^7 \cdot 13$
5, 11

Identify the lesser power.

6. 2^3 and 2^8 2^3
7. 3^6 and 3^4 3^4
8. 11^7 and 11^3 11^3
9. 5 and 5^2 5

Find the GCF.

10. 24 and 56 8
11. 14 and 75 1
12. 45, 60, and 80 5
13. $36m^3$ and $45m^8$ $9m^3$
14. $60r$ and 72 12
15. $20p^7$, $25p$, and $42p^4$ p

Exercises

Find the GCF.

A
1. 2 and 16 2
2. 7 and 14 7
3. 8 and 15 1
4. 54 and 99 9
5. 40 and 100 20
6. 50 and 81 1
7. 16, 24, and 72 8
8. 45, 72, and 108 9
9. $90x$ and $96x$ $6x$

Find the GCF.

10. $48a$ and $51a$ _3a_   **11.** $27bc$ and $49bd$ _b_   **12.** $75mn$ and $90m$ _15m_

13. $12y$, 42, and $44y$ _2_   **14.** 15, $40k$, and 56 _1_

15. $32r^{12}$ and $36r^8$ _4r⁸_   **16.** $18y^5$ and $30y^2$ _6y²_

17. $16n^3$, $28n^2$, and $32n^5$ _4n²_   **18.** $20q^4$, $35q^5$, and $70q^{11}$ _5q⁴_

19. A 48-member band will be marching behind a 54-member band. They must both march in the same number of columns. What is the greatest number of columns in which they can march? _6_

20. Mr. Liu's gym class has 24 students and Mr. Standish's gym class has 28 students. Each class is divided into teams. Teams from Mr. Liu's class play against teams from Mr. Standish's class. Each team must have the same number of students. What is the greatest number of students that each team can have? _4_

THINKING SKILLS 21–24. Application   25–26. Synthesis

Two whole numbers are said to be *relatively prime* if their only common factor is 1. Whole numbers do not have to be prime to be relatively prime.

Tell whether each pair of numbers is relatively prime. Write *Yes* or *No*.

B **21.** 25 and 27 _Yes._ **22.** 12 and 55 _Yes._ **23.** 18 and 21 _No._ **24.** 40 and 99 _Yes._

25. Generate three pairs of composite numbers between 50 and 100 that are relatively prime. _Answers may vary. Examples: 51, 70; 52, 55; 81, 85_

26. Generate three pairs of composite numbers between 100 and 150 that are relatively prime. _Answers may vary. Examples: 108, 121; 143, 144; 110, 119_

LOGICAL REASONING

Classify each statement as *True* or *False*. If the statement is false, give a *counterexample* that demonstrates why it is false.

C **27.** The GCF of an odd and an even number is always an odd number. _True._

28. Two even numbers are never relatively prime. _True._

29. The GCF of a prime number and an odd number is always odd. _True._

30. The GCF of a prime number and an even number is always odd. _False; the GCF of 2 and any other even number is 2._

SPIRAL REVIEW

S **31.** Solve: $5x - 13 = -3$ *(Lesson 4-5)* _2_

32. An angle measures 37°. Find the measure of an angle that is complementary to it. *(Lesson 6-3)* _53°_

33. Estimate the sum: $2.8 + 3.4 + 2.9 + 3.2$ *(Toolbox Skill 4)* _about 12_

34. Find the GCF: $4x^3$ and $22x$ *(Lesson 7-2)* _2x_

Number Theory and Fraction Concepts   **295**

Exercise Notes

Problem Solving

Exercises 19 and 20 can be solved by using diagrams similar to the one in the Exploration of Lesson 7-1. Students can look at various rectangular arrangements until they find the ones with the greatest common width, which represents the GCF.

Reasoning

Exercises 27–30 reinforce the idea that one counterexample is all that is needed to prove a statement false.

Follow-Up

Extension

Have each student choose a composite number and find its prime factorization. Students should then (1) identify three primes that are not factors of the composite numbers they chose, and (2) use these primes to find at least five other composite numbers that are relatively prime to their original numbers.

Enrichment

Enrichment 66 is pictured on page 288E.

Practice

Practice 81, *Resource Book*

Practice 81
Skills and Applications of Lesson 7-2

Find the GCF. 28–39. See below.

1. 4 and 10 2	2. 12 and 18 6	3. 15 and 21 3
4. 14 and 15 1	5. 20 and 30 10	6. 45 and 75 15
7. 8 and 50 2	8. 75 and 90 15	9. 27 and 36 9
10. 35 and 49 7	11. 39 and 52 13	12. 44 and 66 22
13. 18 and 60 6	14. 21 and 28 7	15. 90 and 150 30
16. 180 and 270 90	17. 168 and 441 21	18. 8, 14, and 30 2
19. 27 and 63 9	20. 30 and 70 10	21. 24 and 56 8
22. 42, 70, and 147 7	23. 44, 66, and 77 11	24. 84, 126, and 252 42
25. $14x^2$ and $21x$ 7x	26. $35a^3$ and $42a^6$ 7a³	27. $55b^6$ and $88a^2$ 11a²
28. $30n^2$, $40n^5$, and $50n$	29. $18b^3$ and $27b^2$	30. $40m^2$ and $56m^7$
31. $24p^5$ and $56p^3$	32. $27a^3$ and $50b^3$	33. $25w^3$ and $36w^5$
34. $112x^4$ and $126x^6$	35. $55a^2$ and $99u^2$	36. $35w^3$ and $44y^2$
37. $105m^4$, $135m^2$, and $165m^5$	38. $40a^3$, $120p^3$, and $140x$	39. $112d^3$ and $120d$

40. The ninth grade class has 153 students and the tenth grade class has 180 students. The principal, Mrs. Fernandez, plans to seat the ninth grade class on the left side of the auditorium and the tenth grade on the right side. If each class sits in a rectangular pattern, what is the greatest number of rows in which they can be seated? 9

41. Three schools agree to a debate competition. Students from each school are divided into teams of equal size. What is the greatest number of students that each team can have if Midland has 150 students, Copeland has 162 students, and Abbott has 168 students? 6

28. $10n$	29. $9b^2$	30. $8m^2$
31. $8p^3$	32. 1	33. w^3
34. $14x^4$	35. $11a^2$	36. 1
37. $15m^2$	38. 20	39. $8d$

Teaching the Lesson
- Materials: calculator
- Lesson Starter 61
- Using Technology, p. 288C
- Toolbox Skills 9, 16

Lesson Follow-Up
- Practice 82
- Practice 83: Nonroutine
- Enrichment 67: Application
- Study Guide, pp. 121–122
- Cooperative Activity 14
- Test 31

Warm-Up Exercises

Write the prime factorization of each number.

1. 135 $3^3 \cdot 5$
2. 64 2^6
3. 72 $2^3 \cdot 3^2$

Find the GCF.

4. 32, 56   8
5. 68, 34   34

Lesson Focus

Ask students to imagine a square and an equilateral triangle with the letter K appearing within both in an upright position. Suppose the square and triangle are turned to the right so they are standing on the next side. How many turns will each figure have to make before the K again appears in an upright position on both at the same time? This problem can be solved by using the concept of least common multiple, the focus of today's lesson.

Teaching Notes

Work through the lesson with the students, emphasizing that to find the LCM they should use the *greatest* power of each factor. Stress the difference between the two terms *common multiple* and *least common multiple*.

7-3 Least Common Multiple

Objective: To find least common multiples.

APPLICATION

Terms to Know
- *multiple*
- *common multiple*
- *least common multiple (LCM)*

Two nurses who work at the same hospital get together with their families every time they have the same weekend off. One of the nurses has every fourth weekend off. The other has every sixth weekend off.

QUESTION How often do the nurses and their families get together?

This question can be answered by finding the *least common multiple* of 4 and 6. When a number is multiplied by a nonzero whole number, the product is a **multiple** of the given number. A **common multiple** of two numbers is any number that is a multiple of both numbers. The **least common multiple (LCM)** of two numbers is the least number in the list of their common multiples.

Example 1   Find the LCM of 4 and 6.

Solution 1   the multiples of 4:   4, 8, 12, 16, 20, 24, 28, 32, 36, . . .
the multiples of 6:   6, 12, 18, 24, 30, 36, . . .
the common multiples of 4 and 6:   12, 24, 36, . . .
The LCM of 4 and 6 is 12.

Solution 2   Write the prime factorization of each number.
$4 = 2 \cdot 2 = 2^2$
$6 = 2 \cdot 3$
The prime factors are 2 and 3.

Form a product using the greatest power of each factor.
Greatest power of 2: 2^2
Greatest power of 3: 3
$LCM = 2^2 \cdot 3 = 12$

✓ **Check Your Understanding**

1. In Solution 1 of Example 1, are all the multiples of 4 and 6 listed? Explain.
See *Answers to Check Your Understanding* at the back of the book.

ANSWER The nurses and their families get together every 12 weeks.

As shown in Example 1, you can use prime factorizations to find an LCM. To do this, find all the prime factors that appear. Then multiply using the greatest power of each prime factor.

When you have to find the LCM of variable expressions, include in it the greatest power of each variable factor.

Example 2

Find the LCM.

a. $15a^2$ and $25a^4$

b. $4rs$ and $8s$

Solution

a. $15a^2 = 3 \cdot 5 \cdot a^2$
$25a^4 = 5^2 \cdot a^4$
$\text{LCM} = 3 \cdot 5^2 \cdot a^4 = 75a^4$

b. $4rs = 2^2 \cdot r \cdot s$
$8s = 2^3 \cdot s$
$\text{LCM} = 2^3 \cdot r \cdot s = 8rs$

☑ **Check Your Understanding**

2. In Example 2(a), why was the power 5^2 used instead of 5?

3. In Example 2(b), why was r included as a factor in the LCM?
See *Answers to Check Your Understanding* at the back of the book.

Guided Practice

COMMUNICATION «*Reading*

«**1.** Explain the meaning of the word *multiple* in the following sentence.

There were multiple-choice questions on the test. Questions having more than one suggested answer from which to choose were on the test.

«**2.** Write a sentence using the word *multiple* in a mathematical context.
Answers will vary. Example: A multiple of 3 is 6.

List four multiples of each number. Answers may vary.

3. 7
14, 21, 28, 35

4. 11
22, 33, 44, 55

5. 8
16, 24, 32, 40

6. 15
30, 45, 60, 75

Identify all the prime factors that appear.

7. $3^2 \cdot 5$
$2 \cdot 5^7$ 2, 3, 5

8. $5^2 \cdot 7$
$3 \cdot 7^2$ 3, 5, 7

9. $2 \cdot 7^5 \cdot 11$
$7 \cdot 11^3 \cdot 13$
2, 7, 11, 13

10. $2^2 \cdot 3^5 \cdot 11$
$5 \cdot 7 \cdot 13^2$
2, 3, 5, 7, 11, 13

Identify the greater power.

11. $3^5; 3^2$ 3^5

12. $5^4; 5^3$ 5^4

13. $2^9; 2$ 2^9

14. $13^8; 13^{12}$
13^{12}

Replace each ? with the correct number or expression.

15. $24 = 2^? \cdot$? 3; 3
$50 =$? $\cdot 5^2$ 2
$\text{LCM} = 2^? \cdot$? $\cdot 5^2$ 3; 3
$=$? 600

16. $9xy^3 = 3^? \cdot x \cdot$? 2; y^3
$6x^2 = 2 \cdot$? $\cdot x^2$ 3
$\text{LCM} = 2 \cdot 3^? \cdot x^? \cdot$?
$=$? 2; 2; y^3; $18x^2y^3$

Find the LCM.

17. 3 and 5 15

18. 4, 6, and 15 60

19. $6n$ and 9 $18n$

20. $12x^9$ and $16x^7$ $48x^9$

21. $6xy$ and $39yz$ $78xyz$

22. $5y$, $12y^2$, and y^{10}
$60y^{10}$

Number Theory and Fraction Concepts **297**

Exercise Notes

Communication: Reading
Guided Practice Exercises 1 and 2 contrast the mathematical meaning of *multiple* with its common usage.

Making Connections/Transitions
Exercises 19 and 20 connect mathematics with real world applications. *Exercise 20* points out how the packaging of products often influences purchasing.

Computer
Computer programs may be used to simplify arithmetic computation. Computers can perform operations quickly and then test to see if the necessary conditions are met. *Exercises 21–26* may be solved very quickly by using a computer program.

Communication: Writing
Exercises 27 and 28 ask students to compare and to contrast the concepts in Lessons 7-2 and 7-3. Creating an outline requires students to combine and refine their ideas.

Critical Thinking
Exercises 29–34 require students to *synthesize* concepts from Lessons 7-2 and 7-3 and then to *decide* whether the GCF of two different numbers can ever equal the LCM. These exercises can help students to clarify and understand the concepts involved.

Nonroutine Problem
Greg has some marbles. When he sorted them by 2's, 3's, 4's, and 5's, he had one left over each time. If Greg has fewer than 100 marbles, how many does he have?
61 marbles

Exercises

Find the LCM.

A

1. 2 and 5 10
2. 1 and 9 9
3. 9 and 27 27

4. 6 and 12 12
5. 8 and 10 40
6. 20 and 50 100

7. 4, 5, and 21 420
8. 9, 12, and 15 180
9. 2 and $3x$ $6x$

10. $7r$ and s $7rs$
11. $6k$ and $4k$ $12k$
12. $5x$ and $3x$ $15x$

13. $6rs$ and $15st$ $30rst$
14. $8bz$ and $12bz$ $24bz$
15. $14a^9$ and $21a^5$ $42a^9$

16. $16h^7$ and $24h^4$ $48h^7$
17. $6n$, $15n^4$, and 75 $150n^4$
18. $16m^{10}$, $18m$, and $30m^3$ $720m^{10}$

19. Tracey Donovan can arrange her science class into lab groups of six or eight with no one left out. What is the least number of students Tracey Donovan can have in her science class? 24

20. Paper plates come in packages of 30, paper cups come in packages of 15, and paper napkins come in packages of 20. What is the least number of plates, cups, and napkins that Martha can buy to get an equal number of each? 60

 COMPUTER APPLICATION

Sometimes you can use a BASIC computer program to find an answer quickly. Using a computer program can save you time when you are working with greater numbers. The program below computes the LCM of any two whole numbers.

```
10 PRINT "TO FIND LCM,"
20 PRINT "ENTER TWO NUMBERS."
30 INPUT A, B
40 FOR X = 1 TO B
50 LET A1 = A*X
60 LET Q = A1/B
70 IF Q = INT(Q) THEN 90
80 NEXT X
90 PRINT "LCM OF ";A;" AND ";B
100 PRINT "IS ";A1
110 END
```

Find the LCM. Use the BASIC program above if a computer is available.

B

21. 54 and 72 216
22. 56 and 70 280

23. 45 and 224 10,080
24. 98 and 180 8820

25. 144 and 256 2304
26. 168 and 196 1176

WRITING ABOUT MATHEMATICS Answers to Exercises 27–28 are on p. A20.

27. Describe how you would explain to a friend the difference between a factor and a multiple.

28. For your notebook, outline the steps for finding a GCF and for finding an LCM. Explain any differences.

THINKING SKILLS 29–33. Synthesis 34. Evaluation

Exercises 29–34 require you to combine what you know about the GCF and the LCM.

Find both the GCF and the LCM.

29. 30 and 45 15; 90

30. 63 and 84 21; 252

31. 88 and 132 44; 264

32. 12, 64, and 72 4; 576

C 33. Find two numbers greater than 1 whose GCF is the lesser number and whose LCM is the greater number. Answers may vary.
Example: 4 and 12

34. Is it possible for the GCF of two different numbers to equal the LCM of the numbers? Give a convincing argument to support your answer. See answer on pp. A20–A21.

SPIRAL REVIEW

S 35. Find the product: $5\frac{3}{8} \times 2\frac{2}{3}$ *(Toolbox Skill 20)* $14\frac{1}{3}$

36. Find the LCM: $10mn$ and $12m$ *(Lesson 7-3)* $60mn$

37. Write a variable expression that represents this phrase: seven more than a number y *(Lesson 4-6)* $y + 7$

38. Evaluate $|a - b|$ when $a = -2$ and $b = 5$. *(Lesson 3-6)* 7

Self-Test 1

Tell whether each number is *prime* or *composite*.

1. 23 prime   2. 29 prime   3. 33 composite   4. 37 prime   7-1

Write the prime factorization of each number.

5. 14 $2 \cdot 7$ 6. 12 $2^2 \cdot 3$ 7. 105 $3 \cdot 5 \cdot 7$ 8. 400 $2^4 \cdot 5^2$

Find the GCF.

9. 10 and 22 2   10. 18 and 45 9   7-2
11. $21bc$ and $66c$ $3c$ 12. $24x^5$ and $40x^2$ $8x^2$

Find the LCM.

13. 5 and 15 15   14. 16 and 20 80   7-3
15. $6aw$ and $8ay$ $24awy$ 16. $18n^7$ and $42n$ $126n^7$

Number Theory and Fraction Concepts **299**

Extension (Cooperative Learning)
Divide students into cooperative groups. Ask each group to create a problem that can be solved by finding either the LCM or the GCF. Have each group pass its problem to another group to solve. Have two groups join together and choose one of their problems to share with the class.

Enrichment
Enrichment 67 is pictured on page 288E.

Practice
Practice 82 is shown below.

Quick Quiz 1
See page 333.

Alternative Assessment
See page 333.

Practice 82, *Resource Book*

299

Teaching Notes

Some students may have used fraction bars in previous courses. Fraction bars provide a concrete method of visualizing parts of a whole. The manipulative activities on these pages will serve as a good introduction or review of fraction bars.

Key Questions

1. Does the total number of parts in a whole bar represent the numerator or the denominator?
   denominator
2. What does the shaded part of a fraction bar represent?
   numerator

Activity Notes

Activity I

Activity I serves as an introduction to fraction bars. If students are familiar with fraction bars, you may wish to begin with *Activity II*.

Activities II and III

Activity II on equivalent fractions and *Activity III* on comparing fractions can be used to have students explore these two concepts before their formal introduction in Lessons 7-4 and 7-6. The activities can also be used to reinforce the concepts when they are presented in the lessons.

Activity IV

Activity IV visually demonstrates the concept of common denominator that students must know if they are to add and subtract rational numbers in Chapter 8.

Objective: To use fraction bars to model fractions.

Materials

■ fraction bars:
halves, thirds, fourths, sixths, twelfths

Modeling Fractions Using Fraction Bars

You can use fraction bars to help you visualize fractional parts of a whole. For example, let the entire fraction bar 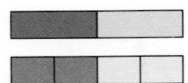 represent the whole. The diagram below then represents the fraction $\frac{2}{3}$.

$$\frac{\text{number of parts shaded}}{\text{total number of parts}} = \frac{2}{3}$$

Activity I *Representing Fractions*

1 Write the fraction that is represented by each diagram.

a. $\frac{1}{2}$ **b.** $\frac{2}{4}$

c. $\frac{3}{6}$ **d.** $\frac{11}{12}$

2 Show how to use a fraction bar to represent each fraction.

a. $\frac{1}{4}$ **b.** $\frac{1}{3}$ **c.** $\frac{3}{4}$ **d.** $\frac{5}{6}$

Answers to Activity I, Step 2, are on p. A21.

Activity II *Equivalent Fractions*

1 Describe the situation that you think is being represented by the fraction bars below. $\frac{1}{2} = \frac{2}{4}$

2 Show how to use fraction bars to illustrate each statement.

a. $\frac{1}{2} = \frac{3}{6}$ **b.** $\frac{4}{12} = \frac{1}{3}$

c. $\frac{3}{4} = \frac{9}{12}$ **d.** $\frac{2}{12} = \frac{1}{6}$

3 Show two different fraction bars that represent the same amount as the given bar.

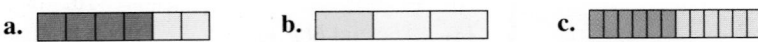

a.   **b.**   **c.**

4 Use your results from Step 3. Complete each statement.

a. $\frac{4}{6} = \underline{\ ?\ } = \underline{\ ?\ }$ **b.** $\frac{1}{3} = \underline{\ ?\ } = \underline{\ ?\ }$ **c.** $\frac{6}{12} = \underline{\ ?\ } = \underline{\ ?\ }$

Answers to Activity II, Steps 2–4, are on p. A21.

Activity III *Comparing Fractions*

1 Which statement do you think the diagram at the right represents?

a. $\frac{3}{3} < \frac{4}{4}$ **(b.)** $\frac{3}{4} > \frac{4}{6}$

c. $\frac{3}{4} < \frac{4}{6}$ **d.** $\frac{3}{4} = \frac{4}{6}$

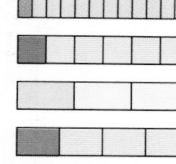

2 Show how to use fraction bars to compare each pair of fractions.
Should the __?__ be replaced by $>$, $<$, or $=$?

a. $\frac{5}{6}$ __?__ $\frac{7}{12}$ $>$

b. $\frac{2}{3}$ __?__ $\frac{3}{4}$ $<$

c. $\frac{1}{4}$ __?__ $\frac{1}{3}$ $<$

d. $\frac{3}{4}$ __?__ $\frac{5}{12}$ $>$

3 Each diagram at the right represents a *unit fraction*.

a. Describe what you think is meant by the term *unit fraction*.
Answers may vary; for example, fraction whose numerator is 1.

b. Write the fractions that the diagrams represent in order from
least to greatest. What pattern do you see? $\frac{1}{12}, \frac{1}{6}, \frac{1}{4}, \frac{1}{3}, \frac{1}{2}$;
the denominators decrease.

c. Do you think $\frac{1}{50}$ is *greater than* or *less than* $\frac{1}{60}$? Explain.
greater than; because 50 < 60

Answers to Activity IV are on p. A21.

Activity IV *Finding a Common Denominator*

1 Compare diagrams A and B, shown below.

a. How are the diagrams alike?

b. How are the diagrams different?

             A B

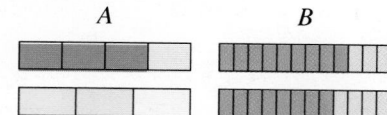

2 For each diagram, show an equivalent pair of fraction bars. The
bars in the new pair should each have the same number of parts.

a.

b.

c.

d.

e.

f.

Cooperative Learning
You may wish to have students
make fraction bars from paper and
work the activities in cooperative
groups; in so doing, they can discuss their solutions with other members of the group.

Teaching the Lesson
• Materials: fraction bars
• Lesson Starter 62
• Visuals, Folder C

Lesson Follow-Up
• Practice 84
• Enrichment 68: Connection
• Study Guide, pp. 123–124
• Manipulative Activity 13

Warm-Up Exercises

Find the GCF.

1. 15, 24   3
2. 32, 48   16
3. 140, 84   28

Find the LCM.

4. 24, 36   72
5. 45, 27   135

Lesson Focus

To motivate today's lesson on equivalent fractions, you may wish to use the application below.

Application

A carpenter is building a set of stairs. Each stair has to be $36\frac{5}{8}$ in. wide. The carpenter's ruler shows inches divided into sixteenths. What must the carpenter do to make a measurement of $36\frac{5}{8}$ in. with the ruler?   The carpenter needs to determine how many sixteenths are equivalent to $\frac{5}{8}$.

Equivalent Fractions

7-4

Objective: To find equivalent fractions and to write fractions in lowest terms.

Terms to Know
• *equivalent fractions*
• *lowest terms*

Ali has a ruler that shows inches divided into fourths. While working on her art project, she draws a line that is three fourths of an inch long. The next day, she borrows Jared's ruler that shows inches divided into eighths. She finds that three fourths of an inch is the same as six eighths of an inch.

Fractions that represent the same amount are called **equivalent fractions.** So $\frac{3}{4}$ and $\frac{6}{8}$ are equivalent fractions.

Example 1

Replace the _?_ with the number that will make the fractions equivalent.

a. $\frac{3}{4} = \frac{?}{12}$

b. $\frac{6}{12} = \frac{?}{6}$

Solution

a.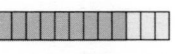

To change 4 parts to 12 parts, multiply by 3.

$\frac{3}{4} = \frac{3 \cdot 3}{4 \cdot 3}$

$= \frac{9}{12}$

b.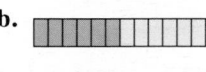

To change 12 parts to 6 parts, divide by 2.

$\frac{6}{12} = \frac{6 \div 2}{12 \div 2}$

$= \frac{3}{6}$

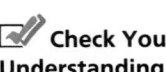 **Check Your Understanding**

1. How are the exercises in Example 1(a) and 1(b) alike? different?

2. In what way are the calculations in Example 1(b) similar to the calculations in Example 1(a)?
See *Answers to Check Your Understanding* at the back of the book.

Generalization: *Finding Equivalent Fractions*

To find a fraction that is equivalent to a given fraction, multiply or divide both the numerator and the denominator of the given fraction by the same nonzero number.

$$\frac{a}{b} = \frac{a \cdot c}{b \cdot c} \qquad \frac{a}{b} = \frac{a \div c}{b \div c}, \quad b \neq 0, c \neq 0$$

A fraction is in **lowest terms** if the GCF of the numerator and the denominator is 1. You can write a fraction in lowest terms by using either the GCF or prime factorization.

Example 2   Write $\frac{12}{18}$ in lowest terms.

Solution 1   Divide both the numerator and the denominator by the GCF, which is 6.

$$\frac{12}{18} = \frac{12 \div 6}{18 \div 6} = \frac{2}{3}$$
⟵ You may find it easier to use this shortcut: $\frac{\overset{2}{\cancel{12}}}{\underset{3}{\cancel{18}}} = \frac{2}{3}$

Solution 2

$$\frac{12}{18} = \frac{\overset{1}{\cancel{2}} \cdot 2 \cdot \overset{1}{\cancel{3}}}{\underset{1}{\cancel{2}} \cdot \underset{1}{\cancel{3}} \cdot 3} = \frac{2}{3}$$
⟵ Write the prime factorization of the numerator and denominator and divide by all common factors.

Guided Practice

COMMUNICATION «*Reading*

«**1.** In a dictionary, find the word *equivalent*. Does it mean the same as *equal*? Yes

«**2.** Explain the meaning of *equivalent* in the following sentence.
By working overtime, Mary earned the equivalent of two days' pay.
Mary earned the same as two days' pay.

COMMUNICATION «*Writing*

5. $\frac{1}{6}, \frac{2}{12}$

Write the pair of equivalent fractions represented by each diagram.

«**3.** 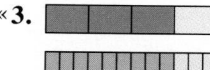 $\frac{3}{4}, \frac{9}{12}$

«**4.** $\frac{1}{3}, \frac{2}{6}$

«**5.**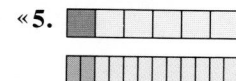

Draw a diagram similar to those in Exercises 3–5 to represent each pair of equivalent fractions. Answers to Guided Practice 6–8 are on p. A21.

«**6.** $\frac{2}{4}, \frac{1}{2}$ «**7.** $\frac{2}{3}, \frac{4}{6}$ «**8.** $\frac{1}{4}, \frac{3}{12}$

Number Theory and Fraction Concepts   **303**

Suggested Assignments

Skills: 1–22, 31–37, 39–42
Average: 1–42
Advanced: 1–19 odd, 21–42,
Challenge

Exercise Notes

Communication: Reading

Guided Practice Exercises 1 and 2 help students understand the meaning of the word *equivalent.*

Communication: Writing

Guided Practice Exercises 3–8 relate the pictorial representation of equivalent fractions to the symbolic representation.

Critical Thinking

Exercises 23–26 require students to *apply* their knowledge of writing fractions in lowest terms to fractions in which the numerator is greater than the denominator. *Exercises 27 and 28* require students to *analyze* their answers and *create* a set of fractions that meet a given condition. *Exercises 29 and 30* ask students to *determine* if a rule is true and then to use the rule to *create* a similar rule that relates to fractions equivalent to 1.

Making Connections/Transitions

Exercises 31–38 connect arithmetic and geometry and integrate the concept of symmetry introduced in Lesson 6-9. The *Challenge* problem connects the concepts of divisibility with the notion of least common multiple.

Is each fraction written in lowest terms? Write *Yes* or *No*.

9. $\frac{10}{19}$ Yes
10. $\frac{14}{25}$ Yes
11. $\frac{9}{27}$ No
12. $\frac{5}{45}$ No
13. $\frac{28}{4}$ No
14. $\frac{30}{17}$ Yes

Replace each ___?___ with the number that will make the fractions equivalent.

15. $\frac{4}{5} = \frac{?}{25}$ 20
16. $\frac{1}{8} = \frac{?}{64}$ 8
17. $\frac{10}{?} = \frac{20}{18}$ 9
18. $\frac{5}{?} = \frac{35}{14}$ 2

Write each fraction in lowest terms.

19. $\frac{8}{14}$ $\frac{4}{7}$
20. $\frac{2}{12}$ $\frac{1}{6}$
21. $\frac{25}{5}$ $\frac{5}{1}$
22. $\frac{150}{60}$ $\frac{5}{2}$

Exercises

Replace each ___?___ with the number that will make the fractions equivalent.

A
1. $\frac{1}{5} = \frac{?}{10}$ 2
2. $\frac{2}{7} = \frac{?}{21}$ 6
3. $\frac{22}{?} = \frac{44}{12}$ 6
4. $\frac{5}{?} = \frac{25}{75}$ 15

5. $\frac{15}{4} = \frac{75}{?}$ 20
6. $\frac{60}{21} = \frac{20}{?}$ 7
7. $\frac{?}{6} = \frac{12}{36}$ 2
8. $\frac{?}{84} = \frac{17}{21}$ 68

Write each fraction in lowest terms.

9. $\frac{3}{9}$ $\frac{1}{3}$
10. $\frac{4}{8}$ $\frac{1}{2}$
11. $\frac{6}{10}$ $\frac{3}{5}$
12. $\frac{15}{24}$ $\frac{5}{8}$

13. $\frac{36}{27}$ $\frac{4}{3}$
14. $\frac{35}{10}$ $\frac{7}{2}$
15. $\frac{16}{18}$ $\frac{8}{9}$
16. $\frac{36}{54}$ $\frac{2}{3}$

17. $\frac{72}{48}$ $\frac{3}{2}$
18. $\frac{100}{86}$ $\frac{50}{43}$
19. $\frac{284}{568}$ $\frac{1}{2}$
20. $\frac{750}{850}$ $\frac{15}{17}$

21. The team won four tenths of its games. Write the fraction in lowest terms. $\frac{2}{5}$

22. Find a fraction that has 16 as its numerator and is equivalent to four fifths. $\frac{16}{20}$

THINKING SKILLS 23–26. Application **27.** Analysis **28.** Synthesis **29–30.** Evaluation

Write each fraction in lowest terms.

B
23. $\frac{18}{3}$ $\frac{6}{1}$
24. $\frac{36}{12}$ $\frac{3}{1}$
25. $\frac{60}{15}$ $\frac{4}{1}$
26. $\frac{45}{9}$ $\frac{5}{1}$

27. Compare your answers to Exercises 23–26. What do they have in common? The denominators are 1; the fractions represent whole numbers.

28. Make a list of ten different fractions that are equivalent to 4.

29. Determine whether this rule is *always*, *sometimes*, or *never* true: *If the numerator of a fraction is 0 and the denominator is not 0, then the fraction is equivalent to 0.* always

30. Make up a rule similar to the one in Exercise 29 to help you determine if a fraction is equivalent to 1. See answer on p. A21.

28. Answers will vary. Examples: $\frac{8}{2}, \frac{12}{3}, \frac{16}{4}$

CONNECTING ARITHMETIC AND GEOMETRY

31. The equilateral triangle at the right is divided into equal parts by all its lines of symmetry. If one part is shaded, what fraction of the triangle is shaded? $\frac{1}{6}$

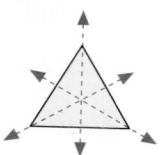

32. Suppose you divide a square into equal parts by drawing all its lines of symmetry. How many parts are there? If one part is shaded, what fraction of the square is shaded? $8; \frac{1}{8}$

33. Using a regular polygon and all its lines of symmetry, model $\frac{1}{12}$.
See answer on p. A21.

The circles below are divided into equal parts by lines of symmetry. Name the fraction represented by each shaded part.

34. 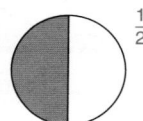 $\frac{1}{2}$

35. 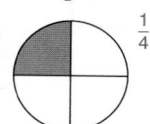 $\frac{1}{4}$

36. 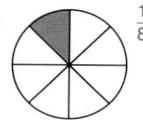 $\frac{1}{8}$

37. 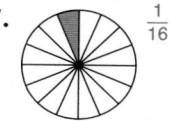 $\frac{1}{16}$

38. Show how you could use the circles in Exercises 34–37 to model the fact $\frac{3}{4} = \frac{12}{16}$. In Exercise 35 shade 3 parts; in Exercise 37 shade 12 parts.

SPIRAL REVIEW

S **39.** Write $\frac{24}{150}$ in lowest terms.   *(Lesson 7-4)* $\frac{4}{25}$

40. Estimate the quotient: $2502 \div 8$ *(Toolbox Skill 3)* about 300

41. Name the figure: $\overset{\bullet\quad\quad\bullet\longrightarrow}{M\quad N}$ *(Lesson 6-1)* $\overrightarrow{MN}$

42. Write an equation to represent this sentence: Eight less than a number x is 17.   *(Lesson 4-7)* $x - 8 = 17$

Challenge

1. Find a number for which all three statements below are true. 59

   a. When the number is divided by 3, the remainder is 2.
   b. When the number is divided by 4, the remainder is 3.
   c. When the number is divided by 5, the remainder is 4.

2. Find a number for which all three statements below are true. 53

   a. When the number is divided by 4, the remainder is 1.
   b. When the number is divided by 5, the remainder is 3.
   c. When the number is divided by 6, the remainder is 5.

Number Theory and Fraction Concepts   **305**

Follow-Up

Extension
As an extension of this lesson, as well as a preview of the next lesson, give students a list of algebraic fractions that do not involve exponents and that are not in lowest terms. Ask students to write each fraction in lowest terms.

Enrichment
Enrichment 68 is pictured on page 288E.

Practice
Practice 84 is shown below.

Practice 84, *Resource Book*

Practice 84
Skills and Applications of Lesson 7-4

Replace each ? with the number that will make the fractions equivalent.

1. $\frac{1}{6} = \frac{?}{18}$ 3
2. $\frac{1}{9} = \frac{?}{36}$ 4
3. $\frac{1}{5} = \frac{3}{?}$ 15
4. $\frac{1}{2} = \frac{?}{16}$ 8
5. $\frac{3}{7} = \frac{?}{21}$ 9
6. $\frac{2}{9} = \frac{4}{?}$ 18
7. $\frac{7}{4} = \frac{15}{20}$ 3
8. $\frac{12}{7} = \frac{?}{28}$ 48
9. $\frac{36}{?} = \frac{9}{4}$ 16
10. $\frac{42}{28} = \frac{6}{?}$ 4
11. $\frac{56}{64} = \frac{7}{8}$ 7
12. $\frac{7}{92} = \frac{32}{23}$ 128
13. $\frac{5}{13} = \frac{?}{52}$ 20
14. $\frac{7}{20} = \frac{9}{2}$ 90
15. $\frac{9}{?} = \frac{108}{84}$ 7

Write each fraction in lowest terms.

16. $\frac{5}{20}$ $\frac{1}{4}$
17. $\frac{14}{21}$ $\frac{2}{3}$
18. $\frac{9}{36}$ $\frac{1}{4}$
19. $\frac{36}{45}$ $\frac{4}{5}$
20. $\frac{63}{27}$ $\frac{7}{3}$
21. $\frac{32}{72}$ $\frac{4}{9}$
22. $\frac{35}{85}$ $\frac{7}{17}$
23. $\frac{180}{30}$ $\frac{6}{1}$
24. $\frac{186}{372}$ $\frac{1}{2}$
25. $\frac{225}{150}$ $\frac{3}{2}$
26. $\frac{225}{375}$ $\frac{3}{5}$
27. $\frac{330}{180}$ $\frac{11}{6}$
28. $\frac{396}{594}$ $\frac{2}{3}$
29. $\frac{315}{294}$ $\frac{15}{14}$
30. $\frac{247}{330}$ $\frac{247}{330}$
31. $\frac{693}{210}$ $\frac{33}{10}$
32. $\frac{385}{462}$ $\frac{5}{6}$
33. $\frac{360}{1350}$ $\frac{4}{15}$

34. Females make up six tenths of a committee. Write the fraction in lowest terms. $\frac{3}{5}$

35. The varsity football team is comprised of 44 members. Seven elevenths of the members are seniors. How many seniors are on the varsity team? 28

36. The fraction $\frac{a}{b}$ is equivalent to $\frac{4}{9}$. If a has the value of 20, find the value of b. 45

Teaching the Lesson
• Lesson Starter 63

Lesson Follow-Up
• Practice 85
• Enrichment 69:
  Communication
• Study Guide, pp. 125–126

Warm-Up Exercises

Write each fraction in lowest terms.

1. $\dfrac{15}{18}$ $\dfrac{5}{6}$

2. $\dfrac{16}{24}$ $\dfrac{2}{3}$

3. $\dfrac{20}{25}$ $\dfrac{4}{5}$

Replace each _?_ with the number that will make the given fractions equivalent.

4. $\dfrac{2}{7} = \dfrac{?}{28}$ 8

5. $\dfrac{9}{14} = \dfrac{?}{42}$ 27

Lesson Focus

Remind students of their earlier work in Chapter 2 with simplifying variable expressions. Today's lesson extends the concepts of Chapter 2 to include algebraic fractions.

Teaching Notes

In working through the lesson with students, point out the similarities between finding equivalent arithmetic fractions and finding equivalent algebraic fractions. Be certain students understand the quotient of powers rule and how it is used in simplifying algebraic fractions.

7-5   Simplifying Algebraic Fractions

Objective: To simplify algebraic fractions.

Terms to Know
• *algebraic fraction*
• *simplify a fraction*
• *quotient of powers rule*

CONNECTION

You have worked with fractions in arithmetic. You can apply what you know to work with fractions in algebra. A fraction that contains a variable is called an **algebraic fraction.**

In Arithmetic		In Algebra
$\dfrac{3}{5}$ ⟵	numerator ⟶	$\dfrac{a}{b}, b \neq 0$
⟵	denominator ⟶	

Any fraction represents a division, so you know that the denominator of a fraction cannot be zero.

Throughout this textbook, you may assume that no denominator equals zero.

You **simplify** a fraction or an algebraic fraction by writing it in lowest terms. To do this, write the prime factorization of both the numerator and the denominator, then divide by all the common factors.

Example 1   **Simplify:**   a. $\dfrac{10xy}{8x}$ b. $\dfrac{n^5}{n^2}$

Solution

a. $\dfrac{10xy}{8x} = \dfrac{2 \cdot 5 \cdot \overset{1}{\cancel{x}} \cdot y}{2 \cdot 2 \cdot 2 \cdot \underset{1}{\cancel{x}}} = \dfrac{5y}{4}$

b. $\dfrac{n^5}{n^2} = \dfrac{\overset{1}{\cancel{n}} \cdot \overset{1}{\cancel{n}} \cdot n \cdot n \cdot n}{\underset{1}{\cancel{n}} \cdot \underset{1}{\cancel{n}}} = \dfrac{n^3}{1} = n^3$

In Example 1(b), the numerator and denominator of the fraction are powers of the same base. You may have noticed that the exponent in the simplified form is the same as the difference of the exponents in the given fraction.

Quotient of Powers Rule

To divide powers having the same base but different exponents, subtract the exponents.

$$\dfrac{a^m}{a^n} = a^{m-n}, a \neq 0$$

Example 2

Simplify: a. $\dfrac{5x^7}{x^3}$ b. $\dfrac{d^3}{4d}$

Solution

a. $\dfrac{5x^7}{x^3} = \dfrac{5x^{7-3}}{1} = 5x^4$

b. $\dfrac{d^3}{4d} = \dfrac{d^3}{4d^1} = \dfrac{d^{3-1}}{4} = \dfrac{d^2}{4}$

✏️ **Check Your Understanding**

1. In Example 2(a), how would the solution be different if the denominator were x^4?
2. In Example 2(b), why is d rewritten as d^1?
   See *Answers to Check Your Understanding* at the back of the book.

Guided Practice

COMMUNICATION « *Reading*

Refer to the text on pages 306–307.

« 1. In this textbook, what can you assume about the denominator of a fraction? It is not 0.

« 2. State the quotient of powers rule in words. To divide powers with the same base but different exponents, subtract the exponents.

Rewrite each fraction using the prime factorization of the numerator and the denominator.

3. $\dfrac{6a}{14}$ 4. $\dfrac{10j}{15j}$ 5. $\dfrac{12s}{18rs}$ 6. $\dfrac{a^9}{a^5}$ 7. $\dfrac{w^6}{7w}$

Simplify.

8. $\dfrac{4b}{16}$ $\dfrac{b}{4}$ 9. $\dfrac{20x}{12x}$ $\dfrac{5}{3}$ 10. $\dfrac{15m}{18mn}$ $\dfrac{5}{6n}$ 11. $\dfrac{n^{14}}{n^{11}}$ n^3 12. $\dfrac{a^4}{8a^3}$ $\dfrac{a}{8}$

Sidebar margin fractions:

3. $\dfrac{2 \cdot 3 \cdot a}{2 \cdot 7}$

4. $\dfrac{2 \cdot 5 \cdot j}{3 \cdot 5 \cdot j}$

5. $\dfrac{2 \cdot 2 \cdot 3 \cdot s}{2 \cdot 3 \cdot 3 \cdot r \cdot s}$

6. $\dfrac{a \cdot a \cdot a \cdot a \cdot a \cdot a \cdot a \cdot a \cdot a}{a \cdot a \cdot a \cdot a \cdot a}$

7. $\dfrac{w \cdot w \cdot w \cdot w \cdot w \cdot w}{7 \cdot w}$

Exercises

Simplify.

A   1. $\dfrac{21b}{28}$ $\dfrac{3b}{4}$ 2. $\dfrac{5}{10n}$ $\dfrac{1}{2n}$ 3. $\dfrac{36w}{45w}$ $\dfrac{4}{5}$ 4. $\dfrac{18x}{4x}$ $\dfrac{9}{2}$ 5. $\dfrac{6r}{3rs}$ $\dfrac{2}{s}$

6. $\dfrac{14vw}{14w}$ v 7. $\dfrac{c^{10}}{c^2}$ c^8 8. $\dfrac{r^7}{r}$ r^6 9. $\dfrac{6a^6}{a^2}$ $6a^4$ 10. $\dfrac{m^{12}}{13m^6}$ $\dfrac{m^6}{13}$

11. $\dfrac{48t}{4t}$ 12 12. $\dfrac{6ab}{27b}$ $\dfrac{2a}{9}$ 13. $\dfrac{30}{6y}$ $\dfrac{5}{y}$ 14. $\dfrac{n^9}{n^3}$ n^6 15. $\dfrac{8x^{24}}{x^8}$ $8x^{16}$

16. Simplify the algebraic fraction whose numerator is $21n$ and whose denominator is $3n$. 7

17. Simplify the algebraic fraction whose numerator is b^6 and whose denominator is $6b$. $\dfrac{b^5}{6}$

Sidebar

Making Connections/Transitions
This lesson connects the concepts of simplifying arithmetic fractions and simplifying algebraic fractions.

Key Question
How do you use the quotient of powers rule to simplify an algebraic fraction? The rule allows you to simplify the fraction by subtracting the exponents.

Reasoning
Students can use inductive reasoning to arrive at the quotient of powers rule from the solution to Example 1(b) and from other examples that are worked in class.

Error Analysis
Students may omit the exponent 1 when working with a variable to the first power. Remind students not to overlook variables raised to the first power.

Additional Examples
Simplify.
1. $\dfrac{20ab}{16a}$ $\dfrac{5b}{4}$
2. $\dfrac{y^8}{y^3}$ y^5
3. $\dfrac{7m^9}{m^3}$ $7m^6$
4. $\dfrac{n^4}{5n}$ $\dfrac{n^3}{5}$

Closing the Lesson
Ask a student to explain why the values of the variables in the denominator of an algebraic fraction must be restricted to nonzero numbers. Ask another student to explain the quotient of powers rule. Have a student simplify an algebraic fraction without exponents and have another student simplify an algebraic fraction with exponents.

Simplify.

B **18.** $\dfrac{9z^{15}}{3z^3}$ $3z^{12}$ **19.** $\dfrac{40n^{16}}{10n^{14}}$ $4n^2$ **20.** $\dfrac{10x^{21}}{15x^{13}}$ $\dfrac{2x^8}{3}$ **21.** $\dfrac{3v^{19}}{21v^{16}}$ $\dfrac{v^3}{7}$

22. $\dfrac{49a^3c^3}{14ac}$ $\dfrac{7a^2c^2}{2}$ **23.** $\dfrac{16v^3w^3}{44v^2}$ $\dfrac{4vw^3}{11}$ **24.** $\dfrac{12m^4n^2}{27m}$ $\dfrac{4m^3n^2}{9}$ **25.** $\dfrac{35x^2y}{5xy}$ $7x$

26. $\dfrac{18x^5y^5}{2y}$ $9x^5y^4$ **27.** $\dfrac{20m^8}{24m^3n^2}$ $\dfrac{5m^5}{6n^2}$ **28.** $\dfrac{3a^4b^3}{15a^2b}$ $\dfrac{a^2b^2}{5}$ **29.** $\dfrac{40s^6t^9}{16s^2t^7}$ $\dfrac{5s^4t^2}{2}$

LOGICAL REASONING

Exercises 30 and 31 refer to these statements.

I. $\dfrac{a^2}{a^3} = \dfrac{a \cdot a}{a \cdot a \cdot a} = \dfrac{1}{a}$ **II.** $\dfrac{a^2}{a^3} = a^{2-3} = a^{-1}$

C **30.** What logical conclusion can you make about the relationship between a^{-1} and $\dfrac{1}{a}$? Give a convincing argument to support your answer.

31. Use your results from Exercise 30. What do you think is another way to write a^{-5}? $\dfrac{1}{a^5}$

30. $a^{-1} = \dfrac{1}{a}$; since $\dfrac{a^2}{a^3} = \dfrac{1}{a}$ and $\dfrac{a^2}{a^3} = a^{-1}$, $\dfrac{1}{a} = a^{-1}$.

SPIRAL REVIEW

S **32.** Last baseball season, Ed had 38 hits, 27 runs batted in, and 4 home runs. How many singles did Ed hit? *(Lesson 5-6)* Not enough information is given.

33. Simplify: $\dfrac{a^{12}}{a^5}$ *(Lesson 7-5)* a^7

34. Complete: A ___?___ triangle has one 90° angle. *(Lesson 6-6)* right

35. Solve: $x - 17 = -9$ *(Lesson 4-3)* 8

36. Evaluate $a^2 + 2b - 15 \div c$ when $a = 4$, $b = 6$, and $c = 5$. *(Lesson 1-10)* 25

37. Find the sum: $-14 + 9 + (-3)$ *(Lesson 3-3)* −8

38. Write 12.5 cm in m. *(Lesson 2-9)* 0.125 m

Historical Note

Recently, computers were used to find the three prime factors of a 155-digit number. Early computers could only do much simpler tasks. The first working computer, Mark I, went into operation at Harvard University in 1944. Rear Admiral Grace Hopper, U.S. Navy (Ret.), worked as a programmer on that computer. She had a long and distinguished career in the computer field. Among her many contributions was the development of computer languages.

Research

A computer language is named Ada in honor of Ada Lovelace. Find out who she was. See answer on p. A21.

Comparing Fractions

Objective: To compare fractions.

Terms to Know
- *least common denominator (LCD)*

DATA ANALYSIS

The table below lists the lengths of different sizes of nails. For a project, a carpenter needs to use nails that are less than $1\frac{1}{3}$ in. long. The carpenter has some 3d nails.

QUESTION Are 3d nails short enough for the carpenter to use on the project?

To find the answer, first find the length of a 3d nail. The table shows that the length is $1\frac{1}{4}$ in. Then compare $1\frac{1}{4}$ to $1\frac{1}{3}$. Because both $1\frac{1}{4}$ and $1\frac{1}{3}$ have 1 as the whole-number part, you need to compare only the fractional parts of $1\frac{1}{4}$ and $1\frac{1}{3}$.

Penny Size (d)	2	3	4	5	6	7
Length (in.)	1	$1\frac{1}{4}$	$1\frac{1}{2}$	$1\frac{3}{4}$	2	$2\frac{1}{4}$

Example 1

Solution

Replace the __?__ with >, <, or =: $\frac{1}{4}$ __?__ $\frac{1}{3}$

To compare fractional parts of a whole, you can use shaded parts of bars to show the relationship.

$\longleftarrow \frac{1}{4}$ One fourth of a whole is less than

$\longleftarrow \frac{1}{3}$ one third of a whole, so $\frac{1}{4} < \frac{1}{3}$.

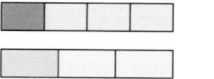 **Check Your Understanding**

1. In Example 1, explain how you know that one fourth of a whole is less than one third of a whole.
   See *Answers to Check Your Understanding* at the back of the book.

ANSWER Because $\frac{1}{4} < \frac{1}{3}$, it follows that $1\frac{1}{4} < 1\frac{1}{3}$. So 3d nails are short enough to use.

You can also compare fractions whose denominators are different, but whose numerators are the same and greater than 1.

Number Theory and Fraction Concepts   **309**

Lesson Planner

Teaching the Lesson
- Materials: fraction bars, clothesline, clothespins, index cards, calculator
- Lesson Starter 64
- Visuals, Folder C
- Using Technology, p. 288C
- Using Manipulatives, p. 288D
- Reteaching/Alternate Approach, p. 288D

Lesson Follow-Up
- Practice 86
- Enrichment 70: Exploration
- Study Guide, pp. 127–128
- Manipulative Activity 14

Warm-Up Exercises

Replace each __?__ with >, <, or =.

1. 3.405 __?__ 3.504   <
2. 0.78 __?__ 0.087   >
3. 1.101 __?__ 10.01   <
4. 6.05 __?__ 6.050   =
5. 0.090 __?__ 0.009   >

Lesson Focus

Ask students to imagine a long-jump competition. Jumps are measured to a fraction of an inch. After each jump, the distances are posted. Suppose two jumps were measured at 16 ft $3\frac{3}{8}$ in. and 16 ft $3\frac{1}{2}$ in. How can you determine which jump was farther? Today's lesson focuses on comparing fractions.

Teaching Notes

Work through the lesson with students, stressing the method of finding the LCD. The LCD is used again in Chapter 8 when students are asked to add and subtract fractions.

Key Questions

1. If the numerators of two fractions are 1, how can you determine which fraction is greater
   without finding a common
   denominator or creating a visual
   model? The denominator indicates the number of parts into
   which a whole is divided. The
   larger the denominator, the
   smaller the part. Thus, the
   greater of the two fractions
   described would have the
   smaller number in the
   denominator.
2. How are equivalent fractions
   used to compare fractions?
   When a common denominator is
   found, the numerators can be
   compared to see which fraction
   is greater.

Error Analysis

When comparing fractions, students
may compare numerators when the
denominators are different. Remind
students that to compare fractions
with unlike numerators and denominators, they need to find a common
denominator.

Life Skills

In sewing and tailoring, fabric measurements are given using fractions.
For instance, seam allowances are
given to a fraction of an inch.

Using Technology

For a suggestion on using a calculator to compare fractions, see page
288C.

Using Manipulatives

For suggestions on using fraction
bars and a fraction clothesline, see
page 288D.

Reteaching/Alternate Approach

For a suggestion on using cross
multiplication to compare fractions,
see page 288D.

Example 2

Replace the __?__ with >, <, or =: $\dfrac{5}{6}$ __?__ $\dfrac{5}{12}$

Solution

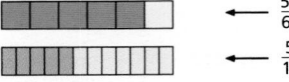

$\longleftarrow \dfrac{5}{6}$

$\longleftarrow \dfrac{5}{12}$

One sixth of a whole is greater than
one twelfth of a whole, so *five* sixths
are greater than *five* twelfths.

$$\dfrac{5}{6} > \dfrac{5}{12}$$

There are times when the best way to compare fractions is to rewrite
them as *equivalent fractions* with a common denominator. Then compare
the numerators. The **least common denominator (LCD)** is the LCM of the
denominators.

Example 3

Replace the __?__ with >, <, or =: $\dfrac{7}{12}$ __?__ $\dfrac{11}{18}$

Solution

The LCM of 12 and 18 is 36, so the LCD is 36.

$$\dfrac{7}{12} = \dfrac{7 \cdot 3}{12 \cdot 3} = \dfrac{21}{36} \qquad\qquad \dfrac{11}{18} = \dfrac{11 \cdot 2}{18 \cdot 2} = \dfrac{22}{36}$$

$$\dfrac{21}{36} < \dfrac{22}{36}, \text{ so } \dfrac{7}{12} < \dfrac{11}{18}.$$

✎ **Check Your Understanding**

2. In Example 3, how do you determine that the LCM is 36?
3. In Example 3, why is the number 3 used to multiply the numerator and
   denominator of $\dfrac{7}{12}$?
   See *Answers to Check Your Understanding* at the back of the book.

Guided Practice

COMMUNICATION «*Writing*

Write the statement represented by each diagram.
$\dfrac{4}{6} = \dfrac{2}{3}$

«**1.** «**2.** «**3.**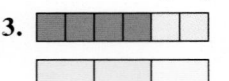

$\dfrac{1}{4} > \dfrac{1}{6}$ $\dfrac{6}{12} < \dfrac{3}{4}$

**Draw a diagram similar to those in Exercises 1–3 to represent each
statement.** Answers to Guided Practice 4–7 are on p. A21.

«**4.** $\dfrac{1}{6} > \dfrac{1}{12}$ «**5.** $\dfrac{1}{4} < \dfrac{1}{3}$ «**6.** $\dfrac{3}{4} = \dfrac{9}{12}$ «**7.** $\dfrac{11}{12} > \dfrac{5}{6}$

Find the LCD of each pair of fractions.

8. $\dfrac{6}{7}, \dfrac{3}{14}$ 14 9. $\dfrac{3}{25}, \dfrac{4}{5}$ 25 10. $\dfrac{2}{3}, \dfrac{3}{5}$ 15 11. $\dfrac{1}{6}, \dfrac{10}{21}$ 42

310 Chapter 7

Replace each __?__ with >, <, or =.

12. $\frac{1}{5}$ __?__ $\frac{1}{15}$ >   **13.** $\frac{7}{9}$ __?__ $\frac{7}{8}$ <   **14.** $\frac{4}{15}$ __?__ $\frac{3}{10}$ <   **15.** $\frac{4}{5}$ __?__ $\frac{3}{4}$ >

Exercises

Replace each __?__ with >, <, or =.

A

1. $\frac{1}{6}$ __?__ $\frac{1}{7}$ >   **2.** $\frac{1}{11}$ __?__ $\frac{1}{8}$ <   **3.** $\frac{15}{22}$ __?__ $\frac{15}{21}$ <   **4.** $\frac{11}{30}$ __?__ $\frac{11}{32}$ >

5. $\frac{5}{6}$ __?__ $\frac{11}{12}$ <   **6.** $\frac{4}{5}$ __?__ $\frac{7}{10}$ >   **7.** $\frac{2}{3}$ __?__ $\frac{5}{9}$ >   **8.** $\frac{1}{3}$ __?__ $\frac{2}{5}$ <

9. $\frac{16}{20}$ __?__ $\frac{4}{5}$ =   **10.** $\frac{21}{30}$ __?__ $\frac{7}{10}$ =   **11.** $\frac{5}{12}$ __?__ $\frac{3}{8}$ >   **12.** $\frac{3}{4}$ __?__ $\frac{7}{9}$ <

13. Marcia's history book is $8\frac{1}{4}$ in. wide. Her science book is $8\frac{1}{2}$ in. wide. Which book is wider? science book

14. Jon's graduation photograph is $3\frac{3}{8}$ in. high and $3\frac{3}{16}$ in. wide. Is the photograph wider than it is high? Explain. No; since $\frac{3}{8} > \frac{3}{16}$, it is higher than it is wide.

Write in order from least to greatest. Use two inequality symbols.

B

15. $\frac{1}{5}, \frac{1}{7}, \frac{1}{4}$ **16.** $\frac{7}{8}, \frac{5}{8}, \frac{2}{3}$ **17.** $\frac{3}{8}, \frac{3}{40}, \frac{7}{16}$ **18.** $2\frac{8}{9}, 2\frac{2}{3}, 2\frac{4}{5}$

15. $\frac{1}{7} < \frac{1}{5} < \frac{1}{4}$

16. $\frac{5}{8} < \frac{2}{3} < \frac{7}{8}$

17. $\frac{3}{40} < \frac{3}{8} < \frac{7}{16}$

18. $2\frac{2}{3} < 2\frac{4}{5} < 2\frac{8}{9}$

DATA ANALYSIS

Use the table at the right.

19. Is the winning height for 1948 *greater than* or *less than* for 1932? less than

20. Is the winning height for 1956 *greater than* or *less than* for 1952? greater than

21. Would you display these data in a *bar graph* or a *line graph*? Explain. bar graph; the data are not continuously changing.

22. In general, did the winning distances *increase* or *decrease* from 1932 to 1956? increase

23. Do you think the winning height for 1960 was *greater than*, *less than*, or *equal to* the winning height for 1956? Explain.

23. greater than; distances are generally increasing.

24. RESEARCH Find the winning height for 1960. Compare it to your answer to Exercise 23. 15 ft $5\frac{1}{8}$ in.

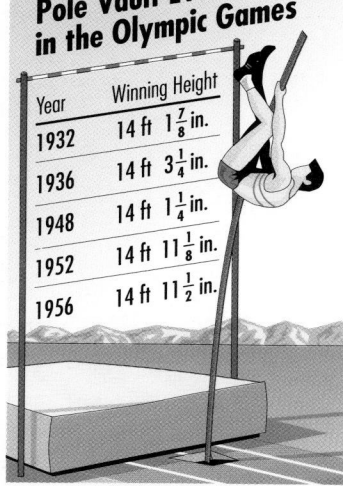

Pole Vault Event in the Olympic Games

Year	Winning Height
1932	14 ft $1\frac{7}{8}$ in.
1936	14 ft $3\frac{1}{4}$ in.
1948	14 ft $1\frac{1}{4}$ in.
1952	14 ft $11\frac{1}{8}$ in.
1956	14 ft $11\frac{1}{2}$ in.

Application

Exercises 25–28 discuss how the shutter speeds of a camera are described by fractions.

Follow-Up

Extension

Give students several sets of three fractions each and ask them to write the fractions in order using inequality symbols.

Enrichment

Enrichment 70 is pictured on page 288E.

Practice

Practice 86 is shown below.

Practice 86, *Resource Book*

CAREER/APPLICATION

A *professional photographer* takes pictures of many different subjects. To control the amount of time that film is exposed to light, the photographer changes the shutter speed. The speeds are marked on a camera with some numbers from 1 to 1000. These numbers represent denominators in fractions, as shown in the table at the right.

A photographer uses a faster shutter speed to photograph something moving quickly and a slower shutter speed for something standing still.

Use the table at the right.

25. List the shutter speeds in order from slowest to fastest: $\frac{1}{500}, \frac{1}{15}, \frac{1}{60}$ $\frac{1}{15}, \frac{1}{60}, \frac{1}{500}$

26. A photographer wants to take a "photo finish" of the cars in a race. Is a shutter speed of $\frac{1}{1000}$ or $\frac{1}{125}$ more appropriate? $\frac{1}{1000}$

27. A photographer has to take pictures of merchandise items for a catalogue. Is a shutter speed of $\frac{1}{250}$ or $\frac{1}{30}$ more appropriate? $\frac{1}{30}$

28. On cloudy days, photographers reduce shutter speed one level to admit more light. A photographer is taking pictures on a sunny day, using a shutter speed of $\frac{1}{125}$. The day suddenly becomes cloudy. What speed should the photographer now use? $\frac{1}{60}$

SPIRAL REVIEW

S 29. Draw the next two figures in the pattern below. Then write the first five numbers in the pattern. *(Lesson 6-8)* See answer on p. A21.

30. Replace the ___?___ with >, <, or =: $\frac{9}{11}$ ___?___ $\frac{4}{5}$ *(Lesson 7-6)* >

31. Find the GCF: 54 and 102 *(Lesson 7-2)* 6

32. Find the difference: $7\frac{4}{9} - 4\frac{5}{6}$ *(Toolbox Skill 19)* $2\frac{11}{18}$

33. Solve: $-24 = \frac{y}{-8}$ *(Lesson 4-4)* 192

34. Simplify: $7n^3 \cdot 9n^5$ *(Lesson 2-1)* $63n^8$

7-7

Fractions and Decimals

Objective: To write fractions as decimals and decimals as fractions.

Terms to Know
• *terminating decimal*
• *repeating decimal*

EXPLORATION

1 Use a calculator to complete each statement.

a. [7.] ÷ [20.] = _?_ 0.35

b. [11.] ÷ [15.] = _?_ 0.733333 . . .

2 Compare your answers to parts (a) and (b) of Step 1. How are they alike? How are they different? Both are decimals; one stops, the other doesn't.

3 Recall that a division bar acts like a division sign. Write each division in Step 1 using a fraction bar. $\frac{7}{20}$; $\frac{11}{15}$

You can write any fraction as a decimal by dividing the numerator by the denominator. When the division results in a remainder of zero, the decimal is called a **terminating decimal.** When the remainder is not zero and a block of digits in the decimal repeats, the decimal is called a **repeating decimal.** You indicate that a block of digits repeats by putting a bar over those digits.

Example 1

Write each fraction or mixed number as a decimal.

a. $\frac{7}{11}$

b. $1\frac{3}{8}$

Solution

a.
$$
\begin{array}{r}
0.6363\ldots \\
11\overline{)7.0000} \\
6\,6 \\
\overline{40} \\
33 \\
\overline{70} \\
66 \\
\overline{40} \\
33 \\
\overline{7}
\end{array}
$$

$\frac{7}{11} = 0.\overline{63}$

b.
$$
\begin{array}{r}
0.375 \\
8\overline{)3.000} \\
2\,4 \\
\overline{60} \\
56 \\
\overline{40} \\
40 \\
\overline{0}
\end{array}
$$

First write $\frac{3}{8}$ as a decimal.

$\frac{3}{8} = 0.375$

$1\frac{3}{8} = 1.375$

✓ Check Your Understanding

1. In Example 1(a), why is the bar over both the 6 and the 3 instead of just over the 3?

2. If you used a calculator in Example 1(a), what number would be shown on an 8-digit display?

3. How could you use Example 1(b) to help you write $6\frac{3}{8}$ as a decimal?
See *Answers to Check Your Understanding* at the back of the book.

Key Questions

1. How can you use a calculator to change a fraction to a decimal? Divide the numerator by the denominator.

2. What types of fractions can be written as terminating decimals? A fraction written in lowest terms that only has powers of 2 and/or 5 as factors of the denominator can be written as a terminating decimal.

Using Technology

The Exploration at the beginning of the lesson shows how to use a calculator to change a fraction to a decimal.

For an additional suggestion on using a calculator or a computer to write a fraction as a repeating decimal, see page 288C.

Error Analysis

When writing a fraction as a repeating decimal, students may incorrectly indicate the digits that repeat. Remind students to place a bar only over the digits that repeat.

Additional Examples

Write each fraction or mixed number as a decimal.

1. $\frac{2}{9}$ $0.\overline{2}$

2. $2\frac{5}{16}$ 2.3125

Write each decimal as a fraction or mixed number in lowest terms.

3. 0.444 $\frac{111}{250}$

4. 5.64 $5\frac{16}{25}$

In order to write a terminating decimal as a fraction in lowest terms, you first write the terminating decimal as a fraction whose denominator is a power of 10. Then you simplify the fraction.

Example 2

Write each decimal as a fraction or mixed number in lowest terms.

a. 0.555 **b.** 4.24

Solution

a. $0.555 = \frac{555}{1000} = \frac{111}{200}$

b. $4.24 = 4 + 0.24 = 4 + \frac{24}{100} = 4\frac{6}{25}$

Check Your Understanding

4. In Example 2(a), how do you know that $\frac{111}{200}$ is written in lowest terms?

See *Answers to Check Your Understanding* at the back of the book.

The chart below shows some of the most commonly used sets of equivalent decimals and fractions. You will find it helpful to become familiar with these.

Equivalent Decimals and Fractions

$0.2 = \frac{1}{5}$	$0.25 = \frac{1}{4}$	$0.125 = \frac{1}{8}$	$0.1\overline{6} = \frac{1}{6}$
$0.4 = \frac{2}{5}$	$0.5 = \frac{1}{2}$	$0.375 = \frac{3}{8}$	$0.\overline{3} = \frac{1}{3}$
$0.6 = \frac{3}{5}$	$0.75 = \frac{3}{4}$	$0.625 = \frac{5}{8}$	$0.\overline{6} = \frac{2}{3}$
$0.8 = \frac{4}{5}$		$0.875 = \frac{7}{8}$	$0.8\overline{3} = \frac{5}{6}$

Guided Practice

«1. COMMUNICATION «*Reading* Replace each __?__ with the correct word or phrase.

When a fraction is written as a __?__, one of two types of decimals results. The two types of decimals are the __?__ and the repeating decimal. In a __?__ decimal, a pattern of digits repeats over and over.
decimal; terminating decimal; repeating

Rewrite each repeating decimal with a bar over the repeating digits.

2. $0.41666\ldots$ $0.41\overline{6}$ **3.** $1.825825\ldots$ $1.\overline{825}$ **4.** $0.32121\ldots$
$0.3\overline{21}$

Write each decimal as a fraction whose denominator is a power of ten.

5. 0.18 $\frac{18}{100}$ **6.** 0.056 $\frac{56}{1000}$ **7.** 0.475 $\frac{475}{1000}$ **8.** 9.44 $9\frac{44}{100}$ **9.** 4.6
$4\frac{6}{10}$

Write each fraction or mixed number as a decimal.

10. $\frac{9}{20}$ 0.45　　11. $\frac{5}{8}$ 0.625　　12. $\frac{5}{11}$ 0.$\overline{45}$　　13. $3\frac{8}{9}$ 3.$\overline{8}$　　14. $7\frac{1}{3}$ 7.$\overline{3}$

Write each decimal as a fraction or mixed number in lowest terms.　7$\frac{3}{4}$

15. 0.205 $\frac{41}{200}$　　16. 0.81 $\frac{81}{100}$　　17. 5.1$\overline{6}$ 5$\frac{1}{6}$　　18. 3.62 3$\frac{31}{50}$　　19. 7.75

Exercises

A

Write each fraction or mixed number as a decimal.　0.$\overline{03}$

1. $\frac{7}{10}$ 0.7　　2. $\frac{3}{5}$ 0.6　　3. $\frac{9}{11}$ 0.$\overline{81}$　　4. $\frac{1}{33}$

5. $6\frac{11}{20}$ 6.55　　6. $9\frac{13}{15}$ 9.8$\overline{6}$　　7. $10\frac{2}{3}$ 10.$\overline{6}$　　8. $4\frac{4}{25}$ 4.16

Write each decimal as a fraction or mixed number in lowest terms.

9. 0.432 $\frac{54}{125}$　　10. 0.525 $\frac{21}{40}$　　11. 0.19 $\frac{19}{100}$　　$\frac{77}{100}$ 12. 0.77

13. 0.1$\overline{6}$ $\frac{1}{6}$　　14. 0.$\overline{3}$ $\frac{1}{3}$　　15. 3.32 3$\frac{8}{25}$　　5$\frac{19}{20}$ 16. 5.95

17. 9.51 9$\frac{51}{100}$　　18. 7.8$\overline{3}$ 7$\frac{5}{6}$　　19. 2.$\overline{6}$ 2$\frac{2}{3}$　　8$\frac{13}{100}$ 20. 8.13

21. Jordan answered $\frac{7}{8}$ of the test questions correctly. Write this fraction as a decimal. 0.875

22. A soccer team won $\frac{6}{11}$ of its games. Write this fraction as a decimal. 0.$\overline{54}$

ESTIMATION

Estimate each decimal as a fraction or mixed number.　about $\frac{3}{4}$

B　23. 0.49 about $\frac{1}{2}$　　24. 0.42 about $\frac{2}{5}$　　25. 0.33 about $\frac{1}{3}$　　26. 0.748

27. 9.58 about 9$\frac{3}{5}$　　28. 4.19 about 4$\frac{1}{5}$　　29. 6.124 about 6$\frac{1}{8}$　　30. 5.168 about 5$\frac{1}{6}$

PATTERNS

Write each fraction as a decimal.

31. $\frac{1}{9}$ 0.$\overline{1}$　　32. $\frac{2}{9}$ 0.$\overline{2}$　　33. $\frac{3}{9}$ 0.$\overline{3}$　　34. $\frac{4}{9}$ 0.$\overline{4}$

35. Use your answers to Exercises 31–34. What pattern do you notice? Use this pattern to make a table of equivalent fractions and decimals for fractions between 0 and 1 that have 9 as a denominator. See answer on p. A21.

Write each fraction as a decimal.　0.$\overline{36}$

36. $\frac{1}{11}$ 0.$\overline{09}$　　37. $\frac{2}{11}$ 0.$\overline{18}$　　38. $\frac{3}{11}$ 0.$\overline{27}$　　39. $\frac{4}{11}$

40. Use your answers to Exercises 36–39. What pattern do you notice? Use this pattern to make a table of equivalent fractions and decimals for fractions between 0 and 1 that have 11 as a denominator. See answer on p. A22.

Number Theory and Fraction Concepts　**315**

Calculator

Exercises 46–48 allow students to explore the conditions under which decimals will terminate or repeat. Students then use the results of their work to make a rule that can be used to determine if a decimal terminates.

Follow-Up

Exploration

Ask students to visit a local delicatessen or meat market where amounts of meat are purchased according to weight. Ask the clerk how an adjustment is made for the weight of a container. Is an adjustment necessary when weighing meats such as sliced turkey? Why or why not? Are any of the meat products priced by a fraction of a pound? Students should write a brief report on the practices they observed.

Enrichment

Enrichment 71 is pictured on page 288F.

Practice

Practice 87, *Resource Book*

Practice 87
Skills and Applications of Lesson 7-7

Write each fraction or mixed number as a decimal.

1. $\frac{2}{5}$ 0.4   2. $\frac{9}{10}$ 0.9   3. $\frac{3}{4}$ 0.75   4. $\frac{2}{3}$ 0.$\overline{6}$

5. $\frac{7}{8}$ 0.875   6. $\frac{1}{8}$ 0.125   7. $\frac{1}{2}$ 0.5   8. $1\frac{3}{5}$ 1.6

9. $1\frac{1}{2}$ 1.5   10. $3\frac{4}{5}$ 3.8   11. $5\frac{1}{4}$ 5.25   12. $1\frac{3}{8}$ 1.375

13. $7\frac{5}{8}$ 7.625   14. $\frac{49}{100}$ 0.49   15. $\frac{28}{70}$ 0.4   16. $12\frac{5}{9}$ 12.$\overline{5}$

17. $\frac{6}{80}$ 0.075   18. $1\frac{24}{25}$ 1.96   19. $9\frac{3}{100}$ 9.03   20. $15\frac{1}{9}$ 15.$\overline{1}$

Write each decimal as a fraction or mixed number in lowest terms. 25–28, 33–36. See below.

21. 0.45 $\frac{9}{20}$ 22. 2.5 $2\frac{1}{2}$ 23. 3.25 $3\frac{1}{4}$ 24. 0.85 $\frac{17}{20}$

25. 0.65   26. 7.50   27. 0.72   28. 0.82

29. 9.36 $9\frac{9}{25}$ 30. 7.3 $7\frac{1}{3}$ 31. .73 $\frac{11}{15}$ 32. 0.48 $\frac{12}{25}$

33. 0.375   34. 3.125   35. 0.185   36. 1.35

37. 6.44 $6\frac{11}{25}$ 38. 6.875 $6\frac{7}{8}$ 39. 9.83 $9\frac{5}{6}$ 40. 15.16 $15\frac{1}{6}$

41. Michiko swam $2\frac{3}{4}$ lengths of the pool before stopping to rest. Write this mixed number as a decimal. 2.75

42. Karl's job paid $4.60 per hour. Write 4.60 as a mixed number. $4\frac{3}{5}$

43. Two thirds of all seniors attended Saturday's football game. Write this fraction as a decimal. 0.$\overline{6}$

44. The guidance counselor announced that $\frac{7}{8}$ of all ninth grade students are taking biology. Write the fraction as a decimal. 0.875

45. One angle of a triangle measures 48.25°. Write 48.25 as a fraction. $48\frac{1}{4}$

25. $\frac{13}{20}$ 26. $7\frac{1}{2}$ 27. $\frac{18}{25}$ 28. $\frac{41}{50}$

33. $\frac{3}{8}$ 34. $3\frac{1}{8}$ 35. $\frac{37}{200}$ 36. $1\frac{7}{20}$

PROBLEM SOLVING/APPLICATION

At delicatessens, customers use fractions when ordering food. However, food items are usually weighed on scales that show weights as decimals.

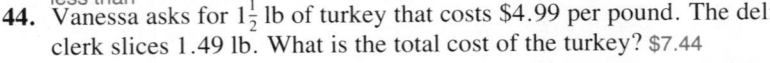

41. A scale shows 0.52 lb. About what fraction of a pound is this? about $\frac{1}{2}$

42. Pete asks for $\frac{1}{4}$ lb of cheese. The deli clerk slices 0.26 lb. Is this *more than* or *less than* $\frac{1}{4}$ lb?
more than

43. Martha asks for $\frac{3}{4}$ lb of roast beef. The deli clerk slices 0.73 lb. Is this *more than* or *less than* $\frac{3}{4}$ lb?
less than

44. Vanessa asks for $1\frac{1}{2}$ lb of turkey that costs $4.99 per pound. The deli clerk slices 1.49 lb. What is the total cost of the turkey? $7.44

45. Lee asks for 1 lb of potato salad. The deli clerk first sets the scale to −0.03 lb. Next the clerk puts a container on the scale. Then the clerk puts potato salad into the container. The scale shows 1.01 lb.
   a. Why did the clerk set the scale to −0.03 lb before weighing?
   b. How much potato salad did the customer buy? 1.01 lb

45a. to account for the weight of the container

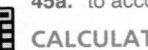 **CALCULATOR** Answers to Exercises 46–48 are on p. A22.

C 46. Using a calculator, find the decimal equivalents for these fractions.

$$\frac{1}{5}, \frac{1}{10}, \frac{1}{15}, \frac{1}{20}, \frac{1}{25}, \frac{1}{30}, \frac{1}{35}, \frac{1}{40}$$

Tell which decimals are repeating and which are terminating.

47. Use your answers to Exercise 46 to predict if $\frac{1}{45}$ and $\frac{1}{50}$ are repeating or terminating decimals. Use a calculator to check the predictions.

48. Using your answers to Exercises 46 and 47, make a rule for determining whether the decimal equivalent of a given fraction will terminate. (*Hint:* Consider the prime factors of the denominators.)

SPIRAL REVIEW

S 49. Measure the angle shown at the right. *(Lesson 6-2)* 60°

50. Write $8\frac{5}{6}$ as a decimal. *(Lesson 7-7)* 8.8$\overline{3}$

51. Estimate the difference: $5827 - 1184$ *(Toolbox Skill 1)* about 4600

52. Solve: $3(m + 5) = -18$ *(Lesson 4-9)* −11

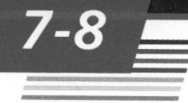

7-8

Strategy:
Drawing a Diagram

Objective: To solve problems by drawing a diagram.

Some problems may be easier to solve if you draw a diagram picturing the facts. Drawing a diagram while you read a problem can help you to understand the facts of the problem and determine what to find for the answer. For example, drawing grids or number lines might help you to solve problems involving distance. Drawing blocks might help solve problems involving lengths.

Problem

During a sightseeing tour, the tour bus travels 2 blocks due north, 3 blocks due east, 6 blocks due south, 11 blocks due west, and 4 blocks due north. At this point, where is the tour bus in relation to the starting point?

Solution

UNDERSTAND The problem is about a tour bus route.
Facts: travels 2 blocks due north,
        3 blocks due east,
        6 blocks due south,
        11 blocks due west,
        4 blocks due north
Find: the location of the bus relative to the starting point

PLAN You can use a piece of graph paper to draw a diagram. Let each square on the graph paper represent one block. Start near the middle of the graph paper and trace the route of the bus. Draw one arrow for each distance traveled. The first arrow will represent 2 blocks due north, the second arrow will represent 3 blocks due east, and so on. Be sure to identify the directions north, south, east, and west.

WORK

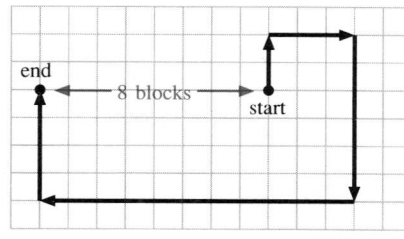

ANSWER Count the blocks between the starting point and the ending point. The tour bus is 8 blocks due west of its starting point.

Look Back What if the tour bus had traveled 11 blocks due east rather than 11 blocks due west? Where would the tour bus have been in relation to the starting point? 14 blocks due east

---

Lesson Planner

Teaching the Lesson
• Materials: graph paper
• Lesson Starter 66

Lesson Follow-Up
• Practice 88
• Practice 89: Nonroutine
• Enrichment 72: Problem Solving
• Study Guide, pp. 131–132
• Test 32

Warm-Up Exercises

1. Hamburger costs $1.58 a pound. Jean buys $2\frac{1}{2}$ pounds. What is the total cost? $3.95
2. Lee drove at 55 mi/h for $3\frac{3}{4}$ h. How many miles did Lee drive? 206.25 mi
3. Maria earns $11.40 per hour. On Monday, she worked $7\frac{1}{4}$ h. How much did Maria earn on Monday? $82.65

Lesson Focus

Ask students to imagine scheduling a bus route. It must be set up so that the bus can pick up and discharge passengers in a minimal amount of time. The route can be shown by using a drawing. Today's lesson focuses on making a diagram to solve problems.

Teaching Notes

Drawing a diagram is a strategy that can be used alone or with other strategies to solve a problem. The Checklist on page 319 gives the other strategies that have been used earlier in the text.

Key Question

How is the four-step plan used to solve the Problem on page 317? The four-step plan is used as follows: (1) the given facts are listed and what is to be found is stated; (2) a plan to draw a diagram is developed; (3) a diagram is made and used to solve the problem; and (4) the answer is stated and checked for reasonableness.

Problem Solving

Drawing a diagram is a strategy that can be used to illustrate the conditions of a problem visually.

Additional Example

On his newspaper route, Josh travels 2 blocks north, 4 blocks west, 5 blocks south, 7 blocks east, and then 3 blocks north. Find Josh's location relative to his starting point.   Josh is 3 blocks east of his original starting point.

Closing the Lesson

You may wish to use the following cooperative learning activity to close the lesson.

Cooperative Learning

Have students work in cooperative groups to create a problem that can be solved by using a diagram. Each group should then create three diagrams: one that solves the problem and two that do not. Groups should exchange problems and diagrams and then decide which diagram solves the problem they have been given.

Suggested Assignments

Skills: 1–19
Average: 1–19
Advanced: 1–19

Guided Practice

COMMUNICATION « *Reading*

Use this paragraph for Exercises 1–5.

A building has 30 floors, including two floors below ground level. An elevator in the building starts at ground level. It rises 20 floors, descends 11 floors, rises 5 floors, and descends 16 floors.

« **1.** How many floors does the building have in all? 30

« **2.** How many floors are above ground? 28

« **3.** Where does the elevator start? ground level

« **4.** In what direction is the elevator traveling when it moves 16 floors? down

« **5.** Draw a diagram to represent the situation. See answer on p. A22.

6. What is the value of *d* in the diagram at the right? 11 m

6 m	|← d →|
10 m	7 m

7. Solve by drawing a diagram: A softball field is located 12 mi due east of Jen's house. A bakery is located halfway between Jen's house and the softball field. The high school is located one third of the way from the bakery to the softball field. Where is the high school in relation to Jen's house? 8 mi due east

Problem Solving Situations

Solve by drawing a diagram.

1. Bernie lives 8 blocks due east of Harry. Harry lives 3 blocks due west of Ian. Where does Ian live in relation to Bernie? 5 blocks due west

2. The capacities of 3 buckets are 3 gal, 5 gal, and 8 gal. How could you use these containers to measure exactly 4 gal of water? See answer on p. A22.

318 Chapter 7

Solve by drawing a diagram.

3. How many diagonals can be drawn in a hexagon? 9

4. The lengths of three steel rods are 8 cm, 14 cm, and 16 cm. How could you use these rods to mark off a length of 10 cm?
See answer on p. A22.

5. An elevator started at ground level. It rose 15 floors, descended 3 floors, rose 8 floors, descended 12 floors, and descended 2 floors. At this point, where was the elevator relative to ground level? 6th floor

6. The exit for Greenville is 18 mi due west of the exit for Clairemont. The exit for Springfield is two thirds of the way from Clairemont to Greenville. The exit for Ashton is halfway between the exits for Springfield and Greenville. Where is the exit for Ashton in relation to the exit for Clairemont? 15 mi due west

7–12. Strategies may vary; likely strategies are given.

Solve using any problem solving strategy.

7. Seaside Gifts has postcards for $.25, $.35, and $.50. In how many different ways could a customer spend $1.35 on postcards? making a table; 3

8. Sonia Garcia bought cans of soup for $1.19 each and cups of yogurt for $.89 each. She spent a total of $13.97. How many of each did Sonia buy?
guess and check; 8 cans of soup, 5 cups of yogurt

9. Sixteen players participated in a single-elimination tennis tournament. In such a tournament, each player is out after one loss. How many games did the winner play? drawing a diagram; 4

10. Mr. Cole's shoe size is one size less than four times his son's shoe size. Mr. Cole wears a size 11 shoe. What size shoe does his son wear? using equations; 3

11. David delivers meals. After he has picked up the meals, he drives 4 blocks due south to make his first delivery. To make his second delivery, he turns left and drives 6 blocks. From there he drives 4 blocks due north to make his third delivery. At this point, how far is David from where he picks up the meals? drawing a diagram; 6 blocks due east

12. What is the least number of magnets that you need to display twelve 4-inch by 6-inch postcards so that they can all be seen? Assume that each corner must have a magnet holding it up. drawing a diagram; 20

13. Rita bought two boxes of computer disks for $10.99 each and paid tax of $1.10. She gave the cashier two $20 bills. How much change did Rita receive? choosing the correct operation; $16.92

PROBLEM SOLVING	
CHECKLIST	

Keep these in mind:
Using a Four-Step Plan
Too Much or Not Enough Information

Consider these strategies:
Choosing the Correct Operation
Making a Table
Guess and Check
Using Equations
Identifying a Pattern
Drawing a Diagram

Exercise Notes

Communication: Reading
Guided Practice Exercises 1–5 ask students to read a paragraph and answer questions related to it.

Life Skills
Directions are often given using landmarks and approximate distances. *Problem Solving Situation 6* uses the idea of drawing a diagram to solve a problem involving road directions.

Communication: Writing
Problem Solving Situations 14 and 15 ask students to examine a diagram and then write a word problem using the diagram.

Follow-Up

Extension

Use Problem Solving Situation 3 as the basis for connecting the concepts of drawing a diagram, making a table or graph, and looking for a pattern.

Ask students to draw a series of polygons having from 3 to 10 sides. In each polygon, ask them to find the number of diagonals. Then have them make a table organizing the information. Using the information developed so far, students should find the number of diagonals in a polygon of 20 sides. Then have them find the number of diagonals in a polygon of n sides.

Enrichment

Enrichment 72 is pictured on page 288F.

Practice

Practice 88 is shown below.

Quick Quiz 2

See pages 336 and 337.

Alternative Assessment

See page 337.

Practice 88, *Resource Book*

WRITING WORD PROBLEMS Answers to Problems 14–15 are on p. A22.

Write a word problem that you could solve by using the given diagram. Then solve the problem.

14.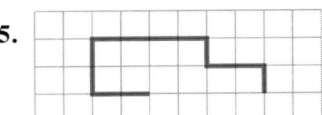
A   B   C

15.

SPIRAL REVIEW

S **16.** Find the quotient: $2\frac{2}{9} \div 4\frac{1}{3}$ *(Toolbox Skill 21)* $\frac{20}{39}$

17. DATA, *pages 92–93* Estimate the record high January temperature in Houston. *(Lesson 5-2)* about 85°F

18. Complete: A __?__ is a quadrilateral with exactly one pair of parallel sides. *(Lesson 6-7)* trapezoid

19. Karen lives 9 blocks due west of Julie and 6 blocks due west of Alice. Where does Alice live in relation to Julie? *(Lesson 7-8)*
3 blocks due west

Self-Test 2

Write each fraction in lowest terms.

1. $\frac{9}{12}$ $\frac{3}{4}$ **2.** $\frac{14}{16}$ $\frac{7}{8}$ **3.** $\frac{20}{25}$ $\frac{4}{5}$ **4.** $\frac{18}{48}$ $\frac{3}{8}$ 7-4

Simplify.

5. $\frac{8c}{24c}$ $\frac{1}{3}$ **6.** $\frac{10x}{15xy}$ $\frac{2}{3y}$ **7.** $\frac{a^8}{a^2}$ a^6 **8.** $\frac{8x^{14}}{x^5}$ $8x^9$ 7-5

Replace each __?__ with >, <, or =.

9. $\frac{1}{8}$ _?_ $\frac{1}{7}$ <   **10.** $\frac{7}{12}$ _?_ $\frac{7}{10}$ <   **11.** $\frac{13}{16}$ _?_ $\frac{17}{24}$ >   7-6

Write each fraction or mixed number as a decimal.

12. $\frac{11}{20}$ 0.55   **13.** $\frac{8}{9}$ $0.\overline{8}$ **14.** $2\frac{17}{25}$ 2.68   **15.** $4\frac{5}{6}$ 4.83   7-7

Write each decimal as a fraction or mixed number in lowest terms.

16. 0.78 $\frac{39}{50}$ **17.** 0.444 $\frac{111}{250}$ **18.** 3.56 $3\frac{14}{25}$ **19.** 8.$\overline{6}$ $8\frac{2}{3}$

20. Solve by drawing a diagram: The lengths of three steel rods are 12 m, 26 m, and 19 m. How could you use these rods to mark off a length of 5 m? See answer on p. A22.   7-8

320   Chapter 7

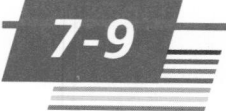

7-9 Rational Numbers

Objective: To recognize rational numbers.

You have worked with whole numbers, integers, and fractions. All these numbers can be written in fractional form. For example, $5 = \frac{5}{1}$ and $-3 = \frac{-6}{2}$. Any number that can be written as a quotient of two integers $\frac{a}{b}$, where b does not equal zero, is called a **rational number.** All whole numbers, integers, and arithmetic fractions as well as many decimals are rational numbers.

Example 1

Express each rational number as a quotient of two integers.

a. -16 **b.** $2\frac{4}{7}$ **c.** $-1.\overline{3}$

Solution

a. $-16 = -\frac{16}{1}$
$= \frac{-16}{1}$

b. $2\frac{4}{7} = \frac{18}{7}$

c. $-1.\overline{3} = -1\frac{1}{3}$
$= -\frac{4}{3}$
$= \frac{-4}{3}$

 **Check Your Understanding**

1. In Example 1(a), would the quotient $\frac{16}{-1}$ also represent -16? Explain.

2. In Example 1(a), would the quotient $\frac{-16}{-1}$ also represent -16? Explain.
See *Answers to Check Your Understanding* at the back of the book.

Warm-Up Exercises

Write each decimal as a fraction in lowest terms.

1. 0.55 $\frac{11}{20}$

2. 0.25 $\frac{1}{4}$

3. 0.888 $\frac{111}{125}$

4. 0.33 $\frac{33}{100}$

5. 0.785 $\frac{157}{200}$

Terms to Know
• *rational number*
• *irrational number*
• *real number*

Numbers that cannot be written as the quotient of two integers are called **irrational numbers.** These numbers are nonrepeating, nonterminating decimals. Any number that is either rational or irrational is called a **real number.** The chart below shows the parts of the *real number system*.

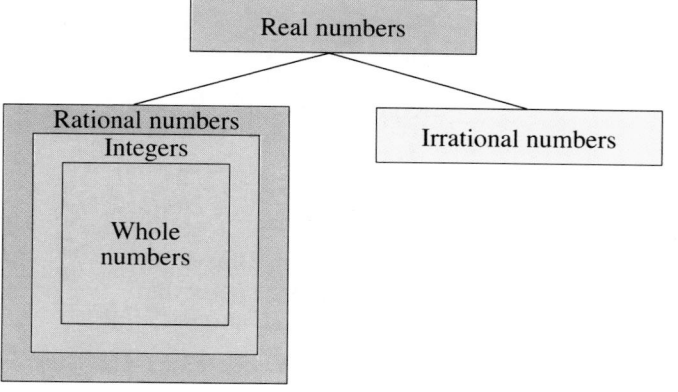

Lesson Focus

Ask students if they have ever wondered how many numbers are on the number line. Ask them to name some different types of numbers that can be placed on a number line. Is there always a number between any two numbers? The number system is made up of many different types of numbers. Today's lesson focuses on the rational numbers.

Teaching Notes

The content of this lesson builds on material that is familiar to students. Stress the terms *rational number*, *irrational number*, and *real number*.

Key Questions

1. How would you describe two nonterminating decimals, one that repeats and one that does not? *The repeating decimal is rational; the nonrepeating decimal is irrational.*

2. Are whole numbers and integers rational numbers? *Yes, any whole number or integer x can be written as $\frac{x}{1}$.*

Error Analysis

Students may make errors when ordering negative rational numbers. Remind them that the greater the absolute value of a number, the farther away it is from zero.

Using Manipulatives

For a suggestion on using a rational number clothesline, see page 288D.

Additional Examples

1. Express each rational number as a quotient of two integers.

   a. -17 $\frac{-17}{1}$

   b. $3\frac{5}{7}$ $\frac{26}{7}$

   c. 5.7 $\frac{57}{10}$

   d. $-2.\overline{6}$ $\frac{-8}{3}$

2. Use the number line below. Name the point that represents each number.

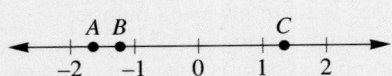

   a. -1.23 *B*

   b. $1\frac{1}{3}$ *C*

   c. -1.65 *A*

Every real number can be represented by a point on a number line.

Example 2   **Use the number line at the right. Name the point that represents each number.**

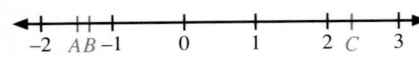

   a. -1.5 b. -1.35 c. $2\frac{1}{3}$

Solution

   a. -1.5 is halfway between -2 and -1. Point A represents -1.5.

   b. -1.35 is between -2 and -1 and is greater than -1.5. Point B represents -1.35.

   c. $2\frac{1}{3}$ is between 2 and 3. Point C represents $2\frac{1}{3}$.

Guided Practice

COMMUNICATION «*Reading*

Refer to the diagram on page 321. To which group(s) of numbers does each number belong? B, C, E

«**1.** 3 A, B, C, E   «**2.** -10 «**3.** 6.23 C, E   **A.** whole numbers

«**4.** $\frac{7}{8}$ C, E   «**5.** $-\frac{14}{9}$ C, E   «**6.** $8.\overline{5}$ C, E   **B.** integers
                                                      C. rational numbers

«**7.** $-0.\overline{27}$ C, E   «**8.** $5\frac{1}{3}$ C, E   «**9.** -18.3 C, E   **D.** irrational numbers
                                                      E. real numbers

10–14. Answers may vary. Examples are given.
Express each rational number as a quotient of two integers.

10. 0.125 $\frac{1}{8}$ **11.** $0\frac{0}{5}$ **12.** $-3\frac{-3}{1}$ **13.** $-2\frac{1}{5}\frac{-11}{5}$ **14.** $0.1\overline{6}$ $\frac{1}{6}$

Use the number line at the right. Name the point that represents each number.

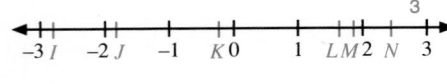

15. $1.\overline{5}$ G   **16.** -2.8 D   **17.** $2\frac{1}{4}$ H   **18.** $-\frac{2}{9}$ F   **19.** -0.654 E

Exercises

1–8. Answers may vary. Examples are given.
Express each rational number as a quotient of two integers.

A   **1.** $12\frac{3}{5}$ $\frac{63}{5}$ **2.** -50 $\frac{-50}{1}$ **3.** $-4\frac{1}{4}$ $\frac{-17}{4}$ **4.** $10\frac{5}{12}$ $\frac{125}{12}$

     5. 20.2 $\frac{202}{10}$ **6.** -0.1 $\frac{-1}{10}$ **7.** $-2.\overline{3}$ $\frac{-7}{3}$ **8.** $8.\overline{6}$ $\frac{26}{3}$

Use the number line at the right. Name the point that represents each number.

9. $2.\overline{4}$ N   **10.** $-1.8\overline{3}$ J   **11.** 1.657 L   **12.** $-\frac{3}{13}$ K   **13.** $-\frac{14}{5}$ I   **14.** $1\frac{7}{8}$ M

15. On a horizontal number line, is the point that represents -1.5 to the *left* or *right* of the point that represents -1? left

16. List five different ways to express 0 as a quotient of two integers.
Answers will vary. Examples: $\frac{0}{1}, \frac{0}{2}, \frac{0}{3}, \frac{0}{4}, \frac{0}{5}$

Replace each __?__ with >, <, or =.

B **17.** $-\frac{1}{2}$ __?__ $-\frac{1}{10}$ < **18.** $-\frac{4}{9}$ __?__ -1 > **19.** $-12.\overline{4}$ __?__ -12.4 <

20. 5.5 __?__ $\frac{16}{3}$ > **21.** -4.315 __?__ -4.35 > **22.** $3\frac{1}{7}$ __?__ $3.\overline{16}$ <

NUMBER SENSE

To graph numbers on a number line, it is helpful to decide whether the number is close to a value whose graph is easily identifiable.

Tell whether each number is closest to -1, $-\frac{1}{2}$, 0, $\frac{1}{2}$, or 1.

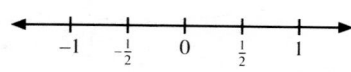

23. -0.705 $-\frac{1}{2}$ **24.** $\frac{8}{9}$ 1 **25.** -0.2 0 **26.** $\frac{1}{7}$ 0 **27.** $-\frac{3}{8}$ $-\frac{1}{2}$ **28.** 0.345 $\frac{1}{2}$

LOGICAL REASONING

Replace each __?__ with the word *All*, *Some*, or *No* to make the statement true.

C **29.** __?__ whole numbers are integers. All

30. __?__ integers are whole numbers. Some

31. __?__ rational numbers are integers. Some

32. __?__ real numbers are rational numbers. Some

SPIRAL REVIEW

S **33.** Identify the polygon at the right. *(Lesson 6-5)* rhombus

34. Express 0.16 as a quotient of two integers. *(Lesson 7-9)*

34. Answers may vary. Example: $\frac{16}{100}$

35. Find the sum: $7\frac{3}{8} + 9\frac{11}{16}$ *(Toolbox Skill 18)* $17\frac{1}{16}$

36. Find the LCM: 24 and 36 *(Lesson 7-3)* 72

Number Theory and Fraction Concepts   **323**

Lesson Planner

Teaching the Lesson
- Lesson Starter 68
- Using Technology, p. 288C
- Reteaching/Alternate Approach, p. 288D
- Cooperative Learning, p. 288D

Lesson Follow-Up
- Practice 91
- Enrichment 74: Connection
- Study Guide, pp. 135–136

Warm-Up Exercises

Write an inequality to describe each of the following.

1. The temperature in Fargo was less than 32°F all day.
   $x < 32°$
2. The price of fruit juice in all stores was greater than or equal to $2.97. $\$2.97 \leq c$
3. The temperature yesterday had a low of 47°F and a high of 63°F. $47° \leq x \leq 63°$
4. When Rebecca stayed home from school, her temperature was above 98.6°F, but it did not go above 102.3°F.
   $98.6° < x \leq 102.3°$
5. Each year, Fresno receives at least 6.4 in. of rain but less than 22.8 in. $6.4 \leq x < 22.8$

Lesson Focus

Ask students to think of various places they have visited that are *more than* fifty miles from home. For some students, there may be many places they recall that fit this description. Today's lesson focuses on graphing open sentences, some of which have many solutions.

Graphing Open Sentences

Objective: To graph solutions of equations and inequalities on a number line.

CONNECTION

> **Terms to Know**
> - open sentence
> - solution of an open sentence
> - graph of an open sentence
> - inequality

In both the English language and mathematics, there are many sentences that can be classified as either true or false. There are other sentences whose truth you cannot determine.

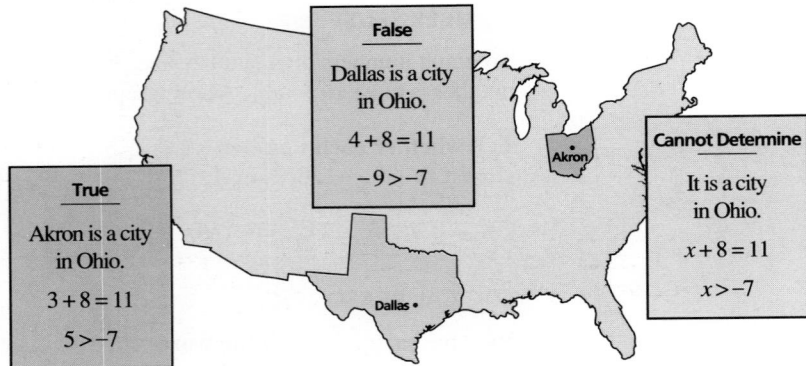

You cannot determine the truth of the English sentence *It is a city in Ohio* because you do not know what city the word *it* represents. You cannot determine the truth of mathematical sentences like $x + 8 = 11$ and $x > -7$ because you do not know what number the variable x represents. For this reason, a mathematical sentence that contains a variable is called an **open sentence.**

An open sentence is by itself neither true nor false. When you substitute a real number for the variable, however, you can determine whether the result is true or false. Any value of the variable that results in a true sentence is called a **solution of the open sentence.** Because a solution is a real number, you can show the **graph of an open sentence** in one variable by graphing all the solutions on a number line.

Example 1

Solution

Graph the equation $8 = 12 + x$.

First solve the equation.
$$8 = 12 + x$$
$$8 - 12 = 12 + x - 12$$
$$-4 = x$$
The solution is -4.

Then graph the solution as a heavy dot on a number line.

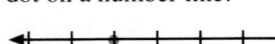

 **Check Your Understanding**

1. Would the graph in Example 1 be different if the equation were $x + 12 = 8$?

See *Answers to Check Your Understanding* at the back of the book.

A mathematical sentence that has an inequality symbol between two numbers or quantities is an **inequality.** When an inequality is an open sentence, like $x > -4.2$, there are infinitely many real-number solutions. To graph an inequality like this on a number line, you use an open dot and an arrow.

Example 2

Solution

Graph the inequality $x > -4.2$.

Because -4.2 is not a solution of the inequality, you place an open dot at -4.2 on a number line. Then shade in a heavy arrow to the right to graph all numbers greater than -4.2.

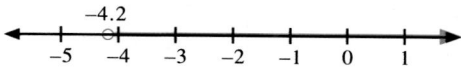

 **Check Your Understanding**

2. Why is -4.2 *not* a solution of $x > -4.2$?
3. Describe how the graph in Example 2 would be different if the inequality were $x < -4.2$.
   See *Answers to Check Your Understanding* at the back of the book.

Two other inequality symbols that are commonly used in mathematics are $\leq$ and $\geq$.

In Words	In Symbols
a is less than or equal to b.	$a \leq b$
a is greater than or equal to b.	$a \geq b$

On a number line, you graph an open sentence that contains one of these inequality symbols by using a closed dot and an arrow.

Example 3

Solution

Graph the inequality $y \leq \frac{3}{4}$.

Because $\frac{3}{4}$ is a solution of the inequality, you place a closed dot at $\frac{3}{4}$ on a number line. Then shade in a heavy arrow to the left to graph all numbers less than $\frac{3}{4}$.

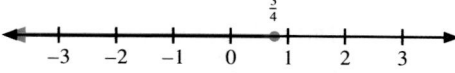

 **Check Your Understanding**

4. Why is $\frac{3}{4}$ a solution of $y \leq \frac{3}{4}$?
5. Describe how the graph in Example 2 would be different if the inequality were $y \geq \frac{3}{4}$.
   See *Answers to Check Your Understanding* at the back of the book.

Discuss the differences in the graphs of the various types of open sentences. Pay particular attention to the open dot and closed dot concepts.

Key Questions
1. When graphing inequalities, how do you know whether to use a closed dot or an open dot? If $\leq$ or $\geq$ are used, the dot is shaded. If the symbols $<$ or $>$ are used, the dot is open.
2. When an inequality is being graphed, how do you know which direction to shade the number line?
   Pick a test number on one side of the open or closed dot. Substitute the number for the variable in the inequality. If the resulting inequality is true, shade the side of the test number. If the resulting inequality is false, shade the other side.

Reasoning
The Check Your Understanding questions rely on students' reasoning abilities. Questions 2 and 4 ask students to reason why a point is or is not a solution to an inequality. Questions 3 and 5 ask students to describe changes in the graphs of inequalities when changes are made in their directions.

Error Analysis
When graphing an inequality, students may incorrectly shade the dot. Remind them that the dot is shaded only if equality is included, as in $\leq$ or $\geq$. In determining the direction of an arrow, students may shade an incorrect portion of a number line. Remind them to check for direction by selecting a test point.

Teaching Notes *(continued)*

Using Technology
For a suggestion on using a computer to find points included in the graph of an open sentence, see page 288C.

Computer graphics may be appropriate for graphing the solutions to equations and inequalities on a number line.

Reteaching/Alternate Approach
For a suggestion on using a cooperative learning activity to graph inequalities, see page 288D.

Additional Examples
1. Graph the equation $7 = 10 + x$.

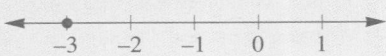

2. Graph the inequality $y > -3.4$.

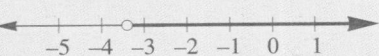

3. Graph the inequality $x \leq \frac{1}{3}$.

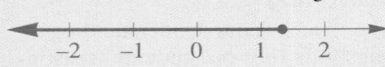

Closing the Lesson
Give students a list of ten related equations and inequalities and ten graphs of these open sentences. Ask them to match each open sentence with its graph.

Suggested Assignments
Skills: 1–19, 33–39
Average: 1–28, 33–39
Advanced: 1–13 odd, 15–39

Exercise Notes
Communication: Writing
Guided Practice Exercises 3–6 ask students to translate sentences into mathematical symbols. *Guided Practice Exercises 7–10* reverse the process by having students translate mathematical symbols into words.

Guided Practice

For each open sentence, choose all the given numbers that are solutions.

1. $z + 8 = -1$ **a.** 7   **b.** -7 (**c.**) -9 **d.** 9

2. $n \geq -\frac{5}{8}$ (**a.**) $\frac{5}{8}$ **b.** -1.3 (**c.**) $-\frac{1}{2}$ (**d.**) $-\frac{5}{8}$

COMMUNICATION «*Writing*

Write each sentence in symbols.

« **3.** A number n is greater than or equal to -4. $n \geq -4$

« **4.** Fifteen is greater than a number x. $15 > x$

« **5.** A number q is less than -2.25. $q < -2.25$

« **6.** A number t is less than or equal to $-\frac{3}{5}$. $t \leq -\frac{3}{5}$

Write each inequality in words.

7. A number a is less than -14.

8. A number d is greater than or equal to $3\frac{2}{9}$.

9. A number k is greater than $-\frac{6}{7}$.

10. A number m is less than or equal to -1.7.

« **7.** $a < -14$ « **8.** $d \geq 3\frac{2}{9}$ « **9.** $k > -\frac{6}{7}$ « **10.** $m \leq -1.7$

Graph each open sentence.

11. $2r - 5 = 3$ **12.** $b < \frac{5}{6}$ **13.** $p > -7\frac{1}{3}$ **14.** $h \geq -3.5$

Answers to Guided Practice 11–14 are on p. A22.

Exercises

Graph each open sentence. Answers to Exercises 1–14 are on p. A22.

A   **1.** $n + 4 = -5$ **2.** $c - 34 = -28$ **3.** $-11 = a - 8$

     4. $69 = 62 + q$ **5.** $17 = 3m + 2$ **6.** $4z - 17 = -25$

     7. $g < 2$ **8.** $y > -1$ **9.** $z > -3\frac{1}{3}$ **10.** $s < 5\frac{2}{5}$

     11. $b \geq -6\frac{1}{7}$ **12.** $m \leq -\frac{2}{3}$ **13.** $a \leq 7.5$ **14.** $v \geq -2.25$

Match each statement with the correct graph.

B   **15.** $n \geq 25$ D   **A.** ⟵—┼╫┼—⟶ 25

     16. $m = 25$ E

     17. $z < 25$ B   **B.** ⟵—┼╫┼—⟶ 25

     18. $p > 25$ A

     19. $x \leq 25$ C   **C.** ⟵—┼╫┼—⟶ 25

     D. ⟵—┼╫┼—⟶ 25

     E. ⟵—┼╫┼—⟶ 25

THINKING SKILLS 20–22. Analysis   23–28. Synthesis

The figure at the right shows the graph of $-3 < x < 1$. For Exercises 20–28, combine what you learned about this type of inequality in Chapter 1 with what you learned about graphing in this lesson.

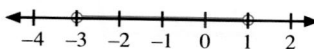

Tell how the graph of each inequality compares with the graph above.

20. $-3 \le x \le 1$
The dots are closed.

21. $1 > x > -3$
no difference

22. $-3 \le x < 1$
The dot at -3 is closed.

Graph each inequality. Answers to Exercises 23–28 are on p. A22.

23. $-4 < t < 2$

24. $-1 > n > -5$

25. $0 \ge a \ge -3$

26. $1.5 \le r \le 6.5$

27. $-4 < x \le 1$

28. $\frac{2}{3} \ge y \ge -\frac{2}{3}$

FUNCTIONS

There is a function of real numbers that is called the *greatest integer function*. This function pairs a real number with the greatest integer less than or equal to that number. The function rule is $x \rightarrow [x]$. For instance, when $x = 2\frac{1}{2}$, $[x] = 2$.

29. Complete the function table at the right. 7; 4; 3; 0

30. List three numbers for which the greatest integer is 5. Answers will vary. Examples: 5.4, 5.1, 5.39

C **31.** Graph on a number line all the numbers for which the greatest integer is 0. See answer on p. A22.

32. Use two inequality symbols to write a statement that represents all the numbers for which the greatest integer is 8. Use n as the variable.
$8 \le n < 9$

x	$[x]$
$9\frac{1}{3}$	9
$7.\overline{6}$	?
4	?
$3.\overline{3}$	?
$\frac{7}{8}$	?

SPIRAL REVIEW

S **33.** Estimate the sum: $157 + 389 + 438$ *(Toolbox Skill 2)* about 1000

34. Graph the inequality $x < -2.5$.   *(Lesson 7-10)* See answer on p. A22.

35. DATA, *pages 288–289*   Write in scientific notation: the number of additions a computer could do in 1964   *(Lesson 2-10)* 1.25×10^6

36. *True* or *False*?: Two squares are always congruent.   *(Lesson 6-9)*
False

37. Use the formula $D = rt$. Let $r = 48$ mi/h and $t = 2.75$ h. Find D.
(Lesson 4-10) 132 mi

38. Simplify: $\frac{30x}{42xy}$ *(Lesson 7-5)* $\frac{5}{7y}$

39. Evaluate $5n^4$ when $n = 3$.   *(Lesson 1-9)* 405

Number Theory and Fraction Concepts   **327**

Critical Thinking
Exercises 20–28 require students to *apply* their knowledge of combined inequalities to *analyze* and to *create* graphs of these inequalities.

Application
Exercises 29–32 introduce the greatest integer function. The notion of greatest integer in the context of an application requires the use of reasoning and common sense. For example, suppose books that measure 1.5 in. thick are placed on a shelf that is 28 in. long. How many such books can be placed on the shelf?   Since $28 \div 1.5 = 18.\overline{6}$, the number of books that can be placed on the shelf is 18.

Follow-Up

Extension
Ask students what would be represented by a graph in which only a single point is *unshaded*. How could Example 1 be changed to create such a graph?

Enrichment
Enrichment 74 is pictured on page 288F.

Practice
Practice 91, *Resource Book*

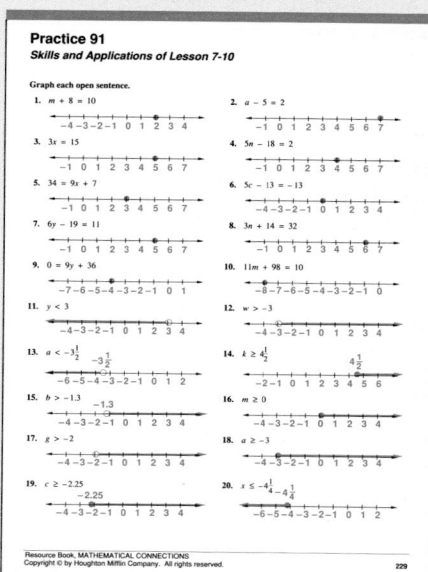

Teaching the Lesson
- Materials: calculator
- Lesson Starter 69

Lesson Follow-Up
- Practice 92
- Practice 93: Nonroutine
- Enrichment 75: Exploration
- Study Guide, pp. 137–138
- Test 33

Warm-Up Exercises

Simplify.

1. 5^4 625
2. 4^5 1024
3. 3^7 2187
4. 2^{10} 1024
5. Write 4,300,000,000 in scientific notation. 4.3×10^9
6. Write 3.45×10^7 in decimal notation.   34,500,000

Lesson Focus

Ask students to simplify 2^5, 2^4, 2^3, 2^2, and 2^1. Then ask them to look for a pattern in what they have written. Ask them what will happen if the pattern continues. How will the exponents and the results change? This lesson focuses on zero and negative exponents.

Teaching Notes

Stress the definition of negative exponent. Point out that $a^0 = 1$ only if $a \neq 0$, since 0^0 is undefined.

Negative and Zero Exponents

Objective: To simplify expressions involving negative and zero exponents, and to use negative exponents to write numbers in scientific notation.

EXPLORATION

1 Replace each __?__ with the number that makes the statement true.

a. $\dfrac{a^2}{a^2} = \dfrac{\cancel{a} \cdot \cancel{a}}{\cancel{a} \cdot \cancel{a}} = $ __?__   1

b. $\dfrac{a^2}{a^2} = a^{2-2} = a^?$ 0

2 Use your results from Step 1. Complete: $1 = a^?$ 0

3 Replace each __?__ with the number that makes the statement true.

a. $\dfrac{a^3}{a^5} = \dfrac{\cancel{a} \cdot \cancel{a} \cdot \cancel{a}}{\cancel{a} \cdot \cancel{a} \cdot \cancel{a} \cdot a \cdot a} = \dfrac{1}{a \cdot a} = \dfrac{1}{a^?}$ 2

b. $\dfrac{a^3}{a^5} = a^{3-5} = a^?$ -2

4 Use your results from Step 3. Complete: $\dfrac{1}{a^2} = a^?$ -2

If an expression has a negative exponent, you can write the expression as a fraction with a positive exponent.
If a is any nonzero number, then

$$a^{-n} = \frac{1}{a^n} \quad \text{and} \quad a^0 = 1.$$

Example 1

Simplify.

a. x^{-7}
b. 3^{-3}
c. $(-2)^{-2}$
d. $(5.67)^0$

Solution

a. $x^{-7} = \dfrac{1}{x^7}$

b. $3^{-3} = \dfrac{1}{3^3} = \dfrac{1}{27}$

c. $(-2)^{-2} = \dfrac{1}{(-2)^2} = \dfrac{1}{4}$

d. $(5.67)^0 = 1$

In Lesson 2-10 you learned to use positive powers of ten to express very large measures in *scientific notation.*

$$3.2 \times 10^{12} = 3.2 \times 1,000,000,000,000 = 3,200,000,000,000$$

Now that you know about negative exponents, you also can use scientific notation to express very small measures.

$$3.2 \times 10^{-4} = 3.2 \times \frac{1}{10^4} = 3.2 \times \frac{1}{10,000} = 3.2 \times 0.0001 = 0.00032$$

Example 2
    a. Write 0.065 in scientific notation.
    b. Write 4.7×10^{-4} in decimal notation.

Solution
    a. Move the decimal point to get a number that is at least 1, but less than 10.

$$0.065 = 6.5 \times 10^{-2}$$

    2 places

    b. Move the decimal point to the left.

$$4.7 \times 10^{-4} = 0.00047$$

    4 places

 **Check Your Understanding**

1. How would Example 2(a) be different if the number were 0.65?
2. How would Example 2(b) be different if the number were 4.7×10^4?
See Answers to Check Your Understanding at the back of the book.

Remember that you can use the [EE] or [EXP] key on your calculator to enter numbers in scientific notation. For instance, to enter 6.3×10^{-7} you use this key sequence.

[6.3] [EE] [7.] [+/-]

Guided Practice

COMMUNICATION « *Reading*

Refer to the text on pages 328–329.

« **1.** What is the main idea of the lesson? Expressions involving negative and zero exponents can be simplified.

« **2.** List three major points that support the main idea.

2. If a is any nonzero number, then (1) $a^{-n} = \frac{1}{a^n}$ and (2) $a^0 = 1$. (3) Numbers in scientific notation with negative exponents can be written in decimal notation.

Simplify.

3. 4^4 256   **4.** 6^3 216   **5.** $(-5)^2$ 25   **6.** $(-3)^5$ −243   **7.** $(-2)^7$ −128

Write each number as a power of ten.

8. 0.01 10^{-2} **9.** 0.0001 10^{-4} **10.** 0.00001 10^{-5} **11.** $\frac{1}{1,000,000}$ 10^{-6}

Simplify.

12. x^{-4} $\frac{1}{x^4}$ **13.** 8^{-2} $\frac{1}{64}$ **14.** $(-3)^{-3}$ $-\frac{1}{27}$ **15.** $(2{,}004{,}627)^0$ 1

Write each number in scientific notation.

16. 0.0704 7.04×10^{-2} **17.** 0.00006 6×10^{-5} **18.** 0.00000057 5.7×10^{-7}

Write each number in decimal notation.

19. 3.295×10^{-2} 0.03295   **20.** 1.7×10^{-5} 0.000017   **21.** 4.44×10^{-7} 0.000000444

Number Theory and Fraction Concepts   **329**

Key Questions

1. How can you write an expression such as a^{-3} so that it contains only a positive exponent? Rewrite the expression as a fraction with 1 as the numerator and a^3 as the denominator.

2. What relationship exists between x^7 and x^{-7}? They are reciprocals of one another.

Additional Examples

Simplify.

1. n^{-8} $\frac{1}{n^8}$ **2.** 4^{-3} $\frac{1}{64}$

3. $(-3)^{-2}$ $\frac{1}{9}$ **4.** $(3.47)^0$ 1

5. Write 0.072 in scientific notation. 7.2×10^{-2}

6. Write 5.6×10^{-5} in decimal notation. 0.000056

Closing the Lesson

Ask a student to explain how to simplify an expression involving negative exponents. Ask another student to explain how to write a number representing a very small measure in scientific notation. Ask a third student to explain how to write a decimal number in scientific notation using a negative exponent.

Suggested Assignments

Skills: 1–24, 31–34
Average: 1–27, 31–34
Advanced: 1–19 odd, 21–34

Exercise Notes

Communication: Reading
Guided Practice Exercises 1 and 2 check students' understanding of the main idea and major points of the lesson.

Calculator

Exercises 23 and 24 assess students' understanding of how to enter a number in scientific notation on a calculator.

Making Connections/Transitions

Exercises 25–30 connect mathematics and biology by showing how scientific notation is actually used in science.

Follow-Up

Project

Exercise 30 can be expanded into a research project. Students can research the history of each type of microscope and find out the circumstances under which each is used. Some students may want to report their findings to the class.

Enrichment

Enrichment 75 is pictured on page 288F.

Practice

Practice 92, *Resource Book*

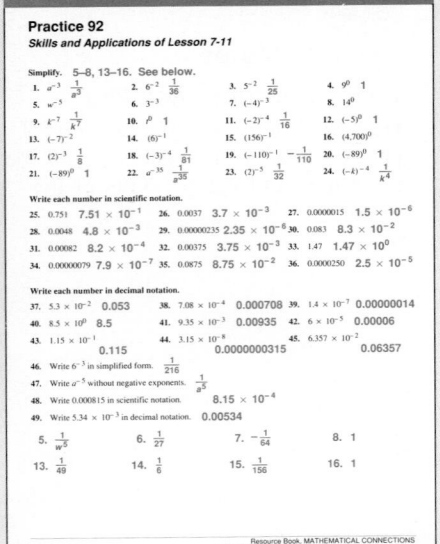

Exercises

Simplify.

A
1. y^{-12} $\frac{1}{y^{12}}$
2. 7^{-2} $\frac{1}{49}$
3. q^0 1
4. 324^0 1
5. 5^{-3} $\frac{1}{125}$
6. w^{-200} $\frac{1}{w^{200}}$
7. $(-4)^{-2}$ $\frac{1}{16}$
8. $(-1)^{-13}$ -1

Write each number in scientific notation.

9. 0.0072 7.2×10^{-3}
10. 0.875 8.75×10^{-1}
11. 0.0006012 6.012×10^{-4}
12. 0.01234 1.234×10^{-2}
13. 0.00000234 2.34×10^{-6}
14. 0.0000005 5×10^{-7}

Write each number in decimal notation.

15. 1.16×10^{-3} 0.00116
16. 5.027×10^{-5} 0.00005027
17. 1×10^{-8} 0.00000001
18. 3.209×10^{-4} 0.0003209
19. 6.1×10^{-9} 0.0000000061
20. 4×10^{-11} 0.00000000004

21. When 0.00562 is written in scientific notation, what is the exponent in the power of 10? -3

22. Write the phrase *three to the negative fourth power* as a number without exponents. $\frac{1}{81}$

 CALCULATOR

Choose the key sequence you would use to enter each number on a calculator.

B
23. 2.18×10^{-6} D
24. 2.18×10^6 B

A. ☐ 2.18 ☐ +/− ☐ EE ☐ 6
B. ☐ 2.18 ☐ EE ☐ 6
C. ☐ 2.18 ☐ +/− ☐ EE ☐ 6 ☐ +/−
D. ☐ 2.18 ☐ EE ☐ 6 ☐ +/−

CONNECTING MATHEMATICS AND SCIENCE

In biology you use a microscope to study organisms. Any organism viewed under a microscope is called a *specimen*. A specimen magnified at 100X appears 100 times as large as its actual size. Some specimens are too small to see with the human eye. The measurements of these specimens are usually written in scientific notation.

25. What is the meaning of 1500X? 1500 times the actual size

26. The diameter of the smallest organism is 1.8×10^{-9} mm. Write this in decimal notation. 0.0000000018

27. The width of a human hair is 0.02 cm. Write this in scientific notation. 2×10^{-2}

28. A bacterial cell appears to be 0.24 cm wide when viewed under a microscope at 500X. What is the actual width of the cell? Write this width in scientific notation. 0.00048 cm; 4.8×10^{-4} cm

C 29. A virus cell with width 6.5×10^{-4} mm is viewed under a microscope at 20,000X. How wide does the cell appear to be? 13 mm

30. **RESEARCH** Two basic types of microscopes are *optical microscopes* and *electron microscopes*. What method does each microscope use to magnify specimens? How powerful is each with regard to magnifying specimens? See answer on p. A22.

SPIRAL REVIEW

S 31. Complete: Two lines in the same plane that do not intersect are __?__. *(Lesson 6-4)*
parallel

32. Write 0.0000074 in scientific notation. *(Lesson 7-11)* 7.4×10^{-6}

33. Tell whether 111 is *prime* or *composite*. *(Lesson 7-1)* composite

34. **DATA,** *pages 140–141* Draw a graph to represent the data in *Average Rates of Trains*. *(Lesson 5-5)* See answer on p. A23.

Self-Test 3

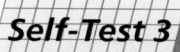

1–4. Answers may vary. Examples are given.

Express each rational number as the quotient of two integers.

1. $3.4 \quad \frac{34}{10}$ 2. $-4\frac{5}{8} \quad \frac{-37}{8}$ 3. $-9 \quad \frac{-9}{1}$ 4. $1.\overline{2} \quad \frac{11}{9}$ 7-9

Graph each open sentence. Answers to Self-Test 5–10 are on p. A23.

5. $x + 5 = -2$ 6. $3y - 4 = 14$ 7. $22 = -5n + 7$ 7-10

8. $p \leq 4$ 9. $x < -2.5$ 10. $z \geq -\frac{1}{3}$

Simplify.

11. $(-8)^{-1} \quad -\frac{1}{8}$ 12. $(-3)^0 \quad 1$ 13. $w^{-3} \quad \frac{1}{w^3}$ 14. $4^{-4} \quad \frac{1}{256}$ 7-11

15. Write 0.00003 in scientific notation. 3×10^{-5}

16. Write 2.43×10^{-7} in decimal notation. 0.000000243

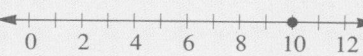
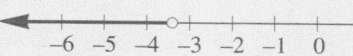
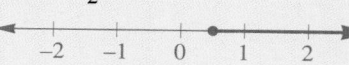
331

Chapter Review

Terms to Know

factor (p. 290)
prime number (p. 290)
composite number (p. 290)
prime factorization (p. 290)
common factor (p. 293)
greatest common factor (GCF)
  (p. 293)
multiple (p. 296)
common multiple (p. 296)
least common multiple (LCM)
  (p. 296)
equivalent fractions (p. 302)
lowest terms (p. 303)
algebraic fraction (p. 306)

simplify a fraction (p. 306)
quotient of powers rule (p. 306)
least common denominator (LCD)
  (p. 310)
terminating decimal (p. 313)
repeating decimal (p. 313)
rational number (p. 321)
irrational number (p. 321)
real number (p. 321)
open sentence (p. 324)
solution of an open sentence
  (p. 324)
graph of an open sentence (p. 324)
inequality (p. 325)

Choose the correct term from the list above to complete each sentence.

1. Fractions that represent the same amount are called __?__.
   equivalent fractions
2. Any number that can be written as a quotient of two integers, where the divisor does not equal zero, is called a(n) __?__.
   rational number
3. A(n) __?__ is a mathematical sentence that contains a variable.
   open sentence
4. A decimal in which a block of digits repeats is a(n) __?__ decimal.
   repeating
5. A number with more than two factors is a(n) __?__ number.
   composite
6. The greatest number in a list of common factors is the __?__.
   greatest common factor

Tell whether each number is *prime* or *composite*. *(Lesson 7-1)*

7. 13 prime
8. 81 composite
9. 77 composite
10. 43 prime

Write the prime factorization of each number. *(Lesson 7-1)*

11. 75 $3 \cdot 5^2$
12. 216 $2^3 \cdot 3^3$
13. 350 $2 \cdot 5^2 \cdot 7$
14. 1200 $2^4 \cdot 3 \cdot 5^2$

Find the GCF. *(Lesson 7-2)*

15. 48 and 64 16
16. 66a and 84a 6a
17. 15xy and 24x 3x
18. 36z^2 and 54z^5
    18z^2

Find the LCM. *(Lesson 7-3)*

19. 6 and 15 30
20. 8 and 12 24
21. 4y and 5x 20xy
22. 2c and 7c^2
    14c^2

Write each fraction in lowest terms. *(Lesson 7-4)*

23. $\frac{6}{12}$ $\frac{1}{2}$
24. $\frac{6}{9}$ $\frac{2}{3}$
25. $\frac{8}{10}$ $\frac{4}{5}$
26. $\frac{3}{21}$ $\frac{1}{7}$
27. $\frac{48}{10}$ $\frac{24}{5}$
28. $\frac{27}{6}$
    $\frac{9}{2}$

Chapter 7

Simplify. *(Lesson 7-5)*

29. $\frac{10ac}{5c}$ $2a$ **30.** $\frac{12a}{15a}$ $\frac{4}{5}$ **31.** $\frac{v^{12}}{v^5}$ v^7 **32.** $\frac{x^{18}}{x^6}$ x^{12} **33.** $\frac{4z^5}{z^2}$ $4z^3$ **34.** $\frac{x^8}{2x}$ $\frac{x^7}{2}$

Replace each __?__ with >, <, or =. *(Lesson 7-6)*

35. $\frac{1}{5}$ _?_ $\frac{1}{6}$ $>$ **36.** $\frac{7}{9}$ _?_ $\frac{17}{18}$ $<$ **37.** $\frac{8}{11}$ _?_ $\frac{8}{13}$ $>$ **38.** $\frac{2}{3}$ _?_ $\frac{7}{12}$ $>$

Write each fraction or mixed number as a decimal. *(Lesson 7-7)*

39. $\frac{1}{5}$ 0.2 **40.** $\frac{3}{10}$ 0.3 **41.** $\frac{4}{9}$ $0.\overline{4}$ **42.** $\frac{5}{11}$ $0.\overline{45}$ **43.** $3\frac{7}{20}$ 3.35 **44.** $5\frac{11}{50}$ 5.22

Write each decimal as a fraction or mixed number in lowest terms.
(Lesson 7-7)

45. 0.37 $\frac{37}{100}$ **46.** 0.255 $\frac{51}{200}$ **47.** 6.875 $6\frac{7}{8}$ **48.** $4.\overline{6}$ $4\frac{2}{3}$

Solve by drawing a diagram. *(Lesson 7-8)*

49. The capacities of three buckets are 10 gal, 4 gal, and 3 gal. How could
you use each bucket once to measure exactly 9 gal of water?
See answer on p. A23.
50. How many diagonals can be drawn in a pentagon? 5
51–56. Answers may vary. Examples are given.
Express each rational number as the quotient of two integers. *(Lesson 7-9)*

51. 3.2 $\frac{32}{10}$ **52.** $6\frac{1}{5}$ $\frac{31}{5}$ **53.** -7 $\frac{-7}{1}$ **54.** 0 $\frac{0}{9}$ **55.** 0.37 $\frac{37}{100}$ **56.** $-9.\overline{6}$ $\frac{-29}{3}$

**Use the number line below. Name the point that represents each
number.** *(Lesson 7-9)*

57. 2.25 D **58.** $\frac{9}{2}$ F **59.** -0.55 B **60.** $-2\frac{3}{5}$ A **61.** $-\frac{1}{3}$ C **62.** 4.34 E

Graph each open sentence. *(Lesson 7-10)*
Answers to Chapter Review 63–66 are on p. A23.
63. $x + 5 = 1$ **64.** $-4n + 5 = -3$ **65.** $z < -1.7$ **66.** $y \geq \frac{2}{3}$

Simplify. *(Lesson 7-11)*

67. m^{-5} $\frac{1}{m^5}$ **68.** 3^{-4} $\frac{1}{81}$ **69.** 12^0 1 **70.** $(-3)^{-2}$ $\frac{1}{9}$

Write each number in scientific notation. *(Lesson 7-11)*

71. 0.000074
7.4×10^{-5}
72. 0.000821
8.21×10^{-4}
73. 0.00000069
6.9×10^{-7}
74. 0.005325
5.325×10^{-3}

Write each number in decimal notation. *(Lesson 7-11)*

75. 2.7×10^{-3}
0.0027
76. 5.5×10^{-6}
0.0000055
77. 9×10^{-9}
0.000000009
78. 3.24×10^{-7}
0.000000324

Number Theory and Fraction Concepts **333**

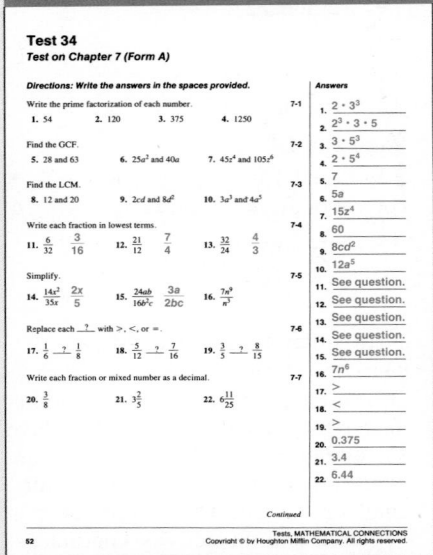

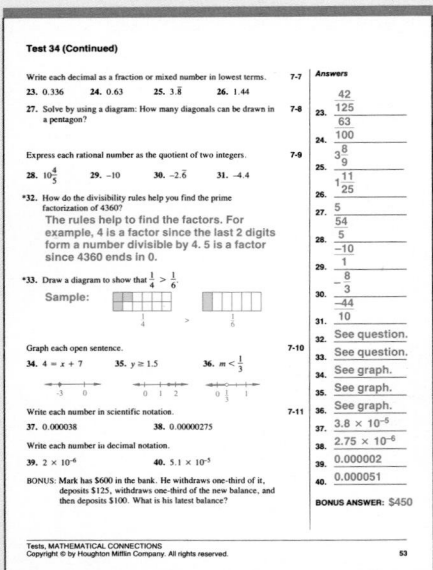

Chapter Test

Write the prime factorization of each number.

1. 84 $2^2 \cdot 3 \cdot 7$ 2. 150 $2 \cdot 3 \cdot 5^2$ 3. 625 5^4 4. 1450 $2 \cdot 5^2 \cdot 29$ **7-1**

Find the GCF.

5. 36 and 64 4   6. $27a$ and $45a$ $9a$ 7. $48z^2$ and $112z^6$ $16z^2$ **7-2**

Find the LCM.

8. 18 and 30 90   9. $3cd$ and $8d$ $24cd$ 10. $5a^2$ and $2a^5$ $10a^5$ **7-3**

Write each fraction in lowest terms.

11. $\frac{8}{12}$ $\frac{2}{3}$ 12. $\frac{9}{15}$ $\frac{3}{5}$ 13. $\frac{18}{8}$ $\frac{9}{4}$ 14. $\frac{21}{6}$ $\frac{7}{2}$ **7-4**

Simplify.

15. $\frac{9x}{15}$ $\frac{3x}{5}$ 16. $\frac{24b}{18bc}$ $\frac{4}{3c}$ 17. $\frac{x^{15}}{x^5}$ x^{10} 18. $\frac{9n^8}{n^2}$ $9n^6$ **7-5**

Replace each __?__ with >, <, or =.

19. $\frac{1}{4}$ _?_ $\frac{1}{6}$ >   20. $\frac{6}{13}$ _?_ $\frac{6}{7}$ <   21. $\frac{3}{8}$ _?_ $\frac{7}{16}$ <   **7-6**

Write each fraction or mixed number as a decimal.

22. $\frac{4}{5}$ 0.8   23. $\frac{4}{15}$ $0.2\overline{6}$ 24. $4\frac{1}{6}$ $4.1\overline{6}$ 25. $7\frac{13}{20}$ 7.65   **7-7**

Write each decimal as a fraction or mixed number in lowest terms.

26. 0.672 $\frac{84}{125}$ 27. 0.27 $\frac{27}{100}$ 28. $7.\overline{3}$ $7\frac{1}{3}$ 29. 2.88 $2\frac{22}{25}$

30. Solve by using a diagram: Eight players participated in a single-elimination tennis tournament. In such a tournament, each player is out after one loss. How many games did the champion play? 3   **7-8**

31–34. Answers may vary. Examples are given.

Express each rational number as the quotient of two integers.

31. 7.4 $\frac{74}{10}$ 32. -4.25 $\frac{-425}{100}$ 33. -8 $\frac{-8}{1}$ 34. $3.\overline{3}$ $\frac{10}{3}$ **7-9**

★ 35. Write the number -1 as a quotient of two integers in three different ways. Answers may vary. Examples: $\frac{-3}{3}$, $\frac{-7}{7}$, $\frac{-12}{12}$

★ 36. Name three real numbers that are between $\frac{1}{3}$ and $\frac{1}{2}$. Answers may vary. Examples: $\frac{3}{8}$, $\frac{5}{12}$, $\frac{11}{24}$

Graph each open sentence. Answers to Chapter Test 37–40 are on p. A23.

37. $6 = x + 9$ 38. $3w - 8 = 1$ 39. $y \geq 5.5$ 40. $m < \frac{7}{8}$ **7-10**

Write each number in scientific notation.   **Write each number in decimal notation.**

41. 0.00076 7.6×10^{-4} 42. 0.00000335 3.35×10^{-6} 43. 3×10^{-5} 0.00003   44. 4.7×10^{-8} 0.000000047   **7-11**

334

CHAPTER EXTENSION

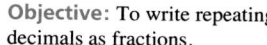

Repeating Decimals as Fractions

Objective: To write repeating decimals as fractions.

You may have memorized fractional equivalents for some common repeating decimals, such as $0.\overline{3} = \frac{1}{3}$. Using algebra, however, it is possible to find the fractional equivalent of any repeating decimal.

Example

Solution

Write $0.\overline{27}$ as a fraction.

Let $n = 0.\overline{27}$.

Then $100n = 27.\overline{27}$. ◀── Multiply n by 10^2 because there are 2 digits in the repeating block.

Subtract n from $100n$.

$$100n = 27.272727\ldots$$
$$-\quad n = -0.272727\ldots$$
$$\overline{99n = 27}$$

Solve for n.

$$\frac{99n}{99} = \frac{27}{99}$$

$$n = \frac{27}{99} = \frac{3}{11}$$

Exercises

Write each repeating decimal as a fraction.

1. $0.\overline{5}$ $\frac{5}{9}$
2. $0.\overline{8}$ $\frac{8}{9}$
3. $0.\overline{35}$ $\frac{35}{99}$
4. $0.\overline{52}$

5. $2.\overline{3}$ $\frac{7}{3}$
6. $6.\overline{7}$ $\frac{61}{9}$
7. $3.\overline{21}$ $\frac{106}{33}$
8. $9.\overline{45}$

9. $1.\overline{123}$ $\frac{374}{333}$
10. $4.\overline{534}$ $\frac{1510}{333}$
11. $5.7\overline{2}$ $\frac{103}{18}$
12. $2.3\overline{21}$

 CALCULATOR   4. $\frac{52}{99}$ 8. $\frac{104}{11}$ 12. $\frac{383}{165}$

13. To multiply $\frac{1}{3} \times 3$ using his four-function calculator, Gustavo entered this key sequence.

    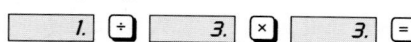

    The result was ⟨ 0.9999999 ⟩ . Francine entered the same

    key sequence on her scientific calculator and got the result ⟨ *1.* ⟩ .
    What conclusion can you make about $0.\overline{9}$ and 1? They are equivalent.

14. Refer to Exercise 13. What do you think the results would be if Gustavo and Francine entered this key sequence on their calculators?

    ⟨ *4.* ⟩ ÷ ⟨ *3.* ⟩ × ⟨ *3.* ⟩ =

    3.9999999; 4.

Number Theory and Fraction Concepts   **335**

Teaching Notes

Some students have difficulty changing repeating decimals to fractions. Work through the Example carefully with students, being certain they understand each step of the procedure.

Making Connections/Transitions

This extension connects the previously learned concepts of solving equations and simplifying fractions. By using a calculator students can verify that a fraction represents a specific decimal.

Additional Example

Write $0.3\overline{8}$ as a fraction. $\frac{7}{18}$

Exercise Note

Calculator

Exercises 13 and 14 demonstrate the rounding differences that occur when doing multiplication of fractions on a four-function calculator and on a scientific calculator. Students are asked to reach an important conclusion about $0.\overline{9}$ and 1.

Write each fraction in lowest terms.

1. $\frac{8}{12}$ $\frac{2}{3}$

2. $\frac{16}{18}$ $\frac{8}{9}$

3. $\frac{30}{36}$ $\frac{5}{6}$

4. $\frac{24}{54}$ $\frac{4}{9}$

Simplify.

5. $\frac{9c}{24c}$ $\frac{3}{8}$

6. $\frac{12a}{18ab}$ $\frac{2}{3b}$

7. $\frac{m^7}{m^3}$ m^4

8. $\frac{9x^{15}}{x^4}$ $9x^{11}$

Replace each ? with >, <, or =.

9. $\frac{1}{5}$? $\frac{1}{4}$ <

10. $\frac{5}{8}$? $\frac{5}{9}$ >

11. $\frac{13}{15}$? $\frac{17}{20}$ >

Write each fraction or mixed number as a decimal.

12. $\frac{13}{20}$ 0.65

13. $\frac{7}{9}$ $0.\overline{7}$

14. $3\frac{19}{25}$ 3.76

15. $6\frac{1}{6}$ $6.1\overline{6}$

Cumulative Review

Standardized Testing Practice

Choose the letter of the correct answer.

1. Reading Preferences of Students

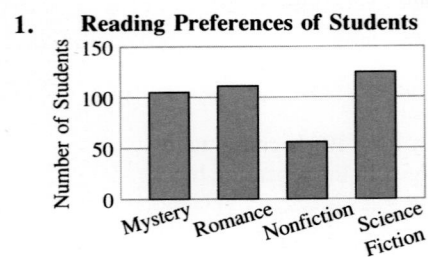

Type of Book

About how many more students prefer mysteries than prefer nonfiction books?

A. about 25   **B.** about 50

C. about 75   D. about 100

2. Evaluate the difference $a - b$ when $a = 51.2$ and $b = 3.43$.

A. 1.69   **B.** 47.77

C. 54.63   D. 8.55

3. Find the LCM: $12xz$ and $18yz$

A. $6z$ B. $216xyz^2$

C. $36xyz^2$ **D.** $36xyz$

4. Find the measure of an angle that is supplementary to an angle with a measure of 83°.

A. 97°   B. 7°

C. 277°   D. 263°

5. Find the GCF: 84 and 140

A. 4   **B.** 28

C. 84   D. 420

6. Average Weights of Infants

During which interval does the greatest increase in weight occur?

A. 0–3 months   B. 3–6 months

C. 6–9 months   D. 9–12 months

7. Solve: $-4n + 8 = 32$

A. $n = -10$ B. $n = 6$

C. $n = -6$ D. $n = 10$

8. Write $\frac{5}{12}$ as a decimal.

A. $0.41\overline{6}$ **B.** $0.41\overline{6}$

C. $0.4\overline{16}$ D. 0.416

9. Evaluate $-xy - 16$ when $x = 3$ and $y = -2$.

A. 10   B. -22

C. -10 D. 22

10. Simplify: $\frac{x^9}{x^3}$

A. x^3 **B.** x^6 C. x^{12} D. x^{-6}

11. **Choose the true statement about the data:** 42, 38, 51, 42, 72
   A. The range is 30.
   B. The mean is less than the mode.
   C.) The mode and the median are equal.
   D. The mean and the median are equal.

12. **Write an equation for the situation.** Ella is 165 cm tall. She is 3 cm shorter than Deion. How tall is Deion?
   A. $3d = 165$
   B. $d + 165 = 3$
   C. $d + 3 = 165$
   D.) $d - 3 = 165$

13. **At noon, the temperature was −9°C. During the next 5 h, it fell 4°C. What was the temperature at 5:00 P.M.?**
   A. 5°C   B. 13°C
   C. −5°C   D.) −13°C

14. **Find the answer mentally:** 6(198)
   A. 1212   B. 1248
   C.) 1188   D. 1152

15. **To which of the sets does the number 333 belong?**
   A. whole numbers
   B. integers
   C. rational numbers
   D.) all of the above

16. **Eva did 10, 15, and 20 sit-ups each day during the first, second, and third weeks of a fitness program. If she continued the pattern, how many sit-ups did she do each day during the fifth week?**
   A.) 30   B. 40
   C. 25   D. 35

17. **Choose the fraction that is *not* equivalent to $\frac{3}{4}$.**
   A. $\frac{39}{52}$ B. $\frac{21}{28}$
   C. $\frac{75}{100}$ D.) $\frac{69}{96}$

18. **Find the sum of the measures of the angles in a 20-sided polygon.**
   A.) 3240°   B. 3600°
   C. 360°   D. 3598°

19. **Simplify:** $8a^6 \cdot 5a^2$
   A. $3a^4$ B. $40a^{12}$
   C. $13a^8$ D.) $40a^8$

20. **Describe the triangle.**
   A.) right, scalene
   B. acute, scalene
   C. equilateral
   D. right, isosceles

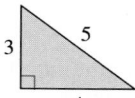

Write each decimal as a fraction or mixed number in lowest terms.

16. 0.86 $\frac{43}{50}$

17. 0.222 $\frac{111}{500}$

18. 4.64 $4\frac{16}{25}$

19. $9.8\overline{3}$ $9\frac{5}{6}$

20. Solve by drawing a diagram: The lengths of three steel rods are 13 m, 27 m, and 19 m. How can you use these rods to mark off a length of 5 m?
   $13 + 19 - 27 = 5$

Alternative Assessment
For Lessons 7-4 through 7-8, you may wish to assess students' knowledge about the material by having groups of students work the Self-Test exercises at the chalkboard.

Number Theory and Fraction Concepts   **337**

Planning Chapter 8

Chapter Overview

In Chapter 8, operations with rational numbers are extended to include algebraic fractions. *Lesson 1* begins by using an area model to explain multiplication of rational numbers. The visual representation of the product of two rational numbers helps students to understand the relative size of the product and thus helps to develop number sense. *Lesson 2* continues the use of the area model in developing the concept of division with rational numbers. Algebraic concepts are interwoven with previously learned ideas on formulas, rational numbers, and problem solving. *Lesson 3* addresses the form of an answer. Students are asked not only to find an answer, but also to determine what form the answer should take so that it makes sense.

The *Focus on Explorations* uses fraction tiles to model equivalent fractions and addition and subtraction of rational numbers. *Lessons 4 and 5* continue the process of adding and subtracting rational numbers, with algebraic fractions included as a natural part of the process. *Lesson 6* presents estimation with rational numbers and continues to develop students' number sense. In *Lesson 7*, instruction is given on how to solve problems with missing information. *Lessons 8 and 9* extend the processes of solving equations to include rational numbers. In *Lesson 10*, formulas are transformed, thus extending a method first presented in Chapter 4.

Background

Chapter 8 continues the strands of estimation, communication, problem solving, connections, and technology begun earlier in the book.

The prominent role of estimation is evident throughout the chapter. Estimation may involve using compatible numbers, rounding, or mental math. As technology becomes more accessible, the need for estimation skills increases.

Communication skills involve discussing mathematics, drawing comparisons, making inferences, and linking mathematical language to the students' own world of experience. Students are frequently asked to explain their reasoning, and their writing can serve as a diagnostic device for the teacher.

Problem solving and decision making involve general knowledge and reasonable answers. Students are asked to call upon their everyday knowledge to supply missing facts in order to solve problems. They are also asked to call upon their common sense to determine whether an answer should be a whole number, a fraction, or a decimal.

Connections are made to varying career areas; students examine problems that may appear in actual career fields. Branches of mathematics are connected and integrated. For example, probability is included, as well as tables and charts. A more cohesive picture of mathematics begins to emerge.

Technology is addressed by the use of the fraction key and the reciprocal key found on many calculators. In some instances, the key sequence is indicated. At other times, a possible result is shown.

Objectives

8-1	To multiply rational numbers and algebraic fractions.
8-2	To find the quotients of rational numbers and of algebraic fractions.
8-3	To determine the correct form of an answer when solving a problem.
☰	**FOCUS ON EXPLORATIONS** To use fraction tiles to model addition and subtraction of fractions.
8-4	To add and subtract rational numbers and algebraic fractions with like denominators.
8-5	To add and subtract rational numbers and algebraic fractions with unlike denominators.
8-6	To estimate the sum, difference, or quotient of rational numbers.
8-7	To solve problems by supplying missing facts.
8-8	To solve equations involving rational numbers.
8-9	To use reciprocals to solve equations.
8-10	To evaluate and transform formulas containing rational numbers.
☰	**FOCUS ON APPLICATIONS** To read and interpret a stock price table.
☰	**CHAPTER EXTENSION** To add and subtract algebraic fractions with unlike denominators that contain variables.

Chapter Planner

Instructional Aids	Manipulatives	Cooperative Learning	Technology	Practice/Reteaching	Assessment

Materials: fraction tiles, calculator
Resource Book:
  Lesson Starters 70–79
Visuals, Folder C
	Lessons 8-1, 8-2,
	Focus on
	Explorations, 8-4, 8-5
	Activities Book:
	  Manipulative
	    Activities 15-16
		Lessons 8-1, 8-2, 8-3,
		8-5, 8-10
		Activities Book:
		  Cooperative
		    Activity 16
			Lessons 8-1, 8-2, 8-3,
			8-5, 8-6, 8-8, 8-9
			Activities Book:
			  Calculator Activity 8
			  Computer Activity 8
			*Connections Plotter
			Plus* Disk
				Toolbox, pp. 753–771
				Extra Practice, p. 741
				Resource Book:
				  Practice 95–108
				Study Guide,
				  pp. 139–158
					Self-Tests 1–3
					Quick Quizzes 1–3
					Chapter Test, p. 382
					Resource Book:
					  Lesson Starters 70–79
					  (Daily Quizzes)
					Tests 36–42

Assignment Guide Chapter 8

Day	Skills Course	Average Course	Advanced Course
1	**8-1:** 1–22, 37–42	**8-1:** 1–22, 23–35 odd, 37–42	**8-1:** 1–19 odd, 21–42
2	**8-2:** 1–22, 31–34	**8-2:** 1–26, 31–34	**8-2:** 1–15 odd, 17–34
3	**8-3:** 1–13 **Exploration:** Activities I–IV	**8-3:** 1–13 **Exploration:** Activities I–IV	**8-3:** 1–13 **Exploration:** Activities I–IV
4	**8-4:** 1–30, 40, 41	**8-4:** 1–30, 40, 41	**8-4:** 1–23 odd, 25–41 **8-5:** 1–45 odd, 47–55
5	**8-5:** 1–26, 27–39 odd, 52–55	**8-5:** 1–45, odd, 52–55	**Mixed Review** **8-6:** 1–24
6	**Mixed Review** **8-6:** 1–16, 21–24	**Mixed Review** **8-6:** 1–24	**8-7:** 1–16, Challenge
7	**8-7:** 1–16	**8-7:** 1–16	**8-8:** 1–32
8	**8-8:** 1–16, 29–32	**8-8:** 1–24, 29–32	**8-9:** 1–43 odd, 45–53, Historical Note
9	**8-9:** 1–14, 15–27 odd, 49–53	**8-9:** 1–43 odd, 49–53	**8-10:** 1–26 **Application**
10	**8-10:** 1–18, 24–26 **Application**	**8-10:** 1–26 **Application**	*Prepare for Chapter Test:* Chapter Review
11	*Prepare for Chapter Test:* Chapter Review	*Prepare for Chapter Test:* Chapter Review	*Administer Chapter 8 Test;* Chapter Extension; Cumulative Review
12	*Administer Chapter 8 Test*	*Administer Chapter 8 Test;* Cumulative Review	

Teacher's Resources

Resource Book
  Chapter 8 Project
  Lesson Starters 70–79
  Practice 95–108
  Enrichment 77–87
  Diagram Masters
  Chapter 8 Objectives
  Family Involvement 8
  Spanish Test 8
  Spanish Glossary

Activities Book
  Cooperative Activity 16
  Manipulative Activities
    15–16
  Calculator Activity 8
  Computer Activity 8

Study Guide, pp. 139–158

Tests
  Tests 36–42

Visuals
  Folder C

*Connections Plotter
Plus* **Disk**

Using Technology

CALCULATORS

This chapter reviews arithmetic operations involving fractions and introduces the use of fractions in solving linear equations. Calculators with a fraction key, usually marked as a b/c , can be used to check answers and rediscover the principles of fraction arithmetic. This approach is likely to be very different from the ones that students are familiar with, which allows them to have fresh insights into fraction arithmetic.

COMPUTERS

Through a simple BASIC program (see below), a computer can be made to act like a calculator with a fraction key. Although this is less convenient than using a calculator in many cases, you may have easier access to computers than to these relatively new calculator models.

Lesson 8-1

CALCULATORS

The rules of fraction multiplication can be reinforced by having students use a calculator equipped with a fraction key. Instructions for using this key are given on page 341 of the text, and are also available in the instruction manual that comes with the calculator. Students using such a calculator can be led to discover alternative algorithms for multiplying mixed numbers and other operations.

COMPUTERS

A computer can be programmed easily to do fraction arithmetic, which may be advantageous if calculators with a fraction key are not readily available. Here is a BASIC program that will multiply mixed numbers.

```
10 INPUT "FIRST WHOLE NUMBER: "; W1
20 INPUT "FIRST NUMERATOR: "; N1
30 INPUT "FIRST DENOMINATOR: "; D1
40 INPUT "SECOND WHOLE NUMBER: "; W2
50 INPUT "SECOND NUMERATOR: "; N2
60 INPUT "SECOND DENOMINATOR: "; D2
70 LET W = W1 * W2 : REM WHOLE PART OF
     RESULT
80 LET N = N1 * W2 * D2 + N2 * W1 * D1 + N1 * N2:
     REM NUMERATOR
```

```
90 LET D = D1 * D2 : REM DENOMINATOR OF
     RESULT
100 IF D > N THEN 130 : REM IF FRACTION IS
     PROPER
110 LET W = W + INT(N/D) : REM MAKE
     FRACTION PROPER
120 LET N = N - D * INT(N/D) : REM CHANGE
     NUMERATOR
130 IF N = 1 THEN 200 : REM IF ALREADY
     REDUCED
140 FOR I = 2 TO N : REM LOOK FOR COMMON
     DIVISORS
150 IF (INT(N/I) <> N/I) OR (INT(D/I) <> D/I)
     THEN 190 : REM IF I IS NOT A DIVISOR
     OF D AND N
160 LET N = N/I : REM REDUCE THE FRACTION
     BY DIVIDING
170 LET D = D/I : REM BOTH N AND D BY I
180 GOTO 150 : REM PERHAPS I IS STILL A
     DIVISOR
190 NEXT I : REM NEXT POSSIBLE DIVISOR
200 PRINT "ANSWER: "; W; " + "; N; "/"; D
210 PRINT
220 GOTO 10
```

Lesson 8-2

CALCULATORS AND COMPUTERS

Using calculators with fraction keys, or a BASIC program like that provided above, students can rediscover the idea of dividing fractions as multiplying by the reciprocal of the divisor. Have them think about division as answering questions such as "What do you multiply $\frac{2}{3}$ by to get 1?" By trial and error, students can find the answer, which will lead them to the rules for dividing fractions in an unusual but very sound way.

Lesson 8-5

CALCULATORS AND COMPUTER

A fraction key or BASIC fraction addition program can be used to explore least common denominators, and therefore greatest common divisors, using the relationship that the GCD of two numbers is the quotient of the product of the numbers and the LCD. The LCD itself can be found by adding two fractions with numerator 1.

Using Manipulatives

The Exploration at the beginning of the lesson can be done by using physical objects rather than drawings. You can create fraction cards by coloring halves, thirds, quarters, and so on, of a set of squares made from index cards. Then use the cards to demonstrate multiplications such as the one in the Exploration. After seeing a demonstration or two, students can use the cards to model their own multiplications.

Lesson 8-2

The fraction cards used in Lesson 8-1 also can be used in this lesson to model division of rational numbers. Students can first find the reciprocal of the divisor and then use the cards to find the product of the dividend and the reciprocal of the divisor.

Lessons 8-4, 8-5

Fraction tiles are very useful in modeling addition and subtraction of rational numbers. Students can use the tiles to perform the activities in the Exploration preceding these two lessons and then use them to model additions and subtractions of their own.

Reteaching/Alternate Approach

Lesson 8-1

COOPERATIVE LEARNING

The Exploration at the beginning of the lesson can be done in cooperative groups. As an extension, ask each group to create three problems involving the product of rational numbers. The groups should draw models to represent the product for each of their problems. Each group should share its problems and models with the class and give an oral explanation of how the models relate to the multiplication process.

Lesson 8-2

COOPERATIVE LEARNING

When dividing with rational numbers, students often find it difficult to understand that division by a number between 0 and 1 results in a greater quotient. To make this concept clearer, have the class work in groups to draw models that demonstrate what happens when a number is divided by a number between 0 and 1. Ask each group to explain what happens to the quotient as the divisor approaches zero. Have each group divide by numbers greater than one and explain how the quotient and the size of the dividend relate to one another.

Lesson 8-3

COOPERATIVE LEARNING

Have students work in cooperative groups to create three word problems. One problem should require a whole number answer, the second a fractional answer, and the third a decimal answer. Each group should present its problems to the class, giving reasons why each type of answer was needed.

Lesson 8-5

COOPERATIVE LEARNING

Have each cooperative group create two exercises involving addition and two exercises involving subtraction of rational numbers with unlike denominators. Each exercise and its solution should be written on an index card. Each group should then present one of its exercises to the other groups. The other groups should then solve the exercise, with the presenting group determining whether the solution is correct. This should continue until all the exercises have been solved.

Lesson 8-7

After introducing the concept of missing facts, you may wish to have students create a ''missing facts'' poster. Students can then fill in the poster with facts that may be missing from problems, such as 36 inches equals 1 yard, or one dozen contains twelve objects. The poster can be organized by type, that is, by length, mass, capacity, general knowledge, and so on.

Teacher's Resources Chapter 8

Enrichment masters from the Resource Book are pictured here. See the Teacher's Resources chart on page 338B for a complete listing of all materials available for this chapter.

8-1 Enrichment
Application

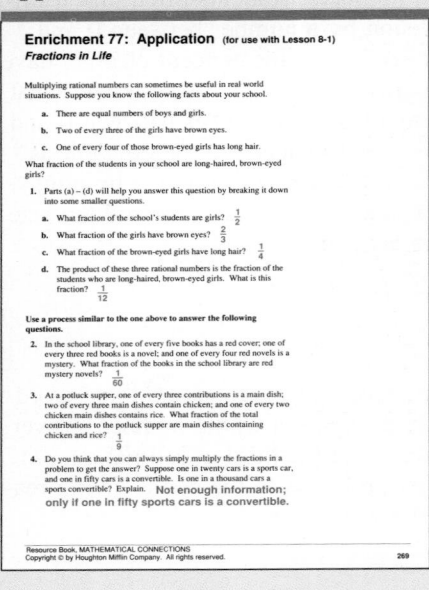

8-2 Enrichment
Problem Solving

8-3 Enrichment
Communication

8-4 Enrichment
Thinking Skills

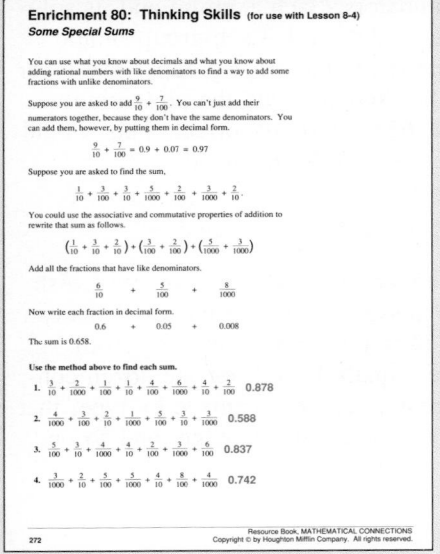

8-5 Enrichment
Connection

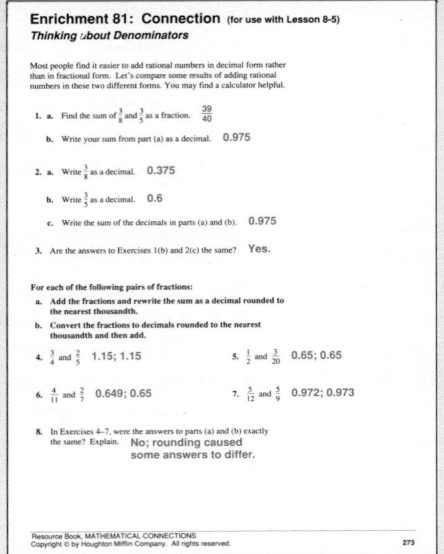

8-6 Enrichment
Data Analysis

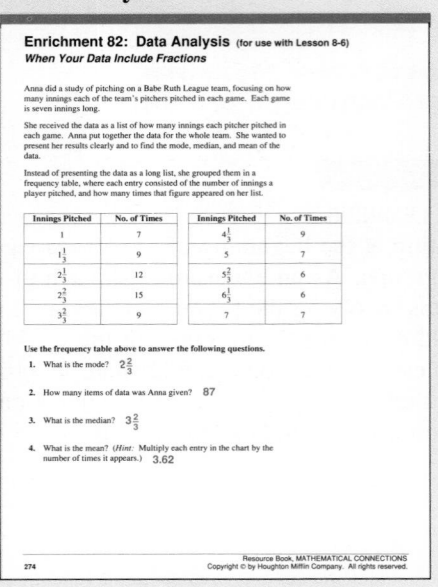

8-7 Enrichment
Thinking Skills

Enrichment 83: Thinking Skills (for use with Lesson 8-7)
Missing Facts That Matter

In some kinds of problems, a fact that you don't know is necessary to solve the problem. These missing facts break down into two types: facts that you can't know because the problem doesn't give you enough information, and facts that you can look up.

Tell what fact you need to know in order to answer each of these questions. If you think you could look up the needed information, say so.

1. How many kilometers did Tina drive if she left at 2 P.M. and drove at a constant rate of 85 kilometers per hour? **What time did she stop driving?**

2. How many Swiss francs did Jeremy receive for 1500 yen at the currency exchange? **How many Swiss francs equal 1 yen? You could look this up if you knew the date of the exchange.**

3. If Earth is 93 million miles from the sun, how many seconds does it take light to travel from the sun to Earth? **What is the speed of light? You can look this up.**

4. If pencils cost $.15 each, and paper costs $1.29 per pad, how much change does Maria receive from a $20 bill? **How many pencils and paper pads did she buy?**

5. a. Fred is $\frac{2}{3}$ of Bob's height and Ann is $\frac{3}{4}$ of Mary's height. Who is taller, Ann or Fred? **How tall are Bob and Mary?**

   b. If you know that Bob is taller than Mary, is that enough information to solve part (a)? What if Mary is taller than Bob? **No; Yes.**

8-8 Enrichment
Exploration

Enrichment 84: Exploration (for use with Lesson 8-8)
Doing and Undoing

1. In the morning, when you're getting ready for school, you must put on both socks, put on both shoes, and tie both shoes. The order in which you do these operations *can* affect the final results. Give an example of how changing the order would affect the results. **Answers will vary.**

2. In the mathematical expression $(5 + 3) \times 7 - 4$, the order in which you perform these operations affects the final result. Give both the right answer and an answer you'd get if you used the wrong order. Explain what you did wrong to get that answer. **52; Answers will vary.**

When it's time to take off your shoes and socks, the order still matters. You must untie shoes, then remove shoes, and then remove socks. Notice that not only is each operation undone, but it is undone in the reverse of the order in which it was done. The same is true in mathematics.

3. Explain how to evaluate the expression $3x + \frac{1}{2}$ for $x = 5$. What is the result? **First, substitute 5 for x. Then multiply 5 by 3 and add $\frac{1}{2}$; $15\frac{1}{2}$**

The equation $3x + \frac{1}{2} = 15\frac{1}{2}$ represents the idea that if x is multiplied by 3, and then $\frac{1}{2}$ is added to the product, the result will be $15\frac{1}{2}$. To *do* the arithmetic, you first multiply by 3, then you add $\frac{1}{2}$. Solving the equation is like *undoing* the arithmetic. Following the "shoes and socks" model, to *undo*, you'd subtract $\frac{1}{2}$, then divide by 3.

4. a. To solve the equation $2x - \frac{3}{2} = 9$, what would you do first? **Add $\frac{3}{2}$ to both sides.**

   b. What would you do next? **Divide both sides by 2.**

   c. What is the result? **$\frac{21}{4}$**

5. Think of a routine that you do every day. Describe each step in the routine. How would changing the order of the steps affect your results? **Answers will vary.**

8-9 Enrichment
Thinking Skills

Enrichment 85: Thinking Skills (for use with Lesson 8-9)
Rational Numbers in Real World Equations

Before you solve each problem, analyze the facts and decide whether the answer will be greater than or less than 18, then solve. Were you right about the size of the answer?

1. In one basketball game, Kele made two thirds of the baskets that he attempted. If he attempted eighteen baskets, how many did he make? **less than; 12**

2. In another game, Kele again made two thirds of his basket attempts, but this time he made eighteen baskets. How many did he attempt? **greater than; 27**

Before you solve each problem, analyze the facts and write down what you can tell about the size of the answer. Then solve. Were you right about the size?

3. Three fourths of the girls in a class play soccer. If twelve more girls in the class played, there would be fifty-seven girls who play soccer. How many girls are in that class? **greater than 57; 60**

4. Four fifths of the people in a certain math class got the right answer to Problem 3. If two more people in the class had gotten the right answer, the number getting it right would have been twenty-six. How many people are in that class? **greater than 26; 30**

5. Jorge is a relief pitcher whose specialty is finishing games. He always pitches exactly one and one third innings per game. If he has pitched twelve innings, in how many games has he pitched? **less than 12; 9**

8-10 Enrichment
Data Analysis

Enrichment 86: Data Analysis (for use with Lesson 8-10)
Mailing a Letter

Sometimes a graph is more useful than a formula. If the formula is long or complicated, a graph can be an easier way to get the information that you need. A good example of this is finding how much it costs to mail a first-class letter.

The postage for a one-ounce letter in 1990 was 25 cents, but each additional full ounce cost only 20 cents. A 1.1-ounce letter cost the same as a 2-ounce letter — they were both 45 cents. Let's look at the graph.

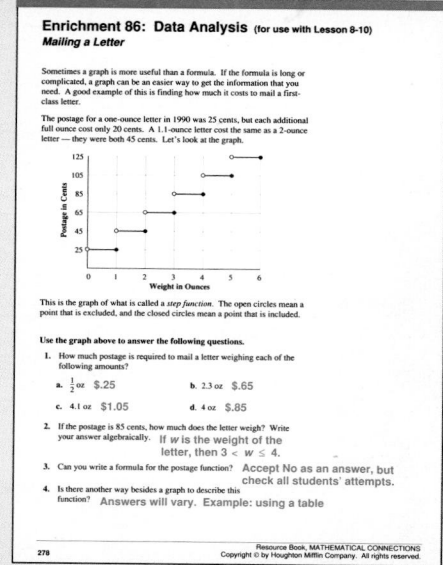

This is the graph of what is called a *step function*. The open circles mean a point that is excluded, and the closed circles mean a point that is included.

Use the graph above to answer the following questions.

1. How much postage is required to mail a letter weighing each of the following amounts?
   a. $\frac{1}{2}$ oz **$.25**   b. 2.3 oz **$.65**
   c. 4.1 oz **$1.05**   d. 4 oz **$.85**

2. If the postage is 85 cents, how much does the letter weigh? Write your answer algebraically. **If w is the weight of the letter, then $3 < w \leq 4$.**

3. Can you write a formula for the postage function? **Accept No as an answer, but check all students' attempts.**

4. Is there another way besides a graph to describe this function? **Answers will vary. Example: using a table**

End of Chapter Enrichment
Extension

Enrichment 87: Extension (for use after Chapter 8)
How It Works

Recall the rule for dividing rational numbers.

$$\frac{a}{b} \div \frac{c}{d} = \frac{a}{b} \cdot \frac{d}{c}$$

Have you ever wondered how or why this works?

For this explanation, it's useful to think of $\frac{a}{b} \div \frac{c}{d}$ as one big fraction. (Mathematicians call it a *complex fraction* because its parts are fractions.) The numerator of the complex fraction is $\frac{a}{b}$, and the denominator is $\frac{c}{d}$.

$$\frac{\frac{a}{b}}{\frac{c}{d}}$$

Just as you can rewrite $\frac{3}{4}$ as $\frac{9}{12}$ by multiplying both the numerator and the denominator by 3, you can also multiply the numerator and the denominator of a complex fraction by the same number. The goal is to end up with a simple fraction rather than a complex fraction, so you want to make the denominator equal to 1. Multiply both numerator and denominator by $\frac{d}{c}$, the reciprocal of $\frac{c}{d}$.

$$\frac{\frac{a}{b} \cdot \frac{d}{c}}{\frac{c}{d} \cdot \frac{d}{c}} = \frac{\frac{a}{b} \cdot \frac{d}{c}}{1} = \frac{a}{b} \cdot \frac{d}{c}$$

Use the method above to do the following division problems.

1. $\frac{3}{4} \div \frac{5}{8}$ **$\frac{6}{5}$**   2. $\frac{11}{32} \div \frac{5}{16}$ **$\frac{11}{10}$**

3. $\frac{2}{9} \div \frac{8}{27}$ **$\frac{3}{4}$**   4. $\frac{5}{9} \div \frac{7}{12}$ **$\frac{20}{21}$**

5. $\frac{4x}{5y} \div \frac{2x}{3y}$ **$\frac{6}{5}$**   6. $\frac{5x}{7y^2} \div \frac{3xy}{4}$ **$\frac{20}{21y^3}$**

For Chapter 8 Opener

Students most likely are aware that the world faces shortages of important fossil fuels in the future. They may be interested in learning about the many efforts under way around the world to develop renewable sources of energy. For example, one such source is geothermal energy—energy obtained from heat stored in rocks, water, and steam under the surface of the earth. Countries that already have geothermal plants include Italy, New Zealand, the Philippines, Japan, and Mexico. The Caribbean Islands, East Africa, Central Asia, the Himalayas, and western Arabia are also thought to be good locations for obtaining geothermal energy.

For Page 374

Fannie Farmer, described in the Historical Note, was physically challenged in her life. Her left leg was partially paralyzed after she contracted polio at the age of sixteen. She was disabled further by strokes suffered later in life, but continued her lectures from a wheelchair.

Research Activities

For Chapter 8 Opener

You may want to have students research and list different kinds of renewable energy. These could include, for example, solar energy, wind energy, and alcohol fuels. Students can work in cooperative groups to find out what countries are using different energy sources. They can also familiarize themselves with the geography and climate of certain countries, to see if those countries are making use of available renewable energy sources. The students can present the results of their research to the class.

Recommended Energy Saving Temperature Settings		
Appliance	Fahrenheit	Celsius
air conditioner	78°F	26°C
freezer	0°F	−18°C
heater (day)	65-68°F	18-20°C
heater (night)	55-60°F	13-16°C
refrigerator	38-42°F	3-6°C
water heater	110-120°F	43-49°C

Recommended Thickness of Glass Fiber Insulation (Inches)		
City	Attic Floors	Basement Ceilings
Atlanta	8-8½	3½-4
San Francisco	8-8½	4-4½
Omaha	9½-10½	6-6½
Grand Rapids	11	6½
Duluth	12-13	6½

338

Rational Numbers

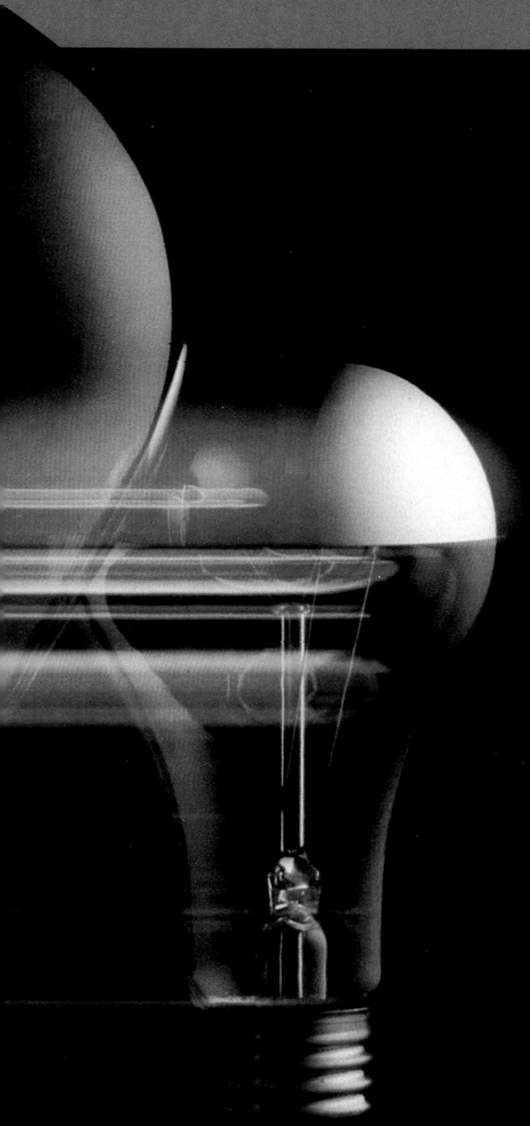

Water Heater
Electric
First hour rating: 58

ENERGY GUIDE

Estimates on the scale are based on a national average electric rate of 8.04¢ per kilowatt-hour.

Only models with first hour ratings of 56-64 gallons are used on this scale.

Model with lowest energy cost
$423

$467

Model with highest energy cost
$547

▼   THIS ▼ MODEL   ▼

Estimated yearly energy cost

Your cost will vary depending on your local energy rate and how you use the product. This energy rate is based on U.S. Government standard tests.

		YEARLY COST
Cost per	4¢	$232
kilowatt-	6¢	$348
hour	8¢	$465
	10¢	$581
	12¢	$697
	14¢	$813

Ask your salesperson or local utility for the energy rate in your area.

Did You Know?

The brightness of a light bulb is indicated by lumens, not wattage. Wattage indicates how much energy a bulb uses. It is most energy-efficient to use bulbs with high lumens and low wattage. Compared to an incandescent light bulb, a fluorescent bulb is 3 to 5 times more efficient. For example, a 40-watt fluorescent bulb gives off 3200 lumens, and a 60-watt incandescent bulb gives off only 882 lumens.

Research Activities
(continued)

For Page 374
Students can research how recipe measurements were given in cookbooks written before Fannie Farmer introduced the idea of standard cooking measurements.

Suggested Resources

Corson, Walter H., ed. *The Global Ecology Handbook: What You Can Do About the Environmental Crisis.* Boston: Beacon Press, 1990.

Our Magnificent Earth: A Rand McNally Atlas of Earth Resources. New York: Rand McNally and Company, 1979.

Teaching the Lesson
- Materials: index cards, calculator
- Lesson Starter 70
- Visuals, Folder C
- Using Technology, p. 338C
- Using Manipulatives, p. 338D
- Reteaching/Alternate Approach, p. 338D
- Cooperative Learning, p. 338D
- Toolbox Skill 20

Lesson Follow-Up
- Practice 95
- Enrichment 77: Application
- Study Guide, pp. 139–140
- Calculator Activity 8
- Manipulative Activity 15

Warm-Up Exercises

Simplify.

1. $\dfrac{24}{30}$ $\quad \dfrac{4}{5}$

2. $\dfrac{32}{48}$ $\quad \dfrac{2}{3}$

3. $\dfrac{12ab}{15b}$ $\quad \dfrac{4a}{5}$

4. $\dfrac{cd}{cd}$ $\quad 1$

5. $\dfrac{40m^3n}{32mn^2}$ $\quad \dfrac{5m^2}{4n}$

Lesson Focus

Ask students if they have ever gone to a store to buy six eggs. Since eggs are usually packaged in cartons of twelve, they probably had to break a carton in half. If they then used two of those eggs in a recipe, they would have used $\frac{1}{3} \cdot \frac{1}{2}$, or $\frac{1}{6}$, of a full carton. Today's lesson focuses on multiplying rational numbers.

8-1 Multiplying Rational Numbers

Objective: To multiply rational numbers and algebraic fractions.

Terms to Know
- *reciprocal*
- *multiplication property of reciprocals*
- *multiplicative inverse*

EXPLORATION

1. What part of the whole is shaded blue? $\frac{1}{2}$

2. What part of the whole is shaded pink? $\frac{1}{3}$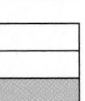

3. What part of the whole is shaded purple? $\frac{1}{6}$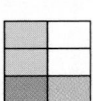

4. Use your results from Step 3. Complete: $\dfrac{1}{2} \cdot \dfrac{1}{3} = \dfrac{?}{?} \dfrac{1}{6}$

5. Draw a model to represent each product. See answer on p. A23.

   a. $\dfrac{1}{3} \cdot \dfrac{1}{4}$ $\qquad$ **b.** $\dfrac{2}{3} \cdot \dfrac{1}{4}$ $\qquad$ **c.** $\dfrac{3}{3} \cdot \dfrac{1}{4}$

To multiply with fractions, mixed numbers, and whole numbers, write all numbers as fractions and then multiply.

> **Generalization:** *Multiplying Rational Numbers*
>
> To multiply rational numbers, multiply the numerators and then multiply the denominators.
>
> **In Arithmetic** $\qquad\qquad$ **In Algebra**
>
> $\dfrac{1}{3} \cdot \dfrac{4}{5} = \dfrac{4}{15}$ $\qquad\qquad$ $\dfrac{a}{b} \cdot \dfrac{c}{d} = \dfrac{ac}{bd}, \ b \neq 0, \ d \neq 0$

Two numbers, like $\frac{5}{9}$ and $\frac{9}{5}$, whose product is 1 are called **reciprocals.** The reciprocal of $-\frac{5}{9}$ is $-\frac{9}{5}$.

> **Multiplication Property of Reciprocals**
>
> The product of a number and its reciprocal is 1.
>
> **In Arithmetic** $\qquad\qquad$ **In Algebra**
>
> $\dfrac{5}{9} \cdot \dfrac{9}{5} = 1$ $\qquad\qquad$ $\dfrac{a}{b} \cdot \dfrac{b}{a} = 1, \ a \neq 0, \ b \neq 0$

When two numbers are reciprocals of each other, they are also called **multiplicative inverses** of each other.

340 Chapter 8

Example 1

Find each product. Simplify if possible.

a. $\dfrac{2}{3} \cdot \dfrac{7}{8}$

b. $-2\left(\dfrac{3}{4}\right)$

Solution

a. $\dfrac{2}{3} \cdot \dfrac{7}{8} = \dfrac{2}{3} \cdot \dfrac{\overset{1}{7}}{\underset{4}{8}} = \dfrac{1 \cdot 7}{3 \cdot 4} = \dfrac{7}{12}$

b. $-2\left(\dfrac{3}{4}\right) = \dfrac{-2}{1} \cdot \dfrac{3}{4}$ ⟵ Write each factor as a fraction.

$= \dfrac{\overset{-1}{-2}}{1} \cdot \dfrac{3}{\underset{2}{4}}$

$= \dfrac{-1 \cdot 3}{1 \cdot 2} = \dfrac{-3}{2} = -1\dfrac{1}{2}$

 **Check Your Understanding**

1. In Example 1(b), why is -2 rewritten as a fraction, $\dfrac{-2}{1}$?
2. Why is the answer to Example 1(b) negative?
3. How would the answer to Example 1(b) be different if $\dfrac{3}{4}$ were $-\dfrac{3}{4}$?
   See Answers to Check Your Understanding at the back of the book.

 You can use the fraction key, [ab/c], on a calculator to perform operations with fractions and mixed numbers. To find the product $\left(3\dfrac{1}{8}\right)\left(\dfrac{1}{3}\right)$, you use the key sequence below. The result is $1\dfrac{1}{24}$.

 [3.] [ab/c] 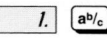 [1.] [ab/c] [8.] [×] [1.] [ab/c]  [3.] [=]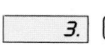

Sometimes you may only need an estimate of a product, or you may want an estimate to check your work. To estimate, first round each factor and then multiply. If one factor is less than 1, you may want to use compatible numbers and then multiply.

Example 2

Estimate each product.

a. $\left(2\dfrac{5}{8}\right)\left(-3\dfrac{1}{3}\right)$

b. $\left(-\dfrac{5}{11}\right)\left(-5\dfrac{1}{12}\right)$

Solution

a. Round.

$\left(2\dfrac{5}{8}\right)\left(-3\dfrac{1}{3}\right) \rightarrow \underbrace{(3)(-3)}_{\text{about } -9}$

b. Use compatible numbers.

$\left(-\dfrac{5}{11}\right)\left(-5\dfrac{1}{12}\right) \rightarrow \underbrace{\left(-\dfrac{1}{2}\right)(-6)}_{\text{about } 3}$

 **Check Your Understanding**

4. In Example 2(a), $2\dfrac{5}{8}$ is rounded to 3 because $\dfrac{5}{8}$ is greater than or equal to $\dfrac{1}{2}$. Explain why $-3\dfrac{1}{3}$ is rounded to -3.
5. In Example 2(b), why is $-5\dfrac{1}{12}$ rewritten as -6?
   See Answers to Check Your Understanding at the back of the book.

Rational Numbers   **341**

Teaching Notes

Emphasize the process of multiplication. Focus on the terms *reciprocal, multiplication property of reciprocals,* and *multiplicative inverse.*

Key Question

What step must you take before multiplying a whole number or a mixed number by a fraction? Write the whole number or mixed number as a fraction.

Using Technology

Many scientific calculators have fraction keys. The text on page 341 gives an example of how this key can be used to multiply fractions (also see page 338C).

For a suggestion on using a BASIC program to multiply mixed numbers, see page 338C.

Using Manipulatives

The Exploration section uses manipulatives to promote the development of number sense and an understanding of the meaning of multiplication with rational numbers.

For a suggestion on using fraction cards to model multiplication of rational numbers, see page 338D.

Reteaching/Alternate Approach

For a suggestion on using a cooperative learning activity to introduce the lesson, see page 338D.

Estimation

When multiplying fractions, an estimate should be made to check the reasonableness of the answer. Example 2 provides two methods of estimating with rational numbers.

Error Analysis

Students may fail to simplify products completely. Remind them to check for common factors, which can be numeric or algebraic.

342

Additional Examples

Find each product. Simplify if possible.

1. $\frac{3}{5} \cdot \frac{8}{9}$ $\frac{8}{15}$

2. $(-3)\frac{5}{6}$ $-2\frac{1}{2}$

Estimate each product.

3. $\left(3\frac{1}{3}\right)\left(2\frac{2}{3}\right)$ about 9

4. $\left(-\frac{4}{7}\right)\left(-4\frac{1}{9}\right)$ about 2

Simplify.

5. $\left(\frac{4x}{5}\right)\left(\frac{b}{8x}\right)$ $\frac{b}{10}$

Closing the Lesson

Demonstrate that the methods used to multiply numeric or algebraic fractions are the same by placing an example of each on the chalkboard. Have two students perform the first step of each example. Point out the similarity in the processes, and continue in this way until both multiplications are complete.

Suggested Assignments

Skills: 1–22, 37–42
Average: 1–22, 23–35 odd, 37–42
Advanced: 1–19 odd, 21–42

Exercise Notes

Communication: Discussion

Guided Practice Exercise 2 requires students to make the connection between fractions and decimals and to discuss the fact that the answer to a single exercise may be presented in different forms, depending on how the exercise is presented.

Mental Math

Exercises 29–32 promote the use of mental math in conjunction with the distributive property.

Calculator

Exercises 33–36 provide practice in using the fraction key by having students write the key sequence they would use to solve an exercise.

Example 3   Simplify $\frac{5x}{3} \cdot \frac{7a}{10x}$.

Solution

$$\frac{5x}{3} \cdot \frac{7a}{10x} = \frac{5\overset{1}{x}}{3} \cdot \frac{7a}{\underset{2}{10x}}$$

$$= \frac{1 \cdot 7a}{3 \cdot 2} \quad \longleftarrow \text{ Multiply the numerators,} \atop \text{and then the denominators.}$$

$$= \frac{7a}{6}$$

When multiplying positive or negative rational numbers that are written as decimals, apply the rules you learned for multiplication of integers.

Guided Practice

« **1. COMMUNICATION** «*Reading*   Replace each _?_ with the correct word or number.

When you multiply a number and its reciprocal, the result is _?_. Another name for reciprocals is _?_. 1; multiplicative inverses

    To estimate the product of two numbers, one of which is less than 1, you can use _?_ numbers and then multiply. compatible

« **2. COMMUNICATION** «*Discussion*   You know that $-\frac{7}{10} \cdot \frac{9}{10} = -\frac{63}{100}$.
Compare $-\frac{7}{10} \cdot \frac{9}{10}$ to $(-0.7)(0.9)$. Is the product $(-0.7)(0.9)$ *positive* or *negative*? Explain. negative; the rule is the same for fractions, whole numbers, and decimals.

Find each product. Simplify if possible.

3. $\frac{5}{12} \cdot \frac{1}{5}$ $\frac{1}{12}$ 4. $-\frac{3}{4} \cdot 3\frac{1}{3}$ $-2\frac{1}{2}$ 5. $(-0.3)(0.8)$ -0.24 6. $(-2.5)(-0.3)$ 0.75

7. $\frac{4m}{7} \cdot \frac{n}{8m}$ $\frac{n}{14}$ 8. $\frac{9a}{10} \cdot \frac{2}{3}$ $\frac{3a}{5}$ 9. $\frac{25}{c} \cdot \frac{6c}{5}$ 30 10. $\frac{7w}{8} \cdot \frac{11}{14}$ $\frac{11w}{16}$

Estimate each product. Estimates may vary. Accept reasonable estimates.

11. $\left(5\frac{1}{8}\right)\left(-5\frac{7}{8}\right)$ 12. $\left(-6\frac{7}{10}\right)\left(-3\frac{2}{5}\right)$ 13. $\left(\frac{7}{20}\right)\left(8\frac{11}{12}\right)$ 14. $\left(-\frac{6}{11}\right)\left(-4\frac{2}{11}\right)$
   about −30   about 21   about 3   about 2

Exercises

Find each product. Simplify if possible.

A   1. $\frac{8}{9} \cdot \frac{4}{5}$ $\frac{32}{45}$ 2. $\frac{2}{3} \cdot \frac{3}{2}$ 1 3. $\frac{5}{6}\left(-2\frac{1}{4}\right)$ $-1\frac{7}{8}$ 4. $\left(-4\frac{1}{3}\right)\left(-\frac{9}{10}\right)$ $3\frac{9}{10}$

5. $(-56)\left(-\frac{1}{56}\right)$ 1 6. $-\frac{3}{20} \cdot 16$ $-2\frac{2}{5}$ 7. $5 \cdot 6\frac{7}{10}$ $33\frac{1}{2}$ 8. $3\frac{3}{4} \cdot 12$ 45

9. $\left(-10\frac{1}{2}\right)\left(-5\frac{1}{3}\right)$ 56   **10.** $\left(7\frac{1}{3}\right)\left(-1\frac{1}{5}\right)$ $-8\frac{4}{5}$ **11.** $(-0.7)(3.5)$ -2.45 **12.** $\left(\frac{3n}{8}\right)\left(\frac{2}{n}\right)$ $\frac{3}{4}$

13. $\left(\frac{z}{10}\right)\left(\frac{5}{3z}\right)$ $\frac{1}{6}$ **14.** $(-2.8)(-0.9)$ 2.52   **15.** $\left(\frac{15a}{4}\right)\left(\frac{b}{3}\right)$ $\frac{5ab}{4}$ **16.** $\left(\frac{b}{7}\right)\left(\frac{14a}{b}\right)$ $2a$

Estimates may vary. Accept reasonable estimates.   **17.** about 10

Estimate each product. **18.** about −12   **19.** about −2   **20.** about 9

17. $\left(-4\frac{1}{2}\right)\left(-1\frac{2}{3}\right)$ **18.** $\left(-17\frac{9}{10}\right)\left(\frac{2}{3}\right)$ **19.** $\left(-\frac{3}{11}\right)\left(7\frac{2}{3}\right)$ **20.** $\left(2\frac{5}{8}\right)\left(3\frac{3}{7}\right)$

21. Leigh made nine new window curtains. How many yards of material did Leigh use if each curtain required $2\frac{2}{3}$ yd? about 27 yd

22. The average weight of eight chickens was $7\frac{3}{4}$ lb. About how much was the total weight of the chickens? about 64 lb

Find each product. Simplify if possible. $\frac{-9}{40}$ or   1.05 or $1\frac{1}{20}$

B **23.** $\frac{1}{2} \cdot 0.75$ $\frac{3}{8}$ or 0.375 **24.** $\frac{3}{8}(-0.6)$ -0.225 **25.** $(-0.25)\left(-4\frac{1}{5}\right)$

26. $\left(3\frac{1}{7}\right)(10)\left(11\frac{1}{5}\right)$ 352   **27.** $\left(-1\frac{1}{7}\right)\left(1\frac{3}{4}\right)(12)$ −24 **28.** $\left(-5\frac{5}{8}\right)\left(\frac{3}{5}\right)(4)$ $-13\frac{1}{2}$

                                                                            or -13.5

MENTAL MATH

You can use the distributive property to multiply a whole number by a mixed number mentally. For example, $6 \cdot 5\frac{2}{3} = 6(5) + 6\left(\frac{2}{3}\right) = 30 + 4 = 34$.

Multiply mentally.

29. $3 \cdot 9\frac{1}{4}$ $27\frac{3}{4}$ **30.** $10 \cdot 5\frac{3}{5}$ 56   **31.** $8 \cdot 4\frac{1}{4}$ 34   **32.** $4 \cdot 5\frac{3}{8}$ $21\frac{1}{2}$

CALCULATOR Answers to Exercises 33–36 are on p. A23.

Write the calculator key sequence you would use to find each product. Assume that the calculator has a fraction key.

33. $4\frac{2}{7} \cdot \frac{2}{7}$ **34.** $8\frac{11}{12} \cdot \frac{19}{20}$ **35.** $\frac{4}{5}\left(17\frac{3}{8}\right)$ **36.** $\left(1\frac{1}{2}\right)\left(111\frac{1}{2}\right)$

SPIRAL REVIEW

S **37.** Estimate the product: 207×92 *(Toolbox Skill 1)* about 18,000

38. Solve by using the guess-and-check strategy: Tim bought some shirts costing $19 each and some ties costing $14 each. He spent $127. How many of each did he buy?   *(Lesson 4-2)* 3 shirts and 5 ties

39. Evaluate $7a - 3b$ when $a = -2$ and $b = -8$.   *(Lesson 3-6)* 10

40. Simplify: $\left(\frac{5c}{9}\right)\left(\frac{d}{10c}\right)$ *(Lesson 8-1)* $\frac{d}{18}$

41. Estimate the sum: $17 + 24 + 22 + 19$ *(Toolbox Skill 4)* about 80

42. The weekly salaries at a store are $315, $295, $325, and $680. Does the mean describe these data well? Explain.   *(Lesson 5-9)* No; the mean is greater than most of the salaries because of the $680.

Rational Numbers   **343**

Nonroutine Problem

Find the numbers x that will yield a product of the following type when multiplied by $3\frac{1}{3}$.

a. less than 1 $x < \frac{3}{10}$

b. equal to 1 $\frac{3}{10}$

c. greater than 1 $x > \frac{3}{10}$

Follow-Up

Exploration (Cooperative Learning)
Tell each group to assume they have a calculator that does not have a fraction key. Have each group make up exercises that can be solved easily with the calculator. Then have each group make up exercises that can be solved easily without the calculator. Have each group explain why they categorized each exercise as they did.

Enrichment
Enrichment 77 is pictured on page 338E.

Practice
Practice 95, *Resource Book*

Practice 95
Skills and Applications of Lesson 8-1

Find each product. Simplify if possible.

1. $\frac{5}{7} \cdot \frac{3}{4}$ $\frac{15}{28}$ 2. $\frac{2}{5} \cdot \frac{3}{10}$ $\frac{3}{25}$ 3. $\frac{7}{8}\left(-1\frac{1}{14}\right)$ $-\frac{15}{16}$

4. $2\frac{6}{7} \cdot 4\frac{1}{5}$ 12   5. $1\frac{3}{8} \cdot \frac{32}{33}$ $1\frac{1}{3}$ 6. $5\frac{1}{4} \cdot 16$ 84

7. $-\frac{7}{15} \cdot \frac{45}{49}$ $-\frac{3}{7}$ 8. $\left(-6\frac{1}{4}\right)\left(-3\frac{1}{5}\right)$ 20   9. $\left(2\frac{1}{5}\right)\left(-1\frac{21}{44}\right)$ $-3\frac{1}{4}$

10. $-1\frac{2}{7} \cdot \frac{28}{45}$ $-\frac{4}{5}$ 11. $2\frac{3}{5} \cdot 2\frac{1}{16}$ $5\frac{1}{2}$ 12. $\frac{7}{9} \cdot \frac{3}{49}$ $\frac{1}{21}$

13. $\left(4\frac{3}{5}\right)\left(4\frac{4}{9}\right)$ 20   14. $\left(\frac{4}{9}\right)\left(-2\frac{1}{4}\right)$ −1   15. $5\frac{1}{4} \cdot 1\frac{3}{4}$ $9\frac{3}{16}$

16. $\frac{5a}{3} \cdot \frac{21}{40a}$ $\frac{7}{8}$ 17. $\frac{6m}{11} \cdot \frac{44}{54}$ $\frac{4m}{9}$ 18. $\frac{35ab}{6} \cdot \frac{18}{25a}$ $\frac{21b}{5}$

19. $\frac{5b}{32a} \cdot \left(-\frac{8a}{15b}\right)$ $-\frac{1}{12}$ 20. $\left(-1\frac{1}{5}\right)\left(-4\frac{1}{12}\right)$ $4\frac{9}{10}$ 21. $\frac{16}{25y} \cdot \frac{35y}{24}$ $\frac{14}{15}$

22. $\left(\frac{12ab}{13}\right)\left(\frac{20}{24a}\right)$ $\frac{3b}{2}$ 23. $\frac{3n}{28} \cdot \frac{7}{18n}$ $\frac{1}{24}$ 24. $-9\frac{3}{8} \cdot 3\frac{1}{5}$ −30

Estimate each product.   Estimates may vary. Examples are given.

25. $\left(-6\frac{1}{7}\right)\left(3\frac{2}{4}\right)$ about −18   26. $\left(8\frac{3}{5}\right)\left(5\frac{1}{7}\right)$ about 45   27. $\left(6\frac{7}{8}\right)\left(\frac{5}{8}\right)$ about 4

28. $\left(4\frac{2}{9}\right)\left(-8\frac{7}{8}\right)$ about −40   29. $\left(2\frac{1}{5}\right)\left(3\frac{3}{4}\right)$ about 8   30. $\left(-\frac{2}{9}\right)\left(-3\frac{3}{8}\right)$ about 1

31. $\left(11\frac{1}{2}\right)\left(\frac{7}{23}\right)$ about 4   32. $\left(-9\frac{1}{9}\right)\left(\frac{36}{41}\right)$ about −9   33. $\left(14\frac{2}{5}\right)\left(1\frac{1}{11}\right)$ about 15

34. It is estimated that $\frac{1}{3}$ of each pound of a lobster is "meat" you can eat. If a bushel of lobsters weighs $32\frac{2}{5}$ lb, about how many pounds of lobster can be eaten?   about 11 lb

35. Dolores' car gets 24.3 miles per gallon of gasoline. Approximately how many miles can she drive on 20 gallons of gasoline? about 480 mi

Resource Book, MATHEMATICAL CONNECTIONS
Copyright © by Houghton Mifflin Company. All rights reserved.   255

Teaching the Lesson
- Materials: index cards, calculator
- Lesson Starter 71
- Using Technology, p. 338C
- Using Manipulatives, p. 338D
- Reteaching/Alternate Approach, p. 338D
- Cooperative Learning, p. 338D
- Toolbox Skill 21

Lesson Follow-Up
- Practice 96
- Practice 97: Nonroutine
- Enrichment 78: Problem Solving
- Study Guide, pp. 141–142
- Test 36

Warm-Up Exercises

Find each reciprocal.

1. $\dfrac{3}{5}$ $\dfrac{5}{3}$

2. 4 $\dfrac{1}{4}$

3. $3\dfrac{1}{5}$ $\dfrac{5}{16}$

4. $\dfrac{a}{b}$ $\dfrac{b}{a}$

5. $\dfrac{7a}{2}$ $\dfrac{2}{7a}$

Lesson Focus

Ask students to imagine a pizza cut into eight slices. This situation could be represented by the expression $1 \div \dfrac{1}{8}$, where 1 represents the pizza and $\dfrac{1}{8}$ represents each slice. Today's lesson focuses on dividing rational numbers.

Teaching Notes

Point out the relationship between division and multiplication of rational numbers. Stress the importance of remembering to use reciprocals.

Dividing Rational Numbers

Objective: To find the quotients of rational numbers and of algebraic fractions.

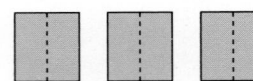

1 How many halves are there in 3? 6

$$3 \div \frac{1}{2} = \underline{}\ 6$$

A related multiplication is $3 \times \underline{} = 6.\ 2$

2 How many two thirds are in 4? 6

$$4 \div \frac{2}{3} = \underline{}\ 6$$

A related multiplication is
$4 \times \underline{} = 6.\ \dfrac{3}{2}$

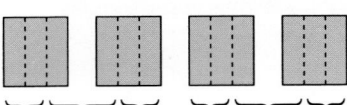

3 When you rewrite a division as a related multiplication, you multiply by the $\underline{}$ of the divisor. reciprocal

---

Generalization: *Dividing Rational Numbers*

To divide rational numbers, multiply by the reciprocal of the divisor.

In Arithmetic

$$\frac{1}{5} \div \frac{2}{3} = \frac{1}{5} \cdot \frac{3}{2} = \frac{3}{10}$$

In Algebra

$$\frac{a}{b} \div \frac{c}{d} = \frac{a}{b} \cdot \frac{d}{c} = \frac{ad}{bc}$$

$$b \neq 0,\ c \neq 0,\ d \neq 0$$

Example 1 **Find each quotient. Simplify if possible.**

a. $5 \div \dfrac{1}{2}$

b. $-\dfrac{5}{6} \div \left(-\dfrac{3}{4}\right)$

Solution

a. $5 \div \dfrac{1}{2} = \dfrac{5}{1} \cdot \dfrac{2}{1}$

$ = 10$

b. $-\dfrac{5}{6} \div \left(-\dfrac{3}{4}\right) = \dfrac{-5}{6} \cdot \dfrac{-4}{3}$

$ = \dfrac{-5}{6} \cdot \dfrac{\overset{-2}{\cancel{-4}}}{3}$

$ = \dfrac{10}{9} = 1\dfrac{1}{9}$

 **Check Your Understanding**

Would the answer to Example 1(a) be different if the quotient were $\dfrac{1}{2} \div 5$? Explain.

See *Answers to Check Your Understanding* at the back of the book.

 Some calculators have a reciprocal key, $\boxed{1/x}$. You can use this key to find a reciprocal, which the calculator displays as a decimal. To find the reciprocal of -8, you can use this key sequence.

$$\boxed{8.}\ \boxed{+/-}\ \boxed{1/x}$$

To find the quotient of variable expressions, write each expression as a fraction. Then multiply by the reciprocal of the divisor.

Example 2

Simplify:

a. $\dfrac{a}{2} \div \dfrac{a}{14}$

b. $\dfrac{3n}{2x} \div 6n$

Solution

a. $\dfrac{a}{2} \div \dfrac{a}{14} = \dfrac{a}{2} \cdot \dfrac{14}{a}$

$= \dfrac{\overset{1}{\cancel{a}}}{\underset{1}{\cancel{2}}} \cdot \dfrac{\overset{7}{\cancel{14}}}{\underset{1}{\cancel{a}}}$

$= 7$

b. $\dfrac{3n}{2x} \div 6n = \dfrac{3n}{2x} \div \dfrac{6n}{1}$

$= \dfrac{3n}{2x} \cdot \dfrac{1}{6n}$

$= \dfrac{\cancel{3n}}{2x} \cdot \dfrac{1}{\underset{2}{\cancel{6n}}}$

$= \dfrac{1}{4x}$

When dividing positive or negative rational numbers that are written as decimals, apply the rules you learned for division of integers.

Guided Practice

«**1.** COMMUNICATION «*Discussion*   You know that $-\frac{1}{4} \div \frac{3}{4} = -\frac{1}{3}$. Compare $-\frac{1}{4} \div \frac{3}{4}$ to $(-0.25) \div (0.75)$. Is the quotient $(-0.25) \div (0.75)$ *positive* or *negative*? Explain. negative; they are different forms of the same numbers.

Replace each ? with the number or variable expression that makes the statement true.

2. $\dfrac{4}{5} \div \left(-\dfrac{1}{8}\right) = \dfrac{4}{5} \cdot \underline{\ ?\ }$ $\dfrac{-8}{1}$

3. $-\dfrac{6}{7} \div 7 = -\dfrac{6}{7} \cdot \underline{\ ?\ }$ $\dfrac{1}{7}$

4. $\dfrac{4z}{9} \div 18z = \dfrac{4z}{9} \cdot \underline{\ ?\ }$ $\dfrac{1}{18z}$

5. $\dfrac{10}{a} \div \dfrac{5a}{6} = \dfrac{10}{a} \cdot \underline{\ ?\ }$ $\dfrac{6}{5a}$

Find each quotient. Simplify if possible.

6. $-\dfrac{2}{3} \div \dfrac{1}{4}$ $-2\frac{2}{3}$

7. $8 \div \dfrac{2}{3}$ 12

8. $-0.1 \div 10$ -0.01

9. $1.5 \div (-0.75)$ -2

10. $\dfrac{c}{5} \div \dfrac{c}{10}$ 2

11. $\dfrac{4n}{5} \div 8n$ $\dfrac{1}{10}$

12. $\dfrac{m}{7} \div 7m$ $\dfrac{1}{49}$

13. $\dfrac{z}{8} \div \dfrac{z}{12}$ $1\frac{1}{2}$

Key Questions
1. When dividing rational numbers, do you use the reciprocal of the dividend or the divisor?   divisor
2. When the divisor is a fraction between 0 and 1, is the quotient greater than or less than the dividend?   greater than

Using Technology
Many scientific calculators have a reciprocal key. The use of this key is discussed on page 345.
   For a suggestion on using a calculator or a BASIC program to divide fractions, see page 338C.

Using Manipulatives
For a suggestion on using fraction cards to model division of rational numbers, see page 338D.

Reteaching/Alternate Approach
For a suggestion on using cooperative learning to explore division by a fraction, see page 338D.

Error Analysis
Students may use the reciprocal of the dividend and thus get an incorrect answer. Remind students that they must use the reciprocal of the divisor to divide rational numbers.

Additional Examples
Find each quotient. Simplify if possible.
1. $6 \div \dfrac{1}{2}$ 12
2. $-\dfrac{1}{6} \div \left(-\dfrac{8}{9}\right)$ $\dfrac{3}{16}$
3. $\dfrac{m}{3} \div \dfrac{m}{24}$ 8
4. $\dfrac{4y}{3} \div 8y$ $\dfrac{1}{6}$

Closing the Lesson

Ask a student to explain how to find the quotient of two rational numbers. Ask another student to explain how to find the quotient of two algebraic fractions. Ask a third student to compare the two methods.

Suggested Assignments

Skills: 1–22, 31–34
Average: 1–26, 31–34
Advanced: 1–15 odd, 17–34

Exercise Notes

Communication: Discussion

Guided Practice Exercise 1 helps students make the connection between fractions and decimals.

Calculator

Exercises 19–22 provide students with practice using the reciprocal key.

Life Skills

Exercises 23–26 present a practical use of multiplying and dividing fractions, namely adjusting recipes.

Application

Exercises 27–30 connect fractions, formulas, and decimals by showing how opticians apply these concepts in their work.

Exercises

Find each quotient. Simplify if possible.

A   **1.** $-\dfrac{5}{8} \div \left(-\dfrac{5}{16}\right)$ 2   **2.** $\dfrac{3}{4} \div \dfrac{1}{5}$ $3\frac{3}{4}$ **3.** $18 \div \left(-\dfrac{1}{3}\right)$ -54 **4.** $-24 \div \dfrac{3}{4}$ -32

5. $\dfrac{w}{5} \div \dfrac{w}{15}$ 3   **6.** $\dfrac{5n}{6} \div \dfrac{d}{5}$ **7.** $\dfrac{2x}{3y} \div 3x$ $\dfrac{2}{9y}$ **8.** $\dfrac{m}{4} \div 8m$ $\dfrac{1}{32}$

9. $3\dfrac{1}{5} \div (-5)$ **10.** $\dfrac{13}{6} \div \dfrac{6}{a}$ **11.** $-1\dfrac{2}{3} \div \left(-\dfrac{1}{5}\right)$ **12.** $-3\dfrac{1}{8} \div \left(-4\dfrac{1}{8}\right)$

13. $\dfrac{11}{5} \div 11y$ $\dfrac{1}{5y}$ **14.** $-\dfrac{14}{15} \div 7$ **15.** $3.4 \div (-1.7)$ **16.** $-0.56 \div (-0.7)$

6. $\dfrac{25n}{6d}$

9. $-\dfrac{16}{25}$

10. $\dfrac{13a}{36}$

11. $8\dfrac{1}{3}$

12. $\dfrac{25}{33}$

14. $-\dfrac{2}{15}$

15. -2

16. 0.8

17. To fasten tents to the ground, the campers cut pieces $5\frac{3}{4}$ ft long from a rope that was 46 ft long. How many pieces could be cut? 8

18. Al is working in a print shop on a page that is $8\frac{1}{2}$ in. wide. The page must be divided into four columns of equal width. What should be the width of each column? $2\frac{1}{8}$ in.

🖩 **CALCULATOR**

Use a calculator reciprocal key [1/x] to find the reciprocal of each number.

B   **19.** -2 -0.5 **20.** -0.50 -2 **21.** $\dfrac{4}{3}$ 0.75   **22.** 1.6 0.625

PROBLEM SOLVING/APPLICATION

APPLE MUFFINS

$1\frac{1}{2}$ c milk   2 tbsp baking soda
2 eggs   2 tsp cinnamon
$\frac{1}{2}$ c shortening $\frac{3}{4}$ c raisins
$4\frac{1}{2}$ c flour   2 c chopped apple

Maria can make two dozen apple muffins using her recipe at the left. Sometimes she changes the amounts of the ingredients because she wants to make more than or less than the two dozen muffins.

23. Maria has $\frac{3}{4}$ c of raisins left in a box. Is this *more than* or *less than* the amount in the recipe? more than

24. To make 72 muffins, Maria triples the recipe. How much milk does she need? $4\frac{1}{2}$ c

See answer on p. A23.

25. Maria decides to make only one dozen muffins, so she divides each amount in the recipe by 2. Rewrite the recipe listing the new amounts.

26. **RESEARCH** Find a recipe in a cookbook or a magazine. How many items does the recipe make or how many people does it serve? Double the recipe and list the new amounts of the ingredients.
Answers will vary.

346 Chapter 8

CAREER/APPLICATION

An *optician* makes and sells lenses and eyeglasses. A lens bends the rays of light entering the eye. The ability to bend light rays is measured in *diopters* (D) and depends on the *focal length* (f) of the lens.

To find the diopter measure, use the formula $D = \frac{1}{f}$, where f is in meters. If the focal length of a lens is 25 cm, first write 25 cm as 0.25 m. Then substitute for f.

$$D = \frac{1}{0.25} = 4 \quad \longleftarrow \text{Use a reciprocal key on a calculator to compute.}$$

Use the formula $D = \frac{1}{f}$ to find each diopter measure.

27. $f = 400$ cm $\frac{1}{4}$ **28.** $f = 15$ cm $6.\overline{6}$ **29.** $f = 0.02$ cm $\underset{5000}{}$ **30.** $f = 800$ cm $\frac{1}{8}$

SPIRAL REVIEW

S **31.** Estimate the sum: $27 + 34 + 54 + 18$ *(Toolbox Skill 2)* about 130

32. Find the quotient: $\frac{z}{9} \div \frac{z}{3}$ *(Lesson 8-2)* $\frac{1}{3}$

33. Write an equation that represents the sentence: A number z divided by nine is fifteen. *(Lesson 4-7)* $\frac{z}{9} = 15$

Self-Test 1

Find each product. Simplify if possible.

1. $(-4)\left(\frac{11}{12}\right)$ $-3\frac{2}{3}$ **2.** $(-0.6)(-1.8)$ 1.08 **3.** $\left(\frac{6a}{7}\right)\left(\frac{c}{9a}\right)$ $\frac{2c}{21}$ **4.** $\left(\frac{4n}{5}\right)\left(\frac{m}{2n}\right)$ $\frac{2m}{5}$ **8-1**

Estimate each product. Estimates may vary.

5. $\left(6\frac{1}{5}\right)\left(-7\frac{7}{8}\right)$ about -48 **6.** $\left(2\frac{1}{2}\right)\left(10\frac{3}{4}\right)$ about 33

7. $\left(-\frac{6}{11}\right)\left(-8\frac{3}{8}\right)$ about 4 **8.** $\left(20\frac{13}{15}\right)\left(-\frac{6}{17}\right)$ about -7

Find each quotient. Simplify if possible.

9. $-\frac{1}{5} \div \frac{1}{8}$ $-1\frac{3}{5}$ **10.** $-6 \div -\frac{1}{3}$ 18 **11.** $3.6 \div (-0.8)$ -4.5 **8-2**

12. $\frac{5}{12} \div 2\frac{1}{12}$ $\frac{1}{5}$ **13.** $\frac{3n}{8} \div \frac{n}{16}$ 6 **14.** $\frac{2c}{5} \div 3c$ $\frac{2}{15}$ **15.** $7 \div \frac{70}{z}$ $\frac{z}{10}$

Rational Numbers **347**

Follow-Up

Enrichment
Enrichment 78 is pictured on page 338E.

Practice
Practice 96 is shown below.

Quick Quiz 1
See page 381.

Alternative Assessment
See page 381.

Practice 96, *Resource Book*

Practice 96
Skills and Applications of Lesson 8-2

Find each quotient. Simplify if possible.

1. $\frac{1}{5} \div \frac{2}{3}$ $\frac{9}{10}$ 2. $-\frac{1}{4} \div \frac{3}{4}$ $-\frac{1}{3}$ 3. $\frac{4}{9} \div \frac{2}{3}$ $\frac{2}{3}$

4. $25 \div \frac{5}{6}$ 30   5. $-18 \div \frac{6}{7}$ -21 6. $\frac{3}{5} \div 9$ $\frac{1}{15}$

7. $\frac{6t}{7} \div \frac{9}{14}$ $\frac{4x}{3}$ 8. $-\frac{5a}{7} \div \frac{10a}{21}$ $-1\frac{1}{2}$ 9. $3\frac{1}{5} \div \left(-1\frac{3}{25}\right)$ $-3\frac{3}{14}$

10. $\frac{3y}{8a} \div \frac{15y}{24a}$ $\frac{3}{5}$ 11. $-2\frac{1}{9} \div \left(2\frac{2}{9}\right)$ $-\frac{19}{20}$ 12. $\frac{7n}{13} \div \left(-\frac{7n}{13}\right)$ -1

13. $-\frac{13}{16} \div \frac{39}{64}$ $-1\frac{1}{3}$ 14. $\frac{8w}{15} \div \frac{32x}{45}$ $\frac{3w}{4y}$ 15. $4\frac{3}{4}m \div 3\frac{1}{9}$ $\frac{3m}{2}$

16. $4\frac{5}{11} \div \left(-1\frac{1}{11}\right)$ -49 17. $-\frac{18a}{19} \div 6n$ $-\frac{3}{19}$ 18. $-4\frac{3}{4} \div \left(-2\frac{1}{4}\right)$ 2

19. $\frac{16m}{36} \div \frac{8m}{21b}$ 14   20. $0.25 \div 0.40$ $\frac{5}{8}$ 21. $-\frac{15}{16} \div \left(-1\frac{1}{15}\right)$ $\frac{225}{256}$

22. $0.35a \div 1\frac{7}{8}$ $\frac{a}{4}$ 23. $5.6 \div 3\frac{11}{15}$ $1\frac{1}{2}$ 24. $-2.8m \div (1.4)$ $-2m$

25. $0.81 \div (-0.09)$ -9 26. $-\frac{27}{56} \div 1\frac{1}{8}$ $-\frac{3}{7}$ 27. $2\frac{22}{25} \div 7\frac{1}{5}$ $\frac{2}{5}$

28. $6\frac{2}{9} \div (-8)$ $-\frac{7}{9}$ 29. $(-36) \div 2\frac{1}{4}$ -16 30. $-\frac{15}{7b} \div \left(-\frac{45}{28b}\right)$ $1\frac{1}{3}$

31. $1.44 \div (-2.4)$ $-\frac{3}{5}$ 32. $(-5.25) \div (-1.05)$ 5   33. $1.03 \div (-7.21)$ $-\frac{1}{7}$

34. $\frac{28w}{27y} \div \frac{14w}{9y}$ $\frac{2}{3}$ 35. $6\frac{2}{25} \div 7\frac{3}{5}$ $\frac{4}{5}$ 36. $\frac{25}{7ab} \div \frac{75}{42a}$ $\frac{2}{b}$

37. Earl's father placed ceramic tile on the wall above the sink. Each tile measures $5\frac{1}{3}$ in. by $5\frac{1}{3}$ in. Each row of tiles measures 96 in. How many tiles are in each row? 18

38. Winter covers for 16 ft boats must be $17\frac{1}{2}$ ft long. A roll of fabric to cover boats is 80 ft long. How many complete covers can be made from 1 roll? 4

256   Resource Book, MATHEMATICAL CONNECTIONS
Copyright © by Houghton Mifflin Company. All rights reserved.

Teaching the Lesson
- Lesson Starter 72
- Reteaching/Alternate Approach, p. 338D
- Cooperative Learning, p. 338D

Lesson Follow-Up
- Practice 98
- Enrichment 79: Communication
- Study Guide, pp. 143–144

Warm-Up Exercises

Solve.

1. $\dfrac{7}{8} \cdot \dfrac{16}{21}$ $\dfrac{2}{3}$

2. $\left(-7\dfrac{1}{2}\right)\left(\dfrac{3}{5}\right)$ $-4\dfrac{1}{2}$

3. $\left(\dfrac{4n}{9}\right)\left(-\dfrac{3}{10n}\right)$ $-\dfrac{2}{15}$

4. $\dfrac{x}{5} \div \dfrac{x}{15}$ 3

5. $\dfrac{h}{9} \div 6h$ $\dfrac{1}{54}$

Lesson Focus

Ask students if they have ever worked a problem and felt that their answer did not make sense. For example, suppose there are 15 people traveling to a concert in cars. If each car holds 4 people, how many cars are needed? The numerical answer may be $3\dfrac{3}{4}$, but $\dfrac{3}{4}$ of a car does not make sense. Today's lesson focuses on the correct form of an answer.

Teaching Notes

Key Question

Is the form of an answer to a problem determined by the form of the numbers in a problem? No; for example, the problem may involve ordering $\dfrac{3}{4}$ lb of roast beef, but the answer may be $3.49.

DECISION MAKING

8-3

Determining the Correct Form of an Answer

Objective: To determine the correct form of an answer when solving a problem.

When you solve a problem, you may have to decide which form of the answer is needed to make the answer appropriate for a given situation.

- A *whole number* is appropriate when the situation involves items that occur only as whole items, such as sweaters or posters.

- A *fraction* is appropriate when the situation involves items that can occur as fractions, such as yards of material.

- A *decimal* is appropriate when the situation involves items that can occur as decimals, such as amounts of money.

Example

Solve. Decide whether a *whole number*, a *fraction*, or a *decimal* is an appropriate answer.

a. Tami worked $22\dfrac{1}{2}$ h in three days. She worked the same number of hours each day. How many hours did she work each day?

b. Asparagus costs $2 per pound. How much does $\dfrac{2}{3}$ lb cost?

c. There will be 75 people at a luncheon. How many tables of 10 seats each are needed to seat everyone?

Solution

a. $22\dfrac{1}{2} \div 3 = 7\dfrac{1}{2}$

The answer is a number of hours, so a fraction is appropriate. Tami worked $7\dfrac{1}{2}$ h each day.

b. $\dfrac{2}{3} \cdot 2 = 1\dfrac{1}{3} = 1.\overline{3}$

The answer is an amount of money, so a decimal is appropriate. The cost of $\dfrac{2}{3}$ lb is $1.33.

c. $75 \div 10 = 7.5$

The answer is a number of tables, so a whole number is appropriate. Eight tables are needed.

 **Check Your Understanding**

1. In part (b), why is $1.\overline{3}$ written as $1.33?

2. In part (c), why is the answer *eight tables*, instead of *seven tables*?
See *Answers to Check Your Understanding* at the back of the book.

Guided Practice

COMMUNICATION «*Discussion*

Decide whether a *whole number*, a *fraction*, or a *decimal* is the appropriate form for each quantity.

«1. a number of shirts whole number

«2. a sale price decimal or whole number

«3. a number of hours fraction, decimal, or whole number

«4. the number of people in class whole number

Solve. Decide whether a *whole number*, a *fraction*, or a *decimal* is an appropriate answer.

5. A teacher wants to divide a class into groups of four. The class has 29 students. How many groups of four can there be? whole number; 7

6. Each week Raphael earns $26 and saves one fourth of his earnings. How much money does he save each week? decimal; $6.50

Exercises

Solve. Decide whether a *whole number*, a *fraction*, or a *decimal* is an appropriate answer. **1.** fraction; $18\frac{3}{4}$ yd **2.** decimal; $7.50

A

1. The junior class is making 25 bows for prom decorations. Each bow uses $\frac{3}{4}$ yd of ribbon. How many yards of ribbon are needed?

2. A watercolor brush costs $3.75. How much do two brushes cost?

3. A tube of caulking can repair $2\frac{2}{3}$ classroom windows. How many tubes of caulking should be bought to repair 50 windows? whole number; 19

4. Jocelyn divides 5 lb of granola evenly between two bags. How many pounds of granola are in each bag? fraction; $2\frac{1}{2}$ lb

5. Mark practices playing his guitar $\frac{3}{4}$ h each day. How many hours does he practice in seven days? fraction; $5\frac{1}{4}$ h

6. Ann has $7\frac{1}{4}$ cups of pecans. The recipe for one pecan pie calls for $1\frac{1}{4}$ cups of pecans. How many pecan pies can she bake? whole number; 5

7. Salmon costs $6 per pound. How much does $2\frac{1}{4}$ lb of salmon cost? decimal; $13.50

💻 COMPUTER APPLICATION   **10.** fraction; $43\frac{3}{4}$

Many businesses use computers to keep a record of inventory and costs. A computer always shows rational numbers in decimal form. Decide whether a *whole number*, a *fraction*, or a *decimal* is the appropriate form of each quantity. Then write the correct form of the quantity.

B

8. The average number of cars sold in a week is 18.125. whole number; 18

9. The base price of a Model X1 car is 10995.54. decimal; $10,995.54

10. The number of hours a salesperson worked in a week is 43.75.

11. The average number of customers in a month is 159.41666667. whole number; 159

SPIRAL REVIEW

S 12. Make a sketch of $\overline{WY}$.   *(Lesson 6-1)* See answer on p. A23.

13. A school bus seats 44 people. How many buses are needed to transport 204 people to the class outing?   *(Lesson 8-3)* 5

Rational Numbers   **349**

Modeling Fractions Using Fraction Tiles

Using Manipulatives

Using fraction tiles to model equivalent fractions and to add and subtract fractions will help students to understand these concepts.

Objective: To use fraction tiles to model addition and subtraction of fractions.

Reasoning

As students work the Activities, you may wish to ask them questions such as the following.

Materials

■ fraction tiles:
$1, \frac{1}{2}, \frac{1}{3}, \frac{1}{4}, \frac{1}{6}, \frac{1}{8}, \frac{1}{12}, \frac{1}{16}$

Activity I

1. In how many different ways can you use like tiles to cover the tile representing $\frac{1}{2}$?   6 ways

2. In how many different ways can you use like tiles to cover the tile representing $\frac{1}{3}$?   3 ways

3. Which size tiles can be used to cover both pieces? How do these pieces relate to the common denominator of $\frac{1}{2}$ and $\frac{1}{3}$?
$\frac{1}{6}$ and $\frac{1}{12}$; both 6 and 12 are common denominators for the fractions.

Activity II

4. In how many different ways can you cover $\frac{1}{2}$ and $\frac{1}{4}$ exactly?
4 ways

Activity III

5. How are equivalent fractions used to find the missing terms in the problems shown?   The empty space represents the fraction equivalent to the fraction bars that were used to fill it.

Activity IV

6. How are equivalent fractions used to complete the subtraction problems shown?   The empty space again represents the fraction equivalent to the fraction bars used to fill it.

Fraction tiles are designed so that a square represents one whole unit. Fractions are represented by rectangular tiles that can be combined to form a whole square. You can use fraction tiles to add and subtract fractions. For example, you can use fraction tiles to show that the sum of $\frac{1}{3}$ and $\frac{1}{4}$ is $\frac{7}{12}$.

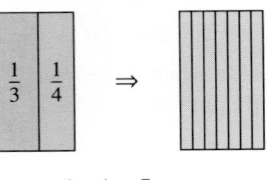

⟵ Seven $\frac{1}{12}$ tiles completely cover both the $\frac{1}{3}$ and $\frac{1}{4}$ tiles.

$$\frac{1}{3} + \frac{1}{4} = \frac{7}{12}$$

Activity I   *Finding Equivalent Fractions*

Suppose you want to cover a fraction tile representing $\frac{1}{2}$ and a fraction tile representing $\frac{1}{3}$. To do this, you want to use fraction tiles that are all the same size.

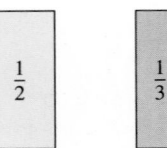

▮ What is the largest size fraction tile you could use? How many of these fraction tiles cover the $\frac{1}{2}$ tile? How many cover the $\frac{1}{3}$ tile? $\frac{1}{6}$; 3; 2

▮ What is the smallest size fraction tile you could use? How many of these fraction tiles cover the $\frac{1}{2}$ tile? How many cover the $\frac{1}{3}$ tile? $\frac{1}{12}$; 6; 4

▮ Use your answers to Steps 1 and 2 to write two fractions equivalent to $\frac{1}{2}$ and two fractions equivalent to $\frac{1}{3}$. $\frac{3}{6}, \frac{6}{12}; \frac{2}{6}, \frac{4}{12}$

Activity II   *More Equivalent Fractions*

Arrange fraction tiles representing $\frac{1}{2}$ and $\frac{1}{4}$ as shown. Cover the tiles using fraction tiles that are all the same size.

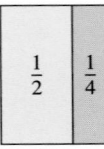

▮ What is the largest size fraction tile you can use? $\frac{1}{4}$

▮ What is the smallest size fraction tile you can use? $\frac{1}{16}$

▮ Use your answers to Steps 1 and 2 to complete the statement.

$$\frac{1}{2} + \frac{1}{4} = \frac{?}{\rule{1cm}{0.4pt}} = \frac{?}{\rule{1cm}{0.4pt}} \quad \frac{3}{4}; \frac{12}{16}$$

Activity III   *Modeling Addition*

1 To complete one whole unit, fill the empty space using fraction tiles that are all the same size. Then replace each __?__ with a fraction in lowest terms that makes the statement true.

a.

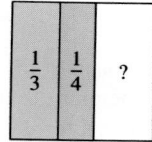

$$\frac{1}{3} + \frac{1}{4} + \underline{\ ?\ } = 1 \qquad \frac{5}{12}$$

b.

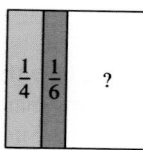

$$\frac{1}{4} + \frac{1}{6} + \underline{\ ?\ } = 1 \qquad \frac{7}{12}$$

c.

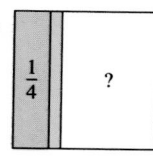

$$\frac{1}{4} + \frac{1}{12} + \underline{\ ?\ } = 1 \qquad \frac{2}{3}$$

d.

$$\frac{1}{6} + \frac{1}{12} + \underline{\ ?\ } = 1 \qquad \frac{3}{4}$$

2 In parts (c) and (d) of Step 2, is there more than one way to fill the empty space using fraction tiles that are all the same size? Explain.
Yes; tiles representing equivalent fractions can be used.

Activity IV   *Modeling Subtraction*

1 Show how to fill the empty space using fraction tiles that are all the same size. Express each answer as a fraction.

a.

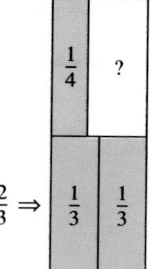

$$\frac{1}{3}$$

b.

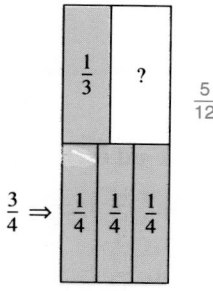

$$\frac{5}{12}$$

c.

$$\frac{5}{12}$$

2 Use your answers to Step 1 to replace each __?__ with the correct fraction in lowest terms.

a. $\frac{1}{2} - \frac{1}{6} = \underline{\ ?\ }\ \frac{1}{3}$ **b.** $\frac{2}{3} - \frac{1}{4} = \underline{\ ?\ }\ \frac{5}{12}$ **c.** $\frac{3}{4} - \frac{1}{3} = \underline{\ ?\ }\ \frac{5}{12}$

3 In Step 1, is there more than one way to fill each empty space using fraction tiles that are all the same size? Explain.
Yes; tiles representing equivalent fractions can be used.

Activity Notes

Activities I and II
Activities I and II prepare students to perform Activities III and IV (and addition and subtraction of fractions in general) by asking them to model equivalent fractions. Both activities allow students to better understand the concept of common denominator.

Activities III and IV
Activities III and IV have students model addition and subtraction of fractions. By physically performing these operations with fraction tiles students will obtain a better understanding of the processes that will be presented in Lessons 8-4 and 8-5.

Rational Numbers   **351**

351

Teaching the Lesson
- Materials: fraction tiles
- Lesson Starter 73
- Using Manipulatives, p. 338D
- Toolbox Skills 17–19

Lesson Follow-Up
- Practice 99
- Enrichment 80: Thinking Skills
- Study Guide, pp. 145–146

Warm-Up Exercises

Write each fraction in lowest terms.

1. $\frac{15}{24}$ $\frac{5}{8}$

2. $\frac{14}{49}$ $\frac{2}{7}$

3. $\frac{18}{36}$ $\frac{1}{2}$

4. $\frac{16}{20}$ $\frac{4}{5}$

5. $\frac{18}{22}$ $\frac{9}{11}$

Lesson Focus

Ask students if they ever have doubled the amounts of a recipe. Did they measure all of the ingredients twice, or did they add the quantity to itself to determine the amount needed? Today's lesson focuses on addition and subtraction of rational numbers with like denominators.

Teaching Notes

When working through the lesson, stress that the denominator indicates the number of parts into which the whole has been divided. Stress also that when adding or subtracting fractions with like denominators, students should add or subtract the numerators, not the denominators.

Adding and Subtracting Rational Numbers with Like Denominators

Objective: To add and subtract rational numbers and algebraic fractions with like denominators.

DATA ANALYSIS

Terry wants to make a bow tie and a sash. She refers to the back of the pattern to determine how many yards of material to buy. She plans to use fabric that is 60 in. wide. She will need $\frac{7}{8}$ yd for the tie and $\frac{5}{8}$ yd for the sash. Terry buys $\frac{12}{8}$ yd, or $1\frac{1}{2}$ yd, of fabric.

Tie	$\frac{7}{8}$ yd
Bow Tie	$\frac{7}{8}$ yd
Scarf	$1\frac{5}{8}$ yd
Sash:	
44/45 in.	$1\frac{1}{4}$ yd
58/60 in.	$\frac{5}{8}$ yd

Generalization: *Adding Rational Numbers with Like Denominators*

To add rational numbers with like denominators, add the numerators and write the sum over the denominator.

In Arithmetic

$$\frac{7}{8} + \frac{5}{8} = \frac{7+5}{8}$$

In Algebra

$$\frac{a}{c} + \frac{b}{c} = \frac{a+b}{c}, \ c \neq 0$$

Example 1

Find each sum. Simplify if possible.

a. $\frac{11}{12} + \frac{5}{12}$

b. $-1\frac{1}{5} + \left(-\frac{3}{5}\right)$

Solution

a. $\frac{11}{12} + \frac{5}{12} = \frac{11+5}{12}$

$= \frac{16}{12}$

$= \frac{4}{3} = 1\frac{1}{3}$

b. $-1\frac{1}{5} + \left(-\frac{3}{5}\right) = \frac{-6}{5} + \left(\frac{-3}{5}\right)$

$= \frac{-6+(-3)}{5}$

$= \frac{-9}{5} = -1\frac{4}{5}$

 **Check Your Understanding**

1. Explain the first step in Example 1(b).

   See *Answers to Check Your Understanding* at the back of the book.

You can also subtract rational numbers with like denominators.

352 Chapter 8

Key Question
Why is it usually a good idea to write each mixed number as a fraction before adding or subtracting? It may be easier to add or subtract the mixed numbers written as fractions since you simply add or subtract the numerators.

Using Manipulatives
For a suggestion on using fraction tiles to model addition and subtraction, see page 338D.

Life Skills
The lesson introduction uses a real-world application of fractions. Fabric patterns use fractional lengths of materials.

Error Analysis
In adding or subtracting rational numbers with like denominators, students may add or subtract the denominators as well as the numerators. Stress that when adding or subtracting rational numbers with like denominators, you add or subtract only the numerators and then write the sum or difference over the denominator.

Additional Examples
Find each sum. Simplify if possible.

1. $\frac{7}{12} + \frac{7}{12}$ $1\frac{1}{6}$

2. $-2\frac{2}{5} + \left(-\frac{1}{5}\right)$ $-2\frac{3}{5}$

Find each difference. Simplify if possible.

3. $3\frac{1}{10} - \frac{7}{10}$ $2\frac{2}{5}$

4. $-4\frac{2}{3} - 7\frac{2}{3}$ $-12\frac{1}{3}$

Simplify.

5. $\frac{7}{y} + \frac{9}{y}$ $\frac{16}{y}$

6. $\frac{11m}{5} - \frac{6m}{5}$ m

Generalization: *Subtracting Rational Numbers with Like Denominators*

To subtract rational numbers with like denominators, subtract the numerators and write the difference over the denominator.

In Arithmetic

$$\frac{7}{8} - \frac{5}{8} = \frac{7-5}{8}$$

In Algebra

$$\frac{a}{c} - \frac{b}{c} = \frac{a-b}{c}, \quad c \neq 0$$

Example 2

Find each difference. Simplify if possible.

a. $2\frac{3}{10} - \frac{7}{10}$

b. $-3\frac{1}{4} - 6\frac{3}{4}$

Solution

a.
$$2\frac{3}{10} - \frac{7}{10} = \frac{23}{10} - \frac{7}{10}$$
$$= \frac{23 - 7}{10}$$
$$= \frac{16}{10}$$
$$= \frac{8}{5} = 1\frac{3}{5}$$

b.
$$-3\frac{1}{4} - 6\frac{3}{4} = \frac{-13}{4} - \frac{27}{4}$$
$$= \frac{-13 - 27}{4}$$
$$= \frac{-40}{4}$$
$$= -10$$

Check Your Understanding

2. In Example 2(b), why does the numerator become -40?

3. What would the numerator in Example 2(b) have become if the problem had been $3\frac{1}{4} - 6\frac{3}{4}$ instead of $-3\frac{1}{4} - 6\frac{3}{4}$?
   See *Answers to Check Your Understanding* at the back of the book.

In Lesson 2-3 you learned to combine like terms. You can use those skills along with what you have just learned about adding and subtracting rational numbers to add and subtract algebraic fractions.

Example 3

Simplify.

a. $\frac{6}{x} + \frac{5}{x}$

b. $\frac{10n}{3} - \frac{4n}{3}$

Solution

a.
$$\frac{6}{x} + \frac{5}{x} = \frac{6 + 5}{x}$$
$$= \frac{11}{x}$$

b.
$$\frac{10n}{3} - \frac{4n}{3} = \frac{10n - 4n}{3}$$
$$= \frac{6n}{3}$$
$$= \frac{2n}{1} = 2n$$

Check Your Understanding

4. In Example 3(b), explain how $\frac{6n}{3}$ was simplified.
   See *Answers to Check Your Understanding* at the back of the book.

Closing the Lesson

Ask a student to explain how to add or subtract rational numbers with like denominators. Ask another student to explain how to add or subtract algebraic fractions with like denominators.

Suggested Assignments

Skills: 1–30, 40, 41
Average: 1–30, 40, 41
Advanced: 1–23 odd, 25–41

Exercise Notes

Communication: Reading

Guided Practice Exercises 1 and 2 require students to interpret information from the text and identify key ideas.

Number Sense

Exercises 27–30 rely on a student's intuition of the relative size of fractions. Students who are having difficulty determining the size of each sum should draw diagrams.

Making Connections/Transitions

Exercises 31–33 connect mathematics with industrial arts through the use of fractional measurements.
Exercises 34–39 connect rational numbers to the probability of an event happening or not happening.

Guided Practice

COMMUNICATION «*Reading* Answers to Guided Practice 1–2 are on p. A23.

Refer to the text on pages 352–353.

«**1.** What is the main idea of the lesson?

«**2.** List two major points that support the main idea.

Express each rational number as a fraction.

3. $-2\frac{1}{4}$ $\frac{-9}{4}$

4. $10\frac{5}{8}$ $\frac{85}{8}$

5. $7\frac{2}{3}$ $\frac{23}{3}$

6. $-1\frac{5}{6}$ $\frac{-11}{6}$

Find each answer. Simplify if possible.

7. $1\frac{5}{9} + 3\frac{1}{9}$ $4\frac{2}{3}$

8. $-\frac{4}{7} - \left(-5\frac{5}{7}\right)$ $5\frac{1}{7}$

9. $-\frac{9}{10} + \left(-\frac{7}{10}\right) + \frac{1}{10}$ $-1\frac{1}{2}$

10. $2\frac{5}{8} + \left(-1\frac{1}{8}\right) + \frac{7}{8}$ $2\frac{3}{8}$

11. $\frac{20}{15a} - \frac{8}{15a}$ $\frac{4}{5a}$

12. $\frac{8b}{9} + \frac{2b}{9} + \frac{5b}{9}$ $\frac{5b}{3}$

Exercises

Find each answer. Simplify if possible.

A **1.** $\frac{14}{15} - \frac{6}{15}$ $\frac{8}{15}$

2. $\frac{1}{6} + \frac{5}{6}$ 1

3. $-\frac{3}{5} - \left(-\frac{3}{5}\right)$ 0

4. $\frac{3}{8} + \left(-\frac{5}{8}\right)$ $-\frac{1}{4}$

5. $8\frac{1}{4} - 5\frac{3}{4}$ $2\frac{1}{2}$

6. $\frac{3}{11} - 2\frac{1}{11}$ $-1\frac{9}{11}$

7. $-\frac{3}{10} - \left(-1\frac{7}{10}\right)$ $1\frac{2}{5}$

8. $-8\frac{1}{10} - 2\frac{3}{10}$ $-10\frac{2}{5}$

9. $-1\frac{11}{14} + \left(-2\frac{3}{14}\right)$ -4

10. $2\frac{3}{4} + \left(-3\frac{1}{4}\right)$ $-\frac{1}{2}$

11. $-4\frac{5}{12} + \frac{11}{12} + 2\frac{7}{12}$ $-\frac{11}{12}$

12. $-3\frac{2}{5} + \frac{3}{5} + \left(-2\frac{4}{5}\right)$ $-5\frac{3}{5}$

13. $\frac{8}{r} + \frac{4}{r}$ $\frac{12}{r}$

14. $\frac{12}{s} - \frac{5}{s}$ $\frac{7}{s}$

15. $\frac{11}{3m} - \frac{5}{3m}$ $\frac{2}{m}$

16. $\frac{3}{2a} + \frac{5}{2a}$ $\frac{4}{a}$

17. $\frac{7}{6x} + \frac{5}{6x}$ $\frac{2}{x}$

18. $\frac{13}{4r} + \frac{9}{4r}$ $\frac{11}{2r}$

19. $\frac{9b}{2} + \frac{7b}{2}$ $8b$

20. $\frac{18h}{5} - \frac{6h}{5}$ $\frac{12h}{5}$

21. $\frac{18x}{25} - \frac{8x}{25}$ $\frac{2x}{5}$

22. $\frac{16b}{5} - \frac{b}{5}$ $3b$

23. $\frac{3c}{7} + \frac{5c}{7} + \frac{6c}{7}$ $2c$

24. $\frac{8}{9n} + \frac{2}{9n} + \frac{11}{9n}$ $\frac{7}{3n}$

25. Ken finished his mathematics assignment in $\frac{3}{4}$ h. Bert finished the same assignment in $1\frac{1}{4}$ h. How much longer did it take Bert to do the assignment than Ken? $\frac{1}{2}$ h

26. Hazel spent $1\frac{3}{8}$ h mowing the lawn, $\frac{7}{8}$ h trimming the hedge, and $1\frac{5}{8}$ h weeding the garden. How many hours in all did Hazel spend doing this work? $3\frac{7}{8}$ h

354 Chapter 8

NUMBER SENSE

Tell if each sum is *greater than* or *less than* 1. Do not find the sum.

B **27.** $\frac{3}{5} + \frac{4}{5}$ **28.** $\frac{3}{7} + \frac{6}{7}$ **29.** $\frac{1}{8} + \frac{3}{8}$ **30.** $\frac{5}{6} + \frac{5}{6}$

greater than   greater than   less than   greater than

CONNECTING MATHEMATICS AND INDUSTRIAL ARTS

Many measurements used in industrial arts involve fractions.

31. From a $33\frac{7}{8}$ in. long board, Jim cut a piece of wood that was $10\frac{5}{8}$ in. long. The saw blade shaved $\frac{1}{8}$ in. off the board. How long was the remaining piece? $23\frac{1}{8}$ in.

32. A drawer front is larger than the drawer opening by $\frac{5}{8}$ in. on each edge. The front of the drawer is $26\frac{3}{8}$ in. long and $8\frac{3}{8}$ in. wide. What are the dimensions of the drawer opening? $25\frac{1}{8}$ in. by $7\frac{1}{8}$ in.

33. Kate is designing a bookshelf. The top of the bookshelf and each of its 4 shelves will be $\frac{3}{4}$ in. thick. The bottom shelf will rest on the floor. The shelves will be $11\frac{1}{4}$ in. apart. Make a sketch of the bookshelf. What will be the overall height of the bookshelf? $48\frac{3}{4}$ in.

CONNECTING ARITHMETIC AND PROBABILITY

The *probability* of an event is a number from 0 to 1 indicating the likelihood that the event will happen. The notation $P(E)$ indicates the probability of an event. The probability that an event will not happen is written $P(\text{not } E) = 1 - P(E)$.

Find the probability that each event will *not* happen.

C **34.** $P(A) = \frac{1}{6}$ $\frac{5}{6}$ **35.** $P(B) = \frac{4}{9}$ $\frac{5}{9}$ **36.** $P(D) = \frac{3}{5}$ $\frac{2}{5}$ **37.** $P(C) = \frac{7}{16}$ $\frac{9}{16}$

38. Eight students run for class president. The probability of being elected is $\frac{1}{8}$. What is the probability of not being elected? $\frac{7}{8}$

39. The probability of choosing a key that will open a door is $\frac{2}{11}$. What is the probability of choosing a key that will not open the door? $\frac{9}{11}$

SPIRAL REVIEW

S **40.** Write 0.0000047 in scientific notation. *(Lesson 7-11)* 4.7×10^{-6}

41. Simplify: $\frac{14m}{5} - \frac{9m}{5}$ *(Lesson 8-4)* m

Rational Numbers **355**

Follow-Up

Extension

To show students the importance of fractions in the real world and to provide motivation for learning to work with them correctly, have students make a list of everyday activities that involve using fractions.

Enrichment

Enrichment 80 is pictured on page 338E.

Practice

Practice 99, *Resource Book*

Practice 99
Skills and Applications of Lesson 8-4

Find each answer. Simplify if possible.

(table of practice problems)

Resource Book, MATHEMATICAL CONNECTIONS
Copyright © by Houghton Mifflin Company. All rights reserved.

259

Teaching the Lesson
- Materials: fraction tiles
- Lesson Starter 74
- Using Technology, p. 338C
- Using Manipulatives, p. 338D
- Reteaching/Alternate Approach, p. 338D
- Cooperative Learning, p. 338D
- Toolbox Skills 17–19

Lesson Follow-Up
- Practice 100
- Enrichment 81: Connection
- Study Guide, pp. 147–148
- Computer Activity 8
- Manipulative Activity 16
- Cooperative Activity 16

Warm-Up Exercises

Simplify.

1. $\dfrac{3n}{4} + \dfrac{2n}{4}$ $\dfrac{5n}{4}$

2. $\dfrac{15}{4x} + \dfrac{1}{4x}$ $\dfrac{4}{x}$

3. $\dfrac{3}{g} - \dfrac{7}{g}$ $-\dfrac{4}{g}$

4. $\dfrac{5a}{b} - \dfrac{9a}{b}$ $-\dfrac{4a}{b}$

5. $\dfrac{7}{k} + \dfrac{7}{k}$ $\dfrac{14}{k}$

Lesson Focus

Ask students to imagine placing two pieces of rope end to end. If one piece is $7\frac{1}{2}$ in. long and the other is $4\frac{1}{8}$ in. long, how long are the two pieces when placed end to end? Today's lesson focuses on adding and subtracting rational numbers with unlike denominators.

Teaching Notes

Remind students that the first step in adding or subtracting rational numbers with unlike denominators is to find the least common denominator (LCD).

8-5

Adding and Subtracting Rational Numbers with Unlike Denominators

Objective: To add and subtract rational numbers and algebraic fractions with unlike denominators.

EXPLORATION

1 Replace each __?__ with a fraction that makes the statement true.

a. $\dfrac{4}{5} \cdot \underline{} = \dfrac{8}{10}$ $\dfrac{2}{2}$ b. $\dfrac{-3}{4} \cdot \underline{} = \dfrac{-15}{20}$ $\dfrac{5}{5}$ c. $\dfrac{7}{2} \cdot \underline{} = \dfrac{21}{6}$ $\dfrac{3}{3}$

2 What do all the answers in Step 1 have in common? They equal 1.

3 In Step 1, what is the relationship between $\frac{4}{5}$ and $\frac{8}{10}$, $\frac{-3}{4}$ and $\frac{-15}{20}$, and $\frac{7}{2}$ and $\frac{21}{6}$? They are equivalent fractions.

You may recall that the product of any number and 1 is the original number. When you multiply a fraction by a fractional form of 1, you obtain an equivalent fraction.

In order to add or subtract rational numbers with unlike denominators, you first write each rational number as a fraction. Then you write equivalent fractions having the LCD. You do this by multiplying each fraction by the form of 1 that will produce the LCD.

Example 1

Find each answer. Simplify if possible.

a. $\dfrac{1}{6} + \dfrac{3}{10}$

b. $-1\dfrac{1}{8} - 2\dfrac{3}{4}$

Solution

a. The LCD is 30.

$$\dfrac{1}{6} + \dfrac{3}{10} = \dfrac{1}{6} \cdot \dfrac{5}{5} + \dfrac{3}{10} \cdot \dfrac{3}{3}$$

$$= \dfrac{5}{30} + \dfrac{9}{30}$$

$$= \dfrac{5+9}{30}$$

$$= \dfrac{14}{30}$$

$$= \dfrac{7}{15}$$

b. The LCD is 8.

$$-1\dfrac{1}{8} - 2\dfrac{3}{4} = \dfrac{-9}{8} - \dfrac{11}{4}$$

$$= \dfrac{-9}{8} - \dfrac{11}{4} \cdot \dfrac{2}{2}$$

$$= \dfrac{-9}{8} - \dfrac{22}{8}$$

$$= \dfrac{-9-22}{8}$$

$$= \dfrac{-31}{8}$$

$$= -3\dfrac{7}{8}$$

✓ Check Your Understanding

1. How would Example 1(b) be different if you used 16 as the common denominator?
   See *Answers to Check Your Understanding* at the back of the book.

The process demonstrated in Example 1 can also be used to add and subtract algebraic fractions.

Example 2

Simplify $\frac{5x}{12} + \frac{3x}{8}$.

Solution

$$\frac{5x}{12} + \frac{3x}{8} = \frac{5x}{12} \cdot \frac{2}{2} + \frac{3x}{8} \cdot \frac{3}{3}$$ ← The LCD is 24.

$$= \frac{10x}{24} + \frac{9x}{24}$$

$$= \frac{10x + 9x}{24} = \frac{19x}{24}$$

To find the sum or difference of rational numbers that are written as decimals, apply the rules that you learned for adding and subtracting integers. Remember to align decimal points.

Example 3

Find each answer: **a.** $-4.25 + (-3.1)$ **b.** $-1.48 - (-3.8)$

Solution

a. $-4.25 + (-3.1) = -7.35$

b. $-1.48 - (-3.8) = -1.48 + 3.8$ ← Write the difference
$$= 2.32$$ as a related addition.

 **Check Your Understanding**

2. In Example 3(a), why is the answer negative?

3. In Example 3(b), why is the answer positive?

See *Answers to Check Your Understanding* at the back of the book.

Guided Practice

«**1.** COMMUNICATION «*Reading*   Refer to the text on pages 356–357. Create an outline for your notebook describing how to add and subtract rational numbers with unlike denominators. See answer on p. A23.

Find the LCD.

2. $\frac{2}{3}$ and $\frac{3}{5}$ 15   **3.** $-\frac{n}{4}$ and $\frac{n}{6}$ 12   **4.** $-1\frac{1}{8}$ and $-\frac{3}{16}$ 16   **5.** 3 and $\frac{1}{2}$ 2

Rewrite each sum or difference with equivalent fractions having the LCD.

6. $\frac{3}{4} + \frac{5}{8}$ $\frac{6}{8} + \frac{5}{8}$ **7.** $1\frac{11}{12} + 5\frac{3}{8}$ $\frac{46}{24} + \frac{129}{24}$ **8.** $-1\frac{5}{9} - 4$ $\frac{-14}{9} - \frac{36}{9}$ **9.** $\frac{10m}{5} - \frac{4m}{11}$ $\frac{110m}{55} - \frac{20m}{55}$

Find each answer. Simplify if possible.

10. $\frac{5}{6} + \frac{3}{4}$ $1\frac{7}{12}$ **11.** $1\frac{1}{4} - 5$ $-3\frac{3}{4}$ **12.** $-3\frac{1}{2} + \left(-4\frac{1}{6}\right) + \frac{2}{3}$ -7

13. $\frac{4a}{5} - \frac{a}{4}$ $\frac{11a}{20}$ **14.** $3x - \frac{x}{3}$ $\frac{8x}{3}$ **15.** $\frac{4r}{5} + \frac{8r}{15} + \frac{r}{3}$ $\frac{5r}{3}$

16. $-10 + 2.66$ -7.34 **17.** $-9.7 - 2.8$ -12.5 **18.** $5.4 + 0.19 + (-3.5)$ 2.09

Rational Numbers   **357**

Key Question
How do you find the least common denominator of two fractions?
The LCD of the two fractions is the LCM of the denominators.

Using Technology
For a suggestion on using a calculator or a BASIC program to find the LCD, see page 338C.

Using Manipulatives
For a suggestion on using fraction tiles to add and subtract fractions, see page 338D.

Reteaching/Alternate Approach
For a suggestion on using a cooperative learning activity to add and subtract fractions, see page 338D.

Error Analysis
When adding or subtracting rational numbers with unlike denominators, students may perform the indicated operations on both the numerators and the denominators. Remind students that they must first find the LCD, write equivalent fractions using the LCD, and then add or subtract.

Additional Examples
Find each answer. Simplify if possible.
1. $\frac{1}{4} + \frac{1}{6}$ $\frac{5}{12}$
2. $-1\frac{1}{6} - 2\frac{2}{3}$ $-3\frac{5}{6}$
3. Simplify: $\frac{7m}{12} + \frac{4m}{15}$ $\frac{17m}{20}$
Find each answer.
4. $-5.65 + (-3.2)$ -8.85
5. $-2.66 - (-3.9)$ 1.24

Closing the Lesson
Ask a student to explain how to find the sum or difference of rational numbers written as fractions. Ask another student to explain how to find the sum or difference of rational numbers written as decimals.

Exercise Notes

Communication: Reading

Guided Practice Exercise 1 requires students to create an outline describing the procedures of the lesson. Rewriting the procedures in their own words will help students better understand them.

Data Analysis

Exercises 41 and 42 help students to compare data involving fractions from a table of values.

Making Connections/Transitions

Exercises 43–46 and 47–51 connect rational numbers to the concepts of *function* and *series.*

Cooperative Learning

Exercises 47–51 may be done in cooperative groups. As an extension of these exercises, each group can create its own fraction series and present it to the class.

Exercises

Find each answer. Simplify if possible.

A **1.** $\frac{2}{3} + \frac{5}{6}$ $1\frac{1}{2}$ **2.** $\frac{7}{12} - \frac{5}{18}$ $\frac{11}{36}$ **3.** $6 - 2\frac{3}{5}$ $3\frac{2}{5}$

4. $5\frac{1}{2} + \frac{1}{6}$ $5\frac{2}{3}$ **5.** $\frac{4}{5} - \left(-\frac{3}{4}\right)$ $1\frac{11}{20}$ **6.** $-3\frac{2}{3} + \frac{3}{4}$ $-2\frac{11}{12}$

7. $-3 + \left(-2\frac{1}{6}\right)$ $-5\frac{1}{6}$ **8.** $-4\frac{1}{2} - 1\frac{11}{18}$ $-6\frac{1}{9}$ **9.** $-1\frac{9}{10} - \left(-2\frac{1}{15}\right)$ $\frac{1}{6}$

10. $1\frac{1}{4} - 3\frac{1}{5}$ $-1\frac{19}{20}$ **11.** $\frac{3}{4} + \frac{5}{18} + \left(-\frac{4}{9}\right)$ $\frac{7}{12}$ **12.** $-3\frac{4}{7} + \left(-5\frac{1}{2}\right) + 4$ $-5\frac{1}{14}$

13. $\frac{5m}{3} - \frac{2m}{7}$ $\frac{29m}{21}$ **14.** $\frac{8x}{5} + \frac{11x}{15}$ $\frac{7x}{3}$ **15.** $3b + \frac{5b}{4}$ $\frac{17b}{4}$

16. $7n - \frac{n}{2}$ $\frac{13n}{2}$ **17.** $\frac{4h}{5} + \frac{3h}{2} + \frac{h}{4}$ $\frac{51h}{20}$ **18.** $\frac{7a}{8} + \frac{5a}{6} + \frac{11a}{12}$ $\frac{21a}{8}$

19. $-5.3 + 1.7$ -3.6 **20.** $8.23 - 10.15$ -1.92

21. $4.72 - (-11.8)$ 16.52 **22.** $-15.4 - (-20)$ 4.6

23. $-6 + 9.03 + (-7.4)$ -4.37 **24.** $0.15 + (-4.8) + (-6.35)$ -11

25. Darla needs $2\frac{3}{4}$ yd of material to make a dress, and $1\frac{2}{3}$ yd to make a matching jacket. How much material does she need in all? $4\frac{5}{12}$ yd

26. Dana could jog $1\frac{1}{4}$ mi without stopping when she joined the track team. After two weeks she could jog $4\frac{1}{3}$ mi without stopping. How much farther could Dana jog after two weeks than when she began? $3\frac{1}{12}$ mi

Find each answer. Simplify if possible.

27. $-20\frac{1}{4}$ or -20.25

31. $-2\frac{3}{8}$ or -2.375

32. $3\frac{19}{40}$ or 3.475

B **27.** $-10.5 - 9\frac{3}{4}$ **28.** $6.35 + 4\frac{3}{5}$ 10.95 or $10\frac{19}{20}$ **29.** $7\frac{1}{9} - 4.\overline{3}$ $2\frac{7}{9}$ or $2.\overline{7}$

30. $-3\frac{5}{6} + 1.\overline{6}$ $-2\frac{1}{6}$ or $-2.1\overline{6}$ **31.** $-\frac{1}{8} + 1.25 + \left(-3\frac{1}{2}\right)$ **32.** $0.6 + 5\frac{3}{8} + (-2.5)$

Evaluate each expression when $a = -2\frac{5}{6}$, $b = 3\frac{1}{2}$, $c = -4.8$, **and** $d = 0.6$.

33. $-ab$ $9\frac{11}{12}$ **34.** $a \div b$ $-\frac{17}{21}$ **35.** $2(a + b)$ $1\frac{1}{3}$ **36.** $|a - b|$ $6\frac{1}{3}$

37. $c \div d$ -8 **38.** $5cd$ -14.4 **39.** $d^2 - c$ 5.16 **40.** $-c + d + 7$ 12.4

DATA, *pages 338–339*

41. In Grand Rapids how much greater is the recommended thickness of insulation for attic floors than for basement ceilings? $4\frac{1}{2}$ in.

42. How much greater is the recommended thickness of insulation for attic floors in Duluth than in Atlanta? (Use the figures for the upper end of each range.) $4\frac{1}{2}$ in.

Complete each function table.   **Find each function rule.**

43.

x	4x
$2\frac{1}{6}$	$8\frac{2}{3}$
0.4	1.6
$-\frac{3}{8}$	?
$-1\frac{1}{4}$	?
-2.5	?

(answers: $-1\frac{1}{2}$, -5, -10)

44.

x	$6x + \frac{1}{2}$
$-\frac{1}{2}$	$-2\frac{1}{2}$
$\frac{1}{2}$	$3\frac{1}{2}$
$\frac{1}{3}$	?
$\frac{1}{4}$	?
$\frac{1}{5}$	?

(answers: $2\frac{1}{2}$, 2, $1\frac{7}{10}$)

45.

x	?
-2	$-1\frac{2}{3}$
$-1\frac{2}{3}$	$-1\frac{1}{3}$
$-\frac{1}{3}$	0
$1\frac{1}{3}$	$1\frac{2}{3}$
$3\frac{2}{3}$	4

$x \rightarrow x + \frac{1}{3}$

46.

x	?
-4	-1
-2	$-\frac{1}{2}$
$\frac{1}{2}$	$\frac{1}{8}$
1	$\frac{1}{4}$
5	$1\frac{1}{4}$

$x \rightarrow \frac{x}{4}$

PATTERNS

The expression shown at the right is called a *series*. There is always a pattern to the numbers in a series.

$$\frac{1}{2} + \frac{1}{4} + \frac{1}{8} + \frac{1}{16} + \frac{1}{32} \ldots$$

C **47.** What is the pattern of the fractions in the series? Each fraction is half the preceding fraction.

48. Find the next two fractions in the series. $\frac{1}{64}, \frac{1}{128}$

49. Find the sum of the fractions indicated.
   a. the first two fractions $\frac{3}{4}$
   b. the first three fractions $\frac{7}{8}$
   c. the first four fractions $\frac{15}{16}$
   d. the first five fractions $\frac{31}{32}$

50. Describe the relationship between the numerator and the denominator of each sum in Exercise 49. The numerator is one less than the denominator.

51. Without adding, predict what will be the sum of the first six fractions in the series. $\frac{63}{64}$

SPIRAL REVIEW

52. Write $\frac{12}{56}$ in lowest terms.   *(Lesson 7-4)* $\frac{3}{14}$

53. Draw a pictograph to display the data.   *(Lesson 5-1)*
See answer on p. A23.

Populations of Large Cities (Millions)

City	New York	London	Tokyo	Mexico City	Beijing
Population	7.3	6.8	8.4	12.9	9.3

54. DATA, *pages 140–141*   What is the maximum speed of an elephant? *(Lesson 5-2)* 25 mi/h

55. Simplify: $\frac{7a}{15} + \frac{4a}{5}$ *(Lesson 8-5)* $\frac{19a}{15}$

Rational Numbers   **359**

Follow-Up

Extension
Have students determine the meaning of the term "relatively prime." Ask them to explore when the least common denominator is the product of the denominators. Ask also if the term "relatively prime" has any relationship to the way the least common denominator is found.

Enrichment
Enrichment 81 is pictured on page 338E.

Practice
Practice 100, *Resource Book*

359

Multicultural Notes

Background

Because our nation is made up of people from so many different backgrounds and cultures, the United States is one of the most musically diverse countries in the world. For example, many groups of Native Americans maintain their identity through traditional songs and dances. Musical traditions transported to America by people from West Africa have evolved into relatively new forms of music—blues and jazz. People from the Spanish-speaking cultures of Latin America and the Caribbean Islands have also introduced new kinds of music, such as calypso and reggae.

As well as creating new kinds of music, Americans have also invented new musical instruments. The mountain dulcimer, which has strings that are plucked and resembles German and Scandinavian instruments, first appeared in southern Appalachia in the 1800s. The ukulele was developed after an instrument called the machete, which Portuguese people brought to Hawaii in the 19th century. The word ukulele may have come from the Hawaiian nickname given to a British officer who helped popularize the instrument. The word banjo can be traced to similar words in a number of African languages.

Music Connection

MIXED REVIEW

Operations with Rational Numbers

Jazz music is played all over the world today, but it has its origins in the experience of African-Americans. Music that would develop into jazz was first played in the early 1900s by African-Americans at memorial parades. Bands of musicians used brass instruments to create a sound that came to be called Dixieland jazz. Musicians have since developed many new forms of jazz, but Dixieland music is still popular today.

To find out what American city is often associated with Dixieland music and early jazz, complete the code boxes. Begin by solving Exercise 1. The answer to Exercise 1 is $\frac{35}{72}$, and the letter associated with Exercise 1 is *W*. Write *W* in the box above $\frac{35}{72}$ as shown below. Continue in this way with Exercises 2–22.

1. $\frac{7}{12} \cdot \frac{5}{6}$ $\frac{35}{72}$ (W)

2. $\frac{5}{8} \div \frac{3}{4}$ $\frac{5}{6}$ (Z)

3. $13\frac{4}{5} + 9\frac{4}{5}$ $23\frac{3}{5}$ (N)

4. $4\frac{7}{10} - 2\frac{4}{5}$ $1\frac{9}{10}$ (R)

5. $-6\left(\frac{7}{8}\right) - 5\frac{1}{4}$ (O)

6. $\left(-4\frac{2}{3}\right)\left(-2\frac{6}{7}\right)$ $13\frac{1}{3}$ (L)

7. $\frac{5}{9} \div \left(-\frac{3}{4}\right)$ $-\frac{20}{27}$ (E)

8. $-6\frac{3}{7} \div 9$ $-\frac{5}{7}$ (S)

9. $\frac{3}{5} + \left(-3\frac{2}{5}\right)$ $-2\frac{4}{5}$ (R)

10. $-\frac{7}{9} + \left(-\frac{5}{12}\right)$ $-1\frac{7}{36}$ (O)

11. $-4\frac{1}{2} - \frac{5}{7}$ $-5\frac{3}{14}$ (N)

12. $9\frac{7}{8} - \left(-4\frac{3}{8}\right)$ $14\frac{1}{4}$ (A)

13. $4\frac{4}{9} + \frac{3}{9} + \left(-2\frac{5}{9}\right)$ $2\frac{2}{9}$ (C)

14. $-6 + \left(-2\frac{5}{6}\right) + \frac{11}{12}$ $-7\frac{11}{12}$ (D)

15. $(-0.4)(-0.9)$ 0.36   (E)

16. $(5.6)(-0.5)$ -2.8 (L)

17. $-15.19 \div 3.1$ -4.9 (A)

18. $-36 \div (-1.8)$ 20   (Z)

19. $-2.3 + (-3.8)$ -6.1 (J)

20. $19 + (-14.6)$ 4.4   (E)

21. $8.95 - 35.7$ -26.75 (A)

22. $5.32 - (-0.74)$ 6.06   (F)

N	E		O	R	L	E	A	N	S
?	?	W	?	?	?	?	?	?	?
$-5\frac{3}{14}$	0.36	$\frac{35}{72}$	$-1\frac{7}{36}$	$-2\frac{4}{5}$	-2.8	$-\frac{20}{27}$	-4.9	$23\frac{3}{5}$	$-\frac{5}{7}$

C	R	A	D	L	E		O	F
?	?	?	?	?	?		?	?
$2\frac{2}{9}$	$1\frac{9}{10}$	$14\frac{1}{4}$	$-7\frac{11}{12}$	$13\frac{1}{3}$	4.4		$-5\frac{1}{4}$	6.06

J	A	Z	Z
?	?	?	?
-6.1	-26.75	$\frac{5}{6}$	20

American music combines the musical traditions of many different peoples and places. To learn more about these musical traditions, solve each exercise. The word printed in blue following the correct answer matches the description printed in blue.

23. $\frac{2}{3} \cdot \frac{3}{2}$

Instrument created in America by African-Americans, based on African instruments.

(a.) 1   banjo   b. $2\frac{1}{6}$ harmonica

25. $-\frac{11}{12} \div \frac{3}{4}$

Hawaiian instrument whose name means *leaping flea*.

(a.) $-1\frac{2}{9}$ ukulele   b. $-\frac{11}{16}$ holoku

27. $\frac{15}{16} + \frac{9}{16} + \frac{13}{16}$

Instrument created by settlers in the Appalachians, based on similar European ones.

(a.) $2\frac{5}{16}$ dulcimer   b. $\frac{37}{48}$ mirliton

29. $\frac{1}{25} - \frac{1}{5}$

Kind of blues music for piano that was popular in the 1930s.

a. $\frac{1}{20}$ bebop   (b.) $-\frac{4}{25}$ boogie-woogie

24. $\frac{1}{3}\left(-3\frac{1}{5}\right)$

City where country music is played at the *Grand Ole Opry*.

(a.) $-1\frac{1}{15}$ Nashville   b. $-1\frac{1}{5}$ Atlanta

26. $-3\frac{3}{5} \div 2\frac{6}{7}$

Kind of Mexican music usually played by a band of violins, guitars, and trumpets.

a. $-10\frac{2}{7}$ flamenco   (b.) $-1\frac{13}{50}$ mariachi

28. $-4\frac{2}{5} + \left(-\frac{3}{5}\right)$

Kind of jazz music that combines aspects of jazz with aspects of rock music.

a. $-3\frac{4}{5}$ synthesizer   (b.) -5 fusion

30. $-\frac{1}{4} + \frac{2}{3} - \left(-1\frac{5}{6}\right)$

Decade sometimes called the Golden Age of Jazz.

(a.) $2\frac{1}{4}$ 1920s   b. $-1\frac{5}{12}$ 1940s

Rational Numbers   361

Teaching the Lesson
- Materials: calculator
- Lesson Starter 75
- Toolbox Skills 1–4

Lesson Follow-Up
- Practice 101
- Practice 102: Nonroutine
- Enrichment 82: Data Analysis
- Study Guide, pp. 149–150
- Test 37

Warm-Up Exercises

Simplify.

1. $2\frac{2}{3} + 3\frac{3}{4}$ $6\frac{5}{12}$

2. $\frac{1}{4} + \frac{1}{5}$ $\frac{9}{20}$

3. $\frac{7}{11} - \frac{9}{11}$ $-\frac{2}{11}$

4. $\left(\frac{2}{5}\right)\left(\frac{7}{3}\right)$ $\frac{14}{15}$

5. $\frac{2}{3} \div \frac{5}{6}$ $\frac{4}{5}$

Lesson Focus

Ask students to imagine sharing a bottle of fruit juice equally with a friend. Would they measure or estimate the amount to pour into each glass? Sometimes, using an estimate is the most reasonable approach to solving a problem. Today's lesson focuses on estimating with rational numbers.

Key Question

Which method is the best to use when estimating with rational numbers? When finding a quotient, compatible numbers may be the best method; when finding a sum in which all the numbers have the same sign, the front-end method may be the best; in other cases, rounding may be the best method.

Estimating with Rational Numbers

Objective: To estimate the sum, difference, or quotient of rational numbers.

APPLICATION

To replace some screens, the Okawas have a piece of screening that is $12\frac{1}{3}$ ft long. They cut off a piece that is $3\frac{3}{4}$ ft long and do not know if they have enough screening left for doors that require a total of 7 ft of screening. Since the Okawas do not need to know the exact footage, they estimate.

QUESTION About how many feet of screening are left?

Example 1

Estimate $12\frac{1}{3} - 3\frac{3}{4}$ by rounding.

Solution

$12\frac{1}{3} - 3\frac{3}{4}$ ⟵ Because $\frac{1}{3}$ is less than one half, round down.

Because $\frac{3}{4}$ is greater than one half, round up.

$\underbrace{12 - 4}$

about 8

✏ Check Your Understanding

How can you tell if a fraction is less than one half or greater than one half?
See *Answers to Check Your Understanding* at the back of the book.

ANSWER About 8 ft of screening are left.

When estimating sums of rational numbers with the same sign, front-end estimation may give you a closer estimate than rounding. Add the integers and then adjust the answer.

Example 2

Estimate $\frac{5}{9} + 2\frac{4}{5} + 3\frac{1}{16}$ by using front-end-and-adjust estimation.

Solution

First, add the integers. $2 + 3 = 5$

Then adjust. $\frac{5}{9}$ is about $\frac{1}{2}$, $\frac{4}{5}$ is closer to 1, $\frac{1}{16}$ is closer to 0

$5 + \frac{1}{2} + 1 + 0 \rightarrow$ about $6\frac{1}{2}$

You can use compatible numbers to estimate the quotient of mixed numbers. You round the divisor to the nearest integer and then use a compatible dividend. To estimate the quotient when the divisor is a fraction, first write a simpler fraction. Then round the dividend and divide.

Example 3   **Estimate:**   a. $13\frac{1}{3} \div 5\frac{9}{10}$ b. $-19\frac{7}{8} \div \frac{7}{15}$

Solution   a. $13\frac{1}{3} \div 5\frac{9}{10}$ b. $-19\frac{7}{8} \div \frac{7}{15}$
$\qquad\qquad\qquad\downarrow\quad\;\;\downarrow$ $\qquad\qquad\qquad\quad\downarrow\quad\;\downarrow$
$\qquad\qquad\underbrace{12 \div 6}$ $\qquad\qquad-20 \div \frac{1}{2} = \underbrace{(-20)(2)}$
$\qquad\qquad\;\;$about 2 $\qquad\qquad\qquad\qquad\quad$about -40

Guided Practice

COMMUNICATION «*Reading*

«**1.** Using $8 - 5$ to estimate the sum $7\frac{2}{3} - 5\frac{1}{4}$ is an example of estimating by _?_.   rounding

«**2.** Estimating by writing $9\frac{1}{16} \div 4\frac{1}{3}$ as $8 \div 4$ is an example of estimating by _?_.   compatible numbers

«**3.** Using $5 + 1 + 0$ to estimate the sum $3\frac{7}{8} + 2\frac{1}{6}$ is an example of estimating by _?_.   front-end-and-adjust

Round to the nearest integer.

4. $\frac{3}{16}$ 0 **5.** $2\frac{7}{12}$ 3 **6.** $-3\frac{1}{4}$ -3 **7.** $-15\frac{7}{8}$ -16 **8.** $34\frac{5}{9}$
$\qquad\qquad\qquad\qquad\qquad\qquad\qquad\qquad\qquad\qquad\qquad\qquad\qquad\qquad\qquad\;\;$ 35

Estimate. Estimates may vary. Accept reasonable estimates.

9. $4\frac{3}{4} - 2\frac{1}{2}$ **10.** $-9\frac{3}{11} + \left(-5\frac{1}{6}\right)$ **11.** $10\frac{7}{8} \div \left(-\frac{5}{14}\right)$ **12.** $18\frac{4}{5} \div 8\frac{3}{4}$
$\;\;$about 2 $\qquad\;$about -14 $\qquad\qquad$about -33 $\qquad\quad$about 2

Exercises

Estimates may vary. Accept reasonable estimates.
Estimate. **11.** about -48 **12.** about 5

1. $2\frac{8}{11} - 8\frac{1}{3}$ about -5 **2.** $7\frac{2}{3} - 2\frac{1}{2}$ about 5 **3.** $6\frac{5}{12} + 5\frac{1}{5}$
$\qquad\qquad\qquad\qquad\qquad\qquad\qquad\qquad\qquad\qquad\qquad\qquad$ about 11

4. $-14\frac{1}{8} + \left(-10\frac{3}{5}\right)$ **5.** $-6\frac{3}{4} + 2\frac{2}{9}$ about -5 **6.** $-2\frac{3}{5} - 7\frac{3}{8}$
$\qquad\qquad\quad$ about -25 $\qquad\qquad\qquad\qquad\qquad\qquad\qquad\;$ about -10

7. $\frac{9}{10} + 7\frac{3}{16} + 2\frac{4}{7}$ about 11 **8.** $3\frac{5}{6} + 9\frac{2}{5} + \frac{7}{8}$ about **9.** $41\frac{1}{4} \div 11\frac{1}{8}$
$\qquad\qquad\qquad\qquad\qquad\qquad\qquad\qquad\qquad\qquad\;$ 14 $\qquad\qquad\;\;$ about 4

10. $19\frac{7}{8} \div \left(-\frac{10}{29}\right)$ about -60 **11.** $-11\frac{3}{5} \div \frac{25}{101}$ **12.** $-28\frac{1}{4} \div \left(-5\frac{3}{4}\right)$

13. Gina makes quilts using stitches that are $\frac{9}{32}$ in. long. About how many stitches will there be in a row of stitches that is $8\frac{1}{2}$ in. long? about 32

14. Melissa tutored for $2\frac{3}{4}$ h last week. This week she tutored for $7\frac{1}{3}$ h. About how many more hours did she tutor this week than last week?
about 4 h

Rational Numbers   **363**

363

15. Pedro cycled $9\frac{1}{4}$ mi one weekend, $10\frac{1}{4}$ mi the next weekend, and $13\frac{7}{8}$ mi the following weekend. About how many miles did he cycle altogether on the three weekends? about 33 mi

16. It takes Sam about $1\frac{2}{3}$ h to install a stereo system in an automobile at the plant where he is employed. About how many stereo systems can he plan to install in $8\frac{1}{4}$ h? about 4

 CALCULATOR

Estimate to tell whether each calculator answer is *reasonable* or *unreasonable*. If an answer is unreasonable, find the correct answer.

B   17. $6\frac{1}{3} - 11\frac{12}{13}$ ⎡ −18 . 10 ⌋ 39. ⎤   unreasonable; $-5\frac{23}{39}$

18. $\frac{9}{10} + 4\frac{1}{8} + 3\frac{1}{6}$ ⎡ 8 . 23 ⌋ 120. ⎤   reasonable

19. $\left(-\frac{24}{49}\right)\left(8\frac{19}{20}\right)$ ⎡ −4 . 94 ⌋ 245. ⎤   reasonable

20. $35\frac{1}{5} \div \frac{4}{23}$ ⎡ 6 . 14 ⌋ 115. ⎤   unreasonable; $202\frac{2}{5}$

SPIRAL REVIEW

S   21. Express 4.7 as the quotient of two integers.   *(Lesson 7-9)* $\frac{47}{10}$

22. Find the sum: $-37 + (-19)$ *(Lesson 3-2)* −56

23. Graph $A(-3, -2)$, $B(4, -1)$, and $C(-2, -3)$ on a coordinate plane. *(Lesson 3-8)* See answer on p. A24.

24. Estimate: $22\frac{1}{6} \div \frac{25}{49}$ *(Lesson 8-6)* about 44

 **Self-Test 2**

Solve. Decide whether a *whole number*, a *fraction*, or a *decimal* is an appropriate answer.

1. A cookbook suggests that a roast be cooked for $\frac{1}{2}$ h per pound. How many hours should you cook a 7 lb roast? fraction; $3\frac{1}{2}$ h   8-3

2. Ribbon costs \$2.90/yd. How much does $1\frac{1}{2}$ yd cost? decimal; \$4.35

Find each answer. Simplify if possible.

3. $-\frac{5}{7} + \frac{6}{7}$ $\frac{1}{7}$ 4. $-12\frac{1}{3} - 4\frac{2}{3}$ −17   5. $\frac{16a}{21} - \frac{a}{21}$ $\frac{5a}{7}$ 8-4

6. $-4\frac{11}{18} - \left(-1\frac{1}{2}\right)$ $-3\frac{1}{9}$ 7. $5b + \frac{7b}{4}$ $\frac{27b}{4}$ 8. $-6.7 + 2.9$ −3.8   8-5

9–11. Estimates may vary. Accept reasonable estimates.

Estimate.   about −19   about 5

9. $20\frac{3}{4} - 6\frac{1}{5}$ about 15   10. $-12\frac{11}{20} + \left(-5\frac{7}{8}\right)$ 11. $23\frac{1}{4} \div 4\frac{11}{12}$ 8-6

364 Chapter 8

8-7

Supplying Missing Facts

Objective: To solve problems by supplying missing facts.

Some problems involve quantities with different units of measure. To solve this type of problem, you may need to supply a fact that is not given in the problem. This fact might be the relationship between the two units of measure. Suppose a problem asks you to compare the number of ounces in two weights that are given in pounds. Before you can solve the problem, you will need to supply the fact that 1 lb = 16 oz.

Problem

Randy decides to spend an equal amount of time on each of his five homework subjects. If he plans to do homework for two hours altogether, how many minutes should he spend on each subject?

Solution

UNDERSTAND The problem is about the amount of time spent on different homework subjects.
Facts: equal time for each subject
2 h for all 5 subjects
Find: number of minutes on each subject

PLAN Divide the 2 h into 5 parts to find the time spent on each subject. This time will be in hours, but the question asks for minutes. The missing fact is 1 h = 60 min. Multiply the number of hours spent on each subject by 60 to find the number of minutes.

WORK

Divide: $2 \div 5 = \frac{2}{5}$　⟵ $\frac{2}{5}$ h for each subject

Multiply: $\frac{2}{5} \cdot 60 = 24$　⟵ 24 min for each subject

ANSWER Randy should spend 24 min on each subject.

Look Back An alternative method for solving this problem is to first find the total number of minutes in two hours, and then divide by five. Solve the problem again using this alternative method. 24 min

Rational Numbers **365**

Lesson Planner

Teaching the Lesson
- Materials: poster board
- Lesson Starter 76
- Reteaching/Alternate Approach, p. 338D
- Toolbox Skill 22

Lesson Follow-Up
- Practice 103
- Enrichment 83: Thinking Skills
- Study Guide, pp. 151–152

Warm-Up Exercises

Estimate.

1. $2\frac{3}{4} - 9\frac{1}{3}$　about −6

2. $3\frac{3}{5} \div \frac{1}{2}$　about 7

3. $41\frac{7}{8} \div 5\frac{1}{3}$　about 8

4. $3\frac{3}{8} + 4\frac{5}{9}$　about 8

5. $32\frac{1}{2} - 23\frac{2}{5}$　about 9

Lesson Focus

Ask students to describe situations in which they had to recall a fact, such as the number of days in a week. To solve many problems in the real world, students will often need to recall facts. Today's lesson focuses on solving problems by supplying missing facts.

Teaching Notes

Work through the Example with students, pointing out that the missing fact, 1 h = 60 min, is a fact of general knowledge.

Key Question

Where can you find information about real-world facts? almanacs, textbooks, encyclopedias, and so on

Reteaching/Alternate Approach

For a suggestion on making a "missing facts" poster, see page 338D.

Error Analysis

When supplying missing facts, particularly ones related to measurements, students may incorrectly remember the conversion. Remind students to check the Table of Measures on page xiii to be certain that they have supplied the correct information.

Additional Example

Michelle wants to make new curtains for her living room. She measures the windows and finds that she needs 58 ft of fabric. The store sells fabric by the yard. How many yards of fabric does Michelle need? $19\frac{1}{3}$ yd

Closing the Lesson

Create a "missing facts" chart with students. Have them supply as many relationships among various units of measure that they can remember or determine.

Suggested Assignments

Skills: 1–16
Average: 1–16
Advanced: 1–16, Challenge

Exercise Notes

Communication: Reading

Guided Practice Exercises 1 and 2 require students to show an understanding of the problem by identifying the missing fact.

Application

Guided Practice Exercises 3–6 and Exercise 11 rely on the use of common formulas. Data must often be converted to different units. For example, in football 10 yd = a first down. Yet, the distance needed for a first down may be given in inches.

Guided Practice

COMMUNICATION «*Reading*

Identify the missing fact. Do not solve the problem.

«**1.** The Evans family uses 21 quarts of milk per week. How many gallons do they use in 4 weeks? the number of quarts in a gallon

«**2.** Ed is $67\frac{1}{2}$ in. tall and still growing. How many more inches must he grow before he is 6 ft tall? the number of inches in a foot

Replace each __?__ with the correct number or word.

3. 1 day = __?__ hours 24

4. 1 __?__ = 128 fluid ounces
gallon

5. __?__ pounds = 1 ton 2000

6. 1000 __?__ = 1 kilometer meters

Write a variable expression for each problem.

7. Amy went to camp for c weeks. How many days was she there? 7c

8. There were d dimes in a parking meter. How many dollars is this? $\frac{d}{10}$

9. Paul types w words in 1 h. How many words can he type in 1 min? $\frac{w}{60}$

10. A classroom is m meters wide. How many centimeters is this? 100m

Solve by supplying a missing fact.

11. Georgia bought 19 yards of rope and Mark bought 11 yards of rope. How many more feet of rope did Georgia buy? 24 ft

Problem Solving Situations

Solve by supplying missing facts.

1. Norman drove due north for 85 min and then due east for 55 min. How many hours did he spend driving? $2\frac{1}{3}$ h

2. The band director needs 5 ft of fabric for each band costume. If there are nine members in the band, how many yards of the fabric should the director buy? 15 yd

3. The town of Middleton celebrated its centennial 40 years ago. How many years old is the town now? 140 years

4. Walt divides 6 lb of birdseed among eight bird feeders. How many ounces of birdseed does he put in each bird feeder? 12 oz

5–10. Strategies may vary. Likely strategies are given.
Solve using any problem solving strategy.

5. Manuel bought some used paperback books for $1.25 each and some used children's books for $.90 each. He paid a total of $7. How many of each type of book did he buy? guess and check; 2 paperbacks, 5 children's books

6. using equations; 5
7. supplying missing facts; 5 lb
8. supplying missing facts; $13\frac{1}{2}$ in.

Solve using any problem solving strategy.

6. Jacob Kosofsky is six years older than five times his son's age. If Jacob Kosofsky is 31, how old is his son?

7. A baker plans to make 8 cheese-cakes, each containing 10 oz of cream cheese. How many pounds of cream cheese are needed?

8. In the 1988 Olympics, the women's high jump event was won by Louise Ritter of the United States who jumped 6 ft 8 in. Guennadi Avdeenko of the USSR won the men's high jump event with a jump of 7 ft $9\frac{1}{2}$ in. How many inches higher did Avdeenko jump than Ritter?

9. Angela Bruegge works 40 h per week for 18 weeks. What is her total salary? not enough information; the hourly rate is needed.

10. A fruit stand sells grapefruits for $.85 each, apples for $.25 each, and lemons for $.15 each. In how many different ways can a customer spend $3.30 on these fruits? making a table; 12

PROBLEM SOLVING CHECKLIST

Keep these in mind:
Using a Four-Step Plan
Too Much or Not Enough Information
Supplying Missing Facts

Consider these strategies:
Choosing the Correct Operation
Making a Table
Guess and Check
Using Equations
Identifying a Pattern
Drawing a Diagram

WRITING WORD PROBLEMS

Write a word problem that you would solve by supplying the missing fact. Then solve the problem. Answers to Problems 11 and 12 are on p. A24.

11. 1 mile = 5280 feet

12. 10 years = 1 decade

SPIRAL REVIEW

S **13.** Replace the __?__ with >, <, or =: $\frac{8}{9}$ __?__ $\frac{22}{27}$ *(Lesson 7-6)* >

14. DATA, *pages 288–289* About how many more senior high schools than junior high schools had microcomputers in 1984? *(Lesson 5-3)* about 5000

15. Solve: $5x + 17 = 2$ *(Lesson 4-5)* −3

16. Sandy earns $2750 per month. How much does Sandy earn in two years? *(Lesson 8-7)* $66,000

Challenge

Find three different whole numbers a, b, and c so that $\frac{1}{a} + \frac{1}{b} + \frac{1}{c}$ is also a whole number. 2, 3, and 6

Rational Numbers **367**

Problem Solving

Problem Solving Situations 1–10 provide problems that can be solved by a variety of strategies. The strategies are reviewed by means of a checklist.

Nonroutine Problem

The *Challenge* problem offers students a chance to use their algebraic skills, problem-solving strategies, and knowledge of rational numbers. Remind students of strategies that might be used, such as drawing a diagram or creating a table.

Follow-Up

Extension

Ask each student to create a problem with missing facts based on something he or she did yesterday. The problems can be presented to the class with other students offering solutions.

Enrichment

Enrichment 83 is pictured on page 338F.

Practice

Practice 103, *Resource Book*

Practice 103
Skills and Applications of Lesson 8-7

Solve by supplying missing facts.

1. Phillipa spends 40 min per school day in mathematics class. How many hours is she scheduled to be in mathematics class if there are 180 days in a school year? 120 h
2. Camillo is 5 ft 8 in. tall. After graduation he wants to be a state policeman. However, the minimum height requirement for policemen is $69\frac{1}{2}$ in. How much taller must Camillo grow in order to qualify? $1\frac{1}{2}$ in.
3. Mrs. Ling's homeroom has 27 students. Each student needs a pencil to complete a questionnaire. If each box contains one dozen pencils, how many boxes does she need to supply each student with a pencil? 3 boxes
4. Each side of square measures 1.5 ft. Find the area in square inches. 324 in.²
5. Joel ran 440 yd. Annie ran 660 yd. How many more feet did Annie run than Joel? 660 ft
6. The dance committee plans to prepare four large bowls of punch for the dance. Each bowl requires 2 quarts of ginger ale. How many 16 fluid ounce bottles of ginger ale are needed for each bowl of punch? 4 bottles
7. Tomas determines a cup of iron filings weighs 4,385 g. The experiment he is doing calls for 3 kg. Does the cup contain enough filings for the experiment? By how much? Yes; 1385 g
8. The Firebrand Paint Company advertises that one quart of their paint will cover 100 ft² of wall space. Lani purchases one gallon of the paint. How many square feet of wall space will she be able to cover? 400 ft²

Lesson Planner

Teaching the Lesson
- Materials: calculator
- Lesson Starter 77
- Toolbox Skills 11–14, 17–21

Lesson Follow-Up
- Practice 104
- Enrichment 84: Exploration
- Study Guide, pp. 153–154

Warm-Up Exercises

Solve.

1. $2x + 9 = -17$ -13
2. $12 - y = 44$ -32
3. $2 + 3n = 32$ 10
4. $19 + 3c = -44$ -21
5. $-y + 15 = -7$ 22

Lesson Focus

Ask students to imagine going to a grocery store. Items are often priced as "2 for 99¢" or "3 for $1," where the cost per item does not equal a whole number. Today's lesson focuses on equations that involve rational numbers.

Teaching Notes

Work through the Examples with students, stressing that the processes used are the same as the ones used in Chapter 4.

Key Question

What is an inverse operation? An inverse operation is an operation that undoes another operation.

Error Analysis

When solving equations involving rational numbers, students may perform the arithmetic with fractions or decimals incorrectly. Remind students to follow techniques learned earlier for finding LCDs, aligning decimal points, and so on.

8-8 Equations Involving Rational Numbers

Objective: To solve equations involving rational numbers.

 CONNECTION

In Chapter 4 you solved equations involving integers. You can use inverse operations to solve questions involving fractions and decimals as well.

Example 1

Solve. Check each solution.

a. $7 = -3x + 5$

b. $2m + 5.1 = -1.3$

Solution

a.
$$7 = -3x + 5$$
$$7 - 5 = -3x + 5 - 5$$
$$2 = -3x$$
$$\frac{2}{-3} = \frac{-3x}{-3}$$
$$-\frac{2}{3} = x$$

The solution is $-\frac{2}{3}$.

 **Check**
$$7 = -3x + 5$$
$$7 \overset{?}{=} (-3)\left(-\frac{2}{3}\right) + 5$$
$$7 \overset{?}{=} 2 + 5$$
$$7 = 7$$

b.
$$2m + 5.1 = -1.3$$
$$2m + 5.1 - 5.1 = -1.3 - 5.1$$
$$2m = -6.4$$
$$\frac{2m}{2} = \frac{-6.4}{2}$$
$$m = -3.2$$

The solution is -3.2.

Check
$$2m + 5.1 = -1.3$$
$$2(-3.2) + 5.1 \overset{?}{=} -1.3$$
$$-6.4 + 5.1 \overset{?}{=} -1.3$$
$$-1.3 = -1.3$$

Example 2

Solve $z - \frac{1}{2} = \frac{3}{4}$. Check the solution.

Solution

$$z - \frac{1}{2} = \frac{3}{4}$$
$$z - \frac{1}{2} + \frac{1}{2} = \frac{3}{4} + \frac{1}{2}$$
$$z = \frac{3}{4} + \frac{2}{4}$$
$$z = \frac{5}{4} = 1\frac{1}{4}$$

Check $z - \frac{1}{2} = \frac{3}{4}$
$$\frac{5}{4} - \frac{1}{2} \overset{?}{=} \frac{3}{4}$$
$$\frac{5}{4} - \frac{2}{4} \overset{?}{=} \frac{3}{4}$$
$$\frac{3}{4} = \frac{3}{4}$$

The solution is $1\frac{1}{4}$.

✓ Check Your Understanding

1. In the Check for Example 1(a), why does $(-3)\left(-\frac{2}{3}\right)$ equal 2?
2. In Example 2, what equivalent fraction replaces $\frac{1}{2}$?
3. In Example 2, explain why the solution, $1\frac{1}{4}$, is substituted in the Check as a fraction, not as a mixed number.

See *Answers to Check Your Understanding* at the back of the book.

368 Chapter 8

Guided Practice

COMMUNICATION « *Reading*

Use this paragraph for Exercises 1–3.

Keith bought some fruit at $.39 per pound. After using $2\frac{1}{2}$ lb of the fruit to make a salad, he had $1\frac{3}{4}$ lb left. How many pounds of fruit did Keith buy?

« **1.** Which equation could you use to solve the problem?

   a. $n + 2\frac{1}{2} = 1\frac{3}{4}$ **b.** $n - 2\frac{1}{2} = 1\frac{3}{4}$ **c.** $n + 1\frac{3}{4} = 2\frac{1}{2}$

« **2.** Identify any facts that are not needed to solve the problem. cost per lb

« **3.** Suppose the question in the problem were "How much did it cost Keith for the fruit he used in his salad?" Identify any facts that are not needed to solve this problem. amount left over

Tell what operation(s) you would use to solve each equation. Do not solve.

4. $4k + 18 = 11$ **5.** $\frac{6}{7} = \frac{3}{5} + a$ **6.** $-1.3 = 7b - 2.7$
subtraction, division   subtraction   addition, division

Solve. Check each solution.

7. $8 + 5j = 12 \;\; \frac{4}{5}$ **8.** $4.6 = 7n - 4.5$ 1.3

9. $x + \frac{3}{8} = -\frac{1}{2} \;\; -\frac{7}{8}$ **10.** $-1\frac{5}{6} + y = \frac{2}{3} \;\; 2\frac{1}{2}$

Exercises

Solve. Check each solution.

A   **1.** $6n + 3 = -1 \;\; -\frac{2}{3}$ **2.** $9 + 4g = 12 \;\; \frac{3}{4}$ **3.** $x + \frac{2}{9} = \frac{1}{3} \;\; \frac{1}{9}$

   4. $s - \frac{1}{8} = -\frac{3}{4} \;\; -\frac{5}{8}$ **5.** $\frac{1}{3} + n = -\frac{1}{2} \;\; -\frac{5}{6}$ **6.** $k + 2\frac{3}{4} = -\frac{5}{8} \;\; -3\frac{3}{8}$

   7. $19.8 = 7.17 + 3x$ **8.** $m - 8 = \frac{3}{5} \;\; 8\frac{3}{5}$ **9.** $-5c - 2.7 = 1.35$
   4.21   -0.81

   10. $3\frac{5}{8} + r = \frac{7}{8} \;\; -2\frac{3}{4}$ **11.** $2a - 1.9 = -4.1$ **12.** $9.5 + 6j = 5.3$
                                              -1.1   -0.7

Solve using an equation.

13. Joey trimmed the length of his new poster by $1\frac{1}{2}$ in. to fit the poster into a frame with length $28\frac{3}{4}$ in. What was the length of the poster before Joey trimmed it? $30\frac{1}{4}$ in.

14. Isaac paid for two tickets to a foreign film with a $20 bill. He received $8.30 in change. What was the price of each ticket? $5.85

Additional Examples

Solve the equation. Check the solution.

1. $9 = -4x + 8$ $-\frac{1}{4}$

2. $2m + 6.7 = -1.6$ -4.15

3. $n - \frac{1}{3} = \frac{3}{4}$ $1\frac{1}{12}$

Closing the Lesson

Ask a student to explain how to solve the equation $x - \frac{2}{3} = \frac{5}{6}$. Ask another student to explain how to check the solution, $1\frac{1}{2}$.

Suggested Assignments

Skills: 1–16, 29–32
Average: 1–24, 29–32
Advanced: 1–32

Exercise Notes

Communication: Reading
Guided Practice Exercises 1–3 ask students to read and interpret a problem, identifying any unnecessary facts. Students are then asked to reconsider the problem when the question is changed.

Calculator
Exercises 17–20 show that equations involving fractions can be solved with a calculator that has a fraction key.

Communication: Writing
Exercises 21–24 require students to translate word sentences involving rational numbers into equations.

Making Connections/Transitions
Exercises 25–28 bring together the concepts of data analysis and solving equations with rational numbers.

Extension

Have students use catalogs or newspaper advertisements as the basis for formulating some problems. Ask each student to design a problem scenario having at least three related problems that can be solved using an advertisement or catalog.

Enrichment

Enrichment 84 is pictured on page 338F.

Practice

Practice 104, *Resource Book*

B

15. Laura bought three identical kites at Flights, Unlimited. She paid a total of $47.23, including $4.48 tax. What was the price of each kite? $14.25

16. Kate jogged $2\frac{1}{2}$ mi farther today than she had jogged yesterday. She jogged $7\frac{2}{3}$ mi today. How far did she jog yesterday? $5\frac{1}{6}$ mi

CALCULATOR

Match each equation with the correct solution as it appears on a calculator with a fraction key.

17. $x + 2\frac{2}{3} = 4\frac{1}{6}$ D

18. $6\frac{3}{8} = 2q - 1\frac{7}{8}$ C

19. $-\frac{2}{3} = \frac{5}{6} + g$ A

20. $10\frac{1}{12} + y = 1\frac{5}{6}$ B

A. $\boxed{-1\_1\_2.}$

B. $\boxed{-8\_1\_4.}$

C. $\boxed{4\_1\_8.}$

D. $\boxed{1\_1\_2.}$

COMMUNICATION « *Writing*

Write an equation that represents each sentence. 22. $n + \frac{2}{3} = -\frac{5}{9}$

«21. Three times a number q decreased by 11.3 is -8.9. $3q - 11.3 = -8.9$

«22. The sum of a number n and two thirds is negative five ninths.

«23. Sixteen is a number t increased by three fifths. $16 = t + \frac{3}{5}$

«24. The difference of a number y and 4.9 is 12.7. $y - 4.9 = 12.7$

DATA ANALYSIS

Find the mean, median, mode(s), and range of each set of data.

C 25. 2.3, 0.8, -0.9, -1.1, 0.8, 1.4 0.55, 0.8, 0.8, 3.4
   26. $-\frac{3}{4}$, $2\frac{7}{8}$, $1\frac{5}{8}$ $1\frac{1}{4}$; $1\frac{5}{8}$; none; $3\frac{5}{8}$

27. What equation would you use to find the greatest number in a group of numbers whose least number is $-\frac{3}{5}$ and whose range is $4\frac{2}{3}$? Find the missing number. $n - \left(-\frac{3}{5}\right) = 4\frac{2}{3}$; $4\frac{1}{15}$

28. The mean of two numbers is -7.2. The greater number is 2.15. Use an equation to find the lesser number. -16.55

SPIRAL REVIEW

S 29. Complete: The measure of an acute angle is between __?__ and __?__. (*Lesson 6-3*) 0°; 90°

30. Use the formula $D = rt$ to find r when $D = 147$ mi and $t = 3.5$ h. (*Lesson 4-10*) 42 mi/h

31. Solve: $x - \frac{4}{5} = \frac{3}{10}$ (*Lesson 8-8*) $1\frac{1}{10}$

32. Find the product: $(-4)\left(\frac{5}{8}\right)$ (*Lesson 8-1*) $-2\frac{1}{2}$

370   Chapter 8

8-9

Using Reciprocals to Solve Equations

Objective: To use reciprocals to solve equations.

To solve equations in which the variable has been multiplied by a fraction, you will need to multiply by the reciprocal of the fraction. Remember that the product of a number and its reciprocal is 1.

Teaching the Lesson
• Materials: calculator
• Lesson Starter 78
• Toolbox Skills 20, 21

Lesson Follow-Up
• Practice 105
• Enrichment 85: Problem Solving
• Study Guide, pp. 155–156

Example 1

Solve $\frac{2}{3}x = -5$. Check the solution.

Solution

$$\frac{2}{3}x = -5$$

$$\frac{3}{2}\left(\frac{2}{3}x\right) = \frac{3}{2}(-5) \quad \longleftarrow \text{Multiply both sides by the reciprocal of } \frac{2}{3}.$$

$$1 \cdot x = \frac{3}{2} \cdot \frac{-5}{1}$$

$$x = \frac{-15}{2} = -7\frac{1}{2}$$

The solution is $-7\frac{1}{2}$.

✔ **Check**

$$\frac{2}{3}x = -5$$

$$\frac{2}{3}\left(\frac{-15}{2}\right) \overset{?}{=} -5$$

$$-5 = -5$$

Warm-Up Exercises

Solve.
1. $3x - 14 = 30$ $14\frac{2}{3}$
2. $5 - 2y = -28$ $16\frac{1}{2}$
3. $2 - 4h = 9$ $-1\frac{3}{4}$
4. $11 - \frac{y}{3} = 8$ 9
5. $\frac{y}{3} - 7 = 4$ 33

Before you multiply by a reciprocal, you may need to use inverse operations to get the variable term alone on one side of the equals sign.

Lesson Focus

Show students eight identical pencils and tell them that these pencils represent $\frac{2}{3}$ of the total number in a box. Ask students to write an equation for finding the total number of pencils in the box. Today's lesson focuses on solving equations by using reciprocals.

Example 2

Solve $4 = -\frac{5}{6}b + 1$. Check the solution.

Solution

$$4 = -\frac{5}{6}b + 1$$

$$4 - 1 = -\frac{5}{6}b + 1 - 1$$

$$3 = -\frac{5}{6}b$$

$$\left(-\frac{6}{5}\right)\left(\frac{3}{1}\right) = \left(-\frac{6}{5}\right)\left(-\frac{5}{6}b\right) \quad \longleftarrow \begin{array}{l}\text{Multiply} \\ \text{both sides} \\ \text{by the} \\ \text{reciprocal} \\ \text{of } -\frac{5}{6}.\end{array}$$

$$\frac{-18}{5} = 1 \cdot b$$

$$-3\frac{3}{5} = b$$

The solution is $-3\frac{3}{5}$.

✔ **Check**

$$4 = -\frac{5}{6}b + 1$$

$$4 \overset{?}{=} \frac{-5}{6}\left(\frac{-18}{5}\right) + 1$$

$$4 \overset{?}{=} 3 + 1$$

$$4 = 4$$

Teaching Notes

Work through the Examples with students, stressing the importance of getting the variable term alone on one side of the equals sign before multiplying by the reciprocal of the coefficient.

 **Check Your Understanding**

1. In Example 2, explain why $-\frac{6}{5}$ is used to multiply both sides of the equation.

2. In Example 2, explain why the number 3 is rewritten as $\frac{3}{1}$.
   See *Answers to Check Your Understanding* at the back of the book.

372

Teaching Notes *(continued)*

Key Questions

1. In Example 2, why is 1 subtracted before multiplying by the reciprocal of $-\frac{5}{6}$? to isolate the variable term alone on one side of the equals sign

2. Is the reciprocal of a negative number positive or negative? The reciprocal is negative because the product of a number and its reciprocal must be 1.

Additional Examples

Solve. Check the solution.

1. $\frac{3}{5}y = -4$ $-6\frac{2}{3}$

2. $5 = -\frac{3}{8}y + 1$ $-10\frac{2}{3}$

Closing the Lesson

Ask a student to explain how to solve an equation such as $\frac{3}{4}x = 15$. Ask another student to explain how to solve an equation such as $\frac{2}{3}x + 4 = -8$.

Suggested Assignments

Skills: 1–14, 15–27 odd, 49–53
Average: 1–43 odd, 49–53
Advanced: 1–43 odd, 45–53, Historical Note

Guided Practice

«**1.** COMMUNICATION «*Reading* Replace each __?__ with the correct word.

To solve $-\frac{4}{5}t = 12$, you __?__ each side of the equation by the __?__ of $-\frac{4}{5}$. multiply; reciprocal

2. Replace each __?__ with the number that makes the statement true.

$$\frac{2}{5}w - 3 = -7$$

$$\frac{2}{5}w - 3 + \underline{\ ?\ } = -7 + \underline{\ ?\ } \quad 3; 3$$

$$\frac{2}{5}w = \underline{\ ?\ } \qquad -4$$

$$\frac{?}{?}\left(\frac{2}{5}w\right) = \frac{?}{?}(-4) \qquad \frac{5}{2}, \frac{5}{2}$$

$$1 \cdot w = \frac{5}{2} \cdot \frac{-4}{?} \qquad 1$$

$$w = \underline{\ ?\ } \qquad -10$$

Solve. Check each solution.

3. $-15 = \frac{2}{3}m$ $-22\frac{1}{2}$ **4.** $\frac{9}{10}j + 11 = 5$ $-6\frac{2}{3}$ **5.** $-\frac{8}{9}q - 5 = -13$ 9

Exercises

Solve. Check each solution.

A   **1.** $\frac{4}{11}j = 16$ 44   **2.** $\frac{3}{4}x = -8$ $-10\frac{2}{3}$ **3.** $-\frac{5}{6}k = 9$ $-10\frac{4}{5}$

4. $-\frac{8}{9}c = 12$ $-13\frac{1}{2}$ **5.** $\frac{1}{2}v - 3 = 8$ 22   **6.** $\frac{8}{15}y + 9 = 15$ $11\frac{1}{4}$

7. $12 + \frac{2}{5}f = 5$ $-17\frac{1}{2}$ **8.** $-1 = \frac{7}{12}z + 6$ -12 **9.** $25 = \frac{14}{25}r + 4$ $37\frac{1}{2}$

10. $-\frac{2}{9}b + 4 = 1$ $13\frac{1}{2}$ **11.** $\frac{6}{7}t - 2 = -10$ $-9\frac{1}{3}$ **12.** $-18 = -\frac{5}{11}x - 8$ 22

13. Jennifer carried home a number of one-half pound packages of sliced deli meats and a 10 lb bag of oranges. Her total load was $13\frac{1}{2}$ lb. How many packages of deli meats did she carry? 7

14. After Marty had eaten some apples from a basket, he still had 14 apples left. This is seven eighths of the number of apples Marty picked. How many apples did Marty pick? 16

You can use reciprocals to solve equations involving fractions when the variable has been multiplied by an integer other than 1. For example, to solve $2x = \frac{4}{5}$, multiply both sides by $\frac{1}{2}$.

Solve. Check each solution.   17. $-\frac{1}{5}$ 20. $-\frac{3}{10}$

B   **15.** $3y = \frac{3}{8}\frac{1}{8}$

16. $-5z = \frac{15}{16}-\frac{3}{16}$

17. $\frac{3}{5} = -2m + \frac{1}{5}$

18. $3a - \frac{2}{5} = 2\frac{4}{5}$

19. $14b + \frac{5}{8} = -\frac{1}{4}-\frac{1}{16}$

20. $-9d + 1\frac{1}{2} = 4\frac{1}{5}$

WRITING ABOUT MATHEMATICS Answers to Exercises 21 and 22 are on p. A24.

21. Write a paragraph describing how you solve an equation involving rational numbers written as fractions.

22. Write a paragraph comparing the methods for solving equations that involve integers, equations that involve decimals, and equations that involve fractions.

MENTAL MATH

Use mental math to solve each equation.

23. $\frac{1}{2}y = 2$ 4

24. $-\frac{7}{8}b = \frac{7}{8}$ -1

25. $\frac{3}{4}x = 1\frac{4}{3}$

26. $m + \frac{6}{11} = 1\frac{6}{11}$ 1

27. $r - \frac{1}{2} = \frac{1}{2}$ 1

28. $4 = \frac{5}{6}g + 4$ 0

 CALCULATOR

31. $5\frac{1}{15}$

Solve. Check the solution. Use a calculator with a fraction key if you have one available to you.

29. $\frac{1}{3}g = \frac{4}{5}$ $2\frac{2}{5}$

30. $\frac{2}{3}v = -\frac{5}{8}$ $-\frac{15}{16}$

31. $-\frac{1}{4}x + \frac{2}{3} = -\frac{3}{5}$

32. $\frac{4}{9} + \frac{11}{6}n = \frac{5}{6}$ $\frac{7}{33}$

33. $\frac{1}{2} = -\frac{7}{12}p - \frac{3}{4}$ $-2\frac{1}{7}$

34. $-\frac{3}{5} = \frac{3}{4} + \frac{15}{2}w$ $-\frac{9}{50}$

COMMUNICATION « *Writing*

Write an equation for each sentence.   **37.** $\frac{1}{2}g - 3 = \frac{5}{6}$

«**35.** Two thirds of a number t is negative fourteen. $\frac{2}{3}t = -14$

«**36.** The sum of 11 and three fifths of a number z is ten. $11 + \frac{3}{5}z = 10$

«**37.** The difference of one half of a number g and three is five sixths.

«**38.** One fourth of a number n is nine and two thirds. $\frac{1}{4}n = 9\frac{2}{3}$

Write each equation in words. Answers to Exercises 39–44 are on p. A24.

«**39.** $\frac{1}{3}g = 14$

«**40.** $\frac{4}{5}c = 16$

«**41.** $-2 = \frac{1}{3}x + 9$

«**42.** $\frac{3}{4}m - 4 = 11$

«**43.** $10 = \frac{3}{8}s + \frac{1}{2}$

«**44.** $5d = -\frac{7}{12}$

Rational Numbers   **373**

Critical Thinking

Exercises 45–48 combine concepts from previous chapters. Students are presented with a method for solving equations that uses common denominators and are asked to *analyze* the procedures involved in applying it and to *create* a plan for solving such equations.

Follow-Up

Extension

Exercises 45–48 reinforce the idea that there may be more than one way to solve a problem. Certain methods may be preferable because of their simplicity, but divergent approaches should be recognized as having value. Ask students to use the method presented in these exercises to solve Exercises 5–12.

Enrichment

Enrichment 85 is pictured on page 338F.

Practice

Practice 105, *Resource Book*

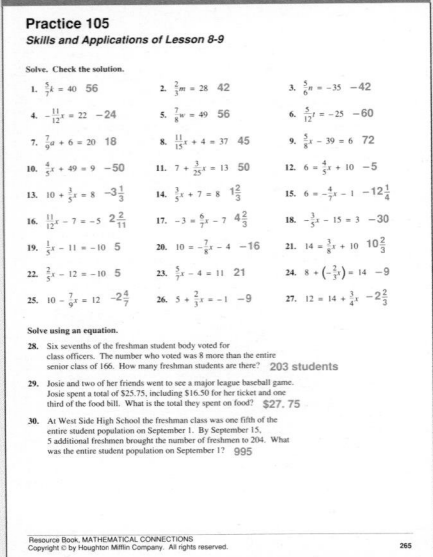

THINKING SKILLS 45–47. Analysis   **48.** Synthesis

The example at the right shows an *alternative method* for solving an equation involving rational numbers.

$$\frac{3}{4}n - \frac{1}{2} = \frac{2}{5}$$

$$20\left(\frac{3}{4}n - \frac{1}{2}\right) = 20\left(\frac{2}{5}\right)$$

$$20\left(\frac{3}{4}n\right) - 20\left(\frac{1}{2}\right) = 20\left(\frac{2}{5}\right)$$

$$15n - 10 = 8$$

$$15n = 18$$

$$n = \frac{18}{15} = 1\frac{1}{5}$$

For Exercises 45–48, analyze the example at the right.

C **45.** How does the number 20 relate to the denominators 4, 2, and 5?
20 is the LCM.

46. What happened to the denominators when each side was multiplied by 20? They became 1; the numbers became integers.

47. Which property allows you to simplify the left side of the equation from $20\left(\frac{3}{4}n - \frac{1}{2}\right)$ to $15n - 10$? distributive property

48. Create a plan for solving an equation using the method illustrated above. Use your plan to solve each equation.

   a. $4 = -\frac{2}{3}y + \frac{1}{2}$ **b.** $\frac{1}{2}m - \frac{3}{5} = \frac{2}{3}$ **c.** $\frac{1}{2} + \frac{3}{4}x = -\frac{1}{8}$
   multiply by 6; $-5\frac{1}{4}$ multiply by 30; $2\frac{8}{15}$ multiply by 8; $-\frac{5}{6}$

SPIRAL REVIEW

S **49.** Solve by drawing a diagram: A telephone installer left the telephone company office and drove 8 blocks due west, 10 blocks due north, 7 blocks due east, and 10 blocks due south. At this point, how far was the installer from the telephone company office? *(Lesson 7-8)*
1 block due west

50. Graph $x \geq 2.5$ on a number line. *(Lesson 7-10)* See answer on p. A24.

51. The measure of an angle is 42°. Find its supplement. *(Lesson 6-3)*
138°

52. Solve: $\frac{3}{4}z = 15$ *(Lesson 8-9)* 20

53. Find the quotient: $\frac{8}{15} \div \left(-\frac{2}{3}\right)$ *(Lesson 8-2)* $-\frac{4}{5}$

Historical Note

F annie Merritt Farmer (1857–1915) introduced the idea of exact measurement of ingredients in her cookbook first published in 1896. She advocated the use of measuring spoons and cups and directed that dry ingredients be leveled off with a knife blade. Earlier cookbooks used only vague terms such as "a pinch of salt." She championed the study of nutrition and gave courses to nurses on preparing diets for sick people. She also lectured on dietetics at Harvard Medical School.

See answer on page A24.

Research

Find out how recipe ingredients are measured in other countries.

374 Chapter 8

8-10

Transforming Formulas

Objective: To evaluate and transform formulas containing rational numbers.

APPLICATION

Alex Gibbons is an exchange student living in France for the school year. He wants to borrow his host-family's kitchen and cook his favorite recipe from home. The recipe suggests a baking temperature of 325°F, but the ovens in France use the Celsius temperature scale. In order to use this recipe, Alex needs to know that 325°F is equivalent to 163°C.

A formula that relates Fahrenheit and Celsius temperatures is $F = \frac{9}{5}C + 32$, where F represents a Fahrenheit temperature and C represents the equivalent Celsius temperature. To find a value for C, you may want to *solve* the formula for C. To solve for a particular variable, you **transform** the formula by using inverse operations to get that variable alone on one side of the equals sign.

Example 1

Solve $F = \frac{9}{5}C + 32$ for C.

Solution

$$F = \frac{9}{5}C + 32$$

$$F - 32 = \frac{9}{5}C + 32 - 32 \qquad \longleftarrow \text{Subtract 32 from both sides.}$$

$$F - 32 = \frac{9}{5}C$$

$$\frac{5}{9}(F - 32) = \frac{5}{9}\left(\frac{9}{5}C\right) \qquad \longleftarrow \begin{array}{l}\text{Multiply both sides by}\\ \text{the reciprocal of } \frac{9}{5}.\end{array}$$

$$\frac{5}{9}(F - 32) = C$$

$$C = \frac{5}{9}(F - 32)$$

Example 2

Find the equivalent Celsius temperature for 325°F.

Solution

Use the formula $C = \frac{5}{9}(F - 32)$.

$$C = \frac{5}{9}(325 - 32) \qquad \longleftarrow \text{Substitute 325 for } F.$$

$$C = \frac{5}{9}(293) = 162.\overline{7} \approx 163$$

325°F is approximately 163°C.

 **Check Your Understanding**

1. In Example 1, why is the *difference* $F - 32$ multiplied by $\frac{5}{9}$?
2. In Example 2, why is 325 substituted for F and not for C?

See *Answers to Check Your Understanding* at the back of the book.

Rational Numbers **375**

Warm-Up Exercises

Use the formula $D = rt$.

1. Let $r = 45$ mi/h and $t = 3$ h. Find D.   135 mi
2. Let $r = 55$ mi/h and $t = 4.5$ h. Find D.   247.5 mi
3. Let $D = 210$ mi and $t = 6$ h. Find r.   35 mi/h
4. Let $D = 190$ mi and $t = 4.75$ h. Find r.   40 mi/h
5. Let $D = 294$ mi and $r = 42$ mi/h. Find t.   7 h

Lesson Focus

Ask students to make a list of some formulas. For each formula, write a transformed version on the chalkboard, and ask students whether the two formulas can be used to find the same quantity. Today's lesson focuses on transforming formulas.

Teaching Notes

Point out to students that when transforming a formula, they use the same techniques as solving for a variable. However, the answer will not be a number, but a variable expression.

Key Question
Do you always need to transform a
formula before using it? No;
sometimes the formula may already
be in the most useful form or sometimes it may be easier to substitute
the given values and then solve for
the missing variable.

Reteaching/Alternate Approach
For a suggestion on making a formula poster, see page 338D.

Reasoning
Transforming a formula requires
students to apply the skills of equation solving in a more general
sense.

Additional Examples
The formula for the perimeter of
a rectangle is $P = 2l + 2w$, where
P is the perimeter, l is the length,
and w is the width.
1. Solve $P = 2l + 2w$ for w.
$w = \frac{P - 2l}{2}$
2. Find the width of a rectangle
with perimeter 48 in. and length
16 in. 8 in.

Closing the Lesson
Use the formulas students provided
in the Lesson Focus and ask them
to transform each formula in as
many ways as possible.

Suggested Assignments
Skills: 1–18, 24–26
Average: 1–26
Advanced: 1–26

Exercise Notes

Making Connections/Transitions
Exercises 17 and 18 connect mathematics and science. Mathematical
formulas are used extensively in
many different branches of science.

Guided Practice

COMMUNICATION «*Reading*

«**1.** Explain the meaning of the word *transformed* in the following
sentence. A caterpillar is changed into a butterfly.
   A caterpillar is transformed into a butterfly.

«**2.** The prefix *trans-*, as in transform, may mean "across," "beyond,"
"from one place to another," or "change." List three other words that
begin with this prefix. Give a definition of each word. Answers will vary.
Examples: translate, transplant, transfer

Choose the letter of a correct transformation of each formula.

3. $P = C - r$ **a.** $C = Pr$ **b.** $C = P + r$ **c.** $C = \frac{P}{r}$

4. $D = \frac{1}{f}$ **a.** $f = D$ **b.** $f = 1 - D$ **c.** $f = \frac{1}{D}$

5. $d = \frac{7}{22}C$

6. $h = \frac{3V}{B}$

7. $b = \frac{2A}{h}$

8. $t = \frac{D}{r}$

Solve for the variable shown in color.

5. $C = \frac{22}{7}d$ **6.** $V = \frac{1}{3}Bh$ **7.** $A = \frac{1}{2}bh$ **8.** $D = rt$

9. Use the formula $d = \frac{7}{22}C$ to find d (in inches) when $C = 11$ in. $3\frac{1}{2}$ in.

10. Use the formula $h = \frac{3V}{B}$ to find h (in feet) when $V = 12$ ft^3 and
$B = 4$ ft^2. 9 ft

Exercises

Solve for the variable shown in color.

4. $b = 2m - a$

6. $r = \frac{C}{6.28}$

11. $a = 2(P - 110)$

12. $l = \frac{1}{2}(P - 2w)$

$b = P - a - c$

A **1.** $A = bh$ $h = \frac{A}{b}$ **2.** $P = 4s$ $s = \frac{P}{4}$.   **3.** $P = a + b + c$

4. $m = \frac{a + b}{2}$ **5.** $F = \frac{W}{d}$ $W = Fd$ **6.** $6.28r = C$

7. $I = Prt$ $P = \frac{I}{rt}$ **8.** $c = np$ $n = \frac{c}{p}$ **9.** $W = IE$ $E = \frac{W}{I}$

10. $l = \frac{A}{w}$ $A = lw$ **11.** $P = \frac{a}{2} + 110$ **12.** $P = 2l + 2w$

13. Use the formula $F = \frac{W}{d}$ to find F (in pounds) when $W = 40$ ft-lb and
$d = 8$ lb. 5 lb

14. Use the formula $A = \frac{1}{2}bh$ to find A (in square feet) when $b = 12$ ft and
$h = 2\frac{1}{2}$ ft. 15 sq ft

15. Use the formula $C = \frac{5}{9}(F - 32)$ to find the equivalent Celsius temperature for each Fahrenheit temperature.
   a. 0°F $-17\frac{7}{9}$°C   **b.** 32°F 0°C   **c.** 212°F 100°C

16. Use the formula $F = \frac{9}{5}C + 32$ to find the equivalent Fahrenheit
temperature for 37°C. 98.6°F

CONNECTING ALGEBRA AND SCIENCE

Ohm's Law is used in science. It states that in an electric circuit $E = IR$, where E represents the force (in volts), I represents the current (in amperes), and R represents the resistance (in ohms).

B **17. a.** Solve Ohm's Law for I. $I = \frac{E}{R}$

    b. What current does a force of 6 volts produce in a circuit with 0.15 ohms of resistance? 40 amperes

18. a. Solve Ohm's Law for R. $R = \frac{E}{I}$

    b. Calculate the resistance of an electric steam iron that takes 8 amperes of current from a source supplying 110 volts. 13.75 ohms

GROUP ACTIVITY

A plumber's fee is a minimum of $30 plus an hourly rate of $22.

19. Write a formula that relates the plumber's fee (f) and the number of hours (h) worked. Put f alone on one side of the equals sign. $f = 22h + 30$

20. Make a table showing both the fees and the number of hours worked for a plumber who works 1, 2, 3, 4, and 5 h. See answer on p. A24.

21. Solve the formula you wrote in Exercise 19 for h. $h = \frac{1}{22}(f - 30)$

22. A plumber's fee is $184. How many hours did the plumber work? 7 h

23. **RESEARCH** Find three formulas that are used in everyday life. (*Hint:* taxicab rates, telephone rates, electric or water rates) See answer on p. A24.

SPIRAL REVIEW

S **24.** Complete: A rhombus has four equal ___?___. (*Lesson 6-7*) sides

25. Solve $A = \frac{1}{2}ps$ for s. (*Lesson 8-10*) $s = \frac{2A}{p}$

26. Find the sum: $-\frac{3}{4} + \frac{5}{16}$ (*Lesson 8-5*) $-\frac{7}{16}$

Self-Test 3

1. Heidi swam 20 ft farther than Kimberly. Kimberly swam 25 yd. How many yards did Heidi swim? $31\frac{2}{3}$ yd   8-7

Solve. Check each solution.

2. $13 = 9 + 20z$ $\frac{1}{5}$ **3.** $4.8n - 7 = 16.4$ 4.875   **4.** $v - \frac{1}{2} = 1\frac{6}{7}$ $2\frac{5}{14}$ 8-8

5. $-\frac{2}{9}x = 24$ −108   **6.** $14 = \frac{4}{7}g - 6$ 35   **7.** $-1 = 5 + \frac{3}{8}a$ −16   8-9

8. Solve $V = lwh$ for h. $h = \frac{V}{lw}$ 8-10

9. Use the formula you found in Exercise 8 to find h (in meters) when $V = 30$ m³, $l = 5$ m, and $w = 3$ m. 2 m

Rational Numbers   **377**

Cooperative Learning

Exercises 19–22 provide a group learning activity on using formulas and creating a table. One problem-solving strategy that might be included is to ask students to extend the table to 10 hours, 25 hours, and *n* hours.

Follow-Up

Exploration (*Cooperative Learning*)
Working in groups, students can use an almanac or other source to research comparative data on weather in countries around the world. Each group should select one country and make a chart showing monthly temperature ranges in both Celsius and Fahrenheit degrees.

Enrichment

Enrichment 86 is pictured on page 338F.

Quick Quiz 3

See page 385.

Alternative Assessment

See page 385.

Practice

Practice 106, *Resource Book*

Practice 106
Skills and Applications of Lesson 8-10

Solve for each equation for the variable requested.

1. $A = \frac{1}{2}bh$ for h $h = \frac{2A}{b}$ 2. $P = 2l + 2w$ for w $w = \frac{P - 2l}{2}$

3. $T = P \cdot V$ for V $V = \frac{T}{P}$ 4. $d = rt$ for t $\frac{d}{r}$

5. $C = 2\pi r$ for r $r = \frac{C}{2\pi}$ 6. $P = I^2R$ for R $R = \frac{P}{I^2}$

7. $A = P + Prt$ for r $r = \frac{A - P}{Pt}$ 8. $V = \frac{1}{3}\pi r^2h$ for h $h = \frac{3V}{\pi r^2}$

9. $IR = E$ for I $I = \frac{E}{R}$ 10. $F = \frac{9}{5}C + 32$ for C $C = \frac{5}{9}(F - 32)$

11. $E = mc^2$ for m $m = \frac{E}{c^2}$ 12. $v = ar + c$ for a $a = \frac{v - c}{r}$

13. $m = \frac{F}{a}$ for F $F = ma$ 14. $P = Fv$ for v $v = \frac{P}{F}$

15. $D = d - 2h$ for h $h = \frac{d - D}{2}$ 16. $A = \pi dh$ for d $d = \frac{A}{\pi h}$

17. Use the formula $C = 2\pi r$ to find the circumference C of a circle if $\pi = \frac{22}{7}$ and $r = 14$ in. 88 in.

18. Use the formula $P = 2l + 2w$ to find the perimeter P of a rectangle when $l = 4$ in. and $w = 7$ in. 22 in.

19. Use the formula $d = rt$ to find the rate r when $d = 95$ ft and $t = 5$ sec. 19 ft/s

20. Use the formula $i = Prt$ to find the simple interest i when $P = 100, $r = 0.055$, and $t = 2$ years. $11

21. Use the formula $E = \frac{O}{I}$ to measure the output O (in watts) if $E = 0.60$ and $I = 800$ watts. 480 watts

22. Use the formula $V = \pi r^2h$ to find the volume V of a cylinder when $\pi = \frac{22}{7}$, $r = 10$ in. and $h = 14$ in. 4400 in.³

Spend time explaining the stock
table on this page, as students may
have trouble discerning what each
entry represents.

Key Question

What do the fractional parts of the
numbers in the table represent?
They represent the fractional parts
of a dollar.

Additional Example

How much did the value of 200
shares of BusiServ change from the
day before? The *Chg* column for
BusiServ has no entry. This means
the value did not change.

Exercise Note

Life Skills

Reading and interpreting financial
information is an important life
skill.

Follow-Up

Extension

Have students bring in the stock
pages from their local newspaper.
Use the tables to find information
on stocks of companies students are
familiar with.

**FOCUS ON
APPLICATIONS**

Using a Stock Price Table

Objective: To read and
interpret a stock price
table.

A shareholder is a person who owns one or more shares of stock. Each
share represents part ownership in a corporation.

Stock price tables like the one shown below can be found in the business
section of a newspaper. They summarize one day of activity on a given
stock exchange. Stock prices in these tables are usually given in eighths,
fourths, and halves of a dollar. A share of stock listed at 27¾ is worth
$27.75.

Stock	Sym	Div	%	PE	Sales 100s	Hi	Lo	Close	Chg
BayOil	BOIL	.50	3.7	10	278	14	13⅝	13⅝	− ⅜
BookClb	BKCL	1.20	4.1	8	1161	29⅝	27¾	29⅜	+1⅝
BusiServ	BUSV	.72	2.4	16	71	30⅛	29⅞	30	…
ByteSys	BTSY	2.80	5.3	13	335	53⅜	52⅞	53	−1¼

The last four columns in a stock price table display information about the
various prices paid for one share of stock on a given day. The columns
marked *Hi* and *Lo* display the highest and lowest prices paid per share that
day. The column marked *Close* displays the price paid per share in the last
transaction of the day for that stock. The column marked *Chg* displays the
change in the closing price from the previous day.

Example

How much did the value of 100 shares of BayOil stock change from the
close of the previous day to the close of the day shown?

Solution

The *Chg* column for BayOil shows −⅜.

This means that the closing price for one share of stock was $⅜ lower than
the closing price on the previous day.

Multiply to find the change in the value of 100 shares of BayOil stock.

$$-\frac{3}{8} \cdot 100 = \frac{-75}{2} = -37\frac{1}{2}$$

Write $-37\frac{1}{2}$ as a decimal: -37.5

The value of 100 shares of BayOil stock decreased by $37.50 from the
close of the previous day to the close of the day shown.

Stocks are sometimes *split*. When a stock splits "*n* for 1," the number of shares is multiplied by *n*, and the price is divided by *n*, so that the investment keeps its value. For example, suppose a shareholder owns 10 shares of stock worth $80 per share. The total value is $800. If the stock splits 2 for 1, the shareholder then owns 10 · 2, or 20, shares of stock at $80 ÷ 2, or $40, per share. The total value is still $800.

Exercises

Use the stock price table on page 378. For Exercises 1–6, name the stock or the category corresponding to the given information.

1. Hi: $29\frac{5}{8}$ BookClb
2. Lo: $13\frac{5}{8}$ BayOil
3. Close: $13\frac{5}{8}$ BayOil
4. BusiServ: $29\frac{7}{8}$ Lo
5. ByteSys: $53\frac{3}{8}$ Hi
6. BookClb: $29\frac{3}{8}$ Close

7. Write the high price for BusiServ stock as a decimal. $30.13

8. Did the price of BookClb stock *increase* or *decrease* from the previous day's price? increase

9. Marion bought 100 shares of BusiServ stock at its lowest price during the day. How much did she pay in all for the stock? $2987.50

10. Find the previous day's closing price for BayOil stock. $14

11. Find the previous day's closing price for BookClb stock. $27.75

12. How much did the value of 400 shares of BookClb stock change from the close of the previous day to the close of the day shown? $650 increase

13. Isabel bought 50 shares of BayOil stock at its highest price. How much would she have saved if she had bought it at its lowest price? $18.75

14. Jen owned 60 shares of BusiServ stock at the close of the day. Then the stock split 3 for 1. How many shares did she own after the split? How much was each share worth? 180; $10

15. Dave owned 200 shares of ByteSys stock at the close of the day. Then the stock split 4 for 1. How many shares did he own after the split? How much was each share worth? 800; $13.25

16. **RESEARCH** Find out what the columns marked *Div*, *%*, *PE*, and *Sales 100s* represent. See answer on p. A24.

Chapter Resources

- Cumulative Practice 108
- Enrichment 87: Extension
- Chapter 8 Project
- Family Involvement 8
- Tests 39, 40
- Cumulative Tests 41, 42
- Spanish Test 8
- Spanish Glossary
- Test Generator

Chapter Review

reciprocal (p. 340)
multiplicative inverse (p. 340)

multiplication property of reciprocals (p. 340)
transform a formula (p. 375)

Choose the correct term from the list above to complete each sentence.

1. Another name for the reciprocal of a number is the __?__. multiplicative inverse

2. The __?__ states that the product of a number and its reciprocal is one. multiplication property of reciprocals

Find each answer. Simplify if possible. *(Lessons 8-1, 8-2, 8-4, 8-5)* 18. $-3\frac{7}{10}$

3. $\frac{5}{8}\cdot\frac{3}{4}$ $\frac{15}{32}$

4. $\frac{5}{6}\div\frac{2}{3}$ $1\frac{1}{4}$

5. $11\frac{5}{6}+2\frac{5}{6}$ $14\frac{2}{3}$

6. $2\frac{1}{6}-\frac{1}{3}$ $1\frac{5}{6}$

7. $-6\left(\frac{3}{4}\right)$ $-4\frac{1}{2}$

8. $\left(-5\frac{1}{4}\right)\left(-2\frac{2}{3}\right)$ 14

9. $\frac{2}{9}\div\left(-\frac{3}{4}\right)$ $-\frac{8}{27}$

10. $-4\frac{4}{9}\div 10$ $-\frac{4}{9}$

11. $\frac{4}{5}+\left(-4\frac{2}{5}\right)$ $-3\frac{3}{5}$

12. $-\frac{8}{9}+\left(-\frac{11}{12}\right)$ $-1\frac{29}{36}$

13. $\left(-4\frac{3}{8}\right)\div\left(-2\frac{3}{16}\right)$ 2

14. $1\frac{3}{10}\div\left(-3\frac{3}{5}\right)$ $-\frac{13}{36}$

15. $-3\frac{1}{2}-\frac{2}{3}$ $-4\frac{1}{6}$

16. $-3\frac{5}{6}-\left(-7\frac{1}{6}\right)$ $3\frac{1}{3}$

17. $3\frac{2}{7}+\frac{1}{7}+\left(-1\frac{3}{7}\right)$ 2

18. $-3+\left(-1\frac{3}{5}\right)+\frac{9}{10}$

19. $(-0.6)(-0.9)$ 0.54

20. $(4.8)(-0.5)$ -2.4

21. $-6.09\div 2.9$ -2.1

22. $-48\div(-1.2)$ 40

23. $-1.6+(-4.7)$ -6.3

24. $12+(-8.1)$ 3.9

25. $3.75-62.5$ -58.75

26. $2.16-(-0.54)$ 2.7

27. $\left(\frac{2a}{5}\right)\left(\frac{3}{8a}\right)$ $\frac{3}{20}$

28. $\left(\frac{x}{4}\right)\left(\frac{8z}{x}\right)$ $2z$

29. $\frac{3c}{2}\div\frac{7c}{3}$ $\frac{9}{14}$

30. $\frac{6a}{11}\div 4a$ $\frac{3}{22}$

31. $\frac{7}{x}+\frac{4}{x}$ $\frac{11}{x}$

32. $\frac{a}{6}+\frac{7a}{12}$ $\frac{3a}{4}$

33. $\frac{3r}{10}-\frac{2r}{15}$ $\frac{r}{6}$

34. $\frac{19m}{12}-\frac{m}{12}$ $\frac{3m}{2}$

35. $8n-\frac{7n}{9}$ $\frac{65n}{9}$

36. $4c+\frac{c}{7}$ $\frac{29c}{7}$

37. $\frac{3}{10n}+\frac{9}{10n}+\frac{13}{10n}$ $\frac{5}{2n}$

38. $\frac{3m}{8}+\frac{m}{4}+\frac{4m}{3}$ $\frac{47m}{24}$

Solve. Decide whether a *whole number*, a *fraction*, or a *decimal* is an appropriate answer. *(Lesson 8-3)*

39. Margaret spends a total of $2\frac{1}{3}$ h commuting to work every day. How many hours does she spend commuting to work in 5 days? fraction; $11\frac{2}{3}$ h

40. A music practice room costs $15 per hour to rent. What is the cost to rent the room for $3\frac{1}{2}$ h? decimal; $52.50

41. John's mother sent him to the grocery store to buy yogurt. She gave him $5.53. Each container of yogurt cost $.79. How many containers of yogurt was John able to buy? whole number; 7

42. The total weight of 12 identical packages is 15 lb. What is the weight of each package? fraction; $1\frac{1}{4}$ lb

Estimate. *(Lessons 8-1, 8-6)* Estimates may vary. Accept reasonable estimates.

43. $\left(3\frac{7}{8}\right)\left(2\frac{1}{9}\right)$ about 8   **44.** $\left(-6\frac{1}{3}\right)\left(-4\frac{3}{4}\right)$ about 30   **45.** $\left(\frac{6}{13}\right)\left(-9\frac{1}{8}\right)$ about −5   **46.** $\left(\frac{8}{27}\right)\left(8\frac{2}{5}\right)$ about 3

47. $13\frac{6}{7} - 8\frac{1}{12}$ about 6   **48.** $4\frac{5}{8} + 3\frac{1}{9} + 2\frac{9}{10}$ about 11   **49.** $-12\frac{5}{6} \div \left(-\frac{7}{16}\right)$ **50.** $-23\frac{9}{11} \div \frac{4}{13}$

51. $-5\frac{5}{6} + \left(-10\frac{2}{9}\right)$ about −16   **52.** $-9\frac{3}{10} - \left(-\frac{4}{5}\right)$ about −8   **53.** $18\frac{1}{2} \div \left(-3\frac{2}{3}\right)$ about −5   **54.** $-16\frac{1}{7} \div \left(-5\frac{1}{4}\right)$ about 3

Solve by supplying missing facts. *(Lesson 8-7)* **49.** about 26   **50.** about −72

55. When Steve reached the age of 30, he was $6\frac{1}{2}$ ft tall. At birth he was 21 in. long. How many inches did Steve grow in 30 years? 57 in.

56. Kiko runs $4\frac{1}{4}$ mi each day. How far does she run in a week? $29\frac{3}{4}$ mi

57. Lisa bought $3\frac{1}{2}$ dozen bran muffins for a party. How many muffins did she buy? 42 muffins

58. Marty swam for $\frac{1}{2}$ h on Monday and $\frac{3}{5}$ h on Tuesday. For how many more minutes did Marty swim on Tuesday than on Monday? 6 min

Solve. Check each solution. *(Lessons 8-8, 8-9)*

59. $5x + 9 = 13$ $\frac{4}{5}$ **60.** $3a - 8 = -9$ $-\frac{1}{3}$ **61.** $-4c - 8.1 = 3.9$ −3

62. $10p - 0.5 = 2.35$ 0.285   **63.** $4 + j = -\frac{2}{5}$ $-4\frac{2}{5}$ **64.** $\frac{1}{8} + h = 4$ $3\frac{7}{8}$

65. $17 + 10x = 5$ $-1\frac{1}{5}$ **66.** $-8b + 13 = 7$ $\frac{3}{4}$ **67.** $-7y + 4.8 = 9$ −0.6

68. $3n - 3.7 = -4.3$ −0.2   **69.** $m - \frac{5}{6} = \frac{2}{3}$ $1\frac{1}{2}$ **70.** $\frac{7}{8} + c = \frac{1}{4}$ $-\frac{5}{8}$

71. $-1.2 = 5b - 4.7$ 0.7   **72.** $-5.9 = -6n + 2.5$ 1.4   **73.** $3\frac{1}{2} + a = 2\frac{4}{5}$ $-\frac{7}{10}$

74. $p + \frac{3}{4} = 2\frac{1}{3}$ $1\frac{7}{12}$ **75.** $\frac{5}{8}x = 30$ 48   **76.** $-\frac{2}{5}n = 5$ $-12\frac{1}{2}$

77. $-\frac{5}{6}a + 9 = 7$ $2\frac{2}{5}$ **78.** $\frac{2}{3}y - 3 = -13$ −15   **79.** $\frac{3}{4}c = -8$ $-10\frac{2}{3}$

80. $-\frac{1}{3}p = -7$ 21   **81.** $19 = -\frac{7}{8}k + 5$ −16   **82.** $11 + \frac{4}{5}m = 5$ $-7\frac{1}{2}$

Solve for the variable shown in color. *(Lesson 8-10)*

83. $d = 2r$ $r = \frac{d}{2}$ **84.** $F = ma$ $m = \frac{F}{a}$ **85.** $A = \frac{1}{2}ps$ $p = \frac{2A}{s}$ **86.** $a = \frac{M - m}{2}$ $M = 2a + m$

87. Use the formula $C = \frac{5}{9}(F - 32)$ to find the equivalent Celsius temperature for 86°F. 30°C

88. Use the formula $P = 2l + 2w$ to find P (in feet) when $l = 5\frac{1}{4}$ ft and $w = 3\frac{1}{2}$ ft. $17\frac{1}{2}$ ft

---

Chapter Test, (Form A)
Test 39, *Tests*

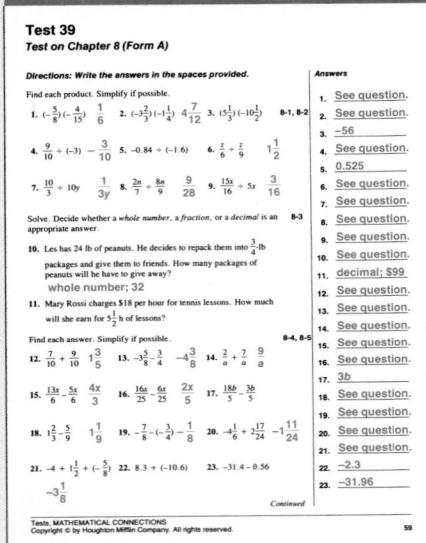

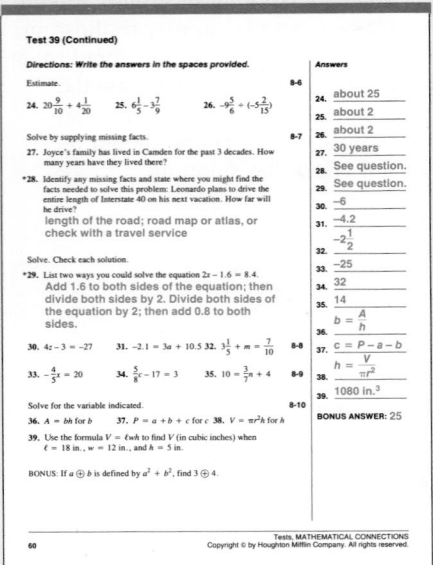

Chapter Test

Find each answer. Simplify if possible.

1. $\left(-\dfrac{5}{9}\right)\left(-\dfrac{3}{10}\right)$ $\frac{1}{6}$ 2. $\left(-2\dfrac{1}{2}\right)\left(4\dfrac{2}{3}\right)$ $-11\frac{2}{3}$ 3. $(2.6)(-3.4)$ -8.84 4. $\left(\dfrac{3a}{5b}\right)\left(\dfrac{10}{9a}\right)$ $\frac{2}{3b}$ **8-1**

5. $\dfrac{15}{16} \div (-3)$ $-\frac{5}{16}$ 6. $-0.846 \div (-1.8)$ 0.47 7. $\dfrac{2n}{9} \div \dfrac{4n}{5}$ $\frac{5}{18}$ 8. $\dfrac{12x}{13} \div 3x$ $\frac{4}{13}$ **8-2**

Solve. Decide whether a *whole number*, a *fraction*, or a *decimal* is an appropriate answer.

9. Lucia has 24 yd of rope to cut into pieces measuring $2\frac{1}{4}$ yd each. How many pieces of this size can Lucia cut from the rope? whole number; 10   **8-3**

10. Mr. Richards charges $20 per hour for tutoring. How much will he earn for $5\frac{3}{4}$ h of tutoring? decimal; $115

Find each answer. Simplify if possible.

11. $\dfrac{7}{12} + \dfrac{11}{12}$ $1\frac{1}{2}$ 12. $-2\dfrac{5}{6} - \dfrac{5}{6}$ $-3\frac{2}{3}$ 13. $\dfrac{9}{a} + \dfrac{7}{a}$ $\frac{16}{a}$ 14. $\dfrac{14x}{9} - \dfrac{8x}{9}$ $\frac{2x}{3}$ **8-4**

15. $1\dfrac{5}{6} - \dfrac{7}{9}$ $1\frac{1}{18}$ 16. $-\dfrac{5}{8} - \left(-\dfrac{2}{3}\right)$ $\frac{1}{24}$ 17. $-5\dfrac{1}{4} + 2\dfrac{11}{12}$ $-2\frac{1}{3}$ 18. $-5 + 2\dfrac{1}{2} + \left(-\dfrac{3}{4}\right)$ $-3\frac{1}{4}$ **8-5**

19. $\dfrac{5a}{9} - \dfrac{5a}{18}$ $\frac{5a}{18}$ 20. $7c + \dfrac{8c}{11}$ $\frac{85c}{11}$ 21. $8.6 + (-9.9)$ -1.3 22. $-42.6 - 0.58$ -43.18

Estimate. Estimates may vary. Accept reasonable estimates.

23. $17\dfrac{1}{10} - 5\dfrac{9}{11}$ about 11 24. $8\dfrac{11}{13} + 4\dfrac{1}{7} + 5\dfrac{7}{12}$ about 19 25. $\left(-9\dfrac{7}{8}\right)\left(\dfrac{7}{16}\right)$ about -5 **8-6**

Solve by supplying missing facts.

26. Leroy earns $350 per week. How much does Leroy earn in a year? $18,200   **8-7**

★27. Identify any missing facts and state where you might find the facts needed to solve this problem: The Amazon river is 3912 mi long. The Nile river is the longest river in the world. How much longer is the Nile river than the Amazon river? length of the Nile; encyclopedia or atlas

Solve. Check each solution.

See answer on p. A24.

★28. List the steps you would use to solve the equation $-2x - 7.5 = -6.2$.   **8-8**

29. $3z - 8 = -10$ $-\frac{2}{3}$ 30. $-2.6 = 4a + 6.6$ -2.3 31. $4\dfrac{3}{5} + m = \dfrac{7}{10}$ $-3\frac{9}{10}$

32. $-\dfrac{3}{5}x = 4$ $-6\frac{2}{3}$ 33. $\dfrac{5}{8}c - 17 = 3$ 32 34. $11 = \dfrac{6}{7}b + 2$ $10\frac{1}{2}$ **8-9**

Solve for the variable shown in color.

35. $D = \dfrac{m}{v}$ $m = Dv$ 36. $A = \dfrac{1}{2}ap$ $a = \dfrac{2A}{p}$ 37. $a + b + c = 180$ $c = 180 - a - b$ **8-10**

38. Use the formula $V = \dfrac{1}{3}Bh$ to find V (in cubic inches) when $B = 4$ in.^{2} and $h = 6$ in. 8 in.^{3}

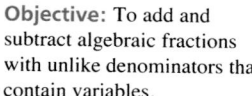
Adding and Subtracting More Algebraic Fractions

Objective: To add and subtract algebraic fractions with unlike denominators that contain variables.

Sometimes the unlike denominators in algebraic fractions contain variables. In order to add and subtract such fractions, you have to find the LCD using the skills you learned in Lesson 7-3 for finding an LCM.

Example

Simplify $\frac{2x}{3y^2} + \frac{5x}{6y^2}$.

Solution

$\frac{2x}{3y^2} + \frac{5x}{6y^2} = \frac{2x}{3y^2} \cdot \frac{2}{2} + \frac{5x}{6y^2}$ ◀— The LCM of $3y^2$ and $6y^2$ is $6y^2$.

$= \frac{4x}{6y^2} + \frac{5x}{6y^2}$

$= \frac{4x + 5x}{6y^2}$

$= \frac{9x}{6y^2} = \frac{3x}{2y^2}$

Exercises

Simplify.

4. $\frac{29b}{7c}$

8. $\frac{4}{3a}$

12. $\frac{5u}{4v^2}$

14. $\frac{29}{6m^2}$

16. $\frac{21m}{10n}$

18. $\frac{19m}{14np}$

20. $\frac{57}{40uv}$

22. $\frac{53x}{36y}$

23. $\frac{31u}{24vw}$

1. $\frac{2}{x} + \frac{1}{5x}$ $\frac{11}{5x}$

2. $\frac{5}{mn} - \frac{5}{6mn}$ $\frac{25}{6mn}$

3. $\frac{9}{2a^2} - \frac{3}{a^2}$ $\frac{3}{2a^2}$

4. $\frac{3b}{c} + \frac{8b}{7c}$

5. $\frac{10x}{3yz} + \frac{4x}{yz}$ $\frac{22x}{3yz}$

6. $\frac{8m}{7n^3} + \frac{2m}{n^3}$ $\frac{22m}{7n^3}$

7. $\frac{5}{6xy} - \frac{7}{12xy}$ $\frac{1}{4xy}$

8. $\frac{9}{5a} - \frac{7}{15a}$

9. $\frac{3}{10n^3} + \frac{6}{5n^3}$ $\frac{3}{2n^3}$

10. $\frac{2b}{9ac} + \frac{5b}{3ac}$ $\frac{17b}{9ac}$

11. $\frac{5r}{4s} - \frac{7r}{8s}$ $\frac{3r}{8s}$

12. $\frac{7u}{4v^2} - \frac{u}{2v^2}$

13. $\frac{7}{3m} - \frac{3}{4m}$ $\frac{19}{12m}$

14. $\frac{7}{2m^2} + \frac{4}{3m^2}$

15. $\frac{11}{4rs} + \frac{1}{5rs}$ $\frac{59}{20rs}$

16. $\frac{5m}{2n} - \frac{2m}{5n}$

17. $\frac{4b}{5c^3} + \frac{8b}{3c^3}$ $\frac{52b}{15c^3}$

18. $\frac{3m}{2np} - \frac{m}{7np}$

19. $\frac{7}{10r} - \frac{1}{6r}$ $\frac{8}{15r}$

20. $\frac{9}{8uv} + \frac{3}{10uv}$

21. $\frac{5r}{4s^3} + \frac{17r}{6s^3}$ $\frac{49r}{12s^3}$

22. $\frac{11x}{12y} + \frac{5x}{9y}$

23. $\frac{13u}{6vw} - \frac{7u}{8vw}$

24. $\frac{4x}{9y^2} - \frac{2x}{15y^2}$ $\frac{14x}{45y^2}$

25. $\frac{9}{mn} + \frac{3}{2mn} + \frac{7}{6mn}$ $\frac{35}{3mn}$

26. $\frac{5u}{2v} + \frac{9u}{10v} + \frac{3u}{5v}$ $\frac{4u}{v}$

27. $\frac{7}{12x^2} + \frac{1}{6x^2} + \frac{9}{8x^2}$ $\frac{15}{8x^2}$

28. $\frac{b}{2c^2} + \frac{2b}{5c^2} + \frac{2b}{3c^2}$ $\frac{47b}{30c^2}$

29. **THINKING SKILLS** A student simplified $\frac{5x + 12y}{8y^2}$ as $\frac{5x + 3}{2y}$. Compare the two expressions. Did the student simplify correctly? Explain.
Analysis: No; $4y$ is not a factor of the numerator.

Solve. Decide whether a *whole number*, a *fraction* or a *decimal* is an appropriate answer.

1. Sue jogs $\frac{3}{4}$ h every day. How many hours does she jog in 5 days? $3\frac{3}{4}$ h; fraction

2. Gasoline costs $1.70/gal. How much does $12\frac{1}{2}$ gal cost? $21.25; decimal

Find each answer. Simplify if possible.

3. $-\frac{7}{9} + \frac{8}{9}$ $\frac{1}{9}$

4. $-11\frac{3}{5} - 2\frac{2}{5}$ -14

5. $\frac{18x}{25} - \frac{x}{25}$ $\frac{17x}{25}$

6. $-5\frac{7}{12} - \left(-1\frac{1}{12}\right)$ $-4\frac{1}{2}$

7. $6x + \frac{7x}{4}$ $\frac{31x}{4}$

8. $-8.5 + 2.8$ -5.7

Estimate.

9. $21\frac{7}{8} - 3\frac{1}{5}$ about 19

10. $-15\frac{13}{25} + \left(-7\frac{8}{9}\right)$ about $-23\frac{1}{2}$

11. $28\frac{1}{3} \div 5\frac{9}{10}$ about 5

Alternative Assessment

Set up cooperative groups of four. Have one student in each group make up a problem involving rational numbers, another student explain how to solve the problem, a third student actually solve the problem, and a fourth student check the work. Rotate roles until each student has served in all four capacities.

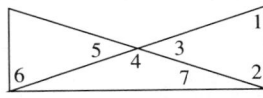

Cumulative Review

Standardized Testing Practice

Choose the letter of the correct answer.

1. **Identify a pair of vertical angles.**

   A. $\angle 2$ and $\angle 7$ B. $\angle 4$ and $\angle 5$
   C. $\angle 3$ and $\angle 5$ D. $\angle 6$ and $\angle 1$

2. **What is the prime factorization of 80?**

   A. $8 \cdot 10$ B. $2 \cdot 5 \cdot 8$
   C. $2^4 \cdot 5$ D. $5 \cdot 16$

3. **Decide which statistical measure best describes the data.**
   4, 3.9, 12.8, 3.7, 3.6, 4.1

   A. mean   B. median
   C. mode   D. range

4. **Decide which is the appropriate form of the answer.**
   Vans hold 12 students each. If 54 students plan to travel in vans, how many vans will be needed?

   A. decimal   B. fraction
   C. dollars   D. whole number

5. **Which equation could you use to solve the problem?**
   Gia's home run total is six more than three times Kelly's home run total. If Gia's total is 15, what is Kelly's total?

   A. $6x + 3 = 15$ B. $6x - 3 = 15$
   C. $3x + 15 = 6$ D. $3x + 6 = 15$

6. **Identify the figure.**

   A. rhombus   B. rectangle
   C. trapezoid   D. hexagon

7. **Find the quotient:** $\frac{t}{12} \div \frac{2t}{3}$

   A. $\frac{1}{2}$ B. 8
   C. $\frac{1}{8}$ D. $\frac{t^2}{18}$

8. **Simplify:** s^{-6}

   A. $-6s$ B. $\frac{s}{6}$
   C. $\frac{1}{s^6}$ D. $s - 6$

9. **A marching band can march in rows of either 8 or 14 with no one left out. What is the least number of members the band could have?**

   A. 22   B. 112
   C. 56   D. 2

10. **Choose the correct relationship:**
    $m = |x - y|, n = |y - x|$

    A. $m = n$
    B. $m > n$
    C. $m < n$
    D. cannot determine

11. Solve: $9x + 2 + 4x = 41$

   Ⓐ $x = 3$

   B. $x = 3\frac{4}{13}$

   C. $x = 2$

   D. $x = 507$

12. Phillip bought $4\frac{1}{2}$ yd of cotton and $1\frac{3}{4}$ yd of wool. How many yards of fabric did he buy altogether?

   Ⓐ $6\frac{1}{4}$ yd   B. $5\frac{4}{6}$ yd

   C. $5\frac{2}{3}$ yd   D. $2\frac{3}{4}$ yd

13. An elevator goes up 6 floors, goes down 8 floors, and then goes down 3 more floors. How much lower is the elevator than it was at its starting point?

   A. 11 floors   B. 2 floors

   Ⓒ 5 floors   D. 17 floors

14. Write 0.0000498 in scientific notation.

   A. 4.98×10^{-4}

   B. 4.98×10^{4}

   C. 4.98×10^{5}

   Ⓓ 4.98×10^{-5}

15. On a trip, Carolina plans to spend 10 days camping, and then 11 days at a resort. How many weeks long is her trip?

   A. 21   B. 2

   C. 7   Ⓓ 3

16. Complete: A triangle with sides that measure 4 cm, 4 cm, and 6 cm is a(n) __?__ triangle.

   A. scalene   Ⓑ isosceles

   C. equilateral   D. rhomboid

17. Complete: An angle that measures 79° is a(n) __?__ angle.

   A. obtuse   Ⓑ acute

   C. right   D. adjacent

18.

```
<—+——+——+——+——+——+——+—>
 -3  -2  -1   0   1   2   3
```

Which of the following inequalities does this graph represent?

   I. $x \geq -1$ II. $x \leq -1$ III. $x > -1$

   Ⓐ I only   B. II only

   C. III only   D. I and II

19. Which of the following can make a bar graph visually misleading?

   A. drawing it horizontally

   B. not shading the bars

   Ⓒ putting a gap in the scale

   D. using a different color

20. Solve: $14 + \frac{15}{16}x = 11$

   Ⓐ $x = -3\frac{1}{5}$ B. $x = 26\frac{2}{3}$

   C. $x = -3$ D. $x = -2\frac{13}{16}$

Quick Quiz 3 (8-7 through 8-10)

1. Nick swam 25 ft farther than Mike. Mike swam 35 yd. How many yards did Nick swim?
$43\frac{1}{3}$ yd

Solve. Check the solution.

2. $23 = 8 + 12z$ $1\frac{1}{4}$

3. $2.4x - 9 = 8.4$ 7.25

4. $w - \frac{1}{3} = 2\frac{4}{5}$ $3\frac{2}{15}$

5. $-\frac{2}{7}c = 36$ -126

6. $15 = \frac{4}{7}h - 5$ 35

7. $-2 = 3 + \frac{5}{8}m$ -8

8. Solve $V = lwh$ for w. $w = \frac{V}{lh}$

9. Use the formula you found in Exercise 8 to find w when $V = 48$ cm^3, $l = 6$ cm, and $h = 4$ cm. 2 cm

Alternative Assessment

Present this situation to students: A clue in a treasure hunt states that to find the buried treasure walk 9 yd and $3\frac{1}{2}$ ft due west of an old gate.

1. What missing fact is needed to relate the two quantities in the situation? 1 yd = 3 ft

2. A "formula" that could be used to find the treasure is $f = 3y + 3\frac{1}{2}$. Solve this formula for y.
$y = \frac{f}{3} - 1\frac{1}{6}$

3. How many feet from the gate is the treasure? How many yards?
$30\frac{1}{2}$ ft; $10\frac{1}{6}$ yd

Chapter Overview

Chapter 9 introduces students to the concepts of ratio, proportion, and percent. *Lesson 1* presents ratios and rates; *Lesson 2* shows how ratios can be combined into proportions; and *Lesson 3* presents an application of ratios and proportions through the study of scale drawings. The problem-solving strategy of using proportions is presented in *Lesson 4*. The *Focus on Explorations* introduces students to the percent concept in a concrete manner by using ten-by-ten grids. The formal presentation of percent begins in *Lesson 5*. The next four lessons, *Lessons 6–9*, cover the standard percent problems: finding a percent of a number, finding the percent one number is of another, finding a number when a percent of it is known, and percent of increase or decrease. *Lesson 10* concludes the chapter by connecting the concepts of percent and proportion.

Background

An understanding of the concepts of ratio, proportion, and percent is fundamentally important to students; proportional reasoning, in particular, is one of the most important ideas students will study in mathematics.

In the Using Manipulatives section on pages 386C and 386D, suggestions are given for activities involving ratios. The notions of ratio and equivalent ratios provide a smooth transition to proportions.

Students are introduced to solving each of the three types of percent problems by using the equation method. This method is followed by using the proportion method. The proportion method usually helps those students who have difficulty solving problems not categorized by type. Technology, of course, can be used to solve each of the three types of percent problems. Calculators with a percent key are very helpful in studying this chapter.

Ample attention should be given to developing the concept of percents greater than 100 and less than 1, since these percents are often troublesome to students. The *Focus on Explorations* on page 402 should be very helpful with these concepts.

Various applications and problem-solving situations are provided throughout the chapter by using scale drawings, equal rates, and percents of increase and decrease. A further application involving compound interest is provided in the *Chapter Extension*.

Objectives

9-1   To write ratios as fractions in lowest terms and to write unit rates.

9-2   To solve proportions.

9-3   To interpret and use scale drawings.

9-4   To use proportions to solve problems.

≡   **FOCUS ON EXPLORATIONS** To use ten-by-ten grids to model percents.

9-5   To write fractions and decimals as percents and to write percents as fractions and decimals.

9-6   To find a percent of a number and to estimate percents of numbers.

9-7   To find the percent one number is of another.

9-8   To find a number when a percent of it is known.

9-9   To find a percent of increase or decrease.

9-10   To use proportions to solve percent problems.

≡   **CHAPTER EXTENSION** To calculate compound interest using the compound interest formula.

Chapter Planner

Instructional Aids	Manipulatives	Cooperative Learning	Technology	Practice/Reteaching	Assessment
Materials: colored tiles, bingo chips, pennies, gumdrops, scale models, grids, calculators *Resource Book:* Lesson Starters 80–89	Lessons 9-1, 9-2, 9-3, Focus on Explorations, 9-5, 9-6, 9-10 *Activities Book:* Manipulative Activities 17–18	Lessons 9-2, 9-3, 9-6, 9-9 *Activities Book:* Cooperative Activities 17–18	Lessons 9-1, 9-2, 9-3, 9-6 *Activities Book:* Calculator Activity 9 Computer Activity 9 *Connections Plotter Plus* Disk	Toolbox, pp. 753–771 Extra Practice, p. 742 *Resource Book:* Practice 109–122 *Study Guide,* pp. 159–178	Self-Tests 1–3 Quick Quizzes 1–3 Chapter Test, p. 430 *Resource Book:* Lesson Starters 80–89 (Daily Quizzes) *Tests 43–47*

Assignment Guide Chapter 9

Day	Skills Course	Average Course	Advanced Course
1	**9-1:** 1–22, 23–29 odd, 35–38	**9-1:** 1–29 odd, 35–38 **9-2:** 1–35 odd, 38–42	**9-1:** 1–33 odd, 35–38 **9-2:** 1–37 odd, 38–42
2	**9-2:** 1–24, 38–42	**9-3:** 1–13, 16–18 **9-4:** 1–14	**9-3:** 1–18 **9-4:** 1–14
3	**9-3:** 1–13, 16–18	**Exploration:** Activities I–IV **9-5:** 1–33, 39–42	**Exploration:** Activities I–IV **9-5:** 1–29 odd, 30–42, Challenge
4	**9-4:** 1–14	**9-6:** 1–32, 45–47	**9-6:** 1–19 odd, 21–47
5	**Exploration:** Activities I–IV **9-5:** 1–33, 39–42	**9-7:** 1–30	**9-7:** 1–15 odd, 17–30, Historical Note **9-8:** 1–15 odd, 17–32
6	**9-6:** 1–24, 25–31 odd, 45–47	**9-8:** 1–32	**9-9:** 1–13 odd, 15–25 **Mixed Review**
7	**9-7:** 1–20, 27–30	**9-9:** 1–25 **Mixed Review**	**9-10:** 1–15 odd, 16–29
8	**9-8:** 1–23, 29–32	**9-10:** 1–21, 25–29	*Prepare for Chapter Test:* Chapter Review
9	**9-9:** 1–19, 22–25 **Mixed Review**	*Prepare for Chapter Test:* Chapter Review	*Administer Chapter 9 Test; Chapter Extension; Cumulative Review*
10	**9-10:** 1–21, 25–29	*Administer Chapter 9 Test; Cumulative Review*	
11	*Prepare for Chapter Test:* Chapter Review		
12	*Administer Chapter 9 Test*		

Teacher's Resources

Resource Book
  Chapter 9 Project
  Lesson Starters 80–89
  Practice 109–122
  Enrichment 88–98
  Diagram Masters
  Chapter 9 Objectives
  Family Involvement 9
  Spanish Test 9
  Spanish Glossary

Activities Book
  Cooperative Activities 17–18
  Manipulative Activities 17–18
  Calculator Activity 9
  Computer Activity 9

Study Guide, pp. 159–178

Tests
  Tests 43–47

***Connections Plotter Plus* Disk**

Using Technology

COMPUTERS

Programs such as *Semantic Calculator* (Sunburst) do calculations with units (such as miles/hour) along with the numbers involved. This design allows students to check the plausibility of calculations involving rates by examining the units that result.

Students can create a spreadsheet to calculate unit prices. This activity can be made more elaborate by including conversions from one unit to another. For example, something like the following could be designed (the formulas will not appear in the cell locations on the screen but will be evaluated by the program).

	A	B	C	D	E	F
		Cost ($)	Weight (kg)	Cost/kg	Cost/lb	Cost/oz
1	Item					
2	Soap	1.12	0.23	+B2/C2	+D2/2.2	+E2/16

CALCULATORS

Using a calculator, students can check the equivalence of two proportions by actually performing the indicated divisions. For example, since $\frac{4}{7} = 0.57142857...$, and $\frac{3.32}{5.81} = 0.57142857...$ as well, these proportions are equivalent, at least to the accuracy this particular calculator allows.

COMPUTERS

Software is available for most computers that allows students to make very good scale drawings. This software works best with a good quality display, and with a graphics input device such as a mouse, trackball, or graphics tablet. Sometimes the software is difficult to use at first, but most students can learn to master the basic skills needed quickly. A very instructive project can be designed that will involve students' geometric intuition, design sense, and knowledge of proportions.

COMPUTERS

Just as most calculators have a percent key, most spreadsheet programs have a percent display option, in which the result of a calculation appears as a percent. This feature can be used to reinforce the conversion of decimals and fractions to percents and vice versa.

Using Manipulatives

Show various ratios by using different arrangements of two colored tiles on an overhead projector. Students should verbalize the ratio shown and then represent the same ratio in another way by using bingo chips of two colors. For example, the ratio of 2:3 can be represented by groups of 2 and 3, 4 and 6, and so on.

Have students use gumdrops and pennies to construct a ratio table. For example, students can do the following.

2 gumdrops for 7 cents,
4 gumdrops for 14 cents,
and so on.

Gumdrops	Pennies
2	7
4	14
6	21

Students should then express the equivalent ratios that have been modeled, $\frac{2}{7} = \frac{4}{14} = \frac{6}{21}$, and confirm the definition of a proportion as a statement that two ratios are equal.

An actual model of a car, airplane, or boat is an excellent manipulative. Students can discuss the scale used to make the model and then calculate real measurements of the full-size object. Blueprints for a house or other building can be used for the same purpose.

Lesson 9-5

The ten-by-ten grid, shown in the Focus on Explorations, is excellent for demonstrating the percent, decimal, and fractional representations of a quantity. Money also can be used effectively to show the relationship between fractions, decimals, and percents. Each cent represents one percent, and a quarter, for example, is $\frac{1}{4}$ of a dollar, 25 cents, $.25, or 25% of a dollar.

Lesson 9-6

Models similar to the ten-by-ten grid can be used to show the relationship of a part to a whole. For example, you can use a ten-by-ten grid to show that $20\% = \frac{20}{100} = \frac{1}{5}$, and then use grids like the ones below to show that 20% of 20 = 4 or that 20% of 40 = 8.

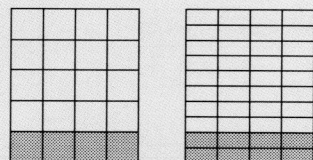

Lesson 9-10

The use of models to show the relationship of a part to a whole, as described above, can be used in this lesson also. For example, to show Example 1 on page 424, you can do the following.

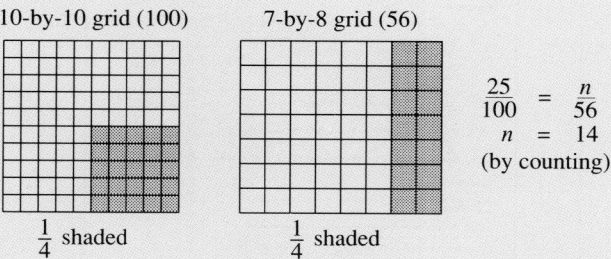

10-by-10 grid (100)   7-by-8 grid (56)

$\frac{1}{4}$ shaded $\frac{1}{4}$ shaded

$$\frac{25}{100} = \frac{n}{56}$$
$$n = 14$$
(by counting)

Reteaching/Alternate Approach

Lesson 9-1

Give students a list of situations to represent by using ratios in various forms. For example, six pairs of socks for $4 can be represented as 6 to 4, 6:4, or $\frac{6}{4}$. Students can then create their own situations and exchange them with a partner, who should use a ratio to represent the situation.

Lesson 9-2

COOPERATIVE LEARNING
Have students working in groups draw a set of rectangles, some of which are similar to others. Each group should examine various combinations of rectangles to see if the sides are in proportion. Each group should then test the proportions to see if they are true. This activity can be extended to compare ratios of perimeters.

Lesson 9-9

COOPERATIVE LEARNING
Have students work in groups. Each group should imagine that members are the management team of a large department store. In order to attract business, the store is having a sale of 10% off on all items. After the sale, the usual prices are reinstated. Have each group consider a jacket that cost $80 and answer the following questions.

1. What is the sale price after the 10% reduction?
2. After the sale is over, does the price of $80 reflect a 10% increase? Why or why not?

Have each group calculate the sale price for other items of their choice. Members of the group should conclude that the percent of decrease and percent of increase are not the same because the base amount changes.

Teacher's Resources Chapter 9

Enrichment masters from the Resource Book are pictured here. See the Teacher's Resources chart on page 386B for a complete listing of all materials available for this chapter.

9-1 Enrichment
Communication

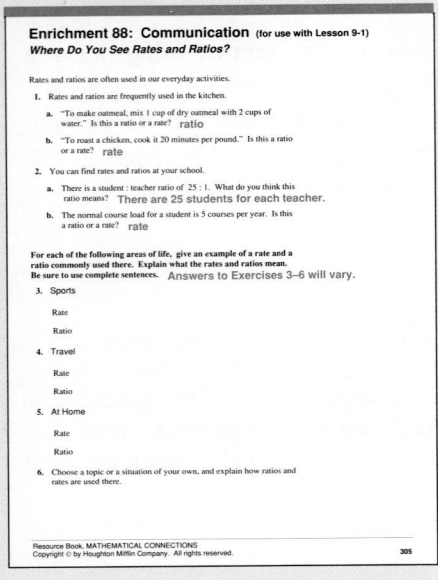

Enrichment 88: Communication (for use with Lesson 9-1)
Where Do You See Rates and Ratios?

Rates and ratios are often used in our everyday activities.

1. Rates and ratios are frequently used in the kitchen.
   a. "To make oatmeal, mix 1 cup of dry oatmeal with 2 cups of water." Is this a ratio or a rate? **ratio**
   b. "To roast a chicken, cook it 20 minutes per pound." Is this a ratio or a rate? **rate**

2. You can find rates and ratios at your school.
   a. There is a student : teacher ratio of 25 : 1. What do you think this ratio means? **There are 25 students for each teacher.**
   b. The normal course load for a student is 5 courses per year. Is this a ratio or a rate? **rate**

For each of the following areas of life, give an example of a rate and a ratio commonly used there. Explain what the rates and ratios mean. Be sure to use complete sentences. **Answers to Exercises 3–6 will vary.**

3. Sports
   Rate
   Ratio

4. Travel
   Rate
   Ratio

5. At Home
   Rate
   Ratio

6. Choose a topic or a situation of your own, and explain how ratios and rates are used there.

Resource Book, MATHEMATICAL CONNECTIONS
Copyright © by Houghton Mifflin Company. All rights reserved.   305

9-2 Enrichment
Connection

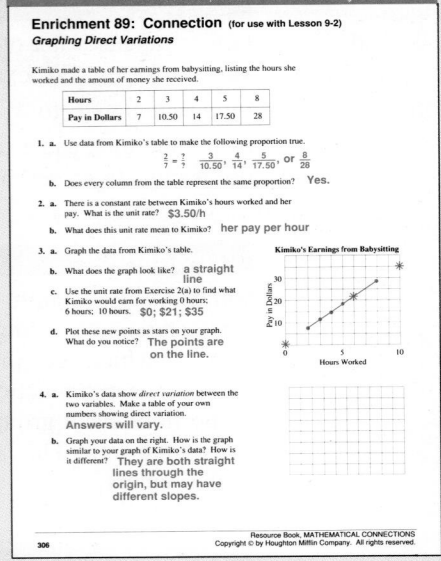

Enrichment 89: Connection (for use with Lesson 9-2)
Graphing Direct Variations

Kimiko made a table of her earnings from babysitting, listing the hours she worked and the amount of money she received.

Hours	2	3	4	5	8
Pay in Dollars	7	10.50	14	17.50	28

1. a. Use data from Kimiko's table to make the following proportion true.
   $\frac{2}{7} = \frac{?}{?}$, $\frac{3}{10.50}$, $\frac{4}{14}$, $\frac{5}{17.50}$, or $\frac{8}{28}$
   b. Does every column from the table represent the same proportion? **Yes.**

2. a. There is a constant rate between Kimiko's hours worked and her pay. What is the unit rate? **$3.50/h**
   b. What does this unit rate mean to Kimiko? **her pay per hour**

3. a. Graph the data from Kimiko's table.
   b. What does the graph look like? **a straight line**
   c. Use the unit rate from Exercise 2(a) to find what Kimiko would earn for working 0 hours; 6 hours; 10 hours. **$0; $21; $35**
   d. Plot these new points as stars on your graph. What do you notice? **The points are on the line.**

   Kimiko's Earnings from Babysitting
   (graph with Pay in Dollars vs Hours Worked)

4. a. Kimiko's data show *direct variation* between the two variables. Make a table of your own numbers showing direct variation. **Answers will vary.**
   b. Graph your data on the right. How is the graph similar to your graph of Kimiko's data? How is it different? **They are both straight lines through the origin, but may have different slopes.**

306   Resource Book, MATHEMATICAL CONNECTIONS
Copyright © by Houghton Mifflin Company. All rights reserved.

9-3 Enrichment
Thinking Skills

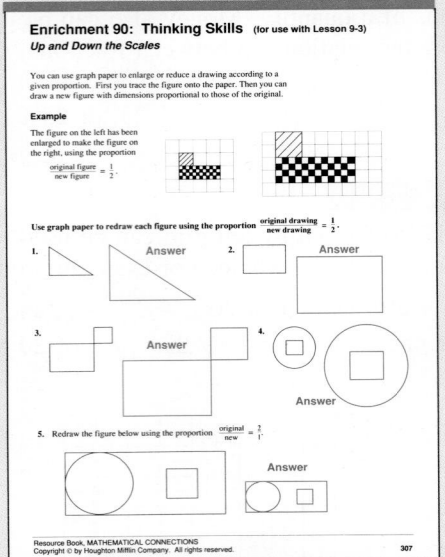

Enrichment 90: Thinking Skills (for use with Lesson 9-3)
Up and Down the Scales

You can use graph paper to enlarge or reduce a drawing according to a given proportion. First you trace the figure onto the paper. Then you can draw a new figure with dimensions proportional to those of the original.

Example
The figure on the left has been enlarged to make the figure on the right, using the proportion
$\frac{\text{original drawing}}{\text{new figure}} = \frac{1}{2}$.

Use graph paper to redraw each figure using the proportion $\frac{\text{original drawing}}{\text{new drawing}} = \frac{1}{2}$.

1. **Answer**
2. **Answer**
3. **Answer**
4. **Answer**

5. Redraw the figure below using the proportion $\frac{\text{original}}{\text{new}} = \frac{2}{1}$. **Answer**

Resource Book, MATHEMATICAL CONNECTIONS
Copyright © by Houghton Mifflin Company. All rights reserved.   307

9-4 Enrichment
Problem Solving

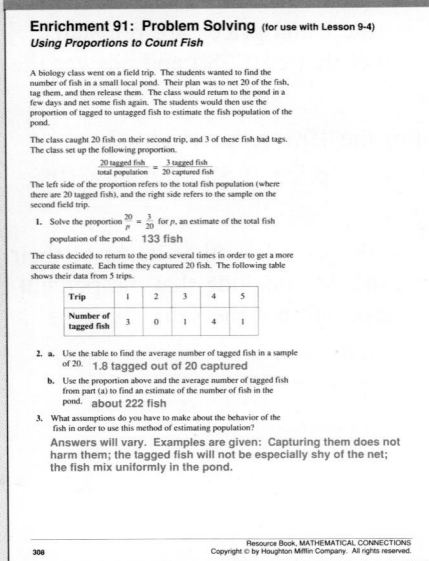

Enrichment 91: Problem Solving (for use with Lesson 9-4)
Using Proportions to Count Fish

A biology class went on a field trip. The students wanted to find the number of fish in a small local pond. Their plan was to net 20 of the fish, tag them, and then release them. The class would return to the pond in a few days and net some fish again. The students would then use the proportion of tagged to untagged fish to estimate the fish population of the pond.

The class caught 20 fish on their second trip, and 3 of these fish had tags. The class set up the following proportion.

$\frac{20 \text{ tagged fish}}{\text{total population}} = \frac{3 \text{ tagged fish}}{20 \text{ population}}$

The left side of the proportion refers to the total fish population (where there are 20 tagged fish), and the right side refers to the sample on the second field trip.

1. Solve the proportion $\frac{20}{p} = \frac{3}{20}$ for p, an estimate of the total fish population of the pond. **133 fish**

The class decided to return to the pond several times in order to get a more accurate estimate. Each time they captured 20 fish. The following table shows their data from 5 trips.

Trip	1	2	3	4	5
Number of tagged fish	3	0	1	4	1

2. a. Use the table to find the average number of tagged fish in a sample of 20. **1.8 tagged out of 20 captured**
   b. Use the proportion above and the average number of tagged fish from part (a) to find an estimate of the number of fish in the pond. **about 222 fish**

3. What assumptions do you have to make about the behavior of the fish in order to use this method of estimating population? **Answers will vary. Examples are given: Capturing them does not harm them; the tagged fish will not be especially shy of the net; the fish mix uniformly in the pond.**

308   Resource Book, MATHEMATICAL CONNECTIONS
Copyright © by Houghton Mifflin Company. All rights reserved.

9-5 Enrichment
Communication

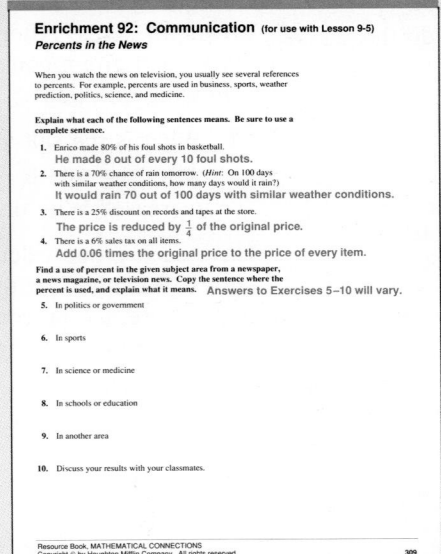

Enrichment 92: Communication (for use with Lesson 9-5)
Percents in the News

When you watch the news on television, you usually see several references to percents. For example, percents are used in business, sports, weather prediction, politics, science, and medicine.

Explain what each of the following sentences means. Be sure to use a complete sentence.

1. Enrico made 80% of his foul shots in basketball.
   He made 8 out of every 10 foul shots.

2. There is a 70% chance of rain tomorrow. (*Hint*: On 100 days with similar weather conditions, how many days would it rain?)
   It would rain 70 out of 100 days with similar weather conditions.

3. There is a 25% discount on records and tapes at the store.
   The price is reduced by $\frac{1}{4}$ of the original price.

4. There is a 6% sales tax on all items.
   Add 0.06 times the original price to the price of every item.

Find a use of percent in the given subject area from a newspaper, a news magazine, or television news. Copy the sentence where the percent is used, and explain what it means. **Answers to Exercises 5–10 will vary.**

5. In politics or government

6. In sports

7. In science or medicine

8. In schools or education

9. In another area

10. Discuss your results with your classmates.

Resource Book, MATHEMATICAL CONNECTIONS
Copyright © by Houghton Mifflin Company. All rights reserved.   309

9-6 Enrichment
Application

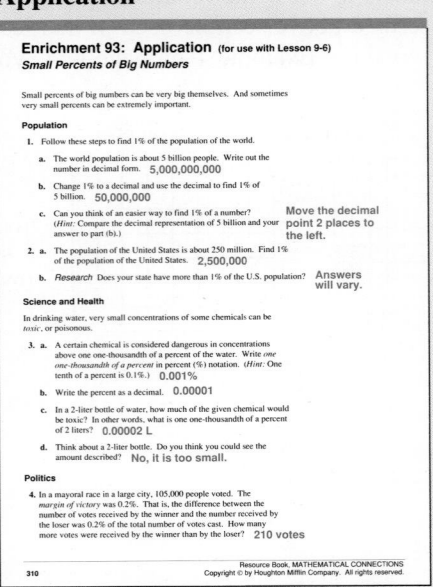

Enrichment 93: Application (for use with Lesson 9-6)
Small Percents of Big Numbers

Small percents of big numbers can be very big themselves. And sometimes very small percents can be extremely important.

Population

1. Follow these steps to find 1% of the population of the world.
   a. The world population is about 5 billion people. Write out the number in decimal form. **5,000,000,000**
   b. Change 1% to a decimal and use the decimal to find 1% of 5 billion. **50,000,000**
   c. Can you think of an easier way to find 1% of a number? (*Hint*: Compare the decimal representation of 5 billion and your answer to part (b).) **Move the decimal point 2 places to the left.**

2. a. The population of the United States is about 250 million. Find 1% of the population of the United States. **2,500,000**
   b. *Research* Does your state have more than 1% of the U.S. population? **Answers will vary.**

Science and Health

In drinking water, very small concentrations of some chemicals can be *toxic*, or poisonous.

3. a. A certain chemical is considered dangerous in concentrations above one one-thousandth of a percent of the water. Write *one one-thousandth of a percent* in percent (%) notation. (*Hint*: One tenth of a percent is 0.1%.) **0.001%**
   b. Write the percent as a decimal. **0.00001**
   c. In a 2-liter bottle of water, how much of the given chemical would be toxic? In other words, what is one one-thousandth of a percent of 2 liters? **0.00002 L**
   d. Think about a 2-liter bottle. Do you think you could see the amount described? **No, it is too small.**

Politics

4. In a mayoral race in a large city, 105,000 people voted. The *margin of victory* was 0.2%. That is, the difference between the number of votes received by the winner and the number received by the loser was 0.2% of the total number of votes cast. How many more votes were received by the winner than by the loser? **210 votes**

310   Resource Book, MATHEMATICAL CONNECTIONS
Copyright © by Houghton Mifflin Company. All rights reserved.

386E

9-7 Enrichment
Data Analysis

Enrichment 94: Data Analysis (for use with Lesson 9-7)
Percents in Politics

Problem

In a recent election Ishana Baweja received 10,493 votes, Quito Pasadas received 11,106 votes, and Adela Canales received 12,003 votes. What percent of the total vote did Adela Canales receive?

Solution

First you need to find the *total vote*. $10,493 + 11,106 + 12,003 = 33,602$

Then the percent of the total vote Adela Canales received is

$$\frac{12,003}{33,602} = 0.357 = 35.7\%.$$

1. a. What percent of the total vote did Quito Pasadas receive? Ishana Baweja? **33.1%; 31.2%**

   b. What is the total of the percents for all three candidates? **100%**

In a student council election, six candidates received most of the votes, and several others received a few votes each. The following table shows the results of the election.

Candidate	Gim	Rini	Ella	Owen	Sara	Jin Ki	Others
Votes	132	121	109	98	91	81	17

2. a. What was the total number of ballots cast? **649**

   b. Rounding to the nearest tenth of a percent, what percent of the total vote did each candidate get?

Candidate	Gim	Rini	Ella	Owen	Sara	Jin Ki	Others
Votes	132	121	109	98	91	81	17
Percent of Total	20.3%	18.6%	16.8%	15.1%	14.0%	12.5%	2.6%

   c. What is the sum of all the percents? **99.9%**

3. Is the sum of all the percents *exactly* what you expected it to be? Explain. **No; round-off error caused it to be not exactly 100%.**

Resource Book, MATHEMATICAL CONNECTIONS
Copyright © by Houghton Mifflin Company. All rights reserved.
311

9-8 Enrichment
Application

Enrichment 95: Application (for use with Lesson 9-8)
Using Percents in Making Budgets

At home and in business, people often use percents to help them predict their income and plan their budgets.

Solve the following problems.

1. Elston Cooper needs to earn $30,000 a year to support his family. He gets an 8% commission on the real estate he sells. (For example, if he sells a $100,000 house, he gets 8% of $100,000, or $8000.) How many dollars' worth of real estate does he need to sell in order to earn $30,000 in commission?

   a. Let n represent the total sales Elston Cooper needs to make. Write an equation showing that 8% of the total sales is $30,000. $0.08n = 30,000$

   b. Solve the equation. How many dollars' worth of real estate does he need to sell? **$375,000**

2. In preparing the state's budget, the treasurer stated that a 5% sales tax on all items sold in the state would provide $10 million in revenue. What did she assume would be the total sales in the state?

   a. Let t = total sales. Write an equation showing that 5% of the total sales is $10 million. $0.05t = 10,000,000$

   b. Solve the equation. How much did the treasurer assume total sales in the state would be? **$200,000,000**

3. To earn more money for the state, the governor is proposing an additional 2.5% sales tax on gasoline. He believes that this will earn $12,500,000 for the state. How many dollars' worth of gasoline will need to be sold? **$500,000,000**

4. A department store needs to raise $1,000,000 to expand its business. The managers feel that raising all prices 7% will raise that amount of money as long as the total sales are at least the same as last year's and other costs do not go up. How much were the total sales last year? **$14,285,714**

5. Explain how you would estimate your family's annual gasoline expenses. What changes would make this projection inaccurate? **Answers may vary. Example: Estimate total mileage, divide by miles per gallon, and multiply by gasoline price per gallon.**

312 Resource Book, MATHEMATICAL CONNECTIONS
Copyright © by Houghton Mifflin Company. All rights reserved.

9-9 Enrichment
Exploration

Enrichment 96: Exploration (for use with Lesson 9-9)
Discounts and Markups

Problem 1 — Discount

Caleb sees an advertisement for 20% off a compact disc that normally sells for $15. What is the new price?

Solution

To find the new price, you can find 20% of $15 and then subtract the result from $15.

$$15 - (15)(20\%) = 15 - (15)(0.20)$$

Or, using the distributive property, you can write,

$$15 - (15)(0.20) = 15(1 - 0.20)$$
$$= 15(0.80)$$
$$= 15 \cdot 80\%.$$

1. a. Find the new price by subtracting 20% of $15 from $15. $15 - \$3 = \$12$

   b. Find the new price by taking 80% of $15. $0.80 \cdot \$15 = \$12$

   c. Do your answers in parts (a) and (b) agree? **Yes.**

Problem 2 — Markup

Most stores buy their merchandise in bulk at wholesale prices. Then, in order to make a profit, they add a *markup*. A markup might be 10% of the wholesale price. The new price is called the retail price. What is the retail price of a CD player with a wholesale price of $150 and a 10% markup?

Solution

You can find the retail price by finding 10% of the wholesale price and then adding the result to the wholesale price. Another way is to see that the retail price is 100% plus 10%, or 110%, of the wholesale price.

2. What is 110% of $150? Is this the same as $150 plus 10% of $150? **$165; Yes**

Use the techniques described above to compute the final prices in the problems below.

3. Original price is $12.50, with a 10% discount. (Find 90% of $12.50.) **$11.25**

4. Wholesale price is $250, with an 11% markup. (Find 111% of $250.) **$277.50**

5. Original price is $134, with a discount of 12%. **$117.92**

6. Wholesale price is $1200, with a markup of 8%. **$1296**

Resource Book, MATHEMATICAL CONNECTIONS
Copyright © by Houghton Mifflin Company. All rights reserved.
313

9-10 Enrichment
Connection

Enrichment 97: Connection (for use with Lesson 9-10)
Percent in Pictures

Draw a figure that is the given percent of the figure shown.

Problem	Solution
50%	Here are two of the possible figures. The area of each of these figures is 50% of the area of the given figure.

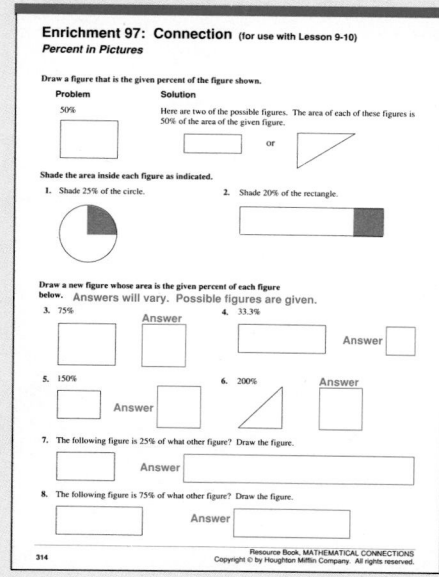

Shade the area inside each figure as indicated.

1. Shade 25% of the circle.

2. Shade 20% of the rectangle.

Draw a new figure whose area is the given percent of each figure below. **Answers will vary. Possible figures are given.**

3. 75% **Answer**

4. 33.3% **Answer**

5. 150% **Answer**

6. 200% **Answer**

7. The following figure is 25% of what other figure? Draw the figure. **Answer**

8. The following figure is 75% of what other figure? Draw the figure. **Answer**

314 Resource Book, MATHEMATICAL CONNECTIONS
Copyright © by Houghton Mifflin Company. All rights reserved.

End of Chapter Enrichment
Extension

Enrichment 98: Extension (for use after Chapter 9)
Inflation — What a Little Percent Can Do

When inflation occurs prices go up and a dollar no longer buys what it once did. Inflation is measured in percent. For example, a 7% annual inflation rate means that the price of an average item goes up 7% in a year. A $1.00 item will cost $1.07 a year later. The new price of an item would be 107% of the old price.

1. If the inflation rate is 7% a year, what will a $10,500 car cost next year? **$11,235**

2. Suppose the inflation rate stays at 7% per year for two years. To find the price of the same car after the second year of inflation, apply the 7% increase to the price after the first year.

   a. $10,500(1.07)(1.07) = \underline{\quad?\quad}$ **$12,021.45**

   b. Suppose the 7% annual inflation lasts for 10 years. Can you think of a way to use exponents to write an expression for the cost of the car after ten years? $10,500(1.07)^{10}$

   c. What would the car cost after 10 years? **$20,655.09**

   d. Are you surprised at the price? Explain. **Answers will vary.**

3. Suppose you graduate from college as a mathematician and get a job for $28,000 a year. Your company says you will get an annual raise equal to 4% plus the inflation rate as long as you work there.

   a. What percent will your annual raise be if inflation is 7%? **11%**

   b. Suppose you stay with the company for 10 years. What will your salary be after 10 years of 7% inflation? **$79,504 a year**

   c. What would your salary be after 20 years? **$225,745 a year**

   d. Are you surprised at the amount of your salary in part (c)? Explain. **Answers will vary.**

Resource Book, MATHEMATICAL CONNECTIONS
Copyright © by Houghton Mifflin Company. All rights reserved.
315

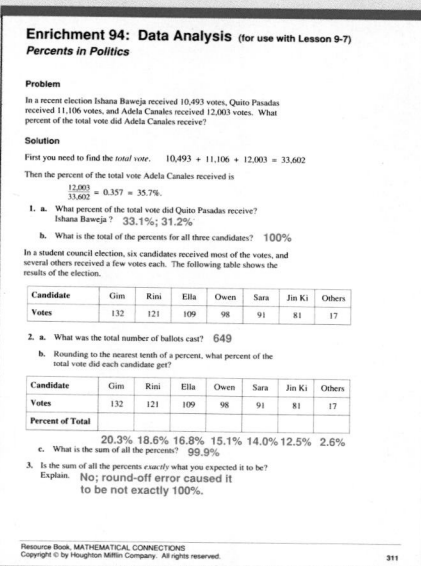
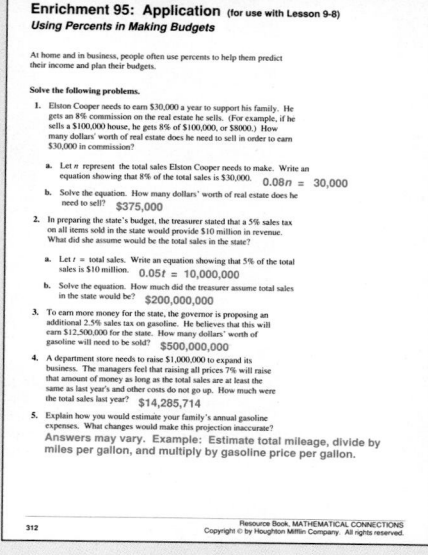

386F

Multicultural Notes

For Chapter 9 Opener
The pleasing dimensions of a golden rectangle show up frequently in architecture. The Italian architect Andrea Palladio (1508–1580) employed the proportions of a golden rectangle in the floor plan of the *Chièsa del Redentóre* (Church of the Redeemer) in Venice, Italy. The French architect Le Corbusier (1887–1965) also used golden rectangle proportions in many of his buildings, including the *Unite d'habitation* in Marseilles, France. The floor plan of the East Treasure House of the Ise Shrine in Ise, Japan, also has the proportions of a golden rectangle.

For Page 394
Students working on Exercise 37 may be interested to learn that *bonsai* means "planted in a tray" in Japanese. The art of training ordinary trees and shrubs to grow in miniature originated in China and Japan over a thousand years ago. Typical bonsai trees can range in size from 2 inches to 2 feet.

Research Activities

For Chapter 9 Opener
Students can estimate to predict whether rectangles found in familiar furniture and objects, or in scale drawings of buildings, are golden rectangles. They can then measure the length and width of the rectangles and calculate the ratio between the sides.

For Page 406
The postage stamps in the illustration on this page feature traditional headdresses of some Native American groups. You may want to have students work cooperatively to research the language, customs, and traditions of one of these Native American groups, the location of the group's traditional homeland, and where members of the group live today. Students can share their research with the entire class.

Construction of a Spiral Based on the Golden Rectangle

Golden Rectangle

x

x y

$$\frac{x+y}{x} = \frac{x}{y} = 1.618$$

Ratio, Proportion, and Percent 9

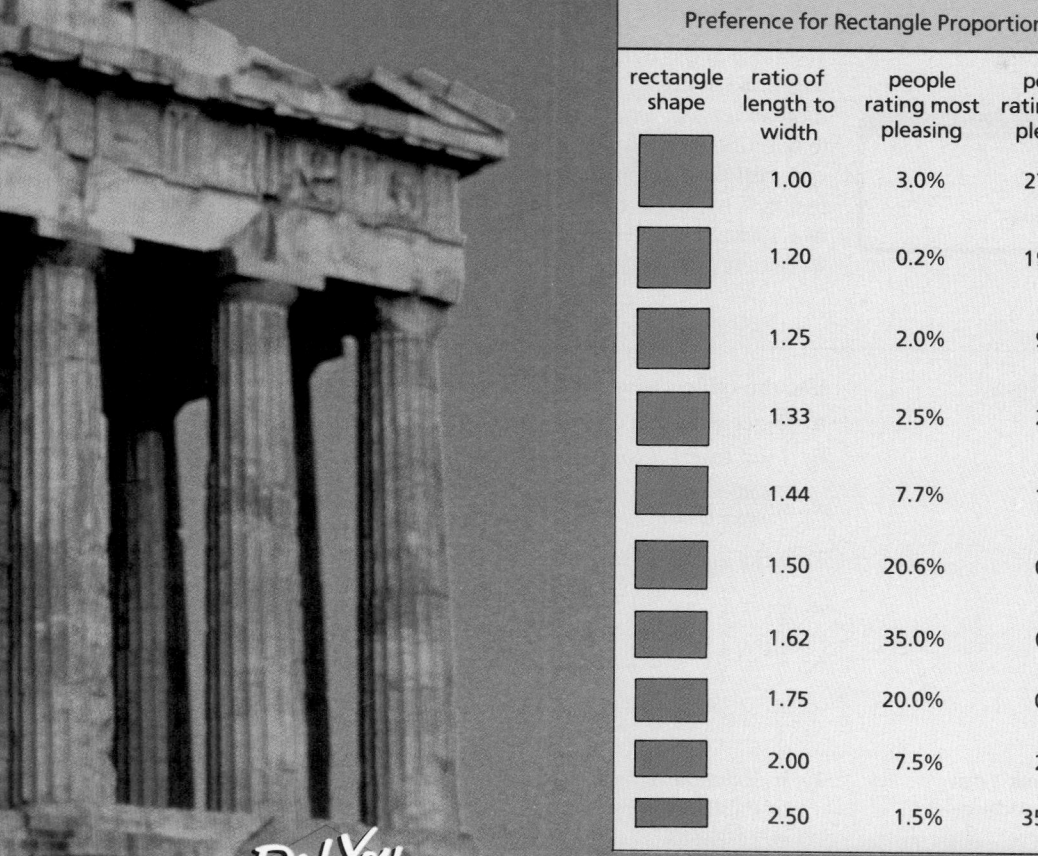

Preference for Rectangle Proportions			
rectangle shape	ratio of length to width	people rating most pleasing	people rating least pleasing
	1.00	3.0%	27.8%
	1.20	0.2%	19.7%
	1.25	2.0%	9.4%
	1.33	2.5%	2.5%
	1.44	7.7%	1.2%
	1.50	20.6%	0.4%
	1.62	35.0%	0.0%
	1.75	20.0%	0.8%
	2.00	7.5%	2.5%
	2.50	1.5%	35.7%

Did You Know?

Ancient philosophers believed, and modern psychologists have demonstrated, that certain shapes are more pleasing to the eye than others. One such shape is the **golden rectangle**, whose ratio of length to width is about 1.618. For thousands of years, the golden rectangle has been used in art and architecture. The front of the Parthenon, a Greek temple built between 447 and 432 B.C., was designed to fit inside a golden rectangle.

387

Suggested Resources

Doczi, Györgi. *The Power of Limits: Proportional Harmonies in Nature, Art and Architecture.* Boulder, CO: Shambhala Press, 1981. (For Teacher)

Waldman, Carl. *Encyclopedia of Native American Tribes.* New York: Facts on File Publications, 1988.

Teaching the Lesson
- Materials: colored tiles, overhead projector, bingo chips
- Lesson Starter 80
- Using Technology, p. 386C
- Using Manipulatives, p. 386C
- Reteaching/Alternate Approach, p. 386D
- Toolbox Skill 10

Lesson Follow-Up
- Practice 109
- Enrichment 88: Communication
- Study Guide, pp. 159–160
- Manipulative Activity 17

Warm-Up Exercises

Write each fraction in lowest terms.

1. $\frac{6}{8}$ $\frac{3}{4}$

2. $\frac{75}{100}$ $\frac{3}{4}$

3. $\frac{10}{12}$ $\frac{5}{6}$

4. $\frac{15}{45}$ $\frac{1}{3}$

Lesson Focus

Ask students if they have ever heard people talking about speed limits. A speed limit, such as 55 mph, is a comparison, or ratio, of a distance to a time; this ratio is called a *rate*. The focus of today's lesson is ratios and rates.

Teaching Notes

Key Questions

1. What is the difference between a ratio and a rate? A ratio is a comparison of two numbers by division. A rate is a special type of ratio that compares two unlike quantities.

9-1   Ratio and Rate

Objective: To write ratios as fractions in lowest terms and to write unit rates.

DATA ANALYSIS

Terms to Know
- *ratio*
- *rate*
- *unit rate*

The table at the right lists the win-loss records of four teams in a football league. You can use *ratios* to describe information in the chart.

A **ratio** is a comparison of two numbers by division. The ratio of two numbers a and b ($b \neq 0$) can be written in three ways.

$$a \text{ to } b \qquad a:b \qquad \frac{a}{b}$$

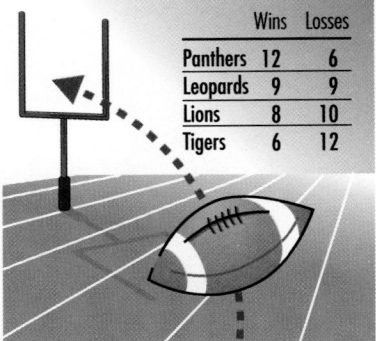

	Wins	Losses
Panthers	12	6
Leopards	9	9
Lions	8	10
Tigers	6	12

Example 1

Use the table above. Write each ratio in lowest terms.

a. Tiger wins to Tiger losses

b. Lion losses : Lion wins

c. $\dfrac{\text{Panther wins}}{\text{Panther losses}}$

Solution

To write a ratio in lowest terms, write it as a fraction in lowest terms.

a. $\dfrac{6}{12} = \dfrac{1}{2}$ or $1:2$ or   1 to 2

b. $\dfrac{10}{8} = \dfrac{5}{4}$ ◄—— Do not write as a mixed number.

c. $\dfrac{12}{6} = \dfrac{2}{1}$ ◄—— Keep the 1 in the denominator.

 **Check Your Understanding**

1. In Example 1, is the ratio of Panther wins to losses the same as the ratio of Panther losses to wins? Explain.
See *Answers to Check Your Understanding* at the back of the book.

When the quantities being compared are in different units, first write the quantities in the same unit and then write the ratio.

Example 2

Write the ratio as a fraction in lowest terms: 21 days to 6 weeks

Solution

Change 21 days to weeks: 7 days = 1 week, so 21 days = 3 weeks

$$\dfrac{3 \text{ weeks}}{6 \text{ weeks}} \rightarrow \dfrac{3}{6} = \dfrac{1}{2}$$ The ratio of 21 days to 6 weeks is $\frac{1}{2}$.

Check Your Understanding

2. In Example 2, is the answer the same when you write the number of weeks as a number of days? Explain.
See *Answers to Check Your Understanding* at the back of the book.

---

388   Chapter 9

A ratio that compares two unlike quantities is called a **rate**. A **unit rate** is a rate for *one* unit of a given quantity. An example of a unit rate is *miles per hour*, which indicates the number of miles for one hour.

Example 3

Write the unit rate.

a. 150 mi in 3 h
b. $42 for 4 shirts

Solution

a. $\dfrac{\text{miles}}{\text{hours}} \rightarrow \dfrac{150}{3} = \dfrac{50}{1}$ The rate is 50 mi in 1 h, or 50 mi/h.

b. $\dfrac{\text{dollars}}{\text{shirts}} \rightarrow \dfrac{42}{4} = \dfrac{10.5}{1}$ The rate is $10.50 for 1 shirt, or $10.50/shirt.

Guided Practice

COMMUNICATION « *Writing*

« **1.** Write the ratio 5 *to* 6 in two other ways. 5:6; $\frac{5}{6}$

« **2.** Write the unit rate in words: 22 mi/gal 22 miles per gallon

« **3.** Write the unit rate in symbols: 53 words per minute 53 words/min

Replace each __?__ with the correct number.

4. 2 ft = __?__ in. 24
5. 3 h = __?__ min 180
6. __?__ years = 730 days 2
7. __?__ yd = 21 ft 7

Write each ratio as a fraction in lowest terms.

8. $\frac{14}{7}$ $\frac{2}{1}$
9. $\frac{36}{15}$ $\frac{12}{5}$
10. 2 h : 1 day $\frac{1}{12}$

11. 5 in. to 2 ft $\frac{5}{24}$
12. $2 to $.30 $\frac{20}{3}$
13. 4 yd : 4 ft $\frac{3}{1}$

Write the unit rate.

14. $45 for 3 lb $15/lb
15. 150 words in 3 min 50 words/min
16. 160 mi in 4 h 40 mi/h

Exercises

Write each ratio as a fraction in lowest terms.

A   **1.** 18 to 24 $\frac{3}{4}$
2. 2:32 $\frac{1}{16}$
3. 35:21 $\frac{5}{3}$
4. 49 to 14 $\frac{7}{2}$

5. $\frac{36}{18}$ $\frac{2}{1}$
6. $\frac{14}{14}$ $\frac{1}{1}$
7. 40 to 25 $\frac{8}{5}$
8. 100:75 $\frac{4}{3}$

9. 28 days to 2 weeks $\frac{2}{1}$
10. 2 m to 40 cm $\frac{5}{1}$
11. 9 in. : 2 ft $\frac{3}{8}$

12. $.50 : $3 $\frac{1}{6}$
13. 12 oz to 1 lb $\frac{3}{4}$
14. 5 yd : 10 ft $\frac{3}{2}$

Ratio, Proportion, and Percent   **389**

Have students write a paragraph
giving definitions and examples of
ratios and rates.

Suggested Assignments

Skills: 1–22, 23–29 odd, 35–38
Average: 1–29 odd, 35–38
Advanced: 1–33 odd, 35–38

Exercise Notes

Data Analysis

*Exercises 27–30 check students'
ability to read and use information
from a pictograph.*

Application

*Exercises 31–34 require students to
apply their knowledge of ratios and
rates to making consumer purchases. Exercise 34 involves comparing actual prices from a grocery
store.*

Follow-Up

Enrichment

Enrichment 88 is pictured on
page 386E.

Practice

Practice 109, *Resource Book*

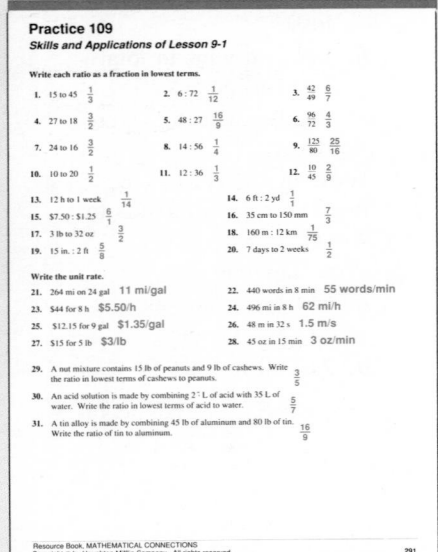

Write the unit rate.

15. 300 mi in 6 h 50 mi/h   **16.** 205 words in 5 min 41 words/min   **17.** 250 mi on 10 gal 25 mi/gal

18. \$42 for 5 h \$8.40/h   **19.** 33 m in 15 s 2.2 m/s   **20.** \$6 for 12 L \$.50/L

21. Sean's chemistry book describes salt water that is made by combining 1 c salt with 9 c water. Write the ratio of water to salt as a fraction in lowest terms. $\frac{9}{1}$

22. DATA, *pages 2–3* Write the ratio of Houston Symphony Pops concerts to Mozart concerts as a fraction in lowest terms. $\frac{7}{6}$

Write each ratio as a fraction in lowest terms.

B **23.** 10x to 20y $\frac{x}{2y}$ **24.** 15y to 5w $\frac{3y}{w}$ **25.** 12n : 4n $\frac{3}{1}$ **26.** $\frac{8t}{6t}$ $\frac{4}{3}$

DATA ANALYSIS

Use the pictograph at the right.

January Car Sales

Compact, Mid Size, Full Size, Luxury

= 2 cars

27. How many cars were sold in January? 27

28. How many more full size cars were sold than luxury cars? 7

29. Write the ratio of *compact cars sold* to *mid size cars sold* as a fraction in lowest terms. $\frac{5}{3}$

30. Write the ratio of *full size cars sold* to *luxury cars sold* as a fraction in lowest terms. $\frac{9}{2}$

PROBLEM SOLVING/APPLICATION
31. 12 pens for \$4.80   **32.** 4 shirts for \$32.50

When shopping, you may find the same items packaged in various sizes or quantities. The item with the lesser unit price is usually the better buy.

C **31.** Find the better buy: 8 pens for \$3.60 or 12 pens for \$4.80.

32. Find the better buy: 3 shirts for \$25 or 4 shirts for \$32.50.

33. Describe a situation in which it may *not* be preferable to buy an item with the lesser unit price. See answer on p. A24.

34. Go to a grocery store and find the prices for different quantities of the same item. Determine the better buy. Answers will vary.

SPIRAL REVIEW

S **35.** Solve: $\frac{2}{3}x = 16$ *(Lesson 8-9)* 24

36. Write the unit rate: \$62.50 for 5 h   *(Lesson 9-1)* \$12.50/h

37. Write the prime factorization of 240.   *(Lesson 7-1)* $2^4 \cdot 3 \cdot 5$

38. Solve: $2x + 7 = -17$ *(Lesson 4-5)* -12

9-2 Proportions

Objective: To solve proportions.

Terms to Know

• *proportion*
• *terms*
• *cross products*
• *solve a proportion*

EXPLORATION

1. Is the statement $\frac{3}{4} = \frac{9}{12}$ *true* or *false*? true

2. Complete: $3(12) =$? ; $4(9) =$? $\frac{3}{4} \bowtie \frac{9}{12}$
   36; 36

3. Is the statement $\frac{3}{12} = \frac{4}{9}$ *true* or *false*? false

4. Complete: $3(9) =$? ; $12(4) =$? $\frac{3}{12} \bowtie \frac{4}{9}$
   27; 48

5. Compare your answers in Steps 1 and 2 with your answers in Steps 3 and 4. Complete.

   Two ratios are equal if the product of the first numerator and the second denominator equals the product of the first ? and the second ? .
   denominator; numerator

A **proportion** is a statement that two ratios are equal.

  You write: $\frac{3}{4} = \frac{9}{12}$ or $3:4 = 9:12$

  You read: *3 is to 4 as 9 is to 12.*

The numbers 3, 4, 9, and 12 are the **terms** of the proportion. If a statement is a true proportion, the **cross products** of the terms are equal.

In Arithmetic	In Algebra
$\frac{3}{4} = \frac{9}{12}$	$\frac{a}{b} = \frac{c}{d}$
$3(12) = 4(9)$	$ad = bc,\ b \neq 0,\ d \neq 0$
$36 = 36$	

Example 1

Tell whether each proportion is *True* or *False*.

a. $\frac{6}{9} = \frac{8}{12}$ b. $\frac{4}{6} = \frac{0.9}{1.3}$

Solution

a. $\frac{6}{9} \bowtie \frac{8}{12}$ b. $\frac{4}{6} \bowtie \frac{0.9}{1.3}$

  $6(12) \stackrel{?}{=} 9(8)$ $4(1.3) \stackrel{?}{=} 6(0.9)$
  $72 \stackrel{?}{=} 72$ $5.2 \stackrel{?}{=} 5.4$
  True   False

 **Check Your Understanding**

1. Would the proportion in Example 1(b) be true if 1.3 were 1.35? Explain.
   See *Answers to Check Your Understanding* at the back of the book.

Ratio, Proportion, and Percent   **391**

Lesson Planner

Teaching the Lesson
• Materials: gumdrops, pennies, algebra tiles
• Lesson Starter 81
• Using Technology, p. 386C
• Using Manipulatives, p. 386C
• Reteaching/Alternate Approach, p. 386D
• Cooperative Learning, p. 386D
• Toolbox Skills 9, 10

Lesson Follow-Up
• Practice 110
• Enrichment 89: Connection
• Study Guide, pp. 161–162

Warm-Up Exercises

Write an equivalent fraction for each of the following. Answers will vary. Examples are given.

1. $\frac{1}{2}$ $\frac{2}{4}$

2. $\frac{3}{4}$ $\frac{6}{8}$

3. $\frac{5}{6}$ $\frac{10}{12}$

4. $\frac{a}{b}$ $\frac{2a}{2b}$

Lesson Focus

You may wish to use the following Application to motivate today's lesson on proportions.

Application

Suppose John is saving all his dimes to buy a new tennis racquet that costs $30. How many dimes must he save? The problem can be solved by using the following proportion.

$$\frac{10 \text{ dimes}}{1 \text{ dollar}} = \frac{? \text{ dimes}}{30 \text{ dollars}}$$

300 dimes

Teaching Notes

Use the Exploration at the beginning of the lesson to lead students to an understanding of proportions. Students should observe that the Exploration exercises used proportions since they involve equivalent ratios.

Key Questions

1. What is the difference between a ratio and a proportion? A ratio is a comparison of two numbers by division; a proportion shows two ratios that are equal.
2. What does ''solving the proportion'' mean? It means finding the value of the variable that makes the proportion true.

Using Technology

For a suggestion on using a calculator, see page 386C.

Using Manipulatives

For a suggestion on using gumdrops and pennies to model ratios, see page 386C.

Reteaching/Alternate Approach

For a suggestion on using a cooperative learning activity, see page 386D.

Additional Examples

1. Tell whether each proportion is *True* or *False*.
   a. $\frac{2}{4} = \frac{0.8}{1.1}$ False
   b. $\frac{2}{4} = \frac{13}{26}$ True

2. Solve each proportion.
   a. $\frac{b}{15} = \frac{5}{3}$ $b = 25$
   b. $\frac{10}{8} = \frac{a}{6}$ $a = 7.5$

Sometimes one of the terms of a proportion is a variable. To **solve a proportion,** you find the value of the variable that makes the proportion true. You can use cross products when you solve a proportion.

Example 2

Solve each proportion.

a. $\frac{n}{6} = \frac{3}{2}$

b. $\frac{6}{4} = \frac{b}{10}$

Solution

a.  $\frac{n}{6} \diagdown \frac{3}{2}$

$2n = 6(3)$
$2n = 18$
$\frac{2n}{2} = \frac{18}{2}$
$n = 9$

b. $\frac{6}{4} \diagdown \frac{b}{10}$

$6(10) = 4b$
$60 = 4b$
$\frac{60}{4} = \frac{4b}{4}$
$15 = b$

✓ **Check Your Understanding**

2. In Example 2(b), would the solution be different if the proportion were $\frac{4}{6} = \frac{10}{b}$? Explain.
   See *Answers to Check Your Understanding* at the back of the book.

A calculator may be helpful in solving proportions. For instance, to solve the proportion $\frac{35}{10} = \frac{n}{7}$, you can use this key sequence.

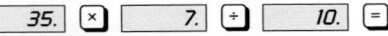

This sequence solves the equation $10n = 35(7)$, which results from the cross products of the proportion. First multiply 35 by 7. Then divide by 10 to find n.

Guided Practice

COMMUNICATION « *Reading*

« **1.** What is the main idea of the lesson? A proportion is a statement that two ratios are equal.
« **2.** List two major points that support the main idea of the lesson. See answer on p. A24.

COMMUNICATION « *Writing*

« **3.** Write *five is to six as ten is to twelve* in symbols. $\frac{5}{6} = \frac{10}{12}$, or 5:6 = 10:12
« **4.** Write *three is to one as fifteen is to five* in symbols. $\frac{3}{1} = \frac{15}{5}$, or 3:1 = 15:5
« **5.** Write the proportion $\frac{6}{9} = \frac{2}{3}$ in words. Answers to Guided Practice 5 and 6 are on p. A24.
« **6.** Write the proportion $\frac{c}{4} = \frac{10}{40}$ in words.

Write the cross products of the terms.

7. $\frac{32}{8} = \frac{20}{5}$
160; 160

8. $\frac{24}{18} = \frac{40}{30}$
720; 720

9. $\frac{25}{40} = \frac{n}{8}$
200; 40n

10. $\frac{c}{2} = \frac{168}{12}$
12c; 336

392 Chapter 9

Tell whether each proportion is *True* or *False*.

11. $\frac{5}{8} = \frac{15}{24}$ True   **12.** $\frac{12}{4} = \frac{4}{12}$ False   **13.** $\frac{18}{9} = \frac{6}{1}$ False   **14.** $\frac{10}{15} = \frac{2.2}{3.3}$
True

Solve each proportion.

15. $\frac{3}{8} = \frac{b}{24}$ 9   **16.** $\frac{16}{8} = \frac{6}{n}$ 3   **17.** $\frac{3}{a} = \frac{5}{10}$ 6   **18.** $\frac{y}{4} = \frac{2.1}{1.4}$ 6

Exercises

Tell whether each proportion is *True* or *False*.

True

A **1.** $\frac{21}{24} = \frac{7}{8}$ True   **2.** $\frac{10}{12} = \frac{3}{4}$ False   **3.** $\frac{8}{6} = \frac{6}{8}$ False   **4.** $\frac{8}{10} = \frac{12}{15}$

5. $\frac{12}{6} = \frac{5}{3}$ False   **6.** $\frac{25}{6} = \frac{8}{2}$ False   **7.** $\frac{1.5}{3} = \frac{10}{20}$ True   **8.** $\frac{50}{100} = \frac{3.2}{6.4}$
True

Solve each proportion.

9. $\frac{3}{2} = \frac{9}{n}$ 6   **10.** $\frac{12}{b} = \frac{4}{6}$ 18   **11.** $\frac{22}{10} = \frac{m}{5}$ 11   **12.** $\frac{a}{5} = \frac{9}{3}$ 15

13. $\frac{1.2}{1.5} = \frac{y}{10}$ 8   **14.** $\frac{2.5}{5} = \frac{c}{8}$ 4   **15.** $\frac{6}{z} = \frac{4}{6}$ 9   **16.** $\frac{7}{p} = \frac{1}{7}$ 49

17. Is $\frac{4}{9} = \frac{36}{81}$ a true proportion? Explain. Yes; (4)(81) = (9)(36)

18. A number n is to 200 as 5 is to 4. Write a proportion and solve for n.
$\frac{n}{200} = \frac{5}{4}$; $n = 250$

Solve each proportion.

B **19.** $\frac{14}{2a} = \frac{6}{9}$ 10.5   **20.** $\frac{3w}{5} = \frac{24}{10}$ 4   **21.** $\frac{n}{25} = \frac{4}{n}$ 10   **22.** $\frac{d}{9} = \frac{4}{d}$
(or −10)   6 (or −6)

23. Solve $\frac{a}{b} = \frac{c}{d}$ for c when $a = 8$, $b = 12$, and $d = 18$. 12

24. Solve $\frac{w}{x} = \frac{y}{z}$ for z when $w = 20$, $x = 8$, and $y = 15$. 6

Solve each proportion. (Use the distributive property.)

25. $\frac{3}{2} = \frac{9}{n+2}$ 4   **26.** $\frac{z-5}{8} = \frac{3}{4}$ 11   **27.** $\frac{6}{y-4} = \frac{3}{5}$ 14   **28.** $\frac{2}{3} = \frac{6+n}{12}$
2

 CALCULATOR

Choose the key sequence you would use to solve each proportion.

29. $\frac{30}{19} = \frac{m}{57}$ C   **A.** [30.] [×] [19.] [÷] [57.]

30. $\frac{30}{57} = \frac{m}{19}$ A   **B.** [30.] [×] [19.] [×] [57.]

   C. [30.] [×] [57.] [÷] [19.]

31. $\frac{30}{57} = \frac{19}{m}$ D   **D.** [57.] [×] [19.] [÷] [30.]

Ratio, Proportion, and Percent **393**

Making Connections/Transitions

Exercises 32–35 connect arithmetic and algebra by making use of variables to solve proportions involving direct variation. In *Exercises 36 and 37*, students solve problems by translating their conditions into a proportion.

Follow-Up

Extension

Have each student create a problem that involves a direct proportion. The problem should relate to something students do each day. For example: I walk to school every day. The school is 0.8 mi from my house. How many miles a week do I walk to and from school?

$$\frac{1.6 \text{ mi}}{1 \text{ day}} = \frac{x \text{ mi}}{5 \text{ days}}; x = 8 \text{ mi}$$

Enrichment

Enrichment 89 is pictured on page 386E.

Practice

Practice 110, *Resource Book*

Practice 110
Skills and Applications of Lesson 9-2

Tell whether each proportion is *True or False.*

1. $\frac{10}{15} = \frac{2}{3}$ True. 2. $\frac{18}{24} = \frac{2}{3}$ False. 3. $\frac{24}{5} = \frac{4}{1}$ False.

4. $\frac{8}{4} = \frac{3}{1}$ False. 5. $\frac{25}{75} = \frac{1}{3}$ True. 6. $\frac{21}{6} = \frac{7}{2}$ True.

7. $\frac{15}{9} = \frac{4}{3}$ False. 8. $\frac{13}{26} = \frac{1}{2}$ True. 9. $\frac{8}{24} = \frac{1}{3}$ True.

10. $\frac{16}{12} = \frac{4}{3}$ True. 11. $\frac{2.5}{0.5} = \frac{5}{2}$ False. 12. $\frac{2.7}{1.8} = \frac{3}{2}$ True.

13. $\frac{26}{14} = \frac{2}{1}$ False. 14. $\frac{24}{21} = \frac{8}{7}$ True. 15. $\frac{12}{16} = \frac{3}{5}$ False.

Solve each proportion.

16. $\frac{4}{3} = \frac{16}{x}$ 12 17. $\frac{8}{x} = \frac{2}{5}$ 20 18. $\frac{5}{35} = \frac{d}{7}$ 1

19. $\frac{h}{6} = \frac{6}{36}$ 1 20. $\frac{5}{3} = \frac{15}{c}$ 9 21. $\frac{21}{m} = \frac{3}{2}$ 14

22. $\frac{2}{9} = \frac{t}{18}$ 4 23. $\frac{r}{24} = \frac{3}{2}$ 36 24. $\frac{1.5}{x} = \frac{4.5}{1.2}$ 0.4

25. $\frac{n}{25} = \frac{0.6}{3}$ 5 26. $\frac{4.8}{0.4} = \frac{120}{p}$ 10 27. $\frac{81}{27} = \frac{r}{0.9}$ 2.7

28. $\frac{6}{15} = \frac{2}{n}$ 5 29. $\frac{a}{65} = \frac{2}{5}$ 26 30. $\frac{36}{27} = \frac{3}{3}$ 4

31. Is $\frac{5}{6} = \frac{25}{30}$ a true proportion? Explain. Yes; 5(30) = 6(25)

32. Is $\frac{4}{5} = \frac{16}{21}$ a true proportion? Explain. No; 4(21) ≠ 5(16)

33. A number b is to 2 as 6 is to 7. Write a proportion and solve for b.

34. A number n is to 9 as 5 is to 45. Write a proportion and solve for n.

33. $\frac{b}{2} = \frac{6}{7}; b = 1\frac{5}{7}$ 34. $\frac{n}{9} = \frac{5}{45}; n = 1$

CONNECTING ARITHMETIC AND ALGEBRA

A *direct variation* is a relationship between two variables in which one variable changes at the same rate as the other. The ratio of the two variables is always the same. If the variables are x and y, you can write this relationship as the proportion

$$\frac{y_1}{x_1} = \frac{y_2}{x_2},$$

where y_1 and y_2 are different values of y, and x_1 and x_2 are different values of x. You say that y is *directly proportional* to x.

Use the proportion above.

32. Find y_2 when $x_1 = 10$, $x_2 = 25$, and $y_1 = 16$. 40

33. Find y_1 when $x_1 = 2$, $x_2 = 3$, and $y_2 = 27$. 18

34. Find x_2 when $x_1 = 14$, $y_1 = 24$, and $y_2 = 60$. 35

35. Find x_1 when $x_2 = 8$, $y_1 = 3.9$, and $y_2 = 6$. 5.2

C 36. An employee's wages are directly proportional to the number of hours worked. One employee earns $120 for working 5 h. How much will that employee earn for working 8 h? $192

37. The length of the shadow of a tree is directly proportional to the height of the tree. A bonsai tree that is 5 in. tall casts a shadow 3.5 in. long. How long is the shadow of a tree that is 20 ft tall? 14 ft

SPIRAL REVIEW

S 38. Find the complement of an angle with measure 62°. *(Lesson 6-3)* 28°

39. Simplify: $\frac{8x^2y^3}{2xyz}$ *(Lesson 7-5)* $\frac{4xy^2}{z}$

40. Solve the proportion: $\frac{9}{16} = \frac{a}{32}$ *(Lesson 9-2)* 18

41. Find the product: $-\frac{8}{27} \cdot \frac{3}{4}$ *(Lesson 8-1)* $-\frac{2}{9}$

42. Estimate the quotient: 3404 ÷ 43 *(Toolbox Skill 3)* about 80

Bonsai is the art of growing small, ornamentally shaped trees in shallow containers.

394 Chapter 9

9-3

Scale Drawings

Objective: To interpret and use scale drawings.

APPLICATION

Scale drawings are drawings that represent real objects. The **scale** is the ratio of the size of the drawing to the actual size of the object. You can use proportions to find the actual measurements of an object when you have a scale drawing.

Below is the scale drawing of a school cafeteria. The scale is 1 in. : 16 ft.

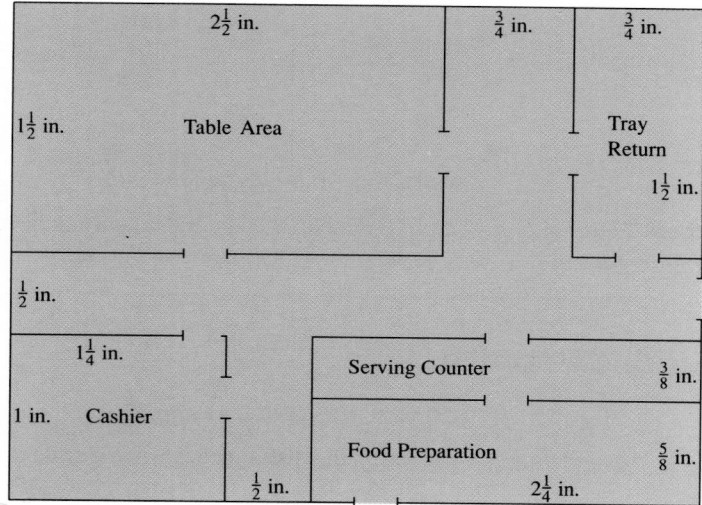

Example 1

Solution

Find the actual length of the tray return area in the scale drawing above.

Let a represent the actual length of the tray return area in feet. Use the scale to write a proportion.

$$\frac{\text{scale drawing length (in.)}}{\text{actual length (ft)}} \;\rightarrow\; \frac{1}{16} = \frac{1\frac{1}{2}}{a}$$

$$1a = 16\left(1\frac{1}{2}\right)$$

$$a = 16\left(\frac{3}{2}\right) = 24$$

The actual length of the tray return area is 24 ft.

 **Check Your Understanding**

1. In Example 1, why is the numerator $1\frac{1}{2}$, not $\frac{3}{4}$?

See *Answers to Check Your Understanding* at the back of the book.

Warm-Up Exercises

Solve each proportion.

1. $\frac{a}{8} = \frac{10}{32}$ 2.5

2. $\frac{1.25}{10} = \frac{x}{40}$ 5

3. $\frac{1.25}{75} = \frac{2.30}{x}$ 138

Lesson Focus

Ask students if they have ever seen a set of blueprints for a building. Many objects are too large to draw their actual size. These objects are drawn using scale drawings, the focus of today's lesson.

Teaching Notes

Use the Application at the beginning of the lesson to illustrate the meaning and purpose of a scale drawing. Work through the Examples, showing how proportions are used to solve for unknown lengths. Stress the need to read the drawings accurately.

Ratio, Proportion, and Percent   **395**

Teaching Notes (continued)

Key Questions

1. What is the "scale" used in a drawing? the ratio of the size of the drawing to the size of the actual object

2. Why are scale drawings useful? Large objects can be represented by a drawing, and their actual measurements can be found by using proportions.

Error Analysis

Students often write the wrong measurement when reading it from a scale drawing. Spend some time with students on how to read drawings accurately.

Critical Thinking

When setting up proportions, students must *associate* the appropriate measurements with the correct units. Also, they must *realize* that while the order chosen for the terms of each ratio does not matter, it must be the same for both ratios.

Using Technology

For a suggestion on using scale-drawing software, see page 386C.

Using Manipulatives

For a suggestion on using scale models, see page 386C.

Additional Examples

1. In the scale drawing on page 395, what is the actual length of table area in the drawing? 40 ft

2. A floor plan for a room is drawn with a scale of 2 in.:12 ft. The length of the room in the drawing is shown as $2\frac{1}{2}$ in. What is the actual length of the room? 15 ft

---

Manufacturers often build models of their products before making the actual product. A **scale model** has the same shape but is usually smaller than the actual object it represents. You can use proportions to find the dimensions of scale models.

Example 2

A car model is being built with a scale of 2 in.:5 ft. The actual length of the car is 12 ft. What is the length of the model?

Solution

Let m represent the length of the model. Use the scale to write a proportion.

$$\frac{\text{model length (in.)}}{\text{actual length (ft)}} \rightarrow \quad \frac{2}{5} = \frac{m}{12}$$

$$2(12) = 5m$$

$$24 = 5m$$

$$\frac{24}{5} = \frac{5m}{5}$$

$$4\frac{4}{5} = m$$

The length of the model is $4\frac{4}{5}$ in.

Check Your Understanding

2. In Example 2, could the proportion have been written $\frac{5}{2} = \frac{12}{m}$? Explain.
See *Answers to Check Your Understanding* at the back of the book.

Guided Practice

COMMUNICATION «Reading

« **1.** Explain what it means when a scale drawing shows a scale of 1 in.:10 ft. 1 in. on the drawing represents 10 ft on the actual object.

« **2.** The following shows the scale on a drawing:
What scale does this represent? 1 in.:1 ft
    1 ft

COMMUNICATION «Writing

« **3.** For a scale drawing of a house, write a scale that shows "every 2 in. represents 15 ft." 2 in.:15 ft

« **4.** For a scale model of a statue, write a scale that shows "every centimeter represents 12 m." 1 cm:12 m

Use the scale drawing of the school cafeteria on page 395. Write a proportion you could use to find each actual dimension. Do not solve.

5. $\frac{1}{16} = \frac{1\frac{1}{4}}{a}$

6. $\frac{1}{16} = \frac{\frac{3}{4}}{a}$

7. $\frac{1}{16} = \frac{\frac{3}{8}}{a}$

8. $\frac{1}{16} = \frac{2\frac{1}{2}}{a}$

5. the length of the cashier's room

6. the width of the tray return area

7. the width of the serving counter

8. the length of the table area

Use the scale drawing of the school cafeteria on page 395. Find each actual dimension.

9. the width of the corridor outside the serving counter 8 ft

10. the length of the cashier's room 20 ft

11. the length of the serving counter 36 ft

The scale of a car model is 3 in.: 10 ft. Find each dimension of the model.

12. The actual width of the seat is 2 ft. $\frac{3}{5}$ in.

13. The actual height of the tire is 3 ft. $\frac{9}{10}$ in.

14. The actual height of the window is $1\frac{1}{4}$ ft. $\frac{3}{8}$ in.

Exercises

Use the scale drawing below of a basketball court. The scale is 1 in.: 25 ft. Find each actual dimension.

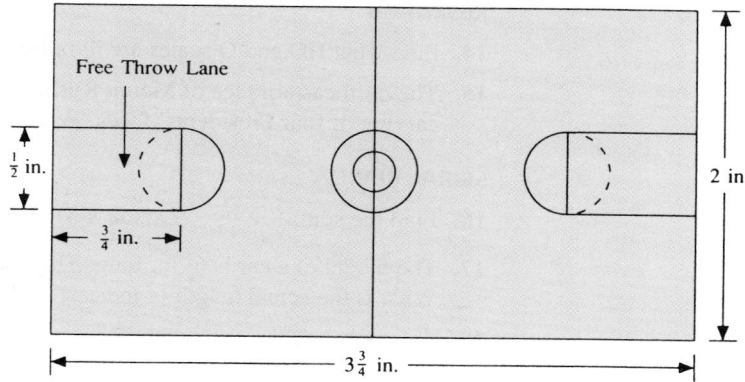

2. $12\frac{1}{2}$ ft

3. $18\frac{3}{4}$ ft

4. $93\frac{3}{4}$ ft

A

1. the width of the court 50 ft

2. the width of the free throw lane

3. the length of the free throw lane

4. the length of the court

5. A kitchen is 13 ft wide and 14 ft long. The scale of a floor plan of the house containing this kitchen is 1 in.: 8 ft. Find the width and the length of the kitchen on the floor plan. $1\frac{5}{8}$ in. wide; $1\frac{3}{4}$ in. long

6. A scale model of a building is constructed using the scale 1 in.: 20 ft. The actual height of the building is 240 ft. Find the height of the scale model. 12 in.

7. A model train is built using the scale 1: 60. The actual length of the dining car is 2100 cm. Find the length of the dining car of the model train. 35 cm

Closing the Lesson

Ask students to summarize the lesson by giving an important consideration when constructing a scale drawing, and when calculating a measurement from a scale drawing.

Suggested Assignments

Skills: 1–13, 16–18
Average: 1–13, 16–18
Advanced: 1–18

Exercise Notes

Application

Guided Practice Exercises 5–11 and Exercises 1–4 have students use scale drawings to write proportions and find actual dimensions.

Cooperative Learning

Exercise 11 provides an opportunity for students to create their own scale drawings by measuring a room.

Research

Exercises 14 and 15 present two situations that can enhance students' understanding of "scales" used in constructions.

Follow-Up

Exploration

Have students use different maps of the United States that have different scales. Using the map scales, students should calculate the approximate distances between at least three given pairs of cities. The results should be displayed in a chart. Students will observe that the calculated distances are approximately the same, even though the scales are different.

Enrichment

Enrichment 90 is pictured on page 386E.

Practice

Practice 111, *Resource Book*

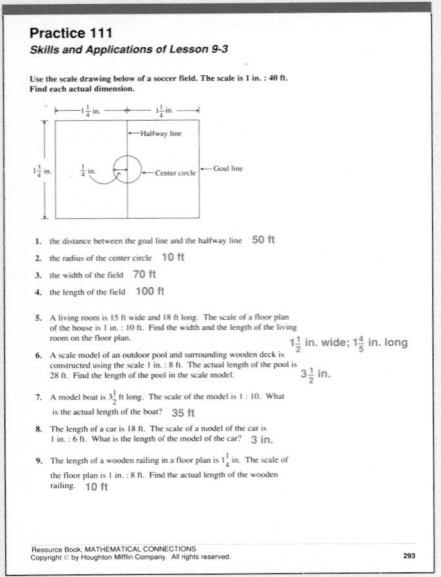

8. The model of a jet plane is $2\frac{1}{2}$ ft long. The scale of the model is $1:40$. What is the actual length of the plane? 100 ft

B 9. The length of the kitchen in a floor plan is 6 in. The actual length of the kitchen is 15 ft. What is the scale of the floor plan? 2 in.:5 ft

10. The wheelbase of a new car is 7 ft. The wheelbase on a model of the car is 2 in. What is the scale of the model? 2 in.:7 ft

GROUP ACTIVITY Answers to Exercises 12–15 are on p. A25.

11. **a.** Measure a room and record the results. Answers will vary.
    b. Choose a convenient scale and make a scale drawing of the room. Draw all the furniture in the room as it would be seen from above.
    c. Sketch a new table that will fit in an empty part of the room. If you were to shop for the table, what size would you look for?

12. What scale might you choose to make a scale model of a ship? Explain your choice.

13. A statue of a person is 21 ft high. Decide what the scale might be. Explain your choice.

RESEARCH

14. Find what HO and O scales are in railroad modeling.

15. The northeastern face of Mount Rushmore in South Dakota has a large carving of four Presidents' faces. What scale did the sculptor use?

SPIRAL REVIEW

S 16. Find the sum: $\frac{5}{9} + \frac{7}{12}$ *(Lesson 8-5)* $1\frac{5}{36}$

17. The model of a car is $6\frac{1}{2}$ in. long. The scale of the model is $1:20$. What is the actual length of the car?   *(Lesson 9-3)* 130 in.

18. Solve by drawing a diagram: On a nature walk, Ed hiked 3 km north, 7 km east, 12 km south, 7 km west, 2 km north. Where was Ed in relation to his original position?   *(Lesson 7-8)* 7 km south

398   Chapter 9

9-4

UNDERSTAND • PLAN • WORK • ANSWER

Strategy:
Using Proportions

Objective: To use proportions to solve problems.

You can solve many problems that involve equal ratios or equal rates by using proportions.

Problem

The Daily Gazette charges $7.20 for 3 weeks of home newspaper delivery. At this rate, what is the cost of 8 weeks of home delivery?

Solution

UNDERSTAND The problem is about home newspaper delivery.
Facts: $7.20 for 3 weeks of home delivery
Find: cost of 8 weeks of home delivery

PLAN Let c = the cost of 8 weeks of home delivery. Write a proportion using the ratio of the number of weeks to the cost of home delivery. Then solve the proportion using cross products.

WORK
$$\frac{3}{7.2} = \frac{8}{c}$$
⟵ number of weeks
⟵ cost of home delivery

$$3c = (7.2)(8)$$
$$3c = 57.6$$
$$\frac{3c}{3} = \frac{57.6}{3}$$
$$c = 19.2$$

ANSWER The cost of 8 weeks of home newspaper delivery is $19.20.

Look Back An alternative method is to solve this problem without using a proportion. Describe how you could do this. Divide $7.20 by 3 to get the weekly price; then multiply by 8.

Guided Practice

COMMUNICATION «*Writing*   1. $\frac{5}{6} = \frac{n}{150}$ 2. $\frac{440}{20} = \frac{n}{35}$ 3. $\frac{3}{1.68} = \frac{7}{n}$

Write the proportion you would use to solve each problem. Use the variable *n*. Do not solve.

«**1.** The ratio of students to parents at a play was 5 to 6. There were 150 parents at the play. How many students were at the play?

«**2.** Last week Mary drove her car 440 mi and used 20 gal of gasoline. At the same rate, how many miles could she drive her car using 35 gal of gasoline?

«**3.** The cost of 3 oranges is $1.68. What is the cost of 7 oranges?

Ratio, Proportion, and Percent   **399**

Lesson Planner

Teaching the Lesson
- Lesson Starter 83
- Toolbox Skills 9, 10, 13, 14

Lesson Follow-Up
- Practice 112
- Practice 113: Nonroutine
- Enrichment 91: Problem Solving
- Study Guide, pp. 165–166
- Test 43

Warm-Up Exercises

Solve the following proportions.

1. $\frac{6}{a} = \frac{3}{9}$ 18

2. $\frac{5}{7} = \frac{10}{b}$ 14

3. $\frac{6}{8} = \frac{12}{d}$ 16

4. $\frac{y}{21} = \frac{3}{7}$ 9

5. $\frac{x}{14} = \frac{6}{2}$ 42

Lesson Focus

Students may wonder why it is necessary to study proportions in detail. Tell them that by using proportions they can solve many problems that occur in everyday life. Using proportions to solve problems is the focus of today's lesson.

Teaching Notes

Spend time on the Examples, making sure that students understand how each of the four steps of the problem-solving model is applied to the problem. Use the Look Back as an opportunity for students to discuss another way to solve the same problem.

Teaching Notes (continued)

Key Question

Is there a best strategy for solving a problem? Not necessarily; it depends on the problem. Many problems can be solved by using more than one strategy. For some problems, using proportions is a good way.

Critical Thinking

In all the problem-solving situations, students must *analyze* a problem carefully before *formulating* a plan to solve it. After the plan is applied and a solution is found, the answer should be *checked* to see if it is reasonable.

Error Analysis

Some students write incorrect proportions because they do not pay attention to the units. Remind students of this fact: Keep the units for the numerators the same and the units for the denominators the same.

Additional Example

Jared earned $90 in the last two weeks at his part-time job. If he continues at this rate, what will be his total earnings for a 12-week period? $540

Closing the Lesson

Review the Problem Solving Checklist on page 400 with students. Have various students describe each strategy in their own words, and identify the type of problem best suited for a particular strategy. Discuss questions such as: Is there a "best" strategy? Are some strategies used more frequently than others? In some problems, can more than one strategy be used?

Suggested Assignments

Skills: 1–14
Average: 1–14
Advanced: 1–14

Solve using a proportion.

4. Josefina sells helium balloons. She charges $9 for 12 balloons. At this rate, what will Josefina charge for 50 balloons? $37.50

5. A photocopy machine copied 50 pages in 1.5 min. At this rate, how long will the machine take to copy 90 pages? 2.7 min

«6. **COMMUNICATION** «*Discussion* Could you solve the Problem on page 399 by using the proportion $\frac{7.2}{3} = \frac{c}{8}$? Explain. Yes; it is an equivalent proportion. The numerators are the costs and the denominators are the numbers of weeks.

Problem Solving Situations

Solve using a proportion.

1. In Plainsville, 5 out of every 7 town meeting members voted to approve the school budget. A total of 190 members voted to approve the school budget. How many town meeting members are there in all? 266

2. Maria spent 3 h addressing 50 wedding invitations. At this rate, how long will it take Maria to address 125 wedding invitations? $7\frac{1}{2}$ h

3. Travel Rent-a-Car charges customers $135 to rent a compact car for 3 days with unlimited mileage. At this rate, what will it cost to rent a compact car for 5 days? $225

4. Shirley paid $12.72 for 8 cans of frozen orange juice to make a punch for a party. Elliot bought 10 cans of the same brand of orange juice to make punch for another party. How much did Elliot pay for the juice? $15.90

Solve using any problem solving strategy.

Answers to Problems 5–8 are on p. A25.

5. Centerville's movie theater is 11 blocks due south of the grocery store. The bank is 5 blocks due north of the movie theater. Where is the bank in relation to the grocery store?

6. Shelly distributed 6 lb of nuts equally among 32 gift baskets. How many ounces of nuts did Shelly put in each basket?

7. There are twice as many geese as ducks in a lake. In all, there are 51 geese and ducks. How many geese are in the lake?

8. There are 6 counselors for every 50 campers at a summer camp. The camp has 21 counselors. Find the number of campers.

> **PROBLEM SOLVING**
>
> **CHECKLIST**
>
> *Keep these in mind:*
> Using a Four-Step Plan
> Too Much or Not Enough Information
> Supplying Missing Facts
>
> *Consider these strategies:*
> Choosing the Correct Operation
> Making a Table
> Guess and Check
> Using Equations
> Identifying a Pattern
> Drawing a Diagram
> Using Proportions

9. The figures below show the first four figures in a pattern of squares. How many squares are in the ninth figure in this pattern? Strategy may vary: identifying a pattern; 83

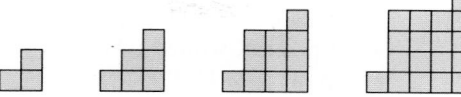

WRITING WORD PROBLEMS

Write a word problem that you could solve using each proportion. Then solve the problem. Answers to Problems 10 and 11 are on p. A25.

10. $\frac{3}{35} = \frac{x}{140}$

11. $\frac{4}{n} = \frac{6}{\$1.44}$

SPIRAL REVIEW

S **12.** Solve using a proportion: In the Springfield Chorus, 2 out of every 7 singers are altos. There are 112 singers in the chorus. How many are altos? *(Lesson 9-4)* 32

13. Graph the inequality $x < 2.5$. *(Lesson 7-10)* See answer on p. A25.

14. Find the mean, median, mode(s), and range:
11, 2, 17, 13, 13, 17, 13 *(Lesson 5-8)* 12.3; 13; 13; 15

Write each ratio as a fraction in lowest terms.

1. $\frac{18}{30}$ $\frac{3}{5}$ **2.** 30 min:2 h $\frac{1}{4}$ **3.** 1 lb to 8 oz $\frac{2}{1}$ **4.** 21:14 $\frac{3}{2}$ 9-1

Write the unit rate.

5. 110 mi in 2 h **6.** \$16 for 4 lb **7.** \$1.95 for 3 combs
55 mi/h   \$4/lb   \$.65/comb

Tell whether each proportion is *True* or *False*.

8. $\frac{5}{10} = \frac{20}{40}$ **9.** $\frac{6.8}{1.7} = \frac{8}{2}$ **10.** $\frac{3}{4} = \frac{20}{15}$ **11.** $\frac{9}{10} = \frac{10}{11}$ 9-2
True   True   False   False

Solve each proportion.

12. $\frac{2}{8} = \frac{b}{40}$ 10 **13.** $\frac{17}{a} = \frac{17}{4}$ 4 **14.** $\frac{10}{3} = \frac{d}{10}$ 33$\frac{1}{3}$ **15.** $\frac{9.2}{n} = \frac{4}{5}$ 11.5

16. The model of a truck is $4\frac{3}{4}$ in. wide. The scale of the model is 9-3
3 in. : 5 ft. What is the actual width of the truck? $7\frac{11}{12}$ ft

17. Kathy pays \$41.25 for 5 dancing lessons. At this rate, how 9-4
much does she pay for 12 dancing lessons? \$99

Ratio, Proportion, and Percent **401**

401

Using a Ten-by-Ten Grid

Objective: To use ten-by-ten grids to model percents.

Materials

■ graph paper

A percent is a ratio that compares a number to 100. For example, 1 percent, written 1%, is the same as the ratio 1:100.

Each square in the ten-by-ten grid at the right represents 1%. You can use such a grid to model ratios and percents. For example, the ratio 48:100, or 48%, is modeled at the right.

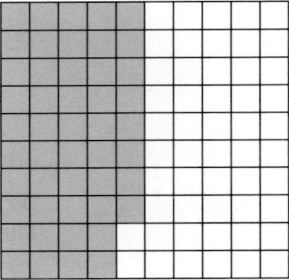

Activity I *Interpreting Models of Percents*

1. What percent of the ten-by-ten grid at the left below is red? 80%

2. What percent of the ten-by-ten grid at the left below is not red? 20%

3. What percent of the ten-by-ten grid at the right below is blue? 65%

4. What percent of the ten-by-ten grid at the right below is not blue? 35%

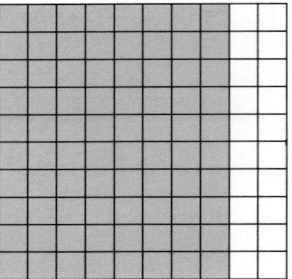
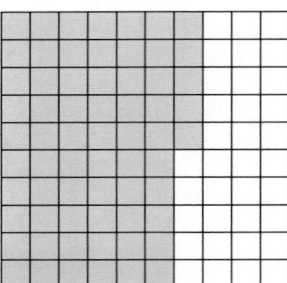

Answers to Activity II, Steps 1–3, are on p. A25.

Activity II *Drawing Models of Percents*

Draw a ten-by-ten grid on a piece of graph paper.

1. Draw $'s in squares to represent 12%.

2. Draw •'s in empty squares to represent 35%.

3. Draw #'s in empty squares to represent 10%.

4. What percent of the grid does not contain any $'s? 88%

5. What percent of the grid does not contain any •'s? 65%

6. What percent of the grid does not contain any #'s? 90%

7. What percent of the grid contains a symbol? 57%

8. What percent of the grid does not contain a symbol? 43%

402 Chapter 9

Activity III *Percents Less than 1 and Greater than 100*

Use the grids below for Steps 1 and 3.

i.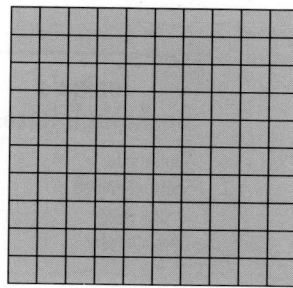

ii.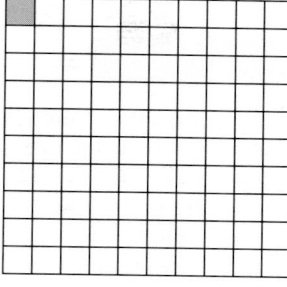

iii.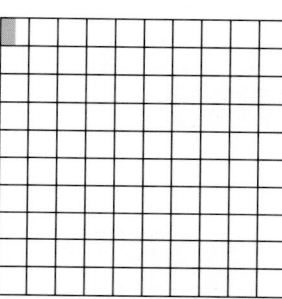

iv.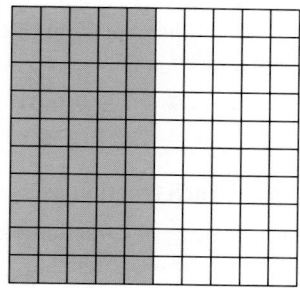

1 Choose the number of the grid that represents each percent.

  a. $\frac{1}{2}$% iii **b.** 1% ii **c.** 50% iv **d.** 100% i

2 Tell how much of a ten-by-ten grid you would have to shade to represent each percent.

$66\frac{2}{3}$ squares

  a. $\frac{1}{5}$% $\frac{1}{5}$ of one square **b.** 20% 20 squares **c.** $\frac{2}{3}$% $\frac{2}{3}$ of one square **d.** $66\frac{2}{3}$%

3 What percent would you obtain if you added the percents represented by grids (i) and (ii)? grids (i) and (iii)? grids (i) and (iv)? 101%; 100.5%; 150%

Activity IV *Converting Fractions and Decimals to Percents*

1 Use the grid at the right. Write each ratio as a fraction in simplest form, as a decimal, and as a percent.

  a. red squares : all squares $\frac{3}{10}$; 0.3; 30%
  b. blue squares : all squares $\frac{1}{4}$; 0.25; 25%
  c. blank squares : all squares $\frac{9}{20}$; 0.45; 45%

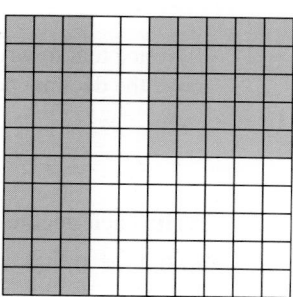

2 Model each number by shading squares on four separate ten-by-ten grids.

  a. $\frac{3}{4}$ **b.** $\frac{2}{5}$ **c.** 0.15 **d.** 0.08

  See answer on p. A25.

3 Use your models to write each number in Step 2 as a percent.

  a. 75% **b.** 40% **c.** 15% **d.** 8%

Ratio, Proportion, and Percent **403**

Teaching the Lesson
- Materials: ten-by-ten grid, coins
- Lesson Starter 84
- Using Manipulatives, p. 386D
- Toolbox Skills 10, 14, 15

Lesson Follow-Up
- Practice 114
- Enrichment 92: Communication
- Study Guide, pp. 167–168
- Manipulative Activity 18
- Cooperative Activity 18

Warm-Up Exercises

Change each ratio to an equivalent ratio with a denominator of 100.

1. $\frac{3}{4} = \frac{x}{100}$ 75

2. $\frac{1}{2} = \frac{a}{100}$ 50

3. $\frac{3}{10} = \frac{b}{100}$ 30

4. $\frac{2}{5} = \frac{c}{100}$ 40

5. $\frac{1}{8} = \frac{d}{100}$ 12.5

Lesson Focus

Many words in English have their origins in another language. This is the case with the word "percent." It comes from the Latin words "per centum", which mean "of a hundred" or "for each one hundred." Can you think of any other words in English that may be linked to percent? century; cents (money) The focus of today's lesson is on percents and how they may be written.

Teaching Notes

Show the students a meter stick and ask them various questions about it to lead to the fact that 1 centimeter is $\frac{1}{100}$, or 1%, of the length.

9-5   Percent

Objective: To write fractions and decimals as percents and to write percents as fractions and decimals.

Terms to Know
- *percent*

DATA ANALYSIS

Jenine attempted twenty shots during the last basketball game. The school paper reported that she made 85% of her shots.

The symbol % is read "percent." A **percent** is a ratio that compares a number to 100. Percent means "per hundred," "hundredths," or "out of every hundred."

$$1\% = \frac{1}{100} = 0.01$$

Jenine's Basketball Record

Shots made	17
Shots attempted	20
$\frac{\text{Shots made}}{\text{Shots attempted}}$	$\frac{17}{20}$

Example 1

Solution

Write each fraction as a percent: a. $\frac{17}{20}$ b. $\frac{3}{8}$

a. Write $\frac{17}{20}$ as a fraction whose denominator is 100.

$$\frac{17}{20} = \frac{17 \cdot 5}{20 \cdot 5} = \frac{85}{100} = 85\%$$

b. The denominator of $\frac{3}{8}$ is not a factor of 100, so use division.

$$\frac{3}{8} = 0.37\frac{1}{2}$$ ⟵ Divide $8\overline{)3.00}$ to the *hundredths* place.

$$= 37\frac{1}{2}\%$$ ⟵ $0.37\frac{1}{2}$ means $37\frac{1}{2}$ *hundredths*.

 **Check Your Understanding**

1. When changing $\frac{17}{20}$ to a percent in Example 1(a), why do you write the ratio as a fraction with a denominator of 100?

2. If you used a calculator in Example 1(b), what number would be displayed for $0.37\frac{1}{2}$?
See Answers to Check Your Understanding at the back of the book.

Notice that to write a decimal as a percent, as in Example 1(b), you move the decimal point two places to the right and insert the symbol %.

$$0.37\tfrac{1}{2} = 3\,7{.}\tfrac{1}{2}\%$$

Example 2

Solution

Write each decimal as a percent: a. 0.43   b. 0.09   c. 2.4

a. $0.43 = 0.43\%$
$= 43\%$

b. $0.09 = 0.09\%$
$= 9\%$

c. $2.4 = 2.40\%$
$= 240\%$

You can also write percents as decimals and as fractions. To write a percent as a decimal, you move the decimal point two places to the left and drop the % symbol.

Notice in Examples 3 and 4 that a percent may be less than 1 or greater than 100.

Example 3

Write each percent as a decimal.

  a. 74%   **b.** 0.25%   **c.** 110%   **d.** $8\frac{1}{2}$%

Solution

  a. 74% = .7 4 % = 0.74

  b. 0.25% = .0 0.25% = 0.0025

  c. 110% = 1.1 0 % = 1.1

  d. $8\frac{1}{2}$% = .0 8.5% = 0.085

Example 4

Write each percent as a fraction or a mixed number in lowest terms.

  a. 125%   **b.** 3.6%   **c.** $\frac{2}{5}$%

Solution

  a. $125\% = \frac{125}{100} = \frac{5}{4} = 1\frac{1}{4}$

  b. $3.6\% = \frac{3.6}{100} = \frac{(3.6)(10)}{(100)(10)} = \frac{36}{1000} = \frac{9}{250}$

  c. $\frac{2}{5}\% = \frac{\frac{2}{5}}{100} = \frac{2}{5} \div 100 = \frac{2}{5} \cdot \frac{1}{100} = \frac{1}{250}$

 **Check Your Understanding**

  3. Describe how the position of the decimal point is changed in writing a percent as a decimal.

  4. In Example 3(d), why is $8\frac{1}{2}$ rewritten as 8.5?

  5. Why are the numerator and denominator in Example 4(b) multiplied by 10?

  See *Answers to Check Your Understanding* at the back of the book.

Guided Practice

COMMUNICATION «*Reading*

«**1.** What is a percent? a ratio that compares a number to 100

«**2.** Describe two methods for changing a fraction to a percent.
See answer on p. A25.

Write each ratio as a percent.

  3. 18 out of 100 18%   **4.** 2 : 100 2%

  5. 56 to 100 56%   **6.** $\frac{100}{100}$ 100%

Ratio, Proportion, and Percent **405**

Closing the Lesson

Have students give the rules for changing a decimal to a percent and a percent to a decimal.

Suggested Assignments

Skills: 1–33, 39–42
Average: 1–33, 39–42
Advanced: 1–29 odd, 30–42,
 Challenge

Exercise Notes

Communication: Reading

Guided Practice Exercises 1 and 2 require students to restate the lesson material on percents in their own words. Describing the methods for changing fractions to percents should help students internalize the rules.

Number Sense

The use of patterns in Exercises 30–33 should improve students' number sense as it relates to percents.

Critical Thinking

Exercises 34–38 help prepare students for Lesson 9-7 as they analyze the progression of steps and create their own plans for finding what percent one number is of another number.

Follow-Up

Exploration

Give students a standard size piece of construction paper. Have them do the following activity.

 Cut the paper in half and label it as "50%; $\frac{1}{2}$." Cut the other half into two equal parts; label one part as "25%; $\frac{1}{4}$," and place it alongside the 50% rectangle. Continue for "$12\frac{1}{2}$%; $\frac{1}{8}$," and then one step further. The figure should begin to look similar to the one on the next page.

---

7. To write $1\frac{5}{8}$ as a percent, replace each __?__ with the number that makes the statement true.

$$1 = \frac{?}{100} = \underline{\ ?\ }\% \quad \text{100; 100}$$

$$\frac{5}{8} = 0.625 = \underline{\ ?\ }\% \quad \text{62.5}$$

$$1\frac{5}{8} = 1 + \frac{5}{8} = \underline{\ ?\ }\% + \underline{\ ?\ }\% = \underline{\ ?\ }\%$$

$$\text{100; 62.5; 162.5}$$

Write each fraction or decimal as a percent.

8. $\frac{11}{50}$ 22%   9. $\frac{3}{5}$ 60%   10. $\frac{7}{8}$ $87\frac{1}{2}$% $\frac{1}{2}$% 11. $\frac{1}{200}$

12. 0.45 45%   13. 0.002 0.2%   14. 1 100%   15. 2.78 278%

Write each percent as a decimal.

16. 65% 0.65   17. 4% 0.04   18. 9.2% 0.092   19. $7\frac{1}{4}$% 0.0725

Write each percent as a fraction or a mixed number in lowest terms.

20. 36% $\frac{9}{25}$ 21. 0.5% $\frac{1}{200}$ 22. $\frac{1}{4}$% $\frac{1}{400}$ 23. 375% $3\frac{3}{4}$

Exercises

Write each fraction, mixed number, or decimal as a percent.

A   1. $\frac{2}{25}$ 8%   2. $\frac{21}{50}$ 42%   3. $\frac{7}{10}$ 70%   4. $\frac{1}{5}$ 20%   5. $\frac{5}{8}$ 62.5%

6. $\frac{9}{16}$ 56.25%   7. 0.21 21%   8. 0.08 8%   9. 0.099 9.9%   10. 1.7 170%

11. $0.15\frac{1}{5}$ $15\frac{1}{5}$%   12. $\frac{7}{12}$ $58\frac{1}{3}$%   13. $\frac{15}{32}$ $46\frac{7}{8}$%   14. $1\frac{4}{15}$ $126\frac{2}{3}$%   15. $2\frac{2}{7}$ $228\frac{4}{7}$%

Write each percent as a decimal.

16. 6% 0.06   17. 49% 0.49   18. 0.3% 0.003   19. 587% 5.87   20. $24\frac{1}{3}$% 0.243

Write each percent as a fraction or mixed number in lowest terms.

21. 50% $\frac{1}{2}$ 22. $\frac{3}{4}$% $\frac{3}{400}$ 23. 2.3% $\frac{23}{1000}$ 24. 475% $4\frac{3}{4}$ 25. 0.15% $\frac{3}{2000}$

26. Christina made 22 out of 25 shots in last week's basketball game. What percent of the shots Christina attempted did she make? 88%

27. The Gray Company receives about 40 letters each day. Sixteen of these letters have postage stamps. The remainder of the letters have been passed through a postage meter. What percent of the letters received each day have postage stamps on them? 40%

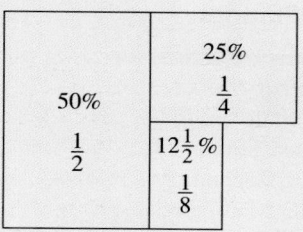

B **28.** Steve got 15 out of 20 problems correct on a math test. What percent of the problems did he get wrong? 25%

29. The Robins won 30 games and lost 20. What percent of the games they played did they win? 60%

PATTERNS

30. Write each percent as a decimal.

   a. 200% 2   **b.** 20% 0.2   **c.** 2% 0.02

31. Use the results of Exercise 30 to find the first six decimals in the pattern.

200%	20%	2%	0.2%	0.02%	0.002%
_?_ 2	_?_ 0.2	_?_ 0.02	_?_ 0.002	_?_ 0.0002	_?_ 0.00002

32. Write each decimal as a percent.

   a. 9 900%   **b.** 0.9 90%   **c.** 0.09 9%

33. Use the results of Exercise 32 to find the first six percents in the pattern.

9	0.9	0.09	0.009	0.0009	0.00009
_?_ % 900	_?_ % 90	_?_ % 9	_?_ % 0.9	_?_ % 0.09	_?_ % 0.009

THINKING SKILLS 34–37. Knowledge   **38.** Analysis; synthesis

Complete to find what percent the first number is of the second number.

C **34.** 97; 100 ⟶ $\frac{?}{100} = \underline{\,?\,}$%   **35.** 19; 19 ⟶ $\frac{19}{?} = \underline{\,?\,}$% 19; 100
    97; 97

36. 100; 200 ⟶ $\frac{?}{200} = \underline{\,?\,}$%   **37.** 200; 100 ⟶ $\frac{200}{?} = \underline{\,?\,}$% 100; 200
    100; 50

38. Analyze your answers to Exercises 34–37. Create a plan for finding what percent one number is of another number. See answer on p. A25.

SPIRAL REVIEW

S **39.** Write 7.5% as a fraction and as a decimal. *(Lesson 9-5)* $\frac{3}{40}$; 0.075

40. Find the GCF: 108 and 60 *(Lesson 7-2)* 12

41. Find the difference: $\frac{29}{37} - \frac{18}{37}$ *(Lesson 8-4)* $\frac{11}{37}$

42. Use a protractor to draw an angle with measure 123°. *(Lesson 6-2)*
See answer on p. A25.

Challenge

A photograph is reduced to four fifths of its original size. What percent of the reduced size is the original size? 125%

Ask: What is the sum of the areas of all the rectangles? What is the sum of the percents shown? What is the sum of the fractions?

Enrichment
Enrichment 92 is pictured on page 386E.

Practice
Practice 114, *Resource Book*

Practice 114
Skills and Applications of Lesson 9-5

Write each fraction, mixed number, or decimal as a percent.

1. $\frac{3}{5}$ 60% 2. $\frac{4}{25}$ 16% 3. $\frac{7}{16}$ 43.75% 4. $\frac{20}{32}$ 62.5%
5. $\frac{11}{12}$ 91$\frac{2}{3}$% 6. $\frac{48}{64}$ 75% 7. $\frac{12}{32}$ 37.5% 8. $\frac{3}{20}$ 15%
9. $\frac{7}{25}$ 28% 10. $\frac{7}{200}$ 3.5% 11. 0.06 6% 12. 0.32 32%
13. 0.064 6.4% 14. 2.2 220% 15. 1.04 104% 16. 0.25 25%
17. 3$\frac{3}{16}$ 318.75% 18. 2$\frac{9}{36}$ 225% 19. 0.42 42% 20. 1$\frac{8}{25}$ 132%

Write each percent as a decimal.

21. 8% 0.08 22. 13% 0.13 23. 89% 0.89 24. 1.2% 0.012
25. 0.25% 0.0025 26. 12.9% 0.129 27. 429% 4.29 28. $\frac{5}{8}$% 0.00625

Write each percent as a fraction or mixed number in lowest terms.

29. 72% $\frac{18}{25}$ 30. $\frac{7}{8}$% $\frac{7}{800}$ 31. $\frac{15}{16}$% $\frac{3}{320}$
32. 3.4% $\frac{17}{500}$ 33. 292% 2$\frac{23}{25}$ 34. 0.16% $\frac{1}{625}$

35. Brian made 5 out of 8 goal attempts in Saturday's soccer game. What percent of the goal attempts did Brian make? 62.5%

36. Mei completed 17 out of 20 problems for a chemistry assignment. What percent of the chemistry assignment did Mei complete? 85%

37. Out of 400 students, 96 prefer basketball to football. What percent of the students prefer basketball? 24%

Teaching the Lesson
- Materials: grids
- Lesson Starter 85
- Using Technology, p. 386C
- Using Manipulatives, p. 386D
- Toolbox Skills 13, 20

Lesson Follow-Up
- Practice 115
- Enrichment 93: Application
- Study Guide, pp. 169–170
- Calculator Activity 9

Warm-Up Exercises

Write the following percents as decimals and fractions.

1. 12% $0.12; \frac{3}{25}$

2. 3% $0.03; \frac{3}{100}$

3. 0.75% $0.0075; \frac{3}{400}$

4. 130% $1.3; \frac{13}{10}$

Lesson Focus

Students most likely have heard expressions such as: "His batting *percentage* is 275 this season," or "out of 150 possible points, how many do I need to get correct in order to score 85%?" The focus of today's lesson is on the interpretation of such expressions and on how to calculate a percent of a number.

9-6

Finding a Percent of a Number

Objective: To find a percent of a number and to estimate percents of numbers.

APPLICATION

Luis Mendez wants to buy a compact disc player. The regular price is $280 but the disc player is on sale today at a 30% discount.

QUESTION How much will Luis save if he buys the compact disc player today?

Example 1

Solution

Find 30% of 280.

To find a percent of a number, you multiply.

$$\begin{array}{ccc} 30\% & \text{of} & 280 \\ \downarrow & \downarrow & \downarrow \\ 0.3 & \cdot & 280 = 84 \end{array}$$

 **Check Your Understanding**

1. In Example 1, 30% was written as a decimal. What is an alternate way to write 30% in order to multiply?
   See *Answers to Check Your Understanding* at the back of the book.

ANSWER Luis will save $84 if he buys the disc player today.

You can use equations to find a percent of a number.

Example 2

Solution

What number is 64% of 75?

Write an equation. Let n be the missing number.

Using a fraction:

$$\text{What number is } \underbrace{64\% \text{ of } 75?}$$
$$\begin{array}{ccc} & \downarrow & \downarrow \quad \downarrow \quad \downarrow \\ n & = & \frac{64}{100} \cdot 75 \\ n & = & \frac{16}{25} \cdot 75 \\ n & = & 48 \end{array}$$

Using a decimal:

$$\text{What number is } \underbrace{64\% \text{ of } 75?}$$
$$\begin{array}{ccc} & \downarrow \quad \downarrow \quad \downarrow \quad \downarrow \\ n & = 0.64 \cdot 75 \\ n & = 48 \end{array}$$

So 48 is 64% of 75.

 **Check Your Understanding**

2. Explain why 64% can be written as either $\frac{64}{100}$ or 0.64.

3. What is $\frac{64}{100}$ as a fraction in lowest terms?
   See *Answers to Check Your Understanding* at the back of the book.

Most calculators have a percent key, $\boxed{\%}$. The way a calculator uses the percent key depends on the type of calculator. To find 64% of 75, you may be able to use one of these two key sequences.

$\boxed{64.}$ $\boxed{\%}$ $\boxed{\times}$ $\boxed{75.}$ $\boxed{=}$ *or*
$\boxed{75.}$ $\boxed{\times}$ $\boxed{64.}$ $\boxed{\%}$

Consult the user's manual to determine how your calculator handles percents.

Even if you have a calculator, you may find it helpful to memorize the equivalent percents, decimals, and fractions in the chart below.

Equivalent Percents, Decimals, and Fractions

$20\% = 0.2 = \frac{1}{5}$	$25\% = 0.25 = \frac{1}{4}$	$12\frac{1}{2}\% = 0.125 = \frac{1}{8}$	$16\frac{2}{3}\% = 0.1\overline{6} = \frac{1}{6}$
$40\% = 0.4 = \frac{2}{5}$	$50\% = 0.5 = \frac{1}{2}$	$37\frac{1}{2}\% = 0.375 = \frac{3}{8}$	$33\frac{1}{3}\% = 0.\overline{3} = \frac{1}{3}$
$60\% = 0.6 = \frac{3}{5}$	$75\% = 0.75 = \frac{3}{4}$	$62\frac{1}{2}\% = 0.625 = \frac{5}{8}$	$66\frac{2}{3}\% = 0.\overline{6} = \frac{2}{3}$
$80\% = 0.8 = \frac{4}{5}$		$87\frac{1}{2}\% = 0.875 = \frac{7}{8}$	$83\frac{1}{3}\% = 0.8\overline{3} = \frac{5}{6}$
$100\% = 1$			

You can use equivalent values from the chart above to estimate percents of numbers.

Example 3

Estimate.

a. 19% of 126　　**b.** $16\frac{2}{3}\%$ of \$358.75　　**c.** 65% of 299

Solution

a. 19% of 126　　← 19% is about 20%.
↓　↓　↓　　You can use the equivalent
$\frac{1}{5}$ · 125　　fraction for 20%.
‿‿‿‿‿
about 25

b. $16\frac{2}{3}\%$ of \$358.75　　**c.** 65% of 299
↓　↓　↓　　　　　　　↓　↓　↓
$\frac{1}{6}$ · \$360　　　　　$\frac{2}{3}$ · 300
‿‿‿‿　　　　　　　‿‿‿‿
about \$60　　　　　　about 200

Check Your Understanding

4. In Example 3(c), explain why 65% is about $\frac{2}{3}$.
See *Answers to Check Your Understanding* at the back of the book.

Teaching Notes

Make certain students understand the procedures used in Examples 1 and 2. Also, if students are using calculators, make certain that they are using the correct key sequence, as this varies from calculator to calculator. In Example 3, stress the importance of checking the reasonableness of an answer.

Key Question

How do you find the percent of a number? You change the percent to a decimal and multiply by the number.

Error Analysis

In percent questions, such as Example 2, students sometimes do not know whether to multiply or divide. The use of an equation will help students follow the correct procedure.

Using Technology

For a suggestion on using a spreadsheet program, see page 386C.

Using Manipulatives

For a suggestion on using grid models, see page 386D.

Additional Examples

1. Find 40% of 360.　144
2. What number is 35% of 120?　42

Estimate.

3. 11% of 475　about 50
4. 34% of 73　about 25
5. 42% of 495　about 200

410

Closing the Lesson

Have students estimate answers to the following exercises and explain their procedure.

1. 36% of 48   about 18
2. 27% of 58   about 15
3. 68% of 290   about 200
4. 5% of 205   about 10

Suggested Assignments

Skills: 1–24, 25–31 odd, 45–47
Average: 1–32, 45–47
Advanced: 1–19 odd, 21–47

Exercise Notes

Communication: Reading

Guided Practice Exercise 1 checks students' comprehension of how to find the percent of a number.

Estimation

Guided Practice Exercises 8–10 and *Exercises 9–14* have students approximate answers by applying their knowledge of equivalent percents and fractions. These exercises should make students understand the importance of memorizing the equivalent values shown in the chart on page 409.

Calculator

Exercises 15–20 ask students to write two different key sequences that could be used to solve the given percent problems.

Guided Practice

« **1.** COMMUNICATION « *Reading*   Replace each ___?___ with the correct word or number.
To find what number is 48% of 64, you ___?___ 64 by ___?___. multiply; 0.48

Choose the letter of the equation you would use to find each answer.

2. 6% of 180
  a. $n = 6 \cdot 180$
  b. $n = 0.6 \cdot 180$
  c. $n = 0.06 \cdot 180$

3. 25% of 488
  a. $n = 25 \cdot 488$
  b. $n = \frac{1}{4} \cdot 488$
  c. $n = 2.5 \cdot 488$

Find each answer.

4. 5% of 320 is what number? 16
5. 50% of 98 is what number? 49
6. $12\frac{1}{2}$% of 84 is what number? 10.5
7. What number is 36% of 70? 25.2

Estimate. Estimates may vary. Accept reasonable estimates.

8. 89% of 51 about 45
9. 17% of $6\frac{1}{4}$ about 1
10. $33\frac{1}{3}$% of $59.99
about $20

Exercises

Find each answer. 1. $94.40 2. 45 3. 240 4. 264 5. 456 6. $1.92

A
1. What number is 40% of $236?
2. What number is $37\frac{1}{2}$% of 120?
3. What number is 2.5% of 9600?
4. 50% of 528 is what number?
5. 100% of 456 is what number?
6. 8% of $24 is what number?

7. Mei bought an electric guitar priced at $350. The sales tax rate is 4.5%. How much sales tax did she pay? $15.75

8. Kwasi scored 85% on a 60-item test. How many items did he get correct? 51

Estimate. Estimates may vary. Accept reasonable estimates.

9. 88% of 41 about 35
10. 60% of 62 about 36
11. 33% of 147 about 50
12. $83\frac{1}{3}$% of 123 about 100
13. 25% of $23.95 about $6
14. 47% of $505 about $250

 CALCULATOR

Write two different calculator key sequences you could use to find each answer. Answers to Exercises 15–20 are on p. A25.

B
15. 5% of 312
16. 2.65% of 180
17. 16% of 975
18. 300% of 159
19. 0.7% of 34.9
20. 26.42% of 100

MENTAL MATH

Finding 10% of a number is the same as multiplying the number by 0.1. You mentally move the decimal point one place to the left. Knowing 10% of a number can help you find other percents of the same number. For example, 5% of a number is one half of 10% of the same number.

Find each answer mentally.

21. a. $2.40   b. $1.20   c. $9.60
22. a. 32   b. 64   c. 96

21. **a.** 10% of $24 **b.** 5% of $24 **c.** 40% of $24
22. **a.** 10% of 320 **b.** 20% of 320 **c.** 30% of 320
23. **a.** 10% of 4.6 0.46 **b.** 5% of 4.6 0.23 **c.** 20% of 4.6 0.92
24. Sam's dinner at a restaurant cost $18. Use the fact that 15% = 10% + 5% to find the 15% tip Sam paid on his dinner. $2.70

NUMBER SENSE

Replace each __?__ with >, <, or =.

25. 40% of 48 __?__ 48 <
26. 120% of 52 __?__ 52 >
27. n __?__ 200% of n, $n > 0$ <
28. x __?__ 90% of x, $x > 0$ >
29. 50 __?__ 100% of 50 =
30. 100% of y __?__ y, $y > 0$ =
31. 175% __?__ $1\frac{3}{4}$ =
32. $\frac{1}{2}$% __?__ $\frac{1}{2}$ <

 COMPUTER APPLICATION

A department store staff uses a spreadsheet to give final prices of items when they are marked up by a given percent. The percents of markup are in column A, the original prices of the items are in row 1, and the final prices appear below the original prices.

	A	B	C	D	E	F
1		$10.00	$20.00	$30.00	$40.00	$50.00
2	40%	$14.00	$28.00	$42.00	$56.00	$70.00
3	50%	$15.00				
4	60%	$16.00				
5	70%	$17.00				

Find the correct value for each cell in the spreadsheet above.

33. C3 $30 34. D3 $45 35. E3 $60 36. F3 $75 37. C4 $32 38. D4 $48
39. E4 $64 40. F4 $80 41. C5 $34 42. D5 $51 43. E5 $68 44. F5 $85

SPIRAL REVIEW

45. Laura pays $75 for four guitar lessons. At this rate, how much would she pay for ten lessons? *(Lesson 9-4)* $187.50

46. Solve $P = 2l + 2w$ for w. *(Lesson 8-10)* $w = \frac{1}{2}(P - 2l)$

47. What number is 15% of 24? *(Lesson 9-6)* 3.6

Ratio, Proportion, and Percent **411**

Mental Math
Exercises 21–24 can be done mentally since they all relate to 10% of an amount. Students should realize that using mental math is the most effective technique for doing problems of this type.

Number Sense
Exercises 25–32 are designed to improve students' number sense in relation to percents that range from less than 1% to greater than 100%.

Computer
Exercises 33–44 require students to calculate final prices and enter the values within a spreadsheet format.

Follow-Up

Enrichment
Enrichment 93 is pictured on page 386E.

Practice
Practice 115, *Resource Book*

Practice 115
Skills and Applications of Lesson 9-6

Find each answer.

1. What number is 30% of $350? $105
2. What number is 80% of $250? $200
3. What number is 3% of 15,000? 450
4. What number is 2% of 2550? 51
5. What number is 45% of 540? 243
6. What number is 15% of 940? 141
7. What number is 90% of 650? 585
8. What number is 60% of 1900? 1140
9. What number is 9% of $4200? $378
10. What number is 40% of 220? 88
11. 70% of 690 is what number? 483
12. 75% of 890 is what number? 667.5
13. 60% of 525 is what number? 315
14. 20% of 1225 is what number? 245
15. 4% of 15,650 is what number? 626
16. 5% of 100 is what number? 5
17. 10% of 7920 is what number? 792
18. 1% of 1860 is what number? 18.6
19. 100% of 789 is what number? 789
20. 30% of 250 is what number? 75

21. Melissa attempted 40 baskets and made 85% of them. How many baskets did she make? 34
22. Masao bought a VCR whose price was $496. The sales tax rate is 7.5%. How much sales tax did he pay? $37.20
23. Of the 680 students at Textor Middle School, 55% are female. How many females attend Textor Middle School? 374

Estimate. Estimates may vary. Examples are given.

24. 68% of 81 about 56
25. 40% of 96 about 40
26. 67% of 598 about 400
27. $62\frac{1}{2}$% of 824 about 500
28. 75% of 608 about 450
29. 53% of 303 about 150

Resource Book, MATHEMATICAL CONNECTIONS
Copyright © by Houghton Mifflin Company. All rights reserved.

297

Teaching the Lesson
- Lesson Starter 86
- Toolbox Skills 10, 14, 15

Lesson Follow-Up
- Practice 116
- Enrichment 94: Data Analysis
- Study Guide, pp. 171–172

Warm-Up Exercises

Find each answer mentally.

1. 20% of 25   5
2. $33\frac{1}{3}$% of 27   9
3. 10% of 38.7   3.87
4. 150% of 40   60

Lesson Focus

Tell students to imagine they are buying an item that costs $12, but their bill is $12.60 because of sales tax. Ask: What percent of the cost is the sales tax? (5%) $.60 is what percent of $12? (5%) Finding what percent one number is of another is the focus of today's lesson.

Teaching Notes

Tell students that finding what percent one number is of another is like finding "what *part* of the *whole*" one number is of the other.

9-7

Finding the Percent One Number Is of Another

Objective: To find the percent one number is of another.

You can use an equation to find out what percent one number is of another number.

Example

a. What percent of 75 is 21?
b. What percent of 16 is 22?
c. 12.5 is what percent of 500?

Solution

Let n = the percent.

a. What percent of 75 is 21?

$$n \cdot 75 = 21$$
$$75n = 21$$
$$\frac{75n}{75} = \frac{21}{75}$$
$$n = \frac{21}{75} = \frac{7}{25} = \frac{28}{100} = 28\%$$

So 28% of 75 is 21.

b. What percent of 16 is 22?

$$n \cdot 16 = 22$$
$$16n = 22$$
$$\frac{16n}{16} = \frac{22}{16}$$
$$n = \frac{22}{16} = \frac{11}{8} = 1.37\frac{1}{2} = 137\frac{1}{2}\%$$

So $137\frac{1}{2}\%$ of 16 is 22.

c. 12.5 is what percent of 500?

$$12.5 = n \cdot 500$$
$$12.5 = 500n$$
$$\frac{12.5}{500} = \frac{500n}{500}$$
$$n = \frac{12.5}{500} = 0.025 = 2.5\%$$

So 12.5 is 2.5% of 500.

✓ **Check Your Understanding**

1. In Example 1(a), why was $\frac{7}{25}$ written with a denominator of 100?

2. Explain how you know that $\frac{12.5}{500} = 0.025$ in Example 1(c).

See *Answers to Check Your Understanding* at the back of the book.

---

412   Chapter 9

Guided Practice

COMMUNICATION « *Writing*

Write an equation that represents each question. Let *n* represent the unknown percent. $100n = 37$

$25n = 19.5$

« **1.** What percent of 100 is 37?

« **2.** What percent of 25 is 19.5?

« **3.** 28 is what percent of 5.6?
$28 = 5.6n$

« **4.** 6 is what percent of 40?
$6 = 40n$

Write a question involving percents that could be represented by each equation. **5.** What percent of 34 is 17? **6.** What percent of $\frac{1}{2}$ is $\frac{3}{4}$?

« **5.** $34n = 17$

« **6.** $\frac{1}{2}n = \frac{3}{4}$

« **7.** $23.7 = 60n$
23.7 is what percent of 60?

« **8.** $78 = 12.5n$
78 is what percent of 12.5?

Write each fraction as a percent.

9. $\frac{16.2}{60}$ 27%

10. $\frac{72}{64.8}$ $111\frac{1}{9}$%

11. $\frac{9.3}{12.4}$
75%

Find each answer.

12. What percent of 50 is 37? 74%

13. 120 is what percent of 150?
80%

14. What percent of 75 is 100?
$133\frac{1}{3}$%

15. 2.4 is what percent of 120?
2%

Exercises

Find each answer. **8.** $66\frac{2}{3}$% **9.** $33\frac{1}{3}$% **10.** 0.5%

A

1. What percent of 20 is 16? 80%

2. What percent of 200 is 68?
34%

3. What percent of 40 is 28? 70%

4. What percent of 32 is 8? 25%

5. 24 is what percent of 25? 96%

6. 3.5 is what percent of 50? 7%

7. 18 is what percent of 18? 100%

8. What percent of 66 is 44?

9. 12.2 is what percent of 36.6?

10. 12 is what percent of 2400?

11. 80 is what percent of 3200? 2.5%

12. What percent of 48 is 2.4? 5%

13. Mike got 35 items correct on a 40-item test. What percent of the items did he get correct? $87\frac{1}{2}$%

14. The sale price of the luggage Caroline bought was $160. The sales tax she paid was $8.80. What was the sales tax rate? 5.5%

15. The goal of a local charity is to raise $400,000. At the present time, it has raised $250,000. What percent of the goal has been reached? $62\frac{1}{2}$%

16. A local charity wishes to raise $500,000. At the end of the fund drive, $600,000 was raised. What percent of its goal did the charity reach?
120%

Ratio, Proportion, and Percent **413**

Key Questions

1. How can you find what percent one number is of another?
   Determine the "part" and the "whole" and then divide the part by the whole. This may be done using an equation in which the unknown represents the percent.

2. In the type of problem presented in this lesson, when will the percent be greater than 100?
   when the part is greater than the whole

Error Analysis

Students may have difficulties with percents greater than 100 because they want to divide the larger number by the smaller. Work several examples of this type of problem to help students overcome their difficulties.

Additional Examples

1. What percent of 60 is 9? 15%
2. What percent of 48 is 72? 150%
3. 19 is what percent of 200? 9.5%

Closing the Lesson

Have students discuss the difference between the following two types of problems:

25% of 40 = __?__

30 is what percent of 90?

Suggested Assignments

Skills: 1–20, 27–30
Average: 1–30
Advanced: 1–15 odd, 17–30, Historical Note

Exercise Notes

Communication: Writing

Guided Practice Exercises 1–8 require students to translate written sentences into algebraic equations and vice versa.

Application

Exercises 21–26 show the application of percents in finding simple interest. This is a very practical use with which all students should become familiar.

Follow-Up

Extension

Have students keep track of how they spend their time for one week and summarize the data in a chart, such as the following.

Activity	Total Hours	%
Sleeping	56	$33\frac{1}{3}$
Watching TV		
Studying		

Enrichment

Enrichment 94 is pictured on page 386F.

Practice

Practice 116, *Resource Book*

B 17. Last November, it snowed on 6 out of 30 days. What percent of the days did it not snow last November? **80%**

18. Jenna's March telephone bill showed $21.50 for local charges and $8.50 for long distance charges. What percent of her total bill was for long distance charges? **$28\frac{1}{3}$%**

19. DATA, *pages 338–339* In Grand Rapids, the recommended basement ceiling insulation is what percent of the attic floor insulation? **$59\frac{1}{11}$%**

20. DATA, *pages 140–141* The maximum speed of a grizzly bear is what percent of the maximum speed of a chicken? **$333\frac{1}{3}$%**

Interest is money paid for the use of money. Money deposited in a bank, on which interest is paid, is called the *principal*. The simple interest formula is $I = Prt$, where I is the interest earned, P is the principal, r is the interest rate per year, and t is the time in years.

Use the formula $I = Prt$.

21. Find I when $P = \$1000$, $r = 5\%$ per year, and $t = 1$ year. **$50**

22. Find I when $P = \$500$, $r = 7.2\%$ per year, and $t = 2$ years. **$72**

23. Find I when $P = \$2500$, $r = 6\frac{1}{2}\%$ per year, and $t = 3$ years. **$487.50**

24. Find I when $P = \$2000$, $r = 5\%$ per year, and $t = 6$ months. **$50**

25. Find r when $I = \$57.50$, $P = \$1250$, and $t = 1$ year. **4.6%**

26. Find t when $I = \$1147.50$, $P = \$6000$, and $r = 8\frac{1}{2}\%$ per year. **$2\frac{1}{4}$ years**

SPIRAL REVIEW

S 27. Replace the __?__ with $>$, $<$, or $=$: $\frac{7}{9}$ __?__ $\frac{7}{10}$ *(Lesson 7-6)* **>**

28. What percent of 40 is 48?   *(Lesson 9-7)* **120%**

29. Find the LCM: 15 and 40   *(Lesson 7-3)* **120**

30. Evaluate $-12 + 5n^2$ when $n = -4$.   *(Lesson 3-6)* **68**

Historical Note

Julio Rey Pastor (1888–1962) was named professor of mathematical analysis at the University of Oviedo in his native Spain at the age of twenty-three. Later he founded a mathematics laboratory in Madrid. In 1912 he began teaching half the year in Spain and the other half in Argentina. During this time he established the journal *Spanish American Mathematical Review*.

See answer on p. A25.

Research

The first mathematics book in the Americas was printed in 1556. Find out the name of its author and the country where it was printed.

9-8

Finding a Number When a Percent of It Is Known

Objective: To find a number when a percent of it is known.

APPLICATION

In an election for senior class president, 126 students voted for Takashi Noma. Takashi received 60% of the total vote.

QUESTION How many students voted in the election?

You can use an equation to find a number when a percent of it is known. In the situation above, 60% of the total vote is 126 students.

Example 1

Solution

60% of what number is 126?

Let n = the missing number.

$$\begin{array}{ccccc} 60\% & \text{of} & \text{what number} & \text{is} & 126? \\ \downarrow & \downarrow & \underline{\quad\quad} & \downarrow & \downarrow \end{array}$$

$$0.6 \cdot \quad n \quad = 126 \quad \longleftarrow \text{Write 60\% as a decimal.}$$

$$0.6n = 126$$

$$\frac{0.6n}{0.6} = \frac{126}{0.6}$$

$$n = 210 \qquad \text{So 60\% of 210 is 126.}$$

Check Your Understanding

1. If 60% of a number is 126, is the missing number *more than* or *less than* 126? Explain.
   See *Answers to Check Your Understanding* at the back of the book.

ANSWER 210 students voted in the election.

Example 2

Solution

120 is $66\frac{2}{3}\%$ of what number?

Let n = the missing number.

$$\begin{array}{cccc} 120 & \text{is} & 66\frac{2}{3}\% & \text{of what number?} \\ \downarrow & \downarrow & \downarrow & \downarrow \underline{\quad\quad} \end{array}$$

$$120 = \frac{2}{3} \quad \cdot \quad n \quad \longleftarrow \text{Write } 66\frac{2}{3}\% \text{ as a fraction.}$$

$$120 = \frac{2}{3}n$$

$$\left(\frac{3}{2}\right)\left(\frac{120}{1}\right) = \left(\frac{3}{2}\right)\left(\frac{2}{3}n\right)$$

$$180 = n \qquad \text{So 120 is } 66\frac{2}{3}\% \text{ of 180.}$$

Check Your Understanding

2. In Example 2, why is $66\frac{2}{3}\%$ rewritten as the fraction $\frac{2}{3}$?
3. In Example 2, why is $\frac{3}{2}$ used to multiply both sides of the equation?
   See *Answers to Check Your Understanding* at the back of the book.

Ratio, Proportion, and Percent **415**

Lesson Planner

Teaching the Lesson
- Lesson Starter 87
- Toolbox Skills 14, 20, 21

Lesson Follow-Up
- Practice 117
- Practice 118: Nonroutine
- Enrichment 95: Application
- Study Guide, pp. 173–174
- Test 44

Warm-Up Exercises

Find each answer.
1. 11 is what percent of 20? 55%
2. What percent of 8 is 7? $87\frac{1}{2}\%$
3. What is 30% of 56? 16.8
4. What is 200% of 15? 30

Lesson Focus

Last season a baseball team won 12 games, which was 60% of the games the team played. What was the total number of games the team played? (20) Today's lesson will focus on how to solve a percent problem of this type.

Teaching Notes

Work through the Application at the beginning of the lesson, encouraging students to set up the equation on their own before referring to the book's solution. Using the "part to whole" concept, stress that the whole is the unknown in this problem and that the part (126) is given.

Key Questions

Using n for the variable, write the equations for the following.
1. 30 is what percent of 40?
   $30 = 40n$
2. What is 30% of 40?
   $n = 0.3(40)$
3. 30 is 40% of what number?
   $30 = 0.4n$

Critical Thinking

The Check Your Understanding exercises require students to *analyze* the components of the problem and *draw a conclusion* about the solution based on their knowledge of percents.

Error Analysis

Students may set up the equations incorrectly. Discussing the problems with a partner often helps eliminate this error.

Additional Examples

1. 42% of what number is 63?
   150

2. 130 is $33\frac{1}{3}$% of what number?
   390

Closing the Lesson

Have students make up an equation that can be used to find a number when a percent of it is known. Students should then exchange their equations with a partner. Each student should create a word problem that can be represented by the equation he or she was given. Have students discuss their equations and problems with one another.

Suggested Assignments

Skills: 1–23, 29–32
Average: 1–32
Advanced: 1–15 odd, 17–32

Exercise Notes

Communication: Writing

Guided Practice Exercises 2–5 ask students to translate written sentences into equations.

Making Connections/Transitions

Exercises 17–24 relate percents to nutritional information and show a practical example of the use of percents.

Guided Practice

1. Which equation would you use to solve the following problem?
   A shoe salesperson earns a commission of 12% of all sales. How much must the salesperson sell to earn $288? a
   a. $0.12 \cdot n = 288$ **b.** $0.12 \cdot 288 = n$ **c.** $0.12 = 288 \cdot n$

COMMUNICATION «*Writing*

Write an equation to represent each question. Use n for the missing number. Do not solve. 2. $24 = 0.5n$ or $24 = \frac{1}{2}n$ 3. $45 = 0.25n$ or $45 = \frac{1}{4}n$

«**2.** 24 is 50% of what number?   «**3.** 45 is 25% of what number?

«**4.** $16\frac{2}{3}$% of what number is 88?   «**5.** $18\frac{1}{2}$% of what number is 160?
                        $\frac{1}{6}n = 88$ $0.185n = 160$

Solve each equation.

6. $0.08n = 60$ 750   **7.** $45 = \frac{3}{5}n$ 75   **8.** $3.4n = 17$
                                                                                5

Find each answer.

9. 75% of what number is 24? 32   **10.** 25% of what number is 36?
                                                    144
11. 2000 is $83\frac{1}{3}$% of what   **12.** 48 is 120% of what number?
     number? 2400   40

Exercises

Find each answer. 5. 2000   11. 36,000

A **1.** 10% of what number is 42?   **2.** 30 is 60% of what number? 50
        420
    3. 17 is 50% of what number? 34   **4.** 40% of what number is 114?
                                                      285
    5. 130 is 6.5% of what number?   **6.** 3% of what number is 54?
                                                      1800
    7. 75% of what number is 60? 80   **8.** 216 is 90% of what number?
                                                      240
    9. 15 is $33\frac{1}{3}$% of what number? 45   **10.** 250% of what number is 50?
                                                      20
    11. 180 is $\frac{1}{2}$% of what number?   **12.** $37\frac{1}{2}$% of what number is 72?
                                                      192

13. Rozlyn got a score of 80% on her last French test. She got 40 items correct. How many items were on the test? 50

14. There are 4230 season ticket holders for the Cougars football games. This represents $56\frac{2}{5}$% of the seating capacity at the stadium. What is the seating capacity at the stadium? 7500

15. Kent earns an 8% commission on all furniture sales. How much must he sell to earn a commission of $336? $4200

16. In last year's Freshman Class election, one candidate received 36 votes. This result represents 45% of the total class vote. How many students voted in the Freshman Class election last year? 80

CONNECTING MATHEMATICS AND HEALTH

The table at the right shows the percent of the Recommended Daily Allowance (RDA) of nutrients contained in a 1 oz serving of a certain cereal.

Use the table at the right.

B **17.** How many calories are in one serving of cereal with whole milk? 180

18. How many more milligrams of potassium are in one serving of cereal with skim milk than with whole milk? 20

19. What percent of the RDA for Vitamin C does one serving of plain cereal give? 100%

20. How much less is the percent of the RDA for Vitamin D for plain cereal than for cereal with skim milk? 10%

NUTRITION INFORMATION
SERVING SIZE: 1 OZ (284 g, ABOUT 1 CUP)
ALONE OR WITH 1/2 CUP SKIM MILK OR WHOLE MILK
SERVINGS PER PACKAGE: 12

	CEREAL	WITH SKIM MILK	WITH WHOLE MILK
CALORIES	110	150	180
PROTEIN	2 g	6 g	6 g
CARBOHYDRATE	24 g	30 g	30 g
FAT	0 g	0 g	4 g
CHOLESTEROL	0 mg	0 mg	15 mg
SODIUM	290 mg	350 mg	350 mg
POTASSIUM	35 mg	240 mg	220 mg

PERCENTAGE OF U.S. RECOMMENDED DAILY ALLOWANCES (U.S. RDA)

	CEREAL	WITH SKIM MILK	WITH WHOLE MILK
PROTEIN	4	15	15
VITAMIN A	100	100	100
VITAMIN C	100	100	100
THIAMIN	100	100	100
RIBOFLAVIN	100	110	110
NIACIN	100	100	100
CALCIUM	*	15	15
IRON	100	100	100
VITAMIN D	50	60	60
VITAMIN E	100	100	100
VITAMIN B_6	100	100	100
FOLIC ACID	100	100	100
VITAMIN B_{12}	100	110	110
PHOSPHORUS	4	15	15
MAGNESIUM	2	6	6
ZINC	100	100	100
COPPER	2	4	10

*CONTAINS LESS THAN 2% OF THE U.S. RDA OF THIS NUTRIENT

21. a. How many grams of protein are in one serving of cereal with skim milk? 6 g
  b. What percent of the RDA for protein does one serving of cereal with skim milk give? 15%
  c. Use your answers to parts (a) and (b) to find the RDA for protein. 40 g

22. The RDA for calcium for teenagers is 1200 mg. How many milligrams of calcium are in one serving of cereal with skim milk? 180 mg

23. The table shows that one serving of cereal with whole milk provides 110% of the RDA for Vitamin B_{12}. Explain what this percent means.
See answer on p. A25.

24. RESEARCH Compare the nutrient percents in the table above with those given on another box of cereal. Do the two cereals provide the same percents? If not, which brand seems more nutritious to you?
Answers will vary.

Ratio, Proportion, and Percent **417**

Quick Quiz 2 *(9–5 through 9–8)*

1. Write 0.58 as a percent. 58%

2. Write $\frac{6}{50}$ as a percent. 12%

3. Write 125% as a decimal.
   1.25

4. Write 36% as a fraction. $\frac{9}{25}$

Find each answer.

5. 43% of 29 is what number?
   12.47

6. What number is $6\frac{1}{2}$% of 210?
   13.65

7. What percent of 84 is 10.5?
   $12\frac{1}{2}$%

8. 33 is what percent of 99?
   $33\frac{1}{3}$%

9. 104 is 40% of what number?
   260

10. 8% of what number is 28?
    350

Alternative Assessment

Imagine you are a real estate agent who receives a 4% commission on the sale of a house.

1. What is your commission on the sale of a $120,000 house?
   $4800

2. How much were your total sales last year if your total commission earnings were $44,000? $1,100,000

CAREER/APPLICATION

A *retail salesperson* sells goods or services directly to consumers. Retail salespeople work in many businesses, such as department stores, car dealerships, insurance agencies, and supermarkets. The work of retail salespeople often involves percents.

25. A salesperson at a clothing store marks an item priced at $32.50 for a 40% discount. What is the final selling price of the item? $19.50

26. A supermarket cashier charges a 5% sales tax on taxable items. A customer buys $14.69 worth of nontaxable items and $8.40 worth of taxable items. What was the total cost of the purchases? $23.51

27. A salesperson at Central Autos earned a $765 commission on the sale of a $9000 car. What is the percent of commission? 8.5%

28. A real estate agent determined that the value of a house had increased by 25%. The owners sold the house for $22,000 more than the price they had originally paid. What did the owners pay for the house originally? $88,000

SPIRAL REVIEW

S 29. Write $\frac{40}{64}$ in lowest terms. *(Lesson 7-4)* $\frac{5}{8}$

30. 36% of what number is 18? *(Lesson 9-8)* 50

31. Find the quotient: $\frac{14}{15} \div \left(-\frac{7}{10}\right)$ *(Lesson 8-2)* $-1\frac{1}{3}$

32. Solve the proportion: $\frac{8}{3} = \frac{n}{9}$ *(Lesson 9-2)* 24

Self-Test 2

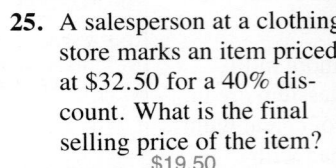

1. Write $\frac{13}{20}$ as a percent. 65% 2. Write 0.45 as a percent. 45% 9-5

3. Write 130% as a decimal. 4. Write 28% as a fraction. $\frac{7}{25}$
   1.3

Find each answer. 5. 18.85 6. 6.75 7. $16\frac{2}{3}$% 9. 380

5. 65% of 29 is what number? 6. What number is $4\frac{1}{2}$% of 150? 9-6

7. What percent of 99 is 16.5? 8. 34 is what percent of 68? 50% 9-7

9. 57 is 15% of what number? 10. 2.5% of what number is 4? 160 9-8

9-9

Percent of Increase or Decrease

Objective: To find a percent of increase or decrease.

Terms to Know
* *percent of increase*
* *percent of decrease*

DATA ANALYSIS

After two years on the soccer team, Matt and Terry compared their records.

QUESTION Which player improved more?

To find the answer, find the *percent of increase* in scoring for each player. The **percent of increase** tells what percent the amount of increase is of the original amount.

$$\text{percent of increase} = \frac{\text{amount of increase}}{\text{original amount}}$$

Soccer Goals Scored

	Matt	Terry
Last Year	10	6
This Year	14	9

Example 1

Find the percent of increase.

a. original amount: 10
   new amount: 14

b. original amount: 6
   new amount: 9

Solution

a. amount of increase: $14 - 10 = 4$

$$\text{percent of increase} = \frac{\text{amount of increase}}{\text{original amount}}$$
$$= \frac{4}{10} = \frac{40}{100} = 40\%$$

b. amount of increase: $9 - 6 = 3$

$$\text{percent of increase} = \frac{\text{amount of increase}}{\text{original amount}}$$
$$= \frac{3}{6} = \frac{1}{2} = 50\%$$

 **Check Your Understanding**

1. In part (a) of Example 1, why is the fraction $\frac{4}{10}$ used, not $\frac{4}{14}$?
   See *Answers to Check Your Understanding* at the back of the book.

ANSWER Matt's percent of increase was 40%, and Terry's percent of increase was 50%. So Terry improved more.

You can use a similar method to find a *percent of decrease*. The **percent of decrease** tells what percent the amount of decrease is of the original amount. When you refer to a sale price of an item, you often call the percent of decrease the *rate of discount*.

Ratio, Proportion, and Percent **419**

Teaching the Lesson
* Lesson Starter 88
* Reteaching/Alternate Approach, p. 386D
* Cooperative Learning, p. 386D
* Toolbox Skills 7–15

Lesson Follow-Up
* Practice 119
* Enrichment 96: Exploration
* Study Guide, pp. 175–176
* Computer Activity 9

Warm-Up Exercises

Find each answer.
1. 8 is what percent of 48? $16\frac{2}{3}\%$
2. 120 is what percent of 80? 150%
3. 6% of 64 is what number? 3.84
4. What is 18% of 500? 90

Lesson Focus

For many businesses, an item that cost $150 five years ago can now cost $250 or more. How can you find the percent of increase? Or, suppose your average grade in mathematics has gone down from 93% to 89%. How can you compute the percent of decrease? The focus of today's lesson is to solve these types of problems.

Teaching Notes

You may wish to use the following Application to introduce the lesson.

Application

Last year, Joanna was able to do 20 laps during the Jog-a-Thon. This year, she did 23 laps in the same amount of time. What was the percent of increase in laps run by Joanna? 15%

419

Key Questions
1. What is meant by ''an increase of more than 100%''? The amount more than doubles.
2. What is the first step in finding the percent of increase or decrease? Find the amount of increase or decrease.

Error Analysis
Students will sometimes compare the amount of increase or decrease to the final amount instead of the original amount. Using the label ''original amount'' in the ratio can help students to avoid this error.

Reteaching/Alternate Approach
For a suggestion on using a cooperative learning activity with this lesson, see page 386D.

Additional Examples
1. Find the percent of increase.
   a. original amount: 16; new amount: 20 25%
   b. original amount: 160; new amount: 168 5%
2. Last month, a train ticket cost $140. This month, a ticket to the same destination costs $126. What is the percent of decrease? 10%

Closing the Lesson
Have students verbalize the steps for finding a percent of increase or decrease.

Suggested Assignments
Skills: 1–19, 22–25
Average: 1–25
Advanced: 1–13 odd, 15–25

Example 2	The original price of a stereo is $150, and the new price is $100. Find the percent of decrease.
Solution	amount of decrease: $150 − $100 = $50

$$\text{percent of decrease} = \frac{\text{amount of decrease}}{\text{original amount}}$$

$$= \frac{50}{150} = \frac{1}{3} = 33\frac{1}{3}\%$$

Guided Practice

«1. **COMMUNICATION** «*Reading* Replace each __?__ with the correct word.
When you find the percent a price has been lowered, you find the rate of __?__. discount
When you find the percent a price has been raised, you find the percent of __?__. increase

Write the fraction you would use to find the percent of increase or decrease. Do not find the percent of increase or decrease.

	original amount	new amount
2.	$40	$30 $\frac{10}{40}$
3.	125 cm	100 cm $\frac{25}{125}$
4.	7.5 ft	12.5 ft $\frac{5}{7.5}$
5.	250 calculators	200 calculators $\frac{50}{250}$

Find the percent of increase.

6. original cost: $200; new cost: $230 15%

7. original population: 1000; new population: 1750 75%

Find the percent of decrease.

8. original price: $20; new price: $19 5%

9. original distance: 84 ft; new distance: 56 ft $33\frac{1}{3}$%

Exercises

Find the percent of increase.

A 1. original score: 100; new score: 118 18%

2. original record: 25 s; new record: 35 s 40%

3. original weight: 150 lb; new weight: 180 lb 20%

4. original price: $15; new price: $25 $66\frac{2}{3}\%$

5. original length: 23 m; new length: 46 m 100%

Find the percent of decrease.

6. original number of boxes: 40; new number of boxes: 14 65%

7. original cost: $100; new cost: $79 21%

8. original population: 500; new population: 460 8%

9. original number of students: 144; new number of students: 20 $86\frac{1}{9}\%$

10. original fare: $250; new fare: $240 4%

11. Last week Suzanne earned $15 doing errands. This week she earned $40. Find the percent of increase in Suzanne's pay. $166\frac{2}{3}\%$

12. Kyle bought a painting for $600 and later sold it for $800. Find the percent of increase in the value of the painting. $33\frac{1}{3}\%$

13. The original price of a jacket is $144. The new price is $54. Find the rate of discount. 62.5%

14. A runner finished a race in 10.5 s. In the next race the runner finished in 9.8 s. Find the percent of decrease in the running time. $6\frac{2}{3}\%$

Tell whether there is an *increase* or a *decrease*. Then find the percent of increase or decrease.

B 15. original number of acres: 40; new number of acres: 90 increase; 125%

16. original weight: 16 kg; new weight: 21 kg increase; $31\frac{1}{4}\%$

17. original population: 3.6 million; new population: 3 million decrease;

18. original length: 2.5 cm; new length: 1.25 cm decrease; 50% $16\frac{2}{3}\%$

19. **WRITING ABOUT MATHEMATICS** Write a paragraph to a friend describing how to find the percent of decrease from $15 to $13.80. See answer on p. A25.

THINKING SKILLS 20–21. Analysis; evaluation

Decide whether each statement is *True* or *False*. Give a convincing argument to support your answer. Answers to Exercises 20–21 are on p. A25.

C 20. Twice an amount is the same as an increase of 200%.

21. Half an amount is the same as a decrease of 50%.

SPIRAL REVIEW

S 22. Solve: $1.4x + 0.4 = 20$ *(Lesson 8-8)* 14 $\frac{18}{1}$

23. Write the ratio 3 min : 10 s as a fraction in lowest terms. *(Lesson 9-1)*

24. Write $\frac{17}{20}$ as a decimal. *(Lesson 7-7)* 0.85

25. The original price of a bracelet is $3. The new price is $2.40. Find the percent of decrease. *(Lesson 9-9)* 20%

Communication: Writing

Exercise 19 requires students to write a paragraph that describes the procedure for finding a specific percent of decrease.

Critical Thinking

Exercises 20 and 21 require students to *consider* statements that typify some common misconceptions about percents. Using an example as part of their *argument* would help students *validate* their thinking process.

Follow-Up

Extension

Ask each student to choose some item that has changed in price during the past year. Students should show at least three different values and then calculate the percent of increase or decrease for each value.

Enrichment

Enrichment 96 is pictured on page 386F.

Practice

Practice 119, *Resource Book*

Practice 119
Skills and Applications of Lesson 9-9

Find the percent of increase.

1. original number of students: 50; new number of students: 58 16%

2. original price: $75; new price: $90 20%

3. original salary: $625; new salary: $750 20%

4. original depth: 8 ft; new depth: 12 ft 50%

5. original length: 20 cm; new length: 24 cm 20%

6. original width: 15 cm; new width: 20 cm $33\frac{1}{3}\%$

7. original fare: $160; new fare: $180 $12\frac{1}{2}\%$

8. original cost: $64; new cost: $80 25%

Find the percent of decrease.

9. original amount: $80; new amount: $68 15%

10. original score: 125; new score: 75 40%

11. original price: $18; new price: $16 $11\frac{1}{9}\%$

12. original salary: $500; new salary: $450 10%

13. original height: 27 in.; new height: 18 in. $33\frac{1}{3}\%$

14. original distance: 400 km; new distance: 320 km 20%

15. original width: 36 mm; new width: 33 mm $8\frac{1}{3}\%$

16. original length: 48 in.; new length: 42 in. $12\frac{1}{2}\%$

17. The original price of a shirt was $32. The new price is $28. Find the percent of decrease. 12.5%

18. In June, Gary's Appliance Store sold 60 telephones. In July the store sold 72 telephones. Find the percent of increase in the number of sales of telephones. 20%

19. Last year Shani made 12 soccer goals. This year she made 10 soccer goals. Find the percent of decrease in the number of goals Shani scored. $16\frac{2}{3}\%$

20. The Gonzalez family increased the size of their family room from 180 sq ft to 240 sq ft. Find the percent of increase in the family room area. $33\frac{1}{3}\%$

Multicultural Notes

Background

Every word in English has a story behind it. For instance, Spanish explorers first tasted *chocolate* when they encountered the Aztec civilization in the early 1500s. The Spaniards flavored the bitter drink with vanilla and cinnamon. The first recorded use of the word chocolate in English took place in 1604.

English settlers adopted Native American words to describe the unfamiliar plants and animals they saw in the New World. The settlers transformed a Cree word, *otchek*, meaning a marten or fisher, into a compound of wood and chuck. But the animal they called the woodchuck was not an *otchek*, nor did it chuck wood.

The word *limousine* originally referred to an early French automobile. In front of its enclosed passenger seat, the limousine had a driver's seat, which was covered by a roof that may have looked like the hood of a cloak.

Language Connection

MIXED REVIEW
Working with Percent

As early as 1709, American colonists were using the word *barbecue* to refer to meat roasted over an open flame. The word *barbecue* came from a Spanish word *barbacoa*. Spanish explorers had learned about barbecuing meat from a group of people native to the West Indies. The Spaniards then borrowed the word *barbacoa*, which meant a stick framework used to roast or smoke meat.

To discover the name of the people from whom the Spanish explorers learned the word *barbacoa*, complete the code boxes below. Begin by solving Exercise 1. The answer to Exercise 1 is 20, and T is associated with Exercise 1. Write T in the box above 20 as shown below. Continue in this way with Exercises 2–21.

1. What number is 25% of 80?   (T)
   20
2. What number is 74% of 50?   (A)
   37
3. What percent of 160 is 56?   (F)
   35
4. What percent of 325 is 13?   (O)
   4
5. $66\frac{2}{3}$% of what number is 80?   (H)
   120
6. 70% of what number is 294?   (L)
   420
7. 60% of 125 is what number?   (T)
   75
8. 15% of 320 is what number?   (O)
   48
9. 42 is what percent of 150?   (I)
   28
10. 32 is what percent of 256?   (H)
    12.5
11. $37\frac{1}{2}$% of what number is 24?   (O)
    64
12. 3% of what number is 6?   (P)
    200
13. What number is 65% of 270?   (I)
    175.5
14. What number is 82% of 200?   (A)
    164
15. What percent of 75 is 66?   (E)
    88
16. What percent of 228 is 76?   (E)
    $33\frac{1}{3}$
17. 50% of what number is 457?   (T)
    914
18. 396 is 75% of what number?   (N)
    528
19. 18 is $\frac{1}{2}$% of what number?   (I)
    3600
20. $33\frac{1}{3}$% of 87 is what number?   (E)
    29
21. What percent of 420 is 378?   (P)
    90

T	H	E		A	I	N	O
?	?	?	T	?	?	?	?
914	120	88	20	164	28	528	4

P	E	O	P	L	E		O	F		H	A	I	T	I
?	?	?	?	?	?		?	?		?	?	?	?	?
90	29	48	200	420	$33\frac{1}{3}$		64	35		12.5	37	175.5	75	3600

422   Chapter 9

Many words in the English language have their origins in other languages. By solving Exercises 22–29, you can learn about the origins of some common English words. Solve each exercise. The word printed in blue following the correct answer matches the description printed in blue.

22. 65 is 13% of what number?

Word that comes from an Arabic word meaning *calendar*.

a. 845   schedule   **b.** 500   almanac

23. What percent of 225 is 75?

Word that comes from a Hindi word for a tie-dying process.

a. $33\frac{1}{3}$ bandanna   **b.** 168   batik

24. 80% of what number is 900?

Word that comes from a French word referring to a type of *cloak*.

a. 720   poncho   **b.** 1125   limousine

25. 40 is 5% of what number?

Word that comes from two Chinese words referring to a fish sauce.

a. 2   tartar   **b.** 800   ketchup

26. What number is $12\frac{1}{2}$% of 88?

Word that comes from an Ojibwa word meaning *head first*.

a. 11   chipmunk   **b.** 1100   woodchuck

27. $37\frac{1}{2}$% of what number is 3000?

Word that comes from an Aztec word meaning *bitter water*.

a. 8000   chocolate   **b.** 1125   coffee

28. What percent of 120 is 90?

Word that comes from a Spanish word meaning *lizard*.

a. 108   crocodile   **b.** 75   alligator

29. 16% of what number is 400?

Word that comes from a Senegalese word meaning *to eat*.

a. 2500   yam   **b.** 64   okra

Ratio, Proportion, and Percent   **423**

Suggested Activities

Students can increase their understanding of other cultures by researching the meanings of words in other languages. An activity that also increases students' understanding of mathematics is to explore the meaning of words used for numbers in other languages. In English, number words are based on groupings of ten. Fourteen is four and ten. Forty is four tens. French number words are also grouped by tens—until the number seventy, that is. The French word for seventy means sixty and ten. Eighty is four twenties. Ninety is four twenties and ten. The words for these numbers are based on grouping by twenties instead of tens.

Many people group numbers by twenties or by fives, as is often reflected in the number words they use. Ask students if they can count in other languages. Have them discuss the meanings of the number words used in these languages.

Suggested Resources

Webster's Word Histories. Springfield, MA: Merriam-Webster Inc., 1989.

Zaslavsky, Claudia. *Africa Counts: Number and Pattern in African Culture*. Boston: Prindle, Weber and Schmidt, Inc., 1973.

————. "Bringing the World Into the Math Class." *Curriculum Review*, January/February 1985, pp. 62–65.

Teaching the Lesson
- Materials: grids
- Lesson Starter 89
- Using Manipulatives, p. 386D
- Toolbox Skills 9, 10, 13–15

Lesson Follow-Up
- Practice 120
- Practice 121: Nonroutine
- Enrichment 97: Connection
- Study Guide, pp. 177–178
- Test 45

Warm-Up Exercises

1. 45% of what number is 63?
   140
2. 10 is 20% of what number?   50
3. What percent of 48 is 16?
   $33\frac{1}{3}\%$
4. What percent of 48 is 6?
   12.5%

Lesson Focus

Out of 550 graduates of a certain high school last year, 375 enrolled in college. What percent of the graduates is this? This type of problem can be solved by using an equation. The focus of today's lesson is to look at another way to solve this problem, namely, by using proportions.

Teaching Notes

Stress that many people have a "mathematical sense" regarding proportional thinking, and that this kind of thinking is used frequently for solving problems.

Percents and Proportions

Objective: To use proportions to solve percent problems.

CONNECTION

You have learned to solve proportions and you know that a percent is a ratio. You can use proportions to solve percent problems. To do this, you write the problem as a proportion.

$$\frac{\text{part}}{\text{whole}} = \frac{\text{part}}{\text{whole}}$$

One of the ratios in the proportion will represent the percent. The *whole* part of this ratio will always be 100.

Example 1

Solution

What number is 25% of 56?

The ratio represented by 25% is $\frac{25}{100}$. The part is 25. The whole is 100. You are looking for a percent of 56, so n is the part and 56 is the whole.

Write the proportion and solve.

$$\frac{\text{unknown part}}{\text{whole}} \longrightarrow \quad \frac{n}{56} = \frac{25}{100} \longleftarrow \quad \frac{\text{part}}{\text{whole}}$$

$$100n = 56(25)$$

$$\frac{100n}{100} = \frac{1400}{100}$$

$$n = 14$$

So 14 is 25% of 56.

Example 2

Solution

What percent of 25 is 20?

You are looking for the percent. So in the ratio representing the percent, n is the part and 100 is the whole. In the other ratio, 20 is the part and 25 is the whole.

$$\frac{\text{part}}{\text{whole}} \longrightarrow \quad \frac{20}{25} = \frac{n}{100} \longleftarrow \quad \frac{\text{unknown part}}{\text{whole}}$$

$$20(100) = 25n$$

$$\frac{2000}{25} = \frac{25n}{25}$$

$$80 = n$$

So 80% of 25 is 20.

✓ **Check Your Understanding**

1. In Example 1, will the result change if you write $\frac{25}{100}$ in simplest form before you find the cross products? Explain.
2. In Example 2, why is the answer 80% rather than 80?
   See *Answers to Check Your Understanding* at the back of the book.

424   Chapter 9

Example 3

Solution

48.6 is 60% of what number?

The ratio represented by 60% is $\frac{60}{100}$. The part is 60. The whole is 100. The number 48.6 is a percent of another number. So in the other ratio, 48.6 is the part and n is the whole.

$$\frac{\text{part}}{\underline{unknown}\text{ whole}} \longrightarrow \frac{48.6}{n} = \frac{60}{100} \longleftarrow \frac{\text{part}}{\text{whole}}$$

$$(48.6)(100) = 60n$$

$$\frac{4860}{60} = \frac{60n}{60}$$

$$81 = n$$

So 48.6 is 60% of 81.

Guided Practice

COMMUNICATION «*Reading*

Tell whether the number shown in color is a *part* or a *whole*.

«**1.** What number is 70% of 220?
whole

«**2.** 25% of what number is 45?
part

«**3.** 30% of 12.4 is what number?
whole

«**4.** 24 is what percent of 36? part

«**5.** What percent of 51 is 34?
whole

«**6.** 0.28 is 80% of what number?
part

COMMUNICATION «*Writing*

«**7.** In symbols, write the proportion described by the statement:
10 is to 25 as an unknown number is to 100. $\frac{10}{25} = \frac{x}{100}$

«**8.** In symbols, write a proportion that represents the situation:
50% of 236 is what number? $\frac{50}{100} = \frac{x}{236}$

«**9.** In words, write the question that is represented by the proportion:
$\frac{9.5}{38} = \frac{n}{100}$ 9.5 is what percent of 38?

«**10.** In words, write the proportion that represents the situation:
36 is 25% of what number? 36 is to an unknown number as 25 is to 100.

Find each answer using a proportion.

11. What number is 45% of 6? 2.7

12. 8 is what percent of 32? 25%

13. 140% of what number is 21? 15

14. 90% of 450 is what number?
405

15. What percent of 24 is 9? 37.5%

16. 2.8 is 5% of what number? 56

17. COMMUNICATION «*Discussion* In Example 2 on p. 424, how can you tell that 25 is the whole? The question asks for a percent, or part, of 25, so 25 must be the whole.

Ratio, Proportion, and Percent **425**

426

Exercises

Find each answer using a proportion.

A
1. What number is 4% of 236?
9.44
2. 18 is 40% of what number? 45
3. 6% of what number is 210?
3500
4. 18% of 350 is what number?
63
5. 18 is what percent of 120?
15%
6. What percent of 40 is 25? $62\frac{1}{2}$%
7. 10% of what number is 24.5?
245
8. 7% of what number is 84?
1200
9. What number is 48% of 2200?
1056
10. 11% of what number is 1.21?
11
11. What percent of 60 is 4.2? 7%
12. 95 is what percent of 95?
100%

13. The Weston Library has 8000 books. Fiction books make up 35% of the total. How many fiction books does the library have? 2800

14. Martin sells stereo equipment and earns a commission on all sales. Last month he earned $840 in commissions. His total monthly sales were $14,000. What percent of Martin's sales was his commission?
6%

15. Last month, General Cable Company made service calls to 135 customers. This represents 3% of all the customers subscribing to General Cable Company. How many customers does General Cable Company have? 4500

PROBLEM SOLVING/APPLICATION

One way that state and local governments raise money is through sales taxes. Tables like the one shown at the right are used to find the sales tax for various purchase prices.

Use the tax table at the right for Exercises 16–20.

B
16. Find the sales tax for an item priced at $8.73. $.48

17. What interval has a sales tax of $.07. $1.18–$1.35

18. What is the total amount, including tax, that you would pay for an item priced at $5.54? $5.85

19. Including tax, how much more would you pay for an item priced at $6.00 than for one priced at $5.00?
$1.05

5.5% Sales Tax Table			
Amount of Sale	Tax	Amount of Sale	Tax
0.00–0.09	0.00	4.96–5.13	0.28
0.10–0.27	0.01	5.14–5.31	0.29
0.28–0.45	0.02	5.32–5.49	0.30
0.46–0.63	0.03	5.50–5.67	0.31
0.64–0.81	0.04	5.68–5.85	0.32
0.82–0.99	0.05	5.86–6.03	0.33
1.00–1.17	0.06	6.04–6.21	0.34
1.18–1.35	0.07	6.22–6.39	0.35
1.36–1.53	0.08	6.40–6.57	0.36
1.54–1.71	0.09	6.58–6.75	0.37
1.72–1.89	0.10	6.76–6.93	0.38
1.90–2.07	0.11	6.94–7.11	0.39
2.08–2.25	0.12	7.12–7.29	0.40
2.26–2.43	0.13	7.30–7.47	0.41
2.44–2.61	0.14	7.48–7.65	0.42
2.62–2.79	0.15	7.66–7.83	0.43
2.80–2.97	0.16	7.84–8.01	0.44
2.98–3.15	0.17	8.02–8.19	0.45
3.16–3.33	0.18	8.20–8.37	0.46
3.34–3.51	0.19	8.38–8.55	0.47
3.52–3.69	0.20	8.56–8.73	0.48
3.70–3.87	0.21	8.74–8.91	0.49
3.88–4.05	0.22	8.92–9.09	0.50
4.06–4.23	0.23	9.10–9.27	0.51
4.24–4.41	0.24	9.28–9.45	0.52
4.42–4.59	0.25	9.46–9.63	0.53
4.60–4.77	0.26	9.64–9.81	0.54
4.78–4.95	0.27	9.82–9.99	0.55

20. Jen has a coupon for 10% off the cost of two mugs. The price of one mug is $4.95. Sales tax is taken on the discounted price. Including tax, what will Jen pay for two mugs? $9.40

21. The price of a compact disc is $13.87. The cost, including tax, is $14.77. What is the sales tax rate to the nearest tenth of a percent? 6.5%

22. **RESEARCH** Find the general sales tax rate for each of the 50 states.

Answers to Exercises 22–24 are on p. A26.

WRITING ABOUT MATHEMATICS

23. Write an outline for your notebook entitled *How to Solve a Percent Problem Using a Proportion*.

24. **a.** Use the method you learned in Lesson 9-8 to solve this problem.
7.5% of what number is 112.5?
Then solve the problem again using a proportion.

   b. Write a paragraph describing how these two methods are alike, and how they are different.

SPIRAL REVIEW

S   25. Write 0.943 as a percent.   *(Lesson 9-5)* 94.3%

26. Solve by supplying missing facts: Will spent 37 min doing mathematics homework and 48 min doing science homework. How many hours did Will spend doing homework?   *(Lesson 8-7)* $1\frac{5}{12}$ h

27. Write 0.000058 in scientific notation.   *(Lesson 7-11)* 5.8×10^{-5}

28. Solve: $3(x + 5) = 9$ *(Lesson 4-9)* -2

29. Find the answer using a proportion: 7 is what percent of 8?
(Lesson 9-10) $87\frac{1}{2}$%

Self-Test 3

Find the percent of increase or decrease.

1. original score: 14; new score: 49 250%   9-9

2. original price: $25; new price: $22.50 10%

3. original length: 75 km; new length: 60 km 20%

4. original population: 3000; new population: 4000 $33\frac{1}{3}$%

Find the answer using a proportion.

5. What number is 20% of 16? 3.2 **6.** 9% of what number is 108?   9-10
1200

7. 2.4 is 6% of what number? 40 **8.** What percent of 45 is 18? 40%

Ratio, Proportion, and Percent   **427**

Chapter Review

Terms to Know

ratio (p. 388)
rate (p. 389)
unit rate (p. 389)
proportion (p. 391)
terms (p. 391)
cross products (p. 391)
solve a proportion (p. 392)

scale drawings (p. 395)
scale (p. 395)
scale model (p. 396)
percent (p. 404)
percent of increase (p. 419)
percent of decrease (p. 419)

Choose the correct term from the list above to complete each sentence.

1. The __?__ is the ratio of the size of a drawing to the actual size of the object it represents. scale

2. A __?__ is a rate for one unit of a given quantity. unit rate

3. A comparison of two numbers by division is a __?__. ratio

4. A __?__ is a statement that two ratios are equal. proportion

5. A __?__ represents a ratio that compares a number to 100. percent

6. In the proportion $\frac{6}{9} = \frac{10}{15}$, the __?__ of the proportion are 6, 9, 10, and 15. terms

Write each ratio as a fraction in lowest terms. *(Lesson 9-1)*

7. $\frac{2}{10}$ $\frac{1}{5}$

8. 60 cm to 3 m $\frac{1}{5}$

9. 8:44 $\frac{2}{11}$

10. $\frac{48}{32}$ $\frac{3}{2}$

Write the unit rate. *(Lesson 9-1)*

11. 21 h in 3 days 7 h/day

12. 108 words in 4 min 27 words/min

13. $21.25 for 5 h $4.25/h

Tell whether each proportion is *True* or *False*. *(Lesson 9-2)*

14. $\frac{12}{12} = \frac{6}{5}$ False

15. $\frac{12}{9} = \frac{4.8}{3.6}$ True

16. $\frac{6}{32} = \frac{3}{48}$ False

17. $\frac{5}{7} = \frac{6}{8}$ False

Solve each proportion. *(Lesson 9-2)*

18. $\frac{n}{5} = \frac{2}{10}$ 1

19. $\frac{14}{7} = \frac{8}{y}$ 4

20. $\frac{w}{1.6} = \frac{5}{8}$ 1

21. $\frac{2}{9.2} = \frac{5}{z}$ 23

Solve. *(Lesson 9-3)*

22. The scale on the floor plan of a museum is 1 in. : 16 ft. On the floor plan, the Far Eastern Art room is $1\frac{1}{4}$ in. wide. Find the actual width of the room. 20 ft

23. The scale of a model stadium is 3 in. : 8 ft. The actual height of the stadium is 40 ft. Find the height of the scale model. 15 in.

Solve using a proportion. *(Lesson 9-4)*

24. Tom can read 34 pages of a novel in 50 min. At this rate, how many minutes will it take Tom to read a 272-page novel? 400 min

25. A department store charges $6.50 for 4 blank cassette tapes. What will the department store charge for 10 blank cassette tapes? $16.25

Write each fraction or mixed number as a percent. *(Lesson 9-5)*

26. $\frac{11}{20}$ 55%   **27.** $\frac{7}{16}$ 43.75%   **28.** $2\frac{3}{4}$ 275%   **29.** $\frac{1}{3}$ $33\frac{1}{3}$%   **30.** $\frac{5}{8}$ 62.5%

Write each decimal as a percent. *(Lesson 9-5)*

31. 0.65 65%   **32.** 0.03 3%   **33.** 1.2 120%   **34.** 0.428 42.8%   **35.** 0.0095 0.95%

Write each percent as a fraction or mixed number in lowest terms and as a decimal. *(Lesson 9-5)*

36. 75% $\frac{3}{4}$; 0.75   **37.** 250% $2\frac{1}{2}$; 2.5   **38.** $5\frac{1}{4}$% $\frac{21}{400}$; 0.0525   **39.** 19.5% $\frac{39}{200}$; 0.195   **40.** 0.05% $\frac{1}{2000}$; 0.0005

Estimate. *(Lesson 9-6)*

41. $33\frac{1}{3}$% of 272 about 90   **42.** 51% of 79 about 40   **43.** 13% of 64 about 8   **44.** 24% of 258 about 60

Find each answer. *(Lessons 9-6, 9-7, 9-8)*

45. What number is 20% of 95? 19

46. 4% of 350 is what number? 14

47. What percent of 18 is 9? 50%

48. What percent of 225 is 180? 80%

49. 22 is what percent of 550? 4%

50. 0.17 is what percent of 6.8? 2.5%

51. 15% of what number is 90? 600

52. 28 is $66\frac{2}{3}$% of what number? 42

53. 16.8 is 24% of what number? 70

54. 75% of what number is 273? 364

Find the percent of increase or percent of decrease. *(Lesson 9-9)*

55. original distance: 50 km; new distance: 200 km 300%

56. original weight: 75 lb; new weight: 80 lb $6\frac{2}{3}$%

57. original time: 3.2 s; new time: 1.6 s 50%

58. original cost: $50; new cost: $40 20%

Find each answer by using a proportion. *(Lesson 9-10)*

59. What number is 48% of 150? 72

60. 60% of what number is 2.7? 4.5

61. What percent of 104 is 39? 37.5%

62. 12% of 24 is what number? 2.88

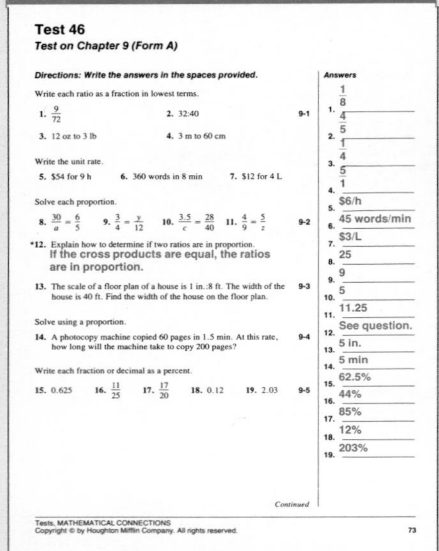

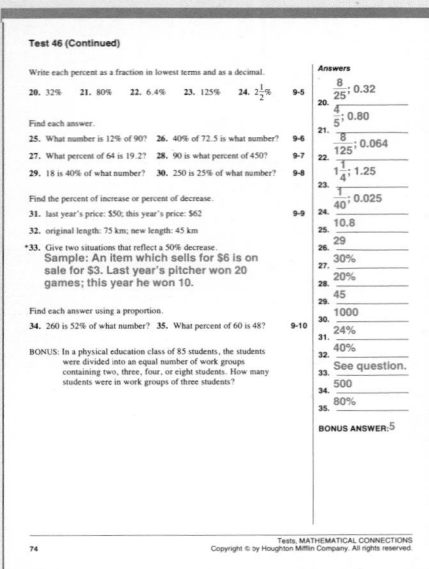

Chapter Test

Write each ratio as a fraction in lowest terms.

1. $\frac{4}{16}$ $\frac{1}{4}$

2. $50:75$ $\frac{2}{3}$

3. $1 to $.40 $\frac{5}{2}$

4. 20 in. to 1 yd $\frac{5}{9}$ **9-1**

Write the unit rate.

5. 80 mi on 2 gal 40 mi/gal

6. $51 for 6 h $8.50/h

7. 135 mi in 3 h 45mi/h

Solve each proportion.

8. $\frac{20}{a} = \frac{5}{6}$ 24

9. $\frac{7}{8} = \frac{y}{16}$ 14

10. $\frac{2.5}{c} = \frac{10}{7}$ 1.75

11. $\frac{1.8}{1.2} = \frac{m}{4}$ 6   **9-2**

★12. Explain the difference between a ratio and a proportion. See answer on p. A26.

13. The scale of a model house is 1 in. : 6 ft. The actual house is 48 ft wide. Find the width of the model house. 8 in.   **9-3**

Solve using a proportion.

14. Isaac Holmes can grade 20 exams in 3 h. At this rate, how long will it take Isaac Holmes to grade 72 exams? 10.8 h   **9-4**

Write each fraction or decimal as a percent.

15. 0.829 82.9%

16. $\frac{2}{5}$ 40%

17. $\frac{7}{25}$ 28%

18. 0.06 6%

19. 1.04 104%   **9-5**

Write each percent as a fraction in lowest terms and as a decimal.

20. 64% $\frac{16}{25}$; 0.64

21. 90% $\frac{9}{10}$; 0.9

22. 7.2% $\frac{9}{125}$; 0.072

23. 350% $3\frac{1}{2}$; 3.5

24. $6\frac{1}{2}$% $\frac{13}{200}$; 0.065

Find each answer.

25. What number is 18% of 70? 12.6

26. 80% of 32.5 is what number? 26   **9-6**

27. What percent of 128 is 38.4? 30%

28. 104 is what percent of 78? $133\frac{1}{3}$%   **9-7**

29. 35 is $87\frac{1}{2}$% of what number? 40

30. 60% of what number is 390? 650   **9-8**

Find the percent of increase or percent of decrease.

31. original number of passengers: 40; new number of passengers: 10 75%   **9-9**

32. original length: 12.5 m; new length: 17.5 m 40%

★33. Give two situations that reflect a 100% increase. See answer on p. A26.

Find each answer using a proportion.

34. 700 is 35% of what number? 2000

35. What percent of 75 is 50? $66\frac{2}{3}$%   **9-10**

The Compound Interest Formula

Warm-Up Exercises

Solve for I in the following problems using the formula $I = prt$.

1. $p = \$6000$; $r = 5\%$; $t = 2$ years; $I = \underline{?}$ $\$600$

2. $p = \$3000$; $r = 6\frac{1}{2}\%$;
   $t = 3$ years; $I = \underline{?}$ $\$585$

3. $p = \$12{,}000$; $r = 8\%$;
   $t = 10$ years; $I = \underline{?}$ $\$9600$

4. $p = \$5000$; $r = 7\%$;
   $t = 6$ months; $I = \underline{?}$ $\$175$

Objective: To calculate compound interest using the compound interest formula.

Banks pay interest for the use of principal that you deposit in your account. If you leave the interest in your account, you will earn **compound interest**, which is interest on the principal *and* any interest previously earned.

Banks pay interest to accounts at specific **interest periods**. Interest that is paid to an account once a year is *compounded annually*.

The total amount (A) in an account at the end of a period of time (t) in years is given by the compound interest formula: $A = P\left(1 + \frac{r}{n}\right)^{nt}$, where P is the principal, r is the annual interest rate, and n is the number of interest periods per year.

Teaching Notes

After this lesson, students should be able to contrast simple interest with compound interest. They also should be able to calculate interest amounts that would be very difficult to find without the use of a scientific calculator. This extension should develop students' appreciation for both the power of mathematics and the power of the calculator.

Example

Solution

Let $P = \$6000$, $r = 8\%$, $n = 4$, and $t = 3$. Find A.

Use the compound interest formula. $A = P\left(1 + \frac{r}{n}\right)^{nt}$

$A = 6000\left(1 + \frac{0.08}{4}\right)^{4(3)} = 6000(1 + 0.02)^{12}$
$= 6000(1.02)^{12}$
$\approx 6000(1.2682418) \approx 7609.4508$ ← Round.

The value of A is $\$7609.45$. (This represents the amount at the end of 3 years when an account earns 8% interest compounded 4 times a year.)

Calculator

Point out the necessity of paying careful attention to the calculator key sequence for the Exercises.

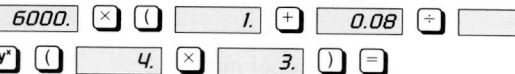

You can use this key sequence on a scientific calculator to find the answer to the Example.

6000. ⊠ ⦅ ⦅ 1. ⊞ 0.08 ⊡ 4. ⦆
y^x ⦅ 4. ⊠ 3. ⦆ ⊟

Additional Example

Let $P = \$5000$, $r = 6\%$, $n = 4$, and $t = 5$. Find A. $\$6734.28$

Exercises

Find each answer using the compound interest formula. Assume that no deposits or withdrawals are made after the first deposit.

1. Let $P = \$2400$, $r = 12\%$, $n = 2$, and $t = 4$. Find A. $\$3825.24$

2. Let $P = \$9000$, $r = 7.2\%$, $n = 4$, and $t = 5$. Find A. $\$12{,}858.73$

3. Let $P = \$4500$, $r = 9\%$, $n = 12$, and $t = 0.5$. Find A. $\$4706.34$

4. Cari deposited $2000 in a savings account. The account earned 8% annual interest compounded semiannually (twice a year). How much money was in the account at the end of 3 years? $\$2530.64$

5. Thrift Savings Bank pays 10% annual interest compounded quarterly (four times a year) on its two year certificates. How much would a $5000 certificate be worth at the end of two years? $\$6092.01$

Ratio, Proportion, and Percent   **431**

Quick Quiz 1 (9–1 through 9–4)

Write each ratio as a fraction in lowest terms.

1. $\frac{16}{34}$ $\frac{8}{17}$

2. 2 gal:6 qt $\frac{4}{3}$

3. 3 ft:2 yd $\frac{1}{2}$

4. 12:8 $\frac{3}{2}$

Write the unit rate.

5. $12 for 3 tickets $4/ticket

6. 120 mi in 2 h   60 mi/h

7. $40 for 4 h $10/h

Tell whether each proportion is *True* or *False*.

8. $\frac{2}{3} = \frac{16}{24}$ True

9. $\frac{3.5}{2.4} = \frac{7.0}{5.0}$ False

10. $\frac{8}{6} = \frac{3}{4}$ False

11. $\frac{0.5}{1.0} = \frac{3}{6}$ True

Solve each proportion.

12. $\frac{4}{5} = \frac{a}{20}$ 16

13. $\frac{4}{3} = \frac{3}{b}$ $2\frac{1}{4}$

14. $\frac{3}{13} = \frac{x}{13}$ 3

15. $\frac{7}{c} = \frac{2}{6.2}$ 21.7

16. A floor plan for a house shows the scale to be 2 in.: 16 ft. If the house is 40 ft wide, what is the width on the floor plan?
    5 in.

17. Carin earns $52 for 8 h work. How much will she earn in 24 h? $156

Cumulative Review

Standardized Testing Practice

Choose the letter of the correct answer.

1.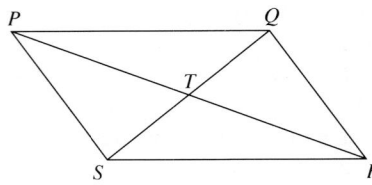

   Which three points lie on the same line?

   A. P, Q, and R B. P, Q, and T
   C. T, Q, and S D. S, R, and P

2. A health club charges an initial fee of $90, plus $30 per month. Doug paid $360 to the health club. For how many months did he sign up to be a member?

   A. 12   B. 9   C. 30   D. 15

3. Which is *not* a rational number?

   A. 1.7   B. $\frac{12}{0}$ C. $\frac{2}{7}$ D. $0.\overline{3}$

4. Find the sum of the measures of the angles of a pentagon.

   A. 900   B. 1260
   C. 540   D. 360

5. Five rulers cost $1.95. What is the cost of 12 rulers?

   A. $4.68   B. $23.64
   C. $9.75   D. $3.90

6.

Hours Spent on Homework	Tally	Frequency
0	⊬⊬	5
1	IIII	4
2	⊬⊬ II	7
3	IIII	4
Total		20

What is the mean of these data?

A. 1 h   B. 3 h
C. 2 h   D. 1.5 h

7. Corrine exercises for about $\frac{3}{4}$ h every day. About how long does she exercise in a week?

   A. 3 h   B. 18 h
   C. $\frac{3}{4}$ h   D. 6 h

8. What is 45% of 120?

   A. 45   B. 54
   C. $2.\overline{6}$ D. 0.45

9. Which fraction is equivalent to $\frac{2}{5}$?

   A. $\frac{12}{15}$ B. $\frac{5}{2}$ C. $\frac{24}{60}$ D. $2\frac{1}{5}$

10. Solve: $z + \frac{1}{3} = -\frac{2}{5}$

    A. $z = -\frac{11}{15}$ B. $z = -\frac{1}{15}$
    C. $z = -\frac{2}{15}$ D. $z = -1\frac{1}{2}$

432   Chapter 9

11. Which property does this statement illustrate?

$\frac{1}{4}(284) = \frac{1}{4} \cdot 200 + \frac{1}{4} \cdot 84$

A. identity property of multiplication
B. additive property of addition
C. associative property of addition
D. distributive property

12. Find the sum. Simplify if possible.

$-2\frac{7}{12} + 3\frac{11}{12}$

A. $-6\frac{1}{2}$ B. $1\frac{1}{3}$

C. $5\frac{1}{2}$ D. $6\frac{1}{2}$

13. A furniture store manager buys lamps for $60 and sells them for $80. What is the percent of increase?

A. 20%   B. $33\frac{1}{3}\%$

C. 25%   D. 140%

14. Which of the following has the greatest value?

$1, |-2|, -14, |0|, |3|-|8|$

A. $|-2|$ B. -14
C. 1 D. $|3| - |8|$

15. Which number is most likely to be estimated?

A. the hourly wage of a cashier
B. the postage for a package
C. the number of frames on a roll of film
D. the number of people who visit an airport in one year

16. A mail carrier walks 18 blocks due east, 26 blocks due west, and 5 blocks due east. Where is the mail carrier in relation to the starting point?

A. 3 blocks due west
B. 3 blocks due east
C. 5 blocks due west
D. 5 blocks due east

17. Solve: $\frac{10}{15} = \frac{x}{36}$

A. $x = \frac{2}{3}$ B. $x = 12$

C. $x = 15$ D. $x = 24$

18. Find the product. Simplify if possible.

$\frac{8c}{15} \cdot \frac{10}{c}$

A. $\frac{18c}{15c}$ B. $5\frac{1}{3}$

C. $\frac{4c^2}{75}$ D. $\frac{18c}{15c}$

19. Write the ratio in lowest terms.

16 in. to 4 ft

A. $\frac{4}{1}$ B. $\frac{1}{3}$

C. $\frac{16}{4}$ D. $\frac{1}{4}$

20. What should be true of a double line graph?

A. the lines may not cross
B. one line should show an increase and one should show a decrease
C. the intervals on the horizontal axis should be equal
D. the bottom line should be red

Alternative Assessment

1. Draw a floor plan of a volleyball court where each half of the court is 30 ft by 30 ft. Use a scale of 1 in.:10 ft.

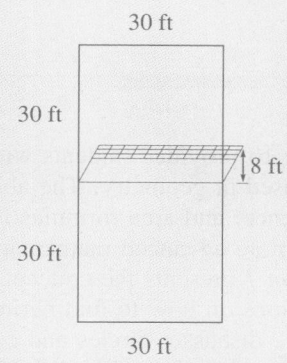

30 ft

30 ft

8 ft

30 ft

30 ft

The scale of this drawing is $\frac{1}{4}$ in.:10 ft.

2. If the actual height of the net were 8 ft, what would be the height shown in the drawing? 0.8 in.

3. The width of the net is 3 ft. How far from the floor is the bottom of the net? 5 ft

4. What is the ratio of the width of one side of the court to the total perimeter? 1:6

5. There are 6 players on each team. What is the ratio of each player to square feet in the court area? 1 player:150 ft^2

Planning Chapter 10

Chapter Overview

Chapter 10 provides students with the basic properties and formulas used in geometry. The ability to apply perimeter, circumference, and area formulas is important in the real world and in more advanced mathematics courses.

Lesson 1 presents the concept of perimeter and gives general instructions on how to find perimeters of various polygons. *Lesson 2* discusses circles and circumference. Area is covered in *Lessons 3 and 4*, with all basic area formulas being discussed. *Lesson 5* brings perimeter, circumference, and area together, as students are asked to decide which of these concepts are needed to solve various problems. *Lesson 6* demonstrates that there are cases in which the solution to a problem may make sense mathematically, but may not make sense in a real-world context. Such problems, therefore, have no solution. In *Lesson 7*, congruent polygons are discussed. *Lessons 8 and 9* treat similar polygons and their application to indirect measurement. *Lesson 10* introduces square roots, and *Lesson 11* applies them through the use of the Pythagorean theorem. The *Chapter Extension* shows methods for simplifying square roots.

Background

Manipulatives can play an important role in presenting and studying geometry. To enhance students' understanding, various models and concrete approaches are recommended for introducing topics, for follow-up, and for exploration. Geoboards can be used extensively to study perimeter, area, congruence, and similarity. Linear and square units can be contrasted effectively by using a geoboard. Transparent mirrors can be used in *Lesson 7* to study congruent figures. The *Focus on Explorations* shows various uses of tangrams to explore area. These experiences will give meaning to the content of the chapter.

In many cases, real-world considerations must be taken into account when working with geometry. The decision-making and problem-solving lessons, *Lessons 5 and 6*, demonstrate this quite clearly. In *Lesson 5*, students decide whether they need

to find perimeter, circumference, or area to solve a problem. In *Lesson 6*, students not only solve a problem mathematically, but they also decide if the answer makes sense physically; for example, does a length of −6 cm have meaning?

A calculator can be a useful tool in Chapter 10 to compute formulas, solve proportions, find square roots, and calculate with π. A full-function calculator is recommended.

Objectives

10-1 To find the perimeter of a polygon.

10-2 To identify and find the radius, diameter, and circumference of a circle.

10-3 To find the area of rectangles, parallelograms, triangles, and trapezoids.

FOCUS ON EXPLORATIONS To explore the concept of area using tangrams.

10-4 To find the area of circles.

10-5 To decide whether perimeter, circumference, or area must be calculated to solve a given problem.

10-6 To recognize when a problem has no solution.

10-7 To identify corresponding parts of congruent figures.

10-8 To recognize similar polygons and use corresponding sides to find unknown lengths.

10-9 To use similar triangles to find an unknown measurement that cannot be measured directly.

10-10 To find the square root of a number using a table or a calculator.

10-11 To use the Pythagorean theorem to find unknown lengths.

CHAPTER EXTENSION To simplify square roots using the product property of square roots.

Chapter Planner

Instructional Aids	Manipulatives	Cooperative Learning	Technology	Practice/Reteaching	Assessment

Materials: geoboards, calculator, metric tape measure, string, tangrams, transparent mirrors
Resource Book:
 Lesson Starters
  90–100
Visuals, Folders E, L, O
	Lessons 10-1, 10-2, Focus on Explorations, 10-7, 10-11
	Activities Book:
	 Manipulative
	  Activities 19–20
		Lessons 10-2, 10-3, 10-6, 10-8, 10-11
		Activities Book:
		 Cooperative
		  Activities 19–20
			Lessons 10-2, 10-4, 10-8, 10-10, 10-11
			Activities Book:
			 Calculator Activity 10
			 Computer Activity 10
			Connections Plotter Plus Disk
				Toolbox, pp. 753–771
				Extra Practice, p. 744
				Resource Book:
				 Practice 123–136
				Study Guide,
				 pp. 179–200
					Self-Tests 1–2
					Quick Quizzes 1–2
					Chapter Test, p. 488
					Resource Book:
					 Lesson Starters
					  90–100
					  (Daily Quizzes)
					Tests 48–52

Assignment Guide Chapter 10

Day	Skills Course	Average Course	Advanced Course
1	**10-1:** 1–10, 11–21 odd, 25–30	**10-1:** 1–21 odd, 25–30 **10-2:** 1–4, 5–15 odd, 17–24, 29–34	**10-1:** 1–21 odd, 23–30 **10-2:** 1–4, 5–15 odd, 17–34
2	**10-2:** 1–22, 29–34	**10-3:** 1–20, 23, 24 **Exploration:** Activities I–IV	**10-3:** 1–24 **Exploration:** Activities I–IV
3	**10-3:** 1–12, 23, 24 **Exploration:** Activities I–IV	**10-4:** 1–34	**10-4:** 1–34, Historical Note
4	**10-4:** 1–24, 31–34	**10-5:** 1–14 **10-6:** 1–16	**10-5:** 1–14 **10-6:** 1–16
5	**10-5:** 1–14	**10-7:** 1–22	**10-7:** 1–22
6	**10-6:** 1–16	**10-8:** 1–20, 25–30	**10-8:** 1–30
7	**10-7:** 1–16, 19–22	**10-9:** 1–16	**10-9:** 1–16
8	**10-8:** 1–12, 25–30	**10-10:** 1–32, 33–47 odd, 53, 54	**10-10:** 1–27 odd, 29–54, Challenge
9	**10-9:** 1–16	**10-11:** 1–28	**10-11:** 1–28
10	**10-10:** 1–32, 33–47 odd, 53, 54	*Prepare for Chapter Test:* Chapter Review	*Prepare for Chapter Test:* Chapter Review
11	**10-11:** 1–16, 25–28	*Administer Chapter 10 Test;* Cumulative Review	*Administer Chapter 10 Test;* Chapter Extension; Cumulative Review
12	*Prepare for Chapter Test:* Chapter Review		
13	*Administer Chapter 10 Test*		

Teacher's Resources

Resource Book
  Chapter 10 Project
  Lesson Starters 90–100
  Practice 123–136
  Enrichment 99–110
  Diagram Masters
  Chapter 10 Objectives
  Family Involvement 10
  Spanish Test 10
  Spanish Glossary

Activities Book
  Cooperative Activities 19–20
  Manipulative Activities 19–20
  Calculator Activity 10
  Computer Activity 10

Study Guide
  pp. 179–200

Tests
  Tests 48–52

Visuals
  Folders E, L, O

***Connections Plotter Plus* Disk**

Using Technology

Lesson 10-2

CALCULATORS

Students can use a protractor, a ruler, and a calculator to find an approximation to the circumference of a circle. They should draw a large accurate circle, and two radii that form a small angle (say, 10°). They can then measure the distance between the intersections of these two radii with the circle, and use their calculators to obtain the approximate circumference (multiply by 36 if the angle formed is 10°, for example). The smaller they make the angle between the radii, the closer their approximation should be to the theoretically predicted circumference.

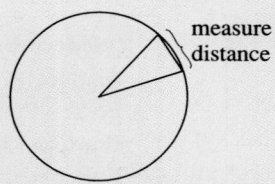

measure distance

Lesson 10-4

COMPUTERS

A simple spreadsheet, such as that illustrated below, could be used to explore what happens to the circumference and area of a circle when a constant is added to the radius, or is multiplied by the radius. (An interesting and surprising result, for example, is that if a string were to encircle Earth, it would take only about six extra inches of string to raise the entire string one inch above the surface of Earth. The area enclosed by the string, however, would be increased considerably by this procedure.)

	A	B	C
1	radius	circumference	area
2	10	2*3.14*A2	3.14*A2*A2
3	20	2*3.14*A3	3.14*A3*A3

Note that the formulas that appear in the sample spreadsheet will not actually appear in the corresponding positions on the screen; their values will be displayed instead.

Lesson 10-8

COMPUTERS

Geometric drawing software can be used by students to draw and manipulate similar figures. Areas and perimeters can then be calculated. A spreadsheet using formulas for perimeters and areas of simple figures also can be used to explore the important effect on perimeter and area of multiplying each linear measurement in a figure by a given constant.

Lesson 10-10

CALCULATORS

Many calculators have a square root key that can be used to explore the properties of square roots (for instance, the fact that $\sqrt{ab} = \sqrt{a}\,\sqrt{b}$, but that in general $\sqrt{a+b} \neq \sqrt{a} + \sqrt{b}$). It is also a useful experience to use a calculator to approximate square roots without using a square root key. This can be done by squaring an initial guess, then changing the guess in a systematic way until the desired degree of accuracy is reached. For example, to approximate $\sqrt{5}$, you might first guess the value 2. But $2^2 = 4$ is too small, so 2.5 might be the next guess. But $2.5^2 = 6.25$ is too big, so 2.25 might be the next guess, and so forth.

Lesson 10-11

CALCULATORS

The Pythagorean theorem can be verified experimentally by drawing an arbitrary right triangle, measuring its sides as accurately as possible, then using a calculator to find the square of the length of each side, and then verifying that the sum of the squares of the two smaller sides is close to the square of the longest side. This is a useful exercise because it reinforces measurement and calculator skills, and underscores the meaning of the Pythagorean theorem itself.

Using Manipulatives

Lesson 10-1

Ask students to form figures with given perimeters on geoboards. The figures will vary in shape, and students should see that shapes and areas can be different for figures with the same perimeter. This activity also can be done by having students draw the figures on graph paper.

Lesson 10-2

COOPERATIVE LEARNING

Divide students into cooperative groups. Distribute metric tape measures and several circular objects of various sizes (plastic lids, pie plates, and so on) to each group. Have each group number each object and measure its circumference and diameter. Students can fill in a chart, such as the one below, and use calculators for this part of the activity.

Circle	Circumference	Diameter	$C \div d$
1			
2			
.			
.			
.			

Have each group share its results with the class. Students should begin to realize that the ratio $\frac{C}{d}$ is slightly more than 3 for each object. This activity is an introduction to the meaning of π.

Lesson 10-7

Students can use transparent mirrors to duplicate geometric figures. By placing the mirror alongside a figure, the image of the figure can be seen and traced on the opposite side of the mirror. Students should discuss why the figures are congruent, and label the corresponding parts.

Lesson 10-11

Ask students to draw a 3-4-5 or 6-8-10 right triangle on graph paper and then draw three squares by using each side of the triangle as the side of a square. Have students cut out the two smaller squares and then make them fit on the largest square, cutting wherever necessary. Students should discuss their observations. This activity leads to an understanding of the Pythagorean theorem.

Reteaching/Alternate Approach

Lesson 10-4

Use an overhead projector or the chalkboard to show a 3 × 3 square divided into nine sections. Superimpose a circle with a diameter of 3 units on the square, as shown below. Ask students to estimate the area covered by the circle (about 7 square units). Use the formula for the area of a circle to calculate the actual area with students (7.065 square units).

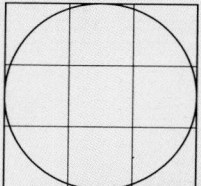

Lesson 10-8

COOPERATIVE LEARNING

Have students working in cooperative groups draw ten figures on graph paper. After exchanging papers with another group, each group should draw figures similar to the ones they received on the same graph paper. The groups should also calculate the ratio of the corresponding sides for each figure.

434D

Teacher's Resources Chapter 10

Enrichment masters from the Resource Book are pictured here. See the Teacher's Resources chart on page 434B for a complete listing of all materials available for this chapter.

10-1 Enrichment
Connection

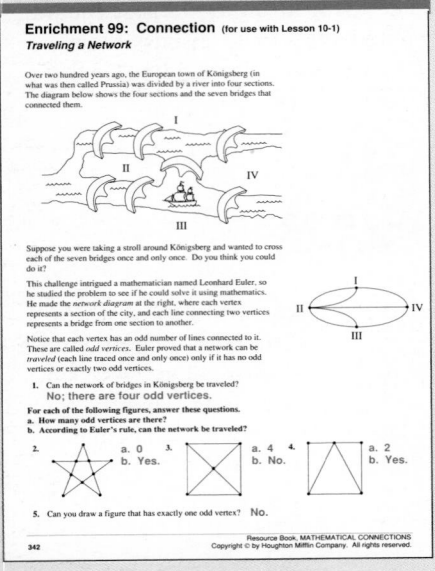

10-2 Enrichment
Data Analysis

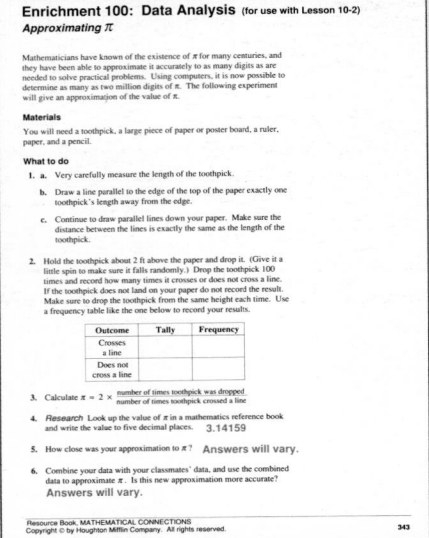

10-3 Enrichment
Exploration

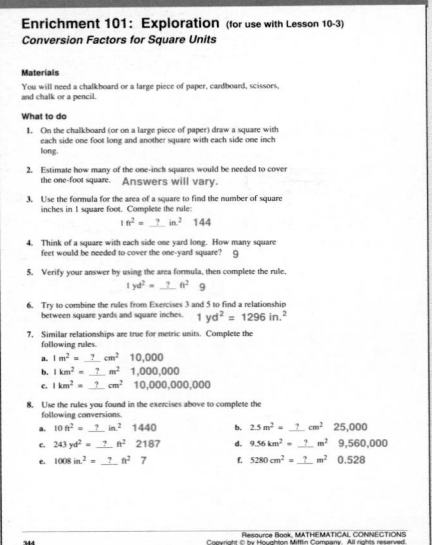

10-4 Enrichment
Thinking Skills

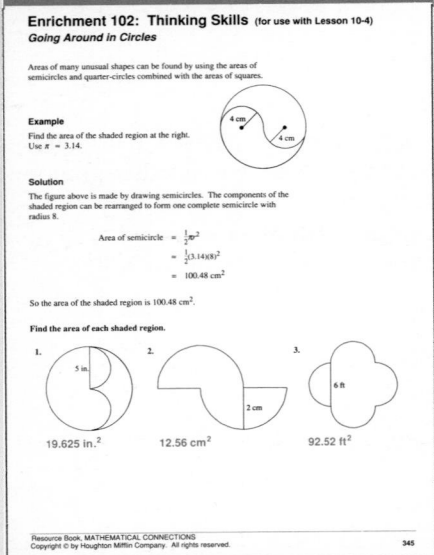

10-5 Enrichment
Data Analysis

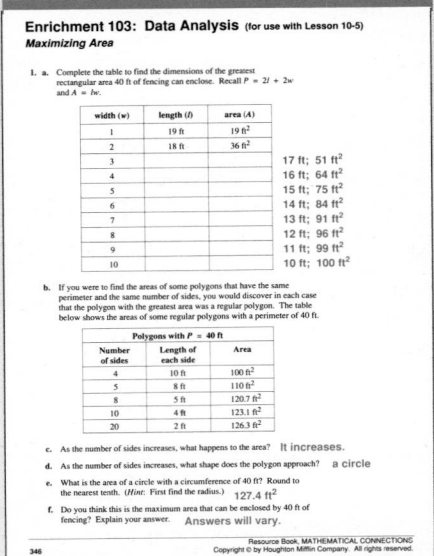

10-6 Enrichment
Thinking Skills

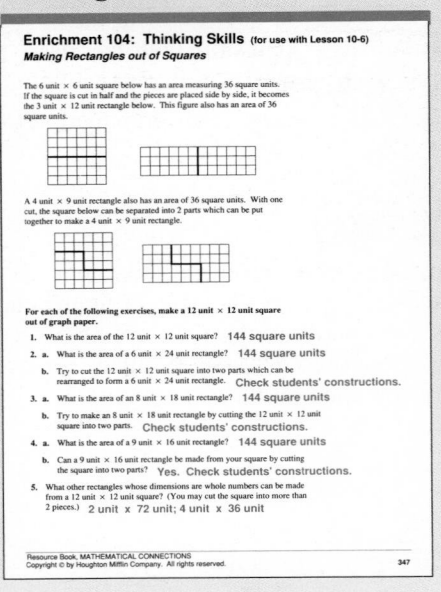

10-7 Enrichment
Exploration

Transformational Geometry

There are three basic ways to transform a geometric figure so that the new figure is congruent to the original one.

A *translation* (or slide) moves the figure from one location to another.

A *reflection* (or flip) changes the orientation of the figure in relation to a line of reflection.

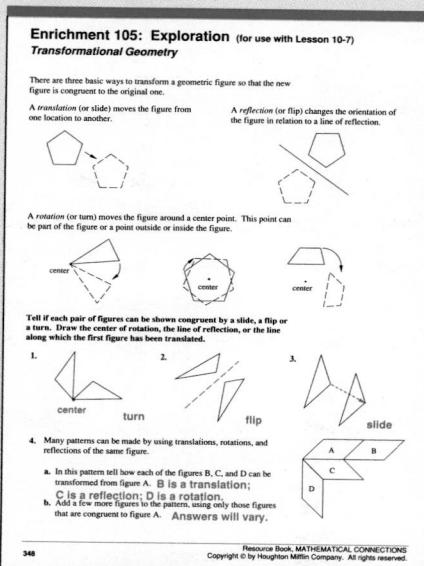

A *rotation* (or turn) moves the figure around a center point. This point can be part of the figure or a point outside or inside the figure.

Tell if each pair of figures can be shown congruent by a slide, a flip or a turn. Draw the center of rotation, the line of reflection, or the line along which the first figure has been translated.

1. center   turn
2. flip
3. slide

4. Many patterns can be made by using translations, rotations, and reflections of the same figure.

   a. In this pattern tell how each of the figures B, C, and D can be transformed from figure A. **B is a translation; C is a reflection; D is a rotation.**
   b. Add a few more figures to the pattern, using only those figures that are congruent to figure A. **Answers will vary.**

10-8 Enrichment
Application

Building Scale Models

Model trains and real trains are examples of similar figures. Railroad models are built to *scale*. The scale is the ratio of the size of each item in the model to that of the corresponding item in the railroad. Model railroads have been built in a wide variety of scales.

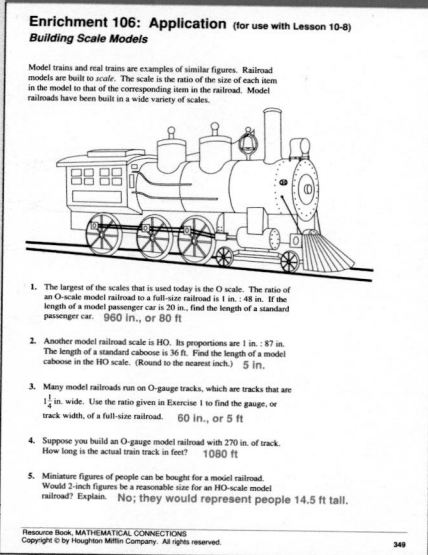

1. The largest of the scales that is used today is the O scale. The ratio of an O-scale model railroad to a full-size railroad is 1 in. : 48 in. If the length of a model passenger car is 20 in., find the length of a standard passenger car. **960 in., or 80 ft**

2. Another model railroad scale is HO. Its proportions are 1 in. : 87 in. The length of a standard caboose is 36 ft. Find the length of a model caboose in the HO scale. (Round to the nearest inch.) **5 in.**

3. Many model railroads run on O-gauge tracks, which are tracks that are $1\frac{1}{4}$ in. wide. Use the ratio given in Exercise 1 to find the gauge, or track width, of a full-size railroad. **60 in., or 5 ft**

4. Suppose you build an O-gauge model railroad with 270 in. of track. How long is the actual train track in feet? **1080 ft**

5. Miniature figures of people can be bought for a model railroad. Would 2-inch figures be a reasonable size for an HO-scale model railroad? Explain. **No; they would represent people 14.5 ft tall.**

10-9 Enrichment
Connection

A Peek at Trigonometry

Surveyors often use indirect measurement techniques. These are part of a branch of mathematics called *trigonometry*. Trigonometry uses certain ratios found in similar right triangles.

For Exercises 1–7 draw a diagram similar to the one below.

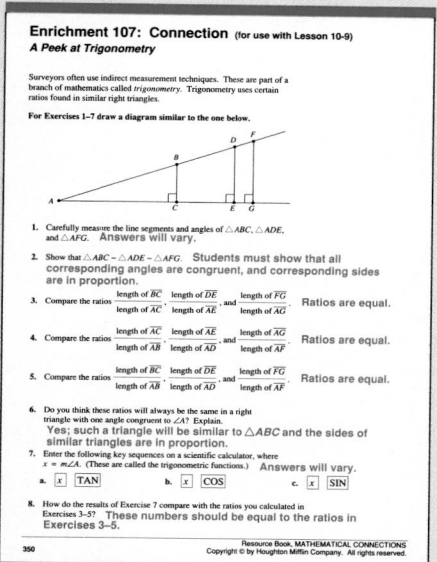

1. Carefully measure the line segments and angles of $\triangle ABC$, $\triangle ADE$, and $\triangle AFG$. **Answers will vary.**

2. Show that $\triangle ABC \sim \triangle ADE \sim \triangle AFG$. **Students must show that all corresponding angles are congruent, and corresponding sides are in proportion.**

3. Compare the ratios $\dfrac{\text{length of } \overline{BC}}{\text{length of } \overline{AC}}$, $\dfrac{\text{length of } \overline{DE}}{\text{length of } \overline{AE}}$, and $\dfrac{\text{length of } \overline{FG}}{\text{length of } \overline{AG}}$ **Ratios are equal.**

4. Compare the ratios $\dfrac{\text{length of } \overline{AC}}{\text{length of } \overline{AB}}$, $\dfrac{\text{length of } \overline{AE}}{\text{length of } \overline{AD}}$, and $\dfrac{\text{length of } \overline{AG}}{\text{length of } \overline{AF}}$ **Ratios are equal.**

5. Compare the ratios $\dfrac{\text{length of } \overline{BC}}{\text{length of } \overline{AB}}$, $\dfrac{\text{length of } \overline{DE}}{\text{length of } \overline{AD}}$, and $\dfrac{\text{length of } \overline{FG}}{\text{length of } \overline{AF}}$ **Ratios are equal.**

6. Do you think these ratios will always be the same in a right triangle with one angle congruent to $\angle A$? Explain. **Yes; such a triangle will be similar to $\triangle ABC$ and the sides of similar triangles are in proportion.**

7. Enter the following key sequences on a scientific calculator, where $x = m\angle A$. (These are called the trigonometric functions.) **Answers will vary.**
   a. x TAN   b. x COS   c. x SIN

8. How do the results of Exercise 7 compare with the ratios you calculated in Exercises 3–5? **These numbers should be equal to the ratios in Exercises 3–5.**

10-10 Enrichment
Exploration

Approximating Square Roots

Suppose you have neither a scientific calculator nor a table of square roots. You can use simple arithmetic to approximate the square root of any number. The approximation can be accurate to as many digits as you wish. Here's a method called the *divide-and-average* method.

Example
Approximate $\sqrt{45}$ to the nearest ten-thousandth.

Solution

Step 1: First estimate the value of $\sqrt{45}$ by locating it between consecutive integers. Select the integer whose square is nearest 45 as the first approximation:

45 is between 36 and 49, two perfect squares.
$\sqrt{45}$ is between $\sqrt{36}$ and $\sqrt{49}$.
$\sqrt{45}$ is between 6 and 7.

So let $a = 7$ since 49 is nearer to 45 than 36 is.

Step 2: Divide 45 by a, carrying out the division to two more digits than there are in the divisor:

$45 \div 7 = 6.42$

Step 3: Find the average of a and $\frac{45}{a}$: $\frac{1}{2}(7 + 6.42) = 6.71$

Step 4: Use the average as the new value for a. Continue repeating Steps 2 and 3 until the divisor and the quotient in Step 2 match in the ten-thousandths' place.

$45 \div 6.71 = 6.7064 \longrightarrow \frac{1}{2}(6.71 + 6.7064) = 6.7082$
$45 \div 6.7082 = 6.708207$

So $\sqrt{45} \approx 6.7082$ to the nearest ten-thousandth.

In this method the approximation is accurate to at least as many digits as match in the divisor and the quotient.

Use the divide-and-average method to approximate each square root to the nearest ten-thousandth.

1. $\sqrt{11}$ **3.3166**   2. $\sqrt{21}$ **4.5825**
3. $\sqrt{38}$ **6.1644**   4. $\sqrt{53}$ **7.2801**

10-11 Enrichment
Communication

Reading and Writing about Pythagoras

Research

You need to write an article about the mathematician Pythagoras for a math column in your school newspaper. Use an encyclopedia or a mathematics reference book to find out about Pythagoras's life, the Pythagorean theorem, and the scholars who taught him. Consider these questions: Who? What? Where? When? and Why?
Answers will vary. A sample answer is given.

Pythagoras was born on the island of Samos in the sixth century B.C. He studied mathematics, philosophy, and religion. Pythagoras was taught by Thales, who did astronomical calculations, and by Anaximander, who drew geographic maps. Pythagoras proved the property that for all right triangles the area of the square on the hypotenuse is equal to the sum of the areas of the squares on the two legs. In his honor, this property is called the Pythagorean theorem. This theorem is still used today as a method of indirect measurement.

End of Chapter Enrichment
Extension

Locating Square Roots on a Number Line

Materials
You will need a straightedge and a compass.

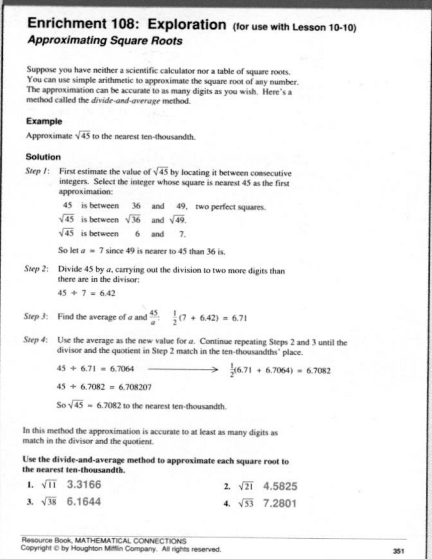

1. Follow these steps to locate $\sqrt{2}$ on a number line, as shown above.
   a. Draw a horizontal line.
   b. Draw a line perpendicular to the first line.
   c. Label the point of intersection 0. Then use your compass to mark off 1, 2, 3, and 4 units to the right of 0 on the horizontal line.
   d. Use your compass to mark off 1 unit from 0 on your vertical line and label the intersection point 1.
   e. Draw a line from the point labeled 1 on the vertical line to the point labeled 1 on the horizontal line. You have just made a right triangle whose legs are each equal to 1 unit. The hypotenuse measures $\sqrt{1^2 + 1^2} = \sqrt{2}$.
   f. Open your compass so that the radius is equal to the hypotenuse of the triangle you just constructed. Put the point of your compass at 0. Mark the radius on the horizontal line and label it $\sqrt{2}$.

2. To locate $\sqrt{3}$ on this number line use the same procedure to make a right triangle with legs measuring 1 unit and $\sqrt{2}$ units. The hypotenuse will measure $\sqrt{1^2 + (\sqrt{2})^2} = \sqrt{3}$.

Locate each number on the same number line.

3. $\sqrt{5}$ 4. $\sqrt{6}$ 5. $\sqrt{7}$ 6. $\sqrt{8}$ 7. $\sqrt{10}$

For Chapter 10 Opener

In all likelihood, the field game with the longest history of being played in North America is lacrosse. Scholars speculate that the Algonquian peoples living in the St. Lawrence valley originated a game that developed into modern lacrosse. By the time Europeans began to settle in North America, the game was played by a number of Native American groups in much of what is now the United States and southern Canada.

The Native American game was played with a deerskin ball and 3- to 4-foot long sticks, curved at the end. A Jesuit missionary who viewed a game being played in 1636 noticed a resemblance between the shape of these sticks and a bishop's crozier, the French word for which is *la crosse*.

In the mid-1800s, European settlers in Canada began to play "lacrosse." The game has become the national sport of Canada, and is played all over the world.

For Page 456

Ptolemy, discussed in the Historical Note, was a Greek scholar who lived in Alexandria, Egypt, around the second century A.D. A discussion of his work as a geographer can be found in the Multicultural Notes on page 580.

Research Activities

For Chapter 10 Opener

Working in cooperative groups, students can find out more about the rules and traditions of the Native American precursor to lacrosse. Groups can also research the history of other sports and report their findings to the class. Encourage the groups to draw diagrams of the various playing fields.

Archery Target

48 in.

9.6 in.

48 in.

434

Circles and Polygons 10

Research Activities
(continued)

For Page 456
As an extension of the Historical Note, students can learn about Ptolemaic astronomy, which dominated western thought until it was eventually replaced by the Copernican system in the sixteenth century.

Suggested Resources

Arlott, John. *The Oxford Companion to Sports and Games*. London: Oxford University Press, 1975.

Twombly, Wells. *200 Years of Sport in America*. New York: McGraw-Hill, 1976.

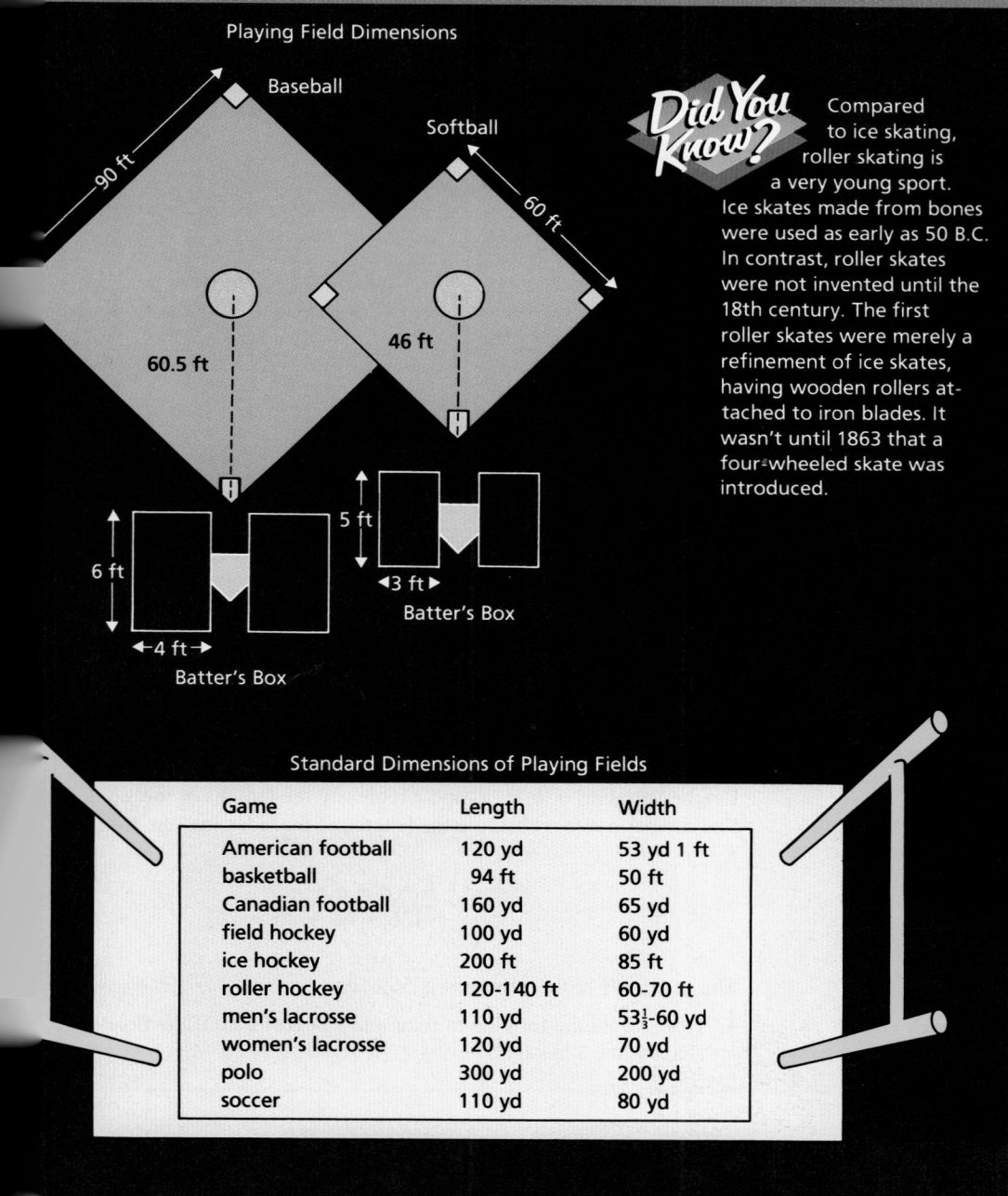

Playing Field Dimensions

Baseball
Softball
90 ft
60 ft
60.5 ft
46 ft

6 ft
◄4 ft►
Batter's Box

5 ft
◄3 ft►
Batter's Box

Did You Know?
Compared to ice skating, roller skating is a very young sport. Ice skates made from bones were used as early as 50 B.C. In contrast, roller skates were not invented until the 18th century. The first roller skates were merely a refinement of ice skates, having wooden rollers attached to iron blades. It wasn't until 1863 that a four-wheeled skate was introduced.

Standard Dimensions of Playing Fields

Game	Length	Width
American football	120 yd	53 yd 1 ft
basketball	94 ft	50 ft
Canadian football	160 yd	65 yd
field hockey	100 yd	60 yd
ice hockey	200 ft	85 ft
roller hockey	120-140 ft	60-70 ft
men's lacrosse	110 yd	53⅓-60 yd
women's lacrosse	120 yd	70 yd
polo	300 yd	200 yd
soccer	110 yd	80 yd

435

Teaching the Lesson
- Materials: geoboards
- Lesson Starter 90
- Using Manipulatives, p. 434D

Lesson Follow-Up
- Practice 123
- Enrichment 99: Connection
- Study Guide, pp. 179–180

Warm-Up Exercises

Find each answer.
1. $13.5 + 10 + 13.5 + 10$ 47
2. $2(14 + 6)$ 40
3. $1.6 + 0.75 + 1.6$ 3.95

Evaluate each expression if $a = 6$, $b = 3$, and $c = 5$.
4. $a + b + c$ 14
5. $2a + 2b$ 18

Lesson Focus

Ask students if they ever had an optometrist test their *peri*pheral vision, or if they have used the word *peri*scope? Discuss the meanings of the words and the prefix *peri*. Finding the perimeter of geometric figures is the focus of today's lesson.

Teaching Notes

In the course of presenting the lesson, use a measuring tape to find the distance around a student desk or some other object in the room. Use the tape to emphasize that perimeter is a linear measure.

Key Question

What is meant by the perimeter of a polygon? The perimeter of a polygon is the sum of the lengths of all its sides.

Perimeter

Objective: To find the perimeter of a polygon.

Terms to Know
- *perimeter of a polygon*

APPLICATION

The athletic field at the new Jay City High School needs to be enclosed by fencing. The figure below is a sketch of the field.

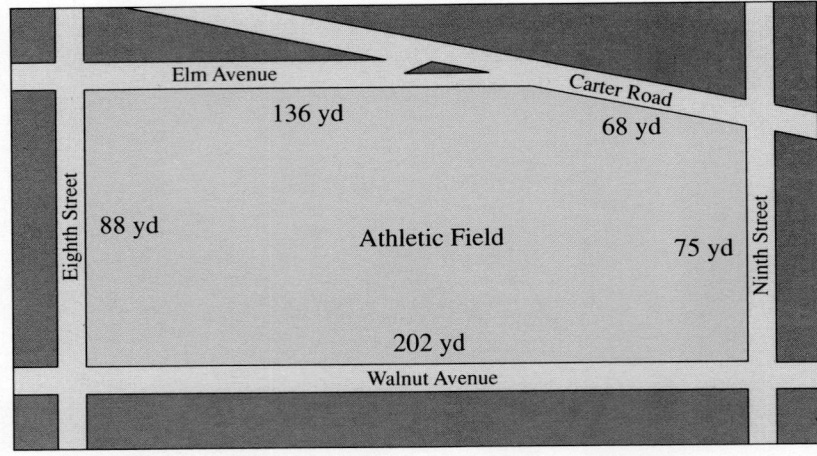

QUESTION What amount of fencing is needed to enclose the field?

The shape of the athletic field is a pentagon. To find the amount of fencing, you need to find the *perimeter* of this pentagon. The **perimeter** of a polygon is the sum of the lengths of all its sides.

Example 1

Solution

Find the perimeter of the pentagon in the sketch above.

Let the variables a, b, c, d, and e represent the lengths of the sides. To find the perimeter, substitute the lengths of the sides into the formula $P = a + b + c + d + e$.

$$P = a + b + c + d + e$$
$$= 88 + 136 + 68 + 75 + 202$$
$$= 569$$

The perimeter of the pentagon is 569 yd.

☑ Check Your Understanding

1. How would the formula in Example 1 be different if the field were shaped like a hexagon?
   See *Answers to Check Your Understanding* at the back of the book.

ANSWER The amount of fencing needed is 569 yd.

To find the perimeter of some figures, you need to apply the properties of polygons that you studied in Chapter 6.

Example 2

The length of one side of a regular pentagon is 4.9 cm. Find the perimeter.

Solution

First make a sketch of the figure. Because it is a regular pentagon, all sides can be marked with the same measure.

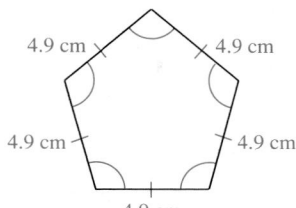

Let n represent the length of one side. To find the perimeter, substitute the length of one side into the formula $P = 5n$.

$$P = 5n = 5(4.9) = 24.5$$

The perimeter of the pentagon is 24.5 cm.

Example 3

The length of a rectangle is $5\frac{1}{2}$ ft and the width is $4\frac{3}{4}$ ft. Find the perimeter.

Solution

First make a sketch of the figure. Because it is a rectangle, opposite sides can be marked with the same measure.

Let l and w represent the length and width, respectively. To find the perimeter, substitute the length and width into the formula $P = 2l + 2w$.

$$P = 2l + 2w = 2(5\tfrac{1}{2}) + 2(4\tfrac{3}{4}) = 11 + 9\tfrac{1}{2} = 20\tfrac{1}{2}$$

The perimeter is $20\frac{1}{2}$ ft.

 **Check Your Understanding**

2. How are Examples 1 and 2 alike? How are they different?
3. How would the solution of Example 2 be different if the figure were a regular octagon?
See Answers to Check Your Understanding at the back of the book.

Guided Practice

COMMUNICATION « *Writing*

Write a formula that can be used to find the perimeter of each figure.

« **1.**

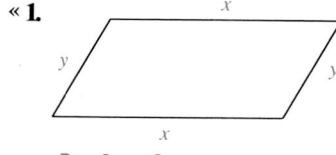

$P = 2x + 2y$

« **2.**

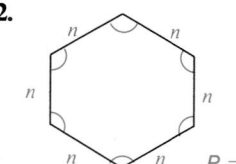

$P = 6n$

« **3.** a rhombus with one side that measures s $P = 4s$

« **4.** a triangle with sides that measure a, b, and c

$P = a + b + c$

Circles and Polygons   **437**

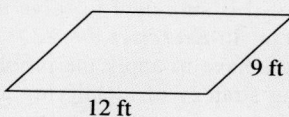

Communication: Discussion
Guided Practice Exercise 8 will help students to understand the necessity of converting all measurements to the same units prior to adding.

Making Connections/Transitions
For *Exercises 11–13*, students have to use simple algebraic equations to find the unknown measurements.

Problem Solving
Exercises 14–18 have students analyze the given information to determine if it is sufficient to solve the problem. In *Exercises 19–22*, students have to apply the problem-solving strategy of identifying a pattern, further demonstrating the usefulness of this strategy.

Critical Thinking
For *Exercises 23 and 24*, students have to *analyze* the given relationships and *apply* their understanding of perimeter to solve the problems.

Find the perimeter of each figure.

5.

8 m

6 m   7.5 m

10.3 m
31.8 m

6.

12 yd

15 yd
54 yd

7. a square with one side that measures $6\frac{1}{3}$ yd
$25\frac{1}{3}$ yd

«**8.** COMMUNICATION «*Discussion* The sides of a triangle measure 1 yd, 60 in., and 4 ft. Describe three ways to find the perimeter.
Convert all the measurements to either yards, inches, or feet. Then add the measurements.

Exercises

Find the perimeter of each figure.

A   1.

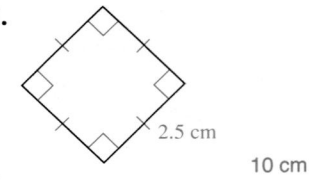

2.5 cm

10 cm

2.

5 in. $5\frac{1}{4}$ in.

$2\frac{1}{2}$ in.

3 in.

4 in.   5 in. $24\frac{3}{4}$ in.

3.

2.31 m

11.55 m

4.

$3\frac{1}{2}$ ft

$7\frac{1}{3}$ ft

$18\frac{1}{6}$ ft

5. a regular decagon with one side that measures 2.8 cm 28 cm

6. a parallelogram with sides that measure 9 ft and $8\frac{1}{4}$ ft $34\frac{1}{2}$ ft

7. an equilateral triangle with one side that measures $5\frac{3}{8}$ in. $16\frac{1}{8}$ in.

8. a rhombus with one side that measures 3.22 m 12.88 m

9. A rectangular playground is 40 yd long and $57\frac{1}{2}$ ft wide. Find the perimeter of the playground. 355 ft or $118\frac{1}{3}$ yd

10. DATA, *pages 434–435* Compare the perimeter of a Canadian football field to the perimeter of an American football field. Which is longer? How much longer? Canadian; $103\frac{1}{3}$ yd

Find the value of x.

B   11.

23.8 m

25.5 m

x

$P = 61.2$ m   11.9 m

12.

9.7 cm

x

$P = 32$ cm   6.3 cm

13.

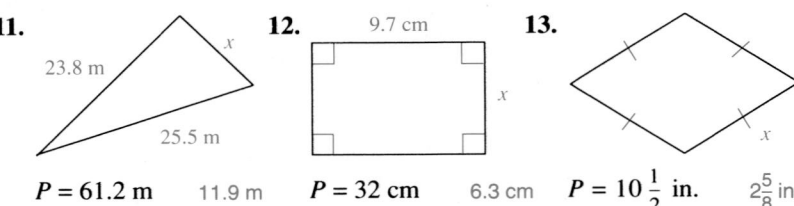

x

x

$P = 10\frac{1}{2}$ in. $2\frac{5}{8}$ in.

15. not enough information; the length of the rectangle is needed.

Solve, if possible. If there is not enough information, tell what additional facts are needed.

14. The perimeter of a regular octagon is 19.6 m. What is the length of one side? 2.45 m

15. The perimeter of a rectangle is 25 cm. What is its width?

16. What is the length of one side of a square with perimeter 14 ft? $3\frac{1}{2}$ ft

17. Four sides of a pentagon each measure $10\frac{1}{2}$ ft. What is the length of the fifth side? not enough information; the perimeter or the fact that the pentagon
is regular is needed.
18. One side of a trapezoid measures 32 in., two other sides each measure 14 in., and the perimeter is 108 in. What is the length of the fourth side? 48 in.

PATTERNS

In the following patterns, the length of one side of a small square (□) is 1 unit. Find the perimeter of the ninth figure in each pattern.

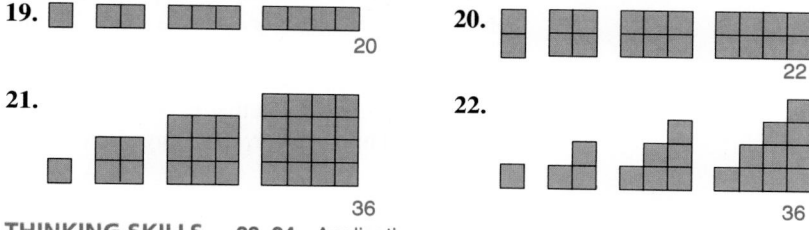

19. 20

20. 22

21. 36

22. 36

THINKING SKILLS   **23, 24.** Application

C   **23.** In triangle ABC, the length of $\overline{BC}$ is 15 mm. The length of $\overline{AC}$ is 4 mm less than the length of $\overline{BC}$, and the length of $\overline{AB}$ is half the length of $\overline{BC}$. What is the perimeter of the triangle? 33.5 mm

24. The length of a rectangle is 5 in. less than twice the width. The perimeter is 62 in. Find the width and the length. 12 in.; 19 in.

SPIRAL REVIEW

S   **25.** Find the mean, median, mode(s), and range of the data: 18, 25, 19, 22, 18, 20, 22, 24, 18   *(Lesson 5-8)* 20.7; 20; 18; 7

26. Simplify: $\frac{4x}{5} + \frac{11x}{5}$ *(Lesson 8-4)* 3x

27. Find the perimeter of a regular hexagon with one side that measures $9\frac{1}{2}$ in.   *(Lesson 10-1)* 57 in.

28. Write 0.00000084 in scientific notation.   *(Lesson 7-11)* 8.4×10^{-7}

29. Find the measure of an angle supplementary to a 30° angle. *(Lesson 6-3)* 150°

30. Solve the proportion: $\frac{15}{45} = \frac{a}{6}$ *(Lesson 9-2)* 2

Circles and Polygons   **439**

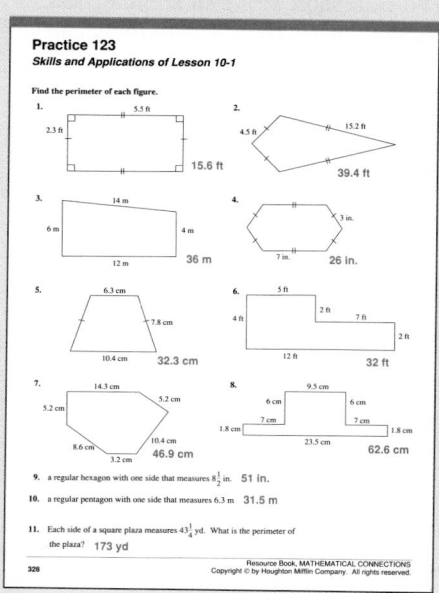

Practice 123
Skills and Applications of Lesson 10-1

Find the perimeter of each figure.

9. a regular hexagon with one side that measures $8\frac{1}{2}$ in. 51 in.

10. a regular pentagon with one side that measures 6.3 m 31.5 m

11. Each side of a square plaza measures $43\frac{1}{4}$ yd. What is the perimeter of the plaza? 173 yd

328

Follow-Up

Exploration
Have students use graph paper to draw three rectangles that enclose 18 squares each. Students should find the perimeter of each rectangle and then write down any conclusions they can reach about the figure.

Enrichment
Enrichment 99 is pictured on page 434E.

Practice
Practice 123, *Resource Book*

Teaching the Lesson
- Materials: protractor, ruler, calculator, metric tape measure, circular objects, string
- Lesson Starter 91
- Using Technology, p. 434C
- Using Manipulatives, p. 434D

Lesson Follow-Up
- Practice 124
- Enrichment 100: Data Analysis
- Study Guide, pp. 181–182
- Manipulative Activity 19

Warm-Up Exercises

Find each answer.

1. $2 \times \frac{22}{7} \times 14$ 88
2. 3.14×4 12.56
3. $2 \times \frac{22}{7} \times \frac{7}{2}$ 22
4. $2 \times 3.14 \times 8.2$ 51.496
5. Use $P = 2(l + w)$ to find P when $l = 5$ and $w = 4$.   18

Lesson Focus

Rotate a piece of chalk tied to the end of a short string in a circle. Ask students what type of path the chalk is traveling and how the total distance it travels in a given number of revolutions can be found. Answering questions of this type is the focus of today's lesson.

Circles and Circumference

Objective: To identify and find the radius, diameter, and circumference of a circle.

APPLICATION

Terms to Know
- circle
- center
- radius (radii)
- chord
- diameter
- circumference

The set of all points in a plane that are a given distance from a given point in the plane is called a **circle.** The given point is the **center** of the circle. For instance the rim of the Ferris wheel in the picture at the right is a circle. A circle is named by its center.

A line segment whose two endpoints are the center of a circle and a point on the circle is called a **radius.** A circle has an infinite number of *radii*, all with the same length.

A line segment whose endpoints are both on a circle is called a **chord** of the circle. Any chord that passes through the center of a circle is a **diameter.** The diameter, d, of any circle is twice its radius, r.

$$d = 2r$$

Example 1

Use the circle at the right.
a. Name the circle.
b. Identify as many radii, chords, and diameters as are shown.
c. Find the radius.
d. Find the diameter.

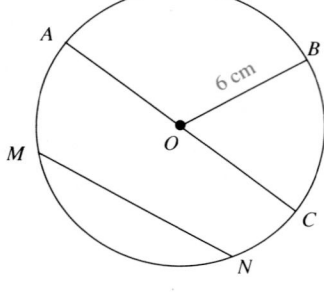

Solution

a. circle O
b. radii: $\overline{OA}$, $\overline{OB}$, and $\overline{OC}$
   chords: $\overline{MN}$ and $\overline{AC}$
   diameter: $\overline{AC}$
c. The length of radius $\overline{OB}$ is labeled as 6 cm. The radius is 6 cm.
d. $r = 6$, so $d = 2r = 2(6) = 12$.
   The diameter is 12 cm.

 Check Your Understanding

1. In Example 1, why is $\overline{AC}$ identified as both a chord and a diameter of circle O?
   See *Answers to Check Your Understanding* at the back of the book.

The distance around a circle is called its **circumference**, C. In all circles, the ratio of the circumference to the diameter is equal to the same number, represented by the Greek letter π (*pi*).

$$\frac{C}{d} = \pi$$

Formulas: *Circumference of a Circle*

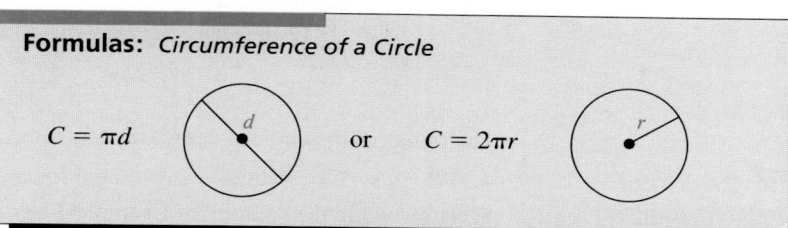

$$C = \pi d \qquad \text{or} \qquad C = 2\pi r$$

Because the number represented by π is irrational, there is no common fraction or decimal that you can use to name it. However, π is so important in mathematics that the following two *approximations* are often used.

$$\pi \approx 3.14 \qquad \text{and} \qquad \pi \approx \frac{22}{7}$$

In general, use 3.14 as an approximation for π. You can use $\frac{22}{7}$ when multiples of 7 make the calculations easier. When you use 3.14 for π, you generally round your answer to the tenths' place.

Example 2
Solution

The diameter of a circle is 21 cm. Find the circumference.

Use the formula $C = \pi d$.

$$C \approx \frac{22}{7} \cdot 21 \quad \longleftarrow \quad \begin{array}{l} \text{Substitute } \frac{22}{7} \text{ for } \pi \\ \text{and 21 for } d. \end{array}$$

$$C \approx 66$$

The circumference of the circle is approximately 66 cm.

☑ **Check Your Understanding**

2. Why was $\frac{22}{7}$ and not 3.14 substituted for π in Example 2?

3. How would the solution of Example 2 be different if the radius of the circle were 21 cm?
   See *Answers to Check Your Understanding* at the back of the book.

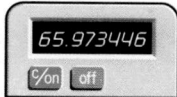

Many calculators have a ⊡ key. When you press this key, the calculator recalls from its memory a decimal approximation of π that has several decimal places. For instance, this key sequence might be used to find the circumference in Example 2.

To the nearest tenth, the answer in the display rounds to 66.0. Notice that this is the same as the result when you substitute $\frac{22}{7}$ for π.

Circles and Polygons **441**

Teaching Notes

Students may not be familiar with some of the terminology in this lesson, such as chord or diameter. To aid their understanding of the terms, you may wish to have students cut out some paper circles and fold them to create chords and diameters.

Key Questions
1. What ratio does the Greek letter π represent? the ratio of the circumference of a circle to its diameter
2. How can you find the circumference of a circle? Multiply the diameter by π.

Error Analysis
Students may use the value for the radius instead of the diameter, or vice versa, when substituting into a circumference formula. Doing some oral exercises requiring substitution only (without solving) should help students to avoid this error.

Estimation
Students can estimate the circumference of a circle easily by multiplying the diameter by three.

Calculators
Students can use calculators in this lesson, since the emphasis of the lesson is not on developing computational proficiency.

Using Technology
For a suggestion on using a calculator, protractor, and ruler to find an approximation to the circumference of a circle, see page 434C.

Using Manipulatives
For a suggestion on finding an approximate value of π using manipulatives, see page 434D.

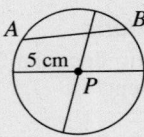

Example 3

Solution

The circumference of a circle is 25.12 m. Find the radius.

Use the formula $C = 2\pi r$.

$$25.12 \approx 2(3.14)r \qquad \leftarrow \text{Substitute 25.12 for } C$$
$$25.12 \approx 6.28r \qquad\qquad \text{and 3.14 for } \pi.$$

$$\frac{25.12}{6.28} \approx \frac{6.28r}{6.28}$$

$$4 \approx r$$

The radius of the circle is approximately 4 m.

✓ **Check Your Understanding**

4. Why was 3.14 and not $\frac{22}{7}$ substituted for π in Example 3?
5. How would the solution of Example 3 be different if you were asked to
   find the diameter of the circle instead of the radius?
   See *Answers to Check Your Understanding* at the back of the book.

Guided Practice

COMMUNICATION «*Reading*

Refer to the text on pages 440–442.   2. the ratio of circumference to diameter

1. Both are line segments
   with one endpoint on the
   circle. A *radius* has its
   other endpoint at the
   center of a circle. A
   diameter passes through
   the center and has its
   other endpoint on the
   circle.

«1. Explain how *a radius* is different from *a diameter*.

«2. What characteristic of a circle is represented by the number π?

«3. Name two commonly used approximations for π. 3.14; $\frac{22}{7}$

«4. Identify two frequently used formulas for circumference. $C = 2\pi r$;
      $C = \pi d$

**Exercises 5 and 6 show a series of steps that apply one of the
circumference formulas to the given figure. In each step, replace the
__?__ with the number that makes the statement true.**

5.

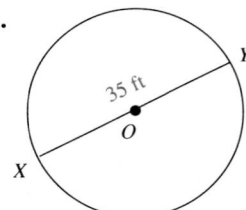

$C = \pi d$
$C \approx (\underline{\ ?\ })(35) \frac{22}{7}$
$C \approx \underline{\ ?\ }$ 110

6.

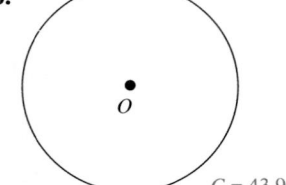

$C = 43.96$ m

$C = 2\pi r$
$43.96 \approx 2(\underline{\ ?\ })r$ 3.14
$43.96 \approx (\underline{\ ?\ })r$ 6.28

$\dfrac{43.96}{?} \approx \dfrac{6.28r}{?}$ 6.28; 6.28

$\underline{\ ?\ } \approx r$ 7

Use the figure at the right.

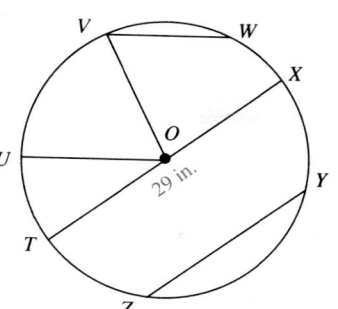

7. Name the circle. circle O

8. radii: $\overline{OT}, \overline{OU}, \overline{OV}, \overline{OX}$
chords: $\overline{TX}, \overline{YZ}, \overline{VW}$
diameter: $\overline{TX}$

8. Identify as many radii, chords, and diameters as are shown.

9. Find the diameter. 29 in.

10. Find the radius. 14.5 in.

Find the circumference of a circle with the given measure.

11. $d = 14$ yd
44 yd

12. $r = 8$ m
50.2 m

Find the radius and diameter of a circle with the given circumference.

13. 5.2 m
0.8 m; 1.7 m

14. 220 ft
35 ft; 70 ft

Exercises

Use the figure at the right.

A

1. Name the circle. circle O

2. radii: $\overline{OP}, \overline{OQ}, \overline{OR}, \overline{OS}$
chords: $\overline{QR}, \overline{RS}, \overline{QS}, \overline{PR}$
diameters: $\overline{PR}, \overline{QS}$

2. Identify as many radii, chords, and diameters as are shown.

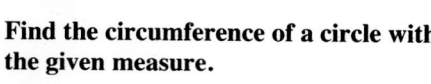

3. Find the radius. 8.6 cm

4. Find the diameter. 17.2 cm

Find the circumference of a circle with the given measure.

5. $d = 2$ m 6.3 m

6. $d = 10.5$ cm 33 cm

7. $r = 2\frac{1}{3}$ yd $14\frac{2}{3}$ yd

8. $r = 2.5$ cm 15.7 cm

9. $r = 7$ mm 44 mm

10. $d = 3.5$ m 11 m

11. $d = 5.2$ m 16.3 m

12. $r = 14$ in. 88 in.

Find the radius and diameter of a circle with the given circumference.

13. 9 m; 18 m
14. $5\frac{1}{4}$ ft; $10\frac{1}{2}$ ft
15. 7 yd; 14 yd
16. 5.5 cm; 11 cm
17. 50 m

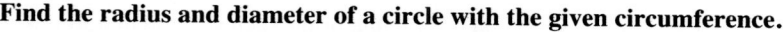

13. 56.52 m　　　**14.** 33 ft　　　**15.** 44 yd　　　**16.** 34.54 cm

17. In a public park, the length of the fence that encloses a circular garden is 157 m. What is the approximate diameter of the garden?

18. The radius of the front wheel of a tricycle is 14 in. The radius of each back wheel is 6 in. How much larger is the circumference of the front wheel than the circumference of a back wheel?  50.2 in.

Exercise Notes

Communication: Reading
Guided Practice Exercises 1–4 check students' understanding of the terms and relationships associated with a circle.

Number Sense
Exercises 19–22 highlight the difference between the exact value of π and its common approximations.

Communication: Writing
Exercises 23 and 24 allow students to express their ideas of the effect that circles have in their lives.

Application
Exercises 25–28 require students to use a formula for finding circumference and to use their knowledge of linear units to find the distances traveled.

Project

Have a few student volunteers research the development of various methods of approximating π and present their reports to the class.

Enrichment

Enrichment 100 is pictured on page 434E.

Practice

Practice 124, *Resource Book*

444

NUMBER SENSE 21. approximately   22. exactly

Replace each __?__ with *exactly* or *approximately* to make a true statement.

B **19.** The product π(3) is __?__ equal to 9.42. approximately

20. The product 2(π)(6) is __?__ equal to the product π(12). exactly

21. The circumference of a circle with diameter 7 in. is __?__ 22 in.

22. The circumference of a circle with radius 5 cm is __?__ 10(π) cm.

WRITING ABOUT MATHEMATICS Answers to Exercises 23–24 are on p. A26.

Write an essay entitled *A World without Circles*. Here are two questions that you should address in this essay.

23. How would nature be different if there were no circular shapes?

24. How would your house and school be different if no circles were used in manufacturing the items that you use?

PROBLEM SOLVING/APPLICATION

Exercises 25–28 refer to a monocycle wheel whose radius is 35 in.

25. What is the circumference of the wheel? 220 in.

26. Approximately how far does the wheel travel in five complete turns? 1100 in.

C **27.** How many complete turns does the wheel make in traveling 1 mi? 288

28. If the wheel has turned one million times, how many miles has it traveled? 3472.2 mi

SPIRAL REVIEW

S **29.** Michiko jogs 4.5 mi/day. How many miles does she jog per week? *(Lesson 8-7)* 31.5 mi

30. Write the prime factorization of 150. *(Lesson 7-1)* $2 \cdot 3 \cdot 5^2$

31. Find the radius of a circle with circumference 785 cm. *(Lesson 10-2)* 125 cm

32. What percent of 60 is 15? *(Lesson 9-7)* 25%

33. The measures of two angles of a triangle are 30° and 60°. Tell whether the triangle is *acute, right,* or *obtuse*. *(Lesson 6-6)* right

34. Evaluate $-3a^2$ when $a = 5$. *(Lesson 3-6)* -75

10-3

Area of Polygons

Objective: To find the area of rectangles, parallelograms, triangles, and trapezoids.

Terms to Know
• *area*
• *base*
• *height*

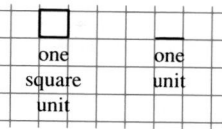

The length of one side of a square on a sheet of grid paper can be thought of as one *unit*. A square is then one *square unit*. The number of square units enclosed by a figure on grid paper is the *area* of the figure.

one square unit   one unit

1 Draw and cut out a rectangle like the one shown below.
   a. Find the area of the rectangle by counting the square units. 28
   b. Can you describe a simpler way to find the area? Count 4 rows of 7 square units each, or 7 columns of 4 square units each.

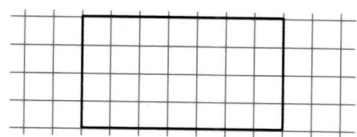

2 Draw and cut out a square that is three units on each side.
   a. Find the area of the square by counting the square units. 9
   b. Can you describe a simpler way to find the area?
   Count 3 rows or columns of 3 square units each.

The region enclosed by a plane figure is called the **area** (*A*) of the figure. When possible, it is far more efficient to find an area using a formula than counting units within a figure.

One area formula that may be familiar to you is the formula for the area of a rectangle.

Formula: *Area of a Rectangle*

Area of a rectangle = length × width

$A = lw$

width (*w*)

length (*l*)

When the length and width of a rectangle are equal, the figure is a square. To find the area of a square, you need to know only one measure, the length of a side.

Formula: *Area of a Square*

Area of a square = (length of side)^2

$A = s^2$

side (*s*)

Circles and Polygons   **445**

Teaching the Lesson
• Lesson Starter 92
• Visuals, Folder E

Lesson Follow-Up
• Practice 125
• Enrichment 101: Exploration
• Study Guide, pp. 183–184
• Cooperative Activity 19

Warm-Up Exercises

1. Find the circumference of a circle with a diameter of 10 cm. Use $\pi = 3.14$.   31.4 cm

2. Find the circumference of a circle with a radius of 28 cm. Use $\pi = \frac{22}{7}$.   176 cm

3. Find the perimeter of a rectangle with a width of 4 in. and a length of 7 in.   22 in.

4. Find the perimeter of a triangle with sides of 6 cm, 8 cm, and 10 cm.   24 cm

Lesson Focus

Ask students to imagine buying a new carpet for a room. The carpet costs $18.99 per square yard. How can the number of square yards of carpet needed be determined, and what would the total cost be? The focus of today's lesson involves finding areas of polygons.

Teaching Notes

This lesson discusses four formulas for finding areas of polygons. A wall chart showing each polygon and its formula for area will be helpful to students.

446

Teaching Notes *(continued)*

Key Questions

1. What type of unit is used to express perimeter? What type is used to express area?
   Perimeter is expressed in linear units. Area is expressed in square units.
2. What are the formulas for the areas of a rectangle, a parallelogram, and a triangle? *$A = lw$, $A = bh$, $A = \frac{1}{2}bh$*

Error Analysis

Student errors may result from using the wrong formula or from substituting incorrectly into the formula. Have students use sketches that are labeled and tell them to write the entire formula before substituting values into it.

Making Connections/Transitions

Using rectangles to develop formulas for finding areas of other polygons will provide students with an understanding of why each formula works.

Additional Examples

1. Find the area of a square with the length of one side equal to 4.5 ft. *20.25 ft²*
2. Find the base of a parallelogram with a height of 10 cm and an area of 95 cm². *9.5 cm*
3. Find the base of a triangle whose area is 36 ft² and whose height is 9 ft. *8 ft*

Example 1

Solution

Find the area of a rectangle with width $3\frac{1}{2}$ ft and length $4\frac{1}{2}$ ft.

Use the formula $A = lw$.

$$A = \left(4\frac{1}{2}\right)\left(3\frac{1}{2}\right)$$  Substitute $4\frac{1}{2}$ for l and $3\frac{1}{2}$ for w.

$$A = \left(\frac{9}{2}\right)\left(\frac{7}{2}\right) = \frac{63}{4} = 15\frac{3}{4}$$ The area is $15\frac{3}{4}$ ft².

Check Your Understanding

1. How would Example 1 be different if the figure were a square with the length of one side equal to $3\frac{1}{2}$ ft?
   See Answers to Check Your Understanding at the back of the book.

Any parallelogram can be "rearranged" to form a rectangle. For this reason, the area formula for a parallelogram is closely related to the rectangle formula.

To use the parallelogram formula, either pair of parallel sides can be the **bases**. The **height** is the perpendicular distance between the bases.

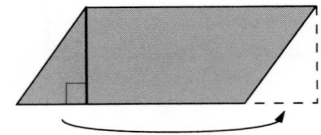

> **Formula:** *Area of a Parallelogram*
>
> Area of a parallelogram = base × height
>
> $$A = bh$$

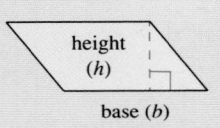

height (h)

base (b)

Example 2

Solution

Find the height of a parallelogram with base 6 yd and area 46 yd².

Use the formula $A = bh$.

$$46 = 6h$$ Substitute 46 for A and 6 for b.

$$\frac{46}{6} = \frac{6h}{6}$$

$$h = \frac{46}{6} = 7\frac{4}{6} = 7\frac{2}{3}$$ The height is $7\frac{2}{3}$ yd.

A diagonal of a parallelogram separates it into identical triangles. You can find the area of a triangle by thinking of it as *one half* the area of a parallelogram with the same base and height.

To use the triangle formula, let any side of the triangle be the **base**. The **height** is the perpendicular distance between the base and the opposite vertex.

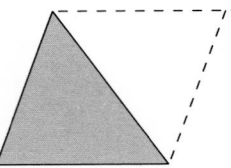

446   Chapter 10

Formula: *Area of a Triangle*

Area of a triangle $= \frac{1}{2} \times$ base $\times$ height

$A = \frac{1}{2}bh$

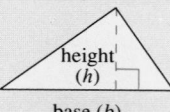

height
(h)

base (b)

Example 3

Solution

Find the area of a triangle with base 11 cm and height 50 mm.

All measures that are substituted into a formula must be expressed in the same unit, so change 50 mm to 5 cm.

Use the formula $A = \frac{1}{2}bh$.

$$A = \frac{1}{2}(11)(5) \quad \longleftarrow \text{Substitute 11 for } b \text{ and 5 for } h.$$

$$A = \frac{1}{2}(55) = 27\frac{1}{2}$$

The area is 27.5 cm².

 **Check Your Understanding**

2. How would the solution to Example 3 be different if 11 cm had been changed to millimeters? Which answer is correct?

3. Would the answer to Example 3 be different if the base were 50 mm and the height were 11 cm?

See *Answers to Check Your Understanding* at the back of the book.

Guided Practice

COMMUNICATION « *Writing*

Make a sketch of each figure. Answers to Guided Practice 1–4 are on p. A26.

« **1.** a rectangle with length 5 cm and width 2.2 cm

« **2.** a triangle with base 15 in. and height 1 ft

« **3.** a parallelogram with base $8\frac{1}{2}$ in. and height 6 in.

« **4.** a square with a side that measures t inches

Find the area of each figure.

5.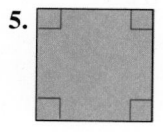

$2\frac{1}{4}$ in.

$2\frac{1}{4}$ in. $5\frac{1}{16}$ in.²

6.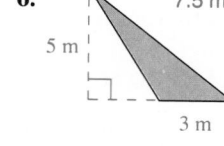

7.5 m²

5 m

3 m

7. a parallelogram with base 220 cm and height 15 m
330,000 cm²
or 33 m²

8. Find the height of a triangle with area 36 ft² and base 8 ft.   9 ft

« **9.** **COMMUNICATION** « *Discussion* How many square inches are in a square foot? How many square centimeters are in a square meter?
144; 10,000

Circles and Polygons   **447**

Exercise Notes *(continued)*

Reasoning
Exercises 11 and 12 have students find areas of irregular quadrilaterals. Students must be able to envision the figures as being made up of triangles before performing any calculations.

Critical Thinking
To work *Exercises 13–16*, students must *synthesize* their knowledge of area and algebraic properties. The exercises lead students step by step to the formula for finding the area of a trapezoid.

Application
Exercises 19–21 have students calculate and compare costs for various floor surfaces. Students must pay careful attention to unit conversions when they use the appropriate formula.

Cooperative Learning
Exercise 22 provides a cooperative learning activity in which groups use real-world data to make decisions regarding flooring.

Exercises

Find the area of each figure.

A **1.** 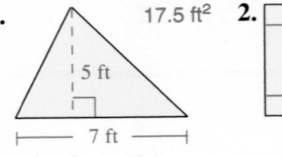 17.5 ft²   **2.** 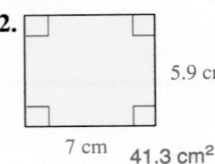   **3.**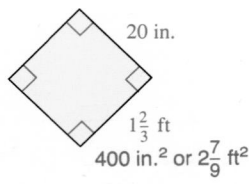

20 in.

5 ft

7 ft

5.9 cm

7 cm   41.3 cm²

$1\frac{2}{3}$ ft

400 in.² or $2\frac{7}{9}$ ft²

4. 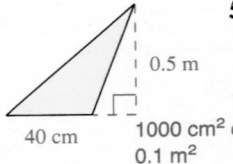   **5.** 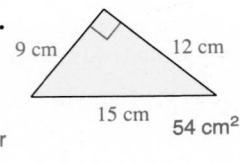   **6.**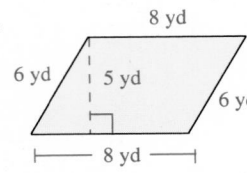

0.5 m

40 cm   1000 cm² or 0.1 m²

9 cm   12 cm

15 cm   54 cm²

8 yd

6 yd   5 yd   6 yd

8 yd

40 yd²

7. a square with length of one side equal to $2\frac{1}{2}$ ft $6\frac{1}{4}$ ft²

8. a parallelogram with height $1\frac{1}{3}$ yd and base $2\frac{1}{2}$ ft   10 ft² or $1\frac{1}{9}$ yd²

9. John Lano's garden is 7 yd wide and 27 ft long. Find the area.   63 yd² or 567 ft²

10. DATA, *pages 434–435* Which of the playing fields listed has the greatest area? polo field

SPATIAL SENSE

Find the area in square units of each quadrilateral. (*Hint:* Use the grid lines to separate each quadrilateral into triangles.)

B **11.** 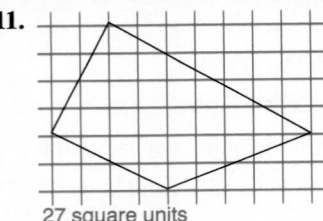   **12.**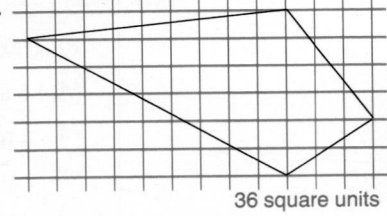

27 square units   36 square units

THINKING SKILLS 13, 14, 16. Knowledge   **15.** Comprehension

17, 18. Application

Use the figure at the right. Replace each ___?___ with the variable or expression that makes the statement true.

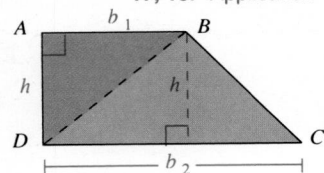

13. The area of triangle *ABD* is $\frac{1}{2}h$ ___?___.

14. The area of triangle *BCD* is $\frac{1}{2}$ ___?___ b_2. *h*

15. Use the results of Exercises 13 and 14. The area of trapezoid *ABCD* is ___?___ + ___?___. $\frac{1}{2}hb_1 + \frac{1}{2}hb_2$

16. By the distributive property, $\frac{1}{2}hb_1 + \frac{1}{2}hb_2 = \frac{1}{2}h($ ___?___ + ___?___ $)$. b_1; b_2

448 Chapter 10

Apply the formula $A = \frac{1}{2}h(b_1 + b_2)$. Find the area of each trapezoid.

17.
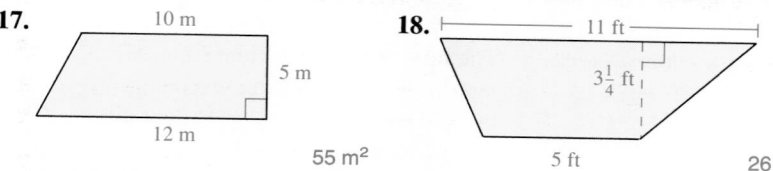

10 m

5 m

12 m

55 m²

18.

11 ft

$3\frac{1}{4}$ ft

5 ft

26 ft²

CAREER/APPLICATION

An *interior designer* plans a room to make it attractive and functional. A designer often calculates and compares the costs of materials.

Family Room

windows

12 ft

windows

door

door

15 ft

Use the floor plan above.

19. A hardwood floor costs $9.50 per square foot to install, including materials and labor. Find the cost of installing a floor in this room.
   $1710

20. Wall-to-wall carpeting costs $28.50 per square yard, including installation. How much does it cost to carpet this room? $570

C 21. Square ceramic tiles that measure 8 in. on each side cost $.89 each. The installation charge is $250. What is the total cost of installing the tiles in this room? $610.45

22. **GROUP ACTIVITY** Using a newspaper, find the costs of several types of flooring. Which type is most expensive? least expensive? Calculate the costs of installing three types of flooring in your classroom. What factors should you consider when choosing a flooring? Answers will vary.

SPIRAL REVIEW

S 23. Find the product: $\frac{2x}{3} \cdot \frac{9}{16}$ *(Lesson 8-1)* $\frac{3x}{8}$

24. Find the area of a triangle with base $1\frac{3}{4}$ yd and height 48 ft. *(Lesson 10-3)* 14 yd² or 126 ft²

Circles and Polygons **449**

Follow-Up

Extension

Give students geo-dot paper and have them make figures meeting a list of specifications.
For example:
1. Draw a figure having four sides and 6 square units of area.
2. Draw a three-sided figure having 3 square units of area.
3. Draw a figure having 4 square units of area and a perimeter of 10 units.   One possible solution:

Enrichment

Enrichment 101 is pictured on page 434E.

Practice

Practice 125, *Resource Book*

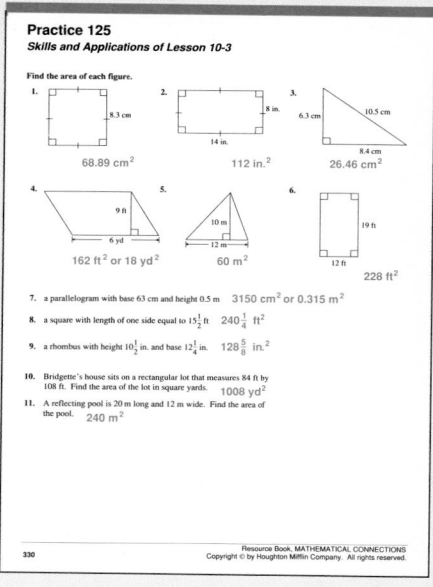

Practice 125
Skills and Applications of Lesson 10-3

Find the area of each figure.

1. 8.3 cm 68.89 cm²
2. 8 in. 14 in. 112 in.²
3. 10.5 cm 6.3 cm 8.4 cm 26.46 cm²
4. 9 ft 6 yd 162 ft² or 18 yd²
5. 10 m 12 m 60 m²
6. 19 ft 12 ft 228 ft²

7. a parallelogram with base 63 cm and height 0.5 m 3150 cm² or 0.315 m²
8. a square with length of one side equal to $15\frac{1}{2}$ ft $240\frac{1}{4}$ ft²
9. a rhombus with height $10\frac{1}{2}$ in. and base $12\frac{1}{4}$ in. $128\frac{5}{8}$ in.²
10. Bridgette's house sits on a rectangular lot that measures 84 ft by 108 ft. Find the area of the lot in square yards. 1008 yd²
11. A reflecting pool is 20 m long and 12 m wide. Find the area of the pool. 240 m²

Exploring Area Using Tangrams

Teaching Note

Using tangrams is an enjoyable way to deepen students' understanding of area. Students who have difficulty with finding ratios of areas pictorially are often more successful with this model because they can fit one piece over the other.

Activity Notes

The purpose of the activities is to enhance students' ability to identify the various tangram pieces and to find ratios of areas among the pieces and to the whole. *Activity IV* allows students to be creative by designing their own shapes.

Critical Thinking

The activities require students to *compare* the pieces to one another in different ways by changing the basis for the square unit. This requires *finding* relationships among the pieces and *drawing conclusions* regarding individual areas.

Objective: To explore the concept of area using tangrams.

Materials

■ tangrams

1. *A, B, C, D, E*; all are isosceles right triangles.

The **tangram** is an ancient Chinese puzzle. It consists of the seven puzzle pieces shown in the figure at the right.

To help you with the activities in this lesson, each tangram piece in the figure has been identified by one of the letters from *A* through *G*. If you have a manufactured set of tangram pieces, you will not see these letters on the pieces.

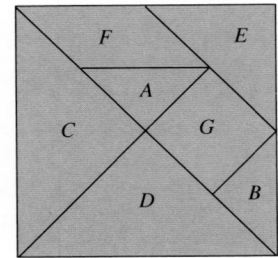

Activity I *Identifying the Tangram Pieces*

■ List the letters of the tangram pieces that are triangles. Next to each letter, classify the triangle by its sides and by its angles.

■ List the letters of the tangram pieces that are quadrilaterals. Next to each letter, give the most appropriate name for the quadrilateral. *G*: square; *F*: parallelogram

Activity II *Calculating Areas*

Suppose that the area of triangle *A* is one square unit. Because triangle *B* is identical to *A*, its area also would be one square unit. The area of triangles *A* and *B* together must be the same as the area of square *G*, so the area of *G* would be two square units.

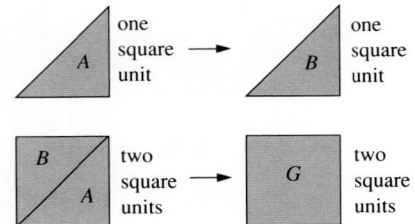

■ Use the reasoning described above. Copy and complete this table.

Tangram piece	A	B	C	D	E	F	G
Area in square units	1	1	? 4	? 4	? 2	? 2	2

■ Now suppose that the area of square *G* is one square unit. Find the areas of the other pieces. Copy and complete this table.

Tangram piece	A	B	C	D	E	F	G
Area in square units	? $\frac{1}{2}$	? $\frac{1}{2}$	? 2	? 2	? 1	? 1	1

■ Next, suppose that the area of triangle *C* is one square unit. Find the areas of the other pieces. Copy and complete this table.

Tangram piece	A	B	C	D	E	F	G
Area in square units	? $\frac{1}{4}$	? $\frac{1}{4}$	1	? 1	? $\frac{1}{2}$	? $\frac{1}{2}$	? $\frac{1}{2}$

Activity III   *Taking a Different Look at Area*

This time, suppose that the area of the entire tangram puzzle is one square unit. It follows that the area of each tangram piece has to be some fractional part of one square unit. For instance, the figure at the right shows that the area of triangle *C* is one fourth the area of the entire tangram puzzle. Therefore, its area is $\frac{1}{4}$ square unit.

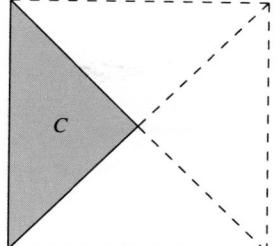

1 Use the reasoning described above to find the areas of the other six tangram pieces. Copy and complete this table.

Tangram piece	A	B	C	D	E	F	G
Area in square units	$?\ \frac{1}{16}$	$?\ \frac{1}{16}$	$\frac{1}{4}$	$?\ \frac{1}{4}$	$?\ \frac{1}{8}$	$?\ \frac{1}{8}$	$?\ \frac{1}{8}$

2 Show how to use four of the tangram pieces to form a square with area equal to $\frac{1}{2}$ square unit. See answer on p. A26.

3 Show how to use triangles *A*, *B*, *C*, and *E* to form each of the following quadrilaterals. What is the area of each quadrilateral? See answer on p. A26.

   a. a square

   b. a rectangle that is not square

   c. a parallelogram that is not a rectangle

   d. a trapezoid

Activity IV   *Creating Tangram Designs*

For many centuries, people have enjoyed the challenge of arranging the tangram pieces to form familiar shapes. The Chinese produced entire books illustrating these designs.

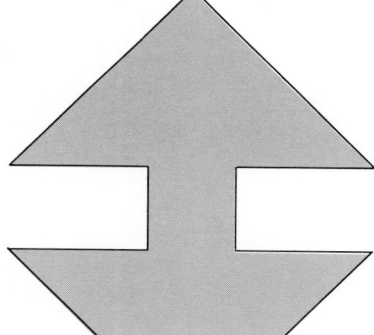

1 Show how the seven tangram pieces can be arranged to form the "sailboat" design shown in the figure at the right. See answer on p. A26.

2 Arrange the seven tangram pieces to form your own design. Trace the outline of your design on a sheet of paper and trade outlines with a partner. Determine how the tangram pieces were arranged to form your partner's design. How does the area of your partner's design compare to the area of your design? The areas are equal.

Circles and Polygons   **451**

Follow-Up

Extension

Have students make their own tangram puzzle. Students can use a file folder and draw a ten centimeter square, as in the figure below. Points *A* and *B* are midpoints; points *E* and *F* are midpoints of segments *AB* and *CD*, respectively. Points *G* and *H* are midpoints of line segments *CF* and *FD*, respectively. After connecting the points, students can cut out the pieces.

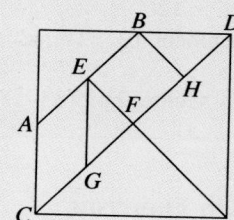

Teaching the Lesson
- Materials: overhead projector
- Lesson Starter 93
- Visuals, Folder E
- Using Technology, p. 434C
- Reteaching/Alternate Approach, p. 434D

Lesson Follow-Up
- Practice 126
- Enrichment 102: Thinking Skills
- Study Guide, pp. 185–186

Warm-Up Exercises

Find each answer.

1. 3.14×15 47.1
2. 3.14×5^2 78.5
3. $3.14 \times (6.2)^2$ 120.7
4. $\frac{22}{7} \times \left(\frac{7}{2}\right)^2$ $38\frac{1}{2}$

Lesson Focus

A store manager advertises that the store will deliver merchandise free of charge to any location within a radius of five miles. How many square miles are included in the free-delivery zone? Finding the area of circular regions is the focus of today's lesson.

Teaching Notes

Estimating areas, as explained in the Exploration at the top of page 452, will help establish students' conceptual understanding of the area of a circle and the formula for calculating circular area, namely $A = \pi r^2$.

Area of Circles

Objective: To find the area of circles.

EXPLORATION

No matter what the shape of a figure is, its area is measured in square units. Although at first it may seem impossible to count the number of square units in a circular region, the following method gives you a fairly good *estimate* of the area of a circle.

1 Estimate the area of a circle with radius 4 units.

a. Count the square units that contain any part of the circular region. Call this number a. 60

b. Count the square units that lie entirely within the circle. Call this number b. 32

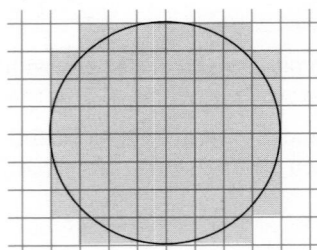 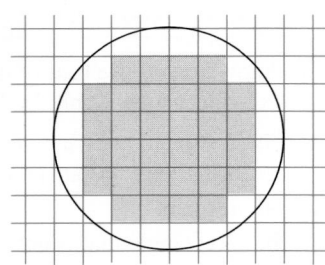

c. Find the average of a and b: $\frac{a+b}{2} = \underline{\quad?\quad}$. 46

d. Complete: The area of the circle is about $\underline{\quad?\quad}$ square units. 46

2 Use the method from Step 1 to estimate the area of these circles.

a.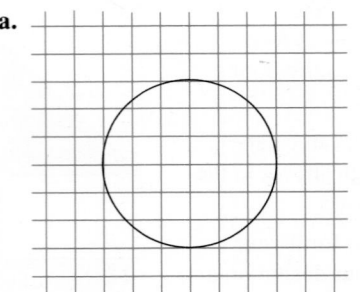

b.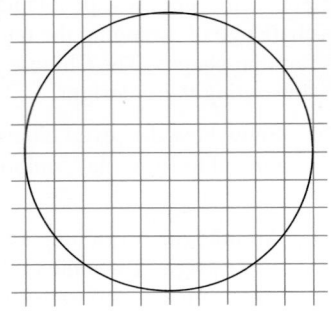

about 26 square units   about 72 square units

In Lesson 10-3, you found the areas of polygons by using their bases and heights. It may seem that finding the area of a circle is totally unrelated, because you describe the size of a circle by its radius or diameter rather than its base and height. However, it is possible to use what you know about polygons to develop an area formula for circles.

Suppose that you could separate a circle into equal parts, then rearrange the parts as shown below. The new figure looks very much like a parallelogram.

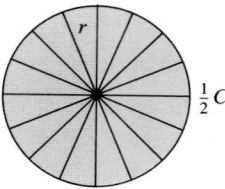 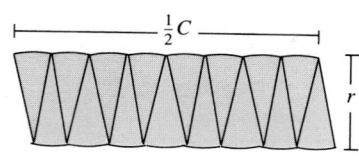

Notice that the base of this "parallelogram" is equal to one half the circumference of the circle, or $\frac{1}{2}C$. The height of the "parallelogram" is just about equal to the radius of the circle, which is r. Now you can use the area formula for a parallelogram to find the area of this figure.

$$A = b \cdot h$$
$$A = \tfrac{1}{2}C \cdot r$$

You know that $C = 2\pi r$, so replace the C in the formula with $2\pi r$.

$$A = \tfrac{1}{2}(2\pi r) \cdot r$$
$$A = \pi r^2$$

Formula: *Area of a Circle*

$$A = \pi r^2$$

Example

Solution

The radius of a circle is 4.4 m. Find the area.

Use the formula $A = \pi r^2$.   ⟵ Substitute 3.14 for π
$$A \approx 3.14(4.4)^2$$ and 4.4 for r.
$$A \approx 3.14(19.36) \approx 60.7904$$

To the nearest tenth, the area of the circle is 60.8 m².

 **Check Your Understanding**

1. Why was 3.14 and not $\frac{22}{7}$ substituted for π in the Example?
2. How would the Example be different if the *diameter* were 4.4 m?
   See *Answers to Check Your Understanding* at the back of the book.

 If you have a calculator with the [π] and [x²] keys, you can use this key sequence to find the area of the circle in the Example.

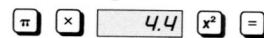

To the nearest tenth, the answer in the display rounds to 60.8.

Circles and Polygons **453**

Key Questions

1. Is there a difference between the units of measure for circumference and for the area of a circle? The circumference is measured in linear units. Area is measured in square units.
2. Which formula for finding area is used as a basis for developing the formula for the area of a circle? formula for area of a parallelogram

Error Analysis

Students may substitute the diameter instead of the radius in the area formula. Sketches with labels are the best approach for avoiding this error.

Using Technology

For a suggestion on using a spreadsheet with this lesson, see page 434C.

Reteaching/Alternate Approach

For a suggestion on using an overhead projector to verify the area formula for a circle, see page 434D.

Additional Example

The radius of a circle is 7 cm. Find the area. 154 cm²

Closing the Lesson

Have students give the formulas for finding the circumference and the area of a circle. Students should give two formulas for circumference.

454

Suggested Assignments

Skills: 1–24, 31–34
Average: 1–34
Advanced: 1–34, Historical Note

Exercise Notes

Communication: Reading

Guided Practice Exercises 1 and 2
check students' understanding of
circumference and area. Students
are required to compare and contrast
the concepts and the formulas.

Communication: Discussion

When discussing *Guided Practice
Exercise 11*, ask students to explain
the difference between linear and
square units of measure. It is important that students understand this
difference.

Making Connections/Transitions

Exercises 9 and 10 give two real-
world problems that can be solved
by using the formula for the area of
a circle. *Exercises 25–28* illustrate
how circumference and area are
both a function of the radius of a
circle.

Calculator

Exercises 11–14 require students to
identify the correct key sequence to
do an exercise.

Application

Exercises 15–24 require students to
find the areas of composite figures
that represent objects such as windows and circular tracks.

Guided Practice

COMMUNICATION «*Reading*

1. The circumference is the distance around the circle; the area is the number of square units enclosed by the circle.

Refer to the text on pages 440–442 and on pages 452–453.

«**1.** Explain how the circumference of a circle is different from its area.

«**2.** Compare the circumference formulas to the area formula for a circle. How are they alike? How are they different? They both use the radius and π. The area formula squares the radius.

Find each answer.

3. $(3.14)(8)(8)$ 200.96

4. $\frac{22}{7}(14)(14)$ 616

5. $\frac{22}{7}\left(\frac{7}{10}\right)^2$ $1\frac{27}{50}$

6. $(3.14)(6.5)^2$ 132.665

Find the area of a circle with the given measure.

7. $d = 2$ mm 3.1 mm²

8. $r = 2.5$ cm 19.6 cm²

9. $r = 21$ ft 1386 ft²

10. $d = 3\frac{1}{2}$ in. $9\frac{5}{8}$ in.²

«**11.** COMMUNICATION «*Discussion* Calculate both the circumference and the area of a circle with radius 2 cm. Are the answers the same or different? Explain. They are the same numerically, but the circumference is 12.6 cm and the area is 12.6 cm².

Exercises

Find the area of a circle with the given measure.

A **1.** $r = 10$ in. 314 in.²

2. $r = 7$ mm 154 mm²

3. $d = 6$ yd 28.3 yd²

4. $r = 1\frac{1}{2}$ ft 7.1 ft²

5. $d = 4\frac{2}{3}$ mi $17\frac{1}{9}$ mi²

6. $d = 3.4$ km 9.1 km²

7.

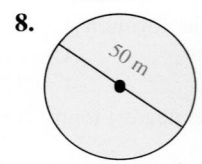

6.2 cm²

8.

1962.5 m²

9. Sometimes the beam from a light-house can be seen for 30 mi in all directions. Over how many square miles of area can the beam be seen? 2826 mi²

10. The diameter of a circular field is 600 ft. One fourth of the field is to be planted with corn. How many square yards are to be planted with corn? 7850 yd²

CALCULATOR

Match each problem with the calculator key sequence that can be used to solve it. (Not every key sequence will be used.)

B 11. Find the area of a circle with radius 6 cm. B

12. Find the circumference of a circle with diameter 6 cm. E

13. Find the circumference of a circle with radius 6 cm. C

14. Find the area of a circle with diameter 6 cm. A

A. π × [3.] x^2 =

B. π × [6.] x^2 =

C. [2.] × π × [6.] =

D. [2.] × π × [12.] =

E. π × [6.] =

F. π × [3.] =

PROBLEM SOLVING/APPLICATION

A **composite** figure is made up, or *composed,* of two or more familiar geometric figures. For instance, the figure at the right is composed of a rectangle and a half-circle. The half-circle is called a **semicircle.**

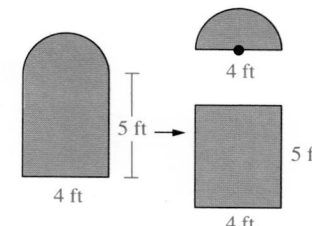

15. How do you think you can find the area of a semicircle? Find half the area of a circle.

16. What is the area of the semicircle in the figure above? 6.3 ft²

17. What is the area of the rectangle in the figure above? 20 ft²

18. What is the total area of the figure above? 26.3 ft²

Find the area of each composite figure.

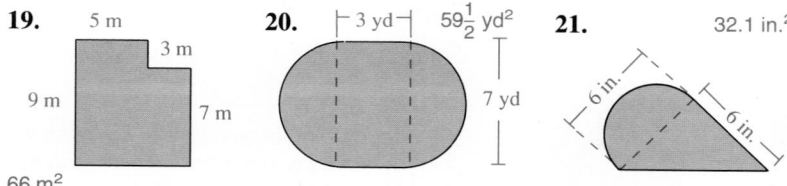

19. 5 m, 3 m, 9 m, 7 m

66 m²

20. 3 yd, 7 yd   59½ yd²

21. 6 in., 6 in.   32.1 in.²

Finding the area of some composite figures involves subtracting areas rather than adding. In each figure, find the area of the shaded region.

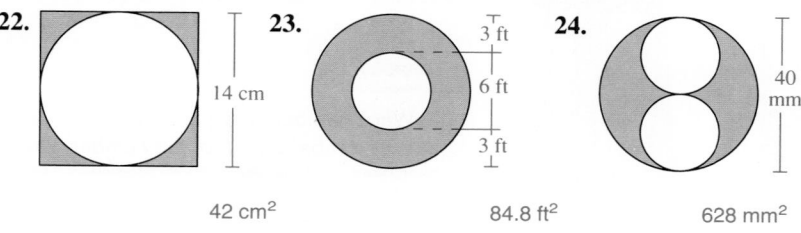

22. 14 cm

42 cm²

23. 3 ft, 6 ft, 3 ft

84.8 ft²

24. 40 mm

628 mm²

Circles and Polygons **455**

Exercises 29 and 30 require students to *apply* their knowledge of circumference and area of circles to *determine* the effect on each of them when the radius is doubled. Students must differentiate between multiplying and squaring a constant.

Follow-Up

Exploration

Students can do the following activity at home and report their results to the class.

Use a round cake pan and measure its radius. Using light cardboard, make a rectangular "box" with bottom dimensions r × 4r, as shown below.

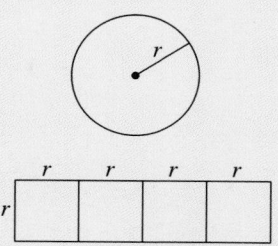

Cover the *bottom* of the cake pan with popcorn or a dry cereal. Pour the contents into the rectangular box, starting at one end and covering the area toward the other end. How much of the area of the box is covered? about ¾

Give a mathematical explanation for the result of this experiment. The area of the bottom of the cake pan is πr^2, which is about $3r^2$. The area of the bottom of the box is $4r^2$.

Enrichment

Enrichment 102 is pictured on page 434E.

FUNCTIONS

Because the area and circumference of a circle are quantities that depend on the radius of the circle, each quantity is a *function* of the radius.

Copy and complete each function table.

25.

Radius (r)	Circumference (2πr)	
1	?	2π
2	?	4π
3	6π	
4	?	8π
5	10π	

26.

Radius (r)	Area (πr²)	
1	?	π
2	4π	
3	?	9π
4	?	16π
5	25π	

27. Each number has π as a factor; for circumference, the radius doubles; for area, the radius is squared.

27. Refer to the tables that you completed for Exercises 25 and 26. Describe the pattern in each table in words.

28. Refer to the tables that you completed for Exercises 25 and 26. Using r as the variable, write a function rule for each table.
$r \rightarrow 2\pi r; r \rightarrow \pi r^2$

THINKING SKILLS **29, 30.** Evaluation

C **29.** Determine what happens to the circumference of a circle when its radius is doubled. The circumference doubles.

30. Determine what happens to the area of a circle when its radius is doubled. The area quadruples.

SPIRAL REVIEW

S **31.** Write 26.7% as a decimal. *(Lesson 9-5)* 0.267

32. Find the area of a circle with diameter 10.6 m. *(Lesson 10-4)* 88.2 m²

33. Estimate: $1\frac{7}{8} + 7\frac{1}{6}$ *(Lesson 8-6)* about 9

34. Solve: $2x + 9 = 17$ *(Lesson 4-5)* 4

Practice

Practice 126, *Resource Book*

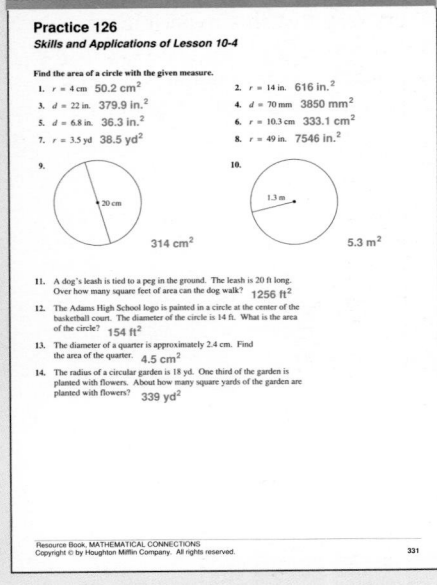

Historical Note

Claudius Ptolemy, a Greek-Egyptian who lived in the second century A.D., wrote a book on astronomy that was the standard in its field for centuries. Much of Ptolemy's work was based on dividing the circumference of the circle into 360 parts. He developed this idea from a Babylonian practice originated nearly 4000 years earlier.

60; in an early system of weights and measures, a larger unit was 60 times a smaller.

Research

What number was the base of the Babylonian number system? Why is it believed that this number was used?

10-5 Choosing Perimeter, Circumference, or Area

Objective: To decide whether perimeter, circumference, or area must be calculated to solve a given problem.

In everyday life, problems that you encounter seldom instruct you to "find the perimeter" or "find the circumference" or "find the area." Usually you must make a decision about which measure is needed.

- Calculate *perimeter* or *circumference* when you need to find the distance around a figure—perimeter for polygons, circumference for circles.

- Calculate *area* when you need to find the measure of the region enclosed by a figure.

Example

Tell whether you would need to calculate *perimeter*, *circumference*, or *area* to find each measure.

a. the amount of artificial turf needed to cover a football field
b. the length of the chalk line around a football field

Solution

a. The artificial turf covers a region enclosed by a rectangle. You need to calculate the *area* of the rectangle.

b. The chalk line outlines the sides of a rectangle. You need to calculate the distance around the rectangle, or its *perimeter*.

 **Check Your Understanding**

In what type of situation do you need to find the *circumference* of a figure?
See *Answers to Check Your Understanding* at the back of the book.

Guided Practice

COMMUNICATION «*Reading*

Refer to the text above. Answers to Guided Practice 1–2 are on p. A26.

«**1.** How do you know when you need to calculate either a perimeter or a circumference? What is the example given?

«**2.** How do you know when you need to calculate area? What is the example given?

Circles and Polygons **457**

Lesson Planner

Teaching the Lesson
- Lesson Starter 94

Lesson Follow-Up
- Practice 127
- Enrichment 103: Data Analysis
- Study Guide, pp. 187–188

Warm-Up Exercises

Match each formula with its description below.

1. $C = \pi d$ b
2. $P = 2(l + w)$ e
3. $A = \pi r^2$ d
4. $P = 4s$ a
5. $A = s^2$ c

(a) perimeter of a square
(b) circumference of a circle
(c) area of a square
(d) area of a circle
(e) perimeter of a rectangle

Lesson Focus

Ask students to imagine that they are going to reseed a lawn. To do so, they need to determine how much grass seed to buy. Ask students if they would calculate the perimeter or area of the lawn. Today's lesson focuses on deciding which measure is appropriate for specific problems.

Teaching Notes

Key Question
When do you need to find the perimeter or circumference of a figure? the area of a figure? when you need to know the distance around the figure; when you need to know the measure of the region enclosed by the figure

Additional Examples

Tell whether you need to calculate *perimeter, circumference,* **or** *area* **to find each measure.**

1. the amount of fertilizer needed for a lawn   area

2. distance walked around 3 blocks   perimeter

Closing the Lesson

Ask students to use examples to explain the meaning of perimeter, circumference, and area.

Suggested Assignments

Skills: 1–14
Average: 1–14
Advanced: 1–14

Exercise Note

Application

Exercises 5–10 help to develop students' ability to apply various measurement formulas.

Follow-Up

Enrichment

Enrichment 103 is pictured on page 434E.

Practice

Practice 127, *Resource Book*

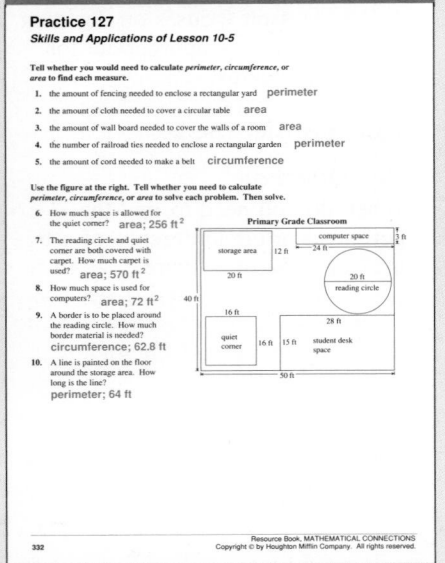

Tell whether you would need to calculate *perimeter, circumference,* **or** *area* **to find each measure.**

3. the amount of grazing space in a square pasture   area

4. the amount of fencing needed for a circular garden   circumference

«**5.** **COMMUNICATION** «*Discussion*   Describe an everyday situation in which you needed to choose perimeter, circumference, or area to solve a problem.   Answers will vary.

Exercises

Tell whether you would need to calculate *perimeter, circumference,* **or** *area* **to find each measure.**

A **1.** the amount of wood needed to frame a circular mirror   circumference

2. the amount of wallpaper needed for a rectangular wall   area

3. the amount of decorative molding needed around a square ceiling   perimeter

4. the amount of floor space covered by a circular rug   area

Use the figure at the right. Tell whether you need to calculate *perimeter, circumference,* **or** *area* **to solve each problem. Then solve.**

5. How much space is there for a swimmer to float in the pool?   area; 706.5 ft²

6. circumference; 150.7 ft   **6.** What amount of fencing is needed to enclose the concrete deck?

7. perimeter; 112 yd   **7.** How much fencing is needed to enclose the entire recreation area?

8. What amount of concrete is needed to cover the play area?   area; 1024 ft²

9. area; 1102.1 ft²   **9.** How much space does the concrete deck provide for chairs?

10. One package of lawn seed covers 100 ft². How many packages are needed to seed the lawn?   area; 41

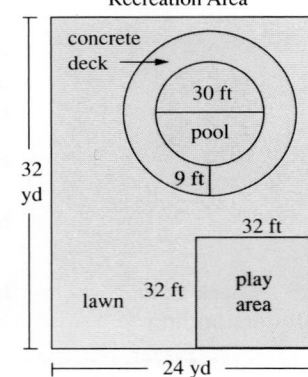

Recreation Area

concrete deck

30 ft

pool

9 ft

32 yd

32 ft

lawn   32 ft   play area

24 yd

SPIRAL REVIEW

S **11.** Solve: $4x + 17 + x = -8$ *(Lesson 4-9)* −5

12. Find the LCM of 154 and 75.   *(Lesson 7-3)* 11,550

13. Tell whether you need to calculate *perimeter, circumference,* or *area* to find the distance around a circular track.   *(Lesson 10-5)* circumference

14. Solve $A = \frac{1}{2}bh$ for h.   *(Lesson 8-10)* $h = \frac{2A}{b}$

10-6

Problems with No Solution

Objective: To recognize when a problem has no solution.

In solving a problem, you must evaluate the answer to your calculations to be sure that it makes sense in a real-world context. You will find that some problems have no solution, even though you have arrived at a correct answer to your calculations.

Problem

The perimeter of a triangular field is 64 ft. The measures of two sides are 16 ft and 14 ft. What is the length of the third side?

Solution

UNDERSTAND The problem is about the perimeter of a field.
Facts: triangular shape
    perimeter 64 ft
    two sides measure 16 ft and 14 ft
Find: length of the third side

PLAN Use the perimeter formula $P = a + b + c$. Substitute 64 for P, 16 for a, and 14 for b. Solve the resulting equation for c.

WORK
$$P = a + b + c$$
$$64 = 16 + 14 + c$$
$$64 = 30 + c$$
$$64 - 30 = 30 + c - 30$$
$$34 = c$$

ANSWER The solution of the equation is 34. However, the sum of the lengths of any two sides of a triangle must be greater than the length of the third side, and $16 + 14 < 34$. The problem has no solution.

Look Back What if the given information were that the lengths of the two sides are 18 ft and 14 ft? Does the problem have a solution?
No; 18 + 14 = 32, which is the length of the third side.

Guided Practice

1. You cannot buy $\frac{3}{4}$ of a sweater.
2. The sum of 35°, 65°, and 90° is greater than 180°.
3. Perimeter must be positive.
4. The number of people must be a whole number.

COMMUNICATION «*Reading*

Explain why each answer has no meaning in the given context.

«**1.** *Context:* How many sweaters were bought? *Answer:* $1\frac{3}{4}$

«**2.** *Context:* What are the measures of the three angles of a triangle? *Answer:* 35°, 65°, and 90°

«**3.** *Context:* What is the perimeter of a pentagon? *Answer:* −60

«**4.** *Context:* How many people attended? *Answer:* 156.3

Circles and Polygons   **459**

Lesson Planner

Teaching the Lesson
- Lesson Starter 95

Lesson Follow-Up
- Practice 128
- Practice 129: Nonroutine
- Enrichment 104: Thinking Skills
- Study Guide, pp. 189–190
- Test 48

Warm-Up Exercises

Give the formula for each of the following.
1. area of a circle $A = \pi r^2$
2. perimeter of a rectangle $P = 2(l + w)$
3. area of a triangle $A = \frac{1}{2}bh$
4. circumference of a circle $C = \pi d = 2\pi r$
5. area of a trapezoid $A = \frac{h}{2}(b_1 + b_2)$

Lesson Focus

Ask students to think of situations in their lives that caused them to say, "This doesn't make sense," or "This just isn't logical." In solving problems, students need to check that an answer does make sense. If there is no sensible or logical answer, the problem has no solution. Today's lesson focuses on problems of this type.

Ask students to solve this problem:

The fifth grade class from East-port Elementary School is going to the zoo. They are traveling by bus, and each bus holds 48 passengers. If 128 persons are going, how many buses are needed?

Discuss why an answer of $2\frac{2}{3}$ buses does not make sense?

Key Question

How do you determine if an answer to a problem is sensible?
You check to see if the answer fits the context of the situation.

Error Analysis

Many students accept answers to problems based solely on calculations. Emphasize that all answers should be checked against the conditions of the problem.

Closing the Lesson

Have students explain what is meant by "problems with no solution."

Suggested Assignments

Skills: 1–16
Average: 1–16
Advanced: 1–16

Exercise Notes

Communication: Reading

Guided Practice Exercises 1–4 are designed to increase students' awareness of the need to check an answer for meaning within a given context.

Problem Solving

In *Exercises 1–4*, students must analyze each situation and explain why there is no solution. *Exercises 5–10* challenge students to select an appropriate problem-solving strategy for each problem, and then apply the strategy.

Explain why each problem has no solution.

5. There are no amounts of dimes and quarters totaling seven that result in $1.20.

6. 68 yd = 204 ft; the two lengths would be greater than the perimeter.

5. A collection of seven coins consists of dimes and quarters and has a total value of $1.20. How many of each type of coin are there?

6. The perimeter of a rectangular auditorium is 306 ft. If the length is 68 yd, what is the width?

«7. **COMMUNICATION** «*Discussion*   Describe how the *problems with no solution* in this lesson are different from the *problems with not enough information* that you studied in Lesson 5-6. See answer on p. A27.

Problem Solving Situations

Explain why each problem has no solution.

1. The number of books must be a whole number.

2. The two sides of the triangle are longer than the perimeter.

3. There are no amounts of dimes and nickels totaling 12 that result in $.60.

4. The number of students must be a whole number.

1. Books at the sale cost $.75 each, including tax. Janeen said she spent $8 at the sale. How many books did she buy?

2. The measure of each of two sides of a triangle is 16 in. The perimeter is measured to be 30 in. What is the measure of the third side?

3. Jake says he has $.60 in dimes and nickels. He has twelve coins altogether. How many of each type of coin does he have?

4. Admission to the school play was $3 for adults and $2 for students. The receipts were recorded as $1400. If 237 adults attended the play, how many students attended?

Solve using any problem solving strategy.

5–10. Strategies may vary. Likely strategies are given.
5. supplying missing facts; $32
6. using equations; 64 cm²
7. choosing the correct operation; 2.5 lb
8. no solution
9. using equations; 22.5 m, 22.5 m
10. making a table; 100 cm²

5. The cost of carpeting a rectangular room with width 12 ft and length 15 ft is $640. What is the cost of the carpeting per square yard?

6. The perimeter of a square is 32 cm. What is its area?

7. Roast beef costs $4.98 per pound. How many pounds can you buy for $12.45?

8. When you double a number and add five, the result is three less than twice the number. What is the number?

9. The perimeter of an isosceles triangle is 60 m. The measure of one side is 15 m. What are the measures of the other two sides?

10. The perimeter of a rectangle is 40 cm. What is the greatest possible area that this rectangle could have?

PROBLEM SOLVING
CHECKLIST
Keep these in mind:
Using a Four-Step Plan
Too Much or Not Enough Information
Supplying Missing Facts
Problems with No Solution
Consider these strategies:
Choosing the Correct Operation
Making a Table
Guess and Check
Using Equations
Identifying a Pattern
Drawing a Diagram
Using Proportions

WRITING WORD PROBLEMS

For each exercise, use the collection of coins that is pictured at the right.

11. Write a word problem about the coins that has exactly one solution.

12. Write a word problem about the coins that has no solution.

SPIRAL REVIEW

S **13.** Write a variable expression that represents the phrase *the sum of a number x and nine.*   *(Lesson 4-6)* $x + 9$

14. Explain why this problem has no solution: Four coins have a total value of $.48. What are the coins?   *(Lesson 10-6)*

14. You need at least 3 pennies; there is no single coin with a value of $.45.

15. Solve: $\frac{5}{8}a = 24$ *(Lesson 8-9)* $38\frac{2}{5}$

16. 18% of 70 is what number?   *(Lesson 9-6)* 12.6

Self-Test 1

8. The sum of the lengths of the two sides given is equal to the perimeter.

Find the perimeter of each figure.

1. a rhombus with one side that measures 11.2 cm 44.8 cm   10-1

2. a rectangle with sides that measure $6\frac{1}{2}$ in. and $9\frac{1}{4}$ in. $31\frac{1}{2}$ in.

3. A circle has a radius of 7 m.   10-2
   a. Find the diameter. 14 m   **b.** Find the circumference.
                                                                    44 m

Find the area of each figure.

4. a square with one side that measures $5\frac{1}{2}$ in. $30\frac{1}{4}$ in.²   10-3

5. a triangle with base 12 cm and height 6 cm 36 cm²

6. a circle with radius 8.4 cm 221.6 cm²   10-4

7. Tell whether you would need to calculate *perimeter*,   10-5
circumference, or *area* to find the amount of fencing needed
to enclose a circular circus tent. circumference

8. Explain why this problem has no solution: A triangle has a   10-6
perimeter of 27 in. Two of the sides measure 18 in. and 9 in.
Find the length of the third side.

Circles and Polygons   **461**

Practice 128
Skills and Applications of Lesson 10-6

Explain why each problem has no solution.   Ex. 1–5. See below.

1. Anthony High School has 7 sophomore homerooms with the same number of students. The total number of sophomores is 170. How many students are in each homeroom?

2. A collection of 5 coins consists of nickels and quarters and has a total value of $1.75. How many of each type of coin are there?

3. A supplement of an angle is measured to be 134°. A complement of the same angle is measured to be 50°. Find the measure of the angle.

4. Beth pays $1.36 for some pencils. The pencils cost $.39 each, including tax. How many pencils did she buy?

5. Two angles of a triangle measure 65° and 117°. Find the measure of the third angle.

Solve using any problem solving strategy.

6. In a recipe for bread, the ratio of yeast to honey is 1 : 2. There are 3 tablespoons of honey in the dough. How many tablespoons of yeast are needed? $1\frac{1}{2}$

7. The area of a circle is 616 in.² The circumference of the same circle is measured to be 94.2 in. Find the radius of the circle.   no solution

8. A roll of wallpaper border is 18 ft long. How many rolls are needed for a rectangular room with width 5 yd and length 7 yd?   4

1. The number of students in each homeroom must be a whole number.
2. There is no amount of nickels and quarters totaling 5 which results in $1.75.
3. The difference of the supplement and complement must be 90°.
4. The number of pencils must be a whole number.
5. The sum of the measures of two angles of a triangle must be less than 180°.

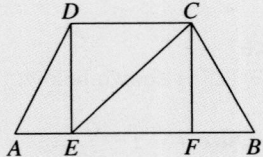

Congruent Polygons

Objective: To identify corresponding parts of congruent figures.

Terms to Know
- *congruent*
- *corresponding angles*
- *corresponding sides*

CONNECTION

In arithmetic, you learned that fractions and decimals that represent the same number are called *equivalent*. For instance, you learned that there are infinitely many ways to represent the number $\frac{1}{2}$.

$$\frac{1}{2} = \frac{2}{4} = \frac{3}{6} = \frac{4}{8} = \ldots \qquad \text{and} \qquad \frac{1}{2} = 0.5 = 0.50 = 0.500 = \ldots$$

Similarly, in geometry there are infinitely many ways to picture a given size and shape. For example, although the four triangles below have different names and are positioned differently, they are identical in size and shape.

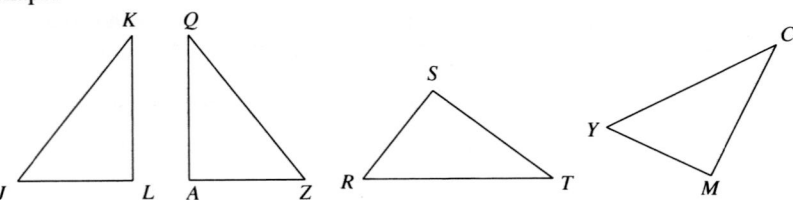

Geometric figures that have the same size and shape are called **congruent** figures. The symbol for congruent is ≅.

Line segments are congruent when they have the same length. In the figure at the right, for example, line segments *AB* and *CD* each measure $1\frac{1}{4}$ in., so they are congruent.

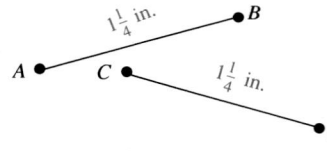

In Words	In Symbols
Line segment AB is congruent to line segment CD.	$\overline{AB} \cong \overline{CD}$

Angles are congruent when they have the same degree measure. In the figure at the right, angles *P* and *Q* are congruent because they each measure 127°.

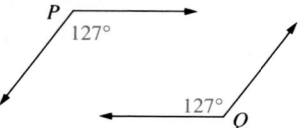

In Words	In Symbols
Angle P is congruent to angle Q.	$\angle P \cong \angle Q$

Polygons are congruent when there is a way to match up their vertices so that all pairs of **corresponding angles** and all pairs of **corresponding sides** are congruent. In the triangles at the right, the red markings indicate that these corresponding parts are congruent.

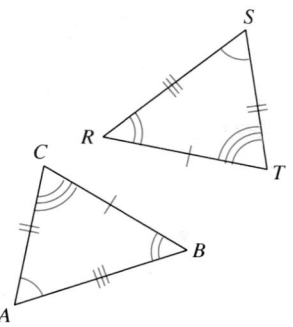

$$\angle A \cong \angle S \qquad \overline{AB} \cong \overline{SR}$$
$$\angle B \cong \angle R \qquad \overline{BC} \cong \overline{RT}$$
$$\angle C \cong \angle T \qquad \overline{AC} \cong \overline{ST}$$

To state that the triangles are congruent, you list the vertices of each triangle in the same order as the corresponding angles.

In Words	In Symbols
Triangle ABC is congruent to triangle SRT.	$\triangle ABC \cong \triangle SRT$

Example

Use the figures at the right to complete each statement.

a. $\overline{SP} \cong \underline{\ ?\ }$

b. $\angle B \cong \underline{\ ?\ }$

c. quadrilateral $PQRS \cong$ quadrilateral $\underline{\ ?\ }$

d. If the length of $\overline{DB}$ is 15 cm, then the length of $\underline{\ ?\ }$ is 15 cm.

e. If $m\angle R = 98°$, then $m\underline{\ ?\ } = 98°$.

Solution

Use the red markings to match up the vertices as follows.

$$P \text{ and } C \qquad Q \text{ and } B$$
$$R \text{ and } D \qquad S \text{ and } A$$

a. $\overline{SP} \cong \overline{AC}$ b. $\angle B \cong \angle Q$

c. quadrilateral $PQRS \cong$ quadrilateral $CBDA$

d. $\overline{DB} \cong \overline{RQ}$, so the length of $\overline{RQ}$ is 15 cm.

e. $\angle R \cong \angle D$, so $m\angle D = 98°$.

✔ **Check Your Understanding**

1. List all the pairs of corresponding angles and corresponding sides in the figure for the Example.

2. In part (c) of the Example, why is the answer not given as "quadrilateral $PQRS \cong$ quadrilateral $ABCD$"?

See *Answers to Check Your Understanding* at the back of the book.

Circles and Polygons   **463**

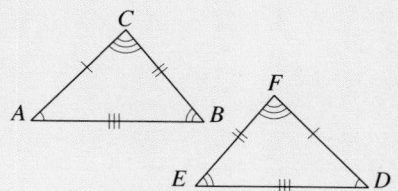

Guided Practice

COMMUNICATION « *Writing*

«**1.** Write the following statement in symbols:
Angle *PQR* is congruent to angle *XYZ*. $\angle PQR \cong \angle XYZ$

«**2.** Write the following statement in words: $\triangle PQR \cong \triangle XYZ$
Triangle *PQR* is congruent to triangle *XYZ*.

Use the figure at the right. Suppose you are given the information that $\triangle DEF \cong \triangle ABC$.

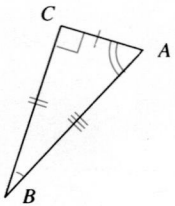

«**3.** Which side of $\triangle DEF$ would have three marks? $\overline{DE}$

«**4.** Which angle of $\triangle DEF$ would have two marks? $\angle D$

«**5.** Which angle of $\triangle DEF$ would be a right angle? $\angle F$

«**6.** Make a sketch of $\triangle DEF$. See answer on p. A27.

Use the figure at the right to complete each statement.

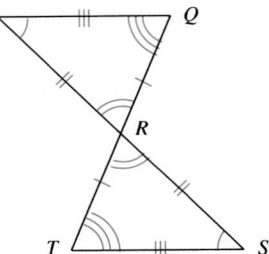

7. $\angle Q \cong$? $\angle T$ **8.** $\overline{TR} \cong$? $\overline{QR}$

9. $\triangle PQR \cong$? $\triangle STR$

10. If the length of $\overline{RQ}$ is 22 in., then the length of ? is 22 in. $\overline{RT}$

11. If $m\angle S = 45°$, then m ? $= 45°$. $\angle P$

12. If $m\angle PRQ = 68°$, then m ? $= 68°$. $\angle SRT$

Exercises

Use the figure at the right to complete each statement.

A   **1.** $\angle F \cong$? $\angle D$ **2.** $\angle CBE \cong$? $\angle ABE$

3. $\overline{BC} \cong$? $\overline{BA}$ **4.** $\overline{FE} \cong$? $\overline{DE}$

5. quadrilateral $ABEF \cong$ quadrilateral ? $CBED$

6. If the length of $\overline{CD}$ is 4.6 m, then the length of ? is 4.6 m. $\overline{AF}$

7. If $m\angle FEB = 70°$, then m ? $= 70°$. $\angle DEB$

8. Which line segment forms a side of both quadrilaterals? $\overline{BE}$

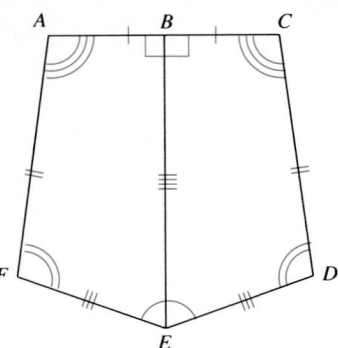

9. If $\triangle PQR \cong \triangle MNO$, which side of $\triangle MNO$ is congruent to $\overline{RP}$? $\overline{OM}$

10. Rectangle $JKLM$ is congruent to rectangle $WXYZ$. List all the segments that are congruent to $\overline{MJ}$. $\overline{ZW}, \overline{KL}, \overline{XY}$

LOGICAL REASONING

Tell whether each statement is *true* or *false*. If the statement is false, give a counterexample to show why it is false.

11. False; a rectangle with sides of lengths 1 and 4 and a rectangle with sides of lengths 2 and 3.

11. Two rectangles are congruent if they have the same perimeter.

12. Two squares are congruent if they have the same perimeter. True.

13. If two circles have the same radius, then they are congruent. True.

15. False; a rectangle with sides of lengths 3 and 4 and a rectangle with sides of lengths 2 and 6.

14. If two circles have the same area, then they are congruent. True.

15. If two rectangles have the same area, then they are congruent.

16. If two rectangles are congruent, then they have the same area. True.

CONNECTING GEOMETRY AND ALGEBRA

C **17.** Give a set of coordinates for point S so that quadrilateral $ABCD \cong$ quadrilateral $PQRS$. (1, −1)

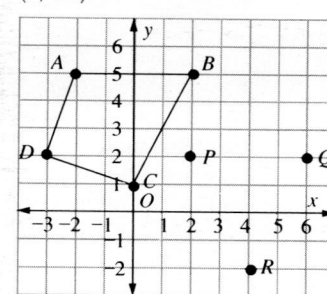

18. Give two different sets of coordinates for point Z so that $\triangle MLN \cong \triangle XYZ$. Answers will vary. Examples: (1, −2); (1, 4)

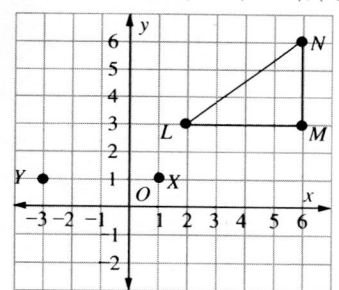

SPIRAL REVIEW

S **19.** Simplify: $\frac{2a}{5} + \frac{3a}{4}$ *(Lesson 8-5)* $\frac{23a}{20}$

20. Solve: $x - 18 = -4$ *(Lesson 4-3)* 14

21. Solve: $\frac{x}{6} = \frac{20}{3}$ *(Lesson 9-2)* 40

22. Use the figures below to complete this statement: If the length of $\overline{WY}$ is 8.3 cm, then the length of ___?___ is 8.3 cm.   *(Lesson 10-7)* $\overline{HJ}$

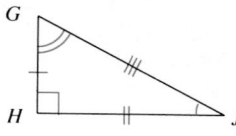
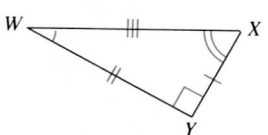

Circles and Polygons   **465**

Extension
Have each student draw an isosceles triangle ABC and cut it out. Next, fold the triangle along the altitude to each side. Then draw in altitudes along each crease. Point out that the altitudes appear to intersect in one point. Have students label this point as D and the intersections of the altitudes with each side as E, F, and G. Finally, have students list all pairs of congruent figures that they see in the drawing.

Enrichment
Enrichment 105 is pictured on page 434F.

Practice
Practice 130, *Resource Book*

Practice 130
Skills and Applications of Lesson 10-7

Use the figure at the right to complete each statement.
1. $\angle BAC \cong$ _?_ $\angle YXZ$
2. $\overline{AB} \cong$ _?_ $\overline{XY}$
3. _?_ $\cong \angle YZX$ $\angle BCA$
4. $\overline{CB} \cong$ _?_ $\overline{ZY}$
5. $\overline{AC} \cong$ _?_ $\overline{XZ}$
6. If $m\angle CBA = 32°$, then $m\angle ZYX =$ _?_ 32°
7. If $m\angle ZXY = 105°$, then $m\angle CAB =$ _?_ 105°
8. If the length of $\overline{AB}$ is 16 cm, then the length of _?_ is 16 cm. $\overline{XY}$

Use the figure at the right to complete each statement.
9. $\overline{AC} \cong$ _?_ $\overline{BC}$
10. _?_ $\cong \angle CBD$ $\angle CAD$
11. $\triangle ACD \cong$ _?_ BCD
12. $\overline{CD} \cong$ _?_ $\overline{CD}$
13. If $m\angle CAD = 44°$, then _?_ $= 44°$. $m\angle CBD$
14. If the length of $\overline{BD}$ is 14 cm, then the length of _?_ is 14 cm. $\overline{AD}$
15. If quadrilateral $ABCD$ is congruent to quadrilateral $MNOP$, which angle of quadrilateral $MNOP$ is congruent to $\angle CDA$? $\angle OPM$
16. If $\triangle HFJ \cong \triangle LKM$, which side of $\triangle HFJ$ is congruent to $\overline{LM}$? $\overline{HJ}$

Teaching the Lesson
- Materials: geoboards
- Lesson Starter 97
- Visuals, Folder L
- Using Technology, p. 434C
- Reteaching/Alternate Approach, p. 434D
- Cooperative Learning, p. 434D

Lesson Follow-Up
- Practice 131
- Enrichment 106: Application
- Study Guide, pp. 193–194
- Computer Activity 10
- Manipulative Activity 20

Warm-Up Exercises

ABCD is a trapezoid with congruent segments and angles as marked.

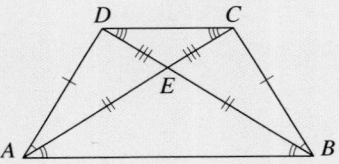

1. List the pairs of triangles that are congruent.
   $\triangle ADE \cong \triangle BEC$;
   $\triangle ADC \cong \triangle BCD$;
   $\triangle ADB \cong \triangle BCA$
2. Identify any triangles that appear to have the same shape but not the same size.
   $\triangle DEC$ and $\triangle BEA$

10-8 Similar Polygons

Objective: To recognize similar polygons and use corresponding sides to find unknown lengths.

Terms to Know
- *similar*

APPLICATION

When you see a new type of car in an automobile showroom, do you realize that the plans for the car were actually begun several years ago? In creating these plans, engineers make a scale model of the car that has the same shape as the actual car but is much smaller in size. Scale models like these are examples of *similar* geometric figures.

Geometric figures that have the same shape but not necessarily the same size are called **similar** figures. For instance, the circles shown at the right are similar. The symbol for similar is $\sim$.

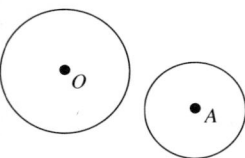

In Words	In Symbols
Circle O is similar to circle A.	$\odot O \sim \odot A$

Polygons are similar when there is a way to match up their vertices so that all pairs of corresponding angles are congruent and all pairs of corresponding sides are *in proportion*. In the triangles at the right, for example, the following pairs of angles are congruent.

$\angle A \cong \angle D \qquad \angle B \cong \angle E \qquad \angle C \cong \angle F$

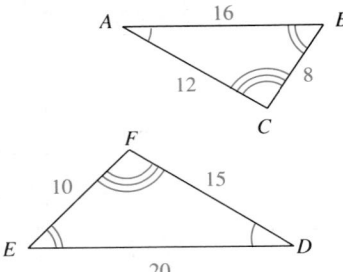

Using the corresponding angles as a guide, you can identify the corresponding sides.

$\overline{AB}$ and $\overline{DE}$ $\qquad$ $\overline{BC}$ and $\overline{EF}$ $\qquad$ $\overline{AC}$ and $\overline{DF}$

Now you can check that each ratio between the lengths of corresponding sides is equal to the same number, $\frac{4}{5}$.

$$\frac{\overline{AB}}{\overline{DE}} \rightarrow \frac{16}{20} = \frac{4}{5} \qquad \frac{\overline{BC}}{\overline{EF}} \rightarrow \frac{8}{10} = \frac{4}{5} \qquad \frac{\overline{AC}}{\overline{DF}} \rightarrow \frac{12}{15} = \frac{4}{5}$$

The corresponding sides are in proportion, so the triangles are similar. You write $\triangle ABC \sim \triangle DEF$, being sure to list the vertices in the same order as the corresponding angles.

When you know that polygons are similar, you can use the fact that corresponding sides are in proportion to find unknown measures.

Example

In the figures at the right, quadrilateral $WXYZ \sim$ quadrilateral $QPRS$. Find the length of $\overline{PR}$.

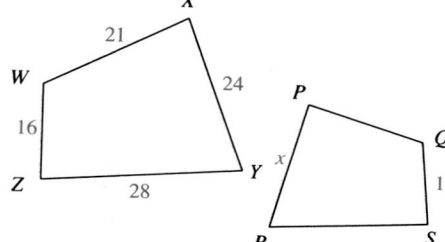

Solution

$\overline{WZ}$ corresponds to $\overline{QS}$. The ratio of these sides is $\frac{16}{12} = \frac{4}{3}$, so the ratio of each pair of corresponding sides is $\frac{4}{3}$.

$\overline{XY}$ corresponds to $\overline{PR}$. Using the ratio $\frac{4}{3}$, write and solve a proportion.

$$\frac{4}{3} = \frac{24}{x} \qquad \begin{array}{l} \longleftarrow \text{length of } \overline{XY} \\ \longleftarrow \text{length of } \overline{PR} \end{array}$$

$$4x = 3(24)$$
$$4x = 72$$
$$x = 18$$

The length of $\overline{PR}$ is 18.

 **Check Your Understanding**

1. Using the information in the Example, what is the length of $\overline{RS}$?
2. How would the solution of the Example be different if the length of $\overline{QS}$ were labeled as 10?

See *Answers to Check Your Understanding* at the back of the book.

Guided Practice

COMMUNICATION «*Reading*

Refer to the text on pages 462–463 and on pages 466–467. Copy and complete each outline in your notebook.

«1. *Congruent Geometric Figures* $\qquad$ *Congruent Polygons*

$\underline{\quad ? \quad}$ size same $\qquad$ congruent corresponding angles are $\underline{\quad ? \quad}$

$\underline{\quad ? \quad}$ shape same $\qquad$ congruent corresponding sides are $\underline{\quad ? \quad}$

Ask students if they have ever seen a photograph that has been enlarged. The objects in the enlarged photo look the same as those in the original, but they are dilated. The objects in the enlarged photo are said to be *similar* to those in the original photo. Applying the concept of similarity to polygons is the focus of today's lesson.

Teaching Notes

Key Questions

1. How are similar figures and congruent figures alike? How are they different? Figures that are congruent and figures that are similar both have the same shape. However, similar figures do not necessarily have the same size.

2. What is true of the corresponding sides of similar polygons? They all are in proportion.

Error Analysis

Students may set up incorrect ratios when working with similar figures. Stress to students that when they form proportions they must match up the corresponding sides.

Using Technology

For a suggestion on using geometric drawing software and a spreadsheet, see page 434C.

Reteaching/Alternate Approach

For a suggestion on using a cooperative learning activity, see page 434D.

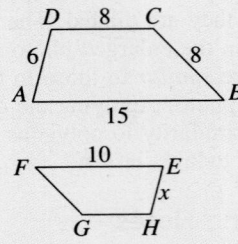
Refer to the text on pages 462–463 and on pages 466–467. Copy and complete each outline in your notebook.

« 2. *Similar Geometric Figures*
    __?__ size not necessarily the same
    __?__ shape same

Similar Polygons congruent
corresponding angles are __?__
corresponding sides are __?__
                   in proportion

Use the figures at the right to write each ratio.

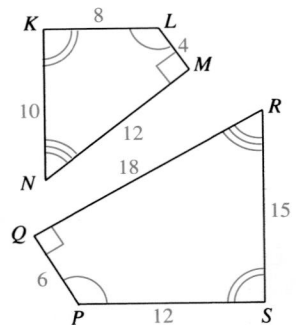

3. $\dfrac{\text{length of } \overline{LM}}{\text{length of } \overline{PQ}}$ $\dfrac{4}{6}$
4. $\dfrac{\text{length of } \overline{MN}}{\text{length of } \overline{QR}}$ $\dfrac{12}{18}$

5. $\dfrac{\text{length of } \overline{NK}}{\text{length of } \overline{RS}}$ $\dfrac{10}{15}$
6. $\dfrac{\text{length of } \overline{KL}}{\text{length of } \overline{SP}}$ $\dfrac{8}{12}$

Use the figures at the right to complete.

7. Written in lowest terms, the ratio of each pair of corresponding sides is __?__. $\dfrac{2}{3}$

8. quadrilateral *KLMN* ~ quadrilateral __?__
    SPQR

9. Two ratios are $\frac{1}{2}$; one ratio is $\frac{3}{5}$.

9. Explain why △*RST* is *not* similar to △*ABC*.

10. Given that △*XYZ* ~ △*EFD*, find the values of *x* and *y*. 6.5; 18

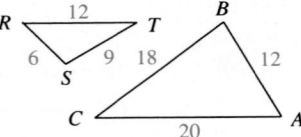

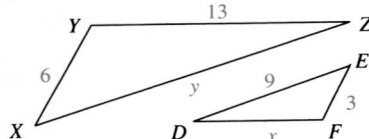

« 11. COMMUNICATION « *Discussion* If two polygons are congruent, are they also similar? Explain. Yes; the ratio of each pair of corresponding sides is 1.

Exercises

Find the values of *x* and *y*.

A   **1.** △*QRS* ~ △*PNM* 12; 12.5
                2. △*XYZ* ~ △*CBA* 6; 6.75

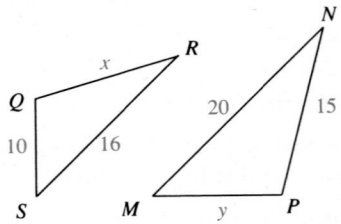

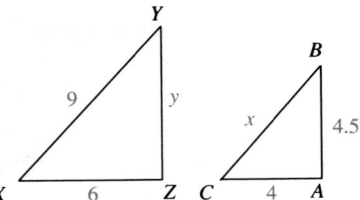

468 Chapter 10

3. quad. *WXYZ* ~ quad. *JKGH*
5; 12.5

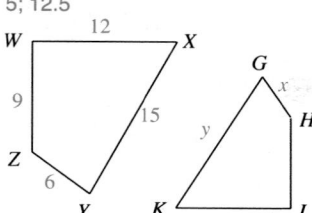

4. quad. *DCBA* ~ quad. *PQRS*
22.5; 36

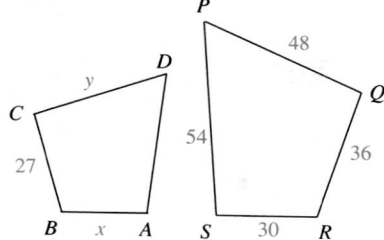

5. The measures of three sides of a triangle are 16 cm, 10 cm, and 12 cm. The shortest side of a similar triangle measures 15 cm. What is the perimeter of the similar triangle? 57 cm

6. The length of a rectangle is 12 ft and its width is 8 ft. The width of a similar rectangle is 18 ft. What is the area of the similar rectangle?
486 ft²

7. *Yes*; the length of $\overline{MN}$ is 10 cm because corresponding sides are congruent.

8. *No*; you need to know the ratio of the sides.

7. Suppose that $\triangle PQR \cong \triangle MNO$, and the length of $\overline{PQ}$ is 10 cm. Can you determine the length of any side of $\triangle MNO$? Explain.

8. Suppose that $\triangle PQR \sim \triangle MNO$, and the length of $\overline{PQ}$ is 10 cm. Can you determine the length of any side of $\triangle MNO$? Explain.

Exercises 9–12 refer to the figures at the right. In the figures, $\triangle RST \sim \triangle LMN$.

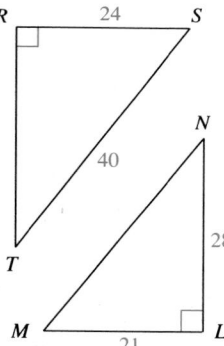

9. Find the length of $\overline{RT}$. 32

10. Find the length of $\overline{NM}$. 35

11. $\frac{8}{7}$; they are equal. **B**

11. Write the ratio $\dfrac{\text{perimeter of } \triangle RST}{\text{perimeter of } \triangle LMN}$ in lowest terms. How is this ratio related to the ratio of the corresponding sides?

12. Suppose that $\triangle ABC \sim \triangle LMN$. You know that the length of $\overline{AB}$ is 4. What is the perimeter of $\triangle ABC$? 16

LOGICAL REASONING

Tell whether two figures of the given type are *always, sometimes,* or *never* similar. **16.** always **18.** always

13. two squares always

14. two rectangles sometimes

15. two isosceles triangles sometimes

16. two equilateral triangles

17. two regular hexagons always

18. two congruent polygons

19. a right triangle and an acute triangle never

20. an obtuse triangle and an equilateral triangle never

Communication: Reading
Guided Practice Exercises 1 and 2 point out the difference between congruent figures and similar figures.

Communication: Discussion
Guided Practice Exercise 11 encourages discussion of the phrase "but not necessarily" as given in the definition on page 466.

Reasoning
Exercises 13–20 test students' ability to visualize many types of polygons and apply their knowledge of similarity to each pair.

Computer
Exercises 21–23 encourage students to use geometric drawing software to explore the properties of a segment that joins the midpoints of two sides of a triangle.

Critical Thinking
Exercise 24 extends *Exercises 21–23* by having students *examine* and *compare* results for additional triangles, and then *make a generalization* about the effect of drawing a segment that joins the midpoints of two sides of a triangle.

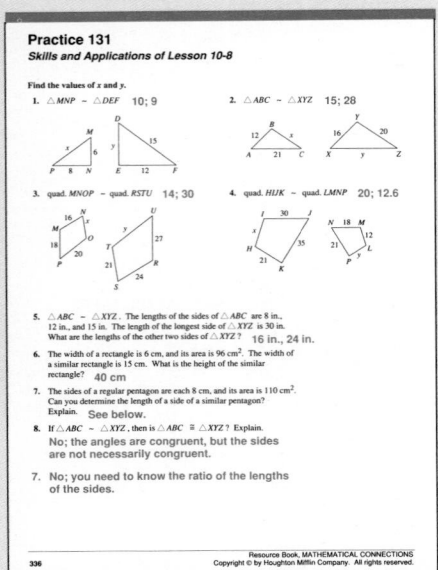
COMPUTER APPLICATION

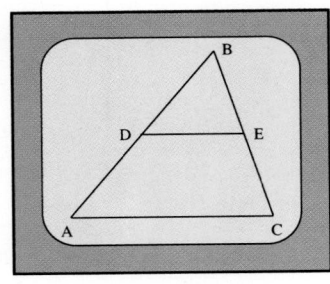

The **midpoint** of a line segment is the point that separates it into two congruent segments. In the figure at the right, point D is the midpoint of $\overline{AB}$ and point E is the midpoint of $\overline{BC}$.

C **21.** Draw the figure at the right, making $\triangle ABC$ any size that you choose. Use geometric drawing software if it is available. Answers will vary.

22. Copy and complete the chart.

22. Answers will vary; corresponding sides will be in proportion and corresponding angles will be congruent.

△*ABC*	△*DBE*
length of $\overline{AB}$ = _?_	length of $\overline{DB}$ = _?_
length of $\overline{BC}$ = _?_	length of $\overline{BE}$ = _?_
length of $\overline{AC}$ = _?_	length of $\overline{DE}$ = _?_
$m\angle BAC$ = _?_	$m\angle BDE$ = _?_
$m\angle B$ = _?_	$m\angle B$ = _?_
$m\angle BCA$ = _?_	$m\angle BED$ = _?_
perimeter = _?_	perimeter = _?_

23. Refer to the chart that you completed for Exercise 22. What is the relationship between $\triangle ABC$ and $\triangle DBE$? They are similar.

24. Answers will vary. Example: the resulting triangle is similar to the original triangle.

24. **THINKING SKILLS** Repeat Exercises 21–23 for three different triangles. Compare the results. Make a generalization by completing this statement: When a line segment is drawn between the midpoints of two sides of a triangle, _____?_____. Synthesis

SPIRAL REVIEW

S **25.** Solve: $2.4x - 10 = 6.2$ *(Lesson 8-8)* 6.75

26. In the figures at the right, $\triangle YJT \sim \triangle ACB$. Find the values of x and y. *(Lesson 10-8)* 10.5, 15

27. Find the area of a circle with radius 14 cm. *(Lesson 10-4)* 616 cm²

28. The scale of the floor plan of a theater is 1 in. : 12 ft. On the floor plan, the stage is $2\frac{3}{4}$ in. wide. Find the actual width of the stage. *(Lesson 9-3)* 33 ft

29. Write 6.89×10^{-4} in decimal notation. *(Lesson 7-11)* 0.000689

30. Tell whether line segments of 5 m, 5 m, and 8 m *can* or *cannot* be the sides of a triangle. If they can, tell whether the triangle would be *scalene*, *isosceles*, or *equilateral*. *(Lesson 6-6)* can; isosceles

Indirect Measurement

Objective: To use similar triangles to find an unknown measurement that cannot be measured directly.

APPLICATION

Jed Carter was standing near a tree on a sunny day. Jed cast a shadow of 12 ft at the same time that the tree cast a shadow of 18 ft. Jed is 6 ft tall.

QUESTION How tall is the tree?

Clearly it would be difficult to find the height of the tree by making a *direct measurement* with a ruler or tape measure. In cases like this, it often is possible to use similar triangles to make an **indirect measurement.**

Terms to Know
• *indirect measurement*

Example 1

In the figure below, the triangles are similar. Find the unknown height h.

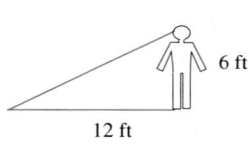 6 ft

12 ft

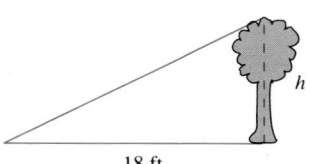 h

18 ft

Solution

Write a proportion involving the corresponding sides of the triangles.

$$\dfrac{\text{length of Jed's shadow}}{\text{length of tree's shadow}} \longrightarrow \quad \dfrac{12}{18} = \dfrac{6}{h} \quad \longleftarrow \text{Jed's height} \atop \longleftarrow \text{tree's height}$$

Write $\dfrac{12}{18}$ as $\dfrac{2}{3}$. $\qquad\qquad \dfrac{2}{3} = \dfrac{6}{h}$

Solve the proportion $\qquad\quad 2h = 3(6)$
using cross products. $\qquad\quad 2h = 18$
$\qquad\qquad\qquad\qquad\qquad h = 9$

The unknown height is 9 ft.

☑ Check Your Understanding

1. Using the solution of Example 1, write the ratio of Jed's height to the tree's height in lowest terms. How does it compare to the ratio of the length of Jed's shadow to the length of the tree's shadow?

2. In Example 1, if the length of Jed's shadow were 18 ft, what would be the length of the shadow cast by the tree at the same time?
See *Answers to Check Your Understanding* at the back of the book.

ANSWER The tree is 9 ft tall.

Circles and Polygons **471**

Lesson Planner

Teaching the Lesson
• Lesson Starter 98

Lesson Follow-Up
• Practice 132
• Enrichment 107: Connection
• Study Guide, pp. 195–196

Warm-Up Exercises

Solve the following proportions.

1. $\dfrac{3}{4} = \dfrac{x}{36}$ 27

2. $\dfrac{3}{9} = \dfrac{4\frac{1}{2}}{x}$ $13\frac{1}{2}$

3. $\dfrac{2.5}{10} = \dfrac{4}{x}$ 16

4. $\dfrac{3.5}{7} = \dfrac{x}{14}$ 7

Lesson Focus

Ask students to imagine that they have to measure the height of a flagpole. Would they measure the pole with a measuring tape? The focus of today's lesson is learning how to use mathematics to measure distances that cannot be measured directly.

After working through the Examples
with students, you may wish to use
the following application.

Application

Take the class outside to the school
flagpole and measure the length of
its shadow. Then measure the
lengths of the shadows of several
students. Use the lengths of the
shadows and the heights of the students to set up ratios, as shown in
Example 2, to find the height of the
flagpole.

Key Questions

1. What is meant by an indirect
   measurement? An indirect measurement is a measurement
   found by using a mathematical
   calculation rather than by making a physical measurement.
2. How is an indirect measurement
   made? An indirect measurement is made by solving a
   proportion based on similar
   triangles.

Error Analysis

Students may set up an incorrect
proportion when solving an indirect
measurement problem. Having students work together to draw sketches
is a good way to help them avoid
this error.

Critical Thinking

The Check Your Understanding
questions of Example 1 have students *examine* the figures and
compare the ratios to validate the
consistency of the answers. Students
also must *create* a new proportion
to find the solution when the conditions of the problem are altered.

Example 2   In the figure below, $\triangle ABE \sim \triangle ACD$. Find the unknown height h.

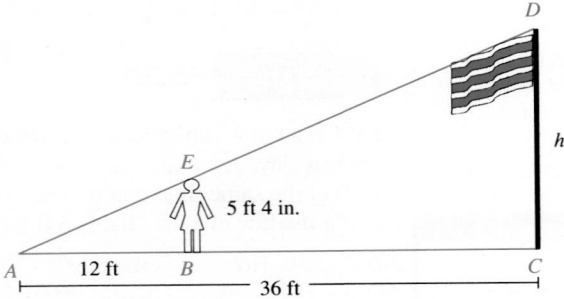

Solution   Write a proportion involving the corresponding sides of the triangles.

$$\frac{\text{length of person's shadow} \longrightarrow}{\text{length of flagpole's shadow} \longrightarrow} \quad \frac{12}{36} = \frac{5 \text{ ft 4 in.}}{h} \quad \begin{array}{l} \longleftarrow \text{person's height} \\ \longleftarrow \text{flagpole's height} \end{array}$$

Write 5 ft 4 in. as $5\frac{1}{3}$ ft. $\dfrac{12}{36} = \dfrac{5\frac{1}{3}}{h}$

Write $\dfrac{12}{36}$ as $\dfrac{1}{3}$. $\dfrac{1}{3} = \dfrac{5\frac{1}{3}}{h}$

Solve the proportion $h = 3\left(5\frac{1}{3}\right)$
using cross products.

$h = \left(\dfrac{3}{1}\right)\left(\dfrac{16}{3}\right) = 16$

The unknown height is 16 ft.

 **Check Your
Understanding**

3. How would the solution of Example 2 be different if you wrote
   5 ft 4 in. as 64 in. and solved the proportion using this number?
4. How would the solution of Example 2 be different if the person's height
   were 5 ft 6 in.?
   See *Answers to Check Your Understanding* at the back of the book.

Guided Practice

COMMUNICATION « *Reading*

Refer to the text on pages 471–472.

« **1.** What is meant by *direct measurement*? A measurement that can be made
directly by applying a ruler or tape measure.
« **2.** Describe the process of indirect measurement that is presented in this
lesson. Use similar triangle relationships to write proportions that can be
solved for the unknown quantity.
« **3.** How is the figure for Example 2 different from the figure for
Example 1? The similar triangles in Example 1 are separate. In Example 2,
they share a vertex.
« **4.** How are the measurements involved in Example 2 different from the
measurements involved in Example 1? In Example 1, the measurements
are all in feet. In Example 2, one measurement is in feet and inches.

COMMUNICATION « Writing

5. See answer on p. A27.

« **5.** Draw a diagram to represent this situation: A fence post that is 2 m tall casts a shadow that is 6 m long. At the same time, a nearby tree that is 6 m tall casts a shadow 18 m long.

6. A person who is 5 ft tall casts a shadow that is 2 ft 6 in. At the same time a tree 9 ft tall casts a shadow that is 4 ft 6 in.

« **6.** Write a description of the situation that is pictured in the figures at the right.

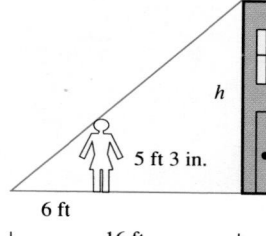

5 ft

9 ft

2 ft 6 in.   4 ft 6 in.

Exercises 7–10 refer to the figure at the right. In this figure, the triangles are similar.

7. Write the ratio of the length of the person's shadow to the length of the building's shadow in lowest terms. $\frac{3}{8}$

8. Write 5 ft 3 in. as a number of feet. $5\frac{1}{4}$ ft

9. $\frac{3}{8} = \frac{5\frac{1}{4}}{h}$

9. Write a proportion involving the corresponding sides of the triangles.

10. Find the unknown height h. 14 ft

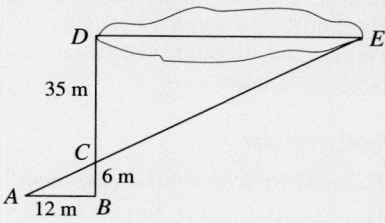

h

5 ft 3 in.

6 ft

|⸻ 16 ft ⸻|

Exercises

In each pair, the triangles are similar. Find the unknown height h.

A **1.**

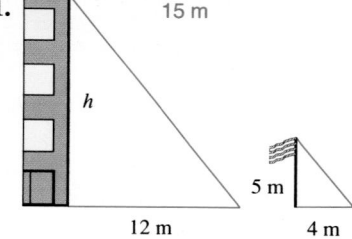

15 m

h

12 m   4 m

5 m

2. 14 ft

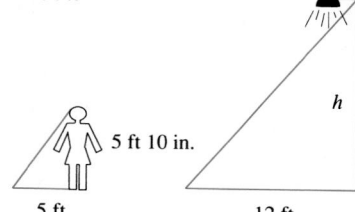

5 ft 10 in.

5 ft   12 ft

h

3.

 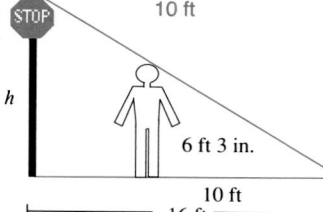

10 ft

h

6 ft 3 in.

|⸻ 10 ft ⸻|
|⸻ 16 ft ⸻|

4. 6 m

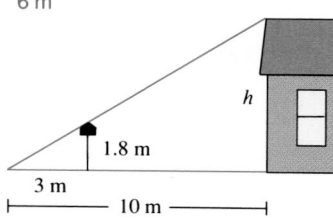

1.8 m

3 m   10 m

h

Circles and Polygons **473**

Additional Examples

1. In the figure below, $\triangle ABC \sim \triangle EDC$. Find the length of $\overline{DE}$.   70 m

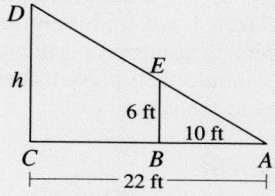

D   E

35 m

C

6 m

A 12 m B

2. In the figure below, $\triangle ABE \sim \triangle ACD$. Find the unknown height h.   13.2 ft

D

h   E

6 ft

10 ft

C   B   A

|⸻ 22 ft ⸻|

Closing the Lesson

Have students summarize the purpose and procedure for finding indirect measurements.

Suggested Assignments

Skills: 1–16
Average: 1–16
Advanced: 1–16

Exercise Notes

Communication: Reading

Guided Practice Exercises 1–4 check students' understanding of concepts in the lesson by having them give their own description of direct and indirect measurement.

Communication: Writing

Guided Practice Exercises 5 and 6 require students to draw a sketch from a description involving indirect measurement, and vice versa.

Application

Exercises 7–11 present a real-world application of how surveyors use indirect measurement. Students verify that the two given triangles are similar and then find the distance across the lake.

Follow-Up

Extension *(Cooperative Learning)*

Have students work in cooperative groups to calculate some indirect measurements. Some examples of objects to measure are the height of the school building, a tree, a flagpole, and so on. Each group should draw a sketch and find the unknown measurement using proportions. Groups should then present their findings to the class.

Enrichment

Enrichment 107 is pictured on page 434F.

Practice

Practice 132, *Resource Book*

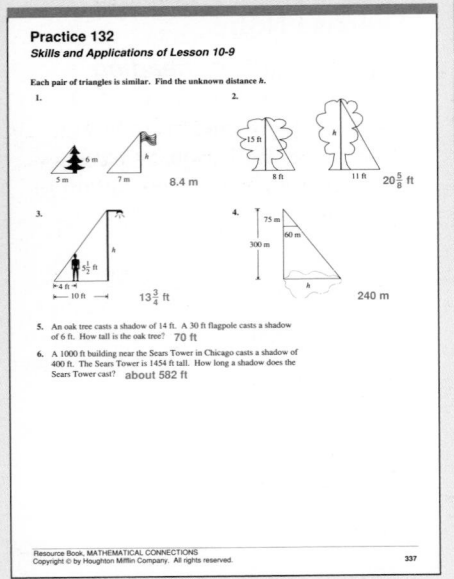

474

B 5. A student who is 5 ft 6 in. tall casts a shadow that is 15 ft long. At the same time a nearby tree casts a shadow that is 45 ft long. What is the height of the tree? 16 ft 6 in.

6. A man who is 1.6 m tall casts a shadow that is 50 cm long. At the same time a nearby television tower casts a shadow that is 8 m long. How tall is the television tower? 25.6 m

PROBLEM SOLVING/APPLICATION

8. ∠A and ∠E; they are alternate interior angles.

The figure at the right illustrates a method of indirect measurement that is used by surveyors. In this case, the unknown distance is the distance across Chosen Lake, represented by the length of line segment $\overline{AB}$. The surveyors create similar triangles by marking off $\overline{DE}$ parallel to $\overline{AB}$.

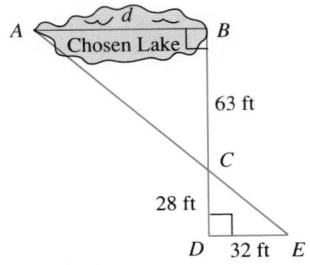

7. Why is it true that $m\angle ACB = m\angle ECD$?
They are vertical angles.

8. Which two angles are equal in measure as a result of the fact that $\overline{DE}$ is parallel to $\overline{AB}$? Explain.

9. Complete this statement: $\triangle ABC \sim \triangle \underline{\ ?\ } EDC$

10. Write a proportion involving corresponding sides of the triangles.
$\frac{d}{63} = \frac{32}{28}$

11. What is the distance across Chosen Lake? 72 ft

C 12. The figure below shows how surveyors laid out similar triangles along the banks of the Sage River. Use these triangles to calculate the distance across the river. 52.5 m

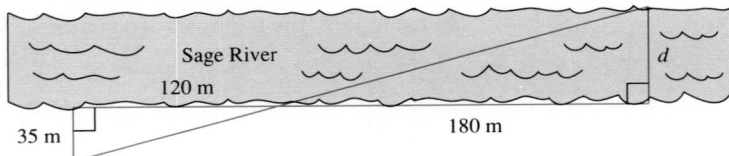

SPIRAL REVIEW

S 13. Find the perimeter of a rectangle with length 16 cm and width 7 cm.
(Lesson 10-1) 46 cm

14. In the figure at the right, the triangles are similar. Find h.   *(Lesson 10-9)* 13 ft

15. The sum of three times a number and four is 52. Find the number.   *(Lesson 4-8)* 16

16. Solve by drawing a diagram: Sara hiked 4 mi south, 2 mi east, 8 mi north, and 2 mi west. Where was she then in relation to her starting point?   *(Lesson 7-8)* 4 mi north

Square Roots

Objective: To find the square root of a number using a table or a calculator.

Lesson Planner

Teaching the Lesson
• Materials: graph paper
• Lesson Starter 99
• Visuals, Folder O
• Using Technology, p. 434C

Lesson Follow-Up
• Practice 133
• Enrichment 108: Exploration
• Study Guide, pp. 197–198
• Calculator Activity 10
• Cooperative Activity 20

EXPLORATION

1 Evaluate x^2 when x has the given value.
  a. 10 100   **b.** -10 100   **c.** 15 225   **d.** -15 225

2 If $x^2 = 144$ and x is a positive number, what is the value of x? 12

3 If $x^2 = 81$ and x is a negative number, what is the value of x? -9

4 If $x^2 = 0$, what is the value of x? 0

5 Do you think there is any real number x for which $x^2 = -100$? Give a convincing argument to support your answer. No; the product of two numbers with the same sign is positive.

If $a^2 = b$, the number a is called a **square root** of b. For instance, 7 is a square root of 49 because $7^2 = 7 \cdot 7 = 49$. Notice that -7 is also a square root of 49 because $(-7)^2 = (-7)(-7) = 49$. The symbol $\sqrt{}$ is used to indicate the positive square root.

In Words	In Symbols
The positive square root of 49 is 7.	$\sqrt{49} = 7$
The negative square root of 49 is -7.	$-\sqrt{49} = -7$

In the real number system, the square root of a negative number does not exist. This is true because there is no real number a for which a^2 is a negative number. Therefore, an expression like $\sqrt{-49}$ has no meaning in the real number system.

Example 1   **Find each square root.**
  a. $-\sqrt{16}$ **b.** $\sqrt{2500}$ **c.** $\sqrt{0}$ **d.** $\sqrt{\dfrac{16}{49}}$ **e.** $-\sqrt{0.16}$

Solution
  a. $4^2 = (4)(4) = 16$, so $-\sqrt{16} = -4$.
  b. $50^2 = 50 \cdot 50 = 2500$, so $\sqrt{2500} = 50$.
  c. $0^2 = 0$, so $\sqrt{0} = 0$.
  d. $\left(\dfrac{4}{7}\right)^2 = \dfrac{4}{7} \cdot \dfrac{4}{7} = \dfrac{16}{49}$, so $\sqrt{\dfrac{16}{49}} = \dfrac{4}{7}$.
  e. $(0.4)^2 = (0.4)(0.4) = 0.16$, so $-\sqrt{0.16} = -0.4$.

✓ Check Your Understanding

1. How would the answer to part (e) of Example 1 be different if you were asked to find $\sqrt{-0.16}$?
See *Answers to Check Your Understanding* at the back of the book.

Warm-Up Exercises

Find each answer.

1. 5^2 25
2. 3^2 9
3. $(-4)^2$ 16
4. $(0.6)^2$ 0.36
5. $-(2^2)$ -4

Lesson Focus

Ask students to imagine purchasing a square rug. The rug is rolled up, but the tag on it says that it covers an area of 100 ft^2. How can the length of the side of the rug be found without unrolling it? Find the square root of 100. Finding square roots is the focus of today's lesson.

Teaching Notes

Allow ample time for students to do the Exploration. Important concepts are presented that should foster student discussion.

Key Questions
1. Which integers have exact square roots? *integers that are perfect squares, such as 1, 4, 9, and so on*

2. Why do negative numbers have no square roots in the real number system? *There is no real number a that once squared gives a negative result.*

3. Does $\sqrt{16} + \sqrt{9} = \sqrt{16 + 9}$? *No; $\sqrt{16} + \sqrt{9} = 4 + 3 = 7$; $\sqrt{16 + 9} = \sqrt{25} = 5$*

Error Analysis
Students sometimes confuse expressions such as $-\sqrt{4}$ and $\sqrt{-4}$. Point out that when the negative sign is outside the square root sign, as in $-\sqrt{4}$, you can find the square root $(-\sqrt{4} = -2)$, but when the negative sign is inside the square root sign, as in $\sqrt{-4}$, the expression has no meaning in the real number system.

Using Technology
For suggestions on using a calculator to explore the properties of square roots and to approximate square roots without using the square root key, see page 434C.

Additional Examples
1. Find each square root.

   a. $\sqrt{400}$ *20*

   b. $-\sqrt{81}$ *−9*

   c. $\sqrt{0.49}$ *0.7*

   d. $\sqrt{\dfrac{25}{36}}$ *$\dfrac{5}{6}$*

   e. $-\sqrt{225}$ *−15*

2. Approximate $\sqrt{73}$ to the nearest thousandth. *8.544*

When $\sqrt{n}$ is an integer, the number n is called a **perfect square.** You could never list all the perfect squares, of course, but you can indicate the set of perfect squares by showing this pattern.

$$\{0^2, 1^2, 2^2, 3^2, 4^2, 5^2, \ldots\} = \{0, 1, 4, 9, 16, 25, \ldots\}$$

If an integer is not a perfect square, then its square root is an irrational number. Therefore, numbers like $\sqrt{2}$, $\sqrt{3}$, $\sqrt{5}$, and $\sqrt{6}$ are irrational. Because an irrational number cannot be expressed as a quotient of two integers, there is no common fraction or decimal that indicates the exact value of these square roots. However, you can *approximate* these numbers either by using a table of squares and square roots like the one on page 480 or by using a calculator.

Example 2

Approximate $\sqrt{46}$ to the nearest thousandth.

Solution 1

Use the table on page 480. Find 46 in the *Number* column. Read the number across from 46 in the *Square Root* column. $\sqrt{46} \approx 6.782$

Solution 2

Use a calculator. Enter ⌊ *46.* ⌋ ⌊√x⌋. The calculator may show ⌊ *6.7823299* ⌋ or ⌊ *6.782329983* ⌋, depending on the number of digits in the display. To the nearest thousandth, either number rounds to 6.782, and $\sqrt{46} \approx 6.782$.

✓ Check Your Understanding

2. In Example 2, why is it incorrect to write $\sqrt{46} = 6.782$?

3. Use a calculator to find $(6.782)^2$. Explain the result.

See *Answers to Check Your Understanding* at the back of the book.

Notice that, in the order of operations, the square root symbol is a grouping symbol.

$$\sqrt{9 + 16} = \sqrt{25} = 5$$

Guided Practice

COMMUNICATION « *Writing*

Write each phrase or sentence in symbols.

«**1.** the positive square root of one and sixty-nine hundredths $\sqrt{1.69}$

«**2.** the square of negative nine $(-9)^2$

«**3.** The square of sixty is three thousand, six hundred. $60^2 = 3600$

«**4.** The negative square root of four ninths is negative two thirds. $-\sqrt{\dfrac{4}{9}} = -\dfrac{2}{3}$

Write each expression or statement in words.

«**5.** $-\sqrt{196}$ «**6.** $\left(\dfrac{1}{16}\right)^2$ «**7.** $(0.2)^2 = 0.04$ «**8.** $-\sqrt{1} = -1$

9. rational number
10. irrational number
11. rational number
12. rational number
13. not a real number
14. rational number
15. irrational number
16. rational number

Tell whether each expression represents a *rational number*, represents an *irrational number*, or is *not a real number*.

9. $\sqrt{81}$ 10. $\sqrt{75}$ 11. $\sqrt{\dfrac{9}{49}}$ 12. $-\sqrt{0.04}$

13. $\sqrt{-36}$ 14. $-\sqrt{36}$ 15. $\sqrt{1000}$ 16. $\sqrt{10{,}000}$

Find each square root.

17. $-\sqrt{64}$ $_{-8}$ 18. $\sqrt{121}$ $_{11}$ 19. $\sqrt{\dfrac{4}{25}}$ $\frac{2}{5}$

20. $\sqrt{0.09}$ $_{0.3}$ 21. $-\sqrt{1.44}$ $_{-1.2}$ 22. $-\sqrt{\dfrac{1}{36}}$ $-\frac{1}{6}$

Use the table on page 480 or use a calculator. Approximate each square root to the nearest thousandth.

23. $\sqrt{116}$ $_{10.770}$ 24. $-\sqrt{8}$ $_{-2.828}$ 25. $\sqrt{40}$ $_{6.325}$

26. $-\sqrt{45}$ $_{-6.708}$ 27. $\sqrt{132}$ $_{11.489}$ 28. $-\sqrt{72}$ $_{-8.485}$

«29. **COMMUNICATION** «*Discussion*   Explain the difference between the expressions $\sqrt{25} + \sqrt{144}$ and $\sqrt{25 + 144}$.
$\sqrt{25} + \sqrt{144} = 5 + 12 = 17;\ \sqrt{25 + 144} = \sqrt{169} = 13$

Exercises

Find each square root.

A   1. $\sqrt{25}$ $_5$ 2. $-\sqrt{4}$ $_{-2}$ 3. $\sqrt{1}$ $_1$ 4. $-\sqrt{81}$ $_{-9}$ 5. $-\sqrt{900}$ $^{-30}$

6. $\sqrt{12{,}100}$ $_{110}$ 7. $\sqrt{\dfrac{9}{64}}$ $\frac{3}{8}$ 8. $-\sqrt{\dfrac{4}{81}}$ $-\frac{2}{9}$ 9. $-\sqrt{0.49}$ $_{-0.7}$ 10. $\sqrt{6.25}$ $_{2.5}$

Use the table on page 480 or use a calculator. Approximate each square root to the nearest thousandth.

11. $-\sqrt{89}$ $_{-9.434}$ 12. $\sqrt{50}$ $_{7.071}$ 13. $\sqrt{140}$ $_{11.832}$ 14. $-\sqrt{131}$ $_{-11.446}$

Find the exact square root if possible. Otherwise, approximate the square root to the nearest thousandth.

15. $\sqrt{400}$ $_{20}$ 16. $\sqrt{40}$ $_{6.325}$ 17. $\sqrt{0.04}$ $_{0.2}$ 18. $\sqrt{\dfrac{1}{25}}$ $\frac{1}{5}$

19. $-\sqrt{33}$ $_{-5.745}$ 20. $\sqrt{225}$ $_{15}$ 21. $\sqrt{1.44}$ $_{1.2}$ 22. $-\sqrt{84}$ $_{-9.165}$

Find each answer. If necessary, round to the nearest thousandth.

23. $\sqrt{36} + \sqrt{64}$ $_{14}$ 24. $\sqrt{36 + 64}$ $_{10}$ 25. $\sqrt{24 + 25}$ ^{7}

26. $\sqrt{24} + \sqrt{25}$ $_{9.899}$ 27. $\sqrt{9} \cdot \sqrt{16}$ $_{12}$ 28. $\sqrt{9 \cdot 16}$ $_{12}$

29. Name three perfect squares between 200 and 300. $_{225,\ 256,\ 289}$

30. Using the table of squares and square roots on page 480, find a number between 5000 and 5100 that is a perfect square. $_{5041}$

Closing the Lesson

You may wish to use the following application to check students' understanding of the concepts in the lesson.

Application

A table top is a square with area 18 ft². What is the exact length of a side of the table? Approximate this length to the nearest tenth of a foot. $\sqrt{18}$ ft; 4.2 ft

Suggested Assignments

Skills: 1–32, 33–47 odd, 53, 54
Average: 1–32, 33–47 odd, 53, 54
Advanced: 1–27 odd, 29–54, Challenge

Exercise Notes

Communication: Discussion
Guided Practice Exercise 29 helps students understand the distinction between the sum of two square roots and the square root of a sum.

Exercise Notes (continued)

Mental Math
Exercises 33–40 show that by knowing the square roots of the perfect squares up to 144, students can do a number of exercises mentally.

Estimation
Exercises 41–48 give students a procedure for estimating irrational square roots without using a calculator or tables.

Making Connections/Transitions
Exercises 49–52 introduce students to the golden rectangle. In *Exercise 51*, students check the ratio against some given rectangles. In *Exercise 52*, students conduct a survey that tests the claim that the golden rectangle is the most pleasing to the eye.

CONNECTING ALGEBRA AND GEOMETRY

The symbol $\sqrt{}$ is read *the square root of* because evaluating $\sqrt{x}$ gives you the length of the side of a square with area x.

B **31.** Suppose that the area of a square is 196 m². What is its perimeter? 56 m

32. The area of a square is 45 ft². Approximate its perimeter to the nearest tenth of a foot.
26.8 ft

Area = 25 square units
Side = $\sqrt{25}$ units
= 5 units

MENTAL MATH

Most people agree that it is a good idea to memorize these simple square roots in order to do calculations more quickly and accurately.

$$\sqrt{1} = 1 \qquad \sqrt{16} = 4 \qquad \sqrt{49} = 7 \qquad \sqrt{100} = 10$$
$$\sqrt{4} = 2 \qquad \sqrt{25} = 5 \qquad \sqrt{64} = 8 \qquad \sqrt{121} = 11$$
$$\sqrt{9} = 3 \qquad \sqrt{36} = 6 \qquad \sqrt{81} = 9 \qquad \sqrt{144} = 12$$

Find each answer mentally.

33. $5 \cdot \sqrt{121}$ 55   **34.** $29 - \sqrt{49}$ 22   **35.** $(0.1)(\sqrt{36})$ 0.6

36. $\sqrt{100} + 87$ 97   **37.** $\frac{1}{2}(\sqrt{25})$ $2\frac{1}{2}$ **38.** $\frac{\sqrt{16}}{8}$ $\frac{1}{2}$

39. $(\sqrt{81} + 1)(\sqrt{144} - \sqrt{4})$ 100   **40.** $\sqrt{1} + \sqrt{64} - \sqrt{9}$ 6

ESTIMATION

When a square root is irrational, you have learned to approximate its value using a table or calculator. However, to check your answers for reasonableness, you also should be able to *estimate* the value by locating it between consecutive integers. Here is an example.

76 is between 64 and 81, two perfect squares.

$\sqrt{76}$ is between $\sqrt{64}$ and $\sqrt{81}$.

$\sqrt{76}$ is between 8 and 9.

Between which two consecutive integers does each square root lie?

41. $\sqrt{38}$ 6 and 7   **42.** $\sqrt{130}$ 11 and 12   **43.** $-\sqrt{19}$ −5 and −4   **44.** $-\sqrt{105}$ −11 and −10

Choose the best estimates *without using a table or a calculator*.

45. $\sqrt{18}$ **a.** about 3.8   **(b.)** about 4.2   **c.** about 4.9

46. $\sqrt{56}$ **(a.)** about 7.5   **b.** about 7.9   **c.** about 8.3

47. $\sqrt{8.1}$ **a.** about 0.9   **(b.)** about 2.8   **c.** about 9.0

48. $\sqrt{0.26}$ **(a.)** about 0.51   **b.** about 1.3   **c.** about 5.1

CONNECTING MATHEMATICS AND ART

Some artists and architects have created designs based on the *golden rectangle*. In a **golden rectangle,** the ratio $\frac{length}{width}$ equals $\frac{1 + \sqrt{5}}{2}$. People seem to find the rectangle's proportions pleasing to the eye. The Triumphal Arch of Constantine in Rome approximates two golden rectangles.

49. Approximate $\frac{1 + \sqrt{5}}{2}$ to the nearest thousandth. 1.618

50. Use your answer to Exercise 49. If the width of a rectangle is 1 m, about how long must it be in order to be a golden rectangle?
about 1.618 m

C 51. Which of these rectangles is most nearly a golden rectangle? II

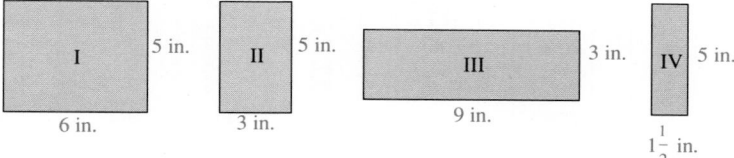

I 5 in. 6 in.   II 5 in. 3 in.   III 3 in. 9 in.   IV 5 in. $1\frac{1}{2}$ in.

52. DATA ANALYSIS Make models of the rectangles in Exercise 51. Then ask each of 30 people to choose the rectangle with the shape that seems most pleasant to look at. Display the results in a frequency table. Did most people choose the golden rectangle? Answers will vary.

SPIRAL REVIEW

S 53. Find the GCF of 96 and 16. *(Lesson 7-2)* 16

54. Find $\sqrt{0.0081}$. *(Lesson 10-10)* 0.09

Challenge

The perimeter of a rectangle is 18 cm. What is its greatest possible area? 20.25 cm²
The area of a rectangle is 36 cm². What is its least possible perimeter? 24 cm

Circles and Polygons   **479**

Follow-Up

Project
Have students do research on the view of irrational numbers during the time of Pythagoras, and the meaning of the square root symbol as a grouping symbol.

Enrichment
Enrichment 108 is pictured on page 434F.

Practice
Practice 133, *Resource Book*

Practice 133
Skills and Applications of Lesson 10-10

Find each square root.

1. $\sqrt{36}$ 6   2. $-\sqrt{49}$ −7   3. $\sqrt{100}$ 10   4. $\sqrt{121}$ 11
5. $-\sqrt{\frac{9}{25}}$ $-\frac{3}{5}$ 6. $\sqrt{\frac{25}{36}}$ $\frac{5}{6}$ 7. $\sqrt{0.01}$ 0.1   8. $\sqrt{\frac{121}{144}}$ $\frac{11}{12}$
9. $\sqrt{400}$ 20   10. $\sqrt{256}$ 16   11. $-\sqrt{144}$ −12   12. $\sqrt{625}$ 25

Use the table on page 480, or use a calculator. Approximate each square root to the nearest thousandth.
13. $\sqrt{150}$ 12.247   14. $\sqrt{40}$ 6.325   15. $\sqrt{85}$ 9.220   16. $-\sqrt{70}$ −8.367

Find the exact square root if possible. Otherwise, approximate the square root to the nearest thousandth.
17. $\sqrt{900}$ 30   18. $\sqrt{80}$ 8.944   19. $\sqrt{169}$ 13   20. $\sqrt{\frac{25}{49}}$ $\frac{5}{7}$
21. $-\sqrt{\frac{36}{49}}$ $-\frac{6}{7}$ 22. $\sqrt{256}$ 16   23. $\sqrt{1.69}$ 1.3   24. $-\sqrt{2.25}$ −1.5

Find each answer. If necessary, round to the nearest thousandth.
25. $\sqrt{16 + 25}$ 6.403   26. $\sqrt{16} + \sqrt{25}$ 9   27. $\sqrt{36} + \sqrt{81}$ 15
28. $\sqrt{50} + \sqrt{14}$ 10.813   29. $\sqrt{25} \cdot \sqrt{64}$ 40   30. $\sqrt{144 + 36}$ 13.416

31. Using the table on page 480 or a calculator, find two numbers between 3000 and 3200 that are perfect squares. 3025; 3136
32. How many perfect squares can you find between 3850 and 3950? none

Table of Squares and Square Roots

NO.	SQUARE	SQUARE ROOT	NO.	SQUARE	SQUARE ROOT	NO.	SQUARE	SQUARE ROOT
1	1	1.000	51	2,601	7.141	101	10,201	10.050
2	4	1.414	52	2,704	7.211	102	10,404	10.100
3	9	1.732	53	2,809	7.280	103	10,609	10.149
4	16	2.000	54	2,916	7.348	104	10,816	10.198
5	25	2.236	55	3,025	7.416	105	11,025	10.247
6	36	2.449	56	3,136	7.483	106	11,236	10.296
7	49	2.646	57	3,249	7.550	107	11,449	10.344
8	64	2.828	58	3,364	7.616	108	11,664	10.392
9	81	3.000	59	3,481	7.681	109	11,881	10.440
10	100	3.162	60	3,600	7.746	110	12,100	10.488
11	121	3.317	61	3,721	7.810	111	12,321	10.536
12	144	3.464	62	3,844	7.874	112	12,544	10.583
13	169	3.606	63	3,969	7.937	113	12,769	10.630
14	196	3.742	64	4,096	8.000	114	12,996	10.677
15	225	3.873	65	4,225	8.062	115	13,225	10.724
16	256	4.000	66	4,356	8.124	116	13,456	10.770
17	289	4.123	67	4,489	8.185	117	13,689	10.817
18	324	4.243	68	4,624	8.246	118	13,924	10.863
19	361	4.359	69	4,761	8.307	119	14,161	10.909
20	400	4.472	70	4,900	8.367	120	14,400	10.954
21	441	4.583	71	5,041	8.426	121	14,641	11.000
22	484	4.690	72	5,184	8.485	122	14,884	11.045
23	529	4.796	73	5,329	8.544	123	15,129	11.091
24	576	4.899	74	5,476	8.602	124	15,376	11.136
25	625	5.000	75	5,625	8.660	125	15,625	11.180
26	676	5.099	76	5,776	8.718	126	15,876	11.225
27	729	5.196	77	5,929	8.775	127	16,129	11.269
28	784	5.292	78	6,084	8.832	128	16,384	11.314
29	841	5.385	79	6,241	8.888	129	16,641	11.358
30	900	5.477	80	6,400	8.944	130	16,900	11.402
31	961	5.568	81	6,561	9.000	131	17,161	11.446
32	1,024	5.657	82	6,724	9.055	132	17,424	11.489
33	1,089	5.745	83	6,889	9.110	133	17,689	11.533
34	1,156	5.831	84	7,056	9.165	134	17,956	11.576
35	1,225	5.916	85	7,225	9.220	135	18,225	11.619
36	1,296	6.000	86	7,396	9.274	136	18,496	11.662
37	1,369	6.083	87	7,569	9.327	137	18,769	11.705
38	1,444	6.164	88	7,744	9.381	138	19,044	11.747
39	1,521	6.245	89	7,921	9.434	139	19,321	11.790
40	1,600	6.325	90	8,100	9.487	140	19,600	11.832
41	1,681	6.403	91	8,281	9.539	141	19,881	11.874
42	1,764	6.481	92	8,464	9.592	142	20,164	11.916
43	1,849	6.557	93	8,649	9.644	143	20,449	11.958
44	1,936	6.633	94	8,836	9.695	144	20,736	12.000
45	2,025	6.708	95	9,025	9.747	145	21,025	12.042
46	2,116	6.782	96	9,216	9.798	146	21,316	12.083
47	2,209	6.856	97	9,409	9.849	147	21,609	12.124
48	2,304	6.928	98	9,604	9.899	148	21,904	12.166
49	2,401	7.000	99	9,801	9.950	149	22,201	12.207
50	2,500	7.071	100	10,000	10.000	150	22,500	12.247

The Pythagorean Theorem

Objective: To use the Pythagorean Theorem to find unknown lengths.

Terms to Know
- *hypotenuse*
- *legs*
- *Pythagorean Theorem*

DATA ANALYSIS

In a right triangle, the side opposite the right angle is always the longest side. It is called the **hypotenuse.** The two shorter sides are called the **legs.** In ancient times, people gathered data about the hypotenuse and legs of many right triangles and began to see a pattern forming. The table below lists these data for several right triangles. Can you see the pattern?

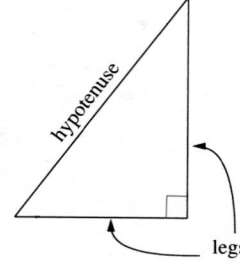

Some Common Measures of Right Triangles

Lengths of Legs	3	5	6	7	8	9
	4	12	8	24	15	12
Length of Hypotenuse	5	13	10	25	17	15

A good way to find the pattern is to work with the simplest of the triangles listed, the "3-4-5" right triangle, and observe what happens when you build a square on each of the legs and on the hypotenuse.

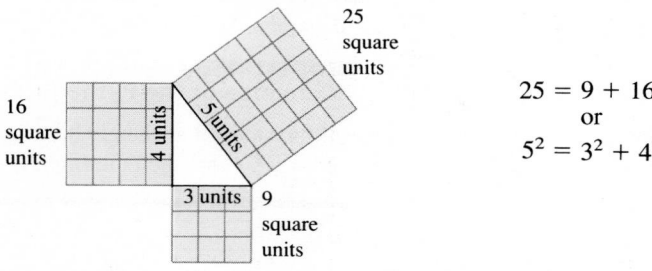

$$25 = 9 + 16$$
or
$$5^2 = 3^2 + 4^2$$

As you can see, the area of the square on the hypotenuse is equal to the sum of the areas of the squares on the two legs. More than 2500 years ago, the Greek mathematician Pythagoras proved that this relationship applies to all right triangles, not just the 3-4-5 right triangle. In his honor, this property is called the **Pythagorean Theorem.**

The Pythagorean Theorem

If the length of the hypotenuse of a right triangle is c and the lengths of the legs are a and b, then the following relationship holds true.

$$c^2 = a^2 + b^2$$

Circles and Polygons **481**

Lesson Planner

Teaching the Lesson
- Materials: graph paper
- Lesson Starter 100
- Visuals, Folder O
- Using Technology, p. 434C
- Using Manipulatives, p. 434D

Lesson Follow-Up
- Practice 134
- Practice 135: Nonroutine
- Enrichment 109: Communication
- Study Guide, pp. 199–200
- Test 49

Warm-Up Exercises

Find each answer. Approximate to the nearest hundredth, if necessary.

1. $x^2 = 16$ 4
2. $x^2 = 56$ 7.48
3. $x^2 = 0.0225$ 0.15
4. $x^2 = 99$ 9.95
5. $x^2 = 40$ 6.32

Lesson Focus

Ask students to imagine a rectangular park with a sidewalk running diagonally across it. The length of the park is 400 yd and the width is 100 yd. How can the length of the sidewalk be found without measuring it? This problem and others like it can be solved by using the Pythagorean theorem, the focus of today's lesson.

If you know the lengths of two sides of a right triangle, you can use the Pythagorean theorem to find the unknown length of the third side.

Example 1

Find the unknown length. If necessary, round to the nearest tenth.

a.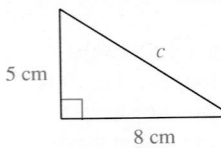

b.

Solution

a.
$$c^2 = a^2 + b^2$$
$$c^2 = 5^2 + 8^2$$
$$c^2 = 25 + 64$$
$$c^2 = 89$$
$$c = \sqrt{89}$$
$$c \approx 9.434$$

The length of the hypotenuse is about 9.4 cm.

b.
$$c^2 = a^2 + b^2$$
$$35^2 = a^2 + 28^2$$
$$1225 = a^2 + 784$$
$$441 = a^2$$
$$\sqrt{441} = a$$
$$21 = a$$

The length of the leg is 21 ft.

✏ Check Your Understanding

1. Why were the squares of the lengths added in part (a) of Example 1, but subtracted in part (b)?
   See *Answers to Check Your Understanding* at the back of the book.

It also has been proved that the *converse* of the Pythagorean theorem is a true statement. You obtain the converse of a statement by interchanging the "if" and "then" parts of the statement.

> **Converse of the Pythagorean Theorem**
>
> If the sides of a triangle have lengths a, b, and c such that $c^2 = a^2 + b^2$, then the triangle is a right triangle.

Example 2

Is a triangle with sides of the given lengths a right triangle?

a. 12 ft, 16 ft, 20 ft
b. 9 m, 15 m, 13 m

Solution

a.
$$c^2 = a^2 + b^2$$
$$20^2 \overset{?}{=} 12^2 + 16^2$$
$$400 \overset{?}{=} 144 + 256$$
$$400 = 400$$

Yes, it is a right triangle.

b.
$$c^2 = a^2 + b^2$$
$$15^2 \overset{?}{=} 9^2 + 13^2$$
$$225 \overset{?}{=} 81 + 169$$
$$225 \neq 250$$

No, it is not a right triangle.

✏ Check Your Understanding

2. In Example 2, how do you know which length to substitute for c in the equation $c^2 = a^2 + b^2$?
   See *Answers to Check Your Understanding* at the back of the book.

Guided Practice

2. The Pythagorean theorem gives the relationship between the sides of a right triangle. The converse states that if the relationship holds, then the triangle is a right triangle.

COMMUNICATION «*Reading*

« **1.** What is the main idea of the lesson? The Pythagorean theorem can be used to find unknown lengths in right triangles.

« **2.** Explain the difference between the Pythagorean theorem and the *converse* of the Pythagorean theorem.

Replace each __?__ with = or ≠ to make a true statement.

3. $7^2 + 3^2$ __?__ 10^2 ≠

4. $5^2 + 12^2$ __?__ 13^2 =

5. $6^2 + 8^2$ __?__ 10^2 =

6. $4^2 + 8^2$ __?__ 9^2 ≠

Find the unknown length. If necessary, round to the nearest tenth. Use the table on page 480 or a calculator as needed.

7.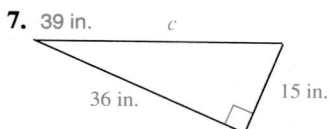

8.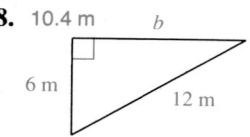

Is a triangle with sides of the given lengths a right triangle? Write *Yes* or *No*.

9. 5 in., 8 in., 11 in. No.

10. 6 cm, 10 cm, 8 cm Yes.

Exercises

Find the unknown length. If necessary, round to the nearest tenth. Use the table on page 480 or a calculator as needed.

A

1.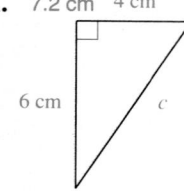

2.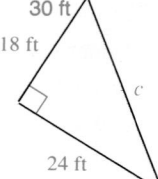

3.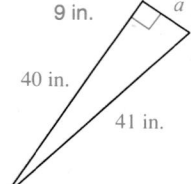

4.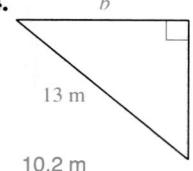

5.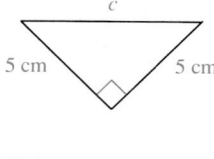

6.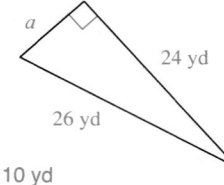

Circles and Polygons **483**

Additional Examples

1. Find the unknown length. If necessary, round to the nearest tenth.

   a.

   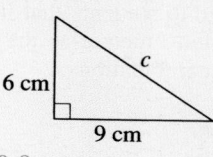

   10.8 cm

   b.

   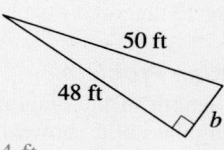

   14 ft

2. Is a triangle with sides of the given lengths a right triangle? Write *Yes* or *No*.

   a. 10 ft, 24 ft, 26 ft   Yes

   b. 7 m, 25 m, 28 m   No

Closing the Lesson

To check students' understanding of the Pythagorean theorem give them diagrams of right triangles whose sides are labeled with letters other than *a*, *b*, and *c*. Ask students to write the Pythagorean equation using these letters.

Suggested Assignments

Skills: 1–16, 25–28
Average: 1–28
Advanced: 1–28

Exercise Notes

Communication: Reading

Guided Practice Exercises 1 and 2 check students' understanding of the lesson objective and the difference between the Pythagorean theorem and its converse.

Application

In *Exercises 13–20*, students find indirect measurements from diagrams or make their own diagrams to help solve problems. These applications will increase students' skills with using the Pythagorean theorem.

483

Cooperative Learning

Exercises 21–23 use the formulas for finding Pythagorean triples. Emphasize to students that it is worthwhile to memorize the more common combinations.

Computer

Exercise 24 challenges students to write their own BASIC program for generating Pythagorean triples.

Nonroutine Problem

Find the length of the diagonal of the rectangular solid shown below.
$d = 13$ cm

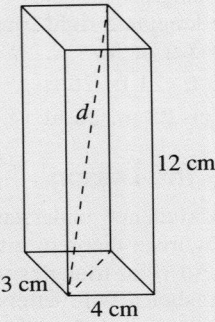

Follow-Up

Extension

Give each student two sheets of paper. Each sheet should contain ten 3 cm-by-6 cm right triangles. Have students cut out the 20 triangles and arrange them to form a square. All of the triangles must be used to form the square.

Is a triangle with sides of the given lengths a right triangle? Write *Yes* or *No*.

7. 3 yd, 4 yd, 5 yd Yes.

8. 8 mm, 10 mm, 12 mm No.

9. 7 m, 24 m, 25 m Yes.

10. 9 ft, 12 ft, 14 ft No.

11. 13 cm, 5 cm, 12 cm Yes.

12. 15 in., 25 in., 20 in. Yes.

PROBLEM SOLVING/APPLICATION

Often the Pythagorean Theorem is used as a method of indirect measurement. In the following figures, use the Pythagorean Theorem to find the unknown distance. If necessary, round to the nearest tenth.

B **13.** 10.3 yd

support wire / l / 5 yd / 9 yd

14.

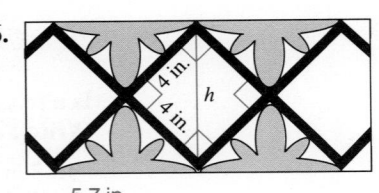

Hidden Lake — d — 40 m — 32 m — 24 m

15.

10 ft / 12 ft / 6.6 ft / d

16.

4 in. / 4 in. / h / 5.7 in.

Solve. If necessary, round the answer to the nearest tenth.

17. Martine Lancois left her house and walked 2 km due east and then 6 km due north. At this point, how far was Martine from her house?
6.3 km

18. A ladder that is 6 m long leans against a wall. The bottom of the ladder rests 2 m from the base of the wall. How far up the wall does the ladder reach? 5.7 m

19. A public garden is to be shaped like a rectangle with length 80 yd and width 60 yd. What is the length of a diagonal walkway across this garden? 100 yd

20. A square gate is reinforced by a wooden brace across its diagonal. The perimeter of the gate is 12 ft. What is the length of the brace?
4.2 ft

1. What is the area of each triangle? 9 cm^2
2. What is the total area of the square? 180 cm^2
3. What is the length of one side of the square? to the nearest hundredth? $\sqrt{180}$ cm; 13.42 cm

GROUP ACTIVITY 21. 3; 4; 5   22. 4; 1

Any three positive integers a, b, and c such that $c^2 = a^2 + b^2$ are said to form a **Pythagorean triple.** For instance, all the integers in the chart on page 481 form Pythagorean triples. To *generate* triples like these, you substitute positive integers x and y, $x > y$, into these formulas.

$$a = x^2 - y^2 \qquad b = 2xy \qquad c = x^2 + y^2$$

21. Find the triple that is generated by the values $x = 2$, $y = 1$.

C 22. Find the values of x and y that generate the triple 8–15–17.

23. Make an organized list of all the Pythagorean triples generated by values of x from 2 through 6. Answers to Exercises 23–24 are on p. A27.

24. **COMPUTER APPLICATION** Write a BASIC program that generates a Pythagorean triple when values are input for x and y.

SPIRAL REVIEW 27. See answer on p. A27.

S 25. Solve: $\frac{1}{4} + n = -\frac{5}{8}$ *(Lesson 8-8)* $-\frac{7}{8}$

26. Find the unknown length in the figure at the right. *(Lesson 10-11)* 11.5 cm

27. Graph the inequality $x \le 3.5$ *(Lesson 7-10)*

28. Find the area of a square with side 7 in. *(Lesson 10-3)* 49 in.²

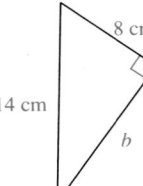

8 cm

14 cm

b

Self-Test 2

Use the congruent quadrilaterals in the Example on page 463 to complete each statement.

1. $\overline{QR} \cong \underline{\ ?\ } \ \overline{BD}$ 2. $\angle A \cong \underline{\ ?\ } \ \angle S$ 10-7

3. Use the similar quadrilaterals in the Example on page 467 to find the length of $\overline{PQ}$. 15.75   10-8

4. A person 6 ft tall casts a shadow 15 ft long. At the same time, a building casts a shadow 50 ft long. How tall is the building? 20 ft   10-9

Find each square root. Use the table on page 480 if necessary.

5. $\sqrt{8100}$ 90   6. $\sqrt{\frac{25}{64}}$ $\frac{5}{8}$ 7. $\sqrt{125}$ 11.180   10-10

8. Is a triangle with sides of 9 m, 12 m, and 13 m a right triangle? Write *Yes* or *No*. No.   10-11

9. Find the unknown length in the triangle at the right. If necessary, round to the nearest tenth. 8.1

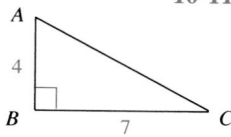

A

4

B 7 C

Enrichment
Enrichment 109 is pictured on page 434F.

Practice
Practice 134 is shown below.

Quick Quiz 2
See page 490.

Alternative Assessment
See page 491.

Practice 134, *Resource Book*

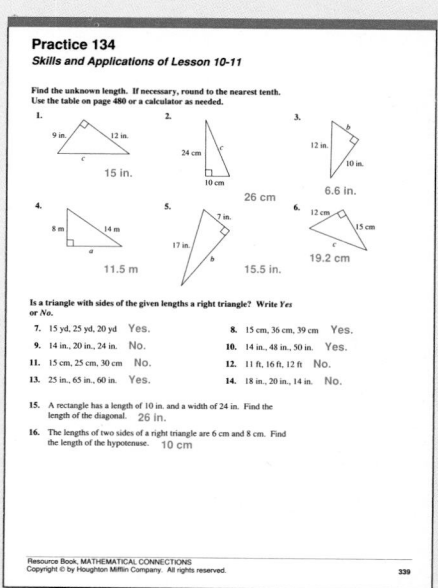

Circles and Polygons **485**

Chapter Review

Terms to Know

perimeter (p. 436)		congruent (p. 462)
circle (p. 440)		corresponding angles (p. 463)
center (p. 440)		corresponding sides (p. 463)
radius (p. 440)		similar (p. 466)
chord (p. 440)		indirect measurement (p. 471)
diameter (p. 440)		square root (p. 475)
circumference (p. 441)		perfect square (p. 476)
area (p. 445)		hypotenuse (p. 481)
base (p. 446)		legs (p. 481)
height (p. 446)		Pythagorean Theorem (p. 481)

Choose the correct term from the list above to complete each sentence. 3. congruent

1. The distance around a circle is called its __?__. circumference

2. The __?__ is the longest side of a right triangle. hypotenuse

3. Two figures that have the same shape and size are called __?__.

4. A __?__ of a circle is a line segment whose endpoints are the center and a point on the circle. radius

5. A __?__ has a square root that is an integer. perfect square

6. The region enclosed by a plane figure is called its __?__. area

---

Find the perimeter of each figure. *(Lesson 10-1)*

7. a regular hexagon with one side that measures 6.2 cm 37.2 cm

8. a parallelogram with sides that measure 7 ft and $10\frac{1}{4}$ ft $34\frac{1}{2}$ ft

Use the circle at the right. *(Lesson 10-2)*

9. Name the circle. circle R

10. Identify as many radii, chords, and diameters as shown. radii: $\overline{RT}$, $\overline{RS}$, $\overline{RW}$; chords: $\overline{WT}$, $\overline{VU}$; diameter: $\overline{WT}$

11. Find the radius. 9 cm

12. Find the diameter. 18 cm

13. Find the circumference. 56.5 cm

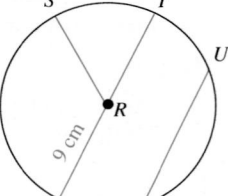

Find the area of each figure. *(Lessons 10-3 and 10-4)*

14. a square with length of one side equal to $5\frac{1}{2}$ ft $30\frac{1}{4}$ ft²

15. a rectangle with length 4.8 m and width 2.6 m 12.48 m²

16. a triangle with base 12 in. and height 9 in. 54 in.²

17. a circle with radius 16 cm 803.8 cm²

Tell whether you would need to calculate *perimeter*, *circumference*, or *area* to find each measure. *(Lesson 10-5)*

18. the amount of asphalt needed to tar a square playground area

19. the fence around the playground perimeter

Explain why each problem has no solution. *(Lesson 10-6)*

20. A collection of five coins has a value of $1 and consists of dimes and quarters. How many of each type of coin are there? There are no amounts of dimes and quarters totaling 5 that result in $1.

21. The perimeter of a rectangle is 24 cm. The length is 12 cm. Find the width. The two lengths together equal the perimeter.

Use the figures at the right to complete each statement. *(Lesson 10-7)*

22. $\overline{AB} \cong$ __?__ $\overline{EF}$ **23.** $\angle C \cong$ __?__ $\angle D$

24. If the length of $\overline{BC}$ is 8 m, then the length of __?__ is 8 m. $\overline{DE}$

25. If $m\angle B = 65°$, then m __?__ $= 65°$. $\angle E$

In the figures at the right
ABCDE ~ UVRST.
(Lesson 10-8)

26. Find x. 2   **27.** Find y. 9

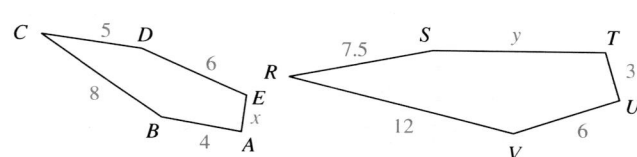

Solve. *(Lesson 10-9)*

28. In the figures at the right, the triangles are similar. Find the unknown height h. 25 ft

Find the exact square root if possible. Otherwise, approximate each square root to the nearest thousandth. *(Lesson 10-10)*

29. $\sqrt{900}$ 30   **30.** $\sqrt{\dfrac{100}{121}}$ $\dfrac{10}{11}$

31. $\sqrt{76}$ 8.718   **32.** $\sqrt{142}$ 11.916

Use the Pythagorean theorem. *(Lesson 10-11)*

33. Is a triangle with sides of 10 ft, 24 ft, and 26 ft a right triangle? Write *Yes* or *No*. Yes.

34. Find the length of the third side of the triangle at the right. 12

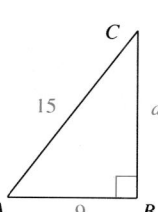

Quick Quiz 1 (10-1 through 10-6)

Find the perimeter of each figure.

1. a square with one side that measures 6.3 in.   25.2 in.

2. a rectangle with sides that measure 7 m and $8\frac{1}{2}$ m   31 m

3. A circle has a diameter of 6 ft.
   a. Find the radius.   3 ft
   b. Find the circumference.
      18.84 ft

Find the area of each figure.

4. a rectangle with sides that measure 9 ft and 11 ft   99 ft²

5. a triangle with base 10 cm and height 5 cm   25 cm²

6. Find the area of a circle with radius 2.5 in.   19.6 in.²

7. Tell whether you would need to calculate *perimeter*, *circumference*, or *area* to find the amount of topsoil needed for a rectangular lawn.   area

8. Explain why this problem has no solution: The base angles of an isosceles triangle each measure 92°. What is the measure of the third angle?   The sum of the three angles cannot exceed 180°.

Alternative Assessment

Tell whether each statement is *True* or *False*.

1. Two circles that have the same circumference are always the same size and shape.   True

2. Two rectangles that have the same perimeter are always the same size and shape.   False

3. If the radius of a circle is doubled, the area is also doubled.   False

487

Alternative Assessment

Exercises 4 and 6 are marked with stars to indicate that they are alternative forms of assessment. Exercise 4 requires students to test various rectangles with a perimeter of 16 ft to find the greatest possible area. Exercise 6 demonstrates what happens to the circumference and the area of a circle when its radius is tripled.

Chapter Test (Form A)

Test 50, *Tests*

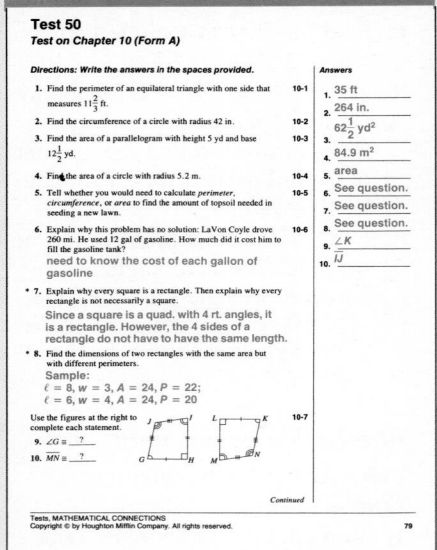

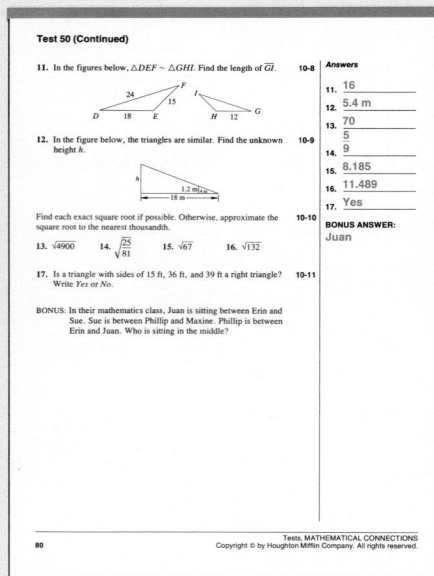

Chapter Test

1. Find the perimeter of a square with one side that measures $7\frac{1}{4}$ in. 29 in. — 10-1

2. Find the circumference of a circle with radius 21 in. 132 in. — 10-2

3. Find the area of a triangle with base 1 yd and height 4 ft. 6 ft² or $\frac{2}{3}$ yd² — 10-3

★ 4. What is the greatest possible area for a rectangle with a perimeter of 16 ft? 16 ft²

5. Find the area of a circle with radius 2.3 m. 16.6 m² — 10-4

★ 6. When the radius of a circle is tripled, what happens to the circumference and area of the circle? The circumference triples; the area is nine times as great.

7. Tell whether you would need to calculate *perimeter*, *circumference*, or *area* to find the amount of sod needed to cover a soccer field. area — 10-5

8. Explain why this problem has no solution: La Toya has $1.10 in nickels and dimes. She has ten coins altogether. How many of each type of coin does she have? Ten dimes have a value of $1.00, which is less than $1.10. — 10-6

Use the figures at the right to complete each statement.

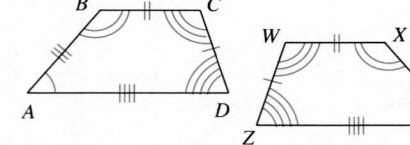

9. $\overline{AB} \cong$ ___?___ $\overline{XY}$ — 10-7

10. $\angle C \cong$ ___?___ $\angle W$

11. In the figures at the right $\triangle ABC \sim \triangle ZYX$. Find the length of $\overline{YZ}$. 16 — 10-8

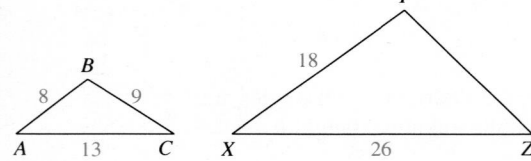

12. In the figure at the right, the triangles are similar. Find the unknown height h. 9 ft — 10-9

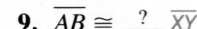

Find each square root if possible. Otherwise, approximate the square root to the nearest thousandth.

13. $\sqrt{3600}$ 60

14. $\sqrt{\frac{9}{49}}$ $\frac{3}{7}$

15. $\sqrt{91}$ 9.539

16. $\sqrt{147}$ 12.124 — 10-10

17. Is a triangle with sides of 16 ft, 30 ft, and 34 ft a right triangle? Write *Yes* or *No*. Yes. — 10-11

488 Chapter 10

488

Simplifying Square Roots

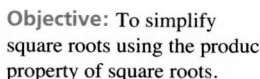

Exercises

Objective: To simplify square roots using the product property of square roots.

You probably know that $\sqrt{144} = 12$ because you remember that $12 \cdot 12 = 144$. However, you could also find $\sqrt{144}$ as follows.

$$\sqrt{144} = \sqrt{4 \cdot 36} = \sqrt{4} \cdot \sqrt{36} = 2 \cdot 6 = 12$$

This method illustrates the following property of square roots.

The Product Property of Square Roots

The square root of the product of two nonnegative real numbers is equal to the product of their square roots.

$$\sqrt{ab} = \sqrt{a} \cdot \sqrt{b} \qquad a \geq 0, b \geq 0$$

You can use the product property to simplify square roots.

Example

Solution

Simplify $\sqrt{80}$.

First write 80 as the product of two factors, one of which is the greatest perfect square possible: $80 = 16 \cdot 5$

Then use the product property of square roots.

$$\sqrt{80} = \sqrt{16 \cdot 5} = \sqrt{16} \cdot \sqrt{5} = 4\sqrt{5}$$

Exercises

Simplify each square root. **4.** $2\sqrt{6}$ **8.** $3\sqrt{7}$ **12.** $4\sqrt{3}$

1. $\sqrt{12}$ $2\sqrt{3}$ 2. $\sqrt{45}$ $3\sqrt{5}$ 3. $\sqrt{50}$ $5\sqrt{2}$ 4. $\sqrt{24}$

5. $\sqrt{8}$ $2\sqrt{2}$ 6. $\sqrt{27}$ $3\sqrt{3}$ 7. $\sqrt{40}$ $2\sqrt{10}$ 8. $\sqrt{63}$

9. $\sqrt{28}$ $2\sqrt{7}$ 10. $\sqrt{44}$ $2\sqrt{11}$ 11. $\sqrt{32}$ $4\sqrt{2}$ 12. $\sqrt{48}$

13. $\sqrt{124}$ $2\sqrt{31}$ 14. $\sqrt{125}$ $5\sqrt{5}$ 15. $\sqrt{60}$ $2\sqrt{15}$ 16. $\sqrt{200}$

17. **b.** 8.944

   c. They are equal.

17. The Example above showed that $\sqrt{80} = 4\sqrt{5}$.
   a. Approximate $\sqrt{80}$ to the nearest thousandth. 8.944
   b. Approximate $\sqrt{5}$ to the nearest thousandth, then multiply by 4.
   c. Compare your answers to (a) and (b). What do you conclude?

16. $10\sqrt{2}$

18. $\sqrt{\dfrac{a}{b}} = \dfrac{\sqrt{a}}{\sqrt{b}}; a \geq 0, b > 0$

$\sqrt{\dfrac{9}{16}} = \dfrac{\sqrt{9}}{\sqrt{16}} = \dfrac{3}{4}$

18. **THINKING SKILLS** Do you think there is a *Quotient Property of Square Roots*? Formulate a possible statement of this property, similar to the statement of the *Product Property* given above. Then create an example to show how this property might be used. Synthesis

Circles and Polygons **489**

Warm-Up Exercises

Find each answer mentally.

1. 9^2 81
2. 11^2 121
3. $3^2 \cdot 2^2$ 36
4. 5^3 125
5. $2^3 \cdot 3^2$ 72

Teaching Note

Students often do these types of problems mechanically without thinking of underlying concepts. To help students develop their number sense with radicals, do several examples in which students estimate each answer before calculating.

Additional Example

Simplify $\sqrt{90}$. $3\sqrt{10}$

Exercise Note

Critical Thinking
Exercise 18 challenges students to *formulate* a quotient property for square roots and to create an example illustrating the property. Testing this *conjecture* provides an opportunity for students to discover relationships on their own.

In the figure below,
$\triangle ABC \cong \triangle DEF$.

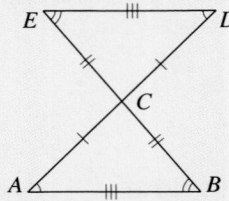

1. $\overline{BC} \cong$ _?_ $\overline{EC}$

2. $\angle E \cong$ _?_ $\angle B$

3. In the figure below, quadrilateral $ABCD \sim EFGH$. Find the length of $\overline{BC}$.   6

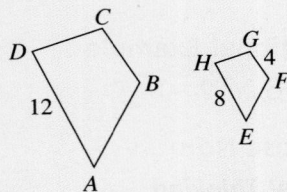

4. A tower casts a shadow 60 ft long. At the same time, a tree 9 ft tall casts a shadow 18 ft long. How tall is the tower?
   30 ft

Find each square root. Use a calculator or the table on page 480 if necessary.

5. $\sqrt{1600}$ 40

6. $\sqrt{\dfrac{9}{100}}$ $\dfrac{3}{10}$

7. $\sqrt{93}$ 9.644

8. Is a triangle with sides of 8 ft, 10 ft, and 12 ft a right triangle? Write *Yes* or *No.*   No

9. Find the unknown length in the triangle below. If necessary, round to the nearest tenth.   4.5

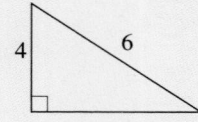

Cumulative Review

Standardized Testing Practice

Choose the letter of the correct answer.

1.

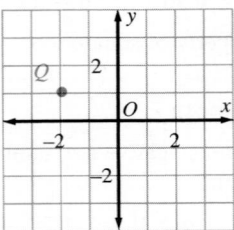

Find the coordinates of point Q.

A. $(-2, -1)$ B. $(-2, 1)$
C. $(1, -2)$ D. $(-1, 2)$

2. Complete: 35% of _?_ is 63.

A. 180   B. 55.6
C. 63   D. 1.8

3. Write the phrase *the sum of three times a number t and seven* as a variable expression.

A. $3t - 7$ B. $3t + 7$
C. $t^3 + 7$ D. $3(t + 7)$

4. The lengths of the legs of a right triangle are 9 in. and 12 in. Find the hypotenuse.

A. 54 in.   B. 15 in.
C. 21 in.   D. 7.9 in.

5. Find the next expression in the pattern: $x + 2,\ 2x + 4,\ 3x + 6,\ \underline{\ ?\ }$

A. $4x + 7$ B. $3x + 8$
C. $4x + 8$ D. $4x + 12$

6.

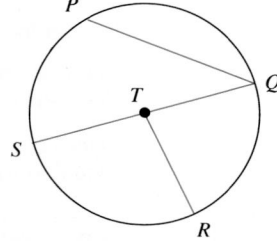

Identify a diameter of this circle.

A. $\overline{TR}$ B. $\overline{PQ}$
C. $\overline{SQ}$ D. $\overline{TQ}$

7. Add: $-36 + (-17)$

A. -53 B. 53
C. 19   D. -19

8. Write 475% as a fraction or mixed number in lowest terms.

A. $\dfrac{475}{100}$ B. $47\frac{1}{2}$

C. $\dfrac{19}{40}$ D. $4\frac{3}{4}$

9. Evaluate $|t + 3| - s$ when $t = -8$ and $s = 4$.

A. 15   B. -9
C. 7   D. 1

10. Simplify: $\sqrt{4900}$

A. 60   B. 2450
C. 70   D. 700

11.

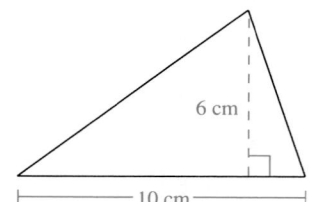

6 cm

10 cm

Find the area of this triangle.

A. 9.5 cm²   (B.) 30 cm²

C. 60 cm²   D. 19 cm²

16.

K

J L

H G M

Complete: $\angle JKG \cong$ ___?___

(A.) $\angle LKG$ B. $\angle LGK$

C. $\angle KGM$ D. $\angle JKL$

---

12. Simplify: $(7x^2)(2x)$

A. $14x^2$ (B.) $14x^3$

C. 7   D. 1

17. Solve: $15 + 2q = 9$

A. $q = \dfrac{9}{17}$ (B.) $q = -3$

C. $q = 12$ D. $q = -12$

---

13. Which type of quadrilateral has exactly two lines of symmetry?

(A.) rhombus

B. square

C. trapezoid

D. none of the above

18. Which of the following is equivalent to 4%?

I. $\dfrac{4}{100}$ II. 0.04   III. $\dfrac{1}{25}$

A. I only

B. I and II only

(C.) I, II, and III

D. II and III only

---

14. The price of a book is marked down from \$18.95 to \$15.16. Find the percent of decrease.

A. 25%   (B.) 20%

C. 37.9%   D. 80%

19. A strip of molding is $\frac{4}{5}$ the length of a room. The room is $12\frac{1}{2}$ ft long. Find the length of the molding.

A. $13\dfrac{3}{10}$ ft   (B.) 10 ft

C. $11\dfrac{7}{10}$ ft   D. $15\dfrac{5}{8}$ ft

---

15. Use the formula $m = \dfrac{a+b}{2}$ to find m when $a = 17$ and $b = 27$.

A. 5   (B.) 22

C. 27   D. 44

20. Which fraction is greater than $-\frac{3}{4}$?

A. $-\dfrac{15}{20}$ B. $-\dfrac{7}{8}$

C. $-\dfrac{13}{14}$ (D.) $-\dfrac{1}{2}$

Alternative Assessment

1. a. State the converse of this statement: If two polygons are congruent, they are also similar.   If two polygons are similar, they are also congruent.

   b. Is the converse true?   No

2. Is the statement

$$\sqrt{a^2 + b^2} = a + b$$ *True* or *False*? Justify your answer. False. The justification will vary, but it should involve substituting numbers for the values of a and b.

Planning Chapter 11

Chapter Overview

Chapter 11 continues the data analysis strand begun earlier in Chapter 5. *Lesson 1* introduces students to frequency tables with intervals, an extension of Lesson 5-10. In *Lesson 2*, the focus is on stem-and-leaf plots. *Lesson 3* concentrates on histograms and frequency polygons, and *Lesson 4* presents box-and-whisker plots. Scattergrams are featured in *Lesson 5* and are used to display relationships between two sets of data. *Lessons 6 and 7* focus on interpreting and constructing circle graphs, which are used to represent data as parts of a whole and may be found in many applications. The *Focus on Computers* shows how technology can be used to create circle graphs. *Lesson 8* reviews the various types of data displays, indicating when each type is most appropriate. *Lesson 9* presents the problem-solving strategy of logical reasoning. The *Focus on Applications* connects the collection of data to the activity of conducting the United States census. Finally, the *Chapter Extension* examines the relative position of data items through percentile rankings.

Background

Students continue their study of data analysis in this chapter. Once again, students have the opportunity to examine data and see why items are displayed in a particular manner. New types of displays are introduced: circle graphs, stem-and-leaf plots, box-and-whisker plots, scattergrams, histograms, and frequency polygons. Pertinent explanations highlight the features of each type of display.

Technology and data analysis are linked in the *Focus on Computers*, where statistical graphing software is used to create circle graphs. Although circle graphs are featured in the *Focus on Computers*, other types of graphs also can be drawn using graphing software. Drawing graphs, a process that is time-consuming and tedious, can be completed quickly and accurately using this type of software.

Problem solving is supported strongly in this chapter through the analysis of tables and graphs. Organizing data in tables and reading and interpreting graphs are important problem-solving strategies.

Communication skills continue to be reinforced throughout the chapter, as students examine, justify, and explain why data items are presented in a particular way.

Objectives

11-1	To organize data in a frequency table with intervals.
11-2	To make and interpret stem-and-leaf plots.
11-3	To interpret histograms and frequency polygons.
11-4	To make and interpret box-and-whisker plots.
11-5	To read and interpret scattergrams.
11-6	To read and interpret circle graphs.
11-7	To construct circle graphs to display data.
▤	**FOCUS ON COMPUTERS** To use statistical graphing software to display and analyze data.
11-8	To decide which method for displaying data is appropriate.
11-9	To solve problems using logical reasoning.
▤	**FOCUS ON APPLICATIONS** To analyze census data.
▤	**CHAPTER EXTENSION** To find a given percentile.

Chapter Planner

Instructional Aids	Manipulatives	Cooperative Learning	Technology	Practice/Reteaching	Assessment

Materials: string, paper, colored beads or cereal, compass or template, calculator
Resource Book:
  Lesson Starters
  101–109
Visuals, Folder J
	Lessons 11-4, 11-7
	Activities Book:
	  Manipulative
	  Activities 21–22
		Lessons 11-1, 11-3, 11-7, 11-9
		Activities Book:
		  Cooperative
		  Activity 21
			Lessons 11-3, 11-4, 11-5, 11-7, Focus on Computers
			Activities Book:
			  Calculator Activity 11
			  Computer Activity 11
			Connections Plotter Plus Disk
				Toolbox, pp. 753–771
				Extra Practice, p. 745
				Resource Book:
				  Practice 137–148
				Study Guide,
				  pp. 201–218
					Self-Tests 1–2
					Quick Quizzes 1–2
					Chapter Test, p. 533
					Resource Book:
					  Lesson Starters
					  (Daily Quizzes)
					Tests 53–56

Assignment Guide Chapter 11

Day	Skills Course	Average Course	Advanced Course
1	**11-1:** 1–11	**11-1:** 1–7 odd, 8–11 **11-2:** 1–20	**11-1:** 1–7 odd, 8–11, Challenge **11-2:** 1–20, Historical Note
2	**11-2:** 1–20	**11-3:** 1–19, 22–26	**11-3:** 1–26
3	**11-3:** 1–17, 22–26	**11-4:** 1–11, 13–16	**11-4:** 1–16
4	**11-4:** 1–8, 13–16	**11-5:** 1–16, 21–28	**11-5:** 1–28
5	**11-5:** 1–16, 23–28	**11-6:** 1–26	**11-6:** 1–17 odd, 19–26 **11-7:** 1–7 odd, 9–18
6	**11-6:** 1–18, 23–26	**11-7:** 1–12, 16–18	**Focus on Computers** **11-8:** 1–6
7	**11-7:** 1–8, 16–18	**Focus on Computers** **11-8:** 1–6	**11-9:** 1–14
8	**11-8:** 1–6	**11-9:** 1–14	*Prepare for Chapter Test:* Chapter Review
9	**11-9** 1–14	*Prepare for Chapter Test:* Chapter Review	*Administer Chapter 11 Test;* Chapter Extension; Cumulative Review
10	*Prepare for Chapter Test:* Chapter Review	*Administer Chapter 11 Test;* Cumulative Review	
11	*Administer Chapter 11 Test*		

Teacher's Resources

Resource Book
  Chapter 11 Project
  Lesson Starters 101–109
  Practice 137–148
  Enrichment 111–120
  Diagram Masters
  Chapter 11 Objectives
  Family Involvement 11
  Spanish Test 11
  Spanish Glossary

Activities Book
  Cooperative Activity 21
  Manipulative Activities 21–22
  Calculator Activity 11
  Computer Activity 11

Study Guide, pp. 201–218

Tests
  Tests 53–56

Visuals
  Folder J

Connections Plotter
  ***Plus* Disk**

Alternate Approaches Chapter 11

Using Technology

COMPUTERS

Much of this chapter concerns methods of analyzing and depicting data informally by the use of tables and diagrams of various kinds. Computer programs that can display data in histograms and scattergrams (see the *Connections Plotter Plus* disk) and that can draw frequency polygons and circle graphs are readily available. Sometimes the programs are designed to analyze data directly, not only providing graphical displays, but also calculating such statistics as the mean, median, standard deviation, and correlation coefficient. In addition, spreadsheet programs often are able to display the contents of cells in graphical form, or stand-alone programs are available that can be used to display the spreadsheet data.

Lesson 11-3
COMPUTERS
Students can verify their histograms and frequency polygons by using appropriate graphing software. They can be given exercises to convert one type of graph to the other by using graphs generated by such a program. In addition, the ease with which such graphs can be generated by the programs allows students to explore how changing data intervals and scales affect the appearance of graphs.

Lesson 11-4
COMPUTERS
Finding quartiles when there is a large amount of unsorted data can be tedious. Statistics and spreadsheet programs that allow sorting on a column of data (arranging the data in increasing or decreasing order) can speed up the process considerably. This and other features of these programs make class statistics projects feasible.

Lesson 11-5
COMPUTERS
Most statistics software will do scattergrams for two columns of related data. This eliminates the tedium of plotting points so that a great many plots can be explored. For any such scattergram, the statistics program will usually provide the correlation coefficient, which is a measure of how closely the data points are clustered around the *trend line,* or *line of best fit.* In addition, an equation of the trend line usually is provided. This equation, or the graph of the trend line itself, can be used by students to predict data given other data (for example, knowing a person's test score, students can predict the number of hours studied).

Lesson 11-7
COMPUTERS
Many statistics and spreadsheet programs can draw circle graphs from a set of data. (Sometimes, you must purchase a separate program to graph data entered into a spreadsheet.) Such programs can be used to check work done by hand, to explore the principles behind circle graphs, and to generate sample circle graphs for analysis by students.

Using Manipulatives

Lesson 11-4
Students themselves can be used to illustrate the principles of developing a box-and-whisker plot. Arrange half the class in a row by height from shortest to tallest. Have the students stand so that they are equally spaced across the front of the room. Ask the rest of the class to identify the student of median height and have that student step forward. Next, find the student of median height of the half that includes the shorter students. If the median is between two students, place a sheet of paper on the floor between them to mark the first quartile. Repeat the process with the half that included the taller students to locate the third quartile. Use string to place a ''box'' on the floor that extends from the first quartile to the third quartile. Place a piece of string that extends from the end of the box at the first quartile to the shortest person and another length that extends from the third quartile to the tallest person.

Lesson 11-7
COOPERATIVE LEARNING
An interesting way to represent data in a circle graph is to use colored beads on a string. (Colored cereal can be used if colored beads are not available.) Give each group a cup of

different colored beads and have the group sort the beads by color. Each group should thread the beads on a string so that like colors are together. After arranging all the beads on one string, each group should arrange the beads in a circle. Have each group locate the center of the circle and draw segments so that sectors of the circle are used to represent each color. Have each group use a protractor to measure the portion represented by each color. The degree measure may be verified computationally by the method presented in the text.

Reteaching/Alternate Approach

Lesson 11-1

COOPERATIVE LEARNING

Have students work in cooperative groups to conduct a survey. The topic of the survey should be determined by each group. A suggested topic might be to determine the average length of time students watched television the previous night or week. Each group should decide how to organize the data and how to record the information in a frequency table.

Lesson 11-7

The degree measure for the sectors of a circle can be found by solving proportions. For instance, in the given Example, the degree measure for each type would be represented by the following proportion.

$$\frac{\text{number sold of each type}}{\text{total sales}} = \frac{\text{number of degrees in sector}}{360°}$$

The degree measure for rock music can be found by solving the following proportion.

$$\frac{240}{540} = \frac{n}{360}$$

Lesson 11-9

COOPERATIVE LEARNING

Have students work in cooperative groups to solve the logic problem below. Write each clue on a separate card and give each group a set of clue cards in random order. Groups should rearrange the cards in the order they feel is best and then work together to arrive at a solution.

Find out who lives next to whom.
1. George, Tomas, Cindy, and Lisa live in the only four houses on a street.
2. Lisa loves her dogs.
3. Cindy is glad she doesn't have dogs living next door.
4. George lives east of Cindy.
5. The two boys are not next-door neighbors.
6. Tomas lives west of Lisa.
7. George has neighbors on both sides of his house.

Tomas, Cindy, George, Lisa

Teacher's Resources Chapter 11

Enrichment masters from the Resource Book are pictured here. See the Teacher's Resources chart on page 492B for a complete listing of all materials available for this chapter.

11-1 Enrichment
Communication

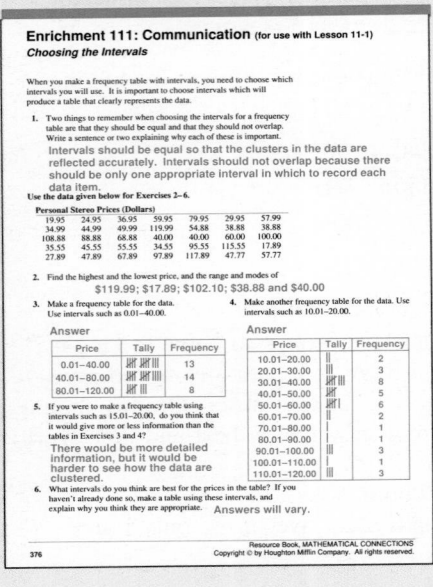

11-2 Enrichment
Connection

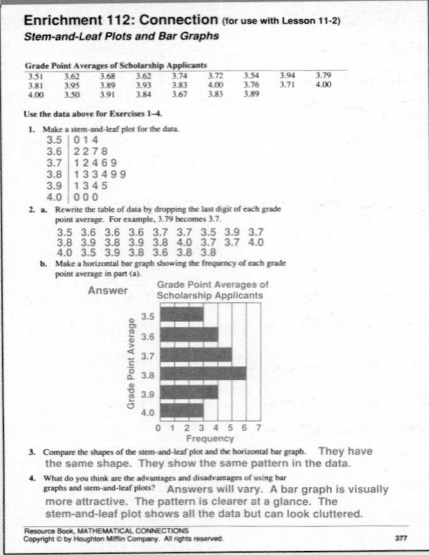

11-3 Enrichment
Application

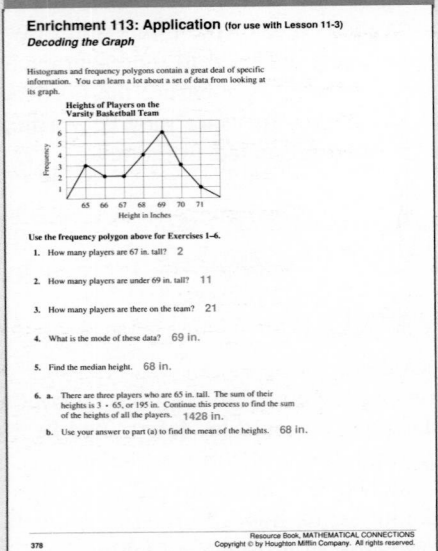

11-4 Enrichment
Data Analysis

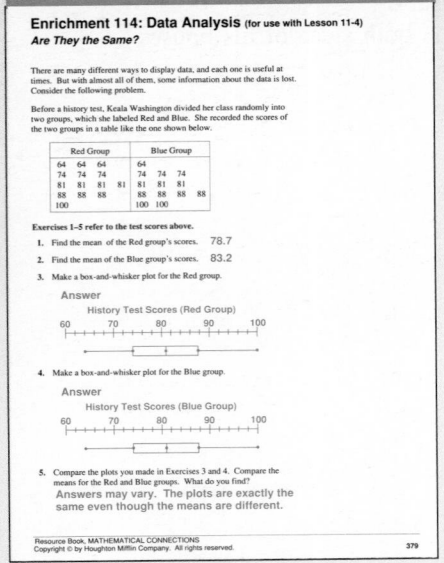

11-5 Enrichment
Thinking Skills

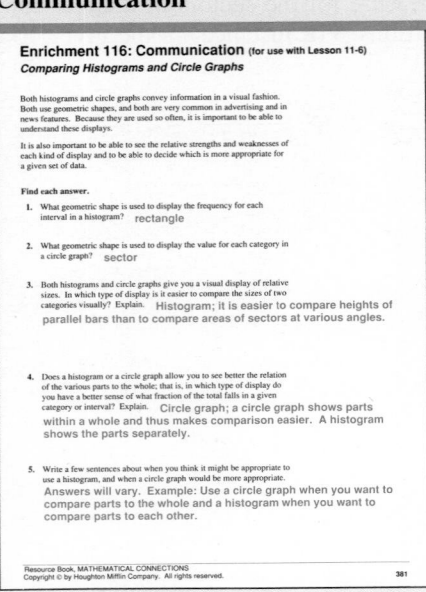

11-6 Enrichment
Communication

11-7 Enrichment
Thinking Skills

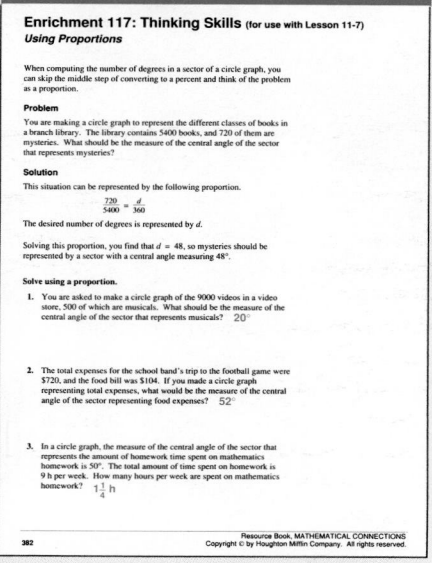

Enrichment 117: Thinking Skills (for use with Lesson 11-7)
Using Proportions

When computing the number of degrees in a sector of a circle graph, you can skip the middle step of converting to a percent and think of the problem as a proportion.

Problem

You are making a circle graph to represent the different classes of books in a branch library. The library contains 5400 books, and 720 of them are mysteries. What should be the measure of the central angle of the sector that represents mysteries?

Solution

This situation can be represented by the following proportion.

$$\frac{720}{5400} = \frac{d}{360}$$

The desired number of degrees is represented by d.

Solving this proportion, you find that $d = 48$, so mysteries should be represented by a sector with a central angle measuring 48°.

Solve using a proportion.

1. You are asked to make a circle graph of the 9000 videos in a video store, 500 of which are musicals. What should be the measure of the central angle of the sector that represents musicals? **20°**

2. The total expenses for the school band's trip to the football game were $720, and the food bill was $104. If you made a circle graph representing total expenses, what would be the measure of the central angle of the sector representing food expenses? **52°**

3. In a circle graph, the measure of the central angle of the sector that represents the amount of homework time spent on mathematics homework is 50°. The total amount of time spent on homework is 9 h per week. How many hours per week are spent on mathematics homework? **$1\frac{1}{4}$ h**

11-8 Enrichment
Exploration

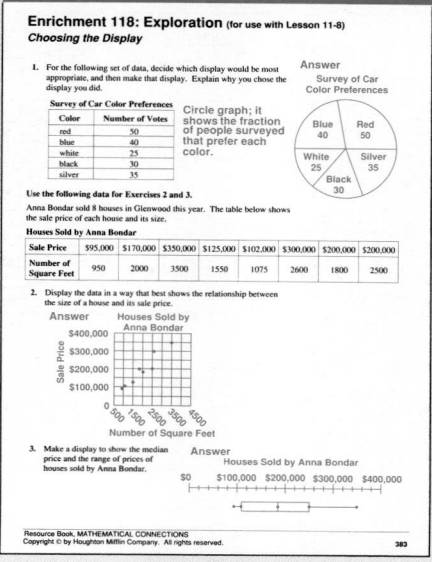

Enrichment 118: Exploration (for use with Lesson 11-8)
Choosing the Display

1. For the following set of data, decide which display would be most appropriate, and then make that display. Explain why you chose the display you did.

Survey of Car Color Preferences

Color	Number of Votes
red	50
blue	40
white	25
black	30
silver	35

Answer Circle graph; it shows the fraction of people surveyed that prefer each color.

Survey of Car Color Preferences
Blue 40, Red 50, White 25, Silver 35, Black 30

Use the following data for Exercises 2 and 3.

Anna Bondar sold 8 houses in Glenwood this year. The table below shows the sale price of each house and its size.

Houses Sold by Anna Bondar

Sale Price	$95,000	$170,000	$350,000	$125,000	$102,000	$300,000	$200,000	$200,000
Number of Square Feet	950	2000	3500	1550	1075	2600	1800	2500

2. Display the data in a way that best shows the relationship between the size of a house and its sale price.

Answer Houses Sold by Anna Bondar (scatter plot, Sale Price vs. Number of Square Feet)

3. Make a display to show the median price and the range of prices of houses sold by Anna Bondar.

Answer Houses Sold by Anna Bondar (box-and-whisker plot, $0 to $400,000)

11-9 Enrichment
Problem Solving

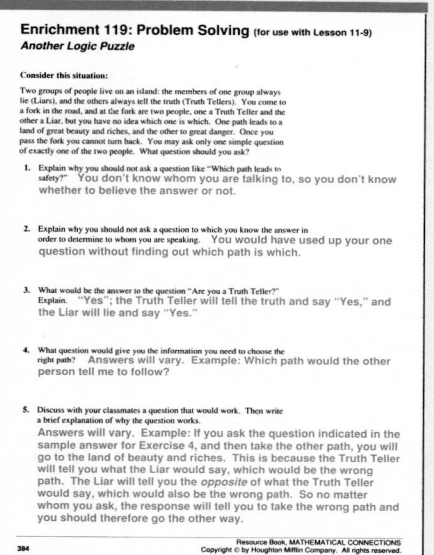

Enrichment 119: Problem Solving (for use with Lesson 11-9)
Another Logic Puzzle

Consider this situation:

Two groups of people live on an island: the members of one group always lie (Liars), and the others always tell the truth (Truth Tellers). You come to a fork in the road, and at the fork are two people, one a Truth Teller and the other a Liar, but you have no idea which one is which. One path leads to a land of great beauty and riches, and the other to great danger. Once you pass the fork you cannot turn back. You may ask only one simple question of exactly one of the two people. What question should you ask?

1. Explain why you should not ask a question like "Which path leads to safety?" **You don't know whom you are talking to, so you don't know whether to believe the answer or not.**

2. Explain why you should not ask a question to which you know the answer in order to determine to whom you are speaking. **You would have used up your one question without finding out which path is which.**

3. What would be the answer to the question "Are you a Truth Teller?" Explain. **"Yes"; the Truth Teller will tell the truth and say "Yes," and the Liar will lie and say "Yes."**

4. What question would give you the information you need to choose the right path? **Answers will vary. Example: Which path would the other person tell me to follow?**

5. Discuss with your classmates a question that would work. Then write a brief explanation of why the question works. **Answers will vary. Example: If you ask the question indicated in the sample answer for Exercise 4, and then take the other path, you will go to the land of beauty and riches. This is because the Truth Teller will tell you what the Liar would say, which would be the wrong path. The Liar will tell you the *opposite* of what the Truth Teller would say, which would also be the wrong path. So no matter whom you ask, the response will tell you to take the wrong path and you should therefore go the other way.**

End of Chapter Enrichment
Extension

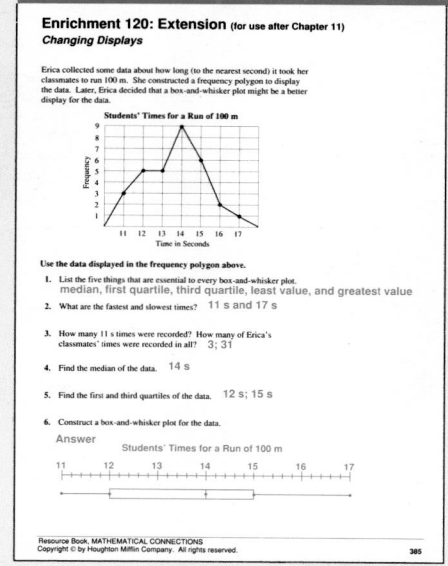

Enrichment 120: Extension (for use after Chapter 11)
Changing Displays

Erica collected some data about how long (to the nearest second) it took her classmates to run 100 m. She constructed a frequency polygon to display the data. Later, Erica decided that a box-and-whisker plot might be a better display for the data.

Students' Times for a Run of 100 m (frequency polygon, Frequency vs. Time in Seconds, 11 to 17)

Use the data displayed in the frequency polygon above.

1. List the five things that are essential to every box-and-whisker plot. **median, first quartile, third quartile, least value, and greatest value**

2. What are the fastest and slowest times? **11 s and 17 s**

3. How many 11 s times were recorded? How many of Erica's classmates' times were recorded in all? **3; 31**

4. Find the median of the data. **14 s**

5. Find the first and third quartiles of the data. **12 s; 15 s**

6. Construct a box-and-whisker plot for the data.

Answer Students' Times for a Run of 100 m (box-and-whisker plot, 11 to 17)

For Chapter 11 Opener

Many Native American groups depended on trees for their living. For example, along the northwest coast of North America, where cedar trees are plentiful, groups such as the Makah used virtually every part of the tree in building houses, canoes, utensils, and tools. The bark was used for clothing, baskets, twine, and matting, and the roots were used for making baskets. The withes, or smaller branches, were used to make ropes, which could be made strong enough to use in whaling.

A typical whaling canoe was quite large, measuring up to thirty-five feet long and able to carry eight whalers. A skilled craftsman and several assistants carved the main body of the canoe from a single cedar log. The bow and stern were carved separately and sewn to the main body with cedar withes.

For Page 500

The Tuskegee Institute, discussed in the Historical Note, was founded by Booker T. Washington in 1881 in Tuskegee, Alabama. The Institute offers courses in arts and sciences, home economics, engineering, architecture, business, education, health and nursing, and veterinary medicine.

Research Activities

For Chapter 11 Opener

Working in cooperative groups, students can conduct research to find out what kinds of trees are indigenous to their local area. They can then conduct research to find out how these trees are used today, and how they may have been used in the past by Native American groups.

492

Statistics and Circle Graphs 11

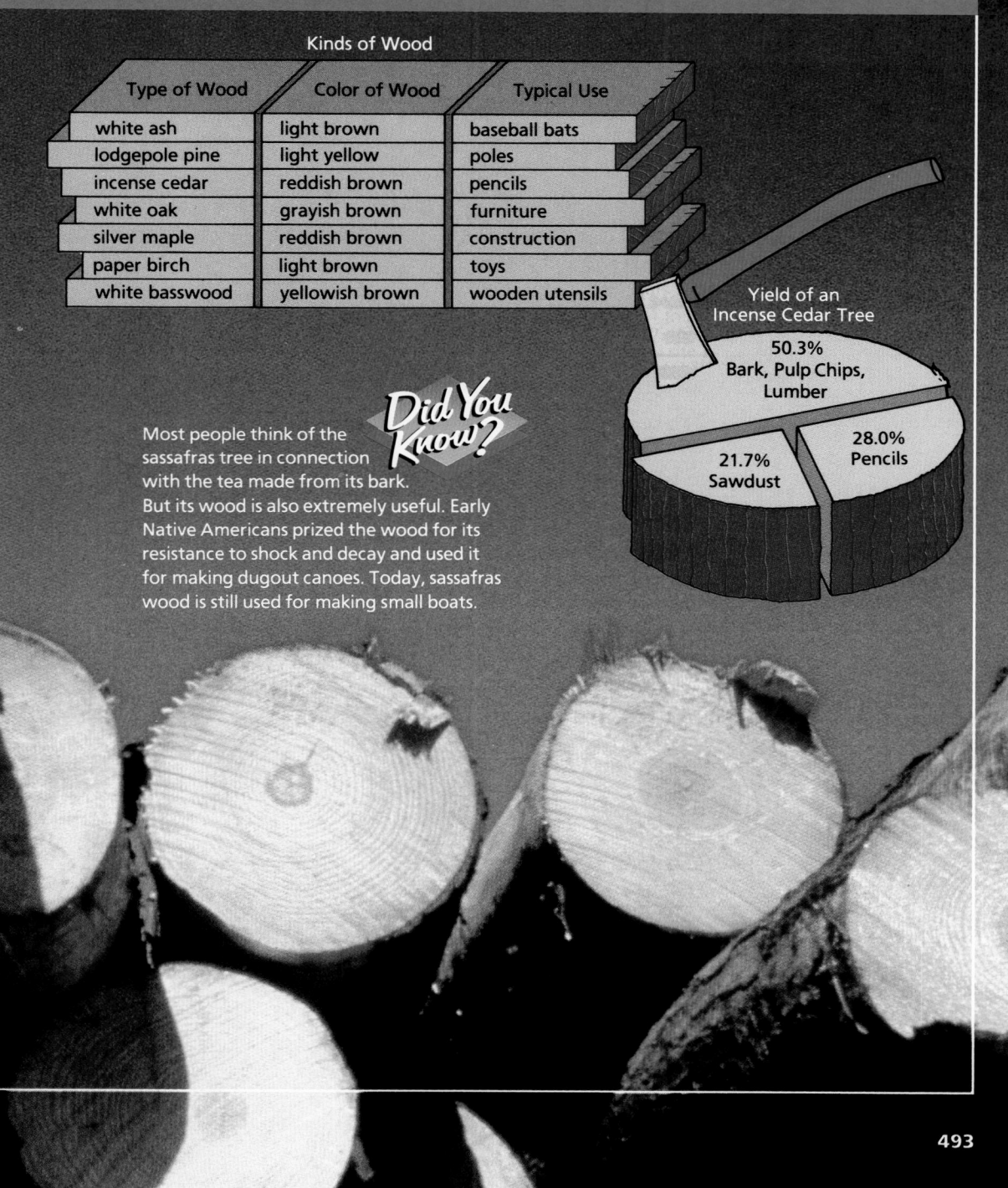

Kinds of Wood

Type of Wood	Color of Wood	Typical Use
white ash	light brown	baseball bats
lodgepole pine	light yellow	poles
incense cedar	reddish brown	pencils
white oak	grayish brown	furniture
silver maple	reddish brown	construction
paper birch	light brown	toys
white basswood	yellowish brown	wooden utensils

Did You Know?

Most people think of the sassafras tree in connection with the tea made from its bark. But its wood is also extremely useful. Early Native Americans prized the wood for its resistance to shock and decay and used it for making dugout canoes. Today, sassafras wood is still used for making small boats.

Yield of an Incense Cedar Tree

50.3% Bark, Pulp Chips, Lumber

28.0% Pencils

21.7% Sawdust

Research Activities (continued)

For Page 517
The table in the example on this page shows four different kinds of music. You may want to discuss the listed categories with students to decide whether or not they feel these are the most representative categories, and to formulate others if necessary. Students can then talk with proprietors of local stores selling compact discs to obtain data to use in constructing a class table.

Suggested Resources

Stewart, Hilary. *Cedar: Tree of Life to the Northwest Coast Indians.* Seattle, WA: University of Washington Press, 1984.

493

Lesson Planner

Teaching the Lesson

- Lesson Starter 101
- Reteaching/Alternate Approach, p. 492D
- Cooperative Learning, p. 492D

Lesson Follow-Up

- Practice 137
- Enrichment 111: Communication
- Study Guide, pp. 201–202
- Manipulative Activity 21

Warm-Up Exercises

Use the table on this page showing bowling scores to answer the following questions.

1. What is the highest score? 158
2. What is the lowest score? 115
3. What is the range of scores? 43
4. How can you determine the median? Arrange the scores in order from lowest to highest; count to find the middle score.
5. What is the median score? 131

Lesson Focus

For the United States census, taken every ten years, enormous amounts of data are collected on census forms. Ask students if they can imagine examining the census data by looking at every form? Today's lesson focuses on organizing data using frequency tables.

11-1 *Frequency Tables with Intervals*

Objective: To organize data in a frequency table with intervals.

CONNECTION

In Chapter 5 you learned that statistics is the branch of mathematics that deals with collecting, organizing, and analyzing data. You also learned that data can be organized in a frequency table. When data are scattered, a frequency table that has the data grouped in equal intervals may be easier to interpret.

Bowling Scores

156	131	124	115	143
158	158	131	121	131
143	152	130	137	143
124	137	131	121	130
152	158	131	124	143

Example Make a frequency table for the data above. Use intervals such as 111–120.

Solution

- Make a table with three columns.
- List the intervals in the first column.
- Make a tally mark in the second column next to each interval for every score that falls within that interval.
- Record the total number of tally marks for each interval in the third column.

Bowling Scores

Score	Tally	Frequency
111–120	|	1
121–130	卌 ||	7
131–140	卌 ||	7
141–150	||||	4
151–160	卌 |	6

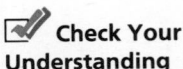

Check Your Understanding What would be the frequency for the last interval in the Example if that interval were 155–160?
See *Answers to Check Your Understanding* at the back of the book.

Guided Practice

«1. **COMMUNICATION** «*Reading* Replace each _?_ with the correct word or phrase. frequency; intervals; frequency

Data can be organized in a _?_ table. A large set of data can be condensed by grouping the data into equal _?_. The number of times that an item occurs is called the _?_ of that item.

2. The first interval in a frequency table is 50–54. What will be the next four intervals? 55–59; 60–64; 65–69; 70–74

Answers to Guided Practice 3–4 are on p. A27.

3. Make a frequency table for the data. Use intervals such as 61–70.

Scores on a Mathematics Test

| 65 | 70 | 70 | 85 | 69 | 85 | 100 | 81 | 70 | 83 | 81 | 90 | 82 | 87 |
| 96 | 68 | 94 | 100 | 95 | 81 | | 83 | 100 | 66 | 81 | 83 | 96 | 84 | 85 |

«4. COMMUNICATION «*Discussion* Can you tell what the individual data items are in a frequency table with intervals? without intervals, as you saw in lesson 5-10? Explain.

Exercises

Make a frequency table for the data. Answers to Exercises 1–6 are on pp. A27–A28.

A 1. **Compact Disc Prices (Dollars)**

| 7.99 | 7.99 | 10.99 | 12.99 | 10.99 | 12.99 | 14.99 | 15.49 | 14.99 | 15.49 |
| 14.99 | 14.49 | 12.49 | 12.99 | 12.99 | 11.99 | 11.99 | 13.99 | 13.99 | 13.49 |

(Use intervals such as 7.00–8.99.)

2. **Ages of Hospital Nurses**

| 23 | 25 | 23 | 29 | 23 | 34 | 31 | 47 | 55 | 57 | 56 | 57 | 45 |
| 37 | 24 | 41 | 59 | 60 | 61 | 45 | 45 | 52 | 39 | 50 | 39 | |

(Use intervals such as 20–29.)

3. **Annual Tuition at Private Colleges (Dollars)**

7800	12000	9550	14000	13750	7500	6400
10900	11200	9600	13400	13200	7000	6900
12300	11200	8500	7400	11500	12200	7600

(Use intervals such as 6000–7999.)

4. **Total Points Scored in Basketball Games**

79	89	75	71	80	83	85	88
93	90	86	74	96	98	105	91
107	82	110	91	88	95	78	86

(Use intervals such as 70–79.)

B 5. Make a frequency table for the data. Choose reasonable intervals.

Cost of a One Week Vacation (Dollars)

890	945	1200	1050	1450	1300
1200	1100	950	1425	1680	1650
1450	1500	1050	1750	1695	1200
1250	1370	980	1575	1625	1300

6. DATA, *pages 186–187* Using intervals such as 1–5, make a frequency table with intervals for the data in the frequency table.

Statistics and Circle Graphs **495**

Teaching Notes

Key Question

Why is the same number of units contained in each interval of a frequency table? Like intervals show how the data are distributed. If the intervals were of different sizes, it would be difficult to compare the data.

Error Analysis

Students may set up intervals that are of different sizes, thus distorting the distribution of the data. Students also may set up intervals that overlap, thus allowing for an item to be placed in either of two intervals. Stress that intervals must be distinct, continuous, and of the same size.

Reteaching/Alternate Approach

For a suggestion on using a cooperative learning activity with this lesson, see page 492D.

Additional Example

Make a frequency table using the data below. Use intervals such as 131–140.

Heights of Students (cm)

142	156	153	167	162
131	165	156	143	152
154	166	173	168	155
134	156	177	165	178
142	175	162	156	155

Heights of Students (cm)

Height	Tally	Frequency
131–140	II	2
141–150	III	3
151–160	HHT IIII	9
161–170	HHT II	7
171–180	IIII	4

Closing the Lesson

Ask a student to explain how to analyze data in order to determine the values for the intervals on a frequency table. Ask another student to explain how the data are then displayed.

Exercise Notes

Critical Thinking

Exercise 7 displays data in a form that is not interpreted easily. Students must *examine* and *analyze* the frequency table to determine that the intervals overlap.

Nonroutine Problem

The Challenge problem requires students to analyze data and use mathematics, such as percents and fractions, to determine the solution.

Follow-Up

Enrichment

Enrichment 111 is pictured on page 492E.

Practice

Practice 137, *Resource Book*

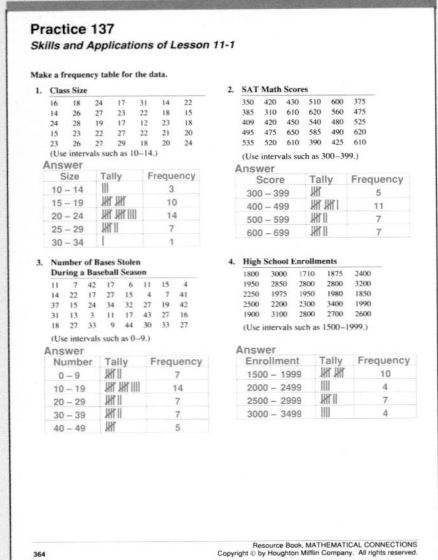

The intervals overlap. It cannot be determined if a score of 70 is in the interval 60–70 or 70–80.

7. THINKING SKILLS Evaluation
Examine the frequency table at the right. Do you notice anything wrong with this table? Determine what is wrong and explain.

Scores on a Biology Test

Scores	Tally	Frequency
60–70	IIII	4
70–80	ʜʜʈ II	7
80–90	ʜʜʈ	5
90–100	ʜʜʈ ʜʜʈ	10

SPIRAL REVIEW

S **8.** The original price is $40 and the new price is $45. Find the percent of increase. *(Lesson 9-9)* 12.5%

9. Write a variable expression that represents this phrase: three less than twice a number *x* *(Lesson 4-6)* $2x - 3$

10. Find $\sqrt{324}$. *(Lesson 10-10)* 18

11. Make a frequency table for the data. Use intervals such as 20–24.
(Lesson 11-1) See answer on p. A28.

Number of Hours Worked per Week

20	26	20	35	29	26	40	20	40
29	40	37	24	20	29	25	24	35
20	20	29	37	37	40	37	37	40
37	24	35	24	26	37	20	37	29

Challenge

The exact *percent* frequency for each data item is shown in the table below. What is the least number of total data items that there could be? 16

Data Item	0	1	2	3	4
Frequency	6.25%	12.5%	12.5%	43.75%	25%

11-2

Stem-and-Leaf Plots

Objective: To make and interpret stem-and-leaf plots.

A frequency table with intervals will not show you the value of each piece of data. An alternative display in which no information is lost is the **stem-and-leaf plot.** In this type of display, each number is represented by a *stem* and a *leaf*. The **leaf** is the last digit on the right of the number. The **stem** is the digit or digits of the number remaining when the leaf is dropped.

<table>
<tr><td>Terms to Know</td></tr>
<tr><td>• stem-and-leaf plot
• stem
• leaf</td></tr>
</table>

4 8 6

48 is the stem. ◄───┘   └──► 6 is the leaf.

Example 1

Make a stem-and-leaf plot for the data.

Scores on a Science Test

78	74	80	84	99	65	81	80
62	95	100	91	75	82	76	92
80	75	100	82	77	70	79	85
90	95	86	82	75	69	83	97

Solution

Write the stems in order from least to greatest to the left of a vertical line. Write each leaf to the right of its stem.

```
 6 | 5  2  9              ◄──── This row contains 65, 62, and 69.
 7 | 8  4  5  6  5  7  0  9  5
 8 | 0  4  1  0  2  0  2  5  6  2  3
 9 | 9  5  1  2  0  5  7
10 | 0  0
```
↑
Stems   Leaves

Rearrange the leaves for each stem in order from least to greatest. Title the stem-and-leaf plot.

Scores on a Science Test

```
 6 | 2  5  9
 7 | 0  4  5  5  5  6  7  8  9
 8 | 0  0  0  1  2  2  2  3  4  5  6
 9 | 0  1  2  5  5  7  9
10 | 0  0
```

 **Check Your Understanding**

1. How would Example 1 be different if a score of 96 were included?
See *Answers to Check Your Understanding* at the back of the book.

Lesson Planner

Teaching the Lesson
• Lesson Starter 102
• Visuals, Folder J
• Toolbox Skill 5

Lesson Follow-Up
• Practice 138
• Enrichment 112: Connection
• Study Guide, pp. 203–204

Warm-Up Exercises

Use the following data to answer the questions below.

Points Scored in Football Games

13	43	27	35	33	20
16	24	19	25	34	22
33	26	54	40	39	43
22	45	56	32	27	19

1. Make a frequency table for the data. Use intervals such as 10–19.

Points Scored in Football Games

Points	Tally	Frequency
10–19	IIII	4
20–29	₩Ⱦ III	8
30–39	₩Ⱦ I	6
40–49	IIII	4
50–59	II	2

2. What is the range of scores?   43

3. Which interval in the frequency table represents the highest number of games?   20–29

4. Which interval in the frequency table represents the number of points least likely to occur?
50–59

497

Lesson Focus

Ask students to imagine a display of data involving the number of home runs hit by a baseball team. If the number of home runs were grouped by intervals, would the display be useful? Or, would showing individual numbers of home runs be more useful? Grouped data can be displayed and yet continue to show individual items. Today's lesson focuses on the type of data display known as a stem-and-leaf plot.

Teaching Notes

Students usually have fewer problems creating a stem-and-leaf plot for two-digit data, as in Example 1, than for three-or-more-digit data, as in Example 2. Point out to students that while the stem can have any number of digits, the leaf must have only one.

Key Question

Can the leaves in a stem-and-leaf plot be two-digit numbers? No; each leaf is the *last* digit at the right of the number.

Reasoning

Ask a student to explain how to relate the display of data in a stem-and-leaf plot to the arrangement used earlier for determining the median of a set of numbers.

Error Analysis

When students read a stem-and-leaf plot, they may inadvertently omit the stem. Remind students that the stem and leaf must be used together to identify the data correctly.

Example 2

Use the stem-and-leaf plot at the right.

a. What is the best free throw average?

b. How many teams have a free throw average of 0.735?

c. Find the median of the free throw averages.

Team Free Throw Averages

0.69	2 4 4
0.70	0 1
0.71	
0.72	5 5
0.73	1 5 5 8 9
0.74	
0.75	2 4 5

Solution

a. The greatest stem is 0.75. The greatest leaf for 0.75 is 5. The best free throw average is 0.755.

b. The stem is 0.73. The number 5 appears twice beside this stem. Two teams have a free throw average of 0.735.

c. The order of the 15 leaves reflects the order of the averages. The middle leaf is 1. The stem corresponding to this leaf is 0.73. The median of the free throw averages is 0.731.

✓ Check Your Understanding

2. Describe how Example 2(b) would be different if you were asked to find the number of teams with a free throw average greater than 0.735.

3. Describe how Example 2(c) would be different if you were asked to find the range of the free throw averages.

See *Answers to Check Your Understanding* at the back of the book.

Guided Practice

COMMUNICATION « *Reading* 1. 3.40, 3.45, 3.45, 3.48, 3.56, 3.58, 3.60, 3.60, 3.61, 3.65

List the data that are represented by each stem-and-leaf plot.

«1.

3.4	0 5 5 8
3.5	6 8
3.6	0 0 1 5

«2.

23	0 8
24	6
25	5 5 8 9

230, 238, 246, 255, 255, 258, 259

Identify the stem and the leaf of each number.

3. 19 1; 9

4. 2.78 2.7; 8

5. 20.57 20.5; 7

6. 1228 122; 8

7. 0.253 0.25; 3

8. 400 40; 0

9. No; no; the stems 0.71 and 0.74 are only placeholders. Each would have a leaf of 0 next to it if the free throw averages 0.710 and 0.740 were part of the stem-and-leaf plot.

«9. COMMUNICATION « *Discussion* Refer to the stem-and-leaf plot in Example 2. Does the stem-and-leaf plot include a free throw average of 0.710? of 0.740? Explain.

10. Make a stem-and-leaf plot for the data. See answer on p. A28.

Prices of Compact Disc Players (Dollars)

190	206	210	228	230	194
209	215	229	236	195	215
238	195	210			

Use the stem-and-leaf plot at the right.

11. Find the least capacity. 165

12. How many theaters have a capacity of 220? 0

13. How many theaters have a capacity less than 200? 8

14. Find the mean, median, mode(s), and range of the data.
202.1; 200; 180, 200, and 238; 73

Capacities of Movie Theaters

16	5	9		
17	5	8		
18	0	0		
19	0	8		
20	0	0	5	
21	5	8	9	
22				
23	4	6	8	8

Exercises

Make a stem-and-leaf plot for the data.
Answers to Exercises 1–2 are on p. A28.

A **1. Earned Run Averages of Leading Pitchers**

2.56 2.80 2.78
2.92 2.62 2.20
2.67 2.90 2.90
2.32 2.25 2.82
2.77 2.84 2.82
2.62

2. Ages of Company Presidents

45 58 60 62
48 50 42 60
56 58 55 48
38 55 47
39 50 65
39 35 44

Use the stem-and-leaf plot at the right.

3. What is the least cost? $25

4. What is the greatest cost? $56

5. How many customers pay $41 per month? 3

6. How many customers pay $27 per month? 0

7. How many customers pay more than $36 per month? 12

8. How many customers pay less than $44 per month? 19

9. How many customers pay from $35 through $50 per month? 13

10. Find the mean, median, mode(s), and range of the data. $37; $36.50; $41; $31

Monthly Cost of Cable (Dollars)

2	5	5	6	8	9	9	
3	0	0	2	5	5	6	7	8
4	0	1	1	1	2	5	5
5	0	2	6				

Statistics and Circle Graphs **499**

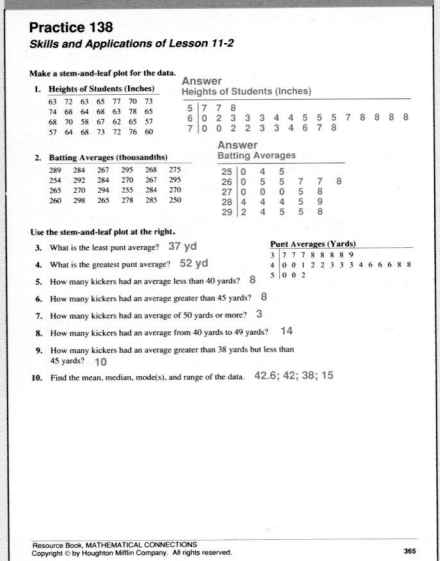
CONNECTING STATISTICS AND SOCIAL STUDIES

B **11.** Make a stem-and-leaf plot for the data below. See answer on p. A28.

Age at First Inauguration of United States Presidents

57	61	57	57	58	57	61	54	68	51	49	64
50	48	65	52	56	46	54	49	50	47	55	54
42	51	56	55	51	54	51	60	62	43	55	56
61	52	69	64								

Use the stem-and-leaf plot that you made for Exercise 11 for Exercises 12–15.

12. How many presidents were 56 years old at their inauguration? 3

13. How many presidents were less than 55 years old at their inauguration? 18

14. How many presidents were 55–64 years old at their inauguration? 19

15. How many presidents were 40–49 years old at their inauguration? 7

SPIRAL REVIEW 17. $5n + 4$; $6n + 4$; $7n + 4$

S **16.** Simplify: $\frac{7x}{4} + \frac{9x}{4}$ *(Lesson 8-4)* $4x$

17. **PATTERNS**   Find the next three expressions in the pattern:
$2n + 4$, $3n + 4$, $4n + 4$, __?__, __?__, __?__   *(Lesson 2-5)*

18. Solve: $4(n + 5) = -24$ *(Lesson 4-9)* -11

19. Use the stem-and-leaf plot on page 499 entitled *Capacities of Movie Theaters*. How many theaters have a capacity greater than 200? *(Lesson 11-2)* 8

20. Solve using an equation: On Sunday, the number of visitors at a crafts fair was 43 less than twice the number of visitors on Saturday. There were 279 visitors on Sunday. How many visitors were there on Saturday?   *(Lesson 4-8)* 161

Historical Note

George Washington Carver (1864–1943), an African-American agricultural chemist, developed ways to diversify crops to prevent soil depletion. Through his research at Tuskegee Institute, he also synthesized hundreds of new products based on sweet potatoes, peanuts, pecans, soybeans, wood shavings, and cotton stalks. Statistics played a role in his research.

Research

Find out what some of the products were that George Washington Carver synthesized. See answer on p. A28.

500   Chapter 11

11-3

Histograms and Frequency Polygons

Objective: To interpret histograms and frequency polygons.

CONNECTION

Data displayed in a frequency table can also be displayed in a *histogram*. A **histogram** is a bar graph like the one shown at the right that is used to show frequencies. In this special type of bar graph, there are no spaces between consecutive bars.

Hours Employees Work per Week

Example 1

Use the histogram above.

a. How many employees work 36–40 h per week?

b. Between which two consecutive intervals does the greatest increase in frequency occur? What is the increase?

c. How many employees work fewer than 31 h per week?

Solution

a. There are 20 employees who work 36–40 h per week.

b. The greatest increase occurs between the intervals of 31–35 and 36–40. The frequencies for these intervals are 12 and 20. The increase is 20 − 12 = 8.

c. There are 12 employees who work 26–30 h, 8 employees who work 21–25 h, and 4 employees who work 16–20 h. Since 12 + 8 + 4 = 24, 24 people work fewer than 31 h.

 **Check Your Understanding**

How would Example 1(c) have been different if you were asked to find how many employees work 31 or more hours per week?
See *Answers to Check Your Understanding* at the back of the book.

Frequencies can also be displayed in a line graph called a **frequency polygon.** One way to construct a frequency polygon is to connect the midpoints of the tops of the bars of a histogram, as shown at the right. You also connect the first point to the origin and the last point to the other end of the horizontal axis.

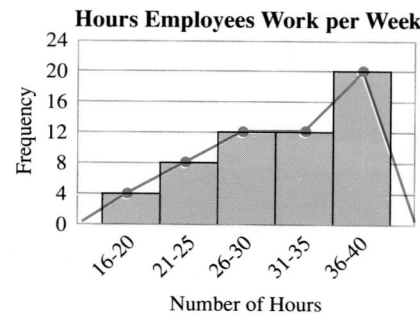

Hours Employees Work per Week

Statistics and Circle Graphs **501**

Key Question

How are histograms and frequency polygons alike and how are they different?   Both graphs display exactly the same information. Histograms use bars with no spaces between them to display data. Frequency polygons use line segments to connect points that represent the heights of the bars on a histogram.

Error Analysis

Students may confuse a bar graph and a histogram. Remind them that a histogram does not leave spaces between the bars of the graph.

Using Technology

For a suggestion on using data graphing software, see page 492C.

Additional Examples

Use the histogram above Example 1 in the text to answer the following questions.

1. How many employees work 26–30 h per week?
   12 employees
2. Between which two consecutive intervals is the least change in frequency? What is the change?   between 26–30 and 31–35; there is no change.
3. How many employees work more than 25 hours per week?
   44 employees

Use the frequency polygon at the right of Example 2 to answer the following questions.

4. How many students are 17 years old?   6 students
5. Exactly four students are a certain age. What is this age?
   16
6. How many students are less than 17 years old?   14 students

Example 2

Use the frequency polygon at the right.

a. How many students are 14 years old?

b. Exactly 3 students are a certain age. What is this age?

c. How many students are there in all?

Ages of Students

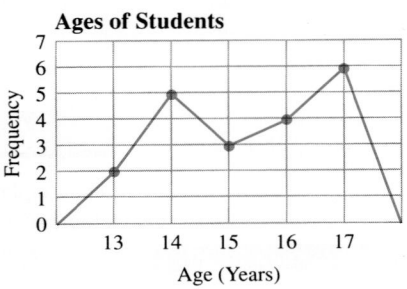

Solution

a. There are 5 students who are 14 years old.

b. The age for which there are exactly 3 students is 15.

c. Find the sum of all the frequencies: $2 + 5 + 3 + 4 + 6 = 20$
   There are 20 students in all.

Guided Practice

1. In a histogram, the vertical axis is scaled for frequencies. Bar graphs have spaces between the bars; histograms do not.
2. In a frequency polygon, the vertical axis is scaled for frequencies, and the line is connected to either end of the horizontal axis.

COMMUNICATION «*Reading*

Refer to the text on pages 501–502.

«**1.** How are an ordinary bar graph and a histogram different?

«**2.** How are an ordinary line graph and a frequency polygon different?

«**3.** COMMUNICATION «*Writing*   On one set of axes, draw a histogram and a frequency polygon for the data in the frequency table below.

See answer on p. A28.

Exam Scores

Score	Tally	Frequency
60–69	I	1
70–79	IIII	4
80–89	⊞ II	7
90–99	⊞ IIII	9

Use the histogram at the top of page 501.

4. How many employees work more than 20 h per week? 52

5. Which interval includes the number of hours worked per week by exactly 8 employees? 21–25

Use the frequency polygon at the top of this page.

6. How many students are more than 15 years old? 10

7. How many students are 13–16 years old? 14

502   Chapter 11

Exercises

Use the histogram at the right.

A 1. How many students ran
9–16 laps? 6

2. How many students ran
9 laps or more? 24

3. How many students ran
16 laps or less? 14

4. How many students
were there in all? 32

5. Did the frequency increase
or decrease between
the intervals 9–16 and
17–24? increase

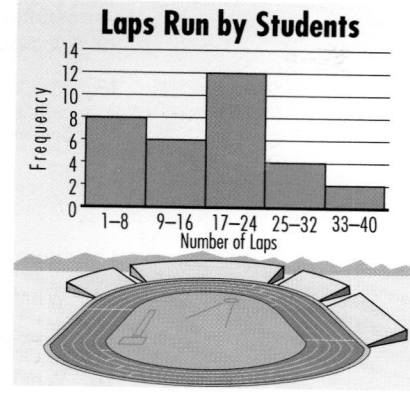

Laps Run by Students

6. What was the greatest decrease between consecutive intervals?
a decrease of 8, between the intervals 17–24 and 25–32

Use the frequency polygon at the right.

7. How many band members
are 64 in. tall? 6

8. What is the height of only
4 band members? 68 in.

9. How many band members
are more than 66 in. tall? 7

10. How many band members
are less than 66 in. tall? 16

11. How many band members
are 65–68 in. tall? 20

12. How many band members are there in all? 29

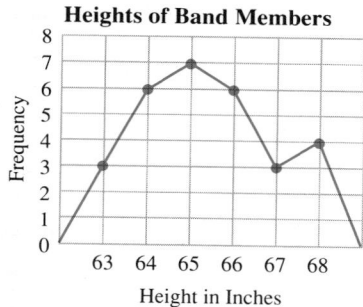

Heights of Band Members

Statistics and Circle Graphs   **503**

Closing the Lesson

Ask a student to explain what a histogram is. Then ask another student to explain how a histogram differs from a bar graph. Ask a third student to explain how a frequency polygon can be made from a histogram.

Suggested Assignments

Skills: 1–17, 22–26
Average: 1–19, 22–26
Advanced: 1–26

Exercise Notes

Communication: Writing

Guided Practice Exercise 3 asks students to create a histogram and a frequency polygon for a set of data on the same axes, thus showing the relationship between these two data displays.

Application

Exercises 13–17 present a use of histograms that is very familiar to students, that is, determining grades.

Cooperative Learning

Exercises 18–21 are designed for use as a group activity. The discussion portion encourages students to verbalize mathematical ideas in order to clarify concepts.

Follow-Up

Exploration *(Cooperative Learning)*

Have students work in cooperative groups to collect data that can be displayed in a histogram. Ask each group to do the following.

1. Make a histogram.

2. Make a frequency polygon using the same data by connecting the midpoint of each bar on the histogram.

3. Make a frequency polygon using the same data by connecting the point on the left-hand side of each bar on the histogram.

4. Make a frequency polygon using the same data by connecting the point on the right-hand side of each bar on the histogram.

5. Compare the frequency polygons made in parts 2, 3, and 4. How are they alike and how do they differ?

6. Explain how a histogram can be constructed from a frequency polygon.

Enrichment

Enrichment 113 is pictured on page 492E.

Practice

Practice 139, *Resource Book*

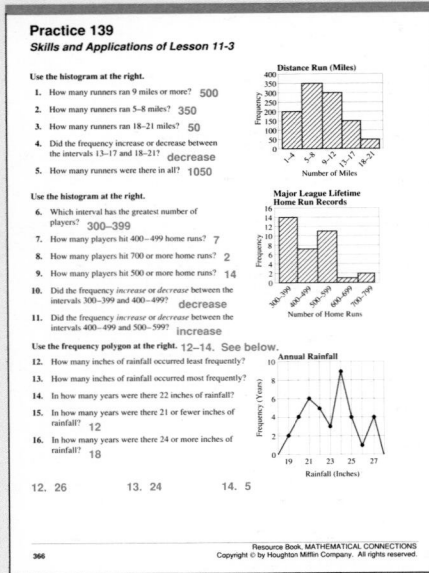

PROBLEM SOLVING/APPLICATION

There are specific ranges of test scores associated with each letter grade. For example, a grade of B is sometimes given for any score in the interval 80–89. For this reason, histograms are often used to display test scores.

B **13.** Make a histogram for the data below using intervals such as 60–69.
See answer on p. A29.

Scores on a Mathematics Test

87	66	74	74	88	97	89	78	68	84	75	85	77
69	93	65	73	76	83	85	79	80	96	84	88	

Use the histogram that you made for Exercise 13 for Exercises 14–17.

14. What letter grade corresponds to each interval in the histogram?

14. 90–99 A
80–89 B
70–79 C
60–69 D

15. How many students received a grade of C on the test? 8

16. What percent of the students earned a grade of B or better on the test? 52%

17. What was the mode of the letter grades on the test? B

GROUP ACTIVITY

Use the data below for Exercises 18 and 19.
Answers to Exercises 18–21 are on p. A29.

Hourly Wages of Part-time Baby Sitters (Dollars)

1.00	1.50	2.25	0.75	0.75	1.50	2.00	2.50	1.75	1.25	0.75
0.75	3.00	2.50	1.75	1.50	1.00	2.50	2.75	2.00	0.75	1.50

18. Make a frequency polygon without intervals for the data.

19. Make a frequency polygon for the data using intervals such as 0.75–1.24.

20. Which of the graphs that you drew for Exercises 18 and 19 displays the individual data items? Explain.

C **21.** Which of the graphs that you drew for Exercises 18 and 19 gives a better picture of the overall trend in the hourly wages of part-time baby sitters? Explain.

SPIRAL REVIEW

S **22.** Use the frequency polygon on page 503. How many band members are 63 in. tall? *(Lesson 11-3)* 3

23. Write $\frac{13}{20}$ as a decimal. *(Lesson 7-7)* 0.65

24. Solve: $\frac{3}{4}x = 18$ *(Lesson 8-9)* 24

25. What number is 4.5% of 2000? *(Lesson 9-6)* 90

26. Find the mean, median, mode(s), and range:
7.9, 7.3, 6.5, 8.4, 7.3, 6.4, 5.7, 6.9 *(Lesson 5-8)* 7.05; 7.1; 7.3; 2.7

11-4 Box-and-Whisker Plots

Objective: To make and interpret box-and-whisker plots.

Lesson Planner

Teaching the Lesson
- Materials: string, paper
- Lesson Starter 104
- Visuals, Folder J
- Using Technology, p. 492C
- Using Manipulatives, p. 492C
- Toolbox Skills 5, 7, 10

Lesson Follow-Up
- Practice 140
- Practice 141: Nonroutine
- Enrichment 114: Data Analysis
- Study Guide, pp. 207–208
- Test 53

EXPLORATION

1 What is the median of the data below?

3   4   10   11   11   14   20   23   23   23   25   27   17

2 What is the median of the items in the above list that are less than the median in Step 1? greater than the median in Step 1?   10.5; 23

3 Into how many sections do the medians you found in Steps 1 and 2 divide the data?   4

The median divides a set of data into two parts. The **first quartile** is the median of the lower part. The **third quartile** is the median of the upper part. You can display these measures of a set of data along with the least and greatest values of the data in a **box-and-whisker plot.**

Terms to Know
- *first quartile*
- *third quartile*
- *box-and-whisker plot*

Example 1

Make a box-and-whisker plot for the data.

Gasoline Mileage (Miles per Gallon)

24   20   18   25   22   31   30   20   28   29   35   24   38

Solution

Arrange the data in order from least to greatest. Find the median, first quartile, third quartile, least value, and greatest value.

18   20   20   22   24   24   25   28   29   30   31   35   38

| ↑ | ↑ | ↑ | ↑ | ↑ |
| least value | first quartile | median | third quartile | greatest value |

$$\frac{20 + 22}{2} = 21 \qquad\qquad \frac{30 + 31}{2} = 30.5$$

Display the five numbers as points below a number line.

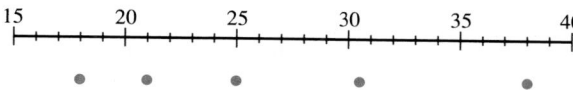

Draw a box with ends at the quartiles. Draw a vertical line through the box at the median. Draw *whiskers* from the ends of the box to the points representing the least and greatest values. Title the graph.

Gasoline Mileage (Miles per Gallon)

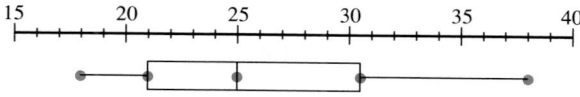

Warm-Up Exercises

1. Explain how to find the median when there are an even number of data items.   Arrange the items in order from least to greatest and then average the middle two numbers.

Use the data given below for Exercises 2–5.

24   35   37   56   71   12   52

2. Find the median.   37
3. Find the range.   59
4. Find the mode(s).   no mode
5. Find the mean.   41

Lesson Focus

Ask students to imagine data tables or displays that include extreme values, that is, values far from the median or mean. These extreme values may distort the measures of central tendency. Today's lesson focuses on making and interpreting a box-and-whisker plot, a visual display of data that shows both measures of central tendency and extreme values.

Statistics and Circle Graphs   **505**

506

Teaching Notes

Box-and-whisker plots may be new to students. Work through the Examples in detail, carefully explaining how to create and interpret this type of data display.

Key Question

When finding the first and third quartiles, do you include the median? No, the quartiles involve only numbers above and below the median.

Using Technology

For a suggestion on using a spreadsheet program or a statistics program, see page 492C.

Using Manipulatives

For a suggestion on using students in the class and string to create a box-and-whisker plot, see page 492C.

Error Analysis

When finding the first and third quartiles, students may incorrectly include the median. Remind students that the first and third quartiles are located by finding the median of the upper or lower part of the data, exclusive of the median value.

Additional Examples

1. Make a box-and-whisker plot for the data.

Temperature Readings in California (°F)

86	87	96	72	66
73	67	83	85	62
92	72	95	97	83

Temperature Readings in California (°F)

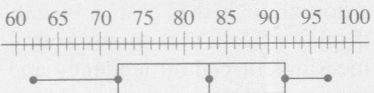

Box-and-whisker plots are often used to compare sets of data.

Example 2

Use the box-and-whisker plot below.

Yearly Stock Summary (Dollars per Share)

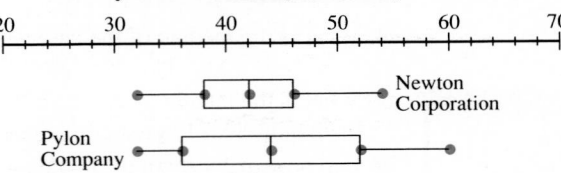

a. Find the median, the first quartile, the third quartile, and the lowest and highest price per share for Newton stock.

b. Which stock had the greater range in price during the year?

Solution

a. Locate the box-and-whisker for Newton Corporation.
The vertical line through the box is at the median, $42.
The left end of the box is at the first quartile, $38.
The right end of the box is at the third quartile, $46.
The lowest price per share is $32.
The highest price per share is $54.

b. The overall length of the box-and-whisker plot is greater for the Pylon Company than for the Newton Corporation. Pylon Company stock had the greater range in price during the year.

✎ **Check Your Understanding**

Describe a different way to find the answer to Example 2(b).
See *Answers to Check Your Understanding* at the back of the book.

Guided Practice

« **1. COMMUNICATION** « *Reading* In statistics, the suffix *-ile* means "a division of a specified size."
a. What is the meaning of the word *quartile*? a division into 4 parts
b. What is the meaning of the word *percentile*? a division into 100 parts

2. No; when finding the first quartile you count only the items below the median. When finding the third quartile you count only the items above the median.

« **2. COMMUNICATION** « *Discussion* Is the median counted when you find the first and third quartiles of a set of data? Explain.

Identify the median and the first and third quartiles for each set of data.

3. 12.4, 12.6, 12.8, 13.0, 13.4, 13.7, 14.0, 14.6, 14.8 13.4; 12.7; 14.3

4. 127, 130, 136, 138, 142, 150, 155, 159, 166, 168 146; 136; 159

5. Draw a box-and-whisker plot using the data in Exercise 3.
See answer on p. A29.

506 Chapter 11

Use the box-and-whisker plot on page 506.

Newton Corp.

6. Which stock had the lesser median price per share for the year?

7. Which stock had the greatest price per share for the year? Pylon Co.

Exercises

Make a box-and-whisker plot for each set of data. Answers to Exercises 1–2 are on p. A29.

A **1. Box Seat Prices at Baseball Games (Dollars)**

| 12.50 | 12.00 | 13.25 | 11.00 | 11.50 | 11.25 | 14.00 | 9.00 |
| 11.75 | 13.50 | 14.50 | | | | | |

2. Prices for 13-inch Color Televisions (Dollars)

| 229 | 299 | 349 | 215 | 250 | 375 | 399 | 415 | 400 | 240 |
| 269 | 329 | 345 | 379 | 411 | | | | | |

Use the box-and-whisker plot below for Exercises 3–8.

Scores on a Chemistry Test

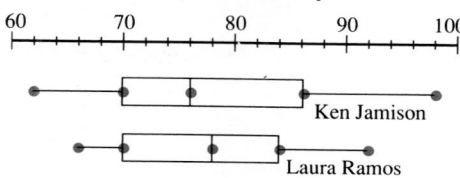

3. Find the median, the first quartile, the third quartile, and the lowest and highest scores for Ken Jamison's class. 76; 70; 86; 62; 98

4. Find the median, the first quartile, the third quartile, and the lowest and highest scores for Laura Ramos's class. 78; 70; 84; 66; 92

5. Ken Jamison's
6. Ken Jamison's
7. Ken Jamison's
8. Laura Ramos's

5. Which teacher's class had the student with the greatest score?

6. Which teacher's class had the student with the least score?

7. Which teacher's class had the greater range of scores on the test?

8. Which teacher's class had the greater median score on the test?

LOGICAL REASONING

Tell whether each statement is *always*, *sometimes*, or *never* true.

B **9.** A box-and-whisker plot displays at least two numbers from a set of data. always

10. When the number of items in a set of data is odd, the median is one of the data items. always

11. Exactly half of the items in a set of data are less than the median. sometimes

2. Use the box-and-whisker plot in Example 2 for the following.

a. Find the median, first quartile, third quartile, and the lowest and highest price per share for Pylon Company. median = $44; first quartile = $36; third quartile = $52; lowest price = $32; highest price = $60

b. Which stock had the lower median price during the year? Newton Corporation

Closing the Lesson

Give students data that represent scores the class achieved on a recent test. Have students organize the data and draw a box-and-whisker plot.

Suggested Assignments

Skills: 1–8, 13–16
Average: 1–11, 13–16
Advanced: 1–16

Exercise Notes

Communication: Discussion

Guided Practice Exercise 2 requires students to demonstrate their understanding of how to find the first and third quartiles of a set of data.

Reasoning

Exercises 9–11 require students to determine whether a statement is always, sometimes, or never true. Students may use examples or counterexamples to verify their thinking.

Application

Exercise 12 has students use information obtained from a box-and-whisker plot to write a paragraph about attendance at plays and musicals. Students also may wish to research how ticket prices affect attendance at an event.

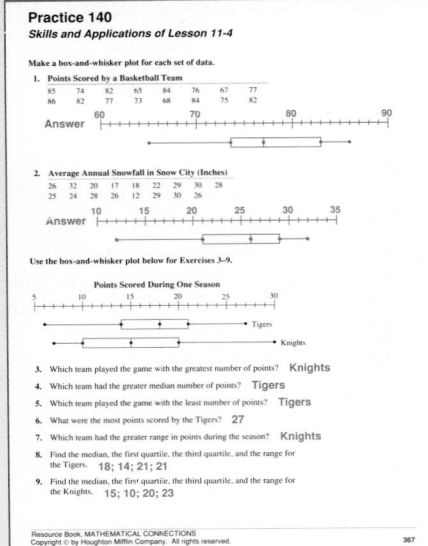

12. WRITING ABOUT MATHEMATICS The box-and-whisker plot below shows data about theater attendance. Use the box-and-whisker plot to write a paragraph describing the relationship between the attendance at plays and musicals. See answer on p. A29.

Theater Attendance

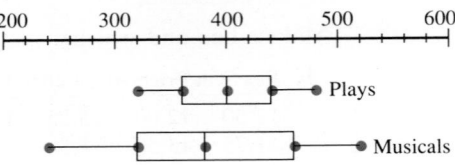

SPIRAL REVIEW

S **13.** Graph the inequality $x \geq -3$. *(Lesson 7-10)* See answer on p. A29.

14. Find the sum of the measures of the angles of a pentagon. *(Lesson 6-5)* 540°

15. Make a box-and-whisker plot to display the following data. *(Lesson 11-4)* See answer on p. A29.

Number of Field Goals Scored

12	20	19	15	24	26
27	23	21	18	14	29

16. Find the quotient: $3\frac{1}{3} \div \left(-\frac{2}{5}\right)$ *(Lesson 8-2)* $-8\frac{1}{3}$

Self-Test 1

Answers to Self-Test 1–2 are on p. A29.

1. Make a frequency table for the data. Use intervals such as 60–69. **11-1**

Scores on an Algebra Test

61	87	69	74	92	77	84	91	85	68
77	83	76	99	77	88	71	92	83	76

2. Make a stem-and-leaf plot for the data in Exercise 1. **11-2**

3. Use the histogram on page 503. How many students ran 17 laps or more? 18 **11-3**

4. Use the frequency polygon on page 503. How many band members are 63–66 in. tall? 22

5. Use the box-and-whisker plot on page 507. Which teacher's class had the greater third quartile score? Ken Jamison's **11-4**

508   Chapter 11

Scattergrams

Objective: To read and interpret scattergrams.

Lesson Planner

> **Teaching the Lesson**
> - Lesson Starter 105
> - Using Technology, p. 492C
> - Toolbox Skill 25
>
> **Lesson Follow-Up**
> - Practice 142
> - Enrichment 115: Thinking Skills
> - Study Guide, pp. 209–210
> - Computer Activity 11

Terms to Know
- *scattergram*
- *trend line*
- *positive correlation*
- *negative correlation*

DATA ANALYSIS

People who analyze data sometimes use a **scattergram** such as the one at the right to display the relationship between two sets of data. In a scattergram, the data are represented by points, but the points are not connected like those on a line graph. Furthermore, a scattergram may have more than one point for a given number on either scale.

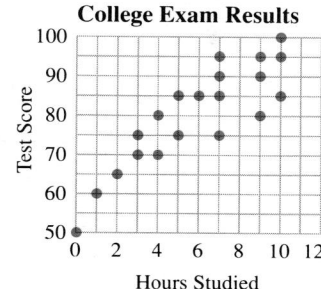

College Exam Results

Example 1

Use the scattergram above.

a. What was the test score of the student who studied for 6 h?

b. How many students studied for 7 h?

c. Find the mode(s) of the test scores.

Solution

a. Locate 6 on the horizontal axis. Move up to the red point, then left to the vertical axis. The student's score was 85.

b. Locate 7 on the horizontal axis. Move up and count the points. A total of 4 students studied for 7 h.

c. Locate the test score with the greatest number of points beside it. The mode of the test scores is 85.

 **Check Your Understanding**

1. In Example 1(b), where on the horizontal axis does the vertical line corresponding to *7 h studied* begin?

2. How would Example 1(c) have been different if you had been asked to find the range of the test scores?
See *Answers to Check Your Understanding* at the back of the book.

The points on a scattergram usually do not lie on a line. However, on some scattergrams the points lie *near* a line. A line that can be drawn near the points on a scattergram is called the **trend line.** It can be used to make predictions about the data.

If the trend line slopes upward to the right, there is a **positive correlation** between the sets of data. If the trend line slopes downward to the right, there is a **negative correlation.**

Warm-Up Exercises

Use the data below.

Number of Sit-Ups in PE Class

20	36	54	67	33
36	50	23	25	34
45	33	21	40	31
36	28	29	37	40

1. Make a frequency table for the data. Use intervals such as 20–29.

Number of Sit-Ups in PE Class

Sit-Ups	Tally	Frequency
20–29	ⅢⅢ I	6
30–39	ⅢⅢ III	8
40–49	III	3
50–59	II	2
60–69	I	1

2. Plot the data using a stem-and-leaf plot.

Number of Sit-Ups in PE Class

2	0 1 3 5 8 9
3	1 3 3 4 6 6 6 7
4	0 0 5
5	0 4
6	7

3. Find the median.   35

4. Find the mode(s).   36

5. Find the range.   47

Lesson Focus

Census takers collect a great amount of data. This data may be displayed by plotting a collection of scatter points, not connected by a line. When the scatter points are analyzed, they may show a trend. Today's lesson focuses on reading and interpreting information on scattergrams.

Teaching Notes

Key Question

Do all scattergrams have a trend line? No; on some scattergrams the data do not show any correlation and therefore no trend exists.

Reasoning

Example 2 shows students how to use a trend line to make a prediction about a data item based on another data item.

Using Technology

For a suggestion on using statistics software, see page 492C.

Additional Examples

For the following, use the scattergram above Example 1 in the text.

1. What was the test score of the student who studied for 2 h? 65
2. How many students studied for 9 h? 3 students
3. Find the median test score. 82.5
4. Is there a positive or negative correlation between the hours studied and the test scores? positive

For the following, use the scattergram to the right of Example 2.

5. Predict the test score for a student who studied 11 h. about 93

Example 2

Use the scattergram at the right.

a. Is there a *positive* or *negative* correlation between the hours studied and the test scores?

b. Predict the test score for a student who studies for 8 h.

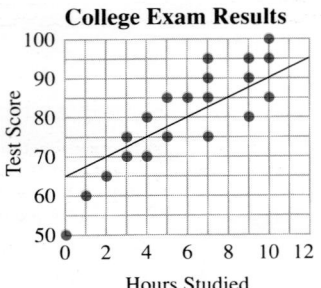

College Exam Results

Solution

a. The trend line slopes upward to the right. As the number of hours studied increases, the test scores tend to increase. There is a positive correlation between the hours studied and the test scores.

b. Locate 8 on the horizontal axis. Move up to the trend line, then left to the vertical axis. The predicted score is about 85.

Guided Practice

COMMUNICATION « *Reading*

Refer to the text on pages 509–510.

« 1. What is the main idea of the lesson? Scattergrams show the relationship between two sets of data.

« 2. List two major points that support the main idea.

2. If the trend line slopes up, there is a positive correlation between the two sets of data. If the trend line slopes down, there is a negative correlation.

3. negative correlation

4. positive correlation

5. no correlation

Identify each scattergram as having a *positive correlation*, a *negative correlation*, or *no correlation*.

3. 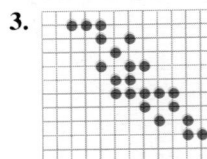   4. 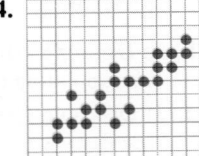   5.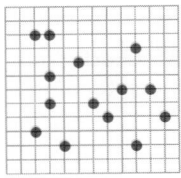

Use the scattergram in Example 2 above.

6. What was the test score of the student who studied for 2 h? 65

7. How many students had a score of 75? 3

8. Find the range of the number of hours studied. 10 h

9. Predict the test score for a student who studied for 12 h. 95

10. What was the test score of the student who studied for 1 h? 60

Exercises

Use the scattergram at the right.

A

1. How much does the woman who is 71 in. tall weigh? 165 lb

2. How tall is the woman who weighs 140 lb? 65 in.

3. How many women are 64 in. tall? 3

4. How many women weigh 155 lb? 2

5. Find the mode(s) of the heights. 64 in.

6. Find the range of the weights. 55 lb

7. Two women are 67 in. tall. How much does each woman weigh? 145, 155

8. As the heights increase, do the weights tend to *increase* or *decrease*? increase

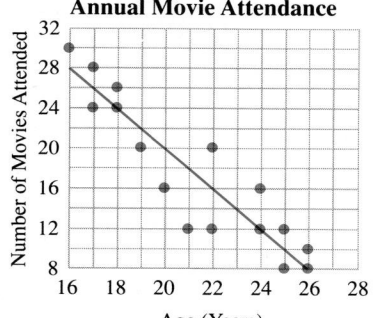

Women's Heights and Weights

Weight in Pounds / Height in Inches

Use the scattergram at the right.

9. Is there a *positive* or a *negative* correlation between the number of movies attended and a person's age? negative

10. Who will probably attend more movies in a year, a 15-year-old or a 25-year-old? 15-year-old

11. Predict the number of movies a 23-year-old will attend in a year. 14

12. Predict the age of a person who attends 18 movies per year. 21

Annual Movie Attendance

Number of Movies Attended / Age (Years)

B

13. Draw a scattergram to display the data. See answer on p. A29.

Jump Shots Made in 25 Attempts from Various Distances

Distance from Basket (ft)	25	15	10	20	15	5	10	5	25
Jump Shots Made	6	10	15	7	9	18	12	20	5

14. Use the scattergram you made in Exercise 13. Tell whether there is a *positive correlation*, a *negative correlation*, or *no correlation* between the two sets of data. negative

DATA, pages 492–493

15. How many trees have branches that spread 32 ft? 3

16. What is the spread of the branches of the tree that is 60 ft tall? 52 ft

Statistics and Circle Graphs **511**

Research

Exercises 21 and 22 reflect the type of approach that a scientist might use. Data is collected, plotted, and studied to determine if any relationships exist.

Follow-Up

Extension *(Cooperative Learning)*

Have students work in cooperative groups to collect data that might be displayed in a scattergram. For example, data on hours studied per week and grade point averages for a semester might be the variables considered in designing the scattergram. Groups may wish to predict the correlation between the number of hours of television viewed each week and grades. Each group should present its scattergram and conclusions to the class.

Enrichment

Enrichment 115 is pictured on page 492E.

Practice

Practice 142, *Resource Book*

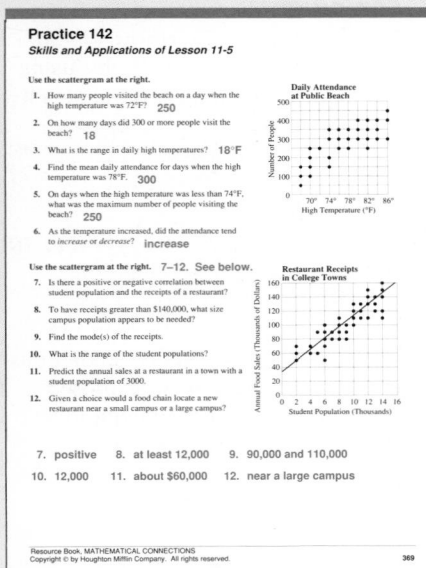

  COMPUTER APPLICATION

Car Registrations and Passenger Train Cars over Ten Years

Cars (Millions)	117	119	122	123	124	126	128	132	135	137
Train Cars	4493	4241	4347	3945	3736	2610	2580	2502	2307	2350

Answers to Exercises 17–18 are on p. A29.

C

17. Using statistical graphing software or graph paper, draw a scattergram displaying the data in the table above.

18. Draw the trend line for the scattergram you drew in Exercise 17.

19. Use the trend line that you drew in Exercise 18. Is there a *positive* or a *negative* correlation between the number of passenger car registrations and the number of passenger train cars? negative

20. Use the trend line that you drew in Exercise 18. If the number of passenger cars registered continues to increase over time, what will probably happen to the number of passenger train cars?
It will continue to decrease.

RESEARCH

21. Use an almanac to find the area and the average depth of the world's ten largest bodies of salt water besides the four oceans. Draw a scattergram showing the relationship between these two sets of data.
See answer on p. A29.
22. Use the scattergram that you drew in Exercise 21. Tell whether there is a *positive correlation*, a *negative correlation*, or *no correlation* between the two sets of data. positive correlation

SPIRAL REVIEW

S 23. Find the mean, median, mode(s), and range of the data:
12, 14, 16, 18, 23, 26, 19, 10, 12, 28, 15, 12   *(Lesson 5-8)*
17.1; 15.5; 12; 18
24. Make a stem-and-leaf plot for the data.   *(Lesson 11-2)*
See answer on p. A29.
Ages of Chorus Members

26	27	28	34	32	36	41	45	48	49	25	29
19	17	16	19	20							

25. Solve: $2x - 2.4 = 3.8$ *(Lesson 8-8)* 3.1

26. Use the scattergram on page 511 entitled *Annual Movie Attendance*. How many movies did the 19-year-old attend?   *(Lesson 11-5)* 20

27. What number is 5% of 720?   *(Lesson 9-6)* 36

28. Solve by making a table: Michael asks the attendant at Village Laundry to give him change for a quarter. In how many different ways could the attendant give Michael change using only dimes and nickels? *(Lesson 3-7)* 3 ways

11-6 Interpreting Circle Graphs

Objective: To read and interpret circle graphs.

APPLICATION

The athletic director at State University analyzed how the annual football budget was spent. The results were presented to the budget director in a *circle graph* like the one at the right.

A **circle graph** is used to represent data expressed as parts of a whole. The entire circle represents the whole, or 100%, of the data. Each wedge or **sector** represents a part of the data.

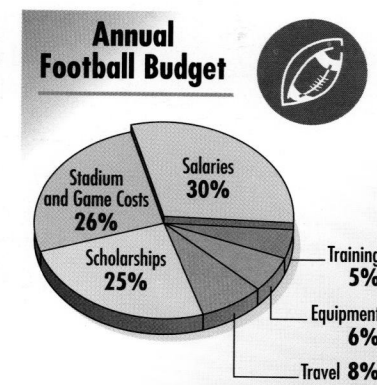

Annual Football Budget

- Stadium and Game Costs 26%
- Salaries 30%
- Training 5%
- Equipment 6%
- Travel 8%
- Scholarships 25%

Terms to Know
- circle graph
- sector

Example 1

Use the circle graph above.

a. If the total budget was $3,500,000, how much was spent on scholarships?

b. If $1,500,000 was spent on salaries, how much was the total budget?

Solution

a. 25% of the budget was spent on scholarships.

Let n = the amount spent on scholarships.

n is 25% of $3,500,000. $\Rightarrow$ $n = 0.25(3,500,000)$
$$= 875,000$$

The amount spent on scholarships was $875,000.

b. 30% of the budget was spent on salaries.

Let n = the total budget.

30% of n is $1,500,000. $\Rightarrow$ $0.3n = 1,500,000$
$$n = 5,000,000$$

The total budget was $5,000,000.

 **Check Your Understanding**

1. In Example 1(b), what had to be done in order for the equation $0.3n = 1,500,000$ to be rewritten as $n = 5,000,000$?
See *Answers to Check Your Understanding* at the back of the book.

The sectors in a circle graph are sometimes labeled with the data items themselves, rather than with percents.

Statistics and Circle Graphs **513**

Lesson Planner

Teaching the Lesson
- Lesson Starter 106
- Toolbox Skills 9, 10, 13, 14

Lesson Follow-Up
- Practice 143
- Enrichment 116: Communication
- Study Guide, pp. 211–212

Warm-Up Exercises

Use the data below.

Number of Minutes Studied for Exams

25	37	72	43	56
38	62	56	44	29
67	55	41	39	33

1. Find the median. 43
2. Find the mode(s). 56
3. Display the data using a frequency table. Use intervals such as 20–29.

Number of Minutes Studied for Exams

Minutes	Tally	Frequency
20–29	||	2
30–39	||||	4
40–49	|||	3
50–59	|||	3
60–69	||	2
70–79	|	1

4. Display the data using a stem-and-leaf plot.

Number of Minutes Studied for Exams

2	5 9
3	3 7 8 9
4	1 3 4
5	5 6 6
6	2 7
7	2

513

5. Display the data using a box-and-whisker plot.

Number of Minutes Studied for Exams

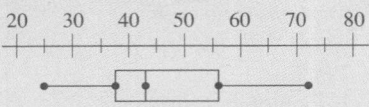

Lesson Focus

Ask students to picture a whole pizza. Each slice of the pizza represents a part of the whole. In mathematics, circle graphs often are used to represent a whole. Today's lesson focuses on interpreting circle graphs.

Teaching Notes

Most students have probably seen circle graphs before. Work through the Examples, pointing out the difference between Example 1, where the sectors represent percents, and Example 2, where the sectors represent actual data.

Key Question

What does an entire circle graph represent? An entire circle graph represents 100% of the data.

Error Analysis

In several of the Exercises, such as Exercise 10, the word *not* is used. Students may overlook this negation and find the percent spent on a particular category, rather than the percent omitted. Remind students to read a problem carefully before beginning to solve it.

Example 2

Use the circle graph at the right.

a. What was the total budget?

b. What percent of the budget was spent on utilities?

c. What percent of the budget was not spent on salaries?

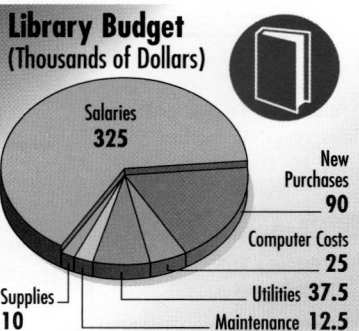

Library Budget (Thousands of Dollars)

Salaries 325
New Purchases 90
Computer Costs 25
Utilities 37.5
Maintenance 12.5
Supplies 10

Solution

a. Add the amounts budgeted for all the expenses.
$325 + 10 + 12.5 + 37.5 + 25 + 90 = 500$
The total budget was $500,000.

b. The amount spent on utilities was $37,500. The total budget was $500,000. Let n = the percent of the budget spent on utilities.

n is $37,500 out of $500,000. $\Rightarrow$ $n = \dfrac{37,500}{500,000}$
$= 0.075$

7.5% of the budget was spent on utilities.

c. Subtract the amount spent on salaries from the total budget.
$500,000 - 325,000 = 175,000$ ←— amount not spent on salaries
Let n = the percent of the budget not spent on salaries.

n is 175,000 out of 500,000. $\Rightarrow$ $n = \dfrac{175,000}{500,000}$
$= 0.35$

35% of the total budget was not spent on salaries.

 **Check Your Understanding**

2. In Example 2(a), why is the total budget $500,000 instead of $500?
See *Answers to Check Your Understanding* at the back of the book.

Guided Practice

COMMUNICATION «*Reading*

Refer to the text on pages 513–514.

«**1.** What percent is represented by the entire circle in a circle graph? 100%

«**2.** What are two ways that you can label the sectors in a circle graph?
percent or data items

Find each answer.

3. What number is 24% of 4500?
1080

4. 30% of what number is 18?
60

Use the circle graph on page 513.

5. If the total budget was $4,000,000, how much was spent on travel?
$320,000

6. If $228,000 was spent on equipment, how much was spent on stadium and game costs? $988,000

Use the circle graph on page 514.

7. What percent of the budget was spent on new purchases? 18%

8. What percent of the budget was spent on maintenance? 2.5%

Exercises

Use the circle graph at the right.

A **1.** What percent of the events were not concerts? 70%

2. If there was a total of 300 events at the stadium, how many were basketball events? 36

3. If there was a total of 240 events at the stadium, how many were baseball events? 108

4. If there were 36 concerts at the stadium, how many events were there in all? 120

5. If there were 12 boxing and wrestling events at the stadium, how many events were there in all? 400

6. If there were 11 football events at the stadium, how many baseball events were there? 99

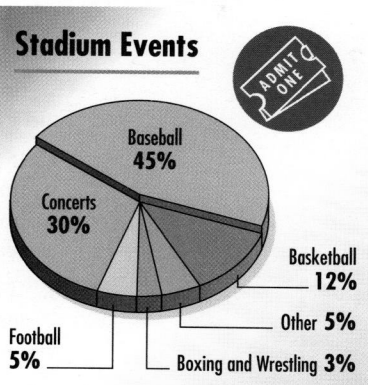

Stadium Events

Baseball 45%
Concerts 30%
Basketball 12%
Other 5%
Boxing and Wrestling 3%
Football 5%

Use the circle graph at the right.

7. What is the total budget for a month? $3750

8. What percent of the budget is spent on transportation? 20%

9. What percent of the budget is saved? 2%

10. What percent of the budget is not spent on housing? 69%

11. What percent of the budget is not spent on clothing? 94%

12. What percent of the budget is spent on food and other expenses? 41%

Monthly Family Budget

Housing $1162.50
Food $712.50
Clothing $225
Transportation $750
Other Expenses $825
Savings $75

ESTIMATION Estimates may vary. Accept reasonable estimates.

B **13.** Estimate the fraction of the monthly budget spent on housing. about $\frac{1}{3}$

14. Estimate the fraction of the monthly budget that is used for savings and other expenses. about $\frac{1}{4}$

Statistics and Circle Graphs **515**

Exercise Notes *(continued)*

Mental Math

Exercises 19–22 can be solved mentally since the percents in the graph are all multiples of 10.

Follow-Up

Nonroutine Problem

Have students examine the information presented in the Application at the beginning of the lesson and make a different type of graph, such as a bar graph, to present the same data. Students should write a paragraph on both the advantages and disadvantages of each type of data representation.

Extension *(Cooperative Learning)*

Using a circle graph from a magazine or newspaper, have students work in groups to write three to five questions to share with the class that can be answered by using the graph.

Enrichment

Enrichment 116 is pictured on page 492E.

Practice

Practice 143, Resource Book

---

15. What fraction of music buyers are 25–29 years old? $\frac{3}{20}$

16. What fraction of music buyers are less than 35 years old? $\frac{3}{4}$

17. What fraction of music buyers are 20–29 years old? $\frac{33}{100}$

18. Out of 3000 customers, how many customers would probably be 15–24 years old? 1260

MENTAL MATH

Use the circle graph at the right. Find each answer mentally.

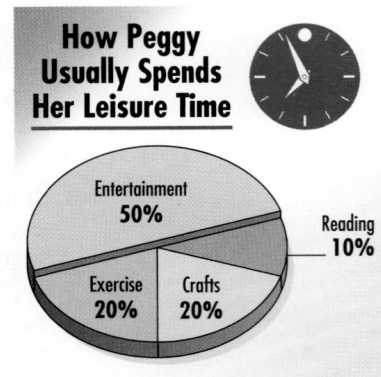

How Peggy Usually Spends Her Leisure Time

Entertainment 50%
Reading 10%
Exercise 20%
Crafts 20%

19. If Peggy had 90 h of leisure time last month, how many hours did she probably spend reading? 9 h

20. If Peggy used 10.5 h of her leisure time last month for entertainment, about how many hours of leisure time did she probably have? 21 h

21. About how many hours of leisure time did Peggy probably have last week if she spent 6 h exercising? 30 h

22. How many hours will Peggy probably spend on crafts if she has 15 h of leisure time? 3 h

SPIRAL REVIEW

S 23. Make a box-and-whisker plot for the data. *(Lesson 11-4)*
    See answer on p. A29.
    Test Scores

65	68	72	75
78	80	82	84
86	91	98	

24. Refer to the circle graph on page 513. If the total budget was $3,250,000, how much was spent on travel? *(Lesson 11-6)* $260,000

25. Find the area of a triangle with base 14 in. and height 9 in. *(Lesson 10-3)* 63 in.^{2}

26. What percent of 30 is 25? *(Lesson 9-7)* $83\frac{1}{3}$%

Constructing Circle Graphs

Objective: To construct circle graphs to display data.

When you construct a circle graph, you divide the circle into sectors. Each sector represents a percent of the data.

The sum of the angles formed by the sectors of a circle is 360°. These angles are called **central angles** because they each have a vertex at the center of the circle. Angle *AOB* in the diagram at the right is a central angle.

Example

Draw a circle graph to display the data.

Sales of Compact Discs

Type	rock	country	jazz	classical
Number Sold	240	180	85	35

Solution

• Find the percent of the total number sold made up by each type of music. Write each percent as a decimal. Then multiply each decimal by 360° to find the number of degrees for each sector.

Find the total number sold: $240 + 180 + 85 + 35 = 540$

rock: $\frac{240}{540} \approx 44.4\%$ $0.444(360°) \approx 160°$

country: $\frac{180}{540} \approx 33.3\%$ $0.333(360°) \approx 120°$

jazz: $\frac{85}{540} \approx 15.7\%$ $0.157(360°) \approx 57°$

classical: $\frac{35}{540} \approx 6.5\%$ $0.065(360°) \approx 23°$

• Use a template or a compass to draw a circle. Draw a radius. Then use a protractor to draw the central angle for each sector.

• Label each sector with the corresponding type of music and the number of compact discs sold.

• Include a title on the circle graph.

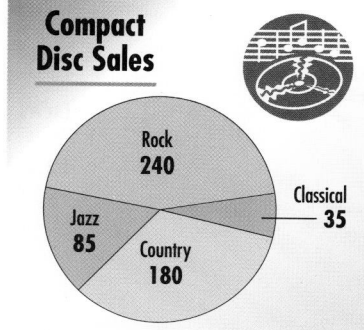

Compact Disc Sales

Rock 240 — Classical 35 — Jazz 85 — Country 180

Check Your Understanding

In the Example, why must you find the total number sold?
See *Answers to Check Your Understanding* at the back of the book.

Warm-Up Exercises

Use the circle below.

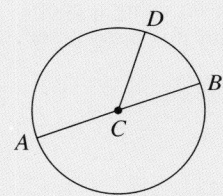

1. Name the center. *C*
2. Name the radii. $\overline{AC}, \overline{BC}, \overline{DC}$
3. Name the diameter. $\overline{AB}$
4. Name the angles.
   $\angle ACD, \angle BCD, \angle ACB$

Lesson Focus

Ask students to picture a pizza cut into slices. Each slice is a sector of the whole pizza. If the slices are taken out and then put back next to one another, the pizza can be reconstructed. Today's lesson focuses on the construction of circle graphs.

Teaching Notes

Work through the Example with students. You may wish to use a board compass to construct the circle graph, thus showing students step by step how the construction is done.

Key Question

How does the measure of a central angle relate to the size of the sector representing the data? The measure of the central angle represents the same percent of 360° as the sector represents of the entire data.

Using Technology

For a suggestion on using a spreadsheet program or a statistics program, see page 492C.

Using Manipulatives

For a suggestion on using manipulatives to create circle graphs, see page 492C.

Reteaching/Alternate Approach

For a suggestion on using a cooperative activity with this lesson, see page 492D.

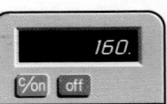

You can use a calculator to find the number of degrees in each sector when you are making a circle graph. For example, this key sequence would be used to find the number of degrees for the rock sector in the Example on page 517.

| 240. | ÷ | 540. | × | 360. | = |

Guided Practice

COMMUNICATION « *Reading*

« **1.** Explain the meaning of the word *sectors* in the following sentence.
At the end of World War II, Berlin was divided into four sectors. parts

« **2.** Write a sentence using the word *sector* in a mathematical context.
Answers will vary. Example: A circle graph may contain many sectors.

Write each fraction as a percent.

3. $\frac{450}{2000}$ 22.5%   **4.** $\frac{75}{375}$ 20%   **5.** $\frac{300}{900}$ 33$\frac{1}{3}$%   **6.** $\frac{30}{513}$ 5.8%

Write each percent as a decimal.

7. 18% 0.18   **8.** 79% 0.79   **9.** 8.5% 0.085   **10.** 39.2% 0.392

11. Copy and complete the chart below as if you were going to draw a circle graph to display the data. Do not draw the graph.

The Movies People Rent

		Type of Movie	Percent	Measure of Central Angle for the Sector
72°	**a.**	comedy	20%	?
144°	**b.**	mystery	40%	?
43°	**c.**	musical	12%	?
90°	**d.**	action/adventure	25%	?
3%; 11°	**e.**	other	?	?

Draw a circle graph to display each set of data. Answers to Guided Practice 12–13 are on p. A30.

12. Protein Source Preferences

Source	pork	chicken	fish	beef	other
Percent	10%	40%	15%	24%	11%

13. Election Results

Candidate	Tyrall	Little	Clemmons	Evans
Number of Votes	1,050,000	1,260,000	510,000	180,000

518   Chapter 11

Exercises

Draw a circle graph to display each set of data. Answers to Exercises 1–4 are on p. A30.

1. Library Inventory

Type of Book	adult fiction	children's fiction	nonfiction	reference	other
Percent	35%	20%	25%	15%	5%

2. Annual Budget for a Four-Year Public University

Budget Item	room and board	tuition and fees	books	travel	other
Percent	41%	24%	11%	7%	17%

3. Cars Rented from Yoko's Rent-A-Car

Type	compact	mid size	full size	mini-vans
Number Rented	90	110	150	50

4. Survey of Favorite Chicken Restaurants

Restaurant	Chicken Hut	Millie's Chicken	Chicken-to-Go	Country Chicken
Number of Votes	50	48	42	60

 CALCULATOR

For Exercises 5–8, choose the letter of the calculator key sequence you would use to find the number of degrees for each sector described.

A. [700.] [÷] [1600.] [×] [360.] [=]

B. [550.] [÷] [2000.] [×] [360.] [=]

C. [700.] [÷] [2000.] [×] [360.] [=]

D. [2000.] [÷] [700.] [×] [360.] [=]

E. [550.] [÷] [1600.] [×] [360.] [=]

F. [360.] [÷] [550.] [×] [1600.] [=]

B

5. Out of a monthly budget of $2000, Kaya spends $700 on rent and household expenses. C

6. Out of a monthly budget of $1600, Faraj spends $550 on rent and household expenses. E

7. Out of a monthly budget of $2000, Anthony spends $550 on car expenses. B

8. Out of a monthly budget of $1600, Mary Kate spends $700 on car expenses. A

Additional Example

Draw a circle graph to display the data.

TV Viewing in Hartford (Thousands)

Type of Program	Number of Viewers
Movies	150
Drama	280
Comedy	95
Action	25

TV Viewing in Hartford (Thousands)

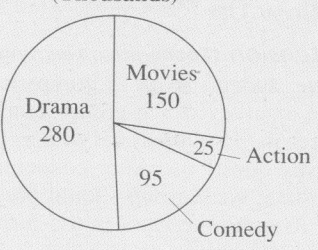

Closing the Lesson

Ask a student to explain how to find the measure of a central angle when constructing a circle graph.

Suggested Assignments

Skills: 1–8, 16–18
Average: 1–12, 16–18
Advanced: 1–7 odd, 9–18

Exercise Notes

Calculator

Exercises 5–8 require students to match the key sequence needed to determine the number of degrees in a sector to the description of the sector.

520

Exercise Notes (continued)

Application
Exercises 9–12 demonstrate the use of percents and circle graphs in a business operation. Many businesses use similar methods for monitoring receipts and expenses.

Making Connections/Transitions
Exercises 13–15 connect the concept of circle graph to the geometric concept of finding the area of a sector of a circle.

Follow-Up

Extension *(Cooperative Learning)*
Have students work in groups to take a survey. Each group should select its own topic and design the survey questions. After collecting the data, each group should present the items to the class in the form of a circle graph.

Enrichment
Enrichment 117 is pictured on page 492F.

Practice
Practice 144, *Resource Book*

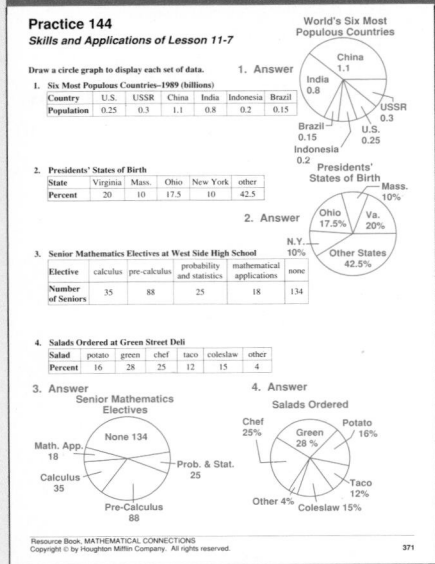

A *Hotel-Restaurant-Travel Administrator* manages hospitality services, such as hotels and motels, amusement parks, campgrounds, and food services. One aspect of an administrator's job is monitoring receipts. The table below shows the annual receipts of a mid-sized motel.

9. What percent of the total receipts came from room rentals? 57.4%

10. Which category contributed the lowest percent of the receipts? other   about $\frac{2}{5}$

11. Estimate the fraction of the total income that came from food and beverage sales.

12. Draw a circle graph to display the data.
See answer on p. A30.

Annual Receipts	
rooms:	$490,560
food:	$236,785
beverages:	$93,584
telephone:	$24,500
other:	$9,499
total:	$854,928

CONNECTING CIRCLE GRAPHS AND GEOMETRY

Finding the area of a sector is like finding the number of degrees in a central angle of a circle graph. You find the fraction of 360° made up by the central angle, and multiply this fraction by the total area. So the area of a sector with a central angle of $n°$ and radius r is $\frac{n}{360}\pi r^2$.

Find the area of each sector.

C **13.**   60°   O 3 cm

4.7 cm²

14.   135°   O 4 in.

18.8 in.²

15.   O   9 m   120°

84.8 m²

SPIRAL REVIEW

S **16.** Write $\frac{5}{8}$ as a percent.   *(Lesson 9-5)* 62.5%

17. Solve: $2(n + 6) = -14$ *(Lesson 4-9)* −13

18. Draw a circle graph to display the data.   *(Lesson 11-7)*
See answer on p. A30.

Commuting Methods of Acme Employees

Method	car	car pool	walk	train	bus
Percent	40%	7%	8%	25%	20%

Using Statistical Graphing Software

Objective: To use statistical graphing software to display and analyze data.

Most large companies have a department whose purpose is to analyze data about the company's performance. People working in such departments have to present summaries to managers in a form that can be quickly and easily understood. Graphs are often included in the summaries, and **statistical graphing software** is used to produce them.

Statistical graphing software can be very flexible. For example, the data below could be presented in either a bar or a circle graph. Graphs developed using statistical graphing software look very professional and can have a strong visual impact.

Personal Computer Sales

Region	Sales (Dollars)
Northeast	45,225,000
North-Central	33,675,000
Northwest	19,950,000
Southeast	31,475,000
South-Central	24,500,000
Southwest	25,175,000
Total	180,000,000

Exercises

Use statistical graphing software to create a circle graph displaying the data in the table above.

1. How many sectors does the circle graph have? 6

2. How are the sectors labeled? Answers will vary. Likely answers are by percent, region, or sales.

3. If each region had the same amount of sales, what fraction of the sales would this be? $\frac{1}{6}$

4. How many regions, if any, had less than 17% of the sales? 3

5. How many regions, if any, had at least 25% of the sales? 1

6. What percent of the sales did the southern regions have? 45%

7. What percent of the sales did the eastern regions have? 43%

8. Suppose that the regions are reorganized so that $10,225,000 in sales from the northeast region are counted as sales for the southeast region. How will this affect the sectors for these two regions? How will this affect the sectors for the other regions?
   See answer on p. A30.

9. What can you visualize in the circle graph of this set of data that you could not visualize in a bar graph of the data?
   the percent of total sales for each region

Statistics and Circle Graphs **521**

Teaching Notes

There are a variety of programs that incorporate graphing with information recorded in a spreadsheet.

You may wish to extend the concepts in this application by having students enter the regions and sales in a spreadsheet and then generate both a circle graph and a bar graph. Also, data could be manipulated so that a circle graph showing percents is generated.

Exercise Note

Exercise 8 could be explored by demonstrating how cells in a program can be combined and/or reconfigured to present data in a different fashion.

Lesson Planner

Teaching the Lesson
• Lesson Starter 108

Lesson Follow-Up
• Practice 145
• Enrichment 118: Exploration
• Study Guide, pp. 215–216

Warm-Up Exercise

List all the methods you know for displaying data.   bar graph, line graph, pictograph, histogram, frequency polygon, scattergram, stem-and-leaf plot, box-and-whisker plot, circle graph

Lesson Focus

Today's lesson focuses on choosing the most appropriate way to display data.

Teaching Notes

Key Question

Why is it necessary to decide on the most appropriate type of display? Different displays serve different functions.

Additional Example

Decide whether it would be more appropriate to draw a scattergram or a circle graph to display the data. Then make the display.   circle graph

Students in Major Courses

Course	Number of Students
English	150
Math	120
Science	90
History	140

DECISION MAKING

11-8

Choosing an Appropriate Data Display

Objective: To decide which method for displaying data is appropriate.

When you have data to display, you must decide which type of display is appropriate.

• A *histogram* or *frequency polygon* may be most appropriate when you want to visually compare the frequencies at which specific data items or groups of data items occur.

• A *circle graph* may be most appropriate when you want to visually compare parts of a set of data to the whole.

• A *scattergram* may be most appropriate when you want to display the correlation between two sets of data.

• A *stem-and-leaf plot* may be most appropriate when you want to display the individual data items in an ordered and concise manner.

• A *box-and-whisker plot* may be most appropriate when you want to display the median, first and third quartiles, and least and greatest values of a set of data, or compare these aspects of two sets of data.

Example

Decide whether it would be appropriate to draw a *scattergram* or a *histogram* to display the data. Then make the display.

Class Grade Point Averages

3.4	2.9	2.9	2.3	3.2	2.5	3.1	3.2	2.9	2.8
2.2	3.8	3.3	2.8	2.7	3.0	2.4	2.6	3.4	2.5
2.9	3.1	4.0	3.7	2.6					

Solution

Histogram. A scattergram is used to compare two sets of data. Only one set of data is given. For this reason, a scattergram is not appropriate.

A histogram is appropriate. It will show how the grade point averages for the class are clustered.

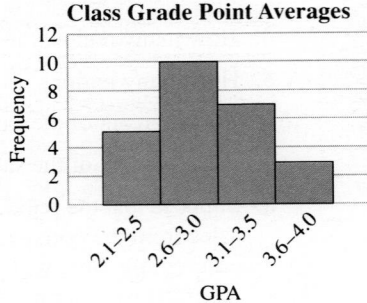

☑ Check Your Understanding

In the Example, would it be appropriate to draw a scattergram if you were also given the grade that each student received in English? Explain.
See *Answers to Check Your Understanding* at the back of the book.

Guided Practice

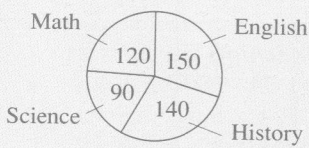
« **1. COMMUNICATION** « *Discussion*   If you are given a set of data in the form of a frequency table with intervals, which displays will automatically be inappropriate? scattergram; stem-and-leaf plot; box-and-whisker plot

2. Decide whether it would be appropriate to draw a *circle graph* or a *stem-and-leaf plot* to display the data. Then make the display.   circle graph; display is on p. A30.

Students in Activities after School

Activity	sports	job	tutoring	other
Students	50	75	34	16

Closing the Lesson

Ask students to explain when it is appropriate to use each type of display discussed in this lesson.

Suggested Assignments

Skills: 1–6
Average: 1–6
Advanced: 1–6

Exercise Note

Communication: Discussion
Guided Practice Exercise 1 shows that data presented in one form cannot always be displayed in a different form.

Follow-Up

Enrichment
Enrichment 118 is pictured on page 492F.

Exercises

Tell which of the two types of displays would be appropriate for each set of data. Then make the display. Displays for Exercises 1–4 and 6 are on p. A30.

A **1.** *box-and-whisker plot* or *circle graph* box-and-whisker plot

Scores on an English Test

100	78	76	78	68	80	89	85	97	75	70	100
98	78	95	84	83	58	93	84	80	90	100	78

2. *scattergram* or *stem-and-leaf plot* stem-and-leaf plot

Weekly Grocery Budgets (Dollars)

125	86	75	52	88	97	78	85	85	116
97	104	88	93	84	71	107	76	96	74

3. *histogram* or *circle graph* circle graph

Survey of the Number of Women in Certain Occupations

Occupation	professional	technical	service	other
Women	125	225	90	60

4. *scattergram* or *box-and-whisker plot* scattergram

Managers' Salaries at Thorn's Department Store

Experience (Years)	1	11	5	6	7	12	6	10
Salary (Thousands)	18	32	24	24	30	35	26	32

SPIRAL REVIEW

S **5.** 18 is 20% of what number?   *(Lesson 9-8)* 90

6. Decide whether it would be appropriate to draw a *circle graph* or a *stem-and-leaf plot* to display the data. Then make the display. *(Lesson 11-8)* stem-and-leaf plot

Price per Gallon of Gasoline (Dollars)

1.29	1.26	1.34	1.38	1.32	1.38	1.35
1.27	1.31	1.40	1.27	1.43	1.41	

Statistics and Circle Graphs   **523**

Practice

Practice 145, *Resource Book*

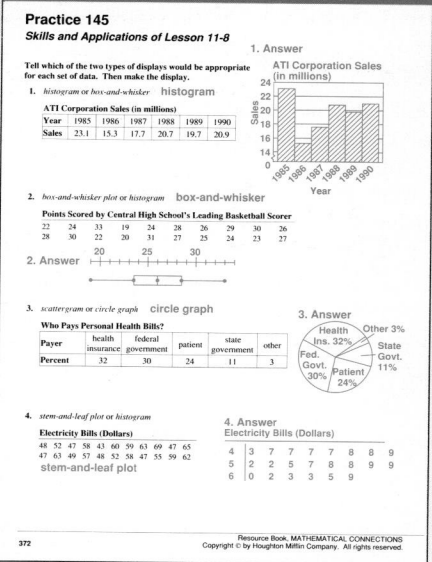

11-9

Strategy:
Using Logical Reasoning

Objective: To solve problems using logical reasoning.

Some problems require you to make conclusions based on the relationships in the facts of the problem. Using a table to organize the facts can help you to reason logically and find the answer to the problem.

Problem

Bob, Frank, and Tran play baseball, football, and tennis. No one plays a sport that begins with the same letter as his name. Tran and the baseball player are cousins. Who plays each sport?

Solution

UNDERSTAND The problem is about three people who play sports.
Facts: first letter of name is not first letter of sport
   the baseball player's cousin is Tran
Find: who plays each sport

PLAN Make a table to show all the possibilities. As you read each fact of the problem, put an × in a space to eliminate a possibility. Write "yes" to indicate a match.

WORK The first letter of the name does not match the first letter of the sport. So eliminate *baseball* from Bob's row, *football* from Frank's row, and *tennis* from Tran's row.
   Tran and the baseball player are cousins, so Tran does not play baseball. Eliminate *baseball* from Tran's row. Tran must play football.
   Because neither Bob nor Tran plays baseball, Frank must play baseball.

	Baseball	Football	Tennis
Bob	×		
Frank	yes	×	
Tran	×	yes	×

Frank plays baseball, so eliminate *tennis* from Frank's row. Tran plays football, so eliminate *football* from Bob's row. Bob must play tennis.

	Baseball	Football	Tennis
Bob	×	×	yes
Frank	yes	×	×
Tran	×	yes	×

ANSWER Bob plays tennis, Frank plays baseball, and Tran plays football.

Look Back What if the baseball player were Frank's cousin, instead of Tran's cousin? What sport would each person play? Bob plays football, Frank plays tennis, and Tran plays baseball.

Warm-Up Exercise

Solve by making a table.

Bill Chung bought three T-shirts for a total of $24. In how many different ways can he pay the $24 with bills of $10, $5, and $1?

8 ways

$10	$5	$1
2	0	4
1	2	4
1	0	14
0	4	4
0	3	9
0	2	14
0	1	19
0	0	24

Lesson Focus

Ask students if they were ever given clues to solve a puzzle. Some board games are based on clues. Today's lesson focuses on using logical reasoning to find "clues" when solving problems.

Guided Practice

COMMUNICATION «*Reading*

Use the problem below for Exercises 1–6.

Art Klein, Jim Pierce, and Sy Fischer teach art, gym, and science. No teacher teaches a subject that sounds like his first name. Art is not interested in athletics. Which teacher teaches each subject?

«**1.** Can Sy Fischer be the science teacher? Explain.
No; it sounds like his first name.

«**2.** Can Art Klein be the art teacher? Explain.
No; it sounds like his first name.

«**3.** What does the second sentence tell you about a subject Jim Pierce does not teach? Jim does not teach gym.

«**4.** What does the third sentence tell you about a subject Art Klein does not teach? Art does not teach gym.

5. Use your answers to Exercises 1–4 to complete a table of the problem. See answer on p. A30.

6. Use your table from Exercise 5 to write an answer to the problem.
Art Klein teaches science, Jim Pierce teaches art, and Sy Fischer teaches gym.

Solve using logical reasoning.

7. Sharon studies painting, Jason studies piano, and Alicia studies ballet.

7. Jason, Sharon, and Alicia study painting, piano, and ballet. The painter painted a portrait of Jason. Sharon and the dancer are sisters. Who studies each art form?

8. Botan, Carl, Adebayo, Dennis

8. Adebayo, Botan, Carl, and Dennis were the first four people to finish a marathon. Adebayo did not finish first or second. Botan did not finish second or third. Carl did not finish first or fourth. Dennis did not finish second. Adebayo finished before Dennis. In what order did the runners finish?

Problem Solving Situations

Solve using logical reasoning.

1. Ann Fernandez teaches typing, Ellen Taylor teaches mathematics, and Sonia Morris teaches French.

1. Ann Fernandez, Ellen Taylor, and Sonia Morris teach French, typing, and mathematics. No one teaches a subject that begins with the same letter as her last name. Ellen Taylor speaks only English. Who teaches each subject?

2. Carmen, Abdul, Ben

2. Abdul, Ben, and Carmen placed first, second, and third in a student council election. Abdul was not first, Carmen was not third, and Ben was not second. Carmen finished ahead of Ben. In what order did the students finish?

3. Marv is the treasurer, Shirley is the president, and Stan is the secretary.

3. Marv, Shirley, and Stan are the officers in a bowling league. The president and Stan are in-laws. Shirley and the treasurer are married. Marv and the secretary are brothers. Who serves in each position?

Communication: Reading

Guided Practice Exercises 1–6 take students step by step through the solution of a problem requiring logical reasoning and, in so doing, help to clarify the solution process.

Problem Solving

Problem Solving Situations 1–4 require students to use logical reasoning to solve each problem. *Problem Solving Situations 5–9* extend the types of situations to include problems that can be solved by using a variety of strategies.

Cooperative Learning

You may wish to have students do *Problem Solving Situations 1–4* in groups. Solving problems in groups encourages students to share ideas and insights.

Communication: Writing

Problem Solving Situations 10 and 11 have students write a problem that fits the conditions given in each of the tables.

Solve using logical reasoning.

4. In a civil case, Jeffries, Daniels, Powers, and Watkins are the judge, defendant, plaintiff, and witness. No one has a role that begins with the first letter of his or her last name. Only Powers went to law school. Jeffries, Watkins, and the plaintiff live in the same town. Who has each role? Jeffries is the witness, Daniels is the plaintiff, Powers is the judge, and Watkins is the defendant.

Solve using any problem solving strategy.

5–8. Strategies may vary; likely strategies are given.

5. Maura has four $5 bills, four $10 bills, and four $20 bills in a box. She takes two bills from the box at random and finds the total value of the bills. How many different totals are possible? making a table; 6

6. A typist can type 100 words in 2 min. How long would it take the typist to type 475 words? using proportions; 9.5 min

7. Aaron, Bianca, and Chris live in Atlanta, Boston, and Chicago. No one lives in a city that begins with the first letter of his or her name. Bianca has never been to the South. Who lives in Boston? using logical reasoning; Aaron

8. Marisa is $62\frac{1}{2}$ in. tall. Marty is $6\frac{1}{2}$ ft tall. Which person is taller? By how many inches? supplying missing facts; Marty is $15\frac{1}{2}$ in. taller.

9. Roger was charged $20.75 for some pumpkins. The price of a pumpkin was $4.95. How many pumpkins did Roger buy? no solution; the number of pumpkins must be a whole number.

> **PROBLEM SOLVING**
> ## CHECKLIST
> **Keep these in mind:**
> Using a Four-Step Plan
> Too Much or Not Enough Information
> Supplying Missing Facts
> Problems with No Solution
> **Consider these strategies:**
> Choosing the Correct Operation
> Making a Table
> Guess and Check
> Using Equations
> Identifying a Pattern
> Drawing a Diagram
> Using Proportions
> Using Logical Reasoning

WRITING WORD PROBLEMS Answers to Problems 10–11 are on p. A31.

Write a word problem that you could solve using logical reasoning. Then solve the problem.

10.

	Cat	Dog	Gerbil
Cindy	×		
Doug	×	×	
Grace			×

11.

	Blue	Brown	Green
Mary			×
Fran		×	
Jim	×		×

SPIRAL REVIEW

S **12.** Solve: $4 = -\frac{3}{5}n + 16$ *(Lesson 8-9)* 20

13. Torres, Wong, and Lipshaw are a doctor, a dentist, and a journalist. The dentist went to college with Torres. Wong and the journalist are neighbors. Lipshaw and Wong are both patients of the doctor. What is each person's profession?   *(Lesson 11-9)* Torres is the doctor, Wong is the dentist, and Lipshaw is the journalist.

14. The lengths of the legs of a right triangle are 7 cm and 24 cm. Find the length of the hypotenuse.   *(Lesson 10-11)* 25 cm

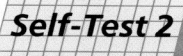
Self-Test 2

Use the scattergram at the top of page 511.

6. Harry majors in physics, Matt majors in history, and Phyllis majors in mathematics. **11-5**

1. How many women are 70 in. tall? 2

2. Is there a *positive* or a *negative* correlation between a woman's weight in pounds and height in inches? positive

3. Use the circle graph on page 513. If the total budget was $4,200,000, how much was spent on training? $210,000 **11-6**

4. Draw a circle graph to display the data. Displays for Self-Test 4–5 are on p. A31. **11-7**

City Budget (Millions of Dollars)

Budget Item	Schools	Police	Fire	Other
Cost	22	11	7	10

5. Decide whether it would be appropriate to draw a *circle graph* or a *scattergram* to display the data. Then make the display. **11-8**

Allowance Record (Dollars) scattergram

Savings	10	5	2	8	5	3	6
Entertainment	2	7	11	2	5	7	4

6. Harry, Matt, and Phyllis are majoring in history, mathematics, and physics. No one majors in a subject that begins with the same letter as his or her first name. Phyllis and the history major take English together. Who majors in each subject? **11-9**

Statistics and Circle Graphs **527**

Follow-Up

Exploration
Ask students to use a textbook from another subject to find data that have been displayed in a graph. Ask them to write a brief report that explains (1) what the graph represents, (2) whether the data represent a sampling or are complete, and (3) why the data were displayed as pictured in the textbook.

Enrichment
Enrichment 119 is pictured on page 492F.

Practice
Practice 146 is shown below.

Quick Quiz 2
See pages 536 and 537.

Alternative Assessment
See page 537.

Practice 146, *Resource Book*

Practice 146
Skills and Applications of Lesson 11-9

Solve using logical reasoning.   1–3. See below.

1. Carlos, Todd, and Michael play baseball, lacrosse, and soccer. Michael and the baseball player are brothers. Carlos and the lacrosse player are best friends. Todd is not related to Michael and does not play soccer. Who plays each sport?

2. Lenny, Ted, and Bob drive a bus, taxi, and limousine. No one drives a vehicle that begins with the same letter as his name. Ted drove the bus driver to the airport. Who drives which vehicle?

3. Anne, Lucille, and Felicia are a lawyer, pharmacist, and teacher. Anne asked the lawyer for advice, and the pharmacist filled Lucille's prescription. Felicia attended the same college as the teacher, and her son is married to the pharmacist. What are their professions?

4. Carmen, Joan, and Cindy sailed yellow, blue, and green boats in a boat race. The yellow boat finished ahead of the green boat but the green boat did not finish ahead of the blue boat. Carmen did not finish first and did not sail the yellow boat. Cindy did not finish ahead of Joan and did not sail the blue boat. Who finished first? in what color boat?   Joan; blue

Solve using any problem solving strategy.

5. An 8-oz can of tomatoes costs $1.45 and a 12-oz can of tomatoes costs $2.05. Which is the better bargain?   12-oz can

6. The mean of five numbers is 33. Four of the numbers are 23, 34, 38, and 42. What is the fifth number?   28

7. The measure of the supplement of an angle is three times the measure of the complement of the angle. Find the measure of the angle.   45°

8. Landscape stone costs $40 for a $3\frac{1}{2}$ ton truckload. Find the cost, to the nearest dollar, of 5 tons of stone. $57

1. Carlos plays baseball, Michael plays soccer, and Todd plays lacrosse.
2. Lenny drives the bus, Ted drives the limo, and Bob drives the taxi.
3. Felicia is the lawyer, Lucille is the teacher, and Anne is the pharmacist.

Make sure that students understand
the difference in the procedures
used to solve parts (a) and (b) of
the Example. A simple way for students to check the reasonableness of
their answers is to look at the percent change: if it is positive, the
1990 population should be greater;
if it is negative, the 1990 population
should be less.

Application

Census data are used by many different businesses and industries.
Marketing and advertising strategies
are developed to take advantage of
shifts in a population, such as a
"baby boom" or the aging of a
population.

Additional Examples

Use the table on this page.

1. What was the population of
   Detroit in 1980?
   about 1,202,176

2. Suppose by the year 2000 the
   population of Phoenix increased
   by another 23%. Predict what
   the population of Phoenix would
   be by the year 2000.
   about 1,195,025

Analyzing Census Data

Objective: To analyze
census data.

Every ten years, the Census Bureau conducts a population census to
determine the number of individuals living in the United States. The 1990
census marked the 200th anniversary of the first census.

The census is important because the distribution of billions of tax dollars
and the redistricting of local and state governments depend on the results.

The table below shows some preliminary figures for the 1990 census.
Cities are ranked according to population.

1990 Rank	1990 Population	Change from 1980	1980 Rank
1. New York	7,033,179	−0.5%	1
2. Los Angeles	3,420,000	15.2%	3
3. Chicago	2,725,979	−9.3%	2
4. Houston	1,609,723	0.9%	5
5. Philadelphia	1,543,313	−8.6%	4
6. San Diego	1,094,524	25%	8
7. Dallas	990,957	9.5%	7
8. Phoenix	971,565	23%	9
9. Detroit	970,156	−19.3%	6
10. San Antonio	926,558	17.9%	11

Example

a. What was the population of Phoenix in 1980?

b. Suppose by the year 2000 the population of San Diego increases from
   the 1990 level by another 25%. Predict what will be the population of
   San Diego by the year 2000.

Solution

a. Let n = the population of Phoenix in 1980.
   Then $0.23n$ = the increase in the population from 1980 to 1990, and
   $n + 0.23n$ = the total population of Phoenix in 1990.

$$n + 0.23n = 971,565$$
$$1.23n = 971,565$$
$$\frac{1.23n}{1.23} = \frac{971,565}{1.23}$$
$$n \approx 789,890$$

The population of Phoenix in 1980 was about 789,890.

b. Find the amount of increase: $0.25(1,094,524) = 273,631$
   Add the amount of increase to the original amount.

$$1,094,524 + 273,631 = 1,368,155$$

The population of San Diego will be about 1,368,155 by the year 2000.

Exercises

Use the table on page 528 for Exercises 1–10.

1. Which city had the greatest decrease in population from 1980 to 1990? Detroit had the greatest percentage decrease, while Chicago had the greatest decrease in number of people.

2. Which city had the least increase in population from 1980 to 1990? Houston

3. Los Angeles, Houston, San Diego, Dallas, Phoenix, San Antonio; Southwest

3. In which cities did the population increase from 1980 to 1990? In what part of the country are these cities located?

4. What was the mean of the populations of the top 10 cities in 1990? How many cities had a population greater than the mean? about 2,128,595; 3

5. What was the median of the populations of the top 10 cities in 1990? about 1,318,919

6. What was the range of the populations of the top 10 cities in 1990? 6,106,621

7. What was the population of Los Angeles in 1980? 2,968,750

8. What was the population of Detroit in 1980? 1,202,176

9. Suppose by the year 2000 the population of the city of New York decreases from the 1990 level by another 0.5%. Predict what will be the population of the city of New York by the year 2000. about 6,998,013

10. Suppose by the year 2000 the population of Dallas increases from the 1990 level by another 9.5%. Predict what will be the population of Dallas by the year 2000. about 1,085,098

RESEARCH Answers to Exercises 11–12 are on p. A31.

11. Find the population of the United States according to every census taken since the census began in 1790. Draw a line graph of the data. Describe the trends you see in the data.

12. Find the 1980 census and 1990 census populations for your state. Did the population increase or decrease from 1980 to 1990? Predict what will be your state's population by the year 2000.

Statistics and Circle Graphs **529**

Exercise Note

Making Connections/Transitions
Exercises 11 and 12 connect data analysis with population trends, an area often explored in social studies.

Chapter Review

Terms to Know

stem-and-leaf plot (p. 497)
stem (p. 497)
leaf (p. 497)
histogram (p. 501)
frequency polygon (p. 501)
first quartile (p. 505)
third quartile (p. 505)
box-and-whisker plot (p. 505)

scattergram (p. 509)
trend line (p. 509)
positive correlation (p. 509)
negative correlation (p. 509)
circle graph (p. 513)
sector (p. 513)
central angle (p. 517)

Choose the correct term from the list above to complete each sentence.

1. A(n) _?_ is a bar graph used to show frequencies. histogram

2. circle graph

2. A(n) _?_ is used to represent data expressed as parts of a whole.

3. If the trend line slopes upward to the right there is a(n) _?_ between the sets of data. positive correlation

4. The _?_ is the median of the upper part of the data. third quartile

5. The _?_ is the last digit on the right of a number. leaf

6. To construct a(n) _?_, you can connect the midpoints of the tops of the bars of a histogram. frequency polygon

---

Use the data at the right for Exercises 7 and 8. *(Lessons 11-1, 11-2)*
Answers to Chapter Review 7–8 are on p. A31.

7. Make a frequency table for the data. Use intervals such as 16–20.

8. Make a stem-and-leaf plot for the data.

Points Scored per Game

17	24	18	31	22	16	22
18	23	24	30	18	25	33
27	23	18	20	22	24	31

Use the stem-and-leaf plot at the right. *(Lesson 11-2)*

9. What is the greatest test score? 100

10. How many test scores of 84 are there? 2

11. How many test scores are less than 90? 17

12. Find the mean, median, mode(s), and range of the test scores. 84.8; 88; 88; 40

Learner's Permit Test Scores

6	0 2 4 8
7	2 6 6 6 8 8
8	4 4 6 8 8 8 8
9	0 0 2 4 4 6 6 6 8 8
10	0 0

Make a box-and-whisker plot for the data. *(Lesson 11-4)*
See answer on p. A31.

13. **Ages of Workers in an Office (Years)**

| 31 | 29 | 34 | 19 | 24 | 42 | 33 |
| 31 | 27 | 28 | 30 | 24 | 30 |

Use the histogram below. *(Lesson 11-3)*

14. How many students are 72–75 in. tall? 4

15. How many students are less than 68 in. tall? 10

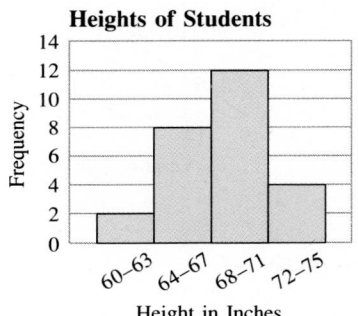

Heights of Students

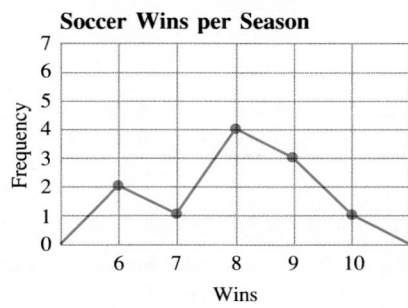

Soccer Wins per Season

Use the frequency polygon above. *(Lesson 11-3)*

16. How many times did the soccer team win 9 games per season? 3

17. How many seasons has the soccer team played? 11

Use the box-and-whisker plot below for Exercises 18–20.
(Lesson 11-4)

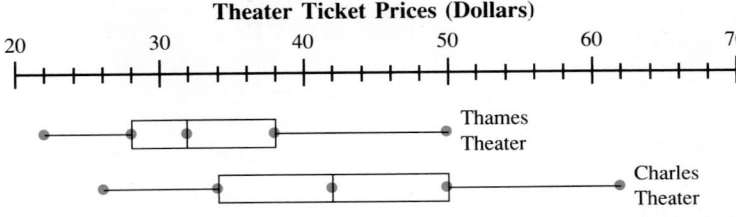

Theater Ticket Prices (Dollars)

18. Find the median, the first quartile, the third quartile, and the highest and lowest ticket prices for the Thames Theater. 32; 28; 38; 50; 22

19. Which theater has the lesser range in ticket prices? Thames

20. Which theater has the greater median ticket price? Charles

Draw a circle graph to display the data. *(Lesson 11-7)*

21. **Colors of Cars Sold**   See answer on p. A31.

Color	Blue	White	Red	Gray	Black
Number Sold	60	80	40	20	40

Quick Quiz 1 *(11-1 through 11-4)*

1. Make a frequency table for the data below. Use intervals such as 70–79.

Scores on a Science Test

71	86	74	92	88
68	74	96	79	83
66	76	92	81	70
75	83	79	76	60

Scores on a Science Test

Score	Tally	Frequency
60–69	|||	3
70–79	++++ ||||	9
80–89	++++	5
90–99	|||	3

2. Make a stem-and-leaf plot for the data in Exercise 1 above.

Scores on a Science Test

6	0 6 8
7	0 1 4 4 5 6 6 9 9
8	1 3 3 6 8
9	2 2 6

3. Use the histogram on page 503. How many students ran 24 or fewer laps?   26 students

4. Use the frequency polygon on page 503. How many band members are 65–68 inches tall? 20 band members

5. Use the box-and-whisker plot on page 507. Which teacher's class had the lower median score? Ken Jamison

Alternative Assessment

1. Have students research the ages of the last twenty Presidents at the time of their first inauguration. Students should then make a frequency table and a stem-and-leaf plot using the age data.

2. Have students use the Presidents' ages to make a box-and-whisker plot.

Chapter 11

Use the scattergram at the right. *(Lesson 11-5)*

22. How much money does the person who went to school for 14 years earn? $30,000

23. How many people earn $30,000 per year? 5

24. Predict the yearly earnings for someone who has had 19 years of schooling. $45,000

25. Is there a *positive* or a *negative* correlation between the number of years of schooling and earnings per year? positive

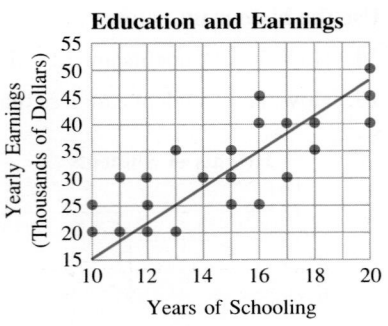

Education and Earnings

Yearly Earnings (Thousands of Dollars) vs. *Years of Schooling*

Use the circle graph at the left below. *(Lesson 11-6)*

26. If the Jones family's total annual budget is $45,000, how much do they spend annually on housing? $13,500

27. If $3500 is spent on entertainment, how much is the total budget? $35,000

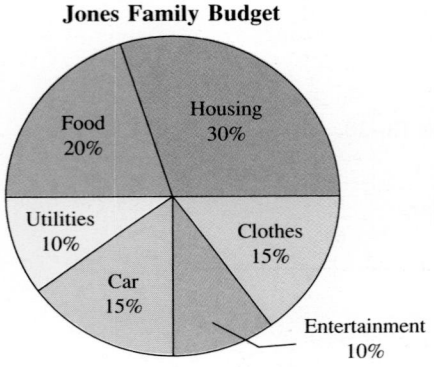

Jones Family Budget

Food 20%
Housing 30%
Utilities 10%
Clothes 15%
Car 15%
Entertainment 10%

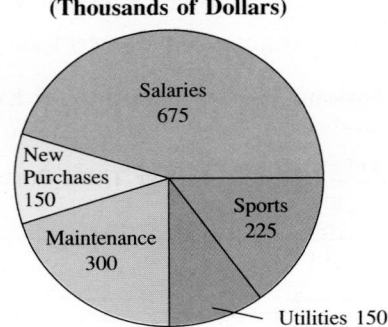

Lake School Budget (Thousands of Dollars)

Salaries 675
New Purchases 150
Maintenance 300
Sports 225
Utilities 150

Use the circle graph at the right above. *(Lesson 11-6)*

28. What was the total budget? $1,500,000

29. What percent of the budget was spent on sports? 15%

30. Decide whether it would be appropriate to draw a *scattergram* or a *stem-and-leaf plot* to display the data at the right. Then make the display. *(Lesson 11-8)* stem-and-leaf plot; see display on p. A31.

Leading Batting Averages

0.322	0.308	0.366	0.304	0.301
0.312	0.311	0.356	0.306	0.305
0.303	0.307	0.312	0.304	0.306

31. Ed, Alice, and Will are an editor, an artist, and a writer. No one has a name that starts with the same letter as his or her job. Alice and the editor had lunch together. Who has each job? *(Lesson 11-9)*
Ed is the artist, Alice is the writer, and Will is the editor.

Chapter Test

Use the data below for Exercises 1, 3, and 4. Answers to Chapter Test 1–3 are on p. A31.

Hours Worked per Week

32	28	40	38	26	42	35	40	38
28	40	41	35	36	30	38	29	44
68	40	37	38	35	32	26	30	40

1. Make a frequency table for the data. Use intervals such as 21–30. **11-1**

★ 2. What are appropriate intervals for a frequency table that displays these prices: $1.49, $1.99, $2.99, $3.50, $3.99, $4.99?

3. Make a stem-and-leaf plot for the data, *Hours Worked per Week*. **11-2**

4. Use the stem-and-leaf plot you made for Exercise 3 to find the median number of hours worked per week. 37 h

Use the histogram at the right for Exercises 5 and 6.

5. How many employees are 31–40 years old? 15 **11-3**

6. How many employees are more than 40 years old? 20

7. Draw a box-and-whisker plot for the data below. See answer on p. A31. **11-4**

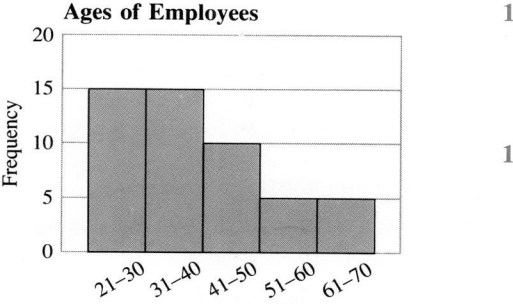

Ages of Employees

Prices of Stereo Systems (Dollars)

800	400	550	700	1200
750	700	600	600	850
750	650	600	700	550

Use the box-and-whisker plot below for Exercises 8–10.

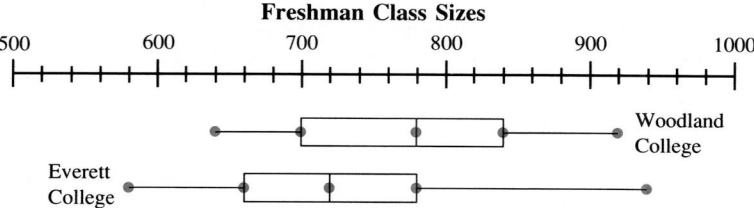

Freshman Class Sizes

8. Which college had the largest freshman class? Everett College

9. Which college had the greater median freshman class size? Woodland College

10. Find the median, the first quartile, the third quartile, and the highest and lowest freshman class size for Woodland College.
780; 700; 840; 920; 640

Alternative Assessment

Exercises 2 and 13 are marked with stars to indicate that they are alternative forms of assessment. *Exercise 2* tests students' ability to chose appropriate intervals for a frequency involving decimal values. *Exercise 13* tests students' understanding of when to use a scattergram.

Chapter Test (Form A)
Test 55, *Tests*

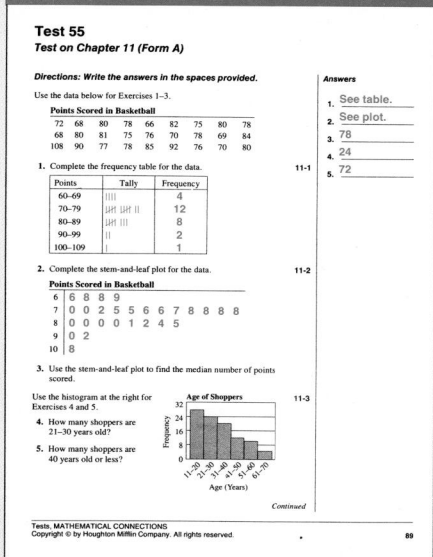

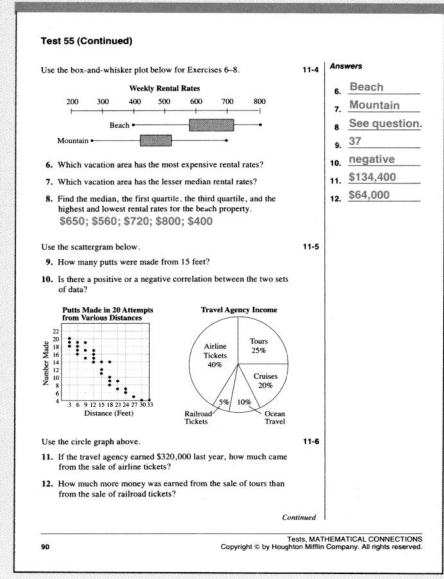

Use the scattergram below.

11. How many people missed 12 days of work due to illness? 3   11-5

12. Is there a *positive* or a *negative* correlation between the two sets of data? negative

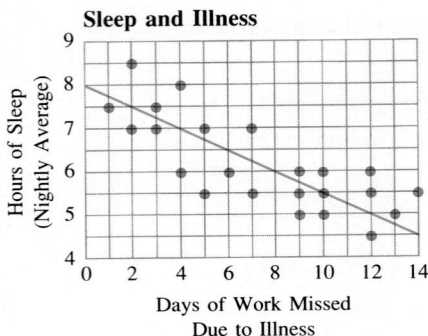

Sleep and Illness

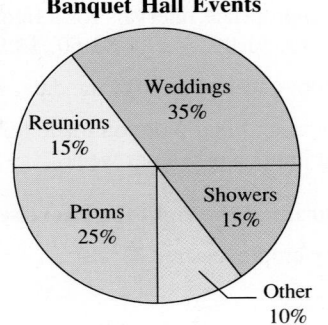

Banquet Hall Events

★13. What is the main purpose for using a scattergram to display data?
to show the relationship between two sets of data

Use the circle graph above.

14. If there was a total of 200 events at the banquet hall, how many events were weddings? 70   11-6

15. If there were 36 reunions at the banquet hall, how many events were there in all? 240

16. Draw a circle graph to display the data. Displays for Chapter Test 16–17 are on p. A31.   11-7

Animals in a Zoo

Class	Mammals	Birds	Reptiles	Amphibians
Animals	210	60	30	20

17. Decide whether it would be appropriate to draw a *circle graph* or a *box-and-whisker plot* for the data below. Then make the display.   11-8

Monthly Commuting Expenses   box-and-whisker plot

| 96 | 36 | 40 | 89 | 36 | 64 | 96 | 25 | 10 | 25 | 50 | 36 | 40 | 89 |

18. Hilda, Wendy, and Cara are a hostess, a waitress, and a chef. No one works at a job that begins with the first letter of her name. Hilda and the chef are sisters. Who has each job? Hilda is the waitress, Wendy is the chef, and Cara is the hostess.   11-9

Percentiles

Objective: To find a given percentile.

When you take a standardized test such as the SAT or the ACT, your score is often reported in terms of percentiles. A *percentile* shows a person's score in relation to other scores in a particular group. For example, if your score on a test is at the 65th percentile, this means approximately 65% of the people who took the test had a lesser score than you while approximately 35% of the people had a greater score.

Example

The following are the test scores of 28 students arranged in order from lowest position (1) to highest position (28).

a. Find the 65th percentile. **b.** Find the 75th percentile.

Position	1	2	3	4	5	6	7	8	9	10	11	12	13	14
Score	62	65	67	69	71	71	72	74	74	75	77	77	78	79

Position	15	16	17	18	19	20	21	22	23	24	25	26	27	28
Score	81	82	84	84	85	88	88	89	90	93	95	96	98	98

Solution

a. Find 65% of 28: $0.65(28) = 18.2$

The answer is not a whole number. When this is the case, round up to the next whole number: $18.2 \rightarrow 19$

Read the score in the 19th position: 85

The 65th percentile is a score of 85.

b. Find 75% of 28: $0.75(28) = 21$

The answer is a whole number. When this is the case, use that number and the next number: 21 and 22

Find the average of the scores in those positions.

$\dfrac{88 + 89}{2} = 88.5$ ◀—— The scores in positions 21 and 22 are 88 and 89, respectively.

The 75th percentile is a score of 88.5.

Exercises

Use the data consisting of the 28 test scores above. Find each percentile.

1. 90th 96   **2.** 77th 89   **3.** 40th 77   **4.** 95th 98

5. 62nd 84   **6.** 25th 73   **7.** 82nd 90   **8.** 80th 90

9. 60th 84   **10.** 55th 82   **11.** 50th 80   **12.** 85th 93

Statistics and Circle Graphs **535**

Quick Quiz 2 (11-5 through 11-9)

Use the scattergram at the top of page 509.

1. How many students studied 7 h?   4 students

2. Is there a positive or negative correlation between the hours studied and the test results?
positive

3. Use the circle graph at the top of page 513. If the total budget were $3,200,000, how much was spent on equipment?
$192,000

4. Draw a circle graph to display the data below.

**County Budget
(Millions of Dollars)**

Budget Items	Cost
Schools	$23
Police	$12
Fire	$5
Other	$10

County Budget
(Millions of Dollars)

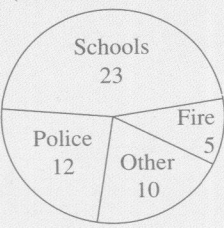

5. Decide whether it would be appropriate to draw a circle graph or a scattergram to display the data below. Then make the display.   scattergram

Allowance Spending (Dollars)

Savings	Entertainment
$10	$3
$6	$8
$3	$12
$7	$1
$6	$6
$4	$8

Cumulative Review

Standardized Testing Practice

Choose the letter of the correct answer.

1. A swimmer, a volleyball player, and a skier are named Arthur, Bernardo, and Carol. Carol's sport only happens outdoors. What can you conclude?
   A. Carol is the skier.
   B. Carol is not the swimmer.
   C. Arthur is not the skier.
   D. all of the above

2. Simplify: $(36x^2)^0$
   A. 1   B. 0
   C. $36x^2$ D. $36x^{20}$

3. Helen has two thirds as many pens as Dave has. Helen has 6 pens. How many pens does Dave have?
   A. $5\frac{1}{3}$ B. $6\frac{2}{3}$
   C. 4   D. 9

4. The radius of a circle is 10 in. What is the area of the circle?
   A. 31.4 in.^{2} B. 62.8 in.^{2}
   C. 100 in.^{2} D. 314 in.^{2}

5. A frequency table shows that 8 out of a total of 20 students are 12–15 years old. How many students are 16–19 years old?
   A. 8   B. 12
   C. 20   D. not enough information

6. The lengths of the legs of a right triangle are 0.3 cm and 0.4 cm. What is the length of the hypotenuse?
   A. 0.84 cm   B. 0.25 cm
   C. 0.5 cm   D. 0.7 cm

7. Solve: $\frac{x}{7} = -21$
   A. $x = 3$ B. $x = -147$
   C. $x = -3$ D. $x = 147$

8. There is exactly one pair of parallel sides in a quadrilateral. What special type of quadrilateral is it?
   A. rhombus   B. trapezoid
   C. rectangle   D. parallelogram

9. Find the range of the data.
   $3, 7, -2, 0, 7, -1, 8, -5, 2$
   A. 2   B. 8
   C. 13   D. 7

10. A sector of a circle graph shows that 35% of consumers prefer wheat bread. Find the number of degrees for that sector.
    A. 35°   B. 126°
    C. 63°   D. 145°

11.

Ages of Car Buyers

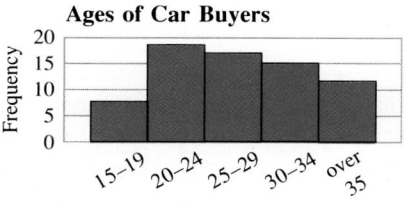

How many buyers are 30–34 years old?

A. 18 B. 8
C. 15 D. 12

12. Write in decimal notation: 6.35×10^5

A. 63.5×10^4 B. 0.0000635
C. 635,000 D. 63.5

13. The correlation in a scattergram is positive. What can you conclude?

A. The trend line slopes up to the right.
B. The trend line slopes down to the right.
C. There is no trend line.
D. The trend line is horizontal.

14. The scale of a statue of a famous citizen is 5 in. : 3 ft. The actual person is 6 ft tall. Find the height of the statue.

A. 6 in. B. 10 in.
C. 2.5 in. D. 3.6 in.

15. Find the LCM: 45 and 66

A. 330 B. 3
C. 990 D. 2970

16.

J 6 cm K   P   Q

10 cm   9 cm

M   L   S   R

Rectangle *JKLM* is similar to rectangle *QRSP*. Find the length of $\overline{RS}$.

A. 13 cm B. 15 cm
C. 5.4 cm D. 12 cm

17. Evaluate $x - y^2$ **when** $x = 8$ **and** $y = 7$.

A. 1 B. −41
C. 15 D. −57

18. Each side of a regular hexagon is 10 mm. Each side of a square is 15 mm. What can you conclude?

A. The hexagon has the greater perimeter.
B. The square has the greater perimeter.
C. The perimeters are equal.
D. The angles are equal.

19. The measures of two angles of a triangle are 63° and 74°. Classify the triangle.

A. right B. acute
C. obtuse D. none of the above

20. What percent of 120 is 72?

A. 52% B. 60%
C. $166\frac{2}{3}\%$ D. 48%

Allowance Spending (Dollars)

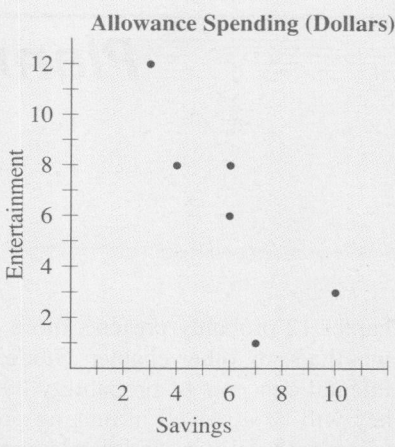

6. Phillipe, Michelle, and Edward are majoring in mathematics, physics, and engineering. None of them majors in a subject that begins with the same letter as his or her first name. Michelle and the physics major take history together. Who majors in each subject? Michelle, engineering; Edward, physics; Phillipe, mathematics

Alternative Assessment

Ask pairs of students to collect some data, identify the topic and source, and give the information to another pair of students to display. The pair that collected the data should then analyze the display and explain whether or not the data have been displayed appropriately.

537

Planning Chapter 12

Chapter Overview

Chapter 12 probably presents more concepts unfamiliar to students than any other chapter. Students will experience the fundamental concepts of probability by performing experiments. They will be involved in making predictions and conjectures, finding probabilities, and displaying and discussing the outcomes of experiments. Many of these activities can be done in cooperative groups, which promote student involvement and discussion.

The chapter opens with a *Focus on Explorations* that involves rolling a number cube. This lesson presents the experimental nature of probability to students in an understandable way. *Lesson 1* continues the experimental approach by discussing probability and odds using a spinner. *Lesson 2* introduces sample spaces and tree diagrams, and *Lesson 3* discusses the counting principle. Independent and dependent events are presented and contrasted in *Lesson 4*. Venn diagrams are presented in *Lesson 5* as a strategy to solve problems. Permutations and combinations are discussed in *Lesson 6*. The difference between theoretical and experimental probability is pointed out in *Lesson 7*. The *Focus on Computers* demonstrates to students how an experiment can be simulated by using a simple program. *Lessons 8 and 9* introduce and discuss the topic of making predictions from a sample. The *Focus on Applications* shows students how to select a sample from a random number table. The *Chapter Extension* on expected value concludes the chapter.

Background

The major topics in Chapter 12 are probability and odds, sample spaces and tree diagrams, the counting principle, independent and dependent events, and permutations and combinations. All of these topics can be studied effectively in cooperative groups and by using manipulatives and models. By performing experiments using number cubes, spinners, and coins, students can discover concepts or verify the mathematics presented in the text.

Technology can play a major role in the teaching and learning of probability. Computers especially are useful for generating random numbers, possible outcomes and permutations, and for simulating experiments, as demonstrated in the *Focus on Computers*. Both calculators and computers can be used to calculate permutations and combinations quickly and efficiently.

There are many connections to the real world in this chapter. In particular, the use of sampling and probability in conducting polls and making business decisions is discussed.

Objectives

FOCUS ON EXPLORATIONS To explore the ratios that occur when number cubes are rolled.

12-1 To find the probability of an event and to determine the odds in favor of an event.

12-2 To use a tree diagram to find sample spaces and probabilities.

12-3 To use the counting principle to find the number of possible outcomes and to find a probability.

12-4 To find the probability of two independent or two dependent events.

12-5 To use Venn diagrams to solve problems.

12-6 To find the number of permutations and the number of combinations of a group of items.

12-7 To find the experimental probability of an event.

FOCUS ON COMPUTERS To estimate the waiting time for an event to happen.

12-8 To make predictions using sampling and probability concepts.

FOCUS ON APPLICATIONS To use a table of random numbers to select a random sample.

CHAPTER EXTENSION To find the expected value of an event.

Chapter Planner

Instructional Aids	Manipulatives	Cooperative Learning	Technology	Practice/Reteaching	Assessment
Materials: red and white cubes, paper bags, spinners, coins, number cubes, index cards, paper cups, calculators *Resource Book:* Lesson Starters 110–117 *Visuals,* Folder K	Lessons 12-1, 12-2, 12-4, 12-6, 12-7 *Activities Book:* Manipulative Activities 23–24	Lessons 12-1, 12-3, 12-4, 12-6, 12-7 *Activities Book:* Cooperative Activity 22	Lessons 12-3, 12-6, 12-7 *Activities Book:* Calculator Activity 12 Computer Activity 12 *Connections Plotter Plus* Disk	Toolbox, pp. 753–771 Extra Practice, p. 747 *Resource Book:* Practice 141–160 *Study Guide,* pp. 219–234	Self-Tests 1–3 Quick Quizzes 1–3 Chapter Test, p. 576 *Resource Book:* Lesson Starters 110–117 (Daily Quizzes) *Tests* 57–63

Assignment Guide Chapter 12

Day	Skills Course	Average Course	Advanced Course
1	**Exploration:** Activities I–III **12-1:** 1–21, 23–39 odd, 46–48	**Exploration:** Activities I–III **12-1:** 1–41, 46–48	**Exploration:** Activities I–III **12-1:** 1–39 odd, 40–48, Challenge
2	**12-2:** 1–26	**12-2:** 1–26	**12-2:** 1–26
3	**12-3:** 1–15, 24–26	**12-3:** 1–19, 24–26	**12-3:** 1–26
4	**12-4:** 1–16, 17–27 odd, 34–38	**12-4:** 1–38	**12-4:** 1–38, Historical Note
5	**12-5:** 1–20	**12-5:** 1–20	**12-5:** 1–20
6	**12-6:** 1–18, 21–27 odd, 37–41	**12-6:** 1–27, 37–41	**12-6:** 1–19 odd, 20–41
7	**12-7:** 1–13, 15, 17, 27–32 **Focus on Computers**	**12-7:** 1–32 **Focus on Computers**	**12-7:** 1–32 **Focus on Computers**
8	**12-8:** 1–19, 24–28 **Focus on Applications**	**12-8:** 1–28 **Focus on Applications**	**12-8:** 1–28 **Focus on Applications**
9	*Prepare for Chapter Test:* Chapter Review	*Prepare for Chapter Test:* Chapter Review	*Prepare for Chapter Test:* Chapter Review
10	*Administer Chapter 12 Test*	*Administer Chapter 12 Test;* Cumulative Review	*Administer Chapter 12 Test;* Chapter Extension; Cumulative Review

Teacher's Resources

Resource Book
  Chapter 12 Project
  Lesson Starters 110–117
  Practice 141–160
  Enrichment 121–129
  Diagram Masters
  Chapter 12 Objectives
  Family Involvement 12
  Spanish Test 12
  Spanish Glossary

Activities Book
  Cooperative Activity 22
  Manipulative Activities 23–24
  Calculator Activity 12
  Computer Activity 12

Study Guide, pp. 219–234

Tests
  Tests 57–63

Visuals, Folder K

Connections Plotter
  Plus Disk

Using Technology

Lesson 12-3

COMPUTERS

A simple BASIC program using nested loops can generate all possible outcomes in a sequence of choices, thus illustrating and reinforcing understanding of the counting principle. For example, if there are five possibilities on a first choice, followed by six on a second, the following program will print all 30 possibilities:

```
10    PRINT "ALL POSSIBILITIES"
20    FOR I = 1 TO 5
30    FOR J = 1 TO 6
40    PRINT "CHOICE: "; I; "; "; J
50    NEXT J
60    NEXT I
```

Students do not have to understand the program, although that is desirable, but their understanding will be enhanced if they try different values in lines 20 and 30 and verify that the counting principle works by actually counting the number of outcomes listed.

Lesson 12-6

COMPUTERS

A BASIC program, such as that given below, can generate all permutations of the numbers 1, 2, 3,..., N. This can allow students to experience the number of possible permutations, and how they look, in situations where writing them out by hand is very tedious.

```
10    INPUT "N: ";N : REM NUMBER OF OBJECTS
20    DIM A(N)
30    FOR I = 1 TO N: LET A(I) = I: NEXT I : REM
         INITIAL VALUES
40    LET CARRY = 1 : REM ADD 1 TO THE NUMBER
         A(1)A(2) ... A(N), WITH CARRY
50    FOR I = N TO 1 STEP − 1
55    LET A(I) = A(I) + CARRY
60    IF A(I) > N THEN LET A(I) = 1: LET CARRY = 1:
         GOTO 75
70    LET CARRY = 0
75    NEXT I
```

```
80    IF CARRY = 1 THEN END : REM IF ALL DONE
90    LET I = 2 : REM LOOK FOR DUPLICATES
100   LET J = 1
110   IF A(I) = A(J) THEN GOTO 40
120   LET J = J + 1
130   IF J < I THEN GOTO 110
140   LET I = I + 1
150   IF I < = N THEN GOTO 100
160   FOR I = 1 TO N : REM IF FOUND A UNIQUE
         PERMUTATION, PRINT IT OUT
170   PRINT A(I);" ";
180   NEXT I
190   PRINT
200   GOTO 40
```

A simple BASIC program can be used to calculate $_nP_r$ and $_nC_r$.

```
10    INPUT "N, R: "; N, R
20    LET P = 1: LET C = 1 : REM # OF PERMUTATIONS,
         COMBINATIONS
30    FOR I = 0 TO R − 1
40    LET P = P * (N − I)
50    LET C = C * (N − I) / (R − I)
60    NEXT I
70    PRINT "P = "; P;" C = "; C
80    END
```

Lesson 12-7

COMPUTERS

The random number generator built into BASIC can be used to simulate the experiments described in the text. For example, to simulate the tossing of a coin, the following program may be used.

```
10    INPUT "# OF TOSSES: "; NTOSS
20    LET H = 0: LET T = 0 : REM # HEADS, TAILS
30    FOR I = 1 TO NTOSS
40    LET R = RND(1) : REM RANDOM NUMBER
         BETWEEN 0 AND 1
50    IF R < .5 THEN LET H = H + 1: PRINT "HEADS":
         GOTO 70
60    LET T = T + 1: PRINT "TAILS"
70    NEXT I
80    PRINT "# HEADS = "; H; "# TAILS = "; T
```

Using Manipulatives

Lesson 12-1

COOPERATIVE LEARNING

Divide students into cooperative groups. Give each group three white cubes, two red cubes, and a paper bag and tell them to place the cubes in the bag. Each member of a group should take turns in drawing cubes until all members can agree on answers to the following:

$P(\text{white}) = \underline{\ ?\ }$ $P(\text{red}) = \underline{\ ?\ }$ $P(\text{red or white}) = \underline{\ ?\ }$
$P(\text{not red}) = \underline{\ ?\ }$ $P(\text{not white}) = \underline{\ ?\ }$ $P(\text{green}) = \underline{\ ?\ }$

Each group's results should be shared with the entire class.

Lesson 12-4

COOPERATIVE LEARNING

Divide students into cooperative groups. Give each group a set of index cards and an envelope as described in the Exploration at the beginning of the lesson. Have each group perform the experiment. Discuss the results with the entire class.

Lesson 12-6

COOPERATIVE LEARNING

Example 1 on page 559 can be performed as a cooperative learning activity. Place students in groups and give each group a set of four index cards with the letters M, A, T, H on them, one letter per card. Each group should display and list all of the possible permutations of the letters.

Lesson 12-7

COOPERATIVE LEARNING

Divide students into cooperative groups. Give each group paper cups and have the groups collect data similar to the data in the chart for Exercises 1–10. Groups should compare results with those in the text.

Reteaching/Alternate Approach

Lesson 12-3

COOPERATIVE LEARNING

Divide students into cooperative groups. Have each group create problems that can be solved using the counting principle. Each group should exchange problems with another group for solving. Problems and solutions should then be shared with the class.

Lesson 12-4

Give students examples of events and have them classify each event as dependent or independent. These events can include items such as grades on tests and the final semester grade, picking counters from a bag with replacement, playing on the basketball team and being selected for the team, or making consecutive decisions based on flipping a coin. Follow this activity with a discussion as to why each event was classified as dependent or independent.

Lesson 12-6

COOPERATIVE LEARNING

Divide students into cooperative groups. Give each group four books and ask the groups to arrange the books on a shelf. Ask each group to determine how many ways the books can be arranged. After groups have done the activity physically and arrived at the answer of 24, point out that after the first book is placed, there are only three possibilities left for the second book, and so on. This activity should help students understand the reason for using 4!.

Lesson 12-8

Begin class by listing randomly selected test grades for ten students (no names) on the chalkboard. Tell students to assume that the next test will be of comparable difficulty. Have students predict the number of grades of A, B, C, and so on for the next test.

Teacher's Resources Chapter 12

Enrichment masters from the Resource Book are pictured here. See the Teacher's Resources chart on page 538B for a complete listing of all supplementary materials available for this chapter.

12-1 Enrichment
Communication

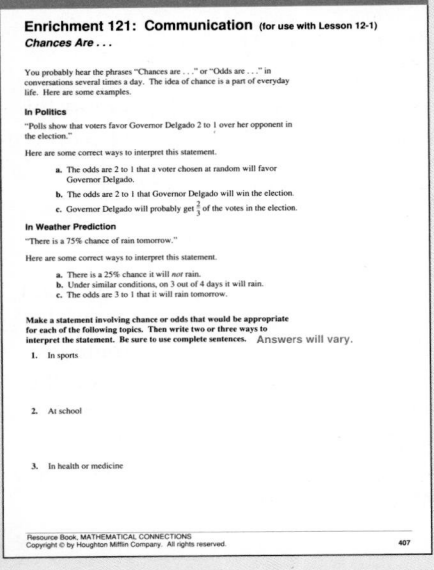

12-2 Enrichment
Application

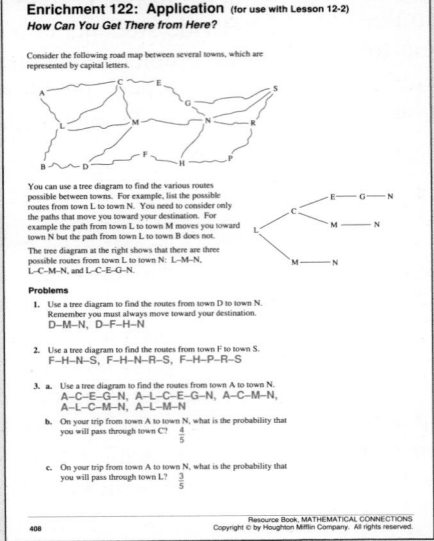

12-3 Enrichment
Connection

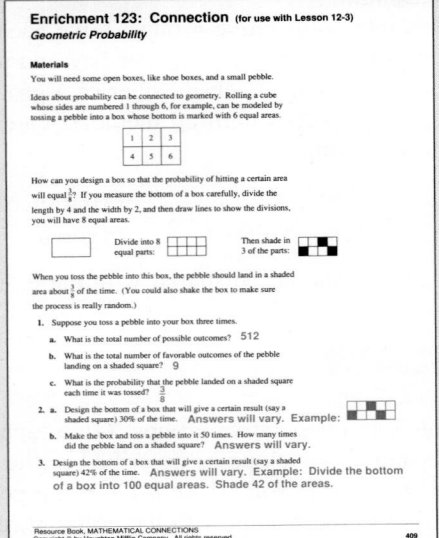

12-4 Enrichment
Exploration

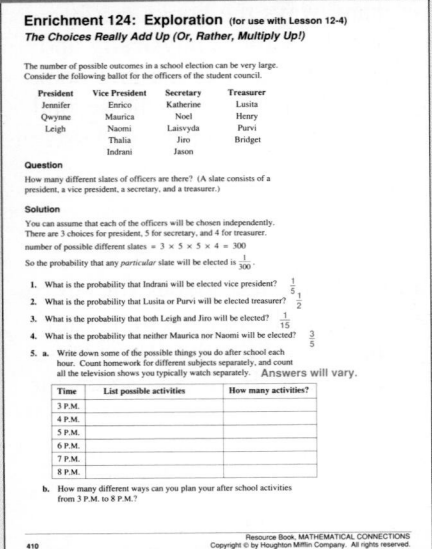

12-5 Enrichment
Problem Solving

12-6 Enrichment
Thinking Skills

12-7 Enrichment
Data Analysis

Enrichment 127: Data Analysis (for use with Lesson 12-7)
You Shouldn't Always Get What You Expect!

Materials

You will need 10 coins and a cup.

If you throw 10 coins, how many heads do you expect to get?

It should not be surprising if you *do not* get exactly what you expect. The following experiment will help establish a reasonable range for the number of heads you get if you toss 10 coins.

Experiment

In this experiment you will toss 10 coins *50 times* (for a total of 500 coin tosses.) Each time you toss the 10 coins, record the number of heads you get with a tally mark. When you have tossed the 10 coins 50 times, add up the tally marks to complete the frequency table. Answers will vary.

Tossing Ten Coins

Number of Heads	Tally	Frequency
0		
1		
2		
3		
4		
5		
6		
7		
8		
9		
10		

1. On what percent of the 50 tosses did you get exactly 5 heads?

2. What percent of the time did you get 3, 4, 5, 6, or 7 heads? (Add up all the frequencies and divide by 50 to get the percent for 50 tosses.)

3. Describe the pattern you see in your data. Compare it with patterns found by your classmates. Expect 5 to be the most frequent result.

4. In what sense is it still reasonable to "expect" 5 heads on each throw of 10 coins? 5 is the average result.

5. What is a reasonable range of the number of heads you might get? 3 to 7

12-8 Enrichment
Application

Enrichment 128: Application (for use with Lesson 12-8)
Expecting Error in the Polling Process

When the results of a political poll are announced, usually there is a comment about possible error in the numbers. For example, the pollster may say the results have a potential error of plus or minus 5 percentage points.

Example

In a poll about an upcoming race for the United States Senate, the results at the right were obtained in a survey of 400 randomly picked voters.

Candidate	Percent
Neilan	48%
Tukey	39%
Undecided	13%

There are 60,000 eligible voters, and the possible error in the poll is plus or minus 5 percentage points. Could Tukey actually be ahead in the race?

Solution

Tukey could have as much as 44% (39% + 5%) of the vote, and Neilan as low as 43% (48% − 5%). So it is possible for Tukey to be slightly ahead.

1. Explain why it is reasonable to expect error in the polling process.
   A small sample cannot fully represent the whole population.
2. Explain what could be done to reduce the degree of error in a poll.
   You could take a larger sample.

In a survey of 200 voters in an upcoming election, the following results were obtained:

Candidate	Number of Votes in Favor
Dumas	78
McMahon	64
Undecided	58

The pollster states that the possible error is plus or minus 4 percentage points. There are 10,000 eligible voters. Use the results of this survey for Exercises 3–8.

Convert the numbers in the survey to percents.

3. Dumas: 78 is what percent of 200? 39%
4. McMahon: 64 is what percent of 200? 32%
5. Undecided: 58 is what percent of 200? 29%
6. Is there an easier way to figure a percent of 200? Explain. Yes; divide the number by 2.
7. With a possible error of plus or minus 4%, what are the maximum and minimum possible vote totals for Dumas and McMahon? Dumas: 3500–4300, McMahon: 2800–3600
8. Could McMahon actually be ahead? Yes.

End of Chapter Enrichment
Extension

Enrichment 129: Extension (for use after Chapter 12)
Acting It Out

If you are having trouble solving a problem, it may help to act it out.

Problem

At an awards ceremony, it takes each student 6 seconds to walk across the stage, 4 seconds to shake hands and receive a certificate, and 6 seconds to leave the stage. How long will it take to present awards to 5 students?

Solution

Ask 6 classmates to help you act out this situation. One student should record the times and another should present the imaginary awards. The remaining students should cross an imaginary stage, shake hands and receive a certificate, and leave the stage. The student recording the times might complete a table like the one shown below.

Student	cross stage	receive certificate	leave stage	
first				6; 4; 6
second				4; 6
third				4; 6
fourth				4; 6
fifth				4; 6

Answer Exercises 1–4 to complete the solution of the problem above.

1. Use the table above. What times would the recorder fill in for the first student? See above.

2. Suppose the second student acting out this situation begins to cross the stage just as the first student has received an award and is turning to leave the stage. Would the recorder need to fill in all three columns of the table for the second student? Complete the table for the second student. No; see above.

3. Suppose each of the remaining students begin to cross the stage just as the preceding student has received an award and is turning to leave the stage. Complete the table for the third, fourth, and fifth students. See above.

4. Add the times in the table to find the total time needed to present awards to five students. 56 s

Multicultural Notes

For Chapter 12 Opener

The sea otter originally flourished along the western coast of North America from Baja California to Alaska, but nearly became extinct as a result of excessive hunting during the eighteenth and nineteenth centuries. After the United States, Russia, and Japan outlawed such hunting in 1911, sea otter populations, particularly those around the Aleutian Islands, began to recover. The Aleutian populations grew so rapidly that the otters experienced overcrowding and could not find enough food. To help assure the continued survival of the sea otter, scientists translocated many otters to new environments along the northwestern coast of North America.

For Page 555

Hilda Geiringer, discussed in the Historical Note, was one of many Jewish academics who were forced to flee Germany in the 1930s. Before coming to the United States, Geiringer sought refuge for several years in Turkey, where she continued her work in mathematics, learning Turkish so that she could give lectures.

In addition to her contributions in the field of probability and statistics, Geiringer also made important contributions to the mathematical theory of plasticity.

For Page 556

John Venn (1834–1923) made important contributions in the field of probability theory.

Probability 12

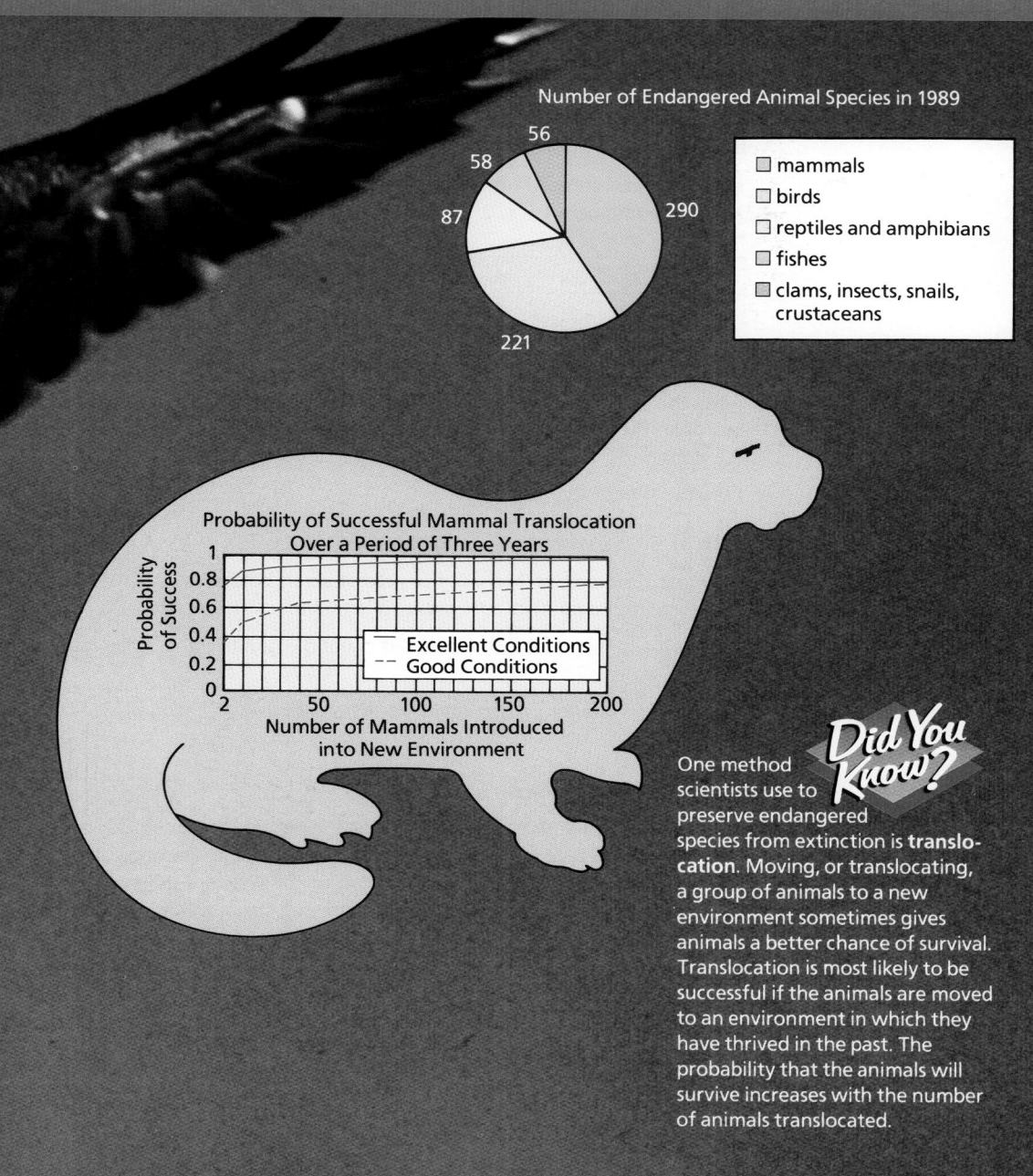

Number of Endangered Animal Species in 1989

- ☐ mammals
- ☐ birds
- ☐ reptiles and amphibians
- ☐ fishes
- ☐ clams, insects, snails, crustaceans

56
58
290
87
221

Probability of Successful Mammal Translocation Over a Period of Three Years

Probability of Success

1
0.8
0.6
0.4
0.2
0

2   50   100   150   200

Number of Mammals Introduced into New Environment

—— Excellent Conditions
- - - Good Conditions

Did You Know?

One method scientists use to preserve endangered species from extinction is **translocation**. Moving, or translocating, a group of animals to a new environment sometimes gives animals a better chance of survival. Translocation is most likely to be successful if the animals are moved to an environment in which they have thrived in the past. The probability that the animals will survive increases with the number of animals translocated.

Research Activities

For Chapter 12 Opener
Students can contact organizations in their own community, such as a local wildlife agency, to find out about endangered animals and what is being done to protect them. Students can also collect data on past and present populations of these animals to use in creating statistical graphs.

For Page 555
Students can research the work of the Nobel Prize-winning American geneticist Barbara McClintock.

Suggested Resources

Kenyon, Karl W. ''Return of the Sea Otter.'' *National Geographic*, October 1971, pp. 520–539.

Shiels, Barbara. *Winners: Women and the Nobel Prize*. Minneapolis: Dillon Press, 1985.

539

Rolling a Number Cube

Objective: To explore the ratios that occur when number cubes are rolled.

Materials
- number cubes

A *number cube* has six sides numbered from 1 through 6. When you roll a number cube, any one of the six sides is equally likely to land face up.

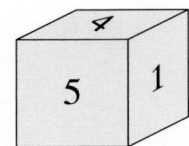

Activity I Answers to Steps 1–2 will vary.

1 Roll a number cube 30 times. Copy and complete the table below to record the number of 1's, 2's, 3's, 4's, 5's, and 6's that land face up.

1's	2's	3's	4's	5's	6's	Total number of rolls
?	?	?	?	?	?	30

2 For each of the six numbers, write a ratio comparing the number of times the number landed face up to the total number of rolls, 30. For example, if twelve 1's are rolled, the ratio is $\frac{12}{30}$. (Do not simplify the fractions.)

3 Find the sum of the six ratios. 1

Activity II Answers to Steps 1–3 will vary.

1 Use your results from Activity I to predict about how many 2's would land face up if a number cube were rolled 60 times.

2 Now roll a number cube 60 times and count the number of 2's that land face up.

3 How close was your prediction to the actual number of 2's that landed face up?

Activity III Answers to Steps 2–3 will vary.

1 Suppose you rolled two number cubes and recorded the sum of the numbers that landed face up. For example, if 3 lands face up on one cube and 5 lands face up on the second cube, you would record the number 8. Make a table similar to the one above, listing all the possible sums in order from least to greatest. See answer on p. A32.

2 Roll a pair of number cubes 30 times. Complete the table that you made in Step 1.

3 For each sum, write a ratio comparing the number of times the sum occurred to the total number of rolls, 30.

4 Find the sum of all the ratios. 1

12-1 Probability and Odds

Objective: To find the probability of an event and to determine the odds in favor of an event.

When the spinner at the right is spun, there are six possible **outcomes.** The spinner may stop on any one of the six numbered sectors of the circle. (Assume that the spinner will not stop on the line between two sectors.) Each outcome is **equally likely** to happen.

An **event** is any group of outcomes. When all possible outcomes are equally likely, the **probability** of an event E, written $P(E)$, is given by the formula below.

$$P(E) = \frac{\text{number of favorable outcomes}}{\text{number of possible outcomes}}$$

A probability is a number from 0 through 1 and is often written as a fraction in lowest terms.

Example 1

Use the spinner shown above. Find each probability.
a. $P(5)$ **b.** $P(\text{red})$ **c.** $P(7)$
d. $P(\text{not white})$ **e.** $P(\text{red, blue, or white})$

Solution

There are 6 possible outcomes when the spinner is spun. Each outcome is equally likely to occur. Use the formula above.

a. $P(5) = \frac{1}{6}$ ← The one favorable outcome is landing on 5.

b. $P(\text{red}) = \frac{3}{6} = \frac{1}{2}$ ← Three sectors are red, so there are 3 favorable outcomes.

c. $P(7) = \frac{0}{6} = 0$ ← The probability of an *impossible* event is 0.

d. $P(\text{not white}) = \frac{5}{6}$ ← Five sectors are not white, so there are 5 favorable outcomes.

e. $P(\text{red, blue, or white}) = \frac{6}{6} = 1$ ← The probability of a *certain* event is 1.

 **Check Your Understanding**

1. Why is it impossible for the event in Example 1(c) to occur?
2. In Example 1(d), describe the five favorable outcomes.
3. Why is the event in Example 1(e) certain to occur?
See *Answers to Check Your Understanding* at the back of the book.

Probability **541**

Lesson Planner

Teaching the Lesson
• Materials: white and red cubes, paper bag
• Lesson Starter 110
• Visuals, Folder K
• Using Manipulatives, p. 538D
• Cooperative Learning, p. 538D

Lesson Follow-Up
• Practice 149
• Enrichment 121: Communication
• Study Guide, pp. 219–220

Warm-Up Exercises
Write each fraction in lowest terms.
1. $\frac{16}{20}$ $\frac{4}{5}$
2. $\frac{12}{48}$ $\frac{1}{4}$
3. $\frac{8}{56}$ $\frac{1}{7}$
4. $\frac{39}{65}$ $\frac{3}{5}$

Lesson Focus

At the beginning of a football game, an official flips a coin to determine which team gets the choice of kicking or receiving the ball. Why is this a fair method to use? Does each team have a 50-50 chance? These questions can be answered by using probability concepts, the focus of today's lesson.

Teaching Notes

Probability and odds may be unfamiliar concepts to many students. Make certain students understand the vocabulary and the methods presented in both Examples.

541

1. What is the probability of an event that cannot occur? of an event that is certain to occur?
   0; 1

2. If there were a 30% chance of rain tomorrow, would the probability of rain be greater than $\frac{9}{10}$ or less than $\frac{9}{10}$? less than $\frac{9}{10}$

3. What is meant by the odds in favor of an event? the ratio of the number of favorable outcomes to the number of unfavorable outcomes

Error Analysis

Some students confuse the probability of an event with the odds in favor of an event. Provide students with several opportunities to relate both concepts to the same problem.

Using Manipulatives

For a suggestion on using colored cubes for a probability experiment, see page 538D.

Additional Examples

Use the spinner shown.

1. Find each probability.
   a. P(an odd number) $\frac{2}{3}$
   b. $P(1)$ $\frac{1}{3}$
   c. P(an even number) $\frac{1}{3}$
   d. P(a number > 2) $\frac{1}{2}$

An event may have favorable outcomes and unfavorable outcomes. The **odds in favor** of an event E are given by the formula below.

$$\text{odds in favor of event } E = \frac{\text{number of favorable outcomes}}{\text{number of unfavorable outcomes}}$$

Example 2

Use the spinner at the right. Find the odds in favor of each event.
a. white   b. number > 2

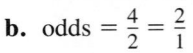

Solution

a. odds $= \frac{1}{5}$ ◄—— 1 favorable outcome
                ◄—— 5 unfavorable outcomes

The odds in favor of spinning white are 1 to 5.

b. odds $= \frac{4}{2} = \frac{2}{1}$

The odds in favor of spinning a number greater than 2 are 2 to 1.

✎ Check Your Understanding

4. Describe the favorable and unfavorable outcomes for the event in Example 2(a).

5. Find the odds in favor of the event *not red*.
   See *Answers to Check Your Understanding* at the back of the book.

Guided Practice

COMMUNICATION «*Writing*

Write each expression in symbols.

1. $P(\text{number} < 5) = \frac{2}{3}$
2. $P(\text{not red}) = \frac{1}{2}$

«**1.** The probability that a spinner lands on a number less than 5 is $\frac{2}{3}$.

«**2.** The probability that a spinner lands on a color that is not red is $\frac{1}{2}$.

Write each expression in words. Assume that a spinner has been spun.

«**3.** P(white or red)   «**4.** P(number > 2)   «**5.** P(not 4)

3. the probability that a spinner lands on white or red
4. the probability that a spinner lands on a number greater than 2
5. the probability that a spinner lands on a number that is not 4

6. The spinner shown is spun. List all the possible outcomes. 1, 2, 3, 4

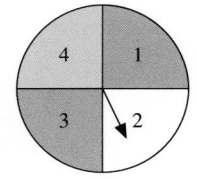

7. Describe the favorable outcomes for the event the spinner shown lands on red. Describe the unfavorable outcomes. favorable: 1, 3; unfavorable: 2, 4

Find each probability for the spinner shown.

8. $P(2)$ $\frac{1}{4}$ 9. P(red) $\frac{1}{2}$ 10. $P(5)$ 0

11. P(white or blue) $\frac{1}{2}$ 12. P(even number) $\frac{1}{2}$ 13. P(number > 0) 1

Find the odds in favor of each event for the spinner shown.

14. red 1 to 1   15. 1 or 2 1 to 1   16. not blue 3 to 1

Exercises

Use the spinner at the right. Find each probability.

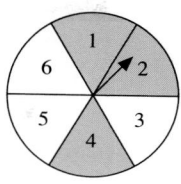

A
1. $P(\text{red})$ $\frac{1}{3}$
2. $P(2 \text{ or } 3)$ $\frac{1}{3}$
3. $P(\text{odd number})$ $\frac{1}{2}$
4. $P(\text{blue})$ $\frac{1}{6}$
5. $P(\text{not red})$ $\frac{2}{3}$
6. $P(\text{number divisible by 3})$ $\frac{1}{3}$
7. $P(\text{not white})$ $\frac{1}{2}$
8. $P(5)$ $\frac{1}{6}$
9. $P(9)$ 0
10. $P(\text{number} > 0)$ 1

Use the spinner at the right. Find the odds in favor of each event.

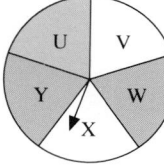

11. W 1 to 4
12. red 2 to 3
13. W, X, or Y 3 to 2
14. blue 1 to 4
15. not white 3 to 2
16. not V 4 to 1
17. red or U 3 to 2
18. white or red 4 to 1
19. X or Y 2 to 3
20. white or not red 3 to 2

21. Assume that there is an equal chance that a coin will show heads or tails facing up when it has been tossed.
    a. Find $P(\text{heads})$. $\frac{1}{2}$
    b. Find the odds in favor of heads. 1 to 1

A number cube that has six sides numbered 1 through 6 is rolled. Find each probability.

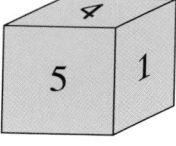

22. $P(1)$ $\frac{1}{6}$
23. $P(6)$ $\frac{1}{6}$

B
24. $P(\text{odd number})$ $\frac{1}{2}$
25. $P(\text{prime number})$ $\frac{1}{2}$
26. $P(3, 4, 5, \text{ or } 6)$ $\frac{2}{3}$
27. $P(\text{number} < 6)$ $\frac{5}{6}$
28. $P(\text{number} > 8)$ 0
29. $P(\text{number} < 9)$ 1

The bag shown at the right contains 3 red, 6 blue, 4 white, and 3 green marbles. One marble is selected *at random*, that is, purely by chance. Find each probability.

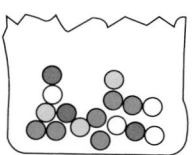

30. $P(\text{blue})$ $\frac{3}{8}$
31. $P(\text{not blue})$ $\frac{5}{8}$
32. $P(\text{red or green})$ $\frac{3}{8}$
33. $P(\text{red, green, or white})$ $\frac{5}{8}$

Use the bag of marbles shown above. One marble is selected at random. Find the odds in favor of each event.

34. red 3 to 13
35. blue or white 5 to 3
36. white 1 to 3
37. not red 13 to 3
38. not blue 5 to 3
39. white or green 7 to 9

Probability 543

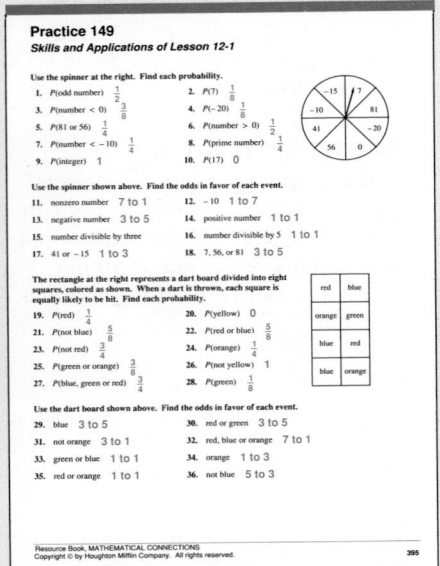
ESTIMATION

You can look at a sample of items chosen at random from a large group to get an idea of what the large group is like. The probability that an event occurs in a sample can be used as an estimate of the probability that the event occurs in the entire group.

40. A sample of 18 film packages is chosen at random from a bin containing 100-speed and 200-speed film packages. Five of the film packages chosen are 100-speed film. Estimate the probability that any package chosen from the entire carton is 100-speed film. about $\frac{5}{18}$

41. A sample of 15 pairs of baby socks is chosen at random from a table of blue pairs of socks and pink pairs of socks at a department store. Six of the pairs of socks in the sample are blue. Estimate the probability that any pair of socks chosen from the entire table is blue. about $\frac{2}{5}$

THINKING SKILLS   42–44. Comprehension   45. Synthesis

42. A number cube is rolled. The probability that a 4 lands face up is $\frac{1}{6}$. What is the probability that a 4 does not land face up? $\frac{5}{6}$

43. A bag contains 9 marbles. Two marbles are red, 4 marbles are blue, and 3 marbles are green. What is the probability that a marble drawn from the bag is blue? What is the probability that a marble drawn from the bag is not blue? $\frac{4}{9}; \frac{5}{9}$

44. a. In Exercise 42, what is $P(4) + P(\text{not } 4)$? 1
   b. In Exercise 43, what is $P(\text{blue}) + P(\text{not blue})$? 1

45. Use your answers to Exercise 44.
   a. Complete the statements to formulate generalizations about the probability that an event E occurs or does not occur.
   For any event E, $P(E) + P(\text{not } E) = \underline{\ ?\ }$. 1
   For any event E, $P(\text{not } E) = 1 - \underline{\ ?\ }$. $P(E)$
   b. A spinner is spun and the probability that the arrow lands on red is $\frac{1}{3}$. Use the generalizations in part (a) to find the probability that the spinner does not land on red. $\frac{2}{3}$

SPIRAL REVIEW

S **46.** A number cube that has sides numbered 1 through 6 is rolled. Find $P(0)$. *(Lesson 12-1)* 0

47. Find the circumference of a circle with radius 9 in. *(Lesson 10-2)* 56.5 in.

48. Solve: $12x - 15x = -15$ *(Lesson 4-9)* 5

Challenge

A box contains 40 pens. Some are red and some are blue. If the odds of selecting a red pen are 3 to 5, how many red pens are in the box? 15

12-2

Sample Spaces and Tree Diagrams

Objective: To use a tree diagram to find sample spaces and probabilities.

A list of all possible outcomes is called the **sample space.** A **tree diagram** is one way of showing a sample space and of finding the number of all possible outcomes.

In the previous lesson, you found probabilities when you rolled one number cube or spun one spinner. Sometimes you perform more than one such activity at the same time or consecutively.

Example

Three coins are tossed. Make a tree diagram to show the sample space. Then find each probability.

a. $P(\text{exactly 2 heads})$ **b.** $P(\text{at least one tail})$

Solution

When each coin is tossed, there are two possibilities, either heads (H) or tails (T).

Terms to Know
- *sample space*
- *tree diagram*

coin 1	coin 2	coin 3	outcomes
H	H	H	HHH
		T	HHT
	T	H	HTH
		T	HTT
T	H	H	THH
		T	THT
	T	H	TTH
		T	TTT

a. There are 3 outcomes that give exactly 2 heads: HHT, HTH, and THH. There are 8 possible outcomes.

$$P(\text{exactly 2 heads}) = \frac{3}{8}$$

b. *At least one tail* means one tail, two tails, or three tails. There are 7 favorable outcomes.

     HHT, HTH, THH   ←   1 tail

     HTT, THT, TTH   ←   2 tails

     TTT   ←   3 tails

$$P(\text{at least one tail}) = \frac{7}{8}$$

✓ Check Your Understanding

1. How would the sample space in the Example be different if only two coins were tossed?
2. If part (b) were changed to $P(\text{at most one tail})$, list the favorable outcomes and then find the probability.
   See *Answers to Check Your Understanding* at the back of the book.

Lesson Planner

Teaching the Lesson
- Materials: spinners, coins, number cubes
- Lesson Starter 111

Lesson Follow-Up
- Practice 150
- Enrichment 122: Application
- Study Guide, pp. 221–222

Warm-Up Exercises

A number cube with faces numbered 2, 4, 6, 8, 10, and 12 is rolled. Find each probability.

1. $P(\text{an even number})$ 1
2. $P(\text{an odd number})$ 0
3. $P(\text{a number} > 6)$ $\frac{1}{2}$
4. $P(10)$ $\frac{1}{6}$
5. $P(4, 8, 12)$ $\frac{1}{2}$

Lesson Focus

Ask students to imagine that they have three posters, four banners, and two pictures, and that they want to select one of each to hang on a wall. Point out that there are many ways to make a selection. How can the number of ways be found without physically making the choices? The focus of today's lesson is to answer questions of this type.

Teaching Notes

Emphasize that a tree diagram is a way to organize the different outcomes of a probability experiment and that it is used to identify specific probabilities.

Teaching Notes (continued)

Key Questions

1. What is a sample space? all possible outcomes of a probability experiment
2. What is an efficient way to list all the possible outcomes of a probability experiment? Draw a tree diagram.

Reasoning

The Check Your Understanding questions test students' ability to examine a sample space and draw conclusions about outcomes and probabilities based on some new conditions.

Additional Example

Use the tree diagram in the Example for the tossing of three coins.

a. Find P(at least one head). $\frac{7}{8}$

b. Find P(exactly 2 tails). $\frac{3}{8}$

Closing the Lesson

Have students explain the definitions of a sample space and a tree diagram. Discuss the advantages of using a tree diagram. Have students list some situations in their lives where this approach could be helpful.

Suggested Assignments

Skills: 1–26
Average: 1–26
Advanced: 1–26

Exercise Notes

Communication: Reading

Guided Practice Exercises 1 and 2 check students' understanding of definitions and their interpretation of phrases.

Guided Practice

COMMUNICATION «*Reading*

«1. Replace each __?__ with the correct word.
The list of all possible outcomes of an experiment is called the __?__. A __?__ diagram can be used to show such a list. sample space; tree

«2. Choose the phrase that means the same as *at least one*.
   a. more than one   (b.) one or more   c. not more than one

3. The spinner at the right is spun and a coin is tossed. Make a tree diagram to show the sample space. See answer on p. A32.

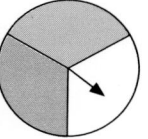

Use the tree diagram from Exercise 3 to find each probability.

4. P(red, heads) $\frac{1}{6}$

5. P(not blue, tails) $\frac{1}{3}$

6. P(green, heads) 0

7. P(red or white, heads) $\frac{1}{3}$

8. P(red or blue, tails) $\frac{1}{3}$

9. P(white, heads or tails) $\frac{1}{3}$

Exercises

A   1. Four coins are tossed. Make a tree diagram to show the sample space. See answer on p. A32.

Use the tree diagram from Exercise 1 to find each probability.

2. Find P(exactly 3 tails). $\frac{1}{4}$

3. Find P(at least 2 heads). $\frac{11}{16}$

4. Find P(5 heads). 0

5. Find P(no tails). $\frac{1}{16}$

6. The number cube below numbered 1 through 6 is rolled and the spinner is spun. Make a tree diagram to show the sample space. See answer on p. A32.

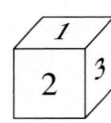 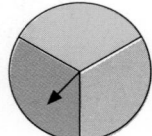

Use the tree diagram from Exercise 6 to find each probability.

7. P(2, blue) $\frac{1}{18}$

8. P(odd number, red) $\frac{1}{6}$

9. P(1 or 3, green) $\frac{1}{9}$

10. P(7, purple) 0

11. P(number < 6, blue) $\frac{5}{18}$

12. P(not 5, red) $\frac{5}{18}$

13. P(even number, red or blue) $\frac{1}{3}$

14. P(3, not green) $\frac{1}{9}$

15. P(8, green) 0

16. P(not 3, not red) $\frac{5}{9}$

Make a tree diagram and solve.

Tree diagrams for Exercises 17–22 are on p. A32.

Dee's Diner offers customers the lunch choices shown at the right. Customers may order a serving of meat, one vegetable, and one soup for one low price. If a customer chooses a meal at random, find each probability.

Meat	chicken, fish, beef
Vegetable	peas, corn
Soup	tomato, onion

B 17. P(chicken, peas, onion soup) $\frac{1}{12}$

18. P(beef, peas or corn, tomato soup) $\frac{1}{6}$

19. P(fish, corn, not tomato soup) $\frac{1}{12}$

20. P(not beef, not corn, tomato soup or onion soup) $\frac{1}{3}$

Whitney can choose an outfit from the following clothes: two pairs of slacks (navy or gray), four blouses (red, blue, striped, or paisley), and three pairs of shoes (white, blue, or black). If Whitney chooses an outfit at random, find each probability.

21. P(navy slacks, paisley blouse, white shoes) $\frac{1}{24}$

22. P(gray slacks, not a red blouse, not white shoes) $\frac{1}{4}$

SPIRAL REVIEW

S 23. Solve the proportion: $\frac{3}{x} = \frac{8}{16}$ *(Lesson 9-2)* 6

24. The radius of a circle is 14 cm. Find the area.   *(Lesson 10-4)*

25. Write 68.4% as a decimal.   *(Lesson 9-5)* 0.684   616 cm²

26. The spinner in Exercise 6 on page 546 is spun and a coin is tossed. Make a tree diagram to show the sample space. Find P(blue, tail). *(Lesson 12-2)* See answer on p. A32.

Probability   **547**

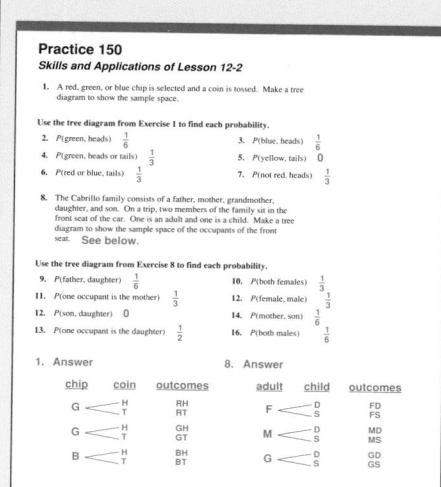

12-3 The Counting Principle

Objective: To use the counting principle to find the number of possible outcomes and to find a probability.

Terms to Know
- *counting principle*

APPLICATION

Susan plans to buy a music system. The table below shows the brands of stereo components that she can afford to buy.

Receiver	Speakers	Compact Disc (CD) player
Supra	Ultra	Star
Majestic	Excel	Carlyle
Heritage	Lobo	
Dynamic		

QUESTION How many different systems can Susan choose?

To answer the question, you can make a tree diagram and count the number of favorable outcomes. You may not find this method convenient when the number of final branches in the tree diagram is too great. Example 1 shows another method you can use to answer the question.

Example 1
●

Solution

Find the number of different music systems that could be chosen from the components in the table above.

There are three stages in choosing a music system from the table above: choosing a receiver, choosing speakers, and choosing a CD player.

Receiver	*Speakers*	*CD player*
4 choices	3 choices	2 choices

You find the number of possible music systems by multiplying the number of choices for each component.

$$4 \cdot 3 \cdot 2 = 24$$

Check Your Understanding

1. Suppose you made a tree diagram to represent the sample space of possible music systems. How many final branches would there be?

2. Suppose there were a fifth brand of receiver listed in the table above. How many different music systems could be chosen?
   See *Answers to Check Your Understanding* at the back of the book.

ANSWER Susan can choose 24 different stereo systems.

Although a tree diagram lists each outcome in a sample space, it is often easier to use the **counting principle** to find the total number of possible outcomes.

The Counting Principle

The number of outcomes for an event is found by multiplying the number of choices for each stage of the event.

Example 2

A whole number from 1 through 20 is chosen at random, then a letter from A through W is chosen at random. Use the counting principle to find P(odd number, C or D).

Solution

First use the counting principle to find the number of favorable outcomes.

$$
\underset{\substack{\text{number of favorable}\\\text{choices for an odd number}}}{10} \cdot \underset{\substack{\text{number of favorable}\\\text{choices for a C or D}}}{2} = 20
$$

There are 20 favorable outcomes.

Then find the number of possible choices.

$$
\underset{\substack{\text{number of choices}\\\text{for a number}}}{20} \cdot \underset{\substack{\text{number of choices}\\\text{for a letter}}}{23} = 460
$$

There are 460 possible outcomes.

$$P(\text{odd number, C or D}) = \frac{20}{460} = \frac{1}{23}$$

 **Check Your Understanding**

3. What would P(odd number, C or D) have been in Example 2 if the letter had been chosen at random from A through T?
See Answers to Check Your Understanding at the back of the book.

Guided Practice

COMMUNICATION « *Reading*

Refer to pages 548–549 of the text.

«**1.** Describe the counting principle in words. To find the number of outcomes for an event, find the number of choices for each stage of the event and multiply them.

«**2.** Complete: According to the counting principle, if there are *m* ways to do one activity, *n* ways to do another activity, and *p* ways to do a third activity, then there are __?__ ways to do the three activities. *mnp*

Use the counting principle to find the number of possible outcomes.

3. rolling a number cube twice 36   **4.** tossing a coin five times 32

Use the counting principle to find each probability.

5. A coin is tossed five times. Find P(all tails). $\frac{1}{32}$

6. A number cube is rolled twice. Find $P(4, 4)$. $\frac{1}{36}$

Probability   **549**

Teaching Notes

Stress to students that the counting principle is very helpful for finding the total number of possible outcomes, but that it does not list each outcome in a sample space.

Key Questions

1. How is the counting principle applied? You multiply the number of choices for each stage of an event to determine the number of outcomes for the event.

2. What is the advantage of using the counting principle over using a tree diagram? When the number of choices is large, a tree diagram is not easy to draw. Using the counting principle is more efficient.

Error Analysis

When using the counting principle, students sometimes add the choices instead of multiplying them. Remind students that the sum of the choices is used with tree diagrams but not with the counting principle.

Reasoning

The first Check Your Understanding question has students examine Example 1, which is solved by the counting principle, and compare it to the tree diagram method.

Using Technology

For a suggestion on using a BASIC program to generate outcomes, see page 538C.

Reteaching/Alternate Approach

For a suggestion on using a cooperative learning activity, see page 538D.

Exercises

Use the counting principle to find the number of possible outcomes.

A 1. selecting one hour of the day and one minute of the hour 1440

2. choosing a letter from A through X and a digit from 0 through 9 240

3. rolling a number cube 3 times 216

4. choosing an outfit from 3 sport coats, 4 shirts, 5 ties, and 3 pairs of pants 180

5. choosing a dinner from 6 main courses, 4 vegetables, 2 salads, and 3 beverages 144

Use the counting principle to find each probability.

6. A coin is tossed four times. Find P(all heads). $\frac{1}{16}$

7. A number cube is rolled twice. Find P(odd number, 4). $\frac{1}{12}$

8. A coin is tossed six times. Find P(all tails). $\frac{1}{64}$

9. One day of the week and one month of the year are selected at random. Find P(Monday or Tuesday, month ending in ''r''). $\frac{2}{21}$

Exercises 10–17 refer to the number cube, spinners, and cards at the right. Determine the number of possible outcomes.

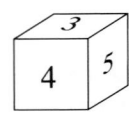

10. spin spinner A and spin spinner B 12

11. draw a card and roll the number cube 30

12. spin spinner A, spin spinner B, and select a card 60

A

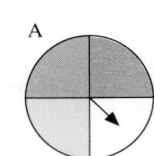

Find each probability.

13. You spin spinner A twice. Find P(red, red). $\frac{1}{16}$

B

14. You draw a card and spin spinner B. Find P(v, not blue). $\frac{2}{15}$

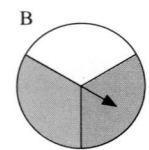

15. You spin spinner A and spin spinner B.
   a. Find P(yellow or white, blue). $\frac{1}{6}$
   b. Find P(not red, not red). $\frac{1}{2}$

B 16. You roll the number cube, spin spinner A, and select a card.
   a. Find P(2, red, x). $\frac{1}{120}$
   b. Find P(even number; not white; x, y, or z). $\frac{9}{40}$

17. You roll the number cube, spin spinner A, spin spinner B, and select a card. Find P(number < 5, red, red, not w). $\frac{2}{45}$

C

PROBLEM SOLVING/APPLICATION

Use this paragraph for Exercises 18–23.

License plates for motor vehicles usually contain six symbols. The first three symbols are often letters from A through Z and the last three symbols are often digits from 0 through 9.

18. How many letters are there from A through Z? 26

19. How many digits are there from 0 through 9? 10

In Exercises 20 and 21, assume that letters and digits can repeat.

20. Use the counting principle to find the number of different license plates possible. 26 · 26 · 26 · 10 · 10 · 10, or 17,576,000

21. Find the probability that the first digit on a license plate is a 5. $\frac{1}{10}$

In Exercises 22 and 23, assume that no letters and no digits repeat.

22. Use the counting principle to find the number of different license plates possible. 26 · 25 · 24 · 10 · 9 · 8, or 11,232,000

23. Find the probability that the third letter on a license plate is a T. $\frac{1}{26}$

SPIRAL REVIEW

S **24.** A number cube is rolled and a letter from A through M is chosen at random. Find P(odd number, A or M). *(Lesson 12-3)* $\frac{1}{13}$

25. Solve the proportion: $\frac{5}{15} = \frac{x}{12}$ *(Lesson 9-2)* 4

26. Tell whether 15 is *prime* or *composite*. *(Lesson 7-1)* composite

For Exercises 1–5, use the spinner shown on page 541.

1. Find P(odd number). $\frac{1}{2}$ **2.** Find P(red or blue). $\frac{5}{6}$ **12-1**

Find the odds in favor of each event.

3. red 1 to 1   **4.** even number 1 to 1   **5.** number < 6   5 to 1

6. Make a tree diagram to show all the possible outcomes when a number cube is rolled and then a coin is tossed. See answer on p. A32.   **12-2**

7. Use the tree diagram from Exercise 6. Find P(odd number, head). $\frac{1}{4}$

8. Use the counting principle to find the number of possible outcomes of rolling a number cube four times. 1296   **12-3**

9. A coin is tossed seven times. Use the counting principle to find P(all tails). $\frac{1}{128}$

Probability   **551**

551

Teaching the Lesson
- Materials: index cards, envelope
- Lesson Starter 113
- Using Manipulatives, p. 538D
- Cooperative Learning, p. 538D
- Reteaching/Alternate Approach, p. 538D
- Toolbox Skill 20

Lesson Follow-Up
- Practice 153
- Enrichment 124: Exploration
- Study Guide, pp. 225–226
- Cooperative Activity 22

Warm-Up Exercises

Find each answer.

1. $\frac{3}{12} \times \frac{5}{6}$ $\frac{5}{24}$

2. $\frac{5}{10} \times \frac{4}{9}$ $\frac{2}{9}$

3. $\frac{7}{8} \times \frac{6}{7}$ $\frac{3}{4}$

4. $\frac{5}{60} \times \frac{4}{25}$ $\frac{1}{75}$

5. $\frac{20}{30} \times \frac{5}{8}$ $\frac{5}{12}$

Lesson Focus

You may wish to use the following application to motivate today's lesson on finding the probability of independent and dependent events.

Application

Suppose you want to wear a different sweater every day this week. You have two brown sweaters, and you wear one of them on Monday. If you randomly select a sweater for Tuesday, would your chance of wearing a brown sweater be affected by Monday's selection? Yes

---

Independent and Dependent Events

Objective: To find the probability of two independent or two dependent events.

EXPLORATION

Write the letter A on each of 3 index cards, the letter B on each of 2 index cards, and the letter C on 1 index card. Place all 6 index cards in a large envelope or a box.

1. Draw a card at random. What is the probability of drawing an A? $\frac{1}{2}$

2. Replace the card you selected and draw again. What is the probability of drawing an A again? $\frac{1}{2}$

3. Remove an A. Do not replace the card, then draw again. What is the probability of drawing an A? $\frac{2}{5}$

4. Was the probability of drawing an A different when you replaced the first card than when you did not? Explain. See answer on p. A32.

Two events may occur either at the same time or one after the other. The two events are **independent** if the occurrence of the first event does not affect that of the second.

Probability of Independent Events

If events A and B are independent, the probability of both events occurring is found by multiplying the probabilities of the events.

$$P(A \text{ and } B) = P(A) \cdot P(B)$$

Example 1

A bag contains 3 red, 5 white, and 4 blue marbles. Two marbles are drawn at random with replacement. Find $P(\text{white, then red})$.

Solution

With replacement means that the first marble is replaced before the second is drawn. The two events are independent.

First find the probability of each event.

$$P(\text{white}) = \frac{5}{12} \qquad P(\text{red}) = \frac{3}{12} = \frac{1}{4}$$

Then multiply. $P(\text{white, then red}) = P(\text{white}) \cdot P(\text{red})$

$$= \frac{5}{12} \cdot \frac{1}{4} = \frac{5}{48}$$

 **Check Your Understanding**

1. Explain why the events in Example 1 are independent.

2. In Example 1, why is $P(\text{white}) = \frac{5}{12}$?

See *Answers to Check Your Understanding* at the back of the book.

Two events are **dependent** if the occurrence of the first event affects that of the second.

Example 2

Use the bag of marbles described in Example 1. Two marbles are drawn at random without replacement. Find P(white, then red).

Solution

Without replacement means that the first marble is not replaced before the second one is drawn. The events are dependent.

P(white, then red) $= P$(white) $\cdot P$(red after removing a white)

First find the probability of each event.

P(white) $= \dfrac{5}{12}$ ◀──── There are 5 white marbles.
◀──── There are 12 marbles altogether.

P(red after white) $= \dfrac{3}{11}$ ◀──── There are still 3 red marbles.
◀──── There are 11 marbles left.

Then multiply. $\qquad P$(white, then red) $= P$(white) $\cdot P$(red after white)

$$= \frac{5}{12} \cdot \frac{3}{11}$$

$$= \frac{15}{132} = \frac{5}{44}$$

 **Check Your Understanding**

3. Explain why the events described in Example 2 are dependent.

4. Consider Examples 1 and 2. Explain why P(white, then red) is different when the marbles are drawn without replacement than when the marbles are drawn with replacement.

See *Answers to Check Your Understanding* at the back of the book.

Guided Practice

1. Independent events have no effect on each other; dependent events have some effect on each other.
2. flipping a coin, then rolling a number cube
3. choosing a sock from a drawer, then choosing a second sock without replacing the first one
4. $P(A) \cdot P(B)$

COMMUNICATION «*Discussion* 1–3. Answers will vary. Examples are given.

«1. Explain in your own words the difference between independent events and dependent events.

«2. Give an example of two events that are independent.

«3. Give an example of two events that are dependent.

«4. **COMMUNICATION** «*Reading* Replace the __?__ with the correct expression.
  If A and B are independent events, then $P(A \text{ and } B) = $ __?__.

5. A number from 1 through 10 is chosen. Find $P(7)$. $\frac{1}{10}$

6. A wallet contains three $5 bills, two $10 bills, and one $20 bill. Two bills are selected at random.
  a. Find $P(\$10, \text{ then } \$5)$ if the bills are selected with replacement. $\frac{1}{6}$
  b. Find $P(\$10, \text{ then } \$5)$ if the bills are selected without replacement. $\frac{1}{5}$

Probability **553**

Teaching Notes

Key Questions

1. What are independent events? events in which the occurrence of the first event does not affect the occurrence of the second event

2. How can you find the probability of independent or dependent events? You multiply the probabilities of the events.

Error Analysis

Students sometimes use incorrect ratios when considering "without replacement." To help students avoid this error, have them do some experiments in groups and discuss the conditions.

Reasoning

The Check Your Understanding questions require students to differentiate between independent and dependent events. Encourage students to verbalize their understanding of these concepts.

Using Manipulatives

For a suggestion on using manipulatives to do the Exploration at the beginning of the lesson, see page 538D.

Reteaching/Alternate Approach

For a suggestion on classifying events as dependent or independent, see page 538D.

Additional Examples

1. A closet contains seven white shirts, three blue shirts, and four striped shirts. If a shirt were selected at random without replacement, what would be the probability that white shirts will be selected on two consecutive days? $\frac{3}{13}$

2. Using the situation in the example above, find P(blue, striped) if two shirts were selected with replacement. $\frac{3}{49}$

Exercises

A drawer contains 10 blue socks, 15 white socks, and 5 black socks. Two socks are selected at random with replacement. Find each probability.

A
1. P(blue, then white) $\frac{1}{6}$
2. P(white, then white) $\frac{1}{4}$
3. P(black, then white) $\frac{1}{12}$
4. P(both blue) $\frac{1}{9}$
5. P(both black) $\frac{1}{36}$
6. P(not blue, white) $\frac{1}{3}$

Exercises 7–16 refer to the cards shown below.

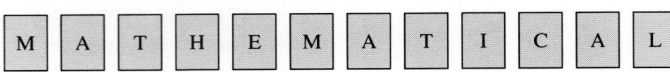

M A T H E M A T I C A L

Two cards are drawn at random with replacement. Find each probability.

7. P(M, then A) $\frac{1}{24}$
8. P(not M, then A) $\frac{5}{24}$
9. P(I, then T) $\frac{1}{72}$
10. P(vowel, then vowel) $\frac{25}{144}$

Two cards are drawn at random without replacement. Find each probability.

11. P(M, then A) $\frac{1}{22}$
12. P(A, then A) $\frac{1}{22}$
13. P(M, then M) $\frac{1}{66}$
14. P(E, then T) $\frac{1}{66}$
15. P(vowel, then vowel) $\frac{5}{33}$
16. P(vowel, then M) $\frac{5}{66}$

A number cube is rolled three times in a row. Find each probability.

B
17. P(all 4's) $\frac{1}{216}$
18. P(odd number, then odd number, then even number) $\frac{1}{8}$

A drawer contains 10 white socks and 6 blue socks. Khaled reaches in the drawer without looking and selects 2 socks.

19. What is the probability that he selects two blue socks? $\frac{1}{8}$
20. What is the probability that he selects first one blue sock and then one white sock? $\frac{1}{4}$
21. What is the probability that he selects no blue socks? $\frac{3}{8}$

A bucket contains 10 yellow, 8 white, and 2 orange tennis balls. Without looking, Trina selects 3 tennis balls.

22. What is the probability that she selects three yellow tennis balls? $\frac{2}{19}$
23. What is the probability that she selects no white tennis balls? $\frac{11}{57}$
24. Trina selects a ball and replaces it. She then selects two more balls without looking. Find P(yellow, then orange, then yellow). $\frac{1}{38}$

Six boys and four girls are finalists in a contest. All of their names are placed in a hat and three names are drawn at random. The prizes are awarded in the order the names are drawn. Find each probability.

25. P(3 girls) $\frac{1}{30}$

26. P(3 boys) $\frac{1}{6}$

27. P(boy, then girl, then boy) $\frac{1}{6}$

28. P(girl, then boy, then girl) $\frac{1}{10}$

The combination for a school locker consists of three one-digit numbers. For example, (0, 5, 9) and (3, 4, 1) are possible combinations. In Exercises 29, 31, and 32, digits may repeat. If combinations are randomly chosen, find each probability.

29. P(2, 2, 2) $\frac{1}{1000}$

30. P(4, 3, 9) $\frac{1}{1000}$

31. P(odd, odd, even) $\frac{1}{8}$

32. P(all digits are the same) $\frac{1}{100}$

C 33. P(all digits are different but none of the digits is 9) $\frac{63}{125}$

SPIRAL REVIEW

S 34. In the figures at the right, $\triangle ABC \sim \triangle XYZ$. Find the values of m and n. *(Lesson 10-8)* 3; 12

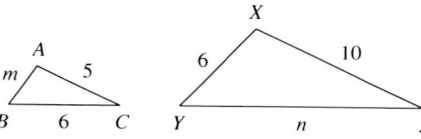

35. A number cube is rolled. Find P(6). *(Lesson 12-1)* $\frac{1}{6}$

36. The Tigers won 40 home games last year and 50 home games this year. Find the percent of increase. *(Lesson 9-9)* 25%

37. A bag contains 4 red marbles and 3 blue marbles. A marble is drawn at random with replacement. Find P(red, then blue). *(Lesson 12-4)* $\frac{12}{49}$

38. **MENTAL MATH** Use mental math to find the solution: $13 = 2a - 7$ *(Lesson 4-1)* 10

Probability **555**

Follow-Up

Enrichment

The diagram below shows where Sato's house *(S)* is in relation to Ben's house *(B)*. Sato plans to walk from her house to Ben's house. Assuming Sato always walks either north or east, in how many ways can she reach Ben's house?
10 ways

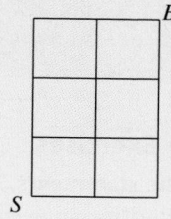

Enrichment

Enrichment 124 is pictured on page 538E.

Practice

Practice 153, *Resource Book*

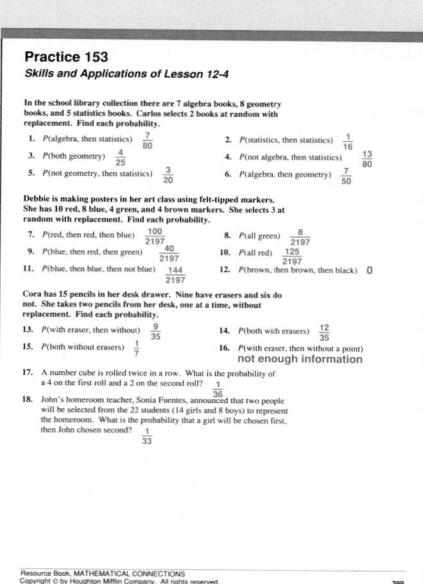

Lesson Planner

Warm-Up Exercises

Complete the statements.

1. Probabilities for an event go from _?_ to _?_. 0; 1
2. When considering two events, if the occurrence of the _?_ event does not affect that of the second, the events are _?_. first; independent
3. Consider independent events A and B. $P(A \text{ and } B) = $ _?_. $P(A) \cdot P(B)$

Lesson Focus

Ask students to consider the group of tall students in the class and the group of students with brown eyes. Ask what relationship exists between the two groups, if any. Then have students consider the group of students who walk to school and the group that eats salad for lunch. Again, ask if there is any relationship between the groups. The focus of today's lesson is on Venn diagrams, which can be used to demonstrate relationships among groups.

Teaching Notes

Recall with students some of the problem-solving strategies used previously in the text, such as making a table, identifying a pattern, and so on. Point out that using Venn diagrams is another useful strategy for solving problems in which relationships among groups need to be shown.

12-5

Strategy:
Using a Venn Diagram

Objective: To use Venn diagrams to solve problems.

The English mathematician John Venn used diagrams to show the relationships between collections of objects. These diagrams are known today as *Venn diagrams*. You can use Venn diagrams as a strategy for solving problems.

Problem

Of 160 students in the senior class, 55 take French, 90 take History, and 25 take both subjects.

a. How many take only French? How many take only History?

b. How many take either French or History?

c. How many seniors take neither of those subjects?

Solution

UNDERSTAND The problem is about the number of students who take different subjects.
Facts: 160 in class; 55 in French, 90 in History, 25 in both French and History
Find: number who take only one, either one, or neither

PLAN A Venn diagram illustrates the situation and can be used to answer the questions. To draw a Venn diagram, first draw a rectangle that represents the 160 students. Then, inside the rectangle, draw 2 overlapping circles. Label one circle F for French and the other H for History. The area where the circles overlap represents the 25 students who take both classes.

WORK

a. number taking only French
$55 - 25 = 30$
number taking only History
$90 - 25 = 65$

b. number taking either subject
$30 + 65 + 25 = 120$

c. number taking neither subject
$160 - 120 = 40$

ANSWER a. 30 students take only French and 65 take only History.

b. 120 students take either French or History.

c. 40 students take neither subject.

Look Back What if the total number of students in the class were 175 instead of 160. Which answer(s) would change? Which answer(s) would remain the same? the number taking neither; the number taking French, History, and either subject

Guided Practice

COMMUNICATION «*Reading*

Use this paragraph for Exercises 1–4.

At Marlin High School, 80 juniors have a part-time job, 110 juniors drive to school, and 35 juniors both drive and have part-time jobs. There are 280 juniors at the school.

« **1.** How many juniors are there at the school? 280

« **2.** How many juniors drive to school? 110

« **3.** Draw a Venn diagram to represent the situation. See answer on p. A32.

« **4.** How many juniors neither drive nor have part-time jobs? 125

Solve using a Venn diagram.

5. Of 50 people at a concert, 32 like rock music, 14 like country music, and 8 like both types of music. How many people like neither rock nor country music? 12

Problem Solving Situations

Solve using a Venn diagram.

Of 250 Wayland residents surveyed, 140 read the *Daily Times*, 80 read the *Evening Tribune*, and 45 read both newspapers.

1. How many residents read only the *Daily Times*? 95

2. How many residents read at least one of the newspapers? 175

3. How many residents read neither newspaper? 75

In a class of 42 people, 23 wear a watch, 12 wear tennis shoes, and 8 wear both tennis shoes and a watch.

4. How many wear tennis shoes but not a watch? 4

5. How many do not wear tennis shoes and do not wear a watch? 15

6. If one person in the class were chosen at random, what is the probability that the person is not wearing a watch? $\frac{19}{42}$

Of the 154 student musicians at Sweetwater High School, 72 are in the marching band, 35 are in the jazz band, and 20 are in both.

7. How many are in neither band? 67

8. How many play in the jazz band, but not the marching band? 15

9. How many play only in the marching band? 52

Probability **557**

Key Question

What do Venn diagrams show?
Venn diagrams show the relationships among groups.

Error Analysis

Students sometimes have difficulties drawing a correct Venn diagram. Show a few examples that demonstrate all possible relationships between two sets of objects. One example can use triangles and quadrilaterals, which have no overlap.

Additional Example

Out of a squad of seven cheerleaders, five take algebra (*A*), four take Spanish (*S*), and two take both subjects.
a. Draw a Venn diagram to illustrate the situation.

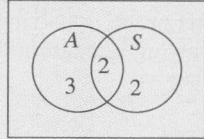

b. How many take algebra only? 3
c. How many take Spanish only? 2

Closing the Lesson

Have students describe Venn diagrams and tell how they are used.

Suggested Assignments

Skills: 1–20
Average: 1–20
Advanced: 1–20

Exercise Notes

Communication: Reading

Guided Practice Exercises 1–4 help students to apply their knowledge of Venn diagrams to a problem situation.

Problem Solving

Exercises 1–9 require students to use Venn diagrams to solve problems. *Exercises 10–14* have students consider the various problem-solving strategies they have learned in order to select an appropriate strategy for solving each problem.

Communication: Writing

For *Exercises 15 and 16,* students create problem-solving situations that can be represented by the Venn diagrams given in the text.

Follow-Up

Extension *(Cooperative Learning)*

Have students work in cooperative groups to create two problems. The problems should be designed to contrast the problem-solving strategies of drawing a diagram and using a Venn diagram.

Enrichment

Enrichment 125 is pictured on page 538E.

Practice

Practice 154, *Resource Book*

Practice 154
Skills and Applications of Lesson 12-5

Solve using a Venn diagram.

Of the 205 seniors at Kennedy High School, 135 applied to colleges out of
state, 120 applied to colleges in state, and 75 applied both to colleges out of
state and in state.

1. How many students applied only to colleges out of state? 60
2. How many students applied only to colleges in state? 45
3. How many seniors did not apply to college? 25

Of 178 ninth grade students, 83 are on a sports team, 111 are in a club, and
48 are involved in both sports and a club.

4. How many ninth grade students are involved in neither sports nor
a club? 32
5. How many ninth grade students are only on a sports team? 35
6. How many ninth grade students are only in a club? 63
7. If one student is chosen at random, what is the probability that the
student is involved in either a sports team or a club? 73/89
8. If a student is chosen at random, what is the probability that the
student is involved in neither a sports team nor a club? 16/89

Of 338 female students at Valley High School, 145 play soccer, 148 play
volleyball, and 60 play both soccer and volleyball.

9. How many students play volleyball only? 88
10. How many students play soccer only? 85
11. If a student is selected at random, what is the probability that the
student plays only volleyball? 44/169
12. How many students play neither sport? 105
13. If a student is chosen at random, what is the probability that the
student does not play either sport? 105/338

400   Resource Book, MATHEMATICAL CONNECTIONS
Copyright © by Houghton Mifflin Company. All rights reserved.

10–14. Strategies will vary; likely strategies are given.
10. not enough information; the amount of time the dentist works in one day is needed.
11. drawing a diagram; Marvin, Kim, Julie, Lance
12. drawing a diagram; 20

Solve using any problem solving strategy.

10. A dentist schedules 15 min for each patient. How many patients can the dentist see in one day?

11. At one point in a race, Julie was $\frac{1}{2}$ mi behind Marvin, Marvin was $\frac{3}{4}$ mi ahead of Lance, and Julie was $\frac{1}{4}$ mi behind Kim. List the order of the runners at that point.

12. How many diagonals can be drawn in an octagon?

13. A spinner has a red, a blue, and a yellow sector, each numbered 1, 2, or 3. The number of the blue sector is an odd number. The red sector is not numbered 1. The number of the yellow sector is one more than the number of the red sector. What number is the red sector? using logical reasoning; 2

14. Of 24 students in a gym class, 12 liked volleyball, 8 liked softball, and 6 liked both. How many students liked neither? using a Venn diagram; 10

WRITING WORD PROBLEMS

Write a word problem that you could solve using each Venn diagram. Then solve the problem. Answers to Problems 15–16 are on p. A32.

15.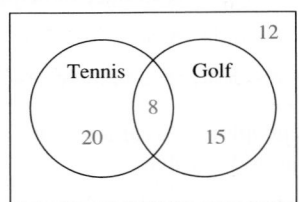

16.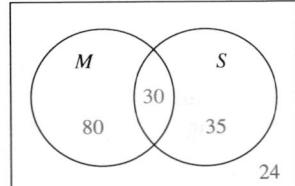

SPIRAL REVIEW

S 17. Solve the proportion: $\frac{6}{9} = \frac{10}{n}$ *(Lesson 9-2)* 15

18. Kinta's math scores were 78, 84, 96, 82, and 80. Find the mean, the median, the mode(s), and the range. *(Lesson 5-8)* 84; 82; none; 18

19. Find the next three numbers in the pattern: 8, 9, 11, 14, __?__, __?__, __?__ *(Lesson 2-5)* 18, 23, 29

20. There are 29 students in a history class. Of these, 16 students take geometry, 18 take Spanish, and 11 take both geometry and Spanish. How many students take neither geometry nor Spanish? *(Lesson 12-5)* 6

12-6

Permutations and Combinations

Objective: To find the number of permutations and the number of combinations of a group of items.

Terms to Know
- *permutation*
- *factorial*
- *combination*

APPLICATION

The personnel director of a computer software company plans to interview 3 people for a job opening. The director can choose any one of the 3 people for the first interview, either of the other 2 for the second, and the remaining person for the last. By the counting principle, the director can arrange the interviews in $3 \cdot 2 \cdot 1 = 6$ different orders.

An arrangement of a group of items in a particular order is called a **permutation.** The number of permutations of 3 items is $3 \cdot 2 \cdot 1 = 6$. The expression $3 \cdot 2 \cdot 1$ can be written 3!, which is read "3 **factorial.**"

Example 1

Find the number of permutations of the letters in the word MATH.

Solution

Use the counting principle. As a letter is chosen for each position in the "word," there is one less letter available for the next position.

position 1		position 2		position 3		position 4	
4	$\cdot$	3	$\cdot$	2	$\cdot$	1	= 24

There are 24 permutations of the letters in the word MATH.

 **Check Your Understanding**

1. In Example 1, list the 6 permutations that start with the letter M.
2. Explain how you could use a factorial to solve Example 1.
   See *Answers to Check Your Understanding* at the back of the book.

You can also make an arrangement using only part of a group of items.

Example 2

In how many different ways can first and second prizes be awarded in a contest among 5 people?

Solution

Find the number of permutations of *5 people taken 2 at a time.*

number of choices for first place		number of choices for second place	
5	$\cdot$	4	= 20

There are 20 ways to award first and second prizes.

Probability   **559**

Lesson Planner

Teaching the Lesson
- Materials: index cards, books, calculators
- Lesson Starter 115
- Using Technology, p. 538C
- Using Manipulatives, p. 538D
- Reteaching/Alternate Approach, p. 538D
- Cooperative Learning, p. 538D
- Toolbox Skills 9, 10

Lesson Follow-Up
- Practice 155
- Practice 156: Nonroutine
- Enrichment 126: Thinking Skills
- Study Guide, pp. 229–230
- Calculator Activity 12
- Test 58

Warm-Up Exercises

Use the counting principle.

1. Sweaters come in three sizes, four colors, and two styles. How many different choices of sweaters are there?   24

2. A number cube with six sides numbered 1 through 6 is rolled twice. Find $P(3, 4)$. $\frac{1}{36}$

Lesson Focus

Suppose a baseball manager has three different players he can use for the first three positions in the batting order. In how many different ways can he arrange the players in those positions? Solving problems of this type is the focus of today's lesson.

Teaching Notes

Work through the Application at the beginning of the lesson using the counting principle. Then relate the solution of the Application to factorial notation.

Key Questions

1. **What is a permutation?** A permutation is an arrangement of a group of objects in a particular order.
2. **How does a combination differ from a permutation?** A combination is a group of objects chosen without regard to order.
3. **What is meant by $_7C_4$?** the combination of 7 objects taken 4 at a time

Error Analysis

Students may confuse permutations with combinations. Stress that they should always consider whether order is important or not in the arrangement.

Using Technology

For a suggestion on using BASIC programs to generate permutations and combinations, see page 538C.

Using Manipulatives

For a suggestion on using index cards to do Example 1, see page 538D.

Reteaching/Alternate Approach

For a suggestion on using a cooperative learning activity, see page 538D.

Additional Examples

1. Find the number of permutations of the letters in the word SANDY. 120
2. In how many ways can a secretary and treasurer be chosen from 6 people? 15
3. Three food items are to be selected from a list of six. Find the number of combinations. 20

Permutations can be described by a brief notation. The number of permutations of 5 items can be written as $_5P_5$. Similarly, the number of permutations of 5 items taken 2 at a time can be written as $_5P_2$.

$$_5P_5 = 5! = 5 \cdot 4 \cdot 3 \cdot 2 \cdot 1 = 120 \qquad _5P_2 = 5 \cdot 4 = 20$$

A **combination** is a group of items chosen from within a group of items. The difference between one combination and another is the items in the combination, not the order of the items.

To understand the difference between a permutation and a combination, consider this example. Suppose there are three candidates, Ling, Jackson, and Garcia, running for two positions on a committee. There are six possible orders in which the candidates can finish in the voting, but only three different possible results of the election.

Possible Orders	Possible Election Results
Ling, Jackson, Garcia	
Jackson, Ling, Garcia	Ling and Jackson are elected.
Ling, Garcia, Jackson	
Garcia, Ling, Jackson	Ling and Garcia are elected.
Jackson, Garcia, Ling	
Garcia, Jackson, Ling	Jackson and Garcia are elected.

The list on the left contains all the permutations for this situation. The list on the right contains all the combinations for this situation.

Notice that there are 6 permutations of 3 people taken 2 at a time.

$$_3P_2 = 3 \cdot 2 = 6$$

To find the number of combinations of items taken 2 at a time, divide by $_2P_2$ to eliminate answers that are the same except for order.

$$_3P_2 \div {_2P_2} = 6 \div 2 = 3$$

There are 3 combinations of 3 people taken 2 at a time.

Example 3

A committee of 3 is to be chosen from a group of 7 people. Find the number of combinations.

Solution

Find the number of combinations of 7 people taken 3 at a time. First find $_7P_3$. Then divide by $_3P_3$.

$$\frac{_7P_3}{_3P_3} = \frac{7 \cdot 6 \cdot 5}{3 \cdot 2 \cdot 1} = \frac{210}{6} = 35$$

There are 35 ways that a committee of 3 can be chosen.

Combinations can also be described by a brief notation. The number of combinations of 7 items taken 3 at a time is written as $_7C_3$.

$$_7C_3 = \frac{_7P_3}{_3P_3} = 35$$

 You can use a calculator to find a number of permutations. If your calculator has a factorial key, you can use this calculator key sequence to find $_4P_4$.

If your calculator does not have a factorial key, you can use this calculator key sequence to find $_4P_4$.

Guided Practice

Tell whether each phrase refers to a *permutation* or a *combination*.

1. the number of ways to arrange 5 books on a shelf permutation

2. the number of ways to select 3 books from a choice of 12 books
combination

COMMUNICATION « *Writing*

Write each permutation or combination in words.

«3. $_2P_2$ 3. the number of permutations of 2 items taken 2 at a time

«4. $_5P_1$ 4. the number of permutations of 5 items taken 1 at a time

«5. $_9C_3$ 5. the number of combinations of 9 items taken 3 at a time

«6. $_7C_4$ 6. the number of combinations of 7 items taken 4 at a time

Use notation as in Exercises 3–6 to express each permutation or combination.

«7. the number of permutations of 8 out of 15 items $_{15}P_8$

«8. the number of combinations of 7 items taken 4 at a time $_7C_4$

Find the value of each expression.

9. 6! 720   **10.** 7! 5040   **11.** $_5P_4$ 120   **12.** $_4C_3$ 4

Find the number of permutations.

13. the letters in the word QUIET
120

14. 3 out of 6 students 120

Find the number of combinations.

15. 5 scarves taken 2 at a time 10

16. 4 out of 6 committee members
15

Exercises

Find the number of permutations.

A **1.** 2 posters on 2 walls 2

2. the digits 1, 2, 3, 4, and 5 120

3. the letters in the word
MARBLE 720

4. the letters in the word
COMPUTE 5040

5. a president and a secretary from 12 students 132

6. first, second, and third prizes among 9 contestants 504

Probability   **561**

Closing the Lesson
Have students summarize the procedures for finding permutations and combinations and how to distinguish between the two.

Suggested Assignments
Skills: 1–18, 21–27 odd, 37–41
Average: 1–27, 37–41
Advanced: 1–19 odd, 20–41

Exercise Notes

Communication: Writing
In *Guided Practice Exercises 3–8*, students translate symbolic expressions into words and vice versa. *Exercise 19* has students write a description of the difference between a permutation and a combination.

Calculator
Exercises 20–24 encourage the use of a calculator, and focus on the process of finding permutations.

Making Connections/Transitions
Exercises 25–27 relate the use of permutations to physical education. These exercises show the usefulness of mathematics to students. *Exercises 32–36* connect algebra and geometry by showing that the number of lines determined by a group of points can be found by taking combinations of the points.

Critical Thinking
Exercises 28–31 ask students to *examine* two combination problems and *compare* their results. Based on their conclusions, students are to *generalize* that $_nC_r = {_n}C_{n-r}$.

Problem Solving
Exercise 36 uses the problem-solving strategy of identifying a pattern to answer other questions.

Extension

Have students place the letters of their first name on index cards, one letter per card. Then have them calculate the number of possible permutations and combinations using two of the letters.

Enrichment

Enrichment 126 is pictured on page 538E.

Practice

Practice 155, *Resource Book*

562

Find the number of combinations.

7. 6 cassette tapes taken 3 at a time   20

8. a committee of 3 selected from a group of 8   56

9. 3 colors selected from 10 choices   120

10. 2 videotapes selected from a group of 6   15

Find the value of each expression.

11. $_4P_4$ 24   12. $_4C_4$ 1   13. $_6C_2$ 15   14. $_8P_2$ 56

15. In how many different ways can 5 people sit in a row of 5 seats?   120

16. In how many different ways can 5 books be chosen from a collection of 5 books?   1

17. In how many different ways can 4 paintings be selected from a group of 10 paintings?   210

18. There are 25 members in the journalism club. In how many different ways can a chairperson and a treasurer be chosen?   600

B   19. **WRITING ABOUT MATHEMATICS**   Write a paragraph describing the difference between a permutation and a combination.
See answer on p. A33.

 CALCULATOR

Use a calculator to find the number of permutations.

20. 7 shirts hanging in a closet   5040

21. the letters in the word PHONE   120

22. 10 charms on a bracelet   3,628,800

23. 9 hats on 9 heads   362,880

24. A calculator displays greater numbers in scientific notation. What is the least whole number with a factorial great enough that it appears in scientific notation on your calculator display?
Answers will vary. For most calculators, the answer is 12! or 14!

CONNECTING MATHEMATICS AND PHYSICAL EDUCATION

25. A school system belongs to a league of 8 cross-country teams. A school plays each of the other schools one time during the season. How many cross-country meets does the league have in a season?   28

26. How many different basketball squads with 5 players can a physical education teacher choose from a class of 12 students?   792

27. In how many different ways can a coach arrange the batting order of the 9 starting players on a baseball team?   362,880

THINKING SKILLS 28–29. Comprehension   **30.** Analysis   **31.** Synthesis

C   28. Find $_7C_2$.   21   29. Find $_7C_5$.   21

30. Compare the results of Exercises 28 and 29. What do you notice?
They are equal.

31. Write a formula comparing $_nC_r$ and $_nC_{n-r}$.
$_nC_r = {_nC_{n-r}}$

CONNECTING ALGEBRA AND GEOMETRY

Through any 2 points there is one and only one line. That is, 2 points determine a line. The number of lines determined by a group of points, no three of which lie on the same line, is the number of combinations of the group taken 2 at a time.

Find the number of lines determined by the given number of points. Assume that no three points lie on the same line.

32. 3 3   **33.** 4 6   **34.** 5 10   **35.** 6 15

36. PATTERNS Use your answers to Exercises 32–35.
  a. What is the pattern that describes the increase in the number of possible lines? Add 3, add 4, add 5, and so on.
  b. Use the pattern to predict the number of lines determined by 7 points, by 8 points, and by 9 points. 21; 28; 36

SPIRAL REVIEW

S   **37.** 25% of what number is 20?   *(Lesson 9-8)* 80

38. The length of a rectangle is 12 cm. The width of the rectangle is 8.5 cm. Find the perimeter.   *(Lesson 10-1)* 41 cm

39. Find the number of permutations of the letters in the word BRIGHT. *(Lesson 12-6)* 720

40. A coin is tossed 4 times. Find P(all heads).   *(Lesson 12-4)* $\frac{1}{16}$

41. DATA, *pages 538–539* In 1989, what percent of endangered species were birds?   *(Lesson 11-6)* about 31%

Self-Test 2

A bag contains 6 brown marbles, 8 white marbles, and 4 black marbles. Two marbles are selected at random with replacement. Find each probability.

1. P(white, then white) $\frac{16}{81}$ **2.** P(brown, then black) $\frac{2}{27}$ **12-4**

3. Two marbles are drawn at random without replacement from the bag described above. Find P(black, then black). $\frac{2}{51}$

4. In a class of 35 people, 21 wore slacks, 10 wore sweaters, and   **12-5**
7 wore slacks and sweaters. Use a Venn diagram to find how many wore slacks but not sweaters. 14

5. Find the number of permutations of the letters in the word   **12-6**
TEAMS. 120

6. In how many different ways can a committee of 2 be selected from a group of 6? 15

Probability   **563**

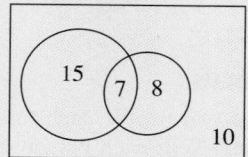

Teaching the Lesson
- Materials: paper cups
- Lesson Starter 116
- Using Technology, p. 538C
- Using Manipulatives, p. 538D
- Cooperative Learning, p. 538D

Lesson Follow-Up
- Practice 157
- Enrichment 127: Data Analysis
- Study Guide, pp. 231–232
- Computer Activity 12

Warm-Up Exercises

Find each answer.

1. $_3P_3$ 6
2. $_2P_2$ 2
3. $_3C_2$ 3
4. $_6P_6$ 720
5. $_4P_4$ 24
6. $_6C_4$ 15

Lesson Focus

Suppose that two people each toss a coin 20 times. Will each person get exactly the same number of heads? Today's lesson focuses on the difference between theoretical and experimental probability, thus providing a better understanding of how events actually occur.

Teaching Notes

To show that the experimental probability will approach the theoretical probability in large samples, have every student toss a coin 20 times and record the results. Contrast this example of experimental probability with one in which the outcomes are not equally likely to occur by using the *Data Analysis* at the beginning of the lesson.

12-7 Experimental Probability

Objective: To find the experimental probability of an event.

DATA ANALYSIS

A nickel is tossed onto a checkerboard 50 times. If the nickel does not land on the checkerboard, the toss is not counted. The results are shown in the frequency table at the right.

In many cases, the possible outcomes are not equally likely to happen. In the situation described above, the nickel is not equally likely to land on red, on black, or on both. So you cannot use the probability formula given on page 541 to determine the probability of the given event. Instead, you find the **experimental probability** based on the collection of actual data.

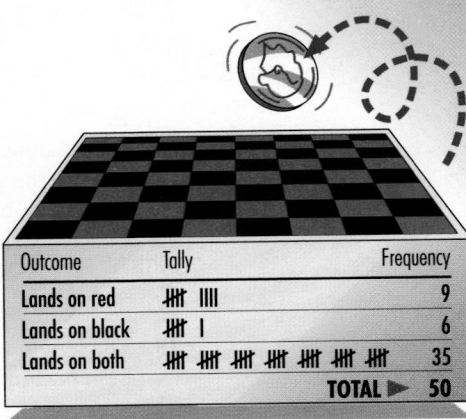

Outcome	Tally	Frequency
Lands on red	ⵏⵏⵏⵏⵏ IIII	9
Lands on black	ⵏⵏⵏⵏⵏ I	6
Lands on both	ⵏⵏⵏⵏⵏ ⵏⵏⵏⵏⵏ ⵏⵏⵏⵏⵏ ⵏⵏⵏⵏⵏ ⵏⵏⵏⵏⵏ ⵏⵏⵏⵏⵏ ⵏⵏⵏⵏⵏ	35
	TOTAL ▶	**50**

Example

Use the results in the table above to find the experimental probability that the nickel lands on both colors.

Solution

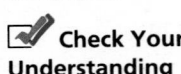 **Check Your Understanding**

$P(\text{both red and black}) = \dfrac{35}{50} = \dfrac{7}{10}$

1. What does 35 represent in the fraction in the Example?
2. If a dime were used instead of a nickel, do you think the results would have been the same? Explain.
See *Answers to Check Your Understanding* at the back of the book.

Guided Practice

COMMUNICATION «*Reading*

«**1.** What is the objective of this lesson? How is it different from the objective of Lesson 12-1? See answer on p. A33.

«**2.** When should experimental probability be used rather than the probability formula? when the outcomes of an experiment are not equally likely to happen

Refer to the frequency table on page 564. Find each probability.

3. P(only black) $\frac{3}{25}$

4. P(only red) $\frac{9}{50}$

5. P(orange) 0

6. P(only red or only black) $\frac{3}{10}$

7. P(only red, only black, or both red and black) 1

Exercises

Matt, Flavio, and Juana each tossed a paper cup 40 times. They recorded whether the cup landed up, down, or on its side. Use their results in the table below to find each experimental probability.

	Up	Down	Side
Matt	8	12	20
Flavio	6	10	24
Juana	7	11	22

6. $\frac{11}{20}$

$\frac{7}{40}$

A **1.** P(up) for Matt $\frac{1}{5}$ **2.** P(up) for Flavio $\frac{3}{20}$ **3.** P(up) for Juana

4. P(down) for Matt $\frac{3}{10}$ **5.** P(side) for Flavio $\frac{3}{5}$ **6.** P(side) for Juana

7. P(down) for Flavio $\frac{1}{4}$ **8.** P(up) for the whole group $\frac{7}{40}$

9. P(down) for the whole group $\frac{11}{40}$ **10.** P(side) for the whole group $\frac{11}{20}$

The table below shows Stan's results when he tossed a bottle cap 80 times. Use the results to find each experimental probability.

Outcome	Frequency	Tally
up	$\cancel{||||}\ \cancel{||||}\ \cancel{||||}\ \cancel{||||}\ \cancel{||||}\ \cancel{||||}\ \cancel{||||}\ \cancel{||||}\ \cancel{||||}$	45
down	$\cancel{||||}\ \cancel{||||}\ \cancel{||||}\ \cancel{||||}\ \cancel{||||}\ \cancel{||||}\ \cancel{||||}$	35

11. P(up) $\frac{9}{16}$ **12.** P(down) $\frac{7}{16}$ **13.** P(up or down) 1

B **14.** Penny also tossed a bottle cap 80 times. It landed down 56 times. Find the probability that the bottle cap does not land down. $\frac{3}{10}$

15. Killian tossed a paper cup. It landed up 20 times, down 26 times, and on its side 34 times. Find P(up), P(down), and P(side). $\frac{1}{4}, \frac{13}{40}, \frac{17}{40}$

16. A baseball player got a hit 36 times in the last 150 times at bat. What is the probability that the player will get a hit the next time at bat? Write the probability as a decimal to the nearest thousandth. 0.240

17. DATA, pages 538–539 The chances of successful mammal translocation are determined by experimental probability. Find the probability that the translocation of 50 mammals into a new environment under excellent conditions will be successful over a period of three years. 0.9

Probability **565**

Exercise Notes

Application
Exercises 1–13 require students to find various probabilities using real data from an experiment.

Cooperative Learning
Exercises 18–26 have students work in cooperative groups to perform an experiment, find experimental probabilities, and make predictions regarding the results.

Follow-Up

Exploration
Have students perform Buffon's Needle Problem. To do so, they will need several 1 in. long toothpicks and a piece of paper with parallel lines 2 in. apart. Students are to drop the toothpicks randomly onto the paper and keep a record of the total number of toothpicks dropped compared to the number of toothpicks falling on a line. Ask students to make a conjecture regarding this ratio.

Enrichment
Enrichment 127 is pictured on page 538F.

Practice
Practice 157, *Resource Book*

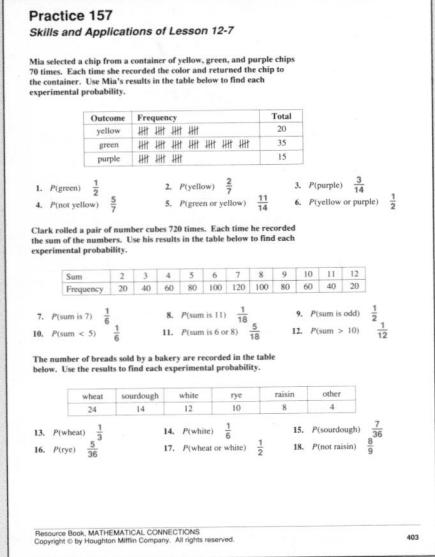

GROUP ACTIVITY 18–25. Answers will vary.

Randomly select a page in this textbook. Count the number of words that contain 1, 2, 3, 4, 5, 6, or more than 6 letters. (*Note:* Numbers should be counted as words. For example, the number 387 should be counted as a 3-letter word. Do not count operation symbols.)

18. Copy and complete the chart to record your data.

Outcome	Tally	Frequency
1 letter		
2 letters		
3 letters		
4 letters		
5 letters		
6 letters		
more than 6 letters		
	Total:	

Use the results recorded in your chart in Exercise 18 to find each experimental probability.

19. $P(4)$ **20.** $P(3)$ **21.** $P(2)$

22. $P(\text{more than } 6)$ **23.** $P(\text{fewer than } 5)$ **24.** $P(\text{more than } 4)$

25. Which word length appeared most often on the page you chose for your experiment? Which word length appeared least often?

26. Would you expect the same results if you chose another page? No.

SPIRAL REVIEW

S   **27.** Solve using a proportion: *The Weekly News* charges $14 for delivery for 8 weeks. At this rate, what is the charge for 26 weeks? (*Lesson 9-4*) $45.50

28. Graph the inequality $x < -1.5$. (*Lesson 7-10*) See answer on p. A33.

29. Alexis tossed a paper cup. It landed up 20 times, down 30 times, and on its side 40 times. Find $P(\text{side})$. (*Lesson 12-7*) $\frac{4}{9}$

30. Find the measure of an angle complementary to 80°. (*Lesson 6-3*) 10°

31. Draw a circle graph to display the data. (*Lesson 11-7*)
See answer on p. A33.

Tickets Sold by Grade Level

Grade	Freshman	Sophomore	Junior	Senior
Number Sold	60	30	90	120

32. Solve by supplying the missing fact: On three separate days, Angela spent 45 min, 55 min, and 50 min studying biology. How many hours in all did she spend studying biology? (*Lesson 8-7*) 2.5 h

Using BASIC to Simulate an Experiment

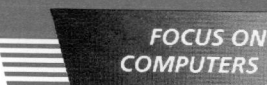

Objective: To estimate the waiting time for an event to happen.

Car license plates in many states have three numbers and three letters, for example 169 SPZ or 343 HAT. A typical *waiting time* problem about license plates is "About how many license plates would you have to inspect to find one whose letter section has the form 'consonant,' 'vowel,' 'consonant'?" A *waiting time* problem asks for the number of experiments you have to perform until you get a particular event.

The computer program below randomly selects three letters and tells you if each letter is a vowel or a consonant. It does this by picking a decimal at random between 0 and 1. If the decimal is less than $\frac{5}{26}$ (the probability of getting a vowel), the word VOWEL is printed. If the random decimal is greater than or equal to $\frac{5}{26}$, the word CONSONANT is printed. The program performs twenty experiments.

```
10   FOR E = 1 TO 20
20   FOR N = 1 TO 3
30   LET D = RND (1)
40   IF D < 5/26 THEN PRINT "VOWEL ";
50   IF D >= 5/26 THEN PRINT "CONSONANT ";
60   NEXT N
70   PRINT
80   NEXT E
90   END
```

Exercises

Use the computer program above. 1–5. Answers will vary.

1. Run the program. Count the number of experiments needed until you first see a CONSONANT VOWEL CONSONANT outcome.

2. Run the program 10 times. Find the average waiting time until you see a VOWEL CONSONANT VOWEL outcome.

3. Find the average waiting time until you see at least one VOWEL.

Change lines 40 and 50 in the program above. Use the new program for Exercises 4 and 5.

```
40   IF D < 1/2 THEN PRINT "ODD ";
50   IF D >= 1/2 THEN PRINT "EVEN ";
```

4. Run the program 10 times. Find the average waiting time until you see the numbers in the order ODD EVEN ODD.

5. Find the average waiting time until you see at least one ODD.

Probability   567

Teaching Note

The application with license plates should be of interest to all students. Be sure to go through each step of the computer program with the students so they understand and appreciate the logical sequence of the steps.

Exercise Note

Exercises 1–3 ask students to run the program and make observations regarding the waiting times for specific results. *Exercises 4 and 5* require students to find other average waiting times for patterns with odd and even numbers, and they provide an opportunity for students to see the effect of changing just two lines in the program.

Follow-Up

Extension

You may wish to have students further alter the program and observe the affect on the results. Students could extend the number of times the program is run and compare results to the original run of ten times, or they may look for other patterns.

Teaching the Lesson
- Lesson Starter 117
- Reteaching/Alternate Approach, p. 538D
- Toolbox Skills 10, 13

Lesson Follow-Up
- Practice 158
- Practice 159: Nonroutine
- Enrichment 128: Application
- Study Guide, pp. 233–234
- Test 59

Warm-Up Exercises

Find the decimal equivalent of each fraction to the nearest thousandth.

1. $\frac{220}{450}$ 0.489

2. $\frac{60}{85}$ 0.706

Find each answer.

3. 100,000(0.125) 12,500
4. (0.52)5000 2600

Lesson Focus

Ask students to imagine building a new restaurant in a community of about 10,000 people. To determine what type of restaurant to build, the owners decide to conduct a random survey of 500 people. The focus of today's lesson is to discuss how random samples can be used to make predictions.

Teaching Notes

You may wish to begin the lesson with the following activity: Give students a questionnaire on which they must indicate their favorite cafeteria food. Tally the results, and discuss with students what they think the results would be if the entire student body responded to the questionnaire.

Using Samples to Make Predictions

Objective: To make predictions using sampling and probability concepts.

Terms to Know
- *random sample*

The Fairtown Tribune conducted a poll before a primary election. The poll takers chose a group of 400 people at random from among the 220,000 registered voters. Each person in this **random sample,** or sample chosen at random from a larger group, was asked, "If the election were held today, which candidate would receive your vote?" The results are shown in the table below.

Candidate	Number of Supporters
Perez	115
Say	90
Wong	109
Wiley	86

Example

Solution

How many votes might Say expect to receive in the election?

Use the sample results to find P(Say). There were 400 people in the random sample. Ninety people in the sample would vote for Say.

$$P(\text{Say}) = \frac{90}{400} = 0.225$$

The number of votes expected is the product of this probability and the total number of registered voters.

P(Say) × number of registered voters = expected number of votes

0.225 × 220,000 = 49,500

Say might expect to receive about 49,500 votes.

Check Your Understanding

1. How does a poll taker determine the number of votes a candidate can expect?
2. About how many voters are not expected to vote for Say?
See *Answers to Check Your Understanding* at the back of the book.

Guided Practice

1. No; random samples must be chosen without bias.

«**1.** COMMUNICATION «*Reading* If the sample described above included 400 of Say's relatives, friends, and business associates, would that be a random sample? Explain.

«**2.** COMMUNICATION «*Discussion* Explain how you would select a random sample to predict the outcome of a class election.
See answer on p. A33.

A television ratings service asked a random sample of 300 of the 44,000 residents of Millsburg to name their favorite program. Of those polled, 60 chose *The Hendersons*, 44 chose *The Lawyers*, 74 chose *Win That Vacation*, 80 chose *Comedy Hour*, and 42 chose *The Detectives*. Use these sample results for Exercises 3–6.

3. Find P(The Hendersons). $\frac{1}{5}$ **4.** Find P(The Detectives). $\frac{7}{50}$

If the whole town were questioned, about how many residents would be expected to select each program as their favorite?

5. The Hendersons about 8800   **6.** The Detectives about 6160

Exercises

Refer to the text on page 568. About how many votes might each candidate expect to receive?

A   **1.** Perez about 63,250   **2.** Wiley about 47,300   **3.** Wong about 59,950

Solve. In Exercises 4–7, assume that each sample is random.

4. When 50 seniors were questioned, 40 said they plan to attend college. About how many of the 480 seniors are expected to attend college? about 384

5. When 75 residents of a senior citizen apartment were questioned, 35 said they favor construction of a new shopping center across the street. About how many of the 450 residents are expected to favor construction of the new shopping center? about 210

6. A random sample of 150 owners of a new Brandon automobile were questioned to determine their satisfaction. Of those polled, 120 stated they were very satisfied, 25 stated they were satisfied, and 5 stated they were dissatisfied. About how many of the 12,000 owners of new Brandon automobiles are expected to be dissatisfied? about 400

Exercise Notes

Communication: Reading

Guided Practice Exercise 1 checks students' understanding of the concept of a random sample.

Communication: Discussion

Guided Practice Exercise 2 promotes student verbalization regarding an effective procedure for obtaining a random sample.

Application

Exercises 1–3 extend the Example on page 568 by having students predict the number of votes for each candidate. Further applications occur in *Exercises 4–7*, which demonstrate real-world situations involving random samples and predictions, and in *Exercises 16–19,* which show applications in several job situations.

Data Analysis

Exercises 8–15 require students to use the data obtained from a random sample to find the probability for each rating and to make predictions about the number of subscribers choosing each rating.

Reasoning

In *Exercises 20–23,* students need to determine if using a poll would be appropriate and, if not, they should suggest a more appropriate method.

B 7. A random survey, or poll, of 400 people showed that 120 listen to radio station KMAT in the mornings and 80 people listen to station KEFA in the evenings. The survey was conducted in an area with a population of about 18,000 people. About how many people in the area are expected to listen to KMAT in the mornings? About how many are expected to listen to KEFA in the evenings? about 5400; about 3600

DATA ANALYSIS

A cable television company polled a random sample of 200 of their 42,000 customers. The customers were asked to rate their service as excellent, good, fair, or poor. The results are shown in the table at the right.

Quality of Service	Number Who Voted
Excellent	104
Good	58
Fair	34
Poor	4

Find the probability that a person polled feels the quality of service is at each level.

8. excellent $\frac{13}{25}$ 9. good $\frac{29}{100}$ 10. fair $\frac{17}{100}$ 11. poor $\frac{1}{50}$

Use the results of the cable company poll to predict how many of the 42,000 subscribers feel their cable service is at each level.

12. excellent
about 21,840

13. good
about 12,180

14. fair
about 7140

15. poor
about 840

CAREER/APPLICATION

A *quality control inspector* checks random samples of certain manufactured goods to find the number of defective goods in the sample. The inspector then predicts the number of defective goods in a larger quantity of the item. Depending on the seriousness of the defect or on the number of defective goods predicted, the manufacturer may decide not to sell certain *production lots* of the item.

16. A quality control inspector selected 40 out of a lot of 6000 light bulbs to inspect and found 3 to be defective. Predict the number of defective light bulbs that might be found in the entire lot. about 450

17. Ten out of a sample of 100 goods inspected were defective. It was later found that 45 goods in the entire production lot were defective. Approximate the number of goods in this production lot. about 450

18. A manufacturer of computer diskettes will not sell a production lot that contains more than 5% defective diskettes. A quality control inspector checks a sample of 50 diskettes in a certain lot and finds that 2 are defective. Would the manufacturer sell this production lot? Yes.

19. Suppose 30 out of the 2400 cartons of milk in a production lot have leaks. What percent of the cartons in the lot have leaks? 1.25%

LOGICAL REASONING

Tell whether it is appropriate to use a poll to predict each of the following. If it is not appropriate, describe a more appropriate method.

20. the winner of a spelling contest

21. the winner of a golf tournament
No; the players' past results should be considered.

22. the favorite television program among students in a school Yes.

23. the favorite book read by students in a school during the summer vacation Yes.

20. No; the contestants' past results and school grades should be considered.

SPIRAL REVIEW

S 24. Find the area of a triangle with base 14 in. and height 10 in. *(Lesson 10-3)* 70 in.²

25. When 40 tenth graders, selected randomly, were questioned, 12 said they walk to school. About how many of the 520 tenth graders in the town walk to school? *(Lesson 12-8)* about 156

26. Simplify: x^{-3} *(Lesson 7-11)* $\frac{1}{x^3}$

27. Estimate: $-15\frac{1}{3} \div \frac{5}{14}$ *(Lesson 8-6)* about -45

28. Tell whether you need to calculate *perimeter*, *circumference*, or *area* to find the amount of grazing land in a rectangular field. *(Lesson 10-5)* area

Self-Test 3

Charlita tossed a paper cup 150 times. It landed up 84 times, down 27 times, and on its side 39 times. Find each experimental probability.

1. $P(\text{up})$ $\frac{14}{25}$ **2.** $P(\text{down})$ $\frac{9}{50}$ **3.** $P(\text{side})$ $\frac{13}{50}$ 12-7

4. The *Daily Newsreporter* polled a random sample of 500 registered voters before the local school board election. The results are shown below. **12-8**

Candidate	Number of Supporters
Bernedez	150
Caldwell	180
Stewart	70
Aluga	100

There are 180,000 registered voters. From about how many voters does Aluga expect support in the election? about 36,000

Probability **571**

Follow-Up

Extension
Have students create a grid such as the one below to display the possible sums when two number cubes are rolled. Then ask students to use the grid to find the probability of rolling a certain sum.

Enrichment
Enrichment 128 is pictured on page 538F.

Practice
Practice 158 is shown below.

Quick Quiz 3
See page 575.

Alternative Assessment
See page 575.

Practice 158, *Resource Book*

Practice 158
Skills and Applications of Lesson 12-8

Solve. Assume that each sample is random.

1. When 30 students were questioned, 18 said they favored changing the school day to nine periods instead of the present eight. About how many of a total of 255 students are expected to favor a nine period day? **about 153**

2. When 40 students at Mountain High School were asked about their favorite pizza crust, 10 said they prefer thick crust. About how many of the total school enrollment of 1280 are expected to prefer thick crust? **about 320**

3. The athletic director asked 80 students at Science High School about football game scheduling. Of those asked, 35 preferred Friday night games, 30 preferred Saturday afternoon games, and 15 had no preference. About how many of the total enrollment of 2960 are expected to prefer Saturday afternoon games? About how many are expected to have no preference? **about 1110; about 555**

4. Ramon asked 30 students the question, "If I ran for class president, would you vote for me?" Of those asked, 14 said yes, 7 said no, and 9 were undecided. Of a 2460 student body, about how many votes might he expect? **about 1148**

5. When 48 seniors were asked if they had taken achievement tests, 26 had taken the mathematics test, 13 had taken the chemistry test, and 6 had taken neither. About how many of 296 seniors are expected to take the mathematics test? About how many of the seniors are expected to take neither? **about 111; about 37**

6. A survey of 45 students showed that 28 students use calculators for computation and 17 use paper and pencil. About how many of a total of 495 students are expected to use a calculator? **about 308**

Resource Book, MATHEMATICAL CONNECTIONS
Copyright © by Houghton Mifflin Company. All rights reserved.
404

Teaching Note

The use of random number tables is a common practice since they can be easily generated by a computer. Have students examine the table on page 573 to confirm the randomness of the numbers. Then have students make predictions concerning how many times a given digit, such as zero, would be expected to appear in a given row. Compare the predictions with the actual results from several rows.

Exercise Note

Application

Students are able to see how various samples can be selected in a random manner by the use of a random number table. The exercises demonstrate real-world situations where this procedure is used. As a result, students' appreciation for this procedure should be increased.

FOCUS ON APPLICATIONS

Using a Table of Random Numbers

Objective: To use a table of random numbers to select a random sample.

Marie DeForge is a marketing executive for a publishing company. She plans to survey 50 students from a university with 20,000 students. She will use a table of random numbers to select the 50 students.

A table of random numbers is a list of numbers in which the digits 0, 1, 2, 3, 4, 5, 6, 7, 8, and 9 are arranged randomly.

Example

a. How can Marie use a table of random numbers to select 50 students at random?

b. What are the first 5 numbers selected by this method?

Solution

a. Marie assigns each student a five-digit number from 00001 to 20000. To select a random sample of 50 students, she randomly selects a starting point and a direction from the table on page 573. She will discard any number that is greater than 20,000. Suppose she starts on column 4, line 1, and reads down the column. When she completes column 4, she will read down column 5, and so on.

b. The first number in column 4, line 1, is 02011. The first student selected has number 2011. The next number is 85393. That number is discarded. The next four numbers Marie selects are 16656, 07972, 10281, and 03427.

Exercises

Use the random number table on page 573 to answer each question. Start with the indicated column and line and read down the column.

1. a. 85393, 97265, 61680, 16656, 42751
   b. They are greater than 15000.
   c. 02011, 07972, 10281, 03427, 08178, 09998, 14346, 07351, 12908, 07391, 07856, 06121, 09172, 13363, 04024

1. The mayor's office intends to interview a random sample of 15 of the town's 15,000 households. Each household is assigned a number from 00001 to 15000. (column 4, line 1)
   a. What are the first five numbers that will be discarded?
   b. Why will the numbers in part (a) be discarded?
   c. What are the 15 numbers selected by this method?

2. A magazine plans to celebrate its twentieth anniversary by randomly selecting 20 of its 97,000 subscribers and awarding them a free lifetime subscription. (column 2, line 6)
   a. To use the table of random numbers, what group of numbers can be assigned to the subscribers? 00001 to 97000
   b. What are the first and last numbers chosen in the random sample?
   06907, 20591
   c. How many numbers will be discarded? 2

3. The customer service department of a bank plans to interview a random sample of 25 of the bank's 85,056 bank card holders to get their reactions to the benefits of the card. Describe how to use a table of random numbers to select 25 card holders at random.

Table of Random Numbers

LINE \ COL.	1	2	3	4	5	6	7	8	9	10
1	10480	15011	01536	02011	81647	91646	69179	14194	62590	36207
2	22368	46573	25595	85393	30995	89198	27982	53402	93965	34095
3	24130	48360	22527	97265	76393	64809	15179	24830	49340	32081
4	42167	93093	06243	61680	07856	16376	39440	53537	71341	57004
5	37570	39975	81837	16656	06121	91782	60468	81305	49684	60672
6	77921	06907	11008	42751	27756	53498	18602	70659	90655	15053
7	99562	72905	56420	69994	98872	31016	71194	18738	44013	48840
8	96301	91977	05463	07972	18876	20922	94595	56869	69014	60045
9	89579	14342	63661	10281	17453	18103	57740	84378	25331	12566
10	85475	36857	53342	53988	53060	59533	38867	62300	08158	17983
11	28918	69578	88231	33276	70997	79936	56865	05859	90106	31595
12	63553	40961	48235	03427	49626	69445	18663	72695	52180	20847
13	09429	93969	52636	92737	88974	33488	36320	17617	30015	08272
14	10365	61129	87529	85689	48237	52267	67689	93394	01511	26358
15	07119	97336	71048	08178	77233	13916	47564	81056	97735	85977
16	51085	12765	51821	51259	77452	16308	60756	92144	49442	53900
17	02368	21382	52404	60268	89368	19885	55322	44819	01188	65255
18	01011	54092	33362	94904	31273	04146	18594	29852	71585	85030
19	52162	53916	46369	58586	23216	14513	83149	98736	23495	64350
20	07056	97628	33787	09998	42698	06691	76988	13602	51851	46104
21	48663	91245	85828	14346	09172	30168	90229	04734	59193	22178
22	54164	58492	22421	74103	47070	25306	76468	26384	58151	06646
23	32639	32363	05597	24200	13363	38005	94342	28728	35806	06912
24	29334	27001	87637	87308	58731	00256	45834	15398	46557	41135
25	02488	33062	28834	07351	19731	92420	60952	61280	50001	67658
26	81525	72295	04839	96423	24878	82651	66566	14778	76797	14780
27	29676	20591	68086	26432	46901	20849	89768	81536	86645	12659
28	00742	57392	39064	66432	84673	40027	32832	61362	98947	96067
29	05366	04213	25669	26422	44407	44048	37937	63904	45766	66134
30	91921	26418	64117	94305	26766	25940	39972	22209	71500	64568
31	00582	04711	87917	77341	42206	35126	74087	99547	81817	42607
32	00725	69884	62797	56170	86324	88072	76222	36086	84637	93161
33	69011	65795	95876	55293	18988	27354	26575	08625	40801	59920
34	25976	57948	29888	88604	67917	48708	18912	82271	65424	69774
35	09763	83473	73577	12908	30883	18317	28290	35797	05998	41688
36	91567	42595	27958	30134	04024	86385	29880	99730	55536	84855
37	17955	56349	90999	49127	20044	59931	06115	20542	18059	02008
38	46503	18584	18845	49618	02304	51038	20655	58727	28168	15475
39	92157	89634	94824	78171	84610	82834	09922	25417	44137	48413
40	14577	62765	35605	81263	39667	47358	56873	56307	61607	49518
41	98427	07523	33362	64270	01638	92477	66969	98420	04880	45585
42	34914	63976	88720	82765	34476	17032	87589	40836	32427	70002
43	70060	28277	39475	46473	23219	53416	94970	25832	69975	94884
44	53976	54914	06990	67245	68350	82948	11398	42878	80287	88267
45	76072	29515	40980	07391	58745	25774	22987	80059	39911	96189
46	90725	52210	83974	29992	65831	38857	50490	83765	55657	14361
47	64364	67412	33339	31926	14883	24413	59744	92351	97473	89286
48	08962	00358	31662	25388	61642	34072	81249	35648	56891	69352
49	95012	68379	93526	70765	10592	04542	76463	54328	02349	17247
50	15664	10493	20492	38391	91132	21999	59516	81652	27195	48223

3. Assign each card holder a number from 00001 to 85056. Randomly select a starting point and direction in the table. Discard each number greater than 85056. Continue until 25 numbers have been selected.

Follow-Up

Extension (Cooperative Learning)
Have students work in cooperative groups to construct their own table of random numbers. Groups can use a number spinner to generate random digits. They should tabulate how many spins are necessary before all ten digits appear. The groups can compare results and the entire class can calculate the average. Discuss with the groups situations for which a random number table could be applied.

Probability   573

Chapter Review

Terms to Know

outcome (p. 541)
equally likely (p. 541)
event (p. 541)
probability (p. 541)
odds in favor (p. 542)
sample space (p. 545)
tree diagram (p. 545)
counting principle (p. 549)

independent events (p. 552)
dependent events (p. 553)
permutation (p. 559)
factorial (p. 559)
combination (p. 560)
experimental probability (p. 564)
random sample (p. 568)

Choose the correct term from the list above to complete each sentence.

1. Two events are __?__ if the occurrence of the first affects the occurrence of the second. dependent events

2. A(n) __?__ is any group of outcomes. event

3. A(n) __?__ is a list of all possible outcomes. sample space

4. An arrangement of a group of items in a particular order is a(n) __?__. permutation

5. The __?__ of an event is based on the collection of actual data. experimental probability

6. The expression 5! is an example of a(n) __?__. factorial

Use the spinner at the right. Find each probability. *(Lesson 12-1)*

7. $P(4)$ $\frac{1}{5}$

8. $P(8)$ 0

9. $P(\text{blue})$ $\frac{2}{5}$

Use the spinner at the right. Find the odds in favor of each event. *(Lesson 12-1)*

10. odd number
3 to 2

11. red 2 to 3

12. not blue
3 to 2

13. The spinner at the right is spun and a coin is tossed. Make a tree diagram to show the sample space. *(Lesson 12-2)*
See answer on p. A33.

Use the tree diagram from Exercise 13 to find each probability.
(Lesson 12-2)

14. $P(4, \text{heads})$ $\frac{1}{10}$

15. $P(2 \text{ or } 5, \text{tails})$ $\frac{1}{5}$

16. $P(\text{not 1, not heads})$ $\frac{2}{5}$

17. $P(6, \text{tails})$ 0

Use the counting principle. *(Lesson 12-3)*

18. Teruko is buying a couch. She has a choice of 3 styles, 5 colors, and 4 different fabrics. Find the number of possible choices. 60

19. A whole number from 1 through 15 is chosen at random and a coin is tossed. Find $P(\text{even, heads})$. $\frac{7}{30}$

Exercises 20–23 refer to a box that contains 7 red plastic chips, 5 blue plastic chips, and 3 white plastic chips.

Two chips are drawn at random with replacement. Find each probability. *(Lesson 12-4)*

20. P(blue, then red) $\frac{7}{45}$

21. P(red, then white) $\frac{7}{75}$

Two chips are drawn at random without replacement. Find each probability. *(Lesson 12-4)*

22. P(red, then red) $\frac{1}{5}$

23. P(blue, then white) $\frac{1}{14}$

24. In a mathematics class of 28 students, 14 are also taking physics, 12 are taking biology, and 5 are taking both physics and biology. Use a Venn diagram to find how many students in the mathematics class are not taking either physics or biology. *(Lesson 12-5)* 7

Find the number of permutations. *(Lesson 12-6)*

25. the letters in the word NUMBER 720

26. the letters in the word TRUCK taken three at a time 60

Find the number of combinations. *(Lesson 12-6)*

27. 4 books selected from a group of 9 126

28. 7 students selected 2 at a time 21

29. A baseball cap is tossed onto a floor made of brown and white tiles. The results are shown in the table at the right. Find P(brown). *(Lesson 12-7)* $\frac{1}{5}$

Outcome	Frequency
Lands on white	12
Lands on brown	10
Lands on both	28

In a poll taken before a recent election, 250 voters were asked their reasons for favoring candidates. *(Lesson 12-8)*

Do you favor a candidate because of the candidate's	Number polled responding "Yes"
television appearance?	65
decisions on key issues?	92
past political voting record?	56
political party affiliation?	37

30. Find the probability that a person polled favors a candidate because of the candidate's television appearance. $\frac{13}{50}$

31. The district where the poll was taken includes 400,000 registered voters. About how many of these voters are expected to favor a candidate because of the candidate's television appearance? about 104,000

Quick Quiz 3 *(12-7 through 12-8)*
Ramon tossed a paper cup 125 times. It landed up 65 times, down 40 times, and on its side 20 times. Find each experimental probability.

1. P(up) $\frac{13}{25}$

2. P(down) $\frac{8}{25}$

3. P(side) $\frac{4}{25}$

4. A random survey of 300 students was conducted from among a student body of 2000 students. The results are shown below.

Favorite Sport	Number Responding
Football	90
Basketball	70
Volleyball	55
Soccer	85

How many students would be expected to choose basketball as their favorite sport if the entire student body were polled? about 467

Alternative Assessment

1. Explain how the size of a sample could affect the accuracy of predictions. If the sample were small, it may not be representative of the total sample space.

2. Give an example of a biased sampling. Answers will vary; for example, asking people who own blue cars only to name their favorite car color.

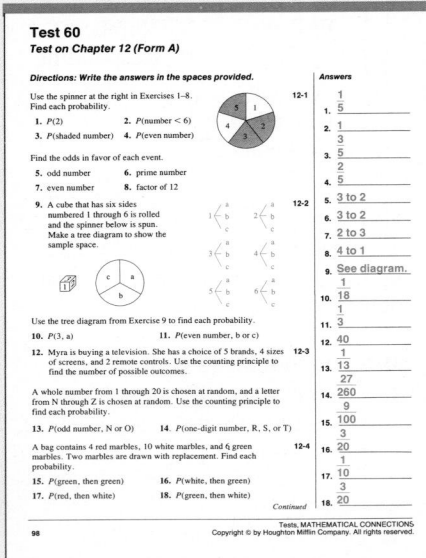
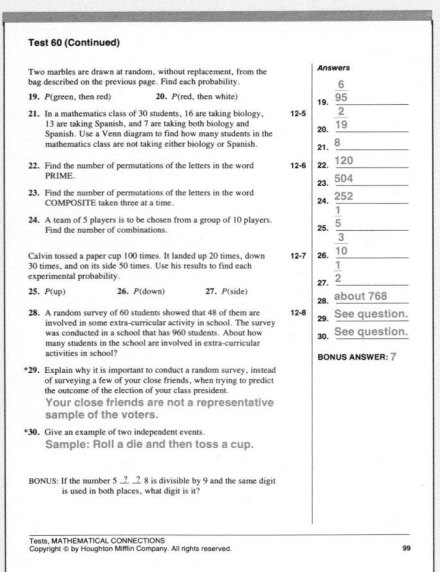
Chapter Test

Use the spinner at the right. Find each probability.

1. $P(4)$ $\frac{1}{6}$ 2. $P(0)$ 0   3. $P(\text{white})$ $\frac{1}{6}$ 12-1

Use the spinner at the right. Find the odds in favor of each event.

4. even number   5. blue   6. not white
    1 to 1   1 to 1   5 to 1

★ 7. Explain why it is impossible for the probability of an event to be greater than 1. See answer on p. A33.

8. A coin is tossed and the spinner above is spun. Make a tree diagram to show the sample space. See answer on p. A33.   12-2

Use the tree diagram from Exercise 8 to find each probability.

9. $P(\text{heads, 5})$ $\frac{1}{12}$ 10. $P(\text{tails, odd number})$ $\frac{1}{4}$

11. Ed is buying a suit. He has a choice of 6 colors, 4 styles, and 2 fabrics. Use the counting principle to find the number of possible outcomes. 48   12-3

12. A whole number from 1 through 10 is chosen at random and a letter from M through Z is chosen at random. Use the counting principle to find $P(\text{prime number, X or Y})$. $\frac{2}{35}$

A bag contains 10 red marbles, 5 blue marbles, and 3 white marbles.

13. Two marbles are chosen with replacement. Find $P(\text{blue, then white})$. $\frac{5}{108}$ 12-4

14. Two marbles are chosen without replacement. Find $P(\text{red, then red})$. $\frac{5}{17}$

★ 15. Write a word problem that involves dependent events. Then solve the problem. See answer on p. A33.

16. Of 100 people, 55 like basketball, 45 like hockey, and 20 like both. Use a Venn diagram to find how many people do not like either hockey or basketball. 20   12-5

17. Find the number of permutations of the letters in the word HISTORY.   12-6
5040
18. Find the number of permutations of the letters in the word LAWYER taken four at a time. 360

19. A committee of 5 people is to be chosen from a group of 8 people. Find the number of combinations. 56

20. A paper cup is tossed 60 times. It lands on its side 38 times, on its top 8 times, and on its bottom 14 times. Find $P(\text{bottom})$. $\frac{7}{30}$ 12-7

21. A random survey of a class of 50 students showed that 12 of the students have dogs as pets. The survey was conducted in a school that has 750 students. About how many students in the school are expected to have dogs as pets? about 180   12-8

Expected Value

Objective: To find the expected value of an event.

Many decisions are based on a knowledge of **expected value.** The expected value of an experiment is the sum of the products of each outcome and its probability. Expected value tells you the result that is obtained *on average* when an experiment is performed repeatedly.

Example

An outdoor theater loses $4000 on concert days when it rains and earns $12,000 on concert days when it does not rain. The probability that it will rain on a concert day is 0.2. Find the amount that the outdoor theater can expect to earn on average for a concert.

Solution

There are two possible outcomes: it will rain and the theater will lose $4000; it will not rain and the theater will earn $12,000.

$P(\text{rain}) = 0.2$; $P(\text{no rain}) = 1 - P(\text{rain}) = 0.8$

$$\text{expected value} = -4000 \cdot P(\text{rain}) + 12{,}000 \cdot P(\text{no rain})$$
$$= -4000(0.2) + 12{,}000(0.8)$$
$$= -800 + 9600 = 8800$$

The outdoor theater can expect to earn $8800 on average for a concert.

Exercises

1. An appliance store offers each customer an envelope that contains a discount. The discounts and their probabilities are shown in the table. Find the discount that a customer can expect on average. $17.40

Discount	Probability
$100	0.02
$50	0.06
$20	0.32
$10	0.60

2. A building contractor bids on a job to build a new school. On such a project, the contractor has a 75% chance of making a profit of $48,000 and a 25% chance of losing $10,000. What can the contractor expect to earn on average on this job? $33,500

3. A game is played by tossing a coin. If you toss heads, you win 10 points. If you toss tails, you lose 5 points. Find the number of points that you can expect to win on average for a given toss. 2.5

4. The players who finish first, second, third, and fourth in a tennis tournament earn $120,000, $80,000, $50,000, and $20,000, respectively. How much can one of the four finalists expect to win on average if all four finalists have an equal chance of winning? $67,500

Use the spinner below. Find each probability.

1. P(an even number) $\frac{1}{2}$

2. P(a number less than 4) $\frac{5}{8}$

Use the spinner above. Find the odds in favor of each event.

3. 2   1 to 7

4. an odd number   1 to 1

5. an even number   1 to 1

6. Make a tree diagram to show all possible outcomes for tossing two coins.

Coin 1	Coin 2	Outcome
	H	HH
H		
	T	HT
	H	TH
T		
	T	TT

7. Use the tree diagram from Exercise 6. Find P(TT). $\frac{1}{4}$

8. Use the counting principle to find the number of possible outcomes of rolling a number cube four times. 1296

9. A coin is tossed six times. Use the counting principle to find P(all heads). $\frac{1}{64}$

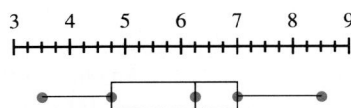

Cumulative Review

Standardized Testing Practice

Choose the letter of the correct answer.

1. There are 5 math teachers, 4 science teachers, 3 administrators, and 7 school board members. A committee must be formed that consists of one member of each group. How many different committees are possible?
   A. 19
   B. 120
   C. 96
   D. 420

2. Find the quotient: $\frac{k}{3} \div k$
   A. $\frac{k^2}{3}$
   B. 3
   C. $\frac{1}{3}$
   D. $\frac{2k}{3}$

3. The price of ABC Company's stock is $6.00 more than twice the price of XYZ Incorporated's stock. ABC Company's stock price is $15.00. What is XYZ Incorporated's stock price?
   A. $13.50
   B. $18.00
   C. $42.00
   D. $4.50

4. A number cube is rolled. What is the probability of rolling an even number?
   A. 0
   B. $\frac{1}{3}$
   C. $\frac{1}{2}$
   D. 1

5. Write $\frac{13}{65}$ as a percent.
   A. 5%
   B. 500%
   C. 2%
   D. 20%

6. **Prices of Audio Tapes (Dollars)**

   What is the median audio tape price?
   A. $6.25
   B. $7.00
   C. $4.75
   D. $8.50

7. Simplify: $9q + 3 + 4q$
   A. $12q + 4$
   B. $9 + 7q$
   C. $16q$
   D. $13q + 3$

8. The number x is 15% of 60. The number y is 75% of 12.

   Which of the following is true?
   A. $x > y$
   B. $x < y$
   C. $x = y$
   D. cannot be determined

9. A coin is flipped 3 times. What is P(heads, then heads, then tails)?
   A. $\frac{1}{2}$
   B. $\frac{1}{3}$
   C. $\frac{1}{8}$
   D. $\frac{1}{16}$

10. The number 14.38 is to be displayed in a stem-and-leaf plot. What is its stem?
    A. 1
    B. 14
    C. 14.3
    D. 14.38

11.

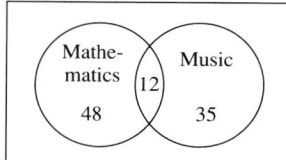

How many students study both mathematics and music?

A. 48   B. 12
C. 83   D. 95

12.

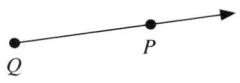

What is the correct name for this figure?

A. $\overrightarrow{PQ}$ B. $\overrightarrow{QP}$
C. $\overleftrightarrow{PQ}$ D. $\overline{PQ}$

13. The base of a parallelogram is 14 cm and the height is 6 cm. Find the area.

A. 40 cm²   B. 20 cm²
C. 84 cm²   D. cannot be
                      determined

14. To compare the incomes from five different sales regions to the total income, which data display is most appropriate?

A. scattergram
B. box-and-whisker plot
C. stem-and-leaf plot
D. circle graph

15. Find the GCF: 210 and 315

A. 3   B. 630
C. 105   D. 7

16. Specialty Books Monthly Sales

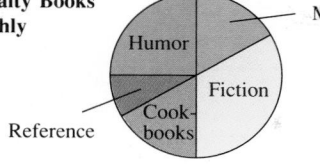

About what fraction of the sales are from fiction books?

A. $\frac{1}{4}$ B. $\frac{1}{3}$ C. $\frac{2}{3}$ D. $\frac{1}{6}$

17. Brad has 5 dimes, 1 quarter, and 6 nickels. In how many different ways can he make $.50?

A. 7   B. 3
C. 5   D. 4

18. James plans to put a wallpaper border around his rectangular bedroom. Which measurement of the room should he find?

A. area   B. circumference
C. diameter   D. perimeter

19. In a random sample of 300 students in a school, 80 rated science as their favorite subject. Predict how many students in the school's total population of 1650 would rate science as their favorite subject.

A. about 80   B. about 440
C. about 1430   D. about 220

20. Simplify: $12 + 4(3 - 5)^2$

A. 64   B. 28
C. −1   D. −64

Alternative Assessment

1. Give an example of an event that would have the given probability. Answers will vary. Examples are given.

a. $\frac{1}{2}$ getting a number greater than 3 when tossing a cube with faces numbered 1 through 6

b. 0 Today is Thursday; tomorrow will be Wednesday.

c. 1 Today is Thursday; tomorrow will be Friday.

2. When is it better to use a tree diagram instead of the counting principle? when you need to see each individual outcome

Probability   **579**

579

Planning Chapter 13

Chapter Overview

Chapter 13 extends skills and concepts that students have learned earlier about solving equations, the coordinate plane, and the solution of inequalities to solving and graphing equations and inequalities with two variables. *Lesson 1* connects the solution of an equation in two variables with the concept of an ordered pair. The problem-solving strategy of making a table helps students visualize how replacing one variable with a value determines the value of the other variable. *Lesson 2* continues to connect equations with their geometric representations by graphing equations on a coordinate plane. *Lesson 3* explores the concepts of slope and intercept, and relates them by presenting the slope-intercept form of the equation of a line. The *Focus on Explorations* utilizes geoboards as a concrete way of exploring changes in slope and finding intercepts. *Lesson 4* extends graphing in the coordinate plane to include the solution of systems of equations, and the *Focus on Computers* shows how function graphing software can be used to solve systems of equations. *Lesson 5* focuses on the problem-solving strategy of solving a simpler problem. This approach encourages students to solve a problem by looking at parts of the problem and solving them first.

Lesson 6 moves from equations to inequalities, as students solve and graph inequalities in one variable on a number line. In *Lesson 7*, the inequalities become more complex in that two steps are required to arrive at a solution. *Lesson 8* connects inequalities to the coordinate plane, as inequalities in two variables are solved and graphed. *Lesson 9* introduces function terminology and notation; students find values of functions and write their results using the proper notation. *Lesson 10* explores the graphing of functions, including parabolic and absolute value functions.

Background

Chapter 13 builds on concepts learned previously. Evaluating expressions is extended to include "evaluating" equations with more than one variable. These equations were introduced earlier in students' work with formulas. In this chapter, concepts are generalized to include equations with two variables.

Equations and inequalities are solved algebraically, and graphing is used to illustrate the solutions and connect the algebraic representations to geometry. Other connections are made with the areas of science, social studies, and carpentry. Thus, students can see that the concepts presented in this chapter are applied in the real world.

Communication skills are enhanced by having students look for similarities and differences in mathematical situations. For example, students try to discern what happens to a graph when certain components are changed. They are asked to explain their reasoning either in writing or verbally as part of a group discussion.

Problem solving is interwoven throughout the chapter. The focus is on solving a simpler problem. The four-step plan serves as the basis for organization of the solution process.

The use of technological tools is encouraged throughout Chapter 13. By graphing various equations or inequalities, students can explore how changing the coefficients affects graphs.

Objectives

13-1   To find solutions of equations with two variables.

13-2   To graph equations with two variables.

13-3   To find the slope and intercepts of a line and to write the equation of a line using the slope and *y*-intercept.

≡   **FOCUS ON EXPLORATIONS**   To use a geoboard to explore slopes and intercepts.

13-4   To solve systems of equations by graphing.

≡   **FOCUS ON COMPUTERS**   To use function graphing software to find solutions to systems of equations.

13-5   To solve problems by using simpler problems.

13-6   To solve and graph inequalities with one variable.

13-7   To solve and graph inequalities with one variable using two steps.

13-8   To determine solutions of inequalities with two variables and to graph inequalities with two variables.

13-9   To use a function rule to find values of a function.

13-10   To graph functions on a coordinate plane.

≡   **CHAPTER EXTENSION**   To graph inequalities with one variable on the coordinate plane.

Chapter Planner

Instructional Aids	Manipulatives	Cooperative Learning	Technology	Practice/Reteaching	Assessment
Materials: chalk, string, graphing calculators, calculators *Resource Book:* Lesson Starters 118–127 *Visuals:* Folders I, M, N	Lessons 13-2, 13-3, Focus on Explorations, 13-4 *Activities Book:* Manipulative Activities 25–26	Lesson 13-8 *Activities Book:* Cooperative Activities 23–24	Lessons 13-2, 13-3, 13-4, Focus on Computers, 13-6, 13-7, 13-8, 13-10 *Activities Book:* Calculator Activity 13 Computer Activity 13 *Connections Plotter Plus* Disk	Toolbox, pp. 753–771 Extra Practice, p. 748 *Resource Book:* Practice 161–174 *Study Guide,* pp. 235–254	Self-Tests 1–3 Quick Quizzes 1–3 Chapter Test, p. 622 *Resource Book:* Lesson Starters 118–127 (Daily Quizzes) *Tests* 64–68

Assignment Guide Chapter 13

Day	Skills Course	Average Course	Advanced Course
1	**13-1:** 1–28	**13-1:** 1–28	**13-1:** 1–11 odd, 13–28 **13-2:** 1–11 odd, 13–24
2	**13-2:** 1–16, 21–24	**13-2:** 1–24	**13-3:** 1–32 **Exploration:** Activities I–IV
3	**13-3:** 1–27, 30–32 **Exploration:** Activities I–IV	**13-3:** 1–32 **Exploration:** Activities I–IV	**13-4:** 1–15 odd, 16–34 **Focus on Computers**
4	**13-4:** 1–20, 32–34	**13-4:** 1–24, 32–34 **Focus on Computers**	**13-5:** 1–13, Challenge
5	**13-5:** 1–13	**13-5:** 1–13	**13-6:** 1–17 odd, 19–33
6	**13-6:** 1–25, 29–33	**13-6:** 1–25, 29–33	**13-7:** 1–27, Historical Note
7	**13-7:** 1–20, 24–27	**13-7:** 1–27	**13-8:** 1–27 odd, 29–42
8	**13-8:** 1–28, 39–42	**13-8:** 1–28, 39–42	**13-9:** 1–22
9	*Prepare for Chapter Test:* Chapter Review	*Prepare for Chapter Test:* Chapter Review	**13-10:** 1–33
10	*Administer Chapter 13 Test*	*Administer Chapter 13 Test;* *Cumulative Review*	*Prepare for Chapter Test:* Chapter Review
11			*Administer Chapter 13 Test;* *Chapter Extension;* *Cumulative Review*

Teacher's Resources

Resource Book
- Chapter 13 Project
- Lesson Starters 118–127
- Practice 161–174
- Enrichment 130–140
- Diagram Masters
- Chapter 13 Objectives
- Family Involvement 13
- Spanish Test 13
- Spanish Glossary

Activities Book
- Cooperative Activities 23–24
- Manipulative Activities 25–26
- Calculator Activity 13
- Computer Activity 13

Study Guide, pp. 235–254

Tests, Tests 64–68

Visuals, Folders I, M, N

Connections Plotter Plus **Disk**

Using Technology

COMPUTERS AND GRAPHING CALCULATORS

This chapter introduces graphing of equations and inequalities. This is a rich area for exploration by students. Using graphing software, such as *Connections Plotter Plus*, or graphing calculators, students can check the work they have done by hand, and they can discover the principles of graphing that are presented in the textbook, along with more advanced principles that go beyond the book.

Lessons 13-2, 13-3

COMPUTERS AND GRAPHING CALCULATORS

Using either of these technologies, students can graph a number of linear equations in a short time and can thereby discover that the slope and y-intercept are evident in the equation itself. All graphing calculators and general graphing programs can graph linear functions written in the form $y = mx + b$. Some can graph lines written in the more general form $ax + by = c$. If students are able to graph both forms at the same time, they can reinforce their understanding of the techniques for transforming an equation from one form to the other.

Lesson 13-4

COMPUTERS AND GRAPHING CALCULATORS

By graphing two functions at the same time, students can discover approximate solutions of systems of equations (see the Focus on Computers). Some programs have a zoom feature that allows students to magnify a small portion of the graphs near an intersection point; this feature is fun to use, and allows students to approximate the solutions to a high degree of accuracy. Note that it is just as easy to have the program or calculator draw nonlinear graphs as linear ones, so that even quite advanced systems of equations can be solved approximately in this way.

Lesson 13-6

COMPUTERS AND GRAPHING CALCULATORS

By graphing two functions at the same time, students can discover a graphical method for solving linear inequalities. For example, to solve $3x + 2 < x - 6$, a student can graph the two functions $y = 3x + 2$ and $y = x - 6$. By solving the inequality by hand or calculating the two expressions for different values of x, the student can discover that the solution is the set of all x-values for which the first function is below the other. This is a very valuable insight and an excellent reason for having students try this activity.

Lesson 13-7

COMPUTERS AND GRAPHING CALCULATORS

By graphing pairs of functions, as described for Lesson 13-6, students can see the meaning of the usual inequality-solving transformations graphically. For example, the graphs of $y = 3x + 2$ and $y = x - 6$ are two lines intersecting at $(-4, -10)$. If you add -2 to both equations, as you might if you were solving $3x + 2 < x - 6$, you get $y = 3x$ and $y = x - 8$, which are two lines intersecting at $(-4, -12)$. The first graph is below the second for exactly the same x-values as was true for the original set of functions; that is, the transformed inequality has the same solution as the original. Two more transformations, subtracting x from both, then dividing both by 2, yield $y = x$ and $y = -4$. These graphs intersect at $(-4, -4)$, and the solution again can be seen to be unchanged.

Lesson 13-8

COMPUTERS

Some graphing software packages, including *Connections Plotter Plus*, contain programs that can graph inequalities in two variables and shade the solution in the plane. Typically, they will solve systems consisting of several inequalities together. Such a program can help students check their work, reinforce their understanding, and make discoveries.

Using Manipulatives

Lesson 13-2

A floor composed of square tiles can be used as a large coordinate plane. Use chalk that can be washed off easily to label an x-axis, a y-axis, and an origin. Have two students stand on points with coordinates that are integers. Call these points A and B. Then have the students face each other holding a stretched string. Ask some students to name other ordered pairs that are on the line segment between points A and B. Then ask a few students to stand on points that are not between A and B but are on the line. Ask students to name the ordered pairs for their points.

Lesson 13-3

Using the coordinate plane model from Lesson 13-2, ask a student to walk from point A to point B in the following manner. From point A, walk north or south until you are directly opposite point B. Count the number of tiles walked; this length is called the *rise*. Now walk east or west until you reach point B. Count the number of tiles walked; this length represents the *run*. The *slope* of the line through points A and B is equal to the ratio of the rise to the run. This activity can be repeated three or four times to show different slopes.

Lesson 13-4

Again, use a floor composed of square tiles to serve as a coordinate plane. Select two students at random and give them a ball of string. Ask these students to stand at two points on the coordinate plane and to stretch the string between them to represent a line. Repeat the process with two other students so that the strings intersect in a point. Ask students to determine the coordinates of the point of intersection.

Now ask the students to stand so that the "lines" are on top of each other. Ask other students how many solutions there are when this occurs. Finally, ask the students to stand so that the "lines" are the same distance apart. Ask them how many solutions there are in this situation.

As an extension, have the students move the strings so that the two "lines" are not in the same plane and not touching. Ask students if there is a point of intersection, if the lines are parallel, and if there is a solution in this situation.

Reteaching/Alternate Approach

Lesson 13-1

Ask students to evaluate expressions with one variable, such as $2x + 3$, using the values -2, -1, 0, 1, and 2 for x. Then designate a name, y, for the expression to make an equation, $y = 2x + 3$. Ask students what the y-value was when x had a value of -2, -1, and so on. This approach will connect the concept of equations in two variables to that of evaluating expressions, which was done earlier in the text.

Lessons 13-6, 13-7

Relate the process of solving inequalities to that of solving equations by solving each of the Examples as an equation first and then as the given inequality. By putting the two solutions next to each other on the chalkboard, students will be able to see the similarities and differences between solving equations and inequalities.

Teacher's Resources Chapter 13

Enrichment masters from the Resource Book are pictured here. See the Teacher's Resources chart on page 580B for a complete listing of all materials available for this chapter.

13-1 Enrichment
Communication

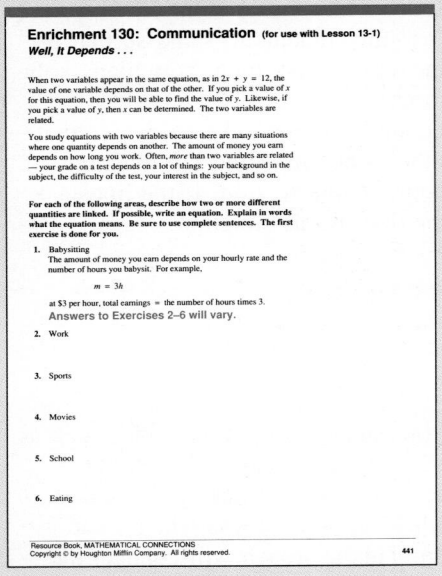

13-2 Enrichment
Thinking Skills

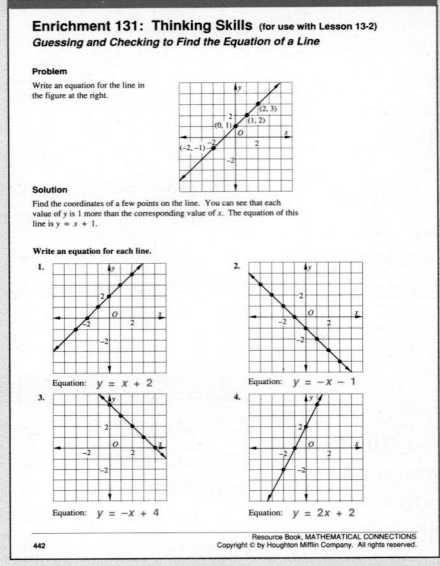

13-3 Enrichment
Data Analysis

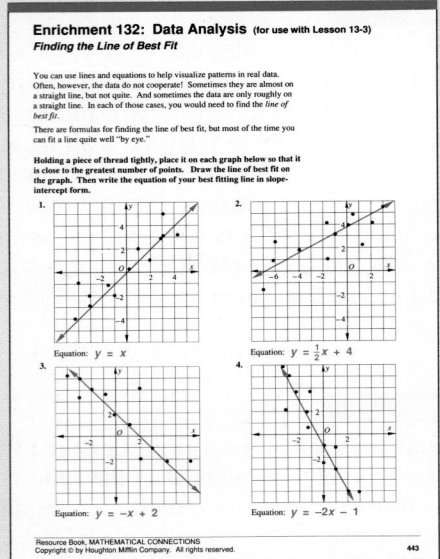

13-4 Enrichment
Application

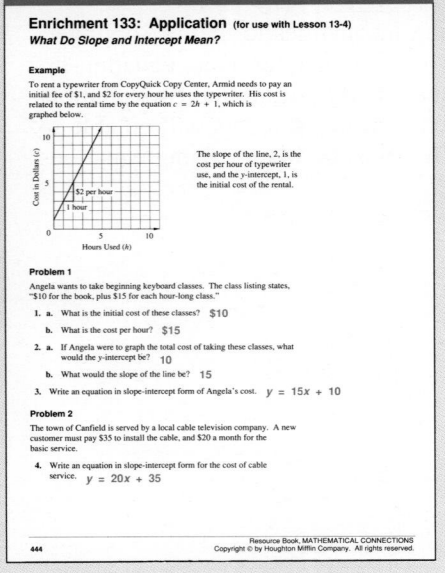

13-5 Enrichment
Problem Solving

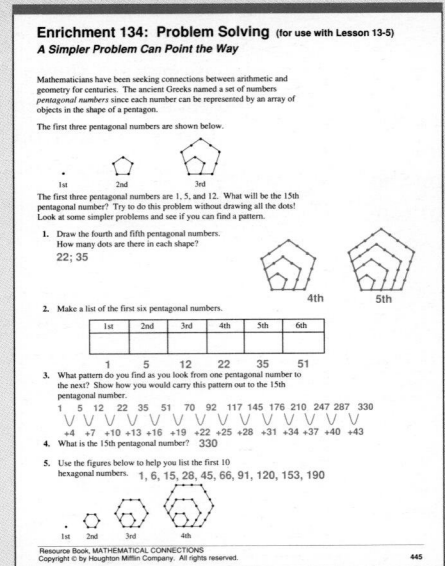

13-6 Enrichment
Application

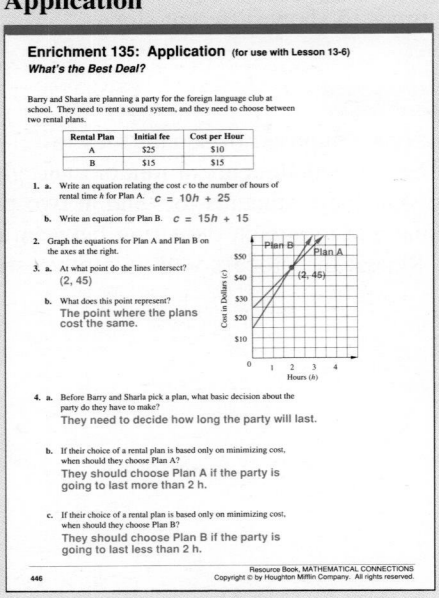

13-7 Enrichment
Connection

Enrichment 136: Connection (for use with Lesson 13-7)
A Geometric Inequality

You have 400 meters of fencing and want to make an enclosure with the greatest area possible. Should you make a circle or a square?

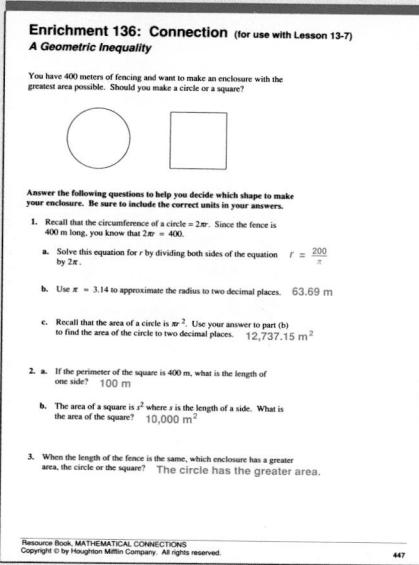

Answer the following questions to help you decide which shape to make your enclosure. Be sure to include the correct units in your answers.

1. Recall that the circumference of a circle is $2\pi r$. Since the fence is 400 m long, you know that $2\pi r = 400$.

   a. Solve this equation for r by dividing both sides of the equation by 2π. $r = \frac{200}{\pi}$

   b. Use $\pi = 3.14$ to approximate the radius to two decimal places.   63.69 m

   c. Recall that the area of a circle is πr^2. Use your answer to part (b) to find the area of the circle to two decimal places.   12,737.15 m²

2. a. If the perimeter of the square is 400 m, what is the length of one side?   100 m

   b. The area of a square is s^2 where s is the length of a side. What is the area of the square?   10,000 m²

3. When the length of the fence is the same, which enclosure has a greater area, the circle or the square?   The circle has the greater area.

13-8 Enrichment
Exploration

Enrichment 137: Exploration (for use with Lesson 13-8)
Using Inequalities to Describe Regions

For Exercises 1–3, refer to the shaded region below.

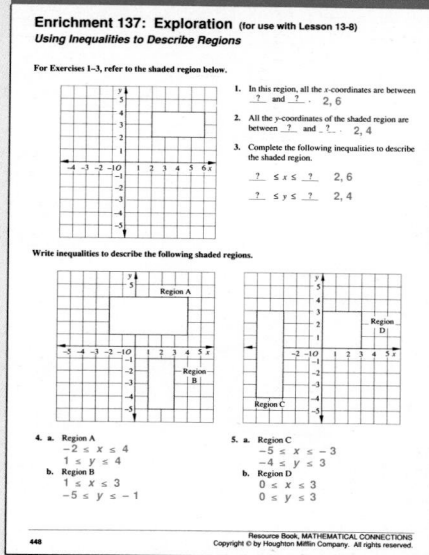

1. In this region, all the x-coordinates are between _?_ and _?_.   2, 6

2. All the y-coordinates of the shaded region are between _?_ and _?_.   2, 4

3. Complete the following inequalities to describe the shaded region.

   _?_ $\leq x \leq$ _?_   2, 6

   _?_ $\leq y \leq$ _?_   2, 4

Write inequalities to describe the following shaded regions.

4. a. Region A
   $-2 \leq x \leq 4$
   $1 \leq y \leq 4$
   b. Region B
   $1 \leq x \leq 3$
   $-5 \leq y \leq -1$

5. a. Region C
   $-5 \leq x \leq -3$
   $-4 \leq y \leq 3$
   b. Region D
   $0 \leq x \leq 3$
   $0 \leq y \leq 3$

13-9 Enrichment
Communication

Enrichment 138: Communication (for use with Lesson 13-9)
Function Machines Are Everywhere

When you turn your television dial to Channel 10 at a certain time of day, you know what program you expect to see. When you put some food in the microwave for 60 seconds, you know what is going to happen to the food. When a driver turns a car's steering wheel a little to the left, she knows where the car is going to go.

Each of these examples is a real-life function. You expect a *specific outcome* from something you do. The outcome should be different for each set of instructions. You do not want just any old television program. You want the food cooked in a certain way. It would be disastrous for a car to go in the wrong direction.

Getting a specific, unique outcome is a characteristic of mathematical functions. The function rule in mathematics is just like the microwave timer, the television dial, and the steering wheel. All of these machines are geared to translate one specific action into another specific action.

Use a complete sentence or two to explain how each of the following machines behaves like a mathematical function.   Answers will vary. Examples are given.

1. A thermostat
   For each spot on the thermostat's dial, there is a specific temperature.

2. A computer keyboard
   For each keystroke, a single symbol is produced.

3. The brakes on a car
   A certain amount of pressure applied to the brake pedal produces a specific amount of pressure on the brake drums.

4. The volume selector on a stereo
   One setting on the selector produces a single level of volume.

5. A kitchen faucet
   A certain setting of the faucet produces a certain flow of water.

6. The accelerator in a car
   A certain amount of pressure on the accelerator produces a certain engine speed.

13-10 Enrichment
Application

Enrichment 139: Application (for use with Lesson 13-10)
Step Functions

In many of the functions that you meet on a day-to-day basis, each value of y is paired with an *interval* of values for x.

Functions like these are called *step functions*. A common example of a step function is a postage-rate function like the one described in Enrichment 86. Another postage-rate function is defined in the table at the right. Its graph appears below.

Postage Rates for Letters and Letter Packages to Canada

Weight (x)		Cost (y)
over	*but not over*	
0 oz	1 oz	$.30
1 oz	2 oz	.52
2 oz	3 oz	.74
3 oz	4 oz	.96
4 oz	5 oz	1.18

Answer these questions about the graph.

1. What does each unit on the x-axis represent? on the y-axis?   1 ounce; $0.20

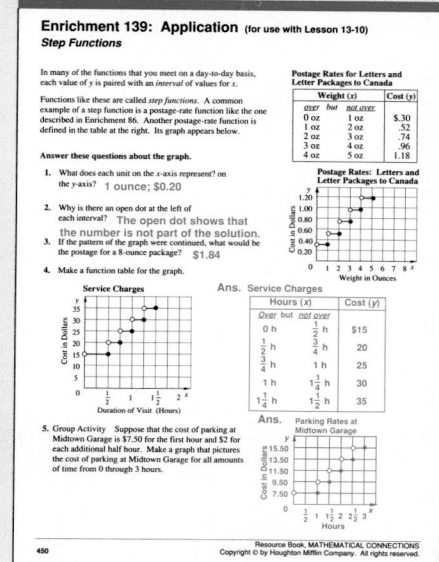

2. Why is there an open dot at the left of each interval?   The open dot shows that the number is not part of the solution.

3. If the pattern of the graph were continued, what would be the postage for a 8-ounce package? $1.84

4. Make a function table for the graph.

Ans. **Service Charges**

Hours (x)		Cost (y)
Over	*but not over*	
0 h	$\frac{1}{4}$ h	$15
$\frac{1}{4}$ h	$\frac{1}{2}$ h	20
$\frac{3}{4}$ h	1 h	25
1 h	$1\frac{1}{4}$ h	30
$1\frac{1}{4}$ h	$1\frac{1}{2}$ h	35

5. **Group Activity** Suppose that the cost of parking at Midtown Garage is $7.50 for the first hour and $2 for each additional half hour. Make a graph that pictures the cost of parking at Midtown Garage for all amounts of time from 0 through 3 hours.

Ans. **Parking Rates at Midtown Garage**

End of Chapter Enrichment
Extension

Enrichment 140: Extension (for use after Chapter 13)
Using Graphs of Inequalities

The shaded region here is bound by the lines

$x + y \leq 5;\ y \leq 2x;\ y \geq 0;\ x \leq 3$

and the labels designate the boundary lines.

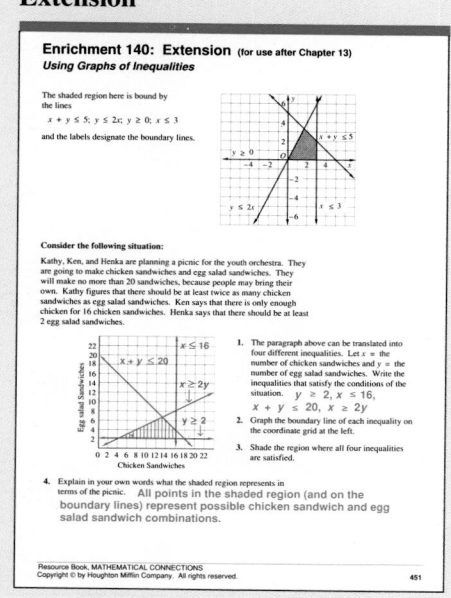

Consider the following situation:

Kathy, Ken, and Henka are planning a picnic for the youth orchestra. They are going to make chicken sandwiches and egg salad sandwiches. They will make no more than 20 sandwiches, because people may bring their own. Kathy figures that there should be at least twice as many chicken sandwiches as egg salad sandwiches. Ken says that there is only enough chicken for 16 chicken sandwiches. Henka says that there should be at least 2 egg salad sandwiches.

1. The paragraph above can be translated into four different inequalities. Let $x =$ the number of chicken sandwiches and $y =$ the number of egg salad sandwiches. Write the inequalities that satisfy the conditions of the situation. $y \geq 2$, $x \leq 16$, $x + y \leq 20$, $x \geq 2y$

2. Graph the boundary line of each inequality on the coordinate grid at the left.

3. Shade the region where all four inequalities are satisfied.

4. Explain in your own words what the shaded region represents in terms of the picnic.   All points in the shaded region (and on the boundary lines) represent possible chicken sandwich and egg salad sandwich combinations.

Multicultural Notes

For Chapter 13 Opener

At the beginning of the fifteenth century, the reappearance of an ancient text on geography excited many scholars throughout Europe. Ptolemy's *Geographia*, which was written in Alexandria, Egypt, in 150 A.D. contained maps that were more accurate than those made by Europeans in the Middle Ages. Ptolemy had worked from the assumption that the world was round, an idea that was still unfamiliar to most Europeans. He marked out latitude and longitude in grids, and gave instructions for making various map projections.

The rediscovery of Ptolemy's work helped cartographers make even more accurate maps of the world. Christopher Columbus was inspired by these new maps. He charted the islands he saw during his 1492 voyage, unknowingly beginning the cartography of the New World.

Research Activities

For Chapter 13 Opener

Students may be interested in looking at maps of the New World made by European cartographers. Students can compare old maps to modern ones, and discuss the many similarities and differences.

Students can also research the history of cartography in other parts of the world. For example, many developments in Chinese cartography predated European developments by more than a thousand years.

For Page 605

Students working on Exercises 26–28 can research other large cats living in various parts of the world, and the maximum speeds that these cats can attain.

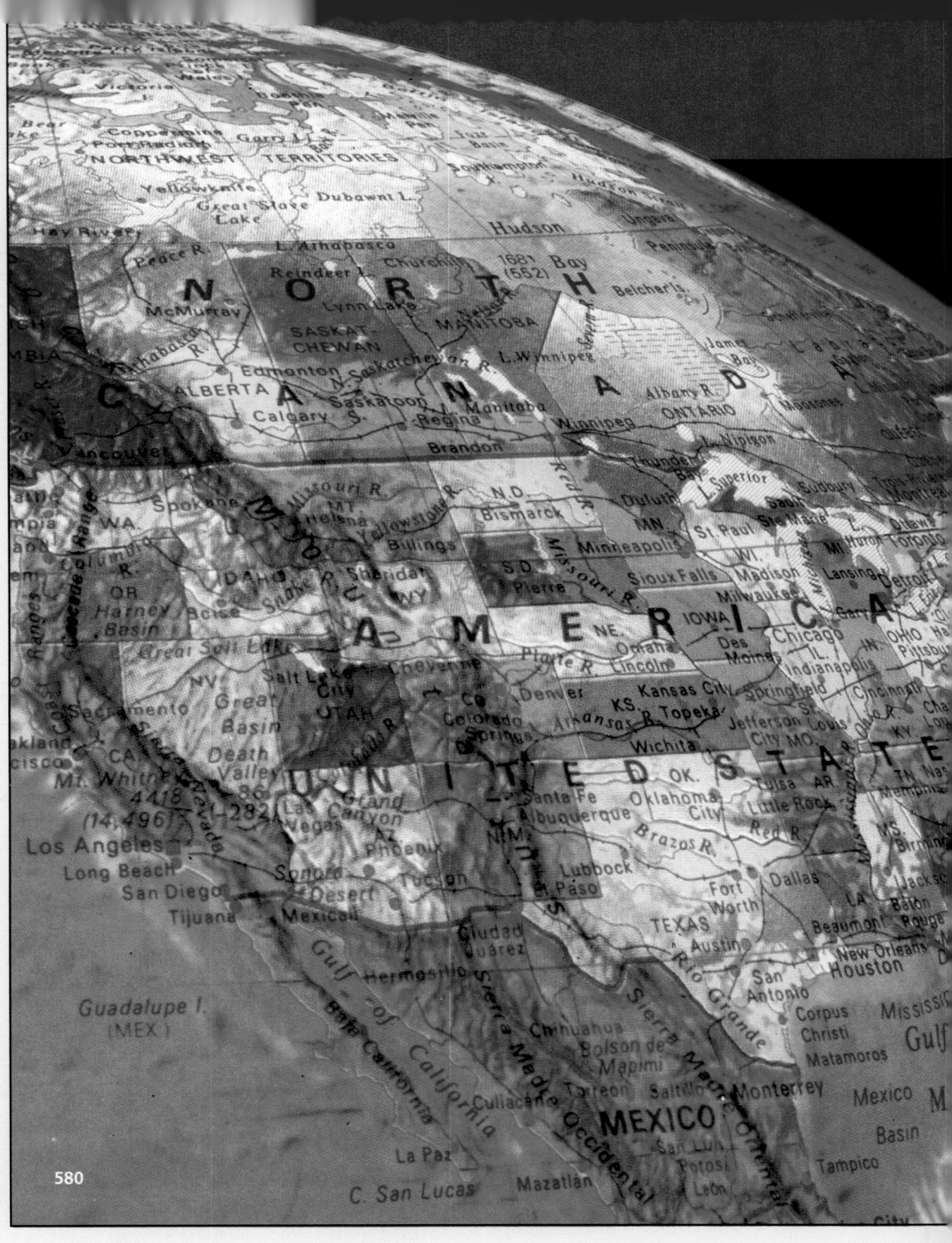

580

Inequalities and Graphing on the Coordinate Plane 13

Suggested Resources

Portinaro, Pierluigi and Knirsch, Franco. *The Cartography of the New World*. New York: Facts on File Publications, 1987.

Wilford, John Noble. *The Mapmakers*. New York: Alfred A. Knopf, 1981.

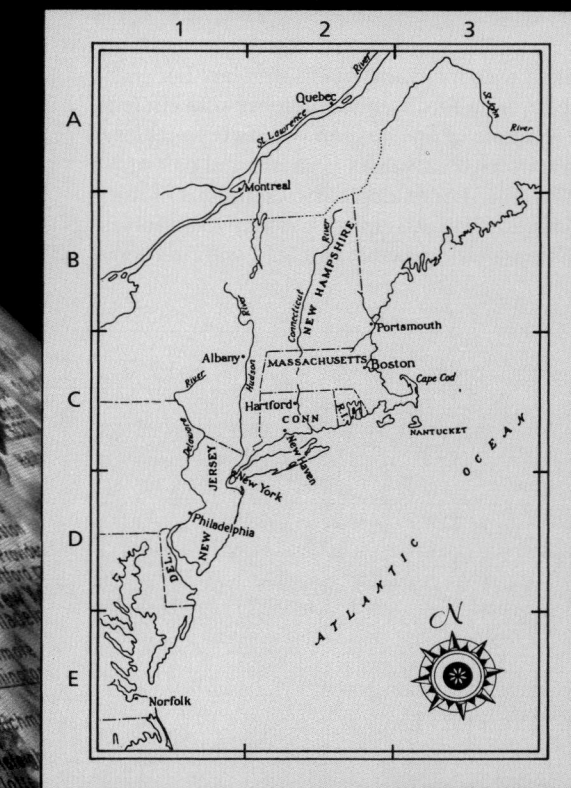

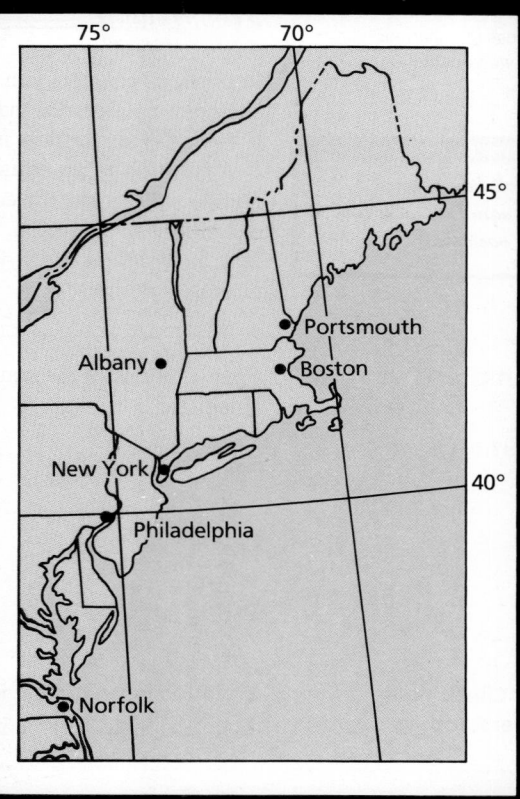

The latitude and longitude lines on a map or globe represent a coordinate system. Latitude is given in degrees north or south of the equator, and longitude is given in degrees east or west of Greenwich, England. By knowing the latitude and longitude of a place, you can locate it on a map or globe. For example, Philadelphia, Pennsylvania, is located at approximately 40° north 75° west.

581

Teaching the Lesson
• Lesson Starter 118
• Reteaching/Alternate
  Approach, p. 580D

Lesson Follow-Up
• Practice 161
• Enrichment 130:
  Communication
• Study Guide, pp. 235–236

Warm-Up Exercises

Solve.

1. $2n - 7 = -11$ -2
2. $3x - 5 = 7 - 11x$ $\frac{6}{7}$
3. $2m - 5 = 12(5 - 2m)$ $\frac{5}{2}$
4. $8 - y = y + 6$ 1
5. $7x + 3 = -2 + 9x$ $\frac{5}{2}$

Lesson Focus

Ask students to think of real-world examples where one quantity depends on another. For example, the distance traveled in a car depends on the time spent driving. Today's lesson focuses on solving equations with two variables, where the value of one variable depends on the value of the other.

Teaching Notes

Key Question

Why is it easier to solve equations with two variables when the variable y is alone on one side? Equations of the form "$y =$" can be solved by direct substitution.

Equations with Two Variables

Objective: To find solutions of equations with two variables.

CONNECTION

In previous chapters you worked with many different formulas. For example, you learned that the formula for the circumference of a circle is $C = 2\pi r$. This formula is an example of an equation with two variables.

A **solution of an equation with two variables** is an ordered pair of numbers that make the equation true. For example, two solutions of the equation $y = \frac{1}{2}x$ are (12, 6) and (20, 10). An equation with two variables may have *infinitely many* solutions. Using a table can help you find some of these solutions.

Terms to Know
• solution of an equation with two variables

Example 1

Find solutions of the equation $y = 5x + 4$. Use $-2, -1, 0, 1$, and 2 as values for x.

Solution

Make a table.

x	$y = 5x + 4$ y	(x, y)
-2	$5(-2) + 4 = -6$	$(-2, -6)$
-1	$5(-1) + 4 = -1$	$(-1, -1)$
0	$5(0) + 4 = 4$	$(0, 4)$
1	$5(1) + 4 = 9$	$(1, 9)$
2	$5(2) + 4 = 14$	$(2, 14)$

✓ Check Your Understanding

1. In Example 1, what is the value of y when $x = 1$?
2. In Example 1, what is the value of x when $y = -6$?
   See *Answers to Check Your Understanding* at the back of the book.

Sometimes it is easier to find solutions of an equation with two variables if you first get one variable alone on one side of the equation.

Example 2

Find solutions of the equation $y - 4x = 7$. Use $-2, -1, 0, 1$, and 2 as values for x.

Solution

Transform the equation so that y is alone on one side.

$$y - 4x = 7$$
$$y - 4x + 4x = 7 + 4x$$
$$y = 7 + 4x$$

Then make a table.

x	$y = 7 + 4x$ y	(x, y)
-2	$7 + 4(-2) = -1$	$(-2, -1)$
-1	$7 + 4(-1) = 3$	$(-1, 3)$
0	$7 + 4(0) = 7$	$(0, 7)$
1	$7 + 4(1) = 11$	$(1, 11)$
2	$7 + 4(2) = 15$	$(2, 15)$

582   Chapter 13

Guided Practice

COMMUNICATION «Reading

Refer to the text on page 582.

«1. What is the solution of an equation with two variables? an ordered pair that makes the equation true

«2. How should you write solutions of an equation with two variables x and y? as ordered pairs in the form (x, y)

Choose the letter of the correct value of y for the given value of x.

3. $y = 3x + 9$; $x = -2$ **a.** 3   **b.** 9   **c.** 15

4. $y = -2x + 4$; $x = -1$ **a.** −2   **b.** 2   **c.** 6

5. $y = -x + 5$; $x = 5$ **a.** 0   **b.** −5   **c.** 10

6. $y = -\frac{1}{3}x + 8$; $x = -9$ **a.** 5   **b.** 11   **c.** 35

Transform each equation to get y alone on one side.

7. $x + y = 8$ $y = 8 - x$ 8. $y - x = -2$ $y = -2 + x$

9. $4x + y = 4$ $y = 4 - 4x$ 10. $y + 5 = x$ $y = x - 5$

11. Find solutions of the equation $y = \frac{3}{4}x + 1$. Use $-8, -1, 0, 2$, and 4 as values for x. $(-8, -5); (-1, \frac{1}{4}); (0, 1); (2, 2\frac{1}{2}); (4, 4)$

«12. **COMMUNICATION** «*Discussion* In Exercise 11, for which values of x was the corresponding value of y an integer? In general, for which values of x will the corresponding value of y in the equation $y = \frac{3}{4}x + 1$ be an integer? −8, 0, 4; multiples of 4 and their opposites.

15. $(-2, -3); (-1, -2\frac{1}{2});$
$(0, -2); (1, -1\frac{1}{2});$
$(2, -1)$

16. $(-2, 6); (-1, 5); (0, 4);$
$(1, 3); (2, 2)$

Find solutions of each equation. Use $-2, -1, 0, 1$, and 2 as values for x.   13. $(-2, -3); (-1, -2); (0, -1); (1, 0); (2, 1)$
14. $(-2, 7); (-1, 5); (0, 3); (1, 1); (2, -1)$

13. $y = x - 1$ 14. $y = -2x + 3$ 15. $y = \frac{1}{2}x - 2$

16. $x + y = 4$ 17. $2x + y = 5$ 18. $y - 3x = 0$

17. $(-2, 9); (-1, 7); (0, 5); (1, 3); (2, 1)$ 18. $(-2, -6); (-1, -3); (0, 0); (1, 3); (2, 6)$

Exercises

1. $(-6, -19); (-2, -3);$
$(0, 5); (2, 13); (6, 29)$

2. $(-6, 5); (-2, 1);$
$(0, -1); (2, -3);$
$(6, -7)$

3. $(-6, 7); (-2, 5\frac{2}{3}); (0, 5);$
$(2, 4\frac{1}{3}); (6, 3)$

A   **Find solutions of each equation. Use $-6, -2, 0, 2$, and 6 as values for x.** Answers to Exercises 4–12 are on p. A33.

1. $y = 4x + 5$ 2. $y = -x - 1$ 3. $y = -\frac{1}{3}x + 5$

4. $y = -6x$ 5. $y = 5x - 9$ 6. $y = \frac{5}{6}x - 3$

7. $x + y = -4$ 8. $x + y = 8$ 9. $2x + y = -1$

10. $y - 4x = 3$ 11. $7x + y = 0$ 12. $y - 2 = 3x$

Inequalities and Graphing on the Coordinate Plane   **583**

Making Connections/Transitions
Students have worked with formulas earlier in this course and also may have done so in science classes. Many formulas are examples of equations in two variables.

Additional Examples

Find solutions for each equation. Use −2, −1, 0, 1, and 2 as values for x.

1. $y = 6x - 5$ $(-2, -17)$, $(-1, -11)$, $(0, -5)$, $(1, 1)$, $(2, 7)$

2. $y - 3x = 8$ $(-2, 2)$, $(-1, 5)$, $(0, 8)$, $(1, 11)$, $(2, 14)$

Closing the Lesson

Ask a student to explain how to use a table to find solutions of equations with two variables when values are given for one of the variables. Ask another student to explain how to isolate one variable of an equation and why this is done.

Suggested Assignments

Skills: 1–28
Average: 1–28
Advanced: 1–11 odd, 13–28

Exercise Notes

Communication: Reading
Guided Practice Exercises 1 and 2 require students to demonstrate their understanding of what a solution to an equation in two variables is and the form it takes.

Communication: Discussion
Guided Practice Exercise 12 provides an opportunity for students to discuss the conditions that yield integer solutions to an equation in two variables. The discussion is likely to include the concepts of multiples or factors.

583

Estimation

Exercises 15–18 require students to estimate a value of *y* for a given value of *x*. Estimation is a useful way of checking the reasonableness of a solution to an equation.

Life Skills

Exercises 19–24 require students to locate cities on a map. Reading coordinates from a map is an important skill used by many people who travel.

Follow-Up

Project *(Cooperative Learning)*

Exercise 24 can be extended and done as a project with students working in groups. Each group can be asked to find the latitude and longitude of two cities in different parts of the country. Each group can write a brief report on how differences in latitude and longitude affect weather, crops, living conditions, and so on.

Enrichment

Enrichment 130 is pictured on page 580E.

Practice

Practice 161, *Resource Book*

Practice 161
Skills and Applications of Lesson 13-1

Find solutions of each equation. Use −5, −3, 0, 3, and 5 as values for *x*.

1. $y = 3x + 2$ (−5,−13); (−3,−7); (0, 2); (3, 11); (5, 17)
2. $y = -x + 3$ (−5, 8); (−3, 6); (0, 3); (3,0); (5, −2)
3. $y = -2x + 5$ (−5, 15); (−3, 11); (0, 5); (3, −1); (5, −5)
4. $y = 5x + 6$ (−5, −19); (−3, −9); (0, 6); (3, 21); (5, 31)
5. $y = -4x + 1$ (−5, 21); (−3, 13); (0, 1); (3, −11); (5, −19)
6. $y = 2x + 7$ (−5, −3); (−3, 1); (0, 7); (3, 13); (5, 17)
7. $y = -2x - 3$ (−5, 7); (−3, 3); (0, −3); (3, −9); (5, −13)
8. $y = -3x - 6$ (−5, 9); (−3, 3); (0, −6);(3, −15); (5, −21)
9. $y = x - 5$ (−5, −10); (−3, −8); (0, −5); (3, −2); (5, 0)
10. $y = 4x - 3$ (−5, −23); (−3, −15); (0, −3); (3, 9); (5, 17)
11. $x + y = -3$ (−5, 2); (−3, 0); (0, −3); (3, −6); (5, −8)
12. $x - y = 4$ (−5, −9); (−3, −7); (0, −4); (3, −1); (5, 1)
13. $2x - y = 4$ (−5, −14); (−3, −10); (0, −4); (3, 2); (5, 6)
14. $y + 5 = 2x$ (−5, −15); (−3, 11); (0, −5); (3, 1); (5, 5)
15. $3x + y = -5$ (−5, 10); (−3, 4); (0, −5); (3, −14); (5, −20)
16. $-2x + y = 3$ (−5, −7); (−3, −3); (0, 3); (3, 9); (5, 13)
17. $3x + 2y = 6$ (−5, 10½); (−3, 7½); (0, 3); (3, −1½); (5, −4½)
18. $5x + 2y = 0$ (−5, 12½); (−3, 7½); (0, 0); (3,−7½); (5, −12½)

19. Lucky Car Rental charges $32 per day for a compact car, with an additional charge of $.15 per mile for each mile traveled over 100 miles.
 a. Write an equation in two variables that relates the cost of the car rental (*c*) to the number of miles (*m*) driven over 100 miles. $c = 0.15m + 32$
 b. What is the cost for renting the car for a day when it is driven 400 miles? $77
 c. How many miles was the car driven on a day when the rental cost was $53? 240

20. The cost for renting skis is $15 for the first day plus $12 for each additional day. The equation $c = 12(d − 1) + 15$ relates the total cost of renting skis (*c*), to the number of rental days (*d*).
 a. Find the cost, in dollars, of renting skis for 7 days. $87
 b. Find the number of days the skis can be rented for $75. 6

---

13. An electrician charges a $20 travel charge and $36 per hour.
 a. Write an equation in two variables that relates what the electrician charges (*C*) to the number of hours worked (*h*). $C = 36h + 20$
 b. What is the charge for a job that takes 3 h? $128
 c. An electrician charges $182. How many hours did the electrician work? 4.5 h

14. A telephone call from Darby to Weston costs 22¢ for the first minute and 14¢ for each additional minute. The equation $C = 14(m − 1) + 22$ relates the total cost of a call in cents (*C*) to the length of the call in minutes (*m*).
 a. Find the cost, in dollars, of a call that lasts 12 min. $1.76
 b. Find the length of a call that costs 78¢. 5 min

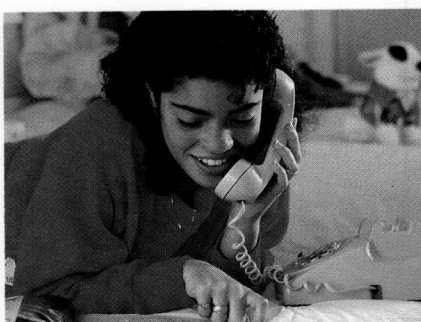

ESTIMATION Estimates may vary. Accept reasonable estimates.

For each equation, estimate a value of *y* for the given value of *x*.

B
15. $y = 1.3x + 6$; $x = 2.8$ about 9
16. $y = 4.6x − 9$; $x = 6.2$ about 21
17. $y = 0.49x + 6.8$; $x = 8$ about 11
18. $y = −0.34x − 1.3$; $x = 6$ about −3

DATA, *pages 580–581*

19. Which city has map coordinates A2? Quebec
20. Name two cities with map coordinates C2. Answers will vary. Example: Hartford and Boston
21. What are the map coordinates of Norfolk? E1
22. Which city is located at approximately 42.5° north, 74° west? Albany
23. Approximate the latitude and longitude of Portsmouth. about 43° north, 71° west
24. **RESEARCH** Find the latitude and longitude of your town or city. Answers will vary.

SPIRAL REVIEW

S
25. Find the number of permutations of the letters in the word RADIO. *(Lesson 12-6)* 120
26. Graph the points $A(3, 1)$, $B(2, −3)$, $C(0, 5)$, and $D(4, 0)$ on a coordinate plane. *(Lesson 3-8)* See answer on p. A33.
27. Find $\sqrt{3600}$. *(Lesson 10-10)* 60
28. Find solutions of the equation $2x + y = −5$. Use −2, −1, 0, 1, and 2 as values for *x*. *(Lesson 13-1)* (−2, −1); (−1, −3); (0, −5); (1, −7); (2, −9)

13-2 Graphing Equations with Two Variables

Objective: To graph equations with two variables.

2. Answers will vary. Examples: (2, 0); (−1, 6)

EXPLORATION

1 Tell whether the coordinates of points A and B are solutions of the equation $y = -2x + 4$. Yes.

2 Find two points on the line other than points A and B whose coordinates are solutions of the equation $y = -2x + 4$.

3 Predict whether it is possible to find a point on the line whose coordinates are not a solution of the equation $y = -2x + 4$. No.

4 How is the line in the figure above related to the equation $y = -2x + 4$? The line contains points whose coordinates are solutions of the equation.

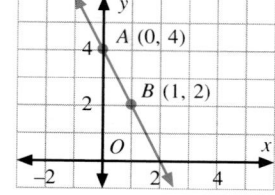

The **graph of an equation with two variables** is all the points whose coordinates are solutions of the equation. The graph of an equation such as $y = -2x + 4$ is a line on a coordinate plane.

Example 1

Solution

Graph the equation $y = \frac{2}{3}x - 2$.

• Make a table.

• Find at least three solutions of the equation. Choose reasonable values for x. Use −3, 0, and 3 as values for x.

$y = \frac{2}{3}x - 2$		
x	y	(x, y)
−3	$\frac{2}{3}(-3) - 2 = -4$	$(-3, -4)$
0	$\frac{2}{3}(0) - 2 = -2$	$(0, -2)$
3	$\frac{2}{3}(3) - 2 = 0$	$(3, 0)$

• Graph each solution as a point on a coordinate plane. Label each point with its coordinates.

• Connect the points with a straight line.

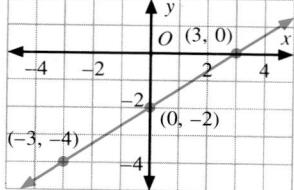

 **Check Your Understanding**

1. In Example 1, why are −3 and 3 good choices as values for x?
2. In Example 1, if you had used other values for x besides −3, 0, and 3, would the graph of the equation have been the same line?
See *Answers to Check Your Understanding* at the back of the book.

Lesson Planner

Teaching the Lesson
• Materials: chalk, string
• Lesson Starter 119
• Visuals, Folder M
• Using Technology, p. 580C
• Using Manipulatives, p. 580D

Lesson Follow-Up
• Practice 162
• Enrichment 131: Thinking Skills
• Study Guide, pp. 237–238

Warm-Up Exercises

Find solutions of each equation. Use −3, −1, 0, 1, and 3 as values for x.

1. $y = 3x - 7$ (−3, −16), (−1, −10), (0, −7), (1, −4), (3, 2)

2. $y = 5x - 11$ (−3, −26), (−1, −16), (0, −11), (1, −6), (3, 4)

3. $y = 2x - 8$ (−3, −14), (−1, −10), (0, −8), (1, −6), (3, −2)

4. $y = 3x + 3$ (−3, −6), (−1, 0), (0, 3), (1, 6), (3, 12)

5. $y = 4x + 8$ (−3, −4), (−1, 4), (0, 8), (1, 12), (3, 20)

Lesson Focus

Ask students to recall the line graphs that they studied earlier. To read information from a line graph, students found the numbers on the axes for the data points. Today's lesson focuses on graphing equations by reversing this process, that is, by first reading coordinates on the axes and plotting the corresponding points.

Key Question
Why should you find at least three solutions before graphing an equation? Although two points determine a line, the third point is a check that no errors were made in determining the first two points.

Using Technology
For a suggestion on using computer graphics or a graphing calculator, see page 580C.

Using Manipulatives
For a suggestion on using a tile floor to represent the coordinate plane and string to represent a line in the plane, see page 580D.

Additional Examples
1. Graph the equation $y = \frac{3}{4}x - 3$.

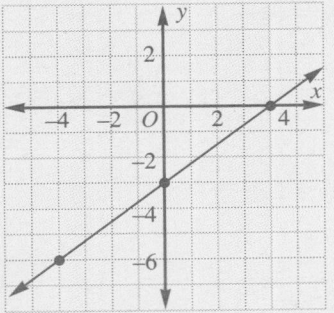

2. Graph the equation $x + y = 2$.

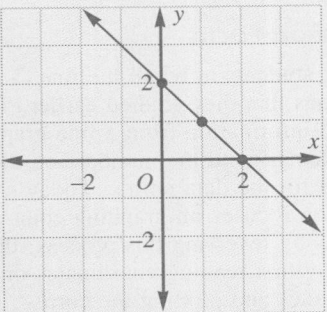

Example 2

Solution

Graph the equation $x + y = 3$.

Transform the equation. Then find at least three solutions. Use $-1, 0,$ and 1 as values for x.

$$x + y = 3$$
$$x + y - x = 3 - x$$
$$y = 3 - x$$

$y = 3 - x$		
x	y	(x, y)
-1	4	$(-1, 4)$
0	3	$(0, 3)$
1	2	$(1, 2)$

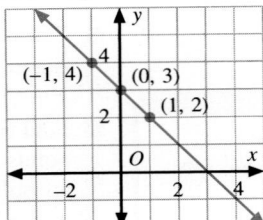

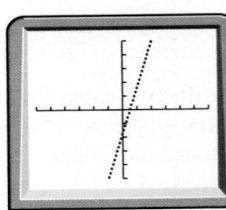

You can use a graphing calculator to graph equations with two variables once you have gotten y alone on one side. Key sequences vary depending on the type of graphing calculator. On some graphing calculators you would use the key sequence below to graph $y = 3x - 2$.

Guided Practice

«1. **COMMUNICATION** «*Reading* Replace each ___?___ with the correct word.

The graph of each solution of an equation with two variables is a(n) ___?___ on a coordinate plane. The graph of all the solutions of an equation such as $3x - y = 6$ is a(n) ___?___ on a coordinate plane. point; line

Tell whether the graph of each ordered pair is on the line shown in the graph at the right. Write *Yes* or *No*.

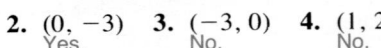

2. $(0, -3)$ 3. $(-3, 0)$ 4. $(1, 2)$
   Yes.   No.   No.

Use the line in the graph at the right to complete each ordered pair.

5. $(2, \underline{?})$ 1 6. $(3, \underline{?})$ 3 7. $(\underline{?}, -1)$
                                                      1

Graph each equation. Answers to Guided Practice 8–13 are on p. A34.

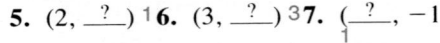

8. $y = x - 3$ 9. $y = -3x + 2$ 10. $y = \frac{1}{2}x - 4$

11. $x + y = 2$ 12. $4x + y = 8$ 13. $5x + y = 0$

Exercises

Graph each equation. Answers to Exercises 1–14 are on p. A34.

A **1.** $y = 4x + 2$ **2.** $y = 3x - 1$ **3.** $y = -4x - 2$

    4. $y = -x + 4$ **5.** $y = \frac{2}{3}x$ **6.** $y = -\frac{1}{3}x + 2$

    7. $x + y = 1$ **8.** $x + y = -5$ **9.** $2x + y = 6$

    10. $3x + y = -1$ **11.** $y - 3 = x$ **12.** $y + 2 = \frac{3}{4}x$

13. Graph the line that passes through the points $(-4, 0)$ and $(0, 1)$.

14. Graph the line that passes through the points $(2, 0)$ and $(0, 3)$.

 CALCULATOR

Choose the letter of the graphing calculator key sequence you would use to graph each equation.

B **15.** $y = 6x - 5$ D

    16. $y = 5x - 6$ B

  A. [GRAPH] [Y=] ▭ 6 [X|T] [−] ▭ 5

  B. [Y=] ▭ 5 [X|T] [−] ▭ 6 [GRAPH]

  C. [Y=] [X|T] ▭ 5 [−] ▭ 6 [GRAPH]

  D. [Y=] ▭ 6 [X|T] [−] ▭ 5 [GRAPH]

THINKING SKILLS 17, 19. Application **18, 20.** Synthesis

C **17.** Think of the equation $y = 2$ as meaning $y = 0x + 2$. Complete each ordered pair.

    a. $(-1, \underline{\ ?\ })$ 2   **b.** $(0, \underline{\ ?\ })$ 2   **c.** $(3, \underline{\ ?\ })$ 2   **d.** $(5, \underline{\ ?\ })$ ²

18. Use the ordered pairs from Exercise 17 to graph the equation $y = 2$ on a coordinate plane. Describe the graph in words. See answer on p. A34.

19. Think of the equation $x = -3$ as meaning $0y - 3 = x$. Complete each ordered pair.

    a. $(\underline{\ ?\ }, -4)$ −3   **b.** $(\underline{\ ?\ }, -2)$ −3   **c.** $(\underline{\ ?\ }, 0)$ −3   **d.** $(\underline{\ ?\ }, 1)$ ⁻³

20. Use the ordered pairs from Exercise 19 to graph the equation $x = -3$ on a coordinate plane. Describe the graph in words. See answer on p. A34.

SPIRAL REVIEW

S **21.** Solve $mx + y = b$ for y.   *(Lesson 8-10)* $y = b - mx$

    22. Evaluate $-3a + b$ when $a = -5$ and $b = 7$.   *(Lesson 3-6)* 22

    23. Find the area of a circle with radius 4 m.   *(Lesson 10-4)* 50.2 m²

    24. Graph the equation $y = 2x - 5$.   *(Lesson 13-2)* See answer on p. A34.

Closing the Lesson

Write the equation $4y - 3x = 8$ on the chalkboard. Ask a student to give the first step necessary to graph it on a coordinate plane. Ask another student for the second step. Continue until all the necessary steps have been given.

Suggested Assignments

Skills: 1–16, 21–24
Average: 1–24
Advanced: 1–11 odd, 13–24

Exercise Note

Calculator

A graphing calculator can be used to graph equations. *Exercises 15 and 16* require students to identify correct key sequences as explained on page 586.

Follow-Up

Enrichment

Enrichment 131 is pictured on page 580E.

Practice

Practice 162, *Resource Book*

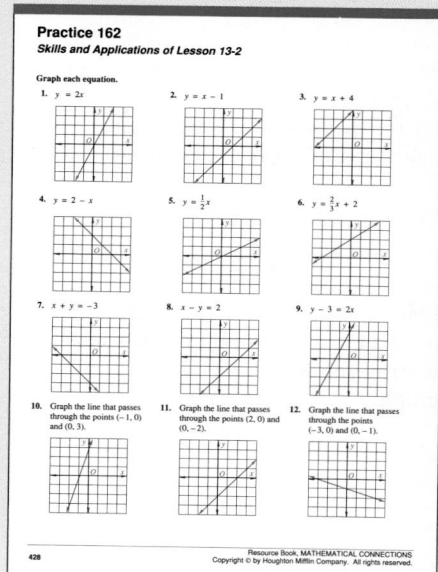

Teaching the Lesson
- Materials: string, chalk
- Lesson Starter 120
- Visuals, Folder M
- Using Technology, p. 580C
- Using Manipulatives, p. 580D

Lesson Follow-Up
- Practice 163
- Enrichment 132: Data Analysis
- Study Guide, pp. 239–240
- Calculator Activity 13
- Manipulative Activity 25

Warm-Up Exercises

Transform each equation to get y alone on one side.

1. $y + 3x = 5$ $y = -3x + 5$
2. $y - 2x = -6$ $y = 2x - 6$
3. $2x + y = 5$ $y = -2x + 5$
4. $\frac{1}{3}x + y = 4$ $y = -\frac{1}{3}x + 4$
5. $\frac{3}{5}x - y = 3$ $y = \frac{3}{5}x - 3$

Lesson Focus

Ask students to picture stairs in front of a building. Then ask them to connect mentally, from top to bottom, points along the right-hand edge of each stair. Can these points be connected by a line? Today's lesson focuses on the slope and intercepts of a linear equation.

Teaching Notes

Be certain students understand the terms introduced in this lesson. Draw the graph of a linear equation on the chalkboard and discuss each term in the lesson in relation to the graph.

13-3 Slope and Intercepts

Objective: To find the slope and intercepts of a line and to write the equation of a line using the slope and y-intercept.

Terms to Know
- *slope*
- *y-intercept*
- *x-intercept*
- *slope-intercept form of an equation*

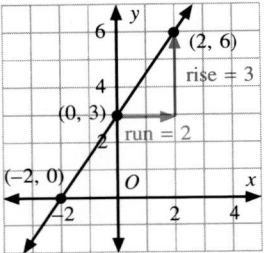

As you move from one point to another on a line, the vertical movement is called the *rise,* and the horizontal movement is called the *run.* The **slope** of a line is the ratio of the *rise* to the *run.* The slope of a line describes the line's steepness and direction. The line at the right has slope $\frac{3}{2}$.

$$\text{slope} = \frac{\text{rise}}{\text{run}} = \frac{3}{2}$$

The slope is the same between any two points on a given line.

The y-coordinate of the point where a graph crosses the y-axis is called the **y-intercept.** Notice that the value of x at this point is 0. The y-intercept of the line above is 3.

The x-coordinate of the point where a graph crosses the x-axis is called the **x-intercept.** Notice that the value of y at this point is 0. The x-intercept of the line above is -2.

The **slope-intercept form of an equation** is $y = mx + b$. In this equation, m represents the slope of the line, and b represents the y-intercept. The slope-intercept form of the equation for the line above is $y = \frac{3}{2}x + 3$.

Example 1

Find the slope and the y-intercept of the line at the right. Then use them to write an equation for the line.

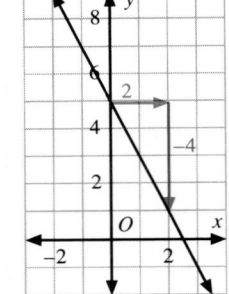

Solution

- Use two points on the line to find the slope. The rise is -4 and the run is 2.

  $$\text{slope} = \frac{\text{rise}}{\text{run}} = \frac{-4}{2} = -2$$

  The slope is -2, or $m = -2$.

- The y-coordinate of the point where the line crosses the y-axis is 5. The y-intercept is 5, or $b = 5$.

- Use the equation $y = mx + b$. An equation for the line is $y = -2x + 5$.

☑ **Check Your Understanding**

1. In Example 1, why is the rise negative?
2. Which two points on the graph in Example 1 were used to find the slope of the line? Would the slope have been the same if two other points on the line had been used?

See *Answers to Check Your Understanding* at the back of the book.

---

588 Chapter 13

When an equation is not written in slope-intercept form, you can transform the equation to find the slope and y-intercept.

Example 2
Solution

Find the slope, the y-intercept, and the x-intercept of the line $y + 6 = -3x$.

• Write the equation in slope-intercept form.

$$y + 6 = -3x$$
$$y + 6 - 6 = -3x - 6$$
$$y = -3x - 6$$
$$y = -3x + (-6)$$

Compare $y = -3x + (-6)$ to $y = mx + b$. $m = -3$ and $b = -6$.

• To find the x-intercept, let $y = 0$. Solve for x.

$$y + 6 = -3x$$
$$0 + 6 = -3x$$
$$6 = -3x$$
$$-2 = x \quad \longleftarrow \text{ Both sides were divided by } -3.$$

The slope is -3, the y-intercept is -6, and the x-intercept is -2.

✔️ **Check Your Understanding**

3. In Example 2, why was the equation $y = -3x - 6$ written as $y = -3x + (-6)$?

4. Where will a graph of the equation in Example 2 cross the y-axis?

See *Answers to Check Your Understanding* at the back of the book.

By looking at the graph of a line, you can tell easily if its slope is *positive* or *negative*. A line with *positive slope* rises as the value of x increases. A line with *negative slope* falls as the value of x increases.

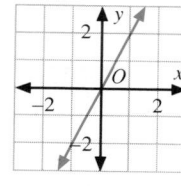
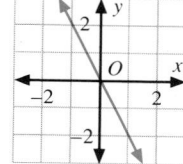

$y = 2x; m = 2$ $y = -2x; m = -2$

Guided Practice

COMMUNICATION « *Reading*

« **1.** Explain the meaning of the word *intercepted* in the sentence below.

The quarterback's pass was intercepted. caught by a player on the other team

« **2.** Explain the meaning of the word *slope* in the sentence below.

The ski instructor brought the class to the beginner slope. hill or incline

« **3.** Write a sentence using each word in a mathematical context.
   a. intercept   **b.** slope

3. Answers will vary. Examples:
a. The y-intercept of the line $y = 6x + 1$ is 1.
b. The slope of the line $y = 6x + 1$ is 6.

Inequalities and Graphing on the Coordinate Plane   **589**

Exercise Notes

Communication: Reading
Guided Practice Exercises 1–3 help students connect mathematical terms with words used in a more familiar context.

Calculator
Exercises 21–24 relate the symbolic form of an equation to its visual form on a graphing calculator. Students may wish to use a graphing calculator to verify their answers.

Application
The concept of slope is used by architects and engineers to draw blueprints. *Exercises 25–27* show how slope is used to plan a roof for a building.

Reasoning
Exercises 28 and 29 require students to use logical reasoning to deduce whether statements are sometimes, always, or never true.

Write an equation for the line with the given slope and y-intercept.

4. slope = 2
   y-intercept = 0 $y = 2x$

5. slope = -1
   y-intercept = $\frac{1}{2}$
   $y = -x + \frac{1}{2}$

6. slope = $\frac{3}{4}$
   y-intercept = -1
   $y = \frac{3}{4}x - 1$

Use the graph at the right.

7. What are the coordinates of the point where the line crosses the x-axis? (3, 0)

8. What are the coordinates of the point where the line crosses the y-axis? (0, −2)

9. Find the slope and the y-intercept of the line. Then use them to write an equation for the line. $\frac{2}{3}$; -2; $y = \frac{2}{3}x - 2$

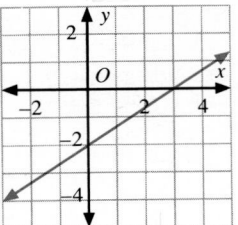

Find the slope, the y-intercept, and the x-intercept of each line. $\frac{4}{5}$; -8; 10

10. $y = 3x - 6$ 3; −6; 2

11. $y = -x - 5$ −1; −5; −5

12. $y = \frac{4}{5}x - 8$

13. $y = \frac{2}{3}x$ $\frac{2}{3}$; 0; 0

14. $y - 3x = 6$ 3; 6; −2

15. $4 + y = x$ 1; −4; 4

Exercises

Find the slope and the y-intercept of each line. Then use them to write an equation for each line.

1. 1; 1; $y = x + 1$
2. -3; -1; $y = -3x - 1$
3. $-\frac{1}{2}$; 0; $y = -\frac{1}{2}x$

A   1.
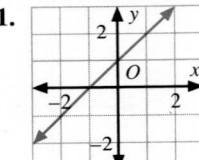

2.
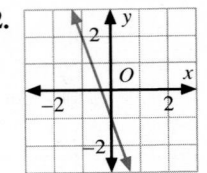

3.
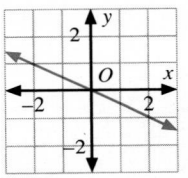

Find the slope, the y-intercept, and the x-intercept of each line. 1; 5; −5

4. $y = 2x - 10$ 2; −10; 5

5. $y = -3x + 9$ −3; 9; 3

6. $y = x + 5$

7. $y = -x - 1$ −1; −1; −1

8. $y = 6x + 5$ 6; 5; $-\frac{5}{6}$

9. -2; 1; $\frac{1}{2}$ 9. $y = -2x + 1$

10. $y = -\frac{1}{3}x + 2$ $-\frac{1}{3}$; 2; 6

11. 1; $-4\frac{1}{2}$; $4\frac{1}{2}$ 11. $y = x - 4\frac{1}{2}$

12. $-\frac{2}{5}$; 0; 0   12. $y = -\frac{2}{5}x$

13. $y = 3x$ 3; 0; 0

14. $y - 8 = 2x$ 2; 8; −4

15. $3 + y = -3x$ −3; −3; −1

16. $y - x = 2$ 1; 2; −2

17. $y - 4x = 6$ 4; 6; $-1\frac{1}{2}$

18. $-\frac{1}{4}$; 7; 28   18. $y - 7 = -\frac{1}{4}x$

19. The equation of a line is $y = -7x + 21$. What are the coordinates of the point where the line crosses the x-axis? the y-axis? (3, 0); (0, 21)

20. The equation of a line is $y - 5x = 10$. What are the coordinates of the point where the line crosses the x-axis? the y-axis? (−2, 0); (0, 10)

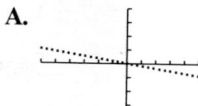 **CALCULATOR**

Choose the letter of the correct graphing calculator display for each equation.

A. B.

C. D.

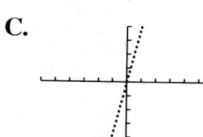
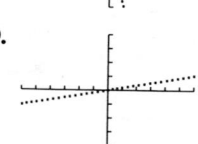

B 21. $y = 4x$ C

22. $y = -4x$ B

23. $y = \frac{1}{4}x$ D

24. $y = -\frac{1}{4}x$ A

CAREER/APPLICATION

Slopes are used in the planning and construction of the roof of a building. The slope of a roof is called the *pitch*. A carpenter must know the pitch of a roof to cut rafters of the correct length.

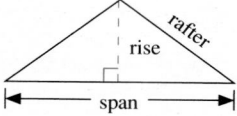

$$\text{Pitch} = \frac{\text{the rise of the roof}}{\text{half the span of the roof}}$$

25. The rise of a roof is 12 ft and the span is 24 ft. Find the pitch. 1

26. The pitch of the roof Tony is building is 5 to 12. Half of the span is 16 ft. What is the rise? $6\frac{2}{3}$ ft

27. About how long should a rafter for the roof described in Exercise 25 be? about 17 ft

LOGICAL REASONING

Tell whether each statement is *always*, *sometimes*, or *never* true.

C 28. The slope of a line with positive *x*- and *y*-intercepts is positive. never

29. The slope of a line with negative *x*- and *y*-intercepts is negative. always

SPIRAL REVIEW

S 30. Solve: $4x + 5.1 = -2.9$ *(Lesson 8-8)* −2

31. In a poll of 30 people, 20 people preferred So-Brite toothpaste. In a similar poll of 87 people, about how many people would be expected to prefer So-Brite toothpaste? *(Lesson 12-8)* about 58

32. Find the slope, the *y*-intercept, and the *x*-intercept of the line $y = 3x - 1$. *(Lesson 13-3)* $3; -1; \frac{1}{3}$

Inequalities and Graphing on the Coordinate Plane **591**

Follow-Up

Exploration

Slope is used by surveyors and civil engineers to construct roads through mountainous areas. Students can research some uses of slope, including the steepest slope allowed in highway construction.

Enrichment

Enrichment 132 is pictured on page 580E.

Practice

Practice 163, *Resource Book*

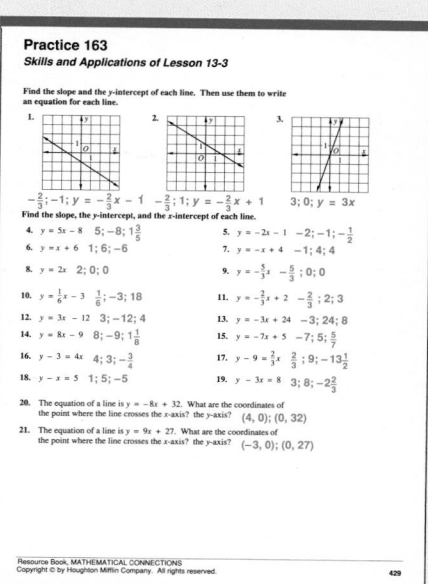

FOCUS ON
EXPLORATIONS

Using Geoboards

In the previous lesson you learned that the slope of a line is represented by the vertical change divided by the horizontal change between any two points on the line. Geoboards can help you visualize the slope of a line.

The geoboard pictured at the right simulates a coordinate plane. The peg at the center represents the origin, and the two rubber bands represent the axes.

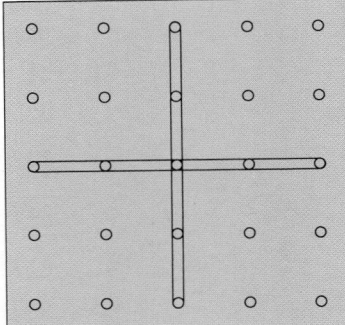

Activity I *Finding Slopes*

Construct a coordinate plane on a geoboard as shown in the diagram above.

1 Stretch a rubber band between the points (2, 2) and (−2, −2). Between these two points, what is the rise? the run? What is the slope of the line? 4; 4; 1

2 Stretch another rubber band between the points (2, 2) and (0, 0). Between these two points, what is the rise? the run? What is the slope of the line? 2; 2; 1

3 Compare the slopes in Steps 1 and 2. What can you conclude about the slope between two different pairs of points on the same line? The slope is the same for both pairs of points.

Activity II *Finding Intercepts*

1 Stretch a rubber band between each pair of points. Name the x-intercept and the y-intercept of each line.
   a. (−1, 2), (1, −2) 0; 0   **b.** (2, 1), (−1, −2) 1; −1
   c. (2, 2), (−2, −2) 0; 0   **d.** (2, −1), (−1, 2) 1; 1

2 You can stretch a rubber band between (0, −2) and two other points to produce a line with an x-intercept of −1. Give the coordinates of these two other points. (−1, 0); (−2, 2)

3 You can stretch a rubber band between (0, −2) and two other points to produce a line with an x-intercept of 1. Give the coordinates of these two other points. (1, 0); (2, 2)

4 You can stretch a rubber band between (−2, 0) and two other points to produce a line with a y-intercept of −1. Give the coordinates of these two other points. (0, −1); (2, −2)

Activity III *Parallel and Perpendicular Lines*

1 The diagram at the right shows one rubber band stretched between the points $(-2, 0)$ and $(0, 2)$, and another stretched between the points $(-1, -2)$ and $(2, 1)$. What is the relationship between the two lines? They are parallel.

2 Find the slope of each line in Step 1. Compare the slopes.
1; 1; they are equal.

3 Stretch one rubber band between the points $(-2, 2)$ and $(0, -2)$, and another between the points $(0, 2)$ and $(2, -2)$. What is the relationship between the two lines? They are parallel.

4 Find the slope of each line in Step 3. Compare the slopes.
−2; −2; they are equal.

5 Use your answers to Steps 2 and 4 to predict the relationship between the slopes of any two parallel lines.
Parallel lines have the same slope.

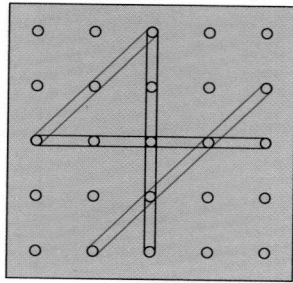

6 The diagram at the right shows one rubber band stretched between the points $(-1, -2)$ and $(2, 1)$, and another stretched between the points $(-2, 1)$ and $(1, -2)$. What is the relationship between the two lines? They are perpendicular.

7 Find the slope of each line in Step 6, and their product.
1; −1; −1

8 Stretch one rubber band between the points $(-1, 2)$ and $(1, -2)$, and another between the points $(-2, -1)$ and $(2, 1)$. What is the relationship between the two lines?
They are perpendicular.

9 Find the slope of each line in Step 8, and their product. $-2; \frac{1}{2}; -1$

10 Use your answers to Steps 7 and 9 to predict the relationship between the slopes of any two perpendicular lines.
The product of the slopes of perpendicular lines is −1.

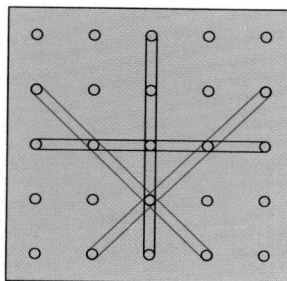

Activity IV *Vertical and Horizontal Lines*

1 The diagram at the right shows a red rubber band stretched between the points $(-2, 1)$ and $(2, 1)$. Between these two points, what is the rise? the run? the slope? 0; 4; 0

2 Stretch a rubber band between the points $(-2, -2)$ and $(2, -2)$. Between these two points, what is the rise? the run? the slope?
0; 4; 0

3 Use your answers to Steps 1 and 2 to predict what will be the slope of any horizontal line. The slope of a horizontal line is 0.

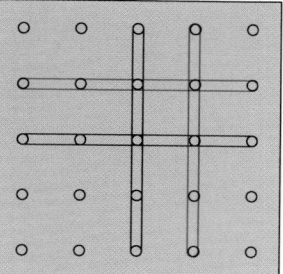

4 The diagram at the right shows a blue rubber band stretched between the points $(1, 2)$ and $(1, -2)$. Between these two points, what is the rise? the run? the slope? 4; 0; undefined

5 Stretch a rubber band between the points $(-2, 2)$ and $(-2, -2)$. Between these two points, what is the rise? the run? the slope?
4; 0; undefined

6 Use your answers to Steps 4 and 5 to predict what will be the slope of any vertical line. The slope of a vertical line is undefined.

Inequalities and Graphing on the Coordinate Plane   **593**

Lesson Planner

Teaching the Lesson
- Materials: string, chalk
- Lesson Starter 121
- Visuals, Folder M
- Using Technology, p. 580C
- Using Manipulatives, p. 580D

Lesson Follow-Up
- Practice 164
- Practice 165: Nonroutine
- Enrichment 133: Application
- Study Guide, pp. 241–242
- Manipulative Activity 26
- Test 64

Warm-Up Exercises

Graph each equation on the same coordinate plane.

1. $y = x + 2$
2. $y = x$
3. $y = -2x - 3$
4. $y = -2x + 4$

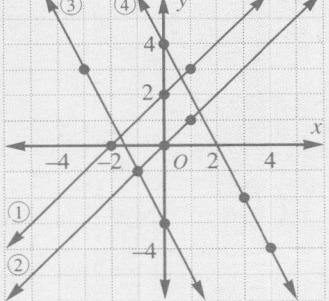

Systems of Equations

Objective: To solve systems of equations by graphing.

APPLICATION

Terms to Know
- system of equations
- solution of a system

Downtown Delivery charges $2 per pound to deliver a package, plus a service fee of $6. ABC Delivery charges $3 per pound, but only a $4 service fee. To find out how much to charge, the companies use the equations $y = 2x + 6$ and $y = 3x + 4$, where y is what a company charges to deliver a package, and x is the weight of a package.

QUESTION For what weight package will the charges be the same?

To answer this question, you find a solution common to both equations.

Two equations with the same variables form a **system of equations.** An ordered pair that is a solution of *both* equations is called a **solution of the system.** You can solve a system of equations by graphing.

Example 1

Solve the system by graphing: $\quad y = 2x + 6$
$\quad\quad\quad\quad\quad\quad\quad\quad\quad\quad\quad\quad y = 3x + 4$

Solution

Make a table for each equation. Then graph both equations on one coordinate plane.

$y = 2x + 6$		
x	y	(x, y)
-1	4	$(-1, 4)$
0	6	$(0, 6)$
1	8	$(1, 8)$

$y = 3x + 4$		
x	y	(x, y)
-1	1	$(-1, 1)$
0	4	$(0, 4)$
1	7	$(1, 7)$

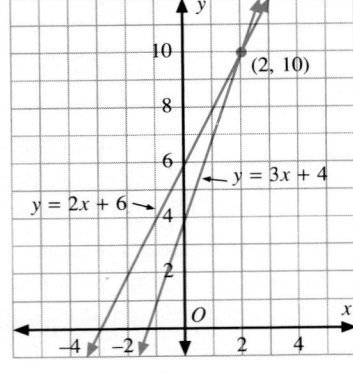

The solution is $(2, 10)$.

 **Check**

$y = 2x + 6$	$y = 3x + 4$
$10 \stackrel{?}{=} 2(2) + 6$	$10 \stackrel{?}{=} 3(2) + 4$
$10 \stackrel{?}{=} 4 + 6$	$10 \stackrel{?}{=} 6 + 4$
$10 = 10$	$10 = 10$

ANSWER Each company will deliver a 2-lb package for $10.

594 Chapter 13

Example 2

Solve the system by graphing:
$$y = -\frac{1}{2}x + 2$$
$$y = -\frac{1}{2}x - 1$$

Solution

$y = -\frac{1}{2}x + 2$		
x	y	(x, y)
-2	3	$(-2, 3)$
0	2	$(0, 2)$
2	1	$(2, 1)$

$y = -\frac{1}{2}x - 1$		
x	y	(x, y)
-2	0	$(-2, 0)$
0	-1	$(0, -1)$
2	-2	$(2, -2)$

The lines are parallel. They do not intersect. The system has no solution.

✔️ **Check Your Understanding**

If you were to use different values of x to graph the system in Example 2, would the lines still be parallel?
See Answers to Check Your Understanding at the back of the book.

Guided Practice

COMMUNICATION « *Reading*

Refer to the text on pages 594–595.

« **1.** When does a system of equations have no solution? when the lines are parallel

« **2.** How can you check if an ordered pair is a solution of the system? check the ordered pair in both equations

Use the graph to name the solution of each system.

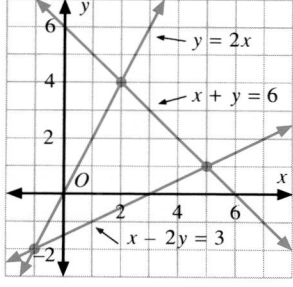

3. $x + y = 6$
   $y = 2x$ (2, 4)

4. $x + y = 6$
   $x - 2y = 3$ (5, 1)

5. $x - 2y = 3$
   $y = 2x$ (−1, −2)

6. No; the lines will be the same no matter what values of x you use.

« **6. COMMUNICATION** « *Discussion* When graphing the equations from a system of equations, do you need to use the same values of x in each table of values? Explain.

Inequalities and Graphing on the Coordinate Plane **595**

Lesson Focus

Ask students to imagine they are renting a car. Suppose each of two companies offers a slightly different deal. Company A charges 15¢ a mile plus $24 a day. Company B charges $35 a day with no charge for mileage. Which car is more economical to rent? Is there a time when both cars would cost the same? If only a few miles will be driven, which car is less expensive? These questions can be answered by solving systems of equations, the focus of today's lesson.

Teaching Notes

Key Question
Do all systems of equations have solutions? No, a system has no solution when the lines representing the equations are parallel.

Error Analysis
Some students make mistakes in graphing a system of equations or they transpose the x- and y-values when writing the solution. Remind students to use three points when graphing a line and to write the coordinates of the solution as (x, y).

Using Technology
For a suggestion on using computer graphics or a graphing calculator, see page 580C.

Using Manipulatives
For a suggestion on using a tile floor to represent the coordinate plane and string to represent a line in the plane, see page 580D.

Additional Examples

1. Solve the system by graphing.
   $y = 2x + 1$
   $y = 3x - 2$ (3, 7)
2. Solve the system by graphing.
   $y = -\frac{1}{3}x + 2$
   $y = -\frac{1}{3}x - 1$ no solution

Closing the Lesson

Ask a student to explain what is meant by a system of equations. Ask another student to explain how to find the solution of a system of equations by graphing.

Suggested Assignments

Skills: 1–20, 32–34
Average: 1–24, 32–34
Advanced: 1–15 odd, 16–34

Exercise Notes

Communication: Discussion

Guided Practice Exercise 6 requires students to think about and discuss the effect of using different values of *x* when graphing equations from a system.

Making Connections/Transitions

Exercises 16–20 connect the graphing of equations with geometric figures, providing a preview of coordinate geometry.

Computer

In *Exercises 21–24,* students can use function graphing software such as *Connections Plotter Plus* to solve systems of three equations.

Critical Thinking

Exercises 25–27 require students to *analyze* situations regarding two lines on a coordinate plane and to *decide* whether the lines will intersect, coincide, or be parallel.

Solve each system by graphing.

7. $y = 3x - 1$
   $y = -x - 5$ (−1, −4)

8. $y = -3x - 2$
   $y = -3x + 1$ no solution

9. $y = \frac{1}{3}x$
   $y = 2x - 5$
   (3, 1)

Exercises

Solve each system by graphing.

A
1. $y = 4x - 2$
   $y = -3x + 5$ (1, 2)

2. $y = -x + 5$
   $y = x + 3$ (1, 4)

3. $y = 3x + 6$
   $y = 3x - 2$
   no solution

4. $y = 4x - 6$
   $y = -2x - 6$ (0, −6)

5. $y = 3x + 3$
   $y = -2x - 2$ (−1, 0)

6. $y = 6x - 4$
   $y = 2x$ (1, 2)

7. $y = -2x$
   $y = -2x + 3$ no solution

8. $y = -x - 4$
   $y = -x + 3$
   no solution

9. $y = -x - 4$
   $y = -2x - 7$
   (−3, −1)

10. $y = -2x + 7$
    $y = 2x - 1$ (2, 3)

11. $y = x - 1$
    $y = -\frac{1}{2}x + 2$ (2, 1)

12. $y = \frac{1}{3}x - 2$
    $y = \frac{1}{3}x + 3$
    no solution

B
13. $y - 3x = -2$
    $y - x = 0$ (1, 1)

14. $y + 7 = 2x$
    $y + 2 = -\frac{1}{2}x$ (2, −3)

15. $y + 2 = -\frac{1}{3}x$
    $y - 1 = \frac{2}{3}x$
    (−3, −1)

CONNECTING ALGEBRA AND GEOMETRY

Use this system of equations for Exercises 16–20: $y = -2x + 8$
$y = 3x + 3$

16. Graph the system. See answer on p. A34.

17. What is the solution of the system? (1, 6)

18. Where does each line intersect the *x*-axis? (−1, 0); (4, 0)

19. What figure is formed by the two lines and the *x*-axis? triangle

20. Find the area of the figure in Exercise 19. 15

 COMPUTER APPLICATION

Answers to Exercises 21–24 are on p. A34.

Use function graphing software or graph paper.

21. Graph this system of three equations on one coordinate plane:
    $y = 3x - 5$, $y = -x + 3$, $y = x - 1$

22. Describe the graph of the system in Exercise 21. Does the system of three equations have a solution? Explain.

23. Graph this system of three equations on one coordinate plane:
    $y = \frac{1}{2}x + 4$, $y = \frac{3}{2}x - 4$, $y = \frac{1}{2}x - 4$

24. Describe the graph of the system in Exercise 23. Does the system of three equations have a solution? Explain.

THINKING SKILLS 25–27. Analysis   28–31. Evaluation

When you graph two equations with two variables on a coordinate plane, the lines will intersect, coincide, or be parallel. Two lines *coincide* when they contain all the same points.

C 25. Tell whether two lines with the same slope and different y-intercepts will *intersect*, *coincide*, or be *parallel*. parallel

26. Tell whether two lines with the same slope and the same y-intercept will *intersect*, *coincide*, or be *parallel*. coincide

27. Tell whether two lines with different slopes will *intersect*, *coincide*, or be *parallel*. intersect

Use your answers to Exercises 25–27. Without graphing, tell whether the two lines *intersect*, *coincide*, or are *parallel*. 28. parallel

29. intersect

28. $y = 5x + 2$, $y = 5x + 7$ 29. $y = -2x$, $y = 4x + 9$

30. $y = 3x + 6$, $y - 3x = 6$ 31. $y - x = 8$, $y + 11 = x$
coincide   parallel

SPIRAL REVIEW

S 32. The lengths of the legs of a right triangle are 10 cm and 24 cm. Find the length of the hypotenuse.   *(Lesson 10-11)* 26 cm

33. Simplify: $\dfrac{10y^3}{y}$ *(Lesson 7-5)* $10y^2$

34. Solve the system by graphing: $y = -2x + 3$, $y = -4x + 7$
(Lesson 13-4) $(2, -1)$

Self-Test 1

Find solutions of each equation. Use -2, -1, 0, 1, and 2 as values for x.

1. $y = 6x - 3$ $(-2, -15)$;
$(-1, -9)$; $(0, -3)$; $(1, 3)$; $(2, 9)$

2. $y - 2x = 5$ $(-2, 1)$;$(-1, 3)$;   **13-1**
$(0, 5)$; $(1, 7)$; $(2, 9)$

Graph each equation. Answers to Self-Test 3–4 are on p. A34.

3. $y = \dfrac{3}{4}x + 2$ 4. $x + y = 4$ **13-2**

Find the slope, the y-intercept, and the x-intercept of each line.

5. $y - 8 = 5x$ $5; 8; -1\dfrac{3}{5}$ 6. $y = 4x$ $4; 0; 0$ **13-3**

Solve by graphing.

7. $y = x + 1$ 8. $y = \dfrac{2}{3}x + 6$ **13-4**
$y = 3x - 5$ $(3, 4)$

$y = \dfrac{2}{3}x - 2$ no solution

Inequalities and Graphing on the Coordinate Plane   **597**

Follow-Up

Exploration *(Cooperative Learning)*
Have students work in groups to write an equation of a line in slope-intercept form. Each group should graph its equation and label a point on the graph. Using a compass, each group should construct a line perpendicular to the graph at the labeled point and then label a second point on the new line. Then the groups should find the slope of the new line. Ask the groups to state a relationship between the slope of the original line and the slope of the line perpendicular to it. The groups should test their conjectures by repeating the process with a different line.

Enrichment
Enrichment 133 is pictured on page 580E.

Practice
Practice 164 is shown below.

Quick Quiz 1
See pages 620 and 621.

Alternative Assessment
See page 621.

Practice 164, *Resource Book*

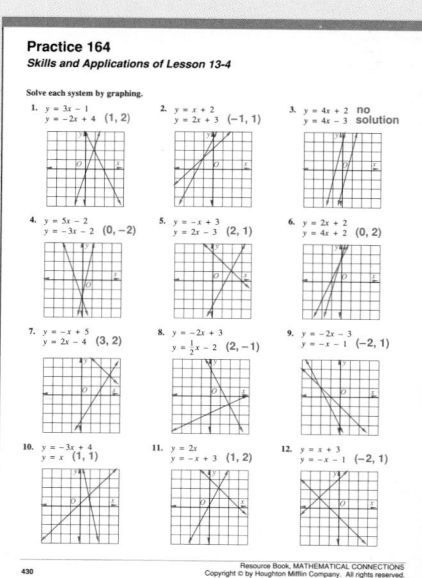

Practice 164
Skills and Applications of Lesson 13-4

Teaching Note

Function graphing software, such as *Connections Plotter Plus*, can be used to graph many types of functions. Although the focus of this lesson is on solving systems of equations, students may wish to use this software to draw other types of equations, inequalities, and functions presented both in this chapter and elsewhere in the textbook.

FOCUS ON COMPUTERS

Using Function Graphing Software

Objective: To use function graphing software to find solutions to systems of equations.

A computer can help you solve a system of equations, especially when the lines do not intersect at a point whose coordinates are integers.

Function graphing software enables you to graph a system of equations and then magnify or zoom in on the solution in order to get closer and closer estimates of the coordinates. The box in the diagram below indicates the section of a graph to be magnified.

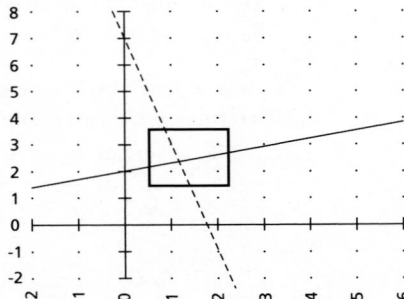

Exercises

The equations of the lines in the diagram above are $y = \frac{3}{10}x + 2$ and $y = -4x + 7$. Use function graphing software to graph the system of equations shown in the diagram.

1. Estimate each coordinate of the solution to the nearest integer. (1, 2)

2. Zoom in on the graph using a box whose center is at the solution of the system. Continue to zoom in until you can accurately estimate each coordinate of the solution to the nearest tenth. What are the coordinates of the solution to the nearest tenth? (1.2, 2.3)

3. Zoom in on the graph again until you can accurately estimate each coordinate of the solution to the nearest hundredth. What are the coordinates of the solution to the nearest hundredth? (1.16, 2.35)

Use function graphing software to graph the system of equations $y = -\frac{1}{2}x + 2$ and $y = 3x + 8$.

4. In what quadrant do the lines intersect? Quadrant II

5. Estimate each coordinate of the solution to the nearest integer. (−2, 3)

6. Zoom in on the graph until you can accurately estimate each coordinate of the solution to the nearest hundredth. What are the coordinates of the solution to the nearest hundredth? (−1.71, 2.86)

13-5

Strategy:
Solving a Simpler Problem

Objective: To solve problems by using simpler problems.

You can solve some complicated problems by solving a simpler problem or by using simpler numbers. Use strategies such as making a table and identifying a pattern to solve the simpler problem, and then solve the original.

Problem

The 10 houses on one side of a street are numbered with the even numbers from 2 to 20. What is the sum of these house numbers?

Solution

UNDERSTAND The problem is about the sum of house numbers.
Facts: houses are numbered with the even numbers from 2 to 20
Find: the sum of the house numbers

PLAN Make a table of partial sums and identify a pattern. Then use the pattern to solve the original problem.

WORK Number of Houses	House Numbers	Sum	Pattern
1	2	2	$1^2 + 1 = 2$
2	2 + 4	6	$2^2 + 2 = 6$
3	2 + 4 + 6	12	$3^2 + 3 = 12$
4	2 + 4 + 6 + 8	20	$4^2 + 4 = 20$

The pattern shows that the sum of the house numbers is the sum of the number of houses and the square of the number of houses.
The number of houses is 10. So $10^2 + 10 = 110$.

ANSWER The sum of the house numbers is 110.

Look Back In the Problem, you made a pattern. You can *generalize* this pattern by using the expression $n^2 + n$. This expression will give you the sum of the first *n* even whole numbers. Use this expression to find the sum of the first 50 even whole numbers. 2550

Inequalities and Graphing on the Coordinate Plane   **599**

Lesson Planner

Teaching the Lesson
• Lesson Starter 122

Lesson Follow-Up
• Practice 166
• Enrichment 134: Problem Solving
• Study Guide, pp. 243–244

Warm-Up Exercises

Solve each system by graphing.
1. $y = -5x + 7$
   $y = 6x - 4$ (1, 2)
2. $y = -4x - 4$
   $y = 5x + 5$ (−1, 0)
3. $y = 2x - 4y = 5x - 13$ (3, 2)
4. $y = \frac{1}{3}x$
   $y = x - 4$ (6, 2)
5. $y = -3x$
   $y = -3x + 5$ no solution

Lesson Focus

Ask students to imagine having to buy all the digits necessary to number the doors in a building with two hundred apartments. How many 1's should be ordered? How many 2's? Many complicated problems can be solved by simpler but similar problems that lead to the solution of the original problem. This strategy is the focus of today's lesson.

Teaching Notes

Point out to students that after identifying a simpler problem, they can then use any strategy introduced earlier in the course to solve it.

Key Question
What problem-solving strategies were used in the Example to solve the problem? the four-step plan, making a table, finding a pattern

Critical Thinking
Students must *analyze* the original problem in order to identify a simpler problem.

Error Analysis
In trying to identify a simpler problem, students may alter the situation and create a problem that is not similar to the original. Emphasize that the new problem must be simpler and similar to the original problem.

Additional Example
The ten houses on the other side of the street are numbered with the odd numbers from 1 to 19. What is the sum of these numbers? 100

Closing the Lesson
Ask students to identify another simpler problem that can be used to solve the Example. Answers will vary. The sum of the first and last house numbers is 22; the sum of the second and ninth house numbers is also 22. There are five such sums. The house numbers add to 5 · 22, or 110.

Suggested Assignments
Skills: 1–13
Average: 1–13
Advanced: 1–13, Challenge

Exercise Notes

Communication: Reading
Guided Practice Exercises 1–4 require students to analyze the information in a problem and, in so doing, create a simpler problem.

Problem Solving
Problem Solving Situations 5–8 require students to decide upon the best strategy to use to solve each problem.

Guided Practice

COMMUNICATION «*Reading*

Use the problem below for Exercises 1–5.

The rooms in a hotel are numbered consecutively from 1 to 185. How many plastic digits are needed to number all the rooms?

«**1.** List the 1-digit numbers that appear on the rooms. How many of them are there? 1, 2, 3, 4, 5, 6, 7, 8, 9; 9

«**2.** What are the least and greatest 2-digit numbers that appear on the rooms? How many 2-digit numbers are there in all? 10, 99; 90

«**3.** What are the least and greatest 3-digit numbers that appear on the rooms? How many 3-digit numbers are there in all? 100, 185; 86

«**4.** Use your answers to Exercises 1–3. How many plastic digits are needed to form the 1-digit numbers? the 2-digit numbers? the 3-digit numbers? 9; 180; 258

5. Use your answers to Exercise 4 to write an answer to the problem.
A total of 447 plastic digits are needed to number all the rooms.

Solve using a simpler problem.

6. A grocer stacks cans in a display using the pattern shown at the right. Each row has 1 fewer can than the row below it. How many cans are there in a stack with 20 rows? (*Hint*: Pair the rows beginning with the top and bottom rows.)
210

7. Find the sum of the first 50 odd whole numbers. (*Hint*: Find partial sums and identify a pattern.) 2500

Problem Solving Situations

Solve using a simpler problem.

1. How many triangles are there in the figure at the right? (*Hint*: There are four different sized triangles in the figure.) 27

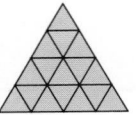

2. A marching band is marching in a triangular formation. There is 1 band member in the first row. Each of the other rows contains 2 more band members than the row in front of it. There are 11 rows in all. How many band members are there? 121

3. Ellen sold tickets to a play. On the first day she sold two tickets. On each of the next twelve days she sold two more tickets than she sold the day before. How many tickets did Ellen sell in all? 182

5. guess and check; 14 quarters and 16 dimes
6. no solution; the number of children must be a whole number.

4. The lockers on one side of a hallway are numbered consecutively from 1 to 100. What is the sum of these locker numbers? 5050

Solve using any problem solving strategy. 5–8. Strategies may vary. Likely strategies are given.

5. Sean has two more dimes than quarters. The coins are worth $5.10. How many of each does he have?

6. Admission to the science museum is $7 for adults and $3 for children. One day, the receipts were recorded as $724. If 80 adults went to the science museum that day, how many children went to the science museum that day?

7. There are 128 sophomores at Lincoln High School. A survey showed that 51 play an instrument, 68 play a sport, and 23 play an instrument and a sport. How many sophomores do neither? using a Venn diagram; 32

8. The houses on River Road are numbered consecutively from 10 to 132. How many brass digits are needed to form all the house numbers? solving a simpler problem; 279

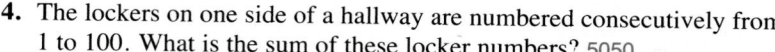

PROBLEM SOLVING

CHECKLIST

Keep these in mind:
Using a Four-Step Plan
Too Much or Not Enough Information
Supplying Missing Facts
Problems with No Solution

Consider these strategies:
Choosing the Correct Operation
Making a Table
Guess and Check
Using Equations
Identifying a Pattern
Drawing a Diagram
Using Proportions
Using Logical Reasoning
Using a Venn Diagram
Solving a Simpler Problem

WRITING WORD PROBLEMS

Using the information given, write a word problem that you could solve using a simpler problem. (You might need to add more information.) Then solve the problem.

Answers to Problems 9–10 are on p. A35.

9. The 84 houses on Oak Street are numbered consecutively.

10. Each row in a chorus line has 2 more dancers than the row in front of it.

SPIRAL REVIEW

S 11. Find the circumference of a circle with radius $5\frac{1}{2}$ in. *(Lesson 10-2)* 34.5 in.

12. Find the sum of the first 200 even whole numbers. *(Lesson 13-5)* 40,200

13. Graph the inequality $x \geq 4\frac{3}{4}$. *(Lesson 7-10)*
See answer on p. A35.

Challenge

What type of figure do you get when you graph the equation $|x| + |y| = 4$ on the coordinate plane? a square with vertices at $(-4, 0)$, $(0, 4)$, $(4, 0)$, and $(0, -4)$

Inequalities and Graphing on the Coordinate Plane **601**

Communication: Writing
Problem Solving Situations 9 and 10 have students create their own word problems that can be solved by using a simpler problem. This helps students to improve both their thinking and writing skills.

Follow-Up

Extension (Cooperative Learning)
Write each of the strategies from the Checklist on page 601 on separate pieces of paper and place them in a box. Form groups and have one student from each group draw a strategy out of the box. Each group should then create a word problem that can be solved by using the strategy selected. Have the groups share their problems and solutions with the entire class.

Enrichment

Enrichment 134 is pictured on page 580E.

Practice

Practice 166, *Resource Book*

Teaching the Lesson
- Lesson Starter 123
- Visuals, Folder I
- Using Technology, p. 580C
- Reteaching/Alternate
  Approach, p. 580D

Lesson Follow-Up
- Practice 167
- Enrichment 135:
  Application
- Study Guide, pp. 245–246
- Cooperative Activity 23

Warm-Up Exercises

Graph each inequality on a
number line.

1. $x > 4$

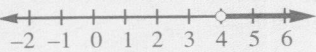

2. $x < -2$

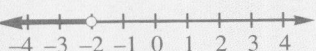

3. $x \leq 0$

4. $x \geq -5$

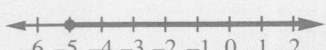

13-6
Inequalities with One Variable

Objective: To solve and
graph inequalities with one
variable.

EXPLORATION

1 You know that $12 > 8$. Replace each __?__ with $>$ or $<$.

a. $12 + 2$ __?__ $8 + 2$ $>$ **b.** $12 - 2$ __?__ $8 - 2$ $>$

c. $2(12)$ __?__ $2(8)$ $>$ **d.** $\frac{12}{2}$ __?__ $\frac{8}{2}$ $>$

e. $-2(12)$ __?__ $-2(8)$ $<$ **f.** $\frac{12}{-2}$ __?__ $\frac{8}{-2}$ $<$

2 Use your answers to Step 1. What are the two situations where
performing a given operation on both sides of an inequality changes the
inequality symbol? multiplying or dividing both sides by a negative number

You solve an inequality involving addition or subtraction the same way you
solve an equation involving addition or subtraction.

Generalization: *Solving Inequalities Using Addition or*
Subtraction

If a number has been *added* to the variable, subtract that number from
both sides of the inequality.
If a number has been *subtracted* from the variable, add that number to
both sides of the inequality.

Example 1

Solution

Solve and graph each inequality.

a. $x + 4 \leq 6$ **b.** $a - 5 > -2$

a. $x + 4 \leq 6$

$x + 4 - 4 \leq 6 - 4$

$x \leq 2$

Check

You cannot check every point on the
graph, but you should check at least
one. Choose $x = 0$.

$x + 4 \leq 6$

$0 + 4 \overset{?}{\leq} 6$

$4 \leq 6$

b. $a - 5 > -2$

$a - 5 + 5 > -2 + 5$

$a > 3$

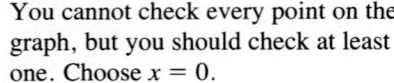

Check

Choose $a = 5$.

$a - 5 > -2$

$5 - 5 \overset{?}{>} -2$

$0 > -2$

Inequalities involving multiplication or division are solved in much the same way as equations involving multiplication or division. The only difference concerns multiplying or dividing by a negative number.

> **Generalization:** *Solving Inequalities Using Multiplication or Division*
>
> If the variable has been multiplied or divided by a *positive* number, solve as you would solve an equation.
> If the variable has been multiplied or divided by a *negative* number, solve as you would solve an equation, and *reverse the direction of the inequality symbol*.

Example 2

Solve and graph each inequality.

a. $5k < -10$

b. $\dfrac{n}{-3} \geq 1$

Solution

a. $5k < -10$

$\dfrac{5k}{5} < \dfrac{-10}{5}$

$k < -2$

Check

Choose $k = -3$.

$5k < -10$

$5(-3) \overset{?}{<} -10$

$-15 < -10$

b. $\dfrac{n}{-3} \geq 1$

$-3\left(\dfrac{n}{-3}\right) \leq -3(1)$

$n \leq -3$

Check

Choose $n = -6$.

$\dfrac{n}{-3} \geq 1$

$\dfrac{-6}{-3} \overset{?}{\geq} 1$

$2 \geq 1$

 **Check Your Understanding**

1. In Example 2(a), why do you not reverse the inequality symbol?
2. In Example 2(b), why do you reverse the inequality symbol?
See *Answers to Check Your Understanding* at the back of the book.

Guided Practice

COMMUNICATION « *Reading*

Refer to the text on pages 602–603.

«**1.** What is the main idea of the lesson? You can solve inequalities with one variable in much the same way as you solve equations with one variable.

«**2.** List three major points that support the main idea. See answer on p. A35.

Lesson Focus

Ask students to imagine they are going shopping. The total amount of money they have with them is $12. If they buy an item that costs $4 and then want to buy something else, what prices can the second item have? The students can buy anything with a price less than or equal to $8. Today's lesson focuses on solving and graphing inequalities with one variable.

Teaching Notes

Stress that the rules used to solve equations involving addition and subtraction can be used to solve inequalities. However, there is one very important fact that students need to remember: the direction of the inequality symbol *must be reversed* when multiplying or dividing both sides of an inequality by a *negative* number.

Key Question

When is a solid dot used in graphing the solution of an inequality? When the inequality symbol is $\leq$ or $\geq$, a solid dot is used.

Reasoning

In Check Your Understanding questions 1 and 2, students must explain the rules for reversing inequality symbols.

Using Technology

For a suggestion on using computer graphics or a graphing calculator, see page 580C.

Additional Examples

Solve and graph each inequality.

1. $x + 3 \le 5 \quad x \le 2$

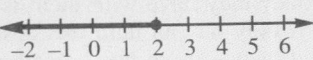

-2 -1 0 1 2 3 4 5 6

2. $b - 6 > -3 \quad b > 3$

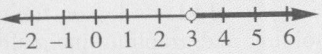

-2 -1 0 1 2 3 4 5 6

3. $6c < -18 \quad c < -3$

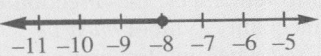

-7 -6 -5 -4 -3 -2 -1 0

4. $\frac{n}{-4} \ge 2 \quad n \le -8$

-11 -10 -9 -8 -7 -6 -5

Closing the Lesson

Ask a student to explain how to solve an inequality with one variable using addition or subtraction. Ask another student to explain how to solve an inequality with one variable using multiplication or division.

Suggested Assignments

Skills: 1–25, 29–33
Average: 1–25, 29–33
Advanced: 1–17 odd, 19–33

Exercise Notes

Communication: Reading

Guided Practice Exercises 1 and 2 require that students read and understand the main idea and major points of the lesson.

Estimation

Exercises 22–25 require students to apply estimation techniques to inequalities.

Making Connections/Transitions

Exercises 26–28 connect the concept of comparing data with inequalities.

Tell whether you would *add*, *subtract*, *multiply*, or *divide* to solve each inequality. Do not solve.

3. $a - 5 > 3$ add **4.** $\frac{1}{2}k \le 5$ multiply **5.** $8p \ge -8$ divide **6.** $4 + b < 9$ subtract

Tell whether you would reverse the inequality symbol when solving. Write *Yes* or *No*. Do not solve.

7. $-2x \le 16$ Yes. **8.** $5y < 30$ No. **9.** $\frac{m}{8} \ge 40$ No. **10.** $n - 9.4 > 3$ No.

Choose the letter of the graph that matches each inequality.

11. $x > 0.6$ C

12. $x < 0.6$ D

13. $x \ge 0.6$ B

14. $x \le 0.6$ A

A.
-1 0 0.6 1 2

B.
-1 0 0.6 1 2

C.
-1 0 0.6 1 2

D.
-1 0 0.6 1 2

Solve and graph each inequality. Graphs for Guided Practice 15–20 are on p. A35.

15. $y - 3 < 1$ $y < 4$ **16.** $6v \ge 12$ $v \ge 2$ **17.** $2x \le -1$ $x \le -\frac{1}{2}$

18. $r + 2.7 > 1.5$ $r > -1.2$ **19.** $-\frac{2}{3}t \le -2$ $t \ge 3$ **20.** $\frac{s}{-4} < 1$ $s > -4$

Exercises

Solve and graph each inequality. Graphs for Exercises 1–18 are on p. A35.

6. $r \ge 7\frac{1}{3}$

12. $y \ge -4\frac{3}{4}$

13. $h \le -0.8$

18. $m < -5\frac{1}{6}$

A **1.** $n - 5 > 12$ $n > 17$ **2.** $b + 9 < 11$ $b < 2$ **3.** $3c < 18$ $c < 6$

4. $\frac{a}{2} \le -4$ $a \le -8$ **5.** $\frac{b}{-3} < -7$ $b > 21$ **6.** $r - 2\frac{2}{3} \ge 4\frac{2}{3}$

7. $p + 2.5 > -5$ $p > -7.5$ **8.** $5x \le -15$ $x \le -3$ **9.** $-4g < 16$ $g > -4$

10. $-\frac{2}{3}y \ge -6$ $y \le 9$ **11.** $-\frac{1}{6}y > 3$ $y < -18$ **12.** $y + 7\frac{3}{4} \ge 3$

13. $h - 1.8 \le -2.6$ **14.** $-3x > 6$ $x < -2$ **15.** $-5h > -30$ $h < 6$

16. $\frac{d}{5} \ge -2$ $d \ge -10$ **17.** $w + 1 < -1$ $w < -2$ **18.** $m + 1\frac{2}{3} < -3\frac{1}{2}$

Write an inequality representing each statement. Then solve each inequality.

19. $n + 2 \ge 4; n \ge 2$

20. $3n \le 15; n \le 5$

21. $\frac{n}{-6} < 12; n > -72$

19. The sum of a number n and 2 is greater than or equal to 4.

20. The product of 3 and a number n is less than or equal to 15.

21. The quotient of a number n divided by -6 is less than 12.

ESTIMATION

To estimate the solution of an inequality, first decide which inverse operation to use. Then use an appropriate estimation method for that operation.

Choose the letter of the best estimate for the given inequality.

B **22.** $x - 3.02 < 1.8$ **(a.)** $x < 5$ **b.** $x < -1$

23. $31t > 894$ **a.** $t > 3$ **(b.)** $t > 30$

24. $\frac{a}{6.2} \le 0.47$ **(a.)** $a \le 3$ **b.** $a \le 12$

25. $m + 12.2 > 5.46$ **a.** $m > 17$ **(b.)** $m > -7$

DATA, *pages 140–141*

Write an inequality to represent each statement. Then use the appropriate data to solve each inequality for g.

C **26.** The maximum speed, g, of a giraffe is greater than 3 times the maximum speed, k, of a chicken. $g > 3k; g > 27$

27. Twice the maximum speed, g, of a giraffe is less than the maximum speed, c, of a cheetah. $2g < c; g < 35$

28. Use your answers to Exercises 26 and 27 to complete this statement with the correct numbers: $\underline{\ ?\ } < g < \underline{\ ?\ }$. Then write the statement in words. 27; 35; The maximum speed of a giraffe is between 27 mi/h and 35 mi/h.

SPIRAL REVIEW 32. $b \ge 9$; see graph on p. A35.

S **29.** Write 345% as a fraction in lowest terms. *(Lesson 9-5)* $3\frac{9}{20}$

30. Solve: $3x + 11 = 29$ *(Lesson 4-5)* 6

31. Solve: $\frac{3}{4}z - 8 = 16$ *(Lesson 8-9)* 32

32. Solve and graph $-9b \le -81$. *(Lesson 13-6)*

33. Write the unit rate: 225 mi in 5 h *(Lesson 9-1)* 45 mi/h

Inequalities and Graphing on the Coordinate Plane **605**

Follow-Up

Extension

Give students a set of combined inequalities such as $4 < x + 2 < 8$. Tell students that these inequalities can be solved by performing the same operation on all three parts. Ask students to solve each inequality and graph its solution on a number line.

Enrichment

Enrichment 135 is pictured on page 580E.

Practice

Practice 167, *Resource Book*

Practice 167
Skills and Applications of Lesson 13-6

Solve and graph each inequality.

1. $x - 3 > 2$ $x > 5$ 2. $c + 8 < 12$ $c < 4$

3. $z + 5 > 3$ $z > -2$ 4. $a - 3 \ge -7$ $a \ge -4$

5. $2c > -6$ $c > -3$ 6. $e + 4 \ge 2$ $e \ge -2$

7. $-5x < 15$ $x > -3$ 8. $4z \le 24$ $z \le 6$

9. $3m \ge -24$ $m \ge -8$ 10. $z + 6 \le 8$ $z \le 2$

11. $4a \le -14$ $a \le -3\frac{1}{2}$ 12. $e - 2 > -2$ $e > 0$

13. $t + 4 \le 6.8$ $t \le 2.8$ 14. $m - 2 < -3.1$ $m < -1.1$

Write an inequality representing each statement. Then solve each inequality.

15. The sum of a number x and 5 is greater than or equal to 3. $x + 5 \ge 3; x \ge -2$

16. The product of 4 and a number n is greater than -16. $4n > -16; n > -4$

17. The quotient of a number a divided by -3 is less than or equal to 2. $\frac{a}{-3} \le 2; n \ge -6$

433

Teaching the Lesson
- Lesson Starter 124
- Using Technology, p. 580C
- Reteaching/Alternate
  Approach, p. 580D

Lesson Follow-Up
- Practice 168
- Enrichment 136: Connection
- Study Guide, pp. 247–248
- Computer Activity 13

Warm-Up Exercises

Solve and graph each inequality.

1. $n - 7 \geq 9$ $n \geq 16$

13 14 15 16 17 18 19

2. $x + 7 < -4$ $x < -11$

−13   −11   −9   −7

3. $y - 2.6 > 4$ $y > 6.6$

6.4   6.6   6.8

4. $\frac{h}{-3} < 4$ $h > -12$

−16   −14   −12   −10

5. $-5m \geq 100$ $m \leq -20$

−22   −20   −18   −16

Lesson Focus

In the Lesson Focus of the previous lesson, students started shopping with $12 and spent $4. If they want to buy four more items all with the same price, what would be the maximum price of each of the items? The answer is that each item would cost no more than $2, since solving $4x + 4 \leq 12$ gives $x \leq 2$. Today's lesson focuses on solving two-step inequalities with one variable.

Two-Step Inequalities with One Variable

Objective: To solve and graph inequalities with one variable using two steps.

APPLICATION

Derek tunes his own car. The engine's danger zone in revolutions per minute (rpm) is 5000 to 6000. As Derek works on the car, he realizes that the idle speed in rpm can be multiplied by 6 and still be more than 200 rpm below the danger zone.

QUESTION What is the idle speed of Derek's car's engine?

To answer the question, write and solve an inequality. Let $r =$ the idle speed in rpm.

six times the rpm	plus	200	is less than	5000
6r	+	200	<	5000

Solving an inequality like this involves using two steps.

Example 1

Solution

Solve the inequality $6r + 200 < 5000$.

$$6r + 200 < 5000$$
$$6r + 200 - 200 < 5000 - 200 \quad \longleftarrow \text{Subtract 200 from both sides of the inequality.}$$
$$6r < 4800$$
$$\frac{6r}{6} < \frac{4800}{6} \quad \longleftarrow \text{Divide both sides of the inequality by 6.}$$
$$r < 800$$

 **Check Your Understanding**

1. In Example 1, which operation was undone first?
2. Was the inequality symbol in Example 1 reversed? Explain.
   See *Answers to Check Your Understanding* at the back of the book.

ANSWER The idle speed of Derek's car's engine is less than 800 rpm.

Generalization: *Solving Two-Step Inequalities*

First undo the addition or subtraction. Then undo the multiplication or division. Remember to reverse the direction of the inequality symbol when you multiply or divide by a negative number.

606   Chapter 13

Example 2
Solution

Solve and graph the inequality $-16a - 12 \leq 36$.

$$-16a - 12 \leq 36$$
$$-16a - 12 + 12 \leq 36 + 12 \quad \longleftarrow \text{Add 12 to both sides.}$$
$$-16a \leq 48$$
$$\frac{-16a}{-16} \geq \frac{48}{-16} \quad \longleftarrow \begin{array}{l}\text{Divide both sides by } -16.\\ \text{Reverse the inequality symbol.}\end{array}$$
$$a \geq -3$$

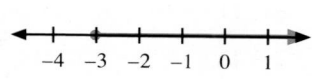

✓ **Check**

Choose $a = 0$.
$$-16a - 12 \leq 36$$
$$-16(0) - 12 \overset{?}{\leq} 36$$
$$0 - 12 \overset{?}{\leq} 36$$
$$-12 \leq 36$$

☑ **Check Your Understanding**

3. In Example 2, what is happening at the step where the inequality symbol gets reversed?

4. Describe how the graph in Example 2 would be different if the inequality were $-16a - 12 < 36$.
See *Answers to Check Your Understanding* at the back of the book.

Guided Practice

COMMUNICATION « *Writing*

Write an inequality that represents each sentence. Do not solve.

« **1.** The sum of twice a number and 5 is less than 17. $2n + 5 < 17$

« **2.** The difference of 9 subtracted from three times a number is greater than or equal to 6. $3n - 9 \geq 6$

« **3.** The sum of 3 and the quotient of a number divided by 2 is less than or equal to 1. $\frac{n}{2} + 3 \leq 1$

Write each inequality in words.

« **4.** $4n - 8 > 10$ « **5.** $5n + 2 \leq -4$ « **6.** $\frac{n}{-5} + 7 < 3$

Tell whether you would first *add* or *subtract* to solve each inequality. Then tell what number you would add or subtract.

7. $3n + 7 < 16$ **8.** $4k - 6 > 14$ **9.** $-5a + 1 \leq -24$
subtract 7   add 6   subtract 1

Solve and graph each inequality. Graphs for Guided Practice 10–15 are on p. A35.

10. $9b - 6 > 21$ **11.** $7h + 5 \geq -9$ **12.** $-4n - 9 < -1$

13. $-\frac{2}{3}x - 2 > 1$ **14.** $4 + \frac{n}{-3} \leq 5$ **15.** $2a - 1.6 < 4.8$

4. The difference of 8 subtracted from four times a number is greater than 10.
5. The sum of five times a number and 2 is less than or equal to -4.
6. The sum of 7 and the quotient of a number divided by -5 is less than 3.
10. $b > 3$
11. $h \geq -2$
12. $n > -2$
13. $x < -4\frac{1}{2}$
14. $n \geq -3$
15. $a < 3.2$

Inequalities and Graphing on the Coordinate Plane   **607**

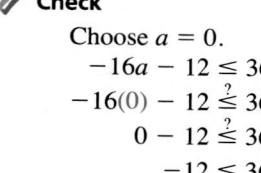

Follow-Up

Enrichment (Cooperative Learning)

Inequalities may occur in pairs linked by the words *and* or *or*. Have students working in groups find and graph the solutions to the inequalities below. Each group should explain why they think its solutions are correct.

1. $x + 3 < 9$ and $x - 3 > -7$
$-4 < x < 6$

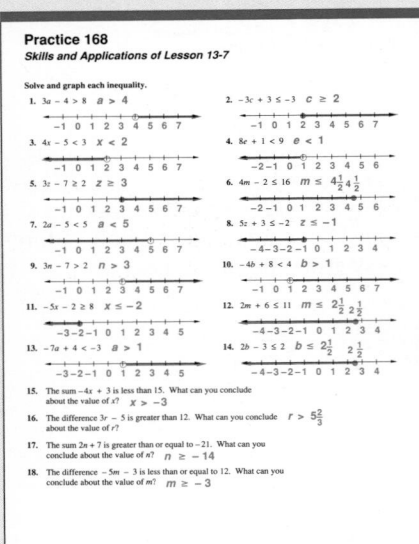

2. $2y - 9 < 15$ and $3y + 11 > 26$
$5 < y < 12$

3. $3m - 8 < 40$ or $2m + 9 > 51$
$m < 16$ or $m > 21$

Enrichment

Enrichment 136 is pictured on page 580F.

Practice

Practice 168, *Resource Book*

Exercises

Solve and graph each inequality. Graphs for Exercises 1–18 are on pp. A35–A36.

A

1. $5t - 8 > 22$ $t > 6$

2. $7h - 4 \leq 24$ $h \leq 4$

3. $11 + 2a \geq 5$ $a \geq -3$

4. $5 + 6x \geq 7$ $x \geq \frac{1}{3}$

5. $\frac{n}{2} - 1 \leq 1$ $n \leq 4$

6. $\frac{z}{3} - 9 < -7$ $z < 6$

7. $2b - 3 < -5$ $b < -1$

8. $5t + 4 \geq 4$ $t \geq 0$

9. $2 + \frac{k}{5} > -1$

10. $5 + \frac{x}{-4} \leq 6$ $x \geq -4$

11. $-8a + 6 \leq 30$

12. $-2q - 7 \leq -9$

13. $\frac{3}{4}y - 2 \geq -4$

14. $-\frac{4}{5}m + 3 < 7$

15. $-4m - 2.7 \geq 4.1$

16. $3b + 4.5 > 6$

17. $\frac{a}{-3} + 2 < 9$ $a > -21$

18. $\frac{t}{-2} - 4 \leq -4$ $t \geq 0$

9. $k > -15$
11. $a \geq -3$
12. $q \geq 1$
13. $y \geq -2\frac{2}{3}$
14. $m > -5$
15. $m \leq -1.7$
16. $b > 0.5$

19. The sum $-2v + 5$ is less than -23. What can you conclude about the value of v? $v > 14$

20. The difference $9j - 6$ is at least 48. What can you conclude about the value of j? $j \geq 6$

Solve each inequality.

B

21. $2(3a - 1) < 16$ $a < 3$

22. $15 \leq 3(t - 4)$ $9 \leq t$

23. $8 > 4(2x + 5)$ $-\frac{3}{2} > x$

SPIRAL REVIEW

S

24. A number cube that has six sides numbered 1 through 6 is rolled. Find $P(\text{number} > 4)$. *(Lesson 12-1)* $\frac{1}{3}$

25. Find solutions of the equation $y = 3x - 5$. Use $-2, -1, 0, 1$, and 2 as values for x. *(Lesson 13-1)* $(-2, -11)$; $(-1, -8)$; $(0, -5)$; $(1, -2)$; $(2, 1)$

26. Quadrilateral $ABCD$ is congruent to quadrilateral $WXYZ$. Which side of $WXYZ$ is congruent to $\overline{AD}$? *(Lesson 10-7)* $\overline{WZ}$

27. Solve and graph the inequality $-2y + 8 < 10$. *(Lesson 13-7)*
$y > -1$; see graph on p. A36.

Historical Note

Jakob Steiner (1796–1863) has been called the greatest geometer since Apollonius of Perga (3rd century B.C.). Yet Steiner did not learn to read or write until he was 14 years old. Born into a poor family in Switzerland, Steiner left home at the age of 18 to study mathematics. He eventually became a professor at the University of Berlin, where he taught many other famous mathematicians including Leopold Kronecker and Bernhard Riemann.

See answer on p. A36.

Research

Both Apollonius of Perga and Jakob Steiner were interested in conic sections. Find out what conic sections are.

608 Chapter 13

13-8 Inequalities with Two Variables

Objective: To determine solutions of inequalities with two variables and to graph inequalities with two variables.

Terms to Know

- *solution of an inequality with two variables*

The line shown in the figure at the right is the graph of $y = -2x + 1$. The line divides the coordinate plane into two parts. The red shaded part *above* the line shows the points for which $y > -2x + 1$. The blue shaded part *below* the line shows the points for which $y < -2x + 1$.

A **solution of an inequality with two variables** is an ordered pair of numbers that make the inequality true. Since the point $(-2, 1)$ is in the blue part, $(-2, 1)$ is a solution of the inequality $y < -2x + 1$.

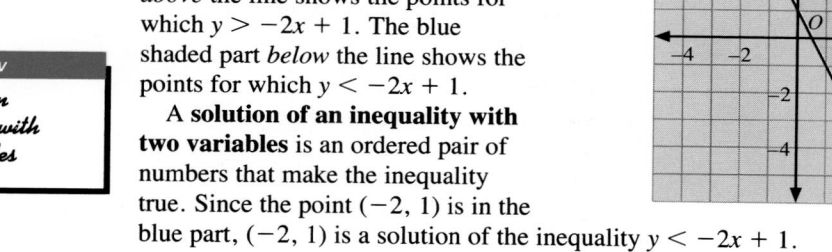

Example 1

Tell whether $(-1, 6)$ is a solution of the inequality $y \geq 3x + 8$. Write *Yes* or *No*.

Solution

Substitute -1 for x and 6 for y.

$$y \geq 3x + 8$$
$$6 \overset{?}{\geq} 3(-1) + 8$$
$$6 \overset{?}{\geq} -3 + 8$$
$$6 \geq 5$$

Yes, $(-1, 6)$ is a solution of the inequality $y \geq 3x + 8$.

Every inequality with two variables has a related equation. For example, the inequality $y < x + 4$ has the related equation $y = x + 4$. The graph of the related equation forms the *boundary* of the graph of the inequality.

> **Generalization:** *Graphing an Inequality with Two Variables*
>
> If an inequality is of the form $y > mx + b$, make a dashed boundary line and shade above.
> If an inequality is of the form $y \geq mx + b$, make a solid boundary line and shade above.
> If an inequality is of the form $y < mx + b$, make a dashed boundary line and shade below.
> If an inequality is of the form $y \leq mx + b$, make a solid boundary line and shade below.

Lesson Planner

Teaching the Lesson
- Lesson Starter 125
- Visuals, Folders I and N
- Using Technology, p. 580C

Lesson Follow-Up
- Practice 169
- Practice 170: Nonroutine
- Enrichment 137: Exploration
- Study Guide, pp. 249–250
- Cooperative Activity 24
- Test 65

Warm-Up Exercises

Graph each equation on the same coordinate plane.

1. $y = 3x + 5$
2. $y = -2x - 5$
3. $y = -\frac{2}{3}x - 2$
4. $y = 2x - 1$

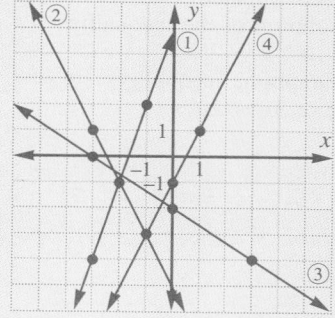

Lesson Focus

The solutions of inequalities with one variable are graphed on a number line. Today's lesson focuses on inequalities whose solutions are graphed on the coordinate plane, that is, inequalities with two variables.

Teaching Notes

Point out to students the similarities in solving inequalities with two variables to solving equations with two variables.

Key Question

When is the boundary of an inequality with two variables a dashed line? when the inequality symbol is < or >

Error Analysis

In graphing the solution to an inequality with two variables, students make errors by shading the wrong part of the graph. Remind students to check their graphs by selecting an ordered pair in the shaded part to see if it is a solution by substituting it in the inequality.

Using Technology

For a suggestion on using computer graphics, see page 580C.

Additional Examples

1. Tell whether $(-2, 7)$ is a solution of the inequality $y \geq 4x + 7$. Yes
2. Graph the inequality $y \geq 3x - 2$.

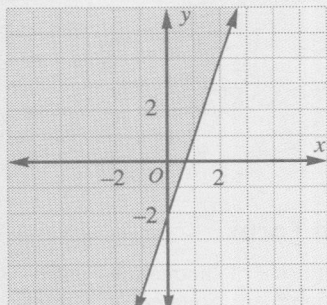

Example 2

Solution

Graph the inequality $y > 2x - 4$.

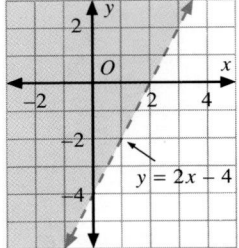

- Begin with the equation $y = 2x - 4$. Three points on this line are $(0, -4)$, $(1, -2)$, and $(2, 0)$. Draw a dashed line through these points to show that points on the line are *not* solutions of the inequality.

- To graph $y > 2x - 4$, shade the region *above* the line.

✔️ **Check**

Choose the point $(0, 0)$.

$$y > 2x - 4$$
$$0 \overset{?}{>} 2(0) - 4$$
$$0 \overset{?}{>} 0 - 4$$
$$0 > -4$$

✔️ **Check Your Understanding**

1. Describe how the graph in Example 2 would be different if the inequality were $y < 2x - 4$.
2. If the inequality in Example 2 were $y \geq 2x - 4$, would the points on the boundary line be solutions of the inequality?
   ʔe *Answers to Check Your Understanding* at the back of the book.

Guided Practice

COMMUNICATION «*Reading*

Refer to the text on pages 609–610.

«**1.** If an inequality contains the inequality symbol "$\geq$," do you use a *solid* or a *dashed* boundary line on the graph? solid

«**2.** If an inequality is of the form $y < mx + b$, do you shade *above* or *below* the boundary line? below

«**3.** When the graph of an inequality has a dashed boundary line, are the points on that line solutions of the inequality? No.

Use the graph at the right. Tell whether each ordered pair is a solution of the inequality $y \leq -\frac{1}{2}x - 2$. Write *Yes* or *No*.

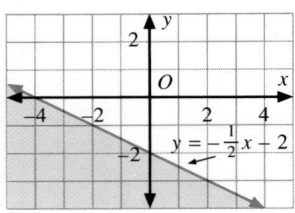

4. $(0, -2)$ Yes.   5. $(1, -1)$ No.

6. $(2, 1)$ No.   7. $(-5, 0)$ Yes.

8. $(-3, -1)$ Yes.   9. $(4, -4)$ Yes.

Tell whether each ordered pair is a solution of each inequality. Write *Yes* or *No*.

10. $(3, -2); y > -3x - 2$ Yes.　　　　11. $(-4, 2); y < \frac{1}{2}x + 4$ No.

Graph each inequality. Answers to Guided Practice 12–14 are on p. A36.

12. $y < -x + 3$　　　　13. $y \geq 2x$　　　　14. $y > \frac{1}{3}x + 1$

Exercises

Tell whether each ordered pair is a solution of each inequality. Write *Yes* or *No*.

A　1. $(2, 1); y > x - 3$ Yes.　　　　2. $(6, 5); y > x + 2$ No.

3. $(-1, 2); y < -2x$ No.　　　　4. $(-2, 1); y \leq -x - 1$ Yes.

5. $(-3, -2); y \geq 2x + 5$ No.　　　　6. $(3, -4); y \leq -3x + 7$ Yes.

7. $(10, -5); y < \frac{3}{5}x - 9$ Yes.　　　　8. $(-8, 1); y > \frac{1}{4}x + 3$ No.

Graph each inequality. Answers to Exercises 9–20 are on p. A36.

9. $y > x$　　　　10. $y < -x$　　　　11. $y \leq 3x - 2$

12. $y \geq -2x + 4$　　　　13. $y > -x - 1$　　　　14. $y \geq x - 2$

15. $y < 3x + 1$　　　　16. $y > -2x$　　　　17. $y \leq 4x - 3$

18. $y < \frac{1}{3}x + 2$　　　　19. $y > -\frac{1}{4}x + 2$　　　　20. $y \geq -\frac{2}{3}x$

Tell whether each ordered pair is a solution of the inequality $y > -x + 3$. Write *Yes* or *No*.

21. $(0, 2)$ No.　　22. $(2, 1)$ No.　　23. $(4, -1)$ No.　24. $(-3, 1)$ No.

25. $(-2, 4.9)$ No.　26. $(1, 2.1)$ Yes.　27. $(0, 3\frac{1}{3})$ Yes.　28. $(2, \frac{3}{4})$ No.

GROUP ACTIVITY

B　29. Graph the inequalities $y > 2x + 3$ and $y \leq -3x - 2$ on the same coordinate plane. See graph on p. A36.
　　　a. Name three points that are solutions of both inequalities.
　　　b. Rewrite one of the two inequalities so that the points $(-1, -4)$, $(0, -6)$, and $(-3, -3)$ are solutions of both inequalities.

30. Graph the inequalities $y \leq x + 3$ and $y \geq x - 3$ on the same coordinate plane. See graph on p. A36.
　　　a. Describe the region common to both inequalities.
　　　b. Rewrite the two inequalities so that the graphs have no points in common and the boundary line of each region is dashed.

29. a. Answers will vary. Examples: $(-3, 0)$; $(-4, 2)$; $(-5, -3)$
　　b. $y \leq 2x + 3$
30. a. The region is a band of points between two parallel lines that slope upward to the right.
　　b. $y > x + 3$; $y < x - 3$

Inequalities and Graphing on the Coordinate Plane　**611**

Closing the Lesson

Ask a student to explain how to determine which part of the graph to shade when graphing an inequality with two variables.

Suggested Assignments

Skills: 1–28, 39–42
Average: 1–28, 39–42
Advanced: 1–27 odd, 29–42

Exercise Notes

Reasoning
Guided Practice Exercises 4–9 and Exercises 1–8 require students to determine if ordered pairs are solutions to inequalities.

Cooperative Learning
Exercises 29 and 30 develop the concept of a system of inequalities in a group setting. Discussion of these problems should help students to clarify their thinking about systems of inequalities.

Communication: Writing
Exercises 31–34 present sentences that students must rewrite as inequalities. In *Exercises 35–38*, the process is reversed, as students must translate the symbols into words.

Follow-Up

Enrichment
Enrichment 137 is pictured on page 580F.

Practice
Practice 169 is shown below.

Quick Quiz 2
See page 624.

Alternative Assessment
See page 625.

Practice 169, *Resource Book*

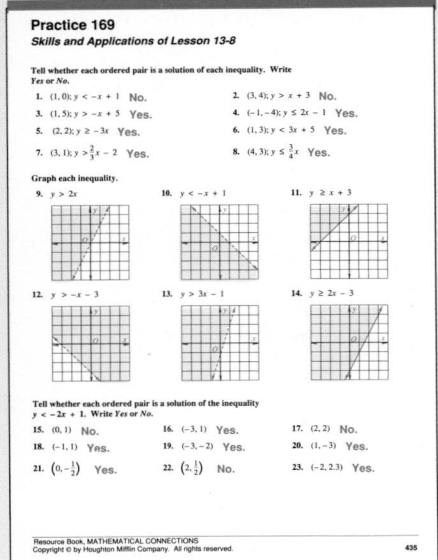

Write an inequality with two variables to represent each statement.

«**31.** The value of a number y is less than or equal to the sum of five times the number x and three. $y \le 5x + 3$

«**32.** The value of a number y is greater than or equal to the difference of seven subtracted from the number x. $y \ge x - 7$

«**33.** The amount of time that Jamie spends doing homework (j) is greater than twice the amount of time that her younger sister Anna spends doing homework (a). $j > 2a$

«**34.** The cost of a house in the Midwest (m) is less than one-third the cost of a similar house in the Northeast (n). $m < \frac{1}{3}n$

Write each inequality in words. Answers to Exercises 35–38 are on p. A36.

«**35.** $y < 3x$ «**36.** $y > 6x + 8$ «**37.** $k \ge m + 1$ «**38.** $c \le \frac{1}{2}t$

SPIRAL REVIEW Answers to Exercises 40–41 are on p. A36.

S **39.** Complete the function table at the right.
(Lesson 2-6)

40. Graph the equation $y = 4x - 2$.
(Lesson 13-2)

41. Draw an angle that measures 48°.
(Lesson 6-2)

42. Tell whether $(-1, 4)$ is a solution of the inequality $y < -x + 5$.
(Lesson 13-8) Yes.

x	$2x + 4$
1	6
4	12
5	?
7	?
10	?

14

18

24

Self-Test 2

1. The 128 apartments in a building are numbered consecutively with metal numbers beginning with the number 1. How many metal digits are needed to number the apartments? 276   13-5

Solve and graph each inequality. Graphs for Self-Test 2–7 are on p. A36.

2. $x - 7 > -4$ $x > 3$ 3. $-2a \ge 8$ $a \le -4$ 13-6

4. $\frac{c}{5} + 7 < 18$ $c < 55$ 5. $-3b - 4 \le -5$ $b \ge \frac{1}{3}$ 13-7

Graph each inequality.

6. $y \le 2x - 3$ 7. $y > 4x + 1$ 13-8

Function Rules

Objective: To use a function rule to find values of a function.

CONNECTION

A function pairs a number in one set of numbers with exactly one number in another set of numbers. In previous lessons you learned to represent a function using function tables, function rules, ordered pairs, and graphs of ordered pairs. Function rules were stated using arrow notation.

Another way to state a function rule is by using an equation with two variables. For example, the function rule $x \rightarrow 4x + 2$ can also be written as $y = 4x + 2$. The set of all possible values of x is called the **domain** of the function, and the set of all possible values of y is called the **range** of the function. The values of y are also referred to as the **values of the function.**

Example 1

Find values of each function. Use -2, 1, and 5 as values for x.

a. $y = -6x - 5$ **b.** $y = x^2 + 1$

Solution

Make function tables.

a.

$y = -6x - 5$	
x	y
-2	$-6(-2) - 5 = 7$
1	$-6(1) - 5 = -11$
5	$-6(5) - 5 = -35$

When $x = -2$, $y = 7$.
When $x = 1$, $y = -11$.
When $x = 5$, $y = -35$.

b.

$y = x^2 + 1$	
x	y
-2	$(-2)^2 + 1 = 5$
1	$(1)^2 + 1 = 2$
5	$(5)^2 + 1 = 26$

When $x = -2$, $y = 5$.
When $x = 1$, $y = 2$.
When $x = 5$, $y = 26$.

 **Check Your Understanding**

1. In Example 1(a), do you first *multiply* or *subtract* when finding each value of the function?

2. In Example 1(b), do you first *find the power* or *add* when finding each value of the function?

See *Answers to Check Your Understanding* at the back of the book.

A third type of notation used to state a function rule is **function notation**. In function notation, the function in Example 1(b) would be written $f(x) = x^2 + 1$, which is read "f of x equals $x^2 + 1$" or "the value of f at x is $x^2 + 1$."

Function notation is a useful shorthand for showing how values are paired. Instead of writing "when $x = 1$, $y = -11$" in Example 1(a), you could use function notation to write $f(1) = -11$.

Inequalities and Graphing on the Coordinate Plane   **613**

Key Question

Given the function $f(x) = 2x - 3$, how can you find $f(7)$? Replace x with 7 and evaluate $2(7) - 3$ to obtain 11.

Error Analysis

A common error students make in finding values of functions involving absolute value is to ignore the placement of the symbol in the function rule. Remind students of the importance of using the correct order of operations when finding values of functions such as $f(x) = |2x - 3|$ and $f(x) = |2x| - 3$.

Additional Examples

1. Find values of each function. Use -3, 1, and 6 as values for x.
   a. $y = -5x - 7$
      when $x = -3$, $y = 8$;
      when $x = 1$, $y = -12$;
      when $x = 6$, $y = -37$
   b. $y = x^2 + 3$
      when $x = -3$, $y = 12$;
      when $x = 1$, $y = 4$;
      when $x = 6$, $y = 39$
2. Find $f(-3)$, $f(0)$, and $f(7)$ for the function $f(x) = |x - 1|$.
   $f(-3) = 4$; $f(0) = 1$; $f(7) = 6$

Closing the Lesson

Ask a student to explain what the terms *domain* and *range* mean. Ask another student to explain how to find the values of a function.

Suggested Assignments

Advanced: 1–22

Exercise Notes

Communication: Writing
Guided Practice Exercises 1–4 require students to write sentences using function notation. *Guided Practice Exercises 5–8* reverse the process.

Example 2   Find $f(-3)$, $f(0)$, and $f(5)$ for the function $f(x) = |x - 2|$.

Solution   Make a function table.

$f(x) = |x - 2|$	
x	$f(x)$
-3	$|-3 - 2| = 5$
0	$|0 - 2| = 2$
5	$|5 - 2| = 3$

$f(-3) = 5$, $f(0) = 2$, and $f(5) = 3$

Check Your Understanding

3. In Example 2, why is the answer $f(-3) = 5$ instead of $f(-3) = -5$?
   See *Answers to Check Your Understanding* at the back of the book.

Guided Practice

COMMUNICATION « *Writing*

Write each sentence using function notation.

«**1.** The function f pairs x with $7x - 1$. $f(x) = 7x - 1$

«**2.** The function f pairs x with $24x$. $f(x) = 24x$

«**3.** The value of the function f at $x = -1$ is 0. $f(-1) = 0$

«**4.** The value of the function f at $x = 3$ is 9. $f(3) = 9$

Write each statement in words.

5. The function f pairs x with $25x$.
6. The function f pairs x with $\frac{x}{2}$.
7. The value of the function f at $x = 2$ is 6.
8. The value of the function f at $x = 1$ is -1.

«**5.** $f(x) = 25x$ «**6.** $f(x) = \frac{x}{2}$ «**7.** $f(2) = 6$ «**8.** $f(1) = -1$

Find each answer.

9. $3(-4)^2$ 48   **10.** $(-6 + 3)^2$ 9   **11.** $|5 - 8|$ 3   **12.** $|-7| + 9$ 16

Find values of each function. Use -1, 3, and 6 as values for x.

13. $y = 3x - 7$ -10; 2; 11   **14.** $y = |x - 5|$ 6; 2; 1   **15.** $y = (x + 4)^2$ 9; 49; 100

Find $f(-2)$, $f(0)$, and $f(4)$ for each function.

16. $f(x) = 4x + 1.5$ **17.** $f(x) = \frac{1}{2}x$ $f(-2) = -1$; **18.** $f(x) = x^2 - 1$
$f(-2) = -6.5$; $f(0) = 1.5$; $f(0) = 0$; $f(4) = 2$ $f(-2) = 3$; $f(0) = -1$;
$f(4) = 17.5$ $f(4) = 15$

Exercises

Find values of each function. Use -6, 2, and 3 as values for x.

A   **1.** $y = \frac{5}{6}x$ -5; $1\frac{2}{3}$; $2\frac{1}{2}$ **2.** $y = -x + 10$ 16; 8; 7   **3.** $y = 2x^2 + 1$ 73; 9; 19

     4. $y = (x - 7)^2$ 169; 25; 16   **5.** $y = |x| + 2.5$ **6.** $y = |x + 6|$
                                                8.5; 4.5; 5.5   0; 8; 9

Find $f(-4)$, $f(-2)$, and $f(5)$ for each function. See answers below.

7. $f(x) = -2x$

8. $f(x) = \frac{1}{2}x - 2$

9. $f(x) = (x + 1)^2$

10. $f(x) = x^2 + 3$

11. $f(x) = 5x^2$

12. $f(x) = |2x - 1|$

CONNECTING MATHEMATICS AND SCIENCE

Functions are used in physics to describe relationships between physical quantities. The function $P(s) = \frac{3}{200}s^3$ shows how the power, in watts, generated by a windmill is related to wind speed (s).

13. Find the power generated by a windmill when the wind speed is 2 mi/h. $\frac{3}{25}$ watts

14. Find the power generated by a windmill when the wind speed is 4 mi/h. $\frac{24}{25}$ watts

15. The wind speed in Exercise 14 is twice the wind speed in Exercise 13. Is the power generated by the windmill in Exercise 14 likewise twice the power generated by the windmill in Exercise 13? Explain. No; the power generated by the windmill in Exercise 14 is 8 times the power generated by the windmill in Exercise 13.

PROBLEM SOLVING/APPLICATION

Engineers use *pulse functions* to make machines perform tasks at specific time intervals. A pulse function can have a value of either 0 or 1. Depending on the machine, the task will be performed either when the value is 1 or when the value is 0.

Let $p(t) = 1$ if t is a multiple of 4, and let $p(t) = 0$ if t is any other number. Use this pulse function for Exercises 16–18.

16. Suppose a machine stamps a package when $p(t) = 1$. Will the machine stamp a package when $t = 48$? Yes.

17. Suppose a machine tightens a bolt when $p(t) = 1$. Will the machine tighten a bolt when $t = 38$? No.

18. Suppose that a machine punches a hole in a metal plate when $p(t) = 0$. How many holes will the machine punch during the interval $1 \leq t \leq 60$? 45

SPIRAL REVIEW

19. Graph the function $x \rightarrow 2x + 3$ when $x = -4, -3, -1, 0,$ and 1. *(Lesson 3-9)* See answer on p. A36.

20. Find the absolute value: $|-4|$ *(Lesson 3-1)* 4

21. Find the slope, the y-intercept, and the x-intercept of the line $y - 2 = -5x$. *(Lesson 13-3)* $-5; 2; \frac{2}{5}$

22. Find $f(-1)$, $f(0)$, and $f(2)$ for the function $f(x) = |x + 8|$. *(Lesson 13-9)* $f(-1) = 7; f(0) = 8; f(2) = 10$

7. $f(-4) = 8; f(-2) = 4;$ $f(5) = -10$

8. $f(-4) = -4;$ $f(-2) = -3; f(5) = \frac{1}{2}$

9. $f(-4) = 9; f(-2) = 1;$ $f(5) = 36$

10. $f(-4) = 19; f(-2) = 7;$ $f(5) = 28$

11. $f(-4) = 80; f(-2) = 20; f(5) = 125$

12. $f(-4) = 9; f(-2) = 5;$ $f(5) = 9$

Inequalities and Graphing on the Coordinate Plane **615**

Making Connections/Transitions
Exercises 13–15 connect mathematics and physics. Functions are used extensively in a variety of scientific applications.

Follow-Up

Extension (*Cooperative Learning*)
Have students work in groups and write a simple function rule. Each group should then compete against another group to guess its function rule. For example, suppose Group A's function rule is $f(x) = 3x - 1$. Group B might ask what $f(2)$ is and Group A would respond that $f(2) = 5$. Group A would then respond to a similar question from Group B. Play would alternate between groups until one group correctly guesses the other group's function rule. An incorrect guess results in that group forfeiting its next opportunity to ask a question.

Enrichment

Enrichment 138 is pictured on page 580F.

Practice

Practice 171, *Resource Book*

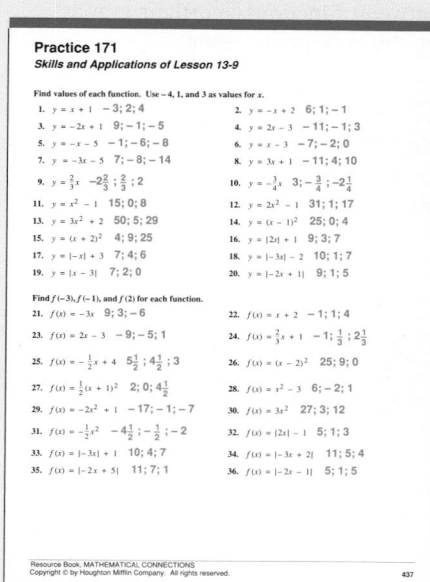

Teaching the Lesson
• Lesson Starter 127

Lesson Follow-Up
• Practice 172
• Practice 173: Nonroutine
• Enrichment 139:
  Application
• Study Guide, pp. 253–254
• Test 66

Warm-Up Exercises

Find values of each function. Use −6, 0, and 4 as values for x.

1. $y = \frac{5}{6}x - 1$ $-6; -1; 2\frac{1}{3}$

2. $y = -x + 6$ 12; 6; 2

3. $y = |x - 1|$ 7; 1; 3

Find $f(-5)$, $f(2)$, and $f(7)$ for each function.

4. $f(x) = (x + 1)^2$
  $f(-5) = 16; f(2) = 9; f(7) = 64$

5. $f(x) = |2x - 5|$
  $f(-5) = 15; f(2) = 1; f(7) = 9$

Lesson Focus

Ask students to recall the connection between functions and equations with two variables. Today's lesson focuses on the connection between functions and their graphs on the coordinate plane.

Teaching Notes

Key Question

Does a function always have exactly one y-value for each x-value?   Yes

---

Function Graphs

Objective: To graph functions on a coordinate plane.

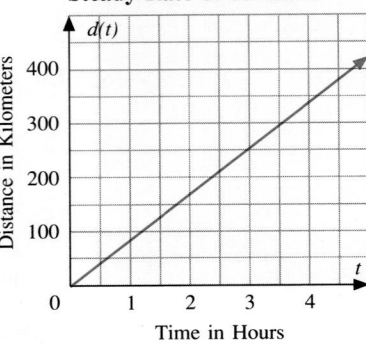

DATA ANALYSIS

The graph at the right shows the function $d(t) = 85t$. This function represents the distance that a car moving at a steady rate of 85 km/h will travel in t hours.

You can graph a function by graphing the equation specified by the function rule. In doing so, you may want to express the rule in terms of x and y, rather than x and $f(x)$. For example, the rule for the graph at the right would be expressed as $y = 85x$.

Distance Traveled at a Steady Rate of 85 km/h

Example

Solution

Graph each function: **a.** $y = x^2$ **b.** $y = |x|$

Make a function table for each function. Then graph the ordered pairs and connect the points.

a.

$y = x^2$		
x	y	
-2	$(-2)^2 = 4$	$(-2, 4)$
-1	$(-1)^2 = 1$	$(-1, 1)$
0	$(0)^2 = 0$	$(0, 0)$
1	$(1)^2 = 1$	$(1, 1)$
2	$(2)^2 = 4$	$(2, 4)$

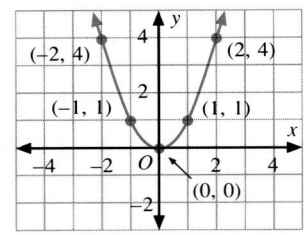

b.

$y = |x|$		
x	y	
-4	$|-4| = 4$	$(-4, 4)$
-2	$|-2| = 2$	$(-2, 2)$
0	$|0| = 0$	$(0, 0)$
2	$|2| = 2$	$(2, 2)$
4	$|4| = 4$	$(4, 4)$

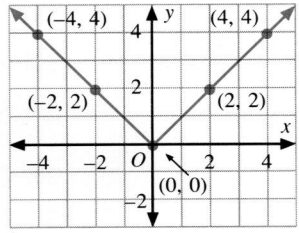

 **Check Your Understanding**

How many values did you find in each function table in the Example before making each graph?
See Answers to Check Your Understanding at the back of the book.

---

616   Chapter 13

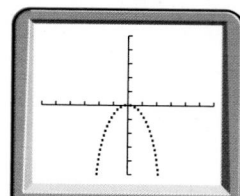

You can use a graphing calculator to graph a variety of functions. On some graphing calculators you would use the key sequence below to graph the function $y = -x^2$.

| Y= | (−) | X|T | ∧ | 2 | GRAPH |

Error Analysis

Students may incorrectly draw the graph of a function even though they have plotted a number of points on the graph correctly. Remind students that not all functions are linear, and that they should plot enough points to be sure their graph is correct.

Guided Practice

COMMUNICATION «*Reading*

Refer to the text on pages 616–617.

«**1.** Describe in words the graph of the function $y = x^2$. a curve, opening up

«**2.** Describe in words the graph of the function $y = |x|$. two lines, forming a right angle, opening up

«**3.** Describe in words the graph of the function $y = -x^2$. a curve, opening down

«**4.** Is the graph of a function always a straight line? No.

Tell whether each ordered pair is on the graph of the function shown at the right. Write *Yes* or *No*.

5. $(1, 0)$ **6.** $(2, 2)$ **7.** $(-3, 2)$
  Yes.   No.   Yes.

Use the graph at the right to complete each ordered pair.

8. $(3, \underline{\ ?\ })$ 2   **9.** $(-2, \underline{\ ?\ })$ 1

10. $(\underline{\ ?\ }, -1)$ 0   **11.** $(2, \underline{\ ?\ })$ 1

Graph each function. Answers to Guided Practice 12–17 are on pp. A36–A37.

12. $y = \frac{1}{2}x - 1$ **13.** $y = 3x$ **14.** $y = \frac{1}{3}x^2$

15. $y = (x + 2)^2$ **16.** $y = |x - 1|$ **17.** $y = -|x|$

Exercises

Graph each function. Answers to Exercises 1–15 are on p. A37.

A   **1.** $y = 2x + 1$ **2.** $y = -4x - 3$ **3.** $y = -3x^2$

4. $y = 3x^2$ **5.** $y = |x - 2|$ **6.** $y = |x| - 2$

7. $y = \frac{3}{5}x$ **8.** $y = -\frac{2}{3}x$ **9.** $y = -\frac{1}{3}x^2$

10. $y = \frac{1}{4}x^2$ **11.** $y = x^2 - 1$ **12.** $y = (x - 1)^2$

13. $y = x^2 + 3$ **14.** $y = |x| + 2$ **15.** $y = -|x| + 1$

Inequalities and Graphing on the Coordinate Plane   **617**

Additional Examples

Graph each function.

1. $y = 2x^2$

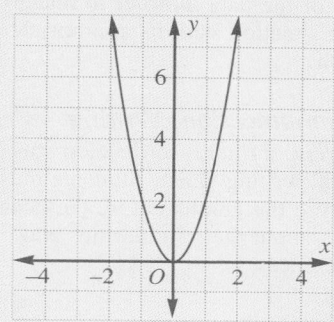

2. $y = |2x|$

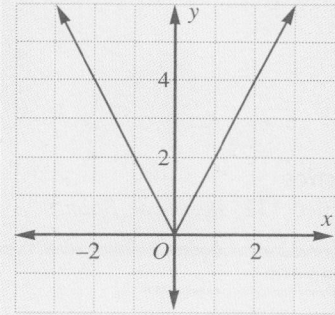

Closing the Lesson

Ask a student to explain how to set up a function table for a function. Ask another student to explain how to graph a function.

Suggested Assignments

Advanced: 1–33

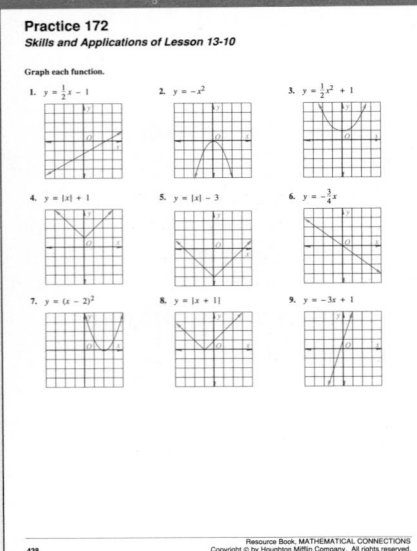
 CALCULATOR

Choose the letter of the correct graphing calculator display for each function.

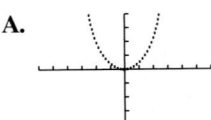

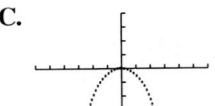 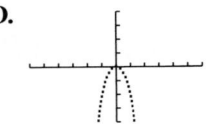

B 16. $y = 2x^2$ B

17. $y = -2x^2$ D

18. $y = \frac{1}{2}x^2$ A

19. $y = -\frac{1}{2}x^2$ C

DATA ANALYSIS

The graph at the top of page 616 shows a relationship between time and distance traveled. Many functions represent a relationship between a physical quantity and time, and these relationships can be displayed in graphs.

Use the graphs below. Write the letter of the graph that most likely represents each situation relating temperature (T) to time (t).

20. Natu turned on the heat in his apartment. C

21. Andrea turned on the air conditioner in her house. A

22. Julieta heated and then sipped a cup of soup for lunch. B

23. Jason put ice in a glass of fruit juice and then drank it while he began his homework. D

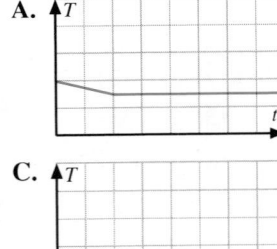

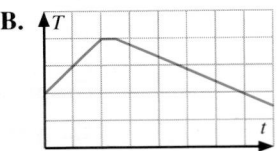

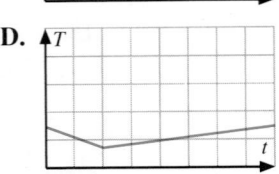

24. **WRITING ABOUT MATHEMATICS** Graph the functions $d(x) = 4x$ and $C(x) = 4x$, where $d(x)$ represents the distance traveled by a person jogging x mi at 4 mi/h, and $C(x)$ represents the cost of buying x yd of material at $4 per yard. Write a paragraph that discusses the similarities and differences in the graphs. See answer on p. A37.

618 Chapter 13

MENTAL MATH

A graph on a coordinate plane represents a function only if each value of x is paired with *exactly one* value of y. You can mentally check whether a graph represents a function. To do this, imagine drawing vertical lines through the graph at various values of x. If any of these lines contain two or more points on the graph, then the graph does not represent a function. This process is called a *vertical-line test*.

Use the vertical-line test to tell mentally whether each graph represents a function. Write Yes or No.

No.
C 25.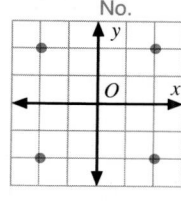

Yes.
26.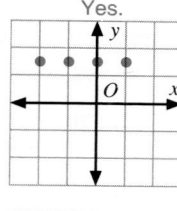

No.
27.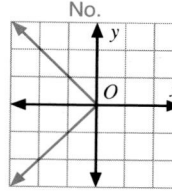

28.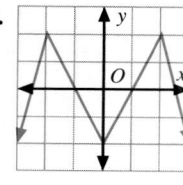

29.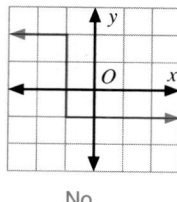

30.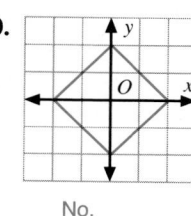

Yes.
No.
No.

SPIRAL REVIEW

S 31. What number is 55% of 250?   *(Lesson 9-6)* 137.5

32. Find the sum of the measures of the angles of a nonagon. *(Lesson 6-5)* 1260°

33. Graph the function $y = -\frac{2}{3}x^2$.   *(Lesson 13-10)*
See answer on p. A37.

Self-Test 3

Find values of each function. Use −2, 0, and 2 as values for x.

1. $y = x^2 - 4$ 0; −4; 0   **2.** $y = |x| - 5$ −3; −5; −3   **13-9**

Find $f(-4)$, $f(0)$, and $f(8)$ for each function.

3. $f(x) = 4x + 7$ $f(-4) = -9$;   **4.** $f(x) = \frac{1}{4}x^2$ $f(-4) = 4$; $f(0) = 0$;
$f(0) = 7$; $f(8) = 39$ $f(8) = 16$

Graph each function. Answers to Self-Test 5–6 are on p. A37.

5. $y = 2x - 1$ **6.** $y = x^2 - 2$ **13-10**

Inequalities and Graphing on the Coordinate Plane   **619**

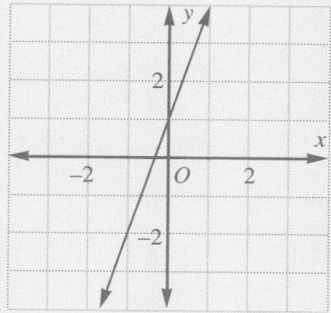
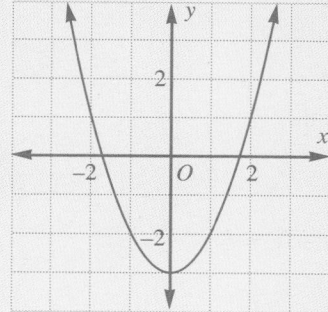

Quick Quiz 1 *(13-1 through 13-4)*
Find solutions of each equation. Use −2, −1, 0, 1, and 2 as values for x.

1. $y = 5x - 3$
   $(-2, -13), (-1, -8), (0, -3),$
   $(1, 2), (2, 7)$

2. $y - 3x = 4$
   $(-2, -2), (-1, 1), (0, 4),$
   $(1, 7), (2, 10)$

Graph each equation.

3. $y = \frac{2}{3}x + 2$

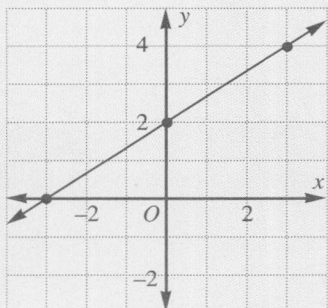

4. $x + y = 5$

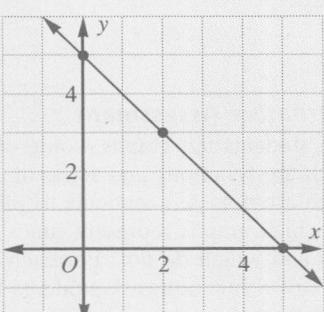

Chapter Review

solution of an equation with two variables (p. 582)
graph of an equation with two variables (p. 585)
slope (p. 588)
y-intercept (p. 588)
x-intercept (p. 588)
slope-intercept form of an equation (p. 588)

system of equations (p. 594)
solution of a system (p. 594)
solution of an inequality with two variables (p. 609)
domain of a function (p. 613)
range of a function (p. 613)
values of a function (p. 613)
function notation (p. 613)

Choose the correct term or phrase from the list above to complete each sentence.

1. The __?__ is the set of all possible x-values of the function.
   domain of a function

2. An ordered pair that is a solution of both equations in a system of equations is a(n) __?__. solution of the system

3. All points whose coordinates are solutions of an equation with two variables represent the __?__. graph of an equation with two variables

4. The __?__ of a line is the ratio of the rise to the run. slope

5. $y = mx + b$ is called the __?__. slope-intercept form of an equation

6. The y-coordinate of the point where a graph crosses the y-axis is the __?__. y-intercept

---

Find solutions of each equation. Use −2, −1, 0, 1, and 2 as values for x. *(Lesson 13-1)*

7. $y = 7x + 4$ $(-2, -10);$
   $(-1, -3); (0, 4); (1, 11); (2, 18)$

8. $y = \frac{1}{2}x + 3$ $(-2, 2); (-1, 2\frac{1}{2});$
   $(0, 3); (1, 3\frac{1}{2}); (2, 4)$

9. $y + 5x = -4$
   $(-2, 6); (-1, 1); (0, -4);$
   $(1, -9); (2, -14)$

Graph each equation. *(Lesson 13-2)* Answers to Chapter Review 10–12 are on p. A37.

10. $y = 2x$

11. $y = \frac{1}{3}x - 3$

12. $y - 2x = 4$

Find the slope and the y-intercept of each line. Then use them to write an equation for each line. *(Lesson 13-3)*
$\frac{2}{3}; 1; y = \frac{2}{3}x + 1$

13.

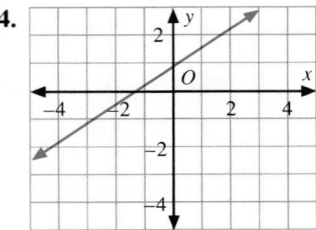

−2; −3; y = −2x − 3

14.

Find the slope, the *y*-intercept, and the *x*-intercept of each line.
(Lesson 13-3)

15. $y - 7 = 8x$ 8; 7; $-\frac{7}{8}$

16. $y - 6x = 18$ 6; 18; −3

17. $y = -4x$ −4; 0; 0

Solve each system by graphing. *(Lesson 13-4)*

18. $y = x + 5$
$y = -2x - 1$ (−2, 3)

19. $y = -3x$
$y = -x - 2$ (1, −3)

20. $y = 2x - 2$
$y = x - 2$ (0, −2)

21. $y = \frac{3}{4}x + 5$
$y = \frac{3}{4}x - 2$ no solution

Solve using a simpler problem. *(Lesson 13-5)*

22. How many squares are in the figure at the right? 30

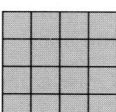

23. Ellen decided to save dimes. On the first day she saved one dime. Every day after that she saved two more dimes than she had saved the day before. How many dimes had Ellen saved after 30 days? 900

Solve and graph each inequality. *(Lessons 13-6, 13-7)* Graphs for Chapter Review 24–32 are on p. A37.

24. $x + 1 \leq 7$ $x \leq 6$

25. $c - 8 > -9$ $c > -1$

26. $4z \geq 16$ $z \geq 4$

27. $\frac{m}{-2} < 2$ $m > -4$

28. $\frac{a}{2} - 1 > -2$ $a > -2$

29. $\frac{2}{3}x - 4 \leq 4$ $x \leq 12$

30. $3a + 7 < 4$ $a < -1$

31. $-4m - 6 \geq 14$ $m \leq -5$

32. $-5y + 4.8 < 1.3$ $y > 0.7$

Tell whether each ordered pair is a solution of each inequality. Write *Yes* or *No*. *(Lesson 13-8)*

33. (2, −4); $y \geq 3x + 1$ No.

34. (5, 7); $y < 2x - 3$ No.

35. (4, 2); $y > -\frac{1}{2}x$ Yes.

Graph each inequality. *(Lesson 13-8)* Answers to Chapter Review 36–38 are on p. A38.

36. $y > 4x + 3$

37. $y \leq -2x - 4$

38. $y \geq \frac{1}{3}x - 2$

Find values of each function. Use −3, 0, and 4 as values for *x*.
(Lesson 13-9)

39. $y = -7x + 11$ 32; 11; −17

40. $y = |x| - 6$ −3; −6; −2

41. $y = (x - 1)^2$ 16; 1; 9

Find $f(-2), f(0), f(3)$ for each function. *(Lesson 13-9)*

42. $f(x) = 5x - 8$
$f(-2) = -18; f(0) = -8; f(3) = 7$

43. $f(x) = x^2 + 5$
$f(-2) = 9; f(0) = 5; f(3) = 14$

44. $f(x) = |x + 4|$
$f(-2) = 2; f(0) = 4; f(3) = 7$

Graph each function. *(Lesson 13-10)* Answers to Chapter Review 45–47 are on p. A38.

45. $y = x + 2$

46. $y = x^2 - 3$

47. $y = |x| + 3$

Find the slope, the *y*-intercept, and the *x*-intercept of each line.

5. $y - 12 = 6x$ 6; 12; −2

6. $y = 3x$ 3; 0; 0

Solve by graphing.

7. $y = x + 3$
$y = 3x - 5$ (4, 7)

8. $y = \frac{3}{4}x + 2$
$y = \frac{3}{4}x - 2$ no solution

Alternative Assessment

1. Write an equation in slope-intercept form using the variables *x* and *y*.

2. Using the equation from (1), find the solutions for *y* using −2, −1, 0, 1, and 2 as values for *x*.

3. Graph the equation.

4. Name the slope, *y*-intercept and *x*-intercept for the line graphed.

5. Write an equation for another line that is parallel to the given line.

6. Graph the parallel line.

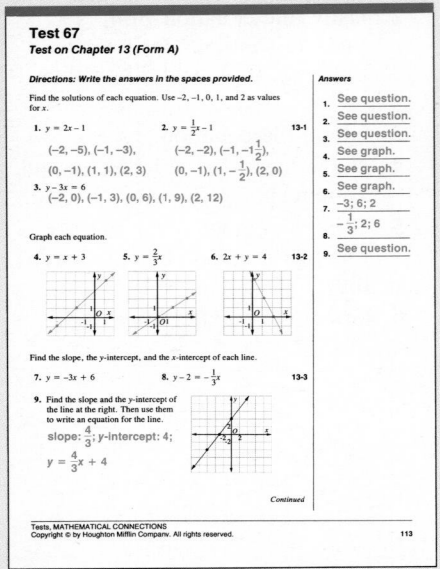
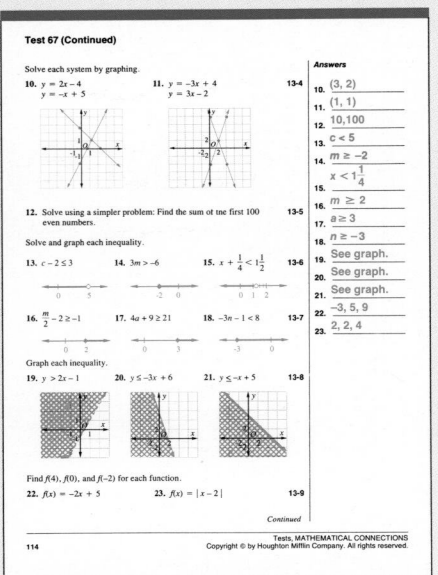
Chapter Test

Find solutions of each equation. Use $-2, -1, 0, 1$, and 2 as values for x.

$(-2, 0); (-1, \frac{1}{2}); (0, 1); (1, 1\frac{1}{2}); (2, 2)$

1. $y = 3x - 2$ $(-2, -8);$
$(-1, -5); (0, -2); (1, 1); (2, 4)$

2. $y = \frac{1}{2}x + 1$

3. $y - 2x = 5$
$(-2, 1); (-1, 3); (0, 5);$
$(1, 7); (2, 9)$

13-1

Graph each equation. Answers to Chapter Test 4–6 are on p. A38.

4. $y = 2x + 3$

5. $y = \frac{3}{4}x$

6. $3x + y = 4$

13-2

Find the slope, the y-intercept, and the x-intercept of each line.
$-\frac{1}{2}; 5; 10$

7. $y = -4x + 12$
$-4; 12; 3$

8. $y - 5 = -\frac{1}{2}x$

13-3

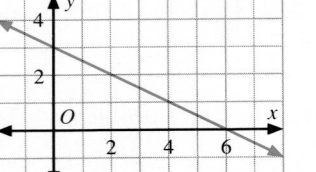

9. Find the slope and the y-intercept of the line at the right. Then use them to write an equation for the line. $-\frac{1}{2}; 3; y = -\frac{1}{2}x + 3$

★ **10.** **a.** Draw a line with positive slope.
b. Draw a line with negative slope. See answer on p. A38.

Solve each system by graphing.

11. $y = 2x + 4$
$y = -x + 7$ $(1, 6)$

12. $y = -5x + 2$
$y = -5x - 2$
no solution

13-4

★ **13.** On a coordinate plane, what represents the solution of a system of two equations? the point of intersection of the graphs of the equations

14. Solve using a simpler problem: Find the sum of the first 100 odd whole numbers. 10,000

13-5

Solve and graph each inequality. Graphs for Chapter Test 15–20 are on p. A38.

15. $c - 3 \leq 5$ $c \leq 8$

16. $4m > -2$ $m > -\frac{1}{2}$

17. $x + 2\frac{1}{2} < 3\frac{3}{4}$ $x < 1\frac{1}{4}$

13-6

18. $\frac{m}{-3} + 2 \geq 1$ $m \leq 3$

19. $5a + 11 \geq 21$ $a \geq 2$

20. $-3n - 7 < 2$
$n > -3$

13-7

Graph each inequality. Answers to Chapter Test 21–23 are on p. A38.

21. $y > 3x + 1$

22. $y \leq -2x - 3$

23. $y < -x + 2$

13-8

Find $f(5), f(0)$, and $f(-3)$ for each function.

24. $f(x) = -3x + 9$
$f(5) = -6; f(0) = 9; f(-3) = 18$

25. $f(x) = |x - 3|$
$f(5) = 2; f(0) = 3; f(-3) = 6$

13-9

Graph each function. Answers to Chapter Test 26–27 are on p. A38.

26. $y = x^2 + 3$

27. $y = |x| - 1$

13-10

Graphing Inequalities with One Variable on the Coordinate Plane

Objective: To graph inequalities with one variable on the coordinate plane.

In Lesson 13-8, you graphed inequalities with two variables. To graph inequalities with one variable on the coordinate plane, you follow a similar procedure. You graph the boundary line and then shade the appropriate region.

A boundary line of the form $y = k$ is a horizontal line. A boundary line of the form $x = k$ is a vertical line.

Example

Solution

Graph $x > 1$ on a coordinate plane.

- Begin with the equation $x = 1$. Three points on this line are $(1, 0)$, $(1, -3)$, and $(1, 6)$. Draw a dashed boundary line through these points.
- To graph $x > 1$, shade the region to the *right* of the line, where every value of x is greater than 1.

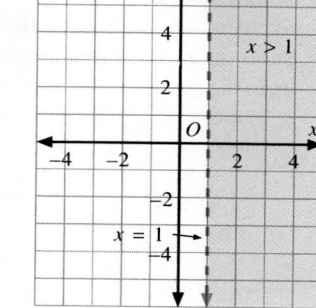

✓ **Check**

Choose the point $(2, 0)$.
$x > 1$
$2 \overset{?}{>} 1$
$2 > 1$

Teaching Note

Stress that $y = k$ is a horizontal line and that $x = k$ is a vertical line. Also, remind students that the symbols $<$ and $>$ result in dashed boundary lines and that $\leq$ and $\geq$ result in solid boundary lines.

Key Question

How do you know which side of the boundary line to shade when graphing inequalities with one variable on a coordinate plane?   Shade above the line for $y > k$ and below the line for $y < k$. Shade to the right of the line for $x > k$ and to the left of the line for $x < k$. The same rules hold for $\leq$ and $\geq$.

Additional Example

Graph $y \leq 3$ on a coordinate plane.

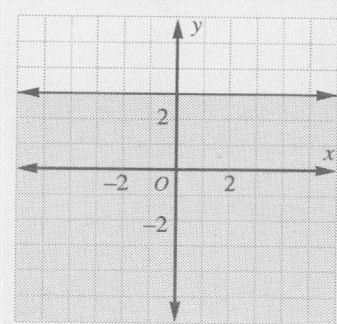

Exercises

Tell whether you shade to the *right* or to the *left* of the boundary line.

1. $x > 4$ right

2. $x \geq -2$ right

3. $x \leq 3$ left

4. $x < -1$ left

5. $x \geq 5$ right

6. $x \leq -7$ left

Tell whether you shade *above* or *below* the boundary line.

7. $y > 0$ above

8. $y < 4$ below

9. $y \leq -6$ below

10. $y \geq -5$ above

11. $y < -2$ below

12. $y > 1$ above

Graph each inequality on a coordinate plane. Answers to Exercises 13–20 are on p. A38.

13. $x < -4$

14. $x > 5$

15. $y \geq 2$

16. $y \leq -2$

17. $y + 3 \leq 6$

18. $2x \geq 4$

19. $3x - 4 < -7$

20. $-2y - 5 > 3$

Inequalities and Graphing on the Coordinate Plane   **623**

Quick Quiz (13-5 through 13-8)

1. There are 125 lockers in a hallway. The lockers are numbered consecutively from 1 to 125 with metal digits. How many metal digits are needed to number all the lockers?   267

Solve and graph each inequality.

2. $x - 8 > -3$ $x > 5$

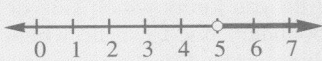

3. $-2x \geq 12$ $x \leq -6$

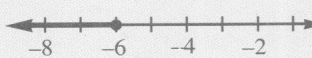

4. $\frac{d}{4} + 6 < 17$ $d < 44$

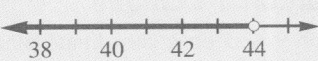

5. $-3k - 5 \leq -7$ $k \geq \frac{2}{3}$

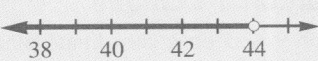

Graph each inequality.

6. $y \leq 2x - 5$

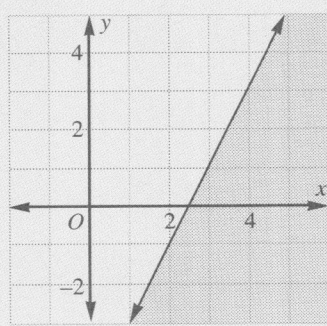

7. $y > 5x + 2$

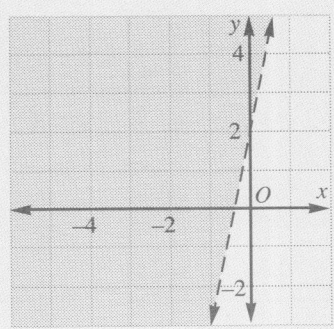

Cumulative Review

Standardized Testing Practice

Choose the letter of the correct answer.

1. Which ordered pair is a solution of $2x + y = 4$?
   I. (2, 2)   II. (7, −10)   III. (0, 4)
   A. I only   B. II only
   C. I and II   D.) II and III

2. The central angle for one sector of a circle graph is 54°. What percent of the total amount is represented by this sector?
   A. 54%   B.) 15%
   C. 36%   D. 30%

3. Simplify: $7x + 9z - 4z + 6x$
   A. $18xz$ B.) $13x + 5z$
   C. $16x + 2z$ D. $x + 5z$

4. Two letters are chosen without replacement from the word MAGNET. What is $P(\text{N, then T})$?
   A.) $\frac{1}{30}$ B. $\frac{1}{36}$
   C. $\frac{11}{30}$ D. $\frac{2}{11}$

5. Find the x-intercept: $y = \frac{2}{3}x - 4$
   A. $\frac{2}{3}$ B. -4
   C.) 6   D. 4

6. Two cocaptains are chosen from a team of 6 swimmers. In how many different ways can this be done?
   A. 36   B. 12
   C.) 15   D. 24

7. The scale of a model house is 2 in. : 15 ft. The actual width of the house is 30 ft. Find the width of the model house.
   A.) 4 in.   B. 30 in.
   C. 1 in.   D. 2.25 in.

8. Solve: $5x + 3x + 9 = 33$
   A. $x = 8$ B.) $x = 3$
   C. $x = 1\frac{16}{17}$ D. $x = 5\frac{1}{4}$

9.
4.3	0 1 1 3 8
4.4	1 4 5 5 5 9
4.5	0 2 3

   How many data items in this stem-and-leaf plot are less than 4.45?
   A.) 7   B. 10
   C. 9   D. 11

10. Simplify: $\frac{15n^2}{n^6}$
    A. $\frac{15}{n^3}$ B. $15n^4$
    C. $15n^8$ D.) $\frac{15}{n^4}$

11.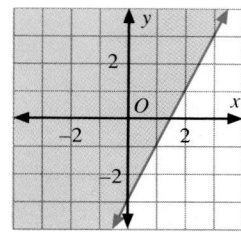

Which inequality does this graph represent?

A. $y > 2x - 3$
B. $y \leq 2x - 3$
C. $y < 2x - 3$
D. $y \geq 2x - 3$

16.

Outcome	Frequency
red	36
yellow	18
blue	26

A three-colored spinner was spun 80 times. The outcomes are recorded in the table above. Find the experimental probability of the spinner landing on red.

A. $\frac{1}{3}$ B. $\frac{9}{20}$

C. $\frac{11}{20}$ D. $\frac{2}{3}$

12. Solve: $17 + j = 6$

A. $j = 9$ B. $j = -9$
C. $j = 11$ D. $j = -11$

17. Simplify: $-5(-4) + 6(-2)$

A. 8   B. 20
C. -32 D. -52

13. A bag contains 5 red, 1 white, and 2 blue marbles. One marble is chosen at random from the bag. Find the odds in favor of selecting a blue marble.

A. $\frac{1}{4}$ B. $\frac{1}{8}$

C. $\frac{1}{3}$ D. $\frac{5}{8}$

18. The length of one side of a regular octagon is 15 cm. Find the perimeter.

A. 90 cm   B. 23 cm
C. 120 cm   D. cannot be determined

14. Write the phrase *the total of four times a number z and fifteen* as a variable expression.

A. $4(z + 15)$ B. $z + 4 \cdot 15$
C. $4z + 15$ D. $z^4 + 15$

19. Find the sum: $\frac{3x}{8} + \frac{5x}{12}$

A. $\frac{11x}{20}$ B. $\frac{x}{3}$

C. $\frac{19x}{24}$ D. $\frac{17x}{12}$

15. Solve: $-4x - 10 < 2$

A. $x < 2$ B. $x > -3$
C. $x > -2$ D. $x < -3$

20. What percent of 120 is 78?

A. 78%   B. 65%
C. 93.6%   D. 154%

Planning Chapter 14

Chapter Overview

Chapter 14 provides students with many opportunities to visualize, examine, and construct various types of space figures. *Lesson 1* develops students' ability to identify different figures. This lesson is followed by the *Focus on Explorations,* which shows students how to construct models for exploring space figures. *Lessons 2 and 3* present methods for finding surface areas of prisms and cylinders; surface areas of pyramids and cones also are included in these lessons. A thorough development of the concepts needed to understand the formulas for finding volumes of prisms and pyramids is given in *Lesson 4,* followed by a related development of the formulas for computing volumes of cylinders and cones in *Lesson 5.* Applications of surface area and the volume of a sphere are presented in *Lesson 6.* The problem-solving strategy of making a model is applied to space figures in *Lesson 7.* The chapter ends with a *Focus on Applications* that discusses the importance of volume in the packaging of products, and a *Chapter Extension* that has students finding volumes of composite space figures.

Background

Chapter 14 continues the development of the geometry strand from Chapters 6 and 10. Students' knowledge of polygons is reinforced as they calculate perimeters and areas of space figures. The chapter develops terminology, characteristics, and procedures for calculating surface areas and volumes of polyhedrons, cylinders, cones, and spheres. The approach taken is to involve students actively in activities that will improve their spatial visualization skills and critical thinking skills. Students also will see how space figures are used in many real-world activities such as construction, art, engineering, and product packaging.

As in earlier geometry chapters, manipulatives and models can play a major role in this chapter. The space figures discussed can be constructed physically and students can actually measure the surface area and the volume. Students may enjoy creating their own models, as described in the activities in the *Focus on Explorations* on pages 632 and 633.

Technology can assist with the learning and discovery of concepts in Chapter 14. With a computer-assisted design program, students can manipulate space figures on screen to explore the effects of various transformations. Calculators can be very useful in carrying out the computations necessary to determine surface areas and volumes, especially when working with the composite space figures presented in the *Chapter Extension.*

Connections and opportunities for communication occur frequently in Chapter 14. Connections to science and the real world are presented in numerous exercises, and the important connection of geometry to algebra is shown many times through the use of formulas. Frequent communication exercises encourage students to explain orally and in writing their perceptions about space figures.

Objectives

14-1　To identify space figures.

≡　FOCUS ON EXPLORATIONS　To explore patterns for space figures.

14-2　To find the surface area of a prism.

14-3　To find the surface area of a cylinder.

14-4　To find the volumes of prisms and pyramids.

14-5　To find the volumes of cylinders and cones.

14-6　To find the surface area and volume of spheres.

14-7　To solve problems by making a model.

≡　FOCUS ON APPLICATIONS　To explore how surface area and volume are used in making consumer-packaging decisions.

≡　CHAPTER EXTENSION　To determine volumes of composite space figures.

Chapter Planner

Instructional Aids	Manipulatives	Cooperative Learning	Technology	Practice/Reteaching	Assessment

Materials: space figure models, centimeter graph paper, orange, cylinders, centimeter cubes, sphere, tiles, calculator
Resource Book:
  Lesson Starters
  128–134
Visuals, Folders
  E, F, G
	Lessons 14-1, Focus on Explorations, 14-3, 14-4, 14-6, 14-7
	Activities Book:
	  Manipulative
	  Activities 27–28
		Lessons 14-1, 14-3, 14-4, 14-5, 14-6, 14-7, Focus on Applications
		Activities Book:
		  Cooperative
		  Activities 25–26
			Lessons 14-1, 14-2, 14-3, 14-4, 14-5, 14-6
			Activities Book:
			  Calculator Activity 14
			  Computer Activity 14
			Connections Plotter Plus Disk
				Toolbox, pp. 753–771
				Extra Practice, p. 749
				Resource Book:
				  Practice 175–184
				Study Guide,
				  pp. 255–268
					Self-Tests 1–2
					Quick Quizzes 1–2
					Chapter Test, p. 662
					Resource Book:
					  Lesson Starters
					  128–134
					  (Daily Quizzes)
					Tests 69–73

Assignment Guide Chapter 14

Day	Skills Course	Average Course	Advanced Course
1	**14-1:** 1–16, 20–22 **Exploration:** Activities I–IV	**14-1:** 1–16, 20–22 **Exploration:** Activities I–IV	**14-1:** 1–22 **Exploration:** Activities I–IV
2	**14-2:** 1–17, 22–26	**14-2:** 1–26	**14-2:** 1–26, Challenge
3	**14-3:** 1–12, 18–22	**14-3:** 1–15, 18–22	**14-3:** 1–22
4	**14-4:** 1–16, 21–23, 26–29	**14-4:** 1–23, 26–29	**14-4:** 1–29
5	**14-5:** 1–14, 23–26	**14-5:** 1–16, 17–21 odd, 23–26	**14-5:** 1–26
6	**14-6:** 1–14, 15–21 odd, 28–33	**14-6:** 1–24, 28–33	**14-6:** 1–33, Historical Note
7	**14-7:** 1–16 **Focus on Applications**	**14-7:** 1–16 **Focus on Applications**	**14-7** 1–16 **Focus on Applications**
8	*Prepare for Chapter Test:* Chapter Review	*Prepare for Chapter Test:* Chapter Review	*Prepare for Chapter Test:* Chapter Review
9	*Administer Chapter 14 Test*	*Administer Chapter 14 Test;* Cumulative Review	*Administer Chapter 14 Test;* Chapter Extension; Cumulative Review

Teacher's Resources

Resource Book
  Chapter 14 Project
  Lesson Starters 128–134
  Practice 175–184
  Enrichment 141–148
  Diagram Masters
  Chapter 14 Objectives
  Family Involvement 14
  Spanish Test 14
  Spanish Glossary

Activities Book
  Cooperative Activities 25–26
  Manipulative Activities 27–28
  Calculator Activity 14
  Computer Activity 14

Study Guide, pp. 255–268

Tests, Tests 69–73

Visuals, Folders E, F, G

Connections Plotter
  Plus **Disk**

Alternate Approaches Chapter 14

Using Technology

CALCULATORS
As usual, calculators can be used in this chapter to provide experience with realistic problems and numbers. As described in the commentary for the lessons, calculators can make practical a process of approximating volumes by imaginatively cutting a figure into simple-shaped slices.

COMPUTERS
The main use for computers in this chapter is to provide a means for students to calculate surface areas and volumes quickly. With this ability, they then can explore the relationships among figures, both similar and dissimilar.

Lesson 14-1

COMPUTERS
If you have a CAD (*C*omputer *A*ssisted *D*esign) program available or a powerful enough computer, you can have students rotate pictures of space figures, project them in any direction, and otherwise manipulate them. This will help students to visualize space figures, and it will give them a more concrete intuition about solids.

Lessons 14-2, 14-3, 14-4, 14-5, 14-6

COMPUTERS
Students can write spreadsheet formulas for surface areas of prisms and cylinders, or for volumes of various space figures. This will give them an opportunity to express important relationships in a mathematically sophisticated and immediately verifiable way. Students can use such a spreadsheet to explore what happens to the surface area and volume, for example, if all three dimensions of a rectangular prism are multiplied by the same constant (thereby creating a similar figure).

Lessons 14-4, 14-5

CALCULATORS OR COMPUTERS
Students can be led to discover or verify the formula for the volume of a pyramid or cone by imagining the process of slicing the figure a number of times parallel to the base. The volume of each slice can be approximated by drawing an accurate cross-section of the figure when cut by a plane perpendicular to the base, finding the dimensions of each slice, and then calculating the area of a face of the slice and multiplying by the thickness of the slice. The total volume of the figure can be approximated by finding the sum of the volumes of all the slices. The process of summing can be done using a calculator, or more efficiently and accurately by using a spreadsheet.

Using Manipulatives

Lesson 14-1

As the various space figures are introduced and discussed, pass models of each figure around the class for students to examine. Students also will find these models helpful when doing the Guided Practice Exercises and Exercises 14–16, which require that figures be sketched from three perspectives.

Focus on Explorations

In this *Focus on Explorations*, students construct their own triangular prisms, square pyramids, and triangular pyramids. Encourage students to construct some of the figures from Activities III and IV for extra credit. You may wish to display the space figures in the classroom.

Lesson 14-3

COOPERATIVE LEARNING
Divide students into cooperative groups. Give each group some centimeter graph paper and a cylinder, and have the students estimate the surface area of the cylinder by ''covering'' it with the graph paper and counting squares. Each group should compare its result with that obtained by using the formula.

Lesson 14-4

COOPERATIVE LEARNING

Have students work in cooperative groups and give each group twenty-four centimeter cubes for building the rectangular prism described in Example 1. Each group should verify the volume of the prism by counting and then should calculate the surface area. Each group should form another prism, such as 2 cm × 2 cm × 6 cm, with the same volume and then compare the surface area of this prism to that of the first prism. The groups should conclude that prisms with equal volumes do not necessarily have equal surface areas.

Lesson 14-6

You can give students an intuitive feeling for the formulas presented in this lesson by using two simple demonstrations.

Surface Area

Slice an orange in half and draw four circles congruent to the great circle created by the cut. Ask students what the total area of the four circles is in terms of r, the radius of the circles. $4\pi r^2$ Peel the orange halves, breaking the peels into smaller pieces, and then fill in the four circles with the peels to verify the formula.

Volume

Find a sphere that fits inside a cylinder exactly, as in the diagram for Exercise 23 on page 653. Fill the cylinder with water and then measure the water displacement of the sphere. Students should notice that the water displacement is two-thirds the volume of the cylinder.

Lesson 14-7

COOPERATIVE LEARNING

Have students work in cooperative groups and give each group a bag of tiles. Ask each group to form as many rectangles as possible with a perimeter of 16 units. The rectangles will vary in their dimensions, for example, 1 by 7, 2 by 6, and so on. Groups should discuss whether the tiles were helpful in arriving at the various arrangements. Stress that a model can be very helpful, particularly in problems involving three dimensions.

Reteaching/Alternate Approach

Lesson 14-1

COOPERATIVE LEARNING

Give students working in cooperative groups a list of the various space figures and have them write down some real-world examples of each type. For example, a real-world example of a cone is a funnel. Follow this activity with a list of questions, such as the following:

1. How are a cylinder and a cone alike? How are they different?
2. How are prisms and pyramids named?

Lesson 14-2

Have each student draw a sketch of a rectangular prism using dimensions of his or her choice. Ask each student to make a two-dimensional sketch of the rectangular faces, label the dimensions, and calculate the surface area. Repeat the procedure for triangular prisms.

Lesson 14-3

Involve several students in the following activity. (Use colored chalk to separate bases and curved surfaces.)

Student 1: Sketch a cylinder on the chalkboard.
Student 2: Sketch the two bases separately.
Student 3: Sketch the curved surface separately.
Student 4: Calculate the area of the bases.
Student 5: Calculate the area of the curved surface.
Student 6: Calculate the total surface area.
Student 7: Respond to any questions regarding any step of the process.

Lesson 14-5

COOPERATIVE LEARNING

Have students working in cooperative groups create problems that involve volumes of cylinders and cones. For example, the problems might involve the amount of corn in a silo, water in a tank, or oil in a funnel. Each group should exchange problems with another group and solve. Problems and solutions can then be discussed with the entire class.

Teacher's Resources Chapter 14

Enrichment masters from the Resource Book are pictured here. See the Teacher's Resources chart on page 626B for a complete listing of all materials available for this chapter.

14-1 Enrichment
Data Analysis

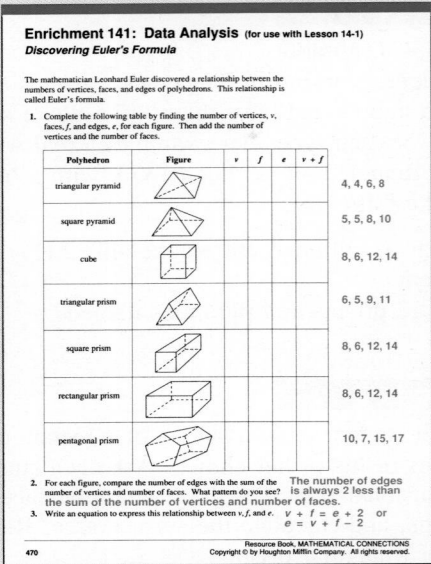

Enrichment 141: Data Analysis (for use with Lesson 14-1)
Discovering Euler's Formula

The mathematician Leonhard Euler discovered a relationship between the numbers of vertices, faces, and edges of polyhedrons. This relationship is called Euler's formula.

1. Complete the following table by finding the number of vertices, v, faces, f, and edges, e, for each figure. Then add the number of vertices and the number of faces.

Polyhedron	Figure	v	f	e	$v + f$
triangular pyramid					4, 4, 6, 8
square pyramid					5, 5, 8, 10
cube					8, 6, 12, 14
triangular prism					6, 5, 9, 11
square prism					8, 6, 12, 14
rectangular prism					8, 6, 12, 14
pentagonal prism					10, 7, 15, 17

2. For each figure, compare the number of edges with the sum of the number of vertices and number of faces. What pattern do you see? The number of edges is always 2 less than the sum of the number of vertices and number of faces.
3. Write an equation to express this relationship between v, f, and e. $v + f = e + 2$ or $e = v + f - 2$

Resource Book, MATHEMATICAL CONNECTIONS
Copyright © by Houghton Mifflin Company. All rights reserved.
470

14-2 Enrichment
Exploration

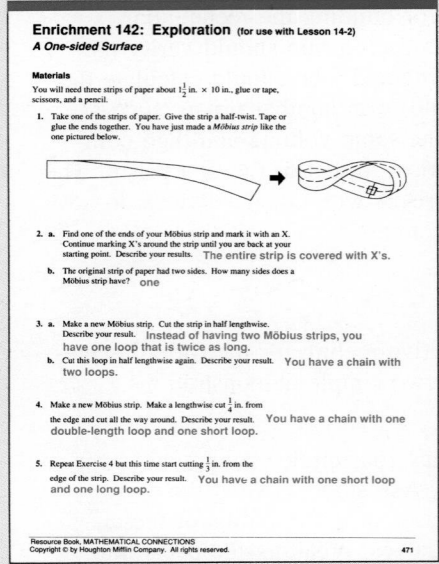

Enrichment 142: Exploration (for use with Lesson 14-2)
A One-sided Surface

Materials
You will need three strips of paper about $1\frac{1}{2}$ in. × 10 in., glue or tape, scissors, and a pencil.

1. Take one of the strips of paper. Give the strip a half-twist. Tape or glue the ends together. You have just made a Möbius strip like the one pictured below.

2. a. Find one of the ends of your Möbius strip and mark it with an X. Continue marking X's around the strip until you are back at your starting point. Describe your results. The entire strip is covered with X's.
   b. The original strip of paper had two sides. How many sides does a Möbius strip have? one

3. a. Make a new Möbius strip. Cut the strip in half lengthwise. Describe your result. Instead of having two Möbius strips, you have one loop that is twice as long.
   b. Cut this loop in half lengthwise again. Describe your result. You have a chain with two loops.

4. Make a new Möbius strip. Make a lengthwise cut $\frac{1}{4}$ in. from the edge and cut all the way around. Describe your result. You have a chain with one double-length loop and one short loop.

5. Repeat Exercise 4 but this time start cutting $\frac{1}{3}$ in. from the edge of the strip. Describe your result. You have a chain with one short loop and one long loop.

Resource Book, MATHEMATICAL CONNECTIONS
Copyright © by Houghton Mifflin Company. All rights reserved.
471

14-3 Enrichment
Exploration

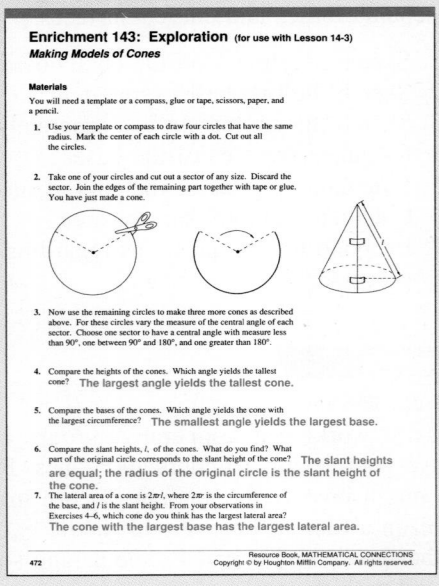

Enrichment 143: Exploration (for use with Lesson 14-3)
Making Models of Cones

Materials
You will need a template or a compass, glue or tape, scissors, paper, and a pencil.

1. Use your template or compass to draw four circles that have the same radius. Mark the center of each circle with a dot. Cut out all the circles.

2. Take one of your circles and cut out a sector of any size. Discard the sector. Join the edges of the remaining part together with tape or glue. You have just made a cone.

3. Now use the remaining circles to make three more cones as described above. For these circles vary the measure of the central angle of each sector. Choose one sector to have a central angle with measure less than 90°, one between 90° and 180°, and one greater than 180°.

4. Compare the heights of the cones. Which angle yields the tallest cone? The largest angle yields the tallest cone.

5. Compare the bases of the cones. Which angle yields the cone with the largest circumference? The smallest angle yields the largest base.

6. Compare the slant heights, l, of the cones. What do you find? What part of the original circle corresponds to the slant height of the cone? The slant heights are equal; the radius of the original circle is the slant height of the cone.
7. The lateral area of a cone is $2\pi r l$, where $2\pi r$ is the circumference of the base, and l is the slant height. From your observations in Exercises 4–6, which cone do you think has the largest lateral area? The cone with the largest base has the largest lateral area.

Resource Book, MATHEMATICAL CONNECTIONS
Copyright © by Houghton Mifflin Company. All rights reserved.
472

14-4 Enrichment
Thinking Skills

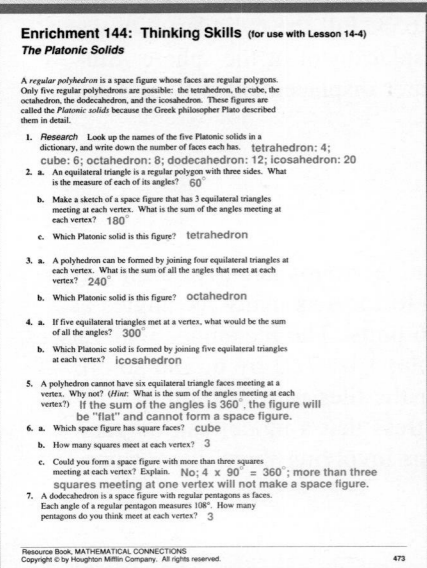

Enrichment 144: Thinking Skills (for use with Lesson 14-4)
The Platonic Solids

A *regular polyhedron* is a space figure whose faces are regular polygons. Only five regular polyhedrons are possible: the tetrahedron, the cube, the octahedron, the dodecahedron, and the icosahedron. These figures are called the *Platonic solids* because the Greek philosopher Plato described them in detail.

1. *Research* Look up the names of the five Platonic solids in a dictionary, and write down the number of faces each has. tetrahedron: 4; cube: 6; octahedron: 8; dodecahedron: 12; icosahedron: 20
2. a. An equilateral triangle is a regular polygon with three sides. What is the measure of each of its angles? 60°
   b. Make a sketch of a space figure that has 3 equilateral triangles meeting at each vertex. What is the sum of the angles meeting at each vertex? 180°
   c. Which Platonic solid is this figure? tetrahedron
3. a. A polyhedron can be formed by joining four equilateral triangles at each vertex. What is the sum of all the angles that meet at each vertex? 240°
   b. Which Platonic solid is this figure? octahedron
4. a. If five equilateral triangles met at a vertex, what would be the sum of all the angles? 300°
   b. Which Platonic solid is formed by joining five equilateral triangles at each vertex? icosahedron
5. A polyhedron cannot have six equilateral triangle faces meeting at a vertex. Why not? (*Hint:* What is the sum of the angles meeting at each vertex?) If the sum of the angles is 360°, the figure will be "flat" and cannot form a space figure.
6. a. Which space figure has square faces? cube
   b. How many squares meet at each vertex? 3
   c. Could you form a space figure with more than three squares meeting at each vertex? Explain. No; 4 × 90° = 360°; more than three squares meeting at one vertex will not make a space figure.
7. A dodecahedron is a space figure with regular pentagons as faces. Each angle of a regular pentagon measures 108°. How many pentagons do you think meet at each vertex? 3

Resource Book, MATHEMATICAL CONNECTIONS
Copyright © by Houghton Mifflin Company. All rights reserved.
473

14-5 Enrichment
Application

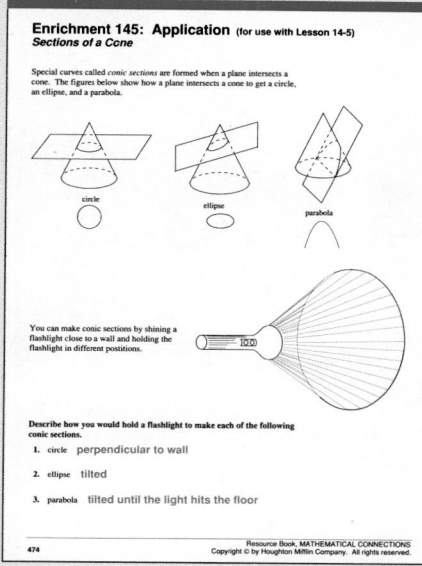

14-6 Enrichment
Communication

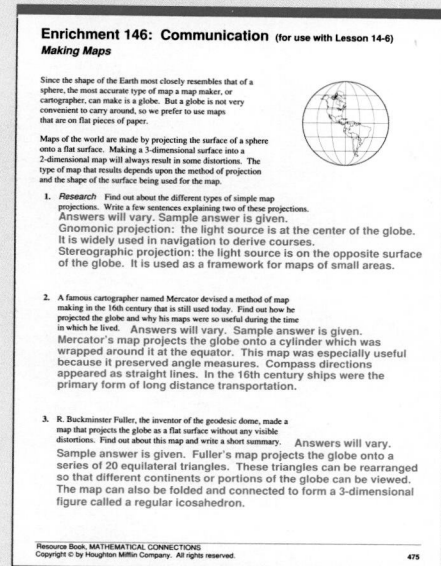

14-7 Enrichment
Problem Solving

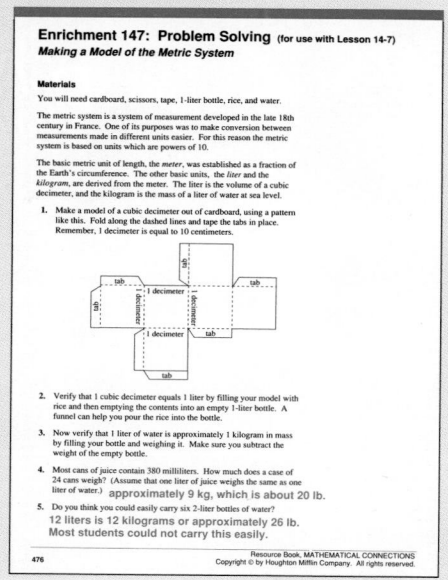

End of Chapter Enrichment
Extension

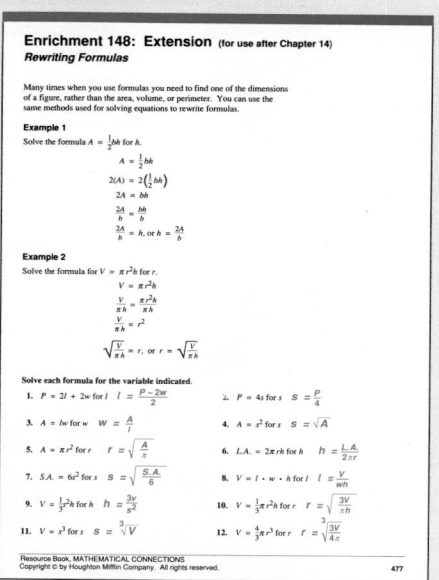

Multicultural Notes

For Chapter 14 Opener

''Give me your tired, your poor, your huddled masses yearning to breathe free... .'' So wrote Emma Lazarus (1849–1887) in ''The New Colossus,'' a sonnet honoring the Statue of Liberty.

''The New Colossus'' was one of many submissions by a number of American writers to a literary auction held in 1883 to raise money for a pedestal for the new Statue of Liberty. It was not until 1903, sixteen years after Lazarus's death, that the poem was engraved on a plaque inside the pedestal. By that time, immigration through Ellis Island had reached a peak; the Statue of Liberty and Lazarus' poem became associated with the American immigrant experience.

For Page 650

With regard to the Application feature, students can inspect a globe to see how it is made. The world map is printed on pieces of paper called gores, which are pasted on a sphere. You may want to discuss with students the advantages and disadvantages of a globe as compared to a flat, projected map of the world.

Research Activities

For Chapter 14 Opener

Students can work in cooperative groups to research the history of the building of the Statue of Liberty and the opening of Ellis Island. Students can also do research on immigration statistics around the turn of the century. They can use the information to create various statistical graphs.

For Page 650

As an extension of the Application feature, students can compare and contrast various kinds of world map projections.

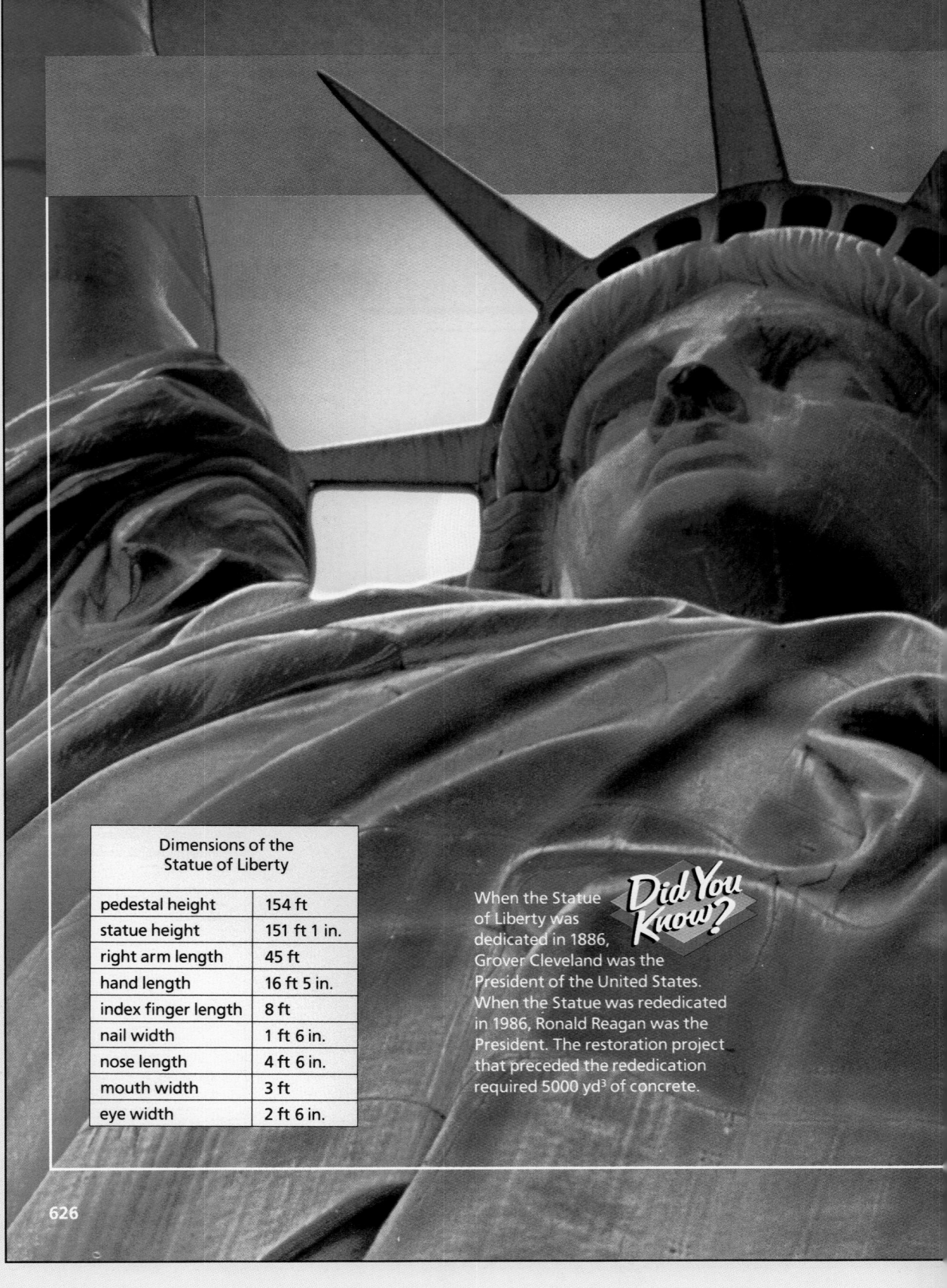

Dimensions of the Statue of Liberty	
pedestal height	154 ft
statue height	151 ft 1 in.
right arm length	45 ft
hand length	16 ft 5 in.
index finger length	8 ft
nail width	1 ft 6 in.
nose length	4 ft 6 in.
mouth width	3 ft
eye width	2 ft 6 in.

When the Statue of Liberty was dedicated in 1886, Grover Cleveland was the President of the United States. When the Statue was rededicated in 1986, Ronald Reagan was the President. The restoration project that preceded the rededication required 5000 yd^3 of concrete.

Did You Know?

626

Surface Area and Volume 14

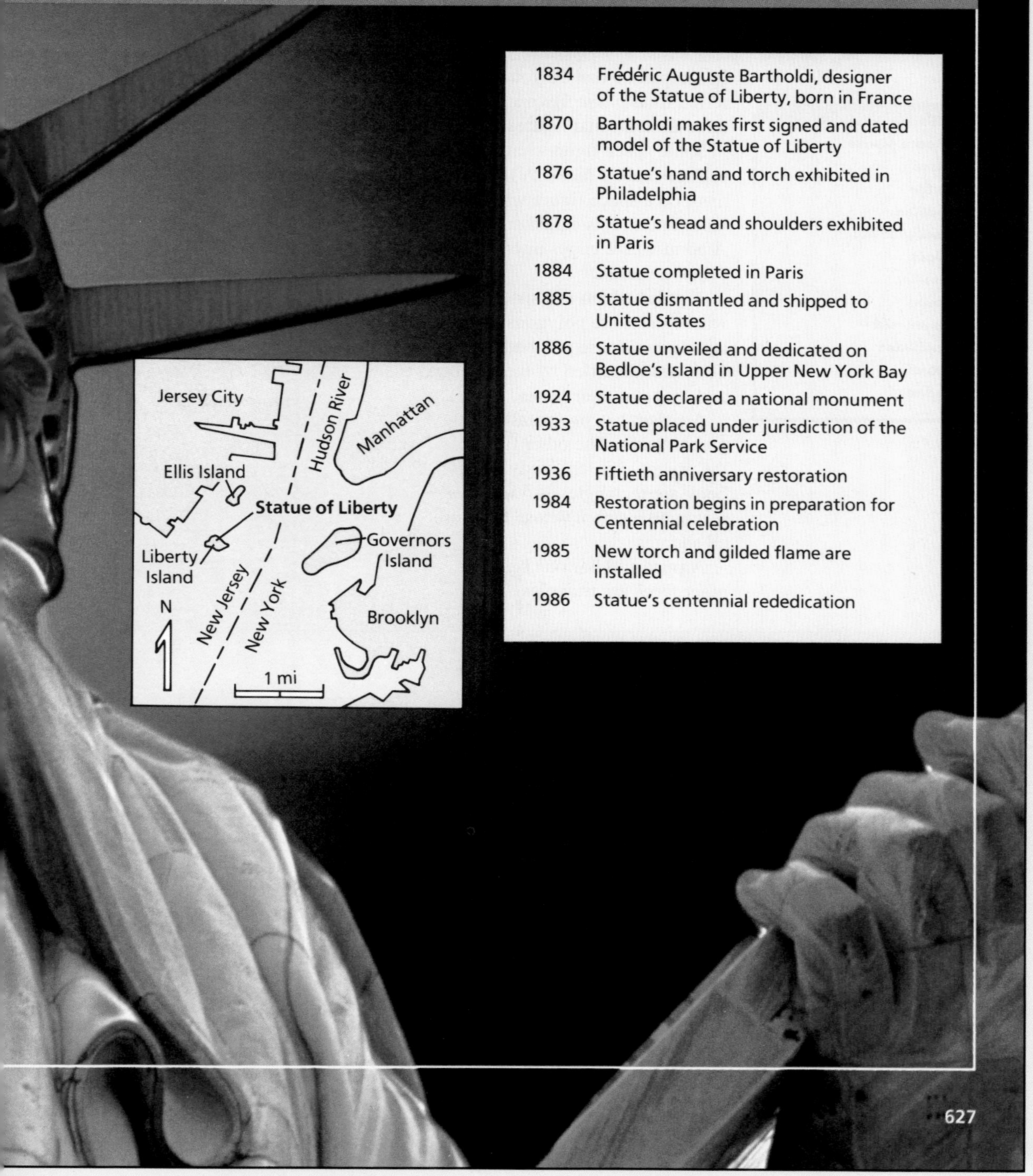

1834	Frédéric Auguste Bartholdi, designer of the Statue of Liberty, born in France
1870	Bartholdi makes first signed and dated model of the Statue of Liberty
1876	Statue's hand and torch exhibited in Philadelphia
1878	Statue's head and shoulders exhibited in Paris
1884	Statue completed in Paris
1885	Statue dismantled and shipped to United States
1886	Statue unveiled and dedicated on Bedloe's Island in Upper New York Bay
1924	Statue declared a national monument
1933	Statue placed under jurisdiction of the National Park Service
1936	Fiftieth anniversary restoration
1984	Restoration begins in preparation for Centennial celebration
1985	New torch and gilded flame are installed
1986	Statue's centennial rededication

Jersey City

Hudson River

Manhattan

Ellis Island

Statue of Liberty

Liberty Island

New Jersey

New York

Governors Island

Brooklyn

N

1 mi

Research Activities (continued)

For Page 653

Students reading the Historical Note can be encouraged to conjecture as to why mathematics was so important to the ancient Greeks, and why they were able to make so many advances in mathematics.

Suggested Resources

Blanchet, Christian, and Dard, Bertrand. *Statue of Liberty: The First Hundred Years*. New York: American Heritage Press, 1985.

Historical Statistics of the United States: Colonial Times to 1970. Washington, D.C.: Bureau of the Census, 1975.

Teaching the Lesson
- Materials: space figure models
- Lesson Starter 128
- Using Technology, p. 626C
- Using Manipulatives, p. 626C
- Reteaching/Alternate Approach, p. 626D
- Cooperative Learning, p. 626D

Lesson Follow-Up
- Practice 175
- Enrichment 141: Data Analysis
- Study Guide, pp. 255–256

Warm-Up Exercises

Identify each of the following figures.

1.

triangle

2.

rectangle

3.

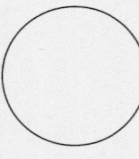

circle

14-1 Space Figures

Objective: To identify space figures.

CONNECTION

Terms to Know
- *space figure*
- *face*
- *edge*
- *vertex*
- *polyhedron*
- *base*
- *prism*
- *cube*
- *pyramid*
- *cylinder*
- *cone*
- *sphere*

In Chapters 6 and 10 you studied polygons. Polygons are sometimes referred to as *plane figures* because they lie in a plane. **Space figures** are three-dimensional figures that enclose part of space.

Some space figures have flat surfaces called **faces.** A line segment on a space figure where two faces intersect is called an **edge.** A point where edges intersect is called a **vertex.**

A **polyhedron** is a space figure whose faces are polygons. *Prisms* and *pyramids* are polyhedrons. They are identified by the number and shape of their **bases.**

A **prism** has *two* parallel congruent bases. The other faces of the prisms you will study in this book are rectangles. A **cube** is a rectangular prism whose faces are all squares.

A **pyramid** has one base. Its other faces are triangles.

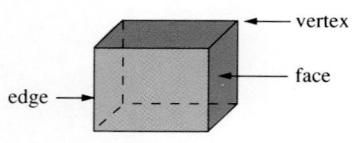

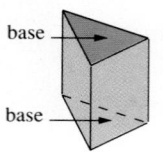

triangular prism

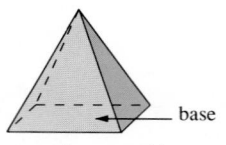

rectangular pyramid

Example 1

Identify each space figure.

a. 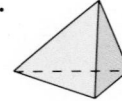 **b.** 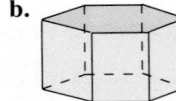 **c.**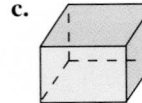

Solution

a. Triangular pyramid. The figure has one triangular base.

b. Hexagonal prism. The figure has two hexagonal bases that are congruent and parallel.

c. Rectangular prism. The figure has two rectangular bases that are congruent and parallel.

☑ **Check Your Understanding**

1. In Example 1(a), how can you tell that the figure is not a prism?
2. Give two reasons why the figure in Example 1(b) is not a pyramid.
   See *Answers to Check Your Understanding* at the back of the book.

Space figures that have curved surfaces are not polyhedrons. Three such figures are a *cylinder*, a *cone*, and a *sphere*.

A **cylinder** has two circular bases that are congruent and parallel.

A **cone** has one circular base and a vertex.

A **sphere** is the set of all points in space that are the same distance from a given point called the center.

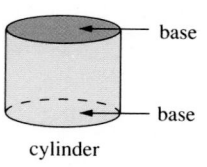
cylinder

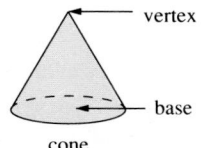
cone

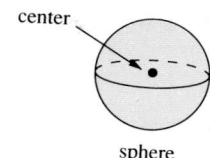
sphere

Example 2

Identify each space figure.

a. 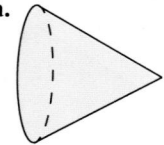   b. 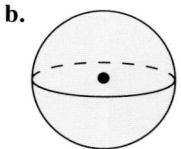   c.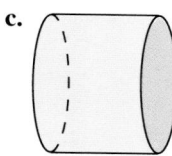

Solution

a. Cone. The figure has one circular base.

b. Sphere.

c. Cylinder. The figure has two circular bases that are congruent and parallel.

Guided Practice

COMMUNICATION «*Reading*

Replace each __?__ with the correct word.

«**1.** A polyhedron with two congruent parallel bases is a(n) __?__. prism

«**2.** A polyhedron with one base is a(n) __?__. pyramid

«**3.** A space figure with a curved surface and two circular bases that are congruent and parallel is a(n) __?__. cylinder

«**4.** A space figure with a curved surface, one circular base, and a vertex is a(n) __?__. cone

5. two congruent and parallel bases

COMMUNICATION «*Discussion* 6. only one base

Name one property that is shared by the two types of figures.

«**5.** cylinders and prisms   «**6.** cones and pyramids

Surface Area and Volume   **629**

Lesson Focus

Show students a picture of an Egyptian pyramid. Ask if anyone knows what the shape is called. Use a model of a pyramid and discuss the shape with students. Geometric shapes and solids are part of the real world. The focus of today's lesson is to learn how to identify space figures.

Teaching Notes

Key Questions

1. What is the difference between a polygon and a polyhedron? A polygon is a plane figure; a polyhedron is a space figure whose faces are polygons.

2. Why is a cylinder not considered a polyhedron? The faces of a polyhedron are polygons, not curved surfaces as in a cylinder.

3. How are prisms identified? by the number and shape of their bases

Error Analysis

Students may confuse the names and related terms for space figures. Have students describe the solids orally, with an emphasis on the correct terminology, to help avoid this error.

Reasoning

The Check Your Understanding questions test students' understanding of the definitions of space figures and their properties.

Using Technology

For a suggestion on using computer-assisted design software, see page 626C.

Using Manipulatives

For a suggestion on using models, see page 626C.

Reteaching/Alternate Approach

For a suggestion on using a cooperative learning activity, see page 626D.

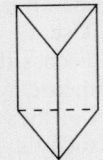
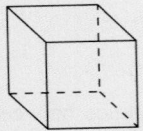

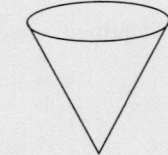
7. hexagonal pyramid
8. cube or rectangular prism
9. cone

Identify each space figure.

7.

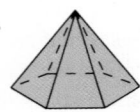

8.

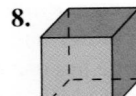

9.

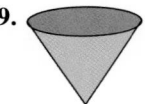

Exercises

Identify each space figure.

1. triangular prism
6. pentagonal prism

sphere   pentagonal pyramid

A   **1.**

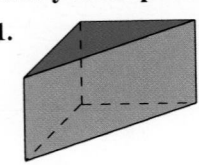

2.

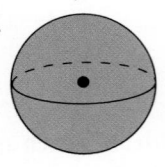

3.

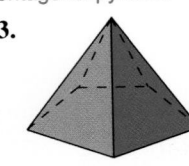

4.   cylinder

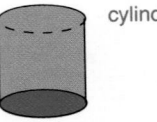

5.
cone

6.

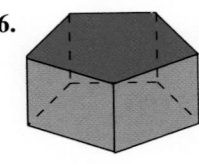

7.

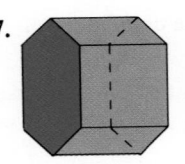

8.

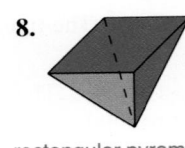

rectangular pyramid

9.

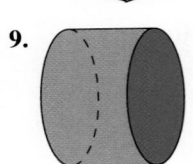

cylinder

hexagonal prism

SPATIAL SENSE

Learning to draw space figures can help you visualize them better. The diagrams below shows the steps involved in drawing a rectangular prism. First draw two congruent rectangles, one to the right of and above the other. Next draw lines connecting the corresponding vertices of the rectangles. Then make dashes for the edges of the prism that should not be visible.

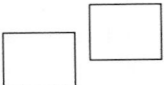 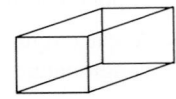 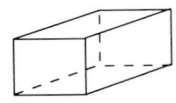

B   **10.** Draw a triangular prism. Answers to Exercises 10–13 are on p. A38.

11. Draw a cylinder. (*Hint:* Make the bases look like ovals.)

12. Draw a cone. (*Hint:* Draw an oval for the base and a point above the center of the base for the vertex.)

13. Draw a rectangular pyramid. (*Hint:* Draw a parallelogram for the base and a point above the base for the vertex.)

CAREER/APPLICATION

An *industrial designer* makes drawings of a product before it is manufactured. The drawings include front, top, and side views, as shown below.

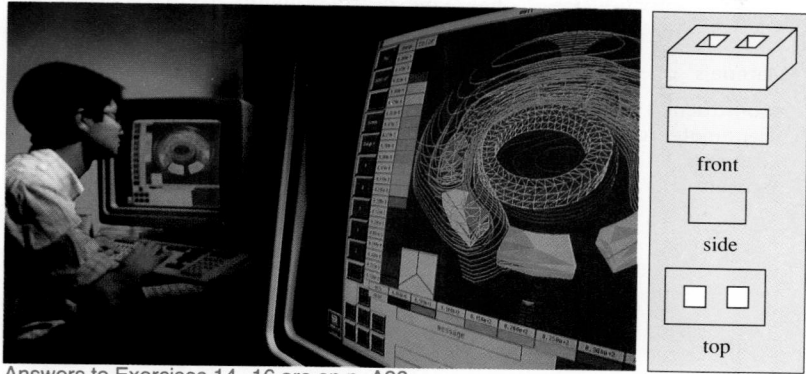

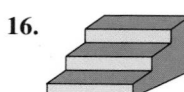

front

side

top

Answers to Exercises 14–16 are on p. A38.

Sketch front, top, and side views of each object.

14. 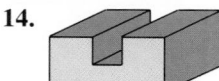   **15.**    **16.**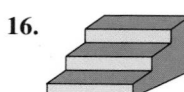

 COMPUTER APPLICATION

Suppose you have drawn a cylinder resting on one of its circular bases. A side view of the cylinder is shown in the diagram at the right. You can use drawing software to rotate it around a horizontal axis or a vertical axis through its center.

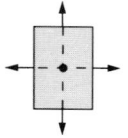

Use computer aided design (CAD) software or spatial sense.

C 17. Rotate the cylinder 90° around the vertical axis. Will the resulting figure look different from the original? No.

18. Rotate the cylinder 90° around the horizontal axis. Make a sketch of the resulting figure. See answer on p. A38.

19. Different amounts of rotation around the horizontal axis will result in different views of the cylinder. Choose another rotation and sketch the resulting figure. Answers will vary.

SPIRAL REVIEW

S 20. Identify the space figure at the right. *(Lesson 14-1)* cone

21. Solve and graph $x - 3 < 2$.   *(Lesson 13-6)*
$x < 5$; see graph on p. A38.

22. Find the area of a triangle with base 10 in. and height 7 in.   *(Lesson 10-3)* 35 in.²

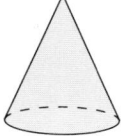

Surface Area and Volume   **631**

Problem Solving

Exercises 10–13 ask students to draw four space figures. Drawing a space figure may be difficult for some students. However, it is a very useful skill in solving problems and should be practiced until mastery is attained.

Computer

Exercises 17–19 demonstrate the effectiveness of computer-aided design software and help to develop students' spatial sense.

Follow-Up

Extension

Have students practice sketching various space figures. Each student should choose a figure and do an enlargement of it for the bulletin board.

Enrichment

Enrichment 141 is pictured on page 626E.

Practice

Practice 175, *Resource Book*

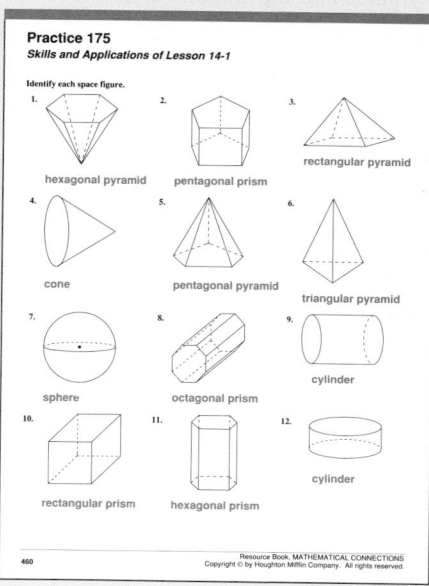

631

In this *Focus on Explorations*, students explore patterns for space figures. The opportunity to construct models is a motivating experience for many students. This lesson will be particularly effective if students work in pairs or cooperative groups.

Error Analysis

Students may have some difficulty associating space figures with two-dimensional patterns. Making a model and testing it is an excellent way to develop spatial reasoning skills.

Using Manipulatives

For a suggestion on using the models constructed in the lesson, see page 626C.

Activity Notes

Activity I

Activity I has students construct a model from a sketch of a triangular prism. Students must also decide what model would result if the triangles were squares.

Activity II

Activity II provides students an opportunity to make their own patterns. This activity leads students through the construction of a square pyramid and a triangular pyramid.

Paper Folding and Cutting

Objective: To explore patterns for space figures.

Materials

■ construction paper
■ centimeter ruler
■ compass
■ scissors

Later in this chapter you may want to make a model of a space figure to help you solve a problem. To make a pattern for a model, you must know the number of faces of the space figure and their shapes.

Activity I *Recognizing a Pattern*

1 A pattern for a polyhedron is shown below. How many rectangular faces does the polyhedron have? How many triangular faces? Do you think the polyhedron is a *prism* or a *pyramid*? 3; 2; prism

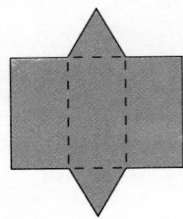

2 On a sheet of construction paper, construct a larger version of the pattern. Include the dashed lines. **2–3.** Check students' models.

3 Cut out the figure you have constructed. Fold along the dashed lines until the edges meet. Tape edges together where they meet.

4 You have built a model of a space figure. Identify it. Was it what you expected? triangular prism; Yes.

5 If the triangles in the pattern were squares, you would have a pattern for a common object. Name a common object that the figure would resemble.
Answers will vary. Example: a box without a lid

Activity II *Making a Pattern*

1 Draw a square with sides 10 cm long. **1–3.** Check students' models.

2 On each side of the square, construct an equilateral triangle with sides 10 cm long.

3 Cut out the figure you have constructed. Fold along the edges of the square until the edges of the triangles meet. Tape the edges together.

4 You have built a model of a space figure. Identify the space figure.
square pyramid

5 Construct an equilateral triangle with sides 4 cm long. On each side of that triangle, construct an isosceles triangle whose two equal sides are 5 cm long. Cut out, fold, and tape the pattern. Identify the space figure.
triangular pyramid

Now that you have cut and folded paper patterns to make models of space figures, you may be able to look at a pattern and visualize the space figure that it would make. You may then want to copy the pattern and construct the model as a check.

Activity III   *Visual Thinking*

1 Which pattern(s) could be folded to form a cube?

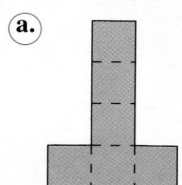 **a.**

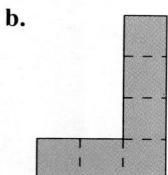 **b.**

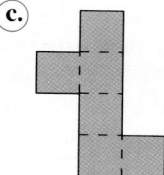 **c.**

2 Which pattern(s) could be folded to make a cylinder?

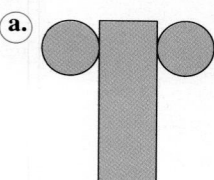 **a.**

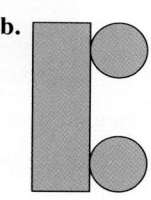 **b.**

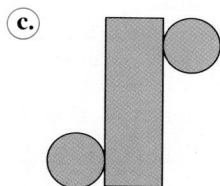 **c.**

3 Identify the space figure that could be formed by folding each pattern.

a.

pentagonal pyramid

b.

cone

c.
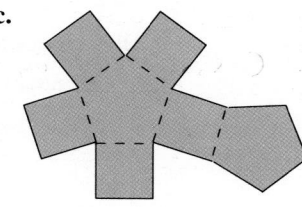
pentagonal prism

You can also develop your visual thinking by looking at a drawing of a space figure and figuring out a pattern that can be used to construct it.

Activity IV   *More Visual Thinking*

Make a pattern for each space figure. See answer on pp. A38–A39.

a.
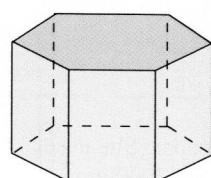

b.
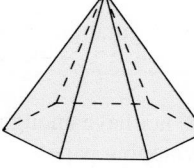

c.
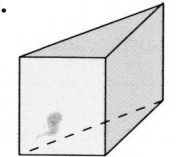

Surface Area and Volume   **633**

Activity III
Activity III requires students to visualize space figures from patterns. Students must determine which pattern will make a given figure.

Activity IV
Activity IV has students create their own patterns for a given space figure.

Follow-Up

Project
Have students do research on the five *Platonic Solids* and write a brief report. You may wish to choose three or four of the reports to be presented orally to the entire class.

Teaching the Lesson
- Lesson Starter 129
- Using Technology, p. 626C
- Reteaching/Alternate
  Approach, p. 626D
- Toolbox Skills 7, 9, 11, 13,
  17, 18, 20

Lesson Follow-Up
- Practice 176
- Enrichment 142:
  Exploration
- Study Guide, pp. 257–258
- Manipulative Activity 27

Warm-Up Exercises

Find the perimeter and area of each figure.

1.

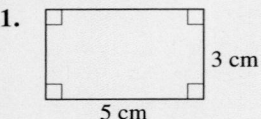

3 cm

5 cm

$P = 16$ cm
$A = 15$ cm^2

2.

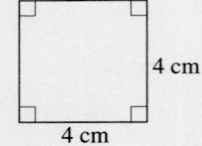

4 cm

4 cm

$P = 16$ cm
$A = 16$ cm^2

3.

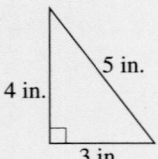

5 in.

4 in.

3 in.

$P = 12$ in.
$A = 6$ in.^{2}

14-2 Surface Area of Prisms

Objective: To find the surface area of a prism.

Terms to Know
- *surface area*

APPLICATION

Megan needs to wrap the box containing her brother John's present. She has a piece of wrapping paper measuring 2000 cm^2. The dimensions of the box are given in the diagram below.

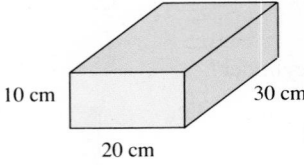

10 cm   30 cm

20 cm

QUESTION Does Megan have enough wrapping paper?

You need to find the *surface area* of the box, or prism.

The **surface area** (S.A.) of a prism is the sum of the areas of the bases and faces of the prism. Surface area is expressed in square units.

Example 1

Solution

Find the surface area of the rectangular prism shown above.

Make a sketch of the rectangular faces and label the dimensions.

top and bottom   front and back   sides

20 cm   10 cm   10 cm

30 cm   20 cm   30 cm

S.A. = top and bottom + front and back + sides
S.A. = 2(30 · 20)   + 2(20 · 10)   + 2(30 · 10)
S.A. = 1200   + 400   + 600
S.A. = 2200

The surface area of the rectangular prism is 2200 cm^2.

ANSWER No, Megan does not have enough wrapping paper. She needs more than 2200 cm^2 of wrapping paper, and she has only 2000 cm^2.

634 Chapter 14

Example 2

Find the surface area of the triangular prism at the right.

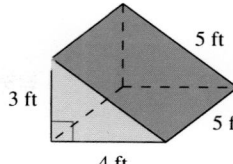
5 ft
3 ft
5 ft
4 ft

Solution

triangular bases

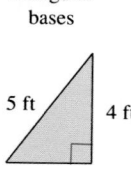
5 ft   4 ft
3 ft

rectangular faces

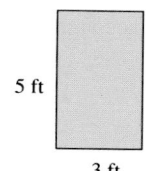
5 ft
3 ft

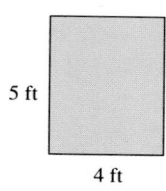
5 ft
4 ft

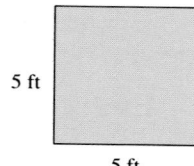
5 ft
5 ft

$$S.A. = 2(\tfrac{1}{2} \cdot 3 \cdot 4) + (3 \cdot 5) + (4 \cdot 5) + (5 \cdot 5)$$

$$S.A. = 12 + 15 + 20 + 25$$

$$S.A. = 72$$

The surface area of the triangular prism is 72 ft^2.

✏️ **Check Your Understanding**

1. In Example 2, why is the area of each base equal to $\frac{1}{2} \cdot 3 \cdot 4$?

2. In Example 2, why can you not triple the area of one rectangular face to find the area of the three rectangular faces?
   See *Answers to Check Your Understanding* at the back of the book.

Guided Practice

COMMUNICATION «*Reading* 1. Clues began to appear.

«**1.** Explain the meaning of the word *surface* in the following sentence.
Clues to the mystery began to surface after some investigation.

2. Answers will vary. Examples: *surcharge*, a charge over the normal amount; *surmount*, to overcome an obstacle.

«**2.** The prefix *sur-*, as in *surface*, means "over." List two other words that begin with this prefix. Write the definition of each word.

Refer to the text on pages 634–635.

3. You find the surface area of a prism by adding the areas of the faces.

«**3.** State the main idea of the lesson.

«**4.** List two major points that support the main idea.

4. The surface area of a rectangular prism is the sum of the areas of 3 pairs of congruent rectangles; the surface area of a triangular prism is the sum of the areas of 2 congruent triangles and 3 rectangles.

5. The cube at the right has edges 2 cm long.

   a. What is the area of each face of the cube? 4 cm²

   b. How many faces does the cube have? 6

   c. What is the surface area of the cube? 24 cm²

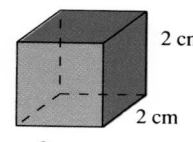
2 cm
2 cm
2 cm

Surface Area and Volume   **635**

Lesson Focus

Suppose a person wants to paint a garage. If a gallon of paint covers 400 ft², how can the number of gallons of paint needed be determined? The focus of today's lesson is on solving problems of this type.

Teaching Notes

Key Question

What is meant by the surface area of a prism? The surface area is the sum of the areas of the bases and faces.

Reasoning

The Check Your Understanding questions in Example 2 require students to understand the relationships among the various faces of the figures in order to apply the correct numerical procedure.

Error Analysis

Students sometimes use the wrong dimensions when calculating the areas of the lateral faces. If they can imagine the figure being unfolded, the number of errors will be reduced.

Using Technology

For a suggestion on using spreadsheets to find surface area, see page 626C.

Reteaching/Alternate Approach

For a suggestion on sketching a rectangular prism and a triangular prism, see page 626D.

Additional Examples

1. Find the surface area of the rectangular prism below.

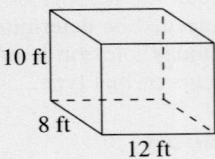

10 ft

8 ft

12 ft

592 ft^2

2. Find the surface area of the triangular prism below.

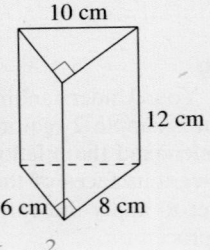

10 cm

12 cm

6 cm   8 cm

336 cm^2

Closing the Lesson

Have students summarize the procedure for finding the surface area of a prism and relate it to the procedure for a square pyramid.

Suggested Assignments

Skills: 1–17, 22–26
Average: 1–26
Advanced: 1–26, Challenge

Exercise Notes

Application

Exercises 10–13 help students to see how the concept of surface area is used in real-world situations.

Mental Math

Exercises 14–17 deal with the surface area of a cube. Students can calculate the surface areas mentally by multiplying the area of one face by 6. *Exercises 16 and 17* show the effect on the surface area when an edge is doubled or tripled.

Draw all the faces of each prism and label their dimensions.

6.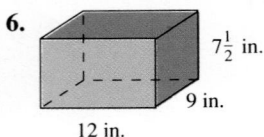

$7\frac{1}{2}$ in.

9 in.

12 in.

7.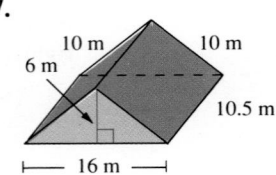

10 m   10 m

6 m

10.5 m

16 m

8. Find the surface area of the prism in Exercise 6 above. 531 in.^{2}

9. Find the surface area of the prism in Exercise 7 above. 474 m^2

Exercises

Find the surface area of each prism.

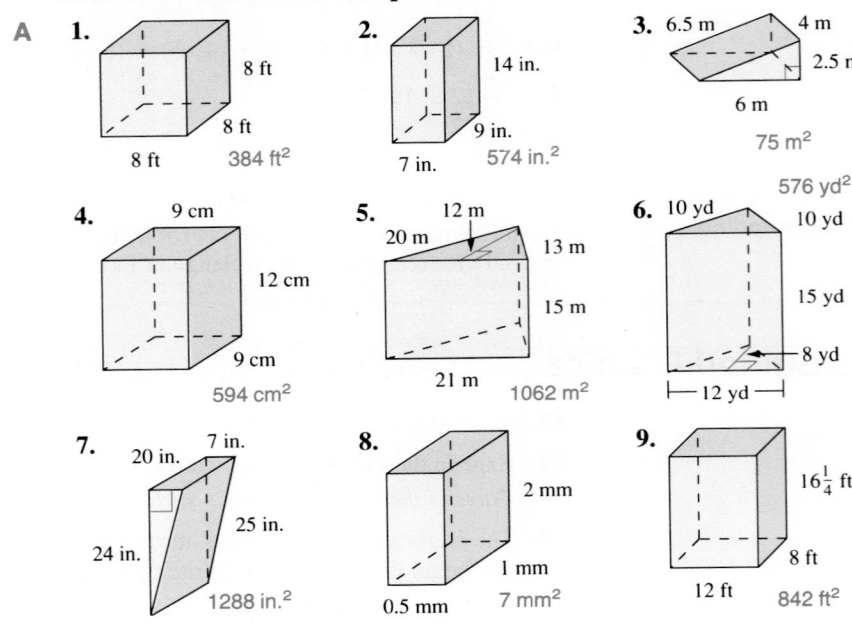

A 1.
8 ft
8 ft
8 ft
384 ft^2

2.
14 in.
9 in.
7 in.
574 in.^{2}

3. 6.5 m   4 m
2.5 m
6 m
75 m^2
576 yd^2

4. 9 cm
12 cm
9 cm
594 cm^2

5. 12 m
20 m   13 m
15 m
21 m
1062 m^2

6. 10 yd   10 yd
15 yd
8 yd
12 yd

7. 20 in. 7 in.
25 in.
24 in.
1288 in.^{2}

8.
2 mm
1 mm
0.5 mm   7 mm^2

9.
$16\frac{1}{4}$ ft
8 ft
12 ft   842 ft^2

10. What is the surface area of a box with length 10 in., width 8 in., and height $4\frac{1}{2}$ in.? 322 in.^{2}

11. What is the surface area of a compact disc case with length 14.3 cm, width 12.5 cm, and height 1 cm? 411.1 cm^2

12. The length of a box of granola is 5 in. The width is 2 in. The height is $7\frac{3}{4}$ in. Find the surface area of the box of granola. 128.5 in.^{2}

13. A perfume bottle is shaped like a rectangular prism with length 4 cm, width 2 cm, and height 7.5 cm. Find the surface area of the bottle of perfume. 106 cm^2

MENTAL MATH

B **14.** What is the surface area of a cube with edges of length 1 cm? 6 cm²

15. What is the surface area of a cube with edges of length 2 ft? 24 ft²

16. One cube has edges of length 5 in. Another cube has edges of length 10 in. Find the ratio of their surface areas. 1:4

17. One cube has edges of length 10 m. Another cube has edges of length 30 m. Find the ratio of their surface areas. 1:9

CONNECTING GEOMETRY AND ALGEBRA

The height of each triangular face of the pyramid at the right is l. The base is a square with a side of length n. If you know the values of l and n, you can find the surface area of this pyramid.

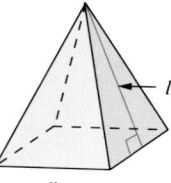

Use the figure above.

C **18.** Choose the letter of the formula for the area of the base.

    a. nl **b.** $4n$ **c.** $\frac{1}{2}nl$ **(d.)** n^2

19. Choose the letter of the formula for the area of a triangular face.

    a. $\frac{1}{2}n^2$ **b.** $\frac{1}{2}l^2$ **(c.)** $\frac{1}{2}nl$ **d.** l^2

20. Use your answers to Exercises 18 and 19 to write a formula for the surface area of the pyramid. S.A. = n^2 + 2nl

21. Use the formula that you wrote in Exercise 20 to find the surface area of the pyramid when $n = 12$ ft and $l = 8$ ft. 336 ft²

SPIRAL REVIEW 23. 1.7

S **22.** Find the surface area of the prism at the right. *(Lesson 14-2)* 285 m²

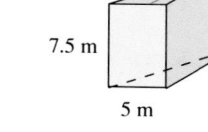

23. Solve: $4x - 7.9 = -1.1$ *(Lesson 8-8)*

24. Find $f(-2), f(0)$, and $f(4)$ for the function
$f(x) = \frac{1}{2}x + 3.$ *(Lesson 13-9)* 2; 3; 5

25. Draw an angle that measures 54°. *(Lesson 6-2)* See answer on p. A39.

26. DATA, *pages 626–627* In lowest terms, write the ratio of the length of the nose on the Statue of Liberty to the width of the mouth. *(Lesson 9-1)* 3:2

Challenge

Number the 8 vertices of a cube from 1 to 8 so that the sum of the 4 numbers at the vertices of each face is 18. See answer on p. A39.

Surface Area and Volume **637**

Making Connections/Transitions
Exercises 18–21 lead students step by step to the formula for the surface area of a square pyramid.

Follow-Up

Enrichment
Suppose a cube is composed of twenty-seven centimeter cubes. Each face of the large cube is painted. How many centimeter cubes have paint on:
a. 1 face? 6
b. 2 faces? 12
c. 3 faces? 8
d. 4 faces? 0
e. no faces? 1

Enrichment
Enrichment 142 is pictured on page 626E.

Practice
Practice 176, *Resource Book*

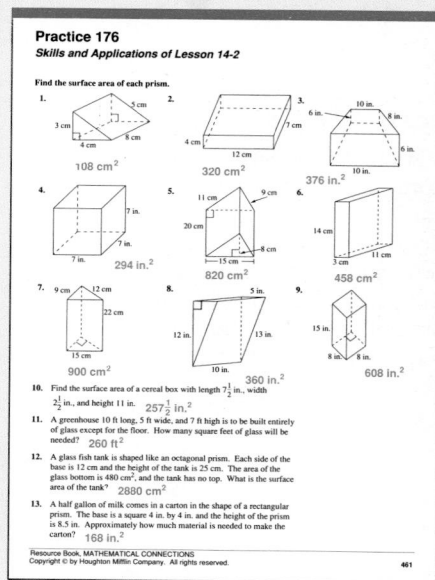

Teaching the Lesson
- Materials: centimeter graph paper, cylinders, colored chalk
- Lesson Starter 130
- Using Technology, p. 626C
- Using Manipulatives, p. 626C
- Cooperative Learning, p. 626C
- Reteaching/Alternate Approach, p. 626D
- Toolbox Skills 7, 9, 11, 13, 17, 18, 20

Lesson Follow-Up
- Practice 177
- Practice 178: Nonroutine
- Enrichment 143: Exploration
- Study Guide, pp. 259–260
- Cooperative Activity 25
- Test 69

Warm-Up Exercises

Find the circumference and the area of circles with the following radii. Use $\frac{22}{7}$ for π.

1. 14 cm   88 cm; 616 cm^2
2. $3\frac{1}{2}$ ft   22 ft; $38\frac{1}{2}$ ft^2
3. 21 m   132 m; 1386 m^2

Lesson Focus

Various foods, such as oatmeal, corn, and soup, are packaged in cans or cylindrical containers. How does the manufacturer determine how much material is needed for the packaging? Finding the surface area of cylinders, the focus of today's lesson, can help solve problems of this type.

14-3   Surface Area of Cylinders

Objective: To find the surface area of a cylinder.

EXPLORATION

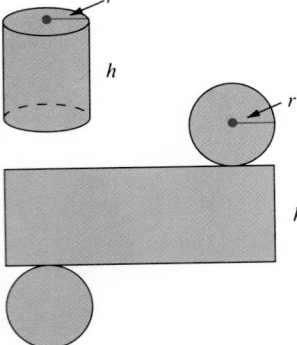

1. How many bases does the cylinder in the diagram at the right have? 2
2. What is the formula for the area of a base of the cylinder? $A = \pi r^2$
3. What is the formula for the circumference of a base of the cylinder? $C = 2\pi r$
4. The curved surface of a cylinder flattens out into a rectangle. How does the length of the rectangle compare with the circumference of a base of the cylinder? They are equal.

The surface area of a cylinder consists of the areas of a rectangle and two congruent circles. The length of the rectangle is the circumference of a base of the cylinder, and the width is the height of the cylinder.

> **Formula:** *Surface Area of a Cylinder*
>
>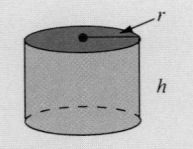
>
> $\text{Surface Area} = \dfrac{\text{area of}}{\text{bases}} + \dfrac{\text{area of curved}}{\text{surface}}$
>
> $\text{S.A.} = 2\pi r^2 + 2\pi rh$

Example

Find the surface area of the cylinder at the right.

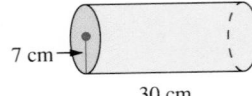

Solution

$\text{S.A.} = 2\pi r^2 + 2\pi rh$

$\text{S.A.} \approx 2\left(\frac{22}{7}\right)(7)^2 + 2\left(\frac{22}{7}\right)(7)(30)$

$\text{S.A.} \approx 308 + 1320 \approx 1628$

The surface area of the cylinder is approximately 1628 cm^2.

 **Check Your Understanding**

1. Why is it more convenient to use $\frac{22}{7}$ as an approximation of π than 3.14 in the Example?
2. Which part of the surface of the cylinder in the Example has an area of 1320 cm^2?

See *Answers to Check Your Understanding* at the back of the book.

You can use the $\boxed{\pi}$ and $\boxed{x^2}$ keys on a calculator to find the surface area of a cylinder. You would use this key sequence to find the surface area of a cylinder with radius 4 and height 15.

Guided Practice

COMMUNICATION « *Reading*

Refer to the text on page 638.

« **1.** Which two measurements must you know to compute the surface area of a cylinder? Answers may vary. Example: height and radius (or diameter or circumference)

« **2.** What is the shape of a pattern for the curved surface of a cylinder? rectangle

« **3.** State the formula for the surface area of a cylinder.
S.A. = $2\pi r^2 + 2\pi rh$

Use the cylinder at the right.

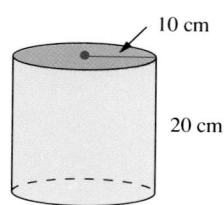

10 cm

20 cm

4. Find the area of a base. 314 cm²

5. Find the circumference of the base. 62.8 cm

6. Find the area of the curved surface. 1256 cm²

7. Use your answers to Exercises 4 and 6 to find the surface area. 1884 cm²

Find the surface area of each cylinder.

8.
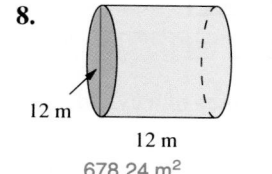
12 m
12 m
678.24 m²

9.
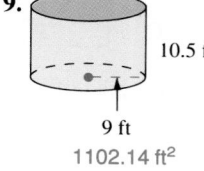
10.5 ft
9 ft
1102.14 ft²

10.
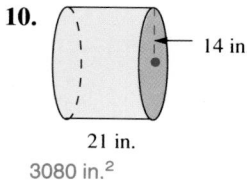
14 in.
21 in.
3080 in.²

Exercises

Find the surface area of each cylinder.

A **1.**
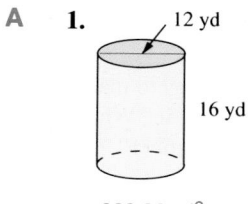
12 yd
16 yd
828.96 yd²

2.
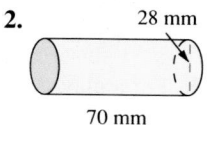
28 mm
70 mm
7392 mm²

3.
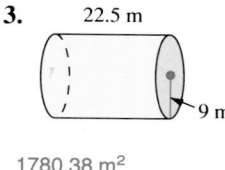
22.5 m
9 m
1780.38 m²

Teaching Notes

Before asking students to do the Exploration, you may wish to demonstrate surface area by cutting up a cardboard cylinder, such as an oatmeal box, so that students can see the resulting rectangle and two circles.

Key Questions
1. What parts make up the surface area of a cylinder? the two circular bases and the curved surface, which is a rectangle when placed in a plane
2. Why is the length of the rectangular part of the cylinder $2\pi r$? The length is the circumference of the circular base.

Error Analysis
Students have several measurements to consider when finding the surface area of a cylinder. Encourage them to draw sketches with labels to avoid using incorrect measurements.

Using Technology
For a suggestion on using spreadsheets to find surface area, see page 626C.

Using Manipulatives
For a suggestion on a cooperative activity using centimeter graph paper to find surface area, see page 626C.

Reteaching/Alternate Approach
For a suggestion on using a step-by-step procedure to calculate the total surface area of a cylinder, see page 626D.

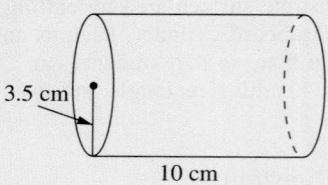

Find the surface area of each cylinder.

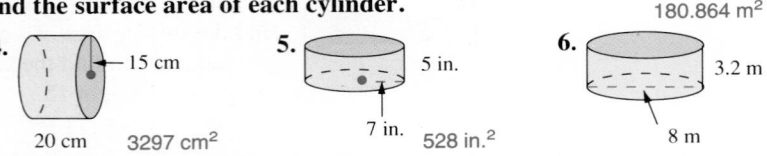

4.

15 cm

20 cm   3297 cm²

5.

5 in.

7 in.   528 in.²

6.   180.864 m²

3.2 m

8 m

7. The diameter of a dime is about 18 mm. The height is about 1.3 mm. Find the approximate surface area of a stack of 15 dimes. 1610.82 mm²

8. A drum is closed on the top and the bottom. The diameter of the drum is 20 in. The height is 27 in. Find the approximate surface area of the drum. 2323.6 in.²

B

9. Find the approximate surface area of the paperweight below. 151.62 cm²

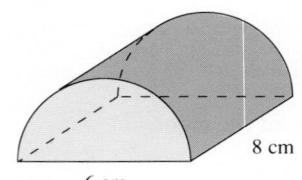

8 cm

6 cm

10. Compare the areas of the two labels. What do you notice? They are equal.

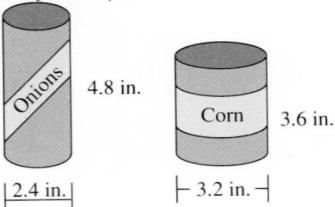

Onions   4.8 in.

Corn   3.6 in.

|2.4 in.| ⊢ 3.2 in. ⊣

CALCULATOR

For Exercises 11 and 12, choose the letter of the key sequence you would use to find each value.

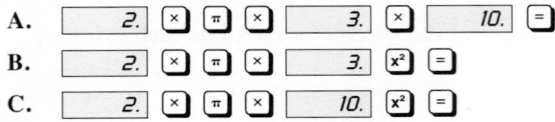

A.   2. × π × 3. × 10. =

B.   2. × π × 3. x² =

C.   2. × π × 10. x² =

11. the area of the two bases of a cylinder with radius 3 and height 10 B

12. the area of the curved surface of a cylinder with radius 3 and height 10
A

PROBLEM SOLVING/APPLICATION

The formula for the surface area of the curved surface of a cone, or *lateral area*, is L.A. = $\pi r l$, where r is the radius of the base of the cone, and l is the *slant height*.

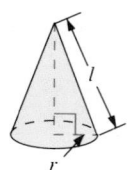

13. For a play, a costume maker has to make a cone-shaped hat with radius 3 in. and slant height 12 in. What will be the lateral area of the hat to the nearest whole number? 113 in.²

14. Suppose the costume maker plans to decorate the hat described in Exercise 13 using 3 sequins per square inch. How many sequins will the costume maker use? 339

15. Suppose the costume maker plans to put ribbon around the bottom edge of the hat described in Exercise 13. Rounding up to the nearest inch, how much ribbon will the costume maker use? 19 in.

THINKING SKILLS 16–17. Evaluation

C

16. Decide which of the following will change when the height of a cylinder is doubled.
   a. area of a base   **(b.)** area of curved surface   **(c.)** surface area

17. Decide which of the following will change when the radius of a can is doubled.
   (a.) area of a base   **(b.)** area of curved surface   **(c.)** surface area

SPIRAL REVIEW

S **18.** Find the surface area of the cylinder at the right.   *(Lesson 14-3)* 904.32 in.²

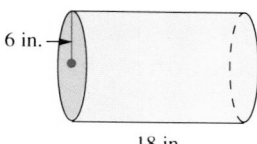
6 in.
18 in.

19. Find the area of a rectangle with length 9 cm and width 5 cm. *(Lesson 10-3)* 45 cm²

20. Graph the function $y = |x| - 3$.   *(Lesson 13-10)* See answer on p. A39.

21. A number cube that has sides numbered 1 through 6 is rolled. Find P(less than 3).   *(Lesson 12-1)* $\frac{1}{3}$

22. Estimate the quotient: $12\frac{1}{2} \div 3\frac{7}{8}$ *(Lesson 8-6)* about 3

Self-Test 1

Identify each space figure.

1.   **2.**   **3.**   14-1

square pyramid   cylinder   sphere

Find the surface area of each space figure.

4. 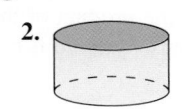   **5.**    14-2

2 mm
$6\frac{3}{4}$ mm
9 mm
184.5 mm²

10 yd
6 yd
9 yd
8 yd
264 yd²

6. 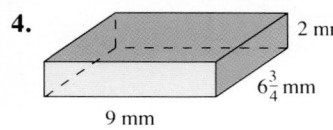   **7.** 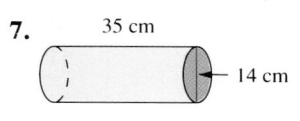   14-3

6 in.
2.5 in.
320.28 in.²

35 cm
14 cm
1848 cm²

Surface Area and Volume   **641**

Critical Thinking
Exercises 16 and 17 require stu-dents to evaluate the effect of doubling the height or radius of a cylinder. Students must make deci-sions based on an analysis of the formula.

Follow-Up

Project
Have students do research on the Leaning Tower of Pisa in order to calculate its surface area.

Enrichment
Enrichment 143 is pictured on page 626E.

Quick Quiz 1
See pages 660 and 661.

Alternative Assessment
See page 661.

Practice
Practice 177, *Resource Book*

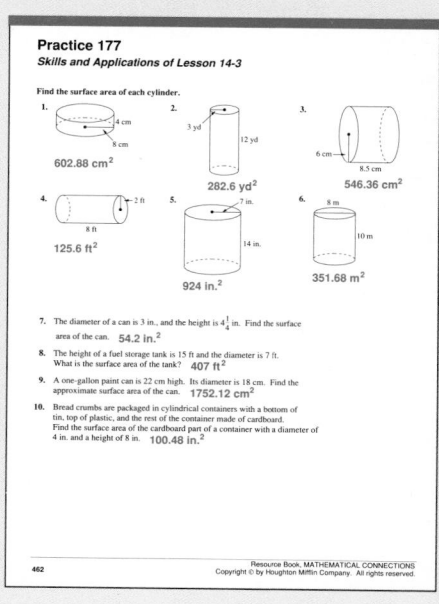

Practice 177
Skills and Applications of Lesson 14-3

Find the surface area of each cylinder.

1. 4 cm, 8 cm 602.88 cm²
2. 3 yd, 12 yd 282.6 yd²
3. 6 cm, 8.5 cm 546.36 cm²
4. 2 ft, 8 ft 125.6 ft²
5. 7 in., 14 in. 924 in.²
6. 8 m, 10 m 351.68 m²

7. The diameter of a can is 3 in., and the height is $4\frac{1}{4}$ in. Find the surface area of the can. 54.2 in.²
8. The height of a fuel storage tank is 15 ft and the diameter is 7 ft. What is the surface area of the tank? 407 ft²
9. A one-gallon paint can is 22 cm high. Its diameter is 18 cm. Find the approximate surface area of the can. 1752.12 cm²
10. Bread crumbs are packaged in cylindrical containers with a bottom of tin, top of plastic, and the rest of the container made of cardboard. Find the surface area of the cardboard part of a container with a diameter of 4 in. and a height of 8 in. 100.48 in.²

462
Resource Book, MATHEMATICAL CONNECTIONS
Copyright © by Houghton Mifflin Company. All rights reserved.

Teaching the Lesson
- Materials: centimeter cubes
- Lesson Starter 131
- Visuals, Folder E
- Using Technology, p. 626C
- Using Manipulatives, p. 626D
- Cooperative Learning, p. 626D
- Toolbox Skills 9, 13, 20

Lesson Follow-Up
- Practice 179
- Enrichment 144: Thinking Skills
- Study Guide, pp. 261–262
- Calculator Activity 14
- Computer Activity 14

Warm-Up Exercises

Find the area of each figure.

1.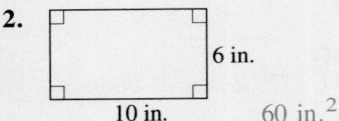

   36 in.^{2}

2.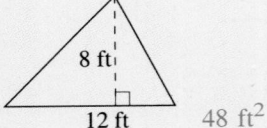

   60 in.^{2}

3.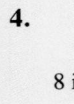

   48 ft^2

4.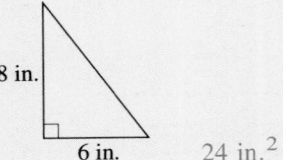

   24 in.^{2}

Volumes of Prisms and Pyramids

Objective: To find the volumes of prisms and pyramids.

Terms to Know
- *volume*

When you pack a suitcase or pour a glass of water, the amount of clothing or the amount of liquid your container can hold depends on its *volume*. The **volume** *V* of a space figure is the amount of space it encloses. To measure volume you use cubic units: for instance, cubic centimeters (cm^3), cubic inches (in.^{3}), and cubic yards (yd^3).

You can find the volume of a rectangular prism by counting the number of unit cubes that can fit inside the prism.

Example 1

Find the volume of the prism at the right.

Solution

To measure the volume, think of the prism as layers of unit cubes that measure 1 cm on each edge.

Number of cubes in 1 layer: $4 \cdot 2 = 8$
Number of cubes in 3 layers: $3 \cdot 8 = 24$

The volume is 24 cm^3.

 **Check Your Understanding**

1. How would the volume change if the height were 5 cm instead of 3 cm?
2. How would the volume change if the length of the base were 6 cm instead of 4 cm?

   See Answers to Check Your Understanding at the back of the book.

In Example 1, notice that you multiplied the area of a base of the prism by the height of the prism to find the volume. In fact, the volume of any prism is the product of the area of a base (*B*) and the height (*h*). The volume of any pyramid is one third the product of the area of the base and the height.

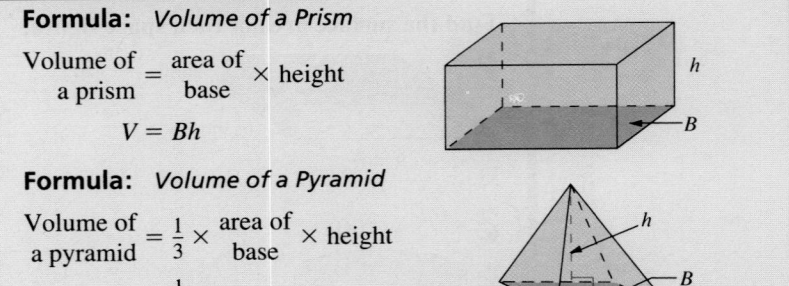

Formula: *Volume of a Prism*

$$\text{Volume of a prism} = \text{area of base} \times \text{height}$$

$$V = Bh$$

Formula: *Volume of a Pyramid*

$$\text{Volume of a pyramid} = \frac{1}{3} \times \text{area of base} \times \text{height}$$

$$V = \frac{1}{3}Bh$$

Example 2
Solution

Find the volume of the pyramid at the right.

The base is a rectangle.
$B = (5.5)(6) = 33$

$V = \frac{1}{3}Bh$

$V = \frac{1}{3}(33)(8) = 88$

The volume is 88 cm³.

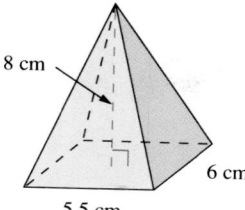
8 cm
6 cm
5.5 cm

Guided Practice

COMMUNICATION « *Reading*

Refer to the text on page 642.

« **1.** What does B represent in the formula $V = Bh$? the area of a base of a prism

« **2.** The volume of a space figure is represented by the formula $V = \frac{1}{3}Bh$. What figure is it? a pyramid

Write the area formula needed to find the base area of each space figure.

$A = \frac{1}{2}bh$

3. rectangular prism   **4.** triangular prism   **5.** square pyramid
$A = lw$ $A = s^2$

6. The lengths of the sides of the base of a given pyramid and the height of the pyramid are measured in feet. What is the most convenient unit of measure for the volume? ft³

The area of the base and the height of each space figure are given. Find the volume.

7.
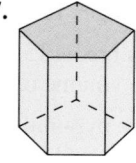

8.
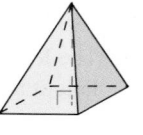

9.
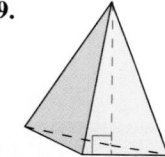

$B = 25$ in.²
$h = 8$ in.   200 in.³

$B = 15$ cm²
$h = 7$ cm   35 cm³

$B = 7.5$ m²
$h = 16.4$ m

41 m³

Find the volume of each space figure.

10.
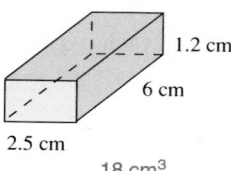
1.2 cm
6 cm
2.5 cm
18 cm³

11.
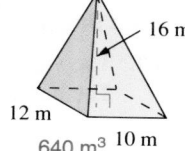
16 m
12 m
10 m
640 m³

12.
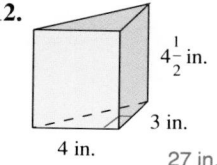
$4\frac{1}{2}$ in.
3 in.
4 in.
27 in.³

Surface Area and Volume   **643**

Lesson Focus

Is it possible to estimate how many gallons of water it takes to fill a swimming pool or how many centimeter cubes it takes to fill a shoe box? The focus of today's lesson is to develop a procedure for calculating how much a container can hold, that is, how to find its volume.

Teaching Notes

Stress the idea that a solid figure is generated by moving a plane figure through space. For instance, a rectangular prism is formed by moving a rectangle through space. The volume of the prism is equal to the area of the rectangle times the distance it moves through space, that is, its height. This description should help students to understand the general approach for finding the volume of a space figure.

Key Questions

1. How do you find the volume of a prism? The volume is found by multiplying the area of a base by the height.

2. What is the ratio of the volume of a pyramid to the volume of a prism having the same height and congruent bases? 1:3

Error Analysis

Students may forget to label volume as cubic units. Spending some time reviewing linear, square, and cubic units should help students to avoid this problem.

Using Technology

For a suggestion on using spreadsheets to find volume, see page 626C.

Using Manipulatives

For a suggestion on a cooperative activity using centimeter cubes, see page 626D.

Additional Examples

1. Find the volume of the cube.

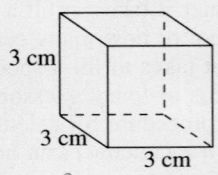

3 cm
3 cm
3 cm

27 cm^3

2. Find the volume of the square pyramid.

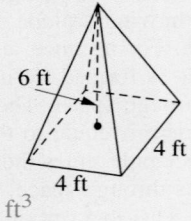

6 ft
4 ft
4 ft

32 ft^3

Closing the Lesson

Have students state the formulas for finding the volumes of prisms and pyramids.

Suggested Assignments

Skills: 1–16, 21–23, 26–29
Average: 1–23, 26–29
Advanced: 1–29

Exercise Notes

Communication: Reading

Guided Practice Exercises 1 and 2 check students' understanding of the volume formulas.

Exercises

1. **a.** 132 in.^{3}
   b. 44 in.^{3}
2. **a.** 228 cm^3
   b. 76 cm^3
3. **a.** 336 ft^3
   b. 112 ft^3
4. **a.** 225 m^3
   b. 75 m^3

Find the volume of each space figure with the given base and height.
a. a prism   b. a pyramid

A **1.** $B = 22$ in.^{2} **2.** $B = 19$ cm^2 **3.** $B = 48$ ft^2 **4.** $B = 25$ m^2
       $h = 6$ in. $h = 12$ cm $h = 7$ ft $h = 9$ m

Find the volume of each space figure.

5.   **6.**   **7.**

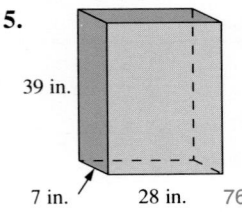   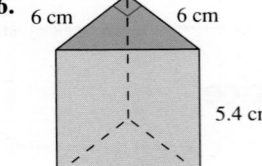   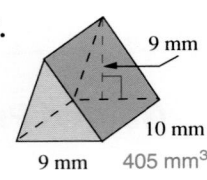

39 in.   6 cm   6 cm   9 mm
                                                                        10 mm
7 in.   28 in.   7644 in.^{3} 5.4 cm   9 mm   405 mm^3
                                97.2 cm^3

8.   **9.**   **10.**

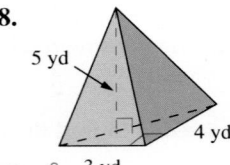   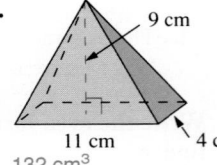   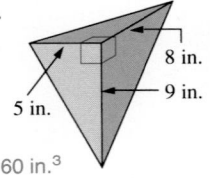

5 yd   9 cm   8 in.
                                                                          9 in.
           4 yd   5 in.
10 yd^3 3 yd   11 cm   4 cm   60 in.^{3}
                               132 cm^3

Find the volume of each space figure described in Exercises 11–14.

11. a cube with sides of length 8 m 512 m^3

12. a rectangular prism with base length $10\frac{1}{2}$ units, base width 8 units, and height 6 units 504 cubic units

13. a square pyramid with base edges 15 ft and height 8 ft 600 ft^3

14. a square pyramid with base edges 11 m and height 7.2 m 290.4 m^3

15. The length of a pancake mix box is 15 cm, the width is 5 cm, and the height is 21 cm. What is the volume of the box? 1575 cm^3

16. A tent is shaped like a square pyramid with height 2 m. If the bottom edges of the tent are 3 m long, find the volume of the tent. 6 m^3

Find the area of the base of each prism described.

B

17. The volume is 102 yd^3.
The height is 12 yd. 8.5 yd^2

18. The volume is 216 cm^3.
The height is 15 cm. 14.4 cm^2

Find the area of the base of each pyramid described.

19. The volume is 98 m^3.
The height is 6 m. 49 m^2

20. The volume is 108 yd^3.
The height is 9 yd. 36 yd^2

DATA ANALYSIS

Many structures are in the shape of rectangular prisms or pyramids. The table below shows the base edge lengths and heights of some structures with square bases.

Structure	Type of figure	Length of each base edge (ft)	Height (ft)
World Trade Center, each tower, New York City	Prism	209	1350
Transamerica Building, San Francisco	Pyramid	145	853
Great Pyramid of Cheops, Giza, Egypt	Pyramid (original measurements)	756	480

Find the volume of each structure to the nearest cubic foot.

21. 58,969,350 ft³
22. 5,978,108 ft³
23. the volume of the Great Pyramid

21. a tower at the World Trade Center   **22.** the Transamerica Building

23. **ESTIMATION** Estimate to decide which is larger, the volume of a World Trade Center tower or the volume of the Great Pyramid.

 CALCULATOR

For Exercises 24 and 25, use the fact that a cubic foot of water is approximately 7.481 gallons. Use a calculator to find the number of gallons of water needed to fill each swimming pool.

C 24.

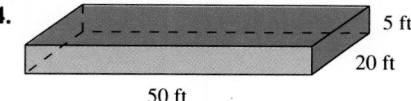

5 ft
20 ft
50 ft

about 37,405 gal

25.

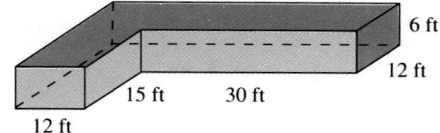

6 ft
12 ft
15 ft   30 ft
12 ft

about 30,702 gal

SPIRAL REVIEW **28.** $x \geq 6$; see graph on p. A39.

S 26. Find the volume of the prism shown at the right. *(Lesson 14-4)* 30 ft³

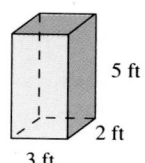

5 ft
2 ft
3 ft

27. What number is 64% of 350? *(Lesson 9-6)*
224

28. Solve and graph: $2x - 7 \geq 5$ *(Lesson 13-7)*

29. Find the area of a circle with radius 5.6 cm. *(Lesson 10-4)* 98.5 cm²

Surface Area and Volume   **645**

Data Analysis
Exercises 21–23 have students compare the volumes of some familiar structures.

Calculator
In *Exercises 24 and 25,* students calculate the volume and the capacity of two swimming pools. *Exercise 25* is a multi-step problem that uses students' visualization skills.

Follow-Up
Extension
Students may enjoy finding the volume of their classroom.

Enrichment
Enrichment 144 is pictured on page 626E.

Practice
Practice 179, *Resource Book*

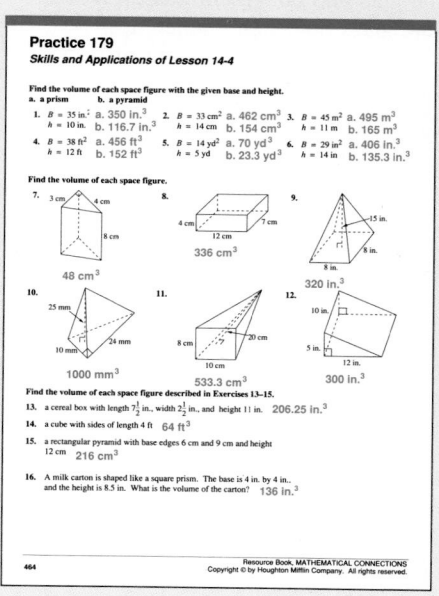

645

Teaching the Lesson
- Lesson Starter 132
- Visuals, Folder E
- Using Technology, p. 626C
- Reteaching/Alternate
  Approach, p. 626D
- Cooperative Learning,
  p. 626D
- Toolbox Skills 9, 13, 20

Lesson Follow-Up
- Practice 180
- Enrichment 145: Application
- Study Guide, pp. 263–264
- Manipulative Activity 28
- Cooperative Activity 26

Warm-Up Exercises

Find the area of each circle with the given radius or diameter. Use 3.14 for π.

1. $r = 5$ in.   78.5 in.^{2}
2. $r = 10$ cm   314 cm^2
3. $d = 8.2$ m   52.78 m^2
4. $d = 3.5$ ft   9.6 ft^2

Lesson Focus

Consider a cylindrical glass and a cone-shaped paper cup of equal heights and congruent bases. Would the glass hold more water, less water, or the same amount of water as *three* of the cones?   They actually hold the same amount.   The focus of today's lesson is finding volumes of cylinders and cones.

Volumes of Cylinders and Cones

Objective: To find the volumes of cylinders and cones.

The formulas for the volumes of a cylinder and a cone are similar to those for a prism and a pyramid. The base of a cylinder or a cone is a circle, so use πr^2 for the area of the base, B, in the formulas.

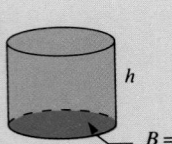

Formula: *Volume of a Cylinder*

$$\text{Volume of a cylinder} = \text{area of the base} \times \text{height}$$

$$V = \pi r^2 h$$

Formula: *Volume of a Cone*

$$\text{Volume of a cone} = \frac{1}{3} \times \text{area of the base} \times \text{height}$$

$$V = \frac{1}{3}\pi r^2 h$$

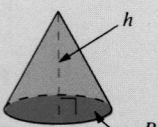

Example 1

The diameter of a cylinder is 30 m and the height is 11 m. Find the volume of the cylinder.

Solution

The radius, r, is $\frac{1}{2}(30) = 15$.

$V = \pi r^2 h$
$V \approx 3.14(15^2)(11)$
$V \approx 7771.5$

The volume is approximately 7771.5 m^3.

Example 2

Find the volume of a cone with radius 14 in. and height 12 in.

Solution

$V = \frac{1}{3}\pi r^2 h$

$V \approx \frac{1}{3}\left(\frac{22}{7}\right)(14^2)(12)$

$V \approx 2464$

The volume is approximately 2464 in.^{3}.

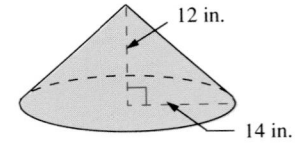

Check Your Understanding

1. How would Example 1 be different if the diameter were 30 in. and the height were 11 in.?
2. How would Example 1 be different if the object were a cone?
3. Why was $\frac{22}{7}$ used for π in Example 2?

See *Answers to Check Your Understanding* at the back of the book.

---

646   Chapter 14

Guided Practice

COMMUNICATION « *Reading*

Replace each __?__ with the correct word, phrase, or expression.

« **1.** To find the volume of a cylinder or cone, you need to know the area of a __?__ and the __?__. base; height

« **2.** The base area of a cone or cylinder with radius *r* is $B =$ __?__. πr^2

« **3.** The volume of a cone is __?__ the volume of a cylinder with the same radius and height. one third

Match each figure with the formula for its volume.

4. cone D

5. prism A

6. pyramid C

7. cylinder B

A. $V = Bh$

B. $V = \pi r^2 h$

C. $V = \frac{1}{3}Bh$

D. $V = \frac{1}{3}\pi r^2 h$

Replace each __?__ with the correct number.

8.

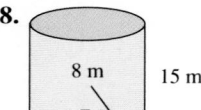

8 m   15 m

$V \approx 3.14 \cdot$ __?__ $\cdot 15$ 8^2

9.

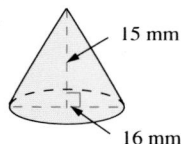

15 mm

16 mm

$V \approx \frac{1}{3}(3.14)(\underline{\;?\;})^2 \cdot$ __?__
  8; 15

Find the volume of each space figure.

10.

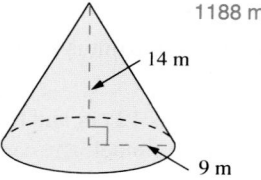

1188 m³

14 m

9 m

11.

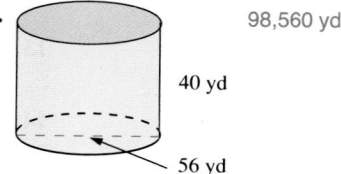

98,560 yd³

40 yd

56 yd

COMMUNICATION « *Discussion*

« **12.** Suppose you know the volume, radius, and height of a cylinder.
 a. How would the volume change if you were to double the height? It would double.
 b. How would the volume change if you were to double the radius? It would quadruple.

« **13.** Which has a greater effect on the volume of a cylinder, doubling the radius or doubling the height? doubling the radius

Teaching Notes

Use models to demonstrate the 1:3 ratio of the volumes of cylinders and cones with equal heights and congruent bases. Pour water or sand from the cone into the cylinder and show students that by filling the cone three times, you can fill the cylinder. Students will understand the formulas after seeing the concept demonstrated physically.

Key Questions

1. What is the ratio of the volume of a cone to the volume of a cylinder that has the same height and radius? 1:3
2. How is the formula for the volume of a cone similar to that for the volume of a pyramid? Both formulas involve taking one-third the area of the base times the height; that is, $V = \frac{1}{3} Bh$.

Error Analysis

Students sometimes make errors by substituting incorrect values, such as a diameter length instead of a radius length, in the formulas. Pointing out possible errors by using several diagrams may help students avoid them.

Using Technology

For a suggestion on using spreadsheets to find surface area, see page 626C.

Reteaching/Alternate Approach

For a suggestion on a cooperative activity involving the writing of volume problems, see page 626D.

Additional Examples

1. The height of a cylinder is 16 in. and the radius is 4 in. Find the volume. 803.8 in.³
2. Find the volume of a cone with diameter 10 ft and height 8 ft. 209.3 ft³

Closing the Lesson

Have students discuss the following questions. How do you find the surface area of a cylinder? How do you find the surface area of a cone? What similarities are there between cylinders and prisms? between pyramids and cones?

Suggested Assignments

Skills: 1–14, 23–26
Average: 1–16, 17–21 odd, 23–26
Advanced: 1–26

Exercise Notes

Communication: Reading

Guided Practice Exercises 1–3 check students' understanding of volume formulas and related concepts for cylinders and cones.

Communication: Discussion

Guided Practice Exercises 12 and 13 encourage students to think about and explain the effect on the volume of a cylinder if its dimensions are altered.

Application

Exercises 13 and 14 relate the concept of volume to real-world objects that are familiar to students.

Communication: Writing

Exercise 17 presents an opportunity for students to compare and contrast the concepts of surface area and volume for a space figure.

Making Connections/Transitions

Exercises 18 and 19 show how the ratio of surface area and volume for cylinders is exemplified in the scientific illustrations with animals. *Exercise 20* extends this concept to a research assignment for students.

Exercises

Find the volume of each space figure.

1. 4019.2 m³
2. 2772 in.³
3. 565.2 yd³
4. 4521.6 in.³
5. 847.8 yd³
6. 110,880 cm³
7. 6867.18 cm³
8. 9504 mm³
9. 792 ft³
10. 753.6 in.³
11. 264,000 in.³
12. 21,195 m³

A

1. 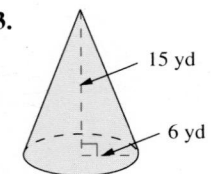 8 m, 20 m

2. 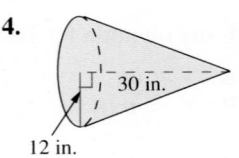 7 in., 18 in.

3. 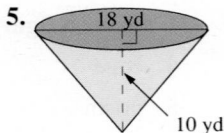 15 yd, 6 yd

4. 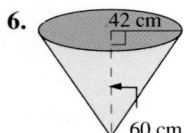 30 in., 12 in.

5. 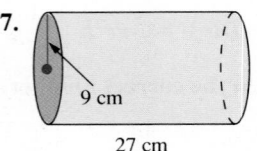 18 yd, 10 yd

6. 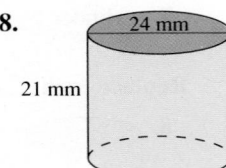 42 cm, 60 cm

7. 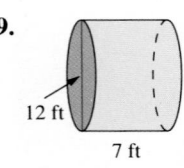 9 cm, 27 cm

8. 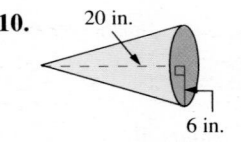 24 mm, 21 mm

9. 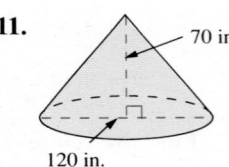 12 ft, 7 ft

10. 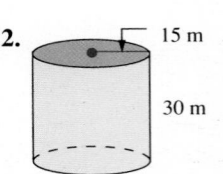 20 in., 6 in.

11. 70 in., 120 in.

12. 15 m, 30 m

13. The diameter of a can of paint is 8 in. and the height is 10 in. Find the volume. 502.4 in.³

14. The height of a funnel is 12 cm and the radius of the base is 7 cm. Find the volume of the funnel. 616 cm³

B 15. The volume of a cylinder with radius 8 cm is approximately 1004.8 cm³. Find the height of the cylinder. Use $\pi \approx 3.14$. 5 cm

16. The volume of a cone with radius 7 yd is approximately 462 yd³. Find the height of the cone. Use $\pi \approx \frac{22}{7}$. 9 yd

17. **WRITING ABOUT MATHEMATICS** Write a paragraph describing how you would explain to a friend the difference between the surface area and the volume of a space figure. See answer on p. A39.

CONNECTING GEOMETRY AND SCIENCE

The relationship between the surface area and the volume of an animal's body is an important factor in the animal's survival. An animal that has more skin area per unit of volume loses body heat faster, so an animal in a warmer climate would benefit from greater skin area.

18. For a cylinder with $r = 5$ cm and $h = 10$ cm, the ratio of surface area to volume is $\frac{3}{5}$. For a cylinder with $r = 15$ cm and $h = 30$ cm, this ratio is $\frac{1}{5}$. Do you think that a larger or a smaller animal has the greater ratio of surface area to volume? a smaller animal

19. The area of the curved surface of a cylinder is S.A. $= 2\pi rh$. Find this area for the following cylinders. 141.3 cm²
  a. $r = 1.5$ cm and $h = 15$ cm
94.2 cm² **b.** $r = 1.5$ cm and $h = 10$ cm
  c. Do you think it would be better for an arctic animal to have long or short legs? short legs

20. RESEARCH Find out how the long ears and legs of a black-tailed jack rabbit help it survive. See answer on p. A39.

 CALCULATOR

Find the volume of the shaded space figure. Use a calculator with a ⬛ key if you have one available to you.

21. Using a π key: 150.79645 cm³; using 3.14 for π: 150.72 cm³
22. Using a π key: 477.52208 ft³; using 3.14 for π: 477.28 ft³

C 21.

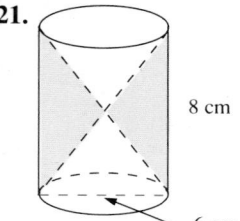
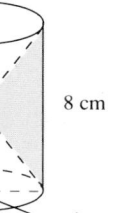

8 cm

6 cm

22.

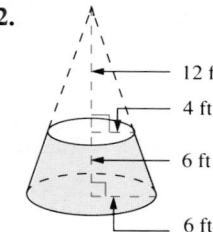

12 ft

4 ft

6 ft

6 ft

SPIRAL REVIEW

S 23. Find the volume of the cone shown at the right. (*Lesson 14-5*) 5652 m³

24. Find $\sqrt{8100}$. (*Lesson 10-10*) 90

25. Solve by graphing: $y = 5x - 3$ (1, 2)
 $y = x + 1$ (*Lesson 13-4*)

26. Solve: $\frac{x}{15} = \frac{9}{5}$ (*Lesson 9-2*) 27

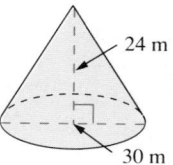

24 m

30 m

Exploration(*Cooperative Learning*)
Have students work in cooperative groups. Provide each group with cone-shaped paper cups and scissors. Ask the groups to imagine a plane cutting through a cone. What geometric figures are formed by the plane passing through the cone at different angles? Have each group cut the cones and examine and discuss the resulting cross-sections. Discuss why the cross-sections are called "conic sections."

Enrichment
Enrichment 145 is pictured on page 626F.

Practice
Practice 180, *Resource Book*

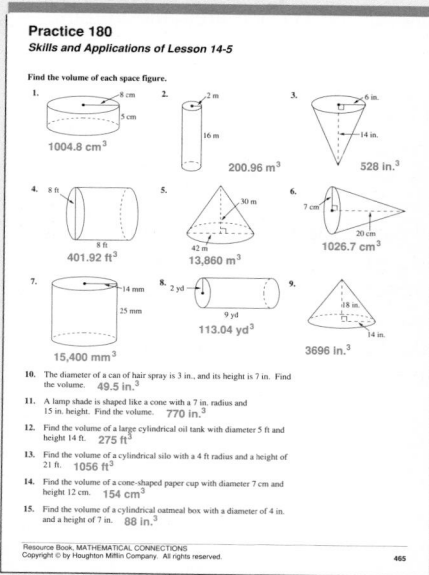

Teaching the Lesson
- Materials: orange, cylinder, sphere
- Lesson Starter 133
- Using Technology, p. 626C
- Using Manipulatives, p. 626D
- Toolbox Skills 9, 13, 20

Lesson Follow-Up
- Practice 181
- Enrichment 146: Communication
- Study Guide, pp. 265–266

Warm-Up Exercises

Find each answer to the nearest tenth.

1. $4 \times 3.14 \times 7$ 87.9
2. $4 \times 3.14 \times 10$ 125.6
3. $\frac{4}{3} \times 3.14 \times 5$ 20.9
4. $\frac{4}{3} \times 3.14 \times 4$ 16.7

Lesson Focus

Ask students to picture a basketball. How much material does it take to make one? How much air does it hold? Today's lesson focuses on answering questions of this type by finding the surface area and volume of spheres.

Teaching Notes

You may wish to begin the lesson by physically performing the Application. A suggestion for materials is given in the Using Manipulatives section on page 626D. As you guide students through the Example, make certain that they understand the difference between the two formulas.

Spheres

Objective: To find the surface area and volume of a sphere.

APPLICATION

Suppose you were to cut open a sphere and lay it flat. The area of the figure formed is the *surface area* of the sphere. This surface area would be four times the area of a circle with the same radius as the sphere.

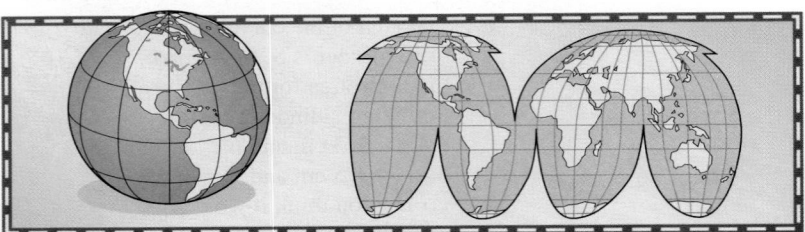

Use these formulas to find the surface area and the volume of a sphere.

Formulas: *Surface Area and Volume of a Sphere*		
Surface Area	**Volume**	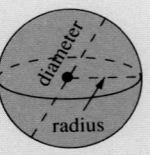
$S.A. = 4\pi r^2$	$V = \frac{4}{3}\pi r^3$	

Example

Find the surface area and volume of a sphere with diameter 12 m.

Solution

$r = \frac{1}{2}d = \frac{1}{2} \cdot 12 = 6$

Substitute 6 for r in each formula.

$S.A. = 4\pi r^2$ $V = \frac{4}{3}\pi r^3$
$S.A. \approx 4(3.14)(6^2)$
$S.A. \approx 4(3.14)(36)$ $V \approx \frac{4}{3}(3.14)(6^3)$
$S.A. \approx 452.16$

$V \approx \frac{4}{3}(3.14)(216)$

$V \approx 904.32$

The surface area of the sphere is approximately 452.16 m² and the volume is approximately 904.32 m³.

 **Check Your Understanding**

1. Why was it necessary to find one half the diameter in the Example?
2. What does 216 represent in the Example?
   See *Answers to Check Your Understanding* at the back of the book.

Guided Practice

COMMUNICATION « *Reading*

Explain the meaning of the word *sphere* or *hemisphere* in each sentence.

« **1.** The courtroom is a judge's sphere. the place where a
judge has control

« **2.** The right hemisphere of the brain largely controls a person's musical ability. the right half of the brain

« **3.** Chile is in the southern hemisphere. the half of Earth south of the equator

Replace each ___?___ with the correct number.

4.

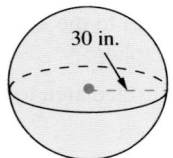
30 in.

S.A. ≈ ___?___ (3.14)(___?___)² 4; 30

$V ≈$ ___?___ (3.14)(___?___)³ $\frac{4}{3}$; 30

5.
42 ft

4; 21

S.A. ≈ ___?___ · $\frac{22}{7}$ · ___?___ ²

$\frac{4}{3}$; 21

$V ≈$ ___?___ · $\frac{22}{7}$ · ___?___ ³

6. Find the approximate surface area and volume of the sphere in Exercise 4. 11,304 in.²; 113,040 in.³

7. Find the approximate surface area and volume of the sphere in Exercise 5. 5544 ft²; 38,808 ft³

Exercises

Find the surface area of each sphere.

A
1.

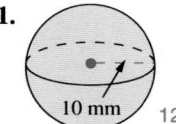
10 mm   1256 mm²

2.

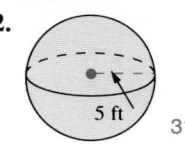
5 ft   314 ft²

3.
16 m

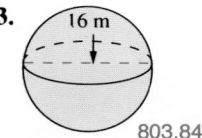
803.84 m²

Find the volume of each sphere.

4.

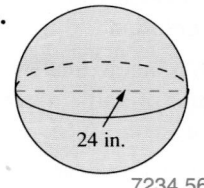
24 in.
7234.56 in.³

5.

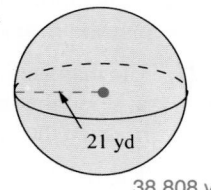
21 yd
38,808 yd³

6.
18 cm

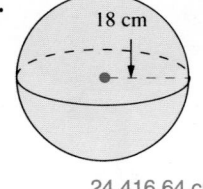
24,416.64 cm³

Key Questions

1. What is the largest possible circle formed by passing a plane through a sphere? a circle with the same radius as the sphere, called a great circle

2. Do you use the diameter or the radius to find the surface area and the volume of a sphere? radius

Error Analysis

Students sometimes confuse the formulas for surface area and volume. Remind students to think about square units for area and cubic units for volume. If they then look at the exponents in the formulas it will be clear which one is used to find area or volume.

Using Technology

For a suggestion on using spreadsheets to find surface area, see page 626C.

Using Manipulatives

For a suggestion on using an orange, a cylinder, and a sphere to demonstrate the formulas, see page 626D.

Additional Example

Find the surface area and volume of a sphere with diameter 8 ft.
S.A. = 201 ft²; $V = 268$ ft³

Closing the Lesson

Have students explain the formulas for finding the surface area and volume of a sphere.

Suggested Assignments

Skills: 1–14, 15–21 odd, 28–33
Average: 1–24, 28–33
Advanced: 1–33, Historical Note

Communication: Reading
Guided Practice Exercises 1–3 give students examples of everyday usage of the terms *sphere* and *hemisphere*.

Making Connections/Transitions
Exercises 21 and 22 connect mathematics and science by applying surface area and volume formulas to planets.

Cooperative Learning
Exercises 23–27 provide an opportunity for students to explore the relationships between a particular sphere and cylinder in a cooperative learning setting.

Critical Thinking
Exercises 23–25 require students to show the relationships between the sphere and the cylinder. Students must *compare* the formulas and realize what substitutions are appropriate for validating the equivalencies. *Exercises 26 and 27* check students' spatial visualization skills and their ability to *apply* a knowledge of volumes of space figures to new situations.

The radius of a basketball is 12 cm.

7. Find the surface area. 1808.64 cm²

8. Find the volume. 7234.56 cm³

The diameter of a soccer ball is 22 cm.

9. Find the surface area. 1519.76 cm²

10. Find the volume. 5572.45 cm³

Find the surface area and the volume of the sphere described.

11. 9156.24 m²;
    82,406.16 m³
12. 49,896 in.²;
    1,047,816 in.³
13. 22,176 cm²;
    310,464 cm³
14. 25,434 ft²;
    381,510 ft³

11. The radius is 27 m.

12. The radius is 63 in.

13. The diameter is 84 cm.

14. The diameter is 90 ft.

The ratio of the surface area to the volume of any sphere with radius r is $3:r$. Use this information for Exercises 15–17.

B 15. The ratio of the surface area to the volume of a given sphere is $3:7$. What is the radius of the sphere? 7 units

16. What is the ratio of the surface area to the volume of a sphere with radius 39 in.? 1:13

17. The surface area of a sphere is 1017.38 cm². The volume is 3052.14 cm³. Using the ratio above, write a proportion to find the radius r. $\frac{1017.38}{3052.14} = \frac{3}{r}$; 9 cm

The hemisphere shown below is one half a sphere with the same radius. The volume of the hemisphere is one half the volume of the sphere.

Volume of a hemisphere $= \frac{1}{2} \cdot$ volume of the sphere

$$V = \frac{1}{2}\left(\frac{4}{3}\pi r^3\right)$$

$$V = \frac{2}{3}\pi r^3$$

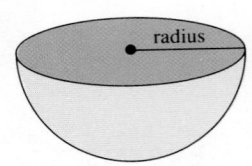
radius

Find the volume of each hemisphere.

18.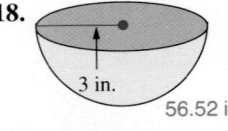
3 in.
56.52 in.³

19.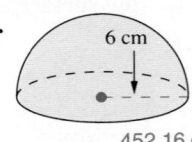
6 cm
452.16 cm³

20.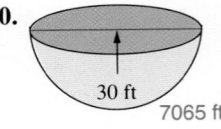
30 ft
7065 ft³

 CALCULATOR

Use a calculator with a $\boxed{\pi}$ key if you have one available to you. Write the answers in scientific notation.

21. The diameter of Earth is approximately 7927 mi.
    a. Find the surface area of Earth. 1.9741×10^8 mi²
    b. Find the volume of Earth. 2.6081×10^{11} mi³

22. The radius of Jupiter is approximately 44,431 mi.
    a. Find the surface area. 2.4807×10^{10} mi²
    b. Find the volume of Jupiter. 3.6741×10^{14} mi³

652 Chapter 14

GROUP ACTIVITY Answers to Exercises 23–24 are on p. A39.

A sphere with radius 3 cm fits inside a cylinder with height 6 cm, as shown.

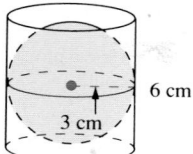

6 cm

3 cm

23. Show that the surface area of the sphere is equal to the area of the curved surface of the cylinder.

24. Show that the volume of the sphere is two thirds the volume of the cylinder.

C 25. Find the volume that is inside the cylinder, but outside the sphere. 56.52 cm³

26. A sphere with radius 3 cm fits inside a cube with edges 6 cm. Find the volume that is inside the cube, but outside the sphere. 102.96 cm³

27. A hemisphere with radius 3 in. fits inside a rectangular prism with length 6 in., width 6 in., and height 3 in. Find the volume that is inside the prism, but outside the hemisphere. 51.48 cm³

SPIRAL REVIEW

S 28. Find the volume of the sphere shown at the right. *(Lesson 14-6)* 3052.08 ft³

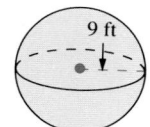

9 ft

29. Write 0.0000081 in scientific notation. *(Lesson 7-11)* 8.1×10^{-6}

30. The lengths of the legs of a right triangle are 7 cm and 24 cm. Find the hypotenuse. *(Lesson 10-11)* 25 cm

31. DATA, *pages 434–435* The length of a field hockey field is how many feet greater than the width of the field? *(Lesson 8-7)* 120 ft

32. Find the slope and the *y*-intercept of the line $y - 5x = 3$. *(Lesson 13-3)* 5; 3

33. Find the number of permutations of 7 objects taken 3 at a time. *(Lesson 12-6)* 210

Archimedes, a Greek mathematician who lived during the third century B.C., is considered to be one of the greatest mathematicians of all time. He discovered many important relationships in geometry. Among them is the fact that the volume of a sphere that fits snugly inside a cylinder is two thirds the volume of the cylinder. He also discovered that a body floating in liquid loses in weight an amount equal to that of the liquid displaced.

Research

Archimedes was also an inventor. Find out about some of his inventions. See answer on p. A39.

Surface Area and Volume **653**

Enrichment
As a follow-up to the manipulative activity given on page 626D, ask students to show how the formula for the volume of a sphere can be obtained from the demonstration.
Cylinder:
$V = \pi r^2 h = \pi r^2(2r) = 2\pi r^3$;
Sphere: $V = \left(\frac{2}{3}\right)2\pi r^3 = \frac{4}{3}\pi r^3$

Enrichment
Enrichment 146 is pictured on page 626F.

Practice
Practice 181, *Resource Book*

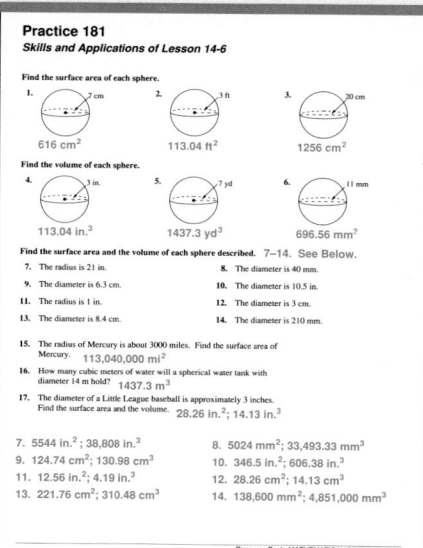

Lesson Planner

Teaching the Lesson
- Materials: tiles
- Lesson Starter 134
- Using Manipulatives, p. 626D
- Cooperative Learning, p. 626D

Lesson Follow-Up
- Practice 182
- Practice 183: Nonroutine
- Enrichment 147: Problem Solving
- Study Guide, pp. 267–268
- Test Master 70

Warm-Up Exercises

1. Sketch a cube with a volume of 27 cubic units.

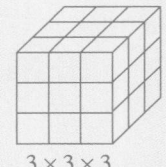

$3 \times 3 \times 3$

2. Sketch a rectangular prism with a volume of 60 cubic units and a base with an area of 15 square units.

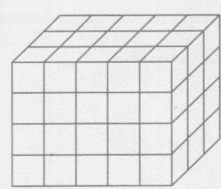

$5 \times 3 \times 4$

14-7

Strategy: Making a Model

Objective: To solve problems by making a model.

In 1990, scientists had difficulty opening a solar panel on the Hubble Space Telescope. To investigate, they built a model. They were able to locate and solve the problem and then fix the telescope.

Sometimes making a model of a problem situation can help you to solve the problem.

Problem

A children's museum has display stands like the one at the right. Each stand consists of 12 cubes. The top and all four sides of the stand are painted red. The bottom is unpainted. The cubes are unpainted on the faces that cannot be seen. How many of the cubes in each stand have exactly 2 red faces and how many have exactly 3 red faces?

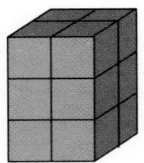

Solution

UNDERSTAND The problem is about the color of the faces of cubes in a display stand.
Facts: 5 sides of a stack of cubes are painted red
Find: the number of cubes with exactly 2 red faces
the number of cubes with exactly 3 red faces

PLAN Make a stack of 12 cubes like the one shown in the figure. Color the appropriate faces red. Take the stack apart and count the number of cubes with 2 red faces, and the number with 3 red faces.

WORK The stack has 3 layers of 4 cubes.

Top layer: Middle layer: Bottom layer:

     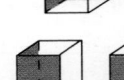 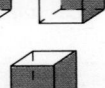   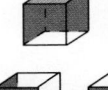

Each cube has 3 red faces.   Each cube has 2 red faces.   Each cube has 2 red faces.

ANSWER In each stand, 8 cubes have exactly 2 red faces and 4 cubes have exactly 3 red faces.

Look Back Suppose the bottom of each stand were also painted red. How would your answer be different? 8 cubes would have 3 red faces and 4 cubes would have 2 red faces.

Guided Practice

COMMUNICATION « *Reading*

Use this problem for Exercises 1–7.

A fence is built using 24 sections, each 1 unit long. What is the greatest possible rectangular area that can be enclosed?

« **1.** What is the shape of the area enclosed by the fence? a rectangle

« **2.** What is the perimeter of the area enclosed by the fence? 24 units

« **3.** What convenient object could you use to represent a section of the fence? Answers will vary. Example: a toothpick

« **4.** If the length of the rectangle enclosed by the fence is 1 unit, what is the width? the area? 11 units; 11 square units

« **5.** How many different rectangles can be built using the 24 sections of the fence? 6

« **6.** Make a sketch of a model showing the arrangement of the sections that has the greatest possible area. See answer on p. A39.

7. Use your model from Exercise 6 to write an answer to the problem.
36 square units

8. Solve by making a model: A rectangular prism with volume 20 cubic units is formed using cubes with edges 1 unit long.

  a. How many such prisms are possible? 4

  b. What are the dimensions of the prism with the greatest surface area? 20 units × 1 unit × 1 unit

Problem Solving Situations

Solve by making a model.

1. All 6 faces of a wooden cube are painted blue. The cube is then cut into 27 smaller cubes. Tell how many of the smaller cubes have the number of painted faces indicated.

  a. exactly 3 blue faces 8   **b.** exactly 2 blue faces 12
  c. exactly 1 blue face 6   **d.** no blue faces 1

2. Suppose a cube is painted so that the top and bottom are painted blue and the other 4 faces are painted red. The cube is then cut into 27 smaller cubes. How many of the smaller cubes have at least 1 red face and 1 blue face? 16

3. A rectangular prism is to be formed using exactly 30 cubes. How many different prisms can be formed? Which prism has the least surface area that can be painted? Which prism has the greatest surface area? 5; 2 × 3 × 5; 30 × 1 × 1

Surface Area and Volume   **655**

Closing the Lesson

You may wish to use the cooperative learning activity below to close the lesson.

Cooperative Learning

Have students work in cooperative groups to create problems that can be solved by using a model. The problems should be shared among the groups to discuss which problem solving strategies are most appropriate to solve each problem.

Suggested Assignments

Skills: 1–16
Average: 1–16
Advanced: 1–16

Exercise Notes

Communication: Reading

Guided Practice Exercises 1–7 have students build a model to solve a problem involving a rectangular fence. Students should conclude that the greatest area occurs with a square lot.

Problem Solving

Problem Solving Situations 1–3 test students' ability to solve problems by making a model. *Problem Solving Situations 4–10* require students to analyze problems and decide which problem solving strategy is most appropriate for a particular problem.

Communication: Writing

In *Problem Solving Situations 11 and 12*, students write their own word problems based on models of rectangular prisms.

Solve using any problem solving strategy. 4–10. Strategies may vary. Likely strategies are given.

4. Of 150 students, 60 take biology, 50 take Spanish, and 40 take both subjects. How many students take neither subject? using a Venn diagram; 80

5. If 3 serving-size bottles of Tree Ripe apple juice cost $1.47, how much will 10 bottles cost? using proportions; $4.90

6. The figures below show the number of sectors resulting when 1, 2, 3, and 4 diameters are drawn in a circle. How many sectors result when 10 diameters are drawn in a circle? identifying a pattern; 20

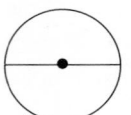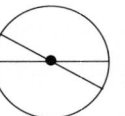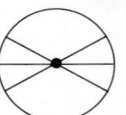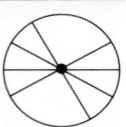

PROBLEM SOLVING

CHECKLIST

Keep these in mind:
Using a Four-Step Plan
Too Much or Not Enough Information
Supplying Missing Facts
Problems with No Solution

Consider these strategies:
Choosing the Correct Operation
Making a Table
Guess and Check
Using Equations
Identifying a Pattern
Drawing a Diagram
Using Proportions
Using Logical Reasoning
Using a Venn Diagram
Solving a Simpler Problem
Making a Model

7. no solution; the sum of the lengths of two sides must be greater than the length of the third side.
8. making a model; 324 square sections

7. An isosceles triangle has perimeter 54 in. The length of one side is 27 in. Find the lengths of the other two sides.

8. A rectangular flower bed is marked off using 72 sections of border fencing. What is the largest possible area of the flower bed?

9. Ted reads 30 min each day after dinner. In a seven-day week, how many hours does he read after dinner? supplying missing facts; $3\frac{1}{2}$ h

10. Ming, Damian, and Cesar are the presidents of the Math, Drama, and Chess Clubs. No person's club begins with the same letter as his or her name. Damian's best friend is president of the Math Club. Which person is president of each club? using logical reasoning; Ming–Drama, Damian–Chess, Cesar–Math

A drama club presents a play using signing for hearing-impaired audience.

656   Chapter 14

WRITING WORD PROBLEMS Answers to Problems 11–12 are on p. A39.

Using any or all of the figures below, write a word problem that you could solve by making a model. Then solve the problem.

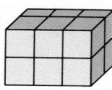 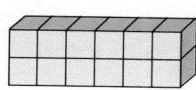 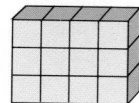

11. Write a problem involving volume.

12. Write a problem involving surface area.

SPIRAL REVIEW 13. 2 units × 3 units × 2 units **14.** 3.375 m³

S **13.** A box is formed to hold 12 cubes with edges 1 unit long. Find the dimensions of the box with the least surface area. *(Lesson 14-7)*

14. Find the volume of a cube with sides of 1.5 m. *(Lesson 14-4)*

15. A number cube is tossed. Find *P*(4). *(Lesson 12-1)* $\frac{1}{6}$

16. Write 0.546 as a fraction in lowest terms. *(Lesson 7-7)* $\frac{273}{500}$

Self-Test 2

Find the volume of each space figure.

1.
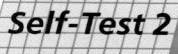
5 in.
4 in.
9 in.
180 in.³

2.

6 m
6 m   6 m
72 m³   **14-4**

3.
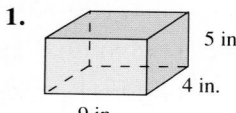
42 cm
50 cm
69,300 cm³

4.
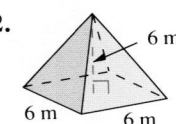
10 ft
8 ft
669.9 ft³   **14-5**

Use the sphere at the right.

5. Find the volume. 3052.08 in.³

6. Find the surface area. 1017.36 in.²

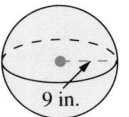
9 in.

14-6

Solve by making a model.

7. Six cylinders of the same size are stacked on top of one another to form one long cylinder that is then painted. How many circular faces of the original cylinders are unpainted? 10   **14-7**

Surface Area and Volume   **657**

Follow-Up

Extension
Have students solve some of the problems written for Problem Solving Situations 11 and 12. They should compare their solutions with the ones supplied by the writers of the problems.

Enrichment
Enrichment 147 is pictured on page 626F.

Practice
Practice 182 is shown below.

Quick Quiz 2
See pages 664 and 665.

Alternative Assessment
See page 665.

Practice 182, *Resource Book*

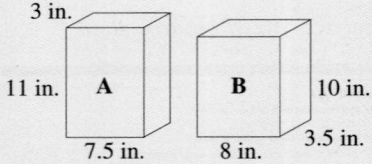

Packaging a Consumer Product

Objective: To explore how surface area and volume are used in making decisions about packaging consumer products.

Surface area and volume are important considerations in packaging a consumer product. Among the many factors affecting a manufacturer's decision in choosing packaging are cost, convenience, storage capacity, and visual appeal.

Example

Family Farm Foods is considering the three metal cans shown for a new Hearty Stew.

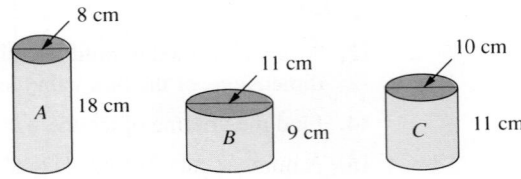

a. Which can holds the most stew?

b. Which can uses the least amount of metal?

Solution

a. The can that holds the most stew is the one with the greatest volume. Find the volume of each can. Use the formula for the volume of a cylinder, $V = \pi r^2 h$.

Can A: $V = \pi r^2 h \approx (3.14)(4)^2(18) \approx 904.32$

Can B: $V = \pi r^2 h \approx (3.14)(5.5)^2(9) \approx 854.865$

Can C: $V = \pi r^2 h \approx (3.14)(5)^2(11) \approx 863.5$

Can A holds the most stew.

b. The can that uses the least amount of metal is the can with the least surface area. Find the surface area of each can. Use the formula for the surface area of a cylinder, S.A. $= 2\pi r^2 + 2\pi rh$.

Can A: S.A. $= 2\pi r^2 + 2\pi rh$
$\approx 2(3.14)(4)^2 + 2(3.14)(4)(18)$
≈ 552.64

Can B: S.A. $= 2\pi r^2 + 2\pi rh$
$\approx 2(3.14)(5.5)^2 + 2(3.14)(5.5)(9)$
≈ 500.83

Can C: S.A. $= 2\pi r^2 + 2\pi rh$
$\approx 2(3.14)(5)^2 + 2(3.14)(5)(11)$
≈ 502.4

Can B uses the least amount of metal.

Exercises

Exercises 1–4 refer to the boxes shown below. Family Farm Foods is considering the three cardboard boxes below for a powdered product.

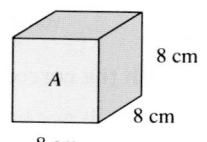 A — 8 cm, 8 cm, 8 cm

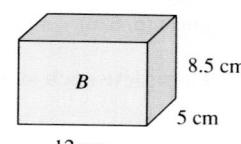 B — 8.5 cm, 5 cm, 12 cm

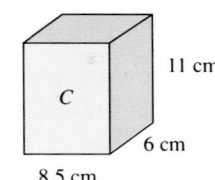 C — 11 cm, 6 cm, 8.5 cm

1. Which box uses the least amount of cardboard? A

2. If the label covers the sides of the box, but not the top and bottom, which box label uses the least amount of paper? A

3. Which box takes up the least area on a shelf? C

4. Which box can hold the most? C

Cans like the one at the right are to be shipped in boxes of 12.

7.5 cm

11.3 cm

5. Find the length, width, and height of the box needed to ship the 12 cans in one layer of three rows, four cans each. 30 cm × 22.5 cm × 11.3 cm

6. Find the dimensions of the box needed to ship the 12 cans in two layers of two rows, three cans each. 15 cm × 22.5 cm × 22.6 cm

7. Find the dimensions of the box needed to ship the 12 cans in one layer of two rows, six cans each. 15 cm × 45 cm × 11.3 cm

8. Which box uses the least cardboard?
the box in Exercise 6

GROUP ACTIVITY

Suppose you had to design a cylindrical container for a new type of juice.

9–12. Answers will vary.

9. Make a sketch of your container. Label the dimensions.

10. What materials might you use to make the container? Why?

11. Find the volume of your container.

12. How many containers would fit on a shelf that is 2 ft wide, 3 ft deep, and 2 ft high?

Surface Area and Volume **659**

Exercise Notes

Critical Thinking

Exercises 1–4 have students *examine* three cardboard boxes and make *decisions* based on the surface areas and volumes. In *Exercises 5–8,* students make similar *judgments* regarding cylinders.

Cooperative Learning

Exercises 9–12 have students working in cooperative groups design their own packages. This is an effective means of assessing students' knowledge of the content related to space figures.

Follow-Up

Extension

Ask students to visit a market and compare various types of packaging for food items. They should select two specific items to discuss in a written paragraph, which includes dimensions, surface areas, and volumes of the items.

Chapter Resources

- Cumulative Practice 184
- Enrichment 148: Extension
- Chapter 14 Project
- Tests 71, 72
- Cumulative Test 73
- Family Involvement 14
- Spanish Test 14
- Spanish Glossary
- Test Generator

Quick Quiz 1 *(14-1 through 14-3)*
Identify each space figure.

1.

triangular prism

2.
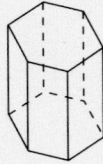
hexagonal prism

3.

cone

Find the surface area of each space figure.

4.
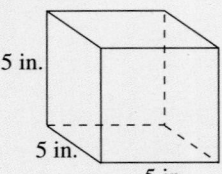
5 in.
5 in.
5 in.
150 in.²

660

Chapter Review

| Terms to Know |

space figure (p. 628)   cube (p. 628)
face (p. 628)   pyramid (p. 628)
edge (p. 628)   cylinder (p. 629)
vertex (p. 628)   cone (p. 629)
polyhedron (p. 628)   sphere (p. 629)
base (p. 628)   surface area (p. 634)
prism (p. 628)   volume (p. 642)

Complete each statement with the correct word or phrase.

1. A(n) __?__ is a space figure with one base whose other faces are triangles. pyramid

2. Any space figure with faces that are polygons is a(n) __?__. polyhedron

3. A(n) __?__ of a space figure is a point where edges intersect. vertex

4. The amount of space that a space figure encloses is called its __?__. volume

5. The set of all points in space that are the same distance from a given point is a(n) __?__. sphere

6. A rectangular prism all of whose edges are the same length is a(n) __?__. cube

---

Identify each space figure. *(Lesson 14-1)*

7.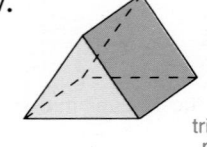
triangular prism

8.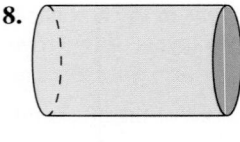
cylinder

9.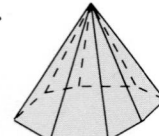
octagonal pyramid

Find the surface area of each prism. *(Lesson 14-2)*

10.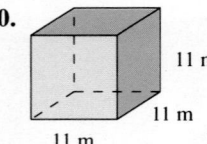
11 m
11 m
11 m
726 m²

11.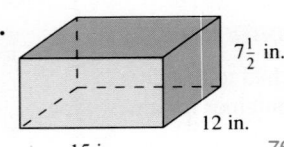
7½ in.
12 in.
15 in.
765 in.²

12.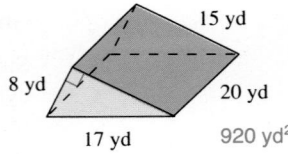
15 yd
8 yd
20 yd
17 yd
920 yd²

Find the surface area of each cylinder. *(Lesson 14-3)*

13.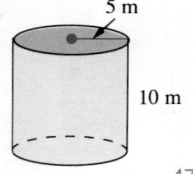
5 m
10 m
471 m²

14.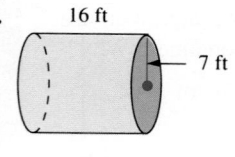
16 ft
7 ft
1012 ft²

15.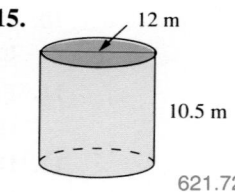
12 m
10.5 m
621.72 m²

Find the volume of each prism or pyramid. *(Lesson 14-4)*

16.

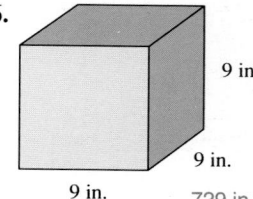

9 in.
9 in.
9 in.
729 in.³

17.

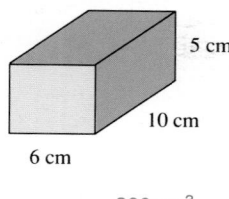

5 cm
10 cm
6 cm
300 cm³

18.

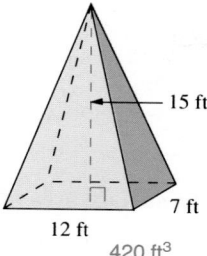

15 ft
7 ft
12 ft
420 ft³

Find the volume of each cylinder or cone. *(Lesson 14-5)*

19.

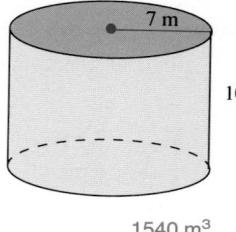

7 m
10 m
1540 m³

20.

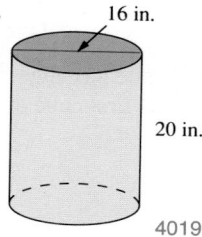

16 in.
20 in.
4019.2 in.³

21.

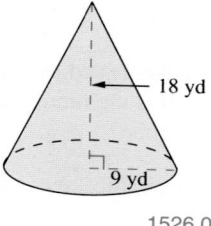

18 yd
9 yd
1526.04 yd³

Find the surface area and volume of each sphere. *(Lesson 14-6)*

22.

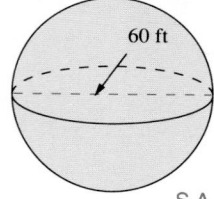

60 ft
S.A. = 11,304 ft²
V = 113,040 ft³

23.

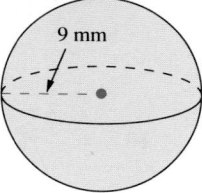

9 mm
S.A. = 1017.36 mm²
V = 3052.08 mm³

24.

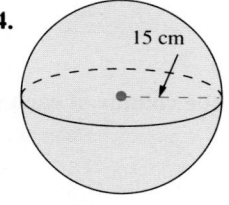

15 cm
S.A. = 2826 cm²
V = 14,130 cm³

Solve by making a model. *(Lesson 14-7)*

25. A rectangular prism like the one shown at the right is painted so that the top and bottom are red and the other 4 faces are blue. The prism is then cut into 18 cubes. How many of the cubes have exactly 1 red face and 1 blue face? 8

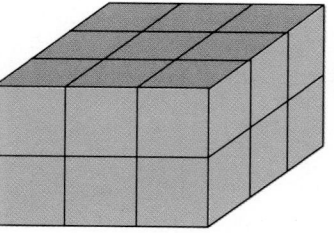

26. A rectangular prism with volume 8 cubic units is formed using cubes with edges 1 unit long. How many such prisms are possible? What are the dimensions of the prism with the greatest surface area? 3; 8 units × 1 unit × 1 unit

5.

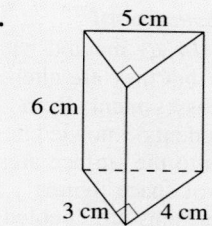

5 cm
6 cm
3 cm   4 cm
84 cm²

6.

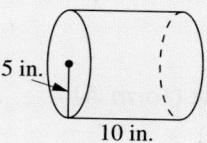

5 in.
10 in.
471 in.²

7.

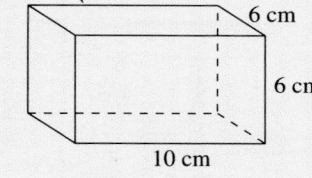

6 cm
6 cm
10 cm
312 cm²

Alternative Assessment

1. Describe the figure that is formed if the rectangle below is rotated about $\overline{AD}$.

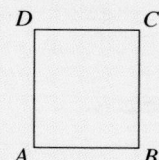

D   C

A   B

a cylinder with a height of *AD* and the radius of the base equal to *AB*

2. Name the geometric shapes that represent the top view of a cylinder and the side view of a cylinder.   circle; rectangle

Alternative Assessment

Exercises 6 and 14 are marked with stars to indicate that they are alternative forms of assessment. Both exercises test students' knowledge of what happens to the surface area and the volume of space figures when their dimensions are doubled.

Chapter Test (Form A)

Test 71, *Tests*

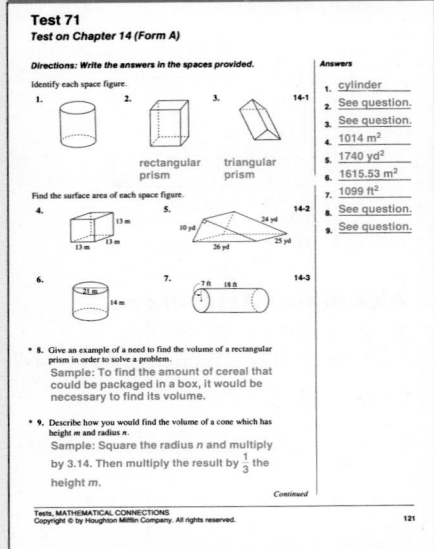
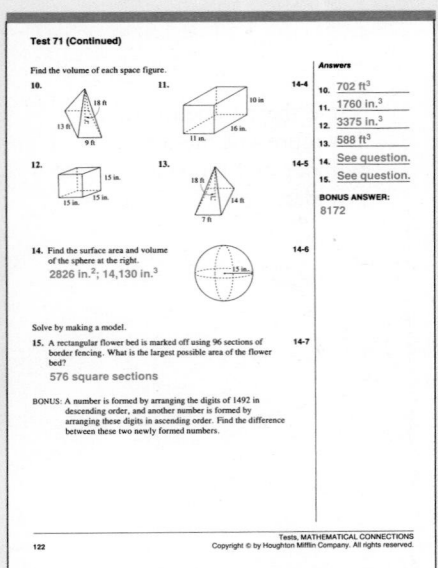
662

Chapter Test

Identify each space figure.   14-1

1.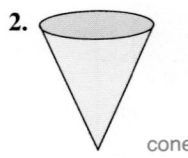
pentagonal prism

2. cone

3. sphere

Find the surface area of each space figure.   14-2

4. 3 cm, 12.6 cm, 5 cm   231.6 cm²

5. 13 ft, 5 ft, 9 ft, 12 ft   330 ft²

★ 6. A cube has edges *n* cm long. A second cube has edges 2*n* cm long. What is the ratio of the surface area of the first cube to that of the second cube? 1 : 4

Find the surface area of each space figure.   14-3

7. 10 yd, 15 yd   1570 yd²

8. 14 m, 10 m   748 m²

Find the volume of each space figure.   14-4

9. 9 ft, 12 ft, 15 ft   1620 ft³

10. 12 yd, 6 yd, 8 yd   192 yd³

11. 50 cm, 50 cm   392,500 cm³   14-5

12. 12 m, 24 m   1808.64 m³

13. Find the surface area and volume of a sphere with radius 24 in.   14-6
7234.56 in.²; 57,876.48 in.³

★ 14. When the radius of a sphere is doubled, what happens to the surface area and volume of the sphere? The surface area is four times as great; the volume is eight times as great.

15. A fence is built using 36 sections, each one unit long. What is the 14-7
greatest possible rectangular area that can be enclosed? 81 square units

Volume of Composite Space Figures

Objective: To determine volumes of composite space figures.

   A **composite space figure** is a combination of two or more space figures. The thumbtack at the right is a composite space figure composed of two cylinders and a cone. The top of the thumbtack is a wide cylinder; the post is a narrow cylinder. The point is a cone.

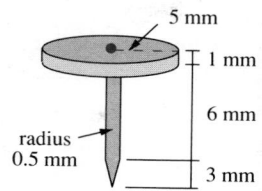

Example

Find the volume of the thumbtack pictured above. Round your answer to the nearest unit.

Solution

The volume of the thumbtack is the sum of the volumes of the three parts.

Top: $V = \pi r^2 h \approx 3.14(5^2)(1) \approx 78.5$

Post: $V = \pi r^2 h \approx 3.14(0.5)^2(6) \approx 4.71$

Point: $V = \frac{1}{3}\pi r^2 h \approx \frac{1}{3}(3.14)(0.5^2)(3) \approx 0.785$

Total Volume $\approx 78.5 + 4.71 + 0.785$
≈ 83.995

The volume is about 84 mm^3.

Exercises

Find the volume of each composite space figure. Round your answer to the nearest unit.

1.
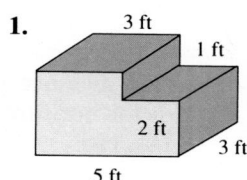
39 ft^3

2.
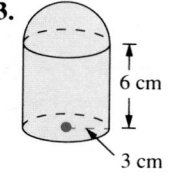
17 in.^{3}

3.
cylinder diagram 6 cm, 3 cm
226 cm^3

4.
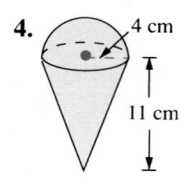
318 cm^3

5.
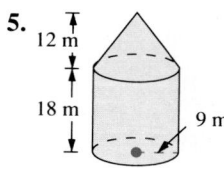
5595 m^3

6.
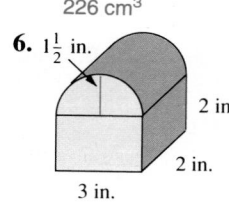
19 in.^{3}

Warm-Up Exercises

Find the volume of each space figure.

1.
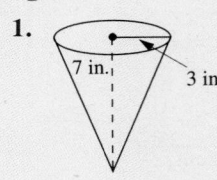
66 in.^{3}

2.
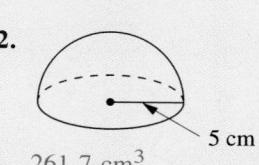
261.7 cm^3

3.
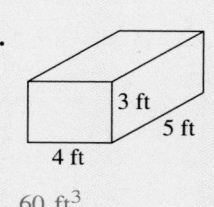
60 ft^3

Teaching Note

Students usually find these types of problems enjoyable. Stress the importance of being organized and of having well-labeled sketches. Drawing the parts of the figures separately is recommended.

Exercise Note

The Exercises require spatial visualization skills for determining the appropriate dimensions in the diagrams. Have models available to assist students who may need to see the actual figures.

Follow-Up

Extension

Find the volume.
120 cubic units

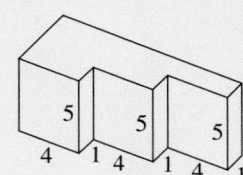

Find the volume of each space figure.

1.

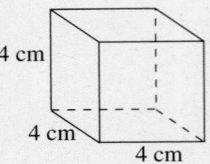

4 cm

4 cm

4 cm

64 cm³

2.

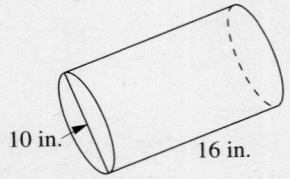

10 in.

16 in.

1256 in.³

3.

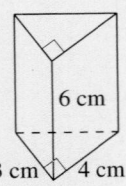

6 cm

3 cm 4 cm

36 cm³

4.

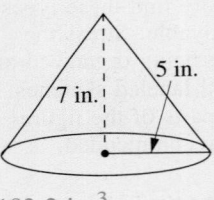

5 in.

7 in.

183.2 in.³

Use the sphere below.

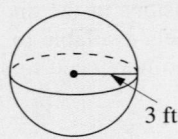

3 ft

5. Find the volume. 113 ft³
6. Find the surface area. 113 ft²

Cumulative Review

Standardized Testing Practice

Choose the letter of the correct answer.

1.

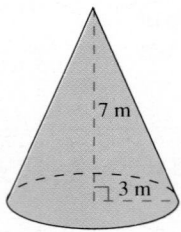

7 m

3 m

What is the volume of this cone?

A. 22 m³   B. 198 m³
C. 44 m³   D. 66 m³

2. Simplify: $\sqrt{25 + 144}$

A. 17   B. 13
C. 84.5   D. 14

3. Write $4y - x = 12$ in slope-intercept form.

A. $4y = x + 12$ B. $x = 4y - 12$

C. $y = \frac{1}{3}x - 4$ D. $y = \frac{1}{4}x + 3$

4. Solve: $r - 13 = -2$

A. $r = 15$ B. $r = -15$
C. $r = 11$ D. $r = -11$

5. Write 0.35 as a fraction.

A. 35%   B. $\frac{9}{20}$

C. $\frac{7}{20}$ D. $\frac{7}{10}$

6.

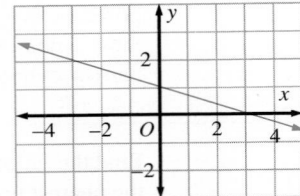

Which equation does this graph represent?

A. $y = -3x + 1$ B. $y = \frac{1}{3}x + 1$

C. $y = 3x + 1$ D. $y = -\frac{1}{3}x + 1$

7. Solve: $8.9 + 3a = 14.6$

A. $a = 7.8$ B. $a = 1.9$
C. $a = 17.1$ D. $a = 1.2$

8. Simplify: $12(3t - 5)$

A. $15t - 17$ B. $36t - 5$
C. $3t - 60$ D. $36t - 60$

9. The measure of an angle is 73°. What is the measure of its complement?

A. 17°   B. 27°
C. 107°   D. 117°

10. Solve: $-4w - 9 < -1$

A. $w < 2$ B. $w < -2$
C. $w > -2$ D. $w < -32$

11.

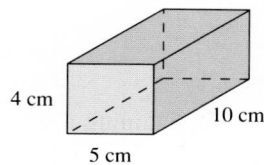

4 cm   10 cm   5 cm

What is the surface area of this prism?

A. 200 cm^2 B. 180 cm^2

C. 220 cm^2 D. 190 cm^2

12. Doris buys $13\frac{1}{2}$ ft of chain and uses $6\frac{7}{8}$ ft for a dog leash. Estimate the amount of chain she has left.

A. about 20 ft   B. about 2 ft

C. about 14 ft   D. about 7 ft

13. In a pictograph, one symbol represents 24 sheep. How many symbols are needed to represent 84 sheep?

A. $3\frac{1}{2}$ B. 60

C. 61   D. 7

14. The height of a cylinder is 8 in. and the radius is 6 in. Find the surface area of the cylinder.

A. 301.44 in.^{2} B. 527.52 in.^{2}

C. 414.48 in.^{2} D. 904.32 in.^{2}

15. The base of a pyramid is a square with sides of length 12 cm. The height of the pyramid is 8 cm. Find the volume of the pyramid.

A. 384 cm^3 B. 32 cm^3

C. 1152 cm^3 D. 128 cm^3

16.

−2   −1   0   1   2

Which inequality does this graph represent?

A. $m + 4 \geq 3$ B. $m - 1 > -2$

C. $m - 2 \geq -1$ D. $4m \leq -4$

17. The price of a carpet is discounted by 15%. The amount of the discount is $39.60. What is the original price of the carpet?

A. $264   B. $45.54

C. $5.94   D. $594

18. The radius of a sphere is 6 mm. What is the surface area of the sphere?

A. 904.32 mm^2 B. 452.16 mm^2

C. 75.36 mm^2 D. 113.04 mm^2

19. There are 3 teachers for every 50 students in a school. The school has 750 students. How many teachers are there at the school?

A. 125   B. 50

C. 703   D. 45

20. One letter from A through Z and one digit from 0 through 9 are selected. How many different outcomes are possible?

A. 234   B. 676

C. 260   D. 36

Solve by making a model.

7. A staircase is built with cubes and has a height of 3 cubes. How many cubes are needed to build the staircase and what is the total surface area of the cubes? Remember to count the faces on the bottom of the staircase.   6 cubes; 24 cubic units

Alternative Assessment

1. What figure is generated if $\triangle ABC$ is rotated about $\overline{BC}$? cone

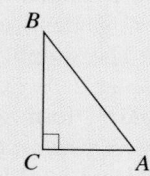

2. The areas of the faces of a rectangular prism are given. Find the volume.   160 ft^3

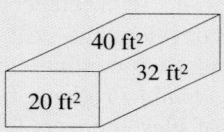

40 ft^2 32 ft^2 20 ft^2

3. Use two strings of equal length to form a circle and a square. Which has the greater area? circle

Planning Chapter 15

Chapter Overview

In this chapter, operations with polynomials are introduced and extended. Connections between mathematical topics are prominent. Concepts that have been introduced earlier are refined, extended, and examined from a more abstract point of view.

In *Lesson 1*, students are introduced to the vocabulary of polynomials. The simplification of variable expressions, introduced in Chapter 2, is extended to include more complex polynomials. Standard form gains importance.

The *Focus on Explorations (Modeling Polynomials Using Tiles)* serves as an introduction to *Lesson 2* on adding polynomials. The manipulative approach is used to illustrate both similarities and differences between terms. The function table in the exercises of *Lesson 2* connects the previously learned skills of evaluating variables, raising a term to a power, and performing operations with integers to polynomial expressions. *Lesson 3* develops the skill of subtracting polynomials.

Manipulatives continue to be used in the *Focus on Explorations (Modeling Products Using Tiles)* and in *Lessons 4 and 5*. The use of tiles connects the multiplication of binomials to the concept of area.

Lesson 6 focuses on the problem-solving strategy of working backward. In this lesson, students may use concepts and strategies from throughout the text.

Lesson 7 extends the concept of the greatest common factor to factoring polynomials. *Lesson 8* addresses division of polynomials. Both of these lessons build on concepts introduced earlier in the book.

When students have completed this chapter, they will have studied many algebraic operations and developed skills that will enable them to get a head start in a first course in algebra.

Background

Mathematical terminology relating to polynomials is used throughout Chapter 15. The language of mathematics is not used in isolation, but rather in the context of a student's personal experiences. Prefixes and meanings can be introduced by asking students to write a sentence containing a word with the same prefix. Students may be asked to define how the same word is used in several different sentences. Meanings change according to context. Rules involving operations with exponents also are re-examined.

Algebra tiles are used to introduce concepts visually. Examples are included that show how manipulatives can be used to illustrate operations with polynomial expressions.

Problem-solving strategies focus on the strategy of working backward. Logical reasoning skills help students relate and understand different concepts.

Communication skills are reinforced by asking students to explain and discuss key ideas and to interpret other information presented in the chapter. It is important also that students be given opportunities to explain their reasoning and how they arrived at a conclusion. Prior knowledge then becomes a meaningful resource.

Objectives

15-1	To simplify polynomials.
≡	**FOCUS ON EXPLORATIONS** To use algebra tiles to model polynomials.
15-2	To add polynomials.
15-3	To subtract polynomials.
≡	**FOCUS ON EXPLORATIONS** To use algebra tiles to model products of monomials and binomials.
15-4	To multiply a polynomial by a monomial.
15-5	To multiply binomials.
15-6	To solve problems by working backward.
15-7	To factor polynomials.
15-8	To divide a polynomial by a monomial.
≡	**CHAPTER EXTENSION** To multiply binomials mentally.

Chapter Planner

Instructional Aids	Manipulatives	Cooperative Learning	Technology	Practice/Reteaching	Assessment
Materials: algebra tiles, calculator *Resource Book:* Lesson Starters 135–142 *Visuals,* Folders Q, R	Lessons Focus on Explorations, 15-2, 15-3, Focus on Explorations, 15-5 *Activities Book:* Manipulative Activities 29–30	Lessons 15-1, 15-2, 15-3, 15-5, 15-6, 15-7 *Activities Book:* Cooperative Activity 27	Lessons 15-2, 15-3, 15-4, 15-5, 15-7, 15-8 *Activities Book:* Calculator Activity 15 Computer Activity 15 *Connections Plotter Plus* Disk	Toolbox, pp. 753–771 Extra Practice, p. 751 *Resource Book:* Practice 185–195 *Study Guide,* pp. 269–284	Self-Tests 1–2 Quick Quizzes 1–2 Chapter Test, p. 702 *Resource Book:* Lesson Starters 135–142 (Daily Quizzes) *Tests 74–77*

Assignment Guide Chapter 15

Day	Skills Course	Average Course	Advanced Course
1		**15-1:** 1–20, 21–33 odd, 41–44 **Exploration:** Activities I–III	**15-1:** 1–33 odd, 35–44 **Exploration:** Activities I–III
2		**15-2:** 1–26, 33–36	**15-2:** 1–27 odd, 28–36 **15-3:** 1–35 odd, 37–46, Challenge
3		**15-3:** 1–22, 23–35 odd, 43–46	**Exploration:** Activities I–II **15-4:** 1–27 odd, 29–44
4		**Exploration:** Activities I–II **15-4:** 1–22, 41–44	**15-5:** 1–19 odd, 21–42
5		**15-5:** 1–26, 39–42	**15-6:** 1–16, Historical Note
6			**15-7:** 1–29 odd, 31–40
7			**15-8:** 1–29 odd, 31–40
8			*Prepare for Chapter Test:* Chapter Review
9			*Administer Chapter 15 Test;* Chapter Extension; Cumulative Review

Teacher's Resources

Resource Book
  Chapter 15 Project
  Lesson Starters 135–142
  Practice 185–195
  Enrichment 149–157
  Diagram Masters
  Chapter 15 Objectives
  Family Involvement 15
  Spanish Test 15
  Spanish Glossary

Activities Book
  Cooperative Activity 27
  Manipulative Activities 29–30
  Calculator Activity 15
  Computer Activity 15

Study Guide, pp. 269–284

Tests, Tests 74–77

Visuals, Folders Q, R

Connections Plotter Plus Disk

Alternate Approaches Chapter 15

Using Technology

CALCULATORS
Since most calculators are useful for numerical calculations rather than symbolic manipulations, they cannot be used directly to simplify, combine, or factor polynomials. They can be used, however, to evaluate polynomials and thereby check the validity of a simplification made algebraically.

COMPUTERS
Like calculators, most computer programs, such as spreadsheets, operate on numbers rather than on the combinations of variables and numbers used in polynomials. However, there are symbol manipulating programs available for computer systems (for example, *Mathematica* for Macintosh computers, and *MuMath* for IBM machines). This software will simplify, combine, factor, or solve just about any algebraic expression or equation. The programs, if available, can provide a means for students to discover some of the principles of polynomial manipulation that appear in the text.

Lessons 15-2, 15-3, 15-4

CALCULATORS
Calculators can be used in these lessons to verify problem solutions experimentally and thus the underlying algebraic ideas. For example, students can make a function table for each of the polynomials $n^2 + 5n + 2$, $3n^2 + 2n + 7$, and for their sum, $4n^2 + 7n + 9$. By choosing several values of n, they can gain confidence in their solution and in the general principle involved. This technique can also generate an interesting discussion of inductive versus deductive reasoning.

COMPUTERS
A spreadsheet or BASIC program can be used to implement the idea mentioned above for calculators. In addition, by using columns of cells to hold the coefficients of different powers of n, students can create a spreadsheet that will add or subtract two polynomials of a single variable by directly manipulating the appropriate coefficients.

Lesson 15-5

COMPUTERS
As an extension to the spreadsheet idea for adding polynomials, students can design a spreadsheet to multiply binomials. To do this, they can reserve a cell for each coefficient, then design a formula for each cell of the resulting trinomial. Such a spreadsheet might look as follows. (The formulas, of course, do not actually appear in the cells' position on the screen; their values appear instead.)

	A	B	C	D
		N^2	N	CONSTANT
1				
2	#1:	0	2	3
3	#2:	0	1	−1
4	RESULT:	+C2*C3	+C2*D3+C3*D2	+D2*D3

Lessons 15-7, 15-8

CALCULATORS AND COMPUTERS
As suggested above for other operations, a calculator or spreadsheet can be used to verify factoring and division solutions by calculating each expression for a number of chosen values of the variable.

Using Manipulatives

Lesson 15-2

You can use algebra tiles to illustrate addition of polynomials with negative terms. Different colored tiles with the same shape can be used to represent opposites. For example, if a light tile represents n^2, then a dark tile of the same size would represent $-n^2$. The zero principle can be used to combine a light tile and a dark tile of the same size.

Consider, for example, the polynomial $x^2 - 2x - 3$. This can be represented by using algebra tiles as follows.

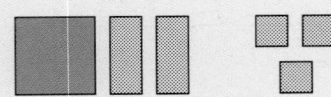

666C

Now, suppose you want to add $x^2 - 2x - 3$ to the polynomial $-x^2 + 3x - 4$. You can build a model to represent the problem.

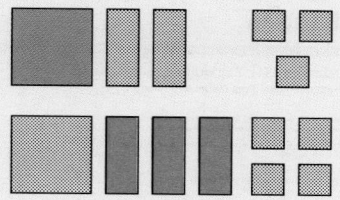

The answer is found by combining light and dark tiles using the zero principle; that is, group zero pairs.

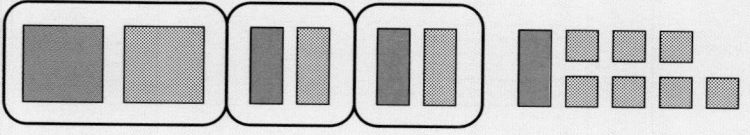

The answer is $x - 7$.

Lesson 15-3

The subtraction of polynomials is very similar to addition. Again, the zero principle can be used. For example, consider $(2x^2 + x - 5) - (x^2 - 2x + 3)$. Subtraction is the same as "taking away" tiles.
Build a model for $2x^2 + x - 5$.

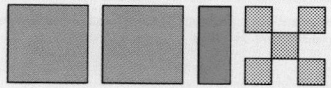

In order to remove the quantity $x^2 - 2x + 3$ from the model, you need to have two $-x$ tiles and three positive unit tiles. Use the zero principle to add these tiles to the model without changing its value. Arrows can indicate which tiles are being removed.

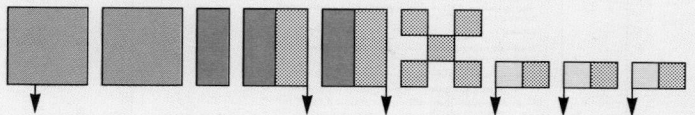

The answer is $x^2 + 3x - 8$.

Reteaching/Alternate Approach

Lesson 15-1

Students may enjoy doing a class activity on simplifying polynomials. Give each student an index card. On the top half of the card have each student write "Who can simplify" followed by a polynomial expression. On the bottom half of the card, have students write "I have" followed by the correct solution. Next cut the cards in half. Shuffle the two decks. Have one student distribute the two decks so that each student has an "I have ..." card and a "Who can ..." card. Any student can be chosen to begin the game by reading a "Who can ..." card aloud. Play continues until the player who asked the original question gets to give an answer.

Lesson 15-5

As an alternate method of multiplying binomials, show students the FOIL method. For example, you can multiply $(2x + 1)(x - 5)$ as follows.

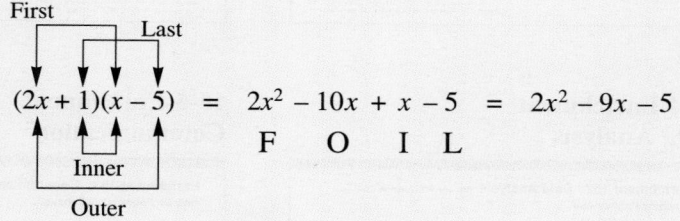

$(2x + 1)(x - 5) = 2x^2 - 10x + x - 5 = 2x^2 - 9x - 5$

F O I L

COOPERATIVE LEARNING

Give students working in groups a large square and a small square cut from construction paper and some extra sheets of the paper. Have each group create a rectangle from the construction paper whose length is that of the large square and whose width is that of the small square. The groups should then cut out more squares of each size. Each group should then make a new square or rectangular shape that uses as many of the squares and rectangles that they have cut out as possible. The groups also should determine the polynomial multiplications that their shapes represent.

Enrichment masters from the Resource Book are pictured here. See the Teacher's Resources chart on page 666B for a complete listing of all materials available for this chapter.

15-1 Enrichment
Application

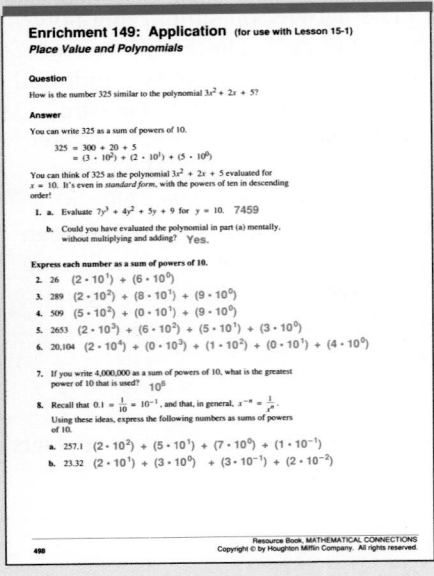

Enrichment 149: Application (for use with Lesson 15-1)
Place Value and Polynomials

Question
How is the number 325 similar to the polynomial $3x^2 + 2x + 5$?

Answer
You can write 325 as a sum of powers of 10.

$325 = 300 + 20 + 5$
$= (3 \cdot 10^2) + (2 \cdot 10^1) + (5 \cdot 10^0)$

You can think of 325 as the polynomial $3x^2 + 2x + 5$ evaluated for $x = 10$. It's even in *standard form*, with the powers of ten in descending order!

1. **a.** Evaluate $7y^3 + 4y^2 + 5y + 9$ for $y = 10$. 7459
   b. Could you have evaluated the polynomial in part (a) mentally, without multiplying and adding? Yes.

Express each number as a sum of powers of 10.

2. 26 $(2 \cdot 10^1) + (6 \cdot 10^0)$
3. 289 $(2 \cdot 10^2) + (8 \cdot 10^1) + (9 \cdot 10^0)$
4. 509 $(5 \cdot 10^2) + (0 \cdot 10^1) + (9 \cdot 10^0)$
5. 2653 $(2 \cdot 10^3) + (6 \cdot 10^2) + (5 \cdot 10^1) + (3 \cdot 10^0)$
6. 20,104 $(2 \cdot 10^4) + (0 \cdot 10^3) + (1 \cdot 10^2) + (0 \cdot 10^1) + (4 \cdot 10^0)$

7. If you write 4,000,000 as a sum of powers of 10, what is the greatest power of 10 that is used? 10^6

8. Recall that $0.1 = \frac{1}{10} = 10^{-1}$, and that, in general, $x^{-n} = \frac{1}{x^n}$. Using these ideas, express the following numbers as sums of powers of 10.
   a. 257.1 $(2 \cdot 10^2) + (5 \cdot 10^1) + (7 \cdot 10^0) + (1 \cdot 10^{-1})$
   b. 23.32 $(2 \cdot 10^1) + (3 \cdot 10^0) + (3 \cdot 10^{-1}) + (2 \cdot 10^{-2})$

15-2 Enrichment
Thinking Skills

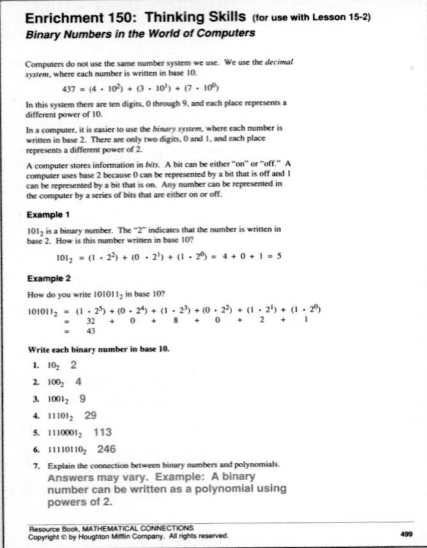

Enrichment 150: Thinking Skills (for use with Lesson 15-2)
Binary Numbers in the World of Computers

Computers do not use the same number system we use. We use the *decimal system*, where each number is written in base 10.

$437 = (4 \cdot 10^2) + (3 \cdot 10^1) + (7 \cdot 10^0)$

In this system there are ten digits, 0 through 9, and each place represents a different power of 10.

In a computer, it is easier to use the *binary system*, where each number is written in base 2. There are only two digits, 0 and 1, and each place represents a different power of 2.

A computer stores information in *bits*. A bit can be either "on" or "off." A computer uses base 2 because 0 can be represented by a bit that is off and 1 can be represented by a bit that is on. Any number can be represented in the computer by a series of bits that are either on or off.

Example 1
101_2 is a binary number. The "2" indicates that the number is written in base 2. How is this number written in base 10?

$101_2 = (1 \cdot 2^2) + (0 \cdot 2^1) + (1 \cdot 2^0) = 4 + 0 + 1 = 5$

Example 2
How do you write 101011_2 in base 10?

$101011_2 = (1 \cdot 2^5) + (0 \cdot 2^4) + (1 \cdot 2^3) + (0 \cdot 2^2) + (1 \cdot 2^1) + (1 \cdot 2^0)$
$= 32 + 0 + 8 + 0 + 2 + 1$
$= 43$

Write each binary number in base 10.

1. 10_2 2
2. 100_2 4
3. 1001_2 9
4. 11101_2 29
5. 1110001_2 113
6. 11101110_2 246

7. Explain the connection between binary numbers and polynomials. Answers may vary. Example: A binary number can be written as a polynomial using powers of 2.

15-3 Enrichment
Exploration

Enrichment 151: Exploration (for use with Lesson 15-3)
Changing Numbers from Decimal to Binary

Computers use the binary system. Binary numbers are sums of powers of 2 in the same way that decimal numbers are sums of powers of 10.

A binary number is a string of 1's and 0's, each indicating the presence or absence of a power of 2.

1. Complete the following table of the powers of 2.

2^0	2^1	2^2	2^3	2^4	2^5	2^6	2^7	2^8	2^9
1	2	4	8	16	32	64	128	256	512

Example
Express the decimal number 11 as a binary number.

Solution
Step 1: Use the table above to find the greatest power of 2 that is less than or equal to 11: $8 = 2^3$
Step 2: Subtract that power of 2 from 11: $11 - 8 = 3$
Repeat Steps 1 and 2 for the number 3.
Step 1: $2 = 2^1$ is the greatest power of 2 that is less than or equal to 3.
Step 2: Subtract that power of 2 from 3: $3 - 2 = 1$
Repeat Steps 1 and 2 for the number 1.
Step 1: $1 = 2^0$ is the largest power of 2 that is less than or equal to 1.
Step 2: Subtract that power of 2 from 1: $1 - 1 = 0$
So, $11 = 8 + 2 + 1$
$= (1 \cdot 2^3) + (0 \cdot 2^2) + (1 \cdot 2^1) + (1 \cdot 2^0)$
$= 1011_2$

Use the method above to express the following decimal numbers as binary numbers.

2. 8 1000_2
3. 9 1001_2
4. 10 1010_2
5. 20 10100_2
6. 30 11110_2
7. 40 101000_2
8. 241 11110001_2
9. 451 111000011_2
10. 635 1001111011_2

11. In the example above, you found that $11 = 1011_2$. Without substituting and evaluating, find the value of $x^3 + x + 1$ when $x = 2$. (*Hint:* How is a binary number like a polynomial?) 11

15-4 Enrichment
Data Analysis

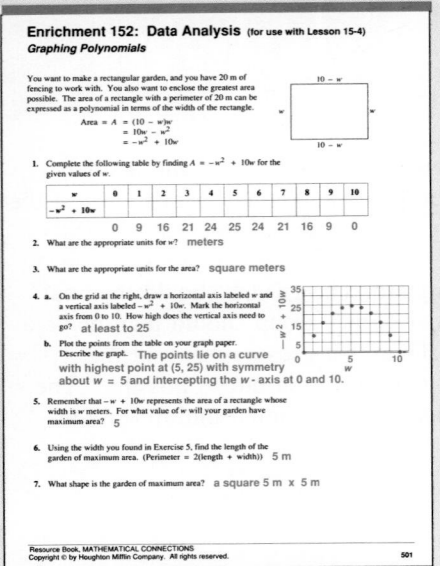

Enrichment 152: Data Analysis (for use with Lesson 15-4)
Graphing Polynomials

You want to make a rectangular garden, and you have 20 m of fencing to work with. You also want to enclose the greatest area possible. The area of a rectangle with a perimeter of 20 m can be expressed as a polynomial in terms of the width of the rectangle.

$Area = A = (10 - w)w$
$= 10w - w^2$
$= -w^2 + 10w$

1. Complete the following table by finding $A = -w^2 + 10w$ for the given values of w.

w	0	1	2	3	4	5	6	7	8	9	10
$-w^2 + 10w$	0	9	16	21	24	25	24	21	16	9	0

2. What are the appropriate units for w? meters

3. What are the appropriate units for the area? square meters

4. **a.** On the grid at the right, draw a horizontal axis labeled w and a vertical axis labeled $-w^2 + 10w$. Mark the horizontal axis from 0 to 10. How high does the vertical axis need to go? at least to 25
   b. Plot the points from the table on your graph paper. Describe the graph. The points lie on a curve with highest point at (5, 25) with symmetry about $w = 5$ and intercepting the w-axis at 0 and 10.

5. Remember that $-w^2 + 10w$ represents the area of a rectangle whose width is w meters. For what value of w will your garden have maximum area? 5

6. Using the width you found in Exercise 5, find the length of the garden of maximum area. (Perimeter = 2(length + width)) 5 m

7. What shape is the garden of maximum area? a square 5 m × 5 m

15-5 Enrichment
Communication

Enrichment 153: Communication (for use with Lesson 15-5)
How Do Polynomials Behave?

Consider the product $(x + 2)(x - 5)$. Multiplying, you get

$(x + 2)(x - 5) = x^2 - 3x - 10$.

How does this polynomial behave? That is, for what values of x is the value of the polynomial positive, negative, or zero?

1. Find the value of $(x + 2)(x - 5)$ for each given value of x.
   a. $x = 10$ 60
   b. $x = -10$ 120
   c. $x = -2$ 0
   d. $x = 5$ 0

2. **a.** Pick a few values for x between -2 and 5. What is the value of $(x + 2)(x - 5)$ for each of these x-values? Answers will vary. Each answer should be negative.
   b. Is the value of the polynomial positive or negative when x is between -2 and 5? negative

3. **a.** Try some values of x that are less than -2. Is the value of the polynomial positive or negative when x is less than -2? positive
   b. Try some values of x that are greater than 5? Is the value of the polynomial positive or negative? positive

4. Explain what happens to the value of $(x + 2)(x - 5)$ as x varies from large negative values, through values of x close to 0, and up to large positive values of x. For $x < -2$, the value of the polynomial is positive; for $-2 < x < 5$, the value of the polynomial is negative; for $x > 5$, it's positive; it is zero when $x = -2$ or 5.

5. **a.** Consider the polynomial product $(x + 6)(x + 3)(x - 2)$. What values of x will this polynomial be equal to 0? $x = -6, -3, $ or 2
   b. Make a table of the polynomial for x-values -8 through 5.

-8	-7	-6	-5	-4	-3	-2	-1	0	1	2	3	4	5
-100	-36	0	14	12	0	-16	-30	-36	-28	0	54	140	264

6. **a.** Describe the behavior of $(x + 6)(x + 3)(x - 2)$ for all values of x. For $x < -6$, it's negative; for $-3 < x < 2$, it's negative; for $x = -6, -3, 2$, it's zero; for $-6 < x < -3$, it's positive; for $x > 2$, it's positive.
   b. Why is it easier to describe the behavior of a polynomial if you have the polynomial in factored form? Because you can see where it is equal to zero.

15-6 Enrichment
Problem Solving

Enrichment 154: Problem Solving (for use with Lesson 15-6)
Working Backward: How Many Ancestors Do You Have?

Problem 1
Kate has a portrait of her great-great-great-great-grandmother on her wall. Kate knows, though, that she must have had more than one great-great-great-great-grandmother. How many others did she have?

Solution
1. How many grandmothers did she have? 2
2. How many great-grandmothers? 4
3. Great-great-grandmothers? 8
4. Great-great-great-grandmothers? 16
5. Great-great-great-great-grandmothers? 32
6. How many portraits of great-great-great-great-grandmothers is Kate missing? 31

Kate's great-great-great-great-grandmother was born in 1812. Kate was born in 1977. That makes seven generations in 165 years. So there were about 25 years to a generation in Kate's family.

Problem 2
Suppose there have been 25 years per generation for several centuries. About how many of Kate's ancestors (male and female) were born around the time of Columbus?

Solution
To solve this problem, you need to work backward from the current generation to the time of Columbus, about 500 years ago.

7. **a.** Go back 1 generation. How many parents did Kate have? 2
   b. Go back 2 generations. How many grandparents did Kate have? 4
   c. What happens to the number of ancestors as you go back each generation? The number of ancestors doubles.

8. Approximately how many generations have there been in Kate's family in 500 years? 20

9. Give an estimate for the number of Kate's ancestors born around the time of Columbus. 2^{20} or 1,048,576

10. Describe some of the hazards in this kind of estimation. How accurate can you expect to be? Are the numbers surprising? Answers will vary. Example: A small error in the number of generations will double or halve the estimate; the generations may have been of different lengths.

15-7 Enrichment Connection

15-8 Enrichment Connection

End of Chapter Enrichment Extension

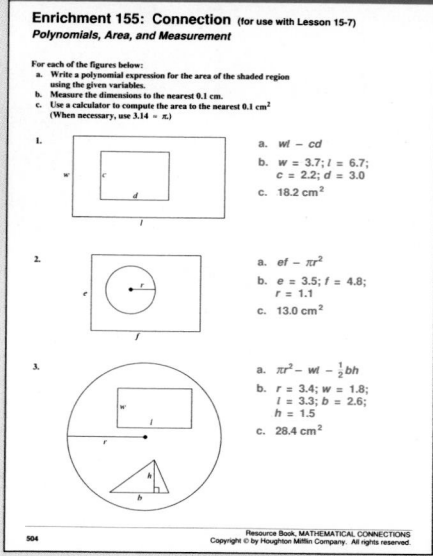

Enrichment 155: Connection (for use with Lesson 15-7)
Polynomials, Area, and Measurement

For each of the figures below:
a. Write a polynomial expression for the area of the shaded region using the given variables.
b. Measure the dimensions to the nearest 0.1 cm.
c. Use a calculator to compute the area to the nearest 0.1 cm² (When necessary, use 3.14 ≈ π.)

1.
  a. $wl - cd$
  b. $w = 3.7; l = 6.7;$ $c = 2.2; d = 3.0$
  c. 18.2 cm²

2.
  a. $ef - \pi r^2$
  b. $e = 3.5; f = 4.8;$ $r = 1.1$
  c. 13.0 cm²

3.
  a. $\pi r^2 - wl - \frac{1}{2}bh$
  b. $r = 3.4; w = 1.8;$ $l = 3.3; b = 2.6;$ $h = 1.5$
  c. 28.4 cm²

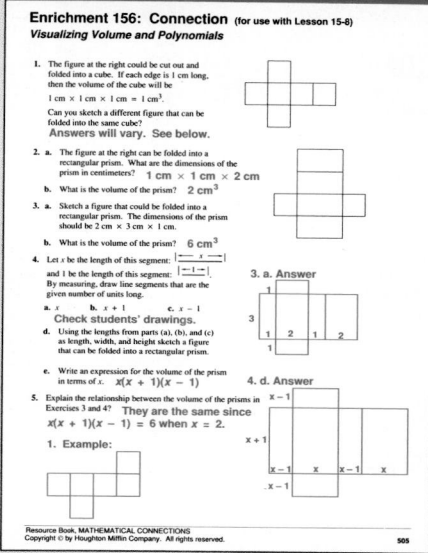

Enrichment 156: Connection (for use with Lesson 15-8)
Visualizing Volume and Polynomials

1. The figure at the right could be cut out and folded into a cube. If each edge is 1 cm long, then the volume of the cube will be
$$1 \text{ cm} \times 1 \text{ cm} \times 1 \text{ cm} = 1 \text{ cm}^3.$$
Can you sketch a different figure that can be folded into the same cube?
Answers will vary. See below.

2. a. The figure at the right can be folded into a rectangular prism. What are the dimensions of the prism in centimeters? 1 cm × 1 cm × 2 cm
   b. What is the volume of the prism? 2 cm³

3. a. Sketch a figure that could be folded into a rectangular prism. The dimensions of the prism should be 2 cm × 3 cm × 1 cm. 6 cm³
   b. What is the volume of the prism? 6 cm³

4. Let x be the length of this segment: |— x —| and 1 be the length of this segment: |—1—|. By measuring, draw line segments that are the given number of units long.
   a. x b. $x + 1$ c. $x - 1$
   Check students' drawings.

   d. Using the lengths from parts (a), (b), and (c) as length, width, and height sketch a figure that can be folded into a rectangular prism.

   e. Write an expression for the volume of the prism in terms of x. $x(x + 1)(x - 1)$

5. Explain the relationship between the volume of the prisms in Exercises 3 and 4? **They are the same since** $x(x + 1)(x - 1) = 6$ **when** $x = 2.$

1. Example:

3. a. Answer

4. d. Answer

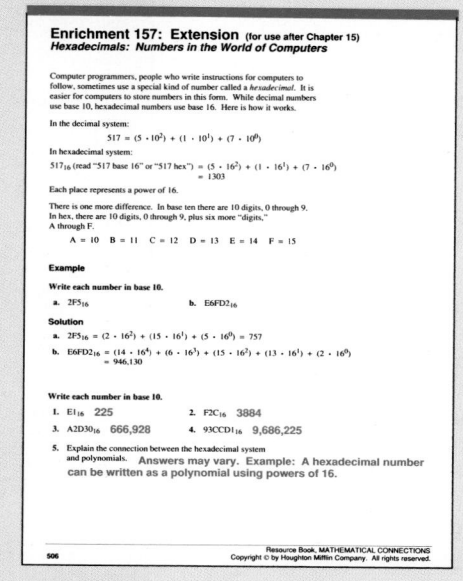

Enrichment 157: Extension (for use after Chapter 15)
Hexadecimals: Numbers in the World of Computers

Computer programmers, people who write instructions for computers to follow, sometimes use a special kind of number called a *hexadecimal*. It is easier for computers to store numbers in this form. While decimal numbers use base 10, hexadecimal numbers use base 16. Here is how it works.

In the decimal system:
$$517 = (5 \cdot 10^2) + (1 \cdot 10^1) + (7 \cdot 10^0)$$

In hexadecimal system:
517_{16} (read "517 base 16" or "517 hex") $= (5 \cdot 16^2) + (1 \cdot 16^1) + (7 \cdot 16^0)$
$= 1303$

Each place represents a power of 16.

There is one more difference. In base ten there are 10 digits, 0 through 9. In hex, there are 10 digits, 0 through 9, plus six more "digits," A through F.
$$A = 10 \quad B = 11 \quad C = 12 \quad D = 13 \quad E = 14 \quad F = 15$$

Example

Write each number in base 10.
a. $2F5_{16}$ b. $E6FD2_{16}$

Solution
a. $2F5_{16} = (2 \cdot 16^2) + (15 \cdot 16^1) + (5 \cdot 16^0) = 757$
b. $E6FD2_{16} = (14 \cdot 16^4) + (6 \cdot 16^3) + (15 \cdot 16^2) + (13 \cdot 16^1) + (2 \cdot 16^0)$
$= 946,130$

Write each number in base 10.
1. $E1_{16}$ 225   2. $F2C_{16}$ 3884
3. $A2D30_{16}$ 666,928   4. $93CCD1_{16}$ 9,686,225

5. Explain the connection between the hexadecimal system and polynomials. **Answers may vary. Example: A hexadecimal number can be written as a polynomial using powers of 16.**

First Nine Rows of Pascal's Triangle

666

Polynomials 15

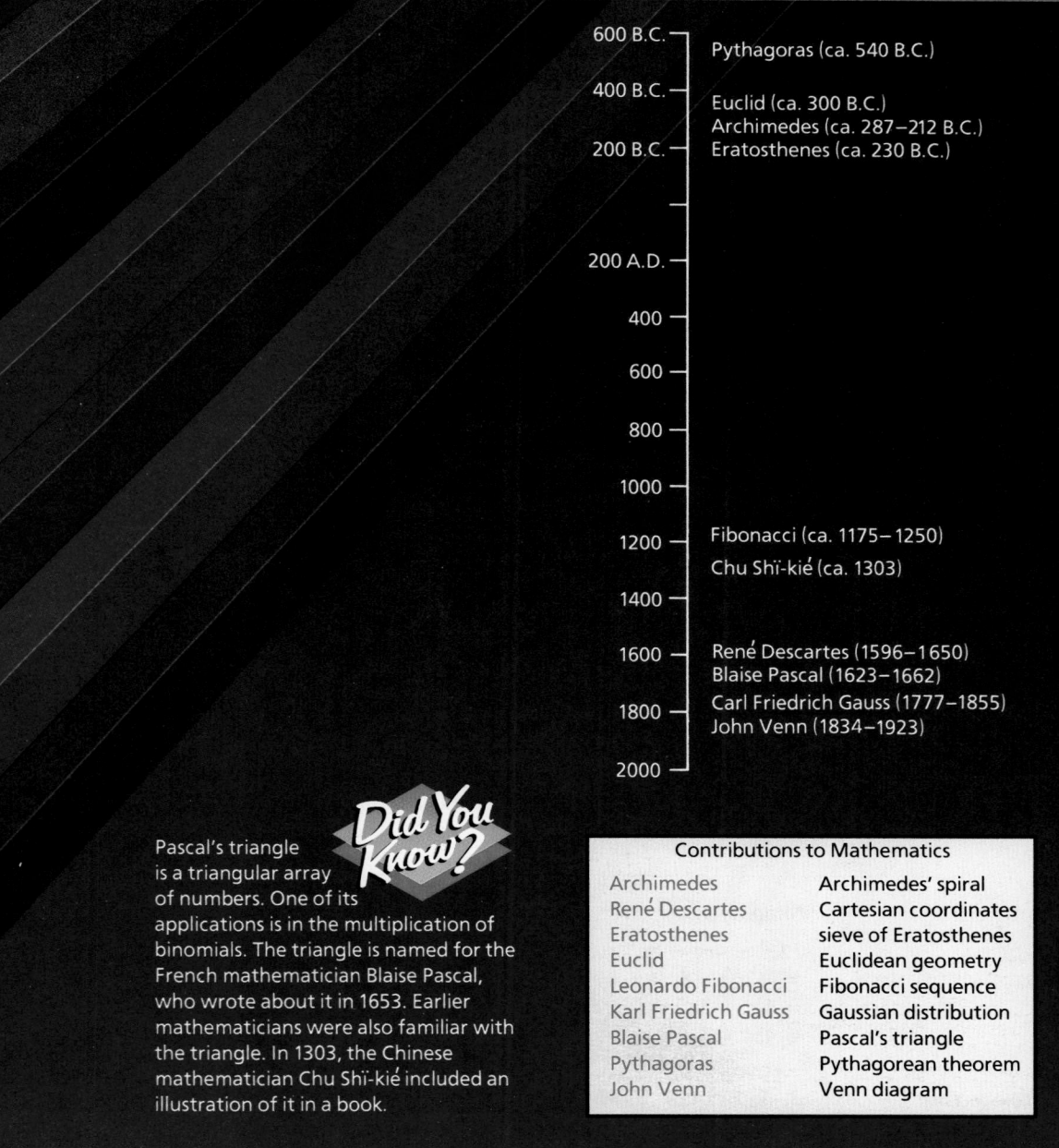

600 B.C. —	Pythagoras (ca. 540 B.C.)
400 B.C. —	Euclid (ca. 300 B.C.)
	Archimedes (ca. 287–212 B.C.)
200 B.C. —	Eratosthenes (ca. 230 B.C.)
200 A.D. —	
400 —	
600 —	
800 —	
1000 —	
1200 —	Fibonacci (ca. 1175–1250)
	Chu Shï-kié (ca. 1303)
1400 —	
1600 —	René Descartes (1596–1650)
	Blaise Pascal (1623–1662)
1800 —	Carl Friedrich Gauss (1777–1855)
	John Venn (1834–1923)
2000 —	

Did You Know?

Pascal's triangle is a triangular array of numbers. One of its applications is in the multiplication of binomials. The triangle is named for the French mathematician Blaise Pascal, who wrote about it in 1653. Earlier mathematicians were also familiar with the triangle. In 1303, the Chinese mathematician Chu Shï-kié included an illustration of it in a book.

Contributions to Mathematics

Archimedes	Archimedes' spiral
René Descartes	Cartesian coordinates
Eratosthenes	sieve of Eratosthenes
Euclid	Euclidean geometry
Leonardo Fibonacci	Fibonacci sequence
Karl Friedrich Gauss	Gaussian distribution
Blaise Pascal	Pascal's triangle
Pythagoras	Pythagorean theorem
John Venn	Venn diagram

667

Research Activities
(continued)

For Page 688

In Exercise 36, students are asked to find out about Blaise Pascal, whose achievements include the invention in 1642 of the first computing machine. You may want to tell students that one of the most widely used computer languages is named after Pascal, and ask them to find out what role Pascal played in the development of computers.

Suggested Resources

Eves, Howard, *An Introduction to the History of Mathematics*. 5th edition. New York: Holt, Rinehart and Winston, 1983. (For Teacher)

Yen, Li, and Dù, Shírán. *Chinese Mathematics: A Concise History*. Oxford, England: Oxford University Press, 1987. (For Teacher)

Teaching the Lesson
- Materials: index cards
- Lesson Starter 135
- Reteaching/Alternate Approach, p. 666D

Lesson Follow-Up
- Practice 185
- Enrichment 149: Application
- Study Guide, pp. 269–270

Warm-Up Exercises

Simplify.

1. $2x - 3y + (-7x)$ $-5x - 3y$
2. $3a + 4b - 4a - (-3b)$
   $-a + 7b$
3. $-11m - (-2m)$ $-9m$
4. $3cd + 4c - 2d + 5cd$
   $8cd + 4c - 2d$
5. $5n^2 - 8n^2 + 9n^2$ $6n^2$

Lesson Focus

Remind students of their earlier work with variable expressions, pointing out the importance of these expressions in algebra. Today's lesson focuses on simplifying polynomials.

Teaching Notes

Stress that monomials, binomials, and trinomials are all polynomials. Also point out to students the similarity between earlier work with variable expressions and the current work with polynomials.

Key Question

Why is it useful to write polynomials in standard form? The use of a consistent form makes it easier to compare results and to see which terms are included or omitted.

15-1

Simplifying Polynomials

Objective: To simplify polynomials.

Terms to Know
- *polynomial*
- *monomial*
- *binomial*
- *trinomial*
- *standard form*

CONNECTION

Since Chapter 1, you have been writing variable expressions and simplifying some expressions. In this lesson you will simplify *polynomials*. Each expression below is a **polynomial,** a variable expression consisting of one or more *terms*.

$$3x^2 \qquad -4t \qquad 2a^2 - 3ab + 2b^2 \qquad x^3 - 1$$

Some polynomials have special names.

A **monomial** has one term. *Examples*: $3x^2$ and $-4t$
A **binomial** has two terms. *Example*: $x^3 - 1$
A **trinomial** has three terms. *Example*: $2a^2 - 3ab + 2b^2$

When you are working with a polynomial, it is often helpful to write the polynomial in **standard form.** To do this, write the terms in order from the highest to the lowest power of one of the variables.

Example 1

Write each polynomial in standard form.

a. $4x^2 + x - 3x^3$

b. $8c^3 + 7 - 9c + 2c^4$

Solution

a. $4x^2 + x - 3x^3$
   $= -3x^3 + 4x^2 + x$

b. $8c^3 + 7 - 9c + 2c^4$
   $= 2c^4 + 8c^3 - 9c + 7$

 **Check Your Understanding**

1. In Example 1(a) and in Example 1(b), list the powers in order.
2. In Example 1(b), why is 7 the last term?
   See *Answers to Check Your Understanding* at the back of the book.

Like terms have the same variables raised to the same powers. To *simplify* a polynomial, you combine like terms and write the resulting polynomial in standard form.

Example 2

Simplify $12c^3 - 4c^2 - 8c^3 - 5 + 7c^2 - 4c$.

Solution

$12c^3 - 4c^2 - 8c^3 - 5 + 7c^2 - 4c$

$= (12c^3 - 8c^3) + (-4c^2 + 7c^2) - 5 - 4c$ ⟵ Group like terms.

$= 4c^3 + 3c^2 - 5 - 4c$ ⟵ Combine like terms.

$= 4c^3 + 3c^2 - 4c - 5$ ⟵ Write in standard notation.

Check Your Understanding

3. In Example 2, why does $4c^3$ come before $3c^2$?
4. In Example 2, why are $4c^3$, $3c^2$, and $4c$ not combined?
   See *Answers to Check Your Understanding* at the back of the book.

---

668   Chapter 15

Guided Practice

COMMUNICATION « *Reading*

Replace each __?__ with the correct word or expression.

1. polynomial;
monomial; binomial;
trinomial

2. standard;
combine like terms

« **1.** A __?__ is a variable expression with one or more terms. A __?__ has one term, a __?__ has two terms, and a __?__ has three terms.

« **2.** A polynomial is in __?__ form when its terms are arranged in order from highest to lowest powers. To simplify a polynomial, you __?__ and write the resulting polynomial in standard form.

Is the polynomial a *monomial*, a *binomial*, or a *trinomial*?

3. $ab + 3$
binomial

4. $x + y - 2xy$
trinomial

5. 5
monomial

6. $-t^6 + s^4$
binomial

Tell whether the terms are *like terms* or *unlike terms*.

7. $3m^3, 5m^3$
like terms

8. $7x^4, 4x^7$
unlike terms

9. xy^3, xy
unlike terms

10. $3ab^2, 5ab^2$
like terms

Write each polynomial in standard form.

11. $4g^4 + 3g^3 - 7g^2 - 3g + 8$

12. $-8k^4 - 9k^3 + 7k^2 + 4k$

11. $3g^3 + 4g^4 - 3g + 8 - 7g^2$

12. $4k - 8k^4 + 7k^2 - 9k^3$

13. $5a + 8a^7 - 2a^3 + 9a^5 - 6$
$8a^7 + 9a^5 - 2a^3 + 5a - 6$

14. $7x^3 + 2x - 5x^8 + 9 - x^5$
$-5x^8 - x^5 + 7x^3 + 2x + 9$

Simplify.

15. $5x^3 + 6x - 2x^3 + 8x - x^2 + 5$ $3x^3 - x^2 + 14x + 5$

16. $7c^2 + 8c + 2c^2 - 9c^3 - 5c - 7$ $-9c^3 + 9c^2 + 3c - 7$

17. $4 - 5a^3 - 2a^2 + 5a - 4a^3 + 8a^2$ $-9a^3 + 6a^2 + 5a + 4$

18. $3n - 7 + 8n^2 + 5n^3 - 3n^2 - 8n$ $5n^3 + 5n^2 - 5n - 7$

Exercises

5. $3z^4 + 8z^3 + 4z^2 - 6z - 7$

6. $2a^4 - 7a^3 + 2a^2 - 13a - 3$

8. $-z^7 - z^5 + 5z^3 + 2z$

Write each polynomial in standard form.

A

1. $x^2 - 2 + x$ $x^2 + x - 2$

2. $5 - 3a^2 + 6a$ $-3a^2 + 6a + 5$

3. $m - 3 + m^3 + 3m^2$
$m^3 + 3m^2 + m - 3$

4. $-x^2 + 8x - 1 + x^3$
$x^3 - x^2 + 8x - 1$

5. $4z^2 - 6z + 3z^4 - 7 + 8z^3$

6. $2a^4 - 3 + 2a^2 - 13a - 7a^3$

7. $c^3 - 5c - 2 + c^4$ $c^4 + c^3 - 5c - 2$

8. $-z^7 + 5z^3 + 2z - z^5$

9. $k^2 - 8k^4 - 6 + 3k^3$
$-8k^4 + 3k^3 + k^2 - 6$

10. $n^3 - 1 + 2n^6 + 4n^4 + 3n^7$
$3n^7 + 2n^6 + 4n^4 + n^3 - 1$

Simplify. Answers to Exercises 11–16 are on p. A40.

11. $2x^2 + x + 2 + 3x - x^2 + 5$

12. $-9d^3 - 4d^2 - d^3 + 4 - 2d^2$

13. $4e - 3e^2 + 4e^2 + 7e - 20$

14. $r^3 - 7r^2 - 6r^3 - 5 + 7$

15. $x^2 + 6x + 7 - 5x^2 + 5 + 3x$

16. $2w^3 - 6w^2 + 7w^3 - 7$

Polynomials   **669**

Reteaching/Alternate Approach

For a suggestion on using a class activity to simplify polynomials, see page 666D.

Error Analysis

In combining like terms, students may incorrectly combine terms that have the same variables but different powers. Remind students that like terms have the *same* variables raised to the *same* powers.

Additional Examples

1. Write $9x^3 - 8 + 7x + 2x^4$ in standard form.
$2x^4 + 9x^3 + 7x - 8$

2. Simplify $14m^3 - 5m^2 - 7m^3 - 6 + 8m^2 - 4m$.
$7m^3 + 3m^2 - 4m - 6$

Closing the Lesson

Have each student write a constant or a term using the variable n raised to some power on an index card. Place these cards in a box. Draw out three to five of the cards and place addition or subtraction signs between the terms on the cards. Ask students to write the resulting expression in standard form.

Suggested Assignments

Average: 1–20, 21–33 odd, 41–44
Advanced: 1–33 odd, 35–44

Exercise Notes

Communication: Reading

Guided Practice Exercises 1 and 2 require students to think about the terms found in the lesson and to use these terms correctly by replacing each blank with the appropriate response.

Critical Thinking

Exercises 39 and 40 require students to *analyze* the concept of standard form for polynomials of more than one variable.

Cooperative Learning

Exercise 40 can be worked on by students in groups. Interaction among students can facilitate understanding of the concept of standard form.

Follow-Up

Extension

Consider the polynomial
$$x^4 + x^3 + x^2 + x.$$
For what values of the variable will the terms be written in

a. increasing value? $0 < x < 1$

b. decreasing value? $x \geq 1$

Enrichment

Enrichment 149 is pictured on page 666E.

Practice

Practice 185, *Resource Book*

Practice 185
Skills and Applications of Lesson 15-1

Simplify. Answers to Exercises 17–34 are on p. A40.

17. $-3x^2 - 6x - 2 - 5x - 4x^2 + 3$ **18.** $2g^2 - 4 + 7g + 5g^2 + 1$

19. $8w^3 + 6w^2 - 12w - 2w^2 + 1$ **20.** $-x^3 + 9x - 3x^3 - 6x^2 + 2$

B **21.** $x^2 - 2x + 1 + 3x^2 + 4x + x^5$ **22.** $-x^2 + 8 + 3x^4 - 9x + x^4$

23. $2a^4 - 6a^3 - 2a^2 - 4a^3 - 5a^2$ **24.** $x^3 - x^2 + 7 + 3x^3 - 11 + x^2$

25. $-x^3 + 2x - 3 - 4x^3 - 2x + 3$ **26.** $6 - 2a^4 + a^2 - 1 + 6a^4 - a^2$

Write each polynomial in standard form for the variable in color.

27. $a^2b + ab^3 + a^2b^2 + 4; b$

28. $m^2n^3 - mn + mn^4 + m^2n^2 - 1; n$

29. $x^3y^3 - 4 + x^2y - xy^2; y$

30. $c^4d - cd^4 + c^2d^2 + 6 - c^3d^3; d$

31. $a^3b^2 + ab^3 - a^2b - 4; a$ **32.** $xz^3 - x^3z + 9 - x^2z^2; x$

33. $a^4c^2 + a^3c^4 - a^2c + ac^3; c$ **34.** $mn^4 - m^2n^3 - m^3n^2 + m^4n; m$

CONNECTING MATHEMATICS AND LANGUAGE ARTS

Explain what the prefix of the word means. For example, the prefix *uni-* in *unicorn* means *one*. Then give an everyday word with the same prefix as the given word. Everyday words may vary. Examples are given.

35. polynomial many; polyester

36. monomial one; monotone

37. binomial two; bicycle

38. trinomial three; tricycle

THINKING SKILLS

39. Application
40. Synthesis
Answers to Exercises 39 and 40 are on p. A40.

C **39.** Write in standard form in three different ways:
$x^3yz^2 + xy^3z^3 + x^2y^2z$

40. Create a polynomial that can be written in standard form in four different ways.

SPIRAL REVIEW

S **41.** Find the volume of a sphere whose radius is 9 cm.   *(Lesson 14-6)*
3052.08 cm^3

42. Simplify: $c^3 - 3c^2 + 8c - 9c^2 + 3c^3 - 6$ *(Lesson 15-1)*
$4c^3 - 12c^2 + 8c - 6$

43. Solve and graph: $3x \leq -18$ *(Lesson 13-6)* See answer on p. A40.

44. Find $\sqrt{529}$.   *(Lesson 10-10)* 23

670   Chapter 15

Modeling Polynomials Using Tiles

Objective: To use algebra tiles to model polynomials.

Materials

■ algebra tiles

You can use algebra tiles to model polynomials. For example, let the

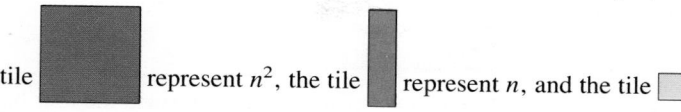

tile represent n^2, the tile represent n, and the tile represent 1. Then you can represent $n^2 + 3n + 2$ as follows.

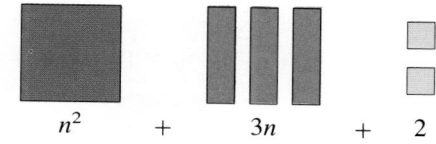

n^2 + $3n$ + 2

Activity I *Representing Polynomials*

1 Write the polynomial represented by the diagram below.

$2n^2 + 5n + 2$

2 Show how to use algebra tiles to represent $3n^2 + 4n + 5$.
See answer on p. A40.

Activity II *Adding Polynomials*

1 Name the polynomial represented by each set of tiles. $n^2 + 2n + 4$;
$n^2 + 4n + 1$

2 How many tiles representing n^2 are there in all? 2

3 How many tiles representing n are there in all? 6

4 How many tiles representing 1 are there in all? 5

5 Write the sum of the polynomials represented by the diagrams. $2n^2 + 6n + 5$

6 Show how to use algebra tiles to represent the sum. See answer on p. A40.
Answers to Activity III, Steps 1–2 are on p. A40.

Activity III *Subtracting Polynomials*

1 Show how to use algebra tiles to represent $2n^2 + 3n + 5$.

2 Remove or cross out one tile representing n^2, two tiles representing n, and three tiles representing 1.

3 Write the polynomial represented by the tiles that are left. $n^2 + n + 2$

4 Complete: When you subtract the polynomial _?_ from the polynomial $2n^2 + 3n + 5$, the difference is _?_. $n^2 + 2n + 3; n^2 + n + 2$

Polynomials **671**

Teaching Notes

Using Manipulatives

Algebra tiles can be used to model the addition and subtraction of polynomials. Models provide students with concrete representations of abstract concepts and operations.

Activity Notes

Activity I

Tiles are used to represent polynomials. Students should be able to identify the polynomial and use tiles or draw a model to represent a polynomial.

Activity II

Students are introduced to the concept of adding polynomials by physically constructing the sum of two polynomials. This helps students to see that there are indeed two n^2's, six n's, and five ones.

Activity III

In this activity, students physically subtract the polynomial $n^2 + 2n + 3$ from $2n^2 + 3n + 5$, resulting in the physical representation of the answer, namely, $n^2 + n + 2$.

Teaching the Lesson
- Materials: algebra tiles
- Lesson Starter 136
- Visuals, Folder Q
- Using Technology, p. 666C
- Using Manipulatives, p. 666C

Lesson Follow-Up
- Practice 186
- Enrichment 150: Thinking Skills
- Study Guide, pp. 271–272

Warm-Up Exercises

Write each polynomial in standard form.

1. $h^2 - 2 + 3h$ $h^2 + 3h - 2$
2. $x^3 - 2 + x^6 + 5x^4 - 3x^7$
   $-3x^7 + x^6 + 5x^4 + x^3 - 2$
3. $-7g^2 + 3g - 8g^2 + g^3 + 15$
   $g^3 - 15g^2 + 3g + 15$
4. $7d^3 + 6d^2 - 12d - 5d^2 + 1$
   $7d^3 + d^2 - 12d + 1$
5. $n^2 - 4n^3 + n - 5n^2 + 42n$
   $-4n^3 - 4n^2 + 43n$

Lesson Focus

Ask students to imagine that they are putting coins in paper rolls; pennies are placed in one roll, nickels in another, and so on. Only like coins are combined in a single roll. Similarly, when adding polynomials, only like terms can be combined. Today's lesson focuses on adding polynomials.

Teaching Notes

Compare aligning like terms with aligning like place values with whole numbers or decimals.

Adding Polynomials

Objective: To add polynomials.

CONNECTION

In previous chapters you performed the four basic operations with rational numbers. You can also add, subtract, multiply, and divide polynomials.

Example 1
Solution

Add: $(n^2 + 3n + 1) + (2n^2 + 2n + 4)$

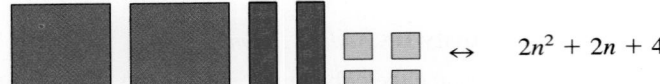

Let the tile ▇ represent n^2, the tile ▌ represent n, and the tile ▫ represent 1.

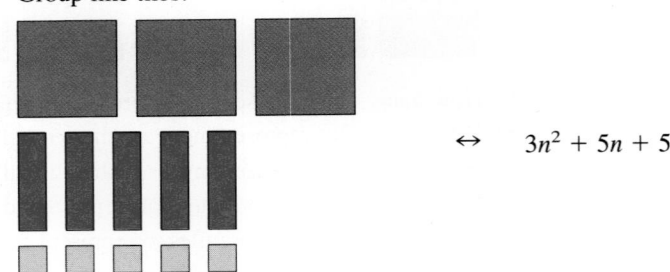

Using tiles: Using symbols:

$\leftrightarrow$ $n^2 + 3n + 1$

$\leftrightarrow$ $2n^2 + 2n + 4$

Group like tiles.

$\leftrightarrow$ $3n^2 + 5n + 5$

Using algebra tiles can help you see how to add polynomials.

Generalization: *Adding Polynomials*

To add polynomials, combine like terms.

When adding polynomials, insert a zero term for a missing power.

672 Chapter 15

Example 2

Solution 1

Add: $(6x^4 - 2x^3 + 7x^2 + x - 6) + (-7x^4 + 2x^3 - 5x + 7)$

Line up like terms vertically.

$$
\begin{array}{r}
6x^4 - 2x^3 + 7x^2 + x - 6 \\
-7x^4 + 2x^3 + 0x^2 - 5x + 7 \\
\hline
-1x^4 + 0x^3 + 7x^2 - 4x + 1
\end{array}
$$

← Insert $0x^2$ since there is no x^2 term.

The sum is written $-x^4 + 7x^2 - 4x + 1$.

Solution 2

$(6x^4 - 2x^3 + 7x^2 + x - 6) + (-7x^4 + 2x^3 - 5x + 7)$
$= (6x^4 - 7x^4) + (-2x^3 + 2x^3) + 7x^2 + (x - 5x) + (-6 + 7)$
$= -1x^4 + 0x^3 + 7x^2 + (-4x) + 1$
$= -x^4 + 7x^2 - 4x + 1$

Check Your Understanding

1. In Solution 2, why does $x + (-5x) = -4x$?
2. In Solutions 1 and 2, why is there no x^3 term in the answer?
   See *Answers to Check Your Understanding* at the back of the book.

Guided Practice

COMMUNICATION « *Reading*

Refer to the text on pages 672–673.

« **1.** Describe two ways to add polynomials. Line up like terms vertically; group like terms horizontally.

« **2.** Why would you insert a zero term for a missing power? To hold the place of a missing term.

COMMUNICATION « *Writing* 3. $2n^2 + n + 3$; $n^2 + 3n + 2$; $3n^2 + 4n + 5$

« **3.** Write the polynomial that is represented by each group of tiles. Then write the sum of the polynomials. Use n as the variable.

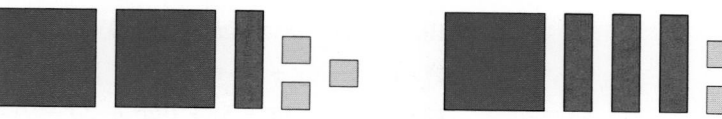

« **4.** Draw a diagram similar to that of Exercise 3 to represent the addition $(3n^2 + 2n + 3) + (2n^2 + 4n + 5)$. See answer on p. A40.

Replace each ? with the term that makes the statement true.

5. $(5x^2 + 3x + 2) + (2x^2 + 6x + 6)$
$= 7x^2 + \underline{?} + 8$ 9x

6. $(3a^3 + 4a^2 - 5a + 1) + (5a^3 - a^2 - 2a + 3)$
$= 8a^3 + \underline{?} - \underline{?} + 4$ $3a^2$; 7a

7. $(-c^4 + 2c^3 + 7c) + (2c^4 + c^3 - 5c^2 + 2c)$
$= \underline{?} + 3c^3 - \underline{?} + 9c$ c^4; $5c^2$

8. $(3m^3 - m^2 + 5m + 2) + (m^3 + m^2 - 3m + 6)$
$= 4m^3 + \underline{?} + 8$ 2m

Polynomials **673**

Suggested Assignments

Average: 1–26, 33–36
Advanced: 1–27 odd, 28–36

Exercise Notes

Communication: Reading
Guided Practice Exercises 1 and 2 require students to interpret material found in the text.

Communication: Writing
Guided Practice Exercises 3 and 4 relate writing skills, the use of manipulatives, and the addition of polynomials.

Life Skills
Exercise 27 asks students to write a paragraph describing the two methods for adding polynomials and to give some advantages of each method. Many real-world situations require an examination of different methods and a written analysis of them.

Critical Thinking
Exercises 28–30 require students to analyze problems and formulate a hypothesis.

Problem Solving
By examining the completed table in *Exercise 31*, students should be able to answer *Exercise 32*: The sum of the values of two polynomial functions is equal to the value of the sum of the two functions.

Calculator
Exercise 31 can be done by using a scientific calculator.

Add. **11.** $-c^3 - 11c^2 - 2c - 12$ **12.** $5z^4 - z^3 + 3z^2 - z - 1$

9. $2x^2 + 3x + 4$
$5x^2 - 4x - 3$ $7x^2 - x + 1$

10. $y^2 + 8y - 2$
$7y^2 - 5y + 8$ $8y^2 + 3y + 6$

11. $-3c^3 - 5c^2 - 11c - 5$
$2c^3 - 6c^2 + 9c - 7$

12. $z^4 - z^3 - 2z^2 - 3z$
$4z^4 \qquad + 5z^2 + 2z - 1$

13. $(6c^3 + 8c^2 - 12c - 4) + (9c^3 + c^2 + 3c + 7)$ $15c^3 + 9c^2 - 9c + 3$

14. $(8w^3 - 9w^2 - 6w + 9) + (5w^3 - 3w^2 - 8w - 11)$
$13w^3 - 12w^2 - 14w - 2$

15. $(z^5 - 2z^3 - 9z + 3) + (-z^4 + 2z^3 - 10)$ $z^5 - z^4 - 9z - 7$

16. $(-6r^4 + 4r^2 + 1) + (-5r^2 - 7r + 4)$ $-6r^4 - r^2 - 7r + 5$

Exercises

Add.

A

1. $5a^2 + 4a + 10$
2. $9c^2 + 11c + 6$
3. $7x^2 + 11x + 11$
4. $11a^2 + 10a + 6$
5. $9v^2 + 6v + 7$
6. $17n^2 + 7n + 11$
8. $12a^4 + 8a^3 + 8a^2 + 8a + 6$
9. $8b^4 + 5b^3 + 12b^2 + 7b + 7$
10. $7d^4 + 7d^3 + 7d^2 + 4d + 5$
19. $5x^5 + 3x^4 + 3x^3 + x^2 - 5$

1. $(2a^2 + 3a + 5) + (3a^2 + a + 5)$

2. $(4c^2 + 3c + 2) + (5c^2 + 8c + 4)$

3. $(3x^2 + 5x + 9) + (4x^2 + 6x + 2)$

4. $(5a^2 + 8a + 1) + (6a^2 + 2a + 5)$

5. $(7v^2 + 3v + 6) + (2v^2 + 3v + 1)$

6. $(9n^2 + 4n + 3) + (8n^2 + 3n + 8)$

7. $(8y^3 + 6y^2 + 3y + 2) + (6y^3 + 6y^2 + 6y + 6)$ $14y^3 + 12y^2 + 9y + 8$

8. $(8a^4 + 6a^3 + 5a^2 + 2a + 4) + (4a^4 + 2a^3 + 3a^2 + 6a + 2)$

9. $(2b^4 + 3b^3 + 4b^2 + 5b + 1) + (6b^4 + 2b^3 + 8b^2 + 2b + 6)$

10. $(2d^4 + 5d^3 + 4d^2 + d + 3) + (5d^4 + 2d^3 + 3d^2 + 3d + 2)$

11. $(7y^2 - 6y + 8) + (-6y^2 + 2y + 1)$ $y^2 - 4y + 9$

12. $(-7a^2 - 4a + 5) + (9a^2 + 2a - 7)$ $2a^2 - 2a - 2$

13. $(4x^2 + 6x + 9) + (2x^2 - 3x - 4)$ $6x^2 + 3x + 5$

14. $(5r^3 + 9r^2 - 3r - 6) + (3r^3 - 7r^2 - 6r - 5)$ $8r^3 + 2r^2 - 9r - 11$

15. $(3k^4 - k^2 + 4k - 6) + (-4k^4 - 7k^3 - 2k^2 - 3k + 4)$
$-k^4 - 7k^3 - 3k^2 + k - 2$

16. $(8k^5 - 3k^3 + 2) + (2k^4 - 3k^2 + 5)$ $8k^5 + 2k^4 - 3k^3 - 3k^2 + 7$

17. $(4k^5 - 1) + (k^5 - k^4 + 1)$ $5k^5 - k^4$

18. $(k^2 - k - 1) + (k^2 - 1)$ $2k^2 - k - 2$

19. Find the sum of $x^5 + 3x^4 - 5x^2 + 4$ and $4x^5 + 3x^3 + 6x^2 - 9$.

20. Find the sum of $n^6 - 5n^4 + 6n^2 - 4$ and $-5n^6 + n^5 - n^4 - n^2 + 2$.
$-4n^6 + n^5 - 6n^4 + 5n^2 - 2$

Add. Write the answer in standard form.

21. $6a^4 + 5a^3 + 9a^2 + 11a + 7$

B **21.** $(a^4 + 6a^2 + 4a^3 + 6 + 7a) + (a^3 + 3a^2 + 5a^4 + 4a + 1)$

22. $(x^2 + 3x^3 + 7x + 8x^4 + 1) + (6x^4 + 3x^3 + 5x + 6 + 3x^2)$
$14x^4 + 6x^3 + 4x^2 + 12x + 7$

23. $(3d^2 - 6d - 2d^3 + 1) + (-d^3 + 2d + 5d^2 - 5)$ $-3d^3 + 8d^2 - 4d - 4$

24. $(b - 4b^2 - b^3 + 1) + (b^2 + 2b^3 - 3b - 5)$ $b^3 - 3b^2 - 2b - 4$

25. $(b^4 - b + 4b^2 - 3) + (-2b^3 - b^2 + b^4 + 2)$ $2b^4 - 2b^3 + 3b^2 - b - 1$

26. $(a^3 - 8a^2 - 2a + 1) + (a^3 + 5a^2 + 3a^4 - 7)$ $3a^4 + 2a^3 - 3a^2 - 2a - 6$

27. **WRITING ABOUT MATHEMATICS** Write a paragraph describing two methods of adding polynomials. Give advantages of each. See answer on p. A40.

THINKING SKILLS

To write the *opposite* of a polynomial, you write the polynomial with all its signs changed to their opposites. For instance, the opposite of the polynomial $2x + 1$ is the polynomial $-2x - 1$.

28–29. Application
30. Synthesis

28. Write the opposite of $2x^2 - 5x - 1$ in standard form. $-2x^2 + 5x + 1$

29. Find the sum of $2x^2 - 5x - 1$ and its opposite. 0

30. Make a generalization about the sum of a polynomial and its opposite.
The sum of a polynomial and its opposite is 0.

FUNCTIONS

C **31.** Complete the function table below.

x	$x^2 - 4x$	$x^3 + 4x + 4$	$x^3 + x^2 + 4$	
-2	?	?	?	12; -12; 0
-1	?	?	?	5; -1; 4
0	?	?	?	0; 4; 4
1	?	?	?	-3; 9; 6
2	?	?	?	-4; 20; 16

32. Use the table in Exercise 31 to complete the following statement:
The sum of the values of two polynomial functions is equal to __?__.
the value of the sum of the two functions

SPIRAL REVIEW

33. Graph $y = 2x - 4$ on a coordinate plane. *(Lesson 13-2)*
See answer on p. A40.

34. Find the surface area of a cube whose edge is 8 in. *(Lesson 14-2)*
384 in.^{2}

35. Add: $(7a^4 + 3a^3 + 2a^2 + 5) + (-6a^3 + 4a^2 - 8)$ *(Lesson 15-2)*
$7a^4 - 3a^3 + 6a^2 - 3$

36. Make a stem-and-leaf plot for the given data. *(Lesson 11-2)*
See answer on p. A40.

Scores on a Mathematics Test

77 68 84 82 79 91 73 88 94 79

Polynomials **675**

Extension (Cooperative Learning)
Have students working in cooperative groups create a function table similar to that in Exercise 31. For functions, the groups should choose a polynomial function and its opposite. After completing the table, each group should reach a conclusion about the sum of the values of a polynomial function and its opposite.

Enrichment
Enrichment 150 is pictured on page 666E.

Practice
Practice 186, *Resource Book*

Teaching the Lesson
- Materials: algebra tiles
- Lesson Starter 137
- Visuals, Folder Q
- Using Technology, p. 666C
- Using Manipulatives, p. 666D

Lesson Follow-Up
- Practice 187
- Enrichment 151: Exploration
- Study Guide, pp. 273–274

Warm-Up Exercises

Add. Write the answer in standard form.

1. $(2x^2 + 3x + 7) + (4x^2 + 6x - 5)$
   $6x^2 + 9x + 2$

2. $(4m^3 - 5m^2 + 8m - 7) + (-7m^3 + 2m^2 - 11m + 3)$
   $-3m^3 - 3m^2 - 3m - 4$

3. $(4k^3 - 6k^2 + 5) + (-5k^3 + 6k^2 + 2k)$
   $-k^3 + 2k + 5$

4. Find the sum of $x^5 + 2x^4 - 7x^2 + 1$ and $2x^5 - 9x^2 + 5$.
   $3x^5 + 2x^4 - 16x^2 + 6$

5. Write the opposite of $3x^4 - 4x - 3$ in standard form.
   $-3x^4 + 4x + 3$

Lesson Focus

Ask students to imagine that they have some quarters, dimes, and nickels in their pockets. To pay for a purchase, they give the clerk a quarter, three dimes, and two nickels. If students want to find out how many of each coin they have left, they must examine each coin individually. Today's lesson focuses on subtracting polynomials, which must be done by examining like terms.

15-3

Subtracting Polynomials

Objective: To subtract polynomials.

EXPLORATION

Let the tile ▇ represent n^2, the tile ▌ represent n, and the tile ☐ represent 1.

The tiles below represent the polynomial $3n^2 + 4n + 8$.

1 Copy the diagram or use algebra tiles. Remove 2 tiles representing n^2, one tile representing n, and five tiles representing 1. *See answer on p. A40.*

2 What polynomial is represented by the tiles that you removed? $2n^2 + n + 5$

3 What polynomial is represented by the tiles that are left? $n^2 + 3n + 3$

4 Complete: $(3n^2 + 4n + 8) - (2n^2 + n + 5) = $ __?__ $n^2 + 3n + 3$

In the Exploration you removed tiles to represent the polynomial that was being subtracted. Recall that subtraction may also be thought of as addition of the opposite. To subtract a rational number, you add the opposite of that number. To subtract a polynomial, you use a similar procedure.

Generalization: *Subtracting Polynomials*

To subtract a polynomial, add the opposite of each term of the polynomial.

Example 1

Solution

Subtract: $(3n^2 + 4n + 8) - (2n^2 + n + 5)$

$(3n^2 + 4n + 8) - (2n^2 + n + 5)$
$= (3n^2 + 4n + 8) + (-2n^2 - n - 5)$ ← The opposite of $2n^2$ is $-2n^2$. The opposite of n is $-n$. The opposite of 5 is -5.
$= (3n^2 - 2n^2) + (4n - n) + (8 - 5)$
$= n^2 + 3n + 3$

✏️ **Check Your Understanding**

1. In Example 1, why was the polynomial $-2n^2 - n - 5$ added to the polynomial $3n^2 + 4n + 8$?
   See Answers to Check Your Understanding at the back of the book.

---

676 Chapter 15

As with addition, you may have to subtract polynomials that have terms missing for some powers.

Example 2

Solution 1

Subtract: $(7a^3 + 3a^2 - 10) - (9a^3 + 4a^2 - 6a - 9)$

Line up like terms. Insert zero terms as needed. Add the opposite.

$$
\begin{array}{r}
7a^3 + 3a^2 + 0a - 10 \\
9a^3 + 4a^2 - 6a - \ 9 \\
\end{array}
\longrightarrow
\begin{array}{r}
7a^3 + 3a^2 + 0a - 10 \\
-9a^3 - 4a^2 + 6a + \ 9 \\
\hline
-2a^3 - \ a^2 + 6a - \ 1 \\
\end{array}
$$

Solution 2

$(7a^3 + 3a^2 - 10) - (9a^3 + 4a^2 - 6a - 9)$
$= (7a^3 + 3a^2 - 10) + (-9a^3 - 4a^2 + 6a + 9)$
$= (7a^3 - 9a^3) + (3a^2 - 4a^2) + 6a + (-10 + 9)$
$= -2a^3 - a^2 + 6a - 1$

 **Check Your Understanding**

2. In Solution 1, why was the term $0a$ inserted in $7a^3 + 3a^2 - 10$?
See *Answers to Check Your Understanding* at the back of the book.

Guided Practice

COMMUNICATION «*Reading*

Replace each __?__ with the correct word.

«**1.** To subtract a number, you add its __?__. opposite

«**2.** To subtract a polynomial, you add the opposite of each __?__. term

Write the opposite of each term of each polynomial.

3. $2a^2 + 5a + 6$ $-2a^2 - 5a - 6$

4. $-6x^3 - 3x^2 - x - 7$ $6x^3 + 3x^2 + x + 7$

5. $4x^4 - 6x^3 + 9x^2 + 4x - 9$ $-4x^4 + 6x^3 - 9x^2 - 4x + 9$

6. $-m^4 + 2m^3 - 8m^2 - m - 8$ $m^4 - 2m^3 + 8m^2 + m + 8$

Subtract.

9. $5b^3 + 2b - 22$ **10.** $16n^3 - 10n^2 - 2n - 2$

7. $\begin{array}{r} -4a + 7 \\ 2a - 3 \\ \hline \end{array}$ $-6a + 10$

8. $\begin{array}{r} 2n^2 - 5n - 1 \\ 3n^2 - \ n + 6 \\ \hline \end{array}$ $-n^2 - 4n - 7$

9. $\begin{array}{r} 7b^3 + 9b^2 \quad \ - 12 \\ 2b^3 + 9b^2 - 2b + 10 \\ \hline \end{array}$

10. $\begin{array}{r} 9n^3 - 6n^2 - n - 3 \\ -7n^3 + 4n^2 + n - 1 \\ \hline \end{array}$

11. $(5a^2 + 7a + 8) - (3a^2 + 4a + 2)$ $2a^2 + 3a + 6$

12. $(v^3 + 8v^2 - 13v + 2) - (v^3 + 5v^2 + 9)$ $3v^2 - 13v - 7$

13. $(15z^3 - 3z^2 + 6z + 13) - (-8z^3 + 7z^2 - 8z - 4)$ $23z^3 - 10z^2 + 14z + 17$

14. $(3w^3 - 5w^2 - 8) - (6w^3 + 2w - 18)$ $-3w^3 - 5w^2 - 2w + 10$

Polynomials **677**

Average: 1–22, 23–35 odd, 43–46
Advanced: 1–35 odd, 37–46,
  Challenge

Exercise Notes

Communication: Writing
*Guided Practice Exercises 1 and 2
require students to interpret information given in the text.*

Making Connections/Transitions
*Exercises 31–36 require students to
use the skills of adding and subtracting polynomials in the same
exercise.*

Critical Thinking
*Exercises 37–42 require students to
analyze polynomial expressions in
order to reach conclusions about
polynomial subtraction.*

Exercises

1. $x^2 + x + 4$
2. $3c^2 + 2c + 3$
3. $3x^2 + 7x + 6$
4. $2a^2 + 6a + 8$
5. $5d^2 - d - 2$
6. $5n^2 - 3n - 6$
7. $-2m^2 - 10m - 4$
8. $-3y^2 - 8y - 9$
14. $-3x^4 - 10x^3 + 2x^2 - x - 2$

Subtract.

A 1. $(2x^2 + 4x + 5) - (x^2 + 3x + 1)$

2. $(5c^2 + 6c + 8) - (2c^2 + 4c + 5)$

3. $(8x^2 + 9x + 9) - (5x^2 + 2x + 3)$

4. $(5a^2 + 8a + 9) - (3a^2 + 2a + 1)$

5. $(8d^2 + 3d + 6) - (3d^2 + 4d + 8)$

6. $(9n^2 + 4n + 3) - (4n^2 + 7n + 9)$

7. $(4m^2 - 3m + 2) - (6m^2 + 7m + 6)$

8. $(6y^2 - 3y - 2) - (9y^2 + 5y + 7)$

9. $(8a^3 - 6a^2 + 2a + 4) - (5a^3 + 7a^2 - 5a - 7)$ $3a^3 - 13a^2 + 7a + 11$

10. $(3b^3 - 4b^2 + b - 4) - (7b^3 + 6b^2 - 4b + 8)$ $-4b^3 - 10b^2 + 5b - 12$

11. $(-6x^2 + 5x - 9) - (-3x^2 - x + 7)$ $-3x^2 + 6x - 16$

12. $(a^2 + 2a - 3) - (-5a^2 + 2a + 4)$ $6a^2 - 7$

13. $(8a^3 - 6a^2 - 2a + 9) - (4a^3 - 2a^2 + 6a - 8)$ $4a^3 - 4a^2 - 8a + 17$

14. $(3x^4 - 8x^3 - 6x^2 + 5x + 2) - (6x^4 + 2x^3 - 8x^2 + 6x + 4)$

15. $(4a^3 - 6a) - (5a^4 - 8a^2 + 2)$ $-5a^4 + 4a^3 + 8a^2 - 6a - 2$

16. $(9y^4 - 6y^2 + 3y - 6) - (4y^3 - 6y - 4)$ $9y^4 - 4y^3 - 6y^2 + 9y - 2$

17. $(7a^3 + 9a^2 - 2) - (5a^4 - 8a^2 - 2)$ $-5a^4 + 7a^3 + 17a^2$

18. $(-6t^4 - 5t^3 - 8t^2 + 4) - (-2t^3 + 8t^2 + t + 1)$ $-6t^4 - 3t^3 - 16t^2 - t + 3$

19. $(z^4 - z^3 - 2z^2 - 3z) - (5z^4 + 7z^2 - z - 8)$ $-4z^4 - z^3 - 9z^2 - 2z + 8$

20. $(-2d^4 + 4d^3 + 3) - (5d^3 - 2d^2 + 3d - 7)$ $-2d^4 - d^3 + 2d^2 - 3d + 10$

21. Find the difference when $4x^3 + 7x^2 - 3x - 6$ is subtracted from
    $8x^3 - 7x^2 + 3x - 9$. $4x^3 - 14x^2 + 6x - 3$

22. Find the difference when $-4x^3 - x^2 + 7x + 3$ is subtracted from
    $-6x^3 + 9x^2 - 5x - 6$. $-2x^3 + 10x^2 - 12x - 9$

Solve.

B 23. $(6a + 9) - (2a + 21) = 0$ 3

24. $(7n + 2) + (3n - 12) = 0$ 1

25. $(3d - 3) - (8d + 5) = -3$ -1

26. $(7y + 1) - (-y + 7) = 3y - 26$ -4

27. $(x^2 - 8x + 3) - (x^2 - 5x - 12) = 0$ 5

28. $(c^2 - 5c + 12) + (-c^2 + 3c - 12) = 0$ 0

29. $(4r + 9r^2) - (9r^2 + 2) = r + 1$ ₁

30. $(n^2 + 2n - 9) - (n^2 - 3n + 16) = n - 5$ ₅

Simplify.

31. $(7a^2 + 4a + 5) + (2a^2 - 5a + 8) - (3a^2 - 6a + 4)$ $6a^2 + 5a + 9$

32. $(x^2 - 6x + 1) + (3x^2 + 7x + 3) - (2x^2 + 3x + 5)$ $2x^2 - 2x - 1$

33. $(6n^2 - 7n + 9) - (3n^2 + 2n + 1) - (n^2 - 4n + 3)$ $2n^2 - 5n + 5$

34. $(c^2 - 3c + 7) - (5c^2 + 3c - 5) - (4c^2 + 7c - 2)$ $-8c^2 - 13c + 14$

35. $(2b^2 + 5b - 8) - (3b^2 + 9b + 5) + (3b^2 - 2b + 7)$ $2b^2 - 6b - 6$

36. $(m^2 + 6m - 2) - (4m^2 - 7m - 3) + (2m^2 + 5m - 3)$ $-m^2 + 18m - 2$

THINKING SKILLS Answers to Exercises 37 and 38 are on p. A41.

37–38. Synthesis
   39. Analysis
40–41. Application
   42. Analysis

C   **37.** Create two polynomials that have a difference of 2.

38. Create two polynomials that have a difference of x.

39. What can you conclude about two polynomials that have a difference of zero? They are the same.

40. Subtract $5x^2 + 2x - 1$ from $3x^2 - 4x + 2$. $-2x^2 - 6x + 3$

41. Subtract $3x^2 - 4x + 2$ from $5x^2 + 2x - 1$. $2x^2 + 6x - 3$

42. Compare your answers from Exercises 40 and 41. What can you conclude about reversing the order of a subtraction? Give another example to support your conclusion. See answer on p. A41.

SPIRAL REVIEW

S   **43.** Make a frequency table for the given data.
   (Lesson 11-1)
   See answer on p. A41.
   Base Hits per Season

   164 173 201 212 183 162 171
   176 160 194 168 185 174 166

44. Find the volume of a pyramid with a base of 144 cm² and a height of 9 cm. *(Lesson 14-4)* 432 cm³

45. Find the difference: $(7d^2 + 8d - 4) - (4d^2 - 5d + 7)$ *(Lesson 15-3)* $3d^2 + 13d - 11$

46. Solve and graph: $2x - 7 > 9$ *(Lesson 13-7)* See answer on p. A41.

Challenge

Find all whole-number values of x such that $(x^3 + 3x^2 - 8x + 1) - (x^3 + 2x^2 - 5x + 1)$ is equal to 0. 0; 3

Polynomials   **679**

Follow-Up

Enrichment *(Cooperative Learning)*
Have students working in groups develop exercises similar to the Challenge. Each group should select one exercise to share with the other groups.

Enrichment
Enrichment 151 is pictured on page 666E.

Practice
Practice 187, *Resource Book*

Teaching Note

Using Manipulatives

This exploration presents a manipulative approach to introduce the idea of multiplying monomials and binomials. The models used in the activities provide a concrete representation to an abstract topic.

Activity Notes

Activity I

The first activity presents students with a model that represents the product of a monomial and a binomial. Then students are asked to draw their own models or use tiles to represent products. Students then use their models to write the products symbolically.

Activity II

The second activity is similar to Activity I but focuses on products of binomials. Students once again examine models of products and then create their own models for products. The students' models then are used to write the products symbolically.

Objective: To use algebra tiles to model products of monomials and binomials.

Materials

■ algebra tiles

Modeling Products Using Tiles

You can use algebra tiles to model products of monomials and binomials or of binomials and binomials. The models use the geometric fact that the area of a rectangle equals the product of the base and the height, or $A = bh$. The base and height represent the two polynomials being multiplied, and the area represents their product.

For example, let the tile  represent n^2, the tile ▮ represent n, and the tile ▪ represent 1. Then you can represent the product $2n(n + 3)$ as shown below.

height $= n + n = 2n$
base $= n + 1 + 1 + 1 = n + 3$
Area $= 2n(n + 3)$

Activity I *Modeling Products of Monomials and Binomials*

1 Write the product represented by each diagram below.

a.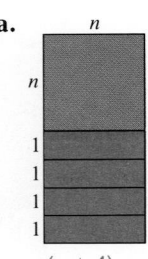

$n(n + 4)$

b.

$2n(2n + 1)$

2 Show how to use algebra tiles to represent each product. See answer on p. A41.
   a. $n(2n + 4)$ **b.** $3n(n + 1)$

3 Use your diagrams from Step 2 to complete each statement.
   a. $n(2n + 4) = \underline{\ ?\ }\ 2n^2 + 4n$

   b. $3n(n + 1) = \underline{\ ?\ }\ 3n^2 + 3n$

You can use algebra tiles to represent the product of a binomial and a binomial. For example, you can represent the product $(2n + 1)(n + 3)$ as shown below.

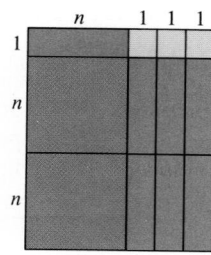

Activity II *Modeling Products of Binomials*

1 Write the product represented by each diagram below.

a. $(n + 4)(n + 2)$

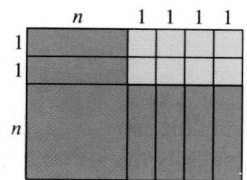

b. $(3n + 2)(n + 2)$

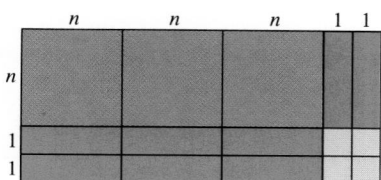

2 Show how to use algebra tiles to represent each product. See answer on p. A41.

a. $(n + 2)(n + 3)$

b. $(n + 3)(n + 3)$

c. $(2n + 1)(n + 1)$

d. $(2n + 1)(2n + 1)$

3 Use your diagrams from Step 2 to complete each statement.

a. $(n + 2)(n + 3) = \underline{\ ?\ }$ a. $n^2 + 5n + 6$

b. $(n + 3)(n + 3) = \underline{\ ?\ }$ b. $n^2 + 6n + 9$
                                                  c. $2n^2 + 3n + 1$
c. $(2n + 1)(n + 1) = \underline{\ ?\ }$ d. $4n^2 + 4n + 1$

d. $(2n + 1)(2n + 1) = \underline{\ ?\ }$

682

Lesson Planner

Teaching the Lesson
- Lesson Starter 138
- Visuals, Folder Q
- Using Technology, p. 666C

Lesson Follow-Up
- Practice 188
- Enrichment 152: Data Analysis
- Study Guide, pp. 275–276
- Calculator Activity 15

Warm-Up Exercises

Find each answer.

1. $(a^2 + 2ab) + (3a^2 + 4b^2)$
   $4a^2 + 2ab + 4b^2$

2. $(3x^3 - 2x^2 + 5) + (2x^2 + 4x - 6)$
   $3x^3 + 4x - 1$

3. $(3x^3 - 2x^2 + 4) - (2x^3 - 3x + 5)$
   $x^3 - 2x^2 + 3x - 1$

4. $(x^3 - 4x + 1) - (-4x^3 - 2x^2 - 4x - 3)$
   $5x^3 + 2x^2 + 4$

5. Write the opposite of the polynomial $2x^3 - 3x^2 - 5$.
   $-2x^3 + 3x^2 + 5$

Lesson Focus

Ask students to give some examples of how they have used the distributive property in this course. Today's lesson focuses on another use, that of multiplying a polynomial by a monomial.

Teaching Notes

Key Question

Why do you add the exponents when multiplying powers with the same base? The product of powers rule states that $a^m \cdot a^n = a^{m+n}$.

 **15-4**

Multiplying a Polynomial by a Monomial

Objective: To multiply a polynomial by a monomial.

 **CONNECTION**

In Chapter 2 you learned to simplify expressions such as $3(x - 7)$ by using the distributive property. Since the expression $x - 7$ is a polynomial, you have already learned how to multiply a number by a polynomial.

Example 1 Multiply: $-4(2x^2 + 5x - 3)$

Solution
$$-4(2x^2 + 5x - 3) = (-4)(2x^2) + (-4)(5x) - (-4)(3)$$
$$= -8x^2 + (-20x) - (-12)$$
$$= -8x^2 - 20x + 12$$

Check Your Understanding

1. In Example 1, why can $-8x^2 + (-20x) - (-12)$ be rewritten as $-8x^2 - 20x + 12$?
   See *Answers to Check Your Understanding* at the back of the book.

Recall that to multiply powers having the same base, you add the exponents. You can use this rule to multiply monomials.

Example 2 Multiply: $(3a^3b^2)(-5a^2b^4)$

Solution
$$(3a^3b^2)(-5a^2b^4) = 3(-5)(a^3 \cdot a^2)(b^2 \cdot b^4)$$
$$= -15(a^{3+2})(b^{2+4})$$
$$= -15a^5b^6$$

← Group integers and powers having the same base.

To multiply a polynomial of two or more terms by a monomial, you use the distributive property and the rule for multiplying powers of the same base.

Example 3 Multiply: $4x^2(7x^3 + 2x^2 - 6x - 4)$

Solution
$$4x^2(7x^3 + 2x^2 - 6x - 4)$$
$$= 4x^2(7x^3) + 4x^2(2x^2) - 4x^2(6x) - 4x^2(4)$$
$$= 28x^5 + 8x^4 - 24x^3 - 16x^2$$

Check Your Understanding

2. Explain how the distributive property was used in Example 3.
   See *Answers to Check Your Understanding* at the back of the book.

---

682 Chapter 15

Generalization: *Multiplying a Polynomial by a Monomial*

To multiply a polynomial by a monomial, multiply each term of the polynomial by the monomial and simplify.

Guided Practice

COMMUNICATION « *Reading*

Refer to the text on pages 682–683.

« **1.** What is the main idea of the lesson? multiplying a polynomial by a
monomial

« **2.** List three major points that support the main idea of the lesson.
See answer on p. A41.

Replace each __?__ with the expression that makes the statement true.

3. $5(2x^3 + x^2 - 4x + 6) = 10x^3 + \underline{\ ?\ } - 20x + 30$ $5x^2$

4. $-3(6x^4 - 7x^3 + 5x^2 - x + 2) = \underline{\ ?\ } + 21x^3 - 15x^2 + \underline{\ ?\ } - 6$
$-18x^4;\ 3x$

5. $7a^3b^4(5a^2b^5) = 35 \cdot \underline{\ ?\ } \cdot b^9$ a^5

6. $4x^5z^2(-8x^3z^7) = \underline{\ ?\ } \cdot x^8z^9$ -32

7. $8n^3(3n^2 + 6n - 5) = 24n^5 + \underline{\ ?\ } - 40n^3$ $48n^4$

8. $-2a^4(-a^3 + 5a^2 - 7a - 8) = \underline{\ ?\ } - \underline{\ ?\ } + 14a^5 + 16a^4$ $2a^7;\ 10a^6$

9. $35c^3 + 20c^2 - 5c - 30$ **10.** $-42x^4 - 35x^3 + 21x^2 + 63x - 35$

Multiply. **13.** $15m^7n^{13}$

9. $5(7c^3 + 4c^2 - c - 6)$ **10.** $-7(6x^4 + 5x^3 - 3x^2 - 9x + 5)$

11. $3x^4y^5(4x^2y^6)$ $12x^6y^{11}$ **12.** $7ab^5(-3a^3b)$ **13.** $(-5m^4n^8)(-3m^3n^5)$
$-21a^4b^6$

14. $7s^4(-5s^3 + 6s^2 - 4s - 8)$ **15.** $-2x^4(3x^3 - 4x^2 - 8x + 3)$
$-35s^7 + 42s^6 - 28s^5 - 56s^4$ $-6x^7 + 8x^6 + 16x^5 - 6x^4$

Exercises

2. $4c^3 - 28c^2 - 24c - 20$

3. $-36a^3 + 12a^2 + 24a - 12$

4. $6b^4 + 21b^3 - 6b^2 - 24$

6. $28a^{11}c^{11}$

7. $-30r^5s^7$

8. $-100x^5y^5$

9. $18a^6x^{11}$

10. $48c^8b^{16}$

13. $15d^5 - 24d^4 + 21d^3 - 18d^2$

14. $-24x^4 + 56x^3 + 32x^2 - 16x$

Multiply.

A **1.** $7(2x^2 + x + 2)$ $14x^2 + 7x + 14$ **2.** $4(c^3 - 7c^2 - 6c - 5)$

3. $-6(6a^3 - 2a^2 - 4a + 2)$ **4.** $-3(-2b^4 - 7b^3 + 2b^2 + 8)$

5. $(4x^2y^3)(6x^4)$ $24x^6y^3$ **6.** $(7a^3c^5)(4a^8c^6)$ **7.** $(6r^3s^3)(-5r^2s^4)$

8. $(-10x^4y)(10xy^4)$ **9.** $(-6a^4x^5)(-3a^2x^6)$ **10.** $(-8c^6b^7)(-6c^2b^9)$

11. $4a(6a^2 + 4a + 5)$ **12.** $-7t(4t^2 + 2t - 9)$
$24a^3 + 16a^2 + 20a$ $-28t^3 - 14t^2 + 63t$

13. $3d^2(5d^3 - 8d^2 + 7d - 6)$ **14.** $-8x(3x^3 - 7x^2 - 4x + 2)$

15. Find the product of $7x^3$ and $4x^3 + 6x^2 - 9x + 7$.
$28x^6 + 42x^5 - 63x^4 + 49x^3$

16. Multiply $3n^4 - n^3 + 5n^2 + 6n - 3$ by $2n^4$.
$6n^8 - 2n^7 + 10n^6 + 12n^5 - 6n^4$

Polynomials **683**

Error Analysis

When multiplying powers with the same base, students may mistakenly multiply the exponents. Remind students that the exponents are *added* when multiplying powers with the same base.

Using Technology

For a suggestion on using a calculator or a spreadsheet program, see page 666C.

Additional Examples

Multiply.

1. $-3(2x^2 + 4x - 5)$
$-6x^2 - 12x + 15$

2. $(4m^3n^2)(-6m^2n^4)$ $-24m^5n^6$

3. $5c^2(9c^3 + 3c^2 - 4c - 5)$
$45c^5 + 15c^4 - 20c^3 - 25c^2$

Closing the Lesson

Make two sets of cards, one containing constants and monomials and the other containing monomials and polynomials. Have each student draw a card at random from each set and explain how to multiply the two expressions shown on the cards.

Suggested Assignments

Average: 1–22, 41–44
Advanced: 1–27 odd, 29–44

Exercise Notes

Communication: Reading

Guided Practice Exercises 1 and 2 require students to analyze information in the text in much the same way they would analyze a reading passage in a language arts class.

Mental Math

Exercises 23–28 encourage students to perform multiplication mentally.

Problem Solving

Exercises 29–32 require students to multiply polynomials by -1. Exercise 33 asks students to look for a pattern in the results of those exercises and use it to complete a statement. Looking for a pattern is an important problem-solving strategy.

683

Critical Thinking

Exercises 34–36 require students to combine previously learned skills to complete statements. Exercise 37 asks students to make a conjecture and to support it by creating an additional example. Exercises 38–40 ask students to use their conjecture to find products involving negative exponents.

Follow-Up

Extension

In Exercise 42, assume that the dimensions of the container are in feet and are whole numbers. What is the greatest possible surface area for the container? Explain how you arrived at your answer. 2896 ft²; the possible dimensions for the base are 1 by 160, 2 by 80, 4 by 40, 5 by 32, and 8 by 20. The 1 ft by 160 ft base produces the greatest surface area (SA = 2lw + 2wh + 2lh).

Enrichment

Enrichment 152 is pictured on page 666E.

Practice

Practice 188, *Resource Book*

Practice 188
Skills and Applications of Lesson 15-4

Multiply. Answers to Exercises 17–22 are on p. A41.

B **17.** $5mn(2mn - 3mn^2 + 4n)$ **18.** $4ac(-3a^2b + 8ab^3 + 5b)$

19. $-xy(5xy + 7x^2y + 8xy^2)$ **20.** $3cd^2(7cd + 4c^2d^2 - c^3d)$

21. $m^3n(-5m^2n^2 - mn + 7m^2n)$

22. $-2a^3b^2(-2a^3 + a^3b - 7ab^4 + 9b^8)$

MENTAL MATH Answers to Exercises 23–28 are on p. A41.

In many cases multiplying a polynomial by a monomial can be done mentally. Do each of the following multiplications mentally.

23. $5(6x^3 + 7x^2 - 9x - 5)$

24. $-3(6c^4 + 8c^3 - 3c^2 - 8c + 2)$

25. $2a(4a^4 + 8a^3 - 7a^2 + a - 9)$ **26.** $-4n(-7n^3 + 2n^2 - 4n + 6)$

27. $2x^2(-9x^3 - 4x^2 + 6x + 8)$ **28.** $-5m^4(5m^3 - 2m^2 - m + 7)$

PATTERNS 29. $-x^3 - 4x^2 - 7x + 5$ **30.** $x^3 + 4x^2 + 7x - 5$

Simplify. 31. $3x^4 - x^3 + 6x^2 + 2x - 7$ **32.** $-3x^4 + x^3 - 6x^2 - 2x + 7$

29. $-(x^3 + 4x^2 + 7x - 5)$ **30.** $-(-x^3 - 4x^2 - 7x + 5)$

31. $-(-3x^4 + x^3 - 6x^2 - 2x + 7)$ **32.** $-(3x^4 - x^3 + 6x^2 + 2x - 7)$

33. Use the pattern in Exercises 29–32 to complete this statement: When you multiply a polynomial by -1, you ___?___. change each sign

THINKING SKILLS 34–37. Synthesis   **38–40.** Application

Combine your knowledge of exponents and fractions to complete.

C **34.** $(a^{-2})(a^{-3}) = \left(\frac{1}{?}\right)\left(\frac{1}{?}\right)$ $a^2; a^3$ **35.** $\left(\frac{1}{a^2}\right)\left(\frac{1}{a^3}\right) = \frac{1}{?}$ a^5 **36.** $\frac{1}{a^5} = a^?$ -5

37. What do the results of Exercises 34–36 seem to indicate about the product of powers rule? Create another example to support your conclusion. See answer on p. A41.

Use the results of Exercise 37 to find each product. Express your answer using positive exponents. 38. $\frac{14}{a^3b^4}$ **39.** $\frac{18r}{s^5}$ **40.** $\frac{-6kr^2}{t^3}$

38. $(7a^2b^{-3})(2a^{-5}b^{-1})$ **39.** $(3r^{-1}s^{-1})(6r^2s^{-4})$ **40.** $(3k^2t^{-6})(-2k^{-1}r^2t^3)$

SPIRAL REVIEW

                                               (Lesson 12-1)

41. A cube with sides numbered 1 through 6 is rolled. Find $P(2 \text{ or } 5)$. $\frac{1}{3}$

42. Find the volume of a rectangular aluminum container with a base area of 160 ft² and a height of 8 ft. *(Lesson 14-4)* 1280 ft³

43. Find the product: $4a(2ab + 5a - 3)$ *(Lesson 15-4)* $8a^2b + 20a^2 - 12a$

44. Solve the system: $x + y = 8$
                        $x - y = 4$ *(Lesson 13-4)* (6, 2)

684 Chapter 15

15-5 Multiplying Binomials

Objective: To multiply binomials.

CONNECTION

The rectangle at the right is $(2n + 3)$ units long and $(n + 2)$ units wide. You can find the area of the rectangle by counting the tiles representing n^2, n, and 1 and writing a polynomial to represent the area.

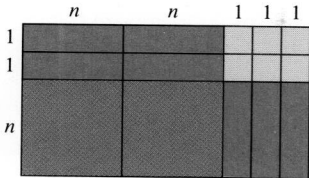

$$\text{Area} = 2n^2 + 7n + 6$$

Recall from geometry that the area of a rectangle is equal to the product of the base and the height, or $A = bh$. Applying this formula to the rectangle shown, you get the following.

$$\text{Area} = (2n + 3)(n + 2) = 2n^2 + 7n + 6$$

To multiply $(2n + 3)(n + 2)$ using algebra, you use the distributive property twice.

Example 1

Solution

Multiply: $(2n + 3)(n + 2)$

$$
\begin{aligned}
(2n + 3)(n + 2) &= 2n(n + 2) + 3(n + 2) \quad \longleftarrow \text{Use the distributive} \\
&= 2n^2 + 4n + 3n + 6 \quad\quad\quad\;\; \text{property.} \\
&= 2n^2 + 7n + 6
\end{aligned}
$$

☑ **Check Your Understanding**

1. In Example 1, why can you add $4n$ and $3n$?
2. In Example 1, will the answer be the same if $(2n + 3)(n + 2)$ is written as $(2n + 3)n + (2n + 3)2$?
   See *Answers to Check Your Understanding* at the back of the book.

Some polynomials involve subtraction. You must be careful to use the correct signs when multiplying such polynomials.

Example 2

Solution

Multiply: $(3x - 2)(7x + 5)$

$$
\begin{aligned}
(3x - 2)(7x + 5) &= 3x(7x + 5) - 2(7x + 5) \\
&= 21x^2 + 15x - 14x - 10 \\
&= 21x^2 + x - 10
\end{aligned}
$$

☑ **Check Your Understanding**

3. In Example 2, why is $2(7x + 5)$ subtracted rather than added?
4. In Example 2, what number is x multiplied by in $21x^2 + x - 10$?
   See *Answers to Check Your Understanding* at the back of the book.

Polynomials **685**

Lesson Planner

Teaching the Lesson
- Materials: algebra tiles, construction paper, scissors
- Lesson Starter 139
- Visuals, Folder Q
- Using Technology, p. 666C
- Reteaching/Alternate Approach, p. 666D
- Cooperative Learning, p. 666D

Lesson Follow-Up
- Practice 189
- Practice 190: Nonroutine
- Enrichment 153: Communication
- Study Guide, pp. 277–278
- Manipulative Activity 29
- Computer Activity 15
- Test 74

Warm-Up Exercises

Multiply.

1. $5k(3k - 4kc + c)$
   $15k^2 - 20k^2c + 5kc$
2. $3j(-3j^2 + 7j - j^3)$
   $-9j^3 + 21j^2 - 3j^4$
3. $5(6h^3 - 3h^2 + 2h)$
   $30h^3 - 15h^2 + 10h$
4. $-3(-3x^2 + 4xy - 5y^2)$
   $9x^2 - 12xy + 15y^2$
5. $(6a^2b^3)(3a^5b^4)$ $18a^7b^7$

Lesson Focus

Ask students if the distributive property can be used twice on the same expression. Most students will tend to say no. However, today's lesson focuses on such a double use of the distributive property in multiplying binomials.

Teaching Notes

Point out the double use of the distributive property while working through the Examples. Also, remind students to be careful about using the correct signs in exercises such as Example 2.

Key Question

When multiplying two binomials, is the order of the binomial factors important? No, multiplication of binomials is commutative.

Error Analysis

In finding the product of two binomials, students often make an error in determining the middle term. Remind students that the middle term is a combination of two like terms.

Using Technology

For a suggestion on using a spreadsheet to multiply binomials, see page 666C.

Reteaching/Alternate Approach

For a suggestion on using the FOIL method to multiply binomials, see page 666D.

Cooperative Learning

For a suggestion on using a cooperative learning activity with this lesson, see page 666D.

Additional Examples

Multiply.

1. $(3n + 2)(n + 4)$ $3n^2 + 14n + 8$
2. $(2x - 3)(5x + 7)$ $10x^2 - x - 21$

Closing the Lesson

Create a set of cards containing binomials. As in the previous lesson, have students randomly select two cards and explain how to find the product of the two binomials shown.

Guided Practice

COMMUNICATION « *Writing*

Write the multiplication represented by each rectangle.

« **1.**

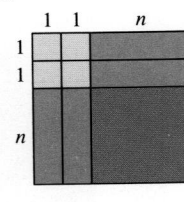

$(n + 2)(n + 2)$

« **2.**

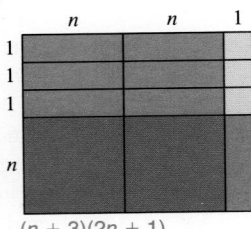

$(n + 3)(2n + 1)$

Draw a diagram similar to the ones in Exercises 1 and 2 to represent each multiplication. Answers to Guided Practice 3 and 4 are on p. A41.

« **3.** $(n + 2)(n + 3)$

« **4.** $(2x + 1)(x + 4)$

Choose the letter of the correct product.

5. $(5x - 3)(x + 2)$ B

6. $(5x + 3)(x - 2)$ A

7. $(5x - 3)(x - 2)$ E

8. $(5x + 3)(x + 2)$ D

A. $5x^2 - 7x - 6$

B. $5x^2 + 7x - 6$

C. $5x^2 + 7x + 6$

D. $5x^2 + 13x + 6$

E. $5x^2 - 13x + 6$

F. $5x^2 - 13x - 6$

Multiply. 9. $x^2 + 3x + 2$ 10. $t^2 - t - 6$ 11. $r^2 - 10r + 24$

9. $(x + 1)(x + 2)$

10. $(t + 2)(t - 3)$

11. $(r - 4)(r - 6)$

12. $(3x + 4)(2x + 4)$

13. $(2y - 1)(3y - 5)$

14. $(6x + 2)(x - 7)$

$6x^2 + 20x + 16$

$6y^2 - 13y + 5$

$6x^2 - 40x - 14$

Exercises

Multiply. Answers to Exercises 1–18 are on p. A41.

A 1. $(x + 4)(x + 2)$

2. $(a + 6)(a + 3)$

3. $(2t + 7)(6t + 5)$

4. $(3x + 1)(5x + 5)$

5. $(4y - 2)(6y + 7)$

6. $(5b + 3)(b - 1)$

7. $(x + 2)(7x - 8)$

8. $(7k - 1)(k + 8)$

9. $(w - 4)(2w - 1)$

10. $(d - 2)(3d - 5)$

11. $(x - 4)(4x - 1)$

12. $(5z - 2)(2z - 5)$

13. $(4x - 6)(3x - 5)$

14. $(6y - 7)(3y - 5)$

15. $(2x + 1)(2x + 1)$

16. $(5c - 3)(5c - 3)$

17. $(2x - 5)(2x + 5)$

18. $(4d + 7)(4d - 7)$

19. Find the product of $3x + 4$ and $5x - 6$. $15x^2 + 2x - 24$

20. Multiply $6m - 5$ by $3m - 9$. $18m^2 - 69m + 45$

Multiply.
21. $a^2 - 2ab - 15b^2$ 22. $2m^2 - 3mn + n^2$ 23. $x^2 - y^2$
24. $r^4 + 2r^2s - 8s^2$ 25. $2x^4 - 11x^2 + 15$ 26. $x^4 + 4x^2 + 3$

B **21.** $(a + 3b)(a - 5b)$ **22.** $(2m - n)(m - n)$ **23.** $(x + y)(x - y)$

24. $(r^2 - 2s)(r^2 + 4s)$ **25.** $(x^2 - 3)(2x^2 - 5)$ **26.** $(x^2 + 1)(x^2 + 3)$

THINKING SKILLS $6x^2 - 11x - 10; 6x^2 - 11x - 10$

27a. Knowledge
27b. Analysis; Synthesis
28a. Comprehension
28b. Synthesis

27. a. Find the products $(3x + 2)(2x - 5)$ and $(2x - 5)(3x + 2)$.
  b. Compare your answers in part (a). Create a mathematical statement relating the two products. What property is illustrated by these results? They are equal; $(3x + 2)(2x - 5) = (2x - 5)(3x + 2)$; commutative property of multiplication

28. a. Determine whether the product of two binomials is *always, sometimes,* or *never* a trinomial. sometimes
  b. Create examples to support your conclusion in part (a). Answers will vary. Examples are $(x + y)(x + y) = x^2 + 2xy + y^2$ and $(x + y)(x - y) = x^2 - y^2$

PROBLEM SOLVING/APPLICATION

A floor of an office building is being remodeled. The floor is currently divided into square work spaces. The new plans call for each work space to be 2 ft less in width and 3 ft less in length.

29. What do $x - 3$ and $x - 2$ represent? The new length and width

30. What does $(x - 3)(x - 2)$ represent? The new area

31. Find the product $(x - 3)(x - 2)$. $x^2 - 5x + 6$

32. What is the original area of each square cubicle in terms of x? x^2

33. Use your answers from Exercises 31 and 32 to find a polynomial that represents the decrease in area in terms of x. $5x - 6$

Polynomials   **687**

Suggested Assignments
Average: 1–26, 39–42
Advanced: 1–19 odd, 21–42

Exercise Notes

Communication: Writing
In *Guided Practice Exercises 1 and 2*, students must interpret the models by writing the multiplication they represent. In *Guided Practice Exercises 3 and 4*, the process is reversed, as students draw a model to illustrate the multiplication.

Critical Thinking
In *Exercises 27 and 28*, students *create* mathematical statements and examples after *analyzing* situations using logical reasoning.

Problem Solving
Exercises 29–33 present a practical application of the multiplication of binomials. Drawing a diagram is a problem-solving strategy that might be used here.

Making Connections/Transitions
Exercises 34–36 connect algebraic skills with an analysis of data in Pascal's triangle.

Computer
Exercises 37 and 38 show how a computer can be used to solve problems involving multiplication of binomials.

Exploration

Use gummed circles to label each side of three cubes. Label one cube 2, -3, x, x^2, x^3, and $-3x^2$; label another cube 1, -2, 4, $-x$, $2x^2$, and $4x^3$; label the third cube $+$, $+$, $+$, $-$, $-$, and $-$. Ask a student to roll all three cubes and write a variable expression using the two terms with the plus or minus sign between them. Have another student roll the cubes a second time and write another variable expression. The entire class should then find the product of the two expressions.

Enrichment

Enrichment 153 is pictured on page 666E.

Practice

Practice 189 is shown below.

Quick Quiz 1

See page 704.

Alternative Assessment

See page 704.

Practice 189, *Resource Book*

Practice 189
Skills and Applications of Lesson 15-5

Multiply.
1. $(c + 5)(x + 4)$ $x^2 + 9x + 20$ 2. $(z + 5)(z + 7)$ $z^2 + 12z + 35$
3. $(a + 5)(a + 3)$ $a^2 + 8a + 15$ 4. $(x + 4)(2x + 1)$ $2x^2 + 9x + 4$
5. $(5k + 2)(4k + 3)$ $20k^2 + 23k + 6$ 6. $(2c + 6)(3c + 1)$ $6c^2 + 20c + 6$
7. $(7e + 2)(2e + 5)$ $14e^2 + 39e + 10$ 8. $(b + 3)(b - 5)$ $b^2 - 2b - 15$
9. $(n + 6)(n - 2)$ $n^2 + 4n - 12$ 10. $(c + 7)(c - 8)$ $c^2 - c - 56$
11. $(t + 2)(t - 5)$ $t^2 - 3t - 10$ 12. $(w + 4)(3w - 1)$ $3w^2 + 11w - 4$
13. $(y - 4)(y + 6)$ $y^2 + 2y - 24$ 14. $(x - 1)(x + 5)$ $x^2 + 4x - 5$
15. $(2z - 9)(7z + 2)$ $14z^2 - 59z - 18$ 16. $(3a - 5)(4a + 3)$ $12a^2 - 11a - 15$
17. $(3z - 2)(4z + 1)$ $12z^2 - 5z - 2$ 18. $(k - 4)(k - 3)$ $k^2 - 7k + 12$
19. $(c - 5)(c - 1)$ $c^2 - 6c + 5$ 20. $(a - 6)(2a - 1)$ $2a^2 - 13a + 6$
21. $(4z - 5)(2z - 1)$ $8z^2 - 14z + 5$ 22. $(3n - 5)(5n - 1)$ $15n^2 - 28n + 5$

23. Find the product of $4z + 5$ and $2z + 3$. $8z^2 + 22z + 15$
24. Find the product of $2x + 3$ and $5x - 2$. $10x^2 + 11x - 6$
25. Find the product of $5a - 6$ and $2a + 3$. $10a^2 + 3a - 18$
26. Multiply $2t - 5$ by $3t - 7$. $6t^2 - 29t + 35$
27. Multiply $3m + 5$ by $4m - 7$. $12m^2 - m - 35$

DATA, *pages 666–667* Answers to Exercises 34–38 are on p. A41.

34. Write the sixth, seventh, and eighth rows of Pascal's triangle.

35. Use your answer to Exercise 34 to find each product.
   a. $(a + b)^5$ **b.** $(a + b)^6$ **c.** $(a + b)^7$

36. **RESEARCH** Find out more about the mathematical and scientific achievements of Blaise Pascal. Then write a brief report.

COMPUTER APPLICATION

You can program a computer in BASIC to print the product $(ax + b)(cx + d)$ as a polynomial in standard form when you input values for a, b, c, and d. You can base your program on this formula.

$$(ax + b)(cx + d) = acx^2 + (ad + bc)x + bd$$

Your print statement in BASIC might be the following.

```
PRINT A*C;"X^2 + ";A*D+B*C;"X + ";B*D
```

37. What do A, B, C, and D represent?

38. Explain how the computer would find $(2x + 3)(4x + 5)$.

SPIRAL REVIEW

S   39. Find the product: $(x + 4)(x - 3)$ *(Lesson 15-5)* $x^2 + x - 12$

40. Find the area of a circle with radius 8.5 cm.   *(Lesson 10-4)* 226.9 cm^2

41. What percent of 60 is 15?   *(Lesson 9-7)* 25%

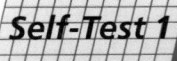

Self-Test 1

Write each polynomial in standard form.

1. $4x + 5x^3 + 6x^4 - 3x^2 - 8$ 2. $5a^2 - 7a + 4a^5 - 2 + 9a^3$ **15-1**
   $6x^4 + 5x^3 - 3x^2 + 4x - 8$ $4a^5 + 9a^3 + 5a^2 - 7a - 2$

Simplify.

3. $4x^3 + 4x^2 + 8x - x^3 + 7$ 4. $7c^3 - 3c - 5c + 4 - 2c^2 + 1$
   $3x^3 + 4x^2 + 8x + 7$ $7c^3 - 2c^2 - 8c + 5$

Find each answer.

5. $(x^2 + 6x + 2) + (x^2 + 2x + 6)$ $2x^2 + 8x + 8$ **15-2**

6. $(z^2 - 4z + 2) + (z^2 + z - 6)$ $2z^2 - 3z - 4$

7. $(2b^3 + b^2 - 4) - (b^3 - b^2 + 2)$ **15-3**

8. $(5n^2 + n - 1) - (n^2 + n + 3)$

9. $5xy^2(-2x^3y^4)$ $-10x^4y^6$ 10. $-3a^2(5a^3 - 3a^5)$ $9a^7 - 15a^5$ **15-4**

11. $(2x + 3)(4x + 5)$ 12. $(5c + 7)(4c - 8)$ $20c^2 - 12c - 56$ **15-5**

7. $b^3 + 2b^2 - 6$
8. $4n^2 - 4$
11. $8x^2 + 22x + 15$

688   Chapter 15

15-6

Strategy:
Working Backward

Objective: To solve problems by working backward.

In some problems, you are given an end result and are asked to find a fact needed to reach that result. One way to solve a problem of this type is to work backward.

Problem

Wendy defeated three opponents to win a chess tournament. At each stage of the tournament, the loser dropped out and the winner continued on to the next stage. How many players were in the tournament?

Solution

UNDERSTAND The problem is about players in a chess tournament.
Facts: winner defeated three players
Find: the number of players in the tournament

PLAN To solve, you work backward from the final match to the first match. Since Wendy needed three wins to complete the tournament, there must have been three stages. Start at the end of the tournament. Draw a diagram that shows the number of players at each stage of the tournament.

WORK To compete in Stage 3, each player had to win in Stage 2. To compete in Stage 2, each player had to win in Stage 1.

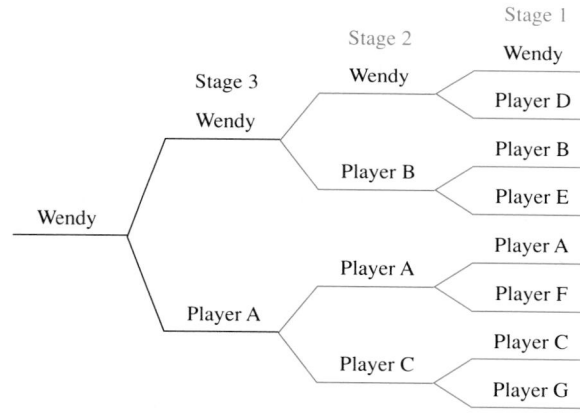

ANSWER There were 8 players in the tournament.

Look Back What if Wendy had defeated four opponents? How many players would have been in the tournament? 16

Lesson Planner

> **Teaching the Lesson**
> • Lesson Starter 140
>
> **Lesson Follow-Up**
> • Practice 191
> • Enrichment 154: Problem Solving
> • Study Guide, pp. 279–280

Warm-Up Exercises

Multiply.
1. $(x + 5)(x + 3)$ $x^2 + 8x + 15$
2. $(3m - 4)(4m - 3)$
   $12m^2 - 25m + 12$
3. $(2k - 7)(2k + 7)$ $4k^2 - 49$
4. $(3x + 2)(4x - 3)$ $12x^2 - x - 6$
5. $(d - 4)(3d + 8)$ $3d^2 - 4d - 32$

Lesson Focus

Ask students if they have ever arrived home from shopping to find they had less money than they thought. One way to account for the missing money is to retrace mentally how much money was spent. Today's lesson focuses on the problem-solving strategy of working backward.

Teaching Notes

When working through the lesson, remind students of the four-step process used to solve problems. Emphasize that a variety of approaches often may be used to solve a single problem.

Key Question

What types of problems are appropriate for the strategy of working backward?   problems in which the final result is known and some earlier information must be found

689

Reasoning

Determining the number of players that would have been in the tournament if Wendy had defeated four opponents requires students to re-examine the Problem and its solution.

Additional Example

Arthur wants to buy a compact disc player that costs $240 plus 5% sales tax. He earns $84 per week from his part-time job. How many weeks will Arthur have to work to buy the compact disc player?   3 weeks

Closing the Lesson

Ask a student to explain how he or she recently used the strategy of working backward to solve a real-world problem.

Suggested Assignments

Advanced: 1–16, Historical Note

Guided Practice

COMMUNICATION « *Reading*

« 1. What is the main idea of this lesson?
   solving problems by working backward

« 2. What other problem solving strategy was used to solve the problem on the previous page? drawing a diagram

Use this problem for Exercises 3–6.

Lawanda had $1125.76 in her checking account at the end of last week. During the week she had deposited $464.55 and had written checks for $129.87, $322.51, and $47.35. She also had made an electronic withdrawal of $100. How much money was in Lawanda's checking account at the beginning of the week?

« 3. What was the total amount of the checks? $499.73

« 4. In the process of solving the problem, do you *add* or *subtract* the total amount of the checks? add

« 5. In the process of solving the problem, do you *add* or *subtract* the deposit? subtract

6. Solve the problem. $1260.94

Solve by working backward.

7. In a tennis tournament, the loser of each match is eliminated. Evan played five matches to win the tournament. How many players were in the tournament? 32

8. Dana wants to buy a turtleneck and jeans. The turtleneck costs $28 and the jeans cost $35. Dana earns $5.25 per hour. How many hours must she work to earn enough money to buy the turtleneck and the jeans?
   12 h

Problem Solving Situations

Solve by working backward.

1. In each stage of a baseball tournament, a team that loses drops out. How many teams compete in the tournament if the winning team plays six games? 64

2. In a round-robin bridge tournament, every player plays every other player. Curtis played seven games and won the tournament. How many players were in the tournament? 8

3. Ming wants to buy a bicycle that costs $140 plus 5% sales tax. He works 4 h per day at a part-time job and earns $4.50 per hour. How many days will he have to work to buy the bicycle? 9

4. Alix has $1042.76 in her checking account at the end of the week. During the week, she wrote a check for $79.09, made a deposit of $243.87, and wrote another check for $120.97. She also transferred $225 from her savings account into her checking account and made an electronic withdrawal of $50. How much money did Alix have in her checking account at the beginning of the week? $823.95

5–9. Strategies may vary; likely strategies are given.

5. working backward; 1.1 mi
6. using equations; 14
7. supplying missing facts; gallons; $.26
8. working backward; 7:35

Solve using any problem solving strategy.

5. Meredith takes a cab to a business meeting. The ride costs $1.20 for the first one-tenth of a mile, and $.90 for each additional one-tenth of a mile. At the end of the ride, Meredith gives the cab driver $15 and asks for $3 back in change, leaving the cab driver with a $1.80 tip. How many miles was the cab ride?

6. The sum of twelve and five times a number is eighty-two. Find the number.

7. Sonia needs 4 qt of oil to change the oil in her car. A gallon of oil costs $4.50, and a quart of oil costs $1.19. To get the best buy, should she buy quarts or gallons? How much can she save?

8. Kristen has a job interview at 9:45 A.M., and she wants to be there 15 min early. She needs 25 min to travel from her home to the interview, and she wants to allow an hour and a half to dress and have breakfast. At what time should Kristen get up?

> **PROBLEM SOLVING**
> ## CHECKLIST
> ### Keep these in mind:
> Using a Four-Step Plan
> Too Much or Not Enough Information
> Supplying Missing Facts
> Problems with No Solution
> ### Consider these strategies:
> Choosing the Correct Operation
> Making a Table
> Guess and Check
> Using Equations
> Identifying a Pattern
> Drawing a Diagram
> Using Proportions
> Using Logical Reasoning
> Using a Venn Diagram
> Using a Simpler Problem
> Making a Model
> Working Backward

Exercise Notes

Communication: Reading
Guided Practice Exercises 1 and 2 require students to state the main idea of the lesson and to list any other strategy used to solve the Problem on page 689.

Problem Solving
The problem-solving strategy of working backward is the focus of this lesson. Other strategies are reviewed in the CHECKLIST and used in *Exercises 5–10.*

Polynomials **691**

Exercise Notes (continued)

Communication: Writing

Exercises 11 and 12 ask students to create their own problems that can be solved by working backward.

Follow-Up

Project (Cooperative Learning)

As a research project, have students work in cooperative groups to investigate the rule of false position mentioned in the Historical Note. Some groups may choose to include information about mathematical reasoning, both inductive and deductive.

Enrichment

Enrichment 154 is pictured on page 666E.

Practice

Practice 191, *Resource Book*

---

9. using equations; 2 h

9. A schedule for a sales meeting showed that Pat was given half the total available time to do her presentation. Donnell was given two thirds of the remaining time, and Cody and Kim were given ten minutes each. How long was the sales meeting expected to last?

10. Anthony's company earned $40,000 in January. The earnings declined 10% each month for the next three months. Following this, the earnings increased for the next two months by 5% each month. In reporting this to the Board of Directors, Anthony decided to use a data display. Which type of display should be used? Make two data displays and compare them. Which shows the information better?
See answer on p. A42.

WRITING WORD PROBLEMS Answers to Problems 11 and 12 are on p. A42.

Write a word problem about the given subject that you could solve by working backward. Then solve the problem.

11. The total number of ancestors Peter has within a given number of generations.

12. The price of a share of stock at the beginning of a week given its price at the end of the week.

SPIRAL REVIEW

S   **13.** A right triangle has legs of 10 cm and 24 cm. Find the length of the hypotenuse.   *(Lesson 10-11)* 26 cm

14. Bill wants to take $120 as spending money on his family's vacation. Bill earns $7.50 per hour from his after-school job. How many hours must Bill work to earn the $120?   *(Lesson 15-6)* 16 h

15. Graph $y \geq 2x$.   *(Lesson 13-8)* See answer on p. A42.

16. Find the sum: $(6x^3 + 9x - 4) + (-2x^3 + 3x^2 - 4x)$ *(Lesson 15-2)*
$4x^3 + 3x^2 + 5x - 4$

Historical Note

Solving problems by working backward is not a new problem solving strategy. It was used at least 1500 years ago in India, most notably by the mathematician Aryabhata. Most of what is known of Aryabhata's work comes from his book on astronomy, the third chapter of which is devoted to mathematics.

See answer on p. A42.

Research

Indian mathematicians solved many problems by the rule of false position. Find out what this rule is and how it was used.

692 Chapter 15

Factoring Polynomials

Objective: To factor polynomials.

EXPLORATION

1 Find the GCF of 24 and 78. 6

2 Find the GCF of $24a^3b^4$ and $78a^2b^5$. $6a^2b^4$

3 Complete: $24a^3b^4 + 78a^2b^5 = \underline{\ ?\ }(4a + 13b)$ $6a^2b^4$

4 Complete: Some polynomials can be written as the product of the $\underline{\ ?\ }$ of their terms and another polynomial. GCF

Terms to Know
- *greatest common monomial factor*
- *factor a polynomial*

The **greatest common monomial factor** of a polynomial is the GCF of its terms. To **factor a polynomial**, you express the polynomial as the product of other polynomials. These polynomials should contain terms involving whole numbers only, no fractions.

Example

Factor each polynomial.

a. $6t^4 + 14t^3 - 24t^2$

b. $3c^4d^2 - 24c^3d^3 - 15c^2d$

Solution

a. The GCF of $6t^4$, $14t^3$, and $24t^2$ is $2t^2$.

$$6t^4 + 14t^3 - 24t^2 = 2t^2(3t^2) + 2t^2(7t) - 2t^2(12)$$
$$= 2t^2(3t^2 + 7t - 12)$$

b. The GCF of $3c^4d^2$, $24c^3d^3$, and $15c^2d$ is $3c^2d$.

$$3c^4d^2 - 24c^3d^3 - 15c^2d = 3c^2d(c^2d) - 3c^2d(8cd^2) - 3c^2d(5)$$
$$= 3c^2d(c^2d - 8cd^2 - 5)$$

✓ Check Your Understanding

1. In part (a), why is $2t^2$ the greatest common monomial factor?
2. In part (b), why is the answer not $3(c^4d^2 - 8c^3d^3 - 5c^2d)$?
3. Describe a way to check the answers in the Example.
   See *Answers to Check Your Understanding* at the back of the book.

Guided Practice

COMMUNICATION « *Reading*

Explain the meaning of the word *factor* **in the following sentences.**

« **1.** The rain was a factor in his poor golf score. reason

« **2.** $6t$ is a factor of $18t^2$ and $30t$. a number that divides into another number

« **3.** Before Camille could state the profit her company made, she had to factor out the production costs.
   consider the effects of production costs on profit

Polynomials **693**

Lesson Planner

Teaching the Lesson
- Lesson Starter 141
- Using Technology, p. 666C

Lesson Follow-Up
- Practice 192
- Enrichment 155: Connection
- Study Guide, pp. 281–282

Warm-Up Exercises

Find the GCF of each pair.

1. 14 and 36   2
2. 56 and 72   8
3. $2x^2$ and $3x^3$ x^2
4. $12n^3$ and $15n^5$ $3n^3$
5. $28xy^2$ and $21x^2y$ $7xy$

Lesson Focus

Work some examples of polynomials multiplied by monomials. Tell students that the process can be reversed. Today's lesson focuses on the reverse process, called factoring.

Teaching Notes

When working through the lesson, point out the similarities between finding the GCF of whole numbers and finding the greatest common monomial factor of polynomials.

Key Question

Why is it important to examine both the coefficients and the variables when determining the greatest common monomial factor? The greatest common monomial factor contains the GCF of the coefficients and the variables.

Error Analysis
Students may not use the greatest common monomial factor when factoring. Stress to students that they must find the *greatest* common factor of the terms before factoring.

Using Technology
For a suggestion on using a calculator or a spreadsheet program, see page 666C.

Additional Examples
Factor each polynomial.
1. $8y^4 + 12y^3 - 24y^2$
   $4y^2(2y^2 + 3y - 6)$
2. $5x^4y^2 - 25x^3y^3 - 20x^2y$
   $5x^2y(x^2y - 5xy^2 - 4)$

Closing the Lesson
Ask a student to explain how to find the GCF of two whole numbers. Ask another student to explain how to find the greatest common monomial factor of two polynomials. Ask a third student to factor a given polynomial.

Suggested Assignments
Advanced: 1–29 odd, 31–40

Exercise Notes
Communication: Reading
Guided Practice Exercises 1–3 ask students to explain various meanings of the term *factor.*

Communication: Discussion
Guided Practice Exercises 22 and 23 have students discuss certain conditions relating to the greatest common monomial factor.

Match each polynomial with its greatest common monomial factor.

4. $3a^2b + 6ab$ C
5. $4a^2b + 6ab^2$ E
6. $12a^2b + 8ab^2$ A
7. $4a^2b^2 + 6a^2b^3$ B
8. $12a^2b^3 + 8a^3b^2$ F
9. $3a^4b^2 + 6a^2b^3$ D

A. $4ab$
B. $2a^2b^2$
C. $3ab$
D. $3a^2b^2$
E. $2ab$
F. $4a^2b^2$

Factor each polynomial. 16. $3z(5z^3 - 6z + 4)$ 17. $6p^2(4p^3 - 3p - 6)$

10. $4c^2 + 6c$ $2c(2c + 3)$
11. $3n^2 + 7n$ $n(3n + 7)$
12. $16e + 2e^2$ $2e(8 + e)$
13. $e^3 + 12e$ $e(e^2 + 12)$
14. $7m^5 - 8m^3$ $m^3(7m^2 - 8)$
15. $14k^5 - 21k^2$ $7k^2(2k^3 - 3)$
16. $15z^4 - 18z^2 + 12z$
17. $24p^5 - 18p^3 - 36p^2$
18. $8n^4 + 4n^2 + 5n$ $n(8n^3 + 4n + 5)$
19. $3t^2 - 15t + 9$ $3(t^2 - 5t + 3)$
20. $6p^2t + 8pt^2 + 12pt$
    $2pt(3p + 4t + 6)$
21. $7rs^3t + 14r^3s^2t^2 + 21rst$
    $7rst(s^2 + 2r^2st + 3)$

COMMUNICATION «*Discussion*

«22. Under what conditions does a polynomial not have a greatest common monomial factor? when its terms are relatively prime

«23. Can the greatest common monomial factor of a polynomial be equal to one of its terms? Explain. Yes; for example, $4x^2 + 2x$

Exercises

Factor each polynomial. Answers to Exercises 1–20 are on p. A42.

A
1. $7x^2 + 14x$
2. $18y^2 - 6y$
3. $4s + 5s^2$
4. $m^4 + 2m^2$
5. $4k^3 + 8k^2$
6. $9n^3 - 3n^2$
7. $5p^5 - 15p^2$
8. $9a^3 - 27a^2$
9. $9xy + 6x^2y$
10. $ab^2 - 3a^2b^2$
11. $4ac - 8bc$
12. $7rst - 5r^2t^2$
13. $12u^4 - 9u^3 - 24u^2$
14. $24t^5 - 18t^2 - 6$
15. $8s^4 - 16s^3 + 32s^2$
16. $18q^5 - 9q^4 - 9q^3 - 18$
17. $25qd^2 + 15q^2d + 30q^2d^2 - 5qd$
18. $35a^5 - 49a^3b^2 - 56a^2b^4$
19. $36y^4z^3 + 45y^3z^4 + 20y^5z^2$
20. $22k^6t^5 + 110k^4t^6 - 55k^7t^5$

B
21. $24r^2t^2u^2 - 12r^3t^3u^3 - 6r^4t^4u^4$ $6r^2t^2u^2(4 - 2rtu - r^2t^2u^2)$
22. $21a^5c^2f - 14ac^2f^2 - 6a^3cf$ $acf(21a^4c - 14cf - 6a^2)$
23. $16a^2b^2c^2 + 4ab^2c + 8a^3b^2c^4 + 12a^2b^2c$ $4ab^2c(4ac + 1 + 2a^2c^3 + 3a)$
24. $9wxy^2 + 15wx^2y^3 + 21wx^2y^2$ $3wxy^2(3 + 5xy + 7x)$

25. $(r - 5)(r + 6)$
26. $(x + 2)(3x - 5)$
27. $(2 - 3d)(7d - 1)$
28. $(8z - 9)(z + 1)$
29. $(y - 3)(y - 1)$
30. $(m + 4)(m + 1)$

Factor each polynomial as the product of two binomials.

25. $r(r - 5) + 6(r - 5)$

26. $3x(x + 2) - 5(x + 2)$

27. $7d(2 - 3d) - (2 - 3d)$

28. $z(8z - 9) + (8z - 9)$

29. $(y - 3)(y - 3) + 2(y - 3)$

30. $(m + 4)(m + 4) - 3(m + 4)$

GROUP ACTIVITY

31. Create the simplest polynomial possible with four terms using the variables r, s, and t and with t as the greatest common monomial factor. $rt + st + t^2 + t$

32. Create the simplest polynomial possible with three terms using the variables w, y, and z and with 1 as the greatest common monomial factor. $w + y + z$

CONNECTING ALGEBRA AND GEOMETRY

Find the area of the shaded part of each figure in terms of π. Give your answer in factored form.

C 33.

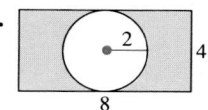

$4(8 - \pi)$

34.

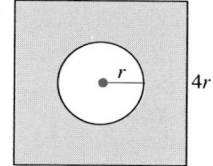

$r^2(16 - \pi)$

35.

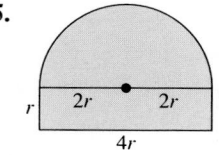

$2r^2(2 + \pi)$

36.

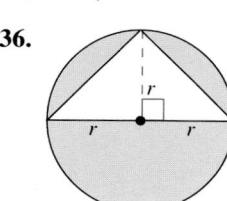

$r^2(\pi - 1)$

SPIRAL REVIEW

S 37. Find the number of permutations of 8 runners in 8 lanes. *(Lesson 12-6)* 8! or 40,320

38. Factor: $16a^2b^3 - 4a^2b^4 + 12a^3b^2$ *(Lesson 15-7)* $4a^2b^2(4b - b^2 + 3a)$

39. Find the difference:
$(11z^4 - 4z^3 + 2z^2 + 6z - 4) - (4z^4 - 3z^2 + 7z - 8)$
(Lesson 15-3) $7z^4 - 4z^3 + 5z^2 - z + 4$

40. Find the surface area of a rectangular prism with length 9 cm, height 4 cm, and width 6 cm. *(Lesson 14-2)* 228 cm²

Polynomials **695**

Cooperative Learning
Exercises 31 and 32 provide an opportunity for students to work in groups to create polynomials that satisfy certain conditions.

Making Connections/Transitions
Exercises 33–36 relate the areas of geometric shapes to algebra through the use of variables.

Follow-Up

Enrichment *(Cooperative Learning)*
Have students work in groups to explore various methods of factoring. They may wish to use first-year algebra textbooks or other resource material. Have each group make a presentation to the class on some aspect of factoring that has not been covered in this course.

Enrichment
Enrichment 155 is pictured on page 666F.

Practice
Practice 192, *Resource Book*

Practice 192
Skills and Applications of Lesson 15-7

Factor each polynomial. Ex. 22, 24–32. See below.
1. $10x^2 + 15x$ $5x(2x + 3)$
2. $6z^2 - 15z$ $3z(2z - 5)$
3. $15n^4 - 3n$ $3n(5n^3 - 1)$
4. $14c^2 - 21c^4$ $7c^2(2 - 3c^2)$
5. $4a^3 + 12a^2$ $4a^2(a^3 + 3)$
6. $12u^3 - 54u$ $6u(2u^2 - 9)$
7. $6t^4 - 4t^3$ $2t^3(3t - 2)$
8. $12s^5 - 3s^4$ $3s^4(4s - 1)$
9. $9y^7 + 27y^4$ $9y^4(y^3 + 3)$
10. $ab - 2a^2b$ $ab(1 - 2a)$
11. $x^3y^3 - x^2y$ $x^2y(xy^2 - x)$
12. $9c^3d^2 - 3c^2d^3$ $3c^2d^2(3c - d)$
13. $4e^2z - e^2u$ $e^2z(4e - f)$
14. $6b^2z^2 - 4b^2z$ $2bz(3s^2z - 2b)$
15. $8x^4y^3 + 20x^3z$ $4x^3y(2ry^2 + 5)$
16. $16m^2x^2 - 10mxt$ $2mx(8mx - 5t)$
17. $10k^4n^4 - 35kn^3$ $5kn^3(2k^3n - 7)$
18. $12x^2z^3 + 20x^2z^2$ $4x^2z^2(3rz + 5)$
19. $5x^5 - 10x + 10$ $5(x^5 - 2x + 2)$
20. $3u^3 + 2u^3 + 5u^2$ $u^2(3u^3 + 2u + 5)$
21. $18x^3 + 9t^2 - 15t$ $3t(6t^2 + 3t - 5)$
22. $18k^4 - 45k^3 + 18c$
23. $5k^8 - 35k^2$ $5k^2(k^6 - 7)$
24. $14x^{12} - 21x^9 + 7x^3$
25. $24m^4z^4 - 16m^2z^6 - 20mz^3$
26. $8k^3z^4 - 16k^4z^3 + 24k^3r^2$
27. $14r^3z^6 - 10r^2z^5$
28. $3x^3yz - 3xy^3z + 3xyz^3$
29. $4a^3bc - 6a^2b^2c + 4a^2bc^2$
30. $63x^3z^2b^3 + 49z^3db^4 - 21x^2u^2b^4$
31. $30w^4z^3z^2 - 12w^3z^2z$
32. $9u^3r^2b^2 - 9z^3z^3z^3 - 9u^2r^3z^3$
22. $9c(2c^3 - 5c + 2)$
24. $7x^3(2x^9 - 3x^6 + 1)$
25. $4mz^3(6m^3z - 4mz^3 - 5)$
26. $8k^3r^2(k^2r^2 - 2kr + 3)$
27. $2r^2t^5(7rst - 5)$
28. $3xyz(x^2 - y^2 + z^2)$
29. $2a^2bc(2a - 3b + 2c)$
30. $7x^2ab^3(9xa + 7xb - 3ab)$
31. $6w^3x^2z(5wxz - 2)$
32. $9a^2r^3t^2(ar - at - rt)$

494
Resource Book, MATHEMATICAL CONNECTIONS
Copyright © by Houghton Mifflin Company. All rights reserved.

Teaching the Lesson
- Lesson Starter 142
- Using Technology, p. 666C

Lesson Follow-Up
- Practice 193
- Practice 194: Nonroutine
- Enrichment 156: Connection
- Study Guide, pp. 283–284
- Manipulative Activity 30
- Test 75

Warm-Up Exercises

Find the greatest common monomial factor for each polynomial.

1. $4x^2y + 8xy$ $4xy$
2. $15a^2b^2 + 25a^2$ $5a^2$

Factor each polynomial.

3. $12m^2n + 16mn + 8mn^2$
   $4mn(3m + 4 + 2n)$
4. $18x^2y^2 + 9x^2y + 3xy$
   $3xy(6xy + 3x + 1)$
5. $3g^2 - 9g + 12gt$ $3g(g - 3 + 4t)$

Lesson Focus

Ask students to recall how to write a fraction in lowest terms. Then ask them to recall how to simplify an algebraic fraction. Today's lesson focuses on the next step in this process, dividing a polynomial by a monomial.

Teaching Notes

Key Question
Is it possible for the addition and subtraction signs in a polynomial to change when it is divided by a monomial?   Yes, if the monomial has a negative sign in front of it.

Dividing a Polynomial by a Monomial

Objective: To divide a polynomial by a monomial.

 **CONNECTION**

To add or subtract fractions with like denominators, you use the following rules.

$$\frac{a}{c} + \frac{b}{c} = \frac{a + b}{c} \quad \text{and} \quad \frac{a}{c} - \frac{b}{c} = \frac{a - b}{c}$$

By using these rules in reverse, you can divide a polynomial by a monomial.

Example 1

Divide: $\dfrac{4a^5 + 8a^4 + 6a^2}{2a}$

Solution

$$\frac{4a^5 + 8a^4 + 6a^2}{2a} = \frac{4a^5}{2a} + \frac{8a^4}{2a} + \frac{6a^2}{2a}$$

$$= \frac{4a^{5-1}}{2} + \frac{8a^{4-1}}{2} + \frac{6a^{2-1}}{2} \quad \longleftarrow \text{Use the quotient of powers rule.}$$

$$= 2a^4 + 4a^3 + 3a$$

☑ **Check Your Understanding**

1. In Example 1, why is 1 being subtracted from each exponent in the numerator?
2. How would the answer to Example 1 be different if the dividend were $4a^5 - 8a^4 + 6a^2$?

See *Answers to Check Your Understanding* at the back of the book.

> **Generalization:** *Dividing a Polynomial by a Monomial*
>
> To divide a polynomial by a monomial, divide each term of the polynomial by the monomial and simplify.

Example 2

Divide: $\dfrac{5x^7y^4 - 35x^5y^5 + 20x^3y^3}{-5x^3y}$

Solution

$$\frac{5x^7y^4 - 35x^5y^5 + 20x^3y^3}{-5x^3y} = \frac{5x^7y^4}{-5x^3y} - \frac{35x^5y^5}{-5x^3y} + \frac{20x^3y^3}{-5x^3y}$$

$$= -x^4y^3 - (-7x^2y^4) + (-4y^2)$$

$$= -x^4y^3 + 7x^2y^4 - 4y^2$$

☑ **Check Your Understanding**

3. In Example 2, why do the addition and subtraction signs change?

See *Answers to Check Your Understanding* at the back of the book.

Guided Practice

Error Analysis
Students may mistakenly divide only one term of a polynomial when dividing it by a monomial. Remind students that each term of the polynomial is divided by the divisor.

Using Technology
For a suggestion on using a calculator or a spreadsheet program, see page 666C.

COMMUNICATION « *Reading*

Replace each __?__ with the correct word or phrase.

« **1.** To divide a polynomial by a monomial, you use the rules for adding and subtracting __?__ in reverse. fractions

« **2.** To divide a polynomial by a monomial, divide __?__ by the monomial and then __?__. each term; simplify

Match each division with the correct quotient.

3. $\dfrac{10ab - 15a^2}{5a}$ C

4. $\dfrac{10ab + 15a^2}{5a}$ A

5. $\dfrac{10ab - 15a^2}{-5a}$ D

6. $\dfrac{10ab + 15a^2}{-5a}$ B

A. $2b + 3a$
B. $-2b - 3a$
C. $2b - 3a$
D. $-2b + 3a$

Divide.

7. $\dfrac{14a^2 - 7a}{7}$ $2a^2 - a$

8. $\dfrac{16c^5 + 3c^3}{c}$ $16c^4 + 3c^2$

9. $\dfrac{24b^6 + 6b^2}{6b^2}$ $4b^4 + 1$

10. $\dfrac{4b^3 + 10b^2}{-2b}$ $-2b^2 - 5b$

11. $\dfrac{6d^6 - 4d^4}{d^3}$ $6d^3 - 4d$

12. $\dfrac{12x^7 - 18x^5}{-3x^4}$ $-4x^3 + 6x$

Exercises

Divide. **5.** $x + 2x^3 - 6x^6$

A **1.** $\dfrac{9x - 12y}{3}$ $3x - 4y$

2. $\dfrac{16r + 8s^2}{-8}$ $-2r - s^2$

3. $\dfrac{7c^3 - c^2}{-c}$ $-7c^2 + c$

4. $\dfrac{5m^7 + 4m^2}{m^2}$ $5m^5 + 4$

5. $\dfrac{4x^2 + 8x^4 - 24x^7}{4x}$

6. $\dfrac{15k^9 - 6k^6}{-3k}$ $-5k^8 + 2k^5$

7. $\dfrac{24t^8 + 64t^3 + 8t^2}{8t^2}$ $3t^6 + 8t + 1$

8. $\dfrac{12s^8 - 36s^6 - 42s^4}{6s^3}$ $2s^5 - 6s^3 - 7s$

9. $\dfrac{-15k^6 - 5k^5 + 60k^4}{-5k^2}$

10. $\dfrac{3m^6 + 4m^5 - 7m^4}{-m}$

11. $\dfrac{21de + 24de^2 + 27d^2e}{3de}$

12. $\dfrac{16k^2t^3 - 24k^3t^2 + 32k^2t^2}{8kt}$

13. $\dfrac{16r^4u^5 - 12r^7u^6}{-4r^4u^5}$ $3r^3u - 4$

14. $\dfrac{48m^2n^2 + 42mn^4 - 54m^3n^5}{-6n^2}$

15. Divide $12x^3y^4 + 21x^4y^2 - 36x^5y^4$ by $3x^2y^2$. $4xy^2 + 7x^2 - 12x^3y^2$

16. Find the quotient when $30a^3b^7 - 15a^7b^5 + 25a^5b^5$ is divided by $5a^2b^4$. $6ab^3 - 3a^5b + 5a^3b$

9. $3k^4 + k^3 - 12k^2$
10. $-3m^5 - 4m^4 + 7m^3$
11. $7 + 8e + 9d$
12. $2kt^2 - 3k^2t + 4kt$
14. $-8m^2 - 7mn^2 + 9m^3n^3$

Additional Examples
Divide.

1. $\dfrac{6x^5 + 8x^4 + 4x^2}{2x}$

$3x^4 + 4x^3 + 2x$

2. $\dfrac{9a^8b^5 - 33a^5b^5 + 21a^3b^3}{-3a^3b}$

$-3a^5b^4 + 11a^2b^4 - 7b^2$

Closing the Lesson
Present students with some worked-out examples that contain incorrect steps, such as not subtracting exponents correctly. Tell students to find all the mistakes in each example.

Suggested Assignments
Advanced: 1–29 odd, 31–40

Polynomials **697**

Divide.

19. $-4xy - \frac{5}{2}x^2y - \frac{2}{3}x^3y$

20. $\frac{4}{3}w^2 + \frac{5}{3}wx + 3x^2$

22. $-5b^2 - \frac{4b^3}{a} - \frac{3b^4}{a^2}$

B 17. $\dfrac{3k^4 - 18k^3}{9k^3}$ $\frac{1}{3}k - 2$

18. $\dfrac{15r^3t^3 - 9r^2t^4}{21t^3}$ $\frac{5}{7}r^3 - \frac{3}{7}r^2t$

19. $\dfrac{24x^2y^2 + 15x^3y^2 + 4x^4y^2}{-6xy}$

20. $\dfrac{12w^4x + 15w^3x^2 + 27w^2x^3}{9w^2x}$

21. $\dfrac{3x^4 + 7x^3 + 9x^2}{2x^3}$ $\frac{3}{2}x + \frac{7}{2} + \frac{9}{2x}$

22. $\dfrac{50a^2b^4 + 40ab^5 + 30b^6}{-10a^2b^2}$

23. $\dfrac{25b^5 + 20b^4 + 5b^2}{15b^3}$ $\frac{5}{3}b^2 + \frac{4}{3}b + \frac{1}{3b}$

24. $\dfrac{8a^5 + 16a^3 + 4a}{6a^2}$ $\frac{4}{3}a^3 + \frac{8}{3}a + \frac{2}{3a}$

Simplify.

25. $\dfrac{8x - 6}{2} + \dfrac{12x + 9}{3}$ $8x$

26. $\dfrac{5a - 10}{5} + \dfrac{6 + 9a}{3}$ $4a$

27. $\dfrac{c^2 + 6c}{c} + \dfrac{4c^2 - 6c}{2c}$ $3c + 3$

28. $\dfrac{n^2 + 2n}{n} - \dfrac{2n^2 - 4n}{2n}$ 4

29. $\dfrac{a^2b + 2a^2b^2}{ab} + \dfrac{2a^2 - 6a^2b}{2a}$ $2a - ab$

30. $\dfrac{m^3n^2 - 2m^2n^3}{m^2} + \dfrac{m^2n^4 - 4m^5n^2}{mn^2}$ $-4m^4 - 2n^3 + 2mn^2$

CAREER APPLICATION

A *physics teacher* uses many formulas, some of which involve polynomials. For example, the height in feet at t seconds of an object thrown vertically upward with a speed of v ft/s is $h = vt - 16t^2$.

C 31. A ball is thrown vertically upward with a speed of 64 ft/s. Write the specific formula that gives its height at t seconds. $h = 64t - 16t^2$

32. Factor the polynomial in the formula that you wrote in Exercise 31. $h = 16t(4 - t)$

33. Make a table using whole-number values of t to find the maximum height of the ball.
See answer on p. A42.

34. Use the table you made in Exercise 33 to find the two times at which the height of the ball is 0 ft. What do these times represent?
0s; 4s; when the ball is about to be thrown and when the ball has fallen back to its starting point.

LOGICAL REASONING

Solve each problem. Apply the reasoning you use to solve Exercise 35 to help you solve Exercise 36.

35. A cardboard box has a height of 12 in. The volume of the box is 3192 in.^{3}. Find the area of the base of the box. Use the formula $V = Bh$, where B is the area of the base. 266 in.^{2}

36. The volume of a rectangular prism is $4x^2y^2 + 8x^2y + 2xy^2 + 4xy$. The height of the prism is $2xy$. Find the area of the base. Use the formula $V = Bh$. $2xy + 4x + y + 2$

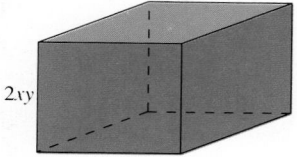
$2xy$

SPIRAL REVIEW

S **37.** Make a box-and-whisker plot to display the given data. *(Lesson 11-4)* See answer on p. A42.

Gasoline Mileage (mi/gal)

26.0	21.8	26.9	26.3	30.5	19.5
23.4	20.8	22.4	24.8	22.5	22.5

38. Find the quotient: $(20b^3c^2 - 10bc^4 + 5b^2c) \div 5bc$ *(Lesson 15-8)* $4b^2c - 2c^3 + b$

39. Find the product: $3x(7x^3 - 2x)$ *(Lesson 15-4)* $21x^4 - 6x^2$

40. Find the area of a parallelogram with height 16 in. and base 7 in. *(Lesson 10-3)* 112 in.^{2}

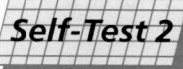

Self-Test 2

Solve by working backward.

1. Shari is saving to buy a new bicycle. The bicycle costs $180 and the sales tax is 5%. Shari earns $4.50 per hour at her weekend job. How many hours will Shari have to work to buy the bicycle? 42 h   15-6

Factor.

2. $12c^2 - 8c - 4cd$
$4c(3c - 2 - d)$

3. $25m^2n + 15mn^2 - 10mn$ 15-7
$5mn(5m + 3m - 2)$

Divide.

4. $\dfrac{18a^3 - 12a^2 + 6a}{6a}$ $3a^2 - 2a + 1$ **5.** $\dfrac{a^2b^2 + 9ab^2 - 6a^2b}{ab}$ 15-8
   $ab + 9b - 6a$

Chapter Review

Terms to Know

polynomial (p. 668)
monomial (p. 668)
binomial (p. 668)
trinomial (p. 668)

standard form of a polynomial
(p. 668)
greatest common monomial factor
(p. 693)
factor a polynomial (p. 693)

Choose the correct term from the list above to complete each sentence.

1. A polynomial with two terms is a __?__. binomial

2. When you write the terms of a polynomial in order from highest to lowest, you have written the __?__. standard form of a polynomial

3. The GCF of the terms of a polynomial is the __?__ of the polynomial. greatest common monomial factor

4. A trinomial is a __?__ with three terms. polynomial

Write each polynomial in standard form. *(Lesson 15-1)*

5. $4x + 7x^3 + 2x^2 + 6x^4 + 7$

6. $8c^2 + 4c^5 - 5c - 9c^3 + 2 + 11c^4$

7. $8 - 4a^3 + a^6 - 2a^4 - 3a$

8. $5m^7 - 2m^3 + 4m - 9m^2 + 6m^5$

Simplify. *(Lesson 15-1)*

9. $4x^3 + 14x^2 + x + 3$
11. $4c^3 + c^2 - 5c + 13$

10. $8b^4 - 6b^2 + b - 5$
12. $-c^4 - 9c^3 + 5c^2 + 3c + 8$

9. $4x^3 + 6x^2 - x + 8x^2 + 3 + 2x$

10. $b^4 + 3b^3 - 6b^2 - 3b^3 + 7b^4 + b - 5$

11. $6c^2 + 3c^3 - 5c^2 + 4 + c^3 - 5c + 9$

12. $c^4 + 4c^2 - 2c^4 + c^2 - 9c^3 + 3c + 8$

13. $-3x^3 + 6x^2 - 7x - 2x^3 + 5 + 4x$
$-5x^3 + 6x^2 - 3x + 5$

14. $-8a^2 - 3a^3 + 7a^2 + a^3 - 5a + 11$
$-2a^3 - a^2 - 5a + 11$

Add. *(Lesson 15-2)* **15.** $5a^2 + 7a + 16$

15. $(3a^2 + 2a + 7) + (2a^2 + 5a + 9)$

16. $(x^3 + x^2 + x + 6) + (x^3 + 3x^2 + x + 1)$
$2x^3 + 4x^2 + 2x + 7$

17. $(n^5 + 4n^3 + 7n) + (2n^5 + 2n + 6)$

18. $(2d^4 + 5d^2 + d + 1) + (3d^2 + 6d + 3)$

19. $(7c^3 + 5c^2 - 4c - 6) + (-2c^3 - c + 1)$
$5c^3 + 5c^2 - 5c - 5$

20. $(n^4 - 2n^3 + n - 1) + (n^3 + 5n^2 - n + 4)$
$n^4 - n^3 + 5n^2 + 3$

Subtract. *(Lesson 15-3)* **21.** $4x^2 + 3x + 5$

21. $(7x^2 + 5x + 9) - (3x^2 + 2x + 4)$

22. $(5a^2 + 8a + 4) - (2a^2 + 4a + 5)$
$3a^2 + 4a - 1$

23. $(8c^2 + 5c - 2) - (3c^2 - 7c + 5)$

24. $(-b^2 - 6b + 2) - (3b^2 + 3b - 6)$

25. $(5x^4 + 2x^3 - x + 6) - (2x^4 + 3x - 4)$

26. $(5a^3 - 6a^2 + 3) - (6a^3 - 2a^2 + 5a + 9)$

5. $6x^4 + 7x^3 + 2x^2 + 4x + 7$
7. $a^6 - 2a^4 - 4a^3 - 3a + 8$
17. $3n^5 + 4n^3 + 9n + 6$
23. $5c^2 + 12c - 7$
25. $3x^4 + 2x^3 - 4x + 10$

6. $4c^5 + 11c^4 - 9c^3 + 8c^2 - 5c + 2$
8. $5m^7 + 6m^5 - 2m^3 - 9m^2 + 4m$
18. $2d^4 + 8d^2 + 7d + 4$
24. $-4b^2 - 9b + 8$
26. $-a^3 - 4a^2 - 5a - 6$

Multiply. *(Lessons 15-4 and 15-5)*   **27.** $6a^4 + 18a^3 - 24a^2 + 21a$
28. $-12x^4 - 28x^3 + 20x^2$

27. $3a(2a^3 + 6a^2 - 8a + 7)$

28. $4x^2(-3x^2 - 7x + 5)$

29. $-6c(4c^4 + 3c^2 - 5c - 8)$
$-24c^5 - 18c^3 + 30c^2 + 48c$

30. $-4n^3(7n^3 + 5n^2 - 8)$
$-28n^6 - 20n^5 + 32n^3$

31. $(x + 5)(2x + 3)$ $2x^2 + 13x + 15$

32. $(4m - 7)(2m + 4)$
$8m^2 + 2m - 28$

33. $(3x - 8)(2x - 5)$ $6x^2 - 31x + 40$

34. $(4d + 3)(7d - 2)$
$28d^2 + 13d - 6$

Solve by working backward.   *(Lesson 15-6)*

35. Elvin Thompson bought three compact discs and two cassettes for a total of $66.73, including sales tax. Each cassette cost $8.99 and the sales tax was $3.78. How much did each compact disc cost? $14.99

36. Lisa Goldberg earns $12.50 per hour for a 40-hour week. She earns double time for overtime. Lisa earned $575 last week. How many hours did she work? 43 h

37. Ysabel Rodriguez wants to buy a new stove. She is able to save $35 per week for that purpose. The stove costs $700, and the sales tax rate is 5%. How many weeks will it take Ysabel to save enough money to buy the stove? 21 weeks

38. Mike Karam has a dental appointment at 10:30 A.M., and he wants to be there 10 min early. He needs 25 min to drive to the dentist's office and wants to allow another 5 min to find a parking space. At what time should Mike leave for his appointment? 9:50

Factor.   *(Lesson 15-7)* Answers to Chapter Review 41–46 are on p. A42.
$3n^2(3n^2 - 2n + 6)$

39. $4x^3 + 8x^2 + 6x$ $2x(2x^2 + 4x + 3)$

40. $9n^4 - 6n^3 + 18n^2$

41. $20d^5 - 10d^4 + 15d^3 - 5d^2$

42. $18m^6 - 30m^4 + 12m^3 - 24m^2$

43. $3c^3d^3 + 5c^2d^2 - 4cd$

44. $4a^3b^2 - 8a^2b^3 + 2ab^2$

45. $25x^4y^3 + 10x^3y^4 - 15x^2y^2$

46. $12mn^3 - 18m^2n^2 + 24m^2n^3$

Divide.   *(Lesson 15-8)* **48.** $2c^4 - 3c^2 + 6c - 1$ **50.** $4m^2n^3 - 2n^2 + 3m^5n$

47. $\dfrac{14x^4 + 8x^3 - 10x^2}{2x}$ $7x^3 + 4x^2 - 5x$

48. $\dfrac{6c^5 - 9c^3 + 18c^2 - 3c}{3c}$

49. $\dfrac{18a^3b^5 + 21a^4b^3 - 6a^2b^2}{3ab}$
$6a^2b^4 + 7a^3b^2 - 2ab$

50. $\dfrac{16m^4n^6 - 8m^2n^5 + 12m^7n^4}{4m^2n^3}$

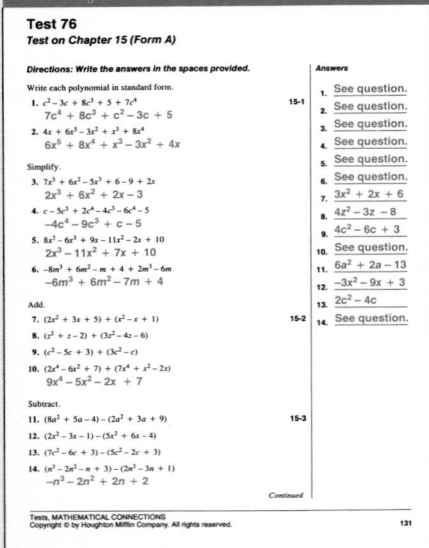
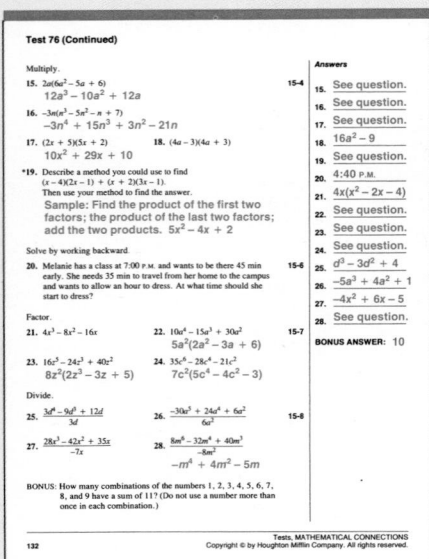
Chapter Test

Write each polynomial in standard form.

1. $c^3 - 2c^2 + 6c^4 - 9c + 7$
   $6c^4 + c^3 - 2c^2 - 9c + 7$

2. $5x - 2x^3 + 7x^5 + 3 - 8x^2$
   $7x^5 - 2x^3 - 8x^2 + 5x + 3$

15-1

Simplify.

3. $6b^3 + 7b^2 - 4b^3 + 5 - 11 + b$
   $2b^3 + 7b^2 + b - 6$

4. $2c - 3c^4 + 2c^3 - 5c - 7c^4 - 7$
   $-10c^4 + 2c^3 - 3c - 7$

5. $5x^3 - 8x + 12x - 9x^3 + 6x - 3$
   $-4x^3 + 10x - 3$

6. $-4m^3 - 6m^3 + 8m^2 - m + 3m^2 - 1$
   $-10m^3 + 11m^2 - m - 1$

Add.

7. $(3x^2 + 5x + 2) + (x^2 - 2x - 1)$
   $4x^2 + 3x + 1$

8. $(z^2 - z - 7) + (3z^2 + 2z - 8)$
   $4z^2 + z - 15$

15-2

9. $(c^2 + 4c - 2) + (2c^2 - 5)$
   $3c^2 + 4c - 7$

10. $(3x^4 - 5x^3 + x) + (4x^4 + 2x^3 + 4)$
    $7x^4 - 3x^3 + x + 4$

Subtract.

11. $(9b^2 + 6b - 5) - (3b^2 + 5b + 7)$
    $6b^2 + b - 12$

12. $(3x^2 - 4x - 2) - (5x^2 + 7x - 2)$
    $-2x^2 - 11x$

15-3

13. $(8c^2 + 7c + 5) - (6c^2 - 7)$
    $2c^2 + 7c + 12$

14. $(n^3 - 4n^2 + n - 1) - (3n^3 - n + 2)$
    $-2n^3 - 4n^2 + 2n - 3$

Multiply.

15. $3b(7b^2 - 4b + 3)$ $21b^3 - 12b^2 + 9b$

16. $-4n(2n^3 - 6n^2 + n - 8)$
    $-8n^4 + 24n^3 - 4n^2 + 32n$

15-4

17. $(4x + 3)(3x + 5)$ $12x^2 + 29x + 15$

18. $(5a - 8)(2a + 7)$ $10a^2 + 19a - 56$

15-5

★19. Show how to use algebra tiles to give a geometric interpretation of $(x + 2)(x + 3)$. See answer on p. A42.

★20. Describe a method you could use to multiply $3n(2n + 5)(4n + 7)$. Then use your method to find the product. See answer on p. A42.

Solve by working backward.

21. Sara Jones has to be at work at 8:30 A.M. She has a 35 min drive to the parking lot near her office and then a 10 min walk to her office. Sara wants to arrive 5 min early. What time should she leave home?
    7:40 A.M.

15-6

Factor.

22. $4x^3 + 8x^2 - 16x$ $4x(x^2 + 2x - 4)$

23. $12a^4 - 6a^3 + 24a^2$
    $6a^2(2a^2 - a + 4)$

15-7

24. $10z^5 - 5z^3 + 15z^2$ $5z^2(2z^3 - z + 3)$

25. $24c^6 - 18c^4 - 12c^3$
    $6c^3(4c^3 - 3c - 2)$

Divide.

26. $\dfrac{7d^3 - 21d^2 + 14d}{7d}$ $d^2 - 3d + 2$

27. $\dfrac{-18b^5 + 12b^4 + 6b^2}{6b^2}$ $-3b^3 + 2b^2 + 1$

15-8

28. $\dfrac{24x^3 - 18x^2 + 12x}{-6x}$ $-4x^2 + 3x - 2$

29. $\dfrac{8m^6 - 24m^5 + 32m^3}{-8m^2}$ $-m^4 + 3m^3 - 4m$

Multiplying Binomials Mentally

Objective: To multiply binomials mentally.

You can find the products $(a + b)(a - b)$, $(a + b)^2$, and $(a - b)^2$ mentally if you learn the following patterns.

$$(a + b)(a - b) = a^2 - b^2$$
$$(a + b)^2 = a^2 + 2ab + b^2$$
$$(a - b)^2 = a^2 - 2ab + b^2$$

Example

Find each product mentally.

a. $(x + 3)(x - 3)$ **b.** $(m + 5)^2$ **c.** $(a - 4)^2$

Solution

a. $(x + 3)(x - 3) = x^2 - 3^2$ ⟵ $(a + b)(a - b) = a^2 - b^2$
$= x^2 - 9$

b. $(m + 5)^2 = m^2 + 2(m)(5) + 5^2$ ⟵ $(a + b)^2 = a^2 + 2ab + b^2$
$= m^2 + 10m + 25$

c. $(a - 4)^2 = a^2 - 2(a)(4) + 4^2$ ⟵ $(a - b)^2 = a^2 - 2ab + b^2$
$= a^2 - 8a + 16$

You can also use the patterns in reverse and work backward to factor certain polynomials as the product of two binomials.

Exercises

6. $m^2 - 24m + 144$
7. $4c^2 - 81$
8. $16b^2 - 121$
9. $9x^2 + 24x + 16$
10. $49a^2 + 42a + 9$
11. $25c^2 - 60c + 36$
12. $81n^2 - 126n + 49$
13. $4x^2 - 9y^2$
14. $25m^2 - 49n^2$
15. $9a^2 + 30ab + 25b^2$
16. $49x^2 + 56xy + 16y^2$
17. $9z^2 - 48xz + 64x^2$
18. $64m^2 - 80mn + 25n^2$

Find each product mentally.

1. $(a + 2)(a - 2)$ $a^2 - 4$ **2.** $(n + 7)(n - 7)$ $n^2 - 49$ **3.** $(x + 8)^2$
$x^2 + 16x + 64$
4. $(z + 6)^2$ $z^2 + 12z + 36$ **5.** $(c - 5)^2$ $c^2 - 10c + 25$ **6.** $(m - 12)^2$

7. $(2c + 9)(2c - 9)$ **8.** $(4b + 11)(4b - 11)$ **9.** $(3x + 4)^2$

10. $(7a + 3)^2$ **11.** $(5c - 6)^2$ **12.** $(9n - 7)^2$

13. $(2x + 3y)(2x - 3y)$ **14.** $(5m + 7n)(5m - 7n)$ **15.** $(3a + 5b)^2$

16. $(7x + 4y)^2$ **17.** $(3z - 8x)^2$ **18.** $(8m - 5n)^2$

Use the patterns to factor each polynomial as the product of two binomials.

19. $x^2 - 36$ $(x - 6)(x + 6)$ **20.** $c^2 - 81$ $(c - 9)(c + 9)$

21. $x^2 + 2x + 1$ $(x + 1)^2$ **22.** $100a^2 - 9b^2$
$(10a - 3b)(10a + 3b)$
23. $c^2 - 18c + 81$ $(c - 9)^2$ **24.** $n^2 + 14n + 49$ $(n + 7)^2$

25. $z^2 - 16z + 64$ $(z - 8)^2$ **26.** $4m^2 - 49n^2$
$(2m - 7n)(2m + 7n)$

Polynomials   **703**

Teaching Notes

Point out to students the reversibility of the equations at the beginning of the lesson. By using these equations from right to left, students can factor certain types of polynomials, as shown in Exercises 19–26.

Mental Math

In this extension, students are asked to look for patterns when multiplying binomials. These patterns can be used to determine the answers mentally.

Problem Solving

The problem-solving strategy of working backward also can be used to factor polynomials as the product of two binomials.

Additional Examples

Find each product mentally.
1. $(c + 9)(c - 9)$ $c^2 - 81$
2. $(a + 6)(a + 6)$ $a^2 + 12a + 36$
3. $(m - 2)(m - 2)$ $m^2 - 4m + 4$

Write each polynomial in standard form.

1. $5y + 5y^3 - 7y^4 + 2y^2 - 9$
   $-7y^4 + 5y^3 + 2y^2 + 5y - 9$
2. $7d^2 - 6d + 5d^4 - 3 + 7d^3$
   $5d^4 + 7d^3 + 7d^2 - 6d - 3$

Simplify.

3. $5m^3 - m^2 + 5m^2 + 9m - m^3 + 4$
   $4m^3 + 4m^2 + 9m + 4$
4. $5b^3 - 4b - 8b + 5 - 3b^2 + 1$
   $5b^3 - 3b^2 - 12b + 6$

Add.

5. $(k^2 + 6k + 3) + (k^2 + 3k + 7)$
   $2k^2 + 9k + 10$
6. $(g^2 - 5g + 3) + (g^2 + g - 3)$
   $2g^2 - 4g$

Subtract.

7. $(4n^3 + n^2 - 5) - (n^3 - n^2 + 3)$
   $3n^3 + 2n^2 - 8$
8. $(6y^2 + y - 2) - (y^2 + y + 3)$
   $5y^2 - 5$

Multiply.

9. $6ab^2(-3a^3b^4)$ $-18a^4b^6$
10. $-4x^2(5x^3 - 3x^5)$
    $-20x^5 + 12x^7$
11. $(3x + 2)(2x + 5)$
    $6x^2 + 19x + 10$
12. $(3d + 5)(4d - 3)$
    $12d^2 + 11d - 15$

Alternative Assessment

1. Write two polynomials whose sum is $2x^2 + 13x + 15$.
   Answers will vary.
2. Write two polynomials whose difference is $2x^2 + 13x + 15$.
   Answers will vary.
3. Write two polynomials whose product is $2x^2 + 13x + 15$.
   $(2x + 3)(x + 5)$

Cumulative Review

Standardized Testing Practice

Choose the letter of the correct answer.

1.

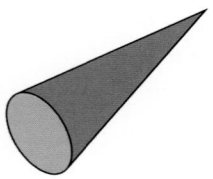

Identify the space figure.
A. cone   B. cylinder
C. sphere   D. pyramid

2. Write $7z^5 + 4z^9 - 6z^3 + 15$ in standard form.
A. $15 + 7z^5 - 6z^3 + 4z^9$
B. $4z^9 + 7z^5 - 6z^3 + 15$
C. $4z^9 - 6z^3 + 7z^5 + 15$
D. $15 - 6z^3 + 7z^5 + 4z^9$

3. 62.5% of what number is 105?
A. 62.5   B. 65.6
C. 168   D. 142.5

4. Divide: $\frac{3x^4 + 12x^2 - 6x}{3x}$
A. $x^4 + 4x^2 - 2x$
B. $x^3 + 4x - 2$
C. $x^3 + 12x^2 - 6x$
D. $9x^5 + 36x^3 - 18x^2$

5. Solve for x: $y = kx$
A. $xy = k$ B. $x = \frac{k}{y}$
C. $x = \frac{y}{k}$ D. $x = ky$

6.

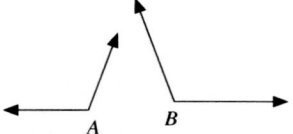

Complete: $\angle A$ and $\angle B$ are both __?__ angles.
A. adjacent   B. right
C. obtuse   D. acute

7. Find the number of permutations of the letters in the word FINE.
A. 24   B. 256
C. 16   D. 120

8. Find the sum: $6 + (-7) + 14 + (-8)$
A. 35   B. -5
C. 5   D. -35

9. Subtract.
$(9y^3 - 3y - 5) - (2y^3 - y + 2)$
A. $7y^3 + 2y + 3$
B. $7y^3 - 2y - 7$
C. $7y^3 - 4y + 3$
D. $7y^3 + 2y + 7$

10. Find the surface area of a cylinder with radius 5 mm and height 40 mm.
A. 1413 mm^2 B. 157 mm^2
C. 1256 mm^2 D. 3140 mm^2

11. Find the product: $(m + 7)(m - 4)$

   A. $m^2 - 28$
   B. $m^2 + 3m - 28$
   C. $m^2 - 11m - 28$
   D. $m^2 + 28$

12. Find the mode of the data.

   18.1, 19.3, 19.7, 22.1, 21.5, 18.1

   A. 19.5   B. 19.8
   C. 4   D. 18.1

13. Factor: $18a^3b^4 + 27a^2b^2$

   A. $9a^2b^2(2ab^2 + 3)$
   B. $9(a^3b^4 + 4a^2b^2)$
   C. $3ab(7a^2b^3 + 8ab)$
   D. $18a^2b^2(ab^2 + 9)$

14. The area of a square is 25 cm². The area of a parallelogram is 36 cm². Which of the following statements is true?

   I. One side of the square is 5 cm.
   II. One side of the parallelogram is 6 cm.

   A. I only   B. II only
   C. I and II   D. Neither I nor II

15. Find the volume of a sphere with a radius of 6 m.

   A. 452.16 m³   B. 904.32 m³
   C. 75.36 m³   D. not enough information

16. Find the slope of the line: $y = 3x - 7$

   A. 3   B. $\frac{1}{3}$
   C. 7   D. -7

17. Find the prime factorization of 240.

   A. $2^4 \cdot 3 \cdot 5$ B. $2^3 \cdot 15$
   C. $15 \cdot 16$ D. $2 \cdot 3 \cdot 5 \cdot 8$

18. Write an equation for the sentence.

   Eight less than half a number x is five.

   A. $8 - \frac{1}{2}x = 5$ B. $8 < \frac{1}{2}x + 5$
   C. $\frac{1}{2}x - 8 = 5$ D. $8 < \frac{1}{2}x - 5$

19.

   Daily Commuting Distances (Miles)

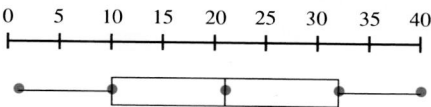

   What is the first quartile of the commuting distances?

   A. 21   B. 32
   C. 10   D. 1

20. Find the surface area of a rectangular prism with length 5 in., height 3 in., and width 4 in.

   A. 54 in.²   B. 45 in.²
   C. 94 in.²   D. 74 in.²

Planning *Looking Ahead*

Chapter Overview

The first three lessons of Looking Ahead give students some experience with transformational geometry. The concept of rigid motion is introduced in the first lesson, which covers translations and reflections. The exercises use pictorial representations initially and end by requiring students to develop the sketches. Rotations, symmetry, and dilations are developed in the next two lessons. In all three lessons, students can perform the various transformations using physical models.

The second part of Looking Ahead gives students a glimpse of trigonometry by introducing the basic functions of sine, cosine, and tangent. The fourth lesson introduces the trigonometric ratios, and the next two lessons require students to use the ratios to find sides and angles of right triangles. Instructions for using a scientific calculator or a trigonometry table to find the values of the trigonometric ratios are provided. It is assumed that students are familiar with the Pythagorean theorem before they do these lessons. The last lesson of the chapter presents a discussion of the special right triangles.

Background

Looking Ahead provides an overview of transformational geometry and trigonometry. The explanatory text and examples provide a strong connection between algebra and geometry. This is especially evident in the lesson on translations and reflections, in which figures are moved about the coordinate plane. The trigonometry lessons continue this connection by having students perform algebraic manipulations to find the missing measures of sides and angles of right triangles.

Real-world applications are stressed throughout the trigonometry lessons, as students find the heights of trees, buildings, towers, and so on. Finding these measures can, of course, be accomplished most easily by using tools such as calculators to determine the correct trigonometric ratios and to perform the necessary computations.

Manipulatives can be used to perform any translation, reflection, or rotation presented in the lessons. Using manipulatives to show these transformations is an important aid to understanding. An overhead projector is quite useful for demonstrating dilations. Computer software for drawing and measuring figures, such as *Connections Plotter Plus,* also can facilitate the exploration of geometric transformations. In the trigonometry lessons, geoboards are useful for learning about sine, cosine, and tangent ratios. Finding the height of the school flagpole or building would be a nice group activity.

Objectives

TRANSLATIONS AND REFLECTIONS To find the image of a figure after a translation or a reflection.

ROTATIONS AND SYMMETRY To find the image of a figure after a rotation and to find rotational symmetries.

ENLARGEMENTS AND REDUCTIONS To find the image of a figure after a dilation.

SINE, COSINE, AND TANGENT To find the sine, cosine, and tangent of acute angles of a right triangle.

USING TRIGONOMETRY To use trigonometric ratios to find the measures of the sides of a right triangle.

FINDING ANGLE MEASURES To use trigonometric ratios to find the measure of an acute angle of a right triangle.

SPECIAL RIGHT TRIANGLES To apply 45°-45°-90° and 30°-60°-90° right triangles.

Suggested Assignments

Suggested assignments can be found in the side columns of each lesson.

Alternate Approaches Looking Ahead

Using Technology

COMPUTERS

A computer program such as Houghton Mifflin's *Connections Plotter Plus,* which is capable of performing geometric transformations, can be used by students to explore this important topic. It is certainly valuable for students to make pencil-and-paper drawings, but the relative ease of using a computer and its accuracy and speed will allow them to work with many more examples. Such explorations can help develop students' intuitions about these new ideas, and can even help them discover some fascinating mathematics.

CALCULATORS

Scientific calculators can be very useful in the trigonometry lessons of this chapter. Students can calculate results quickly and are not restricted to problems that have "neat" answers. The ease with which trigonometric ratios can be found with a scientific calculator allows students to explore their properties effectively.

Translations and Reflections

COMPUTERS

You can have students use a transformation-drawing program to reinforce the ideas in this and the following lessons. Students can be asked to have the program draw a certain shape such as a triangle, and then to translate and reflect the shape in various ways. Exercises 24 and 25 in the text suggest that students can discover what happens when two reflections, each in a different line, are performed on a figure sequentially. Students can discover that if the two lines are parallel, the two reflections are equivalent to a single translation. They also can begin to describe what repeated reflections do when the lines are not parallel.

Rotations and Symmetry

COMPUTERS

With the introduction of rotations in this lesson, students can use a transformation-drawing program to explore the effects of a sequence of any types of rigid motions. At an elementary level, they can discover that two translations are equivalent to a single translation, and they can learn to describe how to find that single equivalent transformation. The same possibility exists for two rotations, or indeed for any combination of any types of transformations.

Enlargements and Reductions

COMPUTERS

A transformation-drawing program again will provide students with an opportunity to try a large number of dilations very quickly. Students can measure segments of dilated figures, compare areas, and explore what happens when two dilations are applied to a figure sequentially.

Trigonometry

CALCULATORS

A scientific calculator is crucial for studying trigonometry. Not only is it important for students to be familiar with their calculators when they come to use trigonometry in later science courses, but a calculator also allows exploration of some of the fascinating properties of the sine, cosine, and tangent functions. For example, the cofunction identity $\sin (90° - x) = \cos(x)$ can be discovered easily by students who are given a list of trigonometric ratios to find. They may even be able, with some guidance, to come up with an explanation of why the identity is true. Another important property that is easily within students' reach is the Pythagorean identity, that is, $\sin^2(x) + \cos^2(x) = 1$.

Using Manipulatives

Models of geometric figures and the use of tangram pieces, pattern blocks, or figures made from stiff paper, would be very helpful in the lessons on transformational geometry. Predicting the new position of a figure as a result of a transformation is made easier by actually seeing the movement.

Translations and Reflections; Rotations and Symmetry

Students should use centimeter graph paper for these activities. Have students place a figure (square, rectangle, triangle, and so on) on graph paper, trace the shape, and label the vertices. Students should then perform a translation, following a given set of instructions, by sliding the figure to its new position. Have students trace the figure in its new position and label the new vertices. Students should check each of the new vertices to see if their coordinates are correct. This activity can be extended to reflections or rotations. For example:

1. Perform the reflection of a triangle across the *y*-axis. Write the new vertices.
2. Repeat Step 1 for a reflection across the *x*-axis.
3. Predict the image of a triangle after a 90° counterclockwise rotation. Lightly sketch your prediction. Perform the rotation of the triangle. How does the image compare with your prediction?

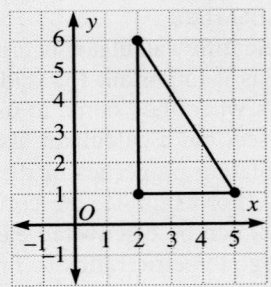

After students are comfortable with single transformations, combining them is facilitated by using models. Marking a vertex with a colored dot helps students follow a transformation more easily.

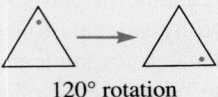

120° rotation

Transparent mirrors are helpful for reflections. The mirror acts as a line of symmetry and students can trace the reflection they see.

Students also can use tracing paper to copy a figure and then perform the transformation by moving the traced figure. Continue to encourage students to sketch their prediction of the transformation before actually doing it.

Enlargements and Reductions

An overhead projector is very useful when working with dilations. For example, you can project the image of a figure onto a screen and then ask students the following types of questions.

1. What is the relationship between the original figure and its projection on the screen? They are similar.
2. What is your estimate of the ratio of the sides? Answers will vary.
3. How can you change the ratio? Move the projector or the screen to change the size of the figure.

Sine, Cosine, and Tangent

A geoboard is a useful manipulative when introducing the trigonometric ratios. Geoboards can be used to review the Pythagorean theorem by having students form a right triangle, construct squares on both legs and the hypotenuse, and then confirm that the area of the square on the hypotenuse is equal to the sum of the areas of the two smaller squares.

Geoboards also can be used by students to form right triangles and list the trigonometric ratios for each triangle. For example, students can do the following exercise by using geoboards.

In right triangle ABC, calculate the length of c by using the Pythagorean theorem and find the sine, cosine, and tangent of angles A and B.

$c = 5$

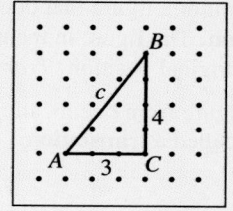

$\sin A = \dfrac{4}{5}$ $\quad\quad$ $\sin B = \dfrac{3}{5}$

$\cos A = \dfrac{3}{5}$ $\quad\quad$ $\cos B = \dfrac{4}{5}$

$\tan A = \dfrac{4}{3}$ $\quad\quad$ $\tan B = \dfrac{3}{4}$

After performing such exercises, students can discuss the relationships they see among the functions.

As an extension of these kinds of exercises, have students change the size of the angle and observe any changes in the function values. For instance, as an angle approaches 90°, what happens to the value of the sine?

This approach can enhance students' comprehension of the trigonometric ratios.

Translations and Reflections; Rotations and Symmetry

COOPERATIVE LEARNING

Have students working in groups draw a polygon on centimeter graph paper. Students should then assign a transformation to their figures and exchange papers with other students in their group. After performing the indicated transformations, the results can be discussed within each group. Groups should pick one or two of their transformations to explain to the class. The use of colored pencils will enhance the drawings.

Sine, Cosine, and Tangent

COOPERATIVE LEARNING

Working in groups students should draw a series of "nested triangles," such as the ones shown below. Students should measure each length and list the trigonometric ratios for each triangle. To get the groups started, show them the following example.

$$\sin A = \frac{BC}{AC} = \underline{\ ?\ }$$

$$\sin A = \frac{DE}{AC} = \underline{\ ?\ }$$

$$\sin A = \frac{FG}{AC} = \underline{\ ?\ }$$

After finding the sine, cosine, and tangent for the common angle in their drawings, students should discuss their conclusions with the other members of their group.

Teaching the Lesson
• Materials: centimeter graph paper, transparent mirrors, tracing paper
• Using Technology, p. 706B
• Using Manipulatives, p. 706C
• Reteaching/Alternate Approach, p. 706D
• Cooperative Learning, p. 706D

Warm-Up Exercises

Draw all the lines of symmetry for each figure.

1.

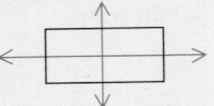

2.

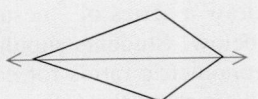

3.

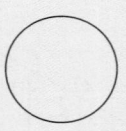

infinitely many

4.

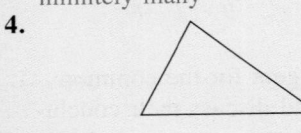

Lesson Focus

Ask students how they would sketch the reflection of a figure. If the figure were a simple one, it would be easy to draw its reflection. If the figure were complex, it would be harder to get the correct reflection. Suppose you wanted to slide a figure from one location to another. How could you give instructions regarding the direction and distance for moving it? This lesson will focus on ways to solve these kinds of problems.

Looking Ahead

Transformations

Translations and Reflections

Objective: To find the image of a figure after a translation or a reflection.

Many patterns in fabric, art, tiling, and nature are the result of sliding or flipping a shape to various positions.

Movement of a figure that does not change its size and shape is called a **rigid motion.** The figure in the new location is called the **image** of the figure in the original position. *Translations* and *reflections* are rigid motions.

Terms to Know
• *rigid motion*
• *translation*
• *reflection*
• *image*

Sliding a figure from one location to another is called a **translation.**

Flipping a figure across a line is called a **reflection.**

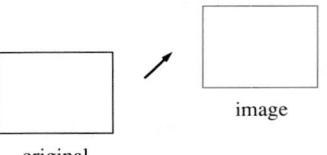

image

original

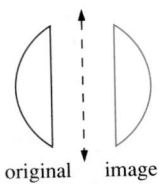

original   image

You can show rigid motions on a coordinate plane.

Example 1

The coordinates of the vertices of △*ABC* are *A*(3, 4), *B*(3, 1), and *C*(7, 1). Give the coordinates of the image of each vertex after the triangle is translated 5 units to the left.

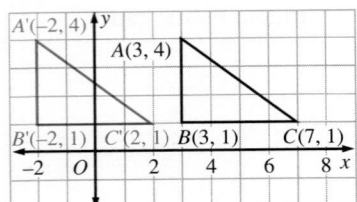

Solution

$A(3, 4) \longrightarrow A'(3 - 5, 4)$ or $A'(-2, 4)$
$B(3, 1) \longrightarrow B'(3 - 5, 1)$ or $B'(-2, 1)$
$C(7, 1) \longrightarrow C'(7 - 5, 1)$ or $C'(2, 1)$

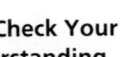 **Check Your Understanding**

1. Explain why the *y*-coordinates did not change.
2. Name a translation that would affect the *y*-coordinates.
   See *Answers to Check Your Understanding* at the back of the book.

You sometimes translate a figure both horizontally and vertically.

706 Looking Ahead

Example 2

A translation moves $\triangle ABC$ 2 units to the right and 4 units down. The image is $\triangle A'B'C'$. Write the coordinates of each vertex of $\triangle A'B'C'$.

Solution

$A(3, 4) \longrightarrow A'(3 + 2, 4 - 4)$
or $A'(5, 0)$

$B(3, 1) \longrightarrow B'(3 + 2, 1 - 4)$
or $B'(5, -3)$

$C(7, 1) \longrightarrow C'(7 + 2, 1 - 4)$
or $C'(9, -3)$

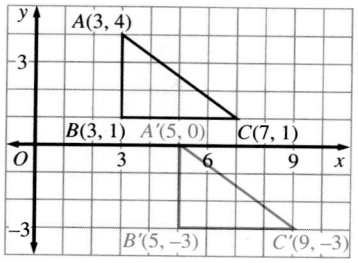

You can use a coordinate plane to show a reflection.

Example 3

Use $\triangle RST$. Write the coordinates of the vertices after a reflection across the x-axis.

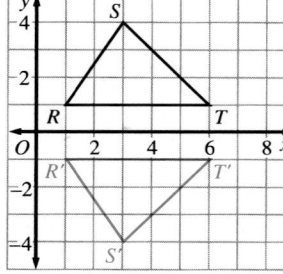

Solution

$R(1, 1) \longrightarrow R'(1, -1)$
$S(3, 4) \longrightarrow S'(3, -4)$
$T(6, 1) \longrightarrow T'(6, -1)$

✏️ Check Your Understanding

3. How far is R above the x-axis? How far is R' below the x-axis?
4. How far is S above the x-axis? How far is S' below the x-axis?
5. If you folded the coordinate graph along the x-axis, would $\triangle RST$ and $\triangle R'S'T'$ match exactly?

See *Answers to Check Your Understanding* at the back of the book.

Exercises

Identify each rigid motion from A to B as a *reflection*, a *translation*, or *neither*.

A

1. translation

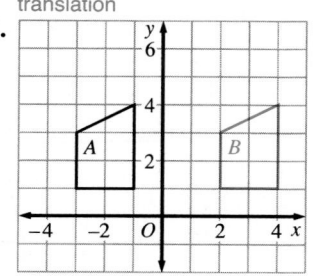

2. neither

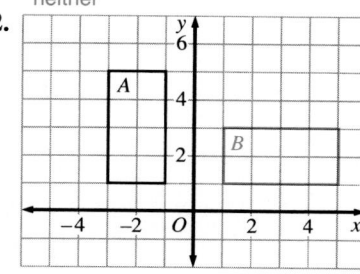

Transformations **707**

Teaching Notes

Translations and reflections may be new to students. Work through the Examples carefully with students, explaining how the coordinates of each vertex are transformed.

Key Questions
1. Why are translations and reflections often called slides and flips? For translations, you actually slide the figure without turning it; for reflections, you turn the figure over or "flip it" across a line.
2. If a figure is translated horizontally, which coordinates change? Why? Only the x-coordinates change because the figure has not moved up or down.

Reasoning
In the Examples, the Check Your Understanding questions require students to apply spatial reasoning skills and explain why certain changes did or did not occur.

Using Technology
For a suggestion on using a transformation drawing program, see page 706B.

Using Manipulatives
For a suggestion on using models to show translations and reflections, see page 706C.

Reteaching/Alternate Approach
For a suggestion on using a cooperative learning activity with this lesson, see page 706D.

Additional Examples

1. The coordinates of the vertices of △ABC are A (0, 0), B (3, 1), and C (2, 3). Give the coordinates of the image of each vertex after it is translated 6 units down.

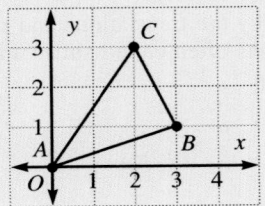

A (0, 0) → A′ (0, −6);
B (3, 1) → B′ (3, −5);
C (2, 3) → C′ (2, −3)

2. A translation moves △ABC 3 units to the left and 2 units up. The image is △A′B′C′. Write the coordinates of each vertex of △A′B′C′.

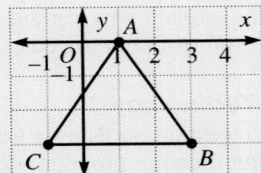

A′ (−2, 2), B′ (0, −1), C′ (−4, −1)

3. Use △DEF. Write the coordinates of the vertices after a reflection across the y-axis.

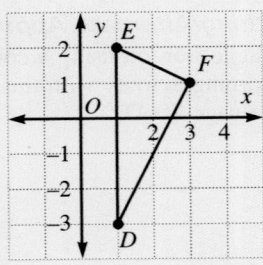

D′ (−1, −3), E′ (−1, 2), F′ (−3, 1)

Closing the Lesson

Have students summarize the main points of the lesson by describing and contrasting translations and reflections.

Identify each rigid motion from A to B as a *reflection*, a *translation*, or *neither*.   reflection   translation

3.
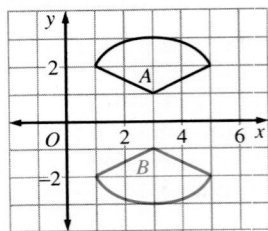

4.
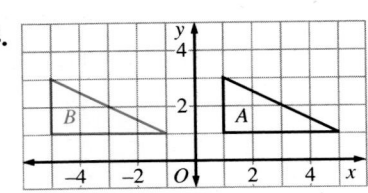

△PQR is the reflection of △ABC across the y-axis. Complete.

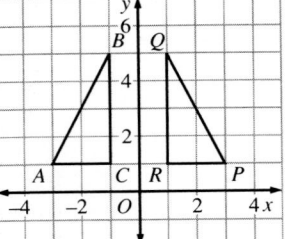

5. A(−3, 1) ⟶ P(_?_, _?_) 3, 1

6. B(−1, 5) ⟶ Q(_?_, _?_) 1, 5

7. C(−1, 1) ⟶ R(_?_, _?_) 1, 1

The vertices of rectangle △ABCD are A(1, 1), B(3, 1), C(3, 4), and D(1, 4).   Answers to Exercises 8–16 are on p. A42–A43.

a. **Sketch the image of rectangle ABCD after each rigid motion.**

b. **State the coordinates of the images of A, B, C, and D.**

8. a translation 1 unit to the right

9. a translation 3 units down

10. a translation 3 units to the left

11. a translation 2 units up

12. a reflection across the y-axis

13. a reflection across the x-axis

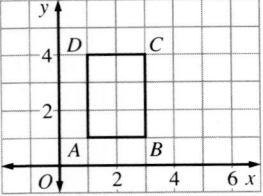

Copy each figure. Then sketch the image when the figure is reflected across the line.

14.
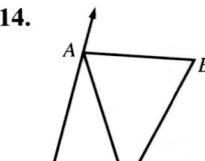

15.
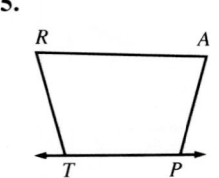

16.
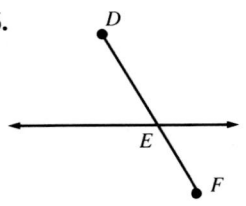

The coordinates of the vertices of △ABC are A(−5, 1), B(−1, 4), and C(−2, 0). Write the coordinates of the images of the vertices after each rigid motion.

17. *A′*(−1, −1); *B′*(3, 2); **B**
    C′(2, −2)
18. *A′*(−2, 2); *B′*(2, 5);
    C′(1, 1)
19. *A′*(5, −1); *B′*(1, −4);
    C′(2, 0)
20. *A′*(−3, −1); *B′*(1, −4);
    C′(0, 0)

17. translation down 2 units, then to the right 4 units

18. translation to the right 3 units, then up 1 unit

19. reflection across the *y*-axis, then across the *x*-axis

20. reflection across the *x*-axis, then translation to the right 2 units

21–23. Answers may vary.
21. translation down 3 units
    and right 2 units
22. translation up 3 units
    and right 2 units
23. reflection in the *x*-axis
    and a translation right
    2 units

Write a description of the rigid motion that takes △DEF to △D′E′F′.

21.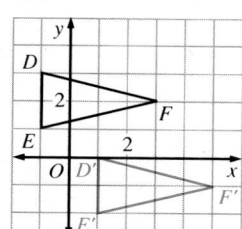

22.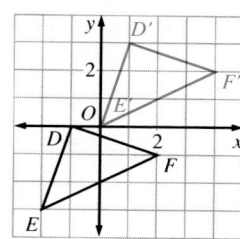

23.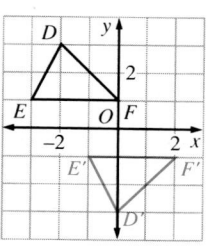

In Exercises 24–25, trace the figure and the two lines.
  a. Sketch the reflection of △ABC in line *m*. Call it △DEF.
  b. Sketch the reflection of △DEF in line *n*. Call it △GHI.
  c. Is the rigid motion from △ABC to △GHI a *translation* or a *reflection*? Answers to Exercises 24 and 25 are on p. A43.

C 24.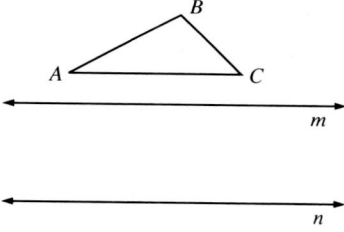

25.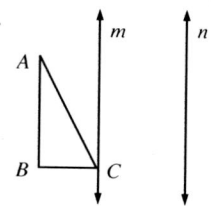

26. The coordinates of the vertices of △RST and the coordinates of the images of the vertices after a rigid motion are given below. Describe the rigid motion from △RST to △R′S′T′.

$$R(-5, 1) \longrightarrow R'(5, 0)$$
$$S(1, 4) \longrightarrow S'(-1, 3)$$
$$T(-1, 1) \longrightarrow T'(1, 0)$$

reflection in the *y*-axis and a translation down one unit

Exercise Notes

Communication: Writing
Exercises 21–23 require students to examine figures that have been moved, and to describe the motion in their own words. *Exercise 23* combines a translation with a reflection.

Critical Thinking
Exercises 24 and 25 require students to perform two reflections on a given figure, *compare* the resulting figure to the original, and state their conclusions.

Follow-Up

Extension
Ask students to describe the result of reflecting a figure twice. Students should validate their conjectures with drawings.

Transformations **709**

Lesson Planner

Teaching the Lesson
- Materials: centimeter graph paper
- Using Technology, p. 706B
- Using Manipulatives, p. 706C
- Reteaching/Alternate Approach, p. 706D
- Cooperative Learning, p. 706D

Warm-Up Exercises

Identify the transformation represented by each action.

1. a checker move   translation
2. lifting a trunk lid   reflection
3. opening a book   reflection
4. climbing a ladder   translation

Lesson Focus

Every day, people encounter examples of objects that rotate around a given point. Consider, for example, the end of a minute hand of a clock. It remains at a constant distance from the center of the face as it rotates. A complete rotation is 360°. This lesson focuses on the notion of rotation as it applies to physical objects and geometric figures.

Teaching Notes

Rotations and rotational symmetry may be unfamiliar to many students. Work through each Example, explaining how these concepts apply to each situation.

Key Question

How does a rotation differ from a translation or a reflection?   A rotation turns a figure through an angle, while a translation does not. A rotation moves a figure about a point, whereas a reflection flips it across a line.

Rotations and Symmetry

Objective: To find the image of a figure after a rotation and to find rotational symmetries.

Imagine that P and Q are points on a bicycle wheel. Suppose the wheel makes a quarter turn. Point P moves to P', and Q moves to Q'. P' and Q' are the images of P and Q. The quarter turn is called a **rotation** of 90° about O. Point O is the **center of rotation.** Notice that a point stays the same distance from the center in a rotation.

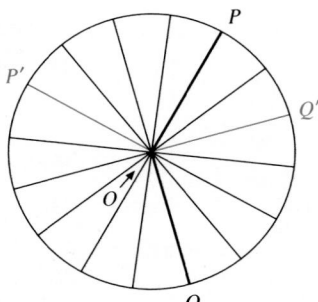

> **Terms to Know**
> - *rotation*
> - *center of rotation*
> - *rotational symmetry*

$$OP = OP' \qquad OQ = OQ'$$

Each point rotates through the same angle.

$$m\angle POP' = 90° \qquad m\angle QOQ' = 90°$$

Example 1

Find the image of square $ABCD$ after a rotation of 45° counterclockwise about P.

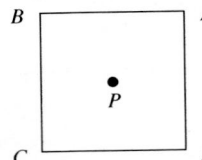

Solution

Rotate each vertex 45° counterclockwise to find its image. Connect the images. Square $A'B'C'D'$ is the image of square $ABCD$.

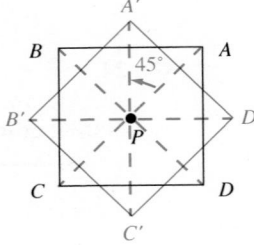

✓ **Check Your Understanding**

1. $m\angle APA' = \underline{\ ?\ }$
2. $m\angle DPD' = \underline{\ ?\ }$
3. If $PA = 3$, then $PA' = \underline{\ ?\ }$

See *Answers to Check Your Understanding* at the back of the book.

In some rotations the figure rotates onto itself. We then say that the figure has **rotational symmetry.** In Example 1, if square $ABCD$ were rotated 90°, the image would fit exactly on the original square. Square $ABCD$ has the following four rotational symmetries.

a rotation of 90° about P	a rotation of 180° about P
a rotation of 270° about P	a rotation of 360° about P

Of course, any figure has 360° rotational symmetry.

Example 2

How many rotational symmetries does the figure have?

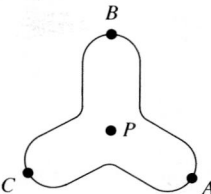

Solution

The image matches the original figure after rotations of 120°, 240°, and 360° about *P*. There are three rotational symmetries.

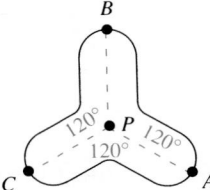

☑ **Check Your Understanding**

Give the image of *A* after each counterclockwise rotation about *P*.

4. 120° **5.** 240° **6.** 360° **7.** 480°

See *Answers to Check Your Understanding* at the back of the book.

A rotation is another kind of rigid motion. The image is the same size and shape as the original figure.

Exercises

Find the image of *A* after each counterclockwise rotation about *P*.

A 1. 120° C **2.** 240° E

3. 600° E **4.** 360° A

5. How many rotational symmetries does the snowflake at the right have? 6

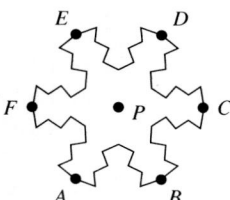

Draw the image of each figure after the given rotation about *P*.

Answers to Exercises 6 and 7 are on p. A43.

6. 90°; counterclockwise **7.** 270°; counterclockwise

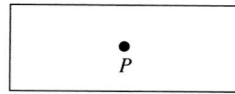 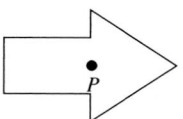

Transformations **711**

Using Technology

For a suggestion on using a transformation drawing program, see page 706B.

Using Manipulatives

For a suggestion on using models to show rotations, see page 706C.

Reteaching/Alternate Approach

For a suggestion on using a cooperative learning activity with this lesson, see page 706D.

Additional Examples

1. Find the image of equilateral triangle *ABC* after a clockwise rotation of 60° about *P*.

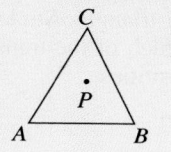 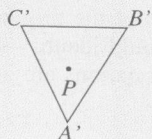

2. How many rotational symmetries does rhombus *ABCD* have?

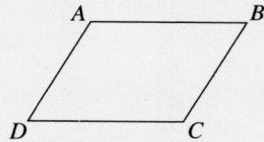

two rotational symmetries at 180° and 360°

Closing the Lesson

Have students write a paragraph that describes the three types of rigid motion: translations, reflections, and rotations. Students should give one example of each type.

Suggested Assignments

Advanced: 1–19 odd

711

Draw the image of each figure after the given rotation about *P*. Answers to Exercises 8 and 9 are on p. A43.

8. 45°; counterclockwise

9. 180°; counterclockwise

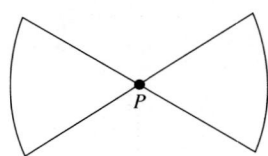

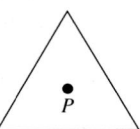

Name all rotational symmetries for each figure. 72°, 144°, 216°, 288°, and 360° about *P*

10. 45°, 90°, 135°, 180°, 225°, 270°, 315°, and 360° about *P*.

B 10.

11.

12.

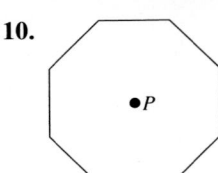

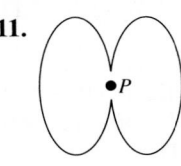

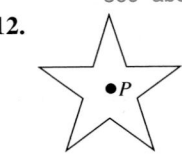

180° and 360° about *P*

Identify the rigid motion as a *translation*, a *reflection*, or a *rotation*.

13. reflection

14. rotation or rotation and translation

15. translation

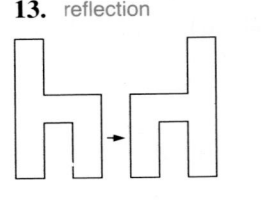

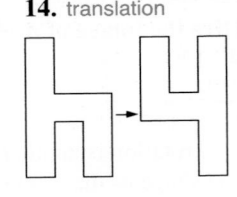

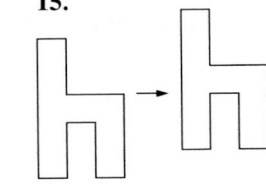

C 16.

17.

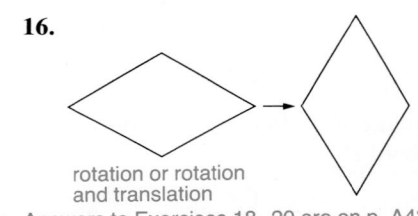

rotation or rotation and translation

translation or rotation and translation

Answers to Exercises 18–20 are on p. A43.

The vertices of a polygon are given below.

    a. Plot the vertices on graph paper. Draw the polygon.

    b. Draw the image of the polygon after the given counterclockwise rotation.

    c. Give the coordinates of the image of each vertex.

18. $A(-5, 6)$, $B(-2, 6)$, $C(-4, 1)$
90° about the origin

19. $A(2, 2)$, $B(5, 2)$, $C(5, 7)$, $D(2, 7)$
270° about D

20. $A(4, 4)$, $B(-4, 4)$, $C(-4, -4)$, $D(4, -4)$
45° about the origin

Enlargements and Reductions

Objective: To find the image of a figure after a dilation.

Terms to Know
- dilation
- enlargement
- reduction
- geometric transformation

Translations, reflections, and rotations are rigid motions; the image and the original are congruent. The image of a figure after a **dilation** is similar to the original. If the image is larger than the original, the dilation is an **enlargement.** If the image is smaller, the dilation is a **reduction.** Because the original figure and its image are similar, corresponding angles are congruent and corresponding sides are in proportion.

Suppose the image of a rectangle $ABCD$ is twice as large as the original. You can say that $A'B' = 2 \cdot AB$. This notation means *the length of segment $\overline{A'B'}$ is two times the length of segment $\overline{AB}$.*

Example 1

Enlarge $\triangle ABC$ so that each side of the image is twice as long as in the original triangle.

Solution

Draw rays through A, B, and C from a point P outside the triangle.
Find A' on $\overrightarrow{PA}$ so that $PA' = 2 \cdot PA$.
Find B' on $\overrightarrow{PB}$ so that $PB' = 2 \cdot PB$.
Find C' on $\overrightarrow{PC}$ so that $PC' = 2 \cdot PC$.
Draw $\triangle A'B'C'$.

$\triangle A'B'C'$ is the image of $\triangle ABC$.

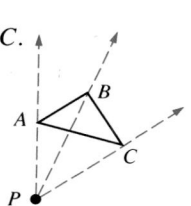 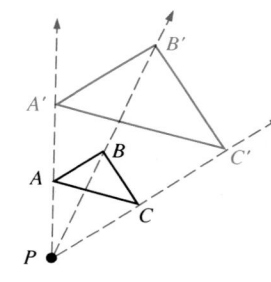

 **Check Your Understanding**

1. If $AB = x$, then $A'B' = $ ___?___ .
2. If $B'C' = a$, then $BC = $ ___?___ .
   See *Answers to Check Your Understanding* at the back of the book.

Example 2

Reduce $\triangle ABC$ so that each side of the image is half as long as in the original triangle.

Solution

Draw rays through A, B, and C from a point P outside the triangle.

Find the points in the middle of $\overline{PA}$, $\overline{PB}$, and $\overline{PC}$. Call them A', B', and C'.
Draw $\triangle A'B'C'$.

$\triangle A'B'C'$ is the image of $\triangle ABC$.

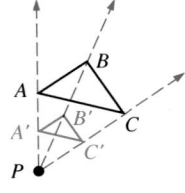

 **Check Your Understanding**

3. If $AB = x$, then $A'B' = $ ___?___ .
4. If $B'C' = a$, then $BC = $ ___?___ .
   See *Answers to Check Your Understanding* at the back of the book.

Transformations **713**

Lesson Planner

Teaching the Lesson
- Materials: overhead projector
- Using Technology, p. 706B
- Using Manipulatives, p. 706C

Warm-Up Exercises

$\triangle ABC \sim \triangle DEF$. **Complete each statement.**

1. $\angle B \cong$ ___?___ $\angle E$
2. $\angle A \cong$ ___?___ $\angle D$
3. $\angle C \cong$ ___?___ $\angle F$
4. $\overline{BC} \sim$ ___?___ $\overline{EF}$
5. $\overline{AC} \sim$ ___?___ $\overline{DF}$

Lesson Focus

Ask students if they have ever had to enlarge or reduce a figure. This can be done by applying the mathematics of similar figures. The focus of today's lesson is on *dilations*, which are transformations that enlarge or reduce figures.

Teaching Notes

Students may have trouble finding the image under a dilation. Make certain they understand the physical procedure for creating dilated images.

Key Question

What is the main difference between dilations and translations, reflections, and rotations? A dilation produces a similar figure; the others produce congruent figures.

Reasoning

The Check Your Understanding questions require students to recognize similar triangles and apply the ratios to corresponding sides. Students also must conclude that the measure of an angle is not affected by a dilation.

Using Technology

For a suggestion on using a transformation drawing program, see page 706B.

Using Manipulatives

For a suggestion on using an overhead projector to show dilations, see page 706C.

Additional Examples

1. Enlarge △*DEF* so that each side of the image is three times as long as in the original triangle.

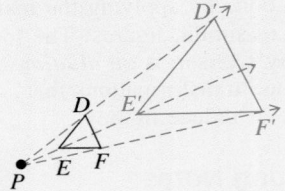

2. Reduce △*ABC* so that each side of the image is one third as long.

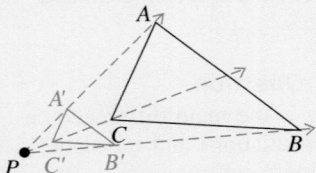

3. Using *R* as the center, enlarge rhombus *RSTU* so that each side of the image is twice as long as the original.

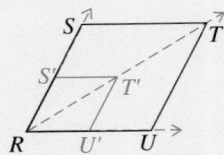

Point *P* in Examples 1 and 2 is called the *center* of the dilation. To enlarge a more complex figure, you may let *P* be a vertex of the original figure rather than a point outside the figure.

Example 3

Using vertex *D* as the center, enlarge quadrilateral *ABCD* so that each side of the image is triple the length of the original.

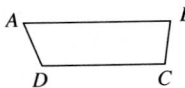

Solution

Draw $\overrightarrow{DA}$, $\overrightarrow{DB}$, and $\overrightarrow{DC}$.
Find A' on $\overrightarrow{DA}$ so that $DA' = 3 \cdot DA$.
Find B' on $\overrightarrow{DB}$ so that $DB' = 3 \cdot DB$.
Find C' on $\overrightarrow{DC}$ so that $DC' = 3 \cdot DC$.
Draw quadrilateral $A'B'C'D$.
Quadrilateral $A'B'C'D$ is the image of quadrilateral *ABCD*.

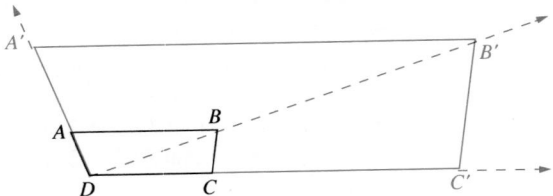

✔ Check Your Understanding

5. Compare the corresponding angles of *ABCD* and $A'B'C'D$.
6. Compare $A'B'$ and *AB*. Compare $B'C'$ and *BC*.
   See *Answers to Check Your Understanding* at the back of the book.

When the position, size, or shape of a figure is changed according to a given rule, it is called a **geometric transformation.** Translations, reflections, rotations, and dilations are examples of transformations.

Exercises

Trace each figure. Enlarge it so that the sides of the image are twice as long. Check students' drawings.

 **1.**

2.

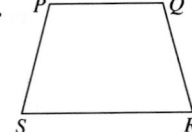

Use a center outside △*ABC*.

Use *R* as the center.

3–10. Check students' drawings.

3. Reduce △ABC in Exercise 1 so that the sides of the image are half as long. Use a center outside △ABC.

4. Reduce quadrilateral PQRS in Exercise 2 so that the sides of the image are half as long. Use point R as the center.

Trace each figure. Enlarge or reduce the figure as indicated.

B

5.
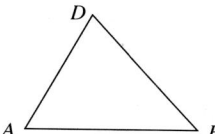
Sides of image are
3 times as long.

6.
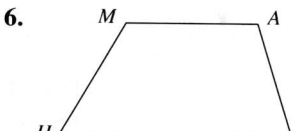
Sides of image are
3 times as long

7.
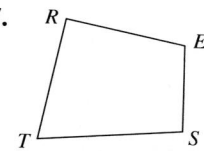
Sides of image are
$\frac{1}{2}$ times as long.

8.
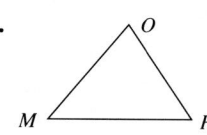
Sides of image are
$2\frac{1}{2}$ times as long.

C **9.** Trace the figure. Enlarge the figure so that each side of the image is twice as long. Use a point P inside the polygon as center.

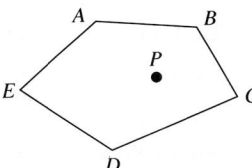

10. Trace the figure. Then reduce it so that each side of the image is one fourth as long.

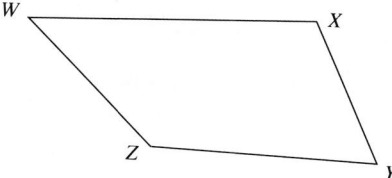

Closing the Lesson

Have students review the procedure for drawing dilations of geometric figures.

Suggested Assignments

Advanced: 1–9 odd

Exercise Note

Making Connections/Transitions

All of the *Exercises* relate numerical ratios to geometric shapes.

Follow-Up

Extension

There are numerous examples of dilations in the real world. Have students look for some examples and describe them to the class.

Lesson Planner

Teaching the Lesson
- Materials: geoboards
- Visuals, Folder P
- Using Technology, p. 706B
- Using Manipulatives, p. 706C
- Reteaching/Alternate Approach, p. 706D
- Cooperative Learning, p. 706D

Warm-Up Exercises

Give the following ratios.

1. the number of days in a week to the number of months in a year
   7 to 12
2. the number of feet in a yard to the number of inches in a foot
   1 to 4
3. the number of hours in a day to the number of minutes in an hour
   2 to 5
4. the number of seconds in a minute to the number of minutes in an hour   1 to 1

Lesson Focus

Ask students if they know what surveyors do. Surveyors measure lengths and angles by using trigonometry. Today's lesson focuses on three important trigonometric ratios: sine, cosine, and tangent.

Teaching Notes

Start the lesson by drawing some right triangles on the chalkboard. Label the angles and have students locate adjacent sides, opposite sides, and the hypotenuse. Introducing the trigonometric ratios will be easier if some time is spent on developing the vocabulary.

Sine, Cosine, and Tangent

Objective: To find the sine, cosine, and tangent of acute angles of a right triangle.

In the figure at the right, $\triangle ABC$ is a right triangle. In relation to $\angle A$, $\overline{BC}$ is called the **opposite side** and $\overline{AC}$ is called the **adjacent side**. AB is the hypotenuse.

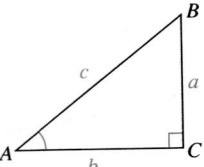

Terms to Know
- *opposite side*
- *adjacent side*
- *trigonometric ratio*
- *sine*
- *cosine*
- *tangent*

The following **trigonometric ratios** are defined for an acute angle A of a right triangle.

sine of $\angle A$: $\sin A = \dfrac{\text{length of side opposite } \angle A}{\text{length of hypotenuse}} = \dfrac{a}{c}$

cosine of $\angle A$: $\cos A = \dfrac{\text{length of side adjacent to } \angle A}{\text{length of hypotenuse}} = \dfrac{b}{c}$

tangent of $\angle A$: $\tan A = \dfrac{\text{length of side opposite } \angle A}{\text{length of side adjacent to } \angle A} = \dfrac{a}{b}$

Sometimes it is useful to use these shortened forms of the definitions.

$$\sin A = \frac{\text{opposite}}{\text{hypotenuse}} \qquad \cos A = \frac{\text{adjacent}}{\text{hypotenuse}} \qquad \tan A = \frac{\text{opposite}}{\text{adjacent}}$$

Notice that $\overline{AC}$ is the side opposite $\angle B$ and $\overline{BC}$ is the side adjacent to $\angle B$.

Example 1

Use the triangle at the right. Find each ratio in lowest terms.

a. $\tan A$ **b.** $\sin B$

c. $\cos A$ **d.** $\cos B$

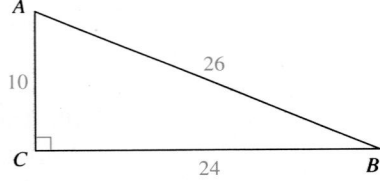

Solution

a. $\tan A = \dfrac{\text{opposite}}{\text{adjacent}} = \dfrac{24}{10} = \dfrac{12}{5}$ **b.** $\sin B = \dfrac{\text{opposite}}{\text{hypotenuse}} = \dfrac{10}{26} = \dfrac{5}{13}$

c. $\cos A = \dfrac{\text{adjacent}}{\text{hypotenuse}} = \dfrac{10}{26} = \dfrac{5}{13}$ **d.** $\cos B = \dfrac{\text{adjacent}}{\text{hypotenuse}} = \dfrac{24}{26} = \dfrac{12}{13}$

 **Check Your Understanding**

1. Which side is longest in a right triangle?
2. How do you know the statement "$\sin A = 2$" can never be true?
3. Find $\tan B$ and $\sin A$.
   See *Answers to Check Your Understanding* at the back of the book.

Example 2

Use the triangle at the right. Find each ratio in lowest terms.

a. $\sin X$

b. $\cos Z$

c. $\tan X$

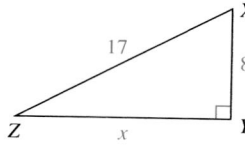

Solution

First use the Pythagorean theorem to find the length of $\overline{YZ}$.

$$x^2 + 8^2 = 17^2$$
$$x^2 + 64 = 289$$
$$x^2 = 225$$
$$x = \sqrt{225}$$
$$x = 15$$

a. $\sin X = \dfrac{\text{opposite}}{\text{hypotenuse}} = \dfrac{15}{17}$

b. $\cos Z = \dfrac{\text{adjacent}}{\text{hypotenuse}} = \dfrac{15}{17}$

c. $\tan X = \dfrac{\text{opposite}}{\text{adjacent}} = \dfrac{15}{8}$

 **Check Your Understanding**

4. In relation to $\angle Z$, what is the length of the opposite side? What is the length of the adjacent side? What is $\tan Z$?

5. Explain how $x^2 = 225$ is obtained from $x^2 + 64 = 289$.

6. The side opposite $\angle Z$ is the side ___?___ to $\angle X$.

See *Answers to Check Your Understanding* at the back of the book.

Exercises

1. **a.** hypotenuse **b.** adjacent; hypotenuse

Replace each ___?___ with the correct word. **c.** opposite; adjacent

A 1. a. $\sin A = \dfrac{\text{opposite}}{?}$ b. $\cos A = \dfrac{?}{?}$ c. $\tan A = \dfrac{?}{?}$

2. Explain the meaning of the word *adjacent* in the following sentence.

The house adjacent to mine was recently painted.
The house next to mine was recently painted.

Find the side indicated in $\triangle DEF$.

3. the hypotenuse *f*

4. the side opposite $\angle D$ *d*

5. the side adjacent to $\angle D$ *e*

6. the side opposite $\angle E$ *e*

7. the side adjacent to $\angle E$ *d*

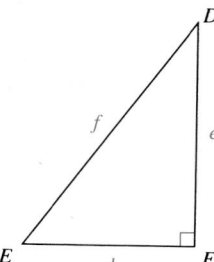

Trigonometry   **717**

Key Questions

1. Why is the sine of a 30° angle always $\frac{1}{2}$, even in right triangles of different sizes? The value of the sine is the ratio of the opposite side to the hypotenuse, which always remains constant.

2. Why is the value of the tangent of a 45° angle always equal to 1? A right triangle with a 45° angle is isosceles, so the ratio of the opposite side to the adjacent side is always equal to 1.

Critical Thinking

In the Check Your Understanding questions, students are asked to explain why $\sin A = 2$ is never true. Students must *apply* their knowledge of ratios and fractions to the sine ratio and *conclude* that the numerator can never be greater than the denominator.

Using Technology

For a suggestion on using a transformation drawing program, see page 706B.

Using Manipulatives

For a suggestion on using geoboards, see page 706C.

Reteaching/Alternate Approach

For a suggestion on using a cooperative learning activity with this lesson, see page 706D.

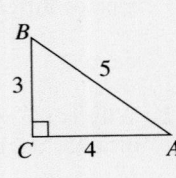
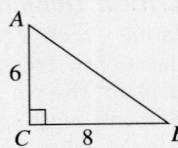
718

Use the triangle at the right.
Find each ratio in lowest terms.

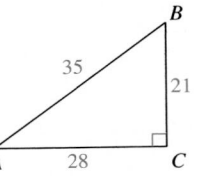

8. $\sin A$ $\frac{3}{5}$ 9. $\cos A$ $\frac{4}{5}$

10. $\tan A$ $\frac{3}{4}$ 11. $\sin B$ $\frac{4}{5}$

12. $\cos B$ $\frac{3}{5}$ 13. $\tan B$ $\frac{4}{3}$

Use the triangle at the right.
Find each ratio in lowest terms.

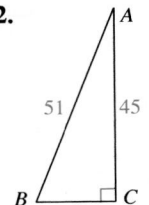

14. $\sin R$ $\frac{12}{13}$ 15. $\cos R$ $\frac{5}{13}$

16. $\tan R$ $\frac{12}{5}$ 17. $\sin S$ $\frac{5}{13}$

18. $\cos S$ $\frac{12}{13}$ 19. $\tan S$ $\frac{5}{12}$

For each right triangle, find the value of x. Then find $\sin A$, $\cos A$, and $\tan A$. Write each ratio in lowest terms.

20. $x = 8; \frac{3}{5}, \frac{4}{5}, \frac{3}{4}$

21. $x = 13; \frac{5}{13}, \frac{12}{13}, \frac{5}{12}$

22. $x = 24; \frac{8}{17}, \frac{15}{17}, \frac{8}{15}$

B   20. 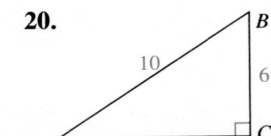   21.   22.

Use the figure at the right. Find each ratio in lowest terms.

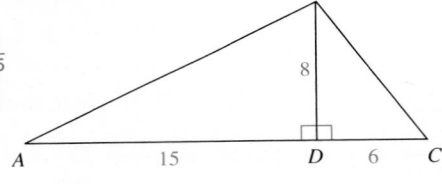

23. $\sin A$ $\frac{8}{17}$ 24. $\tan A$ $\frac{8}{15}$

25. $\sin C$ $\frac{4}{5}$ 26. $\cos C$ $\frac{3}{5}$

27. $\cos \angle ABD$ $\frac{8}{17}$

28. $\tan \angle CBD$ $\frac{3}{4}$

C   29. Is triangle ABC in Exercises 23–28 a right triangle? Explain how you know. No; the Pythagorean theorem does not hold true.

30. K is an acute angle of right triangle JKL and $\tan K = 1$. Give the measures of the angles of $\triangle JKL$. 45°, 45°, 90°

31. Use the triangles in Exercises 20 and 21. Find the value of $(\sin A)^2 + (\cos A)^2$ for each triangle. 1; 1

32. A and B are acute angles of right triangle ABC.
   a. What is the relationship between $\sin A$ and $\cos B$? Explain your answer. $\sin A = \cos B$; the side opposite $\angle A$ is adjacent to $\angle B$.
   b. What is the value of $(\tan A)(\tan B)$? Explain your answer. 1; the values of $\tan A$ and $\tan B$ are reciprocals.

Using Trigonometry

Objective: To use trigonometric ratios to find the lengths of the sides of a right triangle.

The table on page 720 lists the sines, cosines, and tangents of angles with measures from 1° to 89°. To find the value of sin 30°, locate 30° in the *Angle* column. Read the number across from 30 in the *Sine* column: .5000.

$$\sin 30° = 0.5000$$

If you have a scientific calculator, you can find sin 30° by using this key sequence.

You use the TAN and COS keys in a similar manner to find the tangent and cosine of an angle.

You can use trigonometric ratios to find the lengths of the sides of a right triangle.

Example 1

Solution

Find a and b. Round to the nearest tenth.

a represents the length of the side opposite $\angle A$. The length of the hypotenuse is given. Use the sine ratio.

$$\sin A = \frac{\text{opposite}}{\text{hypotenuse}} = \frac{a}{24}$$

$$\sin 52° = \frac{a}{24}$$

$$0.7880 \approx \frac{a}{24}$$

$$24(0.7880) \approx 24\left(\frac{a}{24}\right)$$

$$18.912 \approx a$$

$$a = 18.9 \text{ to the nearest tenth}$$

b represents the length of the side adjacent to $\angle A$. Use the cosine ratio.

$$\cos A = \frac{\text{adjacent}}{\text{hypotenuse}} = \frac{b}{24}$$

$$\cos 52° = \frac{b}{24}$$

$$0.6157 \approx \frac{b}{24}$$

$$24(0.6157) \approx 24\left(\frac{b}{24}\right)$$

$$14.7768 \approx b$$

$$b = 14.8 \text{ to the nearest tenth}$$

 **Check Your Understanding**

1. Use the Pythagorean theorem to check the lengths.

2. How would you find a and b if you were given the measure of $\angle B$?
   See *Answers to Check Your Understanding* at the back of the book.

Trigonometry **719**

719

Error Analysis
If students are using a calculator, they should make certain it is in degree mode before finding trigonometric values.

Number Sense
Remind students that the hypotenuse is always longer than either leg, and that the acute angles total 90°.

Using Technology
For a suggestion on using a transformation drawing program, see page 706B.

Table of Trigonometric Ratios

ANGLE	SINE	COSINE	TANGENT		ANGLE	SINE	COSINE	TANGENT
1°	.0175	.9998	0.175		46°	.7193	.6947	1.0355
2°	.0349	.9994	.0349		47°	.7314	.6820	1.0724
3°	.0523	.9986	.0524		48°	.7431	.6691	1.1106
4°	.0698	.9976	.0699		49°	.7547	.6561	1.1504
5°	.0872	.9962	.0875		50°	.7660	.6428	1.1918
6°	.1045	.9945	.1051		51°	.7771	.6293	1.2349
7°	.1219	.9925	.1228		52°	.7880	.6157	1.2799
8°	.1392	.9903	.1405		53°	.7986	.6018	1.3270
9°	.1564	.9877	.1584		54°	.8090	.5878	1.3764
10°	.1736	.9848	.1763		55°	.8192	.5736	1.4281
11°	.1908	.9816	.1944		56°	.8290	.5592	1.4826
12°	.2079	.9781	.2126		57°	.8387	.5446	1.5399
13°	.2250	.9744	.2309		58°	.8480	.5299	1.6003
14°	.2419	.9703	.2493		59°	.8572	.5150	1.6643
15°	.2588	.9659	.2679		60°	.8660	.5000	1.7321
16°	.2756	.9613	.2867		61°	.8746	.4848	1.8040
17°	.2924	.9563	.3057		62°	.8829	.4695	1.8807
18°	.3090	.9511	.3249		63°	.8910	.4540	1.9626
19°	.3256	.9455	.3443		64°	.8988	.4384	2.0503
20°	.3420	.9397	.3640		65°	.9063	.4226	2.1445
21°	.3584	.9336	.3839		66°	.9135	.4067	2.2460
22°	.3746	.9272	.4040		67°	.9205	.3907	2.3559
23°	.3907	.9205	.4245		68°	.9272	.3746	2.4751
24°	.4067	.9135	.4452		69°	.9336	.3584	2.6051
25°	.4226	.9063	.4663		70°	.9397	.3420	2.7475
26°	.4384	.8988	.4877		71°	.9455	.3256	2.9042
27°	.4540	.8910	.5095		72°	.9511	.3090	3.0777
28°	.4695	.8829	.5317		73°	.9563	.2924	3.2709
29°	.4848	.8746	.5543		74°	.9613	.2756	3.4874
30°	.5000	.8660	.5774		75°	.9659	.2588	3.7321
31°	.5150	.8572	.6009		76°	.9703	.2419	4.0108
32°	.5299	.8480	.6249		77°	.9744	.2250	4.3315
33°	.5446	.8387	.6494		78°	.9781	.2079	4.7046
34°	.5592	.8290	.6745		79°	.9816	.1908	5.1446
35°	.5736	.8192	.7002		80°	.9848	.1736	5.6713
36°	.5878	.8090	.7265		81°	.9877	.1564	6.3138
37°	.6018	.7986	.7536		82°	.9903	.1392	7.1154
38°	.6157	.7880	.7813		83°	.9925	.1219	8.1443
39°	.6293	.7771	.8098		84°	.9945	.1045	9.5144
40°	.6428	.7660	.8391		85°	.9962	.0872	11.4301
41°	.6561	.7547	.8693		86°	.9976	.0698	14.3007
42°	.6691	.7431	.9004		87°	.9986	.0523	19.0811
43°	.6820	.7314	.9325		88°	.9994	.0349	28.6363
44°	.6947	.7193	.9657		89°	.9998	.0175	57.2900
45°	.7071	.7071	1.0000					

Example 2

Find w to the nearest tenth.

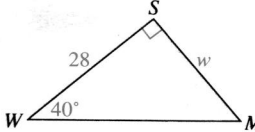

Solution

w represents the length of the side opposite $\angle W$. The length of the adjacent side is given. Use the tangent ratio.

$$\tan W = \frac{\text{opposite}}{\text{adjacent}} = \frac{w}{28}$$

$$\tan 40° = \frac{w}{28}$$

$$0.8391 \approx \frac{w}{28}$$

$$28(0.8391) \approx 28\left(\frac{w}{28}\right)$$

$$23.4948 \approx w$$

$$w = 23.5 \text{ to the nearest tenth}$$

 **Check Your Understanding**

3. Explain how to round 23.4948 to the nearest tenth.
4. Find w if $m\angle W$ had been 60°.
   See *Answers to Check Your Understanding* at the back of the book.

Exercises

Find the value of each trigonometric ratio in the table.

A 1. sin 5° 0.0872   2. tan 59° 1.6643   3. cos 75° 0.2588

   4. cos 45° 0.7071   5. tan 80° 5.6713   6. sin 68° 0.9272

State whether you would use the *sine*, *cosine*, or *tangent* to find x. Do not solve. 7. cosine 10. cosine

7.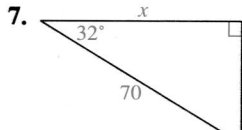

8. tangent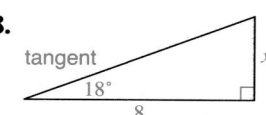

9. sine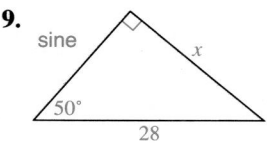

10.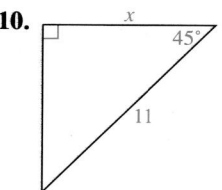

Trigonometry   721

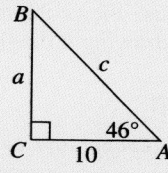
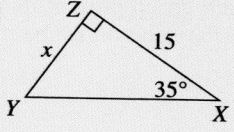
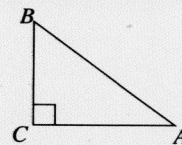

Problem Solving
In *Exercises 18 and 20*, encourage
students to use the problem-solving
strategy of drawing a diagram.

Follow-Up

Extension *(Cooperative Learning)*
Have students work in groups to
create a real-world problem that can
be solved by using trigonometry.
Have each group present its problem and the solution to the class.

Solve for x. Round to the nearest tenth.

11.

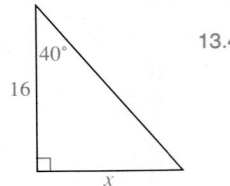

13.4

12.

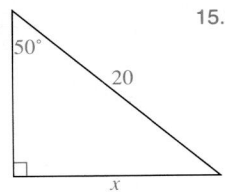

15.3

13.

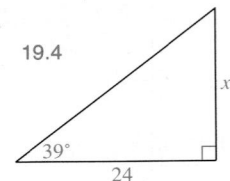

14.

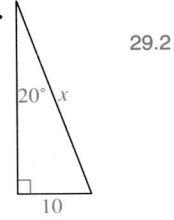

29.2

Find the measures of all the sides and angles of each triangle.

B   15. $a = 11.2$; $b = 27.8$;
$m\angle B = 68°$

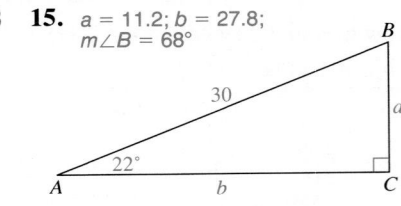

16.
$m\angle B = 46°$;
$a = 29.2$;
$b = 30.2$

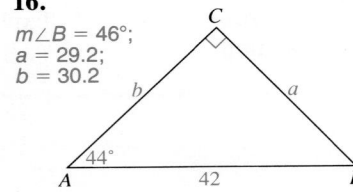

17. What is the height of the
tree, in feet, to the nearest
tenth?
25.0 ft

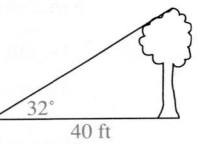

18. A cable 45ft long stretches from the top of an antenna to the ground.
The cable forms an angle of 50° with the ground. What is the height of
the antenna, in feet, to the nearest tenth? 34.5 ft

C   19. Find x to the nearest tenth.
23.7

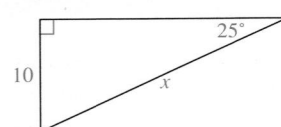

20. The two congruent sides of an isosceles triangle measure 30 cm each.
The two congruent angles measure 50° each. Find the measure of the
third side. (*Hint:* Sketch the triangle and draw the height to the third
side.) 38.6 cm

Finding Angle Measures

Objective: To use trigonometric ratios to find the measure of an angle of a right triangle.

You can use trigonometry to find the measure of an angle in a right triangle. You must know the lengths of two sides.

Example 1

Find the measure of each acute angle to the nearest degree.

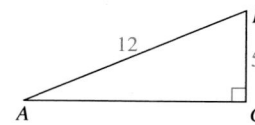

Solution

For $\angle A$, the opposite side and the hypotenuse are given.
First find the sine ratio.

$$\sin A = \frac{\text{opposite}}{\text{hypotenuse}} = \frac{5}{12} \qquad \leftarrow \text{Write } \tfrac{5}{12} \text{ as a decimal.}$$

$$\sin A \approx 0.4167$$

Then use the table on page 720. Look in the *Sine* column.
Find the closest entry to 0.4167.
$\sin 24° \approx 0.4067$, $\sin 25° \approx 0.4226$

Angle	Sine
24°	0.4067
A	0.4167
25°	0.4226

difference: $0.4167 - 0.4067 = 0.0100$
difference: $0.4226 - 0.4167 = 0.0059$

0.4167 is closer to 0.4226 than it is to 0.4067.
To the nearest degree, $m\angle A = 25°$.

Find the measure of $\angle B$.

$$m\angle A + m\angle B + m\angle C = 180°$$
$$25° + m\angle B + 90° = 180°$$
$$115° + m\angle B = 180°$$
$$115° + m\angle B - 115° = 180° - 115°$$
$$m\angle B = 65°$$

 **Check Your Understanding**

1. Explain why 0.4167 is closer to 0.4226 than it is to 0.4067.
2. When finding $m\angle B$ in Example 1, why does the right side of the equation equal 180°?

If you have a scientific calculator, you can find the measure of an acute angle in a triangle given two sides. To find $m\angle A$ in Example 1, you can use this key sequence if your calculator has an **INV** key.

Trigonometry **723**

Warm-Up Exercises

Use $\triangle ABC$ to identify the following.

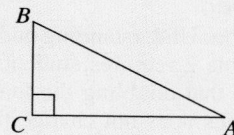

1. side adjacent to $\angle A$ $\overline{AC}$
2. side opposite $\angle B$ $\overline{AC}$
3. hypotenuse $\overline{AB}$
4. complement of $\angle B$ $\angle A$

Lesson Focus

Point out to students that surveyors often have to determine angle measures as well as lengths. Today's lesson focuses on using trigonometry to determine angle measures.

Teaching Notes

The solutions to Example 1 and Example 2 are quite lengthy. Go over the solutions step by step so students can understand the procedures for finding angle measures.

Key Questions

1. What is the relationship of the two acute angles in a right triangle?   complementary
2. What must you know in order to find the measure of an acute angle in a right triangle?   the lengths of any two of the sides

Teaching Notes (continued)

Error Analysis

Students may need help in finding the measures of angles to the nearest degree. They should use the procedures given in the Examples. Caution students not to attempt to do this mentally in order to avoid making errors.

Reasoning

Check Your Understanding question 4 in Example 2 requires students to conclude that doubling the lengths of the sides does not change their ratio. Therefore, the measure of the angle remains the same.

Additional Examples

1. Find the measure of each angle to the nearest degree.

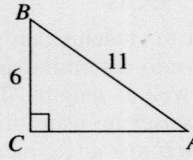

$\angle A = 33°$; $\angle B = 57°$

2. Find the measure of each angle to the nearest degree. Find the length of $\overline{JL}$ to the nearest tenth.

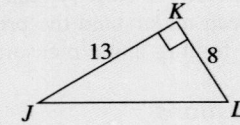

$\angle J = 32°$; $\angle L = 58°$; $JL = 15.3$

Closing the Lesson

$\triangle ABC$ is a right triangle with the right angle at C. Have students explain what values are needed in order to find the measure of $\angle B$. Also, have students draw a conclusion about the relationship between $\sin B$ and $\cos A$.

Example 2

Find the measure of each acute angle to the nearest degree. Find AB to the nearest tenth.

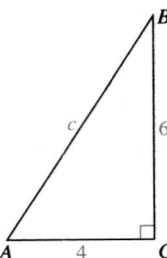

Solution

For $\angle A$, the opposite and adjacent sides are given. First find the tangent ratio.

$$\tan A = \frac{\text{opposite}}{\text{adjacent}} = \frac{6}{4} = 1.5000$$

Then use the table on page 720.

$\tan 56° \approx 1.4826$ $\qquad$ $\tan A = 1.5000$ $\qquad$ $\tan 57° \approx 1.5399$

1.5000 is closer to 1.4826 than it is to 1.5399.

To the nearest degree, $m\angle A = 56°$.

Find $m\angle B$.

$$56° + m\angle B + 90° = 180°$$
$$146° + m\angle B = 180°$$
$$m\angle B = 34°$$

To find the length of the hypotenuse, $\overline{AB}$, use the Pythagorean theorem.

$$a^2 + b^2 = c^2$$
$$6^2 + 4^2 = c^2$$
$$36 + 16 = c^2$$
$$52 = c^2$$
$$c = \sqrt{52} \approx 7.211$$

$AB = 7.2$ to the nearest tenth

☑️ **Check Your Understanding**

3. Explain why 1.5000 is closer to 1.4826 than it is to 1.5399.

4. If $BC = 12$ and $AC = 8$, what would be the measure of $\angle A$? $\angle B$? $\overline{AB}$?

See *Answers to Check Your Understanding* at the back of the book.

Exercises

A 1. Sketch right triangle SDW with the right angle at D. To find $\tan W$, which two sides must have their lengths given? $\overline{DW}$ and $\overline{SD}$

2. How would you find the measure of an acute angle of a right triangle if the measure of the other acute angle is given? subtract the measure of the other acute angle from 90°

Find $m\angle A$ to the nearest degree.

3. $\sin A = 0.3240$ 19°   **4.** $\cos A = 0.8215$ 35°   **5.** $\tan A = 1.4400$
55°

Find the measure of each acute angle to the nearest degree.

6. $m\angle A = 68°$;
$m\angle B = 22°$
7. $m\angle A = 31°$;
$m\angle B = 59°$
8. $m\angle A = 37°$;
$m\angle B = 53°$
9. $m\angle A = 39°$;
$m\angle B = 51°$
10. $m\angle A = 25°$;
$m\angle B = 65°$

6. 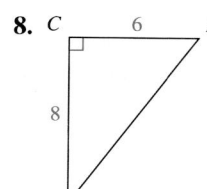   **7.**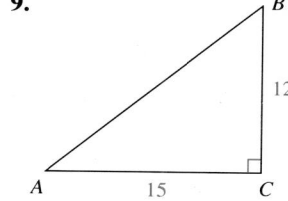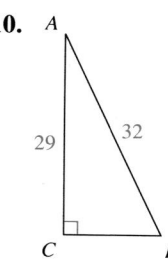

8. 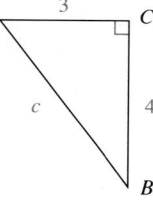   **9.** 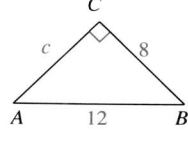   **10.**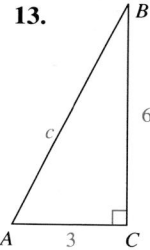

Find the measure of each acute angle to the nearest degree. Find the length of the third side to the nearest tenth.

11. $m\angle A = 53°$;
$m\angle B = 37°$; $c = 5$
12. $m\angle A = 42°$;
$m\angle B = 48°$;
$c = 8.9$
13. $m\angle A = 63°$;
$m\angle B = 27°$; $c = 6.7$

B **11.** 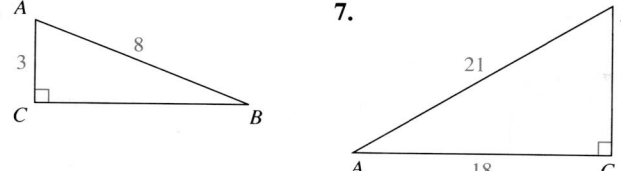   **12.**   **13.**

C **14.** A 25 ft ladder is leaning against a building. The distance from the foot of the ladder to the building is 7 ft. Find the measure of the angle that the ladder makes with the ground to the nearest degree. 74°

15. Find the measures of all three angles of isosceles triangle *TRI* to the nearest degree. (*Hint*: Draw the height from *I* to $\overline{TR}$.)
$m\angle T = 71°; m\angle R = 71°; m\angle I = 38°$

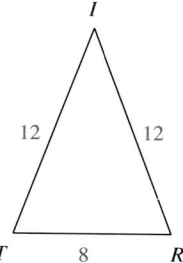

Trigonometry   **725**

Exercise Note

Problem Solving
In *Exercises 14 and 15*, students can make use of the strategy of drawing a diagram to visualize the relationships. In *Exercise 15*, they draw the altitude to form right triangles, thereby making it possible to apply trigonometric ratios.

Follow-Up

Project
Ask students to research the meaning of *angle of elevation* and *angle of depression*. Ask each student to create a problem that involves one of these types of angles. Students should make a sketch to accompany their problems.

Warm-Up Exercises

Find the missing sides and angles in each triangle. Give lengths of sides to the nearest tenth and angle measures to the nearest degree.

1.

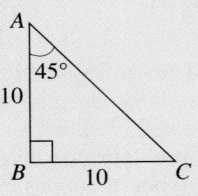

$\angle C = 45°$; $AC = 14.1$

2.

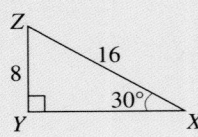

$\angle Z = 60°$; $XY = 13.9$

Lesson Focus

Have you ever heard the expression, "That's a *special* case"? This expression is used to describe a unique situation. For right triangles there are two special cases that are the focus of today's lesson.

Teaching Notes

Use the Application at the beginning of the lesson to make students aware of the fact that the two special right triangles are used frequently. Students should memorize the relationships of the sides.

Work through each Example, paying close attention to how students convert the radical form to a decimal using a calculator.

Special Right Triangles

Objective: To apply 45°-45°-90° and 30°-60°-90° right triangles.

APPLICATION

Terms to Know
• *45°-45°-90° right triangle*
• *30°-60°-90° right triangle*

There are two special right triangles. They are often used by construction and skilled trades workers, architects, draftspersons, and engineers.

One special triangle is the *isosceles right triangle*. In an isosceles right triangle, the two legs are equal. Also, the acute angles are equal in measure and each angle measures 45°. An isosceles right triangle is often called a **45°-45°-90° right triangle.**

In the 45°-45°-90° right triangle shown, each leg is 1 unit long. Find the length of the hypotenuse by using the Pythagorean Theorem.

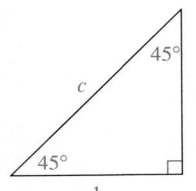

$$c^2 = 1^2 + 1^2$$
$$c^2 = 2$$
$$c = \sqrt{2}$$

The hypotenuse is $\sqrt{2}$ units long.

If each leg of a 45°-45°-90° right triangle has length 3, then the hypotenuse has length $3\sqrt{2}$. This is because all 45°-45°-90° right triangles are similar, and corresponding sides of similar triangles have the same ratio.

45°-45°-90° Right Triangle Property

If each leg of a 45°-45°-90° right triangle has length n, then the hypotenuse has length $n\sqrt{2}$.

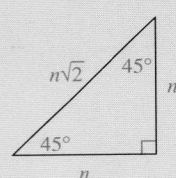

Example 1

Solution

Find the lengths of the missing sides. Round decimal answers to the nearest tenth.

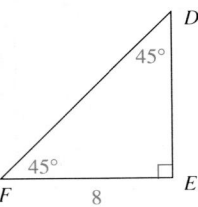

$\triangle DEF$ is a 45°-45°-90° right triangle.

The legs are the same length.
Since $FE = 8$, $DE = 8$.

$\overline{FD}$ is the hypotenuse.
By the 45°-45°-90° right triangle property,
$FD = 8\sqrt{2} \approx 8(1.414) = 11.312$.
$FD = 11.3$ to the nearest tenth

 **Check Your Understanding**

1. Is $8\sqrt{2}$ the *exact length* or the *approximate length* of the hypotenuse?
2. If $FE = 14$, what are the lengths of the other two sides?
3. If $DF = 6\sqrt{2}$, what are the lengths of the other two sides?
   See *Answers to Check Your Understanding* at the back of the book.

 You can use a calculator to find the length of a hypotenuse if you know the length of the shorter leg. To find FD in Example 1, use this key sequence.

$\triangle WXZ$ is an equilateral triangle. Each angle measures 60°. The lengths of the sides are equal. $\triangle WXY$ is half the size of $\triangle WXZ$. $\triangle WXY$ is a **30°-60°-90° right triangle.** If $XY = 1$, then $XW = 2$. You can find WY by using the Pythagorean theorem.

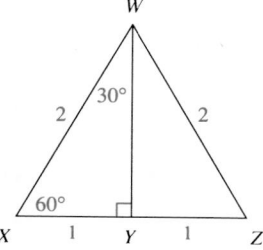

$$(XY)^2 + (WY)^2 = (XW)^2$$
$$1^2 + (WY)^2 = 2^2$$
$$1 + (WY)^2 = 4$$
$$(WY)^2 = 3$$
$$WY = \sqrt{3}$$

For this 30°-60°-90° right triangle we observe that the hypotenuse is twice as long as the shorter leg, and the longer leg is the length of the shorter leg times $\sqrt{3}$.

All 30°-60°-90° right triangles are similar, so corresponding sides have the same ratio.

30°-60°-90° Right Triangle Property

If the shorter leg of a 30°-60°-90° right triangle has length x, then the hypotenuse has length $2x$, and the longer leg has length $x\sqrt{3}$.

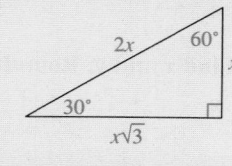

The shorter leg is always opposite the 30° angle. The longer leg is always opposite the 60° angle.

Trigonometry **727**

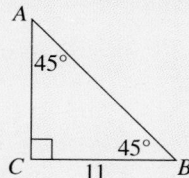
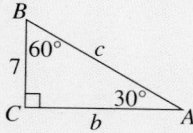

Closing the Lesson

Have students summarize the procedures for finding unknown lengths
in the two special right triangles.

Suggested Assignments

Advanced: 1–23 odd

Exercise Notes

Communication: Reading

Exercises 1–3 check students'
understanding of the relationships
within the special triangles.

Critical Thinking

Exercise 23 requires students to
analyze the sketch, *formulate* a plan
for solving the problem, and provide information needed for the
solution.

Example 2

Find *a* and *b*. Round decimal
answers to the nearest tenth.

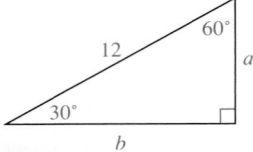

Solution

This is a 30°-60°-90° right triangle. The length of the hypotenuse is twice
the length of the shorter leg.

a is the length of the side opposite the 30° angle.
$$2a = 12$$
$$a = 6$$
b is the length of the side opposite the 60° angle.
$$b = a\sqrt{3}$$
$$b = 6\sqrt{3} \approx 10.392305$$
$$b = 10.4 \text{ to the nearest tenth}$$

 **Check Your**
Understanding

4. If the hypotenuse were 20, what would be the values of *a* and *b*?

5. If *a* = 100, what would be the value of *b*? What would be the length of
   the hypotenuse?
   See *Answers to Check Your Understanding* at the back of the book.

Exercises

Complete.

A   1. If each leg of a 45°-45°-90° right triangle has length *x*, then the
       hypotenuse has length __?__. x √2

2. In a 30°-60°-90° right triangle, the shorter leg is opposite the __?__
   angle, and the longer leg is opposite the __?__ angle. 30°; 60°

3. If the shorter leg of a 30°-60°-90° right triangle has length *b*, then the
   __?__ has length *b* √3, and the __?__ has length 2*b*. longer leg; hypotenuse

Find a decimal approximation to the nearest tenth.

4. 5 √2 7.1   5. 6 √3 10.4   6. 12 √3 20.8   7. 10 √2
                                                        14.1

Find *x* and *y*. Round decimal answers to the nearest tenth.

8. *x* = 7; *y* = 9.9
9. *x* = 6; *y* = 6
10. *x* = 6.9; *y* = 8

8.

9.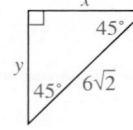

10.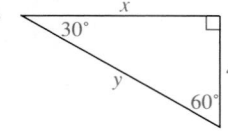

Find *x* and *y*. Round decimal answers to the nearest tenth.

11.
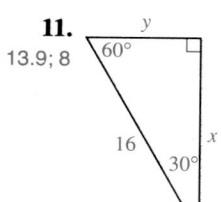
13.9; 8

12.
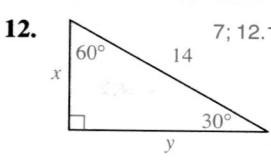
7; 12.1

13. 22; 31.1
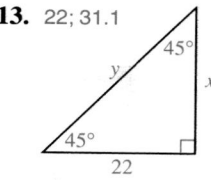

14. 9; 9
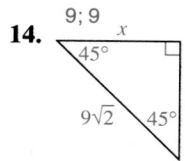

15. 6; 12
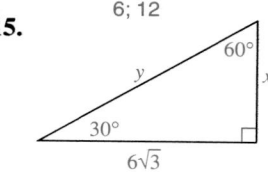

16.
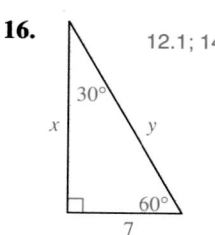
12.1; 14

Find the length marked *x* or *y*. Round decimal answers to the nearest tenth.

B **17.** *ABC* is an equilateral triangle.
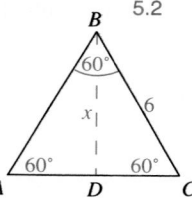

18. *ABCD* is a square. 28.3
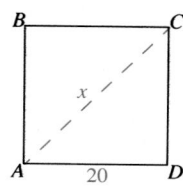

19. *ABCD* is a rectangle. 24; 20.8
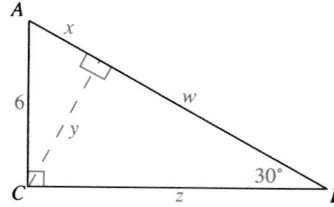

Find the lengths. Round decimal answers to the nearest tenth.

20. A leg of an isosceles right triangle has length 24. Find the lengths of the other two sides. 24; 33.9

21. The shorter leg of a 30°-60°-90° right triangle is 40 cm long. How long are the other two sides? 69.3 cm, 80 cm

C **22.** The hypotenuse of a 45°-45°-90° right triangle has length $16\sqrt{2}$. How long are the other sides? 16; 16

23. a. In $\triangle ABC$ at the right, find *w*, *x*, *y*, and *z*.
   b. What is the area of $\triangle ABC$?
   a. 9; 3; 5.2; 10.4
   b. 31.2 square units

Trigonometry **729**

729

Extra Practice

Chapter 1

Evaluate each expression when $u = 6$, $v = 2$, and $w = 3$.

1. $3 + w$ 6
2. $6v$ 12
3. $17 - u$ 11
4. $9 \div w$ 3

1-1

5. vw 6
6. $u - w$ 3
7. $u \div w$ 2
8. $v + u$ 8

Evaluate each expression when $q = 76$, $r = 47.5$, $s = 103.72$, and $t = 12.4$.

9. $114 + s$ 217.72
10. $92 + q$ 168
11. $r + 12$ 59.5
12. $22 + t$ 34.4

1-2

13. $251 - t$ 238.6
14. $s - r$ 56.22
15. $q - t$ 63.6
16. $q - r$ 28.5

Evaluate each expression when $a = 24$, $b = 18.41$, $c = 7.7$, and $d = 5.2$.

17. $92b$ 1693.72
18. $10.4d$ 54.08
19. $22a$ 528
20. $61c$ 469.7

1-3

21. $20.8 \div d$ 4
22. $b \div 7$ 2.63
23. $a \div 0.12$ 200
24. $b \div 21.04$ 0.875

Use this paragraph for Exercises 25–28.

Chris earns $15 per hour at a chemical factory, where the standard work day is 8 h. He earns double his hourly wage for work on Sundays and holidays. Beginning Monday, July 4, he works 5 days, including July 4.

25. What is the paragraph about? Chris' wages

1-4

26. How many days did Chris work the week of July 4? 5

27. How many hours did he work during that week? 40

28. What facts would not be used to find the wages Chris took home that week? he works at a chemical factory; he earns double his hourly wage on Sundays

Replace each ? with >, <, or =.

29. $7323 \underline{\ ?\ } 4619$ >
30. $14{,}135 \underline{\ ?\ } 14{,}137$ <
31. $5.72 \underline{\ ?\ } 5.720$ =

1-5

32. $110 \underline{\ ?\ } 11.3$ >
33. $9.768 \underline{\ ?\ } 10.315$ <
34. $0.93 \underline{\ ?\ } 0.94$ <

Use the properties to find each sum mentally.

35. $37 + 21 + 313$ 371
36. $45 + 9 + 1025$ 1079
37. $5.6 + 13 + 7.4$ 26

1-6

38. $10.7 + 46 + 3.3$ 60
39. $97 + 0$ 97
40. $83 + 2 + 17$ 102

41. $139 + 4.8 + 91$ 234.8
42. $2.4 + 3 + 0$ 5.4
43. $786 + 12.7 + 94$ 892.7

Use the properties to find each product mentally. 50. 9063   51. 110

44. $5 \cdot 13 \cdot 2$ 130
45. $20(7)(5)$ 700
46. $98 \cdot 17 \cdot 0$ 0
47. $3 \cdot 3 \cdot 1$ 9

1-7

48. $(96)(2)(0.5)$ 96
49. $8(5)(0)$ 0
50. $1 \cdot 9063 \cdot 1$
51. $10(2.2)(5)$

Tell whether it is most efficient to use *mental math*, *paper and pencil*, or a *calculator*. Then find each answer using the method that you chose. 52–59. Answers may vary. Likely answers are given.

52. $168 - 97$
mm; 71

53. $185 - 45$
mm; 140

54. $342 + 977$ *c; 1319*

55. $16.7 + 13.9 + 55.8$ **1-8**
c; 86.4

56. $83 \cdot 42$ *c; 3486*

57. $63.6 \div 0.2$
pp; 318

58. $(0.7)(0.3)$ *pp; 0.21*

59. $32.8 \div 100$
mm; 0.328

Find each answer. 67. 98

60. 2^4 16

61. 0^{17} 0

62. 1^1 1

63. 9^3 729

64. 3^2 9

65. 10^8 100,000,000   **1-9**

66. $3^3 + 3 \cdot 11$ 60

67. $11^2 - 69 \div 3$

68. $97 - 5 \cdot 32 \div 16$
87

69. $9^2 + 8 \div 2 \cdot 7$ 109   **1-10**

70. $6^2 \div (13 - 9)$
4

71. $99 + 3(6 + 17)$
168

72. $2^5 + 9(7 - 3)$ 68

73. $\dfrac{4 \cdot 12}{51 - 3}$ 1

Chapter 2

10. $20a - 24$ **11.** $12 - 18a$ **12.** $40n + 60$ **17.** $8x$

Simplify. 18. $7a + 6$ **19.** $2a + 13c$ **20.** $5n + 3m$

1. $a^3 \cdot a^7$ a^{10}

2. $2b^4 \cdot 7b^3$ $14b^7$

3. $(6x)(11x)$ $66x^2$

4. $(4a)(2b)(5a)$ $40a^2b$ **2-1**

5. $6(a + 7)$
$6a + 42$

6. $5(3b - 8)$
$15b - 40$

7. $2(x + 13)$ $2x + 26$

8. $4(6 - c)$ $24 - 4c$ **2-2**

9. $3(2x + 7)$
$6x + 21$

10. $4(5a - 6)$

11. $6(2 - 3a)$

12. $10(4n + 6)$

13. $5x + 7x$ $12x$

14. $13z - 4z$ $9z$

15. $5x + x + 2x$ $8x$

16. $13y - y$ $12y$ **2-3**

17. $3x + 4x + x$

18. $3a + 4a + 6$

19. $2a + 5c + 8c$

20. $n + 4n + 3m$

21. Mikisha bought two compact discs for $13.50 each and three cassettes for $8.98 each. What was the total cost? $53.94   **2-4**

22. Demiso bought three boxes of herbal tea for $1.59 each and two loaves of bread for $1.39 each. How much more did he pay for the tea than for the loaves of bread? $1.99

23. 39, 45, 51

Find the next three numbers or expressions in each pattern. 24. 20, 10, 5

23. 15, 21, 27, 33, _?_, _?_, _?_

24. 320, 160, 80, 40, _?_, _?_, _?_   **2-5**

25. $x, x + 4, x + 8,$ _?_, _?_, _?_
$x + 12, x + 16, x + 20$

26. $x + 5, 3x + 5, 5x + 5,$ _?_, _?_, _?_
$7x + 5, 9x + 5, 11x + 5$

Complete each function table.   **Find the function rule.**

27.

x	$2x + 5$	
0	?	5
2	?	9
4	?	13
6	?	17
8	?	21

28.

x	$8x$	
10	?	80
11	?	88
12	?	96
13	?	104
14	?	112

29.

x	?
5	14
6	15
7	16
8	17
9	18

$x \rightarrow x + 9$ **2-6**

30. Sharon earns $6.50 per hour at the bike store. She also earns an extra   2-7
$12 for each bike she sells. Last week Sharon worked 30 h and sold
4 bikes. She says that she earned $207 last week. Is this correct?
Explain. No; her earnings are (6.5)(30) + (12)(4) = $243.

Decide whether an *estimate* or an *exact answer* is needed. Then solve.

31. Barbecue meat sells for $1.46 per pound. Alicia needs to buy 40 lb for   2-8
a school picnic. How much money should she take to the store to be
sure that she has enough to pay for the meat? estimate; about $60

32. Charlene charges $4.50 per hour to baby-sit. Last night she baby-sat
for 6 h. How much did she earn? exact answer; $27

Write each measure in the unit indicated.

33. 34 cm; mm 340 mm   **34.** 7.5 kg; g 7500 g   **35.** 4.2 L; mL   2-9
                                                                        4200 mL
36. 120 mm; cm 12 cm   **37.** 4000 mg; g 4 g   **38.** 2.6 m; mm
                                                                        2600 mm

Write each number in scientific notation.

39. 3,500,000 3.5×10^6 **40.** 4210 4.21×10^3 **41.** 80,000   2-10
                                                                        8×10^4

Write each number in decimal notation.

42. 3.7×10^4 37,000   **43.** 2.35×10^5 235,000   **44.** 9.9×10^7
                                                                        99,000,000

Chapter 3

Find each absolute value.

1. $|-15|$ 15   **2.** $|7|$ 7   **3.** $|0|$ 0   **4.** $|-36|$ 36   3-1

Replace each __?__ with >, <, or =.

5. -9 __?__ 7 <   **6.** 0 __?__ -3 >   **7.** -11 __?__ -14 >   **8.** 2 __?__ -5
                                                                                                   >

Find each sum.

9. $-34 + (-9)$ -43 **10.** $57 + 75$ 132   **11.** $24 + 89$ 113   **12.** $-45 + (-37)$ 3-2
                                                                                              -82
13. Thomas withdrew $16 from his savings account last month. He with-
drew $20 and $25 this month. How much has Thomas withdrawn over
the two months? $61

Find each sum.

14. $37 + (-78)$ -41 **15.** $-4 + 25$ 21   **16.** $-6 + 32$ 26   **17.** $8 + (-8)$ 0   3-3

18. Sasha recorded a temperature drop of 13°F followed by an increase of
8°F. Find the total gain or loss in temperature. 5°F loss

Find each answer.

19. $-17 - 6$ -23   **20.** $27 - (-27)$ 54   **21.** $-12 - (-43)$ 31 **22.** $-18 - 31$ -49   3-4

23. $-56 \div (-8)$ 7   **24.** $6(-10)(0)$ 0   **25.** $(-7)(3)(-2)$ 42 **26.** $\dfrac{-90}{15}$ -6   3-5

Evaluate each expression when $a = -4$, $b = 5$, and $c = -2$.

27. $-4c^2$ -16   **28.** $c^4 - b$ 11   **29.** $3a^2 - c$ 50   **30.** $|a| - |c|$ 2   3-6

31. $6 - |a - b|$ -3 **32.** $|a + b|$ 1   **33.** $-ab$ 20   **34.** $-bc - 7$ 3

Solve by making a table.

35. In ice hockey, a win is scored as 2 points, a tie as 1 point, and a loss as 0 points. The Blades have played 4 games in the season. How many different point totals are possible? 9   3-7

Use the coordinate plane at the right. Write the coordinates of each point.

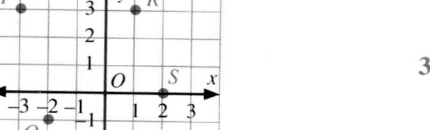

36. P (-3, 3)   **37.** Q (-2, -1)   3-8

38. R (1, 3)   **39.** S (2, 0)

Answers to Exercises 40–47 are on p. A44.
Graph each point on a coordinate plane.

40. $A(-5, -2)$ **41.** $B(3, -4)$

42. $C(0, -3)$ **43.** $D(3, 1)$

Graph each function.

44.

x	$3 - x$
-2	5
-1	4
0	3
1	2

45.

x	$2x + 1$
-3	-5
-2	-3
-1	-1
0	1

3-9

46. $x \rightarrow x^2$, when $x = -3, -2, 0, 2,$ and 3

47. $x \rightarrow |x - 1|$, when $x = -5, -3, -1, 2,$ and 3

Chapter 4

Is the given number a solution of the equation? Write *Yes* or *No*.

1. $n + 7 = 9; 2$ Yes.   **2.** $q - 4 = 18; 21$ No.   **3.** $36 = \dfrac{k}{4}; 4$ No.   **4.** $4p = 24; 6$ Yes.   4-1

Use mental math to find each solution.

5. $3r + 4 = 16$ 4   **6.** $8s - 1 = -25$ -3   **7.** $5t - 6 = 34$ 8   **8.** $\dfrac{y}{-7} + 9 = 5$ 28

Solve by using the guess-and-check strategy.

9. Tickets for a game are $12 and $17. If Harriet spends $300, how many of each type of ticket does she buy? 8 tickets for $12 and 12 tickets for $17 4-2

10. In basketball, there are 3-point goals, 2-point goals, and 1-point goals. Jonathan makes 3 goals and scores 7 points. How many of each type of goal could he make to score the 7 points?
two 2-point goals and one 3-point goal; or two 3-point goals and one 1-point goal

11. In hockey, one point is given for a goal and one point for an assist. Barbara has 28 points. She has four more assists than goals. How many goals and assists does she have? 12 goals and 16 assists

12. Zack bought some pencils for $.15 each. He also bought some books for $1.99 each. He spent $8.41. How many of each item did he buy?
4 books and 3 pencils

Solve. Check each solution.

13. $t + 19 = 31$ 12   14. $n + 6 = -54$ 15. $m - 4 = -11$ 16. $r - 6 = -12$ 4-3
                            −60   −7   −6

17. $-7 = q - 14$ 7   18. $-6 = -6 + s$ 0   19. $14 = h + 6$ 8   20. $-22 = k + 9$
                                                                        −31

21. $8t = 64$ 8   22. $4r = -16$ −4   23. $30 = 6r$ 5   24. $\frac{n}{4} = 16$ 64   4-4

25. $-2 = \frac{k}{-2}$ 4   26. $\frac{a}{-6} = 11$ −66   27. $4 = -n$ −4   28. $-k = -6$ 6

29. $4n + 1 = 21$ 5   30. $6n - 6 = 6$ 2   31. $2b + 7 = -9$ −8   32. $7d - 4 = -18$ 4-5
                                                                        −2

33. $-6 = \frac{t}{-4} + 3$ 34. $7 = \frac{z}{4} - 6$ 52   35. $18 = \frac{v}{-2} + 8$ −20   36. $-4 = \frac{w}{-6} - 3$
    36   6

Write a variable expression that represents each phrase. If necessary, choose a variable to represent the unknown number. 39. $2n + 3$

37. two less than a number x $x - 2$ 38. seven fewer peaches than yesterday $p - 7$ 4-6

39. three more than twice a number n 40. six less than nine times a number k $9k - 6$

41. five times as old as Jennifer $5j$ 42. double the number of points scored $2p$

43. seven less than a number t tripled   44. four more than a number r divided by two
    $3t - 7$ $\frac{r}{2} + 4$

Write an equation that represents the relationship in each sentence.

45. A hockey stick costs $25, which is $21 more than the cost of a hockey puck. $25 = u + 21$ 4-7

46. A laser disk costs $16, which is $9 more than the cost of an audio tape. $16 = c + 9$

47. Mary's total of 20 points was 4 less than the team's points. $20 = t - 4$

48. Jill's time of 47 seconds was 3 seconds less than Serena's time in the track event. $47 = s - 3$

Solve using an equation.

49. Anne pays $7 each for the first two melons she buys and $5 for any others. She pays $64 in all. How many melons does she buy? 12   4-8

50. Sam pays $20 down for a keyboard, and $15 weekly, until he pays the total price of $290. How many weekly payments does he make? 18

51. A car rental agency charges $43 the first two days and $35 each day thereafter. Ted is charged $323. How many days did he rent a car? 10

52. A park charges $12 per person for a group of up to 100 people. After the first 100 people, the park charges $8 per person. A group spends $2000 for an outing at the park. How many people are in the group?
200

Solve. Check each solution.

53. $2x + 5x = 21$ 3   **54.** $25 = 8t - 3t$ 5   **55.** $-9n - n = 100$ −10   4-9

56. $3(2b + 4) = 24$ 2   **57.** $5(k - 4) = 40$ 12   **58.** $3q + 10 + 3q = 130$
20

Use the formula $D = rt$. **59.** 165 mi   **60.** 175 mi   **61.** 47 mi/h   **62.** 4.5 h

59. Let $r = 55$ mi/h and $t = 3$ h. Find D.   **60.** Let $r = 35$ mi/h and $t = 5$ h. Find D.   4-10

61. Let $t = 8$ h and $D = 376$ mi. Find r.   **62.** Let $D = 279$ mi and $r = 62$ mi/h. Find t.

Chapter 5

Draw a pictograph to display the data.

1. Sales of Walking Shoes in the United States (Millions of Dollars)   See answer on p. A44.   5-1

Year	1985	1986	1987	1988	1989
Sales	$263	$368	$512	$752	$888

Use the double bar graph at the right.

2. In 1983, about how many movies made $20 million or more? about 20

3. About how many more movies earned profits of $10–$20 million in 1989 than in 1981?
about 20

5-2

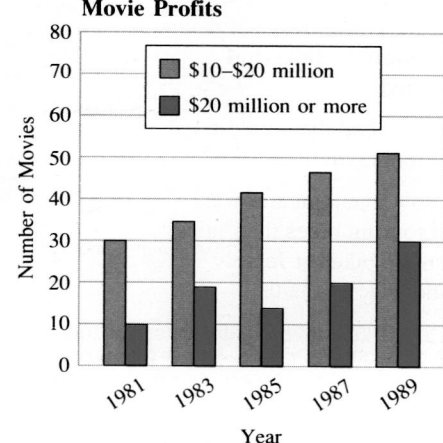

Movie Profits

Use the line graph at the right.

4. Estimate the number of households with telephones in 1885.
   about 180

5. Estimate the increase in households with telephones from 1895
   to 1900. about 1000

6. Describe the overall trend in the
   number of households with telephones. increasing

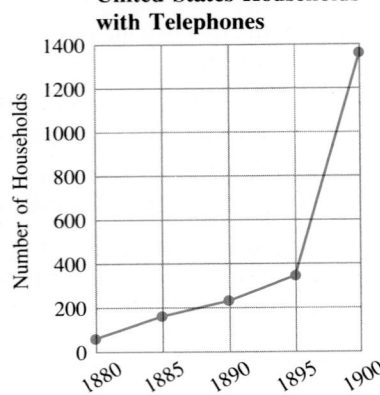

**United States Households
with Telephones**

5-3

Draw a line graph to display the data.

7. **Average Movie Admission Prices in the United States** See answer on p. A44.

5-4

Year	1981	1983	1985	1987	1989
Price	$2.78	$3.15	$3.55	$3.91	$4.45

**Decide whether it would be more appropriate to draw a _bar graph_ or a
line graph to display the data. Then draw the graph.** bar graph; display is
on p. A44.

8. **Time Spent on Volunteer Work (Hours per Week)**

5-5

Name	Jackie	Damon	Rhea	Joel	Samun
Child Care	3.5	5.0	4.2	1.7	2.3
Shopping	3.2	0.6	1.3	5.3	4.5

**Use the bar graph at the right.
Solve, if possible. If there is not
enough information, tell what
facts are needed.**

9. Which store had the greatest
   sales of mountain bikes?
   National Bikes

10. About how much greater were
    sales of touring bikes than sales
    of mountain bikes at Jesse's
    Bike Shop? about $12,000

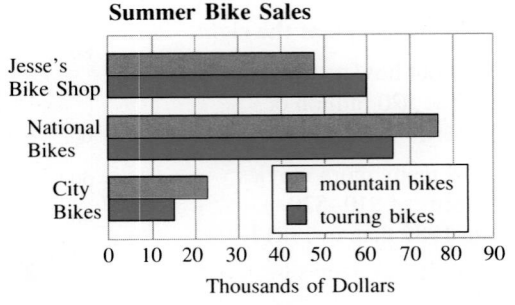

Summer Bike Sales

5-6

11. Does City Bikes have greater sales for touring bikes or racing bikes?
    not enough information; sales figures for racing bikes

Use the bar graph at the right.

12. Describe the visual impression of the results of the class election. See answer on p. A44.

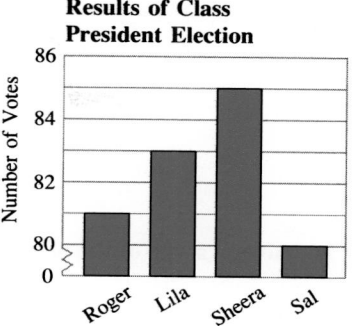

Results of Class President Election

5-7

13. The school newspaper editor wants to use a graph to accompany an article that describes the election as being very close. Explain how she might redraw the graph to support the article.
Remove the gap in the scale; change the intervals on the scale.

14. The distances hiked by mountain club members (in miles) were 50, 16, 25, 23, 23, 16, 39, 43, 21, and 24. Find the mean, median, mode(s), and range of the data. 28; 23.5; 16 and 23; 34

5-8

15. The Yun family's gas bills for the last seven months are $10.56, $9.22, $16.87, $9.36, $9.18, $11.13, and $9.95. Decide whether the mean or the median is the better measure of central tendency for these data. Explain. median; the mean is greater than 5 of the 7 bills.

5-9

16. Find the mean, median, mode(s), and range of the data in the frequency table below. $40.96; $41; $41; $5

5-10

Bicycle Helmet Prices	Tally	Frequency
$39	ℍℍ	5
$40	ℍℍ I	6
$41	ℍℍ II	7
$44	ℍℍ	5
	Total:	23

Chapter 6

Write the name of each figure.

1.
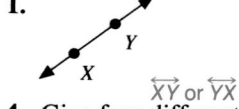
XY or YX

2. A
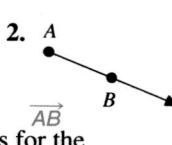
B
AB

3.
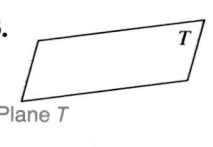
Plane T
T

6-1

4. Give four different names for the angle shown at the right. ∠RST, ∠TSR, ∠S, ∠2

R

S   2

T

Use a protractor to measure each angle.

5. 140° W   X   **6.** 35°   6-2

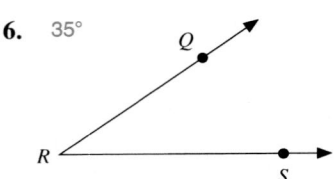

Use a protractor to draw an angle of the given measure.
Answers to Exercises 7 and 8 are on p. A44.

7. 65°   **8.** 138°

Find the measure of a complement of an angle of the given measure.

9. 12° 78°   **10.** 59° 31°   **11.** 87° 3°   **12.** 44°   6-3
                                                                                46°

Find the measure of a supplement of an angle of the given measure.

13. 37° 143°   **14.** 116° 64°   **15.** 135° 45°   **16.** 8°
                                                                           172°

Use the figure at the right. Tell whether each statement is *true* or *false*.

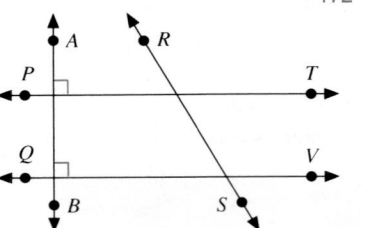

17. $\overrightarrow{PT} \perp \overleftrightarrow{AB}$ True.   **18.** $\overleftrightarrow{PT} \perp \overleftrightarrow{SR}$ False.   6-4

19. $\overleftrightarrow{PT} \parallel \overleftrightarrow{QV}$ True.   **20.** $\overleftrightarrow{AB} \parallel \overleftrightarrow{SR}$ False.

21. $\overleftrightarrow{VQ} \perp \overleftrightarrow{AB}$ True.   **22.** $\overleftrightarrow{QV} \perp \overleftrightarrow{SR}$ False.

23. $\overleftrightarrow{AB} \parallel \overleftrightarrow{PT}$ False.   **24.** $\overleftrightarrow{AB} \perp \overleftrightarrow{SR}$ False.

Identify each polygon. Then list all its diagonals.

quadrilateral; $\overline{QS}$ and $\overline{RP}$

25.   A   triangle; no diagonals   **26.**   P   6-5

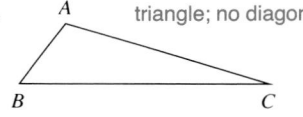
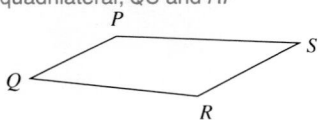

Find each unknown angle measure.

27.   **28.** 85°   **29.** 120°

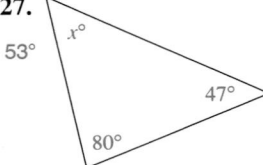
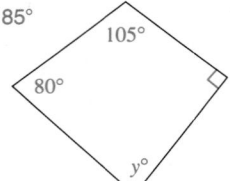

Tell whether line segments of the given lengths *can* or *cannot* be the sides of a triangle. If they can, tell whether the triangle would be *scalene*, *isosceles*, or *equilateral*.

can; isosceles

30. 7 in., 5 in., 12 in. cannot   **31.** 6 m, 7 m, 8 m can; scalene   **32.** 5 ft, 6 ft, 5 ft   6-6

The measures of two angles of a triangle are given. Tell whether the triangle is *acute*, *right*, or *obtuse*.

33. 57°, 33° right

34. 101°, 42° obtuse

35. 29°, 29° obtuse

In the figure at the right, *ABCD* is a rectangle. Find each measure.

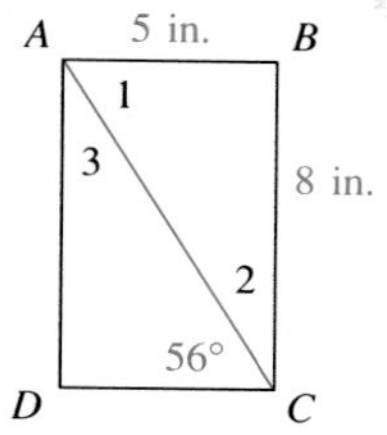

36. $m \angle 2$ 34°

37. $m \angle BAD$ 90°

38. the length of $\overline{AD}$ 8 in.

39. $m \angle 1$ 56°

40. $m \angle 3$ 34°

41. the length of $\overline{DC}$ 5 in.

6-7

Solve by identifying a pattern.

42. The figures below show the first three figures in a pattern of squares. How many triangles are in the sixth figure in this pattern? 24

6-8

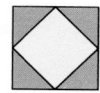

Is $\overleftrightarrow{ST}$ a line of symmetry? Write *Yes* or *No*.

43.

No.

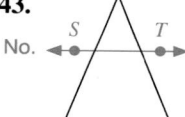

44.

Yes.

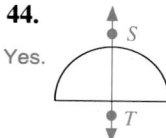

45.

Yes.

6-9

Chapter 7

Tell whether each number is *prime* or *composite*.

1. 7 prime

2. 28 composite

3. 42 composite

4. 53 prime

7-1

Write the prime factorization of each number.

5. 40 $2^3 \cdot 5$

6. 99 $3^2 \cdot 11$

7. 70 $2 \cdot 5 \cdot 7$

8. 156 $2^2 \cdot 3 \cdot 13$

Find the GCF.

9. 16 and 20 4

10. 35 and 49 7

11. 40 and 64 8

12. 75 and 100 25

7-2

13. $4a^2$ and $16a$ 4a

14. $8n^5$ and $32n^3$ $8n^3$

15. $9np$ and $18p$ 9p

16. 24ab, 48ab, and 96ab 24ab

Find the LCM.

17. 8 and 12 24

18. 9 and 15 45

19. 10 and 25 50

20. 6 and 14 42

7-3

21. $9a^4$ and $6a^6$ $18a^6$

22. $4a^5$ and $18a^3$ $36a^5$

23. $8a$ and $4a^6$ $8a^6$

24. $4a$, $8a^5$, and $16a^{10}$ $16a^{10}$

Replace the ? with the number that will make the fractions equivalent.

25. $\frac{6}{8} = \frac{18}{?}$ 24

26. $\frac{9}{18} = \frac{?}{6}$ 3

27. $\frac{?}{7} = \frac{8}{14}$ 4

28. $\frac{8}{?} = \frac{2}{10}$ 40

7-4

Write each fraction in lowest terms.

29. $\frac{4}{14}$ $\frac{2}{7}$

30. $\frac{8}{22}$ $\frac{4}{11}$

31. $\frac{9}{15}$ $\frac{3}{5}$

32. $\frac{12}{32}$ $\frac{3}{8}$

Simplify.

33. $\frac{9nq}{3n}$ $3q$

34. $\frac{12ab}{4b}$ $3a$

35. $\frac{qt^8}{t^3}$ qt^5

36. $\frac{r^{12}}{r^7}$ r^5

7-5

37. $\frac{6p^4}{p^2}$ $6p^2$

38. $\frac{14c^{10}}{c^5}$ $14c^5$

39. $\frac{h^8}{2h^3}$ $\frac{h^5}{2}$

40. $\frac{d^{15}}{5d^{10}}$ $\frac{d^5}{5}$

Replace each ? with >, <, or =.

41. $\frac{2}{3}$ __?__ $\frac{1}{2}$ >

42. $\frac{1}{8}$ __?__ $\frac{1}{3}$ <

43. $\frac{3}{4}$ __?__ $\frac{5}{6}$ <

44. $\frac{1}{6}$ __?__ $\frac{1}{3}$ <

7-6

45. $\frac{3}{5}$ __?__ $\frac{3}{4}$ <

46. $\frac{6}{7}$ __?__ $\frac{6}{11}$ >

47. $\frac{3}{5}$ __?__ $\frac{9}{15}$ =

48. $\frac{7}{12}$ __?__ $\frac{9}{16}$ >

Write each fraction or mixed number as a decimal.

49. $\frac{4}{5}$ 0.8

50. $\frac{3}{11}$ $0.\overline{27}$

51. $2\frac{3}{8}$ 2.375

52. $1\frac{9}{20}$ 1.45

7-7

Write each decimal as a fraction or mixed number in lowest terms.

53. 0.32 $\frac{8}{25}$

54. 0.96 $\frac{24}{25}$

55. 3.25 $3\frac{1}{4}$

56. 4.36 $4\frac{9}{25}$

Solve by drawing a diagram.

57. Sonia's house is 3 blocks east and 8 blocks south of Mary's house. Joan's house is 6 blocks north and 5 blocks west of Mary's house. Where is Joan's house in relation to Sonia's house?
14 blocks north and 8 blocks west

7-8

58. A tour bus starts at a hotel, travels 3 blocks west, 8 blocks north, 2 blocks west, 5 blocks north, and 5 blocks west. Where is the bus in relation to the hotel? 13 blocks north and 10 blocks west

59. How many diagonals can be drawn in an octagon? 20

60. The lengths of three boards are 4 m, 6 m, and 9 m. How could you use these boards to mark off a length of 1 m? See answer on p. A44.

Use the number line below. Name the point that represents each number.

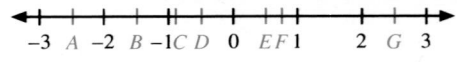

-3 A -2 B -1C D   0   E F 1   2 G 3

61. $2\frac{1}{2}$ G

62. $-1\frac{1}{2}$ B

63. -0.9 C

64. 0.5 E

7-9

Express each rational number as a quotient of two integers. 65–68. Answers may vary. Examples are given.

65. $-9 \frac{-9}{1}$ **66.** $4\frac{2}{3} \frac{14}{3}$ **67.** $-6\frac{1}{5} \frac{-31}{5}$ **68.** $-4.\overline{3} \frac{-13}{3}$

Graph each open sentence. Answers to Exercises 69–76 are on p. A44.

69. $4 = 9 + x$ **70.** $-2 + x = -6$ **71.** $x - 4 = 9$ **72.** $5x + 4 = -1$ **7-10**

73. $x < 3.7$ **74.** $x > -0.5$ **75.** $x \geq -2.5$ **76.** $x \leq -0.5$

Simplify.

77. $x^{-4} \frac{1}{x^4}$ **78.** $2^{-4} \frac{1}{16}$ **79.** $5^{-3} \frac{1}{125}$ **80.** $(-6.782)^0$ 1   **7-11**

Write each number in scientific notation.

81. 0.000082   **82.** 0.0062 6.2×10^{-3} **83.** 0.00007   **84.** 0.000643
     8.2×10^{-5} 7×10^{-5} 6.43×10^{-4}

Write each number in decimal notation.

85. 5×10^{-3} 0.005   **86.** 6.7×10^{-4} 0.00067   **87.** 3.25×10^{-9} **88.** 4.223×10^{-5}
                                              0.00000000325   0.00004223

Chapter 8

Find each product. Simplify if possible.

$-6\frac{2}{13}$

1. $\left(-\frac{9}{10}\right)\left(-\frac{5}{12}\right) \frac{3}{8}$ **2.** $\frac{5}{7} \cdot \frac{7}{5}$ 1   **3.** $\left(2\frac{4}{5}\right)\left(-3\frac{1}{8}\right) -8\frac{3}{4}$ **4.** $\left(-6\frac{2}{3}\right)\left(\frac{12}{13}\right)$ **8-1**

5. $(-4.3)(5.4)$ **6.** $(-2.9)(-0.4)$ 1.16   **7.** $\left(\frac{5a}{7b}\right)\left(\frac{14b}{15}\right) \frac{2a}{3}$ **8.** $\left(\frac{x}{20}\right)\left(\frac{40y}{x}\right)$ 2y
   -23.22

Find each quotient. Simplify if possible.

9. $-\frac{3}{7} \div \left(-\frac{3}{14}\right)$ 2   **10.** $\frac{3}{8} \div \frac{2}{3} \frac{9}{16}$ **11.** $\frac{10}{19} \div (-2) -\frac{5}{19}$ **12.** $-20 \div \frac{4}{5} -25$ **8-2**

13. $-1.44 \div 0.45$ **14.** $-0.81 \div (-2.7)$ **15.** $\frac{4x}{7} \div \frac{8x}{9} \frac{9}{14}$ **16.** $\frac{16a}{17} \div 4a \frac{4}{17}$
   -3.2 0.3

Solve. Decide whether a *whole number*, a *fraction*, or a *decimal* is an appropriate answer.

17. Each month Junko earns $125 and saves one fourth of his earnings. How much money does he save each month? decimal; $31.25   **8-3**

18. There will be 35 people at a luncheon. How many tables of 8 seats each are needed to seat everyone? whole number; 5

Find each answer. Simplify if possible. 22. $-\frac{1}{5}$ 26. $\frac{2n}{5}$

19. $\frac{5}{7} + \frac{6}{7} 1\frac{4}{7}$ **20.** $\frac{5}{8} + \frac{3}{8}$ 1   **21.** $-\frac{4}{9} - \left(-\frac{4}{9}\right)$ 0   **22.** $\frac{7}{10} + \left(-\frac{9}{10}\right)$ **8-4**

23. $-2\frac{1}{3} + \left(-\frac{2}{3}\right) -3$ **24.** $\frac{11}{12} - \left(-\frac{5}{12}\right) 1\frac{1}{3}$ **25.** $\frac{11}{y} + \frac{9}{y} \frac{20}{y}$ **26.** $\frac{12n}{5} - \frac{10n}{5}$

Find each answer. Simplify if possible.

27. $\frac{4}{9} + \frac{1}{6}$ $\frac{11}{18}$ **28.** $-4\frac{4}{7} - \frac{5}{14}$ $-4\frac{13}{14}$ **29.** $-5\frac{2}{3} + 3\frac{11}{12}$ $-1\frac{3}{4}$ **30.** $-6 + 3\frac{1}{3} + \left(-\frac{5}{6}\right)$ $-3\frac{1}{2}$ 8-5

31. $\frac{5a}{7} - \frac{5a}{14}$ $\frac{5a}{14}$ **32.** $3x + \frac{9x}{11}$ $\frac{42x}{11}$ **33.** $6.4 + (-8.7)$ -2.3 **34.** $-37.5 - 0.43$ -37.93

Estimate.

35. $27\frac{2}{13} - 6\frac{11}{12}$ about 20   **36.** $15\frac{1}{3} - 4\frac{5}{8}$ about 10   **37.** $6\frac{27}{28} + 3\frac{1}{5} + 2\frac{7}{12}$ about $12\frac{1}{2}$ 8-6

38. $11\frac{9}{10} \div \left(-\frac{8}{15}\right)$ about -24 **39.** $-17\frac{9}{10} \div \left(-6\frac{1}{5}\right)$ about 3   **40.** $-27\frac{3}{5} \div \frac{25}{99}$ about -112

Solve by supplying missing facts.

41. Elidio's health insurance costs him $85 per month. How much does his insurance cost per year? $1020   8-7

42. It took Roseanne 3 h to complete 5 homework assignments. She spent an equal amount of time on each assignment. How many minutes did she spend on each assignment? 36 min

Solve. Check each solution.

43. $8 = -5b + 3$ -1 **44.** $10 + 5k = 3$ $-1\frac{2}{5}$ **45.** $z - \frac{2}{3} = \frac{5}{6}$ $1\frac{1}{2}$ 8-8

46. $w + 3\frac{1}{8} = -\frac{3}{8}$ $-3\frac{1}{2}$ **47.** $2a - 4.9 = -7.1$ -1.1 **48.** $27.4 = 3.14 + 2x$ 12.13

49. $\frac{8}{9}x = -5$ $-5\frac{5}{8}$ **50.** $-\frac{2}{5}y = 14$ -35 **51.** $11 + \frac{6}{7}m = 8$ $-3\frac{1}{2}$ 8-9

52. $-\frac{2}{9}z + 3 = 7$ -18 **53.** $-1 = \frac{5}{7}f + 4$ -7 **54.** $-9 = -\frac{7}{11}d - 8$ $1\frac{4}{7}$

Solve for the variable shown in color.

55. $A = lw$ $w = \frac{A}{l}$ **56.** $A = \frac{1}{2}bh$ $h = \frac{2A}{b}$ **57.** $A = \frac{1}{2}h(b_1 + b_2)$ $h = \frac{2A}{b_1 + b_2}$ 8-10

58. $p = \frac{m}{a}$ $m = ap$ **59.** $D = rt$ $t = \frac{D}{r}$ **60.** $g = a - h$ $a = g + h$

Chapter 9

Write each ratio as a fraction in lowest terms.

1. $\frac{6}{42}$ $\frac{1}{7}$ **2.** $35 : 50$ $\frac{7}{10}$ **3.** $.40 : 2 $\frac{1}{5}$ **4.** 36 in. to 2 ft $\frac{3}{2}$ 9-1

Write the unit rate.

5. $75 for 6 h $12.50/h   **6.** 165 mi in 3 h 55 mi/h   **7.** 144 mi on 8 gal 18 mi/gal   **8.** 345 words in 5 min 69 words/min

Solve each proportion.

9. $\frac{25}{b} = \frac{5}{7}$ 35 **10.** $\frac{5}{8} = \frac{v}{24}$ 15 **11.** $\frac{8}{a} = \frac{6}{9}$ 12 **12.** $\frac{1.5}{1.2} = \frac{10}{k}$ 8 9-2

13. $\frac{b}{18} = \frac{20}{30}$ 12 **14.** $\frac{2.5}{t} = \frac{10}{9}$ 2.25 **15.** $\frac{0.4}{1.6} = \frac{n}{28}$ 7 **16.** $\frac{x}{7} = \frac{12}{3}$ 28

17. A bedroom is 10 ft wide and 12 ft long. The scale of a floor plan of the bedroom is 1 in. : 12 ft. Find the width and length of the bedroom on the floor plan. $\frac{5}{6}$ in.; 1 in. 9-3

18. The model of a van is $1\frac{1}{2}$ ft long. The scale of the model is 1 : 10. What is the actual length of the van? 15 ft

Solve using a proportion.

19. The ratio of juniors to seniors at a dance was 3 : 4. There were 60 seniors at the dance. How many juniors were at the dance? 45 9-4

20. A computer printer can print 8 pages in 1.5 min. At this rate, how long will it take to print 60 pages? 11.25 min

Write each fraction, mixed number, or decimal as a percent.

21. $\frac{4}{25}$ 16% **22.** 0.09 9% **23.** $1\frac{1}{8}$ 112.5% **24.** 0.765 76.5% 9-5

Write each percent as a fraction in lowest terms and as a decimal.

25. 45% $\frac{9}{20}$; 0.45 **26.** 6.5% $\frac{13}{200}$; 0.065 **27.** $8\frac{1}{5}$% $\frac{41}{500}$; 0.082 **28.** 160% $1\frac{3}{5}$; 1.6

Find each answer.

29. What number is 20% of 48? 9.6 **30.** 6% of 425 is what number? 25.5 9-6

31. What percent of 60 is 4.5? 7.5% **32.** 270 is what percent of 600? 45% 9-7

33. 44 is 25% of what number? 176 **34.** 150 is $66\frac{2}{3}$% of what number? 225 9-8

Find the percent of increase or percent of decrease.

35. original population: 400; new population: 520 30% increase 9-9

36. original cost: $50; new cost: $40 20% decrease

37. original weight: 120 lb; new weight: 100 lb $16\frac{2}{3}$% decrease

Find each answer using a proportion.

38. What number is 75% of 360? 270 **39.** What percent of 120 is 45? 37.5% 9-10

40. 4 is 5% of what number? 80 **41.** 12 is what percent of 18? $66\frac{2}{3}$%

42. What percent of 80 is 4? 5% **43.** What number is 60% of 150? 90

Chapter 10

Find the perimeter of each figure.

1. a rectangle that is $5\frac{2}{3}$ ft long and $3\frac{1}{4}$ ft wide $17\frac{5}{6}$ ft

2. a regular pentagon with one side that measures 3.5 cm 17.5 cm

10-1

Find the radius and diameter of a circle with the given circumference.

3. 75.36 m 12 m; 24 m

4. 439.6 in. 70 in.; 140 in.

5. 40.82 m 6.5 m; 13 m

10-2

Find the area of each figure.

6. a square with length of one side equal to 20 ft 400 ft²

7. a rectangle with length 9.2 m and width 6.6 m 60.72 m²

10-3

8. a circle with radius 6.5 m 132.7 m²

9. a circle with diameter $10\frac{1}{2}$ ft 86.5 ft²

10-4

10. Tell whether you would need to calculate *perimeter*, *circumference*, or *area* to find the amount of fencing needed to enclose a circular park. circumference

10-5

11. Explain why this problem has no solution: Lily said she spent half of her paycheck on rent, one third on food and utilities, and one fourth on clothing. What part of her pay did she save? $\frac{1}{2} + \frac{1}{3} + \frac{1}{4} = 1\frac{1}{12}$; this is more than one whole paycheck.

10-6

Use the figures at the right to complete each statement.

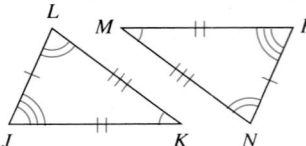

12. $\overline{LK} \cong \underline{\ \ ?\ \ } \overline{NM}$

13. $\angle J \cong \underline{\ \ ?\ \ } \angle P$

10-7

14. In the figures at the right, quadrilateral *MNOP* ~ quadrilateral *STUV*. Find *z*. 18

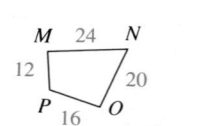

10-8

15. In the figure at the right, the triangles are similar. Find the unknown height *h*. 9 ft

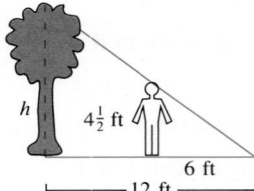

10-9

Find each square root if possible. Otherwise, approximate the square root to the nearest thousandth.

16. $\sqrt{4900}$ 70

17. $\sqrt{\frac{16}{81}}$ $\frac{4}{9}$

18. $\sqrt{87}$ 9.327

19. $\sqrt{139}$ 11.790

10-10

20. Is a triangle with sides of 10 in., 25 in., and 27 in. a right triangle? Write *Yes* or *No*. No.

10-11

Chapter 11

Use the data below for Exercises 1 and 2. Answers to Exercises 1 and 2 are on p. A44.

Test Scores In Math

76 79 86 59 97 94 82 82 80 76 64 60 92 75 98 83 57 72 70 81

1. Make a frequency table for the data. Use intervals such as 50–59.   11-1

2. Make a stem-and-leaf plot for the data.   11-2

Use the stem-and-leaf plot at the right.

Ages of Teachers

3	0	3	3	5	8
4	2	2	6	9	
5	0	0	1	4	7

3. How many teachers are less than 45 years old? 7

4. How many teachers are 42 years old? 2

5. Find the mean, median, mode(s), and range of the data. 43.6; 44; 33, 42, 50; 27

Use the frequency polygon at the right.

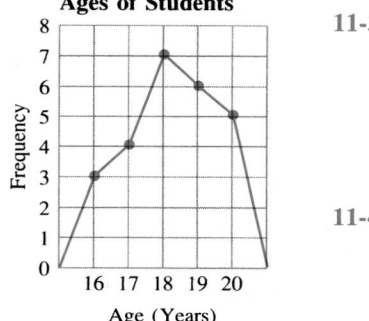

Ages of Students

6. How many students are 19 years old? 6   11-3

7. Exactly 4 students are a certain age. What is this age? 17

8. How many students are less than 19 years old? 14

9. How many students are there in all? 25

10. Make a box-and-whisker plot for the data below.   11-4
See answer on p. A45.

Prices for Compact Disc Players (Dollars)

600 400 300 350 300 500 420 550
375 310 480 500 675 380 480 590

Use the box-and-whisker plot below for Exercises 11–13.

Scores on a Biology Test

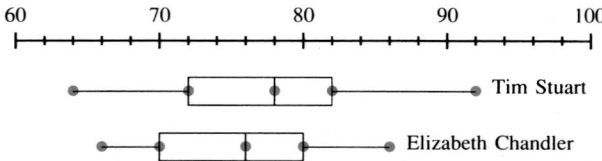

11. Find the median, the first quartile, the third quartile, and the lowest and highest scores for Tim Stuart's class. 78; 72; 82; 64; 92

12. What was the range of test scores for Elizabeth Chandler's class? 20

13. Which teacher's class had the greater median score on the test? Tim Stuart's

Use the scattergram below for Exercises 14–16.

Push-ups and Age

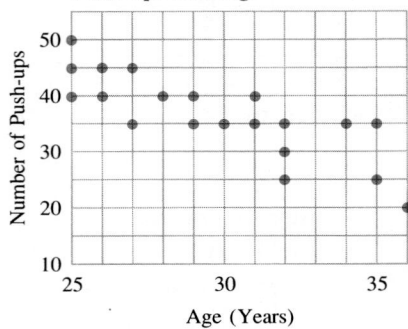

14. How many persons did 45 push-ups? 3   11-5

15. How many persons who were 31 years old did push-ups? 2

16. Is there a *positive* or *negative* correlation between the two sets of data?
    negative

Use the circle graph at the right.   Carson Family Budget

17. If the Carson family's total   11-6
    annual budget is $42,000,
    how much do they plan to
    spend annually on food?
    $10,500

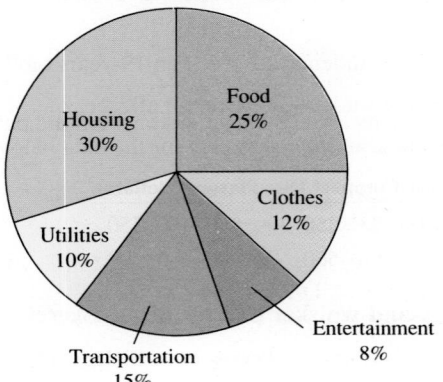

18. If $16,200 is spent annually
    on housing, how much is the
    total budget? $54,000

19. Draw a circle graph to display the data below.   11-7
                                          See answer on p. A45.

    Videos Rented from Victor's Video

Type of Movie	Adventure	Comedy	Family
Number Rented	120	20	60

20. Decide whether it would be appropriate to draw a *circle graph* or a   11-8
    stem-and-leaf plot for the data below. Then make the display.
                                          stem-and-leaf plot; display is on p. A45.
    Scores on a History Test

    75 76 89 98 57 89 73 86 90 65
    67 75 80 98 72 65 83 92 76 79

21. Therese, Lucinda, and Alicia are a teacher, a lawyer, and an accountant. No one works at a job that begins with the first letter of her name. Therese and the accountant went to the same college. Who has each job? Therese is the lawyer, Lucinda is the accountant, and Alicia is the teacher.

11-9

Chapter 12

Use the spinner at the right for Exercises 1–4.

1. Find $P(3)$. $\frac{1}{5}$ **2.** Find P(not white). $\frac{3}{5}$

12-1

Find the odds in favor of each event.

3. an odd number 3 to 2   **4.** not red 4 to 1

5. The spinner above is spun and a coin is tossed. Make a tree diagram to show the sample space. See answer on p. A45.

12-2

6. Use the tree diagram from Exercise 5 to find $P(1,$ heads or tails$)$. $\frac{1}{5}$

7. Use the counting principle to find the number of possible outcomes of choosing an odd number from 1 through 10 and an even number from 11 through 20. 25

12-3

A bag of marbles contains 5 blue, 4 red, and 9 green marbles. Use this information for Exercises 8 and 9.

8. Two marbles are drawn at random from the bag with replacement. Find P(green, then blue). $\frac{5}{36}$

12-4

9. Two marbles are drawn at random from the bag without replacement. Find P(red, then green). $\frac{2}{17}$

10. Of 100 people surveyed, 60 own cars, 35 own trucks, and 20 own both cars and trucks. Use a Venn diagram to find how many of the 100 people do not own either a car or a truck. 25

12-5

11. In how many different ways can 7 pencils all with different colors be arranged in a box? 5040

12-6

12. A committee of 4 is to be chosen from a group of 9 people. Find the number of combinations. 126

13. Yves tossed a thumbtack 60 times. It landed point up 27 times and point down 33 times. Find the experimental probability that the thumbtack lands point up. $\frac{9}{20}$

12-7

14. When 40 employees at a local company were questioned, 24 said that they had worked at the company for at least 10 years. About how many of the 200 employees at the company are expected to have worked at the company for at least 10 years? about 120

12-8

Extra Practice **747**

747

Chapter 13

Answers to Exercises 1–16 are on p. A45.

Find solutions of each equation. Use −2, −1, 0, 1, and 2 as values for x.

13-1

1. $y = 3x + 2$
2. $y = \frac{1}{2}x - 1$
3. $y = -2x + 5$
4. $y = -5x - 2$

5. $y - 3x = 9$
6. $y + 4x = 3$
7. $y - 5x = -7$
8. $y + 3x = -4$

Graph each equation.

13-2

9. $y = \frac{1}{2}x + 4$
10. $y = -\frac{2}{3}x + 1$
11. $y = -\frac{3}{4}x$
12. $y = 2x$

13. $y - 3x = 2$
14. $x + y = -2$
15. $x + y = -1$
16. $2x + y = 1$

Find the slope and the y-intercept of each line. Then use them to write an equation for each line.

17. 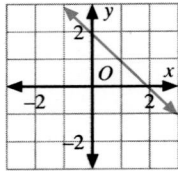 −1; 2; $y = -x + 2$

18. 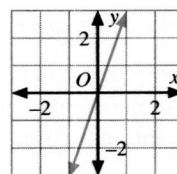 3; 0; $y = 3x$

13-3

19. $\frac{1}{2}$; 0; $y = \frac{1}{2}x$

20. 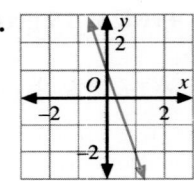 −3; 1; $y = -3x + 1$

Find the slope, the y-intercept, and the x-intercept of each line.

21. $y - 4 = 2x$
    2; 4; −2
22. $y - 2 = 4x$
    4; 2; $-\frac{1}{2}$
23. $y + 6 = -3x$
    −3; −6; −2
24. $y + 1 = x$
    1; −1; 1

Solve each system by graphing.

13-4

25. $y = x + 5$
    $y = 3x + 1$
    (2, 7)
26. $y = -2x + 3$
    $y = 4x - 3$
    (1, 1)
27. $y = -\frac{2}{3}x + 1$
    $y = \frac{1}{6}x - 4$ (6, −3)
28. $y = \frac{1}{2}x + 9$
    $y = -\frac{1}{2}x + 11$
    (2, 10)

29. $y = 3x$
    $y = 4x + 2$
    (−2, −6)
30. $y = -2x$
    $y = 2x$
    (0, 0)
31. $y = -x + 8$
    $y = 2x + 2$
    (2, 6)
32. $y = \frac{1}{2}x + 4$
    $y = x + 2$
    (4,6)

Solve using a simpler problem.

13-5

33. Find the sum of the first 25 positive odd numbers. 625

34. James decided to save quarters. On the first day he saved one quarter. On each of the next thirteen days he saved two more quarters than he saved the day before. How many quarters did James save in all? 196

Solve and graph each inequality. Graphs for Exercises 35–42 are on p. A45.

35. $n + 2 \leq 7$ $n \leq 5$ **36.** $b - 6 \geq 9$ $b \geq 15$ **37.** $p + 4 > -9$ **38.** $q - 3 < -7$ $\quad q < -4$ 13-6
$\qquad\qquad\qquad\qquad\qquad\qquad\qquad\qquad\qquad\; p > -13$

39. $2k < 4$ $k < 2$ **40.** $\dfrac{r}{-6} \geq -2$ $r \leq 12$ **41.** $-5t \leq 15$ $t \geq -3$ **42.** $\dfrac{b}{-4} > 7$ $b < -28$

Solve each inequality.
43. $k \geq 100$ **45.** $q < -7200$
46. $a > -1600$ **47.** $t \geq -600$

43. $6k + 300 \geq 900$ **44.** $3p - 40 \leq 80$ $p \leq 40$ **45.** $-2q - 4600 > 9800$ 13-7

46. $-5a - 800 < 7200$ **47.** $\dfrac{1}{3}t - 500 \geq -700$ **48.** $\dfrac{1}{2}q + 400 < 8000$ $\quad q < 15{,}200$

49. $-\dfrac{2}{3}n + 5 \leq 707$ $n \geq -1053$ **50.** $-1\dfrac{1}{2}w + 300 > 600$ $\quad w < -200$

Tell whether each ordered pair is a solution of each inequality. Write *Yes* or *No*.

51. $(1, 4)$; $y < 6x - 4$ No.   **52.** $y \geq 3x + 2$; $(-2, 8)$ Yes.   13-8

53. $(2, -2)$; $y > -2x - 2$ Yes.   **54.** $y \leq \dfrac{1}{3}x + 2$; $(0, 2)$ Yes.

Graph each inequality. Answers to Exercises 55–58 are on p. A45.

55. $y < 3x - 1$ **56.** $y \geq x + 1$ **57.** $y > -3x$ **58.** $y \leq \dfrac{1}{2}x + 3$

Find values of each function. Use -1, 0, and 3 as values for x.

59. $y = 3x - 7$ $-10, -7, 2$ **60.** $y = \dfrac{2}{3}x - \dfrac{2}{3}$, 0, 2   **61.** $y = x^2 - 4x$ $5, 0, -3$ **62.** $y = |x - 2|$ $3, 2, 1$ 13-9

Find $f(-2)$, $f(1)$, and $f(0)$ for each function. Answers to Exercises 63–68 are on p. A45.

63. $f(x) = -4x$ **64.** $f(x) = \dfrac{1}{8}x$ **65.** $f(x) = 8x^2 - 1$

66. $f(x) = |6x + 1|$ **67.** $f(x) = x^2 - 1$ **68.** $f(x) = (x - 2)^2$

Graph each function. Answers to Exercises 69–74 are on pp. A45–A46.

69. $f(x) = 3x + 1$ **70.** $f(x) = -\dfrac{1}{3}x + 1$ **71.** $f(x) = (x + 1)^2$ 13-10

72. $f(x) = x^2 - 4$ **73.** $f(x) = -|x| + 3$ **74.** $f(x) = |x + 2|$

Chapter 14

Identify each space figure.

1.
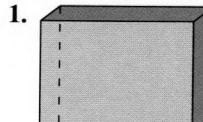
rectangular prism

2.
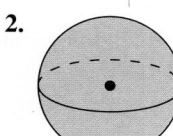
sphere

3.
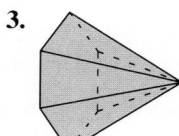
hexagonal pyramid

14-1

Find the surface area of each space figure.

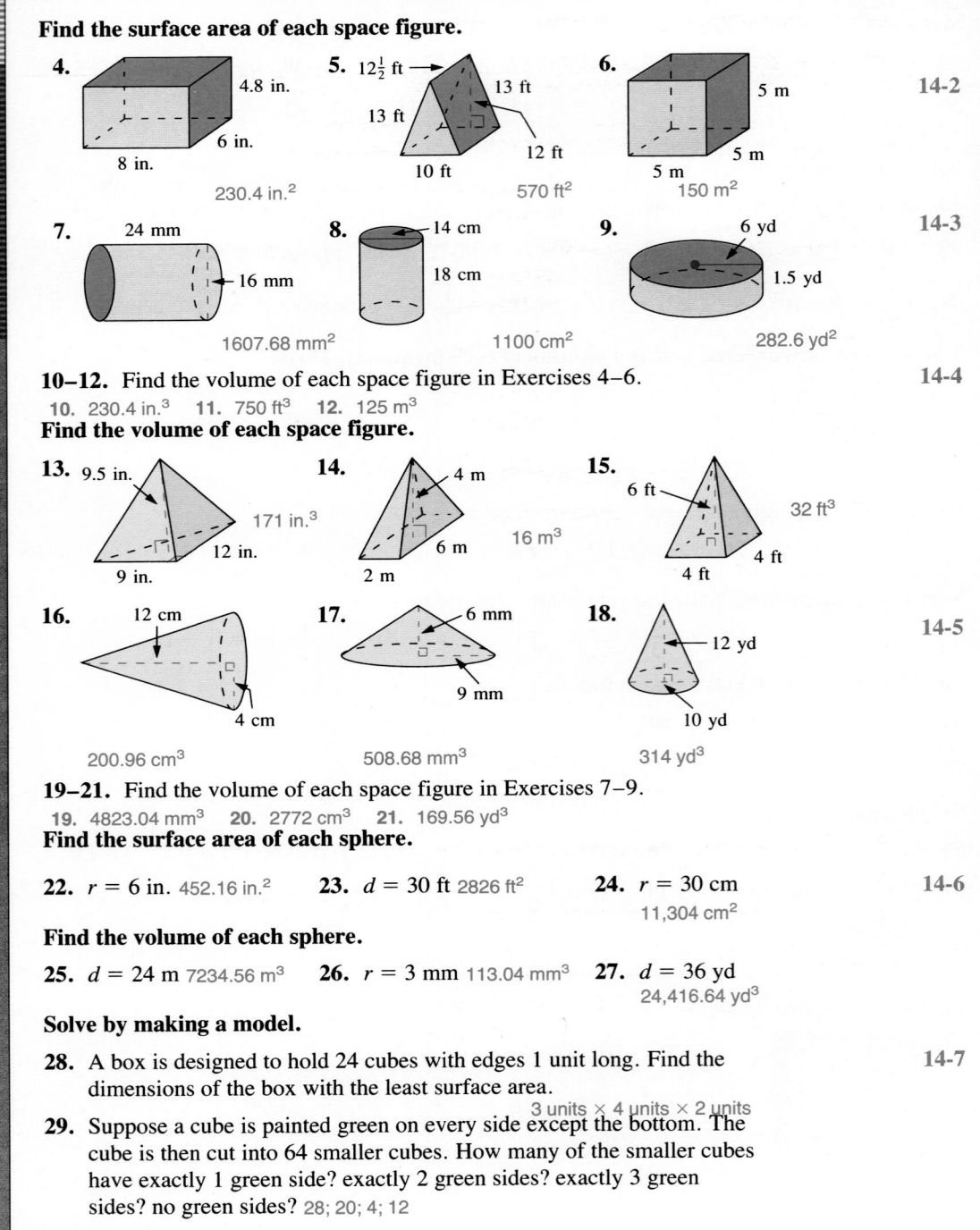

4. 4.8 in. 6 in. 8 in.
230.4 in.²

5. $12\frac{1}{2}$ ft 13 ft 13 ft 12 ft 10 ft
570 ft²

6. 5 m 5 m 5 m
150 m²

14-2

7. 24 mm 16 mm
1607.68 mm²

8. 14 cm 18 cm
1100 cm²

9. 6 yd 1.5 yd
282.6 yd²

14-3

10–12. Find the volume of each space figure in Exercises 4–6.

10. 230.4 in.³   **11.** 750 ft³   **12.** 125 m³

14-4

Find the volume of each space figure.

13. 9.5 in. 12 in. 9 in.
171 in.³

14. 4 m 6 m 2 m
16 m³

15. 6 ft 4 ft 4 ft
32 ft³

16. 12 cm 4 cm
200.96 cm³

17. 6 mm 9 mm
508.68 mm³

18. 12 yd 10 yd
314 yd³

14-5

19–21. Find the volume of each space figure in Exercises 7–9.

19. 4823.04 mm³   **20.** 2772 cm³   **21.** 169.56 yd³

Find the surface area of each sphere.

22. $r = 6$ in. 452.16 in.²   **23.** $d = 30$ ft 2826 ft²   **24.** $r = 30$ cm
11,304 cm²

14-6

Find the volume of each sphere.

25. $d = 24$ m 7234.56 m³   **26.** $r = 3$ mm 113.04 mm³   **27.** $d = 36$ yd
24,416.64 yd³

Solve by making a model.

28. A box is designed to hold 24 cubes with edges 1 unit long. Find the dimensions of the box with the least surface area.
3 units × 4 units × 2 units

14-7

29. Suppose a cube is painted green on every side except the bottom. The cube is then cut into 64 smaller cubes. How many of the smaller cubes have exactly 1 green side? exactly 2 green sides? exactly 3 green sides? no green sides? 28; 20; 4; 12

Chapter 15

Write each polynomial in standard form. $12a^4 + 9a^3 + 8a^2 + 16$

1. $3a^3 + 4a + 9 + 6a^2$
$3a^3 + 6a^2 + 4a + 9$

2. $8a^2 + 9a^3 + 16 + 12a^4$

3. $7a + 4 + 7a^2$
$7a^2 + 7a + 4$

4. $6 + 3a + 9a^2 + 8a^3 + 6a^4$
$6a^4 + 8a^3 + 9a^2 + 3a + 6$

Simplify. $4a^3 - 2a^2 + 4a + 7$

$3n^3 + n^2 + 3n + 8$

5. $6a + 4a^3 - 2a - 2a^2 + 7$

6. $3n^2 - n^3 + 6n - 2n^2 + 4n^3 + 8 - 3n$

7. $4p^2 - 3p + 5p^2 - 6 + 8p + 8$
$9p^2 + 5p + 2$

8. $3x + 4x^2 + 6 - x + 4x - 3 + 8x^2$
$12x^2 + 6x + 3$

Add. Answers to Exercises 9–12 are on p. A46.

9. $(6a^2 + 3a + 9) + (4a^2 + 6a + 1)$

10. $(4b^2 + 18b + 6) + (9b^2 + 9b + 12)$

11. $(12c^2 + 19c + 9) + (6c^2 + 9c + 12)$

12. $(14y^2 + 22y + 12) + (12y^2 + 19y + 6)$

13. $(7x^4 + 3x^3 + 4x^2 - 6x + 2) + (-4x^4 - 6x^3 + 8x - 9)$ $3x^4 - 3x^3 + 4x^2 + 2x - 7$

14. $(3t^4 - 4t^3 + 16t - 4) + (3t^4 + 7t^3 + 3t^2 - 7t - 3)$ $6t^4 + 3t^3 + 3t^2 + 9t - 7$

15. $(8n^4 - 8n^3 - 12n^2 + 5) + (-9n^4 + 9n^3 + 14n^2 - 5n - 2)$ $-n^4 + n^3 + 2n^2 - 5n + 3$

16. $(6q^4 + 7q^3 - 11q^2 - 8q - 12) + (-q^4 + q^3 - q - 1)$ $5q^4 + 8q^3 - 11q^2 - 9q - 13$

Subtract. Answers to Exercises 17–20 are on p. A46.

17. $(5k^2 + 9k + 4) - (3k^2 + 4k + 1)$

18. $(12p^2 + 7p + 8) - (5p^2 + 6p + 5)$

19. $(3h^2 + 4h + 5) - (2h^2 + 3h + 4)$

20. $(10r^2 + 5r + 7) - (8r^2 + 3r + 5)$

21. $(9b^3 + 5b^2 - 8) - (3b^3 - 3b^2 + 5b - 6)$ $6b^3 + 8b^2 - 5b - 2$

22. $(6t^3 - 7t - 4) - (-8t^3 - 6t^2 - 4t + 4)$ $14t^3 + 6t^2 - 3t - 8$

23. $(7x^3 + 10x^2 - 4x + 6) - (10x^3 - 6x^2 + 9)$ $-3x^3 + 16x^2 - 4x - 3$

24. $(12a^3 - 6a^2 + 9a - 4) - (5a^3 - 4a + 8)$ $7a^3 - 6a^2 + 13a - 12$

Multiply. **27.** $12m^3 - 24m^2 + 36m + 36$ **31.** $12a^7 + 24a^6 - 9a^5 - 27a^4$

25. $3(6x^2 + 3x - 2)$ $18x^2 + 9x - 6$

26. $-2(4q^2 - 5q - 8)$ $-8q^2 + 10q + 16$

27. $-4(-3m^3 + 6m^2 - 9m - 9)$

28. $(4x^2y^3)(-2x^5y^2)$ $-8x^7y^5$

29. $(-6c^4d^3)(-5c^3d^7)$ $30c^7d^{10}$

30. $(-8p^4q^7)(9p^5q^6)$ $-72p^9q^{13}$

31. $3a^4(4a^3 + 8a^2 - 3a - 9)$

32. $6d^2(3d^3 - 9d^2 - 6)$
$18d^5 - 54d^4 - 36d^2$

33. $(n + 1)(n + 5)$ $n^2 + 6n + 5$

34. $(4k + 6)(5k + 2)$ $20k^2 + 38k + 12$

35. $(5y + 6)(y + 7)$ $5y^2 + 41y + 42$

36. $(2t + 9)(t + 1)$ $2t^2 + 11t + 9$

37. $(3n + 1)(2n + 5)$ $6n^2 + 17n + 5$

38. $(2g - 3)(8g + 8)$ $16g^2 - 8g - 24$

39. $(4u - 7)(2u + 5)$ $8u^2 + 6u - 35$

40. $(z - 6)(z - 6)$ $z^2 - 12z + 36$

15-1

15-2

15-3

15-4

15-5

Solve by working backward.

15-6

41. In selecting a president, the board began with a pool of people and rejected half of the applicants after each round of interviews. Sara was chosen after 5 rounds. How many people were in the initial pool? 32

42. At the end of January, Adrienne has $1200 in her checking account after writing checks for $350, $73, $82, $625, and making a deposit of $175. The bank charges her $3.50 as a service charge and pays her $2.75 in interest in January. How much money was in Adrienne's account at the beginning of January? $2155.75

Factor. **43.** $2m^2(2m^2 - 8m + 15)$ **47.** $2a^3b^2(2b^3 - 5a + 3a^2b^3)$ **48.** $5c^2d^2(3c^4 - 4 - 9c)$

43. $4m^4 - 16m^3 + 30m^2$

44. $8s^5 - 6s^4 + 12s^3$ $2s^3(4s^2 - 3s + 6)$

15-7

45. $10v^6 - 16v^5 + 12v^4$
$2v^4(5v^2 - 8v + 6)$

46. $14q^2 + 16q + 18$ $2(7q^2 + 8q + 9)$

47. $4a^3b^5 - 10a^4b^2 + 6a^5b^5$

48. $15c^6d^2 - 20c^2d^2 - 45c^3d^2$

49. $24p^4q^5 - 15p^3q^6 + 33p^6q^7$
$3p^3q^5(8p - 5q + 11p^3q^2)$

50. $32r^5t^6 - 18r^4t^4 + 48r^7t^7$
$2r^4t^4(16rt^2 - 9 + 24r^3t^3)$

Divide.

51. $\dfrac{6y^5 + 18y^4 + 9y^3}{3y^2}$ $2y^3 + 6y^2 + 3y$

52. $\dfrac{14x^6 + 21x^4 + 42x^3}{7x^2}$ $2x^4 + 3x^2 + 6x$

15-8

53. $\dfrac{8t^7 + 16t^5 + 24t^4}{4t^3}$ $2t^4 + 4t^2 + 6t$

54. $\dfrac{20r^6 + 24r^4 + 6r^2}{2r}$ $10r^5 + 12r^3 + 3r$

55. $\dfrac{7a^4b^5 - 42a^3b^7 + 14a^5b^2}{7a^2b}$

56. $\dfrac{9s^3t^8 - 24s^5t^6 - 27s^7t^3}{3s^2t^3}$ $3st^5 - 8s^3t^3 - 9s^5$

57. $\dfrac{-22m^2p^5 + 11m^7p^4 - 33m^6p^4}{11m^2p^2}$
$-2p^3 + m^5p^2 - 3m^4p^2$

58. $\dfrac{-63h^7k^8 - 72h^8k^9 - 81h^5k^5}{9h^3k^4}$
$-7h^4k^4 - 8h^5k^5 - 9h^2k$

55. $a^2b^4 - 6ab^6 + 2a^3b$

Estimation Skills

Skill 1 Rounding

1. about 400   2. about 60   3. about 980
4. about 4   5. about 700   6. about 8300
7. about 500   8. about 11   9. about 36,000
10. about 7   11. about 800   12. about 56,000
13. about 4,000,000   14. about 20   15. about 4500

Estimate the answer. Round each number to the place of its leading digit.

$$\begin{array}{r} 56.9 \\ +31.4 \\ \hline \end{array} \longrightarrow \begin{array}{r} 60 \\ +30 \\ \hline \text{about } 90 \end{array}$$

$$\begin{array}{r} 5439 \\ -782 \\ \hline \end{array} \longrightarrow \begin{array}{r} 5000 \\ -800 \\ \hline \text{about } 4200 \end{array}$$

$$\begin{array}{r} 892 \\ \times 4.7 \\ \hline \end{array} \longrightarrow \begin{array}{r} 900 \\ \times 5 \\ \hline \text{about } 4500 \end{array}$$

Estimate the answer. Round each number to the place of its leading digit.

1. $\begin{array}{r} 582 \\ -167 \\ \hline \end{array}$
2. $\begin{array}{r} 17.36 \\ +42.51 \\ \hline \end{array}$
3. $\begin{array}{r} 918 \\ +\ 77 \\ \hline \end{array}$

4. $\begin{array}{r} 4.632 \\ -1.299 \\ \hline \end{array}$
5. $\begin{array}{r} 635 \\ +114 \\ \hline \end{array}$
6. $\begin{array}{r} 7866 \\ +\ 337 \\ \hline \end{array}$

7. $\begin{array}{r} 769 \\ -278 \\ \hline \end{array}$
8. $\begin{array}{r} 2.396 \\ +\ 8.535 \\ \hline \end{array}$
9. $\begin{array}{r} 31,158 \\ +\ 5,676 \\ \hline \end{array}$

10. $\begin{array}{r} 11.784 \\ -\ 3.415 \\ \hline \end{array}$
11. $\begin{array}{r} 36 \\ \times 15 \\ \hline \end{array}$
12. $\begin{array}{r} 809 \\ \times\ 74 \\ \hline \end{array}$

13. $\begin{array}{r} 16,232 \\ \times\ \ \ 198 \\ \hline \end{array}$
14. $\begin{array}{r} 4.83 \\ \times\ 3.7 \\ \hline \end{array}$
15. $\begin{array}{r} 518.8 \\ \times\ 8.63 \\ \hline \end{array}$

16. $\begin{array}{r} 8808 \\ -2220 \\ \hline \text{about } 7000 \end{array}$
17. $\begin{array}{r} 6090 \\ \times\ \ 45 \\ \hline \text{about } 300,000 \end{array}$
18. $\begin{array}{r} 3.579 \\ \times\ \ \ 72 \\ \hline \text{about } 280 \end{array}$

Skill 2 Front-end and Adjust

1. about 1100   2. about 1100   3. about 1400
4. about 7000   5. about 8000   6. about 8000
7. about 30   8. about 300   9. about 5000

Estimate the sum.

$$\begin{array}{r} 70,732 \\ +\ 9,015 \\ \hline 70,000 + 10,000 \\ \text{about } 80,000 \end{array} \quad \text{about } 10,000$$

Estimate the sum.

$$\begin{array}{r} 281.6 \\ +517.1 \\ \hline 700 + 100 \quad \text{about } 800 \end{array} \quad \text{about } 100$$

Estimate the answer by adjusting the sum of the front-end digits.

1. $\begin{array}{r} 534 \\ +607 \\ \hline \end{array}$
2. $\begin{array}{r} 931 \\ +163 \\ \hline \end{array}$
3. $\begin{array}{r} 472 \\ +901 \\ \hline \end{array}$

4. $\begin{array}{r} 5310 \\ +1519 \\ \hline \end{array}$
5. $\begin{array}{r} 2093 \\ +6258 \\ \hline \end{array}$
6. $\begin{array}{r} 1580 \\ +6293 \\ \hline \end{array}$

7. $\begin{array}{r} 24.913 \\ +\ 5.720 \\ \hline \end{array}$
8. $\begin{array}{r} 190.30 \\ +\ 86.10 \\ \hline \end{array}$
9. $\begin{array}{r} 3750.2 \\ +1223.9 \\ \hline \end{array}$

10. $\begin{array}{r} 23,498 \\ 10,976 \\ +\ 8,231 \\ \hline \text{about } 40,000 \end{array}$
11. $\begin{array}{r} 59.226 \\ 14.835 \\ +76.103 \\ \hline \text{about } 150 \end{array}$
12. $\begin{array}{r} 62,018 \\ 13,950 \\ +71,004 \\ \hline \text{about } 150,000 \end{array}$

Skill 3 Compatible Numbers

Estimate the quotient:
493 ÷ 37

First round the divisor to its greatest place value:

493 ÷ 40

Then change the dividend to the closest convenient multiple of the new divisor:

480 ÷ 40

Now divide mentally with the compatible numbers:

480 ÷ 40 = 12

The quotient is about 12.

Use compatible numbers to estimate the quotient.

1. 29)587 about 20
2. 32)617 about 20
3. 26)971 about 30
4. 91)565 about 6
5. 62)1531 about 30
6. 57)1730 about 30
7. 83)4677 about 60
8. 86)2471 about 30
9. 236)7895 about 40
10. 421)2193 about 5
11. 606)38,415 about 60
12. 582)46,478 about 80
13. 156)23,199 about 120
14. 923)82,331 about 90
15. 29.6)2662.7 about 90
16. 463)24,129 about 50
17. 98.7)321.7 about 3
18. 27.82)208.59 about 7
19. 38.7)1648 about 40
20. 9.7)168.9 about 17

Skill 4 Clustering

1. about 60   2. about 240   3. about 120
4. about 200   5. about 630   6. about 120
7. about 1800   8. about 2130

Use clustering to estimate the sum.

Estimate the sum:

62 + 59 + 55 + 67 + 61

Observe that the numbers cluster around 60. Then the sum is about 5 × 60, or about 300.

1. 19 + 26 + 20
2. 85 + 79 + 83
3. 39 + 41 + 43
4. 50 + 48 + 53 + 55
5. 210 + 203 + 211
6. 31 + 26 + 28 + 30
7. 597 + 603 + 609
8. 711 + 709 + 713
9. 130 + 128 + 133 about 390
10. 270 + 272 + 268 about 810

11. 343 + 340 + 337 + 339 + 345 about 1700
12. 1110 + 1119 + 1108 + 1111 + 1102 + 1108 about 6660
13. 7.8 + 7.2 + 8.3 + 8.1 + 8 + 8.9 + 8.6 about 56
14. 10.6 + 9.3 + 9.1 + 10 + 9.4 + 9.9 + 9.8 about 70
15. 7.03 + 6.9 + 6.97 + 7.1 + 7.01 about 35
16. 92 + 90.7 + 89.5 + 90.1 + 88 + 89.8 about 540
17. 1.21 + 1.2 + 1.19 + 1.15 + 1.22 + 1.25 about 7.2

Numerical Skills

Skill 5   Place Value; Writing Numbers

In 7896.513, give the place and value of each digit.

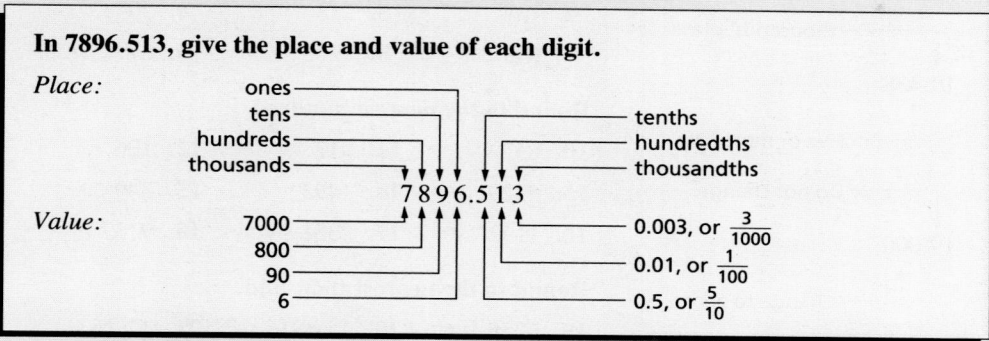

Place:
- ones — tenths
- tens — hundredths
- hundreds — thousandths
- thousands

7896.513

Value:
- 7000 — 0.003, or $\frac{3}{1000}$
- 800 — 0.01, or $\frac{1}{100}$
- 90 — 0.5, or $\frac{5}{10}$
- 6

Give the place and value of the underlined digit.   **15.** thousandths, 0.009

1. 6<u>8</u>9 tens, 80   **2.** 11<u>3</u> ones, 3   **3.** 1<u>0</u>9 tens, 0   **4.** <u>9</u>24 hundreds, 900

5. <u>1</u>562 thousands, 1000   **6.** 3<u>4</u>52 hundreds, 400   **7.** 537<u>1</u> ones, 1   **8.** 4<u>6</u>05 hundreds, 600

9. <u>2</u>638 thousands, 2000   **10.** 0.<u>6</u>2 tenths, 0.6   **11.** 10.1<u>7</u> hundredths, 0.07   **12.** 9.73<u>1</u> thousandths, 0.001

13. 9<u>5</u>.1 tens, 90   **14.** 51.<u>0</u>3 tenths, 0.0   **15.** 146.03<u>9</u>   **16.** 512.<u>9</u>9 ones, 2

17. <u>8</u>.13 ones, 8   **18.** 719.0<u>3</u>3 hundredths, 0.03   **19.** 623.0<u>9</u> hundredths, 0.09   **20.** 1294.31<u>3</u> thousandths, 0.003

Write the word form of each number.

208 ⟶ Two hundred eight   53.29 ⟶ Fifty-three and twenty-nine hundredths

79 ⟶ Seventy-nine   6.418 ⟶ Six and four hundred eighteen thousandths

13 ⟶ Thirteen

Write the word form of each number.   Answers to Exercises 21–50 are on p. A46.

21. 15	**22.** 91	**23.** 162	**24.** 87	**25.** 493
26. 101	**27.** 258	**28.** 911	**29.** 4056	**30.** 3080
31. 32	**32.** 1473	**33.** 7006	**34.** 56	**35.** 385
36. 16.3	**37.** 4.7	**38.** 21.15	**39.** 178.61	**40.** 9.016
41. 83.097	**42.** 492	**43.** 18.315	**44.** 9.036	**45.** 6031.752
46. 24.4	**47.** 193.07	**48.** 8.949	**49.** 307.1	**50.** 12.56

Skill 6 Rounding Whole Numbers and Decimals

Round 19,438 to the nearest thousand.

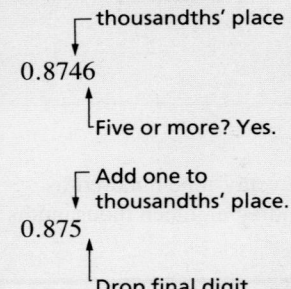

thousands' place

19,438

Five or more? No.

Do not change.

19,000

Change to zeros.

Round to the nearest ten.

1. 18 20	**2.** 52 50	**3.** 99 100
4. 137 140	**5.** 273 270	**6.** 5048 5050
7. 649 650	**8.** 1208 1210	**9.** 8792 8790

Round to the nearest hundred.

10. 652 700	**11.** 912 900	**12.** 106 100
13. 5329 5300	**14.** 1493 1500	**15.** 349 300
16. 1192 1200	**17.** 2351 2400	**18.** 9137 9100

Round to the nearest thousand.

20. 15,000   **21.** 12,000

19. 6358 6000	**20.** 15,491	**21.** 12,253
22. 878 1000	**23.** 7299 7000	**24.** 9602 10,000
25. 2549 3000	**26.** 3197 3000	**27.** 21,840 22,000

Round 0.8746 to the nearest thousandth.

thousandths' place

0.8746

Five or more? Yes.

Add one to thousandths' place.

0.875

Drop final digit.

Round to the nearest tenth.

28. 0.52 0.5	**29.** 4.01 4.0	**30.** 11.93 11.9
31. 0.312 0.3	**32.** 15.11 15.1	**33.** 4.76 4.8
34. 0.08 0.1	**35.** 12.19 12.2	**36.** 92.25 92.3
37. 107.623 107.6	**38.** 22.819 22.8	**39.** 17.853 17.9

Round to the nearest hundredth.

40. 0.739 0.74	**41.** 0.515 0.52	**42.** 127.316 127.32
43. 31.034 31.03	**44.** 0.683 0.68	**45.** 0.299 0.30
46. 2.051 2.05	**47.** 12.994 12.99	**48.** 0.625 0.63
49. 3.408 3.41	**50.** 11.003 11.00	**51.** 71.987 71.99

Round to the nearest thousandth.

52. 0.8376 0.838	**53.** 54.0178	**54.** 0.9003
55. 0.3139 0.314	**56.** 183.5069	**57.** 2.9919
58. 6.0013 6.001	**59.** 13.0297	**60.** 86.9765
61. 4.3482 4.348	**62.** 1.6105 1.611	**63.** 177.0251 177.025

53. 54.018	**54.** 0.900
56. 183.507	**57.** 2.992
59. 13.030	**60.** 86.977

Skill 7 Adding Whole Numbers

Add.

$$
\begin{array}{r} 562 \\ 103 \\ +432 \\ \hline \end{array}
\longrightarrow
\begin{array}{r} 562 \\ 103 \\ +432 \\ \hline 1097 \end{array}
$$

$$
\begin{array}{r} 5738 \\ +6425 \\ \hline \end{array}
\longrightarrow
\begin{array}{r} \overset{1\ \ 1}{5738} \\ +\ 6425 \\ \hline 12{,}163 \end{array}
$$

Add.

1.
$$\begin{array}{r} 57 \\ +31 \\ \hline 88 \end{array}$$

2.
$$\begin{array}{r} 75 \\ +61 \\ \hline 136 \end{array}$$

3.
$$\begin{array}{r} 19 \\ +80 \\ \hline 99 \end{array}$$

4.
$$\begin{array}{r} 274 \\ +615 \\ \hline 889 \end{array}$$

5.
$$\begin{array}{r} 493 \\ +702 \\ \hline 1195 \end{array}$$

6.
$$\begin{array}{r} 128 \\ +931 \\ \hline 1059 \end{array}$$

7.
$$\begin{array}{r} 546 \\ +323 \\ \hline 869 \end{array}$$

8.
$$\begin{array}{r} 948 \\ +\ 51 \\ \hline 999 \end{array}$$

9.
$$\begin{array}{r} 5463 \\ +7124 \\ \hline 12{,}587 \end{array}$$

10.
$$\begin{array}{r} 9807 \\ +5161 \\ \hline 14{,}968 \end{array}$$

11.
$$\begin{array}{r} 5410 \\ +\ 329 \\ \hline 5739 \end{array}$$

12.
$$\begin{array}{r} 8643 \\ +\ 321 \\ \hline 8964 \end{array}$$

13.
$$\begin{array}{r} 8263 \\ 135 \\ +6401 \\ \hline 14{,}799 \end{array}$$

14.
$$\begin{array}{r} 7315 \\ 442 \\ +9211 \\ \hline 16{,}968 \end{array}$$

15.
$$\begin{array}{r} 5471 \\ 206 \\ +\ 112 \\ \hline 5789 \end{array}$$

16.
$$\begin{array}{r} 9641 \\ 105 \\ +\ 230 \\ \hline 9976 \end{array}$$

17.
$$\begin{array}{r} 216 \\ 157 \\ +119 \\ \hline 492 \end{array}$$

Skill 8 Subtracting Whole Numbers

Subtract.

$$
\begin{array}{r} 8635 \\ -5104 \\ \hline \end{array}
\longrightarrow
\begin{array}{r} 8635 \\ -5104 \\ \hline 3531 \end{array}
$$

$$
\begin{array}{r} 6591 \\ -3185 \\ \hline \end{array}
\longrightarrow
\begin{array}{r} \overset{8\,11}{65\cancel{9}\cancel{1}} \\ -3185 \\ \hline 3406 \end{array}
$$

$$
\begin{array}{r} 5073 \\ -1896 \\ \hline \end{array}
\longrightarrow
\begin{array}{r} \overset{4\,9\,16\,13}{5\cancel{0}\cancel{7}\cancel{3}} \\ -1896 \\ \hline 3177 \end{array}
$$

Subtract.

1.
$$\begin{array}{r} 971 \\ -531 \\ \hline 440 \end{array}$$

2.
$$\begin{array}{r} 827 \\ -513 \\ \hline 314 \end{array}$$

3.
$$\begin{array}{r} 694 \\ -572 \\ \hline 122 \end{array}$$

4.
$$\begin{array}{r} 208 \\ -103 \\ \hline 105 \end{array}$$

5.
$$\begin{array}{r} 1965 \\ -\ 740 \\ \hline 1225 \end{array}$$

6.
$$\begin{array}{r} 4835 \\ -\ 625 \\ \hline 4210 \end{array}$$

7.
$$\begin{array}{r} 6679 \\ -4238 \\ \hline 2441 \end{array}$$

8.
$$\begin{array}{r} 8341 \\ -1200 \\ \hline 7141 \end{array}$$

9.
$$\begin{array}{r} 1467 \\ -1354 \\ \hline 113 \end{array}$$

10.
$$\begin{array}{r} 5438 \\ -4216 \\ \hline 1222 \end{array}$$

11.
$$\begin{array}{r} 7032 \\ -5011 \\ \hline 2021 \end{array}$$

12.
$$\begin{array}{r} 4439 \\ -2018 \\ \hline 2421 \end{array}$$

13.
$$\begin{array}{r} 39{,}584 \\ -17{,}160 \\ \hline 22{,}424 \end{array}$$

14.
$$\begin{array}{r} 14{,}761 \\ -10{,}431 \\ \hline 4330 \end{array}$$

15.
$$\begin{array}{r} 73{,}899 \\ -21{,}574 \\ \hline 52{,}325 \end{array}$$

16.
$$\begin{array}{r} 94{,}538 \\ -61{,}527 \\ \hline 33{,}011 \end{array}$$

17.
$$\begin{array}{r} 141 \\ -\ 38 \\ \hline 103 \end{array}$$

18.
$$\begin{array}{r} 116 \\ -\ 89 \\ \hline 27 \end{array}$$

19.
$$\begin{array}{r} 147 \\ -\ 99 \\ \hline 48 \end{array}$$

20.
$$\begin{array}{r} 219 \\ -\ 53 \\ \hline 166 \end{array}$$

21.
$$\begin{array}{r} 289 \\ -193 \\ \hline 96 \end{array}$$

22.
$$\begin{array}{r} 524 \\ -376 \\ \hline 148 \end{array}$$

23.
$$\begin{array}{r} 6843 \\ -1951 \\ \hline 4892 \end{array}$$

24.
$$\begin{array}{r} 3000 \\ -1587 \\ \hline 1413 \end{array}$$

25.
$$\begin{array}{r} 61{,}558 \\ -29{,}709 \\ \hline 31{,}849 \end{array}$$

Skill 9 Multiplying Whole Numbers

Multiply.

```
 332        332
×132   →   ×132
           ────
            664   ←── 2 × 332      ←── Multiply by 2 ones.
           9960   ←── 30 × 332     ←── Multiply by 3 tens.
          33200   ←── 100 × 332    ←── Multiply by 1 hundred.
          ──────
          43,824  ←── 132 × 332
```

Multiply.   **7.** 1,420,868   **8.** 1,361,072   **9.** 264,924

1. 53 ×31 ──── 1643	**2.** 43 ×21 ──── 903	**3.** 714 × 22 ──── 15,708	**4.** 332 × 13 ──── 4316	**5.** 401 ×32 ──── 12,832
6. 832 × 11 ──── 9152	**7.** 7034 × 202	**8.** 4112 × 331	**9.** 6021 × 44	**10.** 641 ×122 ──── 78,202
11. 10,021 × 13 ──── 130,273	**12.** 52,022 × 42 ──── 2,184,924	**13.** 93,122 × 231 ──── 21,511,182	**14.** 71,301 × 121 ──── 8,627,421	**15.** 82,113 × 302 ──── 24,798,126

Multiply.

```
 863        863
×425   →   ×425
           ────
           4315   ←── 5 × 863      ←── Multiply by 5 ones.
          17260   ←── 20 × 863     ←── Multiply by 2 tens.
         345200   ←── 400 × 863    ←── Multiply by 4 hundreds.
         ───────
         366,775  ←── 425 × 863
```

Multiply.   **22.** 100,951   **23.** 232,403   **24.** 317,475   **25.** 638,844

16. 28 ×76 ──── 2128	**17.** 34 ×73 ──── 2482	**18.** 98 ×43 ──── 4214	**19.** 109 × 58 ──── 6322	**20.** 327 × 66 ──── 21,582
21. 429 × 57 ──── 24,453	**22.** 643 ×157	**23.** 839 ×277	**24.** 765 ×415	**25.** 834 ×766
26. 647 ×519 ──── 335,793	**27.** 498 ×306 ──── 152,388	**28.** 6318 × 119 ──── 751,842	**29.** 4265 × 539 ──── 2,298,835	**30.** 2157 × 804 ──── 1,734,228

Divide. Show the remainder in fraction form.

$$24\overline{)762} \longrightarrow \begin{array}{r} 31 \\ 24\overline{)762} \\ \underline{72} \\ 42 \\ \underline{24} \\ 18 \end{array}$$

The answer is $31\frac{18}{24}$, or $31\frac{3}{4}$.

Divide. Round the quotient to the nearest tenth.

$$58\overline{)739} \longrightarrow \begin{array}{r} 12.74 \\ 58\overline{)739.00} \\ \underline{58}\downarrow \\ 159 \\ \underline{116}\downarrow \\ 43\ 0 \\ \underline{40\ 6}\downarrow \\ 2\ 40 \\ \underline{2\ 32} \\ 8 \end{array}$$

To round the quotient to tenths, carry the division to the hundredths' place; then round. To the nearest tenth, the quotient is 12.7.

Divide. If there is a remainder, show it in fraction form.

1. $38\overline{)114}$ 3
2. $54\overline{)78}$ $1\frac{4}{9}$
3. $47\overline{)611}$ 13
4. $81\overline{)702}$ $8\frac{2}{3}$
5. $17\overline{)2975}$ 175

6. $124\overline{)5932}$ $47\frac{26}{31}$
7. $34\overline{)928}$ $27\frac{5}{17}$
8. $763\overline{)28,994}$ 38
9. $328\overline{)63,509}$ $193\frac{5}{8}$
10. $419\overline{)92,195}$ $220\frac{15}{419}$

Divide. Express the quotient as a whole number or a decimal. **12.** 17.6 **15.** 314.8

11. $6\overline{)75}$ 12.5
12. $20\overline{)352}$
13. $35\overline{)665}$ 19
14. $12\overline{)8694}$ 724.5
15. $25\overline{)7870}$

16. $96\overline{)4524}$ 47.125
17. $36\overline{)4122}$ 114.5
18. $90\overline{)31,644}$ 351.6
19. $64\overline{)34,192}$ 534.25
20. $85\overline{)20,111}$ 236.6

Divide. Round the quotient to the nearest tenth. **22.** 25.3 **25.** 143.4

21. $17\overline{)239}$ 14.1
22. $29\overline{)735}$
23. $13\overline{)658}$ 50.6
24. $53\overline{)2947}$ 55.6
25. $61\overline{)8746}$

26. $19\overline{)564}$ 29.7
27. $44\overline{)3593}$ 81.7
28. $25\overline{)9056}$ 362.2
29. $36\overline{)10,362}$ 287.8
30. $19\overline{)24,988}$ 1315.2

Divide. Round the quotient to the nearest hundredth. **31.** 20.11 **32.** 5.91 **33.** 250.94

31. $28\overline{)563}$
32. $78\overline{)461}$
33. $18\overline{)4517}$
34. $92\overline{)7639}$ 83.03
35. $61\overline{)8446}$ 138.46

36. $87\overline{)6398}$ 73.54
37. $21\overline{)4283}$ 203.95
38. $23\overline{)2525}$ 109.78
39. $16\overline{)32,654}$ 2040.88
40. $55\overline{)12,135}$ 220.64

Skill 11 Adding Decimals

1. 767.95 **2.** 1401.15 **3.** 1433.20 **4.** 1088.31
5. 1173.93 **6.** 1127.61 **7.** 11,559.4 **8.** 52.792
9. 1398.898 **10.** 884.58 **11.** 953.23 **12.** 2565.6

Add.

```
 Add.              111 1
351.74            351.74
178.23     ⟶      178.23
+409.17          +409.17
                  939.14
```

Decimal points are aligned. If necessary, use zeros as placeholders.

```
                  21 1
0.67              0.670
5.398      ⟶      5.398
7.2               7.200
+ 29             +29.000
                  42.268
```

Add.

1. 519.76 +248.19	**2.** 748.11 +653.04	**3.** 894.23 +538.97
4. 136.47 +951.84	**5.** 460.09 +713.84	**6.** 384.16 +743.45
7. 2516.3 +9043.1	**8.** 18.289 +34.503	**9.** 957.361 +441.537
10. 14.29 369.51 +500.78	**11.** 29.06 139.24 +784.93	**12.** 851.7 1239.3 +474.6
13. 14 + 7.1 = 21.1	**14.** 25.1 +16.09 = 41.19	**15.** 13.94 + 8.315 = 22.255
16. 541.006 + 39.13 = 580.136	**17.** 93.701 +115.39 = 209.091	**18.** 29.48 150.003 + 4.1 = 183.583	**19.** 29.504 11.06 + 0.013 = 40.577	**20.** 2.006 35.793 + 0.16 = 37.959

Skill 12 Subtracting Decimals

1. 9.38 **2.** 13.53 **3.** 23.73 **4.** 28.043
5. 4.889 **6.** 32.141 **7.** 141.67 **8.** 117.18
9. 288.636

```
 Subtract.        5 9 14 11
605.13     ⟶      6̸0̸5̸.1̸3̸
-298.31          -298.31
                  306.82
```

Decimal points are aligned. Use zero as a placeholder.

```
                  6 12 17 10
73.8       ⟶      7̸3̸.8̸0̸
- 6.93           - 6.93
                  66.87
```

Subtract.

1. 63.87 −54.49	**2.** 33.17 −19.64	**3.** 72.09 −48.36
4. 46.139 −18.096	**5.** 100.036 − 95.147	**6.** 59.218 −27.077
7. 183.76 − 42.09	**8.** 193.29 − 76.11	**9.** 349.001 − 60.365
10. 0.6509 −0.0311 = 0.6198	**11.** 400.31 −278.74 = 121.57	**12.** 112.453 −103.926 = 8.527
13. 53.2 − 7.56 = 45.64	**14.** 29.8 − 5.17 = 24.63	**15.** 96.24 − 7.6 = 88.64	**16.** 713.63 − 5.379 = 708.251	**17.** 11 − 0.03 = 10.97

Skill 13 Multiplying Decimals

7. 19.1205   8. 59.892   9. 7.462
10. 461.3966   11. 2915.415   12. 4479.58

Multiply.

$$23.78 \longleftarrow \text{2 places}$$
$$\times\ 4.6 \longleftarrow +\ \text{1 place}$$
$$\overline{14268}$$
$$95120$$
$$\overline{109.388} \longleftarrow \text{3 places}$$

$$0.07 \longleftarrow \text{2 places}$$
$$\times\ 0.3 \longleftarrow +\ \text{1 place}$$
$$\overline{0.021} \longleftarrow \text{3 places}$$
$$\uparrow$$

Insert one zero as a placeholder.

Multiply.

1. $\begin{array}{r} 5.3 \\ \times\ 2.6 \\ \hline 13.78 \end{array}$ 2. $\begin{array}{r} 7.1 \\ \times\ 9 \\ \hline 63.9 \end{array}$ 3. $\begin{array}{r} 8.4 \\ \times\ 3.9 \\ \hline 32.76 \end{array}$

4. $\begin{array}{r} 12.7 \\ \times\ 4.1 \\ \hline 52.07 \end{array}$ 5. $\begin{array}{r} 13.8 \\ \times\ 9.5 \\ \hline 131.1 \end{array}$ 6. $\begin{array}{r} 17 \\ \times\ 3.9 \\ \hline 66.3 \end{array}$

7. $\begin{array}{r} 6.07 \\ \times\ 3.15 \end{array}$ 8. $\begin{array}{r} 7.13 \\ \times\ 8.4 \end{array}$ 9. $\begin{array}{r} 1.4 \\ \times\ 5.33 \end{array}$

10. $\begin{array}{r} 23.71 \\ \times\ 19.46 \end{array}$ 11. $\begin{array}{r} 38.11 \\ \times\ 76.5 \end{array}$ 12. $\begin{array}{r} 98 \\ \times 45.71 \end{array}$

13. $\begin{array}{r} 0.008 \\ \times\ 0.6 \\ \hline 0.0048 \end{array}$ 14. $\begin{array}{r} 0.56 \\ \times\ 0.3 \\ \hline 0.168 \end{array}$ 15. $\begin{array}{r} 0.019 \\ \times 0.27 \\ \hline 0.00513 \end{array}$

Skill 14 Dividing Decimals

Divide. $3.9\overline{)22.542}$

$3.9\overline{)22.542}$

Move both decimal points one place to the right.

$$\begin{array}{r} 5.78 \\ 39\overline{)225.42} \\ 195 \\ \hline 30\ 4 \\ 27\ 3 \\ \hline 3\ 12 \\ 3\ 12 \\ \hline 0 \end{array}$$

Divide. $0.076\overline{)17.48}$

Annex a zero.

$0.076\overline{)17.480}$

Move three places to the right.

$$\begin{array}{r} 230 \\ 76\overline{)17480} \\ 152 \\ \hline 228 \\ 228 \\ \hline 00 \\ 0 \\ \hline 0 \end{array}$$

Divide. If necessary, round the quotient to the nearest hundredth.   6. 29.32   9. 18.1

1. $3.5\overline{)27.3}$ 7.8   2. $1.9\overline{)7.03}$ 3.7   3. $1.6\overline{)10.384}$ 6.49   4. $2.3\overline{)39.567}$ 17.20

5. $7.82\overline{)148.58}$ 19   6. $19.51\overline{)572.091}$ 7. $2.04\overline{)27.336}$ 13.4   8. $5.06\overline{)47.058}$ 9.3

9. $12.53\overline{)226.793}$ 10. $19.07\overline{)87.722}$ 4.6   11. $0.56\overline{)21.28}$ 38   12. $0.512\overline{)6.656}$ 13

13. $0.381\overline{)24.003}$ 63   14. $0.27\overline{)5.316}$ 19.69   15. $0.63\overline{)73.081}$ 116.00   16. $0.84\overline{)21.84}$ 26

Skill 15 Multiplying and Dividing by 10, 100, and 1000

Multiply.

176.34×10 $= 176.34$ $= 1763.4$
Move one place
to the right.

89.6×100 $= 89.60$ $= 8960$
Move two places
to the right.

0.6043×1000 $= 0.6043$ $= 604.3$
Move three places
to the right.

Divide.

$183.24 \div 10$ $= 183.24$ $= 18.324$
Move one place
to the left.

$0.0319 \div 100$ $= 00.0319$ $= 0.000319$
Move two places
to the left.

$17.83 \div 1000$ $= 017.83$ $= 0.01783$
Move three places
to the left.

Multiply by 10.

1. 0.803 8.03
2. 13.8 138
3. 813.06 8130.6
4. 924 9240
5. 54.1 541
6. 0.519 5.19
7. 7.092 70.92
8. 0.036 0.36

Multiply by 100.

9. 28.4 2840
10. 6.948 694.8
11. 0.3 30
12. 0.092 9.2
13. 127.51 12,751
14. 17.316 1731.6
15. 8.04 804
16. 14.0052 1400.52

Multiply by 1000.

17. 0.0417 41.7
18. 219.3 219,300
19. 0.012 12
20. 16.045 16,045
21. 9.31 9310
22. 0.66 660
23. 82.1 82,100
24. 0.24054 240.54

Divide by 10.

25. 216.35 21.635
26. 0.31 0.031
27. 5.106 0.5106
28. 23 2.3
29. 28.57 2.857
30. 84.19 8.419
31. 0.0792 0.00792
32. 820.51 82.051

Divide by 100.

33. 29.3 0.293
34. 8.47 0.0847
35. 930.1 9.301
36. 0.064 0.00064
37. 1847.5 18.475
38. 506.12 5.0612
39. 0.29 0.0029
40. 605.8 6.058

Divide by 1000.

41. 413 0.413
42. 108.3 0.1083
43. 0.6 0.0006
44. 135.09 0.13509
45. 26.9 0.0269
46. 63,218 63.218
47. 0.025 0.000025
48. 9870 9.87

A number is divisible by a second number if the remainder is zero when the first number is divided by the second.

The table below shows divisibility tests you can use to determine if one number is divisible by another number.

Divisible by	Test
2	The digit in the ones' place is 0, 2, 4, 6, or 8.
5	The digit in the ones' place is 0 or 5.
10	The digit in the ones' place is 0.
3	The sum of the digits is divisible by 3.
9	The sum of the digits is divisible by 9.
4	The number formed by the last two digits is divisible by 4.
8	The number formed by the last three digits is divisible by 8.
6	The number is divisible by both 2 and 3.

Examples:

Is 5435 divisible by 5? ⟶ Yes; the digit in the ones' place is 5.

Is 3741 divisible by 9? ⟶ No; the sum of the digits is $3 + 7 + 4 + 1 = 15$, and 15 is not divisible by 9.

Is 3564 divisible by 8? ⟶ No; the number formed by the last three digits is not divisible by 8.

Is 4938 divisible by 6? ⟶ Yes, the digit in the ones' place is 8 (which is divisible by 2) and the sum of the digits is 24 (which is divisible by 3).

Test the number for divisibility. Write *Yes* or *No*.

By 2:	**1.** 138 Yes.	**2.** 203 No.	**3.** 517 No.
By 5:	**4.** 135 Yes.	**5.** 730 Yes.	**6.** 219 No.
By 10:	**7.** 90 Yes.	**8.** 102 No.	**9.** 305 No.
By 3:	**10.** 216 Yes.	**11.** 735 Yes.	**12.** 889 No.
By 9:	**13.** 258 No.	**14.** 792 Yes.	**15.** 1035 Yes.
By 4:	**16.** 7248 Yes.	**17.** 838 No.	**18.** 2344 Yes.
By 8:	**19.** 7688 Yes.	**20.** 312 Yes.	**21.** 57,680 Yes.
By 6:	**22.** 8324 No.	**23.** 6678 Yes.	**24.** 504 Yes.

Add or subtract. Write the answer in lowest terms.

$\dfrac{5}{7}$ — The fractions have a common denominator. Add the numerators.
$+\dfrac{3}{7}$
$\dfrac{8}{7} = 1\dfrac{1}{7}$

$\dfrac{5}{9}$ — The fractions have a common denominator. Subtract the numerators.
$-\dfrac{1}{9}$
$\dfrac{4}{9}$

Add or subtract. Write the answer in lowest terms.

1. $\dfrac{5}{9}$ $\dfrac{7}{9}$
   $+\dfrac{2}{9}$

2. $\dfrac{1}{5}$ 1
   $+\dfrac{4}{5}$

3. $\dfrac{11}{20}$ $\dfrac{4}{5}$
   $+\dfrac{5}{20}$

4. $\dfrac{2}{5}$ $1\dfrac{2}{5}$
   $\dfrac{1}{5}$
   $+\dfrac{4}{5}$

5. $\dfrac{3}{7}$ 2
   $\dfrac{5}{7}$
   $+\dfrac{6}{7}$

6. $\dfrac{13}{100}$ $\dfrac{1}{4}$
   $\dfrac{3}{100}$
   $+\dfrac{9}{100}$

7. $\dfrac{11}{15}$ $\dfrac{1}{5}$
   $-\dfrac{8}{15}$

8. $\dfrac{11}{12}$ $\dfrac{1}{2}$
   $-\dfrac{5}{12}$

9. $\dfrac{23}{24}$ $\dfrac{1}{2}$
   $-\dfrac{11}{24}$

10. $\dfrac{7}{8}$ $\dfrac{1}{2}$
    $-\dfrac{3}{8}$

11. $\dfrac{9}{10}$ $\dfrac{3}{5}$
    $-\dfrac{3}{10}$

12. $\dfrac{29}{50}$ $\dfrac{2}{5}$
    $-\dfrac{9}{50}$

13. $\dfrac{91}{100}$ $\dfrac{3}{5}$
    $-\dfrac{31}{100}$

14. $\dfrac{5}{6}$ $\dfrac{2}{3}$
    $-\dfrac{1}{6}$

When the fractions have different denominators, rewrite them as equivalent fractions having the least common denominator. Then add or subtract.

$\dfrac{7}{12}$ $\longrightarrow$ $\dfrac{35}{60}$
$+\dfrac{3}{20}$ $\longrightarrow$ $+\dfrac{9}{60}$
$\dfrac{44}{60} = \dfrac{11}{15}$

$\dfrac{5}{9}$ $\longrightarrow$ $\dfrac{10}{18}$
$-\dfrac{1}{6}$ $\longrightarrow$ $-\dfrac{3}{18}$
$\dfrac{7}{18}$

Add or subtract. Write the answer in lowest terms.

15. $\dfrac{3}{16}$ $\dfrac{9}{16}$
    $+\dfrac{3}{8}$

16. $\dfrac{7}{20}$ $1\dfrac{1}{10}$
    $+\dfrac{3}{4}$

17. $\dfrac{7}{9}$ $1\dfrac{25}{36}$
    $+\dfrac{11}{12}$

18. $\dfrac{3}{4}$ $1\dfrac{13}{24}$
    $\dfrac{1}{6}$
    $+\dfrac{5}{8}$

19. $\dfrac{3}{10}$ $\dfrac{26}{45}$
    $\dfrac{7}{45}$
    $+\dfrac{11}{90}$

20. $\dfrac{1}{2}$ $1\dfrac{7}{9}$
    $\dfrac{5}{6}$
    $+\dfrac{4}{9}$

21. $\dfrac{6}{7}$ $\dfrac{11}{21}$
    $-\dfrac{1}{3}$

22. $\dfrac{4}{5}$ $\dfrac{17}{40}$
    $-\dfrac{3}{8}$

23. $\dfrac{9}{10}$ $\dfrac{13}{30}$
    $-\dfrac{7}{15}$

24. $\dfrac{7}{10}$
    $-\dfrac{1}{4}$
    $\dfrac{9}{20}$

25. $\dfrac{11}{50}$
    $-\dfrac{1}{30}$
    $\dfrac{14}{75}$

26. $\dfrac{5}{6}$
    $-\dfrac{5}{8}$
    $\dfrac{5}{24}$

Skill 18 Adding Mixed Numbers

Add.

First add the fractions. Then add the whole numbers.

$$3\frac{4}{5}$$
$$+9\frac{3}{5}$$
$$12\frac{7}{5} = 12 + 1\frac{2}{5} = 13\frac{2}{5}$$

If necessary, rewrite the fractions as equivalent fractions with a common denominator. Then add.

$$16\frac{3}{4} = \quad 16\frac{9}{12}$$
$$+ 7\frac{2}{3} = + \ 7\frac{8}{12}$$
$$23\frac{17}{12} =$$
$$23 + 1\frac{5}{12} = 24\frac{5}{12}$$

Add. Write each sum in lowest terms.

1. $5\frac{7}{11}$ $7\frac{10}{11}$ **2.** 8 $11\frac{1}{7}$ **3.** $6\frac{5}{9}$ $17\frac{5}{9}$
   $+2\frac{3}{11}$ $+3\frac{1}{7}$ $+11$

4. $14\frac{3}{8}$ 20 **5.** $19\frac{5}{12}$ $23\frac{1}{3}$ **6.** $13\frac{5}{16}$ $21\frac{1}{4}$
   $+ 5\frac{5}{8}$ $+ 3\frac{11}{12}$ $+ 7\frac{15}{16}$

7. $23\frac{26}{45}$ $38\frac{11}{15}$ **8.** $16\frac{21}{40}$ $35\frac{1}{4}$ **9.** $9\frac{7}{10}$ $12\frac{3}{5}$
   $+15\frac{7}{45}$ $+18\frac{29}{40}$ $+2\frac{9}{10}$

Add. Write each sum in lowest terms.

10. $7\frac{1}{2}$ $13\frac{1}{4}$ **11.** $9\frac{7}{10}$ $25\frac{19}{20}$ **12.** $13\frac{5}{8}$ $21\frac{11}{24}$
   $+5\frac{3}{4}$ $+16\frac{1}{4}$ $+ 7\frac{5}{6}$

13. $14\frac{7}{12}$ $19\frac{29}{36}$ **14.** $10\frac{3}{7}$ $16\frac{16}{21}$ **15.** $12\frac{9}{10}$ $17\frac{3}{10}$
   $+ 5\frac{2}{9}$ $+ 6\frac{1}{3}$ $+ 4\frac{2}{5}$

16. $18\frac{1}{6}$ $44\frac{13}{18}$ **17.** $4\frac{7}{8}$ $30\frac{5}{8}$ **18.** $3\frac{5}{8}$ $6\frac{5}{24}$
   $+ 26\frac{5}{9}$ $+25\frac{3}{4}$ $+2\frac{7}{12}$

19. $13\frac{5}{12}$ $31\frac{1}{4}$ **20.** $27\frac{9}{10}$ $44\frac{1}{18}$ **21.** $42\frac{11}{15}$ $83\frac{19}{60}$ **22.** $9\frac{91}{100}$ $13\frac{3}{4}$ **23.** $35\frac{5}{6}$ $50\frac{23}{24}$
   $+17\frac{5}{6}$ $+16\frac{7}{45}$ $+40\frac{7}{12}$ $+3\frac{21}{25}$ $+15\frac{1}{8}$

24. $12\frac{11}{32}$ $22\frac{7}{32}$ **25.** $21\frac{4}{5}$ $46\frac{3}{5}$ **26.** $15\frac{13}{20}$ $29\frac{23}{60}$ **27.** $10\frac{1}{4}$ $25\frac{23}{24}$ **28.** $19\frac{3}{10}$ $37\frac{39}{80}$
   $4\frac{15}{32}$ $18\frac{3}{5}$ $6\frac{1}{3}$ $8\frac{1}{3}$ $6\frac{5}{8}$
   $+ 5\frac{13}{32}$ $+ 6\frac{1}{5}$ $+ 7\frac{2}{5}$ $+ 7\frac{3}{8}$ $+11\frac{9}{16}$

Toolbox Skills Practice 765

765

Skill 19 Subtracting Mixed Numbers

Subtract.

Subtract the fractions. Then subtract the whole numbers.

$$17\frac{7}{12}$$
$$-\ 9\frac{1}{12}$$
$$8\frac{6}{12} = 8\frac{1}{2}$$

If necessary, first rewrite the fractions as equivalent fractions with a common denominator.

$$14\frac{9}{16} = \quad 14\frac{9}{16}$$
$$-\ 6\frac{1}{4} = -\ 6\frac{4}{16}$$
$$8\frac{5}{16}$$

If necessary, first rename a whole number or mixed number.

$$4\frac{3}{10} = \quad 4\frac{3}{10} = \quad 3\frac{13}{10}$$
$$-2\frac{4}{5} = -\ 2\frac{8}{10} = -\ 2\frac{8}{10}$$
$$1\frac{5}{10}$$
$$= 1\frac{1}{2}$$

Subtract. Write each difference in lowest terms.

1. $4\frac{6}{7}\ 2\frac{5}{7}$
   $-2\frac{1}{7}$

2. $36\frac{5}{7}\ 23\frac{3}{7}$
   $-13\frac{2}{7}$

3. $21\frac{19}{32}\ 13\frac{1}{8}$
   $-\ 8\frac{15}{32}$

4. $8\frac{3}{4}$
   $-5\frac{1}{4}$
   $3\frac{1}{2}$

5. $6\frac{7}{8}$
   $-1\frac{3}{8}$
   $5\frac{1}{2}$

6. $14\frac{13}{20}$
   $-10\frac{7}{20}$
   $4\frac{3}{10}$

Subtract. Write each difference in lowest terms.

7. $8\frac{3}{4}\ 3\frac{5}{12}$
   $-5\frac{1}{3}$

8. $51\frac{11}{16}\ 15\frac{9}{16}$
   $-36\frac{1}{8}$

9. $26\frac{7}{8}\ 5\frac{5}{8}$
   $-21\frac{1}{4}$

10. $7\frac{1}{2}\ 2\frac{1}{12}$
    $-5\frac{5}{12}$

11. $22\frac{19}{40}\ 13\frac{1}{8}$
    $-\ 9\frac{7}{20}$

12. $13\frac{5}{18}\ 7\frac{7}{36}$
    $-\ 6\frac{1}{12}$

13. $25\frac{5}{7}\ 15\frac{23}{42}$
    $-10\frac{1}{6}$

14. $39\frac{8}{9}\ 9\frac{11}{18}$
    $-30\frac{5}{18}$

15. $16\frac{7}{8}\ 4\frac{27}{40}$
    $-12\frac{1}{5}$

Subtract. Write each difference in lowest terms.

16. $9\frac{11}{20}\ 2\frac{7}{40}$
    $-7\frac{3}{8}$

17. $14\quad 3\frac{7}{10}$
    $-10\frac{3}{10}$

18. $6\quad 3\frac{3}{8}$
    $-2\frac{5}{8}$

19. $19\frac{5}{16}\ 12\frac{11}{16}$
    $-\ 6\frac{5}{8}$

20. $6\frac{2}{9}\ 1\frac{5}{9}$
    $-4\frac{2}{3}$

21. $24\frac{5}{6}\ 13\frac{11}{12}$
    $-10\frac{11}{12}$

22. $11\frac{3}{10}\ 3\frac{17}{30}$
    $-\ 7\frac{11}{15}$

23. $6\frac{3}{14}\ 2\frac{1}{2}$
    $-3\frac{5}{7}$

24. $4\frac{3}{20}\ 2\frac{5}{12}$
    $-1\frac{11}{15}$

Skill 20 Multiplying Fractions and Mixed Numbers

Multiply. Write the product in lowest terms.

$$\frac{7}{8} \times \frac{4}{13} = \frac{7 \times 4}{8 \times 13} = \frac{7}{26} \qquad\qquad 5 \times \frac{3}{4} = \frac{5}{1} \times \frac{3}{4} = \frac{5 \times 3}{1 \times 4} = \frac{15}{4} = 3\frac{3}{4}$$

$$2\frac{1}{5} \times 3\frac{1}{3} = \frac{11}{5} \times \frac{10}{3} = \frac{11 \times 10}{5 \times 3} = \frac{22}{3} = 7\frac{1}{3}$$

Write each product in lowest terms.

1. $\frac{1}{6} \times \frac{2}{5}$ $\frac{1}{15}$

2. $\frac{1}{7} \times \frac{2}{9}$ $\frac{2}{63}$

3. $\frac{5}{11} \times \frac{3}{10}$ $\frac{3}{22}$

4. $16 \times \frac{3}{8}$ 6

5. $\frac{10}{13} \times 26$ 20

6. $\frac{1}{4} \times \frac{5}{9}$ $\frac{5}{36}$

7. $\frac{7}{12} \times \frac{2}{21}$ $\frac{1}{18}$

8. $\frac{5}{8} \times \frac{2}{3}$ $\frac{5}{12}$

9. $\frac{1}{2} \times \frac{10}{13}$ $\frac{5}{13}$

10. $18 \times \frac{7}{9}$ 14

11. $24 \times \frac{5}{6}$ 20

12. $\frac{1}{3} \times \frac{9}{20}$ $\frac{3}{20}$

13. $10 \times 7\frac{3}{5}$ 76

14. $\frac{5}{9} \times 2\frac{7}{10}$ $1\frac{1}{2}$

15. $5\frac{1}{3} \times 1\frac{1}{8}$ 6

16. $5\frac{1}{2} \times \frac{8}{11}$ 4

Skill 21 Dividing Fractions and Mixed Numbers

To divide with fractions and mixed numbers, multiply the dividend by the reciprocal of the divisor.

$$\overbrace{\phantom{\frac{5}{6} \div \frac{2}{5}}}^{\text{reciprocal of } \frac{2}{5}}$$
$$\frac{5}{6} \div \frac{2}{5} = \frac{5}{6} \times \frac{5}{2} = \frac{25}{12} = 2\frac{1}{12} \qquad\qquad \frac{9}{16} \div 3 = \frac{9}{16} \times \frac{1}{3} = \frac{3}{16}$$

$$15\frac{2}{3} \div 1\frac{7}{9} = \frac{47}{3} \div \frac{16}{9} = \frac{47}{3} \times \frac{9}{16} = \frac{141}{16} = 8\frac{13}{16}$$

Write each quotient in lowest terms.

1. $\frac{1}{5} \div \frac{1}{15}$ 3

2. $\frac{5}{6} \div \frac{1}{12}$ 10

3. $\frac{3}{7} \div \frac{6}{11}$ $\frac{11}{14}$

4. $8 \div \frac{4}{7}$ 14

5. $\frac{2}{3} \div \frac{4}{9}$ $1\frac{1}{2}$

6. $\frac{10}{13} \div \frac{25}{26}$ $\frac{4}{5}$

7. $\frac{5}{12} \div \frac{9}{10}$ $\frac{25}{54}$

8. $\frac{5}{9} \div 10$ $\frac{1}{18}$

9. $\frac{3}{8} \div \frac{5}{7}$ $\frac{21}{40}$

10. $2\frac{4}{7} \div \frac{3}{4}$ $3\frac{3}{7}$

11. $1\frac{11}{19} \div 7\frac{1}{2}$ $\frac{4}{19}$

12. $35 \div \frac{7}{10}$ 50

Skill 22 U.S. Customary System of Measurement

Use the Table of Measures on page xiii.

To change from a larger unit to a smaller unit, you *multiply*.

7 ft 8 in. = __?__ in.

```
   12      Think: larger
 × 7  ←    to smaller,
 ────      so multiply.
   84
 + 8       1 ft = 12 in.
 ────          ↘ × 12
   92
```

7 ft 8 in. = 92 in.

To change from a smaller unit to a larger unit, you *divide*.

36 oz = __?__ lb __?__ oz

```
    2     Think: smaller
16)36 ←   to larger,
   32     so divide.
  ───
    4     16 oz = 1 lb
            ↘ ÷ 16
```

32 oz = 2 lb 4 oz

Complete by changing a larger unit to a smaller unit.

1. 12 mi = __?__ yd
   21,120
2. 3 gal = __?__ qt 12
3. 5.5 t = __?__ lb 11,000
4. 2.5 gal = __?__ qt 10
5. $9\frac{1}{3}$ yd = __?__ ft 28
6. 3 t 350 lb = __?__ lb 6350
7. $2\frac{1}{4}$ mi = __?__ ft 11,880
8. 2 lb 13 oz = __?__ oz 45
9. $4\frac{1}{2}$ t = __?__ lb 9000
10. 39 yd = __?__ ft 117
11. 12 qt 1 pt = __?__ pt 25
12. 8.5 lb = __?__ oz 136
13. $\frac{1}{2}$ yd = __?__ in. 18
14. 7 lb 8 oz = __?__ oz 120

Complete by changing a smaller unit to a larger unit.

15. 37 qt = __?__ gal __?__ qt 9; 1
16. 156 in. = __?__ ft 13
17. 110 ft = __?__ yd __?__ ft 36; 2
18. 32 pt = __?__ qt 16
19. 72 in. = __?__ ft 6
20. 6000 lb = __?__ t 3
21. 23 c = __?__ pt __?__ c 11; 1
22. 128 oz = __?__ lb 8
23. 10 pt = __?__ qt 5
24. 18 c = __?__ pt 9
25. 24 fl oz = __?__ pt $1\frac{1}{2}$
26. 41 ft = __?__ yd __?__ ft 13; 2
27. 24 pt = __?__ gal 3
28. 880 ft = __?__ yd __?__ ft 293; 1

Complete.

29. $3\frac{1}{2}$ c = __?__ fl oz 28
30. $7\frac{1}{2}$ ft = __?__ in. 90
31. 64 fl oz = __?__ c 8
32. 108 in. = __?__ yd 3
33. $1\frac{1}{2}$ mi = __?__ yd 2640
34. 97 ft = __?__ yd __?__ ft 32; 1
35. $1\frac{1}{2}$ t = __?__ oz 48,000
36. 19 qt = __?__ gal __?__ qt 4; 3

Graphing Skills

Skill 23 Reading Pictographs

To read a pictograph, find the key to see the amount that each symbol represents. Then multiply that amount by the number of symbols on a line to get the total amount.

About how many letters did each household receive in 1940?

Each symbol represents 50 letters. The 1940 row has 2 symbols.

$$2 \times 50 = 100$$

Each household received about 100 letters in 1940.

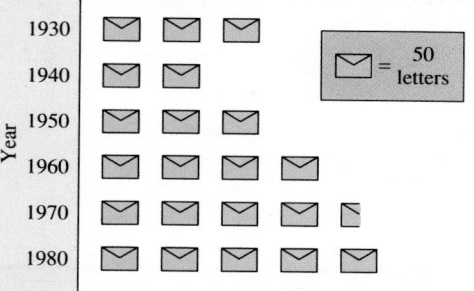

Letters Received per United States Household

Use the pictograph above.

1. In what year was the greatest number of letters received? In what year was the least number of letters received? 1980; 1940

2. About how many letters did each household receive in 1970? about 225

3. About how many more letters were received in 1960 than in 1940?
about 100

4. About how many fewer letters were received in 1930 than in 1960?
about 50

5. During what two years was the number of letters received the same?
1930 and 1950

6. In what year was the number of letters received twice the number received in 1940? 1960

Use the pictograph at the right.

7. About how many pounds of garbage per citizen were collected daily in New York?
about 4 lb

8. In which city was the least amount of garbage per citizen, per day, collected?
Atlanta

9. In which city was the greatest amount of garbage per citizen, per day, collected?
Los Angeles

10. About how many more pounds of garbage per citizen were collected daily in Hartford than in Seattle? about $1\frac{1}{2}$ lb

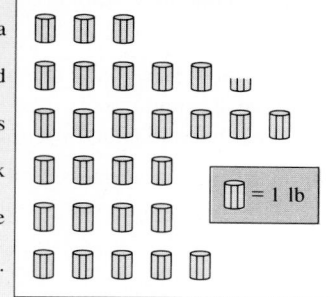

Daily Garbage Collection per Citizen in 1967

Skill 24 Reading Bar Graphs

To read a bar graph, find the bar that represents the information you seek. Trace an imaginary line from the end of the bar to the scale provided. Read the value represented by the bar from the scale.

About how many cars were sold in July?

The bar graph shows that about 16 cars were sold in July.

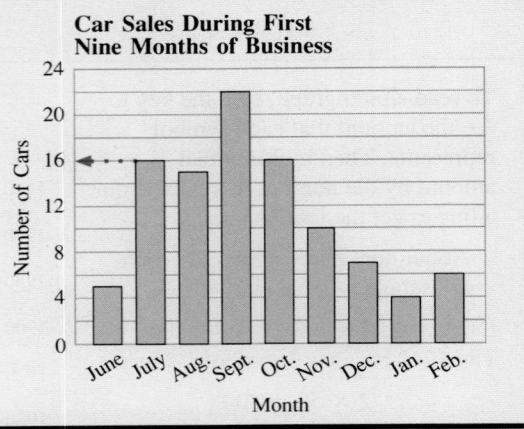

Car Sales During First Nine Months of Business

Use the bar graph above.

1. During which month were the most cars sold? the fewest? September; January

2. During which two months was the same number of cars sold? July and October

3. About how many more cars were sold in November than in February? about 4

4. About how many fewer cars were sold in October than in September? about 6

5. Were car sales increasing or decreasing from September to January? decreasing

6. About how many cars were sold altogether during September, October, and November? about 48

Use the bar graph at the right.

7. In which city were air conditioners used for the greatest number of hours? Dallas

8. For about how many hours were air conditioners used in St. Louis? about 1000 h

9. For about how many fewer hours were air conditioners used in Chicago than in Dallas? about 1000 h

10. Were air conditioners used for more hours in New York or Atlanta? Atlanta

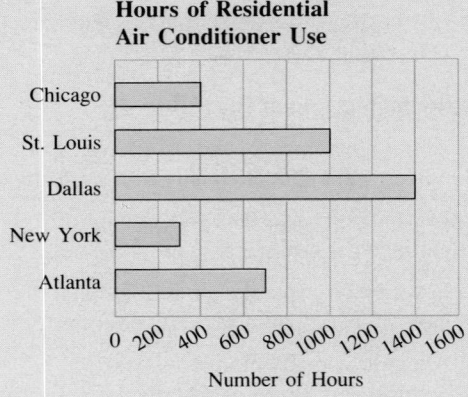

Hours of Residential Air Conditioner Use

Skill 25 Reading Line Graphs

To read a line graph, find the point that represents the information you seek. Trace an imaginary line from the point to the scale on the left. Read the value represented by the point from the scale.

About how many phone calls are placed at 8 P.M.?

The line graph shows that about 30,000 calls are placed at 8 P.M.

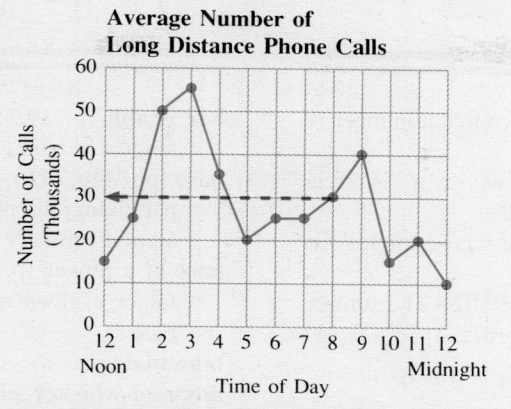

Average Number of Long Distance Phone Calls

Use the line graph above.

1. At what time is the least number of long distance phone calls placed?
   midnight

2. At what time is the greatest number of long distance phone calls placed? 3 P.M.

3. About how many more long distance phone calls are placed at 9 P.M. than at 5 P.M.? about 20,000

4. About how many fewer long distance phone calls are placed at 11 P.M. than at 2 P.M.? about 30,000

5. From noon to 3 P.M., is the number of long distance phone calls increasing or decreasing? increasing

6. At what time is the average number of long distance phone calls placed 50,000? 2 P.M.

Use the line graph at the right.

7. During which month were the fewest homes for sale? January

8. During which month were the most homes for sale? June

9. About how many homes were for sale in August? about 10

10. About how many homes were for sale in October? about 8

11. About how many more homes were for sale in May than in September? about 10

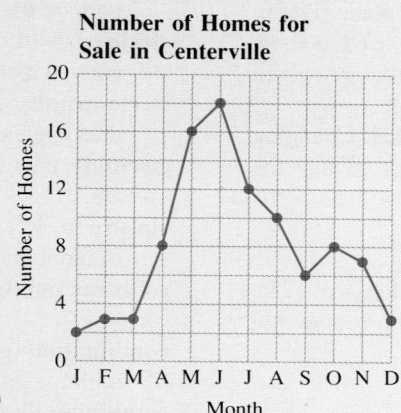

Number of Homes for Sale in Centerville

GLOSSARY

absolute value (p. 95): The distance that a number is from zero on a number line.

acute angle (p. 245): An angle whose measure is greater than 0° and less than 90°.

acute triangle (p. 262): A triangle whose angles are all acute angles.

addition property of opposites (p. 102): The sum of a number and its opposite is zero.

$$a + (-a) = 0 \text{ and } -a + a = 0$$

additive identity (p. 22): The number 0 (zero).

additive inverse (p. 102): The opposite of a number.

adjacent angles (p. 246): Two angles that share a common side but do not overlap each other.

adjacent side (p. 716): In a right triangle, the side of a given angle that is not the hypotenuse.

algebraic fraction (p. 306): A fraction that contains a variable.

alternate interior angles (p. 250): Two angles on alternate sides of a transversal that intersects two lines in the same plane, and interior to the two lines.

angle (p. 237): The figure formed by two rays (called *sides*) which share a common endpoint (called the *vertex*).

area (p. 445): The region enclosed by a plane figure.

arrow notation (p. 66): The symbol → used to show how one number is paired with another in a given function.

associative property of addition (p. 22): Changing the grouping of the terms does not change the sum.

$$(a + b) + c = a + (b + c)$$

associative property of multiplication (p. 25): Changing the grouping of the factors does not change the product.

$$(ab)c = a(bc)$$

average (p. 213): *See* mean.

axes (p. 124): The two number lines that form a coordinate plane. The horizontal number line is the *x-axis*, and the vertical number line is the *y-axis*.

bar graph (p. 192): A type of graph in which the lengths of bars are used to compare data.

base (pp. 446, 628): Either side of a pair of sides of a parallelogram; part of polyhedron used to identify a polyhedron.

base of a power (p. 30): A number that is used as a factor a given number of times. In 5^3, 5 is the base.

binomial (p. 668): A polynomial that has two terms.

box-and-whisker plot (p. 505): A display showing the median of a set of data, the median of each half of the data, and the least and greatest values of the data.

center of a circle (p. 440): The given point in a plane from which all points in a circle are a given distance.

center of rotation (p. 710): The point about which a rotation is made.

central angles (p. 517): Angles having their vertex at the center of the circle.

chord (p. 440): A line segment whose endpoints are both on the circle.

circle (p. 440): The set of all points in a plane that are a given distance from a given point.

circle graph (p. 513): A diagram used to represent data expressed as parts of a whole.

circumference (p. 441): The distance around a circle.

closure (p. 137): A given operation performed on a set of numbers results in a number also in the set.

collinear points (p. 236): Points that lie on the same line.

combination (p. 560): A group of items chosen from within a group of items without regard to order.

combining like terms (p. 56): The process of adding or subtracting like terms.

common factor (p. 293): A number that is a factor of two numbers.

common multiple (p. 296): A number that is a multiple of two numbers.

commutative property of addition (p. 22): Changing the order of the terms does not change the sum.

$$a + b = b + a$$

commutative property of multiplication (p. 25): Changing the order of the factors does not change the product.

$$ab = ba$$

comparison property (p. 18): For any two numbers a and b, exactly one of the following is true: $a > b$, $a < b$, or $a = b$.

complementary angles (p. 245): Two angles the sum of whose measures is 90°.

composite number (p. 290): A number with more than two factors.

composite space figure (p. 663): A combination of two or more space figures.

cone (p. 629): A space figure with one circular base and a vertex.

congruent figures (p. 462): Geometric figures that have the same size and shape.

coordinate plane (p. 124): A grid formed by two number lines that meet at a point.

corresponding angles (pp. 250, 463): Two angles in the same position with respect to two lines and a transversal. *Also* angles in the same position in congruent or similar polygons.

cosine (p. 716): In right triangle ABC,

$$\cos A = \frac{\text{length of side adjacent to } \angle A}{\text{length of hypotenuse}} = \frac{b}{c}.$$

cost (p. 224): The amount of money spent.

counting principle (p. 549): The number of outcomes for an event is found by multiplying the number of choices for each stage of the event.

cross products (p. 391): In a proportion, the product of the first numerator and the second denominator; also, the product of the first denominator and the second numerator.

cube (p. 628): A rectangular prism whose faces are all squares.

cylinder (p. 629): A space figure with two circular bases that are congruent and parallel.

══════════ d ══════════

data (p. 188): A collection of numerical facts.

decimal notation (p. 80): The writing of a number using place values that are powers of ten.

decreasing trend (p. 196): A decrease in the data over a given interval on a line graph, shown by a series of segments that slopes downward.

degree (p. 240): The unit commonly used to measure the size of an angle.

dependent events (p. 553): Events for which the occurrence of the first affects that of the second.

diagonal (p. 256): A line segment that joins two non-consecutive vertices.

diameter (p. 440): Any chord that passes through the center of a circle.

dilation (p. 713): The image of a figure similar to the original figure.

distributive property (p. 52): Each term inside a set of parentheses can be multiplied by a factor outside the parentheses. For example,
$3(80 + 10) = 3(80) + 3(10)$.

domain of a function (p. 613): The set of all possible values of x in a given function.

══════════ e ══════════

edge (p. 628): A line segment on a space figure where two faces intersect.

endpoint (p. 237): Point included in a line segment or ray that defines that line segment or ray.

enlargement (p. 713): The image of a figure after a dilation that is larger than the original.

equally likely (p. 541): Having the same likelihood.

equation (p. 142): A statement that two numbers or two expressions are equal.

equilateral triangle (p. 261): A triangle whose sides all have the same length.

equivalent fractions (p. 302): Fractions that represent the same amount.

evaluate (p. 4): To find the value of a variable expression when a number is substituted for the variable.

event (p. 541): In probability, any group of outcomes.

expected value of an event (p. 577): The sum of the products of each outcome and its probability.

experimental probability (p. 564): Probability based on collection of actual data.

exponent (p. 30): A number used to show how many times another number is used as a factor. In 5^3, 3 is the exponent.

exponential form (p. 30): A shortened form of a multiplication expression in which all the factors are the same. 2^4 is the exponential form of $2 \cdot 2 \cdot 2 \cdot 2$.

face (p. 628): The flat surface of a space figure.

factor (p. 290): When a whole number is divisible by a second whole number, the second number is a *factor* of the first.

factor a polynomial (p. 693): Express a polynomial whose coefficients are whole numbers as the product of other polynomials whose coefficients are whole numbers.

factorial (p. 559): A notation for the product of a number and all nonzero whole numbers less than the given number.

Fibonacci sequence (p. 62): A pattern in nature where, beginning with the third number, 2, each number is the sum of the two numbers immediately preceding it. 1, 1, 2, 3, 5, 8, 13, 21, ...

first quartile of a set of data (p. 505): The median of the lower part of a set of data divided by its median.

formula (p. 176): An equation that states a relationship between two or more quantities.

45°-45°-90° right triangle (p. 726): An isosceles right triangle.

frequency polygon (p. 501): A type of line graph used to show frequencies.

frequency table (p. 221): A form of organizing a set of data to show how often each item in the data occurs.

function (p. 66): A relationship that pairs each number in a given set of numbers with *exactly one* number in a second set of numbers.

function notation (p. 613): A useful shorthand for showing how values are paired in a function.

function rule (p. 66): The description of a function.

function table (p. 66): A table used to show values of the variable expression for a given function.

geometric transformation (p. 714): Process of changing the position, size, or shape of a figure according to a given rule.

gram (p. 76): The basic unit of measure for mass in the metric system.

graph (p. 188): A picture that displays numerical facts, called data.

graph of an equation with two variables (p. 585): All points whose coordinates are solutions of an equation.

graph of a function (p. 128): The points that correspond to all the ordered pairs of a function.

graph of an open sentence (p. 324): A graph of all the solutions on a number line.

graph of a point (p. 124): The point assigned to an ordered pair on a coordinate plane.

greatest common factor (GCF) (p. 293): The greatest number in a list of common factors.

greatest common monomial factor of a polynomial (p. 693): The GCF of the terms of the polynomial.

height (p. 446): The perpendicular distance between bases of a parallelogram.

histogram (p. 501): A type of bar graph that is used to show frequencies.

hypotenuse (p. 481): In a right triangle, the side opposite the right angle.

identity property of addition (p. 23): The sum of any number and zero is the original number.

$$a + 0 = a$$

identity property of multiplication (p. 25): The product of any number and 1 is the original number.

$$a \cdot 1 = a$$

image (p. 706): A figure moved to a new location by rigid motion.

increasing trend (p. 196): An increase in the data over a given interval on a line graph, shown by a series of segments that slopes upward.

independent events (p. 552): Events for which the occurrence of the first event does not affect that of the second.

indirect measurement (p. 471): Finding an unknown measure that cannot be measured easily by direct measurement, by using similar triangles.

inequality (p. 325): A mathematical sentence that has an inequality symbol between two numbers or quantities.

inequality symbols (p. 18): The symbols > (is greater than) and < (is less than).

integer (p. 94): Any number in the set {. . ., -3, -2, -1, 0, 1, 2, 3, . . .}.

intersect (p. 236): To meet. Two lines intersect in a point. Two distinct planes intersect in a line.

inverse operations (p. 149): Operations that undo each other. Addition and subtraction are inverse operations; multiplication and division are inverse operations.

irrational number (p. 321): A number that cannot be written as the quotient of two integers.

isosceles triangle (p. 261): A triangle that has at least two sides of the same length.

k

key to a pictograph (p. 188): The part of the graph that tells how many items one symbol represents.

l

leaf (p. 497): The last digit on the right of a number displayed in a stem-and-leaf plot.

least common denominator (LCD) (p. 310): The least common multiple of the denominators when comparing fractions.

least common multiple (LCM) (p. 296): For two numbers, the least number in the list of their common multiples.

legend (p. 192): An identifying label on a double bar or double line graph.

legs (p. 481): The two sides of a right triangle that are shorter than the hypotenuse.

like terms (p. 56): Terms having identical variable parts.

line (p. 236): A straight arrangement of points that extends forever in opposite directions.

line graph (p. 196): A type of graph using points and line segments to show both amount and direction of change.

line segment (p. 237): A part of a line that consists of two points, called *endpoints*, and all the points between.

line symmetry (p. 275): The property of a pattern or geometric figure that "folds" along a line such that one half fits exactly over the other. The line on which it folds is called a *line of symmetry*.

liter (p. 76): The basic unit of measure for liquid capacity in the metric system.

loss (p. 224): The difference between cost and revenue when cost is greater than revenue.

lowest terms (p. 303): A fraction is in *lowest terms* if the GCF of the numerator and the denominator is 1.

m

mean (p. 213): The sum of the items in a set of data divided by the number of items; also called *average*.

measures of central tendency (p. 218): The mean, median, and mode of a set of data.

median (p. 213): The middle item in a set of data listed in numerical order; for an even number of items, the average of the two middle items.

meter (p. 76): The basic unit of measure for length in the metric system.

mode (p. 213): The item that appears most often in a set of data. A set of data can have more than one mode or no mode.

monomial (p. 668): A polynomial that has only one term.

multiple (p. 296): The product of a number and a non-zero whole number is a multiple of the given number.

multiplication property of −1 (p. 116): The product of any number and −1 is the opposite of the number.

$$-1n = -n \text{ and } -n = -1n$$

multiplication property of reciprocals (p. 340): The product of a number and its reciprocal is 1.

$$\frac{a}{b} \cdot \frac{b}{a} = 1, a \neq 0, b \neq 0$$

multiplication property of zero (p. 26): The product of any number and zero is zero.

$$a \cdot 0 = 0$$

multiplicative identity (p. 25): The number 1.

multiplicative inverse (p. 340): Reciprocal.

n

negative correlation (p. 509): The relation between sets of data in a scattergram when the trendline slopes downward.

negative integer (p. 94): Any integer less than zero.

number cube (p. 540): A cube with sides numbered 1 through 6.

o

obtuse angle (p. 245): An angle whose measure is greater than 90° and less than 180°.

obtuse triangle (p. 262): A triangle that has one obtuse angle.

odds in favor of an event (p. 542): The ratio of the number of favorable outcomes to the number of unfavorable outcomes of an event.

open sentence (p. 324): A mathematical sentence that contains a variable.

opposite side (p. 716): In a right triangle, the side opposite a given angle.

opposites (p. 94): Numbers that are the same distance from zero, but on opposite sides of zero on the number line.

order of operations (p. 36): An agreed-upon order of performing the operations in an expression that involves more than one operation.

ordered pair (p. 124): A pair of numbers assigned to any point on a coordinate plane. The first number is the *x-coordinate,* and the second number is the *y-coordinate.*

origin (p. 124): The point where the axes meet in a coordinate plane.

outcomes (p. 541): In probability, possible happenings, each of which is equally likely to happen.

p

parallel lines (p. 249): Two lines in the same plane that do not intersect.

parallelogram (p. 266): A quadrilateral with two pairs of parallel sides.

percent (p. 404): A ratio that compares a number to 100. The symbol % is read "percent."

percent of decrease (p. 419): The percent the amount of decrease is of the original amount. In business, often called the *discount.*

percent of increase (p. 419): The percent the amount of increase is of the original amount.

perfect square (p. 476): A number whose square root is an integer.

perimeter of a polygon (p. 436): The sum of the lengths of all the sides of the polygon.

permutation (p. 559): An arrangement of a group of things in a particular order.

perpendicular lines (p. 249): Two lines that intersect to form right angles.

pictograph (p. 188): A graph in which a symbol is used to represent a given number of items.

plane (p. 236): A flat surface that extends forever, usually represented by a four-sided figure.

point (p. 236): An exact location in space, represented by a dot.

polygon (p. 256): A closed figure formed by joining three or more line segments in a plane at their endpoints, with each line segment joining exactly two others.

polyhedron (p. 628): A space figure whose faces are polygons.

polynomial (p. 668): A variable expression that consists of one or more terms.

positive correlation (p. 509): The relation between sets of data in a scattergram when the trend line slopes upward.

positive integer (p. 94): Any integer greater than zero.

power (p. 30): The product when a number is multiplied by itself a given number of times. 64, or $4 \cdot 4 \cdot 4$, is the third power of 4.

power of a power rule (p. 89): To find the power of a power, multiply the exponents.

$$(a^m)^n = a^{mn}$$

prime factorization (p. 290): A number written as a product of prime numbers.

prime number (p. 290): A whole number greater than 1 with exactly two factors, 1 and the number itself.

prism (p. 628): A polyhedron with two parallel congruent bases.

probability of an event (p. 541): The ratio of the number of favorable outcomes to the number of possible outcomes of an event.

product of powers rule (p. 48): To multiply powers having the same base, add the exponents.

$$a^m \cdot a^n = a^{m+n}$$

profit (p. 224): The difference between revenue and cost when revenue is greater than cost.

proportion (p. 391): A statement that two ratios are equal.

protractor (p. 240): A geometric tool used to measure an angle.

pyramid (p. 628): A polyhedron with one base and triangular faces.

Pythagorean Theorem (p. 481): If the length of the hypotenuse of a right triangle is c and the lengths of the legs are a and b, then $c^2 = a^2 + b^2$.

<hr>

q

quadrant (p. 124): A section of the coordinate plane.

quotient of powers rule (p. 306): To divide powers having the same base but different exponents, subtract the exponents.

$$\frac{a^m}{a^n} = a^{m-n}, a \neq 0$$

<hr>

r

radius of a circle (p. 440): A line segment whose two endpoints are the center of a circle and a point on the circle.

random sample (p. 568): A sample chosen at random from a larger group.

range (p. 213): The difference between the greatest and least values of the data in a given set of data.

range of a function (p. 613): The set of all possible values of y in a given function. Also referred to as the *values of the function*.

rate (p. 389): A ratio which compares two unlike quantities.

ratio (p. 388): A comparison of two numbers by division.

rational number (p. 321): A number that can be written as a quotient of two integers $\frac{a}{b}$, where b does not equal 0.

ray (p. 237): A part of a line that has one endpoint and extends forever in one direction.

real number (p. 321): Any number that is either rational or irrational.

reciprocals (p. 340): Two numbers whose product is 1.

rectangle (p. 266): A quadrilateral with four right angles.

reduction (p. 713): The image of a figure after a dilation that is smaller than the original.

reflection (p. 706): Flipping a figure across a line.

regular polygon (p. 257): A polygon in which all sides have the same length and all angles have the same measure.

repeating decimal (p. 313): A decimal written by dividing the numerator of a fraction by the denominator and resulting in a remainder that is not zero and a block of digits in the decimal that repeats.

revenue (p. 224): The amount of money collected.

rhombus (p. 266): A quadrilateral with four sides of equal length.

right angle (p. 245): An angle whose measure is equal to 90°.

right triangle (p. 262): A triangle that has one right angle.

rigid motion (p. 706): Movement of a figure that does not change its size and shape.

rotation (p. 710): Movement of a figure in a circular motion around a point.

rotational symmetry (p. 710): A figure that rotates onto itself has rotational symmetry.

<hr>

s

sample space (p. 545): All the possible outcomes of a probability experiment.

scale (p. 395): In a scale drawing, the ratio of the size of the drawing to the actual size of the object.

scale drawing (p. 395): A drawing that represents real objects.

scale of a graph (p. 192): On a graph, numbers along an axis that show what is represented by the distances between the grid lines.

scale model (p. 396): A model of an object with dimensions in proportion to those of the actual object it represents.

scalene triangle (p. 261): A triangle with no sides of the same length.

scattergram (p. 509): A display in which the relationship between two sets of data is shown. The data are represented by unconnected points.

scientific notation (p. 80): A number in scientific notation is written as a number that is at least one but less than ten multiplied by a power of ten.

sector (p. 513): A wedge that represents part of the data in a circle graph.

side of a polygon (p. 256): A line segment that joins two or more line segments to form a polygon.

sides of an angle (p. 237): *See* angle.

sine (p. 716): In right triangle ABC,

$$\sin A = \frac{\text{length of side opposite } \angle A}{\text{length of hypotenuse}} = \frac{a}{c}.$$

similar figures (p. 466): Geometric figures that have the same shape but not necessarily the same size.

simplify an expression (p. 48): To perform as many of the indicated operations as possible.

simplify a fraction (p. 306): Write the fraction in lowest terms.

skew lines (p. 252): Lines that do not lie in the same plane.

slope of a line (p. 588): The ratio of *rise* to *run* of a line, describing the steepness and direction of the line.

slope-intercept form of an equation (p. 588): An equation written in the form $y = mx + b$, where m represents the slope of the line and b represents the *y-intercept*.

solution of an equation (p. 142): A value of the variable that makes an equation true.

solution of an equation with two variables (p. 582): An ordered pair of numbers that makes the equation true.

solution of an inequality with two variables (p. 609): An ordered pair of numbers that makes the inequality true.

solution of an open sentence (p.324): Any value of a variable that results in a true sentence.

solution of a system of equations (p. 594): An ordered pair that is a solution of both equations of a system of equations.

solve (p. 148): To find all values of the variable that make an equation true.

solve a proportion (p. 392): Find the value of the variable that makes the proportion true.

space figure (p. 628): A three-dimensional figure that encloses part of space.

sphere (p. 629): The set of all points in space that are the same distance from a given point called the *center*.

square (p. 266): A quadrilateral with four right angles and four sides of equal length.

square root of a number (p. 475): If $a^2 = b$, the number a is called a *square root* of b.

standard form (p. 668): The form of a polynomial whose terms are written in order from the highest to the lowest power.

statistics (p. 213): The branch of mathematics that deals with collecting, organizing, and analyzing data.

stem (p. 497): The digit or digits of the number remaining in a stem-and-leaf plot when the leaf is dropped.

stem-and-leaf plot (p. 497): A display of data where each number is represented by a *stem* and a *leaf*.

straight angle (p. 245): An angle whose measure is equal to 180°.

supplementary angles (p. 245): Two angles the sum of whose measures is 180°.

surface area of a cylinder (p. 638): The sum of the areas of the bases and the curved surface, expressed in square units.

surface area of a prism (S.A.) (p. 634): The sum of the areas of the bases and faces of the prism, expressed in square units.

system of equations (p. 594): Two equations with the same variables.

---

t

tangent (p. 716): In right triangle ABC,

$$\tan A = \frac{\text{length of side opposite } \angle A}{\text{length of side adjacent to } \angle A} = \frac{a}{b}.$$

terminating decimal (p. 313): A decimal written by dividing the numerator of a fraction by the denominator and resulting in a remainder of zero.

terms of a proportion (p. 391): The numbers or variables in a proportion.

tessellation (p. 285): A pattern in which identical copies of a figure cover a plane without gaps or overlaps.

third quartile of a set of data (p. 505): The median of the upper part of a set of data divided by its median.

30°-60°-90° right triangle (p. 727): A right triangle whose acute angles are 30° and 60°.

transform a formula (p. 375): To solve a formula for a particular variable by using inverse operations.

translation (p. 706): Sliding a figure from one location to another.

transversal (p. 250): A line that intersects two or more lines in the same plane at different points.

trapezoid (p. 266): A quadrilateral with exactly one pair of parallel sides.

tree diagram (p. 545): A representation of all the possible outcomes in a sample space.

trend line (p. 509): A line drawn near the points on a scattergram.

Triangle Inequality (p. 261): In any triangle, the sum of the lengths of any two sides is greater than the length of the third side.

trigonometric ratios (p. 716): Ratios of specific sides of a right triangle. *See* sine, cosine, and tangent.

trinomial (p. 668): A polynomial that has three terms.

u

unit rate (p. 389): A rate for one unit of a given quantity.

unlike terms (p. 56): Terms having different variable parts.

v

values of a function (p. 613): The set of all possible values of y for a function.

value of a variable (p. 4): Any number that is substituted for a variable.

variable (p. 4): A symbol that represents a number.

variable expression (p. 4): An expression that contains a variable.

Venn diagram (p. 556): A diagram that shows the relationships between collections of objects.

vertex of an angle (p. 237): *See* angle.

vertex of a polygon (p. 256): A point where two sides of a polygon meet.

vertex of a space figure (p. 628): A point where edges of a space figure intersect.

vertical angles (p. 246): The angles that are not adjacent to each other, equal in measure, formed by the intersection of two lines.

volume of a space figure (p. 642): The amount of space enclosed by a space figure, measured in cubic units.

w

whole numbers (p. 137): The numbers in the set $\{0, 1, 2, 3, \ldots\}$.

x

x-axis (p. 124): The horizontal number line in a coordinate plane.

x-coordinate (p. 124): The first number of an ordered pair.

x-intercept (p. 588): The x-coordinate of the point where a graph crosses the x-axis.

y

y-axis (p. 124): The vertical number line in a coordinate plane.

y-coordinate (p. 124): The second number of an ordered pair.

y-intercept (p. 588): The y-coordinate of the point where a graph crosses the y-axis.

Credits

Cover: Concept by Martucci Studios; Photographic Special Effect Illustration by VISUAL CONSPIRACY/Martin Stein.

Technical art: Typographic Sales, Inc.; LeGwin Associates

Illustrations:

Neil Pinchin Design: xii, xiii, 122–123, 172, 188, 189, 190, 191, 192, 195, 199, 200, 201, 207, 215, 226, 228, 249, 253, 311, 312, 388, 390, 404, 419, 494, 501, 503, 513, 514, 515, 516, 517, 541, 564, 650

Mark Goldman and Charles Shields: xviii, 1, 6, 7, 9, 11, 21, 33, 61, 94, 97, 120, 121, 132, 159, 206, 215, 221, 324, 346, 352, 586, 591, 598, 618

Victor Ambrus: 10, 55, 127, 175, 208, 239, 308, 374, 414, 456, 500, 555, 608, 653, 692

PHOTOGRAPHS

xiv Comstock.   **xiv** Andrew Sacks/Tony Stone Worldwide.   **xiv** Texas Instruments.   **xv** Texas Instruments.   **xvi** The Telegraph Colour Library/F.P.G.   **2–3** Michael Simpson/F.P.G.   **4** Michael Newman/Photo Edit.   **7** Sepp Seitz/Woodfin Camp and Associates.   **13** Lawrence Migdale/Stock Boston.   **14** © Paul Conklin.   **16** Bill Bachman/Stock Boston.   **20** Hugh Rogers/Monkmeyer Press Photos.   **21** © Dave Schaefer.   **24** © Susan Van Etten.   **29** © Susan Van Etten.   **34** Sekai Bunka.   **36** © Susan Van Etten.   **38** Alvis Upitis/The Image Bank.   **43** Jeffry W. Myers/Stock Boston.   **46–47** Tony Stone Worldwide.   **50** Steve Dunwell/The Image Bank.   **56** Eric Roth/The Picture Cube.   **58** © Bob Daemmrich.   **62** © Susan Van Etten.   **64** © Bob Daemmrich.   **68** © Bob Daemmrich.   **69** E.R. Degginger/Animals, Animals.   **72** © Bob Daemmrich.   **75** Dr. E.R. Degginger.   **79** © Bob Daemmrich.   **82** Scott Berner/Photri.   **85** Comstock.   **92–93** Beryl Bidwell/Tony Stone Worldwide.   **96** Herbert Lanks/Monkmeyer Press Photos.   **100** © Paul Conklin.   **104** N.A.S.A.   **109** Chin Wai-Lan/The Image Bank.   **110** Lawrence Migdale/Photo Edit.   **113** © Bob Daemmrich.   **119** Barrie Rokeach/The Image Bank.   **122–123** L.L.T. Rhodes/earth scenes; The Granger Collection; Gerard Champlong/The Image Bank; The Granger Collection; The Granger Collection.   **125** Minnesota Office of Tourism.   **133** Peter Miller/The Image Bank.   **140–141** Image Works.   **142** © Bob Daemmrich.   **146** MacDonald Photography/Photri.   **151** Lawrence Migdale/Photo Researchers.   **154** N.A.S.A./Peter Arnold, Inc.   **159** Ron Grishaber/Photo Edit.   **160–161** The Bettman Archive; UPI/Bettman; The Granger Collection; The Granger Collection; © Susan Van Etten; N.A.S.A.; Allen Green/Photo Researchers, Inc.; The Granger Collection; The Granger Collection.   **162** Mike Mazzaschi/Stock Boston.   **164** George Zimbel/Monkmeyer Press Photos.   **166** © Bob Daemmrich.   **170** © Bob Daemmrich.   **178** Miro Vintoniv/The Picture Cube.   **186–187** Michael Melford/The Image Bank.   **193** Michal Heron/Monkmeyer Press Photos.   **196** Kindra Clineff.   **217** Kindra Clineff.   **219** Jeffrey M. Myers/Stock Boston.   **221** © Bob Daemmrich.   **223** © Bob Daemmrich.   **225** © Bob Daemmrich.   **231** © Susan Van Etten.   **234–235** Jim Larsen/West Stock.   **237** Camerique.   **240** Bob Daemmrich/Stock Boston.   **257** Bohdan Hrynewych/Stock Boston.   **261** Bill Gallery/Stock Boston.   **264** Bob Crandall/Stock Boston.   **272** Runk/Schoenberger from Grant Heilman.   **276** Kristian Hilsen/Tony Stone Worldwide.   **276** Bill Gallery/Stock Boston.   **278** Odyssey/Frerck/Chicago.   **285** Martin Rogers/Tony Stone Worldwide.   **288–289** Tom Tracy/The Stock Market.   **293** Larry Lefever from Grant Heilman.   **296** Richard Wood/The Picture Cube.   **302** © Bob Daemmrich.   **309** © Susan Van Etten.   **315** Dan Burns/Monkmeyer Press Photos.   **316** Larry Lefever from Grant Heilman.   **318** © Bob Daemmrich.   **319** Spencer Grant/Stock Boston.   **323** The University Museum, University of Pennsylvania.   **331** © Bob Daemmrich.   **338–339** Ted Kurihara/Ted Kurihara Studio.   **343** Larry Lefever from Grant Heilman.   **347** © Bob Daemmrich.   **349** Martin Rogers/Stock Boston.   **352** Jim Harrison/Stock Boston.   **355** Kindra Clineff.   **359** Stan Osolinski/F.P.G.   **360–361** Bob Daemmrich/Stock Boston; Bob Daemmrich; Bob Daemmrich/Uniphoto Picture Agency; The Granger Collection; The Granger Collection.   **363** Cathlyn Melloan/Tony Stone Worldwide.   **365** Mark Scott/F.P.G.   **367** Bob Daemmrich/The Image Works.   **370** Odyssey/Frerck/Chicago.   **372** Barry L. Runk from Grant Heilman.   **377** Joseph Schuyler/Stock Boston.   **379** Steve Leonard/Tony Stone Worldwide.   **386–387** International Stock Photography.   **394** © Susan Van Etten.   **398** Ida Wyman/Monkmeyer Press Photos.   **400** Gans/The Image Works.   **411** Mavournea Hay/Daemmrich Associates.   **417** Dennis Degnan/Uniphoto.   **418** © Bob Daemmrich.   **421** Steven E. Sutton/duomo.   **422–423** John Williamson; Jack Green/Horizon; Patti Murray/Animals Animals; © Susan Van Etten; © Susan Van Etten.   **426** Joseph Nettis/Stock Boston.   **434–435** G.K. & Vicky HA/The Image Bank.   **440** © Bob Daemmrich.   **444** Michael Grecco/Stock Boston.   **449** © Bob Daemmrich.   **454** Don Kelly from Grant Heilman.   **457** © Chris Malazorewicz/Valan Photos.   **461** Martucci Studios.   **463** Stephanie Dinkins/Photo Researchers, Inc.   **466** Peter Menzel/Stock Boston.   **471** Keith Philpott/Stockphotos, Inc.   **474** © Bob Daemmrich.   **479** Scala/Art Resource.   **484** George Holton/Photo Researchers, Inc.   **492–493** Walter Bibikow/The Image Bank.   **496** Tony Freeman/Photo Edit.   **499** © Bob Daemmrich.   **503** Roger Dollarhide/Monkmeyer Press Photos.   **508** Bob Daemmrich/Stock Boston.   **512** Kennedy/TexaStock.   **516** © Bob Daemmrich.   **520** Richard Pasley/Stock Boston.   **523** Freda Leinwand/Monkmeyer Press Photos.   **526** Lee Balterman/The Picture Cube.   **529** Bob Daemmrich/Stock Boston.   **538–539** Stock Imagery/Tom Walker.   **547** © Bob Daemmrich.   **551** Bruce M. Wellman/Stock Boston.   **554** Vernon Doucette/Stock Boston.   **559** © Bob Daemmrich.   **565** © Bob Daemmrich.   **569** Richard Pasley/Stock Boston.   **570** Palmer/Kane/The Stock Market.   **577** Carey/The Image Works.   **580–581** Robert Kristofi/The Image Bank.   **584** Peter Frank/Tony Stone Worldwide.   **591** © Bob Daemmrich.   **599** Robert Frerck/Tony Stone Worldwide.   **601** Lowell J. Georgia/Photo Researchers.   **605** Stan Osolinski/F.P.G.   **606** Stephen Frisch/Stock Boston.   **615** © Susan Van Etten.   **618** Barbara Alper/Stock Boston.   **626–627** Joseph Drivas/The Image Bank.   **631** © M. Greenbar/The Image Works.   **634** © W. Hill/The Image Works.   **640** Bob Daemmrich/Stock Boston.   **644** Paul Mozell/Stock Boston.   **649** E.R. Degginger/Animals, Animals.   **652** Stephen Frisch/Stock Boston.   **656** © Bob Daemmrich.   **659** Larry Lefever from Grant Heilman.   **666–667** John Martucci/Martucci Studios, Inc.   **670** Scala/Art Resource.   **675** © Susan Van Etten.   **679** © Bob Daemmrich.   **684** © Susan Van Etten.   **687** E. Alan McGee/F.P.G.   **689** John Running/Stock Boston.   **691** Jim Pickerell/Tony Stone Worldwide.   **698** DMR/The Image Works.   **706** © Susan Van Etten.   **706** Martin Rogers/Tony Stone Worldwide.

Answers to Check Your Understanding

CHAPTER 1

Page 4 **1.** You would substitute 20 for n. **2.** You would evaluate the expression $6n$. **Page 7** **1.** You estimate before adding to see about what your actual answer should be. **2.** The sum of 87 and 16.49 is close to 100. **3.** You write 287 as 287.00 to give it the same number of decimal places as 116.49, which aids in aligning the decimal points. **Pages 11–12** **1.** The sum of the numbers of the decimal places of the factors is 1. Thus, the number of decimal places in the product is 1. **2.** You place the decimal point in the quotient directly over the decimal point in the dividend. You can also place the decimal point in the quotient by using your estimate of 3. **Pages 18–19** **1.** When you compare the numbers place-by-place from left to right, the digits 7 and 8 are the first digits that differ. **2.** Example 1(b) involves decimals. The numbers in Example 1(a) have the same number of digits, while the numbers in Example 1(b) do not. **3.** When you compare 0.47 and 0.4 place-by-place from left to right, the digits are the same in all places through the tenths' place. You need to write 0.4 as 0.40 so that you can compare the digits in the hundredths' place. **4.** $0.47 > 0.4 > 0.247$ **Page 23** **1.** identity property of addition **2.** The sum of the ones' digits is 10, so the sum of $76 + 14 = 90$ is easy to find. **Page 26** **1.** multiplication property of zero; identity property of multiplication **2.** The product of 25 and 4 is 100, and it is easier to multiply 13×100 mentally. **Page 28** **1.** Paper and pencil; there is only one renaming involved. **Page 30** **1.** The expression 6^3 means that 6 is used as a factor three times. The expression $6 \cdot 3$ means that 6 and 3 are the factors. **2.** The expression 6^3 means that 6 is used as a factor three times. The expression 3^6 means that 3 is used as a factor 6 times. **3.** The number 1 to any power equals 1. **Pages 36–37** **1.** You perform multiplication and division in order from left to right. In Example 1(a) the division occurs before the multiplication when you work from left to right. **2.** You would first add within parentheses ($30 + 24 = 54$), then divide ($54 \div 6 = 9$), and then multiply ($9 \cdot 2 = 18$). **3.** $(24 + 12) \div (13 - 4)$ **4.** You would first add within parentheses ($8 + 4 = 12$), then square the sum ($12^2 = 144$), then divide ($144 \div 3 = 48$), and then add ($48 + 5 = 53$). **5.** You would add within the first set of parentheses ($8 + 4 = 12$), then add within the second set of parentheses ($3 + 5 = 8$), then do the power ($12^2 = 144$), and then divide ($144 \div 8 = 18$).

CHAPTER 2

Page 48 Any number to the first power is equal to that number. **Pages 52–53** **1.** It is easier to use mental math to multiply $7(100 + 8)$ than $7(110 - 2)$. **2.** $7(98) = 7(100 - 2) = 7(100) - 7(2) = 700 - 14 = 686$ **3.** They show that you multiply each term inside a set of parentheses by the factor outside the parentheses when you use the distributive property. **Page 57** **1.** They are the only like terms. **2.** identity property of multiplication **Pages 62–63** **1.** 384, 768 **2.** 8; 40 **3.** 5 **Page 66** **1.** 46 **2.** 496 **Page 74** It is easier to multiply 3×20 than 3×18. **Pages 76–77** **1.** Because a liter is three places above a milliliter in the chart, a liter is $10 \cdot 10 \cdot 10$ times as large as a milliliter. So, you multiply by $10 \cdot 10 \cdot 10$, or 1000, to change from liters to milliliters. **2.** 3 places **Pages 80–81** **1.** You multiply by 10^4 because you need to move the decimal point 4 places to the left. **2.** You move the decimal point 6 places because you are multiplying by 10 to the sixth power. **3.** You would move the decimal point 2 places to the right rather than 5 places to the right. $7.16 \cdot 10^2 = 716$

CHAPTER 3

Page 95 **1.** -3 **2.** 0 **Page 98** **1.** They represent negative numbers. **2.** You would first slide 4 units left and then slide 3 more units left. The sum would be the same. **Page 102** **1.** The positive integer has the greater absolute value, so the sum would be positive. **2.** $|3| = 3$ and $|-3| = 3$. Subtract: $3 - 3 = 0$. The sum is 0. **Pages 107–108** **1.** Two negative chips remain. **2.** Subtracting an integer is the same as adding its opposite. **Page 112** **1.** The factors -20 and -2 have the same sign. The product of two integers with the same sign is positive. **2.** The quotient of two integers with different signs is negative. **Pages 115–116** **1.** $(-4)^3 = (-4)(-4)(-4) = 16(-4) = -64$ **2.** Absolute value signs have priority over addition in the order of operations, so you first find the absolute values of -9 and 4, and then add the absolute values. **3.** The opposite of a number is the product of the number and -1. **Pages 124–125** **1.** Point E is to the left of the origin. **2.** Point F is 0 units up or down from the origin. **3.** Point D is 0 units to the left or right of the origin. **4.** Point A would be below the x-axis. **Pages 128–129** **1.** The x-coordinates are the values of x in the x-column of the table. The y-coordinates are the values obtained from using the function rule. **2.** You use the values given for x.

CHAPTER 4

Pages 142–143 **1.** The two sides of the equation are not equal. **2.** What number minus 6 equals 14? Four times what number equals 20? **3.** The first question would be "What number plus 6 equals 14?" **Pages 148–149** **1.** to keep both sides of the equation equal **2.** You would need to undo the addition. **Page 152** **1.** You need to be able to divide both sides by -1. **2.** You would not need to use the multiplication property of -1. You would just divide both sides by -3. **Pages 156–157** **1.** You would substitute 3 for n in the original equation. **2.** division and subtraction **3.** adding 4 to both sides, then multiplying both sides by 3 **Page 162** **1.** Yes; by the commutative property of addition. **2.** No; subtraction is not commutative. **Page 165 1.** is **2.** The cost of the financial package is given. **Pages 172–173 1.** $4x$ and $3x$ **2.** to simplify the left side of the equation **Page 176 1.** You substitute 240 for D and 40 for r in the formula $D = rt$. **2.** The equation would be solved for r rather than t.

CHAPTER 5

Pages 188–189 1. The number of symbols would be multiplied by 500 instead of by 150. **2.** Find the number of chocolate cones and the number of orange cones sold. Then subtract. **3.** You are rounding to the nearest half-million rather than to the nearest whole million, and 5.7 is closer to 5.5 than to 6. **Page 192** Estimate the heights of the bars for jazz and hard rock. Then subtract. **Page 197** You would locate the point for 1982 on the *red* line. **Page 200** Using intervals of five makes it easier to represent the data accurately. **Page 204** Yes; average attendance changes continuously with each game throughout the baseball season. **Page 209** The attendance in Midway appears to be about 6 times the attendance in Sunville. **Page 213 1.** The 6 represents the number of bowling scores. **2.** You must average the two middle scores because there is an even number of data items. **Page 218** Find the sum of the salaries ($252,000) and then divide by the number of salaries (7). **Page 221 1.** The total of the frequencies is 29. **2.** Count the tally marks until you reach the fifteenth. Then find the rate that corresponds to that tally mark.

CHAPTER 6

Page 237 1. Point B is an endpoint and must come first. **2.** No; point J is not the vertex of the angle. **Pages 240–241 1.** $\overrightarrow{BC}$ coincides with the 0° mark on the bottom scale. **2.** You would use the top scale instead of the bottom scale. **Page 246 1.** You would subtract the given measure from 180°. **2.** No; only acute angles have complements. **3.** $\angle 2$ and $\angle 4$; $\angle 1$ and $\angle 3$ **4.** $\angle 1$ and $\angle 2$; $\angle 2$ and $\angle 3$; $\angle 1$ and $\angle 4$ **Page 250 1.** Four: the right angle, the angle vertical to it, and its two supplements **2.** $\angle 7$; 47° **Page 258 1.** 6; the polygon has six sides. **2.** the sum of the known measures **Pages 262–263 1.** Change the 9 m side to 6 m. **2.** the sum of the measures of the angles of a triangle **3.** It has one obtuse angle, 112°. **Page 267 1.** Yes; $\angle DAC$ and $\angle ACB$ form a pair of alternate interior angles, so they are equal in measure. **2.** The sum of the measures of the angles of any triangle is 180°. **Page 275 1.** When folded, one half does not fit exactly over the other half. **2.** No.

CHAPTER 7

Page 290 1. Two **2.** 2 and 5 are factors of both 10 and 140. **Pages 293–294 1.** $4 > 2$ **2.** 5 and 7 are not common factors. **3.** a^4 is the least power of each com-

mon variable factor. **Pages 296–297 1.** No; the multiples continue infinitely. **2.** You need to use the greatest power. **3.** You need to include the greatest power of each variable factor. **Page 302 1.** Each exercise involves finding equivalent fractions; in 1(a) the equivalent fraction had more parts, while in 1(b) the equivalent fraction has fewer parts. **2.** Both calculations are used to find equivalent fractions. **Page 307 1.** The solution would be $5x^3$. **2.** Writing d as d^1 makes it easier to subtract the exponents. **Pages 309–310 1.** The one-fourth shaded part is smaller than the one-third shaded part. **2.** $12 = 2^2 \cdot 3$ and $18 = 2 \cdot 3^2$, so $2^2 \cdot 3^2$, or 36, is the LCM. **3.** Because you multiply 12 by 3 to get 36. **Pages 313–314 1.** Both numbers repeat. **2.** 0.6363636 **3.** You could write $\frac{3}{8}$ as 0.375 and add 6 to get 6.375. **4.** The GCF of 111 and 200 is 1. **Page 321 1.** Yes; $\frac{16}{-1}$ is equal to -16. **2.** No; $\frac{-16}{-1}$ is equal to 16. **Pages 324–325 1.** No. **2.** Because -4.2 is not greater than -4.2. **3.** You would shade in a heavy arrow to the left to graph all numbers less than -4.2. **4.** Because $\frac{3}{4} = \frac{3}{4}$. **5.** You would shade in a heavy arrow to the right to graph all numbers greater than $\frac{3}{4}$. **Page 329 1.** 6.5×10^{-1} **2.** 47,000.

CHAPTER 8

Page 341 1. All numbers must be written as fractions when multiplying by a fraction. **2.** Multiplying a positive number by a negative number results in a negative number. **3.** The answer would be positive. **4.** $\frac{1}{3}$ is less than $\frac{1}{2}$. **5.** -6 is the closest whole number that is compatible with $\frac{-1}{2}$. **Page 344 1.** Yes; the answer would be $\frac{1}{10}$. **Page 348 1.** Because the question is about money, $1.\overline{3}$ is rounded to the nearest cent. **2.** You need to round up because seven tables aren't enough to seat everyone. **Pages 352–353 1.** The mixed number $-1\frac{1}{5}$ is written as a fraction. **2.** Subtracting the numerators results in -40. **3.** Instead of -40, the numerator would have become -14. **4.** The numerator and the denominator were divided by 3. **Pages 356–357 1.** The fractions would be $\frac{-18}{16}$ and $\frac{44}{16}$; $\frac{-18-44}{16} = \frac{-62}{16} = \frac{-31}{8} = -3\frac{7}{8}$. **2.** Two negatives are added. **3.** $|-3.8| > |-1.48|$ **Page 362** If the denominator is greater than twice the numerator, then the fraction is less than $\frac{1}{2}$. **Page 368 1.** $\frac{(-3)(-2)}{3} = \frac{6}{3} = 2$ **2.** $\frac{2}{4}$ **3.** When subtracting rational numbers you must rewrite mixed numbers as fractions. **Page 371 1.** You multiply the right side, $\frac{-5}{6}b$, by the reciprocal $\frac{-6}{5}$ because you need to solve for b; you multiply the left side by $\frac{-6}{5}$ because you need to balance the equation. **2.** When multiplying by a fraction, you should write each number as a fraction.   **Page 375**   **1.** Each side of the equation must be multiplied by $\frac{5}{9}$, and the left side of the equation is $F - 32$.   **2.** 325° is given in degrees Fahrenheit.

CHAPTER 9
Page 388   **1.** No; the ratio of wins to losses is $\frac{2}{1}$, and the ratio of losses to wins is $\frac{1}{2}$.   **2.** Yes; in lowest terms they are equal.   **Pages 391–392**   **1.** Yes; then the cross products will be equal.   **2.** The cross products would still be 60 and 4b, and $b = 15$.   **Pages 395–396**   **1.** The length of the tray return is asked for.   **2.** Yes; the cross products and solutions would not change.   **Pages 404–405**   **1.** Per cent means "per hundred."   **2.** 0.375   **3.** The decimal point moves two places to the left.   **4.** In order to move the decimal point, you must express the fraction as a decimal.   **5.** so that $\frac{3.6}{100}$ can be expressed as a fraction which can be reduced to lowest terms   **Pages 408–409**   **1.** $\frac{30}{100}$ or $\frac{3}{10}$ **2.** 64% is a ratio that compares 64 to 100, and $\frac{64}{100} = 0.64$.   **3.** $\frac{16}{25}$ **4.** 65% is close to $66\frac{2}{3}$%, which is $\frac{2}{3}$.   **Page 412**   **1.** in order to express it as a percent   **2.** Divide 12.5 by 500; the result is 0.025.   **Page 415**   **1.** more than 126; 60% is part of the whole.   **2.** It is easier to write $66\frac{2}{3}$% as a fraction than as a decimal.   **3.** To solve $120 = \frac{2}{3}n$, multiply both sides by the reciprocal of $\frac{2}{3}$.   **Page 419**   **1.** Percent of increase $= \frac{\text{amount of increase}}{\text{original amount}}$, and 10 is the original amount.   **Page 424**   **1.** No; the cross products will be $4n = 56$, which results in $n = 14$.   **2.** The question asks for a percent.

CHAPTER 10
Pages 436–437   **1.** Another variable would be needed to represent the sixth side; $P = a + b + c + d + e + f$.   **2.** Both figures are pentagons; the pentagon in Example 1 is not a regular pentagon.   **3.** The formula would be 8n.   **Pages 440–442**   **1.** It fulfills both definitions.   **2.** The value for the diameter, 21, is divisible by 7.   **3.** You would use the formula $C = 2\pi r$.   **4.** The circumference is expressed in decimal form.   **5.** You would use the formula $C = \pi d$.   **Pages 446–447**   **1.** You would use the formula $A = s^2$; $A = (3\frac{1}{2})^2 = 12\frac{1}{4}$.   **2.** The solution would be 2750 mm²; both answers are correct.   **3.** No.   **Page 453**   **1.** The radius is expressed as a decimal.   **2.** The radius would be 2.2 m; $A \approx 3.14 \, (2.2)^2 \approx 15.2$ m².   **Page 457**   when finding the distance around something round or circular   **Page 463**   **1.** $\angle P \cong \angle C$, $\angle Q \cong B$, $\angle R \cong \angle D$, $\angle S \cong \angle A$; $\overline{PQ} \cong \overline{CB}$, $\overline{QR} \cong \overline{BD}$, $\overline{RS} \cong \overline{DA}$, $\overline{SP} \cong \overline{AC}$ **2.** The vertices must be listed in the same order as the corresponding congruent angles.

Page 467   **1.** 21   **2.** The ratio would be $\frac{16}{10}$ or $\frac{8}{5}$.
Pages 471–472   **1.** $\frac{2}{3}$; they are equal.   **2.** 27 ft
3. $\frac{12}{36} = \frac{64}{h}$, $\frac{1}{3} = \frac{64}{h}$, $h = 3 \, (64) = 192$ in.   **4.** You would use $5\frac{1}{2}$ instead of $5\frac{1}{3}$ and solve.
Pages 475–476   **1.** The square root of a negative number does not exist.   **2.** $\sqrt{46}$ does not exactly equal 6.782.   **3.** 45.995524; the result is less than 46 because 6.782 is less than $\sqrt{46}$.   **Page 482**   **1.** In part (a), the length of the hypotenuse was unknown; in part (b), the length of a leg was unknown.   **2.** Because c^2 is the sum of a^2 and b^2, c must be greater than both a and b; you should substitute the longest length for c.

CHAPTER 11
Pages 494   4   **Pages 497–498**   **1.** There would be a 6 in the 9 row.   **2.** You would count the numbers greater than 5 on the 0.73 stem and the numbers on stems 0.74 and 0.75. The total would be 5 teams.   **3.** Range equals greatest minus least value: $0.755 - 0.692 = 0.063$
Page 501   12 employees work 31–35 h and 20 employees work 36–40 h; $12 + 20 = 32$ **Page 506**   For each company find the range by subtracting the smallest value from the largest. Then compare the two ranges.
Page 509   **1.** halfway between 6 and 8   **2.** You would find the greatest and least test scores and subtract: $100 - 50$.   **Pages 513–514**   **1.** Both sides had to be divided by 0.3.   **2.** The amounts given in the graph are in thousands.   **Page 517**   The percent for each type equals $\frac{\text{type sales}}{\text{total sales}}$.   **Page 522**   Yes; you would have two sets of data to compare.

CHAPTER 12
Pages 541–542   **1.** None of the sectors is numbered 7.   **2.** 1, 2, 4, 5, 6   **3.** Each of the sectors is red, blue, or white, so one of these colors is certain to occur on each spin.   **4.** favorable: 3; unfavorable: 1, 2, 4, 5, 6   **5.** 1 to 1   **Page 545**   **1.** The sample space would be: HH, HT, TH, TT.   **2.** HHH, HHT, HTH, THH; $\frac{1}{2}$ **Pages 548–549**   **1.** 24   **2.** 30   **3.** $\frac{1}{20}$ **Pages 552–553**   **1.** The first marble drawn is replaced before the second marble is drawn.   **2.** There are twelve marbles in the bag, and five of them are white.   **3.** Drawing the first marble reduces the number of marbles left in the bag for the second drawing.   **4.** There are fewer marbles left in the bag after the first marble is drawn.   **Page 559**   **1.** MATH, MAHT, MTAH, MTHA, MHAT, MHTA   **2.** There are 4 letters in MATH, so find 4!   **Page 564**   **1.** the number of times the nickel landed on both red and black   **2.** A dime is smaller than a nickel, so it might land on a single color more easily.   **Page 568**   **1.** A poll taker finds the probability that a candidate will receive votes by polling a random sample of voters and multiplying this probability by the total number of voters.   **2.** about 170,500 voters

CHAPTER 13

Page 582 **1.** 9 **2.** -2 **Page 585** **1.** The values for y when x equals 3 and -3 are integers, making the solutions easy to graph. **2.** Yes.

Pages 588–589 **1.** The vertical movement is downward. **2.** (2, 1), (0, 5); yes. **3.** The slope-intercept form is $mx + b$. **4.** (0, -6) **Page 595** Yes. **Page 603** **1.** You only reverse the inequality symbol when you multiply or divide both sides by a negative number. **2.** When you multiply both sides of an equality by a negative number, you reverse the inequality symbol.

Pages 606–607 **1.** addition **2.** No; both sides were divided by a positive, not a negative number. **3.** Both sides are being divided by -16. **4.** There would be an open circle at -3. **Page 610** **1.** The region below the line would be shaded. **2.** Yes. **Pages 613–614** **1.** multiply **2.** find the power **3.** $|-3 - 2| = |-5| = 5$

Page 616 5

CHAPTER 14

Page 628 **1.** It has only one base. **2.** It has two bases and its sides are not triangular. **Page 635** **1.** The formula for the area of a triangle is $\frac{1}{2}bh$. **2.** The rectangles are not all the same size. **Page 638** **1.** The radius is a multiple of 7. **2.** the curved surface **Page 642** **1.** There would be 5 layers, so the volume would be $5 \cdot 8 = 40 \text{ cm}^3$. **2.** There would be $6 \cdot 2 = 12$ cubes in one layer, so the volume would be $3 \cdot 12 = 36 \text{ cm}^3$.

Page 646 **1.** If the height were 11 in., the calculation would be the same, but the answer would be approximately 7771.5 in.^3. **2.** The volume formula used would be $V = \frac{1}{3}Bh$, and the volume would be approximately $\frac{1}{3}(7771.5)$, or 2590.5 m^3. **3.** The radius, 14, is a multiple of 7. **Page 650** **1.** The formulas for surface area and volume use the radius, which is one half the diameter. **2.** the radius, 6, cubed

CHAPTER 15

Page 668 **1.** 1(a): 3, 2, 1; 1(b) 4, 3, 1 **2.** It has no variable factor. **3.** 3 is a higher power than 2. **4.** They

are not like terms. **Page 673** **1.** $x = 1x$; therefore, $x + (-5x) = 1x + (-5x) = -4x$ **2.** Adding the x^3 terms for each equation resulted in $0x^3$ or 0.

Pages 676–677 **1.** Because $-2n^2 - n - 5$ is the opposite of $2n^2 + n + 5$ **2.** There is no a term in the first polynomial. **Page 682** **1.** Subtracting is the same as adding the opposite. **2.** Each term of the polynomial was multiplied by $4x^2$. **Page 685** **1.** They are like terms. **2.** Yes. **3.** Subtracting a number is the same as adding its opposite, so $3x - 2 = 3x + (-2)$; when you multiply $3x + (-2)$ by $(7x + 5)$, you get $3x(7x + 5) + (-2)(7x + 5)$, or $3x(7x + 5) - 2(7x + 5)$. **4.** 1

Page 693 **1.** It is the GCF of the terms. **2.** The greatest common monomial factor is $3c^2d$. **3.** Multiply the polynomial by the monomial. **Page 696** **1.** When you divide powers with the same base, you subtract the exponents. **2.** The answer would be $2a^4 - 4a^3 + 3a$. **3.** Adding a number is the same as subtracting its opposite.

LOOKING AHEAD

Pages 706–707 **1.** The translation is horizontal. **2.** For example, move up two units. **3.** 1 unit; 1 unit **4.** 4 units; 4 units **5.** Yes. **Pages 710–711** **1.** $45°$ **2.** $45°$ **3.** 3 **4.** B **5.** C **6.** A **7.** B **Pages 713–714** **1.** $2x$ **2.** $\frac{1}{2}a$ **3.** $\frac{1}{2}x$ **4.** $2a$ **5.** They are congruent. **6.** $A'B' = 3AB$; $B'C' = 3BC$ **Pages 716–717** **1.** the hypotenuse **2.** Because a leg of a triangle is always shorter than the hypotenuse. **3.** $\frac{5}{12}, \frac{12}{13}$ **4.** 8; 15; $\frac{8}{15}$ **5.** Subtract 64 from both sides **6.** adjacent

Pages 719–721 **1.** $(18.9)^2 + (14.8)^2 = 24^2$; result will not be exact due to rounding **2.** You would find sin B and cos B. **3.** The hundredths' digit is 9, and $9 \geq 5$, so you add 1 to the tenths' digit. **4.** 48.5 **Pages 723–724** **1.** The difference between 0.4226 and 0.4167 is less than the difference between 0.4167 and 0.4067. **2.** The sum of the measures of the angles of any triangle is equal to $180°$. **3.** The difference between 1.5000 and 1.4826 is less than the difference between 1.5399 and 1.5000. **4.** $m \angle A \approx 56°$; $m \angle B \approx 34°$; $AB = \sqrt{208} \approx 14.4$

Pages 727–728 **1.** exact **2.** 14, $14\sqrt{2}$ **3.** 6, 6 **4.** $a = 10$, $b = 10\sqrt{3}$ **5.** $b = 100\sqrt{3}$, hypotenuse $= 200$

Answers to Selected Exercises

CHAPTER 1

Pages 5–6 Exercises **1.** 30 **3.** 10 **5.** 12 **7.** 2 **9.** 180 **11.** 20 **13.** 4 **15.** 8 **17.** 19 **19.** $b - 6$ **21.** $7n$ **23.** $n - 1$ **25.** $60 \div n$ **27.** 7 **29.** 20 **31.** $6 * M$ **33.** $J + 12 + K$ **35.** $r = 12; s = 8$ **37.** $r = 30; s = 15$ **39.** nine and two ten-thousandths **41.** 180 **43.** $\frac{2}{5}$

Pages 9–10 Exercises **1.** 39.1 **3.** 204.21 **5.** 81.1 **7.** 101.64 **9.** 200.4 **11.** 85.0 **13.** $576.44 **15.** 1.11 mi **17.** 0.26 mi **19.** about 0.09 mi **21.** $p + 43.61$ **23.** $8.7 - z$ **25.** the sum of 15.8 and a number z **27.** a number b minus 16.4 **29.** The statement $p + q = p - q$ is true only when $q = 0$. **31.** about 35 **33.** $\frac{3}{8}$ **35.** 105,153

Pages 13–14 Exercises **1.** 139.2 **3.** 1693.44 **5.** 0.49 **7.** 1.96 **9.** 100.8 **11.** 0.29 **13.** 187 mi **15.** 42.328 **17.** 0.0532 **19.** 2021.3 **21.** unreasonable; 71.712 **23.** unreasonable; 0.61 **25.** reasonable **27.** $15.4z$ **29.** $986.4 \div n$ **31.** 12.3 times a number a **33.** a number x divided by 5.4 **35.** 60 **37.** 1995 **39.** Answers will vary. **41.** Answers will vary. Example: 840; LEN and DON **43.** about 300 **45.** hundredths

Pages 16–17 Problem Solving Situations **1.** The paragraph is about Cathy's job as a cashier. **3.** the $6.50 per hour that Cathy earns **5.** The paragraph is about Paul's plan to buy a video game system. **7.** No. **9.** The paragraph is about Carol's purchase of a stereo system. **11.** the $100 down payment; the 10 equal payments **13.** 15.9981 **15.** 13,870

Page 17 Self-Test 1 **1.** 32 **2.** 13 **3.** 10 **4.** 64 **5.** 172.4 **6.** 22.71 **7.** 4.77 **8.** 66.36 **9.** 710.84 **10.** 379.68 **11.** 70 **12.** 6.4 **13.** The paragraph is about Jill's movie and video game rentals last year. **14.** 15 **15.** the 15 video games that Jill rented; the $1.50 cost of each video game rental **16.** Multiply 15 times $1.50.

Pages 20–21 Exercises **1.** < **3.** = **5.** > **7.** $0.2 < 0.238 < 0.26$ **9.** $12.03 < 14 < 14.36$ **11.** $25.04 < 25.08 < 25.60$ **13.** Mexico **15.** < **17.** < **19.** < **21.** < **23.** = **25.** > **27.** > **29.** = **31.** < **33.** West; West **35.** 1960: 179.4 million people; 1980: 226.6 million people; about 50 million more people **37.** always **39.** sometimes **41.** always **43.** never **45.** 7.2 **47.** about 900 **49.** $1\frac{3}{7}$ **51.** 327,558

Page 24 Exercises **1.** 37 **3.** 5 **5.** 29 **7.** 114 **9.** 9.9 **11.** 155 **13.** 115 **15.** 19 **17.** 139 **19.** 80 **21.** 12.2 **23.** 60 **25.** 16 **27.** b **29.** 12.35

Page 27 Exercises **1.** 26 **3.** 7 **5.** 2.1 **7.** 320 **9.** 89 **11.** 0 **13.** 120 **15.** 3.3 **17.** 0 **19.** 4300 **21.** 420 **23.** 36 **25.** 0 **27.** 48 **29.** 18 **31.** 36; 8 **33.** 8; 2 **35.** 5 **37.** 0 **39.** 6.6

Page 29 Exercises **1.** pp; 78 **3.** pp; 515 **5.** c; 27,405 **7.** pp; 468.9 **9.** c; 2.4 **11.** mm; 0.28 **13.** mm; $.90 **15.** mm; about $10.50 **17.** 0.002 **19.** =

Pages 32–33 Exercises **1.** 343 **3.** 9 **5.** 1000 **7.** 16 **9.** 128 **11.** 48 **13.** 64 **15.** 16 **17.** 512 **19.** 1024; 40,960 **21.** 25 **23.** 256 **25.** 12 **27.** 0.81 **29.** 0.125 **31.** 0.008 **33.** 0.001 **35.** D **37.** B **39.** 81 **41.** 11 **43.** < **45.** < **47.** $a = 4; b = 2$ **49.** The paragraph is about the amount of money Emily Ling earns and the number of hours that she works. **51.** $1\frac{1}{6}$

Page 33 Challenge The sum of the first and last numbers is 50, the sum of the second and tenth numbers is 50, and so on. There are five of these sums for a total of 250. Add the remaining number, 25, to this sum for a total of 275.

Pages 38–39 Exercises **1.** 96 **3.** 49 **5.** 16 **7.** 82 **9.** 7 **11.** 5 **13.** 23 **15.** 48 **17.** 71 **19.** 4 **21.** 17 **23.** 4,500,000 albums **25.** 36 **27.** 63 **29.** 55 **31.** False; $(24 - 4) 2 = 40$ **33.** True. **35.** False; $3 (4 - 2) 3 = 18$ **37.** False; $(12 - 2^2) \div 4 = 2$ **39.** 33,789 **41.** 42

Page 39 Self-Test 2 **1.** < **2.** = **3.** > **4.** 142 **5.** 10.9 **6.** 106 **7.** 17 **8.** 150 **9.** 0 **10-12.** Answers may vary. Likely answers are given. **10.** c; 73.367 **11.** pp; 66 **12.** mm; 591 **13.** 625 **14.** 192 **15.** 196 **16.** 13 **17.** 8 **18.** 3

Pages 40–41 Chapter Review **1.** variable expression **2.** commutative property of addition **3.** multiplicative identity **4.** variable **5.** base **6.** multiplication property of zero **7.** 21 **8.** 32 **9.** 31 **10.** 28 **11.** 9 **12.** 4 **13.** 16 **14.** 24 **15.** 37.5 **16.** 7.392 **17.** 4.4 **18.** 4.28 **19.** 55.2 **20.** 80.3 **21.** 90.6 **22.** 0.055 **23.** < **24.** = **25.** > **26.** The paragraph is about Yvonne's job typing term papers. **27.** 6 h **28.** the $7 per hour that she earns **29.** You would add the number of hours that Yvonne typed on Monday, Tuesday, Wednesday, and Thursday. **30.** $5642 < 12,375 < 12,456$ **31.** $0.078 < 0.102 < 0.62$ **32.** $0.611 < 6 < 6.10$ **33.** 0.72

34. 44   **35.** 12   **36.** 149   **37.** 0   **38.** 7   **39.** 7
40–43. Answers may vary. Likely answers are given.
40. pp; 143   **41.** c; 94.583   **42.** c; 499.28   **43.** mm; 200
44. 243   **45.** 128   **46.** 1   **47.** 1024   **48.** 10,000
49. 512   **50.** 31   **51.** 7   **52.** 64   **53.** 25   **54.** 243
55. 864   **56.** 486   **57.** 400   **58.** 216   **59.** 7   **60.** 58
61. 2

Page 43 Chapter Extension   1. 13   **3.** 37   **5.** Yes.
Addition is commutative.   **7.** 21   **9.** 70   **11.** No. Answers will vary. An example is 5 ∎ 4 = 21, 4 ∎ 5 = 29.
13. $a \star b = a + 2b$

Pages 44–45 Cumulative Review   1. B   **3.** C   **5.** A
7. D   **9.** C   **11.** D   **13.** D   **15.** B   **17.** C   **19.** C

CHAPTER 2
Page 50 Exercises   1. c^{10} **3.** n^3 **5.** x^{13} **7.** $12d^5$
9. $30b^3$ **11.** $28cd$ **13.** $56w^2y$ **15.** $240wx^2$ **17.** z^5
19. c^4 **21.** $3c$ **23.** $8b^2$ **25.** 8   **27.** 2   **29.** 6   **31.** 2
33. product of powers rule   **35.** Both $2^m \cdot 2^0$ and $2^m \cdot 1$
are equal to 2^m. It follows that the expressions must be
equal to each other.   **37.** The paragraph is about buying
tickets for a basketball game and a hockey game.
39. $30c^3$

Pages 54–55 Exercises   1. 560   **3.** 30   **5.** 436
7. 1773   **9.** 3184   **11.** $90 + 9a$ **13.** $5x - 45$
15. $48m + 72$ **17.** $27 - 12a$ **19.** \$20   **21.** 4   **23.** 3
25. \$240   **27.** 63   **29.** $15x + 25$ **31.** 11.6

Pages 57–58 Exercises   1. $14x$ **3.** $6m$ **5.** $11w$
7. $5x$ **9.** $8z + 7$ **11.** $9k - 6$ **13.** $4w + r$
15. $2a + 9b$ **17.** $18n$ **19.** $10a + 13b$ **21.** $11c + 4w$
23. cannot be simplified   **25.** cannot be simplified
27. $5a + 13b + 3x$ **29.** $14c + 8$ **33.** $88y + 36$
35. $8b + 23$ **37.** 625   **39.** $1\frac{5}{8}$

Page 58 Self-Test 1   1. x^7 **2.** c^8 **3.** a^{16} **4.** $27w^7$
5. $90xz$ **6.** $84a^2b$ **7.** 120   **8.** 535   **9.** 1568
10. $2x + 18$ **11.** $56 - 40x$ **12.** $22 + 14c$ **13.** $7n$
14. $13x$ **15.** $3x - 5y$ **16.** $4a + 11b$ **17.** $8z + 1$
18. $12w$

Pages 60–61 Problem Solving Situations   1. 26 mi
3. \$5.43   **5.** 1688 items   **7.** 163   **9.** 495   **13.** $12a - 32$
15. 119

Pages 64–65 Exercises   1. 52, 62, 72
3. 625; 3125; 15,625   **5.** 20, 26, 33   **7.** 13, 8, 2
9. $56m$, $112m$, $224m$ **11.** $a + 15$, $a + 20$, $a + 25$
13. $4n + 3$, $5n + 3$, $6n + 3$ **15.** January   **17.** 89, 144
19.   **21.** Start with 1, add 2, add

3, add 4, and so on.   **23.** add 3, add 5, add 7, and so on;
square 1, square 2, square 3, and so on   **25.** 14, 16, 20;
20, 22, 44   **29.** 1296; 7776; 46,656   **31.** $3\frac{1}{8}$

Page 65 Challenge   a. A, S, O   **b.** E, N, T
c. ⊻, K, ⊼   **d.** 5⊂ ⊗ ⊼

Pages 68–69 Exercises   1. 40; 60; 80   **3.** 21; 25; 29
5. $x \rightarrow 5x$ **7.** \$57.75; \$74.25; \$99; \$107.25; \$132
9. 110 times   **11.** 21 times   **13.** The book has been kept
out of the library past the day that its return was requested.
15. \$.45   **19.** 23.7   **21.** $5x + 5y$

Page 70 Exercises   1. $2 * A4 + 5$ **3.** add 2   **5.** add 3
7. The pattern will be add 4; each number in the table will
be 4 greater than the number above it.   **9.** The pattern will
be add 5; each number in the table will be 1 greater than
the corresponding number in the table for $x \rightarrow 5x + 2$.

Pages 72–73 Problem Solving Situations   1. Yes.
Estimating gives $20 + 7 + 3(3) = 36$, which is close to
\$37.42.   **3.** \$20.30   **5.** \$45.10   **7.** 80 points   **11.** $<$
13. \$7.11

Page 73 Self-Test 2   1. \$110.01   **2.** 17, 20, 23
3. 256, 1024, 4096   **4.** 16; 24; 28   **5.** $x \rightarrow 7x$ **6.** No.
Marvin multiplied his 5 extra hours by \$7.50 instead of
\$10.75.

Page 75 Exercises   1. estimate; \$6   **3.** exact answer;
9 flights   **5.** estimate; Yes, the coach will spend about
\$210.   **11.** 600

Pages 77–79 Exercises   1. 3 cm   **3.** 0.45 km
5. 25,000 mg   **7.** 0.615 m   **9.** 2.345 L   **11.** 3400 mL
13. 74 cm   **15.** 98 m   **17.** 880 m   **19.** 2 g   **21.** =
23. $<$ **25.** $>$ **27.** 4.36   **29.** 0.578   **31.** 0.809
33. 0.0136   **35.** 0.0245   **37.** b   **39.** c   **41.** A reasonable answer is 0.25 L.   **43.** 17,100 mL   **45.** 2 cm
47. one millionth of a meter   **49.** one billionth of a meter
51. one million liters   **53.** 240 oz   **55.** $10x - 20$

Pages 82–83 Exercises   1. 1.2×10^6 **3.** 5.7×10^3
5. 4.52×10^4 **7.** 8.514×10^8 **9.** 3800   **11.** 1,520,000
13. 500,000,000   **15.** 342,500   **17.** 1.392×10^6 km
19. 36,000,000 mi   **21.** 887,140,000 mi
23. 3.5×10^{11}; 350,000,000,000   **25.** 4.853×10^7;
48,530,000   **27.** D   **29.** $>$ **31.** $<$ **33.** $<$ **35.** 4; 5
37. Both exercises involve multiplying numbers written in
scientific notation. However, an extra step was needed in
Exercise 36 to obtain an answer that was written in scientific notation.   **39.** $7\frac{7}{12}$ **41.** 4.254×10^7 **43.** $3a + 7b$

Page 83 Self-Test 3   1. estimate; Yes, he needs about
\$18.   **2.** 48 cm   **3.** 3200 mL   **4.** 500 g   **5.** 4.4×10^4

6. 9.87×10^6 **7.** 2.13×10^7 **8.** 5600   **9.** 378,000
10. 64,300,000

Pages 84–85 Exercises   1. 2.838 L   **3.** 158.75 cm
5. about 210 mL   **7.** about 50 kg   **9.** meter   **11.** gallon
13–17. Answers may vary.   **13.** 4.0678 L; 3.2637 L
15. 5.1925 m; 4.4175 m   **17.** 6.35 mm; 44.45 mm

Pages 86–87 Chapter Review   1. function   **2.** distributive property   **3.** unlike terms   **4.** gram   **5.** scientific
notation   **6.** arrow notation   **7.** a^6 **8.** c^{13} **9.** $24x + 30$
10. $24 + 2x$ **11.** $24n + 7m$ **12.** $17a$ **13.** $10x + 12$
14. $55ab$ **15.** b^{12} **16.** x^{16} **17.** $54 - 9z$
18. $24c - 72$ **19.** $15x^8$ **20.** $9c$ **21.** $48wx^2$
22. $15 + 12z$ **23.** $7d - 5c$ **24.** $42c^4$ **25.** 180
26. 160   **27.** 832   **28.** 282   **29.** $93.87   **30.** 31, 38, 45
31. 162, 486, 1458   **32.** $a + 9, a + 10, a + 11$
33. $29m, 26m, 23m$ **34.** 12; 16; 20   **35.** 54; 72; 90
36. $x \to \frac{x}{3}$ **37.** $x \to x + 5$ **38.** Yes.   **39.** exact answer;
$49.50   **40.** 1.7 kg   **41.** 0.75 L   **42.** 760 m
43. 190 cm   **44.** 3.5×10^5 **45.** 1.27×10^4
46. 6.55×10^6 **47.** 4.8×10^7 **48.** 130,000
49. 97,000,000   **50.** 26,400   **51.** 588,000,000

Page 89 Chapter Extension   1. b^{10} **3.** c^{28} **5.** x^{36}
7. a. 144   **b.** 144   **9.** Power of a product rule: To find
the power of a product, you find the power of each factor
and then multiply. $(ab)^m = a^m b^m$

Pages 90–91 Cumulative Review   1. B   **3.** A   **5.** B
7. D   **9.** B   **11.** C   **13.** A   **15.** C   **17.** B   **19.** A

CHAPTER 3
Pages 96–97 Exercises   1. 5   **3.** 13   **5.** 1   **7.** >
9. =   **11.** <   **13.** >   **15.** -754 ft   **17.** $-3 < 4 < 9$
19. $-10 < -8 < -6$ **21.** The definition given in the lesson refers to distance from zero. This definition uses a
variable to represent any integer. This definition specifically includes zero.   **23.** $10,000   **25.** Hearty Company
27. never   **29.** never   **31.** always   **33.** sometimes
35. 2203   **37.** 2

Page 100 Exercises   1. 29   **3.** -30 **5.** -63 **7.** 68
9. -45 **11.** 112   **13.** $38   **15.** 33°F   **17.** 18
19. -12 **21.** -22 **23.** -17 **25.** $15y$ **27.** $-10m$
29. $-10r + 4s$ **31.** $8m + 8n$ **33.** about 80
35. $12y - 3$

Pages 103–104 Exercises   1. 0   **3.** -21 **5.** 0   **7.** 5
9. 10   **11.** -7 **13.** -26 **15.** 7   **17.** -14 **19.** less
money; $8 less   **21.** 0   **23.** -8 **25.** 8   **27.** 0   **29.** 12
31. 50 s after liftoff   **33.** exact times; each maneuver begins at an exact time before liftoff.   **35.** T minus 45
37. -8 **39.** -10 **41.** -2 **43.** $-1,000,000,001$

45. infinitely many pairs; for every integer, you can add
the opposite of one greater than the integer.
47. 17, 21, 25   **49.** -28

Pages 109–110 Exercises   1. -9 **3.** 6   **5.** 25   **7.** 0
9. -3 **11.** -4 **13.** 0   **15.** -52 **17.** 4   **19.** 36°F
21. 3   **23.** -16 **25.** -12 **27.** B   **29.** C   **31.** -4,
$-9, -14$ **33.** $-7, -3, 1$ **35.** -33°F   **37.** 17°F
39. about 10°F   **41.** 1070 mm   **43.** about 32

Pages 113–114 Exercises   1. 4   **3.** -11 **5.** 0
7. -4 **9.** 154   **11.** -10 **13.** 100   **15.** -84 **17.** -16
19. -5°F   **21.** -36 **23.** 29   **25.** 9   **27.** -41 **29.** 11
31. 12   **33.** $-12c + 24$ **35.** $14y + 56$ **37.** $-20 + 20b$
39. D   **41.** C   **43.** 1 aluminum, 3 chloride   **47.** -12

Page 114 Self-Test 1   1. >   **2.** <   **3.** =   **4.** -4
5. -22 **6.** -18 **7.** 5   **8.** -6 **9.** 0   **10.** -31
11. -14 **12.** 6   **13.** -32 **14.** 70   **15.** -12

Pages 116–118 Exercises   1. 1296   **3.** -135
5. -166 **7.** 27   **9.** 1   **11.** 11   **13.** 14   **15.** -6
17. -5 **19.** 24   **21.** -1 **23.** 42   **25.** 36   **27.** -15
29. D   **31.** A   **33.** 4; 1; 16   **35.** $x \to -4x$ **37.** 1
39. 1   **41.** -1 **43.** If n is even, $(-1)^n$ is 1. If n is odd,
$(-1)^n$ is -1.   **45.** positive   **47.** negative   **49.** positive
51. 150,000,000   **53.** 16   **55.** No; Julia did not include
the amount that either her brother or her sister paid.   **57.** 0

Page 118 Challenge   Answers will vary. Examples:
$-5 + 7 - 8 \div 4 \cdot 2$ and $-8 \div 2 + 5 - 1 \cdot 3$

Pages 120–121 Problem Solving Situations
1. 12 ways   **3.** 15 totals   **5–9.** Strategies may vary.
Likely strategies are given.   **5.** making a table;
12 totals   **7.** choosing the correct operation; $32
9. making a table; 7 ways   **13.** -12 **15.** -11

Pages 126–127 Exercises   1. $(-4, 3)$ **3.** $(4, -3)$
5. $(0, -3)$ **7.** $(-3, 0)$ **9.** $(2, 3)$ **11.** $(-2, -4)$
13–20.   **21.** I   **23.** II

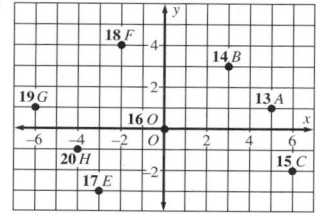

25. y-axis   **27.** x-axis   **29.** The coordinates are all positive.   **31.** The coordinates are all negative.   **33.** The x-
coordinates are all negative.   **35.** The y-coordinates are all
zero.   **37.** quadrilateral

39.

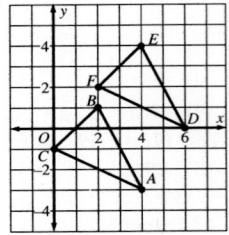

The shapes are the same. △*DEF* is 3 units above and 2 units to the right of △*ABC*.   **41.** $7a - 7b$

43.

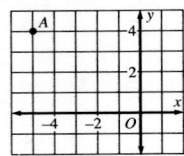

Pages 130–131 Exercises   1.

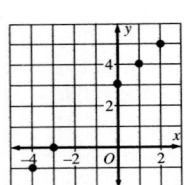

3.

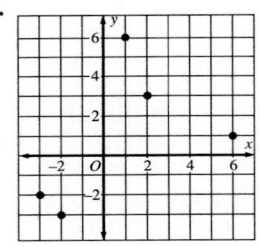

5.

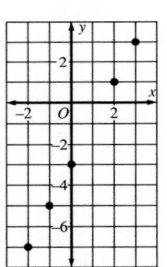

7.

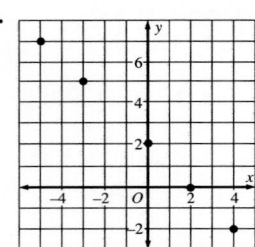

9.

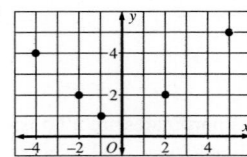

11.

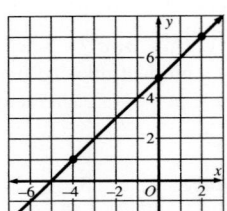

x	$2x$
-2	-4
-1	-2
0	0
1	2
2	4

$x \longrightarrow 2x$

13.

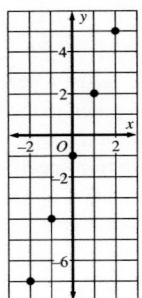

x	$-x$
-2	2
-1	1
0	0
1	-1
2	-2

$x \longrightarrow -x$

15.

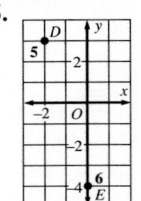

a. Answers will vary.

Examples: $(-2, 3)$, $(-1, 4)$, and $(1, 6)$.   **b.** Yes.

17.   **19.** $68

Page 131 Self-Test 2   1. 31   **2.** 3   **3.** 10   **4.** 6 ways

5–6.   **7.**

Page 133 Exercises   1. B5   **3.** Possible answers: Middletown, Paterson, Newark, and Ramsey   **5.** F3 and F4
7. D3 and D4   **9.** Delaware and Hudson rivers
11. Morristown

13.

Pennsylvania Index	
Bushkill	B5
Cresco	A6
Easton	A3
Quakertown	A2
Portland	B4

15. Answers will vary.

Pages 134–135 Chapter Review   1. opposites
2. absolute value   **3.** integers   **4.** addition property of
opposites   **5.** origin   **6.** x-coordinate   **7.** 11   **8.** 5   **9.** 0
10. 2   **11.** 4   **12.** 3   **13.** <   **14.** =   **15.** <   **16.** <
17.a. 7 ways   **b.** 9 totals   **18.** −5   **19.** 22   **20.** 76
21. −12   **22.** −20   **23.** 48   **24.** 11   **25.** −7   **26.** −77
27. 12   **28.** 108   **29.** −16   **30.** −46   **31.** 0   **32.** −45
33. 0   **34.** 11   **35.** −20   **36.** −8   **37.** 7   **38.** −23
39. −36   **40.** 0   **41.** −42   **42.** 4   **43.** 216   **44.** −24
45. −180   **46.** 40   **47.** 28   **48.** 13   **49.** −5   **50.** 9
51. 8   **52.** 4   **53.** 9   **54.** (−2, 4)   **55.** (1, 0)
56. (3, 3)   **57.** (−1, −2)   **58.** (4, −3)   **59.** (0, −4)
60–63.

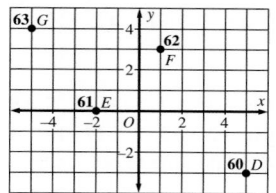

64.

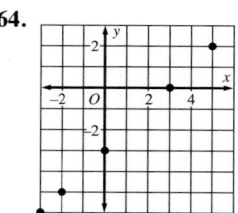

65.   **66.**

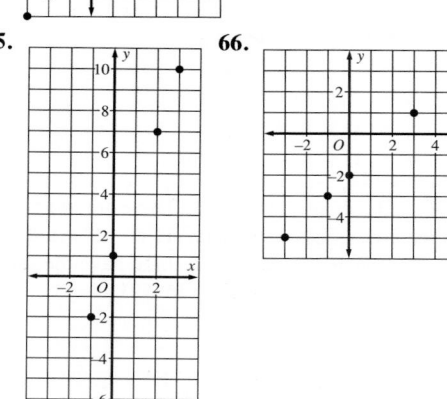

67.

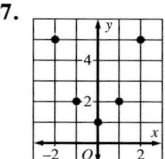

Page 137 Exercises   1.a. closed   **b.** not closed
c. closed   **d.** not closed   **3.a.** closed   **b.** not closed
c. closed   **d.** not closed   **5.a.** not closed   **b.** not closed
c. closed   **d.** closed (exclude division by zero)
7.a. closed   **b.** not closed   **c.** closed   **d.** not closed

Pages 138–139 Cumulative Review   1. D   **3.** A
5. B   **7.** C   **9.** C   **11.** B   **13.** A   **15.** A   **17.** D
19. C

CHAPTER 4

Page 144 Exercises   1. Yes.   **3.** No.   **5.** No.
7. Yes.   **9.** Yes.   **11.** 3   **13.** −45   **15.** −8   **17.** 5
19. No.   **21.** B   **23.** A   **25.** C   **27.** C   **29.** A
31. False.   **33.** True.   **35.** True.   **37.** cannot determine
39. Answers may vary. Examples are given. $m + 3 = 2$;
$9m = −9$ **41.** Answers may vary. Examples are given.
$c − 18 = 6$; $\frac{c}{8} = 3$ **43.** No; the amount $50 was not included in the total.   **45.** 3

Pages 146–147 Problem Solving Situations   1. 24
3. 1320   **5.** 5 sweaters, 7 shirts   **7.** 7 quarters, 8 dimes
9–13. Strategies may vary. Likely strategies are given.
9. making a table; 3 ways   **11.** guess and check; 9
13. choosing the correct operation; $4500   **17.** 23   **19.** 4
compact discs, 3 cassette tapes

Page 150 Exercises   1. 9   **3.** −9   **5.** 4   **7.** −11
9. 18   **11.** 0   **13.** −11   **15.** 396   **17.** −1580
19. −437   **21.** 5; −5   **23.** 3; −3   **25.** 2   **27.** no solution   **29.** 20; −20   **31.** no solution   **33.** 43   **35.** −117

Pages 153–154 Exercises   1. 15   **3.** −8   **5.** 0
7. −56   **9.** −189   **11.** 62   **13.** 15   **15.** −17   **17.** −1
19. 0   **21.** B   **23.** A   **25.** b   **27.** d   **29.** about 1800
31. about 1400   **33.** 23   **35.**

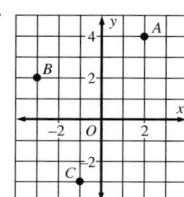

37. about 16

Page 154 Self-Test 1   1. −7   **2.** −9   **3.** 6   **4.** 5
birthday cards, 3 thank you notes   **5.** −3   **6.** −2   **7.** 2
8. −30   **9.** −24   **10.** −7

Pages 158–159 Exercises   1. 4   **3.** −2   **5.** −7
7. −10   **9.** 40   **11.** 39   **13.** −156   **15.** 0   **17.** 5

19. [187.] [−]   [135.] [=] [×]
[8.] [=] ; 416

21. [1505.] [−]   [161.] [=] [÷]
[42.] [+/−] [=] ; −32

23. [88.] [+/−] [−]   [112.] [=] [×]
[15.] [+/−] [=] ; 3000

25. Answers may vary. Example: The first equation requires only one step to solve; $2x + 3 = 15$ requires two steps to solve.   **27.** Two; a chemical reaction has a natural balance.   **29.** 4; 2   **31.** 8,320,000,000   **33.** 0

Pages 163–164 Exercises   1. $x + 5$ **3.** $2w$ **5.** $3c + 8$
7. $6d − 4$ **9.** $a − 12$ **11.** $\frac{s}{6}$ **13.** $t + 17$ **17.** $26 + 12x$
19. $26 + 12x + 7y$ **21.** 23; 31; 39   **23.** $7 + 3x$

Pages 166–167 Exercises   1. $m + 4 = 5$ **3.** $\frac{a}{6} = 12$
5. $15k = 105$ **7.** $7 = \frac{j}{2}$ **9.** $10 = s − 15$ **11.** A   **13.** D
15. C   **17–19.** Answers will vary. Examples are given.
17. The number m decreased by 8 is 16.   **19.** The product of 4 and a number w is 48.   **21.** mental math
23.   **25.** No.

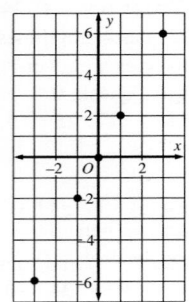

Page 167 Challenge   The weights 11 and 92 on one side balance the weights 47 and 56 on the other side; No.

Pages 169–171 Problem Solving Situations
1. $1000   **3.** 114   **5–11.** Strategies may vary.
5. choosing the correct operation; $91   **7.** making a table; 3 ways   **9.** using equations; $159   **11.** guess and check; 4 books, 4 magazines   **13.** 72 mg   **19.** Yes; Yes; No; No; No; No; No.   **21.** 16

Page 171 Self-Test 2   1. −3   **2.** 24   **3.** 10   **4.** $\frac{z}{14}$
5. $3t$ **6.** $n + 7 = 35$ **7.** $t − 9 = 22$ **8.** 73   **9.** 70 km

Pages 173–175 Exercises   1. 12   **3.** −8   **5.** −5
7. 4   **9.** 3   **11.** 3   **13.** 9   **15.** −28   **17.** 3   **19.** 4

21. 2   **23.** all values of x less than 8   **25.** 8   **27.a.** The paragraph is about Steve working at a job.   **b.** 7 h   **c.** the hourly wage   **d.** Add 7, 8.5, 7, and 2. Multiply the sum by 9.25.   **29.** about 600   **31.** 6, 3, 0

Pages 177–179 Exercises   1. 2200   **3.** 40   **5.** 610
7. 475 mi/h   **9.** 7   **11.** 36   **13.** $P = \frac{a}{2} + 110$ **15.** 80
years old   **17.** 293.4 mi   **19.** $E = IR$ **21.** E = force (in volts), I = current (in amperes), R = resistance (in ohms).
23. 8.4   **25.** 2   **27.** $9a + 4b$ **29.** 1024

Page 179 Self-Test 3   1. 12   **2.** −6   **3.** 12   **4.** 10
5. −2   **6.** 3   **7.** 180 mi   **8.** 350 mi   **9.** 12 h
10. 234 mi/h

Pages 180–181 Chapter Review   1. inverse operations
2. equation   **3.** solve   **4.** formula   **5.** solution of an equation   **6.** Yes.   **7.** No.   **8.** No.   **9.** Yes.   **10.** 5
11. −7   **12.** 50   **13.** 0   **14.** 13 haircuts, 4 perms   **15.** 3 pens, 4 notebooks   **16.** 3   **17.** −2   **18.** −5   **19.** 23
20. −16   **21.** −14   **22.** 45   **23.** 12   **24.** 2   **25.** 9
26. −9   **27.** −88   **28.** $8q − 6$ **29.** $\frac{h}{19}$ **30.** $7b$
31. $2t + 4$ **32.** $z − 15 = 33$ **33.** $m − 5 = 8$
34. $3x = 48$ **35.** $169   **36.** 7 fish   **37.** −2   **38.** −3
39. 16   **40.** 7   **41.** 14   **42.** $47.76   **43.** $126
44. $7.50

Page 183 Chapter Extension   1. −2   **3.** 5   **5.** −26
7. 6   **9.** −2   **11.** 2   **13.** −2   **15.** 3   **17.** 13   **19.** Answers may vary. Example: Add nine to both sides of the equation.

Pages 184–185 Cumulative Review   1. A   **3.** A
5. B   **7.** A   **9.** D   **11.** C   **13.** D   **15.** C   **17.** B   **19.** A

CHAPTER 5
Pages 190–191 Exercises   1. about 3,500,000
3. about 1,000,000   **5.** 1992
7.

Airport Limousine Earnings per Quarter

First	$ $ $ $ $ $ $ $
Second	$ $ $ $ $ $ $ $ $ $
Third	$ $ $ $ $ $ $ $ ᵉ
Fourth	$ $ $ $ $ $ = $10,000

9. about 1550   **11.** about $3600   **13.** There are too few symbols, so the differences in the data are not reflected. For example, Boston and Pittsburgh seem to have the same number of passengers.   **17.** 6   **19.** 136   **21.** 4
23. $40 − 24n$

Pages 194–195 Exercises   1. Estimates may vary; about 2 million.   **3.** Estimates may vary; about 3.5 million.   **5.** bowling, volleyball   **7.** Estimates may vary; about 2 million.   **9.** Western Europe   **11.** Estimates may

vary; about $1 billion. **13.** Estimates may vary; about −35°F. **15.** Estimates may vary; about 40°F
17. 15.4 million **19.** about $1159.2 million **21.** Estimates may vary; about 90.
23.

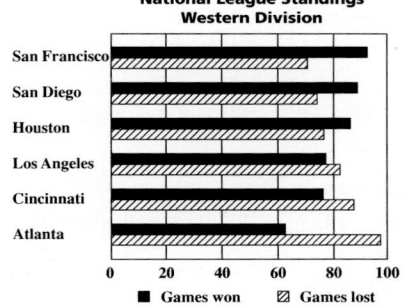

National League Standings
Western Division

■ Games won ▨ Games lost

25. $17u + 5v$

Pages 198–199 Exercises **1.** Estimates may vary; about 8 million **3.** Estimates may vary; about 9.5 million
5. 1977, 1981 **7.** Estimates may vary; about 2.5 million
9. 1973–1977 **11.** Estimates may vary; elementary: about 1.2 million; secondary: about 1.0 million. **13.** Estimates may vary; elementary: about 1.1 million; secondary: about 0.9 million **15.** Estimates may vary; about 0.2 million **17.** Estimates may vary; about 2.2 million **19.** Estimates may vary; about 0.35 million **21.** 5 years
23. 1985 **25.** 50 million **27.** 30 million **29.** 50 million
31.

Year	Number of Sheep
1910	50
1925	40
1940	50
1955	30
1970	20
1985	10

33. $240x^2$ **35.** 2.4

Pages 202–203 Exercises
1.

Adults Participating in
Leisure-Time Activities

Millions of Adults

3.

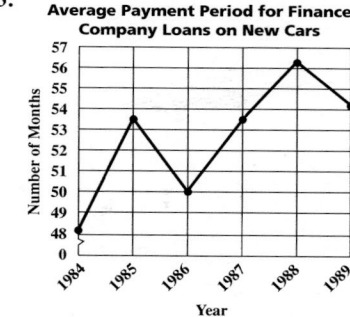

Average Payment Period for Finance
Company Loans on New Cars

5.

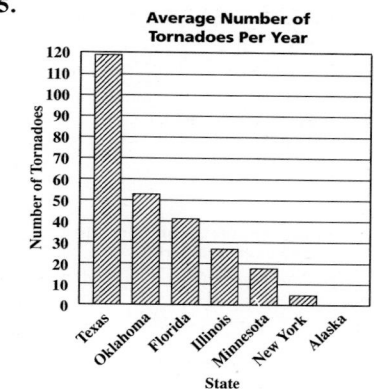

Average Number of
Tornadoes Per Year

7.

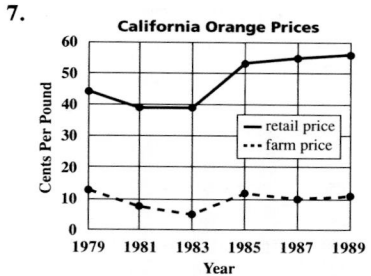

California Orange Prices

— retail price
⋯ farm price

9. Answers may vary. Example: The retail price of oranges is always greater than the farm price. Furthermore, the direction of change in retail and farm prices is generally the same. **11.** −22 **13.** about 8

Page 203 Self-Test 1
1.

Average Annual Snowfall in Four Locations

Blue Canyon	✳✳✳✳✳
Stampede Pass	✳✳✳✳✳✳✳✳
Marquette	✳✳✳
Flagstaff	✳✳

✳ = 50 inches

2. about 15 thousand   **3.** about 35 thousand
4. about 10 million   **5.** increasing
6.

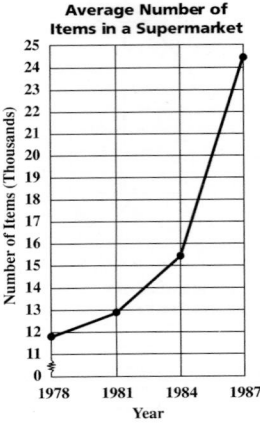

Average Number of Items in a Supermarket

Page 205 Exercises
1. bar graph;

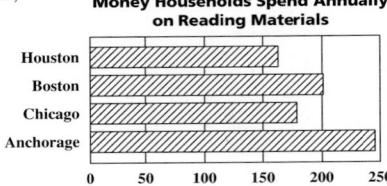

Money Households Spend Annually on Reading Materials

3. bar graph;

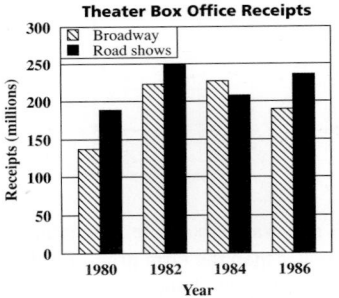

Theater Box Office Receipts

5. $15n + 20$ **7.** Line graph; the population changes continuously.

Pages 207–208 Problem Solving Situations   1. 7
3. not enough information; the number of books sold in Midtown in 1980 and 1990   **5–7.** Strategies may vary. Likely strategies are given.   **5.** making a table; 3
7. choosing the correct operation; $296.52   **11.** $2\frac{5}{12}$
13. 353

Pages 211–212 Exercises   1. about the same   **3.** The gap could be eliminated and intervals of 1 from 0 to 4 could be used.   **5.** *Graph E*. It makes it appear that the advertising campaign did not significantly increase the number of subscribers.   **7.** misleading   **9.** accurate

11.

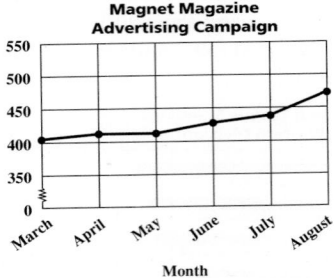

Magnet Magazine Advertising Campaign

13. Changing the scale on a line graph can exaggerate or minimize a trend.   **15.** $9\frac{9}{10}$ **17.** It gives the false visual impression that the level of production at Plant C was about half that at Plant B.

Page 212 Self-Test 2
1. line graph;

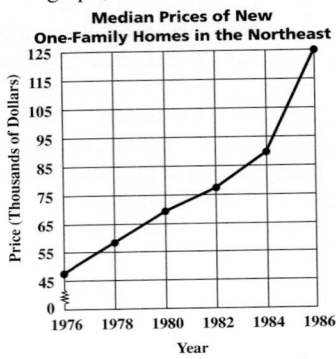

Median Prices of New One-Family Homes in the Northeast

2. $944.76   **3.** not enough information; the number of miles showing on the odometer before the trip   **4.** The level of production at Plant B appears to be five times that at Plant A.

Pages 214–216 Exercises   1. 18.25; 18; 16; 7
3. $43.50; $30.50; none; $467   **5.** $\approx -4.2°C$; $-4°C$; none; 7°C   **7.** 75; 60; 60; 135   **9.** 72   **11.** 64 years
13. unreasonable; ≈ 6.4 **15.** reasonable   **17.** 8.8
19. 8.7   **21.** It is the same as finding the median of the four judges' scores.   **23.** Yes; Yes; Yes; Yes; No; Yes; Yes.   **25.** 250

Page 216 Challenge   10, 10, 15, 34, and 51

Page 219 Exercises   1. Yes; the mean is 72 in. It is not distorted by an extreme value.   **3.** Mode; the data cannot be averaged or listed in numerical order.   **5.** Median; the median is 5 and the mode is 3. The mode is less than most of the family sizes.   **7.** Mean; the mean is 35 h, the median is 29 h, and the mode is 20 h. The mean is equal to the number of hours that the nurse wants to work per week.

9.

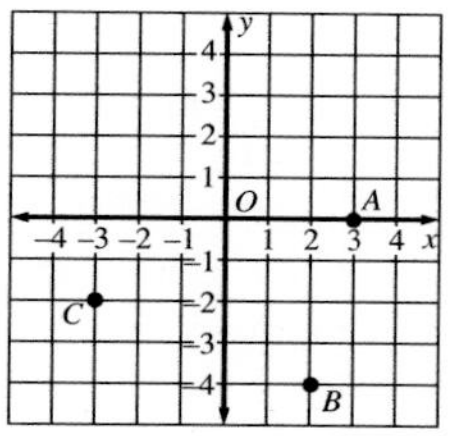

y
4
3
2
1
O
A
−4 −3 −2 −1
1 2 3 4 x
−1
−2
−3
−4
C
B

11. 23, 30, 38

Page 220 Exercises **1.** Metropolitan Area, Region, Food, Housing, Transportation **3.** Atlanta, Dallas/Fort Worth, Washington, D.C. **5.** Washington, D.C. **7.** \$4150.50 **9.** Answers may vary. Example: Washington, D.C.; use the database to sort records in decreasing order by the total cost for food, housing, and transportation. Washington, D.C. appears first, with a total of \$23,208.

Pages 222–223 Exercises **1.** 31; 31; 30; 3

3.

Number of Movies	Tally	Frequency
2	𝍷𝍷	2
3	𝍸 𝍷	6
4	𝍸𝍸	10
5	𝍷𝍷𝍷𝍷	4
12	𝍷	1
15	𝍷	1
		Total 24

5. No; the mean, 4.5, is distorted by the extreme values 12 and 15. **7.** Answers may vary due to possible changes in state laws. **9.** 770 km/h **11.** 36 laps

Page 223 Self-Test 3 **1.** \$43.75; \$40; \$50 and \$20; \$90 **2.** Yes; the mean is 14.5. It is not distorted by an extreme value. **3.** Mode; the data cannot be averaged or listed in numerical order. **4.** 7; 7; 8; 3

Page 225 Exercises **1, 5, 9.** Estimates may vary. **1.** about \$14,000 **3.** 0 up to 80 **5.** about \$2000 **7.** 800 **9.** about \$1.67 **11.** \$10,000 **13.** Answers will vary. Examples: camera, film, rent for shop.

Pages 226–228 Chapter Review **1.** mode **2.** line graph **3.** statistics **4.** data **5.** legend **6.** scale **7.** about 1800 million **8.** about 800 million **9.** not enough information; the amount of corn produced in 1980 **10.** about \$7872 million

11.

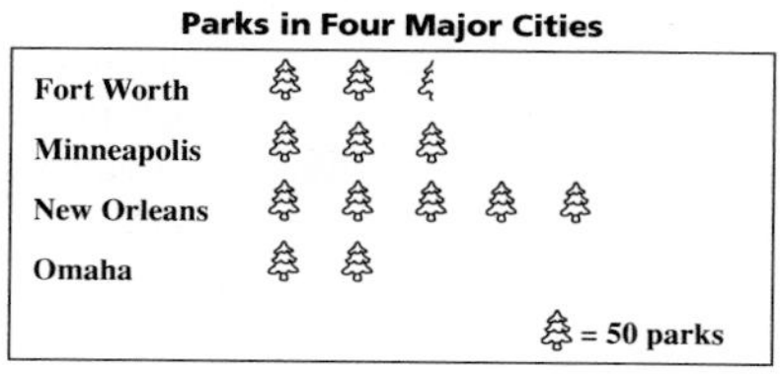

Parks in Four Major Cities
Fort Worth
Minneapolis
New Orleans
Omaha
= 50 parks

12. about \$25 billion **13.** about \$25 billion **14.** direct mail **15.** 1980–1985 **16.** Estimate may vary; about 5.5 visits **17.** 1985 **18.** *Graph A* gives the impression that the average number of annual visits to physicians did not change very dramatically from 1970 to 1985. *Graph B* gives the opposite impression. **19.** *Graph A*

20. bar graph;

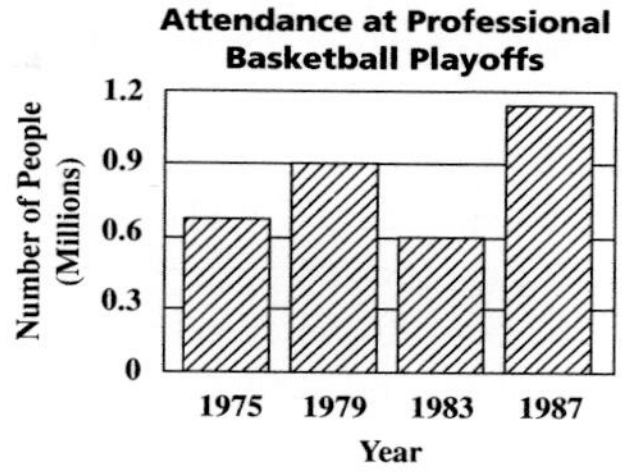

Attendance at Professional Basketball Playoffs
Number of People (Millions)
1.2
0.9
0.6
0.3
0
1975 1979 1983 1987
Year

21. line graph;

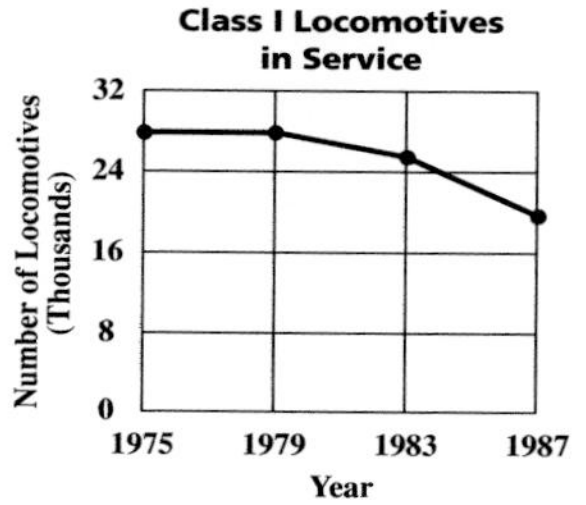

Class I Locomotives in Service
Number of Locomotives (Thousands)
32
24
16
8
0
1975 1979 1983 1987
Year

22. not enough information; the weight of the apples **23.** \$.25 **24.** \$43.31; \$34.60; none; \$53.99 **25.** 1.2; 1; 2; 3 **26.** −4°C; −4°C; none; 12°C **27.** mode; the data cannot be averaged or listed in numerical order.

Page 231 Chapter Extension **1, 3.** Answers are given in the order n, Σx, and $\bar{x}$. **1.** 6; 36; 6 **3.** 5; 94.6; 18.92

Pages 232–233 Cumulative Review **1.** B **3.** D **5.** B **7.** C **9.** B **11.** A **13.** D **15.** D **17.** A **19.** D

CHAPTER 6

Pages 238–239 Exercises **1.** plane *R* **3.** $\angle TRV$, $\angle VRT$, or $\angle R$ **5.** They have different endpoints.

7.

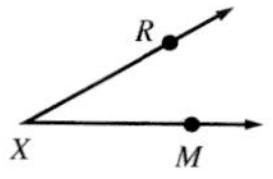

R
X
M

9.

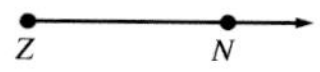

Z
N

11.

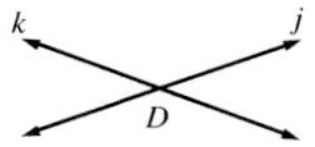

k
j
D

13. *B* **15.** *E* **17.** $\overline{BC}$, $\overline{DC}$, $\overline{GC}$

19. $\overline{BF}$, $\overline{GF}$, $\overline{EF}$ **21.** *M* **23.** $\overline{AB}$, $\overline{BC}$, $\overline{AC}$ **25.** $1\frac{7}{16}$ **27.** 8 adult tickets, 13 student tickets

Page 242 Exercises **1.** 32°

3.

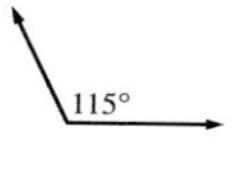

115°

5.

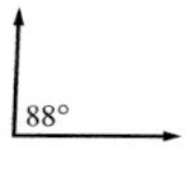

88°

7.

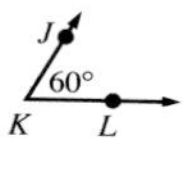

J
60°
K
L

9. D **11.** C **13.** 9 **15.**

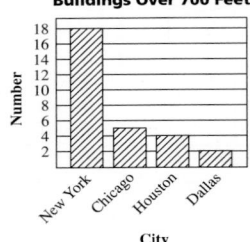

Pages 247–248 Exercises   1. 60° **3.** 61° **5.** 170°
7. 78° **9.** 28° **11.** 28° **13.** 45° **15.** always
17. sometimes **19.** A straight angle has measure 180°.
21. 18 **23.** 60° **25.** 2n + 11 = −31 **27.** −31

Pages 251–253 Exercises   1. False. **3.** True.
5. False. **7.** False. **9.** 37° **11.** 143° **13.** 37° **15.** 90
17. m∠1 = 96°; m∠2 = 84°; m∠3 = 96°; m∠4 = 90°;
m∠5 = 90°; m∠6 = 90°; intersecting **19.** Parallel lines
lie in the same plane, and skew lines do not; both parallel
lines and skew lines do not intersect. **21.** No, because if
two skew lines were perpendicular, they would be in the
same plane. **23.** about 90 **25.** 200
27.

Buildings Over 700 Feet

(bar graph: y-axis "Number" 2–18, x-axis "City" with New York, Chicago, Houston, Dallas)

Page 253 Self-Test 1   1. $\overrightarrow{XY}$ **2.** $\overleftrightarrow{MN}$, $\overleftrightarrow{NM}$ **3.** ∠TSR,
∠RST, or ∠S **4.** (angle diagram 55°)

5. (angle diagram 100°) **6.** (angle diagram 82°) **7.** 41° **8.** 103°

9. 65° **10.** 115° **11.** 65° **12.** 115°

Pages 259–260 Exercises   1. regular triangle; no diago-
nals **3.** quadrilateral; $\overline{LN}$, $\overline{OM}$ **5.** 48° **7.** 33° **9.** B
11. A **13.** 10; 1440; 1440; 10; 144 **15.** 11 **17.** $1\frac{5}{8}$

Pages 263–265 Exercises   1. cannot **3.** can; scalene
5. cannot **7.** can; isosceles **9.** can; equilateral
11. obtuse **13.** acute **15.** right **17.** acute **19.** obtuse
21. No; two sides must be equal in length. **23.** BAE
25. BAE, BED **27.** BED **29.** 38° **31.** 90° **33.** 52°
35. 52° **37.** 142° **39.** 142° **41.** True. **43.** False; the
sum of the angles would be greater than 180°. **45.** True.
47. False; an acute triangle can have three equal sides.
49. FD 40 RT 135 **51.** The three angles are each 65°.
Their sum is greater than 180°.
53. (triangle with 77°, 90, 40, 26°, 77°, 90) **55.** right **57.** 11

Page 265 Challenge   44

Pages 268–269 Exercises   1. 6 m **3.** 12 m **5.** 27°
7. 4 in. **9.** 4 in. **11.** 5 in. **13.** 7 in. **15.** 29°; 151°;
151° **17.** m∠1 = 102°; m∠2 = 33°; m∠3 = 57°
21. quadrilateral, none; parallelogram, four pairs; trapezoid, two pairs; rectangle, four pairs; rhombus, four pairs;
square, four pairs **23.** −51 **25.** False.

Page 270 Exercises   3. Answers will vary. **5.** False.
7. parallelogram, trapezoid, rectangle, rhombus, or square;
true for all the above except trapezoid.

Pages 273–274 Problem Solving Situations   1. 21
3–5. Strategies may vary. Likely strategies are given.
3. using an equation; −21 **5.** making a table; 23 **7.** The
dots represent people; the lines represent handshakes.
9. Count the handshakes between two people, three people, four, five, and so on. Look for a pattern in the number
of handshakes. The solution is 28. **11.** Answers will
vary. How many pieces of pizza result from four cuts
through the center of the pizza? **13.** about 890
15.

(angle diagram 82°)

Pages 277–279 Exercises   1. No. **3.** No.
5. 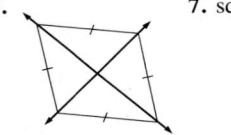 **7.** square **9.**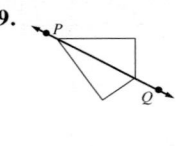

11. 3; 4; 5; 6; 7; 8 **13.** 4 **15.** 14 **17.** rotational
19. both **23.** No.

Page 279 Self-Test 2   1. trapezoid; $\overline{AC}$, $\overline{BD}$ **2.** regular
pentagon: $\overline{VS}$, $\overline{VT}$, $\overline{RU}$, $\overline{RT}$, $\overline{SU}$ **3.** cannot **4.** acute
5. 10 cm **6.** 58° **7.** 28 **8.** (quadrilateral diagram)

Pages 280–282 Chapter Review   1. line segment
2. parallel **3.** straight angle **4.** square **5.** trapezoid
6. equilateral triangle **7.** complementary angles
8. protractor **9.** obtuse angle **10.** acute triangle
11. $\overline{GH}$, $\overline{HG}$ **12.** ∠PQR, ∠RQP, or ∠Q **13.** Plane J
14. 37° **15.** 115° **16.** (angle diagram) **17.** (angle diagrams 25°, 95°)
18. (angle diagram 165°) **19.** (angle diagram 62°) **20.** 66°

21. 45° **22.** 17° **23.** 2° **24.** 150° **25.** 105° **26.** 95°
27. 58° **28.** 40° **29.** 140° **30.** 140° **31.** 140°
32. 40° **33.** 140° **34.** False. **35.** True. **36.** True.
37. False. **38.** True. **39.** False. **40.** rectangle; $\overline{AC}$,
$\overline{BD}$ **41.** triangle; no diagonals **42.** pentagon; $\overline{KH}$, $\overline{KI}$,
$\overline{GJ}$, $\overline{GI}$, $\overline{HJ}$ **43.** 35° **44.** 129° **45.** 130° **46.** can;
equilateral **47.** can; scalene **48.** can; scalene **49.** obtuse **50.** right **51.** acute **52.** acute **53.** 60° **54.** 60°
55. 60° **56.** 7 cm **57.** 60° **58.** 7 cm **59.** 18
60. No. **61.** Yes. **62.** Yes. **63.**

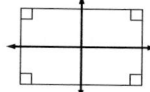

64. none **65.**

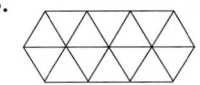

Page 285 Chapter Extension **1.** 4−4−4−4
3.

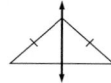

Code: 3−3−3−3−3−3

5. The codes for the combinations: 6−3−6−3,
3−4−6−4, 4−4−3−3, 3−3−4−3−4, 6−3−3−3−3

Pages 286−287 Cumulative Review **1.** C **3.** C
5. B **7.** D **9.** A **11.** C **13.** D **15.** D **17.** D
19. B

CHAPTER 7
Pages 291−292 Exercises **1.** composite **3.** prime
5. composite **7.** prime **9.** 3^3 **11.** $2 \cdot 11$ **13.** $2^2 \cdot 5$
15. $2^2 \cdot 7$ **17.** $2^2 \cdot 3^3$ **19.** $2^2 \cdot 5^3$ **21.** $2^4 \cdot 3^2 \cdot 7$
23. $3 \cdot 5^3 \cdot 7$ **25.** $1 \cdot 60$; $2 \cdot 30$; $3 \cdot 20$; $4 \cdot 15$; $5 \cdot 12$;
$6 \cdot 10$ **27.** y^5 **29.** y^3 **31.** The factors are a^n, a^{n-1},
a^{n-2},..., a^1, 1 **33.** 2, 3, 5, 7, 11, 13, 17, 19, 23, 29, 31,
37, 41, 43, 47, 53, 59, 61, 67, 71, 73, 79, 83, 89, 97
35. **a.** fewer than 21; the number of primes decreases.
b. 211, 223, 227, 229, 233, 239, 241, 251, 257, 263, 269,
271, 277, 281, 283, 293 **c.** 16 **d.** It supports the answer to (a). **37.** 84; 84; none; 24 **39.** $2^5 \cdot 3$ **41.** 0

Pages 294−295 Exercises **1.** 2 **3.** 1 **5.** 20 **7.** 8
9. $6x$ **11.** b **13.** 2 **15.** $4r^8$ **17.** $4n^2$ **19.** 6
21. Yes. **23.** No. **25.** Answers may vary. Examples:
51, 70; 52, 55; 81, 85 **27.** True. **29.** True. **31.** 2
33. about 12

Pages 298−299 Exercises **1.** 10 **3.** 27 **5.** 40
7. 420 **9.** $6x$ **11.** $12k$ **13.** $30rst$ **15.** $42a^9$
17. $150n^4$ **19.** 24 **21.** 216 **23.** 10,080 **25.** 2304
27. A factor is any of two or more numbers multiplied to
form a product. **29.** 15; 90 **31.** 44; 264 **33.** Answers
may vary. Example: 4 and 12 **35.** $14\frac{1}{3}$ **37.** $y + 7$

Page 299 Self-Test 1 **1.** prime **2.** prime
3. composite **4.** prime **5.** $2 \cdot 7$ **6.** $2^2 \cdot 3$ **7.** $3 \cdot 5 \cdot 7$
8. $2^4 \cdot 5^2$ **9.** 2 **10.** 9 **11.** $3c$ **12.** $8x^2$ **13.** 15
14. 80 **15.** $24awy$ **16.** $126n^7$

Pages 304−305 Exercises **1.** 2 **3.** 6 **5.** 20 **7.** 2
9. $\frac{1}{3}$ **11.** $\frac{3}{5}$ **13.** $\frac{4}{3}$ **15.** $\frac{8}{9}$ **17.** $\frac{3}{2}$ **19.** $\frac{1}{2}$ **21.** $\frac{2}{5}$ **23.** $\frac{6}{1}$
25. $\frac{4}{1}$ **27.** The denominators are 1; the fractions represent
whole numbers. **29.** always **31.** $\frac{1}{6}$
33.

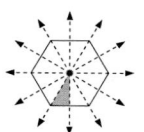

35. $\frac{1}{4}$ **37.** $\frac{1}{16}$ **39.** $\frac{4}{25}$ **41.** $\overrightarrow{MN}$

Page 305 Challenge **1.** 59 **2.** 53

Pages 307−308 Exercises **1.** $\frac{3b}{4}$ **3.** $\frac{4}{5}$ **5.** $\frac{2}{s}$ **7.** c^8
9. $6a^4$ **11.** 12 **13.** $\frac{5}{y}$ **15.** $8x^{16}$ **17.** $\frac{b^5}{6}$ **19.** $4n^2$
21. $\frac{v^3}{7}$ **23.** $\frac{4vw^3}{11}$ **25.** $7x$ **27.** $\frac{5m^5}{6n^2}$ **29.** $\frac{5s^4t^2}{2}$ **31.** $\frac{1}{a^5}$
33. a^7 **35.** 8 **37.** -8

Pages 311−312 Exercises **1.** > **3.** < **5.** < **7.** >
9. = **11.** > **13.** science book **15.** $\frac{1}{7} < \frac{1}{5} < \frac{1}{4}$
17. $\frac{3}{40} < \frac{3}{8} < \frac{7}{16}$ **19.** less than **21.** bar graph; the data
are not continuously changing. **23.** greater than; the distances are generally increasing. **25.** $\frac{1}{15}, \frac{1}{60}, \frac{1}{500}$ **27.** $\frac{1}{30}$
29. [figure] ; 1, 4, 9, 16, 25 **31.** 6
33. 192

Pages 315−316 Exercises **1.** 0.7 **3.** $0.\overline{81}$ **5.** 6.55
7. $10.\overline{6}$ **9.** $\frac{54}{125}$ **11.** $\frac{19}{100}$ **13.** $\frac{1}{6}$ **15.** $3\frac{8}{25}$ **17.** $9\frac{51}{100}$
19. $2\frac{2}{3}$ **21.** 0.875 **23.** about $\frac{1}{2}$ **25.** about $\frac{1}{3}$ **27.** about
$9\frac{3}{5}$ **29.** about $6\frac{1}{8}$ **31.** $0.\overline{1}$ **33.** $0.\overline{3}$ **35.** The numerator
is the repeating part of the decimal.

$\frac{1}{9} = 0.\overline{1}$	$\frac{5}{9} = 0.\overline{5}$
$\frac{2}{9} = 0.\overline{2}$	$\frac{6}{9} = 0.\overline{6}$
$\frac{3}{9} = 0.\overline{3}$	$\frac{7}{9} = 0.\overline{7}$
$\frac{4}{9} = 0.\overline{4}$	$\frac{8}{9} = 0.\overline{8}$

37. $0.\overline{18}$ **39.** $0.\overline{36}$ **41.** about $\frac{1}{2}$ **43.** less than

45. a. to account for the weight of the container
b. 1.01 lb   **47.** $\frac{1}{45}$: repeating; $\frac{1}{50}$: terminating; $\frac{1}{45} = 0.0\overline{2}$;
$\frac{1}{50} = 0.02$ **49.** 60°   **51.** about 4600

Pages 318–320 Problem Solving Situations   1. 5
blocks due west   **3.** 9   **5.** 6th floor   **7–11.** Strategies
may vary. Likely strategies are given.   **7.** making a table;
3   **9.** drawing a diagram; 4   **11.** drawing a diagram;
6 blocks due east   **13.** choosing the correct operation;
$16.92   **17.** about 85°F   **19.** 3 blocks due west

Page 320 Self-Test 2   1. $\frac{3}{4}$ **2.** $\frac{7}{8}$ **3.** $\frac{4}{5}$ **4.** $\frac{3}{8}$ **5.** $\frac{1}{3}$
6. $\frac{2}{3y}$ **7.** a^6 **8.** $8x^9$ **9.** <   **10.** <   **11.** >   **12.** 0.55
13. $0.\overline{8}$ **14.** 2.68   **15.** $4.8\overline{3}$ **16.** $\frac{39}{50}$ **17.** $\frac{111}{250}$ **18.** $3\frac{14}{25}$
19. $8\frac{2}{3}$ **20.** Place the 12 m and 19 m rods end-to-end, cre-
ating a 31 m length. Place the 26 m rod parallel to these
rods, lining up the ends. The difference in length is 5 m.

Pages 322–323 Exercises   1–7. Answers may vary.
Examples are given.   **1.** $\frac{63}{5}$ **3.** $\frac{-17}{4}$ **5.** $\frac{202}{10}$ **7.** $\frac{-7}{3}$
9. N **11.** L **13.** I **15.** left   **17.** <   **19.** <   **21.** >
23. $-\frac{1}{2}$ **25.** 0   **27.** $-\frac{1}{2}$ **29.** All   **31.** Some
33. rhombus   **35.** $17\frac{1}{16}$

Pages 326–327 Exercises
1. (number line −9 to 0)
3. (number line −3 to 3)
5. (number line −3 to 7)
7. (number line −5 to 5)
9. $-3\frac{1}{3}$ (number line −4 to 0)
11. $-6\frac{1}{7}$ (number line −6 to 0)
13. 7.5 (number line 0 to 8)   **15.** D   **17.** B   **19.** C
21. no difference   **23.** (number line −4 to 4)
25. (number line −3 to 3)
27. (number line −5 to 5)   **29.** 7; 4; 3; 0
31. (number line −1 to 1)   **33.** about 1000
35. 1.25×10^6 **37.** 132 mi   **39.** 405

Pages 330–331 Exercises   1. $\frac{1}{y^{12}}$ **3.** 1   **5.** $\frac{1}{125}$ **7.** $\frac{1}{16}$
9. 7.2×10^{-3} **11.** 6.012×10^{-4} **13.** 2.34×10^{-6}
15. 0.00116   **17.** 0.00000001   **19.** 0.0000000061
21. −3   **23.** D   **25.** 1500 times the actual size
27. 2×10^{-2} **29.** 13 mm   **31.** parallel   **33.** composite

Page 331 Self-Test 3   1–4. Answers may vary. Exam-
ples are given.   **1.** $\frac{34}{10}$ **2.** $\frac{-37}{8}$ **3.** $\frac{-9}{1}$ **4.** $\frac{11}{9}$
5. (number line −9 to 1)
6. (number line 0 to 6)
7. (number line −3 to 3)
8. (number line −2 to 6)
9. −2.5 (number line −4 to 0)
10. (number line −2 to 1)   **11.** $-\frac{1}{8}$ **12.** 1
13. $\frac{1}{w^3}$ **14.** $\frac{1}{256}$ **15.** 3×10^{-5} **16.** 0.000000243

Pages 332–333 Chapter Review   1. equivalent fractions   **2.** rational number   **3.** open sentence   **4.** repeating   **5.** composite   **6.** greatest common factor   **7.** prime
8. composite   **9.** composite   **10.** prime   **11.** $3 \cdot 5^2$
12. $2^3 \cdot 3^3$ **13.** $2 \cdot 5^2 \cdot 7$ **14.** $2^4 \cdot 3 \cdot 5^2$ **15.** 16
16. $6a$ **17.** $3x$ **18.** $18z^2$ **19.** 30   **20.** 24   **21.** $20xy$
22. $14c^2$ **23.** $\frac{1}{2}$ **24.** $\frac{2}{3}$ **25.** $\frac{4}{5}$ **26.** $\frac{1}{7}$ **27.** $\frac{24}{5}$ **28.** $\frac{9}{2}$
29. $2a$ **30.** $\frac{4}{5}$ **31.** v^7 **32.** x^{12} **33.** $4z^3$ **34.** $\frac{x^7}{2}$ **35.** >
36. <   **37.** >   **38.** >   **39.** 0.2   **40.** 0.3   **41.** $0.\overline{4}$
42. $0.\overline{45}$ **43.** 3.35   **44.** 5.22   **45.** $\frac{37}{100}$ **46.** $\frac{51}{200}$
47. $6\frac{7}{8}$ **48.** $4\frac{2}{3}$ **49.** Answers may vary. Example: Fill the
10 gal bucket and pour water from it into the 4 gal bucket.
This leaves 6 gal. Fill the 3 gal bucket and pour its water
into the 10 gal bucket. It contains 9 gal of water.   **50.** 5
51–56. Answers may vary. Examples are given.   **51.** $\frac{32}{10}$
52. $\frac{31}{5}$ **53.** $\frac{-7}{1}$ **54.** $\frac{0}{9}$ **55.** $\frac{37}{100}$ **56.** $\frac{-29}{3}$ **57.** D
58. F   **59.** B   **60.** A   **61.** C   **62.** E
63. (number line −8 to 4)
64. (number line −5 to 5)
65. −1.7 (number line −3 to 1)
66. $\frac{2}{3}$ (number line −2 to 3)   **67.** $\frac{1}{m^5}$ **68.** $\frac{1}{81}$
69. 1   **70.** $\frac{1}{9}$ **71.** 7.4×10^{-5} **72.** 8.21×10^{-4}

73. 6.9×10^{-7} **74.** 5.325×10^{-3} **75.** 0.0027
76. 0.0000055 **77.** 0.000000009 **78.** 0.000000324

Page 335 Chapter Extension **1.** $\frac{5}{9}$ **3.** $\frac{35}{99}$ **5.** $\frac{7}{3}$
7. $\frac{106}{33}$ **9.** $\frac{374}{333}$ **11.** $\frac{103}{18}$ **13.** They are equivalent.

Pages 336–337 Cumulative Review **1.** B **3.** D
5. B **7.** C **9.** C **11.** C **13.** D **15.** D **17.** D
19. D

CHAPTER 8

Pages 342–343 Exercises **1.** $\frac{32}{45}$ **3.** $-1\frac{7}{8}$ **5.** 1
7. $33\frac{1}{2}$ **9.** 56 **11.** -2.45 **13.** $\frac{1}{6}$ **15.** $\frac{5ab}{4}$
17–21. Estimates may vary. **17.** about 10 **19.** about
-2 **21.** about 27 yd **23.** $\frac{3}{8}$ or 0.375 **25.** 1.05 or $1\frac{1}{20}$
27. -24 **29.** $27\frac{3}{4}$ **31.** 34

33. $\boxed{4.}$ $\boxed{a^b/c}$ $\boxed{2.}$ $\boxed{a^b/c}$ $\boxed{7.}$ $\boxed{\times}$
$\boxed{2.}$ $\boxed{a^b/c}$ $\boxed{7.}$ $\boxed{=}$

35. $\boxed{4.}$ $\boxed{a^b/c}$ $\boxed{5.}$ $\boxed{\times}$ $\boxed{17.}$ $\boxed{a^b/c}$
$\boxed{3.}$ $\boxed{a^b/c}$ $\boxed{8.}$ $\boxed{=}$

37. about 18,000 **39.** 10 **41.** about 80

Pages 346–347 Exercises **1.** 2 **3.** -54 **5.** 3 **7.** $\frac{2}{9y}$
9. $-\frac{16}{25}$ **11.** $8\frac{1}{3}$ **13.** $\frac{1}{5y}$ **15.** -2 **17.** 8 **19.** -0.5
21. 0.75 **23.** more than **25.** $\frac{3}{4}$ c milk; 1 egg; $\frac{1}{4}$ c shortening; $2\frac{1}{4}$ c flour; 1 tsp cinnamon; $\frac{1}{3}$ c raisins; 1 tbsp baking soda; 1 c chopped apple **27.** $\frac{1}{4}$ **29.** 5000 **31.** about
130 **33.** $\frac{z}{9} = 15$

Page 347 Self-Test 1 **1.** $-3\frac{2}{3}$ **2.** 1.08 **3.** $\frac{2c}{21}$ **4.** $\frac{2m}{5}$
5. about -48 **6.** about 33 **7.** about 4 **8.** about -7
9. $-1\frac{3}{5}$ **10.** 18 **11.** -4.5 **12.** $\frac{1}{5}$ **13.** 6
14. $\frac{2}{15}$ **15.** $\frac{z}{10}$

Page 349 Exercises **1.** fraction; $18\frac{3}{4}$ yd **3.** whole
number; 19 **5.** fraction; $5\frac{1}{4}$ h **7.** decimal; $13.50
9. decimal; $10,995.54 **11.** whole number; 159 **13.** 5

Pages 354–355 Exercises **1.** $\frac{8}{15}$ **3.** 0 **5.** $2\frac{1}{2}$ **7.** $1\frac{2}{5}$
9. -4 **11.** $-\frac{11}{12}$ **13.** $\frac{12}{r}$ **15.** $\frac{2}{m}$ **17.** $\frac{2}{x}$ **19.** $8b$ **21.** $\frac{2x}{5}$
23. $2c$ **25.** $\frac{1}{2}$ h **27.** greater than **29.** less than
31. $23\frac{1}{8}$ in. **33.** $48\frac{3}{4}$ in. **35.** $\frac{5}{9}$ **37.** $\frac{9}{16}$ **39.** $\frac{9}{11}$ **41.** m

Pages 358–359 Exercises **1.** $1\frac{1}{2}$ **3.** $3\frac{2}{5}$ **5.** $1\frac{11}{20}$
7. $-5\frac{1}{6}$ **9.** $\frac{1}{6}$ **11.** $\frac{7}{12}$ **13.** $\frac{29m}{21}$ **15.** $\frac{17b}{4}$ **17.** $\frac{51h}{20}$
19. -3.6 **21.** 16.52 **23.** -4.37 **25.** $4\frac{5}{12}$ yd
27. $-20\frac{1}{4}$ or -20.25 **29.** $2\frac{7}{9}$ or $2.\overline{7}$ **31.** $-2\frac{3}{8}$ or -2.375
33. $9\frac{11}{12}$ **35.** $1\frac{1}{3}$ **37.** -8 **39.** 5.16 **41.** $4\frac{1}{2}$ in.
43. $-1\frac{1}{2}$; -5; -10 **45.** $x \rightarrow x + \frac{1}{3}$ **47.** Each fraction
is half the preceding fraction. **49. a.** $\frac{3}{4}$ **b.** $\frac{7}{8}$ **c.** $\frac{15}{16}$
d. $\frac{31}{32}$ **51.** $\frac{63}{64}$
53.
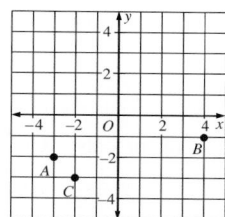

Populations of Large Cities

New York
London
Tokyo
Mexico City
Beijing

$\dagger$ = one million people

55. $\frac{19a}{15}$

Pages 363–364 Exercises **1–15.** Estimates may vary.
1. about -5 **3.** about 11 **5.** about -5 **7.** about 11
9. about 4 **11.** about -48 **13.** about 32 **15.** about
33 mi **17.** unreasonable; $-5\frac{23}{39}$ **19.** reasonable **21.** $\frac{47}{10}$
23.

Page 364 Self-Test 2 **1.** fraction; $3\frac{1}{2}$ h **2.** decimal;
$4.35 **3.** $\frac{1}{7}$ **4.** -17 **5.** $\frac{5a}{7}$ **6.** $-3\frac{1}{9}$ **7.** $\frac{27b}{4}$
8. -3.8 **9–11.** Estimates may vary. **9.** about 15
10. about -19 **11.** about 5

Pages 366–367 Problem Solving Situations **1.** $2\frac{1}{3}$h
3. 140 years **5–9.** Strategies may vary. Likely strategies
are given. **5.** guess and check; 2 paperbacks, 5 children's
books **7.** supplying missing facts; 5 lb **9.** not enough
information; the hourly rate is needed. **13.** $>$ **15.** -3

Page 367 Challenge 2, 3, and 6

Pages 369–370 Exercises **1.** $-\frac{2}{3}$ **3.** $\frac{1}{9}$ **5.** $-\frac{5}{6}$
7. 4.21 **9.** -0.81 **11.** -1.1 **13.** $30\frac{1}{4}$ in. **15.** $14.25
17. D **19.** A **21.** $3q - 11.3 = -8.9$ **23.** $16 = t + \frac{3}{5}$

25. $0.55; 0.8; 0.8; 3.4$ **27.** $n - \left(-\frac{3}{5}\right) = 4\frac{2}{3}; 4\frac{1}{15}$
29. $0°; 90°$ **31.** $1\frac{1}{10}$

Pages 372–374 Exercises   **1.** 44 **3.** $-10\frac{4}{5}$ **5.** 22
7. $-17\frac{1}{2}$ **9.** $37\frac{1}{2}$ **11.** $-9\frac{1}{3}$ **13.** 7 **15.** $\frac{1}{8}$ **17.** $-\frac{1}{5}$
19. $-\frac{1}{16}$ **23.** 4 **25.** $\frac{4}{3}$ **27.** 1 **29.** $2\frac{2}{5}$ **31.** $5\frac{1}{15}$
33. $-2\frac{1}{7}$ **35.** $\frac{2}{3}t = -14$ **37.** $\frac{1}{2}g - 3 = \frac{5}{6}$ **39–43.** Answers will vary.   **39.** One third of a number g is fourteen.
41. Negative two is one third of a number x, increased by nine.   **43.** Ten is three eighths of a number s, increased by one half.   **45.** 20 is the LCM.   **47.** distributive property
49. 1 block due west   **51.** $138°$ **53.** $-\frac{4}{5}$

Pages 376–377 Exercises   **1.** $h = \frac{A}{b}$ **3.** $b = P - a - c$
5. $W = Fd$ **7.** $P = \frac{I}{rt}$ **9.** $E = \frac{W}{I}$ **11.** $a = 2(P - 110)$
13. 5 lb   **15. a.** $-17\frac{7}{9}°$C   **b.** $0°$C   **c.** $100°$C
17. a. $I = \frac{E}{R}$ **b.** 40 amperes   **19.** $f = 22h + 30$
21. $h = \frac{1}{22}(f - 30)$ **25.** $s = \frac{2A}{p}$

Page 377 Self-Test 3   **1.** $31\frac{2}{3}$ yd   **2.** $\frac{1}{5}$ **3.** 4.875
4. $2\frac{5}{14}$ **5.** -108 **6.** 35 **7.** -16 **8.** $h = \frac{V}{lw}$ **9.** 2 m

Page 379 Exercises   **1.** BookClb   **3.** BayOil   **5.** Hi
7. $30.13   **9.** $2987.50   **11.** $27.75   **13.** $18.75
15. $800; $13.25

Pages 380–381 Chapter Review   **1.** multiplicative inverse   **2.** multiplication property of reciprocals   **3.** $\frac{15}{32}$
4. $1\frac{1}{4}$ **5.** $14\frac{2}{3}$ **6.** $1\frac{5}{6}$ **7.** $-4\frac{1}{2}$ **8.** 14 **9.** $-\frac{8}{27}$ **10.** $-\frac{4}{9}$
11. $-3\frac{3}{5}$ **12.** $-1\frac{29}{36}$ **13.** 2 **14.** $-\frac{13}{36}$ **15.** $-4\frac{1}{6}$ **16.** $3\frac{1}{3}$
17. 2 **18.** $-3\frac{7}{10}$ **19.** 0.54 **20.** -2.4 **21.** -2.1
22. 40 **23.** -6.3 **24.** 3.9 **25.** -58.75 **26.** 2.7
27. $\frac{3}{20}$ **28.** $2z$ **29.** $\frac{9}{14}$ **30.** $\frac{3}{22}$ **31.** $\frac{11}{x}$ **32.** $\frac{3a}{4}$ **33.** $\frac{r}{6}$
34. $\frac{3m}{2}$ **35.** $\frac{65n}{9}$ **36.** $\frac{29c}{7}$ **37.** $\frac{5}{2n}$ **38.** $\frac{47m}{24}$
39. fraction; $11\frac{2}{3}$ h   **40.** decimal; $52.50   **41.** whole number; 7   **42.** fraction; $1\frac{1}{4}$ lb   **43–54.** Estimates may vary.   **43.** about 8   **44.** about 30   **45.** about -5
46. about 3   **47.** about 6   **48.** about 11   **49.** about 26
50. about -72 **51.** about -16 **52.** about -8
53. about -5 **54.** about 3   **55.** 57 in.   **56.** $29\frac{3}{4}$ mi
57. 42 muffins   **58.** 6 min   **59.** $\frac{4}{5}$ **60.** $-\frac{1}{3}$ **61.** -3
62. 0.285 **63.** $-4\frac{2}{5}$ **64.** $3\frac{7}{8}$ **65.** $-1\frac{1}{5}$ **66.** $\frac{3}{4}$
67. -0.6 **68.** -0.2 **69.** $1\frac{1}{2}$ **70.** $-\frac{5}{8}$ **71.** 0.7

72. 1.4 **73.** $-\frac{7}{10}$ **74.** $1\frac{7}{12}$ **75.** 48 **76.** $-12\frac{1}{2}$ **77.** $2\frac{2}{5}$
78. -15 **79.** $-10\frac{2}{3}$ **80.** 21 **81.** -16 **82.** $-7\frac{1}{2}$
83. $r = \frac{d}{2}$ **84.** $m = \frac{F}{a}$ **85.** $p = \frac{2A}{s}$ **86.** $M = 2a + m$
87. $30°$C   **88.** $17\frac{1}{2}$ ft

Page 383 Chapter Extension   **1.** $\frac{11}{5x}$ **3.** $\frac{3}{2a^2}$ **5.** $\frac{22x}{3yz}$
7. $\frac{1}{4xy}$ **9.** $\frac{3}{2n^3}$ **11.** $\frac{3r}{8s}$ **13.** $\frac{19}{12m}$ **15.** $\frac{59}{20rs}$ **17.** $\frac{52b}{15c^3}$
19. $\frac{8}{15r}$ **21.** $\frac{49r}{12s^3}$ **23.** $\frac{31u}{24vw}$ **25.** $\frac{35}{3mn}$ **27.** $\frac{15}{8x^2}$
29. No; $4y$ is not a factor of the numerator.

Pages 384–385 Cumulative Review   **1.** C   **3.** B
5. D   **7.** C   **9.** C   **11.** A   **13.** C   **15.** D   **17.** B
19. C

CHAPTER 9
Pages 389–390 Exercises   **1.** $\frac{3}{4}$ **3.** $\frac{5}{3}$ **5.** $\frac{2}{1}$ **7.** $\frac{8}{5}$
9. $\frac{2}{1}$ **11.** $\frac{3}{8}$ **13.** $\frac{3}{4}$ **15.** 50 mi/h   **17.** 25 mi/gal
19. 2.2 m/s   **21.** $\frac{9}{1}$ **23.** $\frac{x}{2y}$ **25.** $\frac{3}{1}$ **27.** 27   **29.** $\frac{5}{3}$
31. 12 pens for $4.80   **33.** Answers will vary. Example: when the item is available only in an inconvenient size
35. 24   **37.** $2^4 \cdot 3 \cdot 5$

Pages 393–394 Exercises   **1.** True.   **3.** False.
5. False.   **7.** True.   **9.** 6   **11.** 11   **13.** 8   **15.** 9
17. Yes; $(4)(81) = (9)(36)$ **19.** 10.5   **21.** 10 or -10
23. 12   **25.** 4   **27.** 14   **29.** C   **31.** D   **33.** 18
35. 5.2   **37.** 14 ft   **39.** $\frac{4xy^2}{z}$ **41.** $-\frac{2}{9}$

Pages 397–398 Exercises   **1.** 50 ft   **3.** $18\frac{3}{4}$ ft
5. $1\frac{5}{8}$ in. wide; $1\frac{3}{4}$ in. long   **7.** 35 cm   **9.** 2 in.: 5 ft
13. Answers will vary. Example: A person is about 6 ft tall, so a 21 ft statue indicates a scale of 6 ft : 21 ft, or 2 ft : 7 ft.   **15.** The heads on Mount Rushmore are about 60 ft high, which makes the scale about 1 in. : 80 in.   **17.** 130 in.

Pages 400–401 Problem Solving Situations   **1.** 266
3. $225   **5–9.** Strategies may vary; likely strategies are given.   **5.** drawing a diagram; 6 blocks due south
7. using equations; 34   **9.** identifying a pattern; 83

Page 401 Self-Test 1   **1.** $\frac{3}{5}$ **2.** $\frac{1}{4}$ **3.** $\frac{2}{1}$ **4.** $\frac{3}{2}$
5. 55 mi/h   **6.** $4/lb   **7.** $.65/comb   **8.** True.   **9.** True.
10. False.   **11.** False.   **12.** 10   **13.** 4   **14.** $33\frac{1}{3}$
15. 11.5   **16.** $7\frac{11}{12}$ ft   **17.** $99

Pages 406–407 Exercises   **1.** 8%   **3.** 70%   **5.** $62\frac{1}{2}$%
7. 21%   **9.** 9.9%   **11.** $15\frac{1}{5}$%   **13.** $46\frac{7}{8}$%   **15.** $228\frac{4}{7}$%
17. 0.49   **19.** 5.87   **21.** $\frac{1}{2}$ **23.** $\frac{23}{1000}$ **25.** $\frac{3}{2000}$

27. 40% **29.** 60% **31.** 2; 0.2; 0.02; 0.002; 0.0002; 0.00002 **33.** 900; 90; 9; 0.9; 0.09; 0.009 **35.** 19; 100 **37.** 100; 200 **39.** $\frac{3}{40}$; 0.075 **41.** $\frac{11}{37}$

Page 407 Challenge 125%

Pages 410–411 Exercises **1.** $94.40 **3.** 240 **5.** 456 **7.** $15.75 **9–13.** Estimates may vary. **9.** about 35 **11.** about 50 **13.** about $6

15. $\boxed{5.}$ $\boxed{\%}$ $\boxed{\times}$ $\boxed{312.}$ $\boxed{=}$;
$\boxed{312.}$ $\boxed{\times}$ $\boxed{5.}$ $\boxed{\%}$

17. $\boxed{16.}$ $\boxed{\%}$ $\boxed{\times}$ $\boxed{975.}$ $\boxed{=}$;
$\boxed{975.}$ $\boxed{\times}$ $\boxed{16.}$ $\boxed{\%}$

19. $\boxed{0.7}$ $\boxed{\%}$ $\boxed{\times}$ $\boxed{34.9}$ $\boxed{=}$;
$\boxed{34.9}$ $\boxed{\times}$ $\boxed{0.7}$ $\boxed{\%}$

21. a. $2.40 **b.** $1.20 **c.** $9.60 **23. a.** 0.46 **b.** 0.23 **c.** 0.92 **25.** < **27.** < **29.** = **31.** = **33.** $30 **35.** $60 **37.** $32 **39.** $64 **41.** $34 **43.** $68 **45.** $187.50 **47.** 3.6

Pages 413–414 Exercises **1.** 80% **3.** 70% **5.** 96% **7.** 100% **9.** $33\frac{1}{3}\%$ **11.** 2.5% **13.** $87\frac{1}{2}\%$ **15.** $62\frac{1}{2}\%$ **17.** 80% **19.** $59\frac{1}{11}\%$ **21.** $50 **23.** $487.50 **25.** 4.6% **27.** > **29.** 120

Pages 416–418 Exercises **1.** 420 **3.** 34 **5.** 2000 **7.** 80 **9.** 45 **11.** 36,000 **13.** 50 **15.** $4200 **17.** 180 **19.** 100% **21. a.** 6 g **b.** 15% **c.** 40 g **23.** Since the cereal alone provides 100% of the RDA for Vitamin B_{12}, and more is provided by the milk, together they provide more than 100% of the RDA for vitamin B_{12}. **25.** $19.50 **27.** $8\frac{1}{2}\%$ **29.** $\frac{5}{8}$ **31.** $-1\frac{1}{3}$

Page 418 Self-Test 2 **1.** 65% **2.** 45% **3.** 1.3 **4.** $\frac{7}{25}$ **5.** 18.85 **6.** 6.75 **7.** $16\frac{2}{3}\%$ **8.** 50% **9.** 380 **10.** 160

Pages 420–421 Exercises **1.** 18% **3.** 20% **5.** 100% **7.** 21% **9.** $86\frac{1}{9}\%$ **11.** $166\frac{2}{3}\%$ **13.** $62\frac{1}{2}\%$ **15.** increase; 125% **17.** decrease; $16\frac{2}{3}\%$ **19.** Answers will vary. **21.** True; half an amount x is $\frac{1}{2}x$, and a decrease of 50% is $x - 0.5x = 0.5x$. **23.** $\frac{18}{1}$ **25.** 20%

Pages 426–427 Exercises **1.** 9.44 **3.** 3500 **5.** 15% **7.** 245 **9.** 1056 **11.** 7% **13.** 2800 **15.** 4500 **17.** $1.18–$1.35 **19.** $1.05 **21.** 6.5% **25.** 94.3% **27.** 5.8×10^{-5} **29.** $87\frac{1}{2}\%$

Page 427 Self-Test 3 **1.** 250% **2.** 10% **3.** 20% **4.** $33\frac{1}{3}\%$ **5.** 3.2 **6.** 1200 **7.** 40 **8.** 40%

Pages 428–429 Chapter Review **1.** scale **2.** unit rate **3.** ratio **4.** proportion **5.** percent **6.** terms **7.** $\frac{1}{5}$ **8.** $\frac{1}{5}$ **9.** $\frac{2}{11}$ **10.** $\frac{3}{2}$ **11.** 7 h/day **12.** 27 words/min **13.** $4.25/h **14.** False. **15.** True. **16.** False. **17.** False. **18.** 1 **19.** 4 **20.** 1 **21.** 23 **22.** 20 ft **23.** 15 in. **24.** 400 min **25.** $16.25 **26.** 55% **27.** $43\frac{3}{4}\%$ **28.** 275% **29.** $33\frac{1}{3}\%$ **30.** $62\frac{1}{2}\%$ **31.** 65% **32.** 3% **33.** 120% **34.** 42.8% **35.** 0.95% **36.** $\frac{3}{4}$; 0.75 **37.** $2\frac{1}{2}$; 2.5 **38.** $\frac{21}{400}$; 0.0525 **39.** $\frac{39}{200}$; 0.195 **40.** $\frac{1}{2000}$; 0.0005 **41.** about 90 **42.** about 40 **43.** about 8 **44.** about 60 **45.** 19 **46.** 14 **47.** 50% **48.** 80% **49.** 4% **50.** $2\frac{1}{2}\%$ **51.** 600 **52.** 42 **53.** 70 **54.** 364 **55.** 300% **56.** $6\frac{2}{3}\%$ **57.** 50% **58.** 20% **59.** 72 **60.** 4.5 **61.** $37\frac{1}{2}\%$ **62.** 2.88

Page 431 Chapter Extension **1.** $3825.24 **3.** $4706.34 **5.** $6092.01

Pages 432–433 Cumulative Review **1.** C **3.** B **5.** A **7.** D **9.** C **11.** D **13.** B **15.** D **17.** D **19.** B

CHAPTER 10
Pages 438–439 Exercises **1.** 10 cm **3.** 11.55 m **5.** 28 cm **7.** $16\frac{1}{8}$ in. **9.** 355 ft or $118\frac{1}{3}$ yd **11.** 11.9 m **13.** $2\frac{5}{8}$ in. **15.** not enough information; the length is needed. **17.** not enough information; the perimeter or if the pentagon is regular is needed. **19.** 20 **21.** 36 **23.** 33.5 mm **25.** 20.7; 20; 18; 7 **27.** 57 in. **29.** 150°

Pages 443–444 Exercises **1.** circle O **3.** 8.6 cm **5.** 6.3 m **7.** $14\frac{2}{3}$ yd **9.** 44 mm **11.** 16.3 m **13.** 9 m; 18 m **15.** 7 yd; 14 yd **17.** 50 m **19.** approximately **21.** approximately **25.** 220 in. **27.** 288 **29.** 31.5 mi **31.** 125 cm **33.** right

Pages 448–449 Exercises **1.** $17\frac{1}{2}$ ft² **3.** 400 in.² or $2\frac{7}{9}$ ft² **5.** 54 cm² **7.** $6\frac{1}{4}$ ft² **9.** 63 yd² or 567 ft² **11.** 27 square units **13.** b_1 **15.** $\frac{1}{2}hb_1 + \frac{1}{2}hb_2$ **17.** 55 m² **19.** $1710 **21.** $610.45 **23.** $\frac{3x}{8}$

Pages 454–456 Exercises **1.** 314 in.² **3.** 28.3 yd² **5.** $17\frac{1}{9}$ mi² **7.** 6.2 cm² **9.** 2826 mi² **11.** B **13.** C

15. Find half the area of a circle.   **17.** 20 ft²   **19.** 66 m²
21. 32.1 in.²   **23.** 84.8 ft²   **25.** 2π; 4π; 8π **27.** Each
number has π as a factor; for circumference, the radius
doubles; for area, the radius is squared.   **29.** The circumference doubles.   **31.** 0.267   **33.** about 9

Page 458 Exercises   1. circumference   **3.** perimeter
5. area; 706.5 ft²   **7.** perimeter; 112 yd   **9.** area;
1102.1 ft²   **11.** −5   **13.** circumference

Pages 460–461 Problem Solving Situations   1. The
number of books must be a whole number.   **3.** There are
no amounts of dimes and nickels totaling 12 that result in
\$.60.   **5–9.** Strategies will vary.   **5.** supplying missing
facts; \$32   **7.** choosing the correct operation; 2.5 lb
9. using equations; 22.5 m, 22.5 m   **13.** $x + 9$ **15.** $38\frac{2}{5}$

Page 461 Self-Test 1   1. 44.8 cm   **2.** $31\frac{1}{2}$ in.

3. a. 14 m   **b.** 44 m   **4.** $30\frac{1}{4}$ in.²   **5.** 36 cm²

6. 221.6 cm²   **7.** circumference   **8.** The sum of the
lengths of the two sides given is equal to the perimeter.

Pages 464–465 Exercises   1. $\angle D$ **3.** $\overline{BA}$ **5.** $CBED$
7. $\angle DEB$ **9.** $\overline{OM}$ **11.** False; a rectangle with sides of
lengths 1 and 4 and a rectangle with sides of lengths 2 and
3.   **13.** True.   **15.** False; a rectangle with sides of lengths
3 and 4 and a rectangle with sides of lengths 2 and 6.
17. $(1, -1)$ **19.** $\frac{23a}{20}$ **21.** 40

Pages 468–470 Exercises   1. $\underline{12}$; 12.5   **3.** 5; 12.5
5. 57 cm   **7.** Yes; the length of $\overline{MN}$ is 10 cm because corresponding sides are congruent.   **9.** 32   **11.** $\frac{8}{7}$; they are
equal.   **13.** always   **15.** sometimes   **17.** always
19. never   **21.** Answers will vary.   **23.** They are similar.
25. 6.75   **27.** 616 cm²   **29.** 0.000689

Pages 473–474 Exercises   1. 15 m   **3.** 10 ft   **5.** 16 ft
6 in.   **7.** They are vertical angles.   **9.** EDC **11.** 72 ft
13. 46 cm   **15.** 16

Pages 477–479 Exercises   1. 5   **3.** 1   **5.** −30   **7.** $\frac{3}{8}$
9. −0.7   **11.** −9.434   **13.** 11.832   **15.** 20   **17.** 0.2
19. −5.745   **21.** 1.2   **23.** 14   **25.** 7   **27.** 12   **29.** 225,
256, 289   **31.** 56 m   **33.** 55   **35.** 0.6   **37.** $2\frac{1}{2}$ **39.** 100
41. 6 and 7   **43.** −5 and −4   **45.** b   **47.** b   **49.** 1.618
51. II   **53.** 16

Page 479 Challenge   20.25 cm²; 24 cm

Pages 483–485 Exercises   1. 7.2 cm   **3.** 9 in.
5. 7.1 cm   **7.** Yes.   **9.** Yes.   **11.** Yes.   **13.** 10.3 yd
15. 6.6 ft   **17.** 6.3 km   **19.** 100 yd   **21.** 3; 4; 5

23.

x	y	a	b	c
2	1	3	4	5
3	1	8	6	10
3	2	5	12	13
4	1	15	8	17
4	2	12	16	20
4	3	7	24	25
5	1	24	10	26
5	2	21	20	29
5	3	16	30	34
5	4	9	40	41
6	1	35	12	37
6	2	32	24	40
6	3	27	36	45
6	4	20	48	52
6	5	11	60	61

25. $-\frac{7}{8}$ **27.**

Page 485 Self-Test 2   1. $\overline{BD}$ **2.** $\angle S$ **3.** 15.75
4. 20 ft   **5.** 90   **6.** $\frac{5}{8}$ **7.** 11.180   **8.** No.   **9.** 8.1

Pages 486–487 Chapter Review   1. circumference
2. hypotenuse   **3.** congruent   **4.** radius   **5.** perfect
square   **6.** area   **7.** 37.2 cm   **8.** $34\frac{1}{2}$ ft   **9.** circle R
10. radii: $\overline{RS}$, $\overline{RT}$, $\overline{RW}$; chords: $\overline{WT}$, $\overline{VU}$; diameter $\overline{WT}$
11. 9 cm   **12.** 18 cm   **13.** 56.5 cm   **14.** $30\frac{1}{4}$ ft²
15. 12.48 m²   **16.** 54 in.²   **17.** 803.8 cm²   **18.** area
19. perimeter   **20.** There are no amounts of dimes and
quarters totaling 5 which result in \$1.   **21.** The two
lengths together equal the perimeter.   **22.** $\overline{EF}$ **23.** $\angle D$
24. $\overline{DE}$ **25.** $\angle E$ **26.** 2   **27.** 9   **28.** 25 ft   **29.** 30
30. $\frac{10}{11}$ **31.** 8.718   **32.** 11.916   **33.** Yes.   **34.** 12

Page 489 Chapter Extension   1. $2\sqrt{3}$ **3.** $5\sqrt{2}$
5. $2\sqrt{2}$ **7.** $2\sqrt{10}$ **9.** $2\sqrt{7}$ **11.** $4\sqrt{2}$ **13.** $2\sqrt{31}$
15. $2\sqrt{15}$ **17. a.** 8.944   **b.** 8.944   **c.** They are equal.

Pages 490–491 Cumulative Review   1. B   **3.** B
5. C   **7.** A   **9.** D   **11.** B   **13.** A   **15, 17, 19.** B

CHAPTER 11
Pages 495–496 Exercises
1.   Compact Disc Prices (Dollars)

Price	Tally	Frequency
7.00-8.99	||	2
9.00-10.99	||	2
11.00-12.99	ⷭ ||	7
13.00-14.99	ⷭ ||	7
15.00-16.99	||	2

3. Annual Tuition at Private Colleges (Dollars)

Tuition	Tally	Frequency
6000-7999	ЖІ ІІ	7
8000-9999	ІІІ	3
10,000-11,999	ІІІІ	4
12,000-13,999	ЖІ І	6
14,000-15,999	І	1

5. Intervals may vary.

Cost of One Week Vacation (Dollars)

Cost	Tally	Frequency
850-1049	ІІІІ	4
1050-1249	ЖІ І	6
1250-1449	ЖІ	5
1450-1649	ЖІ	5
1650-1849	ІІІІ	4

7. The intervals overlap. It cannot be determined if a score of 70 is in the interval 60-70 or 70-80. **9.** 2x − 3

11. Number of Hours Worked Per Week

Hours	Tally	Frequency
20-24	ЖІ ЖІ І	11
25-29	ЖІ ІІІІ	9
30-34		0
35-39	ЖІ ЖІ І	11
40-45	ЖІ	5

Page 496 Challenge   16

Pages 499–500 Exercises
1. Earned Run Averages of Leading Pitchers

2.2	0	5		
2.3	2			
2.4				
2.5	6			
2.6	2	2	7	
2.7	7	8		
2.8	0	2	2	4
2.9	0	0	2	

3. $25   **5.** 3   **7.** 12   **9.** 13
11. Age at Inauguration of United States Presidents

4	2 3 6 7 8 9 9
5	0 0 1 1 1 1 1 2 2 4 4 4 4
	5 5 5 6 6 6 7 7 7 7 8
6	0 1 1 1 2 4 4 5 8 9

13. 19   **15.** 7   **17.** 5n + 4; 6n + 4; 7n + 4   **19.** 8

Pages 503–504 Exercises   **1.** 6   **3.** 14   **5.** increase
7. 6   **9.** 7   **11.** 20   **13.**

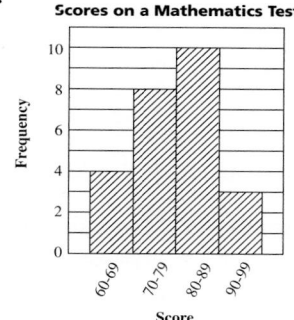
Scores on a Mathematics Test

15. 8   **17.** B   **19.**

Hourly Wages of Part-time Baby Sitters

21. Answers will vary. The graph in Exercise 19 shows the general trend in salaries rather than the individual salaries.
23. 0.65   **25.** 90

Pages 507–508 Exercises
1.

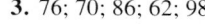
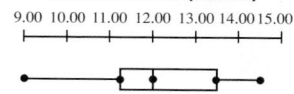
Box Seat Prices at Baseball Games (Dollars)

3. 76; 70; 86; 62; 98

5. Ken Jameson's   **7.** Ken Jameson's   **9.** always
11. sometimes   **13.**

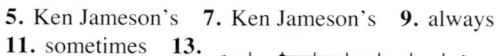

15.

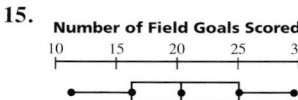
Number of Field Goals Scored

Page 508 Self-Test 1

1.

Scores on an Algebra Test

Score	Tally	Frequency
60-69	|||	3
70-79	||||| ||	7
80-89	|||||		6
90-99	||||	4

2. **Scores on an Algebra Test**

6	1	8	9				
7	1	4	6	6	7	7	7
8	3	3	4	5	7	8	
9	1	2	2	9			

3. 18.　**4.** 22　**5.** Ken Jameson's

Pages 511–512 Exercises　**1.** 165 lb　**3.** 3　**5.** 64 in.
7. 145,155　**9.** negative　**11.** 14
13.　　　　　　　　　　　　　　　　　　　**15.** 3

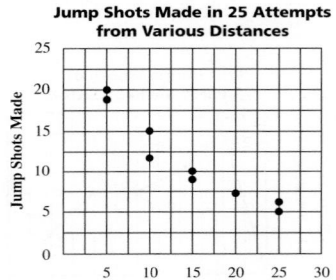

Jump Shots Made in 25 Attempts from Various Distances

17.　　　　　　　　　　　　　　　**19.** negative

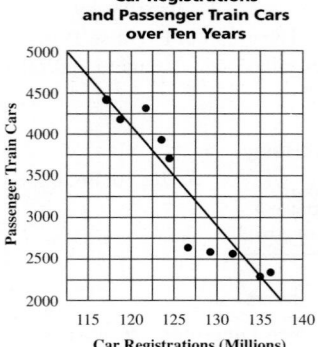

Car Registrations and Passenger Train Cars over Ten Years

23. 17.1; 15.5; 12; 18　**25.** 3.1　**27.** 36

Pages 515–516 Exercises　**1.** 70%　**3.** 108　**5.** 400
7. $3750　**9.** 2%　**11.** 94%　**13.** Estimates may vary.
about $\frac{1}{3}$　**15.** $\frac{3}{20}$　**17.** $\frac{33}{100}$　**19.** 9 h　**21.** 30 h
23.　　　　　　　　　　　　　　　　　**25.** 63 in.^{2}

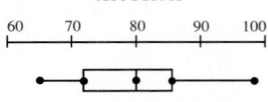

Test Scores

Pages 519–520 Exercises
1.

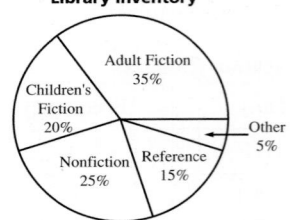

Library Inventory

3.　　　　　　　　　　　**5.** C　**7.** B　**9.** 57.4%

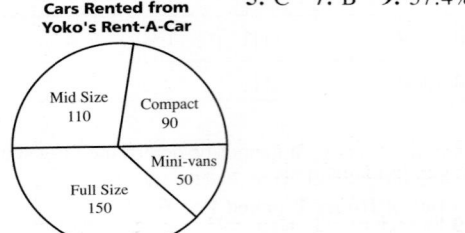

Cars Rented from Yoko's Rent-A-Car

11. about $\frac{2}{5}$　**13.** 4.7 cm^2　**15.** 84.8 m^2　**17.** -13

Page 521 Exercises　**1.** 6　**3.** $\frac{1}{6}$　**5.** 1　**7.** 43%
9. The percent of total sales for each sector.

Page 523 Exercises
1. box-and-whisker plot;

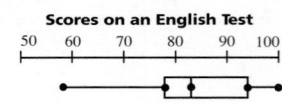

Scores on an English Test

3. circle graph;

Survey of the Number of Women in Certain Occupations

5. 90

Pages 525–527 Problem Solving Situations   1. Ann Fernandez teaches typing, Ellen Taylor teaches mathematics, Sonia Morris teaches French.   **3.** Marv is the treasurer, Shirley is the president, and Stan is the secretary. **5–9.** Strategies may vary; likely strategies are given. **5.** making a table; 6   **7.** using logical reasoning; Aaron **9.** no solution; the number of pumpkins must be a whole number.   **11.** Answers will vary.   **13.** Torres is the doctor, Wong is the dentist, and Lipshaw is the journalist.

Page 527 Self-Test 2   1. 2   **2.** positive   **3.** $210,000 **4.**   **5.** scattergram;

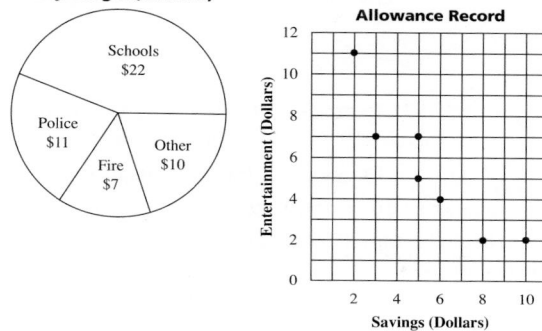

6. Harry majors in physics, Matt majors in history, and Phyllis majors in mathematics.

Page 529 Exercises   1. Detroit had the greatest percentage decrease, while Chicago had the greatest decrease in number of people.   **3.** Los Angeles, Houston, San Diego, Dallas, Phoenix, San Antonio; Southwest   **5.** about 1,318,919   **7.** 2,968,750   **9.** about 6,998,013

Pages 530–532 Chapter Review   1. histogram **2.** circle graph   **3.** positive correlation   **4.** third quartile **5.** leaf   **6.** frequency polygon **7.   Points Scored Per Game**

Points	Tally	Frequency
16-20	⫽⫽⫽ ‖	7
21-25	⫽⫽⫽ ‖‖‖‖	9
26-30	‖	2
31-35	‖‖‖	3

8.   Points Scored per Game

1	6 7 8 8 8 8
2	0 2 2 2 3 3 4 4 4 5 7
3	0 1 1 3

9. 100   **10.** 2   **11.** 17   **12.** 84.8; 88; 88; 40

13.

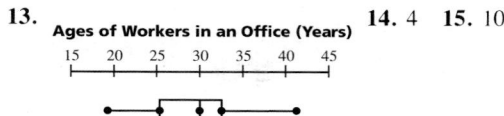

Ages of Workers in an Office (Years)

14. 4   **15.** 10

16. 3   **17.** 11   **18.** 32; 28; 38; 50; 22   **19.** Thames **20.** Charles   **21.**

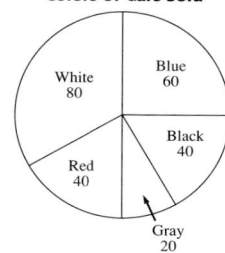

Colors of Cars Sold

22. $30,000

23. 5   **24.** $45,000   **25.** positive **26.** $13,500   **27.** $35,000   **28.** $1,500,000   **29.** 15% **30.** stem-and-leaf plot;

Leading Batting Averages

0.30	1 3 4 4 5 6 6 7 8
0.31	1 2 2
0.32	2
0.33	
0.34	
0.35	6
0.36	6

31. Ed is the artist, Alice is the writer, and Will is the editor.

Page 535 Chapter Extension   1. 96   **3.** 77   **5.** 84 **7.** 90   **9.** 84   **11.** 80

Pages 536–537 Cumulative Review   1. D   **3.** D **5.** D   **7.** B   **9.** C   **11.** C   **13.** A   **15.** C   **17.** B **19.** B

CHAPTER 12
Pages 543–544 Exercises   1. $\frac{1}{3}$ **3.** $\frac{1}{2}$ **5.** $\frac{2}{3}$ **7.** $\frac{1}{2}$ **9.** 0   **11.** 1 to 4   **13.** 3 to 2   **15.** 3 to 2   **17.** 3 to 2 **19.** 2 to 3   **21. a.** $\frac{1}{2}$ **b.** 1 to 1   **23.** $\frac{1}{6}$ **25.** $\frac{1}{2}$ **27.** $\frac{5}{6}$ **29.** 1   **31.** $\frac{5}{8}$ **33.** $\frac{5}{8}$ **35.** 5 to 3   **37.** 13 to 3   **39.** 7 to 9 **41.** about $\frac{2}{5}$ **43.** $\frac{4}{9}$; $\frac{5}{9}$ **45. a.** 1; P(E)   **b.** $\frac{2}{3}$ **47.** 56.5 in.

Page 544 Challenge   15

Pages 546–547 Exercises

1.

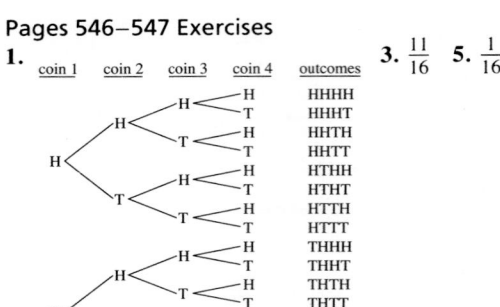

3. $\frac{11}{16}$ **5.** $\frac{1}{16}$

7. $\frac{1}{18}$ **9.** $\frac{1}{9}$ **11.** $\frac{5}{18}$ **13.** $\frac{1}{3}$ **15.** 0   **17.** $\frac{1}{12}$ **19.** $\frac{1}{12}$
21. $\frac{1}{24}$ **23.** 6   **25.** 0.684

Pages 550–551 Exercises   1. 1440   **3.** 216   **5.** 144
7. $\frac{1}{12}$ **9.** $\frac{2}{21}$ **11.** 30   **13.** $\frac{1}{16}$ **15. a.** $\frac{1}{6}$ **b.** $\frac{1}{2}$ **17.** $\frac{2}{45}$
19. 10   **21.** $\frac{1}{10}$ **23.** $\frac{1}{26}$ **25.** 4

Page 551 Self-Test 1   1. $\frac{1}{2}$ **2.** $\frac{5}{6}$ **3.** 1 to 1   **4.** 1 to 1
5. 5 to 1   **6.**   **7.** $\frac{1}{4}$ **8.** 1296   **9.** $\frac{1}{128}$

number cube	coin	outcomes
1	H	1 H
	T	1 T
2	H	2 H
	T	2 T
3	H	3 H
	T	3 T
4	H	4 H
	T	4 T
5	H	5 H
	T	5 T
6	H	6 H
	T	6 T

Pages 554–555 Exercises   1. $\frac{1}{6}$ **3.** $\frac{1}{12}$ **5.** $\frac{1}{36}$ **7.** $\frac{1}{24}$
9. $\frac{1}{72}$ **11.** $\frac{1}{22}$ **13.** $\frac{1}{66}$ **15.** $\frac{5}{33}$ **17.** $\frac{1}{216}$ **19.** $\frac{1}{8}$ **21.** $\frac{3}{8}$
23. $\frac{11}{57}$ **25.** $\frac{1}{30}$ **27.** $\frac{1}{6}$ **29.** $\frac{1}{1000}$ **31.** $\frac{1}{8}$ **33.** $\frac{63}{125}$ **35.** $\frac{1}{6}$
37. $\frac{12}{49}$

Pages 557–558 Problem Solving Situations   1. 95
3. 75   **5.** 15   **7.** 67   **9.** 52   **11–13.** Strategies will
vary.   **11.** drawing a diagram; Marvin, Kim, Julie, Lance
13. using logical reasoning; 2   **17.** 15   **19.** 18, 23, 29

Pages 561–563 Exercises   1. 2   **3.** 720   **5.** 132
7. 20   **9.** 120   **11.** 24   **13.** 15   **15.** 120   **17.** 210
21. 120   **23.** 362,880   **25.** 28   **27.** 362,880   **29.** 21
31. $_nC_r = {_n}C_{n-r}$ **33.** 6   **35.** 15   **37.** 80   **39.** 720
41. about 31%

Page 563 Self-Test 2   1. $\frac{16}{81}$ **2.** $\frac{2}{27}$ **3.** $\frac{2}{51}$ **4.** 14
5. 120   **6.** 15

Pages 565–566 Exercises   1. $\frac{1}{5}$ **3.** $\frac{7}{40}$ **5.** $\frac{3}{5}$ **7.** $\frac{1}{4}$
9. $\frac{11}{40}$ **11.** $\frac{9}{16}$ **13.** 1   **15.** $\frac{1}{4}; \frac{13}{40}; \frac{17}{40}$ **17.** 0.9
19–25. Answers will vary.   **27.** $45.50   **29.** $\frac{4}{9}$
31.

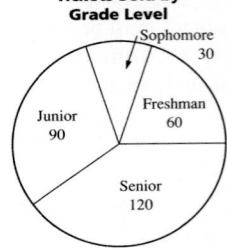

Tickets Sold by Grade Level
Sophomore 30
Freshman 60
Junior 90
Senior 120

Pages 569–571 Exercises   1. about 63,250   **3.** about
59,950   **5.** about 210   **7.** about 5400; about 3600
9. $\frac{29}{100}$ **11.** $\frac{1}{50}$ **13.** about 12,180   **15.** about 840
17. about 450   **19.** 1.25%   **21.** No; the players' past results should be considered.   **23.** Yes.   **25.** about 156
27. about −45

Page 571 Self-Test 3   1. $\frac{14}{25}$ **2.** $\frac{9}{50}$ **3.** $\frac{13}{50}$ **4.** about
36,000

Pages 572–573 Exercises   1. a. 85393, 97265, 61680,
16656, 42751   **b.** They are greater than 15000.
c. 02011, 07972, 10281, 03427, 08178, 09998, 14346,
07351, 12908, 07391, 07856, 06121, 09172, 13363,
04024   **3.** Assign each card holder a number from 00001
to 85056. Randomly select a starting point and direction in
the table. Discard each number greater than 85056. Continue until 25 numbers have been selected.

Pages 574–575 Chapter Review   1. dependent events
2. event   **3.** sample space   **4.** permutation   **5.** experimental probability   **6.** factorial   **7.** $\frac{1}{5}$ **8.** 0   **9.** $\frac{2}{5}$
10. 3 to 2   **11.** 2 to 3   **12.** 3 to 2
13.

spinner	coin	outcomes
1	H	1 (white) H
	T	1 (white) T
2	H	2 (red) H
	T	2 (red) T
3	H	3 (red) H
	T	3 (red) T
4	H	4 (blue) H
	T	4 (blue) T
5	H	5 (blue) H
	T	5 (blue) T

14. $\frac{1}{10}$ **15.** $\frac{1}{5}$ **16.** $\frac{2}{5}$ **17.** 0

18. 60   **19.** $\frac{7}{30}$ **20.** $\frac{7}{45}$ **21.** $\frac{7}{75}$ **22.** $\frac{1}{5}$ **23.** $\frac{1}{14}$ **24.** 7
25. 720   **26.** 60   **27.** 126   **28.** 21   **29.** $\frac{1}{5}$ **30.** $\frac{13}{50}$
31. about 104,000

Page 577 Chapter Extension   1. $17.40   **3.** 2.5

Pages 578–579 Cumulative Review **1.** D **3.** D
5. D **7.** D **9.** C **11.** B **13.** C **15.** C **17.** A
19. B

CHAPTER 13

Pages 583–584 Exercises **1.** (−6, −19); (−2, −3);
(0, 5); (2, 13); (6, 29) **3.** (−6, 7); (−2, 5$\frac{2}{3}$); (0, 5);

(2, 4$\frac{1}{3}$); (6, 3) **5.** (−6, −39); (−2, −19); (0, −9); (2, 1);
(6, 21) **7.** (−6, 2); (−2, −2); (0, −4); (2, −6); (6, −10)
9. (−6, 11); (−2, 3); (0, −1); (2, −5); (6, −13)
11. (−6, 42); (−2, 14); (0, 0); (2, −14); (6, −42)
13. a. $C = 36h + 20$ **b.** $128 **c.** 4.5 h **15.** about 9
17. about 11 **19.** Quebec **21.** E1 **23.** about 43° north,
71° west **25.** 120 **27.** 60

Page 587 Exercises **1.**
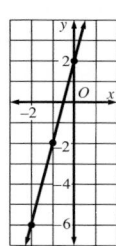

3. 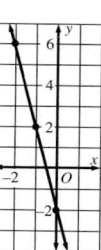 **5.**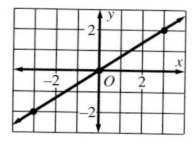

7. 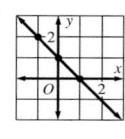 **9.**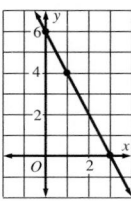

11. 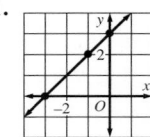 **13.**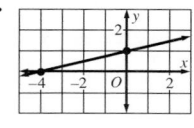

15. D **17. a.** 2 **b.** 2 **c.** 2 **d.** 2 **19. a.** −3 **b.** −3
c. −3 **d.** −3 **21.** $y = b − mx$ **23.** 50.2 m²

Pages 590–591 Exercises **1.** 1; 1; $y = x + 1$
3. $−\frac{1}{2}$; 0; $y = −\frac{1}{2}x$ **5.** −3; 9; 3 **7.** −1; −1; −1
9. −2; 1; $\frac{1}{2}$ **11.** 1; $−4\frac{1}{2}$; $4\frac{1}{2}$ **13.** 3; 0; 0 **15.** −3; −3; −1
17. 4; 6; $−1\frac{1}{2}$ **19.** (3, 0); (0, 21) **21.** C **23.** D **25.** 1

27. about 17 ft **29.** always **31.** about 58

Pages 596–597 Exercises **1.** (1, 2) **3.** no solution
5. (−1, 0) **7.** no solution **9.** (−3, −1) **11.** (2, 1)
13. (1, 1) **15.** (−3, −1) **17.** (1, 6) **19.** triangle
21.
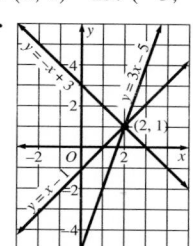

23.
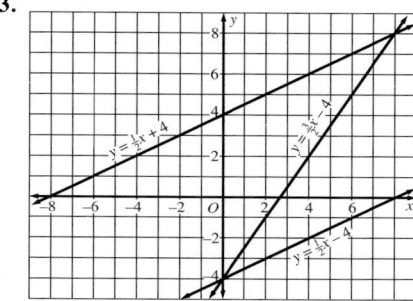

25. parallel **27.** intersect **29.** intersect **31.** parallel
33. $10y^2$

Page 597 Self-Test 1 **1.** (−2, −15); (−1, −9);
(0, −3); (1, 3); (2, 9) **2.** (−2, 1); (−1, 3); (0, 5); (1, 7);
(2, 9) **3.** 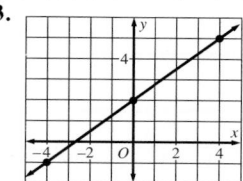 **4.**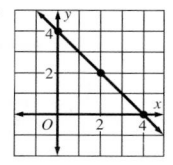

5. 5; 8; $−1\frac{3}{5}$ **6.** 4; 0; 0 **7.** (3, 4) **8.** no solution

Pages 600–601 Problem Solving Situations **1.** 27
3. 182 **5, 7.** Strategies may vary; likely strategies are
given. **5.** guess and check; 14 quarters and 16 dimes
7. using a Venn diagram; 32 **11.** 34.5 in.
13.

Page 601 Challenge a square with vertices at
(−4, 0), (0, 4), (4, 0), and (0, −4)

Pages 604–605 Exercises
1. $n > 17$

3. $c < 6$

5. $b > 21$

7. $p > -7.5$

9. $g > -4$

11. $y < -18$

13. $h \leq -0.8$

15. $h < 6$

17. $w < -2$

19. $n + 2 \geq 4; n \geq 2$ **21.** $\frac{n}{-6} < 12; n > -72$ **23.** b

25. b   **27.** $2g < c; g < 35$ **29.** $3\frac{9}{20}$ **31.** 32   **33.** 45 mi/h

Page 608 Exercises   **1.** $t > 6$

3. $a \geq -3$

5. $n \leq 4$

7. $b < -1$

9. $k > -15$

11. $a \geq -3$

13. $y \geq -2\frac{2}{3}$

15. $m \leq -1.7$

17. $a > -21$

19. $v > 14$ **21.** $a < 3$ **23.** $-1\frac{1}{2} > x$ **25.** $(-2, -11)$; $(-1, -8); (0, -5); (1, -2); (2, 1)$

27. $y > -1$

Pages 611–612 Exercises   **1.** Yes.   **3.** No.   **5.** No.

7. Yes.   **9.** 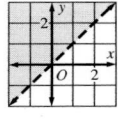   **11.**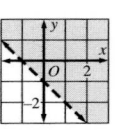

13. 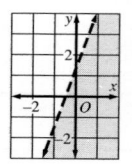   **15.** 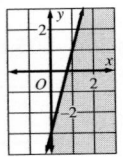   **17.**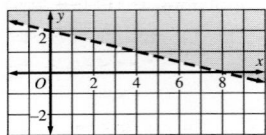

19.   **21.** No.   **23.** No.

25. No.   **27.** Yes.   **29.**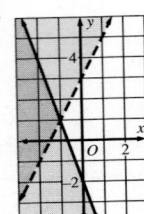

a. Answers will vary. Examples are given. $(-3, 0)$; $(-4, 2); (-5, -3)$ **b.** $y \leq 2x + 3$ **31.** $y \leq 5x + 3$ **33.** $j > 2a$ **35, 37.** Answers will vary. Examples are given.   **35.** The value of a number y is less than three times the number x.   **37.** The value of a number k is greater than or equal to the sum of the number m and 1.   **39.** 14; 18; 24   **41.**

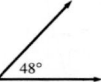

Page 612 Self-Test 2   **1.** 276

2. $x > 3$

3. $a \leq -4$

4. $c < 55$

5. $b \geq \frac{1}{3}$

6. 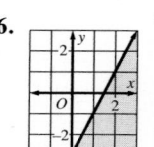   **7.**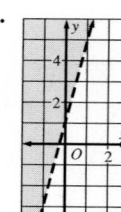

Pages 614–615 Exercises   **1.** When $x = -6$, $y = -5$; when $x = 2$, $y = 1\frac{2}{3}$; when $x = 3$, $y = 2\frac{1}{2}$.   **3.** When $x = -6$, $y = 73$; when $x = 2$, $y = 9$; when $x = 3$, $y = 19$.

5. When $x = -6$, $y = 8.5$; when $x = 2$, $y = 4.5$; when $x = 3$, $y = 5.5$.   **7.** $f(-4) = 8; f(-2) = 4$; $f(5) = -10$ **9.** $f(-4) = 9; f(-2) = 1; f(5) = 36$ **11.** $f(-4) = 80; f(-2) = 20; f(5) = 125$ **13.** $\frac{3}{25}$ watts

15. No; the power generated by the windmill in Exercise 14 is 8 times the power generated by the windmill in Exercise 13. **17.** No

19. 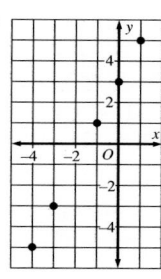 **21.** $-5; 2; \frac{2}{5}$

Pages 617–619 Exercises **1.**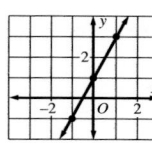

3. 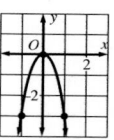 **5.**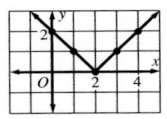

7. 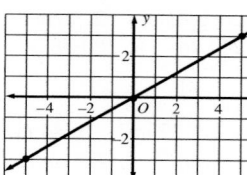 **9.**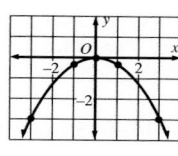

11. 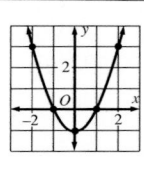 **13.** 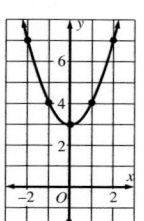 **15.**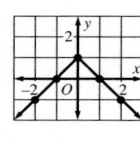

17. D **19.** C **21.** A **23.** D **25.** No. **27.** No. **29.** No. **31.** 137.5 **33.**

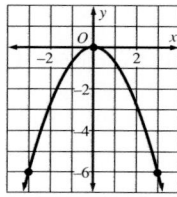

Page 619 Self-Test 3 **1.** When $x = -2$, $y = 0$; when $x = 0$, $y = -4$; when $x = 2$, $y = 0$. **2.** When $x = -2$, $y = -3$; when $x = 0$, $y = -5$; when $x = 2$, $y = -3$. **3.** $f(-4) = -9; f(0) = 7; f(8) = 39$ **4.** $f(-4) = 4$;

$f(0) = 0; f(8) = 16$ **5.** 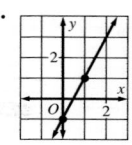 **6.**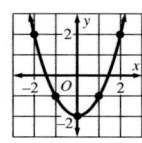

Pages 620–621 Chapter Review **1.** domain of a function **2.** solution of the system **3.** graph of an equation with two variables **4.** slope **5.** slope-intercept form of an equation **6.** y-intercept **7.** $(-2, -10); (-1, -3);$ $(0, 4); (1, 11); (2, 18)$ **8.** $(-2, 2); (-1, 2\frac{1}{2}); (0, 3);$ $(1, 3\frac{1}{2}); (2, 4)$ **9.** $(-2, 6); (-1, 1); (0, -4); (1, -9),$ $(2, -14)$ **10.**

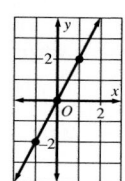

11.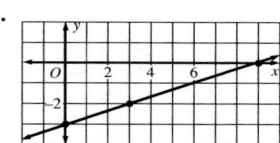

12. 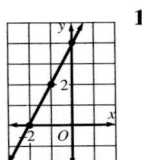 **13.** $-2; -3; y = -2x - 3$

14. $\frac{2}{3}; 1; y = \frac{2}{3}x + 1$ **15.** $8; 7; -\frac{7}{8}$ **16.** $6; 18; -3$
17. $-4; 0; 0$ **18.** $(-2, 3)$ **19.** $(1, -3)$ **20.** $(0, -2)$
21. no solution **22.** 30 **23.** 900
24. $x \le 6$

25. $c > -1$

26. $z \ge 4$

27. $m > -4$

28. $a > -2$

29. $x \le 12$

30. $a < -1$

31. $m \le -5$

32. $y > 0.7$

33. No. **34.** No.

35. Yes. **36.** 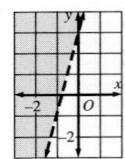 **37.**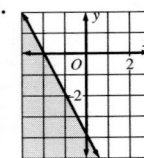

38.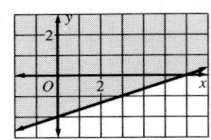

39. When $x = -3$, $y = 32$; when $x = 0$; $y = 11$; when $x = 4$, $y = -17$. **40.** When $x = -3$, $y = -3$; when $x = 0$, $y = -6$; when $x = 4$, $y = -2$. **41.** When $x = -3$, $y = 16$; when $x = 0$, $y = 1$; when $x = 4$, $y = 9$. **42.** $f(-2) = -18; f(0) = -8; f(3) = 7$ **43.** $f(-2) = 9; f(0) = 5; f(3) = 14$ **44.** $f(-2) = 2; f(0) = 4; f(3) = 7$ **45.**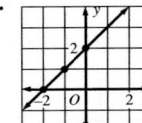

46. 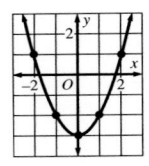 **47.**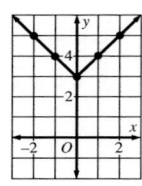

Page 623 Chapter Extension **1.** right **3.** left **5.** right **7.** above **9.** below **11.** below **13.** 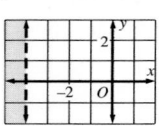 **15.**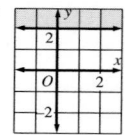

17. 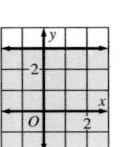 **19.**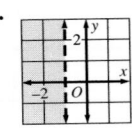

Pages 624–625 Cumulative Review **1.** D **3.** B **5.** C **7.** A **9.** A **11.** D **13.** C **15.** B **17.** A **19.** C

CHAPTER 14
Pages 630–631 Exercises **1.** triangular prism **3.** pentagonal pyramid **5.** cone **7.** hexagonal prism **9.** cylinder **11.**  **13.**

15. 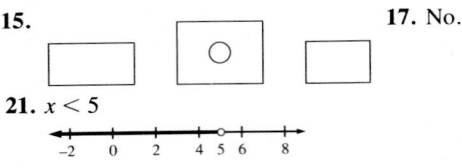 **17.** No.

21. $x < 5$

Pages 636–637 Exercises **1.** 384 ft^2 **3.** 75 m^2 **5.** 1062 m^2 **7.** 1288 in.^{2} **9.** 842 ft^2 **11.** 411.1 cm^2 **13.** 106 cm^2 **15.** 24 ft^2 **17.** 1 : 9 **19.** c **21.** 336 ft^2 **23.** 1.7 **25.**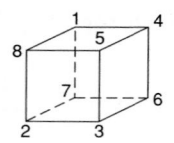

Page 637 Challenge
Answers may vary.

Pages 639–641 Exercises **1.** 828.96 yd^2 **3.** 1780.38 m^2 **5.** 528 in.^{2} **7.** 1610.82 mm^2 **9.** 151.62 cm^2 **11.** B **13.** 113 in.^{2} **15.** 19 in. **17.** a, b, c **19.** 45 cm^2 **21.** $\frac{1}{3}$

Page 641 Self Test 1 **1.** square pyramid **2.** cylinder **3.** sphere **4.** 184.5 mm^2 **5.** 264 yd^2 **6.** 320.28 in.^{2} **7.** 1848 cm^2

Pages 644–645 Exercises **1. a.** 132 in.^{3} **b.** 44 in.^{3} **3. a.** 336 ft^3 **b.** 112 ft^3 **5.** 7644 in.^{3} **7.** 405 mm^3 **9.** 132 cm^3 **11.** 512 m^3 **13.** 600 ft^3 **15.** 1575 cm^3 **17.** 8.5 yd^2 **19.** 49 m^2 **21.** 58,969,350 ft^3 **23.** the volume of the Great Pyramid **25.** about 30,702 gal **27.** 224 **29.** 98.5 cm^2

Pages 648–649 Exercises **1.** 4019.2 m^3 **3.** 565.2 yd^3 **5.** 847.8 yd^3 **7.** 6867.18 cm^3 **9.** 792 ft^3 **11.** 264,000 in.^{3} **13.** 502.4 in.^{3} **15.** 5 cm **19. a.** 141.3 cm^2 **b.** 94.2 cm^2 **c.** short legs **21.** using a π key: 150.79645 cm^3; using 3.14 for π: 150.72 cm^3 **23.** 5652 m^3 **25.** (1, 2)

Pages 651–653 Exercises **1.** 1256 mm^2 **3.** 803.84 m^2 **5.** 38,808 yd^3 **7.** 1808.64 cm^2 **9.** 1519.76 cm^2 **11.** 9156.24 m^2; 82,406.16 m^3 **13.** 22,176 cm^2; 310,464 cm^3 **15.** 7 units **17.** $\frac{1017.38}{3052.14} = \frac{3}{r}$; 9 cm **19.** 452.16 cm^3 **21. a.** 1.9741×10^8 mi^2 **b.** 2.6081×10^{11} mi^3 **23.** S.A. of sphere is $4(3.14)(3^2)$ = 113.04 cm^2. S.A. of curved surface of cylinder is $2(3.14)(3)(6)$ = 113.04 cm^2.

25. 56.52 cm^3 **27.** 51.48 cm^3
29. 8.1×10^{-6} **31.** 120 ft **33.** 210

Pages 655–657 Problem Solving Situations 1. a. 8
b. 12 **c.** 6 **d.** 1 **3.** 5; $2 \times 3 \times 5$; $30 \times 1 \times 1$
5–9. Strategies may vary. **5.** using proportions;
$4.90 **7.** no solution **9.** supplying missing facts; $3\frac{1}{2}$ h
13. 2 units $\times$ 3 units $\times$ 2 units **15.** $\frac{1}{6}$

Page 657 Self Test 2 1. 180 in.^{3} **2.** 72 m^3
3. 69,300 cm^3 **4.** 669.9 ft^3 **5.** 3052.08 in.^{3}
6. 1017.36 in.^{2} **7.** 10

Page 659 Exercises 1. A **3.** C **5.** 30 cm $\times$ 22.5 cm
$\times$ 11.3 cm **7.** 15 cm $\times$ 45 cm $\times$ 11.3 cm **9–11.** Answers will vary.

Pages 660–661 Chapter Review 1. pyramid
2. polyhedron **3.** vertex **4.** volume **5.** sphere
6. cube **7.** triangular prism **8.** cylinder **9.** octagonal
pyramid **10.** 726 m^2 **11.** 765 in.^{2} **12.** 920 yd^2
13. 471 m^2 **14.** 1012 ft^2 **15.** 621.72 m^2 **16.** 729 in.^{3}
17. 300 cm^3 **18.** 420 ft^3 **19.** 1540 m^3 **20.** 4019.2 in.^{3}
21. 1526.04 yd^3 **22.** 11,304 ft^2; 113,040 ft^3
23. 1017.36 mm^2; 3052.08 mm^3 **24.** 2826 cm^2;
14,130 cm^3 **25.** 8 **26.** 3; 8 units $\times$ 1 unit $\times$ 1 unit

Page 663 Chapter Extension 1. 39 ft^3 **3.** 226 cm^3
5. 5595 m^3

Pages 664–665 Cumulative Review 1. D **3.** D
5. C **7.** B **9.** A **11.** C **13.** A **15.** A **17.** A
19. D

CHAPTER 15
Pages 669–670 Exercises 1. $x^2 + x - 2$
3. $m^3 + 3m^2 + m - 3$ **5.** $3z^4 + 8z^3 + 4z^2 - 6z - 7$
7. $c^4 + c^3 - 5c - 2$ **9.** $-8k^4 + 3k^3 + k^2 - 6$
11. $x^2 + 4x + 7$ **13.** $e^2 + 11e - 20$
15. $-4x^2 + 9x + 12$ **17.** $-7x^2 - 11x + 1$
19. $8w^3 + 4w^2 - 12w + 1$ **21.** $x^5 + 4x^2 + 2x + 1$
23. $2a^4 - 10a^3 - 7a^2$ **25.** $-5x^3$
27. $ab^3 + a^2b^2 + a^2b + 4$ **29.** $x^3y^3 - xy^2 + x^2y - 4$
31. $a^3b^2 - a^2b + ab^3 - 4$ **33.** $a^3c^4 + ac^3 + a^4c^2 - a^2c$
35. many; polyester **37.** two; bicycle
39. $x^3yz^2 + x^2y^2z + xy^3z^3$; $xy^3z^3 + x^2y^2z + x^3yz^2$;
$xy^3z^3 + x^3yz^2 + x^2y^2z$ **41.** 3052.08 cm^3
43. $x \le -6$
$$\xleftarrow{\quad\;}{\underset{-18\;\;-12\;\;-6\;\;\;0\;\;\;\;6}{}}$$

Pages 674–675 Exercises 1. $5a^2 + 4a + 10$
3. $7x^2 + 11x + 11$ **5.** $9v^2 + 6v + 7$
7. $14y^3 + 12y^2 + 9y + 8$ **9.** $8b^4 + 5b^3 + 12b^2 + 7b + 7$
11. $y^2 - 4y + 9$ **13.** $6x^2 + 3x + 5$
15. $-k^4 - 7k^3 - 3k^2 + k - 2$ **17.** $5k^5 - k^4$

19. $5x^5 + 3x^4 + 3x^3 + x^2 - 5$
21. $6a^4 + 5a^3 + 9a^2 + 11a + 7$
23. $-3d^3 + 8d^2 - 4d - 4$
25. $2b^4 - 2b^3 + 3b^2 - b - 1$ **29.** 0
31.

12	-12	0
5	-1	4
0	4	4
-3	9	6
-4	20	16

33.

35. $7a^4 - 3a^3 + 6a^2 - 3$

Pages 678–679 Exercises 1. $x^2 + x + 4$
3. $3x^2 + 7x + 6$ **5.** $5d^2 - d - 2$ **7.** $-2m^2 - 10m - 4$
9. $3a^3 - 13a^2 + 7a + 11$ **11.** $-3x^2 + 6x - 16$
13. $4a^3 - 4a^2 - 8a + 17$
15. $-5a^4 + 4a^3 + 8a^2 - 6a - 2$
17. $-5a^4 + 7a^3 + 17a^2$
19. $-4z^4 - z^3 - 9z^2 - 2z + 8$ **21.** $4x^3 - 14x^2 + 6x - 3$
23. 3 **25.** -1 **27.** 5 **29.** 1 **31.** $6a^2 + 5a + 9$
33. $2n^2 - 5n + 5$ **35.** $2b^2 - 6b - 6$ **37.** Answers will
vary. Example: $x^2 + 3x + 7$ and $x^2 + 3x + 5$ **39.** They
are the same. **41.** $2x^2 + 6x - 3$
43. Base Hits Per Season

Hits	Tally	Frequency
160–169	✝✝✝	5
170–179	||||	4
180–189	||	2
190–199	|	1
200–209	|	1
210–219	|	1

45. $3d^2 + 13d - 11$

Page 679 Challenge 0, 3

Pages 683–684 Exercises 1. $14x^2 + 7x + 14$
3. $-36a^3 + 12a^2 + 24a - 12$ **5.** $24x^6y^3$ **7.** $-30r^5s^7$
9. $18a^6x^{11}$ **11.** $24a^3 + 16a^2 + 20a$
13. $15d^5 - 24d^4 + 21d^3 - 18d^2$
15. $28x^6 + 42x^5 - 63x^4 + 49x^3$
17. $10m^2n^2 - 15m^2n^3 + 20mn^2$

19. $-5x^2y^2 - 7x^3y^2 - 8x^2y^3$
21. $-5m^5n^3 - m^4n^2 + 7m^5n^2$
23. $30x^3 + 35x^2 - 45x - 25$
25. $8a^5 + 16a^4 - 14a^3 + 2a^2 - 18a$
27. $-18x^5 - 8x^4 + 12x^3 + 16x^2$
29. $-x^3 - 4x^2 - 7x + 5$ 31. $3x^4 - x^3 + 6x^2 + 2x - 7$
33. change each sign 35. a^5 37. The product of powers rule applies to negative exponents;
$(a^{-4})(a^{-5}) = \frac{1}{a^4}\cdot\frac{1}{a^5} = \frac{1}{a^9} = a^{-9}$ 39. $\frac{18r}{s^5}$ 41. $\frac{1}{3}$
43. $8a^2b + 20a^2 - 12a$

Pages 686–688 Exercises 1. $x^2 + 6x + 8$
3. $12t^2 + 52t + 35$ **5.** $24y^2 + 16y - 14$
7. $7x^2 + 6x - 16$ **9.** $2w^2 - 9w + 4$ **11.** $4x^2 - 17x + 4$
13. $12x^2 - 38x + 30$ **15.** $4x^2 + 4x + 1$ **17.** $4x^2 - 25$
19. $15x^2 + 2x - 24$ **21.** $a^2 - 2ab - 15b^2$ **23.** $x^2 - y^2$
25. $2x^4 - 11x^2 + 15$ **27. a.** $6x^2 - 11x - 10$;
$6x^2 - 11x - 10$ **b.** They are equal; $(3x + 2)(2x - 5) =$
$(2x - 5)(3x + 2)$; commutative property of multiplication
29. the new length and width **31.** $x^2 - 5x + 6$
33. $5x - 6$ **35. a.** $a^5 + 5a^4b + 10a^3b^2 + 10a^2b^3 +$
$5ab^4 + b^5$ **b.** $a^6 + 6a^5b + 15a^4b^2 + 20a^3b^3 + 15a^2b^4 +$
$6ab^5 + b^6$ **c.** $a^7 + 7a^6b + 21a^5b^2 + 35a^4b^3 + 35a^3b^4 +$
$21a^2b^5 + 7ab^6 + b^7$ **37.** *A* and *C* represent the numbers
the variable is multiplied by. *B* and *D* represent the numbers added to the variable. **39.** $x^2 + x - 12$ **41.** 25%

Page 688 Self-Test 1 1. $6x^4 + 5x^3 - 3x^2 + 4x - 8$
2. $4a^5 + 9a^3 + 5a^2 - 7a - 2$ **3.** $3x^3 + 4x^2 + 8x + 7$
4. $7c^3 - 2c^2 - 8c + 5$ **5.** $2x^2 + 8x + 8$
6. $2z^2 - 3z - 4$ **7.** $b^3 + 2b^2 - 6$ **8.** $4n^2 - 4$
9. $-10x^4y^6$ **10.** $9a^7 - 15a^5$ **11.** $8x^2 + 22x + 15$
12. $20c^2 - 12c - 56$

Pages 690–692 Problem Solving Situations 1. 64
3. 9 **5–9.** Strategies will vary. **5.** working backward;
1.1 mi **7.** supplying missing facts; gallons; $.26
9. using equations; 2 h **13.** 26 cm **15.**

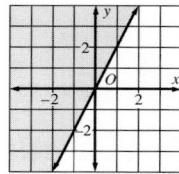

Pages 694–695 Exercises 1. $7x(x + 2)$ **3.** $s(4 + 5s)$
5. $4k^2(k + 2)$ **7.** $5p^2(p^3 - 3)$ **9.** $3xy(3 + 2x)$
11. $4c(a - 2b)$ **13.** $3u^2(4u^2 - 3u - 8)$
15. $8s^2(s^2 - 2s + 4)$ **17.** $5qd(5d + 3q + 6qd - 1)$
19. $y^3z^2(36yz + 45z^2 + 20y^2)$
21. $6r^2t^2u^2(4 - 2rtu - r^2t^2u^2)$
23. $4ab^2c(4ac + 1 + 2a^2c^3 + 3a)$ **25.** $(r - 5)(r + 6)$
27. $(2 - 3d)(7d - 1)$ **29.** $(y - 3)(y - 1)$
31. $rt + st + t^2 + t$ **33.** $4(8 - \pi)$ **35.** $2r^2(2 + \pi)$
37. 8! or 40,320 **39.** $7z^4 - 4z^3 + 5z^2 - z + 4$

Pages 697–699 Exercises 1. $3x - 4y$ **3.** $-7c^2 + c$
5. $x + 2x^3 - 6x^6$ **7.** $3t^6 + 8t + 1$ **9.** $3k^4 + k^3 - 12k^2$
11. $7 + 8e + 9d$ **13.** $3r^3u - 4$
15. $4xy^2 + 7x^2 - 12x^3y^2$ **17.** $\frac{1}{3}k - 2$
19. $-4xy - \frac{5}{2}x^2y - \frac{2}{3}x^3y$ **21.** $\frac{3}{2}x + \frac{7}{2} + \frac{9}{2x}$
23. $\frac{5}{3}b^2 + \frac{4}{3}b + \frac{1}{3b}$ **25.** $8x$ **27.** $3c + 3$ **29.** $2a - ab$
31. $h = 64t - 16t^2$
33.

t (seconds)	$h = 16t(4 - t)$ (feet)
0	0
1	48
2	64
3	48
4	0

35. 266 in.^{2}

37.

Gasoline Mileage (mi/gal)

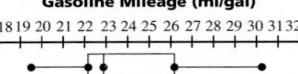

39. $21x^4 - 6x^2$

Page 699 Self-Test 2 1. 42 h **2.** $4c(3c - 2 - d)$
3. $5mn(5m + 3n - 2)$ **4.** $3a^2 - 2a + 1$
5. $ab + 9b - 6a$

Pages 700–701 Chapter Review 1. binomial
2. standard form of a polynomial **3.** greatest common
monomial factor **4.** polynomial **5.** $6x^4 + 7x^3 + 2x^2 +$
$4x + 7$ **6.** $4c^5 + 11c^4 - 9c^3 + 8c^2 - 5c + 2$
7. $a^6 - 2a^4 - 4a^3 - 3a + 8$
8. $5m^7 + 6m^5 - 2m^3 - 9m^2 + 4m$ **9.** $4x^3 + 14x^2 + x + 3$
10. $8b^4 - 6b^2 + b - 5$ **11.** $4c^3 + c^2 - 5c + 13$
12. $-c^4 - 9c^3 + 5c^2 + 3c + 8$ **13.** $-5x^3 + 6x^2 - 3x + 5$
14. $-2a^3 - a^2 - 5a + 11$ **15.** $5a^2 + 7a + 16$
16. $2x^3 + 4x^2 + 2x + 7$ **17.** $3n^5 + 4n^3 + 9n + 6$
18. $2d^4 + 8d^2 + 7d + 4$ **19.** $5c^3 + 5c^2 - 5c - 5$
20. $n^4 - n^3 + 5n^2 + 3$ **21.** $4x^2 + 3x + 5$
22. $3a^2 + 4a - 1$ **23.** $5c^2 + 12c - 7$
24. $-4b^2 - 9b + 8$ **25.** $3x^4 + 2x^3 - 4x + 10$
26. $-a^3 - 4a^2 - 5a - 6$ **27.** $6a^4 + 18a^3 - 24a^2 + 21a$
28. $-12x^4 - 28x^3 + 20x^2$
29. $-24c^5 - 18c^3 + 30c^2 + 48c$
30. $-28n^6 - 20n^5 + 32n^3$ **31.** $2x^2 + 13x + 15$
32. $8m^2 + 2m - 28$ **33.** $6x^2 - 31x + 40$
34. $28d^2 + 13d - 6$ **35.** $14.99 **36.** 43 h
37. 21 weeks **38.** 9:50 A.M. **39.** $2x(2x^2 + 4x + 3)$
40. $3n^2(3n^2 - 2n + 6)$ **41.** $5d^2(4d^3 - 2d^2 + 3d - 1)$
42. $6m^2(3m^4 - 5m^2 + 2m - 4)$ **43.** $cd(3c^2d^2 + 5cd - 4)$
44. $2ab^2(2a^2 - 4ab + 1)$ **45.** $5x^2y^2(5x^2y + 2xy^2 - 3)$
46. $6mn^2(2n - 3m + 4mn)$ **47.** $7x^3 + 4x^2 - 5x$
48. $2c^4 - 3c^2 + 6c - 1$ **49.** $6a^2b^4 + 7a^3b^2 - 2ab$
50. $4m^2n^3 - 2n^2 + 3m^5n$

Page 703 Chapter Extension
1. $a^2 - 4$ **3.** $x^2 + 16x + 64$ **5.** $c^2 - 10c + 25$
7. $4c^2 - 81$ **9.** $9x^2 + 24x + 16$ **11.** $25c^2 - 60c + 36$
13. $4x^2 - 9y^2$ **15.** $9a^2 + 30ab + 25b^2$
17. $9z^2 - 48xz + 64x^2$ **19.** $(x - 6)(x + 6)$
21. $(x + 1)^2$ **23.** $(c - 9)^2$ **25.** $(z - 8)^2$

Pages 704–705 Cumulative Review **1.** A **3.** C
5. C **7.** A **9.** B **11.** B **13.** A **15.** B **17.** A
19. C

LOOKING AHEAD
Pages 707–709 Exercises **1.** translation **3.** reflection
5. (3, 1) **7.** (1, 1) **9. a.**

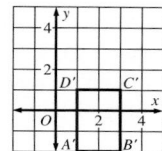

b. A' (1, −2); B' (3, −2); C' (3, 1); D' (1,1)
11. a.

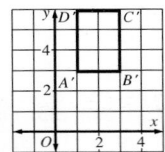

b. A' (1, 3); B' (3, 3); C' (3, 6); D' (1, 6)
13. a.

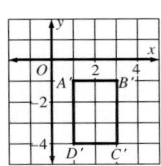

b. A' (1, −1); B' (3, −1); C' (3, −4); D' (1, −4)
15.

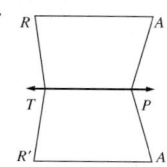

17. A' (−1, −1); B' (3, 2); C' (2, −2) **19.** A' (5, −1);
B' (1, −4); C' (2, 0) **21–23.** Answers may vary.
21. translation down 3 units and right 2 units **23.** reflection across the x-axis and a translation right 2 units
25. a, b. **c.** translation

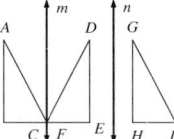

Pages 711–712 Exercises **1.** C **3.** E **5.** 6
7. **9.**

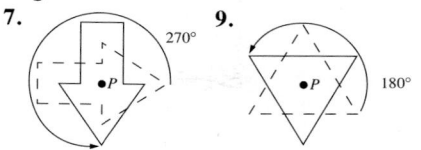

11. 180° and 360° about P **13.** reflection **15.** translation **17.** translation or rotation and translation
19. a, b. **c.** A' (−3, 7);
B' (−3, 4);
C' (2, 4); D' (2, 7)

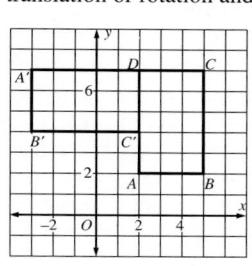

Pages 717–718 Exercises **1. a.** hypotenuse
b. adjacent; hypotenuse **c.** opposite; adjacent **3.** f
5. e **7.** d **9.** $\frac{4}{5}$ **11.** $\frac{4}{5}$ **13.** $\frac{4}{3}$ **15.** $\frac{5}{13}$ **17.** $\frac{5}{13}$ **19.** $\frac{5}{12}$
21. $x = 13$; $\frac{5}{13}, \frac{12}{13}, \frac{5}{12}$ **23.** $\frac{8}{17}$ **25.** $\frac{4}{5}$ **27.** $\frac{8}{17}$
29. No; the Pythagorean theorem does not hold true.
31. 1; 1

Pages 721–722 Exercises **1.** 0.0872 **3.** 0.2588
5. 5.6713 **7.** cosine **9.** sine **11.** 13.4 **13.** 19.4
15. $a = 11.2$; $b = 27.8$; $m\angle B = 68°$ **17.** 25.0 ft
19. 23.7

Pages 724–725 Exercises **1.** $\overline{DW}$ and $\overline{SD}$ **3.** 19°
5. 55° **7.** $m\angle A = 31°$; $m\angle B = 59°$ **9.** $m\angle A = 39°$;
$m\angle B = 51°$ **11.** $m\angle A = 53°$; $m\angle B = 37°$; $c = 5$
13. $m\angle A = 63°$; $m\angle B = 27°$; $c = 6.7$
15. $m\angle T = 71°$; $m\angle R = 71°$; $m\angle I = 38°$

Pages 728–729 Exercises **1.** $x\sqrt{2}$ **3.** longer leg;
hypotenuse **5.** 10.4 **7.** 14.1 **9.** $x = 6$; $y = 6$
11. 13.9; 8 **13.** 22; 31.1 **15.** 6; 12 **17.** 5.2
19. 24; 20.8 **21.** 69.3 cm; 80 cm
23. a. 9; 3; 5.2; 10.4 **b.** 31.2 square units

EXTRA PRACTICE
Pages 730–731 Chapter 1 **1.** 6 **3.** 11 **5.** 6 **7.** 2
9. 217.72 **11.** 59.5 **13.** 238.6 **15.** 63.6 **17.** 1693.72
19. 528 **21.** 4 **23.** 200 **25.** Chris's wages **27.** 40
29. > **31.** = **33.** < **35.** 371 **37.** 26 **39.** 97
41. 234.8 **43.** 892.7 **45.** 700 **47.** 9 **49.** 0 **51.** 110
53–59. Answers may vary. Likely answers are given.
53. mm; 140 **55.** c; 86.4 **57.** pp; 318 **59.** mm; 0.328

61. 0   **63.** 729   **65.** 100,000,000   **67.** 98   **69.** 109
71. 168   **73.** 1

Pages 731–732 Chapter 2   **1.** a^{10} **3.** $66x^2$
5. $6a + 42$ **7.** $2x + 26$ **9.** $6x + 21$ **11.** $12 - 18a$
13. $12x$ **15.** $8x$ **17.** $8x$ **19.** $2a + 13c$ **21.** $53.94
23. 39, 45, 51   **25.** $x + 12, x + 16, x + 20$
27. 5; 9; 13; 17; 21   **29.** $x \rightarrow x + 9$ **31.** estimate; about
$60   **33.** 340 mm   **35.** 4200 mL   **37.** 4 g
39. 3.5×10^6 **41.** 8×10^4 **43.** 235,000

Pages 732–733 Chapter 3   **1.** 15   **3.** 0   **5.** $<$ **7.** $>$
9. -43 **11.** 113   **13.** $61   **15.** 21   **17.** 0   **19.** -23
21. 31   **23.** 7   **25.** 42   **27.** -16 **29.** 50   **31.** -3
33. 20   **35.** 9   **37.** $(-2, -1)$ **39.** $(2, 0)$
40–43.   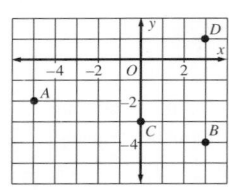   **45.**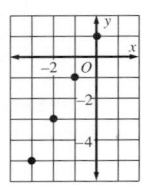

47.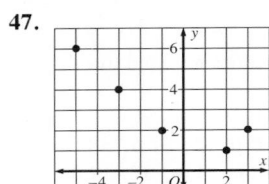

Pages 733–735 Chapter 4   **1.** Yes.   **3.** No.   **5.** 4
7. 8   **9.** 8 tickets for $12 and 12 tickets for $17
11. 12 goals and 16 assists   **13.** 12   **15.** -7 **17.** 7
19. 8   **21.** 8   **23.** 5   **25.** 4   **27.** -4 **29.** 5   **31.** -8
33. 36   **35.** -20 **37.** $x - 2$ **39.** $2n + 3$ **41.** $5j$
43. $3t - 7$ **45.** $25 = u + 21$ **47.** $20 = t - 4$ **49.** 12
51. 10   **53.** 3   **55.** -10 **57.** 12   **59.** 165 mi
61. 47 mi/h

Pages 735–737 Chapter 5
1.

Sales of Walking Shoes in the United States

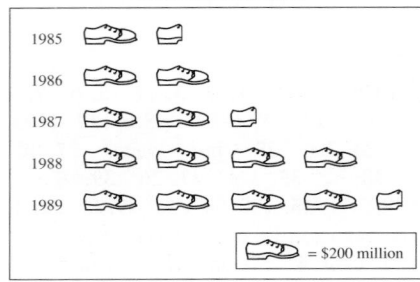

3. about 20   **5.** about 100

7.

Time Spent on Volunteer Work

9. National Bikes   **11.** not enough information; sales figures for racing bikes   **13.** Remove the gap in the scale; change the intervals on the scale.   **15.** median; the mean is greater than 5 of the 7 bills.

Pages 737–739 Chapter 6   **1.** $\overleftrightarrow{XY}$ or $\overleftrightarrow{YX}$ **3.** Plane T
5. 140°   **7.**   **9.** 78°   **11.** 3°   **13.** 143°

65°

15. 45°   **17.** True.   **19.** True.   **21.** True.   **23.** False.
25. triangle; no diagonals   **27.** 53°   **29.** 120°   **31.** can;
scalene   **33.** right   **35.** obtuse   **37.** 90°   **39.** 56°
41. 5 in.   **43.** No.   **45.** Yes.

Pages 739–741 Chapter 7   **1.** prime   **3.** composite
5. $2^3 \cdot 5$ **7.** $2 \cdot 5 \cdot 7$ **9.** 4   **11.** 8   **13.** $4a$ **15.** $9p$
17. 24   **19.** 50   **21.** $18a^6$ **23.** $8a^6$ **25.** 24   **27.** 4
29. $\frac{2}{7}$ **31.** $\frac{3}{5}$ **33.** $3q$ **35.** qt^5 **37.** $6p^2$ **39.** $\frac{h^5}{2}$ **41.** $>$
43. $<$ **45.** $<$ **47.** $=$ **49.** 0.8   **51.** 2.375   **53.** $\frac{8}{25}$
55. $3\frac{1}{4}$ **57.** 14 blocks north and 8 blocks west   **59.** 20
61. G **63.** C **65.** $\frac{-9}{1}$ **67.** $\frac{-31}{5}$
69.
-6 -5 -4 -3 -2 -1 0
71.
10 11 12 13 14 15 16
73.
3.7
2   3   4   5
75.
-3 -2.5 -2 -1.5 -1 -0.5 0   **77.** $\frac{1}{x^4}$ **79.** $\frac{1}{125}$
81. 8.2×10^{-5} **83.** 7×10^{-5} **85.** 0.005
87. 0.00000000325

Pages 741–742 Chapter 8   **1.** $\frac{3}{8}$ **3.** $-8\frac{3}{4}$
5. -23.22 **7.** $\frac{2a}{3}$ **9.** 2   **11.** $-\frac{5}{19}$ **13.** -3.2 **15.** $\frac{9}{14}$
17. decimal; $31.25   **19.** $1\frac{4}{7}$ **21.** 0   **23.** -3 **25.** $\frac{20}{y}$

27. $\frac{11}{18}$ 29. $-1\frac{3}{4}$ 31. $\frac{5a}{14}$ 33. -2.3 35. about 20
37. about $12\frac{1}{2}$ 39. about 3 41. $1020 43. -1 45. $1\frac{1}{2}$
47. -1.1 49. $-5\frac{5}{8}$ 51. $-3\frac{1}{2}$ 53. -7 55. $w = \frac{A}{l}$
57. $h = \frac{2A}{b_1 + b_2}$ 59. $t = \frac{D}{r}$

Pages 742–743 Chapter 9 1. $\frac{1}{7}$ 3. $\frac{1}{5}$ 5. $12.50/h
7. 18 mi/gal 9. 35 11. 12 13. 12 15. 7 17. $\frac{5}{6}$ in.;
1 in. 19. 45 21. 16% 23. 112.5% 25. $\frac{9}{20}$; 0.45
27. $\frac{41}{500}$; 0.082 29. 9.6 31. 7.5% 33. 176 35. 30%
increase 37. $16\frac{2}{3}$% decrease 39. 37.5% 41. $66\frac{2}{3}$%
43. 90

Page 744 Chapter 10 1. $17\frac{5}{6}$ ft 3. 12 m; 24 m
5. 6.5 m; 13 m 7. 60.72 m² 9. 86.5 ft²
11. $\frac{1}{2} + \frac{1}{3} + \frac{1}{4} = 1\frac{1}{12}$; this is more than one whole
paycheck. 13. $\angle P$ 15. 9 ft 17. $\frac{4}{9}$ 19. 11.790

Pages 745–747 Chapter 11
1. Test Scores in Math

Score	Tally	Frequency
50–59	||	2
60–69	||	2
70–79	⊬⊬ |	6
80–89	⊬⊬ |	6
90–99	||||	4

3. 7 5. 43.6; 44; 33, 42, 50; 27 7. 17 9. 25 11. 78;
72; 82; 64; 92 13. Tim Stuart's 15. 2 17. $10,500
19.

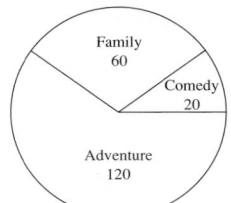

Videos Rented from Victor's Videos
Family 60, Comedy 20, Adventure 120

21. Therese is the lawyer, Lucinda is the accountant, and Alicia is the teacher.

Page 747 Chapter 12 1. $\frac{1}{5}$ 3. 3 to 2
5.

spinner	coin	outcomes
1	H / T	1 (red) H / 1 (red) T
2	H / T	2 (white) H / 2 (white) T
3	H / T	3 (blue) H / 3 (blue) T
4	H / T	4 (blue) H / 4 (blue) T
5	H / T	5 (white) H / 5 (white) T

7. 25 9. $\frac{2}{17}$ 11. 5040 13. $\frac{9}{20}$

Pages 748–749 Chapter 13 1. $(-2, -4)$; $(-1, -1)$; $(0, 2)$; $(1, 5)$; $(2, 8)$ 3. $(-2, 9)$; $(-1, 7)$; $(0, 5)$; $(1, 3)$; $(2, 1)$ 5. $(-2, 3)$; $(-1, 6)$; $(0, 9)$; $(1, 12)$; $(2, 15)$
7. $(-2, -17)$; $(-1, -12)$; $(0, -7)$; $(1, -2)$; $(2, 3)$
9. 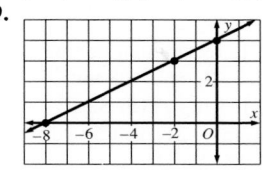 11.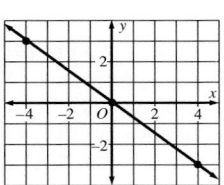
13. 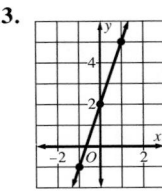 15.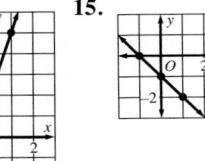
17. -1; 2; $y = -x + 2$ 19. $\frac{1}{2}$; 0; $y = \frac{1}{2}x$ 21. 2; 4; -2
23. -3; -6; -2 25. $(2, 7)$ 27. $(6, -3)$ 29. $(-2, -6)$
31. $(2, 6)$ 33. 625
35. $n \leq 5$
37. $p > -13$
39. $k < 2$
41. $t \geq -3$ 43. $k \geq 100$
45. $q < -7200$ 47. $t \geq -600$ 49. $n \geq -1053$
51. No. 53. Yes. 55. 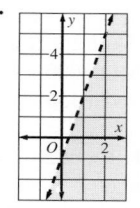 57.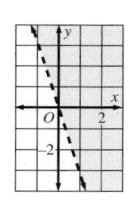
59. When $x = -1$, $y = -10$; when $x = 0$, $y = -7$;
when $x = 3$, $y = 2$. 61. When $x = -1$, $y = 5$;
when $x = 0$, $y = 0$; when $x = 3$, $y = -3$.
63. $f(-2) = 8$; $f(1) = -4$; $f(0) = 0$
65. $f(-2) = 31$; $f(1) = 7$; $f(0) = -1$
67. $f(-2) = 3$; $f(1) = 0$; $f(0) = -1$ 69.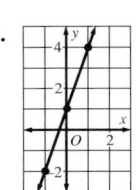

71. 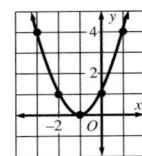 **73.**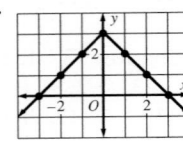

Pages 749–750 Chapter 14 **1.** rectangular prism
3. hexagonal pyramid **5.** 570 ft² **7.** 1607.68 mm²
9. 282.6 yd² **11.** 750 ft³ **13.** 171 in.³ **15.** 32 ft³
17. 508.68 mm³ **19.** 4823.04 mm³ **21.** 169.56 yd³
23. 2826 ft² **25.** 7234.56 m³ **27.** 24,416.64 yd³
29. 28; 20; 4; 12

Pages 751–752 Chapter 15 **1.** $3a^3 + 6a^2 + 4a + 9$
3. $7a^2 + 7a + 4$ **5.** $4a^3 - 2a^2 + 4a + 7$
7. $9p^2 + 5p + 2$ **9.** $10a^2 + 9a + 10$
11. $18c^2 + 28c + 21$ **13.** $3x^4 - 3x^3 + 4x^2 + 2x - 7$
15. $-n^4 + n^3 + 2n^2 - 5n + 3$ **17.** $2k^2 + 5k + 3$
19. $h^2 + h + 1$ **21.** $6b^3 + 8b^2 - 5b - 2$
23. $-3x^3 + 16x^2 - 4x - 3$ **25.** $18x^2 + 9x - 6$
27. $12m^3 - 24m^2 + 36m + 36$ **29.** $30c^7d^{10}$
31. $12a^7 + 24a^6 - 9a^5 - 27a^4$ **33.** $n^2 + 6n + 5$
35. $5y^2 + 41y + 42$ **37.** $6n^2 + 17n + 5$
39. $8u^2 + 6u - 35$ **41.** 32 **43.** $2m^2(2m^2 - 8m + 15)$
45. $2v^4(5v^2 - 8v + 6)$ **47.** $2a^3b^2(2b^3 - 5a + 3a^2b^3)$
49. $3p^3q^5(8p - 5q + 11p^3q^2)$ **51.** $2y^3 + 6y^2 + 3y$
53. $2t^4 + 4t^2 + 6t$ **55.** $a^2b^4 - 6ab^6 + 2a^3b$
57. $-2p^3 + m^5p^2 - 3m^4p^2$

TOOLBOX SKILLS PRACTICE
Page 753 Skill 1 **1.** about 400 **3.** about 980
5. about 700 **7.** about 500 **9.** about 36,000
11. about 800 **13.** about 4,000,000 **15.** about 4500
17. about 300,000 **Skill 2** **1.** about 1100
3. about 1400 **5.** about 8000 **7.** about 30
9. about 5000 **11.** about 150

Page 754 Skill 3 **1.** about 20 **3.** about 30
5. about 30 **7.** about 60 **9.** about 40 **11.** about 60
13. about 120 **15.** about 90 **17.** about 3 **19.** about 40
Skill 4 **1.** about 60 **3.** about 120 **5.** about 630
7. about 1800 **9.** about 390 **11.** about 1700
13. about 56 **15.** about 35 **17.** about 7.2

Page 755 Skill 5 **1.** tens, 80 **3.** tens, 0 **5.** thousands,
1000 **7.** ones, 1 **9.** thousands, 2000 **11.** hundredths,
0.07 **13.** tens, 90 **15.** thousandths, 0.009 **17.** ones,
8 **19.** hundredths, 0.09 **21.** fifteen **23.** one hundred
sixty-two **25.** four hundred ninety-three **27.** two

hundred fifty-eight **29.** four thousand, fifty-six
31. thirty-two **33.** seven thousand, six **35.** three
hundred eighty-five **37.** four and seven tenths
39. one hundred seventy-eight and sixty-one hundredths
41. eighty-three and ninety-seven thousandths
43. eighteen and three hundred fifteen thousandths
45. six thousand thirty-one and seven hundred fifty-two
thousandths **47.** one hundred ninety-three and seven
hundredths **49.** three hundred seven and one tenth

Page 756 Skill 6 **1.** 20 **3.** 100 **5.** 270 **7.** 650
9. 8790 **11.** 900 **13.** 5300 **15.** 300 **17.** 2400
19. 6000 **21.** 12,000 **23.** 7000 **25.** 3000 **27.** 22,000
29. 4.0 **31.** 0.3 **33.** 4.8 **35.** 12.2 **37.** 107.6
39. 17.9 **41.** 0.52 **43.** 31.03 **45.** 0.30 **47.** 12.99
49. 3.41 **51.** 71.99 **53.** 54.018 **55.** 0.314 **57.** 2.992
59. 13.030 **61.** 4.348 **63.** 177.025

Page 757 Skill 7 **1.** 88 **3.** 99 **5.** 1195 **7.** 869
9. 12,587 **11.** 5739 **13.** 14,799 **15.** 5789
17. 492 **Skill 8** **1.** 440 **3.** 122 **5.** 1225 **7.** 2441
9. 113 **11.** 2021 **13.** 22,424 **15.** 52,325 **17.** 103
19. 48 **21.** 96 **23.** 4892 **25.** 31,849

Page 758 Skill 9 **1.** 1643 **3.** 15,708 **5.** 12,832
7. 1,420,868 **9.** 264,924 **11.** 130,273
13. 21,511,182 **15.** 24,798,126 **17.** 2482 **19.** 6322
21. 24,453 **23.** 232,403 **25.** 638,844 **27.** 152,388
29. 2,298,835

Page 759 Skill 10 **1.** 3 **3.** 13 **5.** 175 **7.** $27\frac{5}{17}$
9. $193\frac{5}{8}$ **11.** 12.5 **13.** 19 **15.** 314.8 **17.** 114.5
19. 534.25 **21.** 14.1 **23.** 50.6 **25.** 143.4 **27.** 81.7
29. 287.8 **31.** 20.11 **33.** 250.94 **35.** 138.46
37. 203.95 **39.** 2040.88

Page 760 Skill 11 **1.** 767.95 **3.** 1433.20 **5.** 1173.93
7. 11,559.4 **9.** 1398.898 **11.** 953.23 **13.** 21.1
15. 22.255 **17.** 209.091 **19.** 40.577 **Skill 12** **1.** 9.38
3. 23.73 **5.** 4.889 **7.** 141.67 **9.** 288.636 **11.** 121.57
13. 45.64 **15.** 88.64 **17.** 10.97

Page 761 Skill 13 **1.** 13.78 **3.** 32.76 **5.** 131.1
7. 19.1205 **9.** 7.462 **11.** 2915.415 **13.** 0.0048
15. 0.00513 **Skill 14** **1.** 7.8 **3.** 6.49 **5.** 19 **7.** 13.4
9. 18.1 **11.** 38 **13.** 63 **15.** 116.00

Page 762 Skill 15 **1.** 8.03 **3.** 8130.6 **5.** 541
7. 70.92 **9.** 2840 **11.** 30 **13.** 12,751 **15.** 804
17. 41.7 **19.** 12 **21.** 9310 **23.** 82,100 **25.** 21.635
27. 0.5106 **29.** 2.857 **31.** 0.00792 **33.** 0.293
35. 9.301 **37.** 18.475 **39.** 0.0029 **41.** 0.413
43. 0.0006 **45.** 0.0269 **47.** 0.000025

Page 763 Skill 16   **1.** Yes.   **3.** No.   **5.** Yes.   **7.** Yes.   **9.** No.   **11.** Yes.   **13.** No.   **15.** Yes.   **17.** No.   **19.** Yes.   **21.** Yes.   **23.** Yes.

Page 764 Skill 17   **1.** $\frac{7}{9}$ **3.** $\frac{4}{5}$ **5.** 2   **7.** $\frac{1}{5}$ **9.** $\frac{1}{2}$ **11.** $\frac{3}{5}$ **13.** $\frac{3}{5}$ **15.** $\frac{9}{16}$ **17.** $1\frac{25}{36}$ **19.** $\frac{26}{45}$ **21.** $\frac{11}{21}$ **23.** $\frac{13}{30}$ **25.** $\frac{14}{75}$

Page 765 Skill 18   **1.** $7\frac{10}{11}$ **3.** $17\frac{5}{9}$ **5.** $23\frac{1}{3}$ **7.** $38\frac{11}{15}$ **9.** $12\frac{3}{5}$ **11.** $25\frac{19}{20}$ **13.** $19\frac{29}{36}$ **15.** $17\frac{3}{10}$ **17.** $30\frac{5}{8}$ **19.** $31\frac{1}{4}$ **21.** $83\frac{19}{60}$ **23.** $50\frac{23}{24}$ **25.** $46\frac{3}{5}$ **27.** $25\frac{23}{24}$

Page 766 Skill 19   **1.** $2\frac{5}{7}$ **3.** $13\frac{1}{8}$ **5.** $5\frac{1}{2}$ **7.** $3\frac{5}{12}$ **9.** $5\frac{5}{8}$ **11.** $13\frac{1}{8}$ **13.** $15\frac{23}{42}$ **15.** $4\frac{27}{40}$ **17.** $3\frac{7}{10}$ **19.** $12\frac{11}{16}$ **21.** $13\frac{11}{12}$ **23.** $2\frac{1}{2}$

Page 767 Skill 20   **1.** $\frac{1}{15}$ **3.** $\frac{3}{22}$ **5.** 20   **7.** $\frac{1}{18}$ **9.** $\frac{5}{13}$ **11.** 20   **13.** 76   **15.** 6   **Skill 21**   **1.** 3   **3.** $\frac{11}{14}$ **5.** $1\frac{1}{2}$ **7.** $\frac{25}{54}$ **9.** $\frac{21}{40}$ **11.** $\frac{4}{19}$

Page 768 Skill 22   **1.** 21,120   **3.** 11,000   **5.** 28   **7.** 11,880   **9.** 9000   **11.** 25   **13.** 18   **15.** 9; 1   **17.** 36; 2   **19.** 6   **21.** 11; 1   **23.** 5   **25.** $1\frac{1}{2}$ **27.** 3   **29.** 28   **31.** 8   **33.** 2640   **35.** 48,000

Page 769 Skill 23   **1.** 1980; 1940   **3.** about 100   **5.** 1930 and 1950   **7.** about 4 lb   **9.** Los Angeles

Page 770 Skill 24   **1.** September; January   **3.** about 4   **5.** decreasing   **7.** Dallas   **9.** about 1000 h

Page 771 Skill 25   **1.** midnight   **3.** about 20,000   **5.** increasing   **7.** January   **9.** about 10   **11.** about 10

Additional Answers

Chapter 1

Page 8 *Guided Practice*

2. The objective of this lesson is to evaluate variable expressions involving whole numbers and decimals. It is different from the objective of Lesson 1-1 because Lesson 1-1 involved only whole number values for the variables. Also, Lesson 1-1 involved multiplication and division, not just addition and subtraction. **3–6.** Answers may vary. Examples are given. **3.** Round both 42.1 and 37.7 to 40 and add $40 + 40 = 80$. **4.** Use front-end and adjust. Add $700 + 0 = 700$. Since $14.6 + 81$ is about 100, adjust the sum to $700 + 100 = 800$. **5.** Round 5.14 to 5 and 2.81 to 3 and subtract $5 - 3 = 2$. **6.** Round 6.9037 to 6.9 and 0.751 to 0.8 and subtract $6.9 - 0.8 = 6.1$.

Page 10 *Historical Note*

The ancient Greek use of symbols for unknown quantities has been traced to Diophantus. He merged the Greek letters α and ρ to form a symbol for the unknown. (These are the first two letters of the word "arithmos," or number.) Diophantus also used Δ^Y for unknown squared (from the first two letters of "dunamis," or power) and K^Y for unknown cubed (from the first two letters of "kubos," or cubed).

The ancient Hindus also used abbreviations of words to denote unknowns. The symbol yā (from "yāvattāvat," or so much as) represented the first variable. If additional variables were needed, these were indicated by the first two letters of words for different colors.

Pages 12–13 *Guided Practice*

3. 27×15; $27(15)$; $(27)15$; $(27)(15)$ **4.** $22\overline{)884}$; $\frac{884}{22}$
5. $124.2 \div 6.5$; $\frac{124.2}{6.5}$ **6.** 12.4×26.9; $12.4 \cdot 26.9$; $(12.4)26.9$; $12.4(26.9)$ **7.** Round 3.7 to 4 and 81 to 80 and multiply $4 \cdot 80 = 320$. **8.** Round 4.3 to 4 and 5.1 to 5 and multiply $4 \cdot 5 = 20$. **9.** Use compatible numbers and divide $12 \div 2 = 6$. **10.** Use compatible numbers and divide $15 \div 3 = 5$.

Pages 15–16 *Guided Practice*

4. The paragraph is about clothing that Brett purchased. **5.** four **6.** the date of August 23; the five $20 bills that Brett used to pay **7.** Find the total cost of the clothes and subtract it from $100.

Pages 16–17 *Problem Solving Situations*

1. The paragraph is about Cathy's job as a cashier. **2.** four **3.** the $6.50 per hour that Cathy earns **4.** Find the number of hours she works on Thursday and add one hour to that number. **5.** The paragraph is about Paul's plan to buy a video game system. **6.** $61 **7.** No **8.** the $35.49 cost of additional game cartridges **9.** The paragraph is about Carol's purchase of a stereo system. **10.** $30 **11.** the $100 down payment; the 10 equal payments **12.** Add $600 and $30, subtract $100 from this sum, and then divide the result by 10.

Page 17 *Self-Test 1*

13. The paragraph is about Jill's movie and video game rentals last year. **14.** 15 **15.** the 15 video games that Jill rented; the $1.50 cost of each video game rental **16.** Multiply 15 times $1.50.

Pages 19–20 *Guided Practice*

5. Five thousand one is less than five thousand one hundred **6.** Six tenths is equal to six hundred thousandths. **7.** Nine and three hundredths is greater than nine and seven thousandths. **8.** Five hundred fifty is less than five hundred eighty-one, and five hundred eighty-one is less than six hundred. **19.** Answers will vary. One possible answer is that the smaller end of the inequality symbol always points toward the lesser number. If the symbol is $<$, the lesser number is first and the symbol is read "is less than." If the symbol is $>$, the greater number is first and the symbol is read "is greater than."

35. 1960: 179.4 million people; 1980: 226.6 million people; about 50 million more people **36.** South; Two probable reasons are a favorable climate and more job opportunities.

3. commutative property of addition **4.** identity property of addition **5.** associative property of addition **6.** identity property of addition **7.** commutative property of addition **8.** associative property of addition

26. The commutative and associative properties of addition have a number of differences. One difference is that the commutative property involves two numbers, whereas the associative property involves three numbers. A second difference is that the commutative property switches the order of the numbers, whereas the associative property keeps the numbers in the same order, but changes how they are grouped.

1. The main idea of the lesson is that multiplication has several special properties. **2.** The commutative property of multiplication states that changing the order of the factors does not change the product. The associative property of multiplication states that changing the grouping of the factors does not change the product. The identity property of multiplication states that the product of a number and one is the original number. The multiplication property of zero states that the product of any number and zero is zero. **3.** commutative property of multiplication **4.** identity property of multiplication **5.** multiplication property of zero **6.** associative property of multiplication

1. You only need to multiply 2×3 and annex four zeros.
2. You can add on from 99. **3.** The digits 7 and 3 give you a sum of 10. **4.** Both numbers are multiples of 10, so the answer is the same as $54 \div 6$. **5.** There is no renaming involved.
6. You only need to move the decimal point in 0.9 two places to the right. **7.** Each number has only one decimal place, so the answer is the same as $15 \div 3$. **8.** There is no renaming involved. **9.** $948 + 1003$; There is only one renaming involved.
10. $7.243 - 6.531$; Both numbers have the same number of decimal places and there is only one renaming involved.
11. $(852)(3)$; One of the factors is a single-digit number.
12. $8.208 \div 3$; The divisor is a single-digit number.

The sum of the first and last numbers is 50, the sum of the second and tenth numbers is 50, and so on. There are five of these sums for a total of 250. Add the remaining number, 25, to this sum for a total of 275.

Activity III
1. a. 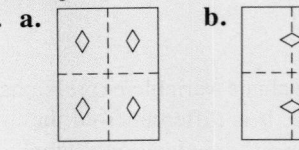 **b.** 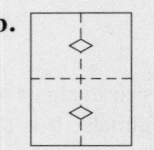 **c.** 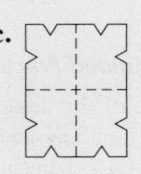 **d.**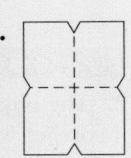

e. 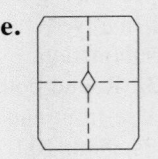 **f.** 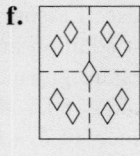 **g.** 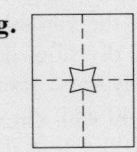 **h.**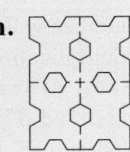

2. a. 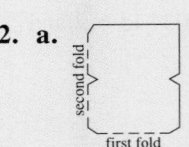 **b.** 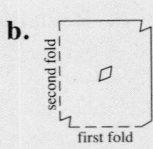 **c.** 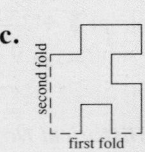 **d.**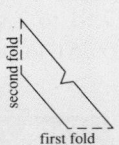

1. First do all work inside parentheses. Then find each power. Then do all multiplications and divisions in order from left to right. Then do all additions and subtractions in order from left to right. **3.** multiply, subtract, add **4.** subtract, divide
5. power, multiply, add **6.** add, subtract, divide

30. False; $4(5 + 6) = 44$ **31.** False; $(24 + 4)2 = 40$ **32.** False; $24 \div (3 + 5) \cdot 2 = 6$ **33.** True **34.** True **35.** False; $3(4 - 2)3 = 18$ **36.** False; $(4 + 4)^2 \div 2 = 32$ **37.** False; $(12 - 2^2) \div 4 = 2$

26. The paragraph is about Yvonne's job typing term papers.
27. 6 h **28.** the $7 per hour that she earns **29.** You would add the number of hours that Yvonne typed on Monday, Tuesday, Wednesday, and Thursday.

13. The paragraph is about Dale's purchase of cassette tapes and a compact disc.   **14.** $13.95   **15.** the $8.95 cost of an album; the two $20 bills Dale used to pay   **16.** Find the total cost of the cassette tapes and the compact disc and subtract it from $40.

Chapter 2

Page 50   *Exercises*

35. Both $2^m \cdot 2^0$ and $2^m \cdot 1$ are equal to 2^m. It follows that the expressions must be equal to each other.   **36.** 1; In Exercise 35, it was shown that $2^m \cdot 2^0 = 2^m \cdot 1$, and the only way that this could be true is if $2^0 = 1$.   **37.** The paragraph is about buying tickets for a basketball game and a hockey game.

Page 51   *Focus on Explorations*

Activity I

2. a.

  b.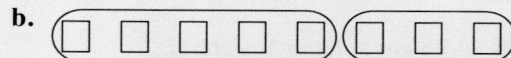

  c.

  d.

  e.

  f.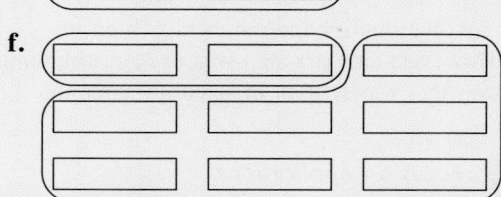

3. Answers will vary. Examples are given.

  a.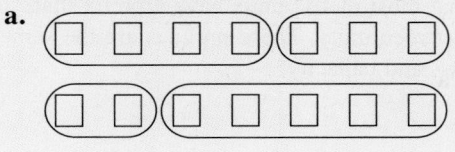

  $4 + 3; 2 + 5$

  b.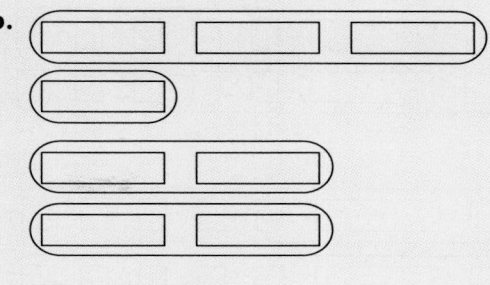

  $3n + n; 2n + 2n$

Activity II

2. a.

  b.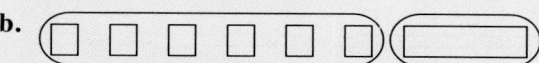

  c.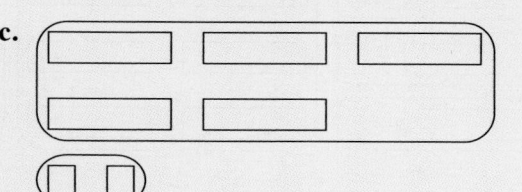

  d.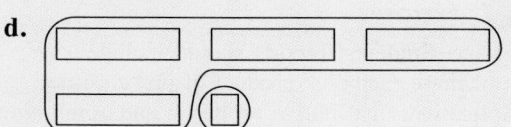

  e.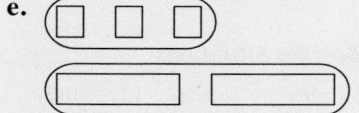

  f.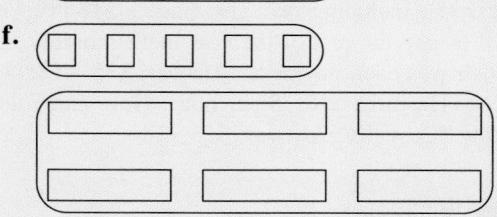

3. Answers will vary. Examples are given.

  a.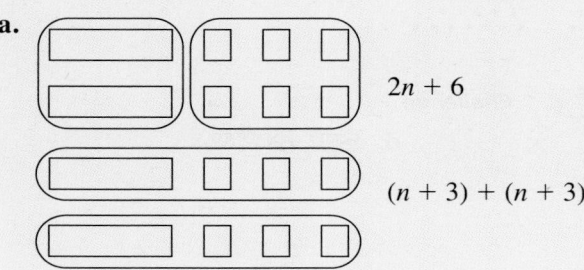

  $2n + 6$

  $(n + 3) + (n + 3)$

b.

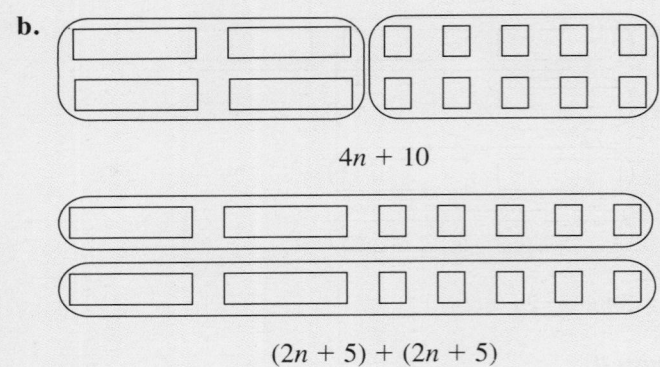

$$4n + 10$$

$$(2n + 5) + (2n + 5)$$

Page 54 **Guided Practice**

4.

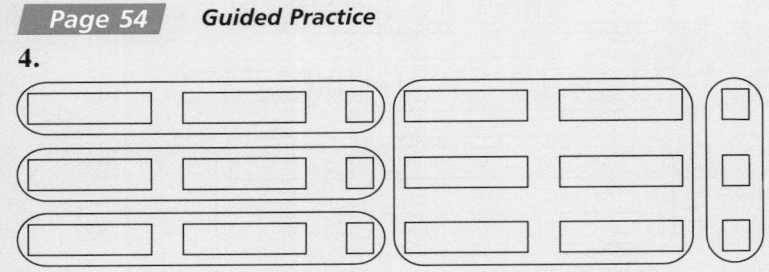

Pages 57–58 **Exercises**

31. Answers will vary. Students' reports should include a discussion of each of these methods: product of powers rule; properties of multiplication; distributive property; and combining like terms.

Pages 60–61 **Problem Solving Situations**

11–12. Answers will vary. Examples are given.   **11.** Susan bought a couch for $600, including tax. She made a $180 down payment and agreed to pay the rest of the cost in 12 monthly payments. How much was each payment? Answer: $35   **12.** Jose works 42 h per week. He earns $10.75 per hour. How much does Jose earn for working 52 weeks? Answer: $23,478

Pages 64–65 **Exercises**

19.

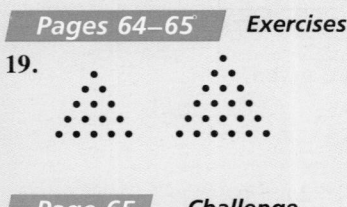

Page 65 **Challenge**

c. ⅄ , K , ⅄   **d.** ⅛ ⅙ ⅄

Page 70 **Focus on Computers**

4. The number being added in the pattern, 2, is the same as the number that multiplies x in the expression.   **6.** The number being added in the pattern, 3, is the same as the number that multiplies x in the expression.   **7.** The pattern will be *add 4*; each number in the table will be 4 greater than the number above it.   **8.** Each function has the pattern *add 5*; each number in the table for $x \rightarrow 5x + 1$ is one less than the corresponding number in the table for $x \rightarrow 5x + 2$.   **9.** The pattern will be *add 5*; each number in the table will be one greater than the corresponding number in the table for $x \rightarrow 5x + 2$.

Pages 71–72 **Guided Practice**

1–4. Answers may vary.   **1.** It is important to check that your answer is correct, or at least reasonable.   **2.** Students' responses should include two of these three ways: check your calculations; solve again using an alternative method; solve again using estimation.   **3.** A person might choose the wrong operation, or perform an operation incorrectly.   **4.** Answering a problem with a complete sentence gives you another opportunity to check that your answer is reasonable.

Pages 72–73 **Problem Solving Situations**

9–10. Answers will vary. Examples are given.   **9.** On a business trip, Marie Chen rented a car for five days. She drove the car the same number of miles each day for a total of 250 mi. How many miles a day did Marie drive the car? Answer: 50 mi
10. Sam Jones bought two cassettes and a compact disc. Cassettes cost $9.95 each and compact discs cost $14.50 each. Excluding tax, how much did Sam spend on cassettes and compact discs? Answer: $34.40

Pages 82–83 **Exercises**

37. Both exercises involve multiplying numbers written in scientific notation. However, an extra step was needed in Exercise 36 to obtain an answer that was written in scientific notation.

Pages 84–85 **Focus on Applications**

19. The metric system was created in the 1790s. One advantage is being able to use multiples of ten to relate one unit to another. Another advantage is that most metric units have a prefix that gives the relationship between units. These prefixes are the same for units of length, mass, and capacity.

Chapter 3

4.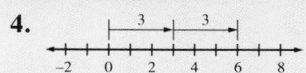

5.

6.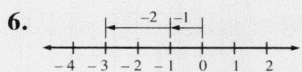

7.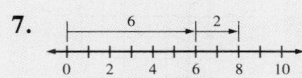

5.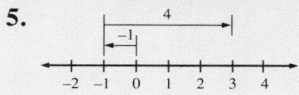

6.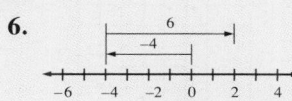

7.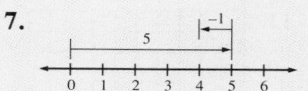

8.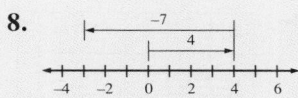

Activity I

2. a. ⊞ ⊞ ⊞ ⊞   **b.** ⊟ ⊟

c. ⊟ ⊟ ⊟ ⊟ ⊟   **d.** ⊞ ⊞ ⊞

3. a. ⊞—⊟   **b.** ⊞—⊟ ⊞—⊟

c. ⊞—⊟ ⊞—⊟
    ⊞—⊟

d. ⊞—⊟ ⊞—⊟
    ⊞—⊟ ⊞—⊟

Activity III

3. a. ⊟ ⊟ ⊟   **b.** ⊟ ⊟ ⊟ ⊟

c. ⊟ ⊟ ⊟ ⊟ ⊟   **d.**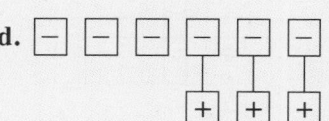

Activity IV

2. a.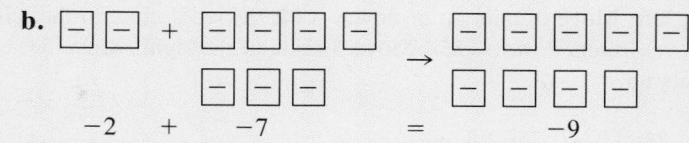

b.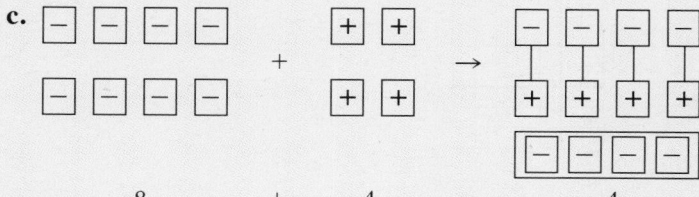

c.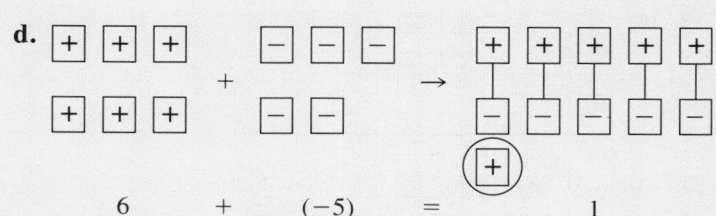

d.

$$6 \quad + \quad (-5) \quad = \quad 1$$

6. Answers may vary. The information may be organized as it appears in the problem, from least value to greatest value, from greatest value to least value, or in any manner that lists all the information. The most efficient method is in order, either least to greatest or greatest to least, as this method allows you to easily check that you have listed all possible cases.

11–12. Answers will vary. Examples are given.   **11.** Joan spends 55¢ in the cafeteria. In how many different ways can Joan pay using only quarters, dimes, and nickels? Answer: 11 ways
12. John is a sales clerk in a small grocery store. The cash register in the store contains only dimes, nickels, and pennies. How many different amounts of change can John make using three coins? Answer: 10 amounts

1. The members of the team agreed to work together.   2. The source, or beginning, of the Mississippi River is Lake Itasca, Minnesota.   10. Move 7 units to the left. Move 2 units up.   11. Move 5 units to the right. Move 3 units down.   12. Move 0 units to the left or right. Move 2 units up.   13. Move 4 units to the left. Move 0 units up or down.   14. Move 2 units to the left. Move 4 units down.   15. Move 3 units to the right. Move 5 units up.

22–25.

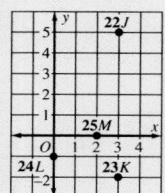

13–20.

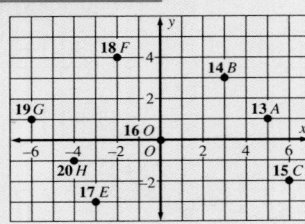

39.

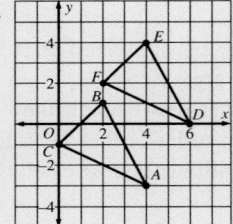

The shapes are the same. $\triangle DEF$ is 3 units up and 2 units to the right of $\triangle ABC$.

40.

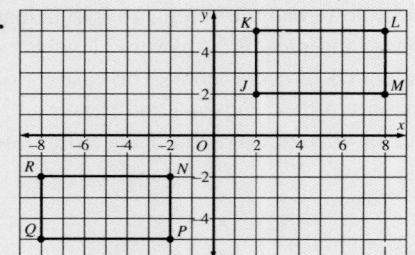

The shapes are the same. Rectangle *NPQR* is lower than and to the left of rectangle *JKLM*.

43.

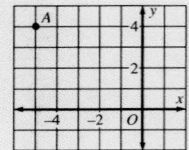

8.

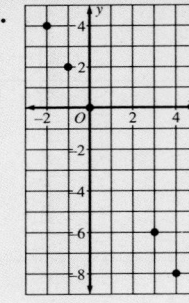

9.

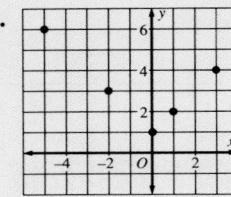

10.

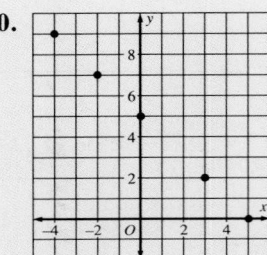

11.

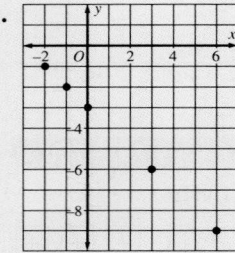

1.

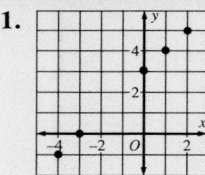

2.

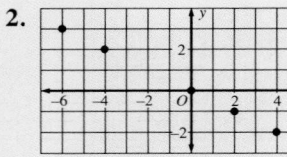

3.

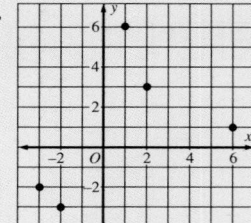

4.

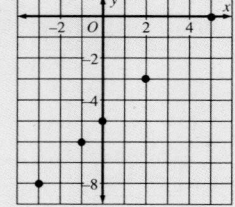

5.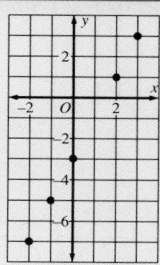

6.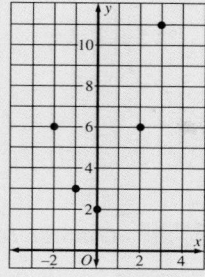

7.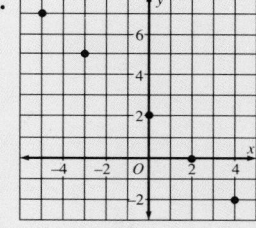

8.

9.

10.

11.

x	2x
−2	−4
−1	−2
0	0
1	2
2	4

$x \rightarrow 2x$

12.

x	x + 6
−2	4
−1	5
0	6
1	7
2	8

$x \rightarrow x + 6$

13.

x	−x
−2	2
−1	1
0	0
1	−1
2	−2

$x \rightarrow -x$

14.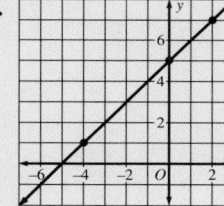

x	x^3
−2	−8
−1	−1
0	0
1	1
2	8

$x \rightarrow x^3$

15.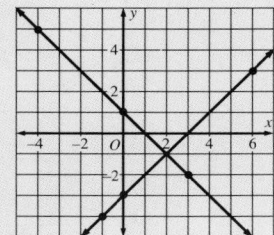

a. Answers will vary. Examples: (−2, 3), (−1, 4), and (1, 6)
b. Yes

16.

a. (2, −1)
b. Both function rules represent the relationship.

17.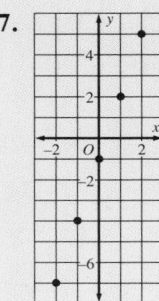

A7

5–6.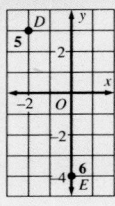

7.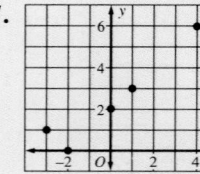

16. Answers will vary. A typical answer is the following. Both a map and a coordinate plane are divided into square sections by horizontal and vertical lines. Locations on both are identified by coordinates and both represent only a portion of a larger space. The differences are that on a map the coordinates are a letter and a number, while on a coordinate plane the coordinates are two numbers. Also the map labels are on the edge of the map, while the coordinate plane labels are on the axes. On a map the coordinates name a square, but on a coordinate plane the coordinates name a point.

60–63.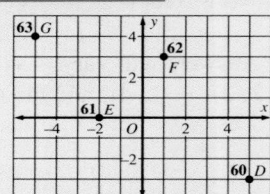

64.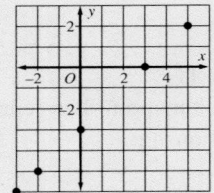

65.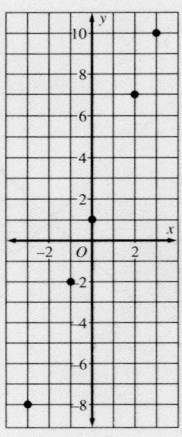

66.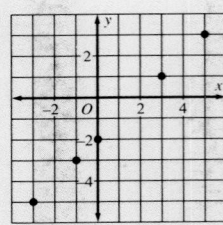

67.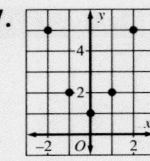

5. If x is negative, $-x$ is positive, so $-x > x$. If x is zero, $-x$ is zero, so $-x = x$. If x is positive, $-x$ is negative, so $x > -x$.

32–34.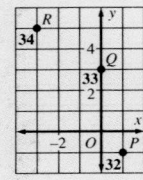

35.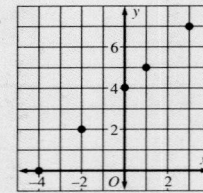

36.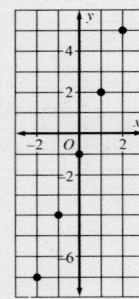

37.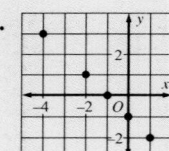

38.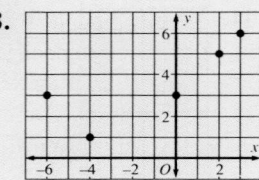

1. a. closed
   b. not closed
   c. closed
   d. not closed

2. a. closed
   b. closed
   c. closed
   d. not closed

3. a. closed
   b. not closed
   c. closed
   d. not closed

4. a. closed
  b. not closed
  c. not closed
  d. not closed

5. a. not closed
  b. not closed
  c. closed
  d. closed (exclude division by zero)

6. a. not closed
  b. not closed
  c. closed
  d. not closed

7. a. closed
  b. not closed
  c. closed
  d. not closed

8. a. not closed
  b. not closed
  c. not closed
  d. not closed

Chapter 4

Page 143   *Guided Practice*

7.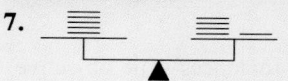

8.

9.

Page 144   *Exercises*

39–42. Answers will vary. Examples are given.
39. $m + 3 = 2$; $9m = -9$ **40.** $-10p = 50$; $p + 12 = 7$
41. $c - 18 = 6$; $\frac{c}{8} = 3$ **42.** $9a = 45$; $a + 11 = 16$

Pages 146–147   *Problem Solving Situations*

14–15. Answers will vary. Examples are given.
14. Germaine bought some apples for 50¢ each and some bananas for 75¢ each. She spent $2.00 altogether. How many of each did she buy? Answer: 1 apple, 2 bananas   **15.** Marc spent $89 on sweaters and shirts. Sweaters cost $17 each and shirts cost $11 each. How many of each did he buy? Answer: 2 sweaters, 5 shirts

Pages 153–154   *Exercises*

35.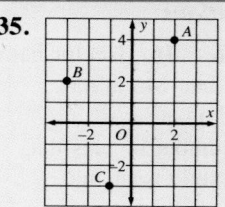

Activity I

2. a. 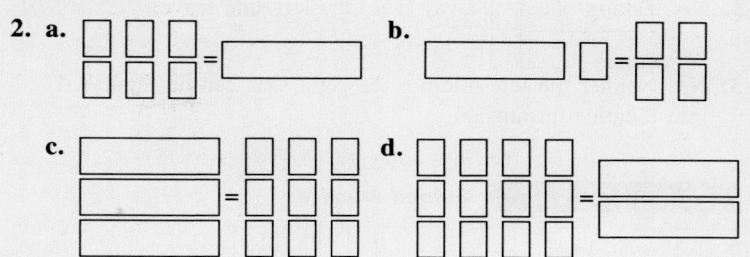   **b.**

  c.   **d.**

Activity II

1. a. The diagram represents the situation of subtracting 1 from each side of the equation $6 = x + 1$.
  b. The diagram represents the situation of subtracting 4 from each side of the equation $x + 4 = 9$.

2. a. 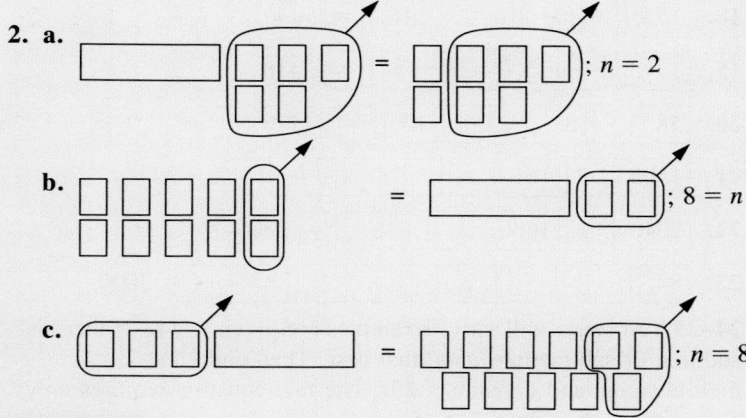   ; $n = 2$

  b.   ; $8 = n$

  c.   ; $n = 8$

Activity III

1. a. The diagram represents the situation of dividing both sides of the equation $3x = 9$ by 3.
  b. The diagram represents the situation of dividing both sides of the equation $2x = 8$ by 2.

2. a. 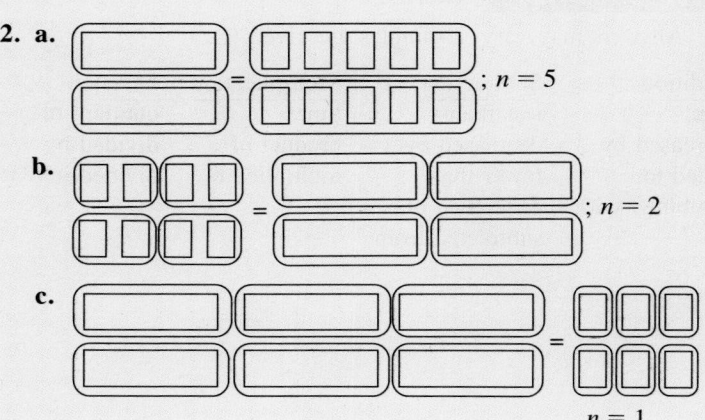   ; $n = 5$

  b.   ; $n = 2$

  c. $n = 1$

1. $2n + 1 = 7$
2. No. Taking one ☐ away from the left side leaves two ☐☐ .
3. No. Neither the left side nor the right side can be separated into identical groups.

2.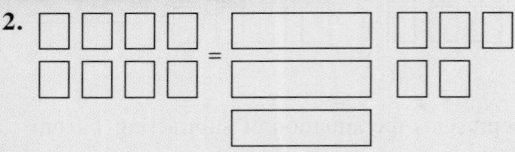

18. │153.│ + │71.│ = │ ÷ │16.│ = │ ; 14
19. │187.│ − │135.│ = │ × │8.│ = │ ; 416
20. │58.│ + │32.│ = │ × │13.│ = │ ; 1170
21. │1505.│ − │161.│ = │ ÷ │42.│ +/− │ = │ ; −32
22. │2040.│ − │1480.│ = │ ÷ │7.│ = │ ; 80
23. │88.│ +/− │ − │112.│ = │ × │15.│ +/− │ = │ ; 3000

24–25. Answers will vary. Examples are given. 24. If there is addition or subtraction, undo that first. Then undo the multiplication and division. 25. The first equation requires only one step to solve; $2x + 3 = 15$ requires two steps to solve.
26. There are the same number of atoms of each element before and after the reaction. 27. Two; a chemical reaction has a natural balance.

15. Answers may vary. Examples are given.

Addition	Subtraction	Multiplication	Division
sum	less than	times	quotient of
increased by	decreased by	product of	divided by
added to	fewer than	multiplied by	divided into
combined with	difference of		
	subtracted from		

20.

x	$26 + 12(x - 1)$
1	26
2	38
3	50
4	62
5	74
6	86
7	98
8	110
9	122
10	134

Rule: $x \rightarrow 26 + 12(x - 1)$

1. The objective is to write equations for sentences. The objective for Lesson 4-6 was to write variable expressions for word phrases. Variable expressions are only parts of equations, as word phrases are only parts of sentences. 2. In Example 2, you must choose the variable.

17–20. Answers will vary. Examples are given. 17. A number m decreased by 8 is 16. 18. Nine more than a number y is 32.
19. The product of 4 and a number w is 48. 20. The quotient of a number u and 9 is 14.
23.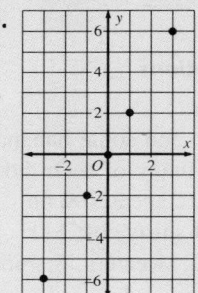

16–17. Answers may vary. Examples are given. 16. Hector has 16 tomato plants in his garden. This is 5 more than Maria has. How many plants does Maria have? Answer: 11 plants
17. Allan made one less than three times the number of pies that Oscar made. If Allan made 26 pies, how many did Oscar make? Answer: 9 pies

1. The main idea is to simplify expressions by combining like terms and by using the distributive property in order to solve equations. **2.** Equations can be simplified before solving by combining like terms. Equations can be simplified before solving by using the distributive property.

Answers may vary. A black hole is a small celestial body with an intense gravitational field that is believed to be a collapsed star.

16. The relationship is expressed as a fraction, with the systolic pressure in the numerator and the diastolic pressure in the denominator.

14. Answers will vary. Example: Inverse operations are operations that undo each other, such as addition and subtraction. **19.** Answers will vary. Example: An equation has an equals sign in it, an expression does not.

19–20. Answers may vary. Examples are given. **19.** Add nine to both sides of the equation. **20.** Subtract $3x$ from both sides; subtract $8x$ from both sides; subtract 5 from both sides; add 10 to both sides

Chapter 5

2–5. Answers may vary. Examples are given.
2. 10 years   **3.** 5 years   **4.** to the nearest multiple of 5

5.

Maximum Life Spans of Certain Animals

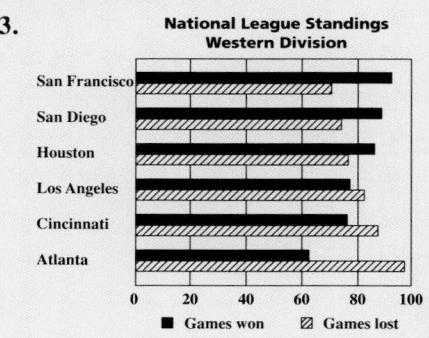

7.

Airport Limousine Earnings per Quarter

First	$ $ $ $ $ $ $ $	
Second	$ $ $ $ $ $ $ $ $ $	
Third	$ $ $ $ $ $ $ $ ᶜ	
Fourth	$ $ $ $ $	$ = \$10,000

8.

Number of Women in State Legislatures

13. There are too few symbols, so the differences in the data are not reflected. For example, Boston and Pittsburgh seem to have the same number of passengers. **14–15.** Answers may vary. Examples are given. **14.** (b); There would be too many symbols if each symbol represented 2 million passengers. If each symbol represented 30 or 40 million passengers, there would be too few symbols to show the differences in the data.

15.

Airport Traffic

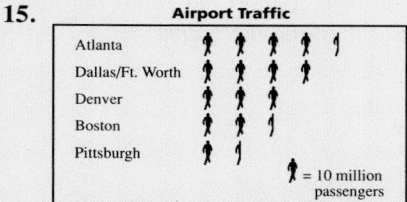

23.

National League Standings Western Division

■ Games won   ▨ Games lost

11–14. Estimates may vary. Answers are given in this order: elementary then secondary.   **11.** about 1.2 million; 1.0 million **12.** about 1.45 million; 1.1 million   **13.** about 1.1 million; 0.9 million   **14.** about 1.35 million; 1.05 million

31.

Year	Number of Sheep (Millions)
1910	50
1925	40
1940	50
1955	30
1970	20
1985	10

1. *How to Draw a Bar Graph:* (1) Draw the axes and position the categories along one axis.   (2) Choose a scale and number the other axis using this scale.   (3) Draw the bars, using the scale as a guide.   (4) Label the axes and title the graph.   *How to Draw a Line Graph:* (1) Draw the axes and position the categories evenly along the horizontal axis.   (2) Choose a scale and number the vertical axis using this scale.   (3) Place a point on the graph for each data item, and then connect the points with line segments.   (4) Label both axes and title the graph.

4.

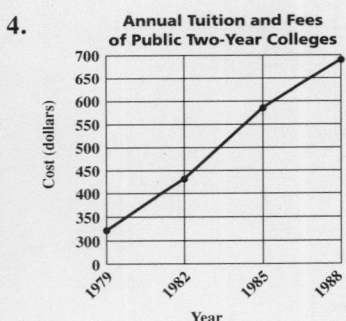

5.

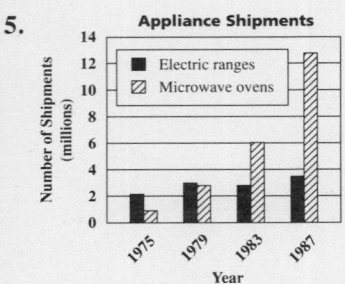

1.

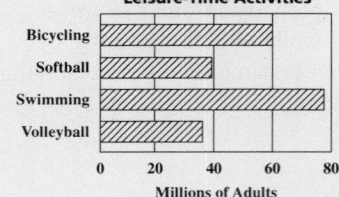

2.

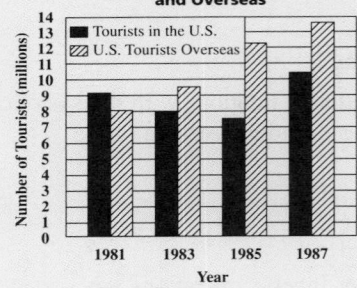

3.

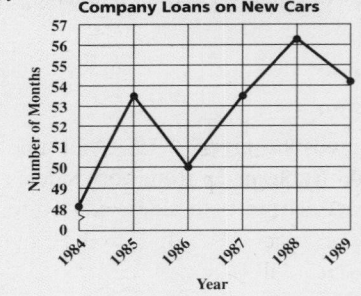

4.

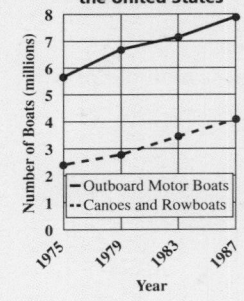

5.

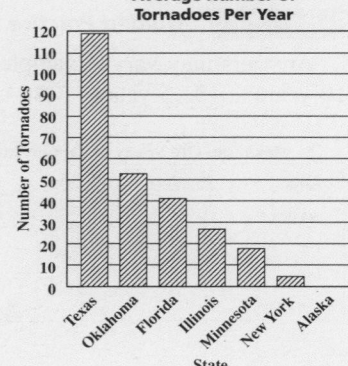

A12

6.

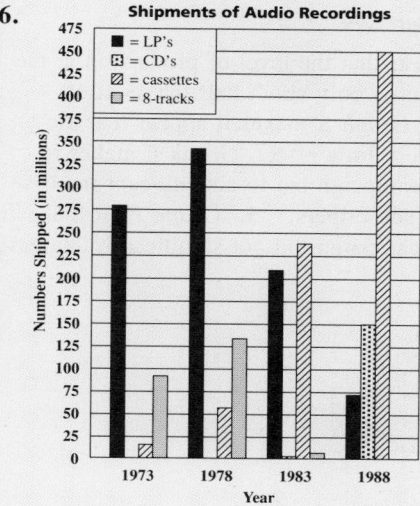

Shipments of Audio Recordings

7.

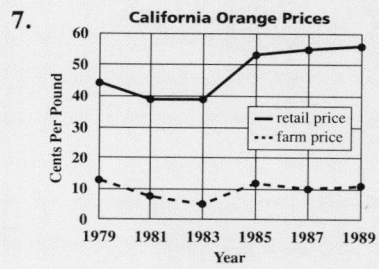

California Orange Prices

8. No; for example, from 1981 to 1983 the retail price did not change, but the farm price dropped by 3¢ per pound.

9. Answers may vary. Example: The retail price of oranges is always greater than the farm price. Furthermore, the direction of change in retail and farm prices is generally the same.

10. Analysis; the graph makes it easier to recognize increases and decreases in the data and to compare trends in the retail and farm prices.

14.

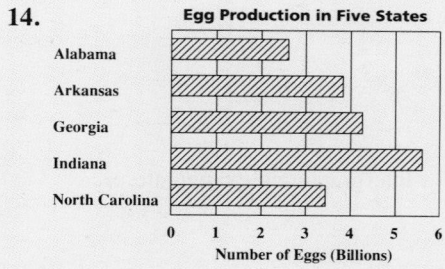

Egg Production in Five States

Page 203 *Self-Test 1*

1.

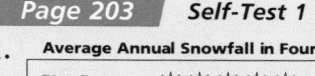

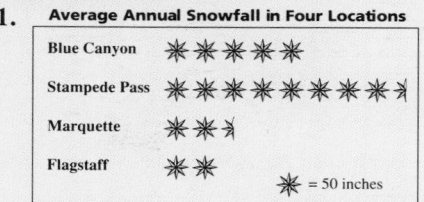

Average Annual Snowfall in Four Locations

6.

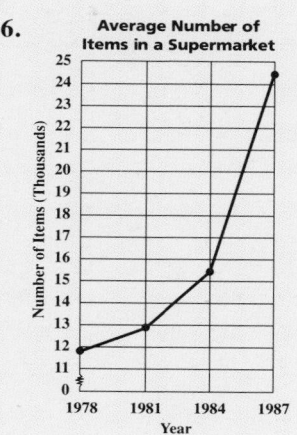

Average Number of Items in a Supermarket

Pages 204–205 *Guided Practice*

5.

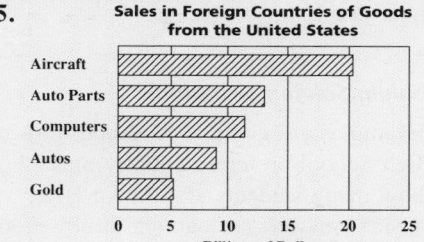

Sales in Foreign Countries of Goods from the United States

Page 205 *Exercises*

1.

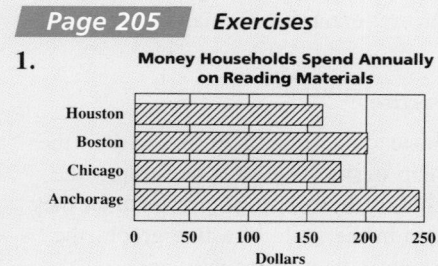

Money Households Spend Annually on Reading Materials

2.

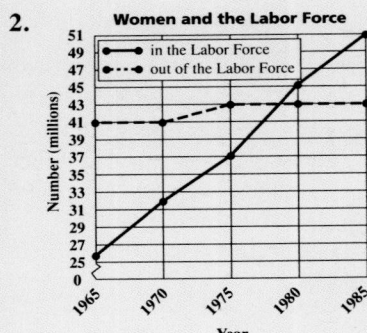

Women and the Labor Force

3.

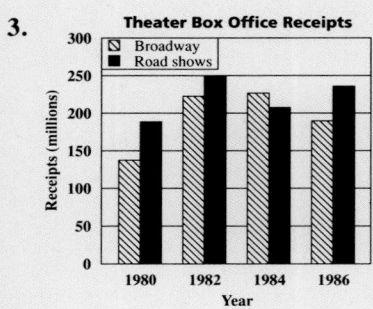

Theater Box Office Receipts

4. Evaluation; Since the data are fairly well dispersed, using a symbol to approximate the data in a pictograph should not significantly distort the data. Furthermore, a pictograph might appear more creative than a bar graph.

Pages 207–208 **Problem Solving Situations**

9–10. Answers will vary. Examples are given. **9.** About how many students at Alltown High School preferred country music? Answer: about 450; About how many students at Alltown High School preferred classical music? Answer: not enough information **10.** About how many people listen to WHMC at 10:00 A.M.? Answer: about 0.1 million; About how many people listen to WHMC at midnight? Answer: not enough information

Page 210 **Guided Practice**

1. Graphs can be visually misleading when there is a gap in the scale. **2.** When there is a gap in the scale of a bar graph, the differences between the heights of the bars look greater than they really are. When there is a gap in the scale of a line graph, the changes may appear more dramatic than they really are.

Pages 211–212 **Exercises**

2. The graph makes it appear that the level of production at the second most productive plant is only about half the level of production at Plant B. **4.** *Graph E* makes it appear that the advertising campaign had very little effect. *Graph F* makes it appear that the advertising campaign led to a significant increase in the number of magazine subscribers. **5.** *Graph E.* It makes it appear that the advertising campaign did not significantly increase the number of subscribers.

10.

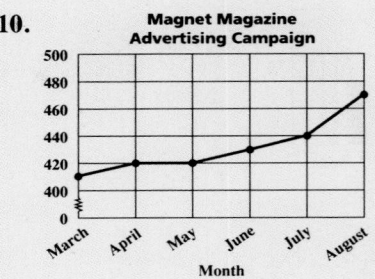

Magnet Magazine Advertising Campaign

11.

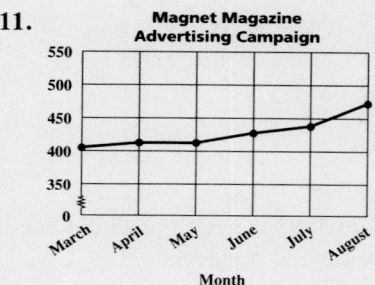

Magnet Magazine Advertising Campaign

12.

Magnet Magazine Advertising Campaign

13. Changing the scale on a line graph can exaggerate or minimize a trend.

1.

**Median Prices of New
One-Family Homes in the Northeast**

Price (Thousands of Dollars)

125
115
105
95
85
75
65
55
45
0

1976   1978   1980   1982   1984   1986

Year

Activity II

4. A set of data will balance around the mean, median, and mode when they are the same. Sets of data will vary. Example: 71, 74, 76, 76, 77, 78, 80.

1. Median; the mean of the salaries for a company will often be distorted by extreme values.   **2.** Mean; test scores generally fall within a fairly narrow range.   **3.** Mode; although shoe sizes are numbers, averaging them is meaningless. The mode tells you what is most in demand.

1. Yes; the mean is 72 in. It is not distorted by an extreme value. **2.** Yes; the mode is $285 and all the payments are reasonably close to it.   **3.** Mode; the data cannot be averaged or listed in numerical order.   **4.** Median; the median is $607.16 and the mean is $1449.34. The mean is distorted by the extreme value $4790.15.   **5.** Median; the median is 5 and the mode is 3. The mode is less than most of the family sizes.   **6.** Mode; the data cannot be averaged or listed in numerical order.   **7.** Mean; the mean is 35 h, the median is 29 h, and the mode is 20 h. The mean is equal to the number of hours that the nurse wants to work per week.

9.

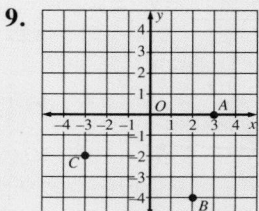

10. Yes; the mean is 83.8 and there are no extreme values.

8. The database software can arrange the data in numerical order so that it is easier to find the middle item(s).   **9.** Answers may vary. Example: Washington, D.C.; use the database to sort the records in decreasing order by the total cost for food, housing, and transportation. Washington, D.C., appears first, with a total of $23,208.

1.

Hours	Tally	Frequency
5	卌 I	6
6	卌 II	7
7	II	2
8	III	3
		Total: 18

3.

Number of Movies	Tally	Frequency
2	II	2
3	卌 I	6
4	卌 卌	10
5	IIII	4
12	I	1
15	I	1
		Total: 24

2. Yes; the mean is 14.5. It is not distorted by an extreme value. **3.** Mode; the data cannot be averaged or listed in numerical order.

14. Answers will vary. Examples: rent, salaries, supplies and equipment, licenses and fees. This general list contains some of the same types of costs as the list in Exercise 13. However, the list in Exercise 13 contains items specific to a passport photo business.

11.

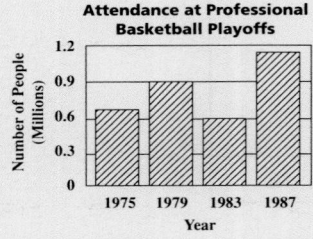

Parks in Four Major Cities

Fort Worth   Minneapolis   New Orleans   Omaha

🌲 = 50 parks

18. *Graph A* gives the impression that the average number of visits to physicians in a year did not change very dramatically from 1970 to 1985. *Graph B* gives the opposite impression.

20.

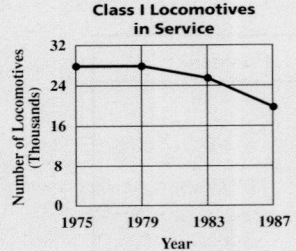

Attendance at Professional Basketball Playoffs

21.

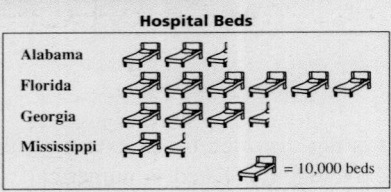

Class I Locomotives in Service

1.

Hospital Beds

Alabama   Florida   Georgia   Mississippi

🛏 = 10,000 beds

4–5. Answers will vary. Examples are given.   **4.** number of students, number of quarters.   **5.** *Class Sizes, Quarters Collected for Charity at Wilson High School.*

8.

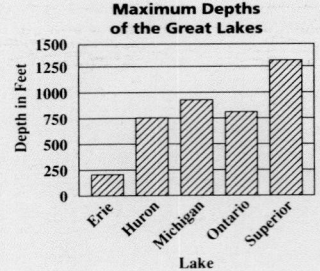

Maximum Depths of the Great Lakes

9.

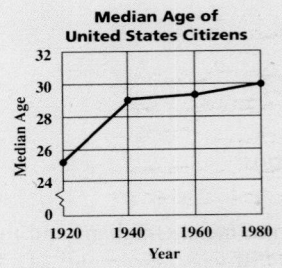

Median Age of United States Citizens

10.

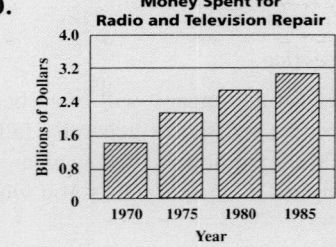

Money Spent for Radio and Television Repair

16. Mode; the data cannot be averaged or listed in numerical order.

Chapter 6

7.

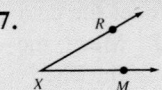

8.

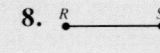

9.

10.

11.

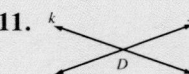

12.

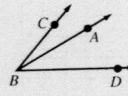

Page 239   Historical Note

Answers will vary. Two examples are given. Truss bridges are supported by a framework of triangles, or trusses. Because of the stability of the triangles, these bridges are quite strong, and are capable of supporting heavy traffic. Truss frameworks of triangles are also used in the construction of cantilever bridges. These consist of two independent beams, called cantilevers, which are supported by piers and a truss framework.

Page 241   Guided Practice

9.

10. 155°

11. 90°

12. 67°

Page 242   Exercises

3. 115°

4. 20°

5. 88°

6. 164°

7. 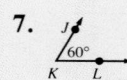 60°

8. 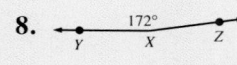 172°

15. 104°

Pages 243–244   Focus on Explorations

Activity I

2. a.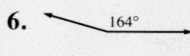

b. Draw a ray. Open the compass to the length of segment *EF*. Put the point of the compass at the endpoint of the ray and mark off a segment equal to *EF*. Put the point on the mark and mark off a second segment equal to *EF*. Repeat the procedure again. Label the endpoint of the ray as *M* and the third mark as *N*. $MN = 3 \cdot EF$.

Activity II

2.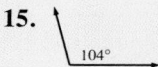

Activity III

2.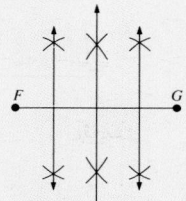

Activity IV

2. The measures of the two angles will always be equal whether the measure of angle *ABC* increases or decreases.

3. a. 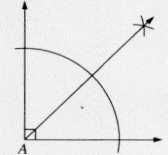   **b.**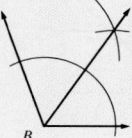

Pages 247–248   Exercises

19. A straight angle has measure 180°.   **20.** They represent the measures of two supplementary angles.

Page 251   Guided Practice

3. Line *AB* is parallel to line *ST*.   **4.** Ray *FG* is perpendicular to line *UV*.   **5.** Segment *XY* is perpendicular to segment *ZW*.

6.

$$G \quad M \quad H$$
with N above M

19. Answers will vary. Examples: to make city planning easier, to make the lots or blocks similar in size, to facilitate orientation.

Pages 251–253   Exercises

19. Parallel lines lie in the same plane, and skew lines do not; both parallel lines and skew lines do not intersect.   **21.** No, because if two skew lines were perpendicular, they would be in the same plane.

27.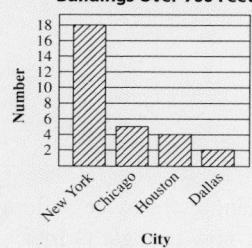

Buildings Over 700 Feet

4. 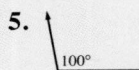   **5.** 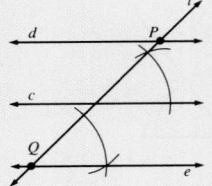 100°   **6.** 82°

55°

Activity III

2. Line *d* is parallel to line *c* because corresponding angles 1 and 2 are equal in measure and when corresponding angles are equal in measure, the lines are parallel.

3.

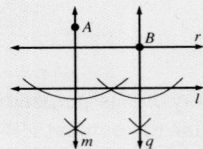

Yes, since all three corresponding angles are equal in measure, all three lines are parallel.

Activity IV

1–2, 4.

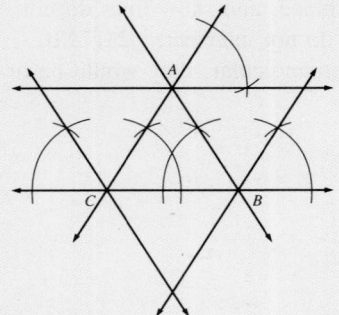

2. Yes, the corresponding angles are 90° and are therefore equal, so the lines are parallel. **3.** Two lines that are perpendicular to the same line are parallel to each other. **4.** yes; yes **5.** Yes, since the lines are parallel, the corresponding angles must be equal in measure. If one angle measures 90°, the other angle measures 90°, but this means that the lines are perpendicular. **6.** The angles formed by the intersection of parallel lines and lines perpendicular to them are all right angles.

7.

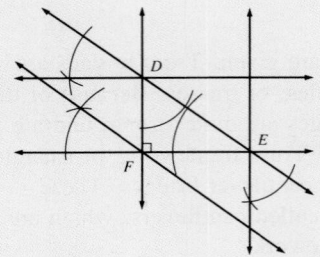

a. triangle **b.** The sides of the new triangle are twice as long as those of triangle *ABC* and the area of the new triangle is four times the area of triangle *ABC*.

8.

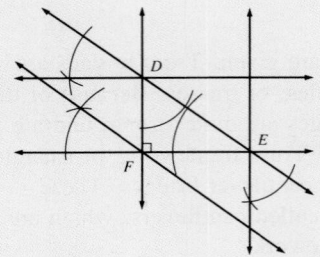

The new shape is a triangle with sides twice as long as those of triangle *DEF* and with four times the area.

4. Answers may vary. An example is given. Strips of lengths 2 in., 3 in., and 5 in. will not form a triangle because the lengths of the two shorter sides added together are not longer than the 5 in. side. If the two shorter strips form an angle less than 180°, their ends will not meet the endpoints of the 5 in. strip.

4.   **5.**   **6.**

52. 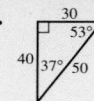   **53.**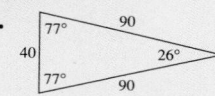

19. Quadrilateral

Definition: A polygon with four sides

	None	One pair	Two pairs
Parallel sides:	(None)	One pair	Two pairs
Opposite sides equal:		yes	(no)
Opposite angles equal:		yes	(no)
Four right angles:		yes	(no)
Four equal sides:		yes	(no)

Parallelogram

Definition: A quadrilateral with two pairs of parallel sides

Parallel sides:	None	One pair	**(Two pairs)**
Opposite sides equal:		**(yes)**	no
Opposite angles equal:		**(yes)**	no
Four right angles:		yes	**(no)**
Four equal sides:		yes	**(no)**

Trapezoid

Definition: A quadrilateral with one pair of parallel sides

Parallel sides:	None	**(One pair)**	Two pairs
Opposite sides equal:		yes	**(no)**
Opposite angles equal:		yes	**(no)**
Four right angles:		yes	**(no)**
Four equal sides:		yes	**(no)**

Rhombus

Definition: A quadrilateral with four sides of equal length

Parallel sides:	None	One pair	**(Two pairs)**
Opposite sides equal:		**(yes)**	no
Opposite angles equal:		**(yes)**	no
Four right angles:		yes	**(no)**
Four equal sides:		**(yes)**	no

Square

Definition: A quadrilateral with four right angles and four sides of equal length

Parallel sides:	None	One pair	**(Two pairs)**
Opposite sides equal:		**(yes)**	no
Opposite angles equal:		**(yes)**	no
Four right angles:		**(yes)**	no
Four equal sides:		**(yes)**	no

20. a. parallelogram, rectangle, rhombus, square

  b. rectangle, square

  c. parallelogram, rectangle, rhombus, square

  d. parallelogram, rectangle, rhombus, square

21.

quadrilateral	none
parallelogram	four pairs
trapezoid	two pairs
rectangle	four pairs
rhombus	four pairs
square	four pairs

Page 270 **Focus on Computers**

7. parallelogram, trapezoid, rectangle, rhombus or square; true for all the figures except trapezoid.
8. The diagonals of an isosceles trapezoid are congruent; true.

Pages 273–274 **Problem Solving Situations**

7. The dots represent people; the lines represent handshakes.
8. Problem Solving Situation 2   **9.** Count the handshakes between two people, three people, four people, and so on. Look for a pattern in the number of handshakes. The solution is 28.
10. Each of the eight people must shake hands with 7 others. Multiply 8 by 7, then divide by 2 to eliminate counting any handshake twice.   **11–12.** Answers will vary. Examples are given.   **11.** How many pieces of pizza result from four cuts through the center of the pizza?   **12.** Find the number of blocks needed to build a "pyramid" with 10 blocks in the bottom row.
15.

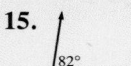

Pages 276–277 **Guided Practice**

8.

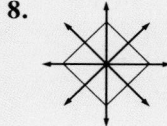

10.

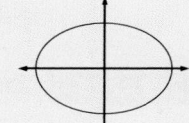

Pages 277–279 **Exercises**

5.

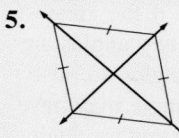

6.

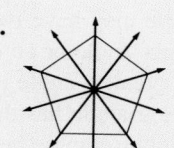

9.

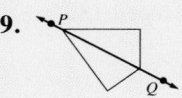

10.

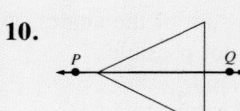

21. Answers will vary, but should include the following points. Trapezoids have no lines of symmetry unless they are isosceles. Parallelograms in general have no lines of symmetry, unless they are one of the following special parallelograms. A rectangle has two lines of symmetry, which connect the midpoints of the opposite sides. A rhombus has two lines of symmetry, which connect the opposite vertices. A square has four lines of symmetry; two as described for a rectangle, and two as described for a rhombus.

24.

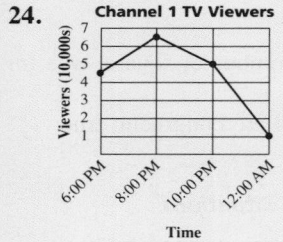

Page 279   *Self-Test 2*

8.

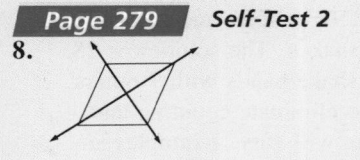

Pages 280–282   *Chapter Review*

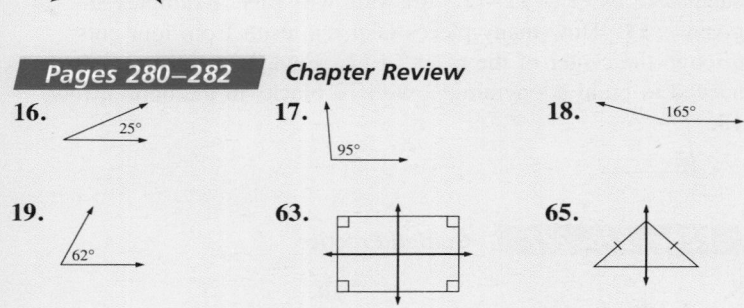

16.   25°   **17.**   95°   **18.**   165°

19.   62°   **63.**   **65.**

Pages 283–284   *Chapter Test*

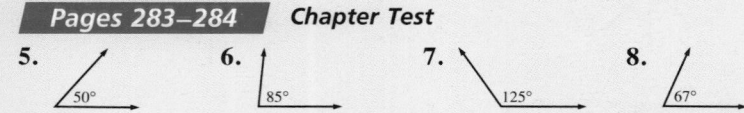

5.   50°   **6.**   85°   **7.**   125°   **8.**   67°

32. A triangle may have two or three acute angles; if the greatest angle is a right angle or an obtuse angle, the other two angles are acute; if the greatest angle is acute, the other two angles are also acute and the triangle has three acute angles. A triangle may have at most one obtuse angle because the measure of an obtuse angle is greater than 90° and if a triangle had more than one obtuse angle the sum of the angles of the triangle would exceed 180°, which is not possible.

40.

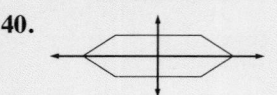

Page 285   *Chapter Extension*

3.

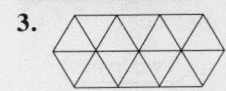

Code: 3-3-3-3-3-3

4.

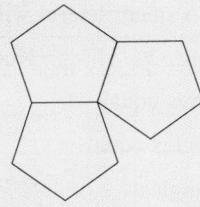

Chapter 7

Pages 291–292   *Exercises*

33. 2, 3, 5, 7, 11, 13, 17, 19, 23, 29, 31, 37, 41, 43, 47, 53, 59, 61, 67, 71, 73, 79, 83, 89, 97   **34. a.** 101, 103, 107, 109, 113, 127, 131, 137, 139, 149, 151, 157, 163, 167, 173, 179, 181, 191, 193, 197, 199   **35. b.** 211, 223, 227, 229, 233, 239, 241, 251, 257, 263, 269, 271, 277, 281, 283, 293   **36.** The sieve of Eratosthenes was used as follows: Write the whole numbers in order, continuing as far as you like. Disregard 1. Retain 2, then cross out every second number after 2. Retain 3, then cross out every third number after 3. The next uncrossed number is 5. Retain 5, then cross out every fifth number after 5. Repeat this procedure for all uncrossed numbers. Eventually, all numbers that are not crossed out are prime numbers.

40.

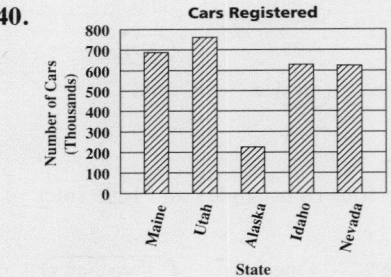

Pages 298–299   *Exercises*

27. A factor is any of two or more numbers multiplied to form a product. A multiple is a product of a given number and any nonzero whole number.   **28.** To find the GCF: (a) Write all the factors of the numbers. (b) Find all the common factors. (c) Find the greatest common factor, which is the GCF. To find the LCM: (a) List the multiples of each number. (b) Find the common multiples. (c) Find the least common multiple, which is the LCM. The difference in the methods is that to find the GCF, you find the greatest common number in the lists, while to find the LCM, you find the least common number.   **34.** No. Since the GCF is a factor of both numbers, it must be less than or equal to the smaller of the two numbers. Since the LCM is a multiple of both numbers, it must be greater than or equal to the larger of the two numbers. Since the two numbers are different, the GCF must be less than the LCM. Therefore, they are not equal.

Pages 300–301 *Focus on Explorations*

Activity I

2. a.   b.

c.   d.

Activity II

2. a.

b.

c.

d.

3. a.

b.

c.

4. Answers may vary. Examples are given.   a. $\frac{2}{3}$; $\frac{8}{12}$ b. $\frac{2}{6}$; $\frac{4}{12}$
c. $\frac{1}{2}$; $\frac{2}{4}$

Activity IV

1. a. The diagram in A represents the same amount as the diagram in B.   b. The fractions represented by A are different from the fractions represented by B.   2. Answers may vary. Examples are given.

2. a.

b.

c.

d.

e.

f.

Pages 303–304 *Guided Practice*

6.

7.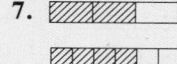

8.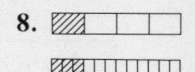

Pages 304–305 *Exercises*

30. If the numerator and denominator of a fraction are the same, then the fraction is equivalent to 1.

33.

Page 308 *Historical Note*

The British mathematician Ada Byron Lovelace (1815–1852) provided the first description of computer programming. This was in relation to Charles Babbage's "Analytical Engine," a machine that could make calculations, store data, and print out results. The daughter of the poet Lord Byron, Lady Lovelace was a gifted writer as well as an accomplished mathematician. In her letters to Babbage she predicted uses and limitations of computers that were not realized until 150 years later.

Pages 310–311 *Guided Practice*

4.

5.

6.

7.

Pages 311–312 *Exercises*

29.    1, 4, 9, 16, 25

Pages 315–316 *Exercises*

35. The numerator is the repeating part of the decimal.

$\frac{1}{9} = 0.\overline{1}$	$\frac{5}{9} = 0.\overline{5}$
$\frac{2}{9} = 0.\overline{2}$	$\frac{6}{9} = 0.\overline{6}$
$\frac{3}{9} = 0.\overline{3}$	$\frac{7}{9} = 0.\overline{7}$
$\frac{4}{9} = 0.\overline{4}$	$\frac{8}{9} = 0.\overline{8}$

40. Nine times the numerator is the repeating part of the decimal.

$\frac{1}{11} = 0.\overline{09}$	$\frac{6}{11} = 0.\overline{54}$
$\frac{2}{11} = 0.\overline{18}$	$\frac{7}{11} = 0.\overline{63}$
$\frac{3}{11} = 0.\overline{27}$	$\frac{8}{11} = 0.\overline{72}$
$\frac{4}{11} = 0.\overline{36}$	$\frac{9}{11} = 0.\overline{81}$
$\frac{5}{11} = 0.\overline{45}$	$\frac{10}{11} = 0.\overline{90}$

46. $\frac{1}{5} = 0.2$; $\frac{1}{25} = 0.04$; $\frac{1}{10} = 0.1$; $\frac{1}{30} = 0.0\overline{3}$;

$\frac{1}{15} = 0.0\overline{6}$; $\frac{1}{35} = 0.0\overline{285714}$; $\frac{1}{20} = 0.05$;

$\frac{1}{40} = 0.025$; Repeating: $\frac{1}{15}, \frac{1}{30}, \frac{1}{35}$;

Terminating: $\frac{1}{5}, \frac{1}{10}, \frac{1}{20}, \frac{1}{25}, \frac{1}{40}$

47. $\frac{1}{45}$: repeating; $\frac{1}{50}$: terminating

$\frac{1}{45} = 0.0\overline{2}$; $\frac{1}{50} = 0.02$

48. If the prime factors of the denominator consist only of 2's and/or 5's, then the decimal terminates.

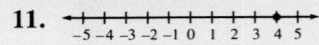

20. Place the 12 m and 19 m rods end-to-end, creating a 31 m length. Place the 26 m rod parallel to them, with one end lined up with one end of the 12 m rod. The distance from the other end of the 26 m rod to the end of the 19 m rod is 5 m.

Page 318   **Guided Practice**

5.

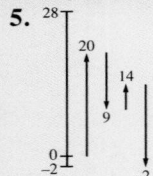

Pages 318–320   **Problem Solving Situations**

Answers will vary. Examples are given.   **2.** Fill the 5 gal bucket and pour into the 3 gal bucket. The 5 gal bucket now contains 2 gal. Pour the 2 gal into the 8 gal bucket. Repeat the procedure. The 8 gal bucket now contains 4 gal.   **4.** Lay the 16 cm rod and the 8 cm rod end-to-end creating a total length of 24 cm. Then lay the 14 cm rod beside them lining up one pair of ends. The difference in lengths is 10 cm.   **14.** Beldenville is between Appleton and Clairemont. The distance from Appleton to Clairemont is 10 mi and the distance from Beldenville to Appleton is 4 mi. How far is it from Clairemont to Beldenville? Answer: 6 mi   **15.** An ant walks 2 cm due west, 2 cm due north, 4 cm due east, 1 cm due south, 2 cm due east, then 1 cm due south. How far is the ant from its starting point? Answer: 4 cm due east

---

Page 326   **Guided Practice**

11. (number line −5 to 5)   **12.** (number line −1 to 1, $\frac{5}{6}$)

13. ($-7\frac{1}{3}$; number line −9 to 1)   **14.** (−3.5; number line −5 to 4)

Pages 326–327   **Exercises**

1. (number line −9 to 0)   **2.** (number line −3 to 6)

3. (number line −3 to 3)   **4.** (number line −1 to 9)

5. (number line −3 to 7)   **6.** (number line −4 to 4)

7. (number line −5 to 5)   **8.** (number line −5 to 5)

9. ($-3\frac{1}{3}$; number line −4 to 0)   **10.** ($5\frac{2}{5}$; number line −2 to 6)

11. ($-6\frac{1}{7}$; number line −6 to 0)   **12.** ($-\frac{2}{3}$; number line −1 to 1)

13. (7.5; number line 0 to 8)   **14.** (−2.25; number line −4 to 1)

23. (number line −4 to 4)   **24.** (number line −5 to 4)

25. (number line −3 to 3)   **26.** (1.5, 6.5; number line −2 to 8)

27. (number line −5 to 5)   **28.** (number line −1 to 1 with thirds)

31. (number line −1 to 1)   **34.** (−2.5; number line −3 to 1)

Pages 330–331   **Exercises**

30. An optical microscope uses light rays bent by lenses to form an enlarged image of a specimen. The most advanced optical microscopes can magnify an object up to 3000 times. An electron microscope uses beams of electrons to magnify objects. The best electron microscopes can magnify objects up to 1,000,000 times.

34.

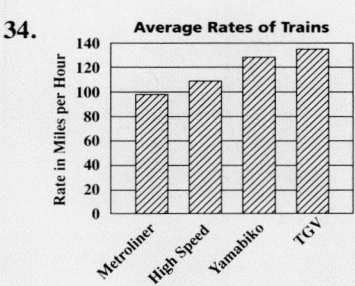

Average Rates of Trains

Page 331   Self-Test 3

5.

$-9\;-8\;-7\;-6\;-5\;-4\;-3\;-2\;-1\;0\;1$

6.

$0\quad 3\quad 6$

7.

$-3\quad 0\quad 3$

8.

$-2\quad 0\quad 2\quad 4\quad 6$

9.

-2.5
$-4\quad -3\quad -2\quad -1\quad 0$

10.

$-2\quad -1\quad -\frac{1}{3}\;0\quad 1$

Pages 332–333   Chapter Review

49. Answers may vary. Example: Fill the 10 gal bucket and pour water from it into the 4 gal bucket. This leaves 6 gal in the 10 gal bucket. Fill the 3 gal bucket and pour the water from it into the 10 gal bucket. The 10 gal bucket now has 9 gal of water in it.

63.

$-8\quad -4\quad 0\quad 4$

64.

$-5\;-4\;-3\;-2\;-1\;0\;1\;2\;3\;4\;5$

65.

-1.7
$-3\quad -2\quad -1\quad 0\quad 1$

66.

$\frac{2}{3}$
$-2\quad -1\quad 0\quad 1\quad 2\quad 3$

Page 334   Chapter Test

37.

$-3\quad 0\quad 3$

38.

$-1\quad 0\quad 1\quad 2\quad 3$

39.

5.5
$-1\;0\;1\;2\;3\;4\;5\;6\;7\;8$

40.

$\frac{7}{8}$
$-1\quad 0\quad 1$

Chapter 8

Page 340   Exploration

5. a. 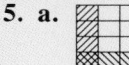   **b.** 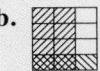   **c.**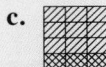

Pages 342–343   Exercises

33. $\boxed{4.}$ $\boxed{a\;b/c}$ $\boxed{2.}$ $\boxed{a\;b/c}$ $\boxed{7.}$ $\boxed{\times}$ $\boxed{2.}$ $\boxed{a\;b/c}$ $\boxed{7.}$ $\boxed{=}$

34. $\boxed{8.}$ $\boxed{a\;b/c}$ $\boxed{11.}$ $\boxed{a\;b/c}$ $\boxed{12.}$ $\boxed{\times}$ $\boxed{19.}$ $\boxed{a\;b/c}$ $\boxed{20.}$ $\boxed{=}$

35. $\boxed{4.}$ $\boxed{a\;b/c}$ $\boxed{5.}$ $\boxed{\times}$ $\boxed{17.}$ $\boxed{a\;b/c}$ $\boxed{3.}$ $\boxed{a\;b/c}$ $\boxed{8.}$ $\boxed{=}$

36. $\boxed{1.}$ $\boxed{a\;b/c}$ $\boxed{1.}$ $\boxed{a\;b/c}$ $\boxed{2.}$ $\boxed{\times}$ $\boxed{111.}$ $\boxed{a\;b/c}$ $\boxed{1.}$ $\boxed{a\;b/c}$ $\boxed{2.}$ $\boxed{=}$

Pages 346–347   Exercises

25.

$\frac{3}{4}$ c milk	1 tbsp baking soda
1 egg	1 tsp cinnamon
$\frac{1}{4}$ c shortening	$\frac{1}{3}$ c raisins
$2\frac{1}{4}$ c flour	1 c chopped apple

Page 349   Exercises

12. $W \quad\quad Y$

Page 354   Guided Practice

1. To add and subtract rational numbers and algebraic fractions with like denominators.   **2.** To add rational numbers or algebraic fractions with like denominators, add the numerators and write the sum over the denominator. To subtract rational numbers or algebraic fractions with like denominators, subtract the numerators and write the difference over the denominator.

Page 357   Guided Practice

1. Answers will vary. The following steps should be included:
a. Write each number as a fraction.   **b.** Find the LCD of the fractions.   **c.** Write equivalent fractions with the LCD.   **d.** Add or subtract the numerators.   **e.** Simplify the result.

Pages 358–359   Exercises

53.

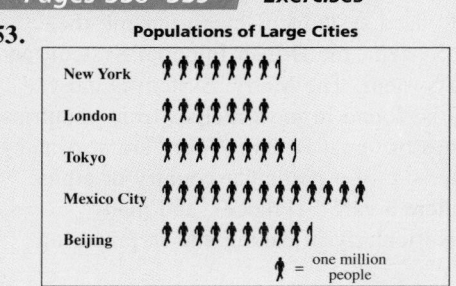

Populations of Large Cities

23.

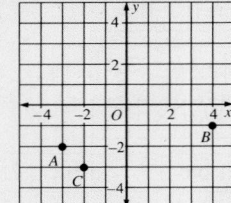

11–12. Answers will vary. Examples are given.   **11.** Angela ran in a 5 mi road race. How many feet did she run? Answer: 26,400 ft   **12.** The Entertainment Palace was a movie theater for 28 years, and has been a music club for the past 12 years. How many decades has it been in business? Answer: 4 decades

21–22. Answers will vary. They should include the following points.   **21.** First undo the addition or subtraction by using the inverse operation. If the denominators of the fractions are different, use the LCD when adding or subtracting. Then, if the variable is multiplied by a number, multiply by the reciprocal of that number to complete the solution.   **22.** The methods for solving equations involving integers and decimals are the same. The method for solving equations involving fractions is the same, unless the variable is multiplied by a number. In that case, multiply by the reciprocal of the number to complete the solution.
39–44. Answers will vary. Examples are given.   **39.** One third of a number g is fourteen.   **40.** Four fifths of a number c is sixteen.   **41.** Negative two is one third of a number x, increased by nine.   **42.** Three fourths of a number m decreased by four is eleven.   **43.** Ten is three eighths of a number s, increased by one half.   **44.** Five times a number d is negative seven twelfths.

50.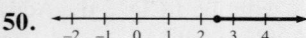

There are three commonly used systems of measurement: the United States Customary System, the British Imperial System, and the International (Metric) System. The Metric System is most commonly used, and will be found in most recipes from countries other than the United States or Great Britain. There are also many measures that are characteristic of a particular country or ethnic group, such as "une cuillière à café" (France), and many measures are estimated, particularly in baking and in preparing traditional foods.

20.

Hours Worked	Fee
1	52
2	74
3	96
4	118
5	140

23. Answers will vary. Examples are given. Taxicab rates: an initial fee, a fee for the first mile, and a fee for each additional mile. If these are \$1.50, \$2.70 and \$1.60, respectively, and the number of additional miles is m, then the formula for the fare, f, is $f = 1.50 + 2.70 + 1.60m$. Telephone rates: a start-up or installation fee, plus a monthly rate. If these are \$37 and \$21 respectively, and the number of months is t, then the formula for the rate r, is $r = 37 + 21t$. Electric rates: a basic monthly charge, an energy charge per kilowatt-hour, and a fuel charge per kilowatt-hour. If these are \$5.78, \$.06882, and \$.022170 respectively, and the number of kilowatt-hours is k, then the formula for the amount, a, is $a = 5.78 + 0.06882k + 0.022170k$.

16. Div is the dividend per share paid by the company. % is the percent of the close used to find the dividend. PE is the price-earnings ratio. Sales 100s is the number of sales on the previous day given in hundreds of shares.

28. Answers may vary. Examples are given.   **a.** Add 7.5 to both sides of the equation.   **b.** Divide both sides of the equation by -2.

Chapter 9

33. Answers will vary. It may not be desirable to buy an item with the lesser unit price if it is available only in an inconvenient size.

2. If a proportion is true, the cross products of the terms are equal; the cross products can be used to solve a proportion.
5. Six is to nine as two is to three.   **6.** The value of c is to four as ten is to forty.

12–13. Answers will vary. Examples are given.   **12.** The scale is dependent on the size of the ship. For a 1500 ft long supertanker, a scale such as 1 in. : 100 ft might be appropriate. For a 235 ft long clipper, a scale such as 1 in. : 20 ft might be appropriate.   **13.** A person is about 6 ft tall, so a 21 ft high statue indicates a scale of 6 ft : 21 ft, or 2 ft : 7 ft.   **14.** The HO scale is used most frequently in model railroads. Each part of the model is $\frac{1}{87}$ the size of the actual part. The O scale is often used for toy trains. Each toy part is $\frac{1}{48}$ actual size.   **15.** The heads on Mount Rushmore are about 60 ft high, which makes the scale about 1 in. : 80 in.

5–8. Strategies will vary. Likely strategies are given.
5. drawing a diagram; 6 blocks due south   **6.** supplying missing facts; 3 oz   **7.** using equations; 34   **8.** using proportions; 175
10–11. Answers will vary. Examples are given.   **10.** An ironing machine can press 35 shirts in 3 h. How long will it take for the machine to press 140 shirts? Answer: 12 h   **11.** The price of 6 pears is $1.44. What is the price of 4 pears? Answer: $.96

13.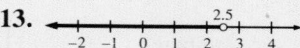

Activity II
1–3.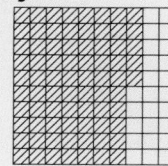

Activity IV
2. a. 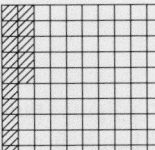   **b.**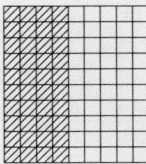

c. 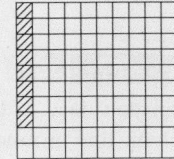   **d.**

2. Method 1: Write the given fraction as an equivalent fraction whose denominator is 100. The numerator is the percent. Method 2: Divide the denominator by the numerator to get a decimal. Move the decimal point two places to the right and insert the % symbol.

38. Answers may vary. An example is given. To find what percent one number is of another, divide the first number by the second number. Convert the result into a percent by moving the decimal point two places to the right and inserting the % symbol.

42.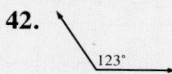

15. $\boxed{5.}\ \boxed{\%}\ \boxed{\times}\ \boxed{312.}\ \boxed{=}$; $\boxed{312.}\ \boxed{\times}\ \boxed{5.}\ \boxed{\%}$

16. $\boxed{2.65}\ \boxed{\%}\ \boxed{\times}\ \boxed{180.}\ \boxed{=}$; $\boxed{180.}\ \boxed{\times}\ \boxed{2.65}\ \boxed{\%}$

17. $\boxed{16.}\ \boxed{\%}\ \boxed{\times}\ \boxed{975.}\ \boxed{=}$; $\boxed{975.}\ \boxed{\times}\ \boxed{16.}\ \boxed{\%}$

18. $\boxed{300.}\ \boxed{\%}\ \boxed{\times}\ \boxed{159.}\ \boxed{=}$; $\boxed{159.}\ \boxed{\times}\ \boxed{300.}\ \boxed{\%}$

19. $\boxed{0.7}\ \boxed{\%}\ \boxed{\times}\ \boxed{34.9}\ \boxed{=}$; $\boxed{34.9}\ \boxed{\times}\ \boxed{0.7}\ \boxed{\%}$

20. $\boxed{26.42}\ \boxed{\%}\ \boxed{\times}\ \boxed{100.}\ \boxed{=}$; $\boxed{100.}\ \boxed{\times}\ \boxed{26.42}\ \boxed{\%}$

The first mathematics book printed in the Americas was a business mathematics book by Juan Diaz. It was printed in Mexico.

23. Since the cereal alone provides 100% of the RDA for Vitamin B_{12}, and more is provided by the milk, together they provide more than 100% of the RDA for Vitamin B_{12}.

19. Answers will vary, but should include the following points. First find the difference between $15 and $13.80, which is $1.20. Then find the percent $1.20 is of the original amount, $15. To do this, divide $1.20 by $15. The result is 0.08, which is 8%.
20. False; twice an amount x is $2x$, but an increase of 200% is $x + 2x = 3x$.   **21.** True; half an amount x is $\frac{1}{2}x$, and a decrease of 50% is $x - 0.5x = 0.5x$.

22. These are the sales tax rates as of October, 1990. All rates shown are percents.

State	Rate	State	Rate	State	Rate	State	Rate	State	Rate
Ala.	4	Haw.	4	Mass.	5	N. Mx.	4.75	S. Dk.	4
Alas.	–	Idaho	5	Mich.	4	N.Y.	4	Tenn.	5.5
Ariz.	5	Ill.	5	Minn.	6	N.C.	3	Texas	6
Ark.	4	Ind.	5	Miss.	6	N. Dk.	6	Utah	5.094
Calif.	4.75	Iowa	4	Mo.	4.225	Ohio	5	Vt.	4
Colo.	3	Kans.	4.25	Mont.	–	Okla.	4	Va.	3.5
Conn.	8	Ky.	5	Nebr.	4	Oreg.	–	Wash.	6.5
Del.	–	La.	4	Nev.	5.75	Pa.	6	W. Va.	6
Fla.	6	Maine	5	N.H.	–	R.I.	6	Wis.	5
Ga.	4	Md.	5	N.J.	6	S.C.	5	Wyo.	3

23. Answers will vary. They should include the following points. Write the problem as a proportion in which one of the ratios represents the percent. This ratio should have 100 in the denominator. The other ratio should have a part in the numerator and a whole in the denominator. Solve the proportion using cross products, and answer the question.

24. a. Method from Lesson 9–8: $0.075\,n = 112.5$

$$n = 1500$$

Method from Lesson 9–10: $\dfrac{7.5}{100} = \dfrac{112.5}{n}$

$$7.5\,n = 11250$$
$$n = 1500$$

b. Paragraphs will vary, but should include the following point. Both methods involve solving equations using division. In the method from Lesson 9–8, 7.5% is converted to a decimal, 0.075. In the method from Lesson 9–10, 7.5% is written as $\dfrac{7.5}{100}$.

12. Answers will vary. An example is given. A ratio is a comparison of two values. A proportion is an equation, each side of which is a ratio.   **33.** original price: $12, new price: $24; original length: 52 yd, new length: 104 yd

Chapter 10

23–24. Answers will vary. Examples are given.   **23.** If there were no circular shapes, common natural objects would have to assume other shapes. These would include objects such as trees,

flower stems, many fruits, and parts of animals such as the iris of the eye.   **24.** Manufactured items with circular shapes include wheels, gears, plates, buckets, clocks, and kitchenware such as pots and pans. If no circles were used in manufacturing, all of these items would have different shapes, and objects such as bicycles and cars that use wheels and gears might not exist.

1.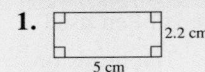

2.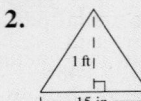

3.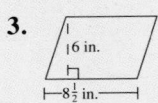

4.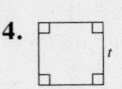

Activities III-IV. Answers will vary. Examples are given.
Activity III
2.

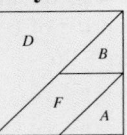

3. The area of each figure is $\frac{1}{2}$ square unit.

a. 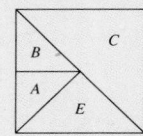   **b.**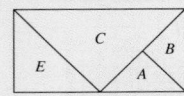

c. 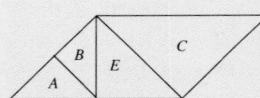   **d.**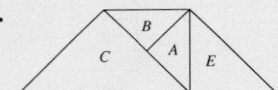

Activity IV
1.

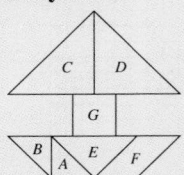

1. When you need to find the distance around a figure; the length of the chalk line around a football field   **2.** When you need to find the measure of the region enclosed by a figure; the amount of artificial turf needed to cover a football field

Guided Practice

7. Answers will vary. An example is given. Problems with no solution contain incorrect information, or lead to a contradiction. Problems with not enough information contain incomplete information and could be solved if more information were given.

Problem Solving Situations

11–12. Answers will vary. Examples are given. **11.** A collection of nine coins consists of quarters, dimes, nickels, and pennies. The total value of the coins is $1.16. How many of each type are there? Answer: 3 quarters, 3 dimes, 2 nickels, 1 penny. **12.** Donna says that she has $1.04 in quarters, dimes, nickels, and pennies. She has nine coins altogether. How many of each type does she have?

Guided Practice

6.

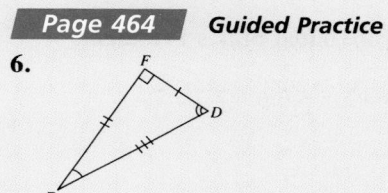

Guided Practice

5.

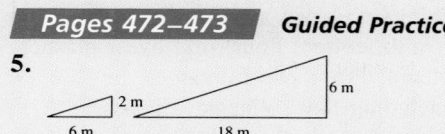

Exercises

23.

x	y	a	b	c
2	1	3	4	5
3	1	8	6	10
3	2	5	12	13
4	1	15	8	17
4	2	12	16	20
4	3	7	24	25
5	1	24	10	26
5	2	21	20	29
5	3	16	30	34
5	4	9	40	41
6	1	35	12	37
6	2	32	24	40
6	3	27	36	45
6	4	20	48	52
6	5	11	60	61

24. Answers will vary. An example is given.

```
10 INPUT X, Y
20 A = X^2 - Y^2
30 B = 2 * X * Y
40 C = X^2 + Y^2
50 PRINT X, Y, A, B, C
```

27.

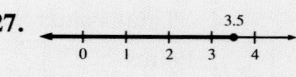

Chapter 11

Guided Practice

3.

Scores on a Mathematics Test		
Score	Tally	Frequency
61–70	卌 II	7
71–80		0
81–90	卌 卌 IIII	14
91–100	卌 II	7

4. No; yes; a frequency table with intervals shows whether a data item lies within a range of values; individual values are lost.

Exercises

1.

Compact Disc Prices (Dollars)		
Price	Tally	Frequency
7.00–8.99	II	2
9.00–10.99	II	2
11.00–12.99	卌 II	7
13.00–14.99	卌 II	7
15.00–16.99	II	2

2.

Ages of Hospital Nurses		
Age	Tally	Frequency
20–29	卌 I	6
30–39	卌	5
40–49	卌	5
50–59	卌 II	7
60–69	II	2

3.

Annual Tuition at Private Colleges (Dollars)		
Tuition	Tally	Frequency
6000–7999	卌 II	7
8000–9999	III	3
10,000–11,999	IIII	4
12,000–13,999	卌 I	6
14,000–15,999	I	1

4.

Total Points Scored in Basketball Games

Points	Tally	Frequency
70–79	ⅢⅡ	5
80–89	ⅢⅡ IIII	9
90–99	ⅢⅡ II	7
100–109	II	2
110–119	I	1

5. Intervals may vary.

Cost of One Week Vacation (Dollars)

Cost	Tally	Frequency
850–1049	IIII	4
1050–1249	ⅢⅡ I	6
1250–1449	ⅢⅡ	5
1450–1649	ⅢⅡ	5
1650–1849	IIII	4

6.

**Olympic Medals Won by Countries
in Track and Field (1988)**

Medals	Tally	Frequency
1–5	ⅢⅡ ⅢⅡ ⅢⅡ II	17
6–10	II	2
11–15		0
16–20		0
21–25		0
26–30	III	3

11.

Number of Hours Worked per Week

Hours	Tally	Frequency
20–24	ⅢⅡ ⅢⅡ I	11
25–29	ⅢⅡ IIII	9
30–34		0
35–39	ⅢⅡ ⅢⅡ I	11
40–44	ⅢⅡ	5

Pages 498–499 *Guided Practice*

10. Prices of Compact Disc Players (Dollars)

19	0 4 5 5
20	6 9
21	0 0 5 5
22	8 9
23	0 6 8

Pages 499–500 *Exercises*

1. Earned Run Averages of Leading Pitchers

2.2	0 5
2.3	2
2.4	
2.5	6
2.6	2 2 7
2.7	7 8
2.8	0 2 2 4
2.9	0 0 2

2. Ages of Company Presidents

3	5 8 9 9
4	2 4 5 7 8 8
5	0 0 5 5 6 8 8
6	0 0 2 5

11. Age at First Inauguration of United States Presidents

4	2 3 6 7 8 9 9
5	0 0 1 1 1 1 2 2 4 4 4 4 5 5 5 5 6 6 6 6 7 7 7 7 8
6	0 1 1 1 2 4 4 5 8 9

Page 500 *Historical Note*

From peanuts, George Washington Carver synthesized 300 products, including cheese, milk, coffee, flour, ink, dyes, plastics, wood stains, soap, and linoleum. From sweet potatoes, he synthesized 118 products, including flour, vinegar, molasses, rubber, ink, and postage stamp glue.

Page 502 *Guided Practice*

3.

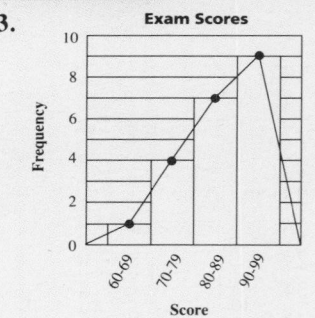

13. Scores on a Mathematics Test
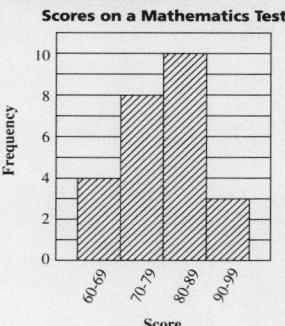

18. Hourly Wages of Part-time Baby Sitters
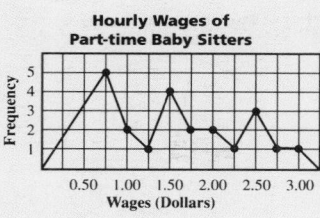

19. Hourly Wages of Part-time Baby Sitters
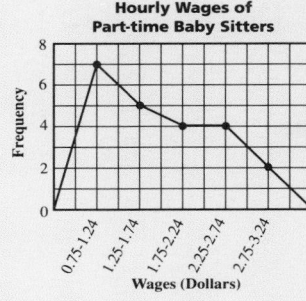

20. The graph for Exercise 18; without intervals, each piece of data is displayed in the graph. **21.** Answers will vary. The graph in Exercise 19 shows the general trend in salaries rather than the individual salaries.

5.
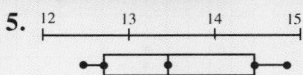

1. Box Seat Prices at Baseball Games (Dollars)
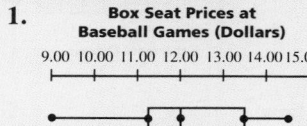

2. Prices for 13-inch Color Televisions (Dollars)
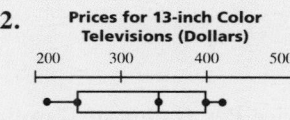

12. Answers will vary, but should include the following points. There is a wider range in attendance at musicals than at plays. The greatest value and third quartile for attendance at musicals are greater than those for attendance at plays. The least value, first quartile, and median for attendance at musicals are less than those for attendance at play.

13.
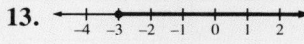

15. Number of Field Goals Scored
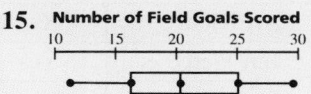

1. Scores on an Algebra Test

Score	Tally	Frequency
60–69	III	3
70–79	IIII II	7
80–89	IIII I	6
90–99	IIII	4

2. Scores on an Algebra Test

6	1 8 9
7	1 4 6 6 7 7 7
8	3 3 4 5 7 8
9	1 2 2 9

13. Jump Shots Made in 25 Attempts from Various Distances
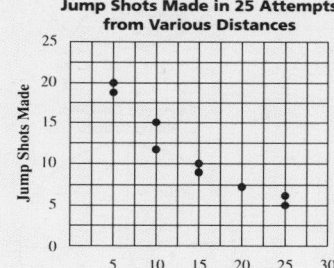

17–18. Car Registrations and Passenger Train Cars over Ten Years
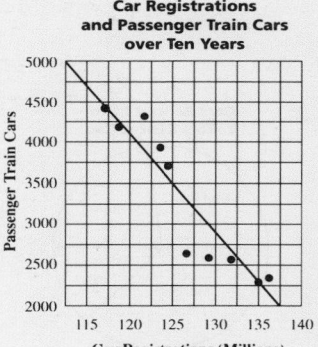

21. Areas and Depths of Bodies of Water
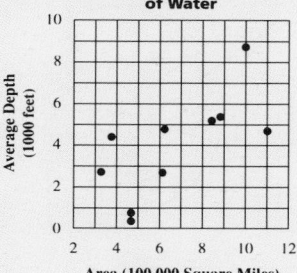

24. Ages of Chorus Members

1	6 7 9 9
2	0 5 6 7 8 9
3	2 4 6
4	1 5 8 9

23. Test Scores
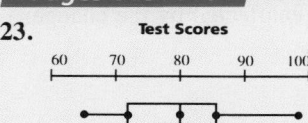

A29

12. Protein Source Preferences

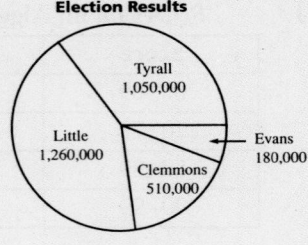

13. Election Results

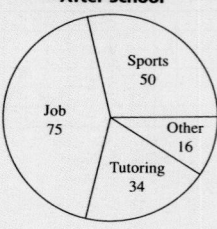

1. Library Inventory

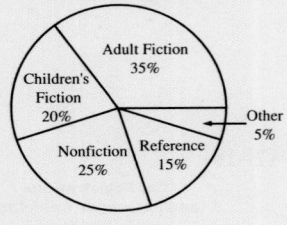

2. Annual Budget for a Four-Year Public University

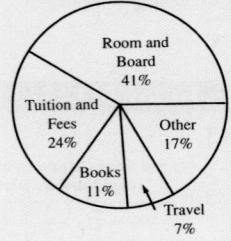

3. Cars Rented from Yoko's Rent-A-Car

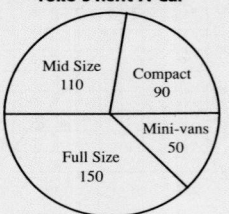

4. Favorite Chicken Restaurants

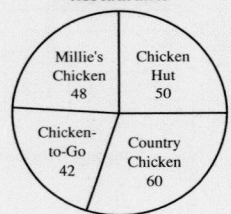

12. Annual Receipts

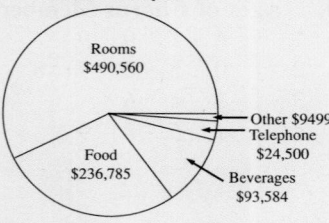

18. Commuting Methods of Acme Employees

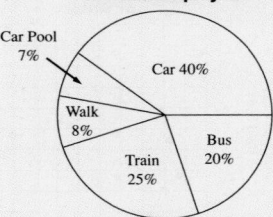

8. Answers will vary. An example is given. The sector for the northeast region will decrease. The sector for the southeast region will increase. The other sectors will be unaffected by the change.

2. Students in Activities After School

1. Scores on an English Test

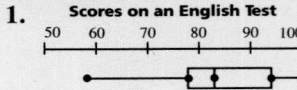

2. Weekly Grocery Budgets (Dollars)

5	2
6	
7	1 4 5 6 8
8	4 5 5 6 8 8
9	3 6 7 7
10	4 7
11	6
12	5

3. Survey of the Number of Women in Certain Occupations

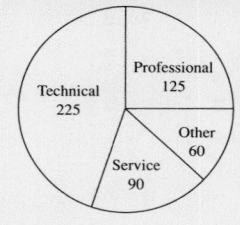

4. Managers' Salaries at Thorn's Department Store

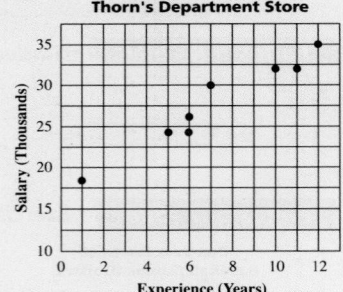

6. Price Per Gallon of Gasoline (Dollars)

1.2	6 7 7 9
1.3	1 2 4 5 8 8
1.4	0 1 3

5.

	Art	Gym	Science
Art Klein	X	X	Yes
Jim Pierce	Yes	X	X
Sy Fischer	X	Yes	X

10–11. Answers will vary. Examples are given. **10.** Cindy, Doug, and Grace own a cat, a dog, and a gerbil. No one owns a pet that begins with the same letter as his or her first name. Doug is allergic to cats. Who owns each pet? Grace owns the cat, Cindy owns the dog, and Doug owns the gerbil. **11.** Mary, Fran, and Jim have blue, brown, and green eyes. Jim does not have blue or green eyes. Fran does not have brown eyes. Mary has either blue or brown eyes. What color are each person's eyes? Mary has blue eyes, Fran has green eyes, and Jim has brown eyes.

4.

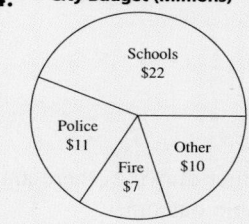

City Budget (Millions)

5.

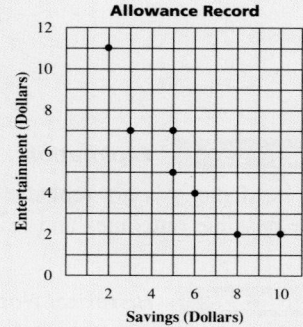

Allowance Record

11.

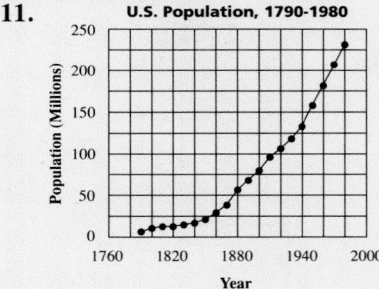

U.S. Population, 1790-1980

There has been an increase in the population of the United States since 1790. The population increased less dramatically in the 19th century than it has in the 20th century.

12. Answers will vary.

7.

Points Scored per Game

Points	Tally	Frequency
16–20	ЖІІ	7
21–25	ЖІІІІ	9
26–30	ІІ	2
31–35	ІІІ	3

8.

Points Scored per Game

1	6 7 8 8 8 8
2	0 2 2 2 3 3 4 4 4 5 7
3	0 1 1 3

13. Ages of Workers in an Office (Years)

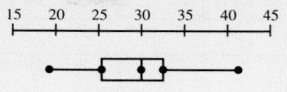

21. Colors of Cars Sold

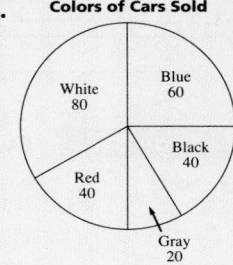

30.

Leading Batting Averages

0.30	1 3 4 4 5 6 6 7 8
0.31	1 2 2
0.32	2
0.33	
0.34	
0.35	6
0.36	6

1.

Hours Worked per Week

Hours	Tally	Frequency
21–30	ЖІІ	7
31–40	ЖЖЖІ	16
41–50	ІІІ	3
51–60		0
61–70	І	1

2. $1.00–$1.99, $2.00–$2.99, $3.00–$3.99, $4.00–$4.99

3.

Hours Worked per Week

2	6 6 8 8 9
3	0 0 2 2 5 5 5 6 7 8 8 8 8
4	0 0 0 0 0 1 2 4
5	
6	8

7. Prices of Stereo Systems (Dollars)

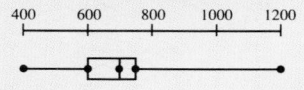

16. Animals in a Zoo

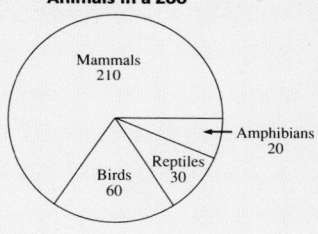

17. Monthly Commuting Expenses

Chapter 12

Focus on Explorations

Activity III

1.

2	3	4	5	6	7	8	9	10	11	12	No. of Rolls
?	?	?	?	?	?	?	?	?	?	?	?

Guided Practice

3.

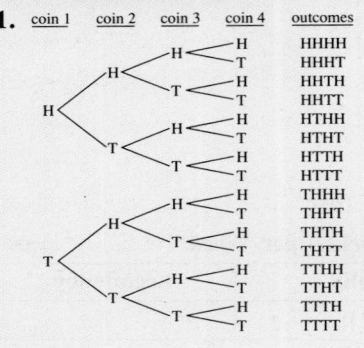

spinner	coin	outcomes
R	H	RH
	T	RT
W	H	WH
	T	WT
B	H	BH
	T	BT

Exercises

1.

coin 1	coin 2	coin 3	coin 4	outcomes
				HHHH
				HHHT
				HHTH
				HHTT
				HTHH
				HTHT
				HTTH
				HTTT
				THHH
				THHT
				THTH
				THTT
				TTHH
				TTHT
				TTTH
				TTTT

6.

number cube	spinner	outcomes
1	R	1 R
	B	1 B
	G	1 G
2	R	2 R
	B	2 B
	G	2 G
3	R	3 R
	B	3 B
	G	3 G
4	R	4 R
	B	4 B
	G	4 G
5	R	5 R
	B	5 B
	G	5 G
6	R	6 R
	B	6 B
	G	6 G

17–20.

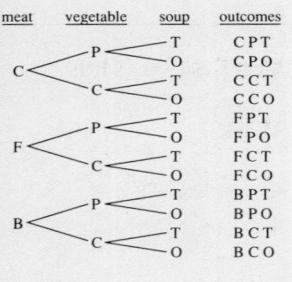

meat	vegetable	soup	outcomes
C	P	T	C P T
		O	C P O
	C	T	C C T
		O	C C O
F	P	T	F P T
		O	F P O
	C	T	F C T
		O	F C O
B	P	T	B P T
		O	B P O
	C	T	B C T
		O	B C O

21–22. K represents black.

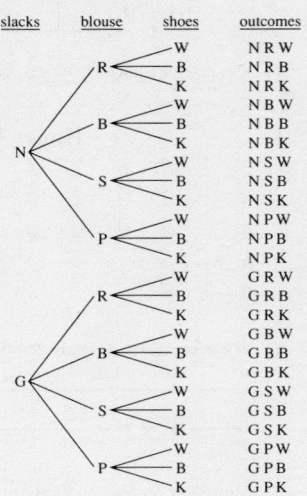

slacks	blouse	shoes	outcomes
N	R	W	N R W
		B	N R B
		K	N R K
	B	W	N B W
		B	N B B
		K	N B K
	S	W	N S W
		B	N S B
		K	N S K
	P	W	N P W
		B	N P B
		K	N P K
G	R	W	G R W
		B	G R B
		K	G R K
	B	W	G B W
		B	G B B
		K	G B K
	S	W	G S W
		B	G S B
		K	G S K
	P	W	G P W
		B	G P B
		K	G P K

26.

spinner	coin	outcomes
R	H	RH
	T	RT
B	H	BH
	T	BT
G	H	GH
	T	GT

$P(\text{blue, tail}) = \dfrac{1}{6}$

Self-Test 1

6.

number cube	coin	outcomes
1	H	1 H
	T	1 T
2	H	2 H
	T	2 T
3	H	3 H
	T	3 T
4	H	4 H
	T	4 T
5	H	5 H
	T	5 T
6	H	6 H
	T	6 T

Exploration

4. Yes; if the A is not replaced after the first drawing, there are fewer possible outcomes and fewer favorable outcomes.

Historical Note

Gregor Mendel is known as the father of the science of genetics. He developed his theories of heredity and genetics by crossbreeding thousands of plants and examining inherited traits.

Guided Practice

3.

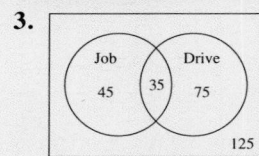

Job 45 | 35 | Drive 75

125

Problem Solving Situations

15–16. Answers will vary. Examples are given. **15.** Of 55 members of a sports club, 28 signed up for tennis lessons, 23 signed up for golf lessons, and 8 signed up for both. How many signed up for neither? Answer: 12 **16.** Of 169 jobs, 110 require a background in mathematics, 30 require backgrounds in both mathematics and science, and 24 have no mathematics or science requirements. How many jobs require only a background in science? Answer: 35

19. Answers will vary. An example is given. A permutation is an arrangement of items in a particular order. To find the number of permutations possible, the order of the items must be considered, so AB and BA are examples of different outcomes. A combination is a group of items chosen from a larger group with no reference to order. When finding the number of combinations possible, the order of the items is not considered, so AB and BA are the same outcome.

1. To find the experimental probability of an event; the objective of Lesson 12-1 was to use the probability formula to find a probability.

28.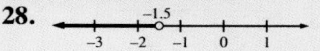

31. **Tickets Sold by Grade Level**

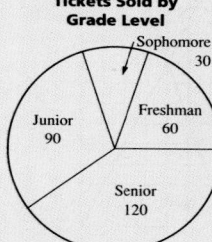

2. Answers will vary. An example is given. Assign each member of the class a number. Choose 10 numbers randomly and ask the corresponding class member for their choice. The candidate with the most votes out of these 10 is the predicted winner.

13.

spinner	coin	outcomes
1	H	1 H
	T	1 T
2	H	2 H
	T	2 T
3	H	3 H
	T	3 T
4	H	4 H
	T	4 T
5	H	5 H
	T	5 T

7. The probability of an event is given by the number of favorable outcomes divided by the total number of outcomes. The number of favorable outcomes can never exceed the total number of outcomes.

8.

coin	spinner	outcomes
H	1	H1
	2	H2
	3	H3
	4	H4
	5	H5
	6	H6
T	1	T1
	2	T2
	3	T3
	4	T4
	5	T5
	6	T6

15. Answers will vary. An example of each is given. Independent: A bag contains 5 blue marbles, 3 red marbles, and 2 white marbles. Two marbles are drawn with replacement. Find P(red, white). Answer. $\frac{3}{50}$ Dependent: A bag contains 5 blue marbles, 3 red marbles, and 2 white marbles. Two marbles are drawn without replacement. Find P (red, white). Answer: $\frac{1}{15}$

Chapter 13

4. $(-6, 36)$; $(-2, 12)$; $(0, 0)$; $(2, -12)$; $(6, -36)$
5. $(-6, -39)$; $(-2, -19)$; $(0, -9)$; $(2, 1)$; $(6, 21)$
6. $(-6, -8)$; $(-2, -4\frac{2}{3})$; $(0, -3)$; $(2, -1\frac{1}{3})$; $(6, 2)$
7. $(-6, 2)$; $(-2, -2)$; $(0, -4)$; $(2, -6)$; $(6, -10)$
8. $(-6, 14)$; $(-2, 10)$; $(0, 8)$; $(2, 6)$; $(6, 2)$
9. $(-6, 11)$; $(-2, 3)$; $(0, -1)$; $(2, -5)$; $(6, -13)$
10. $(-6, -21)$; $(-2, -5)$; $(0, 3)$; $(2, 11)$; $(6, 27)$
11. $(-6, 42)$; $(-2, 14)$; $(0, 0)$; $(2, -14)$; $(6, -42)$
12. $(-6, -16)$; $(-2, -4)$; $(0, 2)$; $(2, 8)$; $(6, 20)$

26.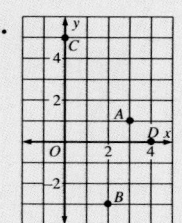

A33

Guided Practice

8. 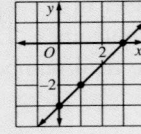 **9.** 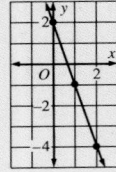 **10.**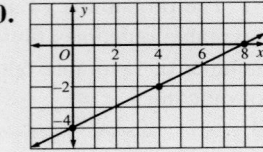

11. 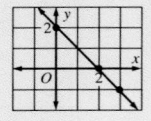 **12.** 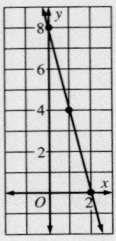 **13.**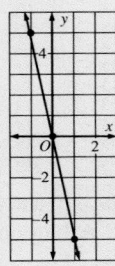

Exercises

1. 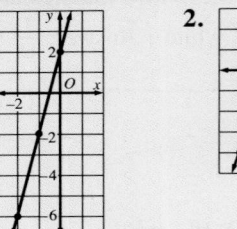 **2.** 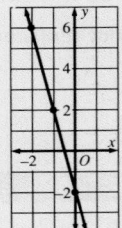 **3.**

4. 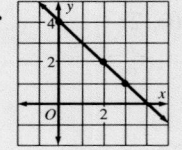 **5.** 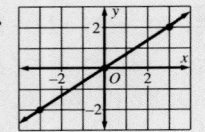 **6.**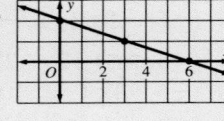

7. 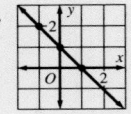 **8.** 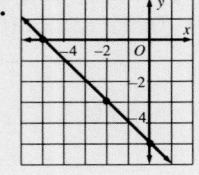 **9.**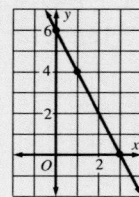

10.  **11.** 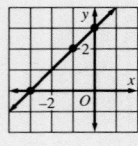 **12.**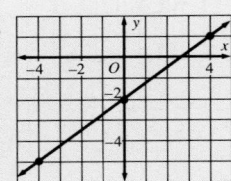

13. 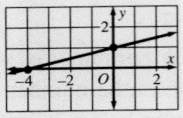 **14.**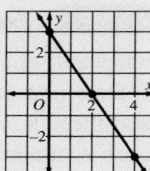

18. 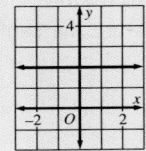 **20.** 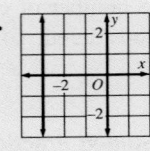 **24.**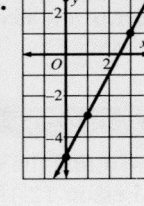

The graph is a horizontal line through (0, 2).

The graph is a vertical line through (−3, 0).

Exercises

16. 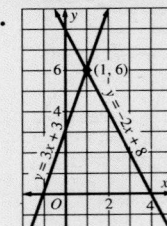 **21.**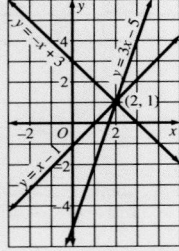

22. Three lines which intersect in one point; yes; the solution is (2, 1), which lies on all three lines.

23.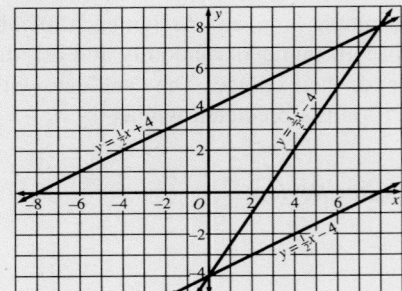

24. Two parallel lines and a third line which intersects both of them; no; there is no single point which lies on all three lines.

Self-Test 1

3. 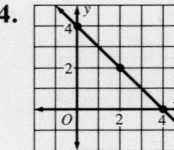 **4.**

A34

9–10. Answers will vary. Examples are given. **9.** The 84 houses on Oak Street are numbered consecutively starting with 1. What is the sum of the house numbers? Answer: 3570 **10.** Each row in a chorus line has 2 more dancers than the row in front of it. There are 24 dancers in the eighth row. How many dancers are in the third row? Answer: 14

13.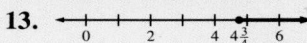

Pages 603–604   *Guided Practice*

2. Inequalities involving addition or subtraction can be solved in the same way equations involving addition or subtraction are solved; inequalities involving multiplication or division by a positive number can be solved in the same way equations involving multiplication or division are solved; in solving an inequality involving multiplication or division by a negative number, reverse the direction of the inequality symbol.

15. 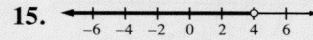   **16.**

17.   **18.**

19.   **20.**

Pages 604–605   *Exercises*

1.   **2.**

3.   **4.**

5.   **6.**

7.   **8.**

9.   **10.**

11.   **12.**

13.   **14.**

15.   **16.**

17. 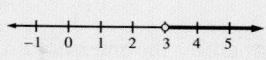   **18.**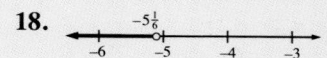

32.

Page 607   *Guided Practice*

10. $b > 3$ **11.** $h \geq -2$

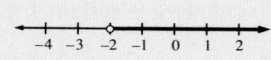

12. $n > -2$ **13.** $x < -4\frac{1}{2}$

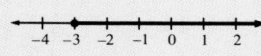

14. $n \geq -3$ **15.** $a < 3.2$

Page 608   *Exercises*

1. $t > 6$ **2.** $h \leq 4$

3. $a \geq -3$ **4.** $x \geq \frac{1}{3}$

5. $n \leq 4$ **6.** $z < 6$

7. $b < -1$ **8.** $t \geq 0$

9. $k > -15$ **10.** $x \geq -4$

11. $a \geq -3$ **12.** $q \geq 1$

13. $y \geq -2\frac{2}{3}$ **14.** $m > -5$

15. $m \leq -1.7$ 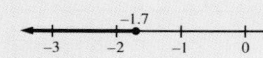 **16.** $b > 0.5$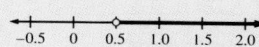

A35

17. $a > -21$
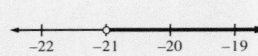

18. $t \geq 0$
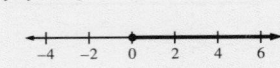

27. $y > -1$
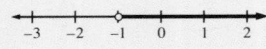

35–38. Answers will vary. Examples are given. **35.** The value of a number y is less than three times the number x. **36.** The value of a number y is greater than the sum of eight and six times the number x. **37.** The value of a number k is greater than or equal to the sum of the number m and 1. **38.** The value of a number c is less than or equal to one half the number t.

40.

41.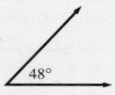

Page 608 **Historical Note**

Conic sections are the cross sections formed by the intersection of a plane and a cone. Depending on the angle of intersection, the conic section may be a circle, a parabola, a hyperbola, or an ellipse.

Pages 610–611 **Guided Practice**

12.  **13.** 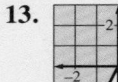 **14.**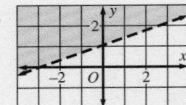

Pages 611–612 **Exercises**

9. 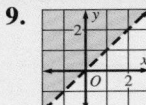 **10.** 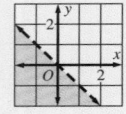 **11.**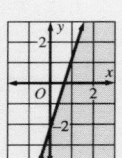

12. 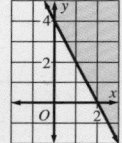 **13.** 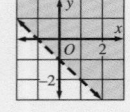 **14.**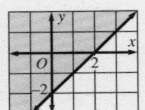

15. 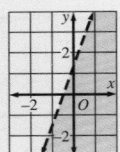 **16.** 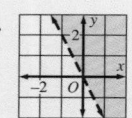 **17.**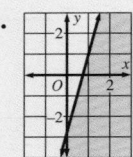

18. 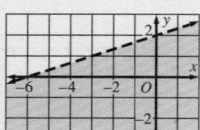 **19.**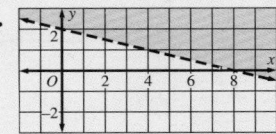

20. 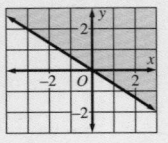 **29.** 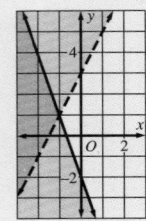 **30.**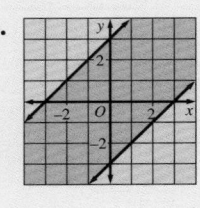

Page 612 **Self-Test 2**

2. 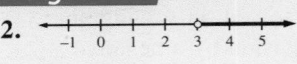 **3.**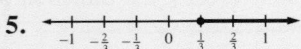

4. 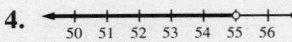 **5.**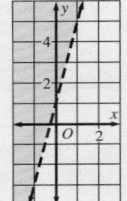

6. 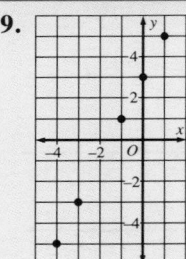 **7.**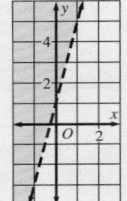

Pages 614–615 **Exercises**

19.

Page 617 **Guided Practice**

12. 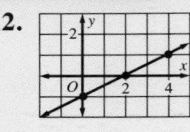 **13.** 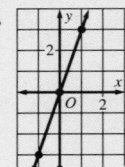 **14.**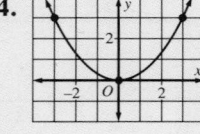

15. 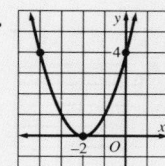   **16.** 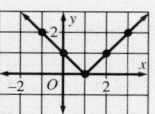   **17.**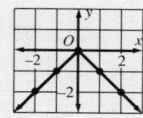

24.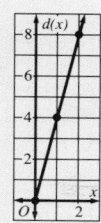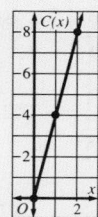

Pages 617–619 *Exercises*

1. 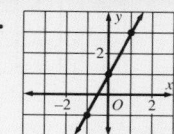   **2.** 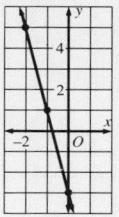   **3.**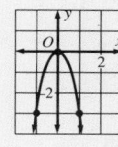

Answers will vary. An example is given. The graphs have the same slope and the same *y*-intercepts. The scale on the *x*-axis of the first graph represents time, and the scale on the *x*-axis of the second graph represents a length of material. The scale on the vertical axis of the first graph represents distance, and the scale on the vertical axis of the second graph represents cost.

4. 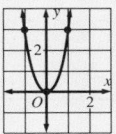   **5.** 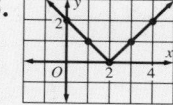   **6.**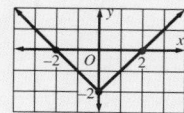

33.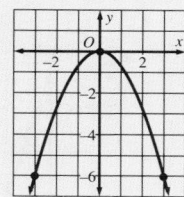

7. 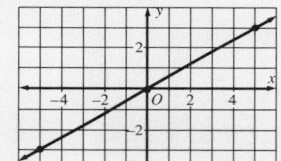   **8.**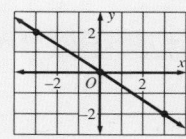

Page 619 *Self-Test 3*

5. 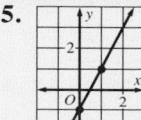   **6.**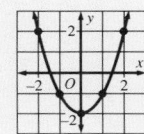

9. 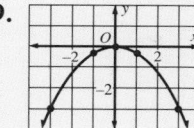   **10.**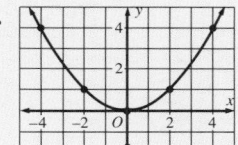

Pages 620–621 *Chapter Review*

10. 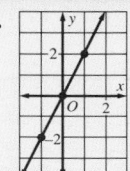   **11.** 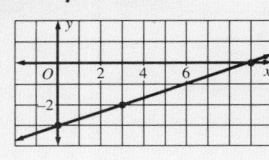   **12.**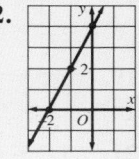

11. 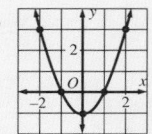   **12.** 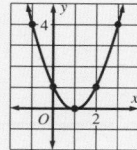   **13.**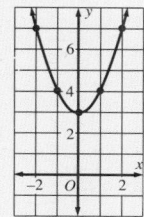

24. ![number line] −2 0 2 4 6 8 10

25. ![number line] −3 −2 −1 0 1 2 3

26. ![number line] −4 −2 0 2 4 6 8

27. ![number line] −6 −4 −2 0 2 4 6

14. 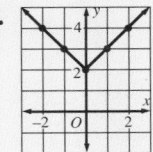   **15.**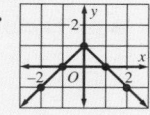

28. ![number line] −3 −2 −1 0 1 2 3

29. ![number line] −8 −4 0 4 8 12 16

30. ![number line] −3 −2 −1 0 1 2 3

31. ![number line] −6 −5 −4 −3 −2 −1 0

32. 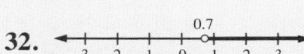 0.7
−3 −2 −1 0 1 2 3

36. 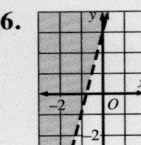 **37.** 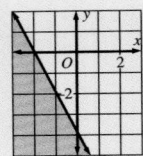 **38.**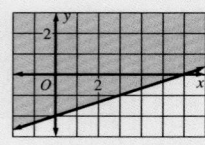

13. 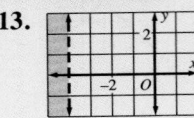 **14.**   **15.**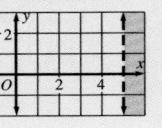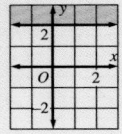

45. 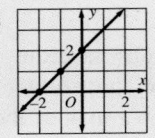 **46.** 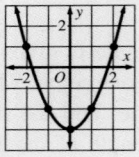 **47.**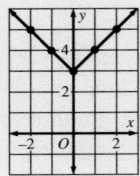

16. 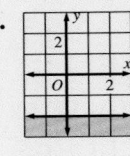 **17.** 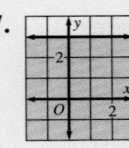 **18.**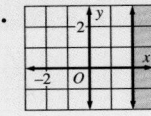

19. 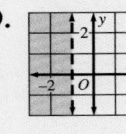 **20.**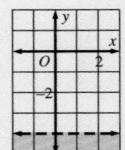

4. 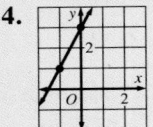 **5.** 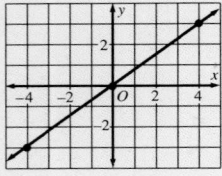 **6.**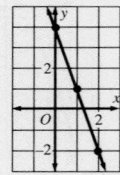

Chapter 14

10. a. 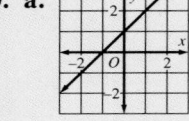 **b.**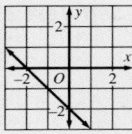

10. 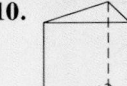 **11.**   **12.**   **13.**

15. 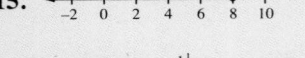 **16.**

14.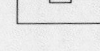

17.   **18.**

15.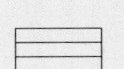

19.   **20.**

16.

21. 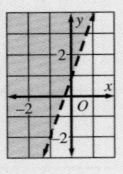 **22.** 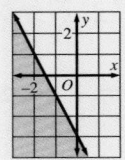 **23.**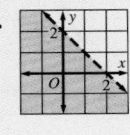

18. **21.**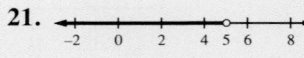

26. 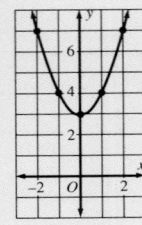 **27.**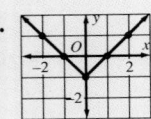

Activity IV

a.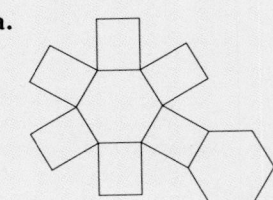

b.

c.

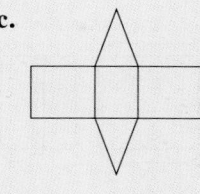

Pages 635–636 *Guided Practice*

6.

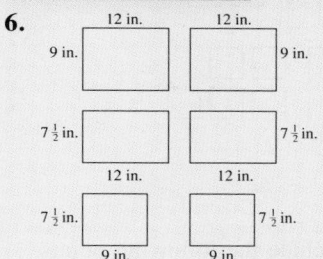

7.

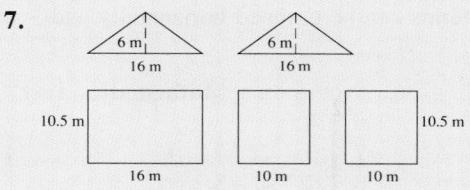

Pages 636–637 *Exercises*

25.

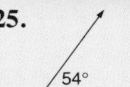

54°

Page 637 *Challenge*

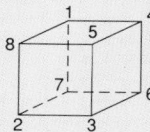

Answers may vary.

Pages 639–641 *Exercises*

20.

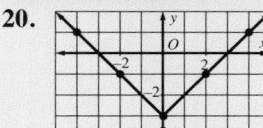

Pages 644–645 *Exercises*

28.

Pages 648–649 *Exercises*

17, 20. Answers will vary. Examples are given. **17.** The surface area of a space figure is a measure of the amount of material needed to form the surface of the space figure. The volume of a space figure is a measure of the amount of space the figure encloses. **20.** The black-tailed jack rabbit is native to the warm climates of Mexico and the southwestern part of the United States. Its long ears and legs provide greater skin area per unit volume, so its body heat dissipates quickly. This helps the jack rabbit survive in hot weather.

Pages 651–653 *Exercises*

23. The surface area of the sphere is $4(3.14)(3^2) = 113.04$ cm^2. The area of the curved surface of the cylinder is $2(3.14)(3)(6) = 113.04$ cm^2. **24.** The volume of the sphere is $\frac{4}{3}(3.14)(3^3) = 113.04$ cm^3. The volume of the cylinder is $(3.14)(3^2)(6) = 169.56$ cm^3. 113.04 cm$^3 = \frac{2}{3}$ (169.56) cm^3

Page 653 *Historical Note*

Archimedes invented a machine for raising water that was used to empty water from the hold of a ship. The same type of machine is still used in Egypt to irrigate fields. He also pioneered the use of levers. Astronomy was one of his major interests, and he invented two astronomical globes, one a star globe, and the other a device for mechanically representing the motions of the sun, moon, and planets.

Page 655 *Guided Practice*

6.

Pages 655–657 *Problem Solving Situations*

11–12. Answers will vary. Examples are given. **11.** A rectangular prism is to be formed using exactly 12 cubes. How many such prisms are possible? Answer: 3 **12.** A rectangular prism is to be formed using exactly 12 cubes. What are the dimensions of the prism with the least possible surface area? Answer: 2 × 3 × 2

Chapter 15

Exercises

11. $x^2 + 4x + 7$ **12.** $-10d^3 - 6d^2 + 4$
13. $e^2 + 11e - 20$ **14.** $-5r^3 - 7r^2 + 2$
15. $-4x^2 + 9x + 12$ **16.** $9w^3 - 6w^2 - 7$
17. $-7x^2 - 11x + 1$ **18.** $7g^2 + 7g - 3$
19. $8w^3 + 4w^2 - 12w + 1$ **20.** $-4x^3 - 6x^2 + 9x + 2$
21. $x^5 + 4x^2 + 2x + 1$ **22.** $4x^4 - x^2 - 9x + 8$
23. $2a^4 - 10a^3 - 7a^2$ **24.** $4x^3 - 4$ **25.** $-5x^3$
26. $4a^4 + 5$ **27.** $ab^3 + a^2b^2 + a^2b + 4$
28. $mn^4 + m^2n^3 + m^2n^2 - mn - 1$
29. $x^3y^3 - xy^2 + x^2y - 4$
30. $-cd^4 - c^3d^3 + c^2d^2 + c^4d + 6$
31. $a^3b^2 - a^2b + ab^3 - 4$ **32.** $-x^3z - x^2z^2 + xz^3 + 9$
33. $a^3c^4 + ac^3 + a^4c^2 - a^2c$
34. $m^4n - m^3n^2 - m^2n^3 + mn^4$
39. $x^3yz^2 + x^2y^2z + xy^3z^3$; $xy^3z^3 + x^2y^2z + x^3yz^2$; $xy^3z^3 + x^3yz^2 + x^2y^2z$ **40.** Answers will vary, but should contain four distinct variables. Example: $a^4b^3c^2d + a^3b^4cd^2 + a^2bc^4d^3 + ab^2c^3d^4$

43.

Focus on Explorations

Activity I

2.

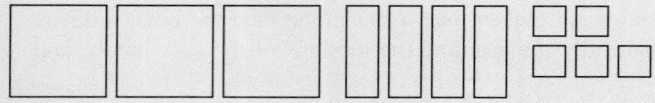

Activity II

6.

Activity III

1.

2.

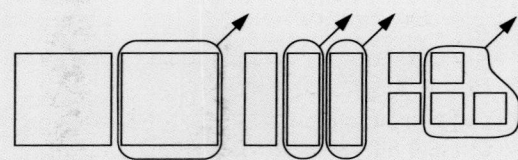

Guided Practice

4.

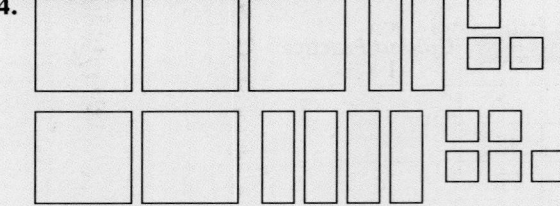

Exercises

27. Answers will vary, but should include the following points: The like terms can be lined up vertically, with zeros inserted when terms are missing. Like terms can be grouped horizontally and then combined.

33.

36. Scores on a Mathematics Test

6	8
7	3 7 9 9
8	2 4 8
9	1 4

Exploration

1.

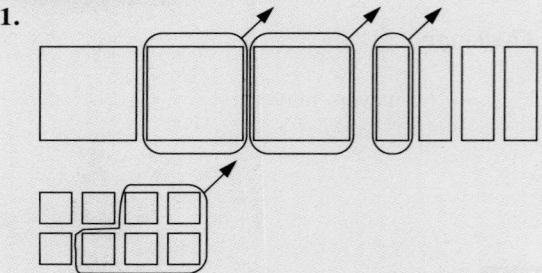

37. Answers will vary. Example: $x^2 + 3x + 7$ and $x^2 + 3x + 5$
38. Answers will vary. Example: $x^2 + 4x + 2$ and $x^2 + 3x + 2$
42. When the order of a subtraction is reversed, the difference is the opposite of the original difference. Examples will vary.
$(6x^2 + 8x - 11) - (2x^2 - 3x + 7) = 4x^2 + 11x - 18$ and $(2x^2 - 3x + 7) - (6x^2 + 8x - 11) = -4x^2 - 11x + 18$

43.   **Base Hits per Season**

Hits	Tally	Frequency
160–169	ⅢⅢ	5
170–179	IIII	4
180–189	II	2
190–199	I	1
200–209	I	1
210–219	I	1

46. $x > 8$

$\begin{array}{cccccc} & & & & \circ & \\ \hline -4 & 0 & 4 & 8 & 12 & \end{array}$

Activity I

2. a.

b.

Activity II

2. a.

b.

c.

d.

2. Answers will vary. Examples: to multiply a polynomial by a monomial, multiply each term of the polynomial by the monomial and simplify; to multiply powers with the same base, add the exponents; use the distributive property.

17. $10m^2n^2 - 15m^2n^3 + 20mn^2$
18. $-12a^3bc + 32a^2b^3c + 20abc$
19. $-5x^2y^2 - 7x^3y^2 - 8x^2y^3$
20. $21c^2d^3 + 12c^3d^4 - 3c^4d^3$ **21.** $-5m^5n^3 - m^4n^2 + 7m^5n^2$
22. $4a^6b^2 - 2a^6b^3 + 14a^4b^6 - 18a^3b^{10}$
23. $30x^3 + 35x^2 - 45x - 25$
24. $-18c^4 - 24c^3 + 9c^2 + 24c - 6$
25. $8a^5 + 16a^4 - 14a^3 + 2a^2 - 18a$
26. $28n^4 - 8n^3 + 16n^2 - 24n$
27. $-18x^5 - 8x^4 + 12x^3 + 16x^2$
28. $-25m^7 + 10m^6 + 5m^5 - 35m^4$
37. The product of powers rule applies to negative exponents; $(a^{-4})(a^{-5}) = \frac{1}{a^4} \cdot \frac{1}{a^5} = \frac{1}{a^9} = a^{-9}$.

3.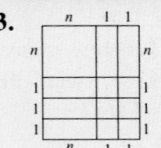

4.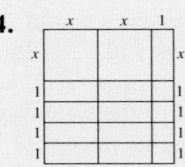

1. $x^2 + 6x + 8$ **2.** $a^2 + 9a + 18$ **3.** $12t^2 + 52t + 35$
4. $15x^2 + 20x + 5$ **5.** $24y^2 + 16y - 14$ **6.** $5b^2 - 2b - 3$
7. $7x^2 + 6x - 16$ **8.** $7k^2 + 55k - 8$
9. $2w^2 - 9w + 4$ **10.** $3d^2 - 11d + 10$ **11.** $4x^2 - 17x + 4$
12. $10z^2 - 29z + 10$ **13.** $12x^2 - 38x + 30$
14. $18y^2 - 51y + 35$ **15.** $4x^2 + 4x + 1$ **16.** $25c^2 - 30c + 9$
17. $4x^2 - 25$ **18.** $16d^2 - 49$

34.   1 5   10   10   5 1
      1 6   15   20   15   6 1
    1 7 21   35   35   21 7 1

35. a. $a^5 + 5a^4b + 10a^3b^2 + 10a^2b^3 + 5ab^4 + b^5$
b. $a^6 + 6a^5b + 15a^4b^2 + 20a^3b^3 + 15a^2b^4 + 6ab^5 + b^6$
c. $a^7 + 7a^6b + 21a^5b^2 + 35a^4b^3 + 35a^3b^4 + 21a^2b^5 + 7ab^6 + b^7$

36. Answers will vary. An example is given. Pascal's law, which states that fluid in a vessel transmits pressures equally in all directions, explains the operations of air compressors, vacuum pumps, and hydraulic elevators, jacks, and presses. Pascal's triangle can be used to calculate probabilities. Pascal also invented a calculating machine which could add and multiply.
37. A, B, C, and D represent the numbers (coefficients and constants).   **38.** The computer would multiply 2 times 4 to get 8. It would then multiply 2 times 5 to get 10 and 3 times 4 to get 12, and then add them to get 22. It would then multiply 3 times 5 to get 15. Finally, the computer would print 8X^2 + 22X + 15.

10. line graph

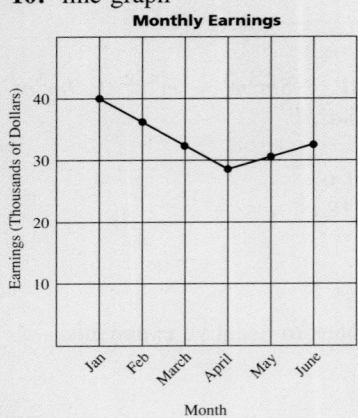

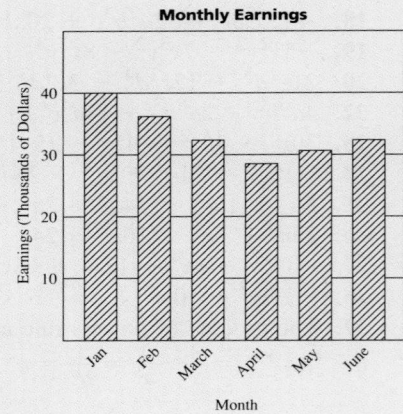

The line graph shows the information better than the bar graph because it shows the trend of the earnings over time.

11–12. Answers will vary. Examples are given.   **11.** How many ancestors does Peter have in the past four generations? Answer: 30

12. At the end of the week, the price of a stock was $16.75. During the week its price rose $2.25, fell $1.75, fell 75¢, rose $1.00, and fell 75¢. What was its price at the beginning of the week? Answer: $16.75.

15.

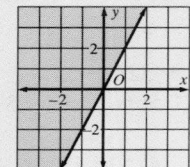

The rule of false position involves substituting a convenient number for the variable in an equation, simplifying, finding the error, and adjusting the substitution accordingly. For example, to solve $x + \frac{x}{5} = 18$, substitute 5 for x. Then $x + \frac{x}{5} = 5 + \frac{5}{5} = 6$. Since 6 must be multiplied by 3 to result in 18, the correct value for x must be 3(5), or 15.

1. $7x(x + 2)$ **2.** $6y(3y - 1)$ **3.** $s(4 + 5s)$

4. $m^2(m^2 + 2)$ **5.** $4k^2(k + 2)$ **6.** $3n^2(3n - 1)$

7. $5p^2(p^3 - 3)$ **8.** $9a^2(a - 3)$ **9.** $3xy(3 + 2x)$

10. $ab^2(1 - 3a)$ **11.** $4c(a - 2b)$ **12.** $rt(7s - 5rt)$

13. $3u^2(4u^2 - 3u - 8)$ **14.** $6(4t^5 - 3t^2 - 1)$

15. $8s^2(s^2 - 2s + 4)$ **16.** $9(2q^5 - q^4 - q^3 - 2)$

17. $5qd(5d + 3q + 6qd - 1)$ **18.** $7a^2(5a^3 - 7ab^2 - 8b^4)$

19. $y^3z^2(36yz + 45z^2 + 20y^2)$

20. $11k^4t^5(2k^2 + 10t - 5k^3)$

33.

t (seconds)	$h = 16t(4 - t)$ (feet)
0	0
1	48
2	64
3	48
4	0

37.   Gasoline Mileage (mi/gal)

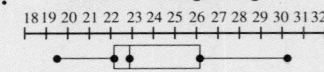

41. $5d^2(4d^3 - 2d^2 + 3d - 1)$ **42.** $6m^2(3m^4 - 5m^2 + 2m - 4)$

43. $cd(3c^2d^2 + 5cd - 4)$ **44.** $2ab^2(2a^2 - 4ab + 1)$

45. $5x^2y^2(5x^2y + 2xy^2 - 3)$ **46.** $6mn^2(2n - 3m + 4mn)$

19.

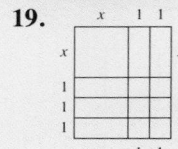

20. Answers will vary. An example is given. Multiply the two binomials together first, then multiply by the monomial. $24n^3 + 102n^2 + 105n$

Looking Ahead

8. a.

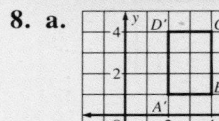

b. $A'(2, 1)$; $B'(4, 1)$; $C'(4, 4)$; $D'(2, 4)$

9. a.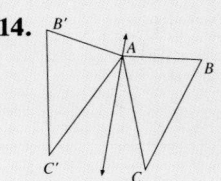

b. $A'(1, -2)$; $B'(3, -2)$; $C'(3, 1)$; $D'(1, 1)$

6.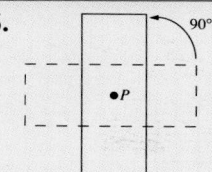

7.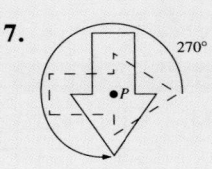

10. a.

b. $A'(-2, 1)$; $B'(0, 1)$; $C'(0, 4)$; $D'(-2, 4)$

8.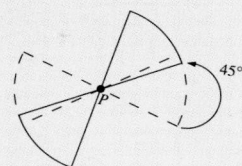

9.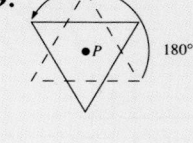

11. a.

b. $A'(1, 3)$; $B'(3, 3)$; $C'(3, 6)$; $D'(1, 6)$

18. a, b.

c. $A'(-6, -5)$; $B'(-6, -2)$; $C'(-1, -4)$

12. a.

b. $A'(-1, 1)$; $B'(-3, 1)$; $C'(-3, 4)$; $D'(-1, 4)$

19. a, b.

c. $A'(-3, 7)$; $B'(-3, 4)$; $C'(2, 4)$; $D'(2, 7)$

13. a.

b. $A'(1, -1)$; $B'(3, -1)$; $C'(3, -4)$; $D'(1, -4)$

14.

15.

16.

20. a, b.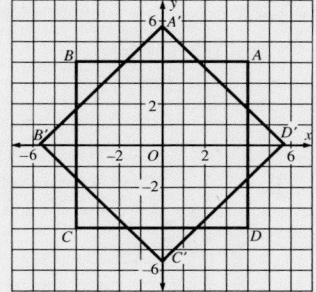

c. $A'(0, 5.7)$; $B'(-5.7, 0)$; $C'(0, -5.7)$; $D'(5.7, 0)$

24. a, b.

25. a, b.

c. translation

c. translation

c. translation

A43

Chapter 3

40–43. 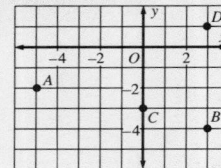 **44.**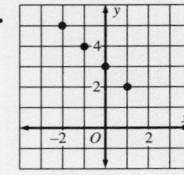

45. 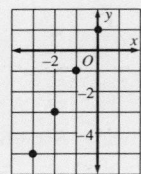 **46.**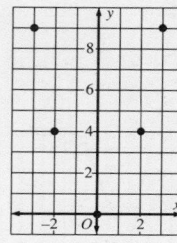

47.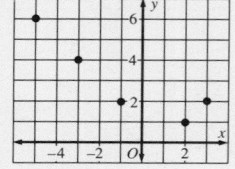

Chapter 5

1.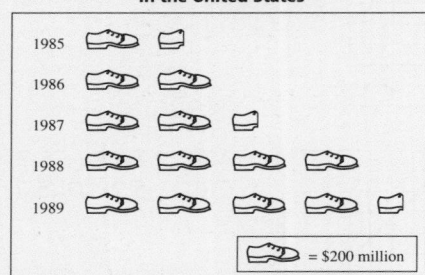

Sales of Walking Shoes in the United States

7. 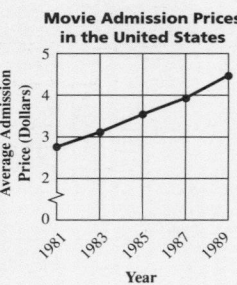 **8.**

12. The graph gives the impression that Sheera won by a large margin.

Chapter 6

7. 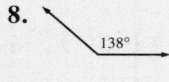 65° **8.** 138°

Chapter 7

60. Lay the 4 m and 6 m boards end-to-end. Then lay the 9 m board beside the other pair, with one pair of ends lined up. The distance between the other pair of ends is 1 m.

69. −6 −5 −4 −3 −2 −1 0 **70.** −5 −4 −3 −2 −1 0 1

71. 10 11 12 13 14 15 16 **72.** −3 −2 −1 0 1 2 3

73. 3.7 2 3 4 5 **74.** −1 −0.5 0 0.5

75. −3 −2.5 −2 −1.5 −1 −0.5 0 **76.** −1 −0.5 0 0.5 1 1.5 2

Chapter 11

1. **Test Scores in Math**

Score	Tally	Frequency
50–59	II	2
60–69	II	2
70–79	IIII I	6
80–89	IIII I	6
90–99	IIII	4

2. **Test Scores in Math**

5	7 9
6	0 4
7	0 2 5 6 6 9
8	0 1 2 2 3 6
9	2 4 7 8

10. Prices for Compact Disc Players (Dollars)

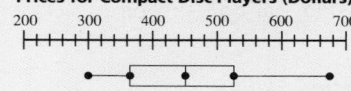

20. Scores on a History Test

5	7
6	5 5 7
7	2 3 5 5 6 6 9
8	0 3 6 9 9
9	0 2 8 8

19. Videos Rented from Victor's Videos

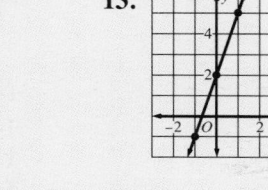

5.

spinner	coin	outcomes
1	H / T	1H / 1T
2	H / T	2H / 2T
3	H / T	3H / 3T
4	H / T	4H / 4T
5	H / T	5H / 5T

---

1. $(-2, -4); (-1, -1); (0, 2); (1, 5); (2, 8)$ **2.** $(-2, -2);$
$(-1, -1\frac{1}{2}); (0, -1); (1, -\frac{1}{2}); (2, 0)$ **3.** $(-2, 9); (-1, 7); (0, 5);$
$(1, 3); (2, 1)$ **4.** $(-2, 8); (-1, 3); (0, -2); (1, -7); (2, -12)$
5. $(-2, 3); (-1, 6); (0, 9); (1, 12); (2, 15)$ **6.** $(-2, 11);$
$(-1, 7); (0, 3); (1, -1); (2, -5)$ **7.** $(-2, -17); (-1, -12);$
$(0, -7); (1, -2); (2, 3)$ **8.** $(-2, 2); (-1, -1); (0, -4);$
$(1, -7); (2, -10)$

9. 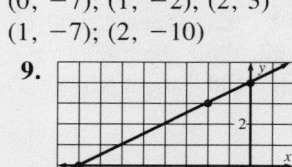 **10.**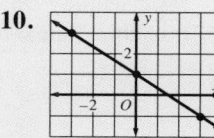

11. 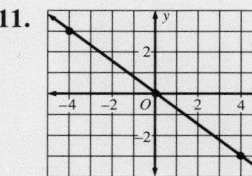 **12.**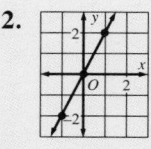

---

13. 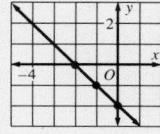 **14.**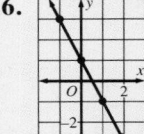

15. 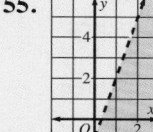 **16.**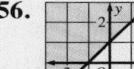

35. 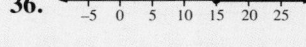 **36.**

37. **38.**

39. **40.**

41. **42.**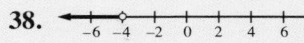

55. **56.**

57. **58.**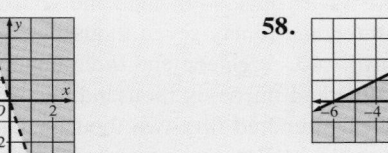

63. $f(-2) = 8; f(1) = -4; f(0) = 0$ **64.** $f(-2) = -\frac{1}{4}; f(1) = \frac{1}{8};$
$f(0) = 0$ **65.** $f(-2) = 31; f(1) = 7; f(0) = -1$ **66.** $f(-2) = 11;$
$f(1) = 7; f(0) = 1$ **67.** $f(-2) = 3; f(1) = 0; f(0) = -1$
68. $f(-2) = 16; f(1) = 1; f(0) = 4$

69. 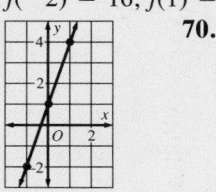 **70.** 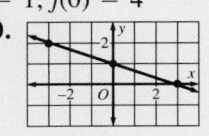 **71.**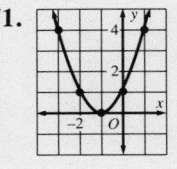

72.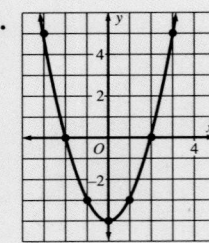

73.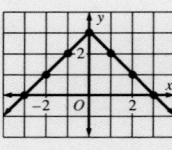

74.

Pages 751–752 **Chapter 15**

9. $10a^2 + 9a + 10$ **10.** $13b^2 + 27b + 18$
11. $18c^2 + 28c + 21$ **12.** $26y^2 + 41y + 18$ **17.** $2k^2 + 5k + 3$ **18.** $7p^2 + p + 3$ **19.** $h^2 + h + 1$ **20.** $2r^2 + 2r + 2$

Toolbox Skills Practice

Page 755 **Skill 5**

21. fifteen **22.** ninety-one **23.** one hundred sixty-two **24.** eighty-seven **25.** four hundred ninety-three **26.** one hundred one **27.** two hundred fifty-eight **28.** nine hundred eleven **29.** four thousand, fifty-six **30.** three thousand, eighty **31.** thirty-two **32.** one thousand, four hundred seventy-three **33.** seven thousand, six **34.** fifty-six **35.** three hundred eighty-five **36.** sixteen and three tenths **37.** four and seven tenths **38.** twenty-one and fifteen hundredths **39.** one hundred seventy-eight and sixty-one hundredths **40.** nine and sixteen thousandths **41.** eighty-three and ninety-seven thousandths **42.** four hundred ninety-two **43.** eighteen and three hundred fifteen thousandths **44.** nine and thirty-six thousandths **45.** six thousand, thirty-one and seven hundred fifty-two thousandths **46.** twenty-four and four tenths **47.** one hundred ninety-three and seven hundredths **48.** eight and nine hundred forty-nine thousandths **49.** three hundred seven and one tenth **50.** twelve and fifty-six hundredths

Diagnostic Test on Toolbox Skills

Page T44 **Test on Estimation Skills**

1-15. Estimates may vary. Accept reasonable estimates.
1. about 400 **2.** about 200 **3.** about 1800 **4.** about 90
5. about 1000 **6.** about 200 **7.** about 500 **8.** about 140,000
9. about 20 **10.** about 30 **11.** about 300 **12.** about 40
13. about 120 **14.** about 240 **15.** about 780 **16.** about 40

Pages T44–T47 **Test on Numerical Skills**

17. tens; 20 **18.** thousandths; 0.001 **19.** five hundred thirty-seven **20.** two hundred nine and three hundredths
21. 800 **22.** 9600 **23.** 15.9 **24.** 543.1 **25.** 157
26. 1697 **27.** 613 **28.** 5536 **29.** 118 **30.** 51,225
31. 136 **32.** 18,297 **33.** 27,993 **34.** 1,168,288 **35.** 2730
36. 89,523 **37.** $2\frac{12}{47}$ **38.** $337\frac{1}{4}$ **39.** 18.4 **40.** 116.9
41. 1198.86 **42.** 500.693 **43.** 55.13 **44.** 64.168
45. 126.92 **46.** 550.119 **47.** 18.27 **48.** 0.0997 **49.** 35.67
50. 140.058 **51.** 0.076 **52.** 0.0472 **53.** 4.5 **54.** 8.13
55. 3.47 **56.** 28.34 **57.** 300 **58.** 172,512.4 **59.** 1.2034
60. 0.0000483 **61.** Yes. **62.** No. **63.** No. **64.** Yes. **65.** $\frac{4}{7}$
66. $\frac{3}{4}$ **67.** $\frac{1}{5}$ **68.** $\frac{5}{6}$ **69.** $12\frac{2}{3}$ **70.** $35\frac{1}{7}$ **71.** $12\frac{5}{14}$
72. $18\frac{71}{80}$ **73.** $5\frac{1}{3}$ **74.** $6\frac{4}{9}$ **75.** $15\frac{9}{14}$ **76.** $10\frac{5}{8}$ **77.** $\frac{1}{25}$
78. $2\frac{1}{2}$ **79.** $1\frac{1}{4}$ **80.** 36 **81.** $4\frac{1}{4}$ **82.** 20 **83.** 121
84. $2\frac{2}{7}$ **85.** $27\frac{3}{4}$; 333 **86.** 44; 88 **87.** 10,000; 5 **88.** 9; 2

Pages T47–T48 **Test on Graphing Skills**

89. September **90.** July **91.** about 4000 **92.** about 22,000
93. Miami **94.** Seattle **95.** about 800 mi **96.** about 600 mi
97. about $100 **98.** January **99.** March **100.** about $100